HANDBOOK OF
Aqueous Solubility Data

SECOND EDITION

HANDBOOK OF

Aqueous
Solubility
Data

SECOND EDITION

Samuel H. Yalkowsky

Yan He

Parijat Jain

CRC Press
Taylor & Francis Group
Boca Raton London New York

CRC Press is an imprint of the
Taylor & Francis Group, an **Informa** business

CRC Press
Taylor & Francis Group
6000 Broken Sound Parkway NW, Suite 300
Boca Raton, FL 33487-2742

First issued in paperback 2019

© 2010 by Taylor & Francis Group, LLC
CRC Press is an imprint of Taylor & Francis Group, an Informa business

No claim to original U.S. Government works

ISBN-13: 978-1-4398-0245-8 (hbk)
ISBN-13: 978-0-367-38417-3 (pbk)

Library of Congress Cataloging-in-Publication Data

Yalkowsky, Samuel H. (Samuel Hyman), 1942-
 Handbook of aqueous solubility data / by Samuel H. Yalkowsky, Yan He, Parijat Jain. -- 2nd ed.
  p. ; cm.
 Includes bibliographical references and index.
 ISBN 978-1-4398-0245-8
 1. Organic compounds--Solubility--Handbooks, manuals, etc. I. He, Yan, 1970- II. Jain, Parijat. III. Title.
 [DNLM: 1. Chemistry, Organic--Handbooks. 2. Solubility--Handbooks. QD 257.7 Y19h 2010]

QD257.7.H33 2010
547'.13422--dc22
                        2009042539

Visit the Taylor & Francis Web site at
http://www.taylorandfrancis.com

and the CRC Press Web site at
http://www.crcpress.com

Contents

Authors

Dr. Samuel Yalkowsky is professor of pharmaceutical sciences at the University of Arizona. He is currently involved in basic research on the relationships between chemical structure and physical phenomena such as solubility, partitioning, and melting. He has also made progress in the development of the state of the art algorithm for the estimation of the melting points, aqueous solubility and other physicochemical properties of organic compounds.

Dr. Yalkowsky is also involved in the alteration of solubility by physical means. This includes the development of formulations for insoluble drugs and the improved dissolution of environmentally important solutes from the soil. The formulation work was extended to include the development of novel dosage forms and the pharmaceutical evaluation of parenteral formulations. This has led to the development of novel methods for screening for hemolysis and for phlebitis.

Dr. Yan He earned her BS in biology from Wuhan University in 1992, her MS and PhD degrees in pharmaceutical sciences from the University of Arizona in 1999 and 2005. She is a senior research investigator in the Pharmaceuticals Sciences Department at Sanofi-Aventis. Her research interests include performing "drugability" assessment, providing formulation for preclinical studies, and preparing preformulation package for preclinical drug candidates. She also conducted basic research on the relationships between chemical structure and physical properties of organic compounds.

Dr. Parijat Jain received his PhD from the University of Arizona in 2008. Currently, he is a formulation scientist in the Pharmaceutical Development Unit at Novartis Pharmaceutical Corporation, East Hanover, NJ.

Acknowledgments

The authors would like to thank the following persons:

Mr. Jingsong Zhang for providing all the information technology support and for extracting and transforming the data to produce the final presentation for the book.

Dr. Julianne M. Braun for her assistance on the alphabetization of chemical names.

Dr. Wei-Youh Kuu and Dr. Rose-Marie Dannenfelser for the compilation of the data and the early development of this database.

Mrs. Piya Jain for assistance in the compilation of the data.

Acknowledgments

The authors would like to thank the following persons:

... Hong, Ong, Zhang, ... for providing all the photographs, examples for important figures, equations and ... to prepare the dissertation into the whole book.

Dr. Johnson, Dr. ... for assistance on the alphanumeric ... of chemical names.

Drs. Vu, Yuen, Lau and Drs. Rose Marie Pangborn ... for the compilation and ... development of the database.

Mrs. ... for assistance in the compilation of the data.

Introduction

The *Handbook of Aqueous Solubility Data* is an extensive compilation of published data for the solubility of a very wide variety of organic nonelectrolytes and unionized weak electrolytes in water. It includes data for pharmaceuticals, pollutants, nutrients, herbicides, pesticides, agricultural, industrial, and energy-related compounds. This handbook contains over sixteen thousand solubility records for more than four thousand compounds. These data were extracted from about eighteen hundred scientific references, contained in the AQUASOL dATAbASE.

Each compound is identified by a sequential number with molecular formula, compound name, synonyms, molecular weight, Chemical Abstracts Service Registry Number, melting point, and boiling point if available. For user convenience, all solubility data are converted to moles per liter and grams per liter. Also, reported numerical temperature values are converted to centigrade. The following symbols are included in the temperature field when non-numerical temperature descriptors are reported:

amb	ambient temperature
c	cold water
h	hot water
rt	room temperature
ns	temperature not stated

Each record has a five-point evaluation for the reporting of the data and a reference code for the citation. Comments are included when necessary. The following alternatives are used in the comments field:

EFG	estimated from graph
LCST	lower critical solution temperature
UCST	upper critical solution temperature

SOLUBILITY DATA

The compounds are sorted by their molecular formula using the Hill system (number of carbons, number of hydrogens, and then alphabetical by element), and then by name. Each compound can contain up to 5 synonyms. This is followed by the Chemical Abstracts Service Registry Number (RN), melting point (MP) in Celsius, molecular weight (MW), and boiling point (BP) in Celsius. Multiple values are presented whenever available. These are sorted by temperature and then by reference source.

CITATIONS

The reference citation is given as a four-character code, in which the first character is alphabetic (referring to the first author's last name) and the next three are numeric. The complete reference citation is provided in the Reference section.

EVALUATION

As listed in the Table of the Explanation of Evaluation Scores, a five-point evaluation is provided for the quality of the reporting of temperature (T), purity of solute (P), equilibration time/agitation (E), analysis (A), and accuracy and/or precision (A).

Explanation of Evaluation Scores

Parameter		Score		
		0	1	2
T	Temperature	Not given, ambient, or room temp	Given with no range	Given with range
P	Purity of solute	Not stated or as received	Stated with no range or as received	Stated with range or altered with range or calculated
E	Equilibration time/agitation	Not stated	Stated briefly	Described in detail
A	Analysis	Not stated	Stated briefly or stated in other paper	Described in detail
A	Accuracy and/or precision	1 significant figure or range > 20%	2 significant figures or range 5–20%	3 significant figures or range 1–5%

INDICES

Entries in the indices are referenced to the sequential number, not to page numbers. The formulas in Index 3 are sorted by the Hill system.

Separate indices are provided for:

Index 1: Molecular Formula
Index 2: Names and Synonyms
Index 3: Chemical Abstracts Service Registry Number (RN)

Solubility Data

1. CHBrCl$_2$

Bromodichloromethane
Dichlorobromomethane
BDCM

RN: 75-27-4   **MP** (°C): −55
MW: 163.83   **BP** (°C): 87

Solubility (Moles/L)	Solubility (Grams/L)	Temp (°C)	Ref (#)	Evaluation (T P E A A)	Comments
1.851E-02	3.032E+00	30	M300	1 1 2 2 2	
1.812E-02	2.968E+00	30	M311	1 1 2 2 2	

2. CHBr$_2$Cl

Chlorodibromomethane
Dibromochloromethane
CDBM

RN: 124-48-1   **MP** (°C): −22
MW: 208.29   **BP** (°C): 119.5

Solubility (Moles/L)	Solubility (Grams/L)	Temp (°C)	Ref (#)	Evaluation (T P E A A)	Comments
5.041E-03	1.050E+00	30	M300	1 1 2 2 2	
1.205E-02	2.509E+00	30	M311	1 1 2 2 2	

3. CHBr$_3$

Bromoform
Tribromomethane
Methyl tribromide

RN: 75-25-2   **MP** (°C): 7.5
MW: 252.75   BP (°C): 149

Solubility (Moles/L)	Solubility (Grams/L)	Temp (°C)	Ref (#)	Evaluation (T P E A A)	Comments
1.187E-02	3.001E+00	15	G029	1 0 2 2 2	
3.957E-03	1.000E+00	20	F300	1 0 0 0 0	
<7.91E-04	<2.00E-01	25	B019	1 0 1 2 0	*sic*
1.262E-02	3.190E+00	30	F300	1 0 0 0 2	
1.258E-02	3.180E+00	30	G029	1 0 2 2 2	
1.555E-02	3.931E+00	30	M311	1 1 2 2 2	
1.256E-02	3.174E+00	30	V009	1 0 0 0 2	
1.227E-02	3.100E+00	ns	O006	0 0 0 0 2	

4. CHClF$_2$

Chlorodifluoromethane
Freon 22
Halocarbon 22

RN: 75-45-6   **MP** (°C):   −146
MW: 86.47   **BP** (°C):   −40.8

Solubility (Moles/L)	Solubility (Grams/L)	Temp (°C)	Ref (#)	Evaluation (T P E A A)	Comments
3.018E-01	2.610E+01	21	M065	1 0 2 1 2	

5. CHCl$_3$

Chloroform
Trichloromethane
Methyl trichloride
Formyl trichloride

RN: 67-66-3   **MP** (°C):   −63
MW: 119.38   **BP** (°C):   61

Solubility (Moles/L)	Solubility (Grams/L)	Temp (°C)	Ref (#)	Evaluation (T P E A A)	Comments
8.896E-02	1.062E+01	0	H101	2 0 0 0 2	
7.077E-02	8.448E+00	15	G029	1 0 2 2 2	
7.134E-02	8.517E+00	15	J036	0 0 0 0 0	
6.648E-02	7.937E+00	20	E019	1 0 1 1 0	
6.785E-02	8.100E+00	20	F300	1 0 0 0 1	
6.886E-02	8.220E+00	20	H101	2 0 0 0 2	
6.869E-02	8.200E+00	20	M133	1 0 0 0 2	
6.827E-02	8.150E+00	20	M368	1 0 0 0 1	
6.648E-02	7.937E+00	20	N034	1 0 0 0 0	
6.869E-02	8.200E+00	20	P046	1 0 0 0 0	
6.750E-02	8.058E+00	20	P073	1 0 0 1 2	
3.504E-02	4.182E+00	22	H072	1 0 1 1 2	
7.472E-02	8.920E+00	25	B019	1 0 1 2 0	
6.050E-02	7.222E+00	25	B173	2 0 2 2 2	
6.660E-02	7.950E+00	25	F071	1 1 2 1 2	
6.648E-02	7.937E+00	25	G056	1 0 0 0 2	
6.813E-02	8.133E+00	25	L319	1 0 2 1 2	
6.618E-02	7.900E+00	25	M037	1 1 0 0 1	
6.648E-02	7.937E+00	25	O026	1 2 0 1 0	
7.472E-02	8.920E+00	25	R321	1 2 1 1 1	
6.236E-02	7.444E+00	25.0	C055	1 2 1 0 1	
6.409E-02	7.651E+00	30	G029	1 0 2 2 2	
6.500E-02	7.760E+00	30	H101	2 0 0 0 2	
2.114E-02	2.524E+00	30	M311	1 1 2 2 2	
6.411E-02	7.653E+00	30	V009	1 0 0 0 2	
6.648E-02	7.937E+00	56.1	C055	2 2 1 0 0	
6.236E-02	7.444E+00	60	R321	1 2 1 1 1	
6.660E-02	7.950E+00	ns	H123	0 0 0 0 0	

(*continued*)

5. $CHCl_3$ (continued)

Solubility (Moles/L)	Solubility (Grams/L)	Temp (°C)	Ref (#)	Evaluation (T P E A A)	Comments
4.168E-02	4.975E+00	ns	I306	0 0 0 0 0	
6.660E-02	7.950E+00	ns	M344	0 0 0 0 2	
6.830E-02	8.153E+00	ns	R028	0 0 0 0 0	

6. CHI_3
Iodoform
Triiodomethane

RN: 75-47-8   **MP** (°C): 121.5
MW: 393.73   **BP** (°C): 218

Solubility (Moles/L)	Solubility (Grams/L)	Temp (°C)	Ref (#)	Evaluation (T P E A A)	Comments
3.000E-04	1.181E-01	25	V009	1 0 0 0 0	
2.540E-04	9.999E-02	rt	D021	0 0 1 1 0	

7. CH_2BrCl
Bromochloromethane
Bromo-chloro-methane
Chlorobromomethane
CBM

RN: 74-97-5   **MP** (°C): −86.5
MW: 129.39   **BP** (°C): 68.1

Solubility (Moles/L)	Solubility (Grams/L)	Temp (°C)	Ref (#)	Evaluation (T P E A A)	Comments
1.290E-01	1.669E+01	25	M342	1 0 1 1 2	
1.142E-01	1.478E+01	ns	O006	0 0 0 0 1	

8. CH_2Br_2
Methylene bromide
Dibrom-methan

RN: 74-95-3   **MP** (°C): −52.7
MW: 173.85   **BP** (°C): 97

Solubility (Moles/L)	Solubility (Grams/L)	Temp (°C)	Ref (#)	Evaluation (T P E A A)	Comments
6.747E-02	1.173E+01	0	H101	2 0 0 0 2	
6.652E-02	1.156E+01	15	G029	1 0 2 2 2	
6.604E-02	1.148E+01	20	H101	2 0 0 0 2	
6.259E-02	1.088E+01	25	O006	1 0 0 0 1	
6.782E-02	1.179E+01	30	G029	1 0 2 2 2	

(continued)

8. CH_2Br_2 (continued)

Solubility (Moles/L)	Solubility (Grams/L)	Temp (°C)	Ref (#)	Evaluation (T P E A A)	Comments
6.765E-02	1.176E+01	30	H101	2 0 0 0 2	
6.779E-02	1.179E+01	30	V009	1 0 0 0 2	
6.558E-02	1.140E+01	ns	F300	0 0 0 0 2	

9. CH_2Cl_2
Methylene chloride
Dichlor-methan
Dichloromethane
Methylene dichloride
Methane dichloride

RN: 75-09-2   **MP** (°C):   −95.1
MW: 84.93   **BP** (°C):   39.8

Solubility (Moles/L)	Solubility (Grams/L)	Temp (°C)	Ref (#)	Evaluation (T P E A A)	Comments
2.782E-01	2.363E+01	0	H101	2 0 0 0 2	
2.309E-01	1.961E+01	20	C057	0 0 0 0 0	
2.355E-01	2.000E+01	20	F300	1 0 0 0 0	
2.355E-01	2.000E+01	20	H101	2 0 0 0 2	
2.263E-01	1.922E+01	20	N034	1 0 0 0 2	
1.887E-01	1.603E+01	20	N038	1 0 0 1 2	
2.309E-01	1.961E+01	25	A094	1 0 0 0 1	
1.534E-01	1.303E+01	25	G056	1 0 0 0 2	
1.554E-01	1.320E+01	25	M037	1 1 0 0 2	
1.554E-01	1.320E+01	25	M133	1 0 0 0 2	
1.554E-01	1.320E+01	25	P046	1 0 0 0 0	
2.275E-01	1.932E+01	30	V009	1 0 0 0 2	
2.284E-01	1.940E+01	ns	H123	0 0 0 0 0	

10. CH_2I_2
Methylene iodide
Diiod-methan

RN: 75-11-6   **MP** (°C):   6.0
MW: 267.84   **BP** (°C):   181

Solubility (Moles/L)	Solubility (Grams/L)	Temp (°C)	Ref (#)	Evaluation (T P E A A)	Comments
3.110E-03	8.330E-01	25	A032	1 2 1 1 2	
4.624E-03	1.238E+00	30	G029	1 0 2 2 2	
4.594E-03	1.231E+00	30	V009	1 0 0 0 1	

11. CH_2N_2
Cyanamide
Cyanamid

RN: 420-04-2	**MP** (°C):
MW: 42.04	**BP** (°C):

Solubility (Moles/L)	Solubility (Grams/L)	Temp (°C)	Ref (#)	Evaluation (T P E A A)	Comments
1.057E+01	4.444E+02	ns	N013	0 0 0 0 1	

12. CH_3Br
Methyl bromide
Bromomethane
Celfume

RN: 74-83-9	**MP** (°C): −94
MW: 94.94	**BP** (°C): 3.56

Solubility (Moles/L)	Solubility (Grams/L)	Temp (°C)	Ref (#)	Evaluation (T P E A A)	Comments
2.748E-01	2.609E+01	10	H081	1 0 2 0 2	
1.893E-01	1.797E+01	17	H081	1 0 2 0 2	
1.893E-01	1.797E+01	17	M061	1 0 0 0 2	
1.933E-01	1.835E+01	19.9	G061	1 2 1 1 2	774.3mm Hg @ 25 °C
1.685E-01	1.600E+01	20	G080	1 0 0 0 1	
1.659E-01	1.575E+01	20	P081	1 0 0 0 1	
1.394E-01	1.323E+01	25	H081	1 0 2 0 2	
1.411E-01	1.340E+01	25	M161	1 0 0 0 2	
1.196E-01	1.136E+01	32	H081	1 0 2 0 2	
9.479E-03	9.000E-01	ns	N013	0 0 0 0 1	

13. $CH_3BrO_6S_2$
Bromomethionic acid
Methanedisulfonic acid, bromo-

RN: 187610-86-2	**MP** (°C):
MW: 255.07	**BP** (°C):

Solubility (Moles/L)	Solubility (Grams/L)	Temp (°C)	Ref (#)	Evaluation (T P E A A)	Comments
3.039E+00	7.752E+02	25	B077	1 2 0 0 2	

14. CH₃Cl

14. CH_3Cl

Methyl chloride
Chloromethane

RN:	74-87-3	MP (°C):	−97.0
MW:	50.49	BP (°C):	−23.7

Solubility (Moles/L)	Solubility (Grams/L)	Temp (°C)	Ref (#)	Evaluation (T P E A A)	Comments
1.531E+01	7.727E+02	0	M061	1 0 0 0 1	*sic*
1.436E-01	7.250E+00	20	M133	1 0 0 0 2	
9.069E-02	4.579E+00	20	N034	1 0 0 0 1	
1.436E-01	7.250E+00	20	P046	1 0 0 0 0	
1.059E-01	5.347E+00	24.9	G061	1 2 1 1 2	756.1mm Hg @ 25 °C
1.455E-01	7.346E+00	30	G056	1 0 0 0 2	
1.466E-01	7.400E+00	30	M037	1 1 0 0 1	

15. CH₃ClO₆S₂

15. $CH_3ClO_6S_2$

Chloromethionic acid
Acide chloromethionique

RN:	74692-14-1	MP (°C):	
MW:	210.61	BP (°C):	

Solubility (Moles/L)	Solubility (Grams/L)	Temp (°C)	Ref (#)	Evaluation (T P E A A)	Comments
1.540E+01	3.243E+03	25	B075	1 2 0 0 2	

16. CH₃F

16. CH_3F

Fluoromethane
Methylfluoride

RN:	593-53-3	MP (°C):	−141.8
MW:	34.03	BP (°C):	−78.2

Solubility (Moles/L)	Solubility (Grams/L)	Temp (°C)	Ref (#)	Evaluation (T P E A A)	Comments
~7.05E-02	~2.40E+00	15	F300	1 0 0 0 0	
5.250E-02	1.787E+00	29.9	G061	1 2 1 1 2	766.8mm Hg @25 °C

17. CH₃I
Iodomethane
Methyl-iodide
Halon 10001
Methyl iodine
Methyliodide

RN: 74-88-4   **MP** (°C): −64
MW: 141.94   **BP** (°C): 42

Solubility (Moles/L)	Solubility (Grams/L)	Temp (°C)	Ref (#)	Evaluation (T P E A A)	Comments
1.103E-01	1.565E+01	0	H101	2 0 0 0 2	
9.997E-02	1.419E+01	20	H101	2 0 0 0 2	
9.727E-02	1.381E+01	20	H127	1 0 0 0 1	
9.727E-02	1.381E+01	20	I316	0 0 0 0 0	
9.600E-02	1.363E+01	20	M171	1 0 0 0 2	
9.590E-02	1.361E+01	22	F001	1 0 1 2 2	
9.511E-02	1.350E+01	22	F300	1 0 0 0 2	
9.590E-02	1.361E+01	22	S006	1 0 0 0 2	
1.007E-01	1.429E+01	30	H101	2 0 0 0 2	
9.957E-02	1.413E+01	30	V009	1 0 0 0 2	
8.725E-03	1.238E+00	ns	O006	0 0 0 0 1	

18. CH₃NO
Formaldehyde oxime
Formaldehyd-oxim

RN: 75-17-2   **MP** (°C):
MW: 45.04   **BP** (°C):

Solubility (Moles/L)	Solubility (Grams/L)	Temp (°C)	Ref (#)	Evaluation (T P E A A)	Comments
3.774E+00	1.700E+02	20	F300	1 0 0 0 1	

19. CH₃NO₂
Nitromethane
Nitrocarbol
NM

RN: 75-52-5   **MP** (°C): −29
MW: 61.04   **BP** (°C): 101

Solubility (Moles/L)	Solubility (Grams/L)	Temp (°C)	Ref (#)	Evaluation (T P E A A)	Comments
1.421E+00	8.676E+01	20	C121	0 0 0 0 1	unit assumed, *sic*
1.627E+00	9.934E+01	25	F049	2 0 2 0 0	
1.802E+00	1.100E+02	25	M136	2 0 0 0 2	
1.802E+00	1.100E+02	25	M139	2 0 0 0 2	
3.039E-01	1.855E+01	ns	D348	0 0 0 0 0	

20. CH_3N_5
5-Aminotetrazole
5-Amino-tetrazol

RN:	4418-61-5	**MP** (°C):	204	
MW:	85.07	**BP** (°C):		

Solubility (Moles/L)	Solubility (Grams/L)	Temp (°C)	Ref (#)	Evaluation (T P E A A)	Comments
1.411E-01	1.200E+01	18	F300	1 0 0 0 1	

21. CH_4
Methane
Methan

RN:	74-82-8	**MP** (°C):	−183	
MW:	16.04	**BP** (°C):	−161	

Solubility (Moles/L)	Solubility (Grams/L)	Temp (°C)	Ref (#)	Evaluation (T P E A A)	Comments
2.468E-03	3.960E-02	0	F300	1 0 0 0 2	
2.210E-03	3.545E-02	4.99	C115	2 0 2 2 2	
1.926E-03	3.090E-02	9.99	C115	2 0 2 2 2	
1.633E-03	2.620E-02	14.99	C115	2 0 2 2 2	
1.567E-03	2.513E-02	19.8	G058	1 0 0 0 2	
1.511E-03	2.424E-02	19.99	C115	2 0 2 2 2	
1.446E-03	2.320E-02	20	F300	1 0 0 0 2	
1.381E-03	2.215E-02	24.99	C115	2 0 2 2 2	
1.521E-03	2.440E-02	25	M001	2 1 2 2 2	
1.521E-03	2.440E-02	25	M002	2 2 1 2 2	
1.502E-03	2.410E-02	25	M040	1 0 0 1 2	
1.550E-03	2.487E-02	25	M102	1 2 2 1 2	
1.266E-03	2.030E-02	29.99	C115	2 0 2 2 2	
1.189E-03	1.907E-02	34.99	C115	2 0 2 2 2	
1.079E-03	1.732E-02	39.99	C115	2 0 2 2 2	
1.056E-03	1.693E-02	40	S212	2 1 2 2 2	
1.055E-03	1.693E-02	44.99	C115	2 0 2 2 2	
8.477E-04	1.360E-02	50	F300	1 0 0 0 2	
9.000E-04	1.444E-02	60	S212	2 1 2 2 2	
8.000E-04	1.283E-02	80	S212	2 1 2 2 2	
1.434E-03	2.300E-02	ns	M091	0 1 0 0 2	
1.378E-03	2.210E-02	ns	S212	2 1 2 2 2	

22. CH₄N₂O

Urea
Harnstoff
Uree

RN: 57-13-6	**MP** (°C): 132.7			
MW: 60.06	**BP** (°C):			

Solubility (Moles/L)	Solubility (Grams/L)	Temp (°C)	Ref (#)	Evaluation (T P E A A)	Comments
6.680E+00	4.012E+02	0	F300	1 0 0 0 2	
4.757E+00	2.857E+02	0	J021	1 0 0 0 2	
6.680E+00	4.012E+02	0	M043	1 0 0 0 1	
6.680E+00	4.012E+02	0	P023	1 2 1 1 2	
7.297E+00	4.382E+02	5	D041	1 0 0 0 1	
5.088E+00	3.056E+02	7	J021	1 0 0 0 2	
5.246E+00	3.151E+02	10	D020	1 2 1 1 2	
5.246E+00	3.151E+02	10	D060	2 2 1 1 2	
7.651E+00	4.595E+02	10	M043	1 0 0 0 1	
7.602E+00	4.565E+02	10	P023	1 2 1 1 2	
5.550E+00	3.333E+02	17	J021	1 0 0 0 2	
7.382E+00	4.433E+02	18.72	S131	2 2 1 1 2	recrystallized
5.536E+00	3.324E+02	20	C052	1 2 1 1 2	
5.617E+00	3.373E+02	20	J021	1 0 0 0 2	
8.529E+00	5.122E+02	20	M043	1 0 0 0 2	
8.517E+00	5.115E+02	20	P023	1 2 1 1 2	
7.594E+00	4.561E+02	21.59	S131	2 2 1 1 2	recrystallized
7.738E+00	4.647E+02	23.85	S131	2 2 1 1 2	recrystallized
5.874E+00	3.528E+02	25	D020	1 2 1 1 2	
9.058E+00	5.440E+02	25	D041	1 0 0 0 2	
5.874E+00	3.528E+02	25	D060	2 2 1 1 2	
8.326E+00	5.000E+02	25	M136	2 0 0 0 2	
7.910E+00	4.750E+02	26.83	S131	2 2 1 1 2	recrystallized
7.966E+00	4.784E+02	27.31	S131	2 2 1 1 2	recrystallized
9.566E+00	5.745E+02	30	M043	1 0 0 0 2	
9.596E+00	5.763E+02	30	P023	1 2 1 1 2	
8.171E+00	4.907E+02	30.38	S131	2 2 1 1 2	recrystallized
6.244E+00	3.750E+02	35	J021	1 0 0 0 2	
1.712E+01	1.028E+03	35	S200	1 0 0 0 2	loc. cit.
8.469E+00	5.086E+02	35.15	S131	2 2 1 1 2	recrystallized
8.465E+00	5.083E+02	35.42	S131	2 2 1 1 2	recrystallized
8.575E+00	5.150E+02	37.36	S131	2 2 1 1 2	recrystallized
1.038E+01	6.232E+02	39.7	P023	1 2 1 1 2	
6.392E+00	3.839E+02	40	D020	1 2 1 1 2	
6.392E+00	3.839E+02	40	D060	2 2 1 1 2	
1.037E+01	6.226E+02	40	M043	1 0 0 0 2	
1.837E+01	1.103E+03	40	S200	1 0 0 0 2	loc. cit.
8.822E+00	5.298E+02	41.11	S131	2 2 1 1 2	recrystallized
8.982E+00	5.394E+02	43.85	S131	2 2 1 1 2	recrystallized
8.967E+00	5.386E+02	43.94	S131	2 2 1 1 2	recrystallized
1.961E+01	1.178E+03	45	S200	1 0 0 0 2	loc. cit.
9.107E+00	5.469E+02	46.56	S131	2 2 1 1 2	recrystallized
1.119E+01	6.721E+02	50	P023	1 2 1 1 2	

(continued)

22. CH_4N_2O (continued)

Solubility (Moles/L)	Solubility (Grams/L)	Temp (°C)	Ref (#)	Evaluation (T P E A A)	Comments
2.109E+01	1.267E+03	50	S200	1 0 0 0 2	loc. cit.
1.122E+01	6.736E+02	50.6	P023	1 2 1 1 2	
9.560E+00	5.742E+02	54.77	S131	2 2 1 1 2	recrystallized
9.584E+00	5.756E+02	54.97	S131	2 2 1 1 2	recrystallized
2.283E+01	1.371E+03	55	S200	1 0 0 0 2	loc. cit.
9.649E+00	5.795E+02	55.88	S131	2 2 1 1 2	recrystallized
9.681E+00	5.814E+02	57.02	S131	2 2 1 1 2	recrystallized
9.806E+00	5.889E+02	59.13	S131	2 2 1 1 2	recrystallized
6.936E+00	4.166E+02	60	J021	1 0 0 0 2	
9.847E+00	5.914E+02	60	K013	1 0 1 1 2	
1.189E+01	7.143E+02	60	M043	1 0 0 0 2	
2.422E+01	1.455E+03	60	S200	1 0 0 0 2	loc. cit.
1.184E+01	7.110E+02	60.0	P023	1 2 1 1 2	
9.930E+00	5.963E+02	61.76	S131	2 2 1 1 2	recrystallized
1.005E+01	6.034E+02	63.79	S131	2 2 1 1 2	recrystallized
1.009E+01	6.060E+02	65	K013	1 0 1 1 2	
2.570E+01	1.543E+03	65	S200	1 0 0 0 2	loc. cit.
1.244E+01	7.468E+02	68.5	P023	1 2 1 1 2	
1.020E+01	6.127E+02	68.50	M059	1 1 2 1 2	
1.270E+01	7.629E+02	70	F300	1 0 0 0 2	
7.206E+00	4.328E+02	70	J021	1 0 0 0 2	
1.033E+01	6.206E+02	70	K013	1 0 1 1 2	
1.263E+01	7.588E+02	70	P023	1 2 1 1 2	
2.730E+01	1.640E+03	70	S200	1 0 0 0 2	loc. cit.
1.038E+01	6.231E+02	70.49	S131	2 2 1 1 2	recrystallized
1.048E+01	6.295E+02	73.11	S131	2 2 1 1 2	recrystallized
1.057E+01	6.345E+02	75	K013	1 0 1 1 2	
1.048E+01	6.296E+02	75.30	M059	1 1 2 1 2	
1.079E+01	6.480E+02	80	K013	1 0 1 1 2	
1.332E+01	8.000E+02	80	M043	1 0 0 0 2	
1.090E+01	6.546E+02	84.40	M059	1 1 2 1 2	
1.101E+01	6.610E+02	85	K013	1 0 1 1 2	
3.229E+01	1.939E+03	85	S200	1 0 0 0 2	loc. cit.
1.122E+01	6.738E+02	90	K013	1 0 1 1 2	
3.426E+01	2.058E+03	90	S200	1 0 0 0 2	loc. cit.
1.131E+01	6.791E+02	93.80	M059	1 1 2 1 2	
1.142E+01	6.858E+02	95	K013	1 0 1 1 2	
3.611E+01	2.169E+03	95	S200	1 0 0 0 2	loc. cit.
1.161E+01	6.975E+02	100	K013	1 0 1 1 2	
1.465E+01	8.795E+02	100	M043	1 0 0 0 2	
3.778E+01	2.269E+03	100	S200	1 0 0 0 2	loc. cit.
1.177E+01	7.066E+02	104.40	M059	1 1 2 1 2	
1.199E+01	7.199E+02	109.90	M059	1 1 2 1 2	
1.219E+01	7.321E+02	115.30	M059	1 1 2 1 2	
1.229E+01	7.383E+02	118.30	M059	1 1 2 1 2	
1.234E+01	7.411E+02	118.70	M059	1 1 2 1 2	
1.245E+01	7.479E+02	121.90	M059	1 1 2 1 2	
1.249E+01	7.503E+02	123.20	M059	1 1 2 1 2	
1.264E+01	7.592E+02	127.50	M059	1 1 2 1 2	

(continued)

22. CH$_4$N$_2$O (continued)

Solubility (Moles/L)	Solubility (Grams/L)	Temp (°C)	Ref (#)	Evaluation (T P E A A)	Comments
1.269E+01	7.619E+02	128.80	M059	1 1 2 1 2	
1.281E+01	7.694E+02	132.60	M059	1 1 2 1 2	
1.665E+01	1.000E+03	ns	B338	0 0 0 0 1	
1.332E+01	8.000E+02	ns	D072	0 0 0 0 0	

23. CH$_4$N$_2$S
Thiourea
Thiouree

RN:	62-56-6	**MP** (°C):	176	
MW:	76.12	**BP** (°C):		

Solubility (Moles/L)	Solubility (Grams/L)	Temp (°C)	Ref (#)	Evaluation (T P E A A)	Comments
6.136E-01	4.671E+01	0	M043	1 0 0 0 1	
9.731E-01	7.407E+01	10	M043	1 0 0 0 1	
1.118E+00	8.507E+01	10	O017	1 0 1 1 2	
1.206E+00	9.180E+01	13	F300	1 0 0 0 2	
1.206E+00	9.179E+01	13	O019	1 0 0 1 2	
1.383E+00	1.053E+02	15	O017	1 0 1 1 2	
1.573E+00	1.197E+02	20	M043	1 0 0 0 2	
1.544E+00	1.175E+02	20	O017	1 0 1 1 2	
1.085E+00	8.257E+01	25	I310	0 0 0 0 0	
1.759E+00	1.339E+02	25	O017	1 0 1 1 2	
2.199E+00	1.674E+02	30	M043	1 0 0 0 2	
3.093E+00	2.355E+02	40	M043	1 0 0 0 2	
5.455E+00	4.152E+02	60	M043	1 0 0 0 1	
7.617E+00	5.798E+02	80	M043	1 0 0 0 2	
9.250E+00	7.041E+02	100	M043	1 0 0 0 2	
7.882E-01	6.000E+01	ns	D072	0 0 0 0 0	

24. CH$_4$N$_4$O$_2$
α-Nitroguanidine
Nitroguanidine
Nitroguanidin

RN:	556-88-7	**MP** (°C):	235	
MW:	104.07	**BP** (°C):		

Solubility (Moles/L)	Solubility (Grams/L)	Temp (°C)	Ref (#)	Evaluation (T P E A A)	Comments
2.597E-02	2.703E+00	19.5	D027	1 2 0 0 2	
1.173E-01	1.221E+01	25	D022	1 1 2 2 2	
4.228E-02	4.400E+00	25	F300	1 0 0 0 1	
4.305E-02	4.480E+00	29.87	M028	1 2 2 1 0	EFG
1.122E-01	1.167E+01	50	D027	1 2 0 0 2	
3.070E-01	3.195E+01	71.67	M028	1 2 2 1 0	EFG
5.695E-01	5.927E+01	83.98	M028	1 2 2 1 0	EFG
9.025E-01	9.392E+01	100	D027	1 2 0 0 2	
7.620E-01	7.930E+01	100	F300	1 0 0 0 2	

25. CH$_4$O
Methanol
Methyl alcohol

RN: 67-56-1   **MP** (°C):   −97.8
MW: 32.04   **BP** (°C):   64.7

Solubility (Moles/L)	Solubility (Grams/L)	Temp (°C)	Ref (#)	Evaluation (T P E A A)	Comments
1.689E+01	5.411E+02	ns	L003	0 0 2 1 2	

26. CH$_4$O$_6$S$_2$
Methionic acid
Acide methionique
Methanedisulfonic acid

RN: 503-40-2   **MP** (°C):   98.0
MW: 176.17   **BP** (°C):

Solubility (Moles/L)	Solubility (Grams/L)	Temp (°C)	Ref (#)	Evaluation (T P E A A)	Comments
1.395E+01	2.458E+03	25	B075	1 2 0 0 2	
4.035E+00	7.108E+02	25	B076	1 2 0 0 2	
4.862E+00	8.566E+02	25	F300	1 0 0 0 2	

27. CH$_4$O$_6$S$_2$.H$_2$O
Methionic acid (monohydrate)

RN: 503-40-2   **MP** (°C):
MW: 194.18   **BP** (°C):

Solubility (Moles/L)	Solubility (Grams/L)	Temp (°C)	Ref (#)	Evaluation (T P E A A)	Comments
4.409E+00	8.562E+02	25	B076	1 2 0 0 2	

28. CH$_5$N
Methylamine
Aminomethane
Carbinamine
Mercurialin

RN: 74-89-5   **MP** (°C):   −93.5
MW: 31.06   **BP** (°C):   −6.3

Solubility (Moles/L)	Solubility (Grams/L)	Temp (°C)	Ref (#)	Evaluation (T P E A A)	Comments
1.906E+01	5.920E+02	4.50	F300	1 0 0 0 2	
2.963E+01	9.202E+02	12.5	D041	1 0 0 0 2	
2.147E+01	6.667E+02	12.50	M081	1 0 0 0 2	
1.916E+01	5.951E+02	20	M081	1 0 0 0 2	

(continued)

28. CH$_5$N (continued)

Solubility (Moles/L)	Solubility (Grams/L)	Temp (°C)	Ref (#)	Evaluation (T P E A A)	Comments
1.789E+01	5.556E+02	25	M081	1 0 0 0 2	
1.664E+01	5.169E+02	30	M081	1 0 0 0 2	
1.380E+01	4.286E+02	40	M081	1 0 0 0 1	
1.143E+01	3.548E+02	50	M081	1 0 0 0 1	
9.034E+00	2.806E+02	60	M081	1 0 0 0 1	

29. CH$_5$N$_5$O$_2$

Nitroaminoguanidine
Hydrazinecarboximidamide, *N*-nitro-
1-Amino-3-nitroguanidine
3-Amino-1-nitroguanidine
1-Amino-2-nitroguanidine
1-Nitro-3-aminoguanidine

RN: 18264-75-0   **MP** (°C): 185
MW: 119.08   **BP** (°C):

Solubility (Moles/L)	Solubility (Grams/L)	Temp (°C)	Ref (#)	Evaluation (T P E A A)	Comments
1.360E-02	1.619E+00	9.33	M047	2 2 1 1 0	EFG
2.254E-02	2.684E+00	20.96	M047	2 2 1 1 0	EFG
3.567E-02	4.248E+00	29.87	M047	2 2 1 1 0	EFG
4.384E-02	5.221E+00	34.53	M047	2 2 1 1 0	EFG
7.087E-02	8.440E+00	44.30	M047	2 2 1 1 0	EFG
9.318E-02	1.110E+01	49.42	M047	2 2 1 1 0	EFG

30. CH$_5$O$_3$As

Methanearsonic acid
MAA
Methylarsonsaeure

RN: 124-58-3   **MP** (°C): 132
MW: 139.97   **BP** (°C):

Solubility (Moles/L)	Solubility (Grams/L)	Temp (°C)	Ref (#)	Evaluation (T P E A A)	Comments
1.456E+00	2.038E+02	20	B200	1 0 0 0 2	
1.563E+00	2.188E+02	25	D305	1 0 0 0 1	

31. CH$_5$As
Methylarsine
Methylarsin
RN: 593-52-2   **MP** (°C):   −143
MW: 91.97   **BP** (°C):   2

Solubility (Moles/L)	Solubility (Grams/L)	Temp (°C)	Ref (#)	Evaluation (T P E A A)	Comments
9.242E-04	8.500E-02	20	F300	1 0 0 0 1	

32. CBrClF$_2$
Bromochlorodifluoromethane
Halon 1211
Chlorodifluorobromomethane
Bromochlorodifluoromethine
RN: 353-59-3   **MP** (°C):
MW: 165.37   **BP** (°C):

Solubility (Moles/L)	Solubility (Grams/L)	Temp (°C)	Ref (#)	Evaluation (T P E A A)	Comments
9.555E-05	1.580E-02	0	G055	1 2 2 2 1	

33. CBr$_3$F
Tribromo-fluoro-methane
Methane, tribromofluoro-
Fluorotribromomethane
RN: 353-54-8   **MP** (°C):
MW: 270.74   **BP** (°C):

Solubility (Moles/L)	Solubility (Grams/L)	Temp (°C)	Ref (#)	Evaluation (T P E A A)	Comments
1.477E-03	3.998E-01	25	O006	1 0 0 0 1	

34. CBr$_4$
Carbon tetrabromide
Tetrabromomethane
RN: 558-13-4   **MP** (°C):   89
MW: 331.65   **BP** (°C):

Solubility (Moles/L)	Solubility (Grams/L)	Temp (°C)	Ref (#)	Evaluation (T P E A A)	Comments
7.235E-04	2.399E-01	30	G029	1 0 2 2 1	
6.998E-04	2.321E-01	30	V009	1 0 0 0 0	

35. CClN
Cyanogen chloride
Chlorcyan

RN: 506-77-4   **MP** (°C):   −6
MW: 61.47   **BP** (°C):   13.8

Solubility (Moles/L)	Solubility (Grams/L)	Temp (°C)	Ref (#)	Evaluation (T P E A A)	Comments
9.761E-01	6.000E+01	0	F300	1 0 0 0 0	

36. CClN$_3$O$_6$
Chlorotrinitromethane
Chlor-trinitro-methan

RN: 1943-16-4   **MP** (°C):
MW: 185.48   **BP** (°C):

Solubility (Moles/L)	Solubility (Grams/L)	Temp (°C)	Ref (#)	Evaluation (T P E A A)	Comments
1.186E-02	2.200E+00	20	F300	1 0 0 0 1	

37. CCl$_2$F$_2$
Dichlorodifluoromethane
Difluorodichloromethane
Freon 12

RN: 75-71-8   **MP** (°C):   −158
MW: 120.91   **BP** (°C):   −29.8

Solubility (Moles/L)	Solubility (Grams/L)	Temp (°C)	Ref (#)	Evaluation (T P E A A)	Comments
1.544E-02	1.867E+00	21	M065	1 0 2 1 2	
2.316E-03	2.800E-01	25	M133	1 0 0 0 2	
2.316E-03	2.800E-01	25	P046	1 0 0 0 0	
2.315E-03	2.799E-01	25	R048	0 0 0 0 0	

38. CCl$_3$F
Trichlorofluoromethane
Fluorotrichloromethane
Freon 11

RN: 75-69-4   **MP** (°C):   −111
MW: 137.37   **BP** (°C):   23.7

Solubility (Moles/L)	Solubility (Grams/L)	Temp (°C)	Ref (#)	Evaluation (T P E A A)	Comments
1.020E-02	1.401E+00	20	H041	0 0 0 0 0	
8.008E-03	1.100E+00	20	M133	1 0 0 0 2	
8.008E-03	1.100E+00	20	P046	1 0 0 0 0	
1.020E-02	1.401E+00	21	H041	0 0 0 0 0	

(continued)

38. CCl₃F (continued)

Solubility (Moles/L)	Solubility (Grams/L)	Temp (°C)	Ref (#)	Evaluation (T P E A A)	Comments
8.013E-03	1.101E+00	25	H041	0 0 0 0 0	
7.999E-03	1.099E+00	25	R048	0 0 0 0 0	
7.997E-03	1.099E+00	27	H041	0 0 0 0 0	
7.853E-03	1.079E+00	30	H041	0 0 0 0 0	
9.892E-03	1.359E+00	31	H041	0 0 0 0 0	
4.152E-03	5.703E-01	50	H041	0 0 0 0 0	
2.258E-03	3.102E-01	75	H041	0 0 0 0 0	

39. CCl₃NO₂
Chloropicrin
Chlorpikrin

RN:	76-06-2	**MP** (°C):	−64	
MW:	164.38	**BP** (°C):	112	

Solubility (Moles/L)	Solubility (Grams/L)	Temp (°C)	Ref (#)	Evaluation (T P E A A)	Comments
1.381E-02	2.270E+00	0	M161	1 0 0 0 2	
1.396E-02	2.295E+00	20	C121	1 0 0 0 1	unit assumed, *sic*
1.186E-02	1.950E+00	20	G080	1 0 0 0 1	
9.718E-03	1.597E+00	20	M061	1 0 0 0 1	
1.214E-02	1.996E+00	20	P081	1 0 0 0 0	
9.874E-03	1.623E+00	25	F300	1 0 0 0 2	
1.217E-02	2.000E+00	ns	N013	0 0 0 0 2	

40. CCl₄
Carbon tetrachloride
Tetrachloromethane
Methane tetrachloride

RN:	56-23-5	**MP** (°C):	−23	
MW:	153.82	**BP** (°C):	76.7	

Solubility (Moles/L)	Solubility (Grams/L)	Temp (°C)	Ref (#)	Evaluation (T P E A A)	Comments
6.306E-03	9.700E-01	0	H101	2 0 0 0 1	
5.002E-03	7.694E-01	15	G029	1 0 2 2 1	
5.002E-03	7.694E-01	15	J036	0 0 0 0 0	
5.197E-03	7.994E-01	20	C121	1 0 0 0 0	unit assumed, *sic*
5.201E-03	8.000E-01	20	H101	2 0 0 0 1	
5.201E-03	8.000E-01	20	M040	1 0 0 1 2	
5.103E-03	7.850E-01	20	M133	1 0 0 0 2	
5.200E-03	7.999E-01	20	M312	1 0 0 0 2	
4.612E-03	7.095E-01	20	N038	1 0 0 1 2	
5.103E-03	7.850E-01	20	P046	1 0 0 0 0	

(continued)

40. CCl₄ (continued)

Solubility (Moles/L)	Solubility (Grams/L)	Temp (°C)	Ref (#)	Evaluation (T P E A A)	Comments
6.494E-03	9.990E-01	25	B019	1 0 1 2 0	
4.920E-03	7.568E-01	25	B173	2 0 2 2 2	
5.000E-03	7.691E-01	25	G038	1 2 2 2 1	
5.000E-03	7.691E-01	25	G053	2 1 2 1 1	
5.197E-03	7.994E-01	25	G056	1 0 0 0 2	
5.197E-03	7.994E-01	25	L319	1 0 2 1 1	
5.201E-03	8.000E-01	25	M037	1 1 0 0 0	
5.197E-03	7.994E-01	25	M061	1 0 0 0 0	
1.820E-03	2.800E-01	25	M161	1 0 0 0 1	
5.006E-03	7.700E-01	25	M368	1 0 0 0 1	
1.038E-02	1.597E+00	25	N034	1 0 0 0 1	*sic*
5.556E-03	8.546E-01	25	S133	1 1 1 1 1	
5.262E-03	8.093E-01	30	G029	1 0 2 2 1	
5.526E-03	8.500E-01	30	H101	2 0 0 0 1	
5.296E-03	8.146E-01	30	V009	1 0 0 0 1	
5.201E-03	8.000E-01	ns	F071	0 1 2 1 2	
5.201E-03	8.000E-01	ns	H080	0 0 0 0 2	
3.249E-03	4.998E-01	ns	I306	0 0 0 0 0	
5.201E-03	8.000E-01	ns	M344	0 0 0 0 2	

41. CF₄
Carbon tetrafluoride
Tetrafluoromethane

RN: 75-73-0 **MP** (°C): −184
MW: 88.00 **BP** (°C): −128

Solubility (Moles/L)	Solubility (Grams/L)	Temp (°C)	Ref (#)	Evaluation (T P E A A)	Comments
2.319E-04	2.041E-02	19.99	C115	2 0 2 2 2	
2.083E-04	1.833E-02	24.99	C115	2 0 2 2 2	
2.111E-04	1.858E-02	25	D055	1 0 0 0 1	
1.940E-04	1.707E-02	29.99	C115	2 0 2 2 2	

42. COS
Carbonyl sulfide
Kohlenoxidsulfid

RN: 463-58-1 **MP** (°C): −138
MW: 60.07 **BP** (°C): −50

Solubility (Moles/L)	Solubility (Grams/L)	Temp (°C)	Ref (#)	Evaluation (T P E A A)	Comments
6.259E-02	3.760E+00	0	F300	1 0 0 0 2	
2.081E-02	1.250E+00	25	F300	1 0 0 0 2	

43. CO_2
Carbon dioxide
Carbonic acid gas
Carbonic anhydride
RN: 124-38-9   **MP** (°C):   –57
MW: 44.01   **BP** (°C):

Solubility (Moles/L)	Solubility (Grams/L)	Temp (°C)	Ref (#)	Evaluation (T P E A A)	Comments
8.641E-02	3.803E+00	16	B109	1 0 0 0 2	unit assumed, *sic*
8.377E-02	3.687E+00	17	B109	1 0 0 0 2	unit assumed, *sic*
8.641E-02	3.803E+00	18	B109	1 0 0 0 2	unit assumed, *sic*
8.123E-02	3.575E+00	18	B109	1 0 0 0 2	unit assumed, *sic*
7.886E-02	3.471E+00	19	B109	1 0 0 0 2	unit assumed, *sic*
7.654E-02	3.369E+00	20	B109	1 0 0 0 2	unit assumed, *sic*
7.432E-02	3.271E+00	21	B109	1 0 0 0 2	unit assumed, *sic*
7.427E-02	3.269E+00	21	B109	1 0 0 0 2	unit assumed, *sic*
7.213E-02	3.174E+00	22	B109	1 0 0 0 2	unit assumed, *sic*
6.582E-02	2.897E+00	25	B109	1 0 0 0 2	unit assumed, *sic*
3.360E-02	1.479E+00	25	H124	1 0 0 1 2	
6.204E-02	2.730E+00	27	B109	1 0 0 0 2	unit assumed, *sic*
6.127E-02	2.696E+00	28	B109	1 0 0 0 2	unit assumed, *sic*
5.714E-02	2.515E+00	30	B109	1 0 0 0 2	unit assumed, *sic*

44. CS_2
Carbon disulfide
Carbon disulphide
Schwefelkohlenstoff
RN: 75-15-0   **MP** (°C):   –112
MW: 76.14   **BP** (°C):   46.5

Solubility (Moles/L)	Solubility (Grams/L)	Temp (°C)	Ref (#)	Evaluation (T P E A A)	Comments
2.679E-02	2.040E+00	0	F300	1 0 0 0 2	
3.257E-02	2.480E+00	0	H101	2 0 0 0 2	
2.883E-02	2.195E+00	20	C121	0 0 0 0 1	unit assumed, *sic*
2.351E-02	1.790E+00	20	F300	1 0 0 0 2	
2.850E-02	2.170E+00	20	G080	1 0 0 0 1	
2.844E-02	2.165E+00	20	M061	1 0 0 0 2	
3.850E-02	2.931E+00	20	N038	1 0 0 1 2	
2.889E-02	2.200E+00	22	P076	1 2 1 1 1	
3.746E-02	2.852E+00	25	L319	1 0 2 1 1	
2.036E-02	1.550E+00	30	F300	1 0 0 0 2	
2.889E-02	2.200E+00	32	M161	1 0 0 0 1	
2.627E-02	2.000E+00	ns	N013	0 0 0 0 2	

45. C₂HBrClF₃

Halothane

2-Bromo-2-chloro-1,1,1-trifluoroethane

Fluothane

RN: 151-67-7   **MP** (°C):   <25

MW: 197.39   **BP** (°C):   50.2

Solubility (Moles/L)	Solubility (Grams/L)	Temp (°C)	Ref (#)	Evaluation (T P E A A)	Comments
1.742E-02	3.438E+00	ns	R028	0 0 0 0 0	

46. C₂HCl₃

Trichloroethylene

Trichloroethene

Trichloro-ethylene

Ethinyl trichloride

Acetylene trichloride

1,1,2-Trichloroethylene

RN: 79-01-6   **MP** (°C):   −87

MW: 131.39   **BP** (°C):   86.7

Solubility (Moles/L)	Solubility (Grams/L)	Temp (°C)	Ref (#)	Evaluation (T P E A A)	Comments
8.372E-03	1.100E+00	20	M133	1 0 0 0 2	
9.654E-03	1.268E+00	20	P041	1 0 0 0 1	
8.372E-03	1.100E+00	20	P046	1 0 0 0 0	
7.603E-03	9.990E-01	25	A094	1 0 0 0 1	
1.120E-02	1.472E+00	25	B173	2 0 2 2 2	
8.363E-03	1.099E+00	25	G056	1 0 0 0 2	
8.372E-03	1.100E+00	25	M037	1 1 0 0 1	
1.040E-02	1.366E+00	25	M342	1 0 1 1 2	
8.372E-03	1.100E+00	25	M368	1 0 0 0 1	
8.363E-03	1.099E+00	25	N034	1 0 0 0 1	
3.032E-02	3.984E+00	25	N309	1 0 0 0 1	*sic*
5.656E-03	7.431E-01	30	M311	1 1 2 2 2	
9.274E-03	1.219E+00	37	P041	1 0 0 0 1	
8.363E-03	1.099E+00	ns	O006	0 0 0 0 1	

47. C₂HCl₃O.H₂O

Chloral (monhydrate)

Chloral-hydrat

RN: 302-17-0   **MP** (°C):   57.0

MW: 165.40   **BP** (°C):

Solubility (Moles/L)	Solubility (Grams/L)	Temp (°C)	Ref (#)	Evaluation (T P E A A)	Comments
2.056E+00	3.400E+02	0	F300	1 0 0 0 2	
4.837E+00	8.000E+02	11.30	F300	1 0 0 0 2	
5.629E+00	9.310E+02	38.10	F300	1 0 0 0 2	
4.794E+00	7.930E+02	rt	D021	0 0 1 1 2	

48. C₂HCl₃O₂
$C_2HCl_3O_2$

Trichloroacetic acid

TCA

RN: 76-03-9   **MP** (°C):   57.5
MW: 163.39   **BP** (°C):   196.5

Solubility (Moles/L)	Solubility (Grams/L)	Temp (°C)	Ref (#)	Evaluation (T P E A A)	Comments
3.338E+00	5.455E+02	25	B185	0 0 0 0 0	
5.685E+00	9.289E+02	25	B200	1 0 0 0 2	
2.146E+00	3.506E+02	25	F018	1 0 0 0 1	
4.024E+00	6.575E+02	25	K040	1 2 1 2 2	
1.000E+01	1.634E+03	ns	M163	0 0 0 0 0	EFG
2.146E+00	3.506E+02	ns	N013	0 0 0 0 1	

49. C₂HCl₅
C_2HCl_5

Pentachloroethane

Pentachloro-ethane

Pentalin

Pentachlorethane

Ethane pentachloride

RN: 76-01-7   **MP** (°C):   −29
MW: 202.30   **BP** (°C):   161

Solubility (Moles/L)	Solubility (Grams/L)	Temp (°C)	Ref (#)	Evaluation (T P E A A)	Comments
2.322E-03	4.698E-01	20	V009	1 0 0 0 1	
2.470E-03	4.998E-01	25	G056	1 0 0 0 2	
2.472E-03	5.000E-01	25	M037	1 1 0 0 1	
2.373E-03	4.800E-01	ns	H123	0 0 0 0 0	
2.322E-03	4.698E-01	ns	O006	0 0 0 0 1	

50. C₂H₂
C_2H_2

Acetylene

Acetylen

RN: 74-86-2   **MP** (°C):   −81
MW: 26.04   **BP** (°C):

Solubility (Moles/L)	Solubility (Grams/L)	Temp (°C)	Ref (#)	Evaluation (T P E A A)	Comments
7.796E-02	2.030E+00	0	F300	1 0 0 0 2	*sic*
4.609E-02	1.200E+00	20	F300	1 0 0 0 2	*sic*
1.862E+01	4.848E+02	25	M101	1 0 0 0 2	
1.959E-02	5.100E-01	60	F300	1 0 0 0 1	*sic*

51. $C_2H_2Br_4$

sym-Tetrabromoethane
1,1,2,2-Tetrabrom-aethan
Acetylene tetrabromide
1,1,2,2-Tetrabromoethane
Tetrabromoacetylene

RN: 79-27-6	**MP** (°C): 0
MW: 345.67	**BP** (°C): 151

Solubility (Moles/L)	Solubility (Grams/L)	Temp (°C)	Ref (#)	Evaluation (T P E A A)	Comments
1.880E-03	6.500E-01	30	F300	1 0 0 0 1	
1.879E-03	6.496E-01	30	O006	1 0 0 0 1	

52. $C_2H_2Cl_2$

Vinylidene chloride
1,1-Dichloroethylene

RN: 75-35-4	**MP** (°C): −122.0
MW: 96.94	**BP** (°C): 31.7

Solubility (Moles/L)	Solubility (Grams/L)	Temp (°C)	Ref (#)	Evaluation (T P E A A)	Comments
2.470E-02	2.394E+00	15	D086	1 0 2 2 1	
2.624E-02	2.544E+00	17	D086	1 0 2 2 2	
4.126E-03	4.000E-01	20	M133	1 0 0 0 2	
4.126E-03	4.000E-01	20	P046	1 0 0 0 0	
2.572E-02	2.494E+00	20.5	D086	1 0 2 2 1	
2.316E-02	2.245E+00	25	D086	1 0 2 2 2	
2.470E-02	2.394E+00	28.5	D086	1 0 2 2 1	
2.624E-02	2.544E+00	29.5	D086	1 0 2 2 2	
2.302E-02	2.232E+00	30	M311	1 1 2 2 2	
2.264E-02	2.195E+00	38.5	D086	1 0 2 2 1	
2.162E-02	2.096E+00	45	D086	1 0 2 2 1	
2.367E-02	2.295E+00	51	D086	1 0 2 2 1	
2.162E-02	2.096E+00	55	D086	1 0 2 2 1	
2.470E-02	2.394E+00	60	D086	1 0 2 2 1	
2.316E-02	2.245E+00	65	D086	1 0 2 2 2	
3.034E-02	2.941E+00	71	D086	1 0 2 2 2	
2.572E-02	2.494E+00	74.5	D086	1 0 2 2 1	
3.034E-02	2.941E+00	81	D086	1 0 2 2 2	
3.803E-02	3.686E+00	85.5	D086	1 0 2 2 1	
3.598E-02	3.488E+00	90.5	D086	1 0 2 2 1	

53. $C_2H_2Cl_2$
cis-Acetylene dichloride
cis-1,2-Dichloroethylene
cis-Dichlorethylene

RN:	156-59-2	**MP** (°C):	−80
MW:	96.94	**BP** (°C):	60

Solubility (Moles/L)	Solubility (Grams/L)	Temp (°C)	Ref (#)	Evaluation (T P E A A)	Comments
3.610E-02	3.500E+00	25	M037	1 1 0 0 1	

54. $C_2H_2Cl_2$
trans-Acetylene dichloride
trans-1,2-Dichloroethylene
trans-Dichlorethylene

RN:	156-60-5	**MP** (°C):	−50
MW:	96.94	**BP** (°C):	48

Solubility (Moles/L)	Solubility (Grams/L)	Temp (°C)	Ref (#)	Evaluation (T P E A A)	Comments
6.499E-02	6.300E+00	25	M037	1 1 0 0 1	

55. $C_2H_2Cl_3As$
Chlorovinyldichloroarsine
Chlorvinylarsin-dichlorid

RN:	541-25-3	**MP** (°C):	
MW:	207.32	**BP** (°C):	

Solubility (Moles/L)	Solubility (Grams/L)	Temp (°C)	Ref (#)	Evaluation (T P E A A)	Comments
2.412E-03	5.000E-01	20	F300	1 0 0 0 0	

56. $C_2H_2Cl_4$
1,1,1,2-Tetrachloroethane
Ethane, 1,1,1,2-tetrachloro-
F 130α
TCA
HCC 130α

RN:	630-20-6	**MP** (°C):	−44
MW:	167.85	**BP** (°C):	

Solubility (Moles/L)	Solubility (Grams/L)	Temp (°C)	Ref (#)	Evaluation (T P E A A)	Comments
7.141E-03	1.199E+00	0	V009	1 0 0 0 2	
6.487E-03	1.089E+00	20	V009	1 0 0 0 2	
1.723E-02	2.892E+00	25	G056	1 0 0 0 2	
<1.66E-02	<2.79E+00	25.50	O005	2 0 2 2 1	
6.843E-03	1.149E+00	35	V009	1 0 0 0 2	
7.438E-03	1.248E+00	50	V009	1 0 0 0 2	

57. C₂H₂Cl₄

1,1,2,2-Tetrachloroethane
sym-Tetrachloroethane

RN:	79-34-5	**MP** (°C):	−36	
MW:	167.85	**BP** (°C):	146.5	

Solubility (Moles/L)	Solubility (Grams/L)	Temp (°C)	Ref (#)	Evaluation (T P E A A)	Comments
1.924E-02	3.230E+00	20	C094	1 0 0 0 2	
1.758E-02	2.951E+00	23.5	S171	2 1 2 2 2	
1.770E-02	2.971E+00	25	B173	2 0 2 2 2	
1.782E-02	2.991E+00	25	F050	1 0 0 0 0	
1.728E-02	2.900E+00	25	M037	1 1 0 0 1	
1.737E-02	2.915E+00	30	M311	1 1 2 2 2	

58. C₂H₂O₄

Oxalic acid
Oxalsaeure

RN:	144-62-7	**MP** (°C):	189	
MW:	90.04	**BP** (°C):		

Solubility (Moles/L)	Solubility (Grams/L)	Temp (°C)	Ref (#)	Evaluation (T P E A A)	Comments
3.683E-01	3.316E+01	0	C066	1 0 1 1 2	
3.665E-01	3.300E+01	0	L041	1 0 0 1 1	
3.756E-01	3.382E+01	0	M043	1 0 0 0 1	
4.907E-01	4.418E+01	4.99	A339	0 0 0 0 0	
6.287E-01	5.660E+01	10	M043	1 0 0 0 1	
5.912E-01	5.323E+01	9.99	A339	0 0 0 0 0	
7.752E-01	6.979E+01	14.99	A339	0 0 0 0 0	
7.441E-01	6.700E+01	15	F066	2 2 2 2 1	
7.464E-01	6.720E+01	15	F300	1 0 0 0 2	
7.775E-01	7.000E+01	15	L041	1 0 0 1 1	
9.468E-01	8.524E+01	19.99	A339	0 0 0 0 0	
9.219E-01	8.300E+01	20	F066	2 2 2 2 1	
9.219E-01	8.300E+01	20	F300	1 0 0 0 1	
9.552E-01	8.600E+01	20	L041	1 0 0 1 1	
9.636E-01	8.676E+01	20	M043	1 0 0 0 1	
8.836E-01	7.956E+01	20	M171	1 0 0 0 1	
1.146E+00	1.032E+02	24.99	A339	0 0 0 0 0	
1.088E+00	9.800E+01	25	F066	2 2 2 2 1	
1.378E+00	1.240E+02	25	F317	2 1 1 1 2	
2.480E+00	2.233E+02	25	H084	1 0 0 0 2	
1.190E+00	1.071E+02	25	H430	0 0 0 0 0	
2.409E+00	2.169E+02	25	K040	1 0 2 1 2	
1.317E+00	1.186E+02	29.99	A339	0 0 0 0 0	
1.407E+00	1.266E+02	30	M043	1 0 0 0 2	
1.623E+00	1.461E+02	34.99	A339	0 0 0 0 0	
1.710E+00	1.540E+02	35	L041	1 0 0 1 2	
1.903E+00	1.713E+02	39.99	A339	0 0 0 0 0	

(*continued*)

58. $C_2H_2O_4$ (continued)

Solubility (Moles/L)	Solubility (Grams/L)	Temp (°C)	Ref (#)	Evaluation (T P E A A)	Comments
1.973E+00	1.776E+02	40	M043	1 0 0 0 2	
2.199E+00	1.979E+02	44.99	A339	0 0 0 0 0	
2.527E+00	2.275E+02	49.99	A339	0 0 0 0 0	
2.150E+00	1.935E+02	50	C066	1 0 1 1 2	
2.821E+00	2.540E+02	50	L041	1 0 0 1 2	
2.867E+00	2.581E+02	54.99	A339	0 0 0 0 0	
3.121E+00	2.810E+02	59.99	A339	0 0 0 0 0	
3.410E+00	3.070E+02	60	M043	1 0 0 0 2	
3.661E+00	3.296E+02	64.99	A339	0 0 0 0 0	
4.121E+00	3.710E+02	65	L041	1 0 0 1 2	
3.583E+00	3.226E+02	80	C066	1 0 1 1 2	
5.084E+00	4.577E+02	80	M043	1 0 0 0 2	
6.059E+00	5.455E+02	90	F300	1 0 0 0 2	

59. $C_2H_2O_4 \cdot 2H_2O$

Oxalic acid dihydrate

Ethanedioic acid, dihydrate

RN: 6153-56-6   **MP (°C):**   101

MW: 126.07   **BP (°C):**

Solubility (Moles/L)	Solubility (Grams/L)	Temp (°C)	Ref (#)	Evaluation (T P E A A)	Comments
1.443E-02	1.820E+00	23	C038	2 2 2 2 0	EFG, 0.1N HCl
1.070E-02	1.349E+00	30	C038	2 2 2 2 0	EFG, 0.1N HCl
7.234E-03	9.120E-01	35	C038	2 2 2 2 0	EFG, 0.1N HCl

60. $C_2H_3Br_3O$

2,2,2-Tribromoethanol

2,2,2-Tribrom-aethanol

RN: 75-80-9   **MP (°C):**   80

MW: 282.77   **BP (°C):**   92

Solubility (Moles/L)	Solubility (Grams/L)	Temp (°C)	Ref (#)	Evaluation (T P E A A)	Comments
1.206E-01	3.410E+01	40	F300	1 0 0 0 2	

61. C$_2$H$_3$Cl
Vinyl chloride
Chloroethylene

RN:	75-01-4	MP (°C):	−153.0
MW:	62.50	BP (°C):	−13.3

Solubility (Moles/L)	Solubility (Grams/L)	Temp (°C)	Ref (#)	Evaluation (T P E A A)	Comments
9.600E-04	6.000E-02	10	M133	1 0 0 0 1	*sic*
9.600E-04	6.000E-02	10	P046	1 0 0 0 0	*sic*
1.506E-01	9.411E+00	15	D086	1 0 2 2 1	
1.576E-01	9.852E+00	16	D086	1 0 2 2 2	
1.081E-01	6.754E+00	20	N034	1 0 0 0 1	
1.451E-01	9.067E+00	20.5	D086	1 0 2 2 2	
<1.76E-02	<1.10E+00	25	I310	0 0 0 0 0	
1.396E-01	8.723E+00	26	D086	1 0 2 2 1	
1.411E-01	8.821E+00	29.5	D086	1 0 2 2 1	
1.490E-01	9.312E+00	35	D086	1 0 2 2 1	
1.411E-01	8.821E+00	41	D086	1 0 2 2 1	
1.396E-01	8.723E+00	46.5	D086	1 0 2 2 1	
6.717E-03	4.198E-01	50	M065	0 0 2 1 1	
1.506E-01	9.411E+00	55	D086	1 0 2 2 1	
1.459E-01	9.116E+00	65	D086	1 0 2 2 1	
1.553E-01	9.705E+00	72.5	D086	1 0 2 2 1	
1.584E-01	9.901E+00	80	D086	1 0 2 2 2	
1.772E-01	1.108E+01	85	D086	1 0 2 2 2	

62. C$_2$H$_3$Cl$_2$NO$_2$
1,1-Dichloro-1-nitroethane
Dichloronitroethane
Ethide

RN:	594-72-9	MP (°C):	
MW:	143.96	BP (°C):	

Solubility (Moles/L)	Solubility (Grams/L)	Temp (°C)	Ref (#)	Evaluation (T P E A A)	Comments
3.456E-02	4.975E+00	20	C121	1 0 0 0 0	unit assumed, *sic*
1.732E-02	2.494E+00	20	M061	1 0 0 0 1	

63. C$_2$H$_3$Cl$_3$

1,1,1-Trichloroethane
1,1,1-Trichloroethane
Trichloroethane
1,1,1-Trichloethane

RN:	71-55-6	**MP** (°C):	−35	
MW:	133.41	**BP** (°C):	74.1	

Solubility (Moles/L)	Solubility (Grams/L)	Temp (°C)	Ref (#)	Evaluation (T P E A A)	Comments
1.190E-02	1.587E+00	0	V009	1 0 0 0 2	
1.342E-02	1.790E+00	3.5	C094	1 0 0 0 2	
1.019E-02	1.360E+00	20	C094	1 0 1 0 2	
3.358E-02	4.480E+00	20	G056	1 0 0 0 2	
3.598E-03	4.800E-01	20	M133	1 0 0 0 2	
9.895E-03	1.320E+00	20	M368	1 0 0 0 1	
3.598E-03	4.800E-01	20	P046	1 0 0 0 0	
9.882E-03	1.318E+00	20	V009	1 0 0 0 2	
8.797E-03	1.174E+00	23.5	S171	2 1 2 2 2	
5.244E-03	6.995E-01	25	A094	1 0 0 0 0	
1.000E-02	1.334E+00	25	B173	2 0 2 2 2	
3.284E-02	4.381E+00	25	N309	1 0 0 0 1	*sic*
9.732E-03	1.298E+00	25	O006	1 0 0 0 1	
3.597E-03	4.798E-01	30	M311	1 1 2 2 2	
9.433E-03	1.258E+00	35	V009	1 0 0 0 2	
9.583E-03	1.278E+00	50	V009	1 0 0 0 2	
5.397E-03	7.200E-01	ns	H123	0 0 0 0 0	

64. C$_2$H$_3$Cl$_3$

1,1,2-Trichloroethane
1,1,2-β-Trichloroethane

RN:	79-00-5	**MP** (°C):	−37	
MW:	133.41	**BP** (°C):	113	

Solubility (Moles/L)	Solubility (Grams/L)	Temp (°C)	Ref (#)	Evaluation (T P E A A)	Comments
3.477E-02	4.638E+00	0	V009	1 0 0 0 2	
3.254E-02	4.341E+00	20	V009	1 0 0 0 2	
3.804E-02	5.074E+00	25	C119	2 2 2 2 2	
3.298E-02	4.400E+00	25	M037	1 1 0 0 1	
3.272E-02	4.365E+00	30	M311	1 1 2 2 2	
3.417E-02	4.559E+00	35	V009	1 0 0 0 2	
3.967E-02	5.292E+00	55	V009	1 0 0 0 2	

65. $C_2H_3FO_2$
Fluoroacetic acid
Essigsaeurefluorid

RN: 144-49-0   **MP** (°C):
MW: 78.04   **BP** (°C):

Solubility (Moles/L)	Solubility (Grams/L)	Temp (°C)	Ref (#)	Evaluation (T P E A A)	Comments
6.407E-04	5.000E-02	20	F300	1 0 0 0 0	

66. C_2H_3N
Acetonitrile
Acetonitril

RN: 75-05-8   **MP** (°C): −45
MW: 41.05   **BP** (°C): 81.6

Solubility (Moles/L)	Solubility (Grams/L)	Temp (°C)	Ref (#)	Evaluation (T P E A A)	Comments
>1.95E+01	>8.00E+02	25	B019	1 0 1 2 0	

67. C_2H_3N
Methylisocyanide
Methyl-isocyanid

RN: 593-75-9   **MP** (°C):
MW: 41.05   **BP** (°C):

Solubility (Moles/L)	Solubility (Grams/L)	Temp (°C)	Ref (#)	Evaluation (T P E A A)	Comments
2.217E+00	9.100E+01	15	F300	1 0 0 0 1	

68. C_2H_3NS
Methyl isothiocyanate
Isothiocyanatomethane

RN: 556-61-6   **MP** (°C): 35
MW: 73.12   **BP** (°C): 119

Solubility (Moles/L)	Solubility (Grams/L)	Temp (°C)	Ref (#)	Evaluation (T P E A A)	Comments
1.039E-01	7.600E+00	20	M161	1 0 0 0 1	
1.032E-01	7.543E+00	20	O300	1 0 0 0 1	
1.085E-01	7.937E+00	20	P081	1 0 0 0 0	

69. C$_2$H$_4$
Ethylene
Ethene

RN: 74-85-1   **MP** (°C):   −169
MW: 28.05   **BP** (°C):   102

Solubility (Moles/L)	Solubility (Grams/L)	Temp (°C)	Ref (#)	Evaluation (T P E A A)	Comments
7.129E+00	2.000E+02	0	R028	0 0 0 0 0	
3.240E+00	9.091E+01	25	R028	0 0 0 0 0	
3.187E+00	8.942E+01	30	C116	0 0 0 0 0	

70. C$_2$H$_4$BrCl
Ethylene chlorobromide
1-Bromo-2-chloroethane

RN: 107-04-0   **MP** (°C):   −17
MW: 143.42   **BP** (°C):   106

Solubility (Moles/L)	Solubility (Grams/L)	Temp (°C)	Ref (#)	Evaluation (T P E A A)	Comments
4.778E-02	6.853E+00	20	C121	1 0 0 0 1	unit assumed, *sic*

71. C$_2$H$_4$Br$_2$
1,2-Dibromoethane
Ethylene dibromide
Curafume
Haltox
1,2-Dibromaethan

RN: 106-93-4   **MP** (°C):   9.97
MW: 187.87   **BP** (°C):   131.7

Solubility (Moles/L)	Solubility (Grams/L)	Temp (°C)	Ref (#)	Evaluation (T P E A A)	Comments
1.777E-02	3.339E+00	0	V009	1 0 0 0 2	
2.078E-02	3.905E+00	15	G029	1 0 2 2 2	
1.874E-02	3.520E+00	20	C094	1 0 1 0 2	
2.279E-02	4.282E+00	20	C121	1 0 0 0 1	unit assumed, *sic*
1.794E-02	3.370E+00	20	G080	1 0 0 0 1	
2.300E-02	4.321E+00	20	M312	1 0 0 0 1	
1.592E-02	2.991E+00	20	P081	1 0 0 0 0	
2.142E-02	4.024E+00	20	V009	1 0 0 0 2	
2.210E-02	4.153E+00	25	O006	1 0 0 0 2	
2.294E-02	4.310E+00	30	F300	1 0 0 0 2	
2.284E-02	4.292E+00	30	G029	1 0 2 2 2	
2.279E-02	4.282E+00	30	M061	1 0 0 0 1	
2.289E-02	4.300E+00	30	M161	1 0 0 0 1	
2.390E-02	4.490E+00	35	V009	1 0 0 0 2	
2.817E-02	5.292E+00	50	V009	1 0 0 0 2	

72. C$_2$H$_4$ClNO

Acetohydroxamic acid chloride
Acethydroximsaeure-chlorid
2-Chloroacetamide
Chloroacetamide
Chloressigsaeureamid
Essigsaeure-N-chloramid

RN:	79-07-2	MP (°C):	119.5		
MW:	93.51	BP (°C):	225		

Solubility (Moles/L)	Solubility (Grams/L)	Temp (°C)	Ref (#)	Evaluation (T P E A A)	Comments
9.624E-01	9.000E+01	24	F300	1 0 0 0 0	

73. C$_2$H$_4$ClNO$_2$

1-Chloro-1-nitroethane
1-Chloronitroethane

RN:	598-92-5	MP (°C):			
MW:	109.51	BP (°C):	124		

Solubility (Moles/L)	Solubility (Grams/L)	Temp (°C)	Ref (#)	Evaluation (T P E A A)	Comments
3.638E-02	3.984E+00	20	C121	1 0 0 0 0	unit assumed, *sic*
3.638E-02	3.984E+00	20	M061	1 0 0 0 0	

74. C$_2$H$_4$Cl$_2$

Ethylidene chloride
1,1-Dichloraethan
1,1-Dichloroethane

RN:	75-34-3	MP (°C):	−97		
MW:	98.96	BP (°C):			

Solubility (Moles/L)	Solubility (Grams/L)	Temp (°C)	Ref (#)	Evaluation (T P E A A)	Comments
6.669E-02	6.600E+00	0	F300	1 0 0 0 1	
6.629E-02	6.560E+00	0	H101	2 0 0 0 2	
5.967E-02	5.905E+00	0	V009	1 0 0 0 2	
5.558E-02	5.500E+00	20	F300	1 0 0 0 1	
5.558E-02	5.500E+00	20	H101	2 0 0 0 2	
5.087E-02	5.035E+00	20	V009	1 0 0 0 2	
5.110E-02	5.057E+00	25	G038	1 2 2 2 2	
5.110E-02	5.057E+00	25	G053	2 2 2 1 2	
5.457E-02	5.400E+00	30	F300	1 0 0 0 1	
4.885E-02	4.834E+00	30	M300	1 1 2 2 2	
4.637E-02	4.589E+00	30	M311	1 1 2 2 2	
5.397E-02	5.341E+00	30	N034	1 0 0 0 2	
4.847E-02	4.797E+00	35	V009	1 0 0 0 2	
5.217E-02	5.163E+00	50	V009	1 0 0 0 2	

75. $C_2H_4Cl_2$
Ethylene dichloride
1,2-Dichloraethan

RN: 107-06-2   **MP** (°C):   −35
MW: 98.96   **BP** (°C):

Solubility (Moles/L)	Solubility (Grams/L)	Temp (°C)	Ref (#)	Evaluation (T P E A A)	Comments
9.095E-02	9.000E+00	0	F300	1 0 0 0 0	
9.317E-02	9.220E+00	0	H101	2 0 0 0 2	
9.232E-02	9.136E+00	0	L103	1 0 0 0 2	unit assumed
8.745E-02	8.654E+00	0	V009	1 0 0 0 2	
8.735E-02	8.645E+00	15	G029	1 0 2 2 2	
8.539E-02	8.450E+00	20	C094	1 0 1 0 2	
8.716E-02	8.625E+00	20	C121	1 0 0 0 1	unit assumed, *sic*
8.716E-02	8.625E+00	20	D052	1 1 0 0 1	
8.716E-02	8.625E+00	20	G056	1 0 0 0 2	
8.781E-02	8.690E+00	20	H101	2 0 0 0 2	
8.706E-02	8.615E+00	20	L103	1 0 0 0 2	unit assumed
8.706E-02	8.615E+00	20	M061	1 0 0 0 2	
8.616E-02	8.527E+00	20	M062	1 0 0 0 1	
8.892E-02	8.800E+00	20	M133	1 0 0 0 2	
8.716E-02	8.625E+00	20	O006	1 0 0 0 1	
8.892E-02	8.800E+00	20	P046	1 0 0 0 0	
8.507E-02	8.419E+00	20	V009	1 0 0 0 2	
8.070E-02	7.986E+00	25	B173	2 0 2 2 2	
1.060E-01	1.049E+01	25	C119	2 2 2 2 2	
8.690E-02	8.600E+00	25	F300	1 0 0 0 2	
8.740E-02	8.649E+00	25	G038	1 2 2 2 2	
8.740E-02	8.649E+00	25	G053	2 1 2 1 2	
8.488E-02	8.400E+00	25	M037	1 1 0 0 1	
9.013E-02	8.920E+00	30	G029	1 0 2 2 1	
8.954E-02	8.861E+00	30	L103	1 0 0 0 2	unit assumed
3.543E-02	3.506E+00	30	M311	1 1 2 2 2	
8.964E-02	8.871E+00	35	V009	1 0 0 0 2	
1.030E-01	1.019E+01	56	V009	1 0 0 0 2	
8.716E-02	8.625E+00	72	B197	0 0 0 0 0	at bp of 72 °C
5.927E-02	5.865E+00	89.3	B197	0 0 0 0 0	at bp of 89.3 °C
4.327E-02	4.282E+00	92.3	B197	0 0 0 0 0	at bp of 92.3 °C
3.324E-02	3.289E+00	94	B197	0 0 0 0 0	at bp of 94 °C
1.312E-02	1.298E+00	98	B197	0 0 0 0 0	at bp of 98 °C
4.345E-02	4.300E+00	rt	M161	0 0 0 0 1	

76. $C_2H_4F_2$
1,1-Difluoroethane
Ethylidene fluoride

RN: 75-37-6   **MP** (°C):   −117
MW: 66.05   **BP** (°C):   −24.7

Solubility (Moles/L)	Solubility (Grams/L)	Temp (°C)	Ref (#)	Evaluation (T P E A A)	Comments
8.132E-02	5.371E+00	0	M065	0 0 2 1 2	

77. C$_2$H$_4$N$_2$O$_2$
Oxamide
Oxalsaeure-diamid
RN: 471-46-5   MP (°C):
MW: 88.07   BP (°C):

Solubility (Moles/L)	Solubility (Grams/L)	Temp (°C)	Ref (#)	Evaluation (T P E A A)	Comments
4.201E-03	3.700E-01	7.30	F300	1 0 0 0 1	
7.040E-02	6.200E+00	100	F300	1 0 0 0 1	

78. C$_2$H$_4$N$_4$
Amitrole
3-Amino-1,2,4-triazole
3-Amino-s-triazole
ATA
Aminotriazole
RN: 61-82-5   MP (°C):   159.0
MW: 84.08   BP (°C):

Solubility (Moles/L)	Solubility (Grams/L)	Temp (°C)	Ref (#)	Evaluation (T P E A A)	Comments
2.602E+00	2.188E+02	23	M061	1 0 0 0 1	
2.602E+00	2.188E+02	25	B185	0 0 0 0 0	
2.602E+00	2.188E+02	25	B200	1 0 0 0 1	
2.602E+00	2.188E+02	25	I310	0 0 0 0 0	
3.330E+00	2.800E+02	25	M161	1 0 0 0 2	
2.602E+00	2.188E+02	ns	B100	0 0 0 0 1	
3.162E+00	2.659E+02	ns	M163	0 0 0 0 0	EFG

79. C$_2$H$_4$N$_4$
Dicyanodiamide
Dicyandiamid
Dicyandiamide
RN: 461-58-5   MP (°C):   210
MW: 84.08   BP (°C):

Solubility (Moles/L)	Solubility (Grams/L)	Temp (°C)	Ref (#)	Evaluation (T P E A A)	Comments
1.526E-01	1.283E+01	0	M043	1 0 0 0 1	
2.218E-01	1.865E+01	10	M043	1 0 0 0 1	
2.617E-01	2.200E+01	13	F300	1 0 0 0 1	
3.688E-01	3.101E+01	20	M043	1 0 0 0 1	
4.876E-01	4.100E+01	25	F300	1 0 0 0 1	
4.717E-01	3.966E+01	25.0	H037	1 2 2 1 2	
5.663E-01	4.762E+01	30	M043	1 0 0 0 1	
8.565E-01	7.201E+01	39.9	H037	1 2 2 1 2	
8.606E-01	7.236E+01	40	M043	1 0 0 0 1	
1.255E+00	1.055E+02	49.8	H037	1 2 2 1 2	

(continued)

79. C₂H₄N₄ (continued)

Solubility (Moles/L)	Solubility (Grams/L)	Temp (°C)	Ref (#)	Evaluation (T P E A A)	Comments
1.899E+00	1.597E+02	60	M043	1 0 0 0 1	
1.878E+00	1.579E+02	60.1	H037	1 2 2 1 2	
2.236E+00	1.880E+02	60.10	F300	1 0 0 0 2	
2.978E+00	2.504E+02	74.5	H037	1 2 2 1 2	
3.275E+00	2.754E+02	80	M043	1 0 0 0 1	
1.492E-01	1.254E+01	.0	H037	1 2 2 1 2	

80. C₂H₄N₄O₂S₂

2-Amino-1,3,4-thiadiazole-5-sulfonamide
5-Amino-1,3,4-thiadiazol-2-sulfonamide
5-Amino-1,3,4-thiadiazole-2-sulfonamide
CL 5343
Tio-urasin

RN: 14949-00-9   **MP** (°C):
MW: 180.21   **BP** (°C):

Solubility (Moles/L)	Solubility (Grams/L)	Temp (°C)	Ref (#)	Evaluation (T P E A A)	Comments
2.630E-02	4.739E+00	15	K024	1 2 1 1 2	

81. C₂H₄O₂

Acetic acid glacial
Acetic acid
Essigsaeure

RN: 64-19-7   **MP** (°C): 16.7
MW: 60.05   **BP** (°C): 118

Solubility (Moles/L)	Solubility (Grams/L)	Temp (°C)	Ref (#)	Evaluation (T P E A A)	Comments
1.004E+01	6.029E+02	25	H084	1 0 0 0 2	

82. C₂H₄O₂

Methyl formate
Methyl methanoate
Formic acid methyl ester

RN: 107-31-3   **MP** (°C): −99.8
MW: 60.05   **BP** (°C): 32

Solubility (Moles/L)	Solubility (Grams/L)	Temp (°C)	Ref (#)	Evaluation (T P E A A)	Comments
+3.80E+00	+2.28E+02	ns	S460	0 0 0 0 0	

83. C$_2$H$_4$O$_3$
Glycolic acid
Glykolsaeure
RN: 79-14-1   MP (°C):   80
MW: 76.05   BP (°C):

Solubility (Moles/L)	Solubility (Grams/L)	Temp (°C)	Ref (#)	Evaluation (T P E A A)	Comments
6.084E+00	4.627E+02	6.99	A340	0 0 0 0 0	
6.913E+00	5.258E+02	10.89	A340	0 0 0 0 0	
7.894E+00	6.004E+02	20.69	A340	0 0 0 0 0	
8.015E+00	6.096E+02	24.99	A340	0 0 0 0 0	
8.168E+00	6.212E+02	30.09	A340	0 0 0 0 0	
8.296E+00	6.309E+02	35.99	A340	0 0 0 0 0	
8.400E+00	6.388E+02	39.99	A340	0 0 0 0 0	
8.533E+00	6.489E+02	47.99	A340	0 0 0 0 0	
8.536E+00	6.492E+02	48.99	A340	0 0 0 0 0	
8.654E+00	6.582E+02	54.99	A340	0 0 0 0 0	
8.721E+00	6.632E+02	59.49	A340	0 0 0 0 0	
8.808E+00	6.698E+02	64.49	A340	0 0 0 0 0	
8.866E+00	6.743E+02	69.99	A340	0 0 0 0 0	
8.932E+00	6.793E+02	74.99	A340	0 0 0 0 0	
8.968E+00	6.820E+02	79.89	A340	0 0 0 0 0	
9.016E+00	6.857E+02	84.49	A340	0 0 0 0 0	
9.043E+00	6.877E+02	88.09	A340	0 0 0 0 0	

84. C$_2$H$_5$Br
Bromoethane
Ethyl bromide
Aethylbromid
RN: 74-96-4   MP (°C):   −119
MW: 108.97   BP (°C):   38.5

Solubility (Moles/L)	Solubility (Grams/L)	Temp (°C)	Ref (#)	Evaluation (T P E A A)	Comments
9.792E-02	1.067E+01	0	H101	2 0 0 0 2	
8.810E-02	9.600E+00	17.5	F001	1 0 1 2 2	
8.810E-02	9.600E+00	17.5	S006	1 0 0 0 2	
8.259E-02	9.000E+00	20	F300	1 0 0 0 0	
8.388E-02	9.140E+00	20	H101	2 0 0 0 2	
8.185E-02	8.920E+00	20	H127	1 0 0 0 0	
8.127E-02	8.856E+00	30	V009	1 0 0 0 1	

85. C$_2$H$_5$Cl

Ethyl chloride
Aethylchlorid
Chloroethane
Monochloroethane

RN: 75-00-3 **MP** (°C): −139.0
MW: 64.52 **BP** (°C): 12.3

Solubility (Moles/L)	Solubility (Grams/L)	Temp (°C)	Ref (#)	Evaluation (T P E A A)	Comments
6.975E-02	4.500E+00	0	M037	1 1 0 0 1	
6.898E-02	4.450E+00	0	V009	1 0 0 0 2	
7.865E-02	5.074E+00	20	G056	1 0 0 0 2	
8.846E-02	5.707E+00	20	N034	1 0 0 0 2	
8.900E-02	5.742E+00	ns	F001	0 0 1 2 2	
8.433E-02	5.440E+00	ns	R028	0 0 0 0 0	

86. C$_2$H$_5$I

Iodoethane
Ethyl iodide
Aethyliodid
Iodaethan

RN: 75-03-6 **MP** (°C): −108
MW: 155.97 **BP** (°C): 71

Solubility (Moles/L)	Solubility (Grams/L)	Temp (°C)	Ref (#)	Evaluation (T P E A A)	Comments
2.828E-02	4.410E+00	0	H101	2 0 0 0 2	
2.571E-02	4.010E+00	20	F300	1 0 0 0 2	
2.584E-02	4.030E+00	20	H101	2 0 0 0 2	
2.510E-02	3.915E+00	20	M171	1 0 0 0 2	
2.510E-02	3.915E+00	22.5	F001	1 0 1 2 2	
2.510E-02	3.915E+00	22.5	S006	1 0 0 0 2	
2.580E-02	4.024E+00	30	G029	1 0 2 2 2	
2.661E-02	4.150E+00	30	H101	2 0 0 0 2	
2.580E-02	4.023E+00	30	V009	1 0 0 0 2	

87. C$_2$H$_5$N

Ethylenimine
Aethylenimin
Aziridine
Ethyleneimine
Dimethyleneimine

RN: 151-56-4 **MP** (°C): −78
MW: 43.07 **BP** (°C): 56

Solubility (Moles/L)	Solubility (Grams/L)	Temp (°C)	Ref (#)	Evaluation (T P E A A)	Comments
2.117E-01	9.116E+00	20	P315	0 0 0 0 0	

88. C₂H₅NO

Acetamide

Acetamid

RN:	60-35-5	**MP** (°C):	81.0		
MW:	59.07	**BP** (°C):	222.0		

Solubility (Moles/L)	Solubility (Grams/L)	Temp (°C)	Ref (#)	Evaluation (T P E A A)	Comments
1.021E+01	6.030E+02	.3	F300	1 0 0 0 2	
8.342E+00	4.927E+02	0	M022	1 0 0 0 2	
9.816E+00	5.798E+02	0	M043	1 0 0 0 2	
1.077E+01	6.364E+02	10	M043	1 0 0 0 2	
1.165E+01	6.880E+02	20	F300	1 0 0 0 2	
9.691E+00	5.724E+02	20	M022	1 0 0 0 2	
1.180E+01	6.970E+02	20	M043	1 0 0 0 2	
1.194E+01	7.050E+02	24.50	F300	1 0 0 0 2	
3.386E+01	2.000E+03	25	I310	0 0 0 0 0	
1.280E+01	7.561E+02	30	M043	1 0 0 0 2	
1.093E+01	6.455E+02	40	M022	1 0 0 0 2	
1.379E+01	8.148E+02	40	M043	1 0 0 0 2	
1.208E+01	7.138E+02	60	M022	1 0 0 0 2	
1.515E+01	8.947E+02	60	M043	1 0 0 0 2	
8.358E+00	4.937E+02	rt	D021	0 0 1 1 2	

89. C₂H₅NO₂

Glycine

Glycin

Glycocoll

RN:	56-40-6	**MP** (°C):	245		
MW:	75.07	**BP** (°C):			

Solubility (Moles/L)	Solubility (Grams/L)	Temp (°C)	Ref (#)	Evaluation (T P E A A)	Comments
1.668E+00	1.252E+02	0	D018	2 2 2 1 2	
1.656E+00	1.243E+02	0	M043	1 0 0 0 2	
1.905E+00	1.430E+02	10	C347	0 0 0 0 0	EFG
2.032E+00	1.525E+02	10	M043	1 0 0 0 2	
3.025E+00	2.271E+02	15	D349	2 1 1 2 2	
1.710E+00	1.284E+02	15	G081	1 0 1 1 2	
3.009E+00	2.259E+02	20	B032	1 2 2 2 2	
2.336E+00	1.754E+02	20	C347	0 0 0 0 0	EFG
3.180E+00	2.387E+02	20	D349	2 1 1 2 2	
2.447E+00	1.837E+02	20	M043	1 0 0 0 2	
2.616E+00	1.964E+02	21	P045	1 0 2 1 2	
2.127E+00	1.597E+02	22.9	Y412	0 0 0 0 0	
2.741E+00	2.058E+02	24.99	C404	2 1 2 2 1	
3.316E+00	2.489E+02	25	B032	1 2 2 2 2	
2.885E+00	2.166E+02	25	C018	0 0 0 0 0	
2.700E-03	2.027E-01	25	C405	2 1 2 2 2	intrinsic zwit

(continued)

89. C$_2$H$_5$NO$_2$ (continued)

Solubility (Moles/L)	Solubility (Grams/L)	Temp (°C)	Ref (#)	Evaluation (T P E A A)	Comments
3.329E+00	2.499E+02	25	D016	1 0 0 0 2	
2.691E+00	2.020E+02	25	D018	2 2 2 1 2	
2.663E+00	1.999E+02	25	D041	1 0 0 0 2	
3.325E+00	2.496E+02	25	D349	2 1 1 2 2	
2.886E+00	2.166E+02	25	E015	1 2 1 1 2	
2.660E+00	1.997E+02	25	F300	1 0 0 0 2	
2.664E+00	2.000E+02	25	G092	2 1 1 1 1	
2.664E+00	2.000E+02	25	G315	0 0 0 0 0	
2.526E+00	1.897E+02	25	K031	2 1 2 1 2	
2.886E+00	2.166E+02	25	M024	1 2 0 1 2	
3.334E+00	2.503E+02	25	M029	2 2 2 2 2	
2.760E+00	2.072E+02	25	N001	0 0 0 0 0	EFG
2.900E+00	2.177E+02	25	N012	2 0 2 1 2	
2.544E+00	1.910E+02	25	O316	1 0 1 2 2	
2.664E+00	2.000E+02	25	O316	1 0 1 2 2	
2.715E+00	2.038E+02	25	O317	1 0 1 2 2	
3.330E+00	2.500E+02	25.1	N024	0 0 0 0 0	
3.352E+00	2.516E+02	25.1	N025	0 0 0 0 0	
3.342E+00	2.509E+02	25.1	N026	0 0 0 0 0	
2.673E+00	2.006E+02	25.1	N027	1 1 2 2 2	
2.220E+00	1.667E+02	25.3	Y412	0 0 0 0 0	
3.144E+00	2.360E+02	27	D036	0 0 0 0 0	
3.074E+00	2.308E+02	27	D036	0 0 0 0 0	
2.312E+00	1.736E+02	29.2	Y412	0 0 0 0 0	
3.630E+00	2.725E+02	29.80	B032	1 2 2 1 2	
2.737E+00	2.054E+02	30	C347	0 0 0 0 0	EFG
2.832E+00	2.126E+02	30	M043	1 0 0 0 1	
3.106E+00	2.332E+02	34.99	C404	2 1 2 2 1	
2.491E+00	1.870E+02	36.8	Y412	0 0 0 0 0	
2.578E+00	1.935E+02	38.2	Y412	0 0 0 0 0	
3.109E+00	2.334E+02	40	C347	0 0 0 0 0	EFG
3.305E+00	2.481E+02	40	M043	1 0 0 0 1	
3.538E+00	2.656E+02	44.99	C404	2 1 2 2 1	
2.749E+00	2.063E+02	45.5	Y412	0 0 0 0 0	
3.547E+00	2.662E+02	50	C347	0 0 0 0 0	EFG
3.816E+00	2.865E+02	50	D018	2 2 2 1 2	
3.745E+00	2.811E+02	50	F300	1 0 0 0 2	
3.921E+00	2.943E+02	60	C347	0 0 0 0 0	EFG
4.134E+00	3.103E+02	60	M043	1 0 0 0 1	
4.215E+00	3.164E+02	70	C347	0 0 0 0 0	EFG
4.863E+00	3.650E+02	75	D018	2 2 2 1 2	
4.693E+00	3.523E+02	75	D041	1 0 0 0 2	
4.693E+00	3.523E+02	75	F300	1 0 0 0 2	
4.517E+00	3.390E+02	80	C347	0 0 0 0 0	EFG
4.836E+00	3.631E+02	80	M043	1 0 0 0 1	
4.753E+00	3.568E+02	90	C347	0 0 0 0 0	EFG
4.911E+00	3.686E+02	100	C347	0 0 0 0 0	EFG
5.353E+00	4.018E+02	100	F300	1 0 0 0 2	

(continued)

89. C$_2$H$_5$NO$_2$ (continued)

Solubility (Moles/L)	Solubility (Grams/L)	Temp (°C)	Ref (#)	Evaluation (T P E A A)	Comments
5.485E+00	4.118E+02	100	M043	1 0 0 0 1	
5.353E+00	4.018E+02	99.99	P349	0 0 0 0 0	
1.612E+00	1.210E+02	—	C347	0 0 0 0 0	EFG
6.661E+00	5.000E+02	ns	D072	0 0 0 0 0	
4.499E+00	3.377E+02	rt	D021	0 0 1 1 2	

90. C$_2$H$_5$NO$_2$
Nitroethane
Nitroetan

RN:	79-24-3	**MP** (°C):	−50
MW:	75.07	**BP** (°C):	114

Solubility (Moles/L)	Solubility (Grams/L)	Temp (°C)	Ref (#)	Evaluation (T P E A A)	Comments
5.736E-01	4.306E+01	20	C121	1 0 0 0 1	unit assumed, *sic*
6.404E-01	4.807E+01	25	M346	2 1 1 1 2	

91. C$_2$H$_5$NO$_2$
Methyl carbamate
Carbamidsaeure-methyl ester
Methyl urethane

RN:	598-55-0	**MP** (°C):	52
MW:	75.07	**BP** (°C):	177

Solubility (Moles/L)	Solubility (Grams/L)	Temp (°C)	Ref (#)	Evaluation (T P E A A)	Comments
9.125E+00	6.850E+02	11	F300	1 0 0 0 2	
9.119E+00	6.845E+02	11	I314	0 0 0 0 0	
9.200E+00	6.906E+02	15.50	F001	1 0 1 0 2	
5.462E+00	4.100E+02	15.50	F300	1 0 0 0 1	

92. C$_2$H$_5$NO$_2$
Glycolamide
2-Hydroxyacetamide
2-Hydroxyacetimidic acid
Glycolic amide
Glycolic acid amide

RN:	598-42-5	**MP** (°C):	
MW:	75.07	**BP** (°C):	

Solubility (Moles/L)	Solubility (Grams/L)	Temp (°C)	Ref (#)	Evaluation (T P E A A)	Comments
5.509E+00	4.135E+02	25	M008	1 0 0 0 2	

93. C₂H₅NS
Thiacetamide
Thioessigsaeureamid
Thioacetamide
Acetothioamide
Ethanethioamide
RN: 62-55-5   **MP** (°C):   113
MW: 75.13   **BP** (°C):

Solubility (Moles/L)	Solubility (Grams/L)	Temp (°C)	Ref (#)	Evaluation (T P E A A)	Comments
1.865E+00	1.402E+02	25	I310	0 0 0 0 0	

94. C₂H₅N.2H₂O
Ethyleneimine (dihydrate)
Aziridine (dihydrate)
RN: 151-56-4   **MP** (°C):
MW: 79.10   **BP** (°C):

Solubility (Moles/L)	Solubility (Grams/L)	Temp (°C)	Ref (#)	Evaluation (T P E A A)	Comments
6.840E-02	5.411E+00	20	P315	0 0 0 0 0	

95. C₂H₅N₃O₂
Methylnitrosourea
MNU
Nitrosomethylurea
RN: 684-93-5   **MP** (°C):   123
MW: 103.08   **BP** (°C):

Solubility (Moles/L)	Solubility (Grams/L)	Temp (°C)	Ref (#)	Evaluation (T P E A A)	Comments
1.400E-01	1.443E+01	24	M031	1 1 1 1 1	
1.413E-01	1.456E+01	ns	R424	0 0 0 0 0	

96. C₂H₅N₃O₂
Biuret
Carbamylurea
RN: 108-19-0   **MP** (°C):
MW: 103.08   **BP** (°C):

Solubility (Moles/L)	Solubility (Grams/L)	Temp (°C)	Ref (#)	Evaluation (T P E A A)	Comments
1.164E-01	1.200E+01	0	F300	1 0 0 0 2	
1.475E-01	1.520E+01	15	F300	1 0 0 0 2	
3.104E+00	3.200E+02	106	F300	1 0 0 0 1	

97. C$_2$H$_5$N$_5$O$_3$

N-Methyl-*N'*-nitro-*N*-nitrosoguanidine
MNNG
1-Methyl-3-nitro-1-nitrosoguanidine

RN: 70-25-7   **MP** (°C): 118
MW: 147.09   **BP** (°C):

Solubility (Moles/L)	Solubility (Grams/L)	Temp (°C)	Ref (#)	Evaluation (T P E A A)	Comments
<3.38E-02	<4.98E+00	ns	I307	0 0 0 0 0	

98. C$_2$H$_5$O$_5$P

Phosphoacetic acid
Phosphor carboxymethyl-phosphonsaeure
Phosphonoacetic acid

RN: 4408-78-0   **MP** (°C): 144.5
MW: 140.03   **BP** (°C):

Solubility (Moles/L)	Solubility (Grams/L)	Temp (°C)	Ref (#)	Evaluation (T P E A A)	Comments
2.799E+00	3.920E+02	0	F300	1 0 0 0 2	
2.800E+00	3.921E+02	0	N028	1 0 0 0 2	

99. C$_2$H$_5$O$_5$As

Arsonoacetic acid
Arsono-essigsaeure

RN: 107-38-0   **MP** (°C): 152
MW: 183.98   **BP** (°C):

Solubility (Moles/L)	Solubility (Grams/L)	Temp (°C)	Ref (#)	Evaluation (T P E A A)	Comments
2.174E+00	4.000E+02	18	F300	1 0 0 0 1	

100. C$_2$H$_6$

Ethane
Aethan

RN: 74-84-0   **MP** (°C): −172
MW: 30.07   **BP** (°C): −88

Solubility (Moles/L)	Solubility (Grams/L)	Temp (°C)	Ref (#)	Evaluation (T P E A A)	Comments
2.587E-01	7.779E+00	0	C075	1 0 1 0 1	
4.157E-03	1.250E-01	0	F300	1 0 0 0 2	
3.601E-03	1.083E-01	4.99	C115	2 0 2 2 2	
2.903E-03	8.730E-02	9.99	C115	2 0 2 2 2	
2.465E-03	7.413E-02	14.99	C115	2 0 2 2 2	

(continued)

100. C$_2$H$_6$ (continued)

Solubility (Moles/L)	Solubility (Grams/L)	Temp (°C)	Ref (#)	Evaluation (T P E A A)	Comments
2.222E-03	6.682E-02	19.8	G058	1 0 0 0 2	
2.129E-03	6.401E-02	19.99	C115	2 0 2 2 2	
1.929E-03	5.800E-02	20	F300	1 0 0 0 1	
1.850E-03	5.563E-02	24.99	C115	2 0 2 2 2	
2.009E-03	6.040E-02	25	M001	2 1 2 2 2	
2.009E-03	6.040E-02	25	M002	2 2 1 2 2	
1.760E-03	5.292E-02	25	M102	1 2 2 1 2	
1.620E-03	4.871E-02	29.99	C115	2 0 2 2 2	
7.981E-04	2.400E-02	60	F300	1 0 0 0 1	

101. C$_2$H$_6$O
Methyl ether
Dimethyl ether
Dimethylaether

RN:	115-10-6	**MP** (°C):	−138
MW:	46.07	**BP** (°C):	−23.6

Solubility (Moles/L)	Solubility (Grams/L)	Temp (°C)	Ref (#)	Evaluation (T P E A A)	Comments
1.476E+00	6.800E+01	18	F300	1 0 0 0 1	
5.669E+00	2.612E+02	24	M065	1 0 2 1 2	

102. C$_2$H$_6$O$_2$
Ethylene glycol
Glycol
1,2-Ethandiol

RN:	107-21-1	**MP** (°C):	−13
MW:	62.07	**BP** (°C):	197.6

Solubility (Moles/L)	Solubility (Grams/L)	Temp (°C)	Ref (#)	Evaluation (T P E A A)	Comments
6.710E+00	4.165E+02	4.50	C022	1 2 0 0 2	
5.562E-01	3.452E+01	25	B004	0 0 0 0 0	

103. C$_2$H$_6$O$_3$S
Methyl methanesulphonate
Methyl mesylate
Methanesulfonic acid methyl ester

RN:	66-27-3	**MP** (°C):	20
MW:	110.13	**BP** (°C):	203

Solubility (Moles/L)	Solubility (Grams/L)	Temp (°C)	Ref (#)	Evaluation (T P E A A)	Comments
1.513E+00	1.667E+02	25	I310	0 0 0 0 0	

104. C₂H₆O₄S

Dimethyl sulfate
Sulfuric acid dimethyl ester

RN:	77-78-1	MP (°C):	−27
MW:	126.13	BP (°C):	188

Solubility (Moles/L)	Solubility (Grams/L)	Temp (°C)	Ref (#)	Evaluation (T P E A A)	Comments
2.220E-01	2.800E+01	18	B078	1 0 0 0 1	
2.159E-01	2.724E+01	18	D049	1 2 0 0 1	

105. C₂H₇N

Ethylamine
Aethylamin

RN:	75-04-7	MP (°C):	−81
MW:	45.08	BP (°C):	16.6

Solubility (Moles/L)	Solubility (Grams/L)	Temp (°C)	Ref (#)	Evaluation (T P E A A)	Comments
2.686E-02	1.211E+00	25	B004	0 0 0 0 0	

106. C₂H₇NO₃S

Taurine
Taurin

RN:	107-35-7	MP (°C):	328
MW:	125.15	BP (°C):	

Solubility (Moles/L)	Solubility (Grams/L)	Temp (°C)	Ref (#)	Evaluation (T P E A A)	Comments
2.999E-01	3.754E+01	0	M043	1 0 0 0 1	
4.523E-01	5.660E+01	10	M043	1 0 0 0 1	
4.842E-01	6.060E+01	12	F300	1 0 0 0 2	
3.919E-01	4.905E+01	15	G081	1 0 1 1 2	
6.448E-01	8.070E+01	20	F300	1 0 0 0 2	
6.463E-01	8.088E+01	20	M043	1 0 0 0 1	
4.700E-01	5.882E+01	24	D031	1 0 0 0 2	
7.580E-01	9.486E+01	25	D041	1 0 0 0 2	
8.815E-01	1.103E+02	30	M043	1 0 0 0 2	
1.149E+00	1.438E+02	40	M043	1 0 0 0 2	
1.719E+00	2.151E+02	60	M043	1 0 0 0 2	
1.985E+00	2.484E+02	70	F300	1 0 0 0 2	
2.105E+00	2.634E+02	75	D041	1 0 0 0 2	
2.217E+00	2.775E+02	80	M043	1 0 0 0 2	
2.506E+00	3.137E+02	100	M043	1 0 0 0 2	

107. C₂H₇O₂As

107. $C_2H_7O_2As$
Cacodylic acid
Dimethylarsinsaeure
Kakodylsaeure
Arsine oxide, hydroxydimethyl-
Cacodylic acid

RN: 75-60-5   **MP** (°C):   195
MW: 138.00   **BP** (°C):

Solubility (Moles/L)	Solubility (Grams/L)	Temp (°C)	Ref (#)	Evaluation (T P E A A)	Comments
2.899E+00	4.001E+02	20	B200	1 0 0 0 2	
3.287E+00	4.536E+02	22	B185	0 0 0 0 0	
3.290E+00	4.540E+02	22	F300	1 0 0 0 2	
4.961E+00	6.845E+02	25	D305	1 0 0 0 2	
1.449E+01	2.000E+03	25	M161	1 0 0 0 0	

108. C₂H₇As

108. C_2H_7As
Ethylarsine
Aethylarsin
Arsen

RN: 593-59-9   **MP** (°C):
MW: 106.00   **BP** (°C):   36

Solubility (Moles/L)	Solubility (Grams/L)	Temp (°C)	Ref (#)	Evaluation (T P E A A)	Comments
1.226E-03	1.300E-01	19	F300	1 0 0 0 1	

109. C₂Cl₂F₄

109. $C_2Cl_2F_4$
1,2-Dichlorotetrafluoroethane
CFC-114
sym-Dichlorotetrafluoroethane
Halon 242
1,2-Dichloro-1,1,2,2-tetrafluoroethane
Cryofluorane

RN: 76-14-2   **MP** (°C):   −94
MW: 170.92   **BP** (°C):   3.8

Solubility (Moles/L)	Solubility (Grams/L)	Temp (°C)	Ref (#)	Evaluation (T P E A A)	Comments
7.605E-04	1.300E-01	25	R048	0 0 0 0 0	

110. C₂Cl₃F₃

1,1,2-Trichloro-1,2,2-trifluoroethane
Freon 113
Fluorocarbon 113
Halocarbon 113

RN: 76-13-1   **MP** (°C):   −36.4
MW: 187.38   **BP** (°C):   47.6

Solubility (Moles/L)	Solubility (Grams/L)	Temp (°C)	Ref (#)	Evaluation (T P E A A)	Comments
9.071E-04	1.700E-01	25	R048	0 0 0 0 0	

111. C₂Cl₄

Tetrachloroethylene
Ethylene tetrachloride
Perchloroethylene
Tetrachloroethene
Tetrachloro-ethylene
PERC

RN: 127-18-4   **MP** (°C):   −22
MW: 165.83   **BP** (°C):   121

Solubility (Moles/L)	Solubility (Grams/L)	Temp (°C)	Ref (#)	Evaluation (T P E A A)	Comments
1.206E-03	2.000E-01	20	C094	1 0 1 0 2	
1.206E-03	2.000E-01	20	C121	0 0 0 0 0	unit assumed, *sic*
9.045E-04	1.500E-01	20	M133	1 0 0 0 2	
9.045E-04	1.500E-01	20	P046	1 0 0 0 0	
9.044E-04	1.500E-01	25	A094	1 0 0 0 1	
2.920E-03	4.842E-01	25	B173	2 0 2 2 2	
1.206E-03	2.000E-01	25	C119	2 2 2 2 2	
2.412E-03	4.000E-01	25	F071	1 1 2 1 2	
9.044E-04	1.500E-01	25	G056	1 0 0 0 2	
9.045E-04	1.500E-01	25	M037	1 1 0 0 1	
9.045E-04	1.500E-01	25	M368	1 0 0 0 1	
9.044E-04	1.500E-01	25	N034	1 0 0 0 1	
2.412E-03	4.000E-01	ns	M344	0 0 0 0 2	
9.044E-04	1.500E-01	ns	O006	0 0 0 0 1	

112. C$_2$Cl$_6$

Hexachloroethane
1,1,1,2,2,2-Hexachloroethane
Avlothane
Distopin
Distopan
Distokal

RN: 67-72-1   **MP** (°C):   187
MW: 236.74   **BP** (°C):   186.8

Solubility (Moles/L)	Solubility (Grams/L)	Temp (°C)	Ref (#)	Evaluation (T P E A A)	Comments
3.253E-05	7.700E-03	20	M339	2 2 2 2 1	
2.112E-04	5.000E-02	22.3	M037	1 1 0 0 0	
1.148E-04	2.718E-02	ns	R427	0 0 0 0 0	

113. C$_2$N$_2$

Cyanogen
Dicyan

RN: 460-19-5   **MP** (°C):
MW: 52.04   **BP** (°C):

Solubility (Moles/L)	Solubility (Grams/L)	Temp (°C)	Ref (#)	Evaluation (T P E A A)	Comments
1.572E+01	8.182E+02	20	F300	1 0 0 0 1	

114. C$_2$N$_4$S$_2$

Cyanogen azidodithiocarbonate

RN: **MP** (°C):
MW: 144.18   **BP** (°C):

Solubility (Moles/L)	Solubility (Grams/L)	Temp (°C)	Ref (#)	Evaluation (T P E A A)	Comments
1.040E-02	1.500E+00	0	A055	0 0 0 0 2	

115. C$_2$N$_6$S$_4$

Thioperoxydicarbonic diazide
Azidoschwefel-kohlenstoff
Azidocarbonicdisulfide

RN: 148832-09-1   **MP** (°C):
MW: 236.32   **BP** (°C):

Solubility (Moles/L)	Solubility (Grams/L)	Temp (°C)	Ref (#)	Evaluation (T P E A A)	Comments
1.269E-03	3.000E-01	25	F300	1 0 0 0 0	

116. C$_3$H$_2$Cl$_2$N$_2$O$_2$
1,3-Dichlorohydantoin
2,4-Imidazolidinedione, 1,3-dichloro-
RN: 2958-99-8   **MP** (°C):
MW: 168.97   **BP** (°C):

Solubility (Moles/L)	Solubility (Grams/L)	Temp (°C)	Ref (#)	Evaluation (T P E A A)	Comments
4.114E-02	6.951E+00	20	B080	1 0 1 1 0	
8.171E-02	1.381E+01	40	B080	1 0 1 1 1	

117. C$_3$H$_2$N$_2$
Malononitrile
Malonsaeure-dinitril
RN: 109-77-3   **MP** (°C):   32
MW: 66.06   **BP** (°C):   218.5

Solubility (Moles/L)	Solubility (Grams/L)	Temp (°C)	Ref (#)	Evaluation (T P E A A)	Comments
1.780E+00	1.176E+02	20	F300	1 0 0 0 2	
1.778E+00	1.175E+02	ns	R424	0 0 0 0 0	

118. C$_3$H$_2$N$_2$O$_3$
Parabanic acid
Parabansaeure
RN: 120-89-8   **MP** (°C):
MW: 114.06   **BP** (°C):

Solubility (Moles/L)	Solubility (Grams/L)	Temp (°C)	Ref (#)	Evaluation (T P E A A)	Comments
3.945E-01	4.500E+01	8	F300	1 0 0 0 1	

119. C$_3$H$_3$Cl$_3$O$_3$
β,β,β-Trichlorolactic acid
β,β,β-Trichlor-milchsaeure
RN: 599-01-9   **MP** (°C):
MW: 193.41   **BP** (°C):

Solubility (Moles/L)	Solubility (Grams/L)	Temp (°C)	Ref (#)	Evaluation (T P E A A)	Comments
2.265E+00	4.380E+02	25	F300	1 0 0 0 2	

120. C₃H₃N

Acrylonitrile
Propenitrile

RN: 107-13-1   **MP** (°C):   −83.5
MW: 53.06   **BP** (°C):   77.3

Solubility (Moles/L)	Solubility (Grams/L)	Temp (°C)	Ref (#)	Evaluation (T P E A A)	Comments
1.266E+00	6.716E+01	0	D046	0 0 0 0 0	
1.266E+00	6.716E+01	0	D046	2 2 0 0 1	EFG
1.282E+00	6.803E+01	20	D046	0 0 0 0 0	
1.282E+00	6.803E+01	20	D046	2 2 0 0 1	EFG
1.298E+00	6.890E+01	25	D046	2 2 0 0 1	EFG
1.298E+00	6.890E+01	25	D046	0 0 0 0 0	
1.298E+00	6.890E+01	25	L096	1 2 0 2 1	
1.413E+00	7.500E+01	25	M161	1 0 0 0 1	
1.315E+00	6.977E+01	28	D046	2 2 0 0 1	EFG
1.347E+00	7.149E+01	36	D046	2 2 0 0 1	EFG
1.364E+00	7.236E+01	39	D046	2 2 0 0 1	EFG
1.388E+00	7.365E+01	41	D046	2 2 0 0 2	EFG
1.508E+00	8.004E+01	49	D046	2 2 0 0 1	EFG
1.508E+00	8.004E+01	53	D046	2 2 0 0 1	EFG
1.540E+00	8.173E+01	59	D046	2 2 0 0 1	EFG
1.603E+00	8.509E+01	63	D046	2 2 0 0 1	EFG
1.760E+00	9.338E+01	65	A324	2 2 2 1 2	
1.651E+00	8.759E+01	68	D046	2 2 0 0 0	EFG
1.721E+00	9.132E+01	72	D046	2 2 0 0 0	EFG
1.869E+00	9.918E+01	80	D046	2 2 0 0 0	EFG
1.974E+00	1.047E+02	85	D046	2 2 1 1 0	EFG
2.124E+00	1.127E+02	90	D046	2 2 1 1 0	EFG

121. C₃H₃NOS₂

Rhodanine
Rhodanin

RN: 141-84-4   **MP** (°C):   170
MW: 133.19   **BP** (°C):

Solubility (Moles/L)	Solubility (Grams/L)	Temp (°C)	Ref (#)	Evaluation (T P E A A)	Comments
1.689E-02	2.250E+00	25	F300	1 0 0 0 2	

122. $C_3H_3N_3O_3$
Cyanuric acid
Cyanursaeure
Isocyanuric acid
Isocyanursaeure

RN: 108-80-5	**MP** (°C):	
MW: 129.08	**BP** (°C):	

Solubility (Moles/L)	Solubility (Grams/L)	Temp (°C)	Ref (#)	Evaluation (T P E A A)	Comments
2.300E-02	2.969E+00	2	B193	1 2 0 0 1	
3.874E-02	5.000E+00	20	F300	1 0 0 0 0	
2.009E-02	2.593E+00	25	B384	0 0 0 0 0	

123. $C_3H_3N_3O_3$
Cyamelide
Cyamelid

RN: 462-02-2	**MP** (°C):	
MW: 129.08	**BP** (°C):	

Solubility (Moles/L)	Solubility (Grams/L)	Temp (°C)	Ref (#)	Evaluation (T P E A A)	Comments
7.747E-04	1.000E-01	15	F300	1 0 0 0 0	

124. $C_3H_3N_3S_3$
Trithiocyanuric acid
s-Triazine-2,4,6-trithiol
Trimercapto-s-triazine

RN: 638-16-4	**MP** (°C):	
MW: 177.27	**BP** (°C):	

Solubility (Moles/L)	Solubility (Grams/L)	Temp (°C)	Ref (#)	Evaluation (T P E A A)	Comments
1.354E-03	2.399E-01	25	B384	0 0 0 0 0	

125. C_3H_4
Propyne
Methyl acetylene
Methylacetylene

RN: 74-99-7	**MP** (°C): −101	
MW: 40.07	**BP** (°C): −23.2	

Solubility (Moles/L)	Solubility (Grams/L)	Temp (°C)	Ref (#)	Evaluation (T P E A A)	Comments
8.085E-02	3.239E+00	21	I011	1 2 2 1 2	
9.085E-02	3.640E+00	25	M001	2 1 2 2 2	
5.488E-02	2.199E+00	38	I011	1 2 2 1 1	
3.606E-02	1.445E+00	54	I011	1 2 2 1 1	
2.220E-02	8.895E-01	71	I011	1 2 2 1 1	
8.886E-03	3.560E-01	88	I011	1 2 2 1 1	

126. $C_3H_4ClN_5$
Desethyl simazine
Amino-2-chloro-6-ethylamino-s-triazine
6-Chloro-N-ethyl-1,3,5-triazine-2,4-diamine
RN: 1007-28-9   **MP** (°C):
MW: 145.55   **BP** (°C):

Solubility (Moles/L)	Solubility (Grams/L)	Temp (°C)	Ref (#)	Evaluation (T P E A A)	Comments
1.200E-03	1.747E-01	2	B193	1 1 0 0 0	

127. $C_3H_4Cl_2$
1,2-Dichloropropene
Dichloropropylene
RN: 26952-23-8   **MP** (°C):
MW: 110.97   **BP** (°C):   92

Solubility (Moles/L)	Solubility (Grams/L)	Temp (°C)	Ref (#)	Evaluation (T P E A A)	Comments
2.427E-02	2.693E+00	20	C121	1 0 0 0 1	unit assumed, *sic*

128. $C_3H_4Cl_2$
trans-1,3-Dichloropropene
1,3-Dichloropropylene (*trans*)
trans-1,3-Dichloropropylene
1,3-Dichloropropene
RN: 542-75-6   **MP** (°C):
MW: 110.97   **BP** (°C):   112

Solubility (Moles/L)	Solubility (Grams/L)	Temp (°C)	Ref (#)	Evaluation (T P E A A)	Comments
2.703E-03	2.999E-01	20	C121	1 0 0 0 0	unit assumed, *sic*
9.011E-03	1.000E+00	20	M161	1 0 0 0 0	
1.071E-02	1.188E+00	30	M300	1 1 2 2 2	

129. $C_3H_4Cl_2$
cis-1,3-Dichloropropene
1,3-Dichloropropropylene (*cis*)
cis-1,3-Dichloropropylene
cis 1,3-Dichloro-propene
cis-1,3-Dichloro-1-propene
(Z)-1,3-Dichloropropene
RN: 10061-01-5   **MP** (°C):
MW: 110.97   **BP** (°C):   108

Solubility (Moles/L)	Solubility (Grams/L)	Temp (°C)	Ref (#)	Evaluation (T P E A A)	Comments
2.433E-02	2.700E+00	20	G080	1 0 0 0 1	
9.651E-03	1.071E+00	30	M300	1 1 2 2 2	
8.211E-03	9.112E-01	30	M311	1 1 2 2 2	

130. C$_3$H$_4$Cl$_2$

trans-1,3-Dichloro-propene
trans-1,3-Dichloro-1-propene
(E)-1,3-Dichloro-1-propene
E-1,3-Dichloropropene

RN:	10061-02-6	MP (°C):	
MW:	110.97	BP (°C):	111

Solubility (Moles/L)	Solubility (Grams/L)	Temp (°C)	Ref (#)	Evaluation (T P E A A)	Comments
2.523E-02	2.800E+00	20	G080	1 0 0 0 1	

131. C$_3$H$_4$Cl$_2$O$_2$

Dalapon
α,α-Dichlor-propionsaeure

RN:	75-99-0	MP (°C):	
MW:	142.97	BP (°C):	187.5

Solubility (Moles/L)	Solubility (Grams/L)	Temp (°C)	Ref (#)	Evaluation (T P E A A)	Comments
3.511E+00	5.020E+02	25	M161	1 0 0 0 2	
3.511E+00	5.020E+02	ns	K138	0 0 0 0 1	

132. C$_3$H$_4$N$_2$O

Cyanoacetamide
Cyanessigsaeure-amid

RN:	107-91-5	MP (°C):	
MW:	84.08	BP (°C):	

Solubility (Moles/L)	Solubility (Grams/L)	Temp (°C)	Ref (#)	Evaluation (T P E A A)	Comments
1.546E+00	1.300E+02	20	F300	1 0 0 0 1	

133. C$_3$H$_4$N$_2$O$_2$

Hydantoin
2,4-Imidazolidinedione

RN:	461-72-3	MP (°C):	220
MW:	100.08	BP (°C):	

Solubility (Moles/L)	Solubility (Grams/L)	Temp (°C)	Ref (#)	Evaluation (T P E A A)	Comments
2.944E+00	2.946E+02	100	F300	1 0 0 0 2	
3.970E-01	3.973E+01	ns	M025	0 2 0 1 2	

134. C$_3$H$_4$N$_2$O$_3$S
2-Imidazole sulfonic acid
Imidazol-sulfosaeure-(2)
RN: 53744-47-1   **MP** (°C):
MW: 148.14   **BP** (°C):

Solubility (Moles/L)	Solubility (Grams/L)	Temp (°C)	Ref (#)	Evaluation (T P E A A)	Comments
5.009E-01	7.420E+01	20	F300	1 0 0 0 2	

135. C$_3$H$_4$N$_4$O$_2$
Ammelide
2,4-Dihydroxy-6-amino-1,3,5-triazine
RN: 645-93-2   **MP** (°C):
MW: 128.09   **BP** (°C):

Solubility (Moles/L)	Solubility (Grams/L)	Temp (°C)	Ref (#)	Evaluation (T P E A A)	Comments
6.000E-04	7.685E-02	2	B193	1 2 0 0 0	

136. C$_3$H$_4$O
Acrolein
2-Propenal
Acrylaldehyde
RN: 107-02-8   **MP** (°C):   −88.0
MW: 56.06   **BP** (°C):   52.5

Solubility (Moles/L)	Solubility (Grams/L)	Temp (°C)	Ref (#)	Evaluation (T P E A A)	Comments
8.690E+00	4.872E+02	0	B111	1 0 0 1 1	Quinol as a stabilizer
3.764E+00	2.110E+02	20	F300	1 0 0 0 2	
3.071E+00	1.722E+02	20	M161	1 0 0 0 1	
8.522E+00	4.778E+02	32.50	B111	1 0 0 1 2	Quinol as a stabilizer
8.429E+00	4.726E+02	44.40	B111	1 0 0 1 2	Quinol as a stabilizer
8.339E+00	4.675E+02	50	B111	1 0 0 1 2	Quinol as a stabilizer
8.288E+00	4.647E+02	53	B111	1 0 0 1 2	Quinol as a stabilizer
7.889E+00	4.423E+02	74.50	B111	1 0 0 1 2	Quinol as a stabilizer
7.338E+00	4.114E+02	82	B111	1 0 0 1 2	Quinol as a stabilizer
7.013E+00	3.932E+02	84	B111	1 0 0 1 2	Quinol as a stabilizer
6.597E+00	3.699E+02	87.80	B111	1 0 0 1 2	Quinol as a stabilizer
6.417E+00	3.598E+02	88	B111	1 0 0 1 2	Quinol as a stabilizer
5.096E+00	2.857E+02	ns	B185	0 0 0 0 0	
3.567E+00	2.000E+02	ns	B200	0 0 0 0 0	

137. $C_3H_4O_4$
Malonic acid
Acide malonique
Malonsaeure

RN: 141-82-2   **MP** (°C):   135
MW: 104.06   **BP** (°C):

Solubility (Moles/L)	Solubility (Grams/L)	Temp (°C)	Ref (#)	Evaluation (T P E A A)	Comments
3.645E+00	3.793E+02	0	F300	1 0 0 0 2	
5.871E+00	6.110E+02	0	L041	1 0 0 1 2	
4.990E+00	5.192E+02	0	M043	1 0 0 0 2	
5.871E+00	6.110E+02	0	M051	1 0 0 0 2	
4.743E+00	4.936E+02	4.99	A339	0 0 0 0 0	
5.427E+00	5.648E+02	10	K077	1 2 2 2 2	average of 3
5.395E+00	5.614E+02	10	M043	1 0 0 0 2	
4.888E+00	5.087E+02	9.99	A339	0 0 0 0 0	
5.034E+00	5.238E+02	14.99	A339	0 0 0 0 0	
5.608E+00	5.836E+02	15	K077	1 2 2 2 2	
6.746E+00	7.020E+02	15	L041	1 0 0 1 2	
6.746E+00	7.020E+02	15	M051	1 0 0 0 2	
5.728E+00	5.961E+02	18	K077	1 2 2 2 2	
5.198E+00	5.409E+02	19.99	A339	0 0 0 0 0	
7.063E+00	7.350E+02	20	L041	1 0 0 1 2	
5.811E+00	6.047E+02	20	M043	1 0 0 0 2	
4.067E+00	4.232E+02	20	M171	1 0 0 0 2	
2.670E+00	2.778E+02	20	S006	1 0 0 0 2	
5.928E+00	6.169E+02	24	K077	1 2 2 2 2	
5.354E+00	5.571E+02	24.99	A339	0 0 0 0 0	
4.221E+00	4.393E+02	25	F300	1 0 0 0 2	
5.990E+00	6.233E+02	25	K077	1 2 2 2 2	
7.332E+00	7.630E+02	25	M051	1 0 0 0 2	
5.494E+00	5.717E+02	29.99	A339	0 0 0 0 0	
6.178E+00	6.429E+02	30	M043	1 0 0 0 2	
5.638E+00	5.867E+02	34.99	A339	0 0 0 0 0	
7.938E+00	8.260E+02	35	L041	1 0 0 1 2	
5.800E+00	6.035E+02	39.99	A339	0 0 0 0 0	
6.530E+00	6.795E+02	40	M043	1 0 0 0 2	
5.913E+00	6.153E+02	44.99	A339	0 0 0 0 0	
6.028E+00	6.273E+02	49.99	A339	0 0 0 0 0	
8.898E+00	9.260E+02	50	L041	1 0 0 1 2	
8.898E+00	9.260E+02	50	M051	1 0 0 0 2	
6.895E+00	7.175E+02	53	K077	1 2 2 2 2	
6.182E+00	6.433E+02	54.99	A339	0 0 0 0 0	
6.328E+00	6.585E+02	59.99	A339	0 0 0 0 0	
7.158E+00	7.449E+02	60	M043	1 0 0 0 2	
6.451E+00	6.713E+02	64.99	A339	0 0 0 0 0	
9.831E+00	1.023E+03	65	L041	1 0 0 1 2	
7.878E+00	8.198E+02	80	M043	1 0 0 0 2	
8.267E+00	8.603E+02	93	K077	1 2 2 2 2	
8.554E+00	8.901E+02	100	M043	1 0 0 0 2	
9.610E+00	1.000E+03	132	K077	1 2 2 2 2	
1.441E+01	1.500E+03	ns	D072	0 0 0 0 1	

138. C$_3$H$_5$Br
Allyl bromide
3-Bromopropene
RN: 106-95-6   **MP** (°C):   −119
MW: 120.98   **BP** (°C):   71.3

Solubility (Moles/L)	Solubility (Grams/L)	Temp (°C)	Ref (#)	Evaluation (T P E A A)	Comments
3.170E-02	3.835E+00	25	M342	1 0 1 1 2	

139. C$_3$H$_5$Bvr$_2$Cl
1,2-Dibromo-3-chloropropane
1-Chloro-2,3-dibromopropane
Nemagon
RN: 96-12-8   **MP** (°C):
MW: 236.34   **BP** (°C):   196

Solubility (Moles/L)	Solubility (Grams/L)	Temp (°C)	Ref (#)	Evaluation (T P E A A)	Comments
5.204E-03	1.230E+00	20	G080	1 0 0 0 1	
4.227E-03	9.990E-01	20	P081	1 0 0 0 0	
4.227E-03	9.990E-01	ns	I316	0 0 0 0 0	
4.227E-03	9.990E-01	ns	M061	0 0 0 0 0	
4.231E-03	1.000E+00	rt	M161	0 0 0 0 0	

140. C$_3$H$_5$Cl
Allyl chloride
3-Chloro-1-propene
RN: 107-05-1   **MP** (°C):   −134
MW: 76.53   **BP** (°C):   44

Solubility (Moles/L)	Solubility (Grams/L)	Temp (°C)	Ref (#)	Evaluation (T P E A A)	Comments
4.687E-02	3.587E+00	20	G056	1 0 0 0 2	
1.305E-02	9.990E-01	ns	N034	0 0 0 0 0	

141. C$_3$H$_5$ClO
Chloroacetone
1-Chloro-2-propanone
Chloraceton
RN: 78-95-5   **MP** (°C):   −44.5
MW: 92.53   **BP** (°C):   119.7

Solubility (Moles/L)	Solubility (Grams/L)	Temp (°C)	Ref (#)	Evaluation (T P E A A)	Comments
8.924E-01	8.257E+01	ns	N034	0 0 0 0 0	

142. C₃H₅ClO

C_3H_5ClO

Epichlorohydrin
Epichloridrina

RN: 106-89-8   **MP** (°C):   −25.6
MW: 92.53   **BP** (°C):   117.9

Solubility (Moles/L)	Solubility (Grams/L)	Temp (°C)	Ref (#)	Evaluation (T P E A A)	Comments
6.577E-01	6.086E+01	0	L061	1 2 2 1 2	
6.615E-01	6.121E+01	10	L061	1 2 2 1 2	
6.501E-01	6.015E+01	20	I313	0 0 0 0 0	
6.692E-01	6.191E+01	30.20	L061	1 2 2 1 2	
7.568E-01	7.003E+01	52	L061	1 2 2 1 2	
8.421E-01	7.792E+01	65	L061	1 2 2 1 2	
9.232E-01	8.542E+01	72	L061	1 2 2 1 2	
1.024E+00	9.478E+01	80.20	L061	1 2 2 1 2	

143. C₃H₅Cl₂NO₂

$C_3H_5Cl_2NO_2$

1,1-Dichloro-1-nitropropane
Propane, 1,1-dichloro-1-nitro-

RN: 595-44-8   **MP** (°C):
MW: 157.98   **BP** (°C):   141

Solubility (Moles/L)	Solubility (Grams/L)	Temp (°C)	Ref (#)	Evaluation (T P E A A)	Comments
3.149E-02	4.975E+00	20	C121	1 0 0 0 0	unit assumed, *sic*

144. C₃H₅Cl₃

$C_3H_5Cl_3$

1,2,3-Trichloropropane
Allyl trichloride
Trichlorohydrin
Glycerol trichlorohydrin

RN: 96-18-4   **MP** (°C):   −14
MW: 147.43   **BP** (°C):   156

Solubility (Moles/L)	Solubility (Grams/L)	Temp (°C)	Ref (#)	Evaluation (T P E A A)	Comments
1.289E-02	1.900E+00	ns	H123	0 0 0 0 0	

145. C₃H₅IO₂

$C_3H_5IO_2$

β-Iodopropionic acid
β-Iod-propionsaeure

RN: 141-76-4   **MP** (°C):   81.5
MW: 199.98   **BP** (°C):

Solubility (Moles/L)	Solubility (Grams/L)	Temp (°C)	Ref (#)	Evaluation (T P E A A)	Comments
3.715E-01	7.430E+01	25	F300	1 0 0 0 2	

146. C$_3$H$_5$N
Propionitrile
Propionsaeure-nitril
n-Propionitrile

RN:	107-12-0	**MP** (°C):	–93	
MW:	55.08	**BP** (°C):	97	

Solubility (Moles/L)	Solubility (Grams/L)	Temp (°C)	Ref (#)	Evaluation (T P E A A)	Comments
6.151E-02	3.388E+00	25	B004	0 0 0 0 0	

147. C$_3$H$_5$N
Ethyl isocyanide
Ethane, isocyano-

RN:	624-79-3	**MP** (°C):	
MW:	55.08	**BP** (°C):	

Solubility (Moles/L)	Solubility (Grams/L)	Temp (°C)	Ref (#)	Evaluation (T P E A A)	Comments
1.814E-02	9.990E-01	ns	L055	0 0 0 0 1	

148. C$_3$H$_5$NO
Acrylamide
2-Propenamide

RN:	79-06-1	**MP** (°C):	84
MW:	71.08	**BP** (°C):	

Solubility (Moles/L)	Solubility (Grams/L)	Temp (°C)	Ref (#)	Evaluation (T P E A A)	Comments
4.299E+00	3.056E+02	0	M147	0 2 1 1 0	EFG
4.690E+00	3.333E+02	10	M147	0 2 1 1 0	EFG
5.220E+00	3.711E+02	20	M147	0 2 1 1 0	EFG
5.695E+00	4.048E+02	30	M147	0 2 1 1 0	EFG
6.075E+00	4.318E+02	40	M147	0 2 1 1 0	EFG
6.253E+00	4.444E+02	50	M147	0 2 1 1 0	EFG
6.625E+00	4.709E+02	60	M147	0 2 1 1 0	EFG
7.034E+00	5.000E+02	80	M147	0 2 1 1 0	EFG

149. C$_3$H$_5$NO$_3$
Formylglycine
N-Formyl glycine

RN:	2491-15-8	**MP** (°C):	
MW:	103.08	**BP** (°C):	

Solubility (Moles/L)	Solubility (Grams/L)	Temp (°C)	Ref (#)	Evaluation (T P E A A)	Comments
1.849E+00	1.906E+02	25	M024	1 2 0 1 2	
1.849E+00	1.906E+02	ns	M025	0 2 0 1 2	

150. C₃H₅N₃O

Ethylnitrosocyanamide

ENC

RN: 38434-77-4   **MP** (°C):

MW: 99.09   **BP** (°C):

Solubility (Moles/L)	Solubility (Grams/L)	Temp (°C)	Ref (#)	Evaluation (T P E A A)	Comments
1.400E-01	1.387E+01	24	M031	1 1 1 1 1	

151. C₃H₅N₃O₉

Nitroglycerin

Nitroglycerol

RN: 55-63-0   **MP** (°C): 13.5

MW: 227.09   **BP** (°C): 256

Solubility (Moles/L)	Solubility (Grams/L)	Temp (°C)	Ref (#)	Evaluation (T P E A A)	Comments
5.629E-03	1.278E+00	15	L063	2 0 1 1 2	
7.926E-03	1.800E+00	20	F300	1 0 0 0 1	
6.069E-03	1.378E+00	20	L063	2 0 1 1 2	
5.504E-03	1.250E+00	25	P312	0 0 0 0 0	
6.595E-03	1.498E+00	30	L063	2 0 1 1 2	
7.342E-03	1.667E+00	40	L063	2 0 1 1 2	
8.570E-03	1.946E+00	50	L063	2 0 1 1 2	
1.041E-02	2.364E+00	60	L063	2 0 1 1 2	
1.265E-02	2.872E+00	70	L063	2 0 1 1 2	
1.518E-02	3.448E+00	80	L063	2 0 1 1 2	

152. C₃H₅N₅O

Ammeline

Ammelin

RN: 645-92-1   **MP** (°C):

MW: 127.11   **BP** (°C):

Solubility (Moles/L)	Solubility (Grams/L)	Temp (°C)	Ref (#)	Evaluation (T P E A A)	Comments
6.000E-04	7.626E-02	2	B193	1 1 0 0 0	
5.901E-04	7.500E-02	23	F300	1 0 0 0 1	
2.486E-03	3.160E-01	100	F300	1 0 0 0 2	

153. C₃H₆

Cyclopropane
Trimethylene

RN: 75-19-4   **MP** (°C):   −127
MW: 42.08   **BP** (°C):   −33

Solubility (Moles/L)	Solubility (Grams/L)	Temp (°C)	Ref (#)	Evaluation (T P E A A)	Comments
2.461E-02	1.036E+00	5.05	Z008	2 1 2 2 2	at 97.26 kPa
1.281E-02	5.390E-01	20	R060	0 0 0 0 0	
1.754E-02	7.382E-01	21	I017	1 2 2 1 2	at 16.9 psia
1.103E-02	4.640E-01	25	R060	0 0 0 0 0	
9.315E-03	3.920E-01	30	R060	0 0 0 0 0	
8.983E-03	3.780E-01	31	R060	0 0 0 0 0	
7.723E-03	3.250E-01	35	R060	0 0 0 0 0	
1.083E-02	4.557E-01	38	I017	1 2 2 1 2	at 17.0 psia
6.844E-03	2.880E-01	39	R060	0 0 0 0 0	
5.917E-03	2.490E-01	45	R060	0 0 0 0 0	
8.386E-03	3.529E-01	71	I017	1 2 2 1 2	at 19.9 psia
3.999E-03	1.683E-01	104	I017	1 2 2 1 2	at 24.9 psia
5.896E+00	2.481E+02	ns	R028	0 0 0 0 0	

154. C₃H₆

Propylene
Methyl ethylene
Propene

RN: 115-07-1   **MP** (°C):   −185
MW: 42.08   **BP** (°C):   −48

Solubility (Moles/L)	Solubility (Grams/L)	Temp (°C)	Ref (#)	Evaluation (T P E A A)	Comments
2.139E-02	9.000E-01	0	F300	1 0 0 0 1	
7.553E-03	3.178E-01	21	A052	1 1 1 2 2	smoothed
7.842E-03	3.300E-01	25	F300	1 0 0 0 1	
4.753E-03	2.000E-01	25	M001	2 1 2 2 2	
4.221E-03	1.776E-01	38	A052	1 1 1 2 1	smoothed
2.333E-03	9.818E-02	54	A052	1 1 1 2 1	smoothed
1.500E-03	6.312E-02	71	A052	1 1 1 2 1	smoothed
7.222E-04	3.039E-02	88	A052	1 1 1 2 1	smoothed

155. C₃H₆BrCl

1-Bromo-3-chloropropane
w-Chlorobromopropane
3-Bromopropyl chloride
3-Chloro-1-bromopropane

RN: 109-70-6   **MP** (°C):   −58.9
MW: 157.44   **BP** (°C):   143.3

Solubility (Moles/L)	Solubility (Grams/L)	Temp (°C)	Ref (#)	Evaluation (T P E A A)	Comments
1.420E-02	2.236E+00	25	M342	1 0 1 1 2	

156. C$_3$H$_6$BrNO$_4$
Bronopol
2-Bromo-2-nitropropane-1,3-diol
RN: 52-51-7   **MP** (°C):   130
MW: 199.99   **BP** (°C):

Solubility (Moles/L)	Solubility (Grams/L)	Temp (°C)	Ref (#)	Evaluation (T P E A A)	Comments
1.000E+00	2.000E+02	22	M161	1 0 0 0 1	

157. C$_3$H$_6$Br$_2$
Trimethylene bromide
1,3-Dibromopropane
RN: 109-64-8   **MP** (°C):   −36
MW: 201.90   **BP** (°C):   167

Solubility (Moles/L)	Solubility (Grams/L)	Temp (°C)	Ref (#)	Evaluation (T P E A A)	Comments
8.406E-03	1.697E+00	20	C121	1 0 0 0 1	unit assumed, *sic*

158. C$_3$H$_6$ClNO$_2$
1-Chloro-1-nitropropane
Propane, 1-chloro-1-nitro-
RN: 600-25-9   **MP** (°C):
MW: 123.54   **BP** (°C):   141

Solubility (Moles/L)	Solubility (Grams/L)	Temp (°C)	Ref (#)	Evaluation (T P E A A)	Comments
6.424E-02	7.937E+00	20	C121	1 0 0 0 0	unit assumed, *sic*
4.027E-02	4.975E+00	20	M061	1 0 0 0 0	

159. C$_3$H$_6$ClNO$_2$
1-Chloro-2-nitropropane
Propane, 1-chloro-2-nitro-
RN: 37809-02-2   **MP** (°C):
MW: 123.54   **BP** (°C):   174

Solubility (Moles/L)	Solubility (Grams/L)	Temp (°C)	Ref (#)	Evaluation (T P E A A)	Comments
6.424E-02	7.937E+00	20	M061	1 0 0 0 0	

160. $C_3H_6Cl_2$
Propylene dichloride
1,2-Dichlor-propan
1,2-Dichloropropane
Propylene chloride
Dichloropropane

RN:	78-87-5	MP (°C):	−100.3
MW:	112.99	BP (°C):	96.8

Solubility (Moles/L)	Solubility (Grams/L)	Temp (°C)	Ref (#)	Evaluation (T P E A A)	Comments
3.160E-02	3.570E+00	20	C094	1 0 1 0 2	
2.383E-02	2.693E+00	20	C121	1 0 0 0 1	unit assumed, *sic*
2.390E-02	2.700E+00	20	F300	1 0 0 0 1	
2.390E-02	2.700E+00	20	M037	1 1 0 0 1	
2.383E-02	2.693E+00	20	M061	1 0 0 0 1	
2.295E-02	2.593E+00	20	M062	1 0 0 0 1	
2.390E-02	2.700E+00	20	M161	1 0 0 0 1	
2.500E-02	2.825E+00	20	M312	1 0 0 0 1	
2.383E-02	2.693E+00	20	N034	1 0 0 0 1	
2.478E-02	2.800E+00	25	F300	1 0 0 0 1	
2.480E-02	2.802E+00	25	G038	1 2 2 2 2	
2.480E-02	2.802E+00	25	G053	2 1 2 1 2	
2.295E-02	2.593E+00	25	G056	1 0 0 0 2	
2.142E-02	2.420E+00	30	M300	1 1 2 2 2	
1.831E-02	2.069E+00	30	M311	1 1 2 2 2	

161. $C_3H_6Cl_2$
1,3-Dichloropropane
1,3-Dichlor-propan

RN:	142-28-9	MP (°C):	−99
MW:	112.99	BP (°C):	120

Solubility (Moles/L)	Solubility (Grams/L)	Temp (°C)	Ref (#)	Evaluation (T P E A A)	Comments
2.559E-02	2.892E+00	20	C121	1 0 0 0 1	unit assumed, *sic*
2.416E-02	2.730E+00	25	F300	1 0 0 0 2	
2.430E-02	2.746E+00	25	G038	1 2 2 2 2	
2.430E-02	2.746E+00	25	G053	2 1 2 1 2	
9.027E-03	1.020E+00	30	M311	1 1 2 2 2	

162. $C_3H_6Cl_2O$
1,3-Dichloro-2-propanol
1,3-Dichlor-propanol-(2)

RN:	96-23-1	MP (°C):	−4
MW:	128.99	BP (°C):	174.3

Solubility (Moles/L)	Solubility (Grams/L)	Temp (°C)	Ref (#)	Evaluation (T P E A A)	Comments
7.675E-01	9.900E+01	19	F300	1 0 0 0 1	
6.984E-01	9.008E+01	19	N034	1 0 0 0 1	
1.124E+00	1.450E+02	72	F300	1 0 0 0 2	

163. C$_3$H$_6$N$_2$O$_2$
Malonic acid diamide
Malonsaeure-diamid
Malonamide
Malonodiamide
Propanediamide

RN: 108-13-4   **MP** (°C): 170
MW: 102.09   **BP** (°C):

Solubility (Moles/L)	Solubility (Grams/L)	Temp (°C)	Ref (#)	Evaluation (T P E A A)	Comments
7.513E-01	7.670E+01	8	F300	1 0 0 0 2	
7.830E-03	7.994E-01	ns	L055	0 0 0 0 1	

164. C$_3$H$_6$N$_2$O$_2$
Methylglyoxime
Methylglyoxim

RN: 1804-15-5   **MP** (°C):
MW: 102.09   **BP** (°C):

Solubility (Moles/L)	Solubility (Grams/L)	Temp (°C)	Ref (#)	Evaluation (T P E A A)	Comments
4.506E-01	4.600E+01	26	F300	1 0 0 0 1	
7.444E-01	7.600E+01	40	F300	1 0 0 0 1	

165. C$_3$H$_6$N$_2$O$_2$
Methylnitrosoacetamide
MNA

RN: 7417-67-6   **MP** (°C):
MW: 102.09   **BP** (°C):

Solubility (Moles/L)	Solubility (Grams/L)	Temp (°C)	Ref (#)	Evaluation (T P E A A)	Comments
1.700E-01	1.736E+01	24	M031	1 1 1 1 1	

166. C$_3$H$_6$N$_2$O$_2$
1-Acetylurea
Acetylharnstoff

RN: 591-07-1   **MP** (°C): 218
MW: 102.09   **BP** (°C): 185

Solubility (Moles/L)	Solubility (Grams/L)	Temp (°C)	Ref (#)	Evaluation (T P E A A)	Comments
1.273E-01	1.300E+01	15	F300	1 0 0 0 1	

167. $C_3H_6N_2O_3$
Hydantoic acid
N-(Carboxymethyl)urea
N-Carbamoylglycine
Carbamoylglycine
Glycoluric acid
RN: 462-60-2   **MP** (°C):
MW: 118.09   **BP** (°C):

Solub ility (Moles/L)	Solubility (Grams/L)	Temp (°C)	Ref (#)	Evaluation (T P E A A)	Comments
2.549E-01	3.010E+01	20	F300	1 0 0 0 2	
3.290E-01	3.885E+01	25	M024	1 2 0 1 2	
3.290E-01	3.885E+01	ns	M025	0 2 0 1 2	

168. $C_3H_6N_2O_7$
Glycerol 1,2-dinitrate
1,2,3-Propanetriol 1,2-dinitrate
1,2-Dinitroglycerol
RN: 131287-51-9   **MP** (°C):
MW: 182.09   **BP** (°C):   106

Solubility (Moles/L)	Solubility (Grams/L)	Temp (°C)	Ref (#)	Evaluation (T P E A A)	Comments
3.386E-01	6.165E+01	20	D013	1 0 1 1 2	

169. $C_3H_6N_2O_7$
Glycerol 1,3-dinitrate
Glycerol-α,α'-dinitrate
Glycerin-α,α'-dinitrate
RN: 623-87-0   **MP** (°C):   26
MW: 182.09   **BP** (°C):   116

Solubility (Moles/L)	Solubility (Grams/L)	Temp (°C)	Ref (#)	Evaluation (T P E A A)	Comments
3.993E-01	7.270E+01	20	D013	1 0 1 1 2	

170. $C_3H_6N_2S$
Ethylenethiourea
Mercaptoimidazoline
Mercozen
RN: 96-45-7   **MP** (°C):   203
MW: 102.16   **BP** (°C):

Solubility (Moles/L)	Solubility (Grams/L)	Temp (°C)	Ref (#)	Evaluation (T P E A A)	Comments
1.919E-01	1.961E+01	30	I310	0 0 0 0 0	
8.082E-01	8.257E+01	60	I310	0 0 0 0 0	
2.991E+00	3.056E+02	90	I310	0 0 0 0 0	

171. C$_3$H$_6$N$_4$Hg
Methylmercuridicyanodiamide
Panogen

RN: 502-39-6   **MP** (°C): 156
MW: 298.70   **BP** (°C):

Solubility (Moles/L)	Solubility (Grams/L)	Temp (°C)	Ref (#)	Evaluation (T P E A A)	Comments
7.265E-02	2.170E+01	20	M061	1 0 0 0 2	
7.265E-02	2.170E+01	rt	M161	0 0 0 0 2	

172. C$_3$H$_6$N$_6$
Melamine
1,3,5-Triazine-2,4,6-triamine
Cymel

RN: 108-78-1   **MP** (°C):
MW: 126.12   **BP** (°C):

Solubility (Moles/L)	Solubility (Grams/L)	Temp (°C)	Ref (#)	Evaluation (T P E A A)	Comments
9.503E-03	1.199E+00	0	M043	1 0 0 0 1	
1.000E-02	1.261E+00	2	B193	1 1 0 0 1	
1.425E-02	1.797E+00	10	M043	1 0 0 0 1	
2.561E-02	3.230E+00	19.90	C023	2 2 0 1 2	
2.135E-02	2.693E+00	20	M043	1 0 0 0 1	
3.316E-02	4.182E+00	30	M043	1 0 0 0 1	
4.651E-02	5.865E+00	34.90	C023	2 2 0 1 2	
5.590E-02	7.050E+00	40	M043	1 0 0 0 1	
8.200E-02	1.034E+01	49.80	C023	2 2 0 1 2	
1.172E-01	1.478E+01	60	M043	1 0 0 0 1	
1.325E-01	1.672E+01	64.10	C023	2 2 0 1 2	
1.836E-01	2.315E+01	74.50	C023	2 2 0 1 2	
2.160E-01	2.724E+01	80	M043	1 0 0 0 1	
2.421E-01	3.054E+01	83.50	C023	2 2 0 1 2	
3.480E-01	4.389E+01	94.80	C023	2 2 0 1 2	
3.812E-01	4.807E+01	99	C023	2 2 0 1 2	
3.776E-01	4.762E+01	100	M043	1 0 0 0 1	

173. C$_3$H$_6$N$_6$O$_6$
Cyclonite
RDX

RN: 121-82-4   **MP** (°C): 205
MW: 222.12   **BP** (°C):

Solubility (Moles/L)	Solubility (Grams/L)	Temp (°C)	Ref (#)	Evaluation (T P E A A)	Comments
2.690E-04	5.975E-02	25	B173	2 0 2 2 2	

174. C$_3$H$_6$O
Propylene oxide
Methyl ethylene oxide
RN: 75-56-9   **MP** (°C): −112
MW: 58.08   **BP** (°C): 34.23

Solubility (Moles/L)	Solubility (Grams/L)	Temp (°C)	Ref (#)	Evaluation (T P E A A)	Comments
4.963E+00	2.883E+02	20	I313	0 0 0 0 0	
2.544E-01	1.478E+01	20	M065	1 0 2 1 1	*sic*
6.389E+00	3.711E+02	25	I313	0 0 0 0 0	

175. C$_3$H$_6$O
Acetone
2-Propanone
Aceton
RN: 67-64-1   **MP** (°C): −94
MW: 58.08   **BP** (°C): 56.5

Solubility (Moles/L)	Solubility (Grams/L)	Temp (°C)	Ref (#)	Evaluation (T P E A A)	Comments
		0	C423	0 0 0 0 0	
		4	C423	0 0 0 0 0	
		10	C423	0 0 0 0 0	

176. C$_3$H$_6$O
Propaldehyde
Propyl aldehyde
Propanal
RN: 123-38-6   **MP** (°C): −81
MW: 58.08   **BP** (°C):

Solubility (Moles/L)	Solubility (Grams/L)	Temp (°C)	Ref (#)	Evaluation (T P E A A)	Comments
2.870E+00	1.667E+02	20	D041	1 0 0 0 0	
2.927E+00	1.700E+02	20	F300	1 0 0 0 1	
5.269E+00	3.060E+02	25	A049	1 0 0 0 2	
3.105E+00	1.803E+02	25	B060	2 0 1 1 1	
2.880E+00	1.673E+02	25	F044	1 0 0 0 2	

177. C$_3$H$_6$O$_2$
Propionic acid
n-Propionic acid
RN: 79-09-4   **MP** (°C): −22
MW: 74.08   **BP** (°C): 141

Solubility (Moles/L)	Solubility (Grams/L)	Temp (°C)	Ref (#)	Evaluation (T P E A A)	Comments
2.733E-01	2.025E+01	25	B004	0 0 0 0 0	

178. C₃H₆O₂

Ethyl formate
Ameisensaeure-aethyl ester
Formic acid ethyl ester

RN:	109-94-4	**MP (°C):**	−80	
MW:	74.08	**BP (°C):**	53	

Solubility (Moles/L)	Solubility (Grams/L)	Temp (°C)	Ref (#)	Evaluation (T P E A A)	Comments
1.094E+00	8.108E+01	5.0	K079	1 0 0 0 2	
1.139E+00	8.437E+01	15.9	K079	1 0 0 0 2	
1.350E+00	1.000E+02	18	F300	1 0 0 0 1	
1.350E+00	1.000E+02	22	S006	1 0 0 0 2	
1.194E+00	8.848E+01	30.2	K079	1 0 0 0 2	
1.239E+00	9.178E+01	38.0	K079	1 0 0 0 2	
1.283E+00	9.507E+01	45.1	K079	1 0 0 0 2	
1.339E+00	9.918E+01	50.0	K079	1 0 0 0 2	
1.383E+00	1.025E+02	55.5	K079	1 0 0 0 2	
1.517E+00	1.124E+02	63.9	K079	1 0 0 0 2	
1.639E+00	1.214E+02	70.0	K079	1 0 0 0 2	
1.778E+00	1.317E+02	75.5	K079	1 0 0 0 2	

179. C₃H₆O₂

Methyl acetate
Essigsaeures methyl
Methylacetat

RN:	79-20-9	**MP (°C):**	−98.0	
MW:	74.08	**BP (°C):**	56.9	

Solubility (Moles/L)	Solubility (Grams/L)	Temp (°C)	Ref (#)	Evaluation (T P E A A)	Comments
3.678E+00	2.725E+02	5.0	K079	1 0 0 0 2	
4.017E+00	2.976E+02	20	E002	1 0 0 0 2	
3.290E+00	2.437E+02	20	F001	1 0 1 2 2	
2.647E+00	1.961E+02	20	F300	1 0 0 0 2	
3.290E+00	2.437E+02	20	M171	1 0 0 0 2	
4.617E+00	3.420E+02	20	P040	0 0 0 0 0	
4.300E+00	3.185E+02	20	S006	1 0 0 0 1	
3.722E+00	2.757E+02	21.0	K079	1 0 0 0 2	
2.772E-02	2.054E+00	25	B004	0 0 0 0 0	*sic*
3.772E+00	2.794E+02	35.0	K079	1 0 0 0 2	
3.889E+00	2.881E+02	58.0	K079	1 0 0 0 2	
3.906E+00	2.893E+02	58.9	K079	1 0 0 0 2	
3.922E+00	2.906E+02	60.1	K079	1 0 0 0 2	
3.950E+00	2.926E+02	61.7	K079	1 0 0 0 2	
4.172E+00	3.091E+02	69.1	K079	1 0 0 0 2	
4.256E+00	3.153E+02	70.5	K079	1 0 0 0 2	
4.294E+00	3.181E+02	71.9	K079	1 0 0 0 2	
4.906E+00	3.634E+02	83.5	K079	1 0 0 0 2	
4.252E-02	3.150E+00	c	L055	0 0 0 0 2	

180. $C_3H_6O_2S_3$
α-Trimethylene trisulphide dioxide
1,3,5-Trithiane, 1,3-dioxide, *trans*-
RN: 60077-04-5   **MP** (°C):
MW: 170.27   **BP** (°C):

Solubility (Moles/L)	Solubility (Grams/L)	Temp (°C)	Ref (#)	Evaluation (T P E A A)	Comments
9.817E-02	1.672E+01	25	B112	1 2 1 1 2	

181. $C_3H_6O_2S_3$
β-Trimethylene trisulphide dioxide
1,3,5-Trithiane, 1,3-dioxide, *cis*-
RN: 60041-48-7   **MP** (°C):
MW: 170.27   **BP** (°C):

Solubility (Moles/L)	Solubility (Grams/L)	Temp (°C)	Ref (#)	Evaluation (T P E A A)	Comments
2.545E-01	4.334E+01	25	B112	1 2 1 1 2	

182. $C_3H_6O_3$
DL-Glyceraldehyde
DL-Glycerin-aldehyd
RN: 56-82-6   **MP** (°C):   145
MW: 90.08   **BP** (°C):   150

Solubility (Moles/L)	Solubility (Grams/L)	Temp (°C)	Ref (#)	Evaluation (T P E A A)	Comments
3.233E-01	2.913E+01	18	D041	1 0 0 0 0	
3.242E-01	2.920E+01	18	F300	1 0 0 0 2	

183. $C_3H_6O_3$
Hydracrylic acid
Hydracrylsaeure
RN: 503-66-2   **MP** (°C):
MW: 90.08   **BP** (°C):

Solubility (Moles/L)	Solubility (Grams/L)	Temp (°C)	Ref (#)	Evaluation (T P E A A)	Comments
2.998E+00	2.701E+02	25	I307	0 0 0 0 0	

184. C₃H₆O₃

s-Trioxane
1,3,5-Trioxan

RN: 110-88-3   **MP** (°C):   64
MW: 90.08   **BP** (°C):   114.5

Solubility (Moles/L)	Solubility (Grams/L)	Temp (°C)	Ref (#)	Evaluation (T P E A A)	Comments
1.715E+00	1.544E+02	20.00	B394	0 0 0 0 0	
1.943E+00	1.750E+02	25	F300	1 0 0 0 2	
2.033E+00	1.831E+02	25.00	B394	0 0 0 0 0	
2.403E+00	2.165E+02	30.10	B394	0 0 0 0 0	
2.741E+00	2.469E+02	34.45	B394	0 0 0 0 0	
4.187E+00	3.772E+02	43.00	B394	0 0 0 0 0	
4.462E+00	4.019E+02	44.00	B394	0 0 0 0 0	
4.606E+00	4.149E+02	44.40	B394	0 0 0 0 0	
4.826E+00	4.348E+02	45.00	B394	0 0 0 0 0	
4.816E+00	4.338E+02	45.10	B394	0 0 0 0 0	
5.355E+00	4.824E+02	46.00	B394	0 0 0 0 0	
5.311E+00	4.784E+02	46.10	B394	0 0 0 0 0	
6.401E+00	5.766E+02	47.10	B394	0 0 0 0 0	
8.161E+00	7.351E+02	47.80	B394	0 0 0 0 0	
8.534E+00	7.687E+02	48.95	B394	0 0 0 0 0	
8.741E+00	7.874E+02	50.20	B394	0 0 0 0 0	
9.095E+00	8.192E+02	55.30	B394	0 0 0 0 0	

185. C₃H₆O₃S₃

α-Trimethylene trisulphoxide
1,3,5-Trithiane, 1,3,5-trioxide, (1α,3α,5α)-

RN: 60102-87-6   **MP** (°C):
MW: 186.27   **BP** (°C):

Solubility (Moles/L)	Solubility (Grams/L)	Temp (°C)	Ref (#)	Evaluation (T P E A A)	Comments
7.184E-03	1.338E+00	25	B112	1 2 1 1 2	

186. C₃H₆O₃S₃

β-Trimethylene trisulphoxide
1,3,5-Trithiane, 1,3,5-trioxide, (1α,3α,5β)-

RN: 60102-88-7   **MP** (°C):
MW: 186.27   **BP** (°C):

Solubility (Moles/L)	Solubility (Grams/L)	Temp (°C)	Ref (#)	Evaluation (T P E A A)	Comments
7.605E-02	1.417E+01	25	B112	1 2 1 1 2	

187. $C_3H_6O_3S$
1,3-Propane sultone
1,2-Oxathiolane 2,2-dioxide
3-Hydroxy-1-propanesulfonic acid g-sultone

RN: 1120-71-4	**MP** (°C):	31
MW: 122.14	**BP** (°C):	112

Solubility (Moles/L)	Solubility (Grams/L)	Temp (°C)	Ref (#)	Evaluation (T P E A A)	Comments
8.187E-01	1.000E+02	ns	I307	0 0 0 0 0	

188. C_3H_7Br
Isopropyl bromide
Isopropylbromid

RN: 75-26-3	**MP** (°C):	−89
MW: 123.00	**BP** (°C):	59

Solubility (Moles/L)	Solubility (Grams/L)	Temp (°C)	Ref (#)	Evaluation (T P E A A)	Comments
3.398E-02	4.180E+00	0	H101	2 0 0 0 2	
2.340E-02	2.878E+00	18	F001	1 0 1 2 2	
2.602E-02	3.200E+00	20	F300	1 0 0 0 1	
2.585E-02	3.180E+00	20	H101	2 0 0 0 2	
2.592E-02	3.188E+00	30	V009	1 0 0 0 1	

189. C_3H_7Br
Propyl bromide
1-Bromopropane
Propylbromid
Bromopropane

RN: 106-94-5	**MP** (°C):	−110
MW: 123.00	**BP** (°C):	71

Solubility (Moles/L)	Solubility (Grams/L)	Temp (°C)	Ref (#)	Evaluation (T P E A A)	Comments
2.415E-02	2.970E+00	0	F300	1 0 0 0 2	
2.423E-02	2.980E+00	0	H101	2 0 0 0 2	
1.850E-02	2.275E+00	19.5	S006	1 0 0 0 2	
1.850E-02	2.275E+00	19.50	F001	1 0 1 0 2	
1.992E-02	2.450E+00	20	H101	2 0 0 0 2	
1.947E-02	2.394E+00	20	H127	1 0 0 0 1	
1.874E-02	2.305E+00	30	G029	1 0 2 2 2	
1.876E-02	2.307E+00	30	V009	1 0 0 0 2	
1.140E-01	1.402E+01	ns	H307	0 0 0 0 0	

190. C$_3$H$_7$BrO
3-Bromo-1-propanol
3-Brom-propanol-(1)
RN: 627-18-9   **MP** (°C):
MW: 139.00   **BP** (°C):

Solubility (Moles/L)	Solubility (Grams/L)	Temp (°C)	Ref (#)	Evaluation (T P E A A)	Comments
1.022E+00	1.420E+02	20	F300	1 0 0 0 2	

191. C$_3$H$_7$Cl
Isopropyl chloride
2-Chloropropane
RN: 75-29-6   **MP** (°C):   −117
MW: 78.54   **BP** (°C):   35

Solubility (Moles/L)	Solubility (Grams/L)	Temp (°C)	Ref (#)	Evaluation (T P E A A)	Comments
5.602E-02	4.400E+00	0	H101	2 0 0 0 2	
4.380E-02	3.440E+00	12.50	F001	1 0 1 0 2	
3.947E-02	3.100E+00	20	F300	1 0 0 0 1	
3.883E-02	3.050E+00	20	H101	2 0 0 0 2	
3.935E-02	3.090E+00	20	N034	1 0 0 0 1	
3.888E-02	3.054E+00	30	V009	1 0 0 0 1	

192. C$_3$H$_7$Cl
Chloropropane
Propyl chloride
1-Chloropropane
RN: 540-54-5   **MP** (°C):   −123
MW: 78.54   **BP** (°C):   43.47

Solubility (Moles/L)	Solubility (Grams/L)	Temp (°C)	Ref (#)	Evaluation (T P E A A)	Comments
4.787E-02	3.760E+00	0	H101	2 0 0 0 2	
2.970E-02	2.333E+00	12.50	F001	1 0 1 0 2	
3.438E-02	2.700E+00	20	F300	1 0 0 0 1	
3.463E-02	2.720E+00	20	H101	2 0 0 0 2	
3.428E-02	2.693E+00	20	N034	1 0 0 0 1	
2.970E-02	2.333E+00	20	S006	1 0 0 0 2	
3.520E-02	2.765E+00	30	V009	1 0 0 0 2	

193. C_3H_7ClO
3-Chloro-1-propanol
3-Chlor-propanol-(1)
RN:　　627-30-5　　　　**MP** (°C):
MW:　　94.54　　　　　**BP** (°C):

Solubility (Moles/L)	Solubility (Grams/L)	Temp (°C)	Ref (#)	Evaluation (T P E A A)	Comments
2.644E+00	2.500E+02	20	F300	1 0 0 0 1	
+2.64E+00	+2.50E+02	ns	S460	0 0 0 0 0	

194. C_3H_7I
Iodopropane
n-Propyl iodide
RN:　　107-08-4　　　　**MP** (°C):　　−101
MW:　　169.99　　　　　**BP** (°C):　　101.5

Solubility (Moles/L)	Solubility (Grams/L)	Temp (°C)	Ref (#)	Evaluation (T P E A A)	Comments
6.706E-03	1.140E+00	0	H101	2 0 0 0 2	
5.100E-03	8.670E-01	20	F001	1 0 1 0 2	
5.118E-03	8.700E-01	20	F300	1 0 0 0 1	
6.294E-03	1.070E+00	20	H101	2 0 0 0 2	
5.100E-03	8.670E-01	20	M171	1 0 0 0 1	
5.100E-03	8.670E-01	20	S006	1 0 0 0 1	
6.258E-03	1.064E+00	23.5	S171	2 1 2 2 2	
6.112E-03	1.039E+00	30	G029	1 0 2 2 2	
6.094E-03	1.036E+00	30	V009	1 0 0 0 1	

195. C_3H_7I
Isopropyl iodide
2-Iodopropane
RN:　　75-30-9　　　　**MP** (°C):　　−90
MW:　　169.99　　　　**BP** (°C):　　89

Solubility (Moles/L)	Solubility (Grams/L)	Temp (°C)	Ref (#)	Evaluation (T P E A A)	Comments
9.824E-03	1.670E+00	0	H101	2 0 0 0 2	
8.236E-03	1.400E+00	20	F300	1 0 0 0 1	
8.236E-03	1.400E+00	20	H101	2 0 0 0 2	
7.889E-03	1.341E+00	30	V009	1 0 0 0 1	

196. C$_3$H$_7$NO$_2$

1-Nitropropane
n-Nitropropane

RN:	108-03-2	**MP** (°C): −108
MW:	89.09	**BP** (°C): 131.6

Solubility (Moles/L)	Solubility (Grams/L)	Temp (°C)	Ref (#)	Evaluation (T P E A A)	Comments
1.550E-01	1.381E+01	20	C121	1 0 0 0 1	unit assumed, *sic*

197. C$_3$H$_7$NO$_2$

2-Nitropropane
Nitroisopropane
Dimethylnitromethane

RN:	79-46-9	**MP** (°C): −93
MW:	89.09	**BP** (°C): 120.3

Solubility (Moles/L)	Solubility (Grams/L)	Temp (°C)	Ref (#)	Evaluation (T P E A A)	Comments
1.876E-01	1.672E+01	20	C121	0 0 0 0 1	unit assumed, *sic*
1.874E-01	1.670E+01	20	F300	1 0 0 0 2	
2.376E-01	2.117E+01	20	H118	1 1 1 1 2	

198. C$_3$H$_7$NO$_2$

α-Alanine
Alanine
2-Aminopropanoic acid
2-Ammoniopropanoate
L-2-Aminopropionic acid

RN:	56-41-7	**MP** (°C): 314.5–316.5
MW:	89.09	**BP** (°C):

Solubility (Moles/L)	Solubility (Grams/L)	Temp (°C)	Ref (#)	Evaluation (T P E A A)	Comments
1.366E+00	1.217E+02	10	C347	0 0 0 0 0	EFG
1.640E+00	1.461E+02	15	D349	2 1 1 2 2	
1.744E+00	1.554E+02	20	B032	1 2 2 1 2	
1.535E+00	1.367E+02	20	C347	0 0 0 0 0	EFG
1.780E+00	1.586E+02	20	D349	2 1 1 2 2	
1.838E+00	1.638E+02	25	B032	1 2 2 1 2	
1.590E+00	1.417E+02	25	D005	2 2 1 1 2	
1.602E+00	1.427E+02	25	D041	1 0 0 0 2	
1.870E+00	1.666E+02	25	D349	2 1 1 2 2	
1.660E+00	1.479E+02	25	E015	1 2 1 1 1	
1.595E+00	1.421E+02	25	G092	2 1 1 1 1	
1.595E+00	1.421E+02	25	G315	0 0 0 0 0	
1.654E+00	1.474E+02	25	G433	0 0 0 0 0	
1.852E+00	1.650E+02	25	J303	0 0 0 0 0	
1.600E+00	1.426E+02	25	N001	0 0 0 0 0	EFG
1.630E+00	1.452E+02	25	N012	2 0 2 1 2	
1.555E+00	1.386E+02	25	O316	1 0 1 2 2	

(continued)

198. C$_3$H$_7$NO$_2$ (continued)

Solubility (Moles/L)	Solubility (Grams/L)	Temp (°C)	Ref (#)	Evaluation (T P E A A)	Comments
1.598E+00	1.424E+02	25	O316	1 0 1 2 2	
1.623E+00	1.446E+02	25	O317	1 0 1 2 2	
1.871E+00	1.667E+02	25.1	N024	0 0 0 0 0	
1.871E+00	1.667E+02	25.1	N026	0 0 0 0 0	
1.606E+00	1.431E+02	25.1	N027	1 1 2 2 2	
1.704E+00	1.518E+02	27	D036	0 0 0 0 0	
1.695E+00	1.510E+02	27	D036	0 0 0 0 0	
1.940E+00	1.728E+02	29.80	B032	1 2 2 1 2	
1.657E+00	1.477E+02	30	C347	0 0 0 0 0	EFG
1.956E+00	1.743E+02	30	J303	0 0 0 0 0	
1.816E+00	1.618E+02	40	C347	0 0 0 0 0	EFG
2.192E+00	1.953E+02	40	J303	0 0 0 0 0	
1.931E+00	1.720E+02	45	F300	1 0 0 0 2	
1.932E+00	1.721E+02	50	C347	0 0 0 0 0	EFG
2.430E+00	2.165E+02	50	J303	0 0 0 0 0	
2.118E+00	1.887E+02	60	C347	0 0 0 0 0	EFG
2.706E+00	2.411E+02	60	J303	0 0 0 0 0	
2.333E+00	2.078E+02	70	C347	0 0 0 0 0	EFG
2.489E+00	2.218E+02	75	D041	1 0 0 0 2	
2.504E+00	2.230E+02	80	C347	0 0 0 0 0	EFG
2.668E+00	2.377E+02	90	C347	0 0 0 0 0	EFG
2.888E+00	2.573E+02	100	C347	0 0 0 0 0	EFG
1.192E+00	1.062E+02	-	C347	0 0 0 0 0	EFG
1.587E+00	1.414E+02	rt	D021	0 0 1 1 2	

199. C$_3$H$_7$NO$_2$
β-Alanine
β-Alanin

RN:	107-95-9	**MP** (°C):	
MW:	89.09	**BP** (°C):	

Solubility (Moles/L)	Solubility (Grams/L)	Temp (°C)	Ref (#)	Evaluation (T P E A A)	Comments
3.959E+00	3.528E+02	25	D041	1 0 0 0 2	
6.123E+00	5.455E+02	25	M024	1 2 0 1 2	

200. C$_3$H$_7$NO$_2$
D-Alanine
D(–)-Alanine

RN:	338-69-2	**MP** (°C):	292
MW:	89.09	**BP** (°C):	

Solubility (Moles/L)	Solubility (Grams/L)	Temp (°C)	Ref (#)	Evaluation (T P E A A)	Comments
1.265E+00	1.127E+02	0	M043	1 0 0 0 2	
1.396E+00	1.243E+02	10	M043	1 0 0 0 2	

(continued)

200. $C_3H_7NO_2$ (continued)

Solubility (Moles/L)	Solubility (Grams/L)	Temp (°C)	Ref (#)	Evaluation (T P E A A)	Comments
1.530E+00	1.363E+02	20	D041	1 0 0 0 2	
1.531E+00	1.364E+02	20	M043	1 0 0 0 2	
1.589E+00	1.416E+02	25	D005	2 2 1 1 2	
1.680E+00	1.497E+02	30	M043	1 0 0 0 2	
1.839E+00	1.639E+02	40	M043	1 0 0 0 2	
2.194E+00	1.955E+02	60	M043	1 0 0 0 2	
2.590E+00	2.308E+02	80	M043	1 0 0 0 2	
3.049E+00	2.717E+02	100	M043	1 0 0 0 2	
3.049E+00	2.717E+02	99.99	P349	0 0 0 0 0	

201. $C_3H_7NO_2$

DL-Alanine
DL-α-Alanine
DL-2-Aminopropionic acid

RN:	302-72-7	**MP (°C):**	289	
MW:	89.09	**BP (°C):**		

Solubility (Moles/L)	Solubility (Grams/L)	Temp (°C)	Ref (#)	Evaluation (T P E A A)	Comments
1.212E+00	1.080E+02	0	D018	2 2 2 1 2	
1.212E+00	1.080E+02	0	F300	1 0 0 0 2	
1.212E+00	1.079E+02	0	M043	1 0 0 0 2	
1.361E+00	1.213E+02	10	M043	1 0 0 0 0	
1.523E+00	1.357E+02	20	M043	1 0 0 0 0	
1.557E+00	1.387E+02	21	P045	1 0 2 1 2	
1.659E+00	1.478E+02	25	C018	0 0 0 0 0	
1.596E+00	1.422E+02	25	D018	2 2 2 1 2	
1.598E+00	1.424E+02	25	D041	1 0 0 0 2	
1.607E+00	1.432E+02	25	F300	1 0 0 0 2	
1.900E+00	1.693E+02	25	J303	0 0 0 0 0	
1.530E+00	1.363E+02	25	K031	2 1 2 1 2	
2.024E+00	1.803E+02	30	J303	0 0 0 0 0	
1.704E+00	1.518E+02	30	M043	1 0 0 0 0	
2.307E+00	2.055E+02	40	J303	0 0 0 0 0	
1.894E+00	1.687E+02	40	M043	1 0 0 0 0	
2.134E+00	1.902E+02	50	D018	2 2 2 1 2	
2.106E+00	1.876E+02	50	F300	1 0 0 0 2	
2.591E+00	2.308E+02	50	J303	0 0 0 0 0	
2.954E+00	2.632E+02	60	J303	0 0 0 0 0	
2.337E+00	2.082E+02	60	M043	1 0 0 0 0	
2.733E+00	2.435E+02	75	D018	2 2 2 1 2	
2.734E+00	2.436E+02	75	D041	1 0 0 0 2	
2.714E+00	2.418E+02	75	F300	1 0 0 0 2	
2.842E+00	2.532E+02	80	M043	1 0 0 0 0	
3.431E+00	3.057E+02	100	F300	1 0 0 0 2	
3.430E+00	3.056E+02	100	M043	1 0 0 0 2	
3.432E+00	3.057E+02	99.99	P349	0 0 0 0 0	

202. C$_3$H$_7$NO$_2$
Lactamide
2-Hydroxypropionamide
RN: 2043-43-8   **MP** (°C):
MW: 89.09   **BP** (°C):

Solubility (Moles/L)	Solubility (Grams/L)	Temp (°C)	Ref (#)	Evaluation (T P E A A)	Comments
8.779E+00	7.822E+02	25	M008	1 0 0 0 2	

203. C$_3$H$_7$NO$_2$
Sarcosine
Sarkosin
RN: 107-97-1   **MP** (°C):   208
MW: 89.09   **BP** (°C):

Solubility (Moles/L)	Solubility (Grams/L)	Temp (°C)	Ref (#)	Evaluation (T P E A A)	Comments
5.151E-01	4.589E+01	20	D041	1 0 0 0 2	
3.367E+00	3.000E+02	20	F300	1 0 0 0 2	
4.807E+00	4.282E+02	20	P045	1 0 2 1 2	

204. C$_3$H$_7$NO$_2$
Urethan
Carbamidsaeure-aethyl ester
Eythyl urethan
Urethane
Ethyl carbamate
Carbamic acid ethyl ester
RN: 51-79-6   **MP** (°C):   49
MW: 89.09   **BP** (°C):   183

Solubility (Moles/L)	Solubility (Grams/L)	Temp (°C)	Ref (#)	Evaluation (T P E A A)	Comments
2.918E+00	2.600E+02	11	F300	1 0 0 0 1	
5.393E+00	4.805E+02	15.5	F001	1 0 1 2 2	
2.245E+01	2.000E+03	25	I310	0 0 0 0 0	
5.074E+00	4.521E+02	25	P065	2 0 1 1 2	
1.800E+01	1.604E+03	37	H006	1 2 2 1 1	
8.901E+00	7.930E+02	40	F300	1 0 0 0 2	

205. C$_3$H$_7$NO$_2$S
Cysteine
2-Amino-3-mercaptopropanoic acid
RN: 3374-22-9   **MP** (°C):   225
MW: 121.16   **BP** (°C):

Solubility (Moles/L)	Solubility (Grams/L)	Temp (°C)	Ref (#)	Evaluation (T P E A A)	Comments
2.773E-02	3.360E+00	20	P045	1 0 2 1 2	

206. C$_3$H$_7$NO$_3$

Serine
2-Amino-3-hydroxypropanoic acid
L(−)-Serin

RN:	56-45-1	MP (°C):	220
MW:	105.09	BP (°C):	

Solubility (Moles/L)	Solubility (Grams/L)	Temp (°C)	Ref (#)	Evaluation (T P E A A)	Comments
1.556E+00	1.635E+02	10.19	J417	0 0 0 0 0	
1.626E+00	1.709E+02	12.69	J417	0 0 0 0 0	
4.530E-01	4.761E+01	15	D349	2 1 1 2 2	
1.864E+00	1.959E+02	16.09	J417	0 0 0 0 0	
1.903E+00	2.000E+02	20	D041	1 0 0 0 1	
4.610E-01	4.845E+01	20	D349	2 1 1 2 2	
9.512E-01	9.997E+01	20	F300	1 0 0 0 2	
3.405E+00	3.578E+02	20.00	B032	1 2 2 1 2	sic
4.700E-01	4.939E+01	25	D349	2 1 1 2 2	
2.807E+00	2.950E+02	25	G315	0 0 0 0 0	sic
4.013E+00	4.217E+02	25	J303	0 0 0 0 0	
4.043E+00	4.249E+02	25.00	B032	1 2 2 0 2	sic
2.228E+00	2.342E+02	25.89	J417	0 0 0 0 0	
3.578E+00	3.760E+02	27	D036	0 0 0 0 0	
2.287E+00	2.404E+02	27.89	J417	0 0 0 0 0	
4.690E+00	4.929E+02	29.80	B032	1 2 2 1 2	sic
5.633E+00	5.920E+02	40	J303	0 0 0 0 0	
2.800E+00	2.943E+02	42.79	J417	0 0 0 0 0	
2.811E+00	2.954E+02	43.79	J417	0 0 0 0 0	
2.861E+00	3.007E+02	44.59	J417	0 0 0 0 0	
2.902E+00	3.050E+02	49.69	J417	0 0 0 0 0	
2.972E+00	3.124E+02	53.89	J417	0 0 0 0 0	
7.574E+00	7.960E+02	60	J303	0 0 0 0 0	

207. C$_3$H$_7$NO$_3$

D-Serine
D-2-Amino-3-hydroxypropanoic acid

RN:	312-84-5	MP (°C):	220
MW:	105.09	BP (°C):	

Solubility (Moles/L)	Solubility (Grams/L)	Temp (°C)	Ref (#)	Evaluation (T P E A A)	Comments
1.903E+00	2.000E+02	20	D041	1 0 0 0 0	
4.010E+00	4.214E+02	25	J303	0 0 0 0 0	
5.709E+00	6.000E+02	40	J303	0 0 0 0 0	
7.631E+00	8.020E+02	60	J303	0 0 0 0 0	

208. $C_3H_7NO_3$
DL-Serine
DL-2-Amino-3-hydroxypropanoic acid

RN: 302-84-1 **MP** (°C): 240
MW: 105.09 **BP** (°C):

Solubility (Moles/L)	Solubility (Grams/L)	Temp (°C)	Ref (#)	Evaluation (T P E A A)	Comments
2.778E-01	2.920E+01	10	F300	1 0 0 0 2	
3.787E-01	3.980E+01	20	F300	1 0 0 0 2	
4.548E-01	4.780E+01	25	D041	1 0 0 0 2	
4.805E-01	5.050E+01	25	J303	0 0 0 0 0	
7.403E-01	7.780E+01	40	J303	0 0 0 0 0	
8.916E-01	9.370E+01	50	F300	1 0 0 0 2	
1.261E+00	1.325E+02	60	J303	0 0 0 0 0	
1.533E+00	1.611E+02	75	D041	1 0 0 0 2	
1.532E+00	1.610E+02	75	F300	1 0 0 0 2	
2.320E+00	2.438E+02	100	F300	1 0 0 0 2	
2.320E+00	2.438E+02	99.99	P349	0 0 0 0 0	

209. $C_3H_7NO_3$
DL-Isoserine
DL-Isoserin

RN: 632-12-2 **MP** (°C): 235
MW: 105.09 **BP** (°C):

Solubility (Moles/L)	Solubility (Grams/L)	Temp (°C)	Ref (#)	Evaluation (T P E A A)	Comments
1.456E-01	1.530E+01	20	F300	1 0 0 0 2	

210. $C_3H_7NO_5$
Glycerol-α-nitrate
Glycerin-α-nitrate

RN: 27321-61-5 **MP** (°C):
MW: 137.09 **BP** (°C):

Solubility (Moles/L)	Solubility (Grams/L)	Temp (°C)	Ref (#)	Evaluation (T P E A A)	Comments
3.004E+00	4.118E+02	15	F300	1 0 0 0 2	

211. $C_3H_7N_3O_2$
Glycocyamine
Guanidin-essigsaeure
Guanidineacetic acid

RN: 352-97-6 **MP** (°C): 280
MW: 117.11 **BP** (°C):

Solubility (Moles/L)	Solubility (Grams/L)	Temp (°C)	Ref (#)	Evaluation (T P E A A)	Comments
3.825E-02	4.480E+00	15	D041	1 0 0 0 1	
3.074E-02	3.600E+00	15	F300	1 0 0 0 1	

212. C₃H₇N₃O₂

Nitrosoethylurea

N-Nitroso-*N*-ethylurea

RN: 759-73-9 **MP** (°C): 103
MW: 117.11 **BP** (°C):

Solubility (Moles/L)	Solubility (Grams/L)	Temp (°C)	Ref (#)	Evaluation (T P E A A)	Comments
1.096E-01	1.283E+01	rt	I306	0 0 0 0 0	

213. C₃H₇O₅P

2-Carboxyethylphosphonic acid

3-Phosphonopropionic acid

RN: 5962-42-5 **MP** (°C):
MW: 154.06 **BP** (°C):

Solubility (Moles/L)	Solubility (Grams/L)	Temp (°C)	Ref (#)	Evaluation (T P E A A)	Comments
1.845E+00	2.842E+02	0	N028	1 0 0 0 2	
2.129E+00	3.280E+02	20	N028	1 0 0 0 2	

214. C₃H₈

Propane

Propan

RN: 74-98-6 **MP** (°C): −187
MW: 44.10 **BP** (°C): −42

Solubility (Moles/L)	Solubility (Grams/L)	Temp (°C)	Ref (#)	Evaluation (T P E A A)	Comments
3.460E-03	1.526E-01	4	K031	2 1 2 1 2	
2.472E-03	1.090E-01	10	F300	1 0 0 0 2	
2.721E-03	1.200E-01	18	M065	0 0 2 1 1	1 atm, *sic*
1.761E-03	7.765E-02	19.8	G058	1 0 0 0 2	
1.746E-03	7.700E-02	20	F300	1 0 0 0 1	
1.420E-03	6.261E-02	25	B342	1 2 1 1 1	
1.530E-03	6.747E-02	25	K031	2 1 2 1 2	
1.415E-03	6.240E-02	25	M001	2 1 2 2 2	
1.415E-03	6.240E-02	25	M002	2 1 2 2 2	
8.400E-04	3.704E-02	50	K031	2 1 2 1 2	
6.123E-04	2.700E-02	60	F300	1 0 0 0 1	

215. $C_3H_8NO_5P$

Glyphosate
N-(Phosphonomethyl)glycine
Bronco

RN:	1071-83-6	**MP** (°C):	230.0		
MW:	169.07	**BP** (°C):			

Solubility (Moles/L)	Solubility (Grams/L)	Temp (°C)	Ref (#)	Evaluation (T P E A A)	Comments
7.097E-02	1.200E+01	25	M161	1 0 0 0 1	
5.856E-02	9.901E+00	ns	B100	0 0 0 0 0	

216. C_3H_8O

n-Propyl alcohol
Propanol

RN:	71-23-8	**MP** (°C):	−127.0
MW:	60.10	**BP** (°C):	97.2

Solubility (Moles/L)	Solubility (Grams/L)	Temp (°C)	Ref (#)	Evaluation (T P E A A)	Comments
3.132E+00	1.882E+02	ns	L003	0 0 2 1 2	
+4.17E+00	+2.51E+02	ns	S460	0 0 0 0 0	

217. C_3H_8O

Isopropyl alcohol
2-Propanol

RN:	67-63-0	**MP** (°C):	−88
MW:	60.10	**BP** (°C):	82.5

Solubility (Moles/L)	Solubility (Grams/L)	Temp (°C)	Ref (#)	Evaluation (T P E A A)	Comments
5.033E+00	3.025E+02	ns	L003	0 0 2 1 1	

218. $C_3H_8OS_2$

2,3-Dimercapto-1-propanol
Dimercaprol

RN:	59-52-9	**MP** (°C):	
MW:	124.22	**BP** (°C):	

Solubility (Moles/L)	Solubility (Grams/L)	Temp (°C)	Ref (#)	Evaluation (T P E A A)	Comments
5.963E-01	7.407E+01	20	D041	1 0 0 0 0	

219. C₃H₈O₂

Methylal

Formaldehyd-dimethyl-acetal

RN:	109-87-5	**MP** (°C):	−105
MW:	76.10	**BP** (°C):	41.5

Solubility (Moles/L)	Solubility (Grams/L)	Temp (°C)	Ref (#)	Evaluation (T P E A A)	Comments
3.208E+00	2.441E+02	16	B117	1 0 0 1 2	
3.022E+00	2.300E+02	20	F300	1 0 0 0 1	

220. C₃H₈O₃

Glycerol

Glycerin

RN:	56-81-5	**MP** (°C):	20
MW:	92.10	**BP** (°C):	

Solubility (Moles/L)	Solubility (Grams/L)	Temp (°C)	Ref (#)	Evaluation (T P E A A)	Comments
5.973E+00	5.501E+02	4.50	C022	1 2 0 0 2	
5.751E-01	5.296E+01	25	B004	0 0 0 0 0	

221. C₃H₉N

Propylamine

Propylamin

n-Propylamine

RN:	107-10-8	**MP** (°C):	−83
MW:	59.11	**BP** (°C):	48

Solubility (Moles/L)	Solubility (Grams/L)	Temp (°C)	Ref (#)	Evaluation (T P E A A)	Comments
2.469E-02	1.459E+00	25	B004	0 0 0 0 0	

222. C₃H₉N

Trimethylamine

N,N-Dimethylmethanamine

RN:	75-50-3	**MP** (°C):	−124.0
MW:	59.11	**BP** (°C):	3.2

Solubility (Moles/L)	Solubility (Grams/L)	Temp (°C)	Ref (#)	Evaluation (T P E A A)	Comments
>6.77E+00	>4.00E+02	20	F300	1 0 0 0 0	
6.936E+00	4.100E+02	25	A049	1 0 0 0 2	

223. C$_3$H$_9$O$_4$P
Trimethyl phosphate
Phosphorsaeure-trimethyl ester
RN: 512-56-1 **MP** (°C):
MW: 140.08 **BP** (°C): 197

Solubility (Moles/L)	Solubility (Grams/L)	Temp (°C)	Ref (#)	Evaluation (T P E A A)	Comments
3.569E+00	5.000E+02	25	F300	1 0 0 0 1	
3.573E+00	5.005E+02	ns	S460	0 0 0 0 0	

224. C$_3$H$_{12}$N$_6$O$_3$
Guanidine carbonate
Guanidin-carbonat
RN: 3425-08-9 **MP** (°C): 198
MW: 180.17 **BP** (°C):

Solubility (Moles/L)	Solubility (Grams/L)	Temp (°C)	Ref (#)	Evaluation (T P E A A)	Comments
1.850E+00	3.333E+02	24	F300	1 0 0 0 2	

225. C$_3$Cl$_3$N$_3$O$_3$
Trichloroisocyanuric acid
Symclosene
RN: 87-90-1 **MP** (°C): 246.5
MW: 232.41 **BP** (°C):

Solubility (Moles/L)	Solubility (Grams/L)	Temp (°C)	Ref (#)	Evaluation (T P E A A)	Comments
3.439E-03	7.994E-01	20	B080	1 0 1 1 0	
2.311E-02	5.371E+00	40	B080	1 0 1 1 1	

226. C$_3$Cl$_6$
Hexachloropropene
Hexachloropropylene
Perchloropropene
Hexachloro-1-propene
1,1,2,3,3,3-Hexachloro-1-propene
1,1,2,3,3,3-Hexachloropropene
RN: 1888-71-7 **MP** (°C):
MW: 248.75 **BP** (°C): 209–210

Solubility (Moles/L)	Solubility (Grams/L)	Temp (°C)	Ref (#)	Evaluation (T P E A A)	Comments
6.026E-04	1.499E-01	ns	S460	0 0 0 0 0	

227. C_4HI_4N
Iodol
2,3,4,5-Tetraiodpyrrol
RN: 87-58-1   **MP** (°C):
MW: 570.68   **BP** (°C):

Solubility (Moles/L)	Solubility (Grams/L)	Temp (°C)	Ref (#)	Evaluation (T P E A A)	Comments
3.505E-04	2.000E-01	15	F300	1 0 0 0 2	

228. C_4H_2
Butadiyne
Diacetylen
RN: 460-12-8   **MP** (°C): −36.4
MW: 50.06   **BP** (°C): 10.3

Solubility (Moles/L)	Solubility (Grams/L)	Temp (°C)	Ref (#)	Evaluation (T P E A A)	Comments
1.998E-03	1.000E-01	25	F300	1 0 0 0 0	

229. $C_4H_2N_2O_4$
Alloxan
Alloxane
RN: 50-71-5   **MP** (°C): 256dec
MW: 142.07   **BP** (°C):

Solubility (Moles/L)	Solubility (Grams/L)	Temp (°C)	Ref (#)	Evaluation (T P E A A)	Comments
5.631E-02	8.000E+00	ns	D072	0 0 0 0 0	
5.623E-02	7.989E+00	ns	R424	0 0 0 0 0	

230. $C_4H_3BrN_2O_2$
5-Bromouracil
5-Bromo-2
4(1H,3H)-Pyrimidinedione
5-Bromo-2,4-dihydroxypyrimidine
Bromouracil
RN: 51-20-7   **MP** (°C): 310
MW: 190.99   **BP** (°C):

Solubility (Moles/L)	Solubility (Grams/L)	Temp (°C)	Ref (#)	Evaluation (T P E A A)	Comments
1.507E-02	2.878E+00	25	S471	0 0 0 0 0	
1.908E-02	3.644E+00	25	S471	0 0 0 0 0	
1.350E-02	2.578E+00	25	Z408	0 0 0 0 0	

231. C₄H₃ClN₂O₂

6-Chlorouracil
4-Chloro-2,6-dihydroxypyrimidine
RN: 4270-27-3   **MP** (°C):
MW: 146.53   **BP** (°C):

Solubility (Moles/L)	Solubility (Grams/L)	Temp (°C)	Ref (#)	Evaluation (T P E A A)	Comments
3.334E-02	4.885E+00	25	S471	0 0 0 0 0	
3.350E-02	4.909E+00	25	S471	0 0 0 0 0	

232. C₄H₃ClN₂O₂

5-Chlorouracil
RN: 1820-81-1   **MP** (°C):
MW: 146.53   **BP** (°C):

Solubility (Moles/L)	Solubility (Grams/L)	Temp (°C)	Ref (#)	Evaluation (T P E A A)	Comments
1.707E-02	2.501E+00	25	S471	0 0 0 0 0	
1.712E-02	2.509E+00	25	S471	0 0 0 0 0	
1.800E-02	2.638E+00	25	Z408	0 0 0 0 0	
9.827E-04	1.440E-01	ns	Y414	0 0 0 0 0	

233. C₄H₃FN₂O₂

5 Fluorouracil
5-Fluorouracil
Fluorouracil
5-Fluoro-2,4(1H,3H)-pyrimidinedione
Fluroblastin
Fluororuracil
RN: 51-21-8   **MP** (°C):   281
MW: 130.08   **BP** (°C):

Solubility (Moles/L)	Solubility (Grams/L)	Temp (°C)	Ref (#)	Evaluation (T P E A A)	Comments
9.600E-02	1.249E+01	21	B416	2 2 1 2 1	
8.533E-02	1.110E+01	22	B321	0 0 0 0 0	pH 4.0
8.533E-02	1.110E+01	22	B332	1 1 0 0 1	pH 4.0
8.533E-02	1.110E+01	22	B388	0 0 0 0 0	
9.379E-02	1.220E+01	22	M317	1 1 1 1 1	
9.379E-02	1.220E+01	25	R023	0 0 0 0 0	
1.356E-01	1.763E+01	25	S471	0 0 0 0 0	
1.382E-01	1.798E+01	25	S471	0 0 0 0 0	
6.940E-02	9.027E+00	25	Z408	0 0 0 0 0	
8.533E-02	1.110E+01	37	B332	0 0 0 0 0	pH 4.0
9.566E-02	1.244E+01	ns	S469	0 0 0 0 0	

234. C$_4$H$_3$IN$_2$O$_2$
5-Iodouracil
5-Iodo-2,4(1H,3H)-pyrimidinedione
5-Iodo-2,4-dihydroxypyrimidine

RN:	696-07-1	MP (°C):	274–276 (°dec)
MW:	237.99	BP (°C):	

Solubility (Moles/L)	Solubility (Grams/L)	Temp (°C)	Ref (#)	Evaluation (T P E A A)	Comments
2.062E-02	4.907E+00	25	S471	0 0 0 0 0	
2.072E-02	4.931E+00	25	S471	0 0 0 0 0	
1.060E-02	2.523E+00	25	Z408	0 0 0 0 0	

235. C$_4$H$_3$N$_2$S
2-Methyl-1,3,4-thiadiazole
Thiodiazolique methyle

RN:	26584-42-9	MP (°C):	
MW:	111.15	BP (°C):	

Solubility (Moles/L)	Solubility (Grams/L)	Temp (°C)	Ref (#)	Evaluation (T P E A A)	Comments
7.918E-03	8.800E-01	37	D084	1 0 1 0 1	

236. C$_4$H$_3$N$_3$O$_4$
5-Nitrouracil

RN:	611-08-5	MP (°C):	
MW:	157.09	BP (°C):	

Solubility (Moles/L)	Solubility (Grams/L)	Temp (°C)	Ref (#)	Evaluation (T P E A A)	Comments
2.300E-02	3.613E+00	25	Z408	0 0 0 0 0	

237. C$_4$H$_3$N$_3$O$_5$
5-Nitrobarbituric acid
Dilitursaeure

RN:	28176-10-5	MP (°C):	176
MW:	173.09	BP (°C):	

Solubility (Moles/L)	Solubility (Grams/L)	Temp (°C)	Ref (#)	Evaluation (T P E A A)	Comments
5.200E-03	9.000E-01	25.60	F300	1 0 0 0 0	

238. $C_4H_4Br_2O_4$

meso-2,3-Dibromosuccinic acid
meso-Dibrom-bernsteinsaeure
DL-2,3-Dibromosuccinic acid
DL-Dibrom-bernsteinsaeure

RN:	526-78-3	MP (°C):	171
MW:	275.89	BP (°C):	

Solubility (Moles/L)	Solubility (Grams/L)	Temp (°C)	Ref (#)	Evaluation (T P E A A)	Comments
7.249E-02	2.000E+01	17	F300	1 0 0 0 2	

239. $C_4H_4Cl_2N_2O_2$

1,3-Dichloro-5-methylhydantoin
2,4-Imidazolidinedione, 1,3-dichloro-5-methyl-
Hydantoin, 1,3-dichloro-5-methyl-

RN:	15216-12-3	MP (°C):	
MW:	182.99	BP (°C):	

Solubility (Moles/L)	Solubility (Grams/L)	Temp (°C)	Ref (#)	Evaluation (T P E A A)	Comments
1.634E-02	2.991E+00	20	B080	1 0 1 1 0	
4.498E-02	8.232E+00	40	B080	1 0 1 1 1	

240. $C_4H_4Cl_2O_4$

L-2,3-Dichlorosuccinic acid
L(–)-Dichlor-bernsteinsaeure
D-2,3-Dichlorosuccinic acid
D(+)-Dichlor-bernsteinsaeure
2,3-Dichlorosuccinic acid
meso-2,3-Dichlorosuccinic acid

RN:	19922-87-3	MP (°C):	168
MW:	186.98	BP (°C):	

Solubility (Moles/L)	Solubility (Grams/L)	Temp (°C)	Ref (#)	Evaluation (T P E A A)	Comments
2.674E+00	5.000E+02	25	H090	0 1 1 1 1	
1.701E-02	3.180E+00	ns	H090	0 2 2 1 2	

241. $C_4H_4N_2$

Succinonitrile
Bersteinsaeure-dinitril

RN:	110-61-2	MP (°C):	57
MW:	80.09	BP (°C):	265

Solubility (Moles/L)	Solubility (Grams/L)	Temp (°C)	Ref (#)	Evaluation (T P E A A)	Comments
1.584E+00	1.269E+02	20	F300	1 0 0 0 2	

242. $C_4H_4N_2O$
4(3H)-Pyrimidone
4-Hydroxypyrimidine
RN: 51953-17-4   MP (°C):   164
MW: 96.09   BP (°C):

Solubility (Moles/L)	Solubility (Grams/L)	Temp (°C)	Ref (#)	Evaluation (T P E A A)	Comments
2.813E+00	2.703E+02	20	B050	1 0 0 0 0	

243. $C_4H_4N_2O$
2-Hydroxypyrimidine
2-Pyrimidinol
RN: 51953-13-0   MP (°C):
MW: 96.09   BP (°C):

Solubility (Moles/L)	Solubility (Grams/L)	Temp (°C)	Ref (#)	Evaluation (T P E A A)	Comments
3.252E+00	3.125E+02	20	B050	1 0 0 0 0	

244. $C_4H_4N_2OS$
2-Thiouracil
Thiouracil
4(1H)-Pyrimidinone
RN: 141-90-2   MP (°C):   340
MW: 128.15   BP (°C):

Solubility (Moles/L)	Solubility (Grams/L)	Temp (°C)	Ref (#)	Evaluation (T P E A A)	Comments
4.679E-03	5.996E-01	20	D041	1 0 0 0 0	
5.530E-03	7.087E-01	25	G016	1 2 1 2 2	intrinsic
3.900E-03	4.998E-01	ns	I310	0 0 0 0 0	

245. $C_4H_4N_2O_2$
Uracil
2,4-Dihydroxypyrimidine
RN: 66-22-8   MP (°C):   335
MW: 112.09   BP (°C):

Solubility (Moles/L)	Solubility (Grams/L)	Temp (°C)	Ref (#)	Evaluation (T P E A A)	Comments
2.964E-02	3.322E+00	20	B050	1 0 0 0 0	
2.500E-02	2.802E+00	20	N019	0 0 0 0 0	
3.200E-02	3.587E+00	25	D041	1 0 0 0 1	
3.212E-02	3.600E+00	25	F300	1 0 0 0 1	
2.380E-02	2.668E+00	25	H061	0 0 0 0 0	
4.109E-02	4.605E+00	25	S471	0 0 0 0 0	
4.125E-02	4.624E+00	25	S471	0 0 0 0 0	

(continued)

245. C$_4$H$_4$N$_2$O$_2$ (continued)

Solubility (Moles/L)	Solubility (Grams/L)	Temp (°C)	Ref (#)	Evaluation (T P E A A)	Comments
2.710E-02	3.038E+00	25	Z408	0 0 0 0 0	
4.015E-02	4.500E+00	37	B390	0 0 0 0 0	
2.676E-02	3.000E+00	ns	B177	0 0 0 0 0	

246. C$_4$H$_4$N$_2$O$_2$
4,6-Dihydroxypyrimidine
4,6-Pyrimidinediol
RN: 1193-24-4   **MP** (°C):   >300
MW: 112.09   **BP** (°C):

Solubility (Moles/L)	Solubility (Grams/L)	Temp (°C)	Ref (#)	Evaluation (T P E A A)	Comments
2.225E-02	2.494E+00	20	B050	1 0 0 0 0	

247. C$_4$H$_4$N$_2$O$_2$
2,4-Dihydroxypyrimidine
RN: 51953-14-1   **MP** (°C):
MW: 112.09   **BP** (°C):

Solubility (Moles/L)	Solubility (Grams/L)	Temp (°C)	Ref (#)	Evaluation (T P E A A)	Comments
2.964E-02	3.322E+00	20	B050	1 0 0 0 0	

248. C$_4$H$_4$N$_2$O$_2$
Maleic hydrazide
Dihydropyridazine-3,6-dione
RN: 123-33-1   **MP** (°C):
MW: 112.09   **BP** (°C):

Solubility (Moles/L)	Solubility (Grams/L)	Temp (°C)	Ref (#)	Evaluation (T P E A A)	Comments
3.554E-02	3.984E+00	20	B185	0 0 0 0 0	
5.321E-02	5.964E+00	25	B185	0 0 0 0 0	
5.353E-02	6.000E+00	25	B200	1 0 0 0 2	
5.321E-02	5.964E+00	25	M061	1 0 0 0 0	
5.353E-02	6.000E+00	25	M161	1 0 0 0 0	
5.321E-02	5.964E+00	ns	B100	0 0 0 0 0	
6.310E-03	7.072E-01	ns	M163	0 0 0 0 0	EFG
3.554E-02	3.984E+00	ns	N013	0 0 0 0 0	

249. $C_4H_4N_2O_3$
Barbituric acid
Barbitursaeure

RN:	67-52-7	MP (°C):	248
MW:	128.09	BP (°C):	

Solubility (Moles/L)	Solubility (Grams/L)	Temp (°C)	Ref (#)	Evaluation (T P E A A)	Comments
1.483E-07	1.900E-05	37	B166	1 0 1 1 1	
5.129E-02	6.569E+00	ns	R424	0 0 0 0 0	
5.129E-02	6.569E+00	ns	R427	0 0 0 0 0	

250. $C_4H_4N_2O_3$
2,4,6-Trihydroxypyrimidine
2,4,6-Pyrimidinetriol

RN:	223674-01-9	MP (°C):	
MW:	128.09	BP (°C):	

Solubility (Moles/L)	Solubility (Grams/L)	Temp (°C)	Ref (#)	Evaluation (T P E A A)	Comments
5.170E-02	6.623E+00	20	B050	1 0 0 0 0	

251. $C_4H_4O_4$
trans-Fumaric acid
Fumaric acid
Fumarsaeure

RN:	110-17-8	MP (°C):	287
MW:	116.07	BP (°C):	

Solubility (Moles/L)	Solubility (Grams/L)	Temp (°C)	Ref (#)	Evaluation (T P E A A)	Comments
1.977E-02	2.295E+00	0	M043	1 0 0 0 1	
3.005E-02	3.488E+00	10	M043	1 0 0 0 1	
4.286E-02	4.975E+00	20	M043	1 0 0 0 1	
5.989E-02	6.951E+00	25	D041	1 0 0 0 1	
6.031E-02	7.000E+00	25	F300	1 0 0 0 0	
5.989E-02	6.951E+00	25	W011	1 2 2 1 1	
6.159E-02	7.149E+00	30	M043	1 0 0 0 1	
9.218E-02	1.070E+01	40	F300	1 0 0 0 2	
9.374E-02	1.088E+01	40	M043	1 0 0 0 1	
9.121E-02	1.059E+01	40	W011	1 2 2 1 2	
1.937E-01	2.248E+01	60	M043	1 0 0 0 1	
2.019E-01	2.344E+01	60	W011	1 2 2 1 1	
4.258E-01	4.943E+01	80	M043	1 0 0 0 1	
7.689E-01	8.925E+01	100	D041	1 0 0 0 1	
8.012E-01	9.300E+01	100	F300	1 0 0 0 1	
7.689E-01	8.925E+01	100	M043	1 0 0 0 1	
7.689E-01	8.925E+01	100	W011	1 2 2 1 1	
5.248E-02	6.092E+00	ns	R424	0 0 0 0 0	

252. $C_4H_4O_4$
Maleic acid
Maleinsaeure

RN: 110-16-7   **MP** (°C):   138
MW: 116.07   **BP** (°C):

Solubility (Moles/L)	Solubility (Grams/L)	Temp (°C)	Ref (#)	Evaluation (T P E A A)	Comments
2.431E+00	2.821E+02	0	M043	1 0 0 0 2	
2.607E+00	3.026E+02	4.99	A339	0 0 0 0 0	
2.872E+00	3.334E+02	10	F300	1 0 0 0 2	
2.872E+00	3.333E+02	10	M043	1 0 0 0 1	
2.880E+00	3.343E+02	9.99	A339	0 0 0 0 0	
3.094E+00	3.591E+02	14.99	A339	0 0 0 0 0	
3.312E+00	3.845E+02	19.99	A339	0 0 0 0 0	
3.547E+00	4.118E+02	20	M043	1 0 0 0 1	
6.789E+00	7.880E+02	22.5	G301	0 0 0 0 0	
3.592E+00	4.170E+02	24.99	A339	0 0 0 0 0	
3.797E+00	4.407E+02	25	D041	1 0 0 0 2	
3.797E+00	4.407E+02	25	F300	1 0 0 0 2	
3.840E+00	4.457E+02	25	H430	0 0 0 0 0	
3.797E+00	4.407E+02	25	W011	1 2 2 1 2	
3.823E+00	4.437E+02	29.99	A339	0 0 0 0 0	
4.081E+00	4.737E+02	30	M043	1 0 0 0 1	
4.117E+00	4.778E+02	34.99	A339	0 0 0 0 0	
4.300E+00	4.991E+02	39.99	A339	0 0 0 0 0	
4.608E+00	5.349E+02	40	M043	1 0 0 0 2	
4.561E+00	5.294E+02	40	W011	1 2 2 1 2	
4.562E+00	5.295E+02	44.99	A339	0 0 0 0 0	
4.677E+00	5.429E+02	49.99	A339	0 0 0 0 0	
4.842E+00	5.620E+02	54.99	A339	0 0 0 0 0	
5.031E+00	5.840E+02	59.99	A339	0 0 0 0 0	
5.516E+00	6.403E+02	60	M043	1 0 0 0 2	
5.151E+00	5.979E+02	60	W011	1 2 2 1 2	
5.166E+00	5.997E+02	64.99	A339	0 0 0 0 0	
6.366E+00	7.389E+02	80	M043	1 0 0 0 2	
6.864E+00	7.967E+02	97.5	D041	1 0 0 0 2	
6.866E+00	7.970E+02	97.5	F300	1 0 0 0 2	
6.866E+00	7.970E+02	97.5	W011	1 2 2 1 2	

253. C_4H_4S
Thiophene
Thiofuran
Thiacyclopentadiene

RN: 110-02-1   **MP** (°C):   −38.3
MW: 84.14   **BP** (°C):   84.4

Solubility (Moles/L)	Solubility (Grams/L)	Temp (°C)	Ref (#)	Evaluation (T P E A A)	Comments
3.583E-02	3.015E+00	25	K119	1 0 0 0 2	
3.583E-02	3.015E+00	25	P051	2 1 1 2 2	
3.583E-02	3.015E+00	25.00	P007	2 1 2 2 2	

254. C₄H₅BrO₄

Bromosuccinic acid
DL-Brombernsteinsaeure

RN:	923-06-8	MP (°C):	
MW:	196.99	BP (°C):	161

Solubility (Moles/L)	Solubility (Grams/L)	Temp (°C)	Ref (#)	Evaluation (T P E A A)	Comments
6.092E-01	1.200E+02	15.5	F300	1 0 0 0 1	

255. C₄H₅ClO₂

2-Chloroisocrotonic acid
α-Chlor-isocrotonsaeure

RN:	24253-33-6	MP (°C):	
MW:	120.54	BP (°C):	

Solubility (Moles/L)	Solubility (Grams/L)	Temp (°C)	Ref (#)	Evaluation (T P E A A)	Comments
5.102E-01	6.150E+01	19	F300	1 0 0 0 2	

256. C₄H₅ClO₂

2-Chlorocrotonic acid
α-Chlor-crotonsaeure

RN:	600-13-5	MP (°C):	
MW:	120.54	BP (°C):	

Solubility (Moles/L)	Solubility (Grams/L)	Temp (°C)	Ref (#)	Evaluation (T P E A A)	Comments
1.742E-01	2.100E+01	19	F300	1 0 0 0 1	

257. C₄H₅ClO₂

3-Chlorocrotonic acid
β-Chlor-crotonsaeure

RN:	6214-28-4	MP (°C):	94
MW:	120.54	BP (°C):	206

Solubility (Moles/L)	Solubility (Grams/L)	Temp (°C)	Ref (#)	Evaluation (T P E A A)	Comments
1.842E-01	2.220E+01	12.5	F300	1 0 0 0 2	
2.481E-01	2.990E+01	19	F300	1 0 0 0 2	

258. C$_4$H$_5$ClO$_2$
3-Chloroisocrotonic acid
β-Chlor-isocrotonsaeure
RN: 6625-00-9   **MP** (°C):
MW: 120.54   **BP** (°C):

Solubility (Moles/L)	Solubility (Grams/L)	Temp (°C)	Ref (#)	Evaluation (T P E A A)	Comments
1.037E-01	1.250E+01	7	F300	1 0 0 0 2	
1.560E-01	1.880E+01	19	F300	1 0 0 0 2	

259. C$_4$H$_5$ClO$_4$
L-Chlorosuccinic acid
L(–)-Chlor-bernsteinsaeure
D-Chlorosuccinic acid
D(+)-Chlor-bernsteinsaeure
RN: 16045-92-4   **MP** (°C):
MW: 152.54   **BP** (°C):

Solubility (Moles/L)	Solubility (Grams/L)	Temp (°C)	Ref (#)	Evaluation (T P E A A)	Comments
1.180E+00	1.800E+02	20	F300	1 0 0 0 1	
1.193E+00	1.820E+02	20	F300	1 0 0 0 2	

260. C$_4$H$_5$F$_3$O
Fluroxene
2,2,2-(Trifluoroethoxy)ethene
Redeptin
Fluoromar
RN: 406-90-6   **MP** (°C):
MW: 126.08   **BP** (°C):   42.7

Solubility (Moles/L)	Solubility (Grams/L)	Temp (°C)	Ref (#)	Evaluation (T P E A A)	Comments
3.173E-05	4.000E-03	ns	R028	0 0 0 0 0	

261. C$_4$H$_5$N
Pyrrole
Azole
Imidole
RN: 109-97-7   **MP** (°C):   –23
MW: 67.09   **BP** (°C):

Solubility (Moles/L)	Solubility (Grams/L)	Temp (°C)	Ref (#)	Evaluation (T P E A A)	Comments
7.098E-01	4.762E+01	rt	B099	0 2 0 0 0	

262. C₄H₅N
Methacrylonitrile
2-Methyl-2-propenenitrile
RN: 126-98-7   **MP** (°C):   −35.8
MW: 67.09   **BP** (°C):   90.3

Solubility (Moles/L)	Solubility (Grams/L)	Temp (°C)	Ref (#)	Evaluation (T P E A A)	Comments
3.692E-01	2.477E+01	25	L096	1 2 0 2 2	

263. C₄H₅NO₂
Hymexazol
3-Hydroxy-5-methyl isoxazole
5-Methyl-3(2H)-isoxazolone
Tachigaren
Isoxazolol, 5-methyl-
RN: 10004-44-1   **MP** (°C):   86
MW: 99.09   **BP** (°C):

Solubility (Moles/L)	Solubility (Grams/L)	Temp (°C)	Ref (#)	Evaluation (T P E A A)	Comments
8.578E-01	8.500E+01	25	M161	1 0 0 0 2	
8.578E-01	8.500E+01	25	N306	1 0 0 0 1	

264. C₄H₅NO₂
Succinimide
2,5-Pyrrolidinedione
Butanimide
RN: 123-56-8   **MP** (°C):   126
MW: 99.09   **BP** (°C):   288

Solubility (Moles/L)	Solubility (Grams/L)	Temp (°C)	Ref (#)	Evaluation (T P E A A)	Comments
9.174E-01	9.091E+01	0	M043	1 0 0 0 1	
1.392E+00	1.379E+02	10	M043	1 0 0 0 1	
2.082E+00	2.063E+02	20	M043	1 0 0 0 1	
1.978E+00	1.960E+02	21	F300	1 0 0 0 2	
3.273E+00	3.243E+02	30	M043	1 0 0 0 1	
4.577E+00	4.536E+02	40	M043	1 0 0 0 1	
5.887E+00	5.833E+02	60	M043	1 0 0 0 2	
6.868E+00	6.805E+02	80	M043	1 0 0 0 2	
1.413E+00	1.400E+02	ns	D072	0 0 0 0 1	
1.995E+00	1.977E+02	ns	R424	0 0 0 0 0	

265. C₄H₅NS

Allyl isothiocyanate
Allyl mustardiol
Allylsenfoel

RN: 57-06-7   **MP** (°C): –8
MW: 99.16   **BP** (°C): 152

Solubility (Moles/L)	Solubility (Grams/L)	Temp (°C)	Ref (#)	Evaluation (T P E A A)	Comments
2.017E-02	2.000E+00	20	F300	1 0 0 0 0	

266. C₄H₅N₃O

Cytosine
2-Oxy-4-amino pyrimidine
2(1H)-Pyrimidinone, 4-amino-

RN: 71-30-7   **MP** (°C): 320
MW: 111.10   **BP** (°C):

Solubility (Moles/L)	Solubility (Grams/L)	Temp (°C)	Ref (#)	Evaluation (T P E A A)	Comments
5.000E-02	5.555E+00	20	C017	2 0 0 1 0	EFG
6.877E-02	7.641E+00	25	D041	1 0 0 0 1	
7.200E-02	8.000E+00	25	F300	1 0 0 0 0	
6.580E-02	7.311E+00	25	H061	0 0 0 0 0	
6.500E-02	7.222E+00	25	R030	0 0 0 0 0	

267. C₄H₅N₃OS

6-Amino-2-thiouracil
2-Mercapto-4-amino-6-hydroxypyrimidine
2-Thio-4-amino-6-hydroxypyrimidine
2-Mercapto-6-aminouracil

RN: 1004-40-6   **MP** (°C):
MW: 143.17   **BP** (°C):

Solubility (Moles/L)	Solubility (Grams/L)	Temp (°C)	Ref (#)	Evaluation (T P E A A)	Comments
1.790E-03	2.563E-01	25	G016	1 2 1 2 2	intrinsic

268. C₄H₅N₃O₂

5-Aminouracil
5-Amino-uracil

RN: 932-52-5   **MP** (°C): >300
MW: 127.10   **BP** (°C):

Solubility (Moles/L)	Solubility (Grams/L)	Temp (°C)	Ref (#)	Evaluation (T P E A A)	Comments
3.934E-03	5.000E-01	20	F300	1 0 0 0 0	
4.700E-03	5.974E-01	25	Z408	0 0 0 0 0	
1.259E-01	1.600E+01	100	F300	1 0 0 0 1	

269. C$_4$H$_5$N$_3$O$_2$

6-Aminouracil
2,4(1H,3H)-Pyrimidinedione, 6-amino
4-Amino-2,6-dihydroxypyrimidine
6-Amino-2,4-pyrimidinediol
4-Amino uracil

RN: 873-83-6 **MP** (°C):
MW: 127.10 **BP** (°C):

Solubility (Moles/L)	Solubility (Grams/L)	Temp (°C)	Ref (#)	Evaluation (T P E A A)	Comments
4.700E-03	5.974E-01	25	Z408	0 0 0 0 0	

270. C$_4$H$_5$N$_3$O$_2$

2-Methyl-4(5)-nitroimidazole
2-Methyl-5-nitroimidazole
Menidazole
RP 8532
L 581490

RN: 696-23-1 **MP** (°C): 257–258
MW: 127.10 **BP** (°C):

Solubility (Moles/L)	Solubility (Grams/L)	Temp (°C)	Ref (#)	Evaluation (T P E A A)	Comments
2.368E-02	3.010E+00	20	D344	0 0 0 0 0	
2.367E-02	3.009E+00	20	D344	0 0 0 0 0	
2.353E-02	2.991E+00	20	D344	0 0 0 0 0	
2.370E-02	3.012E+00	20	D344	0 0 0 0 0	

271. C$_4$H$_6$

1,3-Butadiene
Pyrrolylene

RN: 106-99-0 **MP** (°C): −108.9
MW: 54.09 **BP** (°C): −4.5

Solubility (Moles/L)	Solubility (Grams/L)	Temp (°C)	Ref (#)	Evaluation (T P E A A)	Comments
1.359E-02	7.350E-01	25	M001	2 1 2 2 2	

272. C$_4$H$_6$

1-Butyne
Ethylacetylene
Ethylethyne

RN: 107-00-6 **MP** (°C): −125.7
MW: 54.09 **BP** (°C): 8.1

Solubility (Moles/L)	Solubility (Grams/L)	Temp (°C)	Ref (#)	Evaluation (T P E A A)	Comments
5.306E-02	2.870E+00	25	M001	2 1 2 2 2	

273. C$_4$H$_6$BrNO$_4$

5-Bromo-5-nitro-1,3-dioxane
Bronidox
Microcide I
Bronidox L
1,3-Dioxane, 5-bromo-5-nitro-

RN:	30007-47-7	MP (°C):	49–50
MW:	212.01	BP (°C):	

Solubility (Moles/L)	Solubility (Grams/L)	Temp (°C)	Ref (#)	Evaluation (T P E A A)	Comments
2.706E-02	5.737E+00	25	L013	1 0 2 1 2	

274. C$_4$H$_6$Cl$_2$O$_2$S

3,4-Dichlorotetrahydrothiophene dioxide
3,4-Dichlorotetrahydrothiophene 1,1-dioxide
3,4-Dichlorosulfolane
DAC PRD
3,4-Dichlorothiolane 1,1-dioxide

RN:	3001-57-8	MP (°C):	130
MW:	189.06	BP (°C):	

Solubility (Moles/L)	Solubility (Grams/L)	Temp (°C)	Ref (#)	Evaluation (T P E A A)	Comments
1.161E-02	2.195E+00	20	M061	1 0 0 0 1	

275. C$_4$H$_6$N$_2$O$_2$

2,5-Piperazinedione
Diketopiperazine

RN:	106-57-0	MP (°C):	
MW:	114.10	BP (°C):	

Solubility (Moles/L)	Solubility (Grams/L)	Temp (°C)	Ref (#)	Evaluation (T P E A A)	Comments
1.232E-01	1.406E+01	20	B032	1 2 2 1 2	
1.253E-01	1.430E+01	20	M075	2 0 1 1 2	
1.475E-01	1.683E+01	25	B032	1 2 2 1 2	
1.754E-01	2.001E+01	29.80	B032	1 2 2 1 2	

276. C$_4$H$_6$N$_2$S$_4$Zn

Zineb
Zinc ethylenebis(dithiocarbamate)

RN:	12122-67-7	MP (°C):	
MW:	275.74	BP (°C):	

Solubility (Moles/L)	Solubility (Grams/L)	Temp (°C)	Ref (#)	Evaluation (T P E A A)	Comments
3.627E-06	1.000E-03	20	M061	1 0 0 0 0	
3.627E-05	1.000E-02	rt	M161	0 0 0 0 1	

277. C$_4$H$_6$N$_4$O$_3$
Allantoin
Allantoine

RN:	97-59-6	**MP (°C):**	238	
MW:	158.12	**BP (°C):**		

Solubility (Moles/L)	Solubility (Grams/L)	Temp (°C)	Ref (#)	Evaluation (T P E A A)	Comments
3.303E-02	5.223E+00	20	D041	1 0 0 0 2	
4.755E-02	7.519E+00	c	D004	0 0 0 0 0	
2.040E-01	3.226E+01	h	D004	0 0 0 0 0	
2.530E-02	4.000E+00	ns	D072	0 0 0 0 1	

278. C$_4$H$_6$N$_4$O$_3$S$_2$
Acetazolamide
5-Acetamido-1,3,4-thiadiazole-2-sulfonamide

RN:	59-66-5	**MP (°C):**	258	
MW:	222.25	**BP (°C):**		

Solubility (Moles/L)	Solubility (Grams/L)	Temp (°C)	Ref (#)	Evaluation (T P E A A)	Comments
2.700E-03	6.001E-01	15	K024	1 2 1 1 2	
2.249E-03	4.998E-01	20	D041	1 0 0 0 0	
3.200E-03	7.112E-01	25	C415	1 0 0 1 0	
2.216E-03	4.925E-01	25	F415	0 0 0 0 0	Average
4.409E-03	9.799E-01	30	E049	2 0 2 2 2	
5.174E-03	1.150E+00	37	C054	2 0 2 1 2	
2.880E-03	6.400E-01	amb	L434	0 0 0 0 0	
>2.25E-03	>5.00E-01	ns	B404	0 2 1 1 0	
4.144E-03	9.210E-01	ns	I304	0 0 0 0 0	
4.500E-04	1.000E-01	ns	K444	0 0 0 0 0	
4.365E-03	9.701E-01	ns	R428	0 0 0 0 0	

279. C$_4$H$_6$O
Vinyl ether
1,1'-Oxybisethene
Divinyl ether

RN:	109-93-3	**MP (°C):**		
MW:	70.09	**BP (°C):**	28.4	

Solubility (Moles/L)	Solubility (Grams/L)	Temp (°C)	Ref (#)	Evaluation (T P E A A)	Comments
7.490E-02	5.250E+00	37	R047	0 0 0 0 0	
5.487E-01	3.846E+01	ns	R028	0 0 0 0 0	

280. C₄H₆O
Crotonaldehyde
But-*trans*-enal
RN: 4170-30-3 **MP** (°C): −76.5
MW: 70.09 **BP** (°C):

Solubility (Moles/L)	Solubility (Grams/L)	Temp (°C)	Ref (#)	Evaluation (T P E A A)	Comments
+2.14E+00	+1.50E+02	ns	S460	0 0 0 0 0	

281. C₄H₆O
α-Methylacrolein
α-Methyl-acrolein
RN: 78-85-3 **MP** (°C):
MW: 70.09 **BP** (°C):

Solubility (Moles/L)	Solubility (Grams/L)	Temp (°C)	Ref (#)	Evaluation (T P E A A)	Comments
8.089E-01	5.670E+01	20	F300	1 0 0 0 2	
1.236E+00	8.663E+01	ns	S460	0 0 0 0 0	

282. C₄H₆O
trans-Crotonaldehyde
Crotonaldehyd
RN: 123-73-9 **MP** (°C): −77
MW: 70.09 **BP** (°C):

Solubility (Moles/L)	Solubility (Grams/L)	Temp (°C)	Ref (#)	Evaluation (T P E A A)	Comments
2.140E+00	1.500E+02	20	F300	1 0 0 0 1	

283. C₄H₆O₂
Diacetyl
2,3-Butanedione
RN: 431-03-8 **MP** (°C):
MW: 86.09 **BP** (°C):

Solubility (Moles/L)	Solubility (Grams/L)	Temp (°C)	Ref (#)	Evaluation (T P E A A)	Comments
2.323E+00	2.000E+02	15	F300	1 0 0 0 1	
2.323E+00	2.000E+02	20	D041	1 0 0 0 1	

284. C₄H₆O₂
Methyl acrylate
Acrylic acid methyl ester
2-Propenoic acid methyl ester
RN: 96-33-3 **MP** (°C): −76.5
MW: 86.09 **BP** (°C): 70

Solubility (Moles/L)	Solubility (Grams/L)	Temp (°C)	Ref (#)	Evaluation (T P E A A)	Comments
5.742E-01	4.943E+01	30	L096	1 2 0 2 1	

285. C$_4$H$_6$O$_2$

trans-Crotonic acid
trans-Crotonsaeure

RN:	3724-65-0	MP (°C):	
MW:	86.09	BP (°C):	

Solubility (Moles/L)	Solubility (Grams/L)	Temp (°C)	Ref (#)	Evaluation (T P E A A)	Comments
9.989E-01	8.600E+01	25	F300	1 0 0 0 1	
4.600E+00	3.960E+02	40	F300	1 0 0 0 2	

286. C$_4$H$_6$O$_2$

Vinyl acetate
Vinylacetate

RN:	108-05-4	MP (°C):	−100
MW:	86.09	BP (°C):	72

Solubility (Moles/L)	Solubility (Grams/L)	Temp (°C)	Ref (#)	Evaluation (T P E A A)	Comments
3.136E-01	2.700E+01	50	L097	1 1 1 1 1	

287. C$_4$H$_6$O$_2$

Crotonic acid
2-Butenoic acid
3-Methylacrylic acid

RN:	107-93-7	MP (°C):	73
MW:	86.09	BP (°C):	

Solubility (Moles/L)	Solubility (Grams/L)	Temp (°C)	Ref (#)	Evaluation (T P E A A)	Comments
8.882E-01	7.647E+01	20	D041	1 0 0 0 2	

288. C$_4$H$_6$O$_2$

α-Butyrolacetone
3-Hydroxybutanoic acid β-lactone

RN:	3068-88-0	MP (°C):	
MW:	86.09	BP (°C):	

Solubility (Moles/L)	Solubility (Grams/L)	Temp (°C)	Ref (#)	Evaluation (T P E A A)	Comments
1.541E+00	1.327E+02	18	I313	0 0 0 0 0	

289. C₄H₆O₂S₄

bis(Methylxanthogen) disulfide
Dimethylxanthogen disulfide
Methyl dixanthogen

RN: 1468-37-7   **MP** (°C):   22.75
MW: 214.35   **BP** (°C):

Solubility (Moles/L)	Solubility (Grams/L)	Temp (°C)	Ref (#)	Evaluation (T P E A A)	Comments
1.150E-04	2.465E-02	25	H102	1 2 1 2 2	

290. C₄H₆O₃

Acetic anhydride
Essigsaeure-anhydrid

RN: 108-24-7   **MP** (°C):   −73
MW: 102.09   **BP** (°C):   139

Solubility (Moles/L)	Solubility (Grams/L)	Temp (°C)	Ref (#)	Evaluation (T P E A A)	Comments
1.175E+00	1.200E+02	20	F300	1 0 0 0 2	

291. C₄H₆O₄

Methylmalonic acid
Acide methylmalonique
Methyl-malonsaeure

RN: 516-05-2   **MP** (°C):   129.5
MW: 118.09   **BP** (°C):

Solubility (Moles/L)	Solubility (Grams/L)	Temp (°C)	Ref (#)	Evaluation (T P E A A)	Comments
2.600E+00	3.070E+02	0	F300	1 0 0 0 2	
3.743E+00	4.420E+02	0	M051	1 0 0 0 2	
4.954E+00	5.850E+02	15	M051	1 0 0 0 2	
5.750E+00	6.790E+02	25	M051	1 0 0 0 2	
4.071E+00	4.808E+02	50	F300	1 0 0 0 2	
7.748E+00	9.150E+02	50	M051	1 0 0 0 2	

292. C₄H₆O₄

Succinic acid
Bernsteinsaeure

RN: 110-15-6   **MP** (°C):   185
MW: 118.09   **BP** (°C):   235

Solubility (Moles/L)	Solubility (Grams/L)	Temp (°C)	Ref (#)	Evaluation (T P E A A)	Comments
2.363E-01	2.790E+01	0	L041	1 0 0 1 2	
2.273E-01	2.684E+01	0	M020	1 0 0 1 1	
2.306E-01	2.724E+01	0	M043	1 0 0 0 1	

(continued)

292. C$_4$H$_6$O$_4$ (continued)

Solubility (Moles/L)	Solubility (Grams/L)	Temp (°C)	Ref (#)	Evaluation (T P E A A)	Comments
2.892E-01	3.415E+01	4.99	A339	0 0 0 0 0	
3.569E-01	4.215E+01	10	M043	1 0 0 0 1	
3.616E-01	4.271E+01	9.99	A339	0 0 0 0 0	
3.854E-01	4.551E+01	11.85	L064	2 2 2 1 2	
4.518E-01	5.335E+01	14.99	A339	0 0 0 0 0	
4.102E-01	4.843E+01	15	F055	1 2 2 2 2	
4.149E-01	4.900E+01	15	L041	1 0 0 1 1	
4.149E-01	4.900E+01	15	M051	1 0 0 0 1	
4.912E-01	5.800E+01	17.50	F300	1 0 0 0 1	
4.974E-01	5.874E+01	18	L064	2 2 2 1 2	
5.661E-01	6.685E+01	19.99	A339	0 0 0 0 0	
5.392E-01	6.367E+01	20	D041	1 0 0 0 1	
5.019E-01	5.927E+01	20	F055	1 2 2 2 2	
5.420E-01	6.400E+01	20	F300	1 0 0 0 2	
4.912E-01	5.800E+01	20	L041	1 0 0 1 1	
5.466E-01	6.455E+01	20	M043	1 0 0 0 1	
5.510E-01	6.507E+01	20	M153	1 0 0 0 0	cal. from fitted equation
4.632E-01	5.470E+01	20	M171	1 0 0 0 1	
5.716E-01	6.750E+01	20	W026	1 0 1 1 1	average of 2
6.344E-01	7.492E+01	23.75	L064	2 2 2 1 2	
6.829E-01	8.064E+01	24.99	A339	0 0 0 0 0	
5.930E-01	7.003E+01	25	D061	1 0 0 0 2	
6.032E-01	7.124E+01	25	F055	1 2 2 2 2	
6.849E-01	8.088E+01	25	H430	0 0 0 0 0	
6.518E-01	7.697E+01	25	M020	1 0 0 1 2	
6.634E-01	7.834E+01	25	M153	1 0 0 0 0	cal. from fitted equation
7.402E-01	8.741E+01	28	D050	1 2 1 2 2	
8.003E-01	9.451E+01	29.99	A339	0 0 0 0 0	
8.017E-01	9.502E+01	30	M043	1 0 0 0 2	
8.047E-01	9.502E+01	30	M153	1 0 0 0 0	cal. from fitted equation
8.849E-01	1.045E+02	30	W026	1 0 1 1 2	average of 2
9.508E-01	1.123E+02	34.99	A339	0 0 0 0 0	
8.976E-01	1.060E+02	35	L041	1 0 0 1 2	
9.742E-01	1.150E+02	35	M153	1 0 0 0 0	cal. from fitted equation
1.145E+00	1.353E+02	39.99	A339	0 0 0 0 0	
1.149E+00	1.357E+02	40	B088	1 0 0 0 2	
1.181E+00	1.394E+02	40	M043	1 0 0 0 2	
1.168E+00	1.379E+02	40	M153	1 0 0 0 0	cal. from fitted equation
1.377E+00	1.627E+02	44.99	A339	0 0 0 0 0	
1.600E+00	1.889E+02	49.99	A339	0 0 0 0 0	
1.524E+00	1.800E+02	50	L041	1 0 0 1 2	
1.633E+00	1.929E+02	50	M020	1 0 0 1 2	
1.842E+00	2.175E+02	54.99	A339	0 0 0 0 0	
2.048E+00	2.418E+02	59.99	A339	0 0 0 0 0	
2.232E+00	2.636E+02	60	M043	1 0 0 0 2	
2.398E+00	2.832E+02	64.99	A339	0 0 0 0 0	
2.380E+00	2.810E+02	65	L041	1 0 0 1 2	
3.238E+00	3.824E+02	75	F300	1 0 0 0 2	
3.191E+00	3.768E+02	75	M020	1 0 0 1 2	

(continued)

292. C$_4$H$_6$O$_4$ (continued)

Solubility (Moles/L)	Solubility (Grams/L)	Temp (°C)	Ref (#)	Evaluation (T P E A A)	Comments
3.510E+00	4.145E+02	80	M043	1 0 0 0 2	
8.515E-01	1.006E+02	84.30	B118	1 0 0 0 2	unit assumed
4.636E+00	5.475E+02	100	D041	1 0 0 0 2	
4.738E+00	5.595E+02	100	M043	1 0 0 0 2	
6.821E-01	8.054E+01	rt	H431	0 0 0 0 0	

293. C$_4$H$_6$O$_4$

Methyl oxalate
Oxalic acid ethyl ester
Oxalsaeure-monoaethyl ester

RN:	553-90-2	**MP (°C):**	54.0
MW:	118.09	**BP (°C):**	163.5

Solubility (Moles/L)	Solubility (Grams/L)	Temp (°C)	Ref (#)	Evaluation (T P E A A)	Comments
3.006E-01	3.549E+01	.1	K079	1 0 0 0 2	
6.900E-01	8.148E+01	11.1	K079	1 0 0 0 2	
1.029E+00	1.216E+02	19.5	K079	1 0 0 0 2	
5.106E-01	6.030E+01	25	F300	1 0 0 0 2	
1.489E+00	1.758E+02	27.1	K079	1 0 0 0 2	
1.867E+00	2.204E+02	31.9	K079	1 0 0 0 2	
2.978E+00	3.516E+02	44.4	K079	1 0 0 0 2	
3.372E+00	3.982E+02	49.2	K079	1 0 0 0 2	
3.589E+00	4.238E+02	51.0	K079	1 0 0 0 2	
3.839E+00	4.533E+02	53.0	K079	1 0 0 0 2	
4.783E+00	5.649E+02	75.0	K079	1 0 0 0 2	
4.939E+00	5.832E+02	79.3	K079	1 0 0 0 2	
5.678E+00	6.705E+02	96.1	K079	1 0 0 0 2	
4.929E-01	5.820E+01	rt	D021	0 0 1 1 2	

294. C$_4$H$_6$O$_5$

D-Malic acid
D(–)-Aepfelsaeure

RN:	636-61-3	**MP (°C):**	100
MW:	134.09	**BP (°C):**	

Solubility (Moles/L)	Solubility (Grams/L)	Temp (°C)	Ref (#)	Evaluation (T P E A A)	Comments
3.397E+00	4.555E+02	4.99	A339	0 0 0 0 0	
3.542E+00	4.749E+02	9.99	A339	0 0 0 0 0	
3.695E+00	4.954E+02	14.99	A339	0 0 0 0 0	
3.878E+00	5.200E+02	19.99	A339	0 0 0 0 0	
4.030E+00	5.403E+02	24.99	A339	0 0 0 0 0	
4.146E+00	5.560E+02	29.99	A339	0 0 0 0 0	
4.282E+00	5.742E+02	34.99	A339	0 0 0 0 0	
4.441E+00	5.955E+02	39.99	A339	0 0 0 0 0	
4.544E+00	6.094E+02	44.99	A339	0 0 0 0 0	

(continued)

294. C$_4$H$_6$O$_5$ (continued)

Solubility (Moles/L)	Solubility (Grams/L)	Temp (°C)	Ref (#)	Evaluation (T P E A A)	Comments
4.719E+00	6.328E+02	49.99	A339	0 0 0 0 0	
4.840E+00	6.490E+02	54.99	A339	0 0 0 0 0	
4.976E+00	6.672E+02	59.99	A339	0 0 0 0 0	
5.119E+00	6.865E+02	64.99	A339	0 0 0 0 0	

295. C$_4$H$_6$O$_5$
Diglycolic acid
Di-glykolsaeure

RN: 110-99-6   **MP** (°C):   148
MW: 134.09   **BP** (°C):

Solubility (Moles/L)	Solubility (Grams/L)	Temp (°C)	Ref (#)	Evaluation (T P E A A)	Comments
1.596E+00	2.140E+02	5.09	A340	0 0 0 0 0	
1.932E+00	2.590E+02	10.99	A340	0 0 0 0 0	
2.522E+00	3.382E+02	15.59	A340	0 0 0 0 0	
2.668E+00	3.577E+02	20.59	A340	0 0 0 0 0	
2.834E+00	3.801E+02	23.49	A340	0 0 0 0 0	
3.252E+00	4.361E+02	28.09	A340	0 0 0 0 0	
3.645E+00	4.887E+02	37.49	A340	0 0 0 0 0	
3.794E+00	5.087E+02	39.99	A340	0 0 0 0 0	
4.061E+00	5.445E+02	47.99	A340	0 0 0 0 0	
4.135E+00	5.545E+02	49.99	A340	0 0 0 0 0	
4.353E+00	5.837E+02	54.49	A340	0 0 0 0 0	
4.508E+00	6.044E+02	59.49	A340	0 0 0 0 0	
4.631E+00	6.209E+02	64.99	A340	0 0 0 0 0	
4.776E+00	6.404E+02	69.99	A340	0 0 0 0 0	
4.877E+00	6.540E+02	74.99	A340	0 0 0 0 0	
4.969E+00	6.663E+02	79.89	A340	0 0 0 0 0	
5.067E+00	6.794E+02	83.99	A340	0 0 0 0 0	
5.125E+00	6.872E+02	88.19	A340	0 0 0 0 0	

296. C$_4$H$_6$O$_5$
DL-Malic acid
Malic acid

RN: 6915-15-7   **MP** (°C):   131.5
MW: 134.09   **BP** (°C):

Solubility (Moles/L)	Solubility (Grams/L)	Temp (°C)	Ref (#)	Evaluation (T P E A A)	Comments
3.512E+00	4.709E+02	0	M043	1 0 0 0 1	
3.820E+00	5.122E+02	10	M043	1 0 0 0 2	
4.158E+00	5.575E+02	20	M043	1 0 0 0 2	
4.414E+00	5.918E+02	25	H430	0 0 0 0 0	
4.401E+00	5.902E+02	26	D041	1 0 0 0 2	
4.415E+00	5.920E+02	26	F300	1 0 0 0 2	

(continued)

296. C₄H₆O₅ (continued)

Solubility (Moles/L)	Solubility (Grams/L)	Temp (°C)	Ref (#)	Evaluation (T P E A A)	Comments
5.605E+00	7.516E+02	30	D062	1 0 1 1 0	data given in normality
4.475E+00	6.000E+02	30	M043	1 0 0 0 2	
4.794E+00	6.429E+02	40	M043	1 0 0 0 2	
5.442E+00	7.297E+02	60	M043	1 0 0 0 2	
5.998E+00	8.043E+02	79	D041	1 0 0 0 2	
6.033E+00	8.089E+02	79	F300	1 0 0 0 2	
6.126E+00	8.214E+02	80	M043	1 0 0 0 2	

297. C₄H₆O₆

meso-Tartaric acid
meso-Weinsaeure

RN: 147-73-9   **MP** (°C):   147
MW: 150.09   **BP** (°C):

Solubility (Moles/L)	Solubility (Grams/L)	Temp (°C)	Ref (#)	Evaluation (T P E A A)	Comments
2.239E+00	3.360E+02	0	F300	1 0 0 0 2	
3.702E+00	5.556E+02	15	D041	1 0 0 0 2	
3.731E+00	5.600E+02	15	F300	1 0 0 0 1	
3.731E+00	5.600E+02	20	F300	1 0 0 0 1	

298. C₄H₆O₆

D-(–)-Tartaric acid
D-(–)-Dihydroxysuccinic acid

RN: 147-71-7   **MP** (°C):   173
MW: 150.09   **BP** (°C):   171

Solubility (Moles/L)	Solubility (Grams/L)	Temp (°C)	Ref (#)	Evaluation (T P E A A)	Comments
3.564E+00	5.349E+02	0	M043	1 0 0 0 2	
3.348E+00	5.024E+02	4.99	A339	0 0 0 0 0	
2.350E+00	3.528E+02	10	D020	1 2 1 1 2	
3.715E+00	5.575E+02	10	M043	1 0 0 0 2	
3.431E+00	5.149E+02	9.99	A339	0 0 0 0 0	
3.499E+00	5.251E+02	14.99	A339	0 0 0 0 0	
3.553E+00	5.332E+02	19.99	A339	0 0 0 0 0	
3.875E+00	5.816E+02	20	M043	1 0 0 0 2	
3.629E+00	5.447E+02	24.99	A339	0 0 0 0 0	
2.459E+00	3.691E+02	25	D020	1 2 1 1 2	
3.973E+00	5.963E+02	25	F017	1 0 0 0 2	
3.706E+00	5.562E+02	29.99	A339	0 0 0 0 0	
4.060E+00	6.094E+02	30	M043	1 0 0 0 2	
3.791E+00	5.690E+02	34.99	A339	0 0 0 0 0	
3.846E+00	5.773E+02	39.99	A339	0 0 0 0 0	
4.249E+00	6.377E+02	40	M043	1 0 0 0 2	
3.926E+00	5.892E+02	44.99	A339	0 0 0 0 0	

(continued)

298. $C_4H_6O_6$ (continued)

Solubility (Moles/L)	Solubility (Grams/L)	Temp (°C)	Ref (#)	Evaluation (T P E A A)	Comments
4.021E+00	6.036E+02	49.99	A339	0 0 0 0 0	
4.104E+00	6.160E+02	54.99	A339	0 0 0 0 0	
4.157E+00	6.238E+02	59.99	A339	0 0 0 0 0	
4.581E+00	6.875E+02	60	M043	1 0 0 0 2	
4.232E+00	6.352E+02	64.99	A339	0 0 0 0 0	
4.876E+00	7.319E+02	80	M043	1 0 0 0 2	
5.159E+00	7.743E+02	100	M043	1 0 0 0 2	

299. $C_4H_6O_6$
L-Tartaric acid
L(+)-Weinsaeure
L(+)-Tartaric acid

RN:	87-69-4	**MP (°C):**	169
MW:	150.09	**BP (°C):**	

Solubility (Moles/L)	Solubility (Grams/L)	Temp (°C)	Ref (#)	Evaluation (T P E A A)	Comments
3.565E+00	5.350E+02	0	F300	1 0 0 0 2	
3.564E+00	5.349E+02	0	F302	1 0 0 0 2	
3.634E+00	5.455E+02	5	F302	1 0 0 0 2	
3.702E+00	5.556E+02	10	F302	1 0 0 0 2	
3.791E+00	5.690E+02	15	F302	1 0 0 0 2	
3.878E+00	5.820E+02	20	F300	1 0 0 0 2	
3.875E+00	5.816E+02	20	F302	1 0 0 0 2	
3.965E+00	5.951E+02	25	F302	1 0 0 0 2	
4.060E+00	6.094E+02	30	F302	1 0 0 0 2	
4.158E+00	6.241E+02	35	F302	1 0 0 0 2	
4.249E+00	6.377E+02	40	F302	1 0 0 0 2	
4.325E+00	6.491E+02	45	F302	1 0 0 0 2	
4.397E+00	6.600E+02	50	F300	1 0 0 0 1	
4.404E+00	6.610E+02	50	F302	1 0 0 0 2	
4.485E+00	6.732E+02	55	F302	1 0 0 0 2	
4.568E+00	6.855E+02	60	F302	1 0 0 0 2	
4.644E+00	6.970E+02	65	F302	1 0 0 0 2	
4.726E+00	7.093E+02	70	F302	1 0 0 0 2	
4.802E+00	7.207E+02	75	F302	1 0 0 0 2	
4.876E+00	7.319E+02	80	F302	1 0 0 0 2	
4.954E+00	7.436E+02	85	F302	1 0 0 0 2	
5.026E+00	7.543E+02	90	F302	1 0 0 0 2	
5.095E+00	7.647E+02	95	F302	1 0 0 0 2	
5.157E+00	7.740E+02	100	F300	1 0 0 0 2	
5.159E+00	7.743E+02	100	F302	1 0 0 0 2	

300. $C_4H_6O_6$

DL-Tartaric acid
DL-Weinsaeure
Tartaric acid (racemic)

RN:	133-37-9	**MP (°C):**	206
MW:	150.09	**BP (°C):**	

Solubility (Moles/L)	Solubility (Grams/L)	Temp (°C)	Ref (#)	Evaluation (T P E A A)	Comments
2.279E+00	3.421E+02	0	D039	2 2 1 2 0	EFG
5.630E-01	8.450E+01	0	D041	1 0 0 0 2	
5.084E-01	7.630E+01	0	F300	1 0 0 0 2	
5.049E-01	7.579E+01	0	M043	1 0 0 0 1	
2.333E+00	3.502E+02	10	D039	2 2 1 2 0	EFG
7.298E-01	1.095E+02	10	M043	1 0 0 0 2	
2.350E+00	3.528E+02	20	D039	2 2 1 2 0	EFG
1.138E+00	1.708E+02	20	D041	1 0 0 0 2	
1.139E+00	1.710E+02	20	F300	1 0 0 0 2	
1.016E+00	1.525E+02	20	M043	1 0 0 0 2	
2.459E+00	3.690E+02	25	D039	2 2 1 2 2	EFG
1.179E+00	1.770E+02	25	F017	1 0 0 0 2	
1.026E+01	1.540E+03	25	K040	1 0 2 1 2	
2.483E+00	3.726E+02	30	D039	2 2 1 2 0	EFG
1.341E+00	2.013E+02	30	M043	1 0 0 0 2	
2.563E+00	3.846E+02	40	D039	2 2 1 2 0	EFG
1.799E+00	2.701E+02	40	M043	1 0 0 0 2	
2.612E+00	3.921E+02	50	D039	2 2 1 2 0	EFG
2.687E+00	4.033E+02	60	D039	2 2 1 2 0	EFG
2.612E+00	3.921E+02	60	M043	1 0 0 0 2	
2.750E+00	4.128E+02	70	D039	2 2 1 2 0	EFG
2.811E+00	4.220E+02	80	D039	2 2 1 2 0	EFG
3.299E+00	4.952E+02	80	M043	1 0 0 0 2	
2.860E+00	4.292E+02	90	D039	2 2 1 2 0	EFG
2.920E+00	4.382E+02	100	D039	2 2 1 2 0	EFG
4.324E+00	6.490E+02	100	D041	1 0 0 0 2	
4.331E+00	6.500E+02	100	F300	1 0 0 0 1	
3.863E+00	5.798E+02	100	M043	1 0 0 0 2	

301. C_4H_7Br

4-Bromo-1-butene
1-Bromo-3-butene
Homoallyl bromide
4-Bromobutene-1
3-Butenyl bromide

RN:	5162-44-7	**MP (°C):**	
MW:	135.01	**BP (°C):**	98.5

Solubility (Moles/L)	Solubility (Grams/L)	Temp (°C)	Ref (#)	Evaluation (T P E A A)	Comments
5.660E-03	7.642E-01	25	M342	1 0 1 1 2	

302. C$_4$H$_7$BrN$_2$O$_2$

Propanamide, *N*-(aminocarbonyl)-2-bromo-
(2-Bromopropionyl)urea
α-Bromopropionylurea

RN:	14299-55-9	**MP** (°C):
MW:	195.02	**BP** (°C):

Solubility (Moles/L)	Solubility (Grams/L)	Temp (°C)	Ref (#)	Evaluation (T P E A A)	Comments
2.581E-01	5.033E+01	ns	F056	0 2 2 2 1	

303. C$_4$H$_7$BrO$_2$

α-Bromobutyric acid
DL-2-Bromobutyric acid
DL-Brombuttersaeure

RN:	80-58-0	**MP** (°C):	−4
MW:	167.01	**BP** (°C):	181

Solubility (Moles/L)	Solubility (Grams/L)	Temp (°C)	Ref (#)	Evaluation (T P E A A)	Comments
4.191E-01	7.000E+01	ns	F300	1 0 0 0 0	

304. C$_4$H$_7$Cl

1-Chloro-2-butene
1-Chloro-2-methylpropene-2
α-Methylallyl chloride

RN:	591-97-9	**MP** (°C):
MW:	90.55	**BP** (°C): 84

Solubility (Moles/L)	Solubility (Grams/L)	Temp (°C)	Ref (#)	Evaluation (T P E A A)	Comments
1.103E-02	9.990E-01	ns	M061	0 0 0 0 0	

305. C$_4$H$_7$Cl$_2$O$_4$P

Dichlorvos
O,O-Dimethyl *O*-2-dichlorovinyl phosphate

RN:	62-73-7	**MP** (°C):
MW:	220.98	**BP** (°C): 84

Solubility (Moles/L)	Solubility (Grams/L)	Temp (°C)	Ref (#)	Evaluation (T P E A A)	Comments
4.481E-02	9.901E+00	ns	M061	0 0 0 0 0	
4.525E-02	1.000E+01	rt	M161	0 0 0 0 1	

306. C$_4$H$_7$Cl$_3$O

1,1,1-Trichloro-*tert*-butanol
Acetonchloroform
Chloreton

RN:	57-15-8	MP (°C):	98
MW:	177.46	BP (°C):	167

Solubility (Moles/L)	Solubility (Grams/L)	Temp (°C)	Ref (#)	Evaluation (T P E A A)	Comments
4.508E-02	8.000E+00	20	F300	1 0 0 0 0	
4.467E-02	7.927E+00	ns	R424	0 0 0 0 0	
4.467E-02	7.927E+00	ns	R427	0 0 0 0 0	

307. C$_4$H$_7$N

n-Butyroniitrile
γ-Butyronitrile
Propyl cyanide
1-Cyanopropane
n-Butyronitrile

RN:	109-74-0	MP (°C):	−112
MW:	69.11	BP (°C):	115–117

Solubility (Moles/L)	Solubility (Grams/L)	Temp (°C)	Ref (#)	Evaluation (T P E A A)	Comments
5.446E-02	3.764E+00	25	B004	0 0 0 0 0	

308. C$_4$H$_7$NO$_2$S

4-Thiazolidinecarboxylic acid
Thiazolidine-4-carboxylic acid
γ-Thiaproline
4-Carboxythiazolidine
Detoxepa
Thiaproline

RN:	444-27-9	MP (°C):	196–201
MW:	133.17	BP (°C):	350.3

Solubility (Moles/L)	Solubility (Grams/L)	Temp (°C)	Ref (#)	Evaluation (T P E A A)	Comments
2.200E-01	2.930E+01	21	B414	1 0 0 1 1	

309. C$_4$H$_7$NO$_3$
N-Acetyl glycine
Aceturic acid
Glycin-*N*-acetat
Glycine-*N*-acetate

RN:	543-24-8	**MP** (°C):	206		
MW:	117.11	**BP** (°C):			

Solubility (Moles/L)	Solubility (Grams/L)	Temp (°C)	Ref (#)	Evaluation (T P E A A)	Comments
2.246E-01	2.630E+01	15	F300	1 0 0 0 2	

310. C$_4$H$_7$NO$_4$
Butanoic acid, 4-amino-2-hydroxy-4-oxo-
D-β-Malaminsaeure
r-β-Malaminsaeure

RN:	82310-91-6	**MP** (°C):	149		
MW:	133.10	**BP** (°C):			

Solubility (Moles/L)	Solubility (Grams/L)	Temp (°C)	Ref (#)	Evaluation (T P E A A)	Comments
2.903E-01	3.865E+01	18	L039	1 0 0 0 2	
5.255E-01	6.994E+01	18	L039	1 0 0 0 2	

311. C$_4$H$_7$NO$_4$
DL-Aspartic acid
DL-2-Aminobutanedioic acid

RN:	617-45-8	**MP** (°C):			
MW:	133.10	**BP** (°C):			

Solubility (Moles/L)	Solubility (Grams/L)	Temp (°C)	Ref (#)	Evaluation (T P E A A)	Comments
2.367E-02	3.151E+00	0	D018	2 2 2 1 2	
2.420E-02	3.221E+00	4.99	A405	2 0 1 1 2	
3.140E-02	4.179E+00	9.99	A405	2 0 1 1 2	
3.910E-02	5.204E+00	14.99	A405	2 0 1 1 2	
4.850E-02	6.456E+00	19.99	A405	2 0 1 1 2	
5.900E-02	7.853E+00	24.99	A405	2 0 1 1 2	
6.081E-02	8.094E+00	25	D018	2 2 2 1 2	
6.110E-02	8.133E+00	25	D041	1 0 0 0 1	
7.260E-02	9.663E+00	29.99	A405	2 0 1 1 2	
8.770E-02	1.167E+01	33.99	A405	2 0 1 1 2	
8.950E-02	1.191E+01	34.99	A405	2 0 1 1 2	
1.069E-01	1.423E+01	38.99	A405	2 0 1 1 2	
1.109E-01	1.476E+01	39.99	A405	2 0 1 1 2	
1.293E-01	1.721E+01	44.99	A405	2 0 1 1 2	
1.561E-01	2.078E+01	49.49	A405	2 0 1 1 2	
1.544E-01	2.055E+01	50	D018	2 2 2 1 2	
1.812E-01	2.412E+01	54.99	A405	2 0 1 1 2	
2.170E-01	2.888E+01	58.99	A405	2 0 1 1 2	

(continued)

311. C$_4$H$_7$NO$_4$ (continued)

Solubility (Moles/L)	Solubility (Grams/L)	Temp (°C)	Ref (#)	Evaluation (T P E A A)	Comments
2.543E-01	3.385E+01	61.99	A405	2 0 1 1 2	
3.101E-01	4.128E+01	65.99	A405	2 0 1 1 2	
3.284E-01	4.371E+01	68.99	A405	2 0 1 1 2	
3.646E-01	4.853E+01	70.99	A405	2 0 1 1 2	
3.437E-01	4.575E+01	75	D018	2 2 2 1 2	
3.434E-01	4.571E+01	75	D041	1 0 0 0 2	

312. C$_4$H$_7$NO$_4$
Iminodiacetic acid
Imino-diessigsaeure

RN:	142-73-4	**MP** (°C):	247.5	
MW:	133.10	**BP** (°C):		

Solubility (Moles/L)	Solubility (Grams/L)	Temp (°C)	Ref (#)	Evaluation (T P E A A)	Comments
1.781E-01	2.370E+01	5	F300	1 0 0 0 2	

313. C$_4$H$_7$NO$_4$
L-Aspartic acid
Aspartic acid
L(+)-Asparaginsaeure
L-(+)-Asparaginic acid
L-(+)-Aspartic acid

RN:	56-84-8	**MP** (°C):	270.5	
MW:	133.10	**BP** (°C):		

Solubility (Moles/L)	Solubility (Grams/L)	Temp (°C)	Ref (#)	Evaluation (T P E A A)	Comments
1.675E-02	2.230E+00	0	D018	2 2 2 1 2	
1.780E-02	2.369E+00	4.99	A405	2 0 1 1 2	
2.170E-02	2.888E+00	9.99	A405	2 0 1 1 2	
2.570E-02	3.421E+00	14.99	A405	2 0 1 1 2	
3.160E-02	4.206E+00	19.99	A405	2 0 1 1 2	
3.170E-02	4.220E+00	20	B032	1 2 2 1 2	
3.750E-02	4.991E+00	24.99	A405	2 0 1 1 2	
3.770E-02	5.018E+00	25	B032	1 2 2 1 2	
4.030E-02	5.364E+00	25	D018	2 2 2 1 2	
3.738E-02	4.975E+00	25	D041	1 0 0 0 0	
3.805E-02	5.064E+00	25	G315	0 0 0 0 0	
3.719E-02	4.950E+00	25	J303	0 0 0 0 0	
3.644E-02	4.850E+00	27	D036	0 0 0 0 0	
4.469E-02	5.948E+00	29.80	B032	1 2 2 1 2	
4.550E-02	6.056E+00	29.99	A405	2 0 1 1 2	
5.320E-02	7.081E+00	33.99	A405	2 0 1 1 2	
6.520E-02	8.678E+00	39.99	A405	2 0 1 1 2	
6.348E-02	8.450E+00	40	J303	0 0 0 0 0	

(continued)

313. C$_4$H$_7$NO$_4$ (continued)

Solubility (Moles/L)	Solubility (Grams/L)	Temp (°C)	Ref (#)	Evaluation (T P E A A)	Comments
7.610E-02	1.013E+01	44.99	A405	2 0 1 1 2	
9.304E-02	1.238E+01	50	D018	2 2 2 1 2	
9.110E-02	1.213E+01	50.99	A405	2 0 1 1 2	
1.013E-01	1.348E+01	54.99	A405	2 0 1 1 2	
1.216E-01	1.619E+01	59.99	A405	2 0 1 1 2	
1.232E-01	1.640E+01	60	J303	0 0 0 0 0	
1.316E-01	1.752E+01	62.99	A405	2 0 1 1 2	
1.440E-01	1.917E+01	64.99	A405	2 0 1 1 2	
1.498E-01	1.994E+01	66.99	A405	2 0 1 1 2	
1.725E-01	2.296E+01	69.99	A405	2 0 1 1 2	
1.985E-01	2.642E+01	75	D018	2 2 2 1 2	
2.100E-01	2.795E+01	75	D041	1 0 0 0 2	
2.885E-01	3.840E+01	99	M160	2 1 1 1 0	
3.750E-02	4.991E+00	ns	M025	0 2 0 1 2	
3.738E-02	4.975E+00	rt	H431	0 0 0 0 0	

314. C$_4$H$_7$NO$_4$

L-β-Malamidic acid
L-β-Malaminsaeure

RN:	57229-74-0	**MP** (°C):	149	
MW:	133.10	**BP** (°C):		

Solubility (Moles/L)	Solubility (Grams/L)	Temp (°C)	Ref (#)	Evaluation (T P E A A)	Comments
5.242E-01	6.977E+01	18	L039	1 0 0 0 2	

315. C$_4$H$_7$N$_2$O$_4$

Glycine dipeptide

RN:		**MP** (°C):	
MW:	147.11	**BP** (°C):	

Solubility (Moles/L)	Solubility (Grams/L)	Temp (°C)	Ref (#)	Evaluation (T P E A A)	Comments
1.418E+00	2.086E+02	20	B032	1 2 2 1 2	
1.534E+00	2.257E+02	25	B032	1 2 2 1 2	
1.540E+00	2.266E+02	25.1	N024	0 0 0 0 0	
1.546E+00	2.275E+02	25.1	N026	0 0 0 0 0	
1.647E+00	2.423E+02	29.80	B032	1 2 2 1 2	

316. $C_4H_7N_3O$
Creatinine
Kreatinin
RN: 60-27-5   **MP** (°C):   220.5
MW: 113.12   **BP** (°C):

Solubility (Moles/L)	Solubility (Grams/L)	Temp (°C)	Ref (#)	Evaluation (T P E A A)	Comments
7.075E-01	8.004E+01	16	D041	1 0 0 0 1	
7.081E-01	8.010E+01	16	F300	1 0 0 0 2	

317. C_4H_8
1-Butene
α-Butene
Ethylethylene
α-Butylene
1-Butylene
Butene-1
RN: 106-98-9   **MP** (°C):   −185
MW: 56.11   **BP** (°C):   −6.47

Solubility (Moles/L)	Solubility (Grams/L)	Temp (°C)	Ref (#)	Evaluation (T P E A A)	Comments
3.957E-03	2.220E-01	25	M001	2 1 2 2 2	
1.210E-02	6.791E-01	38	B123	1 2 1 1 2	
1.582E-02	8.876E-01	71	B123	1 2 1 1 2	
2.746E-02	1.541E+00	104	B123	1 2 1 1 2	
3.526E-02	1.979E+00	138	B123	1 2 1 1 2	
3.858E-02	2.165E+00	144.00	B123	1 2 1 1 2	

318. C_4H_8
Isobutylene
2-Methylpropene
RN: 115-11-7   **MP** (°C):   −140.3
MW: 56.11   **BP** (°C):   −6.90

Solubility (Moles/L)	Solubility (Grams/L)	Temp (°C)	Ref (#)	Evaluation (T P E A A)	Comments
4.687E-03	2.630E-01	25	M001	2 1 2 2 2	

319. C$_4$H$_8$Cl$_2$
2,3-Dichlorobutane
Butane, 2,3-dichloro-

RN:	7581-97-7	MP (°C):	−80
MW:	127.01	BP (°C):	117

Solubility (Moles/L)	Solubility (Grams/L)	Temp (°C)	Ref (#)	Evaluation (T P E A A)	Comments
1.430E-02	1.817E+00	0	L103	1 0 0 0 2	unit assumed
4.422E-03	5.617E-01	20	L103	1 0 0 0 2	unit assumed
1.464E-03	1.860E-01	30	L103	1 0 0 0 2	unit assumed
1.755E-03	2.230E-01	40	L103	1 0 0 0 2	unit assumed

320. C$_4$H$_8$Cl$_2$O
sym-Dichloroethyl ether
2,2′-Dichlorodiethylether

RN:	111-44-4	MP (°C):	−50
MW:	143.01	BP (°C):	66

Solubility (Moles/L)	Solubility (Grams/L)	Temp (°C)	Ref (#)	Evaluation (T P E A A)	Comments
7.060E-02	1.010E+01	20	D052	1 1 0 0 0	
7.403E-02	1.059E+01	20	M062	1 0 0 0 2	

321. C$_4$H$_8$Cl$_2$OS
β,β′-Dichlorodiethylsulfoxide
β,β′-Dichlor-diaethylsulfoxid

RN:	5819-08-9	MP (°C):	
MW:	175.08	BP (°C):	

Solubility (Moles/L)	Solubility (Grams/L)	Temp (°C)	Ref (#)	Evaluation (T P E A A)	Comments
6.854E-02	1.200E+01	20	F300	1 0 0 0 1	

322. C$_4$H$_8$Cl$_2$O$_2$S
β,β′-Dichlorodiethylsulfone
β,β′-Dichlor-diaethylsulfon

RN:	471-03-4	MP (°C):	
MW:	191.08	BP (°C):	

Solubility (Moles/L)	Solubility (Grams/L)	Temp (°C)	Ref (#)	Evaluation (T P E A A)	Comments
3.140E-02	6.000E+00	20	F300	1 0 0 0 0	
1.256E-01	2.400E+01	100	F300	1 0 0 0 1	

323. C$_4$H$_8$Cl$_2$S
Mustard gas
Sulfure β′-ethyl dichlore
β,β′-Dichlor-diaethylsulfid

RN: 505-60-2   **MP (°C):**
MW: 159.08   **BP (°C):**

Solubility (Moles/L)	Solubility (Grams/L)	Temp (°C)	Ref (#)	Evaluation (T P E A A)	Comments
4.337E-03	6.900E-01	25	F300	1 0 0 0 1	
3.017E-03	4.800E-01	c	B079	0 0 1 1 1	

324. C$_4$H$_8$Cl$_3$O$_4$P
Trichlorfon
O,O-Dimethyl (1-hydroxy-2,2,2-trichloroethyl)phosphonate

RN: 52-68-6   **MP (°C):**   83.5
MW: 257.44   **BP (°C):**

Solubility (Moles/L)	Solubility (Grams/L)	Temp (°C)	Ref (#)	Evaluation (T P E A A)	Comments
5.982E-01	1.540E+02	25	M161	1 0 0 0 2	
4.255E-01	1.095E+02	ns	M061	0 0 0 0 2	

325. C$_4$H$_8$N$_2$O$_2$
Dimethylglyoxime
Dimethylglyoxim

RN: 95-45-4   **MP (°C):**   240.5
MW: 116.12   **BP (°C):**

Solubility (Moles/L)	Solubility (Grams/L)	Temp (°C)	Ref (#)	Evaluation (T P E A A)	Comments
5.167E-03	6.000E-01	20	F300	1 0 0 0 0	
3.100E-02	3.600E+00	80	F300	1 0 0 0 1	
5.081E-02	5.900E+00	100	F300	1 0 0 0 1	

326. C$_4$H$_8$N$_2$O$_2$
Succinamide
Bersteinsaeure-diamid

RN: 110-14-5   **MP (°C):**   260
MW: 116.12   **BP (°C):**

Solubility (Moles/L)	Solubility (Grams/L)	Temp (°C)	Ref (#)	Evaluation (T P E A A)	Comments
3.858E-02	4.480E+00	15	D041	1 0 0 0 1	
2.842E-02	3.300E+00	15	F300	1 0 0 0 1	
8.534E-01	9.910E+01	100	D041	1 0 0 0 2	
3.445E-04	4.000E-02	c	L055	0 0 0 0 2	
9.463E-03	1.099E+00	h	L055	0 0 0 0 1	
2.818E-02	3.273E+00	ns	R424	0 0 0 0 0	

327. C₄H₈N₂O₃

β-Alanine hydantoic acid
β-Uramidopropionic acid
Glycine, N-(aminocarbonyl)-N-methyl-

RN:	30565-25-4	MP (°C):
MW:	132.12	BP (°C):

Solubility (Moles/L)	Solubility (Grams/L)	Temp (°C)	Ref (#)	Evaluation (T P E A A)	Comments
1.580E-01	2.087E+01	25	M024	1 2 0 1 2	

328. C₄H₈N₂O₃

N-Nitroso-N-methylurethane
N-Nitroso-N-methyl-urethan

RN:	615-53-2	MP (°C):
MW:	132.12	BP (°C):

Solubility (Moles/L)	Solubility (Grams/L)	Temp (°C)	Ref (#)	Evaluation (T P E A A)	Comments
2.800E-01	3.699E+01	24	M031	1 1 1 1 1	

329. C₄H₈N₂O₃

N-Glycylglycine
Diglycine

RN:	556-50-3	MP (°C):	215
MW:	132.12	BP (°C):	

Solubility (Moles/L)	Solubility (Grams/L)	Temp (°C)	Ref (#)	Evaluation (T P E A A)	Comments
1.253E+00	1.656E+02	21	F300	1 0 0 0 2	
1.740E+00	2.299E+02	24.99	B441	0 0 0 0 0	
1.399E+00	1.848E+02	25	G092	2 1 1 1 1	
1.399E+00	1.848E+02	25	G315	0 0 0 0 0	
1.430E+00	1.890E+02	25.1	N027	1 2 2 2 2	
1.512E+00	1.998E+02	ns	M025	0 2 0 1 2	

330. C₄H₈N₂O₃

α-Alanine hydantoic acid
Methylhydantoic acid

RN:	77340-50-2	MP (°C):
MW:	132.12	BP (°C):

Solubility (Moles/L)	Solubility (Grams/L)	Temp (°C)	Ref (#)	Evaluation (T P E A A)	Comments
1.930E-01	2.550E+01	25	M024	1 2 0 1 2	
1.930E-01	2.550E+01	ns	M025	0 2 0 1 2	

331. $C_4H_8N_2O_3$
Asparagine
L-Asparagine
L-Asparagin

RN: 70-47-3   **MP** (°C):   235
MW: 132.12   **BP** (°C):

Solubility (Moles/L)	Solubility (Grams/L)	Temp (°C)	Ref (#)	Evaluation (T P E A A)	Comments
6.509E-02	8.600E+00	0	F300	1 0 0 0 1	
2.180E-01	2.880E+01	15	D349	2 1 1 2 2	
1.759E-01	2.324E+01	20	B032	1 2 2 1 2	
2.210E-01	2.920E+01	20	D349	2 1 1 2 2	
1.589E-01	2.100E+01	20	F300	1 0 0 0 2	
8.477E-02	1.120E+01	21.5	P045	0 0 2 1 2	
2.226E-01	2.941E+01	25	B032	1 2 2 1 2	
2.260E-01	2.986E+01	25	D349	2 1 1 2 2	
1.709E-01	2.258E+01	25	G315	0 0 0 0 0	
1.900E-01	2.510E+01	25.1	N024	0 0 0 0 0	
1.900E-01	2.510E+01	25.1	N025	0 0 0 0 0	
1.900E-01	2.510E+01	25.1	N026	0 0 0 0 0	
1.853E-01	2.449E+01	25.1	N027	1 1 2 2 2	
1.918E-01	2.534E+01	27	D036	0 0 0 0 0	
2.233E-01	2.950E+01	27	D036	0 0 0 0 0	
2.777E-01	3.669E+01	29.80	B032	1 2 2 1 2	
2.604E+00	3.440E+02	98	F300	1 0 0 0 2	
1.817E-01	2.400E+01	ns	D072	0 0 0 0 1	
1.860E-01	2.457E+01	ns	M025	0 2 0 1 2	
1.774E-01	2.344E+01	rt	D021	0 0 1 1 2	

332. $C_4H_8N_2O_3 \cdot H_2O$
L-Asparagine monohydrate
Asparagine, monohydrate, L-

RN: 5794-13-8   **MP** (°C):   234
MW: 150.14   **BP** (°C):

Solubility (Moles/L)	Solubility (Grams/L)	Temp (°C)	Ref (#)	Evaluation (T P E A A)	Comments
1.933E-01	2.902E+01	25	D041	1 0 0 0 2	
1.858E-01	2.790E+01	25	O316	1 0 1 2 2	
1.853E-01	2.781E+01	25	O316	1 0 1 2 2	
1.293E+00	1.941E+02	75	D041	1 0 0 0 2	

333. $C_4H_8N_4O_2$
N,N'-Dinitrosopiperazine
Dinitrosopiperazine

RN: 140-79-4   **MP** (°C):
MW: 144.13   **BP** (°C):

Solubility (Moles/L)	Solubility (Grams/L)	Temp (°C)	Ref (#)	Evaluation (T P E A A)	Comments
4.000E-02	5.765E+00	24	D083	2 0 0 0 1	

334. C₄H₈O

2-Butyraldehyde
Butyraldehyde
Butyraldehyd
n-Butanal

RN: 123-72-8   **MP** (°C):   −96
MW: 72.11   **BP** (°C):   75

Solubility (Moles/L)	Solubility (Grams/L)	Temp (°C)	Ref (#)	Evaluation (T P E A A)	Comments
4.948E-01	3.568E+01	20	D041	1 0 0 0 1	
4.993E-01	3.600E+01	20	F300	1 0 0 0 1	
9.694E-01	6.990E+01	25	A049	1 0 0 0 2	
9.194E-01	6.629E+01	25	B060	2 0 1 1 1	
5.077E-01	3.661E+01	38	J020	2 2 2 1 1	

335. C₄H₈O

Ethyl vinyl ether
Aethyl-vinyl-aether

RN: 109-92-2   **MP** (°C):   −115.0
MW: 72.11   **BP** (°C):   35

Solubility (Moles/L)	Solubility (Grams/L)	Temp (°C)	Ref (#)	Evaluation (T P E A A)	Comments
1.390E-01	1.002E+01	37	E028	1 1 2 2 2	

336. C₄H₈O

Isobutyraldehyde
2-Methyl propanal

RN: 78-84-2   **MP** (°C):   −66
MW: 72.11   **BP** (°C):   64

Solubility (Moles/L)	Solubility (Grams/L)	Temp (°C)	Ref (#)	Evaluation (T P E A A)	Comments
1.167E+00	8.413E+01	20	M146	1 2 2 2 2	
1.234E+00	8.900E+01	25	A049	1 0 0 0 0	

337. C₄H₈O

Methyl ethyl ketone
Butanon-(2)

RN: 78-93-3   **MP** (°C):   −87
MW: 72.11   **BP** (°C):   80

Solubility (Moles/L)	Solubility (Grams/L)	Temp (°C)	Ref (#)	Evaluation (T P E A A)	Comments
		0	C423	0 0 0 0 0	
5.780E+00	4.168E+02	4	C423	0 0 0 0 0	
4.338E+00	3.128E+02	10	C423	0 0 0 0 0	
1.015E+00	7.322E+01	20	A075	1 0 0 0 1	

(continued)

337. C_4H_8O (continued)

Solubility (Moles/L)	Solubility (Grams/L)	Temp (°C)	Ref (#)	Evaluation (T P E A A)	Comments
2.827E+00	2.038E+02	20	D052	1 1 0 0 2	
2.922E+00	2.107E+02	20	E019	1 0 1 1 2	
2.399E+00	1.730E+02	20	F300	1 0 0 0 2	
2.977E+00	2.146E+02	20	G030	1 2 0 0 2	
5.020E+00	3.620E+02	20	P040	0 0 0 0 0	
2.931E+00	2.114E+02	25	A094	1 0 0 0 2	
3.302E+00	2.381E+02	25	A356	0 0 0 0 0	
2.931E+00	2.114E+02	25	B060	2 0 1 1 1	
3.732E+00	2.691E+02	25	C435	0 0 0 0 0	
3.130E+00	2.257E+02	25	F044	1 0 0 0 2	
2.824E+00	2.036E+02	25	G030	1 2 0 0 2	
2.657E+00	1.916E+02	25	J005	1 0 2 1 2	
6.112E+00	4.407E+02	25	K105	2 0 0 0 2	
2.912E+00	2.100E+02	25	M136	2 0 0 0 2	
2.912E+00	2.100E+02	25	M139	2 0 0 0 2	
2.720E+00	1.961E+02	25	N309	1 0 0 0 2	
2.756E+00	1.987E+02	25	O028	2 2 2 2 2	
2.556E+00	1.843E+02	25	P055	1 0 0 0 1	
2.774E+00	2.000E+02	25	R320	1 0 1 1 2	
2.690E+00	1.940E+02	30	G030	1 2 0 0 2	
1.703E+00	1.228E+02	30	R319	2 2 2 1 2	
2.900E+00	2.091E+02	35	A356	0 0 0 0 0	
2.969E+00	2.141E+02	35	C309	2 2 2 2 1	
2.538E+00	1.830E+02	38	J020	2 0 2 1 2	
7.726E-01	5.571E+01	40	A075	1 0 0 0 1	
2.723E+00	1.964E+02	45	A356	0 0 0 0 0	
2.615E+00	1.885E+02	45	C309	2 2 2 2 1	
6.257E+00	4.512E+02	45	K105	2 0 0 0 2	
6.855E-01	4.943E+01	60	A075	1 0 0 0 1	
6.319E+00	4.556E+02	60	K105	2 0 0 0 2	
6.352E-01	4.580E+01	70	A075	1 0 0 0 1	
3.453E+00	2.490E+02	70	P040	0 0 0 0 0	
2.219E+00	1.600E+02	90	F300	1 0 0 0 1	
3.627E+00	2.615E+02	100	P040	0 0 0 0 0	
6.844E+00	4.935E+02	140	P040	0 0 0 0 0	
3.334E+00	2.404E+02	ns	C309	2 2 2 2 1	
+1.89E+00	+1.36E+02	ns	S460	0 0 0 0 0	

338. C_4H_8O
Tetrahydrofuran
1,4-Epoxybutane
Butylene oxide

RN: 109-99-9   **MP** (°C): −108.0
MW: 72.11   **BP** (°C): 66.0

Solubility (Moles/L)	Solubility (Grams/L)	Temp (°C)	Ref (#)	Evaluation (T P E A A)	Comments
4.498E+00	3.243E+02	72.2	M347	2 2 2 1 2	
4.504E+00	3.248E+02	72.25	M347	2 2 2 1 2	

(continued)

338. C_4H_8O (continued)

Solubility (Moles/L)	Solubility (Grams/L)	Temp (°C)	Ref (#)	Evaluation (T P E A A)	Comments
4.536E+00	3.271E+02	72.3	M347	2 2 2 1 2	
4.251E+00	3.065E+02	73.4	M347	2 2 2 1 2	
4.019E+00	2.898E+02	75.4	M347	2 2 2 1 2	
3.678E+00	2.652E+02	78.6	M347	2 2 2 1 2	
3.595E+00	2.593E+02	78.9	M347	2 2 2 1 2	
3.378E+00	2.436E+02	83.3	M347	2 2 2 1 2	
3.257E+00	2.349E+02	87.9	M347	2 2 2 1 2	
3.217E+00	2.320E+02	89.5	M347	2 2 2 1 2	
3.118E+00	2.248E+02	92.9	M347	2 2 2 1 2	
3.042E+00	2.194E+02	102.5	M347	2 2 2 1 2	
3.042E+00	2.194E+02	110.5	M347	2 2 2 1 2	
3.118E+00	2.248E+02	119.3	M347	2 2 2 1 2	
3.257E+00	2.349E+02	127.8	M347	2 2 2 1 2	
3.595E+00	2.593E+02	132.9	M347	2 2 2 1 2	
3.998E+00	2.883E+02	136.1	M347	2 2 2 1 2	
4.067E+00	2.933E+02	136.5	M347	2 2 2 1 2	
4.617E+00	3.329E+02	137.1	M347	2 2 2 1 2	
6.934E+00	5.000E+02	rt	B066	0 2 0 0 2	

339. $C_4H_8O_2$
Ethyl acetate
Athylacetat
Essigsaeureaethyl ester

RN:	141-78-6	**MP** (°C):	−83
MW:	88.11	**BP** (°C):	77

Solubility (Moles/L)	Solubility (Grams/L)	Temp (°C)	Ref (#)	Evaluation (T P E A A)	Comments
9.941E-01	8.759E+01	0	B108	1 2 0 1 1	
1.097E+00	9.666E+01	0	B108	1 2 0 1 2	
1.919E+00	1.691E+02	0	C423	0 0 0 0 0	
1.069E+00	9.420E+01	0	G062	1 2 2 2 2	
1.032E+00	9.091E+01	0	M088	2 0 0 0 1	
1.144E+00	1.008E+02	0	M111	1 0 1 1 2	
1.054E+00	9.290E+01	4	C423	0 0 0 0 0	
8.297E-01	7.310E+01	10	C423	0 0 0 0 0	
9.333E-01	8.223E+01	10	G062	1 2 2 2 2	
1.001E+00	8.817E+01	10	M111	1 0 1 1 2	
9.944E-01	8.762E+01	10.0	K079	1 0 0 0 2	
8.698E-01	7.664E+01	15	M088	2 0 0 0 1	
9.419E-01	8.299E+01	15	M111	1 0 1 1 2	
8.329E-01	7.339E+01	17.0	G101	1 2 1 1 2	
8.718E-01	7.681E+01	20	A016	1 2 1 1 2	
8.212E-01	7.236E+01	20	B108	1 2 0 1 1	
8.795E-01	7.749E+01	20	B108	1 2 0 1 2	
7.346E-01	6.472E+01	20	D052	1 1 0 0 2	
9.556E-01	8.419E+01	20	E002	1 0 0 0 2	
7.310E-01	6.441E+01	20	F001	1 0 1 2 2	

(continued)

339. $C_4H_8O_2$ (continued)

Solubility (Moles/L)	Solubility (Grams/L)	Temp (°C)	Ref (#)	Evaluation (T P E A A)	Comments
8.932E-01	7.870E+01	20	F300	1 0 0 0 2	
8.920E-01	7.860E+01	20	M111	1 0 1 1 2	
7.300E-01	6.432E+01	20	M171	1 0 0 0 1	
7.732E-01	6.812E+01	20	M348	2 2 1 1 2	
9.200E-01	8.106E+01	20	S006	1 0 0 0 1	
8.778E-01	7.734E+01	20.0	K079	1 0 0 0 2	
8.708E-01	7.672E+01	20.40	A016	1 2 1 1 2	
8.417E-01	7.416E+01	25	A016	1 2 1 1 2	
9.084E-01	8.004E+01	25	A094	1 0 0 0 1	
8.243E-01	7.263E+01	25	A326	1 2 0 1 1	
5.396E-02	4.755E+00	25	B004	0 0 0 0 0	*sic*
9.084E-01	8.004E+01	25	B060	2 0 1 1 1	
9.180E-01	8.088E+01	25	B092	2 1 1 1 2	
9.080E-01	8.000E+01	25	B304	2 0 2 2 0	
7.321E-01	6.450E+01	25	C435	0 0 0 0 0	
8.988E-01	7.919E+01	25	D425	0 0 0 0 0	
7.810E-01	6.881E+01	25	G062	1 2 2 2 2	
7.977E-01	7.029E+01	25	L062	2 2 0 1 2	
9.847E-01	8.676E+01	25	L319	1 0 2 1 2	
8.485E-01	7.476E+01	25	M111	1 0 1 1 2	
8.310E-01	7.322E+01	25	P055	1 0 0 0 1	
8.222E-01	7.244E+01	25.0	K079	1 0 0 0 2	
8.436E-01	7.433E+01	25.10	A016	1 2 1 1 2	
7.653E-01	6.743E+01	27.0	G101	1 2 1 1 2	
7.603E-01	6.699E+01	27.5	G101	1 2 1 1 2	
8.124E-01	7.158E+01	30	A016	1 2 1 1 2	
8.115E-01	7.149E+01	30	A016	1 2 1 1 2	
7.524E-01	6.629E+01	30	M088	2 0 0 0 1	
8.124E-01	7.158E+01	30	M111	1 0 1 1 2	
7.524E-01	6.629E+01	30	S357	1 2 1 0 2	
7.889E-01	6.951E+01	30.0	K079	1 0 0 0 2	
7.800E-01	6.873E+01	34	A016	1 2 1 1 2	
7.810E-01	6.881E+01	35	A016	1 2 1 1 2	
7.791E-01	6.864E+01	35	M111	1 0 1 1 2	
8.170E-01	7.198E+01	37	E028	1 0 1 1 2	
7.077E-01	6.235E+01	37	G062	1 2 2 2 2	
7.444E-01	6.559E+01	37.0	K079	1 0 0 0 2	
7.425E-01	6.542E+01	38	J020	2 1 2 1 1	
7.574E-01	6.673E+01	39.90	A016	1 2 1 1 2	
7.504E-01	6.612E+01	40	A016	1 2 1 1 2	
7.395E-01	6.516E+01	40	B108	1 2 0 1 2	
7.524E-01	6.629E+01	40	M111	1 0 1 1 2	
6.696E-01	5.900E+01	40	M348	2 2 1 1 2	
7.278E-01	6.412E+01	40.0	K079	1 0 0 0 2	
6.465E-01	5.696E+01	50	G062	1 2 2 2 2	
6.722E-01	5.923E+01	50.0	K079	1 0 0 0 2	
5.907E-01	5.204E+01	55	M348	2 2 1 1 2	
7.820E-01	6.890E+01	60	B092	2 1 1 1 2	
6.790E-01	5.983E+01	70	A326	1 2 0 1 1	

(continued)

339. C$_4$H$_8$O$_2$ (continued)

Solubility (Moles/L)	Solubility (Grams/L)	Temp (°C)	Ref (#)	Evaluation (T P E A A)	Comments
5.549E-01	4.889E+01	70	M348	2 2 1 1 2	
6.727E-01	5.927E+01	70.4	G101	1 2 1 1 1	
1.156E+00	1.018E+02	.0	K079	1 0 0 0 2	
1.600E-01	1.410E+01	ns	D348	0 0 0 0 0	

340. C$_4$H$_8$O$_2$
Methyl propionate
Methylester propanoic acid
RN: 554-12-1   **MP** (°C):   −87.0
MW: 88.11   **BP** (°C):   79.7

Solubility (Moles/L)	Solubility (Grams/L)	Temp (°C)	Ref (#)	Evaluation (T P E A A)	Comments
1.083E+00	9.545E+01	-2.1	K079	1 0 0 0 2	
1.000E+00	8.811E+01	1.0	K079	1 0 0 0 2	
8.778E-01	7.734E+01	11.5	K079	1 0 0 0 2	
8.500E-01	7.489E+01	14.9	K079	1 0 0 0 2	
8.150E-01	7.181E+01	20	S006	1 0 0 0 2	
8.167E-01	7.195E+01	20.0	K079	1 0 0 0 2	
7.778E-01	6.853E+01	27.1	K079	1 0 0 0 2	
7.667E-01	6.755E+01	32.5	K079	1 0 0 0 2	
7.389E-01	6.510E+01	42.7	K079	1 0 0 0 2	

341. C$_4$H$_8$O$_2$
Isobutyric acid
Isobuttersaeure
RN: 79-31-2   **MP** (°C):   −47
MW: 88.11   **BP** (°C):   153.5

Solubility (Moles/L)	Solubility (Grams/L)	Temp (°C)	Ref (#)	Evaluation (T P E A A)	Comments
1.931E+00	1.701E+02	15.2	P060	1 0 0 0 2	
1.931E+00	1.701E+02	15.2	P060	1 2 0 0 2	
4.171E+00	3.675E+02	17	P060	1 0 0 0 2	
4.171E+00	3.675E+02	17	P060	1 2 0 0 2	
2.619E+00	2.308E+02	17.7	H068	2 0 0 0 1	
1.892E+00	1.667E+02	20	D041	1 0 0 0 0	
1.894E+00	1.669E+02	20	F300	1 0 0 0 2	
3.768E+00	3.320E+02	20.0	P060	1 0 0 0 2	
3.768E+00	3.320E+02	20.0	P060	1 2 0 0 2	
3.732E+00	3.289E+02	20.1	P060	1 0 0 0 2	
3.732E+00	3.289E+02	20.1	P060	1 2 0 0 2	
2.255E+00	1.987E+02	20.2	P060	1 2 0 0 2	
2.255E+00	1.987E+02	20.25	P060	1 0 0 0 2	
2.367E+00	2.085E+02	20.9	P060	1 0 0 0 2	
2.363E+00	2.082E+02	20.9	P060	1 2 0 0 2	
3.363E+00	2.963E+02	21.2	P060	1 2 0 0 2	
3.363E+00	2.963E+02	21.2	P060	1 0 0 0 2	

(continued)

341. C₄H₈O₂ (continued)

Solubility (Moles/L)	Solubility (Grams/L)	Temp (°C)	Ref (#)	Evaluation (T P E A A)	Comments
3.161E+00	2.785E+02	21.5	P060	1 2 0 0 2	
3.161E+00	2.785E+02	21.5	P060	1 0 0 0 2	
2.500E+00	2.203E+02	21.5	P060	1 2 0 0 2	
2.500E+00	2.203E+02	21.5	P060	1 0 0 0 2	
3.240E+00	2.855E+02	21.7	P060	1 2 0 0 2	
3.001E+00	2.644E+02	21.76	P060	1 0 0 0 2	
3.003E+00	2.645E+02	21.79	P060	1 0 0 0 2	
2.831E+00	2.495E+02	21.8	P060	1 2 0 0 2	
2.831E+00	2.495E+02	21.89	P060	1 0 0 0 2	
2.709E+00	2.387E+02	21.9	P060	1 0 0 0 2	
2.709E+00	2.387E+02	21.9	P060	1 2 0 0 2	

342. C₄H₈O₂

3-Hydroxytetrahydrofuran
(RS)-3-Hydroxytetrahydrofuran
Tetrahydro-3-furanol
(±)-3-Hydroxytetrahydrofuran
3-Hydroxyoxolane

RN: 453-20-3   **MP (°C):** <25
MW: 88.11   **BP (°C):**

Solubility (Moles/L)	Solubility (Grams/L)	Temp (°C)	Ref (#)	Evaluation (T P E A A)	Comments
5.675E+00	5.000E+02	rt	B066	0 2 0 0 2	

343. C₄H₈O₂

Butyric acid
Buttersaeure
n-Butyric acid

RN: 107-92-6   **MP (°C):** −7.9
MW: 88.11   **BP (°C):** 163.5

Solubility (Moles/L)	Solubility (Grams/L)	Temp (°C)	Ref (#)	Evaluation (T P E A A)	Comments
2.943E-02	2.593E+00	1.13	H068	2 0 0 0 1	
1.149E-01	1.012E+01	25	B004	0 0 0 0 0	

344. C₄H₈O₂

1,4-Dioxane
1,4-Dioxan
Dioxane

RN: 123-91-1   **MP (°C):** 11.8
MW: 88.11   **BP (°C):** 101

Solubility (Moles/L)	Solubility (Grams/L)	Temp (°C)	Ref (#)	Evaluation (T P E A A)	Comments
>9.08E+00	>8.00E+02	25	B019	1 0 1 2 0	

345. $C_4H_8O_2$
Propyl formate
Ameisensaeure-propylester
Propyl methanoate
n-Propyl formate

RN:	110-74-7	**MP** (°C):	−93		
MW:	88.11	**BP** (°C):	81		

Solubility (Moles/L)	Solubility (Grams/L)	Temp (°C)	Ref (#)	Evaluation (T P E A A)	Comments
4.222E-01	3.720E+01	-1.0	K079	1 0 0 0 2	
3.861E-01	3.402E+01	4.0	K079	1 0 0 0 2	
3.722E-01	3.280E+01	6.0	K079	1 0 0 0 2	
3.444E-01	3.035E+01	12.5	K079	1 0 0 0 2	
3.220E-01	2.837E+01	20	S006	1 0 0 0 2	
3.272E-01	2.883E+01	20.0	K079	1 0 0 0 2	
2.497E-01	2.200E+01	22	F300	1 0 0 0 1	
3.161E-01	2.785E+01	30.0	K079	1 0 0 0 2	
2.880E-01	2.537E+01	32.5	N014	0 0 0 0 0	
3.083E-01	2.717E+01	34.0	K079	1 0 0 0 2	
2.972E-01	2.619E+01	45.0	K079	1 0 0 0 2	

346. C_4H_9Br
Isobutyl bromide
1-Bromo-2-methylpropane

RN:	78-77-3	**MP** (°C):	−119		
MW:	137.03	**BP** (°C):	91.5		

Solubility (Moles/L)	Solubility (Grams/L)	Temp (°C)	Ref (#)	Evaluation (T P E A A)	Comments
3.700E-03	5.070E-01	18	F001	1 0 1 0 2	
3.722E-03	5.100E-01	18	F300	1 0 0 0 1	

347. C_4H_9Br
n-Butyl bromide
Bromobutane

RN:	109-65-9	**MP** (°C):	−112		
MW:	137.03	**BP** (°C):	101.3		

Solubility (Moles/L)	Solubility (Grams/L)	Temp (°C)	Ref (#)	Evaluation (T P E A A)	Comments
4.300E-03	5.892E-01	16	F001	1 0 1 0 2	
4.300E-03	5.892E-01	17	S006	1 0 0 0 1	
<1.46E-03	<2.00E-01	25	B019	1 0 1 2 0	
4.500E-03	6.166E-01	25	K012	1 0 0 0 1	
6.340E-03	8.687E-01	25	M342	1 0 1 1 2	
4.434E-03	6.076E-01	30	G029	1 0 2 2 2	
4.500E-02	6.166E+00	ns	H307	0 0 0 0 0	

348. C$_4$H$_9$Cl
Isobutyl chloride
Isobutylchlorid

RN:	513-36-0	MP (°C):	−131
MW:	92.57	BP (°C):	68

Solubility (Moles/L)	Solubility (Grams/L)	Temp (°C)	Ref (#)	Evaluation (T P E A A)	Comments
1.000E-02	9.257E-01	12.5	F001	1 0 1 2 2	
9.722E-03	9.000E-01	12.50	F300	2 0 0 0 1	

349. C$_4$H$_9$Cl
n-Butyl chloride
1-Chlorobutane

RN:	109-69-3	MP (°C):	−123.0
MW:	92.57	BP (°C):	78.5

Solubility (Moles/L)	Solubility (Grams/L)	Temp (°C)	Ref (#)	Evaluation (T P E A A)	Comments
7.200E-03	6.665E-01	12.5	F001	1 0 1 0 2	
7.130E-02	6.600E+00	12.50	F300	1 0 0 0 1	
8.000E-03	7.406E-01	25	K012	1 0 0 0 0	
9.430E-03	8.729E-01	25	M342	1 0 1 1 2	
7.557E-03	6.995E-01	ns	N034	0 0 0 0 1	

350. C$_4$H$_9$Cl
sec-Butyl chloride
2-Chlorobutane

RN:	78-86-4	MP (°C):	−140
MW:	92.57	BP (°C):	68

Solubility (Moles/L)	Solubility (Grams/L)	Temp (°C)	Ref (#)	Evaluation (T P E A A)	Comments
1.079E-02	9.990E-01	25	N034	1 0 0 0 0	

351. C$_4$H$_9$Cl
tert-Butyl chloride
2-Chloro-2-methylpropane

RN:	507-20-0	MP (°C):	−26.5
MW:	92.57	BP (°C):	51.0

Solubility (Moles/L)	Solubility (Grams/L)	Temp (°C)	Ref (#)	Evaluation (T P E A A)	Comments
8.180E-02	7.572E+00	.99	C064	2 2 1 1 2	
6.620E-02	6.128E+00	5.00	C064	2 2 1 1 2	
3.110E-02	2.879E+00	14.90	C064	2 2 1 1 2	

352. C$_4$H$_9$I
Iodobutane
n-Butyl iodide

RN:	542-69-8	**MP** (°C):	−103
MW:	184.02	**BP** (°C):	130.5

Solubility (Moles/L)	Solubility (Grams/L)	Temp (°C)	Ref (#)	Evaluation (T P E A A)	Comments
1.100E-03	2.024E-01	17.5	F001	1 0 1 0 2	
1.100E-03	2.024E-01	17.5	S006	1 0 0 0 1	
1.100E-03	2.024E-01	20	M171	1 0 0 0 1	
1.700E-03	3.128E-01	25	K012	1 0 0 0 1	

353. C$_4$H$_9$NO
N,N-Dimethylacetamide
Acetdimethylamide
U-5954

RN:	127-19-5	**MP** (°C):	−20
MW:	87.12	**BP** (°C):	163

Solubility (Moles/L)	Solubility (Grams/L)	Temp (°C)	Ref (#)	Evaluation (T P E A A)	Comments
6.071E+00	5.289E+02	4.50	C022	1 2 0 0 2	

354. C$_4$H$_9$NO
Butyramide
n-Butyramide

RN:	541-35-5	**MP** (°C):	116
MW:	87.12	**BP** (°C):	216

Solubility (Moles/L)	Solubility (Grams/L)	Temp (°C)	Ref (#)	Evaluation (T P E A A)	Comments
1.960E+00	1.708E+02	6	H059	0 0 0 0 0	
2.190E+00	1.908E+02	16	H059	0 0 0 0 0	
2.640E+00	2.300E+02	25	H059	0 0 0 0 0	

355. C$_4$H$_9$NO$_2$
γ-Aminobutyric acid
γ-Amino-buttersaeure
γ-Amino-n-butyric acid

RN:	56-12-2	**MP** (°C):	
MW:	103.12	**BP** (°C):	

Solubility (Moles/L)	Solubility (Grams/L)	Temp (°C)	Ref (#)	Evaluation (T P E A A)	Comments
1.261E+01	1.300E+03	25	M029	2 2 2 2 2	

356. C₄H₉NO₂

Propyl carbamate
n-Propyl carbamate

RN:	627-12-3	**MP (°C):**	60
MW:	103.12	**BP (°C):**	196

Solubility (Moles/L)	Solubility (Grams/L)	Temp (°C)	Ref (#)	Evaluation (T P E A A)	Comments
1.940E+00	2.001E+02	37	H006	1 2 2 1 2	

357. C₄H₉NO₂

DL-α-Aminobutyric acid
DL-2-Aminobutyric acid

RN:	2835-81-6	**MP (°C):**	304
MW:	103.12	**BP (°C):**	

Solubility (Moles/L)	Solubility (Grams/L)	Temp (°C)	Ref (#)	Evaluation (T P E A A)	Comments
2.121E+00	2.188E+02	20	D041	1 0 0 0 1	
1.615E+00	1.665E+02	25	K031	2 1 2 1 2	

358. C₄H₉NO₂

β-Aminobutyric acid
β-Amino-*n*-butyric acid

RN:	2835-82-7	**MP (°C):**	193
MW:	103.12	**BP (°C):**	

Solubility (Moles/L)	Solubility (Grams/L)	Temp (°C)	Ref (#)	Evaluation (T P E A A)	Comments
1.212E+01	1.250E+03	25	M029	2 2 2 2 2	

359. C₄H₉NO₂

α-Aminoisobutyric acid
α-Amino-isobuttersaeure
α-Aminoisobutyric acid
2-Methylalanine

RN:	62-57-7	**MP (°C):**	
MW:	103.12	**BP (°C):**	

Solubility (Moles/L)	Solubility (Grams/L)	Temp (°C)	Ref (#)	Evaluation (T P E A A)	Comments
1.330E+00	1.371E+02	25	C018	0 0 0 0 0	
1.170E+00	1.206E+02	25	D041	1 0 0 0 2	
1.482E+00	1.528E+02	25	M029	2 2 2 2 2	
1.759E+00	1.814E+02	25	M097	2 2 2 2 2	

360. $C_4H_9NO_2$
1-Nitrobutane
Butane, 1-nitro-

RN:	627-05-4	**MP** (°C):	−81	
MW:	103.12	**BP** (°C):	152.5	

Solubility (Moles/L)	Solubility (Grams/L)	Temp (°C)	Ref (#)	Evaluation (T P E A A)	Comments
3.500E-02	3.609E+00	25	K012	1 0 0 0 1	

361. $C_4H_9NO_2$
N-Methylurethane
Ethyl methylaminoformate
Ethyl methylcarbamate
Ethyl N-methyl carbamate
Methyl urethane
N-Methylurethane

RN:	105-40-8	**MP** (°C):		
MW:	103.12	**BP** (°C):	170	

Solubility (Moles/L)	Solubility (Grams/L)	Temp (°C)	Ref (#)	Evaluation (T P E A A)	Comments
+3.97E+00	+4.10E+02	ns	S460	0 0 0 0 0	

362. $C_4H_9NO_2$
α-Aminobutyric acid
2-Aminobutanoic acid
α-Amino-n-butyric acid
Butanoic acid

RN:	80 60 4	**MP** (°C):	304	
MW:	103.12	**BP** (°C):		

Solubility (Moles/L)	Solubility (Grams/L)	Temp (°C)	Ref (#)	Evaluation (T P E A A)	Comments
1.845E+00	1.902E+02	25	A048	1 1 1 1 2	form A
1.624E+00	1.674E+02	25	A048	1 1 1 1 2	form B
1.800E+00	1.856E+02	25	C018	0 0 0 0 0	
1.800E+00	1.856E+02	25	E015	1 2 1 1 2	
2.041E+00	2.105E+02	25	M029	2 2 2 2 2	
1.852E+00	1.910E+02	35	A048	1 1 1 1 2	form A
1.771E+00	1.826E+02	35	A048	1 1 1 1 2	form B
1.931E+00	1.991E+02	45	A048	1 1 1 1 2	form A
1.917E+00	1.977E+02	45	A048	1 1 1 1 2	form B

363. C₄H₉NO₃
L-Threonine
Threonine
RN: 72-19-5   **MP** (°C): 270
MW: 119.12   **BP** (°C):

Solubility (Moles/L)	Solubility (Grams/L)	Temp (°C)	Ref (#)	Evaluation (T P E A A)	Comments
7.606E-01	9.060E+01	20	B032	1 2 2 1 2	
8.139E-01	9.695E+01	25	B032	1 2 2 1 2	
7.346E-01	8.751E+01	25	G315	0 0 0 0 0	
8.202E-01	9.770E+01	25.1	N024	0 0 0 0 0	
8.227E-01	9.800E+01	25.1	N026	0 0 0 0 0	
7.493E-01	8.925E+01	25.1	N027	1 1 2 2 2	
8.168E-01	9.730E+01	27	D036	0 0 0 0 0	
8.695E-01	1.036E+02	29.80	B032	1 2 2 1 2	

364. C₄H₉NO₃
DL-*allo*-Threonine
DL-Allothreonine
RN: 144-98-9   **MP** (°C):
MW: 119.12   **BP** (°C):

Solubility (Moles/L)	Solubility (Grams/L)	Temp (°C)	Ref (#)	Evaluation (T P E A A)	Comments
1.024E+00	1.220E+02	25	D041	1 0 0 0 2	
1.987E+00	2.366E+02	80	D041	1 0 0 0 2	

365. C₄H₉NO₃
DL-Threonine
(±)-Threonine
RN: 80-68-2   **MP** (°C): 244
MW: 119.12   **BP** (°C):

Solubility (Moles/L)	Solubility (Grams/L)	Temp (°C)	Ref (#)	Evaluation (T P E A A)	Comments
1.405E+00	1.674E+02	25	D041	1 0 0 0 2	
2.979E+00	3.548E+02	80	D041	1 0 0 0 1	

366. C₄H₉NO₃
Butyl nitrate
N-Butyl nitrate
RN: 928-45-0   **MP** (°C):
MW: 119.12   **BP** (°C):

Solubility (Moles/L)	Solubility (Grams/L)	Temp (°C)	Ref (#)	Evaluation (T P E A A)	Comments
6.500E-03	7.743E-01	25	K012	1 0 0 0 1	

367. C$_4$H$_9$N$_3$O$_2$
Creatine
Kreatin

RN:	57-00-1	**MP** (°C):	219
MW:	131.14	**BP** (°C):	

Solubility (Moles/L)	Solubility (Grams/L)	Temp (°C)	Ref (#)	Evaluation (T P E A A)	Comments
8.222E-02	1.078E+01	10	D041	1 0 0 0 2	
1.016E-01	1.332E+01	18	D041	1 0 0 0 2	
1.014E-01	1.330E+01	18	F300	1 0 0 0 2	

368. C$_4$H$_9$O$_5$P
γ-Phosphono-n-butyric acid
4-Phosphonobutyric acid
Phosphonic acid, (3-carboxypropyl)-
Butyric acid, 4-phosphono-

RN:	4378-43-2	**MP** (°C):	
MW:	168.09	**BP** (°C):	

Solubility (Moles/L)	Solubility (Grams/L)	Temp (°C)	Ref (#)	Evaluation (T P E A A)	Comments
1.739E+00	2.923E+02	0	N028	1 0 0 0 2	
2.068E+00	3.477E+02	20	N028	1 0 0 0 2	

369. C$_4$H$_{10}$
Isobutane
1,1-Dimethylethane
2-Methylpropane
Trimethylmethane
Purifrigor iso 3.5
R 600α

RN:	75-28-5	**MP** (°C):	−159
MW:	58.12	**BP** (°C):	

Solubility (Moles/L)	Solubility (Grams/L)	Temp (°C)	Ref (#)	Evaluation (T P E A A)	Comments
~5.68E-03	~3.30E-01	17	F300	1 0 0 0 0	
8.413E-04	4.890E-02	25	M001	2 1 2 2 2	
8.413E-04	4.890E-02	25	M002	2 1 2 2 2	

370. C$_4$H$_{10}$
Butane
n-Butane
Diethyl
HC 600
Liquefied petroleum gas
R 600 (alkane)

RN: 106-97-8   **MP** (°C): −138
MW: 58.12   **BP** (°C): −0.5

Solubility (Moles/L)	Solubility (Grams/L)	Temp (°C)	Ref (#)	Evaluation (T P E A A)	Comments
3.138E-03	1.824E-01	3	R063	0 0 0 0 0	
3.210E-03	1.866E-01	4	K031	2 1 2 1 2	
2.622E-03	1.524E-01	6	R063	0 0 0 0 0	
2.314E-03	1.345E-01	9	R063	0 0 0 0 0	
1.886E-03	1.096E-01	14	R063	0 0 0 0 0	
1.461E-03	8.492E-02	19.8	G058	1 0 0 0 2	
1.260E-03	7.324E-02	25	K031	2 1 2 1 2	
1.056E-03	6.140E-02	25	M001	2 1 2 2 2	
1.056E-03	6.140E-02	25	M002	2 1 2 2 2	
1.056E-03	6.140E-02	25	M040	1 0 0 1 2	
2.773E-02	1.612E+00	38	R078	0 0 0 0 0	
6.600E-04	3.836E-02	50	K031	2 1 2 1 2	
1.159E-01	6.735E+00	71	R078	0 0 0 0 0	
4.596E-01	2.671E+01	104	R078	0 0 0 0 0	
1.370E+00	7.965E+01	138	R078	0 0 0 0 0	

371. C$_4$H$_{10}$NO$_3$PS
Acephate
Orthene
Acetylphosphoramidothioic acid O,S-dimethyl ester

RN: 30560-19-1   **MP** (°C): 85.5
MW: 183.17   **BP** (°C):

Solubility (Moles/L)	Solubility (Grams/L)	Temp (°C)	Ref (#)	Evaluation (T P E A A)	Comments
2.151E+00	3.939E+02	rt	M161	0 0 0 0 1	

372. C$_4$H$_{10}$N$_2$O
N-Nitrosodiethylamine
Diethyl nitrosamine

RN: 55-18-5   **MP** (°C):
MW: 102.14   **BP** (°C):

Solubility (Moles/L)	Solubility (Grams/L)	Temp (°C)	Ref (#)	Evaluation (T P E A A)	Comments
1.040E+00	1.062E+02	24	D083	2 0 0 0 2	

373. C$_4$H$_{10}$O
Methyl propyl ether
1-Methoxypropane
RN: 557-17-5   **MP** (°C):   <25
MW: 74.12   **BP** (°C):   38.8

Solubility (Moles/L)	Solubility (Grams/L)	Temp (°C)	Ref (#)	Evaluation (T P E A A)	Comments
7.154E-01	5.303E+01	0	B002	2 1 1 2 2	
4.939E-01	3.661E+01	10	B002	2 1 1 2 2	
4.436E-01	3.288E+01	15	B002	2 1 1 2 2	
4.183E-01	3.101E+01	20	B002	2 1 1 2 2	
3.993E-01	2.960E+01	25	B002	2 1 1 2 2	

374. C$_4$H$_{10}$O
tert-Butyl alcohol
2-Methyl-2-propanol
tert-Butanol
RN: 75-65-0   **MP** (°C):   25.6
MW: 74.12   **BP** (°C):   82.41

Solubility (Moles/L)	Solubility (Grams/L)	Temp (°C)	Ref (#)	Evaluation (T P E A A)	Comments
8.712E-02	6.458E+00	79.40	B165	1 0 1 1 2	

375. C$_4$H$_{10}$O
n-Butyl alcohol
Butanol-(1)
n-Butanol
1-Butanol
Butyl alcohol
n-Butyl alcohol
RN: 71-36-3   **MP** (°C):   −90
MW: 74.12   **BP** (°C):   117

Solubility (Moles/L)	Solubility (Grams/L)	Temp (°C)	Ref (#)	Evaluation (T P E A A)	Comments
1.262E+00	9.355E+01	0	E029	1 2 0 1 2	
1.176E+00	8.717E+01	0	M095	2 2 1 2 2	
1.176E+00	8.717E+01	5	H003	1 2 1 1 2	
1.077E+00	7.987E+01	10	E029	1 2 0 1 2	
1.104E+00	8.181E+01	10	H003	1 2 1 1 2	
6.015E+00	4.459E+02	13.0	J012	1 2 0 1 2	
1.024E+00	7.587E+01	15	H003	1 2 1 1 2	
1.034E+00	7.664E+01	15	M095	2 2 1 2 2	
9.190E-01	6.812E+01	18	F001	1 0 1 0 2	
8.634E-01	6.400E+01	18	F300	1 0 0 0 1	
7.396E-01	5.482E+01	20	A075	1 0 0 0 1	
9.762E-01	7.236E+01	20	D040	2 2 1 1 2	

(continued)

375. C$_4$H$_{10}$O (continued)

Solubility (Moles/L)	Solubility (Grams/L)	Temp (°C)	Ref (#)	Evaluation (T P E A A)	Comments
9.993E-01	7.407E+01	20	D052	1 1 0 0 0	
9.482E-01	7.029E+01	20	E029	1 2 0 1 2	
9.773E-01	7.244E+01	20	H003	1 2 1 1 2	
6.302E-01	4.671E+01	20	L084	1 1 1 1 1	
1.040E+00	7.709E+01	20	M312	1 0 0 0 1	
8.270E-01	6.130E+01	23	D063	1 0 1 2 2	
1.021E+00	7.567E+01	23.5	D063	1 0 0 2 2	
9.983E-01	7.400E+01	25	A049	1 0 1 0 0	
1.125E+00	8.341E+01	25	B019	1 0 1 2 0	
9.645E-01	7.149E+01	25	B060	2 0 1 1 1	
1.000E+00	7.412E+01	25	F044	1 0 0 0 0	EFG
8.708E-01	6.455E+01	25	F325	1 2 0 1 1	
9.200E-01	6.819E+01	25	G075	1 0 1 0 1	
9.237E-01	6.847E+01	25	H003	1 2 1 1 2	
9.307E-01	6.899E+01	25	H028	2 0 2 0 2	
1.070E+00	7.931E+01	25	K012	1 0 0 0 2	
9.700E-01	7.190E+01	25	K025	2 2 1 1 1	
8.867E-01	6.572E+01	25	L322	1 1 2 2 1	
8.904E-01	6.600E+01	25	M136	2 0 0 0 1	
8.904E-01	6.600E+01	25	M139	2 0 0 0 1	
8.826E-01	6.542E+01	25.0	P077	1 1 1 1 1	
8.234E-01	6.103E+01	26	O012	1 2 1 1 2	
8.826E-01	6.542E+01	27	R319	2 2 2 1 1	
5.976E+00	4.429E+02	29.82	J012	1 2 0 1 2	
8.944E-01	6.629E+01	30	D040	2 2 1 1 2	
8.897E-01	6.594E+01	30	E029	1 2 0 1 2	
8.920E-01	6.612E+01	30	F053	1 0 2 0 2	
8.920E-01	6.612E+01	30	H003	1 2 1 1 2	
8.838E-01	6.551E+01	30.0	H043	2 2 1 1 2	
8.625E-01	6.393E+01	35	H003	1 2 1 1 2	
9.061E-01	6.716E+01	38	J020	2 0 2 1 1	
8.471E-01	6.279E+01	38	M125	1 1 1 1 1	
5.933E-01	4.398E+01	40	A075	1 0 0 0 1	
8.353E-01	6.191E+01	40	D040	2 2 1 1 2	
8.495E-01	6.297E+01	40	E029	1 2 0 1 2	
8.353E-01	6.191E+01	40	H003	1 2 1 1 2	
8.234E-01	6.103E+01	45	M095	2 2 1 2 2	
8.293E-01	6.147E+01	50	E029	1 2 0 1 2	
8.186E-01	6.068E+01	50	H003	1 2 1 1 2	
7.756E-01	5.749E+01	50	O012	1 2 1 1 2	
5.837E+00	4.327E+02	58.50	J012	1 2 0 1 2	
5.064E-01	3.754E+01	60	A075	1 0 0 0 1	
8.258E-01	6.121E+01	60	E029	1 2 0 1 2	
8.258E-01	6.121E+01	60	H003	1 2 1 1 2	
5.064E-01	3.754E+01	70	A075	1 0 0 0 1	
8.436E-01	6.253E+01	70	E029	1 2 0 1 2	
8.850E-01	6.560E+01	70	F001	1 0 1 0 2	
8.507E-01	6.306E+01	70	H003	1 2 1 1 2	

(continued)

375. $C_4H_{10}O$ (continued)

Solubility (Moles/L)	Solubility (Grams/L)	Temp (°C)	Ref (#)	Evaluation (T P E A A)	Comments
6.669E-01	4.943E+01	75	L084	1 1 1 1 1	
8.590E-01	6.367E+01	75	M095	2 2 1 2 1	
8.708E-01	6.455E+01	80	E029	1 2 0 1 2	
9.460E-01	7.012E+01	80	F001	1 0 1 0 2	
8.696E-01	6.446E+01	80	H003	1 2 1 1 2	
9.412E-01	6.977E+01	90	E029	1 2 0 1 2	
1.054E+00	7.813E+01	90	F001	1 0 1 0 2	
9.762E-01	7.236E+01	90	M095	2 2 1 2 1	
1.084E+00	8.038E+01	97.90	H003	1 2 1 1 2	
1.101E+00	8.164E+01	98.3	R072	2 2 2 1 2	
4.900E+00	3.632E+02	100	E029	1 2 0 1 2	
1.228E+00	9.102E+01	100	F001	1 0 1 0 2	
1.204E+00	8.925E+01	105	M095	2 2 1 2 1	
1.342E+00	9.950E+01	110	E029	1 2 0 1 2	
1.473E+00	1.092E+02	110	F001	1 0 1 0 2	
1.523E+00	1.129E+02	114.50	H003	1 2 1 1 2	
1.600E+00	1.186E+02	116.90	H003	1 2 1 1 2	
1.805E+00	1.338E+02	120	E029	1 2 0 1 2	
2.223E+00	1.648E+02	123.30	H003	1 2 1 1 2	
2.890E+00	2.142E+02	124.80	H003	1 2 1 1 2	
2.567E+00	1.903E+02	125	E029	1 2 0 1 2	
3.334E+00	2.471E+02	125.10	H003	1 2 1 1 2	
3.148E+00	2.334E+02	125.20	H003	1 2 1 1 2	
9.307E-01	6.899E+01	ns	A406	0 0 0 0 1	
7.920E-01	5.871E+01	ns	D348	0 0 0 0 0	
9.744E-01	7.222E+01	ns	L003	0 0 2 1 2	
9.033E+00	6.695E+02	ns	M314	2 1 2 1 2	

376. $C_4H_{10}O$

Methyl isopropyl ether
2-Methoxypropane

RN:	598-53-8	**MP** (°C):	<25	
MW:	74.12	**BP** (°C):	32	

Solubility (Moles/L)	Solubility (Grams/L)	Temp (°C)	Ref (#)	Evaluation (T P E A A)	Comments
1.193E+00	8.842E+01	10	B002	2 1 1 2 2	
1.068E+00	7.919E+01	15	B002	2 1 1 2 2	
9.295E-01	6.890E+01	20	B002	2 1 1 2 2	
8.234E-01	6.103E+01	25	B002	2 1 1 2 2	
8.437E-01	6.254E+01	ns	J300	0 0 0 0 0	

377. C₄H₁₀O

Isobutyl alcohol
2-Methyl-1-propanol

RN: 78-83-1   **MP** (°C): −108
MW: 74.12   **BP** (°C): 108

Solubility (Moles/L)	Solubility (Grams/L)	Temp (°C)	Ref (#)	Evaluation (T P E A A)	Comments
1.351E+00	1.001E+02	18	F001	1 0 1 2 2	
1.228E+00	9.100E+01	18	F300	1 0 0 0 0	
1.278E+00	9.471E+01	20	M146	1 2 2 2 2	
1.280E+00	9.488E+01	20	M312	1 0 0 0 1	
1.000E+00	7.416E+01	25	A037	2 2 2 2 2	
1.226E+00	9.091E+01	25	D052	1 1 0 0 2	
9.529E-01	7.063E+01	25	F050	1 0 0 0 1	
8.967E-01	6.647E+01	25	F317	2 1 1 1 2	
1.045E+00	7.749E+01	29.84	M114	2 2 1 1 1	
9.529E-01	7.063E+01	39.74	M114	2 2 1 1 1	
8.234E-01	6.103E+01	49.64	M114	2 2 1 1 1	
8.590E-01	6.367E+01	59.54	M114	2 2 1 1 1	
9.295E-01	6.890E+01	79.24	M114	2 2 1 1 1	
9.645E-01	7.149E+01	89.14	M114	2 2 1 1 1	
5.168E+00	3.831E+02	90.5	J017	1 0 1 2 2	
5.033E+00	3.730E+02	91.0	J017	1 0 1 2 2	
4.887E+00	3.622E+02	92.0	J017	1 0 1 2 2	
4.871E+00	3.610E+02	92.1	J017	1 0 1 2 2	
4.615E+00	3.421E+02	93.0	J017	1 0 1 2 2	
4.135E+00	3.065E+02	94.3	J017	1 0 1 2 2	
3.820E+00	2.832E+02	95.3	J017	1 0 1 2 2	
1.215E+00	9.008E+01	99.04	M114	2 2 1 1 1	
1.348E+00	9.991E+01	108.94	M114	2 2 1 1 2	
1.708E+00	1.266E+02	118.74	M114	2 2 1 1 2	
2.009E+00	1.489E+02	123.74	M114	2 2 1 1 2	
2.239E+00	1.660E+02	125.64	M114	2 2 1 1 2	
2.415E+00	1.790E+02	128.64	M114	2 2 1 1 2	
2.637E+00	1.955E+02	130.64	M114	2 2 1 1 2	
3.000E+00	2.224E+02	132.64	M114	2 2 1 1 2	
3.527E+00	2.614E+02	134.14	M114	2 2 1 1 2	
1.179E+00	8.740E+01	ns	L003	0 0 2 1 1	

378. C₄H₁₀O

Ethyl ether
Diaethylaether
Diethyl ether

RN: 60-29-7   **MP** (°C): −116
MW: 74.12   **BP** (°C): 34.6

Solubility (Moles/L)	Solubility (Grams/L)	Temp (°C)	Ref (#)	Evaluation (T P E A A)	Comments
1.526E+00	1.131E+02	−3.8	H002	2 0 0 1 2	
1.410E+00	1.045E+02	0	H002	1 0 0 1 2	
1.662E+00	1.232E+02	0	K077	1 2 2 2 2	average of 3
1.338E+00	9.920E+01	7.5	K077	1 2 2 2 2	

(continued)

378. $C_4H_{10}O$ (continued)

Solubility (Moles/L)	Solubility (Grams/L)	Temp (°C)	Ref (#)	Evaluation (T P E A A)	Comments
1.263E+00	9.360E+01	8.5	K077	1 2 2 2 2	
1.118E+00	8.291E+01	10	H002	1 0 0 1 2	
1.115E+00	8.265E+01	10	K002	1 2 1 1 2	
1.105E+00	8.190E+01	12	K077	1 2 2 2 2	
9.796E-01	7.261E+01	15	F055	1 2 2 2 2	
1.133E+00	8.400E+01	15	F300	1 0 0 0 1	
9.893E-01	7.333E+01	15	H002	1 0 0 1 2	
9.843E-01	7.296E+01	15	K002	1 2 1 1 2	
8.430E+00	6.249E+02	15	M069	1 0 0 0 2	
1.137E+00	8.430E+01	15	T033	1 2 1 1 2	
1.029E+00	7.630E+01	16	K077	1 2 2 2 2	
8.837E-01	6.550E+01	19	K077	1 2 2 2 2	average
8.696E-01	6.446E+01	20	F055	1 2 2 2 2	
8.703E-01	6.451E+01	20	H002	1 0 0 1 2	
8.684E-01	6.437E+01	20	K002	1 2 1 1 2	
8.353E-01	6.191E+01	20	M345	2 1 1 1 1	
8.341E-01	6.183E+01	20	N038	1 0 0 1 2	
8.769E-03	6.500E-01	21	H337	1 0 1 0 2	*sic*
1.012E+00	7.502E+01	22	H072	1 0 1 1 2	
9.993E-01	7.407E+01	25	B019	1 0 1 2 0	
7.636E-01	5.660E+01	25	F055	1 2 2 2 2	
8.095E-01	6.000E+01	25	F300	1 0 0 0 1	
7.669E-01	5.684E+01	25	H002	1 0 0 1 2	
7.684E-01	5.696E+01	25	K002	1 2 1 1 2	
8.800E-01	6.523E+01	25	K012	1 0 0 0 1	
6.050E+00	4.484E+02	25	M069	1 0 0 0 2	
8.471E-01	6.279E+01	25	M345	2 1 1 1 1	
8.162E-01	6.050E+01	25	T033	1 2 1 1 2	
1.048E-02	7.770E-01	26	H337	1 0 1 0 2	*sic*
6.839E-01	5.069E+01	30	H002	1 0 0 1 2	
6.839E-01	5.069E+01	30	K002	1 2 1 1 2	
6.799E-01	5.040E+01	30	K077	1 2 2 2 2	
1.073E-02	7.950E-01	32	H337	1 0 1 0 2	*sic*
5.950E-01	4.410E+01	37	E022	1 0 1 1 0	
7.120E-01	5.278E+01	37	E028	1 0 1 1 2	
9.484E-03	7.030E-01	37	H337	1 0 1 0 2	*sic*
6.314E-01	4.680E+01	38	K077	1 2 2 2 2	
9.417E-03	6.980E-01	38.5	H337	1 0 1 0 2	*sic*
9.808E-03	7.270E-01	40	H337	1 0 1 0 2	*sic*
5.545E-01	4.110E+01	49	K077	1 2 2 2 2	
5.491E-01	4.070E+01	51.5	K077	1 2 2 2 2	
4.857E-01	3.600E+01	62.5	K077	1 2 2 2 2	
4.600E-01	3.410E+01	65	K077	1 2 2 2 2	
4.209E-01	3.120E+01	66.5	K077	1 2 2 2 2	
4.020E-01	2.980E+01	71	K077	1 2 2 2 2	
3.912E-01	2.900E+01	72	K077	1 2 2 2 2	
3.643E-01	2.700E+01	82	K077	1 2 2 2 2	
1.770E-01	1.312E+01	ns	D348	0 0 0 0 0	
9.412E-01	6.977E+01	ns	R028	0 0 0 0 0	
8.826E-01	6.542E+01	rt	B066	0 2 0 0 0	

379. $C_4H_{10}O$
sec-Butyl alcohol
DL-*sec*-Butyl alcohol
DL-Butanol-(2)
sec-DL-Butyl alcohol

RN:	78-92-2	**MP** (°C):	−114
MW:	74.12	**BP** (°C):	99.5

Solubility (Moles/L)	Solubility (Grams/L)	Temp (°C)	Ref (#)	Evaluation (T P E A A)	Comments
2.602E+00	1.929E+02	10.04	M119	2 2 2 2 2	
3.222E+00	2.388E+02	20	A070	1 2 1 0 2	
1.499E+00	1.111E+02	20	D052	1 1 0 0 0	
2.106E+00	1.561E+02	20	E019	1 0 1 1 2	
1.497E+00	1.110E+02	20	F300	1 0 0 0 2	
2.230E+00	1.653E+02	20	M112	2 2 1 1 2	
2.267E+00	1.681E+02	20.04	M119	2 2 2 2 2	
1.348E+00	9.991E+01	25	B019	1 0 1 2 0	
1.057E+00	7.834E+01	25	B060	2 0 1 1 1	
1.699E+00	1.260E+02	25	B165	1 0 1 1 1	
2.048E+00	1.518E+02	27.04	M119	2 2 2 2 2	
2.556E+00	1.894E+02	40	A070	1 2 1 0 2	
1.821E+00	1.349E+02	40	M112	2 0 1 1 2	
1.749E+00	1.297E+02	40.04	M119	2 2 2 2 2	
1.573E+00	1.166E+02	50.04	M119	2 2 2 2 2	
2.167E+00	1.606E+02	60	A070	1 2 1 0 2	
1.657E+00	1.228E+02	60	M112	2 0 1 1 2	
1.531E+00	1.135E+02	60.04	M119	2 2 2 2 2	
1.541E+00	1.143E+02	70.04	M119	2 2 2 2 2	
2.167E+00	1.606E+02	80	A070	1 2 1 0 2	
1.657E+00	1.228E+02	80	M112	2 0 1 1 2	
1.636E+00	1.213E+02	80.04	M119	2 2 2 2 2	
1.760E+00	1.304E+02	85	M112	2 0 1 1 2	
5.107E-02	3.786E+00	87.30	B165	1 0 1 1 2	
1.810E+00	1.342E+02	90.04	M119	2 2 2 2 2	
2.087E+00	1.547E+02	100.04	M119	2 2 2 2 2	
2.602E+00	1.929E+02	110.04	M119	2 2 2 2 2	
1.901E+00	1.409E+02	ns	L003	0 0 2 1 2	

380. $C_4H_{10}O_2S$
Diethyl sulfone
Diaethylsulfon

RN:	597-35-3	**MP** (°C):	73
MW:	122.19	**BP** (°C):	248

Solubility (Moles/L)	Solubility (Grams/L)	Temp (°C)	Ref (#)	Evaluation (T P E A A)	Comments
1.105E+00	1.350E+02	16	F300	1 0 0 0 2	

381. C$_4$H$_{10}$O$_4$
DL-Threitol
DL-1,2,3,4-Butanetetrol
RN: 6968-16-7 **MP** (°C): 90
MW: 122.12 **BP** (°C):

Solubility (Moles/L)	Solubility (Grams/L)	Temp (°C)	Ref (#)	Evaluation (T P E A A)	Comments
7.353E+00	8.980E+02	25	C346	0 0 0 0 0	

382. C$_4$H$_{10}$O$_4$
Erythritol
Erythrit
RN: 149-32-6 **MP** (°C): 121.5
MW: 122.12 **BP** (°C): 330

Solubility (Moles/L)	Solubility (Grams/L)	Temp (°C)	Ref (#)	Evaluation (T P E A A)	Comments
3.118E+00	3.808E+02	rt	D021	0 0 1 1 2	
4.995E+00	6.100E+02	rt	F300	0 0 0 0 2	

383. C$_4$H$_{10}$S
Ethyl sulfide
1,1'-Thiobisethane
Diethyl thioether
RN: 352-93-2 **MP** (°C): −100
MW: 90.19 **BP** (°C): 91

Solubility (Moles/L)	Solubility (Grams/L)	Temp (°C)	Ref (#)	Evaluation (T P E A A)	Comments
3.400E-02	3.066E+00	25	K012	1 0 0 0 1	

384. C$_4$H$_{11}$N
sec-Butylamine
DL-sec-Butylamine
DL-sec-Butylamin
RN: 13952-84-6 **MP** (°C):
MW: 73.14 **BP** (°C): 63

Solubility (Moles/L)	Solubility (Grams/L)	Temp (°C)	Ref (#)	Evaluation (T P E A A)	Comments
1.531E+00	1.120E+02	20	F300	1 0 0 0 2	

385. $C_4H_{11}N$
n-Butylamine
n-Butylamin
1-Aminobutane
RN: 109-73-9   **MP** (°C):   −50
MW: 73.14   **BP** (°C):   78

Solubility (Moles/L)	Solubility (Grams/L)	Temp (°C)	Ref (#)	Evaluation (T P E A A)	Comments
3.259E-02	2.384E+00	25	B004	0 0 0 0 0	

386. $C_4H_{11}NO_3$
Tromethamine
tris-(Hydroxymethyl)-amino-methan
tris-(Hydroxymethyl)-aminomethane
2-Amino-2-(hydroxymethyl)-1,3-propanediol
tris(Hydroxymethyl)methylamine
RN: 77-86-1   **MP** (°C):   171.5
MW: 121.14   **BP** (°C):   219.5

Solubility (Moles/L)	Solubility (Grams/L)	Temp (°C)	Ref (#)	Evaluation (T P E A A)	Comments
4.564E+00	5.529E+02	15	E305	0 0 0 0 0	
5.766E+00	6.985E+02	25	E305	0 0 0 0 0	
7.160E+00	8.673E+02	35	E305	0 0 0 0 0	

387. $C_4H_{11}NO_8P_2$
Glyphosine
Polaris
N,N-bis(Phosphonomethyl)glycine
RN: 2439-99-8   **MP** (°C):
MW: 263.08   **BP** (°C):

Solubility (Moles/L)	Solubility (Grams/L)	Temp (°C)	Ref (#)	Evaluation (T P E A A)	Comments
9.427E-01	2.480E+02	20	M161	1 0 0 0 2	

388. C_4Cl_6
Hexachloro-1,3-butadiene
Hexachlorobutadiene
RN: 87-68-3   **MP** (°C):   −19
MW: 260.76   **BP** (°C):   210

Solubility (Moles/L)	Solubility (Grams/L)	Temp (°C)	Ref (#)	Evaluation (T P E A A)	Comments
9.772E-06	2.548E-03	20	C113	1 0 2 1 2	
1.917E-05	5.000E-03	20	M068	1 0 0 0 0	
~7.67E-06	~2.00E-03	20	M133	1 0 0 0 0	
1.240E-05	3.233E-03	25	B173	2 0 2 2 2	
7.668E-04	2.000E-01	ns	M061	0 0 0 0 1	

389. C₅H₂Cl₃NO

2,3,5-Trichloro-4-hydroxypyridine
Daxtrom

RN: 1970-40-7   **MP** (°C):   216
MW: 198.44   **BP** (°C):

Solubility (Moles/L)	Solubility (Grams/L)	Temp (°C)	Ref (#)	Evaluation (T P E A A)	Comments
2.871E-03	5.697E-01	25	M061	1 0 0 0 1	

390. C₅H₂Cl₃NO

3,5,6-Trichloro-2-pyridinol
3,5,6-Trichloropyridinol
Hydroxy-3,5,6-trichloropyridine
Pyridinone, 3,5,6-trichloro-

RN: 6515-38-4   **MP** (°C):
MW: 198.44   **BP** (°C):

Solubility (Moles/L)	Solubility (Grams/L)	Temp (°C)	Ref (#)	Evaluation (T P E A A)	Comments
1.109E-03	2.200E-01	26.70	L095	2 2 1 1 2	
1.109E-03	2.200E-01	ns	K138	0 0 0 0 1	

391. C₅H₃F₃N₂O₂

5-Trifluoromethyl uracil
Trifluorothymine

RN: 54-20-6   **MP** (°C):
MW: 180.09   **BP** (°C):

Solubility (Moles/L)	Solubility (Grams/L)	Temp (°C)	Ref (#)	Evaluation (T P E A A)	Comments
1.451E-01	2.613E+01	25	S471	0 0 0 0 0	
1.492E-01	2.687E+01	25	S471	0 0 0 0 0	

392. C₅H₄ClN₅

2-Chloroadenine
1H-Purin-6-amine, 2-chloro-
6-Amino-2-chloropurine
2-Chloro-6-aminopurine
SQ 22982

RN: 1839-18-5   **MP** (°C):
MW: 169.57   **BP** (°C):

Solubility (Moles/L)	Solubility (Grams/L)	Temp (°C)	Ref (#)	Evaluation (T P E A A)	Comments
4.895E-05	8.300E-03	25	A336	0 0 0 0 0	

393. $C_5H_4N_2O_4$
Orotic acid
Vitamin B13
1,2,3,6-Tetrahydro-2,6-dioxo-4-pyrimidinecarboxylic acid
RN: 65-86-1   **MP** (°C):   345.5
MW: 156.10   **BP** (°C):

Solubility (Moles/L)	Solubility (Grams/L)	Temp (°C)	Ref (#)	Evaluation (T P E A A)	Comments
1.163E-02	1.815E+00	18	B135	1 0 0 0 0	

394. $C_5H_4N_2O_4$
α,β-Imidazoledicarboxylic acid
4,5-Imidazoledicarboxylic acid
Imidazol-di-carbonsaeure-(4,5)
RN: 570-22-9   **MP** (°C):   288
MW: 156.10   **BP** (°C):

Solubility (Moles/L)	Solubility (Grams/L)	Temp (°C)	Ref (#)	Evaluation (T P E A A)	Comments
3.203E-03	5.000E-01	20	F300	1 0 0 0 1	
8.328E-03	1.300E+00	100	F300	1 0 0 0 1	

395. $C_5H_4N_2O_4$
5-Carboxyuracil
5-Uracilcarboxylic acid
2,4-Dihydroxypyrimidine-5-carboxylic acid
Uracil-carbonsaeure-(4)
RN: 23945-44-0   **MP** (°C):   283
MW: 156.10   **BP** (°C):

Solubility (Moles/L)	Solubility (Grams/L)	Temp (°C)	Ref (#)	Evaluation (T P E A A)	Comments
1.153E-02	1.800E+00	20	F300	1 0 0 0 1	
7.000E-03	1.093E+00	20	N019	0 0 0 0 0	

396. $C_5H_4N_4$
Purine
7-Imidazo(4,5-d)pyrimidine
RN: 120-73-0   **MP** (°C):   216
MW: 120.11   **BP** (°C):

Solubility (Moles/L)	Solubility (Grams/L)	Temp (°C)	Ref (#)	Evaluation (T P E A A)	Comments
2.775E+00	3.333E+02	20	A018	1 0 1 1 0	
2.754E+00	3.308E+02	ns	R427	0 0 0 0 0	

397. $C_5H_4N_4O$
Hypoxanthine
Hypoxanthin
RN: 68-94-0   MP (°C):   150dec
MW: 136.11   BP (°C):

Solubility (Moles/L)	Solubility (Grams/L)	Temp (°C)	Ref (#)	Evaluation (T P E A A)	Comments
5.139E-03	6.995E-01	19	D041	1 0 0 0 0	
5.143E-03	7.000E-01	23	F300	1 0 0 0 1	
5.290E-03	7.200E-01	25	A337	0 0 0 0 0	
~1.90E-03	~2.59E-01	39.99	T420	0 0 0 0 0	
1.042E-01	1.418E+01	100	D004	0 0 0 0 0	
1.080E-01	1.470E+01	100	F300	1 0 0 0 2	
5.359E-03	7.294E-01	c	D004	0 0 0 0 0	

398. $C_5H_4N_4O$
Allopurinol
1H-Pyrazolo(3,4-d)pyrimidin-4-ol
Lopurin
RN: 315-30-0   MP (°C):   >350
MW: 136.11   BP (°C):   559.8

Solubility (Moles/L)	Solubility (Grams/L)	Temp (°C)	Ref (#)	Evaluation (T P E A A)	Comments
2.535E-03	3.450E-01	15	C095	1 0 0 1 2	
3.673E-03	5.000E-01	22	B322	0 0 0 0 0	
3.673E-03	5.000E-01	22	B428	1 2 1 2 1	
3.526E-03	4.800E-01	25	B189	1 0 0 0 1	
4.180E-03	5.690E-01	25	C095	1 0 0 1 2	
6.502E-03	8.850E-01	35	C095	1 0 0 1 2	
7.964E-03	1.084E+00	40	C095	1 0 0 1 2	
3.526E-03	4.800E-01	ns	A351	0 0 0 0 0	
2.475E-03	3.369E-01	ns	B404	0 2 1 1 0	
5.730E-03	7.800E-01	ns	H067	0 0 0 0 0	
7.347E-04	1.000E-01	ns	K444	0 0 0 0 0	

399. $C_5H_4N_4O$
8-Hydroxypurine
9H-Purin-8-ol
RN: 51953-05-0   MP (°C):
MW: 136.11   BP (°C):

Solubility (Moles/L)	Solubility (Grams/L)	Temp (°C)	Ref (#)	Evaluation (T P E A A)	Comments
3.048E-02	4.149E+00	20	A022	1 0 0 0 0	

400. $C_5H_4N_4O_2$
Xanthine
2,6-Dioxopurine
1H-Purine-2,6-dione, 3,7-dihydro-

RN:	69-89-6	MP (°C):	>300
MW:	152.11	BP (°C):	

Solubility (Moles/L)	Solubility (Grams/L)	Temp (°C)	Ref (#)	Evaluation (T P E A A)	Comments
3.285E-03	4.998E-01	20	D041	1 0 0 0 0	
3.000E-04	4.563E-02	20.99	T418	0 0 0 0 0	
2.458E-04	3.739E-02	21	L015	1 0 1 1 2	
5.246E-04	7.980E-02	37	L015	1 0 1 1 2	
1.312E-02	1.996E+00	100	D041	1 0 0 0 0	

401. $C_5H_4N_4O_2 \cdot H_2O$
Xanthine (monohydrate)

RN:	69-89-6	MP (°C):	>150dec
MW:	170.13	BP (°C):	

Solubility (Moles/L)	Solubility (Grams/L)	Temp (°C)	Ref (#)	Evaluation (T P E A A)	Comments
4.082E-04	6.944E-02	c	D004	0 0 0 0 0	
3.916E-03	6.662E-01	h	D004	0 0 0 0 0	

402. $C_5H_4N_4O_3$
Uric acid
Harnsaeure

RN:	69-93-2	MP (°C):	
MW:	168.11	BP (°C):	

Solubility (Moles/L)	Solubility (Grams/L)	Temp (°C)	Ref (#)	Evaluation (T P E A A)	Comments
1.190E-04	2.000E-02	0	M043	1 0 0 0 0	
7.110E-05	1.195E-02	2.6	M315	1 0 1 1 2	
1.029E-04	1.730E-02	5	R042	1 2 2 1 2	
1.050E-04	1.765E-02	9.3	M315	1 0 1 1 2	
2.379E-04	4.000E-02	10	M043	1 0 0 0 0	
1.326E-04	2.230E-02	14	B116	2 0 1 1 2	
1.190E-04	2.000E-02	20	D041	1 0 0 0 0	
3.569E-04	6.000E-02	20	M043	1 0 0 0 0	
6.610E-04	1.111E-01	22	M145	1 0 1 2 2	intrinsic
1.862E-04	3.130E-02	25	R042	1 2 2 1 2	
2.070E-04	3.480E-02	25.0	M315	1 0 1 1 2	
5.354E-04	9.000E-02	30	F300	1 0 0 0 2	
5.353E-04	8.999E-02	30	M043	1 0 0 0 0	
3.660E-04	6.153E-02	37.0	M315	1 0 1 1 2	
7.137E-04	1.200E-01	40	M043	1 0 0 0 1	
3.753E-04	6.310E-02	40	R042	1 2 2 1 2	
6.280E-04	1.056E-01	50.0	M315	1 0 1 1 2	
6.960E-04	1.170E-01	54	R042	1 2 2 1 2	
1.368E-03	2.299E-01	60	M043	1 0 0 0 1	

(continued)

402. C$_5$H$_4$N$_4$O$_3$ (continued)

Solubility (Moles/L)	Solubility (Grams/L)	Temp (°C)	Ref (#)	Evaluation (T P E A A)	Comments
1.457E-03	2.450E-01	70	F300	1 0 0 0 2	
2.319E-03	3.898E-01	80	M043	1 0 0 0 1	
2.974E-04	5.000E-02	100	D041	1 0 0 0 0	
4.961E-03	8.340E-01	100	F300	1 0 0 0 0	
3.686E-03	6.196E-01	100	M043	1 0 0 0 1	

403. C$_5$H$_4$N$_4$O$_3$.2H$_2$O
Uric acid (dihydrate)

RN: 69-93-2   **MP** (°C):
MW: 204.14   **BP** (°C):

Solubility (Moles/L)	Solubility (Grams/L)	Temp (°C)	Ref (#)	Evaluation (T P E A A)	Comments
9.620E-05	1.964E-02	2.6	M315	1 0 1 1 2	
1.420E-04	2.899E-02	9.3	M315	1 0 1 1 2	
3.390E-04	6.920E-02	25.0	M315	1 0 1 1 2	
6.560E-04	1.339E-01	37.0	M315	1 0 1 1 2	
1.440E-03	2.940E-01	50.0	M315	1 0 1 1 2	

404. C$_5$H$_4$N$_4$S
6-Mercaptopurine
6-Purinethiol
Mercaptopurine
Purine-6-thiol
Leukeran

RN: 50-44-2   **MP** (°C):
MW: 152.18   **BP** (°C):

Solubility (Moles/L)	Solubility (Grams/L)	Temp (°C)	Ref (#)	Evaluation (T P E A A)	Comments
3.000E-04	4.565E-02	4.62	A034	1 1 2 2 0	EFG
8.148E-04	1.240E-01	25	N063	1 1 1 1 2	
4.500E-02	6.848E+00	29.87	A034	1 1 2 2 1	EFG
1.703E-03	2.591E-01	37	H046	1 1 1 1 2	
2.658E-03	4.045E-01	ns	N050	0 1 1 0 0	

405. C$_5$H$_4$O$_2$
Furfural
2-Furaldehyde
Furfurol

RN: 98-01-1   **MP** (°C):   −36
MW: 96.09   **BP** (°C):   162

Solubility (Moles/L)	Solubility (Grams/L)	Temp (°C)	Ref (#)	Evaluation (T P E A A)	Comments
7.620E-01	7.322E+01	10	M099	1 2 0 1 1	
7.816E-01	7.510E+01	16	M099	1 2 0 1 2	
7.869E-01	7.561E+01	17	M099	1 2 0 1 2	
7.976E-01	7.664E+01	20	D052	1 1 0 0 0	

(continued)

405. C$_5$H$_4$O$_2$ (continued)

Solubility (Moles/L)	Solubility (Grams/L)	Temp (°C)	Ref (#)	Evaluation (T P E A A)	Comments
7.972E-01	7.660E+01	20	F300	1 0 0 0 2	
7.976E-01	7.664E+01	20	M099	1 2 0 1 1	
7.620E-01	7.322E+01	25	C056	1 2 1 1 1	
8.197E-01	7.877E+01	25	C329	1 2 1 1 1	average
7.709E-01	7.407E+01	25	H338	2 2 1 2 2	
7.976E-01	7.664E+01	25	H340	0 0 0 0 0	
7.441E-01	7.149E+01	25	L062	2 2 1 2 1	
7.709E-01	7.407E+01	25	L320	2 2 1 2 1	
8.242E-01	7.919E+01	25	M099	1 2 0 1 1	
8.347E-01	8.021E+01	27	M099	1 2 0 1 2	
8.347E-01	8.021E+01	27.20	M099	1 2 0 1 2	
8.312E-01	7.987E+01	27.50	M099	1 2 0 1 2	
8.418E-01	8.088E+01	30	M099	1 2 0 1 1	
8.488E-01	8.156E+01	35	H338	2 2 1 2 2	
8.506E-01	8.173E+01	35	L320	2 2 1 2 1	
9.029E-01	8.676E+01	38	G050	1 0 2 1 1	
8.619E-01	8.282E+01	39.50	E037	1 2 2 2 2	
9.029E-01	8.676E+01	40	M099	1 2 0 1 1	
9.289E-01	8.925E+01	44	M099	1 2 0 1 2	
9.804E-01	9.420E+01	50	M099	1 2 0 1 2	
1.023E+00	9.829E+01	52	G050	1 0 2 1 2	
9.306E-01	8.942E+01	53.10	E037	1 2 2 2 2	
4.982E+00	4.787E+02	53.30	E037	1 2 2 2 2	
1.090E+00	1.047E+02	60	M099	1 2 0 1 2	
1.107E+00	1.063E+02	61	M099	1 2 0 1 2	
1.156E+00	1.111E+02	66	G050	1 0 2 1 2	
1.156E+00	1.111E+02	66	M099	1 2 0 1 2	
1.214E+00	1.166E+02	70	M099	1 2 0 1 2	
4.895E+00	4.703E+02	73.60	E037	1 2 2 2 2	
1.318E+00	1.266E+02	79	G050	1 0 2 1 2	
1.342E+00	1.289E+02	80	M099	1 2 0 1 2	
1.361E+00	1.307E+02	85.80	E037	1 2 2 2 2	
1.482E+00	1.424E+02	90	M099	1 2 0 1 2	
1.512E+00	1.453E+02	92	M099	1 2 0 1 2	
1.684E+00	1.618E+02	93	G050	1 0 2 1 2	
4.721E+00	4.536E+02	95.90	E037	1 2 2 2 2	
1.617E+00	1.554E+02	97.90	M099	1 2 0 1 2	

406. C$_5$H$_4$O$_2$S

3-Thenoic acid

Thiophen-carbonsaeure-(3)

RN: 88-13-1 **MP** (°C): 137
MW: 128.15 **BP** (°C):

Solubility (Moles/L)	Solubility (Grams/L)	Temp (°C)	Ref (#)	Evaluation (T P E A A)	Comments
3.355E-02	4.300E+00	25	F300	1 0 0 0 1	

407. C$_5$H$_4$O$_3$

2-Furoic acid
Furan-carbon-saeure-(2)

RN:	88-14-2	**MP** (°C):	129.5	
MW:	112.09	**BP** (°C):	231	

Solubility (Moles/L)	Solubility (Grams/L)	Temp (°C)	Ref (#)	Evaluation (T P E A A)	Comments
2.227E-01	2.496E+01	5.99	A341	0 0 0 0 0	
2.243E-01	2.514E+01	6.99	A341	0 0 0 0 0	
2.332E-01	2.614E+01	10.49	A341	0 0 0 0 0	
2.498E-01	2.799E+01	10.99	A341	0 0 0 0 0	
2.543E-01	2.851E+01	11.99	A341	0 0 0 0 0	
3.310E-01	3.710E+01	15	F300	1 0 0 0 2	
2.606E-01	2.921E+01	15.99	A341	0 0 0 0 0	
3.385E-01	3.794E+01	20.99	A341	0 0 0 0 0	
4.216E-01	4.725E+01	24.99	A341	0 0 0 0 0	
4.665E-01	5.229E+01	27.99	A341	0 0 0 0 0	
5.182E-01	5.808E+01	28.99	A341	0 0 0 0 0	
6.448E-01	7.227E+01	33.99	A341	0 0 0 0 0	
6.677E-01	7.484E+01	35.99	A341	0 0 0 0 0	
7.816E-01	8.761E+01	37.99	A341	0 0 0 0 0	
1.120E+00	1.256E+02	41.99	A341	0 0 0 0 0	
1.229E+00	1.378E+02	43.99	A341	0 0 0 0 0	
1.444E+00	1.618E+02	46.64	A341	0 0 0 0 0	
2.159E+00	2.420E+02	49.99	A341	0 0 0 0 0	
2.610E+00	2.926E+02	51.99	A341	0 0 0 0 0	
2.768E+00	3.103E+02	53.99	A341	0 0 0 0 0	
2.815E+00	3.155E+02	54.49	A341	0 0 0 0 0	
3.221E+00	3.610E+02	54.99	A341	0 0 0 0 0	
3.964E+00	4.443E+02	57.49	A341	0 0 0 0 0	
4.219E+00	4.729E+02	60.04	A341	0 0 0 0 0	
4.224E+00	4.735E+02	61.39	A341	0 0 0 0 0	
4.940E+00	5.537E+02	62.99	A341	0 0 0 0 0	
5.529E+00	6.197E+02	67.99	A341	0 0 0 0 0	
1.838E+00	2.060E+02	100	F300	1 0 0 0 2	

408. C$_5$H$_4$O$_3$

Isopyromucic acid
Isobrenzschleimsaeure

RN:	496-64-0	**MP** (°C):		
MW:	112.09	**BP** (°C):		

Solubility (Moles/L)	Solubility (Grams/L)	Temp (°C)	Ref (#)	Evaluation (T P E A A)	Comments
3.845E-01	4.310E+01	0	F300	1 0 0 0 2	

409. $C_5H_5Cl_3N_2OS$
5-Ethoxy-3-trichloromethyl-1,2,4-thiadiazole
RN: 2593-15-9   **MP** (°C):
MW: 247.53   **BP** (°C):

Solubility (Moles/L)	Solubility (Grams/L)	Temp (°C)	Ref (#)	Evaluation (T P E A A)	Comments
4.732E-04	1.171E-01	ns	S460	0 0 0 0 0	

410. C_5H_5NO
3-Hydroxypyridine
3-Pyridinol
RN: 109-00-2   **MP** (°C):   127.5
MW: 95.10   **BP** (°C):   152

Solubility (Moles/L)	Solubility (Grams/L)	Temp (°C)	Ref (#)	Evaluation (T P E A A)	Comments
3.392E-01	3.226E+01	20	B050	1 0 0 0 0	

411. C_5H_5NO
4-Hydroxypyridine
4-Pyridinol
RN: 626-64-2   **MP** (°C):   148
MW: 95.10   **BP** (°C):   232.5

Solubility (Moles/L)	Solubility (Grams/L)	Temp (°C)	Ref (#)	Evaluation (T P E A A)	Comments
5.258E+00	5.000E+02	20	B050	1 0 0 0 0	

412. C_5H_5NO
2-Hydroxypyridine
2-Pyridinol
RN: 72762-00-6   **MP** (°C):   106
MW: 95.10   **BP** (°C):   280.5

Solubility (Moles/L)	Solubility (Grams/L)	Temp (°C)	Ref (#)	Evaluation (T P E A A)	Comments
5.258E+00	5.000E+02	20	B050	1 0 0 0 0	

413. $C_5H_5NO_2$
2,4-Dihydroxypyridine
3-Deazauracil
2,4-Pyridinediol
RN: 626-03-9   **MP** (°C):   278
MW: 111.10   **BP** (°C):

Solubility (Moles/L)	Solubility (Grams/L)	Temp (°C)	Ref (#)	Evaluation (T P E A A)	Comments
5.591E-02	6.211E+00	20	B050	1 0 0 0 0	

414. C$_5$H$_5$N$_3$O
Pyrazinamide
Pyrazine-2-carboxamide
Prazina

RN: 98-96-4   **MP (°C):** 190
MW: 123.12   **BP (°C):**

Solubility (Moles/L)	Solubility (Grams/L)	Temp (°C)	Ref (#)	Evaluation (T P E A A)	Comments
1.413E-01	1.740E+01	25	N041	2 0 1 1 0	EFG
1.218E-01	1.500E+01	ns	K444	0 0 0 0 0	

415. C$_5$H$_5$N$_5$
Adenine
Adenin
6-Aminopurine
1H-Purin-6-amine
Adeninimine
Vitamin B4

RN: 73-24-5   **MP (°C):** 363
MW: 135.13   **BP (°C):**

Solubility (Moles/L)	Solubility (Grams/L)	Temp (°C)	Ref (#)	Evaluation (T P E A A)	Comments
4.719E-03	6.377E-01	17.5	S306	1 0 1 2 2	
6.328E-03	8.551E-01	18.8	S306	1 0 1 2 2	
6.494E-03	8.776E-01	19.2	S306	1 0 1 2 2	
7.382E-03	9.975E-01	19.7	S306	1 0 1 2 2	
7.000E-03	9.459E-01	20	C017	2 0 0 1 0	EFG
6.907E-03	9.333E-01	20.08	D307	0 0 0 0 0	
7.680E-03	1.038E+00	22.36	D307	0 0 0 0 0	
6.586E-03	8.900E-01	25	A337	0 0 0 0 0	
7.200E-03	9.729E-01	25	C416	2 1 1 1 1	
5.476E-03	7.400E-01	25	C437	0 0 0 0 0	Average
6.654E-03	8.992E-01	25	D041	1 0 0 0 0	
7.040E-03	9.513E-01	25	H061	0 0 0 0 0	
7.600E-03	1.027E+00	25	L080	2 1 2 1 2	
8.000E-03	1.081E+00	25	R039	0 0 0 0 0	
8.610E-03	1.163E+00	25.01	D307	0 0 0 0 0	
8.690E-03	1.174E+00	25.03	D307	0 0 0 0 0	
8.250E-03	1.115E+00	25.5	T008	1 1 2 2 2	
7.936E-03	1.072E+00	26.6	S306	1 0 1 2 2	
9.740E-03	1.316E+00	27.47	D307	0 0 0 0 0	
1.087E-02	1.469E+00	29.97	D307	0 0 0 0 0	
9.377E-03	1.267E+00	31.1	S306	1 0 1 2 2	
1.540E-02	2.081E+00	37	L042	2 0 2 2 2	pH 6.47
1.390E-02	1.878E+00	38	T008	1 1 2 2 2	
1.514E-02	2.045E+00	44.0	S306	1 0 1 2 2	
1.707E-02	2.307E+00	45.1	S306	1 0 1 2 2	
1.862E-02	2.516E+00	45.5	S306	1 0 1 2 2	

(continued)

415. C$_5$H$_5$N$_5$ (continued)

Solubility (Moles/L)	Solubility (Grams/L)	Temp (°C)	Ref (#)	Evaluation (T P E A A)	Comments
1.805E-01	2.439E+01	100	D041	1 0 0 0 0	
6.808E-03	9.200E-01	c	D004	0 0 0 0 0	
1.805E-01	2.439E+01	h	D004	0 0 0 0 0	

416. C$_5$H$_5$N$_5$O
Guanine
2-Aminohypoxanthine
2-Amino-6-hydroxypurine
RN: 73-40-5   **MP** (°C):   >300
MW: 151.13   **BP** (°C):

Solubility (Moles/L)	Solubility (Grams/L)	Temp (°C)	Ref (#)	Evaluation (T P E A A)	Comments
1.920E-05	2.902E-03	15.02	D307	0 0 0 0 0	
6.000E-05	9.068E-03	20	C017	2 0 0 1 1	EFG
2.740E-05	4.141E-03	20.05	D307	0 0 0 0 0	
3.290E-05	4.972E-03	22.50	D307	0 0 0 0 0	
3.870E-05	5.849E-03	25.02	D307	0 0 0 0 0	
4.520E-05	6.831E-03	27.54	D307	0 0 0 0 0	
5.350E-05	8.085E-03	30.01	D307	0 0 0 0 0	
7.230E-05	1.093E-02	35.05	D307	0 0 0 0 0	
2.647E-04	4.000E-02	40	D041	1 0 0 0 0	
9.880E-05	1.493E-02	40.22	D307	0 0 0 0 0	
3.311E-04	5.004E-02	ns	R424	0 0 0 0 0	
3.311E-04	5.004E-02	ns	R427	0 0 0 0 0	

417. C$_5$H$_5$N$_5$O
Isoguanine
2-Hydroxy-6-aminopurine
RN: 3373-53-3   **MP** (°C):
MW: 151.13   **BP** (°C):

Solubility (Moles/L)	Solubility (Grams/L)	Temp (°C)	Ref (#)	Evaluation (T P E A A)	Comments
3.970E-04	6.000E-02	25	D041	1 0 0 0 0	
1.654E-03	2.499E-01	100	D041	1 0 0 0 1	

418. C$_5$H$_5$N$_5$O$_2$
2,8-Dioxyadenine
2,8-Dihydroxyadenine
RN: 30377-37-8   **MP** (°C):
MW: 167.13   **BP** (°C):

Solubility (Moles/L)	Solubility (Grams/L)	Temp (°C)	Ref (#)	Evaluation (T P E A A)	Comments
1.316E-05	2.200E-03	25	B049	1 0 1 1 1	
8.556E-06	1.430E-03	37	P068	0 0 0 0 0	

419. C_5H_6
Cyclopentadiene
Pentolex
Pentole
Pyropentylene
R-Pentine
1,3-Cyclopentadiene

RN: 542-92-7 MP (°C): −85
MW: 66.10 BP (°C): 42

Solubility (Moles/L)	Solubility (Grams/L)	Temp (°C)	Ref (#)	Evaluation (T P E A A)	Comments
1.023E-02	6.764E-01	ns	S460	0 0 0 0 0	

420. $C_5H_6Cl_2N_2$
3-Methyluracil
2,4(1H,3H)-Pyrimidinedione, 3-methyl-
Uracil, 3-methyl-

RN: 608-34-4 MP (°C):
MW: 165.02 BP (°C):

Solubility (Moles/L)	Solubility (Grams/L)	Temp (°C)	Ref (#)	Evaluation (T P E A A)	Comments
1.212E+00	2.000E+02	ns	B177	0 0 0 0 2	

421. $C_5H_6Cl_2N_2O_2$
Dantoin
1,3-Dichloro-5,5-dimethyl-2,4-imidazolidinedione
1,3-Dichloro-5,5-dimethylhydantoin

RN: 118-52-5 MP (°C): 132
MW: 197.02 BP (°C):

Solubility (Moles/L)	Solubility (Grams/L)	Temp (°C)	Ref (#)	Evaluation (T P E A A)	Comments
2.537E-03	4.998E-01	20	B080	1 0 1 1 0	
6.590E-03	1.298E+00	40	B080	1 0 1 1 1	

422. $C_5H_6N_2OS$
Methylthiouracil
6-Methyl-2-thiouracil

RN: 56-04-2 MP (°C): 330
MW: 142.18 BP (°C):

Solubility (Moles/L)	Solubility (Grams/L)	Temp (°C)	Ref (#)	Evaluation (T P E A A)	Comments
3.750E-03	5.332E-01	25	G016	1 2 1 2 2	intrinsic
7.026E-03	9.990E-01	c	I310	0 0 0 0 0	
3.715E-03	5.283E-01	ns	R424	0 0 0 0 0	

423. C$_5$H$_6$N$_2$OS
5-Methyl-2-thiouracil
4(1H)-Pyrimidinone, 2,3-dihydro-5-methyl-2-thioxo-
2-Thiothymine

RN: 636-26-0   **MP** (°C):   284
MW: 142.18   **BP** (°C):

Solubility (Moles/L)	Solubility (Grams/L)	Temp (°C)	Ref (#)	Evaluation (T P E A A)	Comments
3.580E-03	5.090E-01	25	G016	1 2 1 2 2	intrinsic

424. C$_5$H$_6$N$_2$O$_2$
Thymine
2,4-Dihydroxy-5-methylpyrimidine
5-Methyluracil

RN: 65-71-4   **MP** (°C):   316
MW: 126.12   **BP** (°C):

Solubility (Moles/L)	Solubility (Grams/L)	Temp (°C)	Ref (#)	Evaluation (T P E A A)	Comments
2.200E-02	2.775E+00	20	C017	2 0 0 1 1	EFG
2.379E-02	3.000E+00	23	F300	1 0 0 0 0	
3.552E-02	4.480E+00	25	D041	1 0 0 0 1	
2.780E-02	3.506E+00	25	H061	0 0 0 0 0	
3.030E-02	3.821E+00	25	L080	2 1 2 1 2	
2.860E-02	3.607E+00	25	R039	0 0 0 0 0	
2.740E-02	3.456E+00	25.5	T008	1 1 2 2 2	
3.500E-02	4.414E+00	30	L080	2 1 2 1 2	

425. C$_5$H$_6$N$_2$O$_2$
1-Methyluracil
2,4(1H,3H)-Pyrimidinedione, 1-methyl-
N1-Methyluracil

RN: 615-77-0   **MP** (°C):   179
MW: 126.12   **BP** (°C):

Solubility (Moles/L)	Solubility (Grams/L)	Temp (°C)	Ref (#)	Evaluation (T P E A A)	Comments
1.586E-01	2.000E+01	ns	B177	0 0 0 0 1	

426. C$_5$H$_6$N$_2$O$_4$
5-Carboxymethylhydantoin
Hydantoin of aspartic acid

RN: 5427-26-9   **MP** (°C):   216
MW: 158.11   **BP** (°C):

Solubility (Moles/L)	Solubility (Grams/L)	Temp (°C)	Ref (#)	Evaluation (T P E A A)	Comments
7.050E-02	1.115E+01	ns	M025	0 2 0 1 2	

427. $C_5H_6O_2$
α-Angelica lactone
α-Angelica-lacton

RN:	591-12-8	MP (°C):	18
MW:	98.10	BP (°C):	56

Solubility (Moles/L)	Solubility (Grams/L)	Temp (°C)	Ref (#)	Evaluation (T P E A A)	Comments
4.689E-01	4.600E+01	15	F300	1 0 0 0 1	

428. $C_5H_6O_4$
Citraconic acid
Citraconsaeure

RN:	498-23-7	MP (°C):	
MW:	130.10	BP (°C):	

Solubility (Moles/L)	Solubility (Grams/L)	Temp (°C)	Ref (#)	Evaluation (T P E A A)	Comments
6.018E+00	7.830E+02	25	F300	1 0 0 0 2	

429. $C_5H_6O_4$
Mesaconic acid
Mesaconsaeure

RN:	498-24-8	MP (°C):	204.5
MW:	130.10	BP (°C):	

Solubility (Moles/L)	Solubility (Grams/L)	Temp (°C)	Ref (#)	Evaluation (T P E A A)	Comments
2.022E-01	2.630E+01	18	F300	1 0 0 0 2	
4.241E+00	5.518E+02	100	F300	1 0 0 0 2	

430. $C_5H_6O_4$
Itaconic acid
Itaconsaeure

RN:	97-65-4	MP (°C):	163
MW:	130.10	BP (°C):	

Solubility (Moles/L)	Solubility (Grams/L)	Temp (°C)	Ref (#)	Evaluation (T P E A A)	Comments
4.281E-01	5.570E+01	10	F300	1 0 0 0 2	
5.891E-01	7.664E+01	20	D041	1 0 0 0 1	
5.903E-01	7.680E+01	20	F300	1 0 0 0 2	

431. C₅H₆S
3-Methylthiophene

RN:	616-44-4	MP (°C):	−69
MW:	98.17	BP (°C):	114 at 738 mm Hg

Solubility (Moles/L)	Solubility (Grams/L)	Temp (°C)	Ref (#)	Evaluation (T P E A A)	Comments
4.074E-03	3.999E-01	ns	S460	0 0 0 0 0	

432. C₅H₇NO₂
Ethyl cyanoacetate
Cyanessigsaeure-aethyl ester

RN:	105-56-6	MP (°C):	
MW:	113.12	BP (°C):	

Solubility (Moles/L)	Solubility (Grams/L)	Temp (°C)	Ref (#)	Evaluation (T P E A A)	Comments
1.768E-01	2.000E+01	25	F300	1 0 0 0 0	
7.072E-01	8.000E+01	80	F300	1 0 0 0 0	

433. C₅H₇NO₄S
2,4-Thiazolidinedicarboxylic acid
Tidiacic acid
Tidiacic
TDCA

RN:	30097-06-4	MP (°C):	
MW:	177.18	BP (°C):	524.3

Solubility (Moles/L)	Solubility (Grams/L)	Temp (°C)	Ref (#)	Evaluation (T P E A A)	Comments
4.300E-02	7.619E+00	21	B414	1 0 0 1 1	

434. C₅H₇N₂O₂
6-Methyluracil
4-Methyl-uracil

RN:	626-48-2	MP (°C):	318dec
MW:	127.12	BP (°C):	

Solubility (Moles/L)	Solubility (Grams/L)	Temp (°C)	Ref (#)	Evaluation (T P E A A)	Comments
5.506E-02	7.000E+00	22	F300	1 0 0 0 0	

435. C₅H₇N₃O
5-Methylcytosine
Mec

RN:	554-01-8	MP (°C):	270
MW:	125.13	BP (°C):	

Solubility (Moles/L)	Solubility (Grams/L)	Temp (°C)	Ref (#)	Evaluation (T P E A A)	Comments
3.441E-01	4.306E+01	25	D041	1 0 0 0 1	

436. $C_5H_7N_3O_2$
Dimetridazole
1,2-Dimethyl-5-nitroimidazole

RN: 551-92-8   **MP** (°C): 137–139
MW: 141.13   **BP** (°C):

Solubility (Moles/L)	Solubility (Grams/L)	Temp (°C)	Ref (#)	Evaluation (T P E A A)	Comments
6.866E-02	9.690E+00	20	D344	0 0 0 0 0	
6.866E-02	9.690E+00	20	D344	0 0 0 0 0	
6.738E-02	9.509E+00	20	D344	0 0 0 0 0	
6.870E-02	9.696E+00	20	D344	0 0 0 0 0	

437. C_5H_8
Isoprene
2-Methyl-1,3-butadiene

RN: 78-79-5   **MP** (°C): −120
MW: 68.12   **BP** (°C): 34.07

Solubility (Moles/L)	Solubility (Grams/L)	Temp (°C)	Ref (#)	Evaluation (T P E A A)	Comments
9.425E-03	6.420E-01	25	M001	2 1 2 2 2	

438. C_5H_8
Cyclopentene

RN: 142-29-0   **MP** (°C): −135
MW: 68.12   **BP** (°C): 44

Solubility (Moles/L)	Solubility (Grams/L)	Temp (°C)	Ref (#)	Evaluation (T P E A A)	Comments
2.411E-02	1.642E+00	24.8	L007	2 1 1 2 2	
7.854E-03	5.350E-01	25	M001	2 1 2 2 2	
2.411E-02	1.642E+00	25.1	L007	2 2 1 1 2	
2.562E-02	1.745E+00	34.8	L007	2 1 1 2 2	

439. C_5H_8
1-Pentyne
Pent-1-yne

RN: 627-19-0   **MP** (°C): −106
MW: 68.12   **BP** (°C): 40

Solubility (Moles/L)	Solubility (Grams/L)	Temp (°C)	Ref (#)	Evaluation (T P E A A)	Comments
2.305E-02	1.570E+00	25	M001	2 1 2 2 2	
1.154E-02	7.861E-01	25	M342	1 0 1 1 2	

440. C₅H₈
1,4-Pentadiene
Penta-1,4-diene
RN:　　591-93-5　　　**MP** (°C):　　−148
MW:　　68.12　　　　**BP** (°C):　　26

Solubility (Moles/L)	Solubility (Grams/L)	Temp (°C)	Ref (#)	Evaluation (T P E A A)	Comments
8.191E-03	5.580E-01	25	M001	2 1 2 2 2	

441. C₅H₈BrNO₄
5-Bromo-2-methyl-5-nitro-1,3-dioxane
Dioxane, 5-bromo-2-methyl-5-nitro-
Nibroxane
RN:　　53983-00-9　　**MP** (°C):　　72
MW:　　226.03　　　　**BP** (°C):

Solubility (Moles/L)	Solubility (Grams/L)	Temp (°C)	Ref (#)	Evaluation (T P E A A)	Comments
2.695E-02	6.093E+00	25	L013	1 0 2 1 2	

442. C₅H₈N₂O₂
5,5′-Dimethylhydantoin
5,5-Dimethylhydantoin
5,5-Dimethyl-2,4-imidazolidinedione
5,5-Dimethylimidazolidine-2,4-dione
RN:　　77-71-4　　　**MP** (°C):　　177
MW:　　128.13　　　**BP** (°C):

Solubility (Moles/L)	Solubility (Grams/L)	Temp (°C)	Ref (#)	Evaluation (T P E A A)	Comments
1.018E+00	1.304E+02	37	F183	1 0 1 1 1	intrinsic

443. C₅H₈N₂O₂
5-Ethylhydantoin
Hydantoin of α-aminobutyric acid
RN:　　15414-82-1　　**MP** (°C):　　119
MW:　　128.13　　　**BP** (°C):

Solubility (Moles/L)	Solubility (Grams/L)	Temp (°C)	Ref (#)	Evaluation (T P E A A)	Comments
8.630E-01	1.106E+02	ns	M025	0 2 0 1 2	

444. C₅H₈N₄O₃S₂

$444.\ C_5H_8N_4O_3S_2$

Methazolamide
Acetamide, N-[5-(aminosulfonyl)-3-methyl-1,3,4-thiadiazol-2(3H)-ylidene]-
N-(4-Methyl-2-sulfamoyl-D2-1,3,4-thiadiazolin-5-ylidene)acetamide
Neptazaneat
Metazolamide
Methenamide

RN:	554-57-4	MP (°C):	213
MW:	236.27	BP (°C):	

Solubility (Moles/L)	Solubility (Grams/L)	Temp (°C)	Ref (#)	Evaluation (T P E A A)	Comments
2.000E-03	4.725E-01	15	K024	1 2 1 1 1	
1.200E-02	2.835E+00	25	C415	1 0 0 1 0	
2.963E-03	7.000E-01	amb	L434	0 0 0 0 0	
1.481E-02	3.500E+00	ns	M032	0 0 0 0 2	
1.479E-02	3.495E+00	ns	R428	0 0 0 0 0	

445. C₅H₈N₄O₁₂

$445.\ C_5H_8N_4O_{12}$

Pentaerythritol tetranitrate
Nitropentaerythritol
1,3-Propanediol, 2,2-*bis*[(nitrooxy)methyl]-, dinitrate (ester)

RN:	78-11-5	MP (°C):	140
MW:	316.14	BP (°C):	

Solubility (Moles/L)	Solubility (Grams/L)	Temp (°C)	Ref (#)	Evaluation (T P E A A)	Comments
6.326E-06	2.000E-03	ns	M013	0 2 0 1 1	

446. C₅H₈O

$446.\ C_5H_8O$

Cyprethylene ether

RN:		MP (°C):	
MW:	84.12	BP (°C):	

Solubility (Moles/L)	Solubility (Grams/L)	Temp (°C)	Ref (#)	Evaluation (T P E A A)	Comments
9.435E-02	7.937E+00	27	K058	1 0 1 1 0	

447. C₅H₈O

$447.\ C_5H_8O$

α-Methylcrotonaldehyde
α-Methyl-crotonaldehyd

RN:	623-36-9	MP (°C):	
MW:	84.12	BP (°C):	

Solubility (Moles/L)	Solubility (Grams/L)	Temp (°C)	Ref (#)	Evaluation (T P E A A)	Comments
2.378E-01	2.000E+01	20	F300	1 0 0 0 1	

448. C₅H₈O₂

Ethyl acrylate
Ethyl propenoate
2-Propenoic acid ethyl ester

RN:	140-88-5	**MP** (°C):	−71
MW:	100.12	**BP** (°C):	99.4

Solubility (Moles/L)	Solubility (Grams/L)	Temp (°C)	Ref (#)	Evaluation (T P E A A)	Comments
1.785E-01	1.787E+01	30	L096	1 2 0 2 2	

449. C₅H₈O₂

Methyl methacrylate
Methacrylic acid methyl ester
Methyl 2-methyl-2-propenoate

RN:	80-62-6	**MP** (°C):	−48
MW:	100.12	**BP** (°C):	100

Solubility (Moles/L)	Solubility (Grams/L)	Temp (°C)	Ref (#)	Evaluation (T P E A A)	Comments
1.563E-01	1.565E+01	20	L096	1 2 0 2 2	

450. C₅H₈O₂

Acetylacetone
2,4-Pentanedione
Acetylaceton

RN:	123-54-6	**MP** (°C):	−23
MW:	100.12	**BP** (°C):	140.5

Solubility (Moles/L)	Solubility (Grams/L)	Temp (°C)	Ref (#)	Evaluation (T P E A A)	Comments
1.678E+00	1.680E+02	19.0	N051	1 2 1 1 2	
1.703E+00	1.705E+02	19.5	N051	1 2 1 1 2	
1.089E+00	1.090E+02	20	F300	1 0 0 0 2	
1.706E+00	1.708E+02	25	B019	1 0 1 2 0	

451. C₅H₈O₃

Dimethylpyruvic acid
DL-Methyl-bernsteinsaeure
α-Ketoisovaleric acid

RN:	759-05-7	**MP** (°C):	
MW:	116.12	**BP** (°C):	

Solubility (Moles/L)	Solubility (Grams/L)	Temp (°C)	Ref (#)	Evaluation (T P E A A)	Comments
3.450E+00	4.006E+02	20	F300	1 0 0 0 2	

452. C$_5$H$_8$O$_3$
Levulinic acid
Laevulinsaeure
4-Oxopentanoic acid
3-Acetyl propionic acid

RN:	123-76-2	MP (°C):	37.2		
MW:	116.12	BP (°C):	245		

Solubility (Moles/L)	Solubility (Grams/L)	Temp (°C)	Ref (#)	Evaluation (T P E A A)	Comments
4.632E+00	5.378E+02	6.99	A340	0 0 0 0 0	
4.990E+00	5.795E+02	9.99	A340	0 0 0 0 0	
5.530E+00	6.422E+02	14.49	A340	0 0 0 0 0	
6.087E+00	7.068E+02	20.79	A340	0 0 0 0 0	
6.400E+00	7.431E+02	24.99	A340	0 0 0 0 0	
6.631E+00	7.700E+02	30.09	A340	0 0 0 0 0	

453. C$_5$H$_8$O$_4$
Methylsuccinic acid
Acide methylsuccinique
1,2-Propanedicarboxylic acid

RN:	498-21-5	MP (°C):	117.5		
MW:	132.12	BP (°C):			

Solubility (Moles/L)	Solubility (Grams/L)	Temp (°C)	Ref (#)	Evaluation (T P E A A)	Comments
5.041E+00	6.660E+02	15	M051	1 0 0 0 2	

454. C$_5$H$_8$O$_4$
Ethylmalonic acid
1,1-Propanedicarboxylic acid
Aethylmalonsaeure
Mono-ethyl malonate
Malonic acid monoethyl ester
Malonsaeure-monoaethyl ester

RN:	601-75-2	MP (°C):	114		
MW:	132.12	BP (°C):	160		

Solubility (Moles/L)	Solubility (Grams/L)	Temp (°C)	Ref (#)	Evaluation (T P E A A)	Comments
2.619E+00	3.460E+02	0	F300	1 0 0 0 2	
3.996E+00	5.280E+02	0	M051	1 0 0 0 2	
4.814E+00	6.360E+02	15	M051	1 0 0 0 2	
5.389E+00	7.120E+02	25	M051	1 0 0 0 2	
3.626E+00	4.790E+02	50	F300	1 0 0 0 2	
6.873E+00	9.080E+02	50	M051	1 0 0 0 2	

455. C$_5$H$_8$O$_4$
Dimethylmalonic acid
Dimethyl-malonsaeure
Dimethyl-propanedioic acid

RN:	595-46-0	MP (°C):	192
MW:	132.12	BP (°C):	

Solubility (Moles/L)	Solubility (Grams/L)	Temp (°C)	Ref (#)	Evaluation (T P E A A)	Comments
6.812E-01	9.000E+01	13	F300	1 0 0 0 0	
1.968E+00	2.600E+02	100	F300	1 0 0 0 1	

456. C$_5$H$_8$O$_4$
Glutaric acid
Glutarsaeure
1,3-Propanedicarboxylic acid

RN:	110-94-1	MP (°C):	96.5
MW:	132.12	BP (°C):	

Solubility (Moles/L)	Solubility (Grams/L)	Temp (°C)	Ref (#)	Evaluation (T P E A A)	Comments
2.272E+00	3.002E+02	0	F300	1 0 0 0 2	
3.247E+00	4.290E+02	0	L041	1 0 0 1 2	
2.410E+00	3.183E+02	3.40	A031	1 2 2 2 2	
2.650E+00	3.501E+02	5.99	A341	0 0 0 0 0	
2.764E+00	3.651E+02	7.99	A341	0 0 0 0 0	
3.127E+00	4.131E+02	10.40	A031	1 2 2 2 2	
2.909E+00	3.843E+02	10.99	A341	0 0 0 0 0	
3.213E+00	4.245E+02	12.99	A341	0 0 0 0 0	
3.433E+00	4.536E+02	14	A031	1 2 2 2 0	
4.443E+00	5.870E+02	15	L041	1 0 0 1 2	
4.443E+00	5.870E+02	15	M051	1 0 0 0 2	
3.521E+00	4.652E+02	15.99	A341	0 0 0 0 0	
3.674E+00	4.854E+02	17.99	A341	0 0 0 0 0	
3.861E+00	5.100E+02	18	A031	1 2 2 2 2	
3.816E+00	5.041E+02	19.99	A341	0 0 0 0 0	
2.954E+00	3.902E+02	20	D041	1 0 0 0 1	
4.837E+00	6.390E+02	20	L041	1 0 0 1 2	
2.952E+00	3.900E+02	20	M171	1 0 0 0 2	
1.340E+00	1.770E+02	20	S006	1 0 0 0 2	
4.278E+00	5.652E+02	23.90	A031	1 2 2 2 2	
4.088E+00	5.401E+02	24.99	A341	0 0 0 0 0	
4.653E+00	6.148E+02	28.30	A031	1 2 2 2 2	
4.394E+00	5.805E+02	28.99	A341	0 0 0 0 0	
4.503E+00	5.949E+02	30.99	A341	0 0 0 0 0	
4.642E+00	6.133E+02	33.99	A341	0 0 0 0 0	
6.033E+00	7.970E+02	35	L041	1 0 0 1 2	
4.796E+00	6.336E+02	36.99	A341	0 0 0 0 0	
4.894E+00	6.466E+02	38.99	A341	0 0 0 0 0	
5.096E+00	6.732E+02	42.99	A341	0 0 0 0 0	
5.131E+00	6.779E+02	43.99	A341	0 0 0 0 0	
5.143E+00	6.795E+02	44.99	A341	0 0 0 0 0	

(continued)

456. C₅H₈O₄ (continued)

Solubility (Moles/L)	Solubility (Grams/L)	Temp (°C)	Ref (#)	Evaluation (T P E A A)	Comments
5.246E+00	6.930E+02	46.99	A341	0 0 0 0 0	
5.341E+00	7.057E+02	49.99	A341	0 0 0 0 0	
7.244E+00	9.570E+02	50	L041	1 0 0 1 2	
5.470E+00	7.227E+02	54.49	A341	0 0 0 0 0	
5.640E+00	7.451E+02	55.99	A341	0 0 0 0 0	
5.713E+00	7.548E+02	58.99	A341	0 0 0 0 0	
5.729E+00	7.569E+02	61.09	A341	0 0 0 0 0	
5.890E+00	7.782E+02	62.99	A341	0 0 0 0 0	
4.032E+00	5.327E+02	65	F300	1 0 0 0 2	
8.462E+00	1.118E+03	65	L041	1 0 0 1 2	
6.038E+00	7.977E+02	68.99	A341	0 0 0 0 0	
4.081E+00	5.392E+02	rt	H431	0 0 0 0 0	

457. C₅H₉BrO₂

α-Bromo-methyl-ethyl-acetate
Ethyl DL-α-bromopropionate
Propanoic acid, 2-bromo-, ethyl ester
Ethyl DL-2-bromopropionate

RN: 535-11-5   **MP (°C):**
MW: 181.04   **BP (°C):**

Solubility (Moles/L)	Solubility (Grams/L)	Temp (°C)	Ref (#)	Evaluation (T P E A A)	Comments
2.780E-01	5.033E+01	ns	F057	0 2 2 2 1	

458. C₅H₉BrO₂

α-Ethyl-β-bromo-propionic ureide

RN: **MP (°C):**
MW: 181.04   **BP (°C):**

Solubility (Moles/L)	Solubility (Grams/L)	Temp (°C)	Ref (#)	Evaluation (T P E A A)	Comments
2.130E-01	3.855E+01	ns	F056	0 2 2 2 1	

459. C₅H₉NO₂

DL-Proline
Pyrrolidine-2-carboxylic acid

RN: 609-36-9   **MP (°C):** 208
MW: 115.13   **BP (°C):**

Solubility (Moles/L)	Solubility (Grams/L)	Temp (°C)	Ref (#)	Evaluation (T P E A A)	Comments
1.217E+01	1.401E+03	20	J303	0 0 0 0 0	
1.146E+01	1.319E+03	25	J303	0 0 0 0 0	
1.425E+01	1.641E+03	40	J303	0 0 0 0 0	
1.708E+01	1.967E+03	50	J303	0 0 0 0 0	
2.082E+01	2.397E+03	60	J303	0 0 0 0 0	

460. C₅H₉NO₂

L-Proline
2-Pyrrolidinecarboxylic acid

RN: 147-85-3   **MP** (°C):
MW: 115.13   **BP** (°C):

Solubility (Moles/L)	Solubility (Grams/L)	Temp (°C)	Ref (#)	Evaluation (T P E A A)	Comments
5.374E+00	6.188E+02	25	D041	1 0 0 0 2	
6.653E+00	7.660E+02	27	D036	0 0 0 0 0	
6.123E+00	7.050E+02	65	D041	1 0 0 0 2	
6.691E+00	7.704E+02	99.99	P349	0 0 0 0 0	

461. C₅H₉NO₂S

2-Methylthiazolidine-4-carboxylic acid
4-Thiazolidinecarboxylic acid, 2-methyl-
Thiazolidine-4-carboxylic acid, 2-methyl-

RN: 4165-32-6   **MP** (°C):   174-175
MW: 147.20   **BP** (°C):   333.0

Solubility (Moles/L)	Solubility (Grams/L)	Temp (°C)	Ref (#)	Evaluation (T P E A A)	Comments
2.100E-01	3.091E+01	21	B414	1 0 0 1 1	partial decomposition

462. C₅H₉NO₃

L-Hydroxyproline
trans-4-Hydroxy-L-proline
L-4-hydroxyproline
(4*S*)-4-Hydroxy-L-proline

RN: 51-35-4   **MP** (°C):
MW: 131.13   **BP** (°C):

Solubility (Moles/L)	Solubility (Grams/L)	Temp (°C)	Ref (#)	Evaluation (T P E A A)	Comments
3.158E+00	4.141E+02	99.99	P349	0 0 0 0 0	

463. C₅H₉NO₃

Formyl-α-aminobutyric acid
Butanoic acid, 2-(formylamino)-

RN: 106873-99-8   **MP** (°C):
MW: 131.13   **BP** (°C):

Solubility (Moles/L)	Solubility (Grams/L)	Temp (°C)	Ref (#)	Evaluation (T P E A A)	Comments
2.560E-01	3.357E+01	25	M024	1 2 0 1 2	
2.560E-01	3.357E+01	ns	M025	0 2 0 1 2	

464. C₅H₉NO₄

D-Glutamic acid

D-2-Aminoglutaric acid

RN:	6893-26-1	**MP** (°C):	201
MW:	147.13	**BP** (°C):	

Solubility (Moles/L)	Solubility (Grams/L)	Temp (°C)	Ref (#)	Evaluation (T P E A A)	Comments
2.337E-02	3.439E+00	0	D018	2 2 2 1 2	
2.303E-02	3.388E+00	0	M043	1 0 0 0 1	
3.381E-02	4.975E+00	10	M043	1 0 0 0 1	
1.004E-01	1.478E+01	20	D041	1 0 0 0 1	
4.859E-02	7.149E+00	20	M043	1 0 0 0 1	
4.472E-02	6.580E+00	21	P045	1 0 2 1 2	
5.981E-02	8.800E+00	25	D018	2 2 2 1 2	
6.729E-02	9.901E+00	30	M043	1 0 0 0 1	
1.004E-01	1.478E+01	40	M043	1 0 0 0 1	
1.481E-01	2.179E+01	50	D018	2 2 2 1 2	
2.107E-01	3.101E+01	60	M043	1 0 0 0 1	
4.148E-01	6.103E+01	80	M043	1 0 0 0 1	
8.347E-01	1.228E+02	100	M043	1 0 0 0 2	
5.850E-02	8.607E+00	ns	M025	0 2 0 1 2	

465. C₅H₉NO₄

DL-Glutamic acid

DL-2-Aminoglutaric acid

RN:	617-65-2	**MP** (°C):	194
MW:	147.13	**BP** (°C):	

Solubility (Moles/L)	Solubility (Grams/L)	Temp (°C)	Ref (#)	Evaluation (T P E A A)	Comments
5.601E-02	8.241E+00	0	D018	2 2 2 1 2	
4.850E-02	7.136E+00	4.99	A405	2 0 1 1 2	
5.990E-02	8.813E+00	9.99	A405	2 0 1 1 2	
6.300E-02	9.269E+00	14.99	A405	2 0 1 1 2	
8.840E-02	1.301E+01	20.99	A405	2 0 1 1 2	
9.370E-02	1.379E+01	24.99	A405	2 0 1 1 2	
1.750E-01	2.575E+01	25	D018	2 2 2 1 2	
1.368E-01	2.013E+01	25	D041	1 0 0 0 2	
1.075E-01	1.582E+01	29.99	A405	2 0 1 1 2	
1.414E-01	2.080E+01	34.99	A405	2 0 1 1 2	
1.684E-01	2.478E+01	39.99	A405	2 0 1 1 2	
2.016E-01	2.966E+01	44.99	A405	2 0 1 1 2	
2.699E-01	3.971E+01	49.99	A405	2 0 1 1 2	
5.131E-01	7.549E+01	50	D018	2 2 2 1 2	
3.502E-01	5.153E+01	54.99	A405	2 0 1 1 2	
3.959E-01	5.825E+01	59.99	A405	2 0 1 1 2	
4.772E-01	7.021E+01	64.99	A405	2 0 1 1 2	
5.621E-01	8.270E+01	69.99	A405	2 0 1 1 2	
6.709E-01	9.871E+01	71.99	A405	2 0 1 1 2	
7.289E-01	1.072E+02	74.99	A405	2 0 1 1 2	
7.206E-01	1.060E+02	75	D041	1 0 0 0 2	

466. C$_5$H$_9$NO$_4$

L-Glutamic acid
L-2-Aminoglutaric acid
L(+)-Glutaminsaeure
Glutamic acid
L(+) Glutaminic acid

RN:	56-86-0	MP (°C):	250
MW:	147.13	BP (°C):	

Solubility (Moles/L)	Solubility (Grams/L)	Temp (°C)	Ref (#)	Evaluation (T P E A A)	Comments
4.866E-02	7.160E+00	20	B032	1 2 2 1 2	
4.486E-02	6.600E+00	21	F302	1 0 0 0 1	
5.825E-02	8.570E+00	25	B032	1 2 2 1 2	
5.822E-02	8.566E+00	25	D041	1 0 0 0 2	
5.845E-02	8.600E+00	25	F300	1 0 0 0 1	
7.262E-02	1.068E+01	25	G315	0 0 0 0 0	
5.614E-02	8.260E+00	27	D036	0 0 0 0 0	
6.980E-02	1.027E+01	29.80	B032	1 2 2 1 2	
1.454E-01	2.140E+01	50	F300	1 0 0 0 2	
3.562E-01	5.240E+01	75	D041	1 0 0 0 2	
3.561E-01	5.240E+01	75	F300	1 0 0 0 2	
8.346E-01	1.228E+02	100	F300	1 0 0 0 2	
4.078E-02	6.000E+00	ns	D072	0 0 0 0 0	
5.802E-02	8.537E+00	rt	H431	0 0 0 0 0	

467. C$_5$H$_{10}$

Cyclopentane
Pentamethylene
Exxsol cyclopentane S
Zeonsolv HP

RN:	287-92-3	MP (°C):	−94.4
MW:	70.14	BP (°C):	49.3

Solubility (Moles/L)	Solubility (Grams/L)	Temp (°C)	Ref (#)	Evaluation (T P E A A)	Comments
4.826E-03	3.385E-01	4.8	L007	2 2 1 2 2	
4.826E-03	3.385E-01	5.1	L007	2 1 1 1 2	
4.870E-03	3.416E-01	14.8	L007	2 2 1 2 2	
4.870E-03	3.416E-01	15.2	L007	2 1 1 1 2	
4.873E-03	3.418E-01	24.8	L007	2 2 1 2 2	
2.338E-03	1.640E-01	25	G313	2 1 1 2 2	
2.281E-03	1.600E-01	25	K119	1 0 0 0 2	
2.224E-03	1.560E-01	25	M001	2 1 2 2 2	
2.224E-03	1.560E-01	25	M002	2 1 2 2 2	
2.281E-03	1.600E-01	25.0	P051	2 1 1 2 2	
2.281E-03	1.600E-01	25.00	P007	2 1 2 2 2	
4.873E-03	3.418E-01	25.1	L007	2 1 1 1 2	
5.252E-03	3.684E-01	34.8	L007	2 2 1 2 2	
5.252E-03	3.684E-01	35.2	L007	2 1 1 1 2	
2.324E-03	1.630E-01	40.1	P051	2 1 1 2 2	

(continued)

467. C₅H₁₀ (continued)

Solubility (Moles/L)	Solubility (Grams/L)	Temp (°C)	Ref (#)	Evaluation (T P E A A)	Comments
2.324E-03	1.630E-01	40.10	P007	2 1 2 2 2	
4.867E-03	3.414E-01	44.8	L007	2 2 1 2 2	
2.566E-03	1.800E-01	55.7	P051	2 1 1 2 2	
2.566E-03	1.800E-01	55.70	P007	2 1 2 2 2	
4.220E-03	2.960E-01	99.1	P051	2 1 1 2 2	
4.220E-03	2.960E-01	99.10	P007	2 1 2 2 2	
5.304E-03	3.720E-01	118.0	P051	2 1 1 2 2	
5.304E-03	3.720E-01	118.00	P007	2 1 2 2 2	
8.712E-03	6.110E-01	137.3	P051	2 1 1 2 2	
8.712E-03	6.110E-01	137.30	P007	2 1 2 2 2	
1.129E-02	7.920E-01	153.1	P051	2 1 1 2 2	
1.129E-02	7.920E-01	153.10	P007	2 1 2 2 2	
2.224E-03	1.560E-01	ns	H123	0 0 0 0 0	

468. C₅H₁₀

3-Methyl-1-butene
2-Methyl-3-butene
3,3-Dimethylpropene
Isopropylethylene

RN: 563-45-1 **MP (°C):** −168
MW: 70.14 **BP (°C):** 20

Solubility (Moles/L)	Solubility (Grams/L)	Temp (°C)	Ref (#)	Evaluation (T P E A A)	Comments
1.854E-03	1.300E-01	25	M001	2 1 2 2 2	

469. C₅H₁₀

2-Pentene
1-Methyl-2-ethylethylene
sym-Methylethylethylene
β-Amylene
β-n-Amylene
3-Pentene

RN: 109-68-2 **MP (°C):** −136
MW: 70.14 **BP (°C):** 36

Solubility (Moles/L)	Solubility (Grams/L)	Temp (°C)	Ref (#)	Evaluation (T P E A A)	Comments
2.894E-03	2.030E-01	25	M001	2 1 2 2 2	

470. C$_5$H$_{10}$

1-Pentene
Propylethylene
α-*n*-Amylene
1-Methyl-3-butene

RN: 109-67-1   **MP** (°C): −165
MW: 70.14   **BP** (°C): 30.1

Solubility (Moles/L)	Solubility (Grams/L)	Temp (°C)	Ref (#)	Evaluation (T P E A A)	Comments
2.609E-03	1.830E-01	23	C332	0 0 0 0 0	
2.110E-03	1.480E-01	25	M001	2 1 2 2 2	

471. C$_5$H$_{10}$Cl$_3$O$_3$P

Diethyl trichloromethyl phosphonate
Phosphonic acid, (trichloromethyl)-, diethyl ester
Ro 3-0658

RN: 866-23-9   **MP** (°C):
MW: 255.47   **BP** (°C):

Solubility (Moles/L)	Solubility (Grams/L)	Temp (°C)	Ref (#)	Evaluation (T P E A A)	Comments
1.761E-02	4.500E+00	25	B070	1 2 0 1 1	

472. C$_5$H$_{10}$N$_2$O

N-Nitrosopiperidine
Pyridine, hexahydro-*N*-nitroso
NPIP

RN: 100-75-4   **MP** (°C): <25
MW: 114.15   **BP** (°C):

Solubility (Moles/L)	Solubility (Grams/L)	Temp (°C)	Ref (#)	Evaluation (T P E A A)	Comments
6.700E-01	7.648E+01	24	D083	2 0 0 0 1	

473. C$_5$H$_{10}$N$_2$O$_2$S

Methomyl
Nudrin
Lannate

RN: 16752-77-5   **MP** (°C): 78.5
MW: 162.21   **BP** (°C):

Solubility (Moles/L)	Solubility (Grams/L)	Temp (°C)	Ref (#)	Evaluation (T P E A A)	Comments
3.576E-01	5.800E+01	25	M161	1 0 0 0 1	

474. $C_5H_{10}N_2O_3$
Glycolylglycineamide
RN: MP (°C):
MW: 146.15   BP (°C):

Solubility (Moles/L)	Solubility (Grams/L)	Temp (°C)	Ref (#)	Evaluation (T P E A A)	Comments
5.820E+00	8.506E+02	25	M008	1 0 0 0 2	

475. $C_5H_{10}N_2O_3$
Glycyl-L-alanine
Glycylalanine
RN: 3695-73-6   MP (°C):
MW: 146.15   BP (°C):

Solubility (Moles/L)	Solubility (Grams/L)	Temp (°C)	Ref (#)	Evaluation (T P E A A)	Comments
4.780E+00	6.986E+02	24.99	B441	0 0 0 0 0	

476. $C_5H_{10}N_2O_3$
D-Glutamine
D-2-Aminoglutaramic acid
RN: 5959-95-5   MP (°C):
MW: 146.15   BP (°C):

Solubility (Moles/L)	Solubility (Grams/L)	Temp (°C)	Ref (#)	Evaluation (T P E A A)	Comments
2.910E-01	4.253E+01	ns	M025	0 2 0 1 2	

477. $C_5H_{10}N_2O_3$
L-Glutamine
L(+)-Glutamin
L(+)-Glutamine
Glutamine
RN: 56-85-9   MP (°C):   185
MW: 146.15   BP (°C):

Solubility (Moles/L)	Solubility (Grams/L)	Temp (°C)	Ref (#)	Evaluation (T P E A A)	Comments
1.184E-01	1.730E+01	0	F300	1 0 0 0 2	
2.378E-01	3.475E+01	18	D041	1 0 0 0 1	
2.444E-01	3.572E+01	20	B032	1 2 2 1 2	
2.829E-01	4.135E+01	25	B032	1 2 2 1 2	
2.789E-01	4.077E+01	25	D041	1 0 0 0 2	
2.701E-01	3.948E+01	25	G315	0 0 0 0 0	
5.891E-02	8.610E+00	25	J303	0 0 0 0 0	
2.997E-01	4.380E+01	25.1	N024	0 0 0 0 0	
2.840E-01	4.150E+01	25.1	N025	0 0 0 0 0	
2.840E-01	4.150E+01	25.1	N026	0 0 0 0 0	

(continued)

477. $C_5H_{10}N_2O_3$ (continued)

Solubility (Moles/L)	Solubility (Grams/L)	Temp (°C)	Ref (#)	Evaluation (T P E A A)	Comments
2.821E-01	4.123E+01	25.1	N027	1 1 2 2 2	
2.737E-01	4.000E+01	27	D036	0 0 0 0 0	
3.285E-01	4.801E+01	29.80	B032	1 2 2 1 2	
3.154E-01	4.610E+01	30	F300	1 0 0 0 2	
1.002E-01	1.464E+01	40	J303	0 0 0 0 0	
2.135E-01	3.120E+01	60	J303	0 0 0 0 0	

478. $C_5H_{10}N_2S_2$

Dazomet
3,5-Dimethyl-1,2,3,5-tetrahydro-1,3,5-thiadiazinethione-2
Thiazone
Thiazon

RN:	533-74-4	**MP** (°C):	106.5	
MW:	162.28	**BP** (°C):		

Solubility (Moles/L)	Solubility (Grams/L)	Temp (°C)	Ref (#)	Evaluation (T P E A A)	Comments
7.386E-03	1.199E+00	25	M061	1 0 0 0 1	
1.169E-02	1.896E+00	30	B185	0 0 0 0 0	
7.395E-03	1.200E+00	30	M161	1 0 0 0 1	

479. $C_5H_{10}N_6O_2$

Dinitrosopentamethylenetetramine
3,7-Dinitroso-1,3,5,7-tetraazabicyclo[3.3.1]nonane

RN:	101-25-7	**MP** (°C):	207	
MW:	186.17	**BP** (°C):		

Solubility (Moles/L)	Solubility (Grams/L)	Temp (°C)	Ref (#)	Evaluation (T P E A A)	Comments
5.318E-02	9.901E+00	ns	I313	0 0 0 0 0	

480. $C_5H_{10}O$

Methy propyl ketone
Methyl propyl ketone
2-Pentanone
Pentan-2-one

RN:	107-87-9	**MP** (°C):	−78	
MW:	86.13	**BP** (°C):	100.5	

Solubility (Moles/L)	Solubility (Grams/L)	Temp (°C)	Ref (#)	Evaluation (T P E A A)	Comments
8.870E-01	7.640E+01	10	G032	1 2 1 1 2	
6.520E-01	5.616E+01	20	G030	1 2 0 0 2	
5.000E-01	4.307E+01	20	M312	1 0 0 0 1	
6.799E-01	5.857E+01	25	A356	0 0 0 0 0	
4.786E-01	4.123E+01	25	B060	2 0 1 1 1	

(continued)

480. C$_5$H$_{10}$O (continued)

Solubility (Moles/L)	Solubility (Grams/L)	Temp (°C)	Ref (#)	Evaluation (T P E A A)	Comments
7.775E-01	6.697E+01	25	C333	0 0 0 0 0	
7.000E-01	6.029E+01	25	F044	1 0 0 0 1	
6.063E-01	5.222E+01	25	G030	1 2 0 0 2	
6.572E-01	5.660E+01	25	P055	1 0 0 0 2	
5.718E-01	4.925E+01	30	G030	1 2 0 0 2	
6.300E-01	5.426E+01	30	G032	1 2 1 1 2	
5.806E-01	5.001E+01	35	A356	0 0 0 0 0	
6.799E-01	5.857E+01	35	C333	0 0 0 0 0	
5.302E-01	4.567E+01	45	A356	0 0 0 0 0	
6.799E-01	5.857E+01	45	C333	0 0 0 0 0	
5.150E-01	4.436E+01	50	G032	1 2 1 1 2	
5.302E-01	4.567E+01	55	A356	0 0 0 0 0	
5.806E-01	5.001E+01	55	C333	0 0 0 0 0	

481. C$_5$H$_{10}$O
Valeraldehyde
n-Valeraldehyde
Valeral
n-Pentanal

RN: 110-62-3   **MP** (°C): −92
MW: 86.13   **BP** (°C): 103

Solubility (Moles/L)	Solubility (Grams/L)	Temp (°C)	Ref (#)	Evaluation (T P E A A)	Comments
1.358E-01	1.170E+01	25	A049	1 0 0 0 2	
2.100E-01	1.809E+01	25	K012	1 0 0 0 1	

482. C$_5$H$_{10}$O
Tetrahydropyran
Pentamethylene oxide

RN: 142-68-7   **MP** (°C): −49.2
MW: 86.13   **BP** (°C): 88

Solubility (Moles/L)	Solubility (Grams/L)	Temp (°C)	Ref (#)	Evaluation (T P E A A)	Comments
1.372E+00	1.182E+02	0	B001	2 0 1 0 0	
1.122E+00	9.666E+01	10	B001	2 0 1 0 0	
1.021E+00	8.792E+01	15	B001	2 0 1 0 0	
9.351E-01	8.054E+01	20	B001	2 0 1 0 0	
8.620E-01	7.425E+01	25	B001	2 0 1 0 0	

483. C₅H₁₀O
Diethyl ketone
3-Pentanone
RN: 96-22-0   **MP** (°C):   −42
MW: 86.13   **BP** (°C):   101.5

Solubility (Moles/L)	Solubility (Grams/L)	Temp (°C)	Ref (#)	Evaluation (T P E A A)	Comments
7.810E-01	6.727E+01	10	G032	1 2 1 1 2	
4.786E-01	4.123E+01	20	D052	1 1 0 0 1	
5.613E-01	4.834E+01	20	G030	1 2 0 0 2	
6.052E-01	5.213E+01	25	B019	1 0 1 2 0	
3.818E-01	3.288E+01	25	B060	2 0 1 1 1	
5.328E-01	4.589E+01	25	G030	1 2 0 0 2	
5.900E-01	5.082E+01	25	K012	1 0 0 0 1	
4.999E-01	4.306E+01	30	G030	1 2 0 0 1	
5.760E-01	4.961E+01	30	G032	1 2 1 1 2	
4.560E-01	3.928E+01	50	G032	1 2 1 1 2	

484. C₅H₁₀O
1-Penten-3-ol
Penten-1-ol-3
RN: 616-25-1   **MP** (°C):
MW: 86.13   **BP** (°C):

Solubility (Moles/L)	Solubility (Grams/L)	Temp (°C)	Ref (#)	Evaluation (T P E A A)	Comments
9.312E-01	8.021E+01	20	G031	1 0 0 0 2	
8.798E-01	7.579E+01	25	G031	1 0 0 0 2	
8.340E-01	7.184E+01	30	G031	1 0 0 0 2	

485. C₅H₁₀O
4-Penten-1-ol
Penten-4-ol-1
RN: 821-09-0   **MP** (°C):
MW: 86.13   **BP** (°C):   135.5

Solubility (Moles/L)	Solubility (Grams/L)	Temp (°C)	Ref (#)	Evaluation (T P E A A)	Comments
6.458E-01	5.562E+01	20	G031	1 0 0 0 2	
6.261E-01	5.393E+01	25	G031	1 0 0 0 2	
6.115E-01	5.267E+01	30	G031	1 0 0 0 2	

486. C$_5$H$_{10}$O
3-Penten-2-ol
Penten-3-ol-2
RN: 1569-50-2   **MP** (°C):
MW: 86.13   **BP** (°C):   120

Solubility (Moles/L)	Solubility (Grams/L)	Temp (°C)	Ref (#)	Evaluation (T P E A A)	Comments
1.003E+00	8.642E+01	20	G031	1 0 0 0 2	
9.508E-01	8.189E+01	25	G031	1 0 0 0 2	
9.075E-01	7.817E+01	30	G031	1 0 0 0 2	

487. C$_5$H$_{10}$O
2-Methyl tetrahydrofuran
2-Methyl oxolane
β-Methyl tetramethylene oxide
RN: 96-47-9   **MP** (°C): −136
MW: 86.13   **BP** (°C):   83

Solubility (Moles/L)	Solubility (Grams/L)	Temp (°C)	Ref (#)	Evaluation (T P E A A)	Comments
1.174E+00	1.011E+02	10	B001	2 0 1 0 0	

488. C$_5$H$_{10}$O
1-Methyl tetrahydrofuran
Methyl oxolane
α-Methyl tetramethylene oxide
RN: 45376-90-7   **MP** (°C):
MW: 86.13   **BP** (°C):   80

Solubility (Moles/L)	Solubility (Grams/L)	Temp (°C)	Ref (#)	Evaluation (T P E A A)	Comments
2.101E+00	1.810E+02	0	B001	2 0 1 0 0	
1.788E+00	1.540E+02	10	B001	2 0 1 0 0	
1.646E+00	1.418E+02	15	B001	2 0 1 0 0	
1.519E+00	1.308E+02	20	B001	2 0 1 0 0	
1.414E+00	1.218E+02	25	B001	2 0 1 0 0	

489. C$_5$H$_{10}$O
Cypreth ether
Cyclopropane, ethoxy-
Ethoxycyclopropane
Ethyl cyclopropyl ether
RN: 5614-38-0   **MP** (°C):
MW: 86.13   **BP** (°C):

Solubility (Moles/L)	Solubility (Grams/L)	Temp (°C)	Ref (#)	Evaluation (T P E A A)	Comments
3.162E-01	2.724E+01	25	K061	1 0 1 1 1	
2.500E-01	2.153E+01	25	K061	1 0 1 1 1	

490. C$_5$H$_{10}$O
3-Methyl-2-butanone
3-Methylbutanone-2

RN: 563-80-4   **MP** (°C): −92
MW: 86.13   **BP** (°C): 94.5

Solubility (Moles/L)	Solubility (Grams/L)	Temp (°C)	Ref (#)	Evaluation (T P E A A)	Comments
8.130E-01	7.003E+01	10	G032	1 2 1 1 2	
7.116E-01	6.130E+01	20	G030	1 2 0 0 2	
6.654E-01	5.732E+01	25	G030	1 2 0 0 2	
6.240E-01	5.375E+01	30	G030	1 2 0 0 2	
6.080E-01	5.237E+01	30	G032	1 2 1 1 2	
5.940E-01	5.116E+01	50	G032	1 2 1 1 2	

491. C$_5$H$_{10}$OS$_2$
Butylxanthogenic acid

RN: **MP** (°C):
MW: 150.26   **BP** (°C):

Solubility (Moles/L)	Solubility (Grams/L)	Temp (°C)	Ref (#)	Evaluation (T P E A A)	Comments
8.000E-04	1.202E-01	25	K012	1 0 0 0 0	

492. C$_5$H$_{10}$O$_2$
Valeric acid
Valeric acid, normal
n-Valeric acid

RN: 109-52-4   **MP** (°C): −34.5
MW: 102.13   **BP** (°C): 185

Solubility (Moles/L)	Solubility (Grams/L)	Temp (°C)	Ref (#)	Evaluation (T P E A A)	Comments
2.295E-01	2.344E+01	25	B060	2 0 1 1 1	
4.636E-01	4.735E+01	25	H028	2 0 2 0 2	
3.697E-01	3.776E+01	25	H122	1 0 0 0 2	
4.055E-01	4.141E+01	25	H338	2 2 1 2 2	
3.750E-01	3.830E+01	25	K012	1 0 0 0 2	
4.893E-01	4.997E+01	35	H338	2 2 1 2 2	
2.936E-03	2.999E-01	c	L055	0 0 0 0 1	
4.636E-01	4.735E+01	ns	A406	0 0 0 0 1	

493. $C_5H_{10}O_2$
Methyl butyrate
Buttersaeure-methyl ester
n-Methyl *n*-butyrate

RN:	623-42-7	MP (°C):	−95
MW:	102.13	BP (°C):	102

Solubility (Moles/L)	Solubility (Grams/L)	Temp (°C)	Ref (#)	Evaluation (T P E A A)	Comments
1.528E-01	1.561E+01	21	F001	1 0 1 2 2	
1.506E-01	1.538E+01	21	F300	1 0 0 0 2	
1.600E-01	1.634E+01	21	S006	1 0 0 0 2	
1.469E-01	1.500E+01	25	A049	1 0 0 0 2	

494. $C_5H_{10}O_2$
3-Hydroxy-2-methyltetrahydrofuran
3-Furanol, tetrahydro-2-methyl-

RN:	29848-44-0	MP (°C):	
MW:	102.13	BP (°C):	

Solubility (Moles/L)	Solubility (Grams/L)	Temp (°C)	Ref (#)	Evaluation (T P E A A)	Comments
1.632E+00	1.667E+02	rt	B066	0 2 0 0 1	
4.896E+00	5.000E+02	rt	B066	0 2 0 0 2	

495. $C_5H_{10}O_2$
Propyl acetate
Essigsaeurepropyl ester

RN:	109-60-4	MP (°C):	−92
MW:	102.13	BP (°C):	101.6

Solubility (Moles/L)	Solubility (Grams/L)	Temp (°C)	Ref (#)	Evaluation (T P E A A)	Comments
2.222E-01	2.270E+01	20	E002	1 0 0 0 2	
1.850E-01	1.889E+01	20	F001	1 0 1 0 2	
1.821E-01	1.860E+01	20	F300	1 0 0 0 2	
1.800E-01	1.838E+01	20	M171	1 0 0 0 1	
2.220E-01	2.267E+01	21	S006	1 0 0 0 2	
1.920E-01	1.961E+01	25	B060	2 0 1 1 1	
1.731E-01	1.768E+01	30	R318	1 2 0 1 1	
1.960E-01	2.002E+01	37	E028	1 0 1 1 2	

496. $C_5H_{10}O_2$

Pivalic acid
Trimethylacetic acid
Trimethylessigsaeure

RN:	75-98-9	**MP** (°C):	35.5
MW:	102.13	**BP** (°C):	163.8

Solubility (Moles/L)	Solubility (Grams/L)	Temp (°C)	Ref (#)	Evaluation (T P E A A)	Comments
2.125E-01	2.170E+01	20	F300	1 0 0 0 2	

497. $C_5H_{10}O_2$

Isopropyl acetate
Essigsaeureisopropyl ester
Iso-propylacetat

RN:	108-21-4	**MP** (°C):	−73
MW:	102.13	**BP** (°C):	89

Solubility (Moles/L)	Solubility (Grams/L)	Temp (°C)	Ref (#)	Evaluation (T P E A A)	Comments
2.556E-01	2.610E+01	20	D052	1 1 0 0 2	average of 2
3.030E-01	3.095E+01	20	F001	1 0 1 2 2	
2.937E-01	3.000E+01	20	F300	1 0 0 0 2	
2.108E-01	2.153E+01	24.6	H121	2 0 0 0 1	
2.759E-01	2.818E+01	25	B060	2 0 1 1 1	
1.930E-01	1.971E+01	37	E028	1 0 1 1 2	

498. $C_5H_{10}O_2$

Butyl formate
Formic acid butyl ester

RN:	592-84-7	**MP** (°C):	
MW:	102.13	**BP** (°C):	106.5

Solubility (Moles/L)	Solubility (Grams/L)	Temp (°C)	Ref (#)	Evaluation (T P E A A)	Comments
9.800E-02	1.001E+01	22	S006	1 0 0 0 1	
6.400E-02	6.537E+00	25	K012	1 0 0 0 1	
7.400E-02	7.558E+00	27	B052	1 0 1 1 2	
7.500E-02	7.660E+00	30.5	N014	0 0 0 0 0	
8.100E-02	8.273E+00	40.0	N014	0 0 0 0 0	

499. C$_5$H$_{10}$O$_2$
Ethyl propionate
Propanoic acid ethyl ester

RN: 105-37-3	**MP** (°C):	−73
MW: 102.13	**BP** (°C):	99

Solubility (Moles/L)	Solubility (Grams/L)	Temp (°C)	Ref (#)	Evaluation (T P E A A)	Comments
1.844E-01	1.884E+01	20	D052	1 1 0 0 2	
2.200E-01	2.247E+01	20	S006	1 0 0 0 1	
2.154E-01	2.200E+01	25	F300	1 0 0 0 1	
1.700E-01	1.736E+01	25	K012	1 0 0 0 1	
2.108E-01	2.153E+01	30	R318	1 1 0 1 1	

500. C$_5$H$_{10}$O$_2$
Isovaleric acid
Isovaleriansaeure

RN: 503-74-2	**MP** (°C):	−29.3
MW: 102.13	**BP** (°C):	176.5

Solubility (Moles/L)	Solubility (Grams/L)	Temp (°C)	Ref (#)	Evaluation (T P E A A)	Comments
3.946E-01	4.031E+01	20	D041	1 0 0 0 1	
3.985E-01	4.070E+01	20	F300	1 0 0 0 2	

501. C$_5$H$_{10}$O$_3$
Methyl β-methoxypropionate
Propionic acid, 3-methoxy-, methyl ester
Methyl 3-methoxypropanoate
Methyl 3-methoxypropionate

RN: 3852-09-3	**MP** (°C):	
MW: 118.13	**BP** (°C):	

Solubility (Moles/L)	Solubility (Grams/L)	Temp (°C)	Ref (#)	Evaluation (T P E A A)	Comments
3.628E+00	4.286E+02	25	R034	1 0 0 0 1	

502. C$_5$H$_{10}$O$_3$
Ethyl carbonate
Diethyl carbonate

RN: 105-58-8	**MP** (°C):	−43
MW: 118.13	**BP** (°C):	126

Solubility (Moles/L)	Solubility (Grams/L)	Temp (°C)	Ref (#)	Evaluation (T P E A A)	Comments
1.562E-01	1.845E+01	20	D052	1 1 0 0 2	

503. C$_5$H$_{10}$O$_5$
D-Xylose
α-Xylose
Wood sugar

RN:	58-86-6	MP (°C):	144.5
MW:	150.13	BP (°C):	

Solubility (Moles/L)	Solubility (Grams/L)	Temp (°C)	Ref (#)	Evaluation (T P E A A)	Comments
2.879E+00	4.322E+02	25	G317	0 0 0 0 0	

504. C$_5$H$_{10}$O$_5$
L-Arabinose
L-Arabinopyranose

RN:	87-72-9	MP (°C):	158
MW:	150.13	BP (°C):	

Solubility (Moles/L)	Solubility (Grams/L)	Temp (°C)	Ref (#)	Evaluation (T P E A A)	Comments
2.482E+00	3.726E+02	10	F300	1 0 0 0 2	

505. C$_5$H$_{11}$Br
n-Amyl bromide
1-Bromopentane
Pentyl bromide
Amylene bromide

RN:	110-53-2	MP (°C):	−87.9
MW:	151.05	BP (°C):	129.6

Solubility (Moles/L)	Solubility (Grams/L)	Temp (°C)	Ref (#)	Evaluation (T P E A A)	Comments
8.380E-04	1.266E-01	25	M342	1 0 1 1 2	
1.800E-02	2.719E+00	ns	H307	0 0 0 0 0	

506. C$_5$H$_{11}$Br
Isoamyl bromide
1-Bromo-3-methylbutane

RN:	107-82-4	MP (°C):	−112
MW:	151.05	BP (°C):	120

Solubility (Moles/L)	Solubility (Grams/L)	Temp (°C)	Ref (#)	Evaluation (T P E A A)	Comments
1.324E-03	2.000E-01	16	F300	1 0 0 0 1	
1.300E-03	1.964E-01	16.5	F001	1 0 1 0 2	

507. C$_5$H$_{11}$NO
Pentanamide
Valeramide

RN: 626-97-1   **MP** (°C): 102–104
MW: 101.15   **BP** (°C):

Solubility (Moles/L)	Solubility (Grams/L)	Temp (°C)	Ref (#)	Evaluation (T P E A A)	Comments
5.530E-01	5.594E+01	6	H059	0 0 0 0 0	
6.360E-01	6.433E+01	16	H059	0 0 0 0 0	
7.880E-01	7.971E+01	25	H059	0 0 0 0 0	
1.108E+00	1.121E+02	37	H059	0 0 0 0 0	

508. C$_5$H$_{11}$NO$_2$
DL-Valine
DL-Valin

RN: 516-06-3   **MP** (°C): 296
MW: 117.15   **BP** (°C):

Solubility (Moles/L)	Solubility (Grams/L)	Temp (°C)	Ref (#)	Evaluation (T P E A A)	Comments
5.593E-01	6.552E+01	0	D018	2 2 2 1 2	
5.711E-01	6.690E+01	25	C018	0 0 0 0 0	
6.035E-01	7.070E+01	25	D016	1 0 0 0 2	
5.912E-01	6.926E+01	25	D018	2 2 2 1 2	
5.614E-01	6.577E+01	25	D041	1 0 0 0 2	
5.975E-01	7.000E+01	25	F300	1 0 0 0 0	
7.352E-01	8.612E+01	50	D018	2 2 2 1 2	
7.170E-01	8.400E+01	50	F300	1 0 0 0 1	
1.003E+00	1.175E+02	75	D018	2 2 2 1 2	
9.559E-01	1.120E+02	75	D041	1 0 0 0 2	
9.560E-01	1.120E+02	75	F300	1 0 0 0 2	
1.349E+00	1.580E+02	100	F300	1 0 0 0 2	
1.351E+00	1.583E+02	99.99	P349	0 0 0 0 0	

509. C$_5$H$_{11}$NO$_2$
L-Norvaline
L-(+)-2-Aminovaleric acid

RN: 6600-40-4   **MP** (°C): >300
MW: 117.15   **BP** (°C):

Solubility (Moles/L)	Solubility (Grams/L)	Temp (°C)	Ref (#)	Evaluation (T P E A A)	Comments
8.286E-01	9.707E+01	15	D041	1 0 0 0 2	

510. C$_5$H$_{11}$NO$_2$
tert-Butyl carbamate
O-t-Butyl carbamate
RN: 4248-19-5   **MP** (°C):   105
MW: 117.15   **BP** (°C):

Solubility (Moles/L)	Solubility (Grams/L)	Temp (°C)	Ref (#)	Evaluation (T P E A A)	Comments
1.250E+00	1.464E+02	37	H006	1 2 2 1 2	
1.259E+00	1.475E+02	ns	R424	0 0 0 0 0	

511. C$_5$H$_{11}$NO$_2$
n-Butyl carbamate
Butyl carbamate
RN: 592-35-8   **MP** (°C):   51
MW: 117.15   **BP** (°C):

Solubility (Moles/L)	Solubility (Grams/L)	Temp (°C)	Ref (#)	Evaluation (T P E A A)	Comments
2.200E-01	2.577E+01	37	H006	1 2 2 1 1	

512. C$_5$H$_{11}$NO$_2$
Isobutyl carbamate
iso-Butyl carbamate
RN: 543-28-2   **MP** (°C):   67
MW: 117.15   **BP** (°C):   206

Solubility (Moles/L)	Solubility (Grams/L)	Temp (°C)	Ref (#)	Evaluation (T P E A A)	Comments
5.000E-01	5.857E+01	37	H006	1 2 2 1 0	

513. C$_5$H$_{11}$NO$_2$
DL-Isovaline
DL-Isovalin
RN: 595-39-1   **MP** (°C):   315
MW: 117.15   **BP** (°C):

Solubility (Moles/L)	Solubility (Grams/L)	Temp (°C)	Ref (#)	Evaluation (T P E A A)	Comments
2.398E+00	2.809E+02	20	F300	1 0 0 0 2	

514. C$_5$H$_{11}$NO$_2$
D-Valine
β-Amino-isovalerian-saeure
β-Aminoisovaleric acid

RN: 640-68-6   **MP** (°C): >295
MW: 117.15   **BP** (°C):

Solubility (Moles/L)	Solubility (Grams/L)	Temp (°C)	Ref (#)	Evaluation (T P E A A)	Comments
1.291E-02	1.512E+00	10	D038	1 0 1 0 0	EFG, unit assumed, *sic*
4.296E-01	5.033E+01	20	D041	1 0 0 0 1	
7.053E-01	8.263E+01	25	C018	0 0 0 0 0	
1.343E-02	1.574E+00	25	D038	1 0 1 0 0	EFG, unit assumed, *sic*
1.384E-02	1.622E+00	33	D038	1 0 1 0 0	EFG, unit assumed, *sic*
1.426E-02	1.671E+00	40	D038	1 0 1 0 0	EFG, unit assumed, *sic*
1.455E-02	1.705E+00	49	D038	1 0 1 0 0	EFG, unit assumed, *sic*
1.500E-02	1.757E+00	57	D038	1 0 1 0 0	EFG, unit assumed, *sic*
1.592E-02	1.865E+00	65	D038	1 0 1 0 0	EFG, unit assumed, *sic*

515. C$_5$H$_{11}$NO$_2$
Betaine
Betain

RN: 107-43-7   **MP** (°C): 296
MW: 117.15   **BP** (°C):

Solubility (Moles/L)	Solubility (Grams/L)	Temp (°C)	Ref (#)	Evaluation (T P E A A)	Comments
5.216E+00	6.110E+02	19.30	F300	1 0 0 0 2	

516. C$_5$H$_{11}$NO$_2$
DL-Norvaline
DL-2-Aminovaleric acid

RN: 760-78-1   **MP** (°C): 303.0
MW: 117.15   **BP** (°C):

Solubility (Moles/L)	Solubility (Grams/L)	Temp (°C)	Ref (#)	Evaluation (T P E A A)	Comments
8.251E-01	9.666E+01	15	D041	1 0 0 0 2	
7.768E-01	9.100E+01	18	F300	1 0 0 0 1	
6.616E-01	7.751E+01	25	K031	2 1 2 1 2	

517. $C_5H_{11}NO_2$

L-Valine
Valine
L-(+)-valine
L-2-Amino-3-methylbutyric acid
2-Amino-3-methylbutyric acid

RN: 72-18-4   **MP** (°C):   315
MW: 117.15   **BP** (°C):

Solubility (Moles/L)	Solubility (Grams/L)	Temp (°C)	Ref (#)	Evaluation (T P E A A)	Comments
7.180E-01	8.411E+01	15	D349	2 1 1 2 2	
4.866E-01	5.701E+01	20	B032	1 2 2 1 2	
7.360E-01	8.622E+01	20	D349	2 1 1 2 2	
4.992E-01	5.848E+01	25	B032	1 2 2 1 2	
6.940E-01	8.130E+01	25	D041	1 0 0 0 2	
7.550E-01	8.845E+01	25	D349	2 1 1 2 2	
4.710E-01	5.518E+01	25	G092	2 1 1 1 1	
4.710E-01	5.518E+01	25	G315	0 0 0 0 0	
5.900E-01	6.912E+01	25	N001	0 0 0 0 0	EFG
4.740E-01	5.553E+01	25	N012	2 0 2 1 2	
5.019E-01	5.880E+01	27	D036	0 0 0 0 0	
5.114E-01	5.991E+01	29.80	B032	1 2 2 1 2	
7.929E-01	9.289E+01	65	D041	1 0 0 0 2	

518. $C_5H_{11}NO_2$

3-Nitropentane
Pentane, 3-nitro-

RN: 551-88-2   **MP** (°C):
MW: 117.15   **BP** (°C):   153

Solubility (Moles/L)	Solubility (Grams/L)	Temp (°C)	Ref (#)	Evaluation (T P E A A)	Comments
1.110E-02	1.300E+00	25	A049	1 0 0 0 1	

519. $C_5H_{11}NO_2S$

DL-Methionine
DL-Methionin
DL-2-Amino-4-(methylthio)butyric acid
Acimetion

RN: 59-51-8   **MP** (°C):   281
MW: 149.21   **BP** (°C):

Solubility (Moles/L)	Solubility (Grams/L)	Temp (°C)	Ref (#)	Evaluation (T P E A A)	Comments
1.200E-01	1.790E+01	0	F300	1 0 0 0 2	
1.905E-01	2.843E+01	19.99	F419	0 0 0 0 0	pH 5.81
2.191E-01	3.269E+01	25	D041	1 0 0 0 2	
2.191E-01	3.270E+01	25	F300	1 0 0 0 2	
3.039E-01	4.535E+01	39.99	F419	0 0 0 0 0	pH 5.56

(continued)

519. C$_5$H$_{11}$NO$_2$S (continued)

Solubility (Moles/L)	Solubility (Grams/L)	Temp (°C)	Ref (#)	Evaluation (T P E A A)	Comments
3.211E-01	4.791E+01	44.99	F419	0 0 0 0 0	pH 5.51
3.833E-01	5.720E+01	50	F300	1 0 0 0 2	
4.241E-01	6.328E+01	54.99	F419	0 0 0 0 0	pH 5.39
5.596E-01	8.350E+01	69.99	F419	0 0 0 0 0	pH 5.24
6.379E-01	9.519E+01	75	D041	1 0 0 0 2	
6.380E-01	9.520E+01	75	F300	1 0 0 0 2	
6.965E-01	1.039E+02	79.99	F419	0 0 0 0 0	pH 5.15
1.003E+00	1.497E+02	100	F300	1 0 0 0 2	
2.212E-01	3.300E+01	ns	K444	0 0 0 0 0	

520. C$_5$H$_{11}$NO$_2$S
Methionine
L-(−)-Methionine
2-Amino-4-(methylthio)butanoic acid

RN: 63-68-3   **MP (°C):** −279
MW: 149.21   **BP (°C):**

Solubility (Moles/L)	Solubility (Grams/L)	Temp (°C)	Ref (#)	Evaluation (T P E A A)	Comments
3.504E-01	5.228E+01	20	B032	1 2 2 1 2	
3.791E-01	5.656E+01	25	B032	1 2 2 1 2	
3.566E-01	5.321E+01	25	G315	0 0 0 0 0	
3.753E-01	5.600E+01	25.1	N024	0 0 0 0 0	
3.746E-01	5.590E+01	25.1	N026	0 0 0 0 0	
3.548E-01	5.294E+01	25.1	N027	1 1 2 2 2	
3.498E-01	5.220E+01	27	D036	0 0 0 0 0	
4.093E-01	6.107E+01	29.80	B032	1 2 2 1 2	

521. C$_5$H$_{11}$NO$_2$S
Penicillamine
3,3-Dimethyl-D-(−)-cysteine
D-3-Mercaptovaline
D-Penicillamine

RN: 52-67-5   **MP (°C):** 198.0
MW: 149.21   **BP (°C):**

Solubility (Moles/L)	Solubility (Grams/L)	Temp (°C)	Ref (#)	Evaluation (T P E A A)	Comments
6.702E-01	1.000E+02	20	C120	0 0 0 0 0	

522. C$_5$H$_{11}$NO$_2$.H$_2$O
Betaine (monohydrate)
Trimethylammonioacetate (monohydrate)
RN: 590-47-6   MP (°C):
MW: 135.16   BP (°C):

Solubility (Moles/L)	Solubility (Grams/L)	Temp (°C)	Ref (#)	Evaluation (T P E A A)	Comments
4.520E+00	6.109E+02	19	D041	1 0 0 0 2	

523. C$_5$H$_{12}$
Pentane
n-Pentane
RN: 109-66-0   MP (°C):   −130
MW: 72.15   BP (°C):

Solubility (Moles/L)	Solubility (Grams/L)	Temp (°C)	Ref (#)	Evaluation (T P E A A)	Comments
9.106E-04	6.570E-02	0	P003	2 2 2 2 2	
5.666E-04	4.088E-02	4.0	N004	1 1 2 2 2	
1.516E-04	1.094E-02	4.8	L007	2 1 1 2 2	
1.516E-04	1.094E-02	5.1	L007	2 0 1 1 2	
5.944E-04	4.289E-02	10.0	N004	1 1 2 2 2	
1.635E-04	1.180E-02	14.8	L007	2 1 1 2 2	
2.425E-04	1.750E-02	20	M337	2 1 2 2 2	
5.444E-04	3.928E-02	20.0	N004	1 1 2 2 2	
1.563E-04	1.128E-02	24.8	L007	2 1 1 2 2	
5.267E-04	3.800E-02	25	A049	1 0 0 0 1	
5.475E-04	3.950E-02	25	K119	1 0 0 0 2	
5.336E-04	3.850E-02	25	M001	2 1 2 2 2	
5.336E-04	3.850E-02	25	M002	2 1 2 2 2	
5.650E-04	4.077E-02	25	M342	1 0 1 1 2	
6.597E-04	4.760E-02	25	P003	2 2 2 2 2	
5.611E-04	4.048E-02	25.0	N004	1 1 2 2 2	
5.475E-04	3.950E-02	25.0	P051	2 1 1 2 2	
5.475E-04	3.950E-02	25.00	P007	2 1 2 2 2	
5.611E-04	4.048E-02	30.0	N004	1 1 2 2 2	
1.509E-04	1.089E-02	34.8	L007	2 1 1 2 2	
5.516E-04	3.980E-02	40.1	P051	2 1 1 2 2	
5.516E-04	3.980E-02	40.10	P007	2 1 2 2 2	
5.793E-04	4.180E-02	55.7	P051	2 1 1 2 2	
5.793E-04	4.180E-02	55.70	P007	2 1 2 2 2	
9.619E-04	6.940E-02	99.1	P051	2 1 1 2 2	
9.619E-04	6.940E-02	99.10	P007	2 1 2 2 2	
1.525E-03	1.100E-01	121.3	P051	2 1 1 2 2	
1.525E-03	1.100E-01	121.30	P007	2 1 2 2 2	
2.786E-03	2.010E-01	137.3	P051	2 1 1 2 2	
2.786E-03	2.010E-01	137.30	P007	2 1 2 2 2	
4.130E-03	2.980E-01	149.5	P051	2 1 1 2 2	
4.130E-03	2.980E-01	149.50	P007	2 1 2 2 2	
1.010E-04	7.287E-03	ns	D348	0 0 0 0 0	

524. C₅H₁₂

524. C_5H_{12}

2-Methylbutane
Isopentane
Izopentan

RN:	78-78-4	MP (°C):	−160
MW:	72.15	BP (°C):	30

Solubility (Moles/L)	Solubility (Grams/L)	Temp (°C)	Ref (#)	Evaluation (T P E A A)	Comments
1.003E-03	7.240E-02	0	P003	2 2 2 2 2	
6.653E-04	4.800E-02	25	K119	1 0 0 0 2	
6.625E-04	4.780E-02	25	M001	2 1 2 2 2	
6.625E-04	4.780E-02	25	M002	2 1 2 2 2	
6.874E-04	4.960E-02	25	P003	2 2 2 2 2	
6.653E-04	4.800E-02	25	P007	2 1 2 2 2	
6.653E-04	4.800E-02	25	P051	2 1 1 2 2	

525. C₅H₁₂

525. C_5H_{12}

Neopentane
2,2-Dimethylpropane

RN:	463-82-1	MP (°C):	
MW:	72.15	BP (°C):	9.5

Solubility (Moles/L)	Solubility (Grams/L)	Temp (°C)	Ref (#)	Evaluation (T P E A A)	Comments
2.220E-04	1.602E-02	25	D346	0 0 0 0 0	
4.601E-04	3.320E-02	25	M001	2 1 2 2 2	
5.611E-04	4.048E-02	25	S212	2 1 2 2 2	
3.833E-04	2.766E-02	40	S212	2 1 2 2 1	
2.667E-04	1.924E-02	60	S212	2 1 2 2 1	
2.389E-04	1.724E-02	80	S212	2 1 2 2 1	

526. C₅H₁₂ClO₂PS₂

526. $C_5H_{12}ClO_2PS_2$

Chlormephos
Dotan
Diethyl *S*-(chloromethyl) dithiophosphate

RN:	24934-91-6	MP (°C):	
MW:	234.70	BP (°C):	83

Solubility (Moles/L)	Solubility (Grams/L)	Temp (°C)	Ref (#)	Evaluation (T P E A A)	Comments
2.556E-04	6.000E-02	20	L303	1 0 0 0 1	
2.556E-04	6.000E-02	20	M161	1 0 0 0 1	
2.559E-04	6.005E-02	ns	S460	0 0 0 0 0	

527. C$_5$H$_{12}$NO$_3$PS$_2$
Dimethoate
O,O-Dimethyl *S*-(*N*-methylcarbamoylmethyl) dithiophosphate
RN: 60-51-5   **MP** (°C):   52.25
MW: 229.26   **BP** (°C):

Solubility (Moles/L)	Solubility (Grams/L)	Temp (°C)	Ref (#)	Evaluation (T P E A A)	Comments
1.096E-01	2.514E+01	20	B179	0 0 0 0 0	
1.309E-01	3.000E+01	20	G319	0 0 0 0 0	
1.090E-01	2.500E+01	21	M161	1 0 0 0 1	
1.701E-01	3.900E+01	ns	M061	0 0 0 0 1	

528. C$_5$H$_{12}$N$_2$
2-Methylpiperazine
2-Methyl-piperazin
RN: 109-07-9   **MP** (°C):   66
MW: 100.16   **BP** (°C):   155

Solubility (Moles/L)	Solubility (Grams/L)	Temp (°C)	Ref (#)	Evaluation (T P E A A)	Comments
4.343E+00	4.350E+02	20	F300	1 0 0 0 2	

529. C$_5$H$_{12}$N$_2$O
Methyl-*n*-butylnitrosamine
MBN
RN: 7068-83-9   **MP** (°C):
MW: 116.16   **BP** (°C):

Solubility (Moles/L)	Solubility (Grams/L)	Temp (°C)	Ref (#)	Evaluation (T P E A A)	Comments
2.000E-01	2.323E+01	24	M031	1 1 1 1 1	

530. C$_5$H$_{12}$O
2-Methyl-1-butanol
DL-2-Methyl-1-butanol
2-Methylbutan-1-ol
RN: 137-32-6   **MP** (°C):   −70
MW: 88.15   **BP** (°C):   128.0

Solubility (Moles/L)	Solubility (Grams/L)	Temp (°C)	Ref (#)	Evaluation (T P E A A)	Comments
4.269E-01	3.763E+01	.5	S307	1 1 0 2 2	
3.720E-01	3.279E+01	9.7	S307	1 1 0 2 2	
3.122E-01	2.752E+01	19.6	S307	1 1 0 2 2	
3.496E-01	3.082E+01	20	G004	2 2 2 2 2	
3.304E-01	2.913E+01	25	C093	2 1 1 1 1	
3.272E-01	2.884E+01	25	G004	2 2 2 2 2	
2.778E-01	2.449E+01	29.6	S307	1 1 0 2 2	
3.122E-01	2.752E+01	30	G004	2 2 2 2 2	

(continued)

530. C₅H₁₂O (continued)

Solubility (Moles/L)	Solubility (Grams/L)	Temp (°C)	Ref (#)	Evaluation (T P E A A)	Comments
2.616E-01	2.306E+01	39.3	S307	1 1 0 2 2	
2.453E-01	2.162E+01	49.6	S307	1 1 0 2 2	
2.301E-01	2.028E+01	59.3	S307	1 1 0 2 2	
2.485E-01	2.191E+01	69.5	S307	1 1 0 2 2	
2.551E-01	2.248E+01	79.7	S307	1 1 0 2 2	
2.724E-01	2.401E+01	90.8	S307	1 1 0 2 2	

531. C₅H₁₂O

tert-Isoamyl alcohol
3-Methyl-1-butanol
Isopentyl alcohol
Isoamyl alcohol

RN:	123-51-3	**MP** (°C):	−117
MW:	88.15	**BP** (°C):	130

Solubility (Moles/L)	Solubility (Grams/L)	Temp (°C)	Ref (#)	Evaluation (T P E A A)	Comments
4.079E-01	3.596E+01	0	S307	1 1 0 2 2	
3.090E-01	2.724E+01	10	A328	0 0 0 0 0	
3.454E-01	3.044E+01	10.1	S307	1 1 0 2 2	
3.347E-01	2.950E+01	15	K002	1 2 1 1 2	
3.130E-01	2.759E+01	18	F001	1 0 1 2 2	
2.918E-01	2.572E+01	19.8	S307	1 1 0 2 2	
3.120E-01	2.750E+01	20	F300	1 0 0 0 2	
3.144E-01	2.771E+01	20	G004	2 2 2 2 2	
3.111E-01	2.743E+01	20	K002	1 2 1 1 2	
9.586E-01	8.450E+01	20	K085	1 0 0 0 2	
2.659E-01	2.344E+01	25	A328	0 0 0 0 0	
3.411E-01	3.007E+01	25	C068	2 2 2 1 2	
2.982E-01	2.629E+01	25	C093	2 1 1 1 1	
3.251E-01	2.865E+01	25	F317	2 1 1 1 2	
2.950E-01	2.601E+01	25	G004	2 2 2 2 2	
2.950E-01	2.601E+01	25	K002	1 2 1 1 2	
2.799E-01	2.468E+01	30	G004	2 2 2 2 2	
2.832E-01	2.496E+01	30	K002	1 2 1 1 2	
2.842E-01	2.506E+01	30.1	H043	2 2 2 2 2	average of 3
2.540E-01	2.239E+01	30.2	S307	1 1 0 2 2	
2.442E-01	2.153E+01	40	A328	0 0 0 0 0	
2.420E-01	2.133E+01	40.0	S307	1 1 0 2 2	
2.257E-01	1.990E+01	49.9	S307	1 1 0 2 2	
2.431E-01	2.143E+01	59.8	S307	1 1 0 2 2	
2.344E-01	2.066E+01	70.0	S307	1 1 0 2 2	
2.442E-01	2.153E+01	80.0	S307	1 1 0 2 2	
2.518E-01	2.220E+01	90.0	S307	1 1 0 2 2	
2.836E-01	2.500E+01	ns	L003	0 0 2 1 2	
2.767E-01	2.439E+01	rt	H111	0 0 0 0 1	

532. C$_5$H$_{12}$O
Neopentyl alcohol
t-Butyl carbinol

RN: 75-84-3   **MP** (°C):   53
MW: 88.15   **BP** (°C):   114

Solubility (Moles/L)	Solubility (Grams/L)	Temp (°C)	Ref (#)	Evaluation (T P E A A)	Comments
4.048E-01	3.568E+01	12.0	S307	1 1 0 2 2	
3.826E-01	3.372E+01	18.8	S307	1 1 0 2 2	
4.090E-01	3.605E+01	20	G004	2 2 2 2 2	
3.836E-01	3.382E+01	25	G004	2 2 2 2 2	
3.603E-01	3.176E+01	30	G004	2 2 2 2 2	
3.229E-01	2.847E+01	30.0	S307	1 1 0 2 2	
2.982E-01	2.629E+01	40.0	S307	1 1 0 2 2	
2.616E-01	2.306E+01	50.0	S307	1 1 0 2 2	
2.778E-01	2.449E+01	60.0	S307	1 1 0 2 2	
2.399E-01	2.114E+01	70.2	S307	1 1 0 2 2	
2.864E-01	2.525E+01	80.0	S307	1 1 0 2 2	
2.637E-01	2.325E+01	90.0	S307	1 1 0 2 2	

533. C$_5$H$_{12}$O
Methyl *tert*-butyl ether
tert-Butyl methyl ether

RN: 1634-04-4   **MP** (°C):   −109
MW: 88.15   **BP** (°C):   54.5

Solubility (Moles/L)	Solubility (Grams/L)	Temp (°C)	Ref (#)	Evaluation (T P E A A)	Comments
6.564E-01	5.786E+01	2.34	S461	0 0 0 0 0	
6.236E-01	5.497E+01	9.99	S461	0 0 0 0 0	
5.196E-01	4.580E+01	20	E019	1 0 1 1 1	
4.738E-01	4.177E+01	24.99	S461	0 0 0 0 0	
5.815E-01	5.126E+01	25	K072	1 0 1 1 1	
5.815E-01	5.126E+01	25	M087	1 1 2 1 2	

534. C$_5$H$_{12}$O
3-Pentanol
Pentan-3-ol
Diethyl carbinol

RN: 584-02-1   **MP** (°C):   <25
MW: 88.15   **BP** (°C):   115.6

Solubility (Moles/L)	Solubility (Grams/L)	Temp (°C)	Ref (#)	Evaluation (T P E A A)	Comments
8.704E-01	7.672E+01	0	S307	1 1 0 2 2	
7.382E-01	6.507E+01	10.2	S307	1 1 0 2 2	
6.026E-01	5.312E+01	20	G004	2 2 2 2 2	

(*continued*)

534. C$_5$H$_{12}$O (continued)

Solubility (Moles/L)	Solubility (Grams/L)	Temp (°C)	Ref (#)	Evaluation (T P E A A)	Comments
6.280E-01	5.536E+01	20.0	S307	1 1 0 2 2	
5.505E-01	4.853E+01	25	C093	2 1 1 1 1	
5.556E-01	4.898E+01	25	G004	2 2 2 2 2	
5.144E-01	4.535E+01	30	G004	2 2 2 2 2	
5.730E-01	5.051E+01	30.0	S307	1 1 0 2 2	
4.510E-01	3.975E+01	40.0	S307	1 1 0 2 2	
4.604E-01	4.058E+01	50.0	S307	1 1 0 2 2	
3.889E-01	3.428E+01	60.0	S307	1 1 0 2 2	
3.783E-01	3.335E+01	70.0	S307	1 1 0 2 2	
3.635E-01	3.204E+01	80.0	S307	1 1 0 2 2	
3.773E-01	3.326E+01	90.0	S307	1 1 0 2 2	
1.392E+00	1.227E+02	ns	L003	0 0 2 1 1	
5.196E-01	4.580E+01	rt	H111	0 0 0 0 1	

535. C$_5$H$_{12}$O
3-Methyl-2-butanol
Methylisopropylcarbinol

RN:	598-75-4	**MP** (°C):	<25
MW:	88.15	**BP** (°C):	113

Solubility (Moles/L)	Solubility (Grams/L)	Temp (°C)	Ref (#)	Evaluation (T P E A A)	Comments
8.771E-01	7.732E+01	0	S307	1 1 0 2 2	
7.609E-01	6.708E+01	10.1	S307	1 1 0 2 2	
6.492E-01	5.723E+01	20	G004	2 2 2 2 2	
6.381E-01	5.625E+01	20.0	S307	1 1 0 2 2	
5.505E-01	4.853E+01	30	G004	2 2 2 2 2	
5.536E-01	4.880E+01	30.0	S307	1 1 0 2 2	
4.833E-01	4.260E+01	40.0	S307	1 1 0 2 2	
4.416E-01	3.892E+01	50.0	S307	1 1 0 2 2	
3.720E-01	3.279E+01	60.0	S307	1 1 0 2 2	
4.005E-01	3.531E+01	70.0	S307	1 1 0 2 2	
3.942E-01	3.475E+01	79.5	S307	1 1 0 2 2	
3.942E-01	3.475E+01	90.0	S307	1 1 0 2 2	

536. C$_5$H$_{12}$O
Ethylisopropyl ether
Propane, 2-ethoxy-

RN:	625-54-7	**MP** (°C):	
MW:	88.15	**BP** (°C):	

Solubility (Moles/L)	Solubility (Grams/L)	Temp (°C)	Ref (#)	Evaluation (T P E A A)	Comments
2.733E-01	2.409E+01	ns	J300	0 0 0 0 0	

537. $C_5H_{12}O$

1-Pentanol
Amyl alcohol
Pentanol
Pentyl alcohol
n-Amyl alcohol

RN:	71-41-0	**MP** (°C):	−79	
MW:	88.15	**BP** (°C):	138	

Solubility (Moles/L)	Solubility (Grams/L)	Temp (°C)	Ref (#)	Evaluation (T P E A A)	Comments
4.321E-01	3.809E+01	−.5	F051	2 1 0 1 2	
3.358E-01	2.960E+01	0	E029	1 2 0 1 2	
3.635E-01	3.204E+01	0	S307	1 1 0 2 2	
3.709E-01	3.269E+01	7	F051	2 1 0 1 2	
2.982E-01	2.629E+01	10	E029	1 2 0 1 2	
2.864E-01	2.525E+01	10.2	S307	1 1 0 2 2	
3.068E-01	2.705E+01	14	F051	2 1 0 1 2	
3.004E-01	2.648E+01	15	F051	2 1 0 1 2	
5.395E+00	4.756E+02	15.5	F051	2 1 0 1 2	
2.875E-01	2.534E+01	16.5	F051	2 1 0 1 2	
2.821E-01	2.487E+01	18	F051	2 1 0 1 2	
2.453E-01	2.162E+01	20	A015	1 2 1 1 2	
1.020E-02	8.992E-01	20	D052	1 1 0 0 0	*sic*
2.605E-01	2.296E+01	20	E029	1 2 0 1 2	
2.616E-01	2.306E+01	20	G004	2 2 2 2 2	
1.676E-01	1.478E+01	20	L049	1 1 2 1 1	
3.070E-01	2.706E+01	20	M312	1 0 0 0 1	
2.496E-01	2.200E+01	20.2	S307	1 1 0 2 2	
3.607E-01	3.180E+01	22	H072	1 0 1 1 2	
2.691E-01	2.372E+01	23	F051	2 1 0 1 2	
3.730E-01	3.288E+01	25	B019	1 0 1 2 0	
2.451E-01	2.160E+01	25	B038	1 0 1 1 2	
1.896E-01	1.672E+01	25	B060	2 0 1 1 1	
2.442E-01	2.153E+01	25	C093	2 1 1 1 1	
1.000E+00	8.815E+01	25	F044	1 0 0 0 0	EFG
2.137E-01	1.884E+01	25	F317	2 1 1 1 2	
2.431E-01	2.143E+01	25	G004	2 2 2 2 2	
2.300E-01	2.027E+01	25	G075	1 0 1 0 1	
2.810E-01	2.477E+01	25	H028	2 0 2 0 2	
2.817E-01	2.483E+01	25	H104	1 0 0 0 1	
2.500E-01	2.204E+01	25	K025	2 2 1 1 1	
2.561E-01	2.258E+01	29	F051	2 1 0 1 2	
2.333E-01	2.057E+01	30	E029	1 2 0 1 2	
2.257E-01	1.990E+01	30	G004	2 2 2 2 2	
2.246E-01	1.980E+01	30.6	S307	1 1 0 2 2	
5.368E+00	4.732E+02	34.0	F051	2 1 0 1 2	
2.475E-01	2.181E+01	36	F051	2 1 0 1 2	
2.130E-01	1.878E+01	37	E028	1 0 1 1 2	
2.115E-01	1.865E+01	40	E029	1 2 0 1 2	
2.082E-01	1.836E+01	40.2	S307	1 1 0 2 2	
2.006E-01	1.768E+01	50	E029	1 2 0 1 2	

(continued)

537. $C_5H_{12}O$ (continued)

Solubility (Moles/L)	Solubility (Grams/L)	Temp (°C)	Ref (#)	Evaluation (T P E A A)	Comments
2.039E-01	1.797E+01	50.0	S307	1 1 0 2 2	
2.475E-01	2.181E+01	58	F051	2 1 0 1 2	
2.006E-01	1.768E+01	60	E029	1 2 0 1 2	
2.039E-01	1.797E+01	60.3	S307	1 1 0 2 2	
5.290E+00	4.664E+02	69.5	F051	2 1 0 1 2	
2.061E-01	1.816E+01	70	E029	1 2 0 1 2	
2.170E-01	1.913E+01	70.0	S307	1 1 0 2 2	
2.561E-01	2.258E+01	72.0	F051	2 1 0 1 2	
2.115E-01	1.865E+01	80	E029	1 2 0 1 2	
2.213E-01	1.951E+01	80.0	S307	1 1 0 2 2	
2.691E-01	2.372E+01	81	F051	2 1 0 1 2	
2.821E-01	2.487E+01	87	F051	2 1 0 1 2	
2.224E-01	1.961E+01	90	E029	1 2 0 1 2	
2.453E-01	2.162E+01	90.7	S307	1 1 0 2 2	
2.875E-01	2.534E+01	91	F051	2 1 0 1 2	
3.004E-01	2.648E+01	95	F051	2 1 0 1 2	
5.180E+00	4.566E+02	97.3	F051	2 1 0 1 2	
3.068E-01	2.705E+01	98	F051	2 1 0 1 2	
2.496E-01	2.200E+01	100	E029	1 2 0 1 2	
2.875E-01	2.534E+01	110	E029	1 2 0 1 2	
3.709E-01	3.269E+01	112	F051	2 1 0 1 2	
3.304E-01	2.913E+01	120	E029	1 2 0 1 2	
5.048E+00	4.450E+02	122.3	F051	2 1 0 1 2	
4.321E-01	3.809E+01	126	F051	2 1 0 1 2	
3.889E-01	3.428E+01	130	E029	1 2 0 1 2	
4.677E-01	4.123E+01	140	E029	1 2 0 1 2	
5.351E-01	4.717E+01	140	F051	2 1 0 1 2	
4.896E+00	4.316E+02	141.6	F051	2 1 0 1 2	
5.853E-01	5.159E+01	145	F051	2 1 0 1 2	
6.290E-01	5.545E+01	148.5	F051	2 1 0 1 2	
5.761E-01	5.078E+01	150	E029	1 2 0 1 2	
4.707E+00	4.149E+02	157.3	F051	2 1 0 1 2	
7.322E-01	6.455E+01	160	E029	1 2 0 1 2	
9.060E-01	7.987E+01	167.0	F051	2 1 0 1 2	
9.889E-01	8.717E+01	170	E029	1 2 0 1 2	
1.001E+00	8.826E+01	171.2	F051	2 1 0 1 2	
4.374E+00	3.856E+02	174.0	F051	2 1 0 1 2	
1.690E+00	1.489E+02	180	E029	1 2 0 1 2	
4.089E+00	3.605E+02	181.3	F051	2 1 0 1 2	
1.435E+00	1.265E+02	182.5	F051	2 1 0 1 2	
3.774E+00	3.327E+02	185.2	F051	2 1 0 1 2	
1.833E+00	1.616E+02	186.0	F051	2 1 0 1 2	
2.270E+00	2.001E+02	186.5	F051	2 1 0 1 2	
3.472E+00	3.061E+02	186.5	F051	2 1 0 1 2	
3.237E+00	2.854E+02	187.4	F051	2 1 0 1 2	
3.040E+00	2.680E+02	187.5	F051	2 1 0 1 2	
2.810E-01	2.477E+01	ns	A406	0 0 0 0 1	
2.538E-01	2.237E+01	ns	L003	0 0 2 1 2	
2.224E-01	1.961E+01	rt	H111	0 0 0 0 1	

538. C$_5$H$_{12}$O

2-Pentanol
iso-Amyl alcohol
sec-Amyl alcohol
Methyl propyl carbinol

RN: 6032-29-7   **MP** (°C):   −50
MW: 88.15   **BP** (°C):   119.3

Solubility (Moles/L)	Solubility (Grams/L)	Temp (°C)	Ref (#)	Evaluation (T P E A A)	Comments
7.708E-01	6.795E+01	0	S307	1 1 0 2 2	
6.189E-01	5.455E+01	10.1	S307	1 1 0 2 2	
5.030E-01	4.434E+01	19.5	S307	1 1 0 2 2	
4.573E-01	4.031E+01	20	C042	0 0 0 0 0	
1.473E-02	1.298E+00	20	D052	1 1 0 0 0	*sic*
4.538E-01	4.000E+01	20	F300	1 0 0 0 1	
5.258E-01	4.635E+01	20	G004	2 2 2 2 2	
3.836E-01	3.382E+01	25	B019	1 0 1 2 0	
4.843E-01	4.270E+01	25	G004	2 2 2 2 2	
4.499E-01	3.966E+01	30	G004	2 2 2 2 2	
4.300E-01	3.791E+01	30.6	S307	1 1 0 2 2	
3.900E-01	3.438E+01	40.0	S307	1 1 0 2 2	
3.645E-01	3.213E+01	50.0	S307	1 1 0 2 2	
3.432E-01	3.026E+01	60.0	S307	1 1 0 2 2	
3.379E-01	2.979E+01	70.1	S307	1 1 0 2 2	
3.443E-01	3.035E+01	79.9	S307	1 1 0 2 2	
3.368E-01	2.969E+01	90.3	S307	1 1 0 2 2	
5.149E-01	4.539E+01	ns	L003	0 0 2 1 2	

539. C$_5$H$_{12}$O

tert-Pentyl alcohol
Dimethylethylcarbinol
tert-Amylalkohol

RN: 75-85-4   **MP** (°C):
MW: 88.15   **BP** (°C):   102.5

Solubility (Moles/L)	Solubility (Grams/L)	Temp (°C)	Ref (#)	Evaluation (T P E A A)	Comments
1.548E+00	1.364E+02	.5	S307	1 1 0 2 2	
1.462E+00	1.289E+02	9.8	S307	1 1 0 2 2	
1.259E+00	1.110E+02	20	F300	1 0 0 0 2	
1.229E+00	1.083E+02	20	G004	2 2 2 2 2	
1.170E+00	1.031E+02	20.8	S307	1 1 0 2 2	
1.124E+00	9.910E+01	25	G004	2 2 2 2 2	
5.965E-01	5.258E+01	25	G004	2 2 2 2 2	
1.026E+00	9.041E+01	29.5	S307	1 1 0 2 2	
1.041E+00	9.173E+01	30	G004	2 2 2 2 2	
8.549E-01	7.536E+01	39.5	S307	1 1 0 2 2	
7.649E-01	6.743E+01	49.0	S307	1 1 0 2 2	
6.673E-01	5.882E+01	60.0	S307	1 1 0 2 2	
6.391E-01	5.634E+01	70.2	S307	1 1 0 2 2	
6.117E-01	5.393E+01	80.1	S307	1 1 0 2 2	
5.883E-01	5.186E+01	90.2	S307	1 1 0 2 2	
1.124E+00	9.910E+01	rt	H111	0 0 0 0 2	

540. $C_5H_{12}O_2$

Formaldehyde diethyl acetal
Diethoxymethane
Diethylacetalformaldehyde
Formaldehyd-diaethyl-acetal

RN: 462-95-3	**MP** (°C):	
MW: 104.15	**BP** (°C):	87.5

Solubility (Moles/L)	Solubility (Grams/L)	Temp (°C)	Ref (#)	Evaluation (T P E A A)	Comments
6.721E-01	7.000E+01	18	F300	1 0 0 0 1	

541. $C_5H_{12}O_4$

Pentaerythritol
2,2-bis(Hydroxymethyl)-1,3-propanediol
PE 200
Tetramethylolmethane

RN: 115-77-5	**MP** (°C):	260
MW: 136.15	**BP** (°C):	

Solubility (Moles/L)	Solubility (Grams/L)	Temp (°C)	Ref (#)	Evaluation (T P E A A)	Comments
2.825E-01	3.846E+01	0	M043	1 0 0 0 0	
3.498E-01	4.762E+01	10	M043	1 0 0 0 0	
3.863E-01	5.260E+01	15	F300	1 0 0 0 2	
4.157E-01	5.660E+01	20	M043	1 0 0 0 0	
5.441E-01	7.407E+01	30	M043	1 0 0 0 0	
8.450E-01	1.150E+02	40	M043	1 0 0 0 1	
1.324E+00	1.803E+02	60	M043	1 0 0 0 1	
2.099E+00	2.857E+02	80	M043	1 0 0 0 1	
3.672E+00	5.000E+02	100	M043	1 0 0 0 2	
3.890E-01	5.297E+01	ns	R424	0 0 0 0 0	

542. $C_5H_{12}O_5$

Adonitol
Adonit
Adonite

RN: 488-81-3	**MP** (°C):	104
MW: 152.15	**BP** (°C):	

Solubility (Moles/L)	Solubility (Grams/L)	Temp (°C)	Ref (#)	Evaluation (T P E A A)	Comments
3.954E+00	6.016E+02	25	C346	0 0 0 0 0	

543. C₅H₁₂O₅
Xylitol

RN:	87-99-0	**MP** (°C):	96 K
MW:	152.15	**BP** (°C):	

Solubility (Moles/L)	Solubility (Grams/L)	Temp (°C)	Ref (#)	Evaluation (T P E A A)	Comments
3.798E+00	5.778E+02	20.12	W414	0 0 0 0 0	
3.963E+00	6.030E+02	25.1	W414	0 0 0 0 0	
4.153E+00	6.319E+02	30.01	W414	0 0 0 0 0	
4.355E+00	6.627E+02	35.05	W414	0 0 0 0 0	
4.550E+00	6.922E+02	40.13	W414	0 0 0 0 0	
4.721E+00	7.183E+02	45.10	W414	0 0 0 0 0	
4.873E+00	7.414E+02	50.09	W414	0 0 0 0 0	
5.001E+00	7.610E+02	55.05	W414	0 0 0 0 0	

544. C₅H₁₂O₅
DL-Arabinitol
(±)-Arabitol

RN:	2152-56-9	**MP** (°C):	103
MW:	152.15	**BP** (°C):	

Solubility (Moles/L)	Solubility (Grams/L)	Temp (°C)	Ref (#)	Evaluation (T P E A A)	Comments
4.459E+00	6.785E+02	25	C346	0 0 0 0 0	

545. C₅H₁₃N
N-Methyldiethylamine
N,N-Diethylmethylamine

RN:	616-39-7	**MP** (°C):	
MW:	87.17	**BP** (°C):	63

Solubility (Moles/L)	Solubility (Grams/L)	Temp (°C)	Ref (#)	Evaluation (T P E A A)	Comments
3.562E+00	3.105E+02	49.40	C086	2 2 2 2 2	average of 5
4.453E+00	3.881E+02	49.50	C086	2 2 2 2 2	
2.236E+00	1.949E+02	49.80	C086	2 2 2 2 2	
5.715E+00	4.982E+02	50.50	C086	2 2 2 2 2	
1.581E+00	1.378E+02	51.20	C086	2 2 2 2 2	
1.413E+00	1.231E+02	52.00	C086	2 2 2 2 2	
6.981E+00	6.085E+02	53.10	C086	2 2 2 2 2	
7.246E+00	6.316E+02	54.00	C086	2 2 2 2 2	

546. C₅H₁₃O₃PS₂

Demephion

O,O-Dimethyl 2-methylmercaptoethyl thiophosphate

Thiolo-tinox

RN:	8065-62-1	MP (°C):	
MW:	216.26	BP (°C):	109

Solubility (Moles/L)	Solubility (Grams/L)	Temp (°C)	Ref (#)	Evaluation (T P E A A)	Comments
2.312E-03	5.000E-01	20	M061	1 0 0 0 2	form II
9.248E-03	2.000E+00	ns	M061	0 0 0 0 2	form I
1.387E-02	3.000E+00	rt	M161	0 0 0 0 0	form II
1.387E-03	3.000E-01	rt	M161	0 0 0 0 2	form I

547. C₅Cl₆

Hexachlorocyclopentadiene

1,2,3,4,5,5-Hexachloro-1,3-cyclopentadiene

Hexachloro-1,3-cyclopentadiene

1,2,3,4,5,5-Hexachlorocyclopentadiene

RN:	77-47-4	MP (°C):	–9.9
MW:	272.77	BP (°C):	239

Solubility (Moles/L)	Solubility (Grams/L)	Temp (°C)	Ref (#)	Evaluation (T P E A A)	Comments
2.951E-06	8.050E-04	22.5	G301	0 0 0 0 0	

548. C₆HCl₃N₂S

4,5,7-Trichloro-2,1,3-benzothiadiazole

PH 40-21

TH 052 H

RN:	1982-55-4	MP (°C):	131.5
MW:	239.51	BP (°C):	

Solubility (Moles/L)	Solubility (Grams/L)	Temp (°C)	Ref (#)	Evaluation (T P E A A)	Comments
6.263E-06	1.500E-03	10	B200	1 0 0 0 1	
1.044E-05	2.500E-03	20	B200	1 0 0 0 1	
1.044E-05	2.500E-03	20	M061	1 0 0 0 1	
1.795E-05	4.300E-03	30	B200	1 0 0 0 1	

549. C₆HCl₄NO₂

2,3,4,5-Tetrachloronitrobenzene

1,2,3,4-Tetrachloro-5-nitrobenzene

2,3,4,5-Tetrachloro-1-nitrobenzene

1-Nitro-2,3,4,5-tetrachlorobenzene

RN:	879-39-0	MP (°C):	66.0
MW:	260.89	BP (°C):	

Solubility (Moles/L)	Solubility (Grams/L)	Temp (°C)	Ref (#)	Evaluation (T P E A A)	Comments
2.800E-05	7.305E-03	20	E308	1 2 2 1 1	

550. C$_6$HCl$_4$NO$_2$
2,3,4,6-Tetrachloronitrobenzene
Benzene, 1,2,3,5-tetrachloro-4-nitro-
RN: 3714-62-3   **MP** (°C):
MW: 260.89   **BP** (°C):

Solubility (Moles/L)	Solubility (Grams/L)	Temp (°C)	Ref (#)	Evaluation (T P E A A)	Comments
2.900E-05	7.566E-03	20	E308	1 2 2 1 1	

551. C$_6$HCl$_4$NO$_2$
2,3,5,6-Tetrachloronitrobenzene
Tecnazene
RN: 117-18-0   **MP** (°C):   99.5
MW: 260.89   **BP** (°C):   304.0

Solubility (Moles/L)	Solubility (Grams/L)	Temp (°C)	Ref (#)	Evaluation (T P E A A)	Comments
8.000E-06	2.087E-03	20	E308	1 2 2 1 0	

552. C$_6$HCl$_5$
Pentachlorobenzene
Penta-chlorobenzene
RN: 608-93-5   **MP** (°C):   82
MW: 250.34   **BP** (°C):   275

Solubility (Moles/L)	Solubility (Grams/L)	Temp (°C)	Ref (#)	Evaluation (T P E A A)	Comments
1.000E-06	2.503E-04	20	K337	1 0 0 0 2	
9.550E-07	2.391E-04	22	K305	1 0 1 1 0	
1.538E-06	3.850E-04	23	C305	1 1 2 2 2	
5.320E-06	1.332E-03	25	B173	2 0 2 2 2	
2.600E-06	6.509E-04	25	B317	0 0 0 0 0	
3.320E-06	8.311E-04	25	M342	1 0 1 1 2	
3.320E-06	8.311E-04	ns	M308	0 0 1 1 2	

553. C$_6$HCl$_5$O
Pentachlorophenol
PCP
2,3,4,5,6-Pentachloro-phenol-
Phenol, 2,3,4,5,6-pentachloro-
Dowicide 7
Fungifen
RN: 87-86-5   **MP** (°C):   174
MW: 266.34   **BP** (°C):   310

Solubility (Moles/L)	Solubility (Grams/L)	Temp (°C)	Ref (#)	Evaluation (T P E A A)	Comments
1.877E-05	5.000E-03	0	C310	0 0 0 0 0	
1.877E-05	5.000E-03	0	G310	1 0 0 0 0	
1.877E-05	5.000E-03	0	M061	1 0 0 0 0	

(continued)

553. C$_6$HCl$_5$O (continued)

Solubility (Moles/L)	Solubility (Grams/L)	Temp (°C)	Ref (#)	Evaluation (T P E A A)	Comments
5.256E-05	1.400E-02	20	B185	0 0 0 0 0	
5.256E-05	1.400E-02	22.5	G301	0 0 0 0 0	
6.195E-05	1.650E-02	25	B183	0 0 0 0 1	
8.260E-05	2.200E-02	25	B185	0 0 0 0 0	
3.600E-05	9.588E-03	25	B316	0 0 0 0 0	
6.908E-05	1.840E-02	25	M373	1 0 2 1 2	
5.256E-05	1.400E-02	25	O320	0 0 0 0 0	
8.035E-05	2.140E-02	25.1	A400	2 1 2 2 2	
5.256E-05	1.400E-02	26.70	L095	2 2 1 1 2	
6.758E-05	1.800E-02	27	C310	0 0 0 0 0	
6.758E-05	1.800E-02	27	G310	1 0 0 0 1	
6.758E-05	1.800E-02	27	M061	1 0 0 0 1	
3.484E-03	9.280E-01	30	A400	2 1 2 2 2	
7.509E-05	2.000E-02	30	M161	1 0 0 0 1	
1.126E-04	3.000E-02	50	B200	1 0 0 0 0	
1.314E-04	3.500E-02	50	C310	0 0 0 0 0	
1.314E-04	3.500E-02	50	G310	1 0 0 0 1	
1.314E-04	3.500E-02	50	M061	1 0 0 0 1	
2.178E-04	5.800E-02	62	C310	0 0 0 0 0	
2.178E-04	5.800E-02	62	G310	1 0 0 0 1	
3.191E-04	8.499E-02	70	C310	0 0 0 0 0	
3.191E-04	8.499E-02	70	G310	1 0 0 0 1	
7.509E-05	2.000E-02	ns	L311	0 0 0 0 1	
7.134E-05	1.900E-02	ns	M110	0 0 0 0 0	EFG
6.007E-06	1.600E-03	ns	N013	0 0 0 0 1	

554. C$_6$HF$_5$O
Pentafluorophenol
PFP
RN: 771-61-9   **MP (°C):** 34–36
MW: 184.07   **BP (°C):** 143

Solubility (Moles/L)	Solubility (Grams/L)	Temp (°C)	Ref (#)	Evaluation (T P E A A)	Comments
3.000E-01	5.522E+01	25	P031	0 0 0 0 0	

555. C$_6$H$_2$Br$_2$ClNO$_2$
2,6-Dibromoquinone-3-chlorimide
2,6-Dibromoquinonechloroimide
RN: **MP (°C):**
MW: 315.36   **BP (°C):**

Solubility (Moles/L)	Solubility (Grams/L)	Temp (°C)	Ref (#)	Evaluation (T P E A A)	Comments
2.000E-04	6.307E-02	20	G043	1 0 1 1 0	

556. $C_6H_2Br_4$

1,2,4,5-Tetrabromobenzene

RN: 636-28-2   **MP** (°C):
MW: 393.72   **BP** (°C):

Solubility (Moles/L)	Solubility (Grams/L)	Temp (°C)	Ref (#)	Evaluation (T P E A A)	Comments
4.724E-08	1.860E-05	10	K440	0 0 0 0 0	
1.105E-07	4.350E-05	25	K440	0 0 0 0 0	
1.976E-07	7.780E-05	35	K440	0 0 0 0 0	

557. $C_6H_2ClN_3O_6$

2,4,6-Trinitro-1-chlorobenzene
Picryl chloride
2-Chlor-1,3,5-trinitrobenzol
Chlorure de picryle

RN: 88-88-0   **MP** (°C):
MW: 247.55   **BP** (°C):

Solubility (Moles/L)	Solubility (Grams/L)	Temp (°C)	Ref (#)	Evaluation (T P E A A)	Comments
7.190E-04	1.780E-01	15	D066	1 2 0 0 2	
7.189E-04	1.780E-01	15	D071	1 2 0 0 2	
7.271E-04	1.800E-01	15	F300	1 0 0 0 1	
2.141E-03	5.300E-01	16	D066	1 2 0 0 2	
2.140E-03	5.297E-01	50	D071	1 2 0 0 1	
1.398E-02	3.460E+00	100	D066	1 2 0 0 2	
1.393E-02	3.448E+00	100	D071	1 2 0 0 2	
1.454E-02	3.600E+00	100	F300	1 0 0 0 1	

558. $C_6H_2Cl_2O_4$

Chloranilic acid
Chloranilsaeure

RN: 87-88-7   **MP** (°C):   283
MW: 208.99   **BP** (°C):

Solubility (Moles/L)	Solubility (Grams/L)	Temp (°C)	Ref (#)	Evaluation (T P E A A)	Comments
9.091E-03	1.900E+00	14	F300	1 0 0 0 1	
6.699E-02	1.400E+01	99	F300	1 0 0 0 1	

559. $C_6H_2Cl_3NO_2$

2,4,5-Trichloronitrobenzene
1,2,4-Trichloro-5-nitrobenzene
2,4,5-Trichloro-1-nitrobenzene
1,4,5-Trichloro-2-nitrobenzene
3,4,6-Trichloronitrobenzene

RN: 89-69-0   **MP** (°C):   57
MW: 226.45   **BP** (°C):   288

Solubility (Moles/L)	Solubility (Grams/L)	Temp (°C)	Ref (#)	Evaluation (T P E A A)	Comments
1.300E-04	2.944E-02	20	E308	1 2 2 1 2	

560. $C_6H_2Cl_3NO_2$
2,3,4-Trichloronitrobenzene
1,2,3-Trichloro-4-nitrobenzene
2,3,4-Trichloro-1-nitrobenzene

RN: 17700-09-3	**MP** (°C): 55.5		
MW: 226.45	**BP** (°C):		

Solubility (Moles/L)	Solubility (Grams/L)	Temp (°C)	Ref (#)	Evaluation (T P E A A)	Comments
1.150E-04	2.604E-02	20	E308	1 2 2 1 2	

561. $C_6H_2Cl_4$
1,2,4,5-Tetrachlorobenzene
s-Tetrachlorobenzene

RN: 95-94-3	**MP** (°C): 139		
MW: 215.89	**BP** (°C): 243		

Solubility (Moles/L)	Solubility (Grams/L)	Temp (°C)	Ref (#)	Evaluation (T P E A A)	Comments
1.445E-06	3.121E-04	20	K337	1 0 0 0 2	
1.349E-06	2.912E-04	22	K305	1 0 1 1 1	
2.154E-06	4.650E-04	25	B304	2 0 2 2 2	
5.900E-06	1.274E-03	25	B317	0 0 0 0 0	
1.090E-05	2.353E-03	25	M342	1 0 1 1 2	
1.600E-06	3.454E-04	25.2	T428	0 0 0 0 0	
1.806E-06	3.900E-04	ns	B393	0 0 0 0 0	
1.090E-05	2.353E-03	ns	M308	0 0 1 1 2	

562. $C_6H_2Cl_4$
Trichlorobenzyl chloride
TCBC

RN: 1344-32-7	**MP** (°C):		
MW: 215.89	**BP** (°C): 93		

Solubility (Moles/L)	Solubility (Grams/L)	Temp (°C)	Ref (#)	Evaluation (T P E A A)	Comments
9.264E-06	2.000E-03	25	B200	1 0 0 0 0	

563. $C_6H_2Cl_4$
1,2,3,4-Tetrachlorobenzene
Benzene, 1,2,3,4-tetrachloro-

RN: 634-66-2	**MP** (°C): 48		
MW: 215.89	**BP** (°C): 254		

Solubility (Moles/L)	Solubility (Grams/L)	Temp (°C)	Ref (#)	Evaluation (T P E A A)	Comments
1.585E-05	3.422E-03	20	K337	1 0 0 0 2	
3.326E-05	7.180E-03	23	C305	1 1 2 2 2	
2.742E-05	5.920E-03	25	B304	2 0 2 2 2	
3.600E-05	7.772E-03	25	B317	0 0 0 0 0	
5.650E-05	1.220E-02	25	M342	1 0 1 1 2	
5.650E-05	1.220E-02	ns	M308	0 0 1 1 2	

564. C$_6$H$_2$Cl$_4$
1,2,3,5-Tetrachlorobenzene
1,2,4,6-Tetrachlorobenzene

RN:	634-90-2	MP (°C):	50
MW:	215.89	BP (°C):	246

Solubility (Moles/L)	Solubility (Grams/L)	Temp (°C)	Ref (#)	Evaluation (T P E A A)	Comments
1.000E-05	2.159E-03	20	K337	1 0 0 0 2	
1.148E-05	2.479E-03	22	K305	1 0 1 1 2	
1.496E-05	3.230E-03	23	C305	1 1 2 2 2	
1.860E-05	4.016E-03	25	B173	2 0 2 2 2	
2.362E-05	5.100E-03	25	B304	2 0 2 2 2	
1.660E-05	3.584E-03	25	B317	0 0 0 0 0	
1.340E-05	2.893E-03	25	M342	1 0 1 1 2	
1.654E-05	3.570E-03	ns	H123	0 0 0 0 0	
1.340E-05	2.893E-03	ns	M308	0 0 1 1 2	

565. C$_6$H$_2$Cl$_4$O
2,3,4,6-Tetrachlorophenol
Phenol, 2,3,4,6-tetrachloro-
1-Hydroxy-2,3,4,6-tetrachlorobenzene
TCP

RN:	58-90-2	MP (°C):	
MW:	231.89	BP (°C):	

Solubility (Moles/L)	Solubility (Grams/L)	Temp (°C)	Ref (#)	Evaluation (T P E A A)	Comments
7.900E-04	1.832E-01	25	B316	0 0 0 0 0	

566. C$_6$H$_2$Cl$_4$O
2,3,4,5-Tetrachlorophenol
Phenol, 2,3,4,5-tetrachloro-

RN:	4901-51-3	MP (°C):	116
MW:	231.89	BP (°C):	

Solubility (Moles/L)	Solubility (Grams/L)	Temp (°C)	Ref (#)	Evaluation (T P E A A)	Comments
7.158E-04	1.660E-01	25	M373	1 0 2 1 2	

567. C$_6$H$_2$Cl$_4$O
2,3,5,6-Tetrachlorophenol
Phenol, 2,3,5,6-tetrachloro-

RN:	935-95-5	MP (°C):	115
MW:	231.89	BP (°C):	

Solubility (Moles/L)	Solubility (Grams/L)	Temp (°C)	Ref (#)	Evaluation (T P E A A)	Comments
4.312E-04	1.000E-01	25	M373	1 0 2 1 2	

568. $C_6H_2Cl_4O_2$

Tetrachlorohydroquinone
2,3,5,6-Tetrachlorohydroquinone

RN: 87-87-6   **MP** (°C):
MW: 247.89   **BP** (°C):

Solubility (Moles/L)	Solubility (Grams/L)	Temp (°C)	Ref (#)	Evaluation (T P E A A)	Comments
8.673E-05	2.150E-02	ns	L311	0 0 0 0 1	

569. $C_6H_2F_4$

1,2,4,5-Tetrafluorobenzene
2,3,5,6-Tetrafluorobenzene
p-Tetrafluorobenzene

RN: 327-54-8   **MP** (°C): 4.5
MW: 150.08   **BP** (°C): 89.5

Solubility (Moles/L)	Solubility (Grams/L)	Temp (°C)	Ref (#)	Evaluation (T P E A A)	Comments
4.215E-03	6.326E-01	25	B349	2 0 2 0 2	

570. $C_6H_2F_4$

1,2,3,5-Tetrafluorobenzene
1,2,4,6-Tetrafluorobenzene
m-Tetrafluorobenzene
1,3,4,5-Tetrafluorobenzene

RN: 2367-82-0   **MP** (°C): −48
MW: 150.08   **BP** (°C): 83

Solubility (Moles/L)	Solubility (Grams/L)	Temp (°C)	Ref (#)	Evaluation (T P E A A)	Comments
4.952E-03	7.431E-01	25	B349	2 0 2 0 2	

571. $C_6H_2F_4O$

2,3,5,6-Tetrafluorophenol
1,2,4,5-Tetrafluoro-3-hydroxybenzene

RN: 769-39-1   **MP** (°C): 38
MW: 166.08   **BP** (°C): 140

Solubility (Moles/L)	Solubility (Grams/L)	Temp (°C)	Ref (#)	Evaluation (T P E A A)	Comments
3.700E-01	6.145E+01	25	P031	0 0 0 0 0	

572. C$_6$H$_3$Br$_2$NO$_2$
2,6-Dibromoquinone oxime

RN: **MP** (°C):
MW: 280.91   **BP** (°C):

Solubility (Moles/L)	Solubility (Grams/L)	Temp (°C)	Ref (#)	Evaluation (T P E A A)	Comments
8.500E-04	2.388E-01	20	G066	1 0 0 0 1	

573. C$_6$H$_3$Br$_3$
1,2,4-Tribromobenzene
Tribromobenzene, 1,2,4-

RN: 615-54-3   **MP** (°C):   43
MW: 314.82   **BP** (°C):

Solubility (Moles/L)	Solubility (Grams/L)	Temp (°C)	Ref (#)	Evaluation (T P E A A)	Comments
1.166E-05	3.670E-03	10	K440	0 0 0 0 0	
2.290E-05	7.210E-03	25	K440	0 0 0 0 0	
3.494E-05	1.100E-02	35	K440	0 0 0 0 0	

574. C$_6$H$_3$Br$_3$O
2,4,6-Tribromobiphenyl
1,1'-Biphenyl, 2,4,6-tribromo-

RN: 59080-33-0   **MP** (°C):   66
MW: 330.82   **BP** (°C):

Solubility (Moles/L)	Solubility (Grams/L)	Temp (°C)	Ref (#)	Evaluation (T P E A A)	Comments
4.111E-02	1.360E+01	26.5	G312	0 0 0 0 0	

575. C$_6$H$_3$Br$_3$O
2,4,6-Tribromophenol
2,4,6-Tribrom-phenol
Tribromophenol
Bromol

RN: 118-79-6   **MP** (°C):   95
MW: 330.82   **BP** (°C):   244

Solubility (Moles/L)	Solubility (Grams/L)	Temp (°C)	Ref (#)	Evaluation (T P E A A)	Comments
2.116E-04	7.000E-02	15	F300	1 0 0 0 1	
2.300E-04	7.609E-02	ns	O310	0 0 0 0 1	

576. $C_6H_3ClN_2O_4$
1-Chloro-2,4-dinitrobenzene
2,4-Dinitro-1-chlorobenzene
4-Chlor-1,3-dinitrobenzol
4-Chloro-1,3-dinitrobenzene

RN:	97-00-7	**MP** (°C):	53		
MW:	202.55	**BP** (°C):	315		

Solubility (Moles/L)	Solubility (Grams/L)	Temp (°C)	Ref (#)	Evaluation (T P E A A)	Comments
3.950E-05	8.000E-03	15	D071	1 2 0 0 0	
3.950E-05	8.000E-03	15	F300	1 0 0 0 0	
4.560E-05	9.236E-03	25	G090	2 2 1 1 1	
2.023E-03	4.098E-01	50	D071	1 2 0 0 1	
7.837E-03	1.587E+00	100	D071	1 2 0 0 2	
8.393E-03	1.700E+00	100	F300	1 0 0 0 1	
7.244E-04	1.467E-01	ns	R427	0 0 0 0 0	

577. $C_6H_3ClN_4$
7-Chloropteridine
Pteridine, 7-chloro-

RN:	1125-84-4	**MP** (°C):	95		
MW:	166.57	**BP** (°C):			

Solubility (Moles/L)	Solubility (Grams/L)	Temp (°C)	Ref (#)	Evaluation (T P E A A)	Comments
1.305E-01	2.174E+01	20	A083	1 2 0 0 0	

578. $C_6H_3Cl_2NO_2$
3,4-Dichloronitrobenzene
1,2-Dichloro-4-nitrobenzene

RN:	99-54-7	**MP** (°C):	41.25		
MW:	192.00	**BP** (°C):	255.5		

Solubility (Moles/L)	Solubility (Grams/L)	Temp (°C)	Ref (#)	Evaluation (T P E A A)	Comments
6.290E-04	1.208E-01	20	E308	1 2 2 1 2	

579. $C_6H_3Cl_2NO_2$
2,5-Dichloronitrobenzene
1,4-Dichloro-2-nitrobenzene

RN:	89-61-2	**MP** (°C):	55.5		
MW:	192.00	**BP** (°C):	267.5		

Solubility (Moles/L)	Solubility (Grams/L)	Temp (°C)	Ref (#)	Evaluation (T P E A A)	Comments
4.800E-04	9.216E-02	20	E308	1 2 2 1 2	

580. C$_6$H$_3$Cl$_2$NO$_2$

2,3-Dichloronitrobenzene
1,2-Dichloro-3-nitrobenzene

RN: 3209-22-1   **MP** (°C): 61.5
MW: 192.00   **BP** (°C): 257.5

Solubility (Moles/L)	Solubility (Grams/L)	Temp (°C)	Ref (#)	Evaluation (T P E A A)	Comments
3.250E-04	6.240E-02	20	E308	1 2 2 1 2	

581. C$_6$H$_3$Cl$_2$NO$_2$

3,6-Dichloropicolinic acid
3,6-Dichloro-2-pyridinecarboxylic acid
Clopyralid
Lontrel
Stinger

RN: 1702-17-6   **MP** (°C): 151.5
MW: 192.00   **BP** (°C):

Solubility (Moles/L)	Solubility (Grams/L)	Temp (°C)	Ref (#)	Evaluation (T P E A A)	Comments
5.208E-03	1.000E+00	20	M161	1 0 0 0 0	
5.208E-03	1.000E+00	ns	K138	0 0 0 0 1	

582. C$_6$H$_3$Cl$_3$

1,2,3-Trichlorobenzene
Benzene, 1,2,3-trichloro-
vic-Trichlorobenzene

RN: 87-61-6   **MP** (°C): 51
MW: 181.45   **BP** (°C): 219

Solubility (Moles/L)	Solubility (Grams/L)	Temp (°C)	Ref (#)	Evaluation (T P E A A)	Comments
7.762E-05	1.408E-02	20	K337	1 0 0 0 2	
6.607E-05	1.199E-02	22	K305	1 0 1 1 2	
8.983E-05	1.630E-02	23	C305	1 1 2 2 2	
9.920E-05	1.800E-02	25	B304	2 0 2 2 2	
1.170E-04	2.123E-02	25	B317	0 0 0 0 0	
9.920E-05	1.800E-02	25	C313	0 0 0 0 0	
6.760E-05	1.227E-02	25	M342	1 0 1 1 2	
9.149E-05	1.660E-02	ns	H123	0 0 0 0 0	
6.760E-05	1.227E-02	ns	M308	0 0 1 1 2	

583. C$_6$H$_3$Cl$_3$
1,3,5-Trichlorobenzene
Benzene, 1,3,5-trichloro-

RN: 108-70-3   **MP** (°C): 64
MW: 181.45   **BP** (°C): 208

Solubility (Moles/L)	Solubility (Grams/L)	Temp (°C)	Ref (#)	Evaluation (T P E A A)	Comments
2.399E-05	4.353E-03	20	K337	1 0 0 0 2	
3.236E-05	5.872E-03	22	K305	1 0 1 1 2	
5.842E-05	1.060E-02	23	C305	1 1 2 2 2	
3.312E-05	6.010E-03	25	B304	2 0 2 2 2	
2.900E-05	5.262E-03	25	B317	0 0 0 0 0	
2.270E-05	4.119E-03	25	M342	1 0 1 1 2	
2.270E-05	4.119E-03	ns	M308	0 0 1 1 2	

584. C$_6$H$_3$Cl$_3$
1,2,4-Trichlorobenzene
Benzene, 1,2,4-trichloro-

RN: 120-82-1   **MP** (°C): 17
MW: 181.45   **BP** (°C): 213

Solubility (Moles/L)	Solubility (Grams/L)	Temp (°C)	Ref (#)	Evaluation (T P E A A)	Comments
1.653E-04	3.000E-02	19	M172	1 0 0 0 0	
1.950E-04	3.538E-02	20	K337	1 0 0 0 2	
1.072E-04	1.944E-02	22	K305	1 0 1 1 2	
1.725E-04	3.130E-02	25	B304	2 0 2 2 2	
2.200E-04	3.992E-02	25	B317	0 0 0 0 0	
2.692E-04	4.884E-02	25	C113	1 0 2 2 2	
2.540E-04	4.609E-02	25	M342	1 0 1 1 2	
3.555E-04	6.451E-02	30	M300	1 1 2 2 2	
3.555E-04	6.450E-02	30	M311	1 1 2 2 2	
2.540E-04	4.609E-02	ns	M308	0 0 1 1 2	

585. C$_6$H$_3$Cl$_3$N$_2$O$_2$
Picloram
4-Amino-3,5,6-trichloropicolinic acid

RN: 1918-02-1   **MP** (°C): 241
MW: 241.46   **BP** (°C):

Solubility (Moles/L)	Solubility (Grams/L)	Temp (°C)	Ref (#)	Evaluation (T P E A A)	Comments
1.967E-03	4.750E-01	10	C031	2 0 2 2 2	pH 2.8
2.260E-03	5.457E-01	20	C031	2 0 2 2 2	pH 2.8
1.781E-03	4.300E-01	25	B185	0 0 0 0 0	
1.781E-03	4.300E-01	25	B200	1 0 0 0 1	
1.781E-03	4.300E-01	25	M161	1 0 0 0 2	
2.830E-03	6.833E-01	30	C031	2 0 2 2 2	pH 2.8
3.290E-03	7.944E-01	40	C031	2 0 2 2 2	pH 2.8
1.781E-03	4.300E-01	ns	K138	0 0 0 0 1	
1.780E-03	4.298E-01	ns	M061	0 0 0 0 1	
3.500E-04	8.451E-02	ns	O025	2 2 2 2 1	intrinsic

586. C$_6$H$_3$Cl$_3$O
2,3,4-Trichlorophenol
2,3,4-Trichlorphenol

RN:	15950-66-0	MP (°C):	80
MW:	197.45	BP (°C):	

Solubility (Moles/L)	Solubility (Grams/L)	Temp (°C)	Ref (#)	Evaluation (T P E A A)	Comments
4.634E-03	9.150E-01	25	M373	1 0 2 1 2	
2.138E-03	4.221E-01	ns	R424	0 0 0 0 0	

587. C$_6$H$_3$Cl$_3$O
2,3,5-Trichlorophenol
2,3,5-Trichlorphenol

RN:	933-78-8	MP (°C):	62
MW:	197.45	BP (°C):	

Solubility (Moles/L)	Solubility (Grams/L)	Temp (°C)	Ref (#)	Evaluation (T P E A A)	Comments
3.905E-03	7.710E-01	25	M373	1 0 2 1 2	

588. C$_6$H$_3$Cl$_3$O
2,3,6-Trichlorophenol
2,3,6-Trichlorphenol

RN:	933-75-5	MP (°C):	58
MW:	197.45	BP (°C):	

Solubility (Moles/L)	Solubility (Grams/L)	Temp (°C)	Ref (#)	Evaluation (T P E A A)	Comments
2.993E-03	5.910E-01	25	M373	1 0 2 1 2	

589. C$_6$H$_3$Cl$_3$O
2,4,6-Trichlorophenol
2,4,6-Trichlorphenol
Dowicide 25

RN:	88-06-2	MP (°C):	69
MW:	197.45	BP (°C):	246

Solubility (Moles/L)	Solubility (Grams/L)	Temp (°C)	Ref (#)	Evaluation (T P E A A)	Comments
2.532E-03	5.000E-01	11.20	F300	1 0 0 0 0	
2.076E-03	4.100E-01	19.5	A400	2 1 2 2 2	
2.163E-03	4.270E-01	20.1	A400	2 1 2 2 2	
4.558E-03	9.000E-01	22.5	G301	0 0 0 0 0	
3.505E-03	6.920E-01	24.9	A400	2 1 2 2 2	
2.200E-03	4.344E-01	25	B316	0 0 0 0 0	
3.586E-03	7.080E-01	25	M373	1 0 2 1 2	
4.554E-03	8.992E-01	25	R041	0 0 0 0 0	
4.558E-03	9.000E-01	25.40	F300	1 0 0 0 0	

(continued)

589. C₆H₃Cl₃O (continued)

589. $C_6H_3Cl_3O$ (continued)

Solubility (Moles/L)	Solubility (Grams/L)	Temp (°C)	Ref (#)	Evaluation (T P E A A)	Comments
3.077E-02	6.075E+00	29.8	A400	2 1 2 2 2	
3.292E-02	6.501E+00	35.1	A400	2 1 2 2 2	
1.266E-02	2.500E+00	96	F300	1 0 0 0 1	
<5.06E-03	<9.99E-01	ns	N034	0 0 0 0 0	
3.981E-03	7.861E-01	ns	R427	0 0 0 0 0	

590. $C_6H_3Cl_3O$
2,4,5-Trichloro-phenol
Phenol, 2,4,5-trichloro-
Dowicide 2
Preventol I
2,4,5-Trichlorophenol
Collunosol

RN:	95-95-4	**MP (°C):**	69	
MW:	197.45	**BP (°C):**		

Solubility (Moles/L)	Solubility (Grams/L)	Temp (°C)	Ref (#)	Evaluation (T P E A A)	Comments
4.800E-03	9.478E-01	25	B316	0 0 0 0 0	
3.287E-03	6.490E-01	25	M373	1 0 2 1 2	

591. $C_6H_3Cl_4N$
Nitrapyrin
2-Chloro-6-(trichloromethyl)pyridine
Donco-163
N-Serve(R)

RN:	1929-82-4	**MP (°C):**	62.5	
MW:	230.91	**BP (°C):**		

Solubility (Moles/L)	Solubility (Grams/L)	Temp (°C)	Ref (#)	Evaluation (T P E A A)	Comments
1.738E-04	4.013E-02	20	B179	0 0 0 0 0	
1.732E-04	4.000E-02	20	G079	1 1 0 0 2	
3.118E-04	7.200E-02	ns	V414	0 0 0 0 0	

592. $C_6H_3FN_2O_4$
1-Fluoro-2,4-dinitrobenzene
FDNB

RN:	70-34-8	**MP (°C):**	26	
MW:	186.10	**BP (°C):**		

Solubility (Moles/L)	Solubility (Grams/L)	Temp (°C)	Ref (#)	Evaluation (T P E A A)	Comments
2.149E-03	4.000E-01	ns	B160	0 0 0 0 2	

593. C$_6$H$_3$F$_3$O
Trifluorophenol
2,3,4-Trifluorophenol

RN: 2822-41-5   **MP** (°C):
MW: 148.09   **BP** (°C):

Solubility (Moles/L)	Solubility (Grams/L)	Temp (°C)	Ref (#)	Evaluation (T P E A A)	Comments
4.200E-01	6.220E+01	25	P031	0 0 0 0 0	

594. C$_6$H$_3$N$_3$O$_6$
sym-Trinitrobenzene
1,3,5-Trinitro-benzol
1,3,5-Trinitrobenzene

RN: 99-35-4   **MP** (°C):   122.5
MW: 213.11   **BP** (°C):

Solubility (Moles/L)	Solubility (Grams/L)	Temp (°C)	Ref (#)	Evaluation (T P E A A)	Comments
1.305E-03	2.780E-01	15	D066	1 2 0 0 2	
1.304E-03	2.779E-01	15	D070	1 2 0 0 2	
1.314E-03	2.800E-01	15	F300	1 0 0 0 1	
1.678E-03	3.577E-01	25	H434	0 0 0 0 0	
4.786E-03	1.020E+00	50	D066	1 2 0 0 2	
4.781E-03	1.019E+00	50	D070	1 2 0 0 2	
2.337E-02	4.980E+00	100	D066	1 2 0 0 2	
2.325E-02	4.955E+00	100	D070	1 2 0 0 2	
2.393E-02	5.100E+00	100	F300	1 0 0 0 1	
1.288E-03	2.745E-01	ns	R427	0 0 0 0 0	

595. C$_6$H$_3$N$_3$O$_7$
Picric acid
2,4,6-Trinitrophenol
Picronitric acid
Pikrinsaeure

RN: 88-89-1   **MP** (°C):   122.5
MW: 229.11   **BP** (°C):

Solubility (Moles/L)	Solubility (Grams/L)	Temp (°C)	Ref (#)	Evaluation (T P E A A)	Comments
2.948E-02	6.754E+00	0	D077	1 0 0 1 1	
4.322E-02	9.901E+00	0	M043	1 0 0 0 1	
4.364E-02	9.999E+00	7.10	E032	1 2 1 2 2	
4.232E-02	9.695E+00	9	D080	1 2 0 0 2	unit assumed
3.507E-02	8.035E+00	10	D077	1 0 0 1 1	
4.749E-02	1.088E+01	10	M043	1 0 0 0 1	
4.407E-02	1.010E+01	18.90	E032	1 2 1 2 2	
4.792E-02	1.098E+01	20	D077	1 0 0 1 2	
5.151E-02	1.180E+01	20	H048	1 0 0 0 2	unit assumed
4.300E-02	9.852E+00	20	K310	1 0 0 1 1	

(continued)

595. $C_6H_3N_3O_7$ (continued)

Solubility (Moles/L)	Solubility (Grams/L)	Temp (°C)	Ref (#)	Evaluation (T P E A A)	Comments
5.176E-02	1.186E+01	20	M043	1 0 0 0 1	
4.932E-02	1.130E+01	23.50	F300	0 0 0 0 2	
5.327E-02	1.220E+01	25	D058	1 0 1 1 2	
5.520E-02	1.265E+01	25	F030	1 0 2 1 2	
5.684E-02	1.302E+01	25	H048	1 0 0 0 2	unit assumed
5.780E-02	1.324E+01	25	K040	1 0 2 1 2	
5.474E-02	1.254E+01	25	M094	1 0 0 1 2	
6.026E-02	1.381E+01	30	D077	1 0 0 1 2	
6.450E-02	1.478E+01	30	M043	1 0 0 0 1	
7.465E-02	1.710E+01	33.30	E032	1 2 1 2 2	
7.633E-02	1.749E+01	40	D077	1 0 0 1 2	
8.138E-02	1.865E+01	40	M043	1 0 0 0 1	
9.396E-02	2.153E+01	44.30	E032	1 2 1 2 2	
9.354E-02	2.143E+01	50	D077	1 0 0 1 2	
9.930E-02	2.275E+01	50	D080	1 2 0 0 2	unit assumed
1.193E-01	2.733E+01	60	D077	1 0 0 1 2	
1.312E-01	3.007E+01	60	M043	1 0 0 0 1	
1.398E-01	3.204E+01	62.90	E032	1 2 1 2 2	
1.464E-01	3.354E+01	70	D077	1 0 0 1 2	
1.703E-01	3.902E+01	72.60	E032	1 2 1 2 2	
1.844E-01	4.224E+01	80	D077	1 0 0 1 2	
1.920E-01	4.398E+01	80	M043	1 0 0 0 1	
1.956E-01	4.481E+01	82	D080	1 2 0 0 2	unit assumed
2.007E-01	4.598E+01	83.90	E032	1 2 1 2 2	
2.362E-01	5.411E+01	90	D077	1 0 0 1 2	
2.160E-01	4.949E+01	90	K310	1 0 0 1 2	
2.244E-01	5.141E+01	90.10	E032	1 2 1 2 2	
2.326E-01	5.330E+01	92.40	E032	1 2 1 2 2	
2.517E-01	5.767E+01	94.80	E032	1 2 1 2 2	
2.947E-01	6.751E+01	100	D077	1 0 0 1 2	
3.083E-01	7.063E+01	100	D080	1 2 0 0 2	unit assumed
3.055E-01	7.000E+01	100	F300	1 0 0 0 1	
2.932E-01	6.716E+01	100	M043	1 0 0 0 1	

596. $C_6H_3N_3O_8$
Styphnic acid
Styphninsaeure

RN:	82-71-3	**MP (°C):**	176	
MW:	245.11	**BP (°C):**		

Solubility (Moles/L)	Solubility (Grams/L)	Temp (°C)	Ref (#)	Evaluation (T P E A A)	Comments
2.393E-02	5.865E+00	6.10	E032	1 2 1 2 2	
2.167E-02	5.312E+00	16.60	E032	1 2 1 2 2	
2.203E-02	5.400E+00	25	F300	1 0 0 0 1	
2.179E-02	5.341E+00	25	K040	1 0 2 1 2	
2.997E-02	7.346E+00	35.70	E032	1 2 1 2 2	
3.471E-02	8.507E+00	47.10	E032	1 2 1 2 2	

(continued)

596. C$_6$H$_3$N$_3$O$_8$ (continued)

Solubility (Moles/L)	Solubility (Grams/L)	Temp (°C)	Ref (#)	Evaluation (T P E A A)	Comments
4.119E-02	1.010E+01	56.90	E032	1 2 1 2 2	
4.692E-02	1.150E+01	62	F300	1 0 0 0 2	
4.758E-02	1.166E+01	63.00	E032	1 2 1 2 2	
6.109E-02	1.497E+01	71.20	E032	1 2 1 2 2	
7.135E-02	1.749E+01	76.20	E032	1 2 1 2 2	
8.000E-02	1.961E+01	80.30	E032	1 2 1 2 2	
9.562E-02	2.344E+01	85.00	E032	1 2 1 2 2	
1.096E-01	2.686E+01	89.80	E032	1 2 1 2 2	
1.357E-01	3.326E+01	95.90	E032	1 2 1 2 2	

597. C$_6$H$_4$BrF

1-Bromo-2-fluorobenzene
2-Bromofluorobenzene

RN: 1072-85-1   **MP** (°C):
MW: 175.01   **BP** (°C):   151.5

Solubility (Moles/L)	Solubility (Grams/L)	Temp (°C)	Ref (#)	Evaluation (T P E A A)	Comments
2.018E-03	3.532E-01	25	B349	2 0 2 0 2	

598. C$_6$H$_4$BrF

1-Bromo-3-fluorobenzene
3-Bromofluorobenzene

RN: 1073-06-9   **MP** (°C):
MW: 175.01   **BP** (°C):   150

Solubility (Moles/L)	Solubility (Grams/L)	Temp (°C)	Ref (#)	Evaluation (T P E A A)	Comments
2.162E-03	3.784E-01	25	B349	2 0 2 0 2	

599. C$_6$H$_4$BrNO$_3$

2-Bromo-4-nitrophenol
2-Brom-4-nitro-phenol

RN: 5847-59-6   **MP** (°C):   114
MW: 218.01   **BP** (°C):

Solubility (Moles/L)	Solubility (Grams/L)	Temp (°C)	Ref (#)	Evaluation (T P E A A)	Comments
1.009E-01	2.200E+01	100	F300	1 0 0 0 1	

600. C$_6$H$_4$Br$_2$

m-Dibromobenzene
1,3-Dibromobenzene

RN: 108-36-1   **MP** (°C): −7
MW: 235.92   **BP** (°C): 218

Solubility (Moles/L)	Solubility (Grams/L)	Temp (°C)	Ref (#)	Evaluation (T P E A A)	Comments
2.860E-04	6.747E-02	35	H077	2 2 2 2 2	

601. C$_6$H$_4$Br$_2$

p-Dibromobenzene
1,4-Dibromobenzene

RN : 106-37-6   **MP** (°C): 87.3
MW: 235.92   **BP** (°C): 220.4

Solubility (Moles/L)	Solubility (Grams/L)	Temp (°C)	Ref (#)	Evaluation (T P E A A)	Comments
4.201E-05	9.910E-03	10	K440	0 0 0 0 0	
8.478E-05	2.000E-02	25	A003	1 0 1 2 1	
5.900E-03	1.392E+00	25	C316	0 0 0 0 0	0.1M NaCl
7.206E-05	1.700E-02	25	K440	0 0 0 0 0	
1.120E-04	2.642E-02	35	H077	2 2 2 2 2	
1.043E-04	2.460E-02	35	K440	0 0 0 0 0	

602. C$_6$H$_4$ClF

1-Chloro-2-fluorobenzene
2-Chlorofluorobenzene

RN: 348-51-6   **MP** (°C): −43
MW: 130.55   **BP** (°C): 137.6

Solubility (Moles/L)	Solubility (Grams/L)	Temp (°C)	Ref (#)	Evaluation (T P E A A)	Comments
3.845E-03	5.019E-01	25	B349	2 0 2 0 2	

603. C$_6$H$_4$ClF

1-Chloro-3-fluorobenzene
3-Chlorofluorobenzene

RN: 625-98-9   **MP** (°C):
MW: 130.55   **BP** (°C): 127.6

Solubility (Moles/L)	Solubility (Grams/L)	Temp (°C)	Ref (#)	Evaluation (T P E A A)	Comments
4.517E-03	5.897E-01	25	B349	2 0 2 0 2	

604. C$_6$H$_4$ClIO$_2$S
Pipsyl chloride
p-Iodobenzenesulfonyl chloride

RN: 98-61-3   **MP** (°C): 81
MW: 302.52   **BP** (°C):

Solubility (Moles/L)	Solubility (Grams/L)	Temp (°C)	Ref (#)	Evaluation (T P E A A)	Comments
5.388E-05	1.630E-02	25	B048	1 0 2 2 2	
8.793E-05	2.660E-02	35	B048	1 0 2 2 2	
1.646E-04	4.980E-02	50	B048	1 0 2 2 2	

605. C$_6$H$_4$ClNO$_2$
6-Chloropicolinic acid
Pyridinecarboxylic acid, 6-chloro-

RN: 4684-94-0   **MP** (°C):
MW: 157.56   **BP** (°C):

Solubility (Moles/L)	Solubility (Grams/L)	Temp (°C)	Ref (#)	Evaluation (T P E A A)	Comments
2.158E-02	3.400E+00	ns	K138	0 0 0 0 1	
2.138E-02	3.369E+00	ns	R427	0 0 0 0 0	

606. C$_6$H$_4$ClNO$_2$
p-Chloronitrobenzene
4-Nitrochlorobenzene
4-CNB
4-Chloronitrobenzene

RN: 100-00-5   **MP** (°C): 82
MW: 157.56   **BP** (°C): 242

Solubility (Moles/L)	Solubility (Grams/L)	Temp (°C)	Ref (#)	Evaluation (T P E A A)	Comments
9.711E-04	1.530E-01	9.99	B403	1 2 2 2 2	
1.777E-04	2.800E-02	17	D071	1 2 0 0 1	
1.777E-04	2.800E-02	17	F300	1 0 0 0 1	
1.327E-03	2.090E-01	19.99	B403	1 2 2 2 2	
2.877E-03	4.533E-01	20	E308	1 2 2 1 2	
1.429E-03	2.251E-01	20	H118	1 1 1 1 2	
1.429E-03	2.251E-01	20	H301	0 0 0 0 0	
<1.27E-03	<2.00E-01	25	B019	1 0 1 2 0	
1.600E-03	2.521E-01	25	G090	2 2 1 1 1	
1.739E-03	2.740E-01	29.99	B403	1 2 2 2 2	
2.348E-03	3.700E-01	39.99	B403	1 2 2 2 2	
7.933E-04	1.250E-01	50	D071	1 2 0 0 2	
9.709E-04	1.530E-01	100	D071	1 2 0 0 2	
1.016E-03	1.600E-01	100	F300	1 0 0 0 2	

607. $C_6H_4ClNO_2$

m-Chloronitrobenzene
1-Chloro-3-nitrobenzene
3-Chloronitrobenzene
m-Nitrochlorobenzene

RN:	121-73-3	**MP** (°C):	46.0	
MW:	157.56	**BP** (°C):	236.0	

Solubility (Moles/L)	Solubility (Grams/L)	Temp (°C)	Ref (#)	Evaluation (T P E A A)	Comments
1.732E-03	2.729E-01	20	E308	1 2 2 1 2	

608. $C_6H_4ClNO_2$

o-Chloronitrobenzene
2-Nitrochlorobenzene
2-CNB
1-Chloro-2-nitrobenzene

RN:	88-73-3	**MP** (°C):	32	
MW:	157.56	**BP** (°C):	245	

Solubility (Moles/L)	Solubility (Grams/L)	Temp (°C)	Ref (#)	Evaluation (T P E A A)	Comments
1.447E-03	2.280E-01	9.99	B403	1 2 2 2 2	
2.133E-03	3.360E-01	19.99	B403	1 2 2 2 2	
2.800E-03	4.412E-01	20	E308	1 2 2 1 2	
<1.27E-03	<2.00E-01	25	B019	1 0 1 2 0	
3.470E-03	5.467E-01	25	G090	2 2 1 1 1	
3.199E-03	5.040E-01	29.99	B403	1 2 2 2 2	
4.271E-03	6.730E-01	39.99	B403	1 2 2 2 2	

609. $C_6H_4Cl_2$

1,2-Dichlorobenzene
o-Dichlorobenzene

RN:	95-50-1	**MP** (°C):	−17	
MW:	147.00	**BP** (°C):	180	

Solubility (Moles/L)	Solubility (Grams/L)	Temp (°C)	Ref (#)	Evaluation (T P E A A)	Comments
9.047E-04	1.330E-01	3.5	C094	1 0 0 0 2	
1.007E-03	1.480E-01	20	C094	1 0 0 0 2	
9.114E-04	1.340E-01	20	K056	1 0 2 2 2	
9.550E-04	1.404E-01	20	K337	1 0 0 0 2	
6.607E-04	9.713E-02	22	K305	1 0 1 1 2	
<1.36E-03	<2.00E-01	25	B019	1 0 1 2 0	
1.060E-03	1.558E-01	25	B173	2 0 2 2 2	
9.864E-04	1.450E-01	25	B185	0 0 0 0 0	
9.319E-04	1.370E-01	25	B304	2 0 2 2 2	
8.000E-04	1.176E-01	25	B317	0 0 0 0 0	
1.047E-03	1.539E-01	25	C113	1 0 2 2 2	
9.864E-04	1.450E-01	25	K056	1 0 2 2 2	
1.156E-03	1.700E-01	25	L319	1 0 2 1 1	
6.280E-04	9.232E-02	25	M342	1 0 1 1 2	
1.163E-03	1.710E-01	30	K056	1 0 2 2 2	

(continued)

609. $C_6H_4Cl_2$ (continued)

Solubility (Moles/L)	Solubility (Grams/L)	Temp (°C)	Ref (#)	Evaluation (T P E A A)	Comments
1.016E-03	1.494E-01	30	M300	1 1 2 2 2	
9.680E-04	1.423E-01	30	M311	1 1 2 2 2	
1.245E-03	1.830E-01	35	K056	1 0 2 2 2	
1.320E-03	1.940E-01	40	K056	1 0 2 2 2	
1.381E-03	2.030E-01	45	K056	1 0 2 2 2	
1.517E-03	2.230E-01	55	K056	1 0 2 2 2	
1.578E-03	2.320E-01	60	K056	1 0 2 2 2	
1.060E+03	1.558E+05	ns	A096	0 0 0 0 2	*sic*
6.280E-04	9.232E-02	ns	M308	0 0 1 1 2	

610. $C_6H_4Cl_2$
1,3-Dichlorobenzene
m-Dichlorobenzene

RN:	541-73-1	**MP** (°C):	−24	
MW:	147.00	**BP** (°C):	172–173	

Solubility (Moles/L)	Solubility (Grams/L)	Temp (°C)	Ref (#)	Evaluation (T P E A A)	Comments
7.551E-04	1.110E-01	20	K056	1 0 2 2 2	
7.943E-04	1.168E-01	20	K337	1 0 0 0 2	
4.677E-04	6.876E-02	22	K305	1 0 1 1 2	
9.080E-04	1.335E-01	25	B173	2 0 2 2 2	
9.728E-04	1.430E-01	25	B304	2 1 2 1 2	
8.300E-04	1.220E-01	25	B317	0 0 0 0 0	
9.120E-04	1.341E-01	25	C113	1 0 2 2 2	
8.367E-04	1.230E-01	25	K056	1 0 2 2 2	
8.470E-04	1.245E-01	25	M342	1 0 1 1 2	
9.523E-04	1.400E-01	30	K056	1 0 2 2 2	
8.537E-04	1.255E-01	30	M300	1 1 2 2 2	
8.537E-04	1.255E-01	30	M311	1 1 2 2 2	
1.020E-03	1.500E-01	35	K056	1 0 2 2 2	
1.136E-03	1.670E-01	40	K056	1 0 2 2 2	
1.204E-03	1.770E-01	45	K056	1 0 2 2 2	
1.333E-03	1.960E-01	55	K056	1 0 2 2 2	
1.367E-03	2.010E-01	60	K056	1 0 2 2 2	
9.080E+02	1.335E+05	ns	A096	0 0 0 0 2	*sic*
8.470E-04	1.245E-01	ns	M308	0 0 1 1 2	

611. $C_6H_4Cl_2$
1,4-Dichlorobenzene
p-Dichlorobenzene

RN:	106-46-7	**MP** (°C):	53.1	
MW:	147.00	**BP** (°C):	173.4	

Solubility (Moles/L)	Solubility (Grams/L)	Temp (°C)	Ref (#)	Evaluation (T P E A A)	Comments
4.680E-04	6.880E-02	20	K056	1 2 2 1 2	average of 4
3.020E-04	4.439E-02	20	K337	1 0 0 0 2	
2.252E-04	3.310E-02	20	T301	1 2 2 2 2	

(continued)

611. C$_6$H$_4$Cl$_2$ (continued)

Solubility (Moles/L)	Solubility (Grams/L)	Temp (°C)	Ref (#)	Evaluation (T P E A A)	Comments
3.311E-04	4.868E-02	22	K305	1 0 1 1 2	
5.292E-04	7.780E-02	22.20	W003	2 2 2 2 2	average of 2
5.673E-04	8.340E-02	24.60	W003	2 2 2 2 2	average of 3
5.170E-04	7.600E-02	25	A003	1 0 1 2 1	
5.928E-04	8.715E-02	25	A058	1 1 1 1 2	
<3.40E-03	<5.00E-01	25	B019	1 0 1 2 0	
5.020E-04	7.380E-02	25	B173	2 0 2 2 2	
4.442E-04	6.530E-02	25	B304	2 0 2 2 2	
5.270E-04	7.747E-02	25	B317	0 0 0 0 0	
3.990E-04	5.865E-02	25	C316	0 0 0 0 0	0.1M NaCl
5.374E-04	7.900E-02	25	F071	1 1 2 1 1	
5.374E-04	7.900E-02	25	H080	1 0 0 0 1	
5.381E-04	7.910E-02	25	K056	1 2 2 2 2	average of 2
5.646E-04	8.300E-02	25	M040	1 0 0 1 1	
5.442E-04	8.000E-02	25	M161	1 0 0 0 1	
2.100E-04	3.087E-02	25	M342	1 0 1 1 2	
6.932E-05	1.019E-02	25	N311	1 0 1 1 2	
4.100E-04	6.027E-02	25.2	T428	0 0 0 0 0	
5.898E-04	8.670E-02	25.50	W003	2 2 2 2 2	average of 2
5.238E-04	7.699E-02	30	G029	1 0 2 2 1	
6.347E-04	9.330E-02	30	K056	1 2 2 2 2	
6.267E-04	9.213E-02	30	M300	1 1 2 2 2	
6.422E-04	9.440E-02	30	M311	1 1 2 2 2	
6.299E-04	9.260E-02	30.00	W003	2 2 2 2 2	average of 2
6.939E-04	1.020E-01	34.50	W003	2 2 2 2 2	average of 3
5.646E-04	8.300E-02	35	K056	1 2 2 2 2	
8.231E-04	1.210E-01	38.40	W003	2 2 2 2 2	
6.857E-04	1.008E-01	40	K056	1 2 2 2 2	average of 2
8.292E-04	1.219E-01	45	K056	1 2 2 2 2	average of 2
1.082E-03	1.590E-01	47.50	W003	2 2 2 2 2	
1.184E-03	1.740E-01	50.10	W003	2 2 2 2 2	average of 3
1.061E-03	1.560E-01	55	K056	1 2 2 2 2	
1.429E-03	2.100E-01	59.20	W003	2 2 2 2 2	
1.109E-03	1.630E-01	60	K056	1 2 2 2 2	
1.483E-03	2.180E-01	60.70	W003	2 2 2 2 2	average of 2
1.565E-03	2.300E-01	65.10	W003	2 2 2 2 2	average of 3
1.612E-03	2.370E-01	65.20	W003	2 2 2 2 2	average of 3
1.912E-03	2.810E-01	73.40	W003	2 2 2 2 2	
2.100E-04	3.087E-02	ns	M308	0 0 1 1 2	
5.374E-04	7.900E-02	ns	M344	0 0 0 0 1	
5.034E-04	7.400E-02	rt	S314	0 0 2 1 1	

612. C$_6$H$_4$Cl$_2$N$_2$O$_2$
Dicloran

RN: 99-30-9   **MP** (°C): 195
MW: 207.02   **BP** (°C):

Solubility (Moles/L)	Solubility (Grams/L)	Temp (°C)	Ref (#)	Evaluation (T P E A A)	Comments
3.020E-05	6.252E-03	ns	R424	0 0 0 0 0	
3.020E-05	6.252E-03	ns	R427	0 0 0 0 0	

613. C$_6$H$_4$Cl$_2$O
2,4-Dichlorophenol
2,4-Dichlor-phenol

RN: 120-83-2   **MP** (°C): 45
MW: 163.00   **BP** (°C):

Solubility (Moles/L)	Solubility (Grams/L)	Temp (°C)	Ref (#)	Evaluation (T P E A A)	Comments
2.390E-02	3.896E+00	15.3	A400	2 1 2 2 2	
2.748E-02	4.480E+00	19	D041	1 0 0 0 1	
~2.76E-02	~4.50E+00	20	F300	1 0 0 0 0	
2.748E-02	4.480E+00	20	N034	1 0 0 0 1	
3.403E-02	5.547E+00	25	M373	1 0 2 1 2	
3.052E-02	4.975E+00	25	R041	0 0 0 0 0	
3.385E-02	5.517E+00	25.2	A400	2 1 2 2 2	
1.748E-01	2.850E+01	34.6	A400	2 1 2 2 2	
2.754E-02	4.490E+00	ns	R427	0 0 0 0 0	

614. C$_6$H$_4$Cl$_2$O
3,5-Dichlorophenol
3,5-DCP

RN: 591-35-5   **MP** (°C): 68
MW: 163.00   **BP** (°C):

Solubility (Moles/L)	Solubility (Grams/L)	Temp (°C)	Ref (#)	Evaluation (T P E A A)	Comments
4.536E-02	7.394E+00	25	M373	1 0 2 1 2	

615. C$_6$H$_4$Cl$_2$O
3,4-Dichlorophenol
4,5-Dichlorophenol
3,4-DCP

RN: 95-77-2   **MP** (°C): 67
MW: 163.00   **BP** (°C):

Solubility (Moles/L)	Solubility (Grams/L)	Temp (°C)	Ref (#)	Evaluation (T P E A A)	Comments
5.678E-02	9.256E+00	25	M373	1 0 2 1 2	

616. C$_6$H$_4$Cl$_2$O
2,6-Dichlorophenol
2,6-DCP

RN: 87-65-0   **MP** (°C): 66.5
MW: 163.00   **BP** (°C):

Solubility (Moles/L)	Solubility (Grams/L)	Temp (°C)	Ref (#)	Evaluation (T P E A A)	Comments
1.610E-02	2.625E+00	25	M373	1 0 2 1 2	

617. $C_6H_4Cl_2O$
2,3-Dichlorophenol
Phenol, 2,3-dichloro-
RN: 576-24-9 **MP** (°C): 59
MW: 163.00 **BP** (°C):

Solubility (Moles/L)	Solubility (Grams/L)	Temp (°C)	Ref (#)	Evaluation (T P E A A)	Comments
5.040E-02	8.215E+00	25	M373	1 0 2 1 2	

618. $C_6H_4Cl_2O$
2,5-Dichlorophenol
2,5-Dichlor-phenol
RN: 583-78-8 **MP** (°C):
MW: 163.00 **BP** (°C):

Solubility (Moles/L)	Solubility (Grams/L)	Temp (°C)	Ref (#)	Evaluation (T P E A A)	Comments
3.800E-02	6.194E+00	25	B316	0 0 0 0 0	

619. C_6H_4FI
1-Fluoro-4-iodobenzene
4-Fluoro-1-iodobenzene
p-Iodofluorobenzene
p-Fluoroiodobenzene
p-Fluorophenyl iodide
RN: 352-34-1 **MP** (°C): −27
MW: 222.00 **BP** (°C): 183

Solubility (Moles/L)	Solubility (Grams/L)	Temp (°C)	Ref (#)	Evaluation (T P E A A)	Comments
7.499E-04	1.665E-01	25	B349	2 0 2 0 2	

620. $C_6H_4I_2$
1,4-Diiodobenzene
p-Diiodobenzene
4-Iodophenyl iodide
RN: 624-38-4 **MP** (°C): 131
MW: 329.91 **BP** (°C): 285

Solubility (Moles/L)	Solubility (Grams/L)	Temp (°C)	Ref (#)	Evaluation (T P E A A)	Comments
4.244E-06	1.400E-03	25	A003	1 2 1 2 1	*sic*
3.100E-02	1.023E+01	25	C316	0 0 0 0 0	0.1M NaCl

621. C₆H₄N₂O₄

p-Dinitrobenzene
1,4-Dinitrobenzene
RN: 100-25-4   **MP** (°C):   173
MW: 168.11   **BP** (°C):

Solubility (Moles/L)	Solubility (Grams/L)	Temp (°C)	Ref (#)	Evaluation (T P E A A)	Comments
4.759E-04	8.000E-02	20	F300	1 0 0 0 0	
2.350E-04	3.951E-02	25	C316	0 0 0 0 0	0.1M NaCl
4.090E-04	6.876E-02	25	I334	2 2 2 1 2	
3.676E-04	6.180E-02	25	L008	2 2 2 1 2	average of 2
6.170E-04	1.037E-01	35	H077	2 2 2 2 2	
1.130E-02	1.900E+00	100	F300	1 0 0 0 1	

622. C₆H₄N₂O₄

m-Dinitrobenzene
1,3-Dinitrobenzene
RN: 99-65-0   **MP** (°C):   89.5
MW: 168.11   **BP** (°C):   301.5

Solubility (Moles/L)	Solubility (Grams/L)	Temp (°C)	Ref (#)	Evaluation (T P E A A)	Comments
4.045E-04	6.800E-02	13	D070	1 2 0 0 1	
4.164E-04	7.000E-02	13	F300	1 0 0 0 0	
3.420E-03	5.749E-01	25	I334	2 2 2 1 2	
3.169E-03	5.328E-01	25	L008	2 2 2 1 2	average of 2
5.116E-03	8.600E-01	25.04	V013	2 2 2 2 2	
3.867E-03	6.500E-01	30	F300	1 0 0 0 1	
3.888E-03	6.536E-01	30	G029	1 0 2 2 2	
4.670E-03	7.851E-01	35	H077	2 2 2 2 2	
2.789E-03	4.688E-01	50	D070	1 2 0 0 2	
1.134E-02	1.906E+00	100	D070	1 2 0 0 2	
1.547E-02	2.600E+00	100	F300	1 0 0 0 1	
2.973E-03	4.998E-01	rt	D021	0 0 1 1 0	

623. C₆H₄N₂O₄

o-Dinitrobenzene
1,2-Dinitrobenzene
RN: 528-29-0   **MP** (°C):   118
MW: 168.11   **BP** (°C):

Solubility (Moles/L)	Solubility (Grams/L)	Temp (°C)	Ref (#)	Evaluation (T P E A A)	Comments
8.328E-04	1.400E-01	20	F300	1 0 0 0 1	
7.910E-04	1.330E-01	25	I334	2 2 2 1 2	
7.418E-04	1.247E-01	25	L008	2 2 2 1 2	average of 3

624. $C_6H_4N_2O_5$
3,5-Dinitrophenol
Phenol, θ-dinitro-
RN: 586-11-8   **MP** (°C):
MW: 184.11   **BP** (°C):

Solubility (Moles/L)	Solubility (Grams/L)	Temp (°C)	Ref (#)	Evaluation (T P E A A)	Comments
7.288E-02	1.342E+01	51.6	S117	1 2 1 1 2	solid hydrate
2.373E+00	4.370E+02	54.1	S117	1 2 1 1 2	anhydrate
2.407E+00	4.431E+02	54.5	S117	1 2 1 1 2	anhydrate
2.442E+00	4.496E+02	55.5	S117	1 2 1 1 2	anhydrate
2.474E+00	4.555E+02	57.9	S117	1 2 1 1 2	anhydrate
2.516E+00	4.633E+02	61.9	S117	1 2 1 1 2	anhydrate
2.583E+00	4.756E+02	69.9	S117	1 2 1 1 2	anhydrate
2.617E+00	4.819E+02	81.3	S117	1 2 1 1 2	anhydrate
5.308E-01	9.772E+01	109.3	S117	1 0 1 1 2	
1.253E+00	2.307E+02	124.6	S117	1 0 1 1 2	

625. $C_6H_4N_2O_5$
2,6-Dinitrophenol
β-Dinitrophenol
RN: 573-56-8   **MP** (°C):
MW: 184.11   **BP** (°C):

Solubility (Moles/L)	Solubility (Grams/L)	Temp (°C)	Ref (#)	Evaluation (T P E A A)	Comments
1.710E-03	3.149E-01	15	D080	1 2 0 0 2	unit assumed
1.629E-03	3.000E-01	15	F300	1 0 0 0 0	
2.805E-02	5.164E+00	50	D080	1 2 0 0 2	unit assumed
6.547E-02	1.205E+01	100	D080	1 2 0 0 2	unit assumed
6.518E-02	1.200E+01	100	F300	1 0 0 0 1	

626. $C_6H_4N_2O_5$
2,4-Dinitrophenol
α-Dinitrophenol
Aldifen
Fenoxyl carbon N
RN: 51-28-5   **MP** (°C): 107.5
MW: 184.11   **BP** (°C): 113

Solubility (Moles/L)	Solubility (Grams/L)	Temp (°C)	Ref (#)	Evaluation (T P E A A)	Comments
1.097E-03	2.020E-01	12.5	D069	1 2 0 0 2	
1.086E-03	2.000E-01	12.50	F300	1 0 0 0 0	
1.629E-03	2.999E-01	15	D079	1 2 0 0 1	
2.254E-03	4.150E-01	15.1	A400	2 1 2 2 2	
3.025E-02	5.569E+00	18	D041	1 0 0 0 1	
2.800E-02	5.155E+00	20	K301	2 2 1 1 1	
2.524E-03	4.647E-01	25	H085	2 0 2 1 2	

(continued)

626. C$_6$H$_4$N$_2$O$_5$ (continued)

Solubility (Moles/L)	Solubility (Grams/L)	Temp (°C)	Ref (#)	Evaluation (T P E A A)	Comments
1.467E-03	2.700E-01	25	P037	2 0 1 1 2	
3.753E-03	6.910E-01	25.0	A400	2 1 2 2 2	
1.901E-01	3.500E+01	35.0	A400	2 1 2 2 2	
4.356E-03	8.020E-01	50	D069	1 2 0 0 2	
9.504E-04	1.750E-01	50	D079	1 2 0 0 2	
7.431E-03	1.368E+00	54.50	E032	1 2 1 2 2	
1.192E-02	2.195E+00	67.60	E032	1 2 1 2 2	
1.630E-02	3.001E+00	75.80	E032	1 2 1 2 2	
3.414E-02	6.286E+00	85	D069	1 2 0 0 2	
3.170E-02	5.836E+00	87.40	E032	1 2 1 2 2	
4.845E-02	8.920E+00	92.40	E032	1 2 1 2 2	
6.547E-02	1.205E+01	96.20	E032	1 2 1 2 2	
7.163E-02	1.319E+01	100	D069	1 2 0 0 2	
8.964E-02	1.650E+01	100	D079	1 2 0 0 2	
7.061E-02	1.300E+01	100	F300	1 0 0 0 1	
2.444E-01	4.500E+01	h	F300	0 0 0 0 1	
2.702E-02	4.975E+00	ns	M061	0 0 0 0 0	

627. C$_6$H$_4$N$_2$O$_6$
2,4-Dinitroresorcinol
2,4-Dinitro-1,3-benzenediol
RN: 519-44-8   **MP** (°C):
MW: 200.11   **BP** (°C):

Solubility (Moles/L)	Solubility (Grams/L)	Temp (°C)	Ref (#)	Evaluation (T P E A A)	Comments
3.129E-02	6.261E+00	57.70	E032	1 2 1 2 2	
4.801E-02	9.607E+00	66.60	E032	1 2 1 2 2	
7.434E-02	1.488E+01	69.50	E032	1 2 1 2 2	
9.895E-02	1.980E+01	76.50	E032	1 2 1 2 2	
1.690E-01	3.382E+01	84.70	E032	1 2 1 2 2	
2.380E-01	4.762E+01	90.00	E032	1 2 1 2 2	
3.495E-01	6.994E+01	93.00	E032	1 2 1 2 2	

628. C$_6$H$_4$N$_2$O$_6$
4,6-Dinitroresorcinol
4,6-Dinitro-1,3-benzenediol
RN: 616-74-0   **MP** (°C):
MW: 200.11   **BP** (°C):

Solubility (Moles/L)	Solubility (Grams/L)	Temp (°C)	Ref (#)	Evaluation (T P E A A)	Comments
1.998E-03	3.998E-01	77.00	E032	1 2 1 2 2	
3.995E-03	7.994E-01	90.50	E032	1 2 1 2 2	
4.992E-03	9.990E-01	96.30	E032	1 2 1 2 2	

629. $C_6H_4N_4$
Pteridine
1,3,5,8-Tetraazanaphthalene
Azinepurine
Pyrimido[4,5-b]pyrazine
Pyrazino[2,3-d]pyrimidine

RN: 91-18-9   **MP** (°C):   138
MW: 132.13   **BP** (°C):

Solubility (Moles/L)	Solubility (Grams/L)	Temp (°C)	Ref (#)	Evaluation (T P E A A)	Comments
9.461E-01	1.250E+02	20	A020	1 2 0 0 1	
9.461E-01	1.250E+02	20	B050	1 0 0 0 0	
9.230E-01	1.220E+02	22.5	A085	1 2 0 0 0	
3.784E+00	5.000E+02	100	B050	1 0 0 0 0	

630. $C_6H_4N_4O$
4-Hydroxypteridine
4-Pteridinol

RN: 700-47-0   **MP** (°C):
MW: 148.12   **BP** (°C):

Solubility (Moles/L)	Solubility (Grams/L)	Temp (°C)	Ref (#)	Evaluation (T P E A A)	Comments
3.359E-02	4.975E+00	20	A020	1 2 0 0 1	
3.359E-02	4.975E+00	20	B050	1 0 0 0 0	
3.359E-02	4.975E+00	22.5	A085	1 2 0 0 0	
2.250E-01	3.333E+01	100	B050	1 0 0 0 0	

631. $C_6H_4N_4O$
6-Hydroxypteridine
6-Pteridinol

RN: 2432-26-0   **MP** (°C):
MW: 148.12   **BP** (°C):

Solubility (Moles/L)	Solubility (Grams/L)	Temp (°C)	Ref (#)	Evaluation (T P E A A)	Comments
1.928E-03	2.856E-01	20	A020	1 2 0 0 1	
1.928E-03	2.856E-01	20	B050	1 0 0 0 0	
2.923E-02	4.329E+00	100	B050	1 0 0 0 0	

632. $C_6H_4N_4O$
7-Hydroxypteridine
7-Pteridinol

RN: 2432-27-1   **MP** (°C):
MW: 148.12   **BP** (°C):

Solubility (Moles/L)	Solubility (Grams/L)	Temp (°C)	Ref (#)	Evaluation (T P E A A)	Comments
7.493E-03	1.110E+00	20	B050	1 0 0 0 0	
8.768E-02	1.299E+01	100	B050	1 0 0 0 0	

633. $C_6H_4N_4O$
2-Hydroxypteridine
2-Pteridinol

RN:	25911-76-6	MP (°C):	240
MW:	148.12	BP (°C):	

Solubility (Moles/L)	Solubility (Grams/L)	Temp (°C)	Ref (#)	Evaluation (T P E A A)	Comments
1.123E-02	1.664E+00	20	A020	1 2 0 0 1	
1.123E-02	1.664E+00	20	B050	1 0 0 0 0	
1.123E-02	1.664E+00	22.5	A085	1 2 0 0 0	
1.324E-01	1.961E+01	100	B050	1 0 0 0 0	

634. $C_6H_4N_4O_2$
2,4-Dihydroxypteridine
2:4-Dihydroxypteridine
Lumazine

RN:	487-21-8	MP (°C):	348.5
MW:	164.12	BP (°C):	

Solubility (Moles/L)	Solubility (Grams/L)	Temp (°C)	Ref (#)	Evaluation (T P E A A)	Comments
7.607E-03	1.248E+00	20	B050	1 0 0 0 0	
7.607E-03	1.248E+00	22.5	A085	1 2 0 0 0	
5.035E-02	8.264E+00	100	B050	1 0 0 0 0	

635. $C_6H_4N_4O_2$
2,7-Dihydroxypteridine
2:7-Dihydroxypteridine

RN:	65882-62-4	MP (°C):	
MW:	164.12	BP (°C):	

Solubility (Moles/L)	Solubility (Grams/L)	Temp (°C)	Ref (#)	Evaluation (T P E A A)	Comments
6.033E-02	9.901E+00	100	A020	1 2 0 0 0	

636. $C_6H_4N_4O_2$
4,6-Dihydroxypteridine
4:6-Dihydroxypteridine

RN:	16310-36-4	MP (°C):	
MW:	164.12	BP (°C):	

Solubility (Moles/L)	Solubility (Grams/L)	Temp (°C)	Ref (#)	Evaluation (T P E A A)	Comments
1.108E-03	1.818E-01	20	A020	1 2 0 0 1	
1.218E-03	2.000E-01	20	B050	1 0 0 0 0	
2.024E-02	3.322E+00	100	B050	1 0 0 0 0	

637. C$_6$H$_4$N$_4$O$_2$
4,7-Dihydroxypteridine
4:7-Dihydroxypteridine
6,7-Dihydroxypteridine
6:7-Dihydroxypteridine

RN: 33669-70-4	**MP** (°C):
MW: 164.12	**BP** (°C):

Solubility (Moles/L)	Solubility (Grams/L)	Temp (°C)	Ref (#)	Evaluation (T P E A A)	Comments
2.030E-03	3.332E-01	20	A020	1 2 0 0 1	
1.523E-03	2.499E-01	20	A020	1 2 0 0 1	
2.030E-03	3.332E-01	20	B050	1 0 0 0 0	
1.523E-03	2.499E-01	20	B050	1 0 0 0 0	
2.094E-02	3.436E+00	100	B050	1 0 0 0 0	
1.014E-02	1.664E+00	100	B050	1 0 0 0 0	

638. C$_6$H$_4$N$_4$O$_2$
2,6-Dihydroxypteridine
2:6-Dihydroxypteridine

RN: 89324-38-9	**MP** (°C):
MW: 164.12	**BP** (°C):

Solubility (Moles/L)	Solubility (Grams/L)	Temp (°C)	Ref (#)	Evaluation (T P E A A)	Comments
1.354E-03	2.222E-01	100	A020	1 2 0 0 1	

639. C$_6$H$_4$N$_4$O$_3$
2,4,7-Trihydroxypteridine
2:4:7-Trihydroxypteridine

RN: 2577-38-0	**MP** (°C):
MW: 180.12	**BP** (°C):

Solubility (Moles/L)	Solubility (Grams/L)	Temp (°C)	Ref (#)	Evaluation (T P E A A)	Comments
4.626E-04	8.333E-02	20	A020	1 2 0 1 1	
4.626E-04	8.333E-02	20	B050	1 0 0 0 0	
3.963E-03	7.138E-01	100	A020	1 2 0 0 1	
3.963E-03	7.138E-01	100	B050	1 0 0 0 0	

640. C$_6$H$_4$N$_4$O$_3$
4,6,7-Trihydroxypteridine
4:6:7-Trihydroxypteridine

RN: 58947-88-9	**MP** (°C):
MW: 180.12	**BP** (°C):

Solubility (Moles/L)	Solubility (Grams/L)	Temp (°C)	Ref (#)	Evaluation (T P E A A)	Comments
2.056E-04	3.704E-02	20	A020	1 2 0 0 1	
2.056E-04	3.704E-02	20	B050	1 0 0 0 0	
7.930E-04	1.428E-01	100	B050	1 0 0 0 0	

641. $C_6H_4N_4O_4$

2,4,6,7-Tetrahydroxypteridine
2,4,6-Trihydroxypteridine
2:4:6-Trihydroxypteridine

RN: 2817-14-3   **MP** (°C):
MW: 196.12   **BP** (°C):

Solubility (Moles/L)	Solubility (Grams/L)	Temp (°C)	Ref (#)	Evaluation (T P E A A)	Comments
8.791E-05	1.724E-02	20	A020	1 2 0 1 1	
6.889E-04	1.351E-01	20	B050	1 0 0 0 0	
8.791E-05	1.724E-02	20	B050	1 0 0 0 0	
1.272E-02	2.494E+00	100	A020	1 2 0 0 0	
7.283E-04	1.428E-01	100	A020	1 2 0 0 0	
1.272E-02	2.494E+00	100	B050	1 0 0 0 0	

642. $C_6H_4N_4O_6$

Picramine
2,4,6-Trinitroaniline
1-Amino-2,4,6-trinitrobenzene
MATB

RN: 489-98-5   **MP** (°C):   192
MW: 228.12   **BP** (°C):

Solubility (Moles/L)	Solubility (Grams/L)	Temp (°C)	Ref (#)	Evaluation (T P E A A)	Comments
8.710E-05	1.987E-02	25	B335	1 2 0 0 1	

643. $C_6H_4N_4S$

4-Mercaptopteridine
4-Pteridinethiol
4(1H)-Pteridinethione
Pteridine-4-thiol

RN: 65882-61-3   **MP** (°C):   176dec
MW: 164.19   **BP** (°C):

Solubility (Moles/L)	Solubility (Grams/L)	Temp (°C)	Ref (#)	Evaluation (T P E A A)	Comments
1.691E-03	2.777E-01	22.5	A085	1 2 0 0 0	

644. $C_6H_4N_4S$

2-Mercaptopteridine
2-Pteridinethiol
2(1H)-Pteridinethione

RN: 16878-76-5   **MP** (°C):   205
MW: 164.19   **BP** (°C):

Solubility (Moles/L)	Solubility (Grams/L)	Temp (°C)	Ref (#)	Evaluation (T P E A A)	Comments
4.347E-03	7.138E-01	22.5	A085	1 2 0 0 0	

645. $C_6H_4N_4S$
7-Mercaptopteridine
7-Pteridinethiol
7(1H)-Pteridinethione
RN: 36653-71-1   **MP** (°C):
MW: 164.19   **BP** (°C):

Solubility (Moles/L)	Solubility (Grams/L)	Temp (°C)	Ref (#)	Evaluation (T P E A A)	Comments
1.964E-03	3.225E-01	20	A083	1 2 0 0 0	
6.760E-03	1.110E+00	100	A083	1 2 0 0 0	

646. $C_6H_4O_2$
Quinone
1,4-Benzoquinone
Benzochinhydrone
p-Quinone
RN: 106-51-4   **MP** (°C):   115.7
MW: 108.10   **BP** (°C):

Solubility (Moles/L)	Solubility (Grams/L)	Temp (°C)	Ref (#)	Evaluation (T P E A A)	Comments
8.630E-02	9.329E+00	11.85	L064	2 2 2 1 2	0.01N HCl
1.013E-01	1.095E+01	17.70	L065	1 0 0 0 2	0.01N HCl
1.021E-01	1.104E+01	17.90	L065	1 0 0 0 2	0.01N HCl
1.030E-01	1.113E+01	17.95	L065	1 0 0 0 2	0.01N HCl
1.030E-01	1.113E+01	18	L064	2 2 2 1 2	0.01N HCl
1.580E-02	1.708E+00	20	B113	1 2 2 1 2	
1.233E-01	1.333E+01	23.85	L064	2 2 2 1 2	0.01N HCl
1.295E-01	1.400E+01	24	F300	1 0 0 0 1	
1.266E-01	1.369E+01	25	G033	1 0 1 1 2	
1.397E-01	1.510E+01	25	K033	1 0 0 1 2	

647. $C_6H_4O_5$
2,5-Dicarboxyfuran
Furan-dicarbon-saeure-(2,5)
RN: 3238-40-2   **MP** (°C):
MW: 156.10   **BP** (°C):

Solubility (Moles/L)	Solubility (Grams/L)	Temp (°C)	Ref (#)	Evaluation (T P E A A)	Comments
6.406E-03	1.000E+00	18	F300	1 0 0 0 0	

648. C$_6$H$_4$O$_5$
2-Carboxy-5-hydroxy-4-pyrone
Komensaeure
Komenic acid
RN: 499-78-5   **MP** (°C):
MW: 156.10   **BP** (°C):

Solubility (Moles/L)	Solubility (Grams/L)	Temp (°C)	Ref (#)	Evaluation (T P E A A)	Comments
3.267E-02	5.100E+00	25	F300	1 0 0 0 1	
3.921E-01	6.120E+01	100	F300	1 0 0 0 2	

649. C$_6$H$_5$Br
Bromobenzene
Phenyl bromide
Monobromobenzene
RN: 108-86-1   **MP** (°C):   −30
MW: 157.02   **BP** (°C):   156.2

Solubility (Moles/L)	Solubility (Grams/L)	Temp (°C)	Ref (#)	Evaluation (T P E A A)	Comments
2.611E-03	4.100E-01	25	A003	1 2 1 2 1	
2.620E-03	4.114E-01	25	W300	2 2 2 2 2	
2.840E-03	4.460E-01	30	F071	1 1 2 1 2	
2.966E-03	4.658E-01	30	G029	1 0 2 2 2	
2.840E-03	4.460E-01	30	H080	1 0 0 0 2	
2.102E-03	3.300E-01	30	M311	1 1 2 2 2	
2.799E-03	4.395E-01	30	V009	1 0 0 0 1	
2.920E-03	4.585E-01	35	H077	2 2 2 2 2	
5.110E-04	8.024E-02	ns	D348	0 0 0 0 0	
2.615E-03	4.106E-01	ns	M344	0 0 0 0 2	

650. C$_6$H$_5$BrO
p-Bromophenol
4-Bromophenol
RN: 106-41-2   **MP** (°C):   66
MW: 173.02   **BP** (°C):   236

Solubility (Moles/L)	Solubility (Grams/L)	Temp (°C)	Ref (#)	Evaluation (T P E A A)	Comments
8.053E-02	1.393E+01	20	R087	0 0 0 0 0	0.15M NaCl
8.542E-02	1.478E+01	25	R041	0 0 0 0 0	
8.128E-02	1.406E+01	ns	R424	0 0 0 0 0	

651. C$_6$H$_5$BrO$_3$S

p-Bromobenzenesulfonic acid
4-Bromobenzenesulfonic acid

RN: 138-36-3   **MP** (°C):
MW: 237.08   **BP** (°C):

Solubility (Moles/L)	Solubility (Grams/L)	Temp (°C)	Ref (#)	Evaluation (T P E A A)	Comments
2.079E+00	4.929E+02	82.3	T023	1 2 2 1 2	
2.088E+00	4.949E+02	89.6	T023	1 2 2 1 2	
2.093E+00	4.961E+02	93.1	T023	1 2 2 1 2	
2.097E+00	4.972E+02	97.6	T023	1 2 2 1 2	

652. C$_6$H$_5$BrO$_3$S.H$_2$O

p-Bromobenzenesulfonic acid (monohydrate)

RN: 138-36-3   **MP** (°C):
MW: 255.09   **BP** (°C):

Solubility (Moles/L)	Solubility (Grams/L)	Temp (°C)	Ref (#)	Evaluation (T P E A A)	Comments
1.799E+00	4.588E+02	43.8	T023	1 2 2 1 2	
1.821E+00	4.644E+02	60.2	T023	1 2 2 1 2	
1.586E+00	4.045E+02	71.2	T023	1 2 2 1 2	
1.924E+00	4.909E+02	76.6	T023	1 2 2 1 2	
1.922E+00	4.903E+02	78.5	T023	1 2 2 1 2	
1.855E+00	4.731E+02	80.3	T023	1 2 2 1 2	
1.868E+00	4.766E+02	86.2	T023	1 2 2 1 2	
1.907E+00	4.865E+02	87.2	T023	1 2 2 1 2	
1.889E+00	4.818E+02	90.2	T023	1 2 2 1 2	

653. C$_6$H$_5$BrO$_3$S.2.5H$_2$O

p-Bromobenzenesulfonic acid (2.5 hydrate)

RN: 138-36-3   **MP** (°C):
MW: 282.12   **BP** (°C):

Solubility (Moles/L)	Solubility (Grams/L)	Temp (°C)	Ref (#)	Evaluation (T P E A A)	Comments
1.375E+00	3.880E+02	−21.0	T023	1 2 2 1 2	
1.409E+00	3.975E+02	−10.5	T023	1 2 2 1 2	
1.495E+00	4.219E+02	12.5	T023	1 2 2 1 2	
1.522E+00	4.294E+02	19.9	T023	1 2 2 1 2	
1.566E+00	4.418E+02	27.6	T023	1 2 2 1 2	
1.613E+00	4.550E+02	34.6	T023	1 2 2 1 2	
1.447E+00	4.081E+02	.0	T023	1 2 2 1 2	

654. C_6H_5Cl
Chlorobenzene
IP Carrier T 40
Phenyl chloride
Tetrosin SP
Monochlorobenzene
MCB

RN: 108-90-7   **MP** (°C):   −45
MW: 112.56   **BP** (°C):   131

Solubility (Moles/L)	Solubility (Grams/L)	Temp (°C)	Ref (#)	Evaluation (T P E A A)	Comments
4.266E-03	4.802E-01	20	K337	1 0 0 0 2	
4.440E-03	4.998E-01	20	M312	1 0 0 0 2	
4.742E-03	5.337E-01	21	C024	2 1 1 2 2	
4.442E-03	5.000E-01	25	A003	1 2 1 2 1	
4.191E-03	4.717E-01	25	A058	1 1 1 1 2	
<1.78E-03	<2.00E-01	25	B019	1 0 1 2 0	
4.460E-03	5.020E-01	25	B304	2 0 2 2 2	
4.300E-03	4.840E-01	25	B317	0 0 0 0 0	
3.108E-03	3.499E-01	25	L319	1 0 2 1 1	
2.620E-03	2.949E-01	25	M342	1 0 1 1 2	
3.540E-02	3.984E+00	25	N309	1 0 0 0 1	*sic*
3.780E-03	4.255E-01	25	S359	2 1 2 2 2	
4.430E-03	4.986E-01	25	W300	2 2 2 2 2	
9.762E-03	1.099E+00	25.50	O005	2 0 2 2 1	*sic*
8.884E-04	1.000E-01	26.70	L095	2 2 1 1 2	
3.980E-03	4.480E-01	30	F071	1 1 2 1 2	
4.353E-03	4.900E-01	30	F300	1 0 0 0 1	
4.333E-03	4.878E-01	30	G029	1 0 2 2 2	
3.980E-03	4.480E-01	30	H080	1 0 0 0 2	
4.000E-03	4.502E-01	30	H332	2 2 2 2 0	
4.351E-03	4.898E-01	30	K065	2 0 2 1 2	
4.211E-03	4.740E-01	30	M300	1 1 2 2 2	
4.211E-03	4.740E-01	30	M311	1 1 2 2 2	
4.298E-03	4.838E-01	30	V009	1 0 0 0 1	
6.259E-03	7.045E-01	40	K065	2 0 2 1 2	
3.560E-03	4.007E-01	45	N043	1 0 2 2 2	
8.521E-03	9.591E-01	50	K065	2 0 2 1 2	
9.762E-03	1.099E+00	60	K065	2 0 2 1 2	
1.424E-02	1.602E+00	70	K065	2 0 2 1 2	
1.601E-02	1.802E+00	80	K065	2 0 2 1 2	
2.216E-02	2.494E+00	90	K065	2 0 2 1 2	
4.185E-03	4.711E-01	ns	H123	0 0 0 0 0	
2.620E-03	2.949E-01	ns	M308	0 0 1 1 2	
4.193E-03	4.720E-01	ns	M344	0 0 0 0 2	

655. C$_6$H$_5$ClN$_2$O$_4$S
4-Chloro-3-nitro-benzenesulfonamide
Benzenesulfonamide, 4-chloro-3-nitro-
RN: 97-09-6   **MP** (°C):
MW: 236.63   **BP** (°C):

Solubility (Moles/L)	Solubility (Grams/L)	Temp (°C)	Ref (#)	Evaluation (T P E A A)	Comments
9.500E-04	2.248E-01	15	K024	1 2 1 1 2	

656. C$_6$H$_5$ClO
m-Chlorophenol
3-Chlorophenol
Chlorophenate
3-Hydroxychlorobenzene
RN: 108-43-0   **MP** (°C):   33
MW: 128.56   **BP** (°C):   214

Solubility (Moles/L)	Solubility (Grams/L)	Temp (°C)	Ref (#)	Evaluation (T P E A A)	Comments
1.945E-01	2.500E+01	20	F300	1 0 0 0 1	
1.919E-01	2.468E+01	20	N034	1 0 0 0 2	
1.726E-01	2.219E+01	25	M373	1 0 2 1 2	
1.995E-01	2.565E+01	ns	R427	0 0 0 0 0	

657. C$_6$H$_5$ClO
p-Chlorophenol
4-Chloro-phenol-
Parachlorophenol
4-Hydroxychlorobenze
4-Chlorophenol
4-Hydroxychlorobenzene
RN: 106-48-9   **MP** (°C):   43.2
MW: 128.56   **BP** (°C):   220

Solubility (Moles/L)	Solubility (Grams/L)	Temp (°C)	Ref (#)	Evaluation (T P E A A)	Comments
1.815E-01	2.334E+01	15.1	A400	2 1 2 2 2	
2.022E-01	2.600E+01	20	F300	1 0 0 0 1	
1.022E-01	1.314E+01	20	H301	0 0 0 0 0	
1.993E-01	2.563E+01	20	N034	1 0 0 0 2	
1.839E-01	2.364E+01	20	R087	0 0 0 0 0	0.15M NaCl
2.100E-01	2.700E+01	25	B316	0 0 0 0 0	
2.053E-01	2.639E+01	25	M373	1 0 2 1 2	
1.823E-01	2.344E+01	25	R041	0 0 0 0 0	
1.987E-01	2.554E+01	25.2	A400	2 1 2 2 2	
1.867E-01	2.401E+01	34.5	A400	2 1 2 2 2	
4.898E+00	6.297E+02	ns	R427	0 0 0 0 0	

658. C₆H₅ClO

o-Chlorophenol
2-Chlorophenol

RN:	95-57-8	**MP** (°C):	9.3	
MW:	128.56	**BP** (°C):	175	

Solubility (Moles/L)	Solubility (Grams/L)	Temp (°C)	Ref (#)	Evaluation (T P E A A)	Comments
1.621E-01	2.084E+01	15.4	A400	2 1 2 2 2	
1.763E-01	2.266E+01	24.6	A400	2 1 2 2 2	
8.830E-02	1.135E+01	25	B173	2 0 2 2 2	
1.809E-01	2.326E+01	25	M373	1 0 2 1 2	
1.674E-01	2.153E+01	25	R041	0 0 0 0 0	
2.097E-01	2.695E+01	ns	N034	0 0 0 0 2	

659. C₆H₅ClO₃S

p-Chlorobenzenesulfonic acid
4-Chlor-benzolsulfosaeure

RN:	98-66-8	**MP** (°C):	67	
MW:	192.62	**BP** (°C):	148	

Solubility (Moles/L)	Solubility (Grams/L)	Temp (°C)	Ref (#)	Evaluation (T P E A A)	Comments
2.583E+00	4.975E+02	59.0	T023	1 2 2 1 2	
2.590E+00	4.988E+02	62.4	T023	1 2 2 1 2	

660. C₆H₅ClO₃S.2.5H₂O

p-Chlorobenzenesulfonic acid (2.5 hydrate)

RN:	98-66-8	**MP** (°C):		
MW:	237.66	**BP** (°C):		

Solubility (Moles/L)	Solubility (Grams/L)	Temp (°C)	Ref (#)	Evaluation (T P E A A)	Comments
1.519E+00	3.609E+02	−26.0	T023	1 2 2 1 2	
1.553E+00	3.690E+02	−20.0	T023	1 2 2 1 2	
1.606E+00	3.816E+02	−11.0	T023	1 2 2 1 2	
1.653E+00	3.929E+02	−2.2	T023	1 2 2 1 2	
1.723E+00	4.095E+02	10.6	T023	1 2 2 1 2	
1.784E+00	4.240E+02	22.9	T023	1 2 2 1 2	
1.817E+00	4.318E+02	27.6	T023	1 2 2 1 2	
1.854E+00	4.406E+02	30.8	T023	1 2 2 1 2	

661. C₆H₅Cl₂NO₂S

3,4-Dichloro-benzenesulfonamide
Benzenesulfonamide, 3,4-dichloro-

RN:	23815-28-3	**MP** (°C):		
MW:	226.08	**BP** (°C):		

Solubility (Moles/L)	Solubility (Grams/L)	Temp (°C)	Ref (#)	Evaluation (T P E A A)	Comments
3.500E-03	7.913E-01	15	K024	1 2 1 1 2	

662. C$_6$H$_5$Cl$_2$PS

Dichlorophenylphosphine sulfide
Benzene phosphorus thiodichloride
Phenylphosphonothioic dichloride
Phenyl phosphorus thiodichloride
DCPPS

RN: 3497-00-5 **MP** (°C):
MW: 211.05 **BP** (°C):

Solubility (Moles/L)	Solubility (Grams/L)	Temp (°C)	Ref (#)	Evaluation (T P E A A)	Comments
7.211E-03	1.522E+00	23	W402	0 0 0 0 0	
2.597E-02	5.481E+00	32	W402	0 0 0 0 0	
4.676E-02	9.868E+00	40	W402	0 0 0 0 0	
7.060E-02	1.490E+01	50	W402	0 0 0 0 0	

663. C$_6$H$_5$F

Fluorobenzene
Fluorbenzol

RN: 462-06-6 **MP** (°C): −42
MW: 96.11 **BP** (°C): 85

Solubility (Moles/L)	Solubility (Grams/L)	Temp (°C)	Ref (#)	Evaluation (T P E A A)	Comments
1.613E-02	1.550E+00	25	A003	1 2 1 2 2	
1.602E-02	1.540E+00	30	F071	1 1 2 1 2	
1.561E-02	1.500E+00	30	F300	1 0 0 0 1	
1.602E-02	1.540E+00	30	H080	1 0 0 0 2	
1.600E-02	1.538E+00	30	J036	0 0 0 0 0	
1.598E-02	1.535E+00	30	V009	1 0 0 0 2	
1.616E-02	1.553E+00	ns	M344	0 0 0 0 2	

664. C$_6$H$_5$FN$_2$O$_3$

3-Acetyl-5-fluoro-2,4(1H,3H)-pyrimidinedi-one
3-Acetyl-5-fluorouracil

RN: 75410-15-0 **MP** (°C): 115–116
MW: 172.12 **BP** (°C):

Solubility (Moles/L)	Solubility (Grams/L)	Temp (°C)	Ref (#)	Evaluation (T P E A A)	Comments
2.487E-01	4.280E+01	22	B321	0 0 0 0 0	pH 4.0
1.660E-01	2.857E+01	22	B416	2 2 1 2 1	

665. C$_6$H$_5$FN$_2$O$_4$

1-Methoxycarbonyl-5-fluorouracil
1(2H)-Pyrimidinecarboxylic acid, 5-fluoro-3,4-dihydro-2,4-dioxo-, methyl ester

RN: 71759-43-8 **MP** (°C):
MW: 188.12 **BP** (°C):

Solubility (Moles/L)	Solubility (Grams/L)	Temp (°C)	Ref (#)	Evaluation (T P E A A)	Comments
1.239E-01	2.330E+01	22	B332	1 1 0 0 1	pH 4.0

666. C$_6$H$_5$FO
2-Fluorophenol
2-Fluor-phenol
o-Fluorophenol

RN:	367-12-4	**MP** (°C):	16.1	
MW:	112.10	**BP** (°C):	171.5	

Solubility (Moles/L)	Solubility (Grams/L)	Temp (°C)	Ref (#)	Evaluation (T P E A A)	Comments
7.200E-01	8.072E+01	25	P031	0 0 0 0 0	

667. C$_6$H$_5$FO
m-Fluorophenol
3-Fluorophenol

RN:	372-20-3	**MP** (°C):	13.7	
MW:	112.10	**BP** (°C):	178	

Solubility (Moles/L)	Solubility (Grams/L)	Temp (°C)	Ref (#)	Evaluation (T P E A A)	Comments
6.900E-01	7.735E+01	25	P031	0 0 0 0 0	

668. C$_6$H$_5$FO
p-Fluorophenol
4-Fluorophenol

RN:	371-41-5	**MP** (°C):	46–48	
MW:	112.10	**BP** (°C):	185–188	

Solubility (Moles/L)	Solubility (Grams/L)	Temp (°C)	Ref (#)	Evaluation (T P E A A)	Comments
5.671E-01	6.357E+01	20	R087	0 0 0 0 0	0.15M NaCl
7.200E-01	8.072E+01	25	P031	0 0 0 0 0	

669. C$_6$H$_5$FO$_3$S.H$_2$O
p-Fluorobenzenesulfonic acid (monohydrate)

RN:	368-88-7	**MP** (°C):		
MW:	194.18	**BP** (°C):		

Solubility (Moles/L)	Solubility (Grams/L)	Temp (°C)	Ref (#)	Evaluation (T P E A A)	Comments
2.243E+00	4.355E+02	22.1	T023	1 2 2 1 2	
2.263E+00	4.394E+02	35.4	T023	1 2 2 1 2	
2.549E+00	4.950E+02	41.4	T023	1 2 2 1 2	
2.306E+00	4.477E+02	54.2	T023	1 2 2 1 2	
2.539E+00	4.930E+02	54.3	T023	1 2 2 1 2	
2.356E+00	4.575E+02	71.2	T023	1 2 2 1 2	
2.509E+00	4.872E+02	74.5	T023	1 2 2 1 2	
2.392E+00	4.644E+02	80.0	T023	1 2 2 1 2	
2.496E+00	4.847E+02	81.0	T023	1 2 2 1 2	
2.463E+00	4.782E+02	85.2	T023	1 2 2 1 2	
2.440E+00	4.739E+02	85.5	T023	1 2 2 1 2	

670. C₆H₅FO₃S.2.5H₂O

p-Fluorobenzenesulfonic acid (2.5 hydrate)

RN: 368-88-7   **MP** (°C):
MW: 221.21   **BP** (°C):

Solubility (Moles/L)	Solubility (Grams/L)	Temp (°C)	Ref (#)	Evaluation (T P E A A)	Comments
1.848E+00	4.088E+02	−15.5	T023	1 2 2 1 2	
1.880E+00	4.160E+02	−3.9	T023	1 2 2 1 2	
1.893E+00	4.187E+02	1.0	T023	1 2 2 1 2	
1.923E+00	4.254E+02	10.1	T023	1 2 2 1 2	
1.966E+00	4.349E+02	21.3	T023	1 2 2 1 2	

671. C₆H₅FO₃S.3H₂O

p-Fluorobenzenesulfonic acid (trihydrate)

RN: 368-88-7   **MP** (°C):
MW: 230.21   **BP** (°C):

Solubility (Moles/L)	Solubility (Grams/L)	Temp (°C)	Ref (#)	Evaluation (T P E A A)	Comments
1.731E+00	3.985E+02	−22.5	T023	1 2 2 1 2	
1.704E+00	3.922E+02	−21.4	T023	1 2 2 1 2	
1.751E+00	4.032E+02	−19.5	T023	1 2 2 1 2	
1.760E+00	4.052E+02	−17.9	T023	1 2 2 1 2	
1.715E+00	3.949E+02	−18.5	T023	1 2 2 1 2	
1.751E+00	4.032E+02	−13.0	T023	1 2 2 1 2	
1.784E+00	4.108E+02	−7.4	T023	1 2 2 1 2	

672. C₆H₅FO₃S.4H₂O

p-Fluorobenzenesulfonic acid (tetrahydrate)

RN: 368-88-7   **MP** (°C):
MW: 248.23   **BP** (°C):

Solubility (Moles/L)	Solubility (Grams/L)	Temp (°C)	Ref (#)	Evaluation (T P E A A)	Comments
1.469E+00	3.648E+02	−38.0	T023	1 2 2 1 2	
1.484E+00	3.684E+02	−35.4	T023	1 2 2 1 2	
1.498E+00	3.719E+02	−34.4	T023	1 2 2 1 2	
1.519E+00	3.771E+02	−32.5	T023	1 2 2 1 2	
1.532E+00	3.803E+02	−30.5	T023	1 2 2 1 2	
1.580E+00	3.922E+02	−26.4	T023	1 2 2 1 2	
1.605E+00	3.985E+02	−24.0	T023	1 2 2 1 2	

673. C₆H₅I

Iodobenzene

RN: 591-50-4   **MP** (°C): −30
MW: 204.01   **BP** (°C): 188

Solubility (Moles/L)	Solubility (Grams/L)	Temp (°C)	Ref (#)	Evaluation (T P E A A)	Comments
8.823E-04	1.800E-01	25	A003	1 2 1 2 1	
9.840E-04	2.007E-01	25	M342	1 0 1 1 2	

(continued)

673. C$_6$H$_5$I (continued)

Solubility (Moles/L)	Solubility (Grams/L)	Temp (°C)	Ref (#)	Evaluation (T P E A A)	Comments
1.667E-03	3.400E-01	30	F071	1 1 2 1 2	
1.667E-03	3.400E-01	30	F300	1 0 0 0 2	
1.667E-03	3.400E-01	30	H080	1 0 0 0 2	
1.667E-03	3.400E-01	30	M344	1 0 0 0 2	
1.699E-03	3.467E-01	30	V009	1 0 0 0 1	

674. C$_6$H$_5$IO
p-Iodophenol
4-Iodophenol

RN:	540-38-5	**MP** (°C):	94
MW:	220.01	**BP** (°C):	138 at 5 mm Hg

Solubility (Moles/L)	Solubility (Grams/L)	Temp (°C)	Ref (#)	Evaluation (T P E A A)	Comments
1.285E-02	2.828E+00	20	R087	0 0 0 0 0	0.15M NaCl

675. C$_6$H$_5$NO$_2$
Nitrobenzene
Nitrobenzol
Benzene, nitro-

RN:	98-95-3	**MP** (°C):	6
MW:	123.11	**BP** (°C):	210

Solubility (Moles/L)	Solubility (Grams/L)	Temp (°C)	Ref (#)	Evaluation (T P E A A)	Comments
1.381E-02	1.700E+00	6	V004	1 0 1 2 2	
1.438E-02	1.770E+00	9.99	B403	1 2 2 2 2	
1.443E-02	1.777E+00	15	G029	1 0 2 2 2	
1.568E-02	1.930E+00	19.99	B403	1 2 2 2 2	
1.549E-02	1.907E+00	20	B179	0 0 0 0 0	
1.543E-02	1.900E+00	20	F300	1 0 0 0 1	
1.600E-02	1.970E+00	20	P073	1 0 0 1 2	
1.543E-02	1.900E+00	22.5	G301	0 0 0 0 0	
1.568E-02	1.930E+00	25	A003	1 2 1 2 2	
1.700E-02	2.093E+00	25	B173	2 0 2 2 2	
1.580E-02	1.945E+00	25	H071	2 2 2 1 2	
1.600E-02	1.970E+00	25	H332	2 2 2 2 1	
1.560E-02	1.921E+00	25	I334	2 2 2 1 2	
1.560E-02	1.921E+00	25	I335	2 2 2 2 2	
1.543E-02	1.900E+00	25	M087	1 1 2 1 2	
1.457E-02	1.794E+00	25.04	V013	2 2 2 2 2	
1.446E-02	1.780E+00	26.70	L095	2 2 1 1 2	
1.673E-02	2.060E+00	29.99	B403	1 2 2 2 2	
1.662E-02	2.046E+00	30	G029	1 0 2 2 2	
1.673E-02	2.060E+00	30	V004	1 0 1 2 2	
1.667E-02	2.052E+00	30	V009	1 0 0 0 2	
1.835E-02	2.259E+00	35	H077	2 2 2 2 2	

(continued)

675. C$_6$H$_5$NO$_2$ (continued)

Solubility (Moles/L)	Solubility (Grams/L)	Temp (°C)	Ref (#)	Evaluation (T P E A A)	Comments
1.787E-02	2.200E+00	39.99	B403	1 2 2 2 2	
2.144E-02	2.640E+00	50	V004	1 0 1 2 2	
2.193E-02	2.700E+00	55	F300	1 0 0 0 1	
2.534E-02	3.120E+00	60	V004	1 0 1 2 2	
2.700E-03	3.324E-01	ns	D348	0 0 0 0 0	

676. C$_6$H$_5$NO$_2$
Nicotinic acid
Niacin

RN: 59-67-6 **MP** (°C): 236
MW: 123.11 **BP** (°C):

Solubility (Moles/L)	Solubility (Grams/L)	Temp (°C)	Ref (#)	Evaluation (T P E A A)	Comments
1.208E-01	1.488E+01	1	H083	1 2 2 1 2	
2.679E-01	3.298E+01	16	C033	1 0 2 1 2	
1.358E-01	1.672E+01	20	D041	1 0 0 0 1	
1.436E-01	1.768E+01	20	H083	1 2 2 1 2	
1.381E-01	1.700E+01	20	M054	1 0 0 0 1	
3.652E-01	4.496E+01	28	C033	1 0 2 1 2	
2.595E-01	3.195E+01	42	H083	1 2 2 1 2	
3.735E-01	4.598E+01	60	H083	1 2 2 1 2	
5.604E-01	6.899E+01	80	H083	1 2 2 1 2	
6.809E-01	8.383E+01	88	H083	1 2 2 1 2	

677. C$_6$H$_5$NO$_3$
o-Nitrophenol
2-Nitrophenol

RN: 88-75-5 **MP** (°C): 44
MW: 139.11 **BP** (°C): 214

Solubility (Moles/L)	Solubility (Grams/L)	Temp (°C)	Ref (#)	Evaluation (T P E A A)	Comments
6.434E-03	8.950E-01	9.99	B403	1 2 2 2 2	
7.735E-03	1.076E+00	15.6	A400	2 1 2 2 2	
9.704E-03	1.350E+00	19.99	B403	1 2 2 2 2	
1.000E-02	1.391E+00	20	H306	1 0 1 2 1	
9.906E-03	1.378E+00	23.10	E032	1 2 1 2 2	
1.220E-02	1.697E+00	24.8	A400	2 1 2 2 2	
1.793E-02	2.494E+00	25	D006	1 2 0 1 2	
1.797E-02	2.500E+00	25	D059	1 2 1 1 1	
1.438E-02	2.000E+00	29.99	B403	1 2 2 2 2	
1.163E-02	1.617E+00	30.40	E032	1 2 1 2 2	
2.110E-02	2.935E+00	34.7	A400	2 1 2 2 2	
1.456E-02	2.026E+00	36.20	E032	1 2 1 2 2	
2.300E-02	3.200E+00	38.40	F300	1 0 0 0 1	
1.936E-02	2.693E+00	39.80	E032	1 2 1 2 2	

(continued)

677. C$_6$H$_5$NO$_3$ (continued)

Solubility (Moles/L)	Solubility (Grams/L)	Temp (°C)	Ref (#)	Evaluation (T P E A A)	Comments
2.042E-02	2.840E+00	39.99	B403	1 2 2 2 2	
2.157E-02	3.000E+00	40	D059	1 2 1 1 0	
2.864E-02	3.984E+00	54.60	E032	1 2 1 2 1	
3.598E-02	5.005E+00	67.20	E032	1 2 1 2 2	
4.429E-02	6.162E+00	72.10	E032	1 2 1 2 2	
5.174E-02	7.198E+00	86.90	E032	1 2 1 2 2	
6.560E-02	9.126E+00	93.80	E032	1 2 1 2 2	
7.979E-02	1.110E+01	100	F300	1 0 0 0 2	

678. C$_6$H$_5$NO$_3$
p-Nitrophenol
4-Nitrophenol

RN: 100-02-7   **MP** (°C):   113
MW: 139.11   **BP** (°C):   279

Solubility (Moles/L)	Solubility (Grams/L)	Temp (°C)	Ref (#)	Evaluation (T P E A A)	Comments
3.576E-02	4.975E+00	0	D006	1 2 0 1 1	
5.787E-02	8.050E+00	9.99	B403	1 2 2 2 2	
7.821E-02	1.088E+01	12.5	D006	1 2 0 1 1	
7.610E-02	1.059E+01	12.60	E032	1 2 1 2 2	
5.780E-02	8.040E+00	15	D069	1 2 0 0 2	
7.305E-02	1.016E+01	15.3	A400	2 1 2 2 2	
1.139E-01	1.584E+01	17.30	E032	1 2 1 2 2	
8.770E-02	1.220E+01	19.99	B403	1 2 2 2 2	
9.700E-02	1.349E+01	20	H306	1 0 1 2 1	
7.188E-02	9.999E+00	20	T301	1 2 2 2 2	
1.078E-01	1.500E+01	22.5	G301	0 0 0 0 0	
1.132E-01	1.575E+01	25	D006	1 2 0 1 1	
1.797E-01	2.500E+01	25	D059	1 2 1 1 1	
8.411E-02	1.170E+01	25	F300	1 0 0 0 2	
9.925E-02	1.381E+01	25	R041	0 0 0 0 0	
1.121E-01	1.560E+01	25.0	A400	2 1 2 2 2	
1.430E-01	1.990E+01	26.60	E032	1 2 1 2 2	
1.794E-01	2.496E+01	27.70	E032	1 2 1 2 2	
2.101E-01	2.922E+01	29.60	E032	1 2 1 2 2	
1.280E-01	1.780E+01	29.99	B403	1 2 2 2 2	
1.409E-01	1.960E+01	30.3	A400	2 1 2 2 2	
1.930E-01	2.685E+01	34.9	A400	2 1 2 2 2	
1.718E-01	2.390E+01	37.99	B403	1 2 2 2 2	
2.026E-01	2.818E+01	40	D006	1 2 0 1 1	
2.085E-01	2.900E+01	40	D059	1 2 1 1 1	
3.021E+00	4.203E+02	40.60	E032	1 2 1 2 2	
2.678E-01	3.726E+01	40.70	E032	1 2 1 2 2	
3.081E+00	4.286E+02	42.50	E032	1 2 1 2 2	
2.961E+00	4.120E+02	42.70	E032	1 2 1 2 2	
3.196E+00	4.447E+02	49.70	E032	1 2 1 2 2	
4.350E-01	6.052E+01	50	D069	1 2 0 0 2	
4.148E-01	5.770E+01	50	F300	1 0 0 0 2	

(continued)

678. $C_6H_5NO_3$ (continued)

Solubility (Moles/L)	Solubility (Grams/L)	Temp (°C)	Ref (#)	Evaluation (T P E A A)	Comments
3.096E-01	4.306E+01	53.30	E032	1 2 1 2 2	
2.900E+00	4.034E+02	54.90	E032	1 2 1 2 2	
3.423E-01	4.762E+01	55.10	E032	1 2 1 2 2	
3.305E+00	4.598E+02	60.70	E032	1 2 1 2 2	
2.834E+00	3.942E+02	65.00	E032	1 2 1 2 2	
3.986E-01	5.545E+01	67.80	E032	1 2 1 2 2	
5.021E-01	6.985E+01	69.40	E032	1 2 1 2 2	
2.768E+00	3.850E+02	73.30	E032	1 2 1 2 2	
3.406E+00	4.739E+02	75.70	E032	1 2 1 2 2	
6.553E-01	9.116E+01	78.30	E032	1 2 1 2 2	
6.837E-01	9.510E+01	79.80	E032	1 2 1 2 2	
2.699E+00	3.754E+02	80.30	E032	1 2 1 2 2	
7.124E-01	9.910E+01	80.70	E032	1 2 1 2 2	
7.987E-01	1.111E+02	82.30	E032	1 2 1 2 2	
9.431E-01	1.312E+02	85.70	E032	1 2 1 2 2	
2.555E+00	3.554E+02	86.00	E032	1 2 1 2 2	
1.076E+00	1.497E+02	88.50	E032	1 2 1 2 2	
2.398E+00	3.336E+02	89.70	E032	1 2 1 2 2	
1.320E+00	1.837E+02	90.70	E032	1 2 1 2 2	
1.438E+00	2.000E+02	91.30	E032	1 2 1 2 2	
2.234E+00	3.107E+02	91.30	E032	1 2 1 2 2	
1.664E+00	2.315E+02	92.10	E032	1 2 1 2 2	
2.056E+00	2.861E+02	92.70	E032	1 2 1 2 2	
1.763E+00	2.453E+02	92.80	E032	1 2 1 2 2	
1.865E+00	2.595E+02	92.90	E032	1 2 1 2 2	
3.503E+00	4.873E+02	93.50	E032	1 2 1 2 2	
5.100E-02	7.095E+00	ns	B157	0 0 0 0 1	
1.148E-01	1.597E+01	ns	R427	0 0 0 0 0	

679. $C_6H_5NO_3$
m-Nitrophenol
3-Nitrophenol

RN:	554-84-7	**MP** (°C):	97
MW:	139.11	**BP** (°C):	194

Solubility (Moles/L)	Solubility (Grams/L)	Temp (°C)	Ref (#)	Evaluation (T P E A A)	Comments
6.412E-02	8.920E+00	0	D006	1 2 0 1 1	
5.176E-02	7.200E+00	9.99	B403	1 2 2 2 2	
8.524E-02	1.186E+01	12.5	D006	1 2 0 1 1	
1.243E-01	1.730E+01	15.90	E032	1 2 1 2 2	
7.764E-02	1.080E+01	19.99	B403	1 2 2 2 2	
8.300E-02	1.155E+01	20	H306	1 0 1 2 1	
1.368E-01	1.903E+01	20.20	E032	1 2 1 2 2	
1.458E-01	2.028E+01	23.40	E032	1 2 1 2 2	
9.575E-02	1.332E+01	25	D006	1 2 0 1 2	
9.740E-02	1.355E+01	25	K040	1 0 2 1 2	
9.225E-02	1.283E+01	25	R041	0 0 0 0 0	

(*continued*)

679. C$_6$H$_5$NO$_3$ (continued)

Solubility (Moles/L)	Solubility (Grams/L)	Temp (°C)	Ref (#)	Evaluation (T P E A A)	Comments
1.685E-01	2.344E+01	29.50	E032	1 2 1 2 2	
1.200E-01	1.670E+01	29.99	B403	1 2 2 2 2	
1.366E-01	1.900E+01	34.99	B403	1 2 2 2 2	
1.944E-01	2.705E+01	35.80	E032	1 2 1 2 2	
2.113E-01	2.940E+01	40	F300	1 0 0 0 2	
2.148E-01	2.988E+01	40.90	E032	1 2 1 2 2	
3.196E+00	4.445E+02	47.10	E032	1 2 1 2 2	
3.046E+00	4.237E+02	49.60	E032	1 2 1 2 2	
3.240E+00	4.507E+02	49.70	E032	1 2 1 2 2	
3.313E+00	4.609E+02	56.50	E032	1 2 1 2 2	
2.979E+00	4.145E+02	58.70	E032	1 2 1 2 2	
2.911E-01	4.049E+01	58.80	E032	1 2 1 2 2	
3.475E-01	4.834E+01	62.70	E032	1 2 1 2 2	
3.387E+00	4.712E+02	62.80	E032	1 2 1 2 2	
2.914E+00	4.054E+02	71.50	E032	1 2 1 2 2	
3.484E+00	4.846E+02	75.10	E032	1 2 1 2 2	
4.703E-01	6.542E+01	77.10	E032	1 2 1 2 2	
2.828E+00	3.935E+02	80.60	E032	1 2 1 2 2	
6.326E-01	8.801E+01	85.30	E032	1 2 1 2 2	
3.549E+00	4.937E+02	85.80	E032	1 2 1 2 2	
2.705E+00	3.762E+02	89.40	E032	1 2 1 2 2	
3.569E+00	4.965E+02	89.80	E032	1 2 1 2 2	
2.649E+00	3.684E+02	92.20	E032	1 2 1 2 2	
9.501E-01	1.322E+02	93.60	E032	1 2 1 2 2	
2.581E+00	3.591E+02	94.20	E032	1 2 1 2 2	
2.475E+00	3.443E+02	95.60	E032	1 2 1 2 2	
1.210E+00	1.683E+02	96.20	E032	1 2 1 2 2	
2.396E+00	3.333E+02	96.60	E032	1 2 1 2 2	
1.440E+00	2.004E+02	97.50	E032	1 2 1 2 2	
2.286E+00	3.181E+02	97.70	E032	1 2 1 2 2	
1.604E+00	2.232E+02	98.10	E032	1 2 1 2 2	
2.341E+00	3.256E+02	98.10	E032	1 2 1 2 2	
1.763E+00	2.453E+02	98.40	E032	1 2 1 2 2	
2.049E+00	2.851E+02	98.50	E032	1 2 1 2 2	
1.965E+00	2.734E+02	98.60	E032	1 2 1 2 2	
3.008E+00	4.184E+02	98.70	F300	1 0 0 0 2	
9.772E-02	1.359E+01	ns	R427	0 0 0 0 0	

680. C$_6$H$_5$NO$_4$
Nitrohydroquinone
2-Nitroquinol
4-Hydroxy-2-nitrophenol

RN:　　16090-33-8　　**MP** (°C):
MW:　　155.11　　　　**BP** (°C):

Solubility (Moles/L)	Solubility (Grams/L)	Temp (°C)	Ref (#)	Evaluation (T P E A A)	Comments
6.888E-02	1.068E+01	30.20	E032	1 2 1 2 2	
1.015E-01	1.575E+01	34.60	E032	1 2 1 2 2	

(continued)

680. C$_6$H$_5$NO$_4$ (continued)

Solubility (Moles/L)	Solubility (Grams/L)	Temp (°C)	Ref (#)	Evaluation (T P E A A)	Comments
1.572E-01	2.439E+01	44.60	E032	1 2 1 2 2	
1.999E-01	3.101E+01	49.60	E032	1 2 1 2 2	
3.128E-01	4.853E+01	54.50	E032	1 2 1 2 2	
4.498E-01	6.977E+01	59.10	E032	1 2 1 2 2	
6.405E-01	9.934E+01	61.70	E032	1 2 1 2 2	
7.163E-01	1.111E+02	64.20	E032	1 2 1 2 2	
8.409E-01	1.304E+02	65.00	E032	1 2 1 2 2	
1.074E+00	1.667E+02	93.80	E032	1 2 1 2 2	

681. C$_6$H$_5$NO$_4$

4-Nitroresorcinol
4-Nitro-1,3-benzenediol

RN: 3163-07-3   **MP** (°C):
MW: 155.11   **BP** (°C):

Solubility (Moles/L)	Solubility (Grams/L)	Temp (°C)	Ref (#)	Evaluation (T P E A A)	Comments
4.354E-02	6.754E+00	18.30	E032	1 2 1 2 2	
5.244E-02	8.133E+00	24.70	E032	1 2 1 2 2	
6.510E-02	1.010E+01	30.80	E032	1 2 1 2 2	
7.959E-02	1.235E+01	36.90	E032	1 2 1 2 2	
1.034E-01	1.604E+01	43.50	E032	1 2 1 2 2	
1.462E-01	2.267E+01	47.50	E032	1 2 1 2 2	
1.817E-01	2.818E+01	49.10	E032	1 2 1 2 2	
2.168E-01	3.363E+01	50.70	E032	1 2 1 2 2	
2.497E-01	3.874E+01	51.20	E032	1 2 1 2 2	
2.776E-01	4.306E+01	52.30	E032	1 2 1 2 2	
3.286E-01	5.096E+01	53.90	E032	1 2 1 2 2	
4.487E-01	6.959E+01	57.80	E032	1 2 1 2 2	
5.951E-01	9.231E+01	62.70	E032	1 2 1 2 2	
8.468E-01	1.313E+02	68.40	E032	1 2 1 2 2	
1.075E+00	1.667E+02	71.90	E032	1 2 1 2 2	
1.209E+00	1.875E+02	72.90	E032	1 2 1 2 2	
1.325E+00	2.055E+02	73.30	E032	1 2 1 2 2	
1.487E+00	2.307E+02	73.40	E032	1 2 1 2 2	

682. C$_6$H$_5$NO$_4$

2-Nitroresorcinol
2-Nitro-1,3-benzenediol

RN: 601-89-8   **MP** (°C): 81
MW: 155.11   **BP** (°C):

Solubility (Moles/L)	Solubility (Grams/L)	Temp (°C)	Ref (#)	Evaluation (T P E A A)	Comments
8.435E-03	1.308E+00	28.40	E032	1 2 1 2 2	
1.306E-02	2.026E+00	36.70	E032	1 2 1 2 2	
2.319E-02	3.597E+00	47.60	E032	1 2 1 2 2	
3.635E-02	5.638E+00	54.90	E032	1 2 1 2 2	

(continued)

682. $C_6H_5NO_4$ (continued)

Solubility (Moles/L)	Solubility (Grams/L)	Temp (°C)	Ref (#)	Evaluation (T P E A A)	Comments
6.276E-02	9.734E+00	67.20	E032	1 2 1 2 2	
8.399E-02	1.303E+01	74.40	E032	1 2 1 2 2	
1.208E-01	1.874E+01	82.90	E032	1 2 1 2 2	
1.529E-01	2.372E+01	92.30	E032	1 2 1 2 2	

683. $C_6H_5NO_4$
3-Nitrocatechol
3-Nitro-1,2-benzenediol

RN: 6665-98-1   **MP** (°C):
MW: 155.11   **BP** (°C):

Solubility (Moles/L)	Solubility (Grams/L)	Temp (°C)	Ref (#)	Evaluation (T P E A A)	Comments
5.377E-02	8.340E+00	14.40	E032	1 2 1 2 2	
6.573E-02	1.019E+01	20.90	E032	1 2 1 2 2	
9.590E-02	1.488E+01	29.50	E032	1 2 1 2 2	
1.277E-01	1.980E+01	35.10	E032	1 2 1 2 2	
1.474E-01	2.286E+01	37.90	E032	1 2 1 2 2	
1.738E-01	2.695E+01	41.00	E032	1 2 1 2 2	
2.372E-01	3.679E+01	45.80	E032	1 2 1 2 2	
2.646E-01	4.104E+01	47.60	E032	1 2 1 2 2	
3.216E-01	4.988E+01	54.50	E032	1 2 1 2 2	
3.615E-01	5.607E+01	61.30	E032	1 2 1 2 2	
4.548E-01	7.055E+01	75.90	E032	1 2 1 2 2	
5.743E-01	8.909E+01	86.80	E032	1 2 1 2 2	
8.164E-01	1.266E+02	96.80	E032	1 2 1 2 2	

684. $C_6H_5NO_4$
4-Nitrocatechol
4-Nitro-1,2-benzenediol

RN: 3316-09-4   **MP** (°C):
MW: 155.11   **BP** (°C):

Solubility (Moles/L)	Solubility (Grams/L)	Temp (°C)	Ref (#)	Evaluation (T P E A A)	Comments
1.211E+00	1.878E+02	24.60	E032	1 2 1 2 2	
1.423E+00	2.208E+02	37.70	E032	1 2 1 2 2	
1.488E+00	2.308E+02	41.30	E032	1 2 1 2 2	
1.664E+00	2.582E+02	51.90	E032	1 2 1 2 2	
1.829E+00	2.837E+02	58.50	E032	1 2 1 2 2	
2.004E+00	3.109E+02	66.50	E032	1 2 1 2 2	
2.049E+00	3.179E+02	67.80	E032	1 2 1 2 2	
2.149E+00	3.334E+02	71.20	E032	1 2 1 2 2	

685. $C_6H_5NO_5S$

p-Nitrobenzenesulfonic acid
4-Nitrobenzenesulfonic acid

RN: 138-42-1   **MP** (°C):
MW: 203.17   **BP** (°C):

Solubility (Moles/L)	Solubility (Grams/L)	Temp (°C)	Ref (#)	Evaluation (T P E A A)	Comments
2.343E+00	4.760E+02	100.5	T023	1 2 2 1 2	
2.412E+00	4.901E+02	105.0	T023	1 2 2 1 2	
2.461E+00	5.000E+02	110.0	T023	1 2 2 1 2	

686. $C_6H_5NO_5S.2H_2O$

p-Nitrobenzenesulfonic acid (dihydrate)

RN: 15481-55-7   **MP** (°C):
MW: 239.21   **BP** (°C):

Solubility (Moles/L)	Solubility (Grams/L)	Temp (°C)	Ref (#)	Evaluation (T P E A A)	Comments
1.667E+00	3.987E+02	36.6	T023	1 2 2 1 2	
1.720E+00	4.113E+02	56.6	T023	1 2 2 1 2	
1.771E+00	4.235E+02	75.5	T023	1 2 2 1 2	
1.822E+00	4.359E+02	90.2	T023	1 2 2 1 2	
1.939E+00	4.638E+02	106.8	T023	1 2 2 1 2	
1.920E+00	4.592E+02	110.2	T023	1 2 2 1 2	

687. $C_6H_5NO_5S.4H_2O$

p-Nitrobenzenesulfonic acid (tetrahydrate)

RN: 15481-55-7   **MP** (°C):
MW: 275.24   **BP** (°C):

Solubility (Moles/L)	Solubility (Grams/L)	Temp (°C)	Ref (#)	Evaluation (T P E A A)	Comments
1.060E+00	2.919E+02	−8.3	T023	1 2 2 1 2	
1.146E+00	3.153E+02	−1.0	T023	1 2 2 1 2	
1.273E+00	3.504E+02	10.8	T023	1 2 2 1 2	
1.318E+00	3.627E+02	16.0	T023	1 2 2 1 2	
1.409E+00	3.877E+02	26.3	T023	1 2 2 1 2	

688. $C_6H_5N_2OS$

Methyl acetylthiodiazole
Thiodiazolique methyle acetyle

RN: **MP** (°C):
MW: 153.18   **BP** (°C):

Solubility (Moles/L)	Solubility (Grams/L)	Temp (°C)	Ref (#)	Evaluation (T P E A A)	Comments
6.528E-04	1.000E-01	37	D084	1 0 1 0 1	

689. $C_6H_5N_3$
Benzotriazole
1,2,3-Benzotriazole
Cobratec 99
1,2,3-triaza-1H-indene
Azimidobenzene
Benzene azimide

RN: 95-14-7   **MP** (°C):   98.5
MW: 119.13   **BP** (°C):   350

Solubility (Moles/L)	Solubility (Grams/L)	Temp (°C)	Ref (#)	Evaluation (T P E A A)	Comments
1.660E-01	1.977E+01	ns	R427	0 0 0 0 0	

690. $C_6H_5N_3O_4$
2,6-Dinitroaniline
2,6-Dinitrobenzenamine

RN: 606-22-4   **MP** (°C):   133
MW: 183.12   **BP** (°C):

Solubility (Moles/L)	Solubility (Grams/L)	Temp (°C)	Ref (#)	Evaluation (T P E A A)	Comments
4.365E-04	7.994E-02	25	B335	1 2 0 0 1	

691. $C_6H_5N_3O_4$
2,4-Dinitroaniline
2,4-Dinitrobenzenamine
2,4-Dinitroaminobenzene
1-Amino-2,4-dinitrobenzene

RN: 97-02-9   **MP** (°C):   176
MW: 183.12   **BP** (°C):

Solubility (Moles/L)	Solubility (Grams/L)	Temp (°C)	Ref (#)	Evaluation (T P E A A)	Comments
4.266E-04	7.812E-02	25	B335	1 2 0 0 1	

692. $C_6H_5N_3O_5$
Picramic acid
2-Amino-4,6-dinitro-phenol

RN: 96-91-3   **MP** (°C):   169
MW: 199.12   **BP** (°C):

Solubility (Moles/L)	Solubility (Grams/L)	Temp (°C)	Ref (#)	Evaluation (T P E A A)	Comments
7.031E-03	1.400E+00	22	F300	1 0 0 0 1	

693. C₆H₅N₅

$C_6H_5N_5$

7-Aminopteridine
7-Pteridinamine

RN: 769-66-4   **MP** (°C):
MW: 147.14   **BP** (°C):

Solubility (Moles/L)	Solubility (Grams/L)	Temp (°C)	Ref (#)	Evaluation (T P E A A)	Comments
4.851E-03	7.138E-01	20	A083	1 2 0 0 0	
3.974E-02	5.848E+00	100	A083	1 2 0 0 0	

694. C₆H₅N₅

$C_6H_5N_5$

4-Aminopteridine
4-Pteridinamine

RN: 6973-01-9   **MP** (°C):   305
MW: 147.14   **BP** (°C):

Solubility (Moles/L)	Solubility (Grams/L)	Temp (°C)	Ref (#)	Evaluation (T P E A A)	Comments
4.851E-03	7.138E-01	22.5	A085	1 2 0 0 0	

695. C₆H₅N₅

$C_6H_5N_5$

2-Aminopteridine
2-Pteridinamine

RN: 700-81-2   **MP** (°C):
MW: 147.14   **BP** (°C):

Solubility (Moles/L)	Solubility (Grams/L)	Temp (°C)	Ref (#)	Evaluation (T P E A A)	Comments
5.031E-03	7.402E-01	22.5	A085	1 2 0 0 0	

696. C₆H₅N₅O

$C_6H_5N_5O$

4-Amino-2-hydroxypteridine
4-Amino-2-oxopteridine
4-Aminopteridin-2-one
4-Amino-2-pteridone

RN: 22005-65-8   **MP** (°C):   >350
MW: 163.14   **BP** (°C):

Solubility (Moles/L)	Solubility (Grams/L)	Temp (°C)	Ref (#)	Evaluation (T P E A A)	Comments
4.378E-04	7.142E-02	20	A019	2 2 1 1 2	
5.104E-03	8.326E-01	100	A019	1 2 1 1 2	

697. C$_6$H$_5$N$_5$O
2-Amino-4-hydroxypteridine
2-Amino-4(1H)-pteridinone
2-Amino-4(3H)-pteridinone
2-Amino-4-pteridone
2-Amino-4-oxopteridine
2-Aminopteridin-4-one

RN: 2236-60-4   **MP** (°C):
MW: 163.14   **BP** (°C):

Solubility (Moles/L)	Solubility (Grams/L)	Temp (°C)	Ref (#)	Evaluation (T P E A A)	Comments
1.075E-04	1.754E-02	22.5	A085	1 2 0 0 0	

698. C$_6$H$_5$N$_5$O
7-Amino-6-hydroxypteridine
7-Amino-6-oxopteridine
7-Aminopteridin-6-one
7-Amino-6-pteridone

RN: 1008-85-1   **MP** (°C):
MW: 163.14   **BP** (°C):

Solubility (Moles/L)	Solubility (Grams/L)	Temp (°C)	Ref (#)	Evaluation (T P E A A)	Comments
1.226E-03	2.000E-01	100	A082	1 2 0 0 0	

699. C$_6$H$_5$N$_5$O$_2$
Xanthopterin
2-Amino-4:6-dihydroxypteridine

RN: 119-44-8   **MP** (°C):
MW: 179.14   **BP** (°C):

Solubility (Moles/L)	Solubility (Grams/L)	Temp (°C)	Ref (#)	Evaluation (T P E A A)	Comments
1.396E-04	2.500E-02	22.5	A085	1 2 0 0 0	

700. C$_6$H$_5$N$_5$O$_3$
Leucopterin
2-Amino-4:6:7-trihydroxypteridine

RN: 492-11-5   **MP** (°C):
MW: 195.14   **BP** (°C):

Solubility (Moles/L)	Solubility (Grams/L)	Temp (°C)	Ref (#)	Evaluation (T P E A A)	Comments
6.833E-06	1.333E-03	22.5	A085	1 2 0 0 0	

701. $C_6H_5N_5O_4S$
3′-Nitrosoniridazole
2-Imidazolidinone, 1-nitroso-3-(5-nitro-2-thiazolyl)-
RN: 34968-90-6   **MP** (°C):   202-203
MW: 243.20   **BP** (°C):

Solubility (Moles/L)	Solubility (Grams/L)	Temp (°C)	Ref (#)	Evaluation (T P E A A)	Comments
3.084E-04	7.500E-02	25	G051	1 0 1 1 0	

702. C_6H_6
Benzene
Benzol
Phenyl hydride
Cyclohexatriene
Benzolene
Phene
RN: 71-43-2   **MP** (°C):   5
MW: 78.11   **BP** (°C):   80

Solubility (Moles/L)	Solubility (Grams/L)	Temp (°C)	Ref (#)	Evaluation (T P E A A)	Comments
2.350E-02	1.836E+00	.20	M151	2 1 2 2 2	
2.347E-02	1.833E+00	.24	M183	1 2 1 1 2	
1.959E-02	1.530E+00	0	F300	1 0 0 0 2	
2.148E-02	1.678E+00	0	P003	2 2 2 2 2	
2.356E-02	1.840E+00	.80	A004	1 2 2 1 2	
2.351E-02	1.837E+00	4.50	B086	2 1 2 2 2	
1.881E-02	1.469E+00	4.62	U013	1 0 0 0 0	EFG
2.646E-02	2.067E+00	4.8	L007	2 1 1 2 2	
1.178E-02	9.200E-01	5	S119	0 0 0 0 1	
2.646E-02	2.067E+00	5.0	L007	2 1 1 1 2	
1.838E-02	1.436E+00	5.39	U010	1 0 0 1 1	EFG
2.310E-02	1.804E+00	6.20	M151	2 1 2 2 2	
2.306E-02	1.802E+00	6.24	M183	1 2 1 1 2	
2.364E-02	1.847E+00	6.30	B086	2 1 2 2 2	
2.313E-02	1.807E+00	7.10	B086	2 1 2 2 2	
2.313E-02	1.807E+00	9	B086	2 1 2 2 2	
2.292E-02	1.790E+00	9.40	A004	1 2 2 1 2	
2.080E-02	1.625E+00	10	B149	2 1 1 2 2	
2.110E-02	1.648E+00	10	J302	2 1 2 2 2	
2.240E-02	1.750E+00	10	M130	1 0 0 0 2	
2.300E-02	1.797E+00	11.00	M151	2 1 2 2 2	
2.300E-02	1.796E+00	11.04	M183	1 2 1 1 2	
2.262E-02	1.767E+00	11.80	B086	2 1 2 2 2	
2.262E-02	1.767E+00	12.10	B086	2 1 2 2 2	
2.270E-02	1.773E+00	14.00	M151	2 1 2 2 2	
2.263E-02	1.767E+00	14.04	M183	1 2 1 1 2	
1.838E-02	1.436E+00	14.20	U013	1 0 0 0 0	EFG
2.655E-02	2.074E+00	14.8	L007	2 1 1 2 2	

(continued)

702. C₆H₆ (continued)

Solubility (Moles/L)	Solubility (Grams/L)	Temp (°C)	Ref (#)	Evaluation (T P E A A)	Comments
2.655E-02	2.074E+00	14.9	L007	2 1 1 1 2	
2.290E-02	1.789E+00	15	I333	1 2 1 1 2	
2.150E-02	1.679E+00	15	S006	1 0 0 0 2	
1.971E-02	1.540E+00	15	S203	1 1 2 1 2	
1.797E-02	1.403E+00	15.02	U010	1 0 0 1 1	EFG
2.287E-02	1.787E+00	15.10	B086	2 1 2 2 2	
2.112E-02	1.650E+00	16	D047	1 0 0 1 2	
2.266E-02	1.770E+00	16.80	A004	1 2 2 1 2	
2.260E-02	1.765E+00	16.90	M151	2 1 2 2 2	
2.253E-02	1.760E+00	16.94	M183	1 2 1 1 2	
2.191E-02	1.711E+00	17	F002	2 2 2 2 2	
2.287E-02	1.787E+00	17.90	B086	2 1 2 2 2	
2.260E-02	1.765E+00	18.60	M151	2 1 2 2 2	
2.259E-02	1.764E+00	18.64	M183	1 2 1 1 2	
2.664E-02	2.081E+00	19.8	L007	2 1 1 2 2	
2.664E-02	2.081E+00	19.9	L007	2 1 1 1 2	
2.220E-02	1.734E+00	20	B149	2 1 1 2 2	
2.180E-02	1.703E+00	20	C006	1 2 1 1 2	
1.023E-02	7.994E-01	20	C121	0 0 0 0 0	unit assumed, *sic*
2.428E-02	1.896E+00	20	D052	1 1 0 0 1	
1.600E-02	1.250E+00	20	E009	1 0 0 0 1	
1.680E-02	1.312E+00	20	E025	1 0 2 2 2	
2.189E-02	1.710E+00	20	F071	1 1 2 1 2	
2.317E-02	1.810E+00	20	F300	1 0 0 0 2	
1.023E-02	7.994E-01	20	I310	0 0 0 0 0	
2.310E-02	1.804E+00	20	I333	1 2 1 1 2	
2.042E-02	1.595E+00	20	K337	1 0 0 0 2	
2.280E-02	1.781E+00	20	M312	1 0 0 0 1	
1.366E-02	1.067E+00	20	M337	2 1 2 2 2	
2.650E-02	2.070E+00	20	P073	1 0 0 1 2	
1.751E-02	1.368E+00	20.0	H043	2 2 2 2 2	
2.249E-02	1.757E+00	20.10	B086	2 1 2 2 2	
2.224E-02	1.737E+00	21	C024	2 1 1 2 2	
2.202E-02	1.720E+00	22	F002	2 2 2 2 2	
2.320E-02	1.812E+00	22.5	I333	1 2 1 1 2	
2.304E-02	1.800E+00	24	A004	1 2 2 1 2	
2.667E-02	2.084E+00	24.8	L007	2 1 1 2 2	
2.227E-02	1.740E+00	25	A001	1 2 2 2 2	
1.917E-02	1.498E+00	25	A037	2 2 2 2 2	
2.292E-02	1.790E+00	25	B003	2 2 2 2 2	
2.045E-02	1.597E+00	25	B019	1 0 1 2 0	
2.279E-02	1.780E+00	25	B060	2 0 1 1 1	
2.292E-02	1.790E+00	25	B090	2 2 2 1 2	
2.292E-02	1.790E+00	25	B151	1 2 2 1 2	
2.330E-02	1.820E+00	25	B153	2 1 1 1 2	
2.240E-02	1.750E+00	25	B173	2 0 2 2 2	
2.300E-02	1.797E+00	25	G323	2 2 2 2 2	
2.300E-02	1.797E+00	25	H332	2 2 2 2 1	
2.330E-02	1.820E+00	25	I333	1 2 1 1 2	

(continued)

702. C₆H₆ (continued)

Solubility (Moles/L)	Solubility (Grams/L)	Temp (°C)	Ref (#)	Evaluation (T P E A A)	Comments
2.310E-02	1.804E+00	25	J302	2 1 2 2 2	
2.390E-02	1.867E+00	25	K001	2 2 2 2 2	
8.961E-03	7.000E-01	25	K072	1 0 1 1 1	
1.300E-02	1.015E+00	25	K123	1 0 2 2 1	
2.170E-02	1.695E+00	25	K316	2 2 2 2 2	
2.259E-02	1.765E+00	25	L002	2 2 2 2 2	
2.313E-02	1.807E+00	25	L319	1 0 2 1 1	
2.166E-02	1.692E+00	25	L322	1 1 2 2 1	
1.770E+00	1.383E+02	25	M021	2 2 2 1 2	*sic*
2.279E-02	1.780E+00	25	M131	1 0 0 0 2	
2.278E-02	1.780E+00	25	M132	2 2 2 1 2	
2.310E-02	1.804E+00	25	M151	2 1 2 2 2	average of 2
2.293E-02	1.791E+00	25	M151	2 1 1 2 2	
2.290E-02	1.789E+00	25	M342	1 0 1 1 2	
1.917E-02	1.498E+00	25	O015	0 0 0 0 0	
2.247E-02	1.755E+00	25	P003	2 2 2 2 2	
2.227E-02	1.740E+00	25	P051	2 1 1 2 2	
2.607E-02	2.036E+00	25	S010	2 1 2 1 2	
2.377E-02	1.857E+00	25	S012	2 0 2 2 2	
2.061E-02	1.610E+00	25	S203	1 1 2 1 2	
2.070E-02	1.617E+00	25	S359	2 1 2 2 2	
2.778E-02	2.170E+00	25	W057	2 0 2 2 2	
2.290E-02	1.789E+00	25	W300	2 2 2 2 2	
2.300E-02	1.797E+00	25.0	H043	2 2 2 2 2	
2.667E-02	2.084E+00	25.0	L007	2 1 1 1 2	
2.227E-02	1.740E+00	25.00	P007	2 1 2 2 2	
2.290E-02	1.789E+00	25.04	M183	1 2 1 1 2	
1.838E-02	1.436E+00	25.35	U010	1 0 0 1 1	EFG
1.881E-02	1.469E+00	25.35	U013	1 0 0 0 0	EFG
2.325E-02	1.816E+00	25.84	M183	1 2 1 1 2	
2.213E-02	1.729E+00	26	F002	2 2 2 2 2	
2.229E-02	1.742E+00	29	F002	2 2 2 2 2	
2.351E-02	1.837E+00	29.99	C349	0 0 0 0 0	
2.368E-02	1.850E+00	30	F300	1 0 0 0 2	
2.364E-02	1.847E+00	30	G029	1 0 2 2 2	
2.350E-02	1.836E+00	30	I333	1 2 1 1 2	
2.343E-02	1.830E+00	31	A004	1 2 2 1 2	
2.285E-02	1.785E+00	32	F002	2 2 2 2 2	
1.970E-02	1.539E+00	34.53	U013	1 0 0 0 0	EFG
2.685E-02	2.098E+00	34.8	L007	2 1 1 2 2	
2.329E-02	1.819E+00	35	F002	2 2 2 2 2	
2.253E-02	1.760E+00	35	S203	1 1 2 1 2	
2.685E-02	2.098E+00	35.1	L007	2 1 1 1 2	
1.925E-02	1.504E+00	35.48	U010	1 0 0 1 1	EFG
2.458E-02	1.920E+00	38	A004	1 2 2 1 2	
2.573E-02	2.010E+00	39.99	C349	0 0 0 0 0	
2.592E-02	2.025E+00	40	B151	1 2 1 1 2	
2.434E-02	1.902E+00	41	F002	2 2 2 2 2	
2.440E-02	1.906E+00	42	F002	2 2 2 2 2	

(continued)

702. C$_6$H$_6$ (continued)

Solubility (Moles/L)	Solubility (Grams/L)	Temp (°C)	Ref (#)	Evaluation (T P E A A)	Comments
2.467E-02	1.927E+00	44	F002	2 2 2 2 2	
2.016E-02	1.574E+00	44.30	U010	1 0 0 1 1	EFG
2.062E-02	1.611E+00	44.30	U013	1 0 0 0 0	EFG
2.599E-02	2.030E+00	44.70	A004	1 2 2 1 2	
2.368E-02	1.850E+00	45	S203	1 1 2 1 2	
2.938E-02	2.295E+00	45.7	L007	2 1 1 1 2	
2.938E-02	2.295E+00	45.8	L007	2 1 1 2 2	
2.534E-02	1.979E+00	46	F002	2 2 2 2 2	
2.827E-02	2.208E+00	49.99	C349	0 0 0 0 0	
2.810E-02	2.195E+00	50	G323	2 2 2 2 1	
2.650E-02	2.070E+00	51	F002	2 2 2 2 2	
2.740E-02	2.140E+00	51.50	A004	1 2 2 1 2	
2.159E-02	1.687E+00	53.64	U010	1 0 0 1 1	EFG
2.210E-02	1.726E+00	54.71	U013	1 0 0 0 0	EFG
5.095E-02	3.980E+00	55.3	P051	2 1 1 2 2	
5.095E-02	3.980E+00	55.30	P007	2 1 2 2 2	
2.788E-02	2.178E+00	56	F002	2 2 2 2 2	
3.162E-02	2.470E+00	57	B124	2 2 2 1 2	
3.776E-02	2.950E+00	57.70	B124	1 2 2 1 2	
2.996E-02	2.340E+00	58.80	A004	1 2 2 1 2	
3.131E-02	2.446E+00	59.99	C349	0 0 0 0 0	
2.938E-02	2.295E+00	60	B126	1 0 1 1 1	
3.101E-02	2.422E+00	60	B151	1 2 1 1 2	
2.943E-02	2.299E+00	61	F002	2 2 2 2 2	
3.004E-02	2.347E+00	63	F002	2 2 2 2 2	
3.290E-02	2.570E+00	65.40	A004	1 2 2 1 2	
2.479E-02	1.936E+00	65.82	U013	1 0 0 0 0	EFG
3.597E-02	2.810E+00	69.20	B124	1 2 2 1 2	
3.587E-02	2.802E+00	69.30	B124	1 0 2 2 2	
3.463E-02	2.705E+00	69.99	C349	0 0 0 0 0	
8.280E-02	6.468E+00	74.7	P051	2 1 1 2 2	
8.280E-02	6.468E+00	74.70	P007	2 1 2 2 2	
3.872E-02	3.024E+00	79.99	C349	0 0 0 0 0	
4.429E-02	3.460E+00	89.99	C349	0 0 0 0 0	
2.560E-02	2.000E+00	100	J023	1 1 2 2 0	
5.256E-02	4.106E+00	99.99	C349	0 0 0 0 0	
7.681E-02	6.000E+00	150	J023	1 1 2 2 0	
2.688E-01	2.100E+01	200	J023	1 1 2 2 1	
9.345E-01	7.300E+01	250	J023	1 1 2 2 1	
1.357E+00	1.060E+02	285	J023	1 1 2 2 2	
1.869E+00	1.460E+02	300	J023	1 1 2 2 2	
2.200E-02	1.719E+00	ns	B059	0 0 1 1 2	
4.000E-03	3.125E-01	ns	D348	0 0 0 0 0	
2.279E-02	1.780E+00	ns	H123	0 0 0 0 0	
3.020E-01	2.359E+01	ns	H307	0 0 0 0 0	
4.500E-02	3.515E+00	ns	H333	0 1 0 1 0	EFG
2.330E-02	1.820E+00	ns	I332	0 0 0 0 2	
2.292E-02	1.790E+00	ns	K304	0 0 0 0 2	
1.933E-02	1.510E+00	ns	M010	0 0 0 0 2	

(continued)

702. C$_6$H$_6$ (continued)

Solubility (Moles/L)	Solubility (Grams/L)	Temp (°C)	Ref (#)	Evaluation (T P E A A)	Comments
2.265E-02	1.769E+00	ns	M175	0 0 2 1 2	
2.279E-02	1.780E+00	ns	M344	0 0 0 0 2	

703. C$_6$H$_6$BrNO$_2$S
4-Bromobenzenesulfonamide
(4-Bromophenyl)sulfonamide
p-Bromobenzenesulfonamide
4-Aminosulfonyl-1-bromobenzene
RN: 701-34-8   **MP** (°C):
MW: 236.09   **BP** (°C):

Solubility (Moles/L)	Solubility (Grams/L)	Temp (°C)	Ref (#)	Evaluation (T P E A A)	Comments
4.200E-03	9.916E-01	15	K024	1 2 1 1 2	

704. C$_6$H$_6$BrNO$_3$S
p-Bromoaniline-o-sulfonic acid
2-Amino-5-bromophenylsulfonic acid
RN: 1576-59-6   **MP** (°C):
MW: 252.09   **BP** (°C):

Solubility (Moles/L)	Solubility (Grams/L)	Temp (°C)	Ref (#)	Evaluation (T P E A A)	Comments
1.107E-02	2.790E+00	8.35	P038	1 0 1 0 2	anhydrate
1.424E-02	3.590E+00	16.75	P038	1 0 1 0 2	anhydrate
1.769E-02	4.460E+00	25.0	P038	1 0 1 0 2	anhydrate
2.578E-02	6.500E+00	40.0	P038	1 0 1 0 2	anhydrate
3.828E-02	9.650E+00	55.0	P038	1 0 1 0 2	anhydrate
5.454E-02	1.375E+01	70.0	P038	1 0 1 0 2	anhydrate
8.013E-02	2.020E+01	85.0	P038	1 0 1 0 2	anhydrate
8.846E-03	2.230E+00	.0	P038	1 0 1 0 2	anhydrate

705. C$_6$H$_6$BrNO$_3$S
p-Bromoaniline-m-sulfonic acid
5-Amino-2-bromobenzenesulfonic acid
RN: 150454-14-1   **MP** (°C):
MW: 252.09   **BP** (°C):

Solubility (Moles/L)	Solubility (Grams/L)	Temp (°C)	Ref (#)	Evaluation (T P E A A)	Comments
3.511E-02	8.850E+00	9.8	P038	1 2 2 2 2	anhydrous rhombic
2.559E-02	6.450E+00	12.55	P038	1 2 2 2 2	anhydrous monoclinic
4.284E-02	1.080E+01	20.0	P038	1 2 2 2 2	anhydrous rhombic
3.419E-02	8.620E+00	25.0	P038	1 2 2 2 2	anhydrous monoclinic
4.740E-02	1.195E+01	25.0	P038	1 2 2 2 2	anhydrous rhombic
5.177E-02	1.305E+01	29.6	P038	1 2 2 2 2	anhydrous rhombic
5.732E-02	1.445E+01	34.7	P038	1 2 2 2 2	anhydrous rhombic
4.820E-02	1.215E+01	40.0	P038	1 2 2 2 2	anhydrous monoclinic

(continued)

705. C₆H₆BrNO₃S (continued)

Solubility (Moles/L)	Solubility (Grams/L)	Temp (°C)	Ref (#)	Evaluation (T P E A A)	Comments
6.387E-02	1.610E+01	40.1	P038	1 2 2 2 2	anhydrous rhombic
6.922E-02	1.745E+01	44.5	P038	1 2 2 2 2	anhydrous rhombic
7.577E-02	1.910E+01	49.7	P038	1 2 2 2 2	anhydrous rhombic
8.330E-02	2.100E+01	54.8	P038	1 2 2 2 2	anhydrous rhombic
7.101E-02	1.790E+01	56.3	P038	1 2 2 2 2	anhydrous monoclinic
9.600E-02	2.420E+01	62.3	P038	1 2 2 2 2	anhydrous rhombic
9.679E-02	2.440E+01	70.0	P038	1 2 2 2 2	anhydrous monoclinic
1.115E-01	2.810E+01	70.4	P038	1 2 2 2 2	anhydrous rhombic
1.329E-01	3.350E+01	85.0	P038	1 2 2 2 2	anhydrous monoclinic
1.452E-01	3.660E+01	85.0	P038	1 2 2 2 2	anhydrous rhombic
2.880E-02	7.260E+00	.0	P038	1 2 2 2 2	anhydrous rhombic
1.884E-02	4.750E+00	.0	P038	1 2 2 2 2	anhydrous monoclinic

706. C₆H₆BrNO₃S.H₂O

p-Bromoaniline-*o*-sulfonic acid (monohydrate)

2-Amino-5-bromophenylsulfonic acid (monohydrate)

RN: 1576-59-6   **MP** (°C):

MW: 270.11   **BP** (°C):

Solubility (Moles/L)	Solubility (Grams/L)	Temp (°C)	Ref (#)	Evaluation (T P E A A)	Comments
1.303E-02	3.520E+00	8.35	P038	1 0 1 0 2	monohydrate
1.751E-02	4.730E+00	16.8	P038	1 0 1 0 2	monohydrate
2.244E-02	6.060E+00	25.0	P038	1 0 1 0 2	monohydrate
9.589E-03	2.590E+00	.0	P038	1 0 1 0 2	monohydrate

707. C₆H₆ClN

p-Chloroaniline

4-Chloroaniline

RN: 106-47-8   **MP** (°C):   72.5

MW: 127.57   **BP** (°C):   232

Solubility (Moles/L)	Solubility (Grams/L)	Temp (°C)	Ref (#)	Evaluation (T P E A A)	Comments
2.157E-02	2.752E+00	20	H118	1 1 1 1 2	
2.157E-02	2.752E+00	20	H301	0 0 0 0 0	
3.057E-02	3.900E+00	22.5	G301	0 0 0 0 0	

708. C₆H₆ClN

o-Chloroaniline

2-Chloroaniline

RN: 95-51-2   **MP** (°C):   −1

MW: 127.57   **BP** (°C):   208.8

Solubility (Moles/L)	Solubility (Grams/L)	Temp (°C)	Ref (#)	Evaluation (T P E A A)	Comments
2.951E-02	3.765E+00	20	C113	1 0 2 1 2	

709. C₆H₆ClN

m-Chloroaniline
3-Chloroaniline

RN:	108-42-9	MP (°C):	−10
MW:	127.57	BP (°C):	230.0

Solubility (Moles/L)	Solubility (Grams/L)	Temp (°C)	Ref (#)	Evaluation (T P E A A)	Comments
4.266E-02	5.442E+00	20	C113	1 0 2 1 2	

710. C₆H₆ClNO₂S

m-Chlorobenzenesulfonamide
MON 5783

RN:	17260-71-8	MP (°C):	
MW:	191.64	BP (°C):	

Solubility (Moles/L)	Solubility (Grams/L)	Temp (°C)	Ref (#)	Evaluation (T P E A A)	Comments
3.500E-03	6.707E-01	15	K024	1 2 1 1 2	

711. C₆H₆ClNO₂S

o-Chlorobenzenesulfonamide
2-Chlorobenzenesulfonamide

RN:	6961-82-6	MP (°C):	
MW:	191.64	BP (°C):	

Solubility (Moles/L)	Solubility (Grams/L)	Temp (°C)	Ref (#)	Evaluation (T P E A A)	Comments
2.600E-03	4.983E-01	15	K024	1 2 1 1 2	

712. C₆H₆ClNO₂S

4-Chlorobenzenesulfonamide
p-Chlorobenzenesulfonamide

RN:	98-64-6	MP (°C):	
MW:	191.64	BP (°C):	

Solubility (Moles/L)	Solubility (Grams/L)	Temp (°C)	Ref (#)	Evaluation (T P E A A)	Comments
6.900E-03	1.322E+00	15	K024	1 2 1 1 2	

713. C₆H₆ClNO₃S

p-Chloroaniline-m-sulfonic acid
1-Amino-4-chlorobenzene-3-sulfonic acid
4-Chloro-3-sulfoaniline
3-Amino-6-chlorobenzenesulfonic acid

RN:	88-43-7	MP (°C):	
MW:	207.64	BP (°C):	

Solubility (Moles/L)	Solubility (Grams/L)	Temp (°C)	Ref (#)	Evaluation (T P E A A)	Comments
5.447E-02	1.131E+01	0	P038	1 0 1 1 2	anhydrate

714. C₆H₆ClNO₃S.H₂O

p-Chloroaniline-*o*-sulfonic acid (monohydrate)
1-Amino-4-chloro-2-benzenesulfonic acid (monohydrate)

RN: 133-74-4   **MP** (°C):
MW: 225.65   **BP** (°C):

Solubility (Moles/L)	Solubility (Grams/L)	Temp (°C)	Ref (#)	Evaluation (T P E A A)	Comments
1.387E-02	3.130E+00	0	P038	1 2 2 1 2	monohydrate

715. C₆H₆ClNO₃S.H₂O

p-Chloroaniline-*m*-sulfonic acid (monohydrate)
1-Amino-4-chlorobenzene-3-sulfonic acid (monohydrate)

RN: 88-43-7   **MP** (°C):
MW: 225.65   **BP** (°C):

Solubility (Moles/L)	Solubility (Grams/L)	Temp (°C)	Ref (#)	Evaluation (T P E A A)	Comments
5.141E-02	1.160E+01	0	P038	1 0 1 1 2	metastable monohydrate

716. C₆H₆Cl₆

β-1,2,3,4,5,6-Hexachlorocyclohexane
β-Benzene hexachloride
β-BHC
β-Hexachlorocyclohexane

RN: 319-85-7   **MP** (°C):   312
MW: 290.83   **BP** (°C):

Solubility (Moles/L)	Solubility (Grams/L)	Temp (°C)	Ref (#)	Evaluation (T P E A A)	Comments
1.719E-05	5.000E-03	20	C099	1 2 0 0 0	
8.252E-07	2.400E-04	25	W025	1 0 2 2 2	
5.501E-07	1.600E-04	28	K120	1 2 2 2 1	average of 2
1.719E-06	5.000E-04	ns	M061	0 0 0 0 0	

717. C₆H₆Cl₆

δ-1,2,3,4,5,6-Hexachlorocyclohexane
δ-Benzene hexachloride

RN: 608-73-1   **MP** (°C):
MW: 290.83   **BP** (°C):

Solubility (Moles/L)	Solubility (Grams/L)	Temp (°C)	Ref (#)	Evaluation (T P E A A)	Comments
3.438E-05	1.000E-02	20	C099	1 2 0 0 1	
1.080E-04	3.140E-02	25	W025	1 0 2 2 2	
4.009E-05	1.166E-02	28	K120	1 2 2 2 2	average of 4

718. $C_6H_6Cl_6$

Lindane

γ-BHC

Benzene hexachloride

RN:	58-89-9	**MP** (°C):	112.5	
MW:	290.83	**BP** (°C):	0	

Solubility (Moles/L)	Solubility (Grams/L)	Temp (°C)	Ref (#)	Evaluation (T P E A A)	Comments
7.393E-06	2.150E-03	15	B083	2 2 1 2 2	
7.393E-06	2.150E-03	15	B162	1 0 0 0 2	
2.816E-05	8.190E-03	19	I018	1 0 0 0 2	
3.438E-05	1.000E-02	20	C099	1 2 0 0 1	
2.709E-05	7.880E-03	22	K137	1 1 2 1 0	
2.706E-05	7.870E-03	24	C313	0 0 0 0 0	
5.845E-05	1.700E-02	24	H116	2 1 0 0 2	
2.338E-05	6.800E-03	25	B083	2 2 1 2 2	
2.338E-05	6.800E-03	25	B162	1 0 0 0 2	
2.586E-05	7.520E-03	25	M060	2 2 1 2 2	
2.510E-05	7.300E-03	25	M130	1 0 0 0 1	
2.682E-05	7.800E-03	25	W025	1 0 2 2 2	
4.126E-05	1.200E-02	27	B161	2 1 2 2 0	EFG
2.235E-05	6.500E-03	28	K120	1 2 2 2 2	average of 4
3.920E-05	1.140E-02	35	B083	2 2 1 2 2	particle size 5 μm
7.221E-05	2.100E-02	35	B161	2 1 2 2 0	EFG
3.920E-05	1.140E-02	35	B162	1 0 0 0 2	
5.226E-05	1.520E-02	45	B083	2 2 1 2 2	particle size 5 μm
9.284E-05	2.700E-02	45	B161	2 1 2 2 0	EFG
1.135E-04	3.300E-02	50	B161	2 1 2 2 0	EFG
1.547E-04	4.500E-02	60	B161	2 1 2 2 0	EFG
2.400E-05	6.980E-03	ns	C318	0 0 0 0 0	
~3.44E-05	~1.00E-02	ns	I308	0 0 0 0 0	
3.138E-07	1.500E-04	ns	K138	0 0 0 0 2	sic
3.438E-06	1.000E-03	ns	M061	0 0 0 0 0	
2.407E-05	7.000E-03	ns	M110	0 0 0 0 0	EFG
2.510E-05	7.300E-03	ns	V414	0 0 0 0 0	
3.438E-05	1.000E-02	rt	M161	0 0 0 0 1	

719. $C_6H_6Cl_6$

α-1,2,3,4,5,6-Hexachlorocyclohexane

α-Benzene hexachloride

α-HCH

α-BHC

α-Hexachlorocyclohexane

RN:	319-84-6	**MP** (°C):	158	
MW:	290.83	**BP** (°C):	288	

Solubility (Moles/L)	Solubility (Grams/L)	Temp (°C)	Ref (#)	Evaluation (T P E A A)	Comments
3.438E-05	1.000E-02	20	C099	1 2 0 0 1	
6.877E-06	2.000E-03	25	W025	1 0 2 2 2	
5.570E-06	1.620E-03	28	K120	1 2 2 2 2	average of 4
3.438E-06	1.000E-03	ns	M061	0 0 0 0 0	

720. $C_6H_6FN_3O_3$

1-Methylcarbamoyl-5-fluorouracil
5-Fluoro-3,4-dihydro-N-methyl-2,4-dioxo-pyrimidinecarboxamide
1-Methylcarbamoyl-5-fluoro-2,4(1H,3H)-pyrimidinedi-one

RN: 56563-18-9 **MP** (°C): 225–228
MW: 187.13 **BP** (°C):

Solubility (Moles/L)	Solubility (Grams/L)	Temp (°C)	Ref (#)	Evaluation (T P E A A)	Comments
3.313E-03	6.200E-01	22	B321	0 0 0 0 0	pH 4.0
3.313E-03	6.200E-01	22	B388	0 0 0 0 0	

721. $C_6H_6INO_3S$

2-Iodoaniline-4-sulphonic acid
Benzenesulfonic acid, 4-amino-2-iodo-

RN: 67877-88-7 **MP** (°C):
MW: 299.09 **BP** (°C):

Solubility (Moles/L)	Solubility (Grams/L)	Temp (°C)	Ref (#)	Evaluation (T P E A A)	Comments
6.781E-02	2.028E+01	25	B107	1 2 1 1 2	

722. $C_6H_6INO_3S$

3-Iodoaniline-4-sulphonic acid
Benzenesulfonic acid, 4-amino-3-iodo-

RN: 25210-30-4 **MP** (°C):
MW: 299.09 **BP** (°C):

Solubility (Moles/L)	Solubility (Grams/L)	Temp (°C)	Ref (#)	Evaluation (T P E A A)	Comments
6.474E-03	1.936E+00	25	B107	1 2 1 1 2	

723. $C_6H_6INO_3S$

4-Iodoaniline-2-sulphonic acid
Benzenesulfonic acid, 2-amino-4-iodo-

RN: 171664-62-3 **MP** (°C):
MW: 299.09 **BP** (°C):

Solubility (Moles/L)	Solubility (Grams/L)	Temp (°C)	Ref (#)	Evaluation (T P E A A)	Comments
1.697E-02	5.074E+00	25	B107	1 2 1 1 1	

724. $C_6H_6INO_3S$

4-Iodoaniline-3-sulphonic acid
Benzenesulfonic acid, 3-amino-4-iodo-

RN: **MP** (°C):
MW: 299.09 **BP** (°C):

Solubility (Moles/L)	Solubility (Grams/L)	Temp (°C)	Ref (#)	Evaluation (T P E A A)	Comments
4.486E-02	1.342E+01	25	B107	1 2 1 1 2	

725. C₆H₆INO₃S

725. C$_6$H$_6$INO$_3$S
5-Iodoaniline-2-sulphonic acid
Benzenesulfonic acid, 2-amino-5-iodo-
RN: **MP** (°C):
MW: 299.09   **BP** (°C):

Solubility (Moles/L)	Solubility (Grams/L)	Temp (°C)	Ref (#)	Evaluation (T P E A A)	Comments
8.671E-03	2.593E+00	25	B107	1 2 1 1 1	

726. C$_6$H$_6$INO$_3$S
6-Iodoaniline-3-sulphonic acid
Benzenesulfonic acid, 3-amino-6-iodo-
RN: **MP** (°C):
MW: 299.09   **BP** (°C):

Solubility (Moles/L)	Solubility (Grams/L)	Temp (°C)	Ref (#)	Evaluation (T P E A A)	Comments
1.597E-02	4.777E+00	25	B107	1 2 1 1 1	

727. C$_6$H$_6$INO$_3$S
5-Iodoaniline-3-sulphonic acid
Benzenesulfonic acid, 3-amino-5-iodo-
RN: **MP** (°C):
MW: 299.09   **BP** (°C):

Solubility (Moles/L)	Solubility (Grams/L)	Temp (°C)	Ref (#)	Evaluation (T P E A A)	Comments
4.323E-02	1.293E+01	25	B107	1 2 1 1 2	

728. C$_6$H$_6$N$_2$O
Nicotiamide
Niacinamide
Nicotinamide
RN: 98-92-0   **MP** (°C):   131
MW: 122.13   **BP** (°C):

Solubility (Moles/L)	Solubility (Grams/L)	Temp (°C)	Ref (#)	Evaluation (T P E A A)	Comments
4.094E+00	5.000E+02	20	D041	1 0 0 0 2	
8.188E+00	1.000E+03	20	M054	1 0 0 0 2	
2.900E-03	3.542E-01	25	A350	0 0 0 0 0	
8.188E+00	1.000E+03	25	D315	0 0 0 0 0	
8.188E-01	1.000E+02	ns	K444	0 0 0 0 0	

729. C$_6$H$_6$N$_2$O$_2$
3-Nitroaniline
1-Amino-3-nitrobenzene
3-Nitrobenzenamine
m-Nitroaminobenzene
m-Nitroaniline
3-Nitro-anilin

RN:	99-09-2	MP (°C):	114
MW:	138.13	BP (°C):	306

Solubility (Moles/L)	Solubility (Grams/L)	Temp (°C)	Ref (#)	Evaluation (T P E A A)	Comments
8.710E-03	1.203E+00	20	B179	0 0 0 0 0	
5.370E-03	7.418E-01	25	B335	1 2 0 0 1	
6.516E-03	9.000E-01	25	F300	1 0 0 0 2	
3.020E-03	4.171E-01	25	L016	1 0 0 0 2	unit assumed
6.582E-03	9.092E-01	25.0	C026	0 0 0 0 0	
1.290E-02	1.782E+00	40.1	C026	0 0 0 0 0	

730. C$_6$H$_6$N$_2$O$_2$
Urocanic acid
Urocaninsaeure

RN:	104-98-3	MP (°C):	225
MW:	138.13	BP (°C):	

Solubility (Moles/L)	Solubility (Grams/L)	Temp (°C)	Ref (#)	Evaluation (T P E A A)	Comments
1.086E-02	1.500E+00	17.40	F300	1 0 0 0 1	
4.318E-02	5.964E+00	37	D041	1 0 0 0 0	
5.575E-02	7.700E+00	50	F300	1 0 0 0 1	
4.098E-01	5.660E+01	100	D041	1 0 0 0 0	

731. C$_6$H$_6$N$_2$O$_2$
p-Nitroaniline
4-Amino-nitrobenzene
Benzenamine
4-Nitroaniline
p-Aminonitrobenzene
4-Nitrobenzenamine

RN:	100-01-6	MP (°C):	146
MW:	138.13	BP (°C):	332

Solubility (Moles/L)	Solubility (Grams/L)	Temp (°C)	Ref (#)	Evaluation (T P E A A)	Comments
5.754E-03	7.948E-01	20	B179	0 0 0 0 0	
2.823E-03	3.900E-01	20	H300	1 2 2 2 1	*sic*
2.815E-03	3.888E-01	20	T301	1 2 2 2 2	
3.020E-03	4.171E-01	25	B335	1 2 0 0 1	
4.344E-03	6.000E-01	25	F300	1 0 0 0 2	*sic*
5.370E-03	7.418E-01	25	L016	1 0 0 0 2	unit assumed
4.110E-03	5.677E-01	25.0	C026	0 0 0 0 0	
5.267E-03	7.275E-01	30	G029	1 0 2 2 2	
8.367E-03	1.156E+00	40.1	C026	0 0 0 0 0	

732. C$_6$H$_6$N$_2$O$_2$
2-Nitroaniline
o-Nitroaniline
1-Amino-2-nitrobenzene
2-Nitro-aniline

RN:	88-74-4	MP (°C):	71.5
MW:	138.13	BP (°C):	284

Solubility (Moles/L)	Solubility (Grams/L)	Temp (°C)	Ref (#)	Evaluation (T P E A A)	Comments
6.467E-03	8.932E-01	20	T301	1 2 2 2 2	
8.764E-03	1.211E+00	25.0	C026	0 0 0 0 0	
1.750E-02	2.417E+00	40.1	C026	0 0 0 0 0	
6.134E-03	8.473E-01	50	T301	1 2 2 2 2	average of 4
6.799E-03	9.391E-01	80	T301	1 2 2 2 2	average of 4

733. C$_6$H$_6$N$_2$O$_3$
5,5-Ethylenebarbituric acid
Spirocyclopropane-1',5-barbituric acid
5,7-Diazaspiro[2.5]octane-4,6,8-trione
Cyclopropane-spirobarbiturate

RN:	6947-77-9	MP (°C):	
MW:	154.13	BP (°C):	

Solubility (Moles/L)	Solubility (Grams/L)	Temp (°C)	Ref (#)	Evaluation (T P E A A)	Comments
1.300E-02	2.004E+00	25	P350	0 0 0 0 0	intrinsic

734. C$_6$H$_6$N$_2$O$_4$
1-Methylorotic acid
4-Pyrimidinecarboxylic acid, 1,2,3,6-tetrahydro-1-methyl-2,6-dioxo-

RN:	705-36-2	MP (°C):	
MW:	170.13	BP (°C):	

Solubility (Moles/L)	Solubility (Grams/L)	Temp (°C)	Ref (#)	Evaluation (T P E A A)	Comments
1.200E-01	2.042E+01	20	N019	0 0 0 0 0	

735. C$_6$H$_6$N$_2$O$_4$S
m-Nitrobenzenesulfonamide
3-Nitrobenzenesulfonamide

RN:	121-52-8	MP (°C):	
MW:	202.19	BP (°C):	

Solubility (Moles/L)	Solubility (Grams/L)	Temp (°C)	Ref (#)	Evaluation (T P E A A)	Comments
2.200E-03	4.448E-01	15	K024	1 2 1 1 2	

736. C$_6$H$_6$N$_2$O$_4$S
4-Nitrobenzenesulfonamide
p-Nitrobenzenesulfonamide
RN: 6325-93-5   **MP** (°C):
MW: 202.19   **BP** (°C):

Solubility (Moles/L)	Solubility (Grams/L)	Temp (°C)	Ref (#)	Evaluation (T P E A A)	Comments
3.000E-03	6.066E-01	15	K024	1 2 1 1 2	

737. C$_6$H$_6$N$_2$O$_4$S
2-Nitrobenzenesulfonamide
o-Nitrobenzenesulfonamide
RN: 5455-59-4   **MP** (°C):
MW: 202.19   **BP** (°C):

Solubility (Moles/L)	Solubility (Grams/L)	Temp (°C)	Ref (#)	Evaluation (T P E A A)	Comments
1.600E-03	3.235E-01	15	K024	1 2 1 1 2	

738. C$_6$H$_6$N$_4$
8-Methylpurine
1H-Purine, 8-methyl-
RN: 934-33-8   **MP** (°C):
MW: 134.14   **BP** (°C):

Solubility (Moles/L)	Solubility (Grams/L)	Temp (°C)	Ref (#)	Evaluation (T P E A A)	Comments
3.924E-01	5.263E+01	20	A022	1 0 0 0 0	

739. C$_6$H$_6$N$_4$O
8-Hydroxymethylpurine
Purine-8-methanol
RN: 6642-26-8   **MP** (°C):
MW: 150.14   **BP** (°C):

Solubility (Moles/L)	Solubility (Grams/L)	Temp (°C)	Ref (#)	Evaluation (T P E A A)	Comments
3.014E-02	4.525E+00	20	A022	1 2 0 0 0	
4.440E-01	6.667E+01	100	A082	1 2 0 0 0	

740. C$_6$H$_6$N$_4$O$_3$
9-Methyluric acid
1H-Purine-2,6,8(3H)-trione, 7,9-dihydro-9-methyl-
*N*9-Methyluric acid
RN: 55441-71-9   **MP** (°C):
MW: 182.14   **BP** (°C):

Solubility (Moles/L)	Solubility (Grams/L)	Temp (°C)	Ref (#)	Evaluation (T P E A A)	Comments
2.999E-03	5.461E-01	ns	B115	0 0 1 1 0	

741. $C_6H_6N_4O_3$
1-Methyluric acid
α-Methyluric acid

RN: 708-79-2   **MP** (°C):   400
MW: 182.14   **BP** (°C):

Solubility (Moles/L)	Solubility (Grams/L)	Temp (°C)	Ref (#)	Evaluation (T P E A A)	Comments
1.153E-02	2.101E+00	ns	B115	0 0 1 1 0	ζ form
8.701E-03	1.585E+00	ns	B115	0 0 1 1 0	γ form
2.731E-02	4.975E+00	ns	B115	0 0 1 1 0	
2.754E-02	5.017E+00	ns	R427	0 0 0 0 0	

742. $C_6H_6N_4O_3S$
Niridazole
Nirodazole

RN: 61-57-4   **MP** (°C):   261
MW: 214.20   **BP** (°C):

Solubility (Moles/L)	Solubility (Grams/L)	Temp (°C)	Ref (#)	Evaluation (T P E A A)	Comments
6.068E-04	1.300E-01	25	A081	1 0 1 1 0	EFG
1.634E-04	3.500E-02	25	G051	1 0 1 1 0	pH 2

743. $C_6H_6N_4O_4$
5-Nitro-2-furaldehyde semicarbazone
Nitrofurazone

RN: 59-87-0   **MP** (°C):   236
MW: 198.14   **BP** (°C):

Solubility (Moles/L)	Solubility (Grams/L)	Temp (°C)	Ref (#)	Evaluation (T P E A A)	Comments
8.225E-04	1.630E-01	ns	B404	0 2 1 1 0	
1.201E-03	2.380E-01	ns	I310	0 0 0 0 0	
8.128E-04	1.611E-01	ns	R427	0 0 0 0 0	

744. $C_6H_6N_6$
2,4-Diaminopteridine
2:4-Diaminopteridine

RN: 1127-93-1   **MP** (°C):
MW: 162.15   **BP** (°C):

Solubility (Moles/L)	Solubility (Grams/L)	Temp (°C)	Ref (#)	Evaluation (T P E A A)	Comments
2.055E-03	3.332E-01	20	A019	2 2 1 1 2	
4.708E-02	7.634E+00	100	A019	1 2 1 1 1	

745. C$_6$H$_6$N$_6$
4,6-Diaminopteridine
4:6-Diaminopteridine
RN: 19167-60-3   **MP** (°C):
MW: 162.15   **BP** (°C):

Solubility (Moles/L)	Solubility (Grams/L)	Temp (°C)	Ref (#)	Evaluation (T P E A A)	Comments
2.569E-04	4.166E-02	20	A020	1 2 0 1 1	
6.554E-03	1.063E+00	100	A020	1 2 0 0 0	

746. C$_6$H$_6$N$_6$
4,7-Diaminopteridine
4:7-Diaminopteridine
RN: 771-41-5   **MP** (°C):
MW: 162.15   **BP** (°C):

Solubility (Moles/L)	Solubility (Grams/L)	Temp (°C)	Ref (#)	Evaluation (T P E A A)	Comments
1.233E-03	2.000E-01	20	A020	1 2 0 0 1	
2.049E-02	3.322E+00	100	A020	1 2 0 0 0	

747. C$_6$H$_6$N$_6$
4-Hydrazinopteridine
4(1H)-Pteridinone, hydrazone
RN: 77632-11-2   **MP** (°C):
MW: 162.15   **BP** (°C):

Solubility (Moles/L)	Solubility (Grams/L)	Temp (°C)	Ref (#)	Evaluation (T P E A A)	Comments
1.367E-02	2.217E+00	20	A083	1 2 0 0 0	
8.686E-02	1.408E+01	100	A083	1 2 0 0 0	

748. C$_6$H$_6$O
Phenol
Carbolic acid
Hydroxybenzene
RN: 108-95-2   **MP** (°C): 40.85
MW: 94.11   **BP** (°C): 182

Solubility (Moles/L)	Solubility (Grams/L)	Temp (°C)	Ref (#)	Evaluation (T P E A A)	Comments
7.164E-01	6.743E+01	0	A056	1 0 1 1 2	
7.136E-01	6.716E+01	0	B031	1 2 2 2 1	
7.164E-01	6.743E+01	0	L059	1 0 1 1 2	
6.858E-01	6.455E+01	8.60	C058	2 0 2 1 1	
7.321E-01	6.890E+01	10	A056	1 0 1 1 2	
7.321E-01	6.890E+01	10	L059	1 0 1 1 2	
8.080E-01	7.604E+01	15.1	A400	2 1 2 2 2	

(continued)

748. C_6H_6O (continued)

Solubility (Moles/L)	Solubility (Grams/L)	Temp (°C)	Ref (#)	Evaluation (T P E A A)	Comments
6.672E-01	6.279E+01	16	D041	1 0 0 0 1	
7.779E-01	7.322E+01	20	B031	1 2 2 2 1	
8.710E-01	8.197E+01	20	B179	0 0 0 0 0	
4.866E+00	4.580E+02	20	C052	1 2 1 1 2	*sic*
8.235E-01	7.750E+01	20	F300	1 0 0 0 2	
8.198E-01	7.715E+01	20	H003	1 2 2 1 2	
1.600E+00	1.506E+02	20	H306	1 0 1 2 1	
8.500E-01	8.000E+01	20	K119	1 0 0 0 2	
7.130E-01	6.710E+01	20	K301	2 2 1 1 2	
6.175E-01	5.811E+01	20	R087	0 0 0 0 0	0.15M NaCl
9.490E-01	8.931E+01	22.70	M135	1 2 1 1 2	
1.000E+00	9.411E+01	25	A021	1 2 1 1 0	
8.930E-01	8.405E+01	25	A400	2 1 2 2 2	
9.882E-01	9.300E+01	25	B060	2 0 1 1 1	
9.400E-01	8.847E+01	25	B316	0 0 0 0 0	
9.000E-01	8.470E+01	25	F044	1 0 0 0 1	
8.468E-01	7.970E+01	25	H003	1 2 2 1 2	
8.245E-01	7.759E+01	25	H028	2 0 2 0 2	
1.527E-01	1.437E+01	25	K129	2 1 2 2 2	
8.854E-01	8.333E+01	25	L022	1 0 0 0 0	
9.000E-01	8.470E+01	25	L088	1 0 0 0 1	
7.413E-01	6.977E+01	25	M041	1 1 0 0 1	
9.300E-01	8.753E+01	25	P031	0 0 0 0 0	
7.688E-01	7.236E+01	25	R041	0 0 0 0 0	
9.900E-01	9.317E+01	26.90	M135	1 2 1 1 2	
8.970E-01	8.442E+01	30	H003	1 2 2 1 2	
8.297E-01	7.809E+01	30	V009	1 0 0 0 1	
1.048E+00	9.863E+01	32.20	M135	1 2 1 1 2	
9.598E-01	9.033E+01	34	B063	1 2 2 1 2	
9.892E-01	9.310E+01	35	A400	2 1 2 2 2	
9.580E-01	9.016E+01	35	H003	1 2 2 1 2	
1.107E+00	1.042E+02	36.00	M135	1 2 1 1 2	
9.130E-01	8.592E+01	40	B031	1 2 2 2 1	
1.158E+00	1.090E+02	43.70	M135	1 2 1 1 2	
1.369E+00	1.288E+02	47.70	M135	1 2 1 1 2	
1.172E+00	1.103E+02	48.00	C058	2 0 2 1 2	
1.138E+00	1.071E+02	50	M041	1 1 0 0 2	
1.476E+00	1.389E+02	50.50	M135	1 2 1 1 2	
1.183E+00	1.113E+02	51.90	B063	1 2 2 1 2	
1.592E+00	1.498E+02	53.50	M135	1 2 1 1 2	
1.725E+00	1.623E+02	55.80	M135	1 2 1 1 2	
1.388E+00	1.306E+02	55.90	B063	1 2 2 1 2	
1.375E+00	1.295E+02	57.30	H003	1 2 2 1 2	
1.856E+00	1.747E+02	57.80	M135	1 2 1 1 2	
1.590E+00	1.497E+02	60	B031	1 2 2 2 2	
2.163E+00	2.036E+02	60.90	M135	1 2 1 1 2	
1.612E+00	1.518E+02	61.70	B063	1 2 2 1 2	
1.723E+00	1.621E+02	62.74	H003	1 2 2 1 2	
1.771E+00	1.667E+02	63.20	B063	1 2 2 1 2	

(continued)

748. C₆H₆O (continued)

Solubility (Moles/L)	Solubility (Grams/L)	Temp (°C)	Ref (#)	Evaluation (T P E A A)	Comments
2.109E+00	1.985E+02	65.40	B063	1 2 2 1 2	
3.064E+00	2.884E+02	65.50	B063	1 2 2 1 2	
2.567E+00	2.416E+02	65.55	B063	1 2 2 1 2	
2.767E+00	2.604E+02	65.60	B063	1 2 2 1 2	
2.388E+00	2.247E+02	65.79	H003	1 2 2 1 2	average of 2
2.590E+00	2.437E+02	65.84	H003	1 2 2 1 2	
2.624E+00	2.469E+02	65.86	H003	1 2 2 1 2	
2.536E+00	2.387E+02	65.90	H003	1 2 2 1 2	
2.818E+00	2.652E+02	66.0	H068	2 0 0 0 2	
2.397E+00	2.256E+02	66.01	H003	1 2 2 1 2	
1.734E+00	1.632E+02	66.30	C058	2 0 2 1 2	
8.243E-01	7.758E+01	ns	A406	0 0 0 0 1	
8.594E-01	8.088E+01	ns	N330	2 2 2 1 2	
8.710E-01	8.197E+01	ns	R427	0 0 0 0 0	
8.043E-01	7.570E+01	rt	N051	0 0 2 1 2	average of 3

749. C₆H₆O₂
Hydroquinone
Hydrochinon
Hydroquinol

RN:	123-31-9	**MP** (°C):	173.5	
MW:	110.11	**BP** (°C):	286	

Solubility (Moles/L)	Solubility (Grams/L)	Temp (°C)	Ref (#)	Evaluation (T P E A A)	Comments
3.493E-01	3.846E+01	0	M043	1 0 0 0 1	
4.653E-01	5.123E+01	10	M043	1 0 0 0 1	
4.904E-01	5.400E+01	15	F300	1 0 0 0 1	
5.077E-01	5.590E+01	17.70	L065	1 0 0 0 2	0.01N HCl
5.087E-01	5.601E+01	17.90	L065	1 0 0 0 2	0.01N HCl
5.101E-01	5.617E+01	17.95	L065	1 0 0 0 2	0.01N HCl
5.103E-01	5.619E+01	18	L064	2 2 2 1 2	0.01N HCl
6.100E-01	6.716E+01	20	M043	1 0 0 0 1	
6.357E-01	7.000E+01	22.5	G301	0 0 0 0 0	
6.180E-01	6.805E+01	23.75	L064	2 2 2 1 2	0.01N HCl
6.450E-01	7.102E+01	25	G033	1 0 1 1 2	
7.283E-01	8.020E+01	25	K033	1 0 0 1 2	
6.660E-01	7.334E+01	25	K040	1 0 2 1 2	
7.955E-01	8.759E+01	30	M043	1 0 0 0 1	
1.045E+00	1.150E+02	40	M043	1 0 0 0 1	
2.354E+00	2.593E+02	60	M043	1 0 0 0 1	
5.694E+00	6.270E+02	75.3	W038	2 2 2 1 2	
4.251E+00	4.681E+02	80	M043	1 0 0 0 1	
7.528E+00	8.289E+02	81.9	W038	2 2 2 1 2	
6.034E+00	6.644E+02	100	M043	1 0 0 0 2	
1.961E+01	2.159E+03	114.6	W038	2 2 2 1 2	
2.180E+01	2.400E+03	120.3	W038	2 2 2 1 2	
2.728E+01	3.004E+03	131.7	W038	2 2 2 1 2	

(continued)

749. $C_6H_6O_2$ (continued)

Solubility (Moles/L)	Solubility (Grams/L)	Temp (°C)	Ref (#)	Evaluation (T P E A A)	Comments
2.942E+01	3.239E+03	136.0	W038	2 2 2 1 2	
3.353E+01	3.692E+03	141.8	W038	2 2 2 1 2	
3.621E+01	3.987E+03	147.2	W038	2 2 2 1 2	
6.026E-01	6.635E+01	ns	R427	0 0 0 0 0	
6.084E-01	6.699E+01	rt	D021	0 0 1 1 2	

750. $C_6H_6O_2$
Pyrocatechol
Brenzkatechin
Catechol

RN: 120-80-9   **MP** (°C): 105
MW: 110.11   **BP** (°C): 245.5

Solubility (Moles/L)	Solubility (Grams/L)	Temp (°C)	Ref (#)	Evaluation (T P E A A)	Comments
2.824E+00	3.110E+02	20	F300	1 0 0 0 2	
2.823E+00	3.108E+02	20	M043	1 0 0 0 2	
4.190E+00	4.614E+02	25	K040	1 0 2 1 2	
5.743E+00	6.324E+02	40	M043	1 0 0 0 2	
1.278E+01	1.408E+03	41.2	W038	2 2 2 1 2	
2.061E+01	2.270E+03	56.7	W038	2 2 2 1 2	
2.068E+01	2.278E+03	57.1	W038	2 2 2 1 2	
7.308E+00	8.047E+02	60	M043	1 0 0 0 2	
2.617E+01	2.882E+03	66.2	W038	2 2 2 1 2	
8.337E+00	9.180E+02	80	M043	1 0 0 0 2	
8.974E+00	9.882E+02	100	M043	1 0 0 0 2	
5.556E+01	6.117E+03	104.5	W038	2 2 2 1 2	
2.823E+00	3.108E+02	rt	D021	0 0 1 1 2	

751. $C_6H_6O_2$
Resorcinol
Resorcin

RN: 108-46-3   **MP** (°C): 110.0
MW: 110.11   **BP** (°C):

Solubility (Moles/L)	Solubility (Grams/L)	Temp (°C)	Ref (#)	Evaluation (T P E A A)	Comments
3.404E+00	3.748E+02	0	M022	1 0 0 0 2	
3.617E+00	3.983E+02	0	M043	1 0 0 0 2	
2.784E+00	3.066E+02	3.70	L090	1 0 0 1 2	
4.173E+00	4.595E+02	10	M043	1 0 0 0 1	
5.413E+00	5.960E+02	12.50	F300	1 0 0 0 2	
3.186E+00	3.508E+02	14.20	L090	1 0 0 1 2	
3.359E+00	3.699E+02	19.50	L090	1 0 0 1 2	
4.576E+00	5.038E+02	20	M022	1 0 0 0 2	
5.009E+00	5.516E+02	20	M043	1 0 0 0 2	
6.515E+00	7.174E+02	25	K040	1 0 2 1 2	

(continued)

751. $C_6H_6O_2$ (continued)

Solubility (Moles/L)	Solubility (Grams/L)	Temp (°C)	Ref (#)	Evaluation (T P E A A)	Comments
6.330E+00	6.970E+02	30	F300	1 0 0 0 2	
5.718E+00	6.296E+02	30	M043	1 0 0 0 2	
3.679E+00	4.051E+02	32.50	L090	1 0 0 1 2	
1.464E+01	1.612E+03	33.61	W038	2 2 2 1 2	
5.641E+00	6.211E+02	40	M022	1 0 0 0 2	
6.287E+00	6.923E+02	40	M043	1 0 0 0 2	
1.843E+01	2.030E+03	44.5	W038	2 2 2 1 2	
2.042E+01	2.249E+03	49.3	W038	2 2 2 1 2	
2.100E+01	2.312E+03	50.4	W038	2 2 2 1 2	
6.465E+00	7.119E+02	60	M022	1 0 0 0 2	
7.228E+00	7.959E+02	60	M043	1 0 0 0 2	
2.701E+01	2.974E+03	64.4	W038	2 2 2 1 2	
2.997E+01	3.300E+03	70.7	W038	2 2 2 1 2	
7.106E+00	7.825E+02	80	M022	1 0 0 0 2	
7.844E+00	8.638E+02	80	M043	1 0 0 0 2	
3.516E+01	3.871E+03	80.5	W038	2 2 2 1 2	
4.008E+01	4.414E+03	88.5	W038	2 2 2 1 2	
7.592E+00	8.360E+02	100	M022	1 0 0 0 2	
8.299E+00	9.138E+02	100	M043	1 0 0 0 2	
5.556E+01	6.117E+03	109.4	W038	2 2 2 1 2	
4.608E+00	5.074E+02	rt	D021	0 0 1 1 2	

752. $C_6H_6O_3$

Maltol
3-Hydroxy-2-methyl-4-pyrone
Hydroxymethylpyrone
Palatone

RN: 118-71-8   **MP** (°C):   161.5
MW: 126.11   **BP** (°C):

Solubility (Moles/L)	Solubility (Grams/L)	Temp (°C)	Ref (#)	Evaluation (T P E A A)	Comments
8.643E-02	1.090E+01	15	F300	1 0 0 0 2	

753. $C_6H_6O_3$

Methyl furoate
5-Methyl-brenzschleimsaeure
5-Methylfuroic acid

RN: 611-13-2   **MP** (°C):
MW: 126.11   **BP** (°C):   181

Solubility (Moles/L)	Solubility (Grams/L)	Temp (°C)	Ref (#)	Evaluation (T P E A A)	Comments
1.475E-01	1.860E+01	20	F300	1 0 0 0 2	

754. C$_6$H$_6$O$_3$
Phloroglucinol
1,3,5-Benzenetriol
1,3,5-Trihydroxybenzene
1,3,5-THB

RN: 108-73-6   **MP** (°C): 218.0
MW: 126.11   **BP** (°C):

Solubility (Moles/L)	Solubility (Grams/L)	Temp (°C)	Ref (#)	Evaluation (T P E A A)	Comments
8.405E-02	1.060E+01	20	F300	1 0 0 0 2	
8.860E-02	1.117E+01	rt	D021	0 0 1 1 2	

755. C$_6$H$_6$O$_3$
Pyrogallol
1,2,3-Trihydroxybenzene
1,2,3-Benzenetriol
Brown AP
Fourrine 85

RN: 87-66-1   **MP** (°C): 131
MW: 126.11   **BP** (°C): 309

Solubility (Moles/L)	Solubility (Grams/L)	Temp (°C)	Ref (#)	Evaluation (T P E A A)	Comments
2.379E+00	3.000E+02	13	F300	1 0 0 0 0	average
3.013E+00	3.800E+02	25	F300	1 0 0 0 1	
4.020E+00	5.070E+02	25	K040	1 0 2 1 2	

756. C$_6$H$_6$O$_3$S
Benzenesulfonic acid
Benzolsulfosaeure

RN: 98-11-3   **MP** (°C): 43
MW: 158.18   **BP** (°C):

Solubility (Moles/L)	Solubility (Grams/L)	Temp (°C)	Ref (#)	Evaluation (T P E A A)	Comments
3.088E+00	4.885E+02	31.4	T023	1 2 2 1 2	
3.109E+00	4.917E+02	42.6	T023	1 2 2 1 2	
3.136E+00	4.960E+02	56.0	T023	1 2 2 1 2	
3.154E+00	4.989E+02	61.3	T023	1 2 2 1 2	

757. C$_6$H$_6$O$_3$S.H$_2$O
Benzenesulfonic acid (monohydrate)

RN: 98-11-3   **MP** (°C):
MW: 176.19   **BP** (°C):

Solubility (Moles/L)	Solubility (Grams/L)	Temp (°C)	Ref (#)	Evaluation (T P E A A)	Comments
2.542E+00	4.478E+02	21.3	T023	1 2 2 1 2	
2.568E+00	4.525E+02	31.0	T023	1 2 2 1 2	
2.770E+00	4.881E+02	32.6	T023	1 2 2 1 2	

(continued)

757. C$_6$H$_6$O$_3$S.H$_2$O (continued)

Solubility (Moles/L)	Solubility (Grams/L)	Temp (°C)	Ref (#)	Evaluation (T P E A A)	Comments
2.598E+00	4.577E+02	39.5	T023	1 2 2 1 2	
2.751E+00	4.846E+02	39.8	T023	1 2 2 1 2	
2.722E+00	4.796E+02	49.0	T023	1 2 2 1 2	
2.641E+00	4.654E+02	49.0	T023	1 2 2 1 2	
2.682E+00	4.726E+02	52.4	T023	1 2 2 1 2	

758. C$_6$H$_6$O$_3$S.2.5H$_2$O
Benzenesulfonic acid (2.5 hydrate)
RN: 98-11-3   **MP** (°C):
MW: 203.22   **BP** (°C):

Solubility (Moles/L)	Solubility (Grams/L)	Temp (°C)	Ref (#)	Evaluation (T P E A A)	Comments
2.107E+00	4.281E+02	−4.0	T023	1 2 2 1 2	
2.122E+00	4.312E+02	−3.3	T023	1 2 2 1 2	
2.150E+00	4.370E+02	−2.3	T023	1 2 2 1 2	
2.131E+00	4.331E+02	−2.5	T023	1 2 2 1 2	

759. C$_6$H$_6$O$_3$S.2H$_2$O
Benzenesulfonic acid (dihydrate)
RN: 98-11-3   **MP** (°C):
MW: 194.21   **BP** (°C):

Solubility (Moles/L)	Solubility (Grams/L)	Temp (°C)	Ref (#)	Evaluation (T P E A A)	Comments
2.250E+00	4.370E+02	2.2	T023	1 2 2 1 2	
2.265E+00	4.399E+02	7.5	T023	1 2 2 1 2	
2.289E+00	4.446E+02	13.7	T023	1 2 2 1 2	
2.297E+00	4.460E+02	15.1	T023	1 2 2 1 2	

760. C$_6$H$_6$O$_3$S.3H$_2$O
Benzenesulfonic acid (trihydrate)
RN: 98-11-3   **MP** (°C):
MW: 212.22   **BP** (°C):

Solubility (Moles/L)	Solubility (Grams/L)	Temp (°C)	Ref (#)	Evaluation (T P E A A)	Comments
1.690E+00	3.586E+02	−40.8	T023	1 2 2 1 2	
1.766E+00	3.748E+02	−29.0	T023	1 2 2 1 2	
1.842E+00	3.909E+02	−18.5	T023	1 2 2 1 2	
1.922E+00	4.078E+02	−10.0	T023	1 2 2 1 2	
1.975E+00	4.191E+02	−5.9	T023	1 2 2 1 2	
2.011E+00	4.267E+02	−4.7	T023	1 2 2 1 2	

761. $C_6H_6O_4$
Muconic acid
Muconsaeure

RN: 505-70-4	**MP** (°C):
MW: 142.11	**BP** (°C):

Solubility (Moles/L)	Solubility (Grams/L)	Temp (°C)	Ref (#)	Evaluation (T P E A A)	Comments
1.407E-03	2.000E-01	20	F300	1 0 0 0 2	

762. $C_6H_7F_3N_4OS$
Thiazafluron
Urea, N,N'-dimethyl-N-[5-(trifluoromethyl)-1,3,4-thiadiazol-2-yl]-

RN: 25366-23-8	**MP** (°C):	136.5
MW: 240.21	**BP** (°C):	

Solubility (Moles/L)	Solubility (Grams/L)	Temp (°C)	Ref (#)	Evaluation (T P E A A)	Comments
8.724E-03	2.096E+00	20	E048	1 2 1 1 2	
8.742E-03	2.100E+00	20	M161	1 0 0 0 1	

763. C_6H_7N
Aniline
Aminobenzene
C.I. Oxidation base 1
Aminophen
Kyanol

RN: 62-53-3	**MP** (°C):	−6.3
MW: 93.13	**BP** (°C):	184

Solubility (Moles/L)	Solubility (Grams/L)	Temp (°C)	Ref (#)	Evaluation (T P E A A)	Comments
3.531E-01	3.288E+01	8.60	C058	2 0 2 1 1	
3.877E-01	3.611E+01	13.8	K119	1 0 0 0 2	
3.747E-01	3.490E+01	18	F300	1 0 0 0 2	
3.818E-01	3.556E+01	18.15	P057	0 0 0 0 0	
3.612E-01	3.364E+01	22	H072	1 0 1 1 2	
3.930E-01	3.660E+01	22.5	G301	0 0 0 0 0	
3.931E-01	3.661E+01	25	B019	1 0 1 2 0	
3.931E-01	3.661E+01	25	B092	2 1 1 1 2	
4.000E-01	3.725E+01	25	F044	1 0 0 0 1	
3.791E-01	3.531E+01	25	G323	2 2 2 2 2	
3.800E-01	3.539E+01	25	H028	2 0 2 0 2	
3.791E-01	3.531E+01	25	H078	1 2 1 0 2	
3.650E-01	3.399E+01	25	M116	2 1 1 1 2	
3.731E-01	3.475E+01	25.40	C058	2 0 2 1 1	
3.930E-01	3.660E+01	26.70	L095	2 2 1 1 2	
4.229E-01	3.939E+01	48.00	C058	2 0 2 1 1	
4.328E-01	4.031E+01	50	G323	2 2 2 2 2	
5.016E-01	4.671E+01	60	B092	2 1 1 1 2	
5.016E-01	4.671E+01	66.30	C058	2 0 2 1 1	
7.025E-01	6.542E+01	96.70	C058	2 0 2 1 1	
3.801E-01	3.540E+01	ns	A406	0 0 0 0 1	

764. C$_6$H$_7$NO
m-Aminophenol
3-Aminophenol

RN:	591-27-5	**MP** (°C):	125
MW:	109.13	**BP** (°C):	164

Solubility (Moles/L)	Solubility (Grams/L)	Temp (°C)	Ref (#)	Evaluation (T P E A A)	Comments
1.797E-01	1.961E+01	10	M043	1 0 0 0 1	
2.291E-01	2.500E+01	20	F300	1 0 0 0 1	
2.409E-01	2.629E+01	20	M043	1 0 0 0 1	
3.355E-01	3.661E+01	30	M043	1 0 0 0 1	
3.261E-01	3.559E+01	32.6	S120	1 2 1 1 2	
4.859E-01	5.303E+01	40	M043	1 0 0 0 1	
6.788E-01	7.407E+01	47.9	S120	1 2 1 1 2	
8.850E-01	9.658E+01	53.0	S120	1 2 1 1 2	
1.590E+00	1.736E+02	60	M043	1 0 0 0 1	
1.406E+00	1.535E+02	60.4	S120	1 2 1 1 2	
2.148E+00	2.344E+02	66.4	S120	1 2 1 1 2	
2.627E+00	2.866E+02	68.9	S120	1 2 1 1 2	
2.927E+00	3.194E+02	70.2	S120	1 2 1 1 2	
3.161E+00	3.450E+02	71.5	S120	1 2 1 1 2	
3.410E+00	3.721E+02	73.2	S120	1 2 1 1 2	
3.737E+00	4.078E+02	77.2	S120	1 2 1 1 2	
6.752E+00	7.368E+02	80	M043	1 0 0 0 2	
4.098E+00	4.472E+02	85.2	S120	1 2 1 1 2	
4.311E+00	4.705E+02	96.0	S120	1 2 1 1 2	
8.291E+00	9.048E+02	100	M043	1 0 0 0 2	

765. C$_6$H$_7$NO
o-Aminophenol
2-Amino-phenol

RN:	95-55-6	**MP** (°C):	172
MW:	109.13	**BP** (°C):	

Solubility (Moles/L)	Solubility (Grams/L)	Temp (°C)	Ref (#)	Evaluation (T P E A A)	Comments
1.558E-01	1.700E+01	0	F300	1 0 0 0 1	
1.532E-01	1.672E+01	0	M043	1 0 0 0 1	
1.709E-01	1.865E+01	10	M043	1 0 0 0 1	
1.797E-01	1.961E+01	20	M043	1 0 0 0 1	
1.973E-01	2.153E+01	30	M043	1 0 0 0 1	
2.148E-01	2.344E+01	40	M043	1 0 0 0 1	
2.409E-01	2.629E+01	60	M043	1 0 0 0 1	
2.669E-01	2.913E+01	80	M043	1 0 0 0 1	
2.686E-01	2.931E+01	80.8	S120	1 2 1 1 1	
3.558E-01	3.883E+01	88.0	S120	1 2 1 1 1	
5.995E-01	6.542E+01	100	M043	1 0 0 0 1	

766. C₆H₇NO

C_6H_7NO

p-Aminophenol
4-Aminophenol

RN:	123-30-8	MP (°C):	190
MW:	109.13	BP (°C):	

Solubility (Moles/L)	Solubility (Grams/L)	Temp (°C)	Ref (#)	Evaluation (T P E A A)	Comments
1.008E-01	1.100E+01	0	F300	1 0 0 0 1	
9.970E-02	1.088E+01	0	M043	1 0 0 0 1	
1.176E-01	1.283E+01	10	M043	1 0 0 0 1	
1.443E-01	1.575E+01	20	M043	1 0 0 0 1	
1.709E-01	1.865E+01	30	M043	1 0 0 0 1	
2.060E-01	2.248E+01	40	M043	1 0 0 0 1	
2.678E-01	2.922E+01	59.0	S120	1 2 1 1 1	
3.184E-01	3.475E+01	60	M043	1 0 0 0 1	
5.544E-01	6.050E+01	77.0	S120	1 2 1 1 1	
6.709E-01	7.322E+01	80	M043	1 0 0 0 1	
8.399E-01	9.165E+01	86.7	S120	1 2 1 1 1	
1.497E+00	1.634E+02	96.6	S120	1 2 1 1 1	
2.475E+00	2.701E+02	100	M043	1 0 0 0 1	

767. C₆H₇NO

C_6H_7NO

Phenylhydroxylamine
Phenylhydroxylamin

RN:	100-65-2	MP (°C):	82
MW:	109.13	BP (°C):	

Solubility (Moles/L)	Solubility (Grams/L)	Temp (°C)	Ref (#)	Evaluation (T P E A A)	Comments
1.833E-01	2.000E+01	5	F300	1 0 0 0 0	
8.247E-01	9.000E+01	100	F300	1 0 0 0 0	

768. C₆H₇NO₂S

$C_6H_7NO_2S$

Benzenesulfonamide
Benzolsulfosaeure-amid

RN:	98-10-2	MP (°C):	151
MW:	157.19	BP (°C):	

Solubility (Moles/L)	Solubility (Grams/L)	Temp (°C)	Ref (#)	Evaluation (T P E A A)	Comments
1.600E-02	2.515E+00	15	K024	1 2 1 1 2	
2.736E-02	4.300E+00	16	F300	1 0 0 0 1	

769. C6H7NO3S
Orthanilic acid
Orthanilsaeure

RN:	88-21-1	**MP** (°C):	325	
MW:	173.19	**BP** (°C):		

Solubility (Moles/L)	Solubility (Grams/L)	Temp (°C)	Ref (#)	Evaluation (T P E A A)	Comments
6.525E-02	1.130E+01	8.25	P038	1 1 2 1 2	monohydrate
7.535E-02	1.305E+01	12.3	P038	1 1 2 1 2	monohydrate
8.459E-02	1.465E+01	15.55	P038	1 1 2 1 2	anhydrate
8.776E-02	1.520E+01	16.75	P038	1 1 2 1 2	anhydrate
1.114E-01	1.930E+01	25	P038	1 1 2 1 2	anhydrate
1.738E-01	3.010E+01	41.3	P038	1 1 2 1 2	anhydrate
2.477E-01	4.290E+01	55.0	P038	1 1 2 1 2	anhydrate
3.672E-01	6.360E+01	70.0	P038	1 1 2 1 2	anhydrate
5.185E-01	8.980E+01	85.0	P038	1 1 2 1 2	anhydrate
4.585E-02	7.940E+00	.0	P038	1 1 2 1 2	monohydrate

770. C6H7NO3S
Sulfanilic acid
4-Aminobenzenesulfonic acid
Sulfanilsaeure

RN:	121-57-3	**MP** (°C):	122	
MW:	173.19	**BP** (°C):		

Solubility (Moles/L)	Solubility (Grams/L)	Temp (°C)	Ref (#)	Evaluation (T P E A A)	Comments
3.672E-02	6.359E+00	0	D077	1 0 0 1 1	
2.587E-02	4.480E+00	0	M043	1 0 0 0 1	
4.810E-02	8.330E+00	10	D077	1 0 0 1 1	
4.850E-02	8.400E+00	10	F300	1 0 0 0 1	
4.583E-02	7.937E+00	10	M043	1 0 0 0 1	
6.169E-02	1.068E+01	20	D077	1 0 0 1 2	
5.774E-02	1.000E+01	20	F300	1 0 0 0 1	
6.395E-02	1.108E+01	20	M043	1 0 0 0 2	
8.477E-02	1.468E+01	30	D077	1 0 0 1 2	
1.115E-01	1.932E+01	40	D077	1 0 0 1 2	
1.109E-01	1.920E+01	40	F300	1 0 0 0 2	
1.149E-01	1.990E+01	40	M043	1 0 0 0 2	
1.414E-01	2.449E+01	50	D077	1 0 0 1 2	
1.736E-01	3.007E+01	60	D077	1 0 0 1 2	
1.687E-01	2.922E+01	60	M043	1 0 0 0 2	
2.159E-01	3.740E+01	69.9	P038	1 0 2 1 2	anhydrate
2.103E-01	3.642E+01	70	D077	1 0 0 1 2	
2.492E-01	4.315E+01	80	D077	1 0 0 1 2	
2.492E-01	4.315E+01	80	M043	1 0 0 0 2	
2.737E-01	4.740E+01	85.0	P038	1 0 2 1 2	anhydrate
3.031E-01	5.249E+01	90	D077	1 0 0 1 2	
3.610E-01	6.253E+01	100	D077	1 0 0 1 2	
3.851E-01	6.670E+01	100	F300	1 0 0 0 2	
3.610E-01	6.253E+01	100	M043	1 0 0 0 2	
6.075E-02	1.052E+01	ns	K076	0 0 0 0 2	

771. C$_6$H$_7$NO$_3$S
Metanilic acid
3-Aminobenzenesulfonic acid
m-Sulfanilic acid

RN:	121-47-1	**MP** (°C):	>300		
MW:	173.19	**BP** (°C):			

Solubility (Moles/L)	Solubility (Grams/L)	Temp (°C)	Ref (#)	Evaluation (T P E A A)	Comments
5.901E-02	1.022E+01	7.75	P038	1 2 2 1 2	anhydrate
7.622E-02	1.320E+01	16.75	P038	1 2 2 1 2	anhydrate
9.440E-02	1.635E+01	24.95	P038	1 2 2 1 2	anhydrate
1.383E-01	2.395E+01	40.0	P038	1 2 2 1 2	anhydrate
1.975E-01	3.420E+01	55.0	P038	1 2 2 1 2	anhydrate
2.714E-01	4.700E+01	70.0	P038	1 2 2 1 2	anhydrate
4.561E-02	7.900E+00	.0	P038	1 2 2 1 2	anhydrate

772. C$_6$H$_7$NO$_3$S.1.5H$_2$O
Metanilic acid (sesquihydrate)
3-Aminobenzenesulfonic acid (sesquihydrate)

RN:	121-47-1	**MP** (°C):	
MW:	200.21	**BP** (°C):	

Solubility (Moles/L)	Solubility (Grams/L)	Temp (°C)	Ref (#)	Evaluation (T P E A A)	Comments
8.041E-02	1.610E+01	8.35	P038	1 2 2 1 2	
1.119E-01	2.240E+01	15.55	P038	1 2 2 1 2	
1.184E-01	2.370E+01	16.8	P038	1 2 2 1 2	
3.247E-01	6.500E+01	85.0	P038	1 2 2 1 2	
5.344E-02	1.070E+01	.0	P038	1 2 2 1 2	

773. C$_6$H$_7$NO$_4$S
2-Aminophenol-4-sulfonic acid
2-Amino-phenol-sulfosaeure-(4)

RN:	98-37-3	**MP** (°C):	>300
MW:	189.19	**BP** (°C):	

Solubility (Moles/L)	Solubility (Grams/L)	Temp (°C)	Ref (#)	Evaluation (T P E A A)	Comments
5.286E-02	1.000E+01	14	F300	1 0 0 0 0	

774. C$_6$H$_7$NO$_4$S
4-Aminophenol-2-sulfonic acid
4-Amino-phenol-sulfosaeure-(2)

RN:	2835-04-3	**MP** (°C):	
MW:	189.19	**BP** (°C):	

Solubility (Moles/L)	Solubility (Grams/L)	Temp (°C)	Ref (#)	Evaluation (T P E A A)	Comments
3.700E-03	7.000E-01	14	F300	1 0 0 0 0	

775. C₆H₇N₃O
Isoniazid
Isonicotinic acid hydrazide
laniazid

RN:	54-85-3	MP (°C):	171
MW:	137.14	BP (°C):	

Solubility (Moles/L)	Solubility (Grams/L)	Temp (°C)	Ref (#)	Evaluation (T P E A A)	Comments
7.813E-01	1.071E+02	20	I307	0 0 0 0 0	
8.955E-01	1.228E+02	25	B187	0 0 0 0 0	
1.458E+00	2.000E+02	37	I307	0 0 0 0 0	
1.505E+00	2.063E+02	40	B187	0 0 0 0 0	
9.115E-01	1.250E+02	ns	K444	0 0 0 0 0	

776. C₆H₇N₃O₃
Orotic acid methylamide
Orotamide, N-methyl-

RN:	1009-04-7	MP (°C):	284–286
MW:	169.14	BP (°C):	

Solubility (Moles/L)	Solubility (Grams/L)	Temp (°C)	Ref (#)	Evaluation (T P E A A)	Comments
3.420E-01	5.785E+01	–4	N018	0 0 0 0 0	
6.840E-01	1.157E+02	16	N018	0 0 0 0 0	
8.340E-01	1.411E+02	25	N018	0 0 0 0 0	

777. C₆H₇N₇
2,4,7-Triaminopteridine
2:4:7-Triaminopteridine

RN:	14439-13-5	MP (°C):	
MW:	177.17	BP (°C):	

Solubility (Moles/L)	Solubility (Grams/L)	Temp (°C)	Ref (#)	Evaluation (T P E A A)	Comments
1.254E-03	2.222E-01	20	A020	1 2 0 0 1	
2.808E-02	4.975E+00	100	A020	1 2 0 0 0	

778. C₆H₇N₇
4,6,7-Triaminopteridine
4:6:7-Triaminopteridine

RN:	19167-62-5	MP (°C):	
MW:	177.17	BP (°C):	

Solubility (Moles/L)	Solubility (Grams/L)	Temp (°C)	Ref (#)	Evaluation (T P E A A)	Comments
4.515E-04	7.999E-02	20	A020	1 2 0 1 1	
1.252E-02	2.217E+00	100	A020	1 2 0 0 1	

779. C$_6$H$_7$O$_2$P
Phenylphosphinic acid
Phenyl-phosphinigsaeure

RN: 1779-48-2   **MP** (°C):   84
MW: 142.10   **BP** (°C):

Solubility (Moles/L)	Solubility (Grams/L)	Temp (°C)	Ref (#)	Evaluation (T P E A A)	Comments
4.757E-01	6.760E+01	14	F300	1 0 0 0 2	
9.460E+00	1.344E+03	24.63	W422	0 0 0 0 0	
1.109E+01	1.576E+03	27.09	W422	0 0 0 0 0	
1.294E+01	1.839E+03	29.24	W422	0 0 0 0 0	
1.593E+01	2.264E+03	32.06	W422	0 0 0 0 0	
2.177E+01	3.093E+03	36.77	W422	0 0 0 0 0	
3.047E+01	4.330E+03	39.68	W422	0 0 0 0 0	
4.843E+00	6.881E+02	100	F300	1 0 0 0 2	

780. C$_6$H$_7$O$_3$P
Phenylphosphonic acid
Phenylphosphonsaeure

RN: 1571-33-1   **MP** (°C):   164.5
MW: 158.09   **BP** (°C):

Solubility (Moles/L)	Solubility (Grams/L)	Temp (°C)	Ref (#)	Evaluation (T P E A A)	Comments
1.202E+00	1.900E+02	15	F300	1 0 0 0 2	
1.202E+00	1.901E+02	ns	R427	0 0 0 0 0	

781. C$_6$H$_7$O$_3$As
Benzenearsonic acid
Phenylarsonsaeure

RN: 98-05-5   **MP** (°C):   160
MW: 202.04   **BP** (°C):

Solubility (Moles/L)	Solubility (Grams/L)	Temp (°C)	Ref (#)	Evaluation (T P E A A)	Comments
1.564E-01	3.160E+01	28	F300	1 0 0 0 2	
9.899E-01	2.000E+02	84	F300	1 0 0 0 1	

782. C$_6$H$_8$
1,4-Cyclohexadiene
1,4-Dihydrobenzene

RN: 628-41-1   **MP** (°C):   −49.2
MW: 80.13   **BP** (°C):   81

Solubility (Moles/L)	Solubility (Grams/L)	Temp (°C)	Ref (#)	Evaluation (T P E A A)	Comments
1.062E-02	8.512E-01	4.8	L007	2 2 1 2 2	
1.062E-02	8.512E-01	5.1	L007	2 1 1 1 2	
1.195E-02	9.576E-01	14.8	L007	2 2 1 2 2	

(continued)

782. C₆H₈ (continued)

Solubility (Moles/L)	Solubility (Grams/L)	Temp (°C)	Ref (#)	Evaluation (T P E A A)	Comments
1.195E-02	9.576E-01	15.2	L007	2 1 1 1 2	
8.002E-03	6.412E-01	20	M337	2 1 2 2 2	
1.167E-02	9.353E-01	24.8	L007	2 2 1 2 2	
8.736E-03	7.000E-01	25	M001	2 1 2 2 2	
1.167E-02	9.353E-01	25.1	L007	2 1 1 1 2	
1.201E-02	9.625E-01	34.8	L007	2 2 1 2 2	
1.201E-02	9.625E-01	35.2	L007	2 1 1 1 2	
1.259E-02	1.009E+00	44.8	L007	2 2 1 2 2	
1.259E-02	1.009E+00	45.2	L007	2 1 1 1 2	

783. C₆H₈ClN₇O
Amiloride

RN: 2609-46-3　　**MP** (°C):
MW: 229.63　　　**BP** (°C):

Solubility (Moles/L)	Solubility (Grams/L)	Temp (°C)	Ref (#)	Evaluation (T P E A A)	Comments
6.531E-04	1.500E-01	22.5	B422	0 0 0 0 0	
2.870E+00	6.590E+02	25	B443	0 0 0 0 0	

784. C₆H₈N₂
2,5-Dimethylpyrazine
2,5-Dimethyl-pyrazin

RN: 123-32-0　　**MP** (°C): 63
MW: 108.14　　　**BP** (°C): 155

Solubility (Moles/L)	Solubility (Grams/L)	Temp (°C)	Ref (#)	Evaluation (T P E A A)	Comments
		25	D425	0 0 0 0 0	

785. C₆H₈N₂
m-Phenylenediamine
m-Phenylendiamin

RN: 108-45-2　　**MP** (°C): 63
MW: 108.14　　　**BP** (°C): 283

Solubility (Moles/L)	Solubility (Grams/L)	Temp (°C)	Ref (#)	Evaluation (T P E A A)	Comments
7.409E-01	8.012E+01	.3	S115	1 2 1 1 2	α form
2.928E-01	3.166E+01	.3	S115	1 2 1 1 2	β form
1.038E+00	1.122E+02	4.6	S115	1 2 1 1 2	α form
1.354E+00	1.465E+02	9.3	S115	1 2 1 1 2	α form
1.618E+00	1.750E+02	11.7	S115	1 2 1 1 2	α form
7.806E-01	8.442E+01	14.3	S115	1 2 1 1 2	β form
2.285E+00	2.472E+02	16.1	S115	1 2 1 1 2	α form
2.671E+00	2.889E+02	17.3	S115	1 2 1 1 2	α form
1.038E+00	1.122E+02	18.3	S115	1 2 1 1 2	β form
3.075E+00	3.326E+02	18.7	S115	1 2 1 1 2	α form
3.339E+00	3.611E+02	19.9	S115	1 2 1 1 2	α form

(continued)

785. C$_6$H$_8$N$_2$ (continued)

Solubility (Moles/L)	Solubility (Grams/L)	Temp (°C)	Ref (#)	Evaluation (T P E A A)	Comments
3.537E+00	3.825E+02	20.8	S115	1 2 1 1 2	α form
1.354E+00	1.465E+02	22.0	S115	1 2 1 1 2	β form
3.796E+00	4.105E+02	22.7	S115	1 2 1 1 2	α form
1.480E+00	1.600E+02	23.1	S115	1 2 1 1 2	β form
1.618E+00	1.750E+02	24.1	S115	1 2 1 1 2	β form
1.918E+00	2.074E+02	25.1	S115	1 2 1 1 2	β form
3.979E+00	4.303E+02	26.0	S115	1 2 1 1 2	α form
2.285E+00	2.472E+02	26.3	S115	1 2 1 1 2	β form
2.671E+00	2.889E+02	27.1	S115	1 2 1 1 2	β form
2.815E+00	3.044E+02	27.1	S115	1 2 1 1 2	β form
3.075E+00	3.326E+02	27.9	S115	1 2 1 1 2	β form
4.085E+00	4.418E+02	28.7	S115	1 2 1 1 2	α form
3.339E+00	3.611E+02	29.0	S115	1 2 1 1 2	β form
3.537E+00	3.825E+02	29.1	S115	1 2 1 1 2	β form
3.796E+00	4.105E+02	30.2	S115	1 2 1 1 2	β form
3.979E+00	4.303E+02	31.5	S115	1 2 1 1 2	β form
4.217E+00	4.560E+02	32.6	S115	1 2 1 1 2	α form
4.085E+00	4.418E+02	32.8	S115	1 2 1 1 2	β form
4.217E+00	4.560E+02	34.4	S115	1 2 1 1 2	β form
4.439E+00	4.800E+02	43.5	S115	1 2 1 1 2	α form
4.549E+00	4.919E+02	53.6	S115	1 2 1 1 2	α form
4.586E+00	4.960E+02	57.6	S115	1 2 1 1 2	α form
4.623E+00	5.000E+02	62.8	S115	1 2 1 1 2	α form

786. C$_6$H$_8$N$_2$

o-Phenylenediamine
o-Phenylendiamin

RN:	95-54-5	MP (°C):	102–103	
MW:	108.14	BP (°C):	257	

Solubility (Moles/L)	Solubility (Grams/L)	Temp (°C)	Ref (#)	Evaluation (T P E A A)	Comments
2.876E-01	3.110E+01	20	T301	1 2 2 2 2	
3.763E-01	4.070E+01	35	F300	1 0 0 0 2	
3.599E-01	3.892E+01	35.1	S115	1 2 1 1 2	
5.110E-01	5.527E+01	45.8	S115	1 2 1 1 2	
9.804E-01	1.060E+02	56.3	S115	1 2 1 1 2	
1.458E+00	1.577E+02	61.3	S115	1 2 1 1 2	
1.755E+00	1.898E+02	62.8	S115	1 2 1 1 2	
2.218E+00	2.398E+02	64.2	S115	1 2 1 1 2	
2.948E+00	3.188E+02	66.1	S115	1 2 1 1 2	
3.558E+00	3.847E+02	67.7	S115	1 2 1 1 2	
3.955E+00	4.277E+02	71.3	S115	1 2 1 1 2	
4.338E+00	4.691E+02	80.8	S115	1 2 1 1 2	
4.476E+00	4.841E+02	88.1	S115	1 2 1 1 2	
4.533E+00	4.902E+02	91.7	S115	1 2 1 1 2	
4.570E+00	4.942E+02	95.5	S115	1 2 1 1 2	

787. C$_6$H$_8$N$_2$

p-Phenylenediamine

1,4-Phenylenediamine

RN:	106-50-3	**MP** (°C):	141
MW:	108.14	**BP** (°C):	267

Solubility (Moles/L)	Solubility (Grams/L)	Temp (°C)	Ref (#)	Evaluation (T P E A A)	Comments
9.880E-02	1.068E+01	3.6	S115	1 2 1 1 2	
3.299E-01	3.568E+01	23.7	S115	1 2 1 1 2	
4.180E-01	4.520E+01	25	F300	1 0 0 0 2	
8.292E-01	8.967E+01	37.8	S115	1 2 1 1 2	
1.460E+00	1.579E+02	49.9	S115	1 2 1 1 2	
1.978E+00	2.140E+02	59.2	S115	1 2 1 1 2	
2.368E+00	2.561E+02	64.6	S115	1 2 1 1 2	
2.724E+00	2.945E+02	69.2	S115	1 2 1 1 2	
3.155E+00	3.412E+02	75.5	S115	1 2 1 1 2	
3.432E+00	3.711E+02	80.3	S115	1 2 1 1 2	
3.809E+00	4.119E+02	88.5	S115	1 2 1 1 2	
4.055E+00	4.385E+02	95.9	S115	1 2 1 1 2	
1.500E-05	1.622E-03	98.59	M180	0 0 2 2 0	EFG
2.500E-05	2.704E-03	111.46	M180	0 0 2 2 0	EFG
4.000E-05	4.326E-03	117.47	M180	0 0 2 2 0	EFG
4.500E-05	4.866E-03	122.10	M180	0 0 2 2 0	EFG
5.000E-05	5.407E-03	126.84	M180	0 0 2 2 0	EFG
7.000E-05	7.570E-03	133.34	M180	0 0 2 2 0	EFG

788. C$_6$H$_8$N$_2$OS

5,6-Dimethyl-2-thiouracil

4(1H)-Pyrimidinone, 2,3-dihydro-5,6-dimethyl-2-thioxo-

5,6-Dimethylthiouracil

RN:	28456-54-4	**MP** (°C):	
MW:	156.21	**BP** (°C):	

Solubility (Moles/L)	Solubility (Grams/L)	Temp (°C)	Ref (#)	Evaluation (T P E A A)	Comments
8.790E-03	1.373E+00	25	G016	1 2 1 2 2	intrinsic

789. C$_6$H$_8$N$_2$O$_2$

N,N-1,3-Dimethyluracil

1,3-Dimethyl-2,4-pyrimidinedione

*N*1,*N*3-Dimethyluracil

N,N'-Dimethyluracil

1,3-Dimethyluracil

RN:	874-14-6	**MP** (°C):	
MW:	140.14	**BP** (°C):	

Solubility (Moles/L)	Solubility (Grams/L)	Temp (°C)	Ref (#)	Evaluation (T P E A A)	Comments
3.568E+00	5.000E+02	ns	B177	0 0 0 0 2	

790. C₆H₈N₂O₂S

o-Aminobenzenesulfonamide
Orthanilamide

RN: 3306-62-5   **MP** (°C):
MW: 172.21   **BP** (°C):

Solubility (Moles/L)	Solubility (Grams/L)	Temp (°C)	Ref (#)	Evaluation (T P E A A)	Comments
3.750E-02	6.458E+00	23	K034	2 2 2 2 1	
3.865E-02	6.655E+00	24	K034	2 2 2 2 1	
4.323E-02	7.444E+00	26	K034	2 2 2 2 1	
4.723E-02	8.133E+00	28	K034	2 2 2 2 1	
5.237E-02	9.018E+00	30.5	K034	2 2 2 2 1	
5.806E-02	9.999E+00	33	K034	2 2 2 2 2	
6.034E-02	1.039E+01	34	K034	2 2 2 2 2	
6.375E-02	1.098E+01	35.5	K034	2 2 2 2 2	
6.886E-02	1.186E+01	37	K034	2 2 2 2 2	
6.829E-02	1.176E+01	37	K034	2 2 2 2 2	
8.356E-02	1.439E+01	42	K034	2 2 2 2 2	
9.707E-02	1.672E+01	46	K034	2 2 2 2 2	
1.139E-01	1.961E+01	50	K034	2 2 2 2 2	

791. C₆H₈N₂O₂S

m-Aminobenzenesulfonamide
Metanilamide
m-Amidobenzenesulfonamide

RN: 98-18-0   **MP** (°C):
MW: 172.21   **BP** (°C):

Solubility (Moles/L)	Solubility (Grams/L)	Temp (°C)	Ref (#)	Evaluation (T P E A A)	Comments
6.545E-02	1.127E+01	23	K034	2 2 2 2 2	
6.942E-02	1.196E+01	24	K034	2 2 2 2 2	
7.678E-02	1.322E+01	26	K034	2 2 2 2 2	
8.469E-02	1.458E+01	28	K034	2 2 2 2 2	
1.077E-01	1.855E+01	33	K034	2 2 2 2 2	
1.244E-01	2.143E+01	35.5	K034	2 2 2 2 2	
1.339E-01	2.306E+01	37	K034	2 2 2 2 2	
1.461E-01	2.515E+01	39	K034	2 2 2 2 2	
1.697E-01	2.922E+01	42	K034	2 2 2 2 2	
2.072E-01	3.568E+01	46	K034	2 2 2 2 2	
2.543E-01	4.379E+01	50	K034	2 2 2 2 2	

792. $C_6H_8N_2O_2S$
Benzenesulfamide
Sulfanilamide
Sulfanilsaeure-amid
p-Aminobenzenesulphonamide

RN:	63-74-1	**MP** (°C):	165	
MW:	172.21	**BP** (°C):		

Solubility (Moles/L)	Solubility (Grams/L)	Temp (°C)	Ref (#)	Evaluation (T P E A A)	Comments
1.159E-02	1.996E+00	1	A047	1 0 0 0 0	EFG
1.057E-02	1.820E+00	4.40	B147	1 2 1 1 2	
1.458E-02	2.510E+00	10.20	B147	1 2 1 1 2	
1.957E-02	3.370E+00	15	B147	1 2 1 1 2	
2.323E-02	4.000E+00	15	F300	1 0 0 0 0	
2.660E-02	4.581E+00	15	K024	1 2 1 1 2	
2.241E-02	3.860E+00	15	S147	1 2 2 2 2	hydrate
2.889E-02	4.975E+00	16	A047	1 0 0 0 0	EFG
2.439E-02	4.200E+00	16	H114	1 0 0 0 2	
2.700E-02	4.650E+00	20	B147	1 2 1 1 2	
3.463E-02	5.964E+00	20	D041	1 0 0 0 0	
4.149E-02	7.145E+00	20	F073	1 2 2 2 2	
2.903E-02	5.000E+00	20	F300	1 0 0 0 0	
3.020E-02	5.200E+00	20	S147	1 2 2 2 2	hydrate
3.693E-02	6.359E+00	23	K034	2 2 2 2 1	
3.979E-02	6.853E+00	24	K034	2 2 2 2 1	
3.484E-02	6.000E+00	25	B147	1 2 1 1 2	
4.855E-02	8.360E+00	25	C102	2 0 2 2 2	
4.550E-02	7.835E+00	25	M116	2 1 1 1 2	
4.274E-02	7.360E+00	25	M440	0 0 0 0 0	
4.820E-02	8.300E+00	25	P015	0 0 0 0 0	
4.216E-02	7.260E+00	25	S147	1 2 2 2 2	hydrate
4.437E-02	7.641E+00	26	K034	2 2 2 2 1	
4.723E-02	8.133E+00	27	K034	2 2 2 2 1	
5.008E-02	8.625E+00	28	K034	2 2 2 2 1	
4.762E-02	8.200E+00	30	B147	1 2 1 1 2	
5.633E-02	9.700E+00	30	S147	1 2 2 2 2	hydrate
5.806E-02	9.999E+00	30.5	K034	2 2 2 2 2	
6.318E-02	1.088E+01	31	A047	1 0 0 0 0	EFG
6.205E-02	1.068E+01	31.7	K034	2 2 2 2 2	
6.829E-02	1.176E+01	33	K034	2 2 2 2 2	
7.282E-02	1.254E+01	34	K034	2 2 2 2 2	
6.388E-02	1.100E+01	35	B147	1 2 1 1 2	
7.543E-02	1.299E+01	35	S147	1 2 2 2 2	β form
7.848E-02	1.351E+01	35.5	K034	2 2 2 2 2	
1.259E-01	2.168E+01	37	A028	1 0 2 1 2	intrinsic
7.375E-02	1.270E+01	37	B147	1 2 1 1 2	
8.478E-02	1.460E+01	37	C102	2 0 2 2 2	
8.594E-02	1.480E+01	37	D084	1 0 1 0 2	
8.018E-02	1.381E+01	37	F072	1 0 0 0 2	
8.710E-02	1.500E+01	37	F300	1 0 0 0 1	
8.920E-02	1.536E+01	37	G028	2 2 1 1 2	δ form, recrystallized

(continued)

792. $C_6H_8N_2O_2S$ (continued)

Solubility (Moles/L)	Solubility (Grams/L)	Temp (°C)	Ref (#)	Evaluation (T P E A A)	Comments
9.070E-02	1.562E+01	37	G028	2 2 1 1 2	β form, recrystallized
9.120E-02	1.571E+01	37	G028	2 2 1 1 2	α form, recrystallized
9.240E-02	1.591E+01	37	G028	2 2 1 1 2	γ form
8.413E-02	1.449E+01	37	K034	2 2 2 2 2	
8.652E-02	1.490E+01	37	K086	1 0 0 0 2	
8.210E-02	1.414E+01	37	K095	2 0 0 0 2	intrinsic
8.710E-02	1.500E+01	37	L091	1 0 0 0 2	pH 5.5
8.469E-02	1.458E+01	37.50	M142	1 0 0 0 2	
9.201E-02	1.584E+01	39	K034	2 2 2 2 2	
8.362E-02	1.440E+01	40	B147	1 2 1 1 2	form II
9.750E-02	1.679E+01	40	G028	2 2 1 1 2	α form, recrystallized
9.640E-02	1.660E+01	40	G028	2 2 1 1 2	γ form
9.640E-02	1.660E+01	40	G028	2 2 1 1 2	δ form, recrystallized
9.680E-02	1.667E+01	40	G028	2 2 1 1 2	β form, recrystallized
9.518E-02	1.639E+01	40	S147	1 2 2 2 2	β form
1.049E-01	1.807E+01	42	K034	2 2 2 2 2	
1.086E-01	1.870E+01	45	B147	1 2 1 1 2	form II
1.201E-01	2.069E+01	45	S147	1 2 2 2 2	β form
1.256E-01	2.162E+01	46	K034	2 2 2 2 2	
1.527E-01	2.629E+01	50	A047	1 0 0 0 0	EFG
1.388E-01	2.390E+01	50	B147	1 2 1 1 2	form II
1.433E-01	2.468E+01	50	G028	2 2 1 1 2	δ form, recrystallized
1.419E-01	2.444E+01	50	G028	2 2 1 1 2	β form, recrystallized
1.430E-01	2.463E+01	50	G028	2 2 1 1 2	γ form
1.435E-01	2.471E+01	50	G028	2 2 1 1 2	α form, recrystallized
1.516E-01	2.610E+01	50	K034	2 2 2 2 2	
1.488E-01	2.562E+01	50	S147	1 2 2 2 2	β form
1.789E-01	3.080E+01	55	B147	1 2 1 1 2	form II
2.294E-01	3.950E+01	60	B147	1 2 1 1 2	form II
2.923E-01	5.033E+01	65	A047	1 0 0 0 0	EFG
2.962E-01	5.100E+01	65	B147	1 2 1 1 2	form II
3.833E-01	6.600E+01	70	B147	1 2 1 1 2	form II
4.599E-01	7.919E+01	75	A047	1 0 0 0 0	EFG
5.168E-01	8.900E+01	75	B147	1 2 1 1 2	form II
5.660E-01	9.747E+01	79	A047	1 0 0 0 0	EFG
6.272E-02	1.080E+01	ns	D035	0 0 0 0 2	
3.050E-02	5.252E+00	ns	L044	0 0 0 0 2	
4.571E-02	7.871E+00	ns	R427	0 0 0 0 0	
4.365E-02	7.517E+00	ns	R428	0 0 0 0 0	

793. $C_6H_8N_2O_2S.H_2O$
Sulfanilamide (monohydrate)
4-Aminobenzenesulfonamide (monohydrate)
p-Anilinesulfonamide (monohydrate)
Bacteramid (monohydrate)

RN: 20203-81-0 **MP** (°C):
MW: 190.22 **BP** (°C):

Solubility (Moles/L)	Solubility (Grams/L)	Temp (°C)	Ref (#)	Evaluation (T P E A A)	Comments
2.200E-02	4.185E+00	15	G028	2 2 1 1 2	
4.320E-02	8.218E+00	26	G028	2 2 1 1 2	
5.600E-02	1.065E+01	30	G028	2 2 1 1 2	
8.420E-02	1.602E+01	37	G028	2 2 1 1 2	

794. $C_6H_8N_2O_3$
5,5-Dimethylbarbituric acid
5,5-Dimethylbarbitursaeure
Barbituric acid, 5,5-dimethyl
2,4,6(1H,3H,5H)-Pyrimidinetrione, 5,5-dimethyl
5,5-Dimethyl barbituric acid
5,5-Dimethylbarbiturate

RN: 24448-94-0 **MP** (°C): 278
MW: 156.14 **BP** (°C):

Solubility (Moles/L)	Solubility (Grams/L)	Temp (°C)	Ref (#)	Evaluation (T P E A A)	Comments
1.812E-02	2.829E+00	25	P350	0 0 0 0 0	intrinsic
1.549E-02	2.419E+00	ns	T003	0 0 0 0 2	

795. $C_6H_8N_2O_3S$
4-Phenylhydrazine sulfonic acid
Phenylhydrazin-sulfosaeure-(4)

RN: 98-71-5 **MP** (°C):
MW: 188.21 **BP** (°C):

Solubility (Moles/L)	Solubility (Grams/L)	Temp (°C)	Ref (#)	Evaluation (T P E A A)	Comments
3.029E-02	5.700E+00	11.50	F300	1 0 0 0 1	
1.860E-01	3.500E+01	100	F300	1 0 0 0 1	

796. C$_6$H$_8$N$_2$O$_8$
Isosorbide dinitrate
1,4:3,6-Dianhydro-D-glucitol dinitrate
Sorbidin
Isogen
Imdur

RN: 87-33-2   **MP (°C):** 70
MW: 236.14   **BP (°C):**

Solubility (Moles/L)	Solubility (Grams/L)	Temp (°C)	Ref (#)	Evaluation (T P E A A)	Comments
2.328E-03	5.497E-01	25	L033	1 0 2 1 2	

797. C$_6$H$_8$N$_4$O
5-Amino-4-carboxymethylaminopyrimidine

RN: **MP (°C):**
MW: 152.16   **BP (°C):**

Solubility (Moles/L)	Solubility (Grams/L)	Temp (°C)	Ref (#)	Evaluation (T P E A A)	Comments
2.120E-01	3.226E+01	100	A082	1 2 0 0 0	

798. C$_6$H$_8$N$_8$
2,4,6,7-Tetraminopteridine
2:4:6:7-Tetraminopteridine

RN: 19167-63-6   **MP (°C):**
MW: 192.18   **BP (°C):**

Solubility (Moles/L)	Solubility (Grams/L)	Temp (°C)	Ref (#)	Evaluation (T P E A A)	Comments
4.002E-04	7.692E-02	20	A020	1 2 0 1 1	

799. C$_6$H$_8$O$_2$
Sorbic acid
2,4-Hexadienoic acid
2-Propenylacrylic acid
Preservastat
Hexadienoic acid
Sorbistat

RN: 110-44-1   **MP (°C):** 134.5
MW: 112.13   **BP (°C):** 228

Solubility (Moles/L)	Solubility (Grams/L)	Temp (°C)	Ref (#)	Evaluation (T P E A A)	Comments
1.700E-02	1.906E+00	30	L069	1 0 1 1 0	EFG

800. C₆H₈O₆

Tricarballylic acid
Tricarballylsaeure
1,2,3-Propanetricarboxylic acid

RN:	99-14-9	**MP** (°C):	166	
MW:	176.13	**BP** (°C):		

Solubility (Moles/L)	Solubility (Grams/L)	Temp (°C)	Ref (#)	Evaluation (T P E A A)	Comments
1.885E+00	3.320E+02	18	F300	1 0 0 0 2	

801. C₆H₈O₆

Ascorbic acid
L-Ascorbic acid
L-Ascorbinsaeure

RN:	50-81-7	**MP** (°C):	193	
MW:	176.13	**BP** (°C):		

Solubility (Moles/L)	Solubility (Grams/L)	Temp (°C)	Ref (#)	Evaluation (T P E A A)	Comments
9.269E-01	1.633E+02	6.99	A341	0 0 0 0 0	
9.509E-01	1.675E+02	7.99	A341	0 0 0 0 0	
9.880E-01	1.740E+02	9.99	A341	0 0 0 0 0	
1.026E+00	1.807E+02	11.99	A341	0 0 0 0 0	
1.142E+00	2.011E+02	15.99	A341	0 0 0 0 0	
1.418E+00	2.498E+02	20	D041	1 0 0 0 2	
1.254E+00	2.208E+02	20	S472	0 0 0 0 0	
1.283E+00	2.260E+02	20.99	A341	0 0 0 0 0	
1.397E+00	2.460E+02	24.99	A341	0 0 0 0 0	
1.891E+00	3.330E+02	25	D315	0 0 0 0 0	
9.757E-01	1.718E+02	25	N003	0 0 0 0 0	
1.388E+00	2.445E+02	25	S472	0 0 0 0 0	
1.551E+00	2.731E+02	28.99	A341	0 0 0 0 0	
1.533E+00	2.699E+02	30	S472	0 0 0 0 0	
1.718E+00	3.025E+02	33.99	A341	0 0 0 0 0	
1.703E+00	2.999E+02	35	S472	0 0 0 0 0	
1.758E+00	3.096E+02	35.99	A341	0 0 0 0 0	
1.856E+00	3.270E+02	38.99	A341	0 0 0 0 0	
1.028E+00	1.810E+02	40	N003	0 0 0 0 0	
1.874E+00	3.301E+02	40	S472	0 0 0 0 0	
2.009E+00	3.539E+02	42.99	A341	0 0 0 0 0	
2.021E+00	3.560E+02	43.99	A341	0 0 0 0 0	
2.066E+00	3.638E+02	44.99	A341	0 0 0 0 0	
2.054E+00	3.618E+02	45	S472	0 0 0 0 0	
2.132E+00	3.755E+02	47.69	A341	0 0 0 0 0	
2.184E+00	3.847E+02	48.49	A341	0 0 0 0 0	
2.235E+00	3.937E+02	49.99	A341	0 0 0 0 0	
2.235E+00	3.936E+02	50	S472	0 0 0 0 0	
2.255E+00	3.972E+02	50.39	A341	0 0 0 0 0	
2.275E+00	4.007E+02	50.99	A341	0 0 0 0 0	
2.373E+00	4.180E+02	52.49	A341	0 0 0 0 0	

(continued)

801. $C_6H_8O_6$ (continued)

Solubility (Moles/L)	Solubility (Grams/L)	Temp (°C)	Ref (#)	Evaluation (T P E A A)	Comments
2.383E+00	4.197E+02	53.99	A341	0 0 0 0 0	
2.413E+00	4.249E+02	54.09	A341	0 0 0 0 0	
2.449E+00	4.314E+02	54.99	A341	0 0 0 0 0	
2.520E+00	4.439E+02	60.02	A341	0 0 0 0 0	
2.551E+00	4.492E+02	61.99	A341	0 0 0 0 0	
2.635E+00	4.641E+02	64.99	A341	0 0 0 0 0	
1.891E+00	3.330E+02	ns	M054	0 0 0 0 2	

802. $C_6H_8O_7$
Citric acid anhydrous
2-Hydroxytricarballylic acid
Citronensaeure
1,2,3-Propanetricarboxylic acid
Citro
Citralite

RN: 77-92-9   **MP** (°C): 153
MW: 192.13   **BP** (°C):

Solubility (Moles/L)	Solubility (Grams/L)	Temp (°C)	Ref (#)	Evaluation (T P E A A)	Comments
2.549E+00	4.898E+02	0	M043	1 0 0 0 1	
1.881E+00	3.613E+02	1.2	K084	1 0 1 0 2	
1.875E+00	3.602E+02	1.6	K084	1 0 1 0 2	
2.562E+00	4.923E+02	4.99	A339	0 0 0 0 0	
1.825E+00	3.506E+02	10	D020	1 2 1 1 2	
2.571E+00	4.940E+02	10	F300	1 0 0 0 2	
1.825E+00	3.506E+02	10	F302	1 0 0 0 1	
2.817E+00	5.413E+02	10	M043	1 0 0 0 2	
1.938E+00	3.723E+02	10.0	K084	1 0 1 0 2	
2.684E+00	5.157E+02	9.99	A339	0 0 0 0 0	
1.927E+00	3.702E+02	10.8	K084	1 0 1 0 2	
2.811E+00	5.400E+02	14.99	A339	0 0 0 0 0	
1.933E+00	3.713E+02	15.0	K084	1 0 1 0 2	
2.918E+00	5.605E+02	19.99	A339	0 0 0 0 0	
3.089E+00	5.935E+02	20	D041	1 0 0 0 2	
2.816E+00	5.410E+02	20	F300	1 0 0 0 2	
1.935E+00	3.719E+02	20	F302	1 0 0 0 2	
3.089E+00	5.935E+02	20	M043	1 0 0 0 2	
3.045E+00	5.851E+02	24.99	A339	0 0 0 0 0	
1.994E+00	3.831E+02	25	D020	1 2 1 1 2	
1.254E+01	2.409E+03	25	K040	1 0 2 1 2	
3.201E+00	6.149E+02	29.99	A339	0 0 0 0 0	
2.037E+00	3.914E+02	30	F302	1 0 0 0 2	
3.366E+00	6.466E+02	30	M043	1 0 0 0 2	
3.296E+00	6.332E+02	34.99	A339	0 0 0 0 0	
2.100E+00	4.034E+02	35.8	D039	2 2 1 2 2	EFG
2.094E+00	4.023E+02	36.6	F302	1 0 0 0 2	
3.201E+00	6.150E+02	36.60	F300	1 0 0 0 2	

(*continued*)

802. C₆H₈O₇ (continued)

Solubility (Moles/L)	Solubility (Grams/L)	Temp (°C)	Ref (#)	Evaluation (T P E A A)	Comments
3.346E+00	6.429E+02	39.99	A339	0 0 0 0 0	
2.118E+00	4.069E+02	40	D020	1 2 1 1 2	
2.116E+00	4.065E+02	40	D039	2 2 1 2 0	EFG
2.118E+00	4.069E+02	40	F302	1 0 0 0 2	
3.553E+00	6.825E+02	40	M043	1 0 0 0 2	
3.438E+00	6.605E+02	44.99	A339	0 0 0 0 0	
3.488E+00	6.702E+02	49.99	A339	0 0 0 0 0	
2.161E+00	4.152E+02	50	D039	2 2 1 2 0	EFG
2.159E+00	4.149E+02	50	F302	1 0 0 0 2	
3.539E+00	6.800E+02	54.99	A339	0 0 0 0 0	
3.601E+00	6.918E+02	59.99	A339	0 0 0 0 0	
2.214E+00	4.253E+02	60	D039	2 2 1 2 0	EFG
2.205E+00	4.236E+02	60	F302	1 0 0 0 2	
3.824E+00	7.347E+02	60	M043	1 0 0 0 2	
3.669E+00	7.050E+02	64.99	A339	0 0 0 0 0	
2.261E+00	4.344E+02	70	D039	2 2 1 2 0	EFG
2.251E+00	4.325E+02	70	F302	1 0 0 0 2	
2.300E+00	4.420E+02	80	D039	2 2 1 2 0	EFG
2.294E+00	4.407E+02	80	F302	1 0 0 0 2	
4.102E+00	7.881E+02	80	M043	1 0 0 0 2	
2.350E+00	4.515E+02	90	D039	2 2 1 2 0	EFG
2.336E+00	4.487E+02	90	F302	1 0 0 0 2	
2.391E+00	4.595E+02	100	D039	2 2 1 2 0	EFG
4.372E+00	8.400E+02	100	D041	1 0 0 0 2	
3.997E+00	7.680E+02	100	F300	1 0 0 0 2	
2.376E+00	4.565E+02	100	F302	1 0 0 0 1	
4.373E+00	8.403E+02	100	M043	1 0 0 0 2	
1.885E+00	3.621E+02	.0	K084	1 0 1 0 2	

803. C₆H₈O₇,H₂O

Citric acid (monohydrate)
2-Hydroxytricarballylic acid (monohydrate)

RN: 5949-29-1　　**MP** (°C):
MW: 210.14　　　**BP** (°C):

Solubility (Moles/L)	Solubility (Grams/L)	Temp (°C)	Ref (#)	Evaluation (T P E A A)	Comments
1.554E+00	3.266E+02	0	D039	2 2 1 2 0	EFG
1.667E+00	3.502E+02	10	D039	2 2 1 2 0	EFG
3.005E+00	6.314E+02	17.20	L031	1 1 2 1 2	average of 2
3.077E+00	6.466E+02	19.80	L031	1 1 2 1 2	
1.771E+00	3.723E+02	20	D039	2 2 1 2 0	EFG
3.080E+00	6.473E+02	20.20	L031	1 1 2 1 2	
3.146E+00	6.610E+02	22.50	L031	1 1 2 1 2	
3.154E+00	6.627E+02	22.90	L031	1 1 2 1 2	
1.822E+00	3.830E+02	25	D039	2 2 1 2 2	EFG
3.214E+00	6.753E+02	25.10	L031	1 1 2 1 2	
3.216E+00	6.759E+02	25.30	L031	1 1 2 1 2	

(continued)

803. $C_6H_8O_7 \cdot H_2O$ (continued)

Solubility (Moles/L)	Solubility (Grams/L)	Temp (°C)	Ref (#)	Evaluation (T P E A A)	Comments
3.272E+00	6.875E+02	27.00	L031	1 1 2 1 2	
3.276E+00	6.885E+02	27.60	L031	1 1 2 1 2	
3.303E+00	6.942E+02	28.60	L031	1 1 2 1 2	
1.864E+00	3.917E+02	30	D039	2 2 1 2 0	EFG
3.359E+00	7.059E+02	30.50	L031	1 1 2 1 2	
3.357E+00	7.054E+02	30.70	L031	1 1 2 1 2	
3.389E+00	7.122E+02	31.80	L031	1 1 2 1 2	
3.440E+00	7.230E+02	33.70	L031	1 1 2 1 2	
3.478E+00	7.308E+02	34.40	L031	1 1 2 1 2	
3.518E+00	7.392E+02	35.40	L031	1 1 2 1 2	

804. C_6H_8S
2-Ethylthiophene
Thiophene, 2-ethyl-
RN: 872-55-9   **MP** (°C):   <25
MW: 112.19   **BP** (°C):   132

Solubility (Moles/L)	Solubility (Grams/L)	Temp (°C)	Ref (#)	Evaluation (T P E A A)	Comments
2.603E-03	2.920E-01	25	K119	1 0 0 0 2	
2.603E-03	2.920E-01	25	P051	2 1 1 2 2	
2.603E-03	2.920E-01	25.00	P007	2 1 2 2 2	

805. $C_6H_9ClO_3$
Ethyl 2-chloroacetoacetate
2-Chloroacetoacetic acid ethyl ester
RN: 609-15-4   **MP** (°C):
MW: 164.59   **BP** (°C):   107 at 14 mm Hg

Solubility (Moles/L)	Solubility (Grams/L)	Temp (°C)	Ref (#)	Evaluation (T P E A A)	Comments
5.407E-02	8.900E+00	30	B433	0 0 0 0 0	

806. $C_6H_9NO_3$
4,6,10-Trioxa-1-azatricyclo[3.3.1.13,7]decane
Trimorpholin
Trimorpholine
RN: 281-36-7   **MP** (°C):
MW: 143.14   **BP** (°C):

Solubility (Moles/L)	Solubility (Grams/L)	Temp (°C)	Ref (#)	Evaluation (T P E A A)	Comments
1.167E+00	1.670E+02	0	F300	1 0 0 0 2	
2.375E+00	3.400E+02	80	F300	1 0 0 0 2	

807. C$_6$H$_9$NO$_3$
Trimethadione
3,5,5-Trimethyl-2,4-diketooxazolidine
3,5,5-Trimethyl-2,4-oxazolidinedione
Tridione

RN:	127-48-0	MP (°C):	46
MW:	143.14	BP (°C):	

Solubility (Moles/L)	Solubility (Grams/L)	Temp (°C)	Ref (#)	Evaluation (T P E A A)	Comments
3.327E-01	4.762E+01	20	D041	1 0 0 0 0	

808. C$_6$H$_9$NO$_6$
Triglycine
Complexon I
N,N-bis(Carboxymethyl)glycine
α,α',α²-Trimethylaminetricarboxylic acid

RN:	139-13-9	MP (°C):	241.5
MW:	191.14	BP (°C):	

Solubility (Moles/L)	Solubility (Grams/L)	Temp (°C)	Ref (#)	Evaluation (T P E A A)	Comments
3.090E-01	5.906E+01	25	M024	1 2 0 1 2	
3.395E-01	6.490E+01	25.1	N024	0 0 0 0 0	
3.374E-01	6.450E+01	25.1	N025	0 0 0 0 0	
3.348E-01	6.400E+01	25.1	N026	0 0 0 0 0	
3.101E-01	5.927E+01	25.1	N027	1 2 2 2 2	

809. C$_6$H$_9$N$_3$
Kyanmethin
6-Amino-2,4-dimethyl-pyrimidin
6-Amino-2,4-dimethylpyrimidine

RN:	461-98-3	MP (°C):	182
MW:	123.16	BP (°C):	

Solubility (Moles/L)	Solubility (Grams/L)	Temp (°C)	Ref (#)	Evaluation (T P E A A)	Comments
5.197E-02	6.400E+00	18	F300	1 0 0 0 1	

810. C$_6$H$_9$N$_3$O$_2$
2-Isopropyl-4(5)-nitroimidazole
1H-Imidazole, 2-(1-methylethyl)-4-nitro-
2-(1-Methylethyl)-4-nitro-1H-imidazole
2-Isopropyl-5-nitroimidazole
2-Isopropyl-4-nitroimidazole

RN: 13373-32-5   **MP** (°C): 182–183
MW: 155.16   **BP** (°C):

Solubility (Moles/L)	Solubility (Grams/L)	Temp (°C)	Ref (#)	Evaluation (T P E A A)	Comments
7.025E-02	1.090E+01	20	D344	0 0 0 0 0	
7.025E-02	1.090E+01	20	D344	0 0 0 0 0	
6.886E-02	1.068E+01	20	D344	0 0 0 0 0	
7.030E-02	1.091E+01	20	D344	0 0 0 0 0	

811. C$_6$H$_9$N$_3$O$_2$
L-Histidine
L-Histidin
Histidine

RN: 71-00-1   **MP** (°C): 287
MW: 155.16   **BP** (°C):

Solubility (Moles/L)	Solubility (Grams/L)	Temp (°C)	Ref (#)	Evaluation (T P E A A)	Comments
2.580E-01	4.003E+01	15	D349	2 1 1 2 2	
2.646E-01	4.106E+01	20	B032	1 2 2 1 2	
2.640E-01	4.096E+01	20	D349	2 1 1 2 2	
2.930E-01	4.546E+01	25	B032	1 2 2 1 2	
2.574E-01	3.994E+01	25	D041	1 0 0 0 2	
2.720E-01	4.220E+01	25	D349	2 1 1 2 2	
2.481E-01	3.850E+01	25	F300	1 0 0 0 2	
2.651E-01	4.114E+01	25	G315	0 0 0 0 0	
2.771E-01	4.300E+01	25.1	N024	0 0 0 0 0	
2.771E-01	4.300E+01	25.1	N025	0 0 0 0 0	
2.771E-01	4.300E+01	25.1	N026	0 0 0 0 0	
2.675E-01	4.150E+01	25.1	N027	1 1 2 2 2	
2.791E-01	4.330E+01	27	D036	0 0 0 0 0	
3.207E-01	4.976E+01	29.80	B032	1 2 2 1 2	
2.834E-01	4.398E+01	30	H062	2 2 2 0 1	EFG
5.213E-01	8.088E+01	50	H062	2 2 2 0 0	EFG
7.915E-01	1.228E+02	70	H062	2 2 2 0 0	EFG

812. C$_6$H$_9$N$_3$O$_2$
6-Amino-1,3-dimethyluracil

RN: 6642-31-5   **MP** (°C):
MW: 155.16   **BP** (°C):

Solubility (Moles/L)	Solubility (Grams/L)	Temp (°C)	Ref (#)	Evaluation (T P E A A)	Comments
4.550E-02	7.060E+00	25	Z408	0 0 0 0 0	

813. C₆H₉N₃O₃

Metronidazole
Flagyl
2-Methyl-5-nitroimidazole-1-ethanol
Metrozine
Rozex
2-Methyl-5-nitro-1-imidazoleethanol

RN: 443-48-1   **MP** (°C): 158
MW: 171.16   **BP** (°C):

Solubility (Moles/L)	Solubility (Grams/L)	Temp (°C)	Ref (#)	Evaluation (T P E A A)	Comments
5.545E-02	9.490E+00	20	D344	0 0 0 0 0	
5.545E-02	9.490E+00	20	D344	0 0 0 0 0	
5.441E-02	9.312E+00	20	D344	0 0 0 0 0	
5.540E-02	9.482E+00	20	D344	0 0 0 0 0	
4.809E-02	8.232E+00	20	H324	0 0 0 0 0	
5.785E-02	9.901E+00	20	I315	0 0 0 0 0	
6.427E-02	1.100E+01	25	C062	1 1 2 1 2	
5.550E-02	9.500E+00	25	C124	2 0 1 1 2	
5.727E-02	9.803E+00	26	H324	0 0 0 0 0	
6.585E-02	1.127E+01	30	H324	0 0 0 0 0	
5.843E-02	1.000E+01	ns	C324	0 0 0 0 0	
5.843E-02	1.000E+01	ns	K444	0 0 0 0 0	

814. C₆H₁₀

1,5-Hexadiene
Biallyl
Diallyl

RN: 592-42-7   **MP** (°C): −141
MW: 82.15   **BP** (°C): 60

Solubility (Moles/L)	Solubility (Grams/L)	Temp (°C)	Ref (#)	Evaluation (T P E A A)	Comments
2.057E-03	1.690E-01	25	M001	2 1 2 2 2	

815. C₆H₁₀

Cyclohexene
1,2,3,4-Tetrahydrobenzene

RN: 110-83-8   **MP** (°C): −104
MW: 82.15   **BP** (°C): 83

Solubility (Moles/L)	Solubility (Grams/L)	Temp (°C)	Ref (#)	Evaluation (T P E A A)	Comments
3.408E-03	2.799E-01	4.8	L007	2 2 1 2 2	
3.408E-03	2.799E-01	5.1	L007	2 0 1 1 2	
3.633E-03	2.984E-01	14.8	L007	2 2 1 2 2	
3.633E-03	2.984E-01	15.2	L007	2 0 1 1 2	
1.583E-03	1.300E-01	20	C008	1 2 2 0 1	
2.769E-03	2.274E-01	20	M337	2 1 2 2 2	

(continued)

815. C$_6$H$_{10}$ (continued)

Solubility (Moles/L)	Solubility (Grams/L)	Temp (°C)	Ref (#)	Evaluation (T P E A A)	Comments
3.450E-03	2.834E-01	23.5	S171	2 1 2 2 2	
3.639E-03	2.989E-01	24.8	L007	2 2 1 2 2	
2.593E-03	2.130E-01	25	M001	2 1 2 2 2	
3.639E-03	2.989E-01	25.1	L007	2 0 1 1 2	
3.681E-03	3.024E-01	34.8	L007	2 2 1 2 2	
3.681E-03	3.024E-01	35.2	L007	2 0 1 1 2	
6.000E-03	4.929E-01	40	P335	0 0 0 0 0	
3.779E-03	3.104E-01	44.8	L007	2 2 1 2 2	
3.779E-03	3.104E-01	45.2	L007	2 0 1 1 2	
1.800E-02	1.479E+00	140	P335	0 0 0 0 0	
1.583E-03	1.300E-01	ns	M010	0 0 0 0 1	

816. C$_6$H$_{10}$
1-Hexyne
Butylacetylene
n-Butylacetylene

RN: 693-02-7   **MP** (°C): −132
MW: 82.15   **BP** (°C): 71

Solubility (Moles/L)	Solubility (Grams/L)	Temp (°C)	Ref (#)	Evaluation (T P E A A)	Comments
4.382E-03	3.600E-01	25	M001	2 1 2 2 2	
8.370E-03	6.876E-01	25	M342	1 0 1 1 2	

817. C$_6$H$_{10}$
3-Hexyne
Diethylacetylene

RN: 928-49-4   **MP** (°C): −103
MW: 82.15   **BP** (°C):

Solubility (Moles/L)	Solubility (Grams/L)	Temp (°C)	Ref (#)	Evaluation (T P E A A)	Comments
6.800E-03	5.586E-01	25	H039	1 2 2 2 1	
6.400E-03	5.257E-01	35	H039	1 2 2 2 1	

818. C$_6$H$_{10}$BrNO$_4$
5-Bromo-2-ethyl-5-nitro-1,3-dioxane
2-Ethyl-5-bromo-5-nitro-1,3-dioxane

RN: 54010-85-4   **MP** (°C): 58–59
MW: 240.06   **BP** (°C):

Solubility (Moles/L)	Solubility (Grams/L)	Temp (°C)	Ref (#)	Evaluation (T P E A A)	Comments
3.205E-03	7.694E-01	25	L013	1 0 2 1 2	

819. $C_6H_{10}BrNO_4$

5-Bromo-2,2-dimethyl-5-nitro-1,3-dioxane
2,2-Dimethyl-5-bromo-5-nitro-1,3-dioxane
m-Dioxane, 5-bromo-2,2-dimethyl-5-nitro-

RN: 60766-57-6   **MP** (°C):   79–81
MW: 240.06   **BP** (°C):

Solubility (Moles/L)	Solubility (Grams/L)	Temp (°C)	Ref (#)	Evaluation (T P E A A)	Comments
4.369E-03	1.049E+00	25	L013	1 0 2 1 2	

820. $C_6H_{10}ClN_5$

Deethylatrazine
2-Amino-4-isopropylamino-6-chloro-*s*-triazine
6-Chloro-*N*-(1-methylethyl)-1,3,5-triazine-2,4-diamine

RN: 6190-65-4   **MP** (°C):
MW: 187.63   **BP** (°C):

Solubility (Moles/L)	Solubility (Grams/L)	Temp (°C)	Ref (#)	Evaluation (T P E A A)	Comments
2.000E-03	3.753E-01	2	B193	1 1 0 0 1	

821. $C_6H_{10}O$

Mesityl oxide
Mesityloxid

RN: 141-79-7   **MP** (°C):   −57
MW: 98.15   **BP** (°C):   130

Solubility (Moles/L)	Solubility (Grams/L)	Temp (°C)	Ref (#)	Evaluation (T P E A A)	Comments
2.862E-01	2.809E+01	20	D052	1 1 0 0 0	
2.975E-01	2.920E+01	ns	F300	0 0 0 0 2	

822. $C_6H_{10}O$

Cyclohexanone
Cyclohexanon

RN: 108-94-1   **MP** (°C):   −47
MW: 98.15   **BP** (°C):

Solubility (Moles/L)	Solubility (Grams/L)	Temp (°C)	Ref (#)	Evaluation (T P E A A)	Comments
1.323E-02	1.298E+00	20	D052	1 1 0 0 1	*sic*
2.485E-01	2.439E+01	25	B060	2 0 1 1 1	
8.975E-01	8.809E+01	25	M323	2 2 1 1 2	

823. $C_6H_{10}OS_2$
Allicin
2-Propene-1-sulfinothioic acid S-2-propenyl ester
RN: 539-86-6   **MP** (°C):   <25
MW: 162.27   **BP** (°C):

Solubility (Moles/L)	Solubility (Grams/L)	Temp (°C)	Ref (#)	Evaluation (T P E A A)	Comments
1.479E-01	2.400E+01	10	F300	1 0 0 0 1	

824. $C_6H_{10}O_2$
Methyl vinyl carbinol acetate
1-Methylallyl acetate
3-Buten-2-yl acetate
RN: 6737-11-7   **MP** (°C):
MW: 114.15   **BP** (°C):

Solubility (Moles/L)	Solubility (Grams/L)	Temp (°C)	Ref (#)	Evaluation (T P E A A)	Comments
1.141E-01	1.303E+01	26	O012	1 2 1 1 2	
6.953E-02	7.937E+00	50	O012	1 2 1 1 2	
1.718E-01	1.961E+01	75	O012	1 2 1 1 2	

825. $C_6H_{10}O_2$
3-Methyl-1,3-pentadione
1,2-Dimethyl-1,3-butadiene
3,4-Dimethylbutadiene
RN: 4549-74-0   **MP** (°C):   −5
MW: 114.15   **BP** (°C):   191

Solubility (Moles/L)	Solubility (Grams/L)	Temp (°C)	Ref (#)	Evaluation (T P E A A)	Comments
9.780E-01	1.116E+02	25	M078	2 0 1 0 2	

826. $C_6H_{10}O_2S_4$
Dixanthogen
Ethyl dixanthogen
RN: 502-55-6   **MP** (°C):   28
MW: 242.40   **BP** (°C):

Solubility (Moles/L)	Solubility (Grams/L)	Temp (°C)	Ref (#)	Evaluation (T P E A A)	Comments
1.300E-05	3.151E-03	22	P076	1 2 1 1 1	
1.140E-05	2.763E-03	25	H102	1 2 1 2 2	
<2.06E-06	<5.00E-04	25	M161	1 0 0 0 0	
1.250E-05	3.030E-03	ns	L083	0 0 0 0 0	EFG, pH 3–9

827. C₆H₁₀O₃

Ethyl acetoacetate
Acetessigsaeure-aethyl ester
Acetoacetic acid ethyl ester

| **RN:** | 141-97-9 | **MP** (°C): | –45 |
| **MW:** | 130.14 | **BP** (°C): | 180.8 |

Solubility (Moles/L)	Solubility (Grams/L)	Temp (°C)	Ref (#)	Evaluation (T P E A A)	Comments
9.613E-01	1.251E+02	10.5	D041	1 0 0 0 2	
8.529E-01	1.110E+02	16.50	F300	1 0 0 0 2	

828. C₆H₁₀O₄

2,2-Dimethylsuccinic acid
α,α-Dimethylbernsteinsaeure

| **RN:** | 597-43-3 | **MP** (°C): | 140.5 |
| **MW:** | 146.14 | **BP** (°C): | |

Solubility (Moles/L)	Solubility (Grams/L)	Temp (°C)	Ref (#)	Evaluation (T P E A A)	Comments
4.790E-01	7.000E+01	14	F300	1 0 0 0 2	

829. C₆H₁₀O₄

sym-Dimethylsuccinic acid
Acide Dimethylsuccinique-*sym*

| **RN:** | 608-40-2 | **MP** (°C): | |
| **MW:** | 146.14 | **BP** (°C): | |

Solubility (Moles/L)	Solubility (Grams/L)	Temp (°C)	Ref (#)	Evaluation (T P E A A)	Comments
2.053E+00	3.000E+02	15	M051	1 0 0 0 2	

830. C₆H₁₀O₄

n-Propylmalonic acid
Acide *n*-propylmalonique

| **RN:** | 616-62-6 | **MP** (°C): | |
| **MW:** | 146.14 | **BP** (°C): | |

Solubility (Moles/L)	Solubility (Grams/L)	Temp (°C)	Ref (#)	Evaluation (T P E A A)	Comments
3.120E+00	4.560E+02	0	M051	1 0 0 0 2	
4.112E+00	6.010E+02	15	M051	1 0 0 0 2	
4.790E+00	7.000E+02	25	M051	1 0 0 0 2	
6.459E+00	9.440E+02	50	M051	1 0 0 0 2	

831. $C_6H_{10}O_4$
Ethylene glycol diacetate
Glycol diacetate
RN: 111-55-7   **MP** (°C):   −31
MW: 146.14   **BP** (°C):   190

Solubility (Moles/L)	Solubility (Grams/L)	Temp (°C)	Ref (#)	Evaluation (T P E A A)	Comments
1.202E+00	1.756E+02	20	D052	1 1 0 0 2	
9.661E-01	1.412E+02	20	M062	1 0 0 0 2	
8.526E-01	1.246E+02	22	F300	1 0 0 0 2	
1.034E+00	1.511E+02	24.50	O005	2 0 2 2 2	
1.070E+00	1.564E+02	25	F064	1 0 0 0 2	
1.220E-01	1.783E+01	ns	F014	0 0 0 0 2	

832. $C_6H_{10}O_4$
DL-2,3-Dimethylsuccinic acid
DL-α,α′-Dimethylbernsteinsaeure
RN: 13545-04-5   **MP** (°C):   120
MW: 146.14   **BP** (°C):

Solubility (Moles/L)	Solubility (Grams/L)	Temp (°C)	Ref (#)	Evaluation (T P E A A)	Comments
2.053E-01	3.000E+01	14	F300	1 0 0 0 0	

833. $C_6H_{10}O_4$
Adipic acid
Adipinsaeure
RN: 124-04-9   **MP** (°C):   152
MW: 146.14   **BP** (°C):   337.5

Solubility (Moles/L)	Solubility (Grams/L)	Temp (°C)	Ref (#)	Evaluation (T P E A A)	Comments
5.431E-02	7.937E+00	0	M043	1 0 0 0 0	
6.766E-02	9.888E+00	4.99	A339	0 0 0 0 0	
6.775E-02	9.901E+00	10	M043	1 0 0 0 1	
7.853E-02	1.148E+01	9.99	A339	0 0 0 0 0	
1.061E-01	1.551E+01	14.99	A339	0 0 0 0 0	
9.580E-02	1.400E+01	15	F300	1 0 0 0 1	
9.580E-02	1.400E+01	15	L041	1 0 0 1 1	
9.580E-02	1.400E+01	15	M051	1 0 0 0 1	
1.303E-01	1.904E+01	19.99	A339	0 0 0 0 0	
1.011E-01	1.478E+01	20	D041	1 0 0 0 1	
1.276E-01	1.865E+01	20	M043	1 0 0 0 1	
9.856E-02	1.440E+01	20	M171	1 0 0 0 1	
9.000E-02	1.315E+01	20	S006	1 0 0 0 1	
4.824E-01	7.050E+01	21	B040	1 0 1 1 2	*sic*
1.664E-01	2.432E+01	24.99	A339	0 0 0 0 0	
2.216E-03	3.239E-01	25	K035	2 0 0 0 2	*sic*
2.053E-01	3.001E+01	29.99	A339	0 0 0 0 0	

(continued)

833. C₆H₁₀O₄ (continued)

Solubility (Moles/L)	Solubility (Grams/L)	Temp (°C)	Ref (#)	Evaluation (T P E A A)	Comments
1.993E-01	2.913E+01	30	M043	1 0 0 0 1	
2.045E-01	2.988E+01	34.10	A031	1 2 2 2 2	
2.546E-01	3.721E+01	34.99	A339	0 0 0 0 0	
2.933E-01	4.287E+01	39.3	G302	2 2 2 2 0	EFG
3.274E-01	4.785E+01	39.99	A339	0 0 0 0 0	
3.333E-01	4.871E+01	40	A031	1 2 2 2 2	
3.382E-01	4.943E+01	40	B088	1 0 0 0 1	
3.258E-01	4.762E+01	40	M043	1 0 0 0 1	
4.383E-01	6.406E+01	44.99	A339	0 0 0 0 0	
5.516E-01	8.062E+01	49.99	A339	0 0 0 0 0	
5.788E-01	8.458E+01	50	A031	1 2 2 2 2	
7.508E-01	1.097E+02	54.99	A339	0 0 0 0 0	
1.011E+00	1.477E+02	59.99	A339	0 0 0 0 0	
1.024E+00	1.497E+02	60	A031	1 2 2 2 2	
1.044E+00	1.525E+02	60	M043	1 0 0 0 1	
1.130E+00	1.652E+02	64.99	A339	0 0 0 0 0	
1.740E+00	2.543E+02	70	A031	1 2 2 2 2	
2.818E+00	4.118E+02	80	M043	1 0 0 0 1	
3.330E+00	4.867E+02	87.10	A031	1 2 2 2 2	
4.277E+00	6.250E+02	100	F300	1 0 0 0 2	
4.211E+00	6.154E+02	100	M043	1 0 0 0 2	
1.662E-01	2.430E+01	rt	H431	0 0 0 0 0	

834. C₆H₁₀O₄

Methyl α-acetoxypropionate
Methyl 2-acetoxypropionate
Methyl O-acetyllactate
Methyl 2-acetyloxypropanoate
RN: 6284-75-9   **MP** (°C):
MW: 146.14   **BP** (°C):

Solubility (Moles/L)	Solubility (Grams/L)	Temp (°C)	Ref (#)	Evaluation (T P E A A)	Comments
5.556E-01	8.120E+01	25	R006	2 2 0 1 2	

835. C₆H₁₀O₅

Propanoic acid, 2-[(methoxycarbonyl)oxy]-, methyl ester
Carbonic acid, methyl ester, ester with methyl lactate
RN: 6288-11-5   **MP** (°C):
MW: 162.14   **BP** (°C):

Solubility (Moles/L)	Solubility (Grams/L)	Temp (°C)	Ref (#)	Evaluation (T P E A A)	Comments
2.412E-01	3.911E+01	25	R007	0 0 0 0 0	

836. C$_6$H$_{10}$O$_8$
D-Talogalactaric acid
D-Taloschleimsaeure
D-Galactaric acid
Galactaric acid
Schleimsaeure

RN: 526-99-8   **MP** (°C):   > 230
MW: 210.14   **BP** (°C):

Solubility (Moles/L)	Solubility (Grams/L)	Temp (°C)	Ref (#)	Evaluation (T P E A A)	Comments
1.565E-02	3.289E+00	14	D041	1 0 0 0 1	
1.570E-02	3.300E+00	14	F300	1 0 0 0 1	
8.090E-02	1.700E+01	100	F300	1 0 0 0 1	

837. C$_6$H$_{11}$Br
Bromocyclohexane
Cyclohexyl bromide

RN: 108-85-0   **MP** (°C):
MW: 163.06   **BP** (°C):   166

Solubility (Moles/L)	Solubility (Grams/L)	Temp (°C)	Ref (#)	Evaluation (T P E A A)	Comments
5.012E-03	8.173E-01	ns	S460	0 0 0 0 0	

838. C$_6$H$_{11}$BrN$_2$O$_2$
α-Methyl-γ-bromo-butanoic ureide

RN: **MP** (°C):
MW: 223.08   **BP** (°C):

Solubility (Moles/L)	Solubility (Grams/L)	Temp (°C)	Ref (#)	Evaluation (T P E A A)	Comments
4.658E-02	1.039E+01	ns	F056	0 2 2 2 1	

839. C$_6$H$_{11}$BrN$_2$O$_2$
α-Bromo-isovaleric ureide
Butanamide, N-(aminocarbonyl)-2-bromo-3-methyl-
Dormigene
Pivadorn
Pivadorm
Isobromyl

RN: 496-67-3   **MP** (°C):
MW: 223.08   **BP** (°C):

Solubility (Moles/L)	Solubility (Grams/L)	Temp (°C)	Ref (#)	Evaluation (T P E A A)	Comments
8.531E-02	1.903E+01	ns	F057	0 2 2 2 2	

840. C₆H₁₁BrN₂O₂

3-Bromo-2-methyl-butanoic ureide
Urea, (2-bromo-2-methylbutyryl)-
DL-*N*-(2-Bromo-2-methylbutanoyl)urea

RN:	14368-76-4	MP (°C):
MW:	223.08	BP (°C):

Solubility (Moles/L)	Solubility (Grams/L)	Temp (°C)	Ref (#)	Evaluation (T P E A A)	Comments
1.390E-01	3.101E+01	ns	F056	0 2 2 2 1	

841. C₆H₁₁BrN₂O₂

β-Bromo-valeric acid ureide

RN:		MP (°C):
MW:	223.08	BP (°C):

Solubility (Moles/L)	Solubility (Grams/L)	Temp (°C)	Ref (#)	Evaluation (T P E A A)	Comments
3.470E-02	7.740E+00	ns	F056	0 2 2 2 1	

842. C₆H₁₁BrN₂O₂

γ-Bromo-valeric acid ureide

RN:		MP (°C):
MW:	223.08	BP (°C):

Solubility (Moles/L)	Solubility (Grams/L)	Temp (°C)	Ref (#)	Evaluation (T P E A A)	Comments
4.307E-02	9.607E+00	ns	F056	0 2 2 2 1	

843. C₆H₁₁BrN₂O₂

α-Bromo-valeric acid ureide
Pentanamide, *N*-(aminocarbonyl)-2-bromo-

RN:	66947-87-3	MP (°C):
MW:	223.08	BP (°C):

Solubility (Moles/L)	Solubility (Grams/L)	Temp (°C)	Ref (#)	Evaluation (T P E A A)	Comments
3.690E-02	8.232E+00	ns	F056	0 2 2 2 1	
3.703E-02	8.261E+00	ns	F057	0 2 2 2 2	

844. C₆H₁₁NO

Caprolactam
ε-Caprolactam

RN:	105-60-2	MP (°C):	70
MW:	113.16	BP (°C):	180

Solubility (Moles/L)	Solubility (Grams/L)	Temp (°C)	Ref (#)	Evaluation (T P E A A)	Comments
3.776E+00	4.273E+02	5.70	B201	2 2 2 1 2	
3.850E+00	4.357E+02	10.30	B201	2 2 2 1 2	

845. C₆H₁₁NO

Cyclohexanone oxime
Antioxidant D
(Hydroxyimino)cyclohexane

RN:	100-64-1	**MP** (°C):	90	
MW:	113.16	**BP** (°C):	208	

Solubility (Moles/L)	Solubility (Grams/L)	Temp (°C)	Ref (#)	Evaluation (T P E A A)	Comments
1.409E-01	1.594E+01	25.5	K087	1 0 0 0 2	
1.580E-01	1.787E+01	32.0	K087	1 0 0 0 2	
1.648E-01	1.865E+01	36.8	K087	1 0 0 0 2	
1.936E-01	2.191E+01	44.0	K087	1 0 0 0 2	
2.155E-01	2.439E+01	48.8	K087	1 0 0 0 2	
2.715E-01	3.073E+01	60.4	K087	1 0 0 0 2	
2.922E-01	3.307E+01	63.7	K087	1 0 0 0 2	
3.194E-01	3.614E+01	76.2	K087	1 0 0 0 2	
3.456E-01	3.911E+01	83.1	K087	1 0 0 0 2	
4.039E-01	4.571E+01	95.2	K087	1 0 0 0 2	
4.939E-01	5.589E+01	110.7	K087	1 0 0 0 2	
5.743E-01	6.498E+01	120	K087	1 0 0 0 2	
7.386E-01	8.358E+01	131	K087	1 0 0 0 2	

846. C₆H₁₁NO₂S

2,2-Dimethylthiazolidine-4-carboxylic acid
4-Thiazolidinecarboxylic acid, 2,2-dimethyl-
Thiazolidine-4-carboxylic acid, 2,2-dimethyl-

RN:	42607-20-5	**MP** (°C):		
MW:	161.22	**BP** (°C):	317.3	

Solubility (Moles/L)	Solubility (Grams/L)	Temp (°C)	Ref (#)	Evaluation (T P E A A)	Comments
3.000E-01	4.837E+01	21	B414	1 0 0 1 1	very fast and extent decompostion, uncertain value

847. C₆H₁₁NO₄

α-Aminoadipic acid
2-Aminohexanedioic acid
α-Amino-adipinsaeure

RN:	542-32-5	**MP** (°C):		
MW:	161.16	**BP** (°C):		

Solubility (Moles/L)	Solubility (Grams/L)	Temp (°C)	Ref (#)	Evaluation (T P E A A)	Comments
1.365E-02	2.200E+00	20	F300	1 0 0 0 1	

848. C$_6$H$_{11}$NO$_4$

Glycine, N-(carboxymethyl)-, 1-ethyl ester

AcGlyOEt

Acetic acid, iminodi-, monoethyl ester

RN: 21885-31-4   **MP** (°C):

MW: 161.16   **BP** (°C):

Solubility (Moles/L)	Solubility (Grams/L)	Temp (°C)	Ref (#)	Evaluation (T P E A A)	Comments
7.074E-03	1.140E+00	27	D036	0 0 0 0 0	

849. C$_6$H$_{11}$N$_2$O$_4$PS$_3$

Methidathion

Supracide

S-((5-Methoxy-2-oxo-1,3,4-thiadiazol-3(2H)-yl)methyl) O,O-dimethyl phosphorodithioate

Ultracide

Somanil

S-2,3-Dihydro-5-methoxy-2-oxo-1,3,4-thiadiazol-3-ylmethyl O,O-dimethylphosphorodithioate

RN: 950-37-8   **MP** (°C):

MW: 302.33   **BP** (°C):

Solubility (Moles/L)	Solubility (Grams/L)	Temp (°C)	Ref (#)	Evaluation (T P E A A)	Comments
6.186E-04	1.870E-01	20	B300	2 2 1 1 2	
8.269E-04	2.500E-01	20	F311	1 2 2 2 1	
7.938E-04	2.400E-01	25	M161	1 0 0 0 2	

850. C$_6$H$_{11}$N$_3$O$_6$

Glycine tripeptide

RN: **MP** (°C):

MW: 221.17   **BP** (°C):

Solubility (Moles/L)	Solubility (Grams/L)	Temp (°C)	Ref (#)	Evaluation (T P E A A)	Comments
2.127E-01	4.705E+01	20	B032	1 2 2 1 2	
2.907E-01	6.430E+01	25	B032	1 2 2 1 2	
3.565E-01	7.884E+01	29.80	B032	1 2 2 1 2	

851. C$_6$H$_{12}$

Methylcyclopentane

MCP

RN: 96-37-7   **MP** (°C):   −142

MW: 84.16   **BP** (°C):   72

Solubility (Moles/L)	Solubility (Grams/L)	Temp (°C)	Ref (#)	Evaluation (T P E A A)	Comments
4.967E-04	4.180E-02	25	K119	1 0 0 0 2	
4.990E-04	4.200E-02	25	M001	2 1 2 2 2	
5.062E-04	4.260E-02	25	M002	2 1 2 2 2	

(continued)

851. C_6H_{12} (continued)

Solubility (Moles/L)	Solubility (Grams/L)	Temp (°C)	Ref (#)	Evaluation (T P E A A)	Comments
4.967E-04	4.180E-02	25	P051	2 1 1 2 2	
4.967E-04	4.180E-02	25.00	P007	2 1 2 2 2	
4.990E-04	4.200E-02	ns	H123	0 0 0 0 0	

852. C_6H_{12}
Cyclohexane
Cyclohexan

RN: 110-82-7   **MP** (°C): 7
MW: 84.16   **BP** (°C):

Solubility (Moles/L)	Solubility (Grams/L)	Temp (°C)	Ref (#)	Evaluation (T P E A A)	Comments
9.734E-04	8.192E-02	4.8	L007	2 1 1 2 2	
9.734E-04	8.192E-02	5.1	L007	2 0 1 1 2	
1.054E-03	8.869E-02	14.8	L007	2 1 1 2 2	
1.054E-03	8.869E-02	15.2	L007	2 0 1 1 2	
9.505E-04	8.000E-02	16	D047	1 0 0 1 1	
<5.94E-04	<5.00E-02	17	F300	1 0 0 0 0	
4.396E-04	3.700E-02	20	M337	2 1 2 2 2	
6.178E-04	5.200E-02	23.5	S171	2 1 2 2 2	
1.055E-03	8.883E-02	24.8	L007	2 1 1 2 2	
9.505E-04	7.999E-02	25	G068	1 0 1 0 0	
6.939E-04	5.840E-02	25	G313	2 1 1 2 2	
1.426E-03	1.200E-01	25	K112	1 0 2 1 1	
7.901E-04	6.650E-02	25	K119	1 0 0 0 2	
6.737E-04	5.670E-02	25	L002	2 2 2 2 2	
6.535E-04	5.500E-02	25	M001	2 1 2 2 2	
6.535E-04	5.500E-02	25	M002	2 1 2 2 2	
6.535E-04	5.500E-02	25	M040	1 0 0 1 1	
6.832E-04	5.750E-02	25	M132	2 2 2 1 2	
7.901E-04	6.650E-02	25	P051	2 1 1 2 2	
6.270E-04	5.277E-02	25	S359	2 1 2 2 2	
7.901E-04	6.650E-02	25.00	P007	2 1 2 2 2	
1.055E-03	8.883E-02	34.8	L007	2 1 1 2 2	
1.055E-03	8.883E-02	35.2	L007	2 0 1 1 2	
5.389E-04	4.535E-02	38	K055	1 2 0 1 1	
1.085E-03	9.131E-02	44.8	L007	2 1 1 2 2	
1.085E-03	9.131E-02	45.2	L007	2 0 1 1 2	
1.426E-03	1.200E-01	50	L097	1 1 1 1 1	
2.020E-03	1.700E-01	56	G068	1 0 1 0 1	
3.222E-04	2.712E-02	71	K055	1 2 0 1 1	
3.326E-03	2.799E-01	94	G068	1 0 1 0 1	
1.200E-04	1.010E-02	ns	D348	0 0 0 0 0	
6.535E-04	5.500E-02	ns	H123	0 0 0 0 0	
5.000E-03	4.208E-01	ns	H333	0 1 0 1 0	EFG
9.505E-04	8.000E-02	ns	M010	0 0 0 0 0	
6.642E-04	5.590E-02	ns	M175	0 0 2 1 2	

853. C$_6$H$_{12}$
4-Methyl-1-pentene
4-Methylpentene
Isohexene

RN:	691-37-2	MP (°C):	−154
MW:	84.16	BP (°C):	53

Solubility (Moles/L)	Solubility (Grams/L)	Temp (°C)	Ref (#)	Evaluation (T P E A A)	Comments
5.703E-04	4.800E-02	25	M001	2 1 2 2 1	

854. C$_6$H$_{12}$
2-Methyl-1-pentene
4-Methyl-4-pentene

RN:	763-29-1	MP (°C):	−136
MW:	84.16	BP (°C):	62

Solubility (Moles/L)	Solubility (Grams/L)	Temp (°C)	Ref (#)	Evaluation (T P E A A)	Comments
9.268E-04	7.800E-02	25	M001	2 1 2 2 2	

855. C$_6$H$_{12}$
1-Hexene
1-n-Hexene
Hexene
Dialen 6

RN:	592-41-6	MP (°C):	−140
MW:	84.16	BP (°C):	64

Solubility (Moles/L)	Solubility (Grams/L)	Temp (°C)	Ref (#)	Evaluation (T P E A A)	Comments
5.822E-04	4.900E-02	23	C332	0 0 0 0 0	
6.583E-04	5.540E-02	25	L002	2 2 2 2 2	
5.941E-04	5.000E-02	25	M001	2 1 2 2 2	
5.941E-04	5.000E-02	25	M040	1 0 0 1 1	
8.280E-04	6.969E-02	25	M342	1 0 1 1 2	

856. C$_6$H$_{12}$ClNO
Acetamide, 2-chloro-N,N-diethyl-
CDEA

RN:	2315-36-8	MP (°C):	
MW:	149.62	BP (°C):	

Solubility (Moles/L)	Solubility (Grams/L)	Temp (°C)	Ref (#)	Evaluation (T P E A A)	Comments
5.264E-01	7.877E+01	25	B185	0 0 0 0 0	

857. C$_6$H$_{12}$Cl$_2$O

Dichloroisopropyl ether
bis(2-Chloro-1-methylethyl) ether
DCIP
β,β′-Dichlorodiisopropyl ether
2,2′-Oxybis[1-chloropropane]
Pichloram

RN:	63283-80-7	**MP** (°C):	
MW:	171.07	**BP** (°C):	187.3

Solubility (Moles/L)	Solubility (Grams/L)	Temp (°C)	Ref (#)	Evaluation (T P E A A)	Comments
9.921E-03	1.697E+00	20	M062	1 0 0 0 1	

858. C$_6$H$_{12}$Cl$_2$O$_2$

1,2-bis(2-Chloroethoxy)ethane
Triglycol dichloride

RN:	112-26-5	**MP** (°C):	121
MW:	187.07	**BP** (°C):	235

Solubility (Moles/L)	Solubility (Grams/L)	Temp (°C)	Ref (#)	Evaluation (T P E A A)	Comments
9.916E-02	1.855E+01	20	M062	1 0 0 0 2	

859. C$_6$H$_{12}$Cl$_3$O$_4$P

tris-(2-Chloroethyl) phosphate
Tri-β-chloroethyl phosphate

RN:	115-96-8	**MP** (°C):	
MW:	285.49	**BP** (°C):	

Solubility (Moles/L)	Solubility (Grams/L)	Temp (°C)	Ref (#)	Evaluation (T P E A A)	Comments
<7.01E-04	<2.00E-01	25	B070	1 2 0 1 0	

860. C$_6$H$_{12}$NO$_3$PS$_2$

Diethyl 1,3-dithietan-2-ylidenephosphoramidate
Nematak
AC 64475
Geofos
Fosthietan
CL 64475

RN:	21548-32-3	**MP** (°C):	
MW:	241.27	**BP** (°C):	

Solubility (Moles/L)	Solubility (Grams/L)	Temp (°C)	Ref (#)	Evaluation (T P E A A)	Comments
2.072E-01	5.000E+01	25	M161	1 0 0 0 1	

861. C$_6$H$_{12}$NO$_4$PS$_2$
Formothion
O,O-Dimethyl S-(N-methyl-N-formylcarbamoylmethyl) dithiophosphate
RN: 2540-82-1   **MP** (°C):
MW: 257.27   **BP** (°C):

Solubility (Moles/L)	Solubility (Grams/L)	Temp (°C)	Ref (#)	Evaluation (T P E A A)	Comments
1.011E-02	2.600E+00	24	M161	1 0 0 0 1	

862. C$_6$H$_{12}$N$_2$O
N-Nitrosohexamethyleneimine
NHMI
RN: 932-83-2   **MP** (°C):
MW: 128.18   **BP** (°C):

Solubility (Moles/L)	Solubility (Grams/L)	Temp (°C)	Ref (#)	Evaluation (T P E A A)	Comments
1.000E-01	1.282E+01	24	M031	1 1 1 1 1	

863. C$_6$H$_{12}$N$_2$O$_2$
2,6-Dimethylnitrosomorpholine
DMNM
RN: 1456-28-6   **MP** (°C):
MW: 144.17   **BP** (°C):

Solubility (Moles/L)	Solubility (Grams/L)	Temp (°C)	Ref (#)	Evaluation (T P E A A)	Comments
8.600E-01	1.240E+02	24	M031	1 1 1 1 1	

864. C$_6$H$_{12}$N$_2$O$_2$
Adipamide
Adipinsaeurediamid
RN: 628-94-4   **MP** (°C):
MW: 144.17   **BP** (°C):

Solubility (Moles/L)	Solubility (Grams/L)	Temp (°C)	Ref (#)	Evaluation (T P E A A)	Comments
3.052E-02	4.400E+00	12.20	F300	1 0 0 0 1	

865. $C_6H_{12}N_2O_3$
Daminozide
N-Dimethylamino-β-carbamyl propionic acid
Succinic acid 2,2-dimethylhydrazide
Alar
DMASA

RN: 1596-84-5   **MP** (°C): 155
MW: 160.17   **BP** (°C):

Solubility (Moles/L)	Solubility (Grams/L)	Temp (°C)	Ref (#)	Evaluation (T P E A A)	Comments
6.243E-01	1.000E+02	25	M161	1 0 0 0 2	

866. $C_6H_{12}N_2O_3$
δ-Aminovaleric hydantoic acid
δ-Uramidovaleric acid

RN: **MP** (°C): 179
MW: 160.17   **BP** (°C):

Solubility (Moles/L)	Solubility (Grams/L)	Temp (°C)	Ref (#)	Evaluation (T P E A A)	Comments
1.740E-02	2.787E+00	25	M024	1 2 0 1 2	

867. $C_6H_{12}N_2O_3S$
Methomyl
Acetamidic acid
N-[(methyl-carbamoyl)oxy]-, methyl ester
Carbamic acid
Lannabait
Nudrin

RN: 16752-77-5   **MP** (°C): 78
MW: 192.24   **BP** (°C): 144

Solubility (Moles/L)	Solubility (Grams/L)	Temp (°C)	Ref (#)	Evaluation (T P E A A)	Comments
3.548E-01	6.821E+01	ns	R424	0 0 0 0 0	

868. $C_6H_{12}N_2O_4S_2$
L-Cystine
3,3′-Dithiobis(2-aminopropanoic acid)

RN: 56-89-3   **MP** (°C):
MW: 240.30   **BP** (°C):

Solubility (Moles/L)	Solubility (Grams/L)	Temp (°C)	Ref (#)	Evaluation (T P E A A)	Comments
2.021E-03	4.858E-01	20	H082	1 2 1 1 2	isomeric
7.905E-04	1.900E-01	20	H082	1 2 1 1 2	plate cystine
6.910E-04	1.660E-01	24.99	C404	2 1 2 2 1	
7.000E-02	1.682E+01	25	C405	2 1 2 2 2	intrinsic zwit

(continued)

868. C$_6$H$_{12}$N$_2$O$_4$S$_2$ (continued)

Solubility (Moles/L)	Solubility (Grams/L)	Temp (°C)	Ref (#)	Evaluation (T P E A A)	Comments
4.536E-04	1.090E-01	25	D017	1 0 0 0 2	
4.577E-04	1.100E-01	25	D041	1 0 0 0 1	
4.661E-04	1.120E-01	25	L001	1 0 1 1 2	pH 6.0
4.910E-04	1.180E-01	27	D036	0 0 0 0 0	
7.100E-04	1.706E-01	34.99	C404	2 1 2 2 1	
8.500E-04	2.043E-01	44.99	C404	2 1 2 2 1	
2.163E-03	5.197E-01	75	D041	1 0 0 0 1	
4.536E-04	1.090E-01	rt	B103	0 0 0 0 2	

869. C$_6$H$_{12}$N$_2$O$_4$S

DL-Lanthionine

L-Cysteine, S-[(2R)-2-amino-2-carboxyethyl]-

RN: 922-55-4   **MP** (°C): 280
MW: 208.24   **BP** (°C):

Solubility (Moles/L)	Solubility (Grams/L)	Temp (°C)	Ref (#)	Evaluation (T P E A A)	Comments
7.193E-03	1.498E+00	25	D041	1 0 0 0 1	

870. C$_6$H$_{12}$N$_2$O$_4$S$_2$

Mesocystine

meso-Cystine

RN: 6020-39-9   **MP** (°C):
MW: 240.30   **BP** (°C):

Solubility (Moles/L)	Solubility (Grams/L)	Temp (°C)	Ref (#)	Evaluation (T P E A A)	Comments
2.330E-04	5.600E-02	25	L001	1 0 1 1 1	pH 6.0

871. C$_6$H$_{12}$N$_2$O$_4$S$_2$

D-Cystine

D-(+)-3,3'-Dithiobis(2-aminopropanoic acid)

RN: 349-46-2   **MP** (°C): 227
MW: 240.30   **BP** (°C):

Solubility (Moles/L)	Solubility (Grams/L)	Temp (°C)	Ref (#)	Evaluation (T P E A A)	Comments
4.577E-04	1.100E-01	25	D041	1 0 0 0 1	
4.702E-04	1.130E-01	25	L001	1 0 1 1 2	pH 6.0

872. $C_6H_{12}N_2O_4S_2$
DL-Cystine
Cystine

RN:	923-32-0	MP (°C):	
MW:	240.30	BP (°C):	

Solubility (Moles/L)	Solubility (Grams/L)	Temp (°C)	Ref (#)	Evaluation (T P E A A)	Comments
2.039E-04	4.900E-02	25	D041	1 0 0 0 1	
2.372E-04	5.700E-02	25	L001	1 0 1 1 1	pH 6.0

873. $C_6H_{12}N_2S_4$
Thiram
Tetramethylthioperoxydicarbonothioic diamine
Tetramethylthiuram disulfide
N,N'-(Dithiodicarbonothioyl)bis(N-methylmethanamine)
Arasan
Nomersan

RN:	137-26-8	MP (°C):	155.5
MW:	240.43	BP (°C):	

Solubility (Moles/L)	Solubility (Grams/L)	Temp (°C)	Ref (#)	Evaluation (T P E A A)	Comments
7.413E-05	1.782E-02	ns	R427	0 0 0 0 0	
1.248E-04	3.000E-02	rt	M161	0 0 0 0 1	

874. $C_6H_{12}N_2S_4Zn$
Ziram
Zinc *bis* dimethyldithiocarbamate
Corozate
Karbam white
Fuklasin
Fuclasin

RN:	137-30-4	MP (°C):	240
MW:	305.81	BP (°C):	

Solubility (Moles/L)	Solubility (Grams/L)	Temp (°C)	Ref (#)	Evaluation (T P E A A)	Comments
2.125E-04	6.500E-02	20	F300	1 0 0 0 1	
1.308E-05	4.000E-03	20	F311	1 2 2 2 1	*sic*
2.125E-04	6.500E-02	25	M161	1 0 0 0 1	

875. C$_6$H$_{12}$N$_4$
Methenamine
Hexamethylen-tetramin

RN: 100-97-0   **MP** (°C):
MW: 140.19   **BP** (°C):

Solubility (Moles/L)	Solubility (Grams/L)	Temp (°C)	Ref (#)	Evaluation (T P E A A)	Comments
2.231E+00	3.128E+02	1.99	B442	0 0 0 0 0	
2.202E+00	3.087E+02	3.99	B442	0 0 0 0 0	
2.183E+00	3.060E+02	5.99	B442	0 0 0 0 0	
2.149E+00	3.012E+02	9.99	B442	0 0 0 0 0	
2.254E+00	3.161E+02	10.99	B442	0 0 0 0 0	
2.250E+00	3.154E+02	11.99	B442	0 0 0 0 0	
3.200E+00	4.486E+02	12	F300	1 0 0 0 2	
2.234E+00	3.131E+02	14.99	B442	0 0 0 0 0	
2.191E+00	3.072E+02	19.99	B442	0 0 0 0 0	
2.156E+00	3.023E+02	24.99	B442	0 0 0 0 0	
2.193E+00	3.074E+02	29.99	B442	0 0 0 0 0	
2.218E+00	3.110E+02	34.99	B442	0 0 0 0 0	
2.233E+00	3.131E+02	39.99	B442	0 0 0 0 0	

876. C$_6$H$_{12}$N$_4$O$_2$
2,6-Dimethyldinitrosopiperazine
DMDNP

RN: 55380-34-2   **MP** (°C):
MW: 172.19   **BP** (°C):

Solubility (Moles/L)	Solubility (Grams/L)	Temp (°C)	Ref (#)	Evaluation (T P E A A)	Comments
1.200E-01	2.066E+01	24	M031	1 1 1 1 1	

877. C$_6$H$_{12}$N$_5$O$_2$PS$_2$
Menazon
O,O-Dimethyl *S*-(4,6-diamino-1,3,5-triazinyl-2-methyl) dithiophosphate

RN: 78-57-9   **MP** (°C):
MW: 281.30   **BP** (°C):

Solubility (Moles/L)	Solubility (Grams/L)	Temp (°C)	Ref (#)	Evaluation (T P E A A)	Comments
8.532E-04	2.400E-01	20	M161	1 0 0 0 1	
3.551E-03	9.990E-01	ns	M061	0 0 0 0 0	

878. $C_6H_{12}O$
Pinacolone
3,3-Dimethyl-2-butanone
3,3-Dimethylbutanone-2

RN:	75-97-8	MP (°C):	−52.5
MW:	100.16	BP (°C):	106.2

Solubility (Moles/L)	Solubility (Grams/L)	Temp (°C)	Ref (#)	Evaluation (T P E A A)	Comments
2.376E-01	2.380E+01	15	F300	1 0 0 0 2	
1.996E-01	1.999E+01	20	G030	1 2 0 0 2	
1.862E-01	1.865E+01	25	G030	1 2 0 0 2	
1.817E-01	1.820E+01	25	K072	1 0 1 1 1	
1.736E-01	1.739E+01	30	G030	1 2 0 0 2	

879. $C_6H_{12}O$
Cyclohexanol
1-Cyclohexanol
Naxol
Cyclohexyl alcoho
Adrona
Hydrophenol

RN:	108-93-0	MP (°C):	23
MW:	100.16	BP (°C):	

Solubility (Moles/L)	Solubility (Grams/L)	Temp (°C)	Ref (#)	Evaluation (T P E A A)	Comments
5.357E-01	5.366E+01	11	F052	1 1 1 0 2	
5.391E-01	5.400E+01	11	F300	1 0 0 0 1	
1.296E-02	1.298E+00	20	D052	1 1 0 0 1	*sic*
3.283E-01	3.288E+01	25	B019	1 0 1 2 0	
3.283E-01	3.288E+01	25	B092	2 1 1 1 2	
3.469E-01	3.475E+01	25	C108	2 2 2 2 2	
3.800E-01	3.806E+01	25	F044	1 0 0 0 1	
3.766E-01	3.772E+01	25	H028	2 0 2 0 2	
3.655E-01	3.661E+01	35	C108	2 2 2 2 2	
3.264E-01	3.269E+01	60	B092	2 1 1 1 2	
3.766E-01	3.772E+01	ns	A406	0 0 0 0 1	

880. $C_6H_{12}O$
Isopropylacetone
4-Methyl-2-pentanone
Methyl isobutyl ketone

RN:	108-10-1	MP (°C):	−80
MW:	100.16	BP (°C):	117

Solubility (Moles/L)	Solubility (Grams/L)	Temp (°C)	Ref (#)	Evaluation (T P E A A)	Comments
3.070E-01	3.075E+01	0	G032	1 2 1 1 2	
2.310E-01	2.314E+01	10	G032	1 2 1 1 2	
1.871E-01	1.874E+01	20	D052	1 1 0 0 2	
1.996E-01	1.999E+01	20	G030	1 2 0 0 2	

(continued)

880. C$_6$H$_{12}$O (continued)

Solubility (Moles/L)	Solubility (Grams/L)	Temp (°C)	Ref (#)	Evaluation (T P E A A)	Comments
1.958E-01	1.961E+01	22.00	O005	2 0 2 2 0	
1.862E-01	1.865E+01	24.6	H121	2 0 0 0 1	
1.862E-01	1.865E+01	25	B060	2 0 1 1 1	
1.717E-01	1.720E+01	25	C329	1 1 1 1 1	average
1.871E-01	1.874E+01	25	G030	1 2 0 0 2	
2.340E-01	2.344E+01	25	K103	1 2 2 2 1	
1.862E-01	1.865E+01	25	L082	1 1 2 1 1	
1.736E-01	1.739E+01	25	L319	1 0 2 1 2	
1.817E-01	1.820E+01	25	M087	1 1 2 1 2	
1.669E-01	1.672E+01	25	R320	1 0 1 1 1	
1.746E-01	1.749E+01	30	G030	1 2 0 0 2	
1.660E-01	1.663E+01	30	G032	1 2 1 1 2	
1.410E-01	1.412E+01	50	G032	1 2 1 1 2	
4.720E+01	4.728E+03	53.0	R308	2 2 1 1 2	
1.669E-01	1.672E+01	70	L082	1 1 2 1 1	
1.370E-01	1.372E+01	75	G032	1 2 1 1 2	
4.300E+01	4.307E+03	97.0	R308	2 2 1 1 2	
4.088E+01	4.094E+03	108.0	R308	2 2 1 1 2	
3.902E+01	3.909E+03	120.0	R308	2 2 1 1 2	
3.333E-01	3.339E+01	125.0	R308	2 2 1 1 1	
5.278E-01	5.286E+01	151.0	R308	2 2 1 1 1	
3.425E+01	3.431E+03	153.0	R308	2 2 1 2 2	

881. C$_6$H$_{12}$O

2-Ethylbutanal
Ethyl butyraldehyde
2-Ethylbutyraldehyde
Diethyl acetaldehyde; ethyl butyraldehyde
Diethyl acetaldehyde
Ethyl butyraldehyde

RN: 97-96-1   **MP** (°C):
MW: 100.16   **BP** (°C):   117

Solubility (Moles/L)	Solubility (Grams/L)	Temp (°C)	Ref (#)	Evaluation (T P E A A)	Comments
3.020E-02	3.025E+00	ns	S460	0 0 0 0 0	

882. C$_6$H$_{12}$O

Caproic aldehyde
Hexaldehyde
n-Hexanal

RN: 66-25-1   **MP** (°C):
MW: 100.16   **BP** (°C):   131

Solubility (Moles/L)	Solubility (Grams/L)	Temp (°C)	Ref (#)	Evaluation (T P E A A)	Comments
5.581E-01	5.590E+01	0	C423	0 0 0 0 0	
4.493E-01	4.500E+01	4	C423	0 0 0 0 0	

(continued)

882. $C_6H_{12}O$ (continued)

Solubility (Moles/L)	Solubility (Grams/L)	Temp (°C)	Ref (#)	Evaluation (T P E A A)	Comments
3.155E-01	3.160E+01	10	C423	0 0 0 0 0	
4.992E-02	5.000E+00	25	A049	1 0 1 0 1	
1.907E-01	1.910E+01	25	C435	0 0 0 0 0	
4.792E-02	4.800E+00	25	J418	0 0 0 0 0	

883. $C_6H_{12}O$
4-Methyl-3-pentanone
4-Methylpentanone-3
RN: 565-69-5 **MP** (°C):
MW: 100.16 **BP** (°C):

Solubility (Moles/L)	Solubility (Grams/L)	Temp (°C)	Ref (#)	Evaluation (T P E A A)	Comments
1.601E-01	1.604E+01	20	G030	1 2 0 0 2	
1.495E-01	1.497E+01	25	G030	1 2 0 0 2	
1.398E-01	1.400E+01	30	G030	1 2 0 0 2	
1.549E-01	1.551E+01	ns	S460	0 0 0 0 0	

884. $C_6H_{12}O$
3-Methyl-2-pentanone
3-Methylpentanone-2
RN: 565-61-7 **MP** (°C): <25
MW: 100.16 **BP** (°C): 118

Solubility (Moles/L)	Solubility (Grams/L)	Temp (°C)	Ref (#)	Evaluation (T P E A A)	Comments
2.206E-01	2.210E+01	20	G030	1 2 0 0 2	
2.044E-01	2.047E+01	25	G030	1 2 0 0 2	
1.890E-01	1.893E+01	30	G030	1 2 0 0 2	

885. $C_6H_{12}O$
3-Hexanone
Hexanone-3
RN: 589-38-8 **MP** (°C): −55.5
MW: 100.16 **BP** (°C): 123

Solubility (Moles/L)	Solubility (Grams/L)	Temp (°C)	Ref (#)	Evaluation (T P E A A)	Comments
1.543E-01	1.546E+01	20	G030	1 2 0 0 2	
1.446E-01	1.449E+01	25	G030	1 2 0 0 2	
1.359E-01	1.361E+01	30	G030	1 2 0 0 2	

886. C$_6$H$_{12}$O
2-Methyl-4-penten-3-ol
2-Methylpenten-4-ol-3
RN: 4798-45-2   **MP** (°C):
MW: 100.16   **BP** (°C):

Solubility (Moles/L)	Solubility (Grams/L)	Temp (°C)	Ref (#)	Evaluation (T P E A A)	Comments
3.180E-01	3.185E+01	20	G031	1 0 0 0 2	
2.964E-01	2.969E+01	25	G031	1 0 0 0 2	
2.804E-01	2.809E+01	30	G031	1 0 0 0 2	

887. C$_6$H$_{12}$O
1-Hexen-3-ol
Hexen-1-ol-3
RN: 4798-44-1   **MP** (°C):
MW: 100.16   **BP** (°C):   134

Solubility (Moles/L)	Solubility (Grams/L)	Temp (°C)	Ref (#)	Evaluation (T P E A A)	Comments
2.644E-01	2.648E+01	20	G031	1 0 0 0 2	
2.454E-01	2.458E+01	25	G031	1 0 0 0 2	
2.302E-01	2.306E+01	30	G031	1 0 0 0 2	

888. C$_6$H$_{12}$O
Methyl butyl ketone
2-Hexanone
Methyl *n*-butyl ketone
RN: 591-78-6   **MP** (°C):   −57
MW: 100.16   **BP** (°C):   127

Solubility (Moles/L)	Solubility (Grams/L)	Temp (°C)	Ref (#)	Evaluation (T P E A A)	Comments
4.323E-01	4.330E+01	0	C423	0 0 0 0 0	
3.335E-01	3.340E+01	4	C423	0 0 0 0 0	
2.386E-01	2.390E+01	10	C423	0 0 0 0 0	
2.040E-01	2.043E+01	10	G032	1 2 1 1 2	
2.192E-02	2.195E+00	20	D052	1 1 0 0 1	*sic*
1.717E-01	1.720E+01	20	G030	1 2 0 0 2	
1.617E-01	1.620E+01	25	C435	0 0 0 0 0	
1.611E-01	1.614E+01	25	G030	1 2 0 0 2	
1.997E-01	2.000E+01	25	J418	0 0 0 0 0	
3.320E-01	3.326E+01	25	P055	1 0 0 0 2	
1.505E-01	1.507E+01	30	G030	1 2 0 0 2	
1.450E-01	1.452E+01	30	G032	1 2 1 1 2	
1.475E-01	1.478E+01	38	J020	2 1 2 1 1	
1.240E-01	1.242E+01	50	G032	1 2 1 1 2	

889. $C_6H_{12}O$
4-Hexen-3-ol
Hexen-4-ol-3
RN: 4798-58-7   **MP** (°C):
MW: 100.16   **BP** (°C):

Solubility (Moles/L)	Solubility (Grams/L)	Temp (°C)	Ref (#)	Evaluation (T P E A A)	Comments
3.895E-01	3.902E+01	20	G031	1 0 0 0 2	
3.664E-01	3.670E+01	25	G031	1 0 0 0 2	
3.451E-01	3.456E+01	30	G031	1 0 0 0 2	

890. $C_6H_{12}O_2$
3-Hydroxy-2,2-dimethyltetrahydrofuran
3-Furanol, tetrahydro-2,2-dimethyl-
2,2-Dimethyltetrahydrofuran-3-ol
RN: 101398-19-0   **MP** (°C):
MW: 116.16   **BP** (°C):

Solubility (Moles/L)	Solubility (Grams/L)	Temp (°C)	Ref (#)	Evaluation (T P E A A)	Comments
7.826E-01	9.091E+01	rt	B066	0 2 0 0 1	

891. $C_6H_{12}O_2$
Diethylacetic acid
2-Ethylbutyric acid
2-Ethyl-butanoic acid
Ethylbutyric acid
RN: 88-09-5   **MP** (°C):   −15
MW: 116.16   **BP** (°C):   194.5

Solubility (Moles/L)	Solubility (Grams/L)	Temp (°C)	Ref (#)	Evaluation (T P E A A)	Comments
2.147E-02	2.494E+00	25	O011	1 0 1 1 1	

892. $C_6H_{12}O_2$
n-Caproic acid
n-Capronsaeure
RN: 142-62-1   **MP** (°C):   −3.4
MW: 116.16   **BP** (°C):   205

Solubility (Moles/L)	Solubility (Grams/L)	Temp (°C)	Ref (#)	Evaluation (T P E A A)	Comments
7.438E-02	8.640E+00	0	B136	1 0 2 1 2	
7.610E-02	8.840E+00	15	F300	1 0 0 0 2	
8.333E-02	9.680E+00	20	B136	1 0 2 1 2	
8.270E-02	9.607E+00	20	D041	1 0 0 0 1	
8.253E-02	9.587E+00	20	R001	1 1 1 1 2	
8.675E-02	1.008E+01	25	H028	2 0 2 0 2	

(continued)

892. $C_6H_{12}O_2$ (continued)

Solubility (Moles/L)	Solubility (Grams/L)	Temp (°C)	Ref (#)	Evaluation (T P E A A)	Comments
8.760E-02	1.018E+01	25	H122	1 0 0 0 2	
8.608E-02	9.999E+00	25	H339	2 2 1 2 2	
9.367E-02	1.088E+01	25	O011	1 0 1 1 1	
8.772E-02	1.019E+01	30	B136	1 0 2 1 2	
8.684E-02	1.009E+01	30	R001	1 1 1 1 2	
9.282E-02	1.078E+01	35	H339	2 2 1 2 2	
9.427E-02	1.095E+01	45	B136	1 0 2 1 2	
9.324E-02	1.083E+01	45	R001	1 1 1 1 2	
1.008E-01	1.171E+01	60	B136	1 0 2 1 2	
9.956E-02	1.156E+01	60	D041	1 0 0 0 2	
9.964E-02	1.157E+01	60	R001	1 1 1 1 2	
7.374E-02	8.566E+00	.0	R001	1 1 1 1 2	
8.692E-02	1.010E+01	ns	A406	0 0 0 0 1	

893. $C_6H_{12}O_2$

n-Butyl acetate
Essigsaeure-n-butyl ester
n-Butylacetat
Butyl acetate
1-Butyl acetate

RN: 123-86-4   **MP (°C):**   −90
MW: 116.16   **BP (°C):**   117.5

Solubility (Moles/L)	Solubility (Grams/L)	Temp (°C)	Ref (#)	Evaluation (T P E A A)	Comments
3.686E-02	4.282E+00	20	D052	1 1 0 0 0	
8.609E-02	1.000E+01	22	F300	1 0 0 0 0	
5.814E-02	6.754E+00	25	B060	2 0 1 1 1	
7.171E-02	8.330E+00	25	L319	1 0 2 1 2	
1.935E-01	2.248E+01	25	P055	1 0 0 0 1	
2.489E-02	2.892E+00	30	N330	2 2 2 1 2	
7.679E-02	8.920E+00	30	R318	1 1 0 1 0	
5.020E-02	5.831E+00	37	E028	1 0 1 1 2	
5.899E-02	6.853E+00	50	O012	1 2 1 1 2	

894. $C_6H_{12}O_2$

Pentyl formate
n-Amyl formate

RN: 638-49-3   **MP (°C):**
MW: 116.16   **BP (°C):**

Solubility (Moles/L)	Solubility (Grams/L)	Temp (°C)	Ref (#)	Evaluation (T P E A A)	Comments
2.500E-02	2.904E+00	22	S006	1 0 0 0 1	

895. C$_6$H$_{12}$O$_2$
Ethyl butyrate
Butanoic acid ethyl ester
Ethyl butanoate
Butyric ether

RN: 105-54-4 **MP** (°C): −135.4
MW: 116.16 **BP** (°C): 120

Solubility (Moles/L)	Solubility (Grams/L)	Temp (°C)	Ref (#)	Evaluation (T P E A A)	Comments
2.410E-02	2.800E+00	0	C423	0 0 0 0 0	
2.763E-02	3.210E+00	4	C423	0 0 0 0 0	
3.151E-02	3.660E+00	10	C423	0 0 0 0 0	
4.198E-02	4.876E+00	20	D052	1 1 0 0 1	
5.310E-02	6.168E+00	22	F001	1 0 1 2 2	
4.300E-02	4.995E+00	22	S006	1 0 0 0 1	
3.702E-02	4.300E+00	25	C435	0 0 0 0 0	
6.832E-02	7.937E+00	30	R318	1 1 0 1 0	

896. C$_6$H$_{12}$O$_2$
sec-Butyl acetate
DL-sec-Butyl acetate

RN: 105-46-4 **MP** (°C):
MW: 116.16 **BP** (°C): 114

Solubility (Moles/L)	Solubility (Grams/L)	Temp (°C)	Ref (#)	Evaluation (T P E A A)	Comments
5.305E-02	6.162E+00	20	D052	1 1 0 0 0	

897. C$_6$H$_{12}$O$_2$
3-Hydroxy-2,5-dimethyltetrahydrofuran
3-Furanol, tetrahydro-2,5-dimethyl-

RN: 30003-26-0 **MP** (°C):
MW: 116.16 **BP** (°C):

Solubility (Moles/L)	Solubility (Grams/L)	Temp (°C)	Ref (#)	Evaluation (T P E A A)	Comments
1.435E+00	1.667E+02	rt	B066	0 2 0 0 1	

898. C$_6$H$_{12}$O$_2$
Propyl propionate
Propionic acid N-propyl ester
n-Propyl propionate

RN: 106-36-5 **MP** (°C):
MW: 116.16 **BP** (°C): 123

Solubility (Moles/L)	Solubility (Grams/L)	Temp (°C)	Ref (#)	Evaluation (T P E A A)	Comments
5.000E-02	5.808E+00	22	S006	1 0 0 0 0	

899. $C_6H_{12}O_2$
Isobutyl acetate
Acetic acid isobutyl ester
Essigsaeureisobutyl ester

RN:	110-19-0	**MP** (°C):	−99	
MW:	116.16	**BP** (°C):	118	

Solubility (Moles/L)	Solubility (Grams/L)	Temp (°C)	Ref (#)	Evaluation (T P E A A)	Comments
6.502E-02	7.553E+00	14.60	L310	2 2 1 1 2	
5.729E-02	6.655E+00	20	D052	1 1 0 0 1	
5.800E-02	6.737E+00	20	F001	1 0 1 2 1	
5.768E-02	6.700E+00	20	F300	1 0 0 0 1	
6.154E-02	7.149E+00	24.90	L310	2 2 1 1 2	
5.390E-02	6.261E+00	25	B060	2 0 1 1 1	
5.967E-02	6.932E+00	47.90	L310	2 2 1 1 2	
6.154E-02	7.149E+00	67.60	L310	2 2 1 1 2	
6.493E-02	7.543E+00	74.90	L310	2 2 1 1 2	
6.502E-02	7.553E+00	75.20	L310	2 2 1 1 2	
6.875E-02	7.986E+00	84.80	L310	2 2 1 1 2	
7.205E-02	8.369E+00	93.20	L310	2 2 1 1 2	
8.253E-02	9.587E+00	111.50	L310	2 2 1 1 2	
8.540E-02	9.921E+00	115.70	L310	2 2 1 1 2	
1.026E-01	1.192E+01	147.10	L310	2 2 1 1 2	

900. $C_6H_{12}O_3$
Paraldehyde
Paraldehyd

RN:	123-63-7	**MP** (°C):	12.6	
MW:	132.16	**BP** (°C):	128	

Solubility (Moles/L)	Solubility (Grams/L)	Temp (°C)	Ref (#)	Evaluation (T P E A A)	Comments
8.853E-01	1.170E+02	8.5	P059	1 1 1 0 1	
8.377E-01	1.107E+02	11.5	P059	1 1 1 0 1	
8.287E-01	1.095E+02	12.0	P059	1 1 1 0 1	
8.323E-01	1.100E+02	13	F300	1 0 0 0 1	
8.047E-01	1.063E+02	13.5	P059	1 1 1 0 1	
7.621E-01	1.007E+02	17.0	P059	1 1 1 0 1	
6.311E-01	8.341E+01	27.0	P059	1 1 1 0 1	
8.475E-01	1.120E+02	30	F300	1 0 0 0 2	
5.377E-01	7.106E+01	40.0	P059	1 1 1 0 1	
5.246E-01	6.933E+01	42.5	P059	1 1 1 0 1	
4.283E-01	5.660E+01	68.0	P059	1 1 1 0 1	
4.148E-01	5.482E+01	75.0	P059	1 1 1 0 1	
4.540E-01	6.000E+01	100	F300	1 0 0 0 0	

901. C$_6$H$_{12}$O$_3$
2-Ethoxyethyl acetate
Cellosolve acetate

RN:	111-15-9	MP (°C):	−61
MW:	132.16	BP (°C):	156

Solubility (Moles/L)	Solubility (Grams/L)	Temp (°C)	Ref (#)	Evaluation (T P E A A)	Comments
1.499E+00	1.981E+02	20	D052	1 1 0 0 2	
1.415E+00	1.870E+02	20	M062	1 0 0 0 2	

902. C$_6$H$_{12}$O$_3$
Methyl β-ethoxypropionate
Methyl 3-ethoxypropionate
3-Ethoxypropionic acid methyl ester

RN:	14144-33-3	MP (°C):	
MW:	132.16	BP (°C):	

Solubility (Moles/L)	Solubility (Grams/L)	Temp (°C)	Ref (#)	Evaluation (T P E A A)	Comments
7.621E-01	1.007E+02	25	D002	1 2 1 1 2	
7.621E-01	1.007E+02	25	R034	0 0 0 0 2	

903. C$_6$H$_{12}$O$_5$
D-Quercitol
D-Quercit

RN:	488-73-3	MP (°C):	234
MW:	164.16	BP (°C):	

Solubility (Moles/L)	Solubility (Grams/L)	Temp (°C)	Ref (#)	Evaluation (T P E A A)	Comments
6.701E-01	1.100E+02	20	F300	1 0 0 0 2	

904. C$_6$H$_{12}$O$_5$
Rhamnose
α-L-Rhamnose
6-Deoxy-L-mannose
L-Mannomethylose
L-Rhamnose

RN:	3615-41-6	MP (°C):	82
MW:	164.16	BP (°C):	

Solubility (Moles/L)	Solubility (Grams/L)	Temp (°C)	Ref (#)	Evaluation (T P E A A)	Comments
2.212E+00	3.631E+02	18	D041	1 0 0 0 1	
3.177E+00	5.215E+02	40	D041	1 0 0 0 1	

905. C₆H₁₂O₆

D-Inositol
D(+)-Inositol
D-Chiro-inositol
(+)-Chiro-inositol
RN: 643-12-9   **MP** (°C):   249.5
MW: 180.16   **BP** (°C):

Solubility (Moles/L)	Solubility (Grams/L)	Temp (°C)	Ref (#)	Evaluation (T P E A A)	Comments
2.239E+00	4.034E+02	11	F300	1 0 0 0 2	

906. C₆H₁₂O₆

D-Mannose
D-(+)-Mannose
Seminose
Carubinose
RN: 3458-28-4   **MP** (°C):   132
MW: 180.16   **BP** (°C):

Solubility (Moles/L)	Solubility (Grams/L)	Temp (°C)	Ref (#)	Evaluation (T P E A A)	Comments
3.956E+00	7.126E+02	17	D041	1 0 0 0 2	
3.957E+00	7.128E+02	17	F300	1 0 0 0 2	
2.399E+00	4.322E+02	25	G317	0 0 0 0 0	

907. C₆H₁₂O₆

Glucose
D-Glucose
D(+)-Glucose
Staleydex 111
Staleydex 333
RN: 50-99-7   **MP** (°C):   146
MW: 180.16   **BP** (°C):

Solubility (Moles/L)	Solubility (Grams/L)	Temp (°C)	Ref (#)	Evaluation (T P E A A)	Comments
1.954E+00	3.520E+02	.50	J019	1 0 1 2 2	
1.749E+00	3.151E+02	0	M043	1 0 0 0 1	
2.286E+00	4.118E+02	10	M043	1 0 0 0 1	
2.271E+00	4.091E+02	10.0	Y020	1 1 2 1 2	
3.365E+00	6.063E+02	15	D041	1 0 0 0 2	
2.660E+00	4.792E+02	20	M043	1 0 0 0 1	
2.314E+00	4.168E+02	20.0	Y020	1 1 2 1 2	
3.033E+00	5.464E+02	30	J019	1 0 1 2 2	
3.031E+00	5.460E+02	30	K122	1 1 1 1 2	
3.028E+00	5.455E+02	30	M043	1 0 0 0 2	
2.355E+00	4.244E+02	30.0	Y020	1 1 2 1 2	
1.901E+00	3.425E+02	30.50	M137	2 1 2 2 2	
2.042E+00	3.678E+02	35	B354	0 0 0 0 0	

(continued)

907. $C_6H_{12}O_6$ (continued)

Solubility (Moles/L)	Solubility (Grams/L)	Temp (°C)	Ref (#)	Evaluation (T P E A A)	Comments
3.416E+00	6.154E+02	40	M043	1 0 0 0 2	
2.396E+00	4.317E+02	40.0	Y020	1 1 2 1 2	
3.936E+00	7.091E+02	50	J019	1 0 1 2 2	
2.436E+00	4.388E+02	50.0	Y020	1 1 2 1 2	
4.090E+00	7.368E+02	60	M043	1 0 0 0 2	
4.005E+00	7.215E+02	70	A420	0 0 0 0 0	
4.523E+00	8.148E+02	80	M043	1 0 0 0 2	
2.227E+00	4.012E+02	.0	Y020	1 1 2 1 2	
2.501E+00	4.505E+02	rt	D021	0 0 1 1 2	

908. $C_6H_{12}O_6$
Fructose
D-Fructose
D-(–)-Fructose
D-(–)-Levulose
Krystar 300
Nevulose

RN: 57-48-7 **MP (°C):** 129
MW: 180.16 **BP (°C):**

Solubility (Moles/L)	Solubility (Grams/L)	Temp (°C)	Ref (#)	Evaluation (T P E A A)	Comments
2.379E+00	4.286E+02	0	M043	1 0 0 0 1	
4.318E+00	7.780E+02	20	F300	1 0 0 0 2	
2.467E+00	4.444E+02	20	M043	1 0 0 0 1	
4.524E+00	8.150E+02	30	K122	1 1 1 1 2	
4.524E+00	8.150E+02	30	K135	1 1 1 1 2	
2.448E+01	4.410E+03	30	K136	1 1 1 1 2	
2.550E+00	4.595E+02	40	M043	1 0 0 0 1	
2.629E+00	4.737E+02	60	M043	1 0 0 0 1	
4.709E+00	8.484E+02	70	A420	0 0 0 0 0	

909. $C_6H_{12}O_6$
Tagatose
Lyxo-2-hexulose
DL-Tagatose

RN: 17598-81-1 **MP (°C):**
MW: 180.16 **BP (°C):**

Solubility (Moles/L)	Solubility (Grams/L)	Temp (°C)	Ref (#)	Evaluation (T P E A A)	Comments
2.084E+00	3.755E+02	22	F300	1 0 0 0 2	

910. C$_6$H$_{12}$O$_6$

D-Galactose
Galactose
(+)-Galactose
D(+)-Galactose

RN: 59-23-4   **MP** (°C):   169
MW: 180.16   **BP** (°C):

Solubility (Moles/L)	Solubility (Grams/L)	Temp (°C)	Ref (#)	Evaluation (T P E A A)	Comments
5.046E-01	9.091E+01	0	D041	1 0 0 0 1	
2.247E+00	4.048E+02	25	D041	1 0 0 0 1	
2.253E+00	4.058E+02	rt	D021	0 0 1 1 2	

911. C$_6$H$_{12}$O$_6$

L-Sorbose
Sorbose
L-1,3,4,5,6-Pentahydroxyhexan-2-one
L-Xylo-2-hexulose

RN: 87-79-6   **MP** (°C):   165
MW: 180.16   **BP** (°C):

Solubility (Moles/L)	Solubility (Grams/L)	Temp (°C)	Ref (#)	Evaluation (T P E A A)	Comments
1.970E+00	3.548E+02	17	D041	1 0 0 0 1	
1.998E+00	3.600E+02	17	F300	1 0 0 0 1	

912. C$_6$H$_{12}$O$_6$

Inositol
Mesoinosit
cis-1,2,3,5-trans-4,6-Cyclohexanehexol
Dambose
Nucite
Phaseomannite

RN: 87-89-8   **MP** (°C):   226
MW: 180.16   **BP** (°C):

Solubility (Moles/L)	Solubility (Grams/L)	Temp (°C)	Ref (#)	Evaluation (T P E A A)	Comments
7.788E-01	1.403E+02	19	F300	1 0 0 0 2	
8.267E-01	1.489E+02	20	D041	1 0 0 0 2	
7.771E-01	1.400E+02	25	M054	1 0 0 0 1	
7.771E-01	1.400E+02	ns	L335	0 0 0 0 2	
7.762E-01	1.398E+02	ns	R424	0 0 0 0 0	

913. $C_6H_{12}O_6$
α-Glucose
α-D-Glucose
D-α-Glucose
Dextrose

RN: 492-62-6   **MP** (°C): 154.5
MW: 180.16   **BP** (°C):

Solubility (Moles/L)	Solubility (Grams/L)	Temp (°C)	Ref (#)	Evaluation (T P E A A)	Comments
1.355E+00	2.441E+02	0	D041	1 0 0 0 2	
2.019E+00	3.638E+02	10.0	Y020	1 1 2 1 2	
2.775E+00	5.000E+02	20	F300	1 0 0 0 0	
2.096E+00	3.775E+02	20.0	Y020	1 1 2 1 2	
2.501E+00	4.505E+02	25	D041	1 0 0 0 2	
2.170E+00	3.909E+02	30.0	Y020	1 1 2 1 2	
2.242E+00	4.040E+02	40.0	Y020	1 1 2 1 2	
2.313E+00	4.168E+02	50.0	Y020	1 1 2 1 2	
2.346E+00	4.227E+02	54.7	Y020	1 1 2 1 2	
1.942E+00	3.498E+02	.0	Y020	1 1 2 1 2	

914. $C_6H_{12}O_6 \cdot H_2O$
Glucose (monohydrate)

RN: 50-99-7   **MP** (°C): 83
MW: 198.17   **BP** (°C):

Solubility (Moles/L)	Solubility (Grams/L)	Temp (°C)	Ref (#)	Evaluation (T P E A A)	Comments
1.449E+00	2.871E+02	10.0	Y020	1 1 2 1 2	
1.619E+00	3.209E+02	20.0	Y020	1 1 2 1 2	
1.781E+00	3.530E+02	30.0	Y020	1 1 2 1 2	
1.933E+00	3.831E+02	40.0	Y020	1 1 2 1 2	
2.072E+00	4.106E+02	50.0	Y020	1 1 2 1 2	
1.784E+00	3.536E+02	73.2	Y020	1 1 2 1 2	
1.274E+00	2.525E+02	.0	Y020	1 1 2 1 2	

915. $C_6H_{12}O_7$
Scyllitol
Scyllit
Quercinitol
Cocositol

RN: 488-59-5   **MP** (°C): 253
MW: 196.16   **BP** (°C):

Solubility (Moles/L)	Solubility (Grams/L)	Temp (°C)	Ref (#)	Evaluation (T P E A A)	Comments
5.149E-02	1.010E+01	18	F300	1 0 0 0 2	

916. C$_6$H$_{13}$Br
1-Bromohexane
Hexyl bromide

RN:	111-25-1	MP (°C):	–84.7
MW:	165.08	BP (°C):	155.3

Solubility (Moles/L)	Solubility (Grams/L)	Temp (°C)	Ref (#)	Evaluation (T P E A A)	Comments
1.560E-04	2.575E-02	25	M342	1 0 1 1 2	

917. C$_6$H$_{13}$N
1-Methylpiperidine
N-Methylpiperidine

RN:	626-67-5	MP (°C):	–18
MW:	99.18	BP (°C):	106

Solubility (Moles/L)	Solubility (Grams/L)	Temp (°C)	Ref (#)	Evaluation (T P E A A)	Comments
+1.70E+00	+1.68E+02	ns	S460	0 0 0 0 0	

918. C$_6$H$_{13}$NO
Caproamide
n-Capronsaeure-amid
Hexanamide
Hexanoic acid, amide

RN:	628-02-4	MP (°C):	99
MW:	115.18	BP (°C):	255

Solubility (Moles/L)	Solubility (Grams/L)	Temp (°C)	Ref (#)	Evaluation (T P E A A)	Comments
1.610E-01	1.854E+01	6	H059	0 0 0 0 0	
2.030E-01	2.338E+01	16	H059	0 0 0 0 0	
2.580E-01	2.972E+01	25	H059	0 0 0 0 0	
2.750E-01	3.167E+01	29	H059	0 0 0 0 0	
3.150E-01	3.628E+01	33	H059	0 0 0 0 0	
3.250E-01	3.743E+01	35	H059	0 0 0 0 0	
3.390E-01	3.904E+01	37	H059	0 0 0 0 0	
3.890E-01	4.480E+01	41	H059	0 0 0 0 0	

919. C$_6$H$_{13}$NO$_2$
L-Norleucine
Norleucine
α-Aminocaproic acid

RN:	327-57-1	MP (°C):	327dec
MW:	131.18	BP (°C):	

Solubility (Moles/L)	Solubility (Grams/L)	Temp (°C)	Ref (#)	Evaluation (T P E A A)	Comments
1.304E-01	1.710E+01	23	K060	1 2 0 0 2	
1.127E-01	1.478E+01	25	D041	1 0 0 0 1	
8.700E-02	1.141E+01	25	E015	1 2 1 1 1	
1.232E-01	1.616E+01	25	K031	2 1 2 1 2	

920. C₆H₁₃NO₂

L-Leucine
L(–)-Leucine
Leucine
2-Amino-4-methylpentanoic acid
L-2-Amino-4-methylpentanoic acid
(2S)-α-Leucine

RN: 61-90-5 **MP** (°C): 286–288
MW: 131.18 **BP** (°C):

Solubility (Moles/L)	Solubility (Grams/L)	Temp (°C)	Ref (#)	Evaluation (T P E A A)	Comments
1.692E-01	2.220E+01	0	F300	1 0 0 0 2	
1.740E-01	2.282E+01	15	D349	2 1 1 2 2	
1.601E-01	2.100E+01	20	B032	1 2 2 1 2	
1.800E-01	2.361E+01	20	D349	2 1 1 2 2	
1.695E-01	2.224E+01	21	P045	1 0 2 1 2	
1.772E-01	2.324E+01	24.99	C404	2 1 2 2 1	
1.640E-01	2.151E+01	25	B032	1 2 2 1 2	
1.712E-01	2.246E+01	25	C018	0 0 0 0 0	
1.851E-01	2.428E+01	25	C018	0 0 0 0 0	
1.700E-04	2.230E-02	25	C405	2 1 2 2 2	intrinsic zwit
1.883E-01	2.470E+01	25	D016	1 0 0 0 2	
1.634E-01	2.143E+01	25	D041	1 0 0 0 2	
1.860E-01	2.440E+01	25	D349	2 1 1 2 2	
1.807E-01	2.370E+01	25	F300	1 0 0 0 2	
1.626E-01	2.133E+01	25	G092	2 1 1 1 1	
1.626E-01	2.133E+01	25	G315	0 0 0 0 0	
1.647E-01	2.160E+01	25.1	N024	0 0 0 0 0	
1.654E-01	2.170E+01	25.1	N025	0 0 0 0 0	
1.647E-01	2.160E+01	25.1	N026	0 0 0 0 0	
1.612E-01	2.114E+01	25.1	N027	1 1 2 2 2	
1.765E-01	2.315E+01	27	D036	0 0 0 0 0	
1.601E-01	2.100E+01	27	D036	0 0 0 0 0	
1.682E-01	2.206E+01	29.80	B032	1 2 2 1 2	
1.907E-01	2.502E+01	34.99	C404	2 1 2 2 1	
2.041E-01	2.677E+01	44.99	C404	2 1 2 2 1	
2.142E-01	2.810E+01	50	F300	1 0 0 0 2	
2.805E-01	3.679E+01	75	D041	1 0 0 0 2	
2.805E-01	3.680E+01	75	F300	1 0 0 0 2	
2.886E-01	3.786E+01	92	M160	2 1 1 1 0	
4.071E-01	5.340E+01	100	F300	1 0 0 0 2	
4.069E-01	5.337E+01	99.99	P349	0 0 0 0 0	
1.830E-01	2.400E+01	ns	D072	0 0 0 0 1	

921. C₆H₁₃NO₂

L-*allo*-Isoleucine
Alloisoleucine

RN: 1509-34-8 **MP** (°C): >280
MW: 131.18 **BP** (°C):

Solubility (Moles/L)	Solubility (Grams/L)	Temp (°C)	Ref (#)	Evaluation (T P E A A)	Comments
2.148E-01	2.818E+01	20	D041	1 0 0 0 1	

922. C₆H₁₃NO₂

D-Leucine
D-2-Amino-4-methylvaleric acid
D-2-Amino-4-methylpentanoic acid

RN:	328-38-1	MP (°C):	>300
MW:	131.18	BP (°C):	

Solubility (Moles/L)	Solubility (Grams/L)	Temp (°C)	Ref (#)	Evaluation (T P E A A)	Comments
1.641E-01	2.153E+01	25	D041	1 0 0 0 2	
1.975E-01	2.591E+01	50	D041	1 0 0 0 2	

923. C₆H₁₃NO₂

D-Norleucine
D-2-Amino-*n*-caproic acid
D-2-Aminohexanoic acid

RN:	327-56-0	MP (°C):	>300
MW:	131.18	BP (°C):	

Solubility (Moles/L)	Solubility (Grams/L)	Temp (°C)	Ref (#)	Evaluation (T P E A A)	Comments
1.201E-01	1.575E+01	19	D041	1 0 0 0 1	

924. C₆H₁₃NO₂

tert-Amyl carbamate
tert-Pentyl carbamate

RN:	590-60-3	MP (°C):	85
MW:	131.18	BP (°C):	

Solubility (Moles/L)	Solubility (Grams/L)	Temp (°C)	Ref (#)	Evaluation (T P E A A)	Comments
1.600E-01	2.099E+01	37	H006	1 2 2 1 1	

925. C₆H₁₃NO₂

n-Amyl carbamate
n-Pentyl carbamate
O-Pentyl carbamate

RN:	638-42-6	MP (°C):	
MW:	131.18	BP (°C):	

Solubility (Moles/L)	Solubility (Grams/L)	Temp (°C)	Ref (#)	Evaluation (T P E A A)	Comments
3.400E-02	4.460E+00	37	H006	1 2 2 1 1	

926. C$_6$H$_{13}$NO$_2$
Isopentyl urethane
Isoamylurethan
Isoamylurethane
RN: 543-86-2   **MP** (°C):
MW: 131.18   **BP** (°C):

Solubility (Moles/L)	Solubility (Grams/L)	Temp (°C)	Ref (#)	Evaluation (T P E A A)	Comments
3.660E-02	4.801E+00	15.5	F001	1 0 1 2 2	

927. C$_6$H$_{13}$NO$_2$
ε-Aminocaproic acid
6-Aminocaproic acid
ε-Amino-capronsaeure
RN: 60-32-2   **MP** (°C):   205
MW: 131.18   **BP** (°C):

Solubility (Moles/L)	Solubility (Grams/L)	Temp (°C)	Ref (#)	Evaluation (T P E A A)	Comments
3.848E+00	5.048E+02	25	M024	1 2 0 1 2	

928. C$_6$H$_{13}$NO$_2$
DL-Norleucine
DL-2-Amino-n-caproic acid
2-Aminohexanoic acid
DL-2-Aminohexanoic acid
RN: 616-06-8   **MP** (°C):   >300
MW: 131.18   **BP** (°C):

Solubility (Moles/L)	Solubility (Grams/L)	Temp (°C)	Ref (#)	Evaluation (T P E A A)	Comments
6.863E-02	9.003E+00	0	D018	2 2 2 1 2	
8.660E-02	1.136E+01	25	C018	0 0 0 0 0	
8.767E-02	1.150E+01	25	D016	1 0 0 0 2	
8.906E-02	1.168E+01	25	D018	2 2 2 1 2	
8.891E-02	1.166E+01	25	D041	1 0 0 0 2	
8.118E-02	1.065E+01	25	K031	2 1 2 1 2	
8.660E-02	1.136E+01	25	M024	1 2 0 1 2	
1.348E-01	1.768E+01	50	D018	2 2 2 1 2	
2.135E-01	2.800E+01	75	D018	2 2 2 1 2	
2.134E-01	2.799E+01	75	D041	1 0 0 0 2	
3.788E-01	4.969E+01	99.99	P349	0 0 0 0 0	

929. C₆H₁₃NO₂

L-Isoleucine
L(+)-Isoleucin
Isoleucine

RN:	73-32-5	MP (°C):	288
MW:	131.18	BP (°C):	

Solubility (Moles/L)	Solubility (Grams/L)	Temp (°C)	Ref (#)	Evaluation (T P E A A)	Comments
2.844E-01	3.730E+01	15.50	F300	1 0 0 0 2	
2.533E-01	3.323E+01	20	B032	1 2 2 1 2	
2.619E-01	3.435E+01	25	B032	1 2 2 1 2	
3.017E-01	3.957E+01	25	D041	1 0 0 0 2	
2.458E-01	3.224E+01	25	G433	0 0 0 0 0	
2.364E-01	3.101E+01	25	O316	1 0 1 2 2	
2.358E-01	3.093E+01	25	O316	1 0 1 2 2	
2.714E-01	3.560E+01	27	D036	0 0 0 0 0	
2.690E-01	3.528E+01	29.80	B032	1 2 2 1 2	
4.369E-01	5.732E+01	75	D041	1 0 0 0 2	
3.801E-01	4.985E+01	84	M160	2 1 1 1 0	

930. C₆H₁₃NO₂

α-Hydroxycaproamide
Hexanamide, 2-hydroxy-
2-Hydroxyhexanamide

RN:	66461-73-2	MP (°C):	
MW:	131.18	BP (°C):	

Solubility (Moles/L)	Solubility (Grams/L)	Temp (°C)	Ref (#)	Evaluation (T P E A A)	Comments
8.300E-02	1.089E+01	25	M008	1 0 0 0 2	

931. C₆H₁₃NO₂

N-Propylurethane
Propylurethan
n-Propyl urethane

RN:	623-85-8	MP (°C):	
MW:	131.18	BP (°C):	

Solubility (Moles/L)	Solubility (Grams/L)	Temp (°C)	Ref (#)	Evaluation (T P E A A)	Comments
7.475E-01	9.805E+01	15.5	F001	1 0 1 2 2	

932. $C_6H_{13}NO_2$

DL-Isoleucine
DL-2-Amino-3-methylpentanoic acid
RN: 443-79-8   **MP** (°C):
MW: 131.18   **BP** (°C):

Solubility (Moles/L)	Solubility (Grams/L)	Temp (°C)	Ref (#)	Evaluation (T P E A A)	Comments
1.311E-01	1.720E+01	0	D018	2 2 2 1 2	
1.632E-01	2.141E+01	25	D018	2 2 2 1 2	
1.662E-01	2.180E+01	25	D041	1 0 0 0 2	
2.235E-01	2.931E+01	50	D018	2 2 2 1 2	
3.510E-01	4.605E+01	75	D018	2 2 2 1 2	
3.357E-01	4.404E+01	75	D041	1 0 0 0 2	
5.517E-01	7.237E+01	99.99	P349	0 0 0 0 0	

933. $C_6H_{13}NO_2$

DL-Leucine
DL-2-Amino-4-methylvaleric acid
DL-2-Amino-4-methylpentanoic acid
RN: 328-39-2   **MP** (°C):   295
MW: 131.18   **BP** (°C):

Solubility (Moles/L)	Solubility (Grams/L)	Temp (°C)	Ref (#)	Evaluation (T P E A A)	Comments
6.659E-02	8.735E+00	0	D018	2 2 2 1 2	
6.022E-02	7.900E+00	0	F300	1 0 0 0 1	
7.433E-02	9.750E+00	25	C018	0 0 0 0 0	
7.517E-02	9.860E+00	25	D016	1 0 0 0 2	
8.898E-02	1.167E+01	25	D018	2 2 2 1 2	
7.481E-02	9.813E+00	25	D041	1 0 0 0 2	
7.471E-02	9.800E+00	25	F300	1 0 0 0 1	
1.321E-01	1.733E+01	50	D018	2 2 2 1 2	
1.060E-01	1.390E+01	50	F300	1 0 0 0 2	
2.105E-01	2.762E+01	75	D018	2 2 2 1 2	
1.696E-01	2.225E+01	75	D041	1 0 0 0 2	
1.700E-01	2.230E+01	75	F300	1 0 0 0 2	
3.080E-01	4.040E+01	100	F300	1 0 0 0 2	
3.077E-01	4.036E+01	99.99	P349	0 0 0 0 0	
7.324E-02	9.607E+00	rt	H431	0 0 0 0 0	average

934. C_6H_{14}

Hexane
Normal hexane
n-Hexane
Skellysolve B
RN: 110-54-3   **MP** (°C):   −95
MW: 86.18   **BP** (°C):   65

Solubility (Moles/L)	Solubility (Grams/L)	Temp (°C)	Ref (#)	Evaluation (T P E A A)	Comments
1.915E-04	1.650E-02	0	P003	2 2 2 2 2	
1.900E-04	1.637E-02	4.0	N004	1 1 2 2 2	

(continued)

934. C_6H_{14} (continued)

Solubility (Moles/L)	Solubility (Grams/L)	Temp (°C)	Ref (#)	Evaluation (T P E A A)	Comments
1.761E-04	1.518E-02	14.0	N004	1 1 2 2 2	
1.600E-03	1.379E-01	15.5	F001	1 0 1 0 2	
6.382E-04	5.500E-02	16	D047	1 0 0 1 1	
1.427E-04	1.230E-02	25	A058	1 1 1 1 2	
1.624E-03	1.400E-01	25	A094	1 0 0 0 1	
1.625E-03	1.400E-01	25	K072	1 0 1 1 1	
1.857E-03	1.600E-01	25	K112	1 0 2 1 1	
1.860E-03	1.603E-01	25	K112	1 0 2 2 2	
1.099E-04	9.470E-03	25	K119	1 0 0 0 2	
1.427E-04	1.230E-02	25	L002	2 2 2 2 2	
1.102E-04	9.500E-03	25	M001	2 1 2 2 2	
1.102E-04	9.500E-03	25	M002	2 1 2 2 2	
1.102E-04	9.500E-03	25	M040	1 0 0 1 1	
1.625E-03v	1.400E-01	25	M087	1 1 2 1 1	
1.430E-04	1.232E-02	25	M342	1 0 1 1 2	
1.439E-04	1.240E-02	25	P003	2 2 2 2 2	
1.624E-03	1.400E-01	25	S012	2 0 2 2 1	
2.128E-04	1.834E-02	25.0	N004	1 1 2 2 2	
1.099E-04	9.470E-03	25.0	P051	2 1 1 2 2	
1.099E-04	9.470E-03	25.00	P007	2 1 2 2 2	
1.494E-04	1.288E-02	35.0	N004	1 1 2 2 2	
4.623E-02	3.984E+00	38	J020	2 0 2 1 0	*sic*
1.172E-04	1.010E-02	40.1	P051	2 1 1 2 2	
1.172E-04	1.010E-02	40.10	P007	2 1 2 2 2	
2.578E-04	2.221E-02	45.0	N004	1 1 2 2 2	
2.553E-03	2.200E-01	50	L097	1 1 1 1 1	
2.456E-04	2.116E-02	55.0	N004	1 1 2 2 2	
1.532E-04	1.320E-02	55.7	P051	2 1 1 2 2	
1.532E-04	1.320E-02	55.70	P007	2 1 2 2 2	
1.775E-04	1.530E-02	69.7	P051	2 1 1 2 2	average of 2
1.764E-04	1.520E-02	69.70	P007	2 1 2 2 2	
1.787E-04	1.540E-02	69.70	P007	2 1 2 2 2	
2.599E-04	2.240E-02	99.1	P051	2 1 1 2 2	
2.599E-04	2.240E-02	99.10	P007	2 1 2 2 2	
3.388E-04	2.920E-02	114.4	P051	2 1 1 2 2	
3.388E-04	2.920E-02	114.40	P007	2 1 2 2 2	
4.363E-04	3.760E-02	121.3	P051	2 1 1 2 2	
4.363E-04	3.760E-02	121.30	P007	2 1 2 2 2	
6.603E-04	5.690E-02	137.3	P051	2 1 1 2 2	
6.603E-04	5.690E-02	137.30	P007	2 1 2 2 2	
1.230E-03	1.060E-01	151.8	P051	2 1 1 2 2	
1.230E-03	1.060E-01	151.80	P007	2 1 2 2 2	
1.102E-04	9.500E-03	ns	H123	0 0 0 0 0	
1.392E-03	1.200E-01	ns	M010	0 0 0 0 1	
1.880E-04	1.620E-02	ns	M175	0 0 2 1 2	

935. C$_6$H$_{14}$
2,2-Dimethylbutane
Neohexane

RN:	75-83-2	MP (°C):	−100
MW:	86.18	BP (°C):	50

Solubility (Moles/L)	Solubility (Grams/L)	Temp (°C)	Ref (#)	Evaluation (T P E A A)	Comments
4.572E-04	3.940E-02	0	P003	2 2 2 2 2	
4.278E-04	3.686E-02	2.34	S461	0 0 0 0 0	
3.444E-04	2.968E-02	9.99	S461	0 0 0 0 0	
2.722E-04	2.346E-02	24.99	S461	0 0 0 0 0	
2.460E-04	2.120E-02	25	K119	1 0 0 0 2	
2.135E-04	1.840E-02	25	M001	2 1 2 2 2	
2.135E-04	1.840E-02	25	M002	2 1 2 2 2	
2.762E-04	2.380E-02	25	P003	2 2 2 2 2	
2.460E-04	2.120E-02	25	P051	2 1 1 2 2	
2.460E-04	2.120E-02	25.00	P007	2 1 2 2 2	
6.600E-04	5.687E-02	ns	J300	0 0 0 0 0	

936. C$_6$H$_{14}$
2,3-Dimethylbutane
Diisopropyl
1,1,2,2-Tetramethylethane

RN:	79-29-8	MP (°C):	−129
MW:	86.18	BP (°C):	58

Solubility (Moles/L)	Solubility (Grams/L)	Temp (°C)	Ref (#)	Evaluation (T P E A A)	Comments
3.818E-04	3.290E-02	0	P003	2 2 2 2 2	
2.216E-04	1.910E-02	25	K119	1 0 0 0 2	
2.611E-04	2.250E-02	25	P003	2 2 2 2 2	
2.216E-04	1.910E-02	25.0	P051	2 1 1 2 2	
2.216E-04	1.910E-02	25.00	P007	2 1 2 2 2	
2.228E-04	1.920E-02	40.1	P051	2 1 1 2 2	
2.228E-04	1.920E-02	40.10	P007	2 1 2 2 2	
2.750E-04	2.370E-02	55.1	P051	2 1 1 2 2	
2.750E-04	2.370E-02	55.10	P007	2 1 2 2 2	
4.653E-04	4.010E-02	99.1	P051	2 1 1 2 2	
4.653E-04	4.010E-02	99.10	P007	2 1 2 2 2	
6.591E-04	5.680E-02	121.3	P051	2 1 1 2 2	
6.591E-04	5.680E-02	121.30	P007	2 1 2 2 2	
1.136E-03	9.790E-02	137.3	P051	2 1 1 2 2	
1.136E-03	9.790E-02	137.30	P007	2 1 2 2 2	
1.984E-03	1.710E-01	149.5	P051	2 1 1 2 2	
1.984E-03	1.710E-01	149.50	P007	2 1 2 2 2	

937. C$_6$H$_{14}$
2-Methylpentane
2-Metylopentan

RN: 107-83-5   **MP** (°C):   −154
MW: 86.18   **BP** (°C):   62

Solubility (Moles/L)	Solubility (Grams/L)	Temp (°C)	Ref (#)	Evaluation (T P E A A)	Comments
2.257E-04	1.945E-02	0	P003	2 2 2 2 2	
5.976E-04	5.150E-02	23	C332	0 0 0 0 0	
1.508E-04	1.300E-02	25	K119	1 0 0 0 2	
1.648E-04	1.420E-02	25	L002	2 2 2 2 2	
1.601E-04	1.380E-02	25	M001	2 1 2 2 2	
1.601E-04	1.380E-02	25	M002	2 1 2 2 2	
1.822E-04	1.570E-02	25	P003	2 2 2 2 2	
1.508E-04	1.300E-02	25.0	P051	2 1 1 2 2	
1.508E-04	1.300E-02	25.00	P007	2 1 2 2 2	
1.601E-04	1.380E-02	40.1	P051	2 1 1 2 2	
1.601E-04	1.380E-02	40.10	P007	2 1 2 2 2	
1.822E-04	1.570E-02	55.7	P051	2 1 1 2 2	
1.822E-04	1.570E-02	55.70	P007	2 1 2 2 2	
3.145E-04	2.710E-02	99.1	P051	2 1 1 2 2	
3.145E-04	2.710E-02	99.10	P007	2 1 2 2 2	
5.210E-04	4.490E-02	118.0	P051	2 1 1 2 2	
5.210E-04	4.490E-02	118.00	P007	2 1 2 2 2	
1.007E-03	8.680E-02	137.3	P051	2 1 1 2 2	
1.007E-03	8.680E-02	137.30	P007	2 1 2 2 2	
1.311E-03	1.130E-01	149.50	P007	2 1 2 2 2	

938. C$_6$H$_{14}$
3-Methylpentane
3-Metylopentan

RN: 96-14-0   **MP** (°C):   −118
MW: 86.18   **BP** (°C):   64

Solubility (Moles/L)	Solubility (Grams/L)	Temp (°C)	Ref (#)	Evaluation (T P E A A)	Comments
2.495E-04	2.150E-02	0	P003	2 2 2 2 2	
1.520E-04	1.310E-02	25	K119	1 0 0 0 2	
1.485E-04	1.280E-02	25	M001	2 1 2 2 2	
2.077E-04	1.790E-02	25	P003	2 2 2 2 2	
1.520E-04	1.310E-02	25	P051	2 1 1 2 2	
1.520E-04	1.310E-02	25.00	P007	2 1 2 2 2	
1.485E-04	1.280E-02	ns	H123	0 0 0 0 0	

939. C₆H₁₄FO₃P

939. $C_6H_{14}FO_3P$

Isofluorphate
Diisopropylfluorophosphate
Phosphorofluoridic acid bis(1-methylethyl) ester
Difluorophate
PF-3
T-1703

RN:	55-91-4	MP (°C):	−82
MW:	184.15	BP (°C):	183

Solubility (Moles/L)	Solubility (Grams/L)	Temp (°C)	Ref (#)	Evaluation (T P E A A)	Comments
8.236E-02	1.517E+01	25	D041	1 0 0 0 2	

940. $C_6H_{14}NO_3PS_2$

Ethoate-methyl
O,O-Dimethyl *S*-(*N*-ethylcarbamoylmethyl) dithiophosphate
Fitios

RN:	116-01-8	MP (°C):	66.1
MW:	243.29	BP (°C):	

Solubility (Moles/L)	Solubility (Grams/L)	Temp (°C)	Ref (#)	Evaluation (T P E A A)	Comments
3.494E-02	8.500E+00	25	M061	1 0 0 0 1	
3.494E-02	8.500E+00	25	M161	1 0 0 0 1	

941. $C_6H_{14}N_2$

trans-2,5-Dimethylpiperazine
trans-2,5-Dimethyl-piperazin

RN:	2815-34-1	MP (°C):	
MW:	114.19	BP (°C):	

Solubility (Moles/L)	Solubility (Grams/L)	Temp (°C)	Ref (#)	Evaluation (T P E A A)	Comments
3.065E+00	3.500E+02	20	F300	1 0 0 0 1	

942. $C_6H_{14}N_2O$

Methyl-*n*-amylnitrosamine
N-Nitroso(methyl)pentylamine

RN:	13256-07-0	MP (°C):	
MW:	130.19	BP (°C):	

Solubility (Moles/L)	Solubility (Grams/L)	Temp (°C)	Ref (#)	Evaluation (T P E A A)	Comments
8.400E-02	1.094E+01	24	D083	2 0 0 0 1	

943. C$_6$H$_{14}$N$_2$O

Di-*n*-propylnitrosamine
N-Nitroso-*N*-propyl-1-propanamine
Dipropylnitrosamine
NDPA
DPNA
Nitrosodipropylamine

RN: 621-64-7   **MP** (°C):
MW: 130.19   **BP** (°C):

Solubility (Moles/L)	Solubility (Grams/L)	Temp (°C)	Ref (#)	Evaluation (T P E A A)	Comments
7.600E-02	9.895E+00	24	D083	2 0 0 0 1	

944. C$_6$H$_{14}$N$_2$O

Ethyl-*n*-butylnitrosamine
Nitroso-*N*-ethyl-*n*-butylamine
N-Nitroso-*N*-butylethylamine
N-Nitroso(ethyl)-*n*-butylamine
NEBA
Butanamine, *N*-ethyl-*N*-nitroso-

RN: 4549-44-4   **MP** (°C):
MW: 130.19   **BP** (°C):

Solubility (Moles/L)	Solubility (Grams/L)	Temp (°C)	Ref (#)	Evaluation (T P E A A)	Comments
9.200E-02	1.198E+01	24	D083	2 0 0 0 1	

945. C$_6$H$_{14}$N$_2$O

Di-isopropylnitrosamine
2-Propanamine, *N*-(1-methylethyl)-*N*-nitroso-
N-Nitrosodiisopropylamine
NdiPA

RN: 601-77-4   **MP** (°C):
MW: 130.19   **BP** (°C):

Solubility (Moles/L)	Solubility (Grams/L)	Temp (°C)	Ref (#)	Evaluation (T P E A A)	Comments
1.000E-01	1.302E+01	24	D083	2 0 0 0 1	

946. C$_6$H$_{14}$N$_2$O$_2$

L(+)-Lysine
L(+)-Lysin
Lysine

RN: 56-87-1   **MP** (°C): 224
MW: 146.19   **BP** (°C):

Solubility (Moles/L)	Solubility (Grams/L)	Temp (°C)	Ref (#)	Evaluation (T P E A A)	Comments
3.995E+00	5.840E+02	27	D036	0 0 0 0 0	

947. $C_6H_{14}N_4O_2$
DL-Arginine
(±)-Arginine

RN:	7200-25-1	MP (°C):
MW:	174.20	BP (°C):

Solubility (Moles/L)	Solubility (Grams/L)	Temp (°C)	Ref (#)	Evaluation (T P E A A)	Comments
1.382E+00	2.407E+02	20	J303	0 0 0 0 0	
1.978E+00	3.445E+02	40	J303	0 0 0 0 0	
2.781E+00	4.844E+02	50	J303	0 0 0 0 0	
3.851E+00	6.709E+02	60	J303	0 0 0 0 0	

948. $C_6H_{14}N_4O_2$
L-Arginine
L(+)-Arginin
Arginine

RN:	74-79-3	MP (°C):	244
MW:	174.20	BP (°C):	

Solubility (Moles/L)	Solubility (Grams/L)	Temp (°C)	Ref (#)	Evaluation (T P E A A)	Comments
6.559E-01	1.143E+02	10	H062	1 2 2 0 0	EFG
8.588E-01	1.496E+02	20	B032	1 2 2 1 2	
7.487E-01	1.304E+02	21	D041	1 0 0 0 1	
8.037E-01	1.400E+02	21	F300	1 0 0 0 0	average
1.044E+00	1.818E+02	25	B032	1 2 2 1 2	
9.230E-01	1.608E+02	25	G315	0 0 0 0 0	
3.060E+00	5.330E+02	27	D036	0 0 0 0 0	
1.241E+00	2.162E+02	29.80	B032	1 2 2 1 2	
1.111E+00	1.935E+02	30	H062	1 2 2 0 0	EFG
1.771E+00	3.084E+02	50	H062	1 2 2 0 0	EFG

949. $C_6H_{14}O$
3-Methyl-3-pentanol
Diethylmethylcarbinol

RN:	77-74-7	MP (°C):	−24
MW:	102.18	BP (°C):	

Solubility (Moles/L)	Solubility (Grams/L)	Temp (°C)	Ref (#)	Evaluation (T P E A A)	Comments
4.286E-01	4.379E+01	9.8	S307	1 1 0 2 2	
3.346E-01	3.419E+01	19.5	S307	1 1 0 2 2	
4.500E-01	4.598E+01	20	G005	1 2 1 1 2	
3.999E-01	4.086E+01	25	G005	1 2 1 1 2	
3.264E-01	3.335E+01	29.8	S307	1 1 0 2 2	
3.592E-01	3.670E+01	30	G005	1 2 1 1 2	
2.647E-01	2.705E+01	39.8	S307	1 1 0 2 2	
2.331E-01	2.382E+01	49.7	S307	1 1 0 2 2	
1.938E-01	1.980E+01	59.5	S307	1 1 0 2 2	
1.834E-01	1.874E+01	70.1	S307	1 1 0 2 2	
1.787E-01	1.826E+01	80.1	S307	1 1 0 2 2	
1.617E-01	1.652E+01	90.4	S307	1 1 0 2 2	

950. C₆H₁₄O
Dipropyl ether
Propyl ether
Dipropylaether
Dipropylether
RN: 111-43-3   **MP** (°C):   –123
MW: 102.18   **BP** (°C):   89

Solubility (Moles/L)	Solubility (Grams/L)	Temp (°C)	Ref (#)	Evaluation (T P E A A)	Comments
5.644E-02	5.767E+00	0	B002	2 1 1 2 2	
3.996E-02	4.083E+00	10	B002	2 1 1 2 2	
3.705E-02	3.786E+00	15	B002	2 1 1 2 2	
2.927E-02	2.991E+00	20	B002	2 1 1 2 2	
2.936E-02	3.000E+00	20	F300	1 0 0 0 0	
6.700E-02	6.846E+00	20	S006	1 0 0 0 1	
2.441E-02	2.494E+00	25	B002	2 1 1 2 2	
1.070E-01	1.093E+01	37	E028	1 0 1 1 2	

951. C₆H₁₄O
tert-Amyl methyl ether
Methyl *tert*-amyl ether
RN: 994-05-8   **MP** (°C):
MW: 102.18   **BP** (°C):   85

Solubility (Moles/L)	Solubility (Grams/L)	Temp (°C)	Ref (#)	Evaluation (T P E A A)	Comments
1.208E-01	1.235E+01	20	E019	1 0 1 1 2	

952. C₆H₁₄O
Propyl isopropyl ether
Propyl-isopropyl-aether
RN: 627-08-7   **MP** (°C):   <25
MW: 102.18   **BP** (°C):   83

Solubility (Moles/L)	Solubility (Grams/L)	Temp (°C)	Ref (#)	Evaluation (T P E A A)	Comments
7.285E-02	7.444E+00	10	B002	2 1 1 2 2	
7.242E-02	7.400E+00	10	F300	1 0 0 0 1	
5.837E-02	5.964E+00	15	B002	2 1 1 2 2	
5.872E-02	6.000E+00	15	F300	1 0 0 0 1	
4.966E-02	5.074E+00	20	B002	2 1 1 2 2	
4.578E-02	4.678E+00	25	B002	2 1 1 2 2	
4.600E-02	4.700E+00	25	F300	1 0 0 0 1	

953. C$_6$H$_{14}$O
Isohexyl alcohol
4-Methyl-1-pentanol

RN:	626-89-1	**MP** (°C):	<25	
MW:	102.18	**BP** (°C):		

Solubility (Moles/L)	Solubility (Grams/L)	Temp (°C)	Ref (#)	Evaluation (T P E A A)	Comments
1.020E-01	1.042E+01	20	H330	0 0 0 0 0	

954. C$_6$H$_{14}$O
4-Methyl-2-pentanol
i-Butylmethylcarbinol
Methyl amyl alcohol

RN:	108-11-2	**MP** (°C):	–90	
MW:	102.18	**BP** (°C):	130	

Solubility (Moles/L)	Solubility (Grams/L)	Temp (°C)	Ref (#)	Evaluation (T P E A A)	Comments
2.684E-01	2.743E+01	0	S307	1 1 0 2 2	
2.004E-01	2.047E+01	9.7	S307	1 1 0 2 2	
1.664E-01	1.701E+01	20	D052	1 1 0 0 2	
1.721E-01	1.759E+01	20	G005	1 2 1 1 2	
1.570E-01	1.604E+01	20.0	S307	1 1 0 2 2	
1.636E-01	1.672E+01	25	C093	2 1 1 1 1	
1.579E-01	1.614E+01	25	G005	1 2 1 1 2	
1.465E-01	1.497E+01	30	G005	1 2 1 1 2	
1.475E-01	1.507E+01	30.0	S307	1 1 0 2 2	
1.246E-01	1.274E+01	40.3	S307	1 1 0 2 2	
1.151E-01	1.176E+01	50.0	S307	1 1 0 2 2	
1.074E-01	1.098E+01	60.1	S307	1 1 0 2 2	
1.094E-01	1.117E+01	70.2	S307	1 1 0 2 2	
1.199E-01	1.225E+01	80.2	S307	1 1 0 2 2	
1.132E-01	1.156E+01	90.2	S307	1 1 0 2 2	

955. C$_6$H$_{14}$O
2,2-Dimethyl-3-butanol
t-Butylmethylcarbinol

RN:	464-07-3	**MP** (°C):		
MW:	102.18	**BP** (°C):		

Solubility (Moles/L)	Solubility (Grams/L)	Temp (°C)	Ref (#)	Evaluation (T P E A A)	Comments
2.517E-01	2.572E+01	20	G005	1 2 1 1 2	
2.322E-01	2.372E+01	25	G005	1 2 1 1 2	
2.163E-01	2.210E+01	30	G005	1 2 1 1 2	

956. C$_6$H$_{14}$O
1-Hexanol
n-Hexanol
Amyl carbinol
Caproic alcohol
n-Hexyl alcohol

RN: 111-27-3   **MP** (°C):
MW: 102.18   **BP** (°C):

Solubility (Moles/L)	Solubility (Grams/L)	Temp (°C)	Ref (#)	Evaluation (T P E A A)	Comments
2.173E-01	2.220E+01	0	C423	0 0 0 0 0	
7.864E-02	8.035E+00	0	E029	1 2 0 1 1	
9.344E-02	9.548E+00	0	S307	1 1 0 2 2	
1.732E-01	1.770E+01	4	C423	0 0 0 0 0	
7.706E-02	7.873E+00	5.54	H110	2 2 2 2 2	
7.487E-02	7.650E+00	6.84	H110	2 2 2 2 2	
7.213E-02	7.370E+00	8.64	H110	2 2 2 2 2	
1.223E-01	1.250E+01	10	C423	0 0 0 0 0	
6.803E-02	6.951E+00	10	E029	1 2 0 1 1	
7.372E-02	7.533E+00	10.2	S307	1 1 0 2 2	
6.906E-02	7.057E+00	11.04	H110	2 2 2 2 2	
6.671E-02	6.816E+00	12.94	H110	2 2 2 2 2	
6.506E-02	6.648E+00	14.64	H110	2 2 2 2 2	
6.287E-02	6.424E+00	17.04	H110	2 2 2 2 2	
6.861E-02	7.011E+00	20	A015	1 2 1 1 2	
6.224E-02	6.359E+00	20	E029	1 2 0 1 1	
6.070E-02	6.202E+00	20	H330	0 0 0 0 0	
4.869E-02	4.975E+00	20	L049	1 1 2 1 0	
5.150E-02	5.262E+00	20	P073	1 0 0 1 2	
6.475E-02	6.616E+00	20.0	S307	1 1 0 2 2	
5.991E-02	6.121E+00	20.74	H110	2 2 2 2 2	
5.854E-02	5.981E+00	22.94	H110	2 2 2 2 2	
6.250E-02	6.386E+00	24	H345	0 0 0 0 0	
6.069E-02	6.201E+00	25	B038	1 2 1 1 2	
5.644E-02	5.767E+00	25	B060	2 0 1 1 1	
5.837E-02	5.964E+00	25	C093	2 1 1 1 1	
7.047E-02	7.200E+00	25	C435	0 0 0 0 0	
1.000E+00	1.022E+02	25	F044	1 0 0 0 0	EFG
8.000E-02	8.174E+00	25	G075	1 0 1 0 0	
5.900E-02	6.028E+00	25	K025	2 2 1 1 2	
8.922E-02	9.116E+00	25	M323	2 2 1 1 2	
5.711E-02	5.835E+00	25.04	H110	2 2 2 2 2	
5.640E-02	5.762E+00	26.94	H110	2 2 2 2 2	
5.579E-02	5.701E+00	28.94	H110	2 2 2 2 2	
5.431E-02	5.549E+00	29.7	S307	1 1 0 2 2	
6.320E-02	6.458E+00	30	C091	1 2 1 1 1	
5.740E-02	5.865E+00	30	E029	1 2 0 1 1	
5.517E-02	5.637E+00	30.94	H110	2 2 2 2 2	
5.440E-02	5.558E+00	33.04	H110	2 2 2 2 2	
5.005E-02	5.114E+00	39.8	S307	1 1 0 2 2	
5.257E-02	5.371E+00	40	E029	1 2 0 1 1	

(continued)

956. $C_6H_{14}O$ (continued)

Solubility (Moles/L)	Solubility (Grams/L)	Temp (°C)	Ref (#)	Evaluation (T P E A A)	Comments
4.869E-02	4.975E+00	50	E029	1 2 0 1 1	
4.840E-02	4.945E+00	50.0	S307	1 1 0 2 2	
5.063E-02	5.173E+00	60	E029	1 2 0 1 1	
5.043E-02	5.153E+00	60.0	S307	1 1 0 2 2	
5.450E-02	5.569E+00	70	E029	1 2 0 1 1	
5.540E-02	5.661E+00	70	F001	1 0 1 0 2	
5.615E-02	5.737E+00	70.3	S307	1 1 0 2 2	
5.934E-02	6.063E+00	80	E029	1 2 0 1 1	
6.080E-02	6.212E+00	80	F001	1 0 1 0 2	
6.079E-02	6.211E+00	80.3	S307	1 1 0 2 2	
6.707E-02	6.853E+00	90	E029	1 2 0 1 1	
6.660E-02	6.805E+00	90	F001	1 0 1 0 2	
6.204E-02	6.340E+00	90.3	S307	1 1 0 2 2	
7.767E-02	7.937E+00	100	E029	1 2 0 1 1	
7.690E-02	7.857E+00	100	F001	1 0 1 0 2	
8.826E-02	9.018E+00	110	E029	1 2 0 1 1	
8.720E-02	8.910E+00	110	F001	1 0 1 0 2	
1.007E-01	1.029E+01	120	E029	1 2 0 1 2	
1.151E-01	1.176E+01	130	E029	1 2 0 1 2	
1.323E-01	1.351E+01	140	E029	1 2 0 1 2	
1.570E-01	1.604E+01	150	E029	1 2 0 1 2	
1.966E-01	2.009E+01	160	E029	1 2 0 1 2	
2.573E-01	2.629E+01	170	E029	1 2 0 1 2	
3.410E-01	3.484E+01	180	E029	1 2 0 1 2	
4.545E-01	4.644E+01	190	E029	1 2 0 1 2	
6.188E-01	6.323E+01	200	E029	1 2 0 1 2	
8.654E-01	8.842E+01	210	E029	1 2 0 1 2	
1.372E+00	1.402E+02	220	E029	1 2 0 1 2	
6.114E-02	6.247E+00	ns	L003	0 0 2 1 2	

957. $C_6H_{14}O$

2-Hexanol

n-Butylmethylcarbinol

1-Methyl pentanol

RN:	626-93-7	**MP** (°C):	<25	
MW:	102.18	**BP** (°C):	136	

Solubility (Moles/L)	Solubility (Grams/L)	Temp (°C)	Ref (#)	Evaluation (T P E A A)	Comments
1.975E-01	2.018E+01	0	S307	1 1 0 2 2	
1.617E-01	1.652E+01	10.1	S307	1 1 0 2 2	
1.246E-01	1.274E+01	19.8	S307	1 1 0 2 2	
1.456E-01	1.488E+01	20	G005	1 2 1 1 2	
1.690E-01	1.727E+01	20	H330	0 0 0 0 0	
1.323E-01	1.351E+01	25	G005	1 2 1 1 2	
1.141E-01	1.166E+01	29.9	S307	1 1 0 2 2	
1.237E-01	1.264E+01	30	G005	1 2 1 1 2	
1.055E-01	1.078E+01	40.0	S307	1 1 0 2 2	
9.306E-02	9.509E+00	50.0	S307	1 1 0 2 2	

(*continued*)

957. C$_6$H$_{14}$O (continued)

Solubility (Moles/L)	Solubility (Grams/L)	Temp (°C)	Ref (#)	Evaluation (T P E A A)	Comments
8.826E-02	9.018E+00	60.2	S307	1 1 0 2 2	
9.498E-02	9.705E+00	70.0	S307	1 1 0 2 2	
1.094E-01	1.117E+01	80.1	S307	1 1 0 2 2	
9.114E-02	9.312E+00	90.2	S307	1 1 0 2 2	

958. C$_6$H$_{14}$O
2,2-Dimethyl-1-butanol
t-Pentylcarbinol

RN: 1185-33-7   **MP** (°C):   −35
MW: 102.18   **BP** (°C):   136

Solubility (Moles/L)	Solubility (Grams/L)	Temp (°C)	Ref (#)	Evaluation (T P E A A)	Comments
7.960E-02	8.133E+00	20	G005	1 2 1 1 1	
7.382E-02	7.543E+00	25	G005	1 2 1 1 1	
6.900E-02	7.050E+00	30	G005	1 2 1 1 1	

959. C$_6$H$_{14}$O
2,3-Dimethyl-1-butanol
Dimethyl-*i*-propylcarbinol
Dimethyl-isopropylcarbinol

RN: 594-60-5   **MP** (°C):   −14
MW: 102.18   **BP** (°C):

Solubility (Moles/L)	Solubility (Grams/L)	Temp (°C)	Ref (#)	Evaluation (T P E A A)	Comments
4.349E-01	4.443E+01	20	G005	1 2 1 1 2	
3.927E-01	4.012E+01	25	G005	1 2 1 1 2	
3.547E-01	3.624E+01	30	G005	1 2 1 1 2	

960. C$_6$H$_{14}$O
Isopropyl ether
Diisopropyl ether

RN: 108-20-3   **MP** (°C):   −60
MW: 102.18   **BP** (°C):   68.5

Solubility (Moles/L)	Solubility (Grams/L)	Temp (°C)	Ref (#)	Evaluation (T P E A A)	Comments
1.351E-01	1.381E+01	24.6	H121	2 0 0 0 1	
8.730E-02	8.920E+00	25	F048	2 0 0 0 0	
7.920E-02	8.092E+00	37	E028	1 0 1 1 2	

961. C$_6$H$_{14}$O
2-Ethyl-1-butanol
2-Ethylbutanol

RN:	97-95-0	MP (°C):	−15
MW:	102.18	BP (°C):	146

Solubility (Moles/L)	Solubility (Grams/L)	Temp (°C)	Ref (#)	Evaluation (T P E A A)	Comments
6.127E-02	6.261E+00	20	D052	1 1 0 0 1	
3.899E-02	3.984E+00	25	C093	2 1 1 1 0	

962. C$_6$H$_{14}$O
3-Methyl-2-pentanol
3-Methyl-2-pentyl alcohol

RN:	565-60-6	MP (°C):	<25
MW:	102.18	BP (°C):	

Solubility (Moles/L)	Solubility (Grams/L)	Temp (°C)	Ref (#)	Evaluation (T P E A A)	Comments
2.004E-01	2.047E+01	20	G005	1 2 1 1 2	
1.863E-01	1.903E+01	25	G005	1 2 1 1 2	
1.721E-01	1.759E+01	30	G005	1 2 1 1 2	

963. C$_6$H$_{14}$O
2-Ethyl-4-butanol
3-Methylpentanol

RN:	105-30-6	MP (°C):	<25
MW:	102.18	BP (°C):	148

Solubility (Moles/L)	Solubility (Grams/L)	Temp (°C)	Ref (#)	Evaluation (T P E A A)	Comments
1.257E-01	1.284E+01	0	S307	1 1 0 2 2	
1.004E-01	1.025E+01	10.0	S307	1 1 0 2 2	
8.518E-02	8.704E+00	19.6	S307	1 1 0 2 2	
5.837E-02	5.964E+00	25	C093	2 1 1 1 1	
7.681E-02	7.848E+00	30.8	S307	1 1 0 2 2	
7.498E-02	7.661E+00	40.3	S307	1 1 0 2 2	
7.295E-02	7.454E+00	50.0	S307	1 1 0 2 2	
7.363E-02	7.523E+00	60.3	S307	1 1 0 2 2	
7.478E-02	7.641E+00	70.1	S307	1 1 0 2 2	
8.133E-02	8.310E+00	80.3	S307	1 1 0 2 2	
8.931E-02	9.126E+00	90.7	S307	1 1 0 2 2	

964. C$_6$H$_{14}$O

2-Methyl-2-pentanol
Dimethyl-*n*-propylcarbinol
1,1-Dimethyl-1-butanol

RN:	590-36-3	**MP** (°C):	−107
MW:	102.18	**BP** (°C):	122

Solubility (Moles/L)	Solubility (Grams/L)	Temp (°C)	Ref (#)	Evaluation (T P E A A)	Comments
3.428E-01	3.503E+01	20	G005	1 2 1 1 2	
3.640E-01	3.719E+01	20	H330	0 0 0 0 0	
3.071E-01	3.138E+01	25	G005	1 2 1 1 2	
2.814E-01	2.875E+01	30	G005	1 2 1 1 2	

965. C$_6$H$_{14}$O

2-Methyl-3-pentanol
i-Propylethylcarbinol

RN:	565-67-3	**MP** (°C):	<25
MW:	102.18	**BP** (°C):	

Solubility (Moles/L)	Solubility (Grams/L)	Temp (°C)	Ref (#)	Evaluation (T P E A A)	Comments
2.144E-01	2.191E+01	20	G005	1 2 0 0 2	
1.928E-01	1.970E+01	25	G005	1 2 1 1 2	
1.749E-01	1.787E+01	30	G005	1 2 1 1 2	

966. C$_6$H$_{14}$O

3-Hexanol
n-Propylethylcarbinol
tert-Hexyl alcohol

RN:	623-37-0	**MP** (°C):	<25
MW:	102.18	**BP** (°C):	134.5

Solubility (Moles/L)	Solubility (Grams/L)	Temp (°C)	Ref (#)	Evaluation (T P E A A)	Comments
2.619E-01	2.676E+01	0	S307	1 1 0 2 2	
1.881E-01	1.922E+01	10.1	S307	1 1 0 2 2	
3.062E-01	3.129E+01	20	A015	1 2 1 1 2	
1.683E-01	1.720E+01	20	G005	1 2 1 1 2	
1.608E-01	1.643E+01	20.0	S307	1 1 0 2 2	
1.551E-01	1.584E+01	25	G005	1 2 1 1 2	
1.437E-01	1.468E+01	30	G005	1 2 1 1 2	
1.342E-01	1.371E+01	30.0	S307	1 1 0 2 2	
1.189E-01	1.215E+01	39.8	S307	1 1 0 2 2	
1.065E-01	1.088E+01	50.0	S307	1 1 0 2 2	
9.882E-02	1.010E+01	60.1	S307	1 1 0 2 2	
9.882E-02	1.010E+01	70.2	S307	1 1 0 2 2	
1.036E-01	1.059E+01	80.2	S307	1 1 0 2 2	
1.065E-01	1.088E+01	90.3	S307	1 1 0 2 2	

967. C$_6$H$_{14}$O
3-Methyl-1-pentanol
3-Methylpentanol
2-Ethyl-4-butanol

RN: 589-35-5   **MP** (°C):
MW: 102.18   **BP** (°C):   151

Solubility (Moles/L)	Solubility (Grams/L)	Temp (°C)	Ref (#)	Evaluation (T P E A A)	Comments
4.190E-02	4.282E+00	25	B060	2 0 1 1 1	

968. C$_6$H$_{14}$O$_2$
Acetal
Acetaldehyd-diaethylacetal
Acetaldehyde diethyl acetal

RN: ' 105-57-7   **MP** (°C):
MW: 118.18   **BP** (°C):   102.7

Solubility (Moles/L)	Solubility (Grams/L)	Temp (°C)	Ref (#)	Evaluation (T P E A A)	Comments
3.723E-01	4.400E+01	25	F300	1 0 0 0 1	

969. C$_6$H$_{14}$O$_2$
Diethyl cellosolve
Ethylene glycol diethyl ether
1,2-Diethoxyethane
3,6-Dioxaoctane
Ethyl glyme
Diethoxyethane

RN: 629-14-1   **MP** (°C):
MW: 118.18   **BP** (°C):   119

Solubility (Moles/L)	Solubility (Grams/L)	Temp (°C)	Ref (#)	Evaluation (T P E A A)	Comments
2.273E-01	2.686E+01	20	D052	1 1 0 0 2	
1.469E+00	1.736E+02	20	M062	1 0 0 0 2	

970. C$_6$H$_{14}$O$_3$
Carbitol
2-(2-Ethoxyethoxy)ethanol

RN: 111-90-0   **MP** (°C):
MW: 134.18   **BP** (°C):   196.0

Solubility (Moles/L)	Solubility (Grams/L)	Temp (°C)	Ref (#)	Evaluation (T P E A A)	Comments
3.610E+00	4.843E+02	4.50	C022	1 2 0 0 2	

971. C₆H₁₄O₆

971. $C_6H_{14}O_6$

D-Mannitol
1,2,3,4,5,6-Hexanehexol
Cordycepic acid
Diosmol
D-Mannite
Manna sugar

RN: 69-65-8	**MP** (°C):	167–170		
MW: 182.17	**BP** (°C):	295		

Solubility (Moles/L)	Solubility (Grams/L)	Temp (°C)	Ref (#)	Evaluation (T P E A A)	Comments
1.148E+00	2.092E+02	ns	R427	0 0 0 0 0	

972. $C_6H_{14}O_6$

Galactitol
Dulcit
Dulcitol

RN: 608-66-2	**MP** (°C):	189.5		
MW: 182.17	**BP** (°C):	277.5		

Solubility (Moles/L)	Solubility (Grams/L)	Temp (°C)	Ref (#)	Evaluation (T P E A A)	Comments
1.599E-01	2.913E+01	14	D041	1 0 0 0 1	
1.702E-01	3.100E+01	15	F300	1 0 0 0 1	
2.086E+00	3.800E+02	100	F300	1 0 0 0 1	

973. $C_6H_{14}O_6$

Sorbitol
D-Sorbitol

RN: 50-70-4	**MP** (°C):	110		
MW: 182.17	**BP** (°C):			

Solubility (Moles/L)	Solubility (Grams/L)	Temp (°C)	Ref (#)	Evaluation (T P E A A)	Comments
3.522E+00	6.416E+02	10	M043	1 0 0 0 2	
3.785E+00	6.894E+02	20	M043	1 0 0 0 2	
4.025E+00	7.333E+02	30	M043	1 0 0 0 2	
4.283E+00	7.802E+02	40	M043	1 0 0 0 2	

974. $C_6H_{14}O_6$

Mannitol
D-Mannit
D-Mannitol

RN: 87-78-5	**MP** (°C):	167		
MW: 182.17	**BP** (°C):	292		

Solubility (Moles/L)	Solubility (Grams/L)	Temp (°C)	Ref (#)	Evaluation (T P E A A)	Comments
5.081E-01	9.256E+01	0	C073	1 2 2 1 2	
5.171E-01	9.420E+01	0	M043	1 0 0 0 2	

(continued)

974. $C_6H_{14}O_6$ (continued)

Solubility (Moles/L)	Solubility (Grams/L)	Temp (°C)	Ref (#)	Evaluation (T P E A A)	Comments
6.614E-01	1.205E+02	10	M043	1 0 0 0 2	
7.734E-01	1.409E+02	15	C073	1 2 2 1 2	
7.740E-01	1.410E+02	15	F300	1 0 0 0 2	
7.408E-01	1.349E+02	18	D041	1 0 0 0 2	
7.936E-01	1.446E+02	19	N051	1 0 2 2 2	
8.609E-01	1.568E+02	20	M043	1 0 0 0 2	
7.571E-01	1.379E+02	21.6	Y412	0 0 0 0 0	
9.762E-01	1.778E+02	25	B106	1 2 2 2 2	
9.732E-01	1.773E+02	25	B106	1 2 2 2 2	
9.739E-01	1.774E+02	25	B106	1 2 2 2 2	
9.639E-01	1.756E+02	25	C073	1 2 2 1 2	
8.255E-01	1.504E+02	25	H087	1 0 2 1 2	
8.373E-01	1.525E+02	26.8	Y412	0 0 0 0 0	
1.000E+00	1.822E+02	30	D011	1 0 1 0 1	
1.105E+00	2.013E+02	30	M043	1 0 0 0 2	
9.149E-01	1.667E+02	30.8	Y412	0 0 0 0 0	
1.254E+00	2.284E+02	35	C073	1 2 2 1 2	
9.899E-01	1.803E+02	35.6	Y412	0 0 0 0 0	
1.062E+00	1.935E+02	38.1	Y412	0 0 0 0 0	
1.411E+00	2.571E+02	40	M043	1 0 0 0 2	
1.133E+00	2.063E+02	41.8	Y412	0 0 0 0 0	
1.760E+00	3.207E+02	50	C073	1 2 2 1 2	
1.827E+00	3.329E+02	51.50	B106	1 2 2 2 2	
2.083E+00	3.794E+02	60	C073	1 2 2 1 2	
2.104E+00	3.833E+02	60	F300	1 0 0 0 2	
2.150E+00	3.917E+02	60	M043	1 0 0 0 2	
2.416E+00	4.401E+02	67.40	B106	1 2 2 2 2	
2.504E+00	4.562E+02	70.50	B106	1 2 2 2 2	
2.936E+00	5.349E+02	80	M043	1 0 0 0 2	
3.015E+00	5.493E+02	82.90	B106	1 2 2 2 2	
3.253E+00	5.927E+02	88.10	B106	1 2 2 2 2	
3.299E+00	6.010E+02	90.10	B106	1 2 2 2 2	
3.590E+00	6.540E+02	98	B106	1 2 2 2 2	
3.628E+00	6.610E+02	99.30	B106	1 2 2 2 2	
3.641E+00	6.633E+02	100	M043	1 0 0 0 2	
8.757E-01	1.595E+02	rt	D021	0 0 1 1 2	

975. $C_6H_{15}N$
Triethylamine
Triaethylamin

RN: 121-44-8   **MP** (°C): −115
MW: 101.19   **BP** (°C): 89

Solubility (Moles/L)	Solubility (Grams/L)	Temp (°C)	Ref (#)	Evaluation (T P E A A)	Comments
1.778E+00	1.799E+02	17.48	K142	1 0 0 0 2	
2.754E+00	2.787E+02	17.59	K142	1 0 0 0 2	
2.754E+00	2.787E+02	17.64	K142	1 0 0 0 2	

(continued)

975. C$_6$H$_{15}$N (continued)

Solubility (Moles/L)	Solubility (Grams/L)	Temp (°C)	Ref (#)	Evaluation (T P E A A)	Comments
1.156E+00	1.170E+02	17.82	K142	1 0 0 0 2	
1.156E+00	1.170E+02	17.85	K142	1 0 0 0 2	
2.791E+00	2.824E+02	18	C088	2 2 2 2 1	
3.434E+00	3.475E+02	18.11	K142	1 0 0 0 2	
3.434E+00	3.475E+02	18.12	K142	1 0 0 0 2	
4.014E+00	4.062E+02	19.12	K142	1 0 0 0 2	
4.014E+00	4.062E+02	19.13	K142	1 0 0 0 2	
8.951E-01	9.058E+01	19.38	K142	1 0 0 0 2	
8.951E-01	9.058E+01	19.43	K142	1 0 0 0 2	
1.403E+00	1.420E+02	20	F300	1 0 0 0 2	
6.780E-01	6.861E+01	25.04	V013	2 2 2 2 2	
1.976E-01	2.000E+01	65	F300	1 0 0 0 1	

976. C$_6$H$_{15}$N

N-Ethyl-sec-butylamine
sec-Butylethylamine
2-Butanamine, N-ethyl-
2-(Ethylamino)butane

RN: 21035-44-9   **MP** (°C):
MW: 101.19   **BP** (°C):

Solubility (Moles/L)	Solubility (Grams/L)	Temp (°C)	Ref (#)	Evaluation (T P E A A)	Comments
8.155E-01	8.253E+01	25	D332	0 0 0 0 0	
6.099E-01	6.172E+01	30	D332	0 0 0 0 0	
4.202E-01	4.252E+01	40	D332	0 0 0 0 0	

977. C$_6$H$_{15}$N

N-Ethyl-n-butylamine
Ethylbutylamine
N-Ethylbutan-1-amine
N-Ethylbutylamine

RN: 13360-63-9   **MP** (°C):   −78
MW: 101.19   **BP** (°C):   108

Solubility (Moles/L)	Solubility (Grams/L)	Temp (°C)	Ref (#)	Evaluation (T P E A A)	Comments
1.003E+00	1.015E+02	10	D332	0 0 0 0 0	
5.310E-01	5.373E+01	20	D332	0 0 0 0 0	
3.793E-01	3.838E+01	30	D332	0 0 0 0 0	
2.859E-01	2.893E+01	40	D332	0 0 0 0 0	

978. C$_6$H$_{15}$N
n-Dipropylamine
Dipropylamine

RN:	142-84-7	MP (°C):	-63
MW:	101.19	BP (°C):	110

Solubility (Moles/L)	Solubility (Grams/L)	Temp (°C)	Ref (#)	Evaluation (T P E A A)	Comments
5.470E-01	5.536E+01	12.2	H038	1 2 1 1 2	
2.794E-01	2.828E+01	36.1	H038	1 2 1 1 2	
2.335E-01	2.363E+01	44.1	H038	1 2 1 1 2	
1.900E-01	1.922E+01	52.6	H038	1 2 1 1 2	

979. C$_6$H$_{15}$O$_2$PS$_3$
Thiometon
O,O-Dimethyl *S*-(2-ethylmercaptoethyl) dithiophosphate

RN:	640-15-3	MP (°C):	
MW:	246.35	BP (°C):	104

Solubility (Moles/L)	Solubility (Grams/L)	Temp (°C)	Ref (#)	Evaluation (T P E A A)	Comments
8.118E-04	2.000E-01	20	M061	1 0 0 0 2	
8.118E-04	2.000E-01	25	M161	1 0 0 0 2	

980. C$_6$H$_{15}$O$_3$PS$_2$
Thiolo-methylmercaptophos
Thiolo-methyl demeton

RN:		MP (°C):	
MW:	230.29	BP (°C):	89

Solubility (Moles/L)	Solubility (Grams/L)	Temp (°C)	Ref (#)	Evaluation (T P E A A)	Comments
1.433E-02	3.300E+00	20	M061	1 0 0 0 2	

981. C$_6$H$_{15}$O$_3$PS$_2$
Thiono-methylmercaptophos
Thiono-methyl demeton

RN:		MP (°C):	
MW:	230.29	BP (°C):	74

Solubility (Moles/L)	Solubility (Grams/L)	Temp (°C)	Ref (#)	Evaluation (T P E A A)	Comments
1.433E-03	3.300E-01	20	M061	1 0 0 0 2	

982. C$_6$H$_{15}$O$_4$P
Triethyl phosphate
Ethyl phosphate
Phosphoric acid, triethyl ester
TEP

RN:	78-40-0	**MP** (°C):	−56.4	
MW:	182.16	**BP** (°C):	215	

Solubility (Moles/L)	Solubility (Grams/L)	Temp (°C)	Ref (#)	Evaluation (T P E A A)	Comments
2.815E+00	5.128E+02	4.50	C022	1 2 0 0 2	
2.745E+00	5.000E+02	25	F300	1 0 0 0 1	
+2.69E+00	+4.90E+02	ns	S460	0 0 0 0 0	

983. C$_6$H$_{16}$FN$_2$OP
Mipafox
N,N'-Diisopropylphosphorodiamidic fluoride

RN:	371-86-8	**MP** (°C):	65	
MW:	182.18	**BP** (°C):		

Solubility (Moles/L)	Solubility (Grams/L)	Temp (°C)	Ref (#)	Evaluation (T P E A A)	Comments
4.066E-01	7.407E+01	ns	M061	0 0 0 0 0	

984. C$_6$H$_{16}$N$_2$
1,6-Hexanediamine
Hexamethylenediamine

RN:	124-09-4	**MP** (°C):	42	
MW:	116.21	**BP** (°C):	205	

Solubility (Moles/L)	Solubility (Grams/L)	Temp (°C)	Ref (#)	Evaluation (T P E A A)	Comments
6.123E+00	7.115E+02	4.50	C022	1 2 0 0 2	

985. C$_6$H$_{17}$N$_3$O$_{10}$S
Glycine sulfate
Triglycine sulfate

RN:	513-29-1	**MP** (°C):		
MW:	323.28	**BP** (°C):		

Solubility (Moles/L)	Solubility (Grams/L)	Temp (°C)	Ref (#)	Evaluation (T P E A A)	Comments
3.314E-01	1.071E+02	0	M043	1 0 0 0 1	
5.155E-01	1.667E+02	10	M043	1 0 0 0 1	
6.576E-01	2.126E+02	20	M043	1 0 0 0 1	
8.188E-01	2.647E+02	30	M043	1 0 0 0 1	
9.600E-01	3.103E+02	40	M043	1 0 0 0 1	
1.326E+00	4.286E+02	60	M043	1 0 0 0 1	

986. $C_6H_{18}N_4$

Triethylenetetramine
N,N'-bis(2-Aminoethyl)-ethylenediamine
1,8-Diamino-3,6-diazaoctane
1,4,7,10-Tetraazadecane
3,6-Diazaoctane-1,8-diamine
Trientine

RN:	112-24-3	**MP (°C):**	12	
MW:	146.24	**BP (°C):**	266	

Solubility (Moles/L)	Solubility (Grams/L)	Temp (°C)	Ref (#)	Evaluation (T P E A A)	Comments
5.655E+00	8.269E+02	4.50	C022	1 2 0 0 2	

987. C_6Br_6

Hexabromobenzene

RN:	87-82-1	**MP (°C):**	327
MW:	551.52	**BP (°C):**	

Solubility (Moles/L)	Solubility (Grams/L)	Temp (°C)	Ref (#)	Evaluation (T P E A A)	Comments
8.558E-11	4.720E-08	10	K440	0 0 0 0 0	
1.994E-10	1.100E-07	25	K440	0 0 0 0 0	
4.207E-10	2.320E-07	35	K440	0 0 0 0 0	

988. $C_6Cl_4O_2$

Chloranil
Tetrachloro-p-benzoquinone
2,3,5,6-Tetrachloro-p-benzoquinone
2,3,5,6-Tetrachloro-2,5-cyclohexadiene-1,4-dione
Vulklor
Coversan

RN:	118-75-2	**MP (°C):**	290
MW:	245.88	**BP (°C):**	

Solubility (Moles/L)	Solubility (Grams/L)	Temp (°C)	Ref (#)	Evaluation (T P E A A)	Comments
1.017E-03	2.500E-01	rt	M161	0 0 0 0 2	

989. C₆Cl₅NO₂

Quintozene
Pentachloronitrobenzene
Avical
Eorthcicle
Quintobenzene

RN: 82-68-8   **MP** (°C):   >139
MW: 295.34   **BP** (°C):   328

Solubility (Moles/L)	Solubility (Grams/L)	Temp (°C)	Ref (#)	Evaluation (T P E A A)	Comments
1.500E-06	4.430E-04	20	E308	1 2 2 1 1	
1.862E-06	5.500E-04	22	K137	1 1 2 1 0	
1.490E-06	4.400E-04	22.5	G301	0 0 0 0 0	

990. C₆Cl₆

Hexachlorobenzene
Benzene hexachloride
HCB
Hexa-chlorobenzene

RN: 118-74-1   **MP** (°C):   228
MW: 284.78   **BP** (°C):   324.5

Solubility (Moles/L)	Solubility (Grams/L)	Temp (°C)	Ref (#)	Evaluation (T P E A A)	Comments
1.259E-07	3.585E-05	20	B179	0 0 0 0 0	
1.721E-08	4.900E-06	20	C113	1 0 1 1 1	
2.598E-08	7.400E-06	20	H300	1 1 2 2 1	
1.896E-08	5.400E-06	20	H300	1 1 2 2 1	
2.042E-08	5.815E-06	20	K337	1 0 0 0 2	
1.380E-08	3.931E-06	22	K305	1 0 1 1 2	
1.756E-08	5.000E-06	22.5	G301	0 0 0 0 0	
1.700E-08	4.841E-06	25	B317	0 0 0 0 0	
1.650E-08	4.699E-06	25	M342	1 0 1 1 2	
2.107E-08	6.000E-06	26.70	L095	2 2 1 1 2	
<3.51E-06	<1.00E-03	30	M311	1 1 2 2 0	
7.023E-08	2.000E-05	ns	L072	0 0 0 0 1	
2.107E-08	6.000E-06	ns	L311	0 0 0 0 1	
1.650E-07	4.699E-05	ns	M308	0 0 1 1 2	
2.458E-05	7.000E-03	rt	H053	0 2 2 2 0	γ isomer

991. C₆F₆

Hexafluorobenzene
Perfluorobenzene

RN:	392-56-3	**MP** (°C):	3.9 C
MW:	186.06	**BP** (°C):	81 C at 743 mm Hg

Solubility (Moles/L)	Solubility (Grams/L)	Temp (°C)	Ref (#)	Evaluation (T P E A A)	Comments
4.186E-03	7.788E-01	8.30	F418	0 0 0 0 0	
3.598E-03	6.694E-01	18.20	F418	0 0 0 0 0	
3.315E-03	6.167E-01	27.81	F418	0 0 0 0 0	
3.198E-03	5.950E-01	37.66	F418	0 0 0 0 0	
3.148E-03	5.857E-01	47.35	F418	0 0 0 0 0	
3.209E-03	5.971E-01	56.61	F418	0 0 0 0 0	
3.420E-03	6.363E-01	66.60	F418	0 0 0 0 0	

992. C₇H₃Br₂NO

Bromoxynil
3,5-Dibromo-4-hydroxybenzonitrile
4-Cyano-2,6-dibromophenol

RN:	1689-84-5	**MP** (°C):	190
MW:	276.93	**BP** (°C):	

Solubility (Moles/L)	Solubility (Grams/L)	Temp (°C)	Ref (#)	Evaluation (T P E A A)	Comments
4.694E-04	1.300E-01	25	M161	1 0 0 0 2	
4.694E-04	1.300E-01	ns	M061	0 0 0 0 2	

993. C₇H₃Br₃O₂

2,4,6-Tribromobenzoic acid
2,4,6-Tribrom-benzoesaeure

RN:	633-12-5	**MP** (°C):	
MW:	358.83	**BP** (°C):	

Solubility (Moles/L)	Solubility (Grams/L)	Temp (°C)	Ref (#)	Evaluation (T P E A A)	Comments
9.754E-03	3.500E+00	15	F300	1 0 0 0 1	
1.533E-02	5.500E+00	100	F300	1 0 0 0 1	

994. C$_7$H$_3$Cl$_2$N
Dichlobenil
2,6-Dichlorobenzonitrile
Benzonitrile, 2,6-dichloro-
RN:	1194-65-6	**MP** (°C):	145
MW:	172.01	**BP** (°C):	270

Solubility (Moles/L)	Solubility (Grams/L)	Temp (°C)	Ref (#)	Evaluation (T P E A A)	Comments
1.046E-04	1.800E-02	20	B185	0 0 0 0 0	
1.046E-04	1.800E-02	20	B200	1 0 0 1 1	
1.046E-04	1.800E-02	20	G319	0 0 0 0 0	
1.046E-04	1.800E-02	20	M161	1 0 0 0 1	
1.163E-04	2.000E-02	25	B185	0 0 0 0 0	
5.813E-05	1.000E-02	25	M061	1 0 0 0 1	
1.046E-04	1.800E-02	ns	V303	0 0 0 0 1	

995. C$_7$H$_3$Cl$_3$O$_2$
2,3,6-Trichlorobenzoic acid
2,3,6-TBA
RN:	50-31-7	**MP** (°C):	125
MW:	225.46	**BP** (°C):	

Solubility (Moles/L)	Solubility (Grams/L)	Temp (°C)	Ref (#)	Evaluation (T P E A A)	Comments
3.726E-02	8.400E+00	20	B200	1 0 0 0 1	
3.415E-02	7.700E+00	22	M161	1 0 0 0 1	

996. C$_7$H$_3$Cl$_5$O
Pentachlorbenzyl alcohol
Blastin
PCBA
RN:	16022-69-8	**MP** (°C):	
MW:	280.37	**BP** (°C):	

Solubility (Moles/L)	Solubility (Grams/L)	Temp (°C)	Ref (#)	Evaluation (T P E A A)	Comments
7.134E-07	2.000E-04	25	M061	0 0 0 0 0	

997. C$_7$H$_3$I$_2$NO
Ioxynil
4-Cyano-2,6-diiodophenol
4-Hydroxy-3,5-diiodobenzonitrile
RN:	1689-83-4	**MP** (°C):	212
MW:	370.92	**BP** (°C):	

Solubility (Moles/L)	Solubility (Grams/L)	Temp (°C)	Ref (#)	Evaluation (T P E A A)	Comments
1.348E-04	5.000E-02	20	F311	1 2 2 2 1	
3.505E-04	1.300E-01	25	B200	1 0 0 0 2	
1.348E-04	5.000E-02	25	M161	1 0 0 0 1	

998. C$_7$H$_3$N$_3$O$_8$

2,4,6-Trinitrobenzoic acid
2,4,6-Trinitrobenzoesaeure
Acide 2,4,6-trinitrobenzoique

RN:	129-66-8	MP (°C):	228.7
MW:	257.12	BP (°C):	

Solubility (Moles/L)	Solubility (Grams/L)	Temp (°C)	Ref (#)	Evaluation (T P E A A)	Comments
7.817E-02	2.010E+01	23	F300	1 0 0 0 2	
7.824E-02	2.012E+01	23.5	D067	1 2 0 0 2	
1.560E-01	4.012E+01	50	D067	1 2 0 0 2	
1.560E-01	4.010E+01	50	F300	1 0 0 0 2	

999. C$_7$H$_4$BrN

4-Bromobenzonitrile
p-Bromobenzonitrile
4-Bromobenzoic acid nitrile

RN:	623-00-7	MP (°C):	111 C
MW:	182.03	BP (°C):	236 C

Solubility (Moles/L)	Solubility (Grams/L)	Temp (°C)	Ref (#)	Evaluation (T P E A A)	Comments
8.635E-04	1.572E-01	22	J420	0 0 0 0 0	pH 6.5

1000. C$_7$H$_4$BrNO$_4$

3-Bromo-2-nitrobenzoic acid
Benzoic acid, 3-bromo-2-nitro-

RN:	116529-61-4	MP (°C):	
MW:	246.02	BP (°C):	

Solubility (Moles/L)	Solubility (Grams/L)	Temp (°C)	Ref (#)	Evaluation (T P E A A)	Comments
3.012E-02	7.410E+00	25	H089	1 2 0 0 2	
1.341E-03	3.300E-01	25	H089	1 2 0 0 1	

1001. C$_7$H$_4$BrNS

4-Bromophenyl isothiocyanate
1-Bromo-4-isothiocyanato-benzene

RN:	1985-12-2	MP (°C):	60.5
MW:	214.09	BP (°C):	

Solubility (Moles/L)	Solubility (Grams/L)	Temp (°C)	Ref (#)	Evaluation (T P E A A)	Comments
5.400E-05	1.156E-02	25	D019	1 1 1 1 1	

1002. C₇H₄BrNS

3-Bromophenyl isothiocyanate
1-Bromo-3-isothiocyanato-benzene

RN: 2131-59-1   **MP** (°C):
MW: 214.09   **BP** (°C):   256.0

Solubility (Moles/L)	Solubility (Grams/L)	Temp (°C)	Ref (#)	Evaluation (T P E A A)	Comments
1.140E-04	2.441E-02	25	D019	1 1 1 1 2	
8.200E-05	1.756E-02	25	K032	2 2 0 1 1	

1003. C₇H₄ClNO₄

3-Chloro-2-nitrobenzoic acid
2-Nitro-3-chlorobenzoic acid

RN: 4771-47-5   **MP** (°C):
MW: 201.57   **BP** (°C):

Solubility (Moles/L)	Solubility (Grams/L)	Temp (°C)	Ref (#)	Evaluation (T P E A A)	Comments
2.332E-03	4.700E-01	25	H089	1 2 0 0 1	

1004. C₇H₄ClNO₄

4-Chloro-3-nitrobenzoic acid
3-Nitro-4-chlorobenzoic acid

RN: 96-99-1   **MP** (°C):   181
MW: 201.57   **BP** (°C):

Solubility (Moles/L)	Solubility (Grams/L)	Temp (°C)	Ref (#)	Evaluation (T P E A A)	Comments
1.700E-03	3.427E-01	ns	C014	0 0 0 1 1	

1005. C₇H₄ClNO₄

5-Chloro-2-nitrobenzoic acid
2-Nitro-5-chlorobenzoic acid

RN: 2516-95-2   **MP** (°C):
MW: 201.57   **BP** (°C):

Solubility (Moles/L)	Solubility (Grams/L)	Temp (°C)	Ref (#)	Evaluation (T P E A A)	Comments
4.797E-02	9.670E+00	25	H089	1 2 0 0 2	

1006. C₇H₄ClNS
3-Chlorophenyl isothiocyanate
1-Chloro-3-isothiocyanato-benzene

RN:	2392-68-9	**MP** (°C):		
MW:	169.63	**BP** (°C):	249.5	

Solubility (Moles/L)	Solubility (Grams/L)	Temp (°C)	Ref (#)	Evaluation (T P E A A)	Comments
2.000E-04	3.393E-02	25	D019	1 1 1 1 0	
1.120E-04	1.900E-02	25	K032	2 2 0 1 2	

1007. C₇H₄Cl₂O₂
3,5-Dichlorobenzoic acid
Benzoic acid, 3,5-dichloro-

RN:	51-36-5	**MP** (°C):	186	
MW:	191.01	**BP** (°C):		

Solubility (Moles/L)	Solubility (Grams/L)	Temp (°C)	Ref (#)	Evaluation (T P E A A)	Comments
7.700E-04	1.471E-01	ns	C014	0 0 0 1 1	

1008. C₇H₄Cl₂O₂
2,6-Dichlorobenzoic acid
2,6-Dichlor-benzoesaeure

RN:	50-30-6	**MP** (°C):		
MW:	191.01	**BP** (°C):		

Solubility (Moles/L)	Solubility (Grams/L)	Temp (°C)	Ref (#)	Evaluation (T P E A A)	Comments
7.400E-02	1.414E+01	ns	C014	0 0 0 1 1	

1009. C₇H₄Cl₂O₂
2,4-Dichlorobenzoic acid
2,4-Dichlor-benzoesaeure

RN:	50-84-0	**MP** (°C):		
MW:	191.01	**BP** (°C):		

Solubility (Moles/L)	Solubility (Grams/L)	Temp (°C)	Ref (#)	Evaluation (T P E A A)	Comments
2.500E-03	4.775E-01	ns	C014	0 2 0 1 1	

1010. C₇H₄Cl₂O₂
3,4-Dichlorobenzoic acid
Benzoic acid, 3,4-dichloro-

RN:	51-44-5	**MP** (°C):	208	
MW:	191.01	**BP** (°C):		

Solubility (Moles/L)	Solubility (Grams/L)	Temp (°C)	Ref (#)	Evaluation (T P E A A)	Comments
3.200E-04	6.112E-02	ns	C014	0 0 0 1 1	

1011. C₇H₄Cl₃NO₃

Triclopyr
Garlon
(3,5,6-Trichloro-2-pyridinyl)oxyacetic acid
Crossbow turflon

RN:	55335-06-3	**MP** (°C):	149
MW:	256.47	**BP** (°C):	290

Solubility (Moles/L)	Solubility (Grams/L)	Temp (°C)	Ref (#)	Evaluation (T P E A A)	Comments
1.677E-03	4.300E-01	ns	K138	0 0 0 0 1	

1012. C₇H₄Cl₄O

2,4,5,6-Tetrachloro-3-methyl-phenol
m-Cresol, 2,4,5,6-tetrachloro-
Phenol, 2,3,4,6-tetrachloro-5-methyl-

RN:	10460-33-0	**MP** (°C):	
MW:	245.92	**BP** (°C):	

Solubility (Moles/L)	Solubility (Grams/L)	Temp (°C)	Ref (#)	Evaluation (T P E A A)	Comments
2.500E-05	6.148E-03	25	B316	0 0 0 0 0	

1013. C₇H₄Cl₄O

2,3,4,5-Tetrachloroanisole
Benzene, 1,2,3,4-tetrachloro-5-methoxy-
Anisole, 2,3,4,5-tetrachloro-

RN:	938-86-3	**MP** (°C):	88
MW:	245.92	**BP** (°C):	

Solubility (Moles/L)	Solubility (Grams/L)	Temp (°C)	Ref (#)	Evaluation (T P E A A)	Comments
5.490E-06	1.350E-03	25	L348	1 2 2 1 2	

1014. C₇H₄INS

4-Iodophenyl isothiocyanate
4-Iodophenylisothiocyanate

RN:	2059-76-9	**MP** (°C):	
MW:	261.09	**BP** (°C):	

Solubility (Moles/L)	Solubility (Grams/L)	Temp (°C)	Ref (#)	Evaluation (T P E A A)	Comments
9.000E-05	2.350E-02	25	D019	1 1 1 1 1	

1015. C₇H₄INS

3-Iodophenyl isothiocyanate
m-Iodophenyl isothiocyanate
RN: 3125-73-3   **MP** (°C):
MW: 261.09   **BP** (°C):

Solubility (Moles/L)	Solubility (Grams/L)	Temp (°C)	Ref (#)	Evaluation (T P E A A)	Comments
2.100E-05	5.483E-03	25	K032	2 2 0 1 0	

1016. C₇H₄I₂O₃

3,5-Diiodosalicylic acid
2-Hydroxy-3,5-diiod-benzoesaeure
RN: 133-91-5   **MP** (°C):   235.5
MW: 389.92   **BP** (°C):

Solubility (Moles/L)	Solubility (Grams/L)	Temp (°C)	Ref (#)	Evaluation (T P E A A)	Comments
4.274E-04	1.666E-01	10	C072	1 2 1 1 2	
1.795E-03	7.000E-01	15	F300	1 0 0 0 1	
4.931E-04	1.923E-01	25	C072	1 2 1 1 2	
3.847E-03	1.500E+00	h	F300	1 0 0 0 1	

1017. C₇H₄N₂O₂S

3-Nitrophenyl isothiocyanate
m-Nitrophenylisothiocyanate
RN: 3529-82-6   **MP** (°C):
MW: 180.19   **BP** (°C):

Solubility (Moles/L)	Solubility (Grams/L)	Temp (°C)	Ref (#)	Evaluation (T P E A A)	Comments
2.800E-04	5.045E-02	25	K032	2 2 0 1 2	

1018. C₇H₄N₂O₆

2,4-Dinitrobenzoic acid
2,4-Dinitrobenzoesaeure
RN: 610-30-0   **MP** (°C):
MW: 212.12   **BP** (°C):

Solubility (Moles/L)	Solubility (Grams/L)	Temp (°C)	Ref (#)	Evaluation (T P E A A)	Comments
8.580E-02	1.820E+01	25	F300	1 0 0 0 2	
4.900E-02	1.039E+01	ns	C014	0 0 0 1 1	

1019. C$_7$H$_4$N$_2$O$_6$
2,6-Dinitrobenzoic acid
2,6-Dinitrobenzoesaeure
RN: 603-12-3   **MP** (°C):
MW: 212.12   **BP** (°C):

Solubility (Moles/L)	Solubility (Grams/L)	Temp (°C)	Ref (#)	Evaluation (T P E A A)	Comments
7.600E-02	1.612E+01	ns	C014	0 2 0 1 1	

1020. C$_7$H$_4$N$_2$O$_6$
3,4-Dinitrobenzoic acid
3,4-Dinitrobenzoesaeure
RN: 528-45-0   **MP** (°C):   166
MW: 212.12   **BP** (°C):

Solubility (Moles/L)	Solubility (Grams/L)	Temp (°C)	Ref (#)	Evaluation (T P E A A)	Comments
3.159E-02	6.700E+00	25	F300	1 0 0 0 1	

1021. C$_7$H$_4$N$_2$O$_6$
3,5-Dinitrobenzoic acid
3,5-Dinitrobenzoesaeure
RN: 99-34-3   **MP** (°C):   205
MW: 212.12   **BP** (°C):

Solubility (Moles/L)	Solubility (Grams/L)	Temp (°C)	Ref (#)	Evaluation (T P E A A)	Comments
6.350E-03	1.347E+00	25	K040	1 0 2 1 2	
2.923E-03	6.200E-01	25	P037	2 0 1 1 1	

1022. C$_7$H$_4$N$_4$O$_9$
2,3,5,6-Tetranitroanisol
RN: **MP** (°C):
MW: 288.13   **BP** (°C):

Solubility (Moles/L)	Solubility (Grams/L)	Temp (°C)	Ref (#)	Evaluation (T P E A A)	Comments
6.941E-04	2.000E-01	50	F300	1 0 0 0 0	
4.165E-03	1.200E+00	100	F300	1 0 0 0 1	

1023. C$_7$H$_4$O$_6$
Chelidonic acid
Chelidonsaeure
RN: 99-32-1   **MP** (°C):
MW: 184.11   **BP** (°C):

Solubility (Moles/L)	Solubility (Grams/L)	Temp (°C)	Ref (#)	Evaluation (T P E A A)	Comments
7.767E-02	1.430E+01	25	F300	1 0 0 0 2	
2.064E-01	3.800E+01	100	F300	1 0 0 0 1	

1024. C$_7$H$_4$O$_7$
Meconic acid
Mekonsaeure
RN: 497-59-6 **MP** (°C):
MW: 200.11 **BP** (°C):

Solubility (Moles/L)	Solubility (Grams/L)	Temp (°C)	Ref (#)	Evaluation (T P E A A)	Comments
4.198E-02	8.400E+00	25	F300	1 0 0 0 1	
1.034E+00	2.070E+02	100	F300	1 0 0 0 2	

1025. C$_7$H$_5$BrO$_2$
p-Bromobenzoic acid
4-Bromobenzoic acid
RN: 586-76-5 **MP** (°C): 252.0
MW: 201.03 **BP** (°C):

Solubility (Moles/L)	Solubility (Grams/L)	Temp (°C)	Ref (#)	Evaluation (T P E A A)	Comments
2.786E-04	5.600E-02	22.5	G301	0 0 0 0 0	
2.985E-04	6.000E-02	ns	B150	0 0 2 2 1	
2.885E-04	5.800E-02	ns	B150	0 0 2 2 1	
2.800E-04	5.629E-02	ns	C014	0 0 0 1 1	

1026. C$_7$H$_5$BrO$_2$
m-Bromobenzoic acid
3-Bromobenzoic acid
RN: 585-76-2 **MP** (°C): 155
MW: 201.03 **BP** (°C):

Solubility (Moles/L)	Solubility (Grams/L)	Temp (°C)	Ref (#)	Evaluation (T P E A A)	Comments
2.000E-03	4.021E-01	ns	C014	0 0 0 1 1	

1027. C$_7$H$_5$ClN$_4$O$_2$
4H-Pyrazolo[3,4-d]pyrimidin-4-one, 1-(chloroacetyl)-1,5-dihydro-
RN: 96448-62-3 **MP** (°C):
MW: 212.60 **BP** (°C):

Solubility (Moles/L)	Solubility (Grams/L)	Temp (°C)	Ref (#)	Evaluation (T P E A A)	Comments
5.174E-04	1.100E-01	22	B428	1 2 1 2 1	

1028. $C_7H_5ClO_2$
meta-Chlorobenzoic acid
3-Chlorobenzoic acid
m-Chlorobenzoic acid
3-Chlor-benzoesaeure

RN:	535-80-8	MP (°C):	154
MW:	156.57	BP (°C):	

Solubility (Moles/L)	Solubility (Grams/L)	Temp (°C)	Ref (#)	Evaluation (T P E A A)	Comments
2.555E-04	4.000E-02	0	F300	1 0 0 0 0	
4.080E-03	6.388E-01	24.99	B391	0 0 0 0 0	
2.555E-03	4.000E-01	25	F300	1 0 0 0 0	
2.543E-03	3.982E-01	25	T066	1 0 0 0 2	
2.555E-03	4.000E-01	37	M360	1 2 1 1 2	
2.460E-03	3.852E-01	ns	O004	0 2 1 1 2	

1029. $C_7H_5ClO_2$
p-Chlorobenzoic acid
4-Chlorobenzoic acid
Chloradracylic
4-Chlor-benzoesaeure

RN:	74-11-3	MP (°C):	235
MW:	156.57	BP (°C):	

Solubility (Moles/L)	Solubility (Grams/L)	Temp (°C)	Ref (#)	Evaluation (T P E A A)	Comments
5.748E-04	9.000E-02	22.5	G301	0 0 0 0 0	
8.000E-04	1.253E-01	24.99	B391	0 0 0 0 0	
7.026E-04	1.100E-01	25	C410	2 0 2 2 1	
4.918E-04	7.700E-02	25	F300	1 0 0 0 1	
4.639E-04	7.263E-02	25	T066	1 0 0 0 2	
7.026E-04	1.100E-01	37	M360	1 2 1 1 2	
4.918E-04	7.700E-02	ns	B150	0 0 2 2 1	
4.350E-04	6.811E-02	ns	O004	0 2 1 1 2	

1030. $C_7H_5ClO_2$
o-Chlorobenzoic acid
2-Chlor-benzoesaeure
2-Chlorobenzoic acid

RN:	118-91-2	MP (°C):	142
MW:	156.57	BP (°C):	

Solubility (Moles/L)	Solubility (Grams/L)	Temp (°C)	Ref (#)	Evaluation (T P E A A)	Comments
2.100E-02	3.288E+00	24.99	B391	0 0 0 0 0	
1.916E-02	3.000E+00	25	C410	2 0 2 2 1	
1.341E-02	2.100E+00	25	F300	1 0 0 0 1	
8.686E-03	1.360E+00	25	P037	2 0 1 1 2	
1.865E-02	2.920E+00	37	M360	1 2 1 1 2	
2.574E-01	4.030E+01	100	F300	1 0 0 0 2	
1.330E-02	2.082E+00	ns	C014	0 0 0 1 2	
1.362E-02	2.132E+00	ns	O004	0 2 1 1 2	

1031. C₇H₅Cl₂NO

1031. $C_7H_5Cl_2NO$

2,6-Dichlorobenzamide
Dichlorobenzamide
BAM

RN:	2008-58-4	**MP** (°C):	198	
MW:	190.03	**BP** (°C):		

Solubility (Moles/L)	Solubility (Grams/L)	Temp (°C)	Ref (#)	Evaluation (T P E A A)	Comments
1.421E-02	2.700E+00	22.5	G301	0 0 0 0 0	

1032. $C_7H_5Cl_2NO_2$

Chloramben
3-Amino-2,5-dichlorobenzoic acid

RN:	133-90-4	**MP** (°C):	201	
MW:	206.03	**BP** (°C):		

Solubility (Moles/L)	Solubility (Grams/L)	Temp (°C)	Ref (#)	Evaluation (T P E A A)	Comments
3.398E-03	7.000E-01	25	B200	1 0 0 0 2	
3.398E-03	7.000E-01	25	M161	1 0 0 0 2	
3.398E-03	7.000E-01	ns	B185	0 0 0 0 0	

1033. $C_7H_5Cl_2NS$

2,6-Dichlorothiobenzamide
Prefix
Chlorthiamid

RN:	1918-13-4	**MP** (°C):	151.5	
MW:	206.09	**BP** (°C):	0	

Solubility (Moles/L)	Solubility (Grams/L)	Temp (°C)	Ref (#)	Evaluation (T P E A A)	Comments
4.561E-03	9.400E-01	20	M061	1 0 0 0 2	
4.610E-03	9.500E-01	21	M161	1 0 0 0 2	

1034. $C_7H_5Cl_3O$

2,3,4-Trichloroanisole
1,2,3-Trichloro-4-methoxy-benzene

RN:	54135-80-7	**MP** (°C):	70	
MW:	211.48	**BP** (°C):		

Solubility (Moles/L)	Solubility (Grams/L)	Temp (°C)	Ref (#)	Evaluation (T P E A A)	Comments
5.107E-05	1.080E-02	25	L348	1 2 2 1 2	

1035. C₇H₅Cl₃O

2,4,6-Trichloro-3-methylphenol
m-Cresol, 2,4,6-trichloro-
2,4,6-Trichloro-*m*-cresol
RN: 551-76-8   **MP** (°C):
MW: 211.48   **BP** (°C):

Solubility (Moles/L)	Solubility (Grams/L)	Temp (°C)	Ref (#)	Evaluation (T P E A A)	Comments
5.300E-04	1.121E-01	25	B316	0 0 0 0 0	

1036. C₇H₅Cl₃O

2,4,6-Trichloroanisole
1-Methoxy-2,4,6-trichlorobenzene
Methyl 2,4,6-trichlorophenyl ether
Tyrene
RN: 87-40-1   **MP** (°C):   61
MW: 211.48   **BP** (°C):

Solubility (Moles/L)	Solubility (Grams/L)	Temp (°C)	Ref (#)	Evaluation (T P E A A)	Comments
6.242E-05	1.320E-02	25	L348	1 2 2 1 2	

1037. C₇H₅FO₂

m-Fluorobenzoic acid
3-Fluor-benzoesaeure
3-Fluorobenzoic acid
RN: 455-38-9   **MP** (°C):   123
MW: 140.12   **BP** (°C):

Solubility (Moles/L)	Solubility (Grams/L)	Temp (°C)	Ref (#)	Evaluation (T P E A A)	Comments
1.071E-02	1.500E+00	25	F300	1 0 0 0 1	

1038. C₇H₅FO₂

o-Fluorobenzoic acid
2-Fluorobenzoic acid
RN: 445-29-4   **MP** (°C):   123
MW: 140.12   **BP** (°C):

Solubility (Moles/L)	Solubility (Grams/L)	Temp (°C)	Ref (#)	Evaluation (T P E A A)	Comments
5.139E-02	7.200E+00	25	F300	1 0 0 0 1	
5.129E-02	7.186E+00	ns	R427	0 0 0 0 0	

1039. C$_7$H$_5$FO$_2$
p-Fluorobenzoic acid
4-Fluor-benzoesaeure
4-Fluorobenzoic acid

RN: 456-22-4   **MP** (°C):   182.6
MW: 140.12   **BP** (°C):

Solubility (Moles/L)	Solubility (Grams/L)	Temp (°C)	Ref (#)	Evaluation (T P E A A)	Comments
8.564E-03	1.200E+00	25	F300	1 0 0 0 1	

1040. C$_7$H$_5$F$_3$N$_2$O$_4$S
3-Trifluoromethyl-4-nitrobenzenesulfonamide
4-Nitro-3-(trifluoromethyl)benzenesulfonamide

RN: 21988-05-6   **MP** (°C):
MW: 270.19   **BP** (°C):

Solubility (Moles/L)	Solubility (Grams/L)	Temp (°C)	Ref (#)	Evaluation (T P E A A)	Comments
6.500E-04	1.756E-01	15	K024	1 2 1 1 2	

1041. C$_7$H$_5$IO$_2$
p-Iodobenzoic acid
4-Iodobenzoic acid

RN: 619-58-9   **MP** (°C):
MW: 248.02   **BP** (°C):

Solubility (Moles/L)	Solubility (Grams/L)	Temp (°C)	Ref (#)	Evaluation (T P E A A)	Comments
1.120E-04	2.778E-02	15	D008	1 0 1 1 2	intrinsic

1042. C$_7$H$_5$IO$_2$
o-Iodobenzoic acid
2-Iodobenzoic acid

RN: 88-67-5   **MP** (°C):   162
MW: 248.02   **BP** (°C):

Solubility (Moles/L)	Solubility (Grams/L)	Temp (°C)	Ref (#)	Evaluation (T P E A A)	Comments
1.860E-03	4.613E-01	15	D008	1 0 1 1 2	0.002N HCl

1043. C$_7$H$_5$IO$_2$
m-Iodobenzoic acid
3-Iodobenzoic acid

RN: 618-51-9   **MP** (°C):   187
MW: 248.02   **BP** (°C):

Solubility (Moles/L)	Solubility (Grams/L)	Temp (°C)	Ref (#)	Evaluation (T P E A A)	Comments
5.380E-04	1.334E-01	15	D008	1 0 1 1 2	0.002N HCl

1044. C$_7$H$_5$I$_2$NO$_3$
3,5-Diiodo-4-pyridone-N-acetic acid
3,5-Diiod-pyridon-(4)-N-essigsaeure
3,5-Diiodo-4-pyridone-1-acetic acid
Diodon
1,4-Dihydro-3,5-diiodo-4-oxopyridine-1-acetic acid

RN:	101-29-1	MP (°C):	244
MW:	404.93	BP (°C):	

Solubility (Moles/L)	Solubility (Grams/L)	Temp (°C)	Ref (#)	Evaluation (T P E A A)	Comments
6.883E-03	2.787E+00	ns	H055	0 0 0 0 0	

1045. C$_7$H$_5$N
Benzonitrile
Benzonitril
Benzenenitrile
Benzoic acid nitrile
Phenyl cyanide
Cyanobenzene

RN:	100-47-0	MP (°C):	−13
MW:	103.12	BP (°C):	190.7

Solubility (Moles/L)	Solubility (Grams/L)	Temp (°C)	Ref (#)	Evaluation (T P E A A)	Comments
1.839E-02	1.896E+00	24.0	P321	0 0 0 0 0	
4.200E-02	4.331E+00	25	M327	1 0 0 1 2	
3.671E-02	3.786E+00	35.5	P321	0 0 0 0 0	
5.400E-02	5.569E+00	50.0	P321	0 0 0 0 0	
4.056E-02	4.182E+00	57.0	P321	0 0 0 0 0	
5.496E-02	5.668E+00	62.5	P321	0 0 0 0 0	
8.268E-02	8.527E+00	85.0	P321	0 0 0 0 0	
8.459E-02	8.723E+00	90.5	P321	0 0 0 0 0	
9.981E-02	1.029E+01	95.5	P321	0 0 0 0 0	
9.697E-02	1.000E+01	100	F300	1 0 0 0 0	
1.065E-01	1.098E+01	101.0	P321	0 0 0 0 0	
1.339E-01	1.381E+01	116.0	P321	0 0 0 0 0	
1.920E-01	1.980E+01	127.5	P321	0 0 0 0 0	
2.171E-01	2.239E+01	142.0	P321	0 0 0 0 0	
2.888E-01	2.979E+01	148.0	P321	0 0 0 0 0	
2.834E-01	2.922E+01	149.0	P321	0 0 0 0 0	
3.873E-01	3.994E+01	160.5	P321	0 0 0 0 0	
5.747E-01	5.927E+01	164.5	P321	0 0 0 0 0	
1.373E+00	1.416E+02	201.0	P321	0 0 0 0 0	
2.937E+00	3.029E+02	211.0	P321	0 0 0 0 0	
9.696E-04	9.999E-02	ns	L055	0 0 0 0 1	

1046. C₇H₅NOS

4-Hydroxyphenyl isothiocyanate
4-Hydroxyphenylisothiocyanate
RN: 2131-60-4 **MP** (°C):
MW: 151.19 **BP** (°C):

Solubility (Moles/L)	Solubility (Grams/L)	Temp (°C)	Ref (#)	Evaluation (T P E A A)	Comments
2.150E-03	3.251E-01	25	D019	1 1 1 1 2	

1047. C₇H₅NOS

3-Hydroxyphenyl isothiocyanate
m-Hydroxyphenyl isothiocyanate
RN: 3125-63-1 **MP** (°C):
MW: 151.19 **BP** (°C):

Solubility (Moles/L)	Solubility (Grams/L)	Temp (°C)	Ref (#)	Evaluation (T P E A A)	Comments
1.020E-02	1.542E+00	25	K032	2 2 0 1 2	
1.023E-02	1.547E+00	ns	R427	0 0 0 0 0	

1048. C₇H₅NO₃

m-Nitrobenzaldehyde
3-Nitrobenzaldehyde
3-Nitro-benzaldehyd
RN: 99-61-6 **MP** (°C): 58
MW: 151.12 **BP** (°C):

Solubility (Moles/L)	Solubility (Grams/L)	Temp (°C)	Ref (#)	Evaluation (T P E A A)	Comments
6.617E-05	1.000E-02	25	F300	1 0 0 0 1	
3.309E+00	5.000E+02	58.0	S118	1 2 0 1 0	
6.292E-02	9.509E+00	75.1	S118	1 2 0 1 1	
3.272E+00	4.945E+02	85.2	S118	1 2 0 1 2	
1.266E-01	1.913E+01	111.9	S118	1 2 0 1 2	
1.934E-01	2.922E+01	136.4	S118	1 2 0 1 2	
3.103E-01	4.689E+01	157.3	S118	1 2 0 1 2	
6.293E-01	9.510E+01	181.0	S118	1 2 0 1 2	
8.142E-01	1.230E+02	191.4	S118	1 2 0 1 2	
1.253E+00	1.893E+02	205.4	S118	1 2 0 1 2	
1.878E+00	2.838E+02	211.8	S118	1 2 0 1 2	

1049. C$_7$H$_5$NO$_3$
o-Nitrobenzaldehyde
2-Nitrobenzaldehyde
2-Nitro-benzaldehyd

RN:	552-89-6	**MP** (°C):	44
MW:	151.12	**BP** (°C):	153

Solubility (Moles/L)	Solubility (Grams/L)	Temp (°C)	Ref (#)	Evaluation (T P E A A)	Comments
1.323E-04	2.000E-02	25	F300	1 0 0 0 1	
4.600E-02	6.951E+00	66.9	S118	1 2 0 1 1	
9.972E-02	1.507E+01	103.1	S118	1 2 0 1 1	
3.001E-01	4.535E+01	166.0	S118	1 2 0 1 1	

1050. C$_7$H$_5$NO$_3$
p-Nitrobenzaldehyde
4-Nitrobenzaldehyde

RN:	555-16-8	**MP** (°C):	106.5
MW:	151.12	**BP** (°C):	

Solubility (Moles/L)	Solubility (Grams/L)	Temp (°C)	Ref (#)	Evaluation (T P E A A)	Comments
1.871E-01	2.828E+01	132.4	S118	1 2 0 1 2	
5.341E-01	8.071E+01	176.5	S118	1 2 0 1 2	
1.133E+00	1.713E+02	205.4	S118	1 2 0 1 2	
1.814E+00	2.742E+02	215.5	S118	1 2 0 1 2	

1051. C$_7$H$_5$NO$_3$S
Saccharin
1,1-Dioxide-1,2-benzisothiazol-3-(2H)-one
3-Benzisothiazolinone 1,1-dioxide
1,2-Benzisothiazol-3(2H)-one-1,1-dioxide
Kandiset
Glucid

RN:	81-07-2	**MP** (°C):	228.8
MW:	183.19	**BP** (°C):	

Solubility (Moles/L)	Solubility (Grams/L)	Temp (°C)	Ref (#)	Evaluation (T P E A A)	Comments
2.347E-02	4.300E+00	25	F300	1 0 0 0 1	
1.880E-01	3.444E+01	30	M015	1 0 2 1 0	EFG

1052. C$_7$H$_5$NO$_4$
Quinolinic acid
2,3-Pyridinedicarboxylic acid
Pyridine-2,3-dicarboxylic acid
Pyridine-2,3-dicarboxylate

RN: 89-00-9   **MP** (°C):   190
MW: 167.12   **BP** (°C):

Solubility (Moles/L)	Solubility (Grams/L)	Temp (°C)	Ref (#)	Evaluation (T P E A A)	Comments
3.291E-02	5.500E+00	7	F300	1 0 0 0 1	
6.600E-02	1.103E+01	25	C104	2 2 1 1 2	
6.400E-02	1.070E+01	25	C104	2 2 1 1 2	

1053. C$_7$H$_5$NO$_4$
p-Nitrobenzoic acid
4-Nitrobenzoic acid

RN: 62-23-7   **MP** (°C):   242.4
MW: 167.12   **BP** (°C):

Solubility (Moles/L)	Solubility (Grams/L)	Temp (°C)	Ref (#)	Evaluation (T P E A A)	Comments
1.197E-03	2.000E-01	15	F300	1 0 0 0 2	
2.525E-03	4.220E-01	24.99	B391	0 0 0 0 0	
1.660E-03	2.774E-01	25	H071	2 2 2 1 2	
3.471E-03	5.800E-01	37	B171	2 0 1 1 2	

1054. C$_7$H$_5$NO$_4$
o-Nitrobenzoic acid
2-Nitrobenzoic acid

RN: 552-16-9   **MP** (°C):   147.5
MW: 167.12   **BP** (°C):

Solubility (Moles/L)	Solubility (Grams/L)	Temp (°C)	Ref (#)	Evaluation (T P E A A)	Comments
3.920E-02	6.551E+00	18	D058	1 0 1 1 2	
3.340E-02	5.582E+00	24.99	B391	0 0 0 0 0	
4.325E-02	7.228E+00	25	D058	1 0 1 1 2	
4.488E-02	7.500E+00	25	F300	1 0 0 0 1	
4.350E-02	7.270E+00	25	H071	2 2 2 1 2	
4.700E-02	7.855E+00	25	K040	1 0 2 1 2	
4.360E-02	7.287E+00	25	K053	2 2 2 2 2	
4.430E-02	7.404E+00	25	L050	2 0 1 2 2	
4.415E-02	7.378E+00	25	R016	0 0 0 0 0	
4.700E-02	7.855E+00	26.4	P043	2 0 1 1 2	

1055. C₇H₅NO₄

m-Nitrobenzoic acid
3-Nitrobenzoic acid

RN:	121-92-6	**MP** (°C):	142.0
MW:	167.12	**BP** (°C):	

Solubility (Moles/L)	Solubility (Grams/L)	Temp (°C)	Ref (#)	Evaluation (T P E A A)	Comments
1.436E-02	2.400E+00	15	F300	1 0 0 0 1	
1.530E-02	2.557E+00	24.99	B391	0 0 0 0 0	
2.121E-02	3.545E+00	25	C076	0 0 0 0 0	
2.140E-02	3.576E+00	25	K040	1 0 2 1 2	
1.227E-02	2.050E+00	25	P037	2 0 1 1 2	
6.582E-02	1.100E+01	37	B171	2 0 1 1 2	
2.334E-02	3.900E+00	ns	B361	0 0 0 0 0	

1056. C₇H₅NO₄

Isocinchomeronic acid
2,5-Pyridinedicarboxylic acid
Pyridine-2,5-dicarboxylic acid

RN:	100-26-5	**MP** (°C):	254
MW:	167.12	**BP** (°C):	

Solubility (Moles/L)	Solubility (Grams/L)	Temp (°C)	Ref (#)	Evaluation (T P E A A)	Comments
7.400E-03	1.237E+00	25	C104	2 2 1 1 2	
7.100E-03	1.187E+00	25	C104	2 2 1 1 2	

1057. C₇H₅NO₄

Cinchomeronic acid
3,4-Pyridinedicarboxylic acid

RN:	490-11-9	**MP** (°C):	256
MW:	167.12	**BP** (°C):	

Solubility (Moles/L)	Solubility (Grams/L)	Temp (°C)	Ref (#)	Evaluation (T P E A A)	Comments
1.400E-02	2.340E+00	25	C104	2 2 1 1 2	
1.380E-02	2.306E+00	25	C104	2 2 1 1 2	

1058. C₇H₅NO₄

3,5-Pyridinedicarboxylic acid
Dinicotinic acid

RN:	499-81-0	**MP** (°C):	
MW:	167.12	**BP** (°C):	

Solubility (Moles/L)	Solubility (Grams/L)	Temp (°C)	Ref (#)	Evaluation (T P E A A)	Comments
6.400E-03	1.070E+00	25	C104	2 2 1 1 2	

1059. C$_7$H$_5$NO$_4$
4-Formyl-2-NO2-phenol

RN: **MP** (°C):
MW: 167.12   **BP** (°C):

Solubility (Moles/L)	Solubility (Grams/L)	Temp (°C)	Ref (#)	Evaluation (T P E A A)	Comments
1.122E-03	1.875E-01	ns	R424	0 0 0 0 0	

1060. C$_7$H$_5$NO$_4$
Lutidinic acid
2,4-Pyridinedicarboxylic acid

RN: 499-80-9   **MP** (°C):   248
MW: 167.12   **BP** (°C):

Solubility (Moles/L)	Solubility (Grams/L)	Temp (°C)	Ref (#)	Evaluation (T P E A A)	Comments
1.490E-02	2.490E+00	25	C104	2 2 1 1 2	
1.480E-02	2.473E+00	25	C104	2 2 1 1 2	

1061. C$_7$H$_5$NO$_5$
3-Nitrosalicylic acid
3-Nitro-salicylsaeure

RN: 85-38-1   **MP** (°C):   128
MW: 183.12   **BP** (°C):

Solubility (Moles/L)	Solubility (Grams/L)	Temp (°C)	Ref (#)	Evaluation (T P E A A)	Comments
7.099E-03	1.300E+00	16	F300	1 0 0 0 1	

1062. C$_7$H$_5$NO$_5$
5-Nitrosalicylic acid
5-Nitrosalicylsaeure

RN: 96-97-9   **MP** (°C):   229–230
MW: 183.12   **BP** (°C):

Solubility (Moles/L)	Solubility (Grams/L)	Temp (°C)	Ref (#)	Evaluation (T P E A A)	Comments
1.092E-02	2.000E+00	45	F300	1 0 0 0 0	

1063. C$_7$H$_5$NS
Benzothiazole
Benzthiazol

RN: 95-16-9   **MP** (°C):   2
MW: 135.19   **BP** (°C):   231

Solubility (Moles/L)	Solubility (Grams/L)	Temp (°C)	Ref (#)	Evaluation (T P E A A)	Comments
3.162E-02	4.275E+00	ns	S460	0 0 0 0 0	

1064. C₇H₅NS

1064. C$_7$H$_5$NS
Phenyl isothiocyanate
Isothiocyanatobenzene
Phenyl mustard oil
PITC

RN:	103-72-0	**MP** (°C):	−21.0
MW:	135.19	**BP** (°C):	221.0

Solubility (Moles/L)	Solubility (Grams/L)	Temp (°C)	Ref (#)	Evaluation (T P E A A)	Comments
6.650E-04	8.990E-02	25	D019	1 1 1 1 2	

1065. C$_7$H$_5$N$_3$O$_6$

2,4,6-Trinitrotoluene
2,4,6-Tronitrotoluol

RN:	118-96-7	**MP** (°C):	80.1
MW:	227.13	**BP** (°C):	

Solubility (Moles/L)	Solubility (Grams/L)	Temp (°C)	Ref (#)	Evaluation (T P E A A)	Comments
4.843E-04	1.100E-01	.3	D065	1 2 2 1 2	
4.842E-04	1.100E-01	.3	T020	1 2 2 2 2	
4.843E-04	1.100E-01	.30	F300	1 0 0 0 1	
4.975E-04	1.130E-01	5.9	D065	1 2 2 1 2	
4.974E-04	1.130E-01	5.9	T020	1 2 2 2 2	
5.283E-04	1.200E-01	20	D065	1 2 2 1 2	
5.283E-04	1.200E-01	20.0	T020	1 2 2 2 2	
8.937E-04	2.030E-01	33.1	D065	1 2 2 1 2	
8.936E-04	2.030E-01	33.1	T020	1 2 2 2 2	
1.497E-03	3.400E-01	44.2	D065	1 2 2 1 2	
1.496E-03	3.399E-01	44.2	T020	1 2 2 2 2	
1.629E-03	3.700E-01	45	D065	1 2 2 1 2	
1.628E-03	3.699E-01	45.0	T020	1 2 2 2 2	
2.351E-03	5.340E-01	53	D065	1 2 2 1 2	
2.350E-03	5.337E-01	53.0	T020	1 2 2 2 2	
2.703E-03	6.140E-01	57.1	D065	1 2 2 1 2	
2.702E-03	6.136E-01	57.1	T020	1 2 2 2 2	
4.240E-03	9.630E-01	73.2	D065	1 2 2 1 2	
4.236E-03	9.621E-01	73.2	T020	1 2 2 2 2	
6.054E-03	1.375E+00	94.4	D065	1 2 2 1 2	
6.045E-03	1.373E+00	94.4	T020	1 2 2 2 2	
6.459E-03	1.467E+00	99.5	D065	1 2 2 1 2	
6.449E-03	1.465E+00	99.5	T020	1 2 2 2 2	
6.459E-03	1.467E+00	99.50	F300	1 0 0 0 2	
6.026E-04	1.369E-01	ns	R427	0 0 0 0 0	

1066. C$_7$H$_5$N$_3$O$_7$
2,4,6-Trinitro-*m*-cresol
2,4,6-Trinitro-*m*-kresol

RN: 3238-38-8   **MP** (°C):
MW: 243.13   **BP** (°C):

Solubility (Moles/L)	Solubility (Grams/L)	Temp (°C)	Ref (#)	Evaluation (T P E A A)	Comments
8.226E-03	2.000E+00	15	F300	1 0 0 0 0	

1067. C$_7$H$_5$N$_3$O$_7$
Methyl picric acid
2,4,6-Trinitro-3-methylphenol
3-Methyl-2,4,6-trinitrophenol
2,4,6-Trinitro-*m*-cresol

RN: 602-99-3   **MP** (°C):
MW: 243.13   **BP** (°C):

Solubility (Moles/L)	Solubility (Grams/L)	Temp (°C)	Ref (#)	Evaluation (T P E A A)	Comments
1.000E-02	2.431E+00	25	K053	2 2 2 2 2	

1068. C$_7$H$_5$N$_3$O$_7$
2,4,6-Trinitroanisole
2-Methoxy-1,3,5-trinitro-benzene
Methyl picrate

RN: 606-35-9   **MP** (°C): 69
MW: 243.13   **BP** (°C):

Solubility (Moles/L)	Solubility (Grams/L)	Temp (°C)	Ref (#)	Evaluation (T P E A A)	Comments
8.224E-04	2.000E-01	15	D079	1 2 0 0 1	
5.627E-03	1.368E+00	50	D079	1 2 0 0 2	
1.594E-02	3.875E+00	100	D079	1 2 0 0 2	

1069. C$_7$H$_5$N$_5$O$_8$
Nitramine
Tetryl
N-Methyl-*N*,2,4,5-tetranitroaniline

RN: 479-45-8   **MP** (°C): 131
MW: 287.15   **BP** (°C):

Solubility (Moles/L)	Solubility (Grams/L)	Temp (°C)	Ref (#)	Evaluation (T P E A A)	Comments
1.776E-04	5.100E-02	.5	T015	1 2 0 1 1	
1.776E-04	5.100E-02	.50	D066	1 2 2 1 2	
1.741E-04	5.000E-02	.50	F300	1 0 0 0 0	
2.403E-04	6.900E-02	9.6	D066	1 2 2 1 2	
2.403E-04	6.900E-02	9.6	T015	1 2 0 1 1	
2.473E-04	7.100E-02	14.8	D066	1 2 2 1 1	

(continued)

1069. C₇H₅N₅O₈ (continued)

Solubility (Moles/L)	Solubility (Grams/L)	Temp (°C)	Ref (#)	Evaluation (T P E A A)	Comments
2.472E-04	7.099E-02	14.8	T015	1 2 0 1 1	
2.577E-04	7.400E-02	20.5	D066	1 2 2 1 1	
2.577E-04	7.399E-02	20.5	T015	1 2 0 1 1	
2.925E-04	8.400E-02	30	D066	1 2 2 1 1	
2.925E-04	8.399E-02	30.0	T015	1 2 0 1 1	
3.274E-04	9.400E-02	35	D066	1 2 2 1 1	
3.273E-04	9.399E-02	35.0	T015	1 2 0 1 1	
3.726E-04	1.070E-01	40	D066	1 2 2 1 2	
3.726E-04	1.070E-01	40.0	T015	1 2 0 1 2	
4.701E-04	1.350E-01	45	D066	1 2 2 1 2	
4.701E-04	1.350E-01	45.0	T015	1 2 0 1 2	
6.965E-04	2.000E-01	50	D066	1 2 2 1 2	
6.964E-04	2.000E-01	50.0	T015	1 2 0 1 2	
1.219E-03	3.500E-01	60	D066	0 0 0 0 0	
1.218E-03	3.499E-01	60.05	T015	1 2 0 1 2	
1.543E-03	4.430E-01	65	D065	1 2 2 1 2	
1.542E-03	4.428E-01	65.05	T015	1 2 0 1 2	
1.849E-03	5.310E-01	69.5	D065	1 2 2 1 2	
1.848E-03	5.307E-01	69.5	T015	1 2 0 1 2	
3.315E-03	9.520E-01	84.2	D065	1 2 2 1 2	
3.312E-03	9.511E-01	84.2	T015	1 2 0 1 2	
5.638E-03	1.619E+00	96.7	D065	1 2 2 1 2	
5.629E-03	1.616E+00	96.7	T015	1 2 0 1 2	
6.112E-03	1.755E+00	98.5	D065	1 2 2 1 2	
6.101E-03	1.752E+00	98.55	T015	1 2 0 1 2	
6.129E-03	1.760E+00	99	F300	1 0 0 0 2	

1070. C₇H₆ClF
2-Fluorobenzyl chloride
o-Fluorobenzyl chloride
RN: 345-35-7   **MP** (°C):
MW: 144.58   **BP** (°C):

Solubility (Moles/L)	Solubility (Grams/L)	Temp (°C)	Ref (#)	Evaluation (T P E A A)	Comments
2.880E-03	4.164E-01	25	M342	1 0 1 1 2	
2.877E-03	4.160E-01	ns	S460	0 0 0 0 0	

1071. C₇H₆ClF
3-Fluorobenzyl chloride
m-Fluorobenzyl chloride
RN: 456-42-8   **MP** (°C):
MW: 144.58   **BP** (°C):

Solubility (Moles/L)	Solubility (Grams/L)	Temp (°C)	Ref (#)	Evaluation (T P E A A)	Comments
2.860E-03	4.135E-01	25	M342	1 0 1 1 2	
2.858E-03	4.131E-01	ns	S460	0 0 0 0 0	

1072. C₇H₆ClF

4-Fluorobenzyl chloride
1-(Chloromethyl)-4-fluoro-benzene
α-Chloro-*p*-fluorotoluene

RN:	352-11-4	MP (°C):	−18
MW:	144.58	BP (°C):	181.2

Solubility (Moles/L)	Solubility (Grams/L)	Temp (°C)	Ref (#)	Evaluation (T P E A A)	Comments
2.884E-03	4.170E-01	ns	S460	0 0 0 0 0	

1073. C₇H₆ClN₃O₄S₂

Chlorothiazide
Diuresal

RN:	58-94-6	MP (°C):	342
MW:	295.72	BP (°C):	

Solubility (Moles/L)	Solubility (Grams/L)	Temp (°C)	Ref (#)	Evaluation (T P E A A)	Comments
9.560E-04	2.827E-01	25	A076	1 0 1 1 2	
9.000E-04	2.662E-01	30	A089	2 0 1 1 0	EFG
9.000E-04	2.662E-01	30	A093	2 0 1 1 0	EFG
6.763E-04	2.000E-01	ns	C114	0 0 0 0 0	
7.439E-04	2.200E-01	rt	A095	0 0 2 2 1	
9.806E-04	2.900E-01	rt	B181	0 0 1 1 2	

1074. C₇H₆ClN₄O₅S₂

4-Nitroso-hydrochlorothiazide

RN:		MP (°C):	155–156
MW:	325.73	BP (°C):	

Solubility (Moles/L)	Solubility (Grams/L)	Temp (°C)	Ref (#)	Evaluation (T P E A A)	Comments
7.368E-04	2.400E-01	25	G051	1 0 1 1 0	

1075. C₇H₆Cl₂N₂O

Chlorambenamide
3,5-Dichloroanthranilamide
Benzamide, 2-amino-3,5-dichloro-

RN:	36765-01-2	MP (°C):	162.5
MW:	205.04	BP (°C):	

Solubility (Moles/L)	Solubility (Grams/L)	Temp (°C)	Ref (#)	Evaluation (T P E A A)	Comments
8.291E-03	1.700E+00	rt	M161	0 0 0 0 1	

1076. C₇H₆Cl₂O

2,6-Dichloroanisole
Benzene, 1,3-dichloro-2-methoxy-
RN: 1984-65-2   **MP** (°C):   31
MW: 177.03   **BP** (°C):

Solubility (Moles/L)	Solubility (Grams/L)	Temp (°C)	Ref (#)	Evaluation (T P E A A)	Comments
7.908E-04	1.400E-01	25	L348	1 2 2 1 2	

1077. C₇H₆Cl₂O

2,3-Dichloroanisole
1,2-Dichloro-3-methoxybenzene
RN: 1984-59-4   **MP** (°C):   32
MW: 177.03   **BP** (°C):

Solubility (Moles/L)	Solubility (Grams/L)	Temp (°C)	Ref (#)	Evaluation (T P E A A)	Comments
4.909E-04	8.690E-02	25	L348	1 2 2 1 2	

1078. C₇H₆Cl₂O

2,6-Dichloro-4-methyl-phenol
2,4-Dichloro-6-methyl-phenol-
RN: 2432-12-4   **MP** (°C):
MW: 177.03   **BP** (°C):

Solubility (Moles/L)	Solubility (Grams/L)	Temp (°C)	Ref (#)	Evaluation (T P E A A)	Comments
1.600E-03	2.833E-01	25	B316	0 0 0 0 0	
3.800E-03	6.727E-01	25	B316	0 0 0 0 0	

1079. C₇H₆N₂O₂S

p-Cyanobenzenesulfonamide
4-Cyanobenzenesulfonamide
RN: 3119-02-6   **MP** (°C):
MW: 182.20   **BP** (°C):

Solubility (Moles/L)	Solubility (Grams/L)	Temp (°C)	Ref (#)	Evaluation (T P E A A)	Comments
6.100E-03	1.111E+00	15	K024	1 2 1 1 2	

1080. C₇H₆N₂O₄

2,4-Dinitrotoluene
2,4-Dinitro-toluol

RN: 121-14-2   **MP** (°C): 71
MW: 182.14   **BP** (°C): 300

Solubility (Moles/L)	Solubility (Grams/L)	Temp (°C)	Ref (#)	Evaluation (T P E A A)	Comments
1.487E-03	2.709E-01	20	T301	1 2 2 2 2	
1.482E-03	2.699E-01	22	D070	1 2 0 0 1	
1.482E-03	2.700E-01	22	F300	1 0 0 0 1	
1.482E-03	2.699E-01	22	L053	1 1 0 0 1	
2.031E-03	3.699E-01	50	D070	1 2 0 0 1	
2.031E-03	3.699E-01	50	L053	1 1 0 0 1	
1.391E-02	2.534E+00	100	D070	1 2 0 0 2	
1.449E-02	2.640E+00	100	F300	1 0 0 0 2	
1.391E-02	2.534E+00	100	L053	1 1 0 0 2	

1081. C₇H₆N₂O₅

2,4-Dinitroanisole
Dinitroanisole
Benzene, 1-methoxy-2,4-dinitro-

RN: 119-27-7   **MP** (°C): 88
MW: 198.14   **BP** (°C):

Solubility (Moles/L)	Solubility (Grams/L)	Temp (°C)	Ref (#)	Evaluation (T P E A A)	Comments
7.822E-04	1.550E-01	15	D079	1 2 0 0 2	
6.863E-04	1.360E-01	50	D079	1 2 0 0 2	
2.401E-02	4.757E+00	100	D079	1 2 0 0 2	

1082. C₇H₆N₂O₅

Dinitrocresol
DNOC
2,4-Dinitro-6-methylphenol
Dinitro-o-cresol

RN: 534-52-1   **MP** (°C): 86
MW: 198.14   **BP** (°C):

Solubility (Moles/L)	Solubility (Grams/L)	Temp (°C)	Ref (#)	Evaluation (T P E A A)	Comments
6.561E-04	1.300E-01	15	M161	1 0 0 0 2	
6.309E-04	1.250E-01	ns	B185	0 0 0 0 0	
6.459E-04	1.280E-01	ns	M061	0 0 0 0 2	
1.000E-03	1.981E-01	ns	M163	0 0 0 0 0	EFG
1.262E-03	2.500E-01	ns	N013	0 0 0 0 2	

1083. C₇H₆N₂S

4-Thiocyanoaniline
Rhodan

RN: 2987-46-4   **MP** (°C):   142
MW: 150.20   **BP** (°C):

Solubility (Moles/L)	Solubility (Grams/L)	Temp (°C)	Ref (#)	Evaluation (T P E A A)	Comments
1.332E-03	2.000E-01	ns	M061	0 0 0 0 0	

1084. C₇H₆N₄

4-Methylpteridine
Pteridine, 4-methyl-

RN: 2432-21-5   **MP** (°C):   151
MW: 146.15   **BP** (°C):

Solubility (Moles/L)	Solubility (Grams/L)	Temp (°C)	Ref (#)	Evaluation (T P E A A)	Comments
3.258E-01	4.762E+01	20	A083	1 2 0 0 0	

1085. C₇H₆N₄

7-Methylpteridine
Pteridine, 7-methyl-

RN: 936-40-3   **MP** (°C):   196.5
MW: 146.15   **BP** (°C):

Solubility (Moles/L)	Solubility (Grams/L)	Temp (°C)	Ref (#)	Evaluation (T P E A A)	Comments
9.775E-01	1.429E+02	20	A083	1 2 0 0 0	

1086. C₇H₆N₄

2-Methylpteridine
Pteridine, 2-methyl-

RN: 2432-20-4   **MP** (°C):   140
MW: 146.15   **BP** (°C):

Solubility (Moles/L)	Solubility (Grams/L)	Temp (°C)	Ref (#)	Evaluation (T P E A A)	Comments
6.842E-01	1.000E+02	20	A083	1 2 0 0 0	

1087. C₇H₆N₄O

2-Methoxypteridine
Pteridine, 2-methoxy-

RN: 102170-44-5   **MP** (°C):   150
MW: 162.15   **BP** (°C):

Solubility (Moles/L)	Solubility (Grams/L)	Temp (°C)	Ref (#)	Evaluation (T P E A A)	Comments
7.614E-02	1.235E+01	20	A019	2 2 1 1 0	
1.233E+00	2.000E+02	100	A019	1 2 1 1 0	

1088. C₇H₆N₄O

1088. $C_7H_6N_4O$

4-Hydroxy-6-methylpteridine
4-Pteridinol, 6-methyl-

RN: 16041-24-0   **MP** (°C):
MW: 162.15   **BP** (°C):

Solubility (Moles/L)	Solubility (Grams/L)	Temp (°C)	Ref (#)	Evaluation (T P E A A)	Comments
2.234E-02	3.623E+00	20	A019	2 2 1 1 2	
1.341E-01	2.174E+01	100	A019	1 2 1 1 1	

1089. $C_7H_6N_4O$

4-Hydroxy-7-methylpteridine
4-Pteridinol, 7-methyl-

RN: 34244-80-9   **MP** (°C):
MW: 162.15   **BP** (°C):

Solubility (Moles/L)	Solubility (Grams/L)	Temp (°C)	Ref (#)	Evaluation (T P E A A)	Comments
2.729E-02	4.425E+00	20	A019	2 2 1 1 2	
1.713E-01	2.778E+01	100	A019	1 2 1 1 1	

1090. $C_7H_6N_4O$

4-Methoxypteridine
Pteridine, 4-methoxy-

RN: 30564-38-6   **MP** (°C):   195
MW: 162.15   **BP** (°C):

Solubility (Moles/L)	Solubility (Grams/L)	Temp (°C)	Ref (#)	Evaluation (T P E A A)	Comments
7.614E-02	1.235E+01	20	A019	2 2 1 1 0	
6.167E-01	1.000E+02	100	A019	1 2 1 1 0	

1091. $C_7H_6N_4O$

7-Methoxypteridine
Pteridine, 7-methoxy-

RN: 204443-27-6   **MP** (°C):
MW: 162.15   **BP** (°C):

Solubility (Moles/L)	Solubility (Grams/L)	Temp (°C)	Ref (#)	Evaluation (T P E A A)	Comments
1.209E-01	1.961E+01	20	A083	1 2 0 0 0	
1.233E+00	2.000E+02	100	A083	1 2 0 0 0	

1092. C₇H₆N₄O

3,4-Dihydro-4-keto-3-methylpteridine
3:4-Dihydro-4-keto-3-methylpteridine
RN: 24851-65-8   **MP** (°C):   286
MW: 162.15   **BP** (°C):

Solubility (Moles/L)	Solubility (Grams/L)	Temp (°C)	Ref (#)	Evaluation (T P E A A)	Comments
8.686E-02	1.408E+01	20	A019	2 2 1 1 0	
6.167E-01	1.000E+02	100	A019	1 2 1 1 0	

1093. C₇H₆N₄O₂

4H-Pyrazolo[3,4-d]pyrimidin-4-one, 1-acetyl-1,5-dihydro-
RN: 96448-60-1   **MP** (°C):
MW: 178.15   **BP** (°C):

Solubility (Moles/L)	Solubility (Grams/L)	Temp (°C)	Ref (#)	Evaluation (T P E A A)	Comments
4.210E-03	7.500E-01	22	B428	1 2 1 2 1	

1094. C₇H₆N₄S

7-Methylthiopteridine
Pteridine, 7-(methylthio)-
Pteridine-7-methyl-thiol
RN: 204443-30-1   **MP** (°C):
MW: 178.22   **BP** (°C):

Solubility (Moles/L)	Solubility (Grams/L)	Temp (°C)	Ref (#)	Evaluation (T P E A A)	Comments
2.792E-02	4.975E+00	20	A083	1 2 0 0 0	
1.439E-01	2.564E+01	100	A083	1 2 0 0 0	

1095. C₇H₆N₄S

4-Methylthiopteridine
Pteridine, 4-(methylthio)-
Pteridine-4-methyl-thiol
RN: 6966-78-5   **MP** (°C):   191
MW: 178.22   **BP** (°C):

Solubility (Moles/L)	Solubility (Grams/L)	Temp (°C)	Ref (#)	Evaluation (T P E A A)	Comments
4.313E-03	7.686E-01	20	A083	1 2 0 0 0	
3.100E-02	5.525E+00	100	A083	1 2 0 0 0	

1096. C$_7$H$_6$N$_4$S
4-Mercapto-7-methylpteridine
4-Pteridinethiol, 7-methyl-

RN: 98550-33-5 **MP** (°C):
MW: 178.22 **BP** (°C):

Solubility (Moles/L)	Solubility (Grams/L)	Temp (°C)	Ref (#)	Evaluation (T P E A A)	Comments
3.738E-03	6.662E-01	100	A083	1 2 0 0 0	

1097. C$_7$H$_6$N$_4$S
2-Methylthiopteridine
Pteridine, 2-(methylthio)-
Pteridine-2-methyl-thiol

RN: 16878-77-6 **MP** (°C): 136
MW: 178.22 **BP** (°C):

Solubility (Moles/L)	Solubility (Grams/L)	Temp (°C)	Ref (#)	Evaluation (T P E A A)	Comments
1.748E-02	3.115E+00	20	A083	1 2 0 0 0	
1.369E-01	2.439E+01	100	A083	1 2 0 0 0	

1098. C$_7$H$_6$O
Benzaldehyde
Benzaldehyd

RN: 100-52-7 **MP** (°C): −55
MW: 106.13 **BP** (°C): 179

Solubility (Moles/L)	Solubility (Grams/L)	Temp (°C)	Ref (#)	Evaluation (T P E A A)	Comments
3.251E-02	3.450E+00	20	C008	1 2 2 0 2	
2.827E-02	3.000E+00	20	F300	1 0 0 0 0	
3.754E-02	3.984E+00	25	B019	1 0 1 2 0	
3.754E-02	3.984E+00	25	B092	2 1 1 1 1	
6.549E-02	6.950E+00	25	C005	2 2 2 2 2	average
3.289E-02	3.490E+00	25	C008	1 2 2 0 2	
6.170E-02	6.548E+00	25	M017	1 2 0 1 2	
3.741E-02	3.970E+00	30	C008	1 2 2 0 2	
2.110E-02	2.239E+00	37	E028	1 0 1 1 2	
8.960E-02	9.509E+00	60	B092	2 0 1 1 1	

1099. C$_7$H$_6$O$_2$
Benzoic acid
Benzenecarboxylic acid
Benzoesaeure
RN: 65-85-0   MP (°C):   122
MW: 122.12   BP (°C):   249

Solubility (Moles/L)	Solubility (Grams/L)	Temp (°C)	Ref (#)	Evaluation (T P E A A)	Comments
1.390E-02	1.697E+00	0	F302	1 0 0 0 2	
1.390E-02	1.697E+00	0	M043	1 0 0 0 1	
1.720E-02	2.100E+00	10	F300	1 0 0 0 1	
1.716E-02	2.096E+00	10	F302	1 0 0 0 2	
1.634E-02	1.996E+00	10	M043	1 0 0 0 1	
2.010E-02	2.455E+00	15	P329	0 0 0 0 0	
1.982E-02	2.421E+00	15.5	K062	2 0 1 1 2	
2.200E-02	2.687E+00	17	B109	1 0 0 0 2	unit assumed, *sic*
2.237E-02	2.732E+00	17.7	K062	2 0 1 1 2	
2.260E-02	2.760E+00	18	B109	1 0 0 0 2	unit assumed, *sic*
2.211E-02	2.700E+00	18	F071	1 1 2 1 2	
2.100E-02	2.565E+00	18	H009	2 1 2 2 0	EFG, 0.01N HCl
2.211E-02	2.700E+00	18	H080	1 0 0 0 2	
2.257E-02	2.756E+00	18	L050	2 0 1 2 2	
2.211E-02	2.700E+00	18	M344	1 0 0 0 2	
2.308E-02	2.819E+00	19.0	K062	2 0 1 1 2	average of 2
2.368E-02	2.892E+00	20	D041	1 0 0 0 1	
2.339E-02	2.857E+00	20	F069	2 2 2 2 2	
2.375E-02	2.900E+00	20	F300	1 0 0 0 1	
2.368E-02	2.892E+00	20	F302	1 0 0 0 2	
2.200E-02	2.686E+00	20	M038	2 2 1 1 2	
2.368E-02	2.892E+00	20	M043	1 0 0 0 1	
2.437E-02	3.000E+00	20	M049	1 0 0 0 1	
2.400E-02	2.931E+00	20	P329	0 0 0 0 0	
2.825E-02	3.450E+00	20	W026	1 0 1 1 1	average of 2
2.540E-02	3.102E+00	22	E045	2 0 1 1 2	
2.605E-02	3.181E+00	23	E045	2 0 1 1 2	
2.807E-02	3.428E+00	24.6	W029	1 2 1 1 2	
2.620E-02	3.200E+00	25	A412	1 0 2 2 1	int
2.449E-02	2.991E+00	25	B019	1 0 1 2 0	
2.751E-02	3.359E+00	25	B085	2 1 1 1 2	
2.683E-02	3.277E+00	25	B097	2 2 1 1 2	0.01M sodium benzoate
2.800E-02	3.420E+00	25	B128	1 0 1 1 2	
2.768E-02	3.381E+00	25	B302	1 0 0 0 0	pH 2.0
2.805E-02	3.426E+00	25	D058	1 0 1 1 2	
2.746E-02	3.354E+00	25	E045	2 0 1 1 2	
2.810E-02	3.432E+00	25	F001	1 0 1 2 2	
2.784E-02	3.400E+00	25	F300	1 0 0 0 1	
2.800E-02	3.419E+00	25	H009	2 1 2 2 0	EFG, 0.01N HCl
2.784E-02	3.400E+00	25	H015	1 0 0 0 1	
2.251E-03	2.749E-01	25	H060	2 0 2 0 2	*sic*
2.760E-02	3.371E+00	25	H071	2 2 2 1 2	
2.800E-02	3.419E+00	25	H084	1 0 0 0 1	

(*continued*)

1099. C$_7$H$_6$O$_2$ (continued)

Solubility (Moles/L)	Solubility (Grams/L)	Temp (°C)	Ref (#)	Evaluation (T P E A A)	Comments
2.760E-02	3.371E+00	25	K005	1 0 0 1 2	
2.727E-02	3.330E+00	25	K047	1 2 1 2 2	
2.760E-02	3.371E+00	25	K057	2 2 1 1 2	
2.775E-02	3.389E+00	25	K064	2 2 2 1 2	
2.781E-02	3.396E+00	25	L048	1 2 2 1 2	
2.780E-02	3.395E+00	25	L050	2 0 1 2 2	
2.596E-02	3.170E+00	25	L338	1 0 1 1 2	
2.619E-02	3.199E+00	25	M038	2 2 1 1 2	
2.702E-02	3.300E+00	25	M049	1 0 0 0 1	
2.790E-02	3.407E+00	25	M116	2 1 1 1 2	
2.160E-02	2.638E+00	25	M149	2 0 2 2 2	intrinsic
2.900E-02	3.542E+00	25	O007	1 0 2 1 2	
2.268E-02	2.770E+00	25	P037	2 0 1 1 2	
2.807E-02	3.428E+00	25	P314	0 0 0 0 0	
8.820E+00	1.077E+03	25	P329	0 0 0 0 0	
2.793E-02	3.411E+00	25	R016	0 0 0 0 0	
2.781E-02	3.396E+00	25.0	K062	2 0 1 1 2	average of 2
2.700E-02	3.297E+00	25.00	M135	1 2 1 1 2	0.01N sodium benzoate
2.781E-02	3.396E+00	25.2	C096	1 0 0 1 2	
2.833E-02	3.460E+00	26	E045	2 0 1 1 2	
2.890E-02	3.529E+00	26.4	P043	2 0 1 1 2	
3.439E-02	4.200E+00	26.70	L095	2 2 1 1 2	
2.936E-02	3.586E+00	27	E045	2 0 1 1 2	
3.146E-02	3.842E+00	28	D050	1 2 1 2 2	
3.147E-02	3.843E+00	30	B109	1 0 0 0 2	unit assumed, *sic*
3.204E-02	3.913E+00	30	B109	1 0 0 0 2	unit assumed, *sic*
3.306E-02	4.037E+00	30	B118	1 0 0 0 2	
3.000E-02	3.664E+00	30	B142	2 0 1 1 0	EFG, 0.1N H$_2$SO$_4$
3.000E-02	3.664E+00	30	C077	0 0 0 0 0	
3.319E-02	4.054E+00	30	D033	2 2 1 2 2	
3.302E-02	4.033E+00	30	D061	1 0 0 0 2	
2.915E-02	3.560E+00	30	F005	1 2 2 2 2	
3.425E-02	4.182E+00	30	F302	1 0 0 0 2	
3.110E-02	3.799E+00	30	M038	2 2 1 1 2	
3.262E-02	3.984E+00	30	M043	1 0 0 0 1	
3.302E-02	4.033E+00	30	S204	2 0 1 0 2	
3.439E-02	4.200E+00	30	W026	1 0 1 1 1	average of 2
3.216E-02	3.927E+00	30.0	K062	2 0 1 1 2	average of 2
3.400E-02	4.152E+00	31	H009	2 1 2 2 0	EFG, 0.01N HCl
3.873E-02	4.730E+00	35	G052	2 1 1 1 2	
3.711E-02	4.532E+00	35	M038	2 2 1 1 2	
4.010E-02	4.897E+00	35	O007	1 0 2 1 2	
3.772E-02	4.607E+00	35	S204	2 0 1 0 2	
3.960E-02	4.836E+00	35.0	K062	2 0 1 1 2	
3.800E-02	4.641E+00	35.00	M135	1 2 1 1 2	0.01N sodium benzoate
4.201E-02	5.131E+00	37	B171	2 0 1 1 2	
3.611E-02	4.410E+00	37	F005	1 2 2 2 2	
4.200E-02	5.129E+00	37	H009	2 1 2 2 0	EFG, 0.01N HCl
3.734E-02	4.560E+00	37	M360	1 2 1 1 2	

(continued)

1099. $C_7H_6O_2$ (continued)

Solubility (Moles/L)	Solubility (Grams/L)	Temp (°C)	Ref (#)	Evaluation (T P E A A)	Comments
4.528E-02	5.529E+00	40	D033	2 2 1 2 2	
4.884E-02	5.964E+00	40	F302	1 0 0 0 1	
4.376E-02	5.345E+00	40	M038	2 2 1 1 2	
4.560E-02	5.569E+00	40	M043	1 0 0 0 1	
4.424E-02	5.403E+00	40	S204	2 0 1 0 2	
5.110E-02	6.241E+00	42.4	W029	1 2 1 1 2	
4.774E-02	5.830E+00	45	F005	1 2 2 2 2	
5.000E-02	6.106E+00	45	H009	2 1 2 2 0	EFG, 0.01N HCl
5.282E-02	6.451E+00	45	M038	2 2 1 1 2	
5.254E-02	6.417E+00	45	S204	2 0 1 0 2	
5.324E-02	6.502E+00	45.0	K062	2 0 1 1 2	
5.500E-02	6.717E+00	45.00	M135	1 2 1 1 2	0.01N sodium benzoate
5.463E-02	6.672E+00	45.3	S124	1 0 0 1 1	
6.878E-02	8.400E+00	50	F300	1 0 0 0 1	
6.901E-02	8.428E+00	50	F302	1 0 0 0 2	
2.107E-02	2.573E+00	50	L006	1 0 0 0 2	
6.237E-02	7.617E+00	50	S204	2 0 1 0 2	
8.032E-02	9.809E+00	53.8	S124	1 0 0 1 2	
7.048E-02	8.607E+00	55	S204	2 0 1 0 2	
8.300E-02	1.014E+01	55.40	M135	1 2 1 1 2	0.01N sodium benzoate
8.853E-02	1.081E+01	57.8	W029	1 2 1 1 2	
9.710E-02	1.186E+01	60	F302	1 0 0 0 2	
9.550E-02	1.166E+01	60	L047	1 1 2 1 2	
9.390E-02	1.147E+01	60	M043	1 0 0 0 2	
1.000E-01	1.221E+01	60.20	M135	1 2 1 1 2	0.01N sodium benzoate
1.129E-01	1.378E+01	62.5	S124	1 0 0 1 2	
1.190E-01	1.453E+01	64.60	M135	1 2 1 1 2	0.01N sodium benzoate
1.390E-01	1.698E+01	68.50	M135	1 2 1 1 2	0.01N sodium benzoate
1.527E-01	1.864E+01	69.4	S124	1 0 0 1 2	
1.424E-01	1.739E+01	70	F302	1 0 0 0 2	
1.658E-01	2.025E+01	74.1	W029	1 2 1 1 2	
1.870E-01	2.284E+01	75.10	M135	1 2 1 1 2	0.01N sodium benzoate
2.242E-01	2.739E+01	79.0	S124	1 0 0 1 2	
2.210E-01	2.699E+01	79.30	M135	1 2 1 1 2	0.01N sodium benzoate
2.192E-01	2.676E+01	80	F302	1 0 0 0 2	
2.168E-01	2.648E+01	80	M043	1 0 0 0 2	
2.540E-01	3.102E+01	82.10	M135	1 2 1 1 2	0.01N sodium benzoate
2.567E-01	3.135E+01	82.3	S124	1 0 0 1 2	
2.485E-01	3.035E+01	83.1	W029	1 2 1 1 2	
3.124E-01	3.815E+01	88.3	W029	1 2 1 1 2	
4.211E-01	5.142E+01	88.6	S124	1 0 0 1 2	
3.550E-01	4.335E+01	88.60	M135	1 2 1 1 2	0.01N sodium benzoate
3.564E-01	4.352E+01	90	F302	1 0 0 0 2	
4.342E-01	5.302E+01	91.5	W029	1 2 1 1 2	average of 3
5.214E-01	6.367E+01	95	D041	1 0 0 0 1	
5.208E-01	6.360E+01	95	F300	1 0 0 0 2	
5.214E-01	6.367E+01	95	F302	1 0 0 0 2	
4.977E-01	6.078E+01	95.3	W029	1 2 1 1 2	
5.493E-01	6.708E+01	98.6	W029	1 2 1 1 2	

(continued)

1099. C$_7$H$_6$O$_2$ (continued)

Solubility (Moles/L)	Solubility (Grams/L)	Temp (°C)	Ref (#)	Evaluation (T P E A A)	Comments
4.547E-01	5.553E+01	100	M043	1 0 0 0 2	
8.241E-01	1.006E+02	109.4	W029	1 2 1 1 2	
1.399E+00	1.709E+02	116.1	W029	1 2 1 1 2	
2.594E+00	3.168E+02	116.3	W029	1 2 1 1 2	
2.001E+00	2.444E+02	117.2	W029	1 2 1 1 2	
9.000E-04	1.099E-01	ns	D037	1 1 1 1 0	pH 3.0, intrinsic

1100. C$_7$H$_6$O$_2$
m-Hydroxybenzaldehyde
3-Hydroxy-benzaldehyd

RN: 100-83-4   **MP** (°C): 104
MW: 122.12   **BP** (°C):

Solubility (Moles/L)	Solubility (Grams/L)	Temp (°C)	Ref (#)	Evaluation (T P E A A)	Comments
2.252E-01	2.750E+01	43	F300	1 0 0 0 2	

1101. C$_7$H$_6$O$_2$
p-Hydroxybenzaldehyde
4-Hydroxy-benzaldehyd

RN: 123-08-0   **MP** (°C): 213.5
MW: 122.12   **BP** (°C):

Solubility (Moles/L)	Solubility (Grams/L)	Temp (°C)	Ref (#)	Evaluation (T P E A A)	Comments
1.056E-01	1.290E+01	30	F300	1 0 0 0 2	

1102. C$_7$H$_6$O$_2$
Salicylaldehyde
Salicylaldehyd

RN: 90-02-8   **MP** (°C): −7
MW: 122.12   **BP** (°C): 197

Solubility (Moles/L)	Solubility (Grams/L)	Temp (°C)	Ref (#)	Evaluation (T P E A A)	Comments
6.614E-04	8.077E-02	25	K129	2 1 2 2 2	
1.392E-01	1.700E+01	86	F300	1 0 0 0 1	

1103. C₇H₆O₃
Salicylic acid
2-Hydroxybenzoic acid
o-Hydroxybenzoic acid

RN:	69-72-7	MP (°C):	158
MW:	138.12	BP (°C):	211

Solubility (Moles/L)	Solubility (Grams/L)	Temp (°C)	Ref (#)	Evaluation (T P E A A)	Comments
6.799E-03	9.391E-01	0	C083	1 2 1 1 2	
5.792E-03	8.000E-01	0	F300	1 0 0 0 0	
9.400E-03	1.298E+00	0	M043	1 0 0 0 0	
9.400E-03	1.298E+00	0	M043	1 0 0 0 1	
9.472E-03	1.308E+00	10	B074	1 2 1 2 2	
8.688E-03	1.200E+00	10	F300	1 0 0 0 1	
1.084E-02	1.498E+00	10	M043	1 0 0 0 0	
1.084E-02	1.498E+00	10	M043	1 0 0 0 1	
8.656E-03	1.196E+00	10	N420	0 0 0 0 0	
9.327E-03	1.288E+00	10	W044	1 0 1 0 2	
1.108E-02	1.531E+00	9.99	A341	0 0 0 0 0	
1.009E-02	1.393E+00	12.1	W044	1 0 1 0 2	
1.207E-02	1.667E+00	14.5	D061	1 0 0 0 2	
1.209E-02	1.670E+00	14.50	B118	1 0 0 0 2	unit assumed
1.028E-02	1.420E+00	15	H022	1 2 2 2 2	
1.520E-03	2.100E-01	15	M461	0 0 0 0 0	
9.875E-03	1.364E+00	15	N420	0 0 0 0 0	
1.258E-02	1.737E+00	17	K046	1 0 0 0 2	spray-dried product
1.330E-02	1.837E+00	20	B074	1 2 1 2 2	
1.303E-02	1.800E+00	20	F071	1 1 2 1 2	
1.303E-02	1.800E+00	20	F300	1 0 0 0 1	
1.303E-02	1.800E+00	20	H080	1 0 0 0 2	
1.296E-02	1.790E+00	20	K047	1 2 1 2 2	
1.445E-02	1.996E+00	20	M043	1 0 0 0 1	
1.445E-02	1.996E+00	20	M043	1 0 0 0 0	
1.445E-02	1.996E+00	20	M107	2 2 1 1 0	EFG
1.303E-02	1.800E+00	20	M344	1 0 0 0 2	
1.154E-02	1.594E+00	20	N420	0 0 0 0 0	
1.593E-02	2.200E+00	20	W026	1 0 1 1 1	average of 2
1.330E-02	1.837E+00	20	W044	1 0 1 0 2	
1.520E-02	2.100E+00	21	B331	1 2 2 1 0	pH 7.4
1.390E-02	1.920E+00	22	E045	2 0 1 1 2	
1.470E-02	2.030E+00	23	E045	2 0 1 1 2	
1.474E-02	2.036E+00	23.0	W044	1 0 1 0 2	
1.550E-02	2.141E+00	24	E045	2 0 1 1 2	
1.847E-02	2.551E+00	24.99	A341	0 0 0 0 0	
1.590E-02	2.196E+00	25	B090	1 1 1 1 2	
1.633E-02	2.255E+00	25	C083	1 2 1 1 2	
1.630E-02	2.251E+00	25	E045	2 0 1 1 2	
1.593E-02	2.200E+00	25	H007	0 0 0 0 0	
1.620E-02	2.238E+00	25	H084	1 0 0 0 2	
1.084E-02	1.498E+00	25	H129	1 0 0 1 0	

(continued)

1103. C$_7$H$_6$O$_3$ (continued)

Solubility (Moles/L)	Solubility (Grams/L)	Temp (°C)	Ref (#)	Evaluation (T P E A A)	Comments
1.613E-02	2.228E+00	25	K040	1 0 2 1 2	
1.634E-02	2.257E+00	25	K053	2 2 2 2 2	
1.620E-02	2.238E+00	25	K057	2 2 1 1 2	
1.601E-02	2.211E+00	25	L050	2 0 1 2 2	
1.665E-03	2.300E-01	25	M461	0 0 0 0 0	
1.370E-02	1.892E+00	25	N420	0 0 0 0 0	
1.680E-02	2.320E+00	25	O007	1 0 2 1 2	
1.621E-02	2.239E+00	25	P314	0 0 0 0 0	
1.491E-02	2.059E+00	25.50	A012	2 2 2 2 2	
1.700E-02	2.348E+00	26	E045	2 0 1 1 2	
1.780E-02	2.459E+00	27	E045	2 0 1 1 2	
1.746E-02	2.411E+00	27	K046	1 0 0 0 2	spray-dried product
1.728E-02	2.387E+00	28	D050	1 2 1 2 2	
1.784E-02	2.464E+00	28.1	W044	1 0 1 0 2	
1.360E-02	1.878E+00	30	A065	2 0 2 2 1	
1.885E-02	2.603E+00	30	B074	1 2 1 2 2	
1.987E-02	2.745E+00	30	B118	1 0 0 0 2	unit assumed
1.750E-02	2.417E+00	30	B142	2 0 1 1 0	EFG, 0.1N H$_2$SO$_4$
1.800E-02	2.486E+00	30	C077	0 0 0 0 0	
1.986E-02	2.743E+00	30	D061	1 0 0 0 2	
1.426E-02	1.970E+00	30	F005	1 2 2 2 2	
1.796E-02	2.481E+00	30	H022	1 2 2 2 2	
1.700E-02	2.348E+00	30	K020	1 0 1 1 0	EFG
1.868E-02	2.580E+00	30	K047	1 2 1 2 2	
2.022E-02	2.792E+00	30	M043	1 0 0 0 0	
2.022E-02	2.792E+00	30	M043	1 0 0 0 1	
2.165E-02	2.991E+00	30	M107	2 2 1 1 0	EFG
2.244E-03	3.100E-01	30	M461	0 0 0 0 0	
1.685E-02	2.327E+00	30	N420	0 0 0 0 0	
2.244E-02	3.100E+00	30	W026	1 0 1 1 2	average of 2
1.906E-02	2.633E+00	30	W044	1 0 1 0 2	
2.172E-02	3.000E+00	30.6	P014	2 1 2 2 0	
2.442E-02	3.373E+00	33.99	A341	0 0 0 0 0	
2.201E-02	3.041E+00	34.4	W044	1 0 1 0 2	
2.273E-02	3.140E+00	35	K047	1 2 1 2 2	
2.039E-02	2.816E+00	35	N420	0 0 0 0 0	
2.390E-02	3.301E+00	35	O007	1 0 2 1 2	
1.332E-02	1.840E+00	37	B171	2 0 1 1 2	
1.861E-02	2.570E+00	37	C079	0 0 0 0 0	
1.897E-02	2.620E+00	37	F005	1 2 2 2 2	
2.452E-02	3.386E+00	37	K046	1 0 0 0 2	spray-dried product
1.303E-02	1.800E+00	37	Y421	0 0 0 0 0	
2.590E-02	3.577E+00	38.7	W044	1 0 1 0 2	
2.848E-02	3.934E+00	40	B074	1 2 1 2 2	
2.679E-02	3.700E+00	40	F300	1 0 0 0 1	
2.672E-02	3.690E+00	40	K047	1 2 1 2 2	
3.028E-02	4.182E+00	40	M043	1 0 0 0 1	

(continued)

1103. C$_7$H$_6$O$_3$ (continued)

Solubility (Moles/L)	Solubility (Grams/L)	Temp (°C)	Ref (#)	Evaluation (T P E A A)	Comments
3.028E-02	4.182E+00	40	M043	1 0 0 0 0	
2.884E-02	3.984E+00	40	M107	2 2 1 1 0	EFG
4.561E-03	6.300E-01	40	M461	0 0 0 0 0	
2.502E-02	3.456E+00	40	N420	0 0 0 0 0	
2.719E-02	3.756E+00	40	W044	1 0 1 0 2	
3.167E-02	4.374E+00	43.99	A341	0 0 0 0 0	
3.743E-02	5.170E+00	44.99	A341	0 0 0 0 0	
2.462E-02	3.400E+00	45	F005	1 2 2 2 2	
3.059E-02	4.226E+00	45	N420	0 0 0 0 0	
3.714E-02	5.130E+00	46.99	A341	0 0 0 0 0	
3.562E-02	4.921E+00	47	K046	1 0 0 0 2	spray-dried product
3.681E-02	5.084E+00	48.6	W044	1 0 1 0 2	
4.102E-02	5.665E+00	49.99	A341	0 0 0 0 0	
4.261E-02	5.885E+00	50	B074	1 2 1 2 2	
6.154E-03	8.500E-01	50	M461	0 0 0 0 0	
3.769E-02	5.206E+00	50	N420	0 0 0 0 0	
3.889E-02	5.371E+00	50	W044	1 0 1 0 2	
4.337E-02	5.991E+00	50.99	A341	0 0 0 0 0	
4.677E-02	6.461E+00	51.99	A341	0 0 0 0 0	
5.151E-02	7.115E+00	53.99	A341	0 0 0 0 0	
5.319E-02	7.347E+00	54.99	A341	0 0 0 0 0	
4.947E-02	6.833E+00	56.0	W044	1 0 1 0 2	
6.104E-02	8.431E+00	57.49	A341	0 0 0 0 0	
6.202E-02	8.566E+00	60	B074	1 2 1 2 2	
6.009E-02	8.300E+00	60	F300	1 0 0 0 1	
6.529E-02	9.018E+00	60	M043	1 0 0 0 1	
6.529E-02	9.018E+00	60	M043	1 0 0 0 0	
5.888E-02	8.133E+00	60	W044	1 0 1 0 2	
7.184E-02	9.922E+00	61.49	A341	0 0 0 0 0	
7.140E-02	9.862E+00	64.0	W044	1 0 1 0 2	
8.184E-02	1.130E+01	65.99	A341	0 0 0 0 0	
8.373E-02	1.156E+01	66.0	W044	1 0 1 0 2	
1.252E-01	1.730E+01	75.0	W044	1 0 1 0 2	
1.499E-01	2.070E+01	80	F300	1 0 0 0 2	
1.600E-01	2.210E+01	80	M043	1 0 0 0 0	
1.600E-01	2.210E+01	80	M043	1 0 0 0 2	
5.437E-01	7.510E+01	100	M043	1 0 0 0 0	
5.437E-01	7.510E+01	100	M043	1 0 0 0 2	
1.598E-02	2.207E+00	ns	O003	0 2 1 1 2	
1.514E-02	2.091E+00	ns	R427	0 0 0 0 0	
1.841E-02	2.544E+00	rt	H431	0 0 0 0 0	

1104. C₇H₆O₃

Protocatechualdehyde
3,4-Dihydroxy-benzaldehyd

RN: 139-85-5 **MP** (°C):
MW: 138.12 **BP** (°C):

Solubility (Moles/L)	Solubility (Grams/L)	Temp (°C)	Ref (#)	Evaluation (T P E A A)	Comments
3.620E-01	5.000E+01	20	F300	1 0 0 0 0	
~1.88E+00	~2.60E+02	100	F300	1 0 0 0 0	

1105. C₇H₆O₃

p-Hydroxybenzoic acid
4-Hydroxy-benzoesaeure
4-Hydroxybenzoic acid
p-Hydroxybenzoicacid
4-Hydroxybenzenecarboxylic acid

RN: 99-96-7 **MP** (°C): 214.5
MW: 138.12 **BP** (°C):

Solubility (Moles/L)	Solubility (Grams/L)	Temp (°C)	Ref (#)	Evaluation (T P E A A)	Comments
1.805E-02	2.494E+00	0	M043	1 0 0 0 1	
1.590E-02	2.196E+00	4.99	A405	2 0 1 1 2	
2.525E-02	3.488E+00	10	M043	1 0 0 0 1	
2.280E-02	3.149E+00	10.99	A405	2 0 1 1 2	
2.216E-02	3.061E+00	12.7	W044	1 0 1 0 2	
5.746E-02	7.937E+00	15	D041	1 0 0 0 0	
3.186E-02	4.400E+00	15	F300	1 0 0 0 1	
2.624E-02	3.624E+00	15	H022	1 2 2 2 2	
2.990E-02	4.130E+00	15.99	A405	2 0 1 1 2	
3.740E-02	5.166E+00	19.99	A405	2 0 1 1 2	
3.470E-02	4.793E+00	20	C006	1 2 1 1 2	
3.817E-02	5.272E+00	20	M043	1 0 0 0 1	
3.602E-02	4.975E+00	20	M107	2 2 1 1 0	EFG
3.545E-02	4.896E+00	20.9	W044	1 0 1 0 2	
4.890E-02	6.754E+00	24.99	A405	2 0 1 1 2	
3.545E-02	4.896E+00	25	D081	1 1 2 1 2	
6.580E-02	9.089E+00	25	D339	0 0 0 0 0	
4.634E-02	6.400E+00	25	H007	0 0 0 0 0	
3.318E-02	4.583E+00	25	M334	1 2 1 1 2	
4.322E-02	5.970E+00	25	N023	1 2 2 1 2	hydrate
6.241E-02	8.620E+00	25	N023	1 2 2 1 2	anhydrate
3.873E-02	5.350E+00	25.50	A012	2 2 2 2 2	
6.340E-02	8.757E+00	29.99	A405	2 0 1 1 2	
5.400E-02	7.459E+00	30	A065	2 0 2 2 1	
4.800E-02	6.630E+00	30	C077	0 0 0 0 0	
5.500E-02	7.597E+00	30	H019	0 0 0 0 0	
5.421E-02	7.488E+00	30	H022	1 2 2 2 2	
5.500E-02	7.597E+00	30	K020	1 0 1 1 0	EFG

(continued)

1105. C$_7$H$_6$O$_3$ (continued)

Solubility (Moles/L)	Solubility (Grams/L)	Temp (°C)	Ref (#)	Evaluation (T P E A A)	Comments
5.746E-02	7.937E+00	30	M043	1 0 0 0 1	
5.746E-02	7.937E+00	30	M107	2 2 1 1 0	EFG
5.538E-02	7.650E+00	30	N023	1 2 2 1 2	hydrate
7.790E-02	1.076E+01	30	N023	1 2 2 1 2	anhydrate
5.496E-02	7.592E+00	30	W044	1 0 1 0 2	
8.120E-02	1.122E+01	33.99	A405	2 0 1 1 2	
7.076E-02	9.774E+00	34.4	W044	1 0 1 0 2	
7.247E-02	1.001E+01	35	N023	1 2 2 1 2	hydrate
9.781E-02	1.351E+01	35	N023	1 2 2 1 2	anhydrate
1.231E-01	1.700E+01	37	B171	2 0 1 1 2	
1.027E-01	1.419E+01	38.99	A405	2 0 1 1 2	
8.663E-02	1.197E+01	39.4	W044	1 0 1 0 2	
8.938E-02	1.235E+01	40	M043	1 0 0 0 2	
9.996E-02	1.381E+01	40	M107	2 2 1 1 0	EFG
1.203E-01	1.662E+01	40	N023	1 2 2 1 2	anhydrate
9.339E-02	1.290E+01	40	N023	1 2 2 1 2	hydrate
1.385E-01	1.913E+01	42.99	A405	2 0 1 1 2	
1.291E-01	1.783E+01	46.0	W044	1 0 1 0 2	
1.804E-01	2.492E+01	47.99	A405	2 0 1 1 2	
2.438E-01	3.367E+01	52.99	A405	2 0 1 1 2	
1.931E-01	2.667E+01	54.6	W044	1 0 1 0 2	
3.330E-01	4.600E+01	56.99	A405	2 0 1 1 2	
2.978E-01	4.114E+01	60	M043	1 0 0 0 2	
4.286E-01	5.920E+01	61.99	A405	2 0 1 1 2	
5.666E-01	7.826E+01	66.99	A405	2 0 1 1 2	
7.269E-01	1.004E+02	71.99	A405	2 0 1 1 2	
1.835E-01	2.534E+01	75	D041	1 0 0 0 1	
8.723E-01	1.205E+02	80	M043	1 0 0 0 2	
1.875E+00	2.590E+02	100	F300	1 0 0 0 2	
2.410E+00	3.329E+02	100	M043	1 0 0 0 2	
3.715E-02	5.132E+00	ns	R427	0 0 0 0 0	
4.854E-02	6.705E+00	rt	H431	0 0 0 0 0	

1106. C$_7$H$_6$O$_3$

β-2-Furyncrylic acid
β-2-Furylacrylic acid
β-Furyl-(2)-acrylsaeure

RN: 539-47-9   **MP** (°C): 143
MW: 138.12   **BP** (°C): 286

Solubility (Moles/L)	Solubility (Grams/L)	Temp (°C)	Ref (#)	Evaluation (T P E A A)	Comments
1.448E-02	2.000E+00	20	F300	1 0 0 0 0	

1107. C$_7$H$_6$O$_3$

m-Hydroxybenzoic acid
3-Hydroxy-benzoesaeure
3-Hydroxybenzoic acid
m-Hydroxybenzoicacid

RN: 99-06-9   MP (°C): 202
MW: 138.12   BP (°C):

Solubility (Moles/L)	Solubility (Grams/L)	Temp (°C)	Ref (#)	Evaluation (T P E A A)	Comments
2.525E-02	3.488E+00	0	M043	1 0 0 0 1	
3.960E-02	5.470E+00	10	M043	1 0 0 0 1	
4.804E-02	6.636E+00	13.3	W044	1 0 1 0 2	
5.068E-02	7.000E+00	15	F300	1 0 0 0 1	
4.477E-02	6.184E+00	15	H022	1 2 2 2 2	
6.052E-02	8.360E+00	18.8	W044	1 0 1 0 2	
6.173E-02	8.527E+00	20	M043	1 0 0 0 1	
4.318E-02	5.964E+00	20	M107	2 2 1 1 0	EFG
7.551E-02	1.043E+01	24.3	W044	1 0 1 0 2	
5.249E-02	7.250E+00	25.50	A012	2 2 2 2 2	
7.800E-03	1.077E+00	30	A065	2 0 2 2 1	
8.600E-02	1.188E+01	30	C077	0 0 0 0 0	
8.800E-02	1.215E+01	30	H019	0 0 0 0 0	
8.300E-02	1.146E+01	30	H021	1 2 1 1 0	EFG
9.291E-02	1.283E+01	30	M043	1 0 0 0 1	
6.813E-02	9.411E+00	30	M107	2 2 1 1 0	EFG
9.552E-02	1.319E+01	30	W044	1 0 1 0 2	
9.855E-02	1.361E+01	30.9	W044	1 0 1 0 2	
1.271E-01	1.756E+01	36.2	W044	1 0 1 0 2	
1.420E-01	1.961E+01	40	M043	1 0 0 0 1	
1.105E-01	1.526E+01	40	M107	2 2 1 1 0	EFG
2.809E-01	3.880E+01	50	F300	1 0 0 0 1	
2.222E-01	3.070E+01	51.0	W044	1 0 1 0 2	
3.118E-01	4.306E+01	60	M043	1 0 0 0 1	
7.987E-01	1.103E+02	80	M043	1 0 0 0 2	
2.678E+00	3.699E+02	100	M043	1 0 0 0 2	
1.810E-02	2.500E+00	ns	B361	0 0 0 0 0	
5.012E-02	6.923E+00	ns	R427	0 0 0 0 0	

1108. C$_7$H$_6$O$_4$

2,6-Dihydroxybenzoic acid
2,6-Dihydroxy-benzoesaeure
γ-Resorcylic acid

RN: 303-07-1   MP (°C):
MW: 154.12   BP (°C):

Solubility (Moles/L)	Solubility (Grams/L)	Temp (°C)	Ref (#)	Evaluation (T P E A A)	Comments
6.200E-02	9.556E+00	ns	C014	0 0 0 1 1	

1109. C₇H₆O₄

Protocatechuic acid
3,4-Dihydroxy-benzoesaeure
3,4-Dihydroxybenzoic acid

RN: 99-50-3   **MP** (°C):
MW: 154.12   **BP** (°C):

Solubility (Moles/L)	Solubility (Grams/L)	Temp (°C)	Ref (#)	Evaluation (T P E A A)	Comments
1.181E-01	1.820E+01	14	F300	1 0 0 0 2	
1.440E+00	2.220E+02	80	F300	1 0 0 0 2	

1110. C₇H₆O₄

β-Resorcyclic acid
2,4-Dihydroxy-benzoesaeure
2,4-Dihydroxybenzoic acid
2,4-Dihydroxybenzoicacid
β-Resorcylic acid
4-Hydroxysalicylic acid

RN: 89-86-1   **MP** (°C):   225
MW: 154.12   **BP** (°C):

Solubility (Moles/L)	Solubility (Grams/L)	Temp (°C)	Ref (#)	Evaluation (T P E A A)	Comments
3.893E-02	6.000E+00	25	H007	0 0 0 0 0	

1111. C₇H₆O₄

Gentisic acid
2,5-Dihydroxy-benzoesaeure
2,5-Dihydroxybenzoic acid
2,5-Dihydroxybenzoicacid
Hydroquinonecarboxylic acid

RN: 490-79-9   **MP** (°C):   205
MW: 154.12   **BP** (°C):

Solubility (Moles/L)	Solubility (Grams/L)	Temp (°C)	Ref (#)	Evaluation (T P E A A)	Comments
1.427E-01	2.200E+01	25	H007	0 0 0 0 0	

1112. C₇H₆O₅

Gallic acid
3,4,5-Trihydroxybenzoesaeure
Gallussaeure

RN: 149-91-7   **MP** (°C):   250
MW: 170.12   **BP** (°C):

Solubility (Moles/L)	Solubility (Grams/L)	Temp (°C)	Ref (#)	Evaluation (T P E A A)	Comments
1.325E+00	2.253E+02	–10.0	L430	0 0 0 0 0	
5.349E-02	9.100E+00	15	M461	0 0 0 0 0	
5.589E-02	9.509E+00	19.99	L430	0 0 0 0 0	

(continued)

1112. $C_7H_6O_5$ (continued)

Solubility (Moles/L)	Solubility (Grams/L)	Temp (°C)	Ref (#)	Evaluation (T P E A A)	Comments
6.995E-02	1.190E+01	20	F300	1 0 0 0 2	
5.820E-02	9.901E+00	24.99	L430	0 0 0 0 0	
8.641E-02	1.470E+01	25	M461	0 0 0 0 0	
8.001E-02	1.361E+01	29.99	L430	0 0 0 0 0	
1.093E-01	1.860E+01	30	M461	0 0 0 0 0	
1.034E-01	1.759E+01	34.99	L430	0 0 0 0 0	
1.355E-01	2.306E+01	39.99	L430	0 0 0 0 0	
1.552E-01	2.640E+01	40	M461	0 0 0 0 0	
1.751E-01	2.979E+01	44.99	L430	0 0 0 0 0	
2.272E-01	3.865E+01	49.99	L430	0 0 0 0 0	
2.240E-01	3.810E+01	50	M461	0 0 0 0 0	
2.879E-01	4.898E+01	54.99	L430	0 0 0 0 0	
3.774E-01	6.420E+01	59.99	L430	0 0 0 0 0	
4.470E-01	7.604E+01	64.99	L430	0 0 0 0 0	
6.044E-01	1.028E+02	69.99	L430	0 0 0 0 0	
7.064E-01	1.202E+02	74.99	L430	0 0 0 0 0	
9.497E-01	1.616E+02	79.99	L430	0 0 0 0 0	
1.198E+00	2.038E+02	84.99	L430	0 0 0 0 0	
1.505E+00	2.561E+02	100	F300	1 0 0 0 2	
4.202E-02	7.149E+00	−.0	L430	0 0 0 0 0	
6.918E-02	1.177E+01	ns	R427	0 0 0 0 0	

1113. $C_7H_6O_5$

2,3,4-Trihydroxybenzoic acid
2,3,4-Trihydroxybenzoesaeure
RN: 610-02-6 **MP** (°C):
MW: 170.12 **BP** (°C):

Solubility (Moles/L)	Solubility (Grams/L)	Temp (°C)	Ref (#)	Evaluation (T P E A A)	Comments
5.878E-03	1.000E+00	12.50	F300	1 0 0 0 0	

1114. C_7H_7Br

m-Bromotoluene
3-Bromotoluene
3-Methyl-1-bromobenzene
1-Bromo-3-methylbenzene
3-Bromo-1-methylbenzene
3-Methylphenyl bromide
RN: 591-17-3 **MP** (°C): −39.8
MW: 171.04 **BP** (°C): 183.7

Solubility (Moles/L)	Solubility (Grams/L)	Temp (°C)	Ref (#)	Evaluation (T P E A A)	Comments
3.000E-04	5.131E-02	ns	O013	0 1 0 1 0	
3.020E-04	5.165E-02	ns	S460	0 0 0 0 0	

1115. C₇H₇Cl

m-Chlorotoluene
3-Chlorotoluene
1-Chloro-3-methylbenzene
m-Tolyl chloride

| **RN:** | 108-41-8 | **MP** (°C): | –48 |
| **MW:** | 126.59 | **BP** (°C): | 161.8 |

Solubility (Moles/L)	Solubility (Grams/L)	Temp (°C)	Ref (#)	Evaluation (T P E A A)	Comments
3.000E-04	3.798E-02	ns	O013	0 1 0 1 0	

1116. C₇H₇Cl

o-Chlorotoluene
2-Chlorotoluene
2-Chloro-1-methylbenzene
2-Methylchlorobenzene
1-Methyl-2-chlorobenzene
OCT

| **RN:** | 95-49-8 | **MP** (°C): | –36 |
| **MW:** | 126.59 | **BP** (°C): | 159.0 |

Solubility (Moles/L)	Solubility (Grams/L)	Temp (°C)	Ref (#)	Evaluation (T P E A A)	Comments
3.000E-04	3.798E-02	ns	O013	0 1 0 1 0	

1117. C₇H₇Cl

p-Chlorotoluene
4-Chlorotoluene
p-Tolyl chloride
4-Chloro-1-methyl-benzene
PCT
1-Chloro-4-methylbenzene

| **RN:** | 106-43-4 | **MP** (°C): | 8 |
| **MW:** | 126.59 | **BP** (°C): | 162.0 |

Solubility (Moles/L)	Solubility (Grams/L)	Temp (°C)	Ref (#)	Evaluation (T P E A A)	Comments
8.415E-04	1.065E-01	20	H118	1 1 1 1 2	
1.084E-03	1.372E-01	20	H301	0 0 0 0 0	
3.000E-04	3.798E-02	ns	O013	0 1 0 1 0	

1118. C$_7$H$_7$ClN$_2$O$_4$S

Saluamine
2-Amino-4-chloro-5-sulfamoylbenzoic acid
4-Chloro-5-sulfamylanthranilic acid
Desfurylmethylfurosemide
4-Chloro-5-sulfamoylanthranilic acid
-Amino-5-aminosulfonyl-4-chlorobenzoic acid

RN: 3086-91-7	**MP** (°C):	
MW: 250.66	**BP** (°C):	549.2

Solubility (Moles/L)	Solubility (Grams/L)	Temp (°C)	Ref (#)	Evaluation (T P E A A)	Comments
2.218E-03	5.560E-01	25	B405	1 1 1 2 2	Buffer pH 2.0
3.008E-03	7.540E-01	25	B405	1 1 1 2 2	

1119. C$_7$H$_7$ClN$_4$O$_2$

8-Chlorotheophylline
8-Chloro-1,3-dimethyl-2,6(1H,3H)-purinedione

RN: 85-18-7	**MP** (°C):	290
MW: 214.61	**BP** (°C):	

Solubility (Moles/L)	Solubility (Grams/L)	Temp (°C)	Ref (#)	Evaluation (T P E A A)	Comments
3.020E-03	6.481E-01	ns	R427	0 0 0 0 0	

1120. C$_7$H$_7$ClO

Chlorocresol
3-Methyl-4-chlorophenol
4-Chloro-3-cresol
6-Chloro-3-hydroxytoluene
3-Methyl-4-chloro-phenol-
Phenol, 4-chloro-3-methyl-

RN: 59-50-7	**MP** (°C):	67
MW: 142.59	**BP** (°C):	

Solubility (Moles/L)	Solubility (Grams/L)	Temp (°C)	Ref (#)	Evaluation (T P E A A)	Comments
2.800E-02	3.992E+00	25	B316	0 0 0 0 0	
3.489E-02	4.975E+00	25	R041	0 0 0 0 0	
3.647E-02	5.200E+00	ns	G024	0 0 0 0 2	

1121. C$_7$H$_7$ClO

4-Chloroanisole
p-Chloroanisole
1-Chloro-4-methoxybenzene

RN: 623-12-1	**MP** (°C):	−18
MW: 142.59	**BP** (°C):	

Solubility (Moles/L)	Solubility (Grams/L)	Temp (°C)	Ref (#)	Evaluation (T P E A A)	Comments
1.662E-03	2.370E-01	25	L348	1 2 2 1 2	

1122. C₇H₇ClO

2-Methyl-6-chloro-phenol
2-Chloro-6-methylphenol
6-Chloro-*o*-cresol
3-Chloro-2-hydroxytoluene
6-Chloro-2-methylphenol

RN: 87-64-9   **MP** (°C):
MW: 142.59   **BP** (°C):

Solubility (Moles/L)	Solubility (Grams/L)	Temp (°C)	Ref (#)	Evaluation (T P E A A)	Comments
2.500E-02	3.565E+00	25	B316	0 0 0 0 0	

1123. C₇H₇ClO

2-Methyl-4-chloro-phenol
4-Chloro-*o*-cresol
4-Chloro-2-methylphenol
5-Chloro-2-hydroxytoluene

RN: 1570-64-5   **MP** (°C):   45–48
MW: 142.59   **BP** (°C):   220–225

Solubility (Moles/L)	Solubility (Grams/L)	Temp (°C)	Ref (#)	Evaluation (T P E A A)	Comments
4.800E-02	6.844E+00	25	B316	0 0 0 0 0	

1124. C₇H₇ClO

2-Chloroanisole
o-Chloroanisole

RN: 766-51-8   **MP** (°C):   −27
MW: 142.59   **BP** (°C):   196

Solubility (Moles/L)	Solubility (Grams/L)	Temp (°C)	Ref (#)	Evaluation (T P E A A)	Comments
3.437E-03	4.900E-01	25	L348	1 2 2 1 2	
3.467E-03	4.944E-01	ns	S460	0 0 0 0 0	

1125. C₇H₇ClO

3-Chloroanisole
m-Chloroanisole
1-Chloro-3-methoxybenzene

RN: 2845-89-8   **MP** (°C):   <25
MW: 142.59   **BP** (°C):

Solubility (Moles/L)	Solubility (Grams/L)	Temp (°C)	Ref (#)	Evaluation (T P E A A)	Comments
1.648E-03	2.350E-01	25	L348	1 2 2 1 2	
1.660E-03	2.366E-01	ns	S460	0 0 0 0 0	

1126. C₇H₇Cl₂NO

Clopidol
3,5-Dichloro-2,6-dimethyl-4-pyridinol
Coyden
Methylchloropindol

RN:	2971-90-6	**MP** (°C):
MW:	192.05	**BP** (°C):

Solubility (Moles/L)	Solubility (Grams/L)	Temp (°C)	Ref (#)	Evaluation (T P E A A)	Comments
2.083E-04	4.000E-02	ns	K138	0 0 0 0 1	

1127. C₇H₇Cl₃NO₃PS

Chlorpyrifos-methyl
Chlorpyrifos-methy

RN:	5598-13-0	**MP** (°C):
MW:	322.54	**BP** (°C):

Solubility (Moles/L)	Solubility (Grams/L)	Temp (°C)	Ref (#)	Evaluation (T P E A A)	Comments
5.581E-06	1.800E-03	10	B324	0 0 0 0 0	
5.581E-06	1.800E-03	10	B324	0 0 0 0 0	
9.922E-06	3.200E-03	20	B300	2 1 1 1 2	
9.922E-06	3.200E-03	20	B324	0 0 0 0 0	
9.921E-06	3.200E-03	20	B324	0 0 0 0 0	
1.476E-05	4.760E-03	20	C053	0 0 0 0 0	
1.240E-05	4.000E-03	24	K069	2 0 0 1 1	
1.240E-05	4.000E-03	25	M161	1 0 0 0 0	
2.139E-05	6.899E-03	30	B324	0 0 0 0 0	
2.139E-05	6.900E-03	30	B324	0 0 0 0 0	
1.476E-05	4.760E-03	ns	F071	0 1 2 1 2	
1.240E-05	4.000E-03	ns	K138	0 0 0 0 1	
1.643E-05	5.300E-03	ns	M110	0 0 0 0 0	EFG

1128. C₇H₇Cl₃NO₄P

Torelle
Dimethyl 3,5,6-trichloro-2-pyridinyl phosphate
DOWCO 217
Fospirate
Phosphoric acid, dimethyl 3,5,6-trichloro-2-pyridyl ester

RN:	5598-52-7	**MP** (°C):
MW:	306.47	**BP** (°C):

Solubility (Moles/L)	Solubility (Grams/L)	Temp (°C)	Ref (#)	Evaluation (T P E A A)	Comments
9.789E-04	3.000E-01	24	K069	2 0 0 1 1	

1129. C₇H₇FN₂O₃

3-Propionyl-5-fluoro-2,4(1H,3H)-pyrimidinedi-one
3-Propionyl-5-fluorouracil

RN: 75410-16-1   **MP** (°C):   113–114
MW: 186.14   **BP** (°C):

Solubility (Moles/L)	Solubility (Grams/L)	Temp (°C)	Ref (#)	Evaluation (T P E A A)	Comments
1.896E-01	3.530E+01	22	B321	0 0 0 0 0	pH 4.0
1.896E-01	3.530E+01	22	B332	1 1 0 0 1	pH 4.0
1.980E-01	3.686E+01	22	B416	2 2 1 2 1	

1130. C₇H₇FN₂O₄

3-Acetoxymethyl-5-fluoro-2,4(1H,3H)-pyrimidinedi-one
3-Acetoxymethyl-5-fluorouracil

RN: 73042-04-3   **MP** (°C):   158–159
MW: 202.14   **BP** (°C):

Solubility (Moles/L)	Solubility (Grams/L)	Temp (°C)	Ref (#)	Evaluation (T P E A A)	Comments
9.894E-02	2.000E+01	22	B321	0 0 0 0 0	pH 4.0

1131. C₇H₇FN₂O₄

1-Acetoxymethyl-5-fluoro-2,4(1H,3H)-pyrimidinedi-one
1-Acetoxymethyl-5-fluorouracil

RN: 62113-41-1   **MP** (°C):   122–123
MW: 202.14   **BP** (°C):

Solubility (Moles/L)	Solubility (Grams/L)	Temp (°C)	Ref (#)	Evaluation (T P E A A)	Comments
2.132E-01	4.310E+01	22	B321	0 0 0 0 0	pH 4.0

1132. C₇H₇FN₂O₄

3-Ethyloxycarbonyl-5-fluoro-2,4(1H,3H)-pyrimidinedi-one
3-Ethyloxycarbonyl-5-fluorouracil
1-Ethyloxycarbonyl-5-fluorouracil

RN: 75410-27-4   **MP** (°C):   126–128
MW: 202.14   **BP** (°C):

Solubility (Moles/L)	Solubility (Grams/L)	Temp (°C)	Ref (#)	Evaluation (T P E A A)	Comments
3.562E-01	7.200E+01	22	B321	0 0 0 0 0	pH 4.0
3.413E-02	6.900E+00	22	B332	1 1 0 0 1	pH 4.0

1133. C₇H₇NO

Benzamide
Benzamid
Phenyl carboxamide
Benzoic acid amide

RN:	55-21-0	MP (°C):	130
MW:	121.14	BP (°C):	288

Solubility (Moles/L)	Solubility (Grams/L)	Temp (°C)	Ref (#)	Evaluation (T P E A A)	Comments
4.923E-02	5.964E+00	10	M043	1 0 0 0 0	
4.750E-02	5.754E+00	12	O019	1 0 0 1 2	
1.000E-01	1.211E+01	20	B139	2 1 1 1 1	
8.173E-02	9.901E+00	20	M043	1 0 0 0 1	
1.100E-01	1.333E+01	22	J037	0 0 0 0 0	
1.106E-01	1.340E+01	25	F300	1 0 0 0 2	
1.059E-01	1.283E+01	30	M043	1 0 0 0 1	
1.300E-01	1.575E+01	40	M043	1 0 0 0 1	
1.651E-01	2.000E+01	50	P064	2 0 1 1 1	
3.931E-01	4.762E+01	60	M043	1 0 0 0 0	
6.191E-01	7.500E+01	70	P064	2 0 1 1 1	
5.503E+00	6.667E+02	80	M043	1 0 0 0 2	
6.686E+00	8.100E+02	90	P064	2 0 1 1 2	
7.338E+00	8.889E+02	100	M043	1 0 0 0 2	
7.842E+00	9.500E+02	110	P064	2 0 1 1 2	
1.100E-01	1.332E+01	rt	D021	0 0 1 1 2	

1134. C₇H₇NO₂

Salicylamide
2-Hydroxybenzoicacidamide
Algamon
Amid-sal
Amidosal
Algiamida

RN:	65-45-2	MP (°C):	140
MW:	137.14	BP (°C):	

Solubility (Moles/L)	Solubility (Grams/L)	Temp (°C)	Ref (#)	Evaluation (T P E A A)	Comments
8.878E-03	1.218E+00	10	N419	0 0 0 0 0	
1.060E-02	1.454E+00	15	D012	1 1 0 1 2	
1.137E-02	1.559E+00	15	N419	0 0 0 0 0	
1.100E-02	1.509E+00	16	D012	1 1 0 1 2	
1.531E-02	2.100E+00	20	E046	1 0 0 0 0	EFG
1.447E-02	1.985E+00	20	N419	0 0 0 0 0	
1.900E-02	2.606E+00	22	J031	0 0 0 0 0	
1.604E-02	2.200E+00	23	B328	1 2 2 1 1	pH 4.0
1.500E-02	2.057E+00	25	D012	1 1 0 1 2	
1.750E-02	2.400E+00	25	E046	1 0 0 0 0	EFG
1.757E-02	2.409E+00	25	N419	0 0 0 0 0	
1.831E-02	2.511E+00	25	P314	0 0 0 0 0	

(continued)

1134. C₇H₇NO₂ (continued)

Solubility (Moles/L)	Solubility (Grams/L)	Temp (°C)	Ref (#)	Evaluation (T P E A A)	Comments
2.115E-02	2.900E+00	30	E046	1 0 0 0 0	EFG
2.166E-02	2.970E+00	30	N419	0 0 0 0 0	
2.771E-02	3.800E+00	35	E046	1 0 0 0 0	EFG
2.685E-02	3.682E+00	35	N419	0 0 0 0 0	
2.900E-02	3.977E+00	37	D012	1 1 0 1 2	
3.427E-02	4.700E+00	40	E046	1 0 0 0 0	EFG
3.285E-02	4.505E+00	40	N419	0 0 0 0 0	
4.280E-02	5.870E+00	45	D012	1 1 0 1 2	
4.181E-02	5.734E+00	45	N419	0 0 0 0 0	
5.323E-02	7.300E+00	50	E046	1 0 0 0 0	EFG
5.371E-02	7.366E+00	50	N419	0 0 0 0 0	
1.677E-03	2.300E-01	ns	B361	0 0 0 0 0	

1135. C₇H₇NO₂

p-Aminobenzoic acid
4-Amino-benzoesaeure
4-Aminobenzoic acid
p-Aminobenzoicacid
1-Amino-4-carboxybenzene

RN: 150-13-0   **MP** (°C):   187.0
MW: 137.14   **BP** (°C):

Solubility (Moles/L)	Solubility (Grams/L)	Temp (°C)	Ref (#)	Evaluation (T P E A A)	Comments
2.479E-02	3.400E+00	12.80	F300	1 0 0 0 1	
3.609E-02	4.950E+00	18	C033	1 0 2 1 2	
3.628E-02	4.975E+00	25	D041	1 0 0 0 0	
3.930E-02	5.390E+00	25	L338	1 0 1 1 2	
3.646E-02	5.000E+00	25	M054	1 0 0 0 0	
3.500E-02	4.800E+00	25	P015	0 0 0 0 0	
4.455E-02	6.110E+00	30	C033	1 0 2 1 2	
4.579E-02	6.280E+00	30	H018	0 0 0 0 0	
4.500E-02	6.171E+00	30	L069	1 0 1 1 0	EFG
6.125E-02	8.400E+00	37	B171	2 0 1 1 2	
6.040E-02	8.283E+00	37	F006	1 1 2 2 2	

1136. C₇H₇NO₂

o-Nitrotoluene
2-Nitro-toluol
2-Nitrotoluene

RN: 88-72-2   **MP** (°C):   –9.5
MW: 137.14   **BP** (°C):   221.7

Solubility (Moles/L)	Solubility (Grams/L)	Temp (°C)	Ref (#)	Evaluation (T P E A A)	Comments
3.872E-03	5.310E-01	9.99	B403	1 2 2 2 2	
4.441E-03	6.090E-01	19.99	B403	1 2 2 2 2	

(*continued*)

1136. C$_7$H$_7$NO$_2$ (continued)

Solubility (Moles/L)	Solubility (Grams/L)	Temp (°C)	Ref (#)	Evaluation (T P E A A)	Comments
5.017E-03	6.880E-01	29.99	B403	1 2 2 2 2	
4.740E-03	6.500E-01	30	F300	1 0 0 0 2	
5.637E-03	7.730E-01	39.99	B403	1 2 2 2 2	

1137. C$_7$H$_7$NO$_2$
o-Aminobenzoic acid
2-Aminobenzoic acid
Anthranilsaeure

RN: 118-92-3   **MP** (°C): 145
MW: 137.14   **BP** (°C):

Solubility (Moles/L)	Solubility (Grams/L)	Temp (°C)	Ref (#)	Evaluation (T P E A A)	Comments
2.181E-02	2.991E+00	10	M043	1 0 0 0 0	
2.543E-02	3.488E+00	14	D041	1 0 0 0 1	
2.552E-02	3.500E+00	14	F300	1 0 0 0 1	
2.543E-02	3.488E+00	20	M043	1 0 0 0 1	
4.349E-02	5.964E+00	30	M043	1 0 0 0 0	
6.504E-02	8.920E+00	40	M043	1 0 0 0 0	
3.552E+00	4.872E+02	100	M043	1 0 0 0 1	

1138. C$_7$H$_7$NO$_2$
m-Nitrotoluene
3-Nitro-toluol
3-Nitrotoluene

RN: 99-08-1   **MP** (°C): 16
MW: 137.14   **BP** (°C): 232.6

Solubility (Moles/L)	Solubility (Grams/L)	Temp (°C)	Ref (#)	Evaluation (T P E A A)	Comments
3.281E-03	4.500E-01	9.99	B403	1 2 2 2 2	
3.580E-03	4.910E-01	19.99	B403	1 2 2 2 2	
3.894E-03	5.340E-01	29.99	B403	1 2 2 2 2	
3.646E-03	5.000E-01	30	F300	1 0 0 0 2	
4.120E-03	5.650E-01	39.99	B403	1 2 2 2 2	

1139. C$_7$H$_7$NO$_2$
m-Aminobenzoic acid
3-Amino-benzoesaeure
3-Aminobenzoic acid

RN: 99-05-8   **MP** (°C): 174
MW: 137.14   **BP** (°C):

Solubility (Moles/L)	Solubility (Grams/L)	Temp (°C)	Ref (#)	Evaluation (T P E A A)	Comments
4.302E-02	5.900E+00	14.90	F300	1 0 0 0 1	
5.830E-02	7.995E+00	30	W007	2 0 2 2 2	

1140. C$_7$H$_7$NO$_2$
Methyl nicotinate
Nicotinsaeure-methyl ester

RN: 93-60-7   **MP** (°C): 39
MW: 137.14   **BP** (°C): 209

Solubility (Moles/L)	Solubility (Grams/L)	Temp (°C)	Ref (#)	Evaluation (T P E A A)	Comments
3.471E-01	4.760E+01	20	F300	1 0 0 0 2	*sic*
8.065E+00	1.106E+03	32	L346	1 0 0 1 0	
3.467E-01	4.755E+01	ns	R424	0 0 0 0 0	

1141. C$_7$H$_7$NO$_2$
p-Nitrotoluene
4-Nitrotoluene

RN: 99-99-0   **MP** (°C): 55
MW: 137.14   **BP** (°C):

Solubility (Moles/L)	Solubility (Grams/L)	Temp (°C)	Ref (#)	Evaluation (T P E A A)	Comments
1.305E-03	1.790E-01	9.99	B403	1 2 2 2 2	
2.917E-04	4.000E-02	14.5	D070	1 2 0 0 1	
2.917E-04	4.000E-02	14.50	F300	1 0 0 0 1	
1.765E-03	2.420E-01	19.99	B403	1 2 2 2 2	
2.100E-03	2.880E-01	20	H306	1 0 1 2 1	
2.150E-03	2.949E-01	20	T301	1 2 2 2 2	
2.348E-03	3.220E-01	29.99	B403	1 2 2 2 2	
3.048E-03	4.180E-01	39.99	B403	1 2 2 2 2	
5.687E-04	7.799E-02	50	D070	1 2 0 0 1	
8.458E-04	1.160E-01	100	D070	1 2 0 0 2	

1142. C$_7$H$_7$NO$_3$
3-Methyl-4-nitrophenol
3-Nitro-*p*-cresol
3-Nitro-*p*-kresol
4-Nitro-5-methylphenol

RN: 2581-34-2   **MP** (°C): 128
MW: 153.14   **BP** (°C):

Solubility (Moles/L)	Solubility (Grams/L)	Temp (°C)	Ref (#)	Evaluation (T P E A A)	Comments
7.769E-03	1.190E+00	25	B104	1 2 1 1 1	
7.762E-03	1.189E+00	ns	R427	0 0 0 0 0	

1143. C₇H₇NO₃

p-Aminosalicylic acid
4-Amino-salicylsaeure
4-Aminosalicylic acid

RN: 65-49-6   **MP** (°C): 150
MW: 153.14   **BP** (°C):

Solubility (Moles/L)	Solubility (Grams/L)	Temp (°C)	Ref (#)	Evaluation (T P E A A)	Comments
1.303E-02	1.996E+00	20	D041	1 0 0 0 0	
1.100E-02	1.685E+00	23	M072	1 2 1 1 0	EFG
2.100E-02	3.216E+00	30	L069	1 0 1 1 0	EFG
1.087E-02	1.664E+00	ns	H125	0 0 0 0 0	

1144. C₇H₇NO₃

p-Nitroanisol
4-Nitro-anisol
4-Nitroanisol

RN: 100-17-4   **MP** (°C): 54
MW: 153.14   **BP** (°C): 260

Solubility (Moles/L)	Solubility (Grams/L)	Temp (°C)	Ref (#)	Evaluation (T P E A A)	Comments
4.571E-04	7.000E-02	15	F300	1 0 0 0 1	
3.853E-03	5.900E-01	30	F300	1 0 0 0 2	

1145. C₇H₇N₂OS

Ethyl acetylthiodiazole
Ethyle acetyle thiodiazolique

RN: **MP** (°C):
MW: 167.21   **BP** (°C):

Solubility (Moles/L)	Solubility (Grams/L)	Temp (°C)	Ref (#)	Evaluation (T P E A A)	Comments
1.196E-03	2.000E-01	37	D084	1 0 1 0 1	

1146. C₇H₇N₅

2-Methylaminopteridine
Pteridine, 2-(methylamino)-

RN: 19167-57-8   **MP** (°C): 219
MW: 161.17   **BP** (°C):

Solubility (Moles/L)	Solubility (Grams/L)	Temp (°C)	Ref (#)	Evaluation (T P E A A)	Comments
1.933E-02	3.115E+00	20	A019	2 2 1 1 1	
1.724E-01	2.778E+01	100	A019	1 2 1 1 1	

1147. C_7H_8
Toluene
Methylbenzene

RN:	108-88-3	MP (°C):	−94
MW:	92.14	BP (°C):	110.6

Solubility (Moles/L)	Solubility (Grams/L)	Temp (°C)	Ref (#)	Evaluation (T P E A A)	Comments
5.819E-03	5.362E-01	.06	U010	1 0 0 1 1	EFG
7.857E-03	7.240E-01	0	P003	2 2 2 2 2	
6.638E-03	6.116E-01	4.50	B086	2 1 2 2 2	
5.557E-03	5.120E-01	4.62	U010	1 0 0 1 1	EFG
5.557E-03	5.120E-01	4.62	U013	1 0 0 0 0	EFG
6.519E-03	6.006E-01	6.30	B086	2 1 2 2 2	
6.356E-03	5.857E-01	7.10	B086	2 1 2 2 2	
6.367E-03	5.867E-01	9	B086	2 1 2 2 2	
6.210E-03	5.722E-01	10	B149	2 1 1 2 2	
6.215E-03	5.727E-01	11.80	B086	2 1 2 2 2	
6.237E-03	5.747E-01	12.10	B086	2 1 2 2 2	
5.307E-03	4.890E-01	14.20	U013	1 0 0 0 0	EFG
5.785E-03	5.330E-01	15	S203	1 1 2 1 2	
6.172E-03	5.687E-01	15.10	B086	2 1 2 2 2	
5.424E-03	4.998E-01	16	D052	1 1 0 0 0	
5.100E-03	4.699E-01	16	F001	1 0 1 2 1	
5.101E-03	4.700E-01	16	F071	1 1 2 1 2	
5.101E-03	4.700E-01	16	F300	1 0 0 0 2	
5.101E-03	4.700E-01	16	H080	1 0 0 0 2	
5.100E-03	4.699E-01	16	S006	1 0 0 0 1	
6.370E-03	5.869E-01	20	B149	2 1 1 2 2	
6.154E-03	5.670E-01	20	B356	0 0 0 0 0	
5.424E-03	4.998E-01	20	C121	1 0 0 0 0	unit assumed, *sic*
5.590E-03	5.151E-01	20	M312	1 0 0 0 2	
4.982E-03	4.591E-01	20	M337	2 1 2 2 2	
6.139E-03	5.657E-01	20.10	B086	2 1 2 2 2	
5.196E-03	4.788E-01	21	C024	2 1 1 2 2	
5.752E-03	5.300E-01	25	A001	1 2 2 2 1	
5.098E-03	4.698E-01	25	A094	1 0 0 0 1	
6.805E-03	6.270E-01	25	B003	2 1 2 2 2	
5.589E-03	5.150E-01	25	B060	2 0 1 1 1	
6.690E-03	6.164E-01	25	B153	2 1 1 1 2	
1.680E-02	1.548E+00	25	B173	2 0 2 2 2	*sic*
5.687E-03	5.240E-01	25	B304	2 0 2 2 2	
8.000E-03	7.371E-01	25	H092	1 1 1 1 0	
6.500E-03	5.989E-01	25	H313	2 1 2 2 1	
6.000E-03	5.529E-01	25	H332	2 2 2 2 0	
6.370E-02	5.869E+00	25	I334	2 2 2 1 2	*sic*
6.370E-03	5.869E-01	25	I335	2 2 2 2 2	
5.430E-03	5.003E-01	25	K001	1 0 2 1 2	
5.318E-03	4.900E-01	25	K072	1 0 1 1 1	
6.290E-03	5.796E-01	25	K316	2 2 2 2 2	
5.641E-03	5.197E-01	25	L319	1 0 2 1 2	

(continued)

1147. C_7H_8 (continued)

Solubility (Moles/L)	Solubility (Grams/L)	Temp (°C)	Ref (#)	Evaluation (T P E A A)	Comments
5.589E-03	5.150E-01	25	M130	1 0 0 0 2	
5.638E-03	5.195E-01	25	M132	2 2 2 1 2	
6.280E-03	5.787E-01	25	M342	1 0 1 1 2	
6.219E-03	5.730E-01	25	P003	2 2 2 2 2	
6.012E-03	5.540E-01	25	P051	2 1 1 2 2	
6.045E-03	5.570E-01	25	S203	1 1 2 1 2	
5.804E-03	5.348E-01	25	S358	2 1 2 2 2	
5.650E-03	5.206E-01	25	S359	2 1 2 2 2	
6.280E-03	5.787E-01	25	W300	2 2 2 2 2	
5.307E-03	4.890E-01	25.35	U010	1 0 0 1 1	EFG
5.307E-03	4.890E-01	25.35	U013	1 0 0 0 0	EFG
3.255E-03	2.999E-01	30	F053	1 0 2 0 2	
6.183E-03	5.697E-01	30	G029	1 0 2 2 1	
5.067E-03	4.669E-01	30	M311	1 1 2 2 2	
1.409E-02	1.298E+00	30	S207	1 0 0 1 1	*sic*
5.557E-03	5.120E-01	34.53	U010	1 0 0 1 1	EFG
5.557E-03	5.120E-01	34.53	U013	1 0 0 0 0	EFG
6.371E-03	5.870E-01	35	S203	1 1 2 1 2	
5.954E-03	5.486E-01	44.30	U010	1 0 0 1 1	EFG
5.819E-03	5.362E-01	44.30	U013	1 0 0 0 0	EFG
6.892E-03	6.350E-01	45	S203	1 1 2 1 2	
1.517E-02	1.398E+00	45	S207	1 0 0 1 1	*sic*
6.529E-03	6.015E-01	54.71	U013	1 0 0 0 0	EFG
1.500E-02	1.382E+00	55	H092	1 1 1 1 1	
6.380E-03	5.879E-01	55.79	U010	1 0 0 1 1	EFG
1.734E-02	1.597E+00	60	S207	1 0 0 1 1	*sic*
7.325E-03	6.749E-01	65.82	U013	1 0 0 0 0	EFG
2.171E-02	2.000E+00	150	J023	1 1 2 2 0	
7.597E-02	7.000E+00	200	J023	1 1 2 2 0	
3.039E-01	2.800E+01	250	J023	1 1 2 2 1	
1.411E+00	1.300E+02	300	J023	1 1 2 2 2	
5.589E-03	5.150E-01	ns	H123	0 0 0 0 0	
1.380E-01	1.272E+01	ns	H307	0 0 0 0 0	*sic*
5.611E-03	5.170E-01	ns	M175	0 0 2 1 2	
5.589E-03	5.150E-01	ns	M344	0 0 0 0 2	

1148. C_7H_8
1,6-Heptadiyne

RN: 2396-63-6   **MP (°C):** −85
MW: 92.14   **BP (°C):** 112

Solubility (Moles/L)	Solubility (Grams/L)	Temp (°C)	Ref (#)	Evaluation (T P E A A)	Comments
1.791E-02	1.650E+00	25	M001	2 1 2 2 2	

1149. C₇H₈

Cycloheptatriene
1,3,5-Cycloheptatriene
Tropilidene
CHT

RN:　　544-25-2　　　**MP** (°C):　　−80
MW:　　92.14　　　　**BP** (°C):　　116.0

Solubility (Moles/L)	Solubility (Grams/L)	Temp (°C)	Ref (#)	Evaluation (T P E A A)	Comments
6.301E-03	5.806E-01	4.8	L007	2 2 1 2 2	
6.301E-03	5.806E-01	5.1	L007	2 1 1 1 2	
7.207E-03	6.641E-01	14.8	L007	2 2 1 2 2	
7.207E-03	6.641E-01	15.2	L007	2 1 1 1 2	
7.260E-03	6.690E-01	24.8	L007	2 2 1 2 2	
6.729E-03	6.200E-01	25	M001	2 1 2 2 2	
7.260E-03	6.690E-01	25.1	L007	2 1 1 1 2	
8.045E-03	7.413E-01	34.8	L007	2 2 1 2 2	
8.045E-03	7.413E-01	35.2	L007	2 1 1 1 2	
8.294E-03	7.642E-01	44.8	L007	2 2 1 2 2	
8.294E-03	7.642E-01	45.2	L007	2 1 1 1 2	

1150. C₇H₈ClN₃O₄S₂

Hydrochlorothiazide
Chlorozide

RN:　　58-93-5　　　**MP** (°C):　　274
MW:　　297.74　　　**BP** (°C):

Solubility (Moles/L)	Solubility (Grams/L)	Temp (°C)	Ref (#)	Evaluation (T P E A A)	Comments
2.425E-03	7.220E-01	25	A076	1 0 1 1 2	
2.032E-03	6.050E-01	25	C437	0 0 0 0 0	Average
2.045E-03	6.090E-01	25	D091	1 0 0 0 2	pH 6.2
2.687E-03	8.000E-01	25	G051	1 0 1 1 0	
2.800E-03	8.337E-01	30	A089	2 0 1 1 0	EFG
2.800E-03	8.337E-01	30	A093	2 0 1 1 0	EFG
2.520E-03	7.503E-01	30	E049	2 0 2 2 2	
3.627E-03	1.080E+00	37	D091	1 0 0 0 2	pH 7.2
7.650E-03	2.278E+00	50	M335	1 0 2 1 2	pH 5
3.359E-03	1.000E+00	ns	K444	0 0 0 0 0	
1.982E-03	5.900E-01	rt	A095	0 0 0 0 0	

1151. $C_7H_8ClN_3O_4S_2$

Hydrochlorothiazide
3,4-Dihydro-6-chloro-7-sulfamoyl-1,2,4-benzothiadiazine-1,1-dioxide
3,4-Dihydrochlorothiazide
6-Chloro-3,4-dihydro-2H-1,2,4-benzothiadiazine-7-sulfonamide-1,1-dioxide
6-Chloro-3,4-dihydro-7-sulfamoyl-2H-1,2,4-benzothiadiazine-1,1-dioxide
Aldactazide

RN: 58-93-5   **MP** (°C):   274
MW: 297.74   **BP** (°C):

Solubility (Moles/L)	Solubility (Grams/L)	Temp (°C)	Ref (#)	Evaluation (T P E A A)	Comments
1.997E-03	5.946E-01	22.5	B422	2 0 2 2 2	
2.351E-06	7.000E-04	25	A408	2 0 1 2 0	
2.115E-03	6.296E-01	25	S450	0 0 0 0 0	

1152. $C_7H_8FN_3O_3$

1-Ethylcarbamoyl-5-fluorouracil
1-Ethylcarbamoyl-5-fluoro-2,4(1H,3H)-pyrimidinedi-one
N-Ethyl-5-fluoro-3,4-dihydro-2,4-dioxo-1-pyrimidinecarboxamide

RN: 58471-47-9   **MP** (°C):   190–196
MW: 201.16   **BP** (°C):

Solubility (Moles/L)	Solubility (Grams/L)	Temp (°C)	Ref (#)	Evaluation (T P E A A)	Comments
7.457E-03	1.500E+00	22	B321	0 0 0 0 0	pH 4.0
7.457E-03	1.500E+00	22	B388	0 0 0 0 0	

1153. $C_7H_8FN_3O_3$

1-(N,N-Dimethylcarbamoyl)-5-fluorouracil
1-Dimethylcarbamoyl-5-fluoro-2,4(1H,3H)-pyrimidinedi-one

RN: 60908-29-4   **MP** (°C):   226–227
MW: 201.16   **BP** (°C):

Solubility (Moles/L)	Solubility (Grams/L)	Temp (°C)	Ref (#)	Evaluation (T P E A A)	Comments
2.983E-02	6.000E+00	22	B321	0 0 0 0 0	pH 4.0
2.983E-02	6.000E+00	22	B388	0 0 0 0 0	

1154. $C_7H_8N_2O_2$

3-Nitro-o-toluidine
3-Nitro-o-toluidin

RN: 603-83-8   **MP** (°C):   92
MW: 152.15   **BP** (°C):   305

Solubility (Moles/L)	Solubility (Grams/L)	Temp (°C)	Ref (#)	Evaluation (T P E A A)	Comments
8.807E-02	1.340E+01	100	F300	1 0 0 0 2	

1155. C₇H₈N₂O₃

5,5-Trimethylenebarbituric acid
6,8-Diazaspiro[3.5]nonane-5,7,9-trione
Cyclobutane-spirobarbiturate

RN: 6128-03-6   **MP** (°C):
MW: 168.15   **BP** (°C):

Solubility (Moles/L)	Solubility (Grams/L)	Temp (°C)	Ref (#)	Evaluation (T P E A A)	Comments
2.213E-02	3.721E+00	25	P350	0 0 0 0 0	intrinsic

1156. C₇H₈N₂O₃

1-Methoxy-2-amino-4-nitrobenzene

RN: 99-59-2   **MP** (°C):   118
MW: 168.15   **BP** (°C):

Solubility (Moles/L)	Solubility (Grams/L)	Temp (°C)	Ref (#)	Evaluation (T P E A A)	Comments
3.388E-03	5.697E-01	rt	N015	0 0 2 2 2	

1157. C₇H₈N₂O₃S

5-Carboethoxy-2-thiouracil
Ethyl 2-thiouracil-5-carboxylate

RN: 38026-46-9   **MP** (°C):   252
MW: 200.22   **BP** (°C):

Solubility (Moles/L)	Solubility (Grams/L)	Temp (°C)	Ref (#)	Evaluation (T P E A A)	Comments
7.970E-03	1.596E+00	25	G016	1 2 1 2 2	intrinsic

1158. C₇H₈N₂O₄

Ethyl orotate
1,2,3,6-Tetrahydro-2,6-dioxo-4-pyrimidine-carboxylic acid, ethyl ester

RN: 1747-53-1   **MP** (°C):
MW: 184.15   **BP** (°C):

Solubility (Moles/L)	Solubility (Grams/L)	Temp (°C)	Ref (#)	Evaluation (T P E A A)	Comments
2.100E-02	3.867E+00	20	N019	0 0 0 0 0	

1159. C₇H₈N₂S

1-Phenyl-2-thiourea
Phenylthioharnstoff

RN: 103-85-5   **MP** (°C):   149
MW: 152.22   **BP** (°C):

Solubility (Moles/L)	Solubility (Grams/L)	Temp (°C)	Ref (#)	Evaluation (T P E A A)	Comments
1.708E-02	2.600E+00	18	F300	1 0 0 0 1	
3.830E-01	5.830E+01	100	F300	1 0 0 0 2	

1160. C₇H₈N₄O₂

Theophylline
1,3-Dimethylxanthine
Aerolate
Bronkotabs
Bronchodid Duracap
Bronkodyl

RN: 58-55-9   **MP (°C):** 272
MW: 180.17   **BP (°C):**

Solubility (Moles/L)	Solubility (Grams/L)	Temp (°C)	Ref (#)	Evaluation (T P E A A)	Comments
3.310E-02	5.964E+00	16	A072	1 0 1 0 1	
2.866E-02	5.164E+00	20	K052	1 1 1 1 2	
1.380E+00	2.486E+02	25	B443	0 0 0 0 0	
3.420E-02	6.162E+00	25	F009	2 2 2 2 0	EFG
3.675E-02	6.621E+00	25	L338	1 0 1 1 2	
4.089E-02	7.366E+00	25	M128	2 0 1 2 2	
4.083E-02	7.356E+00	25	M158	2 0 2 2 2	
3.580E-02	6.450E+00	25	N312	2 1 1 1 1	
4.607E-02	8.300E+00	25	P010	1 0 1 1 1	
4.607E-02	8.300E+00	25	P011	0 0 0 0 0	
4.440E-02	8.000E+00	25	P018	1 0 2 2 1	
4.440E-02	8.000E+00	25	P020	2 0 1 1 1	
4.607E-02	8.300E+00	25	P312	0 0 0 0 0	
4.500E-02	8.108E+00	30	B042	1 2 1 1 1	
4.500E-02	8.108E+00	30	G021	1 0 0 0 2	
4.100E-02	7.387E+00	30	H016	2 2 2 2 0	EFG
4.500E-02	8.108E+00	30	H020	1 0 0 0 1	
5.550E-02	1.000E+01	37	F076	2 0 2 2 0	
2.761E-02	4.975E+00	ns	J025	0 0 0 0 2	
5.550E-03	1.000E+00	ns	K444	0 0 0 0 0	
3.580E-02	6.450E+00	ns	N062	2 0 1 2 2	
2.054E-04	3.700E-02	rt	N015	0 0 2 2 1	*sic*

1161. C₇H₈N₄O₂

Theobromine
Theobromin

RN: 83-67-0   **MP (°C):** 357
MW: 180.17   **BP (°C):**

Solubility (Moles/L)	Solubility (Grams/L)	Temp (°C)	Ref (#)	Evaluation (T P E A A)	Comments
1.665E-03	3.000E-01	18	F300	1 0 0 0 0	
3.328E-03	5.996E-01	19	A072	1 0 1 0 0	
2.419E-03	4.358E-01	20	K052	1 1 1 1 2	
1.830E-03	3.297E-01	25	M158	2 0 2 2 2	
1.832E-03	3.300E-01	25	O302	1 0 0 1 0	
2.775E-03	5.000E-01	25	P010	1 0 1 1 1	
3.330E-03	6.000E-01	25	P011	0 0 0 0 0	
3.386E-03	6.100E-01	25	P018	1 0 2 2 1	

(continued)

1161. C₇H₈N₄O₂ (continued)

Solubility (Moles/L)	Solubility (Grams/L)	Temp (°C)	Ref (#)	Evaluation (T P E A A)	Comments
3.108E-03	5.600E-01	25	P020	2 0 1 1 1	
3.000E-03	5.405E-01	30	B042	1 2 1 1 0	
~3.00E-03	~5.41E-01	30	H020	1 0 0 0 0	
3.830E-02	6.900E+00	100	F300	1 0 0 0 1	
2.774E-03	4.998E-01	c	D004	0 0 0 0 0	
3.676E-02	6.623E+00	h	D004	0 0 0 0 0	
>2.77E-03	>5.00E-01	ns	B404	0 2 1 1 0	

1162. C₇H₈O

p-Cresol
4-Cresol
p-Methylphenol

RN: 106-44-5 **MP** (°C): 35.5
MW: 108.14 **BP** (°C): 201.8

Solubility (Moles/L)	Solubility (Grams/L)	Temp (°C)	Ref (#)	Evaluation (T P E A A)	Comments
1.813E-01	1.961E+01	20	B031	1 0 2 2 1	
1.701E-01	1.840E+01	20	R087	0 0 0 0 0	0.15M NaCl
1.990E-01	2.152E+01	25	A021	1 2 1 1 0	
1.902E-01	2.057E+01	25	B019	1 0 1 2 0	
1.813E-01	1.961E+01	25	L022	1 0 0 0 0	
1.967E-01	2.127E+01	25	P004	0 0 0 0 0	
1.902E-01	2.057E+01	25	R041	0 0 0 0 0	
2.044E-01	2.210E+01	29.5	K119	1 0 0 0 2	
1.999E-01	2.162E+01	29.50	M098	1 2 0 1 2	
2.090E-01	2.260E+01	40	F300	1 0 0 0 2	
3.334E-01	3.605E+01	82.10	M098	1 2 0 1 2	

1163. C₇H₈O

Anisole
Methoxybenzene
Methyl phenyl ether
Phenyl methyl ether

RN: 100-66-3 **MP** (°C): −37.3
MW: 108.14 **BP** (°C): 155.5

Solubility (Moles/L)	Solubility (Grams/L)	Temp (°C)	Ref (#)	Evaluation (T P E A A)	Comments
1.295E-03	1.400E-01	25	A003	1 2 1 2 1	*sic*
9.609E-02	1.039E+01	25	B019	1 0 1 2 0	
1.000E-02	1.081E+00	25	D407	1 0 2 2 2	
1.400E-02	1.514E+00	25	M327	1 0 0 1 2	
1.418E-02	1.533E+00	25.04	V013	2 2 2 2 2	
9.617E-02	1.040E+01	26.70	L095	2 2 1 1 2	

1164. C₇H₈O

2-Cresol
2-Methylphenol
Phenol, 2-methyl-
o-Cresol
o-Methylphenol

RN: 95-48-7   **MP** (°C):   31
MW: 108.14   **BP** (°C):

Solubility (Moles/L)	Solubility (Grams/L)	Temp (°C)	Ref (#)	Evaluation (T P E A A)	Comments
2.519E-01	2.724E+01	20	B031	1 0 2 2 1	
2.276E-01	2.461E+01	20	R087	0 0 0 0 0	0.15M NaCl
2.312E-01	2.500E+01	23	P332	0 0 0 0 0	
2.400E-01	2.595E+01	25	A021	1 2 1 1 0	
1.991E-01	2.153E+01	25	B019	1 0 1 2 0	
2.127E-01	2.300E+01	25	B060	2 0 1 1 1	
2.400E-01	2.595E+01	25	B316	0 0 0 0 0	
2.300E-01	2.487E+01	25	F044	1 0 0 0 1	
2.423E-01	2.620E+01	25	F300	1 0 0 0 2	
2.569E-01	2.778E+01	25	L022	1 0 0 0 0	
2.999E-01	3.244E+01	25	P004	0 0 0 0 0	
2.255E-01	2.439E+01	25	R041	0 0 0 0 0	
1.991E-01	2.153E+01	31	B092	2 1 1 1 2	
2.606E-01	2.818E+01	46.20	M098	1 2 0 1 1	
2.497E-01	2.700E+01	50	K119	1 0 0 0 2	
2.763E-01	2.988E+01	60	B092	2 1 1 1 2	
3.557E-01	3.846E+01	86.70	M098	1 2 0 1 1	
2.291E-01	2.477E+01	ns	R427	0 0 0 0 0	

1165. C₇H₈O

m-Cresol
3-Cresol
m-Methylphenol

RN: 108-39-4   **MP** (°C):   11
MW: 108.14   **BP** (°C):   202

Solubility (Moles/L)	Solubility (Grams/L)	Temp (°C)	Ref (#)	Evaluation (T P E A A)	Comments
1.060E-01	1.147E+01	0	M041	1 1 0 0 2	
2.167E-01	2.344E+01	20	B031	1 2 2 2 1	
2.112E-01	2.284E+01	20	R087	0 0 0 0 0	0.15M NaCl
2.149E-01	2.324E+01	20.3	L339	2 0 2 2 2	
1.420E-01	1.536E+01	25	A021	1 2 1 1 0	
1.991E-01	2.153E+01	25	B019	1 0 1 2 0	
2.053E-01	2.220E+01	25	C060	1 2 1 1 2	
2.099E-01	2.270E+01	25	F300	1 0 0 0 2	
1.946E-01	2.105E+01	25	M041	1 1 0 0 2	
2.255E-01	2.439E+01	25	R041	0 0 0 0 0	
2.292E-01	2.478E+01	40.0	L339	2 0 2 2 2	

(continued)

1165. C₇H₈O (continued)

Solubility (Moles/L)	Solubility (Grams/L)	Temp (°C)	Ref (#)	Evaluation (T P E A A)	Comments
2.682E-01	2.900E+01	46.2	K119	1 0 0 0 2	
2.326E-01	2.515E+01	50	M041	1 1 0 0 2	
2.431E-01	2.629E+01	50.80	M098	1 2 0 1 1	
2.712E-01	2.933E+01	58.4	L339	2 0 2 2 2	
2.693E-01	2.913E+01	60	B031	1 2 2 2 1	
3.331E-01	3.602E+01	77.2	L339	2 0 2 2 2	
3.213E-01	3.475E+01	78.70	M098	1 2 0 1 1	
3.982E-01	4.306E+01	92.20	M098	1 2 0 1 1	
4.387E-01	4.744E+01	98.1	L339	2 0 2 2 2	

1166. C₇H₈O

Benzyl alcohol
Benzylalkohol
Benzenemethanol
Phenylmethanol
Phenylcarbinol
α-Hydroxytoluene

RN:	100-51-6	**MP** (°C):	−15.2
MW:	108.14	**BP** (°C):	204.7

Solubility (Moles/L)	Solubility (Grams/L)	Temp (°C)	Ref (#)	Evaluation (T P E A A)	Comments
3.606E-01	3.900E+01	17	F300	1 0 0 0 1	
3.488E-01	3.772E+01	20	H044	1 0 2 1 2	
3.520E-01	3.807E+01	20	S006	1 0 0 0 2	
3.967E-01	4.290E+01	25	B304	2 0 2 2 2	
3.540E-01	3.828E+01	25	H044	1 0 2 1 2	
4.260E-01	4.607E+01	25	L322	1 1 2 2 1	
3.616E-01	3.911E+01	30	H044	1 0 2 1 2	
3.646E-01	3.943E+01	35	H044	1 0 2 1 2	
3.676E-01	3.975E+01	40	H044	1 0 2 1 2	
3.724E-01	4.027E+01	45	H044	1 0 2 1 2	
3.722E-01	4.025E+01	50	H044	1 0 2 1 2	
3.868E-01	4.182E+01	55	H044	1 0 2 1 2	

1167. C₇H₈O₂

Salicyl alcohol
Salicylalkohol

RN:	90-01-7	**MP** (°C):	86
MW:	124.14	**BP** (°C):	

Solubility (Moles/L)	Solubility (Grams/L)	Temp (°C)	Ref (#)	Evaluation (T P E A A)	Comments
5.075E-01	6.300E+01	22	F300	1 0 0 0 1	

1168. $C_7H_8O_2$
Guaiacol
o-Methoxyphenol
RN: 90-05-1   **MP** (°C):   28
MW: 124.14   **BP** (°C):   205

Solubility (Moles/L)	Solubility (Grams/L)	Temp (°C)	Ref (#)	Evaluation (T P E A A)	Comments
1.506E-01	1.870E+01	15	F300	1 0 0 0 2	
1.880E-01	2.334E+01	24.99	B353	0 0 0 0 0	
1.060E-02	1.316E+00	37	E028	1 0 1 1 2	*sic*
1.288E-03	1.599E-01	ns	R424	0 0 0 0 0	

1169. $C_7H_8O_2$
3-Methoxyphenol
Resorcinol monomethylether
p-Methoxyphenol
RN: 150-19-6   **MP** (°C):
MW: 124.14   **BP** (°C):

Solubility (Moles/L)	Solubility (Grams/L)	Temp (°C)	Ref (#)	Evaluation (T P E A A)	Comments
3.110E-01	3.861E+01	25	B314	0 0 0 0 0	
3.110E-01	3.861E+01	30	B315	0 0 0 0 0	
4.000E-03	4.966E-01	37	E028	1 0 1 1 1	*sic*
4.966E-01	6.165E+01	ns	S460	0 0 0 0 0	

1170. $C_7H_8O_2$
p-Methoxyphenol
p-Hydroxyanisole
Hydroquinone monomethyl ether
4-Methoxyphenol
RN: 150-76-5   **MP** (°C):   52.5
MW: 124.14   **BP** (°C):   243

Solubility (Moles/L)	Solubility (Grams/L)	Temp (°C)	Ref (#)	Evaluation (T P E A A)	Comments
2.073E-01	2.573E+01	20	R087	0 0 0 0 0	0.15M NaCl

1171. C₇H₈O₂

4,6-Dimethyl-1,2-pyrone
4,6-Dimethyl-α-pyrone
2,4-Dimethyl-α-pyrone
Mesitene lactone
4,6-Dimethyl-2-pyranone
4,6-Dimethyl-2H-pyran-2-one

RN: 675-09-2   **MP** (°C):   49
MW: 124.14   **BP** (°C):

Solubility (Moles/L)	Solubility (Grams/L)	Temp (°C)	Ref (#)	Evaluation (T P E A A)	Comments
1.953E+00	2.424E+02	59.7	W022	2 2 1 1 0	EFG
2.088E+00	2.593E+02	86.3	W022	2 2 1 1 0	EFG

1172. C₇H₈O₃S

p-Toluenesulfonic acid
4-Methylbenzenesulfonic acid
Methylbenzenesulfonic acid
Tosic acid
PTSA
Toluene-4-sulfonic acid

RN: 104-15-4   **MP** (°C):   106.5
MW: 172.20   **BP** (°C):

Solubility (Moles/L)	Solubility (Grams/L)	Temp (°C)	Ref (#)	Evaluation (T P E A A)	Comments
2.900E+00	4.993E+02	36.5	T023	1 2 2 1 2	
2.902E+00	4.997E+02	40.5	T023	1 2 2 1 2	
2.903E+00	4.999E+02	42.5	T023	1 2 2 1 2	

1173. C₇H₈O₃S.H₂O

p-Toluenesulfonic acid (monohydrate)

RN: 6192-52-5   **MP** (°C):   104.5
MW: 190.22   **BP** (°C):

Solubility (Moles/L)	Solubility (Grams/L)	Temp (°C)	Ref (#)	Evaluation (T P E A A)	Comments
2.107E+00	4.008E+02	−6.5	T023	1 2 2 1 2	
2.120E+00	4.033E+02	−1.5	T023	1 2 2 1 2	
2.129E+00	4.050E+02	1.5	T023	1 2 2 1 2	
2.168E+00	4.125E+02	20.1	T023	1 2 2 1 2	
2.210E+00	4.203E+02	38.8	T023	1 2 2 1 2	
2.616E+00	4.975E+02	45.3	T023	1 2 2 1 2	
2.257E+00	4.293E+02	55.2	T023	1 2 2 1 2	
2.593E+00	4.933E+02	73.9	T023	1 2 2 1 2	
2.329E+00	4.431E+02	78.4	T023	1 2 2 1 2	
2.566E+00	4.882E+02	89.1	T023	1 2 2 1 2	
2.375E+00	4.517E+02	89.9	T023	1 2 2 1 2	

(continued)

1173. C₇H₈O₃S.H₂O (continued)

Solubility (Moles/L)	Solubility (Grams/L)	Temp (°C)	Ref (#)	Evaluation (T P E A A)	Comments
2.446E+00	4.652E+02	101.1	T023	1 2 2 1 2	
2.525E+00	4.802E+02	102.9	T023	1 2 2 1 2	
2.498E+00	4.751E+02	104.8	T023	1 2 2 1 2	

1174. C₇H₈O₃S.2H₂O
o-Toluenesulfonic acid (dihydrate)
RN: 68066-37-5 **MP** (°C):
MW: 208.23 **BP** (°C):

Solubility (Moles/L)	Solubility (Grams/L)	Temp (°C)	Ref (#)	Evaluation (T P E A A)	Comments
1.718E+00	3.577E+02	−25.0	T023	1 2 2 1 2	
1.773E+00	3.691E+02	−13.0	T023	1 2 2 1 2	
1.823E+00	3.795E+02	.8	T023	1 2 2 1 2	
1.891E+00	3.938E+02	16.8	T023	1 2 2 1 2	
1.954E+00	4.068E+02	31.2	T023	1 2 2 1 2	
2.264E+00	4.715E+02	48.2	T023	1 2 2 1 2	
2.055E+00	4.279E+02	50.0	T023	1 2 2 1 2	
2.243E+00	4.671E+02	54.0	T023	1 2 2 1 2	
2.090E+00	4.353E+02	56.0	T023	1 2 2 1 2	
2.207E+00	4.597E+02	60.4	T023	1 2 2 1 2	
2.148E+00	4.472E+02	61.2	T023	1 2 2 1 2	
2.179E+00	4.538E+02	62.0	T023	1 2 2 1 2	

1175. C₇H₈O₃S.4H₂O
p-Toluenesulfonic acid (tetrahydrate)
RN: 104-15-4 **MP** (°C):
MW: 244.27 **BP** (°C):

Solubility (Moles/L)	Solubility (Grams/L)	Temp (°C)	Ref (#)	Evaluation (T P E A A)	Comments
1.422E+00	3.473E+02	−27.0	T023	1 2 2 1 2	
1.437E+00	3.510E+02	−26.0	T023	1 2 2 1 2	
1.450E+00	3.543E+02	−18.5	T023	1 2 2 1 2	
1.527E+00	3.730E+02	−16.5	T023	1 2 2 1 2	
1.592E+00	3.888E+02	−10.5	T023	1 2 2 1 2	
1.613E+00	3.939E+02	−8.5	T023	1 2 2 1 2	
1.640E+00	4.005E+02	−7.0	T023	1 2 2 1 2	
1.576E+00	3.848E+02	−5.9	T023	1 2 2 1 2	
1.605E+00	3.921E+02	−3.4	T023	1 2 2 1 2	
1.622E+00	3.961E+02	−2.2	T023	1 2 2 1 2	
1.641E+00	4.008E+02	−1.0	T023	1 2 2 1 2	

1176. $C_7H_8O_7$
Methylenecitric acid
Methylen-citronensaeure
RN: 144-16-1   **MP** (°C):
MW: 204.14   **BP** (°C):

Solubility (Moles/L)	Solubility (Grams/L)	Temp (°C)	Ref (#)	Evaluation (T P E A A)	Comments
2.337E-01	4.770E+01	20	F300	1 0 0 0 2	

1177. $C_7H_9ClN_2OS$
TO-2
5-Chloro-4-methyl-2-propionamide-thiazole
CMPT
RN: 13915-79-2   **MP** (°C): 159
MW: 204.68   **BP** (°C):

Solubility (Moles/L)	Solubility (Grams/L)	Temp (°C)	Ref (#)	Evaluation (T P E A A)	Comments
8.794E-04	1.800E-01	ns	M061	0 0 0 0 2	

1178. C_7H_9N
4-Ethylpyridine
4-Aethyl-pyridin
RN: 536-75-4   **MP** (°C): −90.5
MW: 107.16   **BP** (°C): 168.3

Solubility (Moles/L)	Solubility (Grams/L)	Temp (°C)	Ref (#)	Evaluation (T P E A A)	Comments
3.906E+00	4.186E+02	−19	C047	2 2 0 0 1	
2.495E+00	2.674E+02	182	C047	2 2 0 0 2	

1179. C_7H_9N
m-Toluidine
3-Toluidine
4-Methylaniline
p-Toluidine
p-Toluidin
RN: 106-49-0   **MP** (°C): 43
MW: 107.16   **BP** (°C): 203

Solubility (Moles/L)	Solubility (Grams/L)	Temp (°C)	Ref (#)	Evaluation (T P E A A)	Comments
6.066E-02	6.500E+00	15	F300	1 0 0 0 1	
6.026E-02	6.457E+00	20	B179	0 0 0 0 0	
3.890E-01	4.169E+01	20	B179	0 0 0 0 0	
1.403E-01	1.503E+01	20	C113	1 0 2 1 2	
6.200E-02	6.644E+00	20	H306	1 0 1 2 1	
6.119E-02	6.557E+00	20	T301	1 2 2 2 2	

1180. C₇H₉N

Methylaniline

N-Methylaniline

RN: 100-61-8	**MP** (°C):	−57
MW: 107.16	**BP** (°C):	194

Solubility (Moles/L)	Solubility (Grams/L)	Temp (°C)	Ref (#)	Evaluation (T P E A A)	Comments
5.248E-02	5.624E+00	25	C113	1 0 2 1 2	

1181. C₇H₉N

3,4-Lutidine

3,4-Dimethylpyridine

RN: 583-58-4	**MP** (°C):	−12
MW: 107.16	**BP** (°C):	163

Solubility (Moles/L)	Solubility (Grams/L)	Temp (°C)	Ref (#)	Evaluation (T P E A A)	Comments
1.836E+00	1.968E+02	−3.6	C047	2 2 0 0 2	
2.470E+00	2.647E+02	163	C047	2 2 0 0 1	
+2.29E+00	+2.45E+02	ns	S460	0 0 0 0 0	

1182. C₇H₉N

o-Toluidine

2-Toluidine

RN: 95-53-4	**MP** (°C):	−15
MW: 107.16	**BP** (°C):	200

Solubility (Moles/L)	Solubility (Grams/L)	Temp (°C)	Ref (#)	Evaluation (T P E A A)	Comments
1.524E-01	1.633E+01	20	C113	1 0 2 1 2	
1.577E-01	1.690E+01	20	K119	1 0 0 0 2	
1.381E-01	1.480E+01	25	F300	1 0 0 0 2	

1183. C₇H₉N

3-Ethylpyridine

3-Aethyl-pyridin

β-Lutidine

RN: 536-78-7	**MP** (°C):	
MW: 107.16	**BP** (°C):	163

Solubility (Moles/L)	Solubility (Grams/L)	Temp (°C)	Ref (#)	Evaluation (T P E A A)	Comments
2.520E+00	2.701E+02	196	C047	2 2 0 0 1	

1184. C₇H₉N

1184. C_7H_9N
3,5-Lutidine
3,5-Dimethylpyridine

RN:	591-22-0	MP (°C):	−9
MW:	107.16	BP (°C):	169

Solubility (Moles/L)	Solubility (Grams/L)	Temp (°C)	Ref (#)	Evaluation (T P E A A)	Comments
1.896E+00	2.032E+02	−12	C047	2 2 0 0 2	
2.520E+00	2.701E+02	192	C047	2 2 0 0 1	
+2.40E+00	+2.57E+02	ns	S460	0 0 0 0 0	

1185. C₇H₉N

1185. C_7H_9N
2,6-Lutidine
2,6-Dimethyl-pyridin
2,6-Dimethylpyridine

RN:	108-48-5	MP (°C):	−6
MW:	107.16	BP (°C):	144

Solubility (Moles/L)	Solubility (Grams/L)	Temp (°C)	Ref (#)	Evaluation (T P E A A)	Comments
2.154E+00	2.308E+02	34	C047	2 2 0 0 1	
2.714E+00	2.908E+02	231	C047	2 2 0 0 1	
+2.82E+00	+3.02E+02	ns	S460	0 0 0 0 0	

1186. C₇H₉N

1186. C_7H_9N
2,5-Lutidine
2,5-Dimethyl-pyridin
2,5-Dimethylpyridine

RN:	589-93-5	MP (°C):	−15
MW:	107.16	BP (°C):	157

Solubility (Moles/L)	Solubility (Grams/L)	Temp (°C)	Ref (#)	Evaluation (T P E A A)	Comments
1.984E+00	2.126E+02	13.1	C047	2 2 0 0 1	
7.186E-01	7.700E+01	23	F300	1 0 0 0 1	
2.570E+00	2.754E+02	207	C047	2 2 0 0 1	

1187. C₇H₉N

1187. C_7H_9N
2,4-Lutidine
2,4-Dimethyl-pyridin
2,4-Dimethylpyridine

RN:	108-47-4	MP (°C):	−60
MW:	107.16	BP (°C):	159

Solubility (Moles/L)	Solubility (Grams/L)	Temp (°C)	Ref (#)	Evaluation (T P E A A)	Comments
3.961E+00	4.245E+02	23	J007	1 2 0 1 2	average of 2
1.896E+00	2.032E+02	23.4	C047	2 2 0 0 2	

(continued)

1187. C_7H_9N (continued)

Solubility (Moles/L)	Solubility (Grams/L)	Temp (°C)	Ref (#)	Evaluation (T P E A A)	Comments
1.896E+00	2.032E+02	23.40	A009	1 2 1 1 2	LCST
1.287E+00	1.379E+02	24.40	A009	1 2 1 1 2	EFG, LCST
2.419E+00	2.593E+02	25	A009	1 2 1 1 2	EFG, LCST
3.316E+00	3.553E+02	27.2	J007	1 2 0 1 2	
8.484E-01	9.091E+01	30	A009	1 2 1 1 2	EFG, LCST
3.111E+00	3.333E+02	32.50	A009	1 2 1 1 2	EFG, LCST
4.497E+00	4.819E+02	35.0	J007	1 2 0 1 2	
2.902E+00	3.110E+02	39.0	J007	1 2 0 1 2	
6.105E-01	6.542E+01	40	A009	1 2 1 1 2	EFG, LCST
3.500E+00	3.750E+02	50	A009	1 2 1 1 2	EFG, LCST
2.545E+00	2.727E+02	53	J007	1 2 0 1 2	
4.548E+00	4.873E+02	54.3	J007	1 2 0 1 2	
3.777E+00	4.048E+02	62.50	A009	1 2 1 1 2	EFG, LCST
2.204E+00	2.362E+02	68.5	J007	1 2 0 1 2	
6.105E-01	6.542E+01	149	A009	1 2 1 1 2	EFG, UCST
3.794E+00	4.065E+02	165	A009	1 2 1 1 2	EFG, UCST
1.287E+00	1.379E+02	180	A009	1 2 1 1 2	EFG, UCST
3.500E+00	3.750E+02	180	A009	1 2 1 1 2	EFG, UCST
3.111E+00	3.333E+02	186	A009	1 2 1 1 2	EFG, UCST
1.896E+00	2.032E+02	187	A009	1 2 1 1 2	EFG, UCST
2.419E+00	2.593E+02	187	A009	1 2 1 1 2	EFG, UCST
2.520E+00	2.701E+02	189	A009	1 2 1 1 2	UCST
2.520E+00	2.701E+02	189	C047	2 2 0 0 1	

1188. C_7H_9N
2,3-Lutidine
2,3-Dimethylpyridine

RN: 583-61-9   **MP (°C):** −15
MW: 107.16   **BP (°C):** 162

Solubility (Moles/L)	Solubility (Grams/L)	Temp (°C)	Ref (#)	Evaluation (T P E A A)	Comments
1.926E+00	2.063E+02	16.5	C047	2 2 0 0 1	
2.594E+00	2.780E+02	193	C047	2 2 0 0 2	
+2.40E+00	+2.57E+02	ns	S460	0 0 0 0 0	

1189. C_7H_9N
2-Ethylpyridine
α-Lutidine

RN: 100-71-0   **MP (°C):**
MW: 107.16   **BP (°C):** 149

Solubility (Moles/L)	Solubility (Grams/L)	Temp (°C)	Ref (#)	Evaluation (T P E A A)	Comments
2.368E+00	2.537E+02	−5	C047	2 2 0 0 1	
2.760E+00	2.958E+02	231	C047	2 2 0 0 1	
+3.24E+00	+3.47E+02	ns	S460	0 0 0 0 0	

1190. C₇H₉NO

p-Anisidine
4-Methoxybenzenamine
p-Methoxyaniline
4-Methoxy-1-aminobenzene
p-Methoxyphenylamine

RN:	104-94-9	**MP** (°C):	57		
MW:	123.16	**BP** (°C):	246		

Solubility (Moles/L)	Solubility (Grams/L)	Temp (°C)	Ref (#)	Evaluation (T P E A A)	Comments
9.311E-02	1.147E+01	20	T301	1 2 2 2 2	

1191. C₇H₉NO

p-Tolylhydroxylamine
p-Tolylhydroxylamin

RN:	623-10-9	**MP** (°C):	
MW:	123.16	**BP** (°C):	

Solubility (Moles/L)	Solubility (Grams/L)	Temp (°C)	Ref (#)	Evaluation (T P E A A)	Comments
8.120E-02	1.000E+01	5	F300	1 0 0 0 1	
4.027E-01	4.960E+01	100	F300	1 0 0 0 2	

1192. C₇H₉NO

o-Anisidine
2-Anisidine
2-Methoxybenzenamine
o-Methoxyaniline
2-Methoxy-1-aminobenzene
o-Methoxyphenylamine

RN:	90-04-0	**MP** (°C):	5
MW:	123.16	**BP** (°C):	225

Solubility (Moles/L)	Solubility (Grams/L)	Temp (°C)	Ref (#)	Evaluation (T P E A A)	Comments
1.026E-01	1.264E+01	25	B019	1 0 1 2 0	

1193. C₇H₉NO₂

1,2-Dimethyl-3-hydroxy-4-pyridone
DMHP

RN:	30652-11-0	**MP** (°C):	271–273
MW:	139.16	**BP** (°C):	

Solubility (Moles/L)	Solubility (Grams/L)	Temp (°C)	Ref (#)	Evaluation (T P E A A)	Comments
1.130E-01	1.572E+01	25	C340	0 0 0 0 0	pH 9.4

1194. C₇H₉NO₂S

p-Toluenesulfonamide
p-Methylbenzenesulfonamide
4-Methylbenzenesulfonamide

RN:	70-55-3	**MP** (°C):	138	
MW:	171.22	**BP** (°C):		

Solubility (Moles/L)	Solubility (Grams/L)	Temp (°C)	Ref (#)	Evaluation (T P E A A)	Comments
1.110E-02	1.900E+00	9	F300	1 0 0 0 1	
1.180E-02	2.020E+00	15	K024	1 2 1 1 2	
1.843E-02	3.156E+00	25	H105	1 1 0 1 2	

1195. C₇H₉NO₂S

o-Toluenesulfonamide
o-Methylbenzenesulfonamide

RN:	88-19-7	**MP** (°C):	156	
MW:	171.22	**BP** (°C):		

Solubility (Moles/L)	Solubility (Grams/L)	Temp (°C)	Ref (#)	Evaluation (T P E A A)	Comments
5.840E-03	1.000E+00	9	F300	1 0 0 0 0	
1.860E-02	3.185E+00	15	K024	1 2 1 1 2	
9.485E-03	1.624E+00	25	H105	1 1 0 1 2	

1196. C₇H₉NO₂S

m-Toluenesulfonamide
m-Methylbenzenesulfonamide

RN:	1899-94-1	**MP** (°C):		
MW:	171.22	**BP** (°C):		

Solubility (Moles/L)	Solubility (Grams/L)	Temp (°C)	Ref (#)	Evaluation (T P E A A)	Comments
1.750E-02	2.996E+00	15	K024	1 2 1 1 2	
4.563E-02	7.812E+00	25	H105	1 1 0 1 2	

1197. C₇H₉NO₃S

4-Amino-3-methylbenzene sulfonic acid
4-Amino-toluol-sulfosaeure-(3)

RN:	98-33-9	**MP** (°C):		
MW:	187.22	**BP** (°C):		

Solubility (Moles/L)	Solubility (Grams/L)	Temp (°C)	Ref (#)	Evaluation (T P E A A)	Comments
2.671E-02	5.000E+00	20	F300	1 0 0 0 0	

1198. C₇H₉NO₃S
4-Amino-2-methylbenzene sulfonic acid
4-Amino-toluol-sulfosaeure-(2)

RN: 133-78-8   **MP** (°C):
MW: 187.22   **BP** (°C):

Solubility (Moles/L)	Solubility (Grams/L)	Temp (°C)	Ref (#)	Evaluation (T P E A A)	Comments
2.404E-02	4.500E+00	20	F300	1 0 0 0 1	

1199. C₇H₉NO₃S
2-Amino-5-methylbenzene sulfonic acid
2-Amino-toluol-sulfosaeure-(5)

RN: 88-44-8   **MP** (°C): >300
MW: 187.22   **BP** (°C):

Solubility (Moles/L)	Solubility (Grams/L)	Temp (°C)	Ref (#)	Evaluation (T P E A A)	Comments
1.709E-01	3.200E+01	19	F300	1 0 0 0 1	

1200. C₇H₉NO₃S
p-Methoxybenzenesulfonamide
4-Methoxybenzenesulfonamide

RN: 1129-26-6   **MP** (°C):
MW: 187.22   **BP** (°C):

Solubility (Moles/L)	Solubility (Grams/L)	Temp (°C)	Ref (#)	Evaluation (T P E A A)	Comments
1.560E-02	2.921E+00	15	K024	1 2 1 1 2	

1201. C₇H₉N₃O
4-Phenylsemicarbazide
Phenylsemicarbazide

RN: 537-47-3   **MP** (°C): 123.5
MW: 151.17   **BP** (°C):

Solubility (Moles/L)	Solubility (Grams/L)	Temp (°C)	Ref (#)	Evaluation (T P E A A)	Comments
	6.995E-01	15	D068	1 2 0 0 0	

1202. $C_7H_9N_3O_2S_2$

Sulfathiourea
p-Aminobenzenesulfonylthiourea
p-Aminophenylsulfonylthiourea
Badional
Baldinol
Fontamide

RN: 515-49-1   **MP** (°C):   171.5
MW: 231.30   **BP** (°C):

Solubility (Moles/L)	Solubility (Grams/L)	Temp (°C)	Ref (#)	Evaluation (T P E A A)	Comments
2.365E-03	5.470E-01	20	F073	1 2 2 2 2	

1203. $C_7H_9N_3O_3$

Orotic acid ethylamide

RN: 1011-82-1   **MP** (°C):   263–265
MW: 183.17   **BP** (°C):

Solubility (Moles/L)	Solubility (Grams/L)	Temp (°C)	Ref (#)	Evaluation (T P E A A)	Comments
1.940E-01	3.553E+01	–4	N018	0 0 0 0 0	
3.240E-01	5.935E+01	16	N018	0 0 0 0 0	
3.980E-01	7.290E+01	25	N018	0 0 0 0 0	

1204. $C_7H_9N_3O_3S$

Sulfanilylurea
Sulfanilylharnstoff

RN: 547-44-4   **MP** (°C):   146
MW: 215.23   **BP** (°C):

Solubility (Moles/L)	Solubility (Grams/L)	Temp (°C)	Ref (#)	Evaluation (T P E A A)	Comments
1.084E-02	2.333E+00	20	F073	1 2 2 2 2	
5.575E-03	1.200E+00	37	F300	1 0 0 0 1	
5.012E-02	1.079E+01	ns	R427	0 0 0 0 0	

1205. $C_7H_9N_3O_4$

Orotic acid ethanol amide

RN: **MP** (°C):   217–218
MW: 199.17   **BP** (°C):

Solubility (Moles/L)	Solubility (Grams/L)	Temp (°C)	Ref (#)	Evaluation (T P E A A)	Comments
1.800E-01	3.585E+01	–4	N018	0 0 0 0 0	
3.460E-01	6.891E+01	16	N018	0 0 0 0 0	
4.470E-01	8.903E+01	25	N018	0 0 0 0 0	

1206. C$_7$H$_9$O$_3$P
Hydroxymethylphenylphosphinic acid
RN: 61451-78-3 **MP** (°C): 138
MW: 172.12 **BP** (°C):

Solubility (Moles/L)	Solubility (Grams/L)	Temp (°C)	Ref (#)	Evaluation (T P E A A)	Comments
1.166E+02	2.007E+04	0	W422	0 0 0 0 0	
9.900E+00	1.704E+03	34.29	W422	0 0 0 0 0	
2.060E+01	3.546E+03	44.30	W422	0 0 0 0 0	
4.240E+01	7.298E+03	54.41	W422	0 0 0 0 0	
9.660E+01	1.663E+04	64.99	W422	0 0 0 0 0	
1.662E+02	2.861E+04	73.42	W422	0 0 0 0 0	
2.474E+02	4.258E+04	79.6	W422	0 0 0 0 0	
3.120E+02	5.370E+04	83.95	W422	0 0 0 0 0	

1207. C$_7$H$_{10}$
1,3-Cycloheptadiene
RN: 4054-38-0 **MP** (°C):
MW: 94.16 **BP** (°C): 121

Solubility (Moles/L)	Solubility (Grams/L)	Temp (°C)	Ref (#)	Evaluation (T P E A A)	Comments
6.577E-03	6.192E-01	ns	S460	0 0 0 0 0	

1208. C$_7$H$_{10}$N$_2$OS
Propylthiouracil
6-Propyl-2-thiouracil
Propycil
RN: 51-52-5 **MP** (°C): 220.0
MW: 170.23 **BP** (°C):

Solubility (Moles/L)	Solubility (Grams/L)	Temp (°C)	Ref (#)	Evaluation (T P E A A)	Comments
6.520E-03	1.110E+00	20	A091	1 0 0 0 0	
6.455E-03	1.099E+00	20	I310	0 0 0 0 0	
7.070E-03	1.204E+00	25	G016	1 2 1 2 2	intrinsic
5.816E-02	9.901E+00	100	I310	0 0 0 0 0	
5.874E-03	1.000E+00	ns	K444	0 0 0 0 0	

1209. C$_7$H$_{10}$N$_2$O$_2$S
p-Methylaminobenzenesulfonamide
4-Methylaminobenzenesulfonamide
RN: 16891-79-5 **MP** (°C):
MW: 186.23 **BP** (°C):

Solubility (Moles/L)	Solubility (Grams/L)	Temp (°C)	Ref (#)	Evaluation (T P E A A)	Comments
5.000E-03	9.312E-01	15	K024	1 2 1 1 2	

1210. $C_7H_{10}N_2O_2S$
N1-Methylsulfanilamide
4-Amino-N-methylbenzenesulfonamide
N-Methyl-p-aminobenzenesulfonamide
N-Methyl-4-aminobenzenesulfonamide

| RN: | 1709-52-0 | MP (°C): |
| MW: | 186.23 | BP (°C): |

Solubility (Moles/L)	Solubility (Grams/L)	Temp (°C)	Ref (#)	Evaluation (T P E A A)	Comments
9.450E-02	1.760E+01	37	K095	2 0 0 0 2	intrinsic

1211. $C_7H_{10}N_2O_2S$
Toluenesulfamide
Sulfamide, (4-methylphenyl)-
p-Tolylsulfamide

| RN: | 15853-38-0 | MP (°C): |
| MW: | 186.23 | BP (°C): |

Solubility (Moles/L)	Solubility (Grams/L)	Temp (°C)	Ref (#)	Evaluation (T P E A A)	Comments
3.020E-02	5.624E+00	37	A028	1 0 2 1 2	intrinsic

1212. $C_7H_{10}N_2O_3$
Isopropylbarbituric acid
2,4,6(1H,3H,5H)-Pyrimidinetrione, 5-(1-methylethyl)-
Isopropylbarbiturate

| RN: | 7391-69-7 | MP (°C): |
| MW: | 170.17 | BP (°C): |

Solubility (Moles/L)	Solubility (Grams/L)	Temp (°C)	Ref (#)	Evaluation (T P E A A)	Comments
3.482E-02	5.925E+00	20	J030	1 2 2 2 2	
5.905E-02	1.005E+01	37	J030	1 2 2 2 2	

1213. $C_7H_{10}N_2O_3$
5-Ethyl-5-methylbarbituric acid
5-Methyl-5-ethylbarbituric acid

| RN: | 27653-63-0 | MP (°C): |
| MW: | 170.17 | BP (°C): |

Solubility (Moles/L)	Solubility (Grams/L)	Temp (°C)	Ref (#)	Evaluation (T P E A A)	Comments
8.010E-02	1.363E+01	25	M310	2 2 2 2 2	
5.912E-02	1.006E+01	25	P350	0 0 0 0 0	intrinsic

1214. C₇H₁₀N₄O₂S

Sulfanilylguanidine
Sulfaguanidine
Sulfaguanidin
Sulfanilguanidin

RN: 57-67-0   **MP (°C):**   190
MW: 214.25   **BP (°C):**

Solubility (Moles/L)	Solubility (Grams/L)	Temp (°C)	Ref (#)	Evaluation (T P E A A)	Comments
4.131E-03	8.850E-01	20	F073	1 2 2 2 2	
4.663E-03	9.990E-01	25	D041	1 0 0 0 0	
8.868E-03	1.900E+00	37	R045	1 2 1 1 2	
1.025E-02	2.195E+00	37.50	M142	1 2 0 0 2	
4.201E-01	9.000E+01	h	F300	0 0 0 0 0	

1215. C₇H₁₀N₄O₃.H₂O

Theopylline (monohydrate)
1H-Purine-2,6-dione, 3,7-dihydro-1,3-dimethyl-, monohydrate

RN: 5967-84-0   **MP (°C):**   269–272
MW: 216.20   **BP (°C):**

Solubility (Moles/L)	Solubility (Grams/L)	Temp (°C)	Ref (#)	Evaluation (T P E A A)	Comments
3.823E-02	8.264E+00	c	D004	0 0 0 0 0	
		h	D004	0 0 0 0 0	

1216. C₇H₁₀O₄S.H₂O

o-Toluenesulfonic acid (monohydrate)
2-Methyl-benzenesulfonic acid (monohydrate)

RN: 88-20-0   **MP (°C):**
MW: 208.23   **BP (°C):**

Solubility (Moles/L)	Solubility (Grams/L)	Temp (°C)	Ref (#)	Evaluation (T P E A A)	Comments
2.348E+00	4.889E+02	32.5	T023	1 2 2 1 2	
2.335E+00	4.863E+02	38.6	T023	1 2 2 1 2	
2.318E+00	4.827E+02	45.7	T023	1 2 2 1 2	
2.266E+00	4.718E+02	48.5	T023	1 2 2 1 2	
2.302E+00	4.793E+02	48.6	T023	1 2 2 1 2	
2.273E+00	4.733E+02	49.0	T023	1 2 2 1 2	
2.289E+00	4.767E+02	49.6	T023	1 2 2 1 2	

1217. $C_7H_{10}O_5$
Shikimic acid
Shikimisaeure

RN:	138-59-0	MP (°C):	190
MW:	174.15	BP (°C):	

Solubility (Moles/L)	Solubility (Grams/L)	Temp (°C)	Ref (#)	Evaluation (T P E A A)	Comments
8.613E-01	1.500E+02	21	F300	1 0 0 0 1	

1218. $C_7H_{10}O_5$
Mesoxalic acid diethyl ester
Mesooxalsaeure-diethyl ester

RN:	609-09-6	MP (°C):	−30
MW:	174.15	BP (°C):	208

Solubility (Moles/L)	Solubility (Grams/L)	Temp (°C)	Ref (#)	Evaluation (T P E A A)	Comments
3.249E+00	5.658E+02	22	F300	1 0 0 0 2	
+3.25E+00	+5.66E+02	ns	S460	0 0 0 0 0	

1219. $C_7H_{11}NO_2$
Ethosuximide
Zarontin
2-Ethyl-2-methylsuccinimide

RN:	77-67-8	MP (°C):	
MW:	141.17	BP (°C):	

Solubility (Moles/L)	Solubility (Grams/L)	Temp (°C)	Ref (#)	Evaluation (T P E A A)	Comments
1.346E+00	1.900E+02	25	P061	0 0 0 0 0	pH 3-7.9
7.084E-01	1.000E+02	ns	K444	0 0 0 0 0	

1220. $C_7H_{11}N_3O_2$
Ipronidazole
1-Methyl-2-isopropyl-5-nitro-imidazole

RN:	14885-29-1	MP (°C):	58–60
MW:	169.18	BP (°C):	

Solubility (Moles/L)	Solubility (Grams/L)	Temp (°C)	Ref (#)	Evaluation (T P E A A)	Comments
5.556E-02	9.400E+00	20	D344	0 0 0 0 0	
5.550E-02	9.390E+00	20	D344	0 0 0 0 0	
5.446E-02	9.214E+00	20	D344	0 0 0 0 0	
5.560E-02	9.407E+00	20	D344	0 0 0 0 0	

1221. C$_7$H$_{11}$N$_3$O$_2$
1-Methyl-L-histidine
L-1-Methylhistidine

| **RN:** | 15507-76-3 | **MP** (°C): | >254 |
| **MW:** | 169.18 | **BP** (°C): | |

Solubility (Moles/L)	Solubility (Grams/L)	Temp (°C)	Ref (#)	Evaluation (T P E A A)	Comments
9.851E-01	1.667E+02	25	D041	1 0 0 0 0	

1222. C$_7$H$_{11}$N$_7$S
Aziprotryne
2-Azido-4-isopropylamino-6-methylmercapto-*s*-triazine
C-7019

| **RN:** | 4658-28-0 | **MP** (°C): | 95 |
| **MW:** | 225.28 | **BP** (°C): | |

Solubility (Moles/L)	Solubility (Grams/L)	Temp (°C)	Ref (#)	Evaluation (T P E A A)	Comments
2.441E-04	5.500E-02	20	M161	1 0 0 0 1	
3.329E-04	7.500E-02	ns	M061	0 0 0 0 1	

1223. C$_7$H$_{12}$
1,6-Heptadiene

| **RN:** | 3070-53-9 | **MP** (°C): | −129.0 |
| **MW:** | 96.17 | **BP** (°C): | 89 |

Solubility (Moles/L)	Solubility (Grams/L)	Temp (°C)	Ref (#)	Evaluation (T P E A A)	Comments
4.575E-04	4.400E-02	25	M001	2 1 2 2 1	

1224. C$_7$H$_{12}$
1-Heptyne
1-*n*-Heptyne
Pentylacetylene
Amylacetylene

| **RN:** | 628-71-7 | **MP** (°C): | −81 |
| **MW:** | 96.17 | **BP** (°C): | 99 |

Solubility (Moles/L)	Solubility (Grams/L)	Temp (°C)	Ref (#)	Evaluation (T P E A A)	Comments
9.774E-04	9.400E-02	25	M001	2 1 2 2 2	

1225. C₇H₁₂

Cycloheptene

(1Z)-Cycloheptene

cis-Cycloheptene

RN:	628-92-2	MP (°C):	−56
MW:	96.17	BP (°C):	114.7

Solubility (Moles/L)	Solubility (Grams/L)	Temp (°C)	Ref (#)	Evaluation (T P E A A)	Comments
6.863E-04	6.600E-02	25	M001	2 1 2 2 1	

1226. C₇H₁₂

1-Methyl-1-cyclohexene

1-Methylcyclohexene

RN:	591-49-1	MP (°C):	−120
MW:	96.17	BP (°C):	110

Solubility (Moles/L)	Solubility (Grams/L)	Temp (°C)	Ref (#)	Evaluation (T P E A A)	Comments
5.407E-04	5.200E-02	25	M001	2 1 2 2 2	

1227. C₇H₁₂

2-Heptyne

1-Methyl-2-butylacetylene

Butyl(methyl)acetylene

RN:	1119-65-9	MP (°C):	
MW:	96.17	BP (°C):	

Solubility (Moles/L)	Solubility (Grams/L)	Temp (°C)	Ref (#)	Evaluation (T P E A A)	Comments
1.700E-03	1.635E-01	25	H039	1 2 2 2 2	

1228. C₇H₁₂

2-Methyl-3-hexyne

1-Ethyl-2-isopropylacetylene

RN:	36566-80-0	MP (°C):	
MW:	96.17	BP (°C):	

Solubility (Moles/L)	Solubility (Grams/L)	Temp (°C)	Ref (#)	Evaluation (T P E A A)	Comments
1.800E-03	1.731E-01	25	H039	1 2 2 2 2	

1229. C₇H₁₂BrNO₄

5-Bromo-2-propyl-5-nitro-1,3-dioxane
2-Propyl-5-bromo-5-nitro-1,3-dioxane

RN: 53983-01-0   **MP** (°C):   73–75
MW: 254.09   **BP** (°C):

Solubility (Moles/L)	Solubility (Grams/L)	Temp (°C)	Ref (#)	Evaluation (T P E A A)	Comments
1.102E-03	2.799E-01	25	L013	1 0 2 1 2	

1230. C₇H₁₂ClN₅

Norazine
2-Chloro-4-methylamino-6-isopropylamino-s-triazine

RN: 3004-71-5   **MP** (°C):   157–159
MW: 201.66   **BP** (°C):

Solubility (Moles/L)	Solubility (Grams/L)	Temp (°C)	Ref (#)	Evaluation (T P E A A)	Comments
1.289E-03	2.600E-01	20	J033	0 0 0 0 0	
1.289E-03	2.600E-01	21	B192	0 0 0 0 2	

1231. C₇H₁₂ClN₅

Simazine
2-Chloro-4-ethylamino-6-ethylamino-s-triazine
2-Chloro-4,6-bis(ethylamino)-s-triazine
Primatol S

RN: 122-34-9   **MP** (°C):   224
MW: 201.66   **BP** (°C):

Solubility (Moles/L)	Solubility (Grams/L)	Temp (°C)	Ref (#)	Evaluation (T P E A A)	Comments
9.918E-06	2.000E-03	10	B185	0 0 0 0 0	
2.512E-05	5.065E-03	20	B179	0 0 0 0 0	
2.479E-05	5.000E-03	20	B185	0 0 0 0 0	
2.827E-05	5.700E-03	20	C048	2 2 2 2 1	
1.736E-05	3.500E-03	20	F311	1 2 2 2 1	
2.479E-05	5.000E-03	21	B192	0 0 0 0 0	
2.479E-05	5.000E-03	21	G099	2 0 0 1 0	
2.479E-05	5.000E-03	22	M061	1 0 0 0 0	
7.500E-05	1.512E-02	26	G001	1 0 1 1 1	
1.310E-04	2.642E-02	50	G001	1 0 1 1 2	
4.165E-04	8.400E-02	85	B185	0 0 0 0 0	
4.110E-04	8.288E-02	85	B200	1 0 0 0 2	
1.736E-05	3.500E-03	ns	C101	0 0 0 0 1	
2.479E-05	5.000E-03	ns	G041	0 0 0 0 0	
2.479E-05	5.000E-03	ns	H112	0 0 0 0 0	
2.479E-05	5.000E-03	ns	J033	0 0 0 0 0	
3.074E-05	6.200E-03	ns	V414	0 0 0 0 0	
2.479E-05	5.000E-03	rt	M161	0 0 0 0 0	

1232. C$_7$H$_{12}$ClN$_5$
2-Chloro-4-methyl amino-6-propyl amino-s-triazine
1,3,5-Triazine-2,4-diamine, 6-chloro-N-methyl-N'-propyl-
s-Triazine, 2-chloro-4-methylamino-6-propylamino-

RN: 73383-40-1   **MP (°C):**
MW: 201.66   **BP (°C):**

Solubility (Moles/L)	Solubility (Grams/L)	Temp (°C)	Ref (#)	Evaluation (T P E A A)	Comments
1.289E-03	2.600E-01	21	G099	2 0 0 1 0	

1233. C$_7$H$_{12}$N$_2$O$_2$
5-Isobutylhydantoin
Hydantoin of DL-leucine

RN: 67337-73-9   **MP (°C):**   208
MW: 156.19   **BP (°C):**

Solubility (Moles/L)	Solubility (Grams/L)	Temp (°C)	Ref (#)	Evaluation (T P E A A)	Comments
1.240E-02	1.937E+00	ns	M025	0 2 0 1 2	

1234. C$_7$H$_{12}$N$_4$O$_5$
Diglycine hydantoic acid
Carbamidoglycylglycine

RN: **MP (°C):**   194
MW: 232.20   **BP (°C):**

Solubility (Moles/L)	Solubility (Grams/L)	Temp (°C)	Ref (#)	Evaluation (T P E A A)	Comments
1.260E-01	2.926E+01	25	M024	1 2 0 1 2	

1235. C$_7$H$_{12}$N$_4$O$_5$
Carbamidodiglycylglycine
Triglycine hydantoin acid

RN: **MP (°C):**   204
MW: 232.20   **BP (°C):**

Solubility (Moles/L)	Solubility (Grams/L)	Temp (°C)	Ref (#)	Evaluation (T P E A A)	Comments
4.460E-02	1.036E+01	25	M024	1 2 0 1 2	

1236. C₇H₁₂O

3-Methylcyclohexanone
m-Methylcyclohexanone

RN: 591-24-2   **MP** (°C): –75
MW: 112.17   **BP** (°C): 162

Solubility (Moles/L)	Solubility (Grams/L)	Temp (°C)	Ref (#)	Evaluation (T P E A A)	Comments
1.335E-02	1.498E+00	20	D052	1 1 0 0 0	
1.349E-02	1.513E+00	ns	S460	0 0 0 0 0	

1237. C₇H₁₂O

2-Methylcyclohexanone
Methyl anone
o-Methylcyohexanone
Methyl cyclohexanone

RN: 583-60-8   **MP** (°C):
MW: 112.17   **BP** (°C):

Solubility (Moles/L)	Solubility (Grams/L)	Temp (°C)	Ref (#)	Evaluation (T P E A A)	Comments
1.135E-01	1.274E+01	23.50	O005	2 0 2 2 2	

1238. C₇H₁₂O₂

Hexahydrobenzoic acid
Cyclohexanecarboxylic acid
Cyclohexan-carbonsaeure

RN: 98-89-5   **MP** (°C): 31
MW: 128.17   **BP** (°C): 232.5

Solubility (Moles/L)	Solubility (Grams/L)	Temp (°C)	Ref (#)	Evaluation (T P E A A)	Comments
1.565E-02	2.006E+00	15	L006	1 0 0 0 2	
1.560E-02	2.000E+00	21	F300	1 0 0 0 0	

1239. C₇H₁₂O₂

Isobutyl propenoate
2-methylpropyl acrylate
2-Propenoic acid, 2-methylpropyl ester
Acrylic acid isobutyl ester
Isobutyl 2-propenoate
Isobutyl acrylate

RN: 106-63-8   **MP** (°C):
MW: 128.17   **BP** (°C): 132

Solubility (Moles/L)	Solubility (Grams/L)	Temp (°C)	Ref (#)	Evaluation (T P E A A)	Comments
6.166E-02	7.903E+00	ns	S460	0 0 0 0 0	

1240. $C_7H_{12}O_4$
Pimelic acid
Heptanedioc acid

RN: 111-16-0   **MP** (°C): 105.7
MW: 160.17   **BP** (°C): 272

Solubility (Moles/L)	Solubility (Grams/L)	Temp (°C)	Ref (#)	Evaluation (T P E A A)	Comments
1.115E-01	1.786E+01	5.99	A341	0 0 0 0 0	
1.151E-01	1.844E+01	7.99	A341	0 0 0 0 0	
1.334E-01	2.137E+01	10.99	A341	0 0 0 0 0	
1.523E-01	2.439E+01	13	D041	1 0 0 0 1	
1.498E-01	2.400E+01	13.50	F300	1 0 0 0 1	
3.122E-01	5.000E+01	15	M051	1 0 0 0 1	
2.236E-01	3.582E+01	15.99	A341	0 0 0 0 0	
2.527E-01	4.048E+01	17.99	A341	0 0 0 0 0	
3.006E-01	4.815E+01	19.99	A341	0 0 0 0 0	
2.973E-01	4.762E+01	20	D041	1 0 0 0 0	
3.122E-01	5.000E+01	20	L041	1 0 0 1 1	
2.953E-01	4.730E+01	20	M171	1 0 0 0 1	
3.000E-02	4.805E+00	20	S006	1 0 0 0 1	
3.332E+00	5.337E+02	21	B040	1 0 1 1 2	*sic*
3.846E-01	6.160E+01	23.99	A341	0 0 0 0 0	
3.938E-01	6.307E+01	24.99	A341	0 0 0 0 0	
4.660E-01	7.464E+01	28.99	A341	0 0 0 0 0	
5.072E-01	8.124E+01	30.99	A341	0 0 0 0 0	
5.690E-01	9.114E+01	33.99	A341	0 0 0 0 0	
6.545E-01	1.048E+02	36.99	A341	0 0 0 0 0	
8.886E-01	1.423E+02	39.99	A341	0 0 0 0 0	
1.527E+00	2.446E+02	42.99	A341	0 0 0 0 0	
1.824E+00	2.922E+02	44.99	A341	0 0 0 0 0	
2.135E+00	3.420E+02	47.49	A341	0 0 0 0 0	
2.551E+00	4.086E+02	49.99	A341	0 0 0 0 0	
3.460E+00	5.542E+02	54.82	A341	0 0 0 0 0	
3.915E+00	6.270E+02	59.99	A341	0 0 0 0 0	
4.365E+00	6.991E+02	64.49	A341	0 0 0 0 0	
4.649E+00	7.446E+02	68.99	A341	0 0 0 0 0	
3.937E-01	6.306E+01	rt	H431	0 0 0 0 0	

1241. $C_7H_{12}O_4$

Diethyl malonate
Malonic
Malonic ester
Propanedioic acid diethyl ester
Ethyl propanedioate
Ethyl methane dicarboxylate

RN: 105-53-3 **MP** (°C): −50
MW: 160.17 **BP** (°C):

Solubility (Moles/L)	Solubility (Grams/L)	Temp (°C)	Ref (#)	Evaluation (T P E A A)	Comments
3.851E+00	6.169E+02	25	H430	0 0 0 0 0	
1.450E-01	2.322E+01	37	E028	1 0 1 1 2	

1242. $C_7H_{12}O_4$

Ethyl α-acetoxypropionate
Ethyl 2-(acetyloxy)propanoate
Ethyl 2-acetoxypropionate

RN: 2985-28-6 **MP** (°C):
MW: 160.17 **BP** (°C):

Solubility (Moles/L)	Solubility (Grams/L)	Temp (°C)	Ref (#)	Evaluation (T P E A A)	Comments
2.104E-01	3.370E+01	25	R006	2 2 0 1 2	

1243. $C_7H_{12}O_4$

3-Methyladipic acid
3-Methylhexanedioic acid

RN: 3058-01-3 **MP** (°C): 101
MW: 160.17 **BP** (°C): 230

Solubility (Moles/L)	Solubility (Grams/L)	Temp (°C)	Ref (#)	Evaluation (T P E A A)	Comments
3.986E-01	6.385E+01	9.50	A031	1 2 2 2 2	
4.732E-01	7.579E+01	12.80	A031	1 2 2 2 2	
1.241E+00	1.987E+02	25.90	A031	1 2 2 2 2	
1.865E+00	2.987E+02	29.80	A031	1 2 2 2 2	
2.531E+00	4.055E+02	33.20	A031	1 2 2 2 2	
3.707E+00	5.938E+02	41.10	A031	1 2 2 2 2	
4.663E+00	7.468E+02	52.30	A031	1 2 2 2 2	
5.340E+00	8.553E+02	64.30	A031	1 2 2 2 2	

1244. C₇H₁₂O₄

$C_7H_{12}O_4$

n-Butylmalonic acid
Acide n-butylmalonique

RN: 534-59-8   **MP** (°C):   102
MW: 160.17   **BP** (°C):

Solubility (Moles/L)	Solubility (Grams/L)	Temp (°C)	Ref (#)	Evaluation (T P E A A)	Comments
7.242E-01	1.160E+02	0	M051	1 0 0 0 2	
1.898E+00	3.040E+02	15	M051	1 0 0 0 2	
2.735E+00	4.380E+02	25	M051	1 0 0 0 2	
4.951E+00	7.930E+02	50	M051	1 0 0 0 2	

1245. C₇H₁₂O₅

$C_7H_{12}O_5$

Propanoic acid, 2-[(ethoxycarbonyl)oxy]-, methyl ester

RN: **MP** (°C):
MW: 176.17   **BP** (°C):

Solubility (Moles/L)	Solubility (Grams/L)	Temp (°C)	Ref (#)	Evaluation (T P E A A)	Comments
9.214E-02	1.623E+01	25	R007	0 0 0 0 0	

1246. C₇H₁₂O₆

$C_7H_{12}O_6$

Quinic acid
Chinasaeure
D-(–)-Quinic acid
1,3,4,5-Tetrahydroxycyclohexanecarboxylic acid

RN: 77-95-2   **MP** (°C):   162
MW: 192.17   **BP** (°C):

Solubility (Moles/L)	Solubility (Grams/L)	Temp (°C)	Ref (#)	Evaluation (T P E A A)	Comments
1.509E+00	2.900E+02	9	F300	1 0 0 0 1	

1247. C₇H₁₃BrN₂O₂

$C_7H_{13}BrN_2O_2$

Carbromal
Adalin
Bromodiethylacetylurea
N-(Aminocarbonyl)-2-bromo-2-ethylbutanamide
1-Bromo-ethyl-butyryl-urea
Bromodiethylacetylcarbamide

RN: 77-65-6   **MP** (°C):   117
MW: 237.10   **BP** (°C):

Solubility (Moles/L)	Solubility (Grams/L)	Temp (°C)	Ref (#)	Evaluation (T P E A A)	Comments
2.109E-03	5.000E-01	20	F300	1 0 0 0 0	

1248. C$_7$H$_{13}$BrN$_2$O$_2$
Bromo-pivalate ureide
RN: **MP** (°C):
MW: 237.10   **BP** (°C):

Solubility (Moles/L)	Solubility (Grams/L)	Temp (°C)	Ref (#)	Evaluation (T P E A A)	Comments
2.161E-01	5.123E+01	ns	F057	0 2 2 2 1	

1249. C$_7$H$_{13}$NO$_2$S
2-Ethyl-2-methyl-4-thiazolidinecarboxylic acid
4-Thiazolidinecarboxylic acid, 2-ethyl-2-methyl-
Thiazolidine-4-carboxylic acid, 2-ethyl-2-methyl-
RN: 56595-20-1   **MP** (°C):
MW: 175.25   **BP** (°C):   327.7

Solubility (Moles/L)	Solubility (Grams/L)	Temp (°C)	Ref (#)	Evaluation (T P E A A)	Comments
2.600E-01	4.557E+01	21	B414	1 0 0 1 1	very fast and extent decompostion, uncertain value

1250. C$_7$H$_{13}$NO$_2$S
2-Propylthiazolidine-4-carboxylic acid
4-Thiazolidinecarboxylic acid, 2-propyl-
RN: 4165-34-8   **MP** (°C):
MW: 175.25   **BP** (°C):   346.1

Solubility (Moles/L)	Solubility (Grams/L)	Temp (°C)	Ref (#)	Evaluation (T P E A A)	Comments
8.500E-02	1.490E+01	21	B414	1 0 0 1 1	partial decomposition

1251. C$_7$H$_{13}$NO$_2$S$_2$
2,2-(Dimethyl)-4-(methoxycarbamyl)-1,3-dithiolane
1,3-Dithiolane-4-methanol, 2,2-dimethyl-, carbamate
RN: 35801-62-8   **MP** (°C):
MW: 207.32   **BP** (°C):

Solubility (Moles/L)	Solubility (Grams/L)	Temp (°C)	Ref (#)	Evaluation (T P E A A)	Comments
6.000E-03	1.244E+00	rt	B174	0 0 1 0 0	

1252. C₇H₁₃NO₃

N-Formylleucine
N-Formyl-DL-leucine

RN:	6113-61-7	MP (°C):
MW:	159.19	BP (°C):

Solubility (Moles/L)	Solubility (Grams/L)	Temp (°C)	Ref (#)	Evaluation (T P E A A)	Comments
1.850E-01	2.945E+01	ns	M025	0 2 0 1 2	

1253. C₇H₁₃NO₃S

2,2-(Dimethyl)-4-(methoxycarbamyl)-1,3-oxathiolane
1,3-Oxathiolane-5-methanol, 2,2-dimethyl-, carbamate

RN:	78002-88-7	MP (°C):
MW:	191.25	BP (°C):

Solubility (Moles/L)	Solubility (Grams/L)	Temp (°C)	Ref (#)	Evaluation (T P E A A)	Comments
3.000E-02	5.738E+00	rt	B174	0 0 1 0 0	

1254. C₇H₁₃N₃O₃S

Oxamyl
Vydate
Thioxamyl
N',N'-Dimethyl-N-[(methylcarbamoyl)oxy]-1-thiooxamimidic acid methyl ester
N,N-Dimethyl-α-methylcarbamoyloxyimino-α-(methylthio)acetamide
DPX 1410

RN:	23135-22-0	MP (°C):	109
MW:	219.26	BP (°C):	

Solubility (Moles/L)	Solubility (Grams/L)	Temp (°C)	Ref (#)	Evaluation (T P E A A)	Comments
1.288E+00	2.825E+02	20	B179	0 0 0 0 0	
1.277E+00	2.800E+02	25	M161	1 0 0 0 2	
9.977E-01	2.188E+02	ns	H308	0 0 0 0 1	

1255. C₇H₁₃N₅O

Hydroxysimazine
1,3,5-Triazin-2(1H)-one, 4,6-bis(ethylamino)-
2-Hydroxysimazine
4,6-bis(Ethylamino)-s-triazin-2-ol
G 30414

RN:	2599-11-3	MP (°C):
MW:	183.21	BP (°C):

Solubility (Moles/L)	Solubility (Grams/L)	Temp (°C)	Ref (#)	Evaluation (T P E A A)	Comments
1.500E-04	2.748E-02	2	B193	1 1 0 0 1	

1256. C₇H₁₄
1-Heptene
1-*n*-Heptene
n-Hept-1-ene

RN:	592-76-7	MP (°C):	−119
MW:	98.19	BP (°C):	93.6

Solubility (Moles/L)	Solubility (Grams/L)	Temp (°C)	Ref (#)	Evaluation (T P E A A)	Comments
1.850E-04	1.817E-02	25	M342	1 0 1 1 2	

1257. C₇H₁₄
2-Heptene

RN:	592-77-8	MP (°C):	
MW:	98.19	BP (°C):	

Solubility (Moles/L)	Solubility (Grams/L)	Temp (°C)	Ref (#)	Evaluation (T P E A A)	Comments
1.528E-04	1.500E-02	23.5	S171	2 1 2 2 2	
1.528E-04	1.500E-02	25	M001	2 1 2 2 1	

1258. C₇H₁₄
Cycloheptane

RN:	291-64-5	MP (°C):	−12
MW:	98.19	BP (°C):	118.5

Solubility (Moles/L)	Solubility (Grams/L)	Temp (°C)	Ref (#)	Evaluation (T P E A A)	Comments
1.854E-04	1.820E-02	20	M337	2 1 2 2 2	
3.055E-04	3.000E-02	25	M001	2 1 2 2 2	
2.760E-04	2.710E-02	30	G313	2 1 1 2 2	

1259. C₇H₁₄
Methylcyclohexane
Hexahydrotoluene
Methyl cyclohexane

RN:	108-87-2	MP (°C):	−126
MW:	98.19	BP (°C):	101

Solubility (Moles/L)	Solubility (Grams/L)	Temp (°C)	Ref (#)	Evaluation (T P E A A)	Comments
2.222E-04	2.182E-02	2.34	S461	0 0 0 0 0	
2.000E-04	1.964E-02	9.99	S461	0 0 0 0 0	
1.711E-04	1.680E-02	20	B318	0 0 0 0 0	EFG
1.691E-04	1.660E-02	20	B356	0 0 0 0 0	
1.324E-04	1.300E-02	20	M337	2 1 2 2 2	
1.667E-04	1.636E-02	24.99	S461	0 0 0 0 0	
1.701E-04	1.670E-02	25	G313	2 1 1 2 2	
1.629E-04	1.600E-02	25	K119	1 0 0 0 2	

(continued)

1259. C₇H₁₄ (continued)

Solubility (Moles/L)	Solubility (Grams/L)	Temp (°C)	Ref (#)	Evaluation (T P E A A)	Comments
1.426E-04	1.400E-02	25	M001	2 1 2 2 2	
1.426E-04	1.400E-02	25	M002	2 1 2 2 2	
1.629E-04	1.600E-02	25.0	P051	2 1 1 2 2	
1.629E-04	1.600E-02	25.00	P007	2 1 2 2 2	
1.644E-04	1.615E-02	26.1	M447	0 0 0 0 0	
1.375E-04	1.350E-02	28	B348	2 1 2 2 2	
1.833E-04	1.800E-02	40.1	P051	2 1 1 2 2	
1.833E-04	1.800E-02	40.10	P007	2 1 2 2 2	
1.925E-04	1.890E-02	55.7	P051	2 1 1 2 2	
1.925E-04	1.890E-02	55.70	P007	2 1 2 2 2	
2.800E-04	2.749E-02	70.5	M447	0 0 0 0 0	
3.442E-04	3.380E-02	99.1	P051	2 1 1 2 2	
3.442E-04	3.380E-02	99.10	P007	2 1 2 2 2	
5.589E-04	5.487E-02	100.5	M447	0 0 0 0 0	
8.097E-04	7.950E-02	120.0	P051	2 1 1 2 2	
8.097E-04	7.950E-02	120.00	P007	2 1 2 2 2	
1.355E-03	1.331E-01	131.0	M447	0 0 0 0 0	
1.416E-03	1.390E-01	137.3	P051	2 1 1 2 2	
1.416E-03	1.390E-01	137.30	P007	2 1 2 2 2	
2.485E-03	2.440E-01	149.5	P051	2 1 1 2 2	
2.485E-03	2.440E-01	149.50	P007	2 1 2 2 2	
2.349E-03	2.307E-01	151.4	M447	0 0 0 0 0	
1.426E-04	1.400E-02	ns	H123	0 0 0 0 0	

1260. C₇H₁₄N₂O₂S

Aldicarb

Temik

2-Methyl-2-(methylthio)propanal O-[(methylamino)carbonyl]oxime

UC 21149

N-Methylcarbamoyloxime,2-methyl-2-methylsulfenylpropionaldehyde

Methylcarbamic acid

RN: 116-06-3   **MP** (°C): 99

MW: 190.27   **BP** (°C):

Solubility (Moles/L)	Solubility (Grams/L)	Temp (°C)	Ref (#)	Evaluation (T P E A A)	Comments
3.162E-02	6.017E+00	20	B179	0 0 0 0 0	
3.153E-02	6.000E+00	ns	H042	0 0 0 0 2	
3.135E-02	5.964E+00	ns	M061	0 0 0 0 0	
3.153E-02	6.000E+00	rt	M161	0 0 0 0 0	

1261. C$_7$H$_{14}$N$_2$O$_3$
ε-Aminocaproic hydantoic acid
ε-Uramidocaproic acid

RN:		MP (°C):	
MW:	174.20	BP (°C):	

Solubility (Moles/L)	Solubility (Grams/L)	Temp (°C)	Ref (#)	Evaluation (T P E A A)	Comments
6.900E-03	1.202E+00	25	M024	1 2 0 1 2	

1262. C$_7$H$_{14}$N$_2$O$_3$
α-Aminocaproic hydantoic acid
α-Uramidocaproic acid

RN:		MP (°C):	169
MW:	174.20	BP (°C):	

Solubility (Moles/L)	Solubility (Grams/L)	Temp (°C)	Ref (#)	Evaluation (T P E A A)	Comments
6.900E-03	1.202E+00	25	M024	1 2 0 1 2	

1263. C$_7$H$_{14}$N$_2$O$_4$S$_2$
Djenkoic acid
Djenkolsaeure

RN:	498-59-9	MP (°C):	
MW:	254.33	BP (°C):	

Solubility (Moles/L)	Solubility (Grams/L)	Temp (°C)	Ref (#)	Evaluation (T P E A A)	Comments
1.966E-02	5.000E+00	100	F300	1 0 0 0 0	

1264. C$_7$H$_{14}$N$_6$
N2,N2,N4,N4-Tetramethylmelamine
Tetramethylmelamine

RN:	2827-47-6	MP (°C):	227.0
MW:	182.23	BP (°C):	

Solubility (Moles/L)	Solubility (Grams/L)	Temp (°C)	Ref (#)	Evaluation (T P E A A)	Comments
2.052E-03	3.740E-01	25	C051	1 2 1 1 2	pH 7

1265. C$_7$H$_{14}$O
Cycloheptanol

RN:	502-41-0	MP (°C):	
MW:	114.19	BP (°C):	185

Solubility (Moles/L)	Solubility (Grams/L)	Temp (°C)	Ref (#)	Evaluation (T P E A A)	Comments
1.318E-01	1.505E+01	ns	S460	0 0 0 0 0	

1266. C₇H₁₄O

Heptyl aldehyde
Heptanal
Oenanthaldehyd

RN:	111-71-7	**MP** (°C): −43.3
MW:	114.19	**BP** (°C): 152.8

Solubility (Moles/L)	Solubility (Grams/L)	Temp (°C)	Ref (#)	Evaluation (T P E A A)	Comments
2.715E-02	3.100E+00	0	F300	1 0 0 0 1	
1.576E-02	1.800E+00	40	F300	1 0 0 0 1	

1267. C₇H₁₄O

4-Methyl-cyclohexanol

RN:	589-91-3	**MP** (°C): −41
MW:	114.19	**BP** (°C): 171–173

Solubility (Moles/L)	Solubility (Grams/L)	Temp (°C)	Ref (#)	Evaluation (T P E A A)	Comments
1.318E-01	1.505E+01	ns	S460	0 0 0 0 0	

1268. C₇H₁₄O

Dipropyl ketone
4-Heptanone

RN:	123-19-3	**MP** (°C): −32.6
MW:	114.19	**BP** (°C): 144

Solubility (Moles/L)	Solubility (Grams/L)	Temp (°C)	Ref (#)	Evaluation (T P E A A)	Comments
6.430E-02	7.342E+00	0	G032	1 2 1 1 2	
4.660E-02	5.321E+00	10	G032	1 2 1 1 2	
3.750E-02	4.282E+00	20	D052	1 1 0 0 1	
2.793E-02	3.190E+00	25.50	O005	2 0 2 2 1	
3.350E-02	3.825E+00	30	G032	1 2 1 1 2	
2.880E-02	3.289E+00	50	G032	1 2 1 1 2	
2.720E-02	3.106E+00	75	G032	1 2 1 1 2	

1269. C₇H₁₄O

2-Heptanone
Heptan-2-one

RN:	110-43-0	**MP** (°C): −31
MW:	114.19	**BP** (°C): 151.5

Solubility (Moles/L)	Solubility (Grams/L)	Temp (°C)	Ref (#)	Evaluation (T P E A A)	Comments
3.489E-02	3.984E+00	20	D052	1 1 0 0 0	
3.836E-02	4.381E+00	20	G030	1 2 0 0 1	
3.800E-02	4.339E+00	20	M312	1 0 0 0 1	
3.750E-02	4.282E+00	25	G030	1 2 0 0 1	
1.675E-01	1.913E+01	25	P055	1 0 0 0 1	
3.570E-02	4.077E+00	25	W300	2 2 2 2 2	
3.489E-02	3.984E+00	30	G030	1 2 0 0 1	

1270. C₇H₁₄O

1270. C$_7$H$_{14}$O

5-Methyl-2-hexanone
Methyl isoamyl ketone
Isopentyl methyl ketone
Methylhexanone
Methyl isoamyl ketone
MIAK

RN:	110-12-3	**MP** (°C):	−74	
MW:	114.19	**BP** (°C):	144	

Solubility (Moles/L)	Solubility (Grams/L)	Temp (°C)	Ref (#)	Evaluation (T P E A A)	Comments
4.677E-02	5.341E+00	ns	S460	0 0 0 0 0	

1271. C$_7$H$_{14}$O

2,4-Dimethyl-3-pentanone
2,4-Dimethylpentanone-3

RN:	565-80-0	**MP** (°C):	−80	
MW:	114.19	**BP** (°C):	124	

Solubility (Moles/L)	Solubility (Grams/L)	Temp (°C)	Ref (#)	Evaluation (T P E A A)	Comments
5.137E-02	5.865E+00	20	G030	1 2 0 0 1	
4.963E-02	5.668E+00	25	G030	1 2 0 0 1	
4.877E-02	5.569E+00	30	G030	1 2 0 0 1	
4.972E-02	5.677E+00	ns	J300	0 0 0 0 0	

1272. C$_7$H$_{14}$O$_2$

Heptoic acid
Heptanoic acid
n-Heptanoic acid

RN:	111-14-8	**MP** (°C):		
MW:	130.19	**BP** (°C):		

Solubility (Moles/L)	Solubility (Grams/L)	Temp (°C)	Ref (#)	Evaluation (T P E A A)	Comments
1.459E-02	1.900E+00	0	B136	1 0 2 1 2	
1.843E-02	2.400E+00	15	F300	1 0 0 0 1	
1.847E-02	2.404E+00	15	L006	1 0 0 0 2	
1.721E-02	2.240E+00	20	B136	1 0 2 1 2	
1.870E-02	2.434E+00	20.0	R001	1 1 1 1 2	
2.161E-02	2.813E+00	25	H122	1 0 0 0 2	
2.082E-02	2.710E+00	30	B136	1 0 2 1 2	
2.076E-02	2.703E+00	30.0	R001	1 1 1 1 2	
2.389E-02	3.110E+00	45	B136	1 0 2 1 2	
2.381E-02	3.100E+00	45.0	R001	1 1 1 1 2	
2.711E-02	3.530E+00	60	B136	1 0 2 1 2	
2.702E-02	3.518E+00	60.0	R001	1 1 1 1 2	
1.457E-02	1.896E+00	.0	R001	1 1 1 1 2	

1273. C$_7$H$_{14}$O$_2$
Pentyl acetate
Amyl acetate

RN:	628-63-7	**MP** (°C):	−100		
MW:	130.19	**BP** (°C):	142		

Solubility (Moles/L)	Solubility (Grams/L)	Temp (°C)	Ref (#)	Evaluation (T P E A A)	Comments
1.304E-02	1.697E+00	20	D052	1 1 0 0 1	
1.290E-02	1.679E+00	20	S006	1 0 0 0 2	
1.329E-02	1.730E+00	25	K072	1 0 1 1 1	
1.329E-02	1.730E+00	25	M087	1 1 2 1 2	
3.060E-02	3.984E+00	30	R318	1 1 0 1 0	

1274. C$_7$H$_{14}$O$_2$
Isopropyl *N*-butyrate
Isopropyl butyrate
N-Butyric acid isopropyl ester

RN:	638-11-9	**MP** (°C):			
MW:	130.19	**BP** (°C):			

Solubility (Moles/L)	Solubility (Grams/L)	Temp (°C)	Ref (#)	Evaluation (T P E A A)	Comments
1.198E-02	1.560E+00	ns	J300	0 0 0 0 0	

1275. C$_7$H$_{14}$O$_2$
3-Hydroxy-5-methyl-5-ethyltetrahydrofuran
3-Furanol, 5-ethyltetrahydro-5-methyl-

RN:	30010-08-3	**MP** (°C):			
MW:	130.19	**BP** (°C):			

Solubility (Moles/L)	Solubility (Grams/L)	Temp (°C)	Ref (#)	Evaluation (T P E A A)	Comments
6.983E-01	9.091E+01	rt	B066	0 2 0 0 1	

1276. C$_7$H$_{14}$O$_2$
Isoamyl acetate
Acetic acid isoamyl ester
Essigsaeureisoamyl ester

RN:	123-92-2	**MP** (°C):	−79		
MW:	130.19	**BP** (°C):	142		

Solubility (Moles/L)	Solubility (Grams/L)	Temp (°C)	Ref (#)	Evaluation (T P E A A)	Comments
1.920E-02	2.500E+00	15	F300	1 0 0 0 1	
1.222E-02	1.591E+00	20	E002	1 0 0 0 1	
1.227E-02	1.597E+00	23.50	O005	2 0 2 2 1	
1.533E-02	1.996E+00	25	L062	2 2 0 1 0	

1277. C$_7$H$_{14}$O$_2$
Methyl hexanoate
Methyl caproate

RN: 106-70-7   **MP** (°C):   -71.0
MW: 130.19   **BP** (°C):   151.0

Solubility (Moles/L)	Solubility (Grams/L)	Temp (°C)	Ref (#)	Evaluation (T P E A A)	Comments
1.018E-02	1.325E+00	20	M337	2 1 2 2 2	

1278. C$_7$H$_{14}$O$_2$
Ethyl pentanoate
Ethyl *n*-valerate
Ethyl valerianate

RN: 539-82-2   **MP** (°C):
MW: 130.19   **BP** (°C):   145

Solubility (Moles/L)	Solubility (Grams/L)	Temp (°C)	Ref (#)	Evaluation (T P E A A)	Comments
1.710E-02	2.226E+00	ns	S460	0 0 0 0 0	

1279. C$_7$H$_{14}$O$_2$
n-Butyl propionate
Butyl propionate

RN: 590-01-2   **MP** (°C):   -89
MW: 130.19   **BP** (°C):   146.8

Solubility (Moles/L)	Solubility (Grams/L)	Temp (°C)	Ref (#)	Evaluation (T P E A A)	Comments
1.150E-02	1.498E+00	20	D052	1 1 0 0 0	
9.500E-03	1.237E+00	25	K012	1 0 0 0 1	
1.514E-02	1.970E+00	ns	S460	0 0 0 0 0	

1280. C$_7$H$_{14}$O$_2$
Propyl butyrate
Buttersaeure-propyl ester
n-Propyl *n*-butyrate

RN: 105-66-8   **MP** (°C):   -95
MW: 130.19   **BP** (°C):   143

Solubility (Moles/L)	Solubility (Grams/L)	Temp (°C)	Ref (#)	Evaluation (T P E A A)	Comments
1.240E-02	1.614E+00	17	F001	1 0 1 0 2	
1.244E-02	1.620E+00	17	F300	1 0 0 0 2	
1.200E-02	1.562E+00	17	S006	1 0 0 0 1	

1281. C$_7$H$_{14}$O$_2$

sec-Amyl acetate
2-Pentyl acetate
1-Methylbutyl acetate
RN: 53496-15-4   MP (°C):
MW: 130.19   BP (°C):

Solubility (Moles/L)	Solubility (Grams/L)	Temp (°C)	Ref (#)	Evaluation (T P E A A)	Comments
1.457E-02	1.896E+00	20	D052	1 1 0 0 0	

1282. C$_7$H$_{14}$O$_3$

n-Ethyl β-ethoxypropionate
Ethyl β-ethoxypropionate
RN: 763-69-9   MP (°C):
MW: 146.19   BP (°C):   166

Solubility (Moles/L)	Solubility (Grams/L)	Temp (°C)	Ref (#)	Evaluation (T P E A A)	Comments
3.597E-01	5.258E+01	25	D002	1 2 1 1 2	
3.566E-01	5.213E+01	25	R034	0 0 0 0 1	

1283. C$_7$H$_{14}$O$_3$

Butyl lactate
Butyl α-hydroxypropionate
2-Propanoic acid
Lactic acid butyl ester
Butyl 2-hydroxypropanoate
RN: 138-22-7   MP (°C):   −28
MW: 146.19   BP (°C):   185

Solubility (Moles/L)	Solubility (Grams/L)	Temp (°C)	Ref (#)	Evaluation (T P E A A)	Comments
2.631E-01	3.846E+01	20	D052	1 1 0 0 1	
2.982E-01	4.360E+01	25	R006	2 2 0 1 2	

1284. C$_7$H$_{14}$O$_3$

n-Propyl β-methoxypropionate
Propionic acid, 3-methoxy-, propyl ester
RN: 5349-56-4   MP (°C):
MW: 146.19   BP (°C):

Solubility (Moles/L)	Solubility (Grams/L)	Temp (°C)	Ref (#)	Evaluation (T P E A A)	Comments
2.121E-01	3.101E+01	25	R034	0 0 0 0 1	

1285. C$_7$H$_{14}$O$_3$
Methyl β-*n*-propoxypropionate
Propanoic acid, 3-propoxy-, methyl ester
RN: 14144-39-9   **MP** (°C):
MW: 146.19   **BP** (°C):

Solubility (Moles/L)	Solubility (Grams/L)	Temp (°C)	Ref (#)	Evaluation (T P E A A)	Comments
2.249E-01	3.288E+01	25	R034	0 0 0 0 1	

1286. C$_7$H$_{14}$O$_3$
3-Methoxy butyl acetate
3-Methoxy-1-butanol acetate
Methyl-1,3-butylene glycol acetate
3-Methoxybutyl acetate
Butoxyl
Butoxyl (3-methoxy-*N*-butyl acetate)
RN: 4435-53-4   **MP** (°C):
MW: 146.19   **BP** (°C):

Solubility (Moles/L)	Solubility (Grams/L)	Temp (°C)	Ref (#)	Evaluation (T P E A A)	Comments
4.151E-01	6.068E+01	20	D052	1 1 0 0 2	

1287. C$_7$H$_{14}$O$_6$
β-Methyl-D-glucoside
β-Methyl-D-glucosid
RN: 709-50-2   **MP** (°C):
MW: 194.19   **BP** (°C):

Solubility (Moles/L)	Solubility (Grams/L)	Temp (°C)	Ref (#)	Evaluation (T P E A A)	Comments
1.892E+00	3.674E+02	17	F300	1 0 0 0 2	

1288. C$_7$H$_{14}$O$_6$
α-D-Methylglucoside
α-Methyl-D-glucoside
α-Methyl-D-glucosid
RN: 97-30-3   **MP** (°C):   168
MW: 194.19   **BP** (°C):

Solubility (Moles/L)	Solubility (Grams/L)	Temp (°C)	Ref (#)	Evaluation (T P E A A)	Comments
1.992E+00	3.868E+02	17	F300	1 0 0 0 2	
2.543E+00	4.938E+02	17.8	W013	1 2 1 1 2	
2.637E+00	5.120E+02	22.5	W013	1 2 1 1 2	
2.657E+00	5.159E+02	25.5	W013	1 2 1 1 2	
2.696E+00	5.236E+02	26.6	W013	1 2 1 1 2	
2.699E+00	5.241E+02	27.3	W013	1 2 1 1 2	

(continued)

1288. C$_7$H$_{14}$O$_6$ (continued)

Solubility (Moles/L)	Solubility (Grams/L)	Temp (°C)	Ref (#)	Evaluation (T P E A A)	Comments
2.751E+00	5.342E+02	31.8	W013	1 2 1 1 2	
2.806E+00	5.448E+02	33.9	W013	1 2 1 1 2	
2.849E+00	5.533E+02	37.2	W013	1 2 1 1 2	
2.951E+00	5.731E+02	43.2	W013	1 2 1 1 2	
3.060E+00	5.942E+02	49.0	W013	1 2 1 1 2	
3.078E+00	5.978E+02	49.6	W013	1 2 1 1 2	
3.131E+00	6.079E+02	51.8	W013	1 2 1 1 2	
3.166E+00	6.148E+02	54.4	W013	1 2 1 1 2	
3.213E+00	6.240E+02	57.3	W013	1 2 1 1 2	
3.297E+00	6.402E+02	60.6	W013	1 2 1 1 2	
3.332E+00	6.471E+02	62.7	W013	1 2 1 1 2	
3.360E+00	6.525E+02	64.2	W013	1 2 1 1 2	
3.403E+00	6.608E+02	66.2	W013	1 2 1 1 2	
3.435E+00	6.670E+02	67.8	W013	1 2 1 1 2	
3.542E+00	6.878E+02	73.2	W013	1 2 1 1 2	
3.651E+00	7.090E+02	78.0	W013	1 2 1 1 2	

1289. C$_7$H$_{14}$O$_6$
α-Methyl-D-mannoside
α-Methyl-D-mannosid

RN:	617-04-9	**MP** (°C):	
MW:	194.19	**BP** (°C):	

Solubility (Moles/L)	Solubility (Grams/L)	Temp (°C)	Ref (#)	Evaluation (T P E A A)	Comments
1.018E+00	1.976E+02	17	F300	1 0 0 0 2	

1290. C$_7$H$_{14}$O$_7$
D-Mannoheptose
D-Sedoheptose

RN:	7634-39-1	**MP** (°C):	
MW:	210.19	**BP** (°C):	

Solubility (Moles/L)	Solubility (Grams/L)	Temp (°C)	Ref (#)	Evaluation (T P E A A)	Comments
>4.76E-01	>1.00E+02	20	F300	1 0 0 0 0	

1291. C$_7$H$_{14}$O$_7$
D-α-Glucoheptose
Gluco-heptose

RN:	62475-58-5	**MP** (°C):	
MW:	210.19	**BP** (°C):	

Solubility (Moles/L)	Solubility (Grams/L)	Temp (°C)	Ref (#)	Evaluation (T P E A A)	Comments
4.128E-01	8.676E+01	20	D041	1 0 0 0 1	

1292. C₇H₁₅Br

1-Bromoheptane
Heptyl bromide

RN:	629-04-9	**MP** (°C):	−56.1	
MW:	179.11	**BP** (°C):	178.5	

Solubility (Moles/L)	Solubility (Grams/L)	Temp (°C)	Ref (#)	Evaluation (T P E A A)	Comments
3.710E-05	6.645E-03	25	M342	1 0 1 1 2	

1293. C₇H₁₅Cl

1-Chloroheptane
Heptyl chloride

RN:	629-06-1	**MP** (°C):	−69.5	
MW:	134.65	**BP** (°C):	159	

Solubility (Moles/L)	Solubility (Grams/L)	Temp (°C)	Ref (#)	Evaluation (T P E A A)	Comments
1.010E-04	1.360E-02	25	M342	1 0 1 1 2	

1294. C₇H₁₅Cl₂N₂O₂P

Cyclophosphamide
Cyclophosphoramide
2-(bis(2-Chloroethyl)-amino)tetrahydro-2H-1,3,2-oxazaphosphorine 2-oxide
Cycloblastin
Sendoxan
Claphene

RN:	50-18-0	**MP** (°C):	
MW:	261.09	**BP** (°C):	

Solubility (Moles/L)	Solubility (Grams/L)	Temp (°C)	Ref (#)	Evaluation (T P E A A)	Comments
1.532E-01	4.000E+01	ns	K444	0 0 0 0 0	

1295. C₇H₁₅I

1-Iodoheptane
Heptyl iodide

RN:	4282-40-0	**MP** (°C):	−48.2	
MW:	226.10	**BP** (°C):	204	

Solubility (Moles/L)	Solubility (Grams/L)	Temp (°C)	Ref (#)	Evaluation (T P E A A)	Comments
1.550E-05	3.505E-03	25	M342	1 0 1 1 2	

1296. C₇H₁₅NO₂

1296. $C_7H_{15}NO_2$

Isobutyl urethane
Isobutylurethan

RN:	539-89-9	MP (°C):
MW:	145.20	BP (°C):

Solubility (Moles/L)	Solubility (Grams/L)	Temp (°C)	Ref (#)	Evaluation (T P E A A)	Comments
1.709E-01	2.482E+01	15.5	F001	1 0 1 2 2	

1297. $C_7H_{15}NO_2$

n-Hexyl carbamate
Hexyl carbamate

RN:	2114-20-7	MP (°C):	62
MW:	145.20	BP (°C):	

Solubility (Moles/L)	Solubility (Grams/L)	Temp (°C)	Ref (#)	Evaluation (T P E A A)	Comments
1.200E-02	1.742E+00	37	H006	1 2 2 1 1	

1298. $C_7H_{15}NO_2$

tert-Hexyl carbamate
3,3-Dimethyl-1-butanol carbamate

RN:	3124-38-7	MP (°C):	
MW:	145.20	BP (°C):	

Solubility (Moles/L)	Solubility (Grams/L)	Temp (°C)	Ref (#)	Evaluation (T P E A A)	Comments
3.400E-02	4.937E+00	37	H006	1 2 2 1 1	

1299. C_7H_{16}

3,3-Dimethylpentane
3,3-Dwumetylopentan

RN:	562-49-2	MP (°C):	−135
MW:	100.21	BP (°C):	86

Solubility (Moles/L)	Solubility (Grams/L)	Temp (°C)	Ref (#)	Evaluation (T P E A A)	Comments
5.928E-05	5.940E-03	25	K119	1 0 0 0 2	
5.908E-05	5.920E-03	25.0	P051	2 1 1 2 2	
5.908E-05	5.920E-03	25.00	P007	2 1 2 2 2	
6.766E-05	6.780E-03	40.1	P051	2 1 1 2 2	
6.766E-05	6.780E-03	40.10	P007	2 1 2 2 2	
8.153E-05	8.170E-03	55.7	P051	2 1 1 2 2	
8.153E-05	8.170E-03	55.70	P007	2 1 2 2 2	
1.028E-04	1.030E-02	69.7	P051	2 1 1 2 2	
1.028E-04	1.030E-02	69.70	P007	2 1 2 2 2	
1.577E-04	1.580E-02	99.1	P051	2 1 1 2 2	

(continued)

1299. C₇H₁₆ (continued)

Solubility (Moles/L)	Solubility (Grams/L)	Temp (°C)	Ref (#)	Evaluation (T P E A A)	Comments
1.577E-04	1.580E-02	99.10	P007	2 1 2 2 2	
2.724E-04	2.730E-02	118.0	P051	2 1 1 2 2	
2.724E-04	2.730E-02	118.00	P007	2 1 2 2 2	
6.716E-04	6.730E-02	120.4	P051	2 1 1 2 2	
6.716E-04	6.730E-02	120.40	P007	2 1 2 2 2	
8.592E-04	8.610E-02	150.4	P051	2 1 1 2 2	
8.592E-04	8.610E-02	150.40	P007	2 1 2 2 2	

1300. C₇H₁₆
3-Methylhexane
3-Metyloheksan

RN:	589-34-4	**MP** (°C):	−119	
MW:	100.21	**BP** (°C):	91	

Solubility (Moles/L)	Solubility (Grams/L)	Temp (°C)	Ref (#)	Evaluation (T P E A A)	Comments
5.229E-05	5.240E-03	0	P003	2 2 2 2 2	
1.048E-04	1.050E-02	23	C332	0 0 0 0 0	
2.635E-05	2.640E-03	25	K119	1 0 0 0 2	
4.940E-05	4.950E-03	25	P003	2 2 2 2 2	
2.635E-05	2.640E-03	25	P051	2 1 1 2 2	
2.635E-05	2.640E-03	25.00	P007	2 1 2 2 2	

1301. C₇H₁₆
2,4-Dimethylpentane
2,4-Dwumetylopentan

RN:	108-08-7	**MP** (°C):	−123	
MW:	100.21	**BP** (°C):	80	

Solubility (Moles/L)	Solubility (Grams/L)	Temp (°C)	Ref (#)	Evaluation (T P E A A)	Comments
6.487E-05	6.500E-03	0	P003	2 2 2 2 2	
4.401E-05	4.410E-03	25	K119	1 0 0 0 2	
4.052E-05	4.060E-03	25	M001	2 1 2 2 2	
3.613E-05	3.620E-03	25	M002	2 1 2 2 2	
5.489E-05	5.500E-03	25	P003	2 2 2 2 2	
4.401E-05	4.410E-03	25	P051	2 1 1 2 2	
4.401E-05	4.410E-03	25.00	P007	2 1 2 2 2	
4.100E-05	4.108E-03	ns	J300	0 0 0 0 0	

1302. C₇H₁₆

2,3-Dimethylpentane
2,3-Dwumetylopentan
RN: 565-59-3   MP (°C):   <25
MW: 100.21   BP (°C):   89

Solubility (Moles/L)	Solubility (Grams/L)	Temp (°C)	Ref (#)	Evaluation (T P E A A)	Comments
5.239E-05	5.250E-03	25	K119	1 0 0 0 2	
5.239E-05	5.250E-03	25	P051	2 1 1 2 2	
5.239E-05	5.250E-03	25.00	P007	2 1 2 2 2	

1303. C₇H₁₆

2-Methylhexane
2-Metyloheksan
RN: 591-76-4   MP (°C):   −118
MW: 100.21   BP (°C):   90

Solubility (Moles/L)	Solubility (Grams/L)	Temp (°C)	Ref (#)	Evaluation (T P E A A)	Comments
1.397E-04	1.400E-02	23	C332	0 0 0 0 0	
2.535E-05	2.540E-03	25	K119	1 0 0 0 2	
2.535E-05	2.540E-03	25	P051	2 1 1 2 2	
2.535E-05	2.540E-03	25.00	P007	2 1 2 2 2	

1304. C₇H₁₆

2,2-Dimethylpentane
2,2-Dwumetylopentan
RN: 590-35-2   MP (°C):   −123
MW: 100.21   BP (°C):   79.2

Solubility (Moles/L)	Solubility (Grams/L)	Temp (°C)	Ref (#)	Evaluation (T P E A A)	Comments
4.391E-05	4.400E-03	25	K119	1 0 0 0 2	
4.391E-05	4.400E-03	25	P051	2 1 1 2 2	
4.391E-05	4.400E-03	25.00	P007	2 1 2 2 2	
4.100E-05	4.108E-03	ns	J300	0 0 0 0 0	

1305. C₇H₁₆

Heptane
n-Heptane
RN: 142-82-5   MP (°C):   −90.7
MW: 100.21   BP (°C):   98.4

Solubility (Moles/L)	Solubility (Grams/L)	Temp (°C)	Ref (#)	Evaluation (T P E A A)	Comments
4.381E-05	4.390E-03	0	P003	2 2 2 2 2	
8.333E-05	8.350E-03	2.34	S461	0 0 0 0 0	

(continued)

1305. C$_7$H$_{16}$ (continued)

Solubility (Moles/L)	Solubility (Grams/L)	Temp (°C)	Ref (#)	Evaluation (T P E A A)	Comments
1.950E-05	1.954E-03	4.3	N004	1 1 2 2 2	
1.667E-05	1.670E-03	9.99	S461	0 0 0 0 0	
2.017E-05	2.021E-03	13.5	N004	1 1 2 2 2	
4.990E-04	5.000E-02	15	F300	1 0 0 0 1	
5.200E-04	5.211E-02	15.50	F001	1 0 1 0 2	
1.497E-04	1.500E-02	16	D047	1 0 0 1 0	
2.694E-05	2.700E-03	20	M337	2 1 2 2 1	
1.111E-05	1.113E-03	24.99	S461	0 0 0 0 0	
3.990E-03	3.998E-01	25	G323	2 2 2 2 0	
4.990E-04	5.000E-02	25	K072	1 0 1 1 1	
2.235E-05	2.240E-03	25	K119	1 0 0 0 2	
2.924E-05	2.930E-03	25	M001	2 1 2 2 2	
2.924E-05	2.930E-03	25	M002	2 1 2 2 2	
4.990E-04	5.000E-02	25	M087	1 1 2 1 0	
3.050E-05	3.056E-03	25	M342	1 0 1 1 2	
3.363E-05	3.370E-03	25	P003	2 2 2 2 2	
4.989E-04	5.000E-02	25	S012	2 0 2 2 0	
2.656E-05	2.661E-03	25.0	N004	1 1 2 2 2	
2.235E-05	2.240E-03	25.0	P051	2 1 1 2 2	
2.235E-05	2.240E-03	25.00	P007	2 1 2 2 2	
2.261E-05	2.266E-03	35.0	N004	1 1 2 2 2	
2.625E-05	2.630E-03	40.1	P051	2 1 1 2 2	
2.400E-05	2.405E-03	45.0	N004	1 1 2 2 2	
8.973E-03	8.992E-01	50	G323	2 2 2 2 0	
3.104E-05	3.110E-03	55.7	P051	2 1 1 2 2	
3.104E-05	3.110E-03	55.70	P007	2 1 2 2 2	
5.589E-05	5.600E-03	99.1	P051	2 1 1 2 2	
5.589E-05	5.600E-03	99.10	P007	2 1 2 2 2	
1.138E-04	1.140E-02	118	P007	2 1 2 2 2	
1.138E-04	1.140E-02	118.0	P051	2 1 1 2 2	
2.724E-04	2.730E-02	136.6	P051	2 1 1 2 2	
2.724E-04	2.730E-02	136.60	P007	2 1 2 2 2	
4.361E-04	4.370E-02	150.4	P051	2 1 1 2 2	
4.361E-04	4.370E-02	150.40	P007	2 1 2 2 2	
3.692E-05	3.700E-03	ns	B151	0 2 1 1 1	
7.000E-04	7.014E-02	ns	H012	0 2 2 0 0	

1306. C$_7$H$_{16}$O

3-Heptanol
(±)-3-Heptanol
3-Hydroxyheptane
1-Ethyl-1-pentanol

RN:	589-82-2	**MP** (°C):	−70
MW:	116.20	**BP** (°C):	156.0

Solubility (Moles/L)	Solubility (Grams/L)	Temp (°C)	Ref (#)	Evaluation (T P E A A)	Comments
4.100E-02	4.764E+00	20	H330	0 0 0 0 0	
3.428E-02	3.984E+00	25	C093	2 1 1 1 0	

1307. C₇H₁₆O

2-Heptanol
2-Hydroxyheptane
Amylmethylcarbinol

RN: 543-49-7   **MP** (°C):   <25
MW: 116.20   **BP** (°C):   159.00

Solubility (Moles/L)	Solubility (Grams/L)	Temp (°C)	Ref (#)	Evaluation (T P E A A)	Comments
5.532E-02	6.428E+00	0	S307	1 1 0 2 2	
3.966E-02	4.609E+00	10.2	S307	1 1 0 2 2	
3.633E-02	4.222E+00	19.5	S307	1 1 0 2 2	
3.001E-02	3.488E+00	30.7	S307	1 1 0 2 2	
2.813E-02	3.269E+00	40.0	S307	1 1 0 2 2	
2.514E-02	2.921E+00	50.0	S307	1 1 0 2 2	
2.471E-02	2.872E+00	60.3	S307	1 1 0 2 2	
2.754E-02	3.200E+00	70.3	S307	1 1 0 2 2	
2.754E-02	3.200E+00	80.0	S307	1 1 0 2 2	
2.942E-02	3.418E+00	90.2	S307	1 1 0 2 2	

1308. C₇H₁₆O

3-Methyl-3-hexanol
3-Methylhexanol-3

RN: 597-96-6   **MP** (°C):   <25
MW: 116.20   **BP** (°C):

Solubility (Moles/L)	Solubility (Grams/L)	Temp (°C)	Ref (#)	Evaluation (T P E A A)	Comments
1.146E-01	1.332E+01	20	G006	1 2 1 1 2	
1.012E-01	1.176E+01	25	G006	1 2 1 1 2	
9.110E-02	1.059E+01	30	G006	1 2 1 1 2	

1309. C₇H₁₆O

3-Ethyl-3-pentanol
3-Ethyl-pentanol-3
Triethyl carbinol

RN: 597-49-9   **MP** (°C):   −12
MW: 116.20   **BP** (°C):   141.0

Solubility (Moles/L)	Solubility (Grams/L)	Temp (°C)	Ref (#)	Evaluation (T P E A A)	Comments
1.613E-01	1.874E+01	20	G006	1 2 1 1 2	
1.422E-01	1.652E+01	25	G006	1 2 1 1 2	
1.272E-01	1.478E+01	30	G006	1 2 1 1 2	
1.071E-01	1.244E+01	40	G006	1 2 1 1 2	

1310. C$_7$H$_{16}$O
2-Methyl-2-hexanol
2-Methylhexanol-2

RN: 625-23-0   **MP** (°C):   <25
MW: 116.20   **BP** (°C):   141

Solubility (Moles/L)	Solubility (Grams/L)	Temp (°C)	Ref (#)	Evaluation (T P E A A)	Comments
9.195E-02	1.068E+01	20	G006	1 2 1 1 2	
8.267E-02	9.607E+00	25	G006	1 2 1 1 1	
7.422E-02	8.625E+00	30	G006	1 2 1 1 1	

1311. C$_7$H$_{16}$O
2,4-Dimethyl-3-pentanol
2,4-Dimethylpentanol-3
Diisopropyl carbinol

RN: 600-36-2   **MP** (°C):   −70
MW: 116.20   **BP** (°C):

Solubility (Moles/L)	Solubility (Grams/L)	Temp (°C)	Ref (#)	Evaluation (T P E A A)	Comments
1.009E-01	1.172E+01	0	S307	1 1 0 2 2	
8.942E-02	1.039E+01	10.0	S307	1 1 0 2 2	
6.660E-02	7.740E+00	20	G006	1 2 1 1 1	
6.067E-02	7.050E+00	20.2	S307	1 1 0 2 2	
1.935E-01	2.248E+01	24.50	O005	2 0 2 2 1	
5.982E-02	6.951E+00	25	G006	1 2 1 1 1	
5.727E-02	6.655E+00	30	G006	1 2 1 1 1	
5.489E-02	6.379E+00	30.6	S307	1 1 0 2 2	
4.562E-02	5.302E+00	39.5	S307	1 1 0 2 2	
4.332E-02	5.035E+00	49.7	S307	1 1 0 2 2	
3.992E-02	4.638E+00	60.3	S307	1 1 0 2 2	
3.778E-02	4.391E+00	70.2	S307	1 1 0 2 2	
3.667E-02	4.262E+00	80.2	S307	1 1 0 2 2	
3.855E-02	4.480E+00	90.6	S307	1 1 0 2 2	

1312. C$_7$H$_{16}$O
2,4-Dimethyl-2-pentanol
2,4-Dimethylpentanol-2

RN: 625-06-9   **MP** (°C):   <−20
MW: 116.20   **BP** (°C):

Solubility (Moles/L)	Solubility (Grams/L)	Temp (°C)	Ref (#)	Evaluation (T P E A A)	Comments
1.272E-01	1.478E+01	20	G006	1 2 1 1 2	
1.138E-01	1.322E+01	25	G006	1 2 1 1 2	
1.037E-01	1.205E+01	30	G006	1 2 1 1 2	

1313. C₇H₁₆O

1313. $C_7H_{16}O$

2,3-Dimethyl-2-pentanol
2,3-Dimethylpentanol-2
RN: 4911-70-0   **MP** (°C):   <25
MW: 116.20   **BP** (°C):

Solubility (Moles/L)	Solubility (Grams/L)	Temp (°C)	Ref (#)	Evaluation (T P E A A)	Comments
1.430E-01	1.662E+01	20	G006	1 2 1 1 2	
1.305E-01	1.517E+01	25	G006	1 2 1 1 2	
1.188E-01	1.381E+01	30	G006	1 2 1 1 2	

1314. $C_7H_{16}O$

2,3,3-Trimethyl-2-butanol
Dimethyl-*tert*-butylcarbinol
1,1,2,2-Tetramethylpropanol
1,1,2,2-Tetramethylpropyl alcohol
RN: 594-83-2   **MP** (°C):   17
MW: 116.20   **BP** (°C):   131

Solubility (Moles/L)	Solubility (Grams/L)	Temp (°C)	Ref (#)	Evaluation (T P E A A)	Comments
1.852E-01	2.153E+01	40	G006	1 2 1 1 2	

1315. $C_7H_{16}O$

2,2-Dimethyl-3-pentanol
2,2-Dimethylpentanol-3
RN: 3970-62-5   **MP** (°C):   −5
MW: 116.20   **BP** (°C):   132

Solubility (Moles/L)	Solubility (Grams/L)	Temp (°C)	Ref (#)	Evaluation (T P E A A)	Comments
7.507E-02	8.723E+00	20	G006	1 2 1 1 1	
6.999E-02	8.133E+00	25	G006	1 2 1 1 1	
6.745E-02	7.838E+00	30	G006	1 2 1 1 1	

1316. $C_7H_{16}O$

1-Heptanol
1-Hydroxyheptane
Heptan-1-ol
Heptanol-(1)
n-Heptyl alcohol
RN: 111-70-6   **MP** (°C):   −34.6
MW: 116.20   **BP** (°C):   175.8

Solubility (Moles/L)	Solubility (Grams/L)	Temp (°C)	Ref (#)	Evaluation (T P E A A)	Comments
2.916E-02	3.388E+00	0	E029	1 2 0 1 1	
2.026E-02	2.354E+00	0	S307	1 1 0 2 2	
1.897E-02	2.205E+00	6.04	H110	2 2 2 2 2	

(*continued*)

1316. C₇H₁₆O (continued)

Solubility (Moles/L)	Solubility (Grams/L)	Temp (°C)	Ref (#)	Evaluation (T P E A A)	Comments
2.232E-02	2.593E+00	10	E029	1 2 0 1 1	
1.739E-02	2.020E+00	10.24	H110	2 2 2 2 2	
2.172E-02	2.524E+00	10.5	S307	1 1 0 2 2	
1.720E-02	1.999E+00	10.54	H110	2 2 2 2 2	
1.067E-02	1.240E+00	11.4	N042	1 0 2 1 1	
1.608E-02	1.869E+00	15.04	H110	2 2 2 2 2	
1.544E-02	1.795E+00	17.94	H110	2 2 2 2 2	
8.000E-03	9.296E-01	18	F001	1 0 1 0 2	
8.605E-03	1.000E+00	18	F300	1 0 0 0 1	
1.478E-02	1.717E+00	20	A015	1 2 1 1 2	
1.718E-02	1.996E+00	20	E029	1 2 0 1 1	
1.450E-02	1.685E+00	20	H330	0 0 0 0 0	
1.507E-02	1.751E+00	20.04	H110	2 2 2 2 2	
1.581E-02	1.837E+00	20.2	S307	1 1 0 2 2	
1.476E-02	1.716E+00	21.94	H110	2 2 2 2 2	
1.450E-02	1.685E+00	23.94	H110	2 2 2 2 2	
1.443E-02	1.677E+00	24.94	H110	2 2 2 2 2	
1.546E-02	1.797E+00	25	B038	1 2 1 1 2	
1.000E+00	1.162E+02	25	F044	1 0 0 0 0	EFG
1.460E-02	1.697E+00	25	K025	2 1 1 1 1	
1.434E-02	1.666E+00	25.04	H110	2 2 2 2 2	
1.423E-02	1.653E+00	26.04	H110	2 2 2 2 2	
1.411E-02	1.640E+00	28.04	H110	2 2 2 2 2	
1.375E-02	1.597E+00	30	E029	1 2 0 1 1	
1.397E-02	1.624E+00	30.14	H110	2 2 2 2 2	
1.399E-02	1.626E+00	30.14	H110	2 2 2 2 2	
1.323E-02	1.538E+00	30.6	S307	1 1 0 2 2	
1.386E-02	1.611E+00	32.94	H110	2 2 2 2 2	
1.426E-02	1.657E+00	39.8	S307	1 1 0 2 2	
1.117E-02	1.298E+00	40	E029	1 2 0 1 1	
9.456E-03	1.099E+00	50	E029	1 2 0 1 1	
1.392E-02	1.617E+00	50.1	S307	1 1 0 2 2	
9.456E-03	1.099E+00	60	E029	1 2 0 1 1	
1.529E-02	1.777E+00	60.0	S307	1 1 0 2 2	
1.289E-02	1.498E+00	70	E029	1 2 0 1 1	
1.080E-02	1.255E+00	70	F001	1 0 1 0 2	
1.752E-02	2.036E+00	70.1	S307	1 1 0 2 2	
1.632E-02	1.896E+00	80	E029	1 2 0 1 1	
1.460E-02	1.697E+00	80	F001	1 0 1 0 2	
1.863E-02	2.165E+00	80.1	S307	1 1 0 2 2	
1.975E-02	2.295E+00	90	E029	1 2 0 1 1	
1.940E-02	2.254E+00	90	F001	1 0 1 0 2	
2.086E-02	2.424E+00	90.5	S307	1 1 0 2 2	
2.488E-02	2.892E+00	100	E029	1 2 0 1 1	
2.460E-02	2.859E+00	100	F001	1 0 1 0 2	
2.582E-02	3.000E+00	100	F300	1 0 0 0 1	
3.001E-02	3.488E+00	110	E029	1 2 0 1 1	
3.060E-02	3.556E+00	110	F001	1 0 1 0 2	
3.685E-02	4.282E+00	120	E029	1 2 0 1 1	

(continued)

1316. C$_7$H$_{16}$O (continued)

Solubility (Moles/L)	Solubility (Grams/L)	Temp (°C)	Ref (#)	Evaluation (T P E A A)	Comments
4.537E-02	5.272E+00	130	E029	1 2 0 1 1	
5.557E-02	6.458E+00	140	E029	1 2 0 1 1	
6.830E-02	7.937E+00	150	E029	1 2 0 1 1	
8.352E-02	9.705E+00	160	E029	1 2 0 1 1	
1.046E-01	1.215E+01	170	E029	1 2 0 1 2	
1.355E-01	1.575E+01	180	E029	1 2 0 1 2	
1.753E-01	2.038E+01	190	E029	1 2 0 1 2	
2.213E-01	2.572E+01	200	E029	1 2 0 1 2	
2.894E-01	3.363E+01	210	E029	1 2 0 1 2	
3.847E-01	4.471E+01	220	E029	1 1 0 1 2	
5.404E-01	6.279E+01	230	E029	1 2 0 1 2	
7.894E-01	9.173E+01	240	E029	1 2 0 1 2	
1.054E+00	1.225E+02	245	E029	1 2 0 1 2	
1.029E-02	1.195E+00	ns	H012	0 2 2 0 2	
1.558E-02	1.810E+00	ns	L003	0 0 2 1 2	

1317. C$_7$H$_{16}$O

Isopropyl *tert*-butyl ether
2-Methyl-2-(1-methylethoxy)-propane
t-Butyl isopropyl ether

RN:	17348-59-3	**MP** (°C):	−88	
MW:	116.20	**BP** (°C):	87.6	

Solubility (Moles/L)	Solubility (Grams/L)	Temp (°C)	Ref (#)	Evaluation (T P E A A)	Comments
4.303E-03	5.000E-01	25	K072	1 0 1 1 1	
4.303E-03	5.000E-01	25	M087	1 1 2 1 1	

1318. C$_7$H$_{16}$O

Heptanol

RN:	53535-33-4	**MP** (°C):	−36	
MW:	116.20	**BP** (°C):	176	

Solubility (Moles/L)	Solubility (Grams/L)	Temp (°C)	Ref (#)	Evaluation (T P E A A)	Comments
1.009E-01	1.173E+01	20	S006	1 0 0 0 2	
1.240E-02	1.441E+00	24	H345	0 0 0 0 0	

1319. C$_7$H$_{16}$O

4-Heptanol
Dipropyl carbinol

RN:	589-55-9	**MP** (°C):	−42	
MW:	116.20	**BP** (°C):		

Solubility (Moles/L)	Solubility (Grams/L)	Temp (°C)	Ref (#)	Evaluation (T P E A A)	Comments
4.090E-02	4.753E+00	20	H330	0 0 0 0 0	

1320. C₇H₁₆O
2,3-Dimethyl-3-pentanol
2,3-Dimethylpentanol-3

RN: 595-41-5   **MP** (°C): <25
MW: 116.20   **BP** (°C): 140

Solubility (Moles/L)	Solubility (Grams/L)	Temp (°C)	Ref (#)	Evaluation (T P E A A)	Comments
1.580E-01	1.836E+01	20	G006	1 2 1 1 2	
1.389E-01	1.614E+01	25	G006	1 2 1 1 2	
1.213E-01	1.410E+01	30	G006	1 2 1 1 2	

1321. C₇H₁₆O₄S₂
Sulfonmethane
Sulfonal

RN: 115-24-2   **MP** (°C): 125
MW: 228.33   **BP** (°C): 300

Solubility (Moles/L)	Solubility (Grams/L)	Temp (°C)	Ref (#)	Evaluation (T P E A A)	Comments
5.962E-02	1.361E+01	16	A072	1 0 1 0 2	
5.956E-02	1.360E+01	16	F300	1 0 0 0 2	
1.027E-02	2.345E+00	18	F062	1 0 2 2 2	
2.847E-01	6.500E+01	100	F300	1 0 0 0 1	
5.888E-02	1.345E+01	ns	R427	0 0 0 0 0	

1322. C₇H₁₆O₇
(+)-Perseitol
D-Manno-α-heptit

RN: 527-06-0   **MP** (°C): 188
MW: 212.20   **BP** (°C):

Solubility (Moles/L)	Solubility (Grams/L)	Temp (°C)	Ref (#)	Evaluation (T P E A A)	Comments
3.044E-01	6.460E+01	18	F300	1 0 0 0 2	
1.466E+00	3.110E+02	74	F300	1 0 0 0 1	

1323. C₇H₁₇O₂PS₃
Phorate
Thimet
Rampart
Phosphorodithioic acid *O,O*-diethyl *S*-[(ethylthio)methyl] ester
American Cyanamid 3911
CL 35024

RN: 298-02-2   **MP** (°C): –43
MW: 260.38   **BP** (°C):

Solubility (Moles/L)	Solubility (Grams/L)	Temp (°C)	Ref (#)	Evaluation (T P E A A)	Comments
6.874E-05	1.790E-02	20	B169	2 1 1 1 1	
1.905E-04	4.961E-02	20	B179	0 0 0 0 0	

(continued)

1323. $C_7H_{17}O_2PS_3$ (continued)

Solubility (Moles/L)	Solubility (Grams/L)	Temp (°C)	Ref (#)	Evaluation (T P E A A)	Comments
7.681E-05	2.000E-02	24	F179	2 2 2 2 2	
2.688E-04	7.000E-02	ns	M061	0 0 0 0 1	
1.920E-04	5.000E-02	rt	M161	0 0 0 0 1	

1324. $C_7H_{17}O_2PS_3$

S-2-Isopropylthioethyl O,O-dimethyl phosphorodithioate
Isothioate
O,O-Dimethyls-isopropylthioethyl phosphoroditjioate

RN: 36614-38-7   **MP** (°C):
MW: 260.38   **BP** (°C): 55

Solubility (Moles/L)	Solubility (Grams/L)	Temp (°C)	Ref (#)	Evaluation (T P E A A)	Comments
3.725E-04	9.700E-02	25	M161	1 0 0 0 1	
3.725E-04	9.700E-02	25	N304	1 0 0 0 1	

1325. $C_7H_{17}O_4PS_3$

Phorate sulfone
O,O'-Diethyl S-ethylsulfonylmethyl-phosphorodithioate
Thimet sulfone
CL 18161
Phosphorodithioic acid O,O-diethyl S-[(ethylsulfonyl)methyl] ester

RN: 2588-04-7   **MP** (°C):
MW: 292.38   **BP** (°C):

Solubility (Moles/L)	Solubility (Grams/L)	Temp (°C)	Ref (#)	Evaluation (T P E A A)	Comments
2.939E-03	8.593E-01	19	B169	2 0 1 1 2	

1326. $C_8H_2Cl_4N_2$

Chlorquinox
5,6,7,8-Tetrachloroquinoxaline
Lucel
Tetrachloroquinoxaline

RN: 3495-42-9   **MP** (°C): 190
MW: 267.93   **BP** (°C):

Solubility (Moles/L)	Solubility (Grams/L)	Temp (°C)	Ref (#)	Evaluation (T P E A A)	Comments
3.732E-06	1.000E-03	25	M161	1 0 0 0 0	

1327. C₈H₂Cl₄O₄

Tetrachlorophthalic acid
Tetrachlorphthalsaeure
Tetrachloro-1,2-benzenedicarboxylic acid

RN:	632-58-6	MP (°C):
MW:	303.91	BP (°C):

Solubility (Moles/L)	Solubility (Grams/L)	Temp (°C)	Ref (#)	Evaluation (T P E A A)	Comments
1.876E-02	5.700E+00	14	F300	1 0 0 0 1	
1.007E-01	3.060E+01	99	F300	1 0 0 0 2	

1328. C₈H₃Cl₂F₃N₂

Chlorflurazole
4,5-Dichloro-2-(trifluoromethyl)-benzimidazole
Dichloro-2-(trifluoromethyl)benzimidazole
2-Trifluoromethyl-4,5-dichlorobenzimidazole

RN:	3615-21-2	MP (°C):
MW:	255.03	BP (°C):

Solubility (Moles/L)	Solubility (Grams/L)	Temp (°C)	Ref (#)	Evaluation (T P E A A)	Comments
2.353E-04	6.000E-02	ns	B100	0 0 0 0 0	
2.353E-04	6.000E-02	ns	M061	0 0 0 0 1	

1329. C₈H₃Cl₅O₂

Pentachlorophenyl acetate
Pentachlorophenol acetate
Rabcon

RN:	1441-02-7	MP (°C):
MW:	308.38	BP (°C):

Solubility (Moles/L)	Solubility (Grams/L)	Temp (°C)	Ref (#)	Evaluation (T P E A A)	Comments
6.486E-05	2.000E-02	ns	L311	0 0 0 0 1	

1330. C₈H₃Cl₅O₃

2,3,4,5,6-Pentachlorophenoxyacetic acid
Pentachlorophenoxyacetic acid

RN:	2877-14-7	MP (°C):
MW:	324.38	BP (°C):

Solubility (Moles/L)	Solubility (Grams/L)	Temp (°C)	Ref (#)	Evaluation (T P E A A)	Comments
1.800E-04	5.839E-02	25	L030	1 0 2 1 1	

1331. C$_8$H$_4$Cl$_4$O$_3$
2,3,4,6-Tetrachlorophenoxyacetic acid
Acetic acid, (2,3,4,6-tetrachlorophenoxy)-
RN: 10587-37-8   **MP** (°C):
MW: 289.93   **BP** (°C):

Solubility (Moles/L)	Solubility (Grams/L)	Temp (°C)	Ref (#)	Evaluation (T P E A A)	Comments
3.900E-04	1.131E-01	25	L030	1 0 2 1 1	

1332. C$_8$H$_4$N$_2$
1,4-Benzenedicarbonitrile
Terephthalonitrile
1,4-Dicyanobenzene
RN: 623-26-7   **MP** (°C):
MW: 128.13   **BP** (°C):

Solubility (Moles/L)	Solubility (Grams/L)	Temp (°C)	Ref (#)	Evaluation (T P E A A)	Comments
6.970E-04	8.931E-02	25	C316	0 0 0 0 0	0.1M NaCl

1333. C$_8$H$_4$N$_2$S
m-Cyanophenyl isothiocyanate
3-Isothiocyanato-benzonitrile
3-Cyanophenyl isothiocyanate
RN: 3125-78-8   **MP** (°C):
MW: 160.20   **BP** (°C):

Solubility (Moles/L)	Solubility (Grams/L)	Temp (°C)	Ref (#)	Evaluation (T P E A A)	Comments
6.410E-04	1.027E-01	25	K032	2 2 0 1 2	

1334. C$_8$H$_4$N$_2$S$_2$
m-Isothiocyanophenyl isothiocyanate
3-Isothiocyanophenyl isothiocyanate
RN: 3125-77-7   **MP** (°C):
MW: 192.26   **BP** (°C):

Solubility (Moles/L)	Solubility (Grams/L)	Temp (°C)	Ref (#)	Evaluation (T P E A A)	Comments
2.000E-05	3.845E-03	25	K032	2 2 0 1 1	

1335. C₈H₄O₃

Phthalic anhydride
1,2-Benzenedicarboxylic acid anhydride
1,3-Isobenzofurandione
Phthalic acid anhydride
1,3-Dioxophthalan
1,3 Phthalandione

RN: 85-44-9   **MP** (°C):   130.8
MW: 148.12   **BP** (°C):   295.0

Solubility (Moles/L)	Solubility (Grams/L)	Temp (°C)	Ref (#)	Evaluation (T P E A A)	Comments
4.186E-02	6.200E+00	26.70	L095	2 2 1 1 2	
4.027E-02	5.964E+00	rt	D021	0 0 1 1 2	

1336. C₈H₅ClO₄

3-Chlorophthalic acid
3-Chlor-phthalsaeure

RN: 27563-65-1   **MP** (°C):
MW: 200.58   **BP** (°C):

Solubility (Moles/L)	Solubility (Grams/L)	Temp (°C)	Ref (#)	Evaluation (T P E A A)	Comments
1.057E-01	2.120E+01	14	F300	1 0 0 0 2	

1337. C₈H₅Cl₃O₂

Chlorfenac
2,3,6-Trichlorophenylacetic acid
Fenac

RN: 85-34-7   **MP** (°C):   161
MW: 239.49   **BP** (°C):

Solubility (Moles/L)	Solubility (Grams/L)	Temp (°C)	Ref (#)	Evaluation (T P E A A)	Comments
8.351E-04	2.000E-01	28	M161	1 0 0 0 2	
8.351E-04	2.000E-01	30	M061	1 0 0 0 2	

1338. C₈H₅Cl₃O₃

2,4,5-Trichlorophenoxyacetic acid
Acetic acid, (2,4,5-trichlorophenoxy)-
(2,4,5-Trichlorophenoxy)acetic acid
2,4,5-T

RN: 93-76-5   **MP** (°C):   156
MW: 255.49   **BP** (°C):

Solubility (Moles/L)	Solubility (Grams/L)	Temp (°C)	Ref (#)	Evaluation (T P E A A)	Comments
9.316E-04	2.380E-01	20	B185	0 0 0 0 0	
7.398E-04	1.890E-01	20	M061	1 0 0 0 2	
1.090E-03	2.785E-01	24.99	N417	0 0 0 0 0	

(continued)

1338. C$_8$H$_5$Cl$_3$O$_3$ (continued)

Solubility (Moles/L)	Solubility (Grams/L)	Temp (°C)	Ref (#)	Evaluation (T P E A A)	Comments
1.100E-03	2.810E-01	25	B164	1 0 1 1 2	
1.096E-03	2.800E-01	25	B185	0 0 0 0 0	
1.050E-03	2.683E-01	25	L030	1 0 2 1 2	
1.088E-03	2.780E-01	25	M161	1 0 0 0 2	
9.316E-04	2.380E-01	30	B200	1 0 0 0 2	
9.783E-04	2.499E-01	ns	B100	0 0 0 0 1	
7.828E-04	2.000E-01	ns	B185	0 0 0 0 0	
8.000E-04	2.044E-01	ns	F184	0 0 0 0 1	
9.316E-04	2.380E-01	ns	K138	0 0 0 0 1	
9.824E-04	2.510E-01	ns	L024	0 0 0 0 2	
2.512E-04	6.418E-02	ns	M163	0 0 0 0 0	EFG
7.828E-04	2.000E-01	ns	N013	0 0 0 0 2	

1339. C$_8$H$_5$Cl$_3$O$_3$

3,4,5-Trichlorophenoxyacetic acid
Acetic acid, (3,4,5-trichlorophenoxy)-
3,4,5-T

RN: 80496-87-3   **MP** (°C):
MW: 255.49   **BP** (°C):

Solubility (Moles/L)	Solubility (Grams/L)	Temp (°C)	Ref (#)	Evaluation (T P E A A)	Comments
1.150E-03	2.938E-01	25	L030	1 0 2 1 2	

1340. C$_8$H$_5$Cl$_3$O$_3$

2,3,4-Trichlorophenoxyacetic acid
Acetic acid, (2,3,4-trichlorophenoxy)-
2,3,4-T

RN: 25141-27-9   **MP** (°C):
MW: 255.49   **BP** (°C):

Solubility (Moles/L)	Solubility (Grams/L)	Temp (°C)	Ref (#)	Evaluation (T P E A A)	Comments
8.000E-04	2.044E-01	25	L030	1 0 2 1 1	

1341. C$_8$H$_5$Cl$_3$O$_3$

2,4,6-Trichlorophenoxyacetic acid
Acetic acid, (2,4,6-trichlorophenoxy)-
2,4,6-T

RN: 575-89-3   **MP** (°C): 45
MW: 255.49   **BP** (°C):

Solubility (Moles/L)	Solubility (Grams/L)	Temp (°C)	Ref (#)	Evaluation (T P E A A)	Comments
9.700E-04	2.478E-01	25	L030	1 0 2 1 1	

1342. C$_8$H$_5$Cl$_3$O$_3$
2,3,6-Trichlorophenoxyacetic acid
Acetic acid, (2,3,6-trichlorophenoxy)-
2,3,6-T

RN:	4007-00-5	MP (°C):	148
MW:	255.49	BP (°C):	

Solubility (Moles/L)	Solubility (Grams/L)	Temp (°C)	Ref (#)	Evaluation (T P E A A)	Comments
2.400E-03	6.132E-01	25	L030	1 0 2 1 2	

1343. C$_8$H$_5$Cl$_3$O$_3$
2,3,5-Trichlorophenoxyacetic acid
Acetic acid, (2,3,5-trichlorophenoxy)-
2,3,5-T

RN:	33433-95-3	MP (°C):	
MW:	255.49	BP (°C):	

Solubility (Moles/L)	Solubility (Grams/L)	Temp (°C)	Ref (#)	Evaluation (T P E A A)	Comments
1.000E-03	2.555E-01	25	L030	1 0 2 1 2	

1344. C$_8$H$_5$F$_3$O
2,2,2-Trifluoroacetophenone
Trifluoroacetophenone
α,α,α-Trifluoroacetophenone
Phenyl trifluoromethyl ketone
2,2,2-Trifluoro-1-phenylethanone

RN:	434-45-7	MP (°C):	−40
MW:	174.12	BP (°C):	165–166

Solubility (Moles/L)	Solubility (Grams/L)	Temp (°C)	Ref (#)	Evaluation (T P E A A)	Comments
7.007E-02	1.220E+01	30	B433	0 0 0 0 0	

1345. C$_8$H$_5$F$_3$O$_2$
α,α,α-Trifluoro-o-toluic acid
Trifluoro-o-toluic acid
Acide orthotrifluortoluique

RN:	433-97-6	MP (°C):	111
MW:	190.12	BP (°C):	247

Solubility (Moles/L)	Solubility (Grams/L)	Temp (°C)	Ref (#)	Evaluation (T P E A A)	Comments
2.525E-02	4.800E+00	25	D064	1 2 1 1 2	

1346. C$_8$H$_5$NO$_2$
Phthalimide
Phthalimid

RN: 85-41-6   MP (°C): 238.0
MW: 147.13   BP (°C):

Solubility (Moles/L)	Solubility (Grams/L)	Temp (°C)	Ref (#)	Evaluation (T P E A A)	Comments
2.447E-03	3.600E-01	25	F300	1 0 0 0 1	
2.719E-02	4.000E+00	100	F300	1 0 0 0 0	
4.075E-03	5.996E-01	rt	D021	0 0 1 1 0	

1347. C$_8$H$_5$NO$_2$S
3-Carboxyphenylisothiocyanate
m-Isothiocyanobenzoic acid

RN: 2131-63-7   MP (°C):
MW: 179.20   BP (°C):

Solubility (Moles/L)	Solubility (Grams/L)	Temp (°C)	Ref (#)	Evaluation (T P E A A)	Comments
5.600E-04	1.004E-01	25	D019	1 1 1 1 2	
8.000E-04	1.434E-01	25	K032	2 2 0 1 1	

1348. C$_8$H$_5$NO$_2$S
4-Carboxyphenylisothiocyanate
p-Carboxyphenylisothiocyanate

RN: 2131-62-6   MP (°C):
MW: 179.20   BP (°C):

Solubility (Moles/L)	Solubility (Grams/L)	Temp (°C)	Ref (#)	Evaluation (T P E A A)	Comments
1.060E-04	1.900E-02	25	D019	1 1 1 1 2	

1349. C$_8$H$_5$NO$_4$
6-Nitrophthalide
6-Nitro-phthalid

RN: 610-93-5   MP (°C): 145
MW: 179.13   BP (°C):

Solubility (Moles/L)	Solubility (Grams/L)	Temp (°C)	Ref (#)	Evaluation (T P E A A)	Comments
2.233E-03	4.000E-01	25	F300	1 0 0 0 2	

1350. C$_8$H$_5$NO$_6$
3-Nitrophthalic acid
3-Nitro-phthalsaeure

RN: 603-11-2   **MP** (°C): 218
MW: 211.13   **BP** (°C):

Solubility (Moles/L)	Solubility (Grams/L)	Temp (°C)	Ref (#)	Evaluation (T P E A A)	Comments
9.520E-02	2.010E+01	25	F300	1 0 0 0 2	

1351. C$_8$H$_5$NO$_6$
2,3,4-Pyridinetricarboxylic acid
Pyridin-tricarbonsaeure-(2,3,4)

RN: 632-95-1   **MP** (°C): 250
MW: 211.13   **BP** (°C):

Solubility (Moles/L)	Solubility (Grams/L)	Temp (°C)	Ref (#)	Evaluation (T P E A A)	Comments
5.684E-02	1.200E+01	15	F300	1 0 0 0 1	

1352. C$_8$H$_6$
Ethynylbenzene
Phenylacetylene

RN: 536-74-3   **MP** (°C): −44.8
MW: 102.14   **BP** (°C): 142.4

Solubility (Moles/L)	Solubility (Grams/L)	Temp (°C)	Ref (#)	Evaluation (T P E A A)	Comments
4.467E-03	4.562E-01	ns	D001	0 0 0 0 2	

1353. C$_8$H$_6$BrNS
3-Bromobenzyl isothiocyanate
m-Bromobenzyl isothiocyanate

RN: 3845-33-8   **MP** (°C):
MW: 228.12   **BP** (°C):

Solubility (Moles/L)	Solubility (Grams/L)	Temp (°C)	Ref (#)	Evaluation (T P E A A)	Comments
1.070E-04	2.441E-02	25	D014	1 0 0 0 1	

1354. C$_8$H$_6$BrNS
4-Bromobenzyl isothiocyanate
p-Bromobenzyl isothiocyanate

RN: 2076-56-4   **MP** (°C):
MW: 228.12   **BP** (°C):

Solubility (Moles/L)	Solubility (Grams/L)	Temp (°C)	Ref (#)	Evaluation (T P E A A)	Comments
6.500E-05	1.483E-02	25	D014	1 0 0 0 1	
1.500E-04	3.422E-02	25	D019	1 1 1 1 2	

1355. C_8H_6ClNS

3-Chlorobenzyl isothiocyanate
m-Chlorobenzyl isothiocyanate

RN: 3694-58-4   **MP** (°C):
MW: 183.66   **BP** (°C):

Solubility (Moles/L)	Solubility (Grams/L)	Temp (°C)	Ref (#)	Evaluation (T P E A A)	Comments
1.370E-04	2.516E-02	25	D014	1 0 0 0 1	

1356. C_8H_6ClNS

4-Chlorobenzyl isothiocyanate
p-Chlorobenzyl isothiocyanate

RN: 3694-45-9   **MP** (°C):
MW: 183.66   **BP** (°C):

Solubility (Moles/L)	Solubility (Grams/L)	Temp (°C)	Ref (#)	Evaluation (T P E A A)	Comments
1.480E-04	2.718E-02	25	D014	1 0 0 0 1	

1357. $C_8H_6Cl_2O_3$

2,4-Dichlorophenoxyacetic acid
2,4-D
(2,4-Dichlorophenoxy)acetic acid

RN: 94-75-7   **MP** (°C): 138
MW: 221.04   **BP** (°C):

Solubility (Moles/L)	Solubility (Grams/L)	Temp (°C)	Ref (#)	Evaluation (T P E A A)	Comments
2.805E-03	6.200E-01	20	F311	1 2 2 2 1	
2.443E-03	5.400E-01	20	M061	1 0 0 0 2	
2.939E-03	6.496E-01	21.50	B200	1 0 0 0 0	
4.072E-03	9.000E-01	22.5	G301	0 0 0 0 0	
3.060E-03	6.764E-01	24.99	N417	0 0 0 0 0	
3.085E-03	6.820E-01	25	B164	1 0 1 1 2	
3.280E-03	7.250E-01	25	B185	0 0 0 0 0	
4.026E-03	8.900E-01	25	F071	1 1 2 1 2	
2.360E-03	5.217E-01	25	L030	1 0 2 1 2	
2.805E-03	6.200E-01	25	M161	1 0 0 0 2	
2.713E-03	5.996E-01	ns	B100	0 0 0 0 0	
4.072E-03	9.000E-01	ns	B185	0 0 0 0 0	
1.810E-03	4.000E-01	ns	B185	0 0 0 0 0	
2.500E-03	5.526E-01	ns	F184	0 0 0 0 1	
4.072E-03	9.000E-01	ns	K138	0 0 0 0 1	
2.805E-03	6.200E-01	ns	L024	0 0 0 0 2	
4.298E-03	9.500E-01	ns	M110	0 0 0 0 0	EFG
1.259E-03	2.783E-01	ns	M163	0 0 0 0 0	EFG
4.026E-03	8.900E-01	ns	M344	0 0 0 0 2	
2.488E-03	5.500E-01	ns	N013	0 0 0 0 2	

1358. C₈H₆Cl₂O₃
Dicamba
2-Methoxy-3,6-dichlorobenzoic acid
RN: 1918-00-9   **MP** (°C):   98
MW: 221.04   **BP** (°C):

Solubility (Moles/L)	Solubility (Grams/L)	Temp (°C)	Ref (#)	Evaluation (T P E A A)	Comments
2.036E-02	4.500E+00	25	B200	1 0 0 0 1	
2.036E-02	4.500E+00	25	M161	1 0 0 0 1	
3.591E-02	7.937E+00	ns	B100	0 0 0 0 0	

1359. C₈H₆Cl₂O₃
3,5-Dichlorophenoxyacetic acid
3,5-D
RN: 587-64-4   **MP** (°C):
MW: 221.04   **BP** (°C):

Solubility (Moles/L)	Solubility (Grams/L)	Temp (°C)	Ref (#)	Evaluation (T P E A A)	Comments
4.350E-03	9.615E-01	25	L030	1 0 2 1 2	

1360. C₈H₆Cl₂O₃
3,4-Dichlorophenoxyacetic acid
3,4-D
RN: 588-22-7   **MP** (°C):   138
MW: 221.04   **BP** (°C):

Solubility (Moles/L)	Solubility (Grams/L)	Temp (°C)	Ref (#)	Evaluation (T P E A A)	Comments
2.070E-03	4.576E-01	25	L030	1 0 2 1 2	
2.090E-03	4.620E-01	ns	B185	0 0 0 0 0	

1361. C₈H₆Cl₂O₃
2,6-Dichlorophenoxyacetic acid
2,6-D
RN: 575-90-6   **MP** (°C):
MW: 221.04   **BP** (°C):

Solubility (Moles/L)	Solubility (Grams/L)	Temp (°C)	Ref (#)	Evaluation (T P E A A)	Comments
7.050E-03	1.558E+00	25	L030	1 0 2 1 2	

1362. $C_8H_6Cl_2O_3$
2,3-Dichlorophenoxyacetic acid
2,3-D
RN: 2976-74-1   **MP** (°C):   173
MW: 221.04   **BP** (°C):

Solubility (Moles/L)	Solubility (Grams/L)	Temp (°C)	Ref (#)	Evaluation (T P E A A)	Comments
1.550E-03	3.426E-01	25	L030	1 0 2 1 2	

1363. $C_8H_6Cl_2O_3$
2,5-Dichlorophenoxyacetic acid
2,5-D
RN: 582-54-7   **MP** (°C):
MW: 221.04   **BP** (°C):

Solubility (Moles/L)	Solubility (Grams/L)	Temp (°C)	Ref (#)	Evaluation (T P E A A)	Comments
2.420E-03	5.349E-01	25	L030	1 0 2 1 2	

1364. $C_8H_6Cl_4O_2$
Tetrachloroveratrole
3,4,5,6-Tetrachloro-1,2-dimethoxybenzene
RN: 944-61-6   **MP** (°C):
MW: 275.95   **BP** (°C):

Solubility (Moles/L)	Solubility (Grams/L)	Temp (°C)	Ref (#)	Evaluation (T P E A A)	Comments
5.762E-06	1.590E-03	25	L348	1 2 2 1 2	

1365. $C_8H_6Cl_5NO_2$
Penclomedine
Pyridine
3,5-Dichloro-2,4-dimethoxy-6-(trichloromethyl)
NSC 338720
RN: 108030-77-9   **MP** (°C):
MW: 325.41   **BP** (°C):

Solubility (Moles/L)	Solubility (Grams/L)	Temp (°C)	Ref (#)	Evaluation (T P E A A)	Comments
1.229E-06	4.000E-04	25	P325	0 0 0 0 0	
1.229E-06	4.000E-04	25	P336	0 0 0 0 0	

1366. C$_8$H$_6$F$_3$N$_3$O$_4$S$_2$
Flumethiazide
6-(Trifluoromethyl)-2H-1,2,4-benzothiadiazine-7-sulfonamide 1,1-dioxide
6-Trifluoromethyl-7-sulfamoyl-4H-1,2,4-benzothiadiazine 1,1-dioxide
Trifluoromethylthiazide

RN: 148-56-1	**MP** (°C):
MW: 329.28	**BP** (°C):

Solubility (Moles/L)	Solubility (Grams/L)	Temp (°C)	Ref (#)	Evaluation (T P E A A)	Comments
3.189E-03	1.050E+00	rt	A095	0 0 2 2 2	

1367. C$_8$H$_6$INS
3-Iodobenzyl isothiocyanate
m-Iodobenzyl isothiocyanate

RN: 3696-68-2	**MP** (°C):
MW: 275.11	**BP** (°C):

Solubility (Moles/L)	Solubility (Grams/L)	Temp (°C)	Ref (#)	Evaluation (T P E A A)	Comments
5.500E-05	1.513E-02	25	D014	1 0 0 0 1	

1368. C$_8$H$_6$INS
4-Iodobenzyl isothiocyanate
p-Iodobenzyl isothiocyanate

RN: 3694-49-3	**MP** (°C):
MW: 275.11	**BP** (°C):

Solubility (Moles/L)	Solubility (Grams/L)	Temp (°C)	Ref (#)	Evaluation (T P E A A)	Comments
5.100E-05	1.403E-02	25	D014	1 0 0 0 1	

1369. C$_8$H$_6$N$_2$O$_2$S
3-Nitrobenzyl isothiocyanate
m-Nitrobenzyl isothiocyanate

RN: 3696-69-3	**MP** (°C):
MW: 194.21	**BP** (°C):

Solubility (Moles/L)	Solubility (Grams/L)	Temp (°C)	Ref (#)	Evaluation (T P E A A)	Comments
8.200E-05	1.593E-02	25	D014	1 0 0 0 1	

1370. C$_8$H$_6$N$_2$O$_2$S
4-Nitrobenzyl isothiocyanate
p-Nitrobenzyl isothiocyanate

RN: 3694-47-1	**MP** (°C):
MW: 194.21	**BP** (°C):

Solubility (Moles/L)	Solubility (Grams/L)	Temp (°C)	Ref (#)	Evaluation (T P E A A)	Comments
2.330E-04	4.525E-02	25	D014	1 0 0 0 1	

1371. C$_8$H$_6$N$_4$O$_5$
Nitrofurantoin
1-[(5-Nitrofurfurylidene)amino]hydantoin
Furatoin
Macrodantin
Macrobid
Welfurin

RN: 67-20-9   **MP** (°C):   268
MW: 238.16   **BP** (°C):

Solubility (Moles/L)	Solubility (Grams/L)	Temp (°C)	Ref (#)	Evaluation (T P E A A)	Comments
4.619E-04	1.100E-01	22	B154	1 1 1 1 1	pH 3.5
3.338E-04	7.950E-02	24	C034	2 0 2 2 2	
3.338E-04	7.950E-02	24	C118	1 0 0 0 2	
5.207E-04	1.240E-01	25	M457	0 0 0 0 0	
4.753E-04	1.132E-01	30	C011	2 0 2 1 0	EFG
4.761E-04	1.134E-01	30	C034	2 0 2 2 2	
4.761E-04	1.134E-01	30	C118	1 0 0 0 2	
8.264E-04	1.968E-01	37	A330	0 0 0 0 0	
1.142E-03	2.720E-01	37	B044	2 2 2 1 2	pH 7.2
7.310E-04	1.741E-01	37	C011	2 0 2 1 0	EFG
7.310E-04	1.741E-01	37	C034	2 0 2 2 2	
7.310E-04	1.741E-01	37	C118	1 0 0 0 2	
5.878E-04	1.400E-01	37	E044	1 0 1 1 2	
6.508E-04	1.550E-01	37	P034	1 0 0 0 2	pH 5
1.055E-03	2.512E-01	45	C034	2 0 2 2 2	
1.055E-03	2.512E-01	45	C118	1 0 0 0 2	
7.978E-04	1.900E-01	ns	K444	0 0 0 0 0	
5.249E-04	1.250E-01	ns	P033	0 0 0 0 2	
5.248E-04	1.250E-01	ns	R427	0 0 0 0 0	

1372. C$_8$H$_6$N$_4$O$_8$
Alloxantin
Uroxine
Alloxantin hydrate

RN: 76-24-4   **MP** (°C):   254dec
MW: 286.16   **BP** (°C):

Solubility (Moles/L)	Solubility (Grams/L)	Temp (°C)	Ref (#)	Evaluation (T P E A A)	Comments
1.753E-03	5.017E-01	25	B119	1 0 2 2 0	EFG
1.013E-02	2.900E+00	25	F300	1 0 0 0 1	
2.097E-01	6.000E+01	100	F300	1 0 0 0 0	

1373. $C_8H_6N_4S_2$

Methylthiobenzothiazole
Benzothiazole

RN: 76006-86-5 **MP** (°C):
MW: 222.29 **BP** (°C):

Solubility (Moles/L)	Solubility (Grams/L)	Temp (°C)	Ref (#)	Evaluation (T P E A A)	Comments
4.948E-04	1.100E-01	22	P323	0 0 0 0 0	

1374. $C_8H_6O_2$

Phthalic dicarboxaldehyde
o-Phthalaldehyd

RN: 643-79-8 **MP** (°C): 56.5
MW: 134.14 **BP** (°C):

Solubility (Moles/L)	Solubility (Grams/L)	Temp (°C)	Ref (#)	Evaluation (T P E A A)	Comments
1.044E-01	1.400E+01	h	F300	0 0 0 0 1	

1375. $C_8H_6O_2$

Terephthaldicarboxaldehyde
Terephthalaldehyd

RN: 623-27-8 **MP** (°C): 115
MW: 134.14 **BP** (°C): 246.5

Solubility (Moles/L)	Solubility (Grams/L)	Temp (°C)	Ref (#)	Evaluation (T P E A A)	Comments
1.491E-03	2.000E-01	20	F300	1 0 0 0 0	
1.297E-01	1.740E+01	100	F300	1 0 0 0 1	

1376. $C_8H_6O_3$

Piperonal
Heliotropine
3,4-Dihydroxybenzaldehyde methylene ketal
Methylenedioxy procatechuic aldehyde
Protocatechuic aldehyde methylene ether
Piperonyl aldehyde

RN: 120-57-0 **MP** (°C): 37
MW: 150.14 **BP** (°C): 263

Solubility (Moles/L)	Solubility (Grams/L)	Temp (°C)	Ref (#)	Evaluation (T P E A A)	Comments
2.331E-02	3.500E+00	20	F300	1 0 0 0 1	
4.463E-02	6.700E+00	78	F300	1 0 0 0 1	

1377. C₈H₆O₃

Benzoylformic acid
Phenyglyoxilic acid
RN: 611-73-4   **MP** (°C): 67
MW: 150.14   **BP** (°C):

Solubility (Moles/L)	Solubility (Grams/L)	Temp (°C)	Ref (#)	Evaluation (T P E A A)	Comments
6.128E+00	9.200E+02	0	C020	1 2 1 1 1	

1378. C₈H₆O₄

1,4-Benzenedicarboxylic acid
Terephthalic acid
p-Phthalic acid
RN: 100-21-0   **MP** (°C):
MW: 166.13   **BP** (°C):

Solubility (Moles/L)	Solubility (Grams/L)	Temp (°C)	Ref (#)	Evaluation (T P E A A)	Comments
9.029E-05	1.500E-02	20	F300	1 0 0 0 1	
1.920E-03	3.190E-01	25	C316	0 0 0 0 0	0.1M HCL
6.019E-04	9.999E-02	80	A027	1 0 0 0 0	

1379. C₈H₆O₄

1,2-Benzenedicarboxylic acid
o-Phthalic acid
Phthalic acid
Phthalsaeure
Benzene-1,2-dicarboxylic acid
RN: 88-99-3   **MP** (°C): 230
MW: 166.13   **BP** (°C):

Solubility (Moles/L)	Solubility (Grams/L)	Temp (°C)	Ref (#)	Evaluation (T P E A A)	Comments
1.381E-02	2.295E+00	0	M043	1 0 0 0 1	
2.219E-02	3.686E+00	2	A027	1 0 0 0 1	
2.159E-02	3.587E+00	10	M043	1 0 0 0 1	
7.935E-03	1.318E+00	10	S198	2 1 2 2 2	
1.571E-02	2.611E+00	10.49	A341	0 0 0 0 0	
3.471E-02	5.767E+00	20	A027	1 0 0 0 1	
3.435E-02	5.707E+00	20	F069	2 2 2 2 2	
3.431E-02	5.700E+00	20	F300	1 0 0 0 1	
3.352E-02	5.569E+00	20	M043	1 0 0 0 1	
7.214E-03	1.199E+00	20	S198	2 1 2 2 2	
3.915E-02	6.504E+00	22.99	A341	0 0 0 0 0	
4.200E-02	6.978E+00	24.99	A341	0 0 0 0 0	
8.600E-02	1.429E+01	25	H084	1 0 0 0 1	
8.520E-02	1.415E+01	25	K040	1 0 2 1 2	
4.192E-02	6.965E+00	25	M030	2 1 0 1 2	
4.279E-02	7.109E+00	25.8	W029	1 2 1 1 2	
4.808E-02	7.988E+00	28	D050	1 2 1 2 2	
5.152E-02	8.560E+00	29.49	A341	0 0 0 0 0	

(continued)

1379. C$_8$H$_6$O$_4$ (continued)

Solubility (Moles/L)	Solubility (Grams/L)	Temp (°C)	Ref (#)	Evaluation (T P E A A)	Comments
4.900E-02	8.141E+00	30	H019	0 0 0 0 0	
4.777E-02	7.937E+00	30	M043	1 0 0 0 0	
8.235E-03	1.368E+00	30	S198	2 1 2 2 2	
5.865E-02	9.743E+00	33.99	A341	0 0 0 0 0	
6.033E-02	1.002E+01	35	M030	2 1 0 1 2	
6.561E-02	1.090E+01	35.99	A341	0 0 0 0 0	
6.925E-02	1.150E+01	37.99	A341	0 0 0 0 0	
7.137E-02	1.186E+01	40	M043	1 0 0 0 1	
8.274E-02	1.375E+01	41.99	A341	0 0 0 0 0	
7.865E-02	1.307E+01	43.7	W029	1 2 1 1 2	
8.981E-02	1.492E+01	43.99	A341	0 0 0 0 0	
8.991E-02	1.494E+01	44.99	A341	0 0 0 0 0	
8.580E-02	1.425E+01	45	M030	2 1 0 1 2	
9.890E-02	1.643E+01	45.99	A341	0 0 0 0 0	
9.753E-02	1.620E+01	48.9	W029	1 2 1 1 2	
1.212E-01	2.014E+01	49.99	A341	0 0 0 0 0	
1.116E-01	1.854E+01	49.99	A341	0 0 0 0 0	
1.349E-01	2.241E+01	53.99	A341	0 0 0 0 0	
1.277E-01	2.122E+01	55	M030	2 1 0 1 2	
1.339E-01	2.225E+01	58.0	W029	1 2 1 1 2	
1.639E-01	2.724E+01	60	M043	1 0 0 0 1	
1.741E-01	2.892E+01	60.99	A341	0 0 0 0 0	
1.695E-01	2.815E+01	63.7	W029	1 2 1 1 2	
2.145E-01	3.564E+01	64.99	A341	0 0 0 0 0	
1.892E-01	3.144E+01	65	M030	2 1 0 1 2	
2.826E-01	4.695E+01	75	M030	2 1 0 1 2	
3.042E-01	5.053E+01	77.8	W029	1 2 1 1 2	
3.567E-01	5.927E+01	80	M043	1 0 0 0 1	
4.334E-01	7.200E+01	85	F300	1 0 0 0 0	
4.297E-01	7.138E+01	85	M030	2 1 0 1 2	
4.248E-01	7.058E+01	85.7	W029	1 2 1 1 2	
6.377E-01	1.059E+02	94.8	W029	1 2 1 1 2	
9.182E-01	1.525E+02	100	M043	1 0 0 0 2	
8.208E-01	1.364E+02	101.1	W029	1 2 1 1 2	
1.370E+00	2.276E+02	113.8	W029	1 2 1 1 2	
9.015E-03	1.498E+00	ns	F014	0 0 0 0 2	
2.458E-02	4.083E+00	rt	H431	0 0 0 0 0	

1380. C$_8$H$_6$O$_4$

Isophthalic acid
1,3-Benzenedicarboxylic acid
m-Phthalic acid

RN:	121-91-5	**MP** (°C):	345	
MW:	166.13	**BP** (°C):		

Solubility (Moles/L)	Solubility (Grams/L)	Temp (°C)	Ref (#)	Evaluation (T P E A A)	Comments
3.611E-04	6.000E-02	2	A027	1 0 0 0 0	
6.019E-04	9.999E-02	20	A027	1 0 0 0 0	

(continued)

1380. C$_8$H$_6$O$_4$ (continued)

Solubility (Moles/L)	Solubility (Grams/L)	Temp (°C)	Ref (#)	Evaluation (T P E A A)	Comments
1.090E-03	1.811E-01	28.29	L437	0 0 0 0 0	
1.656E-03	2.752E-01	40.99	L437	0 0 0 0 0	
2.535E-03	4.212E-01	51.99	L437	0 0 0 0 0	
4.021E-03	6.681E-01	64.99	L437	0 0 0 0 0	
6.260E-03	1.040E+00	76.49	L437	0 0 0 0 0	
6.013E-03	9.990E-01	80	A027	1 0 0 0 0	
8.300E-03	1.379E+00	83.49	L437	0 0 0 0 0	
9.441E-03	1.568E+00	86.47	L437	0 0 0 0 0	
1.286E-02	2.137E+00	93.42	L437	0 0 0 0 0	
4.610E-04	7.659E-02	rt	H431	0 0 0 0 0	

1381. C$_8$H$_6$O$_5$

2-Hydroxyisophthalic acid
2-Hydroxy-*iso*-phthalsaeure

RN: 606-19-9 **MP** (°C): 244
MW: 182.13 **BP** (°C):

Solubility (Moles/L)	Solubility (Grams/L)	Temp (°C)	Ref (#)	Evaluation (T P E A A)	Comments
1.449E-01	2.640E+01	100	F300	1 0 0 0 2	

1382. C$_8$H$_6$O$_5$

4-Hydroxyisophthalic acid
4-Hydroxy-*iso*-phthasaeure

RN: 636-46-4 **MP** (°C): 310
MW: 182.13 **BP** (°C):

Solubility (Moles/L)	Solubility (Grams/L)	Temp (°C)	Ref (#)	Evaluation (T P E A A)	Comments
1.647E-03	3.000E-01	24	F300	1 0 0 0 1	

1383. C$_8$H$_6$O$_5$

5-Hydroxyisophthalic acid
5-Hydroxy-*iso*-phthalsaeure

RN: 618-83-7 **MP** (°C): 293
MW: 182.13 **BP** (°C):

Solubility (Moles/L)	Solubility (Grams/L)	Temp (°C)	Ref (#)	Evaluation (T P E A A)	Comments
3.294E-03	6.000E-01	15	F300	1 0 0 0 1	
8.889E-01	1.619E+02	99	F300	1 0 0 0 2	

1384. C_8H_6S

Thianaphthene
Benzo[b]thiophene
Benzothiofuran
1-Benzothiophene

RN: 95-15-8   **MP** (°C): 29–32
MW: 134.20   **BP** (°C): 221–222

Solubility (Moles/L)	Solubility (Grams/L)	Temp (°C)	Ref (#)	Evaluation (T P E A A)	Comments
1.611E-03	2.162E-01	59.0	L339	2 0 2 2 2	
2.610E-03	3.503E-01	78.5	L339	2 0 2 2 2	
4.386E-03	5.886E-01	99.0	L339	2 0 2 2 2	

1385. $C_8H_7BrN_2O_3$

o-Nitro-o-bromacetanilide
2-Bromo-5-nitroacetanilide

RN: 245115-83-7   **MP** (°C):
MW: 259.07   **BP** (°C):

Solubility (Moles/L)	Solubility (Grams/L)	Temp (°C)	Ref (#)	Evaluation (T P E A A)	Comments
7.720E-02	2.000E+01	rt	F043	0 0 2 1 1	

1386. $C_8H_7BrN_2O_3$

p-Nitro-o-bromacetanilide
2-Bromo-4-nitroacetanilide

RN: 57045-86-0   **MP** (°C):
MW: 259.07   **BP** (°C):

Solubility (Moles/L)	Solubility (Grams/L)	Temp (°C)	Ref (#)	Evaluation (T P E A A)	Comments
6.832E-02	1.770E+01	rt	F043	0 0 2 1 2	

1387. $C_8H_7ClN_2O_3$

p-Nitro-o-chloracetanilide
2-Chloro-4-nitroacetanilide

RN: 881-87-8   **MP** (°C):
MW: 214.61   **BP** (°C):

Solubility (Moles/L)	Solubility (Grams/L)	Temp (°C)	Ref (#)	Evaluation (T P E A A)	Comments
5.172E-02	1.110E+01	rt	F043	0 0 2 1 2	

1388. C₈H₇ClN₂O₃

o-Nitro-*o*-chloracetanilide
2-Chloro-5-nitroacetanilide

RN: 72487-80-0 **MP** (°C):
MW: 214.61 **BP** (°C):

Solubility (Moles/L)	Solubility (Grams/L)	Temp (°C)	Ref (#)	Evaluation (T P E A A)	Comments
5.172E-02	1.110E+01	rt	F043	0 0 2 1 2	

1389. C₈H₇ClO₃

4-Chlorophenoxyacetic acid
4-CPA
p-Chlorophenoxyacetic acid

RN: 122-88-3 **MP** (°C): 157
MW: 186.60 **BP** (°C):

Solubility (Moles/L)	Solubility (Grams/L)	Temp (°C)	Ref (#)	Evaluation (T P E A A)	Comments
4.545E-03	8.480E-01	25	B164	1 0 1 1 2	
2.042E-03	3.810E-01	25	B185	0 0 0 0 0	
5.130E-03	9.572E-01	25	L030	1 0 2 1 2	

1390. C₈H₇ClO₃

3-Chlorophenoxyacetic acid
m-Chlorophenoxyacetic acid

RN: 588-32-9 **MP** (°C):
MW: 186.60 **BP** (°C):

Solubility (Moles/L)	Solubility (Grams/L)	Temp (°C)	Ref (#)	Evaluation (T P E A A)	Comments
1.265E-02	2.360E+00	25	L030	1 0 2 1 2	

1391. C₈H₇ClO₃

2-Chlorophenoxyacetic acid
o-Chlorophenoxyacetic acid

RN: 614-61-9 **MP** (°C): 146
MW: 186.60 **BP** (°C):

Solubility (Moles/L)	Solubility (Grams/L)	Temp (°C)	Ref (#)	Evaluation (T P E A A)	Comments
6.850E-03	1.278E+00	25	L030	1 0 2 1 2	

1392. C$_8$H$_7$Cl$_2$NO$_2$
Chloramben methyl ester
Vegiben 2E
Methyl 3-amino-2,5-dichlorobenzoate
Amchem 65-81-B
Methyl chloramben
Chloramben methyl

RN: 7286-84-2   **MP (°C):**   63.5
MW: 220.06   **BP (°C):**

Solubility (Moles/L)	Solubility (Grams/L)	Temp (°C)	Ref (#)	Evaluation (T P E A A)	Comments
5.453E-04	1.200E-01	20	M161	1 0 0 0 2	

1393. C$_8$H$_7$Cl$_3$O
2,4,6-Trichloro-3,5-dimethyl-phenol
3,5-Xylenol, 2,4,6-trichloro-

RN: 6972-47-0   **MP (°C):**
MW: 225.50   **BP (°C):**

Solubility (Moles/L)	Solubility (Grams/L)	Temp (°C)	Ref (#)	Evaluation (T P E A A)	Comments
2.200E-05	4.961E-03	25	B316	0 0 0 0 0	

1394. C$_8$H$_7$Cl$_3$O$_2$
3,4,5-Trichloroveratrole
4,5,6-Trichloroveratrole

RN: 16766-29-3   **MP (°C):**   66
MW: 241.50   **BP (°C):**

Solubility (Moles/L)	Solubility (Grams/L)	Temp (°C)	Ref (#)	Evaluation (T P E A A)	Comments
4.265E-05	1.030E-02	25	L348	1 2 2 1 2	

1395. C$_8$H$_7$N
Indole
2,3-Benzopyrrole
Benzopyrrole
1-Benzazole
1-Benzol β pyrrol

RN: 120-72-9   **MP (°C):**   52
MW: 117.15   **BP (°C):**   253

Solubility (Moles/L)	Solubility (Grams/L)	Temp (°C)	Ref (#)	Evaluation (T P E A A)	Comments
9.219E-02	1.080E+01	25	K119	1 0 0 0 2	
3.037E-02	3.558E+00	25	P051	2 1 1 2 2	
3.037E-02	3.558E+00	25.00	P007	2 1 2 2 2	

1396. C₈H₇N
C_8H_7N

p-Toluonitrile
p-Cyanotoluene
p-Methylbenzonitrile
4-Methylbenzenecarbonitrile

RN: 104-85-8   **MP** (°C):
MW: 117.15   **BP** (°C):

Solubility (Moles/L)	Solubility (Grams/L)	Temp (°C)	Ref (#)	Evaluation (T P E A A)	Comments
1.300E-02	1.523E+00	25	M327	1 0 0 1 2	

1397. C₈H₇NOS

m-Methoxyphenyl isothiocyanate
3-Methoxyphenyl isothiocyanate

RN: 3125-64-2   **MP** (°C):
MW: 165.22   **BP** (°C):

Solubility (Moles/L)	Solubility (Grams/L)	Temp (°C)	Ref (#)	Evaluation (T P E A A)	Comments
2.700E-04	4.461E-02	25	K032	2 2 0 1 2	

1398. C₈H₇NOS

p-Methoxyphenyl isothiocyanate
4-Methoxyphenylisothiocyanate

RN: 2284-20-0   **MP** (°C):   18.0
MW: 165.22   **BP** (°C):   280.5

Solubility (Moles/L)	Solubility (Grams/L)	Temp (°C)	Ref (#)	Evaluation (T P E A A)	Comments
2.500E-04	4.130E-02	25	D019	1 1 1 1 2	

1399. C₈H₇NO₃

Oxanilic acid
N-Phenyloxalic acid monoamide
Oxanilsaure

RN: 500-72-1   **MP** (°C):   150
MW: 165.15   **BP** (°C):

Solubility (Moles/L)	Solubility (Grams/L)	Temp (°C)	Ref (#)	Evaluation (T P E A A)	Comments
4.990E-02	8.241E+00	25	D058	1 0 1 1 2	

1400. C$_8$H$_7$NO$_4$
6-Nitro-3-methylbenzoic acid
2-Nitro-5-methylbenzoic acid
5-Methyl-2-nitrobenzoic acid
3-Methyl-6-nitrobenzoic acid

RN: 3113-72-2   **MP** (°C):
MW: 181.15   **BP** (°C):

Solubility (Moles/L)	Solubility (Grams/L)	Temp (°C)	Ref (#)	Evaluation (T P E A A)	Comments
2.043E-02	3.700E+00	10	G063	1 0 0 0 1	
2.595E-02	4.700E+00	20	G063	1 0 0 0 1	
9.385E-02	1.700E+01	40	G063	1 0 0 0 1	
9.937E-02	1.800E+01	50	G063	1 0 0 0 1	
1.490E-01	2.700E+01	60	G063	1 0 0 0 1	
1.932E-01	3.500E+01	65	G063	1 0 0 0 1	
2.484E-01	4.500E+01	70	G063	1 0 0 0 1	
3.643E-01	6.600E+01	80	G063	1 0 0 0 1	
3.699E-01	6.700E+01	100	G063	1 0 0 0 1	

1401. C$_8$H$_7$NO$_4$
2-Nitro-3-methylbenzoic acid
2-Nitro-m-toluic acid
3-Methyl-2-nitrobenzoic acid

RN: 5437-38-7   **MP** (°C):
MW: 181.15   **BP** (°C):

Solubility (Moles/L)	Solubility (Grams/L)	Temp (°C)	Ref (#)	Evaluation (T P E A A)	Comments
2.208E-03	4.000E-01	20	G063	1 0 0 0 1	
8.832E-03	1.600E+00	40	G063	1 0 0 0 1	
3.202E-02	5.800E+00	80	G063	1 0 0 0 1	
3.312E-02	6.000E+00	100	G063	1 0 0 0 0	

1402. C$_8$H$_7$NS
Benzyl isothiocyanate
Benzylisothiocyanate
Isothiocyanatomethylbenzene

RN: 622-78-6   **MP** (°C): 112
MW: 149.22   **BP** (°C): 242

Solubility (Moles/L)	Solubility (Grams/L)	Temp (°C)	Ref (#)	Evaluation (T P E A A)	Comments
7.300E-04	1.089E-01	25	D014	1 0 0 0 2	

1403. C$_8$H$_7$NS
p-Tolyl isothiocyanate
4-Tolylisothiocyanate

RN: 622-59-3 **MP** (°C): 25
MW: 149.22 **BP** (°C): 237

Solubility (Moles/L)	Solubility (Grams/L)	Temp (°C)	Ref (#)	Evaluation (T P E A A)	Comments
1.900E-05	2.835E-03	25	D019	1 1 1 1 1	

1404. C$_8$H$_7$NS
m-Methylphenyl isothiocyanate
3-Methylphenyl isothiocyanate

RN: 614-69-7 **MP** (°C):
MW: 149.22 **BP** (°C):

Solubility (Moles/L)	Solubility (Grams/L)	Temp (°C)	Ref (#)	Evaluation (T P E A A)	Comments
1.420E-04	2.119E-02	25	K032	2 2 0 1 2	

1405. C$_8$H$_7$N$_5$O
7-Acetamidopteridine

RN: **MP** (°C):
MW: 189.18 **BP** (°C):

Solubility (Moles/L)	Solubility (Grams/L)	Temp (°C)	Ref (#)	Evaluation (T P E A A)	Comments
4.035E-02	7.634E+00	100	A083	1 2 0 0 0	

1406. C$_8$H$_7$N$_5$O
2-Acetamidopteridine

RN: **MP** (°C):
MW: 189.18 **BP** (°C):

Solubility (Moles/L)	Solubility (Grams/L)	Temp (°C)	Ref (#)	Evaluation (T P E A A)	Comments
1.705E-01	3.226E+01	100	A083	1 2 0 0 0	

1407. C$_8$H$_7$N$_5$O
4-Acetamidopteridine

RN: **MP** (°C):
MW: 189.18 **BP** (°C):

Solubility (Moles/L)	Solubility (Grams/L)	Temp (°C)	Ref (#)	Evaluation (T P E A A)	Comments
1.762E+00	3.333E+02	100	A083	1 2 0 0 0	

1408. C$_8$H$_7$N$_5$O$_8$

2,4,6-Trinitrophenylethylnitramine
Tetrethyl
Trinitrophenylethylnitramine
Ethyl tetryl

RN: 6052-13-7   **MP** (°C):
MW: 301.17   **BP** (°C):

Solubility (Moles/L)	Solubility (Grams/L)	Temp (°C)	Ref (#)	Evaluation (T P E A A)	Comments
1.992E-04	6.000E-02	22	D067	1 2 0 0 0	
8.633E-04	2.600E-01	50	D067	1 2 0 0 1	
8.998E-03	2.710E+00	100	D067	1 2 0 0 2	

1409. C$_8$H$_8$

Styrene
Phenylethylene
Styrolene
Styrol
Ethenylbenzene
Annamene

RN: 100-42-5   **MP** (°C):   −30
MW: 104.15   **BP** (°C):   145

Solubility (Moles/L)	Solubility (Grams/L)	Temp (°C)	Ref (#)	Evaluation (T P E A A)	Comments
2.784E-03	2.899E-01	7	L028	1 0 1 1 1	
2.400E-03	2.499E-01	15	L028	1 0 1 1 1	
1.152E-03	1.200E-01	20	L096	1 2 0 2 2	
3.167E-03	3.299E-01	24	L028	1 0 1 1 1	
2.880E-03	3.000E-01	25	A002	1 2 1 1 1	
1.540E-03	1.604E-01	25	B173	2 0 2 2 2	
2.975E-03	3.099E-01	25	L028	1 0 1 1 1	
3.455E-03	3.599E-01	32	L028	1 0 1 1 1	
3.839E-03	3.998E-01	40	L028	1 0 1 1 1	
3.839E-03	3.998E-01	44	L028	1 0 1 1 1	
4.319E-03	4.498E-01	49	L028	1 0 1 1 1	
4.319E-03	4.498E-01	51	L028	1 0 1 1 1	
4.798E-03	4.998E-01	56	L028	1 0 1 1 1	
8.658E-02	9.018E+00	65	A324	2 2 2 1 1	
5.566E-03	5.797E-01	65	L028	1 0 1 1 1	

1410. C$_8$H$_8$BrCl$_2$O$_3$PS

Bromophos
O-(4-Bromo-2,5-dichlorophenyl) O,O-dimethyl phosphorothioate
Nexion
Brofene
Brophene
Omexan

RN:	2104-96-3	**MP (°C):**	51		
MW:	366.00	**BP (°C):**			

Solubility (Moles/L)	Solubility (Grams/L)	Temp (°C)	Ref (#)	Evaluation (T P E A A)	Comments
6.557E-07	2.400E-04	10	B324	0 0 0 0 0	
6.558E-07	2.400E-04	10	B324	0 0 0 0 0	
8.197E-07	3.000E-04	20	B169	2 1 1 1 1	*sic*
9.290E-07	3.400E-04	20	B324	0 0 0 0 0	
9.290E-07	3.400E-04	20	B324	0 0 0 0 0	
2.732E-06	1.000E-03	20	F311	1 2 2 2 1	*sic*
1.093E-04	4.000E-02	20	M061	1 0 0 0 1	
1.093E-04	4.000E-02	20	W311	1 0 0 0 1	
2.634E-06	9.641E-04	30	B324	0 0 0 0 0	
2.623E-06	9.600E-04	30	B324	0 0 0 0 0	
1.093E-04	4.000E-02	ns	E050	0 0 0 0 1	
1.093E-04	4.000E-02	rt	M161	0 0 0 0 1	

1411. C$_8$H$_8$BrNO

4'-Bromoacetanilide
Acetamide, N-(4-bromophenyl)-
Acetanilide, 4'-bromo-
Bromoantifebrin

RN:	103-88-8	**MP (°C):**		
MW:	214.07	**BP (°C):**		

Solubility (Moles/L)	Solubility (Grams/L)	Temp (°C)	Ref (#)	Evaluation (T P E A A)	Comments
7.000E-04	1.498E-01	25	D044	0 0 0 0 0	

1412. C$_8$H$_8$ClNO

p-Chloroacetanilide
Acetamide, N-(4-chlorophenyl)-
Acetanilide, 4'-chloro-

RN:	539-03-7	**MP (°C):**		
MW:	169.61	**BP (°C):**		

Solubility (Moles/L)	Solubility (Grams/L)	Temp (°C)	Ref (#)	Evaluation (T P E A A)	Comments
1.000E-03	1.696E-01	25	D044	0 0 0 0 0	

1413. $C_8H_8Cl_2IO_3PS$

Iodofenphos
O-(2,5-Dichloro-4-iodophenyl) O,O-dimethyl phosphorothioate
Nuvanol-N
Dimethyl O-2,5-dichloro-4-iodophenyl thiophosphate
Alfacron
Jodfenphos

RN: 18181-70-9   **MP** (°C):   72
MW: 413.00   **BP** (°C):

Solubility (Moles/L)	Solubility (Grams/L)	Temp (°C)	Ref (#)	Evaluation (T P E A A)	Comments
2.421E-07	1.000E-04	20	B169	2 1 1 1 1	
4.843E-06	2.000E-03	20	M161	1 0 0 0 0	

1414. $C_8H_8Cl_2O$

2,4-Dichloro-6-ethyl-phenol
Phenol, 2,4-dichloro-6-ethyl-

RN: 24539-94-4   **MP** (°C):
MW: 191.06   **BP** (°C):

Solubility (Moles/L)	Solubility (Grams/L)	Temp (°C)	Ref (#)	Evaluation (T P E A A)	Comments
1.300E-03	2.484E-01	25	B316	0 0 0 0 0	

1415. $C_8H_8Cl_2O_2$

Chloroneb
Demosan
Terraneb
Terraneb SP
1,4-Dichloro-2,5-dimethoxybenzene
Terraneb B

RN: 2675-77-6   **MP** (°C):   134.5
MW: 207.06   **BP** (°C):   268

Solubility (Moles/L)	Solubility (Grams/L)	Temp (°C)	Ref (#)	Evaluation (T P E A A)	Comments
3.864E-05	8.000E-03	25	M161	1 0 0 0 0	

1416. $C_8H_8Cl_2O_2$

4,5-Dichloroveratrole
Benzene, 1,2-dichloro-4,5-dimethoxy-

RN: 2772-46-5   **MP** (°C):   83
MW: 207.06   **BP** (°C):

Solubility (Moles/L)	Solubility (Grams/L)	Temp (°C)	Ref (#)	Evaluation (T P E A A)	Comments
3.492E-04	7.230E-02	25	L348	1 2 2 1 2	average of 2

1417. $C_8H_8Cl_3O_3PS$
Ronnel
Fenchlorphos
Dermafos
Dimethyl trichlorophenylthiophosphate

RN: 299-84-3   **MP** (°C): 35
MW: 321.55   **BP** (°C):

Solubility (Moles/L)	Solubility (Grams/L)	Temp (°C)	Ref (#)	Evaluation (T P E A A)	Comments
1.866E-06	6.000E-04	20	B169	2 2 1 1 1	
3.359E-06	1.080E-03	20	C053	0 0 0 0 0	
3.110E-06	1.000E-03	20	E048	1 2 1 1 0	
7.775E-06	2.500E-03	20	F311	1 2 2 2 1	
5.287E-06	1.700E-03	ns	F040	1 2 2 2 1	
3.359E-06	1.080E-03	ns	F071	0 1 2 1 2	
1.866E-05	6.000E-03	ns	K138	0 0 0 0 1	
1.368E-04	4.400E-02	ns	M061	0 0 0 0 1	
1.244E-04	4.000E-02	rt	M161	0 0 0 0 1	

1418. C_8H_8FNO
4'-Fluoroacetanilide
Acetamide, N-(4-fluorophenyl)-
4-Fluoroacetanilide

RN: 351-83-7   **MP** (°C):
MW: 153.16   **BP** (°C):

Solubility (Moles/L)	Solubility (Grams/L)	Temp (°C)	Ref (#)	Evaluation (T P E A A)	Comments
1.630E-02	2.496E+00	25	D044	0 0 0 0 0	

1419. $C_8H_8F_3N_3O_4S_2$
Hydroflumethiazide
Diucardin
Saluron

RN: 135-09-1   **MP** (°C): 272
MW: 331.29   **BP** (°C):

Solubility (Moles/L)	Solubility (Grams/L)	Temp (°C)	Ref (#)	Evaluation (T P E A A)	Comments
1.449E-03	4.800E-01	37	C087	0 0 0 0 0	
2.048E-03	6.785E-01	37	C315	0 0 0 0 0	0.1N HCL, average of 4
5.643E-04	1.870E-01	ns	B404	0 2 1 1 0	
9.958E-04	3.299E-01	rt	K144	0 0 0 0 1	

1420. C$_8$H$_8$INO

p-Iodoaniline-N-acetate
4-Iodanilin-N-acetat
4-Iodoacetanilide
Acetanilide, 4'-iodo-
4-Acetamidophenyl iodide
p-Iodoacetanilide

RN: 622-50-4   **MP** (°C):
MW: 261.06   **BP** (°C):

Solubility (Moles/L)	Solubility (Grams/L)	Temp (°C)	Ref (#)	Evaluation (T P E A A)	Comments
7.000E-04	1.827E-01	25	D044	0 0 0 0 0	

1421. C$_8$H$_8$N$_2$O$_2$

Phthalamide
1,2-Benzenedicarboxamide

RN: 88-96-0   **MP** (°C):   228
MW: 164.17   **BP** (°C):

Solubility (Moles/L)	Solubility (Grams/L)	Temp (°C)	Ref (#)	Evaluation (T P E A A)	Comments
1.218E-03	2.000E-01	20	A027	1 0 0 0 0	*sic*
3.594E-02	5.900E+00	30	K004	1 0 0 0 1	

1422. C$_8$H$_8$N$_2$O$_2$

Ricinine
Ricinin

RN: 524-40-3   **MP** (°C):   201.5
MW: 164.17   **BP** (°C):

Solubility (Moles/L)	Solubility (Grams/L)	Temp (°C)	Ref (#)	Evaluation (T P E A A)	Comments
1.645E-02	2.700E+00	10	F300	1 0 0 0 1	

1423. C$_8$H$_8$N$_2$O$_3$

4-Nitroaniline-N-acetate
4-Nitro-anilin-N-acetat
p-Nitroacetanilide
1-Nitro-4-acetylaminobenzene

RN: 104-04-1   **MP** (°C):   216
MW: 180.16   **BP** (°C):

Solubility (Moles/L)	Solubility (Grams/L)	Temp (°C)	Ref (#)	Evaluation (T P E A A)	Comments
1.221E-02	2.200E+00	20	F300	1 0 0 0 1	
6.000E-04	1.081E-01	25	D044	0 0 0 0 0	
1.221E-02	2.200E+00	rt	F043	0 0 2 1 1	

1424. $C_8H_8N_2O_3$
2-Nitroaniline-*N*-acetate
2-Nitro-anilin-*N*-acetat
o-Nitroacetanilide
RN: 552-32-9   **MP** (°C):
MW: 180.16   **BP** (°C):

Solubility (Moles/L)	Solubility (Grams/L)	Temp (°C)	Ref (#)	Evaluation (T P E A A)	Comments
1.221E-02	2.200E+00	20	F300	1 0 0 0 1	
1.221E-02	2.200E+00	rt	F043	0 0 2 1 1	

1425. $C_8H_8N_2O_6S$
MB 8882
Methyl *N*-(4-nitrobenzenesulphonyl)carbamate
RN: 3337-70-0   **MP** (°C):   151
MW: 260.23   **BP** (°C):

Solubility (Moles/L)	Solubility (Grams/L)	Temp (°C)	Ref (#)	Evaluation (T P E A A)	Comments
3.839E-03	9.990E-01	ns	M061	0 0 0 0 0	

1426. $C_8H_8N_4$
6,7-Dimethylpteridine
6:7-Dimethylpteridine
RN: 704-61-0   **MP** (°C):
MW: 160.18   **BP** (°C):

Solubility (Moles/L)	Solubility (Grams/L)	Temp (°C)	Ref (#)	Evaluation (T P E A A)	Comments
3.468E-01	5.556E+01	20	A083	1 2 0 0 0	

1427. $C_8H_8N_4$
Hydralazine
Apresoline
RN: 86-54-4   **MP** (°C):   172
MW: 160.18   **BP** (°C):

Solubility (Moles/L)	Solubility (Grams/L)	Temp (°C)	Ref (#)	Evaluation (T P E A A)	Comments
2.996E-05	4.800E-03	22.5	B440	0 0 0 0 0	

1428. $C_8H_8N_4O$
4-Hydroxy-6,7-dimethylpteridine
4-Hydroxy-6:7-dimethylpteridine
RN: 14684-54-9   **MP** (°C):
MW: 176.18   **BP** (°C):

Solubility (Moles/L)	Solubility (Grams/L)	Temp (°C)	Ref (#)	Evaluation (T P E A A)	Comments
5.155E-03	9.083E-01	22.5	A085	1 2 0 0 0	

1429. $C_8H_8N_4O_2$
H-Pyrazolo[3,4-d]pyrimidin-4-one, 1,5-dihydro-1-(1-oxopropyl)-
RN: 96448-61-2   **MP** (°C):
MW: 192.18   **BP** (°C):

Solubility (Moles/L)	Solubility (Grams/L)	Temp (°C)	Ref (#)	Evaluation (T P E A A)	Comments
1.561E-03	3.000E-01	22	B428	1 2 1 2 1	

1430. $C_8H_8N_4O_2S_2$
2-Sulfanilamido-1,3,4-thiadiazole
Sulfathiadiazole
Sulfanilamide, *N*1-1,3,4-thiadiazol-2-yl-
RN: 16806-29-4   **MP** (°C):
MW: 256.31   **BP** (°C):

Solubility (Moles/L)	Solubility (Grams/L)	Temp (°C)	Ref (#)	Evaluation (T P E A A)	Comments
2.848E-03	7.300E-01	37	R045	1 2 1 1 1	

1431. $C_8H_8N_4O_3$
1-Acetoxymethyl allopurinol
4H-Pyrazolo[3,4-d]pyrimidin-4-one, 1-[(acetyloxy)methyl]-1,5-dihydro-
RN: 98846-64-1   **MP** (°C): 257-258
MW: 208.18   **BP** (°C):

Solubility (Moles/L)	Solubility (Grams/L)	Temp (°C)	Ref (#)	Evaluation (T P E A A)	Comments
2.786E-03	5.800E-01	22	B322	0 0 0 0 0	

1432. $C_8H_8N_4O_4$
Nifuradene
1-[5-Nitrofurfuryllidene)amino]-2-imidazolidinone
RN: 555-84-0   **MP** (°C): 261.5
MW: 224.18   **BP** (°C):

Solubility (Moles/L)	Solubility (Grams/L)	Temp (°C)	Ref (#)	Evaluation (T P E A A)	Comments
3.925E-04	8.800E-02	ns	I310	0 0 0 0 0	

1433. C₈H₈N₄O₄S₃

CL 11366

RN: **MP** (°C):

MW: 320.37   **BP** (°C):

Solubility (Moles/L)	Solubility (Grams/L)	Temp (°C)	Ref (#)	Evaluation (T P E A A)	Comments
1.405E-03	4.500E-01	ns	M032	0 0 0 0 1	

1434. C₈H₈N₄O₄S₃

Benzolamide

2-Benzenesulfonamide-1,3,4-thiadiazole-5-sulfonamide

5-Benzenesulfonamido-1,3,4-thiadiazole-2-sulfonamide

1,3,4-Thiadiazole-2-sulfonamide, 5-[(phenylsulfonyl)amino]-

1,3,4-Thiadiazole-2-sulfonamide, 5-benzenesulfonamido-

RN: 3368-13-6   **MP** (°C):

MW: 320.37   **BP** (°C):   585.9

Solubility (Moles/L)	Solubility (Grams/L)	Temp (°C)	Ref (#)	Evaluation (T P E A A)	Comments
1.200E-03	3.844E-01	25	C415	1 0 0 1 0	

1435. C₈H₈N₄O₆

2,4,6-Trinitroethylaniline

2-4-6-Trinitromonoethylaniline

RN: 7449-27-6   **MP** (°C):

MW: 256.18   **BP** (°C):

Solubility (Moles/L)	Solubility (Grams/L)	Temp (°C)	Ref (#)	Evaluation (T P E A A)	Comments
3.904E-04	1.000E-01	19	D067	1 2 0 0 2	
1.210E-03	3.100E-01	50	D067	1 2 0 0 2	
5.699E-03	1.460E+00	100	D067	1 2 0 0 2	

1436. C₈H₈O

Acetophenone

Acetophenon

Methyl phenyl ketone

RN: 98-86-2   **MP** (°C):   20.05

MW: 120.15   **BP** (°C):   202

Solubility (Moles/L)	Solubility (Grams/L)	Temp (°C)	Ref (#)	Evaluation (T P E A A)	Comments
4.503E-02	5.411E+00	24	H106	1 0 2 2 2	
4.611E-02	5.540E+00	24	M303	1 0 1 1 2	
5.243E-02	6.300E+00	25	A003	1 2 1 2 2	
4.470E-02	5.371E+00	25	B019	1 0 1 2 0	
4.470E-02	5.371E+00	25	B092	2 1 1 1 1	
9.600E-02	1.153E+01	25	D407	1 0 2 2 2	
5.600E-03	6.729E-01	25	F063	1 1 0 0 1	
6.605E-02	7.937E+00	60	B092	2 1 1 1 1	

1437. C₈H₈O

Styrene oxide
1,2-Epoxyethylbenzene

RN: 96-09-3   **MP** (°C): −36.8
MW: 120.15   **BP** (°C): 194.1

Solubility (Moles/L)	Solubility (Grams/L)	Temp (°C)	Ref (#)	Evaluation (T P E A A)	Comments
2.324E-02	2.792E+00	25	I313	0 0 0 0 0	

1438. C₈H₈O

2,2,3-Trimethyl-3-pentanol
2,2,3-Trimethylpentanol-3

RN: 7294-05-5   **MP** (°C): −6
MW: 120.15   **BP** (°C):

Solubility (Moles/L)	Solubility (Grams/L)	Temp (°C)	Ref (#)	Evaluation (T P E A A)	Comments
4.120E+00	4.950E+02	20	G007	1 2 0 1 2	
4.119E+00	4.949E+02	25	G007	1 2 0 1 2	
4.119E+00	4.949E+02	30	G007	1 2 0 1 2	

1439. C₈H₈O

4-Methylbenzaldehyde
p-Methylbenzaldehyde

RN: 104-87-0   **MP** (°C):
MW: 120.15   **BP** (°C): 204

Solubility (Moles/L)	Solubility (Grams/L)	Temp (°C)	Ref (#)	Evaluation (T P E A A)	Comments
1.890E-02	2.271E+00	25	M017	1 2 0 1 2	

1440. C₈H₈O₂

2′-Hydroxyacetophenone
1-(2-Hydroxyphenyl)ethanone
2-Acetylphenol

RN: 118-93-4   **MP** (°C): 6
MW: 136.15   **BP** (°C): 213 at 717 mm Hg

Solubility (Moles/L)	Solubility (Grams/L)	Temp (°C)	Ref (#)	Evaluation (T P E A A)	Comments
5.000E-02	6.808E+00	30	K441	0 0 0 0 0	
1.100E-01	1.498E+01	40	K441	0 0 0 0 0	
1.400E-01	1.906E+01	50	K441	0 0 0 0 0	

1441. $C_8H_8O_2$

4-Hydroxyacetophenone
4'-Hydroxy-acetophenon

RN: 99-93-4 **MP** (°C): 110
MW: 136.15 **BP** (°C):

Solubility (Moles/L)	Solubility (Grams/L)	Temp (°C)	Ref (#)	Evaluation (T P E A A)	Comments
7.271E-02	9.900E+00	22	F300	1 0 0 0 1	
7.000E-02	9.531E+00	30	K441	0 0 0 0 0	
1.400E-01	1.906E+01	40	K441	0 0 0 0 0	
1.800E-01	2.451E+01	50	K441	0 0 0 0 0	

1442. $C_8H_8O_2$

p-Anisaldehyde
Anisaldehyd
p-Methoxybenzaldehyde

RN: 123-11-5 **MP** (°C): 0
MW: 136.15 **BP** (°C): 249.5

Solubility (Moles/L)	Solubility (Grams/L)	Temp (°C)	Ref (#)	Evaluation (T P E A A)	Comments
1.469E-02	2.000E+00	20	F300	1 0 0 0 0	
3.900E-02	5.310E+00	25	D407	1 0 2 2 2	
3.150E-02	4.289E+00	25	I019	1 0 1 2 2	

1443. $C_8H_8O_2$

m-Toluic acid
3-Methylbenzoic acid
m-Methylbenzoic acid
β-Methylbenzoic acid

RN: 99-04-7 **MP** (°C): 112
MW: 136.15 **BP** (°C): 263

Solubility (Moles/L)	Solubility (Grams/L)	Temp (°C)	Ref (#)	Evaluation (T P E A A)	Comments
7.200E-03	9.803E-01	25	F001	1 0 1 0 2	
7.198E-03	9.800E-01	25	F300	1 0 0 0 2	
7.785E-03	1.060E+00	37	M360	1 2 1 1 2	

1444. C$_8$H$_8$O$_2$

p-Toluic acid
4-Methylbenzoic acid
Toluenecarboxylic acid

RN:	99-94-5	MP (°C):	180
MW:	136.15	BP (°C):	274

Solubility (Moles/L)	Solubility (Grams/L)	Temp (°C)	Ref (#)	Evaluation (T P E A A)	Comments
2.500E-03	3.404E-01	25	F001	1 0 1 0 2	
2.938E-03	4.000E-01	25	F300	1 0 0 0 2	
2.277E-03	3.100E-01	37	M360	1 2 1 1 2	
2.780E-03	3.785E-01	ns	C014	0 0 0 1 2	

1445. C$_8$H$_8$O$_2$

o-Toluic acid
o-Tolylsaeure
o-Toluylic acid
2-Methylbenzoic acid

RN:	118-90-1	MP (°C):	107
MW:	136.15	BP (°C):	258

Solubility (Moles/L)	Solubility (Grams/L)	Temp (°C)	Ref (#)	Evaluation (T P E A A)	Comments
8.700E-03	1.185E+00	25	F001	1 0 1 0 2	
8.780E-03	1.195E+00	25	R016	0 0 0 0 0	
1.014E-02	1.380E+00	37	M360	1 2 1 1 2	

1446. C$_8$H$_8$O$_2$

Phenylacetic acid
Phenylessigsaeure

RN:	103-82-2	MP (°C):	76.5
MW:	136.15	BP (°C):	266

Solubility (Moles/L)	Solubility (Grams/L)	Temp (°C)	Ref (#)	Evaluation (T P E A A)	Comments
1.175E-01	1.600E+01	20	F071	1 1 2 1 2	
1.219E-01	1.660E+01	20	H080	1 0 0 0 2	
1.219E-01	1.660E+01	20	M344	1 0 0 0 2	
1.300E-01	1.770E+01	25	F300	1 0 0 0 2	
1.267E-01	1.725E+01	25	H071	2 2 2 1 2	
1.310E-01	1.784E+01	25	K040	1 0 2 1 2	
1.300E-01	1.770E+01	25.00	M135	1 2 1 1 2	0.01N sodium phenylacetate
1.451E-01	1.975E+01	30	D033	2 2 1 2 2	
1.910E-01	2.600E+01	35.00	M135	1 2 1 1 2	
2.113E-01	2.877E+01	40	D033	2 2 1 2 2	
2.880E-01	3.921E+01	41.50	M135	1 2 1 1 2	
2.900E-01	3.948E+01	45.00	M135	1 2 1 1 2	
3.650E-01	4.970E+01	58.40	M135	1 2 1 1 2	

(continued)

1446. C$_8$H$_8$O$_2$ (continued)

Solubility (Moles/L)	Solubility (Grams/L)	Temp (°C)	Ref (#)	Evaluation (T P E A A)	Comments
4.350E-01	5.923E+01	68.80	M135	1 2 1 1 2	
5.130E-01	6.985E+01	76.50	M135	1 2 1 1 2	
6.110E-01	8.319E+01	83.00	M135	1 2 1 1 2	
6.860E-01	9.340E+01	86.70	M135	1 2 1 1 2	
7.712E-01	1.050E+02	100	F300	1 0 0 0 2	
1.259E-01	1.714E+01	ns	R424	0 0 0 0 0	

1447. C$_8$H$_8$O$_2$

Methyl benzoate

Methyl p-hydroxybenzoate

RN: 93-58-3   **MP** (°C): −12

MW: 136.15   **BP** (°C): 198

Solubility (Moles/L)	Solubility (Grams/L)	Temp (°C)	Ref (#)	Evaluation (T P E A A)	Comments
7.337E-03	9.990E-01	15	G040	1 0 2 0 0	
3.085E-02	4.200E+00	22	N317	1 1 2 1 2	
2.926E-02	3.984E+00	25	G040	1 0 2 0 0	
1.447E-02	1.970E+00	25	L086	1 0 1 1 2	
1.497E-02	2.038E+00	25	M334	1 0 1 1 2	
1.777E-02	2.420E+00	30	L012	2 0 2 2 2	
1.796E-02	2.445E+00	30	L086	1 0 1 1 2	
3.654E-02	4.975E+00	35	G040	1 0 2 0 0	
2.221E-02	3.024E+00	35	L086	1 0 1 1 2	
2.723E-02	3.708E+00	40	L086	1 0 1 1 2	

1448. C$_8$H$_8$O$_2$Hg

Phenylmercuric acetate

Ceresan

PMAC

Acetate, phenylmercuric

PMA

RN: 62-38-4   **MP** (°C): 149

MW: 336.74   **BP** (°C):

Solubility (Moles/L)	Solubility (Grams/L)	Temp (°C)	Ref (#)	Evaluation (T P E A A)	Comments
7.335E-02	2.470E+01	20	M061	1 0 0 0 2	
1.389E-02	4.678E+00	ns	B185	0 0 0 0 0	
1.396E-02	4.700E+00	ns	N013	0 0 0 0 2	
1.298E-02	4.370E+00	rt	M161	0 0 0 0 2	

1449. C₈H₈O₃

Methyl salicylate
Salicylsaeure-methyl ester
Methyl hydroxybenzoate
Betula oil
Panalgesic
Betula

RN:	119-36-8	**MP** (°C):	−8		
MW:	152.15	**BP** (°C):	222		

Solubility (Moles/L)	Solubility (Grams/L)	Temp (°C)	Ref (#)	Evaluation (T P E A A)	Comments
4.206E-03	6.400E-01	21	B331	0 0 0 0 0	
4.000E-03	6.086E-01	25	D407	1 0 2 2 2	
1.312E-02	1.996E+00	25	R041	0 0 0 0 0	
4.601E-03	7.000E-01	30	F300	1 0 0 0 0	
6.244E-03	9.500E-01	30	L012	2 0 2 2 1	

1450. C₈H₈O₃

Vanillin
4-Hydroxy-3-methoxybenzaldehyde
3-Methoxy-4-hydroxybenzaldehyde
Methylprotocatechuic aldehyde
Vanillic aldehyde
Vanillaldehyde

RN:	121-33-5	**MP** (°C):	82		
MW:	152.15	**BP** (°C):	285		

Solubility (Moles/L)	Solubility (Grams/L)	Temp (°C)	Ref (#)	Evaluation (T P E A A)	Comments
4.439E-02	6.754E+00	.2	D073	1 1 2 1 1	
1.972E-02	3.000E+00	4.40	M096	1 1 2 1 1	
3.418E-02	5.200E+00	15.60	M096	1 1 2 1 2	
8.114E-02	1.235E+01	20	D073	1 1 2 1 2	
6.572E-02	1.000E+01	20	F300	1 0 0 0 0	
5.915E-02	9.000E+00	23.90	M096	1 1 2 1 2	
4.800E-02	7.303E+00	25	D407	1 0 2 2 2	
7.240E-02	1.102E+01	25	I019	1 0 1 2 2	
9.713E-02	1.478E+01	30	D073	1 1 2 1 2	
8.500E-02	1.293E+01	30	L069	1 0 1 1 0	EFG
1.697E-01	2.582E+01	40	D073	1 1 2 1 2	
3.010E-01	4.580E+01	50	D073	1 1 2 1 2	
3.160E-01	4.807E+01	60	D073	1 1 2 1 2	
3.286E-01	5.000E+01	80	F300	1 0 0 0 0	

1451. C$_8$H$_8$O$_3$

Methylparaben
Me-paraben
Methyl p-hydroxybenzoic acid
Methyl 4-hydroxybenzoate
Methyl paraben

RN:	99-76-3	MP (°C):	131
MW:	152.15	BP (°C):	275

Solubility (Moles/L)	Solubility (Grams/L)	Temp (°C)	Ref (#)	Evaluation (T P E A A)	Comments
8.310E-03	1.264E+00	15	B355	0 0 0 0 0	
1.026E-02	1.561E+00	15	M352	1 1 1 1 2	
9.970E-03	1.517E+00	20	B355	0 0 0 0 0	
1.334E-02	2.030E+00	20	H056	1 0 2 1 2	
1.441E-02	2.193E+00	25	A059	1 0 1 1 2	
1.140E-02	1.735E+00	25	B355	0 0 0 0 0	
1.639E-02	2.494E+00	25	D081	1 2 2 1 2	
1.600E-02	2.434E+00	25	D339	0 0 0 0 0	
3.162E-02	4.811E+00	25	F322	2 0 1 1 0	EFG
1.364E-02	2.075E+00	25	L075	1 0 1 1 2	
1.393E-02	2.120E+00	25	L338	1 0 1 1 2	
1.460E-02	2.221E+00	25	M014	2 0 1 1 2	
1.585E-02	2.412E+00	25	M352	1 1 1 1 2	
1.643E-02	2.500E+00	25	O027	1 0 1 0 1	
1.485E-02	2.260E+00	25	P013	0 0 0 0 0	
1.446E-02	2.200E+00	25	P053	1 0 1 1 2	
1.600E-02	2.434E+00	27	B129	2 2 2 2 2	
1.500E-02	2.282E+00	27	G078	2 1 0 1 0	EFG
1.600E-02	2.434E+00	27	P019	1 2 1 1 0	EFG
1.450E-02	2.206E+00	27.0	G067	2 0 1 1 2	
1.828E-02	2.782E+00	30	A059	1 0 1 1 2	
1.564E-02	2.380E+00	30	M325	1 0 0 0 1	
2.275E-02	3.462E+00	35	A059	1 0 1 1 2	
2.550E-02	3.880E+00	37	B171	2 0 1 1 2	
2.268E-02	3.451E+00	39.3	G302	2 2 2 2 0	EFG
2.551E-02	3.882E+00	40	A059	1 0 1 1 2	
3.773E-02	5.740E+00	40	M352	1 1 1 1 2	
4.168E-02	6.341E+00	50	M352	1 1 1 1 2	

1452. C$_8$H$_8$O$_3$

D-Mandelic acid
(R)(−)Mandelic acid
(S)-α-Hydroxybenzeneacetic acid
L-Mandelic acid
(S)-(+)-Mandelic acid

RN:	17199-29-0	MP (°C):	132
MW:	152.15	BP (°C):	

Solubility (Moles/L)	Solubility (Grams/L)	Temp (°C)	Ref (#)	Evaluation (T P E A A)	Comments
5.310E-01	8.080E+01	0	A043	1 2 1 1 2	
5.310E-01	8.080E+01	0	L035	1 2 2 1 2	

(continued)

1452. C$_8$H$_8$O$_3$ (continued)

Solubility (Moles/L)	Solubility (Grams/L)	Temp (°C)	Ref (#)	Evaluation (T P E A A)	Comments
6.874E-01	1.046E+02	10	A043	1 2 1 1 2	
6.874E-01	1.046E+02	10	L035	1 2 2 1 2	
7.766E-01	1.182E+02	15	A043	1 2 1 1 2	
7.766E-01	1.182E+02	15	L035	1 2 2 1 2	
9.158E-01	1.393E+02	20	A043	1 2 1 1 2	
9.158E-01	1.393E+02	20	L035	1 2 2 1 2	
5.371E-01	8.173E+01	24.5	L035	1 2 2 1 1	
5.371E-01	8.173E+01	24.50	A043	1 2 1 1 1	
1.183E+00	1.800E+02	25	A043	1 2 1 1 2	
6.503E-01	9.894E+01	25	C045	2 2 0 1 2	
6.705E-01	1.020E+02	25	C045	2 2 0 1 2	
1.183E+00	1.800E+02	25	L035	1 2 2 1 2	
6.460E-01	9.829E+01	27.5	L035	1 2 2 1 2	
6.460E-01	9.829E+01	27.50	A043	1 2 1 1 2	
1.791E+00	2.725E+02	30	A043	1 2 1 1 2	
1.791E+00	2.725E+02	30	L035	1 2 2 1 2	
8.223E-01	1.251E+02	31.5	L035	1 2 2 1 2	
8.223E-01	1.251E+02	31.50	A043	1 2 1 1 2	
2.957E+00	4.499E+02	35	A043	1 2 1 1 2	
2.957E+00	4.499E+02	35	L035	1 2 2 1 2	
3.434E+00	5.224E+02	37	A043	1 2 1 1 2	
1.132E+00	1.722E+02	37	A043	1 2 1 1 2	
3.434E+00	5.224E+02	37	L035	1 2 2 1 2	
1.132E+00	1.722E+02	37	L035	1 2 2 1 2	
4.075E+00	6.201E+02	40	A043	1 2 1 1 2	
4.075E+00	6.201E+02	40	L035	1 2 2 1 2	
1.517E+00	2.308E+02	41.5	L035	1 2 2 1 2	
1.517E+00	2.308E+02	41.50	A043	1 2 1 1 2	
4.325E+00	6.580E+02	42.5	L035	1 2 2 1 2	
4.325E+00	6.580E+02	42.50	A043	1 2 1 1 2	
1.871E+00	2.847E+02	44	A043	1 2 1 1 2	
1.871E+00	2.847E+02	44	L035	1 2 2 1 2	
4.678E+00	7.118E+02	45	L035	1 2 2 1 2	
4.678E+00	7.118E+02	45.50	A043	1 2 1 1 2	
2.351E+00	3.577E+02	46.5	L035	1 2 2 1 2	
2.351E+00	3.577E+02	46.50	A043	1 2 1 1 2	
4.816E+00	7.328E+02	47	L035	1 2 2 1 2	
4.816E+00	7.328E+02	47.50	A043	1 2 1 1 2	
2.795E+00	4.253E+02	48.5	L035	1 2 2 1 2	
2.795E+00	4.253E+02	48.50	A043	1 2 1 1 2	
5.183E+00	7.886E+02	50	A043	1 2 1 1 2	
5.183E+00	7.886E+02	50	L035	1 2 2 1 2	
3.192E+00	4.856E+02	50.5	L035	1 2 2 1 2	
3.192E+00	4.856E+02	50.50	A043	1 2 1 1 2	
3.484E+00	5.301E+02	52.5	L035	1 2 2 1 2	
3.484E+00	5.301E+02	52.50	A043	1 2 1 1 2	
3.704E+00	5.635E+02	54.50	A043	1 2 1 1 2	
3.704E+00	5.635E+02	54.50	L035	1 2 2 1 2	
3.996E+00	6.080E+02	57	A043	1 2 1 1 2	

(continued)

1452. C$_8$H$_8$O$_3$ (continued)

Solubility (Moles/L)	Solubility (Grams/L)	Temp (°C)	Ref (#)	Evaluation (T P E A A)	Comments
3.996E+00	6.080E+02	57	L035	1 2 2 1 2	
4.337E+00	6.599E+02	60.5	L035	1 2 2 1 2	
4.337E+00	6.599E+02	60.50	A043	1 2 1 1 2	
4.884E+00	7.431E+02	68	A043	1 2 1 1 2	
4.884E+00	7.431E+02	68	L035	1 2 2 1 2	

1453. C$_8$H$_8$O$_3$

m-Cresotic acid
2-Hydroxy-p-tolylsaeure-(1)
m-Kresotinsaeure

RN:	50-85-1	MP (°C):	177	
MW:	152.15	BP (°C):		

Solubility (Moles/L)	Solubility (Grams/L)	Temp (°C)	Ref (#)	Evaluation (T P E A A)	Comments
6.638E-02	1.010E+01	100	F300	1 0 0 0 2	

1454. C$_8$H$_8$O$_3$

o-Anisic acid
2-Methoxybenzoic acid
Salicylic acid methyl ether
Salicylsaeure-methylaether
o-Methoxybenzoic acid

RN:	579-75-9	MP (°C):	101	
MW:	152.15	BP (°C):	200	

Solubility (Moles/L)	Solubility (Grams/L)	Temp (°C)	Ref (#)	Evaluation (T P E A A)	Comments
1.030E-02	1.567E+00	4.99	A405	2 0 1 1 2	
1.220E-02	1.856E+00	9.99	A405	2 0 1 1 2	
1.420E-02	2.161E+00	14.99	A405	2 0 1 1 2	
1.710E-02	2.602E+00	19.99	A405	2 0 1 1 2	
2.070E-02	3.150E+00	23.99	A405	2 0 1 1 2	
2.760E-02	4.200E+00	25	H007	0 0 0 0 0	
2.440E-02	3.712E+00	26.99	A405	2 0 1 1 2	
3.286E-02	5.000E+00	30	F300	1 0 0 0 0	
2.760E-02	4.199E+00	30.99	A405	2 0 1 1 2	
3.120E-02	4.747E+00	34.99	A405	2 0 1 1 2	
3.503E-02	5.330E+00	37	M360	1 2 1 1 2	
3.750E-02	5.706E+00	38.99	A405	2 0 1 1 2	
4.390E-02	6.679E+00	41.99	A405	2 0 1 1 2	
4.800E-02	7.303E+00	44.99	A405	2 0 1 1 2	
5.930E-02	9.023E+00	47.99	A405	2 0 1 1 2	
6.930E-02	1.054E+01	52.99	A405	2 0 1 1 2	
8.370E-02	1.274E+01	53.99	A405	2 0 1 1 2	
9.500E-02	1.445E+01	56.99	A405	2 0 1 1 2	
1.261E-01	1.919E+01	60.99	A405	2 0 1 1 2	
1.683E-01	2.561E+01	64.99	A405	2 0 1 1 2	

(continued)

1454. C$_8$H$_8$O$_3$ (continued)

Solubility (Moles/L)	Solubility (Grams/L)	Temp (°C)	Ref (#)	Evaluation (T P E A A)	Comments
2.326E-01	3.539E+01	68.99	A405	2 0 1 1 2	
2.630E-01	4.002E+01	69.99	A405	2 0 1 1 2	
3.467E-01	5.275E+01	72.99	A405	2 0 1 1 2	

1455. C$_8$H$_8$O$_3$
Mandelic acid
Amygdalic acid
α-Hydroxyphenylacetic acid
Uromaline
α-Hydroxy-benzeneacetic acid

RN: 90-64-2   **MP** (°C):   119.0
MW: 152.15   **BP** (°C):

Solubility (Moles/L)	Solubility (Grams/L)	Temp (°C)	Ref (#)	Evaluation (T P E A A)	Comments
1.191E+00	1.812E+02	25	K040	1 0 2 1 2	*sic*
8.795E-03	1.338E+00	25	R049	0 0 0 0 0	
9.120E-01	1.388E+02	ns	R427	0 0 0 0 0	

1456. C$_8$H$_8$O$_3$
3-Hydroxy-*p*-toluic acid
3-Hydroxy-*p*-tolylsaeure-(1)

RN: 586-30-1   **MP** (°C):
MW: 152.15   **BP** (°C):

Solubility (Moles/L)	Solubility (Grams/L)	Temp (°C)	Ref (#)	Evaluation (T P E A A)	Comments
2.859E-01	4.350E+01	100	F300	1 0 0 0 2	

1457. C$_8$H$_8$O$_3$
3-Methoxybenzoic acid
3-Methoxy-benzoesaeure
m-Anisic acid
m-Methoxybenzoic acid

RN: 586-38-9   **MP** (°C):   110
MW: 152.15   **BP** (°C):   170

Solubility (Moles/L)	Solubility (Grams/L)	Temp (°C)	Ref (#)	Evaluation (T P E A A)	Comments
1.282E-02	1.950E+00	37	M360	1 2 1 1 2	
1.183E-03	1.800E-01	ns	B361	0 0 0 0 0	

1458. C₈H₈O₃

1458. $C_8H_8O_3$

DL-Mandelic acid
DL-Mandelsaeure

RN:	611-72-3	**MP** (°C):	122	
MW:	152.15	**BP** (°C):		

Solubility (Moles/L)	Solubility (Grams/L)	Temp (°C)	Ref (#)	Evaluation (T P E A A)	Comments
9.050E-01	1.377E+02	20	F300	1 0 0 0 2	
1.134E+00	1.725E+02	24	F300	1 0 0 0 2	

1459. $C_8H_8O_3$

4-Hydroxy-*m*-toluic acid
4-Hydroxy-*m*-tolylsaeure-(1)
o-Cresotic acid
2-Hydroxy-*m*-toluic acid
2-Hydroxy-*m*-tolylsaeure-(1)

RN:	83-40-9	**MP** (°C):	165.5	
MW:	152.15	**BP** (°C):		

Solubility (Moles/L)	Solubility (Grams/L)	Temp (°C)	Ref (#)	Evaluation (T P E A A)	Comments
7.624E-02	1.160E+01	100	F300	1 0 0 0 2	
3.411E-01	5.190E+01	100	F300	1 0 0 0 2	

1460. $C_8H_8O_3$

Phenoxyacetic acid
Glycolic acid phenyl ether
O-Phenylglycolic acid

RN:	122-59-8	**MP** (°C):	98	
MW:	152.15	**BP** (°C):	285	

Solubility (Moles/L)	Solubility (Grams/L)	Temp (°C)	Ref (#)	Evaluation (T P E A A)	Comments
7.887E-02	1.200E+01	10	F071	1 1 2 1 2	
8.084E-03	1.230E+00	10	F300	1 0 0 0 2	
7.887E-02	1.200E+01	10	H080	1 0 0 0 2	
7.887E-02	1.200E+01	10	M344	1 0 0 0 2	
1.100E-04	1.674E-02	25	L030	1 0 2 1 2	

1461. C$_8$H$_8$O$_3$
p-Methoxybenzoic acid
4-Methoxybenzoic acid
p-Anisic acid
Anissaeure

RN: 100-09-4   **MP** (°C):   184
MW: 152.15   **BP** (°C):   275

Solubility (Moles/L)	Solubility (Grams/L)	Temp (°C)	Ref (#)	Evaluation (T P E A A)	Comments
7.300E-04	1.111E-01	2.99	A405	2 0 1 1 2	
9.400E-04	1.430E-01	4.99	A405	2 0 1 1 2	
1.070E-03	1.628E-01	10.99	A405	2 0 1 1 2	
1.270E-03	1.932E-01	14.99	A405	2 0 1 1 2	
1.775E-02	2.700E+00	19	F300	1 0 0 0 1	
1.330E-03	2.024E-01	19.99	A405	2 0 1 1 2	
1.680E-03	2.556E-01	24.99	A405	2 0 1 1 2	
2.020E-03	3.073E-01	28.99	A405	2 0 1 1 2	
2.300E-03	3.499E-01	33.99	A405	2 0 1 1 2	
3.483E-03	5.300E-01	37	B171	2 0 1 1 2	
1.380E-03	2.100E-01	37	M360	1 2 1 1 2	
3.110E-03	4.732E-01	39.99	A405	2 0 1 1 2	
3.870E-03	5.888E-01	43.99	A405	2 0 1 1 2	
5.130E-03	7.805E-01	50.99	A405	2 0 1 1 2	
6.110E-03	9.296E-01	55.99	A405	2 0 1 1 2	
8.170E-03	1.243E+00	59.99	A405	2 0 1 1 2	
9.000E-03	1.369E+00	64.99	A405	2 0 1 1 2	
1.080E-02	1.643E+00	65.99	A405	2 0 1 1 2	
1.100E-02	1.674E+00	66.99	A405	2 0 1 1 2	
1.460E-02	2.221E+00	71.99	A405	2 0 1 1 2	
1.778E-02	2.706E+00	ns	R427	0 0 0 0 0	

1462. C$_8$H$_8$O$_3$
p-Cresotic acid
6-Hydroxy-m-toluic acid
6-Hydroxy-m-tolylsaeure-(1)

RN: 89-56-5   **MP** (°C):   151
MW: 152.15   **BP** (°C):

Solubility (Moles/L)	Solubility (Grams/L)	Temp (°C)	Ref (#)	Evaluation (T P E A A)	Comments
1.439E-01	2.190E+01	100	F300	1 0 0 0 2	

1463. C$_8$H$_8$O$_4$
Vanillic acid
Vanillinsaeure

RN: 121-34-6   **MP** (°C):   214
MW: 168.15   **BP** (°C):

Solubility (Moles/L)	Solubility (Grams/L)	Temp (°C)	Ref (#)	Evaluation (T P E A A)	Comments
8.921E-03	1.500E+00	14	F300	1 0 0 0 1	
1.546E-01	2.600E+01	100	F300	1 0 0 0 2	

1464. C$_8$H$_8$O$_4$
Homogentisic acid
2,5-Dihydroxyphenylacetic acid
2,5-Dihydroxy-benzeneacetic acid

RN: 451-13-8   **MP** (°C):   151
MW: 168.15   **BP** (°C):

Solubility (Moles/L)	Solubility (Grams/L)	Temp (°C)	Ref (#)	Evaluation (T P E A A)	Comments
2.732E+00	4.595E+02	25	D041	1 0 0 0 1	

1465. C$_8$H$_8$O$_5$
Methyl gallate
Gallussaeuremethyl ester
Methyl-3,4,5-trihydroxybenzoate

RN: 99-24-1   **MP** (°C):   201.5
MW: 184.15   **BP** (°C):

Solubility (Moles/L)	Solubility (Grams/L)	Temp (°C)	Ref (#)	Evaluation (T P E A A)	Comments
5.323E-02	9.803E+00	19.99	L430	0 0 0 0 0	
5.696E-02	1.049E+01	24.99	L430	0 0 0 0 0	
6.757E-02	1.244E+01	29.99	L430	0 0 0 0 0	
9.549E-02	1.759E+01	34.99	L430	0 0 0 0 0	
1.340E-01	2.468E+01	39.99	L430	0 0 0 0 0	
1.704E-01	3.138E+01	44.99	L430	0 0 0 0 0	
2.542E-01	4.680E+01	49.99	L430	0 0 0 0 0	
4.328E-01	7.970E+01	54.99	L430	0 0 0 0 0	
5.879E-01	1.083E+02	59.99	L430	0 0 0 0 0	
7.775E-01	1.432E+02	64.99	L430	0 0 0 0 0	
1.054E+00	1.941E+02	69.99	L430	0 0 0 0 0	
1.624E-02	2.991E+00	–.0	L430	0 0 0 0 0	
5.756E-02	1.060E+01	ns	F300	0 0 0 0 2	

1466. C₈H₉ClNO₅PS

Chlorthion

O,O-Dimethyl O-4-nitro-3-chlorophenyl thiophosphate

RN:	500-28-7	MP (°C):	21
MW:	297.66	BP (°C):	

Solubility (Moles/L)	Solubility (Grams/L)	Temp (°C)	Ref (#)	Evaluation (T P E A A)	Comments
1.344E-04	4.000E-02	20	M061	1 0 0 0 1	

1467. C₈H₉ClNO₅PS

Dicapthon

O-(2-Chloro-4-nitrophenyl) O,O-dimethyl phosphorothioate

Dicaptan

Isochlorthion

RN:	2463-84-5	MP (°C):	
MW:	297.66	BP (°C):	

Solubility (Moles/L)	Solubility (Grams/L)	Temp (°C)	Ref (#)	Evaluation (T P E A A)	Comments
4.233E-05	1.260E-02	10	B324	0 0 0 0 0	
4.233E-05	1.260E-02	10	B324	0 0 0 0 0	
4.939E-05	1.470E-02	20	B300	2 1 1 1 2	
4.939E-05	1.470E-02	20	B324	0 0 0 0 0	
4.939E-05	1.470E-02	20	B324	0 0 0 0 0	
2.100E-05	6.250E-03	20	C053	0 0 0 0 0	
1.485E-04	4.420E-02	30	B324	0 0 0 0 0	
1.485E-04	4.420E-02	30	B324	0 0 0 0 0	
2.100E-05	6.250E-03	ns	F071	0 1 2 1 2	
1.176E-04	3.500E-02	ns	M061	0 0 0 0 1	
2.620E-05	7.800E-03	rt	F040	1 2 2 2 1	

1468. C₈H₉ClO

2,5-Dimethyl-4-chloro-phenol

4-Chloro-2,5-xylenol

4-Chloro-2,5-dimethylphenol

RN:	1124-06-7	MP (°C):	114–116
MW:	156.61	BP (°C):	

Solubility (Moles/L)	Solubility (Grams/L)	Temp (°C)	Ref (#)	Evaluation (T P E A A)	Comments
5.700E-02	8.927E+00	25	B316	0 0 0 0 0	

1469. C₈H₉ClO

2,6-Dimethyl-4-chloro-phenol

4-Chloro-2,6-xylenol

RN:	1123-63-3	MP (°C):	
MW:	156.61	BP (°C):	

Solubility (Moles/L)	Solubility (Grams/L)	Temp (°C)	Ref (#)	Evaluation (T P E A A)	Comments
3.300E-03	5.168E-01	25	B316	0 0 0 0 0	

1470. C$_8$H$_9$ClO
Chloroxylenol
3,5-Dimethyl-4-chloro-phenol-

RN:	88-04-0	MP (°C):	115.5
MW:	156.61	BP (°C):	246

Solubility (Moles/L)	Solubility (Grams/L)	Temp (°C)	Ref (#)	Evaluation (T P E A A)	Comments
1.596E-03	2.500E-01	20	M018	1 2 2 1 0	EFG
1.979E-03	3.099E-01	20	M093	1 0 0 1 1	
2.200E-02	3.445E+00	25	B316	0 0 0 0 0	*sic*
1.915E-03	2.999E-01	25	R041	0 0 0 0 0	
1.585E-03	2.482E-01	ns	R427	0 0 0 0 0	

1471. C$_8$H$_9$FN$_2$O$_3$
2,4(1H,3H)-Pyrimidinedione, 5-fluoro-3-(1-oxobutyl)-

RN:	94452-21-8	MP (°C):	
MW:	200.17	BP (°C):	

Solubility (Moles/L)	Solubility (Grams/L)	Temp (°C)	Ref (#)	Evaluation (T P E A A)	Comments
5.300E-02	1.061E+01	22	B416	2 2 1 2 1	

1472. C$_8$H$_9$FN$_2$O$_3$
Ftorafur
THFFU
1-(2-Tetrahydrofuryl)-5-fluorouracil

RN:	37076-68-9	MP (°C):	167
MW:	200.17	BP (°C):	

Solubility (Moles/L)	Solubility (Grams/L)	Temp (°C)	Ref (#)	Evaluation (T P E A A)	Comments
1.400E-01	2.802E+01	37	N017	0 0 0 0 0	

1473. C$_8$H$_9$FN$_2$O$_4$
1-Propionyloxymethyl-5-fluorouracil
1-Propionyloxymethyl-5-fluoro-2,4(1H,3H)-pyrimidinedi-one

RN:	66542-36-7	MP (°C):	100–102
MW:	216.17	BP (°C):	

Solubility (Moles/L)	Solubility (Grams/L)	Temp (°C)	Ref (#)	Evaluation (T P E A A)	Comments
1.554E-01	3.360E+01	22	B321	0 0 0 0 0	pH 4.0

1474. C$_8$H$_9$FN$_2$O$_4$
1-Isopropyloxyarbonyl-5-fluorouracil
1(2H)-Pyrimidinecarboxylic acid, 5-fluoro-3,4-dihydro-2,4-dioxo-, 1-methylethyl ester

RN: 109232-73-7   **MP** (°C):   180
MW: 216.17   **BP** (°C):

Solubility (Moles/L)	Solubility (Grams/L)	Temp (°C)	Ref (#)	Evaluation (T P E A A)	Comments
2.174E-02	4.700E+00	22	B332	1 1 0 0 1	pH 4.0

1475. C$_8$H$_9$N
Indoline
2,3-Dihydro-1H-indole
2,3-Dihydroindole

RN: 496-15-1   **MP** (°C):   <25
MW: 119.17   **BP** (°C):   220.5

Solubility (Moles/L)	Solubility (Grams/L)	Temp (°C)	Ref (#)	Evaluation (T P E A A)	Comments
2.934E-02	3.497E+00	20.3	L339	2 0 2 2 2	
9.063E-02	1.080E+01	25	P051	2 1 1 2 2	
9.063E-02	1.080E+01	25.00	P007	2 1 2 2 1	
3.651E-02	4.350E+00	40.0	L339	2 0 2 2 2	
4.586E-02	5.465E+00	59.4	L339	2 0 2 2 2	
5.738E-02	6.838E+00	79.0	L339	2 0 2 2 2	
8.142E-02	9.703E+00	100.0	L339	2 0 2 2 2	

1476. C$_8$H$_9$NO
p-Aminoacetophenone
4'-Aminoacetophenone

RN: 99-92-3   **MP** (°C):   106
MW: 135.17   **BP** (°C):   294

Solubility (Moles/L)	Solubility (Grams/L)	Temp (°C)	Ref (#)	Evaluation (T P E A A)	Comments
2.480E-02	3.352E+00	37.5	G002	1 1 1 1 2	

1477. C$_8$H$_9$NO
Acetanilide
Acetanilid

RN: 103-84-4   **MP** (°C):   114
MW: 135.17   **BP** (°C):   304

Solubility (Moles/L)	Solubility (Grams/L)	Temp (°C)	Ref (#)	Evaluation (T P E A A)	Comments
2.652E-02	3.585E+00	0	L029	2 2 2 2 2	
3.534E-02	4.777E+00	10	M043	1 0 0 0 1	
3.251E-02	4.395E+00	10.1	L029	2 2 2 2 2	
2.970E-02	4.014E+00	14	O016	1 0 0 0 2	
3.688E-02	4.985E+00	15	L038	1 0 1 0 2	

(continued)

1477. C$_8$H$_9$NO (continued)

Solubility (Moles/L)	Solubility (Grams/L)	Temp (°C)	Ref (#)	Evaluation (T P E A A)	Comments
3.710E-02	5.015E+00	20	B101	0 0 0 0 0	
3.666E-02	4.955E+00	20	K078	1 0 2 1 2	
4.129E-02	5.581E+00	20	L029	2 2 2 2 2	
3.827E-02	5.173E+00	20	M043	1 0 0 0 1	
3.330E-02	4.501E+00	20	O019	1 0 0 1 2	
3.884E-02	5.250E+00	20	W026	1 0 1 1 1	average of 2
4.142E-02	5.598E+00	25	B101	0 0 0 0 0	
4.450E-02	6.015E+00	25	B434	0 0 0 0 0	
4.786E-02	6.468E+00	25	B434	0 0 0 0 0	
4.160E-02	5.623E+00	25	D044	0 0 0 0 0	
4.143E-02	5.600E+00	25	F300	1 0 0 0 1	
4.697E-02	6.349E+00	25	L029	2 2 2 2 2	
4.486E-02	6.063E+00	25	M094	1 0 0 1 1	
3.699E-02	5.000E+00	25	P016	1 0 0 1 0	
4.887E-02	6.606E+00	30	B101	0 0 0 0 0	
5.262E-02	7.113E+00	30	B434	0 0 0 0 0	
5.240E-02	7.083E+00	30	B434	0 0 0 0 0	
5.351E-02	7.232E+00	30	L029	2 2 2 2 2	
4.632E-02	6.261E+00	30	M043	1 0 0 0 1	
5.253E-02	7.100E+00	30	W026	1 0 1 1 1	average of 2
5.792E-02	7.828E+00	32.6	L038	1 0 1 0 2	
5.930E-02	8.015E+00	35	B101	0 0 0 0 0	
5.799E-02	7.838E+00	35	B434	0 0 0 0 0	
5.760E-02	7.786E+00	35	B434	0 0 0 0 0	
6.787E-02	9.174E+00	40	B434	0 0 0 0 0	
6.730E-02	9.097E+00	40	B434	0 0 0 0 0	
7.134E-02	9.643E+00	40	L029	2 2 2 2 2	
6.381E-02	8.625E+00	40	M043	1 0 0 0 1	
9.682E-02	1.309E+01	50	L029	2 2 2 2 2	
1.349E-01	1.823E+01	60	L029	2 2 2 2 2	
1.522E-01	2.057E+01	60	M043	1 0 0 0 1	
1.928E-01	2.606E+01	70	L029	2 2 2 2 2	
3.321E-01	4.489E+01	80	M043	1 0 0 0 1	
4.047E-02	5.470E+00	rt	D021	0 0 1 1 1	

1478. C$_8$H$_9$NO

m-Aminoacetophenone
3′-Aminoacetophenone

RN:	99-03-6	MP (°C):	97
MW:	135.17	BP (°C):	

Solubility (Moles/L)	Solubility (Grams/L)	Temp (°C)	Ref (#)	Evaluation (T P E A A)	Comments
5.220E-02	7.056E+00	37.5	G002	1 1 1 1 2	pH 6.8

1479. C$_8$H$_9$NO$_2$

Acetaminophen
4-Acetamidophenol
4-Amino-phenol-*N*-acetat
p-Acetaminophen
p-Hydroxyacetanilide

RN:	103-90-2	**MP** (°C):	167
MW:	151.17	**BP** (°C):	

Solubility (Moles/L)	Solubility (Grams/L)	Temp (°C)	Ref (#)	Evaluation (T P E A A)	Comments
7.307E-02	1.105E+01	15	M352	1 1 1 1 2	
5.462E-01	8.257E+01	16.9	Y412	0 0 0 0 0	
6.014E-01	9.091E+01	21.5	Y412	0 0 0 0 0	
1.323E-01	2.000E+01	25	B010	1 1 1 1 0	
1.016E-01	1.536E+01	25	B434	0 0 0 0 0	
9.500E-02	1.436E+01	25	C032	2 2 1 2 0	EFG
7.710E-02	1.165E+01	25	D044	0 0 0 0 0	
9.133E-02	1.381E+01	25	D078	1 2 1 1 2	
5.185E-02	7.838E+00	25	F415	0 0 0 0 0	Average
1.000E-01	1.512E+01	25	K041	1 0 0 0 0	
9.851E-02	1.489E+01	25	M352	1 1 1 1 2	
9.923E-02	1.500E+01	25	P016	1 0 0 1 1	
7.277E-02	1.100E+01	25	P312	0 0 0 0 0	
9.326E-02	1.410E+01	25	W019	1 0 1 1 2	
3.538E-01	5.348E+01	25	Y410	0 0 0 0 0	
9.140E-02	1.382E+01	25	Z408	0 0 0 0 0	
6.556E-01	9.910E+01	26.3	Y412	0 0 0 0 0	
1.241E-01	1.876E+01	30	B434	0 0 0 0 0	
1.240E-01	1.874E+01	30	B434	0 0 0 0 0	
1.120E-01	1.693E+01	30	L069	1 0 1 1 0	EFG
7.088E-01	1.071E+02	31.5	Y412	0 0 0 0 0	
1.684E-01	2.545E+01	35	B434	0 0 0 0 0	
1.684E-01	2.546E+01	35	B434	0 0 0 0 0	
7.610E-01	1.150E+02	35.3	Y412	0 0 0 0 0	
1.323E-01	2.000E+01	37	F076	2 0 2 2 0	
1.442E-01	2.180E+01	37	K086	1 0 0 0 2	
8.124E-01	1.228E+02	37	Y412	0 0 0 0 0	
1.349E-01	2.039E+01	39.3	G302	2 0 2 2 0	EFG
2.234E-01	3.377E+01	40	B434	0 0 0 0 0	
2.238E-01	3.384E+01	40	B434	0 0 0 0 0	
1.440E-01	2.177E+01	40	M352	1 1 1 1 2	
1.800E-01	2.720E+01	50	M352	1 1 1 1 2	
1.019E-01	1.540E+01	c	B434	0 0 0 0 0	
6.615E-04	1.000E-01	ns	K444	0 0 0 0 0	
8.004E-02	1.210E+01	rt	R431	0 0 0 0 0	Average

1480. $C_8H_9NO_2$
Benzyl carbamate
O-Benzyl carbamate
Benzyloxycarbonyl amine

RN: 621-84-1   **MP** (°C): 87
MW: 151.17   **BP** (°C):

Solubility (Moles/L)	Solubility (Grams/L)	Temp (°C)	Ref (#)	Evaluation (T P E A A)	Comments
4.500E-01	6.802E+01	37	H006	1 2 2 1 1	
4.467E-01	6.752E+01	ns	R427	0 0 0 0 0	

1481. $C_8H_9NO_2$
DL-2-Phenylglycine
2-Amino-phenyl-essigsaeure
2-Aminophenylacetic acid

RN: 2835-06-5   **MP** (°C): 255
MW: 151.17   **BP** (°C):

Solubility (Moles/L)	Solubility (Grams/L)	Temp (°C)	Ref (#)	Evaluation (T P E A A)	Comments
7.608E-01	1.150E+02	100	F300	1 0 0 0 2	

1482. $C_8H_9NO_2$
N-Methylanthranilic acid
N-Methyl-anthranilsaeure

RN: 119-68-6   **MP** (°C): 171
MW: 151.17   **BP** (°C):

Solubility (Moles/L)	Solubility (Grams/L)	Temp (°C)	Ref (#)	Evaluation (T P E A A)	Comments
1.323E-03	2.000E-01	20	F300	1 0 0 0 2	
2.646E-03	4.000E-01	100	F300	1 0 0 0 2	

1483. $C_8H_9NO_2$
D-Phenylglycine
D-2-Phenylglycine
D-(–)-α-Aminophenylacetic acid
Benzeneacetic acid, α-amino-

RN: 875-74-1   **MP** (°C): 302 C
MW: 151.17   **BP** (°C):

Solubility (Moles/L)	Solubility (Grams/L)	Temp (°C)	Ref (#)	Evaluation (T P E A A)	Comments
3.034E-02	4.586E+00	25	R419	0 0 0 0 0	

1484. C$_8$H$_9$NO$_2$

Methyl-*p*-aminobenzoate
Methyl *p*-aminobenzoate
4-Aminobenzoic acid methyl ester
RN: 619-45-4   **MP** (°C):
MW: 151.17   **BP** (°C):

Solubility (Moles/L)	Solubility (Grams/L)	Temp (°C)	Ref (#)	Evaluation (T P E A A)	Comments
5.884E-03	8.894E-01	15	M352	1 1 1 1 2	
9.542E-03	1.442E+00	25	M352	1 1 1 1 2	
1.070E-02	1.618E+00	25	P303	0 0 0 0 0	
1.397E-02	2.112E+00	33	P303	0 0 0 0 0	
2.530E-02	3.825E+00	37	F006	1 1 2 2 2	
1.646E-02	2.488E+00	40	M352	1 1 1 1 2	
1.839E-02	2.780E+00	40	P303	0 0 0 0 0	
7.940E-03	1.200E+00	ns	M066	0 0 0 0 2	
7.940E-03	1.200E+00	rt	B016	0 0 1 1 2	pH 7.4

1485. C$_8$H$_9$NO$_2$S$_2$

2-(2-Thienyl)-ʟ-thiazolidine-4-carboxylic acid
4-Thiazolidinecarboxylic acid, 2-(2-thienyl)-
RN: 32451-19-7   **MP** (°C):
MW: 215.29   **BP** (°C):   454.1

Solubility (Moles/L)	Solubility (Grams/L)	Temp (°C)	Ref (#)	Evaluation (T P E A A)	Comments
4.900E-03	1.055E+00	21	B414	1 0 0 1 1	fast decomposition

1486. C$_8$H$_9$NO$_3$

ᴅ-(*p*-hydroxy)phenylglycine
RN: **MP** (°C):
MW: 167.17   **BP** (°C):

Solubility (Moles/L)	Solubility (Grams/L)	Temp (°C)	Ref (#)	Evaluation (T P E A A)	Comments
1.159E-01	1.937E+01	25	R419	0 0 0 0 0	

1487. C$_8$H$_9$NO$_3$S

p-Acetylbenzenesulfonamide
4-Acetylbenzenesulfonamide
RN: 1565-17-9   **MP** (°C):
MW: 199.23   **BP** (°C):

Solubility (Moles/L)	Solubility (Grams/L)	Temp (°C)	Ref (#)	Evaluation (T P E A A)	Comments
2.300E-03	4.582E-01	15	K024	1 2 1 1 2	

1488. C₈H₉NO₄

1488. $C_8H_9NO_4$

Biliverdic acid

Biliverdinsaeure

RN: 487-65-0   **MP** (°C):

MW: 183.17   **BP** (°C):

Solubility (Moles/L)	Solubility (Grams/L)	Temp (°C)	Ref (#)	Evaluation (T P E A A)	Comments
2.129E-01	3.900E+01	20	F300	1 0 0 0 1	

1489. $C_8H_9N_3O_3$

Orotic acid allylamide

4-Pyrimidinecarboxamide, 1,2,3,6-tetrahydro-2,6-dioxo-*N*-2-propenyl-

RN: 292870-71-4   **MP** (°C):   259–262

MW: 195.18   **BP** (°C):

Solubility (Moles/L)	Solubility (Grams/L)	Temp (°C)	Ref (#)	Evaluation (T P E A A)	Comments
1.780E-01	3.474E+01	−4	N018	0 0 0 0 0	
3.000E-01	5.855E+01	16	N018	0 0 0 0 0	
3.710E-01	7.241E+01	25	N018	0 0 0 0 0	

1490. $C_8H_9N_5$

7-Dimethylaminopteridine

7-Pteridinamine, *N,N*-dimethyl-

RN: 204443-26-5   **MP** (°C):

MW: 175.19   **BP** (°C):

Solubility (Moles/L)	Solubility (Grams/L)	Temp (°C)	Ref (#)	Evaluation (T P E A A)	Comments
8.154E-01	1.429E+02	20	A083	1 2 0 0 0	
1.903E+00	3.333E+02	100	A083	1 2 0 0 0	

1491. $C_8H_9N_5$

2-Dimethylaminopteridine

2-Pteridinamine, *N,N*-dimethyl-

RN: 41047-52-3   **MP** (°C):

MW: 175.19   **BP** (°C):

Solubility (Moles/L)	Solubility (Grams/L)	Temp (°C)	Ref (#)	Evaluation (T P E A A)	Comments
1.631E+00	2.857E+02	22.5	A085	1 2 0 0 0	

1492. C₈H₉N₅

1492. $C_8H_9N_5$

4-Dimethylaminopteridine

4-Pteridinamine, *N,N*-dimethyl-

RN: 14131-04-5 **MP** (°C): 165

MW: 175.19 **BP** (°C):

Solubility (Moles/L)	Solubility (Grams/L)	Temp (°C)	Ref (#)	Evaluation (T P E A A)	Comments
9.357E-02	1.639E+01	20	A019	2 2 1 1 0	
1.392E-01	2.439E+01	100	A019	1 2 1 1 0	

1493. $C_8H_9O_3PS$

2-Methoxy-4H-benzo-1,3,2-dioxaphosphorin-2-thione

Dioxabenzofos

Salithion

Fenfosphorin

Dioxabenzophos

RN: 3811-49-2 **MP** (°C): 55.5

MW: 216.20 **BP** (°C):

Solubility (Moles/L)	Solubility (Grams/L)	Temp (°C)	Ref (#)	Evaluation (T P E A A)	Comments
2.683E-04	5.800E-02	30	M161	1 0 0 0 1	

1494. C_8H_{10}

Ethylbenzene

Phenylethane

Ethylenzene

Ethylbenzol

EB

RN: 100-41-4 **MP** (°C): −95

MW: 106.17 **BP** (°C): 136.3

Solubility (Moles/L)	Solubility (Grams/L)	Temp (°C)	Ref (#)	Evaluation (T P E A A)	Comments
1.856E-03	1.970E-01	0	P003	2 2 2 2 2	
1.846E-03	1.960E-01	4.50	B086	2 1 2 2 2	
1.808E-03	1.920E-01	6.30	B086	2 1 2 2 2	
1.677E-03	1.781E-01	7.09	F418	0 0 0 0 0	
1.752E-03	1.860E-01	7.10	B086	2 1 2 2 2	
1.761E-03	1.870E-01	9	B086	2 1 2 2 2	
1.910E-03	2.028E-01	10	B149	2 1 1 2 2	
1.850E-03	1.964E-01	10	O312	2 2 0 2 2	
1.705E-03	1.810E-01	11.80	B086	2 1 2 2 2	
1.723E-03	1.830E-01	12.10	B086	2 1 2 2 2	
1.812E-03	1.924E-01	14	O312	2 2 0 2 2	
1.300E-03	1.380E-01	15	F001	1 0 1 2 1	
1.300E-03	1.380E-01	15	S006	1 0 0 0 1	
1.658E-03	1.760E-01	15	S203	1 1 2 1 2	
1.695E-03	1.800E-01	15.10	B086	2 1 2 2 2	

(continued)

1494. C$_8$H$_{10}$ (continued)

Solubility (Moles/L)	Solubility (Grams/L)	Temp (°C)	Ref (#)	Evaluation (T P E A A)	Comments
1.639E-03	1.740E-01	16.93	F418	0 0 0 0 0	
1.776E-03	1.886E-01	17	O312	2 2 0 2 2	
1.733E-03	1.840E-01	17.90	B086	2 1 2 2 2	
2.901E-03	3.080E-01	18	F185	1 0 0 0 2	
2.788E-03	2.960E-01	18	F185	1 0 0 0 2	
1.725E-03	1.831E-01	18	O312	2 2 0 2 2	
3.080E-03	3.270E-01	19	F185	1 0 0 0 2	
1.676E-03	1.779E-01	19	O312	2 2 0 2 2	
2.000E-03	2.123E-01	20	B149	2 1 1 2 2	
1.695E-03	1.800E-01	20	B356	0 0 0 0 0	
1.770E-03	1.879E-01	20	O312	2 2 0 2 2	
1.695E-03	1.800E-01	20.10	B086	2 1 2 2 1	
1.724E-03	1.830E-01	21	O312	2 2 0 2 2	
3.297E-03	3.500E-01	22	F185	1 0 0 0 2	
1.713E-03	1.819E-01	22	O312	2 2 0 2 2	
3.391E-03	3.600E-01	23	F185	1 0 0 0 2	
1.751E-03	1.859E-01	23.5	O312	2 2 0 2 2	
3.655E-03	3.880E-01	24	F185	1 0 0 0 2	
1.582E-03	1.680E-01	25	A002	1 2 1 1 2	
1.883E-03	2.000E-01	25	A094	1 0 0 0 0	
1.959E-03	2.080E-01	25	B003	2 2 2 2 2	
1.432E-03	1.520E-01	25	B060	2 0 1 1 1	
2.000E-03	2.123E-01	25	B153	2 1 1 1 2	
1.640E-03	1.741E-01	25	K001	1 0 2 1 2	
1.319E-03	1.400E-01	25	K072	1 0 1 1 1	
1.760E-03	1.869E-01	25	M342	1 0 1 1 2	
1.811E-03	1.923E-01	25	O312	2 2 0 2 2	
1.667E-03	1.770E-01	25	P003	2 2 2 2 2	
1.234E-03	1.310E-01	25	P051	2 1 1 2 2	
1.705E-03	1.810E-01	25	S203	1 1 2 1 2	
1.518E-03	1.612E-01	25	S358	2 1 2 2 2	
1.370E-03	1.455E-01	25	S359	2 1 2 2 2	
1.760E-03	1.869E-01	25	W300	2 2 2 2 2	
1.959E-03	2.080E-01	25.0	G035	1 0 0 0 2	
1.753E-03	1.861E-01	25.8	O312	2 2 0 2 2	
1.705E-03	1.810E-01	26.74	F418	0 0 0 0 0	
4.653E-03	4.940E-01	27	F185	1 0 0 0 2	
1.677E-03	1.780E-01	28	B348	2 1 2 2 2	
1.747E-03	1.855E-01	28	O312	2 2 0 2 2	
5.604E-03	5.950E-01	29	F185	1 0 0 0 2	
1.600E-03	1.698E-01	29.99	C350	0 0 0 0 0	
1.391E-03	1.477E-01	30	M311	1 1 2 2 2	
1.777E-03	1.887E-01	30	O312	2 2 0 2 2	
6.103E-03	6.480E-01	31	F185	1 0 0 0 2	
6.395E-03	6.790E-01	32	F185	1 0 0 0 2	
7.017E-03	7.450E-01	34	F185	1 0 0 0 2	
7.319E-03	7.770E-01	35	F185	1 0 0 0 2	
1.818E-03	1.930E-01	35	O312	2 2 0 2 2	
1.827E-03	1.940E-01	35	S203	1 1 2 1 2	

(continued)

1494. C_8H_{10} (continued)

Solubility (Moles/L)	Solubility (Grams/L)	Temp (°C)	Ref (#)	Evaluation (T P E A A)	Comments
7.865E-03	8.350E-01	36	F185	1 0 0 0 2	
1.805E-03	1.917E-01	36.55	F418	0 0 0 0 0	
8.637E-03	9.170E-01	38	F185	1 0 0 0 2	
1.622E-03	1.722E-01	39.99	C350	0 0 0 0 0	
1.928E-03	2.047E-01	40	O312	2 2 0 2 2	
9.466E-03	1.005E+00	41	F185	1 0 0 0 2	
1.991E-03	2.114E-01	45	O312	2 2 0 2 2	
2.025E-03	2.150E-01	45	S203	1 1 2 1 2	
1.994E-03	2.117E-01	46.49	F418	0 0 0 0 0	
1.154E-02	1.225E+00	47	F185	1 0 0 0 2	
1.224E-02	1.300E+00	49	F185	1 0 0 0 2	
1.861E-03	1.976E-01	49.99	C350	0 0 0 0 0	
2.216E-03	2.353E-01	56.73	F418	0 0 0 0 0	
2.261E-03	2.400E-01	59.99	C350	0 0 0 0 0	
2.560E-03	2.718E-01	66.64	F418	0 0 0 0 0	
2.738E-03	2.907E-01	69.99	C350	0 0 0 0 0	
3.327E-03	3.532E-01	79.99	C350	0 0 0 0 0	
3.860E-03	4.098E-01	89.99	C350	0 0 0 0 0	
4.742E-03	5.035E-01	99.99	C350	0 0 0 0 0	
4.829E-03	5.127E-01	115.0	G035	1 0 0 0 2	
1.120E-02	1.189E+00	140.5	G035	1 0 0 0 2	
3.332E-02	3.537E+00	170.5	G035	1 0 0 0 2	
6.185E-02	6.567E+00	210.0	G035	1 0 0 0 2	
1.052E-01	1.116E+01	233.5	G035	1 0 0 0 2	
1.432E-03	1.520E-01	ns	H123	0 0 0 0 0	
6.300E-02	6.689E+00	ns	H307	0 0 0 0 0	
1.432E-03	1.520E-01	ns	M344	0 0 0 0 2	

1495. C_8H_{10}
m-Xylene
1,3-Xylene

RN: 108-38-3   **MP** (°C):   −47.4
MW: 106.17   **BP** (°C):   139.3

Solubility (Moles/L)	Solubility (Grams/L)	Temp (°C)	Ref (#)	Evaluation (T P E A A)	Comments
1.846E-03	1.960E-01	0	P003	2 2 2 2 2	
1.463E-03	1.554E-01	20	M337	2 1 2 2 2	
1.629E-03	1.730E-01	25	A001	1 2 2 2 2	
1.846E-03	1.960E-01	25	B003	2 2 2 2 2	
1.262E-03	1.340E-01	25	K119	1 0 0 0 2	
1.510E-03	1.603E-01	25	M342	1 0 1 1 2	
1.526E-03	1.620E-01	25	P003	2 2 2 2 2	
1.262E-03	1.340E-01	25	P051	2 1 1 2 2	
1.375E-03	1.460E-01	25	S005	2 2 2 2 2	
1.375E-03	1.460E-01	25	S191	1 2 2 2 2	
1.375E-03	1.460E-01	25	S358	2 1 2 2 2	
1.330E-03	1.412E-01	25	S359	2 1 2 2 2	

(continued)

1495. C_8H_{10} (continued)

Solubility (Moles/L)	Solubility (Grams/L)	Temp (°C)	Ref (#)	Evaluation (T P E A A)	Comments
1.510E-03	1.603E-01	25	W300	2 2 2 2 2	
1.262E-03	1.340E-01	25.00	P007	2 1 2 2 2	
1.940E-03	2.059E-01	25.04	V013	2 2 2 2 2	
3.277E-03	3.479E-01	67.7	P005	1 1 2 1 2	
6.257E-03	6.643E-01	107.3	P005	1 1 2 1 2	
9.707E-03	1.031E+00	124.2	P005	1 1 2 1 2	
2.363E-02	2.509E+00	164.2	P005	1 1 2 1 2	
4.327E-02	4.594E+00	186.4	P005	1 1 2 1 2	
4.293E-02	4.557E+00	189.9	P005	1 1 2 1 2	
2.675E-01	2.840E+01	266.6	P005	1 1 2 1 2	
2.698E-01	2.865E+01	270.6	P005	1 1 2 1 2	

1496. C_8H_{10}

o-Xylene
1,2-Dimethylbenzene
1,2-Xylene

RN:	95-47-6	**MP** (°C):	−25
MW:	106.17	**BP** (°C):	144

Solubility (Moles/L)	Solubility (Grams/L)	Temp (°C)	Ref (#)	Evaluation (T P E A A)	Comments
1.337E-03	1.420E-01	0	P003	2 2 2 2 2	
2.000E-03	2.123E-01	10	B149	2 1 1 2 2	
2.260E-03	2.399E-01	20	B149	2 1 1 2 2	
1.605E-03	1.704E-01	20	M337	2 1 2 2 2	
1.921E-03	2.040E-01	25	A001	1 2 2 2 2	
1.648E-03	1.750E-01	25	B060	2 0 1 1 1	
1.573E-03	1.670E-01	25	K119	1 0 0 0 2	
1.648E-03	1.750E-01	25	M001	2 1 2 2 2	
1.648E-03	1.750E-01	25	M002	2 1 2 2 2	
1.648E-03	1.750E-01	25	M040	1 0 0 1 2	
1.648E-03	1.750E-01	25	M130	1 0 0 0 2	
2.080E-03	2.208E-01	25	M342	1 0 1 1 2	
2.006E-03	2.130E-01	25	P003	2 2 2 2 2	
1.573E-03	1.670E-01	25	P051	2 1 1 2 2	
1.606E-03	1.705E-01	25	S005	2 2 2 2 2	
1.606E-03	1.705E-01	25	S191	1 2 2 2 2	
1.606E-03	1.705E-01	25	S358	2 1 2 2 2	
1.680E-03	1.784E-01	25	S359	2 1 2 2 2	
2.080E-03	2.208E-01	25	W300	2 2 2 2 2	
1.573E-03	1.670E-01	25.00	P007	2 1 2 2 2	
1.272E-03	1.350E-01	ns	B150	0 0 2 2 2	
1.648E-03	1.750E-01	ns	M344	0 0 0 0 2	

1497. C$_8$H$_{10}$

p-Xylene
1,4-Dimethylbenzene
1,4-Xylene

RN: 106-42-3 **MP** (°C): 13
MW: 106.17 **BP** (°C): 137

Solubility (Moles/L)	Solubility (Grams/L)	Temp (°C)	Ref (#)	Evaluation (T P E A A)	Comments
1.545E-03	1.640E-01	0	P003	2 2 2 2 2	
1.780E-03	1.890E-01	10	B149	2 1 1 2 2	
1.800E-03	1.911E-01	20	B149	2 1 1 2 2	
1.552E-03	1.648E-01	20	M337	2 1 2 2 2	
1.884E-03	2.000E-01	25	A001	1 2 2 2 2	
1.865E-03	1.980E-01	25	B003	2 2 2 2 2	
1.224E-03	1.300E-01	25	K072	1 0 1 1 1	
1.479E-03	1.570E-01	25	K119	1 0 0 0 2	
1.789E-03	1.900E-01	25	L319	1 0 2 1 1	
1.224E-03	1.300E-01	25	M087	1 1 2 1 1	
2.020E-03	2.145E-01	25	M342	1 0 1 1 2	
1.743E-03	1.850E-01	25	P003	2 2 2 2 2	
1.479E-03	1.570E-01	25	P051	2 1 1 2 2	
1.469E-03	1.560E-01	25	S005	2 2 2 2 2	
1.469E-03	1.560E-01	25	S191	1 2 2 2 2	
1.469E-03	1.560E-01	25	S358	2 1 2 2 2	
1.510E-03	1.603E-01	25	S359	2 1 2 2 2	
2.020E-03	2.145E-01	25	W300	2 2 2 2 2	
1.479E-03	1.570E-01	25.00	P007	2 1 2 2 2	
1.589E-03	1.687E-01	29.99	C350	0 0 0 0 0	
1.766E-03	1.875E-01	39.99	C350	0 0 0 0 0	
2.410E-03	2.559E-01	43.0	P005	1 1 2 1 2	
1.911E-03	2.029E-01	49.99	C350	0 0 0 0 0	
2.832E-03	3.007E-01	56.4	P005	1 1 2 1 2	
2.244E-03	2.382E-01	59.99	C350	0 0 0 0 0	
3.199E-03	3.396E-01	65.0	P005	1 1 2 1 2	
2.683E-03	2.848E-01	69.99	C350	0 0 0 0 0	
3.643E-03	3.868E-01	75.3	P005	1 1 2 1 2	
3.171E-03	3.367E-01	79.99	C350	0 0 0 0 0	
4.326E-03	4.593E-01	87.2	P005	1 1 2 1 2	
3.721E-03	3.950E-01	89.99	C350	0 0 0 0 0	
4.853E-03	5.152E-01	99.99	C350	0 0 0 0 0	
2.363E-02	2.509E+00	162.5	P005	1 1 2 1 2	
4.251E-02	4.513E+00	188.1	P005	1 1 2 1 2	
1.614E-01	1.713E+01	243.2	P005	1 1 2 1 2	
4.053E-01	4.303E+01	282.5	P005	1 1 2 1 2	
4.011E-01	4.258E+01	294.9	P005	1 1 2 1 2	
1.743E-03	1.850E-01	ns	H123	0 0 0 0 0	

1498. C$_8$H$_{10}$

Xylene
Dimethylbenzene
Xylol

RN: 1330-20-7   **MP** (°C):
MW: 106.17   **BP** (°C): 137

Solubility (Moles/L)	Solubility (Grams/L)	Temp (°C)	Ref (#)	Evaluation (T P E A A)	Comments
8.469E-03	8.992E-01	20	C121	1 0 0 0 0	unit assumed, *sic*
1.000E-03	1.062E-01	25	H332	2 2 2 2 0	
<9.41E-03	<9.99E-01	25.50	O005	2 0 2 2 0	
9.419E-03	1.000E+00	150	J023	1 1 2 2 0	
3.297E-02	3.500E+00	200	J023	1 1 2 2 1	
1.036E-01	1.100E+01	250	J023	1 1 2 2 1	

1499. C$_8$H$_{10}$NO$_5$PS

Methyl parathion
Parathion-methyl
Methylparathion

RN: 298-00-0   **MP** (°C): 36
MW: 263.21   **BP** (°C):

Solubility (Moles/L)	Solubility (Grams/L)	Temp (°C)	Ref (#)	Evaluation (T P E A A)	Comments
8.282E-05	2.180E-02	10	B324	0 0 0 0 0	
8.283E-05	2.180E-02	10	B324	0 0 0 0 0	
1.432E-04	3.770E-02	19.50	B169	2 2 1 1 2	
1.444E-04	3.801E-02	20	B324	0 0 0 0 0	
1.444E-04	3.800E-02	20	B324	0 0 0 0 0	
9.498E-05	2.500E-02	20	M040	1 0 0 1 1	
2.090E-04	5.500E-02	25	M061	1 0 0 0 1	
2.185E-04	5.750E-02	25	M161	1 0 0 0 0	
2.089E-04	5.500E-02	25	Z409	0 0 0 0 0	EFG
2.223E-04	5.851E-02	30	B324	0 0 0 0 0	
2.222E-04	5.850E-02	30	B324	0 0 0 0 0	
1.900E-04	5.000E-02	ns	C117	0 0 0 0 0	
1.445E-04	3.805E-02	ns	R427	0 0 0 0 0	
1.432E-04	3.770E-02	ns	V414	0 0 0 0 0	

1500. C$_8$H$_{10}$N$_2$O

p-Phenylenediaminemono-*N*-acetate
p-Phenylendiamin-mono-*N*-acetat

RN: 589-29-7   **MP** (°C):
MW: 150.18   **BP** (°C):

Solubility (Moles/L)	Solubility (Grams/L)	Temp (°C)	Ref (#)	Evaluation (T P E A A)	Comments
4.128E-01	6.200E+01	57	F300	1 0 0 0 1	

1501. C$_8$H$_{10}$N$_2$O
m-Aminoacetanilide
3-Aminoacetanilide
RN: 102-28-3   **MP** (°C):
MW: 150.18   **BP** (°C):

Solubility (Moles/L)	Solubility (Grams/L)	Temp (°C)	Ref (#)	Evaluation (T P E A A)	Comments
5.526E-01	8.299E+01	48.7	S115	1 2 1 1 2	
1.021E+00	1.534E+02	82.9	S115	1 2 1 1 2	

1502. C$_8$H$_{10}$N$_2$O
o-Aminoacetanilide
2-Aminoacetanilide
RN: 34801-09-7   **MP** (°C):
MW: 150.18   **BP** (°C):

Solubility (Moles/L)	Solubility (Grams/L)	Temp (°C)	Ref (#)	Evaluation (T P E A A)	Comments
2.189E-01	3.288E+01	7.2	S115	1 2 1 1 2	
7.161E-01	1.075E+02	22.0	S115	1 2 1 1 2	
1.215E+00	1.825E+02	33.5	S115	1 2 1 1 2	
1.612E+00	2.421E+02	42.1	S115	1 2 1 1 2	
1.958E+00	2.940E+02	50.4	S115	1 2 1 1 2	
2.270E+00	3.409E+02	59.1	S115	1 2 1 1 2	
2.601E+00	3.906E+02	69.9	S115	1 2 1 1 2	
2.781E+00	4.177E+02	78.2	S115	1 2 1 1 2	
2.943E+00	4.420E+02	88.1	S115	1 2 1 1 2	
3.075E+00	4.618E+02	99.0	S115	1 2 1 1 2	
3.213E+00	4.825E+02	115.4	S115	1 2 1 1 2	

1503. C$_8$H$_{10}$N$_2$O
1-(2-Tolyl)urea
o-Tolylurea
RN: 614-77-7   **MP** (°C):
MW: 150.18   **BP** (°C):

Solubility (Moles/L)	Solubility (Grams/L)	Temp (°C)	Ref (#)	Evaluation (T P E A A)	Comments
1.667E-02	2.504E+00	45	W044	1 0 1 0 2	

1504. $C_8H_{10}N_2O$
1-Methyl-3-phenylurea
Desfenuron
N-Phenyl-N'-methylurea
Desphenuron
N-Methyl-N'-phenylurea
IPO 4328

RN:	1007-36-9	MP (°C):
MW:	150.18	BP (°C):

Solubility (Moles/L)	Solubility (Grams/L)	Temp (°C)	Ref (#)	Evaluation (T P E A A)	Comments
4.927E+00	7.400E+02	45	W044	1 0 1 0 2	

1505. $C_8H_{10}N_2O$
1-(4-Tolyl)urea
p-Tolylurea

RN:	622-51-5	MP (°C):
MW:	150.18	BP (°C):

Solubility (Moles/L)	Solubility (Grams/L)	Temp (°C)	Ref (#)	Evaluation (T P E A A)	Comments
2.044E-02	3.070E+00	45	W044	1 0 1 0 2	

1506. $C_8H_{10}N_2O$
p-Aminoacetanilide
4-Aminoacetanilide

RN:	122-80-5	MP (°C):	164.5
MW:	150.18	BP (°C):	

Solubility (Moles/L)	Solubility (Grams/L)	Temp (°C)	Ref (#)	Evaluation (T P E A A)	Comments
1.061E-01	1.593E+01	25	D044	0 0 0 0 0	
4.064E-01	6.103E+01	56.8	S115	1 2 1 1 2	
1.046E+00	1.570E+02	86.3	S115	1 2 1 1 2	
1.441E+00	2.165E+02	92.1	S115	1 2 1 1 2	
1.699E+00	2.552E+02	93.7	S115	1 2 1 1 2	
1.996E+00	2.998E+02	96.5	S115	1 2 1 1 2	
2.193E+00	3.293E+02	98.6	S115	1 2 1 1 2	

1507. C$_8$H$_{10}$N$_2$O
Methylbenzylnitrosamine
N-Nitroso(methyl)benzylamine
N-Nitroso-N-methylbenzylamine
N-Nitroso(benzyl)methylamine
N-Nitroso-N-methylbenzenemethanamine

RN: 937-40-6   **MP** (°C):
MW: 150.18   **BP** (°C):

Solubility (Moles/L)	Solubility (Grams/L)	Temp (°C)	Ref (#)	Evaluation (T P E A A)	Comments
3.000E-02	4.505E+00	24	D083	2 0 0 0 1	

1508. C$_8$H$_{10}$N$_2$O
Benzylurea
Benzyl-harnstoff

RN: 538-32-9   **MP** (°C): 147
MW: 150.18   **BP** (°C):

Solubility (Moles/L)	Solubility (Grams/L)	Temp (°C)	Ref (#)	Evaluation (T P E A A)	Comments
1.132E-01	1.700E+01	45	F300	1 0 0 0 2	
1.139E-01	1.710E+01	45	W044	1 0 1 0 2	

1509. C$_8$H$_{10}$N$_2$O$_3$
5-Methyl-5-allylbarbituric acid
2,4,6(1H,3H,5H)-Pyrimidinetrione, 5-methyl-5-(2-propenyl)
5-Methyl-5-allylbarbiturate

RN: 143585-01-7   **MP** (°C):
MW: 182.18   **BP** (°C):

Solubility (Moles/L)	Solubility (Grams/L)	Temp (°C)	Ref (#)	Evaluation (T P E A A)	Comments
6.920E-02	1.261E+01	25	P350	0 0 0 0 0	intrinsic

1510. C$_8$H$_{10}$N$_2$O$_3$
5,5-Tetramethylenebarbituric acid
7,9-Diazaspiro[4.5]decane-6,8,10-trione
Spirocyclopentabarbituric acid
Cyclopentane-spirobarbiturate

RN: 56209-30-4   **MP** (°C):
MW: 182.18   **BP** (°C):

Solubility (Moles/L)	Solubility (Grams/L)	Temp (°C)	Ref (#)	Evaluation (T P E A A)	Comments
4.476E-03	8.154E-01	25	P350	0 0 0 0 0	intrinsic

1511. C₈H₁₀N₂O₃S
*N*1-Acetylsulfanilamide
Sulfacetamide
Acetyl sulfacetamide

RN:	144-80-9	MP (°C):	183
MW:	214.24	BP (°C):	

Solubility (Moles/L)	Solubility (Grams/L)	Temp (°C)	Ref (#)	Evaluation (T P E A A)	Comments
5.881E-02	1.260E+01	20	F073	1 2 2 2 2	
3.871E-02	8.293E+00	25	M440	0 0 0 0 0	
5.834E-03	1.250E+00	37	B046	1 0 2 2 2	pH 4.5
5.834E-02	1.250E+01	37	B046	1 0 2 2 2	pH 5
6.908E-02	1.480E+01	37	D084	1 0 1 0 2	
5.601E-02	1.200E+01	37	K086	1 0 0 0 2	
5.134E-02	1.100E+01	37	L091	1 0 0 0 2	pH 5.5
2.327E-02	4.985E+00	ns	L044	0 0 0 0 2	
3.090E-02	6.621E+00	ns	R427	0 0 0 0 0	

1512. C₈H₁₀N₂O₃S
*N*4-Acetylsulfanilamide
*N*4-Acetylsulphanilamide

RN:	121-61-9	MP (°C):	216
MW:	214.24	BP (°C):	

Solubility (Moles/L)	Solubility (Grams/L)	Temp (°C)	Ref (#)	Evaluation (T P E A A)	Comments
2.474E-02	5.300E+00	37	L091	1 0 0 0 2	pH 5.5
2.479E-02	5.312E+00	37.50	M142	1 0 0 0 2	

1513. C₈H₁₀N₂O₃S
Tosylurea
Tosyluree

RN:	1694-06-0	MP (°C):	
MW:	214.24	BP (°C):	

Solubility (Moles/L)	Solubility (Grams/L)	Temp (°C)	Ref (#)	Evaluation (T P E A A)	Comments
3.631E-03	7.779E-01	37	A028	1 0 2 1 2	intrinsic

1514. C₈H₁₀N₂O₄S

Asulam

Methyl N-(4-aminobenzenesulphonyl)carbamate

RN:	3337-71-1	**MP** (°C):	144	
MW:	230.24	**BP** (°C):		

Solubility (Moles/L)	Solubility (Grams/L)	Temp (°C)	Ref (#)	Evaluation (T P E A A)	Comments
2.161E-02	4.975E+00	ns	M061	0 0 0 0 0	
2.188E-02	5.037E+00	ns	R427	0 0 0 0 0	
2.172E-02	5.000E+00	rt	M161	0 0 0 0 0	

1515. C₈H₁₀N₄O₂

Caffeine

Coffein

RN:	58-08-2	**MP** (°C):	238	
MW:	194.19	**BP** (°C):		

Solubility (Moles/L)	Solubility (Grams/L)	Temp (°C)	Ref (#)	Evaluation (T P E A A)	Comments
3.887E-02	7.548E+00	0	H023	1 0 2 1 2	
3.800E-02	7.379E+00	1	M116	2 1 1 1 1	
3.757E-02	7.296E+00	2	C074	1 0 0 1 2	
4.786E+00	9.294E+02	5	B429	1 0 1 2 2	
4.859E+00	9.436E+02	15	B429	1 0 1 2 2	
6.603E-02	1.282E+01	15	H023	1 0 2 1 2	
5.800E-02	1.126E+01	15	O017	1 0 1 1 1	
5.770E-02	1.121E+01	15	O018	1 2 1 1 2	
5.770E-02	1.121E+01	15	O019	1 0 0 1 2	
6.859E-02	1.332E+01	16	A072	1 0 1 0 2	
7.415E-02	1.440E+01	20	F300	1 0 0 0 2	
6.779E-02	1.316E+01	20	J009	2 0 2 2 2	
1.242E-01	2.411E+01	25	A068	2 0 0 0 2	
4.931E+00	9.575E+02	25	B429	1 0 1 2 2	
1.066E-01	2.071E+01	25	E016	1 1 1 1 2	
1.081E-01	2.100E+01	25	F300	1 0 0 0 1	
1.080E-01	2.097E+01	25	L329	2 2 1 2 2	
1.110E-01	2.156E+01	25	M116	2 1 1 1 2	
1.244E-01	2.415E+01	25	M158	2 0 2 2 2	
1.000E-01	1.942E+01	25	O017	1 0 1 1 2	
1.002E-01	1.946E+01	25	O018	1 2 1 1 2	
1.098E-02	2.132E+00	25	O019	1 0 0 1 2	
1.272E-01	2.470E+01	25	O302	1 0 0 1 0	
1.107E-01	2.150E+01	25	P010	1 0 1 1 2	
1.123E-01	2.180E+01	25	P011	0 0 0 0 0	
1.195E-01	2.320E+01	25	P018	1 0 2 2 2	
1.081E-01	2.100E+01	25	P020	2 0 1 1 1	
1.330E-01	2.583E+01	30	B042	1 2 1 1 2	
1.330E-01	2.583E+01	30	G021	1 0 0 0 2	
1.330E-01	2.583E+01	30	H020	1 0 0 0 2	
1.333E-01	2.589E+01	30	H023	1 0 2 1 2	

(continued)

1515. C₈H₁₀N₄O₂ (continued)

Solubility (Moles/L)	Solubility (Grams/L)	Temp (°C)	Ref (#)	Evaluation (T P E A A)	Comments
1.330E-01	2.583E+01	30.60	M116	2 1 1 1 2	
4.999E+00	9.707E+02	35	B429	1 0 1 2 2	
1.670E-01	3.243E+01	35	O017	1 0 1 1 2	
1.909E-01	3.707E+01	37	C074	1 0 0 1 2	
1.930E-01	3.748E+01	37	M116	2 1 1 1 2	
5.041E+00	9.789E+02	40	B429	1 0 1 2 2	
2.266E-01	4.400E+01	40	F300	1 0 0 0 1	
5.211E-01	1.012E+02	57	C074	1 0 0 1 2	
1.408E+00	2.735E+02	83	C065	1 0 0 1 2	
1.407E+00	2.733E+02	85	C074	1 0 0 1 2	
1.739E+00	3.377E+02	87	C065	1 0 0 1 2	
2.343E+00	4.550E+02	90	C074	1 0 0 1 2	
1.287E-01	2.500E+01	ns	D035	0 0 0 0 2	
1.104E-01	2.143E+01	rt	D021	0 0 1 1 2	
1.596E-04	3.100E-02	rt	N015	0 0 2 2 1	*sic*
4.892E-02	9.500E+00	rt	R431	0 0 0 0 0	Average

1516. C₈H₁₀N₄O₂.H₂O

Caffeine (monohydrate)

1H-Purine-2,6-dione, 3,7-dihydro-1,3,7-trimethyl-, monohydrate

RN: 5743-12-4   **MP (°C):** 178
MW: 212.21   **BP (°C):**

Solubility (Moles/L)	Solubility (Grams/L)	Temp (°C)	Ref (#)	Evaluation (T P E A A)	Comments
1.011E-01	2.146E+01	25	D004	0 0 0 0 0	

1517. C₈H₁₀N₄O₃

1,3,7-Trimethyluric acid

8-Oxy-caffeine

RN: 5415-44-1   **MP (°C):** 374
MW: 210.19   **BP (°C):**

Solubility (Moles/L)	Solubility (Grams/L)	Temp (°C)	Ref (#)	Evaluation (T P E A A)	Comments
1.142E-04	2.400E-02	rt	N015	0 0 2 2 1	

1518. C₈H₁₀O

$1518.\ C_8H_{10}O$

4-Ethylphenol

p-Ethylphenol

RN: 123-07-9   **MP** (°C):   43.5

MW: 122.17   **BP** (°C):

Solubility (Moles/L)	Solubility (Grams/L)	Temp (°C)	Ref (#)	Evaluation (T P E A A)	Comments
4.854E-02	5.931E+00	20	R087	0 0 0 0 0	0.15M NaCl
2.332E-02	2.849E+00	25	L022	1 0 0 0 0	
4.011E-02	4.900E+00	25	M127	1 0 0 0 1	
4.072E-02	4.975E+00	25	R041	0 0 0 0 0	
4.467E-02	5.457E+00	ns	R427	0 0 0 0 0	

1519. $C_8H_{10}O$

2,3-Xylenol

2,3-Dimethylphenol

RN: 526-75-0   **MP** (°C):   75

MW: 122.17   **BP** (°C):   218

Solubility (Moles/L)	Solubility (Grams/L)	Temp (°C)	Ref (#)	Evaluation (T P E A A)	Comments
3.740E-02	4.569E+00	25	A021	1 2 1 1 2	

1520. $C_8H_{10}O$

Phenylethylalcohol

Phenyl ethyl alcohol

RN: 60-12-8   **MP** (°C):

MW: 122.17   **BP** (°C):

Solubility (Moles/L)	Solubility (Grams/L)	Temp (°C)	Ref (#)	Evaluation (T P E A A)	Comments
1.470E-01	1.796E+01	20	S006	1 0 0 0 2	
1.720E-01	2.101E+01	25	D407	1 0 2 2 2	
1.432E-01	1.749E+01	25	H044	1 0 2 1 2	
1.455E-01	1.778E+01	30	H044	1 0 2 1 2	
1.487E-01	1.816E+01	35	H044	1 0 2 1 2	
1.518E-01	1.855E+01	40	H044	1 0 2 1 2	
1.542E-01	1.884E+01	45	H044	1 0 2 1 2	
1.562E-01	1.908E+01	50	H044	1 0 2 1 2	
1.597E-01	1.951E+01	55	H044	1 0 2 1 2	

1521. $C_8H_{10}O$

Phloral

RN: **MP** (°C):

MW: 122.17   **BP** (°C):   204.52

Solubility (Moles/L)	Solubility (Grams/L)	Temp (°C)	Ref (#)	Evaluation (T P E A A)	Comments
4.072E-02	4.975E+00	25	L022	1 0 0 0 0	

1522. C₈H₁₀O

2,6-Xylenol
1,3,2-Xylenol
2,6-Dimethylphenol
Vic-*m*-xylenol

RN: 576-26-1 **MP** (°C): 49
MW: 122.17 **BP** (°C): 203

Solubility (Moles/L)	Solubility (Grams/L)	Temp (°C)	Ref (#)	Evaluation (T P E A A)	Comments
3.595E-02	4.392E+00	20	R087	0 0 0 0 0	0.15M NaCl
4.950E-02	6.047E+00	25	A021	1 2 1 1 2	
5.100E-02	6.231E+00	25	B316	0 0 0 0 0	

1523. C₈H₁₀O

2,4-Xylenol
2,4-Dimethylphenol
m-Xylenol
2,4-Dimethyl-phenol-
Phenol, 2,4-dimethyl-
1-Hydroxy-2,4-dimethylbenzene

RN: 105-67-9 **MP** (°C): 26
MW: 122.17 **BP** (°C): 211.5

Solubility (Moles/L)	Solubility (Grams/L)	Temp (°C)	Ref (#)	Evaluation (T P E A A)	Comments
4.400E-02	5.375E+00	20	K132	1 0 1 1 1	
4.300E-02	5.253E+00	20	K309	1 0 0 1 1	
5.271E-02	6.440E+00	20	R087	0 0 0 0 0	0.15M NaCl
5.100E-02	6.231E+00	25	A021	1 2 1 1 2	
6.440E-02	7.868E+00	25	B173	2 0 2 2 2	
7.200E-02	8.796E+00	25	B316	0 0 0 0 0	
6.499E-02	7.940E+00	25	M127	1 0 0 0 2	
2.190E-01	2.675E+01	80	K309	1 0 0 1 2	

1524. C₈H₁₀O

α-Methyl-benzenemethanol
α-Methylbenzyl alcohol
1-Phenylethan-1-*o*
Methylphenylcarbinol
β-Hydroxyethylbenzene
(*S*)-1-Phenylethyl alcohol

RN: 98-85-1 **MP** (°C): 20
MW: 122.17 **BP** (°C): 401 at 0 mm

Solubility (Moles/L)	Solubility (Grams/L)	Temp (°C)	Ref (#)	Evaluation (T P E A A)	Comments
6.898E+00	8.427E+02	14.57	L441	0 0 0 0 0	
6.860E+00	8.380E+02	19.84	L441	0 0 0 0 0	

(continued)

1524. C$_8$H$_{10}$O (continued)

Solubility (Moles/L)	Solubility (Grams/L)	Temp (°C)	Ref (#)	Evaluation (T P E A A)	Comments
5.056E-01	6.177E+01	92.71	L441	0 0 0 0 0	
6.491E+00	7.930E+02	94.89	L441	0 0 0 0 0	
6.445E+00	7.874E+02	105.95	L441	0 0 0 0 0	
6.196E+00	7.569E+02	127.92	L441	0 0 0 0 0	

1525. C$_8$H$_{10}$O

Phenetole
Ethoxybenzene

RN: 103-73-1 **MP (°C):** −30
MW: 122.17 **BP (°C):** 171

Solubility (Moles/L)	Solubility (Grams/L)	Temp (°C)	Ref (#)	Evaluation (T P E A A)	Comments
4.500E-03	5.498E-01	25	M327	1 0 0 1 2	
4.657E-03	5.690E-01	25.04	V013	2 2 2 2 2	

1526. C$_8$H$_{10}$O

2,5-Xylenol
2,5-Dimethylphenol
p-Xylenol
2,5-Dimethyl-phenol-
Phenol, 2,5-dimethyl-

RN: 95-87-4 **MP (°C):** 75
MW: 122.17 **BP (°C):** 212

Solubility (Moles/L)	Solubility (Grams/L)	Temp (°C)	Ref (#)	Evaluation (T P E A A)	Comments
2.900E-02	3.543E+00	25	A021	1 2 1 1 2	
2.600E-02	3.176E+00	25	B316	0 0 0 0 0	

1527. C$_8$H$_{10}$O

4-Methylbenzyl alcohol
4-Methyl-benzylalkohol

RN: 589-18-4 **MP (°C):** 60
MW: 122.17 **BP (°C):** 217

Solubility (Moles/L)	Solubility (Grams/L)	Temp (°C)	Ref (#)	Evaluation (T P E A A)	Comments
6.900E-02	8.430E+00	20	B407	1 0 1 2 2	

1528. C$_8$H$_{10}$O
3,4-Xylenol
3,4-Dimethylphenol
As-*o*-xylenol

RN:	95-65-8	**MP** (°C):	62.5	
MW:	122.17	**BP** (°C):	225	

Solubility (Moles/L)	Solubility (Grams/L)	Temp (°C)	Ref (#)	Evaluation (T P E A A)	Comments
3.100E-02	3.787E+00	20	K132	1 0 1 1 1	
3.900E-02	4.765E+00	25	A021	1 2 1 1 2	
4.072E-02	4.975E+00	25	R041	0 0 0 0 0	
2.530E-02	3.091E+00	37	E028	1 0 1 1 2	

1529. C$_8$H$_{10}$O
3,5-Xylenol
3,5-Dimethylphenol

RN:	108-68-9	**MP** (°C):	64	
MW:	122.17	**BP** (°C):	219.5	

Solubility (Moles/L)	Solubility (Grams/L)	Temp (°C)	Ref (#)	Evaluation (T P E A A)	Comments
3.300E-02	4.032E+00	20	K132	1 0 1 1 1	
2.961E-02	3.618E+00	20	R087	0 0 0 0 0	0.15M NaCl
4.000E-02	4.887E+00	25	A021	1 2 1 1 2	
4.000E-02	4.887E+00	25	B316	0 0 0 0 0	
3.981E-02	4.864E+00	ns	R427	0 0 0 0 0	

1530. C$_8$H$_{10}$O$_2$
o-Ethoxyphenol
2-Ethoxyphenol

RN:	94-71-3	**MP** (°C):		
MW:	138.17	**BP** (°C):		

Solubility (Moles/L)	Solubility (Grams/L)	Temp (°C)	Ref (#)	Evaluation (T P E A A)	Comments
6.090E-02	8.414E+00	24.99	B353	0 0 0 0 0	

1531. C$_8$H$_{10}$O$_2$
Veratrole
o-Dimethoxybenzene

RN:	91-16-7	**MP** (°C):	15	
MW:	138.17	**BP** (°C):	207	

Solubility (Moles/L)	Solubility (Grams/L)	Temp (°C)	Ref (#)	Evaluation (T P E A A)	Comments
4.842E-02	6.690E+00	25	L348	1 2 2 1 2	

1532. C$_8$H$_{10}$O$_2$
1,3-Dimethoxybenzene
m-Dimethoxybenzene
Dimethylresorcinol

RN: 151-10-0   **MP** (°C):
MW: 138.17   **BP** (°C):   86

Solubility (Moles/L)	Solubility (Grams/L)	Temp (°C)	Ref (#)	Evaluation (T P E A A)	Comments
8.800E-03	1.216E+00	25	M327	1 0 0 1 2	

1533. C$_8$H$_{10}$O$_2$
2-Phenoxyethanol
Phenoxyethyl alcohol
Ethylene glycol phenyl ether
Arosol
1-Hydroxy-2-phenoxyethane
Phenoxethol

RN: 122-99-6   **MP** (°C):   12
MW: 138.17   **BP** (°C):   237

Solubility (Moles/L)	Solubility (Grams/L)	Temp (°C)	Ref (#)	Evaluation (T P E A A)	Comments
1.882E-01	2.601E+01	20	M062	1 0 0 0 2	
2.610E-01	3.606E+01	37	E028	1 0 1 1 2	

1534. C$_8$H$_{10}$O$_2$
3-Ethoxyphenol
m-Ethoxy phenol
Resorcinol monoethyl ether

RN: 621-34-1   **MP** (°C):
MW: 138.17   **BP** (°C):

Solubility (Moles/L)	Solubility (Grams/L)	Temp (°C)	Ref (#)	Evaluation (T P E A A)	Comments
1.000E-01	1.382E+01	25	B314	0 0 0 0 0	
1.003E-01	1.386E+01	30	B315	0 0 0 0 0	

1535. C$_8$H$_{10}$O$_2$
p-Ethoxyphenol
Hydroquinone monoethyl ether

RN: 622-62-8   **MP** (°C):   64.5–67.5
MW: 138.17   **BP** (°C):   131 at 9 mm Hg

Solubility (Moles/L)	Solubility (Grams/L)	Temp (°C)	Ref (#)	Evaluation (T P E A A)	Comments
5.097E-02	7.043E+00	20	R087	0 0 0 0 0	0.15M NaCl

1536. C$_8$H$_{10}$O$_2$

p-Dimethoxybenzene
4-Dimethoxybenzene

RN: 150-78-7   **MP** (°C):
MW: 138.17   **BP** (°C):

Solubility (Moles/L)	Solubility (Grams/L)	Temp (°C)	Ref (#)	Evaluation (T P E A A)	Comments
5.530E-05	7.641E-03	25	C316	0 0 0 0 0	0.1M NaCl

1537. C$_8$H$_{10}$O$_3$

1,3-Dimethyl ether pyrogallol
Pyrogallol-1,3-dimethylaether
2,6-Dimethoxyphenol

RN: 91-10-1   **MP** (°C): 56
MW: 154.17   **BP** (°C): 262

Solubility (Moles/L)	Solubility (Grams/L)	Temp (°C)	Ref (#)	Evaluation (T P E A A)	Comments
1.116E-01	1.720E+01	13	F300	1 0 0 0 2	

1538. C$_8$H$_{10}$O$_3$S

Benzene sulfonic acid ethyl ester
Ethyl benzenesulfonate
Ethyl phenylsulfonate

RN: 515-46-8   **MP** (°C):
MW: 186.23   **BP** (°C):

Solubility (Moles/L)	Solubility (Grams/L)	Temp (°C)	Ref (#)	Evaluation (T P E A A)	Comments
7.390E-03	1.376E+00	25	K097	2 0 2 2 2	

1539. C$_8$H$_{10}$O$_4$

Cyclohexene-1,4-dicarboxylic acid
Cyclohexen-(1)-dicarbonsaeure-(1,4)

RN: 2205-27-8   **MP** (°C): 312
MW: 170.17   **BP** (°C):

Solubility (Moles/L)	Solubility (Grams/L)	Temp (°C)	Ref (#)	Evaluation (T P E A A)	Comments
1.175E-03	2.000E-01	20	F300	1 0 0 0 0	

1540. C$_8$H$_{10}$O$_4$

2-Cyclohexene-1,2-dicarboxylic acid
Cyclohexen-(2)-dicarbonsaeure-(1,2)

RN: 38765-78-5   **MP (°C):**
MW: 170.17   **BP (°C):**

Solubility (Moles/L)	Solubility (Grams/L)	Temp (°C)	Ref (#)	Evaluation (T P E A A)	Comments
5.113E-02	8.700E+00	10	F300	1 0 0 0 1	

1541. C$_8$H$_{10}$O$_5$

Endothall
Endothal

RN: 145-73-3   **MP (°C):**   144
MW: 186.17   **BP (°C):**

Solubility (Moles/L)	Solubility (Grams/L)	Temp (°C)	Ref (#)	Evaluation (T P E A A)	Comments
4.883E-01	9.091E+01	20	B200	1 0 0 0 2	
5.372E-01	1.000E+02	20	M161	1 0 0 0 2	
4.883E-01	9.091E+01	ns	B100	0 0 0 0 0	
4.883E-01	9.091E+01	ns	C307	0 0 0 0 1	

1542. C$_8$H$_{10}$O$_8$

meso-1,2,3,4-Butanetetracarboxylic acid
1,2,3,4-Butanetetracarboxylic acid
Butanetetracarboxylic acid
1,2,3,4,-Butane tetracarboxylic acid

RN: 1703-58-8   **MP (°C):**   196
MW: 234.16   **BP (°C):**

Solubility (Moles/L)	Solubility (Grams/L)	Temp (°C)	Ref (#)	Evaluation (T P E A A)	Comments
6.606E-01	1.547E+02	25	M370	1 2 2 1 2	

1543. C$_8$H$_{11}$BrN$_2$O$_2$

Isocil
Uracil, 5-bromo-3-isopropyl-6-methyl-

RN: 314-42-1   **MP (°C):**   158–159
MW: 247.10   **BP (°C):**

Solubility (Moles/L)	Solubility (Grams/L)	Temp (°C)	Ref (#)	Evaluation (T P E A A)	Comments
8.701E-03	2.150E+00	25	B185	0 0 0 0 0	

1544. C₈H₁₁Cl₂NO

N,N-Diallyldichloroacetamide
Dichlormid
N,N-Diallyl dichloroacetamide
2,2-Dichloro-N,N-di-2-propenylacetamide
R 25788

RN: 37764-25-3   **MP** (°C):   5
MW: 208.09   **BP** (°C):

Solubility (Moles/L)	Solubility (Grams/L)	Temp (°C)	Ref (#)	Evaluation (T P E A A)	Comments
2.403E-02	5.000E+00	20	M161	1 0 0 0 0	
2.399E-02	4.992E+00	ns	S460	0 0 0 0 0	

1545. C₈H₁₁Cl₃O₆

Chloralose
1,2-O-(2,2,2-Trichloroethylidene)-α-D-glucofuranose
Anhydroglucochloral
Alfamat
Aphosal
Murex

RN: 15879-93-3   **MP** (°C):   187
MW: 309.53   **BP** (°C):

Solubility (Moles/L)	Solubility (Grams/L)	Temp (°C)	Ref (#)	Evaluation (T P E A A)	Comments
1.434E-02	4.440E+00	15	M161	1 0 0 0 2	

1546. C₈H₁₁N

Xylidine
N,N-Dimethylaniline
Dimethylaminobenzene
Benzenamine
Aminodimethylbenzene

RN: 121-69-7   **MP** (°C):   2
MW: 121.18   **BP** (°C):   194

Solubility (Moles/L)	Solubility (Grams/L)	Temp (°C)	Ref (#)	Evaluation (T P E A A)	Comments
9.120E-03	1.105E+00	25	C113	1 0 2 1 2	

1547. C$_8$H$_{11}$NO

Tyramine
Tyramin
4-Hydroxyphenylethylamine
4-(2-Aminoethyl)phenol
2-(p-Hydroxyphenyl)ethylamine

RN: 51-67-2   **MP** (°C):   164.5
MW: 137.18   **BP** (°C):   206

Solubility (Moles/L)	Solubility (Grams/L)	Temp (°C)	Ref (#)	Evaluation (T P E A A)	Comments
7.574E-02	1.039E+01	15	D041	1 0 0 0 2	
7.581E-02	1.040E+01	15	F300	1 0 0 0 2	

1548. C$_8$H$_{11}$NO

Phenylethanolamine
Phenyl ethanolamine
2-Anilinoethanol
β-Hydroxyethyl aniline
N-Phenylethanolamine
PEA

RN: 7568-93-6   **MP** (°C):   56.5
MW: 137.18   **BP** (°C):   286.0

Solubility (Moles/L)	Solubility (Grams/L)	Temp (°C)	Ref (#)	Evaluation (T P E A A)	Comments
3.192E-01	4.379E+01	20	M062	1 0 0 0 2	

1549. C$_8$H$_{11}$N$_2$O$_5$PS

Parathion-amino
Aminoparathion

RN: **MP** (°C):
MW: 278.23   **BP** (°C):

Solubility (Moles/L)	Solubility (Grams/L)	Temp (°C)	Ref (#)	Evaluation (T P E A A)	Comments
1.419E-03	3.948E-01	19.50	B169	2 2 1 1 2	

1550. C$_8$H$_{11}$N$_3$O$_3$S

Lamivudine
2(1H)-Pyrimidinone,4-amino-1-[2-(hydroxymethyl)-1,3-oxathiolan-5-yl]-,(2R-cis)
Epivir
3′-Thia-2′,3′-dideoxcytidine
(−)NGPB-21
(−) 2′-Deoxy-3′-thiacytidine

RN: 134678-17-4   **MP** (°C):
MW: 229.26   **BP** (°C):

Solubility (Moles/L)	Solubility (Grams/L)	Temp (°C)	Ref (#)	Evaluation (T P E A A)	Comments
3.053E-01	7.000E+01	ns	K444	0 0 0 0 0	
3.053E-01	7.000E+01	rt	B435	0 0 0 0 0	

1551. C$_8$H$_{11}$N$_5$O$_3$

Acyclovir
Acycloguanosine
9-(2-Hydroxyethoxymethyl)guanine
6H-Purin-6-one, 2-amino-1,9-dihydro-9-[(2-hydroxyethoxy)methyl]-
Cargosil
Zovirax

RN:	59277-89-3	MP (°C):
MW:	225.21	BP (°C):

Solubility (Moles/L)	Solubility (Grams/L)	Temp (°C)	Ref (#)	Evaluation (T P E A A)	Comments
6.216E-03	1.400E+00	21	B419	1 1 2 2 1	int
7.150E-03	1.610E+00	22	K443	0 0 0 0 0	
7.244E-03	1.631E+00	22	K445	0 0 0 0 0	
5.380E-03	1.212E+00	22.5	B422	2 0 2 2 2	
2.240E+00	5.045E+02	25	B443	0 0 0 0 0	
8.070E-03	1.817E+00	25	Z407	0 0 0 0 0	
4.440E-02	1.000E+01	ns	K444	0 0 0 0 0	
6.166E-03	1.389E+00	ns	R427	0 0 0 0 0	

1552. C$_8$H$_{12}$

4-Vinylcyclohexene
4-Vinyl-1-cyclohexene

RN:	100-40-3	MP (°C):	−101
MW:	108.18	BP (°C):	145

Solubility (Moles/L)	Solubility (Grams/L)	Temp (°C)	Ref (#)	Evaluation (T P E A A)	Comments
4.622E-04	5.000E-02	25	M001	2 1 2 2 1	

1553. C$_8$H$_{12}$ClNO

Allidochlor
CDAA
N,N-Diallyl-2-chloroacetamide
Randox
2-Chloro-N,N-diallylacetamide
CP 6343

RN:	93-71-0	MP (°C):
MW:	173.64	BP (°C):

Solubility (Moles/L)	Solubility (Grams/L)	Temp (°C)	Ref (#)	Evaluation (T P E A A)	Comments
1.113E-01	1.932E+01	22	J008	1 0 0 0 2	
1.113E-01	1.932E+01	25	B185	0 0 0 0 0	
1.135E-01	1.970E+01	25	G319	0 0 0 0 0	
1.135E-01	1.970E+01	25	M161	1 0 0 0 2	
1.129E-01	1.961E+01	ns	B100	0 0 0 0 0	
1.130E-01	1.962E+01	ns	F184	0 0 0 0 2	
1.129E-01	1.961E+01	ns	M061	0 0 0 0 0	
3.162E-01	5.491E+01	ns	M163	0 0 0 0 0	EFG

1554. C$_8$H$_{12}$N$_2$O$_2$S

N1-Dimethylsulfanilamide
p-Amino-N,N-dimethylbenzenesulfonamide
[(4-Aminophenyl)sulfonyl]dimethylamine
p-(Dimethylsulfamoyl)aniline

RN: 1709-59-7   **MP** (°C):
MW: 200.26   **BP** (°C):

Solubility (Moles/L)	Solubility (Grams/L)	Temp (°C)	Ref (#)	Evaluation (T P E A A)	Comments
3.130E-03	6.268E-01	37	K095	2 0 0 0 2	intrinsic

1555. C$_8$H$_{12}$N$_2$O$_2$S

5,5-Diethyl-2-thiobarbituric acid
4,6(1H,5H)-Pyrimidinedione, 5,5-diethyldihydro-2-thioxo
Barbituric acid, 5,5-diethyl-2-thio
Certodorm

RN: 77-32-7   **MP** (°C):
MW: 200.26   **BP** (°C):

Solubility (Moles/L)	Solubility (Grams/L)	Temp (°C)	Ref (#)	Evaluation (T P E A A)	Comments
6.810E-03	1.364E+00	25	P350	0 0 0 0 0	intrinsic

1556. C$_8$H$_{12}$N$_2$O$_3$

Barbital
5,5-Diethylbarbituric acid
Diethylmalonylurea

RN: 57-44-3   **MP** (°C):   190
MW: 184.20   **BP** (°C):

Solubility (Moles/L)	Solubility (Grams/L)	Temp (°C)	Ref (#)	Evaluation (T P E A A)	Comments
1.700E-02	3.131E+00	0	M143	1 2 1 1 0	
1.900E-02	3.500E+00	0	M143	1 2 1 1 2	
2.562E-02	4.720E+00	10	N007	1 2 2 2 2	form I
1.900E-02	3.500E+00	10	N007	1 2 2 2 2	form III
3.100E-02	5.710E+00	14	I006	1 0 0 0 1	
3.187E-02	5.870E+00	15	H018	0 0 0 0 0	
3.500E-02	6.447E+00	19	I006	1 0 0 0 1	
4.522E-02	8.330E+00	20	D041	1 0 0 0 1	
3.637E-02	6.700E+00	20	F300	1 0 0 0 1	
3.415E-02	6.290E+00	20	J030	1 2 2 2 2	
2.839E-02	5.230E+00	20	N007	1 2 2 2 2	form III
3.409E-02	6.280E+00	20	N007	1 2 2 2 2	form I
3.806E-02	7.011E+00	20	S146	2 2 2 1 2	form I
3.752E-02	6.912E+00	20	S146	2 2 2 1 2	form II
3.881E-02	7.149E+00	25	A023	1 0 0 1 2	
3.963E-02	7.300E+00	25	B011	2 0 0 1 0	
3.971E-02	7.314E+00	25	B065	1 1 1 1 1	
3.746E-02	6.900E+00	25	B167	1 1 0 0 1	pH 5.7

(continued)

1556. $C_8H_{12}N_2O_3$ (continued)

Solubility (Moles/L)	Solubility (Grams/L)	Temp (°C)	Ref (#)	Evaluation (T P E A A)	Comments
3.860E-02	7.110E+00	25	G003	1 1 1 1 2	pH 4.7
2.800E-02	5.158E+00	25	M143	1 2 1 1 2	
4.050E-02	7.460E+00	25	M310	2 2 2 2 2	
4.018E-02	7.401E+00	25	P350	0 0 0 0 0	intrinsic
4.239E-02	7.809E+00	25	S146	2 2 2 1 2	form II
4.010E-03	7.386E-01	25	V033	2 0 1 1 2	
4.010E-02	7.386E+00	25.00	T303	1 0 0 0 2	
4.300E-02	7.920E+00	27	I006	1 0 0 0 1	
4.300E-02	7.920E+00	30	G014	1 1 1 1 0	EFG, 0.003N H_2SO_4
2.704E-02	4.980E+00	30	H005	1 0 1 2 2	average of 4
4.408E-02	8.119E+00	30	H018	0 0 0 0 0	
4.400E-02	8.105E+00	30	I001	2 0 2 1 0	EFG, 0.003N H_2SO_4
4.260E-02	7.847E+00	30	K108	1 2 2 0 2	
4.425E-02	8.150E+00	30	N007	1 2 2 2 2	form I
4.207E-02	7.750E+00	30	N007	1 2 2 2 2	form III
4.720E-02	8.694E+00	30	S146	2 2 2 1 2	form I
4.618E-02	8.507E+00	30	S146	2 2 2 1 2	form II
5.162E-02	9.509E+00	35	S146	2 2 2 1 2	form I
5.184E-02	9.548E+00	35	S146	2 2 2 1 2	form II
5.150E-02	9.486E+00	35.00	T303	1 0 0 0 2	
4.843E-02	8.920E+00	36	A023	1 0 0 1 2	
5.152E-02	9.490E+00	37	J030	1 2 2 2 2	
5.300E-02	9.762E+00	37	K121	1 2 1 2 1	0.1N HCl
5.538E-02	1.020E+01	37	N007	1 2 2 2 2	form III
5.277E-02	9.720E+00	37	N007	1 2 2 2 2	form I
5.668E-02	1.044E+01	37	S146	2 2 2 1 2	form II
5.588E-02	1.029E+01	40	A023	1 0 0 1 1	
6.100E-01	1.124E+02	40	N008	1 0 1 1 2	*sic*
6.967E-02	1.283E+01	45	S146	2 2 2 1 2	form II
6.800E-02	1.253E+01	45.00	T303	1 0 0 0 2	
4.343E-01	8.000E+01	100	F300	1 0 0 0 1	
3.257E-02	6.000E+00	ns	T003	0 0 0 0 2	

1557. $C_8H_{12}O_2$

1-Epoxyethyl-3,4-epoxycyclohexane
Vinylcyclohexene dioxide

RN: 106-87-6   **MP** (°C): <−55
MW: 140.18   **BP** (°C): 227

Solubility (Moles/L)	Solubility (Grams/L)	Temp (°C)	Ref (#)	Evaluation (T P E A A)	Comments
1.103E+00	1.547E+02	20	I313	0 0 0 0 0	

1558. C$_8$H$_{12}$O$_4$
trans-Cyclohexane-1,2-dicarboxylic acid
trans-Cyclohexan-dicarbonsaeure-(1,2)

RN: 2305-32-0   **MP** (°C):
MW: 172.18   **BP** (°C):

Solubility (Moles/L)	Solubility (Grams/L)	Temp (°C)	Ref (#)	Evaluation (T P E A A)	Comments
1.162E-02	2.000E+00	20	F300	1 0 0 0 0	

1559. C$_8$H$_{12}$O$_4$
cis-Cyclohexane-1,2-dicarboxylic acid
cis-Cyclohexan-dicarbonsaeure-(1,2)

RN: 610-09-3   **MP** (°C):   193
MW: 172.18   **BP** (°C):

Solubility (Moles/L)	Solubility (Grams/L)	Temp (°C)	Ref (#)	Evaluation (T P E A A)	Comments
>1.16E-02	>2.00E+00	20	F300	1 0 0 0 0	

1560. C$_8$H$_{12}$O$_4$
trans-Cyclohexane-1,4-dicarboxylic acid
trans-Cyclohexan-dicarbonsaeure-(1,4)

RN: 619-82-9   **MP** (°C):
MW: 172.18   **BP** (°C):

Solubility (Moles/L)	Solubility (Grams/L)	Temp (°C)	Ref (#)	Evaluation (T P E A A)	Comments
4.646E-03	8.000E-01	17	F300	1 0 0 0 0	
7.550E-02	1.300E+01	100	F300	1 0 0 0 1	

1561. C$_8$H$_{13}$BrN$_2$O$_2$
α-Bromethylpropylaceturea

RN: **MP** (°C):
MW: 249.11   **BP** (°C):

Solubility (Moles/L)	Solubility (Grams/L)	Temp (°C)	Ref (#)	Evaluation (T P E A A)	Comments
1.645E-03	4.098E-01	20	O021	1 0 0 0 0	

1562. C$_8$H$_{13}$NO
Diaalylacetamide
α,α-Diallylacetamide
2-(2-Propenyl)4-pentenamide

RN: 60730-94-1   **MP** (°C):
MW: 139.20   **BP** (°C):

Solubility (Moles/L)	Solubility (Grams/L)	Temp (°C)	Ref (#)	Evaluation (T P E A A)	Comments
1.257E-01	1.750E+01	ns	H348	0 0 0 0 0	

1563. $C_8H_{13}N_2O_3PS$
Thionazin
O,O-Diethyl O-pyrazinyl thiophosphate
RN: 297-97-2   **MP** (°C):   −1.7
MW: 248.24   **BP** (°C):   80

Solubility (Moles/L)	Solubility (Grams/L)	Temp (°C)	Ref (#)	Evaluation (T P E A A)	Comments
4.592E-03	1.140E+00	25	M061	1 0 0 0 2	
4.592E-03	1.140E+00	27	M161	1 0 0 0 2	

1564. C_8H_{14}
1-Octyne
Hexylacetylene
n-Hexylacetylene
RN: 629-05-0   **MP** (°C):   −80
MW: 110.20   **BP** (°C):   127

Solubility (Moles/L)	Solubility (Grams/L)	Temp (°C)	Ref (#)	Evaluation (T P E A A)	Comments
2.178E-04	2.400E-02	25	M001	2 1 2 2 2	

1565. C_8H_{14}
2,2-Dimethyl-3-hexyne
1-Ethyl-2-tertbutylacetylene
RN: 4911-60-8   **MP** (°C):
MW: 110.20   **BP** (°C):

Solubility (Moles/L)	Solubility (Grams/L)	Temp (°C)	Ref (#)	Evaluation (T P E A A)	Comments
7.200E-04	7.934E-02	25	H039	1 2 2 2 1	

1566. $C_8H_{14}CINS_2$
Carbamic acid, diethyldithio-2chloroallyl ester
2-Chloroallyl diethyldithiocarbamate
CDEC
RN: 95-06-7   **MP** (°C):   <25
MW: 223.79   **BP** (°C):   128

Solubility (Moles/L)	Solubility (Grams/L)	Temp (°C)	Ref (#)	Evaluation (T P E A A)	Comments
4.469E-04	1.000E-01	25	B185	0 0 0 0 0	
4.111E-04	9.200E-02	25	B200	1 0 0 0 1	
4.469E-04	1.000E-01	25	F019	1 0 0 0 2	
4.111E-04	9.200E-02	25	G319	0 0 0 0 0	
4.111E-04	9.200E-02	25	M161	1 0 0 0 1	
4.468E-04	9.999E-02	ns	M061	0 0 0 0 0	approximate

1567. C₈H₁₄ClN₅

Atrazine

2-Chloro-4-ethylamino-6-isopropylamino-*s*-triazine

RN: 1912-24-9 **MP** (°C): 172

MW: 215.69 **BP** (°C):

Solubility (Moles/L)	Solubility (Grams/L)	Temp (°C)	Ref (#)	Evaluation (T P E A A)	Comments
1.020E-04	2.200E-02	0	B185	0 0 0 0 0	
1.390E-04	2.998E-02	1	G091	1 0 1 2 2	pH 6.0
5.000E-04	1.078E-01	2	B193	1 2 0 0 0	
1.410E-04	3.041E-02	8	G091	1 0 1 2 2	pH 6.0
1.530E-04	3.300E-02	20	A314	0 0 0 0 0	
1.345E-04	2.900E-02	20	C048	2 2 2 2 1	
1.391E-04	3.000E-02	20	E048	1 2 1 1 1	
1.391E-04	3.000E-02	20	F311	1 2 2 2 1	
1.580E-04	3.408E-02	20	G091	1 0 1 2 2	pH 6.0
1.298E-04	2.800E-02	20	M161	1 0 0 0 1	
1.391E-04	3.000E-02	20	N333	0 0 0 0 0	
3.245E-04	7.000E-02	21	B192	0 0 0 0 1	
3.245E-04	7.000E-02	21	G099	2 0 0 1 0	
3.245E-04	7.000E-02	22	M061	1 0 0 0 1	
1.530E-04	3.300E-02	25	H024	2 2 2 2 2	
1.386E-04	2.990E-02	25	H073	2 1 1 2 2	
1.530E-04	3.300E-02	25	P434	0 0 0 0 0	
3.245E-04	7.000E-02	27	B185	0 0 0 0 0	
1.530E-04	3.300E-02	27	B200	1 0 0 0 1	
1.970E-04	4.249E-02	29	G091	1 0 1 2 2	pH 6.0
4.530E-04	9.771E-02	50	G001	1 0 0 1 2	
1.484E-03	3.200E-01	85	B185	0 0 0 0 0	
3.245E-04	7.000E-02	ns	C101	0 0 0 0 1	
3.245E-04	7.000E-02	ns	G041	0 0 0 0 1	
3.245E-04	7.000E-02	ns	H112	0 0 0 0 1	
1.530E-04	3.300E-02	ns	J033	0 0 0 0 0	
3.941E-04	8.500E-02	ns	M110	0 0 0 0 0	EFG
1.609E-04	3.470E-02	ns	V414	0 0 0 0 0	

1568. C₈H₁₄N₂O₂

cis-N,N,N′,N′-Tetramethylfumaramide

2-Butenediamide, *N,N,N′,N′*-tetramethyl-, (Z)-

RN: 35075-35-5 **MP** (°C):

MW: 170.21 **BP** (°C):

Solubility (Moles/L)	Solubility (Grams/L)	Temp (°C)	Ref (#)	Evaluation (T P E A A)	Comments
1.730E+00	2.945E+02	30	K019	1 0 0 0 2	

1569. $C_8H_{14}N_4OS$
Metribuzin
4-Amino-6-*tert*-butyl-3-(methylthio)-as-triazin-5(4H)-one
Bayer 6159H
Lexone
Sencor
Sencorex

RN: 21087-64-9   **MP** (°C): 125.8
MW: 214.29   **BP** (°C):

Solubility (Moles/L)	Solubility (Grams/L)	Temp (°C)	Ref (#)	Evaluation (T P E A A)	Comments
5.600E-03	1.200E+00	20	M161	1 0 0 0 1	
5.693E-03	1.220E+00	22.5	G301	0 0 0 0 0	
4.662E-03	9.990E-01	ns	B100	0 0 0 0 0	
7.000E-03	1.500E+00	ns	M110	0 0 0 0 0	EFG

1570. $C_8H_{14}O$
Bicyclo[2.2.1]heptylcarbinol
2-Norcamphanemethanol

RN: 5240-72-2   **MP** (°C):
MW: 126.20   **BP** (°C):

Solubility (Moles/L)	Solubility (Grams/L)	Temp (°C)	Ref (#)	Evaluation (T P E A A)	Comments
7.916E-03	9.990E-01	ns	M061	0 0 0 0 0	

1571. $C_8H_{14}O_2$
2,4-Octadione
Valerylacetone

RN: 14090-87-0   **MP** (°C):
MW: 142.20   **BP** (°C):

Solubility (Moles/L)	Solubility (Grams/L)	Temp (°C)	Ref (#)	Evaluation (T P E A A)	Comments
2.760E-02	3.925E+00	25	M078	2 0 1 0 2	

1572. $C_8H_{14}O_2$
Cyclohexanol acetate
Hexalin acetate
Cyclohexyl acetate

RN: 622-45-7   **MP** (°C): <25
MW: 142.20   **BP** (°C):

Solubility (Moles/L)	Solubility (Grams/L)	Temp (°C)	Ref (#)	Evaluation (T P E A A)	Comments
1.123E-02	1.597E+00	20	D052	1 1 0 0 1	
2.033E-02	2.892E+00	23.50	O005	2 0 2 2 1	
2.138E-02	3.040E+00	ns	S460	0 0 0 0 0	

1573. C$_8$H$_{14}$O$_2$
6-Methyl-2,4-heptadione
2-Methyl-4,6-heptanedione
Isovalerylacetone

RN: 3002-23-1   **MP** (°C):   <25
MW: 142.20   **BP** (°C):

Solubility (Moles/L)	Solubility (Grams/L)	Temp (°C)	Ref (#)	Evaluation (T P E A A)	Comments
2.490E-02	3.541E+00	25	M078	2 0 1 0 2	

1574. C$_8$H$_{14}$O$_2$
3-Propyl-2,4-pentadione
3-Acetyl-2-hexanone

RN: 1540-35-8   **MP** (°C):   <25
MW: 142.20   **BP** (°C):

Solubility (Moles/L)	Solubility (Grams/L)	Temp (°C)	Ref (#)	Evaluation (T P E A A)	Comments
1.330E-01	1.891E+01	25	M078	2 0 1 0 2	

1575. C$_8$H$_{14}$O$_2$
5,5-Dimethyl-2,4-hexadione
Pivaloylacetone
Pivaloylacetylmethane

RN: 7307-04-2   **MP** (°C):
MW: 142.20   **BP** (°C):

Solubility (Moles/L)	Solubility (Grams/L)	Temp (°C)	Ref (#)	Evaluation (T P E A A)	Comments
2.340E-02	3.327E+00	25	M078	2 0 1 0 2	

1576. C$_8$H$_{14}$O$_2$S$_4$
Propyl dixanthogen
bis(1-Propyl) dixanthogen
Propyl xanthogen disulfide
Dipropyl dixanthogen
Dipropyl thioperoxydicarbonate
Dipropyl xanthogen disulfide

RN: 3750-28-5   **MP** (°C):
MW: 270.46   **BP** (°C):

Solubility (Moles/L)	Solubility (Grams/L)	Temp (°C)	Ref (#)	Evaluation (T P E A A)	Comments
1.500E-06	4.057E-04	25	H102	1 2 1 2 1	

1577. C$_8$H$_{14}$O$_4$
Suberic acid
Korksaeure

RN:	505-48-6	MP (°C):	142
MW:	174.20	BP (°C):	279

Solubility (Moles/L)	Solubility (Grams/L)	Temp (°C)	Ref (#)	Evaluation (T P E A A)	Comments
4.592E-03	8.000E-01	0	L041	1 0 0 1 0	
5.301E-03	9.234E-01	6.99	A340	0 0 0 0 0	
7.097E-03	1.236E+00	12.69	A340	0 0 0 0 0	
8.037E-03	1.400E+00	15	F300	1 0 0 0 1	
7.463E-03	1.300E+00	15	L041	1 0 0 1 1	
7.463E-03	1.300E+00	15	M051	1 0 0 0 1	
9.789E-03	1.705E+00	18.69	A340	0 0 0 0 0	
9.185E-03	1.600E+00	20	L041	1 0 0 1 1	
8.986E-03	1.565E+00	20	M171	1 0 0 0 0	
1.206E-01	2.100E+01	21	B040	1 0 1 1 2	*sic*
1.388E-02	2.417E+00	24.99	A340	0 0 0 0 0	
3.387E-02	5.900E+00	25	F300	1 0 0 0 1	
6.800E-02	1.185E+01	25	K040	1 0 2 1 2	*sic*
1.700E-02	2.961E+00	30	H021	1 0 1 1 0	EFG
1.890E-02	3.293E+00	32.49	A340	0 0 0 0 0	
2.045E-02	3.563E+00	34.49	A340	0 0 0 0 0	
2.583E-02	4.500E+00	35	L041	1 0 0 1 1	
2.326E-02	4.051E+00	39.99	A340	0 0 0 0 0	
2.682E-02	4.673E+00	44.49	A340	0 0 0 0 0	
5.626E-02	9.800E+00	50	L041	1 0 0 1 1	
3.198E-02	5.571E+00	50.19	A340	0 0 0 0 0	
3.534E-02	6.156E+00	52.69	A340	0 0 0 0 0	
5.551E-02	9.670E+00	61.49	A340	0 0 0 0 0	
6.422E-02	1.119E+01	63.99	A340	0 0 0 0 0	
1.274E-01	2.220E+01	65	L041	1 0 0 1 2	
8.182E-02	1.425E+01	70.09	A340	0 0 0 0 0	
1.156E-01	2.013E+01	76.49	A340	0 0 0 0 0	
1.386E-02	2.414E+00	rt	H431	0 0 0 0 0	

1578. C$_8$H$_{14}$O$_4$
Diethyl succinate
Butanedioic acid, diethyl ester

RN:	123-25-1	MP (°C):	−20
MW:	174.20	BP (°C):	217

Solubility (Moles/L)	Solubility (Grams/L)	Temp (°C)	Ref (#)	Evaluation (T P E A A)	Comments
1.089E-02	1.896E+00	ns	F014	0 0 0 0 2	

1579. C$_8$H$_{14}$O$_4$
Butylene glycol diacetate
1,4-Diacetoxybutane
Tetramethylene acetate

RN: 628-67-1   **MP** (°C):
MW: 174.20   **BP** (°C):

Solubility (Moles/L)	Solubility (Grams/L)	Temp (°C)	Ref (#)	Evaluation (T P E A A)	Comments
2.005E-01	3.494E+01	26	O012	1 2 1 1 2	
1.602E-01	2.790E+01	50	O012	1 2 1 1 2	
2.048E-01	3.568E+01	75	O012	1 2 1 1 2	

1580. C$_8$H$_{14}$O$_4$
Tetramethyl succinic acid
Tetramethyl-bernsteinsaeure

RN: 630-51-3   **MP** (°C):
MW: 174.20   **BP** (°C):

Solubility (Moles/L)	Solubility (Grams/L)	Temp (°C)	Ref (#)	Evaluation (T P E A A)	Comments
2.755E-02	4.800E+00	13.5	F300	1 0 0 0 1	

1581. C$_8$H$_{14}$O$_4$
Isoamylmalonic acid
Acide isoamylmalonique

RN: 616-87-5   **MP** (°C):
MW: 174.20   **BP** (°C):

Solubility (Moles/L)	Solubility (Grams/L)	Temp (°C)	Ref (#)	Evaluation (T P E A A)	Comments
2.210E+00	3.850E+02	0	M051	1 0 0 0 2	
2.974E+00	5.180E+02	15	M051	1 0 0 0 2	
3.490E+00	6.080E+02	25	M051	1 0 0 0 2	
4.788E+00	8.340E+02	50	M051	1 0 0 0 2	

1582. C$_8$H$_{14}$O$_4$
Propyl α-acetoxypropionate
Hydracrylic acid, propyl ester, acetate

RN: 20473-73-8   **MP** (°C):
MW: 174.20   **BP** (°C):

Solubility (Moles/L)	Solubility (Grams/L)	Temp (°C)	Ref (#)	Evaluation (T P E A A)	Comments
5.683E-02	9.900E+00	25	R006	2 2 0 1 1	

1583. C$_8$H$_{14}$O$_4$

Ethylene glycol dipropionate
1,2-Ethanediol, dipropanoate
1,2-bis(Propionyloxy)ethane

RN: 123-80-8 **MP** (°C):
MW: 174.20 **BP** (°C):

Solubility (Moles/L)	Solubility (Grams/L)	Temp (°C)	Ref (#)	Evaluation (T P E A A)	Comments
9.480E-02	1.651E+01	25	F064	1 0 0 0 2	
9.170E-03	1.597E+00	ns	F014	0 0 0 0 2	

1584. C$_8$H$_{14}$O$_5$

Propanoic acid, 2-[(propoxycarbonyl)oxy]-, methyl ester

RN: **MP** (°C):
MW: 190.20 **BP** (°C):

Solubility (Moles/L)	Solubility (Grams/L)	Temp (°C)	Ref (#)	Evaluation (T P E A A)	Comments
2.720E-02	5.173E+00	25	R007	0 0 0 0 0	

1585. C$_8$H$_{15}$ClN$_5$O

Hydroxyatrazine
4-(Ethylamino)-6-[(1-methylethyl)amino]-1,3,5-triazin-2(1H)-one
2-Hydroxy atrazine

RN: 2163-68-0 **MP** (°C):
MW: 232.69 **BP** (°C):

Solubility (Moles/L)	Solubility (Grams/L)	Temp (°C)	Ref (#)	Evaluation (T P E A A)	Comments
2.400E-04	5.585E-02	2	B193	1 2 0 0 1	

1586. C$_8$H$_{15}$NO

Pelletierine
Pelletierin

RN: 2858-66-4 **MP** (°C): <25
MW: 141.21 **BP** (°C): 195

Solubility (Moles/L)	Solubility (Grams/L)	Temp (°C)	Ref (#)	Evaluation (T P E A A)	Comments
3.541E-01	5.000E+01	20	F300	1 0 0 0 0	
3.372E-01	4.762E+01	25	D004	0 0 0 0 0	

1587. C$_8$H$_{15}$NO
Propylallylacetamide
2-Propyl-4-pentenamide
PAD
RN: 90204-40-3   **MP** (°C):
MW: 141.21   **BP** (°C):

Solubility (Moles/L)	Solubility (Grams/L)	Temp (°C)	Ref (#)	Evaluation (T P E A A)	Comments
6.727E-02	9.500E+00	37	H347	0 0 0 0 0	

1588. C$_8$H$_{15}$NO$_2$S
4-Thiazolidinecarboxylic acid, 2-butyl-
RN: 90205-28-0   **MP** (°C):
MW: 189.28   **BP** (°C):   355.7

Solubility (Moles/L)	Solubility (Grams/L)	Temp (°C)	Ref (#)	Evaluation (T P E A A)	Comments
5.700E-02	1.079E+01	21	B414	1 0 0 1 1	partial decomposition

1589. C$_8$H$_{15}$NO$_2$S
4-Thiazolidinecarboxylic acid, 2-(2-methylpropyl)-
4-Thiazolidine-4-carboxylic acid, 2-(2-isobutyl)-
RN: 215669-71-9   **MP** (°C):
MW: 189.28   **BP** (°C):   347.7

Solubility (Moles/L)	Solubility (Grams/L)	Temp (°C)	Ref (#)	Evaluation (T P E A A)	Comments
4.900E-02	9.275E+00	21	B414	1 0 0 1 1	partial decomposition

1590. C$_8$H$_{15}$N$_3$O$_2$
Isocarbamid
N-(2-Methylpropyl)-2-oxo-1-imidazolidinecarboxamide
RN: 30979-48-7   **MP** (°C):   95.5
MW: 185.23   **BP** (°C):

Solubility (Moles/L)	Solubility (Grams/L)	Temp (°C)	Ref (#)	Evaluation (T P E A A)	Comments
7.018E-03	1.300E+00	20	M161	1 0 0 0 1	

1591. $C_8H_{15}N_3O_7$
Streptozotocin
Streptozocin
D-2-Deoxy-2-(3-methyl-3-nitrosoureido)glucopyranose
RN: 18883-66-4   **MP** (°C):   115
MW: 265.22   **BP** (°C):

Solubility (Moles/L)	Solubility (Grams/L)	Temp (°C)	Ref (#)	Evaluation (T P E A A)	Comments
1.910E-02	5.066E+00	25	I307	0 0 0 0 0	

1592. $C_8H_{15}N_5O$
Simetone
2-Methoxy-4,6-bis(ethylamino)-s-triazine
s-Triazole, 2,4-bis(ethylamine)-6-methoxy-
RN: 673-04-1   **MP** (°C):   118-120
MW: 197.24   **BP** (°C):

Solubility (Moles/L)	Solubility (Grams/L)	Temp (°C)	Ref (#)	Evaluation (T P E A A)	Comments
1.622E-02	3.200E+00	21	B185	0 0 0 0 0	
1.622E-02	3.200E+00	21	B192	0 0 0 0 2	
1.622E-02	3.200E+00	21	G099	2 0 0 1 0	
3.550E-02	7.002E+00	50	G001	1 0 1 1 2	
1.622E-02	3.200E+00	ns	C101	0 0 0 0 1	

1593. $C_8H_{15}N_5O$
2-Methoxy-4-methylamino-6-isopropylamino-s-triazine
Noratone
RN: 3035-45-8   **MP** (°C):
MW: 197.24   **BP** (°C):

Solubility (Moles/L)	Solubility (Grams/L)	Temp (°C)	Ref (#)	Evaluation (T P E A A)	Comments
1.774E-02	3.500E+00	20	J033	0 0 0 0 0	
1.774E-02	3.500E+00	21	B192	0 0 0 0 2	

1594. $C_8H_{15}N_5S$
Desmetryne
N-Methyl-N'-(1-methylethyl)-6-(methylthio)-1,3,5-triazine-2,4-diamine
Semeron
Methylamino-4-methylthio-6-isopropylamino-1,3,5-triazine
Topusyn
Methylthio-4-isopropylamino-6-methylamino-s-triazine
RN: 1014-69-3   **MP** (°C):
MW: 213.31   **BP** (°C):

Solubility (Moles/L)	Solubility (Grams/L)	Temp (°C)	Ref (#)	Evaluation (T P E A A)	Comments
2.813E-03	6.000E-01	20	F311	1 2 2 2 1	
2.719E-03	5.800E-01	20	M161	1 0 0 0 2	
2.811E-03	5.996E-01	ns	B100	0 0 0 0 0	

(*continued*)

1594. C$_8$H$_{15}$N$_5$S (continued)

Solubility (Moles/L)	Solubility (Grams/L)	Temp (°C)	Ref (#)	Evaluation (T P E A A)	Comments
2.719E-03	5.800E-01	ns	J033	0 0 0 0 0	
2.719E-03	5.800E-01	ns	M061	0 0 0 0 2	

1595. C$_8$H$_{15}$N$_5$S
Simetryne
N,N'-Diethyl-6-(methylthio)-1,3,5-triazine-2,4-diamine
G-32911
bis(Ethylamino)-6-(methylthio)-s-triazine
Methylthio-4,6-bis(ethylamino)-s-triazine
Cymetrin

RN:	1014-70-6	**MP** (°C):	82
MW:	213.31	**BP** (°C):	

Solubility (Moles/L)	Solubility (Grams/L)	Temp (°C)	Ref (#)	Evaluation (T P E A A)	Comments
4.700E-03	1.003E+00	50	G001	1 0 1 1 1	
2.110E-03	4.500E-01	ns	C101	0 0 0 0 1	
2.110E-03	4.500E-01	ns	J033	0 0 0 0 0	
2.110E-03	4.500E-01	rt	M161	0 0 0 0 2	

1596. C$_8$H$_{15}$N$_7$O$_2$S$_3$
Famotidine
Amfamox
N'-(Aminosulfonyl)-3-(((2-((diaminomethylene)amino)-4-thiazolyl)methyl)thio)propanimidamide
Pepcid
Pepcidine
Pepcid PM

RN:	76824-35-6	**MP** (°C):	
MW:	337.45	**BP** (°C):	

Solubility (Moles/L)	Solubility (Grams/L)	Temp (°C)	Ref (#)	Evaluation (T P E A A)	Comments
3.260E-06	1.100E-03	25	A408	2 0 1 2 0	
3.311E-03	1.117E+00	ns	R427	0 0 0 0 0	

1597. C$_8$H$_{16}$
Cyclooctane

RN:	292-64-8	**MP** (°C):	10
MW:	112.22	**BP** (°C):	151

Solubility (Moles/L)	Solubility (Grams/L)	Temp (°C)	Ref (#)	Evaluation (T P E A A)	Comments
1.619E-03	1.817E-01	20	M337	2 1 2 2 2	*sic*
7.040E-05	7.900E-03	25	M001	2 1 2 2 1	
7.040E-05	7.900E-03	ns	H123	0 0 0 0 0	

1598. C$_8$H$_{16}$
Caprylene
1-Octene

RN:	111-66-0	MP (°C):	−102
MW:	112.22	BP (°C):	121.0

Solubility (Moles/L)	Solubility (Grams/L)	Temp (°C)	Ref (#)	Evaluation (T P E A A)	Comments
3.208E-05	3.600E-03	23	C332	0 0 0 0 0	
2.406E-05	2.700E-03	25	M001	2 1 2 2 1	
3.650E-05	4.096E-03	25	M342	1 0 1 1 2	

1599. C$_8$H$_{16}$
1,4-Dimethylcyclohexane
p-Dimethylcyclohexane

RN:	589-90-2	MP (°C):	−87
MW:	112.22	BP (°C):	120

Solubility (Moles/L)	Solubility (Grams/L)	Temp (°C)	Ref (#)	Evaluation (T P E A A)	Comments
3.422E-05	3.840E-03	25	K119	1 0 0 0 2	

1600. C$_8$H$_{16}$
cis-1,2-Dimethylcyclohexane
1-cis-2-Dimethylcyclohexane

RN:	2207-01-4	MP (°C):	−50
MW:	112.22	BP (°C):	129

Solubility (Moles/L)	Solubility (Grams/L)	Temp (°C)	Ref (#)	Evaluation (T P E A A)	Comments
6.773E-05	7.600E-03	20	M337	2 1 2 2 1	
5.347E-05	6.000E-03	25	M001	2 1 2 2 1	

1601. C$_8$H$_{16}$
n-Propylcyclopentane
1-Propylcyclopentane

RN:	2040-96-2	MP (°C):	−117
MW:	112.22	BP (°C):	

Solubility (Moles/L)	Solubility (Grams/L)	Temp (°C)	Ref (#)	Evaluation (T P E A A)	Comments
1.818E-05	2.040E-03	25	K119	1 0 0 0 2	
1.818E-05	2.040E-03	25	P051	2 1 1 2 2	
1.818E-05	2.040E-03	25.00	P007	2 1 2 2 2	

1602. C$_8$H$_{16}$
trans-1,2-Dimethylcyclohexane
1,2-*trans*-Dimethylcyclohexane
RN: 6876-23-9   **MP** (°C):   −89
MW: 112.22   **BP** (°C):   123

Solubility (Moles/L)	Solubility (Grams/L)	Temp (°C)	Ref (#)	Evaluation (T P E A A)	Comments
4.634E-05	5.200E-03	20	M337	2 1 2 2 1	
4.444E-05	4.987E-03	30.2	M447	0 0 0 0 0	
1.061E-04	1.191E-02	70.3	M447	0 0 0 0 0	
2.611E-04	2.930E-02	100.7	M447	0 0 0 0 0	
6.000E-04	6.733E-02	131.0	M447	0 0 0 0 0	
1.239E-03	1.390E-01	151.0	M447	0 0 0 0 0	
1.977E-03	2.219E-01	170.1	M447	0 0 0 0 0	

1603. C$_8$H$_{16}$
trans-1,4-Dimethylcyclohexane
1,4-Transdimethylcyclohexane
RN: 2207-04-7   **MP** (°C):   −37
MW: 112.22   **BP** (°C):   115

Solubility (Moles/L)	Solubility (Grams/L)	Temp (°C)	Ref (#)	Evaluation (T P E A A)	Comments
3.422E-05	3.840E-03	25	P051	2 1 1 2 2	
3.422E-05	3.840E-03	25.00	P007	2 1 2 2 2	

1604. C$_8$H$_{16}$
1,2-Dimethylcyclohexane (*cis* + *trans*)
Cyclohexane, 1,2-dimethyl- (*cis/trans*)
1,2-Dimethylcyclohexane
RN: 583-57-3   **MP** (°C):
MW: 112.22   **BP** (°C):   124 C

Solubility (Moles/L)	Solubility (Grams/L)	Temp (°C)	Ref (#)	Evaluation (T P E A A)	Comments
6.056E-05	6.795E-03	30.0	M447	0 0 0 0 0	
1.200E-04	1.347E-02	70.0	M447	0 0 0 0 0	
2.422E-04	2.718E-02	100.2	M447	0 0 0 0 0	
5.483E-04	6.153E-02	130.5	M447	0 0 0 0 0	
1.089E-03	1.222E-01	150.5	M447	0 0 0 0 0	
2.422E-03	2.717E-01	170.5	M447	0 0 0 0 0	

1605. C$_8$H$_{16}$
1,1,3-Trimethylcyclopentane
Cyclopentane, 1,1,3-trimethyl-
RN: 4516-69-2   **MP** (°C):   −142.4
MW: 112.22   **BP** (°C):   104.9

Solubility (Moles/L)	Solubility (Grams/L)	Temp (°C)	Ref (#)	Evaluation (T P E A A)	Comments
3.324E-05	3.730E-03	25	K119	1 0 0 0 2	
3.324E-05	3.730E-03	25	P051	2 1 1 2 2	
3.324E-05	3.730E-03	25.00	P007	2 1 2 2 2	

1606. C$_8$H$_{16}$
Ethyl cyclohexane
Cyclohexane, ethyl-
RN: 1678-91-7   **MP** (°C):
MW: 112.22   **BP** (°C):   131.8

Solubility (Moles/L)	Solubility (Grams/L)	Temp (°C)	Ref (#)	Evaluation (T P E A A)	Comments
5.614E-05	6.300E-03	20	M337	2 1 2 2 1	
3.883E-05	4.358E-03	30.3	M447	0 0 0 0 0	
7.833E-05	8.790E-03	70.4	M447	0 0 0 0 0	
2.511E-04	2.818E-02	100.5	M447	0 0 0 0 0	
6.055E-04	6.795E-02	131.0	M447	0 0 0 0 0	
9.871E-04	1.108E-01	151.2	M447	0 0 0 0 0	
1.633E-03	1.833E-01	170.8	M447	0 0 0 0 0	

1607. C$_8$H$_{16}$Br$_2$
1,8-Dibromooctane
Octamethylene dibromide
RN: 4549-32-0   **MP** (°C):   15–16
MW: 272.03   **BP** (°C):   270–272

Solubility (Moles/L)	Solubility (Grams/L)	Temp (°C)	Ref (#)	Evaluation (T P E A A)	Comments
7.389E-06	2.010E-03	1.0	S464	0 0 0 0 0	
7.278E-06	1.980E-03	1.0	S464	0 0 0 0 0	
7.462E-06	2.030E-03	4.9	S464	0 0 0 0 0	
7.646E-06	2.080E-03	4.9	S464	0 0 0 0 0	
8.565E-06	2.330E-03	10.0	S464	0 0 0 0 0	
8.896E-06	2.420E-03	14.9	S464	0 0 0 0 0	
8.528E-06	2.320E-03	14.9	S464	0 0 0 0 0	
9.374E-06	2.550E-03	19.9	S464	0 0 0 0 0	
1.062E-05	2.890E-03	25	S464	0 0 0 0 0	
1.066E-05	2.900E-03	25.0	S464	0 0 0 0 0	
1.044E-05	2.840E-03	25.0	S464	0 0 0 0 0	
1.209E-05	3.290E-03	30.0	S464	0 0 0 0 0	
1.239E-05	3.370E-03	30.0	S464	0 0 0 0 0	
1.213E-05	3.300E-03	30.1	S464	0 0 0 0 0	

(continued)

1607. C$_8$H$_{16}$Br$_2$ (continued)

Solubility (Moles/L)	Solubility (Grams/L)	Temp (°C)	Ref (#)	Evaluation (T P E A A)	Comments
1.261E-05	3.430E-03	34.9	S464	0 0 0 0 0	
1.309E-05	3.560E-03	35.0	S464	0 0 0 0 0	
1.430E-05	3.890E-03	40.1	S464	0 0 0 0 0	
1.386E-05	3.770E-03	40.1	S464	0 0 0 0 0	

1608. C$_8$H$_{16}$Cl$_2$
1,8-Dichlorooctane

RN: 2162-99-4   **MP** (°C): −8
MW: 183.12   **BP** (°C): 243

Solubility (Moles/L)	Solubility (Grams/L)	Temp (°C)	Ref (#)	Evaluation (T P E A A)	Comments
2.441E-05	4.470E-03	3.6	S464	0 0 0 0 0	
3.047E-05	5.580E-03	5.1	S464	0 0 0 0 0	
3.014E-05	5.520E-03	5.1	S464	0 0 0 0 0	
3.069E-05	5.620E-03	9.9	S464	0 0 0 0 0	
3.233E-05	5.920E-03	15.1	S464	0 0 0 0 0	
3.211E-05	5.880E-03	25.1	S464	0 0 0 0 0	
3.255E-05	5.960E-03	25.1	S464	0 0 0 0 0	
3.222E-05	5.900E-03	25.1	S464	0 0 0 0 0	
3.517E-05	6.440E-03	30.3	S464	0 0 0 0 0	
3.375E-05	6.180E-03	30.3	S464	0 0 0 0 0	
3.823E-05	7.000E-03	35.2	S464	0 0 0 0 0	
3.828E-05	7.010E-03	35.3	S464	0 0 0 0 0	
3.970E-05	7.270E-03	40.1	S464	0 0 0 0 0	

1609. C$_8$H$_{16}$N$_2$O$_2$
N,N,N′,N′-Tetramethylsuccinamide
N,N,N′,N′-Tetramethylbutanediamide

RN: 7334-51-2   **MP** (°C):
MW: 172.23   **BP** (°C):

Solubility (Moles/L)	Solubility (Grams/L)	Temp (°C)	Ref (#)	Evaluation (T P E A A)	Comments
3.188E+00	5.490E+02	30	K004	1 0 0 0 2	

1610. C$_8$H$_{16}$N$_2$O$_4$S$_2$
DL-Homocystine
DL-*meso*-Homocystine
Oxidized DL-homocysteine

RN: 870-93-9   **MP** (°C): 264
MW: 268.36   **BP** (°C):

Solubility (Moles/L)	Solubility (Grams/L)	Temp (°C)	Ref (#)	Evaluation (T P E A A)	Comments
7.451E-04	2.000E-01	25	D041	1 0 0 0 0	

1611. $C_8H_{16}N_6$
Pentamethylmelamine
1-(Methylamino)-3,5-bis(dimethylamino)-s-triazine
RN: 16268-62-5   **MP** (°C):   107.0
MW: 196.26   **BP** (°C):

Solubility (Moles/L)	Solubility (Grams/L)	Temp (°C)	Ref (#)	Evaluation (T P E A A)	Comments
1.679E-04	3.295E-02	25	B386	0 0 0 0 0	
1.010E-02	1.982E+00	25	B386	0 0 0 0 0	
1.101E-02	2.160E+00	25	C051	1 2 1 1 2	pH 7

1612. $C_8H_{16}N_6O$
N2-Hydroxy-N2,N4,N4,N6,N6-pentamethylmelamine
1-(Hydroxylamino)-3,5-bis(dimethylamino)-s-triazine
RN: 64124-14-7   **MP** (°C):   110.0
MW: 212.26   **BP** (°C):

Solubility (Moles/L)	Solubility (Grams/L)	Temp (°C)	Ref (#)	Evaluation (T P E A A)	Comments
4.412E-03	9.365E-01	25	B386	0 0 0 0 0	
4.259E-03	9.040E-01	25	C051	1 2 1 1 2	pH 7

1613. $C_8H_{16}O$
Cyclooctanol
RN: 696-71-9   **MP** (°C):   15
MW: 128.22   **BP** (°C):   106–108 at 22 mm Hg

Solubility (Moles/L)	Solubility (Grams/L)	Temp (°C)	Ref (#)	Evaluation (T P E A A)	Comments
5.129E-02	6.576E+00	ns	S460	0 0 0 0 0	

1614. $C_8H_{16}O$
1-Octen-3-ol
3-Octenol
Flowtron mosquito attractant
Matsuka alcohol
Vinyl hexanol
RN: 3391-86-4   **MP** (°C):
MW: 128.22   **BP** (°C):   174

Solubility (Moles/L)	Solubility (Grams/L)	Temp (°C)	Ref (#)	Evaluation (T P E A A)	Comments
1.557E-02	1.996E+00	25	D425	0 0 0 0 0	

1615. C₈H₁₆O

Hexyl methyl ketone
2-Octanone
Octan-2-one

RN: 111-13-7   **MP** (°C):   −16.0
MW: 128.22   **BP** (°C):   172.5

Solubility (Moles/L)	Solubility (Grams/L)	Temp (°C)	Ref (#)	Evaluation (T P E A A)	Comments
3.276E-02	4.200E+00	0	C423	0 0 0 0 0	
2.574E-02	3.300E+00	4	C423	0 0 0 0 0	
1.716E-02	2.200E+00	10	C423	0 0 0 0 0	
7.013E-03	8.992E-01	20	D052	1 1 0 0 0	
1.014E-02	1.300E+00	25	C435	0 0 0 0 0	

1616. C₈H₁₆O

Caprylic aldehyde
Octaldehyde
n-Octanal

RN: 124-13-0   **MP** (°C):
MW: 128.22   **BP** (°C):   163.4

Solubility (Moles/L)	Solubility (Grams/L)	Temp (°C)	Ref (#)	Evaluation (T P E A A)	Comments
4.368E-03	5.600E-01	25	A049	1 0 0 0 1	
1.887E-03	2.420E-01	25	L450	0 0 0 0 0	

1617. C₈H₁₆O₂

Ethyl hexanoate
Ethyl butyl acetate
Ethyl caproate
Ethyl n-hexanoate
Ethyl caproate (Nat. C-6 ethyl ester)

RN: 123-66-0   **MP** (°C):
MW: 144.22   **BP** (°C):   168

Solubility (Moles/L)	Solubility (Grams/L)	Temp (°C)	Ref (#)	Evaluation (T P E A A)	Comments
3.120E-03	4.500E-01	0	C423	0 0 0 0 0	
3.606E-03	5.200E-01	4	C423	0 0 0 0 0	
3.952E-03	5.700E-01	10	C423	0 0 0 0 0	
4.507E-03	6.500E-01	25	C435	0 0 0 0 0	
4.467E-03	6.442E-01	ns	S460	0 0 0 0 0	

1618. C$_8$H$_{16}$O$_2$
Valproic acid
Vistora
Valporal
Convulex
Depakote
Dalpro

RN: 99-66-1 **MP (°C):** 120–130
MW: 144.22 **BP (°C):** 220

Solubility (Moles/L)	Solubility (Grams/L)	Temp (°C)	Ref (#)	Evaluation (T P E A A)	Comments
9.014E-03	1.300E+00	ns	K444	0 0 0 0 0	
1.380E-02	1.991E+00	ns	S460	0 0 0 0 0	

1619. C$_8$H$_{16}$O$_2$
2-Ethylhexoic acid
2-Ethyl-1-hexanoic acid
3-Heptanecarboxylic acid
Butylethylacetic acid

RN: 149-57-5 **MP (°C):**
MW: 144.22 **BP (°C):** 228

Solubility (Moles/L)	Solubility (Grams/L)	Temp (°C)	Ref (#)	Evaluation (T P E A A)	Comments
1.039E-02	1.498E+00	25	O011	1 0 1 1 1	

1620. C$_8$H$_{16}$O$_2$
3-Hydroxy-2,2,5,5-tetramethyltetrahydrofuran
3-Furanol, 2-ethyltetrahydro-2,2,5,5-tetramethyl-

RN: 29839-74-5 **MP (°C):**
MW: 144.22 **BP (°C):**

Solubility (Moles/L)	Solubility (Grams/L)	Temp (°C)	Ref (#)	Evaluation (T P E A A)	Comments
6.304E-01	9.091E+01	rt	B066	0 2 0 0 1	

1621. C$_8$H$_{16}$O$_2$
n-Butyl n-butyrate
Butyl butyrate

RN: 109-21-7 **MP (°C):**
MW: 144.22 **BP (°C):** 165

Solubility (Moles/L)	Solubility (Grams/L)	Temp (°C)	Ref (#)	Evaluation (T P E A A)	Comments
3.465E-03	4.998E-01	20	D052	1 1 0 0 0	

1622. C₈H₁₆O₂
Pentyl propionate
Propanoic acid pentyl ester
Amyl *n*-propanoate
n-Pentyl propionate
RN: 624-54-4   **MP** (°C):
MW: 144.22   **BP** (°C):

Solubility (Moles/L)	Solubility (Grams/L)	Temp (°C)	Ref (#)	Evaluation (T P E A A)	Comments
4.900E-03	7.067E-01	20	S006	1 0 0 0 1	

1623. C₈H₁₆O₂
3-Hydroxy-2,2-diethyltetrahydrofuran
RN: **MP** (°C):
MW: 144.22   **BP** (°C):

Solubility (Moles/L)	Solubility (Grams/L)	Temp (°C)	Ref (#)	Evaluation (T P E A A)	Comments
1.360E-01	1.961E+01	rt	B066	0 2 0 0 0	

1624. C₈H₁₆O₂
sec-Hexyl acetate
Methyl amyl acetate
RN: 108-84-9   **MP** (°C):   −64
MW: 144.22   **BP** (°C):   140

Solubility (Moles/L)	Solubility (Grams/L)	Temp (°C)	Ref (#)	Evaluation (T P E A A)	Comments
5.543E-03	7.994E-01	20	D052	1 1 0 0 0	

1625. C₈H₁₆O₂
Isobutyl isobutyrate
Isobutyl 2-methylpropanoate
2-Methylpropyl 2-methylpropanoate
IBIB
RN: 97-85-8   **MP** (°C):   −81
MW: 144.22   **BP** (°C):   147

Solubility (Moles/L)	Solubility (Grams/L)	Temp (°C)	Ref (#)	Evaluation (T P E A A)	Comments
3.952E-03	5.700E-01	25	A049	1 0 0 0 1	

1626. C$_8$H$_{16}$O$_2$

3-Hydroxy-2-ethyl-5,5-dimethyltetrahydrofuran
3-Furanol, 2-ethyltetrahydro-5,5-dimethyl-

RN: 29839-59-6   **MP** (°C):
MW: 144.22   **BP** (°C):

Solubility (Moles/L)	Solubility (Grams/L)	Temp (°C)	Ref (#)	Evaluation (T P E A A)	Comments
3.302E-01	4.762E+01	rt	B066	0 2 0 0 0	

1627. C$_8$H$_{16}$O$_2$

3-Hydroxy-5-ethyl-2,5-dimethyltetrahydrofuran
3-Furanol, 2-ethyltetrahydro-2,5-dimethyl-

RN: 29839-60-9   **MP** (°C):
MW: 144.22   **BP** (°C):

Solubility (Moles/L)	Solubility (Grams/L)	Temp (°C)	Ref (#)	Evaluation (T P E A A)	Comments
3.467E+00	5.000E+02	rt	B066	0 2 0 0 2	

1628. C$_8$H$_{16}$O$_2$

3-Hydroxy-5-methyl-5-propyltetrahydrofuran
3-Furanol, 2-ethyltetrahydro-5-methy-5-propyl-

RN: 29839-52-9   **MP** (°C):
MW: 144.22   **BP** (°C):

Solubility (Moles/L)	Solubility (Grams/L)	Temp (°C)	Ref (#)	Evaluation (T P E A A)	Comments
1.360E-01	1.961E+01	rt	B066	0 2 0 0 0	

1629. C$_8$H$_{16}$O$_2$

Hexyl acetate
2-Ethyl butyl acetate

RN: 142-92-7   **MP** (°C):   −80
MW: 144.22   **BP** (°C):   168

Solubility (Moles/L)	Solubility (Grams/L)	Temp (°C)	Ref (#)	Evaluation (T P E A A)	Comments
4.158E-03	5.996E-01	20	D052	1 1 0 0 0	
3.540E-03	5.105E-01	25	M124	2 1 2 2 2	

1630. C$_8$H$_{16}$O$_2$

Caprylic acid

Caprylsaure

RN: 124-07-2   **MP** (°C):   16.7

MW: 144.22   **BP** (°C):   239.7

Solubility (Moles/L)	Solubility (Grams/L)	Temp (°C)	Ref (#)	Evaluation (T P E A A)	Comments
3.051E-03	4.400E-01	0	B136	1 0 2 1 1	
4.993E-03	7.200E-01	15	F300	1 0 0 0 1	
4.715E-03	6.800E-01	20	B136	1 0 2 1 1	
4.712E-03	6.795E-01	20	D041	1 0 0 0 1	
4.712E-03	6.795E-01	20.0	R001	1 1 1 1 1	
5.478E-03	7.900E-01	30	B136	1 0 2 1 1	
5.471E-03	7.890E-01	30	E005	2 1 1 2 2	
5.474E-03	7.894E-01	30.0	R001	1 1 1 1 1	
5.845E-03	8.430E-01	40	E005	2 1 1 2 2	
6.587E-03	9.500E-01	45	B136	1 0 2 1 1	
6.581E-03	9.491E-01	45.0	R001	1 1 1 1 1	
6.539E-03	9.430E-01	50	E005	2 1 1 2 2	
7.835E-03	1.130E+00	60	B136	1 0 2 1 2	
7.426E-03	1.071E+00	60	E005	2 1 1 2 2	
7.827E-03	1.129E+00	60.0	R001	1 1 1 1 2	
1.803E-02	2.600E+00	100	F300	1 0 0 0 1	
3.050E-03	4.398E-01	.0	R001	1 1 1 1 1	

1631. C$_8$H$_{16}$O$_3$

n-Butyl β-methoxypropionate

Propanoic acid, 3-methoxy-, butyl ester

RN: 4195-88-4   **MP** (°C):

MW: 160.21   **BP** (°C):

Solubility (Moles/L)	Solubility (Grams/L)	Temp (°C)	Ref (#)	Evaluation (T P E A A)	Comments
6.117E-02	9.800E+00	25	R034	0 0 0 0 1	

1632. C$_8$H$_{16}$O$_3$

Amyl lactate

n-Pentyl lactate

RN: 6382-06-5   **MP** (°C):

MW: 160.21   **BP** (°C):

Solubility (Moles/L)	Solubility (Grams/L)	Temp (°C)	Ref (#)	Evaluation (T P E A A)	Comments
6.242E-02	1.000E+01	25	R006	2 2 0 1 2	

1633. C$_8$H$_{16}$O$_3$
Methyl β-*n*-butoxypropionate
Butanoic acid, 3-methoxy-3-oxopropyl ester
RN: 40326-33-8   **MP** (°C):
MW: 160.21   **BP** (°C):

Solubility (Moles/L)	Solubility (Grams/L)	Temp (°C)	Ref (#)	Evaluation (T P E A A)	Comments
5.076E-02	8.133E+00	25	R034	0 0 0 0 1	

1634. C$_8$H$_{16}$O$_3$
n-Propyl β-ethoxypropionate
Propionic acid, 3-ethoxy-, propyl ester
RN: 14144-34-4   **MP** (°C):
MW: 160.21   **BP** (°C):

Solubility (Moles/L)	Solubility (Grams/L)	Temp (°C)	Ref (#)	Evaluation (T P E A A)	Comments
9.466E-02	1.517E+01	25	D002	1 2 1 1 2	

1635. C$_8$H$_{16}$O$_3$
Butylcellosolve acetate
Ethylene glycol monobutyl ether acetate
Ektasolve EB acetate
n-Butyl cellosolve acetate
Ethylene glycol mono-*n*-butyl ether acetate
RN: 112-07-2   **MP** (°C):
MW: 160.21   **BP** (°C):

Solubility (Moles/L)	Solubility (Grams/L)	Temp (°C)	Ref (#)	Evaluation (T P E A A)	Comments
5.567E-02	8.920E+00	20	D052	1 1 0 0 0	

1636. C$_8$H$_{16}$O$_3$
2,2,5,5-Tetramethyltetrahydrofuran-3,4-diol
3,4-Furandiol, tetrahydro-2,2,5,5-tetramethyl-
RN: 29839-67-6   **MP** (°C):
MW: 160.21   **BP** (°C):

Solubility (Moles/L)	Solubility (Grams/L)	Temp (°C)	Ref (#)	Evaluation (T P E A A)	Comments
5.674E-01	9.091E+01	rt	B066	0 2 0 0 1	

1637. C$_8$H$_{16}$O$_3$S

1,2-Oxathiolane, 5-pentyl-, 2,2-dioxide
1-Octanesulfonic acid, 3-hydroxy-, γ-sultone

RN: 5633-87-4   **MP** (°C):
MW: 192.28   **BP** (°C):

Solubility (Moles/L)	Solubility (Grams/L)	Temp (°C)	Ref (#)	Evaluation (T P E A A)	Comments
1.300E-03	2.499E-01	20	B058	1 2 0 0 1	
7.938E-02	1.526E+01	100	B058	1 2 0 0 2	

1638. C$_8$H$_{16}$O$_4$

Metaldehyde
Acetaldehyde homopolymer
Acetaldehyde tetramer

RN: 9002-91-9   **MP** (°C):   112
MW: 176.21   **BP** (°C):

Solubility (Moles/L)	Solubility (Grams/L)	Temp (°C)	Ref (#)	Evaluation (T P E A A)	Comments
1.135E-03	2.000E-01	17	M161	1 0 0 0 2	

1639. C$_8$H$_{17}$Cl

1-Chlorooctane
1-Octylchloride
n-Octyl chloride
Octyl chloride

RN: 111-85-3   **MP** (°C):   −61
MW: 148.68   **BP** (°C):   182

Solubility (Moles/L)	Solubility (Grams/L)	Temp (°C)	Ref (#)	Evaluation (T P E A A)	Comments
2.270E+01	3.375E+03	5.0	S454	0 0 0 0 0	
2.210E+01	3.286E+03	10.0	S454	0 0 0 0 0	
2.260E+01	3.360E+03	9.9	S454	0 0 0 0 0	
2.350E+01	3.494E+03	9.9	S454	0 0 0 0 0	
2.370E+01	3.524E+03	9.9	S454	0 0 0 0 0	
2.540E+01	3.776E+03	19.1	S454	0 0 0 0 0	
2.470E+01	3.672E+03	25.0	S454	0 0 0 0 0	
2.620E+01	3.895E+03	25.1	S454	0 0 0 0 0	
2.580E+01	3.836E+03	25.2	S454	0 0 0 0 0	
2.710E+01	4.029E+03	30.0	S454	0 0 0 0 0	
2.700E+01	4.014E+03	34.8	S454	0 0 0 0 0	
2.800E+01	4.163E+03	35.1	S454	0 0 0 0 0	
2.690E+01	3.999E+03	35.1	S454	0 0 0 0 0	
2.750E+01	4.089E+03	40.0	S454	0 0 0 0 0	

1640. C₈H₁₇N

D-Coniine
α-Propylpiperidine
D-Coniin
Coniine

RN:	458-88-8	MP (°C):	–2
MW:	127.23	BP (°C):	166–167

Solubility (Moles/L)	Solubility (Grams/L)	Temp (°C)	Ref (#)	Evaluation (T P E A A)	Comments
1.415E-01	1.800E+01	19.5	F300	1 0 0 0 1	
7.782E-02	9.901E+00	25	D004	0 0 0 0 0	

1641. C₈H₁₇NO

Ethylbutylacetamide
2-Ethylhexanamide
EBD

RN:	4164-92-5	MP (°C):	
MW:	143.23	BP (°C):	

Solubility (Moles/L)	Solubility (Grams/L)	Temp (°C)	Ref (#)	Evaluation (T P E A A)	Comments
3.072E-02	4.400E+00	37	H347	0 0 0 0 0	

1642. C₈H₁₇NO

Ethylisobutylacetamide
2-Ethyl-4-methylpentanamide
EID

RN:	130482-28-9	MP (°C):	
MW:	143.23	BP (°C):	

Solubility (Moles/L)	Solubility (Grams/L)	Temp (°C)	Ref (#)	Evaluation (T P E A A)	Comments
3.002E-02	4.300E+00	ns	H348	0 0 0 0 0	

1643. C₈H₁₇NO

Caprylylamide
Caprylsaeure-amid

RN:	629-01-6	MP (°C):	
MW:	143.23	BP (°C):	

Solubility (Moles/L)	Solubility (Grams/L)	Temp (°C)	Ref (#)	Evaluation (T P E A A)	Comments
3.288E-02	4.710E+00	100	F300	1 0 0 0 2	

1644. C$_8$H$_{17}$NO
Propylisopropylacetamide
2-Isopropyl-2-propylacetamide
2-Isopropylvaleramide
PID

RN: 6098-19-7   **MP** (°C):
MW: 143.23   **BP** (°C):

Solubility (Moles/L)	Solubility (Grams/L)	Temp (°C)	Ref (#)	Evaluation (T P E A A)	Comments
2.444E-02	3.500E+00	37	H347	0 0 0 0 0	

1645. C$_8$H$_{17}$NO
2-Isopropyl-3-methyl-butyramide
3-Methyl-2-(1-methylethyl)butanamide
Diisopropylacetamide

RN: 5440-65-3   **MP** (°C):
MW: 143.23   **BP** (°C):

Solubility (Moles/L)	Solubility (Grams/L)	Temp (°C)	Ref (#)	Evaluation (T P E A A)	Comments
3.002E-02	4.300E+00	ns	H348	0 0 0 0 0	

1646. C$_8$H$_{17}$NO
Dimethylbutylacetamide
2,2-Dimethylhexanamide
DBD

RN: 20923-67-5   **MP** (°C):
MW: 143.23   **BP** (°C):

Solubility (Moles/L)	Solubility (Grams/L)	Temp (°C)	Ref (#)	Evaluation (T P E A A)	Comments
2.374E-02	3.400E+00	ns	H348	0 0 0 0 0	

1647. C$_8$H$_{17}$NO
Valnoctamide
VCD
Valmethamide
2-Ethyl-3-methyl-pentanamide

RN: 4171-13-5   **MP** (°C):
MW: 143.23   **BP** (°C):

Solubility (Moles/L)	Solubility (Grams/L)	Temp (°C)	Ref (#)	Evaluation (T P E A A)	Comments
6.074E-02	8.700E+00	ns	H348	0 0 0 0 0	

1648. C$_8$H$_{17}$NO
Methylpentylacetamide
2-Methyl-heptanamide
MPD

RN: 4164-91-4   **MP** (°C):
MW: 143.23   **BP** (°C):

Solubility (Moles/L)	Solubility (Grams/L)	Temp (°C)	Ref (#)	Evaluation (T P E A A)	Comments
4.957E-02	7.100E+00	37	H347	0 0 0 0 0	

1649. C$_8$H$_{17}$NO$_2$
n-Heptyl carbamate
Heptyl carbamate

RN: 4248-20-8   **MP** (°C):   66
MW: 159.23   **BP** (°C):

Solubility (Moles/L)	Solubility (Grams/L)	Temp (°C)	Ref (#)	Evaluation (T P E A A)	Comments
2.400E-03	3.822E-01	37	H006	1 2 2 1 1	

1650. C$_8$H$_{17}$NO$_3$
N-Isoamylurethane

RN: **MP** (°C):
MW: 175.23   **BP** (°C):

Solubility (Moles/L)	Solubility (Grams/L)	Temp (°C)	Ref (#)	Evaluation (T P E A A)	Comments
2.329E-02	4.082E+00	20	O021	1 0 0 0 0	

1651. C$_8$H$_{18}$
2,3,4-Trimethylpentane
2,3,4-Trojmetylopentan

RN: 565-75-3   **MP** (°C):   −110
MW: 114.23   **BP** (°C):   113

Solubility (Moles/L)	Solubility (Grams/L)	Temp (°C)	Ref (#)	Evaluation (T P E A A)	Comments
2.048E-05	2.340E-03	0	P003	2 2 2 2 2	
1.191E-05	1.360E-03	25	K119	1 0 0 0 2	
2.013E-05	2.300E-03	25	P003	2 2 2 2 2	
1.191E-05	1.360E-03	25	P051	2 1 1 2 2	
1.191E-05	1.360E-03	25.00	P007	2 1 2 2 2	

1652. C$_8$H$_{18}$
3-Methylheptane
3-Metyloheptan

RN: 589-81-1   **MP** (°C): −121
MW: 114.23   **BP** (°C):

Solubility (Moles/L)	Solubility (Grams/L)	Temp (°C)	Ref (#)	Evaluation (T P E A A)	Comments
2.539E-05	2.900E-03	23	C332	0 0 0 0 0	
6.933E-06	7.920E-04	25	K119	1 0 0 0 2	
6.933E-06	7.920E-04	25	P051	2 1 1 2 2	
6.933E-06	7.920E-04	25.00	P007	2 1 2 2 2	

1653. C$_8$H$_{18}$
Isooctane
2:2:4-Trimethylpentane

RN: 540-84-1   **MP** (°C):
MW: 114.23   **BP** (°C):

Solubility (Moles/L)	Solubility (Grams/L)	Temp (°C)	Ref (#)	Evaluation (T P E A A)	Comments
2.153E-05	2.460E-03	0	P003	2 2 2 2 2	
1.226E-05	1.400E-03	20	M337	2 1 2 2 1	
9.980E-06	1.140E-03	25	K119	1 0 0 0 2	
2.136E-05	2.440E-03	25	M001	2 1 2 2 2	
2.136E-05	2.440E-03	25	M002	2 1 2 2 2	
2.136E-05	2.440E-03	25	M130	1 0 0 0 2	
1.795E-05	2.050E-03	25	P003	2 2 2 2 2	
9.980E-06	1.140E-03	25	P051	2 1 1 2 2	
9.980E-06	1.140E-03	25.00	P007	2 1 2 2 2	
7.879E-06	9.000E-04	ns	B170	0 0 0 0 2	
7.500E-05	8.567E-03	ns	J300	0 0 0 0 0	

1654. C$_8$H$_{18}$
3,4-Dimethylhexane

RN: 583-48-2   **MP** (°C):
MW: 114.23   **BP** (°C): 118

Solubility (Moles/L)	Solubility (Grams/L)	Temp (°C)	Ref (#)	Evaluation (T P E A A)	Comments
6.998E-06	7.994E-04	ns	S460	0 0 0 0 0	

1655. C$_8$H$_{18}$
3-Ethylhexane
Ethyl hexane

RN: 619-99-8   **MP** (°C):
MW: 114.23   **BP** (°C): 119

Solubility (Moles/L)	Solubility (Grams/L)	Temp (°C)	Ref (#)	Evaluation (T P E A A)	Comments
3.076E-06	3.514E-04	ns	S460	0 0 0 0 0	

1656. C$_8$H$_{18}$
2,4-Dimethylhexane
RN: 589-43-5   **MP** (°C):
MW: 114.23   **BP** (°C):   109

Solubility (Moles/L)	Solubility (Grams/L)	Temp (°C)	Ref (#)	Evaluation (T P E A A)	Comments
1.132E-05	1.294E-03	ns	S460	0 0 0 0 0	

1657. C$_8$H$_{18}$
2,3-Dimethylhexane
2:3-Dimethylhexane
RN: ↓ 590-73-8   **MP** (°C):
MW: 114.23   **BP** (°C):   115

Solubility (Moles/L)	Solubility (Grams/L)	Temp (°C)	Ref (#)	Evaluation (T P E A A)	Comments
1.751E-06	2.000E-04	ns	B170	0 0 0 0 2	

1658. C$_8$H$_{18}$
2-Methylheptane
RN: 592-27-8   **MP** (°C):   −109
MW: 114.23   **BP** (°C):

Solubility (Moles/L)	Solubility (Grams/L)	Temp (°C)	Ref (#)	Evaluation (T P E A A)	Comments
3.327E-05	3.800E-03	23	C332	0 0 0 0 0	

1659. C$_8$H$_{18}$NO$_4$PS$_2$
Vamidothion
O,O-Dimethyl *S*-2-(1-*N*-methylcarbamoylethylmercapto)ethyl thiophosphate
RN: 2275-23-2   **MP** (°C):   35.5
MW: 287.34   **BP** (°C):

Solubility (Moles/L)	Solubility (Grams/L)	Temp (°C)	Ref (#)	Evaluation (T P E A A)	Comments
1.392E+01	4.000E+03	20	M161	1 0 0 0 0	
1.392E+01	4.000E+03	ns	M061	0 0 0 0 2	

1660. C$_8$H$_{18}$N$_2$O
Di-*n*-butylnitrosamine
N-Nitroso-di-*n*-butylamine
Dibutylnitrosamine
RN: 924-16-3   **MP** (°C):
MW: 158.25   **BP** (°C):   234

Solubility (Moles/L)	Solubility (Grams/L)	Temp (°C)	Ref (#)	Evaluation (T P E A A)	Comments
8.000E-03	1.266E+00	24	D083	2 0 0 0 0	
7.574E-03	1.199E+00	rt	I307	0 0 0 0 0	

1661. C$_8$H$_{18}$O
2-Octanol
sec-Caprylic alcohol
sec-Octyl alcohol
Methyl hexyl carbinol

RN:	123-96-6	**MP** (°C):	−38.6
MW:	130.23	**BP** (°C):	178.5

Solubility (Moles/L)	Solubility (Grams/L)	Temp (°C)	Ref (#)	Evaluation (T P E A A)	Comments
1.158E-02	1.508E+00	15	M073	1 0 2 2 2	
8.131E-03	1.059E+00	20	A015	1 2 1 1 2	
8.600E-03	1.120E+00	20	H330	0 0 0 0 0	
3.059E-02	3.984E+00	25	C093	2 1 1 1 0	
9.829E-03	1.280E+00	25	M073	1 0 2 2 2	
7.892E-03	1.028E+00	ns	J300	0 0 0 0 0	

1662. C$_8$H$_{18}$O
bis(2-Methyl propyl) ether
iso-Butyl ether
Di-isobutyl ether

RN:	628-55-7	**MP** (°C):	
MW:	130.23	**BP** (°C):	

Solubility (Moles/L)	Solubility (Grams/L)	Temp (°C)	Ref (#)	Evaluation (T P E A A)	Comments
1.059E+00	1.379E+02	25	M375	2 2 2 1 1	
1.227E-02	1.597E+00	51	M375	2 2 2 1 1	
1.002E+00	1.304E+02	60	M375	2 2 2 1 1	

1663. C$_8$H$_{18}$O
DL-2-Octanol
DL-Octanol-(2)

RN:	4128-31-8	**MP** (°C):	−31.6
MW:	130.23	**BP** (°C):	180

Solubility (Moles/L)	Solubility (Grams/L)	Temp (°C)	Ref (#)	Evaluation (T P E A A)	Comments
1.152E-02	1.500E+00	15	F300	1 0 0 0 1	
9.214E-03	1.200E+00	25	F300	1 0 0 0 1	

1664. C$_8$H$_{18}$O

2-Ethyl-1-hexanol
Octyl alcohol
Octyl-(2-ethyl hexyl) alcohol
2-Ethyl hexanol
2-Ethylhexanol
2-Ethylhexan-1-ol

RN: 104-76-7   **MP** (°C): −76
MW: 130.23   **BP** (°C):

Solubility (Moles/L)	Solubility (Grams/L)	Temp (°C)	Ref (#)	Evaluation (T P E A A)	Comments
1.012E-02	1.318E+00	10.2	S307	1 1 0 2 2	
9.586E-03	1.248E+00	19.8	S307	1 1 0 2 2	
4.604E-03	5.996E-01	20	D052	1 1 0 0 0	
6.760E-03	8.804E-01	20	H330	0 0 0 0 0	
9.982E-04	1.300E-01	25	K072	1 0 1 1 1	
7.441E-03	9.691E-01	29.6	S307	1 1 0 2 1	
8.437E-03	1.099E+00	40.1	S307	1 1 0 2 2	
5.678E-03	7.395E-01	50.2	S307	1 1 0 2 1	
6.598E-03	8.593E-01	60.3	S307	1 1 0 2 1	
7.594E-03	9.890E-01	70.1	S307	1 1 0 2 1	
8.284E-03	1.079E+00	80.1	S307	1 1 0 2 2	
8.973E-03	1.169E+00	90.3	S307	1 1 0 2 2	

1665. C$_8$H$_{18}$O

1-Octanol
Caprylic alcohol
n-Octyl alcohol
n-Octanol

RN: 111-87-5   **MP** (°C): −16
MW: 130.23   **BP** (°C): 194

Solubility (Moles/L)	Solubility (Grams/L)	Temp (°C)	Ref (#)	Evaluation (T P E A A)	Comments
3.224E-03	4.198E-01	20	A015	1 2 1 1 2	
3.680E-03	4.793E-01	20	H330	0 0 0 0 0	
3.761E-03	4.898E-01	20.5	S307	1 1 0 2 1	
3.236E-03	4.214E-01	20.96	B178	1 1 0 1 2	EFG
3.162E-03	4.118E-01	23.58	B178	1 1 0 1 2	EFG
2.700E-03	3.516E-01	24	H345	0 0 0 0 0	
4.497E-03	5.857E-01	25	B038	1 2 1 1 2	
3.820E-02	4.975E+00	25	C093	2 1 1 1 0	*sic*
1.000E+00	1.302E+02	25	F044	1 0 0 0 0	EFG
1.060E-03	1.380E-01	25	J035	0 0 0 0 0	
3.830E-03	4.988E-01	25	J302	2 1 2 2 2	
3.800E-03	4.949E-01	25	K025	2 2 1 1 2	
4.530E-03	5.900E-01	25	K072	1 0 1 1 1	
3.970E-03	5.170E-01	25	L322	1 1 2 2 1	
4.530E-03	5.900E-01	25	M087	1 1 2 1 1	
4.110E-03	5.353E-01	25	S359	2 1 2 2 2	

(continued)

1665. C$_8$H$_{18}$O (continued)

Solubility (Moles/L)	Solubility (Grams/L)	Temp (°C)	Ref (#)	Evaluation (T P E A A)	Comments
7.671E-03	9.990E-01	30	R067	0 0 0 0 0	
4.911E-03	6.396E-01	30.6	S307	1 1 0 2 1	
3.236E-03	4.214E-01	34.53	B178	1 1 0 1 2	EFG
1.075E-03	1.400E-01	40	J035	0 0 0 0 0	
4.988E-03	6.496E-01	40.1	S307	1 1 0 2 1	
8.054E-03	1.049E+00	50.0	S307	1 1 0 2 2	
3.548E-03	4.621E-01	60	B178	1 1 0 1 2	EFG
6.751E-03	8.792E-01	60.3	S307	1 1 0 2 1	
3.548E-03	4.621E-01	69.31	B178	1 1 0 1 2	EFG
5.908E-03	7.694E-01	70.3	S307	1 1 0 2 1	
6.675E-03	8.692E-01	80.1	S307	1 1 0 2 1	
6.598E-03	8.593E-01	90.3	S307	1 1 0 2 1	
4.514E-03	5.879E-01	ns	L003	0 0 2 1 2	

1666. C$_8$H$_{18}$O
n-Butyl ether
Butyl ether
Dibutyl ether

RN: 142-96-1   **MP (°C):** −98
MW: 130.23   **BP (°C):** 142.5

Solubility (Moles/L)	Solubility (Grams/L)	Temp (°C)	Ref (#)	Evaluation (T P E A A)	Comments
1.418E-02	1.847E+00	24.80	O005	2 0 2 2 2	
2.700E-03	3.516E-01	25	K012	1 0 0 0 1	
6.138E-03	7.994E-01	25.50	O005	2 0 2 2 0	
1.720E-02	2.240E+00	37	E028	1 0 1 1 2	

1667. C$_8$H$_{18}$O$_2$
Ethohexadiol
2-Ethyl-1,3-hexanediol

RN: 94-96-2   **MP (°C):** −40
MW: 146.23   **BP (°C):** 244.2

Solubility (Moles/L)	Solubility (Grams/L)	Temp (°C)	Ref (#)	Evaluation (T P E A A)	Comments
4.103E-02	6.000E+00	20	M161	1 0 0 0 0	
2.756E-01	4.031E+01	25	C093	2 1 1 1 1	
2.756E-01	4.031E+01	ns	M061	0 0 0 0 1	

1668. C$_8$H$_{18}$O$_4$S$_2$
Sulfonethylmethane
Trional

RN:	76-20-0	**MP** (°C):	75	
MW:	242.36	**BP** (°C):		

Solubility (Moles/L)	Solubility (Grams/L)	Temp (°C)	Ref (#)	Evaluation (T P E A A)	Comments
2.053E-02	4.975E+00	16	A072	1 0 1 0 1	
2.063E-02	5.000E+00	16	F300	1 0 0 0 0	
2.042E-02	4.948E+00	ns	R427	0 0 0 0 0	

1669. C$_8$H$_{19}$N
Octylamine
1-Aminooctane
1-Octanamine
Monoctylamine
n-Octylamine

RN:	111-86-4	**MP** (°C):	−5	
MW:	129.25	**BP** (°C):	175	

Solubility (Moles/L)	Solubility (Grams/L)	Temp (°C)	Ref (#)	Evaluation (T P E A A)	Comments
1.547E-03	2.000E-01	25	K072	1 0 1 1 1	
1.547E-03	2.000E-01	25	M087	1 1 2 1 1	

1670. C$_8$H$_{19}$N
n-Dibutylamine
Di-n-butylamine
N,N-Dibutylamine
N-Butyl-1-butanamine

RN:	111-92-2	**MP** (°C):	−62	
MW:	129.25	**BP** (°C):	159	

Solubility (Moles/L)	Solubility (Grams/L)	Temp (°C)	Ref (#)	Evaluation (T P E A A)	Comments
2.500E-02	3.231E+00	25	K012	1 0 0 0 1	

1671. C$_8$H$_{19}$O$_2$PS$_2$
Ethoprop
Ethoprophos
O-Ethyl-S,S-dipropylphosphorodithioate
Holdem
Rovokil
Ethyl S,S-dipropyl phosphorodithioate

RN:	13194-48-4	**MP** (°C):		
MW:	242.34	**BP** (°C):	88.5	

Solubility (Moles/L)	Solubility (Grams/L)	Temp (°C)	Ref (#)	Evaluation (T P E A A)	Comments
3.095E-03	7.500E-01	ns	M161	0 0 0 0 2	
3.097E-03	7.506E-01	ns	S460	0 0 0 0 0	

1672. C$_8$H$_{19}$O$_2$PS$_3$

Disulfoton

Phosphorodithioic acid O,O-diethyl S-[2-(ethylthio)ethyl] ester

Solvirex

Disyston

Thiodemeton

Ethylthiometon

RN: 298-04-4   **MP** (°C):   108

MW: 274.41   **BP** (°C):

Solubility (Moles/L)	Solubility (Grams/L)	Temp (°C)	Ref (#)	Evaluation (T P E A A)	Comments
5.940E-05	1.630E-02	19.50	B169	2 1 1 1 2	
9.111E-05	2.500E-02	20	M061	1 0 0 0 1	
5.888E-05	1.616E-02	ns	S460	0 0 0 0 0	
9.111E-05	2.500E-02	rt	M161	0 0 0 0 1	

1673. C$_8$H$_{19}$O$_3$P

Dibutyl hydrogen phosphonate

Di-n-butyl phosphite

Dibutoxyphosphine oxide

RN: 1809-19-4   **MP** (°C):

MW: 194.21   **BP** (°C):

Solubility (Moles/L)	Solubility (Grams/L)	Temp (°C)	Ref (#)	Evaluation (T P E A A)	Comments
3.759E-02	7.300E+00	25	B070	1 2 0 1 1	

1674. C$_8$H$_{19}$O$_3$PS$_2$

Demetonthione

Thiophosphorsaeure-O,O-diaethyl-O-[2-(aethylthio)-aethyl]-ester

O,O-Diethyl-O-(2-(ethylthio)-ethyl)ester thiophosphoric acid

O,O-Diethyl 2-ethylmercaptoethyl thiophosphate

Systox

Thiolo-demeton

RN: 298-03-3   **MP** (°C):

MW: 258.34   **BP** (°C):   134

Solubility (Moles/L)	Solubility (Grams/L)	Temp (°C)	Ref (#)	Evaluation (T P E A A)	Comments
2.323E-04	6.000E-02	20	M061	1 0 0 0 1	
7.742E-03	2.000E+00	rt	M161	0 0 0 0 0	form II
2.323E-04	6.000E-02	rt	M161	0 0 0 0 1	form I
1.277E-02	3.300E+00	rt	M161	0 0 0 0 1	

1675. C$_8$H$_{19}$O$_3$PS$_2$

Demetonthiol

Thiophosphorsaeure-*O*,*O*-diaethyl-*S*-[2-(aethylthio)-aethyl]-ester

O,*O*-Diethyl-*S*-(2-(ethylthio)-ethyl)ester thiophosphoric acid

RN:	126-75-0	MP (°C):
MW:	258.34	BP (°C):

Solubility (Moles/L)	Solubility (Grams/L)	Temp (°C)	Ref (#)	Evaluation (T P E A A)	Comments
7.742E-03	2.000E+00	20	F300	1 0 0 0 0	

1676. C$_8$H$_{19}$O$_4$P

Diethyl butyl phosphate

Butyl diethyl phosphate

RN:	2737-00-0	MP (°C):
MW:	210.21	BP (°C):

Solubility (Moles/L)	Solubility (Grams/L)	Temp (°C)	Ref (#)	Evaluation (T P E A A)	Comments
7.136E-02	1.500E+01	25	B070	1 2 0 1 1	

1677. C$_8$H$_{19}$O$_4$P

Diethyl isobutyl phosphate

Ethyl isobutyl phosphate

Phosphoric acid, diethyl 2-methylpropyl ester

RN:	26628-97-7	MP (°C):
MW:	210.21	BP (°C):

Solubility (Moles/L)	Solubility (Grams/L)	Temp (°C)	Ref (#)	Evaluation (T P E A A)	Comments
6.660E-02	1.400E+01	25	B070	1 2 0 1 1	

1678. C$_8$H$_{19}$O$_4$PS$_3$

Disulfoton sulfone

Phosphorodithioic acid *O*,*O*-diethyl *S*-[2-(ethylsulfonyl)ethyl] ester

Disulfoton dioxide

Diethyl *S*-(2-ethylsulfonylethyl) phosphorodithioate

Disyston sulfone

Thiodemeton sulfone

RN:	2497-06-5	MP (°C):
MW:	306.40	BP (°C):

Solubility (Moles/L)	Solubility (Grams/L)	Temp (°C)	Ref (#)	Evaluation (T P E A A)	Comments
2.716E-03	8.323E-01	20	B169	2 2 1 1 1	

1679. C$_8$H$_{20}$Si
Tetraethylsilicane
Tetraethylsilane
Tetraethylsilicon

RN:	631-36-7	MP (°C):
MW:	144.33	BP (°C):

Solubility (Moles/L)	Solubility (Grams/L)	Temp (°C)	Ref (#)	Evaluation (T P E A A)	Comments
2.250E-06	3.248E-04	25	D346	0 0 0 0 0	

1680. C$_8$H$_{20}$Sn
Tetraethyltin
Tetraethylstannane

RN:	597-64-8	MP (°C):	−112
MW:	234.94	BP (°C):	181

Solubility (Moles/L)	Solubility (Grams/L)	Temp (°C)	Ref (#)	Evaluation (T P E A A)	Comments
1.140E-06	2.678E-04	25	D346	1 1 2 2 2	

1681. C$_8$H$_{20}$O$_5$P$_2$S$_2$
Sulfotepp
Pirofos
Tetraethyl dithiopyrophosphate

RN:	3689-24-5	MP (°C):	
MW:	322.32	BP (°C):	137.5

Solubility (Moles/L)	Solubility (Grams/L)	Temp (°C)	Ref (#)	Evaluation (T P E A A)	Comments
9.307E-05	3.000E-02	20	F300	1 0 0 0 0	
7.756E-05	2.500E-02	20	M061	1 0 0 0 1	
7.756E-05	2.500E-02	rt	M161	0 0 0 0 1	

1682. C$_8$H$_{23}$N$_5$
Tetraethylenepentamine
1,4,7,10,13-Pentaazatridecane
N-(2-Aminoethyl)-N′-(2-((2-aminoethyl)amino)ethyl)-1,2-ethanediamine
1,11-Diamino-3,6,9-triazaundecane
3,6,9-Triaza-1,11-undecanediamine
3,6,9-Triazaundecane-1,11-diamine

RN:	112-57-2	MP (°C):	−40
MW:	189.31	BP (°C):	340

Solubility (Moles/L)	Solubility (Grams/L)	Temp (°C)	Ref (#)	Evaluation (T P E A A)	Comments
4.582E+00	8.674E+02	4.50	C022	1 2 0 0 2	

1683. C$_8$Cl$_4$N$_2$
Chlorothalonil
2,4,5,6-Tetrachloro-1,3-benzenedicarbonitrile
Forturf
Exotherm
Bravo

RN: 1897-45-6   **MP** (°C):   250.5
MW: 265.91   **BP** (°C):

Solubility (Moles/L)	Solubility (Grams/L)	Temp (°C)	Ref (#)	Evaluation (T P E A A)	Comments
2.256E-06	6.000E-04	25	M161	1 0 0 0 0	

1684. C$_9$H$_4$Cl$_3$NO$_2$S
Folpet
N-(Trichloromethylthio)phthalimide
Folpan
Folpel
Phaltan
Phalton

RN: 133-07-3   **MP** (°C):   177
MW: 296.56   **BP** (°C):

Solubility (Moles/L)	Solubility (Grams/L)	Temp (°C)	Ref (#)	Evaluation (T P E A A)	Comments
3.388E-06	1.005E-03	20	B179	0 0 0 0 0	
3.372E-06	1.000E-03	20	F311	1 2 2 2 1	
3.388E-06	1.005E-03	ns	R427	0 0 0 0 0	

1685. C$_9$H$_5$Cl$_3$N$_4$
Anilazine
4,6-Dichloro-*N*-(2-chlorophenyl)-1,3,5-triazin-2-amine
Triasyn
Direx
Dyrene
Kemate

RN: 101-05-3   **MP** (°C):   159.5
MW: 275.53   **BP** (°C):

Solubility (Moles/L)	Solubility (Grams/L)	Temp (°C)	Ref (#)	Evaluation (T P E A A)	Comments
3.629E-05	1.000E-02	ns	B160	0 0 0 0 1	

1686. C$_9$H$_6$ClNO$_3$S
Benazolin
7-Chloro-2-oxo-3(2H)-benzothiazolacetic acid
Galipan
Herbazolin
Leymin
Metizolin

RN: 3813-05-6   **MP** (°C):   193
MW: 243.67   **BP** (°C):

Solubility (Moles/L)	Solubility (Grams/L)	Temp (°C)	Ref (#)	Evaluation (T P E A A)	Comments
2.462E-03	6.000E-01	20	M161	1 0 0 0 2	

1687. C$_9$H$_6$Cl$_2$N$_2$O$_3$
Methazole
2-(3,4-Dichlorophenyl)-4-methyl-1,2,4-oxadiazolidine-3,5-dione
Tunic
Paxilon
Chlormethazole
Mezopur

RN: 20354-26-1   **MP** (°C):   123
MW: 261.07   **BP** (°C):

Solubility (Moles/L)	Solubility (Grams/L)	Temp (°C)	Ref (#)	Evaluation (T P E A A)	Comments
5.746E-06	1.500E-03	24	C105	2 1 2 2 2	
5.746E-06	1.500E-03	25	M161	1 0 0 0 1	
5.746E-06	1.500E-03	25	W314	1 0 0 0 1	

1688. C$_9$H$_6$Cl$_6$O$_3$S
Endosulfan

RN: 115-29-7   **MP** (°C):   209
MW: 406.93   **BP** (°C):

Solubility (Moles/L)	Solubility (Grams/L)	Temp (°C)	Ref (#)	Evaluation (T P E A A)	Comments
7.987E-07	3.250E-04	ns	V414	0 0 0 0 0	

1689. $C_9H_6Cl_6O_3S$
α-Endosulfan
5-Norbornene-2,3-dimethanol, 1,4,5,6,7,7-hexachloro-, cyclic sulfite, *endo*-
Endosulfan I
Endosulfan A
Hexachloro-5-norbornene-2,3-dimethanol, cyclic sulfite, *endo*-
Thiodan I

RN:	959-98-8	MP (°C):	109
MW:	406.93	BP (°C):	

Solubility (Moles/L)	Solubility (Grams/L)	Temp (°C)	Ref (#)	Evaluation (T P E A A)	Comments
1.253E-06	5.099E-04	20	B300	2 0 1 1 2	
1.302E-06	5.300E-04	25	W025	1 0 2 2 2	
4.030E-07	1.640E-04	ns	A069	0 0 0 0 2	
1.253E-06	5.100E-04	ns	V414	0 0 0 0 0	

1690. $C_9H_6Cl_6O_3S$
β-Endosulfan
5-Norbornene-2,3-dimethanol, 1,4,5,6,7,7-hexachloro-, cyclic sulfite, *exo*-
Endosulfan II
Hexachloro-5-norbornene-2,3-dimethanol, cyclic sulfite, *exo*-
Thiodan II

RN:	33213-65-9	MP (°C):	209
MW:	406.93	BP (°C):	

Solubility (Moles/L)	Solubility (Grams/L)	Temp (°C)	Ref (#)	Evaluation (T P E A A)	Comments
1.106E-06	4.501E-04	20	B300	2 0 1 1 2	
6.881E-07	2.800E-04	25	W025	1 0 2 2 2	
1.720E-07	7.000E-05	ns	A069	0 0 0 0 1	
1.106E-06	4.500E-04	ns	V414	0 0 0 0 0	

1691. $C_9H_6I_3NO_3$
2,4,6-Triiodo-3-acetaminobenzoic acid
Acetrizoic acid

RN:	85-36-9	MP (°C):	
MW:	556.87	BP (°C):	

Solubility (Moles/L)	Solubility (Grams/L)	Temp (°C)	Ref (#)	Evaluation (T P E A A)	Comments
2.299E-03	1.280E+00	25	L025	1 0 0 0 2	
3.232E-03	1.800E+00	50	L025	1 0 0 0 2	
5.387E-03	3.000E+00	100	L025	1 0 0 0 2	
2.442E-03	1.360E+00	ns	H055	0 0 0 0 0	

1692. C$_9$H$_6$N$_2$S
4-Cyanobenzyl isothiocyanate
p-Cyanobenzyl isothiocyanate
Isothiocyanic acid, p-cyanobenzyl ester

RN:	3694-48-2	MP (°C):
MW:	174.23	BP (°C):

Solubility (Moles/L)	Solubility (Grams/L)	Temp (°C)	Ref (#)	Evaluation (T P E A A)	Comments
3.200E-04	5.575E-02	25	D014	1 0 0 0 1	

1693. C$_9$H$_6$O$_2$
Coumarin
Cumarin
1,2-Benzopyrone
2H-1-Benzopyran-2-one
Benzopyran-2-one
Benzopyrone

RN:	91-64-5	MP (°C):	70
MW:	146.15	BP (°C):	

Solubility (Moles/L)	Solubility (Grams/L)	Temp (°C)	Ref (#)	Evaluation (T P E A A)	Comments
6.153E-03	8.992E-01	.2	D073	1 1 2 1 0	
8.211E-03	1.200E+00	0	F300	1 0 0 0 1	
1.298E-02	1.896E+00	20	D073	1 1 2 1 1	
1.368E-02	2.000E+00	22.5	G301	0 0 0 0 0	
1.706E-02	2.494E+00	25	I312	0 0 0 0 0	
1.774E-02	2.593E+00	30	D073	1 1 2 1 1	
1.847E-02	2.700E+00	30	F300	1 0 0 0 1	
3.065E-02	4.480E+00	40	D073	1 1 2 1 1	
4.419E-02	6.458E+00	50	D073	1 1 2 1 1	
4.756E-02	6.951E+00	60	D073	1 1 2 1 1	
1.342E-01	1.961E+01	100	I312	0 0 0 0 0	
1.507E-02	2.203E+00	ns	R082	0 0 0 0 0	
6.842E-04	9.999E-02	rt	D021	0 0 1 1 0	sic

1694. C$_9$H$_6$O$_3$
7-Hydroxycoumarin
Umbelliferone

RN:	93-35-6	MP (°C):	230
MW:	162.15	BP (°C):	

Solubility (Moles/L)	Solubility (Grams/L)	Temp (°C)	Ref (#)	Evaluation (T P E A A)	Comments
1.918E-03	3.110E-01	ns	R082	0 0 0 0 0	

1695. C$_9$H$_6$O$_5$
Phthalonic acid
Phthalonsaeure
RN: 528-46-1   **MP** (°C):
MW: 194.15   **BP** (°C):

Solubility (Moles/L)	Solubility (Grams/L)	Temp (°C)	Ref (#)	Evaluation (T P E A A)	Comments
2.756E+00	5.350E+02	15	F300	1 0 0 0 2	

1696. C$_9$H$_6$O$_6$
Trimesic acid
1,3,5-Benzenetricarboxylic acid
Benzol-tricarbonsaeure-(1,3,5)
RN: 554-95-0   **MP** (°C):
MW: 210.14   **BP** (°C):

Solubility (Moles/L)	Solubility (Grams/L)	Temp (°C)	Ref (#)	Evaluation (T P E A A)	Comments
1.808E-02	3.800E+00	16	F300	1 0 0 0 1	
1.252E-01	2.630E+01	23	F300	1 0 0 0 2	

1697. C$_9$H$_6$O$_6$
1,2,3-Benzenetricarboxylic acid
Benzol-tricarbonsaeure-(1,2,3)
Hemimellitic acid
RN: 569-51-7   **MP** (°C): 223
MW: 210.14   **BP** (°C):

Solubility (Moles/L)	Solubility (Grams/L)	Temp (°C)	Ref (#)	Evaluation (T P E A A)	Comments
1.456E-01	3.060E+01	19	F300	1 0 0 0 2	

1698. C$_9$H$_6$O$_6$
Hydrastic acid
Hydrastsaeure
RN: 490-26-6   **MP** (°C):
MW: 210.14   **BP** (°C):

Solubility (Moles/L)	Solubility (Grams/L)	Temp (°C)	Ref (#)	Evaluation (T P E A A)	Comments
2.855E-02	6.000E+00	15	F300	1 0 0 0 1	

1699. C$_9$H$_7$Cl$_3$O$_3$
Trichloroethyl salicylate
Benzoic acid, 2-hydroxy-, 2,2,2-trichloroethyl ester
RN: 56529-85-2   **MP** (°C):
MW: 269.51   **BP** (°C):

Solubility (Moles/L)	Solubility (Grams/L)	Temp (°C)	Ref (#)	Evaluation (T P E A A)	Comments
4.081E-03	1.100E+00	37	D009	1 2 1 1 1	0.1N HCl

1700. C$_9$H$_7$Cl$_3$O$_3$
Silvex
2-(2,4,5-Trichlorophenoxy)propionic acid
Fenoprop
Propionic acid, 2(2,4,5-trichlorophenoxy)-
RN: 93-72-1   **MP** (°C):   181.6
MW: 269.51   **BP** (°C):   200

Solubility (Moles/L)	Solubility (Grams/L)	Temp (°C)	Ref (#)	Evaluation (T P E A A)	Comments
2.630E-04	7.088E-02	24.99	N417	0 0 0 0 0	
2.634E-04	7.100E-02	25	B164	1 0 1 1 1	
5.195E-04	1.400E-01	25	B185	0 0 0 0 0	
6.678E-04	1.800E-01	25	B200	1 0 0 0 1	
5.195E-04	1.400E-01	25	L024	1 0 0 0 2	
5.194E-04	1.400E-01	25	M061	1 0 0 0 1	
5.195E-04	1.400E-01	25	M161	1 0 0 0 2	
5.194E-04	1.400E-01	ns	B100	0 0 0 0 1	
5.195E-04	1.400E-01	ns	K138	0 0 0 0 1	

1701. C$_9$H$_7$N
Quinoline
Chinolin
1-Azanaphthalene
Benzopyridine
1-Benzazine
Benzo[b]pyridine
RN: 91-22-5   **MP** (°C):   −15
MW: 129.16   **BP** (°C):   237.7

Solubility (Moles/L)	Solubility (Grams/L)	Temp (°C)	Ref (#)	Evaluation (T P E A A)	Comments
4.730E-02	6.110E+00	20	A050	0 0 0 0 0	
4.913E-02	6.346E+00	20.3	L339	2 0 2 2 2	
4.968E-02	6.417E+00	40.0	L339	2 0 2 2 2	
6.337E-02	8.185E+00	64.8	L339	2 0 2 2 2	
8.136E-02	1.051E+01	80.2	L339	2 0 2 2 2	
1.063E-01	1.373E+01	100.0	L339	2 0 2 2 2	

1702. C₉H₇NO

4-Hydroxyquinoline
4-Hydroxy-chinolin
4-Quinolinol

RN: 611-36-9   **MP** (°C): 201
MW: 145.16   **BP** (°C):

Solubility (Moles/L)	Solubility (Grams/L)	Temp (°C)	Ref (#)	Evaluation (T P E A A)	Comments
3.307E-02	4.800E+00	15	F300	1 0 0 0 1	

1703. C₉H₇NO

5-Hydroxyquinoline
5-Quinolinol

RN: 578-67-6   **MP** (°C): 223
MW: 145.16   **BP** (°C):

Solubility (Moles/L)	Solubility (Grams/L)	Temp (°C)	Ref (#)	Evaluation (T P E A A)	Comments
2.869E-03	4.165E-01	20	A035	1 0 2 2 1	
2.884E-03	4.187E-01	ns	R427	0 0 0 0 0	

1704. C₉H₇NO

6-Hydroxyquinoline
6-Quinolinol

RN: 580-16-5   **MP** (°C): 192
MW: 145.16   **BP** (°C):

Solubility (Moles/L)	Solubility (Grams/L)	Temp (°C)	Ref (#)	Evaluation (T P E A A)	Comments
6.882E-03	9.990E-01	20	A035	1 0 2 2 1	

1705. C₉H₇NO

7-Hydroxyquinoline
7-Quinolinol

RN: 580-20-1   **MP** (°C):
MW: 145.16   **BP** (°C):

Solubility (Moles/L)	Solubility (Grams/L)	Temp (°C)	Ref (#)	Evaluation (T P E A A)	Comments
3.130E-03	4.543E-01	20	A035	1 0 2 2 1	
3.162E-03	4.590E-01	ns	R427	0 0 0 0 0	

1706. C$_9$H$_7$NO
8-Hydroxyquinoline
8-Quinolinol
Hydroxybenzopuridine
RN: 148-24-3   **MP** (°C):   76
MW: 145.16   **BP** (°C):   267

Solubility (Moles/L)	Solubility (Grams/L)	Temp (°C)	Ref (#)	Evaluation (T P E A A)	Comments
3.825E-03	5.552E-01	20	A035	1 0 2 2 1	
4.470E-03	6.489E-01	25.2	P024	2 1 1 1 2	
5.380E-03	7.810E-01	30.3	P024	2 1 1 1 2	

1707. C$_9$H$_7$NO
Carbostyril
2-Hydroxyquinoline
2-Quinolinol
RN: 59-31-4   **MP** (°C):   199.0
MW: 145.16   **BP** (°C):

Solubility (Moles/L)	Solubility (Grams/L)	Temp (°C)	Ref (#)	Evaluation (T P E A A)	Comments
7.244E-03	1.052E+00	20	C035	1 0 2 2 1	

1708. C$_9$H$_7$NO
3-Hydroxyquinoline
3-Quinolinol
RN: 580-18-7   **MP** (°C):
MW: 145.16   **BP** (°C):

Solubility (Moles/L)	Solubility (Grams/L)	Temp (°C)	Ref (#)	Evaluation (T P E A A)	Comments
4.050E-03	5.879E-01	20	A035	1 0 2 2 1	

1709.   C$_9$H$_7$NOS
m-Acetylphenyl isothiocyanate
3-Acetylphenyl isothiocyanate
RN: 3125-71-1   **MP** (°C):
MW: 177.23   **BP** (°C):

Solubility (Moles/L)	Solubility (Grams/L)	Temp (°C)	Ref (#)	Evaluation (T P E A A)	Comments
4.700E-05	8.330E-03	25	K032	2 2 0 1 1	

1710. C_9H_7NOS
p-Acetylphenyl isothiocyanate
4-Acetylphenyl isothiocyanate
RN: 2131-57-9 **MP** (°C):
MW: 177.23 **BP** (°C):

Solubility (Moles/L)	Solubility (Grams/L)	Temp (°C)	Ref (#)	Evaluation (T P E A A)	Comments
9.500E-05	1.684E-02	25	D019	1 1 1 1 1	

1711. C_9H_7NOS
Phenacyl thiocyanate
RN: 5399-30-4 **MP** (°C):
MW: 177.23 **BP** (°C):

Solubility (Moles/L)	Solubility (Grams/L)	Temp (°C)	Ref (#)	Evaluation (T P E A A)	Comments
1.971E-03	3.494E-01	22	J420	0 0 0 0 0	pH 6.5

1712. $C_9H_7NO_2S$
m-Acetoxyphenyl isothiocyanate
Methyl *m*-isothiocyanobenzoate
RN: 3530-01-6 **MP** (°C):
MW: 193.23 **BP** (°C):

Solubility (Moles/L)	Solubility (Grams/L)	Temp (°C)	Ref (#)	Evaluation (T P E A A)	Comments
2.720E-04	5.256E-02	25	K032	2 2 0 1 2	
7.700E-04	1.488E-01	25	K032	2 2 0 1 2	

1713. $C_9H_7NO_5$
2-(Oxalylamino)benzoic acid
Oxanil-carbonsaeure-(2)
Oxanil-*o*-carboxylic acid
RN: 5651-01-4 **MP** (°C):
MW: 209.16 **BP** (°C):

Solubility (Moles/L)	Solubility (Grams/L)	Temp (°C)	Ref (#)	Evaluation (T P E A A)	Comments
5.259E-03	1.100E+00	10	F300	1 0 0 0 1	

1714. $C_9H_7N_3S$
Tricyclazole
Methyl-1,2,4-triazolo(3,4-b)benzothiazole
5-Methyl-1,2,4-triazolo[3,4-b]benzothiazole
RN: 41814-78-2 **MP** (°C): 187.5
MW: 189.24 **BP** (°C):

Solubility (Moles/L)	Solubility (Grams/L)	Temp (°C)	Ref (#)	Evaluation (T P E A A)	Comments
8.455E-03	1.600E+00	25	M161	1 0 0 0 1	

1715. C9H7N7O2S

Azathioprine
Cytostatics
Imuran
Azatioprin
6-(1-Methyl-p-nitro-5-imidazolyl)-thiopurine
Ccucol

RN: 446-86-6   **MP** (°C):   243.5
MW: 277.27   **BP** (°C):

Solubility (Moles/L)	Solubility (Grams/L)	Temp (°C)	Ref (#)	Evaluation (T P E A A)	Comments
4.689E-04	1.300E-01	24	N016	0 0 0 0 0	
4.472E-04	1.240E-01	25	N063	1 1 1 1 2	intrinsic
4.689E-04	1.300E-01	25	N063	1 1 1 1 2	
3.607E-05	1.000E-02	ns	K444	0 0 0 0 0	

1716. C9H8Cl2O3

Dichlorprop
Dichloroprop
α-(2,4-Dichlorophenoxy)propionic acid

RN: 120-36-5   **MP** (°C):   117.5
MW: 235.07   **BP** (°C):

Solubility (Moles/L)	Solubility (Grams/L)	Temp (°C)	Ref (#)	Evaluation (T P E A A)	Comments
1.489E-03	3.500E-01	20	L024	1 0 0 0 2	
1.489E-03	3.500E-01	20	M161	1 0 0 0 2	
1.490E-03	3.503E-01	24.99	N417	0 0 0 0 0	
3.527E-03	8.290E-01	25	B164	1 0 1 1 2	
3.020E-03	7.100E-01	28	B200	1 0 0 0 1	
1.484E-02	3.488E+00	ns	B100	0 0 0 0 1	

1717. C9H8Cl2O3

Methyl (2,4-Dichlorophenoxy)acetate
2,4-Dichorophenoxyacetic acid methyl ester

RN: 5335-03-5   **MP** (°C):
MW: 235.07   **BP** (°C):

Solubility (Moles/L)	Solubility (Grams/L)	Temp (°C)	Ref (#)	Evaluation (T P E A A)	Comments
7.657E-04	1.800E-01	ns	B185	0 0 0 0 0	
5.333E-04	1.254E-01	ns	M120	0 0 1 1 2	

1718. $C_9H_8Cl_3NO_2S$
Captan
N-Trichloromethylthio-4-cyclohexene-1,2-dicarboximide
Vancide 89
Merpan 90
Orthocid-83
Pillarcap

RN:	133-06-2	MP (°C):	178
MW:	300.59	BP (°C):	

Solubility (Moles/L)	Solubility (Grams/L)	Temp (°C)	Ref (#)	Evaluation (T P E A A)	Comments
1.660E-06	4.989E-04	20	B179	0 0 0 0 0	
<1.66E-06	<5.00E-04	20	F311	1 2 2 2 1	
1.544E-05	4.642E-03	ns	H322	0 0 0 0 0	
1.660E-06	4.989E-04	ns	R427	0 0 0 0 0	
1.663E-06	5.000E-04	rt	M161	0 0 0 0 0	

1719. $C_9H_8N_2OS$
m-Acetamidophenyl isothiocyanate
3-Acetamidophenyl isothiocyanate

RN:	3137-83-5	MP (°C):	
MW:	192.24	BP (°C):	

Solubility (Moles/L)	Solubility (Grams/L)	Temp (°C)	Ref (#)	Evaluation (T P E A A)	Comments
2.950E-04	5.671E-02	25	K032	2 2 0 1 2	

1720. $C_9H_8N_4O_6$
Nifurtoinol
3-(Hydroxymethyl)nitrofurantoin

RN:	1088-92-2	MP (°C):	
MW:	268.19	BP (°C):	

Solubility (Moles/L)	Solubility (Grams/L)	Temp (°C)	Ref (#)	Evaluation (T P E A A)	Comments
1.230E-03	3.300E-01	22	B154	1 1 1 1 1	0.1M HCl

1721. C_9H_8O
(E)-Cinnamaldehyde
(E)-3-Phenylpropenal;
(2E)-3-Phenyl-2-propenal
(E)-3-Phenylprop-2-enone
(E)-3-Phenylacrolein
(E)-3-Phenylprop-2-enal

RN:	14371-10-9	MP (°C):	
MW:	132.16	BP (°C):	250–253

Solubility (Moles/L)	Solubility (Grams/L)	Temp (°C)	Ref (#)	Evaluation (T P E A A)	Comments
1.400E-02	1.850E+00	25	D407	1 0 2 2 2	

1722. C$_9$H$_8$O
Cinnamaldehyde
3-Phenyl-2-propenal
Phenylacrolein
3-Phenyl-2-propenaldehyde
Zimtaldehyde

RN:	104-55-2	**MP** (°C):
MW:	132.16	**BP** (°C): 246

Solubility (Moles/L)	Solubility (Grams/L)	Temp (°C)	Ref (#)	Evaluation (T P E A A)	Comments
1.020E-02	1.348E+00	25	I019	1 0 1 2 2	
9.100E-03	1.203E+00	37	E028	1 0 1 1 1	

1723. C$_9$H$_8$O$_2$
trans-Cinnamic acid
trans-3-Phenyl-2-propenoic acid
trans-β-Phenylacrylic acid
(E)-3-Phenyl-2-propenoic acid

RN:	140-10-3	**MP** (°C): 133
MW:	148.16	**BP** (°C): 300

Solubility (Moles/L)	Solubility (Grams/L)	Temp (°C)	Ref (#)	Evaluation (T P E A A)	Comments
1.417E-03	2.100E-01	15	M461	0 0 0 0 0	
2.700E-03	4.000E-01	18	F300	1 0 0 0 0	
2.835E-03	4.200E-01	18	M077	1 2 1 1 2	
3.010E-03	4.460E-01	25	C090	1 2 2 2 2	
3.685E-03	5.460E-01	25	M077	1 2 1 1 2	
1.552E-03	2.300E-01	25	M461	0 0 0 0 0	
2.092E-03	3.100E-01	30	M461	0 0 0 0 0	
5.264E-03	7.800E-01	35	M077	1 2 1 1 2	
4.252E-03	6.300E-01	40	M461	0 0 0 0 0	
7.364E-03	1.091E+00	45	M077	1 2 1 1 2	
5.737E-03	8.500E-01	50	M461	0 0 0 0 0	

1724. C$_9$H$_8$O$_2$
Atropic acid
Atropasaeure

RN:	492-38-6	**MP** (°C): 106
MW:	148.16	**BP** (°C):

Solubility (Moles/L)	Solubility (Grams/L)	Temp (°C)	Ref (#)	Evaluation (T P E A A)	Comments
8.774E-03	1.300E+00	20	F300	1 0 0 0 1	

1725. C₉H₈O₂

Cinnamic acid
Phenylacrylic acid
3-Phenylpropenoic acid
2-Propenoic acid, 3-phenyl-

RN:	621-82-9	**MP** (°C):	133	
MW:	148.16	**BP** (°C):	261.9	

Solubility (Moles/L)	Solubility (Grams/L)	Temp (°C)	Ref (#)	Evaluation (T P E A A)	Comments
2.024E-03	2.999E-01	10	M043	1 0 0 0 0	
3.390E-03	5.023E-01	14.3	D061	1 0 0 0 2	
2.642E-03	3.914E-01	16.3	D061	1 0 0 0 2	
2.643E-03	3.916E-01	16.30	B118	1 0 0 0 2	unit assumed
1.515E-02	2.245E+00	20	C092	2 1 0 1 1	sic
2.699E-03	3.998E-01	20	M043	1 0 0 0 0	
3.170E-03	4.697E-01	22	E045	2 0 1 1 2	
3.260E-03	4.830E-01	23	E045	2 0 1 1 2	
3.360E-03	4.978E-01	24	E045	2 0 1 1 2	
3.450E-03	5.112E-01	25	E045	2 0 1 1 2	
3.850E-03	5.704E-01	25	K040	1 0 2 1 2	
3.340E-03	4.949E-01	25	L048	1 2 2 1 2	
3.340E-03	4.949E-01	25	L050	2 0 1 2 2	
3.540E-03	5.245E-01	26	E045	2 0 1 1 2	
3.800E-03	5.630E-01	26.4	P043	2 0 1 1 2	
3.630E-03	5.378E-01	27	E045	2 0 1 1 2	
4.963E-03	7.353E-01	28	D050	1 2 1 2 2	
4.688E-03	6.946E-01	30	B118	1 0 0 0 2	unit assumed
4.682E-03	6.937E-01	30	D061	1 0 0 0 2	
4.047E-03	5.996E-01	30	M043	1 0 0 0 0	
3.959E-02	5.865E+00	100	M043	1 0 0 0 1	

1726. C₉H₈O₂

cis-Cinnamic acid
cis-Zimtsaeure

RN:	102-94-3	**MP** (°C):		
MW:	148.16	**BP** (°C):		

Solubility (Moles/L)	Solubility (Grams/L)	Temp (°C)	Ref (#)	Evaluation (T P E A A)	Comments
4.657E-02	6.900E+00	18	F300	1 0 0 0 1	
4.644E-02	6.880E+00	18	M077	1 2 1 1 2	form III, mp 68 C
5.143E-02	7.620E+00	18	M077	1 2 1 1 2	form II, mp 58 C
6.041E-02	8.950E+00	18	M077	1 2 1 1 2	form I, mp 42 C
5.703E-02	8.450E+00	25	M077	1 2 1 1 2	form III, mp 68 C
6.324E-02	9.370E+00	25	M077	1 2 1 1 2	form II, mp 58 C
7.445E-02	1.103E+01	25	M077	1 2 1 1 2	form I, mp 42 C
7.519E-02	1.114E+01	35	M077	1 2 1 1 2	form III, mp 68 C
8.362E-02	1.239E+01	35	M077	1 2 1 1 2	form II, mp 58 C

(continued)

1726. C$_9$H$_8$O$_2$ (continued)

Solubility (Moles/L)	Solubility (Grams/L)	Temp (°C)	Ref (#)	Evaluation (T P E A A)	Comments
9.861E-02	1.461E+01	35	M077	1 2 1 1 2	form I, mp 42 C
9.760E-02	1.446E+01	45	M077	1 2 1 1 2	form III, mp 68 C
1.086E-01	1.609E+01	45	M077	1 2 1 1 2	form II, mp 58 C
1.245E-01	1.845E+01	55	M077	1 2 1 1 2	form III, mp 68 C

1727. C$_9$H$_8$O$_3$

2-Acetophenone carboxylic acid
Acetophenon-carbonsaeure-(2)
o-Carboxyacetophenone

RN: 577-56-0   **MP** (°C):
MW: 164.16   **BP** (°C):

Solubility (Moles/L)	Solubility (Grams/L)	Temp (°C)	Ref (#)	Evaluation (T P E A A)	Comments
2.427E-02	3.984E+00	rt	H431	0 0 0 0 0	

1728. C$_9$H$_8$O$_4$

Homophthalic acid
Homophthalsaeure

RN: 89-51-0   **MP** (°C):   184.5
MW: 180.16   **BP** (°C):

Solubility (Moles/L)	Solubility (Grams/L)	Temp (°C)	Ref (#)	Evaluation (T P E A A)	Comments
2.542E-02	4.579E+00	rt	H431	0 0 0 0 0	

1729. C$_9$H$_8$O$_4$

Caffeic acid
3,4-Dihydroxy-*trans*-cinnamate
(E)-3-(3,4-Dihydroxyphenyl)-2-propenoic acid

RN: 331-39-5   **MP** (°C):   196 C
MW: 180.16   **BP** (°C):

Solubility (Moles/L)	Solubility (Grams/L)	Temp (°C)	Ref (#)	Evaluation (T P E A A)	Comments
3.053E-03	5.500E-01	15	M461	0 0 0 0 0	
5.440E-03	9.800E-01	25	M461	0 0 0 0 0	
6.827E-03	1.230E+00	30	M461	0 0 0 0 0	
1.132E-02	2.040E+00	40	M461	0 0 0 0 0	
1.621E-02	2.920E+00	50	M461	0 0 0 0 0	

1730. C$_9$H$_8$O$_4$
4-Methylphthalic acid

RN:	4316-23-8	MP (°C):	149
MW:	180.16	BP (°C):	

Solubility (Moles/L)	Solubility (Grams/L)	Temp (°C)	Ref (#)	Evaluation (T P E A A)	Comments
2.211E-02	3.984E+00	rt	H431	0 0 0 0 0	

1731. C$_9$H$_8$O$_4$
Aspirin
Acetyl-salicylsaeure
Acetylsalicylic acid

RN:	50-78-2	MP (°C):	135
MW:	180.16	BP (°C):	

Solubility (Moles/L)	Solubility (Grams/L)	Temp (°C)	Ref (#)	Evaluation (T P E A A)	Comments
3.121E-02	5.623E+00	4.62	M053	1 0 1 1 0	EFG, 0.1N HCl
1.107E-02	1.995E+00	12.55	M053	1 0 1 1 0	EFG, 0.1N HCl
3.200E-02	5.765E+00	14	O019	1 0 0 1 2	
1.998E-02	3.600E+00	15	E017	1 0 0 0 0	EFG
1.388E-02	2.500E+00	15	F300	1 0 0 0 1	
1.716E-02	3.091E+00	15	H022	1 2 2 2 2	
2.109E-02	3.800E+00	20	E017	1 0 0 0 0	EFG
1.460E-02	2.630E+00	20.96	M053	1 0 1 1 0	EFG, 0.1N HCl
1.769E-02	3.188E+00	22.5	B422	2 0 2 2 2	
2.553E-02	4.600E+00	25	E017	1 0 0 0 0	EFG
2.775E-02	5.000E+00	25	S304	1 2 1 2 2	form IV
2.131E-02	3.840E+00	25	S304	1 2 1 2 2	form I
2.442E-02	4.400E+00	25	S304	1 2 1 2 2	form II
1.890E-02	3.405E+00	25.6	G015	1 0 1 1 2	pH 1.00, pka 3.62, intrinsic
2.500E-02	4.504E+00	30	A065	2 0 2 2 1	
2.831E-02	5.100E+00	30	E017	1 0 0 0 0	EFG
2.387E-02	4.300E+00	30	G042	1 1 1 1 1	0.1N HCl
2.851E-02	5.137E+00	30	H022	1 2 2 2 2	
2.000E-02	3.603E+00	30	L069	1 0 1 1 0	EFG
2.637E-02	4.750E+00	30	S304	1 2 1 2 2	form I
3.275E-02	5.900E+00	30	S304	1 2 1 2 2	form IV
3.108E-02	5.600E+00	30	S304	1 2 1 2 2	form II
3.275E-02	5.900E+00	35	E017	1 0 0 0 0	EFG
2.942E-02	5.300E+00	37	D009	1 2 1 1 1	0.1N HCl
3.219E-02	5.800E+00	37	G042	1 1 1 1 1	0.1N HCl
3.641E-02	6.560E+00	37	G430	0 0 0 0 0	pH 4.5
3.569E-02	6.430E+00	37	K086	1 0 0 0 2	
3.031E-02	5.460E+00	37	M115	2 2 1 1 2	
4.052E-02	7.300E+00	37	S304	1 2 1 2 2	form II
3.830E-02	6.900E+00	37	S304	1 2 1 2 2	form I
4.218E-02	7.600E+00	37	S304	1 2 1 2 2	form IV
3.441E-02	6.200E+00	37	Y421	0 0 0 0 0	

(continued)

1731. C$_9$H$_8$O$_4$ (continued)

Solubility (Moles/L)	Solubility (Grams/L)	Temp (°C)	Ref (#)	Evaluation (T P E A A)	Comments
3.830E-02	6.900E+00	40	E017	1 0 0 0 0	EFG
4.385E-02	7.900E+00	40	S304	1 2 1 2 2	form II
4.218E-02	7.600E+00	40	S304	1 2 1 2 2	form I
4.607E-02	8.300E+00	40	S304	1 2 1 2 2	form IV
4.662E-02	8.400E+00	45	E017	1 0 0 0 0	EFG
4.274E-02	7.700E+00	45	G042	1 1 1 1 1	0.1N HCl
5.551E-02	1.000E+01	49.42	M053	1 0 1 1 0	EFG, 0.1N HCl
4.940E-02	8.900E+00	50	G042	1 1 1 1 1	0.1N HCl
6.829E-02	1.230E+01	60.17	M053	1 0 1 1 0	EFG, 0.1N HCl
1.848E-02	3.330E+00	ns	K444	0 0 0 0 0	
1.551E-02	2.795E+00	rt	R431	0 0 0 0 0	Average

1732. C$_9$H$_9$ClO$_3$

DL-2-(2-Chlorophenoxy)propionic acid
2-(*o*-Chlorophenoxy)propionic acid
3-CP

RN: 76466-16-5   **MP** (°C):   113
MW: 200.62   **BP** (°C):

Solubility (Moles/L)	Solubility (Grams/L)	Temp (°C)	Ref (#)	Evaluation (T P E A A)	Comments
5.974E-03	1.199E+00	22	B200	1 0 0 0 1	
9.726E-02	1.951E+01	100	B200	1 0 0 0 2	

1733. C$_9$H$_9$ClO$_3$

DL-2-(4-Chlorophenoxy)propionic acid

RN: 3307-39-9   **MP** (°C):
MW: 200.62   **BP** (°C):

Solubility (Moles/L)	Solubility (Grams/L)	Temp (°C)	Ref (#)	Evaluation (T P E A A)	Comments
7.352E-03	1.475E+00	25	B164	1 0 1 1 2	
7.352E-03	1.475E+00	25	B185	0 0 0 0 0	

1734. C$_9$H$_9$ClO$_3$

(4-Chloro-2-methylphenoxy)acetic acid
MCPA

RN: 94-74-6   **MP** (°C):   120.0
MW: 200.62   **BP** (°C):

Solubility (Moles/L)	Solubility (Grams/L)	Temp (°C)	Ref (#)	Evaluation (T P E A A)	Comments
3.138E-03	6.296E-01	20	M061	1 0 0 0 1	
5.852E-03	1.174E+00	25	B164	1 0 1 1 2	
5.852E-03	1.174E+00	25	B185	0 0 0 0 0	
7.975E-03	1.600E+00	25	B185	0 0 0 0 0	

(continued)

1734. C$_9$H$_9$ClO$_3$ (continued)

Solubility (Moles/L)	Solubility (Grams/L)	Temp (°C)	Ref (#)	Evaluation (T P E A A)	Comments
4.979E-03	9.990E-01	ns	B100	0 0 0 0 0	
3.190E-03	6.400E-01	ns	B185	0 0 0 0 0	
4.112E-03	8.250E-01	ns	L024	0 0 0 0 2	
4.112E-03	8.250E-01	rt	M161	0 0 0 0 2	

1735. C$_9$H$_9$Cl$_2$NO
Propanil
3′,4′-Dichloropropionanilide
DPA

RN: 709-98-8   **MP** (°C): 85
MW: 218.08   **BP** (°C):

Solubility (Moles/L)	Solubility (Grams/L)	Temp (°C)	Ref (#)	Evaluation (T P E A A)	Comments
5.961E-04	1.300E-01	20	F311	1 2 2 2 1	
2.293E-03	5.000E-01	ns	B185	0 0 0 0 0	
2.292E-03	4.998E-01	ns	B200	0 0 0 0 0	
2.293E-03	5.000E-01	ns	H042	0 0 0 0 2	
1.032E-03	2.250E-01	rt	M161	0 0 0 0 2	

1736. C$_9$H$_9$Cl$_2$NO$_2$
Dichlormate
3,4-Dichlorobenzyl N-methylcarbamate
Romate

RN: 1966-58-1   **MP** (°C): 52
MW: 234.08   **BP** (°C):

Solubility (Moles/L)	Solubility (Grams/L)	Temp (°C)	Ref (#)	Evaluation (T P E A A)	Comments
7.262E-04	1.700E-01	25	B200	1 0 0 0 2	

1737. C$_9$H$_9$Cl$_2$NO$_2$
UC 22463
Sirmate 4E
Rowmate
Sirmate

RN: 62046-37-1   **MP** (°C): 52
MW: 234.08   **BP** (°C):

Solubility (Moles/L)	Solubility (Grams/L)	Temp (°C)	Ref (#)	Evaluation (T P E A A)	Comments
7.262E-04	1.700E-01	ns	H042	0 0 0 0 2	

1738. $C_9H_9I_2NO_3$

L-3,5-Diiodotyrosine
3,5-Diiodo-L-tyrosine
DIT

RN:	300-39-0	MP (°C):	213
MW:	432.99	BP (°C):	

Solubility (Moles/L)	Solubility (Grams/L)	Temp (°C)	Ref (#)	Evaluation (T P E A A)	Comments
1.431E-03	6.196E-01	25	D041	1 0 0 0 1	

1739. $C_9H_9I_2NO_3$

3,5-Diiodotyrosine
3,5-Diiod-DL-tyrosin
DL-Thyronin

RN:	66-02-4	MP (°C):	204
MW:	432.99	BP (°C):	

Solubility (Moles/L)	Solubility (Grams/L)	Temp (°C)	Ref (#)	Evaluation (T P E A A)	Comments
1.039E-03	4.500E-01	15	F300	1 0 0 0 1	
7.850E-04	3.399E-01	25	D041	1 0 0 0 1	
1.386E-03	6.000E-01	25	F300	1 0 0 0 0	
1.316E-02	5.700E+00	75	F300	1 0 0 0 1	

1740. C_9H_9N

Skatole
3-Methyl-indol
3-Methylindole

RN:	83-34-1	MP (°C):	95
MW:	131.18	BP (°C):	265.5

Solubility (Moles/L)	Solubility (Grams/L)	Temp (°C)	Ref (#)	Evaluation (T P E A A)	Comments
3.430E-03	4.500E-01	16	F300	1 0 0 0 0	

1741. C_9H_9NOS

m-Ethoxyphenyl isothiocyanate
3-Ethoxyphenyl isothiocyanate

RN:	3701-44-8	MP (°C):	
MW:	179.24	BP (°C):	

Solubility (Moles/L)	Solubility (Grams/L)	Temp (°C)	Ref (#)	Evaluation (T P E A A)	Comments
3.800E-04	6.811E-02	25	K032	2 2 0 1 2	

1742. C$_9$H$_9$NOS
p-Ethoxyphenyl isothiocyanate
4-Ethoxyphenyl isothiocyanate
RN: 25687-50-7 **MP** (°C):
MW: 179.24 **BP** (°C):

Solubility (Moles/L)	Solubility (Grams/L)	Temp (°C)	Ref (#)	Evaluation (T P E A A)	Comments
5.500E-05	9.858E-03	25	D019	1 1 1 1 1	

1743. C$_9$H$_9$NO$_2$
p-Acetamidobenzaldehyde
Acetamide, N-(4-formylphenyl)-
Acetanilide, 4'-formyl-
Micotiazone
RN: 122-85-0 **MP** (°C):
MW: 163.18 **BP** (°C):

Solubility (Moles/L)	Solubility (Grams/L)	Temp (°C)	Ref (#)	Evaluation (T P E A A)	Comments
1.990E-02	3.247E+00	25	D044	0 0 0 0 0	

1744. C$_9$H$_9$NO$_3$
Hippuric acid
Hippursaeure
N-Benzoylglycine
Benzoylaminoacetic acid
RN: 495-69-2 **MP** (°C): 187
MW: 179.18 **BP** (°C):

Solubility (Moles/L)	Solubility (Grams/L)	Temp (°C)	Ref (#)	Evaluation (T P E A A)	Comments
1.836E-02	3.289E+00	20	D041	1 0 0 0 1	
2.177E-02	3.900E+00	20	F300	1 0 0 0 1	
2.050E-02	3.673E+00	25	B028	1 0 0 0 2	
2.048E-02	3.670E+00	25	K053	2 2 2 2 2	
2.095E-02	3.754E+00	25	L048	1 2 2 1 2	
2.095E-02	3.754E+00	25	L050	2 0 1 2 2	
2.048E-02	3.670E+00	25.1	N026	0 0 0 0 0	
3.320E-02	5.949E+00	38	B028	1 0 0 0 2	
2.334E-02	4.182E+00	rt	D021	0 0 1 1 1	

1745. C$_9$H$_9$NO$_3$
Acetamide, 2-(benzoyloxy)-
Glycolamide, benzoate

RN: 64649-43-0   **MP** (°C):   121
MW: 179.18   **BP** (°C):

Solubility (Moles/L)	Solubility (Grams/L)	Temp (°C)	Ref (#)	Evaluation (T P E A A)	Comments
2.288E-02	4.100E+00	22	B427	1 0 0 1 1	in 0.01M HCl
2.288E-02	4.100E+00	22	N317	1 1 2 1 2	

1746. C$_9$H$_9$NO$_4$
Benzadox
((Benzoylamino)oxy)acetic acid
Topcide

RN: 5251-93-4   **MP** (°C):
MW: 195.18   **BP** (°C):

Solubility (Moles/L)	Solubility (Grams/L)	Temp (°C)	Ref (#)	Evaluation (T P E A A)	Comments
8.069E-02	1.575E+01	ns	B100	0 0 0 0 1	

1747. C$_9$H$_9$NS
p-Methylbenzyl isothiocyanate
4-Methylbenzyl isothiocyanate

RN: 3694-46-0   **MP** (°C):
MW: 163.24   **BP** (°C):

Solubility (Moles/L)	Solubility (Grams/L)	Temp (°C)	Ref (#)	Evaluation (T P E A A)	Comments
1.600E-04	2.612E-02	25	D014	1 0 0 0 1	

1748. C$_9$H$_9$N$_3$OS
Benzthiazuron
Benzothiazol-2-yl-3-methylurea
N-2-Benzothiazolyl-*N'*-methylurea
Gatnon

RN: 1929-88-0   **MP** (°C):
MW: 207.26   **BP** (°C):

Solubility (Moles/L)	Solubility (Grams/L)	Temp (°C)	Ref (#)	Evaluation (T P E A A)	Comments
5.790E-05	1.200E-02	20	M161	1 0 0 0 1	

1749. C$_9$H$_9$N$_3$O$_2$

Carbendazim
1H-Benzimidazol-2-ylcarbamic acid methyl ester

RN: 10605-21-7 **MP** (°C): 302
MW: 191.19 **BP** (°C):

Solubility (Moles/L)	Solubility (Grams/L)	Temp (°C)	Ref (#)	Evaluation (T P E A A)	Comments
3.034E-05	5.800E-03	20	A064	1 0 1 1 1	
3.034E-05	5.800E-03	20	M161	1 0 0 0 1	pH 7

1750. C$_9$H$_9$N$_3$O$_2$S$_2$

Sulfathiazole
Sulphathiazole
N1-2-Thiazolyl-
4-Amino-N-2-thiazolyl-

RN: 72-14-0 **MP** (°C): 202
MW: 255.32 **BP** (°C):

Solubility (Moles/L)	Solubility (Grams/L)	Temp (°C)	Ref (#)	Evaluation (T P E A A)	Comments
1.410E-03	3.600E-01	16	H114	1 0 0 0 1	
1.743E-03	4.450E-01	20	F073	1 2 2 2 2	
1.958E-03	5.000E-01	20	F074	1 0 0 0 2	
4.426E-03	1.130E+00	20	K028	2 1 2 1 2	pH 7.3, form I
1.414E-03	3.610E-01	20	K028	2 1 2 1 2	pH 3.8, form II
2.460E-03	6.280E-01	20	K028	2 1 2 1 2	pH 7.3, form II
2.483E-03	6.340E-01	20	K028	2 1 2 1 2	pH 3.8, form I
1.347E-03	3.439E-01	20	L058	1 0 1 1 1	
2.482E-03	6.336E-01	20	M042	1 0 0 0 2	pH 3.8, form I, mp 200–202 C
1.413E-03	3.609E-01	20	M042	1 0 0 0 2	pH 3.8, form II, mp 175 C
1.305E-03	3.332E-01	25	F415	0 0 0 0 0	Average
1.461E-03	3.730E-01	25	H005	1 0 1 2 2	average of 4
1.821E-03	4.650E-01	25	K096	1 2 2 2 2	α form
3.290E-03	8.400E-01	25	K096	1 2 2 2 2	β form
1.796E-03	4.586E-01	25	M440	0 0 0 0 0	
1.966E-03	5.020E-01	26	C102	2 0 2 2 2	
2.350E-03	6.000E-01	26	L052	1 0 0 0 0	
2.270E-03	5.796E-01	30	H018	0 0 0 0 0	
4.308E-03	1.100E+00	30	K096	1 2 2 2 2	β form
2.327E-03	5.940E-01	30	K096	1 2 2 2 2	α form
2.544E-03	6.496E-01	30	M046	1 0 0 0 1	
4.460E-03	1.139E+00	30.0	H010	2 2 1 1 2	
3.564E-03	9.100E-01	35	H114	1 0 0 0 1	
3.094E-03	7.900E-01	35	K096	1 2 2 2 2	α form
5.354E-03	1.367E+00	35	K096	1 2 2 2 2	β form
3.760E-03	9.600E-01	37	C102	2 0 2 2 2	
3.564E-03	9.100E-01	37	D084	1 0 1 0 1	
3.678E-03	9.391E-01	37	F072	1 0 0 0 2	

(continued)

1750. C$_9$H$_9$N$_3$O$_2$S$_2$ (continued)

Solubility (Moles/L)	Solubility (Grams/L)	Temp (°C)	Ref (#)	Evaluation (T P E A A)	Comments
3.686E-03	9.411E-01	37	F075	1 0 2 2 2	
3.443E-03	8.790E-01	37	K091	1 0 0 0 2	
2.560E-03	6.536E-01	37	K095	2 0 0 0 2	intrinsic
3.838E-03	9.800E-01	37	L091	1 0 0 0 1	pH 5.5
3.721E-03	9.500E-01	37	M057	1 0 0 0 2	pH 5.5
3.799E-03	9.700E-01	37	R044	0 0 0 0 0	
3.756E-03	9.591E-01	37.50	M142	1 0 0 0 1	
3.603E-03	9.200E-01	38	K006	1 0 0 0 2	
6.619E-03	1.690E+00	40	K096	1 2 2 2 2	β form
4.073E-03	1.040E+00	40	K096	1 2 2 2 2	α form
8.284E-03	2.115E+00	45	K096	1 2 2 2 2	β form
5.288E-03	1.350E+00	45	K096	1 2 2 2 2	α form
6.592E-03	1.683E+00	49	K096	1 2 2 2 2	α form
9.964E-03	2.544E+00	49	K096	1 2 2 2 2	β form
1.683E-03	4.298E-01	ns	L044	0 0 0 0 2	
3.467E-03	8.853E-01	ns	R427	0 0 0 0 0	
1.918E-03	4.898E-01	rt	N015	0 0 2 2 2	

1751. C$_9$H$_{10}$

Indan
2,3-Dihydroindene
Hydrindane
1H-Indene, 2,3-dihydro-
Hydrindene

RN: 496-11-7   **MP** (°C): −51.4
MW: 118.18   **BP** (°C): 176.5

Solubility (Moles/L)	Solubility (Grams/L)	Temp (°C)	Ref (#)	Evaluation (T P E A A)	Comments
9.232E-04	1.091E-01	25	M064	1 1 2 2 2	
7.522E-04	8.890E-02	25	P051	2 1 1 2 2	
9.232E-04	1.091E-01	ns	M344	0 0 0 0 2	

1752. C$_9$H$_{10}$

α-Methylstyrene
2-Phenyl-1-propene
Isopropenylbenzene
2-Phenylpropene
β-Phenylpropene

RN: 98-83-9   **MP** (°C): −24.0
MW: 118.18   **BP** (°C): 167.0

Solubility (Moles/L)	Solubility (Grams/L)	Temp (°C)	Ref (#)	Evaluation (T P E A A)	Comments
9.772E-04	1.155E-01	ns	D001	0 0 0 0 2	

1753. C$_9$H$_{10}$BrClN$_2$O$_2$

Chlorbromuron

3-(4-Bromo-3-chlorophenyl)-1-methoxy-1-methylurea

N'-(4-Bromo-3-chlorophenyl)-N-methoxy-N-methylurea

Maloran

RN: 13360-45-7 **MP** (°C):

MW: 293.55 **BP** (°C):

Solubility (Moles/L)	Solubility (Grams/L)	Temp (°C)	Ref (#)	Evaluation (T P E A A)	Comments
1.202E-04	3.529E-02	20	B179	0 0 0 0 0	
1.192E-04	3.500E-02	20	M161	1 0 0 0 1	
1.703E-04	5.000E-02	ns	B200	0 0 0 0 1	
1.703E-04	5.000E-02	ns	G036	0 0 0 0 1	

1754. C$_9$H$_{10}$Cl$_2$N$_2$O

Diuron

1,1-Dimethyl-3-(3,4-dichlorophenyl)urea

3-(3,4-Dichlorophenyl)-1,1-dimethylurea

RN: 330-54-1 **MP** (°C): 158

MW: 233.10 **BP** (°C):

Solubility (Moles/L)	Solubility (Grams/L)	Temp (°C)	Ref (#)	Evaluation (T P E A A)	Comments
1.820E-04	4.242E-02	20	B179	0 0 0 0 0	
9.438E-05	2.200E-02	20	E048	1 2 1 1 1	
1.716E-04	4.000E-02	25	A039	1 1 0 0 2	
1.802E-04	4.200E-02	25	B185	0 0 0 0 0	
1.802E-04	4.200E-02	25	B200	1 0 0 0 1	
1.802E-04	4.200E-02	25	G036	1 0 0 0 1	
1.802E-04	4.200E-02	25	G099	1 0 0 1 0	
1.600E-04	3.730E-02	25	H073	2 1 1 2 2	
1.802E-04	4.200E-02	25	M061	1 0 0 0 1	
1.802E-04	4.200E-02	25	M161	1 0 0 0 1	
1.802E-04	4.200E-02	25	N333	0 0 0 0 0	
1.716E-04	4.000E-02	ns	B160	0 0 0 0 1	
1.802E-04	4.200E-02	ns	H042	0 0 0 0 1	
1.000E+02	2.331E+04	ns	H342	0 0 0 0 0	EFG, *sic*
1.802E-04	4.200E-02	ns	K007	0 0 0 0 1	
1.995E-04	4.651E-02	ns	M163	0 0 0 0 0	EFG
1.802E-04	4.200E-02	ns	V414	0 0 0 0 0	

1755. C$_9$H$_{10}$Cl$_2$N$_2$O$_2$

Linuron

3-(3,4-Dichlorophenyl)-1-methoxy-1-methylurea

RN: 330-55-2 **MP** (°C): 93

MW: 249.10 **BP** (°C):

Solubility (Moles/L)	Solubility (Grams/L)	Temp (°C)	Ref (#)	Evaluation (T P E A A)	Comments
3.020E-04	7.523E-02	20	B179	0 0 0 0 0	
3.011E-04	7.500E-02	25	B185	0 0 0 0 0	

(*continued*)

1755. C$_9$H$_{10}$Cl$_2$N$_2$O$_2$ (continued)

Solubility (Moles/L)	Solubility (Grams/L)	Temp (°C)	Ref (#)	Evaluation (T P E A A)	Comments
3.011E-04	7.500E-02	25	B200	1 0 0 0 1	
3.011E-04	7.500E-02	25	M061	1 0 0 0 1	
3.011E-04	7.500E-02	25	M161	1 0 0 0 1	
3.252E-04	8.100E-02	25	M162	1 1 0 0 1	
3.011E-04	7.500E-02	ns	K007	0 0 0 0 1	

1756. C$_9$H$_{10}$Cl$_2$O
2,4-Dichloro-6-propyl-phenol
RN: 91399-12-1   **MP** (°C):
MW: 205.09   **BP** (°C):

Solubility (Moles/L)	Solubility (Grams/L)	Temp (°C)	Ref (#)	Evaluation (T P E A A)	Comments
4.900E-04	1.005E-01	25	B316	0 0 0 0 0	

1757. C$_9$H$_{10}$Cl$_3$O$_3$PS
Trichlormetafos-3
O-Methyl O-ethyl O-2,4,5-trichlorophenyl thiophosphate
RN: 2633-54-7   **MP** (°C):
MW: 335.58   **BP** (°C): 127

Solubility (Moles/L)	Solubility (Grams/L)	Temp (°C)	Ref (#)	Evaluation (T P E A A)	Comments
<1.19E-04	<4.00E-02	ns	M061	0 0 0 0 0	

1758. C$_9$H$_{10}$NO$_3$
2-Oxo-5-indolinyl acetate
5-Acetoxy-2-oxindole
RN: 74973-14-1   **MP** (°C):
MW: 180.18   **BP** (°C):

Solubility (Moles/L)	Solubility (Grams/L)	Temp (°C)	Ref (#)	Evaluation (T P E A A)	Comments
2.900E-02	5.225E+00	25	A066	1 0 1 1 1	

1759. C$_9$H$_{10}$NO$_3$PS
Cyanophos
Dimethyl O-(p-cyanophenyl) phosphorothioate
Ciafos
CYAP
RN: 2636-26-2   **MP** (°C): 14.5
MW: 243.22   **BP** (°C):

Solubility (Moles/L)	Solubility (Grams/L)	Temp (°C)	Ref (#)	Evaluation (T P E A A)	Comments
1.891E-04	4.600E-02	30	M161	1 0 0 0 1	

1760. C$_9$H$_{10}$N$_2$O$_2$
Phenacemide
Phenylacetyl urea

RN:	63-98-9	MP (°C):	215
MW:	178.19	BP (°C):	

Solubility (Moles/L)	Solubility (Grams/L)	Temp (°C)	Ref (#)	Evaluation (T P E A A)	Comments
1.021E-03	1.820E-01	ns	B404	0 2 1 1 0	

1761. C$_9$H$_{10}$N$_2$O$_3$
p-Nitroacetotoluide
4-Nitroacetotoluide

RN:		MP (°C):	
MW:	194.19	BP (°C):	

Solubility (Moles/L)	Solubility (Grams/L)	Temp (°C)	Ref (#)	Evaluation (T P E A A)	Comments
1.133E-02	2.200E+00	rt	F043	0 0 2 1 1	

1762. C$_9$H$_{10}$N$_2$O$_3$
p-Ureidophenyl acetate
4-Ureidophenyl acetate

RN:	59746-11-1	MP (°C):	
MW:	194.19	BP (°C):	

Solubility (Moles/L)	Solubility (Grams/L)	Temp (°C)	Ref (#)	Evaluation (T P E A A)	Comments
3.200E-03	6.214E-01	25	A066	1 0 1 1 1	

1763. C$_9$H$_{10}$N$_2$O$_3$
o-Nitroacetotoluide
2-Nitroacetotoluide

RN:	612-45-3	MP (°C):	
MW:	194.19	BP (°C):	

Solubility (Moles/L)	Solubility (Grams/L)	Temp (°C)	Ref (#)	Evaluation (T P E A A)	Comments
1.133E-02	2.200E+00	rt	F043	0 0 2 1 1	

1764. C₉H₁₀N₂O₃S₂
Ethoxzolamide
6-Ethoxy-2-benzothiazolesulfonamide
Diuretic C
Cardrase

RN: 452-35-7   **MP** (°C):   188
MW: 258.32   **BP** (°C):

Solubility (Moles/L)	Solubility (Grams/L)	Temp (°C)	Ref (#)	Evaluation (T P E A A)	Comments
4.000E-05	1.033E-02	25	C415	1 0 0 1 0	
1.548E-04	4.000E-02	ns	M032	0 0 0 0 0	
1.549E-04	4.001E-02	ns	R428	0 0 0 0 0	

1765. C₉H₁₀N₂S
4-Dimethylaminophenyl isothiocyanate
4-Isothiocyanato-*N,N*-dimethyl-benzenamine

RN: 2131-64-8   **MP** (°C):
MW: 178.26   **BP** (°C):

Solubility (Moles/L)	Solubility (Grams/L)	Temp (°C)	Ref (#)	Evaluation (T P E A A)	Comments
7.500E-05	1.337E-02	25	D019	1 1 1 1 1	

1766. C₉H₁₀N₂S
3-Dimethylaminophenyl isothiocyanate
N',N'-Dimethyl-*m*-aminophenyl isothiocyanate

RN: 2392-67-8   **MP** (°C):
MW: 178.26   **BP** (°C):

Solubility (Moles/L)	Solubility (Grams/L)	Temp (°C)	Ref (#)	Evaluation (T P E A A)	Comments
4.200E-04	7.487E-02	25	D019	1 1 1 1 2	
1.950E-04	3.476E-02	25	K032	2 2 0 1 2	

1767. C₉H₁₀N₄
2,6,7-Trimethylpteridine
2:6:7-Trimethylpteridine

RN: 23767-00-2   **MP** (°C):
MW: 174.21   **BP** (°C):

Solubility (Moles/L)	Solubility (Grams/L)	Temp (°C)	Ref (#)	Evaluation (T P E A A)	Comments
7.087E-02	1.235E+01	20	A083	1 2 0 0 0	

1768. $C_9H_{10}N_4O_2S_2$
Sulfamethizole
Sulfamethylthiadiazole

RN:	144-82-1	MP (°C):	208
MW:	270.33	BP (°C):	

Solubility (Moles/L)	Solubility (Grams/L)	Temp (°C)	Ref (#)	Evaluation (T P E A A)	Comments
1.957E-03	5.290E-01	20	F073	1 2 2 2 2	
3.320E-03	8.975E-01	37	A046	2 0 1 1 2	
3.884E-03	1.050E+00	37	B046	1 0 2 2 2	pH 4.5
3.270E-03	8.840E-01	37	K091	1 0 0 0 2	
3.270E-03	8.840E-01	37	W016	2 0 1 1 2	
2.938E-03	7.943E-01	ns	N057	1 0 2 2 0	EFG, intrinsic

1769. $C_9H_{10}O$
Propiophenone
1-Phenyl-1-propanone
Ethyl phenyl ketone
Propiophenoe

RN:	93-55-0	MP (°C):	19
MW:	134.18	BP (°C):	217

Solubility (Moles/L)	Solubility (Grams/L)	Temp (°C)	Ref (#)	Evaluation (T P E A A)	Comments
1.479E-02	1.985E+00	ns	S460	0 0 0 0 0	

1770. $C_9H_{10}O_2$
Hydrocinnamic acid
Hydrozimtsaeure

RN:	501-52-0	MP (°C):	48
MW:	150.18	BP (°C):	280

Solubility (Moles/L)	Solubility (Grams/L)	Temp (°C)	Ref (#)	Evaluation (T P E A A)	Comments
3.929E-02	5.900E+00	20	F300	1 0 0 0 2	
6.162E-02	9.254E+00	30	D033	2 2 1 2 2	
7.668E-02	1.152E+01	40	D033	2 2 1 2 2	

1771. $C_9H_{10}O_2$
2,5-Dimethylbenzoic acid
2-Carboxy-1,4-dimethylbenzene
Isoxylic acid

RN:	610-72-0	MP (°C):	132.5–134.5
MW:	150.18	BP (°C):	268

Solubility (Moles/L)	Solubility (Grams/L)	Temp (°C)	Ref (#)	Evaluation (T P E A A)	Comments
1.199E-03	1.800E-01	25	H007	0 0 0 0 0	

1772. C$_9$H$_{10}$O$_2$

2,4-Dimethylbenzoic acid
4-Carboxy-1,3-dimethylbenzene

RN: 611-01-8   **MP** (°C):   124–126
MW: 150.18   **BP** (°C):   267

Solubility (Moles/L)	Solubility (Grams/L)	Temp (°C)	Ref (#)	Evaluation (T P E A A)	Comments
1.065E-03	1.600E-01	25	H007	0 0 0 0 0	

1773. C$_9$H$_{10}$O$_2$

Benzyl acetate
Phenylmethyl acetate
Acetic acid phenylmethyl ester
α-Acetoxytoluene

RN: 140-11-4   **MP** (°C):   −51.3
MW: 150.18   **BP** (°C):   206

Solubility (Moles/L)	Solubility (Grams/L)	Temp (°C)	Ref (#)	Evaluation (T P E A A)	Comments
9.973E-03	1.498E+00	25	M350	1 0 1 1 1	

1774. C$_9$H$_{10}$O$_2$

3,4-Dimethylbenzoic acid
1-Carboxy-3,4-dimethylbenzene

RN: 619-04-5   **MP** (°C):   165
MW: 150.18   **BP** (°C):

Solubility (Moles/L)	Solubility (Grams/L)	Temp (°C)	Ref (#)	Evaluation (T P E A A)	Comments
8.600E-04	1.292E-01	ns	C014	0 0 0 1 1	

1775. C$_9$H$_{10}$O$_2$

Ethyl benzoate
Ethyl p-benzoate
Benzoesaeure-aethyl ester

RN: 93-89-0   **MP** (°C):   −34
MW: 150.18   **BP** (°C):   212

Solubility (Moles/L)	Solubility (Grams/L)	Temp (°C)	Ref (#)	Evaluation (T P E A A)	Comments
7.990E-03	1.200E+00	22	N317	1 1 2 1 2	
4.794E-03	7.200E-01	25	A003	1 2 1 2 1	
6.659E-03	1.000E+00	60	F300	1 0 0 0 0	

1776. C$_9$H$_{10}$O$_3$

4-Hydroxy-3-ethoxybenzaldehyde
Ethylprotal; ethylvanillin
Bourbonal
Ethovan
NSC 67240
Ethavan

RN: 121-32-4   **MP** (°C): 65
MW: 166.18   **BP** (°C):

Solubility (Moles/L)	Solubility (Grams/L)	Temp (°C)	Ref (#)	Evaluation (T P E A A)	Comments
1.019E+00	1.693E+02	25	D407	1 0 2 2 2	

1777. C$_9$H$_{10}$O$_3$

Ethyl salicylate
Ethyl *o*-hydroxybenzoate

RN: 118-61-6   **MP** (°C): 1–3
MW: 166.18   **BP** (°C):

Solubility (Moles/L)	Solubility (Grams/L)	Temp (°C)	Ref (#)	Evaluation (T P E A A)	Comments
4.032E-02	6.700E+00	37	D009	1 2 1 1 1	0.1N HCl

1778. C$_9$H$_{10}$O$_3$

Ethylparaben
4-Hydroxybenzoic acid ethyl ester
Ethyl *p*-hydroxybenzoate
Ethyl 4-hydroxybenzoate

RN: 120-47-8   **MP** (°C): 116
MW: 166.18   **BP** (°C): 297

Solubility (Moles/L)	Solubility (Grams/L)	Temp (°C)	Ref (#)	Evaluation (T P E A A)	Comments
2.750E-03	4.570E-01	15	B355	0 0 0 0 0	
3.370E-03	5.600E-01	20	B355	0 0 0 0 0	
4.910E-03	8.159E-01	20	C006	1 2 1 1 2	
5.329E-03	8.855E-01	25	A059	1 0 1 1 1	
4.090E-03	6.797E-01	25	B355	0 0 0 0 0	
4.510E-03	7.494E-01	25	D081	1 2 2 1 2	
5.300E-03	8.807E-01	25	D339	0 0 0 0 0	
6.310E-03	1.049E+00	25	F322	2 0 1 1 0	EFG
9.628E-03	1.600E+00	25	O027	1 0 1 0 1	
6.379E-03	1.060E+00	25	P013	0 0 0 0 0	
9.500E-03	1.579E+00	27	B129	2 2 2 2 1	
5.200E-03	8.641E-01	27	G078	2 1 0 1 0	EFG
5.400E-03	8.974E-01	27.0	G067	2 0 1 1 1	
6.770E-03	1.125E+00	30	A059	1 0 1 1 2	
8.266E-03	1.374E+00	35	A059	1 0 1 1 2	
7.568E-03	1.258E+00	39.3	G302	2 2 2 2 0	EFG
9.540E-03	1.585E+00	40	A059	1 0 1 1 2	

1779. C$_9$H$_{10}$O$_3$
Methyl-4-methoxybenzoate
Methyl anisate

RN:	121-98-2	MP (°C):	49
MW:	166.18	BP (°C):	244

Solubility (Moles/L)	Solubility (Grams/L)	Temp (°C)	Ref (#)	Evaluation (T P E A A)	Comments
3.870E-03	6.431E-01	20	C006	1 0 1 1 2	

1780. C$_9$H$_{10}$O$_3$
DL-Tropic acid
DL-Tropasaeure

RN:	529-64-6	MP (°C):	118.5
MW:	166.18	BP (°C):	

Solubility (Moles/L)	Solubility (Grams/L)	Temp (°C)	Ref (#)	Evaluation (T P E A A)	Comments
1.173E-01	1.950E+01	20	F300	1 0 0 0 2	

1781. C$_9$H$_{10}$O$_4$
3,4-Methoxybenzoic acid
Veratrumsaeure

RN:	93-07-2	MP (°C):	
MW:	182.18	BP (°C):	

Solubility (Moles/L)	Solubility (Grams/L)	Temp (°C)	Ref (#)	Evaluation (T P E A A)	Comments
2.745E-03	5.000E-01	14	F300	1 0 0 0 0	
3.293E-02	6.000E+00	100	F300	1 0 0 0 0	

1782. C$_9$H$_{11}$BrN$_2$O$_2$
Metobromuron
3-(p-Bromophenyl)-1-methoxy-1-methylurea
Patoran
N'-(4-Bromophenyl)-N-methoxy-N-methylurea
Pattonex

RN:	3060-89-7	MP (°C):	
MW:	259.11	BP (°C):	

Solubility (Moles/L)	Solubility (Grams/L)	Temp (°C)	Ref (#)	Evaluation (T P E A A)	Comments
1.288E-03	3.338E-01	20	B179	0 0 0 0 0	
1.274E-03	3.300E-01	20	B200	1 0 0 0 2	
1.274E-03	3.300E-01	20	G036	1 0 0 0 2	
1.274E-03	3.300E-01	20	M061	1 0 0 0 1	
1.274E-03	3.300E-01	20	M161	1 0 0 0 2	
1.157E-03	2.999E-01	ns	B100	0 0 0 0 0	

1783. C$_9$H$_{11}$ClN$_2$O
Monuron
N'-(4-Chlorophenyl)-N,N-dimethyl-urea
1,1-Dimethyl-3-(p-chlorophenyl)urea

RN:	150-68-5	**MP (°C):**	170.5	
MW:	198.65	**BP (°C):**		

Solubility (Moles/L)	Solubility (Grams/L)	Temp (°C)	Ref (#)	Evaluation (T P E A A)	Comments
1.007E-03	2.000E-01	18	F035	1 0 0 0 0	
1.175E-03	2.334E-01	20	B179	0 0 0 0 0	
1.007E-03	2.000E-01	20	E048	1 2 1 1 2	
1.007E-03	2.000E-01	20	F311	1 2 2 2 1	
1.158E-03	2.300E-01	25	A039	1 1 0 0 2	
1.158E-03	2.300E-01	25	B185	0 0 0 0 0	
1.158E-03	2.300E-01	25	B200	1 0 0 0 2	
1.158E-03	2.300E-01	25	G036	1 0 0 0 2	
1.158E-03	2.300E-01	25	G099	1 0 0 1 0	
1.319E-03	2.620E-01	25	H073	2 1 1 2 2	
1.158E-03	2.300E-01	25	M061	1 0 0 0 2	
1.158E-03	2.300E-01	25	M161	1 0 0 0 2	
1.007E-03	2.000E-01	ns	B100	0 0 0 0 0	
1.158E-03	2.300E-01	ns	B160	0 0 0 0 2	
9.000E-04	1.788E-01	ns	F184	0 0 0 0 0	
1.158E-03	2.300E-01	ns	H112	0 0 0 0 2	
1.158E-03	2.300E-01	ns	K007	0 0 0 0 2	
1.158E-03	2.300E-01	ns	N013	0 0 0 0 2	

1784. C$_9$H$_{11}$ClN$_2$O$_2$
Monolinuron
3-(4-Chlorophenyl)-1-methoxy-1-methylurea
Arresin
Afesin
Aresin

RN:	1746-81-2	**MP (°C):**	80	
MW:	214.65	**BP (°C):**		

Solubility (Moles/L)	Solubility (Grams/L)	Temp (°C)	Ref (#)	Evaluation (T P E A A)	Comments
2.692E-03	5.777E-01	20	B179	0 0 0 0 0	
4.333E-03	9.300E-01	20	G036	1 0 0 0 2	
2.702E-03	5.800E-01	20	M061	1 0 0 0 2	
2.702E-03	5.800E-01	22.5	G301	0 0 0 0 0	
3.424E-03	7.350E-01	25	M162	1 1 0 0 2	
2.794E-03	5.996E-01	ns	B100	0 0 0 0 0	
2.702E-03	5.800E-01	rt	M161	0 0 0 0 2	

1785. C$_9$H$_{11}$ClO

3-Methyl-5-ethyl-4-chloro-phenol

m-Cresol, 4-chloro-5-ethyl-

RN: 1125-66-2   **MP** (°C):

MW: 170.64   **BP** (°C):

Solubility (Moles/L)	Solubility (Grams/L)	Temp (°C)	Ref (#)	Evaluation (T P E A A)	Comments
2.200E-03	3.754E-01	25	B316	0 0 0 0 0	

1786. C$_9$H$_{11}$Cl$_2$N$_3$O$_4$S$_2$

Methylclothiazide

2H-1,2,4-Benzothiadiazine-7-sulfonamide, 6-chloro-3-(chloromethyl)-3,4-dihydro-2-methyl-1,1-dioxide

6-Chloro-3-(chloromethyl)-3,4-dihydro-2-methyl-2H-1,2,4-benzothiadiazine-7-sulfonamide 1,1-dioxide

RN: 135-07-9   **MP** (°C):

MW: 360.24   **BP** (°C):

Solubility (Moles/L)	Solubility (Grams/L)	Temp (°C)	Ref (#)	Evaluation (T P E A A)	Comments
1.388E-04	5.000E-02	rt	A095	0 0 2 2 0	

1787. C$_9$H$_{11}$Cl$_3$NO$_3$PS

Chlorpyrifos

O,O-Diethyl O-3,5,6-trichloro-2-pyridyl phosphorothioate

DOWCO 179

RN: 2921-88-2   **MP** (°C):   41.5

MW: 350.59   **BP** (°C):

Solubility (Moles/L)	Solubility (Grams/L)	Temp (°C)	Ref (#)	Evaluation (T P E A A)	Comments
1.284E-06	4.502E-04	10	B324	0 0 0 0 0	
1.284E-06	4.500E-04	10	B324	0 0 0 0 0	
1.997E-06	7.000E-04	19	B169	2 1 1 1 1	
2.082E-06	7.299E-04	20	B300	2 1 1 1 2	
2.082E-06	7.299E-04	20	B324	0 0 0 0 0	
2.082E-06	7.300E-04	20	B324	0 0 0 0 0	
1.141E-06	4.000E-04	23	B096	1 2 0 0 0	
3.195E-06	1.120E-03	24	F179	2 2 2 2 2	
1.141E-06	4.000E-04	24	K069	2 0 0 1 1	
3.708E-06	1.300E-03	30	B324	0 0 0 0 0	
3.708E-06	1.300E-03	30	B324	0 0 0 0 0	
5.705E-06	2.000E-03	35	M161	1 0 0 0 0	
1.141E-06	4.000E-04	ns	F071	0 1 2 1 0	
8.557E-07	3.000E-04	ns	K138	0 0 0 0 1	
5.705E-06	2.000E-03	ns	M110	0 0 0 0 0	EFG
3.195E-06	1.120E-03	ns	V414	0 0 0 0 0	
5.705E-06	2.000E-03	ns	Y414	0 0 0 0 0	

1788. C$_9$H$_{11}$Cl$_3$NO$_4$P
Chlorpyrifos oxon
Chlorpyrifos oxygen analog
Dursban oxygen analog
DOWCO 180
3,5,6-Trichloro-2-pyridyl diethyl phosphate
RN: 5598-15-2   MP (°C):
MW: 334.53   BP (°C):

Solubility (Moles/L)	Solubility (Grams/L)	Temp (°C)	Ref (#)	Evaluation (T P E A A)	Comments
1.554E-03	5.200E-01	24	K069	2 0 0 1 1	

1789. C$_9$H$_{11}$FN$_2$O$_3$
2,4(1H,3H)-Pyrimidinedione, 5-fluoro-3-(1-oxopentyl)-
RN: 145303-99-7   MP (°C):
MW: 214.20   BP (°C):

Solubility (Moles/L)	Solubility (Grams/L)	Temp (°C)	Ref (#)	Evaluation (T P E A A)	Comments
5.000E-03	1.071 E+00	22	B416	2 2 1 2 1	

1790. C$_9$H$_{11}$FN$_2$O$_4$
1-Butyryloxymethyl-5-fluorouracil
Butanoic acid, (5-fluoro-3,4-dihydro-2,4-dioxo-1(2H)-pyrimidinyl)methyl ester
RN: 66542-37-8   MP (°C):
MW: 230.20   BP (°C):

Solubility (Moles/L)	Solubility (Grams/L)	Temp (°C)	Ref (#)	Evaluation (T P E A A)	Comments
4.170E-02	9.600E+00	22	B321	0 0 0 0 0	pH 4.0
4.170E-02	9.600E+00	22	B332	1 1 0 0 1	pH 4.0
4.952E-02	1.140E+01	22	M317	1 1 1 1 1	

1791. C$_9$H$_{11}$FN$_2$O$_4$
1-Isobutyloxycarbonyl-5-fluorouracil
1(2H)-Pyrimidinecarboxylic acid, 5-fluoro-3,4-dihydro-2,4-dioxo-, 2-methylpropyl ester
RN: 71759-45-0   MP (°C):
MW: 230.20   BP (°C):

Solubility (Moles/L)	Solubility (Grams/L)	Temp (°C)	Ref (#)	Evaluation (T P E A A)	Comments
1.303E-02	3.000E+00	22	B332	1 1 0 0 1	pH 4.0

1792. C₉H₁₁FN₂O₄
1-Butyloxycarbonyl-5-fluorouracil
5-Fluoro-1-(butoxycarbonyl)uracil
RN: 85326-32-5 **MP** (°C):
MW: 230.20 **BP** (°C):

Solubility (Moles/L)	Solubility (Grams/L)	Temp (°C)	Ref (#)	Evaluation (T P E A A)	Comments
2.563E-02	5.900E+00	22	B332	1 1 0 0 1	pH 4.0

1793. C₉H₁₁IN₂O₅
2′-Deoxy-5-iodouridine
Idoxuridine
(+)-5-Iodo-2′-deoxyuridine
Herplex
RN: 54-42-2 **MP** (°C): 165
MW: 354.10 **BP** (°C):

Solubility (Moles/L)	Solubility (Grams/L)	Temp (°C)	Ref (#)	Evaluation (T P E A A)	Comments
5.650E+03	2.001E+06	25	N332	0 0 0 0 0	pH 7.4

1794. C₉H₁₁N
1,2,3,4-Tetrahydroquinoline
Kusol
THQ
RN: 635-46-1 **MP** (°C): 15–17
MW: 133.19 **BP** (°C): 249

Solubility (Moles/L)	Solubility (Grams/L)	Temp (°C)	Ref (#)	Evaluation (T P E A A)	Comments
1.054E-02	1.404E+00	20.3	L339	2 0 2 2 2	
1.386E-02	1.847E+00	40.0	L339	2 0 2 2 2	
1.774E-02	2.362E+00	59.8	L339	2 0 2 2 2	
2.326E-02	3.098E+00	79.6	L339	2 0 2 2 2	
2.988E-02	3.980E+00	100.4	L339	2 0 2 2 2	

1795. C₉H₁₁NO
N-Methylacetanilide
Acetamide, N-methyl-N-phenyl-
RN: 579-10-2 **MP** (°C): 102
MW: 149.19 **BP** (°C):

Solubility (Moles/L)	Solubility (Grams/L)	Temp (°C)	Ref (#)	Evaluation (T P E A A)	Comments
1.475E-01	2.200E+01	20	B101	0 0 0 0 0	
1.673E-01	2.496E+01	25	B101	0 0 0 0 0	
1.908E-01	2.847E+01	30	B101	0 0 0 0 0	
2.166E-01	3.232E+01	35	B101	0 0 0 0 0	
1.122E-01	1.674E+01	ns	R424	0 0 0 0 0	

1796. C$_9$H$_{11}$NO
p-Aminopropiophenone
4′-Aminopropiophenone
RN: 70-69-9   **MP** (°C):   140
MW: 149.19   **BP** (°C):

Solubility (Moles/L)	Solubility (Grams/L)	Temp (°C)	Ref (#)	Evaluation (T P E A A)	Comments
2.360E-03	3.521E-01	37.5	G002	1 1 1 1 2	pH 6.8

1797. C$_9$H$_{11}$NO
Propionanilide
Propionsaeure-anilid
Propanilide
RN: 620-71-3   **MP** (°C):   106
MW: 149.19   **BP** (°C):

Solubility (Moles/L)	Solubility (Grams/L)	Temp (°C)	Ref (#)	Evaluation (T P E A A)	Comments
1.206E-02	1.800E+00	18	F300	1 0 0 0 1	
1.204E-02	1.797E+00	20	B101	0 0 0 0 0	*

1798. C$_9$H$_{11}$NO
Methyl, [3-(acetylamino)phenyl]-
m-Toluidin-*N*-acetat
m-Toluidine-*N*-acetate
RN: 113321-22-5   **MP** (°C):
MW: 149.19   **BP** (°C):

Solubility (Moles/L)	Solubility (Grams/L)	Temp (°C)	Ref (#)	Evaluation (T P E A A)	Comments
2.949E-02	4.400E+00	13	F300	1 0 0 0 1	

1799. C$_9$H$_{11}$NO$_2$
Phe
(*S*)-(−)-Phenylalanine
(*S*)-Phenylalanine
2-Amino-3-phenylpropanoic acid
Phenylalanine
RN: 63-91-2   **MP** (°C):   283
MW: 165.19   **BP** (°C):   295

Solubility (Moles/L)	Solubility (Grams/L)	Temp (°C)	Ref (#)	Evaluation (T P E A A)	Comments
6.047E-02	9.989E+00	0	D018	2 2 2 1 2	
1.174E-01	1.940E+01	0	F300	1 0 0 0 2	
1.740E-01	2.874E+01	15	D349	2 1 1 2 2	
1.515E-01	2.502E+01	20	B032	1 2 2 1 2	
1.770E-01	2.924E+01	20	D349	2 1 1 2 2	

(continued)

1799. C$_9$H$_{11}$NO$_2$ (continued)

Solubility (Moles/L)	Solubility (Grams/L)	Temp (°C)	Ref (#)	Evaluation (T P E A A)	Comments
1.637E-01	2.705E+01	25	B032	1 2 2 1 2	
1.740E-01	2.875E+01	25	D041	1 0 0 0 2	
1.800E-01	2.973E+01	25	D349	2 1 1 2 2	
1.816E-01	3.000E+01	25	F300	1 0 0 0 1	
1.649E-01	2.724E+01	25	G092	2 1 1 1 1	
1.649E-01	2.724E+01	25	G315	0 0 0 0 0	
1.625E-01	2.684E+01	25	G433	0 0 0 0 0	
1.589E-01	2.625E+01	25	K031	2 1 2 1 2	
1.200E-01	1.982E+01	25	M097	2 2 2 2 2	
1.494E-01	2.468E+01	25	M374	1 0 2 1 2	
2.100E-01	3.469E+01	25	N001	0 0 0 0 0	EFG
1.720E-01	2.841E+01	25	N012	2 0 2 1 2	
1.574E-01	2.601E+01	25	O316	1 0 1 2 2	
1.575E-01	2.601E+01	25	O316	1 0 1 2 2	
1.689E-01	2.790E+01	25.1	N024	0 0 0 0 0	
1.689E-01	2.790E+01	25.1	N025	0 0 0 0 0	
1.689E-01	2.790E+01	25.1	N026	0 0 0 0 0	
1.649E-01	2.724E+01	25.1	N027	1 1 2 2 2	
1.717E-01	2.837E+01	27	D036	0 0 0 0 0	
1.683E-01	2.780E+01	27	D036	0 0 0 0 0	
1.834E-01	3.030E+01	28	L081	2 1 2 2 2	
1.790E-01	2.957E+01	29.80	B032	1 2 2 1 2	
2.567E-01	4.240E+01	50	F300	1 0 0 0 2	
3.761E-01	6.212E+01	75	D041	1 0 0 0 2	
3.759E-01	6.210E+01	75	F300	1 0 0 0 2	
4.619E-01	7.630E+01	98	M160	2 1 1 1 0	
5.454E-01	9.010E+01	100	F300	1 0 0 0 2	
9.064E-02	1.497E+01	rt	H431	0 0 0 0 0	

1800. C$_9$H$_{11}$NO$_2$
4-(Dimethylamino)benzoic acid
4-Dimethylaminobenzoic acid

RN: 619-84-1   **MP (°C):**   242.5
MW: 165.19   **BP (°C):**

Solubility (Moles/L)	Solubility (Grams/L)	Temp (°C)	Ref (#)	Evaluation (T P E A A)	Comments
4.000E-04	6.608E-02	ns	C014	0 0 0 1 1	

1801. C$_9$H$_{11}$NO$_2$
p-Methoxyacetanilide
p-Acetanisidine
N-(4-Methoxyphenyl)acetamide
N-(4-Methoxyphenyl)acetic acid amide
p-Acetanisidide
Acetamide, N-(4-methoxyphenyl)-
RN: 51-66-1   **MP (°C):**   400.3
MW: 165.19   **BP (°C):**

Solubility (Moles/L)	Solubility (Grams/L)	Temp (°C)	Ref (#)	Evaluation (T P E A A)	Comments
1.029E-02	1.700E+00	15	F300	1 0 0 0 1	
8.820E-03	1.457E+00	15	M352	1 1 1 1 2	
7.090E-02	1.171E+01	25	D044	0 0 0 0 0	
1.353E-02	2.234E+00	25	M352	1 1 1 1 2	
2.131E-02	3.521E+00	40	M352	1 1 1 1 2	
3.249E-02	5.367E+00	50	M352	1 1 1 1 2	

1802. C$_9$H$_{11}$NO$_2$
2-Methyl-4-acetaminophenol
3-Methyl-4-hydroxyacetanilide
3-Methylparacetamol
RN: 16375-90-9   **MP (°C):**
MW: 165.19   **BP (°C):**

Solubility (Moles/L)	Solubility (Grams/L)	Temp (°C)	Ref (#)	Evaluation (T P E A A)	Comments
2.536E-02	4.189E+00	25	D078	1 2 1 1 2	

1803. C$_9$H$_{11}$NO$_2$
DL-Phenylalanine
DL-Phenylalanin
RN: 150-30-1   **MP (°C):**   166.5
MW: 165.19   **BP (°C):**

Solubility (Moles/L)	Solubility (Grams/L)	Temp (°C)	Ref (#)	Evaluation (T P E A A)	Comments
5.993E-02	9.900E+00	0	F300	1 0 0 0 1	
9.080E-02	1.500E+01	21	F300	1 0 0 0 1	
9.008E-02	1.488E+01	21	P045	1 0 2 1 2	
8.464E-02	1.398E+01	25	D018	2 2 2 1 2	
8.476E-02	1.400E+01	25	D041	1 0 0 0 2	
1.304E-01	2.154E+01	50	D018	2 2 2 1 2	
1.295E-01	2.140E+01	50	F300	1 0 0 0 2	
2.158E-01	3.564E+01	75	D018	2 2 2 1 2	
2.164E-01	3.575E+01	75	D041	1 0 0 0 2	
2.167E-01	3.580E+01	75	F300	1 0 0 0 2	
3.898E-01	6.440E+01	100	F300	1 0 0 0 2	

1804. C₉H₁₁NO₂

m-Tolyl methylcarbamate
3-Tolyl methylcarbamate

RN: 1129-41-5 **MP (°C):** 76.5
MW: 165.19 **BP (°C):**

Solubility (Moles/L)	Solubility (Grams/L)	Temp (°C)	Ref (#)	Evaluation (T P E A A)	Comments
1.574E-02	2.600E+00	30	M161	1 0 0 0 1	

1805. C₉H₁₁NO₂

D-Phenylalanine
D-α-Aminohydrocinnamic acid
D-α-Amino-β-phenylpropionic acid
D-β-Phenyl-α-aminopropionic acid
D-PHE

RN: 673-06-3 **MP (°C):** 273
MW: 165.19 **BP (°C):**

Solubility (Moles/L)	Solubility (Grams/L)	Temp (°C)	Ref (#)	Evaluation (T P E A A)	Comments
1.763E-01	2.913E+01	25	D041	1 0 0 0 0	

1806. C₉H₁₁NO₂

Ethyl p-aminobenzoate
4-Aminobenzoic acid ethyl ester
Ethyl p-aminobenzoic acid
Benzocaine

RN: 94-09-7 **MP (°C):** 89.0
MW: 165.19 **BP (°C):**

Solubility (Moles/L)	Solubility (Grams/L)	Temp (°C)	Ref (#)	Evaluation (T P E A A)	Comments
4.308E-03	7.117E-01	15	M352	1 1 1 1 2	
1.513E-02	2.500E+00	20	F300	1 0 0 0 1	
5.800E-03	9.581E-01	25	A418	0 0 0 0 0	
4.840E-03	7.995E-01	25	H008	0 0 0 0 0	
6.493E-03	1.073E+00	25	M352	1 1 1 1 2	
6.216E-03	1.027E+00	25	P303	0 0 0 0 0	
6.980E-03	1.153E+00	30	A418	0 0 0 0 0	
7.930E-03	1.310E+00	30	B071	1 2 1 1 2	
5.150E-03	8.507E-01	30	H018	0 0 0 0 0	
7.500E-03	1.239E+00	30	J018	1 2 0 1 1	0.05N NaOH
7.000E-03	1.156E+00	30	L069	1 0 1 1 0	EFG
7.680E-03	1.269E+00	30	R003	0 0 0 0 0	
8.156E-03	1.347E+00	33	P303	0 0 0 0 0	
8.750E-03	1.445E+00	35	A418	0 0 0 0 0	
1.020E-02	1.685E+00	37	F006	1 1 2 2 2	
1.024E-02	1.692E+00	40	A418	0 0 0 0 0	
1.164E-02	1.924E+00	40	M352	1 1 1 1 2	

(continued)

1806. $C_9H_{11}NO_2$ (continued)

Solubility (Moles/L)	Solubility (Grams/L)	Temp (°C)	Ref (#)	Evaluation (T P E A A)	Comments
1.032E-02	1.704E+00	40	P303	0 0 0 0 0	
1.701E-02	2.810E+00	50	M352	1 1 1 1 2	
>3.03E-03	>5.00E-01	ns	B404	0 2 1 1 0	
4.810E-03	7.946E-01	ns	M066	0 0 0 0 2	
4.810E-03	7.946E-01	rt	B016	0 0 1 1 2	pH 7.4
5.135E-03	8.483E-01	rt	I404	0 0 0 0 0	Average

1807. $C_9H_{11}NO_3$

L-Tyrosine
3-(4-Hydroxyphenyl)-L-alanine
Tyrosine
(S)-(−)-Tyrosine
p-Tyrosine
L-Tyrosin

RN:	60-18-4	**MP (°C):**	342dec	
MW:	181.19	**BP (°C):**		

Solubility (Moles/L)	Solubility (Grams/L)	Temp (°C)	Ref (#)	Evaluation (T P E A A)	Comments
1.241E-03	2.249E-01	0	D018	2 2 2 1 2	
1.104E-03	2.000E-01	0	F300	1 0 0 0 0	
2.042E-03	3.700E-01	20	B032	1 2 2 1 2	
2.495E-03	4.520E-01	21	P045	1 0 2 1 2	
2.285E-03	4.140E-01	22	A045	2 0 2 2 2	
2.800E-03	5.073E-01	24.99	C404	2 1 2 2 1	
7.800E-02	1.413E+01	25	C405	2 1 2 2 2	intrinsic zwit
2.642E-03	4.788E-01	25	D018	2 2 2 1 2	
2.482E-03	4.498E-01	25	D041	1 0 0 0 1	
2.759E-03	5.000E-01	25	F300	1 0 0 0 0	
2.444E-03	4.428E-01	25	G433	0 0 0 0 0	
2.620E-03	4.747E-01	25	H097	2 2 2 2 2	
2.622E-03	4.750E-01	25.1	N024	0 0 0 0 0	
2.495E-03	4.520E-01	25.1	N025	0 0 0 0 0	
2.489E-03	4.510E-01	25.1	N026	0 0 0 0 0	
2.488E-03	4.508E-01	25.1	N027	1 1 2 2 2	
2.753E-03	4.988E-01	27	D036	0 0 0 0 0	
2.677E-03	4.850E-01	27	D036	0 0 0 0 0	
3.195E-03	5.790E-01	28	L081	2 1 2 2 2	
3.800E-03	6.885E-01	34.99	C404	2 1 2 2 1	
5.050E-03	9.150E-01	44.99	C404	2 1 2 2 1	
6.064E-03	1.099E+00	50	D018	2 2 2 1 2	
6.071E-03	1.100E+00	50	F300	1 0 0 0 1	
1.309E-02	2.372E+00	75	D018	2 2 2 1 2	
1.343E-02	2.434E+00	75	D041	1 0 0 0 2	
1.325E-02	2.400E+00	75	F300	1 0 0 0 1	
3.091E-02	5.600E+00	100	F300	1 0 0 0 1	

1808. C$_9$H$_{11}$NO$_3$
D-Tyrosine
3-(4-Hydroxyphenyl)-D-alanine

RN:	556-02-5	**MP** (°C):	>300
MW:	181.19	**BP** (°C):	

Solubility (Moles/L)	Solubility (Grams/L)	Temp (°C)	Ref (#)	Evaluation (T P E A A)	Comments
2.482E-03	4.498E-01	25	D041	1 0 0 0 1	
5.789E-03	1.049E+00	50	D041	1 0 0 0 2	

1809. C$_9$H$_{11}$NO$_3$
DL-Tyrosine
DL-Tyrosin
3-(4-Hydroxyphenyl)-DL-alanine
DL-2-Amino-3-(4-hydroxyphenyl)-propanoic acid

RN:	556-03-6	**MP** (°C):	325
MW:	181.19	**BP** (°C):	

Solubility (Moles/L)	Solubility (Grams/L)	Temp (°C)	Ref (#)	Evaluation (T P E A A)	Comments
5.519E-04	1.000E-01	0	F300	1 0 0 0 0	
2.208E-03	4.000E-01	20	F300	1 0 0 0 0	
1.936E-03	3.509E-01	25	D041	1 0 0 0 2	
4.610E-03	8.353E-01	50	D041	1 0 0 0 2	
4.415E-03	8.000E-01	50	F300	1 0 0 0 0	
3.753E-02	6.800E+00	100	F300	1 0 0 0 1	

1810. C$_9$H$_{11}$NO$_4$
Dopa
DL-3-(3,4-Dihydroxyphenyl)alanine
DL-Dopa

RN:	63-84-3	**MP** (°C):	>270
MW:	197.19	**BP** (°C):	

Solubility (Moles/L)	Solubility (Grams/L)	Temp (°C)	Ref (#)	Evaluation (T P E A A)	Comments
2.523E-02	4.975E+00	20	D041	1 0 0 0 0	
1.237E-01	2.439E+01	100	D041	1 0 0 0 1	

1811. C$_9$H$_{11}$NO$_4$
Levodopa
L-3,4-Dihydroxyphenylalanin

RN:	59-92-7	**MP** (°C):	277
MW:	197.19	**BP** (°C):	

Solubility (Moles/L)	Solubility (Grams/L)	Temp (°C)	Ref (#)	Evaluation (T P E A A)	Comments
2.536E-02	5.000E+00	20	F300	1 0 0 0 0	
1.917E-02	3.780E+00	25	H015	1 0 0 0 2	

(continued)

1811. C$_9$H$_{11}$NO$_4$ (continued)

Solubility (Moles/L)	Solubility (Grams/L)	Temp (°C)	Ref (#)	Evaluation (T P E A A)	Comments
1.927E-02	3.800E+00	25.1	N025	0 0 0 0 0	
1.268E-01	2.500E+01	100	F300	1 0 0 0 1	
5.071E-03	1.000E+00	ns	K444	0 0 0 0 0	

1812. C$_9$H$_{11}$NS$_2$Hg

Phenylmercury dimethyldithiocarbamate
Chipman merbam
Merfenl 51
Phelam DP

RN: 32407-99-1   **MP** (°C):   175
MW: 397.91   **BP** (°C):

Solubility (Moles/L)	Solubility (Grams/L)	Temp (°C)	Ref (#)	Evaluation (T P E A A)	Comments
1.508E-05	6.000E-03	20	M161	1 0 0 0 0	

1813. C$_9$H$_{11}$N$_3$O

Biacetyl mono(2-pyridyl)-hydrazone
BPH
Biacetyl mono(2-pyridyl)hydrazone

RN: 74158-10-4   **MP** (°C):   95
MW: 177.21   **BP** (°C):

Solubility (Moles/L)	Solubility (Grams/L)	Temp (°C)	Ref (#)	Evaluation (T P E A A)	Comments
5.643E-04	9.999E-02	ns	R080	0 0 0 0 0	

1814. C$_9$H$_{11}$N$_3$O$_2$S$_2$

Sulfathiazoline
Benzenesulfonamide, 4-amino-N-(4,5-dihydro-2-thiazolyl)-

RN: 32365-02-9   **MP** (°C):
MW: 257.33   **BP** (°C):

Solubility (Moles/L)	Solubility (Grams/L)	Temp (°C)	Ref (#)	Evaluation (T P E A A)	Comments
5.790E-04	1.490E-01	20	F073	1 2 2 2 2	

1815. C$_9$H$_{11}$N$_3$O$_4$

Orotic acid morpholine

RN: **MP** (°C):   289–291
MW: 225.21   **BP** (°C):

Solubility (Moles/L)	Solubility (Grams/L)	Temp (°C)	Ref (#)	Evaluation (T P E A A)	Comments
4.400E-01	9.909E+01	–4	N018	0 0 0 0 0	
6.500E-01	1.464E+02	16	N018	0 0 0 0 0	
7.450E-01	1.678E+02	25	N018	0 0 0 0 0	

1816. $C_9H_{11}O_4P$
2-Carboxyethylphenylphosphinic acid
CEPPA
RN: **MP** (°C):
MW: 214.16   **BP** (°C):

Solubility (Moles/L)	Solubility (Grams/L)	Temp (°C)	Ref (#)	Evaluation (T P E A A)	Comments
9.694E-02	2.076E+01	25.1	W412	0 0 0 0 0	
1.947E-01	4.171E+01	35.51	W412	0 0 0 0 0	
3.450E-01	7.389E+01	44.92	W412	0 0 0 0 0	
6.388E-01	1.368E+02	54.02	W412	0 0 0 0 0	
1.068E+00	2.287E+02	64.60	W412	0 0 0 0 0	
1.341E+00	2.873E+02	69.60	W412	0 0 0 0 0	
1.536E+00	3.290E+02	71.91	W412	0 0 0 0 0	
1.883E+00	4.034E+02	76.32	W412	0 0 0 0 0	

1817. C_9H_{12}
1,2,3-Trimethylbenzene
Hemimellitene
Hemellitol
RN: 526-73-8   **MP** (°C):   −25
MW: 120.20   **BP** (°C):   175

Solubility (Moles/L)	Solubility (Grams/L)	Temp (°C)	Ref (#)	Evaluation (T P E A A)	Comments
5.450E-04	6.551E-02	25	M342	1 0 1 1 2	
6.256E-04	7.520E-02	25	S005	2 2 2 2 2	
6.256E-04	7.520E-02	25	S191	1 2 2 2 2	
6.256E-04	7.520E-02	25	S358	2 2 2 2 2	

1818. C_9H_{12}
1-Ethyl-2-methylbenzene
2-Ethyltoluene
o-Ethyltoluene
1-Methyl-2-ethylbenzene
RN: 611-14-3   **MP** (°C):   −80.8
MW: 120.20   **BP** (°C):   165.2

Solubility (Moles/L)	Solubility (Grams/L)	Temp (°C)	Ref (#)	Evaluation (T P E A A)	Comments
6.210E-04	7.464E-02	25	M342	1 0 1 1 2	
7.742E-04	9.305E-02	ns	H123	0 0 0 0 0	

1819. C_9H_{12}
1,8-Nonadiyne
RN: 2396-65-8   **MP** (°C):   −21
MW: 120.20   **BP** (°C):   55

Solubility (Moles/L)	Solubility (Grams/L)	Temp (°C)	Ref (#)	Evaluation (T P E A A)	Comments
1.040E-03	1.250E-01	25	M001	2 1 2 2 2	

1820. C$_9$H$_{12}$
Cumene
Isopropylbenzene
Cumol
2-Phenylpropane

RN: 98-82-8	**MP** (°C):	−96
MW: 120.20	**BP** (°C):	152

Solubility (Moles/L)	Solubility (Grams/L)	Temp (°C)	Ref (#)	Evaluation (T P E A A)	Comments
6.694E-04	8.046E-02	24.94	G034	1 2 2 2 2	
6.073E-04	7.300E-02	25	A002	1 2 1 1 1	
4.018E-04	4.830E-02	25	K119	1 0 0 0 2	
4.160E-04	5.000E-02	25	M001	2 1 2 2 2	
4.409E-04	5.300E-02	25	M002	2 2 1 2 1	
4.160E-04	5.000E-02	25	M130	1 0 0 0 1	
4.018E-04	4.830E-02	25	P051	2 1 1 2 2	
5.433E-04	6.530E-02	25	S005	2 2 2 2 2	
5.433E-04	6.530E-02	25	S191	1 2 2 2 2	
5.433E-04	6.530E-02	25	S358	2 1 2 2 2	
4.018E-04	4.830E-02	25.00	P007	2 1 2 2 2	
6.897E-04	8.290E-02	29.94	G034	1 2 2 2 2	
7.124E-04	8.563E-02	34.94	G034	1 2 2 2 2	
7.469E-04	8.978E-02	39.94	G034	1 2 2 2 2	
7.867E-04	9.456E-02	44.94	G034	1 2 2 2 2	
8.353E-04	1.004E-01	49.94	G034	1 2 2 2 2	
8.894E-04	1.069E-01	54.94	G034	1 2 2 2 2	
9.566E-04	1.150E-01	59.94	G034	1 2 2 2 2	
1.035E-03	1.243E-01	65.14	G034	1 2 2 2 2	
1.128E-03	1.355E-01	70.34	G034	1 2 2 2 2	
1.226E-03	1.473E-01	75.04	G034	1 2 2 2 2	
1.345E-03	1.617E-01	80.24	G034	1 2 2 2 2	
4.160E-04	5.000E-02	ns	H123	0 0 0 0 0	
4.160E-04	5.000E-02	ns	M344	0 0 0 0 1	

1821. C$_9$H$_{12}$
n-Propylbenzene
1-Phenylpropane
Propylbenzene
Isocomene

RN: 103-65-1	**MP** (°C):	−99.2
MW: 120.20	**BP** (°C):	159.2

Solubility (Moles/L)	Solubility (Grams/L)	Temp (°C)	Ref (#)	Evaluation (T P E A A)	Comments
4.470E-04	5.373E-02	10	O312	2 2 0 2 2	
5.000E-04	6.010E-02	15	F001	1 0 1 2 0	
4.350E-04	5.229E-02	15	O312	2 2 0 2 2	
4.520E-04	5.433E-02	20	O312	2 2 0 2 2	
4.576E-04	5.500E-02	25	A002	1 2 1 1 1	
1.000E-03	1.202E-01	25	K001	1 0 2 1 2	

(continued)

1821. C$_9$H$_{12}$ (continued)

Solubility (Moles/L)	Solubility (Grams/L)	Temp (°C)	Ref (#)	Evaluation (T P E A A)	Comments
4.340E-04	5.217E-02	25	M342	1 0 1 1 2	
4.430E-04	5.325E-02	25	O312	2 2 0 2 2	
8.319E-04	9.999E-02	25	S012	2 0 2 2 1	
4.150E-04	4.988E-02	25	S359	2 1 2 2 2	
3.920E-04	4.712E-02	25	T067	2 1 2 1 2	
4.340E-04	5.217E-02	25	W300	2 2 2 2 2	
4.370E-04	5.253E-02	30	O312	2 2 0 2 2	
4.710E-04	5.661E-02	35	O312	2 2 0 2 2	
5.320E-04	6.394E-02	40	O312	2 2 0 2 2	
5.540E-04	6.659E-02	45	O312	2 2 0 2 2	
1.098E-03	1.320E-01	85.8	G035	1 0 0 0 2	
1.381E-03	1.660E-01	114.5	G035	1 0 0 0 2	
2.670E-03	3.209E-01	140.5	G035	1 0 0 0 2	
7.232E-03	8.692E-01	188.0	G035	1 0 0 0 1	
2.033E-02	2.444E+00	222.0	G035	1 0 0 0 2	
4.576E-04	5.500E-02	ns	H123	0 0 0 0 0	
2.700E-02	3.245E+00	ns	H307	0 0 0 0 0	
4.576E-04	5.500E-02	ns	M344	0 0 0 0 1	

1822. C$_9$H$_{12}$
1,2,4-Trimethylbenzene
Pseudocumene

RN:	95-63-6	**MP** (°C):	−44
MW:	120.20	**BP** (°C):	169

Solubility (Moles/L)	Solubility (Grams/L)	Temp (°C)	Ref (#)	Evaluation (T P E A A)	Comments
4.318E-04	5.190E-02	25	K119	1 0 0 0 2	
4.742E-04	5.700E-02	25	M001	2 1 2 2 2	
4.318E-04	5.190E-02	25	P051	2 1 1 2 2	
4.909E-04	5.900E-02	25	S005	2 2 2 2 2	
4.909E-04	5.900E-02	25	S191	1 2 2 2 2	
4.909E-04	5.900E-02	25	S358	2 1 2 2 2	
4.318E-04	5.190E-02	25.00	P007	2 1 2 2 2	
4.742E-04	5.700E-02	ns	M344	0 0 0 0 1	

1823. C$_9$H$_{12}$
p-Ethyltoluene
4-Ethyltoluene
1-Ethyl-4-methylbenzene

RN:	622-96-8	**MP** (°C):	−62
MW:	120.20	**BP** (°C):	162

Solubility (Moles/L)	Solubility (Grams/L)	Temp (°C)	Ref (#)	Evaluation (T P E A A)	Comments
7.891E-04	9.485E-02	ns	H123	0 0 0 0 0	

1824. C$_9$H$_{12}$
Mesitylene
1,3,5-Trimethylbenzene
Mesitelene

RN: 108-67-8	**MP** (°C):	−44.8
MW: 120.20	**BP** (°C):	164.7

Solubility (Moles/L)	Solubility (Grams/L)	Temp (°C)	Ref (#)	Evaluation (T P E A A)	Comments
3.794E-04	4.560E-02	15	S203	1 1 2 1 2	
3.111E-04	3.740E-02	20	M337	2 1 2 2 2	
8.070E-04	9.700E-02	25	A002	1 2 1 1 1	
4.010E-04	4.820E-02	25	S005	2 2 2 2 2	
4.010E-04	4.820E-02	25	S191	1 2 2 2 2	
4.118E-04	4.950E-02	25	S203	1 1 2 1 2	
4.010E-04	4.820E-02	25	S358	2 1 2 2 2	
3.280E-04	3.942E-02	25.04	V013	2 2 2 2 2	
5.322E-04	6.397E-02	29.99	C350	0 0 0 0 0	
4.509E-04	5.420E-02	35	S203	1 1 2 1 2	
5.555E-04	6.677E-02	39.99	C350	0 0 0 0 0	
4.701E-04	5.650E-02	45	S203	1 1 2 1 2	
6.166E-04	7.412E-02	49.99	C350	0 0 0 0 0	
7.555E-04	9.081E-02	59.99	C350	0 0 0 0 0	
9.221E-04	1.108E-01	69.99	C350	0 0 0 0 0	
1.161E-03	1.395E-01	79.99	C350	0 0 0 0 0	
1.361E-03	1.636E-01	89.99	C350	0 0 0 0 0	
1.616E-03	1.943E-01	99.99	C350	0 0 0 0 0	

1825. C$_9$H$_{12}$ClN$_5$O
Moxonidine

RN: 75438-57-2	**MP** (°C):	
MW: 241.68	**BP** (°C):	

Solubility (Moles/L)	Solubility (Grams/L)	Temp (°C)	Ref (#)	Evaluation (T P E A A)	Comments
3.311E-03	8.003E-01	ns	R426	0 0 0 0 0	

1826. C$_9$H$_{12}$ClO$_2$PS$_3$
Carbophenothion-methyl
S-p-Chlorophenylthiomethyl O,O-dimethyl phosphorodithioate

RN: 953-17-3	**MP** (°C):	
MW: 314.81	**BP** (°C):	

Solubility (Moles/L)	Solubility (Grams/L)	Temp (°C)	Ref (#)	Evaluation (T P E A A)	Comments
4.669E-06	1.470E-03	10	B324	0 0 0 0 0	
4.670E-06	1.470E-03	10	B324	0 0 0 0 0	
5.178E-06	1.630E-03	20	B300	2 1 1 1 2	
5.083E-06	1.600E-03	20	B324	0 0 0 0 0	
5.082E-06	1.600E-03	20	B324	0 0 0 0 0	
8.958E-06	2.820E-03	30	B324	0 0 0 0 0	
8.958E-06	2.820E-03	30	B324	0 0 0 0 0	
3.176E-06	1.000E-03	rt	M161	0 0 0 0 0	

1827. C$_9$H$_{12}$ClO$_4$P

Heptenophos
7-Chlorobicyclo[3.2.0]hepta-2,6-dien-6-yl dimethyl phosphate
Ragadan
Hostaquick

RN: 23560-59-0   **MP** (°C):
MW: 250.62   **BP** (°C):   64

Solubility (Moles/L)	Solubility (Grams/L)	Temp (°C)	Ref (#)	Evaluation (T P E A A)	Comments
9.975E-03	2.500E+00	23	M161	1 0 0 0 1	

1828. C$_9$H$_{12}$Cl$_2$N$_4$

2,4-Dichloro-6-cyclohexylamino-1,3,5-triazine
2,4-Dichloro-6-(cyclohexylamino)triazine
1,3,5-Triazin-2-amine, 4,6-dichloro-N-cyclohexyl-

RN: 27282-86-6   **MP** (°C):
MW: 247.13   **BP** (°C):

Solubility (Moles/L)	Solubility (Grams/L)	Temp (°C)	Ref (#)	Evaluation (T P E A A)	Comments
4.046E-04	1.000E-01	ns	B160	0 0 0 0 2	

1829. C$_9$H$_{12}$FN$_3$O$_3$

1-Butylcarbamoyl-5-fluorouracil
N-Butyl-5-fluoro-2,4-dioxo-pyrimidinecarboxamide

RN: 64098-82-4   **MP** (°C):   136
MW: 229.21   **BP** (°C):

Solubility (Moles/L)	Solubility (Grams/L)	Temp (°C)	Ref (#)	Evaluation (T P E A A)	Comments
3.577E-03	8.200E-01	22	B321	0 0 0 0 0	pH 4.0
3.577E-03	8.200E-01	22	B388	0 0 0 0 0	

1830. C$_9$H$_{12}$NO$_5$PS

Fenitrothion
Dimethyl O-(4-nitro-m-tolyl) phosphorothioate
Nuvanol
Novathion
Dybar
Metathionine

RN: 122-14-5   **MP** (°C):   3.4
MW: 277.24   **BP** (°C):

Solubility (Moles/L)	Solubility (Grams/L)	Temp (°C)	Ref (#)	Evaluation (T P E A A)	Comments
9.089E-05	2.520E-02	20	B169	2 0 1 1 2	
1.396E-04	3.870E-02	22	K137	1 1 2 1 0	*sic*
1.082E-04	3.000E-02	ns	F071	0 1 2 1 1	
1.082E-04	3.000E-02	ns	M061	0 0 0 0 1	
1.082E-04	3.000E-02	ns	M110	0 0 0 0 0	EFG

1831. C₉H₁₂NO₅PS

O-Methyl O-ethyl O-4-nitrophenyl thiophosphate
Ethylmethylthiophos
Methylethylthiophos
Methylethylthiofos

RN:	2591-57-3	**MP** (°C):		
MW:	277.24	**BP** (°C):	116	

Solubility (Moles/L)	Solubility (Grams/L)	Temp (°C)	Ref (#)	Evaluation (T P E A A)	Comments
1.443E-04	4.000E-02	ns	M061	0 0 0 0 1	

1832. C₉H₁₂N₂O

Fenuron
3-Phenyl-1,1-dimethylurea
N,N-Dimethyl-N-phenylurea
Beet-Klean

RN:	101-42-8	**MP** (°C):	133–134	
MW:	164.21	**BP** (°C):		

Solubility (Moles/L)	Solubility (Grams/L)	Temp (°C)	Ref (#)	Evaluation (T P E A A)	Comments
2.344E-02	3.849E+00	20	B179	0 0 0 0 0	
2.245E-02	3.686E+00	20	E048	1 2 1 1 2	
2.253E-02	3.700E+00	20	F311	1 2 2 2 1	
1.766E-02	2.900E+00	24	B185	0 0 0 0 0	
1.761E-02	2.892E+00	24	M061	1 0 0 0 1	
1.462E-02	2.400E+00	25	A039	1 1 0 0 2	
2.345E-02	3.850E+00	25	B200	1 0 0 0 0	
2.345E-02	3.850E+00	25	G036	1 0 0 0 2	
1.462E-02	2.400E+00	25	G099	1 0 0 1 0	
2.452E-02	4.027E+00	25	I1073	2 1 1 2 2	
2.345E-02	3.850E+00	25	M161	1 0 0 0 2	
2.426E-02	3.984E+00	ns	B100	0 0 0 0 0	
1.462E-02	2.400E+00	ns	B160	0 0 0 0 2	
2.345E-02	3.850E+00	ns	B185	0 0 0 0 0	
1.761E-02	2.892E+00	ns	N013	0 0 0 0 1	

1833. C₉H₁₂N₂O₂

Dulcin
(4-Ethoxyphenyl)urea
4-Aethoxy-phenylharnstoff

RN:	150-69-6	**MP** (°C):	173	
MW:	180.21	**BP** (°C):		

Solubility (Moles/L)	Solubility (Grams/L)	Temp (°C)	Ref (#)	Evaluation (T P E A A)	Comments
6.714E-03	1.210E+00	21	F300	1 0 0 0 2	
7.214E-03	1.300E+00	45	F300	1 0 0 0 1	
1.110E-01	2.000E+01	100	F300	1 0 0 0 0	
6.928E-03	1.248E+00	c	I314	0 0 0 0 0	
1.088E-01	1.961E+01	h	I314	0 0 0 0 0	

1834. C$_9$H$_{12}$N$_2$O$_2$S
3-Thio-2,4-diazaspiro[5.5]undecane-1,3,5-trione
2,4-Diazaspiro[5.5]undecane-1,5-dione, 3-thioxo-
2,4-Diazaspiro[5.5]undecane-1,3,5-trione, 3-thio

RN:	52-45-9	**MP** (°C):			
MW:	212.27	**BP** (°C):			

Solubility (Moles/L)	Solubility (Grams/L)	Temp (°C)	Ref (#)	Evaluation (T P E A A)	Comments
3.450E-04	7.323E-02	25	P350	0 0 0 0 0	intrinsic

1835. C$_9$H$_{12}$N$_2$O$_3$
5-Allyl-5-ethylbarbituric acid
Barbituric acid, 5-allyl-5-ethyl
5-Ethyl-5-allylbarbituric acid
Dormitiv
5-Ethyl-5-allylbarbiturate

RN:	2373-84-4	**MP** (°C):			
MW:	196.21	**BP** (°C):			

Solubility (Moles/L)	Solubility (Grams/L)	Temp (°C)	Ref (#)	Evaluation (T P E A A)	Comments
2.433E-02	4.774E+00	25	P350	0 0 0 0 0	intrinsic

1836. C$_9$H$_{12}$N$_2$O$_3$
2,4-Diazaspiro[5.5]undecane-1,3,5-trione
Spiro[barbituric acid-5,1'-cyclohexane]
Cyclohexane-spirobarbiturate

RN:	52-44-8	**MP** (°C):			
MW:	196.21	**BP** (°C):			

Solubility (Moles/L)	Solubility (Grams/L)	Temp (°C)	Ref (#)	Evaluation (T P E A A)	Comments
8.700E-04	1.707E-01	25	P350	0 0 0 0 0	intrinsic

1837. C$_9$H$_{12}$N$_2$O$_5$
Deoxyuridine

RN:	951-78-0	**MP** (°C):	168		
MW:	228.21	**BP** (°C):			

Solubility (Moles/L)	Solubility (Grams/L)	Temp (°C)	Ref (#)	Evaluation (T P E A A)	Comments
2.053E+00	4.685E+02	25.31	T420	0 0 0 0 0	

1838. $C_9H_{12}N_2O_6$

Uridine

RN: 58-96-8   **MP** (°C):   166.5
MW: 244.21   **BP** (°C):

Solubility (Moles/L)	Solubility (Grams/L)	Temp (°C)	Ref (#)	Evaluation (T P E A A)	Comments
~3.40E+00	~8.30E+02	21.99	T418	0 0 0 0 0	
~3.20E+00	~7.81E+02	22.99	T418	0 0 0 0 0	

1839. $C_9H_{12}N_4O_2$

7-Ethyl theophylline

7-Ethyl-1,3-dimethylxanthine

1H-Purine-2,6-dione, 7-ethyl-3,7-dihydro-1,3-dimethyl-

RN: 23043-88-1   **MP** (°C):
MW: 208.22   **BP** (°C):

Solubility (Moles/L)	Solubility (Grams/L)	Temp (°C)	Ref (#)	Evaluation (T P E A A)	Comments
1.760E-01	3.665E+01	30	B042	1 2 1 1 2	
1.760E-01	3.665E+01	30	G021	1 0 0 0 2	

1840. $C_9H_{12}N_4O_2$

1-Ethyl theobromine

1-Ethyl-3,7-dimethylxanthine

1H-Purine-2,6-dione, 1-ethyl-3,7-dihydro-3,7-dimethyl-

RN: 39832-36-5   **MP** (°C):   156
MW: 208.22   **BP** (°C):

Solubility (Moles/L)	Solubility (Grams/L)	Temp (°C)	Ref (#)	Evaluation (T P E A A)	Comments
1.910E-01	3.977E+01	30	B042	1 2 1 1 2	
1.910E-01	3.977E+01	30	G021	1 0 0 0 2	

1841. $C_9H_{12}N_4O_2$

1H-Pyrazolo[3,4-d]pyrimidine, 4-(1-ethoxyethoxy)-

1-Ethoxyethyl-4-allopurinyl ether

RN: 52717-51-8   **MP** (°C):
MW: 208.22   **BP** (°C):

Solubility (Moles/L)	Solubility (Grams/L)	Temp (°C)	Ref (#)	Evaluation (T P E A A)	Comments
9.173E-03	1.910E+00	ns	H067	0 0 0 0 0	

1842. C$_9$H$_{12}$N$_4$O$_2$
8-Methyl caffeine
1,3,7,8-Tetramethylxanthine
RN: 832-66-6   **MP** (°C):
MW: 208.22   **BP** (°C):

Solubility (Moles/L)	Solubility (Grams/L)	Temp (°C)	Ref (#)	Evaluation (T P E A A)	Comments
1.045E-02	2.175E+00	20	J009	1 0 2 2 2	

1843. C$_9$H$_{12}$N$_4$O$_3$
7-β-Hydroxyethyltheophylline
1H-Purine-2,6-dione, 3,7-Dihydro-7-(2-hydroxyethyl)-1,3-dimethyl-
Dilaphyllin
Etofylline
Corophyllin-*N*
RN: 519-37-9   **MP** (°C):
MW: 224.22   **BP** (°C):

Solubility (Moles/L)	Solubility (Grams/L)	Temp (°C)	Ref (#)	Evaluation (T P E A A)	Comments
1.439E-01	3.226E+01	ns	J025	0 0 0 0 1	

1844. C$_9$H$_{12}$N$_4$O$_3$
8-Methoxycaffeine
1H-Purine-2,6-dione, 3,7-dihydro-8-methoxy-1,3,7-trimethyl-
RN: 569-34-6   **MP** (°C):
MW: 224.22   **BP** (°C):

Solubility (Moles/L)	Solubility (Grams/L)	Temp (°C)	Ref (#)	Evaluation (T P E A A)	Comments
1.140E-02	2.556E+00	25	K008	1 1 0 1 0	EFG
1.115E-04	2.500E-02	rt	N015	0 0 2 2 1	

1845. C$_9$H$_{12}$N$_4$O$_3$
1,3,7,9-Tetramethyluric acid
1H-Purine-2,6,8(3H)-trione, 7,9-dihydro-1,3,7,9-tetramethyl-
Temorine
Temurin
Ba 2750
RN: 2309-49-1   **MP** (°C):
MW: 224.22   **BP** (°C):

Solubility (Moles/L)	Solubility (Grams/L)	Temp (°C)	Ref (#)	Evaluation (T P E A A)	Comments
1.472E-04	3.300E-02	rt	N015	0 0 2 2 1	

1846. C$_9$H$_{12}$N$_4$O$_3$S
N4-Acetylsulfanilylguanidine
Acetamide, N-[4-[[(aminoiminomethyl)amino]sulfonyl]phenyl]-
p-(Guanidinosulfonyl)acetanilide
Sulgin ASG

RN: 19077-97-5   **MP** (°C):
MW: 256.28   **BP** (°C):

Solubility (Moles/L)	Solubility (Grams/L)	Temp (°C)	Ref (#)	Evaluation (T P E A A)	Comments
1.560E-03	3.998E-01	37.50	M142	1 2 0 0 1	
5.766E-02	1.478E+01	h	M142	0 0 0 0 1	

1847. C$_9$H$_{12}$O
2,3,5-Trimethyl-phenol
Isopseudocumenol
1-Hydroxy-2,3,5-trimethylbenzene

RN: 697-82-5   **MP** (°C):
MW: 136.20   **BP** (°C):

Solubility (Moles/L)	Solubility (Grams/L)	Temp (°C)	Ref (#)	Evaluation (T P E A A)	Comments
5.600E-03	7.627E-01	25	B316	0 0 0 0 0	

1848. C$_9$H$_{12}$O
4-Propylphenol
4-Propyphenol
p-n-Propylphenol

RN: 645-56-7   **MP** (°C):
MW: 136.20   **BP** (°C): 232

Solubility (Moles/L)	Solubility (Grams/L)	Temp (°C)	Ref (#)	Evaluation (T P E A A)	Comments
1.047E-02	1.427E+00	25	L022	1 0 0 0 0	

1849. C$_9$H$_{12}$O
2-Propylphenol
2-n-Propylphenol
2-Propyphenol

RN: 644-35-9   **MP** (°C):
MW: 136.20   **BP** (°C): 225

Solubility (Moles/L)	Solubility (Grams/L)	Temp (°C)	Ref (#)	Evaluation (T P E A A)	Comments
1.222E-02	1.664E+00	25	L022	1 0 0 0 0	

1850. C$_9$H$_{12}$O
3-Methyl-5-ethyl-phenol
Phenol, 3-ethyl-5-methyl-
m-Cresol, 5-ethyl-

RN:	698-71-5	MP (°C):
MW:	136.20	BP (°C):

Solubility (Moles/L)	Solubility (Grams/L)	Temp (°C)	Ref (#)	Evaluation (T P E A A)	Comments
1.700E-02	2.315E+00	25	B316	0 0 0 0 0	

1851. C$_9$H$_{12}$O
4-Ethyl-3-methylphenol
3-Methyl-4-ethylphenol
4-Ethyl-*m*-cresol

RN:	1123-94-0	MP (°C):
MW:	136.20	BP (°C):

Solubility (Moles/L)	Solubility (Grams/L)	Temp (°C)	Ref (#)	Evaluation (T P E A A)	Comments
7.335E-03	9.990E-01	25	L020	1 0 0 0 0	

1852. C$_9$H$_{12}$O
2,4,6-Trimethylphenol
2-Hydroxymesitylene
1-Hydroxy-2,4,6-trimethylbenzene
Mesityl alcohol
Hydroxymesitylene

RN:	527-60-6	MP (°C):	72
MW:	136.20	BP (°C):	220

Solubility (Moles/L)	Solubility (Grams/L)	Temp (°C)	Ref (#)	Evaluation (T P E A A)	Comments
7.400E-03	1.008E+00	25	B316	0 0 0 0 0	
4.892E-03	6.662E-01	25	L020	1 0 0 0 0	

1853. C$_9$H$_{12}$O$_2$
o-Propoxyphenol
2-Propoxyphenol

RN:	6280-96-2	MP (°C):
MW:	152.19	BP (°C):

Solubility (Moles/L)	Solubility (Grams/L)	Temp (°C)	Ref (#)	Evaluation (T P E A A)	Comments
1.550E-02	2.359E+00	24.99	B353	0 0 0 0 0	

1854. C$_9$H$_{12}$O$_2$
Cumene hydroperoxide
CHP

RN:	80-15-9	**MP** (°C):		
MW:	152.19	**BP** (°C):	100	

Solubility (Moles/L)	Solubility (Grams/L)	Temp (°C)	Ref (#)	Evaluation (T P E A A)	Comments
9.140E-02	1.391E+01	25	K051	1 2 2 1 2	

1855. C$_9$H$_{12}$O$_2$
3-Propoxyphenol
m-Propoxy phenol
Phenol, 3-propoxy-

RN:	16533-50-9	**MP** (°C):	
MW:	152.19	**BP** (°C):	

Solubility (Moles/L)	Solubility (Grams/L)	Temp (°C)	Ref (#)	Evaluation (T P E A A)	Comments
2.590E-02	3.942E+00	30	B315	0 0 0 0 0	

1856. C$_9$H$_{12}$O$_2$
1-*O*-Benzylethanediol
Benzylcellosolve
Benzyl cellosolve

RN:	622-08-2	**MP** (°C):	
MW:	152.19	**BP** (°C):	

Solubility (Moles/L)	Solubility (Grams/L)	Temp (°C)	Ref (#)	Evaluation (T P E A A)	Comments
2.618E-02	3.984E+00	20	D052	1 1 0 0 0	
2.813E-02	4.282E+00	23	M062	1 0 0 0 1	

1857. C$_9$H$_{13}$BrN$_2$O$_2$
5-Bromo-3-*tert*-butyl-6-methyluracil
Compound 733

RN:	7286-76-2	**MP** (°C):	188
MW:	261.13	**BP** (°C):	

Solubility (Moles/L)	Solubility (Grams/L)	Temp (°C)	Ref (#)	Evaluation (T P E A A)	Comments
1.570E-03	4.100E-01	25	M061	1 0 0 0 0	
3.121E-03	8.150E-01	ns	B185	0 0 0 0 0	

1858. C$_9$H$_{13}$BrN$_2$O$_2$
Bromacil
5-Bromo-6-methyl-3,5-butyluracil
RN: 314-40-9 **MP** (°C): 158.3
MW: 261.13 **BP** (°C):

Solubility (Moles/L)	Solubility (Grams/L)	Temp (°C)	Ref (#)	Evaluation (T P E A A)	Comments
2.719E-03	7.100E-01	25	B200	1 0 0 0 2	
3.119E-03	8.143E-01	25	B200	1 0 0 0 2	
3.121E-03	8.150E-01	25	M061	1 0 0 0 2	
3.121E-03	8.150E-01	25	M161	1 0 0 0 2	
3.061E-03	7.994E-01	ns	B100	0 0 0 0 0	

1859. C$_9$H$_{13}$ClN$_2$O$_2$
Terbacil
3-*tert*-Butyl-5-chloro-6-methyluracil
5-Chloro-3-(1,1-dimethylethyl)-6-methyl-2,4(1H,3H)-pyrimidinedione
Sinbar 80W
Geonter
DPX-D732
RN: 5902-51-2 **MP** (°C): 176.0
MW: 216.67 **BP** (°C):

Solubility (Moles/L)	Solubility (Grams/L)	Temp (°C)	Ref (#)	Evaluation (T P E A A)	Comments
3.277E-03	7.100E-01	25	M061	1 0 0 0 2	
3.277E-03	7.100E-01	25	M161	1 0 0 0 2	
3.277E-03	7.100E-01	25	P307	1 0 0 0 1	
3.228E-03	6.995E-01	ns	B100	0 0 0 0 0	

1860. C$_9$H$_{13}$ClN$_6$
Cyanazine
Bladex
2-[[4-Chloro-6-(ethylamino)-1,3,5-triazin-2-yl]amino]-2-methylpropanenitrile
Fortrol
Payze
SD 45418
RN: 21725-46-2 **MP** (°C): 166.5
MW: 240.70 **BP** (°C):

Solubility (Moles/L)	Solubility (Grams/L)	Temp (°C)	Ref (#)	Evaluation (T P E A A)	Comments
6.647E-04	1.600E-01	23	B200	1 0 0 0 2	
7.104E-04	1.710E-01	25	B200	1 0 0 0 2	
7.104E-04	1.710E-01	25	M061	1 0 0 0 2	
7.104E-04	1.710E-01	25	M161	1 0 0 0 2	
6.647E-04	1.600E-01	25	S309	1 0 0 0 2	
8.309E-04	2.000E-01	ns	M110	0 0 0 0 0	EFG

1861. C₉H₁₃N

2,4,5-Trimethylaniline
2,4,5-Trimethylanilin

RN: 137-17-7 **MP** (°C):
MW: 135.21 **BP** (°C):

Solubility (Moles/L)	Solubility (Grams/L)	Temp (°C)	Ref (#)	Evaluation (T P E A A)	Comments
8.875E-03	1.200E+00	19.40	F300	1 0 0 0 1	
1.109E-02	1.500E+00	28.70	F300	1 0 0 0 1	

1862. C₉H₁₃NO₃

Adrenaline
Adrenalin
Epinephrine
L-1-(3,4-Dihydroxyphenyl)-2-methylaminoethanol
Primatene
Epipen

RN: 51-43-4 **MP** (°C):
MW: 183.21 **BP** (°C):

Solubility (Moles/L)	Solubility (Grams/L)	Temp (°C)	Ref (#)	Evaluation (T P E A A)	Comments
9.825E-04	1.800E-01	20	F300	1 0 0 0 1	

1863. C₉H₁₃N₃O₃

Orotic acid *n*-butylamide
Orotamide, *N*-butyl-

RN: 13156-38-2 **MP** (°C): 276–277
MW: 211.22 **BP** (°C):

Solubility (Moles/L)	Solubility (Grams/L)	Temp (°C)	Ref (#)	Evaluation (T P E A A)	Comments
5.700E-02	1.204E+01	−4	N018	0 0 0 0 0	
9.600E-02	2.028E+01	16	N018	0 0 0 0 0	
1.180E-01	2.492E+01	25	N018	0 0 0 0 0	

1864. C₉H₁₃N₃O₃

Zalcitabine
2′,3′-Dideoxycytidine
Dideoxycytidine
CCRIS 692
Hivid
DDCYD

RN: 7481-89-2 **MP** (°C): 210–214
MW: 211.22 **BP** (°C):

Solubility (Moles/L)	Solubility (Grams/L)	Temp (°C)	Ref (#)	Evaluation (T P E A A)	Comments
3.360E-01	7.098E+01	ns	S469	0 0 0 0 0	

1865. C$_9$H$_{13}$N$_3$O$_3$
Orotic acid diethylamine
Orotamide, N,N-diethyl-
RN: 883-81-8   **MP** (°C):   192–194
MW: 211.22   **BP** (°C):

Solubility (Moles/L)	Solubility (Grams/L)	Temp (°C)	Ref (#)	Evaluation (T P E A A)	Comments
2.939E+00	6.208E+02	25	N018	0 0 0 0 0	

1866. C$_9$H$_{13}$N$_3$O$_4$
Orotic acid isobutanolamine
RN: **MP** (°C):   247–249
MW: 227.22   **BP** (°C):

Solubility (Moles/L)	Solubility (Grams/L)	Temp (°C)	Ref (#)	Evaluation (T P E A A)	Comments
4.200E-01	9.543E+01	–4	N018	0 0 0 0 0	
7.060E-01	1.604E+02	16	N018	0 0 0 0 0	
8.410E-01	1.911E+02	25	N018	0 0 0 0 0	

1867. C$_9$H$_{13}$N$_3$O$_4$
Cytosine deoxyriboside
RN: 951-77-9   **MP** (°C):
MW: 227.22   **BP** (°C):

Solubility (Moles/L)	Solubility (Grams/L)	Temp (°C)	Ref (#)	Evaluation (T P E A A)	Comments
2.780E+00	6.317E+02	25.23	T420	0 0 0 0 0	

1868. C$_9$H$_{13}$N$_3$O$_5$
Cytidine
RN: 65-46-3   **MP** (°C):   > 215
MW: 243.22   **BP** (°C):

Solubility (Moles/L)	Solubility (Grams/L)	Temp (°C)	Ref (#)	Evaluation (T P E A A)	Comments
~9.70E-01	~2.36E+02	21.99	T418	0 0 0 0 0	
~8.00E-01	~1.95E+02	22.99	T418	0 0 0 0 0	

1869. C$_9$H$_{13}$N$_3$O$_5$
Orotic acid 2-amide-2-methyl-1,3-propanediol
RN: **MP** (°C):   214–215
MW: 243.22   **BP** (°C):

Solubility (Moles/L)	Solubility (Grams/L)	Temp (°C)	Ref (#)	Evaluation (T P E A A)	Comments
3.450E-01	8.391E+01	–4	N018	0 0 0 0 0	
5.860E-01	1.425E+02	16	N018	0 0 0 0 0	
6.970E-01	1.695E+02	25	N018	0 0 0 0 0	

1870. $C_9H_{13}N_5O_4$

Ganciclovir

2-Amino-1,9-dihydro-9-((2-hydroxy-1-(hydroxymethyl)ethoxy)methyl)-6H-purin-6-one

DHPG

RN: 82410-32-0   **MP** (°C):   250
MW: 255.24   **BP** (°C):

Solubility (Moles/L)	Solubility (Grams/L)	Temp (°C)	Ref (#)	Evaluation (T P E A A)	Comments
1.410E-02	3.600E+00	25	B360	0 0 0 0 0	
1.230E-02	3.139E+00	25	Z407	0 0 0 0 0	

1871. $C_9H_{13}O_2P$

Mesitylene phosphinous acid

Phosphinic acid, (2,4,6-trimethylphenyl)-

RN: 6781-97-1   **MP** (°C):   147.0
MW: 184.18   **BP** (°C):

Solubility (Moles/L)	Solubility (Grams/L)	Temp (°C)	Ref (#)	Evaluation (T P E A A)	Comments
1.565E-02	2.882E+00	1	C061	2 2 2 1 2	
1.619E-02	2.981E+00	25	C061	2 2 2 1 2	
1.754E-02	3.230E+00	35	C061	2 2 2 1 2	
2.082E-02	3.835E+00	45	C061	2 2 2 1 2	
2.836E-02	5.223E+00	65	C061	2 2 2 1 2	
3.774E-02	6.951E+00	85	C061	2 2 2 1 2	

1872. $C_9H_{13}O_6PS$

Endothion

O,O-Dimethyl *S*-(5-methoxypyronyl-2-methyl) thiophosphate

RN: 2778-04-3   **MP** (°C):   90.5
MW: 280.24   **BP** (°C):

Solubility (Moles/L)	Solubility (Grams/L)	Temp (°C)	Ref (#)	Evaluation (T P E A A)	Comments
2.141E+00	6.000E+02	ns	M061	0 0 0 0 2	
5.353E+00	1.500E+03	ns	M161	0 0 0 0 1	

1873. $C_9H_{14}ClN_5$

Cyprozine

2-Chloro-4-cyclopropylamino-6-isopropylamino-1,3,5-triazine

RN: 22936-86-3   **MP** (°C):   167
MW: 227.70   **BP** (°C):

Solubility (Moles/L)	Solubility (Grams/L)	Temp (°C)	Ref (#)	Evaluation (T P E A A)	Comments
3.030E-05	6.900E-03	25	B200	1 0 0 0 1	
8.582E-04	1.954E-01	40	B200	1 0 0 0 2	

1874. C$_9$H$_{14}$N$_2$O$_3$
5-Ethyl-5-*n*-propylbarbituric acid
2,4,6(1H,3H,5H)-Pyrimidinetrione, 5-ethyl-5-propyl-
5-Ethyl-5-propylbarbiturate

RN:	33376-25-9	MP (°C):	146.5
MW:	198.22	BP (°C):	

Solubility (Moles/L)	Solubility (Grams/L)	Temp (°C)	Ref (#)	Evaluation (T P E A A)	Comments
2.872E-02	5.694E+00	25	B065	1 2 1 1 1	
3.610E-02	7.156E+00	25	M310	2 2 2 2 2	

1875. C$_9$H$_{14}$N$_2$O$_3$
Metharbital
5,5′-Diethyl-1-methylbarbituric acid

RN:	50-11-3	MP (°C):	155
MW:	198.22	BP (°C):	

Solubility (Moles/L)	Solubility (Grams/L)	Temp (°C)	Ref (#)	Evaluation (T P E A A)	Comments
1.009E-02	2.000E+00	25	B011	2 0 0 1 0	
9.980E-03	1.978E+00	25	B065	1 1 1 1 1	
1.150E-02	2.280E+00	25	G003	1 1 1 1 2	pH 4.7
6.054E-03	1.200E+00	25	P061	0 0 0 0 0	
4.979E-03	9.870E-01	rt	M161	0 0 0 0 2	

1876. C$_9$H$_{14}$N$_2$O$_3$
Probarbital
5-Ethyl-5-isopropylbarbituric acid
2,4,6(1H,3H,5H)-Pyrimidinetrione, 5-ethyl-5-(1-methylethyl)

RN:	76-76-6	MP (°C):	197.5
MW:	198.22	BP (°C):	

Solubility (Moles/L)	Solubility (Grams/L)	Temp (°C)	Ref (#)	Evaluation (T P E A A)	Comments
6.104E-03	1.210E+00	25	B065	1 1 1 1 1	
7.111E-03	1.410E+00	25	P350	0 0 0 0 0	intrinsic
1.210E-01	2.399E+01	40	N008	1 0 1 1 2	*sic*

1877. C$_9$H$_{14}$N$_6$
6-Amino-4-(diallylamino)-1,2-dihydro-1-hydroxy-2-imino-*s*-triazine

RN:		MP (°C):	
MW:	206.25	BP (°C):	

Solubility (Moles/L)	Solubility (Grams/L)	Temp (°C)	Ref (#)	Evaluation (T P E A A)	Comments
1.459E-01	3.010E+01	37	H004	0 0 0 0 0	

1878. C₉H₁₄O₆

1878. $C_9H_{14}O_6$

L-Camphoronic acid

L-Camphoronsaeure

RN: 2385-74-2 **MP** (°C):

MW: 218.21 **BP** (°C):

Solubility (Moles/L)	Solubility (Grams/L)	Temp (°C)	Ref (#)	Evaluation (T P E A A)	Comments
5.087E-01	1.110E+02	16	F300	1 0 0 0 2	

1879. $C_9H_{14}O_6$

Triacetin

Propane-1,2,3-triyl triacetate

Enzactin

Vanay

Triacetylglycerol

Glycerol triacetate

RN: 102-76-1 **MP** (°C): −78

MW: 218.21 **BP** (°C): 258

Solubility (Moles/L)	Solubility (Grams/L)	Temp (°C)	Ref (#)	Evaluation (T P E A A)	Comments
3.290E-01	7.180E+01	15	F300	1 0 0 0 2	
2.389E-01	5.213E+01	24.50	O005	1 0 2 2 1	
3.118E-02	6.803E+00	ns	F014	0 0 0 0 2	

1880. $C_9H_{15}Br_6O_4P$

Tris-BP

tris(2,3-Dibromopropyl) phosphate

2,3-Dibromo-1-propanol phosphate (3:1)

2,3-Dibromopropyl phosphate

Flamex T 23P

Anfram 3PB

RN: 126-72-7 **MP** (°C): 5.5

MW: 697.65 **BP** (°C):

Solubility (Moles/L)	Solubility (Grams/L)	Temp (°C)	Ref (#)	Evaluation (T P E A A)	Comments
1.147E-05	8.000E-03	24	H116	2 1 0 0 2	

1881. C₉H₁₅Cl₆O₄P

Fyrol FR-2
tris(1,3-Dichloroisopropyl) phosphate
TCPP
Emulsion 212
TDCPP
PF 38
RN: 13674-87-8   **MP** (°C):
MW: 430.91   **BP** (°C):

Solubility (Moles/L)	Solubility (Grams/L)	Temp (°C)	Ref (#)	Evaluation (T P E A A)	Comments
1.624E-05	7.000E-03	24	H116	2 1 0 0 2	

1882. C₉H₁₅NO₃

Ecgonine
L-Ekgonin
3-Hydroxy-2-tropane carboxylic acid
RN: 481-37-8   **MP** (°C):   198
MW: 185.22   **BP** (°C):

Solubility (Moles/L)	Solubility (Grams/L)	Temp (°C)	Ref (#)	Evaluation (T P E A A)	Comments
9.610E-01	1.780E+02	ns	F300	0 0 0 0 2	

1883. C₉H₁₅NO₃S

Captopril
1-((2S)-3-mercapto-2-methylpropionyl)-L-proline
Acenorm
Capoten
Capozide
(S)-1-(3-Mercapto-2-methyl-1-oxopropyl)-L-proline
RN: 62571-86-2   **MP** (°C):
MW: 217.29   **BP** (°C):

Solubility (Moles/L)	Solubility (Grams/L)	Temp (°C)	Ref (#)	Evaluation (T P E A A)	Comments
4.602E-01	1.000E+02	ns	K444	0 0 0 0 0	
6.348E-01	1.379E+02	ns	S469	0 0 0 0 0	

1884. C₉H₁₆

2,2,5-Trimethyl-3-hexyne
3-Hexyne, 2,2,5-trimethyl-
RN: 17530-23-3   **MP** (°C):
MW: 124.23   **BP** (°C):

Solubility (Moles/L)	Solubility (Grams/L)	Temp (°C)	Ref (#)	Evaluation (T P E A A)	Comments
2.410E-04	2.994E-02	25	H039	1 2 2 2 2	

1885. C₉H₁₆

1-Nonyne

n-Heptylacetylene

Heptylacetylene

RN:	3452-09-3	MP (°C):	-50
MW:	124.23	BP (°C):	150

Solubility (Moles/L)	Solubility (Grams/L)	Temp (°C)	Ref (#)	Evaluation (T P E A A)	Comments
5.796E-05	7.200E-03	25	M001	2 1 2 2 1	

1886. C₉H₁₆ClN₄

G 30451

2-Chloro-4-propylamino-6-isopropylamino-*s*-triazine

RN:	3567-85-9	MP (°C):	
MW:	215.71	BP (°C):	

Solubility (Moles/L)	Solubility (Grams/L)	Temp (°C)	Ref (#)	Evaluation (T P E A A)	Comments
1.947E-04	4.200E-02	21	B192	0 0 0 0 1	

1887. C₉H₁₆ClN₅

Propazine

2-Chloro-4-isopropylamino-6-isopropylamino-*s*-triazine

RN:	139-40-2	MP (°C):	213
MW:	229.71	BP (°C):	

Solubility (Moles/L)	Solubility (Grams/L)	Temp (°C)	Ref (#)	Evaluation (T P E A A)	Comments
3.744E-05	8.600E-03	20	B185	0 0 0 0 0	
4.000E-05	9.189E-03	20	B200	1 0 0 0 0	
2.307E-05	5.300E-03	20	C048	2 2 2 2 1	
2.177E-05	5.000E-03	20	F311	1 2 2 2 1	
3.744E-05	8.600E-03	20	M161	1 0 0 0 1	
3.744E-05	8.600E-03	21	B192	0 0 0 0 1	
3.744E-05	8.600E-03	21	G099	2 0 0 1 0	
3.744E-05	8.600E-03	22	M061	1 0 0 0 1	
7.700E-05	1.769E-02	50	G001	1 0 1 1 1	
3.744E-05	8.600E-03	ns	C101	0 0 0 0 1	
4.353E-05	1.000E-02	ns	G041	0 0 0 0 1	
3.744E-05	8.600E-03	ns	J033	0 0 0 0 0	

1888. C$_9$H$_{16}$ClN$_5$
Terbuthylazine
Terbutylazine
2-Chloro-4-ethylamino-6-*tert*-butylamino-*s*-triazine
Primatol M

RN:	5915-41-3	MP (°C):	178
MW:	229.71	BP (°C):	

Solubility (Moles/L)	Solubility (Grams/L)	Temp (°C)	Ref (#)	Evaluation (T P E A A)	Comments
2.177E-05	5.000E-03	20	F311	1 2 2 2 1	
3.700E-05	8.500E-03	20	M161	1 0 0 0 1	
3.700E-05	8.500E-03	ns	J033	0 0 0 0 0	

1889. C$_9$H$_{16}$ClN$_5$
Trietazine
2-Chloro-4-diethylamino-6-ethylamino-*s*-triazine
2-Chloro-4-ethylamino-6-diethylamino-*s*-triazines

RN:	1912-26-1	MP (°C):	101
MW:	229.71	BP (°C):	

Solubility (Moles/L)	Solubility (Grams/L)	Temp (°C)	Ref (#)	Evaluation (T P E A A)	Comments
8.706E-05	2.000E-02	20	B185	0 0 0 0 0	
8.706E-05	2.000E-02	21	B192	0 0 0 0 1	
8.706E-05	2.000E-02	21	G099	2 0 0 1 0	
8.706E-05	2.000E-02	25	M161	1 0 0 0 1	
8.706E-05	2.000E-02	ns	J033	0 0 0 0 0	

1890. C$_9$H$_{16}$N$_2$O$_4$
Methyl-2,2-diethylmalonurate
Methyl 2,2-diethylmalonurate

RN:	69577-07-7	MP (°C):	112
MW:	216.24	BP (°C):	

Solubility (Moles/L)	Solubility (Grams/L)	Temp (°C)	Ref (#)	Evaluation (T P E A A)	Comments
1.100E-02	2.379E+00	23	B152	1 2 1 1 1	pH 3.5

1891. C$_9$H$_{16}$N$_4$OS
Tebuthiuron
1-(5-*tert*-Butyl-1,3,4-thiadiazol-2-yl)-1,3-dimethylurea
Graslan
Spike
Spike 20P
Perflan

RN:	34014-18-1	MP (°C):	162.2
MW:	228.32	BP (°C):	

Solubility (Moles/L)	Solubility (Grams/L)	Temp (°C)	Ref (#)	Evaluation (T P E A A)	Comments
1.007E-02	2.300E+00	ns	M161	0 0 0 0 1	

1892. C$_9$H$_{16}$N$_8$
2-Azido-4-ethylamino-4-*t*-butylamino-*s*-triazine
WL 9385

RN:	2854-70-8	MP (°C):	102.5
MW:	236.28	BP (°C):	

Solubility (Moles/L)	Solubility (Grams/L)	Temp (°C)	Ref (#)	Evaluation (T P E A A)	Comments
3.047E-04	7.200E-02	20	M061	1 0 0 0 1	

1893. C$_9$H$_{16}$O$_2$
3-Hydroxy-5-spirocyclohexyltetrahydrofuran
1-Oxaspiro[4.5]decan-3-ol

RN:	29839-61-0	MP (°C):	
MW:	156.23	BP (°C):	

Solubility (Moles/L)	Solubility (Grams/L)	Temp (°C)	Ref (#)	Evaluation (T P E A A)	Comments
1.255E-01	1.961E+01	rt	B066	0 2 0 0 0	contains impurity

1894. C$_9$H$_{16}$O$_2$
g-Nonanolactone
4-Hydroxynonanoic acid lactone
g-*n*-Amylbutyrolactone
g-Pentyl-g-butyrolactone
g-Nonanolide

RN:	104-61-0	MP (°C):	
MW:	156.23	BP (°C):	

Solubility (Moles/L)	Solubility (Grams/L)	Temp (°C)	Ref (#)	Evaluation (T P E A A)	Comments
5.900E-02	9.217E+00	25	D407	1 0 2 2 2	
5.902E-02	9.221E+00	ns	S460	0 0 0 0 0	

1895. C$_9$H$_{16}$O$_2$
3-Hydroxy-2-methyl-5-spirocyclopentyltetrahydrofuran
1-Oxaspiro[4.4]nonan-3-ol, 2-methyl-

RN:	29839-62-1	MP (°C):	
MW:	156.23	BP (°C):	

Solubility (Moles/L)	Solubility (Grams/L)	Temp (°C)	Ref (#)	Evaluation (T P E A A)	Comments
1.067E+00	1.667E+02	rt	B066	0 2 0 0 1	

1896. C$_9$H$_{16}$O$_4$
Butyl α-acetoxypropionate
Hydracrylic acid, butyl ester, acetate
RN: 5422-69-5   MP (°C):
MW: 188.23   BP (°C):

Solubility (Moles/L)	Solubility (Grams/L)	Temp (°C)	Ref (#)	Evaluation (T P E A A)	Comments
1.700E-02	3.200E+00	25	R006	2 2 0 1 1	

1897. C$_9$H$_{16}$O$_4$
Azelaic acid
Azelainsaeure
Nonanedioic acid
RN: 123-99-9   MP (°C):   106.5
MW: 188.23   BP (°C):   287

Solubility (Moles/L)	Solubility (Grams/L)	Temp (°C)	Ref (#)	Evaluation (T P E A A)	Comments
5.313E-03	1.000E+00	0	L041	1 0 0 1 1	
3.298E-03	6.208E-01	6.99	A340	0 0 0 0 0	
4.513E-03	8.494E-01	12.69	A340	0 0 0 0 0	
7.969E-03	1.500E+00	15	L041	1 0 0 1 1	
6.475E-03	1.219E+00	18.69	A340	0 0 0 0 0	
1.275E-02	2.400E+00	20	F300	1 0 0 0 1	
1.275E-02	2.400E+00	20	L041	1 0 0 1 1	
1.297E-02	2.441E+00	20	M171	1 0 0 0 1	
2.667E-01	5.020E+01	21	B040	1 0 1 1 2	*sic*
9.461E-03	1.781E+00	24.99	A340	0 0 0 0 0	
1.589E-02	2.990E+00	34.69	A340	0 0 0 0 0	
2.391E-02	4.500E+00	35	L041	1 0 0 1 1	
1.858E-02	3.498E+00	42.99	A340	0 0 0 0 0	
4.356E-02	8.200E+00	50	L041	1 0 0 1 1	
2.662E-02	5.010E+00	52.59	A340	0 0 0 0 0	
3.858E-02	7.263E+00	56.99	A340	0 0 0 0 0	
5.124E-02	9.645E+00	61.49	A340	0 0 0 0 0	
7.023E-02	1.322E+01	64.99	A340	0 0 0 0 0	
1.169E-01	2.200E+01	65	F300	1 0 0 0 1	
1.169E-01	2.200E+01	65	L041	1 0 0 1 1	
7.255E-02	1.366E+01	70.99	A340	0 0 0 0 0	
8.355E-02	1.573E+01	74.49	A340	0 0 0 0 0	
1.048E-01	1.972E+01	79.89	A340	0 0 0 0 0	
9.430E-02	1.775E+01	84.49	A340	0 0 0 0 0	
9.440E-03	1.777E+00	rt	H431	0 0 0 0 0	

1898. C$_9$H$_{16}$O$_5$

Propanoic acid, 2-[(butoxycarbonyl)oxy]-, methyl ester
Propanoic acid, 2-[(methoxycarbonyl)oxy]-, butyl ester

RN: **MP (°C):**
MW: 204.22 **BP (°C):**

Solubility (Moles/L)	Solubility (Grams/L)	Temp (°C)	Ref (#)	Evaluation (T P E A A)	Comments
8.798E-03	1.797E+00	25	R007	0 0 0 0 0	

1899. C$_9$H$_{17}$ClN$_3$O$_3$PS

Isazophos
Diethyl *O*-(5-chloro-1-(1-methylethyl)-1H-1,2,4-triazol-3-yl) phosphorothioate
Miral
Triumph
CGA-12223

RN: 42509-80-8 **MP (°C):**
MW: 313.74 **BP (°C):** 100

Solubility (Moles/L)	Solubility (Grams/L)	Temp (°C)	Ref (#)	Evaluation (T P E A A)	Comments
4.780E-04	1.500E-01	20	E048	1 2 1 1 2	
4.781E-04	1.500E-01	20	M161	1 0 0 0 1	

1900. C$_9$H$_{17}$NOS

Molinate
S-Ethyl hexahydro-1H-azepine-1-carbothioate
Hydram
Carbothialate, ethyl-1-hexa-methylene imine-
Poperidinecarbothioic acid, *S*-ethyl ester

RN: 2212-67-1 **MP (°C):**
MW: 187.31 **BP (°C):**

Solubility (Moles/L)	Solubility (Grams/L)	Temp (°C)	Ref (#)	Evaluation (T P E A A)	Comments
4.271E-03	8.000E-01	20	B200	1 0 0 0 2	
4.271E-03	8.000E-01	21	M161	1 0 0 0 2	
4.698E-03	8.800E-01	22	K137	1 1 2 1 0	
4.698E-03	8.800E-01	25	P434	0 0 0 0 0	
<5.33E-03	<9.99E-01	ns	B185	0 0 0 0 0	
4.869E-03	9.120E-01	ns	F019	0 0 0 0 2	
5.334E-03	9.990E-01	ns	M061	0 0 0 0 0	

1901. C$_9$H$_{17}$NO$_3$
Diethylaceturethane
Detonal
RN: **MP** (°C):
MW: 187.24   **BP** (°C):

Solubility (Moles/L)	Solubility (Grams/L)	Temp (°C)	Ref (#)	Evaluation (T P E A A)	Comments
2.796E-02	5.236E+00	ns	O021	0 2 0 0 0	

1902. C$_9$H$_{17}$NO$_4$
3,3-Dihydroxy-2,2,5,5-tetramethyl-4-carbamyltetrahydrofuran
3-Furamide, tetrahydro-4,4-dihydroxy-2,2,5,5-tetramethyl-
RN: 29839-68-7   **MP** (°C):
MW: 203.24   **BP** (°C):

Solubility (Moles/L)	Solubility (Grams/L)	Temp (°C)	Ref (#)	Evaluation (T P E A A)	Comments
4.473E-01	9.091E+01	rt	B066	0 2 0 0 1	

1903. C$_9$H$_{17}$N$_5$O
Atratone
2-Methoxy-4-ethylamino-6-isopropylamino-s-triazine
2-Methoxy-4-ethylamino-6-isopropylamino-s-triazines
RN: 1610-17-9   **MP** (°C):
MW: 211.27   **BP** (°C):

Solubility (Moles/L)	Solubility (Grams/L)	Temp (°C)	Ref (#)	Evaluation (T P E A A)	Comments
8.520E-03	1.800E+00	20	B185	0 0 0 0 0	
8.520E-03	1.800E+00	20	M061	1 0 0 0 2	
8.520E-03	1.800E+00	21	B192	0 0 0 0 2	
8.520E-03	1.800E+00	21	G099	2 0 0 1 0	
7.905E-03	1.670E+00	25	H073	2 1 1 2 2	
1.240E-02	2.620E+00	50	G001	1 0 1 1 2	
9.448E-03	1.996E+00	ns	B100	0 0 0 0 0	
8.520E-03	1.800E+00	ns	C101	0 0 0 0 1	
7.829E-03	1.654E+00	ns	J033	0 0 0 0 0	

1904. C$_9$H$_{17}$N$_5$S
Ametryn
(2-Methylthio-4-ethylamino-6-isopropylamino-s-triazine
Ametryne
N-Ethyl-N'-(1-methylethyl)-6-(methylthio)-1,3,5-triazine-2,4-diamine
Ametrex

RN:	834-12-8	MP (°C):	84
MW:	227.33	BP (°C):	

Solubility (Moles/L)	Solubility (Grams/L)	Temp (°C)	Ref (#)	Evaluation (T P E A A)	Comments
8.100E-04	1.841E-01	20	B200	1 0 0 0 1	
8.358E-04	1.900E-01	20	F311	1 2 2 2 1	
8.138E-04	1.850E-01	20	M161	1 0 0 0 2	
9.194E-04	2.090E-01	25	H073	2 1 1 2 2	
1.660E-03	3.774E-01	50	G001	1 0 1 1 2	
8.138E-04	1.850E-01	ns	C101	0 0 0 0 1	
8.490E-04	1.930E-01	ns	J033	0 0 0 0 0	

1905. C$_9$H$_{18}$
1-Nonene
α-Nonene
1-n-Nonene
n-Non-1-ene

RN:	124-11-8	MP (°C):	−81
MW:	126.24	BP (°C):	146.9

Solubility (Moles/L)	Solubility (Grams/L)	Temp (°C)	Ref (#)	Evaluation (T P E A A)	Comments
8.850E-06	1.117E-03	25	M342	1 0 1 1 2	

1906. C$_9$H$_{18}$
1,1,3-Trimethylcyclohexane
Cyclogeraniolane

RN:	3073-66-3	MP (°C):	−65.7
MW:	126.24	BP (°C):	136.6

Solubility (Moles/L)	Solubility (Grams/L)	Temp (°C)	Ref (#)	Evaluation (T P E A A)	Comments
1.402E-05	1.770E-03	25	K119	1 0 0 0 2	
1.402E-05	1.770E-03	25	P051	2 1 1 2 2	
1.402E-05	1.770E-03	25.00	P007	2 1 2 2 2	

1907. C$_9$H$_{18}$N$_2$O$_2$S
Thiofanox
3,3-Dimethyl-1-(methylthio)-2-butanone O-((methylamino)carbonyl)oxime
Thiophanox
DS-15647
Dacamox

RN:	39196-18-4	**MP (°C):**	57		
MW:	218.32	**BP (°C):**			

Solubility (Moles/L)	Solubility (Grams/L)	Temp (°C)	Ref (#)	Evaluation (T P E A A)	Comments
2.382E-02	5.200E+00	22	M161	1 0 0 0 1	

1908. C$_9$H$_{18}$N$_2$O$_4$
Meprobamate
2-Methyl-2-propyl-1,3-propanediol dicarbamate
Deprol
Meprospan
Miltown
Pathibamate

RN:	57-53-4	**MP (°C):**	104	
MW:	218.25	**BP (°C):**		

Solubility (Moles/L)	Solubility (Grams/L)	Temp (°C)	Ref (#)	Evaluation (T P E A A)	Comments
2.841E-02	6.200E+00	25	C039	1 2 2 1 1	form II
1.512E-02	3.300E+00	25	C039	1 2 2 1 1	form I
1.512E-02	3.300E+00	25	D082	1 0 1 0 1	
3.757E-02	8.200E+00	30	C039	1 2 2 1 1	form II
1.970E-02	4.300E+00	30	C039	1 2 2 1 1	form I
2.612E-02	5.700E+00	35	C039	1 2 2 1 1	form I
4.857E-02	1.060E+01	35	C039	1 2 2 1 2	form II
3.391E-02	7.400E+00	40	C039	1 2 2 1 1	form I
5.865E-02	1.280E+01	40	C039	1 2 2 1 2	form II

1909. C$_9$H$_{18}$N$_3$S$_6$Fe
Ferbam
tris(Dimethyldithiocarbamate)iron
Knockmate
Ferbeck
Hexaferb
Trifungol

RN:	14484-64-1	**MP (°C):**		
MW:	416.49	**BP (°C):**		

Solubility (Moles/L)	Solubility (Grams/L)	Temp (°C)	Ref (#)	Evaluation (T P E A A)	Comments
2.881E-04	1.200E-01	rt	I314	0 0 0 0 0	
3.121E-04	1.300E-01	rt	M161	0 0 0 0 2	

1910. C₉H₁₈N₆

Altretamine
Hexamethylmelamine
2,4,6-tris(Dimethylamino)-1,3,5-triazine
HMM
Hexastat
Hemel

RN: 645-05-6   **MP** (°C): 172.0
MW: 210.28   **BP** (°C):

Solubility (Moles/L)	Solubility (Grams/L)	Temp (°C)	Ref (#)	Evaluation (T P E A A)	Comments
3.846E-04	8.088E-02	25	B386	0 0 0 0 0	
4.327E-04	9.100E-02	25	C051	1 2 1 1 1	pH 7
4.150E-04	8.727E-02	25	K043	2 0 0 0 0	extrapolated

1911. C₉H₁₈N₆

1,3,5-Triazine-2,4,6-triamine, *N,N',N''*-Triethyl-
*N*2,*N*4,*N*6-Triethylmelamine
tris(Ethylamino)-1,3,5-triazine
2,4,6-tris(Ethylamino)-1,3,5-triazine
2,4,6-tris(Ethylamino)-*s*-triazine
N,N',N''-Triethyl-1,3,5-triazine-2,4,6-triamine

RN: 16268-92-1   **MP** (°C):
MW: 210.28   **BP** (°C):

Solubility (Moles/L)	Solubility (Grams/L)	Temp (°C)	Ref (#)	Evaluation (T P E A A)	Comments
7.318E-03	1.539E+00	25	B386	0 0 0 0 0	

1912. C₉H₁₈N₆O

N-Methylolpentamethylmelamine
N-(Hydroxymethyl)pentamethylmelamine
(Hydroxymethyl)pentamethylmelamine

RN: 16269-01-5   **MP** (°C): 121.0
MW: 226.28   **BP** (°C):

Solubility (Moles/L)	Solubility (Grams/L)	Temp (°C)	Ref (#)	Evaluation (T P E A A)	Comments
3.977E-03	9.000E-01	25	C051	1 2 1 1 0	pH 7, unstable in water

1913. C₉H₁₈N₆O

Ethanol, 2-[[4,6-bis(dimethylamino)-*s*-triazin-2-yl]amino]-
Ethanol, 2-[[4,6-bis(dimethylamino)-1,3,5-triazin-2-yl]amino]-

RN: 31482-09-4   **MP** (°C):
MW: 226.28   **BP** (°C):

Solubility (Moles/L)	Solubility (Grams/L)	Temp (°C)	Ref (#)	Evaluation (T P E A A)	Comments
1.132E-02	2.562E+00	25	B386	0 0 0 0 0	

1914. C$_9$H$_{18}$N$_6$O$_3$

N2,N4,N6-Trimethyl-N2,N4,N6-trimethylolmelamine
N,N',N''-Trimethyl-N,N',N''-trimethylolmelamine
Trimelamol
CB 10-375

RN:	64124-21-6	MP (°C):	129
MW:	258.28	BP (°C):	

Solubility (Moles/L)	Solubility (Grams/L)	Temp (°C)	Ref (#)	Evaluation (T P E A A)	Comments
3.500E-02	9.040E+00	25	C051	1 2 1 1 2	pH 7

1915. C$_9$H$_{18}$O

Nonyl aldehyde
n-Nonanal

RN:	124-19-6	MP (°C):	
MW:	142.24	BP (°C):	93

Solubility (Moles/L)	Solubility (Grams/L)	Temp (°C)	Ref (#)	Evaluation (T P E A A)	Comments
6.749E-04	9.600E-02	25	A049	1 0 0 0 1	

1916. C$_9$H$_{18}$O

5-Nonanone
Dibutyl ketone

RN:	502-56-7	MP (°C):	−50
MW:	142.24	BP (°C):	186.5

Solubility (Moles/L)	Solubility (Grams/L)	Temp (°C)	Ref (#)	Evaluation (T P E A A)	Comments
3.570E-03	5.078E-01	10	G032	1 2 1 1 2	
1.800E-03	2.560E-01	25	K012	1 0 0 0 1	
2.550E-03	3.627E-01	30	G032	1 2 1 1 2	
2.430E-03	3.457E-01	50	G032	1 2 1 1 2	

1917. C$_9$H$_{18}$O

3-Hydroxy-2,3,4,5,5-pentamethyltetrahydrofuran

RN:		MP (°C):	
MW:	142.24	BP (°C):	

Solubility (Moles/L)	Solubility (Grams/L)	Temp (°C)	Ref (#)	Evaluation (T P E A A)	Comments
6.391E-01	9.091E+01	rt	B066	0 2 0 0 1	

1918. C$_9$H$_{18}$O
2,6-Dimethyl-4-heptanone
Diisobutyl ketone
RN: 108-83-8 **MP** (°C):
MW: 142.24 **BP** (°C): 169

Solubility (Moles/L)	Solubility (Grams/L)	Temp (°C)	Ref (#)	Evaluation (T P E A A)	Comments
1.851E-02	2.633E+00	23.50	O005	2 0 2 2 2	

1919. C$_9$H$_{18}$O
2-Nonanone
Nonan-2-one
RN: 821-55-6 **MP** (°C): −21
MW: 142.24 **BP** (°C):

Solubility (Moles/L)	Solubility (Grams/L)	Temp (°C)	Ref (#)	Evaluation (T P E A A)	Comments
1.336E-03	1.900E-01	25	L450	0 0 0 0 0	

1920. C$_9$H$_{18}$O$_2$
3-Hydroxy-2-isopropyl-5,5-dimethyltetrahydrofuran
3-Furanol, tetrahydro-2-isopropyl-5,5-dimethyl-
RN: 29839-66-5 **MP** (°C):
MW: 158.24 **BP** (°C):

Solubility (Moles/L)	Solubility (Grams/L)	Temp (°C)	Ref (#)	Evaluation (T P E A A)	Comments
3.009E-01	4.762E+01	rt	B066	0 2 0 0 0	

1921. C$_9$H$_{18}$O$_2$
Pelargonic acid
1-Octanecarboxylic acid
Nonylic acid
n-Nonanoic acid
RN: 112-05-0 **MP** (°C): 12
MW: 158.24 **BP** (°C):

Solubility (Moles/L)	Solubility (Grams/L)	Temp (°C)	Ref (#)	Evaluation (T P E A A)	Comments
8.847E-04	1.400E-01	0	B136	1 0 2 1 1	
1.795E-03	2.840E-01	20	B136	1 0 2 1 2	
1.643E-03	2.599E-01	20.0	R001	1 1 1 1 1	
2.003E-03	3.170E-01	30	B136	1 0 2 1 2	
1.340E-03	2.120E-01	30	E005	2 1 1 2 2	
2.022E-03	3.199E-01	30.0	R001	1 1 1 1 1	
2.496E-03	3.950E-01	40	B136	1 0 2 1 2	
1.403E-03	2.220E-01	40	E005	2 1 1 2 2	
2.591E-03	4.100E-01	45	B136	1 0 2 1 1	
2.590E-03	4.098E-01	45.0	R001	1 1 1 1 1	

(continued)

1921. C$_9$H$_{18}$O$_2$ (continued)

Solubility (Moles/L)	Solubility (Grams/L)	Temp (°C)	Ref (#)	Evaluation (T P E A A)	Comments
1.668E-03	2.640E-01	50	E005	2 1 1 2 2	
3.223E-03	5.100E-01	60	B136	1 0 2 1 1	
1.890E-03	2.990E-01	60	E005	2 1 1 2 2	
3.221E-03	5.097E-01	60.0	R001	1 1 1 1 1	
8.846E-04	1.400E-01	.0	R001	1 1 1 1 1	

1922. C$_9$H$_{18}$O$_2$
Methyl octanoate
Methyl caprylate
Methyl octylate

RN:	111-11-5	MP (°C):	−37
MW:	158.24	BP (°C):	194.5

Solubility (Moles/L)	Solubility (Grams/L)	Temp (°C)	Ref (#)	Evaluation (T P E A A)	Comments
4.069E-04	6.440E-02	20	M337	2 1 2 2 2	

1923. C$_9$H$_{18}$O$_2$
3-Hydroxy-5-propyl-2,5-dimethyltetrahydrofuran
3-Furanol, 2,5-dimethyltetrahydro-5-propyl-

RN:		MP (°C):	
MW:	158.24	BP (°C):	

Solubility (Moles/L)	Solubility (Grams/L)	Temp (°C)	Ref (#)	Evaluation (T P E A A)	Comments
1.841E-01	2.913E+01	rt	B066	0 2 0 0 0	

1924. C$_9$H$_{18}$O$_2$
3-Hydroxy-5-methyl-5-isobutyltetrahydrofuran
3-Furanol, 5-isobutyltetrahydro-5-methyl-

RN:		MP (°C):	
MW:	158.24	BP (°C):	

Solubility (Moles/L)	Solubility (Grams/L)	Temp (°C)	Ref (#)	Evaluation (T P E A A)	Comments
6.257E-02	9.901E+00	rt	B066	0 2 0 0 0	

1925. C$_9$H$_{18}$O$_2$
3-Hydroxy-5-methyl-5-butyltetrahydrofuran
3-Furanol, 5-butyltetrahydro-5-methyl-

RN:		MP (°C):	
MW:	158.24	BP (°C):	

Solubility (Moles/L)	Solubility (Grams/L)	Temp (°C)	Ref (#)	Evaluation (T P E A A)	Comments
2.518E-02	3.984E+00	rt	B066	0 2 0 0 0	

1926. C$_9$H$_{18}$O$_2$
3-Hydroxy-3-ethyl-2,2,5-trimethyltetrahydrofuranol
3-Furanol, 3-ethyltetrahydro-2,2,5-trimethyl-
RN: 29839-58-5 **MP (°C):**
MW: 158.24 **BP (°C):**

Solubility (Moles/L)	Solubility (Grams/L)	Temp (°C)	Ref (#)	Evaluation (T P E A A)	Comments
4.134E-01	6.542E+01	rt	B066	0 2 0 0 0	

1927. C$_9$H$_{18}$O$_2$
3-Hydroxy-2-methyl-2,5-diethyltetrahydrofuran
3-Furanol, 2,5-diethyltetrahydro-2-methyl-
RN: 29839-64-3 **MP (°C):**
MW: 158.24 **BP (°C):**

Solubility (Moles/L)	Solubility (Grams/L)	Temp (°C)	Ref (#)	Evaluation (T P E A A)	Comments
1.239E-01	1.961E+01	rt	B066	0 2 0 0 0	

1928. C$_9$H$_{18}$O$_2$
3-Hydroxy-2,2,4,5,5-pentamethyltetrahydrofuran
3-Furanol, tetrahydro-2,2,4,5,5-pentamethyl-
RN: 29839-76-7 **MP (°C):**
MW: 158.24 **BP (°C):**

Solubility (Moles/L)	Solubility (Grams/L)	Temp (°C)	Ref (#)	Evaluation (T P E A A)	Comments
6.257E-02	9.901E+00	rt	B066	0 2 0 0 0	

1929. C$_9$H$_{18}$O$_2$
Butyl valerate
n-Butyl pentanoate
Butyl valerianate
RN: 591-68-4 **MP (°C):**
MW: 158.24 **BP (°C):** 186–187

Solubility (Moles/L)	Solubility (Grams/L)	Temp (°C)	Ref (#)	Evaluation (T P E A A)	Comments
5.300E-04	8.387E-02	25	K012	1 0 0 0 1	

1930. C$_9$H$_{18}$O$_2$
Pentyl butyrate
n-Amyl n-butyrate
Pentyl n-butanoate
RN: 540-18-1 **MP (°C):**
MW: 158.24 **BP (°C):**

Solubility (Moles/L)	Solubility (Grams/L)	Temp (°C)	Ref (#)	Evaluation (T P E A A)	Comments
1.100E-03	1.741E-01	20	S006	1 0 0 0 1	

1931. C$_9$H$_{18}$O$_2$
3-Hydroxy-2-methyl-5,5-diethyltetrahydrofuran
3-Furanol, 5,5-diethyltetrahydro-2-methyl-
RN: 6744-54-3   **MP** (°C):
MW: 158.24   **BP** (°C):

Solubility (Moles/L)	Solubility (Grams/L)	Temp (°C)	Ref (#)	Evaluation (T P E A A)	Comments
3.144E-02	4.975E+00	rt	B066	0 2 0 0 0	

1932. C$_9$H$_{18}$O$_3$
2,2-Diethyl-5-methyl-tetrahydrofuran-3,4-diol
3,4-Furandiol, 2,2-diethyltetrahydro-5-methyl-
RN: 31889-35-7   **MP** (°C):
MW: 174.24   **BP** (°C):

Solubility (Moles/L)	Solubility (Grams/L)	Temp (°C)	Ref (#)	Evaluation (T P E A A)	Comments
9.565E-01	1.667E+02	rt	B066	0 2 0 0 1	

1933. C$_9$H$_{18}$O$_3$
n-Propyl β-n-propoxypropionate
Propanoic acid, 3-propoxy-, propyl ester
RN: 14144-41-3   **MP** (°C):
MW: 174.24   **BP** (°C):

Solubility (Moles/L)	Solubility (Grams/L)	Temp (°C)	Ref (#)	Evaluation (T P E A A)	Comments
2.059E-02	3.587E+00	25	R034	0 0 0 0 1	

1934. C$_9$H$_{18}$O$_3$
n-Amyl β-methoxypropionate
Pentyl 3-methoxypropionate
RN: 10500-16-0   **MP** (°C):
MW: 174.24   **BP** (°C):

Solubility (Moles/L)	Solubility (Grams/L)	Temp (°C)	Ref (#)	Evaluation (T P E A A)	Comments
1.660E-02	2.892E+00	25	R034	0 0 0 0 1	

1935. C$_9$H$_{18}$O$_3$
1,3-Dioxolane-4-methanol, 2-butyl-2-methyl
RN: 5694-76-8   **MP** (°C):
MW: 174.24   **BP** (°C):

Solubility (Moles/L)	Solubility (Grams/L)	Temp (°C)	Ref (#)	Evaluation (T P E A A)	Comments
1.940E-01	3.380E+01	25	P342	0 0 0 0 0	0.0001M Na$_2$CO$_3$

1936. C₉H₁₈O₃

n-Butyl β-ethoxypropionate
Propionic acid, 3-ethoxy-, butyl ester
RN: 14144-35-5 **MP** (°C):
MW: 174.24 **BP** (°C):

Solubility (Moles/L)	Solubility (Grams/L)	Temp (°C)	Ref (#)	Evaluation (T P E A A)	Comments
2.287E-02	3.984E+00	25	D002	1 2 1 1 1	

1937. C₉H₁₈O₃

Hexyl lactate
Propanoic acid, 2-hydroxy-, hexyl ester
RN: 20279-51-0 **MP** (°C):
MW: 174.24 **BP** (°C):

Solubility (Moles/L)	Solubility (Grams/L)	Temp (°C)	Ref (#)	Evaluation (T P E A A)	Comments
1.550E-02	2.700E+00	25	R006	2 2 0 1 1	

1938. C₉H₁₉NOS

Eptam
EPTC
Ethyl *N,N'*-di-*n*-propylthiolcarbamate
S-Ethyl dipropylthiocarbamate
S-Ethyl *N,N*-di-*n*-propylthiocarbamate
RN: 759-94-4 **MP** (°C): <25
MW: 189.32 **BP** (°C): 235

Solubility (Moles/L)	Solubility (Grams/L)	Temp (°C)	Ref (#)	Evaluation (T P E A A)	Comments
3.359E-03	6.360E-01	3	G319	0 0 0 0 0	
1.954E-03	3.700E-01	20	B200	1 0 0 0 2	
1.981E+01	3.750E+03	20	F019	1 0 0 0 2	*sic*
1.981E-03	3.750E-01	20	M061	1 0 0 0 2	
1.928E-03	3.650E-01	20	M161	1 0 0 0 2	
4.170E+00	7.895E+02	25	B185	0 0 0 0 0	*sic*
1.981E-03	3.750E-01	25	G319	0 0 0 0 2	
1.981E-03	3.750E-01	25	M131	0 0 0 0 2	
2.123E-03	4.020E-01	28	H109	1 0 0 0 2	
1.981E-03	3.750E-01	ns	V414	0 0 0 0 0	

1939. C_9H_19NO_2
n-Octyl carbamate
Carbamic acid, octyl ester
RN: 2029-64-3   **MP (°C):**   67
MW: 173.26   **BP (°C):**

Solubility (Moles/L)	Solubility (Grams/L)	Temp (°C)	Ref (#)	Evaluation (T P E A A)	Comments
5.000E-04	8.663E-02	37	H006	1 2 2 1 0	

1940. C_9H_19O_3
3-Hydroxy-4-methylol-2,2,5,5-tetramethyltetrahydrofuran
RN: **MP (°C):**
MW: 175.25   **BP (°C):**

Solubility (Moles/L)	Solubility (Grams/L)	Temp (°C)	Ref (#)	Evaluation (T P E A A)	Comments
1.119E-01	1.961E+01	rt	B066	0 2 0 0 0	

1941. C_9H_20
3,3-Diethylpentane
Tetraethylmethane
RN: 1067-20-5   **MP (°C):**
MW: 128.26   **BP (°C):**

Solubility (Moles/L)	Solubility (Grams/L)	Temp (°C)	Ref (#)	Evaluation (T P E A A)	Comments
9.450E-06	1.212E-03	25	D346	0 0 0 0 0	

1942. C_9H_20
2,5-Dimethylheptane
RN: 2216-30-0   **MP (°C):**
MW: 128.26   **BP (°C):**   136

Solubility (Moles/L)	Solubility (Grams/L)	Temp (°C)	Ref (#)	Evaluation (T P E A A)	Comments
2.489E-06	3.192E-04	ns	S460	0 0 0 0 0	

1943. C_9H_20
3-Methyloctane
Octane, 3-methyl-
RN: 2216-33-3   **MP (°C):**
MW: 128.26   **BP (°C):**

Solubility (Moles/L)	Solubility (Grams/L)	Temp (°C)	Ref (#)	Evaluation (T P E A A)	Comments
6.237E-06	8.000E-04	23	C332	0 0 0 0 0	

1944. C$_9$H$_{20}$
2-Methyl-4-ethylhexane

RN:	3074-75-7	**MP** (°C):		
MW:	128.26	**BP** (°C):	134	

Solubility (Moles/L)	Solubility (Grams/L)	Temp (°C)	Ref (#)	Evaluation (T P E A A)	Comments
2.799E-06	3.590E-04	ns	S460	0 0 0 0 0	

1945. C$_9$H$_{20}$
2,2,3-Trimethylhexane

RN:	16747-25-4	**MP** (°C):		
MW:	128.26	**BP** (°C):	134	

Solubility (Moles/L)	Solubility (Grams/L)	Temp (°C)	Ref (#)	Evaluation (T P E A A)	Comments
2.825E-06	3.623E-04	ns	S460	0 0 0 0 0	

1946. C$_9$H$_{20}$
2,4-Dimethylheptane

RN:	2213-23-2	**MP** (°C):		
MW:	128.26	**BP** (°C):	133	

Solubility (Moles/L)	Solubility (Grams/L)	Temp (°C)	Ref (#)	Evaluation (T P E A A)	Comments
2.938E-06	3.768E-04	ns	S460	0 0 0 0 0	

1947. C$_9$H$_{20}$
2,2,5-Trimethylhexane
Hexane, 2,2,5-trimethyl-

RN:	3522-94-9	**MP** (°C):	−120	
MW:	128.26	**BP** (°C):	124.1	

Solubility (Moles/L)	Solubility (Grams/L)	Temp (°C)	Ref (#)	Evaluation (T P E A A)	Comments
6.159E-06	7.900E-04	0	P003	2 2 2 2 1	
8.966E-06	1.150E-03	25	M001	2 1 2 2 2	
4.210E-06	5.400E-04	25	P003	2 2 2 2 1	

1948. C$_9$H$_{20}$
2,2-Dimethylheptane

RN:	1071-26-7	**MP** (°C):		
MW:	128.26	**BP** (°C):		

Solubility (Moles/L)	Solubility (Grams/L)	Temp (°C)	Ref (#)	Evaluation (T P E A A)	Comments
2.800E-06	3.592E-04	ns	S460	0 0 0 0 0	

1949. C₉H₂₀
Nonane
n-Nonan

RN: 111-84-2   **MP** (°C):   –53
MW: 128.26   **BP** (°C):   151

Solubility (Moles/L)	Solubility (Grams/L)	Temp (°C)	Ref (#)	Evaluation (T P E A A)	Comments
<1.72E-05	<2.20E-03	20	M337	2 1 2 2 1	
9.512E-07	1.220E-04	25	K119	1 0 0 0 2	
1.715E-06	2.200E-04	25	M003	1 0 2 2 2	
1.333E-06	1.710E-04	25	T423	0 0 0 0 0	
9.512E-07	1.220E-04	25.0	P051	2 1 1 2 2	
9.512E-07	1.220E-04	25.00	P007	2 1 2 2 2	
2.409E-06	3.090E-04	69.7	P051	2 1 1 2 2	
3.275E-06	4.200E-04	99.1	P051	2 1 1 2 2	
3.275E-06	4.200E-04	99.10	P007	2 1 2 2 2	
1.325E-05	1.700E-03	121.3	P051	2 1 1 2 2	
1.325E-05	1.700E-03	121.30	P007	2 1 2 2 2	
3.953E-05	5.070E-03	136.6	P051	2 1 1 2 2	
3.953E-05	5.070E-03	136.60	P007	2 1 2 2 2	

1950. C₉H₂₀
4,4-Dimethylheptane

RN: 1068-19-5   **MP** (°C):
MW: 128.26   **BP** (°C):   135

Solubility (Moles/L)	Solubility (Grams/L)	Temp (°C)	Ref (#)	Evaluation (T P E A A)	Comments
2.600E-06	3.335E-04	ns	S460	0 0 0 0 0	

1951. C₉H₂₀
2,6-Dimethylheptane

RN: 1072-05-5   **MP** (°C):   –103
MW: 128.26   **BP** (°C):   135

Solubility (Moles/L)	Solubility (Grams/L)	Temp (°C)	Ref (#)	Evaluation (T P E A A)	Comments
2.594E-06	3.327E-04	ns	S460	0 0 0 0 0	

1952. C₉H₂₀
3,5-Dimethylheptane

RN: 926-82-9   **MP** (°C):
MW: 128.26   **BP** (°C):   136

Solubility (Moles/L)	Solubility (Grams/L)	Temp (°C)	Ref (#)	Evaluation (T P E A A)	Comments
2.489E-06	3.192E-04	ns	S460	0 0 0 0 0	

1953. C$_9$H$_{20}$
3-Ethylheptane

RN: 15869-80-4   **MP** (°C):
MW: 128.26   **BP** (°C): 143

Solubility (Moles/L)	Solubility (Grams/L)	Temp (°C)	Ref (#)	Evaluation (T P E A A)	Comments
1.714E-06	2.198E-04	ns	S460	0 0 0 0 0	

1954. C$_9$H$_{20}$
4-Ethylheptane

RN: 2216-32-2   **MP** (°C):
MW: 128.26   **BP** (°C): 141

Solubility (Moles/L)	Solubility (Grams/L)	Temp (°C)	Ref (#)	Evaluation (T P E A A)	Comments
1.888E-06	2.422E-04	ns	S460	0 0 0 0 0	

1955. C$_9$H$_{20}$
2,3-Dimethylheptane

RN: 3074-71-3   **MP** (°C):
MW: 128.26   **BP** (°C): 141

Solubility (Moles/L)	Solubility (Grams/L)	Temp (°C)	Ref (#)	Evaluation (T P E A A)	Comments
1.959E-06	2.512E-04	ns	S420	0 0 0 0 0	

1956. C$_9$H$_{20}$
2,3,4-Trimethylhexane

RN: 921-47-1   **MP** (°C):
MW: 128.26   **BP** (°C): 139

Solubility (Moles/L)	Solubility (Grams/L)	Temp (°C)	Ref (#)	Evaluation (T P E A A)	Comments
2.113E-06	2.711E-04	ns	S460	0 0 0 0 0	

1957. C$_9$H$_{20}$
3-Ethyl-2-methylhexane
2-Methyl-3-ethylhexane

RN: 16789-46-1   **MP** (°C):
MW: 128.26   **BP** (°C):

Solubility (Moles/L)	Solubility (Grams/L)	Temp (°C)	Ref (#)	Evaluation (T P E A A)	Comments
2.239E-06	2.871E-04	ns	S460	0 0 0 0 0	

1958. C_9H_{20}
3,3-Dimethylheptane
RN: 4032-86-4   **MP** (°C):
MW: 128.26   **BP** (°C):   137

Solubility (Moles/L)	Solubility (Grams/L)	Temp (°C)	Ref (#)	Evaluation (T P E A A)	Comments
2.360E-06	3.028E-04	ns	S460	0 0 0 0 0	

1959. C_9H_{20}
4-Methyloctane
4-Metylooktan
RN: 2216-34-4   **MP** (°C):   −113
MW: 128.26   **BP** (°C):

Solubility (Moles/L)	Solubility (Grams/L)	Temp (°C)	Ref (#)	Evaluation (T P E A A)	Comments
8.966E-07	1.150E-04	25	K119	1 0 0 0 2	
8.966E-07	1.150E-04	25	P051	2 1 1 2 2	
8.966E-07	1.150E-04	25.00	P007	2 1 2 2 2	

1960. $C_9H_{20}NO_3PS_2$
Fostion
FAC 20
O,O-Diethyl *S*-(*N*-isopropylcarbamylmethyl) dithiophosphate
Prothoate
RN: 2275-18-5   **MP** (°C):   24.5
MW: 285.37   **BP** (°C):

Solubility (Moles/L)	Solubility (Grams/L)	Temp (°C)	Ref (#)	Evaluation (T P E A A)	Comments
8.761E-03	2.500E+00	20	M161	1 0 0 0 1	

1961. $C_9H_{20}O$
2,6-Dimethyl-4-heptanol
Diisobutylcarbinol
RN: 108-82-7   **MP** (°C):
MW: 144.26   **BP** (°C):

Solubility (Moles/L)	Solubility (Grams/L)	Temp (°C)	Ref (#)	Evaluation (T P E A A)	Comments
6.925E-03	9.990E-01	25	C093	2 1 1 1 1	

1962. C$_9$H$_{20}$O
n-Nonyl alcohol
Nonanol

RN: 143-08-8   **MP** (°C):
MW: 144.26   **BP** (°C):   215

Solubility (Moles/L)	Solubility (Grams/L)	Temp (°C)	Ref (#)	Evaluation (T P E A A)	Comments
9.340E-04	1.347E-01	20	H330	0 0 0 0 0	
9.700E-04	1.399E-01	25	K025	2 2 1 1 2	

1963. C$_9$H$_{20}$O
3-Ethyl-3-heptanol

RN: 19780-41-7   **MP** (°C):
MW: 144.26   **BP** (°C):

Solubility (Moles/L)	Solubility (Grams/L)	Temp (°C)	Ref (#)	Evaluation (T P E A A)	Comments
3.802E-03	5.485E-01	ns	S460	0 0 0 0 0	

1964. C$_9$H$_{20}$O
2,6-Dimethyl-3-heptanol

RN: 19549-73-6   **MP** (°C):
MW: 144.26   **BP** (°C):

Solubility (Moles/L)	Solubility (Grams/L)	Temp (°C)	Ref (#)	Evaluation (T P E A A)	Comments
3.097E-03	4.468E-01	ns	S460	0 0 0 0 0	

1965. C$_9$H$_{20}$O
3-Nonanol
Hexyl ethyl carbinol
Ethyl *n*-hexyl carbinol

RN: 624-51-1   **MP** (°C):
MW: 144.26   **BP** (°C):

Solubility (Moles/L)	Solubility (Grams/L)	Temp (°C)	Ref (#)	Evaluation (T P E A A)	Comments
1.999E-03	2.884E-01	ns	J300	0 0 0 0 0	

1966. C$_9$H$_{20}$O

3,5,5-Trimethylhexanol
3.,5,5-Trimethyl hexanol
Nonylol
3,5,5-Trimethyl-1-hexanol

RN: 3452-97-9   **MP** (°C):
MW: 144.26   **BP** (°C):

Solubility (Moles/L)	Solubility (Grams/L)	Temp (°C)	Ref (#)	Evaluation (T P E A A)	Comments
3.120E-03	4.501E-01	20	H330	0 0 0 0 0	
3.099E-03	4.470E-01	ns	J300	0 0 0 0 0	

1967. C$_9$H$_{20}$O

Methyl-octyl-alcohol
2-Nonanol
Heptylmethylcarbinol
Methyl *n*-heptyl carbinol

RN: 628-99-9   **MP** (°C):
MW: 144.26   **BP** (°C):

Solubility (Moles/L)	Solubility (Grams/L)	Temp (°C)	Ref (#)	Evaluation (T P E A A)	Comments
<4.00E-03	<5.77E-01	25	F044	1 0 0 0 0	

1968. C$_9$H$_{21}$N

Tripropylamine
Tri-*n*-propylamine
N,*N*-Dipropylpropanamine
N,*N*-Dipropyl-1-propanamine

RN: 102-69-2   **MP** (°C):   −93.5
MW: 143.27   **BP** (°C):   155

Solubility (Moles/L)	Solubility (Grams/L)	Temp (°C)	Ref (#)	Evaluation (T P E A A)	Comments
5.216E-03	7.473E-01	25.04	V013	2 2 2 2 2	

1969. C$_9$H$_{21}$O$_2$PS$_3$

Terbufos
O,*O*-Diethyl *S*-(((1,1-dimethylethyl)thio)methyl) phosphorodithoic acid
Counter 15G
Contraven
ST 100

RN: 13071-79-9   **MP** (°C):
MW: 288.43   **BP** (°C):

Solubility (Moles/L)	Solubility (Grams/L)	Temp (°C)	Ref (#)	Evaluation (T P E A A)	Comments
1.907E-05	5.500E-03	19	B169	2 1 1 1 1	
1.758E-05	5.070E-03	24	F179	2 2 2 2 2	
1.907E-05	5.500E-03	ns	B325	0 1 0 0 1	

(continued)

1969. C$_9$H$_{21}$O$_2$PS$_3$ (continued)

Solubility (Moles/L)	Solubility (Grams/L)	Temp (°C)	Ref (#)	Evaluation (T P E A A)	Comments
3.467E-05	1.000E-02	ns	M110	0 0 0 0 0	EFG
4.334E-05	1.250E-02	ns	M161	0 0 0 0 0	

1970. C$_9$H$_{21}$O$_3$P
Dibutyl methyl phosphonate
Di-n-butyl methanephosphonate
RN: 2404-73-1 **MP** (°C):
MW: 208.24 **BP** (°C):

Solubility (Moles/L)	Solubility (Grams/L)	Temp (°C)	Ref (#)	Evaluation (T P E A A)	Comments
3.842E-02	8.000E+00	25	B070	1 2 0 1 0	

1971. C$_9$H$_{21}$O$_3$PS$_3$
S-Ethylsulphinylmethyl O,O-di-isopropyl phosphorodithioate
O,O-Diisopropyl S-[(ethylsulfinyl)methyl] dithiophosphate
Aphidan
PSP 204
IPSP
RN: 5827-05-4 **MP** (°C):
MW: 304.43 **BP** (°C):

Solubility (Moles/L)	Solubility (Grams/L)	Temp (°C)	Ref (#)	Evaluation (T P E A A)	Comments
4.927E-03	1.500E+00	15	M161	1 0 0 0 1	

1972. C$_9$H$_{21}$O$_3$PS$_3$
Terbufos sulfoxide
Phosphorodithioic acid, S-[[(1,1-dimethylethyl)sulfinyl]methyl] O,O-diethyl ester
RN: 10548-10-4 **MP** (°C):
MW: 304.43 **BP** (°C):

Solubility (Moles/L)	Solubility (Grams/L)	Temp (°C)	Ref (#)	Evaluation (T P E A A)	Comments
>3.61E-03	>1.10E+00	ns	B325	0 1 0 0 1	

1973. C$_9$H$_{21}$O$_4$P
Tripropyl phosphate
Tri-n-propyl phosphate
RN: 513-08-6 **MP** (°C):
MW: 224.24 **BP** (°C):

Solubility (Moles/L)	Solubility (Grams/L)	Temp (°C)	Ref (#)	Evaluation (T P E A A)	Comments
3.100E-02	6.951E+00	30	V300	2 2 0 1 0	

1974. C$_9$H$_{21}$O$_4$P

Dibutyl methyl phosphate
Methyl dibutyl phosphate

RN: 7242-59-3   **MP** (°C):
MW: 224.24   **BP** (°C):

Solubility (Moles/L)	Solubility (Grams/L)	Temp (°C)	Ref (#)	Evaluation (T P E A A)	Comments
3.166E-02	7.100E+00	25	B070	1 2 2 1 1	

1975. C$_9$H$_{21}$O$_4$P

Diethyl amyl phosphate
O,O-Diethyl *O*-pentyl phosphate
Diethyl pentyl phosphate

RN: 20195-08-8   **MP** (°C):
MW: 224.24   **BP** (°C):

Solubility (Moles/L)	Solubility (Grams/L)	Temp (°C)	Ref (#)	Evaluation (T P E A A)	Comments
3.345E-02	7.500E+00	25	B070	1 2 0 1 1	

1976. C$_9$H$_{21}$O$_4$PS$_3$

Terbufos sulfone
Phosphorodithioic acid, *S*-[[(1,1-dimethylethyl)sulfonyl]methyl] *O,O*-diethyl ester
Counter sulfone
AC 94320

RN: 56070-16-7   **MP** (°C):
MW: 320.43   **BP** (°C):

Solubility (Moles/L)	Solubility (Grams/L)	Temp (°C)	Ref (#)	Evaluation (T P E A A)	Comments
1.273E-03	4.078E-01	18.50	B169	2 0 1 1 2	
1.273E-03	4.078E-01	ns	B325	0 1 0 0 1	

1977. C$_9$H$_{22}$O$_4$P$_2$S$_4$

Ethion
O,O,O,O-Tetraethyl *S,S*-methylene bisphosphorodithioate
Nialate
Ethanox
Diethion
Hylemox

RN: 563-12-2   **MP** (°C):   −25
MW: 384.48   **BP** (°C):

Solubility (Moles/L)	Solubility (Grams/L)	Temp (°C)	Ref (#)	Evaluation (T P E A A)	Comments
1.483E-06	5.700E-04	10	B324	0 0 0 0 0	
1.483E-06	5.702E-04	10	B324	0 0 0 0 0	
2.861E-06	1.100E-03	19.50	B169	2 2 1 1 1	
1.769E-06	6.801E-04	20	B324	0 0 0 0 0	
1.769E-06	6.800E-04	20	B324	0 0 0 0 0	

(continued)

1977. C₉H₂₂O₄P₂S₄ (continued)

Solubility (Moles/L)	Solubility (Grams/L)	Temp (°C)	Ref (#)	Evaluation (T P E A A)	Comments
1.977E-06	7.601E-04	30	B324	0 0 0 0 0	
1.977E-06	7.600E-04	30	B324	0 0 0 0 0	

1978. C₁₀H₄Cl₂O₂
Dichlone
2,3-Dichloro-1,4-naphthalenedione
Phygon XL
Phygon
Phygon paste
USR 604

RN: 117-80-6   **MP** (°C):
MW: 227.05   **BP** (°C):

Solubility (Moles/L)	Solubility (Grams/L)	Temp (°C)	Ref (#)	Evaluation (T P E A A)	Comments
4.404E-07	1.000E-04	25	M161	1 0 0 0 0	
3.083E-05	7.000E-03	ns	B160	0 0 0 0 0	
4.404E-06	1.000E-03	ns	B185	0 0 0 0 0	

1979. C₁₀H₅ClN₂O₄
1-Chloro-2,4-dinitronaphthalene
2,4-Dinitro-1-naphthyl chloride
2,4-Dinitrochloronaphthalene
2,4-Dinitro-1-chloronaphthalene

RN: 2401-85-6   **MP** (°C):   148
MW: 252.62   **BP** (°C):

Solubility (Moles/L)	Solubility (Grams/L)	Temp (°C)	Ref (#)	Evaluation (T P E A A)	Comments
3.959E-06	1.000E-03	25	M061	1 0 0 0 0	

1980. C₁₀H₅Cl₇
Heptachlor
1,4,5,6,7,8,8-Heptachloro-3α,4,7,7α-tetrahydro-4,7-methano-1H-indene
3-Chlorochlordene
Tetrahydro
Rhodiachlor
3,4,5,6,7,8,8α-Heptachlorodicyclopentadiene

RN: 76-44-8   **MP** (°C):   95.5
MW: 373.32   **BP** (°C):

Solubility (Moles/L)	Solubility (Grams/L)	Temp (°C)	Ref (#)	Evaluation (T P E A A)	Comments
2.679E-07	1.000E-04	15	B083	2 2 1 2 2	particle size 5 μm
4.786E-07	1.787E-04	24.99	K436	0 0 0 0 0	
4.822E-07	1.800E-04	25	B083	2 2 1 2 2	particle size 5 μm

(continued)

1980. $C_{10}H_5Cl_7$ (continued)

Solubility (Moles/L)	Solubility (Grams/L)	Temp (°C)	Ref (#)	Evaluation (T P E A A)	Comments
1.500E-07	5.600E-05	25	I308	0 0 0 0 0	
1.500E-07	5.600E-05	26.5	P027	1 1 2 2 1	
1.500E-07	5.600E-05	27	M161	0 0 0 0 1	
8.438E-07	3.150E-04	35	B083	2 2 1 2 2	particle size 5 μm
1.313E-06	4.900E-04	45	B083	2 2 1 2 2	particle size 5 μm
8.036E-08	3.000E-05	ns	K138	0 0 0 0 2	
1.875E-07	7.000E-05	ns	M110	0 0 0 0 0	EFG
4.822E-07	1.800E-04	ns	V414	0 0 0 0 0	

1981. $C_{10}H_5Cl_7O$
Heptachlor epoxide
1,4,5,6,7,8,8-Heptachloro-2,3-epoxy-3α,4,7,7α-tetrahydro-4,7-methanoindan
Hepachlor epoxide

RN: 1024-57-3   **MP** (°C): 160
MW: 389.32   **BP** (°C):

Solubility (Moles/L)	Solubility (Grams/L)	Temp (°C)	Ref (#)	Evaluation (T P E A A)	Comments
2.825E-07	1.100E-04	15	B083	2 2 1 2 2	particle size 5 μm
5.137E-07	2.000E-04	25	B083	2 2 1 2 2	particle size 5 μm
5.137E-07	2.000E-04	25	I308	0 0 0 0 0	
8.990E-07	3.500E-04	25	W025	1 0 2 2 2	
8.990E-07	3.500E-04	26.5	P027	1 1 2 2 1	
8.990E-07	3.500E-04	35	B083	2 2 1 2 2	particle size 5 μm
1.541E-06	6.000E-04	45	B083	2 2 1 2 2	particle size 5 μm
1.798E-06	7.000E-04	ns	M110	0 0 0 0 0	EFG
5.137E-07	2.000E-04	ns	V414	0 0 0 0 0	

1982. $C_{10}H_5N_3O_6$
1,3,8-Trinitronaphthalene
1,3,8-Trinitronaphthalin

RN: 2364-46-7   **MP** (°C):
MW: 263.17   **BP** (°C):

Solubility (Moles/L)	Solubility (Grams/L)	Temp (°C)	Ref (#)	Evaluation (T P E A A)	Comments
6.840E-05	1.800E-02	15	F300	1 0 0 0 1	

1983. $C_{10}H_5N_3O_6$
1,4,5-Trinitronaphthalene
1,4,5-Trinitronaphthalin

RN: 2243-95-0   **MP** (°C):
MW: 263.17   **BP** (°C):

Solubility (Moles/L)	Solubility (Grams/L)	Temp (°C)	Ref (#)	Evaluation (T P E A A)	Comments
1.520E-04	4.000E-02	15	F300	1 0 0 0 1	

1984. $C_{10}H_6Br_2$
2,3-Dibromonaphthalene
Naphthalene, 2,3-dibromo-
RN: 13214-70-5   **MP** (°C):
MW: 285.98   **BP** (°C):

Solubility (Moles/L)	Solubility (Grams/L)	Temp (°C)	Ref (#)	Evaluation (T P E A A)	Comments
1.922E-07	5.497E-05	4	D351	1 2 1 1 2	
4.778E-07	1.366E-04	25	D351	1 2 1 1 2	
1.222E-06	3.495E-04	40	D351	1 2 1 1 2	

1985. $C_{10}H_6Br_2$
1,4-Dibromonaphthalene
Naphthalene, 1,4-dibromo-
RN: 83-53-4   **MP** (°C):   80–82
MW: 285.98   **BP** (°C):

Solubility (Moles/L)	Solubility (Grams/L)	Temp (°C)	Ref (#)	Evaluation (T P E A A)	Comments
4.333E-07	1.239E-04	4	D351	1 2 1 1 2	
1.217E-06	3.479E-04	25	D351	1 2 1 1 2	
3.006E-06	8.595E-04	40	D351	1 2 1 1 2	

1986. $C_{10}H_6Cl_2$
1,4-Dichloronaphthalene
Naphthalene, 1,4-dichloro-
RN: 1825-31-6   **MP** (°C):
MW: 197.07   **BP** (°C):

Solubility (Moles/L)	Solubility (Grams/L)	Temp (°C)	Ref (#)	Evaluation (T P E A A)	Comments
1.333E-06	2.628E-04	4	D351	1 2 1 1 2	
4.389E-06	8.649E-04	25	D351	1 2 1 1 2	
1.122E-05	2.212E-03	40	D351	1 2 1 1 2	

1987. $C_{10}H_6Cl_4O_3S$
Glenbar
O,S-Dimethyl tetrachlorothioterephthalate
RN: 3765-57-9   **MP** (°C):   161
MW: 348.03   **BP** (°C):

Solubility (Moles/L)	Solubility (Grams/L)	Temp (°C)	Ref (#)	Evaluation (T P E A A)	Comments
1.437E-06	5.000E-04	22	B200	1 0 0 0 0	
1.034E-06	3.600E-04	ns	M061	0 0 0 0 1	

1988. $C_{10}H_6Cl_4O_4$
Dimethyl tetrachloroterephthalate
DCPA
RN:　　1861-32-1　　**MP** (°C):　　156
MW:　　331.97　　　**BP** (°C):

Solubility (Moles/L)	Solubility (Grams/L)	Temp (°C)	Ref (#)	Evaluation (T P E A A)	Comments
1.506E-06	5.000E-04	25	B200	1 0 0 0 0	
<1.51E-06	<5.00E-04	25	M161	1 0 0 0 0	
<1.51E-06	<5.00E-04	ns	B185	0 0 0 0 0	
1.506E-06	5.000E-04	ns	V414	0 0 0 0 0	

1989. $C_{10}H_6Cl_6$
Chlordene
4,5,6,7,8,8-Hexachloro-3α,4,7,7α-tetrahydro-4,7-methanoindene
RN:　　3734-48-3　　**MP** (°C):　　−62
MW:　　338.88　　　**BP** (°C):

Solubility (Moles/L)	Solubility (Grams/L)	Temp (°C)	Ref (#)	Evaluation (T P E A A)	Comments
2.281E-06	7.730E-04	26.70	L071	1 2 0 1 2	

1990. $C_{10}H_6Cl_6O$
1-Hydroxychlordene
1-Hydroxy-4,5,6,7,8,8-hexachloro-3α,4,7,7α-tetra-hydro-4,7-methanoindene
RN:　　2597-11-7　　**MP** (°C):　　194
MW:　　354.88　　　**BP** (°C):

Solubility (Moles/L)	Solubility (Grams/L)	Temp (°C)	Ref (#)	Evaluation (T P E A A)	Comments
3.469E-06	1.231E-03	26.70	L071	1 2 0 1 2	

1991. $C_{10}H_6Cl_6O$
Chlordene epoxide
2,3-Epoxy-4,5,6,7,8,8-hexachloro-3α,4,7,7α-tetrahydro-4,7-methanoindene
Chlordene hydroxide
4,7-Methano-1H-inden-1-ol, 4,5,6,7,8,8-hexachloro-3α,4,7,7α-tetrahydro-
RN:　　6058-23-7　　**MP** (°C):　　215
MW:　　354.88　　　**BP** (°C):

Solubility (Moles/L)	Solubility (Grams/L)	Temp (°C)	Ref (#)	Evaluation (T P E A A)	Comments
3.829E-06	1.359E-03	26.70	L071	1 2 0 1 2	

1992. C$_{10}$H$_6$Cl$_6$O$_2$
1-Hydroxychlordene epoxide
1-Hydroxy-2,3-epoxy-4,5,6,7,8,8-hexachloro-3α,4,7,7α-tetrahydro-4,7-methanoindene

RN: 24009-06-1 **MP** (°C):
MW: 370.88 **BP** (°C):

Solubility (Moles/L)	Solubility (Grams/L)	Temp (°C)	Ref (#)	Evaluation (T P E A A)	Comments
7.391E-06	2.741E-03	26.70	L071	1 1 1 1 2	

1993. C$_{10}$H$_6$Cl$_8$
cis-Chlordane
(1α,2α,3aα,4β,7β,7aα)-1,2,4,5,6,7,8,8-Octachloro-2,3,3a,4,7,7a-hexahydro-4,7-methano-1H-indene
4,7-Methano-1H-indene 1,2,4,5,6,7,8,8-octachloro-2,3,3a,4,7,7a-hexahydro-, (1α,2α,3aα,4β,7β,7aα)
α-Chlordane

RN: 5103-71-9 **MP** (°C):
MW: 409.78 **BP** (°C):

Solubility (Moles/L)	Solubility (Grams/L)	Temp (°C)	Ref (#)	Evaluation (T P E A A)	Comments
1.367E-07	5.600E-05	ns	V414	0 0 0 0 0	

1994. C$_{10}$H$_6$Cl$_8$
trans-Chlordane
(1α,2β,3aα,4β,7β,7aα)-1,2,4,5,6,7,8,8-Octachloro-2,3,3a,4,7,7a-hexahydro-4,7-methano-1H-indene
4,7-Methano-1H-indene 1,2,4,5,6,7,8,8-octachloro-2,3,3a,4,7,7a-hexahydro-, (1α,2β,3aα,4β,7β,7aα)-
β-Chlordane

RN: 5103-74-2 **MP** (°C):
MW: 409.78 **BP** (°C):

Solubility (Moles/L)	Solubility (Grams/L)	Temp (°C)	Ref (#)	Evaluation (T P E A A)	Comments
1.367E-07	5.600E-05	ns	V414	0 0 0 0 0	

1995. C$_{10}$H$_6$Cl$_8$
Chlordane
1,2,4,5,6,7,8,8-Octachloro-4,7-methano-3α,4,7,7α-tetrahydroindane
Octachlor
Velsicol 1068
Toxichlor
Ortho-Klor

RN: 57-74-9 **MP** (°C): 105
MW: 409.78 **BP** (°C):

Solubility (Moles/L)	Solubility (Grams/L)	Temp (°C)	Ref (#)	Evaluation (T P E A A)	Comments
1.380E-07	5.657E-05	24.99	K436	0 0 0 0 0	
4.515E-06	1.850E-03	25	W025	1 0 2 2 2	
1.367E-07	5.600E-05	ns	K138	0 0 0 0 2	

(continued)

1995. C$_{10}$H$_6$Cl$_8$ (continued)

Solubility (Moles/L)	Solubility (Grams/L)	Temp (°C)	Ref (#)	Evaluation (T P E A A)	Comments
1.708E-07	7.000E-05	ns	M110	0 0 0 0 0	EFG
1.367E-07	5.600E-05	ns	S187	0 2 2 1 1	
1.367E-07	5.600E-05	ns	V414	0 0 0 0 0	

1996. C$_{10}$H$_6$FN$_3$O$_3$
3-Nicotinoyl-5-fluorouracil
RN: **MP** (°C):
MW: 235.18   **BP** (°C):

Solubility (Moles/L)	Solubility (Grams/L)	Temp (°C)	Ref (#)	Evaluation (T P E A A)	Comments
1.148E-02	2.700E+00	22	B332	1 1 0 0 1	pH 4.0

1997. C$_{10}$H$_6$N$_2$O$_4$
1,8-Dinitronaphthalene
1,8-Dinitronaphthalin
RN: 602-38-0   **MP** (°C):   107
MW: 218.17   **BP** (°C):

Solubility (Moles/L)	Solubility (Grams/L)	Temp (°C)	Ref (#)	Evaluation (T P E A A)	Comments
1.558E-04	3.400E-02	15	F300	1 0 0 0 1	

1998. C$_{10}$H$_6$N$_2$O$_4$
1,5-Dinitronaphthalene
1,5-Dinitronaphthalin
RN: 605-71-0   **MP** (°C):   216.5
MW: 218.17   **BP** (°C):

Solubility (Moles/L)	Solubility (Grams/L)	Temp (°C)	Ref (#)	Evaluation (T P E A A)	Comments
2.658E-04	5.800E-02	12	F300	1 0 0 0 1	

1999. C$_{10}$H$_6$O$_8$
Pyromellitic acid
1,2,4,5-Benzenetetracarboxylic acid
Benzol-tetracarbonsaeure-(1,2,4,5)
RN: 89-05-4   **MP** (°C):
MW: 254.15   **BP** (°C):

Solubility (Moles/L)	Solubility (Grams/L)	Temp (°C)	Ref (#)	Evaluation (T P E A A)	Comments
5.508E-02	1.400E+01	16	F300	1 0 0 0 2	

2000. C₁₀H₇Br

1-Bromonaphthalene
Naphthalene, 1-bromo-

RN:	90-11-9	MP (°C):	6.2
MW:	207.08	BP (°C):	281.1

Solubility (Moles/L)	Solubility (Grams/L)	Temp (°C)	Ref (#)	Evaluation (T P E A A)	Comments
4.383E-05	9.077E-03	4	D351	1 2 1 1 2	
4.733E-05	9.802E-03	10	D351	1 2 1 1 2	
4.500E-05	9.318E-03	21	A057	2 1 2 2 1	
6.444E-05	1.334E-02	25	D351	1 2 1 1 2	
9.166E-05	1.898E-02	40	D351	1 2 1 1 2	
6.000E-05	1.242E-02	ns	L060	0 0 0 0 0	
9.120E-05	1.889E-02	ns	S460	0 0 0 0 0	

2001. C₁₀H₇Br

2-Bromonaphthalene
Naphthalene, 2-bromo-

RN:	580-13-2	MP (°C):	53.5
MW:	207.08	BP (°C):	281.1

Solubility (Moles/L)	Solubility (Grams/L)	Temp (°C)	Ref (#)	Evaluation (T P E A A)	Comments
1.850E-05	3.831E-03	4	D351	1 2 1 1 2	
3.883E-05	8.041E-03	25	D351	1 2 1 1 2	
7.611E-05	1.576E-02	40	D351	1 2 1 1 2	
4.000E-05	8.283E-03	ns	L060	0 0 0 0 0	

2002. C₁₀H₇Cl

β-Chloronaphthalene
2-Chloronaphthalene

RN:	91-58-7	MP (°C):	59.5
MW:	162.62	BP (°C):	256

Solubility (Moles/L)	Solubility (Grams/L)	Temp (°C)	Ref (#)	Evaluation (T P E A A)	Comments
<6.15E-06	<1.00E-03	30	M311	1 1 2 2 0	
8.000E-05	1.301E-02	ns	L060	0 0 0 0 0	

2003. C₁₀H₇Cl

1-Chloronaphthalene
α-Chloronaphthalene
1-Naphthyl chloride

RN:	90-13-1	MP (°C):	−20
MW:	162.62	BP (°C):	259.3

Solubility (Moles/L)	Solubility (Grams/L)	Temp (°C)	Ref (#)	Evaluation (T P E A A)	Comments
<1.23E-04	<2.00E-02	ns	L060	0 0 0 0 2	
1.164E-04	1.893E-02	ns	S460	0 0 0 0 0	

2004. C$_{10}$H$_7$I
α-Iodonaphthalene
1-Iodonaphthalene

RN: 90-14-2   **MP** (°C):
MW: 254.07   **BP** (°C):

Solubility (Moles/L)	Solubility (Grams/L)	Temp (°C)	Ref (#)	Evaluation (T P E A A)	Comments
2.800E-05	7.114E-03	ns	L060	0 0 0 0 1	average
2.818E-05	7.161E-03	ns	S460	0 0 0 0 0	

2005. C$_{10}$H$_7$NO$_2$
1-Nitronaphthalene
1-Nitro-naphthalin

RN: 86-57-7   **MP** (°C):   59.5
MW: 173.17   **BP** (°C):   304

Solubility (Moles/L)	Solubility (Grams/L)	Temp (°C)	Ref (#)	Evaluation (T P E A A)	Comments
2.887E-04	5.000E-02	18	F300	1 0 0 0 1	

2006. C$_{10}$H$_7$NO$_3$
1-Nitro-2-naphthol
1-Nitro-naphthol-(2)

RN: 550-60-7   **MP** (°C):   104
MW: 189.17   **BP** (°C):   115

Solubility (Moles/L)	Solubility (Grams/L)	Temp (°C)	Ref (#)	Evaluation (T P E A A)	Comments
1.057E-03	2.000E-01	20	F300	1 0 0 0 2	

2007. C$_{10}$H$_7$NO$_3$
Kynurenic acid
4-Hydroxy-chinolin-carbonsaeure-(2)
Kynurensaeure

RN: 492-27-3   **MP** (°C):   282.5
MW: 189.17   **BP** (°C):

Solubility (Moles/L)	Solubility (Grams/L)	Temp (°C)	Ref (#)	Evaluation (T P E A A)	Comments
4.715E-02	8.920E+00	100	D041	1 0 0 0 0	
4.969E-03	9.400E-01	100	F300	1 0 0 0 1	

2008. $C_{10}H_7N_3O_3$
Orotic acid pyridine
RN: **MP (°C):**
MW: 217.19   **BP (°C):**

Solubility (Moles/L)	Solubility (Grams/L)	Temp (°C)	Ref (#)	Evaluation (T P E A A)	Comments
1.200E-01	2.606E+01	16	N018	0 0 0 0 0	

2009. $C_{10}H_7N_3S$
Thiabendazole
2-(Thiazol-4-yl)benzimidazole
Mintezol
Apl-Luster
Mertect
Tecto
RN: 148-79-8   **MP (°C):**   304.5
MW: 201.25   **BP (°C):**

Solubility (Moles/L)	Solubility (Grams/L)	Temp (°C)	Ref (#)	Evaluation (T P E A A)	Comments
2.484E-04	5.000E-02	25	M161	1 0 0 0 1	intrinsic

2010. $C_{10}H_8$
Naphthalene
Napthalene
Mothballs
Camphor tar
RN: 91-20-3   **MP (°C):**   80.2
MW: 128.18   **BP (°C):**   217.9

Solubility (Moles/L)	Solubility (Grams/L)	Temp (°C)	Ref (#)	Evaluation (T P E A A)	Comments
1.350E-04	1.730E-02	4.99	P331	0 0 0 0 0	
1.320E-04	1.692E-02	8.20	M082	1 1 1 2 2	
1.320E-04	1.692E-02	8.20	M151	2 1 2 2 1	
1.320E-04	1.692E-02	8.24	M183	1 2 1 1 2	
1.390E-04	1.782E-02	10	J302	2 1 2 2 2	
1.580E-04	2.025E-02	9.99	P331	0 0 0 0 0	
1.500E-04	1.923E-02	11.50	M082	1 1 1 2 2	
1.500E-04	1.923E-02	11.50	M151	2 1 2 2 2	
1.502E-04	1.925E-02	11.54	M183	1 2 1 1 2	
1.570E-04	2.012E-02	12	S076	2 2 2 2 2	
1.590E-04	2.038E-02	13.40	M082	1 1 1 2 2	
1.590E-04	2.038E-02	13.40	M151	2 1 2 2 2	
1.591E-04	2.039E-02	13.44	M183	1 2 1 1 2	
1.900E-04	2.435E-02	14.99	P331	0 0 0 0 0	
1.716E-03	2.200E-01	15	F300	1 0 0 0 2	sic
1.716E-04	2.200E-02	15	M073	1 0 2 2 1	
1.680E-04	2.153E-02	15.10	M082	1 1 1 2 2	

(continued)

2010. $C_{10}H_8$ (continued)

Solubility (Moles/L)	Solubility (Grams/L)	Temp (°C)	Ref (#)	Evaluation (T P E A A)	Comments
1.680E-04	2.153E-02	15.10	M151	2 1 2 2 2	
1.677E-04	2.150E-02	15.14	M183	1 2 1 1 2	
1.900E-04	2.435E-02	18	S076	2 2 2 2 2	
2.010E-04	2.576E-02	19.30	M082	1 1 1 2 2	
2.010E-04	2.576E-02	19.30	M151	2 1 2 2 2	
2.013E-04	2.581E-02	19.34	M183	1 2 1 1 2	
2.240E-04	2.871E-02	19.99	P331	0 0 0 0 0	
1.748E-04	2.240E-02	20	A050	1 0 1 1 1	
7.412E-04	9.500E-02	20	B318	0 0 0 0 0	EFG
3.000E-04	3.845E-02	20	E009	1 0 0 0 1	
3.000E-04	3.845E-02	20	E025	1 0 2 2 1	
1.900E-04	2.435E-02	20	H306	1 0 1 2 1	
1.272E-04	1.630E-02	20	T301	1 2 2 2 2	
2.400E-04	3.076E-02	22	A413	2 0 2 2 1	
2.645E-04	3.390E-02	22	C413	2 0 2 2 1	
1.638E-04	2.100E-02	22	N311	1 0 1 1 2	
2.255E-04	2.890E-02	22.20	W003	2 2 2 2 2	average of 3
2.341E-04	3.000E-02	23	P332	0 0 0 0 0	
2.341E-04	3.000E-02	23	P339	0 0 0 0 0	
2.300E-04	2.948E-02	23.40	M082	1 1 1 2 2	
2.300E-04	2.948E-02	23.40	M151	2 1 2 2 2	
2.301E-04	2.949E-02	23.44	M183	1 2 1 1 2	
2.380E-04	3.050E-02	24.50	W003	2 2 2 2 2	average of 5
2.630E-04	3.371E-02	24.99	P331	0 0 0 0 0	
2.458E-04	3.150E-02	25	A001	1 2 2 2 2	
2.350E-04	3.012E-02	25	A325	2 1 2 2 2	
2.684E-04	3.440E-02	25	B003	2 2 2 2 2	
2.465E-04	3.160E-02	25	B319	2 0 1 2 2	average of 2
2.442E-04	3.130E-02	25	D337	0 0 0 0 0	
2.715E-04	3.480E-02	25	D406	1 2 2 2 2	
2.442E-04	3.130E-02	25	E004	2 1 2 2 2	
2.620E-04	3.358E-02	25	G047	2 2 2 2 2	
2.520E-04	3.230E-02	25	J302	2 1 2 2 2	
9.750E-05	1.250E-02	25	K001	2 2 2 2 2	
2.300E-04	2.948E-02	25	K123	1 0 2 2 1	
2.497E-04	3.200E-02	25	L332	1 1 1 1 0	
2.653E-04	3.400E-02	25	M040	1 0 0 1 1	
2.550E-04	3.268E-02	25	M058	2 2 2 2 2	
2.473E-04	3.170E-02	25	M064	1 1 2 2 2	
2.472E-04	3.169E-02	25	M071	2 2 2 2 2	
3.121E-04	4.000E-02	25	M073	1 0 2 2 1	
2.620E-04	3.358E-02	25	M123	1 0 0 0 2	
2.575E-04	3.300E-02	25	M130	1 0 0 0 1	
2.390E-04	3.063E-02	25	M342	1 0 1 1 2	
2.497E-04	3.200E-02	25	O320	0 0 0 0 0	
2.575E-05	3.300E-03	25	P340	0 0 0 0 0	
2.356E-04	3.020E-02	25	R042	1 2 2 2 2	

(continued)

2010. C$_{10}$H$_8$ (continued)

Solubility (Moles/L)	Solubility (Grams/L)	Temp (°C)	Ref (#)	Evaluation (T P E A A)	Comments
2.340E-04	2.999E-02	25	S076	2 2 2 2 2	
1.716E-04	2.200E-02	25	S227	1 2 1 1 1	
2.390E-04	3.063E-02	25	W300	2 2 2 2 2	
2.490E-04	3.192E-02	25.00	M082	1 1 1 2 2	
2.490E-04	3.192E-02	25.00	M151	2 1 2 2 2	
2.472E-04	3.169E-02	25.00	M151	2 1 1 2 2	
6.936E-04	8.890E-02	25.00	P007	2 1 2 2 2	
2.492E-04	3.194E-02	25.04	M183	1 2 1 1 2	
2.510E-04	3.217E-02	25.04	V013	2 2 2 2 2	
2.660E-04	3.409E-02	27.00	M082	1 1 1 2 2	
2.660E-04	3.409E-02	27.00	M151	2 1 2 2 2	
2.666E-04	3.417E-02	27.04	M183	1 2 1 1 2	
2.980E-04	3.820E-02	29.90	W003	2 2 2 2 2	average of 3
3.240E-04	4.153E-02	29.99	P331	0 0 0 0 0	
2.949E-04	3.780E-02	30.30	W003	2 2 2 2 2	average of 3
3.448E-04	4.420E-02	34.50	W003	2 2 2 2 2	average of 2
3.710E-04	4.755E-02	34.99	P331	0 0 0 0 0	
4.112E-04	5.270E-02	39.30	W003	2 2 2 2 2	average of 2
4.360E-04	5.588E-02	39.99	P331	0 0 0 0 0	
4.275E-04	5.480E-02	40.10	W003	2 2 2 2 2	
5.118E-04	6.560E-02	44.70	W003	2 2 2 2 2	average of 3
6.132E-04	7.860E-02	50.20	W003	2 2 2 2 2	
8.270E-04	1.060E-01	55.60	W003	2 2 2 2 2	
1.233E-03	1.580E-01	64.50	W003	2 2 2 2 2	average of 3
1.904E-03	2.440E-01	73.40	W003	2 2 2 2 2	average of 3
2.341E-04	3.000E-02	ns	F071	0 1 2 1 1	
2.341E-04	3.000E-02	ns	H080	0 0 0 0 1	
2.473E-04	3.170E-02	ns	H123	0 0 0 0 0	
2.473E-04	3.170E-02	ns	K304	0 0 0 0 2	
2.340E-04	2.999E-02	ns	L060	0 0 0 0 2	average
2.473E-04	3.170E-02	ns	M344	0 0 0 0 2	
2.341E-04	3.000E-02	ns	O009	0 0 0 0 0	
8.129E-04	1.042E-01	ns	R042	1 2 2 2 2	
2.341E-04	3.000E-02	rt	M161	0 0 0 0 1	
2.848E-04	3.650E-02	rt	S314	0 0 2 1 2	

2011. C$_{10}$H$_8$BrN$_3$O
Bropirimine
2-Amino-5-bromo-6-phenyl-py-rimidin-4(3H)-one
ABPP

RN: 56741-95-8   **MP** (°C):
MW: 266.10   **BP** (°C):

Solubility (Moles/L)	Solubility (Grams/L)	Temp (°C)	Ref (#)	Evaluation (T P E A A)	Comments
2.931E-05	7.800E-03	37	A346	0 0 0 0 0	EFG

2012. C$_{10}$H$_8$BrN$_3$O

Brompyrazone
Amino-4-bromo-2-phenyl-3(2H)-pyridazinone
1-Phenyl-4-amino-5-bromo-6-pyridazone
Pyridazinone, 5-amino-4-bromo-2-phenyl-

RN: 3042-84-0   **MP (°C):** 223.5
MW: 266.10   **BP (°C):**

Solubility (Moles/L)	Solubility (Grams/L)	Temp (°C)	Ref (#)	Evaluation (T P E A A)	Comments
7.516E-04	2.000E-01	20	M161	1 0 0 0 2	

2013. C$_{10}$H$_8$ClN$_3$O

Pyrazon
5-Amino-4-chloro-2-phenyl-3(2H)-pyridazinone

RN: 1698-60-8   **MP (°C):** 207
MW: 221.65   **BP (°C):**

Solubility (Moles/L)	Solubility (Grams/L)	Temp (°C)	Ref (#)	Evaluation (T P E A A)	Comments
1.353E-03	3.000E-01	20	B185	0 0 0 0 0	
1.353E-03	2.999E-01	20	B200	1 0 0 0 0	
1.353E-03	2.999E-01	20	M061	1 0 0 0 0	
1.805E-03	4.000E-01	20	M161	1 0 0 0 2	

2014. C$_{10}$H$_8$N$_2$

γ,γ'-Dipyridyl
4,4'-Bipyridyl

RN: 553-26-4   **MP (°C):** 69
MW: 156.19   **BP (°C):**

Solubility (Moles/L)	Solubility (Grams/L)	Temp (°C)	Ref (#)	Evaluation (T P E A A)	Comments
2.887E-02	4.509E+00	25	B095	2 0 1 1 2	

2015. C$_{10}$H$_8$N$_2$

α,α'-Dipyridyl
2,2'-Dipyridyl
α,α'-Bipyridyl
2,2'-Bipyridine
2,2'-Bipyridyl

RN: 366-18-7   **MP (°C):** 71.5
MW: 156.19   **BP (°C):** 273

Solubility (Moles/L)	Solubility (Grams/L)	Temp (°C)	Ref (#)	Evaluation (T P E A A)	Comments
3.201E-02	5.000E+00	20	F300	1 0 0 0 0	
4.276E-02	6.678E+00	24.99	B444	0 0 0 0 0	
3.778E-02	5.900E+00	25	B095	2 0 1 1 2	
4.094E-02	6.394E+00	25	K063	2 2 0 1 2	

2016. C₁₀H₈N₂O₂

4-Phenyluracil
4-Phenyl-uracil
RN: 21321-07-3 **MP** (°C):
MW: 188.19 **BP** (°C):

Solubility (Moles/L)	Solubility (Grams/L)	Temp (°C)	Ref (#)	Evaluation (T P E A A)	Comments
5.314E-02	1.000E+01	100	F300	1 0 0 0 0	

2017. C₁₀H₈O

1-Naphthol
α-Naphthol
RN: 90-15-3 **MP** (°C): 96
MW: 144.17 **BP** (°C): 288

Solubility (Moles/L)	Solubility (Grams/L)	Temp (°C)	Ref (#)	Evaluation (T P E A A)	Comments
6.030E-03	8.694E-01	11	K307	2 0 1 2 2	
7.700E-03	1.110E+00	20	K130	2 1 1 1 2	
7.700E-03	1.110E+00	20	K301	2 2 1 1 1	
7.700E-03	1.110E+00	20	K307	2 0 1 2 2	
6.001E-03	8.653E-01	24	H106	1 0 2 2 2	
6.007E-03	8.660E-01	24	M303	1 0 1 1 2	
3.029E-03	4.367E-01	25	L085	1 2 0 1 2	
9.430E-03	1.360E+00	30	K307	2 0 1 2 2	
1.490E-02	2.148E+00	40	K307	2 0 1 2 2	
2.150E-02	3.100E+00	50	K307	2 0 1 2 2	

2018. C₁₀H₈O

2-Naphthol
β-Naphthol
RN: 135-19-3 **MP** (°C): 121
MW: 144.17 **BP** (°C): 285

Solubility (Moles/L)	Solubility (Grams/L)	Temp (°C)	Ref (#)	Evaluation (T P E A A)	Comments
2.462E-03	3.550E-01	6.90	M026	2 0 1 2 2	
3.378E-03	4.870E-01	13.45	M026	2 0 1 2 2	
3.473E-03	5.007E-01	15.60	M027	1 0 0 2 2	
3.646E-03	5.257E-01	16.20	M027	1 0 0 2 2	
3.891E-03	5.610E-01	17.70	M026	2 0 1 2 2	
4.450E-03	6.416E-01	20	K130	2 1 1 1 2	
4.500E-03	6.488E-01	20	K301	2 2 1 1 1	
4.450E-03	6.416E-01	20	K308	1 0 0 1 2	
5.800E-03	8.362E-01	20	M122	2 0 2 2 2	
4.945E-03	7.130E-01	21.50	M026	2 0 1 2 2	
4.713E-03	6.795E-01	23.20	M027	1 0 0 2 2	
3.954E-03	5.700E-01	25	F300	1 0 0 0 2	

(continued)

2018. C$_{10}$H$_8$O (continued)

Solubility (Moles/L)	Solubility (Grams/L)	Temp (°C)	Ref (#)	Evaluation (T P E A A)	Comments
5.240E-03	7.555E-01	25	K040	1 0 2 1 2	
5.356E-03	7.722E-01	25	L085	1 2 0 1 2	
6.929E-03	9.990E-01	25	R041	0 0 0 0 0	
6.076E-03	8.760E-01	29.50	M026	2 0 1 2 2	
6.431E-03	9.271E-01	31.30	M027	1 0 0 2 2	
6.832E-03	9.850E-01	33.30	M026	2 0 1 2 2	
9.045E-03	1.304E+00	38.70	M026	2 0 1 2 2	
1.116E-02	1.609E+00	44.50	M026	2 0 1 2 2	
1.388E-02	2.001E+00	49.50	M026	2 0 1 2 2	
1.706E-02	2.460E+00	55.20	M026	2 0 1 2 2	
2.104E-02	3.034E+00	60.00	M026	2 0 1 2 2	
2.928E-02	4.222E+00	68.10	M026	2 0 1 2 2	
3.810E-02	5.493E+00	75.00	M026	2 0 1 2 2	
4.670E-02	6.733E+00	80	K308	1 0 0 1 2	
5.129E-03	7.394E-01	ns	R427	0 0 0 0 0	

2019. C$_{10}$H$_8$O$_2$
2,3-Dihydroxynaphthalene
2,3-Dihydroxy-naphthalin
RN: 92-44-4   **MP (°C):**   162
MW: 160.17   **BP (°C):**

Solubility (Moles/L)	Solubility (Grams/L)	Temp (°C)	Ref (#)	Evaluation (T P E A A)	Comments
1.830E-03	2.931E-01	20	M122	2 0 2 2 2	

2020. C$_{10}$H$_8$O$_2$
2,6-Dihydroxynaphthalene
2,6-Dihydroxy-naphthalin
RN: 581-43-1   **MP (°C):**
MW: 160.17   **BP (°C):**

Solubility (Moles/L)	Solubility (Grams/L)	Temp (°C)	Ref (#)	Evaluation (T P E A A)	Comments
6.243E-03	1.000E+00	14	F300	1 0 0 0 0	

2021. C$_{10}$H$_9$ClN$_4$O$_2$S
2-Sulfanilamido-5-chloropyrimidine
Benzenesulfonamide, 4-amino-N-(5-chloro-2-pyrimidinyl)-
RN: 4482-46-6   **MP (°C):**
MW: 284.73   **BP (°C):**

Solubility (Moles/L)	Solubility (Grams/L)	Temp (°C)	Ref (#)	Evaluation (T P E A A)	Comments
6.322E-05	1.800E-02	37	R046	1 2 1 1 1	

2022. C$_{10}$H$_9$ClN$_4$O$_2$S

5-Sulfanilamido-2-chloropyrimidine
Benzenesulfonamide, 4-amino-*N*-(2-chloro-5-pyrimidinyl)-

| RN: | 17103-49-0 | MP (°C): | |
| MW: | 284.73 | BP (°C): | |

Solubility (Moles/L)	Solubility (Grams/L)	Temp (°C)	Ref (#)	Evaluation (T P E A A)	Comments
1.127E-03	3.210E-01	37	R046	1 2 1 1 1	

2023. C$_{10}$H$_9$Cl$_2$NO

Acrylanilide, 3',4'-dichloro-2-methyl-
Dicryl

| RN: | 2164-09-2 | MP (°C): | 127–128 |
| MW: | 230.10 | BP (°C): | |

Solubility (Moles/L)	Solubility (Grams/L)	Temp (°C)	Ref (#)	Evaluation (T P E A A)	Comments
3.477E-05	8.000E-03	ns	B185	0 0 0 0 0	

2024. C$_{10}$H$_9$Cl$_3$O$_3$

2,4,5-Trichlorophenoxy-γ-butyric acid
2,4,5-TB
4-(2,4,5-Trichlorophenoxy)butyric acid
4-(2,4,5-TB)

| RN: | 93-80-1 | MP (°C): | 114.5 |
| MW: | 283.54 | BP (°C): | |

Solubility (Moles/L)	Solubility (Grams/L)	Temp (°C)	Ref (#)	Evaluation (T P E A A)	Comments
1.481E-04	4.200E-02	25	B164	1 0 1 1 1	
1.481E-04	4.200E-02	ns	B185	0 0 0 0 0	

2025. C$_{10}$H$_9$Cl$_3$O$_3$

2,4-Dichlorophenoxyacetic acid β-monochloroethyl ester
Ethanol, 2-chloro-, (2,4-dichlorophenoxy)acetate

| RN: | 19810-30-1 | MP (°C): | |
| MW: | 283.54 | BP (°C): | |

Solubility (Moles/L)	Solubility (Grams/L)	Temp (°C)	Ref (#)	Evaluation (T P E A A)	Comments
1.910E-04	5.415E-02	ns	M120	0 0 1 1 2	

2026. C$_{10}$H$_9$Cl$_4$NO$_2$S
Captafol
cis-3α,4,7,7α-Tetrahydro-2-(1,1,2,2-tetrachloroethyl)thio-1H-isoindole-1,3(2H)-dione
Crisfolatan
Difolatan
Folcid
RN: 2939-80-2   **MP** (°C):   160.5
MW: 349.06   **BP** (°C):

Solubility (Moles/L)	Solubility (Grams/L)	Temp (°C)	Ref (#)	Evaluation (T P E A A)	Comments
4.074E-06	1.422E-03	20	B179	0 0 0 0 0	
4.011E-06	1.400E-03	ns	M161	0 0 0 0 1	

2027. C$_{10}$H$_9$Cl$_4$O$_4$P
Gardona
2-Chloro-1-(2,4,5-trichlorophenyl)vinyldimethylphosphate
RN: 22248-79-9   **MP** (°C):   97.5
MW: 365.97   **BP** (°C):

Solubility (Moles/L)	Solubility (Grams/L)	Temp (°C)	Ref (#)	Evaluation (T P E A A)	Comments
3.006E-05	1.100E-02	20	M061	1 0 0 0 1	

2028. C$_{10}$H$_9$Cl$_4$O$_4$P
Tetrachlorvinphos
2-Chloro-1-(2,4,5-trichlorophenyl)vinyl dimethyl phosphate
Rabon
Gardona
SD 8447
Stirofos
RN: 961-11-5   **MP** (°C):   96
MW: 365.97   **BP** (°C):

Solubility (Moles/L)	Solubility (Grams/L)	Temp (°C)	Ref (#)	Evaluation (T P E A A)	Comments
3.006E-05	1.100E-02	20	M161	1 0 0 0 1	

2029. C$_{10}$H$_9$N
3-Methyl-isoquinoline
Isoquinoline, 3-methyl-
RN: 1125-80-0   **MP** (°C):
MW: 143.19   **BP** (°C):   519.2

Solubility (Moles/L)	Solubility (Grams/L)	Temp (°C)	Ref (#)	Evaluation (T P E A A)	Comments
6.418E-03	9.190E-01	20	A050	1 0 1 1 2	

2030. $C_{10}H_9N$
2-Naphthylamine
Naphthylamine-(2)
β-Naphthylamin
β-Naphthylamine

RN:	91-59-8	**MP** (°C):	113
MW:	143.19	**BP** (°C):	306.1

Solubility (Moles/L)	Solubility (Grams/L)	Temp (°C)	Ref (#)	Evaluation (T P E A A)	Comments
1.320E-03	1.890E-01	rt	N015	0 0 2 2 2	

2031. $C_{10}H_9N$
1-Naphthylamine
1-Aminonaphthalene
α-Naphthoylamine
α-Naphthylamin
α-Naphthylamine

RN:	134-32-7	**MP** (°C):	50
MW:	143.19	**BP** (°C):	300.8

Solubility (Moles/L)	Solubility (Grams/L)	Temp (°C)	Ref (#)	Evaluation (T P E A A)	Comments
1.187E-02	1.700E+00	20	F300	1 0 0 0 1	
3.600E-04	5.155E-02	ns	L060	0 0 0 0 1	average

2032. $C_{10}H_9NO$
8-Hydroxyquinaldine
2-Methyl 8-quinolinol

RN:	826-81-3	**MP** (°C):	72.5
MW:	159.19	**BP** (°C):	

Solubility (Moles/L)	Solubility (Grams/L)	Temp (°C)	Ref (#)	Evaluation (T P E A A)	Comments
2.460E+03	3.916E+05	25.2	P024	2 2 1 1 2	
2.670E+03	4.250E+05	30.3	P024	2 2 1 1 2	

2033. $C_{10}H_9NO$
4-Hydroxy-2-methylquinoline
4-Hydroxy-2-methyl-chinolin

RN:	607-67-0	**MP** (°C):	234
MW:	159.19	**BP** (°C):	

Solubility (Moles/L)	Solubility (Grams/L)	Temp (°C)	Ref (#)	Evaluation (T P E A A)	Comments
6.282E-02	1.000E+01	20	F300	1 0 0 0 1	
5.936E-01	9.450E+01	100	F300	1 0 0 0 2	

2034. C$_{10}$H$_9$NO$_2$S
Ethyl *m*-isothiocyanobenzoate
Ethyl 3-isothiocyanobenzoate
RN: 3137-84-6   **MP** (°C):
MW: 207.25   **BP** (°C):

Solubility (Moles/L)	Solubility (Grams/L)	Temp (°C)	Ref (#)	Evaluation (T P E A A)	Comments
2.500E-04	5.181E-02	25	K032	2 2 0 1 2	

2035. C$_{10}$H$_9$NO$_2$S
Ethyl 4-isothiocyanatobenzoate
4-Carbethoxyphenylisothiocyanate
Ethyl *p*-isothiocyanatobenzoate
RN: 1205-06-7   **MP** (°C):
MW: 207.25   **BP** (°C):

Solubility (Moles/L)	Solubility (Grams/L)	Temp (°C)	Ref (#)	Evaluation (T P E A A)	Comments
9.000E-05	1.865E-02	25	D019	1 1 1 1 1	

2036. C$_{10}$H$_9$NO$_3$S
Badische acid
2-Naphthylamine-8-sulfonic acid
Naphthylamin-(2)-sulfosaeure-(8)
RN: 86-60-2   **MP** (°C):
MW: 223.25   **BP** (°C):

Solubility (Moles/L)	Solubility (Grams/L)	Temp (°C)	Ref (#)	Evaluation (T P E A A)	Comments
2.688E-03	6.000E-01	20	F300	1 0 0 0 2	

2037. C$_{10}$H$_9$NO$_3$S
2-Naphthylamine-1-sulfonic acid
α-Naphthylamine-*o*-monosulfonic acid
RN: 81-16-3   **MP** (°C):
MW: 223.25   **BP** (°C):

Solubility (Moles/L)	Solubility (Grams/L)	Temp (°C)	Ref (#)	Evaluation (T P E A A)	Comments
1.072E-02	2.394E+00	0	D077	1 0 0 1 1	
1.429E-02	3.190E+00	10	D077	1 0 0 1 1	
1.829E-02	4.083E+00	20	D077	1 0 0 1 1	
2.317E-02	5.173E+00	30	D077	1 0 0 1 1	
2.893E-02	6.458E+00	40	D077	1 0 0 1 1	
3.555E-02	7.937E+00	50	D077	1 0 0 1 1	
4.435E-02	9.901E+00	60	D077	1 0 0 1 2	
6.010E-02	1.342E+01	70	D077	1 0 0 1 2	
7.834E-02	1.749E+01	80	D077	1 0 0 1 2	
1.028E-01	2.296E+01	90	D077	1 0 0 1 2	
1.347E-01	3.007E+01	100	D077	1 0 0 1 2	

2038. $C_{10}H_9NO_3S$
1-Naphthylamine-8-sulfonic acid
Naphthylamin-(1)-sulfosaeure-(8)
Peri acid
RN: 82-75-7   **MP** (°C):
MW: 223.25   **BP** (°C):

Solubility (Moles/L)	Solubility (Grams/L)	Temp (°C)	Ref (#)	Evaluation (T P E A A)	Comments
8.958E-04	2.000E-01	21	F300	1 0 0 0 0	
1.971E-02	4.400E+00	100	F300	1 0 0 0 1	

2039. $C_{10}H_9NO_3S$
1-Naphthylamine-5-sulfonic acid
Laurent's acid
Naphthylamin-(1)-sulfosaeure-(5)
RN: 84-89-9   **MP** (°C):
MW: 223.25   **BP** (°C):

Solubility (Moles/L)	Solubility (Grams/L)	Temp (°C)	Ref (#)	Evaluation (T P E A A)	Comments
4.479E-03	1.000E+00	20	F300	1 0 0 0 2	

2040. $C_{10}H_9NO_3S$
1-Naphthylamine-4-sulfonic acid
4-Amino-1-naphthalenesulfonic acid
Naphthionic acid
Naphthylamin-(1)-sulfosaeure-(4)
Pirias acid
RN: 84-86-6   **MP** (°C):   000
MW: 223.25   **BP** (°C):

Solubility (Moles/L)	Solubility (Grams/L)	Temp (°C)	Ref (#)	Evaluation (T P E A A)	Comments
1.209E-03	2.699E-01	0	D077	1 0 0 1 1	
1.299E-03	2.899E-01	10	D077	1 0 0 1 1	
1.388E-03	3.099E-01	20	D077	1 0 0 1 1	
1.344E-03	3.000E-01	20	F300	1 0 0 0 0	
1.657E-03	3.699E-01	30	D077	1 0 0 1 1	
2.149E-03	4.798E-01	40	D077	1 0 0 1 1	
2.641E-03	5.897E-01	50	D077	1 0 0 1 1	
3.357E-03	7.494E-01	60	D077	1 0 0 1 1	
4.341E-03	9.691E-01	70	D077	1 0 0 1 1	
5.815E-03	1.298E+00	80	D077	1 0 0 1 2	
7.825E-03	1.747E+00	90	D077	1 0 0 1 2	
1.021E-03	2.279E-01	100	D077	1 0 0 1 2	
1.075E-02	2.400E+00	100	F300	1 0 0 0 1	

2041. C_{10}H_9NO_3S

1,6-Cleve's acid
1-Naphthylamine-6-sulfonic acid
Naphthylamin-(1)-sulfosaeure-(6)

| **RN:** | 119-79-9 | **MP** (°C): |
| **MW:** | 223.25 | **BP** (°C): |

Solubility (Moles/L)	Solubility (Grams/L)	Temp (°C)	Ref (#)	Evaluation (T P E A A)	Comments
4.479E-03	1.000E+00	16	F300	1 0 0 0 2	

2042. C_{10}H_9NO_3S

Cassella's acid F
2-Naphthylamine-7-sulfonic acid
Naphthylamin-(2)-sulfosaeure-(7)

| **RN:** | 494-44-0 | **MP** (°C): |
| **MW:** | 223.25 | **BP** (°C): |

Solubility (Moles/L)	Solubility (Grams/L)	Temp (°C)	Ref (#)	Evaluation (T P E A A)	Comments
8.958E-04	2.000E-01	20	F300	1 0 0 0 1	
1.389E-02	3.100E+00	100	F300	1 0 0 0 1	

2043. C_{10}H_9NO_3S

Bronner's acid
2-Naphthylamine-6-sulfonic acid
Naphthylamin-(2)-sulfosaeure-(6)

| **RN:** | 93-00-5 | **MP** (°C): |
| **MW:** | 223.25 | **BP** (°C): |

Solubility (Moles/L)	Solubility (Grams/L)	Temp (°C)	Ref (#)	Evaluation (T P E A A)	Comments
5.375E-04	1.200E-01	20	F300	1 0 0 0 1	
7.615E-03	1.700E+00	100	F300	1 0 0 0 1	

2044. C_{10}H_9NO_3S

1-Naphthylamine-2-sulfonic acid
Naphthylamin-(1)-sulfosaeure-(2)

| **RN:** | 81-06-1 | **MP** (°C): |
| **MW:** | 223.25 | **BP** (°C): |

Solubility (Moles/L)	Solubility (Grams/L)	Temp (°C)	Ref (#)	Evaluation (T P E A A)	Comments
1.836E-02	4.100E+00	20	F300	1 0 0 0 1	
1.402E-01	3.130E+01	100	F300	1 0 0 0 2	

2045. C₁₀H₉NO₃S

2-Naphthylamine-5-sulfonic acid
Dahl's acid
Naphthylamin-(2)-sulfosaeure-(5)

RN: 81-05-0 **MP** (°C):
MW: 223.25 **BP** (°C):

Solubility (Moles/L)	Solubility (Grams/L)	Temp (°C)	Ref (#)	Evaluation (T P E A A)	Comments
1.478E-03	3.300E-01	20	F300	1 0 0 0 2	

2046. C₁₀H₉NO₄S

7-Amino-1-naphthol-3-sulfonic acid
7-Amino-naphtol-(1)-sulfosaeure-(3)

RN: 90-51-7 **MP** (°C):
MW: 239.25 **BP** (°C):

Solubility (Moles/L)	Solubility (Grams/L)	Temp (°C)	Ref (#)	Evaluation (T P E A A)	Comments
1.881E-02	4.500E+00	h	F300	0 0 0 0 1	

2047. C₁₀H₉NO₉S₃

1-Naphthylamine-2,4,7-trisulfonic acid
1,3,6-Naphthalenetrisulfonic acid, 4-amino-

RN: 61986-93-4 **MP** (°C):
MW: 383.38 **BP** (°C):

Solubility (Moles/L)	Solubility (Grams/L)	Temp (°C)	Ref (#)	Evaluation (T P E A A)	Comments
4.799E-01	1.840E+02	20	F054	1 2 1 1 2	
8.216E-01	3.150E+02	80	F054	1 2 1 1 2	

2048. C₁₀H₉N₃O₃S

1-Sulfanilyl-3-methyl-5-pyrazolone

RN: **MP** (°C):
MW: 251.27 **BP** (°C):

Solubility (Moles/L)	Solubility (Grams/L)	Temp (°C)	Ref (#)	Evaluation (T P E A A)	Comments
1.827E-03	4.590E-01	37	R045	1 2 1 1 2	

2049. $C_{10}H_9N_4O_5$

Picrolonic acid

Pikrolonsaeure

RN: 550-74-3   **MP** (°C):   116
MW: 265.21   **BP** (°C):

Solubility (Moles/L)	Solubility (Grams/L)	Temp (°C)	Ref (#)	Evaluation (T P E A A)	Comments
3.394E-02	9.000E+00	17	F300	1 0 0 0 0	
3.582E-02	9.500E+00	100	F300	1 0 0 0 1	

2050. $C_{10}H_{10}Fe$

Ferrocene

bis-Cyclopentadienyliron

Ferrotsen

Iron bis(cyclopentadiene)

RN: 102-54-5   **MP** (°C):
MW: 186.04   **BP** (°C):

Solubility (Moles/L)	Solubility (Grams/L)	Temp (°C)	Ref (#)	Evaluation (T P E A A)	Comments
3.388E-05	6.304E-03	25	B335	1 2 0 0 1	

2051. $C_{10}H_{10}BrNO_3S$

4-Thiazolidinecarboxylic acid, 2-(5-bromo-2-hydroxyphenyl)-

Thiazolidine-4-carboxylic acid, 2-(5-bromo-2-hydroxyphenyl)-

RN: 256235-53-7   **MP** (°C):
MW: 304.17   **BP** (°C):   451.3

Solubility (Moles/L)	Solubility (Grams/L)	Temp (°C)	Ref (#)	Evaluation (T P E A A)	Comments
1.200E-03	3.650E-01	21	B414	1 0 0 1 1	fast decomposition

2052. $C_{10}H_{10}BrNO_4$

5-Bromo-2-*p*-phenyl-5-nitro-1,3-dioxane

m-Dioxane, 5-bromo-5-nitro-2-phenyl-

1,3-Dioxane, 5-bromo-5-nitro-2-phenyl-

RN: 58522-87-5   **MP** (°C):   82–84
MW: 288.10   **BP** (°C):

Solubility (Moles/L)	Solubility (Grams/L)	Temp (°C)	Ref (#)	Evaluation (T P E A A)	Comments
1.596E-03	4.598E-01	25	L013	1 0 2 1 2	

2053. C₁₀H₁₀BrNO₅

2053. $C_{10}H_{10}BrNO_5$

5-Bromo-2-p-phenol-5-nitro-1,3-dioxane
m-Dioxane, 5-bromo-5-nitro-2-phenol-

RN: 60766-61-2	**MP (°C):** 142–144	
MW: 304.10	**BP (°C):**	

Solubility (Moles/L)	Solubility (Grams/L)	Temp (°C)	Ref (#)	Evaluation (T P E A A)	Comments
1.413E-03	4.298E-01	25	L013	1 0 2 1 2	

2054. C₁₀H₁₀ClNO₂S

2054. $C_{10}H_{10}ClNO_2S$

4-Thiazolidinecarboxylic acid, 2-(4-chlorophenyl)-
4-Thiazolidinecarboxylic acid, 2-(p-chlorophenyl)-
Thiazolidine-4-carboxylic acid, (2-(4-chlorophenyl)-

RN: 34491-29-7	**MP (C):** 156-185 (°decomp)	
MW: 243.71	**BP (°C):** 458.3	

Solubility (Moles/L)	Solubility (Grams/L)	Temp (°C)	Ref (#)	Evaluation (T P E A A)	Comments
5.900E-03	1.438E+00	21	B414	1 0 0 1 1	fast decomposition

2055. C₁₀H₁₀ClNO₂S

2055. $C_{10}H_{10}ClNO_2S$

4-Thiazolidinecarboxylic acid, 2-(2-chlorophenyl)-
4-Thiazolidinecarboxylic acid, 2-(o-chlorophenyl)-
Thiazolidine-4-carboxylic acid, 2-(2-chlorophenyl)-

RN: 72678-81-0	**MP (°C):** 145–147	
MW: 243.71	**BP (°C):** 439.7	

Solubility (Moles/L)	Solubility (Grams/L)	Temp (°C)	Ref (#)	Evaluation (T P E A A)	Comments
2.100E-03	5.118E-01	21	B414	1 0 0 1 1	fast decomposition, results from gravimetric determination

2056. C₁₀H₁₀ClNO₃

2056. $C_{10}H_{10}ClNO_3$

Chloroacetyl acetaminophen
Acetic acid, chloro-, 4-(acetylamino)phenyl ester
Acetanilide, 4'-hydroxy-, chloroacetate (ester)

RN: 17321-63-0	**MP (°C):** 184.5–185	
MW: 227.65	**BP (°C):**	

Solubility (Moles/L)	Solubility (Grams/L)	Temp (°C)	Ref (#)	Evaluation (T P E A A)	Comments
1.230E-03	2.800E-01	37	D029	0 0 0 0 0	

2057. C₁₀H₁₀Cl₂F₂N₂OS

3-[3-Chloro-4-(chlorodifluoromethylthio)phenyl]-1,1-dimethylurea
N-[3-Chloro-4-(chlorodifluoromethylthiol)phenyl]-N′,N′-dimethylurea
N-(3-Chloro-4-difluorochloromethylthiophenyl)-N′,N′-dimethylurea
Thiochlormethyl
N-[3-Chloro-4-(chlorodifluoromethylthio)phenyl]-N′,N′-dimethylurea

RN: 33439-45-1   **MP** (°C):   113.5
MW: 315.17   **BP** (°C):

Solubility (Moles/L)	Solubility (Grams/L)	Temp (°C)	Ref (#)	Evaluation (T P E A A)	Comments
2.159E-01	6.803E+01	20	M161	1 0 0 0 1	

2058. C₁₀H₁₀Cl₂O₂

Chlorfenprop-methyl
Methyl 2-chloro-3-(p-chlorophenyl)propionate
Methyl α-p-dichlorohydrocinnamate
Bidisin
Fatex

RN: 14437-17-3   **MP** (°C):
MW: 233.10   **BP** (°C):   111.5

Solubility (Moles/L)	Solubility (Grams/L)	Temp (°C)	Ref (#)	Evaluation (T P E A A)	Comments
1.716E-04	4.000E-02	20	M161	1 0 0 0 1	

2059. C₁₀H₁₀Cl₂O₃

4-(2,4-Dichlorophenoxy)propionic acid
2,4-DB

RN: 94-82-6   **MP** (°C):   118
MW: 249.10   **BP** (°C):

Solubility (Moles/L)	Solubility (Grams/L)	Temp (°C)	Ref (#)	Evaluation (T P E A A)	Comments
2.690E-04	6.700E-02	25	B164	1 0 1 1 1	
1.847E-04	4.600E-02	25	M161	1 0 0 0 1	
2.128E-04	5.300E-02	ns	B185	0 0 0 0 0	
1.847E-04	4.600E-02	ns	L024	1 0 0 0 1	
2.128E-04	5.300E-02	rt	M061	0 0 0 0 1	

2060. C₁₀H₁₀Cl₂O₃

Ethyl (2,4-dichlorophenoxy)acetate
2,4-Dichlorophenoxyacetic acid ethyl ester

RN: 533-23-3   **MP** (°C):
MW: 249.10   **BP** (°C):

Solubility (Moles/L)	Solubility (Grams/L)	Temp (°C)	Ref (#)	Evaluation (T P E A A)	Comments
2.529E-04	6.300E-02	ns	M120	0 0 1 1 2	

2061. $C_{10}H_{10}Cl_8$

Toxaphene
Camphechlor
Campheclor
PhenAcide
Toxakil
Chlorinated champhene

RN:	8001-35-2	**MP** (°C):	65		
MW:	413.82	**BP** (°C):			

Solubility (Moles/L)	Solubility (Grams/L)	Temp (°C)	Ref (#)	Evaluation (T P E A A)	Comments
1.329E-06	5.500E-04	20	M336	2 0 2 2 2	
9.666E-07	4.000E-04	25	C100	1 0 2 1 0	
1.208E-06	5.000E-04	25	P085	0 0 0 0 0	
1.788E-06	7.400E-04	25	W025	1 0 2 2 2	
1.450E-06	6.000E-04	ns	M110	0 0 0 0 0	EFG
1.329E-06	5.500E-04	ns	V414	0 0 0 0 0	
7.250E-06	3.000E-03	rt	M161	0 0 0 0 0	

2062. $C_{10}H_{10}N_2O_4S$

4-Thiazolidinecarboxylic acid, 2-(3-nitrophenyl)-
4-Thiazolidinecarboxylic acid, 2-(m-nitrophenyl)-

RN:	69570-81-6	**MP** (°C):	151–153		
MW:	254.27	**BP** (°C):	500.2		

Solubility (Moles/L)	Solubility (Grams/L)	Temp (°C)	Ref (#)	Evaluation (T P E A A)	Comments
5.300E-03	1.348E+00	21	B414	1 0 0 1 1	fast decomposition, results from gravimetric determination

2063. $C_{10}H_{10}N_4O$

Metamitron
3-Methyl-4-amino-6-phenyl-1,2,4-triazin-5(4H)-one
4-Amino-3-methyl-6-phenyl-1,2,4-triazin-5-one
Goltix

RN:	41394-05-2	**MP** (°C):	166.6		
MW:	202.22	**BP** (°C):			

Solubility (Moles/L)	Solubility (Grams/L)	Temp (°C)	Ref (#)	Evaluation (T P E A A)	Comments
8.901E-03	1.800E+00	20	M161	1 0 0 0 1	

2064. C$_{10}$H$_{10}$N$_4$O$_2$S
Sulfadiazine
Sulphadiazine
N1-(2-Pyrimidinyl)-sulfanilamide
Debenal

RN:	68-35-9	**MP** (°C):	254
MW:	250.28	**BP** (°C):	

Solubility (Moles/L)	Solubility (Grams/L)	Temp (°C)	Ref (#)	Evaluation (T P E A A)	Comments
2.360E-04	5.907E-02	20	C006	1 2 1 1 2	
1.814E-04	4.540E-02	20	E003	2 2 1 1 2	
5.993E-04	1.500E-01	20	F073	1 2 2 2 2	
2.917E-04	7.299E-02	20	L058	1 0 1 1 1	
3.077E-04	7.700E-02	25	C102	2 0 2 2 2	
2.637E-03	6.600E-01	25	K048	1 2 2 1 1	pH 1.26
2.682E-04	6.713E-02	25	M440	0 0 0 0 0	
3.036E-04	7.599E-02	30	E003	2 2 1 1 2	
3.640E-04	9.110E-02	30	H018	0 0 0 0 0	
3.200E-04	8.009E-02	30	L069	1 0 1 1 0	EFG
7.192E-04	1.800E-01	35	H114	1 0 0 0 1	
5.074E-04	1.270E-01	37	C102	2 0 2 2 2	
4.914E-04	1.230E-01	37	F072	1 0 0 0 2	
4.794E-04	1.200E-01	37	F075	1 0 2 2 2	
5.114E-04	1.280E-01	37	K091	1 0 0 0 2	
5.194E-04	1.300E-01	37	L091	1 0 0 0 1	pH 5.5
7.192E-04	1.800E-01	37	M057	1 0 0 0 2	pH 5.5
8.790E-04	2.200E-01	37	R044	0 0 0 0 0	EFG, intrinsic
4.914E-04	1.230E-01	37	R045	1 2 1 1 1	
6.712E-04	1.680E-01	37	S192	1 0 1 1 2	pH 6.0
5.074E-04	1.270E-01	37	W016	2 0 1 1 2	
4.914E-04	1.230E-01	37	W053	1 0 0 0 2	
3.956E-04	9.900E-02	38	K006	1 0 0 0 1	
5.154E-04	1.290E-01	40	E003	2 2 1 1 2	
5.194E-04	1.300E-01	ns	G083	0 0 0 0 1	pH 5.5
3.196E-04	8.000E-02	ns	K444	0 0 0 0 0	
3.981E-04	9.964E-02	ns	R427	0 0 0 0 0	

2065. C$_{10}$H$_{10}$N$_4$O$_2$S
Sulfapyrazine
Sulphapyrazine

RN:	116-44-9	**MP** (°C):	255
MW:	250.28	**BP** (°C):	

Solubility (Moles/L)	Solubility (Grams/L)	Temp (°C)	Ref (#)	Evaluation (T P E A A)	Comments
1.998E-04	5.000E-02	37	L091	1 0 0 0 0	pH 5.5

2066. C₁₀H₁₀N₄O₂S

2066. C$_{10}$H$_{10}$N$_{4}$O$_{2}$S

5-Sulfanilamidopyrimidine
5-Sulfapyrimidine
Sulfanilamide, *N*1-5-pyrimidinyl-
RN: 17103-48-9 **MP** (°C):
MW: 250.28 **BP** (°C):

Solubility (Moles/L)	Solubility (Grams/L)	Temp (°C)	Ref (#)	Evaluation (T P E A A)	Comments
3.916E-04	9.800E-02	37	R046	1 2 1 1 1	

2067. C$_{10}$H$_{10}$N$_{4}$O$_{2}$S

4-Sulfanilamidopyrimidine
4-Sulfapyrimidine
Sulfanilamide, *N*1-4-pyrimidinyl-
RN: 599-82-6 **MP** (°C):
MW: 250.28 **BP** (°C):

Solubility (Moles/L)	Solubility (Grams/L)	Temp (°C)	Ref (#)	Evaluation (T P E A A)	Comments
1.414E-02	3.540E+00	37	R045	1 2 1 1 2	

2068. C$_{10}$H$_{10}$N$_{4}$O$_{4}$S

5-Sulfanilamidouracil
Benzenesulfonamide, 4-amino-*N*-(1,2,3,4-tetrahydro-2,4-dioxo-5-pyrimidinyl)-
RN: 6912-98-7 **MP** (°C):
MW: 282.28 **BP** (°C):

Solubility (Moles/L)	Solubility (Grams/L)	Temp (°C)	Ref (#)	Evaluation (T P E A A)	Comments
1.722E-03	4.860E-01	37	R045	1 2 1 1 0	

2069. C$_{10}$H$_{10}$O

Benzalacetone
4-Phenyl-3-buten-2-one
Methyl styryl ketone
RN: 122-57-6 **MP** (°C):
MW: 146.19 **BP** (°C):

Solubility (Moles/L)	Solubility (Grams/L)	Temp (°C)	Ref (#)	Evaluation (T P E A A)	Comments
9.560E-03	1.398E+00	25	R070	0 0 0 0 0	

2070. C$_{10}$H$_{10}$O$_2$
p-Acetylacetophenone
Ethanone, 1,1'-(1,4-phenylene)bis-
RN: 1009-61-6   **MP (°C):**
MW: 162.19   **BP (°C):**

Solubility (Moles/L)	Solubility (Grams/L)	Temp (°C)	Ref (#)	Evaluation (T P E A A)	Comments
3.890E-05	6.309E-03	25	C316	0 0 0 0 0	0.1M NaCl

2071. C$_{10}$H$_{10}$O$_2$
Methyl cinnamate
2-Propenoic acid
3-Phenyl-, methyl ester
RN: 103-26-4   **MP (°C):**
MW: 162.19   **BP (°C):**

Solubility (Moles/L)	Solubility (Grams/L)	Temp (°C)	Ref (#)	Evaluation (T P E A A)	Comments
2.500E-03	4.055E-01	25	R070	0 0 0 0 0	

2072. C$_{10}$H$_{10}$O$_2$
trans-α-Methyl-cinnamic acid
α-Methyl-trans-zimtsaeure
RN: 1895-97-2   **MP (°C):**
MW: 162.19   **BP (°C):**

Solubility (Moles/L)	Solubility (Grams/L)	Temp (°C)	Ref (#)	Evaluation (T P E A A)	Comments
7.399E-03	1.200E+00	h	F300	0 0 0 0 1	

2073. C$_{10}$H$_{10}$O$_4$
Dimethyl o-phthalate
RN: **MP (°C):** 5.5 C
MW: 194.19   **BP (°C):**

Solubility (Moles/L)	Solubility (Grams/L)	Temp (°C)	Ref (#)	Evaluation (T P E A A)	Comments
1.802E-02	3.500E+00	25	S417	0 0 0 0 0	

2074. $C_{10}H_{10}O_4$

Ferulic acid
3-(4-Hydroxy-3-methoxyphenyl)-2-propenoic acid
4-Hydroxy-3-methoxycinnamic acid

RN: 1135-24-6   **MP** (°C):   169 C
MW: 194.19   **BP** (°C):

Solubility (Moles/L)	Solubility (Grams/L)	Temp (°C)	Ref (#)	Evaluation (T P E A A)	Comments
2.935E-03	5.700E-01	15	M461	0 0 0 0 0	
4.017E-03	7.800E-01	25	M461	0 0 0 0 0	
4.738E-03	9.200E-01	30	M461	0 0 0 0 0	
9.063E-03	1.760E+00	40	M461	0 0 0 0 0	
1.128E-02	2.190E+00	50	M461	0 0 0 0 0	

2075. $C_{10}H_{10}O_4$

Acetyl-*r*-mandelic acid
(R)(–)*O*-Acetylmandelic acid
[R]-[–]-α-(Acetoxy)phenylacetic acid
O-Acetylmandelic acid

RN: 5438-68-6   **MP** (°C):
MW: 194.19   **BP** (°C):

Solubility (Moles/L)	Solubility (Grams/L)	Temp (°C)	Ref (#)	Evaluation (T P E A A)	Comments
2.919E-02	5.668E+00	0	A043	1 2 1 1 1	
2.919E-02	5.668E+00	0	L035	1 2 2 1 1	
3.478E-02	6.754E+00	10	A043	1 2 1 1 1	
3.478E-02	6.754E+00	10	L035	1 2 2 1 1	
3.884E-02	7.543E+00	15	A043	1 2 1 1 1	
3.884E-02	7.543E+00	15	L035	1 2 2 1 1	
4.897E-02	9.509E+00	20	A043	1 2 1 1 1	
4.897E-02	9.509E+00	20	L035	1 2 2 1 1	
5.804E-02	1.127E+01	25	A043	1 2 1 1 2	
5.804E-02	1.127E+01	25	L035	1 2 2 1 2	
7.060E-02	1.371E+01	30	A043	1 2 1 1 2	
7.060E-02	1.371E+01	30	L035	1 2 2 1 2	
1.005E-01	1.951E+01	35	A043	1 2 1 1 2	
1.587E-01	3.082E+01	40	A043	1 2 1 1 2	
2.795E-01	5.428E+01	45	A043	1 2 1 1 2	
2.795E-01	5.428E+01	45	L035	1 2 2 1 2	
6.125E-01	1.189E+02	50	A043	1 2 1 1 2	
6.125E-01	1.189E+02	50	L035	1 2 2 1 2	

2076. C$_{10}$H$_{10}$O$_4$
Dimethyl phthalate
1,2-Benzenedicarboxylic acid, dimethyl ester
Fermine
Unimoll DM
Mipax
Palatinol M

RN: 131-11-3   **MP** (°C):   5.5
MW: 194.19   **BP** (°C):   283.7

Solubility (Moles/L)	Solubility (Grams/L)	Temp (°C)	Ref (#)	Evaluation (T P E A A)	Comments
2.210E-02	4.292E+00	20	L300	2 1 0 2 2	
4.087E-02	7.937E+00	20.00	D343	0 0 0 0 0	
2.317E-01	4.500E+01	25	F067	1 0 2 2 2	*sic*
2.307E-02	4.480E+00	c	F070	1 0 0 0 0	
1.566E-02	3.041E+00	ns	F014	0 0 0 0 2	
2.052E-02	3.984E+00	ns	H069	0 0 1 1 1	
2.214E-02	4.300E+00	rt	M161	0 0 0 0 1	

2077. C$_{10}$H$_{10}$O$_4$
Meconin
Mekonin

RN: 569-31-3   **MP** (°C):   102
MW: 194.19   **BP** (°C):

Solubility (Moles/L)	Solubility (Grams/L)	Temp (°C)	Ref (#)	Evaluation (T P E A A)	Comments
1.287E-02	2.500E+00	25	F300	1 0 0 0 0	
2.420E-02	4.700E+00	100	F300	1 0 0 0 1	

2078. C$_{10}$H$_{10}$O$_4$
Acetylsalicylic acid, methyl ester
Methyl 2-acetoxybenzoate
Benzoic acid, 2-(acetyloxy)-, methyl ester

RN: 580-02-9   **MP** (°C):   48
MW: 194.19   **BP** (°C):

Solubility (Moles/L)	Solubility (Grams/L)	Temp (°C)	Ref (#)	Evaluation (T P E A A)	Comments
1.447E-02	2.810E+00	21	N335	0 0 0 0 0	
1.679E-02	3.260E+00	37	G430	0 0 0 0 0	pH 4.5

2079. C$_{10}$H$_{10}$O$_4$

Terephthalate acid dimethyl ester
Terephthalsaeure-dimethyl ester
1,4-Benzenedicarboxylic acid dimethyl ester
Terephthalic acid
Dimethyl terephthalate
Dimethyl 1,4-Benzenedicarboxylate

RN:	120-61-6	**MP** (°C):	140
MW:	194.19	**BP** (°C):	

Solubility (Moles/L)	Solubility (Grams/L)	Temp (°C)	Ref (#)	Evaluation (T P E A A)	Comments
1.690E-04	3.282E-02	25	C316	0 0 0 0 0	0.1M NaCl
1.540E-02	2.991E+00	h	F070	1 0 0 0 1	

2080. C$_{10}$H$_{10}$O$_5$

Opianic acid
Opiansaeure

RN:	519-05-1	**MP** (°C):	150
MW:	210.19	**BP** (°C):	

Solubility (Moles/L)	Solubility (Grams/L)	Temp (°C)	Ref (#)	Evaluation (T P E A A)	Comments
1.189E-02	2.500E+00	20	F300	1 0 0 0 1	
8.088E-02	1.700E+01	h	F300	0 0 0 0 1	

2081. C$_{10}$H$_{11}$ClO$_3$

Mecoprop
2-(4-Chloro-2-methylphenoxy)propionic acid
2-(2-Methyl-4-chlorophenoxy)propionic acid
2-(MCPP)

RN:	93-65-2	**MP** (°C):	93
MW:	214.65	**BP** (°C):	

Solubility (Moles/L)	Solubility (Grams/L)	Temp (°C)	Ref (#)	Evaluation (T P E A A)	Comments
2.888E-03	6.200E-01	20	B185	0 0 0 0 0	
2.795E-03	6.000E-01	20	B200	1 0 0 0 2	
2.887E-03	6.196E-01	20	M061	1 0 0 0 1	
2.888E-03	6.200E-01	20	M161	1 0 0 0 2	
4.170E-03	8.950E-01	25	B164	1 0 1 1 2	
4.170E-03	8.950E-01	25	B185	0 0 0 0 0	
2.794E-03	5.996E-01	ns	B100	0 0 0 0 0	
2.050E-04	4.400E-02	ns	B185	0 0 0 0 0	
2.888E-03	6.200E-01	ns	L024	1 0 0 0 2	
3.802E-03	8.161E-01	ns	R427	0 0 0 0 0	

2082. C$_{10}$H$_{11}$ClO$_3$
4-(4-Chlorophenoxy)butyric acid
4-(4-CPB)
RN: 3547-07-7   **MP** (°C):
MW: 214.65   **BP** (°C):

Solubility (Moles/L)	Solubility (Grams/L)	Temp (°C)	Ref (#)	Evaluation (T P E A A)	Comments
5.125E-04	1.100E-01	25	B164	1 0 1 1 2	

2083. C$_{10}$H$_{11}$Cl$_3$O$_2$
2,3,6-Trichlorobenzyloxypropanol
1-Propanol, 3-[(2,3,6-trichlorobenzyl)oxy]-
RN: 1591-82-8   **MP** (°C):
MW: 269.56   **BP** (°C):

Solubility (Moles/L)	Solubility (Grams/L)	Temp (°C)	Ref (#)	Evaluation (T P E A A)	Comments
2.708E-04	7.300E-02	25	B185	0 0 0 0 0	
2.708E-04	7.300E-02	25	B200	1 0 0 0 1	

2084. C$_{10}$H$_{11}$FN$_2$O$_6$
1,3-bis(Acetoxymethyl)-5-fluoro-2,4(1H,3H)-pyrimidinedi-one
1,3-bis(Acetoxymethyl)-5-fluorouracil
RN: 66542-48-1   **MP** (°C):   105–106
MW: 274.21   **BP** (°C):

Solubility (Moles/L)	Solubility (Grams/L)	Temp (°C)	Ref (#)	Evaluation (T P E A A)	Comments
1.568E-02	4.300E+00	22	B321	0 0 0 0 0	pH 4.0

2085. C$_{10}$H$_{11}$F$_3$N$_2$O
Fluometuron
1,1-Dimethyl-3-(α,α,α-trifluoro-m-tolyl)urea
RN: 2164-17-2   **MP** (°C):   163
MW: 232.21   **BP** (°C):

Solubility (Moles/L)	Solubility (Grams/L)	Temp (°C)	Ref (#)	Evaluation (T P E A A)	Comments
4.571E-04	1.061E-01	20	B179	0 0 0 0 0	
4.522E-04	1.050E-01	20	M161	1 0 0 0 2	
3.661E-04	8.500E-02	24	C105	2 1 2 2 2	
3.876E-04	9.000E-02	25	B200	1 0 0 0 1	
3.876E-04	9.000E-02	25	G036	1 0 0 0 1	
3.876E-04	9.000E-02	25	M061	1 0 0 0 1	

2086. $C_{10}H_{11}F_3N_2O_3S$
Fluoridamid
Acetamide, N-{4-methyl-3-{{(trifluoromethyl)sulfonyl}amino}phenyl}-
Sustar
MBR6033

RN:	47000-92-0	MP (°C):	182–184
MW:	296.27	BP (°C):	

Solubility (Moles/L)	Solubility (Grams/L)	Temp (°C)	Ref (#)	Evaluation (T P E A A)	Comments
4.388E-04	1.300E-01	22	G307	0 0 0 0 0	

2087. $C_{10}H_{11}NO$
N-Methylcinnamide
2-Propenamide, N-methyl-3-phenyl-

RN:	2757-10-0	MP (°C):	
MW:	161.21	BP (°C):	

Solubility (Moles/L)	Solubility (Grams/L)	Temp (°C)	Ref (#)	Evaluation (T P E A A)	Comments
1.310E-02	2.112E+00	ns	H350	0 0 0 0 0	

2088. $C_{10}H_{11}NOS$
m-Isopropoxyphenyl isothiocyanate
3-Isopropoxyphenyl isothiocyanate

RN:	3528-90-3	MP (°C):	
MW:	193.27	BP (°C):	

Solubility (Moles/L)	Solubility (Grams/L)	Temp (°C)	Ref (#)	Evaluation (T P E A A)	Comments
4.700E-04	9.084E-02	25	K032	2 2 0 1 2	

2089. $C_{10}H_{11}NO_2S$
2-Phenylthiazolidine-4-carboxylic acid
4-Thiazolidinecarboxylic acid, 2-phenyl-

RN:	42607-21-6	MP (°C):	166–168
MW:	209.27	BP (°C):	433.6

Solubility (Moles/L)	Solubility (Grams/L)	Temp (°C)	Ref (#)	Evaluation (T P E A A)	Comments
4.500E-03	9.417E-01	21	B414	1 0 0 1 1	partial decomposition

2090. C$_{10}$H$_{11}$NO$_3$

Acetamide, 2-(benzoyloxy)-*N*-methyl-

RN: 106231-50-9   **MP** (°C):   111
MW: 193.20   **BP** (°C):

Solubility (Moles/L)	Solubility (Grams/L)	Temp (°C)	Ref (#)	Evaluation (T P E A A)	Comments
1.915E-02	3.700E+00	22	B427	1 0 0 1 1	in 0.01M HCl
1.915E-02	3.700E+00	22	N317	1 1 2 1 2	

2091. C$_{10}$H$_{11}$NO$_3$

p-Acetoxy-acetanilide
p-Acetoxyacetanilide
Acetaminophen acetate
Acetyl acetaminophen

RN: 2623-33-8   **MP** (°C):   153
MW: 193.20   **BP** (°C):

Solubility (Moles/L)	Solubility (Grams/L)	Temp (°C)	Ref (#)	Evaluation (T P E A A)	Comments
1.656E-03	3.200E-01	25	B010	1 1 1 1 0	
1.237E-02	2.390E+00	25	E016	1 1 1 1 2	
1.139E-02	2.200E+00	25	M333	1 1 0 0 2	
1.760E-02	3.400E+00	37	D029	0 0 0 0 0	

2092. C$_{10}$H$_{11}$NO$_3$S

4-Thiazolidinecarboxylic acid, 2-(4-hydroxyphenyl)-
4-Thiazolidinecarboxylic acid, 2-(*p*-hydroxyphenyl)-

RN: 69588-11-0   **MP** (°C):
MW: 225.27   **BP** (°C):   507.5

Solubility (Moles/L)	Solubility (Grams/L)	Temp (°C)	Ref (#)	Evaluation (T P E A A)	Comments
7.000E-03	1.577E+00	21	B414	1 0 0 1 1	fast decomposition

2093. C$_{10}$H$_{11}$NO$_3$S

2-(2-Hydroxyphenyl)-4-thiazolidinecarboxylic acid
4-Thiazolidinecarboxylic acid, 2-(2-hydroxyphenyl)-
Thiazolidine-4-carboxylic acid, 2-(2-hydroxyphenyl)

RN: 72678-82-1   **MP** (°C):
MW: 225.27   **BP** (°C):   418.9

Solubility (Moles/L)	Solubility (Grams/L)	Temp (°C)	Ref (#)	Evaluation (T P E A A)	Comments
2.100E-03	4.731E-01	21	B414	1 0 0 1 1	fast decomposition, results from gravimetric determination

2094. C$_{10}$H$_{11}$NO$_4$

Methyl acetaminophen
Carbonic acid, 4-(acetylamino)phenyl methyl ester
Acetanilide, 4′-hydroxy-, methyl carbonate (ester)

RN: 17321-62-9 **MP (°C):** 115.5–116.5
MW: 209.20 **BP (°C):**

Solubility (Moles/L)	Solubility (Grams/L)	Temp (°C)	Ref (#)	Evaluation (T P E A A)	Comments
2.868E-02	6.000E+00	37	D029	0 0 0 0 0	

2095. C$_{10}$H$_{11}$NO$_4$

O-(Acetoxymethyl) salicylamide
2-[(Acetyloxy)methoxy]-benzamide
Benzamide, 2-[(acetyloxy)methoxy]-
O-Acetoxymethyl methyl salicylamide

RN: 102273-25-6 **MP (°C):** 92.5
MW: 209.20 **BP (°C):**

Solubility (Moles/L)	Solubility (Grams/L)	Temp (°C)	Ref (#)	Evaluation (T P E A A)	Comments
>2.39E-02	>5.00E+00	23	B328	1 2 2 1 1	pH 4
2.390E-02	5.000E+00	23	B328	0 0 0 0 0	

2096. C$_{10}$H$_{11}$NO$_4$

Carbobenzoxyglycine
N-Carbobenzyloxyglycine
N-CBZ-glycine
Benzyloxycarbonyl glycine

RN: 1138-80-3 **MP (°C):**
MW: 209.20 **BP (°C):**

Solubility (Moles/L)	Solubility (Grams/L)	Temp (°C)	Ref (#)	Evaluation (T P E A A)	Comments
2.180E-02	4.560E+00	25.1	N026	0 0 0 0 0	
2.170E-02	4.539E+00	25.1	N027	1 1 2 2 2	

2097. C$_{10}$H$_{11}$NO$_5$

Acido D-feniltartrammico tartranilico

RN: **MP (°C):** 194
MW: 225.20 **BP (°C):**

Solubility (Moles/L)	Solubility (Grams/L)	Temp (°C)	Ref (#)	Evaluation (T P E A A)	Comments
1.232E-01	2.774E+01	17.40	C070	1 2 2 1 2	

2098. C$_{10}$H$_{11}$NO$_6$
Acido *p*-ossifeniltartrammico

RN: **MP** (°C):   218
MW: 241.20   **BP** (°C):

Solubility (Moles/L)	Solubility (Grams/L)	Temp (°C)	Ref (#)	Evaluation (T P E A A)	Comments
1.677E-01	4.045E+01	14	C071	1 2 0 1 2	

2099. C$_{10}$H$_{11}$N$_3$OS
Methabenzthiazuron
N-2-Benzothiazolyl-*N,N'*-dimethylurea
1,3-Dimethyl-3-(2-benzothiazolyl)urea
Methyl-*N'*-methyl-*N'*-(2-benzothiazolyl)urea
Tribunil
Preparation 5633

RN: 18691-97-9   **MP** (°C):   119.5
MW: 221.28   **BP** (°C):

Solubility (Moles/L)	Solubility (Grams/L)	Temp (°C)	Ref (#)	Evaluation (T P E A A)	Comments
2.666E-04	5.900E-02	20	M161	1 0 0 0 1	

2100. C$_{10}$H$_{11}$N$_3$O$_2$S$_2$
Methyl sulfathiazole
Sulfathiazol methyle

RN: 15251-46-4   **MP** (°C):
MW: 269.35   **BP** (°C):

Solubility (Moles/L)	Solubility (Grams/L)	Temp (°C)	Ref (#)	Evaluation (T P E A A)	Comments
9.653E-04	2.600E-01	37	D084	1 0 1 0 1	

2101. C$_{10}$H$_{11}$N$_3$O$_2$S
Sulfapyrrole

RN: **MP** (°C):
MW: 237.28   **BP** (°C):

Solubility (Moles/L)	Solubility (Grams/L)	Temp (°C)	Ref (#)	Evaluation (T P E A A)	Comments
2.023E-02	4.800E+00	20	F073	1 2 2 2 2	

2102. $C_{10}H_{11}N_3O_2S_2$

Sulfamethylthiazole
4-Methyl-2-sulfanilamidothiazole
2-(p-Aminobenzenesulfonamido)-4-methylthiazole
2-Sulfanilamido-4-methylthiazole
Aseptil 2
Ciba 3753

RN: 515-59-3   **MP** (°C): 239
MW: 269.35   **BP** (°C):

Solubility (Moles/L)	Solubility (Grams/L)	Temp (°C)	Ref (#)	Evaluation (T P E A A)	Comments
4.084E-04	1.100E-01	20	F073	1 2 2 2 2	
4.084E-04	1.100E-01	20	F074	1 0 0 0 2	

2103. $C_{10}H_{11}N_3O_2S_2$

N1-Methyl-N1-2-thiazolyl-sulfanilamide
N1-Methylsulfathiazole

RN: 51203-19-1   **MP** (°C):
MW: 269.35   **BP** (°C):

Solubility (Moles/L)	Solubility (Grams/L)	Temp (°C)	Ref (#)	Evaluation (T P E A A)	Comments
1.150E-03	3.097E-01	37	K095	2 0 0 0 2	intrinsic

2104. $C_{10}H_{11}N_3O_3$

α-Semicarbazono-p-tolyl acetate

RN: **MP** (°C):
MW: 221.22   **BP** (°C):

Solubility (Moles/L)	Solubility (Grams/L)	Temp (°C)	Ref (#)	Evaluation (T P E A A)	Comments
1.400E-03	3.097E-01	25	A066	1 0 1 1 1	

2105. $C_{10}H_{11}N_3O_3S$

Sulfamethoxazole
4-Amino-N-(5-methyl-3-isoxazolyl)benzenesulfonamide
Cotrimoxazole
Septra
Bactrim
Cotrim

RN: 723-46-6   **MP** (°C): 167
MW: 253.28   **BP** (°C):

Solubility (Moles/L)	Solubility (Grams/L)	Temp (°C)	Ref (#)	Evaluation (T P E A A)	Comments
1.109E-03	2.810E-01	25	D308	0 0 0 0 0	pH 3.22
1.730E-03	4.383E-01	25	F415	0 0 0 0 0	Average

(continued)

2105. $C_{10}H_{11}N_3O_3S$ (continued)

Solubility (Moles/L)	Solubility (Grams/L)	Temp (°C)	Ref (#)	Evaluation (T P E A A)	Comments
1.470E-03	3.723E-01	25	M440	0 0 0 0 0	
1.974E-03	5.000E-01	25	R025	0 0 0 0 0	
1.488E-03	3.770E-01	32	D308	0 0 0 0 0	pH 4.0
1.824E-03	4.620E-01	37	D308	0 0 0 0 0	pH 3.43
2.408E-03	6.100E-01	37	H120	1 1 1 1 1	normal saline
2.480E-03	6.281E-01	37	K095	2 0 0 0 2	intrinsic
5.527E-03	1.400E+00	37	M321	1 0 0 0 2	intrinsic
1.540E-03	3.900E-01	amb	L434	0 0 0 0 0	
1.540E-03	3.900E-01	amb	L437	0 0 0 0 0	
3.948E-05	1.000E-02	ns	K444	0 0 0 0 0	

2106. $C_{10}H_{11}N_5O_2S$

5-Sulfanilamido-2-aminopyrimidine

Benzenesulfonamide, 4-amino-N-(2-amino-5-pyrimidinyl)-

RN: 71119-38-5   **MP** (°C):

MW: 265.30   **BP** (°C):

Solubility (Moles/L)	Solubility (Grams/L)	Temp (°C)	Ref (#)	Evaluation (T P E A A)	Comments
3.129E-04	8.300E-02	37	R046	1 2 1 1 1	

2107. $C_{10}H_{12}$

Tetralin

1,2,3,4-Tetrahydronaphthalene

RN: 119-64-2   **MP** (°C):   −31.0

MW: 132.21   **BP** (°C):   207.2

Solubility (Moles/L)	Solubility (Grams/L)	Temp (°C)	Ref (#)	Evaluation (T P E A A)	Comments
3.404E-04	4.500E-02	20	B356	0 0 0 0 0	
3.532E-04	4.670E-02	28	B348	2 1 2 2 2	
1.513E-03	2.000E-01	150	J023	1 1 2 2 0	
3.026E-03	4.000E-01	200	J023	1 1 2 2 0	
3.026E-02	4.000E+00	250	J023	1 1 2 2 0	
3.236E-04	4.278E-02	ns	D001	0 0 0 0 2	

2108. C$_{10}$H$_{12}$BrCl$_2$O$_3$PS
Bromophos-ethyl
O-(4-Bromo-2,5-dichlorophenyl) O,O-diethyl phosphorothioate
Nexagan
Filariol

RN: 4824-78-6	**MP (°C):**	
MW: 394.06	**BP (°C):**	

Solubility (Moles/L)	Solubility (Grams/L)	Temp (°C)	Ref (#)	Evaluation (T P E A A)	Comments
5.329E-07	2.100E-04	10	B324	0 0 0 0 0	
5.329E-07	2.100E-04	10	B324	0 0 0 0 0	
8.629E-07	3.400E-04	20	B324	0 0 0 0 0	
8.628E-07	3.400E-04	20	B324	0 0 0 0 0	
7.613E-06	3.000E-03	20	F311	1 2 2 2 1	
5.075E-06	2.000E-03	20	W312	1 0 0 0 0	
1.269E-06	5.001E-04	30	B324	0 0 0 0 0	
1.269E-06	5.000E-04	30	B324	0 0 0 0 0	
5.075E-06	2.000E-03	ns	E050	0 0 0 0 0	
5.075E-06	2.000E-03	rt	M161	0 0 0 0 0	

2109. C$_{10}$H$_{12}$ClNO$_2$
Chloro-IPC
Furloe
Taterpex
Chlorpropham
Isopropyl m-chlorocarbanilate

RN: 101-21-3	**MP (°C):** 38	
MW: 213.67	**BP (°C):**	

Solubility (Moles/L)	Solubility (Grams/L)	Temp (°C)	Ref (#)	Evaluation (T P E A A)	Comments
5.055E-04	1.080E-01	20	B185	0 0 0 0 0	
3.744E-04	8.000E-02	25	G099	1 0 0 1 0	
3.744E-04	8.000E-02	25	G319	0 0 0 0 0	
4.165E-04	8.900E-02	25	M161	1 0 0 0 1	
3.744E-04	8.000E-02	ns	B185	0 0 0 0 0	
4.119E-04	8.800E-02	ns	B200	0 0 0 0 1	
3.744E-04	8.000E-02	ns	F035	0 0 0 0 0	
4.119E-04	8.800E-02	ns	H042	0 0 0 0 1	
3.744E-04	8.000E-02	ns	M061	0 0 0 0 1	
3.548E-04	7.581E-02	ns	M163	0 0 0 0 0	EFG
5.055E-04	1.080E-01	ns	N013	0 0 0 0 2	

2110. C$_{10}$H$_{12}$ClNO$_2$
Baclofen
Lioresal
β-(Aminomethyl)-p-chlorohydrocinnamic acid
RN: 1134-47-0   **MP** (°C):
MW: 213.67   **BP** (°C):

Solubility (Moles/L)	Solubility (Grams/L)	Temp (°C)	Ref (#)	Evaluation (T P E A A)	Comments
2.129E-02	4.549E+00	25	M374	1 0 2 1 2	

2111. C$_{10}$H$_{12}$ClN$_3$O$_2$
Tranid
3-Chloro-6-cyanonorbornanone-2-oxime-*O,N*-methylcarbamate
RN: 15271-41-7   **MP** (°C):   143.5
MW: 241.68   **BP** (°C):

Solubility (Moles/L)	Solubility (Grams/L)	Temp (°C)	Ref (#)	Evaluation (T P E A A)	Comments
8.259E-03	1.996E+00	ns	M061	0 0 0 0 0	

2112. C$_{10}$H$_{12}$ClN$_3$O$_3$S
Quinethazone
7-Chloro-2-ethyl-1,2,3,4-tetrahydro-4-oxo-6-quinazolinesulfonamide
Hydromox
CL 36010
Aquamox
RN: 73-49-4   **MP** (°C):   251
MW: 289.74   **BP** (°C):

Solubility (Moles/L)	Solubility (Grams/L)	Temp (°C)	Ref (#)	Evaluation (T P E A A)	Comments
5.176E-04	1.500E-01	25	A081	1 0 1 1 0	EFG

2113. C$_{10}$H$_{12}$ClN$_5$O$_2$
2-Chloro-2′,3′-dideoxyadenosine
2-CIDDA
RN: 114849-58-0   **MP** (°C):
MW: 269.69   **BP** (°C):

Solubility (Moles/L)	Solubility (Grams/L)	Temp (°C)	Ref (#)	Evaluation (T P E A A)	Comments
3.745E-03	1.010E+00	25	A336	0 0 0 0 0	

2114. $C_{10}H_{12}Cl_2O$
2,4-Dichloro-6-butyl-phenol
Phenol, 2-butyl-4,6-dichloro-
RN: 91399-13-2 **MP** (°C):
MW: 219.11 **BP** (°C):

Solubility (Moles/L)	Solubility (Grams/L)	Temp (°C)	Ref (#)	Evaluation (T P E A A)	Comments
2.400E-04	5.259E-02	25	B316	0 0 0 0 0	

2115. $C_{10}H_{12}Cl_3O_2PS$
Trichloronate
Trichloronat
Ethyl O-(2,4,5-trichlorophenyl) ethylphosphonothioate
Agritox
Bay 37289
RN: 327-98-0 **MP** (°C):
MW: 333.60 **BP** (°C): 108

Solubility (Moles/L)	Solubility (Grams/L)	Temp (°C)	Ref (#)	Evaluation (T P E A A)	Comments
2.458E-06	8.200E-04	10	B324	0 0 0 0 0	
2.458E-06	8.200E-04	10	B324	0 0 0 0 0	
1.769E-06	5.901E-04	20	B300	2 1 1 1 2	
2.638E-06	8.800E-04	20	B324	0 0 0 0 0	
2.638E-06	8.800E-04	20	B324	0 0 0 0 0	
1.499E-04	5.000E-02	20	M161	1 0 0 0 1	*sic*
3.208E-06	1.070E-03	30	B324	0 0 0 0 0	
3.207E-06	1.070E-03	30	B324	0 0 0 0 0	

2116. $C_{10}H_{12}N_2O_2$
Acetone N-(phenylcarbamoyl)oxime
Acetone oxime N-phenylcarbamate
Proxypham
RN: **MP** (°C): 109.5
MW: 192.22 **BP** (°C):

Solubility (Moles/L)	Solubility (Grams/L)	Temp (°C)	Ref (#)	Evaluation (T P E A A)	Comments
2.601E-03	5.000E-01	ns	M061	0 0 0 0 2	approximate

2117. $C_{10}H_{12}N_2O_3$
Barbituric-2-14C acid, 5,5-diallyl
RN: 112599-90-3 **MP** (°C):
MW: 208.22 **BP** (°C):

Solubility (Moles/L)	Solubility (Grams/L)	Temp (°C)	Ref (#)	Evaluation (T P E A A)	Comments
8.381E-03	1.745E+00	25	P350	0 0 0 0 0	intrinsic

2118. $C_{10}H_{12}N_2O_3$

Allobarbital

5,5-Diallylbarbituric acid

RN:	52-43-7	MP (°C):	171
MW:	208.22	BP (°C):	

Solubility (Moles/L)	Solubility (Grams/L)	Temp (°C)	Ref (#)	Evaluation (T P E A A)	Comments
6.003E-03	1.250E+00	20	J030	1 2 2 2 2	
7.193E-03	1.498E+00	25	A023	1 0 0 1 2	
8.500E-03	1.770E+00	25	G003	1 1 1 1 1	pH 4.7
8.650E-03	1.801E+00	25	V033	2 0 1 1 2	
8.700E-03	1.812E+00	25.00	T303	1 0 0 0 1	
9.250E-03	1.926E+00	30	G014	1 1 1 1 0	EFG
9.200E-03	1.916E+00	30	I001	2 0 2 1 0	EFG, 0.003N H_2SO_4
9.200E-03	1.916E+00	30	K108	1 2 2 0 1	
1.150E-02	2.394E+00	35	A023	1 0 0 1 2	
1.110E-02	2.311E+00	35.00	T303	1 0 0 0 2	
1.215E-02	2.530E+00	37	J030	1 2 2 2 2	
1.200E-02	2.499E+00	37	K121	1 2 1 2 1	0.1N HCl
1.675E-02	3.488E+00	40	A023	1 0 0 1 2	
1.370E-01	2.853E+01	40	N008	1 0 1 1 2	*sic*
1.690E-02	3.519E+00	45.00	T303	1 0 0 0 2	
7.036E-03	1.465E+00	ns	T003	0 0 0 0 2	

2119. $C_{10}H_{12}N_2O_3S$

Bentazon

2,1,3-Benzothiadiazin-4(3H)-one

Thiadiazinol

Basagran 4E

Adagio

BAS 351H

RN:	25057-89-0	MP (°C):	138.0
MW:	240.28	BP (°C):	

Solubility (Moles/L)	Solubility (Grams/L)	Temp (°C)	Ref (#)	Evaluation (T P E A A)	Comments
2.081E-03	5.000E-01	20	M161	1 0 0 0 2	
2.080E-03	4.998E-01	ns	B100	0 0 0 0 0	
3.329E-03	8.000E-01	ns	M110	0 0 0 0 0	EFG

2120. $C_{10}H_{12}N_2O_4$
Stavudine
1-(2,3-Dideoxy-β-D-glycero-pent-2-enofuranosyl)thymine
BMY-27857
d4T
Zerit
3′-Deoxy-2′-thymidinene

RN: 3056-17-5	**MP** (°C):	159–160
MW: 224.22	**BP** (°C):	

Solubility (Moles/L)	Solubility (Grams/L)	Temp (°C)	Ref (#)	Evaluation (T P E A A)	Comments
3.353E-01	7.519E+01	20.5	M439	0 0 0 0 0	
3.791E-01	8.500E+01	24.8	M439	0 0 0 0 0	
4.238E-01	9.502E+01	29.4	M439	0 0 0 0 0	
4.668E-01	1.047E+02	33.2	M439	0 0 0 0 0	
5.563E-01	1.247E+02	38.4	M439	0 0 0 0 0	
3.702E-01	8.300E+01	ns	K444	0 0 0 0 0	
3.418E-01	7.664E+01	ns	S469	0 0 0 0 0	

2121. $C_{10}H_{12}N_2O_4S$
N1,N4-Diacetylsulfanilamide
N4-Acetylsulphacetamide

RN: 5626-90-4	**MP** (°C):	
MW: 256.28	**BP** (°C):	

Solubility (Moles/L)	Solubility (Grams/L)	Temp (°C)	Ref (#)	Evaluation (T P E A A)	Comments
8.389E-03	2.150E+00	37	L091	1 0 0 0 2	pH 5.5

2122. $C_{10}H_{12}N_2O_5$
D-Monofeniltartramide tartranilamide

RN:	**MP** (°C):	226
MW: 240.22	**BP** (°C):	

Solubility (Moles/L)	Solubility (Grams/L)	Temp (°C)	Ref (#)	Evaluation (T P E A A)	Comments
1.958E-02	4.704E+00	21.50	C070	1 2 2 1 2	

2123. C$_{10}$H$_{12}$N$_2$O$_5$

2,4-Dinitro-6-*sec*-butylphenol
Dinoseb
4,6-Dinitro-2-*S*-butylphenol
Phenol, 4,6-dinitro-2-*sec*-butyl-

RN: 88-85-7   **MP** (°C): 38
MW: 240.22   **BP** (°C):

Solubility (Moles/L)	Solubility (Grams/L)	Temp (°C)	Ref (#)	Evaluation (T P E A A)	Comments
2.165E-04	5.200E-02	25	B200	1 0 0 0 1	
2.165E-04	5.200E-02	25	G319	0 0 0 0 0	
3.053E-03	7.335E-01	25	M061	1 0 0 0 2	
4.159E-03	9.990E-01	ns	B100	0 0 0 0 0	
2.081E-04	5.000E-02	ns	B185	0 0 0 0 0	
1.413E-03	3.393E-01	ns	M163	0 0 0 0 0	EFG
2.165E-04	5.200E-02	ns	V414	0 0 0 0 0	
4.163E-04	1.000E-01	rt	M161	0 0 0 0 2	

2124. C$_{10}$H$_{12}$N$_2$O$_5$S

7-Aminocephalosporanic acid
7-ACA

RN: 957-68-6   **MP** (°C):
MW: 272.28   **BP** (°C):

Solubility (Moles/L)	Solubility (Grams/L)	Temp (°C)	Ref (#)	Evaluation (T P E A A)	Comments
9.901E-04	2.696E-01	1.29	W417	0 0 0 0 0	
1.070E-03	2.914E-01	5.19	W417	0 0 0 0 0	
1.130E-03	3.076E-01	8.29	W417	0 0 0 0 0	
1.255E-03	3.416E-01	12.19	W417	0 0 0 0 0	
1.367E-03	3.723E-01	17.59	W417	0 0 0 0 0	
1.504E-03	4.096E-01	22.99	W417	0 0 0 0 0	
1.627E-03	4.429E-01	27.99	W417	0 0 0 0 0	

2125. C$_{10}$H$_{12}$N$_3$O$_3$PS$_2$

Azinphos-methyl
Guthion
S-(3,4-Dihydro-4-oxobenzo[d][1,2,3]triazin-3-ylmethyl) *O,O*-dimethyl phosphorodithioate
Methyl gusathion

RN: 86-50-0   **MP** (°C): 74
MW: 317.33   **BP** (°C):

Solubility (Moles/L)	Solubility (Grams/L)	Temp (°C)	Ref (#)	Evaluation (T P E A A)	Comments
2.994E-05	9.501E-03	10	B324	0 0 0 0 0	
2.994E-05	9.500E-03	10	B324	0 0 0 0 0	
4.412E-05	1.400E-02	15	A087	1 0 0 1 0	
6.587E-05	2.090E-02	20	B300	2 1 1 1 2	

(continued)

2125. $C_{10}H_{12}N_3O_3PS_2$ (continued)

Solubility (Moles/L)	Solubility (Grams/L)	Temp (°C)	Ref (#)	Evaluation (T P E A A)	Comments
6.587E-05	2.090E-02	20	B324	0 0 0 0 0	
6.586E-05	2.090E-02	20	B324	0 0 0 0 0	
9.454E-05	3.000E-02	20	M061	1 0 0 0 1	
9.139E-05	2.900E-02	25	A087	1 0 0 1 0	
1.374E-04	4.360E-02	30	B324	0 0 0 0 0	
1.374E-04	4.360E-02	30	B324	0 0 0 0 0	
1.481E-04	4.700E-02	35	A087	1 0 0 1 0	
8.913E-05	2.828E-02	ns	R427	0 0 0 0 0	
1.040E-04	3.300E-02	rt	M161	0 0 0 0 1	

2126. $C_{10}H_{12}N_4$
6,7-Diethylpteridine

RN: **MP** (°C):   52
MW: 188.23   **BP** (°C):

Solubility (Moles/L)	Solubility (Grams/L)	Temp (°C)	Ref (#)	Evaluation (T P E A A)	Comments
6.641E-01	1.250E+02	20	A019	2 2 1 1 0	

2127. $C_{10}H_{12}N_4O$
2-Hydroxy-6,7-diethylpteridine
2-Hydroxy-6:7-diethylpteridine
4-Hydroxy-6,7-diethylpteridine
4-Hydroxy-6:7-diethylpteridine

RN: 90870-76-1   **MP** (°C):
MW: 204.23   **BP** (°C):

Solubility (Moles/L)	Solubility (Grams/L)	Temp (°C)	Ref (#)	Evaluation (T P E A A)	Comments
1.221E-02	2.494E+00	20	A019	2 2 1 1 2	
5.434E-03	1.110E+00	20	A019	2 2 1 1 2	

2128. $C_{10}H_{12}N_4O_2$
2,4-Dihydroxy-6,7-diethylpteridine
2,4-Dihydroxy-6:7-diethylpteridine

RN: 113222-29-0   **MP** (°C):   218
MW: 220.23   **BP** (°C):

Solubility (Moles/L)	Solubility (Grams/L)	Temp (°C)	Ref (#)	Evaluation (T P E A A)	Comments
4.124E-03	9.083E-01	20	A019	2 2 1 1 2	

2129. C$_{10}$H$_{12}$N$_4$O$_2$

1H-Pyrazolo[3,4-d]pyrimidine, 4-[(tetrahydro-2H-pyran-2-yl)oxy]-
2-Tetrahydropuran-4-allopurinyl ether

RN: 52717-52-9   **MP** (°C):
MW: 220.23   **BP** (°C):

Solubility (Moles/L)	Solubility (Grams/L)	Temp (°C)	Ref (#)	Evaluation (T P E A A)	Comments
1.653E-02	3.640E+00	ns	H067	0 0 0 0 0	

2130. C$_{10}$H$_{12}$N$_4$O$_2$S

Sulfaethidole
Ethyl thiodiazole
Sulfaethylthiadiazole
Thiodiazolique ethyle

RN: 94-19-9   **MP** (°C):   188
MW: 252.30   **BP** (°C):

Solubility (Moles/L)	Solubility (Grams/L)	Temp (°C)	Ref (#)	Evaluation (T P E A A)	Comments
8.522E-04	2.150E-01	20	F073	1 2 2 2 2	
1.288E-02	3.250E+00	37	B046	1 0 2 2 2	pH 5
1.585E-03	4.000E-01	37	D084	1 0 1 0 1	

2131. C$_{10}$H$_{12}$N$_4$O$_3$

1-Butyryloxymethyl allopurinol
Butanoic acid, (4,5-dihydro-4-oxo-1H-pyrazolo[3,4-d]pyrimidin-1-yl)methyl ester

RN: 98827-21-5   **MP** (°C):   224–226
MW: 236.23   **BP** (°C):

Solubility (Moles/L)	Solubility (Grams/L)	Temp (°C)	Ref (#)	Evaluation (T P E A A)	Comments
1.482E-03	3.500E-01	22	B322	0 0 0 0 0	

2132. C$_{10}$H$_{12}$N$_4$O$_3$

2′,3′-Dideoxyinosine
Videx
Didanosine
CCRIS 805
CCRIS 805Didanosine

RN: 69655-05-6   **MP** (°C):   175
MW: 236.23   **BP** (°C):

Solubility (Moles/L)	Solubility (Grams/L)	Temp (°C)	Ref (#)	Evaluation (T P E A A)	Comments
4.614E-02	1.090E+01	4	A337	0 0 0 0 0	
1.156E-01	2.730E+01	25	A337	0 0 0 0 0	
1.270E-01	3.000E+01	ns	A426	0 0 0 0 0	Intrinsic
1.156E-01	2.730E+01	ns	K444	0 0 0 0 0	
1.125E-01	2.657E+01	ns	S469	0 0 0 0 0	

2133. $C_{10}H_{12}N_4O_3$

2-Butyryloxymethyl allopurinol

Butanoic acid, (4,5-dihydro-4-oxo-2H-pyrazolo[3,4-d]pyrimidin-1-yl)methyl ester

RN:	98827-22-6	MP (°C): 182–183
MW:	236.23	BP (°C):

Solubility (Moles/L)	Solubility (Grams/L)	Temp (°C)	Ref (#)	Evaluation (T P E A A)	Comments
6.350E-03	1.500E+00	22	B322	0 0 0 0 0	

2134. $C_{10}H_{12}N_4O_4$

2′-Deoxy-inosine

2[-Deoxyinosine

Deoxyinosine

RN:	890-38-0	MP (°C):
MW:	252.23	BP (°C):

Solubility (Moles/L)	Solubility (Grams/L)	Temp (°C)	Ref (#)	Evaluation (T P E A A)	Comments
3.301E-02	8.326E+00	25.02	T420	0 0 0 0 0	

2135. $C_{10}H_{12}N_4O_5$

Inosine

Inosin

Hypoxanthine ribonucleoside

RN:	58-63-9	MP (°C): 212dec
MW:	268.23	BP (°C):

Solubility (Moles/L)	Solubility (Grams/L)	Temp (°C)	Ref (#)	Evaluation (T P E A A)	Comments
5.871E-02	1.575E+01	20	D041	1 0 0 0 1	
5.890E-02	1.580E+01	20	F300	1 0 0 0 2	
5.888E-02	1.579E+01	ns	R427	0 0 0 0 0	

2136. $C_{10}H_{12}N_4O_6$

2,4,6-Trinitrodiethylaniline

2-4-6-Trinitrodiethylaniline

RN:	106415-21-8	MP (°C):
MW:	284.23	BP (°C):

Solubility (Moles/L)	Solubility (Grams/L)	Temp (°C)	Ref (#)	Evaluation (T P E A A)	Comments
1.759E-04	5.000E-02	50	D067	1 2 0 0 0	
7.037E-04	2.000E-01	100	D067	1 2 0 0 1	

2137. $C_{10}H_{12}N_5O_6P$

Adenosine 3′,5′-monophosphate
Adenosine, cyclic 3′,5′-(hydrogen phosphate)
4H-Furo[3,2-d]-1,3,2-dioxaphosphorin, adenosine deriv

RN:　60-92-4　　　**MP (°C):**
MW:　329.21　　　**BP (°C):**

Solubility (Moles/L)	Solubility (Grams/L)	Temp (°C)	Ref (#)	Evaluation (T P E A A)	Comments
2.360E-02	7.769E+00	20	D034	0 0 0 0 0	pH 7.0

2138. $C_{10}H_{12}N_6O_2S$

2-S-Cysteinyl-4,6-bis-(dimethylamino)-s-triazine

RN:　　　　　　　　**MP (°C):**　173
MW:　280.31　　　**BP (°C):**

Solubility (Moles/L)	Solubility (Grams/L)	Temp (°C)	Ref (#)	Evaluation (T P E A A)	Comments
7.991E-03	2.240E+00	25	C051	1 2 1 1 2	pH 7

2139. $C_{10}H_{12}O$

Estragole
1-Methoxy-4-(2-propen-1-yl)benzene
Chavicyl methyl ether
4-Allylanisole
Tarragon

RN:　140-67-0　　　**MP (°C):**　<25
MW:　148.21　　　**BP (°C):**　216

Solubility (Moles/L)	Solubility (Grams/L)	Temp (°C)	Ref (#)	Evaluation (T P E A A)	Comments
1.200E-03	1.778E-01	25	I019	1 0 1 2 2	

2140. $C_{10}H_{12}O$

Anethole
Methoxy-4-propenylbenzene
Propenylanisole
p-Propenylanisole
Anise camphor
Isoestragole

RN:　104-46-1　　　**MP (°C):**　21.4
MW:　148.21　　　**BP (°C):**

Solubility (Moles/L)	Solubility (Grams/L)	Temp (°C)	Ref (#)	Evaluation (T P E A A)	Comments
1.000E-03	1.482E-01	25	D407	1 0 2 2 2	
7.490E-04	1.110E-01	25	I019	1 0 1 2 2	
7.413E-04	1.099E-01	ns	S460	0 0 0 0 0	

2141. C$_{10}$H$_{12}$O
5,6,7,8-Tetrahydro-2-naphthol
5,6,7,8-Tetrahydro-naphthol-(2)

RN:	1125-78-6	**MP** (°C):	56.5		
MW:	148.21	**BP** (°C):	275.5		

Solubility (Moles/L)	Solubility (Grams/L)	Temp (°C)	Ref (#)	Evaluation (T P E A A)	Comments
1.012E-02	1.500E+00	20	F300	1 0 0 0 1	

2142. C$_{10}$H$_{12}$O$_2$
Eugenol
1-Allyl-3-methoxy-4-hydroxybenzene
2-Methoxy-4-allylphenol
2-Methoxy-4-(2-propenyl)phenol
4-Allylguaiacol
Allylguaiacol

RN:	97-53-0	**MP** (°C):	15		
MW:	164.21	**BP** (°C):			

Solubility (Moles/L)	Solubility (Grams/L)	Temp (°C)	Ref (#)	Evaluation (T P E A A)	Comments
1.500E-02	2.463E+00	25	I019	1 0 1 2 2	
4.020E-02	6.601E+00	37	E028	1 0 1 1 2	

2143. C$_{10}$H$_{12}$O$_2$
Ethyl 2-phenylacetate
Phenylacetic acid ethyl ester
Ethyl benzeneacetate; ethyl phenacetate
NSC 8894
NSC 406259
Ethyl phenylacetate

RN:	101-97-3	**MP** (°C):			
MW:	164.21	**BP** (°C):			

Solubility (Moles/L)	Solubility (Grams/L)	Temp (°C)	Ref (#)	Evaluation (T P E A A)	Comments
9.000E-03	1.478E+00	25	D407	1 0 2 2 2	
8.995E-03	1.477E+00	ns	S460	0 0 0 0 0	

2144. C$_{10}$H$_{12}$O$_2$
β-Phenylbutyric acid
3-Phenyl-n-butyric acid

RN:	4593-90-2	**MP** (°C):	38		
MW:	164.21	**BP** (°C):	171		

Solubility (Moles/L)	Solubility (Grams/L)	Temp (°C)	Ref (#)	Evaluation (T P E A A)	Comments
5.635E-02	9.254E+00	30	D033	2 2 1 2 2	
7.013E-02	1.152E+01	40	D033	2 2 1 2 2	

2145. C$_{10}$H$_{12}$O$_2$
b-Phenylethanol acetate
Phenylethyl ethanoate
b-Phenylethyl acetate
2-Phenethyl acetate; 2-phenylethyl acetate
Benzylcarbinyl acetate
NSC 71927

RN: 103-45-7   **MP** (°C):
MW: 164.21   **BP** (°C):   232

Solubility (Moles/L)	Solubility (Grams/L)	Temp (°C)	Ref (#)	Evaluation (T P E A A)	Comments
1.300E-02	2.135E+00	25	D407	1 0 2 2 2	
1.300E-02	2.135E+00	ns	S460	0 0 0 0 0	

2146. C$_{10}$H$_{12}$O$_2$
2,4,6-Trimethylbenzoic acid
Mesitylenecarboxylic acid

RN: 480-63-7   **MP** (°C):   154
MW: 164.21   **BP** (°C):

Solubility (Moles/L)	Solubility (Grams/L)	Temp (°C)	Ref (#)	Evaluation (T P E A A)	Comments
4.400E-03	7.225E-01	ns	C014	0 2 0 1 1	

2147. C$_{10}$H$_{12}$O$_2$
n-Propyl benzoate
Propyl benzoate
Benzoicacidpropyl ester

RN: 2315-68-6   **MP** (°C):   –51
MW: 164.21   **BP** (°C):

Solubility (Moles/L)	Solubility (Grams/L)	Temp (°C)	Ref (#)	Evaluation (T P E A A)	Comments
1.531E-03	2.514E-01	20	H301	0 0 0 0 0	

2148. C$_{10}$H$_{12}$O$_3$
Anisyl acetate
4-Methoxybenzyl acetate
Benzenemethanol, 4-methoxy-, acetate

RN: 104-21-2   **MP** (°C):
MW: 180.21   **BP** (°C):   235

Solubility (Moles/L)	Solubility (Grams/L)	Temp (°C)	Ref (#)	Evaluation (T P E A A)	Comments
1.100E-02	1.982E+00	25	D407	1 0 2 2 2	

2149. $C_{10}H_{12}O_3$
Propylparaben
Pr-paraben
Propyl *p*-hydroxybenzoic acid
Propyl 4-hydroxybenzoate
Propyl paraben

RN:	94-13-3	**MP** (°C):	96.5	
MW:	180.21	**BP** (°C):		

Solubility (Moles/L)	Solubility (Grams/L)	Temp (°C)	Ref (#)	Evaluation (T P E A A)	Comments
2.050E-03	3.694E-01	15	B355	0 0 0 0 0	
1.172E-03	2.112E-01	15	M352	1 1 1 1 2	
2.410E-03	4.343E-01	20	B355	0 0 0 0 0	
2.055E-03	3.703E-01	25	A059	1 0 1 1 1	
2.570E-03	4.631E-01	25	B355	0 0 0 0 0	
2.773E-03	4.998E-01	25	D081	1 2 2 1 2	
1.990E-03	3.586E-01	25	D339	0 0 0 0 0	
1.778E-03	3.205E-01	25	F322	2 0 1 1 0	EFG
1.844E-03	3.323E-01	25	M352	1 1 1 1 2	
2.775E-03	5.000E-01	25	O027	1 0 1 0 0	
2.863E-03	5.160E-01	25	P013	0 0 0 0 0	
2.300E-03	4.145E-01	27	B129	2 2 2 2 1	
2.443E-03	4.403E-01	30	A059	1 0 1 1 1	
2.053E-03	3.700E-01	30	M325	1 0 0 0 1	
3.054E-03	5.503E-01	35	A059	1 0 1 1 1	
3.403E-03	6.132E-01	39.3	G302	2 2 2 2 0	EFG
4.053E-03	7.303E-01	40	A059	1 0 1 1 1	
3.925E-03	7.073E-01	40	M352	1 1 1 1 2	
6.492E-03	1.170E+00	50	M352	1 1 1 1 2	
1.515E-03	2.729E-01	ns	B404	0 2 1 1 0	

2150. $C_{10}H_{12}O_4$
Cantharidin
Dimethyl-3,6-epoxyperhydrophthalic anhydride
Cantharides
Hexahydro-3α,7α-dimethyl-4β,7β-epoxyisobenzofuran-1,3-dione
Spanish fly

RN:	56-25-7	**MP** (°C):		
MW:	196.20	**BP** (°C):		

Solubility (Moles/L)	Solubility (Grams/L)	Temp (°C)	Ref (#)	Evaluation (T P E A A)	Comments
1.529E-04	3.000E-02	20	F300	1 0 0 0 0	
3.058E-01	6.000E+01	100	F300	1 0 0 0 0	
1.514E-04	2.970E-02	ns	R427	0 0 0 0 0	

2151. C$_{10}$H$_{12}$O$_5$
Propyl gallate
3,4,5-Trihydroxybenzoic acid propyl ester
Gallic acid propyl ester
Progallin P
n-propyl 3,4,5-trihydroxybenzoate
Nipa 49

RN: 121-79-9	**MP** (°C):	150 C
MW: 212.20	**BP** (°C):	

Solubility (Moles/L)	Solubility (Grams/L)	Temp (°C)	Ref (#)	Evaluation (T P E A A)	Comments
1.316E-02	2.792E+00	19.99	L430	0 0 0 0 0	
1.644E-02	3.488E+00	24.99	L430	0 0 0 0 0	
1.784E-02	3.786E+00	29.99	L430	0 0 0 0 0	
3.276E-02	6.951E+00	34.99	L430	0 0 0 0 0	
4.850E-02	1.029E+01	39.99	L430	0 0 0 0 0	
7.010E-02	1.488E+01	44.99	L430	0 0 0 0 0	
2.321E-01	4.925E+01	49.99	L430	0 0 0 0 0	
1.158E-01	2.458E+01	49.99	L430	0 0 0 0 0	
6.751E-01	1.432E+02	59.99	L430	0 0 0 0 0	
1.111E+00	2.357E+02	64.99	L430	0 0 0 0 0	
5.648E-03	1.199E+00	−.0	L430	0 0 0 0 0	

2152. C$_{10}$H$_{12}$O$_8$
Dilactone
α-Oxo-β-methylol-γ-butyrolactone betrachten

RN:	**MP** (°C):	140
MW: 260.20	**BP** (°C):	

Solubility (Moles/L)	Solubility (Grams/L)	Temp (°C)	Ref (#)	Evaluation (T P E A A)	Comments
9.374E-02	2.439E+01	0	F023	1 1 0 0 1	unit assumed
1.900E-01	4.943E+01	25	F023	1 1 0 0 1	unit assumed
5.972E-01	1.554E+02	50	F023	1 1 0 0 2	unit assumed
1.788E+00	4.652E+02	75	F023	1 1 0 0 2	unit assumed
2.451E+00	6.377E+02	100	F023	1 1 0 0 2	unit assumed

2153. C$_{10}$H$_{13}$ClN$_2$

Chlordimeform

N'-(4-Chloro-2-methylphenyl)-N,N-dimethylmethanimidamide

Bermat

Fundex

Galecon

Chlorophenamidine

RN:	6164-98-3	**MP** (°C):	32		
MW:	196.68	**BP** (°C):			

Solubility (Moles/L)	Solubility (Grams/L)	Temp (°C)	Ref (#)	Evaluation (T P E A A)	Comments
1.032E-03	2.030E-01	10	B324	0 0 0 0 0	
1.032E-03	2.030E-01	10	B324	0 0 0 0 0	
1.373E-03	2.700E-01	20	B300	2 0 1 1 2	
1.373E-03	2.700E-01	20	B324	0 0 0 0 0	
1.372E-03	2.699E-01	20	B324	0 0 0 0 0	
1.271E-03	2.500E-01	20	M161	1 0 0 0 2	

2154. C$_{10}$H$_{13}$ClN$_2$O

Trimeturon

N'-4-Chlorophenyl-O,N,N-trimethylisourea

RN:	3050-27-9	**MP** (°C):	147.5		
MW:	212.68	**BP** (°C):			

Solubility (Moles/L)	Solubility (Grams/L)	Temp (°C)	Ref (#)	Evaluation (T P E A A)	Comments
3.289E-03	6.995E-01	ns	M061	0 0 0 0 1	

2155. C$_{10}$H$_{13}$ClN$_2$O

Chlortoluron

N'-(3-Chloro-4-methylphenyl)-N,N-dimethylurea

Dicuran

Chlortokem

Tolurex

RN:	15545-48-9	**MP** (°C):	147.5		
MW:	212.68	**BP** (°C):			

Solubility (Moles/L)	Solubility (Grams/L)	Temp (°C)	Ref (#)	Evaluation (T P E A A)	Comments
3.311E-04	7.043E-02	20	B179	0 0 0 0 0	
3.291E-04	7.000E-02	20	F311	1 2 2 2 1	
3.291E-04	7.000E-02	20	M161	1 0 0 0 1	

2156. C$_{10}$H$_{13}$ClN$_2$O$_2$

Metoxuron
N'-(3-Chloro-4-methoxyphenyl)-N,N-dimethylurea
Purivel
Sulerex
Dosanex
Dosaflo

RN:　19937-59-8　　**MP (°C):**　125
MW:　228.68　　**BP (°C):**

Solubility (Moles/L)	Solubility (Grams/L)	Temp (°C)	Ref (#)	Evaluation (T P E A A)	Comments
3.020E-03	6.906E-01	20	B179	0 0 0 0 0	
2.622E-03	5.996E-01	20	E048	1 2 1 1 2	
2.965E-03	6.780E-01	23	M161	0 0 0 0 2	
3.059E-03	6.995E-01	ns	B100	0 0 0 0 0	

2157. C$_{10}$H$_{13}$ClN$_2$O$_3$S

Chlorpropamide
N3-Butyl-N1-p-chlorobenzenesulfonylurea
Diabinese
Glucamide
Catanil
Diabaril

RN:　94-20-2　　**MP (°C):**　128
MW:　276.74　　**BP (°C):**

Solubility (Moles/L)	Solubility (Grams/L)	Temp (°C)	Ref (#)	Evaluation (T P E A A)	Comments
3.221E-03	8.913E-01	25	F415	0 0 0 0 0	
9.311E-04	2.577E-01	37	A028	1 0 2 1 2	intrinsic
9.250E-04	2.560E-01	37	A046	2 0 1 1 2	
~1.26E-03	~3.50E-01	37	B140	2 2 1 2 0	pH 1.5, form V
1.203E-03	3.330E-01	37	B140	2 2 1 2 2	pH 1.5, form I
1.384E-03	3.830E-01	37	B140	2 2 1 2 2	pH 1.5, form II
8.925E-04	2.470E-01	37	B140	2 2 1 2 2	pH 1.5, form III
1.153E-03	3.190E-01	37	B140	2 2 1 2 2	pH 1.5, form IV
>1.81E-03	>5.00E-01	ns	B404	0 2 1 1 0	
5.192E-04	1.437E-01	rt	I404	0 0 0 0 0	Average

2158. $C_{10}H_{13}Cl_2FN_2O_2S_2$

Tolylfluanid

1,1-Dichloro-N-((dimethylamino)sulfonyl)-1-fluoro-N-(4-methylphenyl)methanesulfenamide

Dichlofluanid-methyl

Euparen M

Bay 5712α

Bay 49854

RN:	731-27-1	MP (°C):	96
MW:	347.26	BP (°C):	

Solubility (Moles/L)	Solubility (Grams/L)	Temp (°C)	Ref (#)	Evaluation (T P E A A)	Comments
2.570E-06	8.926E-04	ns	R427	0 0 0 0 0	
1.152E-02	4.000E+00	rt	M161	0 0 0 0 0	

2159. $C_{10}H_{13}Cl_2O_3PS$

Dichlofenthion

Diethyl O-dichlorophenyl phosphorothioate

Hexanema

Diclophenthion

Nemacide

TRI-VC13

RN:	97-17-6	MP (°C):	
MW:	315.16	BP (°C):	164

Solubility (Moles/L)	Solubility (Grams/L)	Temp (°C)	Ref (#)	Evaluation (T P E A A)	Comments
7.774E-07	2.450E-04	25	M161	1 0 0 0 2	
7.774E-07	2.450E-04	ns	F071	0 1 2 1 2	
7.774E-04	2.450E-01	ns	M061	0 0 0 0 2	*sic*

2160. $C_{10}H_{13}FN_2O_3$

1-Pivaloyloxymethyl-5-fluorouracil

RN:		MP (°C):	
MW:	228.23	BP (°C):	

Solubility (Moles/L)	Solubility (Grams/L)	Temp (°C)	Ref (#)	Evaluation (T P E A A)	Comments
1.095E-02	2.500E+00	22	M317	1 1 1 1 1	

2161. $C_{10}H_{13}FN_2O_4$

1-Pivaloyloxymethyl-5-fluoro-2,4(1H,3H)-pyrimidinedi-one

1-Pivaloyloxymethyl-5-fluorouracil

RN:	62113-42-2	MP (°C):	158-160
MW:	244.22	BP (°C):	

Solubility (Moles/L)	Solubility (Grams/L)	Temp (°C)	Ref (#)	Evaluation (T P E A A)	Comments
9.418E-03	2.300E+00	22	B321	0 0 0 0 0	pH 4.0

2162. $C_{10}H_{13}NO_2$

Phenacetin
p-Ethoxyacetanilide
p-Acetophenetidide

RN:	62-44-2	**MP** (°C):	134.5	
MW:	179.22	**BP** (°C):		

Solubility (Moles/L)	Solubility (Grams/L)	Temp (°C)	Ref (#)	Evaluation (T P E A A)	Comments
3.010E-04	5.395E-02	14	O019	1 0 0 1 2	
2.010E-03	3.603E-01	15	M352	1 1 1 1 2	
3.903E-03	6.995E-01	20	M043	1 0 0 0 0	
5.167E-03	9.261E-01	25	B434	0 0 0 0 0	
5.180E-03	9.284E-01	25	B434	0 0 0 0 0	
4.300E-02	7.706E+00	25	D044	0 0 0 0 0	
4.464E-03	8.000E-01	25	F300	1 0 0 0 0	
2.801E-03	5.020E-01	25	M333	1 1 0 0 2	
2.799E-03	5.016E-01	25	M352	1 1 1 1 2	
6.271E-03	1.124E+00	30	B434	0 0 0 0 0	
6.280E-03	1.126E+00	30	B434	0 0 0 0 0	
8.653E-03	1.551E+00	35	B434	0 0 0 0 0	
8.680E-03	1.556E+00	35	B434	0 0 0 0 0	
1.183E-02	2.120E+00	40	B434	0 0 0 0 0	
1.185E-02	2.124E+00	40	B434	0 0 0 0 0	
5.483E-03	9.828E-01	40	M352	1 1 1 1 2	
7.878E-03	1.412E+00	50	M352	1 1 1 1 2	
6.616E-02	1.186E+01	100	I315	0 0 0 0 0	
7.867E-02	1.410E+01	100	M043	1 0 0 0 2	
4.237E-03	7.594E-01	c	I315	0 0 0 0 0	
6.584E-03	1.180E+00	ns	F059	1 0 2 2 2	0.1N HCl
5.574E-03	9.990E-01	rt	D021	0 0 1 1 1	

2163. $C_{10}H_{13}NO_2$

Propham
Isopropyl carbanilate
Isopropyl-*N*-phenyl carbamate
IPC

RN:	122-42-9	**MP** (°C):	87	
MW:	179.22	**BP** (°C):		

Solubility (Moles/L)	Solubility (Grams/L)	Temp (°C)	Ref (#)	Evaluation (T P E A A)	Comments
5.580E-04	1.000E-01	25	G099	1 0 0 1 0	
1.116E-04	2.000E-02	ns	B185	0 0 0 0 0	
1.786E-04	3.200E-02	ns	B185	0 0 0 0 0	
1.395E-03	2.500E-01	ns	B200	0 0 0 0 2	
5.580E-04	1.000E-01	ns	F035	0 0 0 0 0	
1.395E-03	2.500E-01	ns	H042	0 0 0 0 2	
1.000E-03	1.792E-01	ns	M163	0 0 0 0 0	EFG
1.395E-03	2.500E-01	ns	N013	0 0 0 0 2	

2164. $C_{10}H_{13}NO_2$
Butyl nicotinate
n-Butyl nicotinate
RN: 6938-06-3   **MP** (°C):
MW: 179.22   **BP** (°C):

Solubility (Moles/L)	Solubility (Grams/L)	Temp (°C)	Ref (#)	Evaluation (T P E A A)	Comments
1.367E-02	2.450E+00	32	L346	1 0 0 1 2	

2165. $C_{10}H_{13}NO_2$
Propyl-*p*-aminobenzoate
Risocaine
4-Aminobenzoic acid propyl ester
RN: 94-12-2   **MP** (°C):   75.5
MW: 179.22   **BP** (°C):

Solubility (Moles/L)	Solubility (Grams/L)	Temp (°C)	Ref (#)	Evaluation (T P E A A)	Comments
1.655E-03	2.966E-01	15	M352	1 1 1 1 2	
2.220E-03	3.979E-01	25	H008	0 0 0 0 0	
2.860E-03	5.125E-01	25	M352	1 1 1 1 2	
3.553E-03	6.368E-01	25	P303	0 0 0 0 0	
4.219E-03	7.561E-01	33	P303	0 0 0 0 0	
4.700E-03	8.423E-01	37	F006	1 1 2 2 2	
4.629E-03	8.297E-01	40	M352	1 1 1 1 2	
5.217E-03	9.351E-01	40	P303	0 0 0 0 0	
7.047E-03	1.263E+00	50	M352	1 1 1 1 2	
1.890E-03	3.387E-01	ns	M066	0 0 0 0 2	
1.890E-03	3.387E-01	rt	B016	0 0 1 1 2	pH 7.4

2166. $C_{10}H_{13}NO_2$
3,4-Xylyl methylcarbamate
3,4-Dimethylphenyl methylcarbamate
3,4-Dimethylphenyl *N*-methylcarbamate
MPMC
Meobal
RN: 2425-10-7   **MP** (°C):   79.5
MW: 179.22   **BP** (°C):   126.5

Solubility (Moles/L)	Solubility (Grams/L)	Temp (°C)	Ref (#)	Evaluation (T P E A A)	Comments
7.254E-03	1.300E+00	30	M161	1 0 0 0 1	

2167. C$_{10}$H$_{13}$NO$_2$
2,6-Dimethyl-4-acetaminophenol
4-Acetamido-2,6-dimethylphenol
RN: 22900-79-4   **MP** (°C):
MW: 179.22   **BP** (°C):

Solubility (Moles/L)	Solubility (Grams/L)	Temp (°C)	Ref (#)	Evaluation (T P E A A)	Comments
1.227E-02	2.200E+00	25	D078	1 2 1 1 2	

2168. C$_{10}$H$_{13}$NO$_2$
Methyl *p*-dimethylaminobenzoic acid
Methyl 4-dimethylaminobenzoate
RN: 1202-25-1   **MP** (°C):   371.7
MW: 179.22   **BP** (°C):

Solubility (Moles/L)	Solubility (Grams/L)	Temp (°C)	Ref (#)	Evaluation (T P E A A)	Comments
3.400E-04	6.093E-02	15	M352	1 1 1 1 2	
4.988E-04	8.940E-02	25	M352	1 1 1 1 2	
8.277E-04	1.483E-01	40	M352	1 1 1 1 2	
1.111E-03	1.991E-01	50	M352	1 1 1 1 2	

2169. C$_{10}$H$_{13}$NO$_2$
2,5-Dimethyl-4-acetaminophenol
4-Acetamido-2,5-dimethylphenol
RN: 69477-71-0   **MP** (°C):
MW: 179.22   **BP** (°C):

Solubility (Moles/L)	Solubility (Grams/L)	Temp (°C)	Ref (#)	Evaluation (T P E A A)	Comments
9.694E-03	1.737E+00	25	D078	1 2 1 1 2	

2170. C$_{10}$H$_{13}$NO$_3$
o-Ethoxyphenyl *N*-methylcarbamate
1,2-Ethoxyphenyl *N*-methylcarbamate
RN: 23409-17-8   **MP** (°C):   79.5
MW: 195.22   **BP** (°C):

Solubility (Moles/L)	Solubility (Grams/L)	Temp (°C)	Ref (#)	Evaluation (T P E A A)	Comments
1.178E-02	2.300E+00	30	D089	2 2 0 0 0	

2171. C$_{10}$H$_{13}$NO$_3$
m-Ethoxyphenyl *N*-methylcarbamate
1,3-Ethoxyphenyl *N*-methylcarbamate
RN: 7225-96-9   **MP** (°C):   57
MW: 195.22   **BP** (°C):

Solubility (Moles/L)	Solubility (Grams/L)	Temp (°C)	Ref (#)	Evaluation (T P E A A)	Comments
6.403E-03	1.250E+00	30	D089	2 2 0 0 0	

2172. C$_{10}$H$_{13}$NO$_4$
Methyldopa
α-Methyldopa
Sembrina
Presinol
Sedometil
Presolisin
RN: 555-30-6   **MP** (°C):   ~300
MW: 211.22   **BP** (°C):

Solubility (Moles/L)	Solubility (Grams/L)	Temp (°C)	Ref (#)	Evaluation (T P E A A)	Comments
4.734E-02	1.000E+01	ns	K444	0 0 0 0 0	

2173. C$_{10}$H$_{13}$N$_3$O$_2$S$_2$
3-Methyl-2-sulfanilamide-2,3-dihydrothiazole
Benzenesulfonamide, 4-amino-*N*-(2,3-dihydro-3-methyl-2-thiazolyl)-
RN: 51203-20-4   **MP** (°C):
MW: 271.36   **BP** (°C):

Solubility (Moles/L)	Solubility (Grams/L)	Temp (°C)	Ref (#)	Evaluation (T P E A A)	Comments
5.690E-04	1.544E-01	37	K095	2 0 0 0 2	intrinsic

2174. C$_{10}$H$_{13}$N$_3$O$_5$S
Nifurtimox
4-((5-Nitrofurfurylidene)amino)-3-methylthiomorpholine-1,1-dioxide
RN: 23256-30-6   **MP** (°C):
MW: 287.30   **BP** (°C):

Solubility (Moles/L)	Solubility (Grams/L)	Temp (°C)	Ref (#)	Evaluation (T P E A A)	Comments
1.149E-01	3.300E+01	ns	K444	0 0 0 0 0	

2175. $C_{10}H_{13}N_4O_3$
Spasmolysin
β-Hydroxypropyltheophylline
RN: 603-00-9 **MP** (°C):
MW: 237.24 **BP** (°C):

Solubility (Moles/L)	Solubility (Grams/L)	Temp (°C)	Ref (#)	Evaluation (T P E A A)	Comments
1.204E+00	2.857E+02	ns	J025	0 0 0 0 1	

2176. $C_{10}H_{13}N_5$
4-Amino-6,7-diethylpteridine
RN: **MP** (°C):
MW: 203.25 **BP** (°C):

Solubility (Moles/L)	Solubility (Grams/L)	Temp (°C)	Ref (#)	Evaluation (T P E A A)	Comments
1.171E-03	2.380E-01	20	A019	2 2 1 1 2	

2177. $C_{10}H_{13}N_5$
2-Amino-6,7-diethylpteridine
RN: **MP** (°C):
MW: 203.25 **BP** (°C):

Solubility (Moles/L)	Solubility (Grams/L)	Temp (°C)	Ref (#)	Evaluation (T P E A A)	Comments
9.110E-04	1.852E-01	20	A019	2 2 1 1 2	

2178. $C_{10}H_{13}N_5O$
2-Amino-4-hydroxy-6,7-diethylpteridine
2-Amino-4-hydroxy-6:7-diethylpteridine
RN: **MP** (°C): >350
MW: 219.25 **BP** (°C):

Solubility (Moles/L)	Solubility (Grams/L)	Temp (°C)	Ref (#)	Evaluation (T P E A A)	Comments
5.303E-05	1.163E-02	20	A019	2 2 1 1 2	

2179. $C_{10}H_{13}N_5O$
4-Amino-2-hydroxy-6,7-diethylpteridine
4-Amino-2-hydroxy-6:7-diethylpteridine
RN: **MP** (°C):
MW: 219.25 **BP** (°C):

Solubility (Moles/L)	Solubility (Grams/L)	Temp (°C)	Ref (#)	Evaluation (T P E A A)	Comments
2.850E-04	6.250E-02	20	A019	2 2 1 1 2	

2180. $C_{10}H_{13}N_5O_2$
2',3'-Dideoxyadenosine
DDA

RN:	4097-22-7	**MP** (°C):	181–184
MW:	235.25	**BP** (°C):	

Solubility (Moles/L)	Solubility (Grams/L)	Temp (°C)	Ref (#)	Evaluation (T P E A A)	Comments
1.228E-01	2.890E+01	4	A337	0 0 0 0 0	
1.836E-01	4.320E+01	25	A337	0 0 0 0 0	

2181. $C_{10}H_{13}N_5O_3$
Deoxyadenosine
2'-Deoxyadenosine
dA

RN:	958-09-8	**MP** (°C):	
MW:	251.25	**BP** (°C):	

Solubility (Moles/L)	Solubility (Grams/L)	Temp (°C)	Ref (#)	Evaluation (T P E A A)	Comments
1.422E-02	3.573E+00	14.88	T420	0 0 0 0 0	
1.827E-02	4.590E+00	20.26	T420	0 0 0 0 0	
2.690E-02	6.759E+00	25	H061	0 0 0 0 0	
2.558E-02	6.427E+00	25.23	T420	0 0 0 0 0	
3.683E-02	9.253E+00	29.97	T420	0 0 0 0 0	
4.780E-02	1.201E+01	35.09	T420	0 0 0 0 0	

2182. $C_{10}H_{13}N_5O_4$
Zidovudine
3-Azido-3-deoxythymidine
AZT
Azidodeoxythymidine
Azidothymidine
Retrovir

RN:	30516-87-1	**MP** (°C):	106–112
MW:	267.25	**BP** (°C):	

Solubility (Moles/L)	Solubility (Grams/L)	Temp (°C)	Ref (#)	Evaluation (T P E A A)	Comments
>1.13E+00	>3.02E+02	25	B443	0 0 0 0 0	
7.521E-02	2.010E+01	ns	K444	0 0 0 0 0	
7.373E-02	1.970E+01	ns	S469	0 0 0 0 0	

2183. C$_{10}$H$_{13}$N$_5$O$_4$
Adenosine
Adenosin
9-B-D-Ribofuranosyl-9H-purin-6-amine adenine riboside
Adenocard
9-β-D-Ribofuranosyladenine

RN: 58-61-7 **MP** (°C): 234
MW: 267.25 **BP** (°C):

Solubility (Moles/L)	Solubility (Grams/L)	Temp (°C)	Ref (#)	Evaluation (T P E A A)	Comments
1.920E-02	5.131E+00	25	H061	0 0 0 0 0	
2.000E-02	5.345E+00	ns	R030	0 0 0 0 0	
1.905E-02	5.092E+00	ns	R427	0 0 0 0 0	
8.232E-05	2.200E-02	rt	N015	0 0 2 2 1	*sic*

2184. C$_{10}$H$_{13}$N$_5$O$_4$
Guanine deoxyriboside

RN: 961-07-9 **MP** (°C):
MW: 267.25 **BP** (°C):

Solubility (Moles/L)	Solubility (Grams/L)	Temp (°C)	Ref (#)	Evaluation (T P E A A)	Comments
6.680E-03	1.785E+00	14.88	T420	0 0 0 0 0	
8.790E-03	2.349E+00	20.26	T420	0 0 0 0 0	
1.118E-02	2.988E+00	25.02	T420	0 0 0 0 0	
1.589E-02	4.247E+00	29.97	T420	0 0 0 0 0	
2.072E-02	5.537E+00	35.09	T420	0 0 0 0 0	

2185. C$_{10}$H$_{13}$N$_5$O$_5$
Guanosine
Guanosin
2-Amino-9-β-D-ribofuranosyl-9H-purine-6-(1H)-one
Guanine riboside
rG

RN: 118-00-3 **MP** (°C): 250
MW: 283.25 **BP** (°C):

Solubility (Moles/L)	Solubility (Grams/L)	Temp (°C)	Ref (#)	Evaluation (T P E A A)	Comments
2.471E-03	7.000E-01	18	F300	1 0 0 0 1	
4.300E-03	1.218E+00	25	C416	2 1 1 1 1	
1.820E-03	5.155E-01	25	H061	0 0 0 0 0	
1.073E-01	3.040E+01	100	F300	1 0 0 0 1	

2186. $C_{10}H_{14}$
Isobutylbenzene
2-Methyl-1-phenylpropane
(2-Methylpropyl)-benzene

RN:	538-93-2	MP (°C):	−51
MW:	134.22	BP (°C):	170.5

Solubility (Moles/L)	Solubility (Grams/L)	Temp (°C)	Ref (#)	Evaluation (T P E A A)	Comments
7.525E-05	1.010E-02	25	P051	2 1 1 2 2	
7.525E-05	1.010E-02	25.00	P007	2 1 2 2 2	
7.525E-05	1.010E-02	ns	H123	0 0 0 0 0	

2187. $C_{10}H_{14}$
Durene
1,2,4,5-Tetramethylbenzene
Durol

RN:	95-93-2	MP (°C):	80.0
MW:	134.22	BP (°C):	192.0

Solubility (Moles/L)	Solubility (Grams/L)	Temp (°C)	Ref (#)	Evaluation (T P E A A)	Comments
2.593E-05	3.480E-03	25	K119	1 0 0 0 2	
2.593E-05	3.480E-03	25	P051	2 1 1 2 2	
2.593E-05	3.480E-03	25.00	P007	2 1 2 2 2	
1.445E-04	1.940E-02	ns	D001	0 0 0 0 2	
7.152E-05	9.600E-03	ns	H123	0 0 0 0 0	

2188. $C_{10}H_{14}$
Butylbenzene
1-Phenylbutane
n-Butylbenzene

RN:	68411-44-9	MP (°C):	−88
MW:	134.22	BP (°C):	

Solubility (Moles/L)	Solubility (Grams/L)	Temp (°C)	Ref (#)	Evaluation (T P E A A)	Comments
1.300E-02	1.745E+00	ns	H307	0 0 0 0 0	

2189. C$_{10}$H$_{14}$

n-Butylbenzene
1-Phenylbutane
Butylbenzene

RN: 104-51-8   **MP** (°C):   −88.5
MW: 134.22   **BP** (°C):   183.1

Solubility (Moles/L)	Solubility (Grams/L)	Temp (°C)	Ref (#)	Evaluation (T P E A A)	Comments
9.940E-05	1.334E-02	7	O312	2 2 0 2 2	
9.670E-05	1.298E-02	10	O312	2 2 0 2 2	
9.790E-05	1.314E-02	12.5	O312	2 2 0 2 2	
9.660E-05	1.297E-02	15	O312	2 2 0 2 2	
9.790E-05	1.314E-02	17.5	O312	2 2 0 2 2	
9.909E-05	1.330E-02	20	B356	0 0 0 0 0	
1.018E-04	1.366E-02	20	O312	2 2 0 2 2	
9.387E-06	1.260E-03	25	A002	1 2 1 1 2	*sic*
3.700E-04	4.966E-02	25	K001	1 0 2 1 2	
1.320E-04	1.772E-02	25	M124	2 1 2 2 2	
1.030E-04	1.382E-02	25	M342	1 0 1 1 2	
1.025E-04	1.376E-02	25	O312	2 2 0 2 2	
8.791E-05	1.180E-02	25	S005	2 2 2 2 2	
3.725E-04	5.000E-02	25	S012	2 0 2 2 0	
8.791E-05	1.180E-02	25	S191	1 2 2 2 2	
8.791E-05	1.180E-02	25	S358	2 1 2 2 2	
1.030E-04	1.382E-02	25	W300	2 2 2 2 2	
1.244E-04	1.670E-02	29.99	C350	0 0 0 0 0	
1.086E-04	1.458E-02	30	O312	2 2 0 2 2	
1.147E-04	1.540E-02	35	O312	2 2 0 2 2	
1.328E-04	1.782E-02	39.99	C350	0 0 0 0 0	
1.234E-04	1.656E-02	40	O312	2 2 0 2 2	
1.411E-04	1.894E-02	45	O312	2 2 0 2 2	
1.517E-04	2.036E-02	49.99	C350	0 0 0 0 0	
2.006E-04	2.692E-02	59.99	C350	0 0 0 0 0	
2.389E-04	3.206E-02	69.99	C350	0 0 0 0 0	
3.555E-04	4.772E-02	79.99	C350	0 0 0 0 0	
4.555E-04	6.114E-02	89.99	C350	0 0 0 0 0	
6.222E-04	8.351E-02	99.99	C350	0 0 0 0 0	
9.387E-05	1.260E-02	ns	H123	0 0 0 0 0	

2190. C$_{10}$H$_{14}$

p-Cymene
1-Methyl-4-isopropylbenzene
4-Cymene
Dolcymine

RN: 99-87-6   **MP** (°C):   −68
MW: 134.22   **BP** (°C):   177

Solubility (Moles/L)	Solubility (Grams/L)	Temp (°C)	Ref (#)	Evaluation (T P E A A)	Comments
2.979E-03	3.998E-01	25	B019	1 0 1 2 0	*sic*
1.740E-04	2.335E-02	25	B173	2 0 2 2 2	*sic*

2191. C$_{10}$H$_{14}$
sec-Butylbenzene
1-Methylpropylbenzene
RN:	135-98-8	MP (°C):	−82.7
MW:	134.22	BP (°C):	173.5

Solubility (Moles/L)	Solubility (Grams/L)	Temp (°C)	Ref (#)	Evaluation (T P E A A)	Comments
2.302E-03	3.090E-01	25	A002	1 2 1 1 2	sic
7.525E-05	1.010E-02	25	K119	1 0 0 0 2	
1.311E-04	1.760E-02	25	S005	2 2 2 2 2	
1.311E-04	1.760E-02	25	S191	1 2 2 2 2	
1.311E-04	1.760E-02	25	S358	2 1 2 2 2	

2192. C$_{10}$H$_{14}$
tert-Butylbenzene
1,1-Dimethylethylbenzene
t-Butylbenzene
RN:	98-06-6	MP (°C):	−58
MW:	134.22	BP (°C):	168.5

Solubility (Moles/L)	Solubility (Grams/L)	Temp (°C)	Ref (#)	Evaluation (T P E A A)	Comments
2.533E-04	3.400E-02	25	A002	1 2 1 1 1	
2.198E-04	2.950E-02	25	S005	2 2 2 2 2	
1.311E-04	1.760E-02	25	S191	1 2 2 2 2	
2.198E-04	2.950E-02	25	S358	2 1 2 2 2	

2193. C$_{10}$H$_{14}$
1,2-Diethylbenzene
o-Diethylbenzene
RN:	135-01-3	MP (°C):	−31
MW:	134.22	BP (°C):	183

Solubility (Moles/L)	Solubility (Grams/L)	Temp (°C)	Ref (#)	Evaluation (T P E A A)	Comments
5.300E-04	7.114E-02	10	B149	2 1 1 2 1	
5.300E-04	7.114E-02	20	B149	2 1 1 2 1	

2194. C$_{10}$H$_{14}$
1,4-Diethylbenzene
p-Diethylbenzene
RN:	105-05-5	MP (°C):	−43
MW:	134.22	BP (°C):	184

Solubility (Moles/L)	Solubility (Grams/L)	Temp (°C)	Ref (#)	Evaluation (T P E A A)	Comments
1.850E-04	2.483E-02	10	B149	2 1 1 2 2	
1.850E-04	2.483E-02	20	B149	2 1 1 2 2	

2195. C₁₀H₁₄Cl₂NO₂PS

DMPA

Isopropylphosphoramidothioate

O-(2,4-Dichlorophenyl)-*O*-methyl

Phosphoramidothioic acid, isopropyl-*o*-(2,4-dichlorophenyl)-*o*-methyl ester

RN: 299-85-4   **MP** (°C):   51.4

MW: 314.17   **BP** (°C):

Solubility (Moles/L)	Solubility (Grams/L)	Temp (°C)	Ref (#)	Evaluation (T P E A A)	Comments
1.595E-05	5.010E-03	25	B185	0 0 0 0 0	
1.591E-05	5.000E-03	25	B200	1 0 0 0 0	
1.591E-05	5.000E-03	ns	M061	0 0 0 0 0	

2196. C₁₀H₁₄Cl₆N₄O₂

Triforine

N,N'-[1,4-Piperazinediylbis(2,2,2-trichloroethylidene)] bisformamide

Funginex

Denarin

Biformylchlorazin

Saprol

RN: 26644-46-2   **MP** (°C):   155

MW: 434.97   **BP** (°C):

Solubility (Moles/L)	Solubility (Grams/L)	Temp (°C)	Ref (#)	Evaluation (T P E A A)	Comments
~1.38E-05	~6.00E-03	rt	D303	0 0 0 0 0	
6.437E-05	2.800E-02	rt	M161	0 0 0 0 0	

2197. C₁₀H₁₄NO₅PS

Parathion

O,O-Diethyl *O-p*-nitrophenyl phosphorothioate

Foliclal

Rhodiatox

Alkron

Fosferno

RN: 56-38-2   **MP** (°C):   6

MW: 291.26   **BP** (°C):

Solubility (Moles/L)	Solubility (Grams/L)	Temp (°C)	Ref (#)	Evaluation (T P E A A)	Comments
3.536E-05	1.030E-02	10	B324	0 0 0 0 0	
3.536E-05	1.030E-02	10	B324	0 0 0 0 0	
4.257E-05	1.240E-02	20	B169	2 1 1 1 1	
8.318E-05	2.423E-02	20	B179	0 0 0 0 0	
4.429E-05	1.290E-02	20	B324	0 0 0 0 0	
4.429E-05	1.290E-02	20	B324	0 0 0 0 0	
2.245E-05	6.540E-03	24	F179	2 2 2 2 2	
8.240E-05	2.400E-02	25	M161	1 0 0 0 1	

(continued)

2197. $C_{10}H_{14}NO_5PS$ (continued)

Solubility (Moles/L)	Solubility (Grams/L)	Temp (°C)	Ref (#)	Evaluation (T P E A A)	Comments
5.219E-05	1.520E-02	30	B324	0 0 0 0 0	
5.219E-05	1.520E-02	30	B324	0 0 0 0 0	
4.086E-05	1.190E-02	ns	F071	0 1 2 1 2	
8.240E-05	2.400E-02	ns	M061	0 0 0 0 1	
6.867E-05	2.000E-02	ns	M110	0 0 0 0 0	EFG
8.240E-05	2.400E-02	ns	M344	0 0 0 0 1	

2198. $C_{10}H_{14}NO_6P$
Paraoxon
Diethyl *p*-nitrophenyl phosphate
Fosfacol
Eticol
Ethyl paraoxon
Miotisal

RN: 311-45-5   **MP** (°C):
MW: 275.20   **BP** (°C): 169–170

Solubility (Moles/L)	Solubility (Grams/L)	Temp (°C)	Ref (#)	Evaluation (T P E A A)	Comments
1.318E-02	3.627E+00	20	B169	2 0 1 1 2	
3.634E-03	1.000E+00	20	F300	1 0 0 0 0	

2199. $C_{10}H_{14}N_2O$
N-(Dimethylaminomethyl)benzamide
Benzamide, *N*-[(dimethylamino)methyl]-
RN: 59917-58-7   **MP** (°C):
MW: 178.24   **BP** (°C):

Solubility (Moles/L)	Solubility (Grams/L)	Temp (°C)	Ref (#)	Evaluation (T P E A A)	Comments
2.600E+00	4.634E+02	22	J037	0 0 0 0 0	

2200. $C_{10}H_{14}N_2O$
N-(Ethylaminomethyl)benzamide
Benzamide, *N*-[(ethylamino)methyl]-
RN: 73239-20-0   **MP** (°C):
MW: 178.24   **BP** (°C):

Solubility (Moles/L)	Solubility (Grams/L)	Temp (°C)	Ref (#)	Evaluation (T P E A A)	Comments
7.300E-02	1.301E+01	22	J037	0 0 0 0 0	

2201. $C_{10}H_{14}N_2O_2$

m-N,N-Dimethylaminophenyl *N*-methylcarbamate
1,3-*N,N*-Dimethylaminophenyl *N*-methylcarbamate

RN: 2631-39-2   **MP** (°C):   86
MW: 194.24   **BP** (°C):

Solubility (Moles/L)	Solubility (Grams/L)	Temp (°C)	Ref (#)	Evaluation (T P E A A)	Comments
3.604E-03	7.000E-01	30	D089	2 2 0 0 0	

2202. $C_{10}H_{14}N_2O_3$

5-Methyl-5-(3-methylbut-2-enyl)barbituric acid
2,4,6(1H,3H,5H)-Pyrimidinetrione, 5-methyl-5-(3-methyl-2-butenyl)
5-Methyl-5-(3-methylbut-2-enyl)barbiturate

RN: 66843-01-4   **MP** (°C):
MW: 210.23   **BP** (°C):

Solubility (Moles/L)	Solubility (Grams/L)	Temp (°C)	Ref (#)	Evaluation (T P E A A)	Comments
2.503E-03	5.262E-01	25	P350	0 0 0 0 0	intrinsic

2203. $C_{10}H_{14}N_2O_3$

2,4-Diazaspiro[5.6]dodecane-1,3,5-trione
Cycloheptane-spirobarbiturate

RN: 143288-61-3   **MP** (°C):
MW: 210.23   **BP** (°C):

Solubility (Moles/L)	Solubility (Grams/L)	Temp (°C)	Ref (#)	Evaluation (T P E A A)	Comments
6.790E-04	1.427E-01	25	P350	0 0 0 0 0	intrinsic

2204. $C_{10}H_{14}N_2O_3$

5-Isopropyl-5-allylbarbituric acid
Aprobarbital
5-(1-Methylethyl)-5-(2-propenyl)-2,4,6(1H,3H,5H)-pyrimidinetrione
5-Allyl-5-isopropylbarbituric acid
Aprobarbitone

RN: 77-02-1   **MP** (°C):   141
MW: 210.23   **BP** (°C):

Solubility (Moles/L)	Solubility (Grams/L)	Temp (°C)	Ref (#)	Evaluation (T P E A A)	Comments
1.617E-02	3.400E+00	20	J030	1 2 2 2 2	
1.960E-02	4.121E+00	25	P350	0 0 0 0 0	intrinsic
1.940E-02	4.079E+00	25	V033	2 0 1 1 2	
1.940E-02	4.079E+00	25.00	T303	1 0 0 0 2	
2.600E-02	5.466E+00	35.00	T303	1 0 0 0 2	
2.664E-02	5.600E+00	37	J030	1 2 2 2 2	
3.340E-02	7.022E+00	45.00	T303	1 0 0 0 2	
1.912E-02	4.020E+00	ns	T003	0 0 0 0 2	

2205. C$_{10}$H$_{14}$N$_2$O$_5$
Thymidine
(1-[2-Deoxy-β-D-ribofuranosyl]-5-methyluracil)
Thyminedeoxyriboside
2′-deoxy-5-methyl
Thymine-2-desoxyriboside
Uridine

RN: 50-89-5   **MP** (°C): 187–189
MW: 242.23   **BP** (°C):

Solubility (Moles/L)	Solubility (Grams/L)	Temp (°C)	Ref (#)	Evaluation (T P E A A)	Comments
2.720E-02	6.589E+00	19.99	T418	0 0 0 0 0	
2.790E-02	6.758E+00	24.96	T418	0 0 0 0 0	
2.780E-02	6.734E+00	24.99	T418	0 0 0 0 0	
3.040E-02	7.364E+00	24.99	T418	0 0 0 0 0	
2.790E-02	6.758E+00	24.99	T418	0 0 0 0 0	
2.870E-02	6.952E+00	24.99	T418	0 0 0 0 0	
2.200E-01	5.329E+01	24.99	T418	0 0 0 0 0	
2.710E-02	6.565E+00	25.49	T418	0 0 0 0 0	

2206. C$_{10}$H$_{14}$N$_2$S
Methiuron
N,N-Dimethyl-N′-3-methylphenylthiourea

RN: 21540-35-2   **MP** (°C): 145
MW: 194.30   **BP** (°C):

Solubility (Moles/L)	Solubility (Grams/L)	Temp (°C)	Ref (#)	Evaluation (T P E A A)	Comments
2.059E-03	4.000E-01	ns	M061	0 0 0 0 2	

2207. C$_{10}$H$_{14}$N$_4$O$_2$
7-Propyl theophylline
3,7-Dimethyl-7-propyl-xanthine

RN: 27760-74-3   **MP** (°C):
MW: 222.25   **BP** (°C):

Solubility (Moles/L)	Solubility (Grams/L)	Temp (°C)	Ref (#)	Evaluation (T P E A A)	Comments
1.044E+00	2.320E+02	30	B042	1 2 1 1 2	
1.040E+00	2.311E+02	30	G021	1 0 0 0 2	

2208. C$_{10}$H$_{14}$N$_4$O$_2$
1-Propyl theobromine
3,7-Dimethyl-1-propyl-xanthine

RN: 204443-29-8   **MP** (°C): 99
MW: 222.25   **BP** (°C):

Solubility (Moles/L)	Solubility (Grams/L)	Temp (°C)	Ref (#)	Evaluation (T P E A A)	Comments
6.190E-02	1.376E+01	30	B042	1 2 1 1 2	

2209. C$_{10}$H$_{14}$N$_4$O$_3$
Ethoxycaffeine
1,3,7-Trimethyl-2,6-dioxo-8-ethoxypurine
RN: 577-66-2   **MP** (°C):   143
MW: 238.25   **BP** (°C):

Solubility (Moles/L)	Solubility (Grams/L)	Temp (°C)	Ref (#)	Evaluation (T P E A A)	Comments
1.255E-02	2.991E+00	19	A072	1 2 1 0 1	

2210. C$_{10}$H$_{14}$N$_4$O$_4$
Dyphylline
7-(2,3-Dihydroxypropyl)theophylline
Lufyllin-EPG
Neothylline
Airet
RN: 479-18-5   **MP** (°C):   158
MW: 254.25   **BP** (°C):

Solubility (Moles/L)	Solubility (Grams/L)	Temp (°C)	Ref (#)	Evaluation (T P E A A)	Comments
6.686E-01	1.700E+02	37	F076	2 0 2 2 1	

2211. C$_{10}$H$_{14}$N$_5$O$_7$P
2′-Adenylic acid
2′-Adenylsaeure
RN: 130-49-4   **MP** (°C):
MW: 347.23   **BP** (°C):

Solubility (Moles/L)	Solubility (Grams/L)	Temp (°C)	Ref (#)	Evaluation (T P E A A)	Comments
1.440E-03	5.000E-01	15	F300	1 0 0 0 0	

2212. C$_{10}$H$_{14}$N$_5$O$_7$P
3′-Adenylic acid
3′-Adenylsaeure
RN: 84-21-9   **MP** (°C):   197
MW: 347.23   **BP** (°C):

Solubility (Moles/L)	Solubility (Grams/L)	Temp (°C)	Ref (#)	Evaluation (T P E A A)	Comments
1.440E-03	5.000E-01	15	F300	1 0 0 0 0	

2213. $C_{10}H_{14}O$

L-Carvone
r-(–)-p-Mentha-6,8-dien-2-one
1-Methyl-4-isopropenyl-6-cyclohexen-2-one
p-Mentha-6,8-dien-2-one

RN: 6485-40-1   **MP (°C):**   <25
MW: 150.22   **BP (°C):**   230

Solubility (Moles/L)	Solubility (Grams/L)	Temp (°C)	Ref (#)	Evaluation (T P E A A)	Comments
8.654E-03	1.300E+00	18	F300	1 0 0 0 1	
8.654E-03	1.300E+00	25	A049	1 0 0 0 1	
1.020E-02	1.532E+00	25	A401	1 0 2 2 0	
1.100E-02	1.652E+00	25	D407	1 0 2 2 2	
1.100E-02	1.652E+00	37	E028	1 0 1 1 2	

2214. $C_{10}H_{14}O$

l-Perillaldehyde
4-Isopropenyl-1-cyclohexene-1-carboxaldehyde
para-Mentha-1,8-dien-7-al
L-4-(1-Methylethenyl)-1-cyclohexene-1-carboxaldehyde
L(–)-Perillaldehyde
(S)-4-(1-Methylethenyl)-1-cyclohexene-1-carboxaldehyde

RN: 18031-40-8   **MP (°C):**
MW: 150.22   **BP (°C):**

Solubility (Moles/L)	Solubility (Grams/L)	Temp (°C)	Ref (#)	Evaluation (T P E A A)	Comments
4.200E-03	6.309E-01	25	A401	1 0 2 2 0	

2215. $C_{10}H_{14}O$

p-n-Butylphenol
4-n-Butylphenol

RN: 1638-22-8   **MP (°C):**
MW: 150.22   **BP (°C):**

Solubility (Moles/L)	Solubility (Grams/L)	Temp (°C)	Ref (#)	Evaluation (T P E A A)	Comments
3.038E-03	4.563E-01	20	R087	0 0 0 0 0	0.15M NaCl
2.662E-03	3.998E-01	25	L022	1 0 0 0 0	

2216. $C_{10}H_{14}O$

o-n-Butylphenol
2-n-Butylphenol

RN: 28805-86-9   **MP (°C):**
MW: 150.22   **BP (°C):**

Solubility (Moles/L)	Solubility (Grams/L)	Temp (°C)	Ref (#)	Evaluation (T P E A A)	Comments
2.662E-03	3.998E-01	25	L022	1 0 0 0 0	

2217. C$_{10}$H$_{14}$O
p-tert-Butylphenol
4-*t*-Butylphenol

RN:	98-54-4	**MP** (°C):	99.5	
MW:	150.22	**BP** (°C):	237	

Solubility (Moles/L)	Solubility (Grams/L)	Temp (°C)	Ref (#)	Evaluation (T P E A A)	Comments
4.327E-03	6.500E-01	22.5	G301	0 0 0 0 0	
3.327E-03	4.998E-01	25	L021	1 0 0 0 0	
3.861E-03	5.800E-01	25	M127	1 0 0 0 1	
4.427E-03	6.650E-01	25	P004	0 0 0 0 0	
5.076E-03	7.625E-01	30	P004	0 0 0 0 0	
5.785E-03	8.690E-01	35	P004	0 0 0 0 0	
6.534E-03	9.815E-01	40	P004	0 0 0 0 0	
4.266E-03	6.408E-01	ns	R427	0 0 0 0 0	

2218. C$_{10}$H$_{14}$O
Thymol
6-Isopropyl-*m*-cresol
3-Hydroxy-*p*-cymene
5-Methyl-2-isopropyl-1-phenol
2-Isopropyl-5-methyl phenol
5-Methyl-2-(1-methylethyl)phenol

RN:	89-83-8	**MP** (°C):	48–51	
MW:	150.22	**BP** (°C):		

Solubility (Moles/L)	Solubility (Grams/L)	Temp (°C)	Ref (#)	Evaluation (T P E A A)	Comments
5.991E-03	9.000E-01	20	F300	1 0 0 0 0	
6.000E-03	9.013E-01	25	D407	1 0 2 2 2	
5.700E-03	8.563E-01	25	F044	1 0 0 0 1	
6.046E-03	9.083E-01	25	L021	1 0 0 0 0	
6.650E-03	9.990E-01	25	R041	0 0 0 0 0	
5.990E-02	8.998E+00	37	E028	1 0 1 1 2	*sic*
8.654E-03	1.300E+00	37	F300	1 0 0 0 1	
6.166E-03	9.263E-01	ns	R427	0 0 0 0 0	

2219. C$_{10}$H$_{14}$O
Carvacrol
2-Methyl-5-isopropylphenol

RN:	499-75-2	**MP** (°C):	3	
MW:	150.22	**BP** (°C):		

Solubility (Moles/L)	Solubility (Grams/L)	Temp (°C)	Ref (#)	Evaluation (T P E A A)	Comments
6.650E-03	9.990E-01	25	L021	1 0 0 0 0	
8.321E-03	1.250E+00	25	M127	1 0 0 0 2	

2220. C$_{10}$H$_{14}$O
4-*sec*-Butylphenol
p-sec-Butylphenol

RN: 99-71-8   **MP** (°C):
MW: 150.22   **BP** (°C):

Solubility (Moles/L)	Solubility (Grams/L)	Temp (°C)	Ref (#)	Evaluation (T P E A A)	Comments
6.391E-03	9.600E-01	25	M127	1 0 0 0 1	

2221. C$_{10}$H$_{14}$O$_2$
3-Butoxyphenol
m-Butoxy phenol
Phenol, 3-butoxy-

RN: 18979-72-1   **MP** (°C):
MW: 166.22   **BP** (°C):

Solubility (Moles/L)	Solubility (Grams/L)	Temp (°C)	Ref (#)	Evaluation (T P E A A)	Comments
8.240E-03	1.370E+00	30	B315	0 0 0 0 0	

2222. C$_{10}$H$_{14}$O$_2$
p-Diethoxybenzene
4-Diethoxybenzene

RN: 122-95-2   **MP** (°C):
MW: 166.22   **BP** (°C):

Solubility (Moles/L)	Solubility (Grams/L)	Temp (°C)	Ref (#)	Evaluation (T P E A A)	Comments
4.560E-04	7.580E-02	25	C316	0 0 0 0 0	0.1M NaCl

2223. C$_{10}$H$_{14}$O$_2$
o-Butoxyphenol
2-Butoxyphenol

RN: 39075-90-6   **MP** (°C):
MW: 166.22   **BP** (°C):

Solubility (Moles/L)	Solubility (Grams/L)	Temp (°C)	Ref (#)	Evaluation (T P E A A)	Comments
3.920E-03	6.516E-01	24.99	B353	0 0 0 0 0	

2224. $C_{10}H_{14}O_8$
1,1,2,2-Ethanetetrol, tetraacetate
Glyoxal-tetraacetat
Glyoxal tetraacetate

RN: 59602-16-3 **MP** (°C):
MW: 262.22 **BP** (°C):

Solubility (Moles/L)	Solubility (Grams/L)	Temp (°C)	Ref (#)	Evaluation (T P E A A)	Comments
3.051E-05	8.000E-03	25	F300	1 0 0 0 1	

2225. $C_{10}H_{15}N$
Diethylaniline
2,6-Diethylaniline

RN: 579-66-8 **MP** (°C): −38
MW: 149.24 **BP** (°C): 215

Solubility (Moles/L)	Solubility (Grams/L)	Temp (°C)	Ref (#)	Evaluation (T P E A A)	Comments
4.489E-03	6.700E-01	26.70	L095	2 2·1 1 2	
4.467E-03	6.666E-01	ns	S460	0 0 0 0 0	

2226. $C_{10}H_{15}NO$
Ethyl phenyl ethanolamine
2-(N-Ethylanilino)ethanol
N-Phenyl-N-ethylethanolamine

RN: 92-50-2 **MP** (°C):
MW: 165.24 **BP** (°C): 268

Solubility (Moles/L)	Solubility (Grams/L)	Temp (°C)	Ref (#)	Evaluation (T P E A A)	Comments
3.011E-02	4.975E+00	20	M062	1 0 0 0 1	

2227. $C_{10}H_{15}NO$
Ephedrine
L-Erythro-2-(methylamino)-1-phenylpropan-1-ol
(1R,2S)-(−)-Ephedrine
L-α-(1-Methylaminoethyl)benzyl alcohol

RN: 299-42-3 **MP** (°C): 38–39
MW: 165.24 **BP** (°C):

Solubility (Moles/L)	Solubility (Grams/L)	Temp (°C)	Ref (#)	Evaluation (T P E A A)	Comments
2.882E-01	4.762E+01	25	D004	0 0 0 0 0	
3.442E-01	5.688E+01	25	L338	1 0 1 1 2	
3.850E-01	6.362E+01	30	L069	1 0 1 1 0	EFG
1.160E+00	1.917E+02	ns	F007	0 0 0 0 2	

2228. C₁₀H₁₅NO

(+)-Pseudoephedrine
(+)-Pseudoephedrin

RN:	90-82-4	**MP** (°C):	118	
MW:	165.24	**BP** (°C):		

Solubility (Moles/L)	Solubility (Grams/L)	Temp (°C)	Ref (#)	Evaluation (T P E A A)	Comments
<3.03E-03	<5.00E-01	rt	B435	0 0 0 0 0	

2229. C₁₀H₁₅NO₂

N-Phenyldiethanolamine
Phenyl diethanolamine
N,N-di(Hydroxyethyl)aniline
2,2'-(Phenylimino)diethanol
PDEA

RN:	120-07-0	**MP** (°C):	57	
MW:	181.24	**BP** (°C):		

Solubility (Moles/L)	Solubility (Grams/L)	Temp (°C)	Ref (#)	Evaluation (T P E A A)	Comments
1.783E-01	3.232E+01	20	M062	1 0 0 0 2	

2230. C₁₀H₁₅N₅O₅

Arabinosyladenine
9-β-D-Arabino furanosyl adenine
Vidarabine
β-D-Arabinosyladenine
Spongoadenosine

RN:	24356-66-9	**MP** (°C):	208	
MW:	285.26	**BP** (°C):		

Solubility (Moles/L)	Solubility (Grams/L)	Temp (°C)	Ref (#)	Evaluation (T P E A A)	Comments
1.800E-03	5.135E-01	ns	R030	0 0 0 0 0	

2231. C₁₀H₁₅OPS₂

Fonofos
Ethyl S-phenyl ethylphosphonothiolthionate
Diphonate
Dyfonate®
Stauffer N-2790

RN:	944-22-9	**MP** (°C):		
MW:	246.33	**BP** (°C):		

Solubility (Moles/L)	Solubility (Grams/L)	Temp (°C)	Ref (#)	Evaluation (T P E A A)	Comments
6.373E-05	1.570E-02	20	B169	2 1 1 1 2	
6.089E-05	1.500E-02	ns	M110	0 0 0 0 0	EFG

(*continued*)

2231. C$_{10}$H$_{15}$OPS$_2$ (continued)

Solubility (Moles/L)	Solubility (Grams/L)	Temp (°C)	Ref (#)	Evaluation (T P E A A)	Comments
5.272E-05	1.299E-02	ns	S460	0 0 0 0 0	
6.374E-05	1.570E-02	ns	V414	0 0 0 0 0	

2232. C$_{10}$H$_{15}$O$_3$PS$_2$

Fenthion
4-Methylmercapto-3-methylphenyl dimethyl thiophosphate
Mercaptofos
Thiophos
Baycid
Entex

RN: 55-38-9   **MP (°C):**   7.5
MW: 278.33   **BP (°C):**

Solubility (Moles/L)	Solubility (Grams/L)	Temp (°C)	Ref (#)	Evaluation (T P E A A)	Comments
2.299E-05	6.400E-03	10	B324	0 0 0 0 0	
2.300E-05	6.402E-03	10	B324	0 0 0 0 0	
2.698E-05	7.509E-03	20	B300	2 1 1 1 2	
3.244E-05	9.029E-03	20	B324	0 0 0 0 0	
3.341E-05	9.300E-03	20	B324	0 0 0 0 0	
1.940E-04	5.400E-02	20	M061	1 0 0 0 1	
4.074E-05	1.134E-02	30	B324	0 0 0 0 0	
4.060E-05	1.130E-02	30	B324	0 0 0 0 0	
1.976E-04	5.500E-02	rt	M161	0 0 0 0 0	

2233. C$_{10}$H$_{16}$

Myrcene
7-Methyl-3-methylene-1,6-octadiene
7-Methyl-3-methylene-1,6-octadiene
7-Methyl-3-methyleneocta-1,6-diene
7-Methyl-3-methylene-octadiene
β-Myrcene

RN: 123-35-3   **MP (°C):**
MW: 136.24   **BP (°C):**   167

Solubility (Moles/L)	Solubility (Grams/L)	Temp (°C)	Ref (#)	Evaluation (T P E A A)	Comments
3.000E-05	4.087E-03	25	A401	1 0 2 2 0	
7.560E-05	1.030E-02	25	L450	0 0 0 0 0	

2234. $C_{10}H_{16}$
β-Pinene
(10)-Pinene
Bicyclo[3.1.1]heptane, 6,6-dimethyl-2-methylene-
Nopinene
Pseudopinene

RN: 127-91-3   **MP** (°C):   −61
MW: 136.24   **BP** (°C):   167

Solubility (Moles/L)	Solubility (Grams/L)	Temp (°C)	Ref (#)	Evaluation (T P E A A)	Comments
9.333E-05	1.272E-02	24.99	T424	0 0 0 0 0	
8.808E-05	1.200E-02	25	L450	0 0 0 0 0	

2235. $C_{10}H_{16}$
D-Limonene
D-1,8-p-Menthadiene
(R)-1-Methyl-4-(1-methylethenyl)cyclohexene
(R)-(+)-Limonene
Hemo-sol

RN: 5989-27-5   **MP** (°C):   95
MW: 136.24   **BP** (°C):   176

Solubility (Moles/L)	Solubility (Grams/L)	Temp (°C)	Ref (#)	Evaluation (T P E A A)	Comments
7.080E-01	9.646E+01	0	M124	2 1 2 2 1	
7.670E-01	1.045E+02	5	M124	2 1 2 2 2	
6.973E-05	9.500E-03	25	L450	0 0 0 0 0	
1.011E-04	1.377E-02	25	M124	2 1 2 2 1	

2236. $C_{10}H_{16}$
Limonene
p-Mentha-1,8-diene
Cyclil decene
Acintene DP dipentene

RN: 138-86-3   **MP** (°C):   73.97
MW: 136.24   **BP** (°C):   175.5

Solubility (Moles/L)	Solubility (Grams/L)	Temp (°C)	Ref (#)	Evaluation (T P E A A)	Comments
3.180E-05	4.332E-03	6	P430	0 0 0 0 0	
4.100E-05	5.586E-03	23.5	P430	0 0 0 0 0	
9.055E-05	1.234E-02	24.99	T424	0 0 0 0 0	
6.390E-05	8.706E-03	25	I019	1 0 1 2 2	
2.202E-04	3.000E-02	25	M350	1 0 1 1 1	

2237. C$_{10}$H$_{16}$

γ-Terpinene
1-Methyl-4-(1-methylethyl)-1,4-cyclohexadiene
1,4-p-Menthadiene
1-Isopropyl-4-methyl-1,4-cyclohexadiene
Moslene
Terpinene

RN: 99-85-4   **MP** (°C):
MW: 136.24   **BP** (°C):   182

Solubility (Moles/L)	Solubility (Grams/L)	Temp (°C)	Ref (#)	Evaluation (T P E A A)	Comments
4.470E-05	6.090E-03	6	P430	0 0 0 0 0	
6.370E-05	8.678E-03	23.5	P430	0 0 0 0 0	

2238. C$_{10}$H$_{16}$

Terpinolene
1-Methyl-4-(1-methylethylidene)cyclohexene
1,4(8)-p-Menthadiene
1-Methyl-4-(1-methylethylidene)cyclohexene
Terpinolene 30/35
Terpinolene 90

RN: 586-62-9   **MP** (°C):
MW: 136.24   **BP** (°C):

Solubility (Moles/L)	Solubility (Grams/L)	Temp (°C)	Ref (#)	Evaluation (T P E A A)	Comments
5.670E-05	7.725E-03	6	P430	0 0 0 0 0	
6.960E-05	9.482E-03	23.5	P430	0 0 0 0 0	
5.000E-05	6.812E-03	25	A401	1 0 2 2 0	

2239. C$_{10}$H$_{16}$

α-Pinene
2,6,6-Trimethylbicyclo[3.1.1]hept-2-ene
Acitene A
Cyclic dexadiene
pin-2(3)-ene
2-Pinene

RN: 80-56-8   **MP** (°C):   −64
MW: 136.24   **BP** (°C):   155

Solubility (Moles/L)	Solubility (Grams/L)	Temp (°C)	Ref (#)	Evaluation (T P E A A)	Comments
1.670E-05	2.275E-03	6	P430	0 0 0 0 0	
1.830E-05	2.493E-03	23.5	P430	0 0 0 0 0	
3.867E-05	5.268E-03	24.99	T424	0 0 0 0 0	
3.523E-05	4.800E-03	25	L450	0 0 0 0 0	

2240. C$_{10}$H$_{16}$Cl$_3$NOS
Triallate
S-(2,3,3-Trichloroallyl)diisopropylthiocarbamate

RN:	2303-17-5	MP (°C):	29
MW:	304.67	BP (°C):	

Solubility (Moles/L)	Solubility (Grams/L)	Temp (°C)	Ref (#)	Evaluation (T P E A A)	Comments
1.313E-05	4.000E-03	25	B200	1 0 0 1 0	
1.313E-05	4.000E-03	25	M161	1 0 0 0 0	
1.313E-05	4.000E-03	ns	F019	0 0 0 0 0	

2241. C$_{10}$H$_{16}$NO$_2$S$_2$
2-Cyclopentamethylene-4-methoxycarbamyl-1,3-dithiolane
2-Cyclopentyl-4-methoxycarbamyl-1,3-dithiolane

RN:		MP (°C):	
MW:	246.37	BP (°C):	

Solubility (Moles/L)	Solubility (Grams/L)	Temp (°C)	Ref (#)	Evaluation (T P E A A)	Comments
3.000E-04	7.391E-02	rt	B174	0 0 1 0 0	

2242. C$_{10}$H$_{16}$NO$_3$S
2-Cyclopentamethylene-4-methoxycarbamyl-1,3-oxathiolane

RN:		MP (°C):	
MW:	230.31	BP (°C):	

Solubility (Moles/L)	Solubility (Grams/L)	Temp (°C)	Ref (#)	Evaluation (T P E A A)	Comments
1.500E-03	3.455E-01	rt	B174	0 0 1 0 1	

2243. C$_{10}$H$_{16}$NO$_4$
2-Cyclopentamethylene-4-methoxycarbamyl-1,3-dioxolane

RN:		MP (°C):	
MW:	214.24	BP (°C):	

Solubility (Moles/L)	Solubility (Grams/L)	Temp (°C)	Ref (#)	Evaluation (T P E A A)	Comments
1.300E-02	2.785E+00	rt	B174	0 0 1 0 1	

2244. C$_{10}$H$_{16}$N$_2$O$_3$
5-Ethyl-5-(2-methylpropyl)barbituric acid

RN:	125-40-6	MP (°C):	174.5
MW:	212.25	BP (°C):	

Solubility (Moles/L)	Solubility (Grams/L)	Temp (°C)	Ref (#)	Evaluation (T P E A A)	Comments
3.997E-03	8.483E-01	25	B065	1 2 1 1 1	

2245. $C_{10}H_{16}N_2O_3$
5,5-Dipropylbarbituric acid
5,5-Dipropylbarbitursaeure
Proponal
5,5-Dipropylbarbiturate

RN: 2217-08-5   **MP** (°C): 146
MW: 212.25   **BP** (°C):

Solubility (Moles/L)	Solubility (Grams/L)	Temp (°C)	Ref (#)	Evaluation (T P E A A)	Comments
2.827E-03	6.000E-01	20	F300	1 0 0 0 0	
2.968E-03	6.300E-01	20	J030	1 2 2 2 1	
5.088E-03	1.080E+00	37	J030	1 2 2 2 2	
6.926E-02	1.470E+01	100	F300	1 0 0 0 2	

2246. $C_{10}H_{16}N_2O_3$
5,5-Diisopropylbarbituric acid
Barbituric acid, 5,5-diisopropyl
2,4,6(1H,3H,5H)-Pyrimidinetrione, 5,5-bis(1-methylethyl)
5,5-Di-i-propylbarbiturate

RN: 99167-69-8   **MP** (°C):
MW: 212.25   **BP** (°C):

Solubility (Moles/L)	Solubility (Grams/L)	Temp (°C)	Ref (#)	Evaluation (T P E A A)	Comments
1.715E-03	3.640E-01	25	P350	0 0 0 0 0	intrinsic

2247. $C_{10}H_{16}N_2O_3$
Butabarbital
Butethal
5-Ethyl-5-*n*-butylbarbituric acid
5-Butyl-5-ethylbarbituric acid

RN: 77-28-1   **MP** (°C): 127
MW: 212.25   **BP** (°C):

Solubility (Moles/L)	Solubility (Grams/L)	Temp (°C)	Ref (#)	Evaluation (T P E A A)	Comments
1.602E-02	3.400E+00	0	D089	0 0 0 0 2	form I
1.484E-02	3.150E+00	20	J030	1 2 2 2 2	
1.044E-02	2.215E+00	20	K078	1 0 2 1 2	
4.052E-03	8.600E-01	25	B011	2 0 0 1 0	
4.218E-03	8.954E-01	25	B065	1 1 1 1 1	
1.936E-02	4.110E+00	25	B065	1 1 1 1 1	
8.000E-03	1.698E+00	25	G003	1 1 1 1 1	pH 4.7
2.300E-02	4.882E+00	25	M310	2 2 2 2 2	
2.130E-02	4.521E+00	25	V033	2 0 1 1 2	
4.070E-03	8.639E-01	25	V033	2 0 1 1 2	
2.130E-02	4.521E+00	25.00	T303	1 0 0 0 2	
7.400E-03	1.571E+00	25.00	T303	1 0 0 0 1	

(continued)

2247. C$_{10}$H$_{16}$N$_2$O$_3$ (continued)

Solubility (Moles/L)	Solubility (Grams/L)	Temp (°C)	Ref (#)	Evaluation (T P E A A)	Comments
1.950E-02	4.139E+00	30	I001	2 0 2 1 0	EFG, 0.003N H$_2$SO$_4$
9.900E-03	2.101E+00	35.00	T303	1 0 0 0 1	
2.430E-02	5.158E+00	35.00	T303	1 0 0 0 2	
2.299E-02	4.880E+00	37	J030	1 2 2 2 2	
3.090E-02	6.559E+00	45.00	T303	1 0 0 0 2	
1.370E-02	2.908E+00	45.00	T303	1 0 0 0 2	
1.743E-02	3.700E+00	amb	D092	0 2 2 1 2	form II
1.602E-02	3.400E+00	amb	D092	0 2 2 1 2	0.1N HCl, form III, mp 124 C
1.743E-02	3.700E+00	amb	D092	0 2 2 1 2	form I
9.362E-03	1.987E+00	ns	T003	0 0 0 0 2	
8.952E-03	1.900E+00	ns	T003	0 0 0 0 2	

2248. C$_{10}$H$_{16}$N$_2$O$_3$S
Biotin d
D-Biotin
Biotin

RN: 58-85-5 **MP** (°C): 232
MW: 244.31 **BP** (°C):

Solubility (Moles/L)	Solubility (Grams/L)	Temp (°C)	Ref (#)	Evaluation (T P E A A)	Comments
9.003E-04	2.200E-01	25	D041	1 0 0 0 1	
1.433E-03	3.500E-01	25	D315	0 0 0 0 0	
8.186E-04	2.000E-01	25	M054	1 0 0 0 0	

2249. C$_{10}$H$_{16}$N$_2$O$_4$
Methyl-2,2-diallylmalonurate
Methyl 2,2-diallylmalonurate

RN: 73632-82-3 **MP** (°C): 84
MW: 228.25 **BP** (°C):

Solubility (Moles/L)	Solubility (Grams/L)	Temp (°C)	Ref (#)	Evaluation (T P E A A)	Comments
6.800E-03	1.552E+00	23	B152	1 2 1 1 1	pH 3.5

2250. C$_{10}$H$_{16}$N$_4$O$_2$
7-Butyl theophylline
1H-Purine-2,6-dione, 7-butyl-3,7-dihydro-1,3-dimethyl-
7-Butyl-1,3-dimethylxanthine

RN: 1021-65-4 **MP** (°C):
MW: 224.26 **BP** (°C):

Solubility (Moles/L)	Solubility (Grams/L)	Temp (°C)	Ref (#)	Evaluation (T P E A A)	Comments
1.560E-02	3.499E+00	30	B042	1 2 1 1 2	
1.560E-02	3.499E+00	30	G021	1 0 0 0 2	

2251. C$_{10}$H$_{16}$N$_4$O$_2$S

3-(5-*tert*-Butyl-1,3,4-thiadiazol-2-yl)-4-hydroxy-1
2-Imidazolidinone, 3-[5-(1,1-dimethylethyl)-1,3,4-thiadiazol-2-yl]-4-hydroxy-1-methyl-
Buthidazole
Ravage
VEL 5026

RN:	55511-98-3	MP (°C):	133.5
MW:	256.33	BP (°C):	

Solubility (Moles/L)	Solubility (Grams/L)	Temp (°C)	Ref (#)	Evaluation (T P E A A)	Comments
1.322E-02	3.388E+00	25	M161	1 0 0 0 1	

2252. C$_{10}$H$_{16}$N$_6$S

Cimetidine
2-Cyano-1-methyl-3-(2-(((5-methylimidazol-4-yl)methyl)thio)ethyl)guanidine
N''-Cyano-*N*-methyl-*N'*-(2-(((5-methyl-1H-imidazol-4-yl)methyl)thio)-ethyl)guanidine
N''-Cyano-*N*-methyl-*N'*-(2-(((5-methyl-1H-imidazol-4-yl)methyl)thio)-ethyl)guanidine
Sigmetadine
Peptol

RN:	51481-61-9	MP (°C):	142
MW:	252.34	BP (°C):	

Solubility (Moles/L)	Solubility (Grams/L)	Temp (°C)	Ref (#)	Evaluation (T P E A A)	Comments
2.382E-02	6.010E+00	22.5	B422	2 0 2 2 2	
3.685E-02	9.300E+00	25	A412	1 0 2 2 1	int
3.963E-03	1.000E+00	ns	K444	0 0 0 0 0	

2253. C$_{10}$H$_{16}$O

D-Fenchone
D-1,3,3-Trimethyl-2-norbornanone
Bicyclo[2.2.1]heptan-2-one, 1,3,3-trimethyl-, (1S)-
α-Fenchone
(+)-Fenchone

RN:	4695-62-9	MP (°C):	6.1
MW:	152.24	BP (°C):	193.5

Solubility (Moles/L)	Solubility (Grams/L)	Temp (°C)	Ref (#)	Evaluation (T P E A A)	Comments
1.311E-02	1.996E+00	20	D052	1 1 0 0 0	
1.410E-02	2.147E+00	25	I019	1 0 1 2 2	
1.413E-02	2.150E+00	ns	S460	0 0 0 0 0	

2254. $C_{10}H_{16}O$
D-Camphor
D-Campher
Camphor

RN:	76-22-2	MP (°C):	179.7	
MW:	152.24	BP (°C):		

Solubility (Moles/L)	Solubility (Grams/L)	Temp (°C)	Ref (#)	Evaluation (T P E A A)	Comments
1.095E-02	1.667E+00	15.50	L073	1 2 2 1 2	
6.569E-03	1.000E+00	20	F300	1 0 0 0 0	
1.363E-02	2.076E+00	20	K078	1 0 2 1 2	
1.030E-02	1.568E+00	25	I019	1 0 1 2 2	
1.340E-02	2.040E+00	25	L338	1 0 1 1 2	
1.630E-02	2.481E+00	37	E028	1 0 1 1 2	
1.115E-02	1.697E+00	ns	F014	0 0 0 0 2	
1.023E-02	1.558E+00	ns	R427	0 0 0 0 0	

2255. $C_{10}H_{16}O$
Carvotan acetone
Carvotan-aceton

RN:	499-71-8	MP (°C):		
MW:	152.24	BP (°C):		

Solubility (Moles/L)	Solubility (Grams/L)	Temp (°C)	Ref (#)	Evaluation (T P E A A)	Comments
5.912E-03	9.000E-01	20	F300	1 0 0 0 0	

2256. $C_{10}H_{16}O$
Citral
trans-3,7-dimethyl-2,6-octadienal
Geranialdehyde
Neral
Geranial
Citral A

RN:	5392-40-5	MP (°C):	<10	
MW:	152.24	BP (°C):	92.5	

Solubility (Moles/L)	Solubility (Grams/L)	Temp (°C)	Ref (#)	Evaluation (T P E A A)	Comments
3.800E-03	5.785E-01	25	A401	1 0 2 2 0	
1.583E-03	2.410E-01	25	L450	0 0 0 0 0	
1.970E-03	2.999E-01	25	M350	1 0 1 1 1	
8.800E-03	1.340E+00	37	E028	1 0 1 1 1	
8.710E-03	1.326E+00	ns	S460	0 0 0 0 0	

2257. $C_{10}H_{16}O$
L-Dihydrocarvone
L-Dihydro-carvon

RN:	619-02-3	MP (°C):	
MW:	152.24	BP (°C):	221

Solubility (Moles/L)	Solubility (Grams/L)	Temp (°C)	Ref (#)	Evaluation (T P E A A)	Comments
6.569E-03	1.000E+00	20	F300	1 0 0 0 0	

2258. $C_{10}H_{16}O$
Neral

RN:	106-26-3	MP (°C):	
MW:	152.24	BP (°C):	

Solubility (Moles/L)	Solubility (Grams/L)	Temp (°C)	Ref (#)	Evaluation (T P E A A)	Comments
1.898E-03	2.890E-01	25	L450	0 0 0 0 0	

2259. $C_{10}H_{16}O_2$
3-Hydroxy-3-ethynyl-2,2,5,5-tetramethyltetrahydrofuran
3-Furanol, 3-ethynyltetrahydro-2,2,5,5-tetramethyl-

RN:	24270-82-4	MP (°C):	
MW:	168.24	BP (°C):	

Solubility (Moles/L)	Solubility (Grams/L)	Temp (°C)	Ref (#)	Evaluation (T P E A A)	Comments
1.165E-01	1.961E+01	rt	B066	0 2 0 0 0	

2260. $C_{10}H_{16}O_3$
cis-Pinonic acid
cis-3-Acetyl-2,2-dimethylcyclobutaneacetic acid

RN:	473-72-3	MP (°C):	104–107
MW:	184.24	BP (°C):	

Solubility (Moles/L)	Solubility (Grams/L)	Temp (°C)	Ref (#)	Evaluation (T P E A A)	Comments
2.001E-02	3.686E+00	0	H430	0 0 0 0 0	
3.612E-02	6.655E+00	rt	H431	0 0 0 0 0	average

2261. $C_{10}H_{16}O_4$
L-Isocamphoric acid
L-Isocamphersaeure

RN:	5394-83-2	MP (°C):	173
MW:	200.24	BP (°C):	

Solubility (Moles/L)	Solubility (Grams/L)	Temp (°C)	Ref (#)	Evaluation (T P E A A)	Comments
1.698E-02	3.400E+00	20	F300	1 0 0 0 1	

2262. $C_{10}H_{16}O_4$
D-Camphoric acid
D-Camphersaeure
RN: 124-83-4   **MP** (°C):
MW: 200.24   **BP** (°C):

Solubility (Moles/L)	Solubility (Grams/L)	Temp (°C)	Ref (#)	Evaluation (T P E A A)	Comments
3.796E-02	7.600E+00	25	F300	1 0 0 0 1	

2263. $C_{10}H_{16}O_5$
DL-Cineolic acid
DL-Cineolsaeure
RN: 473-18-7   **MP** (°C):   208
MW: 216.24   **BP** (°C):

Solubility (Moles/L)	Solubility (Grams/L)	Temp (°C)	Ref (#)	Evaluation (T P E A A)	Comments
6.474E-02	1.400E+01	15	F300	1 0 0 0 1	
3.006E-01	6.500E+01	100	F300	1 0 0 0 1	

2264. $C_{10}H_{17}Cl_2NOS$
Diallate
DATC
S-(2,3-Dichloroallyl-N,N-diisopropylthiocarbamate
RN: 2303-16-4   **MP** (°C):   −10
MW: 270.22   **BP** (°C):

Solubility (Moles/L)	Solubility (Grams/L)	Temp (°C)	Ref (#)	Evaluation (T P E A A)	Comments
1.480E-04	4.000E-02	25	B185	0 0 0 0 0	
5.181E-05	1.400E-02	25	B200	1 0 0 1 1	
1.480E-04	4.000E-02	25	M061	1 0 0 0 1	
5.181E-05	1.400E-02	25	M161	1 0 0 0 1	
1.480E-04	4.000E-02	ns	F019	0 0 0 0 1	
1.480E-04	4.000E-02	rt	I314	0 0 0 0 0	

2265. $C_{10}H_{17}NO_2$
Methyprylon
Dimerin
3,3-Diethyl-5-methyl-2,4-piperidinedione
RN: 125-64-4   **MP** (°C):
MW: 183.25   **BP** (°C):

Solubility (Moles/L)	Solubility (Grams/L)	Temp (°C)	Ref (#)	Evaluation (T P E A A)	Comments
4.147E-01	7.600E+01	25	R027	0 0 0 0 0	

2266. C$_{10}$H$_{17}$N$_2$O$_4$PS
Etrimfos
Dimethyl *O*-(2-ethyl-4-ethoxy-pyrimidin-6-yl)thionophosphate
Ekamet G
Ekamet ULV
Etrimphos

RN: 38260-54-7   **MP** (°C):
MW: 292.30   **BP** (°C):

Solubility (Moles/L)	Solubility (Grams/L)	Temp (°C)	Ref (#)	Evaluation (T P E A A)	Comments
3.421E-02	1.000E+01	20	M161	1 0 0 0 1	
1.368E-04	3.998E-02	ns	S460	0 0 0 0 0	

2267. C$_{10}$H$_{17}$N$_3$O$_5$
Orotic acid choline

RN: **MP** (°C):   102–104
MW: 259.26   **BP** (°C):

Solubility (Moles/L)	Solubility (Grams/L)	Temp (°C)	Ref (#)	Evaluation (T P E A A)	Comments
2.697E+00	6.992E+02	25	N018	0 0 0 0 0	

2268. C$_{10}$H$_{17}$N$_3$O$_6$S
Glutathione
Glutathion

RN: 70-18-8   **MP** (°C):   193.4
MW: 307.33   **BP** (°C):

Solubility (Moles/L)	Solubility (Grams/L)	Temp (°C)	Ref (#)	Evaluation (T P E A A)	Comments
2.958E-01	9.090E+01	0	F300	1 0 0 0 2	

2269. C$_{10}$H$_{17}$O$_3$P
Diethyl phenyl phosphonate
Diethyl benzenephosphonate
Diethyl phenylphosphonate

RN: 1754-49-0   **MP** (°C):
MW: 216.22   **BP** (°C):   110

Solubility (Moles/L)	Solubility (Grams/L)	Temp (°C)	Ref (#)	Evaluation (T P E A A)	Comments
<9.25E-04	<2.00E-01	25	B070	1 2 0 1 0	

2270. $C_{10}H_{18}$
2,2,5,5-Tetramethyl-3-hexyne
Di-*tert*-butylacetylene
Di-*tert*-butylethyne

| RN: | 17530-24-4 | MP (°C): | |
| MW: | 138.25 | BP (°C): | |

Solubility (Moles/L)	Solubility (Grams/L)	Temp (°C)	Ref (#)	Evaluation (T P E A A)	Comments
1.470E-04	2.032E-02	25	H039	1 2 2 2 2	
7.700E-05	1.065E-02	35	H039	1 2 2 2 1	

2271. $C_{10}H_{18}$
Pinane
2,6,6-Trimethylbicyclo[3.1.1]heptane
2,7,7-Trimethylbicyclo[3.1.1]heptane
Dihydropinene

| RN: | 473-55-2 | MP (°C): | |
| MW: | 138.25 | BP (°C): | 151 |

Solubility (Moles/L)	Solubility (Grams/L)	Temp (°C)	Ref (#)	Evaluation (T P E A A)	Comments
1.140E-05	1.576E-03	ns	S460	0 0 0 0 0	

2272. $C_{10}H_{18}$
Decalin
Decahydronaphthalene

| RN: | 91-17-8 | MP (°C): | −31 |
| MW: | 138.25 | BP (°C): | |

Solubility (Moles/L)	Solubility (Grams/L)	Temp (°C)	Ref (#)	Evaluation (T P E A A)	Comments
<1.45E-03	<2.00E-01	25	B019	1 0 1 2 0	
6.430E-06	8.890E-04	25	P051	2 1 1 2 2	
6.148E-06	8.500E-04	25	T423	0 0 0 0 0	
6.430E-06	8.890E-04	25.00	P007	2 1 2 2 2	
4.492E-05	6.210E-03	ns	H123	0 0 0 0 0	

2273. $C_{10}H_{18}$
cis-Decalin
cis-Decahydronaphthalene
cis-Bicyclo[4.4.0]decane

| RN: | 493-01-6 | MP (°C): | −43.2 |
| MW: | 138.25 | BP (°C): | 195.7 |

Solubility (Moles/L)	Solubility (Grams/L)	Temp (°C)	Ref (#)	Evaluation (T P E A A)	Comments
6.452E-02	8.920E+00	300	S355	1 1 1 2 0	EFG

2274. C$_{10}$H$_{18}$ClN$_5$

Ipazine
2-Chloro-4-diethylamino-6-isopropylamino-s-triazine
2-Chloro-4-isopropylamino-6-biethylamino-s-triazines

| RN: | 1912-25-0 | MP (°C): |
| MW: | 243.74 | BP (°C): |

Solubility (Moles/L)	Solubility (Grams/L)	Temp (°C)	Ref (#)	Evaluation (T P E A A)	Comments
1.641E-04	4.000E-02	21	B192	0 0 0 0 1	
1.641E-04	4.000E-02	21	G099	2 0 0 1 0	
1.641E-04	4.000E-02	ns	B185	0 0 0 0 0	

2275. C$_{10}$H$_{18}$N$_2$O$_4$

Ethyl-2,2-diethylmalonurate
Ethyl 2,2-diethylmalnurate

| RN: | 73632-76-5 | MP (°C): | 84.5 |
| MW: | 230.27 | BP (°C): |

Solubility (Moles/L)	Solubility (Grams/L)	Temp (°C)	Ref (#)	Evaluation (T P E A A)	Comments
8.400E-03	1.934E+00	23	B152	1 2 1 1 1	pH 3.5

2276. C$_{10}$H$_{18}$N$_2$O$_5$

Methoxymethyl-2,2-diethylmalonurate
Methoxymethyl 2,2-diethylmalonurate

| RN: | 73632-79-8 | MP (°C): | 113 |
| MW: | 246.27 | BP (°C): |

Solubility (Moles/L)	Solubility (Grams/L)	Temp (°C)	Ref (#)	Evaluation (T P E A A)	Comments
6.800E-03	1.675E+00	23	B152	1 2 1 1 1	pH 3.5

2277. C$_{10}$H$_{18}$N$_6$O$_2$

1-(Sarcosino)-3,5-bis(dimethylamino)-s-triazine
N2-Carboxymethyl-N2,N4,N4,N6,N6-pentamethylmelamine

| RN: | 64124-17-0 | MP (°C): |
| MW: | 254.29 | BP (°C): |

Solubility (Moles/L)	Solubility (Grams/L)	Temp (°C)	Ref (#)	Evaluation (T P E A A)	Comments
7.360E-02	1.872E+01	25	B386	0 0 0 0 0	

2278. C₁₀H₁₈O

Borneol
endo-1,7,7-Trimethyl-bicyclo[2.2.1]heptan-2-ol
L-Borneol

RN:	507-70-0	MP (°C):	206
MW:	154.25	BP (°C):	210

Solubility (Moles/L)	Solubility (Grams/L)	Temp (°C)	Ref (#)	Evaluation (T P E A A)	Comments
4.512E-03	6.960E-01	15	M073	1 0 2 2 2	
4.784E-03	7.380E-01	25	M073	1 0 2 2 2	
4.786E-03	7.383E-01	ns	R427	0 0 0 0 0	

2279. C₁₀H₁₈O

D-Borneol
Borneocamphor
Sumatra camphor
endo-2-Bornanol

RN:	464-43-7	MP (°C):	208
MW:	154.25	BP (°C):	212

Solubility (Moles/L)	Solubility (Grams/L)	Temp (°C)	Ref (#)	Evaluation (T P E A A)	Comments
4.797E-03	7.400E-01	25	F300	1 0 0 0 1	

2280. C₁₀H₁₈O

L-Menthone
trans-p-Menthan-3-one
p-Menthan-3-one
(−)-5-Methyl-2-(1-methylethyl)cyclohexanone
(−)-Menthone

RN:	14073-97-3	MP (°C):	−6
MW:	154.25	BP (°C):	207

Solubility (Moles/L)	Solubility (Grams/L)	Temp (°C)	Ref (#)	Evaluation (T P E A A)	Comments
3.220E-03	4.967E-01	25	I019	1 0 1 2 2	

2281. C$_{10}$H$_{18}$O

Linalool
3,7-Dimethylocta-1,6-dien-3-ol
2,6-Dimethylocta-2,7-dien-6-ol
Linalol
3,7-Dimethyl-1,6-octadien-3-ol

| **RN:** | 78-70-6 | **MP** (°C): | <25 |
| **MW:** | 154.25 | **BP** (°C): | 195.5 |

Solubility (Moles/L)	Solubility (Grams/L)	Temp (°C)	Ref (#)	Evaluation (T P E A A)	Comments
3.570E-03	5.507E-01	6	P430	0 0 0 0 0	
5.530E-03	8.530E-01	23.5	P430	0 0 0 0 0	
1.200E-02	1.851E+00	25	D407	1 0 2 2 2	
1.030E-02	1.589E+00	25	I019	1 0 1 2 2	
9.710E-03	1.498E+00	25	M350	1 0 1 1 1	
3.800E-02	5.862E+00	37	E028	1 0 1 1 2	

2282. C$_{10}$H$_{18}$O

Citronellal
D-Citronellal
(R)-(+)-citronellal
3,7-Dimethyl-6-octen-1-al
3,7-Dimethyl-6-octenal
Rhodinal

| **RN:** | 106-23-0 | **MP** (°C): | |
| **MW:** | 154.25 | **BP** (°C): | |

Solubility (Moles/L)	Solubility (Grams/L)	Temp (°C)	Ref (#)	Evaluation (T P E A A)	Comments
9.000E-04	1.388E-01	25	A401	1 0 2 2 0	

2283. C$_{10}$H$_{18}$O

α-Terpineol
1-p-Menthen-8-ol
1-Methyl-4-isopropyl-1-cyclohexen-8-ol
2-(4-Methyl-3-cyclohexenyl)-2-propanol
p-Menth-1-en-8-ol
α,α,4-Trimethyl-3-cyclohexene-1-methanol

| **RN:** | 98-55-5 | **MP** (°C): | 34.5 |
| **MW:** | 154.25 | **BP** (°C): | 218 |

Solubility (Moles/L)	Solubility (Grams/L)	Temp (°C)	Ref (#)	Evaluation (T P E A A)	Comments
2.202E-03	3.397E-01	6	P430	0 0 0 0 0	
4.600E-03	7.096E-01	23.5	P430	0 0 0 0 0	
1.620E-02	2.499E+00	25	A401	1 0 2 2 0	

2284. C₁₀H₁₈O

Nerol
Allerol
cis-3,7-Dimethyl-2,6-octadien-1-ol
Neraniol
Nerosol
Vernol

RN: 106-25-2   **MP** (°C):
MW: 154.25   **BP** (°C):

Solubility (Moles/L)	Solubility (Grams/L)	Temp (°C)	Ref (#)	Evaluation (T P E A A)	Comments
8.500E-03	1.311E+00	25	A401	1 0 2 2 0	

2285. C₁₀H₁₈O

Geraniol
2,6-Dimethyl-2,6-octadien-8-ol
2,6-Dimethyl-*trans*-2,6-octadien-8-ol
2-*trans*-3,7-Dimethyl-2,6-octadiene-1-ol
3,7-Dimethyl-*trans*-2,6-octadien-1-ol
(E)-3,7-Dimethyl-2,6-octadien-1-ol

RN: 106-24-1   **MP** (°C): 15
MW: 154.25   **BP** (°C): 229

Solubility (Moles/L)	Solubility (Grams/L)	Temp (°C)	Ref (#)	Evaluation (T P E A A)	Comments
5.000E-03	7.713E-01	25	A401	1 0 2 2 0	

2286. C₁₀H₁₈O

Menthone
5-Methyl-2-(1-methylethyl)cyclohexanone
DL-Menthone

RN: 10458-14-7   **MP** (°C): –6
MW: 154.25   **BP** (°C): 207

Solubility (Moles/L)	Solubility (Grams/L)	Temp (°C)	Ref (#)	Evaluation (T P E A A)	Comments
2.000E-03	3.085E-01	25	A401	1 0 2 2 0	

2287. C₁₀H₁₈O

Plinol

RN: 72402-00-7   **MP** (°C):
MW: 154.25   **BP** (°C):

Solubility (Moles/L)	Solubility (Grams/L)	Temp (°C)	Ref (#)	Evaluation (T P E A A)	Comments
5.281E-03	8.146E-01	6	P430	0 0 0 0 0	
9.610E-03	1.482E+00	23.5	P430	0 0 0 0 0	

2288. C$_{10}$H$_{18}$O
1,8-Cineole
Eucalyptol
Cineol
Cineole

RN:	470-82-6	MP (°C):	36.5
MW:	154.25	BP (°C):	

Solubility (Moles/L)	Solubility (Grams/L)	Temp (°C)	Ref (#)	Evaluation (T P E A A)	Comments
4.123E-02	6.359E+00	1.5	E036	1 0 1 1 1	
4.187E-02	6.458E+00	4.0	B352	0 0 0 0 0	
3.674E-02	5.668E+00	7.5	E036	1 0 1 1 1	
3.482E-02	5.371E+00	10	E036	1 0 1 1 1	
3.610E-02	5.569E+00	10.0	B352	0 0 0 0 0	
1.297E-02	2.000E+00	15	F300	1 0 0 0 1	
3.097E-02	4.777E+00	15.0	B352	0 0 0 0 0	
2.261E-02	3.488E+00	21	E036	1 0 1 1 1	
2.454E-02	3.786E+00	21.0	B352	0 0 0 0 0	
2.010E-02	3.100E+00	25	A049	1 0 0 0 1	
2.197E-02	3.388E+00	25	B423	1 1 1 2 1	
1.746E-02	2.693E+00	30.0	B352	0 0 0 0 0	
1.552E-02	2.394E+00	35.0	B352	0 0 0 0 0	
9.100E-03	1.404E+00	37	E028	1 0 1 1 1	
1.359E-02	2.096E+00	40	E036	1 0 1 1 1	
1.423E-02	2.195E+00	40.0	B352	0 0 0 0 0	
1.294E-02	1.996E+00	45.0	B352	0 0 0 0 0	
1.229E-02	1.896E+00	50	E036	1 0 1 1 1	
1.100E-02	1.697E+00	50.0	B352	0 0 0 0 0	

2289. C$_{10}$H$_{18}$O$_2$
2,4-Decadione
Acetylmethyl hexyl ketone

RN:	13329-78-7	MP (°C):	
MW:	170.25	BP (°C):	

Solubility (Moles/L)	Solubility (Grams/L)	Temp (°C)	Ref (#)	Evaluation (T P E A A)	Comments
2.600E-03	4.427E-01	25	M078	2 0 1 0 1	

2290. C$_{10}$H$_{18}$O$_2$
3-Pentyl-2,4-pentadione
3-Amyl-2,4-pentanedione

RN:	27970-50-9	MP (°C):	
MW:	170.25	BP (°C):	

Solubility (Moles/L)	Solubility (Grams/L)	Temp (°C)	Ref (#)	Evaluation (T P E A A)	Comments
1.410E-02	2.401E+00	25	M078	2 0 1 0 2	

2291. $C_{10}H_{18}O_2$
Sobrerol
Pinolhydrat

RN:	498-71-5	**MP** (°C):	130	
MW:	170.25	**BP** (°C):	170	

Solubility (Moles/L)	Solubility (Grams/L)	Temp (°C)	Ref (#)	Evaluation (T P E A A)	Comments
1.880E-01	3.200E+01	15	F300	1 0 0 0 1	
1.938E-01	3.300E+01	ns	L335	0 0 0 0 2	

2292. $C_{10}H_{18}O_2$
D-Campholic acid
D-Campholsaeure

RN:	464-88-0	**MP** (°C):		
MW:	170.25	**BP** (°C):		

Solubility (Moles/L)	Solubility (Grams/L)	Temp (°C)	Ref (#)	Evaluation (T P E A A)	Comments
9.398E-04	1.600E-01	19	F300	1 0 0 0 1	

2293. $C_{10}H_{18}O_3$
2,2,5,5-Tetramethyl-tetrahydro-3-hydroxy-3-furanyl methyl ketone
Ketone, methyl tetrahydro-3-hydroxy-2,2,5,5-tetramethyl-3-furyl

RN:	24282-51-7	**MP** (°C):		
MW:	186.25	**BP** (°C):		

Solubility (Moles/L)	Solubility (Grams/L)	Temp (°C)	Ref (#)	Evaluation (T P E A A)	Comments
1.053E-01	1.961E+01	rt	B066	0 2 0 0 0	

2294. $C_{10}H_{18}O_4$
Sebacic acid
Sebacinsaeure

RN:	111-20-6	**MP** (°C):	134.5	
MW:	202.25	**BP** (°C):	294.5	

Solubility (Moles/L)	Solubility (Grams/L)	Temp (°C)	Ref (#)	Evaluation (T P E A A)	Comments
1.978E-04	4.000E-02	0	F300	1 0 0 0 0	
1.978E-04	4.000E-02	0	L041	1 0 0 1 0	
4.944E-03	1.000E+00	20	F300	1 0 0 0 1	
4.944E-03	1.000E+00	20	L041	1 0 0 1 1	
9.889E-03	2.000E+00	21	B040	1 0 1 1 1	*sic*
7.911E-03	1.600E+00	35	L041	1 0 0 1 1	
1.088E-02	2.200E+00	50	L041	1 0 0 1 1	
2.077E-02	4.200E+00	65	F300	1 0 0 0 1	
2.077E-02	4.200E+00	65	L041	1 0 0 1 1	
8.898E-04	1.800E-01	ns	F014	0 0 0 0 1	

2295. C$_{10}$H$_{18}$O$_4$
Amyl α-acetoxypropionate
Hydracrylic acid, pentyl ester, acetate
RN: 20473-77-2   **MP** (°C):
MW: 202.25   **BP** (°C):

Solubility (Moles/L)	Solubility (Grams/L)	Temp (°C)	Ref (#)	Evaluation (T P E A A)	Comments
3.461E-03	7.000E-01	25	R006	2 2 0 1 1	

2296. C$_{10}$H$_{18}$O$_4$
Ethylene glycol dibutyrate
Ethylene glycol di-*N*-butyrate
RN: 105-72-6   **MP** (°C):
MW: 202.25   **BP** (°C):

Solubility (Moles/L)	Solubility (Grams/L)	Temp (°C)	Ref (#)	Evaluation (T P E A A)	Comments
8.220E-03	1.663E+00	25	F064	1 0 0 0 2	
2.471E-03	4.998E-01	ns	F014	0 0 0 0 1	

2297. C$_{10}$H$_{18}$O$_4$
Diethoxyethyl adipate
Diethyl adipate
RN: 141-28-6   **MP** (°C):   −18
MW: 202.25   **BP** (°C):   251

Solubility (Moles/L)	Solubility (Grams/L)	Temp (°C)	Ref (#)	Evaluation (T P E A A)	Comments
2.965E-03	5.996E-01	ns	F014	0 0 0 0 1	
1.223E-02	2.474E+00	ns	F014	0 0 0 0 2	

2298. C$_{10}$H$_{18}$O$_4$
Dimethyl cyclohexyl oxalate
RN: **MP** (°C):
MW: 202.25   **BP** (°C):

Solubility (Moles/L)	Solubility (Grams/L)	Temp (°C)	Ref (#)	Evaluation (T P E A A)	Comments
<9.89E-06	<2.00E-03	15	H069	1 0 1 1 0	

2299. C$_{10}$H$_{18}$O$_5$
Diethylene glycol dipropionate
Ethanol, 2,2′-oxybis-, dipropanoate
RN: 6942-59-2   **MP** (°C):
MW: 218.25   **BP** (°C):

Solubility (Moles/L)	Solubility (Grams/L)	Temp (°C)	Ref (#)	Evaluation (T P E A A)	Comments
1.592E-01	3.475E+01	ns	F014	0 0 0 0 2	

2300. $C_{10}H_{18}O_5$
Propanoic acid, 2-[(ethoxycarbonyl)oxy]-, butyl ester
Propanoic acid, 2-[(amoxycarbonyl)oxy]-, methyl ester
RN: **MP** (°C):
MW: 218.25   **BP** (°C):

Solubility (Moles/L)	Solubility (Grams/L)	Temp (°C)	Ref (#)	Evaluation (T P E A A)	Comments
2.290E-03	4.998E-01	25	R007	0 0 0 0 0	
3.205E-03	6.995E-01	25	R007	0 0 0 0 0	

2301. $C_{10}H_{19}NO_2S$
4-Thiazolidinecarboxylic acid, 2-hexyl-
Thiazolidine-4-carboxylic acid, 2-hexyl-
RN: 14347-74-1   **MP** (°C):
MW: 217.33   **BP** (°C):   378.1

Solubility (Moles/L)	Solubility (Grams/L)	Temp (°C)	Ref (#)	Evaluation (T P E A A)	Comments
2.800E-03	6.085E-01	21	B414	1 0 0 1 1	partial decomposition

2302. $C_{10}H_{19}NO_3$
Ethylpropylaceturethane
RN: **MP** (°C):
MW: 201.27   **BP** (°C):

Solubility (Moles/L)	Solubility (Grams/L)	Temp (°C)	Ref (#)	Evaluation (T P E A A)	Comments
7.088E-03	1.427E+00	c	O021	0 2 0 0 0	

2303. $C_{10}H_{19}NO_3$
Oenanthylylurethane
RN: **MP** (°C):
MW: 201.27   **BP** (°C):

Solubility (Moles/L)	Solubility (Grams/L)	Temp (°C)	Ref (#)	Evaluation (T P E A A)	Comments
1.043E-03	2.100E-01	ns	O021	0 0 0 0 0	

2304. $C_{10}H_{19}NO_4S$
2-Amino-5-naphthol-1-sulfonic acid
RN: **MP** (°C):
MW: 249.33   **BP** (°C):

Solubility (Moles/L)	Solubility (Grams/L)	Temp (°C)	Ref (#)	Evaluation (T P E A A)	Comments
8.503E-03	2.120E+00	c	B125	1 2 0 0 2	

2305. C₁₀H₁₉N₂O₄PS

Cyanthoate

Phosphorothioic acid, *S*-(2-((1-cyano-1-methylethyl)amino)-2-oxoethyl) *O,O*-diethyl ester

Tartran

RN: 3734-95-0   **MP** (°C):

MW: 294.31   **BP** (°C):

Solubility (Moles/L)	Solubility (Grams/L)	Temp (°C)	Ref (#)	Evaluation (T P E A A)	Comments
2.378E-01	7.000E+01	20	M161	1 0 0 0 1	

2306. C₁₀H₁₉N₅O

Prometone

2-Methoxy-4,6-*bis*-isopropylamino-*s*-triazine

Pramitol

Primatol O

Prometon

2-Methoxy-4,6-*bis*-(isopropyl-amino)-*s*-triazine

RN: 1610-18-0   **MP** (°C):   91.5

MW: 225.30   **BP** (°C):   91–92

Solubility (Moles/L)	Solubility (Grams/L)	Temp (°C)	Ref (#)	Evaluation (T P E A A)	Comments
3.330E-03	7.502E-01	20	B200	1 0 0 0 2	
2.752E-03	6.200E-01	20	F311	1 2 2 2 1	
3.329E-03	7.500E-01	20	M161	1 0 0 0 2	
3.329E-03	7.500E-01	21	B192	0 0 0 0 2	
1.554E-02	3.500E+00	21	G099	2 0 0 1 0	
3.329E-03	7.500E-01	21	G099	2 0 0 1 0	
4.680E-03	1.054E+00	50	G001	1 0 1 1 2	
3.548E-03	7.994E-01	ns	B100	0 0 0 0 0	
3.329E-03	7.500E-01	ns	B185	0 0 0 0 0	
3.329E-03	7.500E-01	ns	C101	0 0 0 0 1	
3.329E-03	7.500E-01	ns	G041	0 0 0 0 2	
3.329E-03	7.500E-01	ns	H112	0 0 0 0 2	
3.329E-03	7.500E-01	ns	J033	0 0 0 0 0	

2307. C₁₀H₁₉N₅O

Terebumeton

1,3,5-Triazine-2,4-diamine, *N*-(1,1-dimethylethyl)-*N*′-ethyl-6-methoxy-

2-Methoxy-4-ethylamino-6-*tert*-butylamino-*s*-triazine

Karagard

4-(Ethylamino)-2-methoxy-6-(*tert*-butylamino)-*s*-triazine

Caragard

RN: 33693-04-8   **MP** (°C):   123.5

MW: 225.30   **BP** (°C):

Solubility (Moles/L)	Solubility (Grams/L)	Temp (°C)	Ref (#)	Evaluation (T P E A A)	Comments
5.770E-04	1.300E-01	20	M161	1 0 0 0 2	

2308. $C_{10}H_{19}N_5O$

2-Methoxy-4-ethylamino-6-diethylamino-s-triazine
G 31432

RN: 13532-26-8 **MP** (°C):
MW: 225.30 **BP** (°C):

Solubility (Moles/L)	Solubility (Grams/L)	Temp (°C)	Ref (#)	Evaluation (T P E A A)	Comments
1.775E-04	4.000E-02	20	J033	0 0 0 0 0	

2309. $C_{10}H_{19}N_5O$

Secbumeton
2-sec-Butylamino-4-ethylamino-6-methoxy-s-triazine
GS-14254

RN: 26259-45-0 **MP** (°C): 86
MW: 225.30 **BP** (°C):

Solubility (Moles/L)	Solubility (Grams/L)	Temp (°C)	Ref (#)	Evaluation (T P E A A)	Comments
2.930E-03	6.601E-01	1	G091	1 0 1 2 2	pH 6.0
3.250E-03	7.322E-01	8	G091	1 0 1 2 2	pH 6.0
2.750E-03	6.196E-01	20	B200	1 0 0 0 2	
2.663E-03	6.000E-01	20	F311	1 2 2 2 1	
3.070E-03	6.917E-01	20	G091	1 0 1 2 2	pH 6.0
2.752E-03	6.200E-01	20	M161	1 0 0 0 2	
3.300E-03	7.435E-01	29	G091	1 0 1 2 2	pH 6.0

2310. $C_{10}H_{19}N_5OS$

Hydroxyprometryne
1,3,5-Triazin-2(1H)-one, 4,6-bis[(1-methylethyl)amino]-
bis(Isopropylamino)hydroxy-s-triazine
GS 11526

RN: 7374-53-0 **MP** (°C):
MW: 257.36 **BP** (°C):

Solubility (Moles/L)	Solubility (Grams/L)	Temp (°C)	Ref (#)	Evaluation (T P E A A)	Comments
4.000E-04	1.029E-01	2	B193	1 2 0 0 0	

2311. C$_{10}$H$_{19}$N$_5$S
Terbutryn
2-Methylthio-4-ethylamino-6-*tert*-butylamino-*s*-triazine
Terbutryne
N-(1,1-Dimethylethyl)-*N'*-ethyl-6-(methylthio)-1,3,5-triazine-2,4-diamine
Terbutrex

RN:	886-50-0	MP (°C):	104
MW:	241.36	BP (°C):	157

Solubility (Moles/L)	Solubility (Grams/L)	Temp (°C)	Ref (#)	Evaluation (T P E A A)	Comments
1.090E-04	2.631E-02	1	G091	1 0 1 2 2	pH 6.0
1.100E-04	2.655E-02	8	G091	1 0 1 2 2	pH 6.0
2.400E-04	5.793E-02	20	B200	1 0 0 0 1	
1.036E-04	2.500E-02	20	E048	1 2 1 1 1	
1.036E-04	2.500E-02	20	F311	1 2 2 2 1	
1.460E-04	3.524E-02	20	G091	1 0 1 2 2	pH 6.0
2.403E-04	5.800E-02	20	M161	1 0 0 0 1	
1.660E-04	4.007E-02	29	G091	1 0 1 2 2	pH 6.0
2.403E-04	5.800E-02	ns	J033	0 0 0 0 0	

2312. C$_{10}$H$_{19}$N$_5$S
Prometryne
N,N'-bis(1-Methylethyl)-6-methylthio-1,3,5-triazine-2,4-diamine
Caparol
Primatol Q
Gesagard
Caparol 80W

RN:	7287-19-6	MP (°C):	118
MW:	241.36	BP (°C):	

Solubility (Moles/L)	Solubility (Grams/L)	Temp (°C)	Ref (#)	Evaluation (T P E A A)	Comments
2.400E-04	5.793E-02	2	B193	1 2 0 0 0	
2.000E-04	4.827E-02	20	B200	1 0 0 0 0	
1.657E-04	4.000E-02	20	F311	1 2 2 2 1	
1.988E-03	4.798E-01	20	M061	1 0 0 0 1	
1.989E-04	4.800E-02	20	M161	1 0 0 0 1	
1.989E-04	4.800E-02	24	C105	2 1 2 2 2	
4.200E-04	1.014E-01	50	G001	1 0 1 1 2	
1.989E-04	4.800E-02	ns	C101	0 0 0 0 1	
1.989E-04	4.800E-02	ns	H112	0 0 0 0 1	
1.989E-04	4.800E-02	ns	J033	0 0 0 0 0	

2313. C$_{10}$H$_{19}$N$_5$S
s-Triazole, 2,4-bis(isopropylamine)-6-methylmercapto-
RN: **MP** (°C):
MW: 241.36   **BP** (°C):

Solubility (Moles/L)	Solubility (Grams/L)	Temp (°C)	Ref (#)	Evaluation (T P E A A)	Comments
1.989E-04	4.800E-02	20	B185	0 0 0 0 0	

2314. C$_{10}$H$_{19}$O$_6$PS$_2$
Malathion
Dicarboethoxyethyl O,O-dimethyl phosphorodithioate
Carbofos
Cythion
Mercaptothion
Phosphothion
RN: 121-75-5   **MP** (°C):   3
MW: 330.36   **BP** (°C):

Solubility (Moles/L)	Solubility (Grams/L)	Temp (°C)	Ref (#)	Evaluation (T P E A A)	Comments
4.267E-04	1.410E-01	10	B324	0 0 0 0 0	
4.268E-04	1.410E-01	10	B324	0 0 0 0 0	
4.329E-04	1.430E-01	20	B300	2 1 1 1 2	
4.389E-04	1.450E-01	20	B324	0 0 0 0 0	
4.388E-04	1.450E-01	20	B324	0 0 0 0 0	
4.389E-04	1.450E-01	20	F311	.1 2 2 2 1	
4.389E-04	1.450E-01	20	M061	1 0 0 0 2	
4.389E-04	1.450E-01	20	M344	1 0 0 0 2	
4.964E-04	1.640E-01	30	B324	0 0 0 0 0	
4.963E-04	1.640E-01	30	B324	0 0 0 0 0	
4.389E-04	1.450E-01	rt	M161	0 0 0 0 2	

2315. C$_{10}$H$_{20}$
Cyclodecane
RN: 293-96-9   **MP** (°C):   10
MW: 140.27   **BP** (°C):   202

Solubility (Moles/L)	Solubility (Grams/L)	Temp (°C)	Ref (#)	Evaluation (T P E A A)	Comments
2.353E-06	3.300E-04	25	T423	0 0 0 0 0	

2316. C₁₀H₂₀

n-Pentylcyclopentane

1-Pentylcyclopentane

RN: 3741-00-2   **MP** (°C):
MW: 140.27   **BP** (°C):

Solubility (Moles/L)	Solubility (Grams/L)	Temp (°C)	Ref (#)	Evaluation (T P E A A)	Comments
8.198E-07	1.150E-04	25	K119	1 0 0 0 2	
8.198E-07	1.150E-04	25	P051	2 1 1 2 2	
8.198E-07	1.150E-04	25.00	P007	2 1 2 2 2	

2317. C₁₀H₂₀NO₄PS

Propetamphos

Methylethyl (E)-3-(((ethylamino)methoxyphosphinothioyl)oxy)-2-butenoate

Safrotin

Seraphos

Zoecon

RN: 31218-83-4   **MP** (°C):
MW: 281.31   **BP** (°C):   88

Solubility (Moles/L)	Solubility (Grams/L)	Temp (°C)	Ref (#)	Evaluation (T P E A A)	Comments
3.910E-04	1.100E-01	24	M161	1 0 0 0 2	

2318. C₁₀H₂₀NO₅PS₂

Mecarbam

O,O-Diethyl S-(N-methyl-N-carboethoxycarbamoylmethyl) dithiophosphate

RN: 2595-54-2   **MP** (°C):
MW: 329.38   **BP** (°C):   144

Solubility (Moles/L)	Solubility (Grams/L)	Temp (°C)	Ref (#)	Evaluation (T P E A A)	Comments
3.033E-03	9.990E-01	rt	M061	0 0 0 0 0	
<3.04E-03	<1.00E+00	rt	M161	0 0 0 0 0	

2319. C₁₀H₂₀N₂S₄

Disulfiram

Tetraethylthioperoxydicarbonothioic diamide

Tetraethylthiuram disulfide

Antadix

Antabuse

Esperal

RN: 97-77-8   **MP** (°C):   70
MW: 296.54   **BP** (°C):

Solubility (Moles/L)	Solubility (Grams/L)	Temp (°C)	Ref (#)	Evaluation (T P E A A)	Comments
6.744E-04	2.000E-01	25	I314	0 0 0 0 0	
1.379E-05	4.090E-03	25	L033	1 0 2 1 2	*sic*
1.012E-03	3.000E-01	ns	N061	0 0 0 0 0	

2320. C$_{10}$H$_{20}$N$_6$O
N-(Methoxymethyl)pentamethylmelamine
N-Methylolpentamethylmelamine methyl ether

RN: 64124-15-8   **MP** (°C):   39
MW: 240.31   **BP** (°C):

Solubility (Moles/L)	Solubility (Grams/L)	Temp (°C)	Ref (#)	Evaluation (T P E A A)	Comments
6.242E-03	1.500E+00	25	C051	1 2 1 1 1	pH 7, unstable in water

2321. C$_{10}$H$_{20}$O
Citronellol
3,7-Dimethyl-6-octen-1-ol
Levo-citronellol
β-Citronellol

RN: 106-22-9   **MP** (°C):
MW: 156.27   **BP** (°C):   222

Solubility (Moles/L)	Solubility (Grams/L)	Temp (°C)	Ref (#)	Evaluation (T P E A A)	Comments
1.280E-03	2.000E-01	25	M350	1 0 1 1 1	

2322. C$_{10}$H$_{20}$O
Decanal
Cuprylaldehyde

RN: 112-31-2   **MP** (°C):   7
MW: 156.27   **BP** (°C):   207–209

Solubility (Moles/L)	Solubility (Grams/L)	Temp (°C)	Ref (#)	Evaluation (T P E A A)	Comments
9.983E-05	1.560E-02	25	L450	0 0 0 0 0	

2323. C$_{10}$H$_{20}$O
Menthol
Cyclohexanol, 5-methyl-2-(1-methylethyl)-, (1α,2β,5α)-
3-p-Menthanol

RN: 89-78-1   **MP** (°C):   42
MW: 156.27   **BP** (°C):   212

Solubility (Moles/L)	Solubility (Grams/L)	Temp (°C)	Ref (#)	Evaluation (T P E A A)	Comments
2.560E-03	4.000E-01	20	F300	1 0 0 0 2	
2.920E-03	4.563E-01	25	I019	1 0 1 2 2	
8.600E-03	1.344E+00	37	E028	1 0 1 1 1	

2324. C$_{10}$H$_{20}$O
l-Menthol
1-Isopropyl-4-methyl cyclohexan-2-ol
1-Methyl-4-isopropyl cyclohexan-3-ol
(1R,2S,5R)-(–)-Menthol
5-Methyl-2-isopropyl hexahydrophenol
Cyclohexanol

RN:	2216-51-5	**MP** (°C):	44		
MW:	156.27	**BP** (°C):	212		

Solubility (Moles/L)	Solubility (Grams/L)	Temp (°C)	Ref (#)	Evaluation (T P E A A)	Comments
4.000E-03	6.251E-01	25	A401	1 0 2 2 0	

2325. C$_{10}$H$_{20}$O$_2$
3-Hydroxy-2-ethyl-5-propyl-5-methyltetrahydrofuran
3-Furanol, 2-ethyltetrahydro-5-methyl-5-propyl-

RN:	29839-73-4	**MP** (°C):	
MW:	172.27	**BP** (°C):	

Solubility (Moles/L)	Solubility (Grams/L)	Temp (°C)	Ref (#)	Evaluation (T P E A A)	Comments
5.747E-02	9.901E+00	rt	B066	0 2 0 0 0	

2326. C$_{10}$H$_{20}$O$_2$
3-Hydroxy-2,2-dimethyl-5,5-diethyltetrahydrofuran
3-Furanol, 5,5-diethyltetrahydro-2,2-dimethyl-

RN:	29839-77-8	**MP** (°C):	
MW:	172.27	**BP** (°C):	

Solubility (Moles/L)	Solubility (Grams/L)	Temp (°C)	Ref (#)	Evaluation (T P E A A)	Comments
2.888E-02	4.975E+00	rt	B066	0 2 0 0 0	

2327. C$_{10}$H$_{20}$O$_2$
3-Hydroxy-2,5,5-triethyltetrahydrofuran
3-Furanol, 2,5,5-triethyltetrahydro-

RN:	29839-70-1	**MP** (°C):	
MW:	172.27	**BP** (°C):	

Solubility (Moles/L)	Solubility (Grams/L)	Temp (°C)	Ref (#)	Evaluation (T P E A A)	Comments
5.747E-02	9.901E+00	rt	B066	0 2 0 0 0	

2328. $C_{10}H_{20}O_2$

3-Hydroxy-2,5-dipropyltetrahydrofuran
3-Furanol, 2,5-dipropyltetrahydro-

RN: 30003-27-1   **MP** (°C):
MW: 172.27   **BP** (°C):

Solubility (Moles/L)	Solubility (Grams/L)	Temp (°C)	Ref (#)	Evaluation (T P E A A)	Comments
1.159E-02	1.996E+00	rt	B066	0 2 0 0 0	

2329. $C_{10}H_{20}O_2$

3-Hydroxy-2-butyl-5,5-methyltetrahydrofuran
3-Furanol, 2-butyltetrahydro-5,5-dimethyl-

RN: 29839-71-2   **MP** (°C):
MW: 172.27   **BP** (°C):

Solubility (Moles/L)	Solubility (Grams/L)	Temp (°C)	Ref (#)	Evaluation (T P E A A)	Comments
2.888E-02	4.975E+00	rt	B066	0 2 0 0 0	

2330. $C_{10}H_{20}O_2$

3-Hydroxy-2-pentyl-5-methyltetrahydrofuran
3-Furanol, 5-methyltetrahydro-2-pentyl-

RN: 29848-45-1   **MP** (°C):
MW: 172.27   **BP** (°C):

Solubility (Moles/L)	Solubility (Grams/L)	Temp (°C)	Ref (#)	Evaluation (T P E A A)	Comments
1.159E-02	1.996E+00	rt	B066	0 2 0 0 0	

2331. $C_{10}H_{20}O_2$

3-Hydroxy-2-propyl-5-methyl-5-ethyltetrahydrofuran
3-Furanol, 5-ethyltetrahydro-5-methyl-2-propyl-

RN: 29839-72-3   **MP** (°C):
MW: 172.27   **BP** (°C):

Solubility (Moles/L)	Solubility (Grams/L)	Temp (°C)	Ref (#)	Evaluation (T P E A A)	Comments
5.747E-02	9.901E+00	rt	B066	0 2 0 0 0	

2332. $C_{10}H_{20}O_2$

n-Capric acid
Caprinsaeure
Decanoic acid
Nonanecarboxylic acid

RN:	334-48-5	MP (°C):	31.4
MW:	172.27	BP (°C):	270

Solubility (Moles/L)	Solubility (Grams/L)	Temp (°C)	Ref (#)	Evaluation (T P E A A)	Comments
5.515E-04	9.500E-02	0	B136	1 0 2 1 1	
1.509E-04	2.600E-02	15	F300	1 0 0 0 1	
2.902E-04	5.000E-02	20	A011	1 2 1 1 1	
9.462E-04	1.630E-01	20	B136	1 0 2 1 2	
8.706E-04	1.500E-01	20	D041	1 0 0 0 1	
8.706E-04	1.500E-01	20.0	R001	1 1 1 1 1	
3.590E-04	6.184E-02	25	J001	1 0 2 1 2	
1.115E-03	1.920E-01	30	B136	1 0 2 1 2	
3.715E-04	6.400E-02	30	E005	2 1 1 2 1	
1.045E-03	1.800E-01	30.0	R001	1 1 1 1 1	
1.294E-03	2.230E-01	40	B136	1 0 2 1 2	
4.179E-04	7.200E-02	40	E005	2 1 1 2 1	
1.335E-03	2.300E-01	45	B136	1 0 2 1 1	
1.335E-03	2.299E-01	45.0	R001	1 1 1 1 1	
4.702E-04	8.100E-02	50	E005	2 1 1 2 1	
5.000E-04	8.613E-02	50	J001	1 0 2 1 2	
1.567E-03	2.700E-01	60	B136	1 0 2 1 1	
5.805E-04	1.000E-01	60	E005	2 1 1 2 2	
1.567E-03	2.699E-01	60.0	R001	1 1 1 1 1	
5.514E-04	9.499E-02	.0	R001	1 1 1 1 1	

2333. $C_{10}H_{20}O_2$

3-Hydroxy-5,5-dipropyltetrahydrofuran
3-Furanol, 5,5-dipropyltetrahydro-

RN:	29839-54-1	MP (°C):	
MW:	172.27	BP (°C):	

Solubility (Moles/L)	Solubility (Grams/L)	Temp (°C)	Ref (#)	Evaluation (T P E A A)	Comments
5.747E-02	9.901E+00	rt	B066	0 2 0 0 0	

2334. $C_{10}H_{20}O_2$

3-Hydroxy-5,5-diisopropyltetrahydrofuran
3-Furanol, 5,5-diisopropyltetrahydro-

RN:	29839-55-2	MP (°C):	
MW:	172.27	BP (°C):	

Solubility (Moles/L)	Solubility (Grams/L)	Temp (°C)	Ref (#)	Evaluation (T P E A A)	Comments
5.747E-02	9.901E+00	rt	B066	0 2 0 0 0	

2335. $C_{10}H_{20}O_2$
3-Hydroxy-2,5-dimethyl-2,5-diethyltetrahydrofuran
3-Furanol, 2,5-diethyltetrahydro-2,5-dimethyl-
RN: 30010-09-4   **MP** (°C):
MW: 172.27   **BP** (°C):

Solubility (Moles/L)	Solubility (Grams/L)	Temp (°C)	Ref (#)	Evaluation (T P E A A)	Comments
1.138E-01	1.961E+01	rt	B066	0 2 0 0 0	

2336. $C_{10}H_{20}O_2 \cdot H_2O$
Terpin (monohydrate)
Terpin-hydrat
RN: 2451-01-6   **MP** (°C):   116
MW: 190.29   **BP** (°C):

Solubility (Moles/L)	Solubility (Grams/L)	Temp (°C)	Ref (#)	Evaluation (T P E A A)	Comments
2.102E-02	4.000E+00	15	F300	1 0 0 0 0	
1.799E-02	3.424E+00	25	M012	1 0 2 1 2	
1.661E-01	3.160E+01	100	F300	1 0 0 0 2	

2337. $C_{10}H_{20}O_3$
1,3-Dioxolane-4-methanol, 2-methyl-2-pentyl
2-Heptanone, cyclic (hydroxymethyl)ethylene acetal
2-Methyl-2-*n*-amyl-4-hydroxymethyl-1,3-dioxolane
2-Methyl-2-pentyl-1,3-dioxolane-4-methanol
RN: 4361-59-5   **MP** (°C):
MW: 188.27   **BP** (°C):

Solubility (Moles/L)	Solubility (Grams/L)	Temp (°C)	Ref (#)	Evaluation (T P E A A)	Comments
5.090E-02	9.583E+00	25	P342	0 0 0 0 0	0.0001M Na_2CO_3

2338. $C_{10}H_{20}O_3$
n-Amyl β-ethoxypropionate
Propionic acid, 3-ethoxy-, pentyl ester
RN: 14144-36-6   **MP** (°C):
MW: 188.27   **BP** (°C):

Solubility (Moles/L)	Solubility (Grams/L)	Temp (°C)	Ref (#)	Evaluation (T P E A A)	Comments
6.366E-03	1.199E+00	25	D002	1 2 1 1 1	

2339. C$_{10}$H$_{20}$O$_4$
Butyl carbitol acetate
Diethylene glycol acetate butyl ether
Diethylene glycol butyl ether acetate
Diglykol-monobutylaether-acetat
RN: 124-17-4   **MP** (°C):   −32
MW: 204.27   **BP** (°C):   245

Solubility (Moles/L)	Solubility (Grams/L)	Temp (°C)	Ref (#)	Evaluation (T P E A A)	Comments
7.709E-02	1.575E+01	20	D052	1 1 0 0 1	
1.792E-01	3.661E+01	20	M062	1 0 0 0 1	

2340. C$_{10}$H$_{21}$NOS
Pebulate
S-Propyl butylethylthiocarbamate
RN: 1114-71-2   **MP** (°C):   <25
MW: 203.35   **BP** (°C):

Solubility (Moles/L)	Solubility (Grams/L)	Temp (°C)	Ref (#)	Evaluation (T P E A A)	Comments
2.951E-04	6.000E-02	20	M161	1 0 0 0 1	
4.524E-04	9.200E-02	21	F019	1 0 0 0 1	
4.524E-04	9.200E-02	21	M061	1 0 0 0 1	
2.951E-04	6.000E-02	ns	B200	0 0 0 0 1	
2.951E-04	6.001E-02	ns	S460	0 0 0 0 0	

2341. C$_{10}$H$_{21}$NOS
Vernolate
S-Propyl dipropylthiocarbamate
Carbamic acid, dipropylthio-, *S*-propyl ester
Carbamate, *n*-propyl-di-*n*-propylthio-
Vernam
RN: 1929-77-7   **MP** (°C):   <25
MW: 203.35   **BP** (°C):

Solubility (Moles/L)	Solubility (Grams/L)	Temp (°C)	Ref (#)	Evaluation (T P E A A)	Comments
4.426E-04	9.000E-02	20	B200	1 0 0 0 1	
5.262E-04	1.070E-01	21	F019	1 0 0 0 2	
5.262E-04	1.070E-01	21	M161	1 0 0 0 2	
<4.92E-04	<1.00E-01	ns	B185	0 0 0 0 0	
4.917E-04	9.999E-02	ns	M061	0 0 0 0 0	

2342. $C_{10}H_{22}$
2,2-Dimethyloctane
RN: 15869-87-1   **MP** (°C):
MW: 142.29   **BP** (°C): 157

Solubility (Moles/L)	Solubility (Grams/L)	Temp (°C)	Ref (#)	Evaluation (T P E A A)	Comments
7.499E-07	1.067E-04	ns	S460	0 0 0 0 0	

2343. $C_{10}H_{22}$
n-Decane
Decane
Decyl hydride
RN: 124-18-5   **MP** (°C): −30.0
MW: 142.29   **BP** (°C): 174.0

Solubility (Moles/L)	Solubility (Grams/L)	Temp (°C)	Ref (#)	Evaluation (T P E A A)	Comments
1.389E-07	1.976E-05	20	B165	1 0 1 1 1	
1.124E-07	1.600E-05	25	B069	1 0 1 1 1	
1.389E-07	1.976E-05	25	F004	0 0 0 0 0	
3.655E-07	5.200E-05	25	M003	1 0 2 2 1	
3.655E-07	5.200E-05	25	M040	1 0 0 1 1	
3.233E-07	4.600E-05	25	T423	0 0 0 0 0	
1.546E-07	2.200E-05	ns	B033	0 0 0 0 2	
1.546E-07	2.200E-05	ns	B033	0 0 0 0 0	
3.655E-07	5.200E-05	ns	H123	0 0 0 0 0	

2344. $C_{10}H_{22}$
4,4-Dimethyloctane
RN: 15869-95-1   **MP** (°C):
MW: 142.29   **BP** (°C): 157.5

Solubility (Moles/L)	Solubility (Grams/L)	Temp (°C)	Ref (#)	Evaluation (T P E A A)	Comments
1.546E-05	2.200E-03	20	M337	2 1 2 2 1	
7.278E-07	1.036E-04	ns	S460	0 0 0 0 0	

2345. $C_{10}H_{22}$
2,3-Dimethyloctane
RN: 7146-60-3   **MP** (°C):
MW: 142.29   **BP** (°C): 164

Solubility (Moles/L)	Solubility (Grams/L)	Temp (°C)	Ref (#)	Evaluation (T P E A A)	Comments
5.117E-07	7.281E-05	ns	S460	0 0 0 0 0	

2346. C₁₀H₂₂

2,6-Dimethyloctane

RN:	2051-30-1	MP (°C):	
MW:	142.29	BP (°C):	159

Solubility (Moles/L)	Solubility (Grams/L)	Temp (°C)	Ref (#)	Evaluation (T P E A A)	Comments
6.266E-07	8.916E-05	ns	S460	0 0 0 0 0	

2347. C₁₀H₂₂

3,6-Dimethyloctane

RN:	15869-94-0	MP (°C):	
MW:	142.29	BP (°C):	161

Solubility (Moles/L)	Solubility (Grams/L)	Temp (°C)	Ref (#)	Evaluation (T P E A A)	Comments
6.109E-07	8.693E-05	ns	S460	0 0 0 0 0	

2348. C₁₀H₂₂

3-Ethyloctane

RN:	5881-17-4	MP (°C):	
MW:	142.29	BP (°C):	167

Solubility (Moles/L)	Solubility (Grams/L)	Temp (°C)	Ref (#)	Evaluation (T P E A A)	Comments
4.581E-07	6.519E-05	ns	S460	0 0 0 0 0	

2349. C₁₀H₂₂

4-Methylnonane

4-Methylnonane(DL)

RN:	17301-94-9	MP (°C):	
MW:	142.29	BP (°C):	

Solubility (Moles/L)	Solubility (Grams/L)	Temp (°C)	Ref (#)	Evaluation (T P E A A)	Comments
4.764E-07	6.779E-05	ns	S460	0 0 0 0 0	

2350. C₁₀H₂₂

3,3-Dimethyloctane

RN:	4110-44-5	MP (°C):	
MW:	142.29	BP (°C):	161

Solubility (Moles/L)	Solubility (Grams/L)	Temp (°C)	Ref (#)	Evaluation (T P E A A)	Comments
5.998E-07	8.534E-05	ns	S460	0 0 0 0 0	

2351. C_{10}H_{22}
4-Ethyloctane

RN: 15869-86-0 **MP** (°C):
MW: 142.29 **BP** (°C): 164

Solubility (Moles/L)	Solubility (Grams/L)	Temp (°C)	Ref (#)	Evaluation (T P E A A)	Comments
5.297E-07	7.536E-05	ns	S460	0 0 0 0 0	

2352. C_{10}H_{22}
3,5-Dimethyloctane

RN: 15869-93-9 **MP** (°C):
MW: 142.29 **BP** (°C): 159

Solubility (Moles/L)	Solubility (Grams/L)	Temp (°C)	Ref (#)	Evaluation (T P E A A)	Comments
6.546E-07	9.315E-05	ns	S460	0 0 0 0 0	

2353. C_{10}H_{22}
3-Methylnonane
3-Methylnonane(DL)

RN: 5911-04-6 **MP** (°C):
MW: 142.29 **BP** (°C):

Solubility (Moles/L)	Solubility (Grams/L)	Temp (°C)	Ref (#)	Evaluation (T P E A A)	Comments
4.295E-07	6.112E-05	ns	S460	0 0 0 0 0	

2354. C_{10}H_{22}O
Decanol

RN: 36729-58-5 **MP** (°C):
MW: 158.29 **BP** (°C):

Solubility (Moles/L)	Solubility (Grams/L)	Temp (°C)	Ref (#)	Evaluation (T P E A A)	Comments
2.000E-04	3.166E-02	24	H345	0 0 0 0 0	

2355. C$_{10}$H$_{22}$O
n-Decyl alcohol
Alcohol C-10
Nonyl acarbinol
Capric alcohol
RN: 36729-58-5   **MP** (°C):
MW: 158.29   **BP** (°C):

Solubility (Moles/L)	Solubility (Grams/L)	Temp (°C)	Ref (#)	Evaluation (T P E A A)	Comments
2.690E-04	4.258E-02	20	H330	0 0 0 0 0	
2.000E-04	3.166E-02	24	H345	2 0 2 2 2	
2.340E-04	3.704E-02	25	K025	2 2 1 1 2	
2.527E-05	4.000E-03	40	W305	1 0 0 1 0	EFG
3.000E-04	4.748E-02	ns	H012	0 2 2 0 0	

2356. C$_{10}$H$_{23}$O$_2$PS$_2$
Cadusafos
Ebufos
Taredan
Rugby
Apache
O-ethyl S,S-bis(1-methylpropyl) phosphorodithioate
RN: 95465-99-9   **MP** (°C):
MW: 270.40   **BP** (°C):

Solubility (Moles/L)	Solubility (Grams/L)	Temp (°C)	Ref (#)	Evaluation (T P E A A)	Comments
9.162E-04	2.477E-01	ns	S460	0 0 0 0 0	

2357. C$_{10}$H$_{23}$O$_3$P
Ethyl dibutyl phosphonate
Dibutyl ethyl phosphonate
RN: 2404-58-2   **MP** (°C):
MW: 222.27   **BP** (°C):

Solubility (Moles/L)	Solubility (Grams/L)	Temp (°C)	Ref (#)	Evaluation (T P E A A)	Comments
2.699E-02	6.000E+00	25	B070	1 2 0 1 0	
5.849E-02	1.300E+01	25	B070	1 2 0 1 1	

2358. C$_{10}$H$_{23}$O$_4$P
Dibutyl ethyl phosphate
RN: 7242-58-2   **MP** (°C):
MW: 238.27   **BP** (°C):

Solubility (Moles/L)	Solubility (Grams/L)	Temp (°C)	Ref (#)	Evaluation (T P E A A)	Comments
1.427E-02	3.400E+00	25	B070	1 2 2 1 1	

2359. $C_{10}Cl_{10}O$
Chlordecone
Kepone
1,2,3,5,6,7,8,9,10,10-Decachloropentacyclo[5.2.1.0(2,6).0(3,9).0(5,8)]decano-4-one
Merex
Decachloroketone

RN: 143-50-0   **MP** (°C):
MW: 490.64   **BP** (°C):

Solubility (Moles/L)	Solubility (Grams/L)	Temp (°C)	Ref (#)	Evaluation (T P E A A)	Comments
8.153E-03	4.000E+00	100	M161	1 0 0 0 0	
6.166E-06	3.025E-03	ns	R424	0 0 0 0 0	
6.166E-06	3.025E-03	ns	R427	0 0 0 0 0	

2360. $C_{10}Cl_{12}$
Mirex
1,2,3,4,5,5-Hexachloro-1,3-cyclopentadiene dimer
Bichlorendo
Ferriamicide
Dechlorane 4070

RN: 2385-85-5   **MP** (°C):
MW: 545.55   **BP** (°C):

Solubility (Moles/L)	Solubility (Grams/L)	Temp (°C)	Ref (#)	Evaluation (T P E A A)	Comments
1.549E-07	8.450E-05	24.99	K436	0 0 0 0 0	
1.397E-10	7.619E-08	25	H434	0 0 0 0 0	
1.558E-07	8.500E-05	25	M134	1 2 1 1 1	
1.741E-07	9.500E-05	ns	M110	0 0 0 0 0	EFG
1.660E-07	9.054E-05	ns	R427	0 0 0 0 0	

2361. $C_{11}H_6BrNS$
1-Bromo-2-naphthylisothiocyanate

RN: 2392-80-5   **MP** (°C):
MW: 264.15   **BP** (°C):

Solubility (Moles/L)	Solubility (Grams/L)	Temp (°C)	Ref (#)	Evaluation (T P E A A)	Comments
4.800E-05	1.268E-02	25	D019	1 1 1 1 1	

2362. $C_{11}H_6O_3$
Psoralen
7H-Furo[3,2-g][1]benzopyran-7-one

RN: 66-97-7   **MP** (°C):   158–161
MW: 186.17   **BP** (°C):

Solubility (Moles/L)	Solubility (Grams/L)	Temp (°C)	Ref (#)	Evaluation (T P E A A)	Comments
3.500E-04	6.516E-02	25	A355	0 0 0 0 0	

2363. $C_{11}H_7Cl_2NO_3$
Pyoluteorin
RN: **MP** (°C):
MW: 272.09   **BP** (°C):

Solubility (Moles/L)	Solubility (Grams/L)	Temp (°C)	Ref (#)	Evaluation (T P E A A)	Comments
5.300E-04	1.442E-01	5.0	L451	0 0 0 0 0	
5.600E-04	1.524E-01	10	L451	0 0 0 0 0	
6.300E-04	1.714E-01	15.0	L451	0 0 0 0 0	
7.500E-04	2.041E-01	20.0	L451	0 0 0 0 0	
7.900E-04	2.150E-01	25.0	L451	0 0 0 0 0	
9.600E-04	2.612E-01	30.0	L451	0 0 0 0 0	
9.900E-04	2.694E-01	35.0	L451	0 0 0 0 0	
1.150E-03	3.129E-01	40.0	L451	0 0 0 0 0	
1.290E-03	3.510E-01	45.0	L451	0 0 0 0 0	
1.500E-03	4.081E-01	50.0	L451	0 0 0 0 0	
1.590E-03	4.326E-01	55.0	L451	0 0 0 0 0	
1.730E-03	4.707E-01	60.0	L451	0 0 0 0 0	

2364. $C_{11}H_7FN_2O_3$
3-Benzoyl-5-fluorouracil
RN: 61251-77-2   **MP** (°C):   169–170
MW: 234.19   **BP** (°C):

Solubility (Moles/L)	Solubility (Grams/L)	Temp (°C)	Ref (#)	Evaluation (T P E A A)	Comments
5.551E-03	1.300E+00	22	B321	0 0 0 0 0	pH 4.0
5.551E-03	1.300E+00	22	B332	1 1 0 0 1	pH 4.0

2365. $C_{11}H_7FN_2O_4$
3-Phenyloxycarbonyl-5-fluoro-2,4(1H,3H)-pyrimidinedi-one
3-Phenyloxycarbonyl-5-fluorouracil
RN: 66999-97-1   **MP** (°C):   169–170
MW: 250.19   **BP** (°C):

Solubility (Moles/L)	Solubility (Grams/L)	Temp (°C)	Ref (#)	Evaluation (T P E A A)	Comments
5.995E-04	1.500E-01	22	B321	0 0 0 0 0	pH 4.0

2366. $C_{11}H_7FN_2O_4$
1-Phenyloxycarbonyl-5-fluorouracil
RN: 75410-28-5   **MP** (°C):
MW: 250.19   **BP** (°C):

Solubility (Moles/L)	Solubility (Grams/L)	Temp (°C)	Ref (#)	Evaluation (T P E A A)	Comments
3.597E-03	9.000E-01	22	B332	1 1 0 0 1	pH 4.0

2367. C₁₁H₇NS

2-Naphthyl isothiocyanate
2-Isothiocyanatonaphthalene
β-Naphthyl mustard oil

RN:	1636-33-5	MP (°C):	
MW:	185.25	BP (°C):	

Solubility (Moles/L)	Solubility (Grams/L)	Temp (°C)	Ref (#)	Evaluation (T P E A A)	Comments
3.600E-05	6.669E-03	25	D019	1 1 1 1 1	

2368. C₁₁H₇NS

1-Naphthyl isothiocyanate
1-Isothiocyanatonaphthalene
α-Naphthyl mustard oil
Kesscocide
ANI
ANIT

RN:	551-06-4	MP (°C):	58.0
MW:	185.25	BP (°C):	

Solubility (Moles/L)	Solubility (Grams/L)	Temp (°C)	Ref (#)	Evaluation (T P E A A)	Comments
2.500E-05	4.631E-03	25	D019	1 1 1 1 1	

2369. C₁₁H₈N₂

β-Carboline
β Carbolin
Norharmane
9H-Pyrido(3,4-b)indole

RN:	244-63-3	MP (°C):	199
MW:	168.20	BP (°C):	

Solubility (Moles/L)	Solubility (Grams/L)	Temp (°C)	Ref (#)	Evaluation (T P E A A)	Comments
1.391E+01	2.340E+03	16	B413	1 0 2 2 1	
1.601E+01	2.693E+03	17	B413	1 0 2 2 1	
2.535E+01	4.264E+03	37	B413	1 0 2 2 1	
2.561E+01	4.308E+03	38	B413	1 0 2 2 1	
2.916E+01	4.905E+03	45	B413	1 0 2 2 1	

2370. C₁₁H₈N₄O₄

Orotic acid nicotinmide

RN:		MP (°C):	252–253
MW:	260.21	BP (°C):	

Solubility (Moles/L)	Solubility (Grams/L)	Temp (°C)	Ref (#)	Evaluation (T P E A A)	Comments
6.800E-02	1.769E+01	25	N018	0 0 0 0 0	

2371. $C_{11}H_8O_2$

2-Naphthoic acid

β-Naphthoic acid

2-Naphthalenecarboxylic acid

RN: 93-09-4   **MP** (°C):

MW: 172.19   **BP** (°C):

Solubility (Moles/L)	Solubility (Grams/L)	Temp (°C)	Ref (#)	Evaluation (T P E A A)	Comments
1.300E-04	2.238E-02	25	M149	2 2 2 2 1	intrinsic, *sic*
1.617E-06	2.785E-04	30	K148	1 1 0 0 2	
2.323E-06	4.000E-04	40	K148	1 1 0 0 1	
3.165E-06	5.450E-04	50	K148	1 1 0 0 2	
3.949E-06	6.800E-04	60	K148	1 1 0 0 2	
4.652E-06	8.010E-04	70	K148	1 1 0 0 2	
5.459E-06	9.400E-04	80	K148	1 1 0 0 2	
6.261E-06	1.078E-03	90	K148	1 1 0 0 2	

2372. $C_{11}H_8O_2$

Menadione

2-Methyl-1,4-naphthoquinone

Vitamin K3

Kativ-G

Panosine

Menaphthone

RN: 58-27-5   **MP** (°C):   106

MW: 172.19   **BP** (°C):

Solubility (Moles/L)	Solubility (Grams/L)	Temp (°C)	Ref (#)	Evaluation (T P E A A)	Comments
9.291E-04	1.600E-01	25	P096	0 0 0 0 0	
6.969E-04	1.200E-01	30	K090	1 2 2 2 0	EFG
8.700E-04	1.498E-01	30	O321	0 0 0 0 0	
8.710E-04	1.500E-01	30	O321	0 0 0 0 0	
9.291E-04	1.600E-01	30.00	E033	1 0 2 1 0	EFG
8.888E-04	1.530E-01	33	D404	2 1 2 2 2	
8.768E-04	1.510E-01	33	D404	2 1 2 2 2	
1.161E-03	2.000E-01	37.00	E033	1 0 2 1 0	EFG

2373. $C_{11}H_8O_3$

8-Hydroxypsoralon

RN: **MP** (°C):

MW: 188.18   **BP** (°C):

Solubility (Moles/L)	Solubility (Grams/L)	Temp (°C)	Ref (#)	Evaluation (T P E A A)	Comments
6.100E-04	1.148E-01	25	A355	0 0 0 0 0	

2374. C₁₁H₈O₃

2-Methoxy-1,4-naphthoquinone
1,4-Naphthalenedione, 2-methoxy-
2-Methoxy-1,4-naphthoquinone

RN: 2348-82-5 **MP** (°C):
MW: 188.18 **BP** (°C):

Solubility (Moles/L)	Solubility (Grams/L)	Temp (°C)	Ref (#)	Evaluation (T P E A A)	Comments
1.660E-04	3.123E-02	ns	R427	0 0 0 0 0	

2375. C₁₁H₉ClO₂S

Tianafac

RN: 51527-19-6 **MP** (°C):
MW: 240.71 **BP** (°C):

Solubility (Moles/L)	Solubility (Grams/L)	Temp (°C)	Ref (#)	Evaluation (T P E A A)	Comments
1.444E-04	3.476E-02	25	C314	0 0 0 0 0	
1.442E-04	3.470E-02	25	C314	0 0 0 0 0	

2376. C₁₁H₉Cl₂NO₂

Barban
4-Chloro-2-butynyl-N-(3-chlorophenyl)carbamate
4-Chloro-2-butynyl-m-chlorocarbanilate

RN: 101-27-9 **MP** (°C): 75
MW: 258.11 **BP** (°C):

Solubility (Moles/L)	Solubility (Grams/L)	Temp (°C)	Ref (#)	Evaluation (T P E A A)	Comments
4.262E-05	1.100E-02	25	B200	1 0 0 0 2	
4.262E-05	1.100E-02	25	M161	1 0 0 0 1	
3.874E-05	1.000E-02	ns	H042	0 0 0 0 1	
4.262E-04	1.100E-01	ns	M061	0 0 0 0 2	

2377. C₁₁H₉Cl₄NO₄

OCS-21693
TMMT
Methyl-2,3,5,6-tetrachloro-N-methoxy-N-methylterephthalamate

RN: 14419-01-3 **MP** (°C): 96
MW: 361.01 **BP** (°C):

Solubility (Moles/L)	Solubility (Grams/L)	Temp (°C)	Ref (#)	Evaluation (T P E A A)	Comments
1.385E-05	5.000E-03	25	B200	1 0 0 0 0	

2378. $C_{11}H_9I_3N_2O_4$

3,5-Diacetylamino-2,4,6-triiodobenzoic acid
Iothalamic acid
Diatrazoic acid

RN: 117-96-4   **MP** (°C):
MW: 613.92   **BP** (°C):

Solubility (Moles/L)	Solubility (Grams/L)	Temp (°C)	Ref (#)	Evaluation (T P E A A)	Comments
8.144E-01	5.000E+02	25	L100	1 0 0 0 2	
9.773E-01	6.000E+02	50	L100	1 0 0 0 2	
1.189E+00	7.297E+02	90	L100	1 0 0 0 2	
2.557E-03	1.570E+00	ns	H055	0 0 0 0 0	

2379. $C_{11}H_{10}$

2-Methylnaphthalene
2-Methyl naphthalene
β-Methyl naphthalenes

RN: 91-57-6   **MP** (°C):   35
MW: 142.20   **BP** (°C):   241.5

Solubility (Moles/L)	Solubility (Grams/L)	Temp (°C)	Ref (#)	Evaluation (T P E A A)	Comments
1.730E-04	2.460E-02	25	E004	2 1 2 2 2	
1.828E-04	2.600E-02	25	L332	1 1 1 1 0	
1.786E-04	2.540E-02	25	M064	1 1 2 2 2	
1.800E-04	2.560E-02	25	M342	1 0 1 1 1	
1.758E-04	2.500E-02	25	O320	0 0 0 0 0	
1.786E-04	2.540E-02	ns	H123	0 0 0 0 0	
8.000E-05	1.138E-02	ns	L060	0 0 0 0 0	
1.786E-04	2.540E-02	ns	M344	0 0 0 0 2	

2380. $C_{11}H_{10}$

1-Methylnaphthalene
1-Methyl naphthalene
1-Methyl-napthalene
α-Methyl naphthalenes
α-Methylnaphthalene

RN: 90-12-0   **MP** (°C):   −22
MW: 142.20   **BP** (°C):   244

Solubility (Moles/L)	Solubility (Grams/L)	Temp (°C)	Ref (#)	Evaluation (T P E A A)	Comments
1.739E-04	2.473E-02	4	D351	1 2 1 1 2	
1.600E-04	2.275E-02	10	S076	2 2 2 2 1	
2.000E-04	2.844E-02	14	S076	2 2 2 2 1	
1.195E-04	1.700E-02	20	A050	1 0 1 1 1	
2.145E-04	3.050E-02	20	B318	0 0 0 0 0	EFG
2.124E-04	3.020E-02	20	B356	0 0 0 0 0	
2.000E-04	2.844E-02	20	S076	2 2 2 2 1	
2.100E-04	2.986E-02	21	A057	2 1 2 2 1	

(continued)

2380. $C_{11}H_{10}$ (continued)

Solubility (Moles/L)	Solubility (Grams/L)	Temp (°C)	Ref (#)	Evaluation (T P E A A)	Comments
2.489E-04	3.539E-02	25	D351	1 2 1 1 2	
1.814E-04	2.580E-02	25	E004	2 1 2 2 2	
1.899E-04	2.700E-02	25	L332	1 1 1 1 0	
2.004E-04	2.850E-02	25	M064	1 1 2 2 2	
2.000E-04	2.844E-02	25	M342	1 0 1 1 2	
2.100E-04	2.986E-02	25	S076	2 2 2 2 1	
2.440E-04	3.470E-02	28	B348	2 2 2 2 2	
2.955E-04	4.203E-02	40	D351	1 2 1 1 2	
2.004E-04	2.850E-02	ns	H123	0 0 0 0 0	
1.600E-04	2.275E-02	ns	L060	0 0 0 0 1	
2.004E-04	2.850E-02	ns	M344	0 0 0 0 2	

2381. $C_{11}H_{10}BrN_3O_2S$
5-Sulfanilamido-2-bromopyridine
Benzenesulfonamide, 4-amino-N-(2-bromo-5-pyridinyl)-
RN: 17103-43-4   **MP** (°C):
MW: 328.19   **BP** (°C):

Solubility (Moles/L)	Solubility (Grams/L)	Temp (°C)	Ref (#)	Evaluation (T P E A A)	Comments
3.717E-04	1.220E-01	37	R058	1 2 1 1 2	

2382. $C_{11}H_{10}BrN_3O_2S$
2-Sulfanilamido-5-bromopyridine
Benzenesulfonamide, 4-amino-N-(5-bromo-2-pyridinyl)-
RN: 16805-99-5   **MP** (°C):
MW: 328.19   **BP** (°C):

Solubility (Moles/L)	Solubility (Grams/L)	Temp (°C)	Ref (#)	Evaluation (T P E A A)	Comments
1.158E-04	3.800E-02	37	R058	1 2 1 1 1	

2383. $C_{11}H_{10}ClNO_2$
Chlorbupham
1-Methylpropyn-2-yl N-(m-chlorophenyl)carbamate
Chlorbufam
Bi-PC
RN: 1967-16-4   **MP** (°C):   45.5
MW: 223.66   **BP** (°C):

Solubility (Moles/L)	Solubility (Grams/L)	Temp (°C)	Ref (#)	Evaluation (T P E A A)	Comments
2.414E-03	5.400E-01	20	B185	0 0 0 0 0	
2.414E-03	5.400E-01	20	M161	1 0 0 0 2	

2384. C$_{11}$H$_{10}$ClN$_3$O$_2$S
5-Sulfanilamido-2-chloropyridine
N1-(6-Chloro-3-pyridyl)sulfanilamide
RN: 34392-82-0   **MP** (°C):
MW: 283.74   **BP** (°C):

Solubility (Moles/L)	Solubility (Grams/L)	Temp (°C)	Ref (#)	Evaluation (T P E A A)	Comments
6.344E-04	1.800E-01	37	R058	1 2 1 1 1	

2385. C$_{11}$H$_{10}$Cl$_2$O$_3$
2,4-Dichlorophenoxyacetic acid allyl ester
Allyl 2,4-dichlorophenoxyacetate
RN: 58965-05-2   **MP** (°C):
MW: 261.11   **BP** (°C):

Solubility (Moles/L)	Solubility (Grams/L)	Temp (°C)	Ref (#)	Evaluation (T P E A A)	Comments
1.426E-04	3.722E-02	ns	M120	0 0 1 1 2	

2386. C$_{11}$H$_{10}$IN$_3$O$_2$S
2-Sulfanilamido-5-iodopyridine
Benzenesulfonamide, 4-amino-N-(5-iodo-2-pyridinyl)-
RN: 71119-21-6   **MP** (°C):
MW: 375.19   **BP** (°C):

Solubility (Moles/L)	Solubility (Grams/L)	Temp (°C)	Ref (#)	Evaluation (T P E A A)	Comments
3.465E-05	1.300E-02	37	R058	1 2 1 1 1	

2387. C$_{11}$H$_{10}$N$_2$O
3-o-Toluoxypyridazine
Credazine
3-(2-Methylphenoxy)-pyridazine
RN: 14491-59-9   **MP** (°C):   78
MW: 186.22   **BP** (°C):

Solubility (Moles/L)	Solubility (Grams/L)	Temp (°C)	Ref (#)	Evaluation (T P E A A)	Comments
1.072E-02	1.996E+00	ns	B100	0 0 0 0 0	
1.074E-02	2.000E+00	rt	M161	0 0 0 0 0	

2388. C₁₁H₁₀N₂O

$2388.$ $C_{11}H_{10}N_2O$

Vasicinone
Pyrrolo[2,1-b]quinazolin-9(1H)-one, 2,3-dihydro-3-hydroxy-, (3S)-
(−)-Vasicinone
L-Vasicinone

RN:	486-64-6	MP (°C):	204
MW:	186.22	BP (°C):	

Solubility (Moles/L)	Solubility (Grams/L)	Temp (°C)	Ref (#)	Evaluation (T P E A A)	Comments
8.578E-03	1.597E+00	25	B194	2 2 2 2 1	

2389. C₁₁H₁₀N₂O₃

$2389.$ $C_{11}H_{10}N_2O_3$

Phenylmethylbarbituric acid
Barbituric acid, 5-methyl-5-phenyl
2,4,6(1H,3H,5H)-Pyrimidinetrione, 5-methyl-5-phenyl
2,4,6-Trioxo-5-methyl-5-phenylhexahydropyrimidine
Heptobarbital

RN:	76-94-8	MP (°C):	226
MW:	218.21	BP (°C):	

Solubility (Moles/L)	Solubility (Grams/L)	Temp (°C)	Ref (#)	Evaluation (T P E A A)	Comments
3.480E-03	7.594E-01	20	J030	1 2 2 2 1	
4.170E-03	9.100E-01	25	P350	0 0 0 0 0	intrinsic
6.133E-03	1.338E+00	37	J030	1 2 2 2 2	

2390. C₁₁H₁₀N₂S

$2390.$ $C_{11}H_{10}N_2S$

1-Naphthylthiourea
ANTU

RN:	86-88-4	MP (°C):	198
MW:	202.28	BP (°C):	

Solubility (Moles/L)	Solubility (Grams/L)	Temp (°C)	Ref (#)	Evaluation (T P E A A)	Comments
2.966E-03	6.000E-01	rt	M161	0 0 0 0 2	

2391. C₁₁H₁₀N₄O₄S

$2391.$ $C_{11}H_{10}N_4O_4S$

2-Sulfanilamido-5-nitropyridine
Benzenesulfonamide, 4-amino-N-(5-nitro-2-pyridinyl)-

RN:	39588-36-8	MP (°C):	
MW:	294.29	BP (°C):	

Solubility (Moles/L)	Solubility (Grams/L)	Temp (°C)	Ref (#)	Evaluation (T P E A A)	Comments
1.257E-04	3.700E-02	37	R058	1 2 1 1 1	

2392. C₁₁H₁₁ClO₃

Alclofenac
(4-Allyloxy-3-chlorophenyl)acetic acid
(3-Chloro-4-allyloxyphenyl)acetic acid

RN: 22131-79-9   **MP** (°C):
MW: 226.66   **BP** (°C):

Solubility (Moles/L)	Solubility (Grams/L)	Temp (°C)	Ref (#)	Evaluation (T P E A A)	Comments
4.850E-05	1.099E-02	5	F306	1 0 1 2 2	intrinsic
5.780E-05	1.310E-02	25	C314	0 0 0 0 0	
5.780E-05	1.310E-02	25	C314	0 0 0 0 0	
6.200E-05	1.405E-02	25	F306	1 0 1 2 2	intrinsic
8.000E-05	1.813E-02	37	F306	1 0 1 2 2	intrinsic

2393. C₁₁H₁₁N

2,4-Dimethylquinoline
Quinoline, 2,4-dimethyl-

RN: 1198-37-4   **MP** (°C):   264
MW: 157.22   **BP** (°C):

Solubility (Moles/L)	Solubility (Grams/L)	Temp (°C)	Ref (#)	Evaluation (T P E A A)	Comments
1.142E-02	1.795E+00	25	K119	1 0 0 0 2	

2394. C₁₁H₁₁N

2,7-Dimethylquinoline
Quinoline, 2,7-dimethyl-

RN: 93-37-8   **MP** (°C):   58
MW: 157.22   **BP** (°C):

Solubility (Moles/L)	Solubility (Grams/L)	Temp (°C)	Ref (#)	Evaluation (T P E A A)	Comments
1.142E-02	1.795E+00	25	P051	2 1 1 2 2	
1.142E-02	1.795E+00	25.00	P007	2 1 2 2 2	

2395. C₁₁H₁₁NO

Aziridine, 1-(1-oxo-3-phenyl-2-propenyl)-
N-Cyclopropylcinnamamide

RN: 53162-40-6   **MP** (°C):
MW: 173.22   **BP** (°C):

Solubility (Moles/L)	Solubility (Grams/L)	Temp (°C)	Ref (#)	Evaluation (T P E A A)	Comments
3.150E-03	5.456E-01	ns	H350	0 0 0 0 0	

2396. $C_{11}H_{11}NO_2$

Phensuximide
Milontin
N-Methyl-2-phenyl-succinimide

RN:	86-34-0	MP (°C):	71–73
MW:	189.22	BP (°C):	

Solubility (Moles/L)	Solubility (Grams/L)	Temp (°C)	Ref (#)	Evaluation (T P E A A)	Comments
2.220E-02	4.200E+00	25	P061	0 0 0 0 0	

2397. $C_{11}H_{11}NO_2S$

Butyric acid, p-isothiocyanatophenyl ester

RN:	96933-13-0	MP (°C):	
MW:	221.28	BP (°C):	

Solubility (Moles/L)	Solubility (Grams/L)	Temp (°C)	Ref (#)	Evaluation (T P E A A)	Comments
8.200E-05	1.814E-02	25	K032	2 2 0 1 1	

2398. $C_{11}H_{11}NO_4$

Acetamide, N-acetyl-2-(benzoyloxy)-

RN:	68659-48-3	MP (°C):	104.5
MW:	221.21	BP (°C):	

Solubility (Moles/L)	Solubility (Grams/L)	Temp (°C)	Ref (#)	Evaluation (T P E A A)	Comments
3.978E-03	8.800E-01	22	N317	1 1 2 1 2	

2399. $C_{11}H_{11}NO_4S$

4-Thiazolidinecarboxylic acid, 2-(4-carboxyphenyl)-
Thiazolidine-4-carboxylic acid, 2-(4-carboxyphenyl)-

RN:	118845-10-6	MP (°C):	
MW:	253.28	BP (°C):	551.7

Solubility (Moles/L)	Solubility (Grams/L)	Temp (°C)	Ref (#)	Evaluation (T P E A A)	Comments
6.000E-04	1.520E-01	21	B414	1 0 0 1 1	fast decomposition, results from gravimetric determination

2400. C$_{11}$H$_{11}$NO$_5$
Benzoxydiglycine

RN: **MP** (°C):
MW: 237.21 **BP** (°C):

Solubility (Moles/L)	Solubility (Grams/L)	Temp (°C)	Ref (#)	Evaluation (T P E A A)	Comments
1.391E-02	3.300E+00	25.1	N026	0 0 0 0 0	

2401. C$_{11}$H$_{11}$NO$_5$
Benzoic acid, 2-(acetyloxy)-, 2-amino-2-oxoethyl ester
(O-Acetylsalicyloyloxy)acetamide

RN: 50785-22-3 **MP** (°C): 128.5
MW: 237.21 **BP** (°C):

Solubility (Moles/L)	Solubility (Grams/L)	Temp (°C)	Ref (#)	Evaluation (T P E A A)	Comments
1.619E-02	3.840E+00	21	N335	0 0 0 0 0	

2402. C$_{11}$H$_{11}$N$_3$OS
Seedvax
2-Amino-4-methyl-5-carboxanilidothiazole

RN: 21452-14-2 **MP** (°C): 221
MW: 233.29 **BP** (°C):

Solubility (Moles/L)	Solubility (Grams/L)	Temp (°C)	Ref (#)	Evaluation (T P E A A)	Comments
4.282E-03	9.990E-01	ns	M061	0 0 0 0 0	

2403. C$_{11}$H$_{11}$N$_3$O$_2$S
Sulfapyridine
2-(Aminobenzene-4'-sulfamido)-pyridine
2-[Aminobenzol-4'-sulfamid]-pyridin
Sulphapyridine
2-Sulfapyridine
N-(2-Pyridyl)sulfanilamide

RN: 144-83-2 **MP** (°C): 192
MW: 249.29 **BP** (°C):

Solubility (Moles/L)	Solubility (Grams/L)	Temp (°C)	Ref (#)	Evaluation (T P E A A)	Comments
6.819E-04	1.700E-01	16	H114	1 0 0 0 2	
2.006E-03	5.000E-01	20	C103	1 2 0 0 2	
1.323E-03	3.299E-01	20	D041	1 0 0 0 1	
8.023E-04	2.000E-01	20	F073	1 2 2 2 2	
1.075E-03	2.680E-01	25	C102	2 0 2 2 2	
1.049E-03	2.615E-01	25	M440	0 0 0 0 0	
1.645E-03	4.100E-01	35	H114	1 0 0 0 1	

(continued)

2403. $C_{11}H_{11}N_3O_2S$ (continued)

Solubility (Moles/L)	Solubility (Grams/L)	Temp (°C)	Ref (#)	Evaluation (T P E A A)	Comments
1.950E-03	4.860E-01	37	C102	2 0 2 2 2	
1.805E-03	4.500E-01	37	D084	1 0 1 0 1	
1.985E-03	4.948E-01	37	F072	1 0 0 0 2	
1.985E-03	4.948E-01	37	F075	1 0 2 2 2	
2.006E-03	5.000E-01	37	F300	1 0 0 0 0	
4.047E-03	1.009E+00	37	G037	2 2 2 1 0	EFG, form V
6.128E-03	1.528E+00	37	G073	2 2 2 1 0	EFG, amorphous
3.807E-03	9.491E-01	37	G073	2 2 2 1 0	EFG, form II
3.807E-03	9.491E-01	37	G073	2 2 2 1 0	EFG, form I
2.090E-03	5.210E-01	37	K095	2 0 0 0 2	intrinsic
2.447E-03	6.100E-01	37	M057	1 0 0 0 2	pH 5.5
2.607E-03	6.500E-01	37	R044	0 0 0 0 0	
6.417E-04	1.600E-01	37.50	M142	1 0 0 0 1	
2.165E-03	5.397E-01	37.50	M142	1 0 0 0 1	
2.006E-03	5.000E-01	38	K006	1 0 0 0 2	
4.412E-03	1.100E+00	40	C103	1 2 0 0 2	
4.212E-02	1.050E+01	100	C103	1 2 0 0 2	
3.972E-02	9.901E+00	100	D041	1 0 0 0 0	
1.995E-03	4.974E-01	ns	R427	0 0 0 0 0	
1.484E-03	3.699E-01	rt	N015	0 0 2 2 2	

2404. $C_{11}H_{11}N_3O_3S_2$
Acetyl sulfathiazole
Sulfathiazol acetyle
N4-Acetylsulfathiazole
N4-Acetylsulphathiazole

RN: 127-76-4 **MP** (°C):
MW: 297.36 **BP** (°C):

Solubility (Moles/L)	Solubility (Grams/L)	Temp (°C)	Ref (#)	Evaluation (T P E A A)	Comments
3.363E-04	1.000E-01	37	D084	1 0 1 0 1	
2.186E-04	6.500E-02	37	F075	1 0 2 2 1	
2.354E-04	7.000E-02	37	L091	1 0 0 0 0	pH 5.5
1.951E-04	5.800E-02	37	M057	1 0 0 0 1	pH 5.5
2.018E-04	6.000E-02	37.50	M142	1 0 0 0 0	
2.388E-04	7.100E-02	38	K006	1 0 0 0 1	

2405. $C_{11}H_{11}N_3O_3S$
5-Sulfanilamido-2-hydroxypyridine
RN: 71119-20-5 **MP** (°C):
MW: 265.29 **BP** (°C):

Solubility (Moles/L)	Solubility (Grams/L)	Temp (°C)	Ref (#)	Evaluation (T P E A A)	Comments
9.725E-03	2.580E+00	37	R058	1 2 1 1 1	

2406. C₁₁H₁₁N₅

Phenazopyridine
3-(Phenylazo)-2,6-pyridinediamine
RN: 94-78-0   **MP** (°C):   235
MW: 213.24   **BP** (°C):

Solubility (Moles/L)	Solubility (Grams/L)	Temp (°C)	Ref (#)	Evaluation (T P E A A)	Comments
4.240E+00	9.042E+02	25	B443	0 0 0 0 0	
1.738E-04	3.706E-02	ns	R427	0 0 0 0 0	

2407. C₁₁H₁₂ClNO₄

Chloroethyl acetaminophen
Carbonic acid, 4-(acetylamino)phenyl 2-chloroethyl ester
Acetanilide, 4′-hydroxy-, 2-chloroethyl carbonate (ester)
RN: 17243-29-7   **MP** (°C):   122.5–123
MW: 257.68   **BP** (°C):

Solubility (Moles/L)	Solubility (Grams/L)	Temp (°C)	Ref (#)	Evaluation (T P E A A)	Comments
1.514E-03	3.900E-01	37	D029	0 0 0 0 0	

2408. C₁₁H₁₂Cl₂N₂O₅

Chloramphenicol
D-(–)-Threo-1-(p-nitrophenyl)-2-dichloroacetamido-1,3-propanediol
Amphicol
Leukomycin
Cloramical
Intramyctin
RN: 56-75-7   **MP** (°C):   150.5
MW: 323.13   **BP** (°C):

Solubility (Moles/L)	Solubility (Grams/L)	Temp (°C)	Ref (#)	Evaluation (T P E A A)	Comments
7.717E-03	2.494E+00	20	D041	1 0 0 0 1	
5.570E-03	1.800E+00	23	M168	2 0 0 0 0	EFG
1.200E-02	3.878E+00	25	A352	0 0 0 0 0	
7.717E-03	2.494E+00	25	I312	0 0 0 0 0	
1.156E-02	3.736E+00	25.5	J011	1 0 2 1 2	pH 4.7
1.370E-02	4.427E+00	30	K020	1 0 1 1 0	EFG
1.238E-02	4.000E+00	37	G010	1 0 1 1 0	EFG
7.737E-03	2.500E+00	ns	K444	0 0 0 0 0	

2409. $C_{11}H_{12}Cl_2O_3$

2,4-D Isopropyl ester
2,4-D-Isopropyl ester
2,4-Dichlorophenoxyacetic acid isopropyl ester
2,4-Dichlorophenoxyacetic acid iso-propyl ester

RN: 94-11-1   **MP** (°C):
MW: 263.12   **BP** (°C): 139

Solubility (Moles/L)	Solubility (Grams/L)	Temp (°C)	Ref (#)	Evaluation (T P E A A)	Comments
1.040E-04	2.736E-02	ns	M120	0 0 1 1 2	
1.419E-04	3.734E-02	ns	M120	0 0 1 1 2	

2410. $C_{11}H_{12}I_3NO_2$

Iopanoic acid
β-(3-Amino-2,4,6-triiodophenyl)-α-ethylpropionic acid
Bilijodon
Cholevid
Choladine
Colepax

RN: 96-83-3   **MP** (°C): 155.2
MW: 570.94   **BP** (°C):

Solubility (Moles/L)	Solubility (Grams/L)	Temp (°C)	Ref (#)	Evaluation (T P E A A)	Comments
6.100E-04	3.483E-01	37	J016	1 0 0 0 1	pH 7.4
2.627E-05	1.500E-02	ns	H055	0 0 0 0 0	

2411. $C_{11}H_{12}NO_4PS_2$

Phosmet
Phosphorodithioic scid S-[(1,3-dihydro-1,3-dioxo-2H-isoindol-2-yl)methyl] O,O-dimethyl ester
Decemthion
Smidan
Appa
Imidan

RN: 732-11-6   **MP** (°C):
MW: 317.32   **BP** (°C):

Solubility (Moles/L)	Solubility (Grams/L)	Temp (°C)	Ref (#)	Evaluation (T P E A A)	Comments
7.690E-05	2.440E-02	20	B300	2 1 1 1 2	
7.878E-05	2.500E-02	25	M061	1 0 0 0 1	
7.878E-05	2.500E-02	25	M161	1 0 0 0 1	
7.878E-05	2.500E-02	ns	F071	0 1 2 1 1	
7.943E-05	2.521E-02	ns	R427	0 0 0 0 0	

2412. C₁₁H₁₂N₂O

Antipyrine

Antipyrin

2,3-Dimethyl-1-phenyl-3-pyrazolin-5-one

1,2-Dihydro-1,5-dimethyl-2-phenyl-3H-pyrazol-3-one

Phenazone

RN: 60-80-0   **MP** (°C):   114

MW: 188.23   **BP** (°C):   319

Solubility (Moles/L)	Solubility (Grams/L)	Temp (°C)	Ref (#)	Evaluation (T P E A A)	Comments
1.550E+00	2.918E+02	2.5	K075	1 0 0 0 2	
1.968E+00	3.705E+02	4.62	M109	2 1 1 1 0	EFG
1.472E-01	2.771E+01	5	L089	1 0 0 0 2	sic
1.613E+00	3.036E+02	6.1	K075	1 0 0 0 2	
1.777E-01	3.344E+01	10	L089	1 0 0 0 2	sic
2.084E+00	3.922E+02	11.74	M109	2 1 1 1 0	EFG
2.261E+00	4.256E+02	14.20	M109	2 1 1 1 0	EFG
1.771E+00	3.333E+02	20	D041	1 0 0 0 0	
2.205E-01	4.150E+01	20	L089	1 0 0 0 2	sic
2.472E+00	4.654E+02	20.96	M109	2 1 1 1 0	EFG
2.621E-01	4.934E+01	25	L089	1 0 0 0 2	sic
3.294E+00	6.200E+02	25	P012	0 0 0 0 0	
3.294E+00	6.200E+02	25	P016	1 0 0 1 2	
3.559E+00	6.700E+02	25	P020	2 0 1 1 2	
2.717E+00	5.114E+02	25.35	M109	2 1 1 1 0	EFG
3.020E+00	5.685E+02	29.87	M109	2 1 1 1 0	EFG
2.621E-01	4.934E+01	30	L089	1 0 0 0 2	sic
2.983E-01	5.616E+01	35	L089	1 0 0 0 2	sic
3.968E+00	7.468E+02	39.34	M109	2 1 1 1 0	EFG
3.359E-01	6.323E+01	40	L089	1 0 0 0 2	sic
5.637E-01	1.061E+02	50	L089	1 0 0 0 2	sic
1.493E+00	2.811E+02	.0	K075	1 0 0 0 2	
2.656E+00	5.000E+02	rt	D021	0 0 1 1 2	

2413. C₁₁H₁₂N₂O₂

DL-Tryptophan

1H-Indole-3-alanine

DL-α-Amino-3-indolepropionic acid

RN: 54-12-6   **MP** (°C):   289

MW: 204.23   **BP** (°C):

Solubility (Moles/L)	Solubility (Grams/L)	Temp (°C)	Ref (#)	Evaluation (T P E A A)	Comments
1.020E-02	2.083E+00	20	N006	0 0 0 0 0	
1.140E-02	2.328E+00	25	N006	0 0 0 0 0	
1.221E-02	2.494E+00	30	D041	1 0 0 0 1	
1.250E-02	2.553E+00	30	N006	0 0 0 0 0	
1.200E-02	2.451E+00	30	N009	0 0 0 0 0	
1.640E-02	3.349E+00	40	N006	0 0 0 0 0	
1.570E-02	3.206E+00	40	N009	0 0 0 0 0	
2.150E-02	4.391E+00	50	N006	0 0 0 0 0	

2414. $C_{11}H_{12}N_2O_2$
Tryptophan
2-Amino-3-(lH-indol-3-yl)-propanoic acid
3-Indol-3-ylalanine
L-β-3-indolylalanine
Trp
(S)-(–)-Tryptophan

RN: 73-22-3   **MP (°C):**
MW: 204.23   **BP (°C):**

Solubility (Moles/L)	Solubility (Grams/L)	Temp (°C)	Ref (#)	Evaluation (T P E A A)	Comments
4.015E-02	8.200E+00	0	F300	1 0 0 0 1	
6.042E-02	1.234E+01	20	B032	1 2 2 1 2	
6.395E-02	1.306E+01	22.5	P045	0 0 2 1 2	
6.551E-02	1.338E+01	25	B032	1 2 2 1 2	
5.519E-02	1.127E+01	25	D041	1 0 0 0 2	
5.337E-02	1.090E+01	25	F300	1 0 0 0 2	
6.665E-02	1.361E+01	25	G092	2 1 1 1 1	
6.665E-02	1.361E+01	25	G315	0 0 0 0 0	
5.519E-02	1.127E+01	25	H070	1 0 0 0 2	
6.267E-02	1.280E+01	25.1	N024	0 0 0 0 0	
6.757E-02	1.380E+01	25.1	N025	0 0 0 0 0	
6.757E-02	1.380E+01	25.1	N026	0 0 0 0 0	
6.665E-02	1.361E+01	25.1	N027	1 1 2 2 2	
1.787E-01	3.650E+01	27	D036	0 0 0 0 0	
5.386E-02	1.100E+01	28	L081	2 1 2 2 2	
7.056E-02	1.441E+01	29.80	B032	1 2 2 1 2	
8.100E-02	1.654E+01	30	N009	0 0 0 0 0	
9.480E-02	1.936E+01	40	N009	0 0 0 0 0	
8.226E-02	1.680E+01	50	F300	1 0 0 0 2	
1.122E-01	2.291E+01	50	N009	0 0 0 0 0	
1.200E-01	2.450E+01	70	F300	1 0 0 0 2	
1.334E-01	2.724E+01	75	D041	1 0 0 0 2	
2.448E-01	5.000E+01	100	F300	1 0 0 0 1	

2415. $C_{11}H_{12}N_2O_2$
5-Ethyl-5-phenylhydantoin
2,4-Imidazolidinedione, 5-ethyl-5-phenyl-
Nirvanol
5-Phenyl-5-ethylhydantoin
Normephenytoin

RN: 631-07-2   **MP (°C):**
MW: 204.23   **BP (°C):**

Solubility (Moles/L)	Solubility (Grams/L)	Temp (°C)	Ref (#)	Evaluation (T P E A A)	Comments
3.938E-03	8.044E-01	37	F183	1 0 1 1 1	intrinsic

2416. C$_{11}$H$_{12}$N$_2$O$_4$

Acetamide, N-(2-amino-2-oxoethyl)-2-(benzoyloxy)-

RN: 106231-53-2   **MP** (°C):   151.5
MW: 236.23   **BP** (°C):   568.6

Solubility (Moles/L)	Solubility (Grams/L)	Temp (°C)	Ref (#)	Evaluation (T P E A A)	Comments
3.175E-02	7.500E+00	22	B427	1 0 0 1 1	in 0.01M HCl
3.175E-02	7.500E+00	22	N317	1 1 2 1 2	

2417. C$_{11}$H$_{12}$N$_4$O$_2$S

2-Sulfanilamido-5-aminopyridine
Benzenesulfonamide, 4-amino-N-(5-amino-2-pyridinyl)-

RN: 16840-28-1   **MP** (°C):
MW: 264.31   **BP** (°C):

Solubility (Moles/L)	Solubility (Grams/L)	Temp (°C)	Ref (#)	Evaluation (T P E A A)	Comments
1.581E-02	4.180E+00	37	R058	1 2 1 1 2	

2418. C$_{11}$H$_{12}$N$_4$O$_2$S

4-Sulfanilamido-2-methylpyrimidine
Benzenesulfonamide, 4-amino-N-(2-methyl-4-pyrimidinyl)-

RN: 599-84-8   **MP** (°C):
MW: 264.31   **BP** (°C):

Solubility (Moles/L)	Solubility (Grams/L)	Temp (°C)	Ref (#)	Evaluation (T P E A A)	Comments
2.357E-02	6.230E+00	37	R046	1 2 1 1 2	

2419. C$_{11}$H$_{12}$N$_4$O$_2$S

Sulfamethylpyrimidine
Ulfamerazine
Sulfamerazine

RN: 127-79-7   **MP** (°C):   234
MW: 264.31   **BP** (°C):

Solubility (Moles/L)	Solubility (Grams/L)	Temp (°C)	Ref (#)	Evaluation (T P E A A)	Comments
8.967E-04	2.370E-01	20	F073	1 2 2 2 2	
7.641E-04	2.020E-01	20	L058	1 0 1 1 2	
8.012E-04	2.118E-01	25	M440	0 0 0 0 0	
1.400E-03	3.700E-01	37	L091	1 0 0 0 1	pH 5.5
1.203E-03	3.180E-01	37	R045	1 2 1 1 2	
1.381E-03	3.650E-01	37	S192	1 0 1 1 2	pH 6.0
1.551E-03	4.100E-01	38	K006	1 0 0 0 1	

2420. $C_{11}H_{12}N_4O_3S_2$
N4-Acetyl sulfamethizole
Acetyl sulfamethylthiazole
RN: 39719-87-4 MP (°C):
MW: 312.37 BP (°C):

Solubility (Moles/L)	Solubility (Grams/L)	Temp (°C)	Ref (#)	Evaluation (T P E A A)	Comments
1.313E-03	4.100E-01	37	B046	1 0 2 2 1	pH 4.5

2421. $C_{11}H_{12}N_4O_3S$
Sulfamethoxypyridazine
Sulphamethoxypyridazine
4-Amino-N-(6-methoxy-3-pyridazinyl)-benzenesulfonamide
RN: 80-35-3 MP (°C): 182.5
MW: 280.31 BP (°C):

Solubility (Moles/L)	Solubility (Grams/L)	Temp (°C)	Ref (#)	Evaluation (T P E A A)	Comments
2.067E-03	5.795E-01	25	E314	0 0 0 0 0	intrinsic
2.569E-02	7.200E+00	37	B046	1 0 2 2 2	pH 4.5

2422. $C_{11}H_{12}N_4O_3S$
Sulfameter
Sulphamethoxydiazine
RN: 651-06-9 MP (°C): 213
MW: 280.31 BP (°C):

Solubility (Moles/L)	Solubility (Grams/L)	Temp (°C)	Ref (#)	Evaluation (T P E A A)	Comments
1.677E-03	4.700E-01	30	M113	2 2 2 2 0	form III, EFG, 0.1N HCl
2.604E-03	7.300E-01	30	M113	2 2 2 2 0	form II, EFG, 0.1N HCl
1.891E-03	5.300E-01	30	M113	2 2 2 2 0	form I, EFG, 0.1N HCl
2.462E-03	6.900E-01	30	M113	2 2 2 2 0	EFG, 0.1N HCl, amorphous
3.211E-04	9.000E-02	37.5	C081	1 0 1 0 0	EFG, form III
6.243E-04	1.750E-01	37.5	C081	1 0 1 0 0	EFG, form II
4.281E-04	1.200E-01	37.5	C081	1 0 1 0 0	EFG, form I

2423. $C_{11}H_{12}N_4O_3S$
5-Sulfanilamido-2-methoxypyrimidine
Benzenesulfonamide, 4-amino-N-(2-methoxy-5-pyrimidinyl)-
RN: 71119-37-4 MP (°C):
MW: 280.31 BP (°C):

Solubility (Moles/L)	Solubility (Grams/L)	Temp (°C)	Ref (#)	Evaluation (T P E A A)	Comments
3.282E-04	9.200E-02	37	R046	1 2 1 1 1	

2424. C₁₁H₁₂N₄O₃S

2-Sulfanilamido-4-methoxypyrimidine
Benzenesulfonamide, 4-amino-*N*-(4-methoxy-2-pyrimidinyl)-
RN: 3213-22-7   **MP** (°C):
MW: 280.31   **BP** (°C):

Solubility (Moles/L)	Solubility (Grams/L)	Temp (°C)	Ref (#)	Evaluation (T P E A A)	Comments
6.493E-04	1.820E-01	37	R046	1 2 1 1 2	

2425. C₁₁H₁₂N₄O₅

2,5-Diacetoxymethyl allopurinol
4H-Pyrazolo[3,4-d]pyrimidin-4-one, 2,5-bis[(acetyloxy)methyl]-2,5-dihydro-
RN: 98827-24-8   **MP** (°C):   153–154
MW: 280.24   **BP** (°C):

Solubility (Moles/L)	Solubility (Grams/L)	Temp (°C)	Ref (#)	Evaluation (T P E A A)	Comments
1.035E-02	2.900E+00	22	B322	0 0 0 0 0	

2426. C₁₁H₁₂N₆O₂S

6-Sulfapurine
RN: **MP** (°C):
MW: 292.32   **BP** (°C):

Solubility (Moles/L)	Solubility (Grams/L)	Temp (°C)	Ref (#)	Evaluation (T P E A A)	Comments
4.447E-05	1.300E-02	20	F073	1 2 2 2 1	

2427. C₁₁H₁₂O₂

Cinnamyl acetate
3-Phenylallyl acetate
3-Phenyl-2-propenyl acetate
1-Acetoxy-3-phenyl-2-propene
3-Phenyl-2-propen-1-ol acetate
NSC 46109
RN: 103-54-8   **MP** (°C):
MW: 176.22   **BP** (°C):   170 (°50 torr)

Solubility (Moles/L)	Solubility (Grams/L)	Temp (°C)	Ref (#)	Evaluation (T P E A A)	Comments
1.000E-03	1.762E-01	25	D407	1 0 2 2 2	
1.000E-03	1.762E-01	ns	S460	0 0 0 0 0	

2428. $C_{11}H_{12}O_2$

Ethyl cinnamate
Ethyl (E)-cinnamate
Ethyl 3-phenyl propenoate
Ethyl phenylacrylate

RN: 103-36-6	**MP** (°C):	6
MW: 176.22	**BP** (°C):	271

Solubility (Moles/L)	Solubility (Grams/L)	Temp (°C)	Ref (#)	Evaluation (T P E A A)	Comments
1.010E-03	1.780E-01	25	A002	1 2 1 1 2	

2429. $C_{11}H_{12}O_4$

3,5-Dimethoxycinnamic acid
Predominantly *trans* isomer

RN: 16909-11-8	**MP** (°C):	174.5
MW: 208.22	**BP** (°C):	

Solubility (Moles/L)	Solubility (Grams/L)	Temp (°C)	Ref (#)	Evaluation (T P E A A)	Comments
1.510E-04	3.144E-02	25	R070	0 0 0 0 0	

2430. $C_{11}H_{12}O_4$

Ethyl acetylsalicylate
Acetyl salicylic acid, ethyl ester

RN: 529-68-0	**MP** (°C):	
MW: 208.22	**BP** (°C):	

Solubility (Moles/L)	Solubility (Grams/L)	Temp (°C)	Ref (#)	Evaluation (T P E A A)	Comments
1.594E-02	3.320E+00	37	G430	0 0 0 0 0	pH 4.5

2431. $C_{11}H_{12}O_4$

Propionyl-*r*-mandelic acid

RN:	**MP** (°C):	126
MW: 208.22	**BP** (°C):	

Solubility (Moles/L)	Solubility (Grams/L)	Temp (°C)	Ref (#)	Evaluation (T P E A A)	Comments
1.389E-02	2.892E+00	0	A043	1 2 1 1 1	
1.389E-02	2.892E+00	0	L035	1 2 2 1 1	
1.675E-02	3.488E+00	10	A043	1 2 1 1 1	
1.675E-02	3.488E+00	10	L035	1 2 2 1 1	
1.770E-02	3.686E+00	15	A043	1 2 1 1 1	
1.770E-02	3.686E+00	15	L035	1 2 2 1 1	
1.818E-02	3.786E+00	20	A043	1 2 1 1 1	
1.818E-02	3.786E+00	20	L035	1 2 2 1 1	
2.484E-02	5.173E+00	25	A043	1 2 1 1 1	

(*continued*)

2431. C$_{11}$H$_{12}$O$_4$ (continued)

Solubility (Moles/L)	Solubility (Grams/L)	Temp (°C)	Ref (#)	Evaluation (T P E A A)	Comments
2.484E-02	5.173E+00	25	L035	1 2 2 1 1	
2.817E-02	5.865E+00	30	A043	1 2 1 1 1	
2.817E-02	5.865E+00	30	L035	1 2 2 1 1	
3.528E-02	7.346E+00	35	A043	1 2 1 1 1	
3.528E-02	7.346E+00	35	L035	1 2 2 1 1	
5.789E-02	1.205E+01	40	A043	1 2 1 1 2	
5.789E-02	1.205E+01	40	L035	1 2 2 1 2	
8.724E-02	1.816E+01	45	A043	1 2 1 1 2	
8.724E-02	1.816E+01	45	L035	1 2 2 1 2	
1.606E-01	3.344E+01	50	A043	1 2 1 1 2	
1.606E-01	3.344E+01	50	L035	1 2 2 1 2	

2432. C$_{11}$H$_{12}$O$_4$S
Benzoic acid, 2-(acetyloxy)-, (methylthio)methyl ester
RN: 76432-30-9   **MP** (°C):
MW: 240.28   **BP** (°C):

Solubility (Moles/L)	Solubility (Grams/L)	Temp (°C)	Ref (#)	Evaluation (T P E A A)	Comments
2.289E-03	5.500E-01	21	N335	0 0 0 0 0	

2433. C$_{11}$H$_{12}$O$_5$S
2-(Acetoxy)-benzoic acid, (methylsulfinyl)methyl ester
RN: 76432-33-2   **MP** (°C):   80.5
MW: 256.28   **BP** (°C):

Solubility (Moles/L)	Solubility (Grams/L)	Temp (°C)	Ref (#)	Evaluation (T P E A A)	Comments
1.651E-02	4.230E+00	21	N335	0 0 0 0 0	

2434. C$_{11}$H$_{12}$O$_6$S
2-(Acetoxy)-benzoic acid, (methylsulfonyl)methyl ester
RN: 76432-35-4   **MP** (°C):   150
MW: 272.28   **BP** (°C):

Solubility (Moles/L)	Solubility (Grams/L)	Temp (°C)	Ref (#)	Evaluation (T P E A A)	Comments
4.040E-04	1.100E-01	21	N335	0 0 0 0 0	

2435. C₁₁H₁₃ClO₃

2435. $C_{11}H_{13}ClO_3$

Bexone
4-(2-Methyl-4-chlorophenoxy)butyric acid
4-(MCPB)
MCPB

RN: 94-81-5 **MP** (°C):
MW: 228.68 **BP** (°C):

Solubility (Moles/L)	Solubility (Grams/L)	Temp (°C)	Ref (#)	Evaluation (T P E A A)	Comments
2.099E-04	4.800E-02	25	B164	1 0 1 1 1	
1.924E-04	4.400E-02	ns	L024	0 0 0 0 1	
1.924E-04	4.400E-02	ns	M061	0 0 0 0 1	
1.924E-04	4.400E-02	rt	M161	0 0 0 0 1	

2436. $C_{11}H_{13}FN_2O_4$

1-Cyclohexyloxycarbonyl-5-fluorouracil
1(2H)-Pyrimidinecarboxylic acid, 5-fluoro-3,4-dihydro-2,4-dioxo-, cyclohexyl ester

RN: 109232-74-8 **MP** (°C):
MW: 256.24 **BP** (°C):

Solubility (Moles/L)	Solubility (Grams/L)	Temp (°C)	Ref (#)	Evaluation (T P E A A)	Comments
3.590E-03	9.200E-01	22	B332	1 1 0 0 1	pH 4.0

2437. $C_{11}H_{13}F_3N_2O_3S$

Mefluidide
N-(2,4-Dimethyl-5-(((trifluoromethyl)sulfonyl)amino)phenyl)acetamide
Vistar
Embark
MBR 12325
Methafluoridamid

RN: 53780-34-0 **MP** (°C): 184
MW: 310.30 **BP** (°C):

Solubility (Moles/L)	Solubility (Grams/L)	Temp (°C)	Ref (#)	Evaluation (T P E A A)	Comments
5.801E-04	1.800E-01	23	M161	1 0 0 0 2	

2438. $C_{11}H_{13}F_3N_4O_4$

Dinitramine
1,3-Benzenediamine, N1,N1-diethyl-2,6-dinitro-4-(trifluoromethyl)-
N3,N3-Diethyl-2,4-dinitro-6-(trifluoromethyl)-m-phenylenediamine
N3,N3-Diethyl-2,4-dinitro-6-(trifluoromethyl)-1,3-phenylenediamine
USB 3584

RN: 29091-05-2 **MP** (°C): 98.5
MW: 322.25 **BP** (°C):

Solubility (Moles/L)	Solubility (Grams/L)	Temp (°C)	Ref (#)	Evaluation (T P E A A)	Comments
3.414E-06	1.100E-03	25	M161	1 0 0 0 1	

2439. C$_{11}$H$_{13}$NO

N,N-Dimethylcinnamide
Cinnamic acid dimethylamide
N,N-Dimethyl-3-phenyl-2-propenamide

RN: 13156-74-6   **MP** (°C):
MW: 175.23   **BP** (°C):

Solubility (Moles/L)	Solubility (Grams/L)	Temp (°C)	Ref (#)	Evaluation (T P E A A)	Comments
1.670E-02	2.926E+00	ns	H350	0 0 0 0 0	

2440. C$_{11}$H$_{13}$NO

N-Ethylcinnamamide
N-Ethyl-3-phenyl-2-propenamide

RN: 23784-45-4   **MP** (°C):
MW: 175.23   **BP** (°C):

Solubility (Moles/L)	Solubility (Grams/L)	Temp (°C)	Ref (#)	Evaluation (T P E A A)	Comments
6.390E-03	1.120E+00	ns	H350	0 0 0 0 0	

2441. C$_{11}$H$_{13}$NO$_2$S

2-p-Tolyl-4-thiazolidinecarboxylic acid
4-Thiazolidinecarboxylic acid, 2-(4-methylphenyl)-

RN: 67189-37-1   **MP** (°C):
MW: 223.30   **BP** (°C): 444.2

Solubility (Moles/L)	Solubility (Grams/L)	Temp (°C)	Ref (#)	Evaluation (T P E A A)	Comments
1.800E-03	4.019E-01	21	B414	1 0 0 1 1	very fast and extent decomposition, uncertain value

2442. C$_{11}$H$_{13}$NO$_3$

Acetaminophen propionate
Propionic acid, p-acetamidophenyl ester

RN: 54942-42-6   **MP** (°C): 130
MW: 207.23   **BP** (°C):

Solubility (Moles/L)	Solubility (Grams/L)	Temp (°C)	Ref (#)	Evaluation (T P E A A)	Comments
1.544E-03	3.200E-01	25	B010	1 1 1 1 0	

2443. C$_{11}$H$_{13}$NO$_3$

Acetamide, 2-(benzoyloxy)-N-ethyl-
2-(Benzoyloxy)-N-ethylacetamide

RN:	64649-57-6	MP (°C):	106
MW:	207.23	BP (°C):	

Solubility (Moles/L)	Solubility (Grams/L)	Temp (°C)	Ref (#)	Evaluation (T P E A A)	Comments
5.791E-03	1.200E+00	22	B427	1 0 0 1 1	in 0.01M HCl
5.791E-03	1.200E+00	22	N317	1 1 2 1 2	

2444. C$_{11}$H$_{13}$NO$_3$

Acetamide, 2-(benzoyloxy)-N,N-dimethyl-
2-(Benzoyloxy)-N,N-dimethylacetamide

RN:	106231-54-3	MP (°C):	81.5
MW:	207.23	BP (°C):	351.6

Solubility (Moles/L)	Solubility (Grams/L)	Temp (°C)	Ref (#)	Evaluation (T P E A A)	Comments
4.246E-02	8.800E+00	22	B427	1 0 0 1 1	in 0.01M HCl
4.246E-02	8.800E+00	22	N317	1 1 2 1 2	

2445. C$_{11}$H$_{13}$NO$_3$S

4-Thiazolidinecarboxylic acid, 2-(4-methoxyphenyl)-
Thiazolidine-4-carboxylic acid, 2-(4-methoxyphenyl)-

RN:	65884-40-4	MP (°C):	165–166
MW:	239.30	BP (°C):	466.0

Solubility (Moles/L)	Solubility (Grams/L)	Temp (°C)	Ref (#)	Evaluation (T P E A A)	Comments
4.000E-04	9.572E-02	21	B414	1 0 0 1 1	fast decomposition

2446. C$_{11}$H$_{13}$NO$_4$

Bendiocarb
2,2-Dimethyl-1,3-benzodioxol-4-ol methylcarbamate
Fuam
Multimet
Garvox

RN:	22781-23-3	MP (°C):	129.5
MW:	223.23	BP (°C):	

Solubility (Moles/L)	Solubility (Grams/L)	Temp (°C)	Ref (#)	Evaluation (T P E A A)	Comments
1.792E-04	4.000E-02	25	M161	1 0 0 0 1	
1.792E-04	4.000E-02	25	W310	1 0 0 0 0	

2447. C₁₁H₁₃NO₄

N,N-Dimethyl glycolamide salicylate
2-Hydroxybenzoic acid, 2-(dimethylamino)-2-oxoethyl ester

RN: 114665-08-6   **MP** (°C):   68
MW: 223.23   **BP** (°C):

Solubility (Moles/L)	Solubility (Grams/L)	Temp (°C)	Ref (#)	Evaluation (T P E A A)	Comments
1.971E-02	4.400E+00	21	B331	1 2 2 1 0	pH 7.4
1.971E-02	4.400E+00	21	B331	0 0 0 0 0	

2448. C₁₁H₁₃NO₄

Ethyl acetaminophen
Carbonic acid, 4-(acetylamino)phenyl ethyl ester
Acetanilide, 4'-hydroxy-, ethyl carbonate (ester)

RN: 17243-26-4   **MP** (°C):   121–122
MW: 223.23   **BP** (°C):

Solubility (Moles/L)	Solubility (Grams/L)	Temp (°C)	Ref (#)	Evaluation (T P E A A)	Comments
4.928E-03	1.100E+00	37	D029	0 0 0 0 0	

2449. C₁₁H₁₃NO₄

Dioxacarb
2-(1,3-Dioxolan-2-yl)phenyl methylcarbamate
2-(1,3-Dioxolan-2-yl)-phenyl *N*-methylcarbamate
Elocron
Famid

RN: 6988-21-2   **MP** (°C):   114.5
MW: 223.23   **BP** (°C):

Solubility (Moles/L)	Solubility (Grams/L)	Temp (°C)	Ref (#)	Evaluation (T P E A A)	Comments
2.688E-02	6.000E+00	20	M161	1 0 0 0 0	

2450. C₁₁H₁₃NO₄S

4-Thiazolidinecarboxylic acid, 2-(2-hydroxy-3-methoxyphenyl)-
Thiazolidine-4-carboxylic acid, 2-(2-hydroxy-3-methoxyphenyl)-

RN: 72678-93-4   **MP** (°C):
MW: 255.29   **BP** (°C):   435.5

Solubility (Moles/L)	Solubility (Grams/L)	Temp (°C)	Ref (#)	Evaluation (T P E A A)	Comments
6.000E-03	1.532E+00	2.1	B414	1 0 0 1 1	fast decomposition

2451. C₁₁H₁₃N₃O

Ampyrone
4-Aminoantipyrine
Aminophenazone

RN:	83-07-8	MP (°C):	109
MW:	203.25	BP (°C):	

Solubility (Moles/L)	Solubility (Grams/L)	Temp (°C)	Ref (#)	Evaluation (T P E A A)	Comments
9.053E-01	1.840E+02	5.39	M109	2 1 1 1 0	EFG
1.088E+00	2.211E+02	10.93	M109	2 1 1 1 0	EFG
1.252E+00	2.544E+02	14.20	M109	2 1 1 1 0	EFG
1.527E+00	3.103E+02	20.96	M109	2 1 1 1 0	EFG
2.076E+00	4.218E+02	25.35	M109	2 1 1 1 0	EFG
2.384E+00	4.845E+02	29.87	M109	2 1 1 1 0	EFG
2.400E-01	4.878E+01	30	I010	2 1 2 2 1	EFG, *sic*
2.862E+00	5.816E+02	39.34	M109	2 1 1 1 0	EFG

2452. C₁₁H₁₃N₃O₃S

Sulfamoxole
Sulfuno
N-(4,5-Dimethyloxazol-2-yl)sulfanilamide

RN:	729-99-7	MP (°C):	193
MW:	267.31	BP (°C):	

Solubility (Moles/L)	Solubility (Grams/L)	Temp (°C)	Ref (#)	Evaluation (T P E A A)	Comments
3.595E-03	9.610E-01	20	K028	2 1 2 1 2	pH 6.0, form I
3.430E-03	9.170E-01	20	K028	2 1 2 1 2	pH 3.8, form I
3.277E-03	8.760E-01	20	K028	2 1 2 1 2	pH 6.0, form II
3.165E-03	8.460E-01	20	K028	2 1 2 1 2	pH 3.8, form II
6.274E-03	1.677E+00	20	K028	2 1 2 1 2	pH 7.3, form I
5.447E-03	1.456E+00	20	K028	2 1 2 1 2	pH 7.3, form II
3.427E-03	9.162E-01	20	M042	1 0 0 0 2	pH 3.8, form I, mp 205-211 C
3.162E-03	8.453E-01	20	M042	1 0 0 0 2	pH 3.8, form II, mp 188-195 C

2453. C₁₁H₁₃N₃O₃S

Sulfisoxazole
4-Amino-*N*-(3,4-dimethyl-5-isoxazolyl)benzenesulfonamide
3,4-Dimethyl-5-sulfanilamidoisoxazole
Gantrisin
Urogan
Urisoxin

RN:	127-69-5	MP (°C):	194
MW:	267.31	BP (°C):	

Solubility (Moles/L)	Solubility (Grams/L)	Temp (°C)	Ref (#)	Evaluation (T P E A A)	Comments
1.235E-03	3.300E-01	37	B046	1 0 2 2 1	pH 4.5
3.142E-04	8.400E-02	37	K022	1 0 1 1 0	intrinsic
1.092E-03	2.920E-01	37	K091	1 0 0 0 2	

2454. C$_{11}$H$_{13}$N$_3$O$_3$S

N1-Methyl-N1-(5-methyl-3-isoxazolyl)sulfanilamide
N1-Methylsulfamethoxazole

RN:	51543-31-8	MP (°C):
MW:	267.31	BP (°C):

Solubility (Moles/L)	Solubility (Grams/L)	Temp (°C)	Ref (#)	Evaluation (T P E A A)	Comments
6.280E-04	1.679E-01	37	K095	2 0 0 0 2	intrinsic

2455. C$_{11}$H$_{13}$N$_5$O$_2$

Carbovir
9-[4α-(Hydroxymethyl)-cyclopent-2-ene-1α-yl]guanine

RN:	118353-05-2	MP (°C):
MW:	247.26	BP (°C):

Solubility (Moles/L)	Solubility (Grams/L)	Temp (°C)	Ref (#)	Evaluation (T P E A A)	Comments
5.015E-03	1.240E+00	25	A338	0 0 0 0 0	

2456. C$_{11}$H$_{13}$N$_5$O$_5$

Arabinosyladenine 5'-formate
Arabinosyladenine 5'-O-formate ester
NSC 171240

RN:	55648-40-3	MP (°C):	168–170
MW:	295.26	BP (°C):	

Solubility (Moles/L)	Solubility (Grams/L)	Temp (°C)	Ref (#)	Evaluation (T P E A A)	Comments
1.152E-01	3.400E+01	ns	R030	0 0 0 0 0	

2457. C$_{11}$H$_{14}$ClNO

Propachlor
2-Chloro-N-isopropylacetanilide
N-Isopropyl-2-chloroacetanilide
N-Isopropyl-α-chloroacetanilide

RN:	1918-16-7	MP (°C):	67
MW:	211.69	BP (°C):	110

Solubility (Moles/L)	Solubility (Grams/L)	Temp (°C)	Ref (#)	Evaluation (T P E A A)	Comments
3.307E-03	7.000E-01	20	B200	1 0 0 0 2	
3.307E-03	7.000E-01	20	M161	1 0 0 0 2	
3.304E-03	6.995E-01	ns	J008	0 0 0 0 0	
3.304E-03	6.995E-01	ns	M061	0 0 0 0 0	
2.362E-03	5.000E-01	ns	M110	0 0 0 0 0	EFG

2458. $C_{11}H_{14}N_2O$

Cytisine
Cytisin

RN:	485-35-8	MP (°C):	155
MW:	190.25	BP (°C):	

Solubility (Moles/L)	Solubility (Grams/L)	Temp (°C)	Ref (#)	Evaluation (T P E A A)	Comments
2.308E+00	4.390E+02	16	F300	1 0 0 0 2	

2459. $C_{11}H_{14}N_2O_3S$

Sulfadicramide
2-Butenamide, N-[(4-aminophenyl)sulfonyl]-3-methyl-
N-Sulfanilyl-β,β-dimethylacrylamide
Sulfirgamid
Irgamide
Sulfirgamide

RN:	115-68-4	MP (°C):	184.5
MW:	254.31	BP (°C):	

Solubility (Moles/L)	Solubility (Grams/L)	Temp (°C)	Ref (#)	Evaluation (T P E A A)	Comments
1.026E-03	2.610E-01	20	F073	1 2 2 2 2	

2460. $C_{11}H_{14}N_4O_2S_2$

4-Amino-N-(5-isopropyl-1,3,4-thiadiazol-2-yl)benzenesulfonamide
N1-(5-Isopropyl-1,3,4-thiadiazol-2-yl)sulfanilamide
Sulfaisopropylthiadiazole
Glyprothiazole
PASIT
RP 2254

RN:	80-34-2	MP (°C):	
MW:	298.39	BP (°C):	

Solubility (Moles/L)	Solubility (Grams/L)	Temp (°C)	Ref (#)	Evaluation (T P E A A)	Comments
7.330E-04	2.187E-01	37	A046	2 0 1 1 2	

2461. $C_{11}H_{14}N_4O_2S_2$

4-Amino-N-(5-propyl-1,3,4-thiadiazol-2-yl)benzenesulfonamide
N1-(5-Propyl-1,3,4-thiadiazol-2-yl)sulfanilamide

RN:	71119-32-9	MP (°C):	
MW:	298.39	BP (°C):	

Solubility (Moles/L)	Solubility (Grams/L)	Temp (°C)	Ref (#)	Evaluation (T P E A A)	Comments
8.980E-04	2.680E-01	37	A046	2 0 1 1 2	

2462. C₁₁H₁₄N₄O₃
2-Pivaloyloxymethyl allopurinol
Propanoic acid, 2,2-dimethyl-, (4,5-dihydro-4-oxo-2H-pyrazolo[3,4-d]pyrimidin-2-yl)methyl ester
RN: 98827-15-7   **MP (°C):**   180–181
MW: 250.26   **BP (°C):**

Solubility (Moles/L)	Solubility (Grams/L)	Temp (°C)	Ref (#)	Evaluation (T P E A A)	Comments
6.793E-03	1.700E+00	22	B322	0 0 0 0 0	

2463. C₁₁H₁₄N₄O₃
1-Pivaloyloxymethyl allopurinol
Propanoic acid, 2,2-dimethyl-, (4,5-dihydro-4-oxo-2H-pyrazolo[3,4-d]pyrimidin-1-yl)methyl ester
RN: 98827-18-0   **MP (°C):**   185–187
MW: 250.26   **BP (°C):**

Solubility (Moles/L)	Solubility (Grams/L)	Temp (°C)	Ref (#)	Evaluation (T P E A A)	Comments
2.078E-03	5.200E-01	22	B322	0 0 0 0 0	

2464. C₁₁H₁₄N₄O₅
6-Methoxypurine arabinoside
9H-Purine, 9-β-D-arabinofuranosyl-6-methoxy-
RN: 91969-06-1   **MP (°C):**
MW: 282.26   **BP (°C):**

Solubility (Moles/L)	Solubility (Grams/L)	Temp (°C)	Ref (#)	Evaluation (T P E A A)	Comments
4.980E-02	1.406E+01	37	C348	0 0 0 0 0	pH 7.00

2465. C₁₁H₁₄O
o-2-Pentenylphenol
Phenol, 2-(2-pentenyl)-
RN: 62536-86-1   **MP (°C):**
MW: 162.23   **BP (°C):**

Solubility (Moles/L)	Solubility (Grams/L)	Temp (°C)	Ref (#)	Evaluation (T P E A A)	Comments
2.054E-03	3.332E-01	25	L021	1 0 0 0 0	

2466. C₁₁H₁₄O₂

δ-Phenylvaleric acid
Benzenepentanoic acid
5-Phenylvaleric acid

RN: 2270-20-4   **MP** (°C): 59
MW: 178.23   **BP** (°C):

Solubility (Moles/L)	Solubility (Grams/L)	Temp (°C)	Ref (#)	Evaluation (T P E A A)	Comments
9.969E-03	1.777E+00	30	D033	2 2 1 2 2	
1.159E-02	2.066E+00	40	D033	2 2 1 2 2	

2467. C₁₁H₁₄O₂

4-Butylbenzoic acid

RN: 20651-71-2   **MP** (°C): 100
MW: 178.23   **BP** (°C):

Solubility (Moles/L)	Solubility (Grams/L)	Temp (°C)	Ref (#)	Evaluation (T P E A A)	Comments
8.318E-04	1.482E-01	ns	R427	0 0 0 0 0	

2468. C₁₁H₁₄O₂

Ethyl hydrocinnamate
Ethyl 3-phenylpropionate
Benzenepropanoic acid, ethyl ester

RN: 2021-28-5   **MP** (°C):
MW: 178.23   **BP** (°C): 122

Solubility (Moles/L)	Solubility (Grams/L)	Temp (°C)	Ref (#)	Evaluation (T P E A A)	Comments
1.234E-03	2.200E-01	25	A002	1 2 1 1 1	

2469. C₁₁H₁₄O₃

n-Butyl salicylate
2-Hydroxy-benzoic acid, butyl ester
Salicylic acid n-butyl ester
Butyl salicylate
Benzoic acid, 2-hydroxy-, butyl ester

RN: 2052-14-4   **MP** (°C):
MW: 194.23   **BP** (°C):

Solubility (Moles/L)	Solubility (Grams/L)	Temp (°C)	Ref (#)	Evaluation (T P E A A)	Comments
1.442E-02	2.800E+00	37	D009	1 2 1 1 1	0.1N HCl

2470. C₁₁H₁₄O₃

2-Hydroxy-3-isopropyl-6-methylbenzoic acid

o-Thymotinic acid

RN: 548-51-6   **MP** (°C):
MW: 194.23   **BP** (°C):   316.3

Solubility (Moles/L)	Solubility (Grams/L)	Temp (°C)	Ref (#)	Evaluation (T P E A A)	Comments
>2.57E-03	>5.00E-01	ns	B404	0 2 1 1 0	

2471. C₁₁H₁₄O₃

Butylparaben

Bu-paraben

Butyl 4-hydroxybenzoate

RN: 94-26-8   **MP** (°C):   68.5
MW: 194.23   **BP** (°C):

Solubility (Moles/L)	Solubility (Grams/L)	Temp (°C)	Ref (#)	Evaluation (T P E A A)	Comments
7.040E-04	1.367E-01	15	B355	0 0 0 0 0	
8.350E-04	1.622E-01	20	B355	0 0 0 0 0	
1.065E-03	2.069E-01	20	C006	1 2 1 1 2	
1.277E-03	2.481E-01	25	A059	1 0 1 1 1	
1.050E-03	2.039E-01	25	B355	0 0 0 0 0	
8.751E-04	1.700E-01	25	D081	1 2 2 1 2	
1.130E-03	2.195E-01	25	D339	0 0 0 0 0	
5.623E-04	1.092E-01	25	F322	2 0 1 1 0	EFG
1.030E-03	2.000E-01	25	O027	1 0 1 0 0	
7.465E-04	1.450E-01	25	P013	0 0 0 0 0	
1.200E-03	2.331E-01	27	B129	2 2 2 2 1	
1.200E-03	2.331E-01	27	G078	2 1 0 1 0	EFG
1.777E-03	3.452E-01	30	A059	1 0 1 1 1	
2.221E-03	4.314E-01	35	A059	1 0 1 1 1	
2.064E-03	4.009E-01	39.3	G302	2 2 2 2 0	EFG
2.610E-03	5.069E-01	40	A059	1 0 1 1 1	
7.155E-04	1.390E-01	ns	B404	0 2 1 1 0	
1.100E-03	2.137E-01	ns	G067	2 0 1 1 1	
9.989E-04	1.940E-01	rt	I404	0 0 0 0 0	Intrinsic, Average

2472. C₁₁H₁₄O₃

4-Methoxyphenylbutyric acid

RN: 4521-28-2   **MP** (°C):   57
MW: 194.23   **BP** (°C):   335

Solubility (Moles/L)	Solubility (Grams/L)	Temp (°C)	Ref (#)	Evaluation (T P E A A)	Comments
5.122E+00	9.949E+02	37	A407	2 2 2 2 2	

2473. $C_{11}H_{14}O_4$
Dimethyl carbate
Dimelone

RN:	5826-73-3	MP (°C):	38
MW:	210.23	BP (°C):	

Solubility (Moles/L)	Solubility (Grams/L)	Temp (°C)	Ref (#)	Evaluation (T P E A A)	Comments
6.197E-02	1.303E+01	35	M061	1 0 0 0 2	

2474. $C_{11}H_{15}BrClO_3PS$
Profenofos
O-(4-Bromo-2-chlorophenyl)-O-ethyl-S-propyl phosphorothioate
Selecron
Curacron
Polycron

RN:	41198-08-7	MP (°C):	
MW:	373.64	BP (°C):	110

Solubility (Moles/L)	Solubility (Grams/L)	Temp (°C)	Ref (#)	Evaluation (T P E A A)	Comments
5.353E-05	2.000E-02	20	E048	1 2 1 1 1	
5.353E-05	2.000E-02	20	M161	1 0 0 0 1	
7.499E-05	2.802E-02	ns	S460	0 0 0 0 0	

2475. $C_{11}H_{15}BrN_2O$
Butallylonal
5-(2-Bromoallyl)-5-sec-butylbarbituric acid
Dial

RN:	1142-70-7	MP (°C):	131.5
MW:	271.16	BP (°C):	

Solubility (Moles/L)	Solubility (Grams/L)	Temp (°C)	Ref (#)	Evaluation (T P E A A)	Comments
2.522E-03	6.840E-01	ns	T003	0 0 0 0 2	

2476. $C_{11}H_{15}FN_2O_4$
1-Hexyloxycarbonyl-5-fluorouracil
1(2H)-Pyrimidinecarboxylic acid, 5-fluoro-3,4-dihydro-2,4-dioxo-, hexyl ester

RN:	66999-99-3	MP (°C):	68
MW:	258.25	BP (°C):	

Solubility (Moles/L)	Solubility (Grams/L)	Temp (°C)	Ref (#)	Evaluation (T P E A A)	Comments
5.808E-03	1.500E+00	22	B332	1 1 0 0 1	pH 4.0

2477. C₁₁H₁₅NO₂

m-Isopropylphenyl *N*-methylcarbamate
3-Isopropylphenyl *N*-methylcarbamate
UC-10854

RN:	64-00-6	MP (°C):	53
MW:	193.25	BP (°C):	

Solubility (Moles/L)	Solubility (Grams/L)	Temp (°C)	Ref (#)	Evaluation (T P E A A)	Comments
4.398E-04	8.500E-02	30	D089	2 2 0 0 0	
4.398E-04	8.500E-02	30	M061	1 0 0 0 1	

2478. C₁₁H₁₅NO₂

Butamben
4-Aminobenzoic acid butyl ester
Butyl *p*-aminobenzoate

RN:	94-25-7	MP (°C):	58.0
MW:	193.25	BP (°C):	

Solubility (Moles/L)	Solubility (Grams/L)	Temp (°C)	Ref (#)	Evaluation (T P E A A)	Comments
1.030E-03	1.990E-01	25	H008	0 0 0 0 0	
8.332E-04	1.610E-01	25	P303	0 0 0 0 0	
1.200E-03	2.319E-01	30	J018	1 2 0 1 1	0.05N NaOH
1.200E-03	2.319E-01	30	J022	1 0 2 1 1	
1.200E-03	2.319E-01	30	N045	1 2 2 2 0	EFG
1.389E-03	2.683E-01	33	P303	0 0 0 0 0	
1.720E-03	3.324E-01	37	F006	1 1 2 2 2	
1.700E-03	3.285E-01	37	J026	2 2 2 1 1	
2.221E-03	4.293E-01	40	P303	0 0 0 0 0	
6.468E-04	1.250E-01	ns	B404	0 2 1 1 0	
7.140E-04	1.380E-01	ns	M066	0 0 0 0 2	
7.140E-04	1.380E-01	rt	B016	0 0 1 1 2	pH 7.4
7.784E-04	1.504E-01	rt	I404	0 0 0 0 0	Average

2479. C₁₁H₁₅NO₂S

Ethiofencarb
2-((Ethylthio)methyl)phenyl methylcarbamate
Ethylmercaptomethylphenyl-*N*-methylcarbamate
Ethiophencarp
Croneton
HOX 1901

RN:	29973-13-5	MP (°C):	<25
MW:	225.31	BP (°C):	

Solubility (Moles/L)	Solubility (Grams/L)	Temp (°C)	Ref (#)	Evaluation (T P E A A)	Comments
8.078E-03	1.820E+00	20	M161	1 0 0 0 2	

2480. C₁₁H₁₅NO₃

α,3-o-Isopropylidene pyriridoxine

RN: **MP** (°C):
MW: 209.25 **BP** (°C):

Solubility (Moles/L)	Solubility (Grams/L)	Temp (°C)	Ref (#)	Evaluation (T P E A A)	Comments
1.196E-02	2.503E+00	37	M067	2 0 1 1 2	

2481. C₁₁H₁₅NO₃

Propoxur

o-Isopropoxyphenyl methylcarbamate

Baygon

Blattanex

Blattosep

Suncide

RN: 114-26-1 **MP** (°C): 91
MW: 209.25 **BP** (°C):

Solubility (Moles/L)	Solubility (Grams/L)	Temp (°C)	Ref (#)	Evaluation (T P E A A)	Comments
8.301E-03	1.737E+00	10	B324	0 0 0 0 0	
8.316E-03	1.740E+00	10	B324	0 0 0 0 0	
8.885E-03	1.859E+00	20	B300	2 2 1 1 2	
9.244E-03	1.934E+00	20	B324	0 0 0 0 0	
9.206E-03	1.926E+00	20	B324	0 0 0 0 0	
9.558E-03	2.000E+00	20	M161	1 0 0 0 0	
1.166E-02	2.440E+00	30	B324	0 0 0 0 0	
1.163E-02	2.434E+00	30	B324	0 0 0 0 0	
4.732E-02	9.901E+00	ns	M061	0 0 0 0 0	approximate
4.301E-04	9.000E-02	ns	M110	0 0 0 0 0	EFG

2482. C₁₁H₁₅NO₄

n-Ethyl-6-hydroxynorbornane-2-carboxamide-3,5-lactone

RN: **MP** (°C):
MW: 225.25 **BP** (°C):

Solubility (Moles/L)	Solubility (Grams/L)	Temp (°C)	Ref (#)	Evaluation (T P E A A)	Comments
2.908E-01	6.550E+01	20	K050	1 1 1 1 2	

2483. C₁₁H₁₅N₃O₂

Formetanate

Methylcarbamic acid, ester with N′-(m-hydroxyphenyl)-N,N-dimethylformamidine

RN: 22259-30-9 **MP** (°C): 102.5
MW: 221.26 **BP** (°C):

Solubility (Moles/L)	Solubility (Grams/L)	Temp (°C)	Ref (#)	Evaluation (T P E A A)	Comments
4.520E-03	1.000E+00	rt	M161	0 0 0 0 0	

2484. C$_{11}$H$_{15}$N$_3$O$_3$

Orotic acid cyclohexylamide
Orotamide, N-cyclohexyl-

RN:	4558-58-1	MP (°C):	284–285
MW:	237.26	BP (°C):	

Solubility (Moles/L)	Solubility (Grams/L)	Temp (°C)	Ref (#)	Evaluation (T P E A A)	Comments
7.500E-02	1.779E+01	−4	N018	0 0 0 0 0	
1.100E-01	2.610E+01	16	N018	0 0 0 0 0	
1.330E-01	3.156E+01	25	N018	0 0 0 0 0	

2485. C$_{11}$H$_{15}$N$_3$O$_5$

Triglycidylurazol
Anaxirone

RN:	77658-97-0	MP (°C):	91
MW:	269.26	BP (°C):	

Solubility (Moles/L)	Solubility (Grams/L)	Temp (°C)	Ref (#)	Evaluation (T P E A A)	Comments
7.426E-04	2.000E-01	ns	D319	0 0 0 0 0	

2486. C$_{11}$H$_{15}$O$_3$P

Diethyl benzoyl phosphonate
Methylene, (diethoxyphosphinyl)phenyl-

RN:	105394-75-0	MP (°C):	
MW:	226.21	BP (°C):	

Solubility (Moles/L)	Solubility (Grams/L)	Temp (°C)	Ref (#)	Evaluation (T P E A A)	Comments
<8.84E-04	<2.00E-01	25	B070	1 2 0 1 0	

2487. C$_{11}$H$_{16}$

tert-Amylbenzene
t-Amylbenzene

RN:	2049-95-8	MP (°C):	−57.8
MW:	148.25	BP (°C):	

Solubility (Moles/L)	Solubility (Grams/L)	Temp (°C)	Ref (#)	Evaluation (T P E A A)	Comments
7.083E-05	1.050E-02	25	A002	1 2 1 1 2	

2488. C$_{11}$H$_{16}$
Amylbenzene
n-Pentylbenzene
Pentylbenzene
n-Amylbenzene
n-Pentylbenzene1-phenylpentane

RN:	538-68-1	**MP (°C):**	−75		
MW:	148.25	**BP (°C):**	205.4		

Solubility (Moles/L)	Solubility (Grams/L)	Temp (°C)	Ref (#)	Evaluation (T P E A A)	Comments
2.348E-05	3.481E-03	7	O312	2 2 0 2 2	
2.144E-05	3.178E-03	10	O312	2 2 0 2 2	
2.323E-05	3.444E-03	12.5	O312	2 2 0 2 2	
2.153E-05	3.192E-03	15	O312	2 2 0 2 2	
2.311E-05	3.426E-03	17.5	O312	2 2 0 2 2	
2.142E-05	3.176E-03	20	O312	2 2 0 2 2	
2.590E-05	3.840E-03	25	M342	1 0 1 1 2	
2.276E-05	3.374E-03	25	O312	2 2 0 2 2	
2.433E-05	3.607E-03	30	O312	2 2 0 2 2	
2.642E-05	3.917E-03	35	O312	2 2 0 2 2	
2.868E-05	4.252E-03	40	O312	2 2 0 2 2	
3.163E-05	4.689E-03	45	O312	2 2 0 2 2	
6.000E-03	8.895E-01	ns	H307	0 0 0 0 0	

2489. C$_{11}$H$_{16}$
Pentamethylbenzene
1,2,3,4,5-Pentamethyl benzene

RN:	700-12-9	**MP (°C):**	50.8		
MW:	148.25	**BP (°C):**	231.0		

Solubility (Moles/L)	Solubility (Grams/L)	Temp (°C)	Ref (#)	Evaluation (T P E A A)	Comments
1.047E-04	1.552E-02	ns	D001	0 0 0 0 2	

2490. C$_{11}$H$_{16}$ClO$_2$PS$_3$
Carbophenothion
O,O-Diethyl S-(4-chlorophenylthiomethyl) dithiophosphate
Trithion
Garrathion
Nephocarp
Lethox

RN:	786-19-6	**MP (°C):**	<25		
MW:	342.87	**BP (°C):**			

Solubility (Moles/L)	Solubility (Grams/L)	Temp (°C)	Ref (#)	Evaluation (T P E A A)	Comments
1.779E-06	6.100E-04	10	B324	0 0 0 0 0	
1.779E-06	6.100E-04	10	B324	0 0 0 0 0	
1.838E-06	6.302E-04	20	B300	2 1 1 1 2	

(continued)

2490. $C_{11}H_{16}ClO_2PS_3$ (continued)

Solubility (Moles/L)	Solubility (Grams/L)	Temp (°C)	Ref (#)	Evaluation (T P E A A)	Comments
1.837E-06	6.300E-04	20	B324	0 0 0 0 0	
1.838E-06	6.302E-04	20	B324	0 0 0 0 0	
2.129E-06	7.300E-04	30	B324	0 0 0 0 0	
2.129E-06	7.300E-04	30	B324	0 0 0 0 0	
<1.17E-04	<4.00E-02	ns	M161	0 0 0 0 0	

2491. $C_{11}H_{16}N_2O_2$
4-Aminobenzoic acid-2-(ethyl-amino)ethyl ester
2-(Ethylamino)ethyl 4-aminobenzoate
RN: **MP** (°C):
MW: 208.26   **BP** (°C):

Solubility (Moles/L)	Solubility (Grams/L)	Temp (°C)	Ref (#)	Evaluation (T P E A A)	Comments
2.700E-02	5.623E+00	ns	M066	0 0 0 0 1	

2492. $C_{11}H_{16}N_2O_2$
Aminocarb
Phenol, 4-(dimethylamino)-3-methyl, methylcarbamate (ester)
Carbamic acid, methyl-, 4-(dimethylamino)-*m*-tolyl ester
RN: 2032-59-9   **MP** (°C):   93
MW: 208.26   **BP** (°C):

Solubility (Moles/L)	Solubility (Grams/L)	Temp (°C)	Ref (#)	Evaluation (T P E A A)	Comments
4.187E-03	8.720E-01	10	B324	0 0 0 0 0	
4.183E-03	8.712E-01	10	B324	0 0 0 0 0	
4.394E-03	9.151E-01	20	B300	2 2 1 1 2	
4.389E-03	9.142E-01	20	B324	0 0 0 0 0	
4.394E-03	9.151E-01	20	B324	0 0 0 0 0	
4.393E-03	9.150E-01	20	G300	1 0 0 0 2	
6.521E-03	1.358E+00	30	B324	0 0 0 0 0	
6.540E-03	1.362E+00	30	B324	0 0 0 0 0	

2493. $C_{11}H_{16}N_2O_3$
Vinbarbital
5-Ethyl-5-(1-methyl-1-butenyl)barbituric acid
RN: 125-42-8   **MP** (°C):   161
MW: 224.26   **BP** (°C):

Solubility (Moles/L)	Solubility (Grams/L)	Temp (°C)	Ref (#)	Evaluation (T P E A A)	Comments
3.121E-03	7.000E-01	25	B011	2 0 0 1 0	
3.164E-03	7.097E-01	25	B065	1 1 1 1 1	
4.870E-03	1.092E+00	25	V033	2 0 1 1 2	

(*continued*)

2493. $C_{11}H_{16}N_2O_3$ (continued)

Solubility (Moles/L)	Solubility (Grams/L)	Temp (°C)	Ref (#)	Evaluation (T P E A A)	Comments
4.900E-03	1.099E+00	25.00	T303	1 0 0 0 1	
7.000E-03	1.570E+00	35.00	T303	1 0 0 0 1	
8.000E-03	1.794E+00	45.00	T303	1 0 0 0 1	

2494. $C_{11}H_{16}N_2O_3$
5-Allyl-5-butylbarbituric acid
n-Butylallylbarbitone
n-Butylallylbarbituric acid
Allylbutylbarbituric acid
Idobutal

RN: 3146-66-5 **MP** (°C):
MW: 224.26 **BP** (°C):

Solubility (Moles/L)	Solubility (Grams/L)	Temp (°C)	Ref (#)	Evaluation (T P E A A)	Comments
6.723E-03	1.508E+00	20	J030	1 2 2 2 2	
8.945E-03	2.006E+00	37	J030	1 2 2 2 2	

2495. $C_{11}H_{16}N_2O_3$
Talbutal
Allyl-sec-butyl-barbituric acid
5-Allyl-5-sec-butylbarbituric acid

RN: 115-44-6 **MP** (°C): 109
MW: 224.26 **BP** (°C):

Solubility (Moles/L)	Solubility (Grams/L)	Temp (°C)	Ref (#)	Evaluation (T P E A A)	Comments
9.632E-03	2.160E+00	ns	T003	0 0 0 0 2	

2496. $C_{11}H_{16}N_2O_3$
Butalbital
Itobarbital
5-Allyl-5-isobutylbarbituric acid
Fioricet
Phrenilin
Medigesic

RN: 77-26-9 **MP** (°C): 138
MW: 224.26 **BP** (°C):

Solubility (Moles/L)	Solubility (Grams/L)	Temp (°C)	Ref (#)	Evaluation (T P E A A)	Comments
7.590E-03	1.702E+00	25	V033	2 0 1 1 2	
7.600E-03	1.704E+00	25.00	T303	1 0 0 0 1	
1.030E-02	2.310E+00	35.00	T303	1 0 0 0 2	
1.410E-02	3.162E+00	45.00	T303	1 0 0 0 2	

2497. C$_{11}$H$_{16}$N$_2$O$_3$
2,4-Diazaspiro[5.7]tridecane-1,3,5-trione
RN: 143288-62-4   **MP** (°C):
MW: 224.26   **BP** (°C):

Solubility (Moles/L)	Solubility (Grams/L)	Temp (°C)	Ref (#)	Evaluation (T P E A A)	Comments
1.042E-03	2.337E-01	25	P350	0 0 0 0 0	intrinsic

2498. C$_{11}$H$_{16}$N$_2$O$_3$
Barbituric acid, 5-ethyl-5-(3-methyl-2-butenyl)
5-Ethyl-5-(3'-methylbut-2'-enyl)barbituric acid
2,4,6(1H,3H,5H)-Pyrimidinetrione, 5-ethyl-5-(3-methyl-2-butenyl)-
2,4,6(1H,3H,5H)-Pyrimidinetrione, 5-ethyl-5-(3-methyl-2-butenyl)
5-Ethyl-5-(3-methylbut-2-enyl)barbiturate
RN: 21149-88-2   **MP** (°C):
MW: 224.26   **BP** (°C):

Solubility (Moles/L)	Solubility (Grams/L)	Temp (°C)	Ref (#)	Evaluation (T P E A A)	Comments
5.583E-03	1.252E+00	25	P350	0 0 0 0 0	intrinsic

2499. C$_{11}$H$_{16}$N$_2$O$_3$S
Phenbutamide
N-(Phenylsulfonyl)-N'-butylurea
N-Benzenesulfonyl-N'-n-butylurea
RN: 3149-00-6   **MP** (°C):   131
MW: 256.33   **BP** (°C):

Solubility (Moles/L)	Solubility (Grams/L)	Temp (°C)	Ref (#)	Evaluation (T P E A A)	Comments
8.995E-04	2.306E-01	37	A028	1 0 2 1 2	intrinsic
9.000E-04	2.307E-01	37	A046	2 0 1 1 2	

2500. C$_{11}$H$_{16}$N$_2$O$_4$
Methyl-2-ethyl-2-allylmalonurate
Methyl 2-ethyl-2-allylmalonurate
RN: 73632-83-4   **MP** (°C):   78.5
MW: 240.26   **BP** (°C):

Solubility (Moles/L)	Solubility (Grams/L)	Temp (°C)	Ref (#)	Evaluation (T P E A A)	Comments
1.200E-02	2.883E+00	23	B152	1 2 1 1 1	pH 3.5

2501. C₁₁H₁₆N₂O₅

Methoxycarbonylmethyl-2,2-diethylmalonurate
Methoxycarbonylmethyl 2,2-diethylmalonurate

RN: **MP (°C):**   89
MW: 256.26   **BP (°C):**

Solubility (Moles/L)	Solubility (Grams/L)	Temp (°C)	Ref (#)	Evaluation (T P E A A)	Comments
9.700E-03	2.486E+00	23	B152	1 2 1 1 1	pH 3.5

2502. C₁₁H₁₆N₄O₂

1-Butyl theobromine
1-Butyl-3,7-dimethylxanthine
1-*n*-Butyl-3,7-dimethylxanthine

RN: 1143-30-2   **MP (°C):**   108
MW: 236.28   **BP (°C):**

Solubility (Moles/L)	Solubility (Grams/L)	Temp (°C)	Ref (#)	Evaluation (T P E A A)	Comments
2.370E-02	5.600E+00	30	B042	1 2 1 1 2	

2503. C₁₁H₁₆N₄O₄

2,6-Piperazinedione, 4,4′-(1-methyl-1,2-ethanediyl)*bis*-
1,2-Di(4-piperazine-2,6-dione)propane
2,6-Piperazinedione, 4,4′-(1-methyl-1,2-ethanediyl)*bis*-, (±)-, polymer with 1,3-dibromopropane
Propane, 1,3-dibromo-, polymer with (±)-4,4′-(1-methyl-1,2-ethanediyl)bis[2,6-piperazinedione]

RN: 21416-67-1   **MP (°C):**   192 dec
MW: 268.27   **BP (°C):**   233 dec

Solubility (Moles/L)	Solubility (Grams/L)	Temp (°C)	Ref (#)	Evaluation (T P E A A)	Comments
1.118E-02	3.000E+00	25	P326	0 0 0 0 0	
~5.59E-02	~1.50E+01	25	R017	0 0 0 0 0	enantimer (R)
~1.12E-02	~3.00E+00	25	R017	0 0 0 0 0	

2504. C₁₁H₁₆O

p-sec-Amylphenol
4-*sec*-Amylphenol

RN: 25735-67-5   **MP (°C):**
MW: 164.25   **BP (°C):**

Solubility (Moles/L)	Solubility (Grams/L)	Temp (°C)	Ref (#)	Evaluation (T P E A A)	Comments
6.408E-04	1.053E-01	25	L021	1 0 0 0 0	

2505. C$_{11}$H$_{16}$O
p-n-Amylphenol
4-*n*-Pentylphenol

RN: 14938-35-3   **MP** (°C):
MW: 164.25   **BP** (°C):

Solubility (Moles/L)	Solubility (Grams/L)	Temp (°C)	Ref (#)	Evaluation (T P E A A)	Comments
6.088E-04	9.999E-02	25	L022	1 0 0 0 0	

2506. C$_{11}$H$_{16}$O
p-tert-Pentylphenol
p-(α,α-Dimethylpropyl)phenol
p-(1,1-Dimethylpropyl)phenol
1-Hydroxy-4(2-methyl-2-butyl)benzene
PTAP

RN: 80-46-6   **MP** (°C):
MW: 164.25   **BP** (°C):

Solubility (Moles/L)	Solubility (Grams/L)	Temp (°C)	Ref (#)	Evaluation (T P E A A)	Comments
7.162E-04	1.176E-01	25	L021	1 0 0 0 0	
1.023E-03	1.680E-01	25	M127	1 0 0 0 2	

2507. C$_{11}$H$_{16}$O
4-(1,1-Dimethylethyl)benzenemethanol
4-(1,1-Dimethylethyl)benzyl alcohol
4-*tert*-Butylbenzyl alcohol
4-*tert*-Butylphenylmethanol
p-tert-Butylbenzyl alcohol

RN: 877-65-6   **MP** (°C):
MW: 164.25   **BP** (°C):   250.3

Solubility (Moles/L)	Solubility (Grams/L)	Temp (°C)	Ref (#)	Evaluation (T P E A A)	Comments
6.400E-03	1.051E+00	20	B407	1 0 1 2 2	

2508. C$_{11}$H$_{16}$O
o-n-Amylphenol
2-*n*-Amylphenol

RN: 87-26-3   **MP** (°C):
MW: 164.25   **BP** (°C):

Solubility (Moles/L)	Solubility (Grams/L)	Temp (°C)	Ref (#)	Evaluation (T P E A A)	Comments
9.365E-04	1.538E-01	25	L022	1 0 0 0 0	

2509. $C_{11}H_{16}O$

o-2-Hexenylphenol
2-2-Hexenylphenol

RN: 75121-79-8   **MP** (°C):
MW: 164.25   **BP** (°C):

Solubility (Moles/L)	Solubility (Grams/L)	Temp (°C)	Ref (#)	Evaluation (T P E A A)	Comments
7.162E-04	1.176E-01	25	L021	1 0 0 0 0	

2510. $C_{11}H_{16}O$

2-Methyl-5-*t*-butylphenol
5-*tert*-Butyl-2-methylphenol
5-*tert*-Butyl-*o*-cresol
o-Cresol, 5-*tert*-butyl-

RN: 5781-02-2   **MP** (°C):
MW: 164.25   **BP** (°C):

Solubility (Moles/L)	Solubility (Grams/L)	Temp (°C)	Ref (#)	Evaluation (T P E A A)	Comments
2.533E-03	4.160E-01	25	M127	1 0 0 0 2	

2511. $C_{11}H_{16}O_2$

4-*n*-Amyl resorcinol
4-*n*-Amyl-resorcin

RN: 533-24-4   **MP** (°C):
MW: 180.25   **BP** (°C):

Solubility (Moles/L)	Solubility (Grams/L)	Temp (°C)	Ref (#)	Evaluation (T P E A A)	Comments
1.110E-02	2.000E+00	20	F300	1 0 0 0 0	

2512. $C_{11}H_{16}O_2$

3-Pentoxyphenol
m-Pentoxy phenol
Phenol, 3-pentoxy-

RN: 18979-73-2   **MP** (°C):
MW: 180.25   **BP** (°C):

Solubility (Moles/L)	Solubility (Grams/L)	Temp (°C)	Ref (#)	Evaluation (T P E A A)	Comments
2.130E-03	3.839E-01	30	B315	0 0 0 0 0	

2513. C₁₁H₁₇NO₃

Dimetan

5,5-Dimethyldihydroresorcinyl *N,N*-dimethylcarbamate

RN: 122-15-6   **MP** (°C):   45.5
MW: 211.26   **BP** (°C):

Solubility (Moles/L)	Solubility (Grams/L)	Temp (°C)	Ref (#)	Evaluation (T P E A A)	Comments
1.379E-01	2.913E+01	ns	M061	0 0 0 0 0	approximate

2514. C₁₁H₁₇N₃O₃

Orotic acid triethylamide

RN: **MP** (°C):   200–202
MW: 239.28   **BP** (°C):

Solubility (Moles/L)	Solubility (Grams/L)	Temp (°C)	Ref (#)	Evaluation (T P E A A)	Comments
2.261E+00	5.410E+02	25	N018	0 0 0 0 0	

2515. C₁₁H₁₇N₃O₃S

Carbutamide

4-Amino-*N*-[(butylamino)carbonyl]-benzenesulfonamide

1-Butyl-3-sulfanilyl urea

RN: 339-43-5   **MP** (°C):   144.5
MW: 271.34   **BP** (°C):

Solubility (Moles/L)	Solubility (Grams/L)	Temp (°C)	Ref (#)	Evaluation (T P E A A)	Comments
1.972E-03	5.352E-01	37	A028	1 0 2 1 2	intrinsic
1.950E-03	5.291E-01	37	A046	2 0 1 1 2	
6.634E-03	1.800E+00	37	C054	2 0 2 1 2	0.1N HCl

2516. C₁₁H₁₇N₃O₆

Orotic acid triethanolamide

RN: **MP** (°C):   104–108
MW: 287.27   **BP** (°C):

Solubility (Moles/L)	Solubility (Grams/L)	Temp (°C)	Ref (#)	Evaluation (T P E A A)	Comments
1.315E+00	3.778E+02	−4	N018	0 0 0 0 0	
1.882E+00	5.407E+02	16	N018	0 0 0 0 0	
2.187E+00	6.283E+02	25	N018	0 0 0 0 0	

2517. $C_{11}H_{17}O_3PS$
Kitazin
O,O-Diethyl S-benzyl thiophosphate
EBP
S-Benzyl O,O-di-ethyl phosphorothioate

RN: 13286-32-3   **MP** (°C):
MW: 260.29   **BP** (°C): 115

Solubility (Moles/L)	Solubility (Grams/L)	Temp (°C)	Ref (#)	Evaluation (T P E A A)	Comments
1.537E-03	4.000E-01	22	K137	1 1 2 1 0	

2518. $C_{11}H_{17}O_3PS_2$
Fensulfothion sulfide
O,O-Diethyl O-[p-(methylthio)phenyl] phosphorothioate
Phosphorothioic acid, O,O-diethyl O-[4-(methylthio)phenyl] ester

RN: 3070-15-3   **MP** (°C):
MW: 292.36   **BP** (°C):

Solubility (Moles/L)	Solubility (Grams/L)	Temp (°C)	Ref (#)	Evaluation (T P E A A)	Comments
1.266E-05	3.700E-03	20	M318	2 2 0 0 2	

2519. $C_{11}H_{17}O_4PS_2$
Fensulfothion
O,O-Diethyl O-(4-(methylsulfinyl)phenyl) phosphorothioate
Dasanit
Bay 25141
Agricur
Chemagro 25141

RN: 115-90-2   **MP** (°C): <25
MW: 308.36   **BP** (°C):

Solubility (Moles/L)	Solubility (Grams/L)	Temp (°C)	Ref (#)	Evaluation (T P E A A)	Comments
6.473E-03	1.996E+00	20	B169	2 2 1 1 2	
6.473E-03	1.996E+00	20	F318	2 2 0 0 2	
4.994E-03	1.540E+00	25	M161	1 0 0 0 2	

2520. $C_{11}H_{17}O_5PS_2$

Fensulfothion sulfone
Phosphorothioic acid, O,O-diethyl O-[p-(methylsulfonyl)phenyl] ester
Dasanit sulfone
Dasanit sulphone

RN: 14255-72-2   **MP** (°C):
MW: 324.36   **BP** (°C):

Solubility (Moles/L)	Solubility (Grams/L)	Temp (°C)	Ref (#)	Evaluation (T P E A A)	Comments
1.242E-04	4.030E-02	10	B324	0 0 0 0 0	
1.243E-04	4.032E-02	10	B324	0 0 0 0 0	
2.300E-04	7.459E-02	20	B169	2 2 1 1 2	
2.633E-04	8.540E-02	20	B324	0 0 0 0 0	
2.633E-04	8.539E-02	20	B324	0 0 0 0 0	
2.300E-04	7.459E-02	20	M318	2 2 0 0 2	
3.576E-04	1.160E-01	30	B324	0 0 0 0 0	
3.576E-04	1.160E-01	30	B324	0 0 0 0 0	

2521. $C_{11}H_{18}N_2O_2S$

Thiopental
5-Ethyl-5-(1-methyl-butyl)-2-thiobarbituric acid
5-Ethyl-5-(1-methylbutyl)-2-thiobarbituric acid
Barbituric acid, 5-ethyl-5-(1-methylbutyl)-2-thio
4,6(1H,5H)-Pyrimidinedione, 5-ethyldihydro-5-(1-methylbutyl)-2-thioxo
Pentothiobarbital

RN: 76-75-5   **MP** (°C):   158
MW: 242.34   **BP** (°C):

Solubility (Moles/L)	Solubility (Grams/L)	Temp (°C)	Ref (#)	Evaluation (T P E A A)	Comments
2.063E-04	5.000E-02	25	A023	1 0 0 1 1	
3.301E-04	8.000E-02	25	B011	2 0 0 1 0	
3.333E-04	8.077E-02	25	B065	1 1 1 1 1	
8.200E-04	1.987E-01	25	G003	1 1 1 1 1	pH 4.7
2.094E-04	5.075E-02	25	P350	0 0 0 0 0	intrinsic
3.000E-04	7.270E-02	30	K108	1 2 2 0 0	
3.301E-04	7.999E-02	35	A023	1 0 0 1 1	
4.126E-04	9.999E-02	40	A023	1 0 0 1 1	

2522. $C_{11}H_{18}N_2O_3$

Amobarbital
5-Ethyl-5-isoamylbarbituric acid
Amylobarbitone

RN: 57-43-2   **MP** (°C): 157
MW: 226.28   **BP** (°C):

Solubility (Moles/L)	Solubility (Grams/L)	Temp (°C)	Ref (#)	Evaluation (T P E A A)	Comments
2.828E-03	6.400E-01	20	J030	1 2 2 2 1	
3.533E-03	7.994E-01	25	A023	1 0 0 1 1	
2.475E-03	5.600E-01	25	B011	2 0 0 1 0	
2.665E-03	6.030E-01	25	B065	1 1 1 1 1	
3.900E-03	8.825E-01	25	G003	1 1 1 1 1	pH 4.7
2.170E-03	4.910E-01	25	V033	2 0 1 1 2	
2.200E-03	4.978E-01	25.00	T303	1 0 0 0 1	
3.000E-03	6.788E-01	30	G014	1 1 1 1 0	EFG
3.100E-03	7.015E-01	30	I001	2 0 2 1 0	EFG, 0.003N H_2SO_4
2.846E-03	6.440E-01	30	I015	1 2 2 1 2	pH 6.0, 3 forms
3.200E-03	7.241E-01	30	K108	1 2 2 0 1	
3.300E-03	7.467E-01	35.00	T303	1 0 0 0 1	
4.375E-03	9.900E-01	37	J030	1 2 2 2 1	
4.000E-03	9.051E-01	37	K121	1 2 1 2 0	0.1N HCl
5.517E-03	1.248E+00	40	A023	1 0 0 1 1	
3.820E-02	8.644E+00	40	N008	1 0 1 1 2	sic
4.300E-03	9.730E-01	45.00	T303	1 0 0 0 1	
2.342E-03	5.300E-01	ns	T003	0 0 0 0 2	

2523. $C_{11}H_{18}N_2O_3$

Pentobarbital
5-ethyl-5-(1-methyl-butyl)-barbituric acid

RN: 76-74-4   **MP** (°C): 130
MW: 226.28   **BP** (°C):

Solubility (Moles/L)	Solubility (Grams/L)	Temp (°C)	Ref (#)	Evaluation (T P E A A)	Comments
4.415E-03	9.990E-01	25	A023	1 0 0 1 1	
2.210E-03	5.000E-01	25	B011	2 0 0 1 0	
2.221E-03	5.026E-01	25	B065	1 1 1 1 1	
3.000E-03	6.788E-01	25	G003	1 1 1 1 1	pH 4.7
4.070E-03	9.210E-01	25	V033	2 0 1 1 2	
4.100E-03	9.277E-01	25.00	T303	1 0 0 0 1	
6.000E-03	1.358E+00	30	K108	1 2 2 0 1	
6.178E-03	1.398E+00	35	A023	1 0 0 1 1	
5.700E-03	1.290E+00	35.00	T303	1 0 0 0 1	
7.000E-03	1.584E+00	37	K121	1 2 1 2 0	0.1N HCl
7.060E-03	1.597E+00	40	A023	1 0 0 1 1	
7.640E-02	1.729E+01	40	N008	1 0 1 1 2	sic
6.900E-03	1.561E+00	45.00	T303	1 0 0 0 1	
4.365E-03	9.877E-01	ns	R427	0 0 0 0 0	

2524. $C_{11}H_{18}N_2O_3$
5-*n*-Pentyl-5-ethylbarbituric acid
5-Ethyl-5-pentylbarbituric acid
5-Ethyl-5-pentylbarbiturate

RN:	115-58-2	MP (°C):	135.5
MW:	226.28	BP (°C):	

Solubility (Moles/L)	Solubility (Grams/L)	Temp (°C)	Ref (#)	Evaluation (T P E A A)	Comments
6.657E-03	1.506E+00	25	B065	1 2 1 1 1	
2.448E-03	5.540E-01	ns	T003	0 0 0 0 2	

2525. $C_{11}H_{18}N_2O_3$
Pilocarpic acid
1,2-Secopilocarpin-2-oic acid

RN:	28406-15-7	MP (°C):	
MW:	226.28	BP (°C):	

Solubility (Moles/L)	Solubility (Grams/L)	Temp (°C)	Ref (#)	Evaluation (T P E A A)	Comments
5.303E-04	1.200E-01	23	B340	1 1 2 1 1	pH 9

2526. $C_{11}H_{18}N_4O_2$
Pirimicarb
2-(Dimethylamino)-5,6-dimethyl-4-pyrimidinyl dimethylcarbamate
Abol
Rapid
Fernos
Aphox

RN:	23103-98-2	MP (°C):	90.5
MW:	238.29	BP (°C):	

Solubility (Moles/L)	Solubility (Grams/L)	Temp (°C)	Ref (#)	Evaluation (T P E A A)	Comments
1.133E-02	2.700E+00	25	M161	1 0 0 0 1	

2527. $C_{11}H_{19}N_3O$
Dimethirimol
2-Dimethylamino-4-hyroxy-5-*n*-butyl-6-methylpyrimidine

RN:	5221-53-4	MP (°C):	102
MW:	209.29	BP (°C):	

Solubility (Moles/L)	Solubility (Grams/L)	Temp (°C)	Ref (#)	Evaluation (T P E A A)	Comments
5.734E-03	1.200E+00	25	M161	1 0 0 0 1	
5.727E-03	1.199E+00	ns	M061	0 0 0 0 1	

2528. $C_{11}H_{19}N_3O$
Ethirimol
5-Butyl-2-(ethylamino)-4-hydroxy-6-methylpyrimidine
Milgo
Milcurb super
Milstem

RN: 23947-60-6 **MP** (°C): 159.5
MW: 209.29 **BP** (°C):

Solubility (Moles/L)	Solubility (Grams/L)	Temp (°C)	Ref (#)	Evaluation (T P E A A)	Comments
9.556E-04	2.000E-01	rt	M161	0 0 0 0 0	

2529. $C_{11}H_{20}$
2-Methyldecalin
Decahydro-2-methylnaphthalene

RN: 2958-76-1 **MP** (°C):
MW: 152.28 **BP** (°C):

Solubility (Moles/L)	Solubility (Grams/L)	Temp (°C)	Ref (#)	Evaluation (T P E A A)	Comments
2.666E-07	4.060E-05	25	B069	1 0 1 1 2	

2530. $C_{11}H_{20}ClN_5$
Chlorazine
2-Chloro-4-diethylamino-6-diethylamino-s-triazine
2-Chloro-4,6-*bis*-(diethylamino)-s-triazine chlorazine
1,3,5-Triazine
1,3,5-Triazine-2,4-diamine

RN: 580-48-3 **MP** (°C):
MW: 257.77 **BP** (°C):

Solubility (Moles/L)	Solubility (Grams/L)	Temp (°C)	Ref (#)	Evaluation (T P E A A)	Comments
3.492E-05	9.000E-03	20	J033	0 0 0 0 0	
3.879E-05	1.000E-02	21	B192	0 0 0 0 1	
3.492E-05	9.000E-03	21	G099	2 0 0 1 0	

2531. $C_{11}H_{20}N_2O_4$
Isopropyl-2,2-diethylmalonurate
Isopropyl 2,2-diethylmalonurate

RN: 73632-77-6 **MP** (°C): 99.5
MW: 244.29 **BP** (°C):

Solubility (Moles/L)	Solubility (Grams/L)	Temp (°C)	Ref (#)	Evaluation (T P E A A)	Comments
1.700E-03	4.153E-01	23	B152	1 2 1 1 1	pH 3.5

2532. C$_{11}$H$_{20}$N$_3$O$_3$PS
Pirimiphos-methyl
Pirimiphosmethyl

RN: 29232-93-7   **MP** (°C):   15
MW: 305.34   **BP** (°C):

Solubility (Moles/L)	Solubility (Grams/L)	Temp (°C)	Ref (#)	Evaluation (T P E A A)	Comments
7.139E-05	2.180E-02	10	B324	0 0 0 0 0	
7.946E-05	2.426E-02	10	B324	0 0 0 0 0	
7.363E-05	2.248E-02	20	B300	2 1 1 1 2	
1.119E-04	3.417E-02	20	B324	0 0 0 0 0	
1.005E-04	3.070E-02	20	B324	0 0 0 0 0	
1.640E-04	5.008E-02	30	B324	0 0 0 0 0	
1.474E-04	4.500E-02	30	B324	0 0 0 0 0	
1.638E-05	5.000E-03	30	M161	1 0 0 0 0	*sic*

2533. C$_{11}$H$_{20}$N$_6$
1-(Pyrrolidinyl)-3,5-bis(dimethylamino)-s-triazine
1-Pyrrolidino-3,5-bis(dimethylamino)-s-triazine

RN: 13452-85-2   **MP** (°C):
MW: 236.32   **BP** (°C):

Solubility (Moles/L)	Solubility (Grams/L)	Temp (°C)	Ref (#)	Evaluation (T P E A A)	Comments
1.641E-04	3.878E-02	25	B386	0 0 0 0 0	

2534. C$_{11}$H$_{20}$N$_6$O
1-(Morpholinyl)-3,5-bis(dimethylamino)-s-triazine
s-Triazine, 2,4-bis(dimethylamino)-6-morpholino-

RN: 16269-02-6   **MP** (°C):
MW: 252.32   **BP** (°C):

Solubility (Moles/L)	Solubility (Grams/L)	Temp (°C)	Ref (#)	Evaluation (T P E A A)	Comments
1.303E-03	3.288E-01	25	B386	0 0 0 0 0	

2535. C$_{11}$H$_{20}$N$_6$S
1-(Thiomorpholinyl)-3,5-bis(dimethylamino)-s-triazine
1,3,5-Triazine-2,4-diamine, N,N,N',N'-tetramethyl-6-(4-thiomorpholinyl)-

RN: 41492-69-7   **MP** (°C):
MW: 268.39   **BP** (°C):

Solubility (Moles/L)	Solubility (Grams/L)	Temp (°C)	Ref (#)	Evaluation (T P E A A)	Comments
5.689E-05	1.527E-02	25	B386	0 0 0 0 0	

2536. $C_{11}H_{20}O_2$

Undecylenic acid
10-Undecylenic acid
Hendecenoic acid
RN: 112-38-9   **MP** (°C): 25
MW: 184.28   **BP** (°C):

Solubility (Moles/L)	Solubility (Grams/L)	Temp (°C)	Ref (#)	Evaluation (T P E A A)	Comments
4.000E-04	7.371E-02	30	D051	2 0 0 1 2	
1.074E-04	1.980E-02	30	E005	2 1 1 2 2	
1.248E-04	2.300E-02	40	E005	2 1 1 2 1	
1.411E-04	2.600E-02	50	E005	2 1 1 2 1	
1.000E-03	1.843E-01	60	D051	2 0 0 1 2	
1.736E-04	3.200E-02	60	E005	2 1 1 2 1	

2537. $C_{11}H_{20}O_4$

Hexyl α-acetoxypropionate
Propanoic acid, 2-(acetyloxy)-, hexyl ester
RN: 96884-73-0   **MP** (°C):
MW: 216.28   **BP** (°C):

Solubility (Moles/L)	Solubility (Grams/L)	Temp (°C)	Ref (#)	Evaluation (T P E A A)	Comments
9.247E-04	2.000E-01	25	R006	2 2 0 1 1	

2538. $C_{11}H_{20}O_4$

Undecanedioic acid
1,9-Nonanedicarboxylic acid
Nonan-dicarbonsaeure-(1,9)
RN: 1852-04-6   **MP** (°C):
MW: 216.28   **BP** (°C):

Solubility (Moles/L)	Solubility (Grams/L)	Temp (°C)	Ref (#)	Evaluation (T P E A A)	Comments
2.358E-02	5.100E+00	21	B040	1 0 1 1 1	*sic*
6.473E-04	1.400E-01	ns	F300	0 0 0 0 2	

2539. $C_{11}H_{20}O_5$

Propanoic acid, 2-[(hexthoxycarbonyl)oxy]-, methyl ester
RN: **MP** (°C):
MW: 232.28   **BP** (°C):

Solubility (Moles/L)	Solubility (Grams/L)	Temp (°C)	Ref (#)	Evaluation (T P E A A)	Comments
4.305E-04	9.999E-02	25	R007	0 0 0 0 0	

2540. C_{11}H_{21}BrO_2

11-Bromoundecanoic acid

Bromo-11-undecanoique acide

RN:	2834-05-1	MP (°C):	49.5
MW:	265.20	BP (°C):	173.5

Solubility (Moles/L)	Solubility (Grams/L)	Temp (°C)	Ref (#)	Evaluation (T P E A A)	Comments
2.000E-04	5.304E-02	30	D051	2 0 0 1 2	
7.500E-04	1.989E-01	60	D051	2 0 0 1 2	

2541. C_{11}H_{21}NOS

Cycloate

S-Ethyl N-ethylthiocyclohexanecarbamate

RO-Neet

S-Ethyl N,N-ethylcyclohexylthiocarbamate

RN:	1134-23-2	MP (°C):	12
MW:	215.36	BP (°C):	

Solubility (Moles/L)	Solubility (Grams/L)	Temp (°C)	Ref (#)	Evaluation (T P E A A)	Comments
3.947E-04	8.500E-02	22	B200	1 0 0 0 1	
3.947E-04	8.500E-02	22	F019	1 0 0 0 1	
3.947E-04	8.500E-02	22	M161	1 0 0 0 1	

2542. C_{11}H_{21}NO_3

Dipropylaceturethane

RN:		MP (°C):	
MW:	215.29	BP (°C):	

Solubility (Moles/L)	Solubility (Grams/L)	Temp (°C)	Ref (#)	Evaluation (T P E A A)	Comments
1.857E-03	3.998E-01	20	O021	1 2 0 0 0	

2543. C_{11}H_{21}N_5O

Ipatone

1,3,5-Triazine, 2-(diethylamino)-4-(isopropylamino)-6-methoxy

1,3,5-Triazine-2,4-diamine, N,N-diethyl-6-methoxy-N'-(1-methylethyl)

RN:	3004-70-4	MP (°C):	
MW:	239.32	BP (°C):	

Solubility (Moles/L)	Solubility (Grams/L)	Temp (°C)	Ref (#)	Evaluation (T P E A A)	Comments
4.178E-04	1.000E-01	20	J033	0 0 0 0 0	

2544. C₁₁H₂₁N₅OS

Gesaran
2-Methylthio-4-isopropylamino-6-(3-methoxypropylamino)-s-triazine
Methoprotryne

RN: 841-06-5   **MP** (°C):   69
MW: 271.39   **BP** (°C):

Solubility (Moles/L)	Solubility (Grams/L)	Temp (°C)	Ref (#)	Evaluation (T P E A A)	Comments
1.179E-03	3.200E-01	20	F311	1 2 2 2 1	
1.179E-03	3.200E-01	20	M161	1 0 0 0 2	
1.179E-03	3.200E-01	ns	J033	0 0 0 0 0	
3.681E-03	9.990E-01	ns	M061	0 0 0 0 0	

2545. C₁₁H₂₁N₅S

Dimethametryn
N-(1,2-Dimethylpropyl)-N′-ethyl-6-(methylthio)-1,3,5-triazine-2,4-diamine
Belclene 310

RN: 22936-75-0   **MP** (°C):
MW: 255.39   **BP** (°C):   152

Solubility (Moles/L)	Solubility (Grams/L)	Temp (°C)	Ref (#)	Evaluation (T P E A A)	Comments
1.958E-04	5.000E-02	20	M161	1 0 0 0 I	

2546. C₁₁H₂₁N₅S

Dipropetryn
2-(Ethylthio)-4,6-bis(isopropylamino)-s-triazine
Cotofor
Sancap
Sancap 80W

RN: 4147-51-7   **MP** (°C):   105
MW: 255.39   **BP** (°C):

Solubility (Moles/L)	Solubility (Grams/L)	Temp (°C)	Ref (#)	Evaluation (T P E A A)	Comments
6.265E-05	1.600E-02	rt	M161	0 0 0 0 1	

2547. C₁₁H₂₁N₅S

Ipatryne
2-Methylmercapto-4-isopropylamino-6-diethylamino-s-triazine

RN: **MP** (°C):
MW: 255.39   **BP** (°C):

Solubility (Moles/L)	Solubility (Grams/L)	Temp (°C)	Ref (#)	Evaluation (T P E A A)	Comments
2.100E-05	5.363E-03	26	G001	1 0 1 1 1	

2548. C₁₁H₂₁N₇

1-(1-Piperizinyl)-3,5-bis(dimethylamino)-*s*-triazine
1,3,5-Triazine-2,4-diamine, *N,N,N',N'*-tetramethyl-6-(1-piperazinyl)-
RN: 125867-94-9   **MP** (°C):
MW: 251.34   **BP** (°C):

Solubility (Moles/L)	Solubility (Grams/L)	Temp (°C)	Ref (#)	Evaluation (T P E A A)	Comments
1.081E-02	2.717E+00	25	B386	0 0 0 0 0	

2549. C₁₁H₂₁O₅

Propanoic acid, 2-[(proxycarbonyl)oxy]-, butyl ester
RN: **MP** (°C):
MW: 233.29   **BP** (°C):

Solubility (Moles/L)	Solubility (Grams/L)	Temp (°C)	Ref (#)	Evaluation (T P E A A)	Comments
4.286E-04	9.999E-02	25	R007	0 0 0 0 0	

2550. C₁₁H₂₂N₂O

Cycluron
N'-Cyclooctyl-*N,N*-dimethylurea
Cyclooctyl-1,1-dimethylurea
OMU
RN: 2163-69-1   **MP** (°C):   138
MW: 198.31   **BP** (°C):

Solubility (Moles/L)	Solubility (Grams/L)	Temp (°C)	Ref (#)	Evaluation (T P E A A)	Comments
7.564E-04	1.500E-01	20	B185	0 0 0 0 0	
6.051E-03	1.200E+00	20	G036	1 0 0 0 2	
5.541E-03	1.099E+00	20	M061	1 0 0 0 1	
5.547E-03	1.100E+00	20	M161	1 0 0 0 1	
6.310E-04	1.251E-01	ns	M163	0 0 0 0 0	EFG

2551. C₁₁H₂₂N₆

*N*6,*N*6-Diethyl-*N*2,*N*2,*N*4,*N*4-tetramethylmelamine
1,3,5-Triazine-2,4,6-triamine, *N,N*-diethyl-*N',N',N'',N''*-tetramethyl-
RN: 16268-75-0   **MP** (°C):   42.0
MW: 238.34   **BP** (°C):

Solubility (Moles/L)	Solubility (Grams/L)	Temp (°C)	Ref (#)	Evaluation (T P E A A)	Comments
2.979E-04	7.100E-02	25	C051	1 2 1 1 1	pH 7

2552. $C_{11}H_{22}O_2$
Undecanoic acid
Undecanoique acide

RN:	112-37-8	MP (°C):	28.5
MW:	186.30	BP (°C):	228

Solubility (Moles/L)	Solubility (Grams/L)	Temp (°C)	Ref (#)	Evaluation (T P E A A)	Comments
3.382E-03	6.300E-01	0	B136	1 0 2 1 1	
5.744E-04	1.070E-01	20	B136	1 0 2 1 2	
4.992E-04	9.299E-02	20.0	R001	1 1 1 1 1	
6.978E-04	1.300E-01	30	B136	1 0 2 1 2	
2.800E-04	5.216E-02	30	D051	2 0 0 1 2	
5.904E-04	1.100E-01	30.0	R001	1 1 1 1 1	
7.730E-04	1.440E-01	40	B136	1 0 2 1 2	
6.978E-04	1.300E-01	45	B136	1 0 2 1 1	
6.977E-04	1.300E-01	45.0	R001	1 1 1 1 1	
8.052E-04	1.500E-01	60	B136	1 0 2 1 1	
6.000E-04	1.118E-01	60	D051	2 0 0 1 2	
8.050E-04	1.500E-01	60.0	R001	1 1 1 1 1	
3.381E-04	6.300E-02	.0	R001	1 1 1 1 1	

2553. $C_{11}H_{22}O_2$
Methyl caprate
Capric acid methyl ester
Methyl decanoate

RN:	110-42-9	MP (°C):	−13
MW:	186.30	BP (°C):	223

Solubility (Moles/L)	Solubility (Grams/L)	Temp (°C)	Ref (#)	Evaluation (T P E A A)	Comments
<2.36E-05	<4.40E-03	20	M337	2 1 2 2 1	
2.051E-05	3.821E-03	ns	S460	0 0 0 0 0	

2554. $C_{11}H_{22}O_2$
Ethyl nonanoate
Ethyl nonylate

RN:	123-29-5	MP (°C):	
MW:	186.30	BP (°C):	119 at 23 mm

Solubility (Moles/L)	Solubility (Grams/L)	Temp (°C)	Ref (#)	Evaluation (T P E A A)	Comments
1.585E-04	2.953E-02	ns	S460	0 0 0 0 0	

2555. C₁₁H₂₂O₂

3-Hydroxy-2-propyl-5,5-diethyltetrahydrofuran

RN: **MP** (°C):
MW: 186.30   **BP** (°C):

Solubility (Moles/L)	Solubility (Grams/L)	Temp (°C)	Ref (#)	Evaluation (T P E A A)	Comments
1.053E-01	1.961E+01	rt	B066	0 2 0 0 0	

2556. C₁₁H₂₂O₃

n-Hexyl β-ethoxypropionate
Propionic acid, 3-ethoxy-, hexyl ester

RN: 14144-37-7   **MP** (°C):
MW: 202.30   **BP** (°C):

Solubility (Moles/L)	Solubility (Grams/L)	Temp (°C)	Ref (#)	Evaluation (T P E A A)	Comments
1.483E-03	2.999E-01	25	D002	1 2 1 1 0	

2557. C₁₁H₂₂O₃

1,3-Dioxolane-4-methanol, 2-hexyl-2-methyl
2-Octanone, cyclic (hydroxymethyl)ethylene acetal

RN: 5660-52-6   **MP** (°C):
MW: 202.30   **BP** (°C):

Solubility (Moles/L)	Solubility (Grams/L)	Temp (°C)	Ref (#)	Evaluation (T P E A A)	Comments
1.360E-02	2.751E+00	25	P342	0 0 0 0 0	0.0001M Na₂CO₃

2558. C₁₁H₂₂O₃

Octyl lactate
Propanoic acid, 2-hydroxy-, octyl ester

RN: 5464-71-1   **MP** (°C):
MW: 202.30   **BP** (°C):

Solubility (Moles/L)	Solubility (Grams/L)	Temp (°C)	Ref (#)	Evaluation (T P E A A)	Comments
3.955E-03	8.000E-01	25	R006	2 2 0 1 0	

2559. C₁₁H₂₂O₃

n-Butyl β-n-butoxypropionate
Butyl 3-butoxypropionate
Propanoic acid, 3-butoxy-, butyl ester

RN: 14144-48-0   **MP** (°C):
MW: 202.30   **BP** (°C):

Solubility (Moles/L)	Solubility (Grams/L)	Temp (°C)	Ref (#)	Evaluation (T P E A A)	Comments
3.951E-03	7.994E-01	25	R034	0 0 0 0 0	

2560. $C_{11}H_{22}O_4$
1,3-Dioxolane-4-methanol, 2-(2-butoxyethyl)-2-methyl
RN: 143458-55-3   **MP** (°C):
MW: 218.30   **BP** (°C):

Solubility (Moles/L)	Solubility (Grams/L)	Temp (°C)	Ref (#)	Evaluation (T P E A A)	Comments
2.640E-01	5.763E+01	25	P342	0 0 0 0 0	0.0001M Na_2CO_3

2561. $C_{11}H_{23}NOS$
Butylate
S-Ethyl diisobutylthiocarbamate
RN: 2008-41-5   **MP** (°C):   <25
MW: 217.38   **BP** (°C):

Solubility (Moles/L)	Solubility (Grams/L)	Temp (°C)	Ref (#)	Evaluation (T P E A A)	Comments
2.070E-04	4.500E-02	22	B200	1 0 0 0 1	
2.070E-04	4.500E-02	22	F019	1 0 0 0 1	
1.656E-04	3.599E-02	ns	S460	0 0 0 0 0	
2.070E-04	4.500E-02	rt	M161	0 0 0 0 1	

2562. $C_{11}H_{23}NO_2$
11-Aminoundecanoic acid
Amino-11-undecanoique acide
RN: 2432-99-7   **MP** (°C):   191
MW: 201.31   **BP** (°C):

Solubility (Moles/L)	Solubility (Grams/L)	Temp (°C)	Ref (#)	Evaluation (T P E A A)	Comments
1.986E-03	3.998E-01	20	E039	2 0 1 1 1	smoothed
1.600E-03	3.221E-01	30	D051	2 0 0 1 2	
4.962E-03	9.990E-01	30	E039	2 0 1 1 2	smoothed
8.925E-03	1.797E+00	40	E039	2 0 1 1 2	smoothed
1.486E-02	2.991E+00	50	E039	2 0 1 1 2	smoothed
1.000E-02	2.013E+00	60	D051	2 0 0 1 2	
2.471E-02	4.975E+00	60	E039	2 0 1 1 2	smoothed
3.453E-02	6.951E+00	65	E039	2 0 1 1 2	smoothed
4.431E-02	8.920E+00	70	E039	2 0 1 1 2	smoothed
5.405E-02	1.088E+01	75	E039	2 0 1 1 2	smoothed
6.858E-02	1.381E+01	80	E039	2 0 1 1 2	smoothed
8.183E-02	1.647E+01	85	E039	2 0 1 1 2	smoothed
9.740E-02	1.961E+01	90	E039	2 0 1 1 2	smoothed
1.145E-01	2.306E+01	95	E039	2 0 1 1 2	smoothed
1.259E-01	2.534E+01	100	E039	2 0 1 1 2	smoothed

2563. $C_{11}H_{24}$

Undecane

n-Undecane

n-Hendecane

RN:	1120-21-4	**MP** (°C):	−26	
MW:	156.31	**BP** (°C):	196	

Solubility (Moles/L)	Solubility (Grams/L)	Temp (°C)	Ref (#)	Evaluation (T P E A A)	Comments
<9.60E-06	<1.50E-03	20	M337	2 1 2 2 1	
2.815E-08	4.400E-06	25	M003	1 0 2 2 1	
5.758E-08	9.000E-06	25	T423	0 0 0 0 0	

2564. $C_{12}HCl_7O$

1,2,3,4,6,7,8-Heptachlorodibenzofuran

1,2,3,4,6,7,8-HpCDF

PCDF 131

F 131

RN:	67562-39-4	**MP** (°C):	236	
MW:	409.31	**BP** (°C):		

Solubility (Moles/L)	Solubility (Grams/L)	Temp (°C)	Ref (#)	Evaluation (T P E A A)	Comments
3.310E-12	1.355E-09	22.5	F314	1 1 0 2 2	

2565. $C_{12}HCl_7O_2$

1,2,3,4,6,7,8-Heptachlorodibenzo-p-dioxin

1,2,3,4,6,7,8-HpCDD

PCDD 73

D 73

Heptachlorodibenzo-p-dioxin

RN:	35822-46-9	**MP** (°C):	265	
MW:	425.31	**BP** (°C):		

Solubility (Moles/L)	Solubility (Grams/L)	Temp (°C)	Ref (#)	Evaluation (T P E A A)	Comments
2.200E-12	9.357E-10	7.0	F315	1 2 0 2 2	
2.690E-12	1.144E-09	11.5	F315	1 2 0 2 2	
3.040E-12	1.293E-09	17.0	F315	1 2 0 2 2	
5.400E-12	2.297E-09	21.0	F315	1 2 0 2 2	
6.030E-12	2.565E-09	26.0	F315	1 2 0 2 2	
1.481E-11	6.300E-09	40	F303	1 2 1 2 1	
1.490E-11	6.337E-09	41.0	F315	1 2 0 2 2	

2566. C₁₂HCl₉

2,2′,3,3′,4,4′,5,5′,6-Nonachlorobiphenyl
2,3,4,5,6,2′,3′,4′,5′-Nonachlorbiphenyl

RN: 40186-72-9 **MP** (°C): 204.5
MW: 464.22 **BP** (°C):

Solubility (Moles/L)	Solubility (Grams/L)	Temp (°C)	Ref (#)	Evaluation (T P E A A)	Comments
1.680E-10	7.800E-08	22	O311	2 2 1 2 1	
5.490E-11	2.549E-08	25	D331	2 1 2 2 2	
5.493E-11	2.550E-08	25	D335	1 0 0 0 2	
2.413E-10	1.120E-07	25	W025	1 0 2 2 2	
5.490E-11	2.549E-08	25.0	M324	1 2 1 1 2	
1.100E-10	5.106E-08	32	D331	2 1 2 2 2	
1.100E-10	5.106E-08	32.0	M324	1 2 1 1 2	
1.420E-10	6.592E-08	40	D331	2 1 2 2 2	
1.420E-10	6.592E-08	40.0	M324	1 2 1 1 2	
2.840E-10	1.318E-07	50	D331	2 1 2 2 2	
2.840E-10	1.318E-07	50.0	M324	1 2 1 1 2	

2567. C₁₂HCl₉

2,2′,3,3′,4,5,5′,6,6′-Nonachlorobiphenyl
1,1′-Biphenyl, 2,2′,3,3′,4,5,5′,6,6′-nonachloro-
PCB 208

RN: 52663-77-1 **MP** (°C): 182
MW: 464.22 **BP** (°C):

Solubility (Moles/L)	Solubility (Grams/L)	Temp (°C)	Ref (#)	Evaluation (T P E A A)	Comments
3.880E-11	1.801E-08	25	M342	1 0 1 1 2	

2568. C₁₂H₂Br₈

Octabromobiphenyl
OBBP
Bromkal 80

RN: 27858-07-7 **MP** (°C): 225.0
MW: 785.42 **BP** (°C):

Solubility (Moles/L)	Solubility (Grams/L)	Temp (°C)	Ref (#)	Evaluation (T P E A A)	Comments
3.183E-08	2.500E-05	25	N326	1 0 0 0 1	average

2569. C$_{12}$H$_2$Cl$_6$O
1,2,3,6,7,8-Hexachlorodibenzofuran
1,2,3,6,7,8-HxCDF
F 121
PCDF 121
2,3,4,7,8,9-Hexachlorodibenzofuran

RN: 57117-44-9   **MP (°C):** 233
MW: 374.87   **BP (°C):**

Solubility (Moles/L)	Solubility (Grams/L)	Temp (°C)	Ref (#)	Evaluation (T P E A A)	Comments
4.720E-11	1.769E-08	22.5	F314	1 1 0 2 2	

2570. C$_{12}$H$_2$Cl$_6$O
1,2,3,4,7,8-Hexachlorodibenzofuran
1,2,3,4,7,8-HxCDF
F 118
PCDF 118

RN: 70648-26-9   **MP (°C):** 226
MW: 374.87   **BP (°C):**

Solubility (Moles/L)	Solubility (Grams/L)	Temp (°C)	Ref (#)	Evaluation (T P E A A)	Comments
2.200E-11	8.247E-09	22.5	F314	1 1 0 2 2	

2571. C$_{12}$H$_2$Cl$_6$O$_2$
1,2,3,4,7,8-Hexachlorodibenzo-*p*-dioxin
1,2,3,4,7,8-Hexachlorodibenzo[b,e][1,4]dioxin
1,2,3,4,7,8-Hexachlorodibenzo[1,4]dioxin
1,2,3,4,7,8-HxCDD
D 66
PCDD 66

RN: 39227-28-6   **MP (°C):** 273
MW: 390.87   **BP (°C):**

Solubility (Moles/L)	Solubility (Grams/L)	Temp (°C)	Ref (#)	Evaluation (T P E A A)	Comments
5.910E-12	2.310E-09	7.0	F315	1 2 0 2 2	
7.980E-12	3.119E-09	11.5	F315	1 2 0 2 2	
1.070E-11	4.182E-09	17.0	F315	1 2 0 2 2	
1.126E-11	4.400E-09	20	F303	1 2 1 2 1	
1.250E-11	4.886E-09	21.0	F315	1 2 0 2 2	
2.020E-11	7.896E-09	26.0	F315	1 2 0 2 2	
4.861E-11	1.900E-08	40	F303	1 2 1 2 2	
4.860E-11	1.900E-08	41.0	F315	1 2 0 2 2	

2572. C$_{12}$H$_2$Cl$_8$

2,2′,3,3′,4,4′,5,5′-Octachlorobiphenyl
2,3,4,5,2′,3′,4′,5′-Octachlorbiphenyl
PCB 194

RN:	35694-08-7	**MP** (°C):	156	
MW:	429.77	**BP** (°C):		

Solubility (Moles/L)	Solubility (Grams/L)	Temp (°C)	Ref (#)	Evaluation (T P E A A)	Comments
2.885E-10	1.240E-07	22	O311	2 2 1 2 2	
6.329E-10	2.720E-07	25	W025	1 0 2 2 2	

2573. C$_{12}$H$_2$Cl$_8$

2,2′,3,3′,5,5′,6,6′-Octachlorobiphenyl
2,3,5,6,2′,3′,5′,6′-Octachlorbiphenyl

RN:	2136-99-4	**MP** (°C):	161	
MW:	429.77	**BP** (°C):		

Solubility (Moles/L)	Solubility (Grams/L)	Temp (°C)	Ref (#)	Evaluation (T P E A A)	Comments
2.650E-10	1.139E-07	20	D331	2 1 2 2 2	
2.650E-10	1.139E-07	20.0	M324	1 2 1 1 2	
3.420E-10	1.470E-07	25	D331	2 1 2 2 2	
3.420E-10	1.470E-07	25	D335	1 0 0 0 2	
9.150E-10	3.932E-07	25	M342	1 0 1 1 2	
4.188E-10	1.800E-07	25	W025	1 0 2 2 1	
3.420E-10	1.470E-07	25.0	M324	1 2 1 1 2	
4.930E-10	2.119E-07	32	D331	2 1 2 2 2	
4.930E-10	2.119E-07	32.0	M324	1 2 1 1 2	
1.780E-09	7.650E-07	50	D331	2 1 2 2 2	
1.780E-09	7.650E-07	50.0	M324	1 2 1 1 2	

2574. C$_{12}$H$_3$Cl$_5$O

2,3,4,7,8-Pentachlorodibenzofuran
2,3,4,7,8-P5CDF
PeCDF, 2,3,4,7,8-

RN:	57117-31-4	**MP** (°C):	195.5	
MW:	340.42	**BP** (°C):		

Solubility (Moles/L)	Solubility (Grams/L)	Temp (°C)	Ref (#)	Evaluation (T P E A A)	Comments
6.920E-10	2.356E-07	22.5	F314	1 1 0 2 2	

2575. C$_{12}$H$_3$Cl$_5$O$_2$
1,2,3,4,7-Pentachlorodibenzo-p-dioxin
Dibenzo[b,e][1,4]dioxin, 1,2,3,4,7-pentachloro-
PCDD 50

RN:	39227-61-7	MP (°C):	195
MW:	356.42	BP (°C):	

Solubility (Moles/L)	Solubility (Grams/L)	Temp (°C)	Ref (#)	Evaluation (T P E A A)	Comments
1.420E-10	5.061E-08	7.0	F315	1 2 0 2 2	
1.880E-10	6.701E-08	11.5	F315	1 2 0 2 2	
2.440E-10	8.697E-08	17.0	F315	1 2 0 2 2	
3.367E-10	1.200E-07	20	F303	1 2 1 2 1	
3.450E-10	1.230E-07	21.0	F315	1 2 0 2 2	
4.630E-10	1.650E-07	26.0	F315	1 2 0 2 2	
1.291E-09	4.600E-07	40	F303	1 2 1 2 1	
1.280E-09	4.562E-07	41.0	F315	1 2 0 2 2	

2576. C$_{12}$H$_3$Cl$_7$
2,2′,3,3′,4,4′,5-Heptachlorobiphenyl
1,1′-Biphenyl, 2,2′,3,3′,4,4′,5-heptachloro-
PCB 170
CB 170

RN:	35065-30-6	MP (°C):	134.5
MW:	395.33	BP (°C):	

Solubility (Moles/L)	Solubility (Grams/L)	Temp (°C)	Ref (#)	Evaluation (T P E A A)	Comments
8.778E-09	3.470E-06	20	M336	2 0 2 2 2	

2577. C$_{12}$H$_3$Cl$_7$
2,2′,3,4′,5,5′,6-Heptachlorobiphenyl
1,1′-Biphenyl, 2,2′,3,4′,5,5′,6-heptachloro-
PCB 187

RN:	52663-68-0	MP (°C):	104
MW:	395.33	BP (°C):	

Solubility (Moles/L)	Solubility (Grams/L)	Temp (°C)	Ref (#)	Evaluation (T P E A A)	Comments
1.141E-08	4.510E-06	20	M336	2 0 2 2 2	

2578. C$_{12}$H$_3$Cl$_7$
2,2′,3,3′,5,5′,6-Heptachlorobiphenyl
1,1′-Biphenyl, 2,2′,3,3′,5,5′,6-heptachloro-
PCB 178

RN:	52663-67-9	MP (°C):	
MW:	395.33	BP (°C):	

Solubility (Moles/L)	Solubility (Grams/L)	Temp (°C)	Ref (#)	Evaluation (T P E A A)	Comments
2.236E-08	8.840E-06	20	M336	2 0 2 2 2	

2579. $C_{12}H_3Cl_7$
2,2',3,3',4,6,6'-Heptachlorobiphenyl
1,1'-Biphenyl, 2,2',3,3',4,6,6'-heptachloro-
PCB 176
RN: 52663-65-7   **MP** (°C):
MW: 395.33   **BP** (°C):

Solubility (Moles/L)	Solubility (Grams/L)	Temp (°C)	Ref (#)	Evaluation (T P E A A)	Comments
1.480E-08	5.850E-06	20	M336	2 0 2 2 2	

2580. $C_{12}H_3Cl_7$
2,2',3,3',4,5,6-Heptachlorobiphenyl
1,1'-Biphenyl, 2,2',3,3',4,5,6-heptachloro-
PCB 173
RN: 68194-16-1   **MP** (°C):
MW: 395.33   **BP** (°C):

Solubility (Moles/L)	Solubility (Grams/L)	Temp (°C)	Ref (#)	Evaluation (T P E A A)	Comments
1.052E-08	4.160E-06	20	M336	2 0 2 2 2	

2581. $C_{12}H_3Cl_7$
2,2',3,3',4,5,6'-Heptachlorobiphenyl
1,1'-Biphenyl, 2,2',3,3',4,5,6'-heptachloro-
PCB 174
RN: 38411-25-5   **MP** (°C):
MW: 395.33   **BP** (°C):

Solubility (Moles/L)	Solubility (Grams/L)	Temp (°C)	Ref (#)	Evaluation (T P E A A)	Comments
1.328E-08	5.250E-06	20	M336	2 0 2 2 2	

2582. $C_{12}H_3Cl_7$
2,2',3,4,4',5',6-Heptachlorobiphenyl
1,1'-Biphenyl, 2,2',3,4,4',5',6-heptachloro-
PCB 183
RN: 52663-69-1   **MP** (°C):   83
MW: 395.33   **BP** (°C):

Solubility (Moles/L)	Solubility (Grams/L)	Temp (°C)	Ref (#)	Evaluation (T P E A A)	Comments
1.239E-08	4.900E-06	20	M336	2 0 2 2 2	

2583. C$_{12}$H$_3$Cl$_7$

2,2′,3,3′,4,5′,6-Heptachlorobiphenyl
1,1′-Biphenyl, 2,2′,3,3′,4,5′,6-heptachloro-
PCB 175

RN: 40186-70-7   **MP** (°C):
MW: 395.33   **BP** (°C):

Solubility (Moles/L)	Solubility (Grams/L)	Temp (°C)	Ref (#)	Evaluation (T P E A A)	Comments
2.261E-08	8.940E-06	20	M336	2 0 2 2 2	

2584. C$_{12}$H$_3$Cl$_7$

2,2′,3,3′,4,4′,6-Heptachlorobiphenyl
1,1′-Biphenyl, 2,2′,3,3′,4,4′,6-heptachloro-
PCB 171

RN: 52663-71-5   **MP** (°C):   117
MW: 395.33   **BP** (°C):

Solubility (Moles/L)	Solubility (Grams/L)	Temp (°C)	Ref (#)	Evaluation (T P E A A)	Comments
1.042E-08	4.120E-06	20	M336	2 0 2 2 2	
5.490E-09	2.170E-06	25	M342	1 0 1 1 2	
5.490E-09	2.170E-06	ns	M308	0 0 1 1 2	

2585. C$_{12}$H$_3$Cl$_7$

2,2′,3,3′,4′,5,6-Heptachlorobiphenyl
1,1′-Biphenyl, 2,2′,3,3′,4,5′,6′-heptachloro-
PCB 177

RN: 52663-70-4   **MP** (°C):
MW: 395.33   **BP** (°C):

Solubility (Moles/L)	Solubility (Grams/L)	Temp (°C)	Ref (#)	Evaluation (T P E A A)	Comments
1.219E-08	4.820E-06	20	M336	2 0 2 2 2	

2586. C$_{12}$H$_3$Cl$_7$

2,2′,3,3′,4,5,5′-Heptachlorobiphenyl
1,1′-Biphenyl, 2,2′,3,3′,4,5,5′-heptachloro-
PCB 172

RN: 52663-74-8   **MP** (°C):
MW: 395.33   **BP** (°C):

Solubility (Moles/L)	Solubility (Grams/L)	Temp (°C)	Ref (#)	Evaluation (T P E A A)	Comments
1.088E-08	4.300E-06	20	M336	2 0 2 2 2	

2587. $C_{12}H_3Cl_7$
2,2',3,4,4',5,5'-Heptachlorobiphenyl
1,1'-Biphenyl, 2,2',3,4,4',5,5'-heptachloro-
PCB 180
RN: 35065-29-3   MP (°C):   112
MW: 395.33   BP (°C):

Solubility (Moles/L)	Solubility (Grams/L)	Temp (°C)	Ref (#)	Evaluation (T P E A A)	Comments
9.739E-09	3.850E-06	20	M336	2 0 2 2 2	

2588. $C_{12}H_3Cl_7$
Heptachlorobiphenyl
1,1'-Biphenyl, heptachloro-
Heptachlorodiphenyl
RN: 28655-71-2   MP (°C):
MW: 395.33   BP (°C):

Solubility (Moles/L)	Solubility (Grams/L)	Temp (°C)	Ref (#)	Evaluation (T P E A A)	Comments
1.581E-08	6.250E-06	11.5	D085	0 0 0 0 0	mixed isomers

2589. $C_{12}H_3Cl_7$
2,2',3,4,5,5',6-Heptachlorobiphenyl
2,3,4,5,6,2',5'-Heptachlorbiphenyl
PCB 185
RN: 52712-05-7   MP (°C):   147
MW: 395.33   BP (°C):

Solubility (Moles/L)	Solubility (Grams/L)	Temp (°C)	Ref (#)	Evaluation (T P E A A)	Comments
1.381E-08	5.460E-06	20	M336	2 0 2 2 2	*sic*
1.189E-09	4.700E-07	25	W025	1 0 2 2 1	

2590. $C_{12}H_4Br_6$
FireMaster FF-1 (hexabromobiphenyl mixture)
RN: MP (°C):
MW: 627.62   BP (°C):

Solubility (Moles/L)	Solubility (Grams/L)	Temp (°C)	Ref (#)	Evaluation (T P E A A)	Comments
1.753E-08	1.100E-05	25	H303	1 0 0 0 1	

2591. C$_{12}$H$_4$Br$_6$

2,2',4,4',6,6'-Hexabromobiphenyl
Hexabromobiphenyl
Polybromilated biphenyl

RN:	36355-01-8	MP (°C):	72
MW:	627.62	BP (°C):	

Solubility (Moles/L)	Solubility (Grams/L)	Temp (°C)	Ref (#)	Evaluation (T P E A A)	Comments
9.954E-04	6.247E-01	26.5	G312	0 0 0 0 0	

2592. C$_{12}$H$_4$Br$_6$

Fire Master BP-6 (hexabromophenyl mixture)

RN:	59536-65-1	MP (°C):	
MW:	627.62	BP (°C):	

Solubility (Moles/L)	Solubility (Grams/L)	Temp (°C)	Ref (#)	Evaluation (T P E A A)	Comments
1.753E-08	1.100E-05	25	H303	1 0 0 0 1	

2593. C$_{12}$H$_4$Br$_6$O

2,2',4,4',5,5'-Hexabromodiphenylether

RN:		MP (°C):	
MW:	643.62	BP (°C):	

Solubility (Moles/L)	Solubility (Grams/L)	Temp (°C)	Ref (#)	Evaluation (T P E A A)	Comments
4.723E-11	3.040E-08	10	K431	0 0 0 0 0	
7.831E-11	5.040E-08	25	K431	0 0 0 0 0	
1.896E-10	1.220E-07	35	K431	0 0 0 0 0	

2594. C$_{12}$H$_4$Cl$_4$O

2,3,7,8-Tetrachlorodibenzofuran
2,3,7,8-T4CDF

RN:	51207-31-9	MP (°C):	227
MW:	305.98	BP (°C):	

Solubility (Moles/L)	Solubility (Grams/L)	Temp (°C)	Ref (#)	Evaluation (T P E A A)	Comments
1.370E-09	4.192E-07	22.5	F314	1 1 0 2 2	

2595. $C_{12}H_4Cl_4O_2$

1,2,3,4-Tetrachlorodibenzo-p-dioxin

1,2,3,4-TCDD

1,2,3,4-Tetrachlorodibenzo[b,e][1,4]dioxin

RN: 30746-58-8 **MP** (°C): 184–186

MW: 321.98 **BP** (°C):

Solubility (Moles/L)	Solubility (Grams/L)	Temp (°C)	Ref (#)	Evaluation (T P E A A)	Comments
3.510E-10	1.130E-07	4.0	D330	2 2 1 2 2	
4.007E-11	1.290E-08	4.3	L321	2 0 2 2 2	
1.065E-09	3.430E-07	5	S352	2 2 0 2 2	
1.401E-09	4.510E-07	15	S352	2 2 0 2 2	
1.500E-09	4.830E-07	17.3	L321	2 0 2 2 2	
1.708E-09	5.500E-07	25	S352	2 2 0 2 1	average of 2
1.957E-09	6.300E-07	25	S352	2 2 0 2 1	
1.460E-09	4.701E-07	25.0	D330	2 2 1 2 2	
3.541E-09	1.140E-06	35	S352	2 2 0 2 2	
3.630E-09	1.169E-06	40.0	D330	2 2 1 2 2	
6.476E-09	2.085E-06	45	S352	2 2 0 2 2	

2596. $C_{12}H_4Cl_4O_2$

2,3,7,8-Tetrachlorodibenzo-p-dioxin

TCDD

2,3,7,8-Tetrachlorodibenzodioxin

RN: 1746-01-6 **MP** (°C): 310

MW: 321.98 **BP** (°C):

Solubility (Moles/L)	Solubility (Grams/L)	Temp (°C)	Ref (#)	Evaluation (T P E A A)	Comments
5.994E-11	1.930E-08	22	M340	1 2 2 1 2	
6.212E-10	2.000E 07	ns	C098	0 0 0 0 0	
6.212E-10	2.000E-07	ns	K138	0 0 0 0 2	
6.212E-10	2.000E-07	ns	N320	0 0 0 0 0	
2.457E-11	7.910E-09	rt	A323	0 2 2 1 2	

2597. $C_{12}H_4Cl_4O_2$

1,3,6,8-Tetrachlorodibenzo-p-dioxin

PCDD 42

1,3,6,8-Tetrachlorodibenzo[1,4]dioxin

RN: 33423-92-6 **MP** (°C): 219

MW: 321.98 **BP** (°C):

Solubility (Moles/L)	Solubility (Grams/L)	Temp (°C)	Ref (#)	Evaluation (T P E A A)	Comments
9.939E-10	3.200E-07	20	F303	1 2 1 2 1	
9.939E-10	3.200E-07	20	W319	1 2 1 2 1	
1.211E-09	3.900E-07	40	F303	1 2 1 2 1	
1.211E-09	3.900E-07	40	W319	1 2 1 2 1	
9.845E-10	3.170E-07	ns	W332	0 1 0 2 2	

2598. $C_{12}H_4Cl_4O_2$

1,2,3,7-Tetrachlorodibenzo-p-dioxin
PCDD 29

RN: 67028-18-6   **MP** (°C):   175
MW: 321.98   **BP** (°C):

Solubility (Moles/L)	Solubility (Grams/L)	Temp (°C)	Ref (#)	Evaluation (T P E A A)	Comments
7.560E-10	2.434E-07	7.0	F315	1 2 0 2 2	
8.120E-10	2.614E-07	11.5	F315	1 2 0 2 2	
1.250E-09	4.025E-07	17.0	F315	1 2 0 2 2	
1.336E-09	4.300E-07	20	F303	1 2 1 2 1	
1.490E-09	4.797E-07	21.0	F315	1 2 0 2 2	
2.260E-09	7.277E-07	26.0	F315	1 2 0 2 2	
3.944E-09	1.270E-06	40	F303	1 2 1 2 1	
4.330E-09	1.394E-06	41.0	F315	1 2 0 2 2	

2599. $C_{12}H_4Cl_6$

2,2′,3,4′,5,5′-Hexachlorobiphenyl
1,1′-Biphenyl, 2,2′,3,4′,5,5′-hexachloro-
PCB 146

RN: 51908-16-8   **MP** (°C):
MW: 360.88   **BP** (°C):

Solubility (Moles/L)	Solubility (Grams/L)	Temp (°C)	Ref (#)	Evaluation (T P E A A)	Comments
2.103E-08	7.590E-06	20	M336	2 0 2 2 2	

2600. $C_{12}H_4Cl_6$

2,2′,3,3′,4,4′-Hexachlorobiphenyl
2,3,4,2′,3′,4′-Hexachlorbiphenyl
PCB 128
1,1′-Biphenyl, 2,2′,3,3′,4,4′-hexachloro-

RN: 38380-07-3   **MP** (°C):   150
MW: 360.88   **BP** (°C):

Solubility (Moles/L)	Solubility (Grams/L)	Temp (°C)	Ref (#)	Evaluation (T P E A A)	Comments
1.857E-08	6.700E-06	20	M336	2 0 2 2 2	*sic*
9.690E-10	3.497E-07	25	D306	2 1 2 2 2	
7.840E-10	2.829E-07	25	M342	1 0 1 1 2	
1.219E-09	4.400E-07	25	W025	1 0 2 2 1	

2601. $C_{12}H_4Cl_6$

2,2',3,3',4,5-Hexachlorobiphenyl
2,3,4,5,2',3'-Hexachlorbiphenyl
2,2',3,3',4,5'-Hexachlorobiphenyl
PCB 129

RN:	55215-18-4	MP (°C):	101
MW:	360.88	BP (°C):	

Solubility (Moles/L)	Solubility (Grams/L)	Temp (°C)	Ref (#)	Evaluation (T P E A A)	Comments
1.577E-08	5.690E-06	20	M336	2 0 2 2 2	
1.610E-08	5.810E-06	25	D306	2 1 2 2 2	
2.355E-09	8.500E-07	25	W025	1 0 2 2 1	

2602. $C_{12}H_4Cl_6$

2,3,3',4,4',5'-Hexachlorobiphenyl
2,3,3',4,4',5-Hexachlorobiphenyl

RN:	38380-08-4	MP (°C):	127
MW:	360.88	BP (°C):	

Solubility (Moles/L)	Solubility (Grams/L)	Temp (°C)	Ref (#)	Evaluation (T P E A A)	Comments
1.477E-08	5.330E-06	20	M336	2 0 2 2 2	

2603. $C_{12}H_4Cl_6$

2,2',3,3',6,6'-Hexachlorobiphenyl
1,1'-Biphenyl, 2,2',3,3',6,6'-hexachloro-, (+)-
(+)-PCB 136

RN:	207004-30-6	MP (°C):	114
MW:	360.88	BP (°C):	

Solubility (Moles/L)	Solubility (Grams/L)	Temp (°C)	Ref (#)	Evaluation (T P E A A)	Comments
3.050E-09	1.101E-06	4	D331	2 1 2 2 2	
3.050E-09	1.101E-06	4.0	M324	1 2 1 1 2	
9.010E-09	3.252E-06	20	D331	2 1 2 2 2	
5.586E-08	2.016E-05	20	M336	2 0 2 2 2	
9.010E-09	3.252E-06	20.0	M324	1 2 1 1 2	
1.250E-08	4.511E-06	25	D331	2 1 2 2 2	
1.250E-08	4.510E-06	25	D335	1 0 0 0 2	
1.670E-08	6.027E-06	25	M342	1 0 1 1 2	
1.250E-08	4.511E-06	25.0	M324	1 2 1 1 2	
1.850E-08	6.676E-06	32	D331	2 1 2 2 2	
1.850E-08	6.676E-06	32.0	M324	1 2 1 1 2	
1.670E-08	6.027E-06	ns	M308	0 0 1 1 2	

2604. $C_{12}H_4Cl_6$
2,2′,3,3′,5,6′-Hexachlorobiphenyl
1,1′-Biphenyl, 2,2′,3,3′,5,6′-hexachloro-
PCB 135

RN:	52744-13-5	**MP** (°C):		
MW:	360.88	**BP** (°C):		

Solubility (Moles/L)	Solubility (Grams/L)	Temp (°C)	Ref (#)	Evaluation (T P E A A)	Comments
3.586E-08	1.294E-05	20	M336	2 0 2 2 2	

2605. $C_{12}H_4Cl_6$
2,3,3′,4′,5,6-Hexachlorobiphenyl
1,1′-Biphenyl, 2,3,3′,4′,5,6-hexachloro-
PCB 163

RN:	74472-44-9	**MP** (°C):	122	
MW:	360.88	**BP** (°C):		

Solubility (Moles/L)	Solubility (Grams/L)	Temp (°C)	Ref (#)	Evaluation (T P E A A)	Comments
1.469E-08	5.300E-06	25	B319	2 0 1 2 1	
1.471E-08	5.310E-06	25	H341	1 0 0 0 2	

2606. $C_{12}H_4Cl_6$
2,3,3′,4,4′,6-Hexachlorobiphenyl
1,1′-Biphenyl, 2,3,3′,4,4′,6-hexachloro-
PCB 158

RN:	74472-42-7	**MP** (°C):	107	
MW:	360.88	**BP** (°C):		

Solubility (Moles/L)	Solubility (Grams/L)	Temp (°C)	Ref (#)	Evaluation (T P E A A)	Comments
2.236E-08	8.070E-06	20	M336	2 0 2 2 2	

2607. $C_{12}H_4Cl_6$
Hexachlorobiphenyl
1,1′-Biphenyl, hexachloro-

RN:	26601-64-9	**MP** (°C):		
MW:	360.88	**BP** (°C):		

Solubility (Moles/L)	Solubility (Grams/L)	Temp (°C)	Ref (#)	Evaluation (T P E A A)	Comments
2.754E-08	9.940E-06	11.5	D085	0 0 0 0 0	mixed isomers

2608. C$_{12}$H$_4$Cl$_6$
Aroclor 1260
Arochlor 1260

RN: 11096-82-5   **MP** (°C):
MW: 360.88   **BP** (°C):   402.5

Solubility (Moles/L)	Solubility (Grams/L)	Temp (°C)	Ref (#)	Evaluation (T P E A A)	Comments
3.879E-08	1.400E-05	4	M336	2 0 2 2 1	
3.990E-08	1.440E-05	20	M336	2 0 2 2 2	
6.927E-08	2.500E-05	20	N326	1 0 0 0 1	

2609. C$_{12}$H$_4$Cl$_6$
2,2',3,5,5',6-Hexachlorobiphenyl
1,1'-Biphenyl, 2,2',3,5,5',6-hexachloro-
PCB 151

RN: 52663-63-5   **MP** (°C):   100
MW: 360.88   **BP** (°C):

Solubility (Moles/L)	Solubility (Grams/L)	Temp (°C)	Ref (#)	Evaluation (T P E A A)	Comments
3.755E-08	1.355E-05	20	M336	2 0 2 2 2	

2610. C$_{12}$H$_4$Cl$_6$
2,2',3,3',4,6-Hexachlorobiphenyl
2,2',3,4',5',6'-Hexachlorobiphenyl
PCB 131

RN: 61798-70-7   **MP** (°C):
MW: 360.88   **BP** (°C):

Solubility (Moles/L)	Solubility (Grams/L)	Temp (°C)	Ref (#)	Evaluation (T P E A A)	Comments
3.358E-08	1.212E-05	20	M336	2 0 2 2 2	

2611. C$_{12}$H$_4$Cl$_6$
2,2',3,3',5,6-Hexachlorobiphenyl
2,3,5,6,2',3'-Hexachlorbiphenyl

RN: 52704-70-8   **MP** (°C):   132
MW: 360.88   **BP** (°C):

Solubility (Moles/L)	Solubility (Grams/L)	Temp (°C)	Ref (#)	Evaluation (T P E A A)	Comments
3.588E-08	1.295E-05	20	M336	2 0 2 2 2	*sic*
2.522E-09	9.100E-07	25	W025	1 0 2 2 1	

2612. C$_{12}$H$_4$Cl$_6$

2,2′,3,4,5,5′-Hexachlorobiphenyl
1,1′-Biphenyl, 2,2′,3,4,5,5′-hexachloro-
PCB 141

RN: 52712-04-6   **MP** (°C): 85
MW: 360.88   **BP** (°C):

Solubility (Moles/L)	Solubility (Grams/L)	Temp (°C)	Ref (#)	Evaluation (T P E A A)	Comments
2.092E-08	7.550E-06	20	M336	2 0 2 2 2	

2613. C$_{12}$H$_4$Cl$_6$

2,2′,3,4,5′,6-Hexachlorobiphenyl
1,1′-Biphenyl, 2,2′,3,4,5′,6-hexachloro-
PCB 144

RN: 68194-14-9   **MP** (°C):
MW: 360.88   **BP** (°C):

Solubility (Moles/L)	Solubility (Grams/L)	Temp (°C)	Ref (#)	Evaluation (T P E A A)	Comments
3.586E-08	1.294E-05	20	M336	2 0 2 2 2	

2614. C$_{12}$H$_4$Cl$_6$

2,2′,4,4′,5,5′-Hexachlorobiphenyl
2,4,5,2′,4′,5′-PCB
2,4,5,2′,4′,5′-Hexachlorobiphenyl
PCB 129
PCB 153

RN: 35065-27-1   **MP** (°C): 103
MW: 360.88   **BP** (°C):

Solubility (Moles/L)	Solubility (Grams/L)	Temp (°C)	Ref (#)	Evaluation (T P E A A)	Comments
1.280E-08	4.619E-06	4.0	D330	2 2 1 2 2	
2.533E-08	9.140E-06	20	M336	2 0 2 2 2	*sic*
7.759E-09	2.800E-06	22	C413	2 0 2 2 1	
3.187E-09	1.150E-06	22	O311	2 2 1 2 2	
2.632E-09	9.500E-07	24	C053	0 0 0 0 0	
2.632E-09	9.500E-07	24	F071	1 1 2 1 1	
2.632E-09	9.500E-07	24	M344	1 0 0 0 1	
2.390E-09	8.625E-07	25	D306	2 1 2 2 2	
3.325E-09	1.200E-06	25	W025	1 0 2 2 1	
2.340E-08	8.445E-06	25.0	D330	2 2 1 2 2	
3.540E-08	1.278E-05	40	D330	2 2 1 2 2	
2.641E-09	9.530E-07	ns	H058	0 1 2 1 2	

2615. C₁₂H₄Cl₆

2,2',3,4,4',5'-Hexachlorobiphenyl
1,1'-Biphenyl, 2,2',3,4,4',5'-hexachloro-
PCB 138
CB 138
K 138

RN:	35065-28-2	MP (°C):	80.5
MW:	360.88	BP (°C):	

Solubility (Moles/L)	Solubility (Grams/L)	Temp (°C)	Ref (#)	Evaluation (T P E A A)	Comments
2.020E-08	7.290E-06	20	M336	2 0 2 2 2	

2616. C₁₂H₄Cl₆

2,2',3,4,4',6-Hexachlorobiphenyl
1,1'-Biphenyl, 2,2',3,4,4',6-hexachloro-
PCB 139

RN:	56030-56-9	MP (°C):	73
MW:	360.88	BP (°C):	

Solubility (Moles/L)	Solubility (Grams/L)	Temp (°C)	Ref (#)	Evaluation (T P E A A)	Comments
3.372E-08	1.217E-05	20	M336	2 0 2 2 2	

2617. C₁₂H₄Cl₆

2,2',3,4,4',5-Hexachlorobiphenyl
1,1'-Biphenyl, 2,2',3,4,4',5-hexachloro-
PCB 137

RN:	35694-06-5	MP (°C):	77
MW:	360.88	BP (°C):	

Solubility (Moles/L)	Solubility (Grams/L)	Temp (°C)	Ref (#)	Evaluation (T P E A A)	Comments
2.328E-08	8.400E-06	20	M336	2 0 2 2 2	

2618. C₁₂H₄Cl₆

2,2',4,4',6,6'-Hexachlorobiphenyl
1,1'-Biphenyl, 2,2',4,4',6,6'-hexachloro-
PCB 155

RN:	33979-03-2	MP (°C):	112.5
MW:	360.88	BP (°C):	

Solubility (Moles/L)	Solubility (Grams/L)	Temp (°C)	Ref (#)	Evaluation (T P E A A)	Comments
3.020E-09	1.090E-06	22	O311	2 2 1 2 2	
6.280E-09	2.266E-06	25	D306	2 1 2 2 2	
9.120E-09	3.291E-06	25	L322	1 1 2 2 2	
1.130E-09	4.078E-07	25	M342	1 0 1 1 2	
2.494E-09	9.000E-07	25	W025	1 0 2 2 1	
1.130E-09	4.078E-07	ns	M308	0 0 1 1 2	

2619. C$_{12}$H$_5$Br$_5$

2,2',4,5,5'-Pentabromobiphenyl
1,1'-Biphenyl, 2,2',4,5,5'-pentabromo-
PBB 101

RN: 67888-96-4   **MP** (°C):
MW: 548.72   **BP** (°C):

Solubility (Moles/L)	Solubility (Grams/L)	Temp (°C)	Ref (#)	Evaluation (T P E A A)	Comments
1.880E-10	1.032E-07	4.0	D330	2 2 1 2 2	
8.060E-10	4.423E-07	25	D330	2 2 1 2 2	
1.790E-09	9.822E-07	40.0	D330	2 2 1 2 2	

2620. C$_{12}$H$_5$Br$_5$O

2,2',4,4',5-Pentabromodiphenyl ether

RN: **MP** (°C):
MW: 564.72   **BP** (°C):

Solubility (Moles/L)	Solubility (Grams/L)	Temp (°C)	Ref (#)	Evaluation (T P E A A)	Comments
4.108E-09	2.320E-06	10	K431	0 0 0 0 0	
7.738E-09	4.370E-06	25	K431	0 0 0 0 0	
1.186E-08	6.700E-06	35	K431	0 0 0 0 0	

2621. C$_{12}$H$_5$Cl$_3$O$_2$

1,2,4-Trichlorodibenzo-p-dioxin
Dibenzo[b,e][1,4]dioxin, 1,2,4-trichloro-
PCDD 14

RN: 39227-58-2   **MP** (°C):   129
MW: 287.53   **BP** (°C):

Solubility (Moles/L)	Solubility (Grams/L)	Temp (°C)	Ref (#)	Evaluation (T P E A A)	Comments
7.617E-09	2.190E-06	5	S352	2 2 0 2 2	
1.659E-08	4.770E-06	15	S352	2 2 0 2 2	
2.925E-08	8.410E-06	25	S352	2 2 0 2 2	
2.925E-08	8.410E-06	25	S352	2 2 0 2 2	
5.801E-08	1.668E-05	35	S352	2 2 0 2 2	
9.815E-08	2.822E-05	45	S352	2 2 0 2 2	

2622. C$_{12}$H$_5$Cl$_5$

2,2',3,4,4'-Pentachlorobiphenyl
1,1'-Biphenyl, 2,2',3,4,4'-pentachloro-
PCB 85

RN: 65510-45-4   **MP** (°C):
MW: 326.44   **BP** (°C):

Solubility (Moles/L)	Solubility (Grams/L)	Temp (°C)	Ref (#)	Evaluation (T P E A A)	Comments
6.712E-08	2.191E-05	20	M336	2 0 2 2 2	

2623. $C_{12}H_5Cl_5$
2,2',3,4',6-Pentachlorobiphenyl
2,2',4,6,6'-Pentachlorobiphenyl
PCB 104

RN:	56558-16-8	MP (°C):	85
MW:	326.44	BP (°C):	

Solubility (Moles/L)	Solubility (Grams/L)	Temp (°C)	Ref (#)	Evaluation (T P E A A)	Comments
1.208E-07	3.945E-05	20	M336	2 0 2 2 2	
4.770E-08	1.557E-05	25	D306	2 1 2 2 2	

2624. $C_{12}H_5Cl_5$
2,2',3,3',6-Pentachlorobiphenyl
1,1'-Biphenyl, 2,2',3,3',6-pentachloro-
PCB 84

RN:	52663-60-2	MP (°C):	
MW:	326.44	BP (°C):	

Solubility (Moles/L)	Solubility (Grams/L)	Temp (°C)	Ref (#)	Evaluation (T P E A A)	Comments
1.440E-07	4.702E-05	20	M336	2 0 2 2 2	

2625. $C_{12}H_5Cl_5$
2,2',3,3',5-Pentachlorobiphenyl
1,1'-Biphenyl, 2,2',3,3',5-pentachloro-
PCB 83

RN:	60145-20-2	MP (°C):	
MW:	326.44	BP (°C):	

Solubility (Moles/L)	Solubility (Grams/L)	Temp (°C)	Ref (#)	Evaluation (T P E A A)	Comments
8.648E-08	2.823E-05	20	M336	2 0 2 2 2	

2626. $C_{12}H_5Cl_5$
2',3,4,5,5'-Pentachlorobiphenyl
1,1'-Biphenyl, 2,3',4',5,5'-pentachloro-
PCB 124

RN:	70424-70-3	MP (°C):	105
MW:	326.44	BP (°C):	

Solubility (Moles/L)	Solubility (Grams/L)	Temp (°C)	Ref (#)	Evaluation (T P E A A)	Comments
4.843E-08	1.581E-05	20	M336	2 0 2 2 2	

2627. C$_{12}$H$_5$Cl$_5$

2,2',3',4,5-Pentachlorobiphenyl
1,1'-Biphenyl, 2,2',3,4',5'-pentachloro-
PCB 87

RN: 41464-51-1 **MP** (°C): 81
MW: 326.44 **BP** (°C):

Solubility (Moles/L)	Solubility (Grams/L)	Temp (°C)	Ref (#)	Evaluation (T P E A A)	Comments
8.703E-08	2.841E-05	20	M336	2 0 2 2 2	

2628. C$_{12}$H$_5$Cl$_5$

2,2',3,4,5'-Pentachlorobiphenyl
2,3,4,2',5'-Pentachlorbiphenyl
PCB 87

RN: 38380-02-8 **MP** (°C): 112
MW: 326.44 **BP** (°C):

Solubility (Moles/L)	Solubility (Grams/L)	Temp (°C)	Ref (#)	Evaluation (T P E A A)	Comments
9.009E-08	2.941E-05	20	M336	2 0 2 2 2	
1.379E-08	4.500E-06	25	W025	1 0 2 2 1	

2629. C$_{12}$H$_5$Cl$_5$

2,3,3',4',6-Pentachlorobiphenyl
1,1'-Biphenyl, 2,3,3',4',6-pentachloro-
PCB 110

RN: 38380-03-9 **MP** (°C):
MW: 326.44 **BP** (°C):

Solubility (Moles/L)	Solubility (Grams/L)	Temp (°C)	Ref (#)	Evaluation (T P E A A)	Comments
8.829E-08	2.882E-05	20	M336	2 0 2 2 2	

2630. C$_{12}$H$_5$Cl$_5$

2',3,3',4,5-Pentachlorobiphenyl
1,1'-Biphenyl, 2,3,3',4',5'-pentachloro-
PCB 122

RN: 76842-07-4 **MP** (°C):
MW: 326.44 **BP** (°C):

Solubility (Moles/L)	Solubility (Grams/L)	Temp (°C)	Ref (#)	Evaluation (T P E A A)	Comments
3.933E-08	1.284E-05	20	M336	2 0 2 2 2	

2631. $C_{12}H_5Cl_5$
2,2',3,3',4-Pentachlorobiphenyl
1,1'-Biphenyl, 2,2',3,3',4-pentachloro-
PCB 82
RN: 52663-62-4 **MP** (°C): 119
MW: 326.44 **BP** (°C):

Solubility (Moles/L)	Solubility (Grams/L)	Temp (°C)	Ref (#)	Evaluation (T P E A A)	Comments
8.908E-08	2.908E-05	20	M336	2 0 2 2 2	

2632. $C_{12}H_5Cl_5$
2,2',3,4,5-Pentachlorobiphenyl
2,3,4,5,2'-Pentachlorbiphenyl
PCB 86
RN: 55312-69-1 **MP** (°C): 112
MW: 326.44 **BP** (°C):

Solubility (Moles/L)	Solubility (Grams/L)	Temp (°C)	Ref (#)	Evaluation (T P E A A)	Comments
7.046E-08	2.300E-05	23	W024	0 0 0 0 0	
1.042E-07	3.400E-05	25	B319	2 0 1 2 1	
1.069E-07	3.490E-05	25	H341	1 0 0 0 2	
3.002E-08	9.800E-06	25	W025	1 0 2 2 2	

2633. $C_{12}H_5Cl_5$
2,2',3,4,6-Pentachlorobiphenyl
2,3,4,6,2'-Pentachlorbiphenyl
PCB 88
RN: 55215-17-3 **MP** (°C): 63
MW: 326.44 **BP** (°C):

Solubility (Moles/L)	Solubility (Grams/L)	Temp (°C)	Ref (#)	Evaluation (T P E A A)	Comments
3.676E-08	1.200E-05	25	W025	1 0 2 2 2	

2634. $C_{12}H_5Cl_5$
2,2',3,5',6-Pentachlorobiphenyl
1,1'-Biphenyl, 2,2',3,5',6-pentachloro-
PCB 95
RN: 38379-99-6 **MP** (°C): 94
MW: 326.44 **BP** (°C):

Solubility (Moles/L)	Solubility (Grams/L)	Temp (°C)	Ref (#)	Evaluation (T P E A A)	Comments
1.658E-07	5.413E-05	20	M336	2 0 2 2 2	

2635. C$_{12}$H$_5$Cl$_5$

2,2′,4,4′,5-Pentachlorobiphenyl
1,1′-Biphenyl, 2,2′,4,4′,5-pentachloro-
PCB 99

RN: 38380-01-7   **MP** (°C):
MW: 326.44   **BP** (°C):

Solubility (Moles/L)	Solubility (Grams/L)	Temp (°C)	Ref (#)	Evaluation (T P E A A)	Comments
6.798E-08	2.219E-05	20	M336	2 0 2 2 2	

2636. C$_{12}$H$_5$Cl$_5$

2,3′,4,4′,5-Pentachlorobiphenyl
1,1′-Biphenyl, 2,3′,4,4′,5-pentachloro-
PCB 118
CB 118

RN: 31508-00-6   **MP** (°C):   109
MW: 326.44   **BP** (°C):

Solubility (Moles/L)	Solubility (Grams/L)	Temp (°C)	Ref (#)	Evaluation (T P E A A)	Comments
4.117E-08	1.344E-05	20	M336	2 0 2 2 2	

2637. C$_{12}$H$_5$Cl$_5$

2,3,3′,4′,5-Pentachlorobiphenyl
1,1′-Biphenyl, 2,3,3′,4′,5-pentachloro-
PCB 107

RN: 70424-68-9   **MP** (°C):
MW: 326.44   **BP** (°C):

Solubility (Moles/L)	Solubility (Grams/L)	Temp (°C)	Ref (#)	Evaluation (T P E A A)	Comments
4.546E-08	1.484E-05	20	M336	2 0 2 2 2	

2638. C$_{12}$H$_5$Cl$_5$

2,3,4,4′,5-Pentachlorobiphenyl
1,1′-Biphenyl, 2,3,4,4′,5-pentachloro-
PCB 114

RN: 74472-37-0   **MP** (°C):   98
MW: 326.44   **BP** (°C):

Solubility (Moles/L)	Solubility (Grams/L)	Temp (°C)	Ref (#)	Evaluation (T P E A A)	Comments
4.895E-08	1.598E-05	20	M336	2 0 2 2 2	

2639. C$_{12}$H$_5$Cl$_5$

2,3,4,5,6-Pentachlorobiphenyl
1,1'-Biphenyl, 2,3,4,5,6-pentachloro-
PCB 116

RN:	18259-05-7	**MP** (°C):	123		
MW:	326.44	**BP** (°C):			

Solubility (Moles/L)	Solubility (Grams/L)	Temp (°C)	Ref (#)	Evaluation (T P E A A)	Comments
4.166E-08	1.360E-05	22	O311	2 2 1 2 2	
1.230E-08	4.015E-06	25	D306	2 1 2 2 2	
1.680E-08	5.484E-06	25	M342	1 0 1 1 2	
2.083E-08	6.800E-06	25	W025	1 0 2 2 1	
1.680E-08	5.484E-06	ns	M308	0 0 1 1 2	

2640. C$_{12}$H$_5$Cl$_5$

2,2',4,5,5'-Pentachlorobiphenyl
2,4,5,2',5'-PCB
2,2',4,5,5'-PCB

RN:	37680-73-2	**MP** (°C):	77		
MW:	326.44	**BP** (°C):			

Solubility (Moles/L)	Solubility (Grams/L)	Temp (°C)	Ref (#)	Evaluation (T P E A A)	Comments
1.880E-08	6.137E-06	4	D331	2 1 2 2 2	
1.880E-08	6.137E-06	4.0	M324	1 2 1 1 2	
3.710E-08	1.211E-05	20	D331	2 1 2 2 2	
8.044E-08	2.626E-05	20	M336	2 0 2 2 2	
3.710E-08	1.211E-05	20.0	M324	1 2 1 1 2	
3.063E-08	1.000E-05	24	C053	0 0 0 0 0	
3.370E-08	1.100E-05	24	C311	0 0 0 0 0	EFG
3.063E-08	1.000E-05	24	F071	1 1 2 1 1	
3.063E-08	1.000E-05	24	M344	1 0 0 0 1	
3.370E-08	1.100E-05	25	C313	0 0 0 0 0	
2.070E-08	6.757E-06	25	D306	2 1 2 2 2	
4.720E-08	1.541E-05	25	D331	2 1 2 2 2	
4.718E-08	1.540E-05	25	D335	1 0 0 0 2	
5.920E-08	1.933E-05	25	M342	1 0 1 1 2	
1.287E-08	4.200E-06	25	W025	1 0 2 2 1	
4.720E-08	1.541E-05	25.0	M324	1 2 1 1 2	
6.830E-08	2.230E-05	32	D331	2 1 2 2 2	
6.830E-08	2.230E-05	32.0	M324	1 2 1 1 2	
3.155E-08	1.030E-05	ns	H058	0 1 2 1 2	
5.820E-08	1.900E-05	ns	M118	0 1 1 1 1	
5.920E-08	1.933E-05	ns	M308	0 0 1 1 2	

2641. C$_{12}$H$_5$Cl$_5$
Pentachlorobiphenyl
2,2′,4,4′,6-Pentachlorobiphenyl
Kanekrol 500
RN: 25429-29-2   **MP** (°C):
MW: 326.44   **BP** (°C):

Solubility (Moles/L)	Solubility (Grams/L)	Temp (°C)	Ref (#)	Evaluation (T P E A A)	Comments
6.341E-08	2.070E-05	11.5	D085	0 0 0 0 0	mixed isomers
9.496E-08	3.100E-05	22.5	G301	0 0 0 0 0	

2642. C$_{12}$H$_5$N$_5$O$_{11}$
Pentanitrophenylether
Benzene, 2-(2,4-dinitrophenoxy)-1,3,5-trinitro-
RN: 5950-87-8   **MP** (°C):
MW: 395.20   **BP** (°C):

Solubility (Moles/L)	Solubility (Grams/L)	Temp (°C)	Ref (#)	Evaluation (T P E A A)	Comments
1.771E-04	7.000E-02	27	D067	1 2 0 0 0	
4.302E-04	1.700E-01	50	D067	1 2 0 0 1	
2.404E-03	9.500E-01	100	D067	1 2 0 0 1	

2643. C$_{12}$H$_5$N$_7$O$_{12}$
Hexanitrodiphenylamine
Benzenamine, 2,4,6-trinitro-*N*-(2,4,6-trinitrophenyl)-
RN: 131-73-7   **MP** (°C):
MW: 439.21   **BP** (°C):

Solubility (Moles/L)	Solubility (Grams/L)	Temp (°C)	Ref (#)	Evaluation (T P E A A)	Comments
1.366E-04	6.000E-02	17	D070	1 2 0 0 0	
4.325E-04	1.900E-01	50	D070	1 2 0 0 1	
7.738E-04	3.399E-01	100	D070	1 2 0 0 1	

2644. C$_{12}$H$_6$Br$_4$
2,2′,5,5′-Tetrabromobiphenyl
Tetrabromobiphenyl
RN: 59080-37-4   **MP** (°C):   143
MW: 469.82   **BP** (°C):

Solubility (Moles/L)	Solubility (Grams/L)	Temp (°C)	Ref (#)	Evaluation (T P E A A)	Comments
8.630E-03	4.054E+00	26.5	G312	0 0 0 0 0	

2645. C$_{12}$H$_6$Br$_4$O
2,2',4,4'-Tetrabromodiphenylether
RN: **MP (°C):**
MW: 485.82   **BP (°C):**

Solubility (Moles/L)	Solubility (Grams/L)	Temp (°C)	Ref (#)	Evaluation (T P E A A)	Comments
1.661E-08	8.070E-06	10	K431	0 0 0 0 0	
3.026E-08	1.470E-05	25	K431	0 0 0 0 0	
5.105E-09	2.480E-06	35	K431	0 0 0 0 0	

2646. C$_{12}$H$_6$Cl$_2$O$_2$
2,7-Dichlorodibenzo-p-dioxin
2,7-DCDD
2,8-Dichlorodibenzodioxin
RN: 33857-26-0   **MP (°C):**   201
MW: 253.09   **BP (°C):**

Solubility (Moles/L)	Solubility (Grams/L)	Temp (°C)	Ref (#)	Evaluation (T P E A A)	Comments
4.307E-09	1.090E-06	5	S352	2 2 0 2 2	
7.942E-09	2.010E-06	15	S352	2 2 0 2 2	
1.482E-08	3.750E-06	25	S352	2 2 0 2 2	
1.482E-08	3.750E-06	25	S352	2 2 0 2 2	
2.873E-08	7.270E-06	35	S352	2 2 0 2 2	
5.295E-08	1.340E-05	45	S352	2 2 0 2 2	

2647. C$_{12}$H$_6$Cl$_2$O$_2$
2,8-Dichlorodibenzo-p-dioxin
2,8-Dichlorodibenzodioxin
PCDD 12
3,6-Dichloro-9,10-dioxaanthracene
RN: 38964-22-6   **MP (°C):**   151
MW: 253.09   **BP (°C):**

Solubility (Moles/L)	Solubility (Grams/L)	Temp (°C)	Ref (#)	Evaluation (T P E A A)	Comments
1.746E-08	4.420E-06	5	S352	2 2 0 2 2	
3.394E-08	8.590E-06	15	S352	2 2 0 2 2	
6.599E-08	1.670E-05	25	S352	2 2 0 2 2	
6.614E-08	1.674E-05	25	S352	2 2 0 2 2	
1.088E-07	2.753E-05	35	S352	2 2 0 2 2	
2.035E-07	5.150E-05	45	S352	2 2 0 2 2	

2648. C$_{12}$H$_6$Cl$_2$O$_2$
2,3-Dichlorodibenzo-p-dioxin
2,3-Dichlorodibenzodioxin
PCDD 10
RN: 29446-15-9   **MP** (°C):   160
MW: 253.09   **BP** (°C):

Solubility (Moles/L)	Solubility (Grams/L)	Temp (°C)	Ref (#)	Evaluation (T P E A A)	Comments
1.454E-08	3.680E-06	5	S352	2 2 0 2 2	
2.829E-08	7.160E-06	15	S352	2 2 0 2 2	
5.887E-08	1.490E-05	25	S352	2 2 0 2 2	
5.887E-08	1.490E-05	25	S352	2 2 0 2 2	
1.201E-07	3.040E-05	35	S352	2 2 0 2 2	
2.315E-07	5.860E-05	45	S352	2 2 0 2 2	

2649. C$_{12}$H$_6$Cl$_3$NO$_3$
Quinonamid
2-(Dichloroacetamido)-3-chloro-1,4-naphthoquinone
HOE 13465OH
Chinonamid
2-[(Dichloroacetyl)amino]-3-chloro-1,4-naphthoquinone
RN: 27541-88-4   **MP** (°C):   212.5
MW: 318.55   **BP** (°C):

Solubility (Moles/L)	Solubility (Grams/L)	Temp (°C)	Ref (#)	Evaluation (T P E A A)	Comments
9.418E-06	3.000E-03	23	M161	1 0 0 0 0	pH 4.6

2650. C$_{12}$H$_6$Cl$_3$NO$_3$
Chlornitrofen
4-Nitrophenyl 2,4,6-trichlorophenyl ether
1,3,5-Trichloro-2-(4-nitrophenoxy)benzene
1′,3′,5′-Trichlorophenyl-4-nitrophenyl ether
RN: 1836-77-7   **MP** (°C):
MW: 318.55   **BP** (°C):

Solubility (Moles/L)	Solubility (Grams/L)	Temp (°C)	Ref (#)	Evaluation (T P E A A)	Comments
2.398E-06	7.640E-04	22	K137	1 1 2 1 0	

2651. $C_{12}H_6Cl_4$
2,2',4,5-Tetrachlorobiphenyl
1,1'-Biphenyl, 2,2',4,5-tetrachloro-
PCB 48

RN:	70362-47-9	MP (°C):	63.9
MW:	291.99	BP (°C):	

Solubility (Moles/L)	Solubility (Grams/L)	Temp (°C)	Ref (#)	Evaluation (T P E A A)	Comments
1.026E-07	2.995E-05	20	M336	2 0 2 2 2	
5.630E-08	1.644E-05	25	M342	1 0 1 1 2	

2652. $C_{12}H_6Cl_4$
2,3',4,6-Tetrachlorobiphenyl
1,1'-Biphenyl, 2,3',4,6-tetrachloro-
PCB 69

RN:	60233-24-1	MP (°C):	46
MW:	291.99	BP (°C):	

Solubility (Moles/L)	Solubility (Grams/L)	Temp (°C)	Ref (#)	Evaluation (T P E A A)	Comments
7.004E-08	2.045E-05	20	M336	2 0 2 2 2	

2653. $C_{12}H_6Cl_4$
Aroclor 1254
Arochlor 1254

RN:	11097-69-1	MP (°C):	
MW:	291.99	BP (°C):	

Solubility (Moles/L)	Solubility (Grams/L)	Temp (°C)	Ref (#)	Evaluation (T P E A A)	Comments
1.336E-07	3.900E-05	4	M336	2 0 2 2 1	
8.288E-08	2.420E-05	11.5	D085	0 0 0 0 0	
9.623E-08	2.810E-05	16.50	W033	1 0 2 2 2	
8.459E-08	2.470E-05	16.50	W033	1 0 2 2 2	
1.473E-07	4.300E-05	20	M336	2 0 2 2 1	
1.712E-07	5.000E-05	20	N326	1 0 0 0 1	
~1.92E-07	~5.60E-05	ns	H117	0 2 2 2 0	
1.541E-07	4.500E-05	ns	L106	0 0 2 1 1	
1.370E-07	4.000E-05	ns	M184	0 0 0 0 0	

2654. $C_{12}H_6Cl_4$
Aroclor 1248
Arochlor 1248

RN:	12672-29-6	MP (°C):	
MW:	291.99	BP (°C):	357.5

Solubility (Moles/L)	Solubility (Grams/L)	Temp (°C)	Ref (#)	Evaluation (T P E A A)	Comments
3.425E-07	1.000E-04	20	N326	1 0 0 0 2	

2655. C$_{12}$H$_6$Cl$_4$
3,3′,5,5′-Tetrachlorobiphenyl
1,1′-Biphenyl, 3,3′,5,5′-tetrachloro-
PCB 80

RN: 33284-52-5   **MP** (°C):   164
MW: 291.99   **BP** (°C):

Solubility (Moles/L)	Solubility (Grams/L)	Temp (°C)	Ref (#)	Evaluation (T P E A A)	Comments
4.220E-09	1.232E-06	25	D306	2 1 2 2 2	

2656. C$_{12}$H$_6$Cl$_4$
3,3′,4,4′-Tetrachlorobiphenyl
3,4,3′,4′-Tetrachlorbiphenyl

RN: 32598-13-3   **MP** (°C):   183
MW: 291.99   **BP** (°C):

Solubility (Moles/L)	Solubility (Grams/L)	Temp (°C)	Ref (#)	Evaluation (T P E A A)	Comments
5.000E-10	1.460E-07	4	D331	2 1 2 2 2	
5.000E-10	1.460E-07	4.0	M324	1 2 1 1 2	
1.490E-09	4.351E-07	20	D331	2 1 2 2 2	
1.490E-09	4.351E-07	20.0	M324	1 2 1 1 2	
6.165E-09	1.800E-06	22	O311	2 2 1 2 1	
1.404E-07	4.100E-05	23	W024	0 0 0 0 0	*sic*
1.880E-09	5.489E-07	25	D306	2 1 2 2 2	
1.950E-09	5.694E-07	25	D331	2 1 2 2 2	
1.949E-09	5.690E-07	25	D335	1 0 0 0 2	
2.569E-09	7.500E-07	25	W025	1 0 2 2 1	
1.950E-09	5.694E-07	25.0	M324	1 2 1 1 2	
4.040E-09	1.180E-06	32	D331	2 1 2 2 2	
4.040E-09	1.180E-06	32.0	M324	1 2 1 1 2	

2657. C$_{12}$H$_6$Cl$_4$
2,4,4′,6-Tetrachlorobiphenyl
1,1′-Biphenyl, 2,4,4′,6-tetrachloro-
PCB 75

RN: 32598-12-2   **MP** (°C):   65
MW: 291.99   **BP** (°C):

Solubility (Moles/L)	Solubility (Grams/L)	Temp (°C)	Ref (#)	Evaluation (T P E A A)	Comments
3.120E-07	9.110E-05	25	D306	2 1 2 2 2	

2658. $C_{12}H_6Cl_4$
2,4,4′,5-Tetrachlorobiphenyl
1,1′-Biphenyl, 2,4,4′,5-tetrachloro-
PCB 74
RN: 32690-93-0 **MP** (°C):
MW: 291.99 **BP** (°C):

Solubility (Moles/L)	Solubility (Grams/L)	Temp (°C)	Ref (#)	Evaluation (T P E A A)	Comments
1.049E-07	3.064E-05	20	M336	2 0 2 2 2	

2659. $C_{12}H_6Cl_4$
2,3,4,5-Tetrachlorobiphenyl
1,1′-Biphenyl, 2,3,4,5-tetrachloro-
PCB 61
RN: 33284-53-6 **MP** (°C): 92
MW: 291.99 **BP** (°C):

Solubility (Moles/L)	Solubility (Grams/L)	Temp (°C)	Ref (#)	Evaluation (T P E A A)	Comments
3.390E-08	9.900E-06	25	B319	2 0 1 2 1	
4.780E-08	1.396E-05	25	D306	2 1 2 2 2	
4.677E-08	1.366E-05	25	L322	1 1 2 2 2	
7.170E-08	2.094E-05	25	M342	1 0 1 1 2	
6.575E-08	1.920E-05	25	W025	1 0 2 2 2	
7.170E-08	2.094E-05	ns	M308	0 0 1 1 2	

2660. $C_{12}H_6Cl_4$
2,3,4,4′-Tetrachlorobiphenyl
1,1′-Biphenyl, 2,3,4,4′-tetrachloro-
PCB 60
RN: 33025-41-1 **MP** (°C): 142
MW: 291.99 **BP** (°C):

Solubility (Moles/L)	Solubility (Grams/L)	Temp (°C)	Ref (#)	Evaluation (T P E A A)	Comments
1.333E-07	3.893E-05	20	M336	2 0 2 2 2	

2661. $C_{12}H_6Cl_4$
2,3,4′,6-Tetrachlorobiphenyl
1,1′-Biphenyl, 2,3,4′,6-tetrachloro-
PCB 64
RN: 52663-58-8 **MP** (°C):
MW: 291.99 **BP** (°C):

Solubility (Moles/L)	Solubility (Grams/L)	Temp (°C)	Ref (#)	Evaluation (T P E A A)	Comments
3.207E-07	9.365E-05	20	M336	2 0 2 2 2	

2662. C$_{12}$H$_6$Cl$_4$
2,3,3′,4′-Tetrachlorobiphenyl
1,1′-Biphenyl, 2,3,3′,4′-tetrachloro-
PCB 56

RN: 41464-43-1   **MP** (°C):
MW: 291.99   **BP** (°C):

Solubility (Moles/L)	Solubility (Grams/L)	Temp (°C)	Ref (#)	Evaluation (T P E A A)	Comments
1.334E-07	3.894E-05	20	M336	2 0 2 2 2	

2663. C$_{12}$H$_6$Cl$_4$
2,3′,4,4′-Tetrachlorobiphenyl
1,1′-Biphenyl, 2,3′,4,4′-tetrachloro-
PCB 66

RN: 32598-10-0   **MP** (°C):   128.0
MW: 291.99   **BP** (°C):

Solubility (Moles/L)	Solubility (Grams/L)	Temp (°C)	Ref (#)	Evaluation (T P E A A)	Comments
1.259E-07	3.676E-05	20	M336	2 0 2 2 2	

2664. C$_{12}$H$_6$Cl$_4$
2,3′,4′,5-Tetrachlorobiphenyl
1,1′-Biphenyl, 2,3′,4′,5-tetrachloro-
PCB 70

RN: 32598-11-1   **MP** (°C):   106
MW: 291.99   **BP** (°C):

Solubility (Moles/L)	Solubility (Grams/L)	Temp (°C)	Ref (#)	Evaluation (T P E A A)	Comments
1.239E-07	3.618E-05	20	M336	2 0 2 2 2	
2.055E-07	6.000E-05	23	W024	0 0 0 0 0	
7.534E-08	2.200E-05	ns	B301	0 2 1 1 1	

2665. C$_{12}$H$_6$Cl$_4$
2,2′,6,6′-Tetrachlorobiphenyl
1,1′-Biphenyl, 2,2′,6,6′-tetrachloro-
PCB 54

RN: 15968-05-5   **MP** (°C):   198.0
MW: 291.99   **BP** (°C):

Solubility (Moles/L)	Solubility (Grams/L)	Temp (°C)	Ref (#)	Evaluation (T P E A A)	Comments
9.247E-09	2.700E-06	22	O311	2 2 1 2 1	
4.070E-08	1.188E-05	25	D306	2 1 2 2 2	

2666. $C_{12}H_6Cl_4$

2,2′,5,5′-Tetrachlorobiphenyl
1,1′-Biphenyl, 2,2′,5,5′-tetrachloro-
PCB 52

RN:	35693-99-3	MP (°C):	87
MW:	291.99	BP (°C):	

Solubility (Moles/L)	Solubility (Grams/L)	Temp (°C)	Ref (#)	Evaluation (T P E A A)	Comments
3.855E-07	1.126E-04	20	M336	2 0 2 2 2	
5.240E-08	1.530E-05	22	O311	2 2 1 2 2	
1.575E-07	4.600E-05	23	W024	0 0 0 0 0	
5.822E-07	1.700E-04	25	B319	2 0 1 2 2	
3.750E-07	1.095E-04	25	D306	2 1 2 2 2	
1.250E-07	3.650E-05	25	H341	1 0 0 0 2	
1.884E-07	5.500E-05	ns	B301	0 2 1 1 1	
9.076E-08	2.650E-05	ns	H058	0 1 2 1 2	
5.480E-08	1.600E-05	ns	M118	0 1 1 1 1	

2667. $C_{12}H_6Cl_4$

2,2′,4,4′-Tetrachlorobiphenyl
1,1′-Biphenyl, 2,2′,4,4′-tetrachloro-
PCB 47

RN:	2437-79-8	MP (°C):	42.0
MW:	291.99	BP (°C):	

Solubility (Moles/L)	Solubility (Grams/L)	Temp (°C)	Ref (#)	Evaluation (T P E A A)	Comments
2.260E-07	6.600E-05	22	C413	2 0 2 2 1	
1.853E-07	5.410E-05	22	O311	2 2 1 2 2	
5.993E-07	1.750E-04	23	W024	0 0 0 0 0	
7.534E-07	2.200E-04	25	B351	1 0 0 1 1	

2668. $C_{12}H_6Cl_4$

2,3,4′,5-Tetrachlorobiphenyl
1,1′-Biphenyl, 2,3,4′,5-tetrachloro-
PCB 63

RN:	74472-34-7	MP (°C):	
MW:	291.99	BP (°C):	

Solubility (Moles/L)	Solubility (Grams/L)	Temp (°C)	Ref (#)	Evaluation (T P E A A)	Comments
8.997E-08	2.627E-05	20	M336	2 0 2 2 2	

2669. C₁₂H₆Cl₄

2,2′,5,6′-Tetrachlorobiphenyl
1,1′-Biphenyl, 2,2′,5,6′-tetrachloro-
PCB 53

RN: 41464-41-9   **MP** (°C):   103
MW: 291.99   **BP** (°C):

Solubility (Moles/L)	Solubility (Grams/L)	Temp (°C)	Ref (#)	Evaluation (T P E A A)	Comments
3.717E-07	1.085E-04	20	M336	2 0 2 2 2	
1.630E-07	4.759E-05	25	D306	2 1 2 2 2	

2670. C₁₂H₆Cl₄

2,2′,3,5′-Tetrachlorobiphenyl
1,1′-Biphenyl, 2,2′,3,5′-tetrachloro-
PCB 44

RN: 41464-39-5   **MP** (°C):   47
MW: 291.99   **BP** (°C):

Solubility (Moles/L)	Solubility (Grams/L)	Temp (°C)	Ref (#)	Evaluation (T P E A A)	Comments
3.426E-07	1.001E-04	20	M336	2 0 2 2 2	
2.226E-07	6.500E-05	23	W024	0 0 0 0 0	
2.740E-07	8.000E-05	25	B319	2 0 1 2 0	

2671. C₁₂H₆Cl₄

2,2′,3,4′-Tetrachlorobiphenyl
1,1′-Biphenyl, 2,2′,3,4′-tetrachloro-
PCB 42

RN: 36559-22-5   **MP** (°C):   68
MW: 291.99   **BP** (°C):

Solubility (Moles/L)	Solubility (Grams/L)	Temp (°C)	Ref (#)	Evaluation (T P E A A)	Comments
2.083E-07	6.083E-05	20	M336	2 0 2 2 2	

2672. C₁₂H₆Cl₄

2,2′,3,6′-Tetrachlorobiphenyl
1,1′-Biphenyl, 2,2′,3,6′-tetrachloro-
PCB 46

RN: 41464-47-5   **MP** (°C):
MW: 291.99   **BP** (°C):

Solubility (Moles/L)	Solubility (Grams/L)	Temp (°C)	Ref (#)	Evaluation (T P E A A)	Comments
3.628E-07	1.059E-04	20	M336	2 0 2 2 2	

2673. $C_{12}H_6Cl_4$

Tetrachlorobiphenyl
1,1'-Biphenyl, tetrachloro-
Pyralene 1498

RN: 26914-33-0 **MP** (°C):
MW: 291.99 **BP** (°C):

Solubility (Moles/L)	Solubility (Grams/L)	Temp (°C)	Ref (#)	Evaluation (T P E A A)	Comments
1.825E-07	5.330E-05	11.5	D085	0 0 0 0 0	mixed isomers

2674. $C_{12}H_6Cl_4$

2',3,4,5-Tetrachlorobiphenyl
1,1'-Biphenyl, 2,3',4',5'-tetrachloro-
PCB 76

RN: 70362-48-0 **MP** (°C): 92.0
MW: 291.99 **BP** (°C):

Solubility (Moles/L)	Solubility (Grams/L)	Temp (°C)	Ref (#)	Evaluation (T P E A A)	Comments
1.888E-07	5.513E-05	20	M336	2 0 2 2 2	

2675. $C_{12}H_6Cl_4$

2,2',3,4-Tetrachlorobiphenyl
1,1'-Biphenyl, 2,2',3,4-tetrachloro-
PCB 41

RN: 52663-59-9 **MP** (°C):
MW: 291.99 **BP** (°C):

Solubility (Moles/L)	Solubility (Grams/L)	Temp (°C)	Ref (#)	Evaluation (T P E A A)	Comments
2.219E-07	6.480E-05	20	M336	2 0 2 2 2	

2676. $C_{12}H_6Cl_4$

2,2',4,5'-Tetrachlorobiphenyl
2,2',4',5-Tetrachlorobiphenyl
PCB 49

RN: 41464-40-8 **MP** (°C): 67
MW: 291.99 **BP** (°C):

Solubility (Moles/L)	Solubility (Grams/L)	Temp (°C)	Ref (#)	Evaluation (T P E A A)	Comments
2.676E-07	7.814E-05	20	M336	2 0 2 2 2	
5.630E-08	1.644E-05	ns	M308	0 0 1 1 2	

2677. C₁₂H₆Cl₄

2,2′,3,3′-Tetrachlorobiphenyl
1,1′-Biphenyl, 2,2′,3,3′-tetrachloro-
PCB 40

RN:	38444-93-8	MP (°C):	121.0
MW:	291.99	BP (°C):	

Solubility (Moles/L)	Solubility (Grams/L)	Temp (°C)	Ref (#)	Evaluation (T P E A A)	Comments
2.764E-07	8.070E-05	20	M336	2 0 2 2 2	
5.822E-07	1.700E-04	23	W024	0 0 0 0 0	
5.340E-08	1.559E-05	25	D306	2 1 2 2 2	

2678. C₁₂H₆Cl₄O₂S

Tetradifon
2,4,5,4′-Tetrachlorodiphenyl sulfone
Tedion
Aracnol K
Akaritox
Rotetra

RN:	116-29-0	MP (°C):	148.5
MW:	356.06	BP (°C):	

Solubility (Moles/L)	Solubility (Grams/L)	Temp (°C)	Ref (#)	Evaluation (T P E A A)	Comments
1.404E-07	5.000E-05	10	V301	1 0 0 0 0	
5.617E-04	2.000E-01	50	M161	1 0 0 0 0	
9.549E-07	3.400E-04	50	V301	1 0 0 0 1	

2679. C₁₂H₇BrClNO₂

Halacrinate
7-Bromo-5-chloro-8-quinolinyl 2-propenoate
Halocrinate

RN:	34462-96-9	MP (°C):	100.5
MW:	312.56	BP (°C):	

Solubility (Moles/L)	Solubility (Grams/L)	Temp (°C)	Ref (#)	Evaluation (T P E A A)	Comments
1.920E-05	6.000E-03	20	M161	1 0 0 0 0	

2680. C$_{12}$H$_7$ClO$_2$
1-Chlorodibenzo-p-dioxin
1-Monochlorodibenzodioxin
PCDD 1

RN: 39227-53-7   **MP** (°C):   98
MW: 218.64   **BP** (°C):

Solubility (Moles/L)	Solubility (Grams/L)	Temp (°C)	Ref (#)	Evaluation (T P E A A)	Comments
6.220E-07	1.360E-04	5	S352	2 2 0 2 2	
1.066E-06	2.330E-04	15	S352	2 2 0 2 2	
1.907E-06	4.170E-04	25	S352	2 2 0 2 2	
1.907E-06	4.170E-04	25	S352	2 2 0 2 2	
3.316E-06	7.250E-04	35	S352	2 2 0 2 2	
5.671E-06	1.240E-03	45	S352	2 2 0 2 2	

2681. C$_{12}$H$_7$ClO$_2$
2-Chlorodibenzo-p-dioxin
2-Monochlorodibenzo-p-dioxin
PCDD 2

RN: 39227-54-8   **MP** (°C):   89
MW: 218.64   **BP** (°C):

Solubility (Moles/L)	Solubility (Grams/L)	Temp (°C)	Ref (#)	Evaluation (T P E A A)	Comments
6.100E-07	1.334E-04	3.90	D330	2 2 1 2 2	
2.904E-07	6.350E-05	5	S352	2 2 0 2 2	
6.266E-07	1.370E-04	15	S352	2 2 0 2 2	
1.363E-06	2.980E-04	25	S352	2 2 0 2 2	average of 2
1.271E-06	2.780E-04	25	S352	2 2 0 2 2	
1.460E-06	3.192E-04	25.0	D330	2 2 1 2 2	
2.987E-06	6.530E-04	35	S352	2 2 0 2 2	
3.430E-06	7.499E-04	39.0	D330	2 2 1 2 2	
5.072E-06	1.109E-03	45	S352	2 2 0 2 2	

2682. C$_{12}$H$_7$Cl$_2$NO$_3$
Nitrofen
2,4-Dichlorophenyl-4-nitrophenyl ether

RN: 1836-75-5   **MP** (°C):   70.5
MW: 284.10   **BP** (°C):

Solubility (Moles/L)	Solubility (Grams/L)	Temp (°C)	Ref (#)	Evaluation (T P E A A)	Comments
3.520E-06	1.000E-03	22	M061	1 0 0 0 0	
3.344E-05	9.500E-03	22	M161	1 0 0 0 0	
3.520E-06	1.000E-03	ns	B100	0 0 0 0 0	
2.144E-06	6.090E-04	ns	H322	0 0 0 0 0	

2683. $C_{12}H_7Cl_3$
2,2',4-Trichlorobiphenyl
1,1'-Biphenyl, 2,2',4-trichloro-
RN: 37680-66-3   **MP** (°C):
MW: 257.55   **BP** (°C):

Solubility (Moles/L)	Solubility (Grams/L)	Temp (°C)	Ref (#)	Evaluation (T P E A A)	Comments
1.006E-06	2.592E-04	20	M336	2 0 2 2 2	

2684. $C_{12}H_7Cl_3$
2,2',6-Trichlorobiphenyl
1,1'-Biphenyl, 2,2',6-trichloro-
RN: 38444-73-4   **MP** (°C):
MW: 257.55   **BP** (°C):

Solubility (Moles/L)	Solubility (Grams/L)	Temp (°C)	Ref (#)	Evaluation (T P E A A)	Comments
1.741E-06	4.483E-04	20	M336	2 0 2 2 2	

2685. $C_{12}H_7Cl_3$
2,3',6-Trichlorobiphenyl
1,1'-Biphenyl, 2,3',6-trichloro-
RN: 38444-76-7   **MP** (°C):
MW: 257.55   **BP** (°C):

Solubility (Moles/L)	Solubility (Grams/L)	Temp (°C)	Ref (#)	Evaluation (T P E A A)	Comments
1.498E-07	3.858E-05	20	M336	2 0 2 2 2	

2686. $C_{12}H_7Cl_3$
2,4,5-Trichlorobiphenyl
1,1'-Biphenyl, 2,4,5-trichloro-
RN: 15862-07-4   **MP** (°C):   77
MW: 257.55   **BP** (°C):

Solubility (Moles/L)	Solubility (Grams/L)	Temp (°C)	Ref (#)	Evaluation (T P E A A)	Comments
3.300E-07	8.500E-05	23	W024	0 0 0 0 0	
5.436E-07	1.400E-04	25	B319	2 0 1 2 1	
5.514E-07	1.420E-04	25	H341	1 0 0 0 2	
6.320E-07	1.628E-04	25	M342	1 0 1 1 2	
3.572E-07	9.200E-05	25	W025	1 0 2 2 1	
6.320E-07	1.628E-04	ns	M308	0 0 1 1 2	

2687. $C_{12}H_7Cl_3$
2,3,4'-Trichlorobiphenyl
1,1'-Biphenyl, 2,3,4'-trichloro-
RN: 38444-85-8 **MP** (°C): 69
MW: 257.55 **BP** (°C):

Solubility (Moles/L)	Solubility (Grams/L)	Temp (°C)	Ref (#)	Evaluation (T P E A A)	Comments
5.500E-07	1.417E-04	20	M336	2 0 2 2 2	

2688. $C_{12}H_7Cl_3$
2,3,6-Trichlorobiphenyl
1,1'-Biphenyl, 2,3,6-trichloro-
RN: 55702-45-9 **MP** (°C): 49
MW: 257.55 **BP** (°C):

Solubility (Moles/L)	Solubility (Grams/L)	Temp (°C)	Ref (#)	Evaluation (T P E A A)	Comments
5.126E-07	1.320E-04	20	M336	2 0 2 2 2	

2689. $C_{12}H_7Cl_3$
2,2',3-Trichlorobiphenyl
1,1'-Biphenyl, 2,2',3-trichloro-
RN: 38444-78-9 **MP** (°C): 28.1
MW: 257.55 **BP** (°C):

Solubility (Moles/L)	Solubility (Grams/L)	Temp (°C)	Ref (#)	Evaluation (T P E A A)	Comments
1.138E-06	2.930E-04	20	M336	2 0 2 2 2	

2690. $C_{12}H_7Cl_3$
2,2',5-Trichlorobiphenyl
1,1'-Biphenyl, 2,2',5-trichloro-
PCB 18
RN: 37680-65-2 **MP** (°C): 44
MW: 257.55 **BP** (°C):

Solubility (Moles/L)	Solubility (Grams/L)	Temp (°C)	Ref (#)	Evaluation (T P E A A)	Comments
1.160E-06	2.986E-04	20	M336	2 0 2 2 2	
1.980E-06	5.099E-04	25	D306	2 1 2 2 2	
2.485E-06	6.400E-04	25	W025	1 0 2 2 2	
4.271E-07	1.100E-04	ns	B301	0 2 1 1 2	
9.629E-07	2.480E-04	ns	H058	0 1 2 1 2	
6.212E-08	1.600E-05	ns	M118	0 1 1 1 1	

2691. C$_{12}$H$_7$Cl$_3$
3,4,4'-Trichlorobiphenyl
3,4,4'-Trichlorbiphenyl
RN: 38444-90-5   **MP** (°C):   88
MW: 257.55   **BP** (°C):

Solubility (Moles/L)	Solubility (Grams/L)	Temp (°C)	Ref (#)	Evaluation (T P E A A)	Comments
2.791E-07	7.189E-05	20	M336	2 0 2 2 2	
3.106E-07	8.000E-05	23	W024	0 0 0 0 0	
5.902E-08	1.520E-05	25	W025	1 0 2 2 2	

2692. C$_{12}$H$_7$Cl$_3$
2,4',5-Trichlorobiphenyl
2,5,4'-Trichlorobiphenyl
PCB 31
RN: 16606-02-3   **MP** (°C):   67
MW: 257.55   **BP** (°C):

Solubility (Moles/L)	Solubility (Grams/L)	Temp (°C)	Ref (#)	Evaluation (T P E A A)	Comments
5.559E-07	1.432E-04	20	M336	2 0 2 2 2	
3.494E-07	9.000E-05	22	O311	2 2 1 2 1	
4.271E-07	1.100E-04	22.5	G301	0 0 0 0 0	
2.912E-07	7.500E-05	ns	B301	0 2 1 1 1	

2693. C$_{12}$H$_7$Cl$_3$
2,4,6-Trichlorobiphenyl
1,1'-Biphenyl, 2,4,6-trichloro-
RN: 35693-92-6   **MP** (°C):   62.5
MW: 257.55   **BP** (°C):

Solubility (Moles/L)	Solubility (Grams/L)	Temp (°C)	Ref (#)	Evaluation (T P E A A)	Comments
3.120E-07	8.036E-05	4.0	D330	2 2 1 2 2	
9.800E-07	2.524E-04	25	D306	2 1 2 2 2	
9.333E-07	2.404E-04	25	L322	1 1 2 2 2	
8.760E-07	2.256E-04	25	M342	1 0 1 1 2	
7.250E-07	1.867E-04	25.0	D330	2 2 1 2 2	
1.690E-06	4.353E-04	40.0	D330	2 2 1 2 2	
8.760E-07	2.256E-04	ns	M308	0 0 1 1 2	

2694. C₁₂H₇Cl₃

2,4,4′-Trichlorobiphenyl

2,4,4′-PCB

RN:	7012-37-5	**MP** (°C):	57		
MW:	257.55	**BP** (°C):			

Solubility (Moles/L)	Solubility (Grams/L)	Temp (°C)	Ref (#)	Evaluation (T P E A A)	Comments
4.465E-07	1.150E-04	20	C302	1 1 2 2 2	
5.559E-07	1.432E-04	20	M336	2 0 2 2 2	
2.601E-07	6.700E-05	22	O311	2 2 1 2 1	
4.271E-07	1.100E-04	24	C311	0 0 0 0 0	EFG
4.504E-07	1.160E-04	25	C313	0 0 0 0 0	
4.530E-07	1.167E-04	25	D306	2 1 2 2 2	
1.010E-06	2.600E-04	25	W025	1 0 2 2 2	

2695. C₁₂H₇Cl₃

Aroclor 1242

Arochlor 1242

RN:	53469-21-9	**MP** (°C):			
MW:	257.55	**BP** (°C):			

Solubility (Moles/L)	Solubility (Grams/L)	Temp (°C)	Ref (#)	Evaluation (T P E A A)	Comments
7.377E-07	1.900E-04	4	M336	2 0 2 2 2	
5.160E-07	1.329E-04	11.5	D085	0 0 0 0 0	
1.076E-06	2.770E-04	20	M336	2 0 2 2 2	
7.766E-07	2.000E-04	20	N326	1 0 0 0 2	
1.747E-07	4.500E-05	ns	L106	0 0 2 1 1	
7.766E-07	2.000E-04	ns	M184	0 0 0 0 0	

2696. C₁₂H₇Cl₃

2′,3,4-Trichlorobiphenyl

1,1′-Biphenyl, 2′,3,4-trichloro-

RN:	38444-86-9	**MP** (°C):	60.0		
MW:	257.55	**BP** (°C):			

Solubility (Moles/L)	Solubility (Grams/L)	Temp (°C)	Ref (#)	Evaluation (T P E A A)	Comments
5.147E-07	1.326E-04	20	M336	2 0 2 2 2	
1.165E-07	3.000E-05	23	W024	0 0 0 0 0	

2697. C$_{12}$H$_7$Cl$_3$

2,3',5-Trichlorobiphenyl
1,1'-Biphenyl, 2,3',5-trichloro-
RN: 38444-81-4   **MP** (°C):   40
MW: 257.55   **BP** (°C):

Solubility (Moles/L)	Solubility (Grams/L)	Temp (°C)	Ref (#)	Evaluation (T P E A A)	Comments
5.374E-07	1.384E-04	20	M336	2 0 2 2 2	
9.810E-07	2.527E-04	25	D306	2 1 2 2 2	

2698. C$_{12}$H$_7$Cl$_3$

Trichlorobiphenyl
Apirolio 1431C
Pyranol 1499
Pyralene 3011
RN: 25323-68-6   **MP** (°C):
MW: 257.55   **BP** (°C):

Solubility (Moles/L)	Solubility (Grams/L)	Temp (°C)	Ref (#)	Evaluation (T P E A A)	Comments
4.620E-07	1.190E-04	11.5	D085	0 0 0 0 0	mixed isomers

2699. C$_{12}$H$_7$Cl$_3$O$_2$

Triclosan
5-Chloro-2-(2,4-dichlorophenoxy)-phenol
RN: 3380-34-5   **MP** (°C):   55.2
MW: 289.55   **BP** (°C):

Solubility (Moles/L)	Solubility (Grams/L)	Temp (°C)	Ref (#)	Evaluation (T P E A A)	Comments
3.454E-05	1.000E-02	20	A067	1 0 0 0 0	
		amb	L434	0 0 0 0 0	
3.467E-05	1.004E-02	ns	R427	0 0 0 0 0	

2700. C$_{12}$H$_7$NO$_2$

1,8-Naphthalimide
1,8-Naphthalenedicarboximide
Naphthalimide
1,8-Naphthalenedicarboxylic acid imide
1H-Benz[de]isoquinoline-1,3(2H)-dione
RN: 81-83-4   **MP** (°C):   292-300
MW: 197.20   **BP** (°C):   428.8

Solubility (Moles/L)	Solubility (Grams/L)	Temp (°C)	Ref (#)	Evaluation (T P E A A)	Comments
3.000E-05	5.916E-03	23	B410	2 1 2 2 2	

2701. $C_{12}H_7N_3O_2$
5-Nitro-1,10-phenanthroline
5-Nitro-o-phenanthroline
RN: 4199-88-6   **MP** (°C):
MW: 225.21   **BP** (°C):

Solubility (Moles/L)	Solubility (Grams/L)	Temp (°C)	Ref (#)	Evaluation (T P E A A)	Comments
1.210E-04	2.725E-02	25.04	B094	1 2 1 2 2	

2702. $C_{12}H_7N_5O_8$
2,4,5,6-Tetranitrodiphenylamine
RN: **MP** (°C):
MW: 349.22   **BP** (°C):

Solubility (Moles/L)	Solubility (Grams/L)	Temp (°C)	Ref (#)	Evaluation (T P E A A)	Comments
2.348E-04	8.199E-02	13.5	D070	1 2 0 0 1	
2.949E-04	1.030E-01	50	D070	1 2 0 0 2	
5.783E-04	2.020E-01	100	D070	1 2 0 0 2	

2703. $C_{12}H_7N_5O_8$
2,4,2′,4′-Tetranitrodiphenylamine
2,4,2′,4-Tetranitro-diphenylamin
RN: 2908-76-1   **MP** (°C):
MW: 349.22   **BP** (°C):

Solubility (Moles/L)	Solubility (Grams/L)	Temp (°C)	Ref (#)	Evaluation (T P E A A)	Comments
5.727E-04	2.000E-01	100	F300	1 0 0 0 2	

2704. $C_{12}H_8$
Acenaphthylene
1,2-Dehydroacenaphthalene
Acenaphthalene
RN: 208-96-8   **MP** (°C): 93.5–94.5
MW: 152.20   **BP** (°C):

Solubility (Moles/L)	Solubility (Grams/L)	Temp (°C)	Ref (#)	Evaluation (T P E A A)	Comments
2.582E-05	3.930E-03	25	L332	1 1 1 1 2	

2705. C$_{12}$H$_8$Br$_2$
4,4'-Dibromobiphenyl
p,p'-Dibromobiphenyl

RN:	92-86-4	MP (°C):	170
MW:	312.02	BP (°C):	357

Solubility (Moles/L)	Solubility (Grams/L)	Temp (°C)	Ref (#)	Evaluation (T P E A A)	Comments
1.841E-02	5.743E+00	26.5	G312	0 0 0 0 0	

2706. C$_{12}$H$_8$Br$_2$O
4,4'-Dibromodiphenylether
bis-p-Bromophenyl ether
Dibromodiphenyl ether, p,p'-

RN:	2050-47-7	MP (°C):	59 C
MW:	328.01	BP (°C):	357 C

Solubility (Moles/L)	Solubility (Grams/L)	Temp (°C)	Ref (#)	Evaluation (T P E A A)	Comments
2.878E-07	9.440E-05	10	K431	0 0 0 0 0	
6.585E-07	2.160E-04	25	K431	0 0 0 0 0	
1.171E-06	3.840E-04	35	K431	0 0 0 0 0	

2707. C$_{12}$H$_8$Cl$_2$
2,5-Dichlorobiphenyl
1,1'-Biphenyl, 2,5-dichloro-

RN:	34883-39-1	MP (°C):	23
MW:	223.10	BP (°C):	

Solubility (Moles/L)	Solubility (Grams/L)	Temp (°C)	Ref (#)	Evaluation (T P E A A)	Comments
6.454E-06	1.440E-03	23	W024	0 0 0 0 0	
5.000E-06	1.116E-03	25	D306	2 1 2 2 2	
8.700E-06	1.941E-03	25	M342	1 0 1 1 1	
2.600E-06	5.800E-04	25	W025	1 0 2 2 2	
8.516E-07	1.900E-04	ns	B301	0 2 1 1 2	
2.680E-05	5.979E-03	ns	M308	0 0 1 1 2	

2708. C$_{12}$H$_8$Cl$_2$
2,4-Dichlorobiphenyl
1,1'-Biphenyl, 2,4-dichloro-

RN:	33284-50-3	MP (°C):	25.0
MW:	223.10	BP (°C):	

Solubility (Moles/L)	Solubility (Grams/L)	Temp (°C)	Ref (#)	Evaluation (T P E A A)	Comments
2.747E-06	6.129E-04	20	M336	2 0 2 2 2	
3.138E-07	7.000E-05	23	W024	0 0 0 0 0	sic
5.065E-06	1.130E-03	25	B319	2 0 1 2 2	
5.065E-06	1.130E-03	25	B350	1 0 0 0 2	
5.150E-06	1.149E-03	25	D306	2 1 2 2 2	

2709. $C_{12}H_8Cl_2$
2,4'-Dichlorobiphenyl
2,4'-PCB
RN: 34883-43-7   **MP** (°C): 43
MW: 223.10   **BP** (°C):

Solubility (Moles/L)	Solubility (Grams/L)	Temp (°C)	Ref (#)	Evaluation (T P E A A)	Comments
2.855E-06	6.370E-04	20	C302	1 1 2 2 2	
2.413E-06	5.383E-04	20	M336	2 0 2 2 2	
2.241E-06	5.000E-04	24	H100	2 0 2 2 0	
2.779E-06	6.200E-04	25	W025	1 0 2 2 2	
2.855E-06	6.370E-04	ns	H058	0 1 2 1 2	

2710. $C_{12}H_8Cl_2$
2,3'-Dichlorobiphenyl
1,1'-Biphenyl, 2,3'-dichloro-
RN: 25569-80-6   **MP** (°C):
MW: 223.10   **BP** (°C):

Solubility (Moles/L)	Solubility (Grams/L)	Temp (°C)	Ref (#)	Evaluation (T P E A A)	Comments
2.599E-06	5.798E-04	20	M336	2 0 2 2 2	

2711. $C_{12}H_8Cl_2$
2,6-Dichlorobiphenyl
1,1'-Biphenyl, 2,6-dichloro-
PCB 10
RN: 33146-45-1   **MP** (°C): 35
MW: 223.10   **BP** (°C):

Solubility (Moles/L)	Solubility (Grams/L)	Temp (°C)	Ref (#)	Evaluation (T P E A A)	Comments
2.420E-06	5.400E-04	22	O311	2 2 1 2 2	
1.080E-05	2.410E-03	25	D306	2 1 2 2 2	
6.230E-06	1.390E-03	25	M342	1 0 1 1 2	
6.230E-06	1.390E-03	ns	M308	0 0 1 1 2	

2712. $C_{12}H_8Cl_2$
2,2'-Dichlorobiphenyl
2,2'-PCB
RN: 13029-08-8   **MP** (°C): 61
MW: 223.10   **BP** (°C):

Solubility (Moles/L)	Solubility (Grams/L)	Temp (°C)	Ref (#)	Evaluation (T P E A A)	Comments
3.214E-06	7.170E-04	20	C302	1 1 2 2 2	
5.038E-06	1.124E-03	20	M336	2 0 2 2 2	
3.541E-06	7.900E-04	22.5	G301	0 0 0 0 0	

(continued)

2712. $C_{12}H_8Cl_2$ (continued)

Solubility (Moles/L)	Solubility (Grams/L)	Temp (°C)	Ref (#)	Evaluation (T P E A A)	Comments
6.275E-06	1.400E-03	23	W024	0 0 0 0 0	
4.034E-06	9.000E-04	24	H100	2 0 2 2 0	
5.410E-06	1.207E-03	25	D306	2 1 2 2 2	
3.541E-06	7.900E-04	25	W025	1 0 2 2 2	

2713. $C_{12}H_8Cl_2$
3,4-Dichlorobiphenyl
1,1'-Biphenyl, 3,4-dichloro-

RN: 2974-92-7   **MP** (°C): 49.5
MW: 223.10   **BP** (°C): 197.5

Solubility (Moles/L)	Solubility (Grams/L)	Temp (°C)	Ref (#)	Evaluation (T P E A A)	Comments
3.550E-08	7.920E-06	25	D306	2 1 2 2 2	
4.074E-07	9.089E-05	ns	R424	0 0 0 0 0	

2714. $C_{12}H_8Cl_2$
4,4'-Dichlorobiphenyl
4,4'-PCB
Dichlorobiphenyl

RN: 2050-68-2   **MP** (°C): 149
MW: 223.10   **BP** (°C): 317

Solubility (Moles/L)	Solubility (Grams/L)	Temp (°C)	Ref (#)	Evaluation (T P E A A)	Comments
1.488E-06	3.320E-04	11.5	D085	0 0 0 0 0	mixed isomers
2.779E-07	6.200E-05	20	C053	0 0 0 0 0	
2.779E-07	6.200E-05	20	F071	1 1 1 1 1	
2.779E-07	6.200E-05	20	M344	1 0 0 0 1	
2.689E-07	6.000E-05	24	H100	2 0 2 2 0	
2.376E-07	5.300E-05	25	B319	2 0 1 2 2	average of 2
2.062E-07	4.600E-05	25	B350	1 0 0 0 1	
1.630E-07	3.637E-05	25	D306	2 1 2 2 2	
2.913E-07	6.500E-05	25	H341	1 0 0 0 1	
2.510E-07	5.600E-05	25	W025	1 0 2 2 1	

2715. $C_{12}H_8Cl_2$
3,3'-Dichlorobiphenyl
1,1'-Biphenyl, 3,3'-dichloro-

RN: 2050-67-1   **MP** (°C): 29
MW: 223.10   **BP** (°C): 323.0

Solubility (Moles/L)	Solubility (Grams/L)	Temp (°C)	Ref (#)	Evaluation (T P E A A)	Comments
1.590E-06	3.547E-04	25	D306	2 1 2 2 2	

2716. $C_{12}H_8Cl_2O_2S$
bis(4-Chlorophenyl) sulfone
4,4′-Dichlorodiphenyl sulfone
1,1′-Sulfonylbis(4-chlorobenzene)
p-Chlorophenyl sulfone

RN:	80-07-9	MP (°C):	149 C
MW:	287.17	BP (°C):	397 C

Solubility (Moles/L)	Solubility (Grams/L)	Temp (°C)	Ref (#)	Evaluation (T P E A A)	Comments
1.741E-07	5.000E-05	22	J420	0 0 0 0 0	pH 6.5

2717. $C_{12}H_8Cl_6$
Aldrin
1,2,3,4,10,10-Hexachloro-1,4,4α,5,8,8α-hexahydro-1,4:5,8-dimethanonaphthalene
Aldrite
Seedrin
Aldrosol
HHDN

RN:	309-00-2	MP (°C):	104.3
MW:	364.92	BP (°C):	

Solubility (Moles/L)	Solubility (Grams/L)	Temp (°C)	Ref (#)	Evaluation (T P E A A)	Comments
2.877E-07	1.050E-04	15	B083	2 2 1 2 2	particle size 5 μm
7.413E-08	2.705E-05	20	B179	0 0 0 0 0	
4.659E-08	1.700E-05	22.5	G301	0 0 0 0 0	
4.898E-07	1.787E-04	24.99	K436	0 0 0 0 0	
4.933E-07	1.800E-04	25	B083	2 2 1 2 2	particle size 5 μm
5.481E-07	2.000E-04	25	M130	1 0 0 0 0	
4.659E-08	1.700E-05	25	W025	1 0 2 2 2	
7.399E-08	2.700E-05	26.5	P027	1 1 2 2 1	
5.481E-07	2.000E-04	26.70	L095	2 2 1 1 2	
7.399E-08	2.700E-05	27	M161	0 0 0 0 1	
9.591E-07	3.500E-04	35	B083	2 2 1 2 2	particle size 5 μm
1.644E-06	6.000E-04	45	B083	2 2 1 2 2	particle size 5 μm
7.399E-08	2.700E-05	ns	I308	0 0 0 0 0	
3.562E-08	1.300E-05	ns	K138	0 0 0 0 2	
1.096E-07	4.000E-05	ns	M110	0 0 0 0 0	EFG

2718. C$_{12}$H$_8$Cl$_6$O
Endrin
1,2,3,4,10,10-Hexachloro-6,7-epoxy-1,4,4α,5,6,7,8,8α-octahydro-1,4-*endo-endo*-5,8-dimethano-
   naphthalene
Mendrin
Nendrin

RN:	72-20-8	**MP** (°C):	228.0		
MW:	380.91	**BP** (°C):			

Solubility (Moles/L)	Solubility (Grams/L)	Temp (°C)	Ref (#)	Evaluation (T P E A A)	Comments
3.413E-07	1.300E-04	15	B083	2 2 1 2 2	particle size 5 μm
6.607E-07	2.517E-04	24.99	K436	0 0 0 0 0	
6.563E-07	2.500E-04	25	B083	2 2 1 2 2	particle size 5 μm
6.826E-07	2.600E-04	25	W025	1 0 2 2 2	
1.103E-06	4.200E-04	35	B083	2 2 1 2 2	particle size 5 μm
1.641E-06	6.250E-04	45	B083	2 2 1 2 2	particle size 5 μm
6.301E-08	2.400E-05	ns	K138	0 0 0 0 2	
1.050E-06	4.000E-04	ns	M110	0 0 0 0 0	EFG
<2.63E-07	<1.00E-04	ns	N034	0 0 0 0 0	
6.563E-07	2.500E-04	ns	V414	0 0 0 0 0	

2719. C$_{12}$H$_8$Cl$_6$O
Dieldrin
3,4,5,6,9,9-Hexachloro-1α,2,2α,3,6,6α,7,7α-octahydro-2,7:3,6-dimethanonaphth[2,3-b]oxirene
Alvit
Quintox
Oxralox

RN:	60-57-1	**MP** (°C):	175.5		
MW:	380.91	**BP** (°C):			

Solubility (Moles/L)	Solubility (Grams/L)	Temp (°C)	Ref (#)	Evaluation (T P E A A)	Comments
2.100E-07	7.999E-05	10	B324	0 0 0 0 0	
2.100E-07	8.000E-05	10	B324	0 0 0 0 0	
2.363E-07	9.000E-05	15	B083	2 2 1 2 1	particle size 5 μm
4.898E-07	1.866E-04	20	B179	0 0 0 0 0	
3.675E-07	1.400E-04	20	B324	0 0 0 0 0	
3.676E-07	1.400E-04	20	B324	0 0 0 0 0	
1.229E-06	4.680E-04	22	K137	1 1 2 1 0	
5.129E-07	1.954E-04	24.99	K436	0 0 0 0 0	
5.119E-07	1.950E-04	25	B083	2 2 1 2 2	particle size 5 μm
4.883E-07	1.860E-04	25	I308	0 0 0 0 0	
6.563E-07	2.500E-04	25	M130	1 0 0 0 1	
5.251E-07	2.000E-04	25	W025	1 0 2 2 2	
1.313E-07	5.000E-05	26	M061	1 0 0 0 0	
4.883E-07	1.860E-04	26.5	P027	1 1 2 2 2	
5.251E-07	2.000E-04	27	B161	2 1 2 2 0	EFG
4.883E-07	1.860E-04	27	M161	0 0 0 0 2	
5.251E-07	2.000E-04	30	B324	0 0 0 0 0	
5.251E-07	2.000E-04	30	B324	0 0 0 0 0	

(continued)

2719. C$_{12}$H$_8$Cl$_6$O (continued)

Solubility (Moles/L)	Solubility (Grams/L)	Temp (°C)	Ref (#)	Evaluation (T P E A A)	Comments
1.050E-06	4.000E-04	35	B083	2 2 1 2 2	particle size 5 µm
1.313E-06	5.000E-04	40	B161	2 1 2 2 0	EFG
1.706E-06	6.500E-04	45	B083	2 2 1 2 2	particle size 5 µm
2.363E-06	9.000E-04	50	B161	2 1 2 2 0	EFG
3.544E-06	1.350E-03	60	B161	2 1 2 2 0	EFG
6.511E-06	2.480E-03	70	B161	2 1 2 2 0	EFG
6.563E-07	2.500E-04	ns	H322	0 0 0 0 0	
5.776E-08	2.200E-05	ns	K138	0 0 0 0 2	
7.876E-07	3.000E-04	ns	M110	0 0 0 0 0	EFG
<2.63E-07	<1.00E-04	ns	N034	0 0 0 0 0	
5.119E-07	1.950E-04	ns	V414	0 0 0 0 0	

2720. C$_{12}$H$_8$N$_2$
p-Phenanthroline
p-Phenanthrolin

RN: 230-07-9 **MP** (°C):
MW: 180.21 **BP** (°C):

Solubility (Moles/L)	Solubility (Grams/L)	Temp (°C)	Ref (#)	Evaluation (T P E A A)	Comments
8.000E-03	1.442E+00	ns	K114	0 0 0 0 0	

2721. C$_{12}$H$_8$N$_2$
Phenazine
Dibenzopyrazine

RN: 92-82-0 **MP** (°C): 175.5
MW: 180.21 **BP** (°C):

Solubility (Moles/L)	Solubility (Grams/L)	Temp (°C)	Ref (#)	Evaluation (T P E A A)	Comments
1.400E-04	2.523E-02	25	K009	1 2 1 1 0	EFG

2722. C$_{12}$H$_8$N$_2$
o-Phenanthroline
1,10-Phenanthroline
o-Phenanthrolin

RN: 66-71-7 **MP** (°C): 115
MW: 180.21 **BP** (°C): >300

Solubility (Moles/L)	Solubility (Grams/L)	Temp (°C)	Ref (#)	Evaluation (T P E A A)	Comments
1.854E-02	3.340E+00	24.99	B444	0 0 0 0 0	
1.526E-02	2.750E+00	25	M155	1 0 1 1 0	EFG
1.490E-02	2.685E+00	25.04	B094	1 2 1 2 2	
1.850E-02	3.334E+00	31	B094	1 2 1 2 2	

(continued)

2722. C$_{12}$H$_8$N$_2$ (continued)

Solubility (Moles/L)	Solubility (Grams/L)	Temp (°C)	Ref (#)	Evaluation (T P E A A)	Comments
2.090E-02	3.766E+00	35	B094	1 2 1 2 2	
2.550E-02	4.595E+00	40.04	B094	1 2 1 2 2	
2.880E-02	5.190E+00	45.44	B094	1 2 1 2 2	
3.410E-02	6.145E+00	50.04	B094	1 2 1 2 2	

2723. C$_{12}$H$_8$N$_2$
m-Phenanthroline
m-Phenanthrolin
RN: 230-46-6   **MP** (°C):
MW: 180.21   **BP** (°C):

Solubility (Moles/L)	Solubility (Grams/L)	Temp (°C)	Ref (#)	Evaluation (T P E A A)	Comments
4.000E-03	7.208E-01	ns	K114	0 0 0 0 0	

2724. C$_{12}$H$_8$N$_4$O$_2$
4H-Pyrazolo[3,4-d]pyrimidin-4-one, 1-benzoyl-1,5-dihydro-
RN: 96448-63-4   **MP** (°C):
MW: 240.22   **BP** (°C):

Solubility (Moles/L)	Solubility (Grams/L)	Temp (°C)	Ref (#)	Evaluation (T P E A A)	Comments
5.828E-05	1.400E-02	22	B428	1 2 1 2 1	

2725. C$_{12}$H$_8$N$_4$O$_6$
Picrylaniline
2,4,6-Trinitrodiphenyllamine
RN: 2919-12-2   **MP** (°C):
MW: 304.22   **BP** (°C):

Solubility (Moles/L)	Solubility (Grams/L)	Temp (°C)	Ref (#)	Evaluation (T P E A A)	Comments
5.888E-05	1.791E-02	25	B335	1 2 0 0 1	

2726. C$_{12}$H$_8$O
Dibenzofuran
Diphenylene oxide
DBF
RN: 132-64-9   **MP** (°C):   83
MW: 168.20   **BP** (°C):   154

Solubility (Moles/L)	Solubility (Grams/L)	Temp (°C)	Ref (#)	Evaluation (T P E A A)	Comments
9.820E-06	1.652E-03	4.0	D330	2 2 1 2 2	
5.960E-05	1.002E-02	25	B173	2 0 2 2 2	

(continued)

2726. $C_{12}H_8O$ (continued)

Solubility (Moles/L)	Solubility (Grams/L)	Temp (°C)	Ref (#)	Evaluation (T P E A A)	Comments
1.850E-05	3.112E-03	25	L301	1 1 2 2 2	
2.592E-05	4.360E-03	25	O406	0 0 0 0 0	
2.812E-05	4.730E-03	25	O406	0 0 0 0 0	
2.510E-05	4.222E-03	25.0	D330	2 2 1 2 2	
4.140E-05	6.963E-03	39.8	D330	2 2 1 2 2	

2727. $C_{12}H_8O_2$
Dibenzo-*p*-dioxin
Dibenzo[1,4]dioxin
Oxanthrene
Phenodioxin

RN: 262-12-4   **MP** (°C):   119
MW: 184.20   **BP** (°C):

Solubility (Moles/L)	Solubility (Grams/L)	Temp (°C)	Ref (#)	Evaluation (T P E A A)	Comments
1.150E-06	2.118E-04	4.10	D330	2 2 1 2 2	
1.113E-06	2.050E-04	5	S352	2 2 0 2 2	
2.497E-06	4.600E-04	15	S352	2 2 0 2 2	
7.601E-06	1.400E-03	25	O406	0 0 0 0 0	
6.841E-06	1.260E-03	25	O406	0 0 0 0 0	
4.729E-06	8.710E-04	25	S352	2 2 0 2 2	average of 2
4.571E-06	8.420E-04	25	S352	2 2 0 2 2	
4.890E-06	9.007E-04	25.0	D330	2 2 1 2 2	
9.566E-06	1.762E-03	35	S352	2 2 0 2 2	
1.300E-05	2.395E-03	40.0	D330	2 2 1 2 2	
1.771E-05	3.262E-03	45	S352	2 2 0 2 2	

2728. $C_{12}H_8O_4$
Methoxsalen
Ammoidin
8-Methoxy-2′,3′,6,7-furocoumarin
Methoxalen
8-Methoxyfuranocoumarin
Oxypsoralen

RN: 298-81-7   **MP** (°C):   148
MW: 216.20   **BP** (°C):

Solubility (Moles/L)	Solubility (Grams/L)	Temp (°C)	Ref (#)	Evaluation (T P E A A)	Comments
2.200E-04	4.756E-02	30	E012	1 2 1 1 0	

2729. C$_{12}$H$_8$S
Dibenzothiophene
Diphenylene sulfide

RN: 132-65-0   **MP** (°C):   97
MW: 184.26   **BP** (°C):   332

Solubility (Moles/L)	Solubility (Grams/L)	Temp (°C)	Ref (#)	Evaluation (T P E A A)	Comments
7.978E-06	1.470E-03	24	H106	1 0 2 2 2	
7.978E-06	1.470E-03	24	M303	1 0 1 1 2	
2.871E-06	5.291E-04	25	L301	1 1 2 2 2	
7.978E-06	1.470E-03	ns	H107	0 0 0 0 2	

2730. C$_{12}$H$_9$Br
4-Bromobiphenyl
1,1'-Biphenyl, 4-bromo-
Bromodiphenyl

RN: 92-66-0   **MP** (°C):   91.5
MW: 233.11   **BP** (°C):   310.0

Solubility (Moles/L)	Solubility (Grams/L)	Temp (°C)	Ref (#)	Evaluation (T P E A A)	Comments
1.010E-06	2.354E-04	4.0	D330	2 2 1 2 2	
2.800E-06	6.527E-04	25.0	D330	2 2 1 2 2	
3.740E-06	8.718E-04	40.0	D330	2 2 1 2 2	

2731. C$_{12}$H$_9$Cl
2-Chlorobiphenyl
2-PCB

RN: 2051-60-7   **MP** (°C):   32
MW: 188.66   **BP** (°C):   274

Solubility (Moles/L)	Solubility (Grams/L)	Temp (°C)	Ref (#)	Evaluation (T P E A A)	Comments
1.993E-05	3.760E-03	20	C302	1 1 2 2 2	
3.074E-05	5.800E-03	23	W024	0 0 0 0 0	
4.771E-06	9.000E-04	24	H100	2 0 2 2 0	
4.134E-05	7.800E-03	25	B351	1 0 0 1 1	
2.680E-05	5.056E-03	25	M342	1 0 1 1 2	
2.189E-05	4.130E-03	25	W025	1 0 2 2 2	
2.680E-05	5.056E-03	ns	M308	0 0 1 1 2	

2732. C$_{12}$H$_9$Cl

4-Chlorobiphenyl
1-Chloro-4-phenyl benzene
4-Monochloro-biphenyl

RN:	2051-62-9	MP (°C):	77
MW:	188.66	BP (°C):	291

Solubility (Moles/L)	Solubility (Grams/L)	Temp (°C)	Ref (#)	Evaluation (T P E A A)	Comments
6.202E-06	1.170E-03	23	W024	0 0 0 0 0	
2.120E-06	4.000E-04	24	H100	2 0 2 2 0	
7.103E-06	1.340E-03	25	B319	2 0 1 2 2	average of 2
6.891E-06	1.300E-03	25	B350	1 0 0 0 2	
6.361E-06	1.200E-03	25	B351	1 0 0 1 1	
6.361E-06	1.200E-03	25	H341	1 0 0 0 2	
7.087E-06	1.337E-03	25	L322	1 1 2 2 2	average of 2
7.079E-06	1.336E-03	25	L322	1 1 2 2 2	average of 2
4.771E-06	9.000E-04	25	W025	1 0 2 2 2	

2733. C$_{12}$H$_9$Cl

3-Chlorobiphenyl
3-Chlorbiphenyl

RN:	2051-61-8	MP (°C):	16
MW:	188.66	BP (°C):	285

Solubility (Moles/L)	Solubility (Grams/L)	Temp (°C)	Ref (#)	Evaluation (T P E A A)	Comments
1.908E-05	3.600E-03	23	W024	0 0 0 0 0	
9.806E-06	1.850E-03	23	W024	0 0 0 0 0	
1.924E-05	3.630E-03	25	B319	2 0 1 2 2	
6.891E-06	1.300E-03	25	W025	1 0 2 2 2	

2734. C$_{12}$H$_9$Cl

Aroclor 1221
Arochlor 1221

RN:	11104-28-2	MP (°C):	
MW:	188.66	BP (°C):	

Solubility (Moles/L)	Solubility (Grams/L)	Temp (°C)	Ref (#)	Evaluation (T P E A A)	Comments
>1.06E-06	>2.00E-04	ns	M184	0 0 0 0 0	

2735. C$_{12}$H$_9$ClF$_3$N$_3$O

Norflurazon
4-Chloro-5-(methylamino)-2-(α,α,α-trifluoro-*m*-tolyl)-3(2H)-pyridazinone
Zorial

RN: 27314-13-2 **MP** (°C): 177
MW: 303.67 **BP** (°C):

Solubility (Moles/L)	Solubility (Grams/L)	Temp (°C)	Ref (#)	Evaluation (T P E A A)	Comments
9.220E-05	2.800E-02	23	M161	1 0 0 0 1	
9.220E-05	2.800E-02	24	C105	2 1 2 2 2	
9.220E-05	2.800E-02	25	B310	1 1 0 0 1	

2736. C$_{12}$H$_9$ClN$_2$

4-Chloroazobenzene
Diazene, (4-chlorophenyl)phenyl-, (E)-

RN: 4340-77-6 **MP** (°C): 88
MW: 216.67 **BP** (°C):

Solubility (Moles/L)	Solubility (Grams/L)	Temp (°C)	Ref (#)	Evaluation (T P E A A)	Comments
2.000E-06	4.333E-04	25	B333	0 0 0 0 0	

2737. C$_{12}$H$_9$ClO

4-Chlorophenyl phenyl ether
1-Chloro-4-phenoxybenzene
p-Chlorodiphenyl oxide

RN: 7005-72-3 **MP** (°C):
MW: 204.66 **BP** (°C):

Solubility (Moles/L)	Solubility (Grams/L)	Temp (°C)	Ref (#)	Evaluation (T P E A A)	Comments
1.612E-05	3.300E-03	25	B131	1 0 0 0 1	

2738. C$_{12}$H$_9$Cl$_2$NO$_2$S

N-(2,3-Chlorophenyl)-benzene-sulfonamide

RN: **MP** (°C):
MW: 302.18 **BP** (°C):

Solubility (Moles/L)	Solubility (Grams/L)	Temp (°C)	Ref (#)	Evaluation (T P E A A)	Comments
1.922E-05	5.809E-03	20	P433	0 0 0 0 0	
2.256E-05	6.816E-03	25	P433	0 0 0 0 0	
2.717E-05	8.209E-03	30	P433	0 0 0 0 0	
3.511E-05	1.061E-02	37	P433	0 0 0 0 0	
4.300E-05	1.299E-02	42	P433	0 0 0 0 0	

2739. C$_{12}$H$_9$Cl$_2$NO$_3$

Vinclozolin
3-(3,5-Dichlorophenyl)-5-ethenyl-5-methyl-2,4-oxazolidinedione
Ornalin
Vinclozalin
Ronilan

RN:	50471-44-8	**MP (°C):**	108		
MW:	286.12	**BP (°C):**			

Solubility (Moles/L)	Solubility (Grams/L)	Temp (°C)	Ref (#)	Evaluation (T P E A A)	Comments
3.495E-03	1.000E+00	20	M161	1 0 0 0 0	
9.120E-06	2.609E-03	ns	R427	0 0 0 0 0	

2740. C$_{12}$H$_9$Cl$_3$NO$_2$S

Reserptyl
4′-[Chlorophenyl]-3,4-dichlorophenylbenzene-sulphonamide

RN:		**MP (°C):**	127–129		
MW:	337.63	**BP (°C):**			

Solubility (Moles/L)	Solubility (Grams/L)	Temp (°C)	Ref (#)	Evaluation (T P E A A)	Comments
1.066E-04	3.600E-02	25	L014	1 0 1 1 1	

2741. C$_{12}$H$_9$FN$_2$O$_4$

1-Benzyloxycarbonyl-5-fluorouracil
1(2H)-Pyrimidinecarboxylic acid, 5-fluoro-3,4-dihydro-2,4-dioxo-, phenylmethyl ester

RN:	66999-98-2	**MP (°C):**			
MW:	264.21	**BP (°C):**			

Solubility (Moles/L)	Solubility (Grams/L)	Temp (°C)	Ref (#)	Evaluation (T P E A A)	Comments
3.028E-04	8.000E-02	22	B332	1 1 0 0 1	pH 4.0

2742. C$_{12}$H$_9$N

Carbazole
9-Azafluorene
Dibenzo[b,d]pyrrole
Diphenylenimine
9H-Carbazole
Dibenzopyrrole

RN:	86-74-8	**MP (°C):**	245		
MW:	167.21	**BP (°C):**	355		

Solubility (Moles/L)	Solubility (Grams/L)	Temp (°C)	Ref (#)	Evaluation (T P E A A)	Comments
7.177E-06	1.200E-03	20	H300	1 1 2 2 1	
5.427E-06	9.075E-04	25	L301	1 1 2 2 2	

2743. C$_{12}$H$_9$NO$_3$
Furo[3,4-b]quinolin-3(1H)-one, 9-hydroxy-1-methyl-
RN: 74103-11-0   **MP** (°C):
MW: 215.21   **BP** (°C):

Solubility (Moles/L)	Solubility (Grams/L)	Temp (°C)	Ref (#)	Evaluation (T P E A A)	Comments
3.253E-07	7.000E-05	25	P089	0 0 0 0 0	
4.321E-07	9.300E-05	37	P089	0 0 0 0 0	
5.529E-07	1.190E-04	51	P089	0 0 0 0 0	

2744. C$_{12}$H$_9$NS
Phenothiazine
Dibenzo-1,4-thiazine
Thiodiphenylamine
RN: 92-84-2   **MP** (°C):   185.1
MW: 199.28   **BP** (°C):

Solubility (Moles/L)	Solubility (Grams/L)	Temp (°C)	Ref (#)	Evaluation (T P E A A)	Comments
6.000E-06	1.196E-03	20	M177	2 2 2 2 0	EFG
8.000E-06	1.594E-03	25	M177	2 2 2 2 0	EFG
1.000E-05	1.993E-03	30	M177	2 2 2 2 0	EFG

2745. C$_{12}$H$_9$N$_3$O$_2$
4-Nitroazobenzene
Diazene, (*p*-nitrophenyl)phenyl-, (E)-
RN: 2491-52-3   **MP** (°C):
MW: 227.22   **BP** (°C):

Solubility (Moles/L)	Solubility (Grams/L)	Temp (°C)	Ref (#)	Evaluation (T P E A A)	Comments
2.800E-06	6.362E-04	25	B333	0 0 0 0 0	

2746. C$_{12}$H$_9$N$_3$O$_3$
Dis. A. 3
4-[(4-Nitrophenyl)azo]phenol
p-Nitrophenylazophenol
p-Hydroxy-*p*′-nitroazobenzene
RN: 1435-60-5   **MP** (°C):   216
MW: 243.22   **BP** (°C):

Solubility (Moles/L)	Solubility (Grams/L)	Temp (°C)	Ref (#)	Evaluation (T P E A A)	Comments
1.600E-05	3.892E-03	25	B333	0 0 0 0 0	

2747. $C_{12}H_9N_3O_4$

2,4-Dinitrodiphenylamine
2,4-Dinitrodiphenylamin
C.I. Disperse yellow 14

RN: 961-68-2   **MP** (°C): 160
MW: 259.22   **BP** (°C):

Solubility (Moles/L)	Solubility (Grams/L)	Temp (°C)	Ref (#)	Evaluation (T P E A A)	Comments
1.466E-04	3.800E-02	15	D070	1 2 0 0 1	
1.543E-04	4.000E-02	15	F300	1 0 0 0 0	
5.100E-06	1.322E-03	25	B333	0 0 0 0 0	*sic*
3.240E-04	8.399E-02	50	D070	1 2 0 0 1	
5.516E-04	1.430E-01	100	D070	1 2 0 0 2	

2748. $C_{12}H_9N_3O_5$

C.I. Disperse yellow 1
C.I. Disperse yellow 1
p-(2,4-Dinitroanilino)
2,4-Dinitro-4'-hydroxydiphenylamine
4-Hydroxy-2',4'-dinitrodiphenylamine

RN: 119-15-3   **MP** (°C): 194
MW: 275.22   **BP** (°C):

Solubility (Moles/L)	Solubility (Grams/L)	Temp (°C)	Ref (#)	Evaluation (T P E A A)	Comments
9.000E-06	2.477E-03	25	B333	0 0 0 0 0	
6.195E-05	1.705E-02	60	P313	0 0 0 0 0	average of 2
1.546E-04	4.255E-02	70	P313	0 0 0 0 0	average of 2
2.954E-04	8.130E-02	80	P313	0 0 0 0 0	average of 2
5.559E-04	1.530E-01	90	P313	0 0 0 0 0	average of 2
1.163E-03	3.200E-01	100	P313	0 0 0 0 0	

2749. $C_{12}H_9N_5O_3$

1-Nicotinoyloxymethyl allopurinol
3-Pyridinecarboxylic acid, (4,5-dihydro-4-oxo-1H-pyrazolo[3,4-d]pyrimidin-1-yl)methyl ester

RN: 98846-66-3   **MP** (°C): 242–243
MW: 271.24   **BP** (°C):

Solubility (Moles/L)	Solubility (Grams/L)	Temp (°C)	Ref (#)	Evaluation (T P E A A)	Comments
3.429E-04	9.300E-02	22	B322	0 0 0 0 0	

2750. C₁₂H₁₀

Diphenyl
Biphenyl
Phenylbenzene
1,1′-Biphenyl
Lemonene

RN:	92-52-4	**MP** (°C):	69.1
MW:	154.21	**BP** (°C):	254

Solubility (Moles/L)	Solubility (Grams/L)	Temp (°C)	Ref (#)	Evaluation (T P E A A)	Comments
1.718E-05	2.650E-03	–.7	N053	1 0 0 1 0	EFG
1.973E-05	3.042E-03	4.62	N053	1 0 0 1 0	EFG
2.670E-05	4.118E-03	10	J302	2 1 2 2 2	
2.372E-05	3.658E-03	10.13	N053	1 0 0 1 0	EFG
2.918E-05	4.500E-03	14.20	N053	1 0 0 1 0	EFG
3.800E-05	5.860E-03	20	H306	1 0 1 2 1	
4.182E-05	6.450E-03	20	T301	1 2 2 2 2	
3.590E-05	5.536E-03	20.10	N053	1 0 0 1 0	EFG
4.100E-05	6.323E-03	21	A057	2 1 2 2 1	
4.533E-05	6.990E-03	22	C413	2 0 2 2 1	
4.850E-05	7.480E-03	22.5	G301	0 0 0 0 0	
1.187E-04	1.830E-02	23.5	S171	2 1 2 2 2	
2.983E-05	4.600E-03	24	H100	2 0 2 2 1	
5.512E-05	8.500E-03	24	H116	2 1 0 0 2	
4.708E-05	7.260E-03	24.60	W003	2 2 2 2 2	average of 3
3.852E-05	5.940E-03	25	A001	1 0 2 2 2	
4.570E-05	7.048E-03	25	A325	2 1 2 2 2	
4.850E-05	7.480E-03	25	B003	2 2 2 2 2	
3.910E-05	6.030E-03	25	B173	2 0 2 2 2	
4.799E-05	7.400E-03	25	B319	2 0 1 2 1	average of 2
4.409E-05	6.800E-03	25	B351	1 0 0 1 1	
4.831E-05	7.450E-03	25	E004	2 1 2 2 2	
4.850E-05	7.479E-03	25	J302	2 1 2 2 2	
4.863E-05	7.500E-03	25	M040	1 0 0 1 1	
4.539E-05	7.000E-03	25	M064	1 1 2 2 1	
4.850E-05	7.480E-03	25	M130	1 0 0 0 2	
4.350E-05	6.708E-03	25	M342	1 0 1 1 2	
4.540E-05	7.001E-03	25	M342	1 0 1 1 2	
4.234E-04	6.530E-02	25	S005	2 2 2 2 2	
4.910E-05	7.572E-03	25.04	V013	2 2 2 2 2	
4.416E-05	6.811E-03	25.35	N053	1 0 0 1 0	EFG
5.689E-05	8.774E-03	28.95	N053	1 0 0 1 0	EFG
5.700E-05	8.790E-03	29.90	W003	2 2 2 2 2	average of 3
5.525E-05	8.520E-03	30.30	W003	2 2 2 2 2	average of 3
8.624E-05	1.330E-02	38.40	W003	2 2 2 2 2	average of 3
8.624E-05	1.330E-02	40.10	W003	2 2 2 2 2	average of 3
1.219E-04	1.880E-02	47.50	W003	2 2 2 2 2	average of 3
1.381E-04	2.130E-02	50.10	W003	2 2 2 2 2	average of 3
1.381E-04	2.130E-02	50.20	W003	2 2 2 2 2	average of 2
1.855E-04	2.860E-02	54.70	W003	2 2 2 2 2	average of 3

(continued)

2750. $C_{12}H_{10}$ (continued)

Solubility (Moles/L)	Solubility (Grams/L)	Temp (°C)	Ref (#)	Evaluation (T P E A A)	Comments
2.347E-04	3.620E-02	59.20	W003	2 2 2 2 2	average of 3
2.620E-04	4.040E-02	60.50	W003	2 2 2 2 2	
2.918E-04	4.500E-02	64.50	W003	2 2 2 2 2	average of 3
4.539E-05	7.000E-03	ns	H123	0 0 0 0 0	
4.350E-05	6.708E-03	ns	M308	0 0 1 1 2	
4.539E-05	7.000E-03	ns	M344	0 0 0 0 1	

2751. $C_{12}H_{10}$

Acenaphthene
1,2-Dihydroacenaphthene
1,8-Ethylenenaphthalene
peri-Ethylenenaphthalene

RN:	83-32-9	**MP** (°C):	95
MW:	154.21	**BP** (°C):	279

Solubility (Moles/L)	Solubility (Grams/L)	Temp (°C)	Ref (#)	Evaluation (T P E A A)	Comments
2.315E-05	3.570E-03	22.20	W003	2 2 2 2 2	
4.780E-05	7.371E-03	25	B173	2 0 2 2 2	
2.250E-05	3.470E-03	25	E004	2 1 2 2 2	
2.218E-05	3.420E-03	25	L332	1 1 1 1 2	
2.548E-05	3.930E-03	25	M064	1 1 2 2 2	
2.550E-05	3.932E-03	25	M342	1 0 1 1 2	
8.889E-07	1.371E-04	25	R084	2 2 2 2 1	*sic*
2.330E-05	3.593E-03	25.04	V013	2 2 2 2 2	
3.041E-05	4.690E-03	30.00	W003	2 2 2 2 2	average of 3
3.761E-05	5.800E-03	34.50	W003	2 2 2 2 2	average of 3
4.520E-05	6.970E-03	39.30	W003	2 2 2 2 1	average of 3
6.076E-05	9.370E-03	44.70	W003	2 2 2 2 1	average of 3
8.060E-05	1.243E-02	50.10	W003	2 2 2 2 2	average of 3
1.038E-04	1.600E-02	55.60	W003	2 2 2 2 2	average of 3
1.741E-04	2.685E-02	64.50	W003	2 2 2 2 2	average of 3
1.511E-04	2.330E-02	65.20	W003	2 2 2 2 2	average of 3
2.118E-04	3.267E-02	69.80	W003	2 2 2 2 2	average of 3
2.283E-04	3.520E-02	71.90	W003	2 2 2 2 2	
2.568E-04	3.960E-02	73.40	W003	2 2 2 2 2	average of 2
2.597E-04	4.005E-02	74.70	W003	2 2 2 2 2	average of 2
3.981E-05	6.139E-03	ns	D001	0 0 0 0 2	
2.248E-05	3.467E-03	ns	I332	0 0 0 0 1	
2.000E-05	3.084E-03	ns	L060	0 0 0 0 0	average
2.548E-05	3.930E-03	ns	M344	0 0 0 0 2	
2.344E-05	3.615E-03	ns	R424	0 0 0 0 0	

2752. C$_{12}$H$_{10}$ClN
4-Amino-4′-chlorodiphenyl
4-Chloro-4′-aminobiphenyl
p-Amino-p′-chlorobiphenyl
p′-Chloro-p-phenylaniline

RN:	135-68-2	MP (°C):
MW:	203.67	BP (°C):

Solubility (Moles/L)	Solubility (Grams/L)	Temp (°C)	Ref (#)	Evaluation (T P E A A)	Comments
2.300E-05	4.684E-03	ns	B305	0 2 0 0 1	

2753. C$_{12}$H$_{10}$ClNO$_2$S
N-(2-Chlorophenyl)-benzene-sulfonamide

RN:		MP (°C):
MW:	267.74	BP (°C):

Solubility (Moles/L)	Solubility (Grams/L)	Temp (°C)	Ref (#)	Evaluation (T P E A A)	Comments
4.178E-05	1.119E-02	20	P433	0 0 0 0 0	
4.800E-05	1.285E-02	25	P433	0 0 0 0 0	
5.239E-05	1.403E-02	30	P433	0 0 0 0 0	
5.667E-05	1.517E-02	37	P433	0 0 0 0 0	
6.444E-05	1.725E-02	42	P433	0 0 0 0 0	

2754. C$_{12}$H$_{10}$ClNO$_2$S
N-(4-Chlorophenyl)-benzene-sulfonamide

RN:		MP (°C):
MW:	267.74	BP (°C):

Solubility (Moles/L)	Solubility (Grams/L)	Temp (°C)	Ref (#)	Evaluation (T P E A A)	Comments
7.611E-05	2.038E-02	20	P433	0 0 0 0 0	
9.333E-05	2.499E-02	25	P433	0 0 0 0 0	
1.244E-04	3.332E-02	30	P433	0 0 0 0 0	
1.789E-04	4.789E-02	37	P433	0 0 0 0 0	
2.189E-04	5.860E-02	42	P433	0 0 0 0 0	

2755. C$_{12}$H$_{10}$Cl$_2$N$_2$
3,3′-Dichlorobenzidine
3,3′-Dichloro-4,4′-biphenyldiamine
o,o′-Dichlorobenzidine
4,4′-Diamino-3,3′-dichlorobiphenyl

RN:	91-94-1	MP (°C):	132
MW:	253.13	BP (°C):	

Solubility (Moles/L)	Solubility (Grams/L)	Temp (°C)	Ref (#)	Evaluation (T P E A A)	Comments
1.230E-05	3.114E-03	25	B173	2 0 2 2 2	
<3.95E-06	<1.00E-03	30	M311	1 1 2 2 0	

2756. $C_{12}H_{10}N_2$

Harmane
1-Methyl-9H-pyrido[3,4-b]indole
Aribine

RN:	486-84-0	MP (°C):	235–238
MW:	182.23	BP (°C):	

Solubility (Moles/L)	Solubility (Grams/L)	Temp (°C)	Ref (#)	Evaluation (T P E A A)	Comments
6.010E+00	1.095E+03	15	B413	1 0 2 2 1	
6.250E+00	1.139E+03	16	B413	1 0 2 2 1	
6.710E+00	1.223E+03	17	B413	1 0 2 2 1	
8.360E+00	1.523E+03	20	B413	1 0 2 2 1	
1.364E+01	2.486E+03	37	B413	1 0 2 2 1	
1.434E+01	2.613E+03	38	B413	1 0 2 2 1	
1.617E+01	2.947E+03	45	B413	1 0 2 2 1	

2757. $C_{12}H_{10}N_2O$

4-Phenylazophenol
4-Hydroxyazobenzene
p-Hydroxyazobenzene
C.I. Solvent yellow 7

RN:	1689-82-3	MP (°C):	150
MW:	198.23	BP (°C):	220

Solubility (Moles/L)	Solubility (Grams/L)	Temp (°C)	Ref (#)	Evaluation (T P E A A)	Comments
4.540E-04	9.000E-02	20	F300	1 0 0 0 1	
1.100E-04	2.180E-02	25	B333	0 0 0 0 0	
1.715E-04	3.400E-02	37	H120	1 1 1 1 1	normal saline
4.036E-03	8.000E-01	100	F300	1 0 0 0 1	

2758. $C_{12}H_{10}N_2O$

Diphenylnitrosamine
Redax
N-Nitroso-N-phenylaniline

RN:	86-30-6	MP (°C):	67
MW:	198.23	BP (°C):	

Solubility (Moles/L)	Solubility (Grams/L)	Temp (°C)	Ref (#)	Evaluation (T P E A A)	Comments
1.770E-04	3.509E-02	25	B173	2 0 2 2 2	

2759. $C_{12}H_{10}N_2O_2$
2,4-Dihydroxyazobenzene
2,4-Dihydroxy-azobenzol

RN: 2051-85-6   **MP** (°C):   170
MW: 214.23   **BP** (°C):

Solubility (Moles/L)	Solubility (Grams/L)	Temp (°C)	Ref (#)	Evaluation (T P E A A)	Comments
9.336E-04	2.000E-01	20	F300	1 0 0 0 0	

2760. $C_{12}H_{10}N_2O_3$
3-Hydroxyazobenzene
3-Hydroxy-azobenzol

RN: 40038-46-8   **MP** (°C):
MW: 230.23   **BP** (°C):

Solubility (Moles/L)	Solubility (Grams/L)	Temp (°C)	Ref (#)	Evaluation (T P E A A)	Comments
3.475E-03	8.000E-01	100	F300	1 0 0 0 1	

2761. $C_{12}H_{10}N_4O_2$
C.I. Disperse orange 3
4′-Nitro-4-aminoazobenzene
4-Amino-4′-nitroazobenzene
4-(4-Nitrophenylazo)aniline

RN: 730-40-5   **MP** (°C):   211
MW: 242.24   **BP** (°C):

Solubility (Moles/L)	Solubility (Grams/L)	Temp (°C)	Ref (#)	Evaluation (T P E A A)	Comments
1.200E-06	2.907E-04	25	B333	0 0 0 0 0	

2762. $C_{12}H_{10}N_4O_4$
C.I. Disperse yellow 9
2,4-Dinitro-4′-aminodiphenylamine
4-Amino-2′,4′-dinitrodiphenylamine
C.I. 10375

RN: 6373-73-5   **MP** (°C):   188
MW: 274.24   **BP** (°C):

Solubility (Moles/L)	Solubility (Grams/L)	Temp (°C)	Ref (#)	Evaluation (T P E A A)	Comments
6.000E-06	1.645E-03	25	B333	0 0 0 0 0	

2763. C₁₂H₁₀O

$\text{2763. } C_{12}H_{10}O$

p-Phenylphenol
p-Hydroxybiphenyl

RN:	92-69-3	MP (°C):	164.5
MW:	170.21	BP (°C):	306.5

Solubility (Moles/L)	Solubility (Grams/L)	Temp (°C)	Ref (#)	Evaluation (T P E A A)	Comments
3.300E-04	5.617E-02	25	E014	2 2 2 1 2	pH 7.2
5.875E-05	1.000E-02	25	L021	1 0 0 0 0	

2764. C₁₂H₁₀O

$\text{2764. } C_{12}H_{10}O$

o-Phenylphenol
2-Phenylphenol

RN:	90-43-7	MP (°C):	56.5
MW:	170.21	BP (°C):	282

Solubility (Moles/L)	Solubility (Grams/L)	Temp (°C)	Ref (#)	Evaluation (T P E A A)	Comments
9.790E-04	1.666E-01	25	L021	1 0 0 0 0	
4.110E-03	6.995E-01	25	M061	0 0 0 0 0	
4.112E-03	7.000E-01	25	M161	1 0 0 0 0	
3.162E-04	5.383E-02	rt	D056	0 1 1 1 0	EFG, pH 6–8, *sic*

2765. C₁₂H₁₀O

$\text{2765. } C_{12}H_{10}O$

Phenyl ether
Diphenyl ether

RN:	101-84-8	MP (°C):	28
MW:	170.21	BP (°C):	259

Solubility (Moles/L)	Solubility (Grams/L)	Temp (°C)	Ref (#)	Evaluation (T P E A A)	Comments
2.341E-02	3.984E+00	25	B019	1 0 1 2 0	*sic*
1.060E-04	1.804E-02	25	B173	2 0 2 2 2	
1.234E-04	2.100E-02	25	F071	1 1 2 1 1	
1.100E-04	1.872E-02	25.04	V013	2 2 2 2 2	

2766. C₁₂H₁₀O₂

$\text{2766. } C_{12}H_{10}O_2$

1-Naphthaleneacetic acid
NAA

RN:	86-87-3	MP (°C):	134
MW:	186.21	BP (°C):	

Solubility (Moles/L)	Solubility (Grams/L)	Temp (°C)	Ref (#)	Evaluation (T P E A A)	Comments
2.040E-03	3.799E-01	17	B200	1 0 0 0 1	
2.255E-03	4.198E-01	20	B200	1 0 0 0 1	
1.179E-02	2.195E+00	20	C092	2 2 0 1 2	
2.228E-03	4.148E-01	25	M061	1 0 0 0 2	average of 2

2767. $C_{12}H_{10}O_2$
2-Hydroxydiphenyl ether
2-Hydroxy-diphenyl-aether
RN: 2417-10-9   **MP** (°C):
MW: 186.21   **BP** (°C):

Solubility (Moles/L)	Solubility (Grams/L)	Temp (°C)	Ref (#)	Evaluation (T P E A A)	Comments
5.907E-04	1.100E-01	20	F300	1 0 0 0 1	

2768. $C_{12}H_{10}O_3$
β-Naphthoxyacetic acid
(2-Naphthoxy)acetic acid
Phyomone
BNOA
RN: 120-23-0   **MP** (°C):   155–157
MW: 202.21   **BP** (°C):

Solubility (Moles/L)	Solubility (Grams/L)	Temp (°C)	Ref (#)	Evaluation (T P E A A)	Comments
4.330E-04	8.756E-02	25	D088	0 0 0 0 0	
8.100E-04	1.638E-01	35	D088	0 0 0 0 0	
1.100E-05	2.224E-03	45	D088	0 0 0 0 0	

2769. $C_{12}H_{10}O_4$
Quinhydrone
Chinhydron
RN: 106-34-3   **MP** (°C):   171
MW: 218.21   **BP** (°C):

Solubility (Moles/L)	Solubility (Grams/L)	Temp (°C)	Ref (#)	Evaluation (T P E A A)	Comments
1.861E-02	4.061E+00	25	B121	1 2 2 1 2	average of 4

2770. $C_{12}H_{11}ClN_2O_5S$
Furosemide
Frusemide
RN: 54-31-9   **MP** (°C):   206
MW: 330.75   **BP** (°C):

Solubility (Moles/L)	Solubility (Grams/L)	Temp (°C)	Ref (#)	Evaluation (T P E A A)	Comments
5.593E-05	1.850E-02	20	B405	1 1 1 2 2	
1.814E-05	6.000E-03	22.5	C438	0 0 0 0 0	
1.784E-05	5.900E-03	25	A408	2 0 1 2 0	
2.691E-05	8.900E-03	25	B405	1 1 1 2 2	Buffer pH 2.0
7.559E-05	2.500E-02	25	B405	1 1 1 2 2	
1.875E-05	6.200E-03	25	F415	0 0 0 0 0	Average
2.210E-04	7.310E-02	30	E049	2 0 2 2 2	

(continued)

2770. $C_{12}H_{11}ClN_2O_5S$ (continued)

Solubility (Moles/L)	Solubility (Grams/L)	Temp (°C)	Ref (#)	Evaluation (T P E A A)	Comments
3.023E-05	1.000E-02	ns	K444	0 0 0 0 0	
1.778E-05	5.882E-03	ns	R427	0 0 0 0 0	

2771. $C_{12}H_{11}Cl_2NO$
Propyzamide
3,5-Dichloro-N-(1,1-dimethyl-2-propynyl)benzamide
Pronamide
Kerb 50W
RH-315

RN: 23950-58-5   **MP** (°C): 155.5
MW: 256.13   **BP** (°C):

Solubility (Moles/L)	Solubility (Grams/L)	Temp (°C)	Ref (#)	Evaluation (T P E A A)	Comments
5.856E-05	1.500E-02	25	M161	1 0 0 0 1	

2772. $C_{12}H_{11}I_3N_2O_4$
Iodamide
3-Acetamido-5-acetamidomethyl-2,4,6-triiodobenzoic acid
3-Acetylamino-5-acetylaminomethyl-2,4,6-triiodobenzoic acid
Jodomiron 380
Uromiro
Uromiron

RN: 440-58-4   **MP** (°C):
MW: 627.95   **BP** (°C):

Solubility (Moles/L)	Solubility (Grams/L)	Temp (°C)	Ref (#)	Evaluation (T P E A A)	Comments
4.777E-03	3.000E+00	20	F045	1 2 2 2 1	
5.096E-03	3.200E+00	40	F045	1 2 2 2 1	
6.211E-03	3.900E+00	60	F045	1 2 2 2 1	

2773. $C_{12}H_{11}N$
Diphenylamine
4-Aminobiphenyl

RN: 122-39-4   **MP** (°C): 53.5
MW: 169.23   **BP** (°C): 302.0

Solubility (Moles/L)	Solubility (Grams/L)	Temp (°C)	Ref (#)	Evaluation (T P E A A)	Comments
1.820E-03	3.079E-01	20	B179	0 0 0 0 0	
3.132E-04	5.300E-02	20	H300	1 2 2 2 1	
3.274E-04	5.540E-02	20	T301	1 2 2 2 2	
2.765E-04	4.680E-02	25	F029	1 0 0 0 2	
3.415E-04	5.780E-02	50	T301	1 2 2 2 2	average of 5
3.557E-04	6.020E-02	80	T301	1 2 2 2 2	average of 5
1.772E-03	2.999E-01	rt	D021	0 0 1 1 0	

2774. C$_{12}$H$_{11}$NO$_2$

Fenfuram

2-Methyl-*N*-phenyl-3-furancarboxamide

Pano-ram

RN: 24691-80-3   **MP** (°C):   109.5

MW: 201.23   **BP** (°C):

Solubility (Moles/L)	Solubility (Grams/L)	Temp (°C)	Ref (#)	Evaluation (T P E A A)	Comments
4.970E-04	1.000E-01	20	M161	1 0 0 0 0	

2775. C$_{12}$H$_{11}$NO$_2$

Carbaryl

1-Naphthyl *N*-methylcarbamate

Devicarb

Hexavin

Karbaspray

Murvin

RN: 63-25-2   **MP** (°C):   142

MW: 201.23   **BP** (°C):

Solubility (Moles/L)	Solubility (Grams/L)	Temp (°C)	Ref (#)	Evaluation (T P E A A)	Comments
2.710E-04	5.453E-02	5	H343	0 0 0 0 0	
3.598E-04	7.239E-02	10	B324	0 0 0 0 0	
3.444E-04	6.930E-02	10	B324	0 0 0 0 0	
3.150E-04	6.339E-02	10	H343	0 0 0 0 0	
3.740E-04	7.526E-02	15	H343	0 0 0 0 0	
1.995E-04	4.015E-02	20	B179	0 0 0 0 0	
5.164E-04	1.039E-01	20	B300	2 1 1 1 2	
4.947E-04	9.955E-02	20	B324	0 0 0 0 0	
5.168E-04	1.040E-01	20	B324	0 0 0 0 0	
2.485E-04	5.000E-02	20	F311	1 2 2 2 1	
4.450E-04	8.955E-02	20	H343	0 0 0 0 0	
1.690E-04	3.400E-02	22	K137	1 1 2 1 0	
1.988E-04	4.000E-02	22.5	G301	0 0 0 0 0	
5.210E-04	1.048E-01	25	H343	0 0 0 0 0	
6.184E-04	1.244E-01	30	B324	0 0 0 0 0	
6.460E-04	1.300E-01	30	B324	0 0 0 0 0	
1.988E-04	4.000E-02	30	D089	2 2 0 0 0	
6.520E-04	1.312E-01	30	H343	0 0 0 0 0	
1.988E-04	4.000E-02	30	M161	1 0 0 0 1	
7.860E-04	1.582E-01	35	H343	0 0 0 0 0	
8.990E-04	1.809E-01	40	H343	0 0 0 0 0	
1.006E-03	2.024E-01	45	H343	0 0 0 0 0	
1.988E-04	4.000E-02	ns	H042	0 0 0 0 1	
2.783E-04	5.600E-02	ns	M110	0 0 0 0 0	EFG

2776. C₁₂H₁₁N₃

C.I. Solvent yellow 1
p-Aminoazobenzene
4-Aminoazobenzene
4-Amino-azobenzol

RN:	60-09-3	**MP** (°C):	125
MW:	197.24	**BP** (°C):	>360

Solubility (Moles/L)	Solubility (Grams/L)	Temp (°C)	Ref (#)	Evaluation (T P E A A)	Comments
6.591E-04	1.300E-01	18	F300	1 0 0 0 1	
1.500E-04	2.959E-02	25	B333	0 0 0 0 0	
2.484E-04	4.900E-02	37	H120	1 1 1 1 1	normal saline
5.510E-04	1.087E-01	60	B198	1 2 1 1 2	
1.041E-03	2.053E-01	71.80	B198	1 2 1 1 2	
1.907E-03	3.761E-01	84.10	B198	1 2 1 1 2	
3.431E-03	6.767E-01	97.40	B198	1 2 1 1 2	

2777. C₁₂H₁₁N₃

Diazoaminobenzene
1,3-Diphenyltriazene
Anilinoazobenzene
N-(Phenylazo)aniline

RN:	136-35-6	**MP** (°C):	98.0
MW:	197.24	**BP** (°C):	

Solubility (Moles/L)	Solubility (Grams/L)	Temp (°C)	Ref (#)	Evaluation (T P E A A)	Comments
2.534E-03	4.998E-01	rt	D021	0 0 1 1 0	

2778. C₁₂H₁₁N₃O₃

Orotic acid benzylamide
Orotamide, *N*-benzyl-

RN:	13156-36-0	**MP** (°C):	260–263
MW:	245.24	**BP** (°C):	

Solubility (Moles/L)	Solubility (Grams/L)	Temp (°C)	Ref (#)	Evaluation (T P E A A)	Comments
4.600E-02	1.128E+01	–4	N018	0 0 0 0 0	
8.700E-02	2.134E+01	16	N018	0 0 0 0 0	
1.180E-01	2.894E+01	25	N018	0 0 0 0 0	

2779. C$_{12}$H$_{11}$O$_4$P
Diphenyl phosphate
Phosphoric acid, diphenyl ester
RN: 838-85-7   **MP** (°C):   63
MW: 250.19   **BP** (°C):

Solubility (Moles/L)	Solubility (Grams/L)	Temp (°C)	Ref (#)	Evaluation (T P E A A)	Comments
>1.08E-03	>2.70E-01	24	H116	2 1 0 0 0	

2780. C$_{12}$H$_{12}$
1,5-Dimethylnaphthalene
RN: 571-61-9   **MP** (°C):   81
MW: 156.23   **BP** (°C):   265.5

Solubility (Moles/L)	Solubility (Grams/L)	Temp (°C)	Ref (#)	Evaluation (T P E A A)	Comments
1.754E-05	2.740E-03	25	E004	2 1 2 2 2	
2.163E-05	3.380E-03	25	M064	1 1 2 2 2	
2.160E-05	3.375E-03	25	M342	1 0 1 1 2	
2.163E-05	3.380E-03	ns	M344	0 0 0 0 2	

2781. C$_{12}$H$_{12}$
1-Ethylnaphthalene
RN: 1127-76-0   **MP** (°C):   −15
MW: 156.23   **BP** (°C):   258

Solubility (Moles/L)	Solubility (Grams/L)	Temp (°C)	Ref (#)	Evaluation (T P E A A)	Comments
5.200E-05	8.124E-03	10	S076	2 2 2 2 1	
5.200E-05	8.124E-03	14	S076	2 2 2 2 1	
6.400E-05	9.999E-03	20	S076	2 2 2 2 1	
6.849E-05	1.070E-02	25	M064	1 1 2 2 2	
6.850E-05	1.070E-02	25	M342	1 0 1 1 2	
6.400E-05	9.999E-03	25	S076	2 2 2 2 1	
6.849E-05	1.070E-02	ns	M344	0 0 0 0 2	

2782. C$_{12}$H$_{12}$
2,3-Dimethylnaphthalene
RN: 581-40-8   **MP** (°C):   103
MW: 156.23   **BP** (°C):   269

Solubility (Moles/L)	Solubility (Grams/L)	Temp (°C)	Ref (#)	Evaluation (T P E A A)	Comments
1.274E-05	1.990E-03	25	E004	2 1 2 2 2	
1.920E-05	3.000E-03	25	M064	1 1 2 2 1	
1.920E-05	3.000E-03	25	M342	1 0 1 1 2	
1.920E-05	3.000E-03	ns	M344	0 0 0 0 1	

2783. $C_{12}H_{12}$
1,3-Dimethylnaphthalene
RN: 575-41-7   **MP** (°C):   −5
MW: 156.23   **BP** (°C):   263

Solubility (Moles/L)	Solubility (Grams/L)	Temp (°C)	Ref (#)	Evaluation (T P E A A)	Comments
5.121E-05	8.000E-03	25	M064	1 1 2 2 1	
5.120E-05	7.999E-03	25	M342	1 0 1 1 2	
5.121E-05	8.000E-03	ns	M344	0 0 0 0 1	

2784. $C_{12}H_{12}$
2,6-Dimethylnaphthalene
RN: 581-42-0   **MP** (°C):   109
MW: 156.23   **BP** (°C):

Solubility (Moles/L)	Solubility (Grams/L)	Temp (°C)	Ref (#)	Evaluation (T P E A A)	Comments
1.280E-05	2.000E-03	25	M064	1 1 2 2 1	
1.280E-05	2.000E-03	25	M342	1 0 1 1 2	
1.280E-05	2.000E-03	ns	M344	0 0 0 0 1	

2785. $C_{12}H_{12}$
2-Ethylnaphthalene
RN: 939-27-5   **MP** (°C):   −7.4
MW: 156.23   **BP** (°C):   251.5

Solubility (Moles/L)	Solubility (Grams/L)	Temp (°C)	Ref (#)	Evaluation (T P E A A)	Comments
5.895E-05	9.210E-03	20	B356	0 0 0 0 0	
5.121E-05	8.000E-03	25	E004	2 1 2 2 2	

2786. $C_{12}H_{12}$
1,4-Dimethylnaphthalene
RN: 571-58-4   **MP** (°C):   7.6
MW: 156.23   **BP** (°C):   262

Solubility (Moles/L)	Solubility (Grams/L)	Temp (°C)	Ref (#)	Evaluation (T P E A A)	Comments
4.544E-05	7.100E-03	4	D351	1 2 1 1 2	
4.744E-05	7.412E-03	10	D351	1 2 1 1 2	
6.081E-05	9.500E-03	20	B318	0 0 0 0 0	EFG
6.062E-05	9.470E-03	20	B356	0 0 0 0 0	
6.167E-05	9.634E-03	25	D351	1 2 1 1 2	
7.297E-05	1.140E-02	25	M064	1 1 2 2 2	
7.300E-05	1.140E-02	25	M342	1 0 1 1 1	
7.944E-05	1.241E-02	40	D351	1 2 1 1 2	
7.297E-05	1.140E-02	ns	M344	0 0 0 0 2	

2787. C$_{12}$H$_{12}$ClNO
2-Chloro-*N*-(1-methyl-2-propynyl)acetanilide
Basamaize
RN: 35846-47-0   **MP** (°C):   40
MW: 221.69   **BP** (°C):

Solubility (Moles/L)	Solubility (Grams/L)	Temp (°C)	Ref (#)	Evaluation (T P E A A)	Comments
2.255E-03	5.000E-01	20	B200	1 0 0 0 0	

2788. C$_{12}$H$_{12}$N$_2$
Benzidine
Benzidin
p-Diaminobiphenyl
RN: 92-87-5   **MP** (°C):   117
MW: 184.24   **BP** (°C):   400

Solubility (Moles/L)	Solubility (Grams/L)	Temp (°C)	Ref (#)	Evaluation (T P E A A)	Comments
1.953E-03	3.599E-01	24	H106	1 0 2 2 2	
1.954E-03	3.600E-01	24	M303	1 0 1 1 2	pH 5.9
2.712E-03	4.998E-01	25	B019	1 0 1 2 0	
2.822E-03	5.200E-01	25	B068	2 0 1 1 1	
2.700E-04	4.975E-02	25	H091	1 2 2 2 1	*sic*
1.465E-03	2.699E-01	rt	N015	0 0 2 2 2	

2789. C$_{12}$H$_{12}$N$_2$
m-Benzidine
3-Benzidine
RN: 2050-89-7   **MP** (°C):   117
MW: 184.24   **BP** (°C):   400

Solubility (Moles/L)	Solubility (Grams/L)	Temp (°C)	Ref (#)	Evaluation (T P E A A)	Comments
5.970E-02	1.100E+01	100	F300	1 0 0 0 1	

2790. C$_{12}$H$_{12}$N$_2$OS
2,4-Dimethyl-5-carboxanilidothiazole
G-696
RN: 21452-18-6   **MP** (°C):   141
MW: 232.31   **BP** (°C):

Solubility (Moles/L)	Solubility (Grams/L)	Temp (°C)	Ref (#)	Evaluation (T P E A A)	Comments
1.056E-02	2.454E+00	25	M061	1 0 0 0 2	

2791. $C_{12}H_{12}N_2O_2S$
Dapsone
4,4'-Diaminodiphenyl sulphone
RN: 80-08-0   **MP** (°C): 175
MW: 248.31   **BP** (°C):

Solubility (Moles/L)	Solubility (Grams/L)	Temp (°C)	Ref (#)	Evaluation (T P E A A)	Comments
5.638E-04	1.400E-01	25	P351	0 0 0 0 0	pH 7.4
6.444E-04	1.600E-01	25	P351	0 0 0 0 0	
1.530E-03	3.800E-01	37	L037	1 2 2 1 1	
4.027E-04	1.000E-01	ns	K444	0 0 0 0 0	

2792. $C_{12}H_{12}N_2O_2S$
Sulfabenz
Sulfanilid
RN: 127-77-5   **MP** (°C):
MW: 248.31   **BP** (°C):

Solubility (Moles/L)	Solubility (Grams/L)	Temp (°C)	Ref (#)	Evaluation (T P E A A)	Comments
2.819E-02	7.000E+00	100	F300	1 0 0 0 0	

2793. $C_{12}H_{12}N_2O_3$
Nalidixic acid
NegGRAM
1-Ethyl-1,4-dihydro-7-methyl-4-oxo-1,8-naphthyridine-3-carboxylic acid
Nalidic acid
RN: 389-08-2   **MP** (°C): 228
MW: 232.24   **BP** (°C):

Solubility (Moles/L)	Solubility (Grams/L)	Temp (°C)	Ref (#)	Evaluation (T P E A A)	Comments
4.306E-04	1.000E-01	23	G098	1 0 0 0 0	
7.079E-01	1.644E+02	37	O307	1 0 1 2 1	pH 2, EFG
4.306E-04	1.000E-01	ns	K444	0 0 0 0 0	

2794. $C_{12}H_{12}N_2O_3$
Phenobarbital
5-Ethyl-5-phenylbarbituric acid
Phenylethylmalonylurea
RN: 50-06-6   **MP** (°C): 176
MW: 232.24   **BP** (°C):

Solubility (Moles/L)	Solubility (Grams/L)	Temp (°C)	Ref (#)	Evaluation (T P E A A)	Comments
3.980E-03	9.243E-01	15	H018	0 0 0 0 0	
3.680E-03	8.546E-01	15	S149	1 2 2 1 2	anhydrate

(continued)

2794. $C_{12}H_{12}N_2O_3$ (continued)

Solubility (Moles/L)	Solubility (Grams/L)	Temp (°C)	Ref (#)	Evaluation (T P E A A)	Comments
3.180E-03	7.385E-01	15	S149	1 2 2 1 2	hydrate
4.736E-03	1.100E+00	20	I009	1 2 2 1 1	EFG, 0.005M HCl
3.789E-03	8.800E-01	20	J030	1 2 2 2 1	
4.521E-03	1.050E+00	20	K143	1 2 2 2 2	form II
5.081E-03	1.180E+00	20	K143	1 2 2 2 2	form III
3.143E-03	7.300E-01	20	N023	1 2 2 1 1	hydrate
4.866E-03	1.130E+00	20	N023	1 2 2 1 2	anhydrate
3.920E-03	9.104E-01	20	S149	1 2 2 1 2	hydrate
4.510E-03	1.047E+00	20	S149	1 2 2 1 2	anhydrate
4.731E-03	1.099E+00	25	A023	1 0 0 1 1	
5.167E-03	1.200E+00	25	B011	2 0 0 1 0	
4.994E-03	1.160E+00	25	B065	1 1 1 1 0	
5.590E-03	1.298E+00	25	E011	2 1 1 2 1	
7.737E-03	1.797E+00	25	E011	2 1 1 2 1	pH 7.0
3.078E-02	7.149E+00	25	E011	2 1 1 2 1	pH 8.0
4.731E-03	1.099E+00	25	F009	2 2 2 2 0	EFG
4.600E-03	1.068E+00	25	G003	1 1 1 1 1	pH 4.7
2.734E-03	6.350E-01	25	H005	1 0 1 2 2	
5.161E-03	1.199E+00	25	K010	2 0 0 1 1	
6.114E-03	1.420E+00	25	K143	1 2 2 2 2	form III
5.512E-03	1.280E+00	25	K143	1 2 2 2 2	form II
4.650E-03	1.080E+00	25	L032	2 1 2 0 2	
4.790E-03	1.112E+00	25	M056	2 2 2 2 2	
5.684E-03	1.320E+00	25	N023	1 2 2 1 2	anhydrate
4.995E-03	1.160E+00	25	N023	1 2 2 1 2	hydrate
6.020E-03	1.398E+00	25	P006	2 0 2 2 1	
4.306E-03	1.000E+00	25	P015	0 0 0 0 0	
4.761E-03	1.106E+00	25	P350	0 0 0 0 0	intrinsic
4.830E-03	1.122E+00	25	S149	1 2 2 1 2	hydrate
5.320E-03	1.236E+00	25	S149	1 2 2 1 2	anhydrate
5.170E-03	1.201E+00	25	V033	2 0 1 1 2	
5.200E-03	1.208E+00	25.00	T303	1 0 0 0 1	
6.700E-03	1.556E+00	30	A065	2 0 2 2 1	
6.310E-03	1.465E+00	30	H018	0 0 0 0 0	
6.000E-03	1.393E+00	30	I001	2 0 2 1 0	EFG, 0.003N H_2SO_4
6.100E-03	1.417E+00	30	K108	1 2 2 0 1	
6.502E-03	1.510E+00	30	K143	1 2 2 2 2	form II
7.148E-03	1.660E+00	30	K143	1 2 2 2 2	form III
6.071E-03	1.410E+00	30	N023	1 2 2 1 2	hydrate
6.502E-03	1.510E+00	30	N023	1 2 2 1 2	anhydrate
6.020E-03	1.398E+00	30	O321	0 0 0 0 0	
6.000E-03	1.393E+00	30	O321	0 0 0 0 0	
8.612E-03	2.000E+00	32	M157	2 0 1 1 0	EFG
7.737E-03	1.797E+00	35	A023	1 0 0 1 2	
7.700E-03	1.788E+00	35	S149	1 2 2 1 2	hydrate
7.750E-03	1.800E+00	35	S149	1 2 2 1 2	anhydrate
8.500E-03	1.974E+00	35.00	T303	1 0 0 0 1	

(continued)

2794. $C_{12}H_{12}N_2O_3$ (continued)

Solubility (Moles/L)	Solubility (Grams/L)	Temp (°C)	Ref (#)	Evaluation (T P E A A)	Comments
7.923E-03	1.840E+00	37	J030	1 2 2 2 2	
8.000E-03	1.858E+00	37	K121	1 2 1 2 0	0.1N HCl
9.023E-03	2.096E+00	40	A023	1 0 0 1 2	
9.000E-02	2.090E+01	40	N008	1 0 1 1 2	*sic*
1.055E-02	2.450E+00	45	S149	1 2 2 1 2	anhydrate
1.108E-02	2.573E+00	45	S149	1 2 2 1 2	hydrate
1.130E-02	2.624E+00	45.00	T303	1 0 0 0 2	
1.266E-02	2.940E+00	50	S149	1 2 2 1 2	anhydrate
1.506E-02	3.498E+00	50	S149	1 2 2 1 2	hydrate
1.698E-02	3.943E+00	55	S149	1 2 2 1 2	hydrate
1.499E-02	3.481E+00	55	S149	1 2 2 1 2	anhydrate
1.033E-02	2.400E+00	60	I009	1 2 2 1 1	EFG, 0.005M HCl
4.306E-03	1.000E+00	ns	K444	0 0 0 0 0	
4.177E-03	9.700E-01	ns	T003	0 0 0 0 2	

2795. $C_{12}H_{12}N_2O_6S_2$
Benzidine-2,2'-disulfonic acid
Benzidin-disulfosaeure-(2,2')
RN: 117-61-3 **MP** (°C):
MW: 344.37 **BP** (°C):

Solubility (Moles/L)	Solubility (Grams/L)	Temp (°C)	Ref (#)	Evaluation (T P E A A)	Comments
2.323E-03	8.000E-01	25	F300	1 0 0 0 0	

2796. $C_{12}H_{12}N_2S$
Thiopyrine
1-Phenyl-2,3-dimethyl-3-pyrazoline-5-thione
RN: 5702-69-2 **MP** (°C): 166
MW: 216.31 **BP** (°C):

Solubility (Moles/L)	Solubility (Grams/L)	Temp (°C)	Ref (#)	Evaluation (T P E A A)	Comments
6.600E-02	1.428E+01	ns	D087	0 2 0 0 2	

2797. $C_{12}H_{12}N_4O_3$
Benznidazole
2-Nitro-*N*-(phenylmethyl)-imidazole-1-acetamide
RN: 22994-85-0 **MP** (°C):
MW: 260.25 **BP** (°C):

Solubility (Moles/L)	Solubility (Grams/L)	Temp (°C)	Ref (#)	Evaluation (T P E A A)	Comments
1.537E-03	4.000E-01	ns	K444	0 0 0 0 0	

2798. C₁₂H₁₂N₄O₃S

$C_{12}H_{12}N_4O_3S$

N4-Acetylsulfapyrazine
N4-Acetylsulphapyrazine

RN: 5433-91-0   **MP** (°C):
MW: 292.32   **BP** (°C):

Solubility (Moles/L)	Solubility (Grams/L)	Temp (°C)	Ref (#)	Evaluation (T P E A A)	Comments
1.710E-04	5.000E-02	37	L091	1 0 0 0 0	pH 5.5

2799. C₁₂H₁₂N₄O₃S

$C_{12}H_{12}N_4O_3S$

N4-Acetyl sulfadiazine
N4-Acetylsulfadiazine
Acetyl sulfadiazine
2-N4-Acetylsulfanilamidopyrimidine

RN: 127-74-2   **MP** (°C):
MW: 292.32   **BP** (°C):

Solubility (Moles/L)	Solubility (Grams/L)	Temp (°C)	Ref (#)	Evaluation (T P E A A)	Comments
	1.500E-01	37	F075	1 0 2 2 2	
7.200E-04	2.105E-01	37	G026	1 0 1 1 0	EFG, pH 5.4
6.842E-04	2.000E-01	37	L091	1 0 0 0 1	pH 5.5
8.723E-04	2.550E-01	37	M057	1 0 0 0 2	pH 5.5
5.131E-04	1.500E-01	37	R045	1 2 1 1 1	

2800. C₁₂H₁₂N₆O₆

$C_{12}H_{12}N_6O_6$

TMPPT
1,3,7,9-Tetramethylpyrimido(5,4-γ) pteridine-2,4,6,8(1H,3H,7H,9H)-tetrone

RN: **MP** (°C):
MW: 336.27   **BP** (°C):

Solubility (Moles/L)	Solubility (Grams/L)	Temp (°C)	Ref (#)	Evaluation (T P E A A)	Comments
3.860E-04	1.298E-01	25	K008	1 1 0 1 0	EFG
3.900E-04	1.311E-01	25	K009	1 2 1 1 0	EFG

2801. C₁₂H₁₂O₆

$C_{12}H_{12}O_6$

Benzoic acid, 2-(acetyloxy)-, (acetyloxy)methyl ester
Salicylic acid acetate, hydroxymethyl ester acetate

RN: 32620-68-1   **MP** (°C):   oil
MW: 252.23   **BP** (°C):

Solubility (Moles/L)	Solubility (Grams/L)	Temp (°C)	Ref (#)	Evaluation (T P E A A)	Comments
9.634E-03	2.430E+00	21	N335	0 0 0 0 0	

2802. C$_{12}$H$_{13}$ClN$_2$O

Buturon

3-(*para*-Chlorophenyl)-1-methyl-1-(1-methyl-2-propynyl) urea

Urea, *N'*-(4-chlorophenyl)-*N*-methyl-*N*-(1-methyl-2-propynyl)

Eptapur

RN:	3766-60-7	**MP** (°C):	145.5
MW:	236.70	**BP** (°C):	

Solubility (Moles/L)	Solubility (Grams/L)	Temp (°C)	Ref (#)	Evaluation (T P E A A)	Comments
1.267E-04	3.000E-02	20	G036	1 0 0 0 1	
1.267E-04	3.000E-02	20	M161	1 0 0 0 1	

2803. C$_{12}$H$_{13}$ClN$_4$

Pyrimethamine

RN:	58-14-0	**MP** (°C):	238
MW:	248.72	**BP** (°C):	

Solubility (Moles/L)	Solubility (Grams/L)	Temp (°C)	Ref (#)	Evaluation (T P E A A)	Comments
4.021E-05	1.000E-02	ns	K444	0 0 0 0 0	

2804. C$_{12}$H$_{13}$I$_3$N$_2$O$_2$

Iopodic acid

Ipodic acid

RN:	5587-89-3	**MP** (°C):	
MW:	597.96	**BP** (°C):	

Solubility (Moles/L)	Solubility (Grams/L)	Temp (°C)	Ref (#)	Evaluation (T P E A A)	Comments
3.027E-03	1.810E+00	ns	H055	0 0 0 0 0	

2805. C$_{12}$H$_{13}$I$_3$N$_2$O$_3$

Iocetamic acid

N-(3-Amino-2,4,6-triiodophenyl)-3-acetamido-2-methylpropionic acid

Cholebrine

MP 620

DRC 1201

RN:	16034-77-8	**MP** (°C):	224
MW:	613.96	**BP** (°C):	

Solubility (Moles/L)	Solubility (Grams/L)	Temp (°C)	Ref (#)	Evaluation (T P E A A)	Comments
8.610E-03	5.286E+00	37	J016	1 0 0 0 2	pH 7.4

2806. C$_{12}$H$_{13}$NO$_2$
Methsuximide
Celontin
N-Methyl-α-methyl-α-phenylsuccinimide

RN:	77-41-8	MP (°C):	52–53
MW:	203.24	BP (°C):	

Solubility (Moles/L)	Solubility (Grams/L)	Temp (°C)	Ref (#)	Evaluation (T P E A A)	Comments
1.378E-02	2.800E+00	25	P061	0 0 0 0 0	

2807. C$_{12}$H$_{13}$NO$_2$S
Carboxin
2,3-Dihydro-5-carboxanilido-6-methyl-1,4-oxathiin
Vitavax

RN:	5234-68-4	MP (°C):	94
MW:	235.31	BP (°C):	

Solubility (Moles/L)	Solubility (Grams/L)	Temp (°C)	Ref (#)	Evaluation (T P E A A)	Comments
7.225E-04	1.700E-01	25	M061	1 0 0 0 2	
7.225E-04	1.700E-01	25	M161	1 0 0 0 2	

2808. C$_{12}$H$_{13}$NO$_2$S
4-Thiazolidinecarboxylic acid, 2-(4-ethenylphenyl)-

RN:	256235-52-6	MP (°C):	
MW:	235.31	BP (°C):	464.3

Solubility (Moles/L)	Solubility (Grams/L)	Temp (°C)	Ref (#)	Evaluation (T P E A A)	Comments
3.700E-03	8.706E-01	21	B414	1 0 0 1 1	partial decomposition

2809. C$_{12}$H$_{13}$NO$_3$
Azetidine, 1-[(benzoyloxy)acetyl]-

RN:	115178-66-0	MP (°C):	74.5
MW:	219.24	BP (°C):	

Solubility (Moles/L)	Solubility (Grams/L)	Temp (°C)	Ref (#)	Evaluation (T P E A A)	Comments
2.463E-02	5.400E+00	22	N317	1 1 2 1 2	

2810. C₁₂H₁₃NO₃

Crotonyl acetaminophen
Crotonic acid, ester with 4'-hydroxyacetanilide
Acetanilide, 4'-hydroxy-, crotonate (ester)

RN:	20675-24-5	MP (°C):	146-147
MW:	219.24	BP (°C):	

Solubility (Moles/L)	Solubility (Grams/L)	Temp (°C)	Ref (#)	Evaluation (T P E A A)	Comments
1.961E-03	4.300E-01	37	D029	0 0 0 0 0	

2811. C₁₂H₁₃NO₄

Acetamide, N-acetyl-2-(benzoyloxy)-N-methyl-

RN:	115178-80-8	MP (°C):	
MW:	235.24	BP (°C):	

Solubility (Moles/L)	Solubility (Grams/L)	Temp (°C)	Ref (#)	Evaluation (T P E A A)	Comments
1.360E-03	3.200E-01	22	N317	1 1 2 1 2	

2812. C₁₂H₁₃NO₄S

Plantvax
2,3-Dihydro-5-carboxanilido-6-methyl-1,4-oxathiin-4,4-dioxide
Oxycarboxin

RN:	5259-88-1	MP (°C):	128.7
MW:	267.31	BP (°C):	

Solubility (Moles/L)	Solubility (Grams/L)	Temp (°C)	Ref (#)	Evaluation (T P E A A)	Comments
3.741E-03	1.000E+00	25	M161	1 0 0 0 0	
3.741E-03	1.000E+00	ns	M061	0 0 0 0 2	

2813. C₁₂H₁₃NO₄S₂

4-Ethylsulfonylnaphthalene-1-sulfonamide
ENS
4-ENS

RN:	842-00-2	MP (°C):	
MW:	299.37	BP (°C):	

Solubility (Moles/L)	Solubility (Grams/L)	Temp (°C)	Ref (#)	Evaluation (T P E A A)	Comments
3.775E-04	1.130E-01	c	K042	2 2 2 2 2	

2814. $C_{12}H_{13}NO_5$
Glycine, N-[(benzoyloxy)acetyl]-N-methyl-
RN: 106231-64-5   **MP** (°C):   160.5
MW: 251.24   **BP** (°C):   475.6

Solubility (Moles/L)	Solubility (Grams/L)	Temp (°C)	Ref (#)	Evaluation (T P E A A)	Comments
5.572E-03	1.400E+00	22	B427	1 0 0 1 1	in 0.01M HCl
5.572E-03	1.400E+00	22	N317	1 1 2 1 2	

2815. $C_{12}H_{13}NO_5$
Succinyl acetaminophen
Butanedioic acid, mono[4-(acetylamino)phenyl] ester
Acetanilide, 4′-hydroxy-, hydrogen succinate ester
RN: 20675-25-6   **MP** (°C):   145.5-146.5
MW: 251.24   **BP** (°C):

Solubility (Moles/L)	Solubility (Grams/L)	Temp (°C)	Ref (#)	Evaluation (T P E A A)	Comments
2.587E-02	6.500E+00	37	D029	0 0 0 0 0	

2816. $C_{12}H_{13}NO_6$
Carbobenzoxydiglycine
RN: **MP** (°C):
MW: 267.24   **BP** (°C):

Solubility (Moles/L)	Solubility (Grams/L)	Temp (°C)	Ref (#)	Evaluation (T P E A A)	Comments
2.432E-03	6.500E-01	25.1	N026	0 0 0 0 0	
2.804E-03	7.494E-01	25.1	N027	1 1 2 2 2	

2817. $C_{12}H_{13}N_3O_2$
Isocarboxazid
Marplan
RN: 59-63-2   **MP** (°C):
MW: 231.26   **BP** (°C):

Solubility (Moles/L)	Solubility (Grams/L)	Temp (°C)	Ref (#)	Evaluation (T P E A A)	Comments
3.459E-03	8.000E-01	25	R024	0 0 0 0 0	

2818. $C_{12}H_{13}N_3O_2S$
N1-Methyl-N1-2-pyridyl-sulfanilamide
N1-Methyl-N1-(2-pyridyl)sulfanilamide
RN: 51543-29-4   **MP** (°C):
MW: 263.32   **BP** (°C):

Solubility (Moles/L)	Solubility (Grams/L)	Temp (°C)	Ref (#)	Evaluation (T P E A A)	Comments
4.740E-03	1.248E+00	37	K095	2 0 0 0 2	intrinsic

2819. C$_{12}$H$_{13}$N$_3$O$_3$S$_2$
Methyl acetyl sulfathiazole
Sulfathiazol methyle acetyle
RN: **MP** (°C):
MW: 311.38   **BP** (°C):

Solubility (Moles/L)	Solubility (Grams/L)	Temp (°C)	Ref (#)	Evaluation (T P E A A)	Comments
2.248E-04	7.000E-02	37	D084	1 0 1 0 0	

2820. C$_{12}$H$_{13}$N$_3$O$_4$S
Acetylsulfamethoxazole
Acetanilide, 4′-[(5-methyl-3-isoxazolyl)sulfamoyl]-
4′-Acetyl-3-sulfa-5-methylisoxazole
RN: 21312-10-7   **MP** (°C):
MW: 295.32   **BP** (°C):

Solubility (Moles/L)	Solubility (Grams/L)	Temp (°C)	Ref (#)	Evaluation (T P E A A)	Comments
2.573E-04	7.600E-02	37	H120	1 1 1 1 1	normal saline

2821. C$_{12}$H$_{14}$ClNO$_2$
Clomazone
Command
Dimethazone
Fenoxan
FMC 57020
Gamit
RN: 81777-89-1   **MP** (°C):   25
MW: 239.70   **BP** (°C):   275.4

Solubility (Moles/L)	Solubility (Grams/L)	Temp (°C)	Ref (#)	Evaluation (T P E A A)	Comments
4.592E-03	1.101E+00	ns	S460	0 0 0 0 0	

2822. C$_{12}$H$_{14}$Cl$_2$O$_3$
2,4-Dichlorophenoxyacetic acid n-butyl ester
2,4-Dichlorophenoxyacetic acid butyl ester
RN: 94-80-4   **MP** (°C):
MW: 277.15   **BP** (°C):

Solubility (Moles/L)	Solubility (Grams/L)	Temp (°C)	Ref (#)	Evaluation (T P E A A)	Comments
5.495E-05	1.523E-02	ns	M120	0 0 1 1 2	

2823. C₁₂H₁₄Cl₂O₃

$2823.\ C_{12}H_{14}Cl_2O_3$

2,4-Dichlorophenoxyacetic acid *sec*-butyl ester

RN: 94-79-1   **MP** (°C):
MW: 277.15   **BP** (°C):

Solubility (Moles/L)	Solubility (Grams/L)	Temp (°C)	Ref (#)	Evaluation (T P E A A)	Comments
6.252E-05	1.733E-02	ns	M120	0 0 1 1 2	

2824. $C_{12}H_{14}Cl_3O_4P$

Chlorfenvinphos

2-Chloro-1-(2,4-dichlorophenyl)ethenyl phosphoric acid, diethyl ester

Dermaton
Birlanex
Birlane
Steladone

RN: 470-90-6   **MP** (°C):
MW: 359.58   **BP** (°C):

Solubility (Moles/L)	Solubility (Grams/L)	Temp (°C)	Ref (#)	Evaluation (T P E A A)	Comments
3.476E-04	1.250E-01	10	B324	0 0 0 0 0	
3.476E-04	1.250E-01	10	B324	0 0 0 0 0	
4.074E-04	1.465E-01	20	B179	0 0 0 0 0	
3.449E-04	1.240E-01	20	B300	2 1 1 1 2	
3.449E-04	1.240E-01	20	B324	0 0 0 0 0	
3.448E-04	1.240E-01	20	B324	0 0 0 0 0	
3.893E-04	1.400E-01	20	F311	1 2 2 2 1	
4.033E-04	1.450E-01	20	M061	1 0 0 0 2	
4.033E-04	1.450E-01	23	M161	1 0 0 0 2	
2.976E-04	1.070E-01	30	B324	0 0 0 0 0	
2.975E-04	1.070E-01	30	B324	0 0 0 0 0	

2825. $C_{12}H_{14}NO_4PS$

Ditalimfos

O,O-Diethyl (1,3-dihydro-1,3-dioxo-2H-isoindol-2-yl) phosphonothioate

Laptran
Plondrel

RN: 5131-24-8   **MP** (°C): 83.5
MW: 299.29   **BP** (°C):

Solubility (Moles/L)	Solubility (Grams/L)	Temp (°C)	Ref (#)	Evaluation (T P E A A)	Comments
4.444E-04	1.330E-01	rt	M161	0 0 0 0 2	

2826. $C_{12}H_{14}N_2O_2$

Primidone
5-Ethyldihydro-5-phenyl-4,6(1H,5H)-pyrimidinedione
Desoxyphenobarbitone
2-Deoxyphenobarbital

RN: 125-33-7   **MP** (°C): 281.5
MW: 218.26   **BP** (°C):

Solubility (Moles/L)	Solubility (Grams/L)	Temp (°C)	Ref (#)	Evaluation (T P E A A)	Comments
2.153E-03	4.700E-01	25	C437	0 0 0 0 0	Average
2.200E-03	4.802E-01	30	K108	1 2 2 0 1	
2.747E-03	5.996E-01	37	P061	0 0 0 0 0	
2.291E-03	5.000E-01	rt	D025	0 0 0 0 0	

2827. $C_{12}H_{14}N_2O_4$

Acetamide, N-(2-amino-2-oxoethyl)-2-(benzoyloxy)-N-methyl-

RN: 106231-62-3   **MP** (°C): 101.5
MW: 250.26   **BP** (°C): 496.6

Solubility (Moles/L)	Solubility (Grams/L)	Temp (°C)	Ref (#)	Evaluation (T P E A A)	Comments
1.207E-01	3.020E+01	22	B427	1 0 0 1 1	in 0.01M HCl
1.207E-01	3.020E+01	22	N317	1 1 2 1 2	

2828. $C_{12}H_{14}N_2O_4$

Propanamide, 2-[[(benzoyloxy)acetyl]amino]-

RN: 115193-30-1   **MP** (°C): 201.5
MW: 250.26   **BP** (°C):

Solubility (Moles/L)	Solubility (Grams/L)	Temp (°C)	Ref (#)	Evaluation (T P E A A)	Comments
1.918E-03	4.800E-01	22	N317	1 1 2 1 2	

2829. $C_{12}H_{14}N_2O_5$

2-Cyclohexyl-4,6-dinitrophenol
Dinex
4,6-Dinitro-2-cyclohexylphenol
2,4-Dinitro-6-cyclohexylphenol

RN: 131-89-5   **MP** (°C): 106
MW: 266.26   **BP** (°C):

Solubility (Moles/L)	Solubility (Grams/L)	Temp (°C)	Ref (#)	Evaluation (T P E A A)	Comments
5.634E-05	1.500E-02	25	M061	1 0 0 0 1	pH 6.5
6.760E-06	1.800E-03	25	M061	1 0 0 0 1	pH 1

2830. $C_{12}H_{14}N_2O_6$
Dinoseb acetate
Aretit

RN: 2813-95-8   **MP** (°C):   26.5
MW: 282.26   **BP** (°C):

Solubility (Moles/L)	Solubility (Grams/L)	Temp (°C)	Ref (#)	Evaluation (T P E A A)	Comments
7.794E-03	2.200E+00	rt	M161	0 0 0 0 1	

2831. $C_{12}H_{14}N_4O_2S$
6-Sulfanilamido-2,4-dimethylpyrimidine
6-Sulfanilamido-2,4-dimethylpyrimidin
Sulfisomidine
Sulphasomidine

RN: 515-64-0   **MP** (°C):   243.0
MW: 278.33   **BP** (°C):

Solubility (Moles/L)	Solubility (Grams/L)	Temp (°C)	Ref (#)	Evaluation (T P E A A)	Comments
4.965E-03	1.382E+00	25	M319	2 1 1 1 2	
6.862E-03	1.910E+00	37	K086	1 0 0 0 2	
5.802E-03	1.615E+00	ns	B133	0 2 0 1 2	pH 7.4
1.075E-02	2.991E+00	ns	M141	0 0 0 0 0	

2832. $C_{12}H_{14}N_4O_2S$
Sulfamethazine
Sulfadimezine
2-Sulfanilamido-4,6,-dimethylpyrimidine

RN: 57-68-1   **MP** (°C):   176
MW: 278.33   **BP** (°C):

Solubility (Moles/L)	Solubility (Grams/L)	Temp (°C)	Ref (#)	Evaluation (T P E A A)	Comments
5.317E-03	1.480E+00	20	F073	1 2 2 2 2	
1.544E-03	4.298E-01	20	L058	1 0 1 1 2	
1.893E-03	5.269E-01	20	O032	1 0 0 0 2	
1.424E-03	3.963E-01	24	N021	2 0 1 2 2	pH 5.6
1.600E-03	4.453E-01	25	M440	0 0 0 0 0	
5.389E-03	1.500E+00	29	C049	0 0 0 0 0	
2.695E-03	7.500E-01	37	L091	1 0 0 0 1	pH 5.5
6.862E-03	1.910E+00	37	M057	1 0 0 0 2	pH 5.5
2.414E-03	6.720E-01	37	S192	1 0 1 1 2	pH 6.0
2.299E-03	6.400E-01	38	K006	1 0 0 0 1	
1.185E-03	3.299E-01	ns	L044	0 0 0 0 2	

2833. $C_{12}H_{14}N_4O_2S.0.5H_2O$
Sulphamethazine (hemihydrate)
Sulfamethazine hemihydrate
RN: 57-68-1   MP (°C):
MW: 287.34   BP (°C):

Solubility (Moles/L)	Solubility (Grams/L)	Temp (°C)	Ref (#)	Evaluation (T P E A A)	Comments
6.786E-03	1.950E+00	37	R044	0 0 0 0 0	

2834. $C_{12}H_{14}N_4O_2S$
2-Sulfanilamido-4,5-dimethylpyrimidine
RN: 4462-43-5   MP (°C):   225.7
MW: 278.33   BP (°C):

Solubility (Moles/L)	Solubility (Grams/L)	Temp (°C)	Ref (#)	Evaluation (T P E A A)	Comments
7.186E-04	2.000E-01	29	C049	0 0 0 0 0	

2835. $C_{12}H_{14}N_4O_2S$
2-Sulfanilylamino-4-ethylpyrimidine
RN: 2276-96-2   MP (°C):
MW: 278.33   BP (°C):

Solubility (Moles/L)	Solubility (Grams/L)	Temp (°C)	Ref (#)	Evaluation (T P E A A)	Comments
6.180E-04	1.720E-01	37	R076	1 2 0 0 2	

2836. $C_{12}H_{14}N_4O_3S_2$
Acetyl sulfaethylthiadiazole
Acetamide, N-[4-[[(5-ethyl-1,3,4-thiadiazol-2-yl)amino]sulfonyl]phenyl]-
RN: 1037-51-0   MP (°C):
MW: 326.40   BP (°C):

Solubility (Moles/L)	Solubility (Grams/L)	Temp (°C)	Ref (#)	Evaluation (T P E A A)	Comments
4.963E-03	1.620E+00	37	B046	1 0 2 2 2	pH 4.6

2837. $C_{12}H_{14}N_4O_3S$
Sulfamethomidine
Sulphamethomidine
RN: 3772-76-7   MP (°C):   146.0
MW: 294.33   BP (°C):

Solubility (Moles/L)	Solubility (Grams/L)	Temp (°C)	Ref (#)	Evaluation (T P E A A)	Comments
2.864E-03	8.430E-01	ns	B133	0 2 0 1 2	pH 7.4
2.884E-03	8.489E-01	ns	R427	0 0 0 0 0	

2838. $C_{12}H_{14}N_4O_3S$
2-Sulfanilamido-4-ethoxypyrimidine
RN: 71138-72-2 **MP** (°C):
MW: 294.33 **BP** (°C):

Solubility (Moles/L)	Solubility (Grams/L)	Temp (°C)	Ref (#)	Evaluation (T P E A A)	Comments
1.801E-04	5.300E-02	37	R046	1 2 1 1 2	

2839. $C_{12}H_{14}N_4O_4S$
Sulfadimethoxine
Sulphadimethoxine
RN: 122-11-2 **MP** (°C): 202.0
MW: 310.33 **BP** (°C):

Solubility (Moles/L)	Solubility (Grams/L)	Temp (°C)	Ref (#)	Evaluation (T P E A A)	Comments
1.492E-04	4.630E-02	37	W055	1 2 0 1 2	
1.105E-03	3.430E-01	ns	B133	0 2 0 1 2	pH 7.4

2840. $C_{12}H_{14}N_4O_4S$
Sulfadoxine
Sulformethoxine
Sulforthomidine
4-Amino-*N*-(5,6-dimethoxy-4-pyrimidinyl)benzenesulfonamide
Fanzil
Fanasil
RN: 2447-57-6 **MP** (°C): 190–194
MW: 310.33 **BP** (°C):

Solubility (Moles/L)	Solubility (Grams/L)	Temp (°C)	Ref (#)	Evaluation (T P E A A)	Comments
6.761E-04	2.098E-01	ns	R427	0 0 0 0 0	

2841. $C_{12}H_{14}O_4$
Diethyl phthalate
Ethyl phthalate
Di-ethyl phthalate
Phthalic acid ethyl ester
Phthalsaeure-diethyl ester
RN: 84-66-2 **MP** (°C): −40.5
MW: 222.24 **BP** (°C): 296.1

Solubility (Moles/L)	Solubility (Grams/L)	Temp (°C)	Ref (#)	Evaluation (T P E A A)	Comments
4.495E-03	9.990E-01	20	F070	1 0 0 0 0	
4.180E-03	9.290E-01	20	L300	2 1 0 2 2	
1.793E-02	3.984E+00	20.00	D343	0 0 0 0 0	
5.399E-03	1.200E+00	25	F067	1 0 2 2 2	
4.500E-03	1.000E+00	25	F300	1 0 0 0 0	

2842. C₁₂H₁₄O₄

Trimethylacetyl salicylate
Salicylic acid, pivalate
2-Carboxyphenyl pivalate

RN: 2704-58-7   **MP** (°C):
MW: 222.24   **BP** (°C):

Solubility (Moles/L)	Solubility (Grams/L)	Temp (°C)	Ref (#)	Evaluation (T P E A A)	Comments
9.730E-04	2.162E-01	25.6	G015	1 0 1 1 2	pH 1.00, pka 3.74, intrinsic

2843. C₁₂H₁₄O₄

Diethyl o-phthalate

RN: **MP** (°C):   −40 C
MW: 222.24   **BP** (°C):

Solubility (Moles/L)	Solubility (Grams/L)	Temp (°C)	Ref (#)	Evaluation (T P E A A)	Comments
3.618E-03	8.040E-01	25	S417	0 0 0 0 0	

2844. C₁₂H₁₅ClNO₄PS₂

Phosalone
Diethyl S-((6-chloro-2-oxobenzoxazolin-3-yl)methyl) phosphorodithioate
Rubitox
Benzophosphate

RN: 2310-17-0   **MP** (°C):
MW: 367.81   **BP** (°C):

Solubility (Moles/L)	Solubility (Grams/L)	Temp (°C)	Ref (#)	Evaluation (T P E A A)	Comments
3.263E-06	1.200E-03	10	B324	0 0 0 0 0	
3.263E-06	1.200E-03	10	B324	0 0 0 0 0	
7.069E-06	2.600E-03	20	B300	2 2 1 1 2	
7.069E-06	2.600E-03	20	B324	0 0 0 0 0	
7.069E-06	2.600E-03	20	B324	0 0 0 0 0	
5.845E-06	2.150E-03	20	C053	0 0 0 0 0	
1.006E-05	3.700E-03	30	B324	0 0 0 0 0	
1.006E-05	3.700E-03	30	B324	0 0 0 0 0	
5.845E-06	2.150E-03	ns	F071	0 1 2 1 2	
2.719E-05	1.000E-02	rt	M161	0 0 0 0 1	

2845. C$_{12}$H$_{15}$ClO$_3$

Clofibrate

2-(p-Chlorophenoxy)-2-methylpropionic acid ethyl ester

Abitrate

Atromid S

RN: 637-07-0   **MP** (°C):

MW: 242.70   **BP** (°C):

Solubility (Moles/L)	Solubility (Grams/L)	Temp (°C)	Ref (#)	Evaluation (T P E A A)	Comments
4.000E-04	9.708E-02	rt	G093	0 1 1 1 2	

2846. C$_{12}$H$_{15}$IN$_2$O$_6$

Uridine, 2'-deoxy-5-iodo-, 5'-propanoate

5'-Propionyl 5-iodo-2'-deoxyuridine

5-Iodo-2'-deoxyuridine 5'-propionate

RN: 84043-25-4   **MP** (°C):   167.5

MW: 410.17   **BP** (°C):

Solubility (Moles/L)	Solubility (Grams/L)	Temp (°C)	Ref (#)	Evaluation (T P E A A)	Comments
3.480E+03	1.427E+06	25	N332	0 0 0 0 0	pH 7.4

2847. C$_{12}$H$_{15}$NO

n-Propylcinnamamide

Cinnamamide, N-propyl-

2-Propenamide, 3-phenyl-N-propyl-

RN: 6329-15-3   **MP** (°C):

MW: 189.26   **BP** (°C):

Solubility (Moles/L)	Solubility (Grams/L)	Temp (°C)	Ref (#)	Evaluation (T P E A A)	Comments
2.300E-03	4.353E-01	ns	H350	0 0 0 0 0	

2848. C$_{12}$H$_{15}$NO$_3$

Acetamide, N-[2-(benzoyloxy)ethyl]-N-methyl-

RN: 57440-16-1   **MP** (°C):

MW: 221.26   **BP** (°C):

Solubility (Moles/L)	Solubility (Grams/L)	Temp (°C)	Ref (#)	Evaluation (T P E A A)	Comments
1.415E-01	3.130E+01	22	N317	1 1 2 1 2	

2849. $C_{12}H_{15}NO_3$
Acetamide, 2-(benzoyloxy)-N-propyl-
RN: 106231-51-0   **MP** (°C):   89.5
MW: 221.26   **BP** (°C):

Solubility (Moles/L)	Solubility (Grams/L)	Temp (°C)	Ref (#)	Evaluation (T P E A A)	Comments
2.893E-03	6.400E-01	22	B427	1 0 0 1 1	in 0.01M HCl
2.893E-03	6.400E-01	22	N317	1 1 2 1 2	

2850. $C_{12}H_{15}NO_3$
Propanamide, 3-(benzoyloxy)-N,N-dimethyl-
RN: 115178-77-3   **MP** (°C):
MW: 221.26   **BP** (°C):

Solubility (Moles/L)	Solubility (Grams/L)	Temp (°C)	Ref (#)	Evaluation (T P E A A)	Comments
7.955E-02	1.760E+01	22	N317	1 1 2 1 2	

2851. $C_{12}H_{15}NO_3$
Carbofuran
2,3-Dihydro-2,2-dimethyl-7-benzofuranol methylcarbamate
Crisfuran
Furadanx
Curaterr
RN: 1563-66-2   **MP** (°C):   152
MW: 221.26   **BP** (°C):

Solubility (Moles/L)	Solubility (Grams/L)	Temp (°C)	Ref (#)	Evaluation (T P E A A)	Comments
1.315E-03	2.909E-01	10	B324	0 0 0 0 0	
1.315E-03	2.910E-01	10	B324	0 0 0 0 0	
1.446E-03	3.199E-01	19	B169	2 1 1 1 1	
1.446E-03	3.199E-01	20	B324	0 0 0 0 0	
1.446E-03	3.199E-01	20	B324	0 0 0 0 0	
3.164E-03	7.000E-01	25	M161	1 0 0 0 2	
1.695E-03	3.750E-01	30	B324	0 0 0 0 0	
1.694E-03	3.749E-01	30	B324	0 0 0 0 0	
1.446E-03	3.200E-01	ns	V414	0 0 0 0 0	

2852. C$_{12}$H$_{15}$NO$_3$
Acetaminophen butyrate
Butyryl acetaminophen
Butanoic acid, 4-(acetylamino)phenyl ester
Acetanilide, 4'-hydroxy-, butyrate

RN: 14771-98-3   **MP** (°C): 140
MW: 221.26   **BP** (°C):

Solubility (Moles/L)	Solubility (Grams/L)	Temp (°C)	Ref (#)	Evaluation (T P E A A)	Comments
1.491E-03	3.300E-01	25	B010	1 1 1 1 0	
2.441E-03	5.400E-01	37	D029	0 0 0 0 0	

2853. C$_{12}$H$_{15}$NO$_3$
Acetamide, 2-(benzoyloxy)-N-(1-methylethyl)-

RN: 115193-27-6   **MP** (°C): 129.5
MW: 221.26   **BP** (°C):

Solubility (Moles/L)	Solubility (Grams/L)	Temp (°C)	Ref (#)	Evaluation (T P E A A)	Comments
1.853E-03	4.100E-01	22	N317	1 1 2 1 2	

2854. C$_{12}$H$_{15}$NO$_4$
Isopropyl acetaminophen
Carbonic acid, 4-(acetylamino)phenyl 1-methylethyl ester
Acetanilide, 4'-hydroxy-, isopropyl carbonate

RN: 17239-27-9   **MP** (°C): 131.5–132
MW: 237.26   **BP** (°C):

Solubility (Moles/L)	Solubility (Grams/L)	Temp (°C)	Ref (#)	Evaluation (T P E A A)	Comments
4.636E-03	1.100E+00	37	D029	0 0 0 0 0	

2855. C$_{12}$H$_{15}$NO$_4$
Acetamide, 2-(benzoyloxy)-N-(2-hydroxyethyl)-N-methyl-

RN: 106231-59-8   **MP** (°C): 79
MW: 237.26   **BP** (°C): 428.2

Solubility (Moles/L)	Solubility (Grams/L)	Temp (°C)	Ref (#)	Evaluation (T P E A A)	Comments
8.135E-02	1.930E+01	22	B427	1 0 0 1 1	in 0.01M HCl
8.135E-02	1.930E+01	22	N317	1 1 2 1 2	

2856. C₁₂H₁₅NO₄

O-(Butyryloxymethyl) salicylamide
O-Butyryloxymethyl salicylamide
Butanoic acid, [2-(aminocarbonyl)phenoxy]methyl ester
RN: 103951-39-9 **MP** (°C): 57
MW: 237.26 **BP** (°C):

Solubility (Moles/L)	Solubility (Grams/L)	Temp (°C)	Ref (#)	Evaluation (T P E A A)	Comments
1.054E-02	2.500E+00	23	B328	1 2 2 1 1	pH 4.0
1.054E-02	2.500E+00	23	B328	0 0 0 0 0	

2857. C₁₂H₁₅NO₅

Benzoic acid, 2-hydroxy-, 2-[(2-hydroxyethyl)methylamino]-2-oxoethyl ester
N-Methyl-N-carbamoylmethyl glycolamide salicylate
RN: 114665-09-7 **MP** (°C): 92.5
MW: 253.26 **BP** (°C):

Solubility (Moles/L)	Solubility (Grams/L)	Temp (°C)	Ref (#)	Evaluation (T P E A A)	Comments
2.488E-02	6.300E+00	21	B331	1 2 2 1 0	pH 7.4
2.488E-02	6.300E+00	21	B331	0 0 0 0 0	

2858. C₁₂H₁₅NO₆

Ethonyphenyl tartramic acid
RN: **MP** (°C): 201
MW: 269.26 **BP** (°C):

Solubility (Moles/L)	Solubility (Grams/L)	Temp (°C)	Ref (#)	Evaluation (T P E A A)	Comments
1.481E-02	3.989E+00	14	C069	1 2 0 1 2	

2859. C₁₂H₁₅N₂O₃PS

Phoxim
4-Ethoxy-7-phenyl-3,5-dioxa-6-aza-4-phosphaoct-6-ene-8-nitrile 4-sulfide
Baythion
Sebacil
Volation
RN: 14816-18-3 **MP** (°C):
MW: 298.30 **BP** (°C):

Solubility (Moles/L)	Solubility (Grams/L)	Temp (°C)	Ref (#)	Evaluation (T P E A A)	Comments
1.106E-05	3.300E-03	10	B324	0 0 0 0 0	
1.106E-05	3.299E-03	10	B324	0 0 0 0 0	
1.374E-05	4.099E-03	20	B300	2 1 1 1 2	
1.374E-05	4.099E-03	20	B324	0 0 0 0 0	
1.374E-05	4.100E-03	20	B324	0 0 0 0 0	

(continued)

2859. C₁₂H₁₅N₂O₃PS (continued)

Solubility (Moles/L)	Solubility (Grams/L)	Temp (°C)	Ref (#)	Evaluation (T P E A A)	Comments
2.347E-05	7.000E-03	20	M161	1 0 0 0 0	
1.643E-05	4.901E-03	30	B324	0 0 0 0 0	
1.643E-05	4.900E-03	30	B324	0 0 0 0 0	
1.374E-05	4.099E-03	ns	S460	0 0 0 0 0	

2860. C₁₂H₁₅N₂O₃PS

Quinalphos
Diethyl *O*-(2-quinoxalyl) phosphorothioate
Diethquinalphion
Bayrusil
Ekalux
RN: 13593-03-8   **MP (°C):**   33.5
MW: 298.30   **BP (°C):**

Solubility (Moles/L)	Solubility (Grams/L)	Temp (°C)	Ref (#)	Evaluation (T P E A A)	Comments
7.375E-05	2.200E-02	24	M161	1 0 0 0 1	

2861. C₁₂H₁₅N₃O₂S

1-Methyl-2-sulfanilamide-1,2-dihydropyridine
Benzenesulfonamide, 4-amino-*N*-(1,2-dihydro-1-methyl-2-pyridinyl)-
RN: 51543-30-7   **MP (°C):**
MW: 265.34   **BP (°C):**

Solubility (Moles/L)	Solubility (Grams/L)	Temp (°C)	Ref (#)	Evaluation (T P E A A)	Comments
3.690E-03	9.791E-01	37	K095	2 0 0 0 2	intrinsic

2862. C₁₂H₁₅N₃O₂S

Albendazole
Bilutac
Eskazole
Proftril
Valbazan
Zentel
RN: 54965-21-8   **MP (°C):**
MW: 265.34   **BP (°C):**

Solubility (Moles/L)	Solubility (Grams/L)	Temp (°C)	Ref (#)	Evaluation (T P E A A)	Comments
2.827E-06	7.500E-04	209	D426	0 0 0 0 0	
3.769E-05	1.000E-02	ns	K444	0 0 0 0 0	

2863. $C_{12}H_{15}N_3O_3$
Triallyl cyanurate
Cyanursaeure-triallylaether

RN:	101-37-1	**MP** (°C):	26–28
MW:	249.27	**BP** (°C):	119–120

Solubility (Moles/L)	Solubility (Grams/L)	Temp (°C)	Ref (#)	Evaluation (T P E A A)	Comments
2.407E-02	6.000E+00	20	F300	1 0 0 0 0	

2864. $C_{12}H_{15}N_3O_3S$
Albendazole sulphoxide
Ricobendazole
Albendazole oxide
Methoxy-*N*-[5-(propylsulfinyl)benzimidazol-2-yl]carboxamide
Albendazole oxide [BAN:INN]
Carbamic acid

RN:	54029-12-8	**MP** (°C):	
MW:	281.34	**BP** (°C):	

Solubility (Moles/L)	Solubility (Grams/L)	Temp (°C)	Ref (#)	Evaluation (T P E A A)	Comments
2.204E-04	6.200E-02	25	W416	0 0 0 0 0	
1.094E-03	3.079E-01	94.1	D426	0 0 0 0 0	

2865. $C_{12}H_{15}N_3O_6$
1,3,5-Triglycidyl-*S*-triazinetrione
α-TGT

RN:	2451-62-9	**MP** (°C):	
MW:	297.27	**BP** (°C):	

Solubility (Moles/L)	Solubility (Grams/L)	Temp (°C)	Ref (#)	Evaluation (T P E A A)	Comments
4.373E-02	1.300E+01	0	A088	0 0 1 1 1	

2866. $C_{12}H_{15}N_5O_5$
9-[5′-(*O*-Acetyl)-β-D-arabinofuranosyl]adenine ester
Vidarabine 5′-acetate

RN:	65926-28-5	**MP** (°C):	198.0
MW:	309.28	**BP** (°C):	

Solubility (Moles/L)	Solubility (Grams/L)	Temp (°C)	Ref (#)	Evaluation (T P E A A)	Comments
2.134E-02	6.600E+00	ns	B134	0 1 1 1 1	

2867. $C_{12}H_{15}N_5O_5$
Pivaloyl salicylate
9-(2-O-Acetyl-β-D-arabinofuranosyl)adenine
RN: 87970-03-4   **MP** (°C):   195
MW: 309.28   **BP** (°C):

Solubility (Moles/L)	Solubility (Grams/L)	Temp (°C)	Ref (#)	Evaluation (T P E A A)	Comments
3.026E-01	9.360E+01	37	B306	1 2 0 1 2	pH 7.3

2868. $C_{12}H_{15}O_3P$
Diallyl phenyl phosphonate
Phosphonic acid, phenyl-, di-2-propenyl ester
RN: 2948-89-2   **MP** (°C):
MW: 238.23   **BP** (°C):

Solubility (Moles/L)	Solubility (Grams/L)	Temp (°C)	Ref (#)	Evaluation (T P E A A)	Comments
1.259E-03	3.000E-01	25	B070	1 2 0 1 0	

2869. $C_{12}H_{16}ClNOS$
Orbencarb
Lanray
S-((2-Chlorophenyl)methyl) diethylcarbamothioate
RN: 34622-58-7   **MP** (°C):
MW: 257.78   **BP** (°C):

Solubility (Moles/L)	Solubility (Grams/L)	Temp (°C)	Ref (#)	Evaluation (T P E A A)	Comments
9.311E-05	2.400E-02	ns	S460	0 0 0 0 0	

2870. $C_{12}H_{16}ClNOS$
Thiobencarb
S-4-Chlorobenzyl diethylthiocarbamate
Diethylcarbamothioic acid S-[(4-chlorophenyl)methyl] ester
4-Chlorobenzyl N,N-diethylthiocarbamate
RN: 28249-77-6   **MP** (°C):
MW: 257.78   **BP** (°C):   127.5

Solubility (Moles/L)	Solubility (Grams/L)	Temp (°C)	Ref (#)	Evaluation (T P E A A)	Comments
1.164E-04	3.000E-02	22	K137	1 1 2 1 0	
1.164E-04	3.001E-02	ns	S460	0 0 0 0 0	

2871. $C_{12}H_{16}Cl_2N_2O$
Neburon
1-Butyl-3-(3,4-dichlorophenyl)-1-methylurea

RN:	555-37-3	MP (°C):	101.5
MW:	275.18	BP (°C):	

Solubility (Moles/L)	Solubility (Grams/L)	Temp (°C)	Ref (#)	Evaluation (T P E A A)	Comments
1.744E-05	4.800E-03	20	F311	1 2 2 2 1	
1.744E-05	4.800E-03	24	B185	0 0 0 0 0	
1.744E-05	4.800E-03	24	G036	1 0 0 0 1	
1.744E-05	4.800E-03	24	M061	1 0 0 0 1	
1.744E-05	4.800E-03	24	M161	1 0 0 0 1	
1.744E-05	4.800E-03	25	A039	1 1 0 0 1	
1.744E-05	4.800E-03	25	G099	1 0 0 1 0	
1.744E-05	4.800E-03	ns	K007	0 0 0 0 1	

2872. $C_{12}H_{16}N_2$
Etryptamine
α-Ethyltryptamine

RN:	2235-90-7	MP (°C):	97
MW:	188.27	BP (°C):	

Solubility (Moles/L)	Solubility (Grams/L)	Temp (°C)	Ref (#)	Evaluation (T P E A A)	Comments
2.709E-03	5.100E-01	rt	M011	0 0 2 1 1	intrinsic

2873. $C_{12}H_{16}N_2O$
N-(Piperidinomethyl)benzamide
Benzamide, N-(1-pyrrolidinylmethyl)-

RN:	92788-60-8	MP (°C):	
MW:	204.27	BP (°C):	

Solubility (Moles/L)	Solubility (Grams/L)	Temp (°C)	Ref (#)	Evaluation (T P E A A)	Comments
7.100E-03	1.450E+00	22	J037	0 0 0 0 0	

2874. $C_{12}H_{16}N_2O_2$
N,N,N',N'-Tetramethylterephthalamide
1,4-Benzenedicarboxamide, N,N,N',N'-tetramethyl-

RN:	13158-31-1	MP (°C):	
MW:	220.27	BP (°C):	

Solubility (Moles/L)	Solubility (Grams/L)	Temp (°C)	Ref (#)	Evaluation (T P E A A)	Comments
1.843E+00	4.060E+02	30	K004	1 0 0 0 2	
1.840E+00	4.053E+02	30	K019	1 0 0 0 2	

2875. C$_{12}$H$_{16}$N$_2$O$_2$
N,N,N',N'-Tetramethylphthalamide
1,2-Benzenedicarboxamide, *N,N,N',N'*-tetramethyl-
RN: 6329-16-4   **MP** (°C):
MW: 220.27   **BP** (°C):

Solubility (Moles/L)	Solubility (Grams/L)	Temp (°C)	Ref (#)	Evaluation (T P E A A)	Comments
3.223E+00	7.100E+02	30	K004	1 0 0 0 2	

2876. C$_{12}$H$_{16}$N$_2$O$_2$
N,N,N',N'-Tetramethylisophthalamide
1,3-Benzenedicarboxamide, *N,N,N',N'*-tetramethyl-
RN: 14334-36-2   **MP** (°C):
MW: 220.27   **BP** (°C):

Solubility (Moles/L)	Solubility (Grams/L)	Temp (°C)	Ref (#)	Evaluation (T P E A A)	Comments
3.069E+00	6.760E+02	30	K004	1 0 0 0 2	
3.070E+00	6.762E+02	30	K019	1 0 0 0 2	

2877. C$_{12}$H$_{16}$N$_2$O$_2$S
4-Thiazolidinecarboxylic acid, 2-[4-(dimethylamino)phenyl]-
4-Thiazolidinecarboxylic acid, 2-(*p*-dimethylaminophenyl)-
RN: 72678-86-5   **MP** (°C):
MW: 252.34   **BP** (°C):   481.4

Solubility (Moles/L)	Solubility (Grams/L)	Temp (°C)	Ref (#)	Evaluation (T P E A A)	Comments
2.700E-03	6.813E-01	21	B414	1 0 0 1 1	fast decomposition

2878. C$_{12}$H$_{16}$N$_2$O$_3$
Hexobarbital
5-(1-Cyclohexen-1-yl)-1,5-dimethylbarbituric acid
5-(1-Cyclohexenyl)-1,5-dimethylbarbituric acid
Hexabarital
RN: 56-29-1   **MP** (°C):   146
MW: 236.27   **BP** (°C):

Solubility (Moles/L)	Solubility (Grams/L)	Temp (°C)	Ref (#)	Evaluation (T P E A A)	Comments
1.227E-03	2.900E-01	20	J030	1 2 2 2 1	
1.840E-03	4.347E-01	25	M056	2 2 2 2 2	
2.000E-03	4.725E-01	30	K108	1 2 2 0 1	
2.709E-03	6.400E-01	37	J030	1 2 2 2 1	

2879. C₁₂H₁₆N₂O₃

2879. $C_{12}H_{16}N_2O_3$

Carbetamide

N-Ethyl-2-((((phenylamino)carbonyl)oxy)propanamide

Leguarme

RN: 16118-49-3	**MP (°C):** >110	
MW: 236.27	**BP (°C):**	

Solubility (Moles/L)	Solubility (Grams/L)	Temp (°C)	Ref (#)	Evaluation (T P E A A)	Comments
1.481E-02	3.500E+00	20	M161	1 0 0 0 1	

2880. $C_{12}H_{16}N_2O_3$

Cyclobarbital

Phanodorm

RN: 52-31-3	**MP (°C):** 173	
MW: 236.27	**BP (°C):**	

Solubility (Moles/L)	Solubility (Grams/L)	Temp (°C)	Ref (#)	Evaluation (T P E A A)	Comments
6.772E-03	1.600E+00	20	F300	1 0 0 0 1	
6.941E-03	1.640E+00	20	J030	1 2 2 2 2	
3.500E-02	8.270E+00	25	G003	1 1 1 1 1	pH 4.7
8.000E-03	1.890E+00	30	G014	1 1 1 1 0	EFG
7.800E-03	1.843E+00	30	I001	2 0 2 1 0	EFG, 0.003N H₂SO₄
8.000E-03	1.890E+00	30	K108	1 2 2 0 1	
9.735E-03	2.300E+00	37	F300	1 0 0 0 1	
9.523E-03	2.250E+00	37	J030	1 2 2 2 2	
9.140E-02	2.160E+01	40	N008	1 2 1 1 2	*sic*

2881. $C_{12}H_{16}N_3O_3PS$

Triazophos

O,O-Diethyl *O*-(1-phenyl-1H-1,2,4-triazol-3-yl) phosphorothioate

Hostathion

RN: 24017-47-8	**MP (°C):**	
MW: 313.32	**BP (°C):**	

Solubility (Moles/L)	Solubility (Grams/L)	Temp (°C)	Ref (#)	Evaluation (T P E A A)	Comments
7.884E-05	2.470E-02	20	B300	2 1 1 1 2	
1.245E-04	3.900E-02	23	M161	1 0 0 0 1	
1.245E-04	3.900E-02	23	T305	1 0 0 0 1	
1.245E-04	3.899E-02	ns	S460	0 0 0 0 0	

2882. $C_{12}H_{16}N_3O_3PS_2$
Azinphos-ethyl
O,O-Diethyl S-[(4-oxo-1,2,3-benzotriazin-3(4H)-yl)methyl] phosphorodithioate
Azinos
Ethyl guthion

RN:	2642-71-9	MP (°C):	
MW:	345.38	BP (°C):	

Solubility (Moles/L)	Solubility (Grams/L)	Temp (°C)	Ref (#)	Evaluation (T P E A A)	Comments
1.940E-05	6.700E-03	10	B324	0 0 0 0 0	
1.940E-05	6.700E-03	10	B324	0 0 0 0 0	
3.040E-05	1.050E-02	20	B300	2 2 1 1 2	
3.040E-05	1.050E-02	20	B324	0 0 0 0 0	
3.040E-05	1.050E-02	20	B324	0 0 0 0 0	
7.152E-05	2.470E-02	30	B324	0 0 0 0 0	
7.151E-05	2.470E-02	30	B324	0 0 0 0 0	
3.020E-05	1.043E-02	ns	R427	0 0 0 0 0	

2883. $C_{12}H_{16}N_4O_2$
2,5-Diaziridinyl-3,6-bis(methylamino)-1,4-benzoquinone
Benzoquinone-2,5-bisaziridinyl-3,6-bismethyl amino

RN:	59886-52-1	MP (°C):	220
MW:	248.29	BP (°C):	

Solubility (Moles/L)	Solubility (Grams/L)	Temp (°C)	Ref (#)	Evaluation (T P E A A)	Comments
<4.03E-04	<1.00E-01	rt	C317	0 0 0 0 0	

2884. $C_{12}H_{16}N_4O_2S_2$
Glybuthiazole
p-Aminobenzenesulfamido-$tert$-butylthiodiazole
Glipasol
Glypasol

RN:	535-65-9	MP (°C):	222
MW:	312.41	BP (°C):	

Solubility (Moles/L)	Solubility (Grams/L)	Temp (°C)	Ref (#)	Evaluation (T P E A A)	Comments
1.820E-04	5.686E-02	37	A046	2 0 1 1 2	

2885. $C_{12}H_{16}N_4O_2S_2$
4-Amino-N-(5-butyl-1,3,4-thiadiazol-2-yl)benzenesulfonamide
Sulfanilamide, N1-(5-butyl-1,3,4-thiadiazol-2-yl)-

RN:	71119-31-8	MP (°C):	
MW:	312.41	BP (°C):	

Solubility (Moles/L)	Solubility (Grams/L)	Temp (°C)	Ref (#)	Evaluation (T P E A A)	Comments
2.710E-04	8.466E-02	37	A046	2 0 1 1 2	

2886. $C_{12}H_{16}N_4O_7S$
2′-Methylsulfonyl-6-methoxypurine arabinoside
9H-Purine, 6-methoxy-9-[2-O-(methylsulfonyl)-β-D-arabinofuranosyl]-

| RN: | 145913-48-0 | MP (°C): | 188-190 |
| MW: | 360.35 | BP (°C): | |

Solubility (Moles/L)	Solubility (Grams/L)	Temp (°C)	Ref (#)	Evaluation (T P E A A)	Comments
1.720E-02	6.198E+00	37	C348	0 0 0 0 0	pH 7.00

2887. $C_{12}H_{16}N_5O_3PS_2$
Azinphos-ethyl O-analog

| RN: | | MP (°C): | |
| MW: | 373.39 | BP (°C): | |

Solubility (Moles/L)	Solubility (Grams/L)	Temp (°C)	Ref (#)	Evaluation (T P E A A)	Comments
1.017E-02	3.797E+00	10	B300	2 2 1 1 2	

2888. $C_{12}H_{16}O$
o-Cyclohexylphenol
2-Cyclohexylphenol

| RN: | 119-42-6 | MP (°C): | |
| MW: | 176.26 | BP (°C): | |

Solubility (Moles/L)	Solubility (Grams/L)	Temp (°C)	Ref (#)	Evaluation (T P E A A)	Comments
4.727E-04	8.333E-02	25	L021	1 0 0 0 0	

2889. $C_{12}H_{16}O$
p-Cyclohexylphenol
4-Cyclohexylphenol

| RN: | 1131-60-8 | MP (°C): | |
| MW: | 176.26 | BP (°C): | |

Solubility (Moles/L)	Solubility (Grams/L)	Temp (°C)	Ref (#)	Evaluation (T P E A A)	Comments
3.782E-04	6.666E-02	25	L021	1 0 0 0 0	

2890. $C_{12}H_{16}O_2$
ε-Phenylcaproic acid
6-Phenylcaproic acid
6-Phenylhexanoic acid

| RN: | 5581-75-9 | MP (°C): | |
| MW: | 192.26 | BP (°C): | |

Solubility (Moles/L)	Solubility (Grams/L)	Temp (°C)	Ref (#)	Evaluation (T P E A A)	Comments
2.495E-03	4.798E-01	30	D033	2 2 1 2 2	
4.002E-03	7.694E-01	40	D033	2 2 1 2 2	

2891. $C_{12}H_{16}O_2$

4-Cyclohexylresorcinol

p-Cyclohexylresorcinol

RN: 2138-20-7   **MP** (°C):

MW: 192.26   **BP** (°C):

Solubility (Moles/L)	Solubility (Grams/L)	Temp (°C)	Ref (#)	Evaluation (T P E A A)	Comments
2.599E-03	4.998E-01	25	L021	1 0 0 0 0	

2892. $C_{12}H_{16}O_3$

Isoamyl salicylate

Isoamyl *o*-hydroxybenzoate

3-Methylbutyl salicylate

3-Methylbutyl *o*-hydroxybenzoate

RN: 87-20-7   **MP** (°C):

MW: 208.26   **BP** (°C):

Solubility (Moles/L)	Solubility (Grams/L)	Temp (°C)	Ref (#)	Evaluation (T P E A A)	Comments
6.961E-04	1.450E-01	25	D081	1 2 2 1 2	
6.918E-04	1.441E-01	ns	S460	0 0 0 0 0	

2893. $C_{12}H_{16}O_7 \cdot H_2O$

Arbutin (monohydrate)

Hydroquinone-β-D-glucopyranoside monohydrate

RN: 6058-77-1   **MP** (°C): 195–200

MW: 290.27   **BP** (°C):

Solubility (Moles/L)	Solubility (Grams/L)	Temp (°C)	Ref (#)	Evaluation (T P E A A)	Comments
3.828E-01	1.111E+02	c	D004	0 0 0 0 0	
1.723E+00	5.000E+02	h	D004	0 0 0 0 0	

2894. $C_{12}H_{17}NO_2$

2,6-Diethyl-4-acetaminophenol

3,5-Diethylparacetamol

4-Acetamido-2,6-diethylphenol

RN: 55205-89-5   **MP** (°C):

MW: 207.27   **BP** (°C):

Solubility (Moles/L)	Solubility (Grams/L)	Temp (°C)	Ref (#)	Evaluation (T P E A A)	Comments
2.943E-03	6.101E-01	25	D078	1 2 1 1 2	

2895. $C_{12}H_{17}NO_2$
Promecarb
5-Isopropyl-*m*-tolyl methylcarbamate
Carbamult

RN:	2631-37-0	MP (°C):	87.5
MW:	207.27	BP (°C):	117

Solubility (Moles/L)	Solubility (Grams/L)	Temp (°C)	Ref (#)	Evaluation (T P E A A)	Comments
4.439E-04	9.200E-02	rt	M161	0 0 0 0 1	

2896. $C_{12}H_{17}NO_2$
Pentyl *p*-aminobenzoate
4-Aminobenzoic acid pentyl ester

RN:	13110-37-7	MP (°C):	
MW:	207.27	BP (°C):	

Solubility (Moles/L)	Solubility (Grams/L)	Temp (°C)	Ref (#)	Evaluation (T P E A A)	Comments
3.900E-04	8.084E-02	37	F006	1 1 2 2 1	
1.890E-04	3.917E-02	ns	M066	0 0 0 0 2	
1.890E-04	3.917E-02	rt	B016	0 0 1 1 2	pH 7.4

2897. $C_{12}H_{17}NO_2$
2-*sec*-Butylphenyl methylcarbamate
BPMC
2-(1-Methylpropyl)phenol methylcarbamate
N-Methyl *O*-*sec*-butylphenylcarbamate

RN:	3766-81-2	MP (°C):	32
MW:	207.27	BP (°C):	112.5

Solubility (Moles/L)	Solubility (Grams/L)	Temp (°C)	Ref (#)	Evaluation (T P E A A)	Comments
4.294E-04	8.900E-02	22	K137	1 1 2 1 0	
3.184E-03	6.600E-01	30	M161	1 0 0 0 2	

2898. $C_{12}H_{17}NO_2$
Hexyl nicotinate
n-Hexyl nicotinoate
Nicotinic acid *n*-hexyl ester

RN:	23597-82-2	MP (°C):	
MW:	207.27	BP (°C):	

Solubility (Moles/L)	Solubility (Grams/L)	Temp (°C)	Ref (#)	Evaluation (T P E A A)	Comments
8.202E-04	1.700E-01	32	L346	1 0 0 1 2	

2899. C$_{12}$H$_{17}$NO$_2$
m-tert-Butylphenyl *N*-methylcarbamate
3-*tert*-Butylphenyl *N*-methylcarbamate

RN: 780-11-0 **MP** (°C): 144.0
MW: 207.27 **BP** (°C):

Solubility (Moles/L)	Solubility (Grams/L)	Temp (°C)	Ref (#)	Evaluation (T P E A A)	Comments
<2.41E-06	<5.00E-04	30	D089	2 2 0 0 0	

2900. C$_{12}$H$_{17}$NO$_3$
m-sec-Butoxyphenyl *N*-methylcarbamate
3-*sec*-Butoxyphenyl *N*-methylcarbamate

RN: 13538-22-2 **MP** (°C): 53
MW: 223.27 **BP** (°C):

Solubility (Moles/L)	Solubility (Grams/L)	Temp (°C)	Ref (#)	Evaluation (T P E A A)	Comments
3.583E-04	8.000E-02	30	D089	2 2 0 0 0	

2901. C$_{12}$H$_{17}$NO$_3$
m-n-Butoxyphenyl *N*-methylcarbamate
3-*n*-Butoxyphenyl *N*-methylcarbamate

RN: 3978-68-5 **MP** (°C): 54.5
MW: 223.27 **BP** (°C):

Solubility (Moles/L)	Solubility (Grams/L)	Temp (°C)	Ref (#)	Evaluation (T P E A A)	Comments
4.031E-04	9.000E-02	30	D089	2 2 0 0 0	

2902. C$_{12}$H$_{17}$NO$_3$
Acetamide, *N*-[4-(1-ethoxyethoxy)phenyl]-
1-(*p*-Acetaminophenoxy)-1-ethoxyethane

RN: 51736-24-4 **MP** (°C):
MW: 223.27 **BP** (°C):

Solubility (Moles/L)	Solubility (Grams/L)	Temp (°C)	Ref (#)	Evaluation (T P E A A)	Comments
3.000E-03	6.698E-01	ns	H076	0 0 0 0 0	

2903. C$_{12}$H$_{17}$NO$_4$
3,5-Dimethoxy-acetophenetide

RN: **MP** (°C):
MW: 239.27 **BP** (°C):

Solubility (Moles/L)	Solubility (Grams/L)	Temp (°C)	Ref (#)	Evaluation (T P E A A)	Comments
4.904E-01	1.173E+02	21.80	B102	2 0 1 1 1	solid hydrate
3.344E+00	8.000E+02	35.60	B102	2 0 1 1 2	liquid hydrate
8.778E-01	2.100E+02	39.40	B102	2 0 1 1 1	solid hydrate

(continued)

2903. $C_{12}H_{17}NO_4$ (continued)

Solubility (Moles/L)	Solubility (Grams/L)	Temp (°C)	Ref (#)	Evaluation (T P E A A)	Comments
3.233E+00	7.736E+02	45.60	B102	2 0 1 1 2	liquid hydrate
1.586E+00	3.795E+02	57	B102	2 0 1 1 1	solid hydrate
3.172E+00	7.591E+02	58.10	B102	2 0 1 1 2	liquid hydrate
3.172E+00	7.591E+02	68.50	B102	2 0 1 1 2	liquid hydrate
2.100E+00	5.026E+02	69.50	B102	2 0 1 1 1	solid hydrate
2.288E+00	5.474E+02	72.80	B102	2 0 1 1 1	solid hydrate
2.569E+00	6.147E+02	77.10	B102	2 0 1 1 2	solid hydrate
2.790E+00	6.675E+02	80.20	B102	2 0 1 1 2	solid hydrate
2.947E+00	7.053E+02	82.60	B102	2 0 1 1 2	solid hydrate
3.049E+00	7.296E+02	84.20	B102	2 0 1 1 2	solid hydrate
3.233E+00	7.736E+02	84.30	B102	2 0 1 1 2	liquid hydrate
3.172E+00	7.591E+02	86	B102	2 0 1 1 2	solid hydrate
3.233E+00	7.736E+02	86.90	B102	2 0 1 1 2	solid hydrate
3.348E+00	8.011E+02	99.80	B102	2 0 1 1 2	liquid hydrate
3.459E+00	8.275E+02	111.10	B102	2 0 1 1 2	liquid hydrate
3.527E+00	8.440E+02	118.40	B102	2 0 1 1 2	liquid hydrate
3.632E+00	8.690E+02	129.20	B102	2 0 1 1 2	liquid hydrate
4.031E+00	9.645E+02	173.60	B102	2 0 1 1 2	liquid hydrate

2904. $C_{12}H_{17}N_2O_2$
4-Aminobenzoic acid-2-(propyl-amino)ethyl ester
2-(Propylamino)ethyl 4-aminobenzoate
4-Aminobenzoic acid 2-(propyl-amino)ethyl ester
RN: **MP** (°C):
MW: 221.28   **BP** (°C):

Solubility (Moles/L)	Solubility (Grams/L)	Temp (°C)	Ref (#)	Evaluation (T P E A A)	Comments
3.000E-04	6.638E-02	ns	M066	0 0 0 0 0	

2905. $C_{12}H_{17}N_3O_4S$
3'-Nitroso-tolbutamide
RN: **MP** (°C):
MW: 299.35   **BP** (°C):

Solubility (Moles/L)	Solubility (Grams/L)	Temp (°C)	Ref (#)	Evaluation (T P E A A)	Comments
3.341E-04	1.000E-01	25	G051	1 0 1 1 0	

2906. $C_{12}H_{17}N_5O_3$
N,N-Diethylglycyloxymethyl-1-allopurinol
Glycine, N,N-diethyl-, (4,5-dihydro-4-oxo-1H-pyrazolo[3,4-d]pyrimidin-1-yl)methyl ester
RN: 98204-08-1   **MP** (°C):
MW: 279.30   **BP** (°C):

Solubility (Moles/L)	Solubility (Grams/L)	Temp (°C)	Ref (#)	Evaluation (T P E A A)	Comments
1.611E-02	4.500E+00	22	B323	0 0 0 0 0	

2907. $C_{12}H_{17}O_4PS_2$
Phenthoate
Dimethyl-S-(α-ethoxycarbonylbenzyl) phosphorodithioate
Elsan
Fenthoate
Phent
Cidial

RN:	2597-03-7	MP (°C):
MW:	320.37	BP (°C):

Solubility (Moles/L)	Solubility (Grams/L)	Temp (°C)	Ref (#)	Evaluation (T P E A A)	Comments
6.243E-04	2.000E-01	20	M161	1 0 0 0 2	
3.434E-05	1.100E-02	22	K137	1 1 2 1 0	
3.119E-05	9.992E-03	ns	S460	0 0 0 0 0	

2908. $C_{12}H_{18}$
1-Phenylhexane
Hexylbenzene
n-Hexylbenzene

RN:	1077-16-3	MP (°C):	−61
MW:	162.28	BP (°C):	226

Solubility (Moles/L)	Solubility (Grams/L)	Temp (°C)	Ref (#)	Evaluation (T P E A A)	Comments
5.678E-06	9.214E-04	5.04	M183	1 2 1 1 2	
5.678E-06	9.214E-04	6.04	M183	1 2 1 1 2	
5.140E-06	8.341E-04	7	O312	2 2 0 2 2	
5.667E-06	9.196E-04	8.04	M183	1 2 1 1 2	
5.583E-06	9.060E-04	9.04	M183	1 2 1 1 2	
5.150E-06	8.357E-04	10	O312	2 2 0 2 2	
5.572E-06	9.042E-04	10.04	M183	1 2 1 1 2	
5.717E-06	9.277E-04	11.04	M183	1 2 1 1 2	
5.733E-06	9.304E-04	12.04	M183	1 2 1 1 2	
5.667E-06	9.196E-04	13.04	M183	1 2 1 1 2	
5.700E-06	9.250E-04	14.04	M183	1 2 1 1 2	
5.090E-06	8.260E-04	15	O312	2 2 0 2 2	
5.594E-06	9.079E-04	15.04	M183	1 2 1 1 2	
5.661E-06	9.187E-04	16.04	M183	1 2 1 1 2	
5.606E-06	9.097E-04	17.04	M183	1 2 1 1 2	
5.678E-06	9.214E-04	18.04	M183	1 2 1 1 2	
5.811E-06	9.430E-04	19.04	M183	1 2 1 1 2	
5.860E-06	9.509E-04	20	O312	2 2 0 2 2	
5.850E-06	9.493E-04	20.04	M183	1 2 1 1 2	
5.889E-06	9.556E-04	21.04	M183	1 2 1 1 2	
5.872E-06	9.529E-04	22.04	M183	1 2 1 1 2	
6.056E-06	9.827E-04	23.04	M183	1 2 1 1 2	
6.133E-06	9.953E-04	24.04	M183	1 2 1 1 2	
6.270E-06	1.017E-03	25	M342	1 0 1 1 2	
5.560E-06	9.023E-04	25	O312	2 2 0 2 2	

(continued)

2908. C$_{12}$H$_{18}$ (continued)

Solubility (Moles/L)	Solubility (Grams/L)	Temp (°C)	Ref (#)	Evaluation (T P E A A)	Comments
6.156E-06	9.989E-04	25.04	M183	1 2 1 1 2	
6.156E-06	9.989E-04	26.04	M183	1 2 1 1 2	
6.239E-06	1.012E-03	27.04	M183	1 2 1 1 2	
6.261E-06	1.016E-03	29.04	M183	1 2 1 1 2	
6.140E-06	9.964E-04	30	O312	2 2 0 2 2	
6.590E-06	1.069E-03	35	O312	2 2 0 2 2	
6.590E-06	1.069E-03	40	O312	2 2 0 2 2	
8.000E-06	1.298E-03	45	O312	2 2 0 2 2	
2.000E-03	3.246E-01	ns	H307	0 0 0 0 0	

2909. C$_{12}$H$_{18}$N$_2$O
Isoproturon
N,N-Dimethyl-*N′*-(4-(1-methylethyl)phenyl)urea
3-(4-Isopropylphenyl)-1,1-dimethylurea
Tolkan
DPX 6774
RN: 34123-59-6   **MP (°C):**   158.5
MW: 206.29   **BP (°C):**

Solubility (Moles/L)	Solubility (Grams/L)	Temp (°C)	Ref (#)	Evaluation (T P E A A)	Comments
2.909E-04	6.000E-02	20	M161	1 0 0 0 1	

2910. C$_{12}$H$_{18}$N$_2$O$_2$
Zectran
4-Dimethylamino-3,5-dimethylphenol methylcarbamate ester
Mexacarbole
Mexacarbate
RN: 315-18-4   **MP (°C):**   85
MW: 222.29   **BP (°C):**

Solubility (Moles/L)	Solubility (Grams/L)	Temp (°C)	Ref (#)	Evaluation (T P E A A)	Comments
4.498E-04	9.999E-02	25	I314	0 0 0 0 0	

2911. C$_{12}$H$_{18}$N$_2$O$_2$S
Thiamylal
5-Allyl-5-(1-methyl-butyl)-barbituric acid
5-Allyl-5-(1-methylbutyl)-2-thiobarbituric acid
RN: 77-27-0   **MP (°C):**   132
MW: 254.35   **BP (°C):**

Solubility (Moles/L)	Solubility (Grams/L)	Temp (°C)	Ref (#)	Evaluation (T P E A A)	Comments
5.104E-03	1.298E+00	25	A023	1 0 0 1 1	
1.966E-04	5.000E-02	25	B011	2 0 0 1 0	
1.944E-04	4.946E-02	25	B065	1 1 1 1 2	

(continued)

2911. C$_{12}$H$_{18}$N$_2$O$_2$S (continued)

Solubility (Moles/L)	Solubility (Grams/L)	Temp (°C)	Ref (#)	Evaluation (T P E A A)	Comments
3.480E-04	8.852E-02	25	G003	1 1 1 1 1	pH 4.7
7.500E-03	1.908E+00	30	G014	1 1 1 1 0	EFG
6.600E-03	1.679E+00	30	I001	2 0 2 1 0	EFG, 0.003N H$_2$SO$_4$
8.630E-03	2.195E+00	40	A023	1 0 0 1 1	
3.750E-03	9.538E-01	40	N008	1 2 1 1 2	*sic*
8.792E-03	2.236E+00	ns	G039	0 0 0 0 0	EFG

2912. C$_{12}$H$_{18}$N$_2$O$_3$

5-Isopropyl-5-(3-methylbut-2-enyl)barbituric acid
2,4,6(1H,3H,5H)-Pyrimidinetrione, 5-(3-methyl-2-butenyl)-5-(1-methylethyl)
5-*i*-Propyl-5-(3-methylbut-2-enyl)barbiturate

RN: 67051-26-7   **MP** (°C):
MW: 238.29   **BP** (°C):

Solubility (Moles/L)	Solubility (Grams/L)	Temp (°C)	Ref (#)	Evaluation (T P E A A)	Comments
2.555E-03	6.088E-01	25	P350	0 0 0 0 0	intrinsic

2913. C$_{12}$H$_{18}$N$_2$O$_3$

Secobarbital
5-Allyl-5-(1-methylbutyl)barbituric acid
Seconal

RN: 76-73-3   **MP** (°C): 98
MW: 238.29   **BP** (°C):

Solubility (Moles/L)	Solubility (Grams/L)	Temp (°C)	Ref (#)	Evaluation (T P E A A)	Comments
7.250E-03	1.728E+00	25	G003	1 1 1 1 2	pH 7
4.410E-03	1.051E+00	25	V033	2 0 1 1 2	
4.400E-03	1.048E+00	25.00	T303	1 0 0 0 1	
6.300E-03	1.501E+00	35.00	T303	1 0 0 0 1	
7.900E-02	1.882E+01	40	N008	1 0 1 1 2	*sic*
9.400E-03	2.240E+00	45.00	T303	1 0 0 0 1	

2914. C$_{12}$H$_{18}$N$_2$O$_3$S

Tolbutamide
1-Butyl-3-(*para*-tolylsulfonyl) urea
Oramide
Orinase

RN: 64-77-7   **MP** (°C): 129
MW: 270.35   **BP** (°C):

Solubility (Moles/L)	Solubility (Grams/L)	Temp (°C)	Ref (#)	Evaluation (T P E A A)	Comments
5.178E-04	1.400E-01	25	G051	1 0 1 1 0	
4.068E-04	1.100E-01	25	P096	0 0 0 0 0	

(continued)

2914. $C_{12}H_{18}N_2O_3S$ (continued)

Solubility (Moles/L)	Solubility (Grams/L)	Temp (°C)	Ref (#)	Evaluation (T P E A A)	Comments
3.900E-04	1.054E-01	30	G318	0 0 0 0 0	EFG
4.027E-04	1.089E-01	37	A028	1 0 2 1 2	intrinsic
4.030E-04	1.090E-01	37	A046	2 0 1 1 2	
5.659E-04	1.530E-01	37	B138	1 2 0 0 2	pH 1.5, form II
5.289E-04	1.430E-01	37	B138	1 2 0 0 2	pH 1.5, form III
5.067E-04	1.370E-01	37	B138	1 2 0 0 2	pH 1.5, form I
3.699E-04	1.000E-01	37.0	H033	1 0 2 1 0	pH 1.4, intrinsic
3.031E-03	8.193E-01	37.5	F015	1 0 2 2 1	pH 6.0, pKa 5.32
2.535E-02	6.853E+00	37.5	F015	1 0 2 2 2	pH 7.0, pKa 5.32

2915. $C_{12}H_{18}N_2O_4S$
Anisylbutamide
Methoxyphenylbutazolamide
Methoxytolbutamide
RN: 24535-67-9   **MP** (°C):
MW: 286.35   **BP** (°C):

Solubility (Moles/L)	Solubility (Grams/L)	Temp (°C)	Ref (#)	Evaluation (T P E A A)	Comments
4.236E-04	1.213E-01	37	A028	1 0 2 1 2	intrinsic
4.260E-04	1.220E-01	37	A046	2 0 1 1 2	

2916. $C_{12}H_{18}N_2O_5$
D-Mannosephenylhydrazone
D-Mannose-phenylhydrazon
RN: 6147-14-4   **MP** (°C): 195.5
MW: 270.29   **BP** (°C):

Solubility (Moles/L)	Solubility (Grams/L)	Temp (°C)	Ref (#)	Evaluation (T P E A A)	Comments
3.811E-02	1.030E+01	100	F300	1 0 0 0 2	

2917. $C_{12}H_{18}N_4O_6S$
Oryzalin
3,5-Dinitro-N4,N4-dipropylsulfanilamide
RN: 19044-88-3   **MP** (°C): 137
MW: 346.36   **BP** (°C):

Solubility (Moles/L)	Solubility (Grams/L)	Temp (°C)	Ref (#)	Evaluation (T P E A A)	Comments
2.454E-04	8.500E-02	25	B200	1 0 0 0 1	
6.929E-06	2.400E-03	25	M161	1 0 0 0 1	

2918. C$_{12}$H$_{18}$O
Propofol
2,6-Diisopropylphenol
Diisopropylphenol
Diprivan
RN: 2078-54-8   **MP** (°C):
MW: 178.28   **BP** (°C):

Solubility (Moles/L)	Solubility (Grams/L)	Temp (°C)	Ref (#)	Evaluation (T P E A A)	Comments
8.975E-04	1.600E-01	amb	L434	0 0 0 0 0	

2919. C$_{12}$H$_{18}$O
2-Butyl-4-ethylphenol
Phenol, 2-butyl-4-ethyl-
RN: 3781-74-6   **MP** (°C):
MW: 178.28   **BP** (°C):

Solubility (Moles/L)	Solubility (Grams/L)	Temp (°C)	Ref (#)	Evaluation (T P E A A)	Comments
1.402E-04	2.500E-02	25	L020	1 0 0 0 0	

2920. C$_{12}$H$_{18}$O
2-Butyl-4,6-dimethylphenol
2,6-Xylenol, 2-butyl-
RN: 6483-60-9   **MP** (°C):
MW: 178.28   **BP** (°C):

Solubility (Moles/L)	Solubility (Grams/L)	Temp (°C)	Ref (#)	Evaluation (T P E A A)	Comments
1.603E-04	2.857E-02	25	L020	1 0 0 0 0	

2921. C$_{12}$H$_{18}$O
o-n-Hexylphenol
2-n-Hexylphenol
RN: 3226-32-2   **MP** (°C):
MW: 178.28   **BP** (°C):

Solubility (Moles/L)	Solubility (Grams/L)	Temp (°C)	Ref (#)	Evaluation (T P E A A)	Comments
2.244E-04	4.000E-02	25	L022	1 0 0 0 0	

2922. C$_{12}$H$_{18}$O
2-Butyl-4,5-dimethylphenol
Phenol, 2-butyl-4,5-dimethyl-
RN: **MP** (°C):
MW: 178.28   **BP** (°C):

Solubility (Moles/L)	Solubility (Grams/L)	Temp (°C)	Ref (#)	Evaluation (T P E A A)	Comments
1.870E-04	3.333E-02	25	L020	1 0 0 0 0	

2923. C$_{12}$H$_{18}$O
2-Butyl-6-ethylphenol
Phenol, 2-butyl-6-ethyl-
RN: 22496-45-3 **MP** (°C):
MW: 178.28 **BP** (°C):

Solubility (Moles/L)	Solubility (Grams/L)	Temp (°C)	Ref (#)	Evaluation (T P E A A)	Comments
1.870E-04	3.333E-02	25	L020	1 0 0 0 0	

2924. C$_{12}$H$_{18}$O
2,6-Dipropylphenol
Phenol, 2,6-dipropyl-
RN: 6626-32-0 **MP** (°C):
MW: 178.28 **BP** (°C):

Solubility (Moles/L)	Solubility (Grams/L)	Temp (°C)	Ref (#)	Evaluation (T P E A A)	Comments
1.402E-04	2.500E-02	25	L020	1 0 0 0 0	

2925. C$_{12}$H$_{18}$O
4-Butyl-2,5-dimethylphenol
2,5-Xylenol, 4-butyl-
RN: 91763-77-8 **MP** (°C):
MW: 178.28 **BP** (°C):

Solubility (Moles/L)	Solubility (Grams/L)	Temp (°C)	Ref (#)	Evaluation (T P E A A)	Comments
2.244E-04	4.000E-02	25	L020	1 0 0 0 0	

2926. C$_{12}$H$_{18}$O
4-Butyl-2,6-dimethylphenol
Phenol, 4-butyl-2,6-dimethyl-
2,6-Xylenol, 4-butyl-
RN: 6676-26-2 **MP** (°C):
MW: 178.28 **BP** (°C):

Solubility (Moles/L)	Solubility (Grams/L)	Temp (°C)	Ref (#)	Evaluation (T P E A A)	Comments
2.244E-04	4.000E-02	25	L020	1 0 0 0 0	

2927. C$_{12}$H$_{18}$O
p-n-Hexylphenol
4-n-Hexylphenol
RN: 2446-69-7 **MP** (°C):
MW: 178.28 **BP** (°C):

Solubility (Moles/L)	Solubility (Grams/L)	Temp (°C)	Ref (#)	Evaluation (T P E A A)	Comments
1.603E-04	2.857E-02	25	L022	1 0 0 0 0	

2928. $C_{12}H_{18}O$
2,4-Dipropylphenol
Phenol, 2,4-dipropyl-

RN: 23167-99-9   **MP** (°C):
MW: 178.28   **BP** (°C):

Solubility (Moles/L)	Solubility (Grams/L)	Temp (°C)	Ref (#)	Evaluation (T P E A A)	Comments
1.402E-04	2.500E-02	25	L020	1 0 0 0 0	

2929. $C_{12}H_{18}O_2$
4-Hexylresorcinol
4-n-Hexylresorcin

RN: 136-77-6   **MP** (°C):   68
MW: 194.28   **BP** (°C):   334

Solubility (Moles/L)	Solubility (Grams/L)	Temp (°C)	Ref (#)	Evaluation (T P E A A)	Comments
2.574E-03	5.000E-01	18	F300	1 0 0 0 1	

2930. $C_{12}H_{18}O_4S_2$
Di-isopropyl 1,3-dithiolan-2-ylidinemalonate
Isoprothiolane
Fuji-one
bis(1-Methylethyl) 1,3-dithiolan-2-ylidenepropanedioate

RN: 50512-35-1   **MP** (°C):   52.25
MW: 290.40   **BP** (°C):   168

Solubility (Moles/L)	Solubility (Grams/L)	Temp (°C)	Ref (#)	Evaluation (T P E A A)	Comments
1.653E-04	4.800E-02	20	H309	0 0 0 0 0	
1.653E-04	4.800E-02	20	M161	1 0 0 0 1	

2931. $C_{12}H_{19}BrN_2O_2$
Neostigmine bromide
Neostigmine bromide
Neostigmine;
Prostigmin

RN: 114-80-7   **MP** (°C):
MW: 303.21   **BP** (°C):

Solubility (Moles/L)	Solubility (Grams/L)	Temp (°C)	Ref (#)	Evaluation (T P E A A)	Comments
3.298E+00	1.000E+03	ns	K444	0 0 0 0 0	

2932. C$_{12}$H$_{19}$ClNO$_3$P

Crufomate

O-Methyl O-2-chloro-4-tert-butyphenyl N-methylamidophosphate

RN: 299-86-5   **MP** (°C):   60.25

MW: 291.72   **BP** (°C):   117.5

Solubility (Moles/L)	Solubility (Grams/L)	Temp (°C)	Ref (#)	Evaluation (T P E A A)	Comments
1.705E-02	4.975E+00	ns	M061	0 0 0 0 0	

2933. C$_{12}$H$_{19}$N$_3$O$_8$

Orotic acid methylglucamide

RN: **MP** (°C):   184–186

MW: 333.30   **BP** (°C):

Solubility (Moles/L)	Solubility (Grams/L)	Temp (°C)	Ref (#)	Evaluation (T P E A A)	Comments
4.470E-01	1.490E+02	–4	N018	0 0 0 0 0	
7.090E-01	2.363E+02	16	N018	0 0 0 0 0	
8.150E-01	2.716E+02	25	N018	0 0 0 0 0	

2934. C$_{12}$H$_{19}$N$_6$OP

Triamiphos

5-Amino-1-(bis(dimethylamino)phosphoryl)-3-phenyl-1,2,4-triazole

Triamifos

Wepsyn 155

Wepsyn

bis(Dimethylamino)-(3-amino-5-phenyl-1,2,4-triazol-1-yl)-phosphine oxide

RN: 1031-47-6   **MP** (°C):   167.5

MW: 294.30   **BP** (°C):

Solubility (Moles/L)	Solubility (Grams/L)	Temp (°C)	Ref (#)	Evaluation (T P E A A)	Comments
8.495E-04	2.500E-01	20	M161	1 0 0 0 2	

2935. C$_{12}$H$_{19}$O$_2$PS$_3$

Sulprofos

O-Ethyl O-[4-(methylthio)phenyl]phosphorodithioic acid S-propyl ester

Morpafos

Bolstar

Heliothion

Merdafos

RN: 35400-43-2   **MP** (°C):

MW: 322.45   **BP** (°C):   155–158

Solubility (Moles/L)	Solubility (Grams/L)	Temp (°C)	Ref (#)	Evaluation (T P E A A)	Comments
9.616E-07	3.101E-04	ns	S460	0 0 0 0 0	

2936. C$_{12}$H$_{20}$
Triisobutene
1,8-Nonadiene, 2,8-dimethyl-5-methylene-
RN: 36370-80-6   **MP** (°C):
MW: 164.29   **BP** (°C):

Solubility (Moles/L)	Solubility (Grams/L)	Temp (°C)	Ref (#)	Evaluation (T P E A A)	Comments
4.944E-08	8.123E-06	20	B165	1 0 1 1 1	
5.838E-03	9.591E-01	97.30	B165	1 0 1 1 1	

2937. C$_{12}$H$_{20}$N$_2$O$_3$
5-Ethyl-5-*n*-hexylbarbituric acid
2,4,6(1H,3H,5H)-Pyrimidinetrione, 5-ethyl-5-hexyl-
Hexethal
Ortal
Ortol
RN: 77-30-5   **MP** (°C):
MW: 240.30   **BP** (°C):

Solubility (Moles/L)	Solubility (Grams/L)	Temp (°C)	Ref (#)	Evaluation (T P E A A)	Comments
8.930E-04	2.146E-01	25	M310	2 2 2 2 2	

2938. C$_{12}$H$_{20}$N$_4$O$_2$
3-Cyclohexyl-6-dimethylamino-1-methyl-1,3,5-triazine-2,4-dione
1,3,5-Triazine-2,4(1H,3H)-dione, 3-cyclohexyl-6-(dimethylamino)-1-methyl-
Hexazinone
Pronone
DPX 3674
RN: 51235-04-2   **MP** (°C):   116
MW: 252.32   **BP** (°C):

Solubility (Moles/L)	Solubility (Grams/L)	Temp (°C)	Ref (#)	Evaluation (T P E A A)	Comments
1.308E-01	3.300E+01	25	M161	1 0 0 0 1	

2939. C$_{12}$H$_{20}$N$_4$O$_6$
Acetyltetraglycine ethyl ester
Glycine, *N*-acetylglycylglycylglycyl-, ethyl ester
RN: 637-83-2   **MP** (°C):   264
MW: 316.32   **BP** (°C):

Solubility (Moles/L)	Solubility (Grams/L)	Temp (°C)	Ref (#)	Evaluation (T P E A A)	Comments
8.220E-04	2.600E-01	0	R036	0 0 0 0 0	
2.466E-03	7.800E-01	25	R036	0 0 0 0 0	
5.216E-03	1.650E+00	40	R036	0 0 0 0 0	

2940. $C_{12}H_{20}O_2$
Linalyl acetate
Bergamol
3,7-Dimethyl-1,6-octadien-3-yl acetate
Linalyl
RN: 115-95-7   **MP** (°C):
MW: 196.29   **BP** (°C):   220

Solubility (Moles/L)	Solubility (Grams/L)	Temp (°C)	Ref (#)	Evaluation (T P E A A)	Comments
2.546E-03	4.998E-01	25	M350	1 0 1 1 1	

2941. $C_{12}H_{20}O_4$
Dibutyl maleate
Di-*n*-butyl maleate
RN: 105-76-0   **MP** (°C):
MW: 228.29   **BP** (°C):

Solubility (Moles/L)	Solubility (Grams/L)	Temp (°C)	Ref (#)	Evaluation (T P E A A)	Comments
1.073E-03	2.450E-01	25	F067	1 0 2 2 2	

2942. $C_{12}H_{20}O_6$
Tripropionin
1,2,3-Propanetriol, tripropanoate
1,2,3-Propanetriyl tripropionate
Tripropionylglycerol
Tripropanoylglycerol
RN: 139-45-7   **MP** (°C):
MW: 260.29   **BP** (°C):

Solubility (Moles/L)	Solubility (Grams/L)	Temp (°C)	Ref (#)	Evaluation (T P E A A)	Comments
1.199E-02	3.120E+00	ns	F014	0 0 0 0 2	

2943. $C_{12}H_{21}NO_8S$
Topiramate
2,3:4,5-di-*O*-isopropylidene-β-D-fructopyranose sulfamate
Topamax
Tracrium
RN: 97240-79-4   **MP** (°C):
MW: 339.37   **BP** (°C):

Solubility (Moles/L)	Solubility (Grams/L)	Temp (°C)	Ref (#)	Evaluation (T P E A A)	Comments
2.860E-02	9.705E+00	ns	S469	0 0 0 0 0	

2944. C$_{12}$H$_{21}$N$_2$O$_3$PS

Diazinon
O,O-Diethyl *O*-(2-isopropyl-6-methyl-4-pyrimidinyl), phosphorothioate
Dimpylate
Basudin
Spectracide
Fezudin

RN: 333-41-5	**MP** (°C):	>120
MW: 304.35	**BP** (°C):	

Solubility (Moles/L)	Solubility (Grams/L)	Temp (°C)	Ref (#)	Evaluation (T P E A A)	Comments
2.336E-04	7.109E-02	10	B324	0 0 0 0 0	
2.336E-04	7.110E-02	10	B324	0 0 0 0 0	
1.318E-04	4.012E-02	20	B179	0 0 0 0 0	
2.261E-04	6.881E-02	20	B300	2 1 1 1 2	
1.758E-04	5.350E-02	20	B324	0 0 0 0 0	
1.758E-04	5.350E-02	20	B324	0 0 0 0 0	
1.314E-04	4.000E-02	20	M061	1 0 0 0 1	
2.260E-04	6.880E-02	22	B169	2 1 1 1 2	
1.331E-04	4.050E-02	22	K137	1 1 2 1 0	
1.436E-04	4.370E-02	30	B324	0 0 0 0 0	
1.436E-04	4.370E-02	30	B324	0 0 0 0 0	
1.314E-04	4.000E-02	rt	M161	0 0 0 0 1	

2945. C$_{12}$H$_{21}$N$_5$O$_2$S$_2$

Nizatidine
Axid
N-(2-(((2-((Dimethylamino)methyl)-4-thiazolyl)methyl)thio)ethyl)-*N'*-methyl-2-
　nitro-1,1-ethenediamine

RN: 76963-41-2	**MP** (°C):	
MW: 331.46	**BP** (°C):	

Solubility (Moles/L)	Solubility (Grams/L)	Temp (°C)	Ref (#)	Evaluation (T P E A A)	Comments
6.457E-02	2.140E+01	ns	R427	0 0 0 0 0	

2946. C$_{12}$H$_{21}$N$_7$O

1-(4'-Formyl-1-piperizinyl)-3,5-bis(dimethylamino)-*s*-triazine
1-Piperazinecarboxaldehyde, 4-[4,6-bis(dimethylamino)-1,3,5-triazin-2-yl]-

RN: 126974-79-6	**MP** (°C):	
MW: 279.35	**BP** (°C):	

Solubility (Moles/L)	Solubility (Grams/L)	Temp (°C)	Ref (#)	Evaluation (T P E A A)	Comments
3.670E-03	1.025E+00	25	B386	0 0 0 0 0	

2947. $C_{12}H_{22}N_2O_2$
N,N,N',N'-Tetraethylfumaramide
2-Butenediamide, *N,N,N',N'*-tetraethyl-
RN: 111328-65-5   **MP** (°C):
MW: 226.32   **BP** (°C):

Solubility (Moles/L)	Solubility (Grams/L)	Temp (°C)	Ref (#)	Evaluation (T P E A A)	Comments
6.900E-01	1.562E+02	30	K019	1 0 0 0 1	

2948. $C_{12}H_{22}N_6$
1-(Piperidinyl)-3,5-bis(dimethylamino)-*s*-triazine
s-Triazine, 2,4-bis(dimethylamino)-6-piperidino-
RN: 16268-79-4   **MP** (°C):
MW: 250.35   **BP** (°C):

Solubility (Moles/L)	Solubility (Grams/L)	Temp (°C)	Ref (#)	Evaluation (T P E A A)	Comments
1.758E-04	4.402E-02	25	B386	0 0 0 0 0	

2949. $C_{12}H_{22}O_2$
Arbanol
RN: 7070-15-7   **MP** (°C):
MW: 198.31   **BP** (°C):

Solubility (Moles/L)	Solubility (Grams/L)	Temp (°C)	Ref (#)	Evaluation (T P E A A)	Comments
1.523E-03	3.020E-01	6	P430	0 0 0 0 0	
2.911E-03	5.773E-01	23.5	P430	0 0 0 0 0	

2950. $C_{12}H_{22}O_4$
Ethylene glycol divalerate
RN: **MP** (°C):
MW: 230.31   **BP** (°C):

Solubility (Moles/L)	Solubility (Grams/L)	Temp (°C)	Ref (#)	Evaluation (T P E A A)	Comments
6.460E-04	1.488E-01	25	F064	1 0 0 0 2	

2951. C$_{12}$H$_{22}$O$_4$
1,10-Decanedicarboxylic acid
Decan-dicarbonsaeure-(1,10)
Dodecanedioc acid
RN: 693-23-2   **MP** (°C):   128
MW: 230.31   **BP** (°C):

Solubility (Moles/L)	Solubility (Grams/L)	Temp (°C)	Ref (#)	Evaluation (T P E A A)	Comments
1.737E-04	4.000E-02	20	F300	1 0 0 0 0	
3.039E-03	7.000E-01	21	B040	1 0 1 1 0	*sic*
5.124E-03	1.180E+00	100	F300	1 0 0 0 2	

2952. C$_{12}$H$_{22}$O$_4$
Dibutyl succinate
Succinic acid di-*n*-butyl ester
Tabutrex
RN: 141-03-7   **MP** (°C):   –29
MW: 230.31   **BP** (°C):   108

Solubility (Moles/L)	Solubility (Grams/L)	Temp (°C)	Ref (#)	Evaluation (T P E A A)	Comments
9.984E-04	2.299E-01	ns	F014	0 0 0 0 1	

2953. C$_{12}$H$_{22}$O$_6$
Triethylene glycol dipropionate
Ethanol, 2,2′-[1,2-ethanediylbis(oxy)]*bis*-, dipropanoate
RN: 141-34-4   **MP** (°C):
MW: 262.31   **BP** (°C):

Solubility (Moles/L)	Solubility (Grams/L)	Temp (°C)	Ref (#)	Evaluation (T P E A A)	Comments
2.394E-01	6.279E+01	ns	F014	0 0 0 0 2	

2954. C$_{12}$H$_{22}$O$_6$
Dibutyl tartrate
(2R,3R)-Di-*n*-butyl tartrate
ENT 396
RN: 87-92-3   **MP** (°C):   21
MW: 262.31   **BP** (°C):

Solubility (Moles/L)	Solubility (Grams/L)	Temp (°C)	Ref (#)	Evaluation (T P E A A)	Comments
1.840E-02	4.827E+00	ns	F014	0 0 0 0 2	

2955. C$_{12}$H$_{22}$O$_6$
Dimethoxyethyl adipate

RN: **MP (°C):**
MW: 262.31   **BP (°C):**

Solubility (Moles/L)	Solubility (Grams/L)	Temp (°C)	Ref (#)	Evaluation (T P E A A)	Comments
5.338E-02	1.400E+01	ns	F014	0 0 0 0 2	

2956. C$_{12}$H$_{22}$O$_{11}$
Maltose
D-Glucose, 4-*O*-α-D-glucopyranosyl-
α-Maltose
Malt sugar

RN: 69-79-4   **MP (°C):**   102.5
MW: 342.30   **BP (°C):**

Solubility (Moles/L)	Solubility (Grams/L)	Temp (°C)	Ref (#)	Evaluation (T P E A A)	Comments
7.166E-01	2.453E+02	0	C401	1 0 0 0 0	EFG
1.061E+00	3.631E+02	0	M043	1 0 0 0 1	
1.151E+00	3.939E+02	10	M043	1 0 0 0 1	
9.066E-01	3.103E+02	20	C401	1 0 0 0 0	EFG
1.517E+00	5.192E+02	20	D041	1 0 0 0 2	
1.280E+00	4.382E+02	20	M043	1 0 0 0 1	
1.408E+00	4.819E+02	30	M043	1 0 0 0 1	
1.124E+00	3.846E+02	40	C401	1 0 0 0 0	EFG
1.037E+00	3.548E+02	40	C401	1 0 0 0 0	EFG
1.530E+00	5.238E+02	40	M043	1 0 0 0 2	
1.252E+00	4.286E+02	60	C401	1 0 0 0 0	EFG
1.859E+00	6.364E+02	60	M043	1 0 0 0 2	
1.298E+00	4.444E+02	80	C401	1 0 0 0 0	EFG
2.191E+00	7.500E+02	80	M043	1 0 0 0 2	
1.298E+00	4.444E+02	90	C401	1 0 0 0 0	EFG
1.321E+00	4.521E+02	100	C401	1 0 0 0 0	EFG
1.517E+00	5.192E+02	rt	D021	0 0 1 1 2	

2957. C$_{12}$H$_{22}$O$_{11}$
β-Lactose
B-Lactose
Milchzucker
4-*O*-β-D-Galactopyranosyl-D-glucose

RN: 5965-66-2   **MP (°C):**   253
MW: 342.30   **BP (°C):**

Solubility (Moles/L)	Solubility (Grams/L)	Temp (°C)	Ref (#)	Evaluation (T P E A A)	Comments
1.525E-01	5.220E+01	20	F300	1 0 0 0 2	
7.303E-02	2.500E+01	h	F300	0 0 0 0 1	

2958. C$_{12}$H$_{22}$O$_{11}$
Cellobiose
4-O-β-D-Glucopyranosyl-D-glucose
4-β-D-Glucopyransoyl-D-glucopyranose
D-(+)-Cellobiose

RN: 528-50-7   **MP** (°C):
MW: 342.30   **BP** (°C):

Solubility (Moles/L)	Solubility (Grams/L)	Temp (°C)	Ref (#)	Evaluation (T P E A A)	Comments
3.243E-01	1.110E+02	15	F300	1 0 0 0 2	
3.475E-01	1.189E+02	30.50	M137	2 1 2 2 2	
1.198E+00	4.100E+02	h	F300	0 0 0 0 1	

2959. C$_{12}$H$_{22}$O$_{11}$
Lactose
4-O-B-D-Galactopyranosyl-D-glucose
Milk sugar

RN: 63-42-3   **MP** (°C):   201
MW: 342.30   **BP** (°C):

Solubility (Moles/L)	Solubility (Grams/L)	Temp (°C)	Ref (#)	Evaluation (T P E A A)	Comments
2.656E-01	9.091E+01	0	C401	1 0 0 0 0	EFG
3.177E-01	1.087E+02	0	M043	1 0 0 0 2	
3.116E-01	1.067E+02	0	P052	1 0 2 2 2	
4.701E-01	1.609E+02	1	P049	1 0 1 1 1	
3.811E-01	1.304E+02	10	M043	1 0 0 0 2	
4.351E-01	1.489E+02	20	C401	1 0 0 0 0	EFG
4.767E-01	1.632E+02	20	M043	1 0 0 0 2	
5.189E-01	1.776E+02	25	D041	1 0 0 0 2	
5.470E-01	1.873E+02	25	P049	1 0 1 1 1	
6.000E-01	2.054E+02	30	D011	1 0 1 0 1	
5.880E-01	2.013E+02	30	M043	1 0 0 0 2	
5.843E-01	2.000E+02	40	C401	1 0 0 0 0	EFG
7.298E-01	2.498E+02	40	M043	1 0 0 0 2	
7.574E-01	2.593E+02	60	C401	1 0 0 0 0	EFG
1.067E+00	3.651E+02	60	M043	1 0 0 0 2	
9.738E-01	3.333E+02	80	C401	1 0 0 0 0	EFG
1.475E+00	5.050E+02	80	M043	1 0 0 0 2	
1.699E+00	5.816E+02	89	D041	1 0 0 0 2	
1.096E+00	3.750E+02	95	C401	1 0 0 0 0	EFG
1.124E+00	3.846E+02	100	C401	1 0 0 0 0	EFG
1.767E+00	6.047E+02	100	M043	1 0 0 0 2	
4.775E-01	1.635E+02	rt	D021	0 0 1 1 2	

2960. $C_{12}H_{22}O_{11}$

Sucrose

Saccharose

β-D-Fructofuranosyl-α-D-glucopyranoside

α-D-Glucopyranosyl β-D-fructofuranoside

Beet sugar

Cane sugar

RN: 57-50-1   **MP** (°C): 191

MW: 342.30   **BP** (°C):

Solubility (Moles/L)	Solubility (Grams/L)	Temp (°C)	Ref (#)	Evaluation (T P E A A)	Comments
1.140E+00	3.902E+02	0	C401	1 0 0 0 0	EFG
1.878E+00	6.429E+02	0	D041	1 0 0 0 2	
1.876E+00	6.421E+02	0	G046	1 0 1 1 2	
1.142E+00	3.909E+02	0	H094	1 0 0 0 2	
1.874E+00	6.416E+02	0	M043	1 0 0 0 2	
1.884E+00	6.450E+02	0	P052	1 0 2 2 2	
1.880E+00	6.435E+02	.90	M074	1 0 0 0 2	average of 3
1.157E+00	3.961E+02	10	H094	1 0 0 0 2	
1.914E+00	6.552E+02	10	M043	1 0 0 0 2	
1.943E+00	6.650E+02	12.5	F300	1 0 0 0 2	
1.938E+00	6.633E+02	15	D041	1 0 0 0 2	
1.934E+00	6.622E+02	15.80	M074	1 0 0 0 2	average of 3
1.931E+00	6.609E+02	18.5	W013	1 2 1 1 2	
1.177E+00	4.030E+02	20	C401	1 0 0 0 0	EFG
1.203E+00	4.118E+02	20	C401	1 0 0 0 0	EFG
1.946E+00	6.660E+02	20	F300	1 0 0 0 2	
1.170E+00	4.005E+02	20	G060	1 0 0 0 2	
1.173E+00	4.015E+02	20	H094	1 0 0 0 2	
1.960E+00	6.711E+02	20	M043	1 0 0 0 2	
1.956E+00	6.697E+02	23.9	W013	1 2 1 1 2	
1.954E+00	6.689E+02	24.4	W013	1 2 1 1 2	
1.964E+00	6.723E+02	24.9	W013	1 2 1 1 2	
1.986E+00	6.798E+02	25	G046	1 0 1 1 2	
1.179E+00	4.036E+02	25	G060	1 0 0 0 2	
1.981E+00	6.779E+02	25.60	M074	1 0 0 0 2	average of 3
1.963E+00	6.721E+02	25.9	W013	1 2 1 1 2	
1.188E+00	4.067E+02	30	G060	1 0 0 0 2	
1.190E+00	4.072E+02	30	H094	1 0 0 0 2	
2.006E+00	6.865E+02	30	M043	1 0 0 0 2	
1.997E+00	6.836E+02	30.0	W013	1 2 1 1 2	
1.996E+00	6.831E+02	30.5	W013	1 2 1 1 2	
2.003E+00	6.855E+02	30.50	M074	1 0 0 0 2	average of 3
2.008E+00	6.873E+02	31.5	W013	1 2 1 1 2	
2.005E+00	6.862E+02	33.1	W013	1 2 1 1 2	
2.025E+00	6.932E+02	34.5	W013	1 2 1 1 2	
2.030E+00	6.950E+02	35	G046	1 0 1 1 2	
1.198E+00	4.100E+02	35	G060	1 0 0 0 2	
2.028E+00	6.942E+02	36.0	W013	1 2 1 1 2	
2.028E+00	6.941E+02	36.4	W013	1 2 1 1 2	
1.252E+00	4.286E+02	40	C401	1 0 0 0 0	EFG

(continued)

2960. $C_{12}H_{22}O_{11}$ (continued)

Solubility (Moles/L)	Solubility (Grams/L)	Temp (°C)	Ref (#)	Evaluation (T P E A A)	Comments
1.207E+00	4.133E+02	40	G060	1 0 0 0 2	
1.207E+00	4.132E+02	40	H094	1 0 0 0 2	
2.057E+00	7.041E+02	40	M043	1 0 0 0 2	
2.050E+00	7.017E+02	40.2	W013	1 2 1 1 2	
2.052E+00	7.023E+02	40.7	W013	1 2 1 1 2	
2.055E+00	7.035E+02	41.0	W013	1 2 1 1 2	
2.061E+00	7.055E+02	42.2	W013	1 2 1 1 2	
2.067E+00	7.074E+02	42.3	W013	1 2 1 1 2	
2.080E+00	7.120E+02	45	F300	1 0 0 0 2	
1.217E+00	4.167E+02	45	G060	1 0 0 0 2	
2.093E+00	7.163E+02	46.1	W013	1 2 1 1 2	
2.107E+00	7.212E+02	49.6	W013	1 2 1 1 2	
2.111E+00	7.225E+02	50	G046	1 0 1 1 2	
1.228E+00	4.202E+02	50	G060	1 0 0 0 2	
7.596E+00	2.600E+03	50	H063	1 0 0 0 2	
1.225E+00	4.194E+02	50	H094	1 0 0 0 2	
2.101E+00	7.191E+02	50.2	W013	1 2 1 1 2	
2.118E+00	7.251E+02	51.1	W013	1 2 1 1 2	
2.124E+00	7.272E+02	52.2	W013	1 2 1 1 2	
2.126E+00	7.276E+02	52.6	W013	1 2 1 1 2	
2.134E+00	7.304E+02	53.6	W013	1 2 1 1 2	
2.134E+00	7.305E+02	53.8	W013	1 2 1 1 2	
2.126E+00	7.278E+02	54.1	W013	1 2 1 1 2	
1.237E+00	4.235E+02	55	G060	1 0 0 0 2	
2.137E+00	7.316E+02	55.8	W013	1 2 1 1 2	
2.147E+00	7.350E+02	56.1	W013	1 2 1 1 2	
2.154E+00	7.372E+02	56.4	W013	1 2 1 1 2	
2.151E+00	7.364E+02	57.5	W013	1 2 1 1 2	
2.154E+00	7.374E+02	57.8	W013	1 2 1 1 2	
2.152E+00	7.368E+02	58.4	W013	1 2 1 1 2	
2.165E+00	7.410E+02	58.6	W013	1 2 1 1 2	
2.166E+00	7.415E+02	59.7	W013	1 2 1 1 2	
1.252E+00	4.286E+02	60	C401	1 0 0 0 0	EFG
1.248E+00	4.273E+02	60	G060	1 0 0 0 2	
1.244E+00	4.259E+02	60	H094	1 0 0 0 2	
2.167E+00	7.416E+02	60	M043	1 0 0 0 2	
2.176E+00	7.448E+02	61.1	W013	1 2 1 1 2	
2.176E+00	7.447E+02	61.4	W013	1 2 1 1 2	
2.182E+00	7.469E+02	62.6	W013	1 2 1 1 2	
2.189E+00	7.493E+02	62.9	W013	1 2 1 1 2	
2.193E+00	7.505E+02	64.6	W013	1 2 1 1 2	
1.258E+00	4.307E+02	65	G060	1 0 0 0 2	
2.204E+00	7.543E+02	65.5	W013	1 2 1 1 2	
2.214E+00	7.580E+02	66.4	W013	1 2 1 1 2	
2.219E+00	7.595E+02	66.5	W013	1 2 1 1 2	
2.222E+00	7.607E+02	68.2	W013	1 2 1 1 2	
2.221E+00	7.603E+02	69.0	W013	1 2 1 1 2	
1.269E+00	4.344E+02	70	G060	1 0 0 0 2	

(continued)

2960. C$_{12}$H$_{22}$O$_{11}$ (continued)

Solubility (Moles/L)	Solubility (Grams/L)	Temp (°C)	Ref (#)	Evaluation (T P E A A)	Comments
2.230E+00	7.632E+02	70.1	W013	1 2 1 1 2	
2.233E+00	7.645E+02	70.4	W013	1 2 1 1 2	
2.251E+00	7.706E+02	72.8	W013	1 2 1 1 2	
2.249E+00	7.698E+02	73.8	W013	1 2 1 1 2	
2.267E+00	7.760E+02	74.5	W013	1 2 1 1 2	
2.265E+00	7.752E+02	74.6	W013	1 2 1 1 2	
2.256E+00	7.724E+02	75	G046	1 0 1 1 2	
1.280E+00	4.380E+02	75	G060	1 0 0 0 2	
2.266E+00	7.758E+02	75.1	W013	1 2 1 1 2	
2.290E+00	7.840E+02	79.5	W013	1 2 1 1 2	
1.276E+00	4.366E+02	80	C401	1 0 0 0 0	EFG
1.291E+00	4.417E+02	80	G060	1 0 0 0 2	
1.090E+01	3.730E+03	80	H063	1 0 0 0 2	
2.289E+00	7.835E+02	80	M043	1 0 0 0 2	
2.304E+00	7.886E+02	82.3	W013	1 2 1 1 2	
2.333E+00	7.985E+02	85.1	W013	1 2 1 1 2	
2.335E+00	7.994E+02	85.3	W013	1 2 1 1 2	
2.337E+00	7.999E+02	85.5	W013	1 2 1 1 2	
2.344E+00	8.022E+02	86.6	W013	1 2 1 1 2	
2.346E+00	8.032E+02	88.0	W013	1 2 1 1 2	
1.298E+00	4.444E+02	90	C401	1 0 0 0 0	EFG
2.355E+00	8.061E+02	90	G046	1 0 1 1 2	
2.363E+00	8.087E+02	90.2	W013	1 2 1 1 2	
2.388E+00	8.176E+02	95	G046	1 0 1 1 2	
2.409E+00	8.247E+02	98	G046	1 0 1 1 2	
1.321E+00	4.521E+02	100	C401	1 0 0 0 0	EFG
2.424E+00	8.296E+02	100	D041	1 0 0 0 2	
2.424E+00	8.296E+02	100	G046	1 0 1 1 2	
2.424E+00	8.296E+02	100	M043	1 0 0 0 2	

2961. C$_{12}$H$_{23}$NO$_3$
Propylbutylaceturethane
RN: **MP** (°C):
MW: 229.32 **BP** (°C):

Solubility (Moles/L)	Solubility (Grams/L)	Temp (°C)	Ref (#)	Evaluation (T P E A A)	Comments
1.395E-03	3.199E-01	20	O021	1 2 0 0 0	

2962. C$_{12}$H$_{23}$N$_7$
1-(4′-Methyl-1-piperizinyl)-3,5-bis(dimethylamino)-s-triazine
2-(4-Methyl-1-piperazinyl)-4,6-bis(dimethylamino)-s-triazine
RN: 5512-05-0 **MP** (°C):
MW: 265.36 **BP** (°C):

Solubility (Moles/L)	Solubility (Grams/L)	Temp (°C)	Ref (#)	Evaluation (T P E A A)	Comments
4.514E-03	1.198E+00	25	B386	0 0 0 0 0	

2963. $C_{12}H_{24}N_2O_2$

N,N,N',N'-Tetramethylsuberamide
Octanediamide, N,N,N',N'-tetramethyl-

RN:	27397-05-3	MP (°C):
MW:	228.34	BP (°C):

Solubility (Moles/L)	Solubility (Grams/L)	Temp (°C)	Ref (#)	Evaluation (T P E A A)	Comments
2.520E+00	5.754E+02	30	D010	1 2 1 1 2	

2964. $C_{12}H_{24}N_3O_3PS$

Thiophosphoryl trimorpholide
Morpholine, 4,4',4''-phosphinothioylidynetris-
Phosphine sulfide, trimorpholino-

RN:	14129-98-7	MP (°C):
MW:	321.38	BP (°C):

Solubility (Moles/L)	Solubility (Grams/L)	Temp (°C)	Ref (#)	Evaluation (T P E A A)	Comments
9.987E-03	3.210E+00	25	A040	1 0 0 0 2	

2965. $C_{12}H_{24}N_3O_4P$

Phosphoryl trimorpholide
Morpholine, 4,4',4''-phosphinylidynetris-
Phosphine oxide, trimorpholino-

RN:	4441-12-7	MP (°C):
MW:	305.32	BP (°C):

Solubility (Moles/L)	Solubility (Grams/L)	Temp (°C)	Ref (#)	Evaluation (T P E A A)	Comments
1.989E+00	6.072E+02	25	A040	1 0 0 0 2	

2966. $C_{12}H_{24}N_6$

$N2,N4,N6$-Triethyl-$N2,N4,N6$-trimethylmelamine
1,3,5-Triazine-2,4,6-triamine, N,N',N''-triethyl-N,N',N''-trimethyl-

RN:	64124-20-5	MP (°C):
MW:	252.37	BP (°C):

Solubility (Moles/L)	Solubility (Grams/L)	Temp (°C)	Ref (#)	Evaluation (T P E A A)	Comments
1.981E-04	5.000E-02	25	C051	1 2 1 1 0	pH 7

2967. $C_{12}H_{24}N_9P_3$

Hexaziridinocyclotriphosphazene
2,2,4,4,6,6-Hexahydro-2,2,4,4,6,6-hexakis(1-aziridinyl)-1,3,5,2,4,6-triazatriphosphorine
2,2,4,4,6,6-Hexakis(1-aziridinyl)cyclotriphosphaza-1,3,5-triene
Apholate
APN
ENT 26316

RN: 52-46-0   **MP** (°C):
MW: 387.31   **BP** (°C):

Solubility (Moles/L)	Solubility (Grams/L)	Temp (°C)	Ref (#)	Evaluation (T P E A A)	Comments
2.582E-01	1.000E+02	ns	L076	0 1 0 0 0	approximate

2968. $C_{12}H_{24}O_2$

Lauric acid
Dodecanoic acid
Laurostearic acid

RN: 143-07-7   **MP** (°C):   44
MW: 200.32   **BP** (°C):

Solubility (Moles/L)	Solubility (Grams/L)	Temp (°C)	Ref (#)	Evaluation (T P E A A)	Comments
1.847E-04	3.700E-02	0	B136	1 0 2 1 1	
2.895E-04	5.800E-02	20	B136	1 0 2 1 1	
2.745E-04	5.500E-02	20	D041	1 0 0 0 1	
2.745E-04	5.500E-02	20.0	R001	1 1 1 1 1	
2.400E-05	4.808E-03	25	J001	1 0 2 1 2	
8.486E-06	1.700E-03	25	M083	1 0 0 1 1	
1.150E-05	2.304E-03	25	R002	0 0 0 0 0	intrinsic
2.080E-05	4.167E-03	25	R002	0 0 0 0 0	
3.345E-04	6.700E-02	30	B136	1 0 2 1 1	
3.145E-04	6.300E-02	30.0	R001	1 1 1 1 1	
3.494E-04	7.000E-02	40	B136	1 0 2 1 1	
3.844E-05	7.700E-03	40	E005	2 1 1 2 1	
3.744E-04	7.500E-02	45	B136	1 0 2 1 1	
3.744E-04	7.499E-02	45.0	R001	1 1 1 1 1	
4.593E-05	9.200E-03	50	E005	2 1 1 2 1	
5.470E-05	1.096E-02	50	J001	1 0 2 1 2	
4.343E-04	8.700E-02	60	B136	1 0 2 1 1	
5.791E-05	1.160E-02	60	E005	2 1 1 2 2	
4.343E-04	8.699E-02	60.0	R001	1 1 1 1 1	
1.847E-04	3.700E-02	.0	R001	1 1 1 1 1	

2969. C$_{12}$H$_{24}$O$_2$
3-Hydroxy-2,2,5,5-tetraethyltetrahydrofuran
3-Furanol, 2,2,5,5-tetraethyltetrahydro-
RN: 29839-78-9   **MP** (°C):
MW: 200.32   **BP** (°C):

Solubility (Moles/L)	Solubility (Grams/L)	Temp (°C)	Ref (#)	Evaluation (T P E A A)	Comments
1.493E-02	2.991E+00	rt	B066	0 2 0 0 0	

2970. C$_{12}$H$_{24}$O$_3$
1,3-Dioxolane-4-methanol, 2-heptyl-2-methyl
2-Heptyl-4-hydroxymethyl-2-methyl-1,3-dioxolane
RN: 5660-50-4   **MP** (°C):
MW: 216.32   **BP** (°C):

Solubility (Moles/L)	Solubility (Grams/L)	Temp (°C)	Ref (#)	Evaluation (T P E A A)	Comments
3.560E-03	7.701E-01	25	P342	0 0 0 0 0	0.0001M Na$_2$CO$_3$

2971. C$_{12}$H$_{24}$O$_4$
1,3-Dioxolane-4-methanol, 2-methyl-2-[2-(pentyloxy)ethyl]
RN: 143458-56-4   **MP** (°C):
MW: 232.32   **BP** (°C):

Solubility (Moles/L)	Solubility (Grams/L)	Temp (°C)	Ref (#)	Evaluation (T P E A A)	Comments
6.250E-02	1.452E+01	25	P342	0 0 0 0 0	0.0001M Na$_2$CO$_3$

2972. C$_{12}$H$_{26}$
2-Methylundecane
Isododecane
RN: 31807-55-3   **MP** (°C):
MW: 170.34   **BP** (°C):

Solubility (Moles/L)	Solubility (Grams/L)	Temp (°C)	Ref (#)	Evaluation (T P E A A)	Comments
1.174E-08	2.000E-06	25	T423	0 0 0 0 0	

2973. C$_{12}$H$_{26}$
3,3,6,6-Tetramethyloctane
RN: 62199-46-6   **MP** (°C):
MW: 170.34   **BP** (°C):

Solubility (Moles/L)	Solubility (Grams/L)	Temp (°C)	Ref (#)	Evaluation (T P E A A)	Comments
1.233E-07	2.100E-05	25	T423	0 0 0 0 0	

2974. C$_{12}$H$_{26}$
Dodecane
N-Dodecane
Alkane C(12)
Duodecane
Bihexyl
Adakane 12

| **RN:** | 112-40-3 | **MP** (°C): | −9.6 |
| **MW:** | 170.34 | **BP** (°C): | 216.3 |

Solubility (Moles/L)	Solubility (Grams/L)	Temp (°C)	Ref (#)	Evaluation (T P E A A)	Comments
4.931E-08	8.400E-06	22.5	G301	0 0 0 0 0	
2.055E-08	3.500E-06	23	C332	0 0 0 0 0	
1.068E-08	1.820E-06	25	B156	1 0 2 2 2	
4.944E-08	8.422E-06	25	F004	0 0 0 0 0	
5.871E-09	1.000E-06	25	T423	0 0 0 0 0	
3.900E-09	6.643E-07	ns	D348	0 0 0 0 0	
2.231E-08	3.800E-06	ns	H123	0 0 0 0 0	

2975. C$_{12}$H$_{26}$
2,2,4,6,6-Pentamethylheptane

| **RN:** | 13475-82-6 | **MP** (°C): | |
| **MW:** | 170.34 | **BP** (°C): | |

Solubility (Moles/L)	Solubility (Grams/L)	Temp (°C)	Ref (#)	Evaluation (T P E A A)	Comments
1.468E-07	2.500E-05	25	T423	0 0 0 0 0	

2976. C$_{12}$H$_{26}$O
Dodecanol
Dodecyl alcohol
Lauryl alcohol
Undecyl carbinol

| **RN:** | 112-53-8 | **MP** (°C): | 24 |
| **MW:** | 186.34 | **BP** (°C): | 261 |

Solubility (Moles/L)	Solubility (Grams/L)	Temp (°C)	Ref (#)	Evaluation (T P E A A)	Comments
9.100E-06	1.696E-03	16	K011	1 2 1 1 2	
2.300E-05	4.286E-03	25	R002	0 0 0 0 0	
1.560E-05	2.907E-03	34	K011	1 2 1 1 2	
1.930E-05	3.596E-03	49	K011	1 2 1 1 2	

2977. C$_{12}$H$_{27}$N
Tributylamine
tris-n-Butylamine
N,N-Dibutyl-1-butanamine

RN:	102-82-9	MP (°C):	−70
MW:	185.36	BP (°C):	216

Solubility (Moles/L)	Solubility (Grams/L)	Temp (°C)	Ref (#)	Evaluation (T P E A A)	Comments
7.649E-04	1.418E-01	25.04	V013	2 2 2 2 2	

2978. C$_{12}$H$_{27}$N.4H$_2$O
Dodecylamine (tetrahydrate)

RN:	124-22-1	MP (°C):	
MW:	257.42	BP (°C):	

Solubility (Moles/L)	Solubility (Grams/L)	Temp (°C)	Ref (#)	Evaluation (T P E A A)	Comments
2.776E-03	7.145E-01	ns	R037	0 2 2 1 0	

2979. C$_{12}$H$_{27}$OP
Tributyl phosphine oxide
Tributylphosphine oxide
TBPO

RN:	814-29-9	MP (°C):	64
MW:	218.32	BP (°C):	

Solubility (Moles/L)	Solubility (Grams/L)	Temp (°C)	Ref (#)	Evaluation (T P E A A)	Comments
1.035E+00	2.260E+02	13.20	H031	1 2 2 2 2	
8.794E-01	1.920E+02	13.40	H031	1 2 2 2 2	
4.718E-01	1.030E+02	16.30	H031	1 2 2 2 2	
1.832E-01	4.000E+01	25	B070	1 2 0 1 1	
2.551E-01	5.570E+01	25.00	H031	1 2 2 2 2	
2.299E-01	5.020E+01	27.00	H032	1 1 2 1 2	
2.244E-01	4.900E+01	27.8	H032	1 1 2 1 2	
2.125E-01	4.640E+01	29.0	H032	1 1 2 1 2	
2.020E-01	4.410E+01	30.2	H032	1 1 2 1 2	
1.974E-01	4.310E+01	31.1	H032	1 1 2 1 2	
1.892E-01	4.130E+01	32.0	H032	1 1 2 1 2	
1.818E-01	3.970E+01	32.5	H032	1 1 2 1 2	
1.626E-01	3.550E+01	34.50	H031	1 2 2 2 2	
1.530E-01	3.340E+01	36.0	H032	1 1 2 1 2	
1.205E-01	2.630E+01	42.6	H032	1 1 2 1 2	
1.063E-01	2.320E+01	46.0	H032	1 1 2 1 2	
1.035E-01	2.260E+01	46.70	H031	1 2 2 2 2	
8.932E-02	1.950E+01	50.4	H032	1 1 2 1 2	
7.466E-02	1.630E+01	56.00	H031	1 2 2 2 2	
5.176E-02	1.130E+01	76.50	H031	1 2 2 2 2	
4.306E-02	9.400E+00	99.00	H031	1 2 2 2 2	

2980. $C_{12}H_{27}O_2P$
Butyl dibutyl phosphinate
Butoxydibutylphosphine oxide
Dibutylbutoxyphosphine oxide
Butyl dibutylphosphinate
RN: 2950-47-2 **MP (°C):**
MW: 234.32 **BP (°C):**

Solubility (Moles/L)	Solubility (Grams/L)	Temp (°C)	Ref (#)	Evaluation (T P E A A)	Comments
1.920E-02	4.500E+00	25	B070	1 2 0 1 1	

2981. $C_{12}H_{27}O_3P$
Diethyl octyl phosphonate
Diethyl octanephosphonate
RN: 1068-07-1 **MP (°C):**
MW: 250.32 **BP (°C):**

Solubility (Moles/L)	Solubility (Grams/L)	Temp (°C)	Ref (#)	Evaluation (T P E A A)	Comments
<7.99E-04	<2.00E-01	25	B070	1 2 0 1 0	

2982. $C_{12}H_{27}O_3P$
Dibutyl butyl phosphonate
Dibutoxybutylphosphine oxide
Dibutyl butanephosphonate
Dibutyl butylphosphonate
TC 44
RN: 78-46-6 **MP (°C):**
MW: 250.32 **BP (°C):**

Solubility (Moles/L)	Solubility (Grams/L)	Temp (°C)	Ref (#)	Evaluation (T P E A A)	Comments
1.997E-03	5.000E-01	25	B070	1 2 0 1 0	

2983. $C_{12}H_{27}O_4P$
Tributyl phosphate
Tri-*n*-butyl phosphate
RN: 126-73-8 **MP (°C):**
MW: 266.32 **BP (°C):** 289.0

Solubility (Moles/L)	Solubility (Grams/L)	Temp (°C)	Ref (#)	Evaluation (T P E A A)	Comments
4.036E-03	1.075E+00	3.4	H027	2 1 2 2 2	
3.800E-03	1.012E+00	4.0	H027	2 1 2 2 2	
3.593E-03	9.570E-01	5.0	H027	2 1 2 2 2	
2.403E-03	6.400E-01	13.0	H027	2 1 2 2 2	
1.500E-03	3.995E-01	25	B070	1 2 0 1 2	
1.464E-03	3.900E-01	25	B070	1 2 0 1 1	

(continued)

2983. C$_{12}$H$_{27}$O$_4$P (continued)

Solubility (Moles/L)	Solubility (Grams/L)	Temp (°C)	Ref (#)	Evaluation (T P E A A)	Comments
2.253E-02	6.000E+00	25	F300	1 0 0 0 0	
1.585E-03	4.220E-01	25.0	H027	2 1 2 2 2	
1.570E-03	4.180E-01	25.0	H032	2 2 2 1 1	EFG
1.070E-03	2.850E-01	50.0	H027	2 1 2 2 2	
1.239E-03	3.299E-01	ns	F014	0 0 0 0 1	

2984. C$_{12}$H$_{28}$Ge
Tetrapropylgermanium
Tetra-*n*-propylgermane
RN:　994-65-0　　**MP** (°C):
MW:　244.96　　**BP** (°C):

Solubility (Moles/L)	Solubility (Grams/L)	Temp (°C)	Ref (#)	Evaluation (T P E A A)	Comments
3.320E-08	8.133E-06	25	D346	1 1 2 2 2	

2985. C$_{12}$Br$_{10}$O
Decabromodiphenyl ether
DBDPO
Decabromodiphenyl oxide
RN:　1163-19-5　　**MP** (°C):　298.0
MW:　959.22　　**BP** (°C):

Solubility (Moles/L)	Solubility (Grams/L)	Temp (°C)	Ref (#)	Evaluation (T P E A A)	Comments
2.606E-08	2.500E-05	25	N326	1 0 0 0 1	average

2986. C$_{12}$Cl$_8$O$_2$
Octachlorodibenzo-*p*-dioxin
OCDD
1,2,3,4,6,7,8,9-Octachlorodibenzodioxin
O8CDD
Octachlorodibenzo[b,e][1,4]dioxin
RN:　3268-87-9　　**MP** (°C):　330
MW:　459.76　　**BP** (°C):

Solubility (Moles/L)	Solubility (Grams/L)	Temp (°C)	Ref (#)	Evaluation (T P E A A)	Comments
8.700E-13	4.000E-10	20	F303	1 2 1 2 0	
8.700E-13	4.000E-10	20	W319	1 2 1 2 1	
1.610E-13	7.400E-11	25	S352	2 2 0 2 1	
1.610E-13	7.402E-11	25.0	D330	2 2 1 2 2	
4.350E-12	2.000E-09	40	F303	1 2 1 2 1	
4.350E-12	2.000E-09	40	W319	1 2 1 2 1	
6.750E-13	3.103E-10	40.0	D330	2 2 1 2 2	
3.960E-12	1.821E-09	60.0	D330	2 2 1 2 2	
1.710E-12	7.862E-10	80.0	D330	2 2 1 2 2	
8.374E-13	3.850E-10	ns	W332	0 1 0 2 2	

2987. C₁₂Cl₁₀

Decachlorobiphenyl

Decachlorbiphenyl

2,2',3,3',4,4',5,5',6,6'-Decachlorobiphenyl

RN:	2051-24-3	**MP** (°C):	305	
MW:	498.66	**BP** (°C):		

Solubility (Moles/L)	Solubility (Grams/L)	Temp (°C)	Ref (#)	Evaluation (T P E A A)	Comments
4.211E-11	2.100E-08	22	O311	2 2 1 2 1	
1.300E-12	6.483E-10	25	D331	2 1 2 2 2	
1.303E-11	6.500E-09	25	D335	1 0 0 0 1	
1.490E-11	7.430E-09	25	M342	1 0 1 1 2	
3.209E-11	1.600E-08	25	W025	1 0 2 2 1	
1.300E-12	6.483E-10	25.0	M324	1 2 1 1 2	
1.680E-11	8.378E-09	60	D331	2 1 2 2 2	
1.680E-11	8.378E-09	60.0	M324	1 2 1 1 2	
3.530E-11	1.760E-08	70	D331	2 1 2 2 2	
3.530E-11	1.760E-08	70.0	M324	1 2 1 1 2	
9.930E-11	4.952E-08	80	D331	2 1 2 2 2	
9.930E-11	4.952E-08	80.0	M324	1 2 1 1 2	

2988. C₁₃H₆Cl₅NO₃

Oxyclozanide

3,5,6,3',5'-Pentachloro-2,2'-dihydroxybenzanilide

Zanilox

Diplin

ICI 46638

Zanil

RN:	2277-92-1	**MP** (°C):		
MW:	401.46	**BP** (°C):		

Solubility (Moles/L)	Solubility (Grams/L)	Temp (°C)	Ref (#)	Evaluation (T P E A A)	Comments
7.224E-05	2.900E-02	25	P036	0 0 0 0 0	average of 3, form III
2.665E-06	1.070E-03	25	P036	0 0 0 0 0	average of 3, form II
6.227E-07	2.500E-04	25	P036	0 0 0 0 0	average of 3, form I

2989. C₁₃H₆Cl₆O₂

Hexachlorophene

2,2'-Methylenebis[3,4,6-trichlorophenol]

Bilevon

AT-7

Dermadex

Exofene

RN:	70-30-4	**MP** (°C):	164.5	
MW:	406.91	**BP** (°C):		

Solubility (Moles/L)	Solubility (Grams/L)	Temp (°C)	Ref (#)	Evaluation (T P E A A)	Comments
6.142E-04	2.499E-01	22	M048	1 0 1 1 0	EFG
4.669E-05	1.900E-02	25	A008	1 0 0 0 0	EFG
3.441E-04	1.400E-01	25	A010	2 2 2 1 1	0.003N HCl
7.373E-07	3.000E-04	ns	V302	0 0 0 0 0	*sic*

2990. C₁₃H₇Br₂N₃O₆

Bromofenoxim
3,5-Dibromo-4-hydroxybenzaldehyde-2,4-dinitrophenyloxime
Faneron
Bromfenim

RN:	13181-17-4	**MP (°C):**	196.5
MW:	461.04	**BP (°C):**	

Solubility (Moles/L)	Solubility (Grams/L)	Temp (°C)	Ref (#)	Evaluation (T P E A A)	Comments
2.169E-07	1.000E-04	20	M161	1 0 0 0 0	
1.288E-06	5.939E-04	ns	R427	0 0 0 0 0	

2991. C₁₃H₇F₃N₂O₅

Fluorodifen
p-Nitrophenyl α,α,α-trifluoro-2-nitro-p-tolyl ether

RN:	15457-05-3	**MP (°C):**	90
MW:	328.21	**BP (°C):**	

Solubility (Moles/L)	Solubility (Grams/L)	Temp (°C)	Ref (#)	Evaluation (T P E A A)	Comments
6.094E-06	2.000E-03	20	E048	1 2 1 1 0	
6.094E-06	2.000E-03	20	M161	1 0 0 0 0	
<6.09E-06	<2.00E-03	ns	B200	0 0 0 0 0	
6.094E-06	2.000E-03	ns	M061	0 0 0 0 0	

2992. C₁₃H₈ClFO₂

4'-Chloro-5-fluoro-2-hydroxy benzophenone
SL 79182

RN:	62433-26-5	**MP (°C):**	
MW:	250.66	**BP (°C):**	

Solubility (Moles/L)	Solubility (Grams/L)	Temp (°C)	Ref (#)	Evaluation (T P E A A)	Comments
3.590E-05	8.999E-03	37	F309	1 0 2 2 2	

2993. C₁₃H₈ClNO

CP 31675
2-Chloro-N-(2-methyl-6-t-butylphenyl)acetamide

RN:	3785-20-4	**MP (°C):**	115
MW:	229.67	**BP (°C):**	

Solubility (Moles/L)	Solubility (Grams/L)	Temp (°C)	Ref (#)	Evaluation (T P E A A)	Comments
1.306E-03	3.000E-01	ns	M061	0 0 0 0 2	

2994. C₁₃H₈ClN₃O

RJ-64

3,4-Pyridyl-(5)-2-chlorophenyl-1,2,4-oxadiazole

RN: 27199-40-2 **MP** (°C):

MW: 257.68 **BP** (°C):

Solubility (Moles/L)	Solubility (Grams/L)	Temp (°C)	Ref (#)	Evaluation (T P E A A)	Comments
5.045E-03	1.300E+00	37	C054	2 2 2 1 2	0.1N HCl

2995. C₁₃H₈Cl₂N₂O₄

Niclosamide

2′,5-Dichloro-4′-nitrosalicylanilide

2-Chloro-4-nitrophenylamide-6-chlorosalicylic acid

Cestocid

Devermine

Bayluscid

RN: 50-65-7 **MP** (°C): 230

MW: 327.13 **BP** (°C):

Solubility (Moles/L)	Solubility (Grams/L)	Temp (°C)	Ref (#)	Evaluation (T P E A A)	Comments
4.072E-05	1.332E-02	25	T426	0 0 0 0 0	
1.987E-05	6.500E-03	rt	M161	0 0 0 0 0	

2996. C₁₃H₈F₂O₃

Diflunisal

5-(2,4-Difluorophenyl) salicylic acid

Dolobid

RN: 22494-42-4 **MP** (°C):

MW: 250.20 **BP** (°C):

Solubility (Moles/L)	Solubility (Grams/L)	Temp (°C)	Ref (#)	Evaluation (T P E A A)	Comments
2.472E-05	6.186E-03	24.99	K447	0 0 0 0 0	pH 2.0
1.199E-05	3.000E-03	37	Y421	0 0 0 0 0	

2997. C₁₃H₈N₂O₂

Phenazine-1-carboxylic acid

PCA

RN: **MP** (°C):

MW: 224.22 **BP** (°C):

Solubility (Moles/L)	Solubility (Grams/L)	Temp (°C)	Ref (#)	Evaluation (T P E A A)	Comments
2.300E-04	5.157E-02	5.0	Y409	0 0 0 0 0	
2.300E-04	5.157E-02	10.0	Y409	0 0 0 0 0	
2.400E-04	5.381E-02	15.0	Y409	0 0 0 0 0	
2.500E-04	5.606E-02	20.0	Y409	0 0 0 0 0	

(continued)

2997. C$_{13}$H$_8$N$_2$O$_2$ (continued)

Solubility (Moles/L)	Solubility (Grams/L)	Temp (°C)	Ref (#)	Evaluation (T P E A A)	Comments
2.700E-04	6.054E-02	25.0	Y409	0 0 0 0 0	
2.900E-04	6.502E-02	30.0	Y409	0 0 0 0 0	
3.200E-04	7.175E-02	35.0	Y409	0 0 0 0 0	
3.500E-04	7.848E-02	40.0	Y409	0 0 0 0 0	
3.900E-04	8.745E-02	45.0	Y409	0 0 0 0 0	
4.400E-04	9.866E-02	50.0	Y409	0 0 0 0 0	
5.100E-04	1.144E-01	55.0	Y409	0 0 0 0 0	

2998. C$_{13}$H$_8$N$_2$O$_2$S

m-Pyridine carboxyphenylisothiocyanate
Picolinic acid, m-isothiocyanatophenyl ester
RN: 5174-37-8   **MP** (°C):
MW: 256.28   **BP** (°C):

Solubility (Moles/L)	Solubility (Grams/L)	Temp (°C)	Ref (#)	Evaluation (T P E A A)	Comments
5.000E-05	1.281E-02	25	K032	2 2 0 1 1	

2999. C$_{13}$H$_9$ClN$_2$O$_4$

4'-Chloro-2-hydroxy-3-nitrobenzanilide
Salicylanilide, 4'-chloro-5-nitro-
Benzamide, N-(4-chlorophenyl)-2-hydroxy-5nitro-
RN: 6490-98-8   **MP** (°C):   253–254
MW: 292.68   **BP** (°C):

Solubility (Moles/L)	Solubility (Grams/L)	Temp (°C)	Ref (#)	Evaluation (T P E A A)	Comments
7.551E-06	2.210E-03	25	D400	2 0 0 1 2	

3000. C$_{13}$H$_9$ClN$_2$O$_4$

4'-Chloro-2-hydroxy-3-nitrobenzanilide
Benzamide, N-(4-chlorophenyl)-2-hydroxy-3-nitro-
Salicylanilide, 4'-chloro-3-nitro-
NSC 22899
4'-Chloro-3-nitrosalicylanilide
RN: 6490-99-9   **MP** (°C):
MW: 292.68   **BP** (°C):

Solubility (Moles/L)	Solubility (Grams/L)	Temp (°C)	Ref (#)	Evaluation (T P E A A)	Comments
2.851E-05	8.344E-03	25	D400	2 0 0 1 2	

3001. C$_{13}$H$_9$Cl$_2$NO$_4$

2,4-Dichlorophenyl 3-methoxy-4-nitrophenyl ether
Chlomethoxyfen
Chlomethoxynil

RN:	32861-85-1	MP (°C):	113.5
MW:	314.13	BP (°C):	

Solubility (Moles/L)	Solubility (Grams/L)	Temp (°C)	Ref (#)	Evaluation (T P E A A)	Comments
9.550E-07	3.000E-04	15	M161	1 0 0 0 0	

3002. C$_{13}$H$_9$F$_3$N$_2$O$_2$

Niflumic acid
2-[3-(Trifluoromethyl)anilino]nicotinic acid
Actol
Flogovital
Donalgin
Landruma

RN:	4394-00-7	MP (°C):	204
MW:	282.22	BP (°C):	378.0

Solubility (Moles/L)	Solubility (Grams/L)	Temp (°C)	Ref (#)	Evaluation (T P E A A)	Comments
2.733E-04	7.714E-02	10	B429	1 0 1 2 2	
2.805E-04	7.917E-02	15	B429	1 0 1 2 2	
2.916E-04	8.231E-02	20	B429	1 0 1 2 2	
3.028E-04	8.544E-02	25	B429	1 0 1 2 2	
3.128E-04	8.827E-02	30	B429	1 0 1 2 2	
3.261E-04	9.203E-02	35	B429	1 0 1 2 2	
6.732E-05	1.900E-02	rt	H302	0 0 2 1 1	intrinsic
1.400E-04	3.950E-02	rt	R431	0 0 0 0 0	Average

3003. C$_{13}$H$_9$N

Phenanthridine
Phenanthridin
9-Azaphenanthrene
3,4-Benzoisoquinoline
5-Azaphenanthrene

RN:	229-87-8	MP (°C):	106.5
MW:	179.22	BP (°C):	349

Solubility (Moles/L)	Solubility (Grams/L)	Temp (°C)	Ref (#)	Evaluation (T P E A A)	Comments
1.674E-03	3.000E-01	20	F300	1 0 0 0 1	

3004. C₁₃H₉N

Acridine

2,3,5,6-Dibenzopyridine

Acridin

RN:	260-94-6	MP (°C):	107
MW:	179.22	BP (°C):	346

Solubility (Moles/L)	Solubility (Grams/L)	Temp (°C)	Ref (#)	Evaluation (T P E A A)	Comments
3.200E-04	5.735E-02	24	A029	2 0 0 0 1	0.01N KOH
2.142E-04	3.840E-02	24	H106	1 0 2 2 2	
2.143E-04	3.840E-02	24	M303	1 0 1 1 2	
3.348E-04	6.000E-02	30	K090	1 2 2 2 0	EFG
3.348E-04	6.000E-02	30	K090	1 2 2 2 0	

3005. C₁₃H₉NO

2-Hydroxyacridine

o-Hydroxyacridine

RN:	22817-17-0	MP (°C):	
MW:	195.22	BP (°C):	

Solubility (Moles/L)	Solubility (Grams/L)	Temp (°C)	Ref (#)	Evaluation (T P E A A)	Comments
2.000E-05	3.904E-03	20	A029	1 0 0 0 0	

3006. C₁₃H₉NS

p-Biphenyl isothiocyanate

4-Biphenyl isothiocyanate

RN:	25687-48-3	MP (°C):	
MW:	211.29	BP (°C):	

Solubility (Moles/L)	Solubility (Grams/L)	Temp (°C)	Ref (#)	Evaluation (T P E A A)	Comments
1.400E-05	2.958E-03	25	D019	1 1 1 1 1	

3007. C₁₃H₉NS

m-Biphenyl isothiocyanate

3-Biphenyl isothiocyanate

RN:	1510-25-4	MP (°C):	
MW:	211.29	BP (°C):	

Solubility (Moles/L)	Solubility (Grams/L)	Temp (°C)	Ref (#)	Evaluation (T P E A A)	Comments
3.000E-05	6.339E-03	25	K032	2 2 0 1 1	

3008. $C_{13}H_{10}$

Fluorene
o-Biphenylmethane
2,3-Benzindene
o-Biphenylenemethane
Diphenylenemethane
2,2'-Methylenebiphenyl

RN:	86-73-7	**MP** (°C):	116	
MW:	166.22	**BP** (°C):	295	

Solubility (Moles/L)	Solubility (Grams/L)	Temp (°C)	Ref (#)	Evaluation (T P E A A)	Comments
4.320E-06	7.181E-04	6.60	M082	1 1 1 2 2	
4.320E-06	7.181E-04	6.60	M151	2 1 2 2 2	
4.326E-06	7.190E-04	6.64	M183	1 2 1 1 2	
5.820E-06	9.674E-04	13.20	M082	1 1 1 2 2	
5.820E-06	9.674E-04	13.20	M151	2 1 2 2 2	
5.822E-06	9.678E-04	13.24	M183	1 2 1 1 2	
7.240E-06	1.203E-03	18.00	M082	1 1 1 2 2	
7.240E-06	1.203E-03	18.00	M151	2 1 2 2 2	
7.244E-06	1.204E-03	18.04	M183	1 2 1 1 2	
9.012E-06	1.498E-03	20	V416	0 0 0 0 0	
9.720E-06	1.616E-03	24.00	M082	1 1 1 2 2	
9.720E-06	1.616E-03	24.00	M151	2 1 2 2 2	
9.728E-06	1.617E-03	24.04	M183	1 2 1 1 2	
1.137E-05	1.890E-03	24.60	W003	2 2 2 2 2	average of 3
1.179E-05	1.960E-03	25	B319	2 0 1 2 2	
2.790E-05	4.638E-03	25	L301	1 1 2 2 2	
1.143E-05	1.900E-03	25	L332	1 1 1 1 1	
1.191E-05	1.980E-03	25	M064	1 1 2 2 2	
1.014E-05	1.685E-03	25	M071	2 2 2 2 2	
1.190E-05	1.978E-03	25	M342	1 0 1 1 2	
1.010E-05	1.679E-03	25	W300	2 2 2 2 2	
1.014E-05	1.685E-03	25.00	M151	2 1 1 2 2	
1.110E-05	1.845E-03	27.00	M082	1 1 1 2 2	
1.110E-05	1.845E-03	27.00	M151	2 1 2 2 2	
1.111E-05	1.847E-03	27.04	M183	1 2 1 1 2	
1.420E-05	2.360E-03	29.90	W003	2 2 2 2 2	average of 3
1.317E-05	2.190E-03	30.30	W003	2 2 2 2 2	average of 3
1.350E-05	2.244E-03	31.10	M082	1 1 1 2 2	
1.350E-05	2.244E-03	31.10	M151	2 1 2 2 2	
1.353E-05	2.250E-03	31.14	M183	1 2 1 1 2	
2.244E-05	3.730E-03	38.40	W003	2 2 2 2 2	average of 2
2.223E-05	3.695E-03	40	V416	0 0 0 0 0	
2.322E-05	3.860E-03	40.10	W003	2 2 2 2 2	average of 3
3.387E-05	5.630E-03	47.50	W003	2 2 2 2 2	average of 3
3.862E-05	6.420E-03	50.10	W003	2 2 2 2 2	average of 3
3.772E-05	6.270E-03	50.20	W003	2 2 2 2 2	
5.071E-05	8.430E-03	54.70	W003	2 2 2 2 2	average of 3
6.317E-05	1.050E-02	59.20	W003	2 2 2 2 2	
5.298E-05	8.806E-03	60	V416	0 0 0 0 0	
6.678E-05	1.110E-02	60.50	W003	2 2 2 2 2	average of 3

(continued)

3008. C$_{13}$H$_{10}$ (continued)

Solubility (Moles/L)	Solubility (Grams/L)	Temp (°C)	Ref (#)	Evaluation (T P E A A)	Comments
8.543E-05	1.420E-02	65.10	W003	2 2 2 2 2	average of 3
1.119E-04	1.860E-02	70.70	W003	2 2 2 2 2	average of 3
1.131E-04	1.880E-02	71.90	W003	2 2 2 2 2	
1.293E-04	2.150E-02	73.40	W003	2 2 2 2 2	
1.191E-05	1.980E-03	ns	M344	0 0 0 0 2	

3009. C$_{13}$H$_{10}$BrCl$_2$O$_2$PS

Leptophos

Phenylphosphonothioic acid *O*-(4-bromo-2,5-dichlorophenyl) *O*-methyl ester

Phosvel

NK 711

Velsicol 506

Oleophosvel

RN: 21609-90-5   **MP** (°C):   60
MW: 412.08   **BP** (°C):

Solubility (Moles/L)	Solubility (Grams/L)	Temp (°C)	Ref (#)	Evaluation (T P E A A)	Comments
7.280E-09	3.000E-06	10	B324	0 0 0 0 0	
8.707E-09	3.588E-06	10	B324	0 0 0 0 0	
1.699E-07	7.000E-05	20	B169	2 2 1 1 0	
6.095E-08	2.512E-05	20	B300	2 2 1 1 2	
6.095E-08	2.512E-05	20	B324	0 0 0 0 0	
5.096E-08	2.100E-05	20	B324	0 0 0 0 0	
1.141E-08	4.700E-06	20	C053	0 0 0 0 0	
1.213E-08	5.000E-06	22	K137	1 1 2 1 0	
7.280E-08	3.000E-05	24	C105	2 1 2 2 2	
5.824E-06	2.400E-03	25	M161	1 0 0 0 1	*sic*
1.306E-07	5.382E-05	30	B324	0 0 0 0 0	
1.092E-07	4.500E-05	30	B324	0 0 0 0 0	
2.184E-08	9.000E-06	ns	F040	1 2 2 2 0	
1.141E-08	4.700E-06	ns	F071	0 1 2 1 1	
1.699E-07	7.000E-05	ns	M110	0 0 0 0 0	EFG

3010. C$_{13}$H$_{10}$BrCl$_2$O$_3$P

Leptophos oxon

O-(4-Bromo-2,5-dichlorophenyl) *O*-methyl phenylphosphonate

Phosvel oxon

RN: 25006-32-0   **MP** (°C):
MW: 396.01   **BP** (°C):

Solubility (Moles/L)	Solubility (Grams/L)	Temp (°C)	Ref (#)	Evaluation (T P E A A)	Comments
8.586E-06	3.400E-03	20.50	B169	2 2 1 1 2	

3011. $C_{13}H_{10}ClNO_2$

4'-Chloro salicylanilide
N-(p-Chlorophenyl)-o-hydroxybenzamide
N-(p-Chlorophenyl)salicylamide

RN: 3679-63-8	**MP** (°C):	
MW: 247.68	**BP** (°C):	

Solubility (Moles/L)	Solubility (Grams/L)	Temp (°C)	Ref (#)	Evaluation (T P E A A)	Comments
4.885E-08	1.210E-05	ns	N336	0 0 0 0 0	intrinsic

3012. $C_{13}H_{10}Cl_2O$

2,4,-Dichloro-6-benzyl-phenol
o-Cresol, 4,6-dichloro-α-phenyl-
2-Benzyl-4,6-dichlorophenol

RN: 19578-81-5	**MP** (°C):	
MW: 253.13	**BP** (°C):	

Solubility (Moles/L)	Solubility (Grams/L)	Temp (°C)	Ref (#)	Evaluation (T P E A A)	Comments
2.300E-05	5.822E-03	25	B316	0 0 0 0 0	

3013. $C_{13}H_{10}Cl_2O_2$

Dichlorophen
2,2'-Dihydroxy-5,5'-dichlorodiphenylmethane
G-4

RN: 97-23-4	**MP** (°C): 177–178	
MW: 269.13	**BP** (°C):	

Solubility (Moles/L)	Solubility (Grams/L)	Temp (°C)	Ref (#)	Evaluation (T P E A A)	Comments
1.115E-04	3.000E-02	25	M061	1 0 0 0 0	
1.115E-04	3.000E-02	25	M161	1 0 0 0 1	
1.122E-04	3.020E-02	ns	R427	0 0 0 0 0	

3014. $C_{13}H_{10}INO$

Benodanil
2-Iodo-N-phenylbenzamide
Iodobenzanilide
Calirus

RN: 15310-01-7	**MP** (°C): 137	
MW: 323.14	**BP** (°C):	

Solubility (Moles/L)	Solubility (Grams/L)	Temp (°C)	Ref (#)	Evaluation (T P E A A)	Comments
6.189E-05	2.000E-02	20	M161	1 0 0 0 1	

3015. C$_{13}$H$_{10}$N$_2$
9-Aminoacridine
10-Amino-5-azaanthracene
Monacrin
Izoacridina
Aminacrine
9AA

RN:	90-45-9	**MP** (°C):	241
MW:	194.24	**BP** (°C):	

Solubility (Moles/L)	Solubility (Grams/L)	Temp (°C)	Ref (#)	Evaluation (T P E A A)	Comments
6.000E-05	1.165E-02	24	A029	2 0 0 1 0	0.01N KOH

3016. C$_{13}$H$_{10}$N$_2$
4-Aminoacridine
4-Acridinamine

RN:	578-07-4	**MP** (°C):	108.5
MW:	194.24	**BP** (°C):	346

Solubility (Moles/L)	Solubility (Grams/L)	Temp (°C)	Ref (#)	Evaluation (T P E A A)	Comments
7.000E-05	1.360E-02	24	A029	2 0 0 1 0	0.01N KOH

3017. C$_{13}$H$_{10}$N$_2$
3-Aminoacridine
3-Acridinamine

RN:	581-29-3	**MP** (°C):	108.5
MW:	194.24	**BP** (°C):	346

Solubility (Moles/L)	Solubility (Grams/L)	Temp (°C)	Ref (#)	Evaluation (T P E A A)	Comments
1.500E-04	2.914E-02	24	A029	2 0 0 1 1	0.01N KOH

3018. C$_{13}$H$_{10}$N$_2$
2-Aminoacridine
2-Acridinamine

RN:	581-28-2	**MP** (°C):	108.5
MW:	194.24	**BP** (°C):	346

Solubility (Moles/L)	Solubility (Grams/L)	Temp (°C)	Ref (#)	Evaluation (T P E A A)	Comments
5.000E-05	9.712E-03	24	A029	2 0 0 1 0	0.01N KOH

3019. $C_{13}H_{10}N_2$
1-Aminoacridine
1-Acridinamine
RN: 578-06-3   **MP** (°C): 183
MW: 194.24   **BP** (°C): 346

Solubility (Moles/L)	Solubility (Grams/L)	Temp (°C)	Ref (#)	Evaluation (T P E A A)	Comments
6.000E-05	1.165E-02	24	A029	2 0 0 0 1	intrinsic

3020. $C_{13}H_{10}N_4O_3$
1-Benzoyloxymethyl allopurinol
4H-Pyrazolo[3,4-d]pyrimidin-4-one, 1-[(benzoyloxy)methyl]-1,5-dihydro-
RN: 98846-65-2   **MP** (°C): 217–219
MW: 270.25   **BP** (°C):

Solubility (Moles/L)	Solubility (Grams/L)	Temp (°C)	Ref (#)	Evaluation (T P E A A)	Comments
8.881E-05	2.400E-02	22	B322	0 0 0 0 0	
8.913E-05	2.409E-02	ns	R427	0 0 0 0 0	

3021. $C_{13}H_{10}O$
Benzophenone
α-Oxodiphenylmethane
Diphenylmethanone
Benzoylbenzene
α-Oxoditane
Oxoditane
RN: 119-61-9   **MP** (°C): 48.5
MW: 182.22   **BP** (°C): 305.4

Solubility (Moles/L)	Solubility (Grams/L)	Temp (°C)	Ref (#)	Evaluation (T P E A A)	Comments
4.121E-04	7.510E-02	20	H301	0 0 0 0 0	
7.500E-04	1.367E-01	25	F063	1 1 0 0 1	
3.292E-04	6.000E-02	ns	F014	0 0 0 0 0	

3022. $C_{13}H_{10}O_3$
2,4-Dihydroxybenzophenone
RN: 131-56-6   **MP** (°C):
MW: 214.22   **BP** (°C):

Solubility (Moles/L)	Solubility (Grams/L)	Temp (°C)	Ref (#)	Evaluation (T P E A A)	Comments
3.291E-02	7.050E+00	19.99	L452	0 0 0 0 0	
4.255E-02	9.116E+00	24.99	L452	0 0 0 0 0	
4.805E-02	1.029E+01	29.99	L452	0 0 0 0 0	
5.672E-02	1.215E+01	34.99	L452	0 0 0 0 0	
7.396E-02	1.584E+01	39.99	L452	0 0 0 0 0	

(continued)

3022. $C_{13}H_{10}O_3$ (continued)

Solubility (Moles/L)	Solubility (Grams/L)	Temp (°C)	Ref (#)	Evaluation (T P E A A)	Comments
8.659E-02	1.855E+01	44.99	L452	0 0 0 0 0	
1.174E-01	2.515E+01	49.99	L452	0 0 0 0 0	
1.500E-01	3.213E+01	54.99	L452	0 0 0 0 0	
1.925E-01	4.123E+01	59.99	L452	0 0 0 0 0	
2.559E-01	5.482E+01	64.99	L452	0 0 0 0 0	
3.498E-01	7.493E+01	69.99	L452	0 0 0 0 0	

3023. $C_{13}H_{10}O_3$
Phenyl salicylate
Salol
2-Hydroxybenzoic acid phenyl ester

RN: 118-55-8   **MP (°C):** 42.0
MW: 214.22   **BP (°C):** 173.0

Solubility (Moles/L)	Solubility (Grams/L)	Temp (°C)	Ref (#)	Evaluation (T P E A A)	Comments
7.002E-04	1.500E-01	25	F300	1 0 0 0 1	
7.469E-05	1.600E-02	ns	B404	0 2 1 1 0	
1.866E-03	3.998E-01	rt	D021	0 0 1 1 0	

3024. $C_{13}H_{10}O_4$
2,3,4-Trihydroxybenzophenone
2,3,4-Trihydroxy-benzophenon

RN: 1143-72-2   **MP (°C):**
MW: 230.22   **BP (°C):**

Solubility (Moles/L)	Solubility (Grams/L)	Temp (°C)	Ref (#)	Evaluation (T P E A A)	Comments
4.811E-02	1.108E+01	19.99	L452	0 0 0 0 0	
5.743E-02	1.322E+01	24.99	L452	0 0 0 0 0	
8.057E-02	1.855E+01	29.99	L452	0 0 0 0 0	
1.051E-01	2.420E+01	34.99	L452	0 0 0 0 0	
1.392E-01	3.204E+01	39.99	L452	0 0 0 0 0	
1.831E-01	4.215E+01	44.99	L452	0 0 0 0 0	
2.574E-01	5.927E+01	49.99	L452	0 0 0 0 0	
3.440E-01	7.919E+01	54.99	L452	0 0 0 0 0	
4.723E-01	1.087E+02	59.99	L452	0 0 0 0 0	
6.152E-01	1.416E+02	64.99	L452	0 0 0 0 0	
7.804E-01	1.797E+02	69.99	L452	0 0 0 0 0	

3025. $C_{13}H_{10}O_4$
2,4,6-Trihydroxybenzophenone
2,4,6-Trihydroxy-benzophenon
RN: 3555-86-0   **MP** (°C):
MW: 230.22   **BP** (°C):

Solubility (Moles/L)	Solubility (Grams/L)	Temp (°C)	Ref (#)	Evaluation (T P E A A)	Comments
1.347E-02	3.100E+00	22	F300	1 0 0 0 1	

3026. $C_{13}H_{10}O_5$
2,2′,4,4′-Tetrahydroxybenzophenone
RN: **MP** (°C):
MW: 246.22   **BP** (°C):

Solubility (Moles/L)	Solubility (Grams/L)	Temp (°C)	Ref (#)	Evaluation (T P E A A)	Comments
2.863E-02	7.050E+00	19.99	L452	0 0 0 0 0	
3.583E-02	8.821E+00	24.99	L452	0 0 0 0 0	
4.538E-02	1.117E+01	29.99	L452	0 0 0 0 0	
6.199E-02	1.526E+01	34.99	L452	0 0 0 0 0	
8.431E-02	2.076E+01	39.99	L452	0 0 0 0 0	
1.079E-01	2.657E+01	44.99	L452	0 0 0 0 0	
1.487E-01	3.661E+01	49.99	L452	0 0 0 0 0	
2.190E-01	5.393E+01	54.99	L452	0 0 0 0 0	
3.285E-01	8.088E+01	59.99	L452	0 0 0 0 0	
4.448E-01	1.095E+02	64.99	L452	0 0 0 0 0	
5.572E-01	1.372E+02	69.99	L452	0 0 0 0 0	

3027. $C_{13}H_{10}O_5$
2,3,4,4′-Tetrahydroxybenzophenone
RN: 31127-54-5   **MP** (°C):
MW: 246.22   **BP** (°C):

Solubility (Moles/L)	Solubility (Grams/L)	Temp (°C)	Ref (#)	Evaluation (T P E A A)	Comments
4.578E-02	1.127E+01	19.99	L452	0 0 0 0 0	
6.120E-02	1.507E+01	24.99	L452	0 0 0 0 0	
8.820E-02	2.172E+01	29.99	L452	0 0 0 0 0	
1.202E-01	2.960E+01	34.99	L452	0 0 0 0 0	
1.712E-01	4.215E+01	39.99	L452	0 0 0 0 0	
2.299E-01	5.660E+01	44.99	L452	0 0 0 0 0	
3.216E-01	7.919E+01	49.99	L452	0 0 0 0 0	
4.768E-01	1.174E+02	54.99	L452	0 0 0 0 0	
6.166E-01	1.518E+02	59.99	L452	0 0 0 0 0	
8.432E-01	2.076E+02	64.99	L452	0 0 0 0 0	
1.084E+00	2.669E+02	69.99	L452	0 0 0 0 0	

3028. C₁₃H₁₀O₆
Maclurin
2,4,6,3′,4′-Penta-hydroxy-benzophenol
2,4,6,3′,4′-Pentahydroxybenzophenon

RN:	519-34-6	MP (°C):	222.5
MW:	262.22	BP (°C):	

Solubility (Moles/L)	Solubility (Grams/L)	Temp (°C)	Ref (#)	Evaluation (T P E A A)	Comments
1.907E-02	5.000E+00	14	F300	1 0 0 0 0	

3029. C₁₃H₁₁ClF₃N₃O
San 6706
4-Chloro-5-(dimethylamino)-2-(α,α,α-trifluoro-m-tolyl)-3(2H)-pyridazinone

RN:	23576-23-0	MP (°C):	151
MW:	317.70	BP (°C):	

Solubility (Moles/L)	Solubility (Grams/L)	Temp (°C)	Ref (#)	Evaluation (T P E A A)	Comments
3.305E-05	1.050E-02	23.50	B200	2 0 0 0 2	

3030. C₁₃H₁₁ClN₄O
6H-Dipyrido[3,2-b:2′,3′-e][1,4]diazepin-6-one, 2-chloro-11-ethyl-5,11-dihydro-

RN:	134698-40-1	MP (°C):	
MW:	274.71	BP (°C):	

Solubility (Moles/L)	Solubility (Grams/L)	Temp (°C)	Ref (#)	Evaluation (T P E A A)	Comments
4.365E-06	1.199E-03	ns	M381	0 1 1 1 2	pH 7.0

3031. C₁₃H₁₁ClO
Chlorophene
5-Chloro-2-hydroxydiphenylmethane
Benzylchlorophenol

RN:	120-32-1	MP (°C):	48.5
MW:	218.69	BP (°C):	

Solubility (Moles/L)	Solubility (Grams/L)	Temp (°C)	Ref (#)	Evaluation (T P E A A)	Comments
1.900E-02	4.155E+00	20	A008	1 0 0 0 0	EFG
1.100E-01	2.406E+01	ns	B047	0 0 0 0 0	EFG

3032. $C_{13}H_{11}N$
2-Aminofluorene
9H-Fluoren-2-amine
2-Fluorenamine

RN: 153-78-6 **MP** (°C): 129
MW: 181.24 **BP** (°C):

Solubility (Moles/L)	Solubility (Grams/L)	Temp (°C)	Ref (#)	Evaluation (T P E A A)	Comments
1.710E-04	3.100E-02	rt	N015	0 0 2 2 1	

3033. $C_{13}H_{11}NO_2$
Salicylanilide
2-Hydroxy-N-phenylbenzamide
2-Hydroxybenzanilide

RN: 87-17-2 **MP** (°C): 136
MW: 213.24 **BP** (°C):

Solubility (Moles/L)	Solubility (Grams/L)	Temp (°C)	Ref (#)	Evaluation (T P E A A)	Comments
2.579E-04	5.500E-02	23	M061	1 0 0 0 1	
2.579E-04	5.500E-02	25	M161	1 0 0 0 1	

3034. $C_{13}H_{11}NO_3$
Furo[3,4-b]quinolin-3(1H)-one, 9-hydroxy-1,7-dimethyl-

RN: 74103-12-1 **MP** (°C):
MW: 229.24 **BP** (°C):

Solubility (Moles/L)	Solubility (Grams/L)	Temp (°C)	Ref (#)	Evaluation (T P E A A)	Comments
9.597E-08	2.200E-05	25	P089	0 0 0 0 0	
1.527E-07	3.500E-05	37	P089	0 0 0 0 0	
2.116E-07	4.850E-05	51	P089	0 0 0 0 0	

3035. $C_{13}H_{11}NO_3$
Furo[3,4-b]quinolin-3(1H)-one, 9-hydroxy-1,6-dimethyl-

RN: **MP** (°C):
MW: 229.24 **BP** (°C):

Solubility (Moles/L)	Solubility (Grams/L)	Temp (°C)	Ref (#)	Evaluation (T P E A A)	Comments
2.530E-07	5.800E-05	25	P089	0 0 0 0 0	
3.054E-07	7.000E-05	37	P089	0 0 0 0 0	
3.817E-07	8.750E-05	51	P089	0 0 0 0 0	

3036. C$_{13}$H$_{11}$NO$_5$
Oxolinic acid
5-Ethyl-5,8-dihydro-8-oxo-1,3-dioxolo(4,5-g)quinoline-7-carboxylic acid
Dioxacin
Gramurin
Starner
S-0208
RN: 14698-29-4   **MP** (°C):
MW: 261.24   **BP** (°C):

Solubility (Moles/L)	Solubility (Grams/L)	Temp (°C)	Ref (#)	Evaluation (T P E A A)	Comments
1.230E-05	3.214E-03	ns	R427	0 0 0 0 0	

3037. C$_{13}$H$_{11}$N$_3$O$_2$
Benquinox
Cerenox
Seredon
Benzoylhydrazone of quinone oxime
RN: 495-73-8   **MP** (°C):
MW: 241.25   **BP** (°C):

Solubility (Moles/L)	Solubility (Grams/L)	Temp (°C)	Ref (#)	Evaluation (T P E A A)	Comments
2.073E-05	5.000E-03	ns	M061	0 0 0 0 0	

3038. C$_{13}$H$_{11}$N$_3$O$_2$S$_2$
2-Sulfanilamidobenzothiazole
RN: **MP** (°C):
MW: 305.38   **BP** (°C):

Solubility (Moles/L)	Solubility (Grams/L)	Temp (°C)	Ref (#)	Evaluation (T P E A A)	Comments
3.275E-06	1.000E-03	37	R045	1 2 1 1 1	

3039. C$_{13}$H$_{11}$N$_3$O$_4$S$_2$
Tenoxicam
Mobiflex
RN: 59804-37-4   **MP** (°C):
MW: 337.38   **BP** (°C):

Solubility (Moles/L)	Solubility (Grams/L)	Temp (°C)	Ref (#)	Evaluation (T P E A A)	Comments
1.835E-04	6.190E-02	32	C411	2 1 1 2 1	

3040. C13H11N7O4S

5-p-Nitrobenzenesulfonamidotetrazole

RN: **MP** (°C):
MW: 361.34   **BP** (°C):

Solubility (Moles/L)	Solubility (Grams/L)	Temp (°C)	Ref (#)	Evaluation (T P E A A)	Comments
2.214E-05	8.000E-03	37	R045	1 2 1 1 0	

3041. C13H11O3P

4-Carboxyethylphenylphenylphosphinic acid
CPPPA

RN: **MP** (°C):
MW: 246.20   **BP** (°C):

Solubility (Moles/L)	Solubility (Grams/L)	Temp (°C)	Ref (#)	Evaluation (T P E A A)	Comments
1.399E-02	3.443E+00	-239.0	W412	0 0 0 0 0	
1.242E-02	3.059E+00	26.7	W412	0 0 0 0 0	
1.676E-02	4.127E+00	45.08	W412	0 0 0 0 0	
1.931E-02	4.754E+00	54.4	W412	0 0 0 0 0	
2.609E-02	6.424E+00	64.15	W412	0 0 0 0 0	
3.477E-02	8.561E+00	75.71	W412	0 0 0 0 0	
4.371E-02	1.076E+01	84.38	W412	0 0 0 0 0	
3.780E+00	9.307E+02	94.52	W412	0 0 0 0 0	

3042. C13H12

Diphenylmethane
1,1'-Methylenebis-benzene
Phenylbenzyl
Benzylbenzene

RN: 101-81-5   **MP** (°C):   25.9
MW: 168.24   **BP** (°C):   264.5

Solubility (Moles/L)	Solubility (Grams/L)	Temp (°C)	Ref (#)	Evaluation (T P E A A)	Comments
1.783E-05	3.000E-03	24	H116	2 1 0 0 2	
8.381E-05	1.410E-02	25	A001	1 2 2 2 2	
8.381E-05	1.410E-02	25	A017	1 0 0 0 2	
8.710E-05	1.465E-02	25	D001	1 0 0 0 2	

3043. C$_{13}$H$_{12}$

4-Methylbiphenyl
4-Phenyltoluene

RN: 644-08-6   **MP** (°C): 49.5
MW: 168.24   **BP** (°C): 267.5

Solubility (Moles/L)	Solubility (Grams/L)	Temp (°C)	Ref (#)	Evaluation (T P E A A)	Comments
1.090E-05	1.834E-03	4.9	D330	2 2 1 2 2	
2.410E-05	4.055E-03	25.0	D330	2 2 1 2 2	
4.180E-05	7.032E-03	40.0	D330	2 2 1 2 2	

3044. C$_{13}$H$_{12}$F$_2$N$_6$O

Fluconazole
1H-1,2,4-Triazole-1-ethanol, α(2,4-difluorophenyl)-α-(1H-1,2,4-triazol-1-ylmethyl)
2,4-Difluoro-α,α1-bis(1H-1,2,4-triazol-1-ylmethyl)benzyl alcohol
Diflucan
Triflucan

RN: 86386-73-4   **MP** (°C): 138–140
MW: 306.28   **BP** (°C):

Solubility (Moles/L)	Solubility (Grams/L)	Temp (°C)	Ref (#)	Evaluation (T P E A A)	Comments
3.265E-03	1.000E+00	ns	K444	0 0 0 0 0	

3045. C$_{13}$H$_{12}$N$_2$O

Carbanilide
Diphenylurea
N,N'-Diphenylurea

RN: 102-07-8   **MP** (°C): 238.0
MW: 212.25   **BP** (°C):

Solubility (Moles/L)	Solubility (Grams/L)	Temp (°C)	Ref (#)	Evaluation (T P E A A)	Comments
7.079E-04	1.503E-01	ns	R427	0 0 0 0 0	
7.066E-04	1.500E-01	rt	D021	0 0 1 1 1	

3046. C$_{13}$H$_{12}$N$_2$O$_3$

Phenallymal
5-Allyl-5-phenylbarbituric acid
2,4,6(1H,3H,5H)-Pyrimidinetrione, 5-phenyl-5-(2-propenyl)
Barbituric acid, 5-allyl-5-phenyl
5-Allyl-5-phenylbarbiturate

RN: 115-43-5   **MP** (°C): 156.5
MW: 244.25   **BP** (°C):

Solubility (Moles/L)	Solubility (Grams/L)	Temp (°C)	Ref (#)	Evaluation (T P E A A)	Comments
4.499E-03	1.099E+00	20	J030	1 2 2 2 2	
4.272E-03	1.043E+00	25	P350	0 0 0 0 0	intrinsic
7.764E-03	1.896E+00	37	J030	1 2 2 2 2	

3047. C$_{13}$H$_{12}$N$_2$O$_5$S
Nimesulide
N-(4-Nitro-2-phenoxyphenyl)-methanesulfonamide
RN: 51803-78-2 **MP** (°C):
MW: 308.31 **BP** (°C):

Solubility (Moles/L)	Solubility (Grams/L)	Temp (°C)	Ref (#)	Evaluation (T P E A A)	Comments
4.541E-05	1.400E-02	25	S415	0 0 0 0 0	
7.395E-05	2.280E-02	37	P432	0 0 0 0 0	

3048. C$_{13}$H$_{12}$O
p-Benzylphenol
4-Benzylphenol
RN: 101-53-1 **MP** (°C): 81.5
MW: 184.24 **BP** (°C): 322

Solubility (Moles/L)	Solubility (Grams/L)	Temp (°C)	Ref (#)	Evaluation (T P E A A)	Comments
5.427E-04	9.999E-02	25	L021	1 0 0 0 0	

3049. C$_{13}$H$_{12}$O
o-Benzylphenol
2-Benzylphenol
RN: 28994-41-4 **MP** (°C): 53.5
MW: 184.24 **BP** (°C): 312

Solubility (Moles/L)	Solubility (Grams/L)	Temp (°C)	Ref (#)	Evaluation (T P E A A)	Comments
1.085E-03	2.000E-01	25	L021	1 0 0 0 0	

3050. C$_{13}$H$_{12}$O
Benzhydrol
Diphenylmethanol
RN: 91-01-0 **MP** (°C): 69
MW: 184.24 **BP** (°C): 298

Solubility (Moles/L)	Solubility (Grams/L)	Temp (°C)	Ref (#)	Evaluation (T P E A A)	Comments
2.714E-03	5.000E-01	20	F300	1 0 0 0 0	
2.800E-03	5.159E-01	25	D007	2 0 1 1 1	

3051. C₁₃H₁₂O₅

bis(4-Hydroxy-3-coumarin) acetic acid ethyl ester

RN: 548-00-5 **MP** (°C):
MW: 248.24 **BP** (°C):

Solubility (Moles/L)	Solubility (Grams/L)	Temp (°C)	Ref (#)	Evaluation (T P E A A)	Comments
2.188E-04	5.431E-02	ns	R427	0 0 0 0 0	

3052. C₁₃H₁₃Cl₂N₃O₃

Glycophen
1-Imidazolidinecarboxamide, 3-(3,5-dichlorophenyl)-*N*-(1-methylethyl)-2,4-dioxo-
Iprodial
LFA 2043
Iprodione

RN: 36734-19-7 **MP** (°C): 136
MW: 330.17 **BP** (°C):

Solubility (Moles/L)	Solubility (Grams/L)	Temp (°C)	Ref (#)	Evaluation (T P E A A)	Comments
3.937E-05	1.300E-02	20	M161	1 0 0 0 1	

3053. C₁₃H₁₃NO₂

α-(β-Naphthyl)-α-alanine
Alanine, 3-(1(4H)-naphthylidene)-

RN: 13913-40-1 **MP** (°C):
MW: 215.25 **BP** (°C):

Solubility (Moles/L)	Solubility (Grams/L)	Temp (°C)	Ref (#)	Evaluation (T P E A A)	Comments
2.260E-03	4.865E-01	25	M097	2 2 2 2 2	

3054. C₁₃H₁₃NO₅

2-Azetidinecarboxylic acid, 1-[(benzoyloxy)acetyl]-

RN: 115178-74-0 **MP** (°C): 149.5
MW: 263.25 **BP** (°C):

Solubility (Moles/L)	Solubility (Grams/L)	Temp (°C)	Ref (#)	Evaluation (T P E A A)	Comments
7.217E-03	1.900E+00	22	N317	1 1 2 1 2	

3055. $C_{13}H_{13}N_3O_3S$
*N*4-Acetyl sulfapyridine
Acetylsulfapyridine
Sulfapyridine acetylee
RN: 19077-98-6   **MP** (°C):
MW: 291.33   **BP** (°C):

Solubility (Moles/L)	Solubility (Grams/L)	Temp (°C)	Ref (#)	Evaluation (T P E A A)	Comments
1.098E-03	3.200E-01	37	D084	1 0 1 0 1	
7.207E-04	2.100E-01	37	F075	1 0 2 2 2	
1.119E-03	3.260E-01	37	M057	1 0 0 0 2	pH 5.5

3056. $C_{13}H_{13}N_3O_5S_2$
Succinylsulfathiazole
2-(*N*(4)-Succinylsulfanilamido)thiazole
p-2-Thiazolylsulfamoylsuccinanilic acid
Kaoxidin
Colistatin
Cremosuxidine
RN: 116-43-8   **MP** (°C):
MW: 355.39   **BP** (°C):

Solubility (Moles/L)	Solubility (Grams/L)	Temp (°C)	Ref (#)	Evaluation (T P E A A)	Comments
1.379E-03	4.900E-01	38	K006	1 0 0 0 1	

3057. $C_{13}H_{13}O_4P$
Diphenyl methyl phosphate
Methyl diphenyl phosphate
RN: 115-89-9   **MP** (°C):
MW: 264.22   **BP** (°C):

Solubility (Moles/L)	Solubility (Grams/L)	Temp (°C)	Ref (#)	Evaluation (T P E A A)	Comments
3.633E-06	9.600E-04	24	H116	2 1 0 0 2	*sic*
7.569E-03	2.000E+00	25	A044	1 0 0 0 0	*sic*

3058. $C_{13}H_{14}$
1,4,5-Trimethylnaphthalene
Naphthalene, 1,4,5-trimethyl-
RN: 2131-41-1   **MP** (°C):   58
MW: 170.26   **BP** (°C):

Solubility (Moles/L)	Solubility (Grams/L)	Temp (°C)	Ref (#)	Evaluation (T P E A A)	Comments
1.233E-05	2.100E-03	25	M064	1 1 2 2 1	
1.190E-05	2.026E-03	25	M342	1 0 1 1 2	
1.233E-05	2.100E-03	ns	M344	0 0 0 0 1	

3059. C$_{13}$H$_{14}$F$_3$N$_3$O$_4$

Ethalfluralin

N-Ethyl-N-(2-methyl-2-propenyl)-2,6-dinitro-4-(trifluoromethyl)benzenamine

Buvilan

Solanan

RN:	55283-68-6	**MP** (°C):	55.5	
MW:	333.27	**BP** (°C):		

Solubility (Moles/L)	Solubility (Grams/L)	Temp (°C)	Ref (#)	Evaluation (T P E A A)	Comments
6.001E-07	2.000E-04	25	M161	1 0 0 0 0	pH 7
9.002E-07	3.000E-04	ns	D304	1 0 0 0 0	

3060. C$_{13}$H$_{14}$N$_2$

4,4'-Methylenedianiline

4,4'-Methylenebisbenzeneamine

Tonox

HT 972

RN:	101-77-9	**MP** (°C):	93	
MW:	198.27	**BP** (°C):	398	

Solubility (Moles/L)	Solubility (Grams/L)	Temp (°C)	Ref (#)	Evaluation (T P E A A)	Comments
5.044E-03	1.000E+00	19	I307	0 0 0 0 0	

3061. C$_{13}$H$_{14}$N$_2$O$_3$

Mephobarbital

5-Ethyl-1-methyl-5-phenylbarbituric acid

5-Ethyl-N-methyl-5-phenylbarbituric acid

Mebaral

Prominal

Methylphenobarbital

RN:	115-38-8	**MP** (°C):	176	
MW:	246.27	**BP** (°C):		

Solubility (Moles/L)	Solubility (Grams/L)	Temp (°C)	Ref (#)	Evaluation (T P E A A)	Comments
6.090E-04	1.500E-01	20	J030	1 2 2 2 1	
4.872E-04	1.200E-01	37	J030	1 2 2 2 1	

3062. C$_{13}$H$_{14}$N$_2$O$_6$

Benzoic acid, 2-(acetyloxy)-, 2-[(2-amino-2-oxoethyl)amino]-2-oxoethyl ester

RN:	118247-02-2	**MP** (°C):	186	
MW:	294.27	**BP** (°C):		

Solubility (Moles/L)	Solubility (Grams/L)	Temp (°C)	Ref (#)	Evaluation (T P E A A)	Comments
2.990E-03	8.800E-01	21	N335	0 0 0 0 0	

3063. C$_{13}$H$_{14}$N$_4$

Pyridine-2-azo-p-dimethylaniline
PADA
2-(p-N,N-Dimethylaminophenylazo)pyridine
p-(2-Pyridylazo)-N,N-dimethylaniline
N,N-Dimethyl-4-(2-pyridylazo)aniline
2-(p-N,N-Dimethylaminophenylazo)pyridine

RN: 13103-75-8 **MP** (°C):
MW: 226.28 **BP** (°C): 392.8

Solubility (Moles/L)	Solubility (Grams/L)	Temp (°C)	Ref (#)	Evaluation (T P E A A)	Comments
9.400E-05	2.127E-02	ns	B418	0 2 1 1 2	

3064. C$_{13}$H$_{14}$N$_4$O$_3$S

N4-Acetylsulfamerazine
N4-Acetylsulphamerazine
2-N4-Acetylsulfanilamido-4-methylpyrimidine

RN: 127-73-1 **MP** (°C):
MW: 306.35 **BP** (°C):

Solubility (Moles/L)	Solubility (Grams/L)	Temp (°C)	Ref (#)	Evaluation (T P E A A)	Comments
1.200E-03	3.676E-01	37	G026	1 0 1 1 0	EFG, pH 5.4
2.579E-03	7.900E-01	37	L091	1 0 0 0 1	pH 5.5
9.140E-04	2.800E-01	37	R045	1 2 1 1 2	
9.140E-04	2.800E-01	37	R045	1 2 1 1 1	
1.234E-03	3.780E-01	37	S192	1 0 1 1 2	pH 6.0
2.611E-03	8.000E-01	38	K006	1 0 0 0 1	

3065. C$_{13}$H$_{14}$N$_4$O$_4$S

Acetyl sulfamethoxypyridazine
3-(N1-Acetylsulfanilamido)-6-methoxypyridazine
Acetylmidicel

RN: 127-75-3 **MP** (°C):
MW: 322.34 **BP** (°C):

Solubility (Moles/L)	Solubility (Grams/L)	Temp (°C)	Ref (#)	Evaluation (T P E A A)	Comments
6.825E-04	2.200E-01	37	B046	1 0 2 2 1	pH 4.5

3066. C$_{13}$H$_{14}$O$_6$

Salicylic acid acetate, hydroxymethyl ester propionate

RN: 32620-70-5 **MP** (°C): 51.5
MW: 266.25 **BP** (°C):

Solubility (Moles/L)	Solubility (Grams/L)	Temp (°C)	Ref (#)	Evaluation (T P E A A)	Comments
2.629E-03	7.000E-01	21	N335	0 0 0 0 0	

3067. C$_{13}$H$_{14}$O$_6$
Methylphthalyl ethyl glycolate
2-Ethoxy-2-oxoethyl methyl ester
RN: 85-71-2   **MP** (°C):   <−35
MW: 266.25   **BP** (°C):   189

Solubility (Moles/L)	Solubility (Grams/L)	Temp (°C)	Ref (#)	Evaluation (T P E A A)	Comments
1.990E-03	5.297E-01	20	F070	1 0 0 0 2	

3068. C$_{13}$H$_{15}$NO$_2$
Glutethimide
Doriden
Noxyron
RN: 77-21-4   **MP** (°C):   84
MW: 217.27   **BP** (°C):

Solubility (Moles/L)	Solubility (Grams/L)	Temp (°C)	Ref (#)	Evaluation (T P E A A)	Comments
4.372E-03	9.500E-01	27	B043	1 0 1 2 0	EFG
4.600E-03	9.994E-01	30	D010	1 2 1 1 2	
4.603E-03	1.000E+00	32	B043	1 0 1 2 0	EFG
5.753E-03	1.250E+00	37	B043	1 0 1 2 0	EFG
5.523E-05	1.200E-02	37	B045	1 0 1 1 2	
4.603E-03	1.000E+00	ns	A090	0 0 0 0 1	*sic*
4.600E-03	9.994E-01	ns	R010	0 1 0 0 2	

3069. C$_{13}$H$_{15}$NO$_2$
Pyracarbolid
3,4-Dihydro-6-methyl-*N*-phenyl-2H-pyran-5-carboxamide
Sicarol
RN: 24691-76-7   **MP** (°C):   110.5
MW: 217.27   **BP** (°C):

Solubility (Moles/L)	Solubility (Grams/L)	Temp (°C)	Ref (#)	Evaluation (T P E A A)	Comments
2.762E-03	6.000E-01	40	M161	1 0 0 0 0	

3070. C$_{13}$H$_{15}$NO$_2$S
m-Carboxylpentylphenylisothiocyanate
RN: **MP** (°C):
MW: 249.33   **BP** (°C):

Solubility (Moles/L)	Solubility (Grams/L)	Temp (°C)	Ref (#)	Evaluation (T P E A A)	Comments
7.300E-05	1.820E-02	25	K032	2 2 0 1 1	

3071. C₁₃H₁₅NO₃

Pyrrolidine, 1-[(benzoyloxy)acetyl]-

RN: 115178-67-1	**MP** (°C):	58
MW: 233.27	**BP** (°C):	

Solubility (Moles/L)	Solubility (Grams/L)	Temp (°C)	Ref (#)	Evaluation (T P E A A)	Comments
2.701E-02	6.300E+00	22	N317	1 1 2 1 2	

3072. C₁₃H₁₅NO₄

Morpholine, 4-[(benzoyloxy)acetyl]-

RN: 106231-68-9	**MP** (°C):	103.5
MW: 249.27	**BP** (°C):	453.9

Solubility (Moles/L)	Solubility (Grams/L)	Temp (°C)	Ref (#)	Evaluation (T P E A A)	Comments
1.685E-02	4.200E+00	22	B427	1 0 0 1 1	
1.685E-02	4.200E+00	22	N317	1 1 2 1 2	

3073. C₁₃H₁₅NO₅

Benzoic acid, 2-(acetyloxy)-, 2-(dimethylamino)-2-oxoethyl ester

RN: 118247-04-4	**MP** (°C):	75.5
MW: 265.27	**BP** (°C):	

Solubility (Moles/L)	Solubility (Grams/L)	Temp (°C)	Ref (#)	Evaluation (T P E A A)	Comments
2.827E-02	7.500E+00	21	N335	0 0 0 0 0	

3074. C₁₃H₁₅NO₅

Benzoic acid, 2-(acetyloxy)-, 2-(ethylamino)-2-oxoethyl ester

RN: 118247-01-1	**MP** (°C):	80.5
MW: 265.27	**BP** (°C):	

Solubility (Moles/L)	Solubility (Grams/L)	Temp (°C)	Ref (#)	Evaluation (T P E A A)	Comments
2.081E-02	5.520E+00	21	N335	0 0 0 0 0	

3075. C₁₃H₁₅N₃O₂

Pyrolan

1-Phenyl-3-methylpyrazolyl-5-dimethylcarbamate

RN: 87-47-8	**MP** (°C):	50
MW: 245.28	**BP** (°C):	161

Solubility (Moles/L)	Solubility (Grams/L)	Temp (°C)	Ref (#)	Evaluation (T P E A A)	Comments
8.138E-03	1.996E+00	ns	M061	0 0 0 0 0	

3076. $C_{13}H_{15}N_3O_3S$
2-Sulfanilamido-3-ethoxypyridine
Benzenesulfonamide, 4-amino-N-(3-ethoxy-2-pyridinyl)-
RN: 71119-19-2   **MP** (°C):
MW: 293.35   **BP** (°C):

Solubility (Moles/L)	Solubility (Grams/L)	Temp (°C)	Ref (#)	Evaluation (T P E A A)	Comments
8.011E-04	2.350E-01	37	R058	1 2 1 1 2	

3077. $C_{13}H_{15}N_3O_3S$
5-Sulfanilamido-2-ethoxypyridine
Benzenesulfonamide, 4-amino-N-(6-ethoxy-3-pyridinyl)-
RN: 71720-65-5   **MP** (°C):
MW: 293.35   **BP** (°C):

Solubility (Moles/L)	Solubility (Grams/L)	Temp (°C)	Ref (#)	Evaluation (T P E A A)	Comments
1.227E-04	3.600E-02	37	R058	1 2 1 1 1	

3078. $C_{13}H_{15}N_3O_4S$
Acetyl sulfisoxazole
N1-Acetyl-sulfaisoxazole
RN: 80-74-0   **MP** (°C): 193.5
MW: 309.35   **BP** (°C):

Solubility (Moles/L)	Solubility (Grams/L)	Temp (°C)	Ref (#)	Evaluation (T P E A A)	Comments
2.586E-04	8.000E-02	37	B046	1 0 2 2 0	pH 4.5
1.199E-04	3.710E-02	37	M117	2 1 1 1 2	pH 6.0

3079. $C_{13}H_{15}N_3O_4S$
N1-(3,4-Dimethyl-5-isoxazolyl)-N4-acetylsulfanilamide
Acetylsulfadimethylisoxazole
N4-Acetylsulfisoxazole
4-N-Acetylsulfisoxazole
N-Acetylsulfisoxazole
RN: 4206-74-0   **MP** (°C):
MW: 309.35   **BP** (°C):

Solubility (Moles/L)	Solubility (Grams/L)	Temp (°C)	Ref (#)	Evaluation (T P E A A)	Comments
2.450E-02	7.579E+00	37	B110	1 0 2 2 2	pH 6.7

3080. $C_{13}H_{16}Cl_2O_3$
2,4-Dichlorophenoxyacetic acid 1-ethylpropyl ester
RN: 65267-94-9   **MP** (°C):
MW: 291.18   **BP** (°C):

Solubility (Moles/L)	Solubility (Grams/L)	Temp (°C)	Ref (#)	Evaluation (T P E A A)	Comments
1.667E-05	4.855E-03	ns	M120	0 0 1 1 2	

3081. $C_{13}H_{16}Cl_2O_3$
2,4-Dichlorophenoxyacetic acid *n*-pentyl ester
2,4-D Pentyl ester
Pentyl 2,4-dichlorophenoxyacetate
Amyl 2,4-dichlorophenoxyacetate
RN: 1917-92-6   **MP** (°C):
MW: 291.18   **BP** (°C):

Solubility (Moles/L)	Solubility (Grams/L)	Temp (°C)	Ref (#)	Evaluation (T P E A A)	Comments
2.897E-05	8.436E-03	ns	M120	0 0 1 1 2	

3082. $C_{13}H_{16}Cl_2O_3$
2,4-Dichlorophenoxyacetic acid 2-methylbutyl ester
RN: **MP** (°C):
MW: 291.18   **BP** (°C):

Solubility (Moles/L)	Solubility (Grams/L)	Temp (°C)	Ref (#)	Evaluation (T P E A A)	Comments
1.291E-05	3.760E-03	ns	M120	0 0 1 1 2	

3083. $C_{13}H_{16}F_3N_3O_4$
Benefin
Benfluralin
RN: 1861-40-1   **MP** (°C):   65
MW: 335.29   **BP** (°C):

Solubility (Moles/L)	Solubility (Grams/L)	Temp (°C)	Ref (#)	Evaluation (T P E A A)	Comments
<2.98E-06	<1.00E-03	25	B200	1 0 0 0 0	
<2.98E-06	<1.00E-03	25	M161	1 0 0 0 0	
<2.98E-06	<1.00E-03	25	P028	0 0 0 0 0	
2.088E-04	7.000E-02	ns	M061	0 0 0 0 1	

3084. C₁₃H₁₆F₃N₃O₄
Trifluralin
α,α,α-Trifluoro-2,6-dinitro-*N*,*N*-dipropyl-*p*-toluidine
RN: 1582-09-8   **MP** (°C):   48.5
MW: 335.29   **BP** (°C):

Solubility (Moles/L)	Solubility (Grams/L)	Temp (°C)	Ref (#)	Evaluation (T P E A A)	Comments
1.193E-05	4.000E-03	20	F311	1 2 2 2 1	
2.419E-05	8.110E-03	22	K137	1 1 2 1 0	
1.730E-06	5.800E-04	25	G319	0 0 0 0 0	
<2.98E-06	<1.00E-03	27	B200	1 0 0 0 0	
<2.98E-06	<1.00E-03	27	M161	1 0 0 0 0	
<2.98E-06	<1.00E-03	27	P028	0 0 0 0 0	
7.158E-05	2.400E-02	ns	B185	0 0 0 0 0	
1.193E-04	4.000E-02	ns	M061	0 0 0 0 1	
2.088E-06	7.000E-04	ns	M110	0 0 0 0 0	EFG
5.488E-07	1.840E-04	ns	V414	0 0 0 0 0	

3085. C₁₃H₁₆NO₄PS
Isoxathion
O,O-Diethyl *O*-5-phenylisoxazol-3-yl phosphorothioate
E-48
Karphos
SI-6711
RN: 18854-01-8   **MP** (°C):
MW: 313.31   **BP** (°C):   160

Solubility (Moles/L)	Solubility (Grams/L)	Temp (°C)	Ref (#)	Evaluation (T P E A A)	Comments
6.064E-06	1.900E-03	25	N305	1 0 0 0 1	

3086. C₁₃H₁₆N₂
3-(1-Methyl-2-pyrrolidinyl)-indole
RN: **MP** (°C):
MW: 200.29   **BP** (°C):

Solubility (Moles/L)	Solubility (Grams/L)	Temp (°C)	Ref (#)	Evaluation (T P E A A)	Comments
3.510E-03	7.030E-01	37	H004	0 0 0 0 0	
3.510E-03	7.030E-01	37	H011	0 0 0 0 0	

3087. C₁₃H₁₆N₂O₂
Melatonin
Prime-X
RN: 8041-44-9   **MP** (°C):
MW: 232.28   **BP** (°C):

Solubility (Moles/L)	Solubility (Grams/L)	Temp (°C)	Ref (#)	Evaluation (T P E A A)	Comments
1.870E-03	4.344E-01	25	B426	1 1 2 2 2	

3088. $C_{13}H_{16}N_2O_4$

N-Acetyl-L-tyrosinamide acetate

RN: **MP** (°C):
MW: 264.28   **BP** (°C):

Solubility (Moles/L)	Solubility (Grams/L)	Temp (°C)	Ref (#)	Evaluation (T P E A A)	Comments
1.300E-02	3.436E+00	25	A066	1 0 1 1 1	

3089. $C_{13}H_{16}N_2O_4$

Methyl-2-ethyl-2-phenylmalonurate
Methyl 2-ethyl-2-phenylmalonurate

RN: 73632-81-2   **MP** (°C):   105
MW: 264.28   **BP** (°C):

Solubility (Moles/L)	Solubility (Grams/L)	Temp (°C)	Ref (#)	Evaluation (T P E A A)	Comments
1.800E-03	4.757E-01	23	B152	1 2 1 1 1	pH 3.5

3090. $C_{13}H_{16}N_2O_6$

Medinoterb acetate
m-Cresol, 6-*tert*-butyl-2,4-dinitro-, acetate
MC 1488

RN: 2487-01-6   **MP** (°C):   86.5
MW: 296.28   **BP** (°C):

Solubility (Moles/L)	Solubility (Grams/L)	Temp (°C)	Ref (#)	Evaluation (T P E A A)	Comments
3.375E-05	1.000E-02	rt	M161	0 0 0 0 1	

3091. $C_{13}H_{16}N_4O_2S$

2-Sulfanilylamino-4-ethyl-5-methylpyrimidine

RN: **MP** (°C):
MW: 292.36   **BP** (°C):

Solubility (Moles/L)	Solubility (Grams/L)	Temp (°C)	Ref (#)	Evaluation (T P E A A)	Comments
8.551E-04	2.500E-01	37	R076	1 2 0 0 1	

3092. $C_{13}H_{16}N_4O_2S$

2-*p*-Aminobenzenesulphonamido-4,5,6-trimethylpyrimidine
Sulfanilamide, *N*1-(4,5,6-trimethyl-2-pyrimidinyl)-

RN: 5433-64-7   **MP** (°C):
MW: 292.36   **BP** (°C):

Solubility (Moles/L)	Solubility (Grams/L)	Temp (°C)	Ref (#)	Evaluation (T P E A A)	Comments
5.131E-04	1.500E-01	37	R075	1 0 0 0 1	

3093. C$_{13}$H$_{16}$N$_4$O$_6$.0.5H$_2$O
9-[5-O-(Acetate-β-D-arabinofuranosyl)]-6-methoxy-9H-purine (hemihydrate)
2′-Acetyl-6-methoxypurine arabinoside (hemihydrate)

RN: 121032-43-7   **MP** (°C):   174-176
MW: 333.30   **BP** (°C):

Solubility (Moles/L)	Solubility (Grams/L)	Temp (°C)	Ref (#)	Evaluation (T P E A A)	Comments
3.250E-02	1.083E+01	37	C348	0 0 0 0 0	pH 7.00
5.310E-02	1.770E+01	37	M378	1 2 1 1 2	pH 7.2

3094. C$_{13}$H$_{16}$O$_4$
Diethylacetyl salicylate
Salicylic acid, 2-ethylbutyrate

RN: 100613-21-6   **MP** (°C):
MW: 236.27   **BP** (°C):

Solubility (Moles/L)	Solubility (Grams/L)	Temp (°C)	Ref (#)	Evaluation (T P E A A)	Comments
2.800E-03	6.616E-01	25.6	G015	1 0 1 1 2	pH 1.00, pka 4.00, intrinsic

3095. C$_{13}$H$_{16}$O$_6$
Methyl phthalyl ethyl glycollate

RN: **MP** (°C):
MW: 268.27   **BP** (°C):

Solubility (Moles/L)	Solubility (Grams/L)	Temp (°C)	Ref (#)	Evaluation (T P E A A)	Comments
4.096E-03	1.099E+00	15	H069	1 0 1 1 1	
1.975E-03	5.297E-01	ns	F014	0 0 0 0 1	

3096. C$_{13}$H$_{16}$O$_7$.0.75H$_2$O
Helicin (0.75 hydrate)
Salicylaldehyde β-D-glucoside
Benzaldehyde, 2-(β-D-glucopyranosyloxy)-, hydrate (4:3)

RN: 618-65-5   **MP** (°C):
MW: 297.78   **BP** (°C):

Solubility (Moles/L)	Solubility (Grams/L)	Temp (°C)	Ref (#)	Evaluation (T P E A A)	Comments
5.505E-02	1.639E+01	c	D004	0 0 0 0 0	
		h	D004	0 0 0 0 0	

3097. C$_{13}$H$_{17}$ClO$_3$
MCPB-ethyl

| RN: | 10443-70-6 | MP (°C): | |
| MW: | 256.73 | BP (°C): | |

Solubility (Moles/L)	Solubility (Grams/L)	Temp (°C)	Ref (#)	Evaluation (T P E A A)	Comments
3.899E-05	1.001E-02	ns	S460	0 0 0 0 0	

3098. C$_{13}$H$_{17}$IN$_2$O$_6$
Uridine, 2′-deoxy-5-iodo-, 5′-butanoate
5′-Butyryl 5-iodo-2′-deoxyuridine
5-Iodo-2′-deoxyuridine 5′-butyrate

| RN: | 84043-26-5 | MP (°C): | 145.5 |
| MW: | 424.19 | BP (°C): | |

Solubility (Moles/L)	Solubility (Grams/L)	Temp (°C)	Ref (#)	Evaluation (T P E A A)	Comments
1.450E+03	6.151E+05	25	N332	0 0 0 0 0	pH 7.4

3099. C$_{13}$H$_{17}$IN$_2$O$_6$
Uridine, 2′-deoxy-5-iodo-, 5′-(2-methylpropanoate)
5′-Isobutyryl 5-iodo-2′-deoxyuridine
5-Iodo-2′-deoxyuridine 5′-isobutyrate

| RN: | 84043-27-6 | MP (°C): | 144.5 |
| MW: | 424.19 | BP (°C): | |

Solubility (Moles/L)	Solubility (Grams/L)	Temp (°C)	Ref (#)	Evaluation (T P E A A)	Comments
1.750E+03	7.423E+05	25	N332	0 0 0 0 0	pH 7.4

3100. C$_{13}$H$_{17}$NO
N-Butylcinnamamide
N-Butyl-3-phenyl-2-propenamide

| RN: | 6299-56-5 | MP (°C): | |
| MW: | 203.29 | BP (°C): | |

Solubility (Moles/L)	Solubility (Grams/L)	Temp (°C)	Ref (#)	Evaluation (T P E A A)	Comments
9.700E-04	1.972E-01	ns	H350	0 0 0 0 0	

3101. C$_{13}$H$_{17}$NO
N,N-Diethylcinnamamide
N,N-Diethyl-3-phenyl-2-propenamide

| RN: | 3680-04-4 | MP (°C): | |
| MW: | 203.29 | BP (°C): | |

Solubility (Moles/L)	Solubility (Grams/L)	Temp (°C)	Ref (#)	Evaluation (T P E A A)	Comments
7.450E-03	1.514E+00	ns	H350	0 0 0 0 0	

3102. C$_{13}$H$_{17}$NO$_3$
Acetamide, 2-(benzoyloxy)-N-butyl-
RN: 115193-28-7 **MP** (°C): 69.5
MW: 235.29 **BP** (°C):

Solubility (Moles/L)	Solubility (Grams/L)	Temp (°C)	Ref (#)	Evaluation (T P E A A)	Comments
1.743E-03	4.100E-01	22	N317	1 1 2 1 2	

3103. C$_{13}$H$_{17}$NO$_3$
N-Acetyl-L-phenylalanine ethyl ester
RN: 2361-96-8 **MP** (°C):
MW: 235.29 **BP** (°C):

Solubility (Moles/L)	Solubility (Grams/L)	Temp (°C)	Ref (#)	Evaluation (T P E A A)	Comments
1.084E-02	2.550E+00	5	L081	2 1 2 2 2	
1.755E-02	4.130E+00	28	L081	2 1 2 2 2	
2.814E-02	6.620E+00	40	L081	2 1 2 2 2	
3.417E-02	8.040E+00	55	L081	2 1 2 2 2	
7.268E-02	1.710E+01	65	L081	2 1 2 2 2	

3104. C$_{13}$H$_{17}$NO$_3$
2-(p-Acetaminophenoxy)tetrahydropyran
RN: 51453-65-7 **MP** (°C): 60
MW: 235.29 **BP** (°C):

Solubility (Moles/L)	Solubility (Grams/L)	Temp (°C)	Ref (#)	Evaluation (T P E A A)	Comments
3.000E-03	7.059E-01	ns	H076	0 0 0 0 0	

3105. C$_{13}$H$_{17}$NO$_3$
Pivalyl acetaminophen
Propanoic acid, 2,2-dimethyl-, 4-(acetylamino)phenyl ester
Acetanilide, 4'-hydroxy-, pivalate (ester)
RN: 20675-23-4 **MP** (°C): 162.5–163
MW: 235.29 **BP** (°C):

Solubility (Moles/L)	Solubility (Grams/L)	Temp (°C)	Ref (#)	Evaluation (T P E A A)	Comments
4.675E-04	1.100E-01	37	D029	0 0 0 0 0	

3106. C$_{13}$H$_{17}$NO$_3$
Acetamide, 2-(benzoyloxy)-N-(1,1-dimethylethyl)-
RN: 106231-52-1 **MP** (°C): 112–113
MW: 235.29 **BP** (°C): 418.2

Solubility (Moles/L)	Solubility (Grams/L)	Temp (°C)	Ref (#)	Evaluation (T P E A A)	Comments
1.360E-03	3.200E-01	22	B427	1 0 0 1 1	

3107. C₁₃H₁₇NO₃

Acetamide, 2-(benzoyloxy)-*N*,*N*-diethyl-

RN: 64649-63-4   **MP** (°C):   72.5
MW: 235.29   **BP** (°C):   377.0

Solubility (Moles/L)	Solubility (Grams/L)	Temp (°C)	Ref (#)	Evaluation (T P E A A)	Comments
8.500E-03	2.000E+00	22	B427	1 0 0 1 1	in 0.01M HCl
8.500E-03	2.000E+00	22	N317	1 1 2 1 2	

3108. C₁₃H₁₇NO₃

Butanamide, 4-(benzoyloxy)-*N*,*N*-dimethyl-

RN: 115178-78-4   **MP** (°C):   40.5
MW: 235.29   **BP** (°C):

Solubility (Moles/L)	Solubility (Grams/L)	Temp (°C)	Ref (#)	Evaluation (T P E A A)	Comments
5.908E-02	1.390E+01	22	N317	1 1 2 1 2	

3109. C₁₃H₁₇NO₄

Benzoic acid, 2-hydroxy-, 2-(diethylamino)-2-oxoethyl ester
N,*N*-Diethylglycolamide salicylate
N,*N*-Diethyl glycolamide salicylate

RN: 65783-69-9   **MP** (°C):   74–75
MW: 251.28   **BP** (°C):

Solubility (Moles/L)	Solubility (Grams/L)	Temp (°C)	Ref (#)	Evaluation (T P E A A)	Comments
2.786E-03	7.000E-01	21	B331	1 2 2 1 1	pH 7.4
2.786E-03	7.000E-01	21	B331	0 0 0 0 0	

3110. C₁₃H₁₇NO₄

Butyl acetaminophen
Carbonic acid, butyl ester, ester with 4′-hydroxyacetanilide
Acetanilide, 4′-hydroxy-, butyl carbonate (ester)

RN: 19872-68-5   **MP** (°C):   119.5–120
MW: 251.28   **BP** (°C):

Solubility (Moles/L)	Solubility (Grams/L)	Temp (°C)	Ref (#)	Evaluation (T P E A A)	Comments
6.367E-04	1.600E-01	37	D029	0 0 0 0 0	

3111. C$_{13}$H$_{17}$NO$_4$

Isobutyl acetaminophen
Carbonic acid, isobutyl ester, ester with 4'-hydroxyacetanilide
Acetanilide, 4'-hydroxy-, isobutyl carbonate (ester)

RN: 20460-96-2   **MP (°C):**   119–121
MW: 251.28   **BP (°C):**

Solubility (Moles/L)	Solubility (Grams/L)	Temp (°C)	Ref (#)	Evaluation (T P E A A)	Comments
1.512E-03	3.800E-01	37	D029	0 0 0 0 0	

3112. C$_{13}$H$_{17}$NO$_4$

O-(Pivaloyloxymethyl) salicylamide

RN: **MP (°C):**   95
MW: 251.28   **BP (°C):**

Solubility (Moles/L)	Solubility (Grams/L)	Temp (°C)	Ref (#)	Evaluation (T P E A A)	Comments
2.428E-03	6.100E-01	23	B328	1 2 2 1 1	pH 4

3113. C$_{13}$H$_{17}$NO$_4$

Propanoic acid, 2,2-dimethyl-, [2-(aminocarbonyl)phenoxy]methyl ester
O-Pivaloyloxymethyl salicylamide

RN: 103951-40-2   **MP (°C):**   94–96
MW: 251.28   **BP (°C):**

Solubility (Moles/L)	Solubility (Grams/L)	Temp (°C)	Ref (#)	Evaluation (T P E A A)	Comments
2.428E-03	6.100E-01	23	B328	0 0 0 0 0	

3114. C$_{13}$H$_{17}$NO$_4$

Acetamide, 2-(benzoyloxy)-*N*-ethyl-*N*-(2-hydroxyethyl)-

RN: 106231-60-1   **MP (°C):**   79.5
MW: 251.28   **BP (°C):**   437.5

Solubility (Moles/L)	Solubility (Grams/L)	Temp (°C)	Ref (#)	Evaluation (T P E A A)	Comments
4.298E-02	1.080E+01	22	B427	1 0 0 1 1	in 0.01M HCl
4.298E-02	1.080E+01	22	N317	1 1 2 1 2	

3115. C$_{13}$H$_{17}$NO$_4$

N-Acetyl-L-tyrosine ethyl ester
Ethyl *N*-acetyl-L-tyrosinate

RN: 840-97-1   **MP (°C):**
MW: 251.28   **BP (°C):**

Solubility (Moles/L)	Solubility (Grams/L)	Temp (°C)	Ref (#)	Evaluation (T P E A A)	Comments
5.571E-03	1.400E+00	5	L081	2 1 2 2 2	
1.385E-02	3.480E+00	28	L081	2 1 2 2 2	

3116. C$_{13}$H$_{17}$NO$_5$

Acetamide, 2-(benzoyloxy)-*N,N*-bis(2-hydroxyethyl)-

RN:	106231-61-2	MP (°C):	81
MW:	267.28	BP (°C):	497.5

Solubility (Moles/L)	Solubility (Grams/L)	Temp (°C)	Ref (#)	Evaluation (T P E A A)	Comments
2.694E+00	7.200E+02	22	B427	1 0 0 1 1	in 0.01M HCl
2.694E+00	7.200E+02	22	N317	1 1 2 1 2	

3117. C$_{13}$H$_{17}$NO$_6$

Acetamide, 2-(benzoyloxy)-*N*-[2-hydroxy-1,1-bis(hydroxymethyl)ethyl]-

RN:	115193-31-2	MP (°C):	126.5
MW:	283.28	BP (°C):	

Solubility (Moles/L)	Solubility (Grams/L)	Temp (°C)	Ref (#)	Evaluation (T P E A A)	Comments
5.401E-02	1.530E+01	22	N317	1 1 2 1 2	

3118. C$_{13}$H$_{17}$N$_3$O

Aminopyrine
Amidopyrine
4-Dimethylaminoantipyrine
Febrinina
Febron
Itamidone

RN:	58-15-1	MP (°C):	108
MW:	231.30	BP (°C):	

Solubility (Moles/L)	Solubility (Grams/L)	Temp (°C)	Ref (#)	Evaluation (T P E A A)	Comments
2.827E-01	6.540E+01	0	C025	0 0 0 0 2	form A
5.607E-01	1.297E+02	4.62	M109	2 1 1 1 0	EFG
5.463E-01	1.264E+02	10.93	M109	2 1 1 1 0	EFG
5.430E-01	1.256E+02	15.02	M109	2 1 1 1 0	EFG
2.291E-01	5.300E+01	20	C025	0 0 0 0 2	form A
5.452E-01	1.261E+02	20.96	M109	2 1 1 1 0	EFG
2.291E-01	5.300E+01	25	P012	0 0 0 0 0	
2.162E-01	5.000E+01	25	P016	1 0 0 1 1	
2.075E-01	4.800E+01	25	P020	2 0 1 1 1	
1.773E+00	4.100E+02	25	P020	2 0 1 1 2	
5.618E-01	1.300E+02	25.35	M109	2 1 1 1 0	EFG
5.965E-01	1.380E+02	29.87	M109	2 1 1 1 0	EFG
2.350E-01	5.436E+01	30	A078	2 1 2 1 0	EFG
2.291E-01	5.300E+01	37	C025	0 0 0 0 2	form A
6.329E-01	1.464E+02	38.37	M109	2 1 1 1 0	EFG
6.646E-01	1.537E+02	49.42	M109	2 1 1 1 0	EFG
3.415E-01	7.900E+01	55	C025	0 0 0 0 2	form A
5.638E-01	1.304E+02	65	C025	0 0 0 0 2	form A
2.162E+00	5.000E+02	69.50	C025	0 0 0 0 2	form A

(continued)

3118. $C_{13}H_{17}N_3O$ (continued)

Solubility (Moles/L)	Solubility (Grams/L)	Temp (°C)	Ref (#)	Evaluation (T P E A A)	Comments
1.729E+00	4.000E+02	70	C025	0 0 0 0 2	form B
1.167E+00	2.700E+02	70.50	C025	0 0 0 0 2	form B
2.879E+00	6.660E+02	74.40	C025	0 0 0 0 2	form B
8.647E-01	2.000E+02	77.50	C025	0 0 0 0 2	form B
6.485E-01	1.500E+02	81	C025	0 0 0 0 2	form B
3.243E+00	7.500E+02	84	C025	0 0 0 0 2	form B
3.359E+00	7.770E+02	92	C025	0 0 0 0 2	form B

3119. $C_{13}H_{17}N_5O_5$

9-(2-O-Propionyl-β-D-arabinofuranosyl)adenine

RN: 65174-99-4 **MP** (°C):
MW: 323.31 **BP** (°C):

Solubility (Moles/L)	Solubility (Grams/L)	Temp (°C)	Ref (#)	Evaluation (T P E A A)	Comments
3.618E-04	1.170E-01	37	B306	1 2 0 1 2	pH 7.3

3120. $C_{13}H_{17}N_5O_5$

9-[5′-(O-Propionyl)-β-D-arabinofuranosyl]adenine ester

RN: 14000-32-9 **MP** (°C): 202.0
MW: 323.31 **BP** (°C):

Solubility (Moles/L)	Solubility (Grams/L)	Temp (°C)	Ref (#)	Evaluation (T P E A A)	Comments
2.846E-02	9.200E+00	ns	B134	0 1 1 1 1	

3121. $C_{13}H_{17}N_5O_6$

9-(1,3-Diacetate-2-propoxymethyl)guanine

RN: 86357-19-9 **MP** (°C): 238
MW: 339.31 **BP** (°C):

Solubility (Moles/L)	Solubility (Grams/L)	Temp (°C)	Ref (#)	Evaluation (T P E A A)	Comments
1.709E-03	5.800E-01	25	B360	0 0 0 0 0	

3122. $C_{13}H_{17}N_5O_8$

9-(1,3-Dimethoxycarbonyl-2-propoxymethyl)guanine

RN: 91625-66-0 **MP** (°C): 178
MW: 371.31 **BP** (°C):

Solubility (Moles/L)	Solubility (Grams/L)	Temp (°C)	Ref (#)	Evaluation (T P E A A)	Comments
3.851E-04	1.430E-01	25	B360	0 0 0 0 0	

3123. C₁₃H₁₈ClNO

Monalide

N-(4-Chlorophenyl)-2,2-dimethylvaleramide

RN: 7287-36-7 **MP (°C):** 87.5
MW: 239.75 **BP (°C):**

Solubility (Moles/L)	Solubility (Grams/L)	Temp (°C)	Ref (#)	Evaluation (T P E A A)	Comments
9.510E-05	2.280E-02	23	M161	1 0 0 0 2	
9.510E-05	2.280E-02	ns	M061	0 0 0 0 2	

3124. C₁₃H₁₈ClNO

Pentanochlor

Solan

Pentamide, N-(3-chloro-4-methylphenyl)-2-methyl-

RN: 2307-68-8 **MP (°C):** 84
MW: 239.75 **BP (°C):**

Solubility (Moles/L)	Solubility (Grams/L)	Temp (°C)	Ref (#)	Evaluation (T P E A A)	Comments
3.337E-05	8.000E-03	ns	B185	0 0 0 0 0	
3.545E-05	8.500E-03	rt	M161	0 0 0 0 0	

3125. C₁₃H₁₈ClN₃O₄S₂

Cyclopenthiazide

6-Chloro-3-cyclopentylmethyl-3,4-dihydro-2H-1,2,4-benzothiadiazine-7-sulphonamide 1,1-
  dioxide

RN: 742-20-1 **MP (°C):** 235
MW: 379.89 **BP (°C):**

Solubility (Moles/L)	Solubility (Grams/L)	Temp (°C)	Ref (#)	Evaluation (T P E A A)	Comments
1.316E-04	5.000E-02	rt	A095	0 0 2 2 0	

3126. C₁₃H₁₈Cl₂N₂O₂

Melphalan

4-[bis(2-Chloroethyl)amino]-L-phenylalanine

RN: 148-82-3 **MP (°C):**
MW: 305.21 **BP (°C):**

Solubility (Moles/L)	Solubility (Grams/L)	Temp (°C)	Ref (#)	Evaluation (T P E A A)	Comments
1.442E-02	4.400E+00	30	L343	2 1 1 1 0	EFG
5.561E-03	1.697E+00	ns	S469	0 0 0 0 0	

3127. $C_{13}H_{18}N_2O_2$

Lenacil
3-Cyclohexyl-5,6-trimethyleneuracil

RN: 2164-08-1   **MP** (°C):   290
MW: 234.30   **BP** (°C):

Solubility (Moles/L)	Solubility (Grams/L)	Temp (°C)	Ref (#)	Evaluation (T P E A A)	Comments
2.561E-05	6.000E-03	25	M061	1 0 0 0 0	
2.561E-05	6.000E-03	25	M161	1 0 0 0 0	

3128. $C_{13}H_{18}N_2O_3$

Heptabarbital
5-(1-Cyclohepten-1-yl)-5-ethyl-2,4,6(1H,3H,5H)-pyrimidinetrione
5-(1-Cyclohepten-1-yl)-5-ethylbarbituric acid
Heptabarbitone

RN: 509-86-4   **MP** (°C):   174
MW: 250.30   **BP** (°C):

Solubility (Moles/L)	Solubility (Grams/L)	Temp (°C)	Ref (#)	Evaluation (T P E A A)	Comments
1.000E-03	2.503E-01	25	V033	2 0 1 1 2	
1.000E-03	2.503E-01	25.00	T303	1 0 0 0 1	
1.400E-03	3.504E-01	35.00	T303	1 0 0 0 1	
1.170E-02	2.929E+00	40	N008	1 0 1 1 2	*sic*
1.800E-03	4.505E-01	45.00	T303	1 0 0 0 1	

3129. $C_{13}H_{18}N_2O_3S$

Tosylcyclopentylurea
Tosylcyclopentyluree

RN: 1027-87-8   **MP** (°C):
MW: 282.36   **BP** (°C):

Solubility (Moles/L)	Solubility (Grams/L)	Temp (°C)	Ref (#)	Evaluation (T P E A A)	Comments
2.649E-04	7.478E-02	37	A028	1 0 2 1 2	intrinsic
2.650E-04	7.483E-02	37	A046	2 0 1 1 2	

3130. $C_{13}H_{18}N_2O_4$

Methyl-2-methyl-2-cyclohexenyl-6-methylmalonurate
Methyl 2-methyl-2-cyclohexenyl-6-methylmalonurate

RN: **MP** (°C):   94
MW: 266.30   **BP** (°C):

Solubility (Moles/L)	Solubility (Grams/L)	Temp (°C)	Ref (#)	Evaluation (T P E A A)	Comments
2.100E-03	5.592E-01	23	B152	1 2 1 1 1	pH 3.5

3131. $C_{13}H_{18}N_4O_2S_2$

4-Amino-N-(5-pentyl-1,3,4-thiadiazol-2-yl)benzenesulfonamide
Benzenesulfonamide, 4-amino-N-(5-pentyl-1,3,4-thiadiazol-2-yl)-

RN: 71119-30-7 **MP** (°C):
MW: 326.44 **BP** (°C):

Solubility (Moles/L)	Solubility (Grams/L)	Temp (°C)	Ref (#)	Evaluation (T P E A A)	Comments
1.120E-04	3.656E-02	37	A046	2 0 1 1 2	

3132. $C_{13}H_{18}N_4O_2S_2$

4-Amino-N-(5-isopentyl-1,3,4-thiadiazol-2-yl)benzenesulfonamide
Benzenesulfonamide, 4-amino-N-[5-(3-methylbutyl)-1,3,4-thiadiazol-2-yl]-

RN: 71119-29-4 **MP** (°C):
MW: 326.44 **BP** (°C):

Solubility (Moles/L)	Solubility (Grams/L)	Temp (°C)	Ref (#)	Evaluation (T P E A A)	Comments
9.000E-05	2.938E-02	37	A046	2 0 1 1 2	

3133. $C_{13}H_{18}O_2$

Ibuprofen
2-(4-Isobutylphenyl)propionic acid
Advil
Ebufac
Rufen
RS-Ibuprofen

RN: 15687-27-1 **MP** (°C): 75
MW: 206.29 **BP** (°C): 319.6

Solubility (Moles/L)	Solubility (Grams/L)	Temp (°C)	Ref (#)	Evaluation (T P E A A)	Comments
2.320E-04	4.786E-02	4	B411	1 1 1 2 2	
3.340E-05	6.890E-03	5	F306	1 0 1 2 2	intrinsic
1.080E-04	2.228E-02	12	B411	1 1 1 2 2	
1.460E-04	3.012E-02	20	B411	1 1 1 2 2	
7.271E-05	1.500E-02	20	N316	1 0 1 1 0	EFG
3.102E-04	6.400E-02	21	B331	1 2 2 1 2	pH 7.4
2.375E-04	4.900E-02	25	A408	2 0 1 2 0	int
1.018E-04	2.100E-02	25	A427	0 0 0 0 0	
5.478E-05	1.130E-02	25	C314	0 0 0 0 0	
5.560E-05	1.147E-02	25	C314	0 0 0 0 0	
9.430E-04	1.945E-01	25	D345	0 0 0 0 0	
4.300E-05	8.870E-03	25	F301	1 1 0 0 1	pH 2.0, *sic*
4.300E-05	8.870E-03	25	F306	1 0 1 2 2	intrinsic
5.520E-05	1.139E-02	25	G431	0 0 0 0 0	
2.424E-04	5.000E-02	25	S450	0 0 0 0 0	Intrinsic
2.090E-04	4.311E-02	29	B411	1 1 1 2 2	
7.505E-05	1.548E-02	30	G431	0 0 0 0 0	
1.212E-04	2.500E-02	30	N316	1 0 1 1 0	EFG

(continued)

3133. C$_{13}$H$_{18}$O$_2$ (continued)

Solubility (Moles/L)	Solubility (Grams/L)	Temp (°C)	Ref (#)	Evaluation (T P E A A)	Comments
9.970E-05	2.057E-02	35	G431	0 0 0 0 0	
5.210E-05	1.075E-02	37	F306	1 0 1 2 2	intrinsic
1.551E-04	3.200E-02	37	N316	1 0 1 1 0	EFG
5.332E-05	1.100E-02	37	P432	0 0 0 0 0	
2.909E-04	6.000E-02	37	Y421	0 0 0 0 0	
3.040E-04	6.271E-02	38	B411	1 1 1 2 2	
1.281E-04	2.643E-02	40	G431	0 0 0 0 0	
4.760E-04	9.819E-02	47	B411	1 1 1 2 2	
1.600E-04	3.301E-02	50	M335	1 0 2 1 2	pH 5
2.036E-04	4.200E-02	50	N316	1 0 1 1 0	EFG
2.327E-04	4.800E-02	60	N316	1 0 1 1 0	EFG
2.600E-04	5.363E-02	ns	F327	0 0 1 2 2	
4.848E-05	1.000E-02	ns	K444	0 0 0 0 0	
1.018E-04	2.100E-02	rt	H302	0 0 2 1 2	intrinsic
4.096E-04	8.450E-02	rt	R431	0 0 0 0 0	Average

3134. C$_{13}$H$_{18}$O$_2$
S-Ibuprofen
(S)-(+)-2-(4-Isobutylphenyl)propionic acid
D-Ibuprofen
Seractil
Dexibuprofen
RN: 51146-56-6   **MP** (°C):
MW: 206.29   **BP** (°C):

Solubility (Moles/L)	Solubility (Grams/L)	Temp (°C)	Ref (#)	Evaluation (T P E A A)	Comments
2.320E-04	4.786E-02	4	B411	1 1 1 2 2	
2.560E-04	5.281E-02	12	B411	1 1 1 2 2	
3.390E-04	6.993E-02	20	B411	1 1 1 2 2	
1.790E-03	3.693E-01	25	D345	0 0 0 0 0	
4.670E-04	9.634E-02	29	B411	1 1 1 2 2	
6.090E-04	1.256E-01	38	B411	1 1 1 2 2	

3135. C$_{13}$H$_{18}$O$_2$
r-Ibuprofen
(R)-2-(4-Isobutylphenyl)propanoic acid
r-(−)-p-Isobutylhydratropic acid
l-Ibuprofen
RN: 51146-57-7   **MP** (°C):
MW: 206.29   **BP** (°C):

Solubility (Moles/L)	Solubility (Grams/L)	Temp (°C)	Ref (#)	Evaluation (T P E A A)	Comments
1.790E-03	3.693E-01	25	D345	0 0 0 0 0	

3136. C₁₃H₁₈O₃

Hexyl *p*-hydroxybenzoate

4-Hydroxybenzoic acid *N*-hexyl ester

RN:	1083-27-8	**MP** (°C):
MW:	222.29	**BP** (°C):

Solubility (Moles/L)	Solubility (Grams/L)	Temp (°C)	Ref (#)	Evaluation (T P E A A)	Comments
3.680E-04	8.180E-02	15	B355	0 0 0 0 0	
3.810E-04	8.469E-02	20	B355	0 0 0 0 0	
6.190E-04	1.376E-01	25	B355	0 0 0 0 0	
1.704E-03	3.789E-01	25	D081	1 2 2 1 2	
3.162E-04	7.029E-02	25	F322	2 0 1 1 0	EFG

3137. C₁₃H₁₈O₃

n-Hexyl salicylate

n-Hexyl 2-hydroxybenzoate

RN:	6259-76-3	**MP** (°C):
MW:	222.29	**BP** (°C):

Solubility (Moles/L)	Solubility (Grams/L)	Temp (°C)	Ref (#)	Evaluation (T P E A A)	Comments
1.260E-03	2.800E-01	37	D009	1 2 1 1 1	0.1N HCl

3138. C₁₃H₁₈O₅S

Ethofumesate

2-Ethoxy-2,3-dihydro-3,3-dimethyl-5-benzofuranyl methanesulfonate

Nortran

Tramat

RN:	26225-79-6	**MP** (°C):	71
MW:	286.35	**BP** (°C):	

Solubility (Moles/L)	Solubility (Grams/L)	Temp (°C)	Ref (#)	Evaluation (T P E A A)	Comments
1.758E-04	5.034E-02	25	H434	0 0 0 0 0	
3.841E-04	1.100E-01	25	M161	1 0 0 0 2	
3.841E-04	1.100E-01	25	W313	1 0 0 0 1	

3139. C₁₃H₁₈O₇

Salicin

2-(Hydroxymethyl)phenyl-β-D-glucopyranoside

Salicoside

RN:	138-52-3	**MP** (°C):	199
MW:	286.28	**BP** (°C):	

Solubility (Moles/L)	Solubility (Grams/L)	Temp (°C)	Ref (#)	Evaluation (T P E A A)	Comments
1.397E-01	4.000E+01	25	F300	1 0 0 0 0	
9.082E-01	2.600E+02	100	F300	1 0 0 0 1	
1.455E-01	4.167E+01	c	D004	0 0 0 0 0	
8.733E-01	2.500E+02	h	D004	0 0 0 0 0	

3140. C₁₃H₁₉NO₂

Hexyl *p*-aminobenzoate
4-Aminobenzoic acid hexyl ester
RN: 55791-76-9   **MP** (°C):
MW: 221.30   **BP** (°C):

Solubility (Moles/L)	Solubility (Grams/L)	Temp (°C)	Ref (#)	Evaluation (T P E A A)	Comments
1.040E-04	2.302E-02	37	F006	1 1 2 2 2	
4.500E-05	9.959E-03	ns	M066	0 0 0 0 1	
4.300E-05	9.516E-03	rt	B016	0 0 1 1 1	pH 7.4

3141. C₁₃H₁₉NO₂

Ibuproxam
2-(4-Isobutylphenyl)propionohydroxamic acid
Ibudros
RN: 53648-05-8   **MP** (°C):   123
MW: 221.30   **BP** (°C):

Solubility (Moles/L)	Solubility (Grams/L)	Temp (°C)	Ref (#)	Evaluation (T P E A A)	Comments
9.037E-04	2.000E-01	ns	M148	0 2 0 0 0	

3142. C₁₃H₁₉NO₄

N,N-Diethyl-6-hydroxynorbornane-2-carboxamide-3,5-lactone
RN: **MP** (°C):
MW: 253.30   **BP** (°C):

Solubility (Moles/L)	Solubility (Grams/L)	Temp (°C)	Ref (#)	Evaluation (T P E A A)	Comments
1.153E-01	2.920E+01	20	K050	1 1 1 1 2	

3143. C₁₃H₁₉NO₄S

Probenecid
Parabenem
4-((Dipropylamino)sulfonyl)benzoic acid
p-(Dipropylsulfamoyl)benzoic
RN: 57-66-9   **MP** (°C):   195
MW: 285.36   **BP** (°C):

Solubility (Moles/L)	Solubility (Grams/L)	Temp (°C)	Ref (#)	Evaluation (T P E A A)	Comments
1.262E-05	3.600E-03	22.5	B422	2 0 2 2 2	
2.089E-06	5.962E-04	ns	R427	0 0 0 0 0	

3144. $C_{13}H_{19}N_3O_4$
N-(1-Ethylpropyl)-2,6-dinitro-3,4-xylidine
Pendimethalin

RN:	40487-42-1	**MP** (°C):	56.5
MW:	281.31	**BP** (°C):	

Solubility (Moles/L)	Solubility (Grams/L)	Temp (°C)	Ref (#)	Evaluation (T P E A A)	Comments
1.066E-06	3.000E-04	20	M161	1 0 0 0 0	
1.081E-03	3.040E-01	ns	B185	0 0 0 0 0	
1.066E-06	3.000E-04	ns	V414	0 0 0 0 0	

3145. $C_{13}H_{19}N_3O_6S$
Nitralin
4-(Methylsulfonyl)-2,6-dinitro-N,N-dipropylaniline

RN:	4726-14-1	**MP** (°C):	151
MW:	345.38	**BP** (°C):	

Solubility (Moles/L)	Solubility (Grams/L)	Temp (°C)	Ref (#)	Evaluation (T P E A A)	Comments
1.737E-06	6.000E-04	22	M161	1 0 0 0 0	
1.737E-06	6.000E-04	25	B200	1 0 0 0 0	
1.737E-07	6.000E-05	25	P028	0 0 0 0 0	
1.737E-06	6.000E-04	ns	M061	0 0 0 0 0	

3146. $C_{13}H_{20}N_2O$
Prilocaine

RN:	721-50-6	**MP** (°C):	
MW:	220.32	**BP** (°C):	

Solubility (Moles/L)	Solubility (Grams/L)	Temp (°C)	Ref (#)	Evaluation (T P E A A)	Comments
2.800E-02	6.169E+00	25	D402	1 2 2 2 0	EFG
2.900E-02	6.389E+00	37	D402	1 2 2 2 0	EFG

3147. $C_{13}H_{20}N_2O_2$
Procaine
Novacaine
Novokain

RN:	59-46-1	**MP** (°C):	60
MW:	236.32	**BP** (°C):	

Solubility (Moles/L)	Solubility (Grams/L)	Temp (°C)	Ref (#)	Evaluation (T P E A A)	Comments
4.000E-02	9.453E+00	30	L068	1 0 0 1 0	EFG
4.200E-02	9.925E+00	37.5	L034	2 2 0 1 2	pH 7.4
5.494E-03	1.298E+00	ns	E031	0 0 2 1 2	
2.700E-02	6.381E+00	ns	M066	0 0 0 0 1	

3148. C$_{13}$H$_{20}$N$_2$O$_2$

N,N′-Diethyl-bicyclo(2.2.1)hept-5-ene-2,3-*trans*-dicarboxamide

RN: **MP** (°C):
MW: 236.32   **BP** (°C):

Solubility (Moles/L)	Solubility (Grams/L)	Temp (°C)	Ref (#)	Evaluation (T P E A A)	Comments
3.216E-02	7.600E+00	20	K050	1 1 1 1 2	

3149. C$_{13}$H$_{20}$N$_2$O$_2$

4-Aminobenzoic acid-2-(butyl-amino)ethyl ester
2-(Butylamino)ethyl 4-aminobenzoate

RN: **MP** (°C):
MW: 236.32   **BP** (°C):

Solubility (Moles/L)	Solubility (Grams/L)	Temp (°C)	Ref (#)	Evaluation (T P E A A)	Comments
1.700E-04	4.017E-02	ns	M066	0 0 0 0 1	

3150. C$_{13}$H$_{20}$N$_2$O$_3$

5-Allyl-5-ethylbutylbarbituric acid

RN: **MP** (°C):
MW: 252.32   **BP** (°C):

Solubility (Moles/L)	Solubility (Grams/L)	Temp (°C)	Ref (#)	Evaluation (T P E A A)	Comments
1.587E-02	4.004E+00	20	J030	1 2 2 2 2	
2.579E-02	6.507E+00	37	J030	1 2 2 2 2	

3151. C$_{13}$H$_{20}$N$_2$O$_3$

2,4,6(1H,3H,5H)-Pyrimidinetrione, 5-(1,1-dimethylethyl)-5-(3-methyl-2-butenyl)
5-*t*-Butyl-5-(3-methylbut-2-enyl)barbiturate

RN: 143585-02-8   **MP** (°C):
MW: 252.32   **BP** (°C):

Solubility (Moles/L)	Solubility (Grams/L)	Temp (°C)	Ref (#)	Evaluation (T P E A A)	Comments
2.810E-04	7.090E-02	25	P350	0 0 0 0 0	intrinsic

3152. C$_{13}$H$_{20}$O

2-Hexyl-6-methylphenol
o-Cresol, 6-hexyl-

RN: 106593-25-3   **MP** (°C):
MW: 192.30   **BP** (°C):

Solubility (Moles/L)	Solubility (Grams/L)	Temp (°C)	Ref (#)	Evaluation (T P E A A)	Comments
2.600E-05	5.000E-03	25	L020	1 0 0 0 0	

3153. $C_{13}H_{20}O$
2-Hexyl-4-methylphenol
2-Hexyl-p-cresol

RN: 54612-53-2   **MP** (°C):
MW: 192.30   **BP** (°C):

Solubility (Moles/L)	Solubility (Grams/L)	Temp (°C)	Ref (#)	Evaluation (T P E A A)	Comments
3.467E-05	6.667E-03	25	L020	1 0 0 0 0	

3154. $C_{13}H_{20}O$
b-Damascone
b-Damascone, *trans*-
trans-2,6,6-Trimethyl-1-crotonylcyclohex-1-ene
trans-b-Damascone
Damascone β

RN: 23726-91-2   **MP** (°C):
MW: 192.30   **BP** (°C):

Solubility (Moles/L)	Solubility (Grams/L)	Temp (°C)	Ref (#)	Evaluation (T P E A A)	Comments
1.000E-03	1.923E-01	25	D407	1 0 2 2 2	

3155. $C_{13}H_{20}O$
β-Damascone
4-(2,6,6-Trimethyl cyclohex-1-enyl)but-2-en-4-one
Damasione

RN: 23726-92-3   **MP** (°C):
MW: 192.30   **BP** (°C):

Solubility (Moles/L)	Solubility (Grams/L)	Temp (°C)	Ref (#)	Evaluation (T P E A A)	Comments
1.000E-03	1.923E-01	ns	S460	0 0 0 0 0	

3156. $C_{13}H_{20}O$
o-n-Heptylphenol
2-n-Heptylphenol

RN: 5284-22-0   **MP** (°C):
MW: 192.30   **BP** (°C):

Solubility (Moles/L)	Solubility (Grams/L)	Temp (°C)	Ref (#)	Evaluation (T P E A A)	Comments
6.118E-05	1.176E-02	25	L022	1 0 0 0 0	

3157. C₁₃H₂₀O

4-Hexyl-2-methylphenol
o-Cresol, 4-hexyl-
RN: 3280-61-3   **MP** (°C):
MW: 192.30   **BP** (°C):

Solubility (Moles/L)	Solubility (Grams/L)	Temp (°C)	Ref (#)	Evaluation (T P E A A)	Comments
2.600E-05	5.000E-03	25	L020	1 0 0 0 0	

3158. C₁₃H₂₀O

α-Ionone
α-Irisone
Cyclocitrylideneacetone
Ionone α
Buten-2-one, 4-(2,6,6-trimethyl-2-cyclohexen-1-yl)-
RN: 127-41-3   **MP** (°C):
MW: 192.30   **BP** (°C):   229

Solubility (Moles/L)	Solubility (Grams/L)	Temp (°C)	Ref (#)	Evaluation (T P E A A)	Comments
5.508E-04	1.059E-01	ns	S460	0 0 0 0 0	

3159. C₁₃H₂₀O

p-n-Heptylphenol
4-*n*-Heptylphenol
RN: 1987-50-4   **MP** (°C):
MW: 192.30   **BP** (°C):

Solubility (Moles/L)	Solubility (Grams/L)	Temp (°C)	Ref (#)	Evaluation (T P E A A)	Comments
5.778E-05	1.111E-02	25	L022	1 0 0 0 0	

3160. C₁₃H₂₁NO₃

Salbutamol
Albuterol
Ventolin
RN: 18559-94-9   **MP** (°C):   151
MW: 239.32   **BP** (°C):

Solubility (Moles/L)	Solubility (Grams/L)	Temp (°C)	Ref (#)	Evaluation (T P E A A)	Comments
7.400E-02	1.771E+01	20	M380	1 0 2 1 0	EFG
7.500E-02	1.795E+01	25	M380	1 0 2 1 0	EFG
7.400E-02	1.771E+01	37	M380	1 0 2 1 0	EFG
5.885E-02	1.408E+01	ns	A092	0 0 0 0 0	

3161. C₁₃H₂₁O₃PS

3161. $C_{13}H_{21}O_3PS$

S-Benzyl *O,O*-di-isopropyl phosphorothioate
Isokitazine
Kitazin P
IBP
Iprobenfos
Kitazin L

RN: 26087-47-8 **MP** (°C):
MW: 288.35 **BP** (°C): 126

Solubility (Moles/L)	Solubility (Grams/L)	Temp (°C)	Ref (#)	Evaluation (T P E A A)	Comments
3.468E-03	1.000E+00	18	M161	1 0 0 0 0	

3162. $C_{13}H_{21}O_4PS$

4-(Methylthio)phenyl dipropyl phosphate
O,O-Dipropyl *O*-4-methylthiophenyl phosphate
Propaphos
Kayaphos
Kayphosnac

RN: 7292-16-2 **MP** (°C):
MW: 304.35 **BP** (°C): 176

Solubility (Moles/L)	Solubility (Grams/L)	Temp (°C)	Ref (#)	Evaluation (T P E A A)	Comments
4.107E-04	1.250E-01	25	M161	1 0 0 0 2	

3163. $C_{13}H_{22}NO_3PS$

Fenamiphos
1-(Methylethyl)-*O*-ethyl-*O*-(3-methyl-4-(methylthio)phenyl)phosphoramidate
Nemacur
Bay 68138

RN: 22224-92-6 **MP** (°C):
MW: 303.36 **BP** (°C):

Solubility (Moles/L)	Solubility (Grams/L)	Temp (°C)	Ref (#)	Evaluation (T P E A A)	Comments
1.008E-03	3.059E-01	10	B324	0 0 0 0 0	
1.009E-03	3.061E-01	10	B324	0 0 0 0 0	
2.291E-03	6.950E-01	20	B179	0 0 0 0 0	
1.084E-03	3.288E-01	20	B300	2 1 1 1 2	
1.085E-03	3.291E-01	20	B324	0 0 0 0 0	
1.084E-03	3.289E-01	20	B324	0 0 0 0 0	
1.381E-03	4.189E-01	30	B324	0 0 0 0 0	
1.381E-03	4.188E-01	30	B324	0 0 0 0 0	
2.307E-03	7.000E-01	rt	M161	0 0 0 0 2	

3164. $C_{13}H_{22}N_2O$

Isonoruron

Urea, 3-[hexahydro-4,7-methanoindan-1(or 2)-yl]-1,1-dimethyl-

Tricuron

BAS 2103H

RN:	28346-65-8	**MP** (°C):	165	
MW:	222.33	**BP** (°C):		

Solubility (Moles/L)	Solubility (Grams/L)	Temp (°C)	Ref (#)	Evaluation (T P E A A)	Comments
9.895E-04	2.200E-01	20	M161	1 0 0 0 2	

3165. $C_{13}H_{22}N_2O$

Noruron

3-(Hexahydro-4,7-methanoindan-5-yl)-1,1-dimethylurea

Norea

RN:	18530-56-8	**MP** (°C):	171	
MW:	222.33	**BP** (°C):		

Solubility (Moles/L)	Solubility (Grams/L)	Temp (°C)	Ref (#)	Evaluation (T P E A A)	Comments
6.747E-04	1.500E-01	20	M061	1 0 0 0 2	
6.747E-04	1.500E-01	25	B200	1 0 0 0 2	
6.747E-04	1.500E-01	ns	G036	0 0 0 0 2	

3166. $C_{13}H_{22}N_2O_3$

5-Ethyl-5-n-heptylbarbituric acid

5-Ethyl-5-heptylbarbituric acid

5-Ethyl-5-heptylbarbiturate

RN:	60784-70-5	**MP** (°C):		
MW:	254.33	**BP** (°C):		

Solubility (Moles/L)	Solubility (Grams/L)	Temp (°C)	Ref (#)	Evaluation (T P E A A)	Comments
6.050E-04	1.539E-01	25	M310	2 2 2 2 2	

3167. $C_{13}H_{22}O_3$

Methyl dihydrojasmonate

Hedione

Methyl 3-oxo-2-pentylcyclopentaneacetate

Claigeon

RN:	24851-98-7	**MP** (°C):		
MW:	226.32	**BP** (°C):		

Solubility (Moles/L)	Solubility (Grams/L)	Temp (°C)	Ref (#)	Evaluation (T P E A A)	Comments
1.767E-03	3.998E-01	25	M350	1 0 1 1 1	

3168. C$_{13}$H$_{24}$N$_3$O$_3$PS
Pirimiphos-ethyl
Diethyl *O*-(2-(diethylamino)-6-methyl-4-pyrimidinyl) phosphorothioate
Fernex
Primotec
Solgard

RN: 23505-41-1	**MP** (°C):			
MW: 333.39	**BP** (°C):			

Solubility (Moles/L)	Solubility (Grams/L)	Temp (°C)	Ref (#)	Evaluation (T P E A A)	Comments
1.190E-05	3.967E-03	20	B300	2 1 1 1 2	
<3.00E-06	<1.00E-03	30	M161	1 0 0 0 0	

3169. C$_{13}$H$_{24}$N$_4$O$_3$S
Bupirimate
5-Butyl-2-(ethylamino)-6-methyl-4-pyrimidinyl dimethylsulfamate
Nimrod

RN: 41483-43-6	**MP** (°C): 50.5			
MW: 316.43	**BP** (°C):			

Solubility (Moles/L)	Solubility (Grams/L)	Temp (°C)	Ref (#)	Evaluation (T P E A A)	Comments
6.918E-05	2.189E-02	ns	R427	0 0 0 0 0	
6.953E-05	2.200E-02	rt	M161	0 0 0 0 1	

3170. C$_{13}$H$_{24}$N$_6$
1-(Hexamethyleneiminel)-3,5-bis(dimethylamino)-*s*-triazine
1,3,5-Triazine-2,4-diamine, 6-(hexahydro-1H-azepin-1-yl)-*N,N,N′,N′*-tetramethyl-

RN: 125867-92-7	**MP** (°C):			
MW: 264.38	**BP** (°C):			

Solubility (Moles/L)	Solubility (Grams/L)	Temp (°C)	Ref (#)	Evaluation (T P E A A)	Comments
2.265E-05	5.988E-03	25	B386	0 0 0 0 0	

3171. C$_{13}$H$_{24}$O$_4$
Octyl α-acetoxypropionate
Propanoic acid, 2-(acetyloxy)-, octyl ester

RN: 6283-90-5	**MP** (°C):			
MW: 244.33	**BP** (°C):			

Solubility (Moles/L)	Solubility (Grams/L)	Temp (°C)	Ref (#)	Evaluation (T P E A A)	Comments
4.093E-04	1.000E-01	25	R006	2 2 0 1 1	

3172. C$_{13}$H$_{24}$O$_4$
1,11-Undecanedicarboxylic acid
1,13-Tridecanedioic acid
Brassylic acid

RN:	505-52-2	**MP** (°C):	111
MW:	244.33	**BP** (°C):	

Solubility (Moles/L)	Solubility (Grams/L)	Temp (°C)	Ref (#)	Evaluation (T P E A A)	Comments
6.139E-03	1.500E+00	21	B040	1 0 1 1 1	*sic*
1.637E-04	4.000E-02	24	F300	1 0 0 0 0	*sic*

3173. C$_{13}$H$_{25}$NO$_3$
Dibutylaceturethane

RN:		**MP** (°C):	
MW:	243.35	**BP** (°C):	

Solubility (Moles/L)	Solubility (Grams/L)	Temp (°C)	Ref (#)	Evaluation (T P E A A)	Comments
3.287E-04	7.999E-02	44	O021	1 2 0 0 0	

3174. C$_{13}$H$_{26}$N$_2$O$_2$
N,N,N′,N′-Tetramethylazelamide
Nonanediamide, *N,N,N′,N′*-tetramethyl-

RN:	13424-87-8	**MP** (°C):	
MW:	242.36	**BP** (°C):	

Solubility (Moles/L)	Solubility (Grams/L)	Temp (°C)	Ref (#)	Evaluation (T P E A A)	Comments
3.900E+00	9.452E+02	30	D010	1 2 1 1 2	

3175. C$_{13}$H$_{26}$O$_2$
n-Tridecanoic acid
Tridecanoic acid

RN:	638-53-9	**MP** (°C):	41.5
MW:	214.35	**BP** (°C):	236

Solubility (Moles/L)	Solubility (Grams/L)	Temp (°C)	Ref (#)	Evaluation (T P E A A)	Comments
9.797E-05	2.100E-02	0	B136	1 0 2 1 1	
1.540E-04	3.300E-02	20	B136	1 0 2 1 1	
1.539E-04	3.300E-02	20.0	R001	1 1 1 1 1	
1.773E-04	3.800E-02	30	B136	1 0 2 1 1	
1.773E-04	3.800E-02	30.0	R001	1 1 1 1 1	
2.053E-04	4.400E-02	45	B136	1 0 2 1 1	
2.053E-04	4.400E-02	45.0	R001	1 1 1 1 1	
2.519E-04	5.400E-02	60	B136	1 0 2 1 1	
2.519E-04	5.400E-02	60.0	R001	1 1 1 1 1	
9.797E-05	2.100E-02	.0	R001	1 1 1 1 1	

3176. C₁₃H₂₆O₂

Methyl laurate
Dodecanoic acid methyl ester
Methyl dodecanoate

RN:	111-82-0	MP (°C):	41
MW:	214.35	BP (°C):	261

Solubility (Moles/L)	Solubility (Grams/L)	Temp (°C)	Ref (#)	Evaluation (T P E A A)	Comments
<2.05E-05	<4.40E-03	20	M337	2 1 2 2 1	

3177. C₁₃H₂₆O₃

n-Octyl β-ethoxypropionate

RN:		MP (°C):	
MW:	230.35	BP (°C):	

Solubility (Moles/L)	Solubility (Grams/L)	Temp (°C)	Ref (#)	Evaluation (T P E A A)	Comments
<4.34E-04	<10.00E-02	25	D002	1 2 1 1 0	

3178. C₁₃H₂₆O₃

Decyl lactate
2-Hydroxypropionic acid decyl ester

RN:	42175-34-8	MP (°C):	
MW:	230.35	BP (°C):	

Solubility (Moles/L)	Solubility (Grams/L)	Temp (°C)	Ref (#)	Evaluation (T P E A A)	Comments
8.682E-04	2.000E-01	25	R006	2 2 0 1 0	

3179. C₁₃H₂₆O₄

1,3-Dioxolane-4-methanol, 2-[2-(hexyloxy)ethyl]-2-methyl

RN:	124485-63-8	MP (°C):	
MW:	246.35	BP (°C):	

Solubility (Moles/L)	Solubility (Grams/L)	Temp (°C)	Ref (#)	Evaluation (T P E A A)	Comments
1.600E-02	3.942E+00	25	P342	0 0 0 0 0	0.0001M Na₂CO₃

3180. C₁₃H₂₈

Tridecane

RN:	629-50-5	MP (°C):	−5.5
MW:	184.37	BP (°C):	235.4

Solubility (Moles/L)	Solubility (Grams/L)	Temp (°C)	Ref (#)	Evaluation (T P E A A)	Comments
2.170E-09	4.000E-07	25	T423	0 0 0 0 0	

3181. C₁₄H₄N₂O₂S₂

Dithianon
1,4-Dithiaanthraquinone-2,3-dinitrile
2,3-Dicyano-1,4-dithiaanthraquinone
RN: 3347-22-6   **MP (°C):**   225
MW: 296.33   **BP (°C):**

Solubility (Moles/L)	Solubility (Grams/L)	Temp (°C)	Ref (#)	Evaluation (T P E A A)	Comments
1.687E-06	5.000E-04	ns	A305	0 0 0 0 0	
4.677E-07	1.386E-04	ns	R427	0 0 0 0 0	

3182. C₁₄H₆Cl₂F₄N₂O₂

Teflubenzuron
Nomolt
RN: 83121-18-0   **MP (°C):**
MW: 381.12   **BP (°C):**

Solubility (Moles/L)	Solubility (Grams/L)	Temp (°C)	Ref (#)	Evaluation (T P E A A)	Comments
2.466E-08	9.400E-06	20	M402	0 0 0 0 0	

3183. C₁₄H₆N₂O₄

1,4,5,8-Naphthalenediimide
1,4,5,8-Naphthalenetetracarboxylic 1,8:4,5-diimide
1,4,5,8-Naphthalenetetracarboxylic 1,8:4,5-diimide
1,4,5,8-Naphthalenetetracarboxylic acid diimide
1,4,5,8-Naphthalenetetracarboxylic diimide
Benzo[lmn][3,8]phenanthroline-1,3,6,8(2H,7H)-tetrone
RN: 5690-24-4   **MP (°C):**
MW: 266.21   **BP (°C):**   656.2

Solubility (Moles/L)	Solubility (Grams/L)	Temp (°C)	Ref (#)	Evaluation (T P E A A)	Comments
		23	B410	2 1 2 2 2	

3184. C₁₄H₆O₈

Ellagic acid
2,3,7,8-Tetrahydroxy(1)benzopyrano(5,4,3-cde)(1)benzopyran-5,10-dione
Elagostasine
Benzoaric acid
Alizarine yellow
4,4′,5,5′,6,6′-Hexahydrodiphenic acid 2,6,2′,6′-dilactone
RN: 476-66-4   **MP (°C):**   >360
MW: 302.20   **BP (°C):**

Solubility (Moles/L)	Solubility (Grams/L)	Temp (°C)	Ref (#)	Evaluation (T P E A A)	Comments
3.210E-05	9.700E-03	37	B438	0 0 0 0 0	

3185. C$_{14}$H$_7$ClO$_5$S
1,5-Chloroanthraquinone sulfonic acid
1-Anthracenesulfonic acid, 5-chloro-9,10-dihydro-9,10-dioxo-
RN: **MP** (°C):
MW: 322.73   **BP** (°C):

Solubility (Moles/L)	Solubility (Grams/L)	Temp (°C)	Ref (#)	Evaluation (T P E A A)	Comments
1.033E+00	3.333E+02	18	F047	1 2 1 1 0	

3186. C$_{14}$H$_7$ClO$_5$S
1,7-Chloroanthraquinone sulfonic acid
RN: **MP** (°C):
MW: 322.73   **BP** (°C):

Solubility (Moles/L)	Solubility (Grams/L)	Temp (°C)	Ref (#)	Evaluation (T P E A A)	Comments
6.197E-01	2.000E+02	18	F047	1 2 1 1 0	

3187. C$_{14}$H$_7$ClO$_5$S
1,6-Chloroanthraquinone sulfonic acid
2-Anthracenesulfonic acid, 5-chloro-9,10-dihydro-9,10-dioxo-
RN: 300360-23-0   **MP** (°C):
MW: 322.73   **BP** (°C):

Solubility (Moles/L)	Solubility (Grams/L)	Temp (°C)	Ref (#)	Evaluation (T P E A A)	Comments
6.197E-01	2.000E+02	18	F047	1 2 1 1 0	

3188. C$_{14}$H$_8$Cl$_2$N$_4$
Clofentezine
3,6-bis(2-Chlorophenyl)-1,2,4,5-tetrazine
Apollo
Acaristop
bis(2-Chlorophenyl)-1,2,4,5-tetrazine
Panatac
RN: 74115-24-5   **MP** (°C):   182.3
MW: 303.15   **BP** (°C):

Solubility (Moles/L)	Solubility (Grams/L)	Temp (°C)	Ref (#)	Evaluation (T P E A A)	Comments
8.318E-09	2.522E-06	ns	R424	0 0 0 0 0	
8.318E-09	2.522E-06	ns	R427	0 0 0 0 0	

3189. C$_{14}$H$_8$Cl$_4$

2,4'-Dichlorodiphenyldichloroethylene
1-(2-Chlorophenyl)-1-(4-chlorophenyl)-2,2-dichloroethylene
o,p'-DDE

RN: 3424-82-6   **MP (°C):**   76.5
MW: 318.03   **BP (°C):**

Solubility (Moles/L)	Solubility (Grams/L)	Temp (°C)	Ref (#)	Evaluation (T P E A A)	Comments
4.402E-07	1.400E-04	25	B083	2 2 1 2 2	particle size 5 µm

3190. C$_{14}$H$_8$Cl$_4$

p,p'-Dichlorodiphenyldichloroethylene
2,2-bis(4-Chlorophenyl)-1,1-dichloroethylene
p,p'-DDE

RN: 72-55-9   **MP (°C):**   89.0
MW: 318.03   **BP (°C):**

Solubility (Moles/L)	Solubility (Grams/L)	Temp (°C)	Ref (#)	Evaluation (T P E A A)	Comments
1.729E-07	5.500E-05	15	B083	2 2 1 2 1	particle size 5 µm
1.258E-07	4.000E-05	20	C053	0 0 0 0 0	
1.258E-07	4.000E-05	20	F071	1 1 2 1 1	
3.773E-07	1.200E-04	25	B083	2 2 1 2 2	particle size 5 µm
3.773E-07	1.200E-04	25	I308	0 0 0 0 0	
4.088E-09	1.300E-06	25	M134	1 2 1 1 1	
4.402E-08	1.400E-05	25	W025	1 0 1 1 1	
7.389E-07	2.350E-04	35	B083	2 2 1 2 2	particle size 5 µm
1.415E-06	4.500E-04	45	B083	2 2 1 2 2	particle size 5 µm
4.717E-09	1.500E-06	ns	M110	0 0 0 0 0	EFG
4.088E-09	1.300E-06	ns	M118	0 1 1 1 1	

3191. C$_{14}$H$_8$O$_2$

Anthraquinone
9,10-Anthraquinone
9,10-Dioxoanthracene
Corbit
Morkit
Hoelite

RN: 84-65-1   **MP (°C):**   286
MW: 208.22   **BP (°C):**   377

Solubility (Moles/L)	Solubility (Grams/L)	Temp (°C)	Ref (#)	Evaluation (T P E A A)	Comments
6.500E-06	1.353E-03	25	E014	2 2 2 1 1	pH 7.3
3.000E-06	6.247E-04	ns	G077	0 0 0 0 1	

3192. C$_{14}$H$_8$O$_4$
Alizarin
Alizarine
C.I. Mordant red 11

RN:	72-48-0	MP (°C):	290
MW:	240.22	BP (°C):	430

Solubility (Moles/L)	Solubility (Grams/L)	Temp (°C)	Ref (#)	Evaluation (T P E A A)	Comments
1.300E-05	3.123E-03	25	B333	0 0 0 0 0	*sic*
1.664E-03	3.998E-01	rt	D021	0 0 1 1 1	*sic*

3193. C$_{14}$H$_8$O$_4$
Quinizarin
1,4-Dihydroxyanthraquinone
C.I. Pigment violet 12

RN:	81-64-1	MP (°C):	192
MW:	240.22	BP (°C):	

Solubility (Moles/L)	Solubility (Grams/L)	Temp (°C)	Ref (#)	Evaluation (T P E A A)	Comments
4.000E-07	9.609E-05	25	B333	0 0 0 0 0	
6.000E-05	1.441E-02	98.59	M180	0 0 2 2 0	EFG
9.200E-05	2.210E-02	111.46	M180	0 0 2 2 0	EFG
1.100E-04	2.642E-02	117.47	M180	0 0 2 2 0	EFG
1.800E-04	4.324E-02	123.67	M180	0 0 2 2 0	EFG
2.000E-04	4.804E-02	126.84	M180	0 0 2 2 0	EFG
2.100E-04	5.045E-02	135.00	M180	0 0 2 2 0	EFG
4.900E-04	1.177E-01	141.78	M180	0 0 2 2 0	EFG
7.500E-04	1.802E-01	152.37	M180	0 0 2 2 0	EFG

3194. C$_{14}$H$_8$O$_5$
Purpurin
1,2,4-Trihydroxy-anthrachinon

RN:	81-54-9	MP (°C):	
MW:	256.22	BP (°C):	

Solubility (Moles/L)	Solubility (Grams/L)	Temp (°C)	Ref (#)	Evaluation (T P E A A)	Comments
2.500E-05	6.405E-03	25	B333	0 0 0 0 0	

3195. C₁₄H₈O₆

Quinalizarin
1,2,5,8-Tetrahydroxyanthraquinone
9,10-Anthracenedione
Alizarine Bordeaux B
Mordant violet 26

RN: 81-61-8   **MP** (°C):
MW: 272.22   **BP** (°C):

Solubility (Moles/L)	Solubility (Grams/L)	Temp (°C)	Ref (#)	Evaluation (T P E A A)	Comments
9.500E-06	2.586E-03	25	B333	0 0 0 0 0	

3196. C₁₄H₈O₈S₂

Anthraquinone-1,8-disulfonic acid
1,8-Disulfonic acid anthraquinone
Anthrachinon-disulfosaeure-(1,8)
1,8-Anthraquinone disulfonic acid

RN: 82-48-4   **MP** (°C):   293
MW: 368.34   **BP** (°C):

Solubility (Moles/L)	Solubility (Grams/L)	Temp (°C)	Ref (#)	Evaluation (T P E A A)	Comments
1.086E+00	4.000E+02	18	F047	1 2 1 1 1	

3197. C₁₄H₈O₈S₂

1,6-Anthraquinone disulfonic acid
Anthraquinone-1,6-disulfonic acid

RN: 14486-58-9   **MP** (°C):   216
MW: 368.34   **BP** (°C):

Solubility (Moles/L)	Solubility (Grams/L)	Temp (°C)	Ref (#)	Evaluation (T P E A A)	Comments
1.357E+00	5.000E+02	18	F047	1 2 1 1 0	

3198. C₁₄H₈O₈S₂

1,5-Anthraquinone disulfonic acid
Anthraquinone-1,5-disulfonic acid

RN: 252967-17-2   **MP** (°C):   310.0
MW: 368.34   **BP** (°C):

Solubility (Moles/L)	Solubility (Grams/L)	Temp (°C)	Ref (#)	Evaluation (T P E A A)	Comments
1.086E+00	4.000E+02	18	F047	1 2 1 1 1	

3199. $C_{14}H_9ClF_2N_2O_2$
Difluron
Diflubenzuron
TH 6040

RN: 35367-38-5 **MP (°C):** 239
MW: 310.69 **BP (°C):**

Solubility (Moles/L)	Solubility (Grams/L)	Temp (°C)	Ref (#)	Evaluation (T P E A A)	Comments
6.437E-07	2.000E-04	20	M161	1 0 0 0 0	
2.865E-07	8.900E-05	20	M402	0 0 0 0 0	
6.437E-07	2.000E-04	20	R303	1 0 0 0 0	
9.656E-07	3.000E-04	24	C105	2 1 2 2 2	
1.609E-06	5.000E-04	ns	M110	0 0 0 0 0	EFG
2.570E-07	7.986E-05	ns	R427	0 0 0 0 0	

3200. $C_{14}H_9ClF_3NO_2$
Efavirenz
8-Chloro-5-(2-cyclopropylethynyl)-5-(trifluoromethyl)-4-oxa-2-azabicyclo [4.4.0]deca-7,9,11-
    trien-3-one

RN: 154598-52-4 **MP (°C):**
MW: 315.68 **BP (°C):**

Solubility (Moles/L)	Solubility (Grams/L)	Temp (°C)	Ref (#)	Evaluation (T P E A A)	Comments
2.534E-05	8.000E-03	ns	A426	0 0 0 0 0	intrinsic
3.168E-05	1.000E-02	ns	K444	0 0 0 0 0	

3201. $C_{14}H_9Cl_2NO_5$
Bifenox
5-(2,4-Dichlorphenoxy)-2-nitro-benzoic acid methyl ester
Modown 4 flowable
Modown

RN: 42576-02-3 **MP (°C):** 85
MW: 342.14 **BP (°C):**

Solubility (Moles/L)	Solubility (Grams/L)	Temp (°C)	Ref (#)	Evaluation (T P E A A)	Comments
1.461E-06	5.000E-04	ns	M110	0 0 0 0 0	EFG
1.023E-06	3.500E-04	ns	M161	0 0 0 0 1	
1.023E-06	3.501E-04	ns	R427	0 0 0 0 0	

3202. C$_{14}$H$_9$Cl$_5$
o,p'-DDT
1-(2-Chlorophenyl)-1-(4-chlorophenyl)-2,2,2-trichloroethane
2,4'-DDT
2-(2-Chlorophenyl)-2-(4-chlorophenyl)-1,1,1-trichloroethane

RN:	789-02-6	**MP** (°C):	74.0
MW:	354.49	**BP** (°C):	

Solubility (Moles/L)	Solubility (Grams/L)	Temp (°C)	Ref (#)	Evaluation (T P E A A)	Comments
1.410E-07	5.000E-05	15	B083	2 2 1 2 1	particle size 5 μm
2.398E-07	8.500E-05	25	B083	2 2 1 2 1	particle size 5 μm
2.398E-07	8.500E-05	25	I308	0 0 0 0 0	
7.334E-08	2.600E-05	25	W025	1 0 2 2 1	
3.808E-07	1.350E-04	35	B083	2 2 1 2 2	particle size 5 μm
5.642E-07	2.000E-04	45	B083	2 2 1 2 2	particle size 5 μm

3203. C$_{14}$H$_9$Cl$_5$
p,p'-DDT
2,2-bis(*p*-Chlorophenyl)-1,1,1-trichloroethane
p,p'-TDEE

RN:	50-29-3	**MP** (°C):	108.5
MW:	354.49	**BP** (°C):	260

Solubility (Moles/L)	Solubility (Grams/L)	Temp (°C)	Ref (#)	Evaluation (T P E A A)	Comments
3.385E-09	1.200E-06	0	G319	0 0 0 0 0	
1.664E-08	5.900E-06	2	B186	2 0 2 2 2	
4.796E-08	1.700E-05	15	B083	2 2 1 2 1	particle size 5 μm
1.834E-07	6.500E-05	15	B083	2 2 1 2 1	particle size 5 μm
2.800E-07	9.926E-05	18	G054	1 0 1 0 1	
1.410E-08	5.000E-06	20	C111	1 0 0 0 0	
1.410E-08	5.000E-06	20	C113	1 0 2 1 1	
1.128E-07	4.000E-05	20	E048	1 2 1 1 0	
2.172E-08	7.700E-06	20	F303	1 2 1 2 1	
2.172E-08	7.700E-06	20	W319	1 2 1 2 1	
1.552E-08	5.500E-06	24	C311	0 0 0 0 0	EFG
1.523E-08	5.400E-06	24	C313	0 0 0 0 0	
2.821E-09	1.000E-06	24	K069	2 0 0 1 1	
7.079E-08	2.510E-05	24.99	K436	0 0 0 0 0	
3.385E-09	1.200E-06	25	B036	1 1 0 1 1	
3.949E-07	1.400E-04	25	B083	2 2 1 2 2	particle size 5 μm
7.052E-08	2.500E-05	25	B083	2 2 1 2 1	particle size 5 μm
4.796E-09	1.700E-06	25	B093	2 2 2 2 1	
1.055E-07	3.740E-05	25	B186	2 0 2 2 2	
9.168E-09	3.250E-06	25	F071	1 1 2 1 1	
3.385E-09	1.200E-06	25	M040	1 0 0 1 1	
3.385E-09	1.200E-06	25	M130	1 0 0 0 1	
2.821E-09	1.000E-06	25	P085	0 0 0 0 0	
1.552E-08	5.500E-06	25	W025	1 0 2 2 1	
3.385E-09	1.200E-06	26.70	L095	2 2 1 1 2	

(continued)

3203. $C_{14}H_9Cl_5$ (continued)

Solubility (Moles/L)	Solubility (Grams/L)	Temp (°C)	Ref (#)	Evaluation (T P E A A)	Comments
1.044E-07	3.700E-05	35	B083	2 2 1 2 1	particle size 5 μm
7.334E-07	2.600E-04	35	B083	2 2 1 2 2	particle size 5 μm
1.269E-07	4.500E-05	37.50	B186	2 0 2 2 2	
1.269E-07	4.500E-05	45	B083	2 2 1 2 1	particle size 5 μm
1.439E-06	5.100E-04	45	B083	2 2 1 2 2	particle size 5 μm
1.552E-08	5.500E-06	ns	C318	0 0 0 0 0	
3.385E-09	1.200E-06	ns	I300	0 0 0 0 1	
4.796E-09	1.700E-06	ns	K138	0 0 0 0 2	
2.821E-09	1.000E-06	ns	M061	0 0 0 0 0	
3.103E-09	1.100E-06	ns	M110	0 0 0 0 0	EFG
5.642E-09	2.000E-06	ns	M138	0 0 0 0 0	
8.745E-09	3.100E-06	ns	M344	0 0 0 0 1	
2.821E-08	1.000E-05	ns	V414	0 0 0 0 0	
2.539E-07	9.000E-05	ns	V414	0 0 0 0 0	

3204. $C_{14}H_9Cl_5O$

Dicofol

4-Chloro-α-(4-chlorophenyl)-α-(trichloromethyl)benzenemethanol

4,4′-Dichloro-α-(trichloromethyl)benzhydrol

Acarin

Carbox

Cekudifol

RN: 115-32-2   **MP** (°C):   79

MW: 370.49   **BP** (°C):

Solubility (Moles/L)	Solubility (Grams/L)	Temp (°C)	Ref (#)	Evaluation (T P E A A)	Comments
3.563E-06	1.320E-03	25	W025	1 0 2 2 2	

3205. $C_{14}H_9F$

1-Fluoroanthracene

RN: 7651-80-1   **MP** (°C):

MW: 196.23   **BP** (°C):

Solubility (Moles/L)	Solubility (Grams/L)	Temp (°C)	Ref (#)	Evaluation (T P E A A)	Comments
1.325E-06	2.600E-04	ns	M344	0 0 0 0 2	

3206. C$_{14}$H$_9$NO$_2$
2-Aminoanthraquinone
2-Amino-9,10-anthracenedione
2-Amino-9,10-anthraquinone
Aminoanthraquinone
AAQ

RN: 117-79-3   **MP** (°C): 310
MW: 223.23   **BP** (°C):

Solubility (Moles/L)	Solubility (Grams/L)	Temp (°C)	Ref (#)	Evaluation (T P E A A)	Comments
7.300E-07	1.630E-04	25	B333	0 0 0 0 0	

3207. C$_{14}$H$_9$NO$_2$
1-Aminoanthraquinone
1-Amino-9,10-anthracenedione
1-Amino-9,10-anthraquinone

RN: 82-45-1   **MP** (°C): 254
MW: 223.23   **BP** (°C):

Solubility (Moles/L)	Solubility (Grams/L)	Temp (°C)	Ref (#)	Evaluation (T P E A A)	Comments
1.400E-06	3.125E-04	25	B333	0 0 0 0 0	

3208. C$_{14}$H$_9$NO$_2$
2-Phenyl-3,1-benzoxazin-4-one
Bentranil
Linarotox
Linurotox

RN: 1022-46-4   **MP** (°C): 123.5
MW: 223.23   **BP** (°C):

Solubility (Moles/L)	Solubility (Grams/L)	Temp (°C)	Ref (#)	Evaluation (T P E A A)	Comments
2.464E-05	5.500E-03	20	M161	1 0 0 0 0	

3209. C$_{14}$H$_9$NO$_2$S
4-Benzoyl phenylisothiocyanate
4-Isothiocyanatobenzophenone

RN: 26328-59-6   **MP** (°C):
MW: 255.30   **BP** (°C):

Solubility (Moles/L)	Solubility (Grams/L)	Temp (°C)	Ref (#)	Evaluation (T P E A A)	Comments
1.400E-05	3.574E-03	25	K032	2 2 0 1 1	

3210. $C_{14}H_9NO_3$

1-Amino-4-hydroxyanthraquinone
C.I. Disperse red 15
Disperse red 15
Celliton fast pink B

RN:	116-85-8	**MP** (°C):	208
MW:	239.23	**BP** (°C):	

Solubility (Moles/L)	Solubility (Grams/L)	Temp (°C)	Ref (#)	Evaluation (T P E A A)	Comments
1.200E-06	2.871E-04	25	B333	0 0 0 0 0	
1.129E-05	2.700E-03	60	P313	0 0 0 0 0	average of 2
1.797E-05	4.300E-03	70	P313	0 0 0 0 0	average of 2
2.320E-05	5.550E-03	80	P313	0 0 0 0 0	average of 2
4.828E-05	1.155E-02	90	P313	0 0 0 0 0	average of 2
1.500E-04	3.588E-02	98.59	M180	0 0 2 2 0	EFG
2.500E-04	5.981E-02	111.46	M180	0 0 2 2 0	EFG
3.000E-04	7.177E-02	114.44	M180	0 0 2 2 0	EFG
4.500E-04	1.077E-01	122.10	M180	0 0 2 2 0	EFG
6.000E-04	1.435E-01	126.84	M180	0 0 2 2 0	EFG
6.500E-04	1.555E-01	130.07	M180	0 0 2 2 0	EFG
1.500E-03	3.588E-01	152.37	M180	0 0 2 2 0	EFG

3211. $C_{14}H_{10}$

Phenanthrene
Phenanthracene

RN:	85-01-8	**MP** (°C):	100
MW:	178.24	**BP** (°C):	340

Solubility (Moles/L)	Solubility (Grams/L)	Temp (°C)	Ref (#)	Evaluation (T P E A A)	Comments
1.462E-06	2.607E-04	-.7	N053	1 0 0 1 0	EFG
1.970E-06	3.511E-04	4.00	M082	1 1 1 2 2	
1.970E-06	3.511E-04	4.00	M151	2 1 2 2 2	
2.027E-06	3.613E-04	4.04	M183	1 2 1 1 2	
2.265E-06	4.037E-04	4.62	N053	1 0 0 1 0	EFG
2.373E-06	4.230E-04	8.50	M063	2 1 2 2 2	
2.370E-06	4.224E-04	8.50	M082	1 1 1 2 2	
2.370E-06	4.224E-04	8.50	M151	2 1 2 2 2	
2.375E-06	4.233E-04	8.54	M183	1 2 1 1 2	
2.626E-06	4.680E-04	10.00	M063	2 1 2 2 2	
2.630E-06	4.688E-04	10.00	M082	1 1 1 2 2	
2.630E-06	4.688E-04	10.00	M151	2 1 2 2 2	
2.628E-06	4.684E-04	10.04	M183	1 2 1 1 2	
3.055E-06	5.446E-04	10.13	N053	1 0 0 1 0	EFG
2.873E-06	5.120E-04	12.50	M063	2 1 2 2 2	
2.870E-06	5.115E-04	12.50	M082	1 1 1 2 2	
2.870E-06	5.115E-04	12.50	M151	2 1 2 2 2	
2.875E-06	5.124E-04	12.54	M183	1 2 1 1 2	
3.759E-06	6.700E-04	14.20	N053	1 0 0 1 0	EFG
3.372E-06	6.010E-04	15.00	M063	2 1 2 2 2	

(continued)

3211. C$_{14}$H$_{10}$ (continued)

Solubility (Moles/L)	Solubility (Grams/L)	Temp (°C)	Ref (#)	Evaluation (T P E A A)	Comments
3.370E-06	6.007E-04	15.00	M082	1 1 1 2 2	
3.370E-06	6.007E-04	15.00	M151	2 1 2 2 2	
3.375E-06	6.015E-04	15.04	M183	1 2 1 1 2	
1.500E-05	2.674E-03	20	E025	1 0 2 2 2	
6.200E-06	1.105E-03	20	H306	1 0 1 2 1	
5.061E-06	9.020E-04	20	V416	0 0 0 0 0	
4.420E-06	7.878E-04	20.00	M082	1 1 1 2 2	
4.420E-06	7.878E-04	20.00	M151	2 1 2 2 2	
4.419E-06	7.877E-04	20.04	M183	1 2 1 1 2	
4.578E-06	8.160E-04	21.00	M063	2 1 2 2 2	
4.580E-06	8.163E-04	21.00	M082	1 1 1 2 2	
4.580E-06	8.163E-04	21.00	M151	2 1 2 2 2	
4.582E-06	8.167E-04	21.04	M183	1 2 1 1 2	
7.200E-06	1.283E-03	22	A413	2 0 2 2 1	
5.582E-06	9.950E-04	24.30	M063	2 1 2 2 2	
5.360E-06	9.553E-04	24.30	M082	1 1 1 2 2	
5.360E-06	9.553E-04	24.30	M151	2 1 2 2 2	
5.363E-06	9.558E-04	24.34	M183	1 2 1 1 2	
6.284E-06	1.120E-03	24.60	W003	2 2 2 2 2	average of 2
5.577E-06	9.940E-04	25	A001	1 2 2 2 2	
6.059E-06	1.080E-03	25	B319	2 0 1 2 1	
4.617E-06	8.230E-04	25	D406	1 2 2 2 2	
6.003E-06	1.070E-03	25	E004	2 1 2 2 2	
9.000E-06	1.604E-03	25	K001	2 2 2 2 0	
5.611E-06	1.000E-03	25	L332	1 1 1 1 1	
7.238E-06	1.290E-03	25	M064	1 1 2 2 2	
6.620E-06	1.180E-03	25	M342	1 0 1 1 2	
3.815E-06	6.800E-04	25	P340	0 0 0 0 0	
7.278E-06	1.297E-03	25	T066	1 0 0 0 2	
5.610E-06	9.999E-04	25	W300	2 2 2 2 2	
5.622E-06	1.002E-03	25.00	M151	2 1 1 2 2	
6.800E-06	1.212E-03	25.04	V013	2 2 2 2 2	
5.690E-06	1.014E-03	25.35	N053	1 0 0 1 0	EFG
8.977E-06	1.600E-03	27	D003	1 0 0 1 1	
9.257E-06	1.650E-03	27	D043	2 0 0 0 2	average of 2
7.854E-06	1.400E-03	28.95	N053	1 0 0 1 0	EFG
6.845E-06	1.220E-03	29	M071	2 2 2 2 2	
6.845E-06	1.220E-03	29.00	M151	2 1 1 2 2	
7.165E-06	1.277E-03	29.90	M063	2 1 2 2 2	
7.160E-06	1.276E-03	29.90	M082	1 1 1 2 2	
7.160E-06	1.276E-03	29.90	M151	2 1 2 2 2	
8.360E-06	1.490E-03	29.90	W003	2 2 2 2 2	
6.867E-06	1.224E-03	29.94	M183	1 2 1 1 2	
8.304E-06	1.480E-03	30.30	W003	2 2 2 2 2	average of 2
1.035E-05	1.845E-03	34.53	N053	1 0 0 1 0	EFG
1.375E-05	2.450E-03	38.40	W003	2 2 2 2 2	average of 2
1.440E-05	2.566E-03	40	V416	0 0 0 0 0	
1.274E-05	2.270E-03	40.10	W003	2 2 2 2 2	average of 3
2.171E-05	3.870E-03	47.50	W003	2 2 2 2 2	average of 3

(continued)

3211. $C_{14}H_{10}$ (continued)

Solubility (Moles/L)	Solubility (Grams/L)	Temp (°C)	Ref (#)	Evaluation (T P E A A)	Comments
2.429E-05	4.330E-03	50.10	W003	2 2 2 2 2	average of 3
2.289E-05	4.080E-03	50.20	W003	2 2 2 2 2	average of 3
3.164E-05	5.640E-03	54.70	W003	2 2 2 2 2	average of 3
4.034E-05	7.190E-03	59.20	W003	2 2 2 2 2	average of 3
3.559E-05	6.344E-03	60	V416	0 0 0 0 0	
4.096E-05	7.300E-03	60.50	W003	2 2 2 2 1	average of 3
5.498E-05	9.800E-03	65.10	W003	2 2 2 2 1	average of 3
7.013E-05	1.250E-02	70.70	W003	2 2 2 2 2	average of 3
7.238E-05	1.290E-02	71.90	W003	2 2 2 2 2	
8.528E-05	1.520E-02	73.40	W003	2 2 2 2 2	
7.238E-06	1.290E-03	ns	H123	0 0 0 0 0	
7.238E-06	1.290E-03	ns	K304	0 0 0 0 2	
7.238E-06	1.290E-03	ns	M344	0 0 0 0 2	
1.500E-05	2.674E-03	ns	W005	0 0 1 2 1	

3212. $C_{14}H_{10}$

Anthracene
Paranaphthalene
Anthracin
Green oil
Anthraxcene

RN: 120-12-7   **MP** (°C): 218
MW: 178.24   **BP** (°C): 342

Solubility (Moles/L)	Solubility (Grams/L)	Temp (°C)	Ref (#)	Evaluation (T P E A A)	Comments
7.125E-08	1.270E-05	5.20	M063	2 1 2 2 2	
7.100E-08	1.265E-05	5.20	M082	1 1 1 2 1	
7.100E-08	1.265E-05	5.20	M151	2 1 2 2 1	
7.133E-08	1.271E-05	5.24	M183	1 2 1 1 2	
9.818E-08	1.750E-05	10.00	M063	2 1 2 2 2	
9.800E-08	1.747E-05	10.00	M082	1 1 1 2 1	
9.800E-08	1.747E-05	10.00	M151	2 1 2 2 1	
9.828E-08	1.752E-05	10.04	M183	1 2 1 1 2	
9.094E-08	1.621E-05	9.74	M183	1 2 1 1 2	
1.246E-07	2.220E-05	14.10	M063	2 1 2 2 2	
1.250E-07	2.228E-05	14.10	M082	1 1 1 2 2	
1.250E-07	2.228E-05	14.10	M151	2 1 2 2 2	
1.247E-07	2.223E-05	14.14	M183	1 2 1 1 2	
1.212E-07	2.160E-05	15	B385	0 0 0 0 0	
1.409E-07	2.512E-05	16.64	M183	1 2 1 1 2	
1.633E-07	2.910E-05	18.30	M063	2 1 2 2 2	
1.630E-07	2.905E-05	18.30	M082	1 1 1 2 2	
1.630E-07	2.905E-05	18.30	M151	2 1 2 2 2	
1.634E-07	2.912E-05	18.34	M183	1 2 1 1 2	
2.400E-07	4.278E-05	20	E009	1 0 0 0 1	
2.240E-07	3.992E-05	20	E025	1 0 2 2 2	
1.851E-07	3.300E-05	20	H300	1 1 2 2 1	

(continued)

3212. $C_{14}H_{10}$ (continued)

Solubility (Moles/L)	Solubility (Grams/L)	Temp (°C)	Ref (#)	Evaluation (T P E A A)	Comments
2.087E-07	3.720E-05	22.40	M063	2 1 2 2 2	
2.090E-07	3.725E-05	22.40	M082	1 1 1 2 2	
2.090E-07	3.725E-05	22.40	M151	2 1 2 2 2	
2.089E-07	3.723E-05	22.44	M183	1 2 1 1 2	
2.974E-07	5.300E-05	22.5	G301	0 0 0 0 0	
3.927E-07	7.000E-05	23	P332	0 0 0 0 0	
3.927E-07	7.000E-05	23	P339	0 0 0 0 0	
2.123E-07	3.784E-05	23.24	M183	1 2 1 1 2	
2.435E-07	4.340E-05	24.60	M063	2 1 2 2 2	
2.440E-07	4.349E-05	24.60	M082	1 1 1 2 2	
2.440E-07	4.349E-05	24.60	M151	2 1 2 2 2	
2.437E-07	4.344E-05	24.64	M183	1 2 1 1 2	
2.500E-07	4.456E-05	25	A325	2 1 2 2 1	
2.188E-07	3.900E-05	25	B319	2 0 1 2 1	average of 2
2.174E-07	3.875E-05	25	B385	0 0 0 0 0	
5.218E-07	9.300E-05	25	D406	1 2 2 2 2	
4.470E-07	7.967E-05	25	K001	2 2 2 2 2	
3.800E-07	6.773E-05	25	K123	1 0 2 2 1	
4.152E-07	7.400E-05	25	L301	1 1 2 2 2	
3.927E-07	7.000E-05	25	L332	1 1 1 1 2	
4.096E-07	7.300E-05	25	M064	1 1 2 2 1	
4.100E-06	7.308E-04	25	M342	1 0 1 1 2	
1.683E-07	3.000E-05	25	S227	1 2 1 1 1	
4.211E-07	7.506E-05	25	T066	1 0 0 0 2	
2.500E-07	4.456E-05	25	W300	2 2 2 2 2	
2.502E-07	4.460E-05	25.00	M151	2 1 1 2 2	
4.208E-07	7.500E-05	27	D003	1 0 0 1 1	
3.125E-07	5.570E-05	28.70	M063	2 1 2 2 2	
3.130E-07	5.579E-05	28.70	M082	1 1 1 2 2	
3.130E-07	5.579E-05	28.70	M151	2 1 2 2 2	
3.128E-07	5.575E-05	28.74	M183	1 2 1 1 2	
3.198E-07	5.700E-05	29	M071	2 2 2 2 2	
3.198E-07	5.700E-05	29.00	M151	2 1 1 2 2	
3.212E-07	5.724E-05	29.34	M183	1 2 1 1 2	
3.512E-07	6.260E-05	35	B385	0 0 0 0 0	
6.845E-07	1.220E-04	35.40	W003	2 2 2 2 2	average of 3
8.416E-07	1.500E-04	39.30	W003	2 2 2 2 2	average of 3
1.167E-06	2.080E-04	44.70	W003	2 2 2 2 2	average of 3
1.565E-06	2.790E-04	47.50	W003	2 2 2 2 2	
1.683E-06	3.000E-04	50.10	W003	2 2 2 2 2	average of 3
2.211E-06	3.940E-04	54.70	W003	2 2 2 2 2	average of 3
2.794E-06	4.980E-04	59.20	W003	2 2 2 2 2	average of 3
3.703E-06	6.600E-04	64.50	W003	2 2 2 2 1	average of 3
3.703E-06	6.600E-04	65.10	W003	2 2 2 2 1	average of 3
5.162E-06	9.200E-04	69.80	W003	2 2 2 2 1	
5.274E-06	9.400E-04	70.70	W003	2 2 2 2 1	average of 3
5.106E-06	9.100E-04	71.90	W003	2 2 2 2 2	
6.677E-06	1.190E-03	74.70	W003	2 2 2 2 2	average of 3
2.356E-07	4.200E-05	ns	H123	0 0 0 0 0	

(continued)

3212. $C_{14}H_{10}$ (continued)

Solubility (Moles/L)	Solubility (Grams/L)	Temp (°C)	Ref (#)	Evaluation (T P E A A)	Comments
1.800E-07	3.208E-05	ns	H306	1 0 1 2 1	
4.096E-07	7.300E-05	ns	K304	0 0 0 0 1	
4.096E-07	7.300E-05	ns	M344	0 0 0 0 2	
5.000E-07	8.912E-05	ns	W005	0 0 1 2 0	

3213. $C_{14}H_{10}Cl_2O_3$
Fenclofenac
Benzeneacetic acid, 2-(2,4-dichlorophenoxy)-
RX 67408

RN: 34645-84-6   **MP (°C):** 136
MW: 297.14   **BP (°C):**

Solubility (Moles/L)	Solubility (Grams/L)	Temp (°C)	Ref (#)	Evaluation (T P E A A)	Comments
2.840E-05	8.439E-03	25	C314	0 0 0 0 0	
2.827E-05	8.400E-03	25	C314	0 0 0 0 0	

3214. $C_{14}H_{10}Cl_4$
DDD
1,1-Dichloro-2,2-bis(*p*-chlorophenyl)ethane
p,p'-TDE
Dichlorodiphenyldichloroethane

RN: 72-54-8   **MP (°C):** 109.5
MW: 320.05   **BP (°C):** 193

Solubility (Moles/L)	Solubility (Grams/L)	Temp (°C)	Ref (#)	Evaluation (T P E A A)	Comments
1.562E-07	5.000E-05	15	B083	2 2 1 2 1	particle size 5 μm
2.812E-07	9.000E-05	25	B083	2 2 1 2 1	particle size 5 μm
6.249E-08	2.000E-05	25	W025	1 0 2 2 1	
4.687E-07	1.500E-04	35	B083	2 2 1 2 2	particle size 5 μm
7.499E-07	2.400E-04	45	B083	2 2 1 2 2	particle size 5 μm
9.374E-09	3.000E-06	ns	M110	0 0 0 0 0	EFG

3215. $C_{14}H_{10}Cl_4$
1-(2-Chlorophenyl)-1-(4-chlorophenyl)-2,2-dichloroethane
o,p'-DDD

RN: 53-19-0   **MP (°C):** 76
MW: 320.05   **BP (°C):**

Solubility (Moles/L)	Solubility (Grams/L)	Temp (°C)	Ref (#)	Evaluation (T P E A A)	Comments
1.875E-07	6.000E-05	15	B083	2 2 1 2 1	particle size 5 μm
3.125E-07	1.000E-04	25	B083	2 2 1 2 2	particle size 5 μm
8.749E-07	2.800E-04	35	B083	2 2 1 2 2	particle size 5 μm
9.842E-07	3.150E-04	45	B083	2 2 1 2 2	particle size 5 μm

3216. C$_{14}$H$_{10}$F$_3$NO$_2$

Flufenamic acid

N-(α,α,α-Trifluoro-m-tolyl)anthranilic acid

N-(3-Trifluoromethylphenyl)anthranilic acid

RN:	530-78-9	MP (°C):	132–135
MW:	281.24	BP (°C):	

Solubility (Moles/L)	Solubility (Grams/L)	Temp (°C)	Ref (#)	Evaluation (T P E A A)	Comments
3.890E-06	1.094E-03	25	G085	2 0 0 0 0	EFG
4.000E-05	1.125E-02	25	I007	1 2 2 2 0	EFG
1.031E-04	2.900E-02	30	D015	2 0 1 1 0	EFG
6.670E-06	1.876E-03	35	G085	2 0 0 0 0	EFG
6.200E-04	1.744E-01	35	H091	1 2 2 2 1	sic
2.133E-04	6.000E-02	37	D015	2 0 1 1 0	EFG
3.556E-05	1.000E-02	rt	H302	0 0 2 1 2	intrinsic

3217. C$_{14}$H$_{10}$N$_2$O$_2$

C.I. Disperse violet 1

1,4-Diamino-9,10-anthraquinone

Acetate red violet R

Acetoquinone light heliotrope NL

Supracet brilliant violet 3R

Violet 14447

RN:	128-95-0	MP (°C):	275
MW:	238.25	BP (°C):	

Solubility (Moles/L)	Solubility (Grams/L)	Temp (°C)	Ref (#)	Evaluation (T P E A A)	Comments
9.600E-07	2.287E-04	25	B333	0 0 0 0 0	

3218. C$_{14}$H$_{10}$N$_2$O$_6$

Dipentum

Olsalazine

RN:	15722-48-2	MP (°C):	
MW:	302.25	BP (°C):	

Solubility (Moles/L)	Solubility (Grams/L)	Temp (°C)	Ref (#)	Evaluation (T P E A A)	Comments
3.800E-08	1.149E-05	25	D311	0 0 0 0 0	0.1M NaCl

3219. C$_{14}$H$_{10}$O

2-Anthranol

2-Anthrol

RN:	613-14-9	MP (°C):	
MW:	194.24	BP (°C):	

Solubility (Moles/L)	Solubility (Grams/L)	Temp (°C)	Ref (#)	Evaluation (T P E A A)	Comments
4.720E-04	9.167E-02	25	L085	1 2 0 1 2	

3220. C$_{14}$H$_{10}$O
1-Anthranol
1-Anthrol
Anthranol

| **RN:** | 529-86-2 | **MP** (°C): | 152 |
| **MW:** | 194.24 | **BP** (°C): | |

Solubility (Moles/L)	Solubility (Grams/L)	Temp (°C)	Ref (#)	Evaluation (T P E A A)	Comments
1.850E-04	3.593E-02	25	L085	1 2 0 1 2	

3221. C$_{14}$H$_{10}$O$_3$
Diphenyleneglycollic acid

| **RN:** | | **MP** (°C): | |
| **MW:** | 226.23 | **BP** (°C): | |

Solubility (Moles/L)	Solubility (Grams/L)	Temp (°C)	Ref (#)	Evaluation (T P E A A)	Comments
1.082E-02	2.448E+00	25	K040	1 0 2 1 2	

3222. C$_{14}$H$_{10}$O$_4$
Diphenic acid
1,1′-Biphenyl-2,2′-dicarboxylic acid
2,2′-Biphenyldicarboxylic acid

| **RN:** | 482-05-3 | **MP** (°C): | 228 |
| **MW:** | 242.23 | **BP** (°C): | |

Solubility (Moles/L)	Solubility (Grams/L)	Temp (°C)	Ref (#)	Evaluation (T P E A A)	Comments
5.200E-03	1.260E+00	25	K040	1 0 2 1 2	

3223. C$_{14}$H$_{10}$O$_4$
Benzoyl peroxide
Benzoyl-peroxid

| **RN:** | 94-36-0 | **MP** (°C): | 105 |
| **MW:** | 242.23 | **BP** (°C): | |

Solubility (Moles/L)	Solubility (Grams/L)	Temp (°C)	Ref (#)	Evaluation (T P E A A)	Comments
6.399E-07	1.550E-04	rt	C342	0 0 0 0 0	

3224. C$_{14}$H$_{10}$O$_5$

Gentisin
9H-Xanthen-9-one, 1,7-dihydroxy-3-methoxy-
Gentianic acid
Gentianin

RN:	437-50-3	MP (°C):	266.5
MW:	258.23	BP (°C):	

Solubility (Moles/L)	Solubility (Grams/L)	Temp (°C)	Ref (#)	Evaluation (T P E A A)	Comments
1.162E-03	3.000E-01	16	F300	1 0 0 0 2	

3225. C$_{14}$H$_{10}$O$_9$

Digallic acid
m-Digallic acid
m-Digallussaeure

RN:	536-08-3	MP (°C):	
MW:	322.23	BP (°C):	

Solubility (Moles/L)	Solubility (Grams/L)	Temp (°C)	Ref (#)	Evaluation (T P E A A)	Comments
1.552E-03	5.000E-01	25	F300	1 0 0 0 0	
5.896E-02	1.900E+01	100	F300	1 0 0 0 1	

3226. C$_{14}$H$_{11}$ClNO$_2$

7-Chloro-5,11-dihydrodibenz[b,e][1,4]oxazepine-5-carboxamide

RN:		MP (°C):	
MW:	260.70	BP (°C):	

Solubility (Moles/L)	Solubility (Grams/L)	Temp (°C)	Ref (#)	Evaluation (T P E A A)	Comments
1.534E-04	4.000E-02	37	G020	1 0 0 0 1	

3227. C$_{14}$H$_{11}$ClN$_2$O$_4$

2′-Methyl-3′chloro-2-hydroxy-5-nitrobenzanilide
Benzamide, *N*-(3-chloro-2-methyphenyl)-2-hydroxy-5-nitro-

RN:	213460-66-3	MP (°C):	
MW:	306.71	BP (°C):	

Solubility (Moles/L)	Solubility (Grams/L)	Temp (°C)	Ref (#)	Evaluation (T P E A A)	Comments
1.102E-05	3.379E-03	25	D400	2 0 0 1 2	

3228. C$_{14}$H$_{11}$ClN$_2$O$_4$

2'-Methyl-5'-chloro-2-hydroxy-5-nitrobenzanilide
Benzamide, N-(5-chloro-2-methylphenyl)-2-hydroxy-5-nitro-
RN: 213460-65-2   **MP** (°C):
MW: 306.71   **BP** (°C):

Solubility (Moles/L)	Solubility (Grams/L)	Temp (°C)	Ref (#)	Evaluation (T P E A A)	Comments
7.534E-06	2.311E-03	25	D400	2 0 0 1 2	

3229. C$_{14}$H$_{11}$ClN$_2$O$_4$

2'-Methyl-3'-chloro-2-hydroxy-3-nitrobenzanilide
Benzamide, N-(3-chloro-2methylphenyl)-2-hydroxy-3-nitro-
RN: 73544-88-4   **MP** (°C):
MW: 306.71   **BP** (°C):   324.7–408.7

Solubility (Moles/L)	Solubility (Grams/L)	Temp (°C)	Ref (#)	Evaluation (T P E A A)	Comments
1.528E-05	4.685E-03	25	D400	2 0 0 1 2	

3230. C$_{14}$H$_{11}$ClN$_2$O$_4$

2'-Methyl-3'-chloro-2-hydroxy-3nitrobenzanilide
Benzamide, N-(5-chloro-2methylphenyl)-2-hydroxy-3-nitro-
RN: 213460-62-9   **MP** (°C):
MW: 306.71   **BP** (°C):

Solubility (Moles/L)	Solubility (Grams/L)	Temp (°C)	Ref (#)	Evaluation (T P E A A)	Comments
1.528E-05	4.685E-03	25	D400	2 0 0 1 2	

3231. C$_{14}$H$_{11}$ClN$_2$O$_4$S

Chlorthalidone
2-Chloro-5-(2,3-dihydro-1-hydroxy-3-oxo-1H-isoindol-1-yl)benzenesulfonamide
Hygroton
Thalitone
Chlortalidone
RN: 77-36-1   **MP** (°C):
MW: 338.77   **BP** (°C):

Solubility (Moles/L)	Solubility (Grams/L)	Temp (°C)	Ref (#)	Evaluation (T P E A A)	Comments
3.542E-04	1.200E-01	25	P312	0 0 0 0 0	
4.510E-04	1.528E-01	ns	I304	0 0 0 0 0	

3232. $C_{14}H_{11}Cl_2NO_2$
Diclofenac
2-[(2,6-Dichlorophenyl)amino]benzeneacetic acid
RN: 15307-86-5   **MP** (°C):   157
MW: 296.16   **BP** (°C):   412

Solubility (Moles/L)	Solubility (Grams/L)	Temp (°C)	Ref (#)	Evaluation (T P E A A)	Comments
4.317E-06	1.278E-03	30	P438	0 0 0 0 0	pH 2.0
1.182E-05	3.500E-03	32	C411	2 1 1 2 1	
4.478E-06	1.326E-03	33	P438	0 0 0 0 0	pH 2.0
5.117E-06	1.515E-03	37	P438	0 0 0 0 0	pH 2.0
5.389E-06	1.596E-03	39.5	P438	0 0 0 0 0	pH 2.0
5.822E-06	1.724E-03	42	P438	0 0 0 0 0	pH 2.0

3233. $C_{14}H_{11}Cl_3O_2$
2,2-bis(-*p*-Hydroxyphenyl)-1,1,1-trichloroethylene
Hydroxychlor
p,p'-Hydroxy-DDT
RN: 2971-36-0   **MP** (°C):   194
MW: 317.60   **BP** (°C):

Solubility (Moles/L)	Solubility (Grams/L)	Temp (°C)	Ref (#)	Evaluation (T P E A A)	Comments
2.393E-04	7.600E-02	ns	K117	0 1 2 1 1	

3234. $C_{14}H_{11}FN_2O_5$
1-Acetoxymethyl-3-benzoyl-5-fluoro-2,4(1H,3H)-pyrimidinedi-one
1-Acetoxymethyl-3-benzoyl-5-fluorouracil
RN: 97096-67-8   **MP** (°C):   127–128
MW: 306.25   **BP** (°C):

Solubility (Moles/L)	Solubility (Grams/L)	Temp (°C)	Ref (#)	Evaluation (T P E A A)	Comments
4.571E-04	1.400E-01	22	B321	0 0 0 0 0	pH 4.0

3235. $C_{14}H_{11}N$
2-Aminoanthracene
2-Anthrylamine
β-Aminoanthracene
2-Anthracenamine
2-Anthramine
Anthracene amine
RN: 613-13-8   **MP** (°C):   238
MW: 193.25   **BP** (°C):

Solubility (Moles/L)	Solubility (Grams/L)	Temp (°C)	Ref (#)	Evaluation (T P E A A)	Comments
6.727E-06	1.300E-03	24	H106	1 0 2 2 2	
6.727E-09	1.300E-06	ns	M349	0 2 1 1 2	

3236. C$_{14}$H$_{11}$N
Acetonitrile, diphenyl-
Diphenatrile

RN:	86-29-3	MP (°C):	74
MW:	193.25	BP (°C):	

Solubility (Moles/L)	Solubility (Grams/L)	Temp (°C)	Ref (#)	Evaluation (T P E A A)	Comments
1.138E-03	2.200E-01	ns	B185	0 0 0 0 0	

3237. C$_{14}$H$_{11}$NO$_2$
N-Benzoylbenzamide
Dibenzamid

RN:	614-28-8	MP (°C):	152
MW:	225.25	BP (°C):	

Solubility (Moles/L)	Solubility (Grams/L)	Temp (°C)	Ref (#)	Evaluation (T P E A A)	Comments
5.327E-03	1.200E+00	15	F300	1 0 0 0 1	

3238. C$_{14}$H$_{11}$N$_3$O$_2$
Salicylolhydrazone of picolinealdehyde

RN:		MP (°C):	
MW:	253.26	BP (°C):	

Solubility (Moles/L)	Solubility (Grams/L)	Temp (°C)	Ref (#)	Evaluation (T P E A A)	Comments
7.897E-04	2.000E-01	ns	G089	0 1 2 0 1	

3239. C$_{14}$H$_{12}$
1-Methylfluorene
1-Methyl-9H-fluorene

RN:	1730-37-6	MP (°C):	87
MW:	180.25	BP (°C):	

Solubility (Moles/L)	Solubility (Grams/L)	Temp (°C)	Ref (#)	Evaluation (T P E A A)	Comments
6.047E-06	1.090E-03	25	B319	2 0 1 2 2	
6.060E-06	1.092E-03	25	M342	1 0 1 1 2	

3240. C$_{14}$H$_{12}$
1,1-Diphenylethene
1,1-Diphenylethylene

RN:	530-48-3	MP (°C):	8.2
MW:	180.25	BP (°C):	277

Solubility (Moles/L)	Solubility (Grams/L)	Temp (°C)	Ref (#)	Evaluation (T P E A A)	Comments
3.662E-05	6.600E-03	25	A002	1 0 1 1 1	

3241. $C_{14}H_{12}$

9,10-Dihydroanthracene

RN: 613-31-0   **MP** (°C):   104–107
MW: 180.25   **BP** (°C):   312

Solubility (Moles/L)	Solubility (Grams/L)	Temp (°C)	Ref (#)	Evaluation (T P E A A)	Comments
2.578E-06	4.646E-04	4.96	R423	0 0 0 0 0	
2.622E-06	4.727E-04	5.85	R423	0 0 0 0 0	
2.917E-06	5.257E-04	7.95	R423	0 0 0 0 0	
3.317E-06	5.978E-04	10.95	R423	0 0 0 0 0	
3.556E-06	6.409E-04	12.05	R423	0 0 0 0 0	
4.261E-06	7.681E-04	14.95	R423	0 0 0 0 0	
4.961E-06	8.942E-04	18.00	R423	0 0 0 0 0	
5.811E-06	1.047E-03	20.96	R423	0 0 0 0 0	
7.389E-06	1.332E-03	24.59	R423	0 0 0 0 0	
8.011E-06	1.444E-03	26.59	R423	0 0 0 0 0	
9.400E-06	1.694E-03	29.05	R423	0 0 0 0 0	
1.114E-05	2.009E-03	32.66	R423	0 0 0 0 0	
1.288E-05	2.321E-03	36.28	R423	0 0 0 0 0	
1.498E-05	2.701E-03	40.01	R423	0 0 0 0 0	

3242. $C_{14}H_{12}$

trans-Stilbene
trans-Diphenylethylene
1,2-Diphenylethene
trans-1,2-Diphenylethylene
trans-α, β-Diphenylethylene
Toluylene

RN: 103-30-0   **MP** (°C):   124
MW: 180.25   **BP** (°C):   306

Solubility (Moles/L)	Solubility (Grams/L)	Temp (°C)	Ref (#)	Evaluation (T P E A A)	Comments
1.609E-06	2.900E-04	25	A002	1 0 1 1 1	

3243. $C_{14}H_{12}F_3NO_4S_2$

Perfluidone
Methyl-4-(phenylsulfonyl)trifluoromethanesulfonanilide
1,1,1-Trifluoro-*N*-(2-methyl-4-(phenylsulfonyl)phenyl)methanesulfonamide
Destun
MBR 8251
Trifluoro-*N*-(2-methyl-4-(phenylsulfonyl)phenyl)methanesulfonamide

RN: 37924-13-3   **MP** (°C):   143
MW: 379.38   **BP** (°C):

Solubility (Moles/L)	Solubility (Grams/L)	Temp (°C)	Ref (#)	Evaluation (T P E A A)	Comments
1.582E-04	6.000E-02	22	G306	1 0 0 0 1	
1.582E-04	6.000E-02	22	M161	1 0 0 0 1	

3244. C$_{14}$H$_{12}$N$_2$O$_4$
4'-Methyl-2-hydroxy-5-nitrobenzanilide
Benzamide, 2-hydroxy-N-(4-methylphenyl)--nitro-
RN: 68507-96-0   **MP** (°C):
MW: 272.26   **BP** (°C):

Solubility (Moles/L)	Solubility (Grams/L)	Temp (°C)	Ref (#)	Evaluation (T P E A A)	Comments
1.413E-05	3.846E-03	25	D400	2 0 0 1 2	

3245. C$_{14}$H$_{12}$N$_2$O$_4$
4'-Methyl-2-hydroxy-3-nitrobenzanilide
Benzamide,2-hydroxy-N-(4-methylpheyl)-3-nitro-
RN: 68507-90-4   **MP** (°C):
MW: 272.26   **BP** (°C):   305.7–389.7

Solubility (Moles/L)	Solubility (Grams/L)	Temp (°C)	Ref (#)	Evaluation (T P E A A)	Comments
3.069E-05	8.356E-03	25	D400	2 0 0 1 2	

3246. C$_{14}$H$_{12}$N$_2$O$_4$
2'-Methyl-2-hydroxy-3-nitrobenzanilide
Benzamide, 2-hydroxy-N-(2-methylphenyl)-3-nitro-
RN: 68507-89-1   **MP** (°C):
MW: 272.26   **BP** (°C):   302–384.5

Solubility (Moles/L)	Solubility (Grams/L)	Temp (°C)	Ref (#)	Evaluation (T P E A A)	Comments
2.818E-05	7.673E-03	25	D400	2 0 0 1 2	

3247. C$_{14}$H$_{12}$N$_2$O$_5$
4'-Methoxy-2-hydroxy-5nitrobenzanilide
p-Salicylanisidide, 5-nitro-
Benzamide, 2-hydroxy-N-(4-methoxyphenyl)-5-nitro-
RN: 68507-94-8   **MP** (°C):
MW: 288.26   **BP** (°C):

Solubility (Moles/L)	Solubility (Grams/L)	Temp (°C)	Ref (#)	Evaluation (T P E A A)	Comments
1.928E-05	5.556E-03	25	D400	2 0 0 1 2	

3248. C$_{14}$H$_{12}$N$_2$O$_5$
4'-Methoxy-2-hydroxy-3nitrobenzanilide
Benzamide, 2-hydroxy-N-(4-methoxyphenyl)-3-nitro-
RN: 68507-88-0   **MP** (°C):
MW: 288.26   **BP** (°C):

Solubility (Moles/L)	Solubility (Grams/L)	Temp (°C)	Ref (#)	Evaluation (T P E A A)	Comments
3.532E-05	1.018E-02	25	D400	2 0 0 1 2	

3249. $C_{14}H_{12}N_2S$
2-(4-Aminophenyl)-6-methyl-benzothiazole
Dehydrothio-*N*-toluidin
Dehydrothio-*N*-toluidine

RN: 92-36-4	**MP (°C):** 194.8	
MW: 240.33	**BP (°C):** 434	

Solubility (Moles/L)	Solubility (Grams/L)	Temp (°C)	Ref (#)	Evaluation (T P E A A)	Comments
2.080E-04	5.000E-02	100	F300	1 0 0 0 0	

3250. $C_{14}H_{12}N_4O_2$
C.I. Disperse blue 1
9,10-Anthracenedione, 1,4,5,8-tetraamino-

RN: 2475-45-8	**MP (°C):** 332	
MW: 268.28	**BP (°C):**	

Solubility (Moles/L)	Solubility (Grams/L)	Temp (°C)	Ref (#)	Evaluation (T P E A A)	Comments
1.000E-07	2.683E-05	25	B333	0 0 0 0 0	

3251. $C_{14}H_{12}O_2$
4-Biphenylacetic acid
Felbinac

RN: 5728-52-9	**MP (°C):**	
MW: 212.25	**BP (°C):**	

Solubility (Moles/L)	Solubility (Grams/L)	Temp (°C)	Ref (#)	Evaluation (T P E A A)	Comments
1.850E-04	3.927E-02	25	P344	0 0 0 0 0	EFG

3252. $C_{14}H_{12}O_2$
Benzoin
2-Hydroxy-1,2-diphenylethanone
Benzoylphenylcarbinol
2-Hydroxy-2-phenylacetophenone
Hydroxy-2-phenyl acetophenone

RN: 579-44-2	**MP (°C):** 137	
MW: 212.25	**BP (°C):** 344	

Solubility (Moles/L)	Solubility (Grams/L)	Temp (°C)	Ref (#)	Evaluation (T P E A A)	Comments
1.413E-03	3.000E-01	25	F300	1 0 0 0 0	
1.413E-03	2.999E-01	rt	D021	0 0 1 1 0	

3253. C$_{14}$H$_{12}$O$_2$
Benzyl benzoate
Ascabin
Scabagen
Benzoic acid phenylmethyl ester
Benylate
Phenylmethyl benzoate

RN: 120-51-4   **MP** (°C): 19
MW: 212.25   **BP** (°C): 323

Solubility (Moles/L)	Solubility (Grams/L)	Temp (°C)	Ref (#)	Evaluation (T P E A A)	Comments
1.225E-04	2.600E-02	15	H069	1 0 1 1 1	
6.960E-03	1.477E+00	30	M444	0 0 0 0 0	
7.020E-03	1.490E+00	40	M444	0 0 0 0 0	
7.150E-03	1.518E+00	50	M444	0 0 0 0 0	
7.230E-03	1.535E+00	60	M444	0 0 0 0 0	

3254. C$_{14}$H$_{12}$O$_2$
Diphenylacetic acid
Diphenyl-essigsaeure

RN: 117-34-0   **MP** (°C): 148
MW: 212.25   **BP** (°C):

Solubility (Moles/L)	Solubility (Grams/L)	Temp (°C)	Ref (#)	Evaluation (T P E A A)	Comments
6.000E-04	1.274E-01	25	K040	1 0 2 1 2	

3255. C$_{14}$H$_{12}$O$_3$
Benzilic acid
2,2-Diphenyl-2-hydroxyacetic acid
Diphenylglycolic acid
Benzeneacetic acid, α-hydroxy-α-phenyl-
2-Hydroxy-2,2-diphenylethanoic acid

RN: 76-93-7   **MP** (°C): 150
MW: 228.25   **BP** (°C):

Solubility (Moles/L)	Solubility (Grams/L)	Temp (°C)	Ref (#)	Evaluation (T P E A A)	Comments
7.690E-03	1.755E+00	25	K040	1 0 2 1 2	
6.190E-03	1.413E+00	25	L050	2 0 1 2 2	

3256. C₁₄H₁₂O₃

Benzylparaben
Benzyl 4-hydroxybenzoate
Phenylmethyl ester
RN: 94-18-8   **MP** (°C):
MW: 228.25   **BP** (°C):

Solubility (Moles/L)	Solubility (Grams/L)	Temp (°C)	Ref (#)	Evaluation (T P E A A)	Comments
4.031E-04	9.200E-02	25	P013	0 0 0 0 0	

3257. C₁₄H₁₂O₅

Khellin
Amicardine
RN: 82-02-0   **MP** (°C):   154.5
MW: 260.25   **BP** (°C):

Solubility (Moles/L)	Solubility (Grams/L)	Temp (°C)	Ref (#)	Evaluation (T P E A A)	Comments
9.500E-01	2.472E+02	25	E312	0 0 0 0 0	EFG, *sic*
1.153E-04	3.000E-02	25	J028	1 2 0 2 0	
7.000E-04	1.822E-01	30	E012	1 2 1 1 0	
1.300E-03	3.383E-01	42	E012	1 2 1 1 0	

3258. C₁₄H₁₃ClN₄O

6H-Dipyrido[3,2-b:2′,3′-e][1,4]diazepin-6-one, 2-chloro-11-ethyl-5,11-dihydro-5-methyl-
RN: 133627-12-0   **MP** (°C):
MW: 288.74   **BP** (°C):

Solubility (Moles/L)	Solubility (Grams/L)	Temp (°C)	Ref (#)	Evaluation (T P E A A)	Comments
7.691E-05	2.221E-02	ns	M381	0 1 1 1 2	pH 7.0

3259. C₁₄H₁₃NO₆

Benzoic acid, 2-(acetyloxy)-, (2,5-dioxo-1-pyrrolidinyl)methyl ester
Salicylic acid acetate, ester with *N*-(hydroxymethyl)succinimide
RN: 32620-72-7   **MP** (°C):   117.5
MW: 291.26   **BP** (°C):

Solubility (Moles/L)	Solubility (Grams/L)	Temp (°C)	Ref (#)	Evaluation (T P E A A)	Comments
1.717E-03	5.000E-01	21	N335	0 0 0 0 0	

3260. C₁₄H₁₃N₂

4,7-Dimethyl-1,10-phenanthroline
4,7-Dimethyl-o-phenanthroline

RN: 3248-05-3   **MP** (°C): 193
MW: 209.27   **BP** (°C):

Solubility (Moles/L)	Solubility (Grams/L)	Temp (°C)	Ref (#)	Evaluation (T P E A A)	Comments
1.070E-04	2.239E-02	25.04	B094	1 2 1 2 2	

3261. C₁₄H₁₃N₃O₂

Pyrido[2,3-b][1,5]benzoxazepin-5(6H)-one, 3-amino-6,9-dimethyl-

RN: 134894-45-4   **MP** (°C):
MW: 255.28   **BP** (°C):

Solubility (Moles/L)	Solubility (Grams/L)	Temp (°C)	Ref (#)	Evaluation (T P E A A)	Comments
9.057E-04	2.312E-01	ns	M381	0 1 1 1 2	pH 7.0

3262. C₁₄H₁₃N₃O₄S₂

Meloxicam

RN: 71125-38-7   **MP** (°C):
MW: 351.41   **BP** (°C):

Solubility (Moles/L)	Solubility (Grams/L)	Temp (°C)	Ref (#)	Evaluation (T P E A A)	Comments
6.500E-05	2.284E-02	25	C434	0 0 0 0 0	pH 6.0
3.415E-05	1.200E-02	25	S415	0 0 0 0 0	
9.500E-05	3.338E-02	30	C434	0 0 0 0 0	pH 6.0
1.550E-05	5.447E-03	37	C434	0 0 0 0 0	pH 6.0
3.699E-06	1.300E-03	37	Y421	0 0 0 0 0	
2.800E-05	9.839E-03	45	C434	0 0 0 0 0	pH 6.0

3263. C₁₄H₁₄

4,4′-Dimethylbiphenyl
4,4′-Dimethyl-1,1′-biphenyl
p,p′-Bitoluene

RN: 613-33-2   **MP** (°C): 125.0
MW: 182.27   **BP** (°C): 295.0

Solubility (Moles/L)	Solubility (Grams/L)	Temp (°C)	Ref (#)	Evaluation (T P E A A)	Comments
3.770E-07	6.871E-05	4.0	D330	2 2 1 2 2	
9.590E-07	1.748E-04	25.0	D330	2 2 1 2 2	
2.420E-06	4.411E-04	40.0	D330	2 2 1 2 2	

3264. C$_{14}$H$_{14}$
Bibenzyl
1,2-Diphenylethane
Benzene, 1,1'-(1,2-ethanediyl)*bis*-
RN: 103-29-7   **MP** (°C):   52.0
MW: 182.27   **BP** (°C):   284

Solubility (Moles/L)	Solubility (Grams/L)	Temp (°C)	Ref (#)	Evaluation (T P E A A)	Comments
2.359E-05	4.300E-03	25	A002	1 0 1 1 1	

3265. C$_{14}$H$_{14}$NO$_4$PS
EPN
Ethyl *O*-(*p*-nitrophenyl) phenylphosphonothionate
O-Ethyl *O-p*-nitrophenyl benzenephosphonothioate
Ethyl *O*-(*p*-nitrophenyl) benzenethiophosphonate
RN: 2104-64-5   **MP** (°C):   36
MW: 323.31   **BP** (°C):

Solubility (Moles/L)	Solubility (Grams/L)	Temp (°C)	Ref (#)	Evaluation (T P E A A)	Comments
9.629E-06	3.113E-03	22	K137	1 1 2 1 0	

3266. C$_{14}$H$_{14}$N$_4$O
6H-Dipyrido[3,2-b:2',3'-e][1,4]diazepin-6-one, 11-ethyl-5,11-dihydro-5-methyl
RN: 132312-85-7   **MP** (°C):
MW: 254.29   **BP** (°C):

Solubility (Moles/L)	Solubility (Grams/L)	Temp (°C)	Ref (#)	Evaluation (T P E A A)	Comments
2.399E-03	6.100E-01	ns	M381	0 1 1 1 2	pH 7.0

3267. C$_{14}$H$_{14}$N$_4$O$_2$
Dis. A. 7
RN: 2491-74-9   **MP** (°C):   236
MW: 270.29   **BP** (°C):

Solubility (Moles/L)	Solubility (Grams/L)	Temp (°C)	Ref (#)	Evaluation (T P E A A)	Comments
2.000E-09	5.406E-07	25	B333	0 0 0 0 0	

3268. C$_{14}$H$_{14}$N$_4$O$_2$
Dye II
4-[[(4-Dimethylamino)phenyl]azo]nitrobenzene
RN: **MP** (°C):
MW: 270.29   **BP** (°C):

Solubility (Moles/L)	Solubility (Grams/L)	Temp (°C)	Ref (#)	Evaluation (T P E A A)	Comments
7.800E-07	2.108E-04	84.10	B198	1 2 1 1 1	
2.040E-06	5.514E-04	97.40	B198	1 2 1 1 2	

3269. $C_{14}H_{14}N_4O_4$
β,γ-Dihydroxypropyltheophylline
RN: 180262-60-6   **MP** (°C):
MW: 302.29   **BP** (°C):

Solubility (Moles/L)	Solubility (Grams/L)	Temp (°C)	Ref (#)	Evaluation (T P E A A)	Comments
3.007E-01	9.091E+01	ns	J025	0 0 0 0 1	

3270. $C_{14}H_{14}N_4S$
6H-Dipyrido[3,2-b:2',3'-e][1,4]diazepine-6-thione, 11-ethyl-5,11-dihydro-5-methyl
RN: 134698-27-4   **MP** (°C):
MW: 270.36   **BP** (°C):

Solubility (Moles/L)	Solubility (Grams/L)	Temp (°C)	Ref (#)	Evaluation (T P E A A)	Comments
2.323E-05	6.280E-03	ns	M381	0 1 1 1 2	pH 7.0

3271. $C_{14}H_{14}O$
6-Benzyl-*m*-cresol
Phenol, 5-methyl-2-(phenylmethyl)-
RN: 30091-04-4   **MP** (°C):
MW: 198.27   **BP** (°C):

Solubility (Moles/L)	Solubility (Grams/L)	Temp (°C)	Ref (#)	Evaluation (T P E A A)	Comments
1.441E-04	2.857E-02	25	L021	1 0 0 0 0	

3272. $C_{14}H_{14}O$
DL-1,2-Diphenylethanol
DL-1,2-Diphenyl-aethanol
RN: 614-29-9   **MP** (°C):   67
MW: 198.27   **BP** (°C):

Solubility (Moles/L)	Solubility (Grams/L)	Temp (°C)	Ref (#)	Evaluation (T P E A A)	Comments
3.026E-03	6.000E-01	100	F300	1 0 0 0 0	

3273. $C_{14}H_{14}O_2$
DL-Hydrobenzoin
Hydrobenzoin
RN: 27134-24-3   **MP** (°C):   139
MW: 214.27   **BP** (°C):

Solubility (Moles/L)	Solubility (Grams/L)	Temp (°C)	Ref (#)	Evaluation (T P E A A)	Comments
1.167E-02	2.500E+00	15	F300	1 0 0 0 1	
8.867E-03	1.900E+00	15	F300	1 0 0 0 1	
6.021E-02	1.290E+01	100	F300	1 0 0 0 2	

3274. C$_{14}$H$_{14}$O$_3$
Pindone
2-Pivaloylindandione-1,3
RN: 83-26-1   **MP** (°C):   109
MW: 230.27   **BP** (°C):

Solubility (Moles/L)	Solubility (Grams/L)	Temp (°C)	Ref (#)	Evaluation (T P E A A)	Comments
7.817E-05	1.800E-02	25	M061	1 0 0 0 1	
7.817E-05	1.800E-02	25	M161	1 0 0 0 1	

3275. C$_{14}$H$_{14}$O$_3$
Naproxen
6-Methoxy-α-methyl-2-naphthaleneacetic acid
(S)-6-Methoxy-α-methyl-2-naphthaleneacetic acid
Laraflex
RN: 22204-53-1   **MP** (°C):   155.3
MW: 230.27   **BP** (°C):

Solubility (Moles/L)	Solubility (Grams/L)	Temp (°C)	Ref (#)	Evaluation (T P E A A)	Comments
4.310E-05	9.924E-03	5	F306	1 0 1 2 2	intrinsic
6.948E-05	1.600E-02	21	B331	1 2 2 1 2	pH 7.4
6.080E-05	1.400E-02	25	A408	2 0 1 2 0	int
6.905E-05	1.590E-02	25	A427	0 0 0 0 0	
6.905E-05	1.590E-02	25	C059	1 2 1 1 2	
6.900E-05	1.589E-02	25	F306	1 0 1 2 2	intrinsic
1.146E-04	2.639E-02	37	F306	1 0 1 2 2	intrinsic
2.171E-05	5.000E-03	37	Y421	0 0 0 0 0	
5.211E-04	1.200E-01	amb	L434	0 0 0 0 0	
5.646E-05	1.300E-02	rt	H302	0 0 2 1 2	intrinsic

3276. C$_{14}$H$_{14}$O$_3$S
o-Cresyl-p-toluene sulfonate
2-Methylphenyl tosylate
o-Tolyl tosylate
2-Tolyl tosylate
RN: 599-75-7   **MP** (°C):
MW: 262.33   **BP** (°C):

Solubility (Moles/L)	Solubility (Grams/L)	Temp (°C)	Ref (#)	Evaluation (T P E A A)	Comments
1.144E-04	3.000E-02	ns	F014	0 0 0 0 0	

3277. $C_{14}H_{14}O_4$
Diallyl *m*-phthalate
RN: **MP** (°C):
MW: 246.27 **BP** (°C):

Solubility (Moles/L)	Solubility (Grams/L)	Temp (°C)	Ref (#)	Evaluation (T P E A A)	Comments
1.990E-04	4.900E-02	25	S417	0 0 0 0 0	

3278. $C_{14}H_{14}O_4$
Diallyl phthalate
Di-2-propenyl phthalate
RN: 131-17-9 **MP** (°C): −70
MW: 246.27 **BP** (°C): 165

Solubility (Moles/L)	Solubility (Grams/L)	Temp (°C)	Ref (#)	Evaluation (T P E A A)	Comments
<4.06E-04	<10.00E-02	20	F070	1 0 0 0 1	
7.390E-04	1.820E-01	20	L300	2 1 0 2 2	
7.413E-04	1.826E-01	ns	S460	0 0 0 0 0	

3279. $C_{14}H_{15}N$
p-Aminostilbene
4-Aminostilbene
RN: 834-24-2 **MP** (°C):
MW: 197.28 **BP** (°C):

Solubility (Moles/L)	Solubility (Grams/L)	Temp (°C)	Ref (#)	Evaluation (T P E A A)	Comments
2.534E-05	5.000E-03	rt	N015	0 0 2 2 0	

3280. $C_{14}H_{15}NO_5$
L-Proline, 1-[(benzoyloxy)acetyl]-
RN: 115178-75-1 **MP** (°C): 72.5
MW: 277.28 **BP** (°C):

Solubility (Moles/L)	Solubility (Grams/L)	Temp (°C)	Ref (#)	Evaluation (T P E A A)	Comments
2.561E-02	7.100E+00	22	N317	1 1 2 1 2	

3281. $C_{14}H_{15}N_3$
o-Aminoazotoluene
2-Amino-5-azotoluene
RN: 97-56-3 **MP** (°C): 101
MW: 225.30 **BP** (°C):

Solubility (Moles/L)	Solubility (Grams/L)	Temp (°C)	Ref (#)	Evaluation (T P E A A)	Comments
3.107E-05	7.000E-03	37	H120	1 1 1 1 1	normal saline

3282. C₁₄H₁₅N₃

p-Dimethylaminoazobenzene
4-Dimethylaminoazobenzol
Dimethylaminoazobenzene
Methylgelb
C. I. Solvent yellow 2

RN:	60-11-7	**MP** (°C):	116
MW:	225.30	**BP** (°C):	

Solubility (Moles/L)	Solubility (Grams/L)	Temp (°C)	Ref (#)	Evaluation (T P E A A)	Comments
8.877E-04	2.000E-01	20	F300	1 0 0 0 0	
6.214E-06	1.400E-03	20	J027	1 0 0 0 1	
1.700E-06	3.830E-04	25	B333	0 0 0 0 0	*sic*
1.775E-06	4.000E-04	30	R430	0 0 0 0 0	
7.101E-04	1.600E-01	rt	D021	0 0 1 1 1	*sic*

3283. C₁₄H₁₅N₃O₃S

Gly-dapsone
Acetamide, 2-amino-N-[4-[(4-aminophenyl)sulfonyl]phenyl]

RN:	160349-02-0	**MP** (°C):	
MW:	305.36	**BP** (°C):	

Solubility (Moles/L)	Solubility (Grams/L)	Temp (°C)	Ref (#)	Evaluation (T P E A A)	Comments
2.849E-03	8.700E-01	25	P351	0 0 0 0 0	pH 7.4
>4.91E-02	>1.50E+01	25	P351	0 0 0 0 0	

3284. C₁₄H₁₅N₅O₅

9-(2-O-Butyryl-β-D-arabinofuranosyl)adenine
9H-Purin-6-amine, 9-[3,5-*bis*-O-[(1,1-dimethylethyl)dimethylsilyl]-2-O-(1-oxobutyl)-β-D-arabinofuranosyl]-

RN:	87970-05-6	**MP** (°C):	
MW:	333.31	**BP** (°C):	

Solubility (Moles/L)	Solubility (Grams/L)	Temp (°C)	Ref (#)	Evaluation (T P E A A)	Comments
1.023E-04	3.410E-02	37	B306	1 2 0 1 2	pH 7.3

3285. C₁₄H₁₅N₅O₆S

Metasulfron-methyl
Metsulfuron methyl ester
Allie
Escort
DPX-T6376
Ally

RN:	74223-64-6	**MP** (°C):	158
MW:	381.37	**BP** (°C):	

Solubility (Moles/L)	Solubility (Grams/L)	Temp (°C)	Ref (#)	Evaluation (T P E A A)	Comments
7.079E-05	2.700E-02	ns	R427	0 0 0 0 0	

3286. C$_{14}$H$_{15}$O$_2$PS$_2$

Edifenphos

RN: 17109-49-8 **MP** (°C):
MW: 310.38 **BP** (°C): 154

Solubility (Moles/L)	Solubility (Grams/L)	Temp (°C)	Ref (#)	Evaluation (T P E A A)	Comments
1.803E-04	5.596E-02	ns	S460	0 0 0 0 0	

3287. C$_{14}$H$_{16}$ClN$_3$O$_2$

Triadimefon

1-(4-Chlorophenoxy)-3,3-dimethyl-1-(1H-1,2,4-triazol-1-yl)-2-butanone

Triamefon

Bayleton

RN: 43121-43-3 **MP** (°C): 82.3
MW: 293.76 **BP** (°C):

Solubility (Moles/L)	Solubility (Grams/L)	Temp (°C)	Ref (#)	Evaluation (T P E A A)	Comments
8.851E-04	2.600E-01	20	M161	1 0 0 0 2	

3288. C$_{14}$H$_{16}$ClO$_5$PS

Coumaphos

O,O-Diethyl *O*-(3-chloro-4-methylcoumarinyl-7) thiophosphate

RN: 56-72-4 **MP** (°C): 91
MW: 362.77 **BP** (°C):

Solubility (Moles/L)	Solubility (Grams/L)	Temp (°C)	Ref (#)	Evaluation (T P E A A)	Comments
4.135E-06	1.500E-03	20	M061	1 0 0 0 1	
4.169E-06	1.512E-03	ns	R427	0 0 0 0 0	

3289. C$_{14}$H$_{16}$Cl$_2$O$_3$

2,4-Dichlorophenoxyacetic acid cyclohexyl ester

Cyclohexyl 2,4-dichlorophenoxyacetate

RN: 65267-97-2 **MP** (°C):
MW: 303.19 **BP** (°C):

Solubility (Moles/L)	Solubility (Grams/L)	Temp (°C)	Ref (#)	Evaluation (T P E A A)	Comments
1.811E-05	5.492E-03	ns	M120	0 0 1 1 2	

3290. C$_{14}$H$_{16}$FN$_3$O$_3$

2,5-Diaziridinyl-3-floro-6-morpholino-1,4-benzoquinone

2,5-Cyclohexadiene-1,4-dione, 2,5-bis(1-aziridinyl)-3-fluoro-6-(4-morpholinyl)-

RN: 59886-45-2 **MP** (°C): 157
MW: 293.30 **BP** (°C):

Solubility (Moles/L)	Solubility (Grams/L)	Temp (°C)	Ref (#)	Evaluation (T P E A A)	Comments
6.819E-03	2.000E+00	rt	C317	0 0 0 0 0	

3291. C$_{14}$H$_{16}$F$_3$N$_3$O$_4$
Profluralin
N-(Cyclopropylmethyl)-2,6-dinitro-N-propyl-4-(trifluoromethyl)benzenamine
Pregard
Tolban
ER-5461

RN: 26399-36-0 **MP** (°C): 32
MW: 347.30 **BP** (°C):

Solubility (Moles/L)	Solubility (Grams/L)	Temp (°C)	Ref (#)	Evaluation (T P E A A)	Comments
2.879E-07	1.000E-04	20	E048	1 2 1 1 0	
2.879E-07	1.000E-04	20	M161	1 0 0 0 0	
2.879E-07	1.000E-04	27	K315	1 0 0 0 1	

3292. C$_{14}$H$_{16}$N$_2$
o-Tolidine
3,3'-Dimethylbenzidine

RN: 119-93-7 **MP** (°C): 130.0
MW: 212.30 **BP** (°C):

Solubility (Moles/L)	Solubility (Grams/L)	Temp (°C)	Ref (#)	Evaluation (T P E A A)	Comments
6.123E-03	1.300E+00	25	B068	2 0 1 1 1	

3293. C$_{14}$H$_{16}$N$_2$O$_2$
3,3'-Dimethoxybenzidine
o-Dianisidine
Dianisidine

RN: 119-90-4 **MP** (°C): 137
MW: 244.30 **BP** (°C):

Solubility (Moles/L)	Solubility (Grams/L)	Temp (°C)	Ref (#)	Evaluation (T P E A A)	Comments
2.456E-04	6.000E-02	25	B068	2 0 1 1 0	

3294. C$_{14}$H$_{16}$N$_2$O$_4$
2-Pyrrolidinecarboxamide, 1-[(benzoyloxy)acetyl]-

RN: 116482-82-7 **MP** (°C): 194.5
MW: 276.29 **BP** (°C):

Solubility (Moles/L)	Solubility (Grams/L)	Temp (°C)	Ref (#)	Evaluation (T P E A A)	Comments
5.429E-03	1.500E+00	22	N317	1 1 2 1 2	

3295. C₁₄H₁₆N₂O₄

2-Pyrrolidinecarboxamide, 1-[(benzoyloxy)acetyl]-

RN: 106231-69-0   **MP** (°C):
MW: 276.29   **BP** (°C):   570.7

Solubility (Moles/L)	Solubility (Grams/L)	Temp (°C)	Ref (#)	Evaluation (T P E A A)	Comments
5.429E-03	1.500E+00	22	B427	1 0 0 1 1	

3296. C₁₄H₁₆N₂O₄S₂

4-Thiazolidinecarboxylic acid, 2,2′-(1,4-phenylene)*bis*-
4-Thiazolidinecarboxylic acid, 2,2′-*p*-phenylenebis-

RN: 83690-84-0   **MP** (°C):
MW: 340.42   **BP** (°C):   697.9

Solubility (Moles/L)	Solubility (Grams/L)	Temp (°C)	Ref (#)	Evaluation (T P E A A)	Comments
1.800E-03	6.128E-01	21	B414	1 0 0 1 1	fast decomposition

3297. C₁₄H₁₆N₄

Disperse black 3
N,N-Dimethyl-4,4′-azodian
4-Amino-4′-(dimethylamino)azobenzene
C.I. 11025

RN: 539-17-3   **MP** (°C):
MW: 240.31   **BP** (°C):

Solubility (Moles/L)	Solubility (Grams/L)	Temp (°C)	Ref (#)	Evaluation (T P E A A)	Comments
5.000E-07	1.202E-04	25	B333	0 0 0 0 0	

3298. C₁₄H₁₆N₄O₂S

2-Sulfanilamido-5,6,7,8-tetrahydroquinazoline
2-Sulfanilamido-5,6,7,8,-tetrahydroquinazoline

RN: 71119-34-1   **MP** (°C):   255
MW: 304.37   **BP** (°C):

Solubility (Moles/L)	Solubility (Grams/L)	Temp (°C)	Ref (#)	Evaluation (T P E A A)	Comments
2.234E-04	6.800E-02	29	C049	0 0 0 0 0	

3299. C$_{14}$H$_{16}$N$_4$O$_3$S
N4-Acetylsulfamethazine
N4-Acetylsulfamezathine
N4-Acetylsulphamethazine
Acetylsulfamethazine
2-p-Acetamidobenzenesulphonamido-4:6-dimethylpyri-
RN: 100-90-3   **MP** (°C):
MW: 320.37   **BP** (°C):

Solubility (Moles/L)	Solubility (Grams/L)	Temp (°C)	Ref (#)	Evaluation (T P E A A)	Comments
2.900E-03	9.291E-01	37	G026	1 0 1 1 0	EFG, pH 5.4
3.590E-03	1.150E+00	37	L091	1 0 0 0 2	pH 5.5
3.590E-03	1.150E+00	37	M057	1 0 0 0 2	pH 5.5
3.590E-03	1.150E+00	37	R075	1 2 0 0 2	
2.197E-03	7.040E-01	37	S192	1 0 1 1 2	pH 6.0
2.622E-03	8.400E-01	38	K006	1 0 0 0 1	

3300. C$_{14}$H$_{16}$N$_4$O$_3$S
N4-Acetylsulphasomidine
Acetamide, N-[4-[[(2,6-dimethyl-4-pyrimidinyl)amino]sulfonyl]phenyl]-
RN: 3163-31-3   **MP** (°C):
MW: 320.37   **BP** (°C):

Solubility (Moles/L)	Solubility (Grams/L)	Temp (°C)	Ref (#)	Evaluation (T P E A A)	Comments
1.373E-04	4.400E-02	ns	B133	0 2 0 0 1	pH 7.4

3301. C$_{14}$H$_{16}$N$_4$O$_3$S
2-(N4-Acetylsulfanilylamino)-4-ethylpyrimidine
RN: **MP** (°C):
MW: 320.37   **BP** (°C):

Solubility (Moles/L)	Solubility (Grams/L)	Temp (°C)	Ref (#)	Evaluation (T P E A A)	Comments
2.435E-05	7.800E-03	37	R076	1 2 0 0 2	

3302. C$_{14}$H$_{16}$N$_4$O$_4$S
N4-Acetylsulphamethomidine
RN: **MP** (°C):
MW: 336.37   **BP** (°C):

Solubility (Moles/L)	Solubility (Grams/L)	Temp (°C)	Ref (#)	Evaluation (T P E A A)	Comments
7.730E-04	2.600E-01	ns	B133	0 2 0 0 2	pH 7.4

3303. C$_{14}$H$_{16}$N$_4$O$_5$S

N4-Acetylsulphadimethoxine
N4-Acetyl-2,4-dimethoxy-6-sulfanilamidopyrimidine
N4-Acetylsulfadimethoxypyrimidine
Sulfadimethoxine N4-acetate

| RN: | 555-25-9 | MP (°C): | |
| MW: | 352.37 | BP (°C): | |

Solubility (Moles/L)	Solubility (Grams/L)	Temp (°C)	Ref (#)	Evaluation (T P E A A)	Comments
5.392E-04	1.900E-01	ns	B133	0 2 0 0 2	pH 7.4

3304. C$_{14}$H$_{16}$O$_6$

Benzoic acid, 2-(acetyloxy)-, (1-oxobutoxy)methyl ester

| RN: | 118247-07-7 | MP (°C): | Oil |
| MW: | 280.28 | BP (°C): | |

Solubility (Moles/L)	Solubility (Grams/L)	Temp (°C)	Ref (#)	Evaluation (T P E A A)	Comments
1.249E-03	3.500E-01	21	N335	0 0 0 0 0	

3305. C$_{14}$H$_{16}$O$_6$

Ethylphthalyl ethyl glycolate
Ethoxycarbonylmethyl ethyl phthalate
Ethylphthalyl ethylglycolate

| RN: | 84-72-0 | MP (°C): | 20 |
| MW: | 280.28 | BP (°C): | 320 |

Solubility (Moles/L)	Solubility (Grams/L)	Temp (°C)	Ref (#)	Evaluation (T P E A A)	Comments
<2.85E-03	<7.99E-01	20	F070	1 0 0 0 1	

3306. C$_{14}$H$_{17}$ClNO$_4$PS$_2$

Dialifos
Dialifor
Diethyl S-(2-chloro-1-phthalimidoethyl) phosphorodithioate
Torak
Hercules 14503

| RN: | 10311-84-9 | MP (°C): | 67 |
| MW: | 393.85 | BP (°C): | |

Solubility (Moles/L)	Solubility (Grams/L)	Temp (°C)	Ref (#)	Evaluation (T P E A A)	Comments
4.570E-07	1.800E-04	ns	F071	0 1 2 1 1	
4.571E-07	1.800E-04	ns	R427	0 0 0 0 0	

3307. C₁₄H₁₇NO

1-Cinnamoylpiperidine
N,N-Pentamethylenecinnamamide
1-(1-Oxo-3-phenyl-2-propenyl)-piperidine
RN: 5422-81-1   **MP** (°C):
MW: 215.30   **BP** (°C):

Solubility (Moles/L)	Solubility (Grams/L)	Temp (°C)	Ref (#)	Evaluation (T P E A A)	Comments
9.600E-04	2.067E-01	ns	H350	0 0 0 0 0	

3308. C₁₄H₁₇NO

N-Cyclopentylcinnamamide
2-Propenamide, *N*-cyclopentyl-3-phenyl-
RN: 59831-97-9   **MP** (°C):
MW: 215.30   **BP** (°C):

Solubility (Moles/L)	Solubility (Grams/L)	Temp (°C)	Ref (#)	Evaluation (T P E A A)	Comments
2.280E-04	4.909E-02	ns	H350	0 0 0 0 0	

3309. C₁₄H₁₇NO₂S

m-Carboxylhexylphenylisothiocyanate
3-Carboxylhexylphenylisothiocyanate
RN: **MP** (°C):
MW: 263.36   **BP** (°C):

Solubility (Moles/L)	Solubility (Grams/L)	Temp (°C)	Ref (#)	Evaluation (T P E A A)	Comments
7.000E-05	1.844E-02	25	K032	2 2 0 1 1	

3310. C₁₄H₁₇NO₃

Piperidine, 1-[(benzoyloxy)acetyl]-
RN: 106231-67-8   **MP** (°C):   88
MW: 247.30   **BP** (°C):   433.2

Solubility (Moles/L)	Solubility (Grams/L)	Temp (°C)	Ref (#)	Evaluation (T P E A A)	Comments
3.154E-03	7.800E-01	22	B427	1 0 0 1 1	
3.154E-03	7.800E-01	22	N317	1 1 2 1 2	

3311. C₁₄H₁₇NO₄

4-Piperidinol, 1-[(benzoyloxy)acetyl]-
RN: 115178-71-7   **MP** (°C):   121.5
MW: 263.30   **BP** (°C):

Solubility (Moles/L)	Solubility (Grams/L)	Temp (°C)	Ref (#)	Evaluation (T P E A A)	Comments
4.482E-02	1.180E+01	22	N317	1 1 2 1 2	

3312. $C_{14}H_{17}NO_5$
Glycine, N-[(benzoyloxy)acetyl]-N-methyl-, ethyl ester

RN:	106231-63-4	MP (°C):	39.5
MW:	279.30	BP (°C):	426.4

Solubility (Moles/L)	Solubility (Grams/L)	Temp (°C)	Ref (#)	Evaluation (T P E A A)	Comments
2.148E-02	6.000E+00	22	B427	1 0 0 1 1	in 0.01M HCl
2.148E-02	6.000E+00	22	N317	1 1 2 1 2	

3313. $C_{14}H_{17}N_5O_3$
Pipemidic acid
Pipemidique acide

RN:	51940-44-4	MP (°C):	253
MW:	303.32	BP (°C):	

Solubility (Moles/L)	Solubility (Grams/L)	Temp (°C)	Ref (#)	Evaluation (T P E A A)	Comments
1.060E-03	3.215E-01	25	D051	2 0 0 1 2	0.05N NaCl
1.160E-03	3.519E-01	37	D051	2 0 0 1 2	0.05N NaCl

3314. $C_{14}H_{18}ClN_3S$
Chlorothen
N,N-Dimethyl-N'-(2-pyridyl)-N'-(5-chloro-2-thenyl)ethylenediamine
Chloromethapyrilene
5-Chloro-N-(2-(dimethylamino)ethyl)-N-(2-pyridyl)-2-thenylamine
Chloropyrilene

RN:	148-65-2	MP (°C):	
MW:	295.84	BP (°C):	155.5

Solubility (Moles/L)	Solubility (Grams/L)	Temp (°C)	Ref (#)	Evaluation (T P E A A)	Comments
6.800E-03	2.012E+00	37.5	L034	2 2 0 1 2	pH 7.4

3315. $C_{14}H_{18}Cl_2O_3$
2,4-Dichlorophenoxyacetic acid n-hexyl ester
Chloroxone
Agrotect
Amoxone
BH 2,4-D

RN:	1917-95-9	MP (°C):	
MW:	305.20	BP (°C):	

Solubility (Moles/L)	Solubility (Grams/L)	Temp (°C)	Ref (#)	Evaluation (T P E A A)	Comments
1.941E-05	5.924E-03	ns	M120	0 0 1 1 2	

3316. C₁₄H₁₈N₂O

Propyphenazone

Isopropylantipyrine

1,2-Dihydro-1,5-dimethyl-4-(isopropyl)-2-phenyl-pyrazol-3-one

4-Isopropyl-2,3-dimethyl-5-oxo-1-phenyl-3-pyrazoline

RN: 479-92-5   **MP** (°C):   103

MW: 230.31   **BP** (°C):

Solubility (Moles/L)	Solubility (Grams/L)	Temp (°C)	Ref (#)	Evaluation (T P E A A)	Comments
3.383E+00	7.791E+02	4.62	M109	2 1 1 1 0	EFG
3.330E+00	7.670E+02	10.93	M109	2 1 1 1 0	EFG
3.257E+00	7.501E+02	15.02	M109	2 1 1 1 0	EFG
3.238E+00	7.458E+02	20.96	M109	2 1 1 1 0	EFG
3.229E+00	7.436E+02	25.35	M109	2 1 1 1 0	EFG
3.238E+00	7.458E+02	29.87	M109	2 1 1 1 0	EFG
3.257E+00	7.501E+02	38.37	M109	2 1 1 1 0	EFG
3.348E+00	7.711E+02	40.32	M109	2 1 1 1 0	EFG

3317. C₁₄H₁₈N₂O₃

Reposal

5-Bicyclo[3.2.1]oct-2-en-3-yl-5-ethyl-2,4,6(1H,3H,5H)-pyrimidinetri-one

5-Bicyclo[3.2.1]oct-2-en-3-yl-5-ethylbarbituric acid

RN: 3625-25-0   **MP** (°C):   213

MW: 262.31   **BP** (°C):

Solubility (Moles/L)	Solubility (Grams/L)	Temp (°C)	Ref (#)	Evaluation (T P E A A)	Comments
1.680E-03	4.407E-01	25	V033	2 0 1 1 2	
1.700E-03	4.459E-01	25.00	T303	1 0 0 0 1	
2.300E-03	6.033E-01	35.00	T303	1 0 0 0 1	
2.500E-03	6.558E-01	45.00	T303	1 0 0 0 1	

3318. C₁₄H₁₈N₂O₃

Piperazine, 1-[(benzoyloxy)acetyl]-4-methyl-

RN: 106231-70-3   **MP** (°C):

MW: 262.31   **BP** (°C):   438.1

Solubility (Moles/L)	Solubility (Grams/L)	Temp (°C)	Ref (#)	Evaluation (T P E A A)	Comments
>7.62E-01	>2.00E+02	22	B427	1 0 0 1 1	

3319. C₁₄H₁₈N₄O₂S

2-Sulfanilylamino-4-isobutylpyrimidine

RN: 106596-34-3   **MP** (°C):

MW: 306.39   **BP** (°C):

Solubility (Moles/L)	Solubility (Grams/L)	Temp (°C)	Ref (#)	Evaluation (T P E A A)	Comments
3.264E-04	1.000E-01	37	R076	1 2 0 0 1	

3320. $C_{14}H_{18}N_4O_3$

Benomyl

(1-(Butylamino)carbonyl)-1H-benzimidazol-2-yl)carbamic acid methyl ester

RN: 17804-35-2   **MP** (°C):

MW: 290.32   **BP** (°C):

Solubility (Moles/L)	Solubility (Grams/L)	Temp (°C)	Ref (#)	Evaluation (T P E A A)	Comments
1.309E-05	3.800E-03	20	A064	1 0 1 1 1	
1.309E-05	3.800E-03	20	M161	1 0 0 0 1	pH 7
~6.89E-06	~2.00E-03	ns	B309	0 0 0 0 0	

3321. $C_{14}H_{18}N_4O_3$

Trimethoprim

5-(3,4,5-Trimethoxybenzyl)-2,4-diaminopyrimidine

Monotrim

Syraprim

Proloprim

Trimpex

RN: 738-70-5   **MP** (°C): 201

MW: 290.32   **BP** (°C):

Solubility (Moles/L)	Solubility (Grams/L)	Temp (°C)	Ref (#)	Evaluation (T P E A A)	Comments
6.034E-05	1.752E-02	22.5	B440	0 0 0 0 0	
1.396E-03	4.053E-01	25	H434	0 0 0 0 0	
1.378E-03	4.000E-01	25	M167	1 0 0 0 0	
1.722E-03	5.000E-01	32	D308	0 0 0 0 0	pH 8.54
2.711E-03	7.870E-01	37	G086	1 0 0 0 1	
1.378E-03	4.000E-01	37	M321	1 0 0 0 2	intrinsic
>1.72E-03	>5.00E-01	ns	B404	0 2 1 1 0	
1.378E-03	4.000E-01	ns	K444	0 0 0 0 0	

3322. $C_{14}H_{18}N_4O_6.0.5H_2O$

2′-Propionyl-6-methoxypurine arabinoside (hemihydrate)

RN: 145913-38-8   **MP** (°C): 60–65

MW: 347.33   **BP** (°C):

Solubility (Moles/L)	Solubility (Grams/L)	Temp (°C)	Ref (#)	Evaluation (T P E A A)	Comments
1.100E-01	3.821E+01	37	C348	0 0 0 0 0	pH 7.00

3323. $C_{14}H_{18}N_4O_7.0.5H_2O$

9-[5-O-(Methoxyacetate-β-D-arabinofuranosyl)]-6-methoxy-9H-purine (hemihydrate)

RN: 121032-38-0   **MP** (°C): 137–139

MW: 363.33   **BP** (°C):

Solubility (Moles/L)	Solubility (Grams/L)	Temp (°C)	Ref (#)	Evaluation (T P E A A)	Comments
7.810E-02	2.838E+01	37	M378	1 2 1 1 2	pH 7.2

3324. C$_{14}$H$_{18}$N$_4$O$_7$.0.9H$_2$O
2′-Methoxyacetyl-6-methoxypurine arabinoside (0.9 hydrate)
RN: 145913-47-9   **MP** (°C):
MW: 370.54   **BP** (°C):

Solubility (Moles/L)	Solubility (Grams/L)	Temp (°C)	Ref (#)	Evaluation (T P E A A)	Comments
9.090E-02	3.368E+01	37	C348	0 0 0 0 0	pH 7.00

3325. C$_{14}$H$_{18}$N$_6$O
(1S,4R)-4-[2-Amino-6-(cyclopropylamino)-9H-purin-9-yl]-2-cyclopentene-1-methanol sulfate
   (salt)
ABC sulfate[47]
ABC[48]
Abacavir
RN: 188062-50-2   **MP** (°C):
MW: 286.34   **BP** (°C):

Solubility (Moles/L)	Solubility (Grams/L)	Temp (°C)	Ref (#)	Evaluation (T P E A A)	Comments
5.867E-09	1.680E-06	32	M458	0 0 0 0 0	

3326. C$_{14}$H$_{18}$N$_6$O$_4$
2,5-Diaziridinyl-3,6-bis(glycinamide)-1,4-benzoquinone
RN: 59886-49-6   **MP** (°C):   200
MW: 334.34   **BP** (°C):

Solubility (Moles/L)	Solubility (Grams/L)	Temp (°C)	Ref (#)	Evaluation (T P E A A)	Comments
1.495E-03	5.000E-01	rt	C317	0 0 0 0 0	

3327. C$_{14}$H$_{18}$O$_4$
Diisopropyl phthalate
bis(1-Methyl-ethyl) phthalate
RN: 605-45-8   **MP** (°C):
MW: 250.30   **BP** (°C):

Solubility (Moles/L)	Solubility (Grams/L)	Temp (°C)	Ref (#)	Evaluation (T P E A A)	Comments
1.330E-03	3.329E-01	20	L300	2 1 0 2 2	

3328. $C_{14}H_{18}O_4$
Di-*n*-propyl phthalate
Dipropyl phthalate

RN: 131-16-8   **MP** (°C):
MW: 250.30   **BP** (°C):

Solubility (Moles/L)	Solubility (Grams/L)	Temp (°C)	Ref (#)	Evaluation (T P E A A)	Comments
4.320E-04	1.081E-01	20	L300	2 1 0 2 2	

3329. $C_{14}H_{18}O_4$
Diisopropyl *o*-phthalate

RN: **MP** (°C):
MW: 250.30   **BP** (°C):

Solubility (Moles/L)	Solubility (Grams/L)	Temp (°C)	Ref (#)	Evaluation (T P E A A)	Comments
6.672E-04	1.670E-01	25	S417	0 0 0 0 0	

3330. $C_{14}H_{18}O_6$
Methyl glycol phthalate
bis(2-Methoxyethyl) phthalate

RN: 117-82-8   **MP** (°C):
MW: 282.30   **BP** (°C):

Solubility (Moles/L)	Solubility (Grams/L)	Temp (°C)	Ref (#)	Evaluation (T P E A A)	Comments
3.090E-02	8.723E+00	15	H069	1 0 1 1 1	

3331. $C_{14}H_{18}O_6$
Ethyl phthalyl ethyl glycollate

RN: **MP** (°C):
MW: 282.30   **BP** (°C):

Solubility (Moles/L)	Solubility (Grams/L)	Temp (°C)	Ref (#)	Evaluation (T P E A A)	Comments
1.770E-03	4.998E-01	15	H069	1 0 1 1 0	
1.770E-03	4.998E-01	ns	F014	0 0 0 0 1	

3332. $C_{14}H_{18}O_6$
Dimethoxyethyl phthalate
1,2-Benzenedicarboxylic acid, di(2-methoxyethyl) ester

RN: 34006-76-3   **MP** (°C):
MW: 282.30   **BP** (°C):

Solubility (Moles/L)	Solubility (Grams/L)	Temp (°C)	Ref (#)	Evaluation (T P E A A)	Comments
2.986E-02	8.428E+00	20	F070	1 0 0 0 1	
2.944E-02	8.310E+00	ns	F014	0 0 0 0 2	

3333. C$_{14}$H$_{19}$Cl$_2$NO$_2$
Chlorambucil
N,N-di-(2-Chloroethyl)-γ-(p-aminophenyl)butyric acid
Linfolysin
Elcoril
Linfolizin
Leukersan

RN: 305-03-3   **MP** (°C):   64
MW: 304.22   **BP** (°C):

Solubility (Moles/L)	Solubility (Grams/L)	Temp (°C)	Ref (#)	Evaluation (T P E A A)	Comments
3.000E-04	9.127E-02	3	G434	0 0 0 0 0	pH 4.13
<3.29E-03	<1.00E+00	30	L343	2 1 1 1 0	EFG

3334. C$_{14}$H$_{19}$IN$_2$O$_6$
Uridine, 2'-deoxy-5-iodo-, 5'-pentanoate
5'-Valeryl 5-iodo-2'-deoxyuridine

RN: 84052-69-7   **MP** (°C):   142.5
MW: 438.22   **BP** (°C):

Solubility (Moles/L)	Solubility (Grams/L)	Temp (°C)	Ref (#)	Evaluation (T P E A A)	Comments
4.000E+02	1.753E+05	25	N332	0 0 0 0 0	pH 7.4

3335. C$_{14}$H$_{19}$IN$_2$O$_6$
Uridine, 2'-deoxy-5-iodo-, 5'-(2,2-dimethylpropanoate)
5'-Pivaloyl 5-iodo-2'-deoxyuridine
5-Iodo-2'-deoxyuridine 5'-pivalate

RN: 84043-28-7   **MP** (°C):   106.5
MW: 438.22   **BP** (°C):

Solubility (Moles/L)	Solubility (Grams/L)	Temp (°C)	Ref (#)	Evaluation (T P E A A)	Comments
4.400E+02	1.928E+05	25	N332	0 0 0 0 0	pH 7.4

3336. C$_{14}$H$_{19}$NO
n-Pentylcinnamamide
2-Propenamide, N-pentyl-3-phenyl-

RN: 23784-51-2   **MP** (°C):
MW: 217.31   **BP** (°C):

Solubility (Moles/L)	Solubility (Grams/L)	Temp (°C)	Ref (#)	Evaluation (T P E A A)	Comments
8.200E-05	1.782E-02	ns	H350	0 0 0 0 0	

3337. $C_{14}H_{19}NO_3$

Acetaminophen hexanoate
Hexanyl acetaminophen
Hexanoic acid, 4-(acetylamino)phenyl ester
4′-Hydroxyacetanilide hexanoate

RN:	20675-21-2	**MP** (°C):	107	
MW:	249.31	**BP** (°C):		

Solubility (Moles/L)	Solubility (Grams/L)	Temp (°C)	Ref (#)	Evaluation (T P E A A)	Comments
7.220E-05	1.800E-02	25	B010	1 1 1 1 0	
2.286E-04	5.700E-02	37	D029	0 0 0 0 0	

3338. $C_{14}H_{19}NO_3$

Propanamide, 2-(benzoyloxy)-N,N-diethyl-

RN:	115178-79-5	**MP** (°C):	53.5	
MW:	249.31	**BP** (°C):		

Solubility (Moles/L)	Solubility (Grams/L)	Temp (°C)	Ref (#)	Evaluation (T P E A A)	Comments
5.214E-03	1.300E+00	22	N317	1 1 2 1 2	

3339. $C_{14}H_{19}NO_4$

Anisomycin
(2R,3R,4R)-2-(4-Methoxybenzyl)-3,4-pyrrolidinediol-3-acetate

RN:	22862-76-6	**MP** (°C):	140.5	
MW:	265.31	**BP** (°C):		

Solubility (Moles/L)	Solubility (Grams/L)	Temp (°C)	Ref (#)	Evaluation (T P E A A)	Comments
2.469E-02	6.550E+00	28	A038	2 0 1 1 2	

3340. $C_{14}H_{19}N_3S$

Methapyrilene
N,N-Dimethyl-N',2-pyridinyl-N'-(2-thienylmethyl)-1,2-ethanediamine
Cope
A 3322
AH-42
Semiken

RN:	91-80-5	**MP** (°C):	<25	
MW:	261.39	**BP** (°C):		

Solubility (Moles/L)	Solubility (Grams/L)	Temp (°C)	Ref (#)	Evaluation (T P E A A)	Comments
2.300E-03	6.012E-01	30	L068	1 0 0 1 0	EFG
1.700E-02	4.444E+00	37.5	L034	2 2 0 1 2	pH 7.4

3341. C$_{14}$H$_{19}$N$_3$S

Thenyldiamine

1,2-Ethanediamine, N,N-dimethyl-N'-2-pyridinyl-N'-(3-thienylmethyl)-

N-(2-Dimethylaminoethyl)-N-2-pyridyl-3-thenylamine

Thefanil

Thenfadil

Tenfidil

RN:	91-79-2	MP (°C):
MW:	261.39	BP (°C):

Solubility (Moles/L)	Solubility (Grams/L)	Temp (°C)	Ref (#)	Evaluation (T P E A A)	Comments
1.700E-02	4.444E+00	37.5	L034	2 2 0 1 2	pH 7.4

3342. C$_{14}$H$_{19}$N$_5$O$_4$

N,N-Diethylsuccinamyloxymethyl-1-allopurinol

Butanoic acid, 4-(diethylamino)-4-oxo-, (4,5-dihydro-4-oxo-1H-pyrazolo[3,4-d]pyrimidin-1-yl) methyl ester

RN:	98827-27-1	MP (°C):	138-140
MW:	321.34	BP (°C):	

Solubility (Moles/L)	Solubility (Grams/L)	Temp (°C)	Ref (#)	Evaluation (T P E A A)	Comments
1.027E-01	3.300E+01	22	B322	0 0 0 0 0	

3343. C$_{14}$H$_{19}$N$_5$O$_5$

9-[5'-(O-Butyryl)-β-D-arabinofuranosyl]adenine ester

Vidarabine 5'-butyrate

RN:	65926-30-9	MP (°C):
MW:	337.34	BP (°C):

Solubility (Moles/L)	Solubility (Grams/L)	Temp (°C)	Ref (#)	Evaluation (T P E A A)	Comments
4.773E-02	1.610E+01	ns	B134	0 1 1 1 2	

3344. C$_{14}$H$_{19}$O$_6$P

Crotoxyphos

Dimethylphosphate of α-methylbenzyl-3-hydroxy-cis-crotonate

RN:	7700-17-6	MP (°C):
MW:	314.28	BP (°C): 135

Solubility (Moles/L)	Solubility (Grams/L)	Temp (°C)	Ref (#)	Evaluation (T P E A A)	Comments
3.179E-03	9.990E-01	ns	M061	0 0 0 0 0	
3.182E-03	1.000E+00	rt	M161	0 0 0 0 0	

3345. $C_{14}H_{20}ClNO_2$

Alachlor

2-Chloro-2′,6′-diethyl-N-(methoxymethyl)acetanilide

RN: 15972-60-8 **MP** (°C): 39.5

MW: 269.77 **BP** (°C):

Solubility (Moles/L)	Solubility (Grams/L)	Temp (°C)	Ref (#)	Evaluation (T P E A A)	Comments
8.896E-04	2.400E-01	23	M161	1 0 0 0 2	
5.486E-04	1.480E-01	25	B200	1 0 0 0 2	
5.486E-04	1.480E-01	ns	M061	0 0 0 0 2	
5.560E-04	1.500E-01	ns	M110	0 0 0 0 0	EFG
8.896E-04	2.400E-01	ns	V414	0 0 0 0 0	

3346. $C_{14}H_{20}ClNO_2$

Acetochlor

Doubleplay

Harness

Topnotch

Top Hand

Acenit

RN: 34256-82-1 **MP** (°C):

MW: 269.77 **BP** (°C):

Solubility (Moles/L)	Solubility (Grams/L)	Temp (°C)	Ref (#)	Evaluation (T P E A A)	Comments
8.260E-04	2.228E-01	ns	S460	0 0 0 0 0	

3347. $C_{14}H_{20}N_2O$

Siduron

1-(2-Methylcyclohexyl)-3-phenylurea

RN: 1982-49-6 **MP** (°C): 133

MW: 232.33 **BP** (°C):

Solubility (Moles/L)	Solubility (Grams/L)	Temp (°C)	Ref (#)	Evaluation (T P E A A)	Comments
7.748E-05	1.800E-02	25	B200	1 0 0 0 1	
7.748E-05	1.800E-02	25	G036	1 0 0 0 1	
7.748E-05	1.800E-02	25	M161	1 0 0 0 1	

3348. $C_{14}H_{20}N_2O_2$
Pindolol
Barbloc
Visken
2-Propanol, 1-(1H-indol-4-yloxy)-3-[(-methylethyl)amino]-

RN: 13523-86-9 **MP** (°C):
MW: 248.33 **BP** (°C):

Solubility (Moles/L)	Solubility (Grams/L)	Temp (°C)	Ref (#)	Evaluation (T P E A A)	Comments
1.329E-04	3.300E-02	22.5	B422	2 0 2 2 2	

3349. $C_{14}H_{20}N_2O_3S$
Tolcyclamide
1-Cyclohexyl-3-*para*-tolylsulfonylurea
Glycyclamide

RN: 664-95-9 **MP** (°C): 175
MW: 296.39 **BP** (°C):

Solubility (Moles/L)	Solubility (Grams/L)	Temp (°C)	Ref (#)	Evaluation (T P E A A)	Comments
6.194E-05	1.836E-02	37	A028	1 0 2 1 2	intrinsic
6.200E-05	1.838E-02	37	A046	2 0 1 1 2	

3350. $C_{14}H_{20}N_3O_5PS$
Pyrazophos
2-[(Diethoxyphosphinothioyl)oxy]-5-methylpyrazolo[1,5-a]pyrimidine-6-carboxylic acid ethyl
Ester
Afugan
Curamil

RN: 13457-18-6 **MP** (°C): 50.5
MW: 373.37 **BP** (°C):

Solubility (Moles/L)	Solubility (Grams/L)	Temp (°C)	Ref (#)	Evaluation (T P E A A)	Comments
1.125E-05	4.200E-03	20	A306	0 0 0 0 0	
1.125E-05	4.200E-03	20	M161	1 0 0 0 1	

3351. $C_{14}H_{20}N_4O_2$
2,5-bis(Methylaziridinyl)-3,6-bis(methylamino)-1,4-benzoquinone

RN: 64947-06-4 **MP** (°C): 179
MW: 276.34 **BP** (°C):

Solubility (Moles/L)	Solubility (Grams/L)	Temp (°C)	Ref (#)	Evaluation (T P E A A)	Comments
<3.62E-04	<1.00E-01	rt	C317	0 0 0 0 0	

3352. C$_{14}$H$_{20}$N$_4$O$_2$

2,5-Diaziridinyl-3,6-bis(dimethylamino)-1,4-benzoquinone

RN: 59886-50-9   **MP** (°C): 112
MW: 276.34   **BP** (°C):

Solubility (Moles/L)	Solubility (Grams/L)	Temp (°C)	Ref (#)	Evaluation (T P E A A)	Comments
3.619E-02	1.000E+01	rt	C317	0 0 0 0 0	

3353. C$_{14}$H$_{20}$N$_4$O$_2$

2,5-Diaziridinyl-3,6-bis(ethylamino)-1,4-benzoquinone

RN: 59886-53-2   **MP** (°C): 157
MW: 276.34   **BP** (°C):

Solubility (Moles/L)	Solubility (Grams/L)	Temp (°C)	Ref (#)	Evaluation (T P E A A)	Comments
1.809E-03	5.000E-01	rt	C317	0 0 0 0 0	

3354. C$_{14}$H$_{20}$N$_4$O$_4$

2,5-Diaziridinyl-3,6-bis(hydroxyethylamino)-1,4-benzoquinone

RN: 59886-54-3   **MP** (°C): 188
MW: 308.34   **BP** (°C):

Solubility (Moles/L)	Solubility (Grams/L)	Temp (°C)	Ref (#)	Evaluation (T P E A A)	Comments
6.486E-03	2.000E+00	rt	C317	0 0 0 0 0	

3355. C$_{14}$H$_{20}$O$_3$

Heptyl p-hydroxybenzoate
n-Heptyl 4-hydroxybenzoate

RN: 1085-12-7   **MP** (°C): 48
MW: 236.31   **BP** (°C):

Solubility (Moles/L)	Solubility (Grams/L)	Temp (°C)	Ref (#)	Evaluation (T P E A A)	Comments
2.630E-04	6.215E-02	−244	B355	0 0 0 0 0	
2.010E-04	4.750E-02	15	B355	0 0 0 0 0	
2.520E-04	5.955E-02	20	B355	0 0 0 0 0	
5.827E-03	1.377E+00	25	D081	1 2 2 1 2	sic
1.259E-04	2.975E-02	25	F322	2 0 1 1 0	EFG

3356. $C_{14}H_{21}NO_2$
Heptyl p-aminobenzoate
Heptyl 4-aminobenzoate

RN: 14309-40-1   **MP** (°C):
MW: 235.33   **BP** (°C):

Solubility (Moles/L)	Solubility (Grams/L)	Temp (°C)	Ref (#)	Evaluation (T P E A A)	Comments
2.000E-05	4.707E-03	37	F006	1 1 2 2 1	
3.300E-05	7.766E-03	ns	M066	0 0 0 0 1	

3357. $C_{14}H_{21}NO_2$
2,6-Diisopropyl-4-acetaminophenol
3,5-Diisopropylparacetamol
4-Acetamido-2,6-diisopropylphenol

RN: 1988-14-3   **MP** (°C):
MW: 235.33   **BP** (°C):

Solubility (Moles/L)	Solubility (Grams/L)	Temp (°C)	Ref (#)	Evaluation (T P E A A)	Comments
5.844E-04	1.375E-01	25	D078	1 2 2 1 2	

3358. $C_{14}H_{21}NO_2$
Octyl nicotinate
Nicotinic acid n-octyl ester

RN: 70136-02-6   **MP** (°C):
MW: 235.33   **BP** (°C):

Solubility (Moles/L)	Solubility (Grams/L)	Temp (°C)	Ref (#)	Evaluation (T P E A A)	Comments
4.249E-05	1.000E-02	32	L346	1 0 0 1 2	

3359. $C_{14}H_{21}NO_2$
Benzeneacetamide, N-hydroxy-α-dipropyl

RN: 60631-09-6   **MP** (°C):
MW: 235.33   **BP** (°C):

Solubility (Moles/L)	Solubility (Grams/L)	Temp (°C)	Ref (#)	Evaluation (T P E A A)	Comments
1.300E-03	3.059E-01	26	G076	1 0 0 0 1	

3360. $C_{14}H_{21}NO_2$
Benzenepropanamide, N-hydroxy-α2,4,6-pentamethyl

RN: 60631-10-9   **MP** (°C):
MW: 235.33   **BP** (°C):

Solubility (Moles/L)	Solubility (Grams/L)	Temp (°C)	Ref (#)	Evaluation (T P E A A)	Comments
3.000E-04	7.060E-02	26	G076	1 0 0 0 1	

3361. C$_{14}$H$_{21}$NO$_3$
4-Methoxybenzoic acid-2-(diethylamino)ethyl ester
Diethylaminoethyl *p*-anisate

RN:	10367-84-7	**MP** (°C):
MW:	251.33	**BP** (°C):

Solubility (Moles/L)	Solubility (Grams/L)	Temp (°C)	Ref (#)	Evaluation (T P E A A)	Comments
5.300E-03	1.332E+00	ns	M066	0 0 0 0 1	

3362. C$_{14}$H$_{21}$NO$_4$P
Phenyl(di-morpholido)-phosphate

RN:		**MP** (°C):
MW:	298.30	**BP** (°C):

Solubility (Moles/L)	Solubility (Grams/L)	Temp (°C)	Ref (#)	Evaluation (T P E A A)	Comments
2.583E+00	7.706E+02	25	A040	1 0 0 0 2	

3363. C$_{14}$H$_{21}$N$_3$O$_3$
Karbutilate
m-(3,3-Dimethylureido)phenyl-*tert*-butylcarbamate
Tandex

RN:	4849-32-5	**MP** (°C):	176.3
MW:	279.34	**BP** (°C):	

Solubility (Moles/L)	Solubility (Grams/L)	Temp (°C)	Ref (#)	Evaluation (T P E A A)	Comments
1.163E-03	3.250E-01	20	B200	1 0 0 0 2	
1.163E-03	3.250E-01	rt	M161	0 0 0 0 2	

3364. C$_{14}$H$_{21}$N$_3$O$_3$S
Tolazamide
N-((((Hexahydro-1H-azepin-1-yl)amino)carbonyl)-4-methylbenzenesulfonamide
Tolinase
N-(*p*-Toluenesulfonyl)-*N'*-hexamethyleniminourea
U 17835

RN:	1156-19-0	**MP** (°C):	170
MW:	311.41	**BP** (°C):	

Solubility (Moles/L)	Solubility (Grams/L)	Temp (°C)	Ref (#)	Evaluation (T P E A A)	Comments
2.100E-04	6.540E-02	30	H025	1 0 2 1 1	intrinsic
1.124E-03	3.499E-01	ns	B404	0 2 1 1 0	

3365. C$_{14}$H$_{22}$
2-Octylbenzene
(1-Methylheptyl)benzene
RN: 777-22-0   **MP** (°C):
MW: 190.33   **BP** (°C):

Solubility (Moles/L)	Solubility (Grams/L)	Temp (°C)	Ref (#)	Evaluation (T P E A A)	Comments
1.585E-06	3.017E-04	ns	D001	0 0 0 0 2	

3366. C$_{14}$H$_{22}$N$_2$O
Lidocaine
2-(Diethylamino)-N-(2,6-dimethylphenyl)acetamide
2-Diethylamino-2′,6′-acetoxylidide
Lignocaine
Leostesin
Xylocaine
RN: 137-58-6   **MP** (°C):   68
MW: 234.34   **BP** (°C):

Solubility (Moles/L)	Solubility (Grams/L)	Temp (°C)	Ref (#)	Evaluation (T P E A A)	Comments
1.850E-02	4.335E+00	14.5	N046	2 0 1 1 1	intrinsic
5.460E-05	1.279E-02	22.5	B440	0 0 0 0 0	
1.550E-02	3.632E+00	25	D402	1 2 2 2 0	EFG
1.643E-02	3.850E+00	25	L338	1 0 1 1 2	
1.630E-02	3.820E+00	25	N046	2 0 1 1 1	intrinsic
1.488E-02	3.488E+00	25	S450	0 0 0 0 0	Intrinsic
1.750E-02	4.101E+00	30	L068	1 0 0 1 0	EFG
1.460E-02	3.421E+00	34.5	N046	2 0 1 1 1	intrinsic
1.700E-02	3.984E+00	37	D402	1 2 2 2 0	
1.440E-02	3.375E+00	37	N044	2 1 1 2 2	intrinsic

3367. C$_{14}$H$_{22}$N$_2$O$_2$
4-Methylaminobenzoic acid-2-(diethyl-amino)ethyl ester
Benzoic acid, 4-(methylamino)-, 2-(diethylamino)ethyl ester
Benzoic acid, p-(methylamino)-, 2-(diethylamino)ethyl ester
RN: 16488-52-1   **MP** (°C):
MW: 250.34   **BP** (°C):

Solubility (Moles/L)	Solubility (Grams/L)	Temp (°C)	Ref (#)	Evaluation (T P E A A)	Comments
7.750E-03	1.940E+00	ns	M066	0 0 0 0 2	

3368. C$_{14}$H$_{22}$N$_2$O$_2$

4-Aminobenzoic acid-2-(diethyl-amino)propyl ester
2-Diethylamino)propyl 4-aminobenzoate

RN: 5878-13-7   **MP** (°C):
MW: 250.34   **BP** (°C):

Solubility (Moles/L)	Solubility (Grams/L)	Temp (°C)	Ref (#)	Evaluation (T P E A A)	Comments
1.290E-02	3.229E+00	ns	M066	0 0 0 0 2	

3369. C$_{14}$H$_{22}$N$_2$O$_3$

2,4-Diazaspiro[5.10]hexadecane-1,3,5-trione

RN: 143288-63-5   **MP** (°C):
MW: 266.34   **BP** (°C):

Solubility (Moles/L)	Solubility (Grams/L)	Temp (°C)	Ref (#)	Evaluation (T P E A A)	Comments
2.600E-05	6.925E-03	25	P350	0 0 0 0 0	intrinsic

3370. C$_{14}$H$_{22}$N$_2$O$_3$

Atenolol
Anselol
Apo-atenolol
Benzeneacetamide
4-(2′-Hydroxy-3′-((1-methylethyl)amino)propoxy)-
Noten

RN: 29122-68-7   **MP** (°C):
MW: 266.34   **BP** (°C):

Solubility (Moles/L)	Solubility (Grams/L)	Temp (°C)	Ref (#)	Evaluation (T P E A A)	Comments
5.069E-05	1.350E-02	25	A408	2 0 1 2 0	int
7.134E-10	1.900E-07	32	M458	0 0 0 0 0	
9.950E-02	2.650E+01	ns	K444	0 0 0 0 0	

3371. C$_{14}$H$_{22}$N$_2$O$_4$

Ethyl-2-methyl-2-cyclohexenyl-6-methylmalonurate
Ethyl 2-methyl-2-cyclohexenyl-6-methylmalonurate

RN: **MP** (°C):   97.5
MW: 282.34   **BP** (°C):

Solubility (Moles/L)	Solubility (Grams/L)	Temp (°C)	Ref (#)	Evaluation (T P E A A)	Comments
1.000E-03	2.823E-01	23	B152	1 2 1 1 1	pH 3.5

3372. C₁₄H₂₂N₂O₅

Methoxymethyl-2-methyl-2-cyclohexenyl-6-methylmalonurate
RN: **MP** (°C):   73
MW: 298.34   **BP** (°C):

Solubility (Moles/L)	Solubility (Grams/L)	Temp (°C)	Ref (#)	Evaluation (T P E A A)	Comments
3.800E-03	1.134E+00	23	B152	1 2 1 1 1	pH 3.5

3373. C₁₄H₂₂O

Methyl ionone
6-Methylionone
RN: 1335-46-2   **MP** (°C):
MW: 206.33   **BP** (°C):

Solubility (Moles/L)	Solubility (Grams/L)	Temp (°C)	Ref (#)	Evaluation (T P E A A)	Comments
9.693E-05	2.000E-02	25	M350	1 0 1 1 1	

3374. C₁₄H₂₂O

o-n-Octylphenol
2-*n*-Octylphenol
RN: 949-13-3   **MP** (°C):
MW: 206.33   **BP** (°C):

Solubility (Moles/L)	Solubility (Grams/L)	Temp (°C)	Ref (#)	Evaluation (T P E A A)	Comments
1.385E-05	2.857E-03	25	L022	1 0 0 0 0	

3375. C₁₄H₂₂O

p-n-Octylphenol
4-Octylphenol
RN: 1806-26-4   **MP** (°C):   44.5
MW: 206.33   **BP** (°C):

Solubility (Moles/L)	Solubility (Grams/L)	Temp (°C)	Ref (#)	Evaluation (T P E A A)	Comments
6.107E-05	1.260E-02	20.5	A335	0 0 0 0 0	
6.120E-05	1.263E-02	20.5	A335	0 0 0 0 0	
8.812E-06	1.818E-03	25	L022	1 0 0 0 0	

3376. C₁₄H₂₃O₃P

Dibutyl phenyl phosphonate
Dibutoxyphenylphosphine oxide
Dibutyl phenylphosphonate

RN: 1024-34-6	**MP** (°C):	
MW: 270.31	**BP** (°C):	

Solubility (Moles/L)	Solubility (Grams/L)	Temp (°C)	Ref (#)	Evaluation (T P E A A)	Comments
<7.40E-04	<2.00E-01	25	B070	1 2 0 1 0	

3377. C₁₄H₂₄NO₄PS₃

Bensulide
O,O-bis(1-Methylethyl) *S*-(2-((phenylsulfonyl)amino)ethyl) phosphorodithioate
Betasan
Betamec
Exporsan
Benzulfide

RN: 741-58-2	**MP** (°C):	34.4
MW: 397.52	**BP** (°C):	

Solubility (Moles/L)	Solubility (Grams/L)	Temp (°C)	Ref (#)	Evaluation (T P E A A)	Comments
6.289E-05	2.500E-02	20	B200	1 2 0 0 1	
6.289E-05	2.500E-02	rt	M161	0 0 0 0 1	

3378. C₁₄H₂₄N₂O₃

5-Ethyl-5-*n*-octylbarbituric acid
2,4,6(1H,3H,5H)-Pyrimidinetrione, 5-ethyl-5-octyl-
5-Ethyl-5-octylbarbiturate

RN: 64810-90-8	**MP** (°C):	
MW: 268.36	**BP** (°C):	

Solubility (Moles/L)	Solubility (Grams/L)	Temp (°C)	Ref (#)	Evaluation (T P E A A)	Comments
1.140E-04	3.059E-02	25	M310	2 2 2 2 2	

3379. C₁₄H₂₄N₂O₃

p-5-Ethyl-5-methylhexylcarbinylbarbituric acid

RN:	**MP** (°C):	
MW: 268.36	**BP** (°C):	

Solubility (Moles/L)	Solubility (Grams/L)	Temp (°C)	Ref (#)	Evaluation (T P E A A)	Comments
1.543E-03	4.140E-01	ns	T003	0 0 0 0 2	

3380. $C_{14}H_{24}O_2$
3-Hydroxy-2,5-dispirocyclohexyltetrahydrofuran
7-Oxadispiro[5.1.5.2]pentadecan-14-ol
RN: 29839-63-2   **MP** (°C):
MW: 224.35   **BP** (°C):

Solubility (Moles/L)	Solubility (Grams/L)	Temp (°C)	Ref (#)	Evaluation (T P E A A)	Comments
3.098E-02	6.951E+00	rt	B066	0 2 0 0 0	contains impurity

3381. $C_{14}H_{26}O_4$
1,12-Dodecanedicarboxylic acid
Tetradecanedioic acid
RN: 821-38-5   **MP** (°C):   127
MW: 258.36   **BP** (°C):

Solubility (Moles/L)	Solubility (Grams/L)	Temp (°C)	Ref (#)	Evaluation (T P E A A)	Comments
7.741E-04	2.000E-01	21	B040	1 0 1 1 0	*sic*

3382. $C_{14}H_{27}NO_2$
Pentanamide,*N*-hydroxy-α,α-dipropyl
RN: **MP** (°C):
MW: 241.38   **BP** (°C):

Solubility (Moles/L)	Solubility (Grams/L)	Temp (°C)	Ref (#)	Evaluation (T P E A A)	Comments
5.000E-04	1.207E-01	26	G076	1 0 0 0 1	

3383. $C_{14}H_{28}NO_3PS_2$
Piperophos
S-(2-(2-Methyl-1-piperidinyl)-2-oxoethyl) *O,O*-dipropyl phosphorodithioate
RN: 24151-93-7   **MP** (°C):
MW: 353.49   **BP** (°C):

Solubility (Moles/L)	Solubility (Grams/L)	Temp (°C)	Ref (#)	Evaluation (T P E A A)	Comments
7.072E-05	2.500E-02	20	M161	1 0 0 0 1	

3384. $C_{14}H_{28}N_2O_2$
N,N,N′,N′-Tetramethylsebacamide
Decanediamide, *N,N,N′,N′*-tetramethyl-
RN: 13424-83-4   **MP** (°C):
MW: 256.39   **BP** (°C):

Solubility (Moles/L)	Solubility (Grams/L)	Temp (°C)	Ref (#)	Evaluation (T P E A A)	Comments
5.270E-01	1.351E+02	30	D010	1 2 1 1 2	

3385. $C_{14}H_{28}O_2$

Myristic acid
Tetradecanoic acid
Crodacid
1-Tridecanecarboxylic acid

RN:	544-63-8	**MP** (°C):	54		
MW:	228.38	**BP** (°C):			

Solubility (Moles/L)	Solubility (Grams/L)	Temp (°C)	Ref (#)	Evaluation (T P E A A)	Comments
5.692E-05	1.300E-02	0	B136	1 0 2 1 1	
8.757E-05	2.000E-02	20	B136	1 0 2 1 1	
8.757E-05	2.000E-02	20	D041	1 0 0 0 0	
8.757E-05	2.000E-02	20	R001	1 1 1 1 1	
4.700E-06	1.073E-03	25	J001	1 0 2 1 1	average of 2
8.000E-07	1.827E-04	25	R002	0 0 0 0 0	intrinsic
3.710E-06	8.473E-04	25	R002	0 0 0 0 0	
9.633E-05	2.200E-02	30	B136	1 0 2 1 1	
1.051E-04	2.400E-02	30	R001	1 1 1 1 1	
1.270E-04	2.900E-02	40	B136	1 0 2 1 1	
1.270E-04	2.900E-02	45	B136	1 0 2 1 1	
1.270E-04	2.900E-02	45	R001	1 1 1 1 1	
1.839E-05	4.200E-03	50	E005	2 1 1 2 1	
9.700E-06	2.215E-03	50	J001	1 0 2 1 1	
1.489E-04	3.400E-02	60	B136	1 0 2 1 1	
2.452E-05	5.600E-03	60	E005	2 1 1 2 1	
1.489E-04	3.400E-02	60	R001	1 1 1 1 1	
5.692E-05	1.300E-02	.0	R001	1 1 1 1 1	

3386. $C_{14}H_{28}O_4$

1,3-Dioxolane-4-methanol, 2-[2-(heptyloxy)ethyl]-2-methyl

RN:	143458-57-5	**MP** (°C):			
MW:	260.38	**BP** (°C):			

Solubility (Moles/L)	Solubility (Grams/L)	Temp (°C)	Ref (#)	Evaluation (T P E A A)	Comments
4.440E-03	1.156E+00	25	P342	0 0 0 0 0	0.0001M Na_2CO_3

3387. $C_{14}H_{29}NO_2$

Benzenepropanamide, N-hydroxy-α2,3-pentamethyl
Octanamide, N-hydroxy-2,2-dipropyl

RN:	60631-08-5	**MP** (°C):			
MW:	243.39	**BP** (°C):			

Solubility (Moles/L)	Solubility (Grams/L)	Temp (°C)	Ref (#)	Evaluation (T P E A A)	Comments
4.500E-04	1.095E-01	26	G076	1 0 0 0 1	
1.500E-03	3.651E-01	26	G076	1 0 0 0 1	

3388. C$_{14}$H$_{29}$NO$_2$
Octanamide, 2,2,4-triethyl-N-hydroxy
RN: 60631-07-4   **MP** (°C):
MW: 243.39   **BP** (°C):

Solubility (Moles/L)	Solubility (Grams/L)	Temp (°C)	Ref (#)	Evaluation (T P E A A)	Comments
4.500E-04	1.095E-01	26	G076	1 0 0 0 1	

3389. C$_{14}$H$_{29}$NO$_2$
Decanamide, 2,2-diethyl-N-hydroxy
RN: 60631-06-3   **MP** (°C):
MW: 243.39   **BP** (°C):

Solubility (Moles/L)	Solubility (Grams/L)	Temp (°C)	Ref (#)	Evaluation (T P E A A)	Comments
6.000E-06	1.460E-03	26	G076	1 0 0 0 1	

3390. C$_{14}$H$_{29}$NO$_2$
Dodecanamide, N-hydroxy-2,2-dimethyl
RN: 60631-05-2   **MP** (°C):
MW: 243.39   **BP** (°C):

Solubility (Moles/L)	Solubility (Grams/L)	Temp (°C)	Ref (#)	Evaluation (T P E A A)	Comments
1.600E-05	3.894E-03	26	G076	1 0 0 0 1	

3391. C$_{14}$H$_{29}$NO$_2$
Pentanamide, N-hydroxy-4-methyl-2,2-bis(2-methylpropyl)
RN: 60469-53-6   **MP** (°C):
MW: 243.39   **BP** (°C):

Solubility (Moles/L)	Solubility (Grams/L)	Temp (°C)	Ref (#)	Evaluation (T P E A A)	Comments
1.000E+01	2.434E+03	26	G076	1 0 0 0 1	

3392. C$_{14}$H$_{29}$NO$_2$
Hexanamide, 2,2-dibutyl-N-hydroxy
2,2-Dibutyl-N-hydroxyhexanamide
Tri-n-butylacetohydroxamic acid
RN: 52061-82-2   **MP** (°C):
MW: 243.39   **BP** (°C):

Solubility (Moles/L)	Solubility (Grams/L)	Temp (°C)	Ref (#)	Evaluation (T P E A A)	Comments
7.000E-05	1.704E-02	26	G076	1 0 0 0 1	

3393. C$_{14}$H$_{29}$NO$_2$
Tetradecanamide, *N*-hydroxy
Myristohydroxamic acid
N-Hydroxytetradecanamide

RN: 17698-03-2 **MP** (°C):
MW: 243.39 **BP** (°C):

Solubility (Moles/L)	Solubility (Grams/L)	Temp (°C)	Ref (#)	Evaluation (T P E A A)	Comments
1.000E-04	2.434E-02	26	G076	1 0 0 0 1	

3394. C$_{14}$H$_{30}$
n-Tetradecane
Tetradecane

RN: 629-59-4 **MP** (°C): 5.89
MW: 198.40 **BP** (°C): 253.7

Solubility (Moles/L)	Solubility (Grams/L)	Temp (°C)	Ref (#)	Evaluation (T P E A A)	Comments
1.663E-09	3.300E-07	23	C332	0 0 0 0 0	
3.500E-08	6.944E-06	25	F004	0 0 0 0 0	
1.159E-08	2.300E-06	ns	H123	0 0 0 0 0	

3395. C$_{14}$H$_{30}$O
Tetradecanol

RN: 27196-00-5 **MP** (°C):
MW: 214.39 **BP** (°C):

Solubility (Moles/L)	Solubility (Grams/L)	Temp (°C)	Ref (#)	Evaluation (T P E A A)	Comments
1.460E-06	3.130E-04	25	R002	0 0 0 0 0	

3396. C$_{14}$H$_{30}$O
Myristyl alcohol
Tetradecanol

RN: 112-72-1 **MP** (°C): 38
MW: 214.39 **BP** (°C): 289

Solubility (Moles/L)	Solubility (Grams/L)	Temp (°C)	Ref (#)	Evaluation (T P E A A)	Comments
9.049E-08	1.940E-05	4	H030	2 2 2 2 2	
9.049E-08	1.940E-05	4	H103	1 2 2 2 2	
8.909E-07	1.910E-04	25	H103	1 2 2 2 2	
5.737E-07	1.230E-04	32	H030	2 2 2 2 2	
5.737E-07	1.230E-04	32	H103	1 2 2 2 2	
1.105E-06	2.370E-04	45	H030	2 2 2 2 2	
1.105E-06	2.370E-04	45	H103	1 2 2 2 2	
2.094E-06	4.490E-04	61	H030	2 2 2 2 2	
2.094E-06	4.490E-04	61	H103	1 2 2 2 2	

3397. C$_{14}$H$_{31}$O$_2$P
Ethyl dihexyl phosphinate
Phosphinic acid, dihexyl-, ethyl ester
RN: 113977-19-8   **MP** (°C):
MW: 262.38   **BP** (°C):

Solubility (Moles/L)	Solubility (Grams/L)	Temp (°C)	Ref (#)	Evaluation (T P E A A)	Comments
<3.81E-04	<1.00E-01	25	B070	1 2 0 1 0	

3398. C$_{14}$H$_{31}$O$_3$P
Dibutyl hexyl phosphonate
Phosphinic acid, hexyl-, dibutyl ester
RN: 5929-66-8   **MP** (°C):
MW: 278.38   **BP** (°C):

Solubility (Moles/L)	Solubility (Grams/L)	Temp (°C)	Ref (#)	Evaluation (T P E A A)	Comments
<7.18E-04	<2.00E-01	25	B070	1 2 0 1 0	

3399. C$_{14}$H$_{31}$O$_3$P
Diethyl hexyl phosphonate
Phosphinic acid, hexyl-, diethyl ester
RN: 16165-66-5   **MP** (°C):
MW: 278.38   **BP** (°C):

Solubility (Moles/L)	Solubility (Grams/L)	Temp (°C)	Ref (#)	Evaluation (T P E A A)	Comments
2.155E-03	6.000E-01	25	B070	1 2 0 1 0	

3400. C$_{14}$H$_{31}$O$_4$P
Dibutyl hexyl phosphate
Phosphoric acid, dibutyl hexyl ester
RN: 80421-90-5   **MP** (°C):
MW: 294.37   **BP** (°C):

Solubility (Moles/L)	Solubility (Grams/L)	Temp (°C)	Ref (#)	Evaluation (T P E A A)	Comments
<3.40E-04	<1.00E-01	25	B070	1 2 0 1 0	

3401. C$_{14}$H$_{31}$O$_4$P
Diethyl decyl phosphate
Phosphoric acid, decyl ester
RN: 20195-16-8   **MP** (°C):
MW: 294.37   **BP** (°C):

Solubility (Moles/L)	Solubility (Grams/L)	Temp (°C)	Ref (#)	Evaluation (T P E A A)	Comments
<3.40E-04	<1.00E-01	25	B070	1 2 0 1 0	

3402. C$_{14}$H$_{31}$O$_5$P
Dibutyl ethoxybutyl phosphate
RN: 100888-67-3   **MP** (°C):
MW: 310.37   **BP** (°C):

Solubility (Moles/L)	Solubility (Grams/L)	Temp (°C)	Ref (#)	Evaluation (T P E A A)	Comments
2.255E-03	7.000E-01	25	B070	1 2 0 1 0	

3403. C$_{15}$H$_{10}$
4,5-Methylenephenanthrene
4H-Cyclopenta[def]phenanthrene
RN: 203-64-5   **MP** (°C):   76
MW: 190.25   **BP** (°C):

Solubility (Moles/L)	Solubility (Grams/L)	Temp (°C)	Ref (#)	Evaluation (T P E A A)	Comments
5.782E-06	1.100E-03	27	D003	1 0 0 1 1	

3404. C$_{15}$H$_{10}$Cl$_2$N$_2$O$_2$
Lorazepam
Alzapam
Ativan
Apo-lorazepam
7-Chloro-5-(o-chlorophenyl)-1,3-dihydro-3-hydroxy-2H-1,4-benzodiazepin-2-one
RN: 846-49-1   **MP** (°C):   167
MW: 321.17   **BP** (°C):

Solubility (Moles/L)	Solubility (Grams/L)	Temp (°C)	Ref (#)	Evaluation (T P E A A)	Comments
1.681E-04	5.400E-02	ns	N315	0 2 2 1 2	pH 7.09

3405. C$_{15}$H$_{10}$O$_2$
9-Anthracenecarboxylic acid
Anthracene-9-carboxylic acid
RN: 723-62-6   **MP** (°C):   214
MW: 222.25   **BP** (°C):

Solubility (Moles/L)	Solubility (Grams/L)	Temp (°C)	Ref (#)	Evaluation (T P E A A)	Comments
3.824E-04	8.499E-02	24	H106	1 0 2 2 2	
3.825E-07	8.500E-05	ns	M349	0 2 1 1 2	

3406. C₁₅H₁₀O₄S
7-Methylthio-2-xanthonecarboxylic acid
RN: 40363-76-6   **MP** (°C):
MW: 286.31   **BP** (°C):

Solubility (Moles/L)	Solubility (Grams/L)	Temp (°C)	Ref (#)	Evaluation (T P E A A)	Comments
9.081E-07	2.600E-04	25	C059	1 2 1 1 1	

3407. C₁₅H₁₀O₅S
7-Methylsulfinyl-2-xanthonecarboxylic acid
RN: 40691-50-7   **MP** (°C):
MW: 302.31   **BP** (°C):

Solubility (Moles/L)	Solubility (Grams/L)	Temp (°C)	Ref (#)	Evaluation (T P E A A)	Comments
9.064E-06	2.740E-03	25	C059	1 2 1 1 2	

3408. C₁₅H₁₀O₆
Eriodictyol
5,7,3′,4′-Tetra-hydroxyflavon
RN: 552-58-9   **MP** (°C):   257dec
MW: 286.24   **BP** (°C):

Solubility (Moles/L)	Solubility (Grams/L)	Temp (°C)	Ref (#)	Evaluation (T P E A A)	Comments
2.445E-04	7.000E-02	20	F300	1 0 0 0 1	
6.987E-04	2.000E-01	100	F300	1 0 0 0 2	

3409. C₁₅H₁₀O₇
Morin
3,5,7,2′,4′,-Penta-hydroxyflavon
RN: 480-16-0   **MP** (°C):   299.5
MW: 302.24   **BP** (°C):

Solubility (Moles/L)	Solubility (Grams/L)	Temp (°C)	Ref (#)	Evaluation (T P E A A)	Comments
8.271E-04	2.500E-01	20	F300	1 0 0 0 1	
2.978E-03	9.000E-01	100	F300	1 0 0 0 0	

3410. C₁₅H₁₀O₇

Quarcetin
2-(3,4-Dihydroxyphenyl)-3,5,7-trihydroxy-4H-1-benzopyran-4-one
3,3′,4′,5,7-Pentahydroxyflavone
3′,4′,5,7-Tetrahydroxyflavon-3-ol
Xanthaurine
Meletin

RN: 117-39-5	**MP** (°C): 316–317			
MW: 302.24	**BP** (°C):			

Solubility (Moles/L)	Solubility (Grams/L)	Temp (°C)	Ref (#)	Evaluation (T P E A A)	Comments
1.985E-02	6.000E+00	ns	Z411	0 0 0 0 0	

3411. C₁₅H₁₀O₇.H₂O

Morin hydrate
4H-1-Benzopyran-4-one, 2-(2,4-dihydroxyphenyl)-3,5,7-trihydroxy-, monohydrate
Flavone, 2′,3,4′,5,7-pentahydroxy-, monohydrate
Morin monohydrate

RN: 6202-27-3	**MP** (°C):			
MW: 320.26	**BP** (°C):			

Solubility (Moles/L)	Solubility (Grams/L)	Temp (°C)	Ref (#)	Evaluation (T P E A A)	Comments
5.994E-04	1.920E-01	ns	B404	0 2 1 1 0	

3412. C₁₅H₁₁ClF₃NO₄

Oxyfluorfen
Oxyfluorofen
Koltar
Goal
2-Chloro-1-(3-ethoxy-4-nitrophenoxy)-4-(trifluoromethyl)benzene
Goal 1.6E

RN: 42874-03-3	**MP** (°C): 83–84			
MW: 361.71	**BP** (°C): >240			

Solubility (Moles/L)	Solubility (Grams/L)	Temp (°C)	Ref (#)	Evaluation (T P E A A)	Comments
3.236E-07	1.170E-04	ns	R427	0 0 0 0 0	

3413. $C_{15}H_{11}ClN_2O_2$

Oxazepam
Serax
7-Chloro-1,3-dihydro-3-hydroxy-5-phenyl-2H-1,4-benzodiazepin-2-one
Apo-oxazepam
Abboxampam

RN:　604-75-1　　**MP** (°C):　205.5
MW:　286.72　　**BP** (°C):

Solubility (Moles/L)	Solubility (Grams/L)	Temp (°C)	Ref (#)	Evaluation (T P E A A)	Comments
6.975E-05	2.000E-02	22	N319	0 0 0 0 0	
1.744E-04	5.000E-02	amb	L434	0 0 0 0 0	
7.673E-05	2.200E-02	c	B362	0 0 0 0 0	

3414. $C_{15}H_{11}ClO_3$

Chlorflurecol-methyl
Chlorflurenol
Methyl-2-chloro-9-hydroxyfluorene-9-carboxylate

RN:　2536-31-4　　**MP** (°C):　152
MW:　274.71　　**BP** (°C):

Solubility (Moles/L)	Solubility (Grams/L)	Temp (°C)	Ref (#)	Evaluation (T P E A A)	Comments
6.552E-05	1.800E-02	20	A308	1 0 0 0 1	
7.936E-05	2.180E-02	20	B200	1 0 0 0 2	
6.552E-05	1.800E-02	20	M161	1 0 0 0 1	

3415. $C_{15}H_{11}NO_2$

C.I. Disperse orange 11
1-Amino-2-methylanthraquinone
2-Methyl-1-anthraquinonylamine
Acetate fast orange R

RN:　82-28-0　　**MP** (°C):　208
MW:　237.26　　**BP** (°C):

Solubility (Moles/L)	Solubility (Grams/L)	Temp (°C)	Ref (#)	Evaluation (T P E A A)	Comments
1.400E-06	3.322E-04	25	B333	0 0 0 0 0	

3416. C₁₅H₁₁NO₂

C.I. Disperse red 9
1-(Methylamino)-9,10-anthraquinone
Serilene fast pink BT
Smoke red M

RN: 82-38-2   **MP** (°C):   161
MW: 237.26   **BP** (°C):

Solubility (Moles/L)	Solubility (Grams/L)	Temp (°C)	Ref (#)	Evaluation (T P E A A)	Comments
3.100E-07	7.355E-05	25	B333	0 0 0 0 0	

3417. C₁₅H₁₁NO₃

N-epoxymethyl-1,8-naphthamilide
ENA

RN: **MP** (°C):
MW: 253.26   **BP** (°C):

Solubility (Moles/L)	Solubility (Grams/L)	Temp (°C)	Ref (#)	Evaluation (T P E A A)	Comments
4.580E-05	1.160E-02	ns	D428	0 0 0 0 0	

3418. C₁₅H₁₁N₃

2,2′,6,2″-terpyridine
Terpyridine
Tripyridyl

RN: 1148-79-4   **MP** (°C):
MW: 233.28   **BP** (°C):

Solubility (Moles/L)	Solubility (Grams/L)	Temp (°C)	Ref (#)	Evaluation (T P E A A)	Comments
6.310E-03	1.472E+00	24.99	B444	0 0 0 0 0	

3419. C₁₅H₁₁N₃O₃

Nitrazepam
1,3-Dihydro-7-nitro-5-phenyl-2H-1,4-benzodiazepin-2-one
Mogadon
Unisomnia

RN: 146-22-5   **MP** (°C):   224
MW: 281.27   **BP** (°C):

Solubility (Moles/L)	Solubility (Grams/L)	Temp (°C)	Ref (#)	Evaluation (T P E A A)	Comments
1.529E-04	4.300E-02	30	O321	0 0 0 0 0	

3420. C₁₅H₁₂
1-Methylphenanthrene

RN:	832-69-9	**MP** (°C):	118
MW:	192.26	**BP** (°C):	358

Solubility (Moles/L)	Solubility (Grams/L)	Temp (°C)	Ref (#)	Evaluation (T P E A A)	Comments
4.952E-07	9.520E-05	6.60	M063	2 1 2 2 2	
4.950E-07	9.517E-05	6.60	M082	1 1 1 2 2	
4.950E-07	9.517E-05	6.60	M151	2 1 2 2 2	
4.956E-06	9.529E-04	6.64	M183	1 2 1 1 2	
5.929E-07	1.140E-04	8.90	M063	2 1 2 2 2	
5.940E-07	1.142E-04	8.90	M082	1 1 1 2 2	
5.940E-07	1.142E-04	8.90	M151	2 1 2 2 2	
5.933E-07	1.141E-04	8.94	M183	1 2 1 1 2	
7.646E-07	1.470E-04	14.00	M063	2 1 2 2 2	
7.650E-07	1.471E-04	14.00	M082	1 1 1 2 2	
7.650E-07	1.471E-04	14.00	M151	2 1 2 2 2	
7.650E-07	1.471E-04	14.04	M183	1 2 1 1 2	
1.004E-06	1.930E-04	19.20	M063	2 1 2 2 2	
1.010E-06	1.942E-04	19.20	M082	1 1 1 2 2	
1.010E-06	1.942E-04	19.20	M151	2 1 2 2 2	
1.004E-06	1.931E-04	19.24	M183	1 2 1 1 2	
1.326E-06	2.550E-04	24.10	M063	2 1 2 2 2	
1.320E-06	2.538E-04	24.10	M082	1 1 1 2 2	
1.320E-06	2.538E-04	24.10	M151	2 1 2 2 2	
1.327E-06	2.552E-04	24.14	M183	1 2 1 1 2	
1.399E-06	2.690E-04	25.00	M151	2 1 2 2 2	
1.581E-06	3.040E-04	26.90	M063	2 1 2 2 2	
1.580E-06	3.038E-04	26.90	M082	1 1 1 2 2	
1.580E-06	3.038E-04	26.90	M151	2 1 2 2 2	
1.583E-06	3.043E-04	26.94	M183	1 2 1 1 2	
1.846E-06	3.550E-04	29.90	M063	2 1 2 2 2	
1.850E-06	3.557E-04	29.90	M082	1 1 1 2 2	
1.850E-06	3.557E-04	29.90	M151	2 1 2 2 2	
1.848E-06	3.553E-04	29.94	M183	1 2 1 1 2	

3421. C₁₅H₁₂
2-Methylanthracene

RN:	613-12-7	**MP** (°C):	204
MW:	192.26	**BP** (°C):	

Solubility (Moles/L)	Solubility (Grams/L)	Temp (°C)	Ref (#)	Evaluation (T P E A A)	Comments
3.672E-08	7.060E-06	6.30	M063	2 1 2 2 2	
3.670E-08	7.056E-06	6.30	M082	1 1 1 2 2	
3.670E-08	7.056E-06	6.30	M151	2 1 2 2 2	
3.675E-08	7.066E-06	6.34	M183	1 2 1 1 2	
4.411E-08	8.480E-06	9.10	M063	2 1 2 2 2	
4.410E-08	8.479E-06	9.10	M082	1 1 1 2 2	
4.410E-08	8.479E-06	9.10	M151	2 1 2 2 2	
4.414E-08	8.487E-06	9.14	M183	1 2 1 1 2	

(continued)

3421. $C_{15}H_{12}$ (continued)

Solubility (Moles/L)	Solubility (Grams/L)	Temp (°C)	Ref (#)	Evaluation (T P E A A)	Comments
4.905E-08	9.430E-06	10.80	M063	2 1 2 2 2	
4.900E-08	9.421E-06	10.80	M082	1 1 1 2 2	
4.900E-08	9.421E-06	10.80	M151	2 1 2 2 2	
4.909E-08	9.438E-06	10.84	M183	1 2 1 1 2	
5.773E-08	1.110E-05	13.90	M063	2 1 2 2 2	
5.750E-08	1.106E-05	13.90	M082	1 1 1 2 2	
5.750E-08	1.106E-05	13.90	M151	2 1 2 2 2	
5.778E-08	1.111E-05	13.94	M183	1 2 1 1 2	
7.542E-08	1.450E-05	18.30	M063	2 1 2 2 2	
7.540E-08	1.450E-05	18.30	M082	1 1 1 2 2	
7.540E-08	1.450E-05	18.30	M151	2 1 2 2 2	
7.550E-08	1.452E-05	18.34	M183	1 2 1 1 2	
9.934E-08	1.910E-05	23.10	M063	2 1 2 2 2	
9.940E-08	1.911E-05	23.10	M082	1 1 1 2 2	
9.940E-08	1.911E-05	23.10	M151	2 1 2 2 2	
9.944E-08	1.912E-05	23.14	M183	1 2 1 1 2	
2.028E-07	3.900E-05	25	M064	1 1 2 2 1	
1.108E-07	2.130E-05	25.00	M151	2 1 1 2 2	
1.259E-07	2.420E-05	27.00	M063	2 1 2 2 2	
1.260E-07	2.423E-05	27.00	M082	1 1 1 2 2	
1.260E-07	2.423E-05	27.00	M151	2 1 2 2 2	
1.260E-07	2.423E-05	27.04	M183	1 2 1 1 2	
1.670E-07	3.210E-05	31.10	M063	2 1 2 2 2	
1.670E-07	3.211E-05	31.10	M082	1 1 1 2 2	
1.670E-07	3.211E-05	31.10	M151	2 1 2 2 2	
1.671E-07	3.213E-05	31.14	M183	1 2 1 1 2	

3422. $C_{15}H_{12}$
9-Methylanthracene

RN:	779-02-2	MP (°C):	79
MW:	192.26	BP (°C):	196

Solubility (Moles/L)	Solubility (Grams/L)	Temp (°C)	Ref (#)	Evaluation (T P E A A)	Comments
1.358E-06	2.610E-04	25	M064	1 1 2 2 2	
1.330E-06	2.557E-04	25	M342	1 0 1 1 2	
1.358E-06	2.610E-04	ns	M344	0 0 0 0 2	

3423. $C_{15}H_{12}Cl_2O_3$
2,4-Dichlorophenoxyacetic acid benzyl ester
Benzyl 2,4-dichlorophenoxyacetate
2,4-DBE

RN:	13246-97-4	MP (°C):	
MW:	311.17	BP (°C):	

Solubility (Moles/L)	Solubility (Grams/L)	Temp (°C)	Ref (#)	Evaluation (T P E A A)	Comments
4.955E-05	1.542E-02	ns	M120	0 0 1 1 2	

3424. C$_{15}$H$_{12}$Cl$_2$O$_3$

Ethanol, 2-(2,4-dicholrophenoxy)-, benzoate
Benzoate, 2-(2,4-dichlorophenoxy)ethyl-
2,4-DEB

| **RN:** | 94-83-7 | **MP** (°C): | 74 |
| **MW:** | 311.17 | **BP** (°C): | |

Solubility (Moles/L)	Solubility (Grams/L)	Temp (°C)	Ref (#)	Evaluation (T P E A A)	Comments
1.543E-04	4.800E-02	ns	B185	0 0 0 0 0	

3425. C$_{15}$H$_{12}$I$_3$NO$_4$

Liothyronine
3,3′,5-Triiodothyronine

| **RN:** | 6893-02-3 | **MP** (°C): | 236dec |
| **MW:** | 650.98 | **BP** (°C): | |

Solubility (Moles/L)	Solubility (Grams/L)	Temp (°C)	Ref (#)	Evaluation (T P E A A)	Comments
6.080E-06	3.958E-03	37	L094	2 0 0 1 2	pH 4-5, zwitterion

3426. C$_{15}$H$_{12}$N$_2$O

5H-Dibenz[b,f]azepine-5-carboxamide
Carbazepine
5-Carbamoyl-5H-dibenz[b,f]azepine
Iminostilbene
Carbamazepine
Epitol

| **RN:** | 298-46-4 | **MP** (°C): | 190–193 |
| **MW:** | 236.28 | **BP** (°C): | |

Solubility (Moles/L)	Solubility (Grams/L)	Temp (°C)	Ref (#)	Evaluation (T P E A A)	Comments
4.655E-04	1.100E-01	20	B196	0 0 0 0 0	
4.700E-04	1.110E-01	20	B196	0 0 0 0 0	
6.349E-04	1.500E-01	25	C437	0 0 0 0 0	Average
1.864E-03	4.404E-01	32	F425	0 0 0 0 0	pH 7.4
1.100E-03	2.600E-01	amb	L434	0 0 0 0 0	
4.232E-05	1.000E-02	ns	K444	0 0 0 0 0	
4.000E-03	9.451E-01	rt	B397	0 0 0 0 0	EFG

3427. $C_{15}H_{12}N_2O_2$

Phenytoin
5,5-Diphenyl-2,4-imidazolidinedione
Dilantin
5,5-Diphenylhydantoin
Ekko
Zentropil

RN: 57-41-0   **MP** (°C):   296.5
MW: 252.28   **BP** (°C):

Solubility (Moles/L)	Solubility (Grams/L)	Temp (°C)	Ref (#)	Evaluation (T P E A A)	Comments
3.765E-04	9.499E-02	0	B114	1 1 1 2 1	pH 6-7
1.268E-04	3.200E-02	22	B154	1 1 1 1 1	0.1M HCl
7.531E-05	1.900E-02	25	A408	2 0 1 2 0	int
5.549E-05	1.400E-02	25	P061	0 0 0 0 0	pH 1-7
1.526E-04	3.850E-02	37	F183	1 0 1 1 2	intrinsic
2.600E-04	6.559E-02	50	M335	1 0 2 1 2	pH 5
2.323E-04	5.860E-02	ns	K446	0 0 0 0 0	
7.650E-05	1.930E-02	rt	I404	0 0 0 0 0	Average

3428. $C_{15}H_{12}N_2O_2$

Disperse violet 4
1-Amino-4-(N-methylamino)anthraquinone
Interchem acetate violet 6B

RN: 1220-94-6   **MP** (°C):   193
MW: 252.28   **BP** (°C):

Solubility (Moles/L)	Solubility (Grams/L)	Temp (°C)	Ref (#)	Evaluation (T P E A A)	Comments
2.300E-06	5.802E-04	25	B333	0 0 0 0 0	

3429. $C_{15}H_{12}N_2O_3$

5-Phenyl-5-(p-hydroxy)phenyl-hydantoin
DL-5-(p-Hydroxyphenyl-5-phenylhydantoin
p-Hydroxyphenytoin
Hydroxydiphenylhydantoin
p-Hydroxydiphenylhydantoin

RN: 2784-27-2   **MP** (°C):
MW: 268.27   **BP** (°C):

Solubility (Moles/L)	Solubility (Grams/L)	Temp (°C)	Ref (#)	Evaluation (T P E A A)	Comments
1.342E-04	3.600E-02	37	F183	1 0 1 1 2	intrinsic

3430. C$_{15}$H$_{12}$N$_2$O$_3$
Furfurin
1H-Imidazole, 2,4,5-tri-2-furanyl-4,5-dihydro-
2-Imidazoline, 2,4,5-tri-2-furyl-
RN: 550-23-2   **MP** (°C):
MW: 268.27   **BP** (°C):

Solubility (Moles/L)	Solubility (Grams/L)	Temp (°C)	Ref (#)	Evaluation (T P E A A)	Comments
7.455E-04	2.000E-01	8	F300	1 0 0 0 0	
2.870E-02	7.700E+00	100	F300	1 0 0 0 1	

3431. C$_{15}$H$_{12}$O$_4$
Benzoyl-r-mandelic acid
p-Benzoylmandelic acid
RN: 100915-04-6   **MP** (°C):   177
MW: 256.26   **BP** (°C):

Solubility (Moles/L)	Solubility (Grams/L)	Temp (°C)	Ref (#)	Evaluation (T P E A A)	Comments
1.980E-02	5.074E+00	0	A043	1 2 1 1 1	
1.980E-02	5.074E+00	0	L035	1 2 2 1 1	
2.327E-02	5.964E+00	10	A043	1 2 1 1 1	
2.327E-02	5.964E+00	10	L035	1 2 2 1 1	
2.520E-02	6.458E+00	15	A043	1 2 1 1 1	
2.520E-02	6.458E+00	15	L035	1 2 2 1 1	
2.828E-02	7.247E+00	20	A043	1 2 1 1 1	
2.828E-02	7.247E+00	20	L035	1 2 2 1 1	
3.059E-02	7.838E+00	25	A043	1 2 1 1 1	
3.059E-02	7.838E+00	25	L035	1 2 2 1 1	
3.557E-02	9.116E+00	30	A043	1 2 1 1 1	
3.557E-02	9.116E+00	30	L035	1 2 2 1 1	
4.017E-02	1.029E+01	35	A043	1 2 1 1 2	
4.017E-02	1.029E+01	35	L035	1 2 2 1 2	
4.894E-02	1.254E+01	40	A043	1 2 1 1 2	
4.894E-02	1.254E+01	40	L035	1 2 2 1 2	
6.032E-02	1.546E+01	45	A043	1 2 1 1 2	
6.032E-02	1.546E+01	45	L035	1 2 2 1 2	
7.201E-02	1.845E+01	50	A043	1 2 1 1 2	
7.201E-02	1.845E+01	50	L035	1 2 2 1 2	

3432. $C_{15}H_{12}O_4$
Benzoic acid, 2-(acetyloxy)-, phenyl ester
Phennin
Phenyl 2-acetoxybenzoate
Vesipyrin
Spiroform
Phenyl acetylsalicylate
RN: 134-55-4   MP (°C): 97.5
MW: 256.26   BP (°C):

Solubility (Moles/L)	Solubility (Grams/L)	Temp (°C)	Ref (#)	Evaluation (T P E A A)	Comments
7.805E-05	2.000E-02	21	N335	0 0 0 0 0	

3433. $C_{15}H_{13}Cl_3O_2$
2-*p*-Methoxyphenyl-2-*p*-hydroxyphenyl-1,1,1-trichloro-ethane
Phenol, 4-[2,2,2-trichloro-1-(4-methoxyphenyl)ethyl]-
RN: 28463-03-8   MP (°C): 112–114
MW: 331.63   BP (°C):

Solubility (Moles/L)	Solubility (Grams/L)	Temp (°C)	Ref (#)	Evaluation (T P E A A)	Comments
2.412E-06	8.000E-04	ns	K117	0 1 2 1 1	

3434. $C_{15}H_{13}FO_2$
Flurbiprofen
3-Fluoro-4-phenylhydratropic acid
Froben
Ansaid
RN: 5104-49-4   MP (°C): 110
MW: 244.27   BP (°C):

Solubility (Moles/L)	Solubility (Grams/L)	Temp (°C)	Ref (#)	Evaluation (T P E A A)	Comments
2.530E-05	6.180E-03	5	F306	1 0 1 2 2	intrinsic
2.761E-05	6.744E-03	24.99	K447	0 0 0 0 0	pH 2.0
4.339E-05	1.060E-02	25	A408	2 0 1 2 0	int
5.000E-05	1.221E-02	25	A411	1 0 0 1 0	int
1.332E-04	3.254E-02	25	C314	0 0 0 0 0	
1.331E-04	3.250E-02	25	C314	0 0 0 0 0	
3.870E-05	9.453E-03	25	F306	1 0 1 2 2	intrinsic
1.940E-04	4.739E-02	25	O303	1 0 0 1 0	EFG
4.600E-05	1.124E-02	37	F306	1 0 1 2 2	intrinsic
2.866E-05	7.000E-03	37	Y421	0 0 0 0 0	
>2.05E-03	>5.00E-01	ns	B404	0 2 1 1 0	
2.700E-04	6.595E-02	ns	O304	0 0 1 2 2	
3.275E-05	8.000E-03	rt	H302	0 0 2 1 2	intrinsic

3435. C₁₅H₁₃F₃N₄O

6H-Dipyrido[3,2-b:2′,3′-e][1,4]diazepin-6-one, 11-ethyl-5,11-dihydro-2-methyl-4-
   (trifluoromethyl)-

RN: 135794-72-8   **MP** (°C):

MW: 322.29   **BP** (°C):

Solubility (Moles/L)	Solubility (Grams/L)	Temp (°C)	Ref (#)	Evaluation (T P E A A)	Comments
6.209E-05	2.001E-02	ns	M381	0 1 1 1 2	pH 7.0

3436. C₁₅H₁₃NO

7-Benzoylindoline

U-26,952

RN: 33244-57-4   **MP** (°C):   124

MW: 223.28   **BP** (°C):

Solubility (Moles/L)	Solubility (Grams/L)	Temp (°C)	Ref (#)	Evaluation (T P E A A)	Comments
1.026E-05	2.290E-03	25	C046	0 0 0 0 0	

3437. C₁₅H₁₃NO₂

Dibenz[b,f][1,4]oxazepin-11(10H)-one, 10-ethyl-

RN: 17296-50-3   **MP** (°C):

MW: 239.28   **BP** (°C):

Solubility (Moles/L)	Solubility (Grams/L)	Temp (°C)	Ref (#)	Evaluation (T P E A A)	Comments
2.089E-04	4.999E-02	ns	M381	0 1 1 1 2	pH 7.0

3438. C₁₅H₁₃NO₂S

Metiazinic acid

Methiazinic acid

RN: 13993-65-2   **MP** (°C):   146

MW: 271.34   **BP** (°C):

Solubility (Moles/L)	Solubility (Grams/L)	Temp (°C)	Ref (#)	Evaluation (T P E A A)	Comments
1.142E-04	3.100E-02	30	D015	2 0 1 1 0	EFG
2.211E-04	6.000E-02	37	D015	2 0 1 1 0	EFG

3439. C₁₅H₁₃NO₃

Ketorolac

RN: 74103-06-3   **MP** (°C):

MW: 255.28   **BP** (°C):

Solubility (Moles/L)	Solubility (Grams/L)	Temp (°C)	Ref (#)	Evaluation (T P E A A)	Comments
7.167E-04	1.830E-01	32	C411	2 1 1 2 1	
4.309E-04	1.100E-01	37	Y421	0 0 0 0 0	

3440. C₁₅H₁₃NO₃

Benzoyl acetaminophen
Acetamide, N-[4-(benzoyloxy)phenyl]-
Acetanilide, 4′-hydroxy-, benzoate (ester)

RN: 537-52-0   **MP** (°C): 170.5–171.5
MW: 255.28   **BP** (°C):

Solubility (Moles/L)	Solubility (Grams/L)	Temp (°C)	Ref (#)	Evaluation (T P E A A)	Comments
6.659E-05	1.700E-02	37	D029	0 0 0 0 0	

3441. C₁₅H₁₃NO₄

Phenyl acetaminophen
Carbonic acid, 4-(acetylamino)phenyl phenyl ester
Acetanilide, 4′-hydroxy-, phenyl carbonate (ester)

RN: 17239-23-5   **MP** (°C): 139–140.5
MW: 271.28   **BP** (°C):

Solubility (Moles/L)	Solubility (Grams/L)	Temp (°C)	Ref (#)	Evaluation (T P E A A)	Comments
2.322E-04	6.300E-02	37	D029	0 0 0 0 0	

3442. C₁₅H₁₃N₃O₄S

Piroxicam
2H-1,2-Benzothiazine-3-carboxamide, 4-hydroxy-2-methyl-N-2-pyridinyl-, 1,1-dioxide
Fensaid
Feldene
Candyl
Mobilis

RN: 36322-90-4   **MP** (°C): 198
MW: 331.35   **BP** (°C):

Solubility (Moles/L)	Solubility (Grams/L)	Temp (°C)	Ref (#)	Evaluation (T P E A A)	Comments
2.535E-04	8.400E-02	25	M457	0 0 0 0 0	
1.608E-04	5.330E-02	32	C411	2 1 1 2 1	
<3.02E-04	<1.00E-01	rt	B435	0 0 0 0 0	
6.941E-05	2.300E-02	rt	H302	0 0 2 1 2	intrinsic

3443. C₁₅H₁₄ClN₃O₄S

Cefaclor

5-Thia-1-azabicyclo[4.2.0]oct-2-ene-2-carboxylic acid, 7-[[(2*R*)-aminophenylacetyl]amino]-3-
   chloro-8-oxo-, (6*R*,7*R*)-

Ceclor

Alfacet

Cephaclor

RN: 53994-73-3   **MP** (°C):

MW: 367.81   **BP** (°C):

Solubility (Moles/L)	Solubility (Grams/L)	Temp (°C)	Ref (#)	Evaluation (T P E A A)	Comments
2.719E-02	1.000E+01	ns	L099	0 0 0 0 0	

3444. C₁₅H₁₄ClN₃O₄S₃

Benzthiazide

Exna

Hydrex

RN: 91-33-8   **MP** (°C):

MW: 431.94   **BP** (°C):

Solubility (Moles/L)	Solubility (Grams/L)	Temp (°C)	Ref (#)	Evaluation (T P E A A)	Comments
2.547E-05	1.100E-02	ns	B404	0 2 1 1 0	
6.482E-06	2.800E-03	rt	I404	0 0 0 0 0	Intrinsic, Average

3445. C₁₅H₁₄Cl₂F₃N₃O₃

Carfentrazone-ethyl

Df herbicide

Benzenepropanoic acid, α-2-dichloro-5-{4-(difluoromethyl)-4,5-dihydro-3-methyl-5-oxo-1H-
   1,2,4-triazol-1-yl}-4-fluoro-, ethyl ester

Ethyl 2-chloro-3-{2-chloro-4-fluoro-5-{4-(difluoromethyl)-4,5-dihydro-3-methyl-5-oxo-1H-1,2,4-
   triazol-1-yl}phenyl}propanoate

F 8426

RN: 128639-02-1   **MP** (°C):

MW: 412.20   **BP** (°C):

Solubility (Moles/L)	Solubility (Grams/L)	Temp (°C)	Ref (#)	Evaluation (T P E A A)	Comments
5.333E-05	2.198E-02	ns	S460	0 0 0 0 0	

3446. $C_{15}H_{14}Cl_2N_4O_3$

C.I. Disperse orange 5
Ethanol, 2-[[4-[(2,6-dichloro-4-nitrophenyl)azo]phenyl]methylamino]
Amacel fast brown 3R
Celliton fast brown 3R

RN:	6232-56-0	**MP** (°C):	127		
MW:	369.21	**BP** (°C):			

Solubility (Moles/L)	Solubility (Grams/L)	Temp (°C)	Ref (#)	Evaluation (T P E A A)	Comments
4.300E-07	1.588E-04	25	B333	0 0 0 0 0	
8.938E-06	3.300E-03	60	P313	0 0 0 0 0	average of 2
1.530E-05	5.650E-03	70	P313	0 0 0 0 0	average of 2
2.939E-05	1.085E-02	80	P313	0 0 0 0 0	average of 2
6.378E-05	2.355E-02	90	P313	0 0 0 0 0	average of 2
1.354E-04	5.000E-02	100	P313	0 0 0 0 0	

3447. $C_{15}H_{14}F_3N_3O_4S_2$

Bendroflumethiazide
Corzide
Rauzide
Naturetin

RN:	73-48-3	**MP** (°C):	222		
MW:	421.42	**BP** (°C):			

Solubility (Moles/L)	Solubility (Grams/L)	Temp (°C)	Ref (#)	Evaluation (T P E A A)	Comments
1.200E-04	5.057E-02	20	A080	1 0 2 1 2	
2.570E-04	1.083E-01	25	A076	1 0 1 1 2	
2.847E-05	1.200E-02	ns	B404	0 2 1 1 0	
9.492E-05	4.000E-02	rt	A095	0 0 2 2 0	
3.631E-05	1.530E-02	rt	I404	0 0 0 0 0	Intrinsic, Average

3448. $C_{15}H_{14}NO_2PS$

Cyanofenphos
O-(4-Cyanophenyl) O-ethyl phenylphosphonothioate
Surecide

RN:	13067-93-1	**MP** (°C):	83		
MW:	303.32	**BP** (°C):			

Solubility (Moles/L)	Solubility (Grams/L)	Temp (°C)	Ref (#)	Evaluation (T P E A A)	Comments
1.978E-06	6.000E-04	30	M161	1 0 0 0 0	

3449. $C_{15}H_{14}N_2O_2$
Dibenz[b,f][1,4]oxazepin-11(10H)-one, 8-amino-2-methyl-
RN:　　155206-47-6　　**MP** (°C):
MW:　　254.29　　　　**BP** (°C):

Solubility (Moles/L)	Solubility (Grams/L)	Temp (°C)	Ref (#)	Evaluation (T P E A A)	Comments
1.180E-04	3.001E-02	ns	M381	0 1 1 1 2	pH 7.0

3450. $C_{15}H_{14}N_2O_3$
p-(3-Phenylureido)phenyl acetate
Benzeneacetic acid, 4-[[(phenylamino)carbonyl]amino]-
RN:　　181518-40-1　　**MP** (°C):
MW:　　270.29　　　　**BP** (°C):

Solubility (Moles/L)	Solubility (Grams/L)	Temp (°C)	Ref (#)	Evaluation (T P E A A)	Comments
3.600E-05	9.730E-03	25	A066	1 0 1 1 1	

3451. $C_{15}H_{14}N_2O_5$
2′-Ethoxy-2hydroxy-5-nitrobenzanilide
Benzamide, N-(2-ethoxyphenyl)-2-hydroxy-5-nitro-
RN:　　213460-67-4　　**MP** (°C):
MW:　　302.29　　　　**BP** (°C):

Solubility (Moles/L)	Solubility (Grams/L)	Temp (°C)	Ref (#)	Evaluation (T P E A A)	Comments
1.687E-05	5.098E-03	25	D400	2 0 0 1 2	

3452. $C_{15}H_{14}N_2O_5$
4′-Ethoxy-2-hydroxy-3-nitrobenzanilide
Benzamide, N-(4-ethoxyphenyl)-2-hydroxy-3-nitro-
RN:　　213460-61-8　　**MP** (°C):
MW:　　302.29　　　　**BP** (°C):　342.2–426.2

Solubility (Moles/L)	Solubility (Grams/L)	Temp (°C)	Ref (#)	Evaluation (T P E A A)	Comments
3.119E-05	9.428E-03	25	D400	2 0 0 1 2	

3453. $C_{15}H_{14}N_2O_5$
2′-Ethoxy-2-hydroxy-3-nitrobenzanilide
Benzamide, N-(2-ethoxyphenyl)-2-hydroxy-3-nitro-
RN:　　213460-63-0　　**MP** (°C):
MW:　　302.29　　　　**BP** (°C):

Solubility (Moles/L)	Solubility (Grams/L)	Temp (°C)	Ref (#)	Evaluation (T P E A A)	Comments
2.432E-05	7.352E-03	25	D400	2 0 0 1 2	

3454. C₁₅H₁₄N₄O

Nevarapine

6H-Dipyrido[3,2-b:2′,3′-e][1,4]diazepin-6-one, 11-cyclopropyl-5,11-dihydro-4-methyl

Nevirapine

BI-RG 587

RN: 129618-40-2   **MP** (°C):   248
MW: 266.31   **BP** (°C):

Solubility (Moles/L)	Solubility (Grams/L)	Temp (°C)	Ref (#)	Evaluation (T P E A A)	Comments
3.755E-04	1.000E-01	ns	K444	0 0 0 0 0	
6.412E-04	1.708E-01	ns	M381	0 1 1 1 2	pH 7.0

3455. C₁₅H₁₄O₃

Methyl benzoyl benzoate

Benzoic acid, 4-hydroxy-, (4-methylphenyl)methyl ester

RN: 84833-58-9   **MP** (°C):
MW: 242.28   **BP** (°C):

Solubility (Moles/L)	Solubility (Grams/L)	Temp (°C)	Ref (#)	Evaluation (T P E A A)	Comments
2.064E-04	5.000E-02	ns	F014	0 0 0 0 0	

3456. C₁₅H₁₄O₃

[4-(Benzyloxy)phenyl]acetic acid

(4-Boph)

RN: 6547-53-1   **MP** (°C):
MW: 242.28   **BP** (°C):

Solubility (Moles/L)	Solubility (Grams/L)	Temp (°C)	Ref (#)	Evaluation (T P E A A)	Comments
2.711E-04	6.568E-02	20	K437	0 0 0 0 0	pH 2.0
3.711E-04	8.990E-02	25	K437	0 0 0 0 0	pH 2.0
6.338E-04	1.536E-01	30	K437	0 0 0 0 0	pH 2.0
7.293E-04	1.767E-01	37	K437	0 0 0 0 0	pH 2.0

3457. C₁₅H₁₄O₃

Fenoprofen

Fenoporfen

Progesic

Fenpron

Nalfon

Fenopron

RN: 31879-05-7   **MP** (°C):
MW: 242.28   **BP** (°C):

Solubility (Moles/L)	Solubility (Grams/L)	Temp (°C)	Ref (#)	Evaluation (T P E A A)	Comments
4.128E-04	1.000E-01	37	Y421	0 0 0 0 0	

3458. C$_{15}$H$_{15}$ClF$_3$N$_3$O
Triflumizole
RN: 99387-89-0   **MP** (°C): 63.5
MW: 345.75   **BP** (°C):

Solubility (Moles/L)	Solubility (Grams/L)	Temp (°C)	Ref (#)	Evaluation (T P E A A)	Comments
4.480E-05	1.549E-02	25	V410	0 0 0 0 0	

3459. C$_{15}$H$_{15}$ClN$_2$O$_2$
Chlorooxuron
(N′-4-(4-Chlorophenoxy)phenyl-N,N-dimethylurea)
3-[p-(p′-Chlorophenoxy)phenyl]-1,1-dimethylurea
N-4-(4′-Chlorophenoxy)phenyl-N′,N′-dimethylurea
Tenoran
RN: 1982-47-4   **MP** (°C): 151
MW: 290.75   **BP** (°C):

Solubility (Moles/L)	Solubility (Grams/L)	Temp (°C)	Ref (#)	Evaluation (T P E A A)	Comments
1.273E-05	3.700E-03	20	B185	0 0 0 0 0	
1.273E-05	3.700E-03	20	G036	1 0 0 0 1	
1.273E-05	3.700E-03	20	M161	1 0 0 0 1	pH 7
9.286E-06	2.700E-03	ns	B200	0 0 0 0 1	
1.273E-04	3.700E-02	ns	M061	0 0 0 0 1	

3460. C$_{15}$H$_{15}$ClN$_2$O$_4$S
Xipamide
2′,6′-Salicyloxylidide, 4-chloro-5-sulfamoyl-
Aquaphor
Aquaphor (diuretic)
BEI 1293
Diurex
RN: 14293-44-8   **MP** (°C): 256
MW: 354.81   **BP** (°C):

Solubility (Moles/L)	Solubility (Grams/L)	Temp (°C)	Ref (#)	Evaluation (T P E A A)	Comments
1.635E-04	5.800E-02	25	H074	1 2 2 1 1	

3461. $C_{15}H_{15}ClN_4O_6S$

Chlorimuron-ethyl
Chlorimuron ethyl ester
Classic 75DF
Classic
Chlorimuron Et
2-[[[[(4-Chloro-6-methoxy-2-pyrimidinyl)amino]carbonyl]amino]sulfonyl]benzoic acid ethyl ester

RN:	90982-32-4	MP (°C):	180–182
MW:	414.83	BP (°C):	

Solubility (Moles/L)	Solubility (Grams/L)	Temp (°C)	Ref (#)	Evaluation (T P E A A)	Comments
2.455E-06	1.018E-03	ns	R427	0 0 0 0 0	

3462. $C_{15}H_{15}ClO$

2-Benzyl-3,5-dimethyl-4-chloro-phenol

RN:	1867-85-2	MP (°C):	
MW:	246.74	BP (°C):	

Solubility (Moles/L)	Solubility (Grams/L)	Temp (°C)	Ref (#)	Evaluation (T P E A A)	Comments
5.000E-05	1.234E-02	25	B316	0 0 0 0 0	

3463. $C_{15}H_{15}NO_2$

Mefenamic acid
2′,3′-Dimethyl-N-phenyl-anthranilic acid
Forte mefenamic acid
N-(2,3-Xylyl)anthranilic acid
Ponstel
Ponstan

RN:	61-68-7	MP (°C):	230.5
MW:	241.29	BP (°C):	

Solubility (Moles/L)	Solubility (Grams/L)	Temp (°C)	Ref (#)	Evaluation (T P E A A)	Comments
8.289E-05	2.000E-02	30	D015	2 0 1 1 0	EFG
2.800E-05	6.756E-03	35	H091	1 2 2 2 1	sic
1.658E-04	4.000E-02	37	D015	2 0 1 1 0	EFG
1.658E-06	4.000E-04	37	P432	0 0 0 0 0	
1.227E-04	2.960E-02	37	P432	0 0 0 0 0	
8.289E-07	2.000E-04	37	Y421	0 0 0 0 0	
1.100E-04	2.654E-02	ns	O304	0 0 1 2 2	

3464. C₁₅H₁₅NO₃

Tolmetin
Tolectin

RN: 26171-23-3   **MP** (°C):
MW: 257.29   **BP** (°C):

Solubility (Moles/L)	Solubility (Grams/L)	Temp (°C)	Ref (#)	Evaluation (T P E A A)	Comments
7.773E-05	2.000E-02	37	Y421	0 0 0 0 0	

3465. C₁₅H₁₅N₃O

5H-Pyrido[2,3-b][1,5]benzodiazepine-5-one, 11-ethyl-6,11-dihydro-6-methyl-

RN: 132686-75-0   **MP** (°C):
MW: 253.31   **BP** (°C):

Solubility (Moles/L)	Solubility (Grams/L)	Temp (°C)	Ref (#)	Evaluation (T P E A A)	Comments
1.782E-05	4.515E-03	ns	M381	0 1 1 1 2	pH 7.0
4.742E-04	1.201E-01	ns	M381	0 1 1 1 2	pH 7.0

3466. C₁₅H₁₅N₃O₂

Pyrido[2,3-b][1,5]benzoxazepin-5(6H)-one, 3-amino-6,7,9-trimethyl-

RN: **MP** (°C):
MW: 269.31   **BP** (°C):

Solubility (Moles/L)	Solubility (Grams/L)	Temp (°C)	Ref (#)	Evaluation (T P E A A)	Comments
1.730E-04	4.658E-02	ns	M381	0 1 1 1 2	pH 7.0

3467. C₁₅H₁₅N₃O₂

C.I. Disperse yellow 3
Acetamide, N-[4-[(2-hydroxy-5-methylphenyl)azo]phenyl]-

RN: 2832-40-8   **MP** (°C): 195
MW: 269.31   **BP** (°C):

Solubility (Moles/L)	Solubility (Grams/L)	Temp (°C)	Ref (#)	Evaluation (T P E A A)	Comments
1.200E-07	3.232E-05	25	B333	0 0 0 0 0	

3468. C₁₅H₁₅N₃S

5H-Pyrido[2,3-b][1,5]benzodiazepine-5-thione, 11-ethyl-6,11-dihydro-6-methyl-

RN: 132686-95-4   **MP** (°C):
MW: 269.37   **BP** (°C):

Solubility (Moles/L)	Solubility (Grams/L)	Temp (°C)	Ref (#)	Evaluation (T P E A A)	Comments
1.968E-05	5.301E-03	ns	M381	0 1 1 1 2	pH 7.0

3469. $C_{15}H_{16}N_2O_2$
Ancymidol
α-Cyclopropyl-α-(4-methoxyphenyl)-5-pyrimidinemethanol
A-Rest

RN:	12771-68-5	MP (°C):	110.5
MW:	256.31	BP (°C):	

Solubility (Moles/L)	Solubility (Grams/L)	Temp (°C)	Ref (#)	Evaluation (T P E A A)	Comments
2.536E-03	6.500E-01	25	M161	1 0 0 0 2	

3470. $C_{15}H_{16}N_4O$
6H-Dipyrido[3,2-b:2′,3′-e][1,4]diazepin-6-one, 5,11-dihydro-5-methyl-11-propyl-

RN:	132312-81-3	MP (°C):	
MW:	268.32	BP (°C):	

Solubility (Moles/L)	Solubility (Grams/L)	Temp (°C)	Ref (#)	Evaluation (T P E A A)	Comments
1.327E-03	3.562E-01	ns	M381	0 1 1 1 2	pH 7.0

3471. $C_{15}H_{16}N_4O$
6H-Dipyrido[3,2-b:2′,3′-e][1,4]diazepin-6-one, 11-ethyl-5,11-dihydro-2,4-dimethyl-

RN:	134698-31-0	MP (°C):	
MW:	268.32	BP (°C):	

Solubility (Moles/L)	Solubility (Grams/L)	Temp (°C)	Ref (#)	Evaluation (T P E A A)	Comments
2.793E-05	7.493E-03	ns	M381	0 1 1 1 2	pH 7.0

3472. $C_{15}H_{16}N_4O$
6H-Dipyrido[3,2-b:2′,3′-e][1,4]diazepin-6-one, 5,11-diethyl-5,11-dihydro-

RN:	132312-82-4	MP (°C):	
MW:	268.32	BP (°C):	

Solubility (Moles/L)	Solubility (Grams/L)	Temp (°C)	Ref (#)	Evaluation (T P E A A)	Comments
1.380E-03	3.704E-01	ns	M381	0 1 1 1 2	pH 7.0

3473. $C_{15}H_{16}N_4O_2$
6H-Dipyrido[3,2-b:2′,3′-e][1,4]diazepin-6-one, 11-ethyl-5,11-dihydro-2-methoxy-4-methyl-

RN:	135794-75-1	MP (°C):	
MW:	284.32	BP (°C):	

Solubility (Moles/L)	Solubility (Grams/L)	Temp (°C)	Ref (#)	Evaluation (T P E A A)	Comments
7.031E-06	1.999E-03	ns	M381	0 1 1 1 2	pH 7.0

3474. C$_{15}$H$_{16}$N$_4$O$_2$
1H-Purine-2,6-dione, 1,3-diethyl-3,7-dihydro-8-phenyl-
1,3-Diethyl-8-phenylxanthine
8-Phenyl-1,3-diethylxanthine

RN: 75922-48-4 **MP** (°C):
MW: 284.32 **BP** (°C):

Solubility (Moles/L)	Solubility (Grams/L)	Temp (°C)	Ref (#)	Evaluation (T P E A A)	Comments
3.517E-06	1.000E-03	ns	H316	0 0 0 0 0	0.1N HCL
2.110E-05	6.000E-03	ns	H316	0 0 0 0 0	pH 7.4

3475. C$_{15}$H$_{16}$N$_4$O$_5$S
Benzenesulfonic acid, 4-(1,3-Diethyl-2,3,6,7-tetrahydro-2,6-dioxo-1H-purin-8-yl)-

RN: 89073-47-2 **MP** (°C): >360
MW: 364.38 **BP** (°C):

Solubility (Moles/L)	Solubility (Grams/L)	Temp (°C)	Ref (#)	Evaluation (T P E A A)	Comments
>1.56E-01	>5.70E+01	ns	H316	0 0 0 0 0	pH 7.4
>2.20E-02	>8.00E+00	ns	H316	0 0 0 0 0	0.1N HCL

3476. C$_{15}$H$_{16}$O$_2$
Bisphenol A
2,2-*bis*-[4-Hydroxyphenyl]-propan
2,2-*bis*-(4-Hydroxypheny)-propane

RN: 80-05-7 **MP** (°C):
MW: 228.29 **BP** (°C):

Solubility (Moles/L)	Solubility (Grams/L)	Temp (°C)	Ref (#)	Evaluation (T P E A A)	Comments
1.533E-03	3.500E-01	20	F300	1 0 0 0 1	
4.775E-04	1.090E-01	22	Y419	0 0 0 0 0	
1.314E-03	3.000E-01	23	S448	0 0 0 0 0	*Temperature 20-25
5.256E-04	1.200E-01	23	S448	0 0 0 0 0	*Temperature 20-25
5.256E-04	1.200E-01	25	D415	1 0 0 0 0	
5.256E-04	1.200E-01	25	D416	0 0 0 0 0	
1.314E-03	3.000E-01	25	S468	0 0 0 0 0	

3477. $C_{15}H_{16}O_2$

Bisphenol A

RN: 80-05-7   **MP** (°C):

MW: 228.29   **BP** (°C):

Solubility (Moles/L)	Solubility (Grams/L)	Temp (°C)	Ref (#)	Evaluation (T P E A A)	Comments
5.256E-04	1.200E-01	25	D416	0 0 0 0 0	

3478. $C_{15}H_{16}O_2$

Nabumetone

RN: 42924-53-8   **MP** (°C):

MW: 228.29   **BP** (°C):

Solubility (Moles/L)	Solubility (Grams/L)	Temp (°C)	Ref (#)	Evaluation (T P E A A)	Comments
2.628E-05	6.000E-03	22.5	C438	0 0 0 0 0	

3479. $C_{15}H_{16}O_3$

Osthole

2H-1-Benzopyran-2-one, 7-methoxy-8-(3-methyl-2-butenyl)-

RN: 484-12-8   **MP** (°C): 83.5

MW: 244.29   **BP** (°C):

Solubility (Moles/L)	Solubility (Grams/L)	Temp (°C)	Ref (#)	Evaluation (T P E A A)	Comments
4.912E-05	1.200E-02	30	B144	1 0 1 0 1	

3480. $C_{15}H_{16}O_9 \cdot 2H_2O$

Aesculin (dihydrate)

Esculin

6,7-Dihydroxycoumarin 6-glucoside

2H-1-Benzopyran-2-one, 6-(β-D-glucopyranosyloxy)-7-hydroxy-

RN: 531-75-9   **MP** (°C): 205dec

MW: 376.32   **BP** (°C):

Solubility (Moles/L)	Solubility (Grams/L)	Temp (°C)	Ref (#)	Evaluation (T P E A A)	Comments
4.605E-03	1.733E+00	c	D004	0 0 0 0 0	
		h	D004	0 0 0 0 0	

3481. C$_{15}$H$_{17}$FN$_4$O$_2$

Flupirtine

Carbamic acid, [2-amino-6-[[(4-fluorophenyl)methyl]amino]-3-pyridinyl]-, ethyl ester

RN: 56995-20-1 **MP** (°C): 175.8–177.7
MW: 304.33 **BP** (°C):

Solubility (Moles/L)	Solubility (Grams/L)	Temp (°C)	Ref (#)	Evaluation (T P E A A)	Comments
3.286E-03	1.000E+00	ns	D321	0 0 0 0 0	

3482. C$_{15}$H$_{17}$NO$_3$

Acetamide, 2-(benzoyloxy)-*N,N*-di-acetamide, 2-(benzoyloxy)-*N,N*-di-2-propenyl-

RN: 106231-58-7 **MP** (°C): 42.5
MW: 259.31 **BP** (°C): 401.1

Solubility (Moles/L)	Solubility (Grams/L)	Temp (°C)	Ref (#)	Evaluation (T P E A A)	Comments
2.738E-03	7.100E-01	22	B427	1 0 0 1 1	in 0.01M HCl
2.738E-03	7.100E-01	22	N317	1 1 2 1 2	

3483. C$_{15}$H$_{17}$NO$_5$

L-Proline, 1-[(benzoyloxy)acetyl]-, methyl ester

RN: 115178-76-2 **MP** (°C): 72.5
MW: 291.31 **BP** (°C):

Solubility (Moles/L)	Solubility (Grams/L)	Temp (°C)	Ref (#)	Evaluation (T P E A A)	Comments
8.239E-03	2.400E+00	22	N317	1 1 2 1 2	

3484. C$_{15}$H$_{17}$NO$_7$

Glycine, *N*-[[[2-(acetyloxy)benzoyl]oxy]acetyl]-, ethyl ester

RN: 118247-03-3 **MP** (°C): 68.5
MW: 323.31 **BP** (°C):

Solubility (Moles/L)	Solubility (Grams/L)	Temp (°C)	Ref (#)	Evaluation (T P E A A)	Comments
1.336E-02	4.320E+00	21	N335	0 0 0 0 0	

3485. C₁₅H₁₇N₃O₃S

L-Ala-dapsone
2-Amino-N-[4-[(4-aminophenyl)sulfonyl]phenyl]-, (S)-propanamide
RN: 160348-99-2 **MP** (°C):
MW: 319.39 **BP** (°C):

Solubility (Moles/L)	Solubility (Grams/L)	Temp (°C)	Ref (#)	Evaluation (T P E A A)	Comments
2.066E-02	6.600E+00	25	P351	0 0 0 0 0	pH 7.4
>9.39E-02	>3.00E+01	25	P351	0 0 0 0 0	

3486. C₁₅H₁₈Cl₂N₂O₃

Oxadiazon
3-[2,4-Dichloro-5-(1-methylethoxy)phenyl]-5-(1,1-dimethylethyl)-1,3,4-oxadiazol-2(3H)-one
Ronstar
Scotts OH I
RP-17623
RN: 19666-30-9 **MP** (°C): 88
MW: 345.23 **BP** (°C):

Solubility (Moles/L)	Solubility (Grams/L)	Temp (°C)	Ref (#)	Evaluation (T P E A A)	Comments
2.028E-06	7.000E-04	20	M161	1 0 0 0 0	
2.028E-06	7.000E-04	24	C105	2 1 2 2 2	

3487. C₁₅H₁₈I₃NO₅

Iopronic acid
Butanoic acid, 2-[[2-[3-(acetylamino)-2,4,6-triiodophenoxy]ethoxy]methyl]-
RN: 37723-78-7 **MP** (°C): 130
MW: 673.03 **BP** (°C):

Solubility (Moles/L)	Solubility (Grams/L)	Temp (°C)	Ref (#)	Evaluation (T P E A A)	Comments
2.984E-02	2.008E+01	37	J016	1 0 0 0 2	pH 7.4
1.456E-04	9.799E-02	50	F013	1 0 1 1 1	

3488. C₁₅H₁₈N₂O₃

N-Acetyl-L-tryptophan ethyl ester
RN: 2382-80-1 **MP** (°C): 106
MW: 274.32 **BP** (°C):

Solubility (Moles/L)	Solubility (Grams/L)	Temp (°C)	Ref (#)	Evaluation (T P E A A)	Comments
1.896E-03	5.200E-01	5	L081	2 2 2 2 1	
5.359E-03	1.470E+00	28	L081	2 1 2 2 2	

3489. C$_{15}$H$_{18}$N$_4$O$_3$S
2-(*N*4-Acetylsulfanilylamino)-4-ethyl-5-methylpyrimidine

RN: **MP** (°C):
MW: 334.40   **BP** (°C):

Solubility (Moles/L)	Solubility (Grams/L)	Temp (°C)	Ref (#)	Evaluation (T P E A A)	Comments
1.077E-05	3.600E-03	37	R076	1 2 0 0 1	

3490. C$_{15}$H$_{18}$N$_4$O$_3$S
2-(*N*4-Acetylsulfanilylamino)-4-*n*-propylpyrimidine

RN: **MP** (°C):
MW: 334.40   **BP** (°C):

Solubility (Moles/L)	Solubility (Grams/L)	Temp (°C)	Ref (#)	Evaluation (T P E A A)	Comments
1.914E-05	6.400E-03	37	R076	1 2 0 0 2	

3491. C$_{15}$H$_{18}$N$_4$O$_5$
Mitomycin C
MMC
6-Amino-8-[[(aminocarbonyl)oxy]methyl]-1,1α,2,8,8α,8β-hexahydro-8α-methoxy-5-methyl,[1aS-
(1α,8β,8aα,8bα)]-azirino[2′,3′:3,4]pyrrolo[1,2a]indole-4,7-dione
Mitomycinum

RN: 50-07-7   **MP** (°C):   >360
MW: 334.33   **BP** (°C):

Solubility (Moles/L)	Solubility (Grams/L)	Temp (°C)	Ref (#)	Evaluation (T P E A A)	Comments
2.730E-03	9.127E-01	25	M316	1 1 1 1 2	
8.500E-01	2.842E+02	ns	B406	0 0 2 2 0	EFG

3492. C$_{15}$H$_{18}$O$_3$
Santonin
Naphtho[1,2-b]furan-2,8(3H,4H)-dione, 3α,5,5α,9β-tetrahydro-3,5α,9-trimethyl-,
(3S,3αS,5αS,9βS)-

RN: 481-06-1   **MP** (°C):   170
MW: 246.31   **BP** (°C):

Solubility (Moles/L)	Solubility (Grams/L)	Temp (°C)	Ref (#)	Evaluation (T P E A A)	Comments
8.120E-04	2.000E-01	17.5	F300	1 0 0 0 0	
1.624E-02	4.000E+00	100	F300	1 0 0 0 0	

3493. C₁₅H₁₈O₄

β-Cyclopentylpropionyl salicylate

RN: **MP** (°C):
MW: 262.31   **BP** (°C):

Solubility (Moles/L)	Solubility (Grams/L)	Temp (°C)	Ref (#)	Evaluation (T P E A A)	Comments
1.060E-04	2.780E-02	25.6	G015	1 0 1 1 2	pH 1.00, pka 3.91, intrinsic

3494. C₁₅H₁₉ClO₂

1,1-Drichloro-1-methyl-2,2-bis(p-methoxylphenyl)ethane

RN: 56288-27-8   **MP** (°C):
MW: 266.77   **BP** (°C):

Solubility (Moles/L)	Solubility (Grams/L)	Temp (°C)	Ref (#)	Evaluation (T P E A A)	Comments
6.373E-06	1.700E-03	rt	C122	0 0 0 0 0	

3495. C₁₅H₁₉NO

N,N-Hexamethylenecinnamamide

Hexahydro-1-(1-oxo-3-phenyl-2-propenyl)1H-azepine

RN: 59832-05-2   **MP** (°C):
MW: 229.32   **BP** (°C):

Solubility (Moles/L)	Solubility (Grams/L)	Temp (°C)	Ref (#)	Evaluation (T P E A A)	Comments
2.460E-04	5.641E-02	ns	H350	0 0 0 0 0	

3496. C₁₅H₁₉NO

N-Cyclohexylcinnamamide

2-Propenamide, N-cyclohexyl-3-phenyl-

RN: 6750-98-7   **MP** (°C):
MW: 229.32   **BP** (°C):

Solubility (Moles/L)	Solubility (Grams/L)	Temp (°C)	Ref (#)	Evaluation (T P E A A)	Comments
4.040E-05	9.265E-03	ns	H350	0 0 0 0 0	

3497. C$_{15}$H$_{19}$NO$_2$
Tropacocaine
RN: 537-26-8   **MP** (°C):   49
MW: 245.32   **BP** (°C):

Solubility (Moles/L)	Solubility (Grams/L)	Temp (°C)	Ref (#)	Evaluation (T P E A A)	Comments
4.300E-03	1.055E+00	15	K059	2 2 2 0 1	

3498. C$_{15}$H$_{19}$NO$_3$
1H-Azepine, 1-[(benzoyloxy)acetyl]hexahydro-
RN: 115178-68-2   **MP** (°C):   107.5
MW: 261.32   **BP** (°C):

Solubility (Moles/L)	Solubility (Grams/L)	Temp (°C)	Ref (#)	Evaluation (T P E A A)	Comments
2.870E-03	7.500E-01	22	N317	1 1 2 1 2	

3499. C$_{15}$H$_{19}$NO$_5$
Benzoic acid, 2-(acetyloxy)-, 2-(diethylamino)-2-oxoethyl ester
RN: 116482-56-5   **MP** (°C):   76.5
MW: 293.32   **BP** (°C):

Solubility (Moles/L)	Solubility (Grams/L)	Temp (°C)	Ref (#)	Evaluation (T P E A A)	Comments
7.773E-03	2.280E+00	21	N335	0 0 0 0 0	

3500. C$_{15}$H$_{20}$N$_2$O$_4$
Benzyl-2,2-diethylmalonurate
Benzyl 2,2-diethylmalonurate

RN: 73632-78-7   **MP** (°C):   107
MW: 292.34   **BP** (°C):

Solubility (Moles/L)	Solubility (Grams/L)	Temp (°C)	Ref (#)	Evaluation (T P E A A)	Comments
2.200E-04	6.431E-02	23	B152	1 2 1 1 1	pH 3.5

3501. $C_{15}H_{20}N_2O_4S$
Acetohexamide
Acetohexamid
1-(p-Acetylbenzenesulfonyl)-3-cyclohexylurea
Dymelor
Dimelin

RN: 968-81-0   **MP** (°C): 189
MW: 324.40   **BP** (°C):

Solubility (Moles/L)	Solubility (Grams/L)	Temp (°C)	Ref (#)	Evaluation (T P E A A)	Comments
7.706E-04	2.500E-01	25	K023	1 0 2 2 1	EFG, pH 6.5, average of 2
3.483E-05	1.130E-02	37	B130	1 2 1 1 2	pH 1.5, form II
4.963E-05	1.610E-02	37	B130	1 2 1 1 2	pH 1.5, form III
8.015E-05	2.600E-02	37	K106	1 2 2 2 0	EFG, form I
9.556E-05	3.100E-02	37	K106	1 2 2 2 0	EFG, form II

3502. $C_{15}H_{20}N_4O_2S$
2-Sulfanilylamino-4-amylpyrimidine
RN: 107203-72-5   **MP** (°C):
MW: 320.42   **BP** (°C):

Solubility (Moles/L)	Solubility (Grams/L)	Temp (°C)	Ref (#)	Evaluation (T P E A A)	Comments
6.242E-04	2.000E-01	37	R076	1 2 0 0 1	

3503. $C_{15}H_{20}N_4O_5$
1,5-Dibutyryloxymethyl allopurinol
RN: 98827-19-1   **MP** (°C): 122–123
MW: 336.35   **BP** (°C):

Solubility (Moles/L)	Solubility (Grams/L)	Temp (°C)	Ref (#)	Evaluation (T P E A A)	Comments
1.487E-04	5.000E-02	22	B322	0 0 0 0 0	

3504. $C_{15}H_{20}N_4O_5$
2,5-Dibutyryloxymethyl allopurinol
RN: 98827-20-4   **MP** (°C): 133–135
MW: 336.35   **BP** (°C):

Solubility (Moles/L)	Solubility (Grams/L)	Temp (°C)	Ref (#)	Evaluation (T P E A A)	Comments
2.795E-04	9.400E-02	22	B322	0 0 0 0 0	

3505. $C_{15}H_{20}N_4O_6$

9-[5-O-(Butyrate-β-D-arabinofuranosyl)]-6-methoxy-9H-purine

RN: 121032-41-5   **MP** (°C):   108–110

MW: 352.35   **BP** (°C):

Solubility (Moles/L)	Solubility (Grams/L)	Temp (°C)	Ref (#)	Evaluation (T P E A A)	Comments
9.680E-03	3.411E+00	37	M378	1 2 1 1 2	pH 7.2

3506. $C_{15}H_{20}N_4O_6 \cdot 0.3H_2O$

2′-Butyryl-6-methoxypurine arabinoside (0.3 hydrate)

RN: 121032-41-5   **MP** (°C):

MW: 357.75   **BP** (°C):

Solubility (Moles/L)	Solubility (Grams/L)	Temp (°C)	Ref (#)	Evaluation (T P E A A)	Comments
2.310E-01	8.264E+01	37	C348	0 0 0 0 0	pH 7.00

3507. $C_{15}H_{20}N_4O_6$

2′-Isobutyryl-6-methoxypurine arabinoside

RN: 121032-44-8   **MP** (°C):

MW: 352.35   **BP** (°C):

Solubility (Moles/L)	Solubility (Grams/L)	Temp (°C)	Ref (#)	Evaluation (T P E A A)	Comments
6.700E-01	2.361E+02	37	C348	0 0 0 0 0	pH 7.00

3508. $C_{15}H_{20}N_4O_6 \cdot 0.25H_2O$

9-[5-O-(Isobutyrate-β-D-arabinofuranosyl)]-6-methoxy-9H-purine (0.25 hydrate)

RN: 121032-44-8   **MP** (°C):   glass

MW: 356.85   **BP** (°C):

Solubility (Moles/L)	Solubility (Grams/L)	Temp (°C)	Ref (#)	Evaluation (T P E A A)	Comments
3.830E-02	1.367E+01	37	M378	1 2 1 1 2	pH 7.2

3509. $C_{15}H_{21}NO$

N,N-Dipropylcinnamamide

Cinnamamide, N,N-dipropyl-

RN: 23784-56-7   **MP** (°C):

MW: 231.34   **BP** (°C):

Solubility (Moles/L)	Solubility (Grams/L)	Temp (°C)	Ref (#)	Evaluation (T P E A A)	Comments
2.890E-03	6.686E-01	ns	H350	0 0 0 0 0	

3510. C₁₅H₂₁NO₂

Meperidine
Ethyl 1-methyl-4-phenylpiperidine-4-carboxylate
Demerol
Dolantin
Pethidine

RN: 57-42-1 **MP (°C):** 30
MW: 247.34 **BP (°C):**

Solubility (Moles/L)	Solubility (Grams/L)	Temp (°C)	Ref (#)	Evaluation (T P E A A)	Comments
2.648E-02	6.550E+00	25	R338	0 0 0 0 0	
1.300E-02	3.215E+00	30	L068	1 0 0 1 0	EFG

3511. C₁₅H₂₁NO₂S₂

2-(p-Isopropylphenyl)-2-methyl-4-(methoxycarbamyl)-1,3-dithiolane

RN: 35801-67-3 **MP (°C):**
MW: 311.47 **BP (°C):**

Solubility (Moles/L)	Solubility (Grams/L)	Temp (°C)	Ref (#)	Evaluation (T P E A A)	Comments
2.500E-05	7.787E-03	rt	B174	0 0 1 0 1	

3512. C₁₅H₂₁NO₃

Acetamide, 2-(benzoyloxy)-N,N-bis(1-methylethyl)-

RN: 106231-56-5 **MP (°C):** 105.5
MW: 263.34 **BP (°C):** 391.6

Solubility (Moles/L)	Solubility (Grams/L)	Temp (°C)	Ref (#)	Evaluation (T P E A A)	Comments
4.557E-04	1.200E-01	22	B427	1 0 0 1 1	in 0.01M HCl
4.557E-04	1.200E-01	22	N317	1 1 2 1 2	

3513. C₁₅H₂₁NO₃

Acetamide, 2-(benzoyloxy)-N-hexyl-

RN: 115193-29-8 **MP (°C):** 130.5
MW: 263.34 **BP (°C):**

Solubility (Moles/L)	Solubility (Grams/L)	Temp (°C)	Ref (#)	Evaluation (T P E A A)	Comments
1.253E-04	3.300E-02	22	N317	1 1 2 1 2	

3514. C₁₅H₂₁NO₃

Acetamide, 2-(benzoyloxy)-N,N-dipropyl-

RN: 106231-55-4 **MP (°C):** 20
MW: 263.34 **BP (°C):** 402.9

Solubility (Moles/L)	Solubility (Grams/L)	Temp (°C)	Ref (#)	Evaluation (T P E A A)	Comments
4.177E-03	1.100E+00	22	B427	1 0 0 1 1	in 0.01M HCl
4.177E-03	1.100E+00	22	N317	1 1 2 1 2	

3515. C₁₅H₂₁NO₃S

$C_{15}H_{21}NO_3S$

2-(p-Isopropylphenyl)-2-methyl-4-(methoxycarbamyl)-1,3-oxathiolane

RN: 24606-94-8   **MP** (°C):
MW: 295.40   **BP** (°C):

Solubility (Moles/L)	Solubility (Grams/L)	Temp (°C)	Ref (#)	Evaluation (T P E A A)	Comments
6.000E-05	1.772E-02	rt	B174	0 0 1 0 0	

3516. C₁₅H₂₁NO₄

$C_{15}H_{21}NO_4$

Metalaxyl
Methyl N-(2,6-dimethyl-phenyl)-N-(2′-methoxyacetyl)-DL-alaninate
Apron
Ridomil
Subdue
Fubol

RN: 57837-19-1   **MP** (°C): 72
MW: 279.34   **BP** (°C):

Solubility (Moles/L)	Solubility (Grams/L)	Temp (°C)	Ref (#)	Evaluation (T P E A A)	Comments
2.488E-02	6.951E+00	20	E048	1 2 1 1 2	

3517. C₁₅H₂₁NO₄

$C_{15}H_{21}NO_4$

Hexyl acetaminophen
Carbonic acid, 4-(acetylamino)phenyl hexyl ester
Acetanilide, 4′-hydroxy-, hexyl carbonate (ester)

RN: 17239-22-4   **MP** (°C): 112.5-113.5
MW: 279.34   **BP** (°C):

Solubility (Moles/L)	Solubility (Grams/L)	Temp (°C)	Ref (#)	Evaluation (T P E A A)	Comments
1.325E-04	3.700E-02	37	D029	0 0 0 0 0	

3518. C₁₅H₂₁NO₄

$C_{15}H_{21}NO_4$

2-(p-Isopropylphenyl)-2-methyl-4-(methoxycarbamyl)-1,3-dioxolane

RN: 35858-24-3   **MP** (°C):
MW: 279.34   **BP** (°C):

Solubility (Moles/L)	Solubility (Grams/L)	Temp (°C)	Ref (#)	Evaluation (T P E A A)	Comments
9.000E-04	2.514E-01	rt	B174	0 0 1 0 0	

3519. C₁₅H₂₁NO₅

$C_{15}H_{21}NO_5$

Acetamide, 2-(benzoyloxy)-N,N-bis(2-methoxyethyl)-

RN: 115178-64-8   **MP** (°C): 57.5
MW: 295.34   **BP** (°C):

Solubility (Moles/L)	Solubility (Grams/L)	Temp (°C)	Ref (#)	Evaluation (T P E A A)	Comments
2.672E-02	7.890E+00	22	N317	1 1 2 1 2	

3520. $C_{15}H_{21}NO_5$

Acetamide, 2-(benzoyloxy)-*N,N*-bis(2-hydroxypropyl)-

RN: 115178-63-7 **MP** (°C): 105.5
MW: 295.34 **BP** (°C):

Solubility (Moles/L)	Solubility (Grams/L)	Temp (°C)	Ref (#)	Evaluation (T P E A A)	Comments
6.636E-02	1.960E+01	22	N317	1 1 2 1 2	

3521. $C_{15}H_{21}N_2O_3$

C.I. Disperse red 11

RN: 2872-48-2 **MP** (°C): 242
MW: 277.35 **BP** (°C):

Solubility (Moles/L)	Solubility (Grams/L)	Temp (°C)	Ref (#)	Evaluation (T P E A A)	Comments
2.500E-06	6.934E-04	25	B333	0 0 0 0 0	

3522. $C_{15}H_{21}N_3O$

Primaquine
Primaquine phosphate
Neo-quipenyl
8-(4-Amino-1-methylbutylamino)-6-methoxyquinoline
8-((4-Amino-1-methylbutyl)amino)-6-methoxyquinoline phosphate
Palum

RN: 90-34-6 **MP** (°C):
MW: 259.35 **BP** (°C):

Solubility (Moles/L)	Solubility (Grams/L)	Temp (°C)	Ref (#)	Evaluation (T P E A A)	Comments
2.770E+00	7.184E+02	25	B443	0 0 0 0 0	

3523. $C_{15}H_{21}N_5O_5$

9-(2-*O*-Valeryl-β-D-arabinofuranosyl)adenine

RN: 87984-85-8 **MP** (°C):
MW: 351.37 **BP** (°C):

Solubility (Moles/L)	Solubility (Grams/L)	Temp (°C)	Ref (#)	Evaluation (T P E A A)	Comments
2.960E-04	1.040E-01	37	B306	1 2 0 1 2	pH 7.3

3524. $C_{15}H_{21}N_5O_5$

9-[5′-(*O*-Isovaleryl)-β-D-arabinofuranosyl]adenine ester

RN: 65926-32-1 **MP** (°C):
MW: 351.37 **BP** (°C):

Solubility (Moles/L)	Solubility (Grams/L)	Temp (°C)	Ref (#)	Evaluation (T P E A A)	Comments
5.635E-02	1.980E+01	ns	B134	0 1 1 1 2	

3525. C₁₅H₂₁N₅O₅

3525. $C_{15}H_{21}N_5O_5$

9-[5′-(O-Valeryl)-β-D-arabinofuranosyl]adenine ester

RN: 65926-31-0   **MP** (°C):
MW: 351.37   **BP** (°C):

Solubility (Moles/L)	Solubility (Grams/L)	Temp (°C)	Ref (#)	Evaluation (T P E A A)	Comments
2.391E-02	8.400E+00	ns	B134	0 1 1 1 1	

3526. $C_{15}H_{21}N_5O_5$

9-[5′-(O-Pivaloyl)-β-D-arabinofuranosyl]adenine ester

RN: 65926-33-2   **MP** (°C):
MW: 351.37   **BP** (°C):

Solubility (Moles/L)	Solubility (Grams/L)	Temp (°C)	Ref (#)	Evaluation (T P E A A)	Comments
1.992E-02	7.000E+00	ns	B134	0 1 1 1 1	

3527. $C_{15}H_{21}N_5O_6$

9-(1,3-Dipropionate-2-propoxymethyl)guanine

RN: 86357-20-2   **MP** (°C): 192
MW: 367.36   **BP** (°C):

Solubility (Moles/L)	Solubility (Grams/L)	Temp (°C)	Ref (#)	Evaluation (T P E A A)	Comments
7.622E-03	2.800E+00	25	B360	0 0 0 0 0	

3528. $C_{15}H_{22}ClNO_2$

Metolachlor
2-Chloro-N-(2-ethyl-6-methylphenyl)-N-(2-methoxy-1-methylethyl)acetamide
Dual
Cotoran Multi
Ontrack 8E
Bicep 6L

RN: 51218-45-2   **MP** (°C): <25
MW: 283.80   **BP** (°C): 100

Solubility (Moles/L)	Solubility (Grams/L)	Temp (°C)	Ref (#)	Evaluation (T P E A A)	Comments
1.867E-03	5.297E-01	20	E048	1 2 1 1 2	
1.868E-03	5.300E-01	20	M161	1 0 0 0 2	
1.866E-03	5.297E-01	ns	S460	0 0 0 0 0	
1.868E-03	5.300E-01	ns	V414	0 0 0 0 0	

3529. $C_{15}H_{22}ClNO_2$
CP 52223
2-Chloro-N-(2,6-dimethyl)phenyl-N-isopropoxymethylacetamide
RN: 24353-58-0 **MP** (°C):
MW: 283.80 **BP** (°C): 137.5

Solubility (Moles/L)	Solubility (Grams/L)	Temp (°C)	Ref (#)	Evaluation (T P E A A)	Comments
2.079E-04	5.900E-02	ns	M061	0 0 0 0 1	

3530. $C_{15}H_{22}N_2O$
DL-Mepivacaine
Carbocaine
1-Methyl-2′,6′-pipecoloxylidide
Carbocain
RN: 96-88-8 **MP** (°C): 150
MW: 246.36 **BP** (°C):

Solubility (Moles/L)	Solubility (Grams/L)	Temp (°C)	Ref (#)	Evaluation (T P E A A)	Comments
1.360E-02	3.350E+00	14.9	N046	2 0 1 1 1	intrinsic
3.653E-02	9.000E+00	23	F176	2 0 0 2 0	EFG, pH 7.4, intrinsic
2.841E-02	7.000E+00	23	F176	2 0 0 2 0	EFG, pH 7.4, intrinsic
9.000E-03	2.217E+00	25	D402	1 2 2 2 0	EFG
1.020E-02	2.513E+00	25	N046	2 0 1 1 1	intrinsic
9.910E-03	2.441E+00	34.5	N046	2 0 1 1 1	intrinsic
1.000E-02	2.464E+00	37	D402	1 2 2 2 0	EFG
7.970E-03	1.963E+00	37	N044	2 1 1 2 2	intrinsic

3531. $C_{15}H_{22}O_3$
Gemfibrozil
2,2-Dimethyl-5-(2,5-xylyloxy)valeric acid
Jezil
Lobid
Lopid
RN: 25812-30-0 **MP** (°C):
MW: 250.34 **BP** (°C):

Solubility (Moles/L)	Solubility (Grams/L)	Temp (°C)	Ref (#)	Evaluation (T P E A A)	Comments
>2.00E-03	>5.00E-01	ns	B404	0 2 1 1 0	

3532. C$_{15}$H$_{22}$O$_3$
Octyl p-hydroxybenzoate
n-Octyl 4-hydroxybenzoate

RN:	1219-38-1	MP (°C):	54
MW:	250.34	BP (°C):	

Solubility (Moles/L)	Solubility (Grams/L)	Temp (°C)	Ref (#)	Evaluation (T P E A A)	Comments
1.470E-05	3.680E-03	15	B355	0 0 0 0 0	
2.300E-04	5.758E-02	20	B355	0 0 0 0 0	
4.650E-04	1.164E-01	25	B355	0 0 0 0 0	
3.273E-03	8.193E-01	25	D081	1 2 2 1 2	
3.162E-04	7.916E-02	25	F322	2 0 1 1 0	EFG

3533. C$_{15}$H$_{22}$O$_5$
Octyl gallate
Octyl 3,4,5-trihydroxybenzoate
n-Octyl gallate

RN:	1034-01-1	MP (°C):	
MW:	282.34	BP (°C):	

Solubility (Moles/L)	Solubility (Grams/L)	Temp (°C)	Ref (#)	Evaluation (T P E A A)	Comments
7.084E-05	2.000E-02	29.99	L430	0 0 0 0 0	
8.500E-05	2.400E-02	34.99	L430	0 0 0 0 0	
1.133E-04	3.200E-02	39.99	L430	0 0 0 0 0	
1.806E-04	5.100E-02	44.99	L430	0 0 0 0 0	
3.152E-04	8.899E-02	49.99	L430	0 0 0 0 0	
4.214E-04	1.190E-01	54.99	L430	0 0 0 0 0	
4.710E-04	1.330E-01	59.99	L430	0 0 0 0 0	
5.064E-04	1.430E-01	64.99	L430	0 0 0 0 0	

3534. C$_{15}$H$_{23}$NO$_2$
Octyl m-aminobenzoate
Octyl 3-aminobenzoate

RN:	52222-35-2	MP (°C):	
MW:	249.36	BP (°C):	

Solubility (Moles/L)	Solubility (Grams/L)	Temp (°C)	Ref (#)	Evaluation (T P E A A)	Comments
3.000E-05	7.481E-03	ns	M066	0 0 0 0 0	

3535. C$_{15}$H$_{23}$NO$_2$
Octyl p-aminobenzoate
4-Aminobenzoic acid octyl ester

RN:	14309-41-2	MP (°C):	
MW:	249.36	BP (°C):	

Solubility (Moles/L)	Solubility (Grams/L)	Temp (°C)	Ref (#)	Evaluation (T P E A A)	Comments
3.200E-06	7.979E-04	37	F006	1 1 2 2 1	

3536. $C_{15}H_{23}NO_2$
Alprenolol
Aptin

RN:	13655-52-2
MW:	249.36

MP (°C):	
BP (°C):	

Solubility (Moles/L)	Solubility (Grams/L)	Temp (°C)	Ref (#)	Evaluation (T P E A A)	Comments
1.471E-03	3.669E-01	22.5	B422	0 0 0 0 0	

3537. $C_{15}H_{23}NO_3$
Parethoxycaine
4-Ethoxybenzoic acid-2-(diethylamino)ethyl ester

RN:	94-23-5
MW:	265.36

MP (°C):	173.0
BP (°C):	

Solubility (Moles/L)	Solubility (Grams/L)	Temp (°C)	Ref (#)	Evaluation (T P E A A)	Comments
1.930E-03	5.121E-01	ns	M066	0 0 0 0 2	

3538. $C_{15}H_{23}NO_4$
Cycloheximide
3-((R)-2-((1S,3S,5S)-3,5-Dimethyl-2-oxocyclohexyl)-2-hydroxyethyl)glutarimide
Actidione
Actispray
Naramycin
Kaken

RN:	66-81-9
MW:	281.35

MP (°C):	116.3
BP (°C):	

Solubility (Moles/L)	Solubility (Grams/L)	Temp (°C)	Ref (#)	Evaluation (T P E A A)	Comments
7.464E-02	2.100E+01	2	M161	1 0 0 0 1	

3539. $C_{15}H_{23}N_3O_4$
Isopropalin
2,6-Dinitro-N,N-dipropylcumidene
4-Isopropyl-2,6-dinitro-N,N-dipropylaniline
2,6-Dinitro-N,N-dipropylcumidine
Paarlan
Paarlan EC

RN:	33820-53-0
MW:	309.37

MP (°C):	
BP (°C):	

Solubility (Moles/L)	Solubility (Grams/L)	Temp (°C)	Ref (#)	Evaluation (T P E A A)	Comments
3.232E-07	1.000E-04	25	M161	1 0 0 0 0	

3540. C$_{15}$H$_{23}$N$_3$O$_4$S
Sulpiride
N-[(1-Ethyl-2-pyrrolidinyl)methyl]-2-methoxy-5-sulfamoylbenzamide
RN: 15676-16-1   **MP** (°C):
MW: 341.43   **BP** (°C):

Solubility (Moles/L)	Solubility (Grams/L)	Temp (°C)	Ref (#)	Evaluation (T P E A A)	Comments
<6.15E-04	<2.10E-01	25	P312	0 0 0 0 0	

3541. C$_{15}$H$_{23}$N$_3$O$_4$S
Cyclacillin
Anhydrous 6-(1-aminocyclohexanecarboxamido)penicillanic acid
RN: 3485-14-1   **MP** (°C):
MW: 341.43   **BP** (°C):

Solubility (Moles/L)	Solubility (Grams/L)	Temp (°C)	Ref (#)	Evaluation (T P E A A)	Comments
1.611E-01	5.500E+01	7	P035	0 0 0 0 0	EFG
1.054E-01	3.600E+01	20	P035	0 0 0 0 0	EFG
9.372E-02	3.200E+01	25	P035	0 0 0 0 0	EFG
7.908E-02	2.700E+01	30	P035	0 0 0 0 0	EFG
6.736E-02	2.300E+01	40	P035	0 0 0 0 0	EFG
6.151E-02	2.100E+01	50	P035	0 0 0 0 0	EFG
5.858E-02	2.000E+01	60	P035	0 0 0 0 0	EFG

3542. C$_{15}$H$_{23}$N$_3$O$_4$S.2H$_2$O
Cyclacillin (dihydrate)
Dihydrate 6-(1-aminocyclohexanecarboxamido)penicillanic acid
RN: 3485-14-1   **MP** (°C):
MW: 377.46   **BP** (°C):

Solubility (Moles/L)	Solubility (Grams/L)	Temp (°C)	Ref (#)	Evaluation (T P E A A)	Comments
3.709E-02	1.400E+01	10	P035	0 0 0 0 0	EFG
3.709E-02	1.400E+01	20	P035	0 0 0 0 0	EFG
3.656E-02	1.380E+01	25	P035	0 0 0 0 0	EFG
3.656E-02	1.380E+01	30	P035	0 0 0 0 0	EFG
3.682E-02	1.390E+01	40	P035	0 0 0 0 0	EFG
3.762E-02	1.420E+01	50	P035	0 0 0 0 0	EFG
4.504E-02	1.700E+01	60	P035	0 0 0 0 0	EFG

3543. C₁₅H₂₄NO₄PS

Isofenphos
Methylethyl 2-((ethoxy((1-methylethyl)amino)phosphinothioyl)oxy)benzoate
Amaze
Oftanol
Pryfon

| RN: | 25311-71-1 | MP (°C): |
| MW: | 345.40 | BP (°C): |

Solubility (Moles/L)	Solubility (Grams/L)	Temp (°C)	Ref (#)	Evaluation (T P E A A)	Comments
6.399E-05	2.210E-02	20	B300	2 1 1 1 2	*sic*
6.891E-02	2.380E+01	20	M161	1 0 0 0 2	*sic*

3544. C₁₅H₂₄N₂O₂

N,N,N′-Triethyl-bicyclo(2.2.1)hept-5-ene-2,3-*trans*-dicarboxamide

| RN: | 62249-37-0 | MP (°C): |
| MW: | 264.37 | BP (°C): |

Solubility (Moles/L)	Solubility (Grams/L)	Temp (°C)	Ref (#)	Evaluation (T P E A A)	Comments
2.232E-01	5.900E+01	20	K050	1 1 1 1 2	

3545. C₁₅H₂₄N₂O₂

Tetracaine
Pantocaine
Cetacaine

| RN: | 94-24-6 | MP (°C): |
| MW: | 264.37 | BP (°C): |

Solubility (Moles/L)	Solubility (Grams/L)	Temp (°C)	Ref (#)	Evaluation (T P E A A)	Comments
5.900E-04	1.560E-01	ns	E031	0 0 2 1 2	

3546. C₁₅H₂₄N₂O₂

4-Ethylaminobenzoic acid-2-(diethyl-amino)ethyl ester

| RN: | 16488-53-2 | MP (°C): |
| MW: | 264.37 | BP (°C): |

Solubility (Moles/L)	Solubility (Grams/L)	Temp (°C)	Ref (#)	Evaluation (T P E A A)	Comments
4.600E-03	1.216E+00	ns	M066	0 0 0 0 1	

3547. C$_{15}$H$_{24}$N$_2$O$_2$
4-Aminobenzoic acid-2-(diethyl-amino)butyl ester
2-(Diethyl(amino)butyl 4-aminobenzoate

RN:	5878-14-8	**MP** (°C):			
MW:	264.37	**BP** (°C):			

Solubility (Moles/L)	Solubility (Grams/L)	Temp (°C)	Ref (#)	Evaluation (T P E A A)	Comments
4.300E-03	1.137E+00	ns	M066	0 0 0 0 1	

3548. C$_{15}$H$_{24}$N$_2$O$_3$
2,4-Diazaspiro[5.11]heptadecane-1,3,5-trione

RN:	143288-64-6	**MP** (°C):			
MW:	280.37	**BP** (°C):			

Solubility (Moles/L)	Solubility (Grams/L)	Temp (°C)	Ref (#)	Evaluation (T P E A A)	Comments
1.600E-06	4.486E-04	25	P350	0 0 0 0 0	intrinsic

3549. C$_{15}$H$_{24}$O
Butylated hydroxytoluene
2,6-Di-*tert*-butyl-*p*-cresol
2,6-Di-*tert*-butyl-1-hydroxy-4-methylbenzene
4-Hydroxy-3,5-di-*tert*-butyltoluene

RN:	128-37-0	**MP** (°C):	71		
MW:	220.36	**BP** (°C):	265		

Solubility (Moles/L)	Solubility (Grams/L)	Temp (°C)	Ref (#)	Evaluation (T P E A A)	Comments
<4.54E-05	<1.00E-02	25	P312	0 0 0 0 0	

3550. C$_{15}$H$_{24}$O
4-Nonylphenol
4-*t*-Nonylphenol

RN:	104-40-5	**MP** (°C):			
MW:	220.36	**BP** (°C):			

Solubility (Moles/L)	Solubility (Grams/L)	Temp (°C)	Ref (#)	Evaluation (T P E A A)	Comments
2.090E-05	4.605E-03	2	A335	0 0 0 0 0	
2.088E-05	4.600E-03	2	A335	0 0 0 0 0	
2.230E-05	4.914E-03	10	A335	0 0 0 0 0	
2.233E-05	4.920E-03	10	A335	0 0 0 0 0	
2.380E-05	5.245E-03	14	A335	0 0 0 0 0	
2.378E-05	5.240E-03	14	A335	0 0 0 0 0	
2.470E-05	5.443E-03	20.5	A335	0 0 0 0 0	
2.464E-05	5.430E-03	20.5	A335	0 0 0 0 0	
2.882E-05	6.350E-03	25	A335	0 0 0 0 0	
2.890E-05	6.368E-03	25	A335	0 0 0 0 0	
3.177E-05	7.000E-03	25	M127	1 0 0 0 0	

3551. $C_{15}H_{24}O$

Nonylphenol

RN:	25154523	MP (°C):	
MW:	220.36	BP (°C):	293–297

Solubility (Moles/L)	Solubility (Grams/L)	Temp (°C)	Ref (#)	Evaluation (T P E A A)	Comments
2.224E-05	4.900E-03	25	B420	1 1 1 1 1	

3552. $C_{15}H_{26}N_2$

Sparteine

(–)-Spartein

RN:	90-39-1	MP (°C):	30
MW:	234.39	BP (°C):	

Solubility (Moles/L)	Solubility (Grams/L)	Temp (°C)	Ref (#)	Evaluation (T P E A A)	Comments
1.297E-02	3.040E+00	22	F300	1 0 0 0 2	
1.297E-02	3.040E+00	25	D004	0 0 0 0 0	

3553. $C_{15}H_{26}N_2O_3$

5-Allyl-5-methylhexylcarbinylbarbituric acid

RN:		MP (°C):	
MW:	282.39	BP (°C):	

Solubility (Moles/L)	Solubility (Grams/L)	Temp (°C)	Ref (#)	Evaluation (T P E A A)	Comments
1.084E-02	3.060E+00	ns	T003	0 0 0 0 2	

3554. $C_{15}H_{26}N_2O_3$

5-Ethyl-5-*n*-nonylbarbituric acid

5-Ethyl-5-nonylbarbiturate

RN:	64810-91-9	MP (°C):	
MW:	282.39	BP (°C):	

Solubility (Moles/L)	Solubility (Grams/L)	Temp (°C)	Ref (#)	Evaluation (T P E A A)	Comments
3.450E-04	9.742E-02	25	M310	2 2 2 2 2	

3555. $C_{15}H_{26}O_6$

Tributyrin

Glyceryl tributyrate

Tributanoylglycerol

1,2,3-Propanetriyl tributyrate

RN:	60-01-5	MP (°C):	173
MW:	302.37	BP (°C):	287.5

Solubility (Moles/L)	Solubility (Grams/L)	Temp (°C)	Ref (#)	Evaluation (T P E A A)	Comments
3.307E-04	9.999E-02	ns	F014	0 0 0 0 1	

3556. C₁₅H₂₈O₄

1,13-Tridecanedicarboxylic acid
1,15-Pentadecandioic acid

RN: 1460-18-0   **MP** (°C):
MW: 272.39   **BP** (°C):

Solubility (Moles/L)	Solubility (Grams/L)	Temp (°C)	Ref (#)	Evaluation (T P E A A)	Comments
1.285E-03	3.500E-01	21	B040	1 0 1 1 1	*sic*

3557. C₁₅H₃₀

1-Pentadecene

RN: 13360-61-7   **MP** (°C):
MW: 210.41   **BP** (°C):

Solubility (Moles/L)	Solubility (Grams/L)	Temp (°C)	Ref (#)	Evaluation (T P E A A)	Comments
1.778E-09	3.740E-07	23	C332	0 0 0 0 0	

3558. C₁₅H₃₀O₂

Pentadecylic acid
Pentadecanoic acid

RN: 1002-84-2   **MP** (°C):   52
MW: 242.41   **BP** (°C):

Solubility (Moles/L)	Solubility (Grams/L)	Temp (°C)	Ref (#)	Evaluation (T P E A A)	Comments
3.135E-05	7.600E-03	0	B136	1 0 2 1 1	
4.950E-05	1.200E-02	20	B136	1 0 2 1 1	
4.950E-05	1.200E-02	20.0	R001	1 1 1 1 1	
5.775E-05	1.400E-02	30	B136	1 0 2 1 1	
5.775E-05	1.400E-02	30.0	R001	1 1 1 1 1	
7.013E-05	1.700E-02	45	B136	1 0 2 1 1	
7.013E-05	1.700E-02	45.0	R001	1 1 1 1 1	
8.251E-05	2.000E-02	60	B136	1 0 2 1 1	
8.250E-05	2.000E-02	60.0	R001	1 1 1 1 1	
3.135E-05	7.600E-03	.0	R001	1 1 1 1 1	

3559. C₁₅H₃₀O₃

Dodecyl lactate
Propanoic acid, 2-hydroxy-, dodecyl ester

RN: 6283-92-7   **MP** (°C):
MW: 258.40   **BP** (°C):

Solubility (Moles/L)	Solubility (Grams/L)	Temp (°C)	Ref (#)	Evaluation (T P E A A)	Comments
3.870E-04	1.000E-01	25	R006	2 2 0 1 0	

3560. C₁₅H₃₂

Pentadecane
n-Pentadecane
Pentadecane-d32
Pentadecane (n)

RN:	629-62-9	**MP** (°C):	9.9
MW:	212.42	**BP** (°C):	270.63

Solubility (Moles/L)	Solubility (Grams/L)	Temp (°C)	Ref (#)	Evaluation (T P E A A)	Comments
1.883E-10	4.000E-08	25	T423	0 0 0 0 0	

3561. C₁₅H₃₂O

Pentadecanol
Pentadecan-1-ol
1-Pentadecanol

RN:	629-76-5	**MP** (°C):	46
MW:	228.42	**BP** (°C):	

Solubility (Moles/L)	Solubility (Grams/L)	Temp (°C)	Ref (#)	Evaluation (T P E A A)	Comments
4.500E-07	1.028E-04	25	R002	0 0 0 0 0	

3562. C₁₆H₈Cl₂F₆N₂O₃

Hexaflumuron

RN:	86479-06-3	**MP** (°C):	
MW:	461.15	**BP** (°C):	

Solubility (Moles/L)	Solubility (Grams/L)	Temp (°C)	Ref (#)	Evaluation (T P E A A)	Comments
3.513E-08	1.620E-05	20	M402	0 0 0 0 0	

3563. C₁₆H₁₀

Fluoranthene
1,2-Benzacenaphthene
1,2-(1,8-Naphthalenediyl)benzene
Benzo[j,k]fluorene
Idryl
FA

RN:	206-44-0	**MP** (°C):	107
MW:	202.26	**BP** (°C):	384

Solubility (Moles/L)	Solubility (Grams/L)	Temp (°C)	Ref (#)	Evaluation (T P E A A)	Comments
4.050E-07	8.191E-05	8.10	M082	1 1 1 2 2	
4.050E-07	8.191E-05	8.10	M151	2 1 2 2 2	
4.058E-07	8.207E-05	8.14	M183	1 1 1 1 2	
5.290E-07	1.070E-04	13.20	M082	1 1 1 2 2	
5.290E-07	1.070E-04	13.20	M151	2 1 2 2 2	

(continued)

3563. $C_{16}H_{10}$ (continued)

Solubility (Moles/L)	Solubility (Grams/L)	Temp (°C)	Ref (#)	Evaluation (T P E A A)	Comments
5.295E-07	1.071E-04	13.24	M183	1 1 1 1 2	
7.330E-07	1.483E-04	19.70	M082	1 1 1 2 2	
7.330E-07	1.483E-04	19.70	M151	2 1 2 2 2	
7.339E-07	1.484E-04	19.74	M183	1 2 1 1 2	
1.190E-06	2.407E-04	20	E009	1 0 0 1 2	
9.394E-07	1.900E-04	20	H300	1 1 2 2 1	
8.850E-07	1.790E-04	20	V416	0 0 0 0 0	
5.933E-07	1.200E-04	24	H116	2 1 0 0 2	
1.000E-06	2.023E-04	24.60	M082	1 1 1 2 2	
1.000E-06	2.023E-04	24.60	M151	2 1 2 2 2	
1.003E-06	2.028E-04	24.64	M183	1 2 1 1 2	
1.400E-06	2.832E-04	25	A325	2 1 2 2 1	
1.023E-06	2.070E-04	25	D406	1 2 2 2 2	
1.320E-06	2.670E-04	25	K001	2 2 2 2 2	
1.335E-06	2.700E-04	25	L332	1 1 1 1 2	
1.285E-06	2.600E-04	25	M064	1 1 2 2 1	
1.019E-06	2.060E-04	25	M071	2 2 2 2 2	
1.300E-06	2.629E-04	25	M342	1 0 1 1 1	
1.167E-06	2.360E-04	25	S227	1 2 1 1 2	
1.019E-06	2.060E-04	25.00	M151	2 1 1 2 2	
1.187E-06	2.400E-04	27	D003	1 0 0 1 1	
1.305E-06	2.640E-04	29	M071	2 2 2 2 2	
1.305E-06	2.640E-04	29.00	M151	2 1 1 2 2	
1.380E-06	2.791E-04	29.90	M082	1 1 1 2 2	
1.380E-06	2.791E-04	29.90	M151	2 1 2 2 2	
1.382E-06	2.796E-04	29.94	M183	1 2 1 1 2	
2.947E-06	5.960E-04	40	V416	0 0 0 0 0	
8.464E-06	1.712E-03	60	V416	0 0 0 0 0	
1.300E-06	2.630E-04	ns	I332	0 0 0 0 1	

3564. $C_{16}H_{10}$
Pyrene
Benzo[def]phenanthrene

RN: 129-00-0   **MP** (°C):   156
MW: 202.26   **BP** (°C):   404

Solubility (Moles/L)	Solubility (Grams/L)	Temp (°C)	Ref (#)	Evaluation (T P E A A)	Comments
<1.00E-07	<2.02E-05	4	K049	1 2 1 1 0	
2.430E-07	4.915E-05	4.70	M082	1 1 1 2 2	
2.430E-07	4.915E-05	4.70	M151	2 1 2 2 2	
2.434E-07	4.924E-05	4.74	M183	1 2 1 1 2	
2.890E-07	5.845E-05	9.50	M082	1 1 1 2 2	
2.890E-07	5.845E-05	9.50	M151	2 1 2 2 2	
2.895E-07	5.855E-05	9.54	M183	1 2 1 1 2	
3.560E-07	7.200E-05	14.30	M082	1 1 1 2 2	
3.560E-07	7.200E-05	14.30	M151	2 1 2 2 2	
3.563E-07	7.206E-05	14.34	M183	1 2 1 1 2	

(continued)

3564. C₁₆H₁₀ (continued)

Solubility (Moles/L)	Solubility (Grams/L)	Temp (°C)	Ref (#)	Evaluation (T P E A A)	Comments
3.588E-07	7.258E-05	15	B385	0 0 0 0 0	
4.610E-07	9.324E-05	18.70	M082	1 1 1 2 2	
4.610E-07	9.324E-05	18.70	M151	2 1 2 2 2	
4.617E-07	9.338E-05	18.74	M183	1 2 1 1 2	
5.200E-07	1.052E-04	20	E009	1 0 0 0 1	
5.200E-07	1.052E-04	20	E025	1 0 1 2 1	
4.700E-07	9.506E-05	20	H306	1 0 1 2 1	
5.370E-07	1.086E-04	21.20	M082	1 1 1 2 2	
5.370E-07	1.086E-04	21.20	M151	2 1 2 2 2	
5.394E-07	1.091E-04	21.24	M183	1 2 1 1 2	
7.200E-07	1.456E-04	22	A413	2 0 2 2 1	
6.279E-07	1.270E-04	22.20	W003	2 1 2 2 2	average of 3
6.675E-07	1.350E-04	24	H106	1 0 2 2 2	
1.582E-07	3.200E-05	24	H116	2 1 0 0 1	
6.675E-07	1.350E-04	24	M129	1 2 1 1 2	
5.834E-07	1.180E-04	25	B319	2 0 1 2 2	
6.490E-07	1.313E-04	25	B385	0 0 0 0 0	
7.700E-07	1.557E-04	25	K001	1 0 2 1 2	
4.700E-07	9.506E-05	25	K123	1 0 2 2 1	
7.911E-07	1.600E-04	25	L332	1 1 1 1 2	
6.675E-07	1.350E-04	25	M064	1 1 2 2 2	
6.526E-07	1.320E-04	25	M071	2 2 2 2 2	
6.675E-07	1.350E-04	25	M156	1 2 1 1 2	
6.670E-07	1.349E-04	25	M342	1 0 1 1 2	
3.955E-07	8.000E-05	25	P340	0 0 0 0 0	
3.556E-08	7.191E-06	25	R084	2 2 2 2 1	*sic*
7.400E-07	1.497E-04	25	R302	1 2 1 2 1	
8.455E-07	1.710E-04	25	S227	1 2 1 1 2	
6.526E-07	1.320E-04	25.00	M151	2 1 1 2 2	
6.730E-07	1.361E-04	25.50	M082	1 1 1 2 2	
6.730E-07	1.361E-04	25.50	M151	2 1 2 2 2	
6.728E-07	1.361E-04	25.54	M183	1 2 1 1 2	
8.158E-07	1.650E-04	27	D003	1 0 0 1 1	
8.010E-07	1.620E-04	29	M071	2 2 2 2 2	
8.010E-07	1.620E-04	29.00	M151	2 1 1 2 2	
8.390E-07	1.697E-04	29.90	M082	1 1 1 2 2	
8.390E-07	1.697E-04	29.90	M151	2 1 2 2 2	
8.411E-07	1.701E-04	29.94	M183	1 2 1 1 2	
1.147E-06	2.320E-04	34.50	W003	2 1 2 2 2	average of 2
9.888E-07	2.000E-04	35	B385	0 0 0 0 0	
1.973E-06	3.990E-04	44.70	W003	2 1 2 2 2	average of 3
2.784E-06	5.630E-04	50.10	W003	2 1 2 2 2	average of 3
3.758E-06	7.600E-04	55.60	W003	2 1 2 2 1	average of 3
3.659E-06	7.400E-04	56.00	W003	2 1 2 2 1	
4.648E-06	9.400E-04	60.70	W003	2 1 2 2 1	average of 3
6.329E-06	1.280E-03	65.20	W003	2 1 2 2 2	average of 2
9.196E-06	1.860E-03	71.90	W003	2 1 2 2 2	average of 3
1.093E-05	2.210E-03	74.70	W003	2 1 2 2 2	
6.675E-07	1.350E-04	ns	H123	0 0 0 0 0	

(continued)

3564. C$_{16}$H$_{10}$ (continued)

Solubility (Moles/L)	Solubility (Grams/L)	Temp (°C)	Ref (#)	Evaluation (T P E A A)	Comments
6.675E-07	1.350E-04	ns	K304	· 0 0 0 0 2	
6.675E-07	1.350E-04	ns	M344	0 0 0 0 2	
5.000E-07	1.011E-04	ns	M383	0 2 1 1 0	
1.000E-06	2.023E-04	ns	W005	0 0 1 2 0	

3565. C$_{16}$H$_{10}$N$_2$O$_8$S$_2$
C.I. Acid blue 74(free acid)
Indigo-disulfosaeure-(5,5′)
Indigotinsulfonic acid

RN:	860-22-0	**MP** (°C):		
MW:	422.39	**BP** (°C):		

Solubility (Moles/L)	Solubility (Grams/L)	Temp (°C)	Ref (#)	Evaluation (T P E A A)	Comments
~2.37E-02	~1.00E+01	25	F300	1 0 0 0 0	

3566. C$_{16}$H$_{11}$NO$_2$
Cinchophen
2-Phenyl-4-quinolinecarboxylic acid
2-Phenylcinchoninic acid

RN:	132-60-5	**MP** (°C):	213	
MW:	249.27	**BP** (°C):		

Solubility (Moles/L)	Solubility (Grams/L)	Temp (°C)	Ref (#)	Evaluation (T P E A A)	Comments
6.418E-04	1.600E-01	25	L074	2 2 1 1 2	

3567. C$_{16}$H$_{12}$F$_3$NO
6H-Dibenz[b,e]azepin-6-one, 5,11-dihydro-5-(2,2,2-trifluoroethyl)-

RN:	155206-49-8	**MP** (°C):		
MW:	291.28	**BP** (°C):		

Solubility (Moles/L)	Solubility (Grams/L)	Temp (°C)	Ref (#)	Evaluation (T P E A A)	Comments
1.589E-05	4.627E-03	ns	M381	0 1 1 1 2	pH 7.0

3568. C$_{16}$H$_{12}$N$_2$O$_3$
5,5-Diphenylbarbituric acid
2,4,6(1H,3H,5H)-Pyrimidinetrione, 5,5-diphenyl
Barbituric acid, 5,5-diphenyl
5,5-Diphenylbarbiturate

RN:	21914-07-8	**MP** (°C):		
MW:	280.29	**BP** (°C):		

Solubility (Moles/L)	Solubility (Grams/L)	Temp (°C)	Ref (#)	Evaluation (T P E A A)	Comments
6.370E-05	1.785E-02	25	P350	0 0 0 0 0	intrinsic

3569. $C_{16}H_{12}N_2O_4S$

Sulfanaphthoquinone

RN: **MP** (°C):
MW: 328.35 **BP** (°C):

Solubility (Moles/L)	Solubility (Grams/L)	Temp (°C)	Ref (#)	Evaluation (T P E A A)	Comments
1.370E-04	4.500E-02	20	F073	1 2 2 2 1	

3570. $C_{16}H_{12}O_6$

Hematein

Haematein

Benz[b]indeno[1,2-d]pyran-9(6H)-one, 6α,7-dihydro-3,4,6α,10-tetrahydroxy-

RN: 475-25-2 **MP** (°C): >200
MW: 300.27 **BP** (°C):

Solubility (Moles/L)	Solubility (Grams/L)	Temp (°C)	Ref (#)	Evaluation (T P E A A)	Comments
1.998E-03	6.000E-01	20	F300	1 0 0 0 1	

3571. $C_{16}H_{12}O_6$

Benzoic acid, 2-(acetyloxy)-, 2-carboxyphenyl ester

RN: 530-75-6 **MP** (°C): 166.5
MW: 300.27 **BP** (°C):

Solubility (Moles/L)	Solubility (Grams/L)	Temp (°C)	Ref (#)	Evaluation (T P E A A)	Comments
6.661E-05	2.000E-02	21	N335	0 0 0 0 0	

3572. $C_{16}H_{13}ClN_2O$

Diazepam

7-Chloro-1-methyl-5-phenyl-2H-1,4-benzodiazepin-2-one

Valium

Valrelease

Vazepam

Diazemuls

RN: 439-14-5 **MP** (°C): 125
MW: 284.75 **BP** (°C):

Solubility (Moles/L)	Solubility (Grams/L)	Temp (°C)	Ref (#)	Evaluation (T P E A A)	Comments
1.475E-04	4.200E-02	20	N059	2 0 2 2 2	average of 2
1.756E-04	5.000E-02	25	G084	2 0 2 2 1	
1.756E-04	5.000E-02	25	G095	2 1 2 2 1	
1.756E-04	5.000E-02	25	M159	1 0 2 2 0	EFG, pH 7.0
2.320E-04	6.606E-02	25	M320	2 2 1 1 2	
1.089E-04	3.100E-02	25	M457	0 0 0 0 0	
1.510E-04	4.300E-02	25	N055	2 0 2 2 1	
1.580E-04	4.500E-02	25	N055	2 0 2 1 2	

(continued)

3572. $C_{16}H_{13}ClN_2O$ (continued)

Solubility (Moles/L)	Solubility (Grams/L)	Temp (°C)	Ref (#)	Evaluation (T P E A A)	Comments
1.721E-04	4.900E-02	25	N055	2 0 2 1 2	
1.405E-04	4.000E-02	30	R081	1 2 2 2 0	
2.900E-04	8.258E-02	50	M335	1 0 2 1 2	pH 6.0
1.200E-04	3.417E-02	ns	F327	0 0 1 2 2	
3.512E-05	1.000E-02	ns	K444	0 0 0 0 0	
1.756E-04	5.000E-02	ns	M036	0 0 0 0 0	

3573. $C_{16}H_{13}Cl_2NO_4$
Aceclofenac

RN: 89796-99-6   **MP** (°C):
MW: 354.19   **BP** (°C):

Solubility (Moles/L)	Solubility (Grams/L)	Temp (°C)	Ref (#)	Evaluation (T P E A A)	Comments
9.034E-05	3.200E-02	32	C411	2 1 1 2 1	

3574. $C_{16}H_{13}I_3N_2O_3$
Iobenzamic acid
N-(3-Amino-2,4,6-triiodobenzoyl)-N-phenyl-β-alanine
Orbil
Osbiland
Razebil
Osbil

RN: 3115-05-7   **MP** (°C):
MW: 662.01   **BP** (°C):

Solubility (Moles/L)	Solubility (Grams/L)	Temp (°C)	Ref (#)	Evaluation (T P E A A)	Comments
1.737E-04	1.150E-01	ns	H055	0 0 0 0 0	

3575. $C_{16}H_{13}NO_3$
C.I. Disperse red 3
N-(2-Hydroxyethyl)-1-aminoanthraquinone
Disperse red 3
Disperse red 66

RN: 4465-58-1   **MP** (°C):   168
MW: 267.29   **BP** (°C):

Solubility (Moles/L)	Solubility (Grams/L)	Temp (°C)	Ref (#)	Evaluation (T P E A A)	Comments
1.600E-05	4.277E-03	25	B333	0 0 0 0 0	

3576. $C_{16}H_{13}N_3$
Yellow AB
1-Phenylazo-2-naphthylamine
RN: 85-84-7   **MP** (°C): 102
MW: 247.30   **BP** (°C):

Solubility (Moles/L)	Solubility (Grams/L)	Temp (°C)	Ref (#)	Evaluation (T P E A A)	Comments
1.213E-06	3.000E-04	37	H120	1 1 1 1 0	normal saline

3577. $C_{16}H_{13}N_3O_3$
Mebendazole
Methyl 5-benzoyl benzimidazole-2-carbamate
Pantelmin
Methyl 5-benzoyl-2-benzimidazolecarbamate
RN: 31431-39-7   **MP** (°C): 288.5
MW: 295.30   **BP** (°C):

Solubility (Moles/L)	Solubility (Grams/L)	Temp (°C)	Ref (#)	Evaluation (T P E A A)	Comments
1.693E-06	5.000E-04	21	N337	0 0 0 0 0	pH 5
1.700E-06	5.020E-04	21	N337	0 0 0 0 0	pH 5
1.199E-04	3.540E-02	25	H075	1 0 2 1 2	polymorph C
2.414E-04	7.130E-02	25	H075	1 0 2 1 2	polymorph B
3.332E-05	9.840E-03	25	H075	1 0 2 1 2	polymorph A
3.725E-06	1.100E-03	288.5	D426	0 0 0 0 0	
1.318E-04	3.893E-02	ns	R427	0 0 0 0 0	

3578. $C_{16}H_{14}$
9,10-Dimethylanthracene
RN: 781-43-1   **MP** (°C): 182
MW: 206.29   **BP** (°C):

Solubility (Moles/L)	Solubility (Grams/L)	Temp (°C)	Ref (#)	Evaluation (T P E A A)	Comments
2.715E-07	5.600E-05	25	M064	1 1 2 2 1	
2.700E-07	5.570E-05	25	M342	1 0 1 1 1	
2.715E-07	5.600E-05	ns	M344	0 0 0 0 2	

3579. C$_{16}$H$_{14}$ClN$_3$O

Chlordiazepoxide

7-Chloro-2-(methylamino)-5-phenyl-3H-1,4-benzodiazepine-4-oxide

Librium

Menrium

Tropium

SK-Lygen

RN: 58-25-3   **MP** (°C):   236

MW: 299.76   **BP** (°C):

Solubility (Moles/L)	Solubility (Grams/L)	Temp (°C)	Ref (#)	Evaluation (T P E A A)	Comments
6.607E-03	1.981E+00	ns	R427	0 0 0 0 0	
6.672E-03	2.000E+00	rt	M035	0 0 0 0 0	

3580. C$_{16}$H$_{14}$Cl$_2$N$_2$O$_2$

Phenobenzuron

Benzoyl-1-(3,4-dichlorophenyl)-3,3-dimethylurea

Benzomarc

Urea, N-benzoyl-N-(3,4-dichlorophenyl)-N',N'-dimethyl-

RN: 3134-12-1   **MP** (°C):   119

MW: 337.21   **BP** (°C):

Solubility (Moles/L)	Solubility (Grams/L)	Temp (°C)	Ref (#)	Evaluation (T P E A A)	Comments
4.745E-05	1.600E-02	22	M161	1 0 0 0 1	

3581. C$_{16}$H$_{14}$Cl$_2$O$_3$

Chlorobenzilate

Ethyl 4,4'-dichlorobenzilate

Acaraben

Benzilen

Folbex

Kopmite

RN: 510-15-6   **MP** (°C):   36

MW: 325.19   **BP** (°C):   157

Solubility (Moles/L)	Solubility (Grams/L)	Temp (°C)	Ref (#)	Evaluation (T P E A A)	Comments
3.998E-05	1.300E-02	20	F311	1 2 2 2 1	

3582. C$_{16}$H$_{14}$Cl$_2$O$_4$

Diclotop-methyl

Methyl (+/−)-2-[4-(2,4-dichlorophenoxy)phenoxy]propionate

RN: 51338-27-3   **MP** (°C):   40

MW: 341.19   **BP** (°C):

Solubility (Moles/L)	Solubility (Grams/L)	Temp (°C)	Ref (#)	Evaluation (T P E A A)	Comments
1.465E-04	5.000E-02	22	M161	1 0 0 0 1	

3583. C$_{16}$H$_{14}$FNO
6H-Dibenz[b,e]azepin-6-one, 5-(2-fluoroethyl)-5,11-dihydro-
RN: 155206-48-7 **MP** (°C):
MW: 255.29 **BP** (°C):

Solubility (Moles/L)	Solubility (Grams/L)	Temp (°C)	Ref (#)	Evaluation (T P E A A)	Comments
2.917E-04	7.448E-02	ns	M381	0 1 1 1 2	pH 7.0

3584. C$_{16}$H$_{14}$N$_2$O
Methaqualone
Quaalude
Mandrax
Somnafac
RN: 72-44-6 **MP** (°C): 114–117
MW: 250.30 **BP** (°C):

Solubility (Moles/L)	Solubility (Grams/L)	Temp (°C)	Ref (#)	Evaluation (T P E A A)	Comments
1.198E-03	2.999E-01	23	P094	0 0 0 0 0	

3585. C$_{16}$H$_{14}$N$_2$O$_2$
C.I. Disperse blue 14
9,10-Anthracenedione, 1,4-bis(methylamino)-
RN: 2475-44-7 **MP** (°C): 226
MW: 266.30 **BP** (°C):

Solubility (Moles/L)	Solubility (Grams/L)	Temp (°C)	Ref (#)	Evaluation (T P E A A)	Comments
1.400E-07	3.728E-05	25	B333	0 0 0 0 0	

3586. C$_{16}$H$_{14}$N$_2$O$_3$
3-(Hydroxymethyl)phenytoin
3-(Hydroxymethyl)-5,5-diphenyl-2,4-imidazolidinedione
RN: 21616-46-6 **MP** (°C):
MW: 282.30 **BP** (°C):

Solubility (Moles/L)	Solubility (Grams/L)	Temp (°C)	Ref (#)	Evaluation (T P E A A)	Comments
4.959E-04	1.400E-01	22	B154	1 1 1 1 1	0.1M HCl

3587. C₁₆H₁₄N₂O₄

C.I. Disperse blue 26
9,10-Anthracenedione, 1,5-dihydroxy-4,8-bis(methylamino)-
Resiren blue TG
Navilene blue GL
PTB 31

RN:	3860-63-7	**MP** (°C):	217
MW:	298.30	**BP** (°C):	

Solubility (Moles/L)	Solubility (Grams/L)	Temp (°C)	Ref (#)	Evaluation (T P E A A)	Comments
6.800E-08	2.028E-05	25	B333	0 0 0 0 0	

3588. C₁₆H₁₄O₃

Ketoprofen
2-(*meta*-Benzoylphenyl) propionic acid
Orudis
Alrheumat
Oruvail

RN:	22071-15-4	**MP** (°C):	94
MW:	254.29	**BP** (°C):	

Solubility (Moles/L)	Solubility (Grams/L)	Temp (°C)	Ref (#)	Evaluation (T P E A A)	Comments
2.509E-04	6.380E-02	5	F306	1 0 1 2 2	intrinsic
9.045E-04	2.300E-01	21	B331	1 2 2 1 1	pH 7.4
3.696E-04	9.399E-02	22.5	B422	2 0 2 2 2	
4.640E-04	1.180E-01	25	A408	2 0 1 2 0	int
2.006E-04	5.100E-02	25	A427	0 0 0 0 0	
5.646E-04	1.436E-01	25	F306	1 0 1 2 2	intrinsic
1.156E-03	2.939E-01	32	C411	2 1 1 2 1	
8.066E-04	2.051E-01	37	F306	1 0 1 2 2	intrinsic
5.112E-04	1.300E-01	37	Y421	0 0 0 0 0	
3.933E-05	1.000E-02	amb	L434	0 0 0 0 0	
2.006E-04	5.100E-02	rt	H302	0 0 2 1 2	intrinsic
8.219E-04	2.090E-01	rt	R431	0 0 0 0 0	Average

3589. C₁₆H₁₄O₃

Fenbufen
3-(4-Biphenylylcarbonyl) propionic acid
Lederfen

RN:	36330-85-5	**MP** (°C):	185
MW:	254.29	**BP** (°C):	

Solubility (Moles/L)	Solubility (Grams/L)	Temp (°C)	Ref (#)	Evaluation (T P E A A)	Comments
3.700E-06	9.409E-04	5	F306	1 0 1 2 2	intrinsic
1.000E-05	2.543E-03	24.99	K447	0 0 0 0 0	pH 2.0
6.430E-05	1.635E-02	25	C314	0 0 0 0 0	
6.410E-05	1.630E-02	25	C314	0 0 0 0 0	
8.700E-06	2.212E-03	25	F301	1 1 0 0 1	pH 2.0, *sic*

(continued)

3589. $C_{16}H_{14}O_3$ (continued)

Solubility (Moles/L)	Solubility (Grams/L)	Temp (°C)	Ref (#)	Evaluation (T P E A A)	Comments
8.700E-06	2.212E-03	25	F306	1 0 1 2 2	intrinsic
1.800E-05	4.577E-03	37	F306	1 0 1 2 2	intrinsic
7.865E-06	2.000E-03	rt	H302	0 0 2 1 1	intrinsic

3590. $C_{16}H_{15}ClN_2$
Medazepam
7-Chloro-2,3-dihydro-1-methyl-5-phenyl-1H-1,4-benzodiazepine
Nobrium

RN:	2898-12-6	**MP (°C):**
MW:	270.76	**BP (°C):**

Solubility (Moles/L)	Solubility (Grams/L)	Temp (°C)	Ref (#)	Evaluation (T P E A A)	Comments
4.000E-05	1.083E-02	37	L011	1 0 2 1 1	

3591. $C_{16}H_{15}Cl_2NO_2$
Clomeprop
2-(2,4-Dichloro-3-methylphenoxy)-N-phenylpropanamide

RN:	84496-56-0	**MP (°C):**
MW:	324.21	**BP (°C):**

Solubility (Moles/L)	Solubility (Grams/L)	Temp (°C)	Ref (#)	Evaluation (T P E A A)	Comments
9.772E-08	3.168E-05	ns	R427	0 0 0 0 0	

3592. $C_{16}H_{15}Cl_3OS_2$
2-(p-Methylthiophenyl)-2-(p-methylsulfinylphenyl)-1,1,1-trichloroethane

RN:	28463-05-0	**MP (°C):**	133-136
MW:	393.78	**BP (°C):**	

Solubility (Moles/L)	Solubility (Grams/L)	Temp (°C)	Ref (#)	Evaluation (T P E A A)	Comments
3.174E-06	1.250E-03	ns	K117	0 1 2 1 1	

3593. $C_{16}H_{15}Cl_3O_2$
Methoxychlor
1,1'-(2,2,2-Trichloroethylidene)-bis[4-methoxybenzene]
Maralate
Methoxy DDT
Marlate
Chemform

RN:	72-43-5	**MP (°C):**	82.5
MW:	345.66	**BP (°C):**	

Solubility (Moles/L)	Solubility (Grams/L)	Temp (°C)	Ref (#)	Evaluation (T P E A A)	Comments
5.786E-08	2.000E-05	15	B083	2 2 1 2 1	particle size 5 µm
1.302E-07	4.500E-05	25	B083	2 2 1 2 1	particle size 5 µm

(continued)

3593. C₁₆H₁₅Cl₃O₂ (continued)

Solubility (Moles/L)	Solubility (Grams/L)	Temp (°C)	Ref (#)	Evaluation (T P E A A)	Comments
1.447E-07	5.000E-05	25	P085	0 0 0 0 0	
2.893E-07	1.000E-04	25	W025	1 0 2 2 2	
2.748E-07	9.500E-05	35	B083	2 2 1 2 1	particle size 5 μm
5.352E-07	1.850E-04	45	B083	2 2 1 2 2	particle size 5 μm
1.794E-06	6.200E-04	ns	K117	0 1 2 1 1	
8.679E-09	3.000E-06	ns	K138	0 0 0 0 2	
2.314E-06	8.000E-04	ns	M110	0 0 0 0 0	EFG
1.794E-06	6.200E-04	ns	M138	0 1 0 0 1	
3.472E-07	1.200E-04	ns	M344	0 0 0 0 1	
2.089E-07	7.222E-05	ns	R427	0 0 0 0 0	

3594. C₁₆H₁₅Cl₃O₂S₂

2,2-bis(p-Methylsulfinylphenyl)-1,1,1-trichloroethane
2-(p-Methylsulfoxidephenyl)-1,1,1-trichloroethane

RN: 28396-87-4   **MP** (°C):   150–153
MW: 409.78   **BP** (°C):

Solubility (Moles/L)	Solubility (Grams/L)	Temp (°C)	Ref (#)	Evaluation (T P E A A)	Comments
7.077E-05	2.900E-02	ns	K117	0 1 2 1 1	

3595. C₁₆H₁₅Cl₃O₄S₂

2,2-bis(p-Methylsulfonylphenyl)-1,1,1-trichloroethane

RN: 30665-94-2   **MP** (°C):   236.0
MW: 441.78   **BP** (°C):

Solubility (Moles/L)	Solubility (Grams/L)	Temp (°C)	Ref (#)	Evaluation (T P E A A)	Comments
3.395E-06	1.500E-03	ns	K117	0 1 2 1 1	

3596. C₁₆H₁₅Cl₃S₂

2,2-*bis*-(p-Methylthiophenyl)-1,1,1-trichloroethane

RN: 19679-38-0   **MP** (°C):   115-117
MW: 377.78   **BP** (°C):

Solubility (Moles/L)	Solubility (Grams/L)	Temp (°C)	Ref (#)	Evaluation (T P E A A)	Comments
1.509E-06	5.700E-04	ns	K117	0 1 2 1 1	

3597. C₁₆H₁₅FN₂O₅

1-Butyryloxymethyl-3-benzoyl-5-fluoro-2,4(1H,3H)-pyrimidinedi-one
1-Butyryloxymethyl-3-benzoyl-5-fluorouracil

RN: 97108-48-0   **MP** (°C):   81–82
MW: 334.31   **BP** (°C):

Solubility (Moles/L)	Solubility (Grams/L)	Temp (°C)	Ref (#)	Evaluation (T P E A A)	Comments
1.855E-04	6.200E-02	22	B321	0 0 0 0 0	pH 4.0

3598. C$_{16}$H$_{15}$NO
4-Cyano-4′-propyloxybiphenyl
3 COB

RN: 52709-86-1   **MP** (°C):
MW: 237.30   **BP** (°C):

Solubility (Moles/L)	Solubility (Grams/L)	Temp (°C)	Ref (#)	Evaluation (T P E A A)	Comments
9.000E-07	2.136E-04	21	D300	2 2 1 1 2	

3599. C$_{16}$H$_{15}$NO$_2$
N-Butyl-1,8-naphthalimide
Naphthalimide, N-butyl-
1H-Benz[de]isoquinoline-1,3(2H)-dione, 2-butyl-

RN: 6914-62-1   **MP** (°C): 95
MW: 253.30   **BP** (°C): 412.3

Solubility (Moles/L)	Solubility (Grams/L)	Temp (°C)	Ref (#)	Evaluation (T P E A A)	Comments
2.000E-05	5.066E-03	23	B410	2 1 2 2 2	

3600. C$_{16}$H$_{15}$NO$_2$
Cinnamyl anthranilate
2-Propen-1-ol, 3-phenyl-, 2-aminobenzoate
2-Aminobenzoic acid 3-phenyl-2-propenyl ester
3-Phenyl-2-propen-1-yl anthranilate
3-Phenyl-2-propenyl 2-aminobenzoate
Cinnamyl alcohol

RN: 87-29-6   **MP** (°C): 60
MW: 253.30   **BP** (°C): 332

Solubility (Moles/L)	Solubility (Grams/L)	Temp (°C)	Ref (#)	Evaluation (T P E A A)	Comments
9.080E-07	2.300E-04	ns	B338	0 0 0 0 1	

3601. C$_{16}$H$_{15}$NO$_3$
Benzoylphenylalanine
N-Benzoyl-DL-phenylalanine

RN: 2901-76-0   **MP** (°C):
MW: 269.30   **BP** (°C):

Solubility (Moles/L)	Solubility (Grams/L)	Temp (°C)	Ref (#)	Evaluation (T P E A A)	Comments
3.156E-03	8.500E-01	25.1	N026	0 0 0 0 0	

3602. C₁₆H₁₅NO₄

Benzoyltyrosine
N-Benzoyl-L-tyrosine

RN:	2566-23-6	**MP** (°C):	
MW:	285.30	**BP** (°C):	

Solubility (Moles/L)	Solubility (Grams/L)	Temp (°C)	Ref (#)	Evaluation (T P E A A)	Comments
1.290E-02	3.680E+00	25.1	N026	0 0 0 0 0	

3603. C₁₆H₁₅N₅

6H-Dipyrido[3,2-b:2',3'-e][1,4]diazepin-6-nitrile, 11-cyclopropyl-5,11-dihydro-4-methyl

RN:		**MP** (°C):	
MW:	277.33	**BP** (°C):	

Solubility (Moles/L)	Solubility (Grams/L)	Temp (°C)	Ref (#)	Evaluation (T P E A A)	Comments
1.816E-05	5.035E-03	ns	M381	0 1 1 1 2	pH 7.0

3604. C₁₆H₁₅N₅O₄S

2,5-Disulfanilamidopyridine

RN:		**MP** (°C):	
MW:	373.39	**BP** (°C):	

Solubility (Moles/L)	Solubility (Grams/L)	Temp (°C)	Ref (#)	Evaluation (T P E A A)	Comments
1.326E-03	4.950E-01	37	R058	1 2 1 1 2	

3605. C₁₆H₁₆

1,2,3,6,7,8-Hexahydropyrene

RN:	1732-13-4	**MP** (°C):	133
MW:	208.31	**BP** (°C):	

Solubility (Moles/L)	Solubility (Grams/L)	Temp (°C)	Ref (#)	Evaluation (T P E A A)	Comments
1.100E-06	2.291E-04	4	K049	1 0 2 1 1	

3606. C₁₆H₁₆ClN₃O₃S

Metolazone
2-Methyl-3-(*o*-tolyl)-6-sulfamyl-7-chloro-1,2,3,4-tetrahydro-4-quinazolinone
Zaroxolyn
Mykrox
Diulo

RN:	17560-51-9	**MP** (°C):	256.0
MW:	365.84	**BP** (°C):	

Solubility (Moles/L)	Solubility (Grams/L)	Temp (°C)	Ref (#)	Evaluation (T P E A A)	Comments
9.321E-05	3.410E-02	10	B030	1 0 1 1 2	
1.339E-04	4.900E-02	20	B030	1 0 1 1 2	

(continued)

3606. $C_{16}H_{16}ClN_3O_3S$ (continued)

Solubility (Moles/L)	Solubility (Grams/L)	Temp (°C)	Ref (#)	Evaluation (T P E A A)	Comments
1.648E-04	6.030E-02	25	B030	1 0 1 1 2	
1.971E-04	7.210E-02	30	B030	1 0 1 1 2	
2.236E-04	8.180E-02	35	B030	1 0 1 1 2	
2.733E-04	1.000E-01	36	B030	1 0 1 1 2	
1.640E-04	6.000E-02	37	H013	1 0 0 0 0	
2.952E-04	1.080E-01	40	B030	1 0 1 1 2	
3.799E-04	1.390E-01	45	B030	1 0 1 1 2	
4.155E-04	1.520E-01	50	B030	1 0 1 1 2	

3607. $C_{16}H_{16}N_2$
3,4,7,8-Tetramethyl-1,10-phenanthroline
RN: 1660-93-1 **MP (°C):** 278.5
MW: 236.32 **BP (°C):**

Solubility (Moles/L)	Solubility (Grams/L)	Temp (°C)	Ref (#)	Evaluation (T P E A A)	Comments
6.400E-06	1.512E-03	25.04	B094	1 2 1 2 1	

3608. $C_{16}H_{16}N_2O_4$
Phenmedipham
Methyl *m*-hydroxycarbanilate *m*-methylcarbanilate
RN: 13684-63-4 **MP (°C):** 143
MW: 300.32 **BP (°C):**

Solubility (Moles/L)	Solubility (Grams/L)	Temp (°C)	Ref (#)	Evaluation (T P E A A)	Comments
<3.33E-05	<1.00E-02	20	B200	1 0 0 0 0	
3.330E-06	1.000E-03	20	F311	1 2 2 2 1	
1.397E-05	4.194E-03	25	H434	0 0 0 0 0	
3.330E-05	1.000E-02	ns	M061	0 0 0 0 1	
9.989E-06	3.000E-03	rt	M161	0 0 0 0 0	

3609. $C_{16}H_{16}N_2O_4$
Desmedipham
Ethyl *m*-hydroxycarbanilate carbanilate
Carbamic acid, *N*-phenyl-, 3-((ethoxycarbonyl)amino)phenyl ester
Betanex
Betanal-475
Betamix 70 WP
RN: 13684-56-5 **MP (°C):** 120
MW: 300.32 **BP (°C):**

Solubility (Moles/L)	Solubility (Grams/L)	Temp (°C)	Ref (#)	Evaluation (T P E A A)	Comments
2.331E-05	7.000E-03	rt	M161	0 0 0 0 0	
2.331E-05	7.000E-03	rt	R304	0 0 0 0 0	

3610. C$_{16}$H$_{16}$N$_4$
Disperse black 1
RN: 6054-48-4   **MP** (°C):
MW: 264.33   **BP** (°C):

Solubility (Moles/L)	Solubility (Grams/L)	Temp (°C)	Ref (#)	Evaluation (T P E A A)	Comments
3.000E-07	7.930E-05	25	B333	0 0 0 0 0	

3611. C$_{16}$H$_{16}$N$_4$O
6H-Dipyrido[3,2-b:2′3′-e][1,4]diazepin-6-one, 11-cyclopropyl-5,11-dihydro-2,4-dimethyl-
RN: 135794-77-3   **MP** (°C):
MW: 280.33   **BP** (°C):

Solubility (Moles/L)	Solubility (Grams/L)	Temp (°C)	Ref (#)	Evaluation (T P E A A)	Comments
5.346E-05	1.499E-02	ns	M381	0 1 1 1 2	pH 7.0

3612. C$_{16}$H$_{16}$N$_4$O
6H-Dipyrido[3,2-b:2′,3′-e][1,4]diazepin-6-one, 11-cyclobutyl-5,11-dihydro-5-methyl-
RN: 135794-88-6   **MP** (°C):
MW: 280.33   **BP** (°C):

Solubility (Moles/L)	Solubility (Grams/L)	Temp (°C)	Ref (#)	Evaluation (T P E A A)	Comments
2.911E-04	8.160E-02	ns	M381	0 1 1 1 2	pH 7.0

3613. C$_{16}$H$_{16}$N$_6$O$_4$S
2,5-Disulfanilamidopyrimidine
RN: **MP** (°C):
MW: 388.41   **BP** (°C):

Solubility (Moles/L)	Solubility (Grams/L)	Temp (°C)	Ref (#)	Evaluation (T P E A A)	Comments
5.664E-05	2.200E-02	37	R046	1 2 1 1 1	

3614. C$_{16}$H$_{16}$O$_2$
4-Methoxy-3,3′-dimethylbenzophenone
RN: 41295-28-7   **MP** (°C):   62.25
MW: 240.30   **BP** (°C):

Solubility (Moles/L)	Solubility (Grams/L)	Temp (°C)	Ref (#)	Evaluation (T P E A A)	Comments
8.323E-06	2.000E-03	20	M161	1 0 0 0 0	

3615. C$_{16}$H$_{16}$O$_3$
Ethyl benzoyl benzoate

RN:	106396-19-4	MP (°C):
MW:	256.30	BP (°C):

Solubility (Moles/L)	Solubility (Grams/L)	Temp (°C)	Ref (#)	Evaluation (T P E A A)	Comments
3.901E-04	9.999E-02	ns	F014	0 0 0 0 1	

3616. C$_{16}$H$_{16}$O$_3$
Anisyl phenylacetate
p-Methoxybenzyl phenylacetate
Phenylacetic acid, *p*-methoxybenzyl ester

RN:	102-17-0	MP (°C):
MW:	256.30	BP (°C):

Solubility (Moles/L)	Solubility (Grams/L)	Temp (°C)	Ref (#)	Evaluation (T P E A A)	Comments
2.000E-03	5.126E-01	25	D407	1 0 2 2 2	
2.000E-03	5.126E-01	ns	S460	0 0 0 0 0	

3617. C$_{16}$H$_{17}$ClN$_2$S
Chlorphenethazine
2-Chloro-*N*,*N*-dimethyl-10H-phenothiazine-10-ethanamide

RN:	2095-24-1	MP (°C):
MW:	304.84	BP (°C):

Solubility (Moles/L)	Solubility (Grams/L)	Temp (°C)	Ref (#)	Evaluation (T P E A A)	Comments
1.500E-05	4.573E-03	ns	G023	0 0 1 1 1	

3618. C$_{16}$H$_{17}$ClN$_4$O$_3$
C.I. Disperse red 13
4-Nitro-2-chloro-4′-[ethyl(2-hydroxyethyl)amino]azobenzene
Acetoquinone light rubine BLZ
Acetamine rubine B
Acetate fast rubine B

RN:	3180-81-2	MP (°C):	133
MW:	348.79	BP (°C):	

Solubility (Moles/L)	Solubility (Grams/L)	Temp (°C)	Ref (#)	Evaluation (T P E A A)	Comments
3.300E-08	1.151E-05	25	B333	0 0 0 0 0	

3619. C$_{16}$H$_{17}$ClN$_4$O$_4$

C.I. Disperse red 7

Ethanol, 2,2′-[[3-chloro-4-[(4-nitrophenyl)azo]phenyl]imino]*bis*-

RN: 4540-00-5   **MP** (°C):   190
MW: 364.79   **BP** (°C):

Solubility (Moles/L)	Solubility (Grams/L)	Temp (°C)	Ref (#)	Evaluation (T P E A A)	Comments
1.100E-06	4.013E-04	25	B333	0 0 0 0 0	

3620. C$_{16}$H$_{17}$NO

Diphenamid

Dyamid

Enide

N,N-Dimethyl-α-phenylbenzeneacetamide

N,N-Dimethyldiphenylacetamide

Diherbid

RN: 957-51-7   **MP** (°C):   132
MW: 239.32   **BP** (°C):

Solubility (Moles/L)	Solubility (Grams/L)	Temp (°C)	Ref (#)	Evaluation (T P E A A)	Comments
1.003E-03	2.399E-01	25	M061	1 0 0 0 1	
1.086E-03	2.600E-01	25	M161	1 0 0 0 2	
1.090E-03	2.609E-01	27	B200	1 0 0 0 2	
1.086E-03	2.600E-01	ns	B185	0 0 0 0 0	
2.079E-02	4.975E+00	ns	B200	0 0 0 0 0	
1.086E-03	2.600E-01	ns	H042	0 0 0 0 2	

3621. C$_{16}$H$_{17}$NO$_4$

2-Naphthaleneacetic acid, 6-methoxy-α-methyl-, 2-amino-2-oxoethyl ester, (*S*)

Naproxen, *N,N*-glycolamide ester

2-Naphthaleneacetic acid, 6-methoxy-α-methyl-, 2-amino-2-oxoethyl ester

Naproxen *N,N*-glycolamide ester

RN: 114665-17-7   **MP** (°C):   139.5
MW: 287.32   **BP** (°C):

Solubility (Moles/L)	Solubility (Grams/L)	Temp (°C)	Ref (#)	Evaluation (T P E A A)	Comments
1.183E-04	3.400E-02	21	B331	1 2 2 1·2	pH 7.4
1.183E-04	3.400E-02	21	B331	0 0 0 0 0	

3622. C$_{16}$H$_{17}$N$_3$O$_4$S

Cephalexin
Cefanex
C-Lexin
Keflex
Cefalexin

RN: 15686-71-2   **MP** (°C):
MW: 347.40   **BP** (°C):

Solubility (Moles/L)	Solubility (Grams/L)	Temp (°C)	Ref (#)	Evaluation (T P E A A)	Comments
1.724E-02	5.990E+00	10	O305	2 2 1 2 2	noncrystalline
1.569E-01	5.450E+01	15	O305	2 2 1 2 2	noncrystalline
1.416E-01	4.920E+01	20	O305	2 2 1 2 2	noncrystalline
3.598E-02	1.250E+01	25	P311	0 0 0 0 0	EFG
1.330E-02	4.620E+00	25	U001	0 0 0 0 0	
3.500E-03	1.216E+00	35	E311	0 0 0 0 0	

3623. C$_{16}$H$_{17}$N$_3$O$_4$S.H$_2$O

Cephalexin (monohydrate)

RN: 23325-78-2   **MP** (°C):
MW: 365.41   **BP** (°C):

Solubility (Moles/L)	Solubility (Grams/L)	Temp (°C)	Ref (#)	Evaluation (T P E A A)	Comments
3.694E-02	1.350E+01	25	M165	1 0 0 0 2	

3624. C$_{16}$H$_{17}$N$_5$O$_5$

Dis. A. 12
Ethanol, 2-[[4-[(2,4-dinitrophenyl)azo]phenyl]ethylamino]-

RN: 62570-20-1   **MP** (°C):
MW: 359.34   **BP** (°C):

Solubility (Moles/L)	Solubility (Grams/L)	Temp (°C)	Ref (#)	Evaluation (T P E A A)	Comments
2.000E-06	7.187E-04	25	B333	0 0 0 0 0	

3625. C$_{16}$H$_{17}$N$_5$O$_6$

Dis. A. 14
4-[bis(2-Hydroxyethyl)amino]-2',4'-dinitroazobenzene

RN: 60129-67-1   **MP** (°C):
MW: 375.34   **BP** (°C):

Solubility (Moles/L)	Solubility (Grams/L)	Temp (°C)	Ref (#)	Evaluation (T P E A A)	Comments
6.000E-06	2.252E-03	25	B333	0 0 0 0 0	

3626. $C_{16}H_{18}ClNO_4S$
Oxathiin carboxanilide
Benzoic acid, 2-chlloro-5-[[(5,6-dihydro-2-methyl-1,4-oxathiin-3-yl)carnonyl]amino]isopropyl
Ester

RN: 135812-04-3   **MP** (°C):   130
MW: 355.84   **BP** (°C):

Solubility (Moles/L)	Solubility (Grams/L)	Temp (°C)	Ref (#)	Evaluation (T P E A A)	Comments
3.653E-06	1.300E-03	25	O319	0 0 0 0 0	

3627. $C_{16}H_{18}FN_3O_3$
Norfloxacin
Noroxin

RN: 70458-96-7   **MP** (°C):
MW: 319.34   **BP** (°C):

Solubility (Moles/L)	Solubility (Grams/L)	Temp (°C)	Ref (#)	Evaluation (T P E A A)	Comments
8.455E-04	2.700E-01	6	Y421	0 0 0 0 0	
6.576E-04	2.100E-01	25	A414	1 0 1 1 1	pH 8.5 bicarbonate buffer (0.05 M)
6.263E-04	2.000E-01	25	A414	1 0 1 1 1	pH 7.4 phosphate buffer
2.505E-02	8.000E+00	25	A414	1 0 1 1 1	pH 5 citrate buffer (0.1 M)
5.950E-04	1.900E-01	25	A414	1 0 1 1 1	
1.159E-03	3.700E-01	25	Y421	0 0 0 0 0	
2.662E-03	8.500E-01	40	Y421	0 0 0 0 0	

3628. $C_{16}H_{18}NO_5P$
Diphenylmorpholidophosphate

RN: **MP** (°C):
MW: 335.30   **BP** (°C):

Solubility (Moles/L)	Solubility (Grams/L)	Temp (°C)	Ref (#)	Evaluation (T P E A A)	Comments
6.844E-03	2.295E+00	25	A040	1 0 0 0 2	

3629. $C_{16}H_{18}N_2O_3$
Difenoxuron
N-4-(4'-Methoxyphenoxy)phenyl-N',N'-dimethylurea
C-3470

RN: 14214-32-5   **MP** (°C):   138.5
MW: 286.33   **BP** (°C):

Solubility (Moles/L)	Solubility (Grams/L)	Temp (°C)	Ref (#)	Evaluation (T P E A A)	Comments
6.985E-05	2.000E-02	20	M161	1 0 0 0 1	
6.985E-05	2.000E-02	ns	M061	0 0 0 0 1	

3630. C₁₆H₁₈N₂O₄S

$3630.\ C_{16}H_{18}N_2O_4S$

Penicillin G
Benzylpenicillin
Pfizerpen

RN:	61-33-6	**MP** (°C):		
MW:	334.40	**BP** (°C):		

Solubility (Moles/L)	Solubility (Grams/L)	Temp (°C)	Ref (#)	Evaluation (T P E A A)	Comments
8.000E-03	2.675E+00	25	U001	0 0 0 0 0	

3631. C₁₆H₁₈N₄O

$3631.\ C_{16}H_{18}N_4O$

6H-Dipyrido[3,2-b:2′,3′-e][1,4]diazepin-6-one, 11-(1,1-dimethylethyl)-5,11-dihydro-5-methyl-

RN:	135794-80-8	**MP** (°C):		
MW:	282.35	**BP** (°C):		

Solubility (Moles/L)	Solubility (Grams/L)	Temp (°C)	Ref (#)	Evaluation (T P E A A)	Comments
1.416E-05	3.997E-03	ns	M381	0 1 1 1 2	pH 7.0

3632. C₁₆H₁₈N₄O₂

$3632.\ C_{16}H_{18}N_4O_2$

Dye III
4[[(4-Diethylamino)phenyl]azo]nitrobenzene

RN:		**MP** (°C):		
MW:	298.35	**BP** (°C):		

Solubility (Moles/L)	Solubility (Grams/L)	Temp (°C)	Ref (#)	Evaluation (T P E A A)	Comments
9.100E-07	2.715E-04	97.40	B198	1 2 1 1 1	

3633. C₁₆H₁₈N₄O₂

$3633.\ C_{16}H_{18}N_4O_2$

Dis. A. 5
4-Nitro-4′-diethylaminoazobenzene
4-Nitro-4′-N,N-diethylaminoazobenzene
DEANAB

RN:	3025-52-3	**MP** (°C):	152	
MW:	298.35	**BP** (°C):		

Solubility (Moles/L)	Solubility (Grams/L)	Temp (°C)	Ref (#)	Evaluation (T P E A A)	Comments
4.000E-11	1.193E-08	25	B333	0 0 0 0 0	

3634. C$_{16}$H$_{18}$N$_4$O$_3$
Disperse red 1
Dye IV
C.I. Disperse red 1
1-[N-Ethyl-N-(2-hydroxyethyl)amino]-4-(4-nitrophenylazo)benzene
4-Nitro-4′-[ethyl(2-hydroxyethyl)amino]azobenzene
RN: 2872-52-8   **MP** (°C):   161
MW: 314.35   **BP** (°C):

Solubility (Moles/L)	Solubility (Grams/L)	Temp (°C)	Ref (#)	Evaluation (T P E A A)	Comments
5.400E-07	1.697E-04	25	B333	0 0 0 0 0	
5.400E-06	1.697E-03	60	B198	1 2 1 1 1	
6.521E-06	2.050E-03	60	P313	0 0 0 0 0	average of 2
1.082E-05	3.400E-03	70	P313	0 0 0 0 0	average of 2
1.310E-05	4.118E-03	71.80	B198	1 2 1 1 2	
1.797E-05	5.650E-03	80	P313	0 0 0 0 0	average of 2
3.120E-05	9.808E-03	84.10	B198	1 2 1 1 2	
3.388E-05	1.065E-02	90	P313	0 0 0 0 0	average of 2
7.130E-05	2.241E-02	97.40	B198	1 2 1 1 2	

3635. C$_{16}$H$_{18}$N$_4$O$_4$
Disperse red 19
Dye V
C.I. Disperse red 19
2-[(2-Hydroxyethyl)[4-(4-nitrophenylazo)phenyl]amino]ethanol
4′-[(N,N-Dihydroxyethyl)amino]-4-nitroazobenzene
RN: 2734-52-3   **MP** (°C):   209
MW: 330.35   **BP** (°C):

Solubility (Moles/L)	Solubility (Grams/L)	Temp (°C)	Ref (#)	Evaluation (T P E A A)	Comments
7.100E-07	2.345E-04	25	B333	0 0 0 0 0	
1.170E-05	3.865E-03	60	B198	1 2 1 1 2	
3.030E-05	1.001E-02	71.80	B198	1 2 1 1 2	
8.330E-05	2.752E-02	84.10	B198	1 2 1 1 2	
2.100E-04	6.937E-02	97.40	B198	1 2 1 1 2	

3636. C$_{16}$H$_{18}$O$_3$
Naproxen ethyl esterv
2-Naphthaleneacetic acid, 6-methoxy-α-methyl-, ethyl ester, (alphaS)-
RN: 31220-35-6   **MP** (°C):
MW: 258.32   **BP** (°C):

Solubility (Moles/L)	Solubility (Grams/L)	Temp (°C)	Ref (#)	Evaluation (T P E A A)	Comments
4.645E-06	1.200E-03	21	B331	1 2 2 1 2	pH 7.4
4.645E-06	1.200E-03	21	B331	0 0 0 0 0	

3637. C$_{16}$H$_{19}$ClN$_2$
Chlorpheniramine
1-(*p*-Chlorophenyl)-1-(2-pyridyl)-3-dimethylaminopropane
RN: 132-22-9   **MP** (°C):   <25
MW: 274.80   **BP** (°C):   142

Solubility (Moles/L)	Solubility (Grams/L)	Temp (°C)	Ref (#)	Evaluation (T P E A A)	Comments
2.000E-02	5.496E+00	37.5	L034	2 2 0 1 2	pH 7.4

3638. C$_{16}$H$_{19}$NO$_7$
Benzoic acid, 2-(acetyloxy)-, 2-[(2-ethoxy-2-oxoethyl)methylamino]-2-oxoethyl ester
RN: 116482-77-0   **MP** (°C):   47.5
MW: 337.33   **BP** (°C):

Solubility (Moles/L)	Solubility (Grams/L)	Temp (°C)	Ref (#)	Evaluation (T P E A A)	Comments
2.846E-03	9.600E-01	21	N335	0 0 0 0 0	

3639. C$_{16}$H$_{19}$N$_3$O$_2$
C.I. Solvent yellow 58
p-[bis(2-Hydroxyethyl)amino]azobenzene
4-[bis(2-Hydroxyethyl)amino]azobenzene
RN: 2452-84-8   **MP** (°C):   134
MW: 285.35   **BP** (°C):

Solubility (Moles/L)	Solubility (Grams/L)	Temp (°C)	Ref (#)	Evaluation (T P E A A)	Comments
1.100E-04	3.139E-02	25	B333	0 0 0 0 0	

3640. C$_{16}$H$_{19}$N$_3$O$_4$S
Cephradine
Anspor
Velosef
RN: 38821-53-3   **MP** (°C):   140
MW: 349.41   **BP** (°C):

Solubility (Moles/L)	Solubility (Grams/L)	Temp (°C)	Ref (#)	Evaluation (T P E A A)	Comments
6.096E-02	2.130E+01	ns	F181	0 0 0 0 2	

3641. $C_{16}H_{19}N_3O_4S$

Ampicillin

$(2S,5R,6R)$-6-[(R)-2-Amino-2-phenylacetamido]-3,3-dimethyl-7-oxo-4-thia-1-azabicyclo[3.2.0]
  heptane-2-carboxylic acid

Aminobenzylpenicillin

Unasyn

Wymox

Totacillin

RN: 69-53-4   **MP** (°C):
MW: 349.41   **BP** (°C):

Solubility (Moles/L)	Solubility (Grams/L)	Temp (°C)	Ref (#)	Evaluation (T P E A A)	Comments
4.293E-02	1.500E+01	7.5	P009	1 0 2 1 0	EFG
3.721E-02	1.300E+01	20	P009	1 0 2 1 0	EFG
2.890E-02	1.010E+01	21	M044	2 0 2 2 2	
3.978E-02	1.390E+01	25	H051	1 2 2 2 2	
6.600E-03	2.306E+00	25	K444	0 0 0 0 0	
3.434E-02	1.200E+01	30	P009	1 0 2 1 0	EFG
3.291E-02	1.150E+01	40	P009	1 0 2 1 0	EFG

3642. $C_{16}H_{19}N_3O_4S.3H_2O$

Ampicillin (trihydrate)

RN: 7177-48-2   **MP** (°C):   198
MW: 403.46   **BP** (°C):

Solubility (Moles/L)	Solubility (Grams/L)	Temp (°C)	Ref (#)	Evaluation (T P E A A)	Comments
1.413E-02	5.700E+00	7.5	P009	1 0 2 1 0	EFG
1.487E-02	6.000E+00	20	P009	1 0 2 1 0	EFG
1.873E-02	7.558E+00	21	M044	2 0 2 2 2	
1.983E-02	8.000E+00	30	P009	1 0 2 1 0	EFG
2.479E-02	1.000E+01	40	P009	1 0 2 1 0	EFG

3643. $C_{16}H_{19}N_3O_5S$

Amoxicillin

RN: 61336-70-7   **MP** (°C):
MW: 365.41   **BP** (°C):

Solubility (Moles/L)	Solubility (Grams/L)	Temp (°C)	Ref (#)	Evaluation (T P E A A)	Comments
1.095E-02	4.000E+00	ns	K444	0 0 0 0 0	

3644. $C_{16}H_{19}N_3O_5S\cdot3H_2O$

Amoxicillin (trihydrate)

4-Thia-1-azabicyclo(3,2,0)heptane-2-carboxylic acid (trihydrate)

RN: 61336-70-7 **MP** (°C):

MW: 419.46 **BP** (°C):

Solubility (Moles/L)	Solubility (Grams/L)	Temp (°C)	Ref (#)	Evaluation (T P E A A)	Comments
~9.54E-03	~4.00E+00	ns	B188	0 0 0 0 0	

3645. $C_{16}H_{19}N_5O$

6H-Dipyrido[3,2-b:2′,3′-e][1,4]diazepin-6-one, 2-dimethylamino)-11-ethyl-5,11-dihydro-4-methyl-

RN: 135795-08-3 **MP** (°C):

MW: 297.36 **BP** (°C):

Solubility (Moles/L)	Solubility (Grams/L)	Temp (°C)	Ref (#)	Evaluation (T P E A A)	Comments
1.346E-05	4.002E-03	ns	M381	0 1 1 1 2	pH 7.0

3646. $C_{16}H_{19}N_5O_2$

6H-Dipyrido[3,2-b:2′,3′-e][1,4]diazepin-6-one, 11-ethyl-5,11-dihydro-2-[(2-hydroxyethyl) methylamino

RN: 155206-46-5 **MP** (°C):

MW: 313.36 **BP** (°C):

Solubility (Moles/L)	Solubility (Grams/L)	Temp (°C)	Ref (#)	Evaluation (T P E A A)	Comments
4.365E-04	1.368E-01	ns	M381	0 1 1 1 2	pH 7.0

3647. $C_{16}H_{19}O_4P$

Butyl diphenyl phosphate

RN: 2752-95-6 **MP** (°C):

MW: 306.30 **BP** (°C):

Solubility (Moles/L)	Solubility (Grams/L)	Temp (°C)	Ref (#)	Evaluation (T P E A A)	Comments
<6.53E-04	<2.00E-01	25	B070	1 2 0 1 0	

3648. $C_{16}H_{20}I_3N_3O_7$

1,3-Benzenedicarboxamide, N-(2-hydroxyethyl)-N′-[2-hydroxy-1-(hydroxymethyl)ethyl]-5-[(2-hydroxy-1-oxopropyl)amino]-2,4,6-triiodo-, (S)-

RN: 77868-44-1 **MP** (°C):

MW: 747.07 **BP** (°C):

Solubility (Moles/L)	Solubility (Grams/L)	Temp (°C)	Ref (#)	Evaluation (T P E A A)	Comments
2.625E-02	1.961E+01	25	P091	0 0 0 0 0	

3649. $C_{16}H_{20}I_3N_3O_7$

1,3-Benzenedicarboxamide, N-(2,3-dihydroxypropyl)-N'-(2-hydroxyethyl)-5-[(2-hydroxy-1-oxopropyl)amino]-2,4,6-triiodo-(RS)

RN: 77868-43-0   **MP** (°C):
MW: 747.07   **BP** (°C):

Solubility (Moles/L)	Solubility (Grams/L)	Temp (°C)	Ref (#)	Evaluation (T P E A A)	Comments
6.374E-02	4.762E+01	25	P091	0 0 0 0 0	

3650. $C_{16}H_{20}I_3N_3O_8$

1,3-Benzenedicarboxamide, N,N'-bis(2,3-dihydroxypropyl)-5S-[(hydroxyacetyl)amino]-2,4,6-triiodo- [RS-(RS^*,RS^*)]-

RN: 77868-40-7   **MP** (°C):
MW: 763.07   **BP** (°C):

Solubility (Moles/L)	Solubility (Grams/L)	Temp (°C)	Ref (#)	Evaluation (T P E A A)	Comments
2.317E-02	1.768E+01	25	P091	0 0 0 0 0	

3651. $C_{16}H_{20}I_3N_3O_8$

1,3-Benzenedicarboxamide, 5-[(hydroxyacetyl)amino]-N,N'-bis[2-hydroxy-1-(hydroxymethyl) ethyl]-2,4,6-triiodo-

RN: 77868-41-8   **MP** (°C):
MW: 763.07   **BP** (°C):

Solubility (Moles/L)	Solubility (Grams/L)	Temp (°C)	Ref (#)	Evaluation (T P E A A)	Comments
5.282E-02	4.031E+01	25	P091	0 0 0 0 0	

3652. $C_{16}H_{20}N_4O_2$

Apazone
APZ
Azapropazone

RN: 13539-59-8   **MP** (°C):   247
MW: 300.36   **BP** (°C):

Solubility (Moles/L)	Solubility (Grams/L)	Temp (°C)	Ref (#)	Evaluation (T P E A A)	Comments
4.900E-04	1.472E-01	35	H091	1 2 2 2 1	*sic*
2.896E-04	8.700E-02	rt	H302	0 0 2 1 1	intrinsic

3653. $C_{16}H_{20}N_4O_3S$

2-(N4-Acetylsulfanilylamino)-4-isobutylpyrimidine

RN: **MP** (°C):
MW: 348.43   **BP** (°C):

Solubility (Moles/L)	Solubility (Grams/L)	Temp (°C)	Ref (#)	Evaluation (T P E A A)	Comments
1.091E-05	3.800E-03	37	R076	1 2 0 0 1	

3654. C$_{16}$H$_{20}$N$_8$O$_2$S
6-[D-2-Amino-2-(4-aminophenyl)-acetamido]-3,3-dimethyl-7-oxo-4-thia-1-azabicyclo[3,2,0]hept-
   2-yl-5-t

RN: **MP** (°C):
MW: 388.45   **BP** (°C):

Solubility (Moles/L)	Solubility (Grams/L)	Temp (°C)	Ref (#)	Evaluation (T P E A A)	Comments
5.277E-03	2.050E+00	25	B148	2 2 2 1 2	

3655. C$_{16}$H$_{20}$O$_6$P$_2$S$_3$
Temephos
O,O'-(Thiodi-4,1-phenylene)bis(O,O'-dimethylphosphorothioate)
Abate
Tetramethyl O,O'-thiodi-p-phenylene phosphorothioate
Abaphos
Tetrafenphos

RN: 3383-96-8   **MP** (°C):
MW: 466.47   **BP** (°C):

Solubility (Moles/L)	Solubility (Grams/L)	Temp (°C)	Ref (#)	Evaluation (T P E A A)	Comments
1.929E-08	9.000E-06	10	B324	0 0 0 0 0	
1.929E-08	8.998E-06	10	B324	0 0 0 0 0	
5.788E-07	2.700E-04	20	B300	2 1 1 1 2	
5.788E-07	2.700E-04	20	B324	0 0 0 0 0	
5.788E-07	2.700E-04	20	B324	0 0 0 0 0	
1.501E-06	7.002E-04	30	B324	0 0 0 0 0	
1.501E-06	7.000E-04	30	B324	0 0 0 0 0	

3656. C$_{16}$H$_{21}$ClN$_3$S
Methylene blue
Methylenblau
C.I. 52015

RN: 61-73-4   **MP** (°C):
MW: 322.88   **BP** (°C):

Solubility (Moles/L)	Solubility (Grams/L)	Temp (°C)	Ref (#)	Evaluation (T P E A A)	Comments
~1.02E-01	~3.30E+01	20	F300	1 0 0 0 0	

3657. C$_{16}$H$_{21}$NO
N,N-Heptamethylenecinnamamide
Octahydro-1-(1-oxo-3-phenyl-2-propenyl) azocine

RN: 59832-06-3   **MP** (°C):
MW: 243.35   **BP** (°C):

Solubility (Moles/L)	Solubility (Grams/L)	Temp (°C)	Ref (#)	Evaluation (T P E A A)	Comments
2.560E-04	6.230E-02	ns	H350	0 0 0 0 0	

3658. C$_{16}$H$_{21}$NO

N-Cycloheptylcinnamamide
N-Cycloheptyl-3-phenyl-2-propenamide
RN: 59831-98-0   **MP** (°C):
MW: 243.35   **BP** (°C):

Solubility (Moles/L)	Solubility (Grams/L)	Temp (°C)	Ref (#)	Evaluation (T P E A A)	Comments
3.570E-06	8.688E-04	ns	H350	0 0 0 0 0	

3659. C$_{16}$H$_{21}$NO$_2$

Propranolol
2-Propanol, 1-[(1-methylethyl)amino]-3-(1-naphthalenyloxy)-
RN: 525-66-6   **MP** (°C):
MW: 259.35   **BP** (°C):

Solubility (Moles/L)	Solubility (Grams/L)	Temp (°C)	Ref (#)	Evaluation (T P E A A)	Comments
1.195E-04	3.100E-02	22.5	B422	0 0 0 0 0	
3.123E-04	8.099E-02	25	S450	0 0 0 0 0	
3.092E-08	8.020E-06	32	M458	0 0 0 0 0	

3660. C$_{16}$H$_{21}$NO$_2$S

m-Carboxyloctylphenylisothiocyanate
3-Carboxyloctylphenylisothiocyanate
RN: **MP** (°C):
MW: 291.42   **BP** (°C):

Solubility (Moles/L)	Solubility (Grams/L)	Temp (°C)	Ref (#)	Evaluation (T P E A A)	Comments
6.000E-05	1.748E-02	25	K032	2 2 0 1 1	

3661. C$_{16}$H$_{21}$NO$_3$

Piperidine, 1-[(benzoyloxy)acetyl]-2-ethyl-
RN: 115178-69-3   **MP** (°C):   54.5
MW: 275.35   **BP** (°C):

Solubility (Moles/L)	Solubility (Grams/L)	Temp (°C)	Ref (#)	Evaluation (T P E A A)	Comments
1.889E-03	5.200E-01	22	N317	1 1 2 1 2	

3662. C$_{16}$H$_{21}$NO$_3$

Piperidine, 1-[(benzoyloxy)acetyl]-2,6-dimethyl-
RN: 115178-70-6   **MP** (°C):   118
MW: 275.35   **BP** (°C):

Solubility (Moles/L)	Solubility (Grams/L)	Temp (°C)	Ref (#)	Evaluation (T P E A A)	Comments
5.448E-04	1.500E-01	22	N317	1 1 2 1 2	

3663. $C_{16}H_{21}NO_3$
Acetamide, 2-(benzoyloxy)-*N*-cyclohexyl-*N*-methyl-
RN: 106231-65-6   **MP** (°C):
MW: 275.35   **BP** (°C):   439.6

Solubility (Moles/L)	Solubility (Grams/L)	Temp (°C)	Ref (#)	Evaluation (T P E A A)	Comments
5.084E-04	1.400E-01	22	B427	1 0 0 1 1	

3664. $C_{16}H_{21}NO_5$
Benzoic acid, 2-(acetyloxy)-, 2-(diethylamino)-1-methyl-2-oxoethyl ester
RN: 118247-09-9   **MP** (°C):   40.5
MW: 307.35   **BP** (°C):

Solubility (Moles/L)	Solubility (Grams/L)	Temp (°C)	Ref (#)	Evaluation (T P E A A)	Comments
2.499E-02	7.680E+00	21	N335	0 0 0 0 0	

3665. $C_{16}H_{21}N_3$
Tripelennamine
N-Benzyl-*N'*,*N'*-dimethyl-*N*-2-pyridylethylenediamine
PBZ
Pelamine
RN: 91-81-6   **MP** (°C):   <25
MW: 255.37   **BP** (°C):

Solubility (Moles/L)	Solubility (Grams/L)	Temp (°C)	Ref (#)	Evaluation (T P E A A)	Comments
2.300E-03	5.873E-01	30	L068	1 0 0 1 0	EFG
1.500E-02	3.830E+00	37.5	L034	2 2 0 1 2	pH 7.4

3666. $C_{16}H_{22}Cl_2O_3$
2,4-Dichlorophenoxyacetic acid *n*-octyl ester
2,4-Dichlorophenoxyacetic acid capryl ester
RN: 1928-44-5   **MP** (°C):
MW: 333.26   **BP** (°C):

Solubility (Moles/L)	Solubility (Grams/L)	Temp (°C)	Ref (#)	Evaluation (T P E A A)	Comments
2.128E-05	7.092E-03	ns	M120	0 0 1 1 2	

3667. C$_{16}$H$_{22}$N$_4$O
Neohetramine
N,N-Dimethyl-N'-(p-methoxybenzyl)-N'-(2-pyrimidyl)ethylenediamine
Tonzilamine
RN: 91-85-0   **MP** (°C):
MW: 286.38   **BP** (°C):

Solubility (Moles/L)	Solubility (Grams/L)	Temp (°C)	Ref (#)	Evaluation (T P E A A)	Comments
1.900E-02	5.441E+00	37.5	L034	2 2 0 1 2	pH 7.4

3668. C$_{16}$H$_{22}$N$_4$O$_2$S
2-Sulfanilamido-4-methyl-5-n-amylpyrimidine
RN: 71119-35-2   **MP** (°C):   188-190
MW: 334.44   **BP** (°C):

Solubility (Moles/L)	Solubility (Grams/L)	Temp (°C)	Ref (#)	Evaluation (T P E A A)	Comments
8.372E-05	2.800E-02	29	C049	0 0 0 0 0	

3669. C$_{16}$H$_{22}$N$_4$O$_6$
2′-Valeryl-6-methoxypurine arabinoside
2′-Trimethylacetyl-6-methoxypurine arabinoside
RN: 121032-22-2   **MP** (°C):   118–120
MW: 366.38   **BP** (°C):

Solubility (Moles/L)	Solubility (Grams/L)	Temp (°C)	Ref (#)	Evaluation (T P E A A)	Comments
2.400E-01	8.793E+01	37	C348	0 0 0 0 0	pH 7.00
1.070E-01	3.920E+01	37	C348	0 0 0 0 0	pH 7.00

3670. C$_{16}$H$_{22}$N$_4$O$_6$.0.5H$_2$O
6-Methoxy-9-(5-O-pivalate-β-D-arabinofuranosyl)]-9H-purine (hemihydrate)
RN: 121032-42-6   **MP** (°C):   glass
MW: 375.38   **BP** (°C):

Solubility (Moles/L)	Solubility (Grams/L)	Temp (°C)	Ref (#)	Evaluation (T P E A A)	Comments
3.560E-02	1.336E+01	37	M378	1 2 1 1 2	pH 7.2

3671. C$_{16}$H$_{22}$N$_4$O$_6$.0.5H$_2$O
6-Methoxy-9-(5-O-valerate-β-D-arabinosyl)]-6-methoxy-9H-purine (hemihydrate)
RN: 142963-77-7   **MP** (°C):   foam
MW: 375.38   **BP** (°C):

Solubility (Moles/L)	Solubility (Grams/L)	Temp (°C)	Ref (#)	Evaluation (T P E A A)	Comments
1.720E-03	6.457E-01	37	M378	1 2 1 1 2	pH 7.2

3672. C$_{16}$H$_{22}$O$_4$
Dibutyl phthalate
n-Butyl phthalate

RN:	84-74-2	MP (°C):	−35
MW:	278.35	BP (°C):	430

Solubility (Moles/L)	Solubility (Grams/L)	Temp (°C)	Ref (#)	Evaluation (T P E A A)	Comments
4.455E-05	1.240E-02	10	S198	2 1 2 2 2	
3.952E-05	1.100E-02	15	H069	1 0 1 1 1	
3.630E-05	1.010E-02	20	L300	2 1 0 2 2	
3.880E-05	1.080E-02	20	S198	2 1 2 2 2	
3.593E-04	1.000E-01	22	N311	1 0 1 1 2	
3.377E-05	9.400E-03	22	Y419	0 0 0 0 0	
6.574E-05	1.830E-02	23.5	S171	2 1 2 2 2	
3.126E-05	8.700E-03	25	D336	0 0 0 0 0	
3.449E-05	9.600E-03	25	D336	0 0 0 0 0	
4.670E-05	1.300E-02	25	F067	1 0 2 2 2	
1.609E-02	4.480E+00	25	F070	1 0 0 0 2	*sic*
4.095E-05	1.140E-02	30	S198	2 1 2 2 2	
1.437E-03	4.000E-01	rt	M161	0 0 0 0 2	

3673. C$_{16}$H$_{22}$O$_4$
Diisobutyl phthalate
1,2-Benzenedicarboxylic acid, bis(2-methylpropyl) esterpalatinol
Phthalic acid diisobutyl ester
Palatinolic

RN:	84-69-5	MP (°C):	
MW:	278.35	BP (°C):	

Solubility (Moles/L)	Solubility (Grams/L)	Temp (°C)	Ref (#)	Evaluation (T P E A A)	Comments
3.592E-04	9.999E-02	20	F070	1 0 0 0 2	
7.300E-05	2.032E-02	20	L300	2 1 0 2 2	
2.227E-05	6.200E-03	24	H116	2 1 0 0 2	
5.030E-06	1.400E-03	25	D336	0 0 0 0 0	

3674. C$_{16}$H$_{22}$O$_4$
tere-Butyl phthalate

RN:	30448-43-2	MP (°C):	
MW:	278.35	BP (°C):	

Solubility (Moles/L)	Solubility (Grams/L)	Temp (°C)	Ref (#)	Evaluation (T P E A A)	Comments
3.952E-06	1.100E-03	25	D336	0 0 0 0 0	

3675. C$_{16}$H$_{22}$O$_4$
Di-*n*-butyl *o*-phthalate
RN: **MP** (°C):   −35 C
MW: 278.35   **BP** (°C):

Solubility (Moles/L)	Solubility (Grams/L)	Temp (°C)	Ref (#)	Evaluation (T P E A A)	Comments
3.593E-05	1.000E-02	25	S417	0 0 0 0 0	

3676. C$_{16}$H$_{22}$O$_6$
Diethoxyethyl phthalate
bis(2-Ethoxyethyl) phthalate
RN: 605-54-9   **MP** (°C):
MW: 310.35   **BP** (°C):

Solubility (Moles/L)	Solubility (Grams/L)	Temp (°C)	Ref (#)	Evaluation (T P E A A)	Comments
6.271E-03	1.946E+00	ns	F014	0 0 0 0 2	

3677. C$_{16}$H$_{22}$O$_8$·2H$_2$O
Coniferin (dihydrate)
4-Hydroxy-3-methoxy-1-(γ-hydroxypropenyl)benzene-4-D-glucoside (dihydrate)
Abietin(dihydrate)
Coniferosi(dihydrate)
RN: 531-29-3   **MP** (°C):   185
MW: 378.38   **BP** (°C):

Solubility (Moles/L)	Solubility (Grams/L)	Temp (°C)	Ref (#)	Evaluation (T P E A A)	Comments
1.315E-02	4.975E+00	c	D004	0 0 0 0 0	
		h	D004	0 0 0 0 0	

3678. C$_{16}$H$_{22}$O$_{11}$
β-D-Glucose pentaacetate
β-Glucose-penta-acetat
RN: 604-69-3   **MP** (°C):   131
MW: 390.35   **BP** (°C):

Solubility (Moles/L)	Solubility (Grams/L)	Temp (°C)	Ref (#)	Evaluation (T P E A A)	Comments
2.306E-03	9.000E-01	18	F300	1 0 0 0 0	

3679. C$_{16}$H$_{22}$O$_{11}$

α-D-Glucose pentaacetate
1,2,3,4,6-Penta-*O*-acetyl-α-D-glucose
Pentaacetate
Glucopyranose pentaacetate
Glucose pentaacetate;
α-D-Glucopyranose

RN: 604-68-2   **MP** (°C): 109–111
MW: 390.35   **BP** (°C):

Solubility (Moles/L)	Solubility (Grams/L)	Temp (°C)	Ref (#)	Evaluation (T P E A A)	Comments
3.802E-03	1.484E+00	ns	R427	0 0 0 0 0	

3680. C$_{16}$H$_{22}$O$_{11}$

α-Glucose pentaacetate
α-Glucose-penta-acetat

RN: 3891-59-6   **MP** (°C): 110
MW: 390.35   **BP** (°C):

Solubility (Moles/L)	Solubility (Grams/L)	Temp (°C)	Ref (#)	Evaluation (T P E A A)	Comments
3.843E-03	1.500E+00	18	F300	1 0 0 0 1	

3681. C$_{16}$H$_{23}$FN$_2$O$_6$

1,3-bis(Pivaloyloxymethyl)-5-fluoro-2,4(1H,3H)-pyrimidinedi-one
1,3-bis(Pivaloyloxymethyl)-5-fluorouracil

RN: 66542-50-5   **MP** (°C): 102–104
MW: 358.37   **BP** (°C):

Solubility (Moles/L)	Solubility (Grams/L)	Temp (°C)	Ref (#)	Evaluation (T P E A A)	Comments
1.256E-04	4.500E-02	22	B321	0 0 0 0 0	pH 4.0

3682. C$_{16}$H$_{23}$NO

n-Heptylcinnamamide
2-Propenamide, *N*-heptyl-3-phenyl-

RN: 59831-99-1   **MP** (°C):
MW: 245.37   **BP** (°C):

Solubility (Moles/L)	Solubility (Grams/L)	Temp (°C)	Ref (#)	Evaluation (T P E A A)	Comments
7.600E-06	1.865E-03	ns	H350	0 0 0 0 0	

3683. C$_{16}$H$_{23}$NO$_2$
Etoxadrol
(+)-2-(2-Ethyl-2-phenyl-1,3-dioxolan-4-yl)piperidine
RN: 28189-85-7   **MP** (°C):
MW: 261.37   **BP** (°C):

Solubility (Moles/L)	Solubility (Grams/L)	Temp (°C)	Ref (#)	Evaluation (T P E A A)	Comments
2.487E-03	6.500E-01	20	K017	1 2 2 2 2	pH 10, intrinsic
1.098E-02	2.870E+00	30	K017	1 2 2 2 2	pH 10, intrinsic
4.668E-02	1.220E+01	40	K017	1 2 2 2 2	pH 10, intrinsic

3684. C$_{16}$H$_{23}$NO$_3$
Acetaminophen octanoate
Octanoic acid, 4-(acetylamino)phenyl ester
RN: 54942-41-5   **MP** (°C):   103
MW: 277.37   **BP** (°C):

Solubility (Moles/L)	Solubility (Grams/L)	Temp (°C)	Ref (#)	Evaluation (T P E A A)	Comments
3.605E-05	1.000E-02	25	B010	1 1 1 1 0	

3685. C$_{16}$H$_{23}$NO$_3$S$_2$
N-[2-(3,4-Dihydroxyphenyl)ethyl]-5-[(3R)-1,2-dithiolan-3-yl]-pentanamide
RN: **MP** (°C):
MW: 341.49   **BP** (°C):

Solubility (Moles/L)	Solubility (Grams/L)	Temp (°C)	Ref (#)	Evaluation (T P E A A)	Comments
1.054E-06	3.600E-04	ns	S453	0 0 0 0 0	

3686. C$_{16}$H$_{23}$NO$_6$
Monocrotaline
(−)-Monocrotaline
RN: 315-22-0   **MP** (°C):   202
MW: 325.36   **BP** (°C):

Solubility (Moles/L)	Solubility (Grams/L)	Temp (°C)	Ref (#)	Evaluation (T P E A A)	Comments
3.644E-02	1.186E+01	ns	I312	0 0 0 0 0	

3687. C$_{16}$H$_{23}$N$_5$O$_5$
9-[5′-(O-Caproyl)-β-D-arabinofuranosyl]adenine ester
RN: 65926-34-3   **MP** (°C):
MW: 365.39   **BP** (°C):

Solubility (Moles/L)	Solubility (Grams/L)	Temp (°C)	Ref (#)	Evaluation (T P E A A)	Comments
6.842E-03	2.500E+00	ns	B134	0 1 1 1 1	

3688. $C_{16}H_{23}N_5O_5$

9-[5′-(O-tert-Butylacetyl)-β-D-arabinofuranosyl]adenine ester

RN: 68325-42-8   **MP** (°C):
MW: 365.39   **BP** (°C):

Solubility (Moles/L)	Solubility (Grams/L)	Temp (°C)	Ref (#)	Evaluation (T P E A A)	Comments
2.135E-02	7.800E+00	ns	B134	0 1 1 1 1	

3689. $C_{16}H_{24}N_2O_2$

N,N,N′,N′-Tetraethylterephthalamide

RN: 15394-30-6   **MP** (°C):
MW: 276.38   **BP** (°C):

Solubility (Moles/L)	Solubility (Grams/L)	Temp (°C)	Ref (#)	Evaluation (T P E A A)	Comments
2.000E-02	5.528E+00	30	K019	1 0 0 0 1	

3690. $C_{16}H_{24}N_2O_2$

N,N,N′,N′-Tetraethylisophthalamide

RN: 13698-87-8   **MP** (°C):
MW: 276.38   **BP** (°C):

Solubility (Moles/L)	Solubility (Grams/L)	Temp (°C)	Ref (#)	Evaluation (T P E A A)	Comments
7.200E-01	1.990E+02	30	K019	1 0 0 0 2	

3691. $C_{16}H_{24}N_4O_2$

2,5-Diaziridinyl-3,6-bis(propylamino)-1,4-benzoquinone

RN: 59886-47-4   **MP** (°C): 140
MW: 304.40   **BP** (°C):

Solubility (Moles/L)	Solubility (Grams/L)	Temp (°C)	Ref (#)	Evaluation (T P E A A)	Comments
<3.29E-04	<1.00E-01	rt	C317	0 0 0 0 0	

3692. $C_{16}H_{24}N_4O_6$

2,5-Diaziridinyl-3,6-bis(2′-hydroxyl-3′-hydroxylpropylamino)-1,4-benzoquinone
2,5-Diaziridinyl-3,6-bis(hydroxylethylmethylamino)-1,4-benzoquinone

RN: 59886-55-4   **MP** (°C): 273
MW: 368.39   **BP** (°C):

Solubility (Moles/L)	Solubility (Grams/L)	Temp (°C)	Ref (#)	Evaluation (T P E A A)	Comments
1.629E-01	6.000E+01	rt	C317	0 0 0 0 0	
8.143E-02	3.000E+01	rt	C317	0 0 0 0 0	

3693. C₁₆H₂₄N₆

1-(Methylphenethylamino)-3,5-bis(dimethylamino)-s-triazine

RN: 125867-93-8 **MP** (°C):
MW: 300.41 **BP** (°C):

Solubility (Moles/L)	Solubility (Grams/L)	Temp (°C)	Ref (#)	Evaluation (T P E A A)	Comments
2.427E-05	7.291E-03	25	B386	0 0 0 0 0	

3694. C₁₆H₂₄O₃

Nonyl *p*-hydroxybenzoate
Nonyl 4-hydroxybenzoate

RN: 38713-56-3 **MP** (°C):
MW: 264.37 **BP** (°C):

Solubility (Moles/L)	Solubility (Grams/L)	Temp (°C)	Ref (#)	Evaluation (T P E A A)	Comments
4.824E-03	1.275E+00	25	D081	1 2 2 1 2	

3695. C₁₆H₂₄O₄

3,4-Epoxy-6-methylcyclohexylmethyl-3,4-epoxy-6-methylcyclohexane carboxylate
EP 201

RN: 141-37-7 **MP** (°C):
MW: 280.37 **BP** (°C):

Solubility (Moles/L)	Solubility (Grams/L)	Temp (°C)	Ref (#)	Evaluation (T P E A A)	Comments
1.067E-02	2.991E+00	ns	I313	0 0 0 0 0	

3696. C₁₆H₂₅NOS

S-Benzyl di-*sec*-butylthiocarbamate
Thiocarbazil
Tiocarbazil

RN: 36756-79-3 **MP** (°C):
MW: 279.45 **BP** (°C):

Solubility (Moles/L)	Solubility (Grams/L)	Temp (°C)	Ref (#)	Evaluation (T P E A A)	Comments
8.946E-06	2.500E-03	30	M161	1 0 0 0 1	

3697. C$_{16}$H$_{25}$NO$_2$
Butacarb
Carbamic acid, *N*-methyl-, 3,5-di-*tert*-butylphenyl ester
3,5-Di-*tert*-butylphenyl methylcarbamate

RN:	2655-19-8	MP (°C):	102.9
MW:	263.38	BP (°C):	

Solubility (Moles/L)	Solubility (Grams/L)	Temp (°C)	Ref (#)	Evaluation (T P E A A)	Comments
5.695E-05	1.500E-02	20	M161	1 0 0 0 1	

3698. C$_{16}$H$_{25}$NO$_2$
Nonyl *p*-aminobenzoate
Nonyl 4-aminobenzoate

RN:	37139-21-2	MP (°C):	
MW:	263.38	BP (°C):	

Solubility (Moles/L)	Solubility (Grams/L)	Temp (°C)	Ref (#)	Evaluation (T P E A A)	Comments
1.020E-06	2.687E-04	37	F006	1 1 2 2 2	

3699. C$_{16}$H$_{25}$NO$_3$
4-Propoxybenzoic acid-2-(diethyl-amino)ethyl ester

RN:	15788-85-9	MP (°C):	
MW:	279.38	BP (°C):	

Solubility (Moles/L)	Solubility (Grams/L)	Temp (°C)	Ref (#)	Evaluation (T P E A A)	Comments
4.500E-04	1.257E-01	ns	M066	0 0 0 0 1	

3700. C$_{16}$H$_{26}$
2-Phenyldecane

RN:		MP (°C):	
MW:	218.39	BP (°C):	

Solubility (Moles/L)	Solubility (Grams/L)	Temp (°C)	Ref (#)	Evaluation (T P E A A)	Comments
2.600E-08	5.678E-06	25	S377	0 0 0 0 0	

3701. C$_{16}$H$_{26}$
3-Phenyldecane

RN:		MP (°C):	
MW:	218.39	BP (°C):	

Solubility (Moles/L)	Solubility (Grams/L)	Temp (°C)	Ref (#)	Evaluation (T P E A A)	Comments
3.800E-08	8.299E-06	25	S377	0 0 0 0 0	

3702. C$_{16}$H$_{26}$
4-Phenyldecane

RN: **MP** (°C):
MW: 218.39   **BP** (°C):

Solubility (Moles/L)	Solubility (Grams/L)	Temp (°C)	Ref (#)	Evaluation (T P E A A)	Comments
3.600E-08	7.862E-06	25	S377	0 0 0 0 0	

3703. C$_{16}$H$_{26}$
5-Phenyldecane

RN: **MP** (°C):
MW: 218.39   **BP** (°C):

Solubility (Moles/L)	Solubility (Grams/L)	Temp (°C)	Ref (#)	Evaluation (T P E A A)	Comments
3.500E-08	7.643E-06	25	S377	0 0 0 0 0	

3704. C$_{16}$H$_{26}$N$_2$O$_2$
4-Propylaminobenzoic acid-2-(diethyl-amino)ethyl ester

RN: 16488-54-3   **MP** (°C):
MW: 278.40   **BP** (°C):

Solubility (Moles/L)	Solubility (Grams/L)	Temp (°C)	Ref (#)	Evaluation (T P E A A)	Comments
1.030E-03	2.867E-01	ns	M066	0 0 0 0 2	

3705. C$_{16}$H$_{26}$O$_2$
4-Octylphenol monoethoxylate

RN: 51437-89-9   **MP** (°C):
MW: 250.38   **BP** (°C):

Solubility (Moles/L)	Solubility (Grams/L)	Temp (°C)	Ref (#)	Evaluation (T P E A A)	Comments
3.195E-05	8.000E-03	20.5	A335	0 0 0 0 0	
3.200E-05	8.012E-03	20.5	A335	0 0 0 0 0	

3706. C$_{16}$H$_{26}$O$_5$
Artemether

RN: 71963-77-4   **MP** (°C):
MW: 298.38   **BP** (°C):

Solubility (Moles/L)	Solubility (Grams/L)	Temp (°C)	Ref (#)	Evaluation (T P E A A)	Comments
		ns	K444	0 0 0 0 0	

3707. C$_{16}$H$_{26}$O$_6$
Triethylene glycol dibutyrate
RN: 26962-26-5   **MP** (°C):
MW: 314.38   **BP** (°C):

Solubility (Moles/L)	Solubility (Grams/L)	Temp (°C)	Ref (#)	Evaluation (T P E A A)	Comments
2.524E-02	7.937E+00	ns	F014	0 0 0 0 2	

3708. C$_{16}$H$_{28}$N$_3$O$_2$
Dioxyethylaminoazobenzene
RN: **MP** (°C):
MW: 294.42   **BP** (°C):

Solubility (Moles/L)	Solubility (Grams/L)	Temp (°C)	Ref (#)	Evaluation (T P E A A)	Comments
2.945E-04	8.670E-02	0	K036	1 0 0 0 2	
4.212E-04	1.240E-01	25	K036	1 0 0 0 2	
2.819E-03	8.300E-01	90	K036	1 0 0 0 2	

3709. C$_{16}$H$_{32}$O$_2$
Palmitic acid
Hexadecanoic acid
RN: 57-10-3   **MP** (°C):   56
MW: 256.43   **BP** (°C):

Solubility (Moles/L)	Solubility (Grams/L)	Temp (°C)	Ref (#)	Evaluation (T P E A A)	Comments
1.794E-05	4.600E-03	0	B136	1 0 2 1 1	
2.808E-05	7.200E-03	20	B136	1 0 2 1 1	
2.808E-05	7.200E-03	20.0	R001	1 1 1 1 1	
3.200E-06	8.206E-04	25	J001	1 0 2 1 1	
1.200E-07	3.077E-05	25	R002	0 0 0 0 0	intrinsic
2.680E-06	6.872E-04	25	R002	0 0 0 0 0	
3.237E-05	8.300E-03	30	B136	1 0 2 1 1	
3.237E-05	8.300E-03	30.0	R001	1 1 1 1 1	
3.900E-05	1.000E-02	45	B136	1 0 2 1 1	
3.900E-05	1.000E-02	45.0	R001	1 1 1 1 1	
4.000E-06	1.026E-03	50	J001	1 0 2 1 1	
4.680E-05	1.200E-02	60	B136	1 0 2 1 1	
4.680E-05	1.200E-02	60.0	R001	1 1 1 1 1	
1.794E-05	4.600E-03	.0	R001	1 1 1 1 1	

3710. C$_{16}$H$_{34}$
2,2,4,4,6,8,8-Heptamethylnonane

RN:	4390-04-9	MP (°C):	
MW:	226.45	BP (°C):	240

Solubility (Moles/L)	Solubility (Grams/L)	Temp (°C)	Ref (#)	Evaluation (T P E A A)	Comments
1.369E-09	3.100E-07	25	T423	0 0 0 0 0	

3711. C$_{16}$H$_{34}$
3-Methylpentadecane

RN:	2882-96-4	MP (°C):	−22
MW:	226.45	BP (°C):	282

Solubility (Moles/L)	Solubility (Grams/L)	Temp (°C)	Ref (#)	Evaluation (T P E A A)	Comments
4.328E-10	9.800E-08	23	C332	0 0 0 0 0	

3712. C$_{16}$H$_{34}$
Hexadecane
n-Hexadecane
Cetane

RN:	544-76-3	MP (°C):	18.17
MW:	226.45	BP (°C):	

Solubility (Moles/L)	Solubility (Grams/L)	Temp (°C)	Ref (#)	Evaluation (T P E A A)	Comments
2.778E-08	6.290E-06	25	F004	0 0 0 0 0	

3713. C$_{16}$H$_{34}$
2-Methylpentadecane

RN:	1560-93-6	MP (°C):	−7
MW:	226.45	BP (°C):	282

Solubility (Moles/L)	Solubility (Grams/L)	Temp (°C)	Ref (#)	Evaluation (T P E A A)	Comments
4.681E-10	1.060E-07	23	C332	0 0 0 0 0	

3714. C$_{16}$H$_{34}$O
Hexadecanol
Cetyl alcohol

RN:	36653-82-4	MP (°C):	49
MW:	242.45	BP (°C):	344

Solubility (Moles/L)	Solubility (Grams/L)	Temp (°C)	Ref (#)	Evaluation (T P E A A)	Comments
1.699E-07	4.120E-05	22.5	G301	0 0 0 0 0	
1.700E-07	4.122E-05	25	R002	0 0 0 0 0	
3.300E-08	8.001E-06	34	K011	1 2 1 1 2	

(continued)

3714. C₁₆H₃₄O (continued)

Solubility (Moles/L)	Solubility (Grams/L)	Temp (°C)	Ref (#)	Evaluation (T P E A A)	Comments
6.393E-08	1.550E-05	43	H030	2 2 2 2 2	
6.393E-08	1.550E-05	43	H103	1 2 2 2 2	
1.270E-07	3.079E-05	55	K011	1 2 1 1 2	
1.675E-07	4.060E-05	61	H030	2 2 2 2 2	
1.675E-07	4.060E-05	61	H103	1 2 2 2 2	

3715. C₁₆H₃₅O₃P
Dibutyl isooctyl phosphonate
RN: 108979-58-4   **MP** (°C):
MW: 306.43   **BP** (°C):

Solubility (Moles/L)	Solubility (Grams/L)	Temp (°C)	Ref (#)	Evaluation (T P E A A)	Comments
<6.53E-04	<2.00E-01	25	B070	1 2 0 1 0	

3716. C₁₆H₃₅O₄P
Dibutyl octyl phosphate
RN: 25786-28-1   **MP** (°C):
MW: 322.43   **BP** (°C):

Solubility (Moles/L)	Solubility (Grams/L)	Temp (°C)	Ref (#)	Evaluation (T P E A A)	Comments
<3.10E-04	<1.00E-01	25	B070	1 2 0 1 0	

3717. C₁₇H₁₁NO₃
Furo[3,4-b]quinolin-3(1H)-one, 9-hydroxy-1-phenyl-
RN: 74103-09-6   **MP** (°C):
MW: 277.28   **BP** (°C):

Solubility (Moles/L)	Solubility (Grams/L)	Temp (°C)	Ref (#)	Evaluation (T P E A A)	Comments
1.190E-07	3.300E-05	25	P089	0 0 0 0 0	
1.388E-07	3.850E-05	37	P089	0 0 0 0 0	
1.677E-07	4.650E-05	51	P089	0 0 0 0 0	

3718. C₁₇H₁₂
1,2-Benzofluorene
Benzo[a]fluorene
11H-Benzo[a]fluorene
RN: 238-84-6   **MP** (°C):   187
MW: 216.29   **BP** (°C):   407

Solubility (Moles/L)	Solubility (Grams/L)	Temp (°C)	Ref (#)	Evaluation (T P E A A)	Comments
2.081E-07	4.500E-05	25	M064	1 1 2 2 1	
2.100E-07	4.542E-05	25	M342	1 0 1 1 1	
2.081E-07	4.500E-05	ns	M344	0 0 0 0 2	

3719. C₁₇H₁₂

$C_{17}H_{12}$

2,3-Benzofluorene
Benzo[b]fluorene
11H-Benzo[b]fluorene

RN: 243-17-4 **MP** (°C): 209
MW: 216.29 **BP** (°C):

Solubility (Moles/L)	Solubility (Grams/L)	Temp (°C)	Ref (#)	Evaluation (T P E A A)	Comments
1.849E-08	4.000E-06	25	B319	2 0 1 2 0	
9.247E-09	2.000E-06	25	M064	1 1 2 2 1	
9.250E-09	2.001E-06	25	M342	1 0 1 1 2	

3720. C₁₇H₁₂ClFN₂O

$C_{17}H_{12}ClFN_2O$

Nuarimol
Triminol
Trimidal
Gauntlet
2-Chloro-4′-fluoro-α-(5-pyrimidinyl)benzhydryl alcohol
α-(2-Chlorophenyl)-α-(4-fluorophenyl)-5-pyrimidinemethanol

RN: 63284-71-9 **MP** (°C):
MW: 314.75 **BP** (°C):

Solubility (Moles/L)	Solubility (Grams/L)	Temp (°C)	Ref (#)	Evaluation (T P E A A)	Comments
8.318E-05	2.618E-02	ns	R427	0 0 0 0 0	

3721. C₁₇H₁₂ClFN₃O₂

$C_{17}H_{12}ClFN_3O_2$

α-(4-Chlorophenyl)-α-(1-2-(2-chloro)phenylethenyl)-1H-1,2,4-triazole-1-ethanol
X-7801
DuP 860

RN: **MP** (°C):
MW: 344.76 **BP** (°C):

Solubility (Moles/L)	Solubility (Grams/L)	Temp (°C)	Ref (#)	Evaluation (T P E A A)	Comments
4.612E-06	1.590E-03	22	M362	1 1 2 1 1	

3722. C₁₇H₁₂ClNO₂S

$C_{17}H_{12}ClNO_2S$

Fentiazac
4-(p-Chlorophenyl)-2-phenyl-5-thiazoleacetic acid

RN: 18046-21-4 **MP** (°C): 161.1
MW: 329.81 **BP** (°C):

Solubility (Moles/L)	Solubility (Grams/L)	Temp (°C)	Ref (#)	Evaluation (T P E A A)	Comments
9.400E-06	3.100E-03	5	F306	1 0 1 2 2	intrinsic
9.600E-05	3.166E-02	25	C314	0 0 0 0 0	
9.612E-05	3.170E-02	25	C314	0 0 0 0 0	

(continued)

3722. $C_{17}H_{12}ClNO_2S$ (continued)

Solubility (Moles/L)	Solubility (Grams/L)	Temp (°C)	Ref (#)	Evaluation (T P E A A)	Comments
1.080E-05	3.562E-03	25	F306	1 0 1 2 2	intrinsic
1.310E-05	4.320E-03	37	F306	1 0 1 2 2	intrinsic
1.072E-05	3.534E-03	ns	R427	0 0 0 0 0	

3723. $C_{17}H_{12}Cl_2N_2O$
Fenarimol
2,4'-Dichloro-α-(5-pyrimidinyl)benzhydryl alcohol
α-(2-Chlorophenyl)-α-(4-chlorophenyl)-5-pyrimidinemethanol
Tebulan
Rubigan 4AS
Rimidin

RN: 60168-88-9 **MP** (°C): 118
MW: 331.20 **BP** (°C):

Solubility (Moles/L)	Solubility (Grams/L)	Temp (°C)	Ref (#)	Evaluation (T P E A A)	Comments
4.136E-05	1.370E-02	25	M161	1 0 0 0 2	pH 7

3724. $C_{17}H_{12}Cl_2N_4$
Triazolam
8-Chloro-6-(o-chlorophenyl)-1-methyl-4H-s-triazolo[4,3-a][1,4]benzodiazepine
Apo-Triazo
Gen-Triazolam
Halcion
Novo-Triolam

RN: 28911-01-5 **MP** (°C):
MW: 343.22 **BP** (°C):

Solubility (Moles/L)	Solubility (Grams/L)	Temp (°C)	Ref (#)	Evaluation (T P E A A)	Comments
8.741E-05	3.000E-02	amb	L434	0 0 0 0 0	

3725. $C_{17}H_{12}Cl_{10}O_3$
Kelevan
Allied GC 9160
Despirol

RN: 4234-79-1 **MP** (°C): 91
MW: 618.81 **BP** (°C):

Solubility (Moles/L)	Solubility (Grams/L)	Temp (°C)	Ref (#)	Evaluation (T P E A A)	Comments
8.888E-06	5.500E-03	20	M164	1 0 0 0 1	

3726. C₁₇H₁₂I₂O₃

Benziodarone
Algocor
Amplivix
Dilafurane
RN: 68-90-6   **MP** (°C):
MW: 518.09   **BP** (°C):

Solubility (Moles/L)	Solubility (Grams/L)	Temp (°C)	Ref (#)	Evaluation (T P E A A)	Comments
1.135E-05	5.881E-03	20	H301	0 0 0 0 0	

3727. C₁₇H₁₂O₆

Aflatoxin B1
AFB1
RN: 1162-65-8   **MP** (°C):   268
MW: 312.28   **BP** (°C):

Solubility (Moles/L)	Solubility (Grams/L)	Temp (°C)	Ref (#)	Evaluation (T P E A A)	Comments
4.803E-05	1.500E-02	ns	I306	0 0 0 0 0	

3728. C₁₇H₁₂O₇

Aflatoxin G1
RN: 1165-39-5   **MP** (°C):   244
MW: 328.28   **BP** (°C):

Solubility (Moles/L)	Solubility (Grams/L)	Temp (°C)	Ref (#)	Evaluation (T P E A A)	Comments
4.569E-05	1.500E-02	ns	I306	0 0 0 0 0	

3729. C₁₇H₁₃ClN₄

Alprazolam
8-Chloro-1-methyl-6-phenyl-4H-s-triazolo[4,3-a][1,4]benzodiazepine
Apo-Alpraz
Kalma
Novo-Alprazol
Nu-Alpraz
RN: 28981-97-7   **MP** (°C):
MW: 308.77   **BP** (°C):

Solubility (Moles/L)	Solubility (Grams/L)	Temp (°C)	Ref (#)	Evaluation (T P E A A)	Comments
2.267E-04	7.000E-02	amb	L434	0 0 0 0 0	
3.239E-04	1.000E-01	amb	L445	0 0 0 0 0	intrinsic

3730. C$_{17}$H$_{13}$ClO$_3$
Itanoxone
2'-Chloro-α-methylene-γ-oxo[1,1'-biphenyl]-4-butanoic acid
F 1379
RN: 58182-63-1   **MP** (°C): 212
MW: 300.74   **BP** (°C):

Solubility (Moles/L)	Solubility (Grams/L)	Temp (°C)	Ref (#)	Evaluation (T P E A A)	Comments
6.318E-04	1.900E-01	20	C112	2 0 1 1 2	

3731. C$_{17}$H$_{13}$Cl$_2$N$_3$O$_2$
α-(2,4-Difluorophenyl)-α-(1-2-(2-chloro)phenylethenyl)-1H-1,2,4-triazole-1-ethanol
A-9991
DuP 991
RN: **MP** (°C):
MW: 362.22   **BP** (°C):

Solubility (Moles/L)	Solubility (Grams/L)	Temp (°C)	Ref (#)	Evaluation (T P E A A)	Comments
1.933E-05	7.000E-03	22	M362	1 1 2 1 1	

3732. C$_{17}$H$_{14}$F$_3$N$_3$O$_2$S
Celecoxib
4-[5-(4-Methylphenyl)-3-(trifluoromethyl)
Celebrex
SC-58635
YM-177
RN: 169590-42-5   **MP** (°C):
MW: 381.38   **BP** (°C):

Solubility (Moles/L)	Solubility (Grams/L)	Temp (°C)	Ref (#)	Evaluation (T P E A A)	Comments
1.835E-05	7.000E-03	25	S415	0 0 0 0 0	
7.866E-06	3.000E-03	37	Y412	0 0 0 0 0	

3733. C$_{17}$H$_{14}$N$_2$O
1-o-Tolylazo-2-naphthol
Orange OT
Oil orange SS
1-(o-Tolylazo)-2-naphthol
RN: 2646-17-5   **MP** (°C): 131
MW: 262.31   **BP** (°C):

Solubility (Moles/L)	Solubility (Grams/L)	Temp (°C)	Ref (#)	Evaluation (T P E A A)	Comments
7.624E-07	2.000E-04	30	R430	0 0 0 0 0	
1.000E-07	2.623E-05	rt	M163	0 0 0 0 1	

3734. C₁₇H₁₄O₄S

Rofecoxib
4-(4-Methylsulfonylphenyl)-3-phenyl-5H-furan-2-one
RN: 162011-90-7   **MP** (°C):
MW: 314.36   **BP** (°C):

Solubility (Moles/L)	Solubility (Grams/L)	Temp (°C)	Ref (#)	Evaluation (T P E A A)	Comments
2.605E-05	8.190E-03	24.99	D414	0 0 0 0 0	
2.863E-05	9.000E-03	25	S415	0 0 0 0 0	
2.977E-05	9.360E-03	29.99	D414	0 0 0 0 0	
3.556E-05	1.118E-02	34.99	D414	0 0 0 0 0	
2.545E-06	8.000E-04	37	Y421	0 0 0 0 0	

3735. C₁₇H₁₄O₆

Aflatoxin B2
RN: 7220-81-7   **MP** (°C):   286
MW: 314.30   **BP** (°C):

Solubility (Moles/L)	Solubility (Grams/L)	Temp (°C)	Ref (#)	Evaluation (T P E A A)	Comments
4.773E-05	1.500E-02	ns	I306	0 0 0 0 0	

3736. C₁₇H₁₄O₇

Aflatoxin G2
RN: 7241-98-7   **MP** (°C):   237
MW: 330.30   **BP** (°C):

Solubility (Moles/L)	Solubility (Grams/L)	Temp (°C)	Ref (#)	Evaluation (T P E A A)	Comments
4.541E-05	1.500E-02	ns	I306	0 0 0 0 0	

3737. C₁₇H₁₅NO₃

Cinnamyl acetaminophen
Cinnamic acid, ester with 4'-hydroxyacetanilide
Acetanilide, 4'-hydroxy-, cinnamate (ester)
RN: 20682-28-4   **MP** (°C):   200–201
MW: 281.31   **BP** (°C):

Solubility (Moles/L)	Solubility (Grams/L)	Temp (°C)	Ref (#)	Evaluation (T P E A A)	Comments
4.977E-06	1.400E-03	37	D029	0 0 0 0 0	

3738. $C_{17}H_{15}NO_5$

Benzoic acid, 2-(acetyloxy)-, 4-(acetylamino)phenyl ester
RN: 5003-48-5 **MP** (°C): 174.5
MW: 313.31 **BP** (°C):

Solubility (Moles/L)	Solubility (Grams/L)	Temp (°C)	Ref (#)	Evaluation (T P E A A)	Comments
6.383E-05	2.000E-02	21	N335	0 0 0 0 0	

3739. $C_{17}H_{16}Br_2O_3$

Bromopropylate
1-Methylethyl-4-bromo-α-(4-bromophenyl)-α-hydroxybenzeneacetate
Neoron
GS-19851
Phenisobromolate
RN: 18181-80-1 **MP** (°C): 77
MW: 428.13 **BP** (°C):

Solubility (Moles/L)	Solubility (Grams/L)	Temp (°C)	Ref (#)	Evaluation (T P E A A)	Comments
<1.17E-06	<5.00E-04	20	F311	1 2 2 2 1	
1.168E-05	5.000E-03	20	M161	1 0 0 0 0	

3740. $C_{17}H_{16}ClFN_2O_2$

Progabide
Butanamide, 4-[[(4-chlorophenyl)(5-fluoro-2-hydroxyphenyl)methylene]amino]-
Gabrene
SL 76-002
RN: 62666-20-0 **MP** (°C):
MW: 334.78 **BP** (°C):

Solubility (Moles/L)	Solubility (Grams/L)	Temp (°C)	Ref (#)	Evaluation (T P E A A)	Comments
1.110E-04	3.716E-02	37	F309	1 0 2 2 2	
1.110E-04	3.716E-02	37	F318	2 2 0 0 2	

3741. $C_{17}H_{16}Cl_2O_3$

Chloropropylate
1-Methylethyl-4-chloro-α-(4-chlorophenyl)-α-hydroxybenzenacetate
Chlormite
Acaralate
G-24163
Rospin
RN: 5836-10-2 **MP** (°C): 74
MW: 339.22 **BP** (°C):

Solubility (Moles/L)	Solubility (Grams/L)	Temp (°C)	Ref (#)	Evaluation (T P E A A)	Comments
4.422E-06	1.500E-03	20	F311	1 2 2 2 1	
2.948E-05	1.000E-02	rt	M161	0 0 0 0 1	

3742. C$_{17}$H$_{16}$N$_2$O$_2$S
1-Sulfamethylnaphthalene
RN: **MP** (°C):
MW: 312.39   **BP** (°C):

Solubility (Moles/L)	Solubility (Grams/L)	Temp (°C)	Ref (#)	Evaluation (T P E A A)	Comments
3.201E-05	1.000E-02	20	F073	1 2 2 2 1	

3743. C$_{17}$H$_{16}$N$_2$O$_3$
C.I. Disperse blue 3
1-[(2-Hydroxyethyl)amino]-4-(methylamino)-9,10-anthracenedione
C.I. 61505
RN: 2475-46-9   **MP** (°C):   187
MW: 296.33   **BP** (°C):

Solubility (Moles/L)	Solubility (Grams/L)	Temp (°C)	Ref (#)	Evaluation (T P E A A)	Comments
1.200E-07	3.556E-05	25	B333	0 0 0 0 0	

3744. C$_{17}$H$_{16}$N$_2$O$_3$S
4-Sulfahydroxymethylnaphthalene
RN: **MP** (°C):
MW: 328.39   **BP** (°C):

Solubility (Moles/L)	Solubility (Grams/L)	Temp (°C)	Ref (#)	Evaluation (T P E A A)	Comments
1.675E-04	5.500E-02	20	F073	1 2 2 2 1	

3745. C$_{17}$H$_{16}$N$_2$O$_4$
p-(p-Acetamidobenzamido)phenyl acetate
RN: 74973-19-6   **MP** (°C):
MW: 312.33   **BP** (°C):

Solubility (Moles/L)	Solubility (Grams/L)	Temp (°C)	Ref (#)	Evaluation (T P E A A)	Comments
3.900E-05	1.218E-02	25	A066	1 0 1 1 1	

3746. C$_{17}$H$_{16}$N$_2$O$_4$S
1-Benzenesulfonyl-5-ethyl-5-phenyl-hydantoin
5-Ethyl-5phenyl-1(phenylsulfonyl)-2,4-imidazolidinedione
RN: 21413-25-2   **MP** (°C):
MW: 344.39   **BP** (°C):

Solubility (Moles/L)	Solubility (Grams/L)	Temp (°C)	Ref (#)	Evaluation (T P E A A)	Comments
9.782E-04	3.369E-01	37	F183	1 0 1 1 1	intrinsic

3747. C$_{17}$H$_{16}$N$_2$O$_5$

p-4-Acetaminophenyl acetaminophen
Acetamide, *N,N'*-[carbonylbis(oxy-4,1-phenylene)]*bis*-
Acetanilide, 4'-hydroxy-, carbonate (2:1) (ester)

RN: 19872-72-1   **MP** (°C):   219.5–220
MW: 328.33   **BP** (°C):

Solubility (Moles/L)	Solubility (Grams/L)	Temp (°C)	Ref (#)	Evaluation (T P E A A)	Comments
1.827E-04	6.000E-02	37	D029	0 0 0 0 0	

3748. C$_{17}$H$_{17}$ClO$_6$

Griseofulvin
(2*S-trans*)-7-Chloro-2',4,6-trimethoxy-6'-methylspiro[benzofuran-2(3H),1'-[2]cyclohexene]-3,4'-
dione
Fulvicin
Grisactin
Grifulvin
Griseostatin

RN: 126-07-8   **MP** (°C):   220.0
MW: 352.77   **BP** (°C):

Solubility (Moles/L)	Solubility (Grams/L)	Temp (°C)	Ref (#)	Evaluation (T P E A A)	Comments
1.830E-05	6.456E-03	15	E010	2 2 2 2 2	
2.466E-05	8.700E-03	20	N322	0 0 0 0 0	
3.260E-05	1.150E-02	21	E316	0 0 0 0 0	
3.175E-04	1.120E-01	21	M044	2 0 2 2 2	*sic*
4.025E-04	1.420E-01	21	M044	2 0 2 2 2	microsize, *sic*
2.126E-05	7.500E-03	22	C040	2 0 2 2 0	EFG
2.076E-05	7.325E-03	22	M382	2 1 1 1 1	average of 2
1.474E-05	5.200E-03	22.5	B422	2 0 2 2 2	
2.523E-05	8.900E-03	23	B362	0 0 0 0 0	
2.268E-05	8.000E-03	25	C037	2 1 2 2 2	
2.450E-05	8.643E-03	25	E010	2 2 2 2 2	
3.685E-05	1.300E-02	25	H015	1 0 0 0 1	
2.835E-05	1.000E-02	25	L033	1 0 2 1 1	
2.268E-05	8.000E-03	25	M457	0 0 0 0 0	
2.750E-05	9.700E-03	25	P096	0 0 0 0 0	
2.551E-05	9.000E-03	27	B043	1 0 1 2 0	EFG
2.835E-05	1.000E-02	30	M045	2 0 0 0 0	
4.000E-05	1.411E-02	30	O321	0 0 0 0 0	
4.252E-05	1.500E-02	30	O321	0 0 0 0 0	
3.510E-05	1.238E-02	35	E010	2 2 2 2 2	
3.969E-05	1.400E-02	37	B039	2 1 1 1 0	EFG
4.252E-05	1.500E-02	37	B043	1 0 1 2 0	EFG
3.969E-05	1.400E-02	37	B045	1 0 1 1 1	
4.054E-05	1.430E-02	37	F033	2 0 2 0 2	
3.968E-05	1.400E-02	37	G011	1 0 1 1 0	EFG
4.252E-05	1.500E-02	37	K018	1 0 0 0 1	
5.669E-05	2.000E-02	45	B043	1 0 1 2 0	EFG

(continued)

3748. C$_{17}$H$_{17}$ClO$_6$ (continued)

Solubility (Moles/L)	Solubility (Grams/L)	Temp (°C)	Ref (#)	Evaluation (T P E A A)	Comments
6.140E-05	2.166E-02	45	E010	2 2 2 2 2	
3.798E-05	1.340E-02	ns	D340	0 0 0 0 0	
2.835E-04	1.000E-01	ns	K444	0 0 0 0 0	
2.466E-05	8.700E-03	ns	N323	0 0 0 0 0	

3749. C$_{17}$H$_{17}$Cl$_2$N
Sertraline
(1S-cis)-4-(3,4-dichlorophenyl)-1,2,3,4-tetrahydro-N-methyl-1-naphthalenamine
RN: 79617-96-2   **MP** (°C):
MW: 306.24   **BP** (°C):

Solubility (Moles/L)	Solubility (Grams/L)	Temp (°C)	Ref (#)	Evaluation (T P E A A)	Comments
<3.27E-04	<1.00E-01	rt	B435	0 0 0 0 0	

3750. C$_{17}$H$_{17}$NO$_2$
Apomorphine
Apomorphin
RN: 58-00-4   **MP** (°C):
MW: 267.33   **BP** (°C):

Solubility (Moles/L)	Solubility (Grams/L)	Temp (°C)	Ref (#)	Evaluation (T P E A A)	Comments
4.000E-04	1.069E-01	15	K059	2 2 2 0 0	
7.481E-02	2.000E+01	25	P312	0 0 0 0 0	

3751. C$_{17}$H$_{17}$NO$_5$
N-Benzyloxycarbonyl-L-tyrosine
Carbobenzoxytyrosine
RN: 1164-16-5   **MP** (°C):
MW: 315.33   **BP** (°C):

Solubility (Moles/L)	Solubility (Grams/L)	Temp (°C)	Ref (#)	Evaluation (T P E A A)	Comments
4.852E-03	1.530E+00	25.1	N026	0 0 0 0 0	

3752. C$_{17}$H$_{17}$N$_5$O$_5$
9-[5′-(O-Benzoyl)-β-D-arabinofuranosyl]adenine ester
RN: 42782-57-0   **MP** (°C):   223.0
MW: 371.36   **BP** (°C):

Solubility (Moles/L)	Solubility (Grams/L)	Temp (°C)	Ref (#)	Evaluation (T P E A A)	Comments
2.154E-04	8.000E-02	ns	B134	0 1 1 1 0	

3753. $C_{17}H_{18}ClNO_6$
Griseofulvin-4'-oxime
Spiro[benzofuran-2(3H),1'-[2]cyclohexene]-3,4'-dione, 7-chloro-2',4,6-trimethoxy-6'-methyl-, 4'-oxime
RN: 13215-54-8 **MP** (°C):
MW: 367.79 **BP** (°C):

Solubility (Moles/L)	Solubility (Grams/L)	Temp (°C)	Ref (#)	Evaluation (T P E A A)	Comments
3.589E-04	1.320E-01	37	F033	2 0 2 0 2	

3754. $C_{17}H_{18}ClN_5O_6$
Dis. A. 8
Ethanol, 2,2'-[[4-[(2-chloro-4,6-dinitrophenyl)azo]-3-methylphenyl]imino]bis-
RN: 65125-87-3 **MP** (°C):
MW: 423.82 **BP** (°C):

Solubility (Moles/L)	Solubility (Grams/L)	Temp (°C)	Ref (#)	Evaluation (T P E A A)	Comments
5.000E-07	2.119E-04	25	B333	0 0 0 0 0	

3755. $C_{17}H_{18}Cl_2N_4O_4$
Dis. A. 10
Ethanol, 2,2'-[4-(2,6-dichloro-4-nitrophenylazo)-m-tolylimino]di-
RN: 58528-60-2 **MP** (°C):
MW: 413.26 **BP** (°C):

Solubility (Moles/L)	Solubility (Grams/L)	Temp (°C)	Ref (#)	Evaluation (T P E A A)	Comments
1.100E-06	4.546E-04	25	B333	0 0 0 0 0	

3756. $C_{17}H_{18}FN_3O_3$
Ciprofloxacin
1-Cyclopropyl-6-fluoro-1,4-dihydro-4-oxo-7-(1-piperazinyl)-3-quinolinecarboxylic
Baycip
Velmonit
RN: 85721-33-1 **MP** (°C):
MW: 331.35 **BP** (°C):

Solubility (Moles/L)	Solubility (Grams/L)	Temp (°C)	Ref (#)	Evaluation (T P E A A)	Comments
1.117E-04	3.700E-02	6	Y421	0 0 0 0 0	
1.630E-04	5.400E-02	22.5	B422	2 0 2 2 2	
2.595E-04	8.600E-02	25	Y421	0 0 0 0 0	
4.225E-04	1.400E-01	30	Y421	0 0 0 0 0	
5.131E-04	1.700E-01	40	Y421	0 0 0 0 0	
3.730E+00	1.236E+03	c	B443	0 0 0 0 0	

3757. C$_{17}$H$_{18}$N$_2$O$_6$
Nifedipine
3,5-Pyridinedicarboxylicacid
RN: 21829-25-4   **MP** (°C):   172–174
MW: 346.34   **BP** (°C):

Solubility (Moles/L)	Solubility (Grams/L)	Temp (°C)	Ref (#)	Evaluation (T P E A A)	Comments
1.675E-05	5.800E-03	25	B387	0 0 0 0 0	
2.887E-05	1.000E-02	ns	K444	0 0 0 0 0	
1.738E-05	6.019E-03	ns	R427	0 0 0 0 0	

3758. C$_{17}$H$_{18}$N$_4$O$_3$S
4-Sulfanilamido-1-phenyl-2,3-dimethyl-5-pyrazolone
RN: 71119-16-9   **MP** (°C):
MW: 358.42   **BP** (°C):

Solubility (Moles/L)	Solubility (Grams/L)	Temp (°C)	Ref (#)	Evaluation (T P E A A)	Comments
4.352E-04	1.560E-01	37	R045	1 2 1 1 2	

3759. C$_{17}$H$_{19}$ClN$_2$S
4-Chloropromazine
4-Chloro-*N,N*-dimethyl-10H-phenothiazine-10-propanamide
RN: 13094-24-1   **MP** (°C):
MW: 318.87   **BP** (°C):

Solubility (Moles/L)	Solubility (Grams/L)	Temp (°C)	Ref (#)	Evaluation (T P E A A)	Comments
1.100E-05	3.508E-03	ns	G023	0 0 1 1 1	

3760. C$_{17}$H$_{19}$ClN$_2$S
3-Chloropromazine
3-Chloro-*N,N*-dimethyl-10H-phenothiazine-10-propanamide
RN: 484-19-5   **MP** (°C):
MW: 318.87   **BP** (°C):

Solubility (Moles/L)	Solubility (Grams/L)	Temp (°C)	Ref (#)	Evaluation (T P E A A)	Comments
1.000E-05	3.189E-03	ns	G023	0 0 1 1 1	

3761. C$_{17}$H$_{19}$ClN$_2$S
1-Chloropromazine
1-Chloro-*N,N*-dimethyl-10H-phenothiazine-10-propanamide
RN: 13100-13-5   **MP** (°C):
MW: 318.87   **BP** (°C):

Solubility (Moles/L)	Solubility (Grams/L)	Temp (°C)	Ref (#)	Evaluation (T P E A A)	Comments
1.200E-05	3.826E-03	0	G023	0 0 0 0 1	

3762. C₁₇H₁₉ClN₄O₄
C.I. Disperse red 5
Ethanol, 2,2′-[[4-[(2-chloro-4-nitrophenyl)azo]-3-methylphenyl]imino]*bis*-
RN: 3769-57-1   **MP** (°C):   192
MW: 378.82   **BP** (°C):

Solubility (Moles/L)	Solubility (Grams/L)	Temp (°C)	Ref (#)	Evaluation (T P E A A)	Comments
3.800E-07	1.440E-04	25	B333	0 0 0 0 0	

3763. C₁₇H₁₉ClO₆
Griseofulvin-4′-ol
Spiro[benzofuran-2(3H),1′-[2]cyclohexen]-3-one, 7-chloro-4′-hydroxy-2′,4,6-trimethoxy-6′-methyl-
RN: 13215-53-7   **MP** (°C):
MW: 354.79   **BP** (°C):

Solubility (Moles/L)	Solubility (Grams/L)	Temp (°C)	Ref (#)	Evaluation (T P E A A)	Comments
7.129E-04	2.529E-01	37	F033	2 0 2 0 2	average of 2

3764. C₁₇H₁₉NO₃
Piperine
Piperidine, 1-[5-(1,3-benzodioxol-5-yl)-1-oxo-2,4-pentadienyl]-, (E,E)-
N-[(E,E)-Piperoyl]piperidine
RN: 94-62-2   **MP** (°C):   130.0
MW: 285.35   **BP** (°C):

Solubility (Moles/L)	Solubility (Grams/L)	Temp (°C)	Ref (#)	Evaluation (T P E A A)	Comments
1.400E-04	3.995E-02	15	K059	2 2 2 0 1	
1.402E-04	4.000E-02	18	F300	1 0 0 0 0	
3.504E-04	9.999E-02	rt	D021	0 0 1 1 0	

3765. C₁₇H₁₉NO₃
1-Methyl-1-nitro-2-(*p*-methylphenyl)-2-*p*-ethoxylphenyl)ethane
RN: 53982-07-3   **MP** (°C):
MW: 285.35   **BP** (°C):

Solubility (Moles/L)	Solubility (Grams/L)	Temp (°C)	Ref (#)	Evaluation (T P E A A)	Comments
8.060E-06	2.300E-03	rt	C122	0 0 0 0 0	

3766. C₁₇H₁₉NO₃
Hydromorphone
Dilaudid
PMS-Hydromorphone
Dihydromorphinone
RN: 466-99-9   **MP** (°C):
MW: 285.35   **BP** (°C):

Solubility (Moles/L)	Solubility (Grams/L)	Temp (°C)	Ref (#)	Evaluation (T P E A A)	Comments
6.767E-03	1.931E+00	25	R338	0 0 0 0 0	

3767. C₁₇H₁₉NO₃
Morphine
Morphin
7,8-Didehydro-4,5-epoxy-17-methylmorphinan-3,6-diol
RN: 57-27-2   **MP** (°C):   254dec
MW: 285.35   **BP** (°C):

Solubility (Moles/L)	Solubility (Grams/L)	Temp (°C)	Ref (#)	Evaluation (T P E A A)	Comments
5.000E-04	1.427E-01	15	K059	2 2 2 0 0	
5.222E-04	1.490E-01	20	B061	1 0 1 1 2	
5.257E-04	1.500E-01	20	F300	1 0 0 0 0	
1.209E-03	3.450E-01	25	R338	0 0 0 0 0	
7.200E-04	2.054E-01	30	L068	1 0 0 1 0	EFG
1.000E-03	2.853E-01	30	L069	1 0 1 1 0	EFG
8.761E-04	2.500E-01	35	R418	0 0 0 0 0	Intrinsic
1.051E-03	2.999E-01	rt	D021	0 0 1 1 0	

3768. C₁₇H₁₉NO₃.H₂O
Morphine (monohydrate)
Morphinan-3,6-diol, 7,8-didehydro-4,5-epoxy-17-methyl-(5α,6α)-, monohydrate
RN: 6009-81-0   **MP** (°C):   254dec
MW: 303.36   **BP** (°C):

Solubility (Moles/L)	Solubility (Grams/L)	Temp (°C)	Ref (#)	Evaluation (T P E A A)	Comments
9.328E-04	2.830E-01	c	D004	0 0 0 0 0	
3.064E-03	9.294E-01	h	D004	0 0 0 0 0	

3769. C₁₇H₁₉NO₄
1-Methyl-1-nitro-2,2-bis(p-methoxylphenyl)ethane
RN: 34197-26-7   **MP** (°C):
MW: 301.35   **BP** (°C):

Solubility (Moles/L)	Solubility (Grams/L)	Temp (°C)	Ref (#)	Evaluation (T P E A A)	Comments
2.854E-05	8.600E-03	rt	C122	0 0 0 0 0	

3770. C₁₇H₁₉N₃

Antazoline

Albalon-A

RN: 91-75-8	**MP** (°C): 120
MW: 265.36	**BP** (°C):

Solubility (Moles/L)	Solubility (Grams/L)	Temp (°C)	Ref (#)	Evaluation (T P E A A)	Comments
2.500E-03	6.634E-01	30	L068	1 0 0 1 0	EFG
1.900E-02	5.042E+00	37.5	L034	2 2 0 1 2	pH 7.4

3771. C₁₇H₁₉N₅O₆

Dis. A. 1

Ethanol, 2,2′-[4-(2,4-dinitrophenylazo)-*m*-tolylimino]di-

Disperse violet 4K

Terasil violet P 4RT

RN: 41541-13-3	**MP** (°C): 190
MW: 389.37	**BP** (°C):

Solubility (Moles/L)	Solubility (Grams/L)	Temp (°C)	Ref (#)	Evaluation (T P E A A)	Comments
7.000E-07	2.726E-04	25	B333	0 0 0 0 0	

3772. C₁₇H₂₀ClN₅O₂

1H-Purine-2,6-dione, 8-(2-amino-4-chlorophenyl)-3,7-dihydro-1,3-dipropyl-

1,3-Dipropyl-8-(2-amino-4-chlorophenyl)xanthine

PACPX

RN: 85872-51-1	**MP** (°C):
MW: 361.83	**BP** (°C):

Solubility (Moles/L)	Solubility (Grams/L)	Temp (°C)	Ref (#)	Evaluation (T P E A A)	Comments
<2.76E-07	<1.00E-04	ns	H316	0 0 0 0 0	pH 7.4
1.105E-06	4.000E-04	ns	H316	0 0 0 0 0	0.1N HCL

3773. C₁₇H₂₀N₂O

Michler's ketone

Tetramethyldiaminobenzophenone

bis[4-(Dimethylamino)phenyl]-methanone

p,p′-bis(*N,N*-Dimethylamino)benzophenone

4,4[-bis(Dimethylamino)benzophenone

RN: 90-94-8	**MP** (°C): 172.0
MW: 268.36	**BP** (°C):

Solubility (Moles/L)	Solubility (Grams/L)	Temp (°C)	Ref (#)	Evaluation (T P E A A)	Comments
1.490E-03	3.998E-01	rt	D021	0 0 1 1 0	

3774. $C_{17}H_{20}N_2O_2$
Tropicamide
RN: 1508-75-4 **MP** (°C):
MW: 284.36 **BP** (°C):

Solubility (Moles/L)	Solubility (Grams/L)	Temp (°C)	Ref (#)	Evaluation (T P E A A)	Comments
7.032E-04	2.000E-01	25	C414	1 0 1 1 0	EFG

3775. $C_{17}H_{20}N_2S$
Promethazine
10-(2-Dimethylaminopropyl)phenothiazine
10-(2-Dimethylamino-2-methylethyl)phenothiazine
Fenergan
Protazine
Thiergan
RN: 60-87-7 **MP** (°C): 60
MW: 284.43 **BP** (°C): 191

Solubility (Moles/L)	Solubility (Grams/L)	Temp (°C)	Ref (#)	Evaluation (T P E A A)	Comments
1.350E-06	3.839E-04	22.5	B440	0 0 0 0 0	
5.500E-05	1.564E-02	24	G023	2 0 1 1 1	
4.390E+00	1.249E+03	c	B443	0 0 0 0 0	

3776. $C_{17}H_{20}N_2S$
Promazine
Primazine
Sparine
Prozine
RN: 58-40-2 **MP** (°C): 32
MW: 284.43 **BP** (°C):

Solubility (Moles/L)	Solubility (Grams/L)	Temp (°C)	Ref (#)	Evaluation (T P E A A)	Comments
5.000E-05	1.422E-02	24	G023	2 0 1 1 1	
5.000E-05	1.422E-02	ns	G023	0 0 0 0 1	

3777. $C_{17}H_{20}N_4O_4$
C.I. Disperse red 17
Ethanol, 2,2′-[[3-methyl-4-[(4-nitrophenyl)azo]phenyl]imino]*bis*-
RN: 3179-89-3 **MP** (°C): 160
MW: 344.37 **BP** (°C):

Solubility (Moles/L)	Solubility (Grams/L)	Temp (°C)	Ref (#)	Evaluation (T P E A A)	Comments
1.800E-06	6.199E-04	25	B333	0 0 0 0 0	

3778. C₁₇H₂₀N₄O₅

Dis. A. 13
4-Nitro-2-methoxy-4′-di(β-hydroxyethyl)-aminoazobenzene
Ethanol, 2,2′-[[4-[(2-methoxy-4-nitrophenyl)azo]phenyl]imino]bis
Ethanol, 2,2′-[p-(2-methoxy-4-nitrophenylazo)phenylimino]di-

RN: 41541-14-4 **MP** (°C):
MW: 360.37 **BP** (°C):

Solubility (Moles/L)	Solubility (Grams/L)	Temp (°C)	Ref (#)	Evaluation (T P E A A)	Comments
2.000E-05	7.207E-03	25	B333	0 0 0 0 0	
6.826E-04	2.460E-01	100	P313	0 0 0 0 0	

3779. C₁₇H₂₀N₄O₅S

Benzenesulfonic acid, 4-(2,3,6,7-tetrahydro-2,6-dioxo-1,3-dipropyl-1H-purin-8-yl)-

RN: 89073-57-4 **MP** (°C):
MW: 392.44 **BP** (°C):

Solubility (Moles/L)	Solubility (Grams/L)	Temp (°C)	Ref (#)	Evaluation (T P E A A)	Comments
3.313E-03	1.300E+00	ns	H316	0 0 0 0 0	0.1N HCL
>6.12E-02	>2.40E+01	ns	H316	0 0 0 0 0	pH 7.4

3780. C₁₇H₂₀N₄O₆

Riboflavine
Riboflavin
Robiflavine
7,8-Dimethyl-10-ribitylisoalloxazine
Zinvit-G
E-101

RN: 83-88-5 **MP** (°C): 290
MW: 376.37 **BP** (°C):

Solubility (Moles/L)	Solubility (Grams/L)	Temp (°C)	Ref (#)	Evaluation (T P E A A)	Comments
2.657E-04	9.999E-02	20	A022	1 0 0 0 0	
2.250E-04	8.468E-02	25	A079	1 0 1 1 2	
2.657E-04	9.999E-02	25	D041	1 0 0 0 0	
1.754E-04	6.600E-02	25	D315	0 0 0 0 0	
2.192E-04	8.250E-02	30	C409	2 0 1 2 2	
3.959E-04	1.490E-01	37	E018	1 0 2 1 2	
2.089E-04	7.864E-02	ns	R427	0 0 0 0 0	

3781. C₁₇H₂₀O₆

3781. $C_{17}H_{20}O_6$

Mycophenolic acid
6-(1,3-Dihydro-7-hydroxy-5-methoxy-4-methyl-1-oxoisobenzofuran-6-yl)-4-methyl-4-hexanoic
    acid

RN: 24280-93-1   **MP** (°C):
MW: 320.35   **BP** (°C):

Solubility (Moles/L)	Solubility (Grams/L)	Temp (°C)	Ref (#)	Evaluation (T P E A A)	Comments
4.058E-05	1.300E-02	25	L333	1 1 1 1 0	

3782. $C_{17}H_{21}NO_2$

Napropamide
N,N-Diethyl-2-(1-naphthyloxy)propanamide
Devrinol 50W
Devrinol
Devrinol 10G
Devrinol 2E

RN: 15299-99-7   **MP** (°C):   75.1
MW: 271.36   **BP** (°C):

Solubility (Moles/L)	Solubility (Grams/L)	Temp (°C)	Ref (#)	Evaluation (T P E A A)	Comments
2.690E-04	7.300E-02	20	M161	1 0 0 0 1	

3783. $C_{17}H_{21}NO_3$

Etodolac
(+/-)-1,8-Diethyl-1,3,4,9-tetrahydropyrano-(3,4-b)indole-1-acetic acid
Lodine

RN: 41340-25-4   **MP** (°C):
MW: 287.36   **BP** (°C):

Solubility (Moles/L)	Solubility (Grams/L)	Temp (°C)	Ref (#)	Evaluation (T P E A A)	Comments
1.392E-04	4.000E-02	37	Y421	0 0 0 0 0	

3784. $C_{17}H_{21}NO_4$

Scopolamine
Scopolamin
Hyoscine
Murocoll
Plexonal
Transderm-SCOP

RN: 51-34-3   **MP** (°C):   59
MW: 303.36   **BP** (°C):

Solubility (Moles/L)	Solubility (Grams/L)	Temp (°C)	Ref (#)	Evaluation (T P E A A)	Comments
3.132E-01	9.500E+01	15	F300	1 0 0 0 1	
3.296E-01	1.000E+02	ns	C109	0 0 0 0 1	

3785. C₁₇H₂₁NO₄

Cocaine
L-Cocaine
L-Cocain

RN:	50-36-2	**MP** (°C):	98	
MW:	303.36	**BP** (°C):		

Solubility (Moles/L)	Solubility (Grams/L)	Temp (°C)	Ref (#)	Evaluation (T P E A A)	Comments
4.000E-03	1.213E+00	15	K059	2 2 2 0 0	
5.934E-03	1.800E+00	22	F300	1 0 0 0 1	
5.485E-03	1.664E+00	25	D004	0 0 0 0 0	
5.266E-03	1.597E+00	25	D041	1 0 0 0 1	
1.248E-02	3.786E+00	80	D041	1 0 0 0 1	

3786. C₁₇H₂₁N₃O₂

Dis. A. 2
Ethanol, 2,2′-[[3-methyl-4-(phenylazo)phenyl]imino]*bis*-
4-[bis(2-Hydroxyethyl)amino]-2-methylazobenzene

RN:	3771-38-8	**MP** (°C):	111	
MW:	299.38	**BP** (°C):		

Solubility (Moles/L)	Solubility (Grams/L)	Temp (°C)	Ref (#)	Evaluation (T P E A A)	Comments
7.600E-05	2.275E-02	25	B333	0 0 0 0 0	

3787. C₁₇H₂₁N₅O₂

1H-Purine-2,6-dione, 8-(2-aminophenyl)-3,7-dihydro-1,3-dipropyl-

RN:	96445-34-0	**MP** (°C):	276dec
MW:	327.39	**BP** (°C):	

Solubility (Moles/L)	Solubility (Grams/L)	Temp (°C)	Ref (#)	Evaluation (T P E A A)	Comments
<3.05E-06	<1.00E-03	ns	H316	0 0 0 0 0	pH 7.4
1.222E-05	4.000E-03	ns	H316	0 0 0 0 0	0.1N HCL

3788. C₁₇H₂₁N₅O₁₀

9-(1,3-Dihemisuccinate-2-propoxymethyl)guanine

RN:	88110-76-3	**MP** (°C):	167
MW:	455.38	**BP** (°C):	

Solubility (Moles/L)	Solubility (Grams/L)	Temp (°C)	Ref (#)	Evaluation (T P E A A)	Comments
1.039E-01	4.730E+01	25	B360	0 0 0 0 0	

3789. C$_{17}$H$_{22}$I$_3$N$_3$O$_8$

1,3-Benzenedicarboxamide, *N,N'*-bis(2,3-dihydroxypropyl)-5*RS*-[(2-hydroxy-1-oxopropyl)amino]-
2,4,6-triiodo-[*RS*-(*RS**,*RS**)]-

RN: 60166-94-1   **MP** (°C):
MW: 777.09   **BP** (°C):

Solubility (Moles/L)	Solubility (Grams/L)	Temp (°C)	Ref (#)	Evaluation (T P E A A)	Comments
1.379E-01	1.071E+02	25	P091	0 0 0 0 0	

3790. C$_{17}$H$_{22}$I$_3$N$_3$O$_8$

1,3-Benzenedicarboxamide, *N,N'*-bis(2,3-dihydroxypropyl)-5*S*-[(2-hydroxy-1-oxopropyl)amino]-
2,4,6-triiodo-[*RS*-(*RS**,*RS**)]-

RN: 77942-93-9   **MP** (°C):
MW: 777.09   **BP** (°C):

Solubility (Moles/L)	Solubility (Grams/L)	Temp (°C)	Ref (#)	Evaluation (T P E A A)	Comments
1.480E-01	1.150E+02	25	P091	0 0 0 0 0	

3791. C$_{17}$H$_{22}$I$_3$N$_3$O$_8$

1,3-Benzenedicarboxamide, *N*-(2,3-dihydroxypropyl)-*N'*-[2-hydroxy-1-(hydroxymethyl)ethyl]-5-
[(2-hydroxy-1-oxopropyl)amino]-2,4,6-triiodo-

RN: 77868-45-2   **MP** (°C):
MW: 777.09   **BP** (°C):

Solubility (Moles/L)	Solubility (Grams/L)	Temp (°C)	Ref (#)	Evaluation (T P E A A)	Comments
1.379E-01	1.071E+02	25	P091	0 0 0 0 0	

3792. C$_{17}$H$_{22}$I$_3$N$_3$O$_8$

1,3-Benzenedicarboxamide, *N,N'*-bis(2,3-dihydroxypropyl)-5*S*-[(2-hydroxy-1-oxopropyl)amino]-
2,4,6-triiodo-[*S*-(*S**,*S**)]-

RN: **MP** (°C):
MW: 777.09   **BP** (°C):

Solubility (Moles/L)	Solubility (Grams/L)	Temp (°C)	Ref (#)	Evaluation (T P E A A)	Comments
1.379E-01	1.071E+02	25	P091	0 0 0 0 0	

3793. $C_{17}H_{22}I_3N_3O_8$

DL-Iopamidol

1,3-Benzenedicarboxamide, N,N'-bis[2-hydroxy-1-(hydroxymethyl)ethyl]-5-[(2-hydroxy-1-oxopropyl)amino]-2,4,6-triiodo-, (S)-

L-Iopamidol

1,3-Benzenedicarboxamide, N,N'-bis(2,3-dihydroxypropyl)-5RS-[(2-hydroxy-1-oxopropyl)amino]-2,4,6-triiodo-[RS-(S*,S*)]-

RN: 60166-93-0   **MP** (°C):

MW: 777.09   **BP** (°C):

Solubility (Moles/L)	Solubility (Grams/L)	Temp (°C)	Ref (#)	Evaluation (T P E A A)	Comments
6.096E-01	4.737E+02	20	F178	1 0 0 0 1	EFG
1.580E-01	1.228E+02	20	F178	1 0 0 0 1	EFG
6.096E-01	4.737E+02	25	P091	0 0 0 0 0	
1.580E-01	1.228E+02	25	P091	0 0 0 0 0	
5.798E-01	4.505E+02	40	F178	1 0 0 0 1	EFG
1.963E-01	1.525E+02	40	F178	1 0 0 0 1	EFG
5.679E-01	4.413E+02	60	F178	1 0 0 0 1	EFG
3.120E-01	2.424E+02	60	F178	1 0 0 0 1	EFG
6.235E-01	4.845E+02	80	F178	1 0 0 0 1	EFG
5.209E-01	4.048E+02	80	F178	1 0 0 0 1	EFG
6.911E-01	5.370E+02	100	F178	1 0 0 0 1	EFG
7.098E-01	5.516E+02	100	F178	1 0 0 0 1	EFG

3794. $C_{17}H_{22}I_3N_3O_8$

1,3-Benzenedicarboxamide, N,N'-bis(2,3-dihydroxypropyl)-5RS-[(2-hydroxy-1-oxopropyl)amino]-2,4,6-triiodo-[RS-(RS*,S*)]-

RN: **MP** (°C):

MW: 777.09   **BP** (°C):

Solubility (Moles/L)	Solubility (Grams/L)	Temp (°C)	Ref (#)	Evaluation (T P E A A)	Comments
1.379E-01	1.071E+02	25	P091	0 0 0 0 0	

3795. $C_{17}H_{22}I_3N_3O_8$

1,3-Benzenedicarboxamide, N,N'-bis[2-hydroxy-1-(hydroxymethyl)ethyl]-5-[(2-hydroxy-1-oxopropyl)amino]-2,4,6-triiodo-, (RS)-

RN: 60208-45-9   **MP** (°C):

MW: 777.09   **BP** (°C):

Solubility (Moles/L)	Solubility (Grams/L)	Temp (°C)	Ref (#)	Evaluation (T P E A A)	Comments
1.775E-01	1.379E+02	25	P091	0 0 0 0 0	

3796. C$_{17}$H$_{22}$I$_3$N$_3$O$_8$
1,3-Benzenedicarboxamide, N,N'-bis(2,3-dihydroxypropyl)-5S-[(2-hydroxy-1-oxopropyl)amino]-
2,4,6-triiodo-(RS,S)-
RN: **MP** (°C):
MW: 777.09   **BP** (°C):

Solubility (Moles/L)	Solubility (Grams/L)	Temp (°C)	Ref (#)	Evaluation (T P E A A)	Comments
1.379E-01	1.071E+02	25	P091	0 0 0 0 0	

3797. C$_{17}$H$_{22}$I$_3$N$_3$O$_9$
1,3-Benzenedicarboxamide, 5-[(2,3-dihydroxy-1-oxopropyl)amino]-N,N'-bis[2-hydroxy-1-
(hydroxymethyl)ethyl]-2,4,6-triiodo-
RN: 69698-47-1   **MP** (°C):
MW: 793.09   **BP** (°C):

Solubility (Moles/L)	Solubility (Grams/L)	Temp (°C)	Ref (#)	Evaluation (T P E A A)	Comments
6.573E-02	5.213E+01	25	P091	0 0 0 0 0	

3798. C$_{17}$H$_{22}$I$_3$N$_3$O$_9$
1,3-Benzenedicarboxamide, 5-[(2,3-dihydroxy-1-oxobutyl)amino]-N,N'-bis[2-hydroxy-1-
(hydroxymethyl)ethyl]-2,4,6-triiodo-
RN: 129968-26-9   **MP** (°C):
MW: 793.09   **BP** (°C):

Solubility (Moles/L)	Solubility (Grams/L)	Temp (°C)	Ref (#)	Evaluation (T P E A A)	Comments
5.430E-02	4.306E+01	25	P091	0 0 0 0 0	

3799. C$_{17}$H$_{22}$N$_4$O$_3$S
2-(N4-Acetylsulfanilylamino)-4-n-amylpyrimidine
RN: **MP** (°C):
MW: 362.45   **BP** (°C):

Solubility (Moles/L)	Solubility (Grams/L)	Temp (°C)	Ref (#)	Evaluation (T P E A A)	Comments
1.214E-05	4.400E-03	37	R076	1 2 0 0 1	

3800. C$_{17}$H$_{22}$N$_4$O$_7$.0.75H$_2$O
2'-(2-Methyl-3-one-pentanyl)-6-methoxypurine arabinoside (0.75 hydrate)
RN: 145913-50-4   **MP** (°C):   55–60
MW: 407.90   **BP** (°C):

Solubility (Moles/L)	Solubility (Grams/L)	Temp (°C)	Ref (#)	Evaluation (T P E A A)	Comments
8.770E-02	3.577E+01	37	C348	0 0 0 0 0	pH 7.00

3801. $C_{17}H_{23}NO$

N,N-Octamethylenecinnamamide
Octahydro-1-(1-oxo-3-phenyl-2-propenyl)1H-azonine
RN: 59832-07-4 **MP** (°C):
MW: 257.38 **BP** (°C):

Solubility (Moles/L)	Solubility (Grams/L)	Temp (°C)	Ref (#)	Evaluation (T P E A A)	Comments
2.460E-04	6.332E-02	ns	H350	0 0 0 0 0	

3802. $C_{17}H_{23}NO$

N-Cyclooctylcinnamamide
2-Propenamide, *N*-cyclooctyl-3-phenyl-
RN: 59832-00-7 **MP** (°C):
MW: 257.38 **BP** (°C):

Solubility (Moles/L)	Solubility (Grams/L)	Temp (°C)	Ref (#)	Evaluation (T P E A A)	Comments
2.660E-06	6.846E-04	ns	H350	0 0 0 0 0	

3803. $C_{17}H_{23}NO_3$

Hyoscyamine
Hyoscyamin
Benzeneacetic acid, α-(hydroxymethyl)-, 8-methyl-8-azabicyclo[3.2.1]oct-3-yl ester, [3(*S*)-*endo*]-
Daturine
Duboisine
L-Hyoscyamine
RN: 101-31-5 **MP** (°C): 108.5
MW: 289.38 **BP** (°C):

Solublllty (Moles/L)	Solubility (Grams/L)	Temp (°C)	Ref (#)	Evaluation (T P E A A)	Comments
1.244E-02	3.600E+00	20	F300	1 0 0 0 2	
1.225E-02	3.546E+00	c	D004	0 0 0 0 0	

3804. $C_{17}H_{23}NO_3$

Atropine
Atropin
8-Methyl-8-azabicyclo[3.2.1]octan-3-yl 3-hydroxy-2-phenylpropionate
Neo-diophen
Minims
RN: 51-55-8 **MP** (°C): 115
MW: 289.38 **BP** (°C):

Solubility (Moles/L)	Solubility (Grams/L)	Temp (°C)	Ref (#)	Evaluation (T P E A A)	Comments
5.500E-03	1.592E+00	15	K059	2 2 2 0 1	
5.529E-03	1.600E+00	18	F300	1 0 0 0 1	

(continued)

3804. C$_{17}$H$_{23}$NO$_3$ (continued)

Solubility (Moles/L)	Solubility (Grams/L)	Temp (°C)	Ref (#)	Evaluation (T P E A A)	Comments
6.898E-03	1.996E+00	20	D041	1 0 0 0 0	
1.032E-02	2.987E+00	20	K052	1 1 1 1 2	
1.610E+00	4.659E+02	25	B443	0 0 0 0 0	
1.148E-02	3.322E+00	25	D004	0 0 0 0 0	
7.586E-03	2.195E+00	rt	D021	0 0 1 1 1	

3805. C$_{17}$H$_{23}$NO$_5$

Benzoic acid, 2-(acetyloxy)-, 2-[bis(1-methylethyl)amino]-2-oxoethyl ester

RN: 116482-76-9   **MP** (°C): 108.9

MW: 321.38   **BP** (°C):

Solubility (Moles/L)	Solubility (Grams/L)	Temp (°C)	Ref (#)	Evaluation (T P E A A)	Comments
5.601E-04	1.800E-01	21	N335	0 0 0 0 0	

3806. C$_{17}$H$_{23}$NO$_5$

Benzoic acid, 2-(acetyloxy)-, 2-(dipropylamino)-2-oxoethyl ester

RN: 116482-75-8   **MP** (°C): 50.5

MW: 321.38   **BP** (°C):

Solubility (Moles/L)	Solubility (Grams/L)	Temp (°C)	Ref (#)	Evaluation (T P E A A)	Comments
2.240E-03	7.200E-01	21	N335	0 0 0 0 0	

3807. C$_{17}$H$_{23}$N$_3$O

Aeo-antergan

1,2-Ethanediamine, N-[(4-methoxyphenyl)methyl]-N',N'-dimethyl-N-2-pyridinyl-

Dorantamin

Anthisan

Dipane

Copsamine

RN: 91-84-9   **MP** (°C):

MW: 285.39   **BP** (°C):

Solubility (Moles/L)	Solubility (Grams/L)	Temp (°C)	Ref (#)	Evaluation (T P E A A)	Comments
1.200E-02	3.425E+00	37.5	L034	2 2 0 1 2	pH 7.4

3808. C$_{17}$H$_{23}$N$_3$O$_2$

2-Methoxy-*N*-[2-(diethyl-amino)ethyl]-4-quinoline carboxamide
N-[2-(Diethylamino)ethyl]-2-methoxyquinoline-4-carboxamide

RN: 2716-98-5 **MP** (°C):
MW: 301.39   **BP** (°C):

Solubility (Moles/L)	Solubility (Grams/L)	Temp (°C)	Ref (#)	Evaluation (T P E A A)	Comments
3.000E-03	9.042E-01	ns	B018	0 0 0 0 1	
3.000E-03	9.042E-01	ns	M066	0 0 0 0 0	

3809. C$_{17}$H$_{24}$N$_4$O$_5$

1,5-Dipivaloyloxymethyl allopurinol

RN: 98827-16-8 **MP** (°C): 136–137
MW: 364.40   **BP** (°C):

Solubility (Moles/L)	Solubility (Grams/L)	Temp (°C)	Ref (#)	Evaluation (T P E A A)	Comments
5.488E-05	2.000E-02	22	B322	0 0 0 0 0	
5.495E-05	2.003E-02	ns	R427	0 0 0 0 0	

3810. C$_{17}$H$_{24}$N$_4$O$_5$

2,5-Dipivaloyloxymethyl allopurinol

RN: 98827-17-9 **MP** (°C): 145–146
MW: 364.40   **BP** (°C):

Solubility (Moles/L)	Solubility (Grams/L)	Temp (°C)	Ref (#)	Evaluation (T P E A A)	Comments
1.235E-04	4.500E-02	22	B322	0 0 0 0 0	

3811. C$_{17}$H$_{24}$N$_4$O$_6$

2′-Hexanyl-6-methoxypurine arabinoside

RN: 145913-39-9 **MP** (°C):
MW: 380.40   **BP** (°C):

Solubility (Moles/L)	Solubility (Grams/L)	Temp (°C)	Ref (#)	Evaluation (T P E A A)	Comments
1.890E-02	7.190E+00	37	C348	0 0 0 0 0	pH 7.00

3812. C$_{17}$H$_{25}$NO

N-Octylcinnamamide
2-Propenamide, *N*-octyl-3-phenyl-

RN: 55030-48-3 **MP** (°C):
MW: 259.39   **BP** (°C):

Solubility (Moles/L)	Solubility (Grams/L)	Temp (°C)	Ref (#)	Evaluation (T P E A A)	Comments
1.390E-06	3.606E-04	ns	H350	0 0 0 0 0	

3813. C₁₇H₂₅NO₃

Acetamide, 2-(benzoyloxy)-*N*,*N*-bis(2-methylpropyl)-

RN: 115193-33-4 **MP** (°C): 44.5
MW: 291.39 **BP** (°C):

Solubility (Moles/L)	Solubility (Grams/L)	Temp (°C)	Ref (#)	Evaluation (T P E A A)	Comments
2.745E-04	8.000E-02	22	N317	1 1 2 1 2	

3814. C₁₇H₂₅NO₃

Acetamide, 2-(benzoyloxy)-*N*,*N*-acetamide, 2-(benzoyloxy)-*N*,*N*-dibutyl-

RN: 106231-57-6 **MP** (°C): 25
MW: 291.39 **BP** (°C): 428.6

Solubility (Moles/L)	Solubility (Grams/L)	Temp (°C)	Ref (#)	Evaluation (T P E A A)	Comments
2.745E-04	8.000E-02	22	B427	1 0 0 1 1	in 0.01M HCl
2.745E-04	8.000E-02	22	N317	1 1 2 1 2	

3815. C₁₇H₂₅NO₄

Octyl acetaminophen
Carbonic acid, octyl ester, ester with 4′-hydroxyacetanilide
Acetanilide, 4′-hydroxy-, octyl carbonate (ester)

RN: 19872-70-9 **MP** (°C): 82.5–83
MW: 307.39 **BP** (°C):

Solubility (Moles/L)	Solubility (Grams/L)	Temp (°C)	Ref (#)	Evaluation (T P E A A)	Comments
1.431E-05	4.400E-03	37	D029	0 0 0 0 0	

3816. C₁₇H₂₅N₅O₆

9-(1,3-Dibutyrate-2-propoxymethyl)guanine

RN: 88110-71-8 **MP** (°C): 200
MW: 395.42 **BP** (°C):

Solubility (Moles/L)	Solubility (Grams/L)	Temp (°C)	Ref (#)	Evaluation (T P E A A)	Comments
3.541E-04	1.400E-01	25	B360	0 0 0 0 0	

3817. C₁₇H₂₆ClNO₂

Butachlor
N-(Butoxymethyl)-2-chloro-N-(2,6-diethylphenyl)acetamide
N-(Butoxymethyl)-2-chloro-2′,6′-diethylacetanilide
Machete
Butanex
Hiltachlor

RN: 23184-66-9   **MP** (°C): <–5
MW: 311.86   **BP** (°C): 196

Solubility (Moles/L)	Solubility (Grams/L)	Temp (°C)	Ref (#)	Evaluation (T P E A A)	Comments
6.413E-05	2.000E-02	20	M161	1 0 0 0 1	
6.412E-05	2.000E-02	ns	S460	0 0 0 0 0	
7.055E-05	2.200E-02	ns	Y414	0 0 0 0 0	
7.055E-02	2.200E+01	ns	Y414	0 0 0 0 0	

3818. C₁₇H₂₆O₃

Decyl-p-hydroxybenzoate
Decyl p-hydroxybenzoate
n-Decyl p-hydroxybenzoate

RN: 69679-30-7   **MP** (°C): 58
MW: 278.39   **BP** (°C):

Solubility (Moles/L)	Solubility (Grams/L)	Temp (°C)	Ref (#)	Evaluation (T P E A A)	Comments
3.200E-05	8.909E-03	15	B355	0 0 0 0 0	
3.710E-05	1.033E-02	20	B355	0 0 0 0 0	
8.800E-05	2.450E-02	25	B355	0 0 0 0 0	
1.303E-03	3.629E-01	25	D081	1 2 2 1 2	sic
7.943E-05	2.211E-02	25	F322	2 0 1 1 0	EFG

3819. C₁₇H₂₇NO₂

Terbutol
2,6-Di-tert-butyl-p-tolyl methylcarbamate

RN: 1918-11-2   **MP** (°C): 185
MW: 277.41   **BP** (°C):

Solubility (Moles/L)	Solubility (Grams/L)	Temp (°C)	Ref (#)	Evaluation (T P E A A)	Comments
2.343E-05	6.500E-03	25	B200	1 0 0 0 0	
2.523E-05	7.000E-03	ns	H042	0 0 0 0 0	

3820. C$_{17}$H$_{27}$NO$_2$
Venlafaxine
RN: 93413-69-5   MP (°C):
MW: 277.41   BP (°C):

Solubility (Moles/L)	Solubility (Grams/L)	Temp (°C)	Ref (#)	Evaluation (T P E A A)	Comments
<3.60E-04	<1.00E-01	rt	B435	0 0 0 0 0	

3821. C$_{17}$H$_{27}$NO$_3$
Pramoxine
Pramocaine
RN: 140-65-8   MP (°C):
MW: 293.41   BP (°C):

Solubility (Moles/L)	Solubility (Grams/L)	Temp (°C)	Ref (#)	Evaluation (T P E A A)	Comments
1.218E-05	3.574E-03	22.5	B440	0 0 0 0 0	

3822. C$_{17}$H$_{27}$NO$_3$
Stadacain
4-Butoxybenzoic acid 2-(diethyl-amino)ethyl ester
RN: 2350-32-5   MP (°C):   146
MW: 293.41   BP (°C):

Solubility (Moles/L)	Solubility (Grams/L)	Temp (°C)	Ref (#)	Evaluation (T P E A A)	Comments
1.300E-04	3.814E-02	ns	M066	0 0 0 0 1	

3823. C$_{17}$H$_{27}$NO$_4$
Nadolol
Corgard
Nadolol
RN: 42200-33-9   MP (°C):
MW: 309.41   BP (°C):

Solubility (Moles/L)	Solubility (Grams/L)	Temp (°C)	Ref (#)	Evaluation (T P E A A)	Comments
2.683E-02	8.300E+00	25	A412	1 0 2 2 1	int

3824. C$_{17}$H$_{28}$
4-Phenylundecane
RN: MP (°C):
MW: 232.41   BP (°C):

Solubility (Moles/L)	Solubility (Grams/L)	Temp (°C)	Ref (#)	Evaluation (T P E A A)	Comments
9.000E-09	2.092E-06	25	S377	0 0 0 0 0	

3825. $C_{17}H_{28}$
6-Phenylundecane
RN: **MP** (°C):
MW: 232.41   **BP** (°C):

Solubility (Moles/L)	Solubility (Grams/L)	Temp (°C)	Ref (#)	Evaluation (T P E A A)	Comments
1.100E-08	2.557E-06	25	S377	0 0 0 0 0	

3826. $C_{17}H_{28}$
3-Phenylundecane
RN: **MP** (°C):
MW: 232.41   **BP** (°C):

Solubility (Moles/L)	Solubility (Grams/L)	Temp (°C)	Ref (#)	Evaluation (T P E A A)	Comments
1.200E-08	2.789E-06	25	S377	0 0 0 0 0	

3827. $C_{17}H_{28}$
2-Phenylundecane
RN: **MP** (°C):
MW: 232.41   **BP** (°C):

Solubility (Moles/L)	Solubility (Grams/L)	Temp (°C)	Ref (#)	Evaluation (T P E A A)	Comments
8.000E-09	1.859E-06	25	S377	0 0 0 0 0	

3828. $C_{17}H_{28}$
5-Phenylundecane
RN: **MP** (°C):
MW: 232.41   **BP** (°C):

Solubility (Moles/L)	Solubility (Grams/L)	Temp (°C)	Ref (#)	Evaluation (T P E A A)	Comments
1.000E-08	2.324E-06	25	S377	0 0 0 0 0	

3829. $C_{17}H_{28}N_2O_2$
4-Butylaminobenzoic acid 2-(diethyl-amino)ethyl ester
RN: 3772-42-7   **MP** (°C):
MW: 292.42   **BP** (°C):

Solubility (Moles/L)	Solubility (Grams/L)	Temp (°C)	Ref (#)	Evaluation (T P E A A)	Comments
4.100E-04	1.199E-01	ns	M066	0 0 0 0 1	

3830. C₁₇H₂₈N₂O₂

Endomid

N,N,N',N'-Tetraethyl-bicyclo(2.2.1)hept-5-ene-2,3-dicarboxamide

RN: 4582-18-7   **MP** (°C):

MW: 292.42   **BP** (°C):

Solubility (Moles/L)	Solubility (Grams/L)	Temp (°C)	Ref (#)	Evaluation (T P E A A)	Comments
5.916E-02	1.730E+01	20	K050	1 1 1 1 2	

3831. C₁₇H₂₈O₂

4-Nonylphenol monoethoxylate

Ethanol, 2-(4-nonylphenoxy)-

RN: 104-35-8   **MP** (°C):

MW: 264.41   **BP** (°C):

Solubility (Moles/L)	Solubility (Grams/L)	Temp (°C)	Ref (#)	Evaluation (T P E A A)	Comments
1.048E-05	2.770E-03	2	A335	0 0 0 0 0	
1.050E-05	2.776E-03	2	A335	0 0 0 0 0	
1.063E-05	2.810E-03	10	A335	0 0 0 0 0	
1.060E-05	2.803E-03	10	A335	0 0 0 0 0	
1.074E-05	2.840E-03	14	A335	0 0 0 0 0	
1.080E-05	2.856E-03	14	A335	0 0 0 0 0	
1.140E-05	3.014E-03	20.5	A335	0 0 0 0 0	
1.142E-05	3.020E-03	20.5	A335	0 0 0 0 0	
1.280E-05	3.384E-03	25	A335	0 0 0 0 0	
1.275E-05	3.370E-03	25	A335	0 0 0 0 0	

3832. C₁₇H₃₄O₂

Margaric acid

Heptadecanoic acid

RN: 506-12-7   **MP** (°C):

MW: 270.46   **BP** (°C):

Solubility (Moles/L)	Solubility (Grams/L)	Temp (°C)	Ref (#)	Evaluation (T P E A A)	Comments
1.035E-05	2.800E-03	0	B136	1 0 2 1 1	
1.553E-05	4.200E-03	20	B136	1 0 2 1 1	
1.553E-05	4.200E-03	20.0	R001	1 1 1 1 1	
1.997E-05	5.400E-03	30	B136	1 0 2 1 1	
2.034E-05	5.500E-03	30.0	R001	1 1 1 1 1	
2.551E-05	6.900E-03	45	B136	1 0 2 1 1	
2.551E-05	6.900E-03	45.0	R001	1 1 1 1 1	
2.995E-05	8.100E-03	60	B136	1 0 2 1 1	
2.995E-05	8.100E-03	60.0	R001	1 1 1 1 1	
1.035E-05	2.800E-03	.0	R001	1 1 1 1 1	

3833. C₁₇H₃₆N₂Ge

Spirogermanium

2-[3-(Dimethylamino)propyl]-8,8-diethyl-2-aza-8-germaspiro[4.5]decane

RN: 41992-23-8   **MP** (°C):

MW: 341.10   **BP** (°C):

Solubility (Moles/L)	Solubility (Grams/L)	Temp (°C)	Ref (#)	Evaluation (T P E A A)	Comments
2.463E-05	8.400E-03	22	M456	0 0 0 0 0	pH 12.5

3834. C₁₇H₃₆O

Heptadecanol

1-Heptadecanol

RN: 1454-85-9   **MP** (°C): 58

MW: 256.48   **BP** (°C):

Solubility (Moles/L)	Solubility (Grams/L)	Temp (°C)	Ref (#)	Evaluation (T P E A A)	Comments
<=1E-7	<=2.56E-5	25	R002	0 0 0 0 0	

3835. C₁₈H₁₀Cl₄

2,4,4″,6-Tetrachloro-p-terphenyl

2,4,4″,6-Tetrachloro-1,1′:4′,1″-terphenyl

RN: 61576-97-4   **MP** (°C):

MW: 368.09   **BP** (°C):

Solubility (Moles/L)	Solubility (Grams/L)	Temp (°C)	Ref (#)	Evaluation (T P E A A)	Comments
1.606E-10	5.910E-08	4	D351	1 2 1 1 2	
4.728E-10	1.740E-07	25	D351	1 2 1 1 2	
1.106E-09	4.069E-07	40	D351	1 2 1 1 2	

3836. C₁₈H₁₀I₆N₂O₇

Ioglycamic acid

N,N′-bis(3-Carboxy-2,4,6-triiodophenyl)-diglycolamide

BE 419

RN: 2618-25-9   **MP** (°C):

MW: 1127.72   **BP** (°C):

Solubility (Moles/L)	Solubility (Grams/L)	Temp (°C)	Ref (#)	Evaluation (T P E A A)	Comments
1.773E-04	2.000E-01	ns	H055	0 0 0 0 0	

3837. C$_{18}$H$_{10}$N$_2$O$_2$S
Disperse brightener
2,2′-(2,5-Thiophenediyl)bisbenzoxazole
Unitex OB
Uvitex EBF

RN: 2866-43-5   **MP** (°C):   219
MW: 318.36   **BP** (°C):

Solubility (Moles/L)	Solubility (Grams/L)	Temp (°C)	Ref (#)	Evaluation (T P E A A)	Comments
3.000E-08	9.551E-06	25	B333	0 0 0 0 0	

3838. C$_{18}$H$_{11}$Cl$_3$
2,4″,5-Trichloro-p-terphenyl
2,4″,5-Trichloro-1,1′:4′,1″-terphenyl

RN: 61576-93-0   **MP** (°C):
MW: 333.65   **BP** (°C):

Solubility (Moles/L)	Solubility (Grams/L)	Temp (°C)	Ref (#)	Evaluation (T P E A A)	Comments
3.028E-10	1.010E-07	4	D351	1 2 1 1 2	
1.233E-09	4.115E-07	25	D351	1 2 1 1 2	
2.567E-09	8.564E-07	39	D351	1 2 1 1 2	

3839. C$_{18}$H$_{11}$NO$_3$
Samaron yellow
Supra light yellow GGL(IG)

RN: 1326-08-5   **MP** (°C):
MW: 289.29   **BP** (°C):

Solubility (Moles/L)	Solubility (Grams/L)	Temp (°C)	Ref (#)	Evaluation (T P E A A)	Comments
4.000E-06	1.157E-03	98.59	M180	0 0 2 2 0	EFG
8.000E-06	2.314E-03	111.46	M180	0 0 2 2 0	EFG
1.000E-05	2.893E-03	112.94	M180	0 0 2 2 0	EFG
1.100E-05	3.182E-03	119.00	M180	0 0 2 2 0	EFG
1.300E-05	3.761E-03	125.25	M180	0 0 2 2 0	EFG
1.400E-05	4.050E-03	128.45	M180	0 0 2 2 0	EFG
2.200E-05	6.364E-03	152.37	M180	0 0 2 2 0	EFG

3840. C$_{18}$H$_{11}$NO$_3$
Disperse yellow 54
C.I. Disperse yellow 54

RN: 7576-65-0   **MP** (°C):
MW: 289.29   **BP** (°C):

Solubility (Moles/L)	Solubility (Grams/L)	Temp (°C)	Ref (#)	Evaluation (T P E A A)	Comments
1.000E-07	2.893E-05	25	B333	0 0 0 0 0	
2.400E-07	6.943E-05	60.0	D093	1 2 1 2 0	EFG

(continued)

3840. $C_{18}H_{11}NO_3$ (continued)

Solubility (Moles/L)	Solubility (Grams/L)	Temp (°C)	Ref (#)	Evaluation (T P E A A)	Comments
6.500E-07	1.880E-04	71.8	D093	1 2 1 2 0	EFG
1.600E-06	4.629E-04	84.1	D093	1 2 1 2 0	EFG
4.000E-06	1.157E-03	97.4	D093	1 2 1 2 0	EFG

3841. $C_{18}H_{12}$
Tetracene
Naphthacene
2,3-Benzanthracene

RN: 92-24-0 **MP** (°C): 341
MW: 228.30 **BP** (°C):

Solubility (Moles/L)	Solubility (Grams/L)	Temp (°C)	Ref (#)	Evaluation (T P E A A)	Comments
1.580E-08	3.607E-06	20	E009	1 0 0 1 2	
6.600E-09	1.507E-06	25	K001	2 2 2 2 1	
2.497E-09	5.700E-07	25	M064	1 1 2 2 1	
2.500E-09	5.707E-07	25	M342	1 0 1 1 1	
4.380E-09	1.000E-06	27	D003	1 0 0 1 1	
2.497E-09	5.700E-07	ns	M344	0 0 0 0 2	
2.754E-09	6.288E-07	ns	R424	0 0 0 0 0	

3842. $C_{18}H_{12}$
Triphenylene
9,10-Benzphenanthrene
Isochrysene

RN: 217-59-4 **MP** (°C): 199
MW: 228.30 **BP** (°C): 425

Solubility (Moles/L)	Solubility (Grams/L)	Temp (°C)	Ref (#)	Evaluation (T P E A A)	Comments
1.180E-08	2.694E-06	8	M082	1 1 1 2 2	
1.180E-08	2.694E-06	8	M151	2 1 2 2 2	
1.311E-08	2.992E-06	8.04	M183	1 2 1 1 2	
1.330E-08	3.036E-06	12.00	M082	1 1 1 2 2	
1.330E-08	3.036E-06	12.00	M151	2 1 2 2 2	
1.328E-08	3.033E-06	12.04	M183	1 2 1 1 2	
1.490E-08	3.402E-06	14.80	M082	1 1 1 2 2	
1.490E-08	3.402E-06	14.80	M151	2 1 2 2 2	
2.500E-07	5.707E-05	20	E009	1 0 0 1 1	
2.140E-08	4.886E-06	20.50	M082	1 1 1 2 2	
2.140E-08	4.886E-06	20.50	M151	2 1 2 2 2	
2.144E-08	4.894E-06	20.54	M183	1 2 1 1 2	
1.800E-07	4.109E-05	25	A325	2 1 2 2 1	
1.880E-07	4.292E-05	25	K001	1 0 2 1 2	
1.884E-07	4.300E-05	25	M064	1 1 2 2 1	
2.891E-08	6.600E-06	25.00	M151	2 1 1 2 1	
1.665E-07	3.800E-05	27	D003	1 0 0 1 1	
3.350E-08	7.648E-06	27.30	M082	1 1 1 2 2	

(continued)

3842. $C_{18}H_{12}$ (continued)

Solubility (Moles/L)	Solubility (Grams/L)	Temp (°C)	Ref (#)	Evaluation (T P E A A)	Comments
3.350E-08	7.648E-06	27.30	M151	2 1 2 2 2	
3.354E-08	7.657E-06	27.34	M183	1 2 1 1 2	
3.550E-08	8.105E-06	28.20	M082	1 1 1 2 2	
3.550E-08	8.105E-06	28.20	M151	2 1 2 2 2	
3.556E-08	8.117E-06	28.24	M183	1 2 1 1 2	
1.486E-08	3.393E-06	114.84	M183	1 2 1 1 2	
1.884E-07	4.300E-05	ns	M344	0 0 0 0 2	

3843. $C_{18}H_{12}$
1,2-Benzanthracene
Benzanthracene
1,2-Benzoanthracene

RN:	56-55-3	**MP** (°C):	155	
MW:	228.30	**BP** (°C):		

Solubility (Moles/L)	Solubility (Grams/L)	Temp (°C)	Ref (#)	Evaluation (T P E A A)	Comments
1.310E-08	2.991E-06	6.90	M082	1 1 1 2 2	
1.310E-08	2.991E-06	6.90	M151	2 1 2 2 2	
1.311E-08	2.992E-06	6.94	M183	1 2 1 1 2	
1.660E-08	3.790E-06	10.70	M082	1 1 1 2 2	
1.660E-08	3.790E-06	10.70	M151	2 1 2 2 2	
1.657E-08	3.783E-06	11.14	M183	1 2 1 1 2	
2.100E-08	4.794E-06	14.24	M183	1 2 1 1 2	
2.100E-08	4.794E-06	14.30	M082	1 1 1 2 2	
2.100E-08	4.794E-06	14.30	M151	2 1 2 2 2	
1.583E-08	3.613E-06	14.34	M183	1 2 1 1 2	
2.365E-08	5.400E-06	15	B385	0 0 0 0 0	
2.446E-08	5.584E-06	18.14	M183	1 2 1 1 2	
2.770E-08	6.324E-06	19.30	M082	1 1 1 2 2	
2.770E-08	6.324E-06	19.30	M151	2 1 2 2 2	
2.775E-08	6.335E-06	19.34	M183	1 2 1 1 2	
3.670E-08	8.378E-06	23.10	M082	1 1 1 2 2	
3.670E-08	8.378E-06	23.10	M151	2 1 2 2 2	
3.669E-08	8.377E-06	23.14	M183	1 2 1 1 2	
3.507E-08	8.007E-06	23.64	M183	1 2 1 1 2	
1.927E-07	4.400E-05	24	H116	2 1 0 0 1	
4.117E-08	9.400E-06	25	B319	2 0 1 2 1	
4.056E-08	9.260E-06	25	B385	0 0 0 0 0	
5.694E-08	1.300E-05	25	D406	1 2 2 2 2	
4.310E-08	9.840E-06	25	K001	2 2 2 2 2	
3.900E-09	8.904E-07	25	K123	1 0 2 2 1	*sic*
2.497E-08	5.700E-06	25	L332	1 1 1 1 2	
6.132E-08	1.400E-05	25	M064	1 1 2 2 1	
4.117E-08	9.400E-06	25	M071	2 2 2 2 2	
6.130E-08	1.399E-05	25	M342	1 0 1 1 2	
4.117E-08	9.400E-06	25.00	M151	2 1 1 2 1	
3.774E-08	8.617E-06	25.04	M183	1 2 1 1 2	
4.818E-08	1.100E-05	27	D003	1 0 0 1 1	

(continued)

3843. C₁₈H₁₂ (continued)

Solubility (Moles/L)	Solubility (Grams/L)	Temp (°C)	Ref (#)	Evaluation (T P E A A)	Comments
5.344E-08	1.220E-05	29	M071	2 2 2 2 2	
5.344E-08	1.220E-05	29.00	M151	2 1 1 2 2	
5.436E-08	1.241E-05	29.54	M183	1 2 1 1 2	
5.580E-08	1.274E-05	29.70	M082	1 1 1 2 2	
5.580E-08	1.274E-05	29.70	M151	2 1 2 2 2	
5.567E-08	1.271E-05	29.74	M183	1 2 1 1 2	
7.635E-08	1.743E-05	35	B385	0 0 0 0 0	
6.132E-08	1.400E-05	ns	M344	0 0 0 0 2	

3844. C₁₈H₁₂

Chrysene
1,2-Benzphenanthrene

RN:	218-01-9	MP (°C):	254
MW:	228.30	BP (°C):	448

Solubility (Moles/L)	Solubility (Grams/L)	Temp (°C)	Ref (#)	Evaluation (T P E A A)	Comments
3.100E-09	7.077E-07	6.50	M082	1 1 1 2 2	
3.100E-09	7.077E-07	6.50	M151	2 1 2 2 2	
3.500E-09	7.990E-07	11.00	M082	1 1 1 2 2	
3.500E-09	7.990E-07	11.00	M151	2 1 2 2 2	
6.130E-09	1.399E-06	20.40	M082	1 1 1 2 2	
6.130E-09	1.399E-06	20.40	M151	2 1 2 2 2	
6.139E-09	1.401E-06	20.44	M183	1 2 1 1 2	
9.199E-09	2.100E-06	23	P339	0 0 0 0 0	
7.446E-08	1.700E-05	24	H116	2 1 0 0 1	
7.360E-09	1.680E-06	24.00	M082	1 1 1 2 2	
7.360E-09	1.680E-06	24.00	M151	2 1 2 2 2	
7.367E-09	1.682E-06	24.04	M183	1 2 1 1 2	
4.818E-09	1.100E-06	25	B319	2 0 1 2 1	average of 2
6.570E-09	1.500E-06	25	D406	1 2 2 2 2	
2.760E-08	6.301E-06	25	K001	2 2 2 2 2	
2.628E-08	6.000E-06	25	L332	1 1 1 1 2	
8.761E-09	2.000E-06	25	M064	1 1 2 2 1	
7.884E-09	1.800E-06	25	M071	2 2 2 2 2	
8.760E-09	2.000E-06	25	M342	1 0 1 1 2	
7.884E-09	1.800E-06	25.00	M151	2 1 1 2 1	
8.280E-09	1.890E-06	25.30	M082	1 1 1 2 2	
8.280E-09	1.890E-06	25.30	M151	2 1 2 2 2	
8.283E-09	1.891E-06	25.34	M183	1 2 1 1 2	
6.570E-09	1.500E-06	27	D003	1 0 0 1 1	
9.680E-09	2.210E-06	28.70	M082	1 1 1 2 2	
9.680E-09	2.210E-06	28.70	M151	2 1 2 2 2	
9.689E-09	2.212E-06	28.74	M183	1 2 1 1 2	
9.637E-09	2.200E-06	29	M071	2 2 2 2 2	
9.637E-09	2.200E-06	29.00	M151	2 1 1 2 1	
8.761E-09	2.000E-06	ns	M344	0 0 0 0 2	
8.710E-09	1.988E-06	ns	R424	0 0 0 0 0	
3.400E-06	7.762E-04	ns	W005	0 0 1 2 1	sic

3845. C$_{18}$H$_{12}$N$_2$
2,2′-Biquinoline
2,2′-Biquinolyl
RN: 119-91-5   **MP** (°C):   193
MW: 256.31   **BP** (°C):

Solubility (Moles/L)	Solubility (Grams/L)	Temp (°C)	Ref (#)	Evaluation (T P E A A)	Comments
3.980E-06	1.020E-03	24	H106	1 0 2 2 2	
3.980E-06	1.020E-03	24	M303	1 0 1 1 2	

3846. C$_{18}$H$_{12}$N$_4$O
4-Hydroxy-6,7-diphenylpteridine
4-Hydroxy-6:7-diphenylpteridine
RN: 102943-71-5   **MP** (°C):
MW: 300.32   **BP** (°C):

Solubility (Moles/L)	Solubility (Grams/L)	Temp (°C)	Ref (#)	Evaluation (T P E A A)	Comments
6.658E-04	2.000E-01	20	A019	2 2 1 1 2	

3847. C$_{18}$H$_{13}$ClFN$_3$
Midazolam
8-Chloro-6-(o-fluorophenyl)-1-methyl-4H-imidazo[1,5-a][1,4]benzodiazepine
RN: 59467-70-8   **MP** (°C):
MW: 325.78   **BP** (°C):

Solubility (Moles/L)	Solubility (Grams/L)	Temp (°C)	Ref (#)	Evaluation (T P E A A)	Comments
1.658E-04	5.400E-02	24	A404	2 0 2 2 2	intrinsic pH = 9.5

3848. C$_{18}$H$_{13}$ClF$_3$NO$_7$
Fluoroglycofen-ethyl
Super Blazer
Fluoroglycofen ethyl ester
Ethoxycarbonylmethyl-5-(2-chloro-4-trifluoromethylphenoxy)-2-nitrobenzoate-
    hyphen-ethoxy-2-oxoethyl 5-(2-chloro-4-(trifluoromethyl)phenoxy)-2-nitrobenzoate
5-[2-Chloro-4-(trifluoromethyl)-phenoxy]-2-nitro-benzoic acid 2-ethoxy-2-oxoethyl ester
RN: 77501-90-7   **MP** (°C):
MW: 447.76   **BP** (°C):

Solubility (Moles/L)	Solubility (Grams/L)	Temp (°C)	Ref (#)	Evaluation (T P E A A)	Comments
1.349E-06	6.040E-04	ns	R427	0 0 0 0 0	

3849. C$_{18}$H$_{13}$N
6-Aminochrysene
6-Chrysenamine

RN: 2642-98-0 **MP** (°C): 210
MW: 243.31 **BP** (°C):

Solubility (Moles/L)	Solubility (Grams/L)	Temp (°C)	Ref (#)	Evaluation (T P E A A)	Comments
6.370E-07	1.550E-04	24	H106	1 0 2 2 2	
6.370E-10	1.550E-07	ns	M349	0 2 1 1 2	

3850. C$_{18}$H$_{13}$NO$_3$
Furo[3,4-b]quinolin-3(1H)-one, 9-hydroxy-6-methyl-1-phenyl-

RN: 74103-08-5 **MP** (°C):
MW: 291.31 **BP** (°C):

Solubility (Moles/L)	Solubility (Grams/L)	Temp (°C)	Ref (#)	Evaluation (T P E A A)	Comments
4.463E-08	1.300E-05	25	P089	0 0 0 0 0	
1.270E-07	3.700E-05	37	P089	0 0 0 0 0	
2.163E-07	6.300E-05	51	P089	0 0 0 0 0	

3851. C$_{18}$H$_{13}$NO$_3$
N-1-Naphthylphthalamic acid
Naptalam
2-((1-Naphthylamino)carbonyl)benzoic acid
Naphthylphthalamic acid
ALANAP-1
NPA

RN: 132-66-1 **MP** (°C): 185
MW: 291.31 **BP** (°C):

Solubility (Moles/L)	Solubility (Grams/L)	Temp (°C)	Ref (#)	Evaluation (T P E A A)	Comments
6.866E-04	2.000E-01	25	B200	1 0 0 0 2	
6.866E-04	2.000E-01	ns	B185	0 0 0 0 0	
6.866E-04	2.000E-01	ns	N013	0 0 0 0 2	
6.866E-04	2.000E-01	rt	M161	0 0 0 0 2	

3852. C$_{18}$H$_{14}$
o-Terphenyl
1,2-Diphenyl benzene

RN: 84-15-1 **MP** (°C): 58
MW: 230.31 **BP** (°C): 332

Solubility (Moles/L)	Solubility (Grams/L)	Temp (°C)	Ref (#)	Evaluation (T P E A A)	Comments
5.380E-06	1.239E-03	25	A325	2 1 2 2 2	

3853. C$_{18}$H$_{14}$
m-Terphenyl
1,3-Diphenyl benzene
RN: 92-06-8　　**MP** (°C): 89
MW: 230.31　　**BP** (°C): 365

Solubility (Moles/L)	Solubility (Grams/L)	Temp (°C)	Ref (#)	Evaluation (T P E A A)	Comments
6.560E-06	1.511E-03	25	A325	2 1 2 2 2	

3854. C$_{18}$H$_{14}$
p-Terphenyl
1,4-Diphenyl benzene
RN: 92-94-4　　**MP** (°C): 213
MW: 230.31　　**BP** (°C):

Solubility (Moles/L)	Solubility (Grams/L)	Temp (°C)	Ref (#)	Evaluation (T P E A A)	Comments
7.800E-08	1.796E-05	25	A325	2 1 2 2 1	

3855. C$_{18}$H$_{14}$Cl$_4$N$_2$O
Miconazole
1-[2-(2,4-Dichlorophenyl)-2-[(2,4-dichlorophenyl)methoxy]ethyl]-1H-imidazole
1-[2,4-Dichloro-β-[(2,4-dichlorobenzyl)oxy]phenethyl]imidazole
Conoderm
RN: 22916-47-8　　**MP** (°C):
MW: 416.14　　**BP** (°C):

Solubility (Moles/L)	Solubility (Grams/L)	Temp (°C)	Ref (#)	Evaluation (T P E A A)	Comments
<4.80E-09	<2.00E-06	25	P348	0 0 0 0 0	
2.163E-04	9.000E-02	amb	L434	0 0 0 0 0	

3856. C$_{18}$H$_{14}$N$_4$O
Disperse yellow 23
Phenol, 4-[[4-(phenylazo)phenyl]azo]-
p-Hydroxy-*p*-bis(azobenzene)
RN: 6250-23-3　　**MP** (°C):
MW: 302.34　　**BP** (°C):

Solubility (Moles/L)	Solubility (Grams/L)	Temp (°C)	Ref (#)	Evaluation (T P E A A)	Comments
2.000E-10	6.047E-08	25	B333	0 0 0 0 0	
1.300E-07	3.930E-05	71.8	D093	1 2 1 2 0	EFG
5.500E-07	1.663E-04	84.1	D093	1 2 1 2 0	EFG
2.300E-06	6.954E-04	97.4	D093	1 2 1 2 0	EFG

3857. C₁₈H₁₄N₄O₂

Disperse orange 1
Dye VI
C.I. Disperse orange 1
4-(*p*-Nitrophenylazo)diphenylamine
4-Anilino-4′-nitroazobenzene
4-(4-Nitrophenylazo)diphenylamine

RN:	2581-69-3	**MP** (°C):	157		
MW:	318.34	**BP** (°C):			

Solubility (Moles/L)	Solubility (Grams/L)	Temp (°C)	Ref (#)	Evaluation (T P E A A)	Comments
1.500E-09	4.775E-07	25	B333	0 0 0 0 0	
3.000E-07	9.550E-05	84.10	B198	1 2 1 1 0	
1.420E-06	4.520E-04	97.40	B198	1 2 1 1 2	
4.900E-06	1.560E-03	111.60	B198	1 2 1 1 1	
1.950E-05	6.208E-03	127	B198	1 2 1 1 2	

3858. C₁₈H₁₄N₄O₅S

Sulfasalazine
Salicylazosulfapyridine
SASP
Sulcolon
Salazosulfapyridine
Salicylazosulfapyridine

RN:	599-79-1	**MP** (°C):	240–245
MW:	398.40	**BP** (°C):	

Solubility (Moles/L)	Solubility (Grams/L)	Temp (°C)	Ref (#)	Evaluation (T P E A A)	Comments
2.510E-05	1.000E-02	ns	K444	0 0 0 0 0	

3859. C₁₈H₁₅Cl₃N₂O

Econazole
1-[2-[(4-Chlorophenyl)methoxy]-2-(2,4-dichlorophenyl)ethyl]-1H-imidazole

RN:	27220-47-9	**MP** (°C):	
MW:	381.69	**BP** (°C):	

Solubility (Moles/L)	Solubility (Grams/L)	Temp (°C)	Ref (#)	Evaluation (T P E A A)	Comments
9.694E-04	3.700E-01	amb	L434	0 0 0 0 0	

3860. C₁₈H₁₅Cl₄N₃O₄

Miconazole nitrate-β cyclidextrin complexant

RN:	22832-87-7	**MP** (°C):	
MW:	479.15	**BP** (°C):	

Solubility (Moles/L)	Solubility (Grams/L)	Temp (°C)	Ref (#)	Evaluation (T P E A A)	Comments
3.700E-04	1.773E-01	25	P348	0 0 0 0 0	

3861. C$_{18}$H$_{15}$NO$_3$

Oxaprozin
4,5-Diphenyl-2-oxazolepropanoic acid
4,5-Diphenyl-2-oxazole-propionic acid
Choledyl
Daypro
Oxaprozin

RN:	21256-18-8	**MP** (°C):
MW:	293.33	**BP** (°C):

Solubility (Moles/L)	Solubility (Grams/L)	Temp (°C)	Ref (#)	Evaluation (T P E A A)	Comments
1.364E-05	4.000E-03	37	Y421	0 0 0 0 0	

3862. C$_{18}$H$_{15}$N$_3$O$_5$

1H-Benzimidazole-1-carboxylic acid, 6-benzoyl-2-[(methoxycarbonyl)amino]-, methyl ester

RN:	104663-14-1	**MP** (°C):	156.5
MW:	353.34	**BP** (°C):	

Solubility (Moles/L)	Solubility (Grams/L)	Temp (°C)	Ref (#)	Evaluation (T P E A A)	Comments
1.981E-05	7.000E-03	21	N337	0 0 0 0 0	pH 5
1.900E-05	6.713E-03	21	N337	0 0 0 0 0	pH 5

3863. C$_{18}$H$_{15}$O$_4$P

Triphenyl phosphate
Phosphoric acid triphenyl ester
Triphenyl phosphoric acid ester
Phenyl phosphate
TPP

RN:	115-86-6	**MP** (°C):	49
MW:	326.29	**BP** (°C):	245

Solubility (Moles/L)	Solubility (Grams/L)	Temp (°C)	Ref (#)	Evaluation (T P E A A)	Comments
2.237E-06	7.300E-04	24	H116	2 1 0 0 2	
6.129E-05	2.000E-02	ns	F014	0 0 0 0 0	

3864. C$_{18}$H$_{16}$ClNO$_5$

Fenoxaprop-*p*-ethyl
Fenoxaprop-*p* ethyl ester
Propanoic acid
2-{4-{(6-Chloro-2-benzoxazolyl)oxy}phenoxy}-ethyl ester

RN:	71283-80-2	**MP** (°C):
MW:	361.78	**BP** (°C):

Solubility (Moles/L)	Solubility (Grams/L)	Temp (°C)	Ref (#)	Evaluation (T P E A A)	Comments
1.950E-06	7.054E-04	ns	R427	0 0 0 0 0	

3865. C$_{18}$H$_{16}$Cl$_3$N$_3$O$_4$
Econazole nitrate
Pevaryl
Spectazole
R 14827

RN: 68797-31-9   **MP** (°C):
MW: 444.70   **BP** (°C):

Solubility (Moles/L)	Solubility (Grams/L)	Temp (°C)	Ref (#)	Evaluation (T P E A A)	Comments
1.600E-03	7.115E-01	25	P348	0 0 0 0 0	

3866. C$_{18}$H$_{16}$N$_2$O$_3$
Benzoyltryptophan
N-Benzoyl-DL-tryptophan

RN: 2901-79-3   **MP** (°C):
MW: 308.34   **BP** (°C):

Solubility (Moles/L)	Solubility (Grams/L)	Temp (°C)	Ref (#)	Evaluation (T P E A A)	Comments
1.816E-03	5.600E-01	25.1	N026	0 0 0 0 0	

3867. C$_{18}$H$_{16}$N$_4$O$_3$S
2-(*N*4-Acetylsulfanilylamino)-4-phenylpyrimidine

RN: **MP** (°C):
MW: 368.42   **BP** (°C):

Solubility (Moles/L)	Solubility (Grams/L)	Temp (°C)	Ref (#)	Evaluation (T P E A A)	Comments
9.772E-06	3.600E-03	37	R076	1 2 0 0 1	

3868. C$_{18}$H$_{17}$ClN$_4$O$_6$.0.5H$_2$O
9-[5-*O*-(4-Chlorobenzoyl-β-D-arabinofuranosyl)]-6-methoxy-9H-purine (hemihydrate)

RN: 121032-34-6   **MP** (°C): 122–124
MW: 429.82   **BP** (°C):

Solubility (Moles/L)	Solubility (Grams/L)	Temp (°C)	Ref (#)	Evaluation (T P E A A)	Comments
1.880E-04	8.081E-02	37	M378	1 2 1 1 2	pH 7.2

3869. $C_{18}H_{17}Cl_2NO_3$

Benzoylprop-ethyl
Ethyl N-benzoyl-N-(3,4-dichlorophenyl)-2-aminopropionate
FX 2182
N-Benzoyl-N-(3,4-dichlorophenyl)-DL-alanine ethyl ester
Enaven
Suffix

RN:	22212-55-1	**MP** (°C):	70.5		
MW:	366.25	**BP** (°C):			

Solubility (Moles/L)	Solubility (Grams/L)	Temp (°C)	Ref (#)	Evaluation (T P E A A)	Comments
5.461E-05	2.000E-02	25	M161	1 0 0 0 1	

3870. $C_{18}H_{17}N_5O_8$

6-Methoxy-9-(5-O-[4-nitrobenzoyl]-β-D-arabinofuranosyl)-9H-purine

RN:	121032-21-1	**MP** (°C):	202–203
MW:	431.36	**BP** (°C):	

Solubility (Moles/L)	Solubility (Grams/L)	Temp (°C)	Ref (#)	Evaluation (T P E A A)	Comments
3.400E-05	1.467E-02	37	M378	1 2 1 1 2	pH 7.2

3871. $C_{18}H_{18}ClNO_4$

Clanobutin
Butanoic acid, 4-[(4-chlorobenzoyl)(4-methoxyphenyl)amino]-
Bykahepar

RN:	30544-61-7	**MP** (°C):	
MW:	347.80	**BP** (°C):	

Solubility (Moles/L)	Solubility (Grams/L)	Temp (°C)	Ref (#)	Evaluation (T P E A A)	Comments
1.270E-04	4.417E-02	37	K093	1 2 1 1 2	pH 3.0

3872. $C_{18}H_{18}ClNO_5$

Etofibrate
3-Pyridinecarboxylic acid, 2-[2-(4-chlorophenoxy)-2-methyl-1-oxopropoxy]ethyl ester
Tricerol
Lipo-Merz

RN:	31637-97-5	**MP** (°C):	
MW:	363.80	**BP** (°C):	

Solubility (Moles/L)	Solubility (Grams/L)	Temp (°C)	Ref (#)	Evaluation (T P E A A)	Comments
2.000E-05	7.276E-03	rt	G093	0 1 1 1 2	pH4

3873. C₁₈H₁₈ClNO₅

Benzoximate

RN: 29104-30-1	**MP** (°C):	73
MW: 363.80	**BP** (°C):	

Solubility (Moles/L)	Solubility (Grams/L)	Temp (°C)	Ref (#)	Evaluation (T P E A A)	Comments
8.318E-05	3.026E-02	ns	R427	0 0 0 0 0	

3874. C₁₈H₁₈ClNS

Chlorprothixene

Taractan

1-Propanamine, 3-(2-chloro-9H-thioxanthen-9-ylidene)-N,N-dimethyl-, (3Z)-

Rentovet

RN: 113-59-7	**MP** (°C):	
MW: 315.87	**BP** (°C):	

Solubility (Moles/L)	Solubility (Grams/L)	Temp (°C)	Ref (#)	Evaluation (T P E A A)	Comments
3.936E-05	1.243E-02	20	H301	0 0 0 0 0	
1.221E-06	3.858E-04	22.5	B440	0 0 0 0 0	

3875. C₁₈H₁₈N₂O₄

C.I. Disperse blue 23

1,4-bis[(2-Hydroxyethyl)amino]anthraquinone

Acetoquinone blue BF

RN: 4471-41-4	**MP** (°C):	248
MW: 326.36	**BP** (°C):	

Solubility (Moles/L)	Solubility (Grams/L)	Temp (°C)	Ref (#)	Evaluation (T P E A A)	Comments
2.400E-06	7.833E-04	25	B333	0 0 0 0 0	

3876. C₁₈H₁₈N₄O₆

9-[5-O-(Benzoyl-β-D-arabinofuranosyl)]-6-methoxy-9H-purine

RN: 121032-31-3	**MP** (°C):	202–204
MW: 386.37	**BP** (°C):	

Solubility (Moles/L)	Solubility (Grams/L)	Temp (°C)	Ref (#)	Evaluation (T P E A A)	Comments
7.400E-05	2.859E-02	37	M378	1 2 1 1 2	pH 7.2

3877. C₁₈H₁₈N₄O₆.0.75H₂O

2′-Benzoyl-6-methoxypurine arabinoside (0.75 hydrate)

RN: 145913-44-6 **MP** (°C): 84–86
MW: 399.88 **BP** (°C):

Solubility (Moles/L)	Solubility (Grams/L)	Temp (°C)	Ref (#)	Evaluation (T P E A A)	Comments
1.780E-02	7.118E+00	37	C348	0 0 0 0 0	pH 7.00

3878. C₁₈H₁₈N₈O₆

7,7′-Succinyldritheophylline

RN: 58447-18-0 **MP** (°C):
MW: 442.39 **BP** (°C):

Solubility (Moles/L)	Solubility (Grams/L)	Temp (°C)	Ref (#)	Evaluation (T P E A A)	Comments
1.630E-03	7.211E-01	25	L067	1 0 1 1 2	

3879. C₁₈H₁₈O₂

Dienestrol
3,4-bis(4-Hydroxyphenyl)-2,4-hexadiene
Dehydrostilbestrol

RN: 84-17-3 **MP** (°C): 227.5
MW: 266.34 **BP** (°C):

Solubility (Moles/L)	Solubility (Grams/L)	Temp (°C)	Ref (#)	Evaluation (T P E A A)	Comments
1.126E-05	3.000E-03	37	B039	2 1 1 1 0	EFG
1.122E-05	2.988E-03	ns	R427	0 0 0 0 0	

3880. C₁₈H₁₈O₂

Equilenin
3-Hydroxy-17-keto-δ(1,3,5-10,6,8)estrapentaene
1,3,5-10,6,8-Estrapentaen-3-ol-17-one

RN: 517-09-9 **MP** (°C): 258
MW: 266.34 **BP** (°C):

Solubility (Moles/L)	Solubility (Grams/L)	Temp (°C)	Ref (#)	Evaluation (T P E A A)	Comments
5.707E-06	1.520E-03	25	L033	1 0 2 1 2	

3881. C₁₈H₁₈O₃

$C_{18}H_{18}O_3$

Flurecol-butyl

Flurenol-*n*-butyl ester

n-Butyl-9-hydroxyfluorene-(9)-carboxylate

RN: 2314-09-2 **MP (°C):** 70

MW: 282.34 **BP (°C):**

Solubility (Moles/L)	Solubility (Grams/L)	Temp (°C)	Ref (#)	Evaluation (T P E A A)	Comments
1.293E-02	3.650E+00	20	B200	1 0 0 0 2	*sic*
1.293E-04	3.650E-02	20	M161	1 0 0 0 2	*sic*

3882. C₁₈H₁₉Cl₂NO₄

$C_{18}H_{19}Cl_2NO_4$

Felodipine

3,5-Pyridinedicarboxylic acid, 4-(2,3-dichlorophenyl)-1,4-dihydro-2,6-dimethyl-, ethyl methyl ester

Plendil

RN: 72509-76-3 **MP (°C):**

MW: 384.26 **BP (°C):**

Solubility (Moles/L)	Solubility (Grams/L)	Temp (°C)	Ref (#)	Evaluation (T P E A A)	Comments
1.301E-06	5.000E-04	20	N322	0 0 0 0 0	
1.179E-05	4.530E-03	22	M382	2 1 1 1 1	

3883. C₁₈H₁₉F₃N₂S

$C_{18}H_{19}F_3N_2S$

4-Trifluoromethyl-*N,N*-dimethyl-10H-phenothiazine-10-propanamide

RN: 3852-94-6 **MP (°C):**

MW: 352.42 **BP (°C):**

Solubility (Moles/L)	Solubility (Grams/L)	Temp (°C)	Ref (#)	Evaluation (T P E A A)	Comments
7.000E-06	2.467E-03	ns	G023	0 0 1 1 0	

3884. C₁₈H₁₉F₃N₂S

$C_{18}H_{19}F_3N_2S$

Fluopromazine

Triflupromazine

RN: 146-54-3 **MP (°C):** <25

MW: 352.42 **BP (°C):**

Solubility (Moles/L)	Solubility (Grams/L)	Temp (°C)	Ref (#)	Evaluation (T P E A A)	Comments
5.000E-06	1.762E-03	24	G022	2 0 1 1 1	
5.000E-06	1.762E-03	ns	F027	0 0 0 0 0	

3885. C$_{18}$H$_{19}$NO
Desmethyldoxepin
1-Propanamine, 3-dibenz[b,e]oxepin-11(6H)-ylidene-N-methyl-

| RN: | 1225-56-5 | MP (°C): |
| MW: | 265.36 | BP (°C): |

Solubility (Moles/L)	Solubility (Grams/L)	Temp (°C)	Ref (#)	Evaluation (T P E A A)	Comments
3.950E-04	1.048E-01	25	E051	1 0 2 1 2	

3886. C$_{18}$H$_{19}$N$_2$O$_4$
N-Benzoyl-L-tyrosinamide acetate

| RN: | | MP (°C): |
| MW: | 327.36 | BP (°C): |

Solubility (Moles/L)	Solubility (Grams/L)	Temp (°C)	Ref (#)	Evaluation (T P E A A)	Comments
1.300E-04	4.256E-02	25	A066	1 0 1 1 1	

3887. C$_{18}$H$_{19}$N$_3$O$_6$S
Cephaloglycin
5-Thia-1-azabicyclo[4.2.0]oct-2-ene-2-carboxylic acid

| RN: | 3577-01-3 | MP (°C): |
| MW: | 405.43 | BP (°C): |

Solubility (Moles/L)	Solubility (Grams/L)	Temp (°C)	Ref (#)	Evaluation (T P E A A)	Comments
2.590E-02	1.050E+01	25	P311	0 0 0 0 0	EFG

3888. C$_{18}$H$_{19}$N$_5$O$_3$
C.I. Disperse dye
Propanenitrile, 3-[(2-hydroxyethyl)[3-methyl-4-[(4-nitrophenyl)azo]phenyl]amino]-
Celliton discharge scarlet RNL
Celliton fast scarlet RN

| RN: | 6054-58-6 | MP (°C): | 156 |
| MW: | 353.38 | BP (°C): |

Solubility (Moles/L)	Solubility (Grams/L)	Temp (°C)	Ref (#)	Evaluation (T P E A A)	Comments
1.900E-07	6.714E-05	25	B333	0 0 0 0 0	

3889. C$_{18}$H$_{19}$N$_5$O$_6$.0.3H$_2$O
9-[5-O-(4-Aminobenzoyl-β-D-arabinofuranosyl)]-6-methoxy-9H-purine (0.3 hydrate)

| RN: | 121032-39-1 | MP (°C): | 198–200 |
| MW: | 406.79 | BP (°C): |

Solubility (Moles/L)	Solubility (Grams/L)	Temp (°C)	Ref (#)	Evaluation (T P E A A)	Comments
3.400E-05	1.383E-02	37	M378	1 2 1 1 2	pH 7.2

3890. $C_{18}H_{19}N_5O_6$

2'-(o-Aminobenzoyl)-6-methoxypurine arabinoside

RN: 121032-55-1 **MP** (°C):
MW: 401.38 **BP** (°C):

Solubility (Moles/L)	Solubility (Grams/L)	Temp (°C)	Ref (#)	Evaluation (T P E A A)	Comments
2.060E-02	8.268E+00	37	C348	0 0 0 0 0	pH 7.00

3891. $C_{18}H_{20}$

2,4-Diphenyl-4-methyl-2-pentene

RN: 6362-80-7 **MP** (°C):
MW: 236.36 **BP** (°C):

Solubility (Moles/L)	Solubility (Grams/L)	Temp (°C)	Ref (#)	Evaluation (T P E A A)	Comments
1.047E-07	2.475E-05	ns	D001	0 0 0 0 2	

3892. $C_{18}H_{20}Cl_2O_2$

1-Dichloro-2,2-bis(p-ethoxylphenyl)ethane

RN: 7388-32-1 **MP** (°C):
MW: 339.26 **BP** (°C):

Solubility (Moles/L)	Solubility (Grams/L)	Temp (°C)	Ref (#)	Evaluation (T P E A A)	Comments
7.664E-08	2.600E-05	rt	C122	0 0 0 0 0	

3893. $C_{18}H_{20}N_4O_7S$

2'-(p-Methylbenzenesulfonyl)-6-methoxypurine arabinoside

RN: 145913-49-1 **MP** (°C): 214–215
MW: 436.45 **BP** (°C):

Solubility (Moles/L)	Solubility (Grams/L)	Temp (°C)	Ref (#)	Evaluation (T P E A A)	Comments
1.240E-04	5.412E-02	37	C348	0 0 0 0 0	pH 7.00

3894. $C_{18}H_{20}O_2$

Equilin
3-Hydroxy-17-keto-δ(1,3,5-10,7)estratetraene
1,3,5(10),7-Estratetraen-3-ol-17-one

RN: 474-86-2 **MP** (°C): 238
MW: 268.36 **BP** (°C):

Solubility (Moles/L)	Solubility (Grams/L)	Temp (°C)	Ref (#)	Evaluation (T P E A A)	Comments
5.217E-06	1.400E-03	25	H049	0 0 0 0 0	
5.254E-06	1.410E-03	25	L033	1 0 2 1 2	

3895. C$_{18}$H$_{20}$O$_2$
Diethylstilbestrol
Diethylstilboestrol
Destrol
4,4′-(1,2-Diethyl-1,2-ethenediyl)bisphenol
Tylosterone
Vagestrol

RN: 56-53-1   **MP** (°C):   169
MW: 268.36   **BP** (°C):

Solubility (Moles/L)	Solubility (Grams/L)	Temp (°C)	Ref (#)	Evaluation (T P E A A)	Comments
4.472E-05	1.200E-02	25	G009	1 0 1 1 1	
9.316E-05	2.500E-02	30	M007	2 2 1 2 2	average of 6
		amb	L434	0 0 0 0 0	

3896. C$_{18}$H$_{21}$ClN$_2$
Chlorocyclizine
Chlorcyclizine

RN: 82-93-9   **MP** (°C):
MW: 300.83   **BP** (°C):

Solubility (Moles/L)	Solubility (Grams/L)	Temp (°C)	Ref (#)	Evaluation (T P E A A)	Comments
1.000E-03	3.008E-01	37.5	L034	2 2 0 1 2	pH 7.4

3897. C$_{18}$H$_{21}$ClN$_2$S
2-Chloro-N,N-dimethyl-10H-phenothiazine-10-butanamine

RN: 13094-23-0   **MP** (°C):
MW: 332.90   **BP** (°C):

Solubility (Moles/L)	Solubility (Grams/L)	Temp (°C)	Ref (#)	Evaluation (T P E A A)	Comments
5.000E-06	1.664E-03	ns	G023	0 0 1 1 0	

3898. C$_{18}$H$_{21}$ClO
1-Chloro-1-methyl-2-(p-methylphenyl)-2-p-ethoxylphenyl)ethane

RN: 56265-27-1   **MP** (°C):
MW: 288.82   **BP** (°C):

Solubility (Moles/L)	Solubility (Grams/L)	Temp (°C)	Ref (#)	Evaluation (T P E A A)	Comments
5.540E-06	1.600E-03	rt	C122	0 0 0 0 0	

3899. C$_{18}$H$_{21}$NO$_3$

Codeine

Codein

Methylmorphin

7,8-Didehydro-4,5-α-epoxy-3-methoxy-17-methylmorphinan-6-α-ol

Nucofed

Robitussin AC

RN: 76-57-3   **MP** (°C):   155

MW: 299.37   **BP** (°C):

Solubility (Moles/L)	Solubility (Grams/L)	Temp (°C)	Ref (#)	Evaluation (T P E A A)	Comments
3.006E-02	9.000E+00	20	A073	1 1 1 1 0	
2.672E-02	8.000E+00	20	F300	1 0 0 0 0	
2.760E-02	8.264E+00	20	K052	1 1 1 1 2	
1.591E-01	4.762E+01	25	E041	2 2 2 2 0	EFG, form III, recrystallized
3.242E-02	9.705E+00	25	E041	2 2 2 2 0	EFG, form II, recrystallized
3.176E-02	9.509E+00	25	E041	2 2 2 2 0	EFG, form I, recrystallized
3.571E-02	1.069E+01	25	R338	0 0 0 0 0	
3.340E-02	1.000E+01	30	A073	1 1 1 1 1	
3.674E-02	1.100E+01	40	A073	1 1 1 1 1	
4.342E-02	1.300E+01	50	A073	1 1 1 1 1	
5.010E-02	1.500E+01	60	A073	1 1 1 1 1	
6.013E-02	1.800E+01	70	A073	1 1 1 1 1	
6.347E-02	1.900E+01	80	A073	1 1 1 1 1	
5.578E-02	1.670E+01	80	F300	1 0 0 0 2	
8.017E-02	2.400E+01	90	A073	1 1 1 1 1	
1.069E-01	3.200E+01	100	A073	1 1 1 1 1	

3900. C$_{18}$H$_{21}$NO$_3$

Thebainone A

Morphinan-6-one, 7,8-didehydro-4-hydroxy-3-methoxy-17-methyl-

Thebainon

RN: 467-98-1   **MP** (°C):   146

MW: 299.37   **BP** (°C):

Solubility (Moles/L)	Solubility (Grams/L)	Temp (°C)	Ref (#)	Evaluation (T P E A A)	Comments
1.336E-02	4.000E+00	20	F300	1 0 0 0 0	
2.839E-02	8.500E+00	100	F300	1 0 0 0 1	

3901. C₁₈H₂₁NO₃,H₂O

Codeine (monohydrate)

Morphinan-6-ol, 7,8-didehydro-4,5-epoxy-3-methoxy-17-methyl-, monohydrate, (5α,6α)

RN: 6059-47-8   **MP** (°C):   155
MW: 317.39   **BP** (°C):

Solubility (Moles/L)	Solubility (Grams/L)	Temp (°C)	Ref (#)	Evaluation (T P E A A)	Comments
2.604E-02	8.264E+00	c	D004	0 0 0 0 0	

3902. C₁₈H₂₁NO₄

2-Naphthaleneacetic acid, 6-methoxy-α-methyl-, 2-(dimethylamino)-2-oxoethyl ester, (S)

Naproxen, N,N-dimethyl glycolamide ester

2-Naphthaleneacetic acid, 6-methoxy-α-methyl-, 2-(dimethylamino)-2-oxoethyl ester

Naproxen N,N-dimethyl glycolamide ester

RN: 114665-18-8   **MP** (°C):   150.5
MW: 315.37   **BP** (°C):

Solubility (Moles/L)	Solubility (Grams/L)	Temp (°C)	Ref (#)	Evaluation (T P E A A)	Comments
1.268E-05	4.000E-03	21	B331	1 2 2 1 2	pH 7.4
1.268E-05	4.000E-03	21	B331	0 0 0 0 0	

3903. C₁₈H₂₂ClNO₄

Oxycodone hydrochloride

4,5-Epoxy-14-hydroxy-3-methoxy-17-methylmorphinan-6-one hydrochloride

Endocet

Percocet

Supeudol

Roxicet

RN: 124-90-3   **MP** (°C):   270–271
MW: 351.83   **BP** (°C):

Solubility (Moles/L)	Solubility (Grams/L)	Temp (°C)	Ref (#)	Evaluation (T P E A A)	Comments
4.060E-01	1.429E+02	ns	S469	0 0 0 0 0	

3904. C₁₈H₂₂N₂

1-(Diphenylmethyl)-4-methylpiperazine

RN: **MP** (°C):
MW: 266.39   **BP** (°C):

Solubility (Moles/L)	Solubility (Grams/L)	Temp (°C)	Ref (#)	Evaluation (T P E A A)	Comments
6.962E-04	1.855E-01	25	M438	0 0 0 0 0	

3905. C$_{18}$H$_{22}$N$_4$O$_5$

Dis. A. 9

Ethanol, 2,2′-[[4-[(2-methoxy-4-nitrophenyl)azo]-3-methylphenyl]imino]*bis*-
4-[bis(2-Hydroxyethyl)amino]-2′-methoxy-2-methyl-4′-nitroazobenzene

RN:	41541-11-1	**MP** (°C):			
MW:	374.40	**BP** (°C):			

Solubility (Moles/L)	Solubility (Grams/L)	Temp (°C)	Ref (#)	Evaluation (T P E A A)	Comments
4.500E-06	1.685E-03	25	B333	0 0 0 0 0	

3906. C$_{18}$H$_{22}$O$_2$

Hexestrol

4,4′-(1,2-Diethylethylene)diphenol

Dihydrodiethylstilbestrol

Esestrolo

RN:	5635-50-7	**MP** (°C):	186.5	
MW:	270.37	**BP** (°C):		

Solubility (Moles/L)	Solubility (Grams/L)	Temp (°C)	Ref (#)	Evaluation (T P E A A)	Comments
4.438E-05	1.200E-02	37	B039	2 1 1 1 0	EFG
3.699E-05	1.000E-02	37	B045	1 0 1 1 1	
4.365E-05	1.180E-02	ns	R427	0 0 0 0 0	

3907. C$_{18}$H$_{22}$O$_2$

Estrone

Oestrone

Folliculin

1,3,5(10)-Estratrien-3-ol-17-one

Estra-1,3,5(10)-Trien-17-one, 3-hydroxy-

Oestrin

RN:	53-16-7	**MP** (°C):	252.5	
MW:	270.37	**BP** (°C):		

Solubility (Moles/L)	Solubility (Grams/L)	Temp (°C)	Ref (#)	Evaluation (T P E A A)	Comments
5.659E-06	1.530E-03	22	Y419	0 0 0 0 0	
2.959E-06	8.000E-04	25	H049	0 0 0 0 0	
1.110E-04	3.000E-02	25	I309	0 0 0 0 0	*sic*
2.959E-06	8.000E-04	25	L033	1 0 2 1 1	
1.109E-03	2.999E-01	25	P324	0 0 0 0 0	
4.808E-06	1.300E-03	25	S468	0 0 0 0 0	
8.200E-06	2.217E-03	37	H034	1 0 2 1 1	pH 7.4
1.184E-05	3.200E-03	37	L010	2 0 2 1 1	
3.162E-06	8.550E-04	ns	A074	0 0 0 0 0	EFG

3908. C₁₈H₂₃NO

Orphenadrine
Disipal
Marflex
Noradex
Orflagen
Norflex

RN: 83-98-7   **MP** (°C):
MW: 269.39   **BP** (°C):   195 at 12 mm Hg

Solubility (Moles/L)	Solubility (Grams/L)	Temp (°C)	Ref (#)	Evaluation (T P E A A)	Comments
6.686E-06	1.801E-03	22.5	B440	0 0 0 0 0	

3909. C₁₈H₂₃N₃O₃S

L-Leu-dapsone
2-Amino-N-[4-[(4-aminophenyl)sulfonyl]phenyl]-4-methyl-, (S)-
Pentanamide

RN: 160349-00-8   **MP** (°C):
MW: 361.47   **BP** (°C):

Solubility (Moles/L)	Solubility (Grams/L)	Temp (°C)	Ref (#)	Evaluation (T P E A A)	Comments
8.576E-04	3.100E-01	25	P351	0 0 0 0 0	pH 7.4
>6.92E-02	>2.50E+01	25	P351	0 0 0 0 0	

3910. C₁₈H₂₃N₃O₄S

Phentolamine methanesulfonate
Vasomax
Regitine mesylate
Regitine methanesulfonate

RN: 65-28-1   **MP** (°C):   177
MW: 377.47   **BP** (°C):

Solubility (Moles/L)	Solubility (Grams/L)	Temp (°C)	Ref (#)	Evaluation (T P E A A)	Comments
3.979E+00	1.502E+03	30	D011	1 0 1 0 2	

3911. C₁₈H₂₄I₃N₃O₉

1,3-Benzenedicarboxamide, 5RS-[(2,3-dihydroxy-1-oxobutyl)amino]-N,N'-bis(2,3-
dihydroxypropyl)-2,4,6-triiodo-[RS-(RS*,S*)]-

RN: 77868-48-5   **MP** (°C):
MW: 807.12   **BP** (°C):

Solubility (Moles/L)	Solubility (Grams/L)	Temp (°C)	Ref (#)	Evaluation (T P E A A)	Comments
1.327E-01	1.071E+02	25	P091	0 0 0 0 0	

3912. $C_{18}H_{24}N_4O_2$
2,5-Diaziridinyl-3,6-dipyrrolidino-1,4-benzoquinone
RN: 59886-43-0 **MP** (°C): 160
MW: 328.42 **BP** (°C):

Solubility (Moles/L)	Solubility (Grams/L)	Temp (°C)	Ref (#)	Evaluation (T P E A A)	Comments
1.522E-03	5.000E-01	rt	C317	0 0 0 0 0	

3913. $C_{18}H_{24}N_4O_2S$
2-Sulfanilamido-5,6,7,8,-tetrahydro-8-isopropyl-5-methyl-quinazoline
RN: 71119-36-3 **MP** (°C): 185-187
MW: 360.48 **BP** (°C):

Solubility (Moles/L)	Solubility (Grams/L)	Temp (°C)	Ref (#)	Evaluation (T P E A A)	Comments
6.658E-05	2.400E-02	29	C049	0 0 0 0 0	

3914. $C_{18}H_{24}N_4O_2S$
2-Sulfanilamidobornylenepyrimidine
RN: **MP** (°C): 276
MW: 360.48 **BP** (°C):

Solubility (Moles/L)	Solubility (Grams/L)	Temp (°C)	Ref (#)	Evaluation (T P E A A)	Comments
8.322E-05	3.000E-02	29	C049	0 0 0 0 0	

3915. $C_{18}H_{24}N_4O_3S$
L-Lys-dapsone
Hexanamide, 2,6-diamino-N-[4-[(4-aminophenyl)sulfonyl]phenyl]-, (S)
RN: 160349-03-1 **MP** (°C):
MW: 376.48 **BP** (°C):

Solubility (Moles/L)	Solubility (Grams/L)	Temp (°C)	Ref (#)	Evaluation (T P E A A)	Comments
>1.73E-01	>6.50E+01	25	P351	0 0 0 0 0	pH 7.4
>1.73E-01	>6.50E+01	25	P351	0 0 0 0 0	

3916. $C_{18}H_{24}O_2$
Estradiol
17-β-Estradiol
Estradiol-17β
RN: 50-28-2 **MP** (°C): 176
MW: 272.39 **BP** (°C):

Solubility (Moles/L)	Solubility (Grams/L)	Temp (°C)	Ref (#)	Evaluation (T P E A A)	Comments
1.652E-05	4.500E-03	20	G072	1 2 2 1 2	

(*continued*)

3916. C$_{18}$H$_{24}$O$_2$ (continued)

Solubility (Moles/L)	Solubility (Grams/L)	Temp (°C)	Ref (#)	Evaluation (T P E A A)	Comments
6.200E-06	1.689E-03	20	L077	1 2 2 2 1	
1.413E-05	3.850E-03	22	Y419	0 0 0 0 0	
2.566E-05	6.990E-03	23	B014	0 0 1 2 2	
7.413E-06	2.019E-03	25	B041	1 0 2 2 0	EFG
6.000E-07	1.634E-04	25	E014	2 2 2 1 1	pH 7.3
1.432E-05	3.900E-03	25	H049	0 0 0 0 0	
1.836E-05	5.000E-03	25	K003	2 1 1 1 1	
5.544E-06	1.510E-03	25	S468	0 0 0 0 0	
1.320E-05	3.596E-03	27.34	L077	1 2 2 2 2	
2.060E-05	5.611E-03	35	L077	1 2 2 2 2	
1.500E-05	4.086E-03	37	H034	1 0 2 1 2	pH 7.4
2.350E-05	6.401E-03	37	H035	1 1 1 1 2	pH 7.4
1.430E-05	3.895E-03	37	H054	0 0 0 0 0	
1.880E-05	5.120E-03	37	R069	0 0 0 0 0	pH 7.4
1.000E-05	2.724E-03	37.50	B041	1 0 2 2 0	EFG
2.830E-05	7.709E-03	42	L077	1 2 2 2 2	
3.560E-05	9.697E-03	50	L077	1 2 2 2 2	

3917. C$_{18}$H$_{24}$O$_2$
α-Estradiol
17-α-Estradiol

RN: 57-91-0   **MP** (°C):   220
MW: 272.39   **BP** (°C):

Solubility (Moles/L)	Solubility (Grams/L)	Temp (°C)	Ref (#)	Evaluation (T P E A A)	Comments
1.432E-05	3.900E-03	25	L033	1 0 2 1 2	

3918. C$_{18}$H$_{24}$O$_3$
Estriol
Oestriol
Drihydroxyestrin

RN: 50-27-1   **MP** (°C):   284.5
MW: 288.39   **BP** (°C):

Solubility (Moles/L)	Solubility (Grams/L)	Temp (°C)	Ref (#)	Evaluation (T P E A A)	Comments
1.047E-04	3.020E-02	22	Y419	0 0 0 0 0	
1.110E-05	3.200E-03	25	H049	0 0 0 0 0	
1.000E-04	2.884E-02	30	O321	0 0 0 0 0	
1.006E-04	2.900E-02	30	O321	0 0 0 0 0	

3919. C₁₈H₂₄O₆

Butylphthalyl butyl glycolate
1,2-Benzenedicarboxylic acid 2-butoxy-2-oxoethyl butyl ester
Butyl carbobutoxymethyl phthalate

RN: 85-70-1   **MP** (°C):   <-35
MW: 336.39   **BP** (°C):   219

Solubility (Moles/L)	Solubility (Grams/L)	Temp (°C)	Ref (#)	Evaluation (T P E A A)	Comments
3.567E-05	1.200E-02	20	F070	1 0 0 0 2	

3920. C₁₈H₂₅I₃N₃O₉

3,5-Diacetylamino-2,4,6-triiodobenzoic acid methyl-glucamide

RN: **MP** (°C):   191
MW: 808.13   **BP** (°C):

Solubility (Moles/L)	Solubility (Grams/L)	Temp (°C)	Ref (#)	Evaluation (T P E A A)	Comments
1.101E+00	8.900E+02	20	L100	1 0 0 0 1	

3921. C₁₈H₂₅NO

Racemethorphan
Dextromethorphan HBr

RN: 510-53-2   **MP** (°C):
MW: 271.41   **BP** (°C):

Solubility (Moles/L)	Solubility (Grams/L)	Temp (°C)	Ref (#)	Evaluation (T P E A A)	Comments
1.326E-01	3.600E+01	37	F008	1 1 2 2 2	0.1N HCl

3922. C₁₈H₂₅NO

Dextromethorphan
(+)-*cis*-1,3,4,9,10,10a-Hexahydro-6-methoxy-11-methyl-2H-10,4a-iminoethanophenanthrene
Romilar CF
DXM Free Base
3-Methoxy-17-methyl-(9α,13α,14α)-morphinan
Benylin DM

RN: 125-71-3   **MP** (°C):
MW: 271.41   **BP** (°C):

Solubility (Moles/L)	Solubility (Grams/L)	Temp (°C)	Ref (#)	Evaluation (T P E A A)	Comments
3.316E-04	9.000E-02	amb	L434	0 0 0 0 0	

3923. $C_{18}H_{25}NO_5S_2$

Methyl N-{5-[(3R)-1,2-dithiolan-3-yl]-pentanoyl}-L-tyrosinate

RN: **MP** (°C):

MW: 399.53   **BP** (°C):

Solubility (Moles/L)	Solubility (Grams/L)	Temp (°C)	Ref (#)	Evaluation (T P E A A)	Comments
3.003E-05	1.200E-02	ns	S453	0 0 0 0 0	

3924. $C_{18}H_{25}N_3O_2$

2-Ethoxy-N-[2-(diethyl-amino)ethyl]-4-quinoline carboxamide

N-[2-(Diethylamino)ethyl]-2-ethoxyquinoline-4-carboxamide

RN: 2716-99-6   **MP** (°C):

MW: 315.42   **BP** (°C):

Solubility (Moles/L)	Solubility (Grams/L)	Temp (°C)	Ref (#)	Evaluation (T P E A A)	Comments
6.600E-04	2.082E-01	ns	M066	0 0 0 0 1	

3925. $C_{18}H_{26}NO_4$

Ibuprofen N-methyl-N-carbamoyl methyl glycolamide ester

RN: **MP** (°C):   100.5

MW: 320.41   **BP** (°C):

Solubility (Moles/L)	Solubility (Grams/L)	Temp (°C)	Ref (#)	Evaluation (T P E A A)	Comments
4.057E-04	1.300E-01	0	B331	1 2 2 1 1	pH 7.4

3926. $C_{18}H_{26}N_2O_4$

Benzeneacetic acid, β-methyl-4-(2-methylpropyl)-, 2-[(2-amino-2-oxoethyl)methylamino]-2-oxo-ethyl ester

Ibuprofen N-methyl-N-carbamoyl methyl glycolamide ester

RN: 114665-11-1   **MP** (°C):   100–101

MW: 334.42   **BP** (°C):

Solubility (Moles/L)	Solubility (Grams/L)	Temp (°C)	Ref (#)	Evaluation (T P E A A)	Comments
3.887E-04	1.300E-01	21	B331	0 0 0 0 0	

3927. $C_{18}H_{26}N_4O_6$

9-[5-O-(Heptylate-β-D-arabinofuranosyl)]-6-methoxy-9H-purine

RN: 142963-79-9   **MP** (°C):   foam

MW: 394.43   **BP** (°C):

Solubility (Moles/L)	Solubility (Grams/L)	Temp (°C)	Ref (#)	Evaluation (T P E A A)	Comments
2.120E-04	8.362E-02	37	M378	1 2 1 1 2	pH 7.2

3928. C$_{18}$H$_{26}$N$_4$O$_6$.0.5H$_2$O
2'-Heptanyl-6-methoxypurine arabinoside (hemihydrate)
RN: 145913-40-2   **MP** (°C):   83–85
MW: 403.44   **BP** (°C):

Solubility (Moles/L)	Solubility (Grams/L)	Temp (°C)	Ref (#)	Evaluation (T P E A A)	Comments
2.780E-03	1.122E+00	37	C348	0 0 0 0 0	pH 7.00

3929. C$_{18}$H$_{26}$O
Acetyl ethyl tetramethyl tetralin
1-(3-Ethyl-5,6,7,8-tetrahydro-5,5,8,8-tetramethyl-2-naphthalenyl)ethanone
AETT
1,1,4,4-Tetramethyl-6-ethyl-7-acetyl-1,2,3,4-tetrahydronaphthalene
Ethanone, 1-(3-ethyl-5,6,7,8-tetrahydro-5,5,8,8-tetramethyl-2-naphthyl)-
RN: 88-29-9   **MP** (°C):
MW: 258.41   **BP** (°C):

Solubility (Moles/L)	Solubility (Grams/L)	Temp (°C)	Ref (#)	Evaluation (T P E A A)	Comments
4.644E-08	1.200E-05	ns	B338	0 0 0 0 1	

3930. C$_{18}$H$_{26}$O$_2$
Nortestosterone
Estr-4-en-3-one, 17-hydroxy-, (17β)
RN: 434-22-0   **MP** (°C):
MW: 274.41   **BP** (°C):

Solubility (Moles/L)	Solubility (Grams/L)	Temp (°C)	Ref (#)	Evaluation (T P E A A)	Comments
1.126E-02	3.090E+00	25	P324	0 0 0 0 0	

3931. C$_{18}$H$_{26}$O$_4$
Dipentyl phthalate
Diamyl phthalate
RN: 131-18-0   **MP** (°C):   <–55
MW: 306.41   **BP** (°C):   342

Solubility (Moles/L)	Solubility (Grams/L)	Temp (°C)	Ref (#)	Evaluation (T P E A A)	Comments
1.450E-06	4.443E-04	20	L300	2 1 0 2 2	
9.791E-07	3.000E-04	25	F067	1 0 2 2 0	
3.263E-04	9.999E-02	ns	F014	0 0 0 0 0	

3932. C$_{18}$H$_{26}$O$_6$
Butyl phthalyl butyl glycollate
RN: **MP** (°C):
MW: 338.40   **BP** (°C):

Solubility (Moles/L)	Solubility (Grams/L)	Temp (°C)	Ref (#)	Evaluation (T P E A A)	Comments
2.955E-05	1.000E-02	15	H069	1 0 1 1 0	
5.318E-04	1.800E-01	ns	F014	0 0 0 0 1	

3933. C$_{18}$H$_{27}$NO
N-Nonylcinnamamide
2-Propenamide, N-nonyl-3-phenyl-
RN: 59832-01-8   **MP** (°C):
MW: 273.42   **BP** (°C):

Solubility (Moles/L)	Solubility (Grams/L)	Temp (°C)	Ref (#)	Evaluation (T P E A A)	Comments
2.220E-06	6.070E-04	ns	H350	0 0 0 0 0	

3934. C$_{18}$H$_{27}$NO$_3$
p-Acetamidophenyl decanoate
Acetaminophen decanoate
RN: 54942-37-9   **MP** (°C):   107
MW: 305.42   **BP** (°C):

Solubility (Moles/L)	Solubility (Grams/L)	Temp (°C)	Ref (#)	Evaluation (T P E A A)	Comments
2.947E-05	9.000E-03	25	B010	1 1 1 1 0	

3935. C$_{18}$H$_{27}$NO$_3$
Capsaicin
Nonenamide, N-((4-hydroxy-3-methoxyphenyl)methyl)-8-methyl-, (E)-
Zostrix
RN: 404-86-4   **MP** (°C):   63 C
MW: 305.42   **BP** (°C):

Solubility (Moles/L)	Solubility (Grams/L)	Temp (°C)	Ref (#)	Evaluation (T P E A A)	Comments
1.040E-04	3.176E-02	27	Z412	0 0 0 0 0	

3936. C$_{18}$H$_{27}$N$_5$O$_5$
9-[5′-(O-Caprylyl)-β-D-arabinofuranosyl]adenine ester
RN: 66460-51-3   **MP** (°C):
MW: 393.45   **BP** (°C):

Solubility (Moles/L)	Solubility (Grams/L)	Temp (°C)	Ref (#)	Evaluation (T P E A A)	Comments
2.542E-04	1.000E-01	ns	B134	0 1 1 1 0	

3937. C₁₈H₂₈N₂O

DL-Bupivacaine
Bupivacaine
Marcaine
Bupivicaine
Marcaine (hydrochloride monohydrate)

RN:	2180-92-9	**MP** (°C):	107	
MW:	288.44	**BP** (°C):		

Solubility (Moles/L)	Solubility (Grams/L)	Temp (°C)	Ref (#)	Evaluation (T P E A A)	Comments
3.750E-04	1.082E-01	14.9	N046	2 0 1 1 2	intrinsic
9.025E-06	2.603E-03	22.5	B440	0 0 0 0 0	
1.733E-03	5.000E-01	23	F176	2 0 0 2 0	EFG, pH 7.4, intrinsic
3.520E-04	1.015E-01	25	D401	1 2 2 2 2	
3.180E-04	9.172E-02	25	N046	2 0 1 1 2	intrinsic
3.130E-04	9.028E-02	34.5	N046	2 0 1 1 2	intrinsic
4.170E-04	1.203E-01	37	N044	2 1 1 2 2	intrinsic

3938. C₁₈H₂₈N₄O₂

2,5-Diaziridinyl-3,6-bis(butylamino)-1,4-benzoquinone

RN:	59886-48-5	**MP** (°C):	95	
MW:	332.45	**BP** (°C):		

Solubility (Moles/L)	Solubility (Grams/L)	Temp (°C)	Ref (#)	Evaluation (T P E A A)	Comments
<3.01E-04	<1.00E-01	rt	C317	0 0 0 0 0	

3939. C₁₈H₂₈O₃

Undecyl *p*-hydroxybenzoate
Undecyl 4-hydroxybenzoate

RN:	69679-31-8	**MP** (°C):		
MW:	292.42	**BP** (°C):		

Solubility (Moles/L)	Solubility (Grams/L)	Temp (°C)	Ref (#)	Evaluation (T P E A A)	Comments
8.079E-03	2.362E+00	25	D081	1 2 2 1 2	

3940. C₁₈H₂₉NO₂

Penbutolol
Levatol
2-Propanol, 1-(2-cyclopentylphenoxy)-3-[(1,1-dimethylethyl)amino]-, (*S*)-

RN:	38363-40-5	**MP** (°C):	70	
MW:	291.44	**BP** (°C):		

Solubility (Moles/L)	Solubility (Grams/L)	Temp (°C)	Ref (#)	Evaluation (T P E A A)	Comments
2.402E-02	7.000E+00	rt	H096	1 0 0 0 0	

3941. C$_{18}$H$_{29}$NO$_3$
4-Pentoxybenzoic acid-2-(diethyl-amino)ethyl ester
RN: 38973-73-8   **MP** (°C):
MW: 307.44   **BP** (°C):

Solubility (Moles/L)	Solubility (Grams/L)	Temp (°C)	Ref (#)	Evaluation (T P E A A)	Comments
6.000E-05	1.845E-02	ns	M066	0 0 0 0 1	

3942. C$_{18}$H$_{30}$
2-Phenyldodecane
RN: **MP** (°C):
MW: 246.44   **BP** (°C):

Solubility (Moles/L)	Solubility (Grams/L)	Temp (°C)	Ref (#)	Evaluation (T P E A A)	Comments
4.000E-09	9.858E-07	25	S377	0 0 0 0 0	

3943. C$_{18}$H$_{30}$
4-Phenyldodecane
RN: **MP** (°C):
MW: 246.44   **BP** (°C):

Solubility (Moles/L)	Solubility (Grams/L)	Temp (°C)	Ref (#)	Evaluation (T P E A A)	Comments
5.000E-09	1.232E-06	25	S377	0 0 0 0 0	

3944. C$_{18}$H$_{30}$
5-Phenyldodecane
RN: **MP** (°C):
MW: 246.44   **BP** (°C):

Solubility (Moles/L)	Solubility (Grams/L)	Temp (°C)	Ref (#)	Evaluation (T P E A A)	Comments
5.000E-09	1.232E-06	25	S377	0 0 0 0 0	

3945. C$_{18}$H$_{30}$
3-Phenyldodecane
RN: **MP** (°C):
MW: 246.44   **BP** (°C):

Solubility (Moles/L)	Solubility (Grams/L)	Temp (°C)	Ref (#)	Evaluation (T P E A A)	Comments
7.000E-09	1.725E-06	25	S377	0 0 0 0 0	

3946. C$_{18}$H$_{30}$
6-Phenyldodecane

RN: **MP** (°C):
MW: 246.44 **BP** (°C):

Solubility (Moles/L)	Solubility (Grams/L)	Temp (°C)	Ref (#)	Evaluation (T P E A A)	Comments
4.000E-09	9.858E-07	25	S377	0 0 0 0 0	

3947. C$_{18}$H$_{30}$N$_2$O$_2$
4-Pentylaminobenzoic acid-2-(diethylamino)ethyl ester

RN: 16488-56-5 **MP** (°C):
MW: 306.45 **BP** (°C):

Solubility (Moles/L)	Solubility (Grams/L)	Temp (°C)	Ref (#)	Evaluation (T P E A A)	Comments
2.100E-04	6.435E-02	ns	M066	0 0 0 0 1	

3948. C$_{18}$H$_{30}$O$_3$
4-Octylphenol diethoxylate
2-[2-(p-Octylphenoxy)ethoxy]ethanol

RN: 51437-90-2 **MP** (°C):
MW: 294.44 **BP** (°C):

Solubility (Moles/L)	Solubility (Grams/L)	Temp (°C)	Ref (#)	Evaluation (T P E A A)	Comments
4.483E-05	1.320E-02	20.5	A335	0 0 0 0 0	
4.490E-05	1.322E-02	20.5	A335	0 0 0 0 0	

3949. C$_{18}$H$_{30}$O$_{15}$.4H$_2$O
Triamylose (tetrahydrate)

RN: **MP** (°C):
MW: 558.49 **BP** (°C):

Solubility (Moles/L)	Solubility (Grams/L)	Temp (°C)	Ref (#)	Evaluation (T P E A A)	Comments
2.298E-02	1.283E+01	20	P048	1 2 1 1 1	

3950. C$_{18}$H$_{31}$NO$_4$
Bisoprolol
1-[Isopropylamino]-3-[isopropoxyethoxymethylphenoxy]-2-propanol
ZEβ
Ziac

RN: 66722-44-9 **MP** (°C):
MW: 325.45 **BP** (°C):

Solubility (Moles/L)	Solubility (Grams/L)	Temp (°C)	Ref (#)	Evaluation (T P E A A)	Comments
1.690E-08	5.500E-06	100	M418	0 0 0 0 0	

3951. C$_{18}$H$_{31}$O$_4$P
Butyl octyl phenyl phosphate
RN: 110459-55-7   **MP** (°C):
MW: 342.42   **BP** (°C):

Solubility (Moles/L)	Solubility (Grams/L)	Temp (°C)	Ref (#)	Evaluation (T P E A A)	Comments
<5.84E-04	<2.00E-01	25	B070	1 2 0 1 0	

3952. C$_{18}$H$_{32}$O$_7$
Tributyl citrate
Tri-*n*-butyl citrate
Butyl citrate
RN: 77-94-1   **MP** (°C):   −20
MW: 360.45   **BP** (°C):

Solubility (Moles/L)	Solubility (Grams/L)	Temp (°C)	Ref (#)	Evaluation (T P E A A)	Comments
1.664E-04	6.000E-02	15	H069	1 0 1 1 0	
2.219E-04	7.999E-02	ns	F014	0 0 0 0 0	

3953. C$_{18}$H$_{32}$O$_{16}$
Raffinose
6G-α-D-Galactosylsucrose
Melitose
Gossypose
Melitriose
RN: 512-69-6   **MP** (°C):   80.0
MW: 504.45   **BP** (°C):

Solubility (Moles/L)	Solubility (Grams/L)	Temp (°C)	Ref (#)	Evaluation (T P E A A)	Comments
6.556E-02	3.307E+01	.02	H040	1 2 2 2 2	
1.227E-01	6.191E+01	10.00	H040	1 2 2 2 1	
1.879E-01	9.478E+01	16.38	H040	1 2 2 2 2	
1.937E-01	9.772E+01	16.90	H040	1 2 2 2 2	
2.480E-01	1.251E+02	20	D041	1 0 0 0 2	
2.373E-01	1.197E+02	20.00	H040	1 2 2 2 2	
3.192E-01	1.610E+02	24.80	H040	1 2 2 2 2	
4.555E-01	2.298E+02	25	P049	1 0 1 1 1	
3.228E-01	1.628E+02	25.05	H040	1 2 2 2 2	
3.340E-01	1.685E+02	25.50	H040	1 2 2 2 2	
4.227E-01	2.132E+02	30.00	H040	1 2 2 2 2	
6.398E-01	3.227E+02	39.38	H040	1 2 2 2 2	
6.599E-01	3.329E+02	40.00	H040	1 2 2 2 2	
9.217E-01	4.650E+02	50.00	H040	1 2 2 2 2	
1.016E+00	5.125E+02	53.20	H040	1 2 2 2 2	
1.201E+00	6.060E+02	60.00	H040	1 2 2 2 2	
1.239E+00	6.250E+02	61.60	H040	1 2 2 2 2	
1.473E+00	7.430E+02	70.00	H040	1 2 2 2 2	

(continued)

3953. $C_{18}H_{32}O_{16}$ (continued)

Solubility (Moles/L)	Solubility (Grams/L)	Temp (°C)	Ref (#)	Evaluation (T P E A A)	Comments
1.682E+00	8.484E+02	78.00	H040	1 2 2 2 2	
6.518E-02	3.288E+01	.00	H040	1 2 2 2 1	
2.480E-01	1.251E+02	rt	D021	0 0 1 1 2	

3954. $C_{18}H_{32}O_{16} \cdot 5H_2O$
Raffinose (pentahydrate)
6G-α-D-Galactosylsucrose (pentahydrate)

RN: 17629-30-0 **MP** (°C): 80
MW: 594.52 **BP** (°C):

Solubility (Moles/L)	Solubility (Grams/L)	Temp (°C)	Ref (#)	Evaluation (T P E A A)	Comments
5.531E-02	3.288E+01	0	M043	1 0 0 0 1	
1.041E-01	6.191E+01	10	M043	1 0 0 0 1	
2.014E-01	1.197E+02	20	M043	1 0 0 0 2	
3.586E-01	2.132E+02	30	M043	1 0 0 0 2	
5.599E-01	3.329E+02	40	M043	1 0 0 0 2	
7.821E-01	4.650E+02	60	M043	1 0 0 0 2	
1.019E+00	6.060E+02	80	M043	1 0 0 0 2	

3955. $C_{18}H_{34}OSn$
Cyhexatin
Tricyclohexylhydroxystannane
Tricyclohexyltin hydroxide
Plictran
Dowco 213

RN: 13121-70-5 **MP** (°C): 196.5
MW: 385.16 **BP** (°C):

Solubility (Moles/L)	Solubility (Grams/L)	Temp (°C)	Ref (#)	Evaluation (T P E A A)	Comments
<2.60E-06	<1.00E-03	25	M161	1 0 0 0 0	
<2.60E-06	<1.00E-03	ns	K138	0 0 0 0 1	

3956. $C_{18}H_{34}O_4$
Dibutyl sebacate
Di-*n*-butyl sebacate
Decanedioic acid dibutyl ester
Dibutyl decanedioate

RN: 109-43-3 **MP** (°C):
MW: 314.47 **BP** (°C):

Solubility (Moles/L)	Solubility (Grams/L)	Temp (°C)	Ref (#)	Evaluation (T P E A A)	Comments
1.590E-04	5.000E-02	ns	F014	0 0 0 0 0	

3957. $C_{18}H_{36}O_2$
Stearic acid
Stearinsaeure
Octadecanoic acid

RN:	57-11-4	**MP** (°C):	70
MW:	284.49	**BP** (°C):	

Solubility (Moles/L)	Solubility (Grams/L)	Temp (°C)	Ref (#)	Evaluation (T P E A A)	Comments
6.327E-06	1.800E-03	0	B136	1 0 2 1 1	
9.842E-06	2.800E-03	20	B136	1 0 2 1 1	
1.055E-05	3.000E-03	20	F300	1 0 0 0 0	
1.019E-05	2.900E-03	20.0	R001	1 1 1 1 1	
2.100E-06	5.974E-04	25	J001	1 0 2 1 1	
1.970E-06	5.604E-04	25	R002	0 0 0 0 0	
1.195E-05	3.400E-03	30	B136	1 0 2 1 1	
1.195E-05	3.400E-03	30.0	R001	1 1 1 1 1	
1.700E-05	4.836E-03	35	M004	2 0 0 0 2	
1.476E-05	4.200E-03	45	B136	1 0 2 1 1	
1.476E-05	4.200E-03	45.0	R001	1 1 1 1 1	
2.700E-06	7.681E-04	50	J001	1 0 2 1 1	
5.770E-05	1.641E-02	50	M004	2 0 0 0 2	
1.758E-05	5.000E-03	60	B136	1 0 2 1 1	
1.758E-05	5.000E-03	60	F300	1 0 0 0 0	
1.758E-05	5.000E-03	60.0	R001	1 1 1 1 1	
1.145E-05	3.257E-03	62.5	M004	1 0 0 0 2	
6.327E-06	1.800E-03	.0	R001	1 1 1 1 1	

3958. $C_{18}H_{38}$
n-Octadecane
Octadecane

RN:	593-45-3	**MP** (°C):	29.5
MW:	254.50	**BP** (°C):	317.0

Solubility (Moles/L)	Solubility (Grams/L)	Temp (°C)	Ref (#)	Evaluation (T P E A A)	Comments
4.715E-07	1.200E-04	10	C331	0 0 0 0 0	
2.358E-08	6.000E-06	25	B069	1 0 1 1 1	
2.240E-08	5.700E-06	25	B069	1 0 1 1 1	
5.894E-07	1.500E-04	30	C331	0 0 0 0 0	
6.680E-07	1.700E-04	60	C331	0 0 0 0 0	
3.045E-08	7.750E-06	ns	B003	0 0 0 0 0	
3.045E-08	7.750E-06	ns	B033	0 0 0 0 2	

3959. C$_{18}$H$_{38}$O
Octadecanol
Stearyl alcohol
Octadecyl alcohol
Steraffine

RN:	112-92-5	**MP** (°C):	61
MW:	270.50	**BP** (°C):	336

Solubility (Moles/L)	Solubility (Grams/L)	Temp (°C)	Ref (#)	Evaluation (T P E A A)	Comments
4.000E-09	1.082E-06	34	K011	1 2 1 1 1	
2.200E-08	5.951E-06	65	K011	1 2 1 1 1	

3960. C$_{18}$H$_{39}$N.2H$_2$O
Octadecylamine (dihydrate)
1-Aminooctadecane (dihydrate)

RN:	124-30-1	**MP** (°C):	
MW:	305.55	**BP** (°C):	

Solubility (Moles/L)	Solubility (Grams/L)	Temp (°C)	Ref (#)	Evaluation (T P E A A)	Comments
5.891E-09	1.800E-06	ns	R037	0 2 2 1 1	

3961. C$_{18}$H$_{39}$O$_3$P
Dibutyl decyl phosphonate

RN:	36378-71-9	**MP** (°C):	
MW:	334.48	**BP** (°C):	

Solubility (Moles/L)	Solubility (Grams/L)	Temp (°C)	Ref (#)	Evaluation (T P E A A)	Comments
<5.98E-04	<2.00E-01	25	B070	1 2 0 1 0	

3962. C$_{18}$H$_{39}$O$_4$P
Dibutyl decyl phosphate

RN:	111440-78-9	**MP** (°C):	
MW:	350.48	**BP** (°C):	

Solubility (Moles/L)	Solubility (Grams/L)	Temp (°C)	Ref (#)	Evaluation (T P E A A)	Comments
<2.85E-04	<1.00E-01	25	B070	1 2 0 1 0	

3963. C$_{18}$H$_{39}$O$_7$P
Tributoxyethyl phosphate

RN:	78-51-3	**MP** (°C):	–70
MW:	398.48	**BP** (°C):	

Solubility (Moles/L)	Solubility (Grams/L)	Temp (°C)	Ref (#)	Evaluation (T P E A A)	Comments
2.760E-03	1.100E+00	25	B070	1 2 0 1 1	

3964. $C_{19}H_{12}O_6$
Dicumarol
3,3'-Methylene-bis(4-hydroxycoumarin)
Dicoumarol

RN:	66-76-2	**MP (°C):**	290
MW:	336.30	**BP (°C):**	

Solubility (Moles/L)	Solubility (Grams/L)	Temp (°C)	Ref (#)	Evaluation (T P E A A)	Comments
5.352E-05	1.800E-02	25	M457	0 0 0 0 0	
<4.46E-04	<1.50E-01	25	P312	0 0 0 0 0	

3965. $C_{19}H_{13}Cl$
6-Chloro-10-methyl-1,2-benzanthracene

RN:	188124-97-2	**MP (°C):**	
MW:	276.77	**BP (°C):**	

Solubility (Moles/L)	Solubility (Grams/L)	Temp (°C)	Ref (#)	Evaluation (T P E A A)	Comments
3.613E-08	1.000E-05	27	D003	1 0 0 1 0	

3966. $C_{19}H_{13}Cl$
4-Fluoro-10-methyl-1,2-benzanthracene
4-FMBA

RN:	2990-70-7	**MP (°C):**	
MW:	276.77	**BP (°C):**	

Solubility (Moles/L)	Solubility (Grams/L)	Temp (°C)	Ref (#)	Evaluation (T P E A A)	Comments
1.900E-08	5.259E-06	22	B062	0 0 0 0 0	

3967. $C_{19}H_{13}Cl$
3-Fluoro-10-methyl-1,2-benzanthracene
3-FMBA

RN:	20629-50-9	**MP (°C):**	
MW:	276.77	**BP (°C):**	

Solubility (Moles/L)	Solubility (Grams/L)	Temp (°C)	Ref (#)	Evaluation (T P E A A)	Comments
1.900E-08	5.259E-06	22	B062	0 0 0 0 0	

3968. $C_{19}H_{14}$
10-Methyl-1,2-benzanthracene

RN:	2541-69-7	**MP (°C):**	141
MW:	242.32	**BP (°C):**	

Solubility (Moles/L)	Solubility (Grams/L)	Temp (°C)	Ref (#)	Evaluation (T P E A A)	Comments
4.539E-08	1.100E-05	24	H116	2 1 0 0 1	

3969. $C_{19}H_{14}$
1′-Methyl-1,2-benzanthracene
RN: 2498-77-3 **MP** (°C): 138
MW: 242.32 **BP** (°C):

Solubility (Moles/L)	Solubility (Grams/L)	Temp (°C)	Ref (#)	Evaluation (T P E A A)	Comments
2.270E-07	5.500E-05	27	D003	1 0 0 1 2	

3970. $C_{19}H_{14}$
5-Methylchrysene
RN: 3697-24-3 **MP** (°C): 117.1
MW: 242.32 **BP** (°C):

Solubility (Moles/L)	Solubility (Grams/L)	Temp (°C)	Ref (#)	Evaluation (T P E A A)	Comments
2.559E-07	6.200E-05	27	D003	1 0 0 1 1	

3971. $C_{19}H_{14}$
9-Methyl-1,2-benzanthracene
RN: 2381-16-0 **MP** (°C): 138
MW: 242.32 **BP** (°C):

Solubility (Moles/L)	Solubility (Grams/L)	Temp (°C)	Ref (#)	Evaluation (T P E A A)	Comments
1.527E-07	3.700E-05	24	H116	2 1 0 0 1	

3972. $C_{19}H_{14}$
6-Methylchrysene
RN: 1705-85-7 **MP** (°C): 149
MW: 242.32 **BP** (°C):

Solubility (Moles/L)	Solubility (Grams/L)	Temp (°C)	Ref (#)	Evaluation (T P E A A)	Comments
2.682E-07	6.500E-05	27	D003	1 0 0 1 1	

3973. $C_{19}H_{14}O_3$
Aurin
Rosolic acid
4-[bis-(p-Hydroxyphenyl)methylene]-2,5-cyclohexadien-1-one
RN: 603-45-2 **MP** (°C):
MW: 290.32 **BP** (°C):

Solubility (Moles/L)	Solubility (Grams/L)	Temp (°C)	Ref (#)	Evaluation (T P E A A)	Comments
4.128E-03	1.199E+00	rt	D021	0 0 1 1 1	

3974. C$_{19}$H$_{14}$O$_5$S
Phenolsulfonaphthalein
Phenolrot

RN:	143-74-8	**MP** (°C):	>300
MW:	354.38	**BP** (°C):	

Solubility (Moles/L)	Solubility (Grams/L)	Temp (°C)	Ref (#)	Evaluation (T P E A A)	Comments
8.748E-04	3.100E-01	100	F300	1 0 0 0 2	

3975. C$_{19}$H$_{16}$O
Triphenylcarbinol
Triphenylmethanol

RN:	76-84-6	**MP** (°C):	164.2
MW:	260.34	**BP** (°C):	

Solubility (Moles/L)	Solubility (Grams/L)	Temp (°C)	Ref (#)	Evaluation (T P E A A)	Comments
5.500E-03	1.432E+00	25	D007	2 0 1 1 2	

3976. C$_{19}$H$_{17}$ClN$_2$O
Prazepam
Centrax
7-Chloro-1-(cyclopropylmethyl)-1,3-dihydro-5-phenyl-2H-1,4-benzodiazepin-2-one
Demetrin
Verstran

RN:	2955-38-6	**MP** (°C):	
MW:	324.81	**BP** (°C):	

Solubility (Moles/L)	Solubility (Grams/L)	Temp (°C)	Ref (#)	Evaluation (T P E A A)	Comments
2.800E-05	9.095E-03	25	M320	2 2 1 1 2	
		amb	L434	0 0 0 0 0	

3977. C$_{19}$H$_{17}$ClN$_2$O$_4$
Quizalofop-ethyl
Quizalofop-et
Quizalofop ethyl ester
Targa
Pilot
NC 302

RN:	76578-14-8	**MP** (°C):	91.7–92.1
MW:	372.81	**BP** (°C):	220 at 0.2 mm Hg

Solubility (Moles/L)	Solubility (Grams/L)	Temp (°C)	Ref (#)	Evaluation (T P E A A)	Comments
8.128E-07	3.030E-04	ns	R427	0 0 0 0 0	

3978. $C_{19}H_{17}ClN_2O_4$
Glafenine
N-(7-Chloro-4-quinolyl)anthranilate
2,3-Dihydroxypropyl-N-(7-chloro-4-quinolinyl)anthranilate
RN: 3820-67-5   **MP** (°C):   169.5
MW: 372.81   **BP** (°C):

Solubility (Moles/L)	Solubility (Grams/L)	Temp (°C)	Ref (#)	Evaluation (T P E A A)	Comments
1.032E-01	3.846E+01	ns	M152	0 0 0 0 0	pH 1.0, intrinsic

3979. $C_{19}H_{17}ClN_4$
Fenbuconazole
α-(2-(4-Chlorophenyl)ethyl)-α-phenyl-1H-1,2,4-triazole-1-propanenitrile
Enable
RH-7592
Fenethanil
1,2,4-Triazole-1-propanenitrile, α-{2-(4-chlorophenyl)ethyl}-α-phenyl
RN: 114369-43-6   **MP** (°C):
MW: 336.83   **BP** (°C):

Solubility (Moles/L)	Solubility (Grams/L)	Temp (°C)	Ref (#)	Evaluation (T P E A A)	Comments
5.888E-07	1.983E-04	ns	R427	0 0 0 0 0	

3980. $C_{19}H_{17}N_3O_4S_2$
Sugordomycin
RN: 1405-50-1   **MP** (°C):
MW: 415.49   **BP** (°C):

Solubility (Moles/L)	Solubility (Grams/L)	Temp (°C)	Ref (#)	Evaluation (T P E A A)	Comments
2.304E-02	9.572E+00	21	M044	2 0 2 2 2	

3981. $C_{19}H_{17}N_3O_4S_2$
Cephaloridine
Glaxoridin
Keflodin
Loridine
RN: 50-59-9   **MP** (°C):   184
MW: 415.49   **BP** (°C):

Solubility (Moles/L)	Solubility (Grams/L)	Temp (°C)	Ref (#)	Evaluation (T P E A A)	Comments
>4.81E-02	>2.00E+01	21	M044	2 0 2 2 0	

3982. C$_{19}$H$_{17}$N$_3$O$_5$

1H-Benzimidazole-1-carboxylic acid, 6-benzoyl-2-[(methoxycarbonyl)amino]-, ethyl ester

RN: 153474-30-7   **MP** (°C):   165.5
MW: 367.36   **BP** (°C):

Solubility (Moles/L)	Solubility (Grams/L)	Temp (°C)	Ref (#)	Evaluation (T P E A A)	Comments
2.722E-05	1.000E-02	21	N337	0 0 0 0 0	pH 5
2.700E-05	9.919E-03	21	N337	0 0 0 0 0	pH 5

3983. C$_{19}$H$_{18}$

1,2,3,4-Tetrahydro-10-methyl-1,2-benzanthracene
10-Methyl-1,2-cyclohexane anthracene

RN: 6366-18-3   **MP** (°C):   117
MW: 246.36   **BP** (°C):

Solubility (Moles/L)	Solubility (Grams/L)	Temp (°C)	Ref (#)	Evaluation (T P E A A)	Comments
1.786E-07	4.400E-05	27	D003	1 0 0 1 1	

3984. C$_{19}$H$_{18}$Cl$_2$N$_2$O$_2$

G-20

p,p-Dichlorophenylbutazone

RN: 4047-57-8   **MP** (°C):
MW: 377.27   **BP** (°C):

Solubility (Moles/L)	Solubility (Grams/L)	Temp (°C)	Ref (#)	Evaluation (T P E A A)	Comments
2.386E-04	9.000E-02	ns	B158	0 0 0 0 1	pH 7.0

3985. C$_{19}$H$_{18}$N$_2$O$_3$

G-23

1-Oxybutylphenylbutazone
3,5-Pyrazolidinedione, 4-butyryl-1,2-diphenyl-

RN: 13167-98-1   **MP** (°C):
MW: 322.37   **BP** (°C):

Solubility (Moles/L)	Solubility (Grams/L)	Temp (°C)	Ref (#)	Evaluation (T P E A A)	Comments
3.722E-04	1.200E-01	ns	B158	0 0 0 0 1	pH 7.0

3986. C$_{19}$H$_{18}$N$_2$O$_3$

Kebuzone
3,5-Pyrazolidinedione

RN: 853-34-9   **MP** (°C):   128
MW: 322.37   **BP** (°C):

Solubility (Moles/L)	Solubility (Grams/L)	Temp (°C)	Ref (#)	Evaluation (T P E A A)	Comments
5.402E-04	1.742E-01	20	M140	2 0 1 1 1	

3987. $C_{19}H_{19}ClFNO_3$

Flamprop-isopropyl

Flufenprop-isopropyl

Isopropyl N-benzoyl-N-(3-chloro-4-fluorophenyl)alanine

1-Methylethyl N-benzoyl-N-(3-chloro-4-fluorophenyl)-DL-alanine

RN:	52756-22-6	MP (°C):	56.5
MW:	363.82	BP (°C):	

Solubility (Moles/L)	Solubility (Grams/L)	Temp (°C)	Ref (#)	Evaluation (T P E A A)	Comments
4.948E-05	1.800E-02	20	M161	1 0 0 0 0	

3988. $C_{19}H_{19}N_7O_6$

Folic acid

N-(p-(((2-Amino-4-hydroxy-6-pteridinyl)methyl)amino)benzoyl)-L-glutamic acid

Vitamin M

Pteroylglutamic acid

Folcysteine

Folacin

RN:	59-30-3	MP (°C):	
MW:	441.41	BP (°C):	

Solubility (Moles/L)	Solubility (Grams/L)	Temp (°C)	Ref (#)	Evaluation (T P E A A)	Comments
3.619E-03	1.597E+00	25	D041	1 0 0 0 1	*sic*
3.625E-06	1.600E-03	25	D315	0 0 0 0 0	
2.243E-02	9.901E+00	100	D041	1 0 0 0 0	*sic*
2.265E-04	1.000E-01	ns	K444	0 0 0 0 0	

3989. $C_{19}H_{20}ClNO_9$

Griseofulvin-4-carboxy-methoxime

RN:		MP (°C):	
MW:	441.83	BP (°C):	

Solubility (Moles/L)	Solubility (Grams/L)	Temp (°C)	Ref (#)	Evaluation (T P E A A)	Comments
1.704E-04	7.529E-02	37	F033	2 0 2 0 2	

3990. $C_{19}H_{20}F_3NO_4$

Fluazifop-butyl

Butyl 2-(4-((5-trifluoromethyl-2-pyridinyl)oxy)phenoxy)propanoate

Onecide

Fluazifop-butyl

Fluazifop butyl ester

Hache uno super

RN:	69806-50-4	MP (°C):	13
MW:	383.37	BP (°C):	165 at 2.02 mm Hg

Solubility (Moles/L)	Solubility (Grams/L)	Temp (°C)	Ref (#)	Evaluation (T P E A A)	Comments
1.400E-06	5.366E-04	ns	S460	0 0 0 0 0	

3991. C$_{19}$H$_{20}$N$_2$O

Cinchoninone
Cinchoninon
9-Deoxy-9-oxocinchonine

RN: 14509-68-3   **MP** (°C):
MW: 292.38   **BP** (°C):

Solubility (Moles/L)	Solubility (Grams/L)	Temp (°C)	Ref (#)	Evaluation (T P E A A)	Comments
6.498E-04	1.900E-01	20	F300	1 0 0 0 1	

3992. C$_{19}$H$_{20}$N$_2$O$_2$

Phenylbutazone
1,2-Diphenyl-4-butyl-3,5-dioxopyrazolidine
Butazolidin
Equiphen
Butazone

RN: 50-33-9   **MP** (°C): 107
MW: 308.38   **BP** (°C):

Solubility (Moles/L)	Solubility (Grams/L)	Temp (°C)	Ref (#)	Evaluation (T P E A A)	Comments
8.415E-05	2.595E-02	20	H301	0 0 0 0 0	
4.864E-05	1.500E-02	20	P026	1 0 1 1 1	
1.102E-04	3.400E-02	25	P096	0 0 0 0 0	
1.540E-04	4.750E-02	30	D015	2 0 1 1 0	EFG
1.000E-03	3.084E-01	35	H091	1 2 2 2 1	*sic*
9.076E-03	2.799E+00	36	I002	2 2 1 1 2	pH 6.95, recrystallized
7.575E-03	2.336E+00	36	I002	2 2 1 1 2	pH 6.95, recrystallized
9.362E-03	2.887E+00	36	I002	2 2 1 1 2	pH 6.95, recrystallized
6.907E-03	2.130E+00	36	I002	2 2 1 1 2	pH 6.95, recrystallized
2.108E-04	6.500E-02	37	D015	2 0 1 1 0	EFG
1.816E-04	5.600E-02	37	E047	1 0 1 1 1	
7.134E-03	2.200E+00	ns	B158	0 0 0 0 1	pH 7.0
1.037E-03	3.199E-01	ns	B404	0 2 1 1 0	
1.300E-04	4.009E-02	ns	O304	0 0 1 2 2	
2.594E-05	8.000E-03	rt	H302	0 0 2 1 2	intrinsic
1.310E-01	4.040E+01	rt	N056	0 0 1 1 2	average of 2

3993. C₁₉H₂₀N₂O₂

G-21

p,p-Dimethylphenylbutazone

RN: 745-27-7   **MP** (°C):
MW: 308.38   **BP** (°C):

Solubility (Moles/L)	Solubility (Grams/L)	Temp (°C)	Ref (#)	Evaluation (T P E A A)	Comments
3.891E-04	1.200E-01	ns	B158	0 0 0 0 1	pH 7.0

3994. C₁₉H₂₀N₂O₃

Oxyphenbutazone

p-Hydroxyphenylbutazone

RN: 129-20-4   **MP** (°C):   124
MW: 324.38   **BP** (°C):

Solubility (Moles/L)	Solubility (Grams/L)	Temp (°C)	Ref (#)	Evaluation (T P E A A)	Comments
1.850E-04	6.000E-02	30	D015	2 0 1 1 0	EFG
2.497E-04	8.100E-02	37	D015	2 0 1 1 0	EFG
3.083E-02	1.000E+01	ns	B158	0 0 0 0 1	pH 7.0, *sic*
>1.54E-03	>5.00E-01	ns	B404	0 2 1 1 0	
6.166E-05	2.000E-02	rt	H302	0 0 2 1 2	intrinsic

3995. C₁₉H₂₀N₄O₆·0.5H₂O

6-Methoxy-9-(5-*O*-[4-methylbenzoyl]-β-D-arabinofuranosyl)-9H-purine (hemihydrate)

RN: 121032-20-0   **MP** (°C):   127–128
MW: 409.40   **BP** (°C):

Solubility (Moles/L)	Solubility (Grams/L)	Temp (°C)	Ref (#)	Evaluation (T P E A A)	Comments
3.500E-05	1.433E-02	37	M378	1 2 1 1 2	pH 7.2

3996. C₁₉H₂₀N₄O₆

2′-(*p*-Toluylyl)-6-methoxypurine arabinoside

2′-Phenylacetyl-6-methoxypurine arabinoside

RN: 121032-52-8   **MP** (°C):   69–73
MW: 400.39   **BP** (°C):

Solubility (Moles/L)	Solubility (Grams/L)	Temp (°C)	Ref (#)	Evaluation (T P E A A)	Comments
5.870E-02	2.350E+01	37	C348	0 0 0 0 0	pH 7.00
5.840E-03	2.338E+00	37	C348	0 0 0 0 0	pH 7.00

3997. C$_{19}$H$_{20}$N$_4$O$_6$.0.1H$_2$O
9-[5-O-(Benzyl formyl-β-D-arabinofuranosyl)]-6-methoxy-9H-purine (0.1 hydrate)
RN: 121032-36-8   **MP** (°C):   foam
MW: 402.20   **BP** (°C):

Solubility (Moles/L)	Solubility (Grams/L)	Temp (°C)	Ref (#)	Evaluation (T P E A A)	Comments
1.050E-02	4.223E+00	37	M378	1 2 1 1 2	pH 7.2

3998. C$_{19}$H$_{20}$N$_4$O$_7$
2′-(p-Methoxybenzoyl)-6-methoxypurine arabinoside
RN: 121032-51-7   **MP** (°C):   71–75
MW: 416.39   **BP** (°C):

Solubility (Moles/L)	Solubility (Grams/L)	Temp (°C)	Ref (#)	Evaluation (T P E A A)	Comments
6.660E-03	2.773E+00	37	C348	0 0 0 0 0	pH 7.00

3999. C$_{19}$H$_{20}$N$_4$O$_7$.0.5H$_2$O
2′-Phenoxyacetyl-6-methoxypurine arabinoside (hemihydrate)
RN: 145913-46-8   **MP** (°C):   123–125
MW: 425.40   **BP** (°C):

Solubility (Moles/L)	Solubility (Grams/L)	Temp (°C)	Ref (#)	Evaluation (T P E A A)	Comments
>2.21E-02	>9.40E+00	37	C348	0 0 0 0 0	pH 7.00

4000. C$_{19}$H$_{20}$N$_4$O$_7$.0.25H$_2$O
9-[5-O-(4-Methoxybenzoyl-β-D-arabinofuranosyl)]-6-methoxy-9H-purine (0.25 hydrate)
RN: 121032-35-7   **MP** (°C):   195–197
MW: 420.90   **BP** (°C):

Solubility (Moles/L)	Solubility (Grams/L)	Temp (°C)	Ref (#)	Evaluation (T P E A A)	Comments
1.960E-04	8.250E-02	37	M378	1 2 1 1 2	pH 7.2

4001. C$_{19}$H$_{20}$N$_4$O$_7$.0.05H$_2$O
9-[5-O-(Benzyl acetate-β-D-arabinofuranosyl)]-6-methoxy-9H-purine (0.05 hydrate)
RN: 121032-37-9   **MP** (°C):   193-195
MW: 417.29   **BP** (°C):

Solubility (Moles/L)	Solubility (Grams/L)	Temp (°C)	Ref (#)	Evaluation (T P E A A)	Comments
3.930E-04	1.640E-01	37	M378	1 2 1 1 2	pH 7.2

4002. C$_{19}$H$_{20}$O$_4$

Butylbenzyl phthalate
Butyl phenyl-methyl phthalate
Benzylbutyl phthalate
Phthalate butyl benzyl ester
Butyl benzyl phthalate
1,2-Benzenedicarboxylic acid butyl phenylmethyl ester

| RN: | 85-68-7 | MP (°C): | <−35 |
| MW: | 312.37 | BP (°C): | 370 |

Solubility (Moles/L)	Solubility (Grams/L)	Temp (°C)	Ref (#)	Evaluation (T P E A A)	Comments
9.020E-06	2.818E-03	20	L300	2 1 0 2 2	
3.778E-05	1.180E-02	22	Y419	0 0 0 0 0	
2.273E-06	7.100E-04	24	H116	2 1 0 0 2	
8.644E-06	2.700E-03	25	F067	1 0 2 2 1	

4003. C$_{19}$H$_{21}$ClO$_4$

Isobutyl (+/−)-2-[4-(4-chlorophenoxy)phenoxy]propionate

| RN: | 51337-71-4 | MP (°C): | 39.5 |
| MW: | 348.83 | BP (°C): | |

Solubility (Moles/L)	Solubility (Grams/L)	Temp (°C)	Ref (#)	Evaluation (T P E A A)	Comments
5.160E-04	1.800E-01	22	M161	1 0 0 0 2	

4004. C$_{19}$H$_{21}$F$_3$N$_2$S

2-Trifluoromethyl-N,N-dimethyl-10H-phenothiazine-10-propanamide

| RN: | 2340-66-1 | MP (°C): | |
| MW: | 366.45 | BP (°C): | |

Solubility (Moles/L)	Solubility (Grams/L)	Temp (°C)	Ref (#)	Evaluation (T P E A A)	Comments
5.000E-06	1.832E-03	ns	G023	0 0 1 1 0	

4005. C$_{19}$H$_{21}$NO

Doxepin
Adapin
Deptran
Sinequan

| RN: | 1668-19-5 | MP (°C): | 120 |
| MW: | 279.39 | BP (°C): | |

Solubility (Moles/L)	Solubility (Grams/L)	Temp (°C)	Ref (#)	Evaluation (T P E A A)	Comments
1.130E-04	3.157E-02	25	E051	1 0 2 1 2	

4006. C₁₉H₂₁NO₃

Thebaine

Paramorphine

Morphinan, 6,7,8,14-tetradehydro-4,5α-epoxy-3,6-dimethoxy-17-methyl-

RN: 115-37-7   **MP** (°C):

MW: 311.38   **BP** (°C):

Solubility (Moles/L)	Solubility (Grams/L)	Temp (°C)	Ref (#)	Evaluation (T P E A A)	Comments
2.200E-03	6.850E-01	15	K059	2 2 2 0 1	

4007. C₁₉H₂₁N₃O

Zolpidem

RN: 82626-48-0   **MP** (°C):

MW: 307.40   **BP** (°C):

Solubility (Moles/L)	Solubility (Grams/L)	Temp (°C)	Ref (#)	Evaluation (T P E A A)	Comments
<3.25E-05	<1.00E-02	rt	B435	0 0 0 0 0	

4008. C₁₉H₂₁N₅O₂

Dis. A. 6

Propanenitrile, 3-[butyl[4-[(4-nitrophenyl)azo]phenyl]amino]-

RN: 69472-19-1   **MP** (°C):   118

MW: 351.41   **BP** (°C):

Solubility (Moles/L)	Solubility (Grams/L)	Temp (°C)	Ref (#)	Evaluation (T P E A A)	Comments
2.000E-08	7.028E-06	25	B333	0 0 0 0 0	

4009. C₁₉H₂₁N₅O₂

Dye VII

4-[[(4-(N-Butyl-N-ethylnitrile)amino)phenyl]azo]nitrobenzene

RN: **MP** (°C):

MW: 351.41   **BP** (°C):

Solubility (Moles/L)	Solubility (Grams/L)	Temp (°C)	Ref (#)	Evaluation (T P E A A)	Comments
4.800E-07	1.687E-04	71.80	B198	1 2 1 1 1	
9.700E-07	3.409E-04	84.10	B198	1 2 1 1 1	
2.020E-06	7.099E-04	97.40	B198	1 2 1 1 2	

4010. C$_{19}$H$_{21}$N$_5$O$_4$
Prazosin
Minipress
Pressin

RN: 19216-56-9   **MP** (°C):
MW: 383.41   **BP** (°C):

Solubility (Moles/L)	Solubility (Grams/L)	Temp (°C)	Ref (#)	Evaluation (T P E A A)	Comments
8.346E-06	3.200E-03	22.5	B422	0 0 0 0 0	

4011. C$_{19}$H$_{21}$N$_5$O$_5$
9-[5′-(O-Hydrocinnamoyl)-β-D-arabinofuranosyl]adenine ester

RN: 68325-41-7   **MP** (°C):
MW: 399.41   **BP** (°C):

Solubility (Moles/L)	Solubility (Grams/L)	Temp (°C)	Ref (#)	Evaluation (T P E A A)	Comments
3.756E-03	1.500E+00	ns	B134	0 1 1 1 1	

4012. C$_{19}$H$_{22}$Cl$_2$O$_2$
1-Methyl-1,1-dichloro-2,2-bis(p-ethoxylphenyl)ethane

RN: 56265-23-7   **MP** (°C):
MW: 353.29   **BP** (°C):

Solubility (Moles/L)	Solubility (Grams/L)	Temp (°C)	Ref (#)	Evaluation (T P E A A)	Comments
1.415E-07	5.000E-05	rt	C122	0 0 0 0 0	

4013. C$_{19}$H$_{22}$N$_2$O
Cinchonidine
Cinchonidin
(8α,9R)-Cinchonan-9-ol
L-Cinchonidine

RN: 485-71-2   **MP** (°C): 210
MW: 294.40   **BP** (°C):

Solubility (Moles/L)	Solubility (Grams/L)	Temp (°C)	Ref (#)	Evaluation (T P E A A)	Comments
9.000E-04	2.650E-01	15	K059	2 2 2 0 0	
9.511E-05	2.800E-02	22	M459	0 0 0 0 0	
6.793E-04	2.000E-01	25	F300	1 0 0 0 0	
1.970E-03	5.800E-01	100	F300	1 0 0 0 1	
6.792E-04	2.000E-01	c	D004	0 0 0 0 0	
8.490E-04	2.499E-01	rt	D021	0 0 1 1 1	

4014. $C_{19}H_{22}N_2O$

Cinchonine
Cinchonan-9-ol
(+)-Cinchonine
(9S)-Cinchonan-9-ol

RN: 118-10-5　　**MP** (°C): 265
MW: 294.40　　　**BP** (°C):

Solubility (Moles/L)	Solubility (Grams/L)	Temp (°C)	Ref (#)	Evaluation (T P E A A)	Comments
4.800E-06	1.413E-03	15	K059	2 2 2 0 1	
2.378E-05	7.000E-03	22	M459	0 0 0 0 0	
9.253E-04	2.724E-01	25	D004	0 0 0 0 0	
9.171E-04	2.700E-01	100	F300	1 0 0 0 1	
8.150E-04	2.399E-01	rt	D021	0 0 1 1 1	

4015. $C_{19}H_{22}N_2OS$

Acetylpromazine
3-Acetyl-10-(3-dimethylaminopropyl)phenothiazine
Plegicil
Vetranquil
Notensil
Plivafen

RN: 61-00-7　　**MP** (°C):
MW: 326.46　　**BP** (°C):

Solubility (Moles/L)	Solubility (Grams/L)	Temp (°C)	Ref (#)	Evaluation (T P E A A)	Comments
4.901E-05	1.600E-02	25	L045	1 1 1 1 2	intrinsic

4016. $C_{19}H_{22}N_2O_5$

2-Naphthaleneacetic acid, 6-methoxy-α-methyl-, 2-[(2-amino-2-oxoethyl)methylamino]-2-
　　oxoethyl ester
Naproxen N-methyl-N-carbamoyl methyl glycolamide ester

RN: 114681-69-5　**MP** (°C): 179
MW: 358.40　　　**BP** (°C):

Solubility (Moles/L)	Solubility (Grams/L)	Temp (°C)	Ref (#)	Evaluation (T P E A A)	Comments
1.646E-04	5.900E-02	21	B331	0 0 0 0 0	

4017. $C_{19}H_{22}N_2S$

Mepazine
Pecazine

RN: 60-89-9　　**MP** (°C): 80
MW: 310.46　　**BP** (°C): 233

Solubility (Moles/L)	Solubility (Grams/L)	Temp (°C)	Ref (#)	Evaluation (T P E A A)	Comments
1.800E-05	5.588E-03	24	G022	2 0 1 1 1	

4018. C$_{19}$H$_{23}$ClO$_2$

1-Chloro-1-methyl-2,2-bis(p-ethoxylphenyl)ethane

RN: 56265-22-6   **MP** (°C):

MW: 318.85   **BP** (°C):

Solubility (Moles/L)	Solubility (Grams/L)	Temp (°C)	Ref (#)	Evaluation (T P E A A)	Comments
2.760E-06	8.800E-04	rt	C122	0 0 0 0 0	

4019. C$_{19}$H$_{23}$NO$_3$

Ethylmorphine

7,8-Didehydro-4,5-epoxy-3-ethoxy-17-methylmorphinan-6-ol

RN: 76-58-4   **MP** (°C):

MW: 313.40   **BP** (°C):

Solubility (Moles/L)	Solubility (Grams/L)	Temp (°C)	Ref (#)	Evaluation (T P E A A)	Comments
8.916E-03	2.794E+00	20	K052	1 1 1 1 2	

4020. C$_{19}$H$_{23}$NO$_4$

1-Methyl-1-nitro-2,2-bis(p-ethoxylphenyl)ethane

RN: 26258-70-8   **MP** (°C):

MW: 329.40   **BP** (°C):

Solubility (Moles/L)	Solubility (Grams/L)	Temp (°C)	Ref (#)	Evaluation (T P E A A)	Comments
1.093E-06	3.600E-04	rt	C122	0 0 0 0 0	

4021. C$_{19}$H$_{23}$NO$_5$

2-Naphthaleneacetic acid, 6-methoxy-α-methyl-, 2-[(2-hydroxyethyl)methylamino]-2-oxoethyl ester

Naproxen N-methyl-N-ethanol glycolamide ester

RN: 114665-19-9   **MP** (°C): 110

MW: 345.40   **BP** (°C):

Solubility (Moles/L)	Solubility (Grams/L)	Temp (°C)	Ref (#)	Evaluation (T P E A A)	Comments
4.053E-04	1.400E-01	21	B331	0 0 0 0 0	

4022. C₁₉H₂₃N₃

Amitraz
1,5-Di(2,4-dimethylphenyl)-3-methyl-1,3,5-triazapenta-1,4-diene
Ovasyn
Mitac
Triazid
Baam

RN: 33089-61-1 **MP** (°C): 86.5
MW: 293.42 **BP** (°C):

Solubility (Moles/L)	Solubility (Grams/L)	Temp (°C)	Ref (#)	Evaluation (T P E A A)	Comments
3.408E-06	1.000E-03	rt	M161	0 0 0 0 0	

4023. C₁₉H₂₃N₃O₂

Ergonovine
9,10-Didehydro-*N*-(2-hydroxy-1-methylethyl)-6-methylergoline-8-carboxamide
Ergometrine

RN: 60-79-7 **MP** (°C):
MW: 325.41 **BP** (°C):

Solubility (Moles/L)	Solubility (Grams/L)	Temp (°C)	Ref (#)	Evaluation (T P E A A)	Comments
>1.21E+00	>3.94E+02	25	B443	0 0 0 0 0	

4024. C₁₉H₂₃N₅O₄

Benzoic acid, 4-[(dimethylamino)methyl]-, 2-[(6-amino-4,5-dihydro-4-oxo-1H-imidazo[4,5-c]
    pyridin-1-yl)methoxy]ethyl ester
1H-Imidazo[4,5-c]pyridine, benzoic acid deriv.

RN: 137605-68-6 **MP** (°C):
MW: 385.43 **BP** (°C): 651.2

Solubility (Moles/L)	Solubility (Grams/L)	Temp (°C)	Ref (#)	Evaluation (T P E A A)	Comments
3.373E-03	1.300E+00	21	B419	1 1 2 2 1	int

4025. C₁₉H₂₄N₂

1-(Diphenylmethyl)-4-ethylpiperazine

RN: **MP** (°C):
MW: 280.42 **BP** (°C):

Solubility (Moles/L)	Solubility (Grams/L)	Temp (°C)	Ref (#)	Evaluation (T P E A A)	Comments
2.030E-03	5.693E-01	25	M438	0 0 0 0 0	

4026. $C_{19}H_{24}N_2$

Imipramine
10,11-Dihydro-*N*,*N*-dimethyl-5H-dibenz[b,f]azepine-5-propanamine
5-[3-(Dimethylamino)propyl]-10,11-dihydro-5H-dibenz[b,f]azepine

RN: 50-49-7	**MP** (°C): 174			
MW: 280.42	**BP** (°C):			

Solubility (Moles/L)	Solubility (Grams/L)	Temp (°C)	Ref (#)	Evaluation (T P E A A)	Comments
6.500E-05	1.823E-02	24	G022	2 0 1 1 1	

4027. $C_{19}H_{24}N_2O$

Hydrocinchonine
Hydrocinchonin
Cinchotine

RN: 485-65-4	**MP** (°C): 268			
MW: 296.42	**BP** (°C):			

Solubility (Moles/L)	Solubility (Grams/L)	Temp (°C)	Ref (#)	Evaluation (T P E A A)	Comments
2.362E-03	7.000E-01	16	F300	1 0 0 0 1	
2.593E-03	7.686E-01	25	D004	0 0 0 0 0	

4028. $C_{19}H_{24}N_2OS$

Methotrimeprazine
Levomepromazine

RN: 60-99-1	**MP** (°C): 117			
MW: 328.48	**BP** (°C):			

Solubility (Moles/L)	Solubility (Grams/L)	Temp (°C)	Ref (#)	Evaluation (T P E A A)	Comments
6.089E-05	2.000E-02	25	A081	1 0 1 1 0	EFG

4029. $C_{19}H_{24}N_2O_2$

Praziquantel
2-Cyclohexyl-carbonyl-1,3,4,6,7,11b-hexahydro-2H-pyrazine(2,1-a)isoquinoline-4-one
Biltricide
Droncit

RN: 55268-74-1	**MP** (°C):			
MW: 312.42	**BP** (°C):			

Solubility (Moles/L)	Solubility (Grams/L)	Temp (°C)	Ref (#)	Evaluation (T P E A A)	Comments
9.000E-04	2.812E-01	30	B402	2 0 1 1 0	EFG
1.280E-03	4.000E-01	ns	K444	0 0 0 0 0	

4030. C$_{19}$H$_{24}$N$_2$O$_2$S

Cyclohexyl-*p*-toluene sulfonamide
Cyclohexyl-4-toluene sulfonamide

RN: **MP** (°C):
MW: 344.48 **BP** (°C):

Solubility (Moles/L)	Solubility (Grams/L)	Temp (°C)	Ref (#)	Evaluation (T P E A A)	Comments
1.742E-04	6.000E-02	ns	F014	0 0 0 0 0	

4031. C$_{19}$H$_{24}$N$_4$O$_7$

Propyloxycarbonyl-mitomycin C

RN: **MP** (°C):
MW: 420.43 **BP** (°C):

Solubility (Moles/L)	Solubility (Grams/L)	Temp (°C)	Ref (#)	Evaluation (T P E A A)	Comments
3.300E-04	1.387E-01	25	M316	1 1 1 1 2	

4032. C$_{19}$H$_{24}$O

1,1-Dimethyl-2-(*p*-methylphenyl)-2-*p*-ethoxylphenyl)ethane

RN: 56265-26-0 **MP** (°C):
MW: 268.40 **BP** (°C):

Solubility (Moles/L)	Solubility (Grams/L)	Temp (°C)	Ref (#)	Evaluation (T P E A A)	Comments
6.706E-07	1.800E-04	rt	C122	0 0 0 0 0	

4033. C$_{19}$H$_{24}$O$_2$

1,1,1-Trimethyl-2,2-bis(*p*-methyloxylphenyl)ethane

RN: 4741-74-6 **MP** (°C):
MW: 284.40 **BP** (°C):

Solubility (Moles/L)	Solubility (Grams/L)	Temp (°C)	Ref (#)	Evaluation (T P E A A)	Comments
2.426E-06	6.900E-04	rt	C122	0 0 0 0 0	

4034. C$_{19}$H$_{24}$O$_3$

Adrenosterone
Androstene-3,11,17-trione

RN: 382-45-6 **MP** (°C): 220
MW: 300.40 **BP** (°C):

Solubility (Moles/L)	Solubility (Grams/L)	Temp (°C)	Ref (#)	Evaluation (T P E A A)	Comments
3.279E-04	9.849E-02	23.5	J003	2 0 2 1 2	average of 2
2.610E-04	7.840E-02	37	H004	0 0 0 0 0	
5.059E-04	1.520E-01	37	J003	1 0 2 1 2	

4035. C$_{19}$H$_{25}$NO

N,N-Dicyclopentylcinnamamide
2-Propenamide, *N,N*-dicyclopentyl-3-phenyl-

RN: 59832-08-5 **MP** (°C):
MW: 283.42 **BP** (°C):

Solubility (Moles/L)	Solubility (Grams/L)	Temp (°C)	Ref (#)	Evaluation (T P E A A)	Comments
7.750E-07	2.196E-04	ns	H350	0 0 0 0 0	

4036. C$_{19}$H$_{26}$I$_3$N$_3$O$_{10}$

1,3-Benzenedicarboxamide, *N,N'*-bis[2-hydroxy-1,1-bis(hydroxymethyl)ethyl]-5-[(2-hydroxy-1-oxopropyl)amino]-2,4,6-triiodo-, (*S*)-

RN: 77868-46-3 **MP** (°C):
MW: 837.15 **BP** (°C):

Solubility (Moles/L)	Solubility (Grams/L)	Temp (°C)	Ref (#)	Evaluation (T P E A A)	Comments
2.342E-02	1.961E+01	25	P091	0 0 0 0 0	

4037. C$_{19}$H$_{26}$N$_6$O$_4$S

Benzenesulfonamide, 4-(1,3-diethyl-2,3,6,7-tetrahydro-2,6-dioxo-1H-purin-8-yl)-*N*-[2-(dimethylamino)ethyl]-

RN: 89073-49-4 **MP** (°C): 264
MW: 434.52 **BP** (°C):

Solubility (Moles/L)	Solubility (Grams/L)	Temp (°C)	Ref (#)	Evaluation (T P E A A)	Comments
2.532E-04	1.100E-01	ns	H316	0 0 0 0 0	pH 7.4
2.647E-02	1.150E+01	ns	H316	0 0 0 0 0	0.1N HCL

4038. C$_{19}$H$_{26}$O

δ-4-Androstene-3-one

RN: **MP** (°C):
MW: 270.42 **BP** (°C):

Solubility (Moles/L)	Solubility (Grams/L)	Temp (°C)	Ref (#)	Evaluation (T P E A A)	Comments
<1.00E-06	<2.70E-04	25	E014	2 2 2 1 0	pH 7.3

4039. $C_{19}H_{26}O_2$
Androstenedione
4-Androstene-3,17-dione
Androst-4-en-3,17-dion
RN: 63-05-8 **MP** (°C):
MW: 286.42 **BP** (°C):

Solubility (Moles/L)	Solubility (Grams/L)	Temp (°C)	Ref (#)	Evaluation (T P E A A)	Comments
2.000E-04	5.728E-02	25	E014	2 2 2 1 2	pH 7.3
2.840E-02	8.133E+00	25	P324	0 0 0 0 0	
1.399E-04	4.007E-02	37	H034	1 0 2 1 2	pH 7.4
1.700E-04	4.870E-02	37	L010	2 0 2 1 1	

4040. $C_{19}H_{27}N_3O$
Doxylamine ethanamine
RN: **MP** (°C):
MW: 313.45 **BP** (°C):

Solubility (Moles/L)	Solubility (Grams/L)	Temp (°C)	Ref (#)	Evaluation (T P E A A)	Comments
3.000E-02	9.403E+00	37.5	L034	2 2 0 1 2	pH 7.4

4041. $C_{19}H_{27}N_3O_2$
2-Propoxy-N-[2-(diethyl-amino)ethyl]-4-quinoline carboxamide
N-[2-(Diethylamino)ethyl]-2-propoxyquinoline-4-carboxamide
RN: 2717-00-2 **MP** (°C):
MW: 329.45 **BP** (°C):

Solubility (Moles/L)	Solubility (Grams/L)	Temp (°C)	Ref (#)	Evaluation (T P E A A)	Comments
3.980E-04	1.311E-01	ns	B018	0 0 0 0 2	
3.980E-04	1.311E-01	ns	M066	0 0 0 0 2	

4042. $C_{19}H_{28}Cl_2O_3$
2,4-Dichlorophenoxyacetic acid n-undecyl ester
RN: 65267-95-0 **MP** (°C):
MW: 375.34 **BP** (°C):

Solubility (Moles/L)	Solubility (Grams/L)	Temp (°C)	Ref (#)	Evaluation (T P E A A)	Comments
1.977E-05	7.420E-03	ns	M120	0 0 1 1 2	

4043. $C_{19}H_{28}N_4O_6$
2′-Octanyl-6-methoxypurine arabinoside
RN: 145913-41-3   **MP** (°C):   75–77
MW: 408.46   **BP** (°C):

Solubility (Moles/L)	Solubility (Grams/L)	Temp (°C)	Ref (#)	Evaluation (T P E A A)	Comments
6.110E-04	2.496E-01	37	C348	0 0 0 0 0	pH 7.00

4044. $C_{19}H_{28}O$
7α-Methyl-19-nortestosterone
Trestolone
19-Nor-7α-methyltestosterone
RN: 3764-87-2   **MP** (°C):
MW: 272.43   **BP** (°C):

Solubility (Moles/L)	Solubility (Grams/L)	Temp (°C)	Ref (#)	Evaluation (T P E A A)	Comments
3.377E-04	9.200E-02	37	H004	0 0 0 0 0	

4045. $C_{19}H_{28}O_2$
Androstanedione
5α-Androstane-3,17-dione
RN: 846-46-8   **MP** (°C):   142
MW: 288.43   **BP** (°C):

Solubility (Moles/L)	Solubility (Grams/L)	Temp (°C)	Ref (#)	Evaluation (T P E A A)	Comments
1.141E-04	3.290E-02	23.5	J003	1 0 2 1 2	average of 2
2.200E-04	6.346E-02	25	E014	2 2 2 1 2	pH 7.3
1.685E-04	4.860E-02	37	J003	1 0 2 1 2	average of 2

4046. $C_{19}H_{28}O_2$
Testosterone
17β-Hydroxyandrost-4-en-3-one
Halotensin
Virilon
Oreton
Testex
RN: 58-22-0   **MP** (°C):   155
MW: 288.43   **BP** (°C):

Solubility (Moles/L)	Solubility (Grams/L)	Temp (°C)	Ref (#)	Evaluation (T P E A A)	Comments
5.600E-05	1.615E-02	10	B012	2 0 1 1 0	
6.390E-05	1.843E-02	10	L017	2 2 2 2 2	
2.254E-04	6.500E-02	15	F042	2 2 2 2 1	
7.550E-05	2.178E-02	15	L017	2 2 2 2 2	
7.900E-05	2.279E-02	20	B012	2 0 1 1 0	

(*continued*)

4046. C$_{19}$H$_{28}$O$_2$ (continued)

Solubility (Moles/L)	Solubility (Grams/L)	Temp (°C)	Ref (#)	Evaluation (T P E A A)	Comments
2.430E-04	7.009E-02	20	F012	1 0 1 1 1	
2.392E-04	6.900E-02	20	F042	2 2 2 2 1	
8.460E-05	2.440E-02	20	G072	1 2 2 1 2	
7.790E-05	2.247E-02	20	L017	2 2 2 2 2	
8.000E-05	2.307E-02	20	L070	1 2 0 2 0	EFG
6.870E-05	1.982E-02	20	L077	1 2 2 2 2	
8.000E-04	2.307E-01	20	L087	1 1 2 1 0	EFG
6.240E-05	1.800E-02	22.5	B422	2 0 2 2 2	
8.100E-05	2.336E-02	25	B012	2 0 1 1 0	
9.500E-05	2.740E-02	25	B041	1 0 2 2 1	
2.531E-04	7.300E-02	25	F042	2 2 2 2 1	
8.321E-05	2.400E-02	25	K003	2 1 1 1 1	
1.664E-04	4.800E-02	25	L009	1 0 0 1 1	
8.480E-05	2.446E-02	25	L017	2 2 2 2 2	
6.934E-05	2.000E-02	25	L338	1 0 1 1 2	
1.040E-04	3.000E-02	27.34	L077	1 2 2 2 2	
1.060E-04	3.057E-02	30	B012	2 0 1 1 0	
2.670E-04	7.700E-02	30	F042	2 2 2 2 1	
9.790E-05	2.824E-02	30	L017	2 2 2 2 2	
1.100E-04	3.173E-02	30	L068	1 0 0 1 0	EFG
2.500E-04	7.211E-02	30	L344	2 0 1 1 0	
1.040E-04	3.000E-02	30	M007	2 2 1 2 2	average of 8
8.876E-05	2.560E-02	30	T005	2 0 2 2 2	
1.096E-04	3.163E-02	31	A025	2 2 2 2 0	EFG
1.300E-04	3.750E-02	35	L017	2 2 2 2 2	
1.397E-04	4.029E-02	35	L077	1 2 2 2 2	
1.950E-04	5.624E-02	37	B013	1 0 2 2 0	average
1.250E-04	3.605E-02	37	E014	2 2 2 1 2	pH 7.3
1.013E-04	2.922E-02	37	H034	1 0 2 1 2	pH 7.4
1.259E-04	3.631E-02	37.50	B041	1 0 2 2 0	EFG
1.260E-04	3.634E-02	37.50	B041	1 0 2 2 2	
1.400E-04	4.038E-02	40	B012	2 0 1 1 0	
1.570E-04	4.528E-02	40	L017	2 2 2 2 2	
3.000E-04	8.653E-02	40	L070	1 2 0 2 0	EFG
1.702E-04	4.909E-02	42.34	L077	1 2 2 2 2	
1.870E-04	5.394E-02	45	L017	2 2 2 2 2	
2.100E-04	6.057E-02	50	B012	2 0 1 1 0	
2.350E-04	6.778E-02	50	L017	2 2 2 2 2	
2.053E-04	5.922E-02	50	L077	1 2 2 2 2	
6.795E-05	1.960E-02	ns	B057	0 2 1 1 2	
3.814E-05	1.100E-02	ns	B338	0 0 0 0 1	

4047. $C_{19}H_{28}O_2$

5,6-Dehydroisoandrosterone
Prasterone
Dehydroepiandrosterone
Dehydroisoandrosterone

RN:	53-43-0	**MP** (°C):	140.5
MW:	288.43	**BP** (°C):	

Solubility (Moles/L)	Solubility (Grams/L)	Temp (°C)	Ref (#)	Evaluation (T P E A A)	Comments
7.558E-05	2.180E-02	23.5	J003	2 0 2 1 2	average of 6
1.000E-04	2.884E-02	37	E014	2 2 2 1 2	pH 7.3
1.040E-04	3.000E-02	37	H034	1 0 2 1 2	pH 7.4
1.144E-04	3.300E-02	37	J003	1 0 2 1 2	average of 4
8.633E-05	2.490E-02	ns	B057	0 2 1 1 2	

4048. $C_{19}H_{28}O_2.H_2O$

Testosterone (monohydrate)
Testosterone monohydrate -I

RN:	58-22-0	**MP** (°C):	
MW:	306.45	**BP** (°C):	

Solubility (Moles/L)	Solubility (Grams/L)	Temp (°C)	Ref (#)	Evaluation (T P E A A)	Comments
6.265E-05	1.920E-02	15	F042	2 2 2 2 2	crystal-II
5.352E-05	1.640E-02	15	F042	2 2 2 2 2	crystal-I
7.081E-05	2.170E-02	20	F042	2 2 2 2 2	crystal-II
6.265E-05	1.920E-02	20	F042	2 2 2 2 2	crystal-I
8.256E-05	2.530E-02	25	F042	2 2 2 2 2	crystal-II
7.310E-05	2.240E-02	25	F042	2 2 2 2 2	crystal-I
9.333E-05	2.860E-02	30	F042	2 2 2 2 2	crystal-II
8.484E-05	2.600E-02	30	F042	2 2 2 2 2	crystal-I

4049. $C_{19}H_{28}O_3$

11-Ketoetiocholanolone
3α-Hydroxy-5β-androstane-11,17-dione
Etiocholanol-11-one
Ba 2684

RN:	739-27-5	**MP** (°C):	
MW:	304.43	**BP** (°C):	

Solubility (Moles/L)	Solubility (Grams/L)	Temp (°C)	Ref (#)	Evaluation (T P E A A)	Comments
7.455E-04	2.269E-01	23	J003	2 0 2 1 2	average of 4
9.457E-04	2.879E-01	37	J003	1 0 2 1 2	average of 2

4050. $C_{19}H_{29}ClN_5O_6$
Terazosin
Hytrin
1-(4-Amino-6,7-dimethoxy-2-quinazolinyl)-4-((tetra-hydro-2-furanyl)carbonyl)-,
  monohydrochloride, dihydrate
(RS)-Piperazine

RN: 63590-64-7   **MP** (°C):
MW: 458.93   **BP** (°C):

Solubility (Moles/L)	Solubility (Grams/L)	Temp (°C)	Ref (#)	Evaluation (T P E A A)	Comments
6.669E-05	3.060E-02	22.5	B440	0 0 0 0 0	

4051. $C_{19}H_{29}NO$
n-Decylcinnamamide
2-Propenamide, N-decyl-3-phenyl-

RN: 59832-02-9   **MP** (°C):
MW: 287.45   **BP** (°C):

Solubility (Moles/L)	Solubility (Grams/L)	Temp (°C)	Ref (#)	Evaluation (T P E A A)	Comments
2.530E-06	7.272E-04	ns	H350	0 0 0 0 0	

4052. $C_{19}H_{29}NO$
Procyclidine
Kemadrin

RN: 77-37-2   **MP** (°C):
MW: 287.45   **BP** (°C):

Solubility (Moles/L)	Solubility (Grams/L)	Temp (°C)	Ref (#)	Evaluation (T P E A A)	Comments
3.669E-06	1.055E-03	22.5	B440	0 0 0 0 0	

4053. $C_{19}H_{29}N_5O_6$
9-(1,3-Dipivaloate-2-propoxymethyl)guanine

RN: 88110-72-9   **MP** (°C):   231
MW: 423.47   **BP** (°C):

Solubility (Moles/L)	Solubility (Grams/L)	Temp (°C)	Ref (#)	Evaluation (T P E A A)	Comments
1.653E-05	7.000E-03	25	B360	0 0 0 0 0	

4054. C$_{19}$H$_{30}$O
Androstane-17-one

RN: 36378-49-1   **MP** (°C): 119
MW: 274.45   **BP** (°C):

Solubility (Moles/L)	Solubility (Grams/L)	Temp (°C)	Ref (#)	Evaluation (T P E A A)	Comments
<2.00E-07	<5.49E-05	25	E014	2 2 2 1 0	pH 7.3

4055. C$_{19}$H$_{30}$OS
Epitiostanol

RN: 2363-58-8   **MP** (°C): 127
MW: 306.51   **BP** (°C):

Solubility (Moles/L)	Solubility (Grams/L)	Temp (°C)	Ref (#)	Evaluation (T P E A A)	Comments
3.915E-06	1.200E-03	37	H120	1 1 1 1 1	normal saline

4056. C$_{19}$H$_{30}$O$_2$
Epiandrosterone
Isoandrosterone

RN: 481-29-8   **MP** (°C): 161
MW: 290.45   **BP** (°C):

Solubility (Moles/L)	Solubility (Grams/L)	Temp (°C)	Ref (#)	Evaluation (T P E A A)	Comments
6.955E-05	2.020E-02	23.5	J003	2 0 2 1 2	average of 5
8.160E-05	2.370E-02	37	J003	1 0 2 1 2	average of 3

4057. C$_{19}$H$_{30}$O$_2$
Androsterone
3α-Hydroxy-17-androstanone
3α-Hydroxy-5α-androstan-17-one
Hydroxy-5α-androstan-17-one
Epihydroxyetioallocholan-17-one
Hydroxy-17-androstanone

RN: 53-41-8   **MP** (°C): 185
MW: 290.45   **BP** (°C):

Solubility (Moles/L)	Solubility (Grams/L)	Temp (°C)	Ref (#)	Evaluation (T P E A A)	Comments
3.959E-05	1.150E-02	23.5	J003	2 0 2 1 2	average of 2
4.300E-05	1.249E-02	37	E014	2 2 2 1 1	pH 7.3
6.163E-05	1.790E-02	37	J003	1 0 2 1 2	average of 2

4058. C$_{19}$H$_{30}$O$_2$
Stanolone
Androstanolone
RN: 521-18-6   **MP** (°C):   181.0
MW: 290.45   **BP** (°C):

Solubility (Moles/L)	Solubility (Grams/L)	Temp (°C)	Ref (#)	Evaluation (T P E A A)	Comments
1.185E+00	3.443E+02	ns	B057	0 2 1 1 2	

4059. C$_{19}$H$_{30}$O$_2$
Etiocholanolone
3α-Hydroxy-5β-androstane-17-one
5-Isoandrosterone
RN: 53-42-9   **MP** (°C):
MW: 290.45   **BP** (°C):

Solubility (Moles/L)	Solubility (Grams/L)	Temp (°C)	Ref (#)	Evaluation (T P E A A)	Comments
1.002E-04	2.910E-02	23.5	J003	2 0 2 1 2	average of 2
7.000E-05	2.033E-02	25	E014	2 2 2 1 1	pH 7.3, pyrogen

4060. C$_{19}$H$_{30}$O$_3$
p-(Dodecyloxy)benzoic acid
Dodecyl p-hydroxybenzoate
RN: 2312-15-4   **MP** (°C):   95
MW: 306.45   **BP** (°C):

Solubility (Moles/L)	Solubility (Grams/L)	Temp (°C)	Ref (#)	Evaluation (T P E A A)	Comments
3.569E-03	1.094E+00	25	D081	1 2 2 1 2	

4061. C$_{19}$H$_{30}$O$_3$
Androstane-3-β,11-β-diol-17-one
Hydroxyisoandrosterone
RN: 514-17-0   **MP** (°C):   235
MW: 306.45   **BP** (°C):

Solubility (Moles/L)	Solubility (Grams/L)	Temp (°C)	Ref (#)	Evaluation (T P E A A)	Comments
2.552E-04	7.819E-02	23.5	J003	1 0 2 1 2	average of 2

4062. C₁₉H₃₀O₃

$C_{19}H_{30}O_3$

11-Hydroxyetiocholanolone

5β-Androstan-17-one, 3α,11-dihydroxy-

RN: 3272-49-9 **MP** (°C):

MW: 306.45 **BP** (°C):

Solubility (Moles/L)	Solubility (Grams/L)	Temp (°C)	Ref (#)	Evaluation (T P E A A)	Comments
1.400E-04	4.290E-02	23.5	J003	2 0 1 1 2	average of 2

4063. C₁₉H₃₁NO₂

$C_{19}H_{31}NO_2$

Dodecyl p-aminobenzoate

p-Aminobenzoic acid dodecyl ester

RN: 20043-94-1 **MP** (°C):

MW: 305.46 **BP** (°C):

Solubility (Moles/L)	Solubility (Grams/L)	Temp (°C)	Ref (#)	Evaluation (T P E A A)	Comments
1.600E-08	4.887E-06	37	F006	1 1 2 2 1	

4064. C₁₉H₃₁NO₃

$C_{19}H_{31}NO_3$

4-Hexoxybenzoic acid-2-(diethyl-amino)ethyl ester

RN: 38973-74-9 **MP** (°C):

MW: 321.46 **BP** (°C):

Solubility (Moles/L)	Solubility (Grams/L)	Temp (°C)	Ref (#)	Evaluation (T P E A A)	Comments
4.000E-05	1.286E-02	ns	M066	0 0 0 0 1	

4065. C₁₉H₃₁NO₉

$C_{19}H_{31}NO_9$

Metoprolol tartrate

1-(Isopropylamino)-3-(p-(2-methoxyethyl)phenoxy)-2-propanol (2:1)

RN: 56392-17-7 **MP** (°C):

MW: 417.46 **BP** (°C):

Solubility (Moles/L)	Solubility (Grams/L)	Temp (°C)	Ref (#)	Evaluation (T P E A A)	Comments
1.030E-01	4.300E+01	25	A412	1 0 2 2 1	int

4066. C₁₉H₃₂

$C_{19}H_{32}$

2-Phenyltridecane

RN: **MP** (°C):

MW: 260.47 **BP** (°C):

Solubility (Moles/L)	Solubility (Grams/L)	Temp (°C)	Ref (#)	Evaluation (T P E A A)	Comments
4.000E-09	1.042E-06	25	S377	0 0 0 0 0	

4067. C$_{19}$H$_{32}$
6-Phenyltridecane
RN: **MP** (°C):
MW: 260.47   **BP** (°C):

Solubility (Moles/L)	Solubility (Grams/L)	Temp (°C)	Ref (#)	Evaluation (T P E A A)	Comments
4.000E-09	1.042E-06	25	S377	0 0 0 0 0	

4068. C$_{19}$H$_{32}$
5-Phenyltridecane
RN: **MP** (°C):
MW: 260.47   **BP** (°C):

Solubility (Moles/L)	Solubility (Grams/L)	Temp (°C)	Ref (#)	Evaluation (T P E A A)	Comments
4.000E-09	1.042E-06	25	S377	0 0 0 0 0	

4069. C$_{19}$H$_{32}$
4-Phenyltridecane
RN: **MP** (°C):
MW: 260.47   **BP** (°C):

Solubility (Moles/L)	Solubility (Grams/L)	Temp (°C)	Ref (#)	Evaluation (T P E A A)	Comments
4.000E-09	1.042E-06	25	S377	0 0 0 0 0	

4070. C$_{19}$H$_{32}$
3-Phenyltridecane
RN: **MP** (°C):
MW: 260.47   **BP** (°C):

Solubility (Moles/L)	Solubility (Grams/L)	Temp (°C)	Ref (#)	Evaluation (T P E A A)	Comments
4.000E-09	1.042E-06	25	S377	0 0 0 0 0	

4071. C$_{19}$H$_{32}$N$_2$O$_2$
4-Hexylaminobenzoic acid-2-(diethyl-amino)ethyl ester
RN: 16488-57-6   **MP** (°C):
MW: 320.48   **BP** (°C):

Solubility (Moles/L)	Solubility (Grams/L)	Temp (°C)	Ref (#)	Evaluation (T P E A A)	Comments
1.900E-04	6.089E-02	ns	M066	0 0 0 0 1	

4072. $C_{19}H_{32}O_3$
4-Nonylphenol diethoxylate
RN: 20427-84-3   **MP** (°C):
MW: 308.47   **BP** (°C):

Solubility (Moles/L)	Solubility (Grams/L)	Temp (°C)	Ref (#)	Evaluation (T P E A A)	Comments
1.180E-05	3.640E-03	2	A335	0 0 0 0 0	
1.080E-05	3.331E-03	10	A335	0 0 0 0 0	
1.096E-05	3.380E-03	10	A335	0 0 0 0 0	
9.700E-06	2.992E-03	14	A335	0 0 0 0 0	
9.726E-06	3.000E-03	14	A335	0 0 0 0 0	
1.100E-05	3.393E-03	20.5	A335	0 0 0 0 0	
1.096E-05	3.380E-03	20.5	A335	0 0 0 0 0	
1.200E-05	3.702E-03	25	A335	0 0 0 0 0	
1.196E-05	3.690E-03	25	A335	0 0 0 0 0	

4073. $C_{19}H_{34}O_3$
Methoprene
Isopropyl (2E,4E)-11-methoxy-3,7,11-trimethyl-2,4-dodecadienoate
Kabat
Precor
Dianex
Pharorid
RN: 40596-69-8   **MP** (°C):   164
MW: 310.48   **BP** (°C):   100

Solubility (Moles/L)	Solubility (Grams/L)	Temp (°C)	Ref (#)	Evaluation (T P E A A)	Comments
4.477E-06	1.390E-03	25	D302	1 0 0 0 2	
6.442E-06	2.000E-03	ns	M110	0 0 0 0 0	EFG

4074. $C_{19}H_{40}$
2,6,10,14-Tetramethylpentadecane
Pristane
RN: 1921-70-6   **MP** (°C):
MW: 268.53   **BP** (°C):   296

Solubility (Moles/L)	Solubility (Grams/L)	Temp (°C)	Ref (#)	Evaluation (T P E A A)	Comments
3.724E-11	1.000E-08	25	T423	0 0 0 0 0	

4075. $C_{20}H_9Cl_3F_5N_3O_3$
Chlorfluazuron
Atabron
Benzamide, N-[4-(3-chloro-5-trifluoromethyl-2-pyridinyl-oxy)-3,5-dichloro-phenyl-
   aminocarbonyl]-2,6-difluoro
Jupiter

| **RN:** | 71422-67-8 | **MP** (°C): | |
| **MW:** | 540.66 | **BP** (°C): | |

Solubility (Moles/L)	Solubility (Grams/L)	Temp (°C)	Ref (#)	Evaluation (T P E A A)	Comments
2.219E-09	1.200E-06	20	M402	0 0 0 0 0	

4076. $C_{20}H_{12}$
Benzo(a)pyrene
1,2-Benzopyrene
3,4-Benzpyrene
Benzo[a]pyrene
Benz[a]pyrene

| **RN:** | 50-32-8 | **MP** (°C): | 179 |
| **MW:** | 252.32 | **BP** (°C): | 310 |

Solubility (Moles/L)	Solubility (Grams/L)	Temp (°C)	Ref (#)	Evaluation (T P E A A)	Comments
3.309E-09	8.350E-07	15	B385	0 0 0 0 0	
2.000E-09	5.046E-07	20	E009	1 0 0 0 1	
2.972E-05	7.500E-03	23	T025	1 2 0 1 1	*sic*
6.341E-09	1.600E-06	25	B319	2 0 1 2 1	
5.667E-09	1.430E-06	25	B385	0 0 0 0 0	
7.213E-09	1.820E-06	25	D406	1 2 2 2 2	
4.400E-10	1.110E-07	25	K123	1 0 2 2 1	
1.506E-08	3.800E-06	25	L332	1 1 1 1 2	
1.506E-08	3.800E-06	25	M064	1 1 2 2 1	
1.500E-08	3.785E-06	25	M342	1 0 1 1 1	
6.428E-09	1.622E-06	25.04	M183	1 2 1 1 2	
1.585E-08	4.000E-06	27	D003	1 0 0 1 1	
9.083E-09	2.292E-06	30.04	M183	1 2 1 1 2	
1.098E-08	2.770E-06	35	B385	0 0 0 0 0	
1.506E-08	3.800E-06	ns	M344	0 0 0 0 2	
2.400E-08	6.056E-06	ns	W005	0 0 1 2 1	
4.756E-09	1.200E-06	ns	W302	0 0 0 0 1	

4077. C₂₀H₁₂

Benzo(k)fluoranthene
11,12-Benzo[k]fluoranthene
11,12-Benzofluoranthene
8,9-Benzofluoranthene
2,3,1′,8′-Binaphthylene
B[K]F

RN:	207-08-9	**MP** (°C):	216	
MW:	252.32	**BP** (°C):		

Solubility (Moles/L)	Solubility (Grams/L)	Temp (°C)	Ref (#)	Evaluation (T P E A A)	Comments
4.320E-09	1.090E-06	25	D406	1 2 2 2 2	
3.171E-09	8.000E-07	ns	W302	0 0 0 0 0	

4078. C₂₀H₁₂

Benzo(j)fluoranthene
Benzo[l]fluoranthene
Benzo-12,13-fluoranthene
10,11-Benzofluoranthene

RN:	205-82-3	**MP** (°C):	165	
MW:	252.32	**BP** (°C):		

Solubility (Moles/L)	Solubility (Grams/L)	Temp (°C)	Ref (#)	Evaluation (T P E A A)	Comments
9.908E-09	2.500E-06	ns	W302	0 0 0 0 1	

4079. C₂₀H₁₂

Benzo(e)pyrene
4,5-Benzopyrene
B[E]P

RN:	192-97-2	**MP** (°C):	178.5	
MW:	252.32	**BP** (°C):		

Solubility (Moles/L)	Solubility (Grams/L)	Temp (°C)	Ref (#)	Evaluation (T P E A A)	Comments
3.900E-09	9.840E-07	25	K123	1 0 2 2 1	
~1.59E-08	~4.00E-06	25	S227	1 2 1 1 0	
6.625E-02	1.672E+01	318	S355	1 1 1 2 0	EFG
1.192E-01	3.007E+01	330	S355	1 1 1 2 0	EFG
1.524E-01	3.846E+01	335	S355	1 1 1 2 0	EFG
2.066E-01	5.213E+01	342	S355	1 1 1 2 0	EFG
4.246E-01	1.071E+02	361	S355	1 1 1 2 0	EFG
4.559E-01	1.150E+02	365	S355	1 1 1 2 0	EFG

4080. $C_{20}H_{12}$
Perylene
Dibenz[de,kl]anthracene
peri-Dinaphthalene
RN: 198-55-0   **MP** (°C):   273
MW: 252.32   **BP** (°C):

Solubility (Moles/L)	Solubility (Grams/L)	Temp (°C)	Ref (#)	Evaluation (T P E A A)	Comments
4.200E-10	1.060E-07	20	E009	1 0 0 1 1	
1.585E-09	4.000E-07	25	M064	1 1 2 2 0	
1.600E-09	4.037E-07	25	M342	1 0 1 1 1	
<1.98E-09	<5.00E-07	27	D003	1 0 0 1 0	
1.585E-09	4.000E-07	ns	M344	0 0 0 0 1	

4081. $C_{20}H_{12}$
Benzo(b)fluoranthene
3,4-Benzofluoranthene
2,3-Benzofluoranthene
B[B]F
RN: 205-99-2   **MP** (°C):   108
MW: 252.32   **BP** (°C):

Solubility (Moles/L)	Solubility (Grams/L)	Temp (°C)	Ref (#)	Evaluation (T P E A A)	Comments
5.945E-09	1.500E-06	ns	W302	0 0 0 0 1	

4082. $C_{20}H_{13}N$
13H-Dibenzo(a,i)carbazole
1:2,7:8-Dibenzocarbazole
RN: 239-64-5   **MP** (°C):   220
MW: 267.33   **BP** (°C):

Solubility (Moles/L)	Solubility (Grams/L)	Temp (°C)	Ref (#)	Evaluation (T P E A A)	Comments
<5.00E-08	<1.34E-05	22	B175	1 0 1 1 0	*sic*
3.890E-08	1.040E-05	24	H106	1 0 2 2 2	
3.890E-08	1.040E-05	24	M303	1 0 1 1 2	

4083. $C_{20}H_{13}N$
3,4,5,6-Dibenzocarbazole
3:4,5:6-Dibenzocarbazole
RN: 194-59-2   **MP** (°C):   158
MW: 267.33   **BP** (°C):

Solubility (Moles/L)	Solubility (Grams/L)	Temp (°C)	Ref (#)	Evaluation (T P E A A)	Comments
2.000E-07	5.347E-05	22	B175	1 0 1 1 0	

4084. $C_{20}H_{13}N$
1,2,5,6-Dibenzocarbazole
1:2,5:6-Dibenzocarbazole
RN: 207-84-1 **MP** (°C):
MW: 267.33 **BP** (°C):

Solubility (Moles/L)	Solubility (Grams/L)	Temp (°C)	Ref (#)	Evaluation (T P E A A)	Comments
5.000E-08	1.337E-05	22	B175	1 0 1 1 0	

4085. $C_{20}H_{14}$
3,4'-Ace-1,2-benzanthracene
Benz[k]acephenanthrene
RN: 5779-79-3 **MP** (°C):
MW: 254.33 **BP** (°C):

Solubility (Moles/L)	Solubility (Grams/L)	Temp (°C)	Ref (#)	Evaluation (T P E A A)	Comments
1.062E-08	2.700E-06	27	D003	1 0 0 1 1	

4086. $C_{20}H_{14}$
Cholanthrene
1,2-Dihydroxybenz[j]aceanthrylene
RN: 479-23-2 **MP** (°C): 173
MW: 254.33 **BP** (°C):

Solubility (Moles/L)	Solubility (Grams/L)	Temp (°C)	Ref (#)	Evaluation (T P E A A)	Comments
1.376E-08	3.500E-06	27	D003	1 0 0 1 1	

4087. $C_{20}H_{14}I_6N_2O_6$
Di(3-carboxy-2,4,6-triiodoanilido)adipic acid
Iodipamide
RN: 606-17-7 **MP** (°C): 306
MW: 1139.77 **BP** (°C):

Solubility (Moles/L)	Solubility (Grams/L)	Temp (°C)	Ref (#)	Evaluation (T P E A A)	Comments
4.036E-04	4.600E-01	20	N035	1 1 2 1 1	
>4.38E-04	>5.00E-01	ns	B404	0 2 1 1 0	
1.404E-04	1.600E-01	ns	H055	0 0 0 0 0	

4088. C₂₀H₁₄N₂O₂
Disperse blue 19
C.I. Disperse blue 19

RN: 4395-65-7   **MP** (°C):   194
MW: 314.35   **BP** (°C):

Solubility (Moles/L)	Solubility (Grams/L)	Temp (°C)	Ref (#)	Evaluation (T P E A A)	Comments
6.100E-10	1.918E-07	25	B333	0 0 0 0 0	
2.100E-07	6.601E-05	60.0	D093	1 2 1 2 0	EFG
5.000E-07	1.572E-04	71.8	D093	1 2 1 2 0	EFG
1.700E-06	5.344E-04	81.4	D093	1 2 1 2 0	EFG
4.200E-06	1.320E-03	97.4	D093	1 2 1 2 0	EFG

4089. C₂₀H₁₄O₂
3,3-Diphenylphthalide
3,3-Diphenyl-phthalid

RN: 596-29-2   **MP** (°C):
MW: 286.33   **BP** (°C):

Solubility (Moles/L)	Solubility (Grams/L)	Temp (°C)	Ref (#)	Evaluation (T P E A A)	Comments
1.397E-04	4.000E-02	25	F300	1 0 0 0 0	

4090. C₂₀H₁₄O₄
Phenolphthalein
2-[bis(4-Hydroxyphenyl)methyl]benzoic acid
Espotabs
Alophen
Figsen
Laxettes

RN: 77-09-8   **MP** (°C):   260.0
MW: 318.33   **BP** (°C):

Solubility (Moles/L)	Solubility (Grams/L)	Temp (°C)	Ref (#)	Evaluation (T P E A A)	Comments
6.283E-06	2.000E-03	25	H064	1 2 2 0 2	
7.476E-04	2.380E-01	100	H064	1 2 2 0 2	
1.256E-03	3.998E-01	rt	D021	0 0 1 1 0	

4091. C₂₀H₁₄O₄
Phenyl phthalate
Diphenyl phthalate

RN: 84-62-8   **MP** (°C):   71
MW: 318.33   **BP** (°C):

Solubility (Moles/L)	Solubility (Grams/L)	Temp (°C)	Ref (#)	Evaluation (T P E A A)	Comments
2.576E-07	8.200E-05	24	H116	2 1 0 0 1	

4092. C$_{20}$H$_{14}$O$_4$

Diphenyl o-phthalate

RN: **MP (°C):**   72 C
MW: 318.33   **BP (°C):**

Solubility (Moles/L)	Solubility (Grams/L)	Temp (°C)	Ref (#)	Evaluation (T P E A A)	Comments
9.424E-06	3.000E-03	25	S417	0 0 0 0 0	

4093. C$_{20}$H$_{15}$O$_5$P

bis(4-Carboxyphenyl)phenylphosphine oxide
BCPPO

RN: 803-19-0   **MP (°C):**
MW: 366.31   **BP (°C):**

Solubility (Moles/L)	Solubility (Grams/L)	Temp (°C)	Ref (#)	Evaluation (T P E A A)	Comments
2.000E-04	7.326E-02	23	W402	0 0 0 0 0	
3.166E-04	1.160E-01	32	W402	0 0 0 0 0	
4.666E-04	1.709E-01	40	W402	0 0 0 0 0	
6.943E-04	2.543E-01	50	W402	0 0 0 0 0	
1.011E-03	3.702E-01	60	W402	0 0 0 0 0	
1.638E-03	6.000E-01	70	W402	0 0 0 0 0	
1.987E-03	7.280E-01	75	W402	0 0 0 0 0	

4094. C$_{20}$H$_{16}$

5,6-Dimethylchrysene
Chrysene, 5,6-dimethyl-

RN: 3697-27-6   **MP (°C):**   127
MW: 256.35   **BP (°C):**   200

Solubility (Moles/L)	Solubility (Grams/L)	Temp (°C)	Ref (#)	Evaluation (T P E A A)	Comments
9.752E-08	2.500E-05	27	D003	1 0 0 1 1	

4095. C$_{20}$H$_{16}$

9,10-Dimethyl-1,2-benzanthracene
7,12-Dimethyl-1,2-benzanthracene
7,12-Dimethylbenz[a]anthracene
9,10-Dimethyl-benz[a]anthracene

RN: 56-56-4   **MP (°C):**   122
MW: 256.35   **BP (°C):**

Solubility (Moles/L)	Solubility (Grams/L)	Temp (°C)	Ref (#)	Evaluation (T P E A A)	Comments
9.518E-08	2.440E-05	24	H106	1 0 2 2 2	
2.145E-07	5.500E-05	24	H116	2 1 0 0 1	
9.752E-08	2.500E-05	24	M129	1 2 1 1 1	
2.380E-07	6.100E-05	25	M064	1 1 2 2 1	
9.518E-08	2.440E-05	25	M156	1 2 1 1 2	
1.677E-07	4.300E-05	27	D003	1 0 0 1 1	

4096. C$_{20}$H$_{16}$
10-Ethyl-1,2-benzanthracene
10-Ethylbenz[a]anthracene
RN: 14854-08-1   **MP** (°C):   114
MW: 256.35   **BP** (°C):

Solubility (Moles/L)	Solubility (Grams/L)	Temp (°C)	Ref (#)	Evaluation (T P E A A)	Comments
1.755E-07	4.500E-05	27	D003	1 0 0 1 1	
1.560E-07	4.000E-05	27	D043	2 0 0 0 0	average of 2

4097. C$_{20}$H$_{16}$O$_4$
Phenolphthalin
Benzoic acid, 2-[bis(4-hydroxyphenyl)methyl]-
RN: 81-90-3   **MP** (°C):   237
MW: 320.35   **BP** (°C):

Solubility (Moles/L)	Solubility (Grams/L)	Temp (°C)	Ref (#)	Evaluation (T P E A A)	Comments
5.463E-04	1.750E-01	20	F300	1 0 0 0 2	

4098. C$_{20}$H$_{17}$FO$_3$S
Sulindac
Aclin
Clinoril
Clusinol
Saldac
RN: 38194-50-2   **MP** (°C):
MW: 356.42   **BP** (°C):

Solubility (Moles/L)	Solubility (Grams/L)	Temp (°C)	Ref (#)	Evaluation (T P E A A)	Comments
1.964E-05	7.000E-03	37	Y421	0 0 0 0 0	

4099. C$_{20}$H$_{18}$O$_2$Sn
Triphenyltin hydroxide acetate
Fentin acetate
RN: 900-95-8   **MP** (°C):   120
MW: 409.06   **BP** (°C):

Solubility (Moles/L)	Solubility (Grams/L)	Temp (°C)	Ref (#)	Evaluation (T P E A A)	Comments
6.845E-05	2.800E-02	20	M161	1 0 0 0 1	

4100. $C_{20}H_{18}O_{10}$
Biphenyl dimethyl dicarboxylate
DDB
RN: **MP** (°C):
MW: 418.36   **BP** (°C):

Solubility (Moles/L)	Solubility (Grams/L)	Temp (°C)	Ref (#)	Evaluation (T P E A A)	Comments
1.004E-05	4.200E-03	ns	K446	0 0 0 0 0	

4101. $C_{20}H_{19}NO_3$
Acronine
3,12-Dihydro-6-methoxy-3,3,12-trimethyl-7H-pyrano(2,3-c)acridin-7-one
Acronycine
RN: 7008-42-6   **MP** (°C):   175–176
MW: 321.38   **BP** (°C):

Solubility (Moles/L)	Solubility (Grams/L)	Temp (°C)	Ref (#)	Evaluation (T P E A A)	Comments
7.779E-06	2.500E-03	22	B064	1 0 1 1 0	
8.401E-06	2.700E-03	25	R071	0 0 0 0 0	

4102. $C_{20}H_{19}NO_5 \cdot 6H_2O$
Berberine (hexahydrate)
Berberine
RN: 2086-83-1   **MP** (°C):   145dec
MW: 461.47   **BP** (°C):

Solubility (Moles/L)	Solubility (Grams/L)	Temp (°C)	Ref (#)	Evaluation (T P E A A)	Comments
9.422E-02	4.348E+01	25	D004	0 0 0 0 0	

4103. $C_{20}H_{19}N_3$
Rosaniline
Basic violet 14
C.I. 42510
Calcozine magenta xx
Cerise B
RN: 632-99-5   **MP** (°C):
MW: 301.39   **BP** (°C):

Solubility (Moles/L)	Solubility (Grams/L)	Temp (°C)	Ref (#)	Evaluation (T P E A A)	Comments
9.951E-04	2.999E-01	rt	D021	0 0 1 1 0	

4104. C$_{20}$H$_{19}$N$_3$O$_5$
1H-Benzimidazole-1-carboxylic acid, 6-benzoyl-2-[(methoxycarbonyl)amino]-, propyl ester
RN: 153474-31-8   **MP** (°C):   113.5
MW: 381.39   **BP** (°C):

Solubility (Moles/L)	Solubility (Grams/L)	Temp (°C)	Ref (#)	Evaluation (T P E A A)	Comments
1.311E-05	5.000E-03	21	N337	0 0 0 0 0	pH 5
1.311E-05	5.000E-03	21	N337	0 0 0 0 0	pH 5

4105. C$_{20}$H$_{20}$ClNO$_7$
BTA-243
1,3-Benzodioxole-2,2-dicarboxylic acid, 5-[2-[[2-(3-chlorophenyl)-2-hydroxyethyl]amino] propyl]-
RN: **MP** (°C):
MW: 421.84   **BP** (°C):

Solubility (Moles/L)	Solubility (Grams/L)	Temp (°C)	Ref (#)	Evaluation (T P E A A)	Comments
8.890E-03	3.750E+00	25	B421	0 0 1 1 0	Zwitterion, EFG

4106. C$_{20}$H$_{20}$N$_2$O$_6$
Succinyl acetaminophen
Butanedioic acid, bis[4-(acetylamino)phenyl] ester
Acetanilide, 4'-hydroxy-, succinate
Acetanilide, 4'-hydroxy-, succinate (2:1) (ester)
RN: 2725-63-5   **MP** (°C):   229–230
MW: 384.39   **BP** (°C):

Solubility (Moles/L)	Solubility (Grams/L)	Temp (°C)	Ref (#)	Evaluation (T P E A A)	Comments
1.769E-05	6.800E-03	37	D029	0 0 0 0 0	

4107. C$_{20}$H$_{20}$N$_6$O$_6$S$_2$
2,5-Di-(N4-acetylsulfanilylamino)pyrimidine
RN: **MP** (°C):
MW: 504.55   **BP** (°C):

Solubility (Moles/L)	Solubility (Grams/L)	Temp (°C)	Ref (#)	Evaluation (T P E A A)	Comments
9.910E-06	5.000E-03	37	R076	1 2 0 0 1	

4108. C$_{20}$H$_{21}$ClO$_4$
Fenofibrate
Proctofene
Sedufen

RN:	49562-28-9	MP (°C):
MW:	360.84	BP (°C):

Solubility (Moles/L)	Solubility (Grams/L)	Temp (°C)	Ref (#)	Evaluation (T P E A A)	Comments
2.217E-06	8.000E-04	25	J415	0 0 0 0 0	

4109. C$_{20}$H$_{21}$NO$_4$
Papaverine
Pantoyl taurine

RN:	58-74-2	MP (°C):	147
MW:	339.39	BP (°C):	

Solubility (Moles/L)	Solubility (Grams/L)	Temp (°C)	Ref (#)	Evaluation (T P E A A)	Comments
1.100E-04	3.733E-02	37.5	L034	2 2 0 1 2	pH 7.4

4110. C$_{20}$H$_{21}$NO$_5$
Aspirin phenylalanine ethyl ester
L-Phenylalanine, N-[2-(acetyloxy)benzoyl]-, ethyl ester

RN:	76748-72-6	MP (°C):
MW:	355.39	BP (°C):

Solubility (Moles/L)	Solubility (Grams/L)	Temp (°C)	Ref (#)	Evaluation (T P E A A)	Comments
4.700E-04	1.670E-01	25	B182	2 2 1 1 1	

4111. C$_{20}$H$_{21}$NO$_5$
Repirinast
Isoamyl 5,6-dihydro-7,8-dimethyl-4,5-dioxo-4H-pyrano(3,2-c)quinoline-2-carboxylate

RN:	73080-51-0	MP (°C):
MW:	355.39	BP (°C):

Solubility (Moles/L)	Solubility (Grams/L)	Temp (°C)	Ref (#)	Evaluation (T P E A A)	Comments
3.377E-06	1.200E-03	ns	S470	0 0 0 0 0	

4112. C₂₀H₂₂ClN

Pyrrobutamine
Pyrrolidine, 1-[4-(4-chlorophenyl)-3-phenyl-2-butenyl]-

RN: 91-82-7   **MP** (°C):
MW: 311.86   **BP** (°C):

Solubility (Moles/L)	Solubility (Grams/L)	Temp (°C)	Ref (#)	Evaluation (T P E A A)	Comments
8.700E-04	2.713E-01	37.5	L034	2 2 0 1 2	pH 7.4

4113. C₂₀H₂₂FN₃O₇

3-Quinolinecarboxylic acid
7-[4-[[(acetyloxy)methoxy]carbonyl]-1-piperazinyl]-1-ethyl-6-fluoro-1,4-dihydro-4-oxo-

RN: 99106-30-6   **MP** (°C):
MW: 435.41   **BP** (°C): 636.3

Solubility (Moles/L)	Solubility (Grams/L)	Temp (°C)	Ref (#)	Evaluation (T P E A A)	Comments
3.445E-04	1.500E-01	25	A414	1 0 1 1 1	pH 8.5 bicarbonate buffer (0.05 M)
1.378E-04	6.000E-02	25	A414	1 0 1 1 1	pH 7.4 phosphate buffer (0.1 M)
6.890E-05	3.000E-02	25	A414	1 0 1 1 1	pH 5 phosphate buffer (0.1 M)

4114. C₂₀H₂₂N₂O₂

Quininone
Chininon
Cinchonan-9-one, 6′-methoxy-, (8α)-

RN: 84-31-1   **MP** (°C): 212
MW: 322.41   **BP** (°C):

Solubility (Moles/L)	Solubility (Grams/L)	Temp (°C)	Ref (#)	Evaluation (T P E A A)	Comments
9.305E-06	3.000E-03	20	F300	1 0 0 0 0	

4115. C₂₀H₂₂N₈O₅

Methotrexate
(+)-4-Amino-10-methylfolic acid
Metatrexan
Methoblastin
Maxtrex
Ledertrexate

RN: 59-05-2   **MP** (°C): 195
MW: 454.45   **BP** (°C):

Solubility (Moles/L)	Solubility (Grams/L)	Temp (°C)	Ref (#)	Evaluation (T P E A A)	Comments
4.290E+00	1.950E+03	c	B443	0 0 0 0 0	
2.200E-05	1.000E-02	ns	K444	0 0 0 0 0	

4116. $C_{20}H_{23}N$
Maprotiline
Maprotyline
RN: 10262-69-8   **MP** (°C):
MW: 277.41   **BP** (°C):

Solubility (Moles/L)	Solubility (Grams/L)	Temp (°C)	Ref (#)	Evaluation (T P E A A)	Comments
3.004E-06	8.334E-04	22.5	B440	0 0 0 0 0	

4117. $C_{20}H_{23}NO_2$
Dexoxadrol
(+)-2-(2,2-Diphenyl-1,3-dioxolan-4-yl)piperidine
Relane
CL 911C
RN: 4741-41-7   **MP** (°C):
MW: 309.41   **BP** (°C):

Solubility (Moles/L)	Solubility (Grams/L)	Temp (°C)	Ref (#)	Evaluation (T P E A A)	Comments
2.262E-04	7.000E-02	rt	K017	0 2 2 2 2	intrinsic

4118. $C_{20}H_{23}N_7O_7$
N5-Formyltetrahydropteroylglutamic acid
RN: 58-05-9   **MP** (°C):
MW: 473.45   **BP** (°C):

Solubility (Moles/L)	Solubility (Grams/L)	Temp (°C)	Ref (#)	Evaluation (T P E A A)	Comments
>2.85E+00	>1.35E+03	25	B443	0 0 0 0 0	

4119. $C_{20}H_{24}ClN_3S$
Prochlorperazine
Compazine
Ultrazine
Cotranzine
Compa-Z
RN: 58-38-8   **MP** (°C):   228
MW: 373.95   **BP** (°C):

Solubility (Moles/L)	Solubility (Grams/L)	Temp (°C)	Ref (#)	Evaluation (T P E A A)	Comments
4.000E-05	1.496E-02	24	G022	2 0 1 1 1	

4120. C$_{20}$H$_{24}$N$_2$
Dimethindene
Dimetindene
Pyridine, 2-[1-[2-[2-(dimethylamino)ethyl]inden-3-yl]ethyl]-
RN: 5636-83-9   **MP** (°C):
MW: 292.43   **BP** (°C):

Solubility (Moles/L)	Solubility (Grams/L)	Temp (°C)	Ref (#)	Evaluation (T P E A A)	Comments
8.160E-04	2.386E-01	37	L094	2 0 0 1 2	pH>10.03, intrinsic

4121. C$_{20}$H$_{24}$N$_2$O$_2$
Quinine
Chinin
Quinine alkaloid
RN: 130-95-0   **MP** (°C):   177
MW: 324.43   **BP** (°C):

Solubility (Moles/L)	Solubility (Grams/L)	Temp (°C)	Ref (#)	Evaluation (T P E A A)	Comments
1.541E-03	5.000E-01	15	F300	1 0 0 0 0	
9.555E-05	3.100E-02	22	M459	0 0 0 0 0	
1.760E-03	5.711E-01	25	D004	0 0 0 0 0	
9.247E-04	3.000E-01	25	P015	0 0 0 0 0	
4.007E-03	1.300E+00	100	F300	1 0 0 0 1	
<3.08E-04	<1.00E-01	rt	B435	0 0 0 0 0	
1.756E-03	5.697E-01	rt	D021	0 0 1 1 1	

4122. C$_{20}$H$_{24}$N$_2$O$_2$
Quinidine
Chinidin
Cinchonan-9-ol, 6′-methoxy-, (9S)-
RN: 56-54-2   **MP** (°C):   174
MW: 324.43   **BP** (°C):

Solubility (Moles/L)	Solubility (Grams/L)	Temp (°C)	Ref (#)	Evaluation (T P E A A)	Comments
7.200E-04	2.336E-01	15	K059	2 2 2 0 1	
1.110E-04	3.600E-02	22	M459	0 0 0 0 0	
4.315E-04	1.400E-01	25	F300	1 0 0 0 1	
1.540E-03	4.998E-01	c	D004	0 0 0 0 0	
3.848E-03	1.248E+00	h	D004	0 0 0 0 0	
1.549E-03	5.025E-01	ns	R427	0 0 0 0 0	

4123. $C_{20}H_{24}N_2O_2.3H_2O$

Quinine (trihydrate)

Quinine, compd. with valeric acid (1:1), hydrate

Cinchonan-9-ol, 6'-methoxy-, trihydrate, (8α,9R)-

RN:	6151-51-5	MP (°C):	57
MW:	378.47	BP (°C):	

Solubility (Moles/L)	Solubility (Grams/L)	Temp (°C)	Ref (#)	Evaluation (T P E A A)	Comments
1.693E-03	6.406E-01	c	D004	0 0 0 0 0	
3.299E-03	1.248E+00	h	D004	0 0 0 0 0	

4124. $C_{20}H_{24}N_2O_4$

Pheniramine maleate

1-Phenyl-1-(2-pyridyl)-3-dimethylaminopropane maleate

Prophenpyridamine maleate

RN:	132-20-7	MP (°C):	
MW:	356.43	BP (°C):	

Solubility (Moles/L)	Solubility (Grams/L)	Temp (°C)	Ref (#)	Evaluation (T P E A A)	Comments
3.100E-02	1.105E+01	37.5	L034	2 2 0 1 2	pH 7.4

4125. $C_{20}H_{24}N_2O_5$

Naproxen, N-methyl-N-carbamoyl methyl-glycolamide ester

RN:		MP (°C):	179.5
MW:	372.42	BP (°C):	

Solubility (Moles/L)	Solubility (Grams/L)	Temp (°C)	Ref (#)	Evaluation (T P E A A)	Comments
1.584E-04	5.900E-02	21	B331	1 2 2 1 1	pH 7.4

4126. $C_{20}H_{24}O_3$

Methylsecodione

RN:	80702-24-5	MP (°C):	
MW:	312.41	BP (°C):	

Solubility (Moles/L)	Solubility (Grams/L)	Temp (°C)	Ref (#)	Evaluation (T P E A A)	Comments
1.919E-03	5.996E-01	25	P324	0 0 0 0 0	

4127. C$_{20}$H$_{24}$O$_4$
3,11-Dioxo-4,17(20)-*cis*-pregnadien-21-oic acid methyl ester
U-2726
RN: **MP** (°C):
MW: 328.41   **BP** (°C):

Solubility (Moles/L)	Solubility (Grams/L)	Temp (°C)	Ref (#)	Evaluation (T P E A A)	Comments
1.309E-05	4.300E-03	ns	K029	0 0 2 1 1	

4128. C$_{20}$H$_{24}$O$_6$
Dibenzo-18-crown-6
DBC
RN: 14187-32-7   **MP** (°C):
MW: 360.41   **BP** (°C):

Solubility (Moles/L)	Solubility (Grams/L)	Temp (°C)	Ref (#)	Evaluation (T P E A A)	Comments
2.025E-05	7.300E-03	25	M127	1 2 1 1 1	
9.000E-05	3.244E-02	26	P029	0 0 0 0 0	

4129. C$_{20}$H$_{25}$ClN$_2$O$_2$
Quinine hydrochloride
Inchonan-9-ol, 6'-methoxy-, monohydrochloride, (8α,9R)-
RN: 130-89-2   **MP** (°C):
MW: 360.89   **BP** (°C):

Solubility (Moles/L)	Solubility (Grams/L)	Temp (°C)	Ref (#)	Evaluation (T P E A A)	Comments
1.523E-03	5.497E-01	25	A412	1 0 2 2 1	int

4130. C$_{20}$H$_{25}$ClO$_2$
1-Chloro-1,1-dimethyl-2,2-bis(*p*-ethoxylphenyl)ethane
RN: 56265-24-8   **MP** (°C):
MW: 332.87   **BP** (°C):

Solubility (Moles/L)	Solubility (Grams/L)	Temp (°C)	Ref (#)	Evaluation (T P E A A)	Comments
5.708E-07	1.900E-04	rt	C122	0 0 0 0 0	

4131. C$_{20}$H$_{25}$NO$_2$

Adiphenine
2-Diethylaminoethyl diphenylacetate
Tranzetil
Patrovine
SKF 962A

RN: 64-95-9 **MP** (°C): 113.5
MW: 311.43 **BP** (°C):

Solubility (Moles/L)	Solubility (Grams/L)	Temp (°C)	Ref (#)	Evaluation (T P E A A)	Comments
1.000E-02	3.114E+00	30	L068	1 0 0 1 0	EFG

4132. C$_{20}$H$_{25}$NO$_4$

2-Naphthaleneacetic acid, 6-methoxy-α-methyl-, 2-(diethylamino)-2-oxoethyl ester, (S)
Naproxen, N,N-diethyl glycolamide ester
2-Naphthaleneacetic acid, 6-methoxy-α-methyl-, 2-(diethylamino)-2-oxoethyl ester
Naproxen N,N-diethyl glycolamide ester

RN: 106231-74-7 **MP** (°C): 89
MW: 343.43 **BP** (°C):

Solubility (Moles/L)	Solubility (Grams/L)	Temp (°C)	Ref (#)	Evaluation (T P E A A)	Comments
3.494E-05	1.200E-02	21	B331	1 2 2 1 1	pH 7.4
3.494E-05	1.200E-02	21	B331	0 0 0 0 0	

4133. C$_{20}$H$_{25}$NO$_4$

3,11-Dioxo-4,17(20)-cis-pregnadien-20-oic acid methyl ester 3-oxime
RN: **MP** (°C):
MW: 343.43 **BP** (°C):

Solubility (Moles/L)	Solubility (Grams/L)	Temp (°C)	Ref (#)	Evaluation (T P E A A)	Comments
1.543E-05	5.300E-03	ns	K029	0 0 2 1 1	

4134. C$_{20}$H$_{25}$NO$_5$

Naproxen, N-methyl-N-hydroxyethyl glycolamide ester
RN: **MP** (°C): 110
MW: 359.43 **BP** (°C):

Solubility (Moles/L)	Solubility (Grams/L)	Temp (°C)	Ref (#)	Evaluation (T P E A A)	Comments
3.895E-04	1.400E-01	21	B331	1 2 2 1 1	pH 7.4

4135. $C_{20}H_{25}NO_6$

2-Naphthaleneacetic acid, 6-methoxy-α-methyl-, 2-[bis(2-hydroxyethyl)amino]-2-oxoethyl ester

Naproxen *N,N*-diethanol glycolamide ester

Naproxen,*N,N*-dihydroxyethyl glycolamide ester

RN:　114665-20-2　**MP** (°C):　113
MW:　375.43　**BP** (°C):

Solubility (Moles/L)	Solubility (Grams/L)	Temp (°C)	Ref (#)	Evaluation (T P E A A)	Comments
1.092E-03	4.100E-01	21	B331	1 2 2 1 1	pH 7.4
1.092E-03	4.100E-01	21	B331	0 0 0 0 0	

4136. $C_{20}H_{26}N_2$

1-(Diphenylmethyl)-4-propylpiperazine

RN:　　　　　　**MP** (°C):
MW:　294.44　**BP** (°C):

Solubility (Moles/L)	Solubility (Grams/L)	Temp (°C)	Ref (#)	Evaluation (T P E A A)	Comments
6.448E-04	1.899E-01	25	M438	0 0 0 0 0	

4137. $C_{20}H_{26}N_2O_2$

Ajmaline

Rauwolfine

Ajmalan-17,21-diol, (17R,21α)-

Merabitol

Raugalline

RN:　4360-12-7　**MP** (°C):　159
MW:　326.44　**BP** (°C):

Solubility (Moles/L)	Solubility (Grams/L)	Temp (°C)	Ref (#)	Evaluation (T P E A A)	Comments
1.100E-03	3.591E-01	0	M106	2 1 1 1 0	EFG
1.300E-03	4.244E-01	15	M106	2 1 1 1 0	EFG
1.500E-03	4.897E-01	30	M106	2 1 1 1 0	EFG

4138. $C_{20}H_{26}N_2O_2$

Hydroquinine

Cinchonan-9-ol, 10,11-dihydro-6′-methoxy-, (8α,9R)-

10,11-Dihydroquinine

RN:　522-66-7　**MP** (°C):　173.5
MW:　326.44　**BP** (°C):

Solubility (Moles/L)	Solubility (Grams/L)	Temp (°C)	Ref (#)	Evaluation (T P E A A)	Comments
3.063E-04	9.999E-02	20	K059	2 2 2 0 1	
>1.53E-03	>5.00E-01	ns	B404	0 2 1 1 0	

4139. $C_{20}H_{26}O_2$
Norethindrone
Norethisterone
RN: 68-22-4 **MP** (°C): 203
MW: 298.43 **BP** (°C):

Solubility (Moles/L)	Solubility (Grams/L)	Temp (°C)	Ref (#)	Evaluation (T P E A A)	Comments
1.334E-05	3.981E-03	10	L078	1 0 1 2 0	EFG
1.679E-05	5.012E-03	20	L078	1 0 1 2 0	EFG
2.360E-05	7.043E-03	25	H099	1 0 2 2 2	
1.884E-05	5.623E-03	25	L078	1 0 1 2 2	
8.377E-03	2.500E+00	25	P312	0 0 0 0 0	
2.114E-05	6.310E-03	30	L078	1 0 1 2 0	EFG
3.610E-05	1.077E-02	37	C004	0 0 0 0 0	EFG
2.986E-05	8.912E-03	40	L078	1 0 1 2 0	EFG
4.218E-05	1.259E-02	50	L078	1 0 1 2 0	EFG
3.351E-05	1.000E-02	ns	K444	0 0 0 0 0	

4140. $C_{20}H_{26}O_2$
1,1-Dimethyl-2,2-bis(p-ethoxylphenyl)ethane
RN: 56265-21-5 **MP** (°C):
MW: 298.43 **BP** (°C):

Solubility (Moles/L)	Solubility (Grams/L)	Temp (°C)	Ref (#)	Evaluation (T P E A A)	Comments
1.441E-07	4.300E-05	rt	C122	0 0 0 0 0	

4141. $C_{20}H_{26}O_4$
Dicyclohexyl phthalate
1,2-Benzenedicarboxylic acid, dicyclohexyl ester
RN: 84-61-7 **MP** (°C): 66
MW: 330.43 **BP** (°C):

Solubility (Moles/L)	Solubility (Grams/L)	Temp (°C)	Ref (#)	Evaluation (T P E A A)	Comments
1.211E-05	4.000E-03	24	H116	2 1 0 0 2	

4142. $C_{20}H_{27}NO_5S_2$
2-(Acetyloxy)-4-[2-({5-[(3R)-1,2-dithiolan-3-yl]-pentanoyl}-amino)ehtyl]phenyl acetate
RN: **MP** (°C):
MW: 425.57 **BP** (°C):

Solubility (Moles/L)	Solubility (Grams/L)	Temp (°C)	Ref (#)	Evaluation (T P E A A)	Comments
1.151E-04	4.900E-02	ns	S453	0 0 0 0 0	

4143. C$_{20}$H$_{27}$NO$_{11}$
Amygdalin
(R)-Amygdalin
(R)-Laenitrile
(R)-Amygdaloside

RN: 29883-15-6   **MP** (°C):   223
MW: 457.44   **BP** (°C):

Solubility (Moles/L)	Solubility (Grams/L)	Temp (°C)	Ref (#)	Evaluation (T P E A A)	Comments
1.705E-01	7.800E+01	10	F300	1 0 0 0 1	
1.698E-01	7.768E+01	ns	R427	0 0 0 0 0	

4144. C$_{20}$H$_{27}$NO$_{11}$.3H$_2$O
Amygdalin (trihydrate)
D-(−)-Amygdalin
(R)-Amygdalin

RN: 29883-15-6   **MP** (°C):   214–216
MW: 511.48   **BP** (°C):

Solubility (Moles/L)	Solubility (Grams/L)	Temp (°C)	Ref (#)	Evaluation (T P E A A)	Comments
1.504E-01	7.692E+01	c	D004	0 0 0 0 0	
		h	D004	0 0 0 0 0	

4145. C$_{20}$H$_{27}$O$_4$P
Octyldiphenyl phosphate
Disflamoll DPO

RN: 115-88-8   **MP** (°C):
MW: 362.41   **BP** (°C):

Solubility (Moles/L)	Solubility (Grams/L)	Temp (°C)	Ref (#)	Evaluation (T P E A A)	Comments
3.863E-07	1.400E-04	24	H116	2 1 0 0 2	

4146. C$_{20}$H$_{28}$O
Vitamin A aldehyde
Retinal
All-*trans*-retinal
All-*trans* vitamin A aldehyde
Retinene

RN: 116-31-4   **MP** (°C):   63
MW: 284.45   **BP** (°C):

Solubility (Moles/L)	Solubility (Grams/L)	Temp (°C)	Ref (#)	Evaluation (T P E A A)	Comments
<2.46E-04	<7.00E-02	25	P312	0 0 0 0 0	

4147. $C_{20}H_{28}O_2$
19-Norprogesterone
19-Norpregn-4-ene-3,20-dione

RN: 472-54-8   **MP** (°C):
MW: 300.44   **BP** (°C):

Solubility (Moles/L)	Solubility (Grams/L)	Temp (°C)	Ref (#)	Evaluation (T P E A A)	Comments
1.202E-04	3.610E-02	37	L010	2 0 2 1 1	

4148. $C_{20}H_{28}O_2$
Retinoic acid
All-*trans* retinoic acid
3,7-Dimethyl-9-(2,6,6-trimethyl-1-cyclohexen-1-yl)-2,4,6,8-nonatetraenoic acid
β-All-*trans*-retinoic acid

RN: 302-79-4   **MP** (°C): 180-181
MW: 300.44   **BP** (°C):

Solubility (Moles/L)	Solubility (Grams/L)	Temp (°C)	Ref (#)	Evaluation (T P E A A)	Comments
<2.33E-04	<7.00E-02	25	P312	0 0 0 0 0	

4149. $C_{20}H_{28}O_3$
5,6-Dehydroisoandrosterone formate
Androst-5-en-17-one, 3α-hydroxy-, formate

RN: 4589-84-8   **MP** (°C):
MW: 316.44   **BP** (°C):

Solubility (Moles/L)	Solubility (Grams/L)	Temp (°C)	Ref (#)	Evaluation (T P E A A)	Comments
4.424E-05	1.400E-02	ns	B057	0 2 1 1 2	

4150. $C_{20}H_{28}O_3$
Testosterone formate
Androst-4-en-17β-ol-3-one formate
Testosterone 17-formate

RN: 3129-42-8   **MP** (°C):
MW: 316.44   **BP** (°C):

Solubility (Moles/L)	Solubility (Grams/L)	Temp (°C)	Ref (#)	Evaluation (T P E A A)	Comments
1.389E-05	4.395E-03	25	J004	1 0 1 1 2	
1.390E-05	4.400E-03	ns	B057	0 2 1 1 1	

4151. C$_{20}$H$_{29}$N$_3$O$_2$
Dibucaine
Cinchocaine

RN:	85-79-0	**MP** (°C):	64
MW:	343.47	**BP** (°C):	

Solubility (Moles/L)	Solubility (Grams/L)	Temp (°C)	Ref (#)	Evaluation (T P E A A)	Comments
1.980E-04	6.801E-02	ns	B018	0 0 0 0 2	
1.980E-04	6.801E-02	ns	M066	0 0 0 0 2	

4152. C$_{20}$H$_{30}$N$_4$O$_6$
2′-Nonyl-6-methoxypurine arabinoside
4-Quinolinecarboxamide, 2-butoxy-N-[2-(diethylamino)ethyl]-

RN:	145913-42-4	**MP** (°C):	88-90
MW:	422.49	**BP** (°C):	

Solubility (Moles/L)	Solubility (Grams/L)	Temp (°C)	Ref (#)	Evaluation (T P E A A)	Comments
1.030E-04	4.352E-02	37	C348	0 0 0 0 0	pH 7.00

4153. C$_{20}$H$_{30}$O
D 263
4,6-Diisopropyl-1,1-dimethyl-7-propionylindan

RN:	290294-31-4	**MP** (°C):	117
MW:	286.46	**BP** (°C):	

Solubility (Moles/L)	Solubility (Grams/L)	Temp (°C)	Ref (#)	Evaluation (T P E A A)	Comments
3.491E-06	1.000E-03	ns	M061	0 0 0 0 0	

4154. C$_{20}$H$_{30}$O
Vitamin A
Retinol
Afaxin
α-Sterol

RN:	68-26-8	**MP** (°C):	62
MW:	286.46	**BP** (°C):	137–138

Solubility (Moles/L)	Solubility (Grams/L)	Temp (°C)	Ref (#)	Evaluation (T P E A A)	Comments
<3.49E-05	<1.00E-02	25	P312	0 0 0 0 0	

4155. $C_{20}H_{30}O_2$
Abietic acid
13-Isopropylpodocarpa-7,13-dien-15-oic acid
Sylvic acid

RN: 514-10-3   **MP** (°C):   172
MW: 302.46   **BP** (°C):

Solubility (Moles/L)	Solubility (Grams/L)	Temp (°C)	Ref (#)	Evaluation (T P E A A)	Comments
1.600E-04	4.839E-02	20	B009	2 2 1 2 0	

4156. $C_{20}H_{30}O_2$
17-Methyltestosterone
17-α-Methyltestosterone
Methyltestosterone
Methyl-testosterone

RN: 58-18-4   **MP** (°C):   161
MW: 302.46   **BP** (°C):

Solubility (Moles/L)	Solubility (Grams/L)	Temp (°C)	Ref (#)	Evaluation (T P E A A)	Comments
1.230E-04	3.720E-02	20	F012	1 0 1 1 1	
1.120E-04	3.388E-02	25	H099	1 0 2 2 2	
1.058E-04	3.200E-02	25	K003	2 1 1 1 1	
4.400E-02	1.331E+01	25	M379	1 0 1 1 0	EFG,*sic*
<5.62E-04	<1.70E-01	25	P312	0 0 0 0 0	
2.313E-03	6.995E-01	25	P324	0 0 0 0 0	
1.018E-04	3.080E-02	30	T005	2 0 2 2 2	
1.200E-04	3.630E-02	37	E014	2 2 2 1 2	pH 7.3
7.472E-05	2.260E-02	ns	B057	0 2 1 1 2	
9.918E 05	3.000E-02	rt	N302	0 2 1 2 1	

4157. $C_{20}H_{30}O_3$
Androstanolone formate
5α-Androstan-3-one, 17-hydroxy-, formate

RN: 4589-90-6   **MP** (°C):
MW: 318.46   **BP** (°C):

Solubility (Moles/L)	Solubility (Grams/L)	Temp (°C)	Ref (#)	Evaluation (T P E A A)	Comments
4.679E-06	1.490E-03	ns	B057	0 2 1 1 2	

4158. C$_{20}$H$_{30}$O$_6$

Butyl glycol phthalate
bis(2-Butoxyethyl) phthalate
Dibutoxyethyl phthalate
bis(2-N-Butoxyethyl) phthalate

| **RN:** | 117-83-9 | **MP** (°C): | 230 |
| **MW:** | 366.46 | **BP** (°C): | 210 |

Solubility (Moles/L)	Solubility (Grams/L)	Temp (°C)	Ref (#)	Evaluation (T P E A A)	Comments
5.458E-05	2.000E-02	15	H069	1 0 1 1 0	
<8.18E-04	<3.00E-01	20	F070	1 0 0 0 1	

4159. C$_{20}$H$_{31}$NO

Trihexyphenidyl
1-Phenyl-1-cyclohexyl-3-piperidyl-1-propanol hydrochloride
Artane
Benzhexol chloride
Trihexyphenidyl-D,L hydrochloride
Tremin

| **RN:** | 52-49-3 | **MP** (°C): | |
| **MW:** | 301.48 | **BP** (°C): | |

Solubility (Moles/L)	Solubility (Grams/L)	Temp (°C)	Ref (#)	Evaluation (T P E A A)	Comments
2.226E-06	6.709E-04	22.5	B440	0 0 0 0 0	

4160. C$_{20}$H$_{31}$NO$_3$

Acetaminophen laurate
Acetaminophen dodecanoate

| **RN:** | 54942-38-0 | **MP** (°C): | 111 |
| **MW:** | 333.47 | **BP** (°C): | |

Solubility (Moles/L)	Solubility (Grams/L)	Temp (°C)	Ref (#)	Evaluation (T P E A A)	Comments
1.799E-05	6.000E-03	25	B010	1 1 1 1 0	

4161. C$_{20}$H$_{32}$O$_3$

Tridecyl p-hydroxybenzoate
p-Hydroxybenzoic acid tridecyl ester

| **RN:** | 69679-32-9 | **MP** (°C): | |
| **MW:** | 320.48 | **BP** (°C): | |

Solubility (Moles/L)	Solubility (Grams/L)	Temp (°C)	Ref (#)	Evaluation (T P E A A)	Comments
1.135E-03	3.639E-01	25	D081	1 2 2 1 2	

4162. $C_{20}H_{32}O_5$
Dinoprostone
Prostaglandin E2

RN:	363-24-6	**MP** (°C):	66–68
MW:	352.48	**BP** (°C):	

Solubility (Moles/L)	Solubility (Grams/L)	Temp (°C)	Ref (#)	Evaluation (T P E A A)	Comments
3.123E-03	1.101E+00	8.53	F068	0 0 2 2 0	
4.022E-03	1.418E+00	19.24	F068	0 0 2 2 0	
4.173E-03	1.471E+00	25.35	F068	0 0 2 2 0	
4.575E-03	1.613E+00	29.9	F068	0 0 2 2 0	

4163. $C_{20}H_{33}NO$
Fenpropimorph
4-(3-(4-(1,1-Dimethylethyl)phenyl)-2-methylpropyl)-2,6-dimethylmorpholine
Corbe
Mistral

RN:	67306-03-0	**MP** (°C):	
MW:	303.49	**BP** (°C):	

Solubility (Moles/L)	Solubility (Grams/L)	Temp (°C)	Ref (#)	Evaluation (T P E A A)	Comments
1.417E-05	4.300E-03	ns	V414	0 0 0 0 0	

4164. $C_{20}H_{33}NO_3$
4-Heptoxybenzoic acid-2-(diethyl-amino)ethyl ester

RN:	38973-75-0	**MP** (°C):	
MW:	335.49	**BP** (°C):	

Solubility (Moles/L)	Solubility (Grams/L)	Temp (°C)	Ref (#)	Evaluation (T P E A A)	Comments
5.000E-05	1.677E-02	ns	M066	0 0 0 0 1	

4165. $C_{20}H_{33}N_3O_4$
Celiprolol

RN:	56980-93-9	**MP** (°C):	
MW:	379.50	**BP** (°C):	

Solubility (Moles/L)	Solubility (Grams/L)	Temp (°C)	Ref (#)	Evaluation (T P E A A)	Comments
6.034E-05	2.290E-02	22.5	B440	0 0 0 0 0	
6.008E-09	2.280E-06	200	M418	0 0 0 0 0	

4166. C$_{20}$H$_{34}$
5-Phenyltetradecane
RN: **MP** (°C):
MW: 274.49   **BP** (°C):

Solubility (Moles/L)	Solubility (Grams/L)	Temp (°C)	Ref (#)	Evaluation (T P E A A)	Comments
5.000E-09	1.372E-06	25	S377	0 0 0 0 0	

4167. C$_{20}$H$_{34}$
2-Phenyltetradecane
RN: **MP** (°C):
MW: 274.49   **BP** (°C):

Solubility (Moles/L)	Solubility (Grams/L)	Temp (°C)	Ref (#)	Evaluation (T P E A A)	Comments
4.000E-09	1.098E-06	25	S377	0 0 0 0 0	

4168. C$_{20}$H$_{34}$
4-Phenyltetradecane
RN: **MP** (°C):
MW: 274.49   **BP** (°C):

Solubility (Moles/L)	Solubility (Grams/L)	Temp (°C)	Ref (#)	Evaluation (T P E A A)	Comments
4.000E-09	1.098E-06	25	S377	0 0 0 0 0	

4169. C$_{20}$H$_{34}$
3-Phenyltetradecane
RN: **MP** (°C):
MW: 274.49   **BP** (°C):

Solubility (Moles/L)	Solubility (Grams/L)	Temp (°C)	Ref (#)	Evaluation (T P E A A)	Comments
5.000E-09	1.372E-06	25	S377	0 0 0 0 0	

4170. C$_{20}$H$_{34}$
6-Phenyltetradecane
RN: **MP** (°C):
MW: 274.49   **BP** (°C):

Solubility (Moles/L)	Solubility (Grams/L)	Temp (°C)	Ref (#)	Evaluation (T P E A A)	Comments
4.000E-09	1.098E-06	25	S377	0 0 0 0 0	

4171. C$_{20}$H$_{34}$N$_2$O$_2$
4-Heptylaminobenzoic acid-2-(diethyl-amino)ethyl ester
RN: **MP** (°C):
MW: 334.51 **BP** (°C):

Solubility (Moles/L)	Solubility (Grams/L)	Temp (°C)	Ref (#)	Evaluation (T P E A A)	Comments
2.100E-04	7.025E-02	ns	M066	0 0 0 0 1	

4172. C$_{20}$H$_{34}$O$_4$
4-Octylphenol triethoxylate
RN: 51437-91-3 **MP** (°C):
MW: 338.49 **BP** (°C):

Solubility (Moles/L)	Solubility (Grams/L)	Temp (°C)	Ref (#)	Evaluation (T P E A A)	Comments
5.436E-05	1.840E-02	20.5	A335	0 0 0 0 0	
5.440E-05	1.841E-02	20.5	A335	0 0 0 0 0	

4173. C$_{20}$H$_{34}$O$_8$
Acetyl tributyl citrate
1,2,3-Propanetricarboxylic acid
Tributyl acetylcitrate
RN: 77-90-7 **MP** (°C):
MW: 402.49 **BP** (°C):

Solubility (Moles/L)	Solubility (Grams/L)	Temp (°C)	Ref (#)	Evaluation (T P E A A)	Comments
4.224E-06	1.700E-03	25	F067	1 0 2 2 1	

4174. C$_{20}$H$_{36}$O$_4$
Dioctyl maleate
2-Butenedioic acid (Z)-
Dioctyl ester
RN: 2915-53-9 **MP** (°C):
MW: 340.51 **BP** (°C):

Solubility (Moles/L)	Solubility (Grams/L)	Temp (°C)	Ref (#)	Evaluation (T P E A A)	Comments
1.762E-06	6.000E-04	25	F067	1 0 2 2 2	

4175. $C_{20}H_{36}O_6$

Dicyclohexyl-18-crown-6
Dibenzo[b,k][1,4,7,10,13,16]hexaoxacyclooctadecin, icosahydro-
Dicyclohexano-18-crown-6
cis-Dicyclohexano-18-crown-6

RN: 16069-36-6　　**MP** (°C):
MW: 372.51　　　　**BP** (°C):

Solubility (Moles/L)	Solubility (Grams/L)	Temp (°C)	Ref (#)	Evaluation (T P E A A)	Comments
3.600E-02	1.341E+01	26	P029	0 0 0 0 0	
2.200E-02	8.195E+00	53	P029	0 0 0 0 0	
1.000E-02	3.725E+00	82	P029	0 0 0 0 0	

4176. $C_{20}H_{40}$

1-Eicosene
n-Eicosene

RN: 3452-07-1　　**MP** (°C):
MW: 280.54　　　　**BP** (°C):

Solubility (Moles/L)	Solubility (Grams/L)	Temp (°C)	Ref (#)	Evaluation (T P E A A)	Comments
1.907E-12	5.350E-10	23	C332	0 0 0 0 0	

4177. $C_{21}H_{11}ClF_6N_2O_3$

Flufenoxuron

RN: 101463-69-8　　**MP** (°C):
MW: 488.78　　　　　**BP** (°C):

Solubility (Moles/L)	Solubility (Grams/L)	Temp (°C)	Ref (#)	Evaluation (T P E A A)	Comments
7.775E-09	3.800E-06	20	M402	0 0 0 0 0	

4178. $C_{21}H_{13}N$

1:2,6:7-Dibenzacridine

RN: 226-92-6　　**MP** (°C):
MW: 279.34　　　**BP** (°C):

Solubility (Moles/L)	Solubility (Grams/L)	Temp (°C)	Ref (#)	Evaluation (T P E A A)	Comments
5.000E-08	1.397E-05	22	B175	1 0 1 1 0	

4179. $C_{21}H_{13}N$

1:2,8:9-Dibenzacridine

RN: 224-53-3　　**MP** (°C):
MW: 279.34　　　**BP** (°C):

Solubility (Moles/L)	Solubility (Grams/L)	Temp (°C)	Ref (#)	Evaluation (T P E A A)	Comments
7.000E-08	1.955E-05	22	B175	1 0 1 1 0	

4180. C$_{21}$H$_{13}$N
3:4,6:7-Dibenzacridine

RN:	226-97-1	MP (°C):
MW:	279.34	BP (°C):

Solubility (Moles/L)	Solubility (Grams/L)	Temp (°C)	Ref (#)	Evaluation (T P E A A)	Comments
2.500E-07	6.984E-05	22	B175	1 0 1 1 1	

4181. C$_{21}$H$_{14}$
5-Methyl-3,4-benzpyrene

RN:	31647-36-6	MP (°C):	216
MW:	266.35	BP (°C):	

Solubility (Moles/L)	Solubility (Grams/L)	Temp (°C)	Ref (#)	Evaluation (T P E A A)	Comments
3.004E-09	8.000E-07	27	D003	1 0 0 1 0	

4182. C$_{21}$H$_{15}$ClN$_2$O$_4$S
1-(p-Chlorobenzenesulfonyl)-5,5-diphenyl-hydantoin

RN:	24759-38-4	MP (°C):
MW:	426.88	BP (°C):

Solubility (Moles/L)	Solubility (Grams/L)	Temp (°C)	Ref (#)	Evaluation (T P E A A)	Comments
7.965E-07	3.400E-04	37	F183	1 0 1 1 2	intrinsic

4183. C$_{21}$H$_{15}$N$_3$O$_6$S
1-(p-Nitrobenzenesulfonyl)-5,5-diphenyl-hydantoin

RN:	21413-53-6	MP (°C):
MW:	437.43	BP (°C):

Solubility (Moles/L)	Solubility (Grams/L)	Temp (°C)	Ref (#)	Evaluation (T P E A A)	Comments
1.486E-06	6.500E-04	37	F183	1 0 1 1 2	intrinsic

4184. C$_{21}$H$_{16}$
3-Methylcholanthrene
1,2-Dihydro-3-methyl-benz[j]aceanthrylene
20-Methylcholanthrene

RN:	56-49-5	MP (°C):	179
MW:	268.36	BP (°C):	280

Solubility (Moles/L)	Solubility (Grams/L)	Temp (°C)	Ref (#)	Evaluation (T P E A A)	Comments
1.204E-08	3.230E-06	24	H106	1 0 2 2 2	
1.081E-08	2.900E-06	25	M064	1 1 2 2 1	
1.204E-08	3.230E-06	25	M156	1 2 1 1 2	
1.100E-08	2.952E-06	25	M342	1 0 1 1 1	
5.589E-09	1.500E-06	27	D003	1 0 0 1 1	
1.081E-08	2.900E-06	ns	M344	0 0 0 0 1	

4185. $C_{21}H_{16}N_2O_2$

C.I. Disperse blue 24
9,10-Anthracenedione, 1-amino-4-hydroxy-2-phenoxy-
Serilene red 2BL
Sumikaron red E-FBL
Solvent red 146

RN: 17418-58-5　**MP** (°C): 151
MW: 328.37　**BP** (°C):

Solubility (Moles/L)	Solubility (Grams/L)	Temp (°C)	Ref (#)	Evaluation (T P E A A)	Comments
5.000E-08	1.642E-05	25	B333	0 0 0 0 0	

4186. $C_{21}H_{16}N_2O_4S$

1-Benzenesulfonyl-5,5-diphenyl-hydantoin

RN: 21413-28-5　**MP** (°C):
MW: 392.44　**BP** (°C):

Solubility (Moles/L)	Solubility (Grams/L)	Temp (°C)	Ref (#)	Evaluation (T P E A A)	Comments
4.587E-06	1.800E-03	37	F183	1 0 1 1 2	intrinsic

4187. $C_{21}H_{16}N_2O_5S$

1-(p-Hydroxylbenzenesulfonyl)-5,5-diphenyl-hydantoin

RN: 24759-35-1　**MP** (°C):
MW: 408.44　**BP** (°C):

Solubility (Moles/L)	Solubility (Grams/L)	Temp (°C)	Ref (#)	Evaluation (T P E A A)	Comments
8.080E-06	3.300E-03	37	F183	1 0 1 1 2	intrinsic

4188. $C_{21}H_{17}N_3O_2S_2$

2-Sulfanilamido-4-p-diphenylthiazole

RN:　**MP** (°C):
MW: 407.52　**BP** (°C):

Solubility (Moles/L)	Solubility (Grams/L)	Temp (°C)	Ref (#)	Evaluation (T P E A A)	Comments
2.454E-06	1.000E-03	37	R045	1 2 1 1 0	

4189. $C_{21}H_{17}N_3O_4S$

1-(p-Aminobenzenesulfonyl)-5,5-diphenyl-hydantoin

RN: 24759-34-0　**MP** (°C):
MW: 407.45　**BP** (°C):

Solubility (Moles/L)	Solubility (Grams/L)	Temp (°C)	Ref (#)	Evaluation (T P E A A)	Comments
3.436E-06	1.400E-03	37	F183	1 0 1 1 2	intrinsic

4190. $C_{21}H_{19}NO_4$

Cinmetacin

1-Cinnamoyl-2-methyl-5-methoxyindolyl-3-acetic acid

Indolacin

RN:	20168-99-4	**MP** (°C):	170		
MW:	349.39	**BP** (°C):			

Solubility (Moles/L)	Solubility (Grams/L)	Temp (°C)	Ref (#)	Evaluation (T P E A A)	Comments
<2.86E-06	<1.00E-03	25	K027	2 0 2 2 0	

4191. $C_{21}H_{20}Cl_2O_3$

Permethrin

3-(2,2-Dichloroethenyl)-2,2-dimethylcyclopropanecarboxylic acid (3-phenoxyphenyl)methyl

Ester

Ambush

Pounce

Ectiban

RN:	52645-53-1	**MP** (°C):	36.5		
MW:	391.30	**BP** (°C):	200		

Solubility (Moles/L)	Solubility (Grams/L)	Temp (°C)	Ref (#)	Evaluation (T P E A A)	Comments
5.111E-07	2.000E-04	ns	M161	0 0 0 0 0	
~5.11E-07	~2.00E-04	ns	Y418	0 0 0 0 0	

4192. $C_{21}H_{20}O_9$

Puerarin

8-ß-D-Glucopyransyl-7-hydroxy-3-(4-hydroxyphenyl)-4H-1benzopyran-4-one

RN:	3681-99-0	**MP** (°C):			
MW:	416.39	**BP** (°C):			

Solubility (Moles/L)	Solubility (Grams/L)	Temp (°C)	Ref (#)	Evaluation (T P E A A)	Comments
8.100E-01	3.373E+02	15.0	W418	0 0 0 0 0	
9.500E-01	3.956E+02	20.0	W418	0 0 0 0 0	
1.100E+00	4.580E+02	25.0	W418	0 0 0 0 0	
1.260E+00	5.246E+02	30.0	W418	0 0 0 0 0	
1.420E+00	5.913E+02	35.0	W418	0 0 0 0 0	
1.710E+00	7.120E+02	40.0	W418	0 0 0 0 0	
2.020E+00	8.411E+02	45.0	W418	0 0 0 0 0	
2.430E+00	1.012E+03	50.0	W418	0 0 0 0 0	
2.840E+00	1.183E+03	55.0	W418	0 0 0 0 0	

4193. C$_{21}$H$_{21}$ClN$_2$O$_8$
Demeclocycline
Declomycin
Methylchlorotetracycline
Demethylchlortetracycline
RN: 127-33-3   **MP** (°C):
MW: 464.86   **BP** (°C):

Solubility (Moles/L)	Solubility (Grams/L)	Temp (°C)	Ref (#)	Evaluation (T P E A A)	Comments
3.259E-03	1.515E+00	21	M044	2 0 2 2 2	
3.012E-03	1.400E+00	25	B191	1 0 0 0 1	neutral pH

4194. C$_{21}$H$_{21}$N
Cyproheptadine
RN: 129-03-3   **MP** (°C):
MW: 287.41   **BP** (°C):

Solubility (Moles/L)	Solubility (Grams/L)	Temp (°C)	Ref (#)	Evaluation (T P E A A)	Comments
1.105E-06	3.176E-04	22.5	B440	0 0 0 0 0	

4195. C$_{21}$H$_{21}$NO$_6$
Rhoeadine
[1,3]Dioxolo[4,5-h]-1,3-dioxolo[7,8][2]benzopyrano[3,4-a][3]benzazepine, 5β,6,7,8,13β,15-
   hexahydro-15-methoxy-6-methyl-, (5bR,13bR,15S)
8-Methoxy-16-methyl-2,3:10,11-bis[methylenebis(oxy)]-, (8β)-
RN: 2718-25-4   **MP** (°C):   245–247dec
MW: 383.40   **BP** (°C):

Solubility (Moles/L)	Solubility (Grams/L)	Temp (°C)	Ref (#)	Evaluation (T P E A A)	Comments
2.172E-03	8.326E-01	25	D004	0 0 0 0 0	

4196. C$_{21}$H$_{21}$NO$_6$
Hydrastine
Hydrastin
(1R,9S)-β-Hydrastine
RN: 118-08-1   **MP** (°C):   132
MW: 383.40   **BP** (°C):

Solubility (Moles/L)	Solubility (Grams/L)	Temp (°C)	Ref (#)	Evaluation (T P E A A)	Comments
8.200E-04	3.144E-01	15	K059	2 2 2 0 1	
7.825E-05	3.000E-02	20	F300	1 0 0 0 1	

4197. $C_{21}H_{21}N_3O_3S$

L-Phe-dapsone

Benzenepropanamide, α-amino-N-[4-[(4-aminophenyl)sulfonyl]phenyl]-, (S)-

RN: 160349-01-9   **MP** (°C):

MW: 395.48   **BP** (°C):

Solubility (Moles/L)	Solubility (Grams/L)	Temp (°C)	Ref (#)	Evaluation (T P E A A)	Comments
5.057E-06	2.000E-03	25	P351	0 0 0 0 0	pH 7.4
3.287E-03	1.300E+00	25	P351	0 0 0 0 0	

4198. $C_{21}H_{21}O_4P$

Tricresyl phosphate

Tritolyl phosphate

Tri-p-cresyl phosphate

RN: 1330-78-5   **MP** (°C):

MW: 368.37   **BP** (°C):   265

Solubility (Moles/L)	Solubility (Grams/L)	Temp (°C)	Ref (#)	Evaluation (T P E A A)	Comments
2.009E-07	7.400E-05	24	H116	2 1 0 0 1	
2.715E-07	1.000E-04	25	F067	1 0 2 2 1	
2.172E-04	7.999E-02	ns	F014	0 0 0 0 0	

4199. $C_{21}H_{22}N_2O_2$

Strychnine

Strychnidin-10-one

Gopher Getter

L-Strychnine

Gopher Bait

RN: 57-24-9   **MP** (°C):   275

MW: 334.42   **BP** (°C):

Solubility (Moles/L)	Solubility (Grams/L)	Temp (°C)	Ref (#)	Evaluation (T P E A A)	Comments
2.700E-04	9.029E-02	15	K059	2 2 2 0 1	
4.186E-04	1.400E-01	20.0	N002	2 1 2 2 1	
5.980E-04	2.000E-01	30.0	N002	2 1 2 2 1	
1.017E-03	3.400E-01	40.0	N002	2 1 2 2 1	
1.196E-03	4.000E-01	50.0	N002	2 1 2 2 1	
1.346E-03	4.500E-01	60.0	N002	2 1 2 2 1	
1.794E-03	6.000E-01	75.0	N002	2 1 2 2 1	
4.672E-04	1.562E-01	c	D004	0 0 0 0 0	
9.643E-04	3.225E-01	h	D004	0 0 0 0 0	
4.276E-04	1.430E-01	rt	M161	0 0 0 0 2	

4200. C$_{21}$H$_{22}$N$_2$O$_5$

Benzeneacetic acid, 4-benzoyl-α-methyl-, 2-[(2-amino-2-oxoethyl)methylamino]-2-oxoethyl
  ester
N-Methyl-N-carbamoyl methyl glycolamide salicylate
RN: 114665-16-6   **MP** (°C):   83
MW: 382.42   **BP** (°C):

Solubility (Moles/L)	Solubility (Grams/L)	Temp (°C)	Ref (#)	Evaluation (T P E A A)	Comments
3.792E-03	1.450E+00	21	B331	0 0 0 0 0	

4201. C$_{21}$H$_{22}$N$_2$O$_5$

Ketoprofen, N-methyl-N-carbamoylmethyl glycolamide ester
Benzeneacetic acid, 3-benzoyl-α-methyl-, 2-[(2-amino-2-oxoethyl)methylamino]-2-oxoethyl
  ester
RN: 116482-84-9   **MP** (°C):   83.5
MW: 382.42   **BP** (°C):

Solubility (Moles/L)	Solubility (Grams/L)	Temp (°C)	Ref (#)	Evaluation (T P E A A)	Comments
3.792E-03	1.450E+00	21	B331	1 2 2 1 1	pH 7.4

4202. C$_{21}$H$_{23}$ClFNO$_2$

Haloperidol
Haldol
4-[4-(p-Chlorophenyl)-4-hydroxypiperidino]-4′-fluorobutyrophenone
Serenace
RN: 52-86-8   **MP** (°C):   148
MW: 375.87   **BP** (°C):

Solubility (Moles/L)	Solubility (Grams/L)	Temp (°C)	Ref (#)	Evaluation (T P E A A)	Comments
1.623E-04	6.100E-02	22	J420	0 0 0 0 0	pH6.5
5.474E-06	2.058E-03	22.5	B440	0 0 0 0 0	
7.981E-06	3.000E-03	30	P044	0 0 0 0 0	
2.660E-05	1.000E-02	ns	K444	0 0 0 0 0	
<2.66E-05	<1.00E-02	rt	B435	0 0 0 0 0	

4203. C$_{21}$H$_{23}$N$_3$OS

Pericyazine
2-Cyano-10-[3′-(4″-hydroxypiperidino)propyl]phenothiazine
Periciazine
RN: 2622-26-6   **MP** (°C):   116
MW: 365.50   **BP** (°C):

Solubility (Moles/L)	Solubility (Grams/L)	Temp (°C)	Ref (#)	Evaluation (T P E A A)	Comments
1.040E-04	3.801E-02	37	F011	1 0 1 1 2	pH 7.4

4204. $C_{21}H_{24}FN_3O_7$

3-Quinolinecarboxylic acid

7-[4-[[1-(Acetyloxy)ethoxy]carbonyl]-1-piperazinyl]-1-ethyl-6-fluoro-1,4-dihydro-4-oxo-

RN: 99106-35-1   **MP** (°C):   216
MW: 449.44   **BP** (°C):   636.7

Solubility (Moles/L)	Solubility (Grams/L)	Temp (°C)	Ref (#)	Evaluation (T P E A A)	Comments
9.122E-04	4.100E-01	25	A414	1 0 1 1 1	pH 8.5 bicarbonate buffer (0.05 M)
1.112E-04	5.000E-02	25	A414	1 0 1 1 1	pH 7.4 phosphate buffer (0.1 M)
1.112E-05	5.000E-03	25	A414	1 0 1 1 1	pH 5 citrate buffer (0.1 M)
1.335E-04	6.000E-02	25	A414	1 0 1 1 1	

4205. $C_{21}H_{24}F_3N_3S$

Trifluoperazine

Stelazine

RN: 117-89-5   **MP** (°C):   232
MW: 407.50   **BP** (°C):   206

Solubility (Moles/L)	Solubility (Grams/L)	Temp (°C)	Ref (#)	Evaluation (T P E A A)	Comments
3.000E-05	1.223E-02	24	G022	2 0 1 1 1	
3.600E-05	1.467E-02	37	F011	1 0 1 1 1	pH 7.4

4206. $C_{21}H_{25}NO$

4-Cyano-4'-octyloxybiphenyl

8 COB

RN: **MP** (°C):
MW: 307.44   **BP** (°C):

Solubility (Moles/L)	Solubility (Grams/L)	Temp (°C)	Ref (#)	Evaluation (T P E A A)	Comments
2.700E-07	8.301E-05	21	D300	2 2 1 1 2	

4207. $C_{21}H_{25}N_5O_5$

Benzoic acid, 4-(4-morpholinylmethyl)-, 2-[(6-amino-4,5-dihydro-4-oxo-1H-imidazo[4,5-c] pyridin-1-yl)methoxy]ethyl ester

1H-Imidazo[4,5-c]pyridine, benzoic acid deriv.

RN: 137605-75-5   **MP** (°C):
MW: 427.46   **BP** (°C):   712.9

Solubility (Moles/L)	Solubility (Grams/L)	Temp (°C)	Ref (#)	Evaluation (T P E A A)	Comments
1.801E-03	7.700E-01	21	B419	1 1 2 2 1	int

4208. $C_{21}H_{26}ClN_3OS$
Perphenazine
4-(3-(2-Chlorophenothiazin-10-YL)propyl)-1-piperazineethanol
Etrafon
Trilafon

RN:	58-39-9	MP (°C):	97
MW:	403.98	BP (°C):	280

Solubility (Moles/L)	Solubility (Grams/L)	Temp (°C)	Ref (#)	Evaluation (T P E A A)	Comments
7.000E-05	2.828E-02	24	G022	2 0 1 1 1	

4209. $C_{21}H_{26}FN_3O_4$
Permafloxacin

RN:	143383-65-7	MP (°C):	
MW:	403.46	BP (°C):	

Solubility (Moles/L)	Solubility (Grams/L)	Temp (°C)	Ref (#)	Evaluation (T P E A A)	Comments
1.853E-02	7.477E+00	25	F415	0 0 0 0 0	Average

4210. $C_{21}H_{26}N_2O_3$
1-(2,3-Dihydro-5-methoxybenzo[b]furan-2-ylmethyl)-4-(*o*-methoxyphenyl)piperazine

RN:		MP (°C):	
MW:	354.45	BP (°C):	

Solubility (Moles/L)	Solubility (Grams/L)	Temp (°C)	Ref (#)	Evaluation (T P E A A)	Comments
5.642E-05	2.000E-02	37	L079	1 0 1 1 0	intrinsic

4211. $C_{21}H_{26}N_2S_2$
Thioridazine
10H-Phenothiazine
10-[2-(1-Methyl-2-piperidyl)ethyl]-2-methylthio
Aldazine
Mellaril
Melleril

RN:	50-52-2	MP (°C):	
MW:	370.58	BP (°C):	

Solubility (Moles/L)	Solubility (Grams/L)	Temp (°C)	Ref (#)	Evaluation (T P E A A)	Comments
3.004E-06	1.113E-03	22.5	B440	0 0 0 0 0	

4212. $C_{21}H_{26}O_4$

Lifibrol

Benzoic acid, 4-[4-[4-(1,1-dimethylethyl)phenyl]-2-hydroxybutoxy]-

RN: 96609-16-4   **MP** (°C):

MW: 342.44   **BP** (°C):   536.8

Solubility (Moles/L)	Solubility (Grams/L)	Temp (°C)	Ref (#)	Evaluation (T P E A A)	Comments
3.800E-07	1.301E-04	12.0	B412	1 0 2 2 1	mod 2 crystal
8.600E-07	2.945E-04	12.0	B412	1 0 2 2 1	mod I crystal
7.000E-07	2.397E-04	20.0	B412	1 0 2 2 1	mod 2 crystal
1.110E-06	3.801E-04	20.0	B412	1 0 2 2 1	mod 1 crystal
1.070E-06	3.664E-04	29.0	B412	1 0 2 2 1	mod 2 crystal
1.640E-06	5.616E-04	29.0	B412	1 0 2 2 1	mod 1 crystal
2.090E-06	7.157E-04	38.0	B412	1 0 2 2 1	mod 2 crystal
2.740E-06	9.383E-04	38.0	B412	1 0 2 2 1	mod 1 crystal
3.080E-06	1.055E-03	47.0	B412	1 0 2 2 1	mod 2 crystal
4.890E-06	1.675E-03	47.0	B412	1 0 2 2 1	mod 1 crystal
4.690E-06	1.606E-03	54.0	B412	1 0 2 2 1	mod 2 crystal
5.900E-06	2.020E-03	54.0	B412	1 0 2 2 1	mod 1 crystal

4213. $C_{21}H_{26}O_4$

17-Hydroxy-6-methyl-16-methylenepregna-4,6-diene-3,20-dione acetate

RN: **MP** (°C):

MW: 342.44   **BP** (°C):

Solubility (Moles/L)	Solubility (Grams/L)	Temp (°C)	Ref (#)	Evaluation (T P E A A)	Comments
8.469E-06	2.900E-03	37	H004	0 0 0 0 0	

4214. $C_{21}H_{26}O_5$

Prednisone

1,4-Pregnadiene-17α,21-diol-3,11,20-trione

1,4-Pregnadiene-17x,21-diol-3,11,20-trione

Delcortin

Metocorten

Panasol

RN: 53-03-2   **MP** (°C):   234

MW: 358.44   **BP** (°C):

Solubility (Moles/L)	Solubility (Grams/L)	Temp (°C)	Ref (#)	Evaluation (T P E A A)	Comments
3.208E-04	1.150E-01	25	K003	2 1 1 1 1	
2.734E-04	9.799E-02	ns	B404	0 2 1 1 0	

4215. $C_{21}H_{27}FO_5$
Fluprednisolone
6α-Fluoro-11β,17,21-trihydroxypregna-1,4-diene-3,20-dione17,21-trihydroxypregna-1,4-diene-
  3,20-dione
Alphadrol
RN: 53-34-9   **MP** (°C):
MW: 378.44   **BP** (°C):

Solubility (Moles/L)	Solubility (Grams/L)	Temp (°C)	Ref (#)	Evaluation (T P E A A)	Comments
2.748E-03	1.040E+00	37	H004	0 0 0 0 0	

4216. $C_{21}H_{27}FO_5 \cdot H_2O$
Fluprednisolone (monohydrate)
RN: 53-34-9   **MP** (°C):
MW: 396.46   **BP** (°C):

Solubility (Moles/L)	Solubility (Grams/L)	Temp (°C)	Ref (#)	Evaluation (T P E A A)	Comments
1.478E-03	5.860E-01	37	H004	0 0 0 0 0	

4217. $C_{21}H_{27}FO_6$
Triamcinolone
9α-Fluoro-11β,16α,17α,21-tetrahydroxy-1,4-pregnadiene-3,20-dione
9α-Fluoro-16α-hydroxyprednisolone
Aristocort
RN: 124-94-7   **MP** (°C):   269
MW: 394.44   **BP** (°C):

Solubility (Moles/L)	Solubility (Grams/L)	Temp (°C)	Ref (#)	Evaluation (T P E A A)	Comments
2.028E-04	7.999E-02	25	F024	1 0 0 0 0	
4.260E-04	1.680E-01	37	C400	2 0 2 2 2	

4218. $C_{21}H_{27}NO_3$
Propafenone
RN: 54063-53-5   **MP** (°C):
MW: 341.45   **BP** (°C):

Solubility (Moles/L)	Solubility (Grams/L)	Temp (°C)	Ref (#)	Evaluation (T P E A A)	Comments
2.226E-06	7.599E-04	22.5	B440	0 0 0 0 0	

4219. $C_{21}H_{28}N_2$

1-(Diphenylmethyl)-4-butylpiperazine

RN: MP (°C):
MW: 308.47   BP (°C):

Solubility (Moles/L)	Solubility (Grams/L)	Temp (°C)	Ref (#)	Evaluation (T P E A A)	Comments
1.561E-03	4.816E-01	25	M438	0 0 0 0 0	

4220. $C_{21}H_{28}N_4O_7$

Pentyloxycarbonyl-mitomycin C

RN: MP (°C):
MW: 448.48   BP (°C):

Solubility (Moles/L)	Solubility (Grams/L)	Temp (°C)	Ref (#)	Evaluation (T P E A A)	Comments
5.900E-04	2.646E-01	25	M316	1 1 1 1 2	

4221. $C_{21}H_{28}O_2$

Norgestrel
Microlut
Microval

RN: 797-63-7   MP (°C):   206
MW: 312.46   BP (°C):

Solubility (Moles/L)	Solubility (Grams/L)	Temp (°C)	Ref (#)	Evaluation (T P E A A)	Comments
3.200E-05	1.000E-02	ns	K444	0 0 0 0 0	

4222. $C_{21}H_{28}O_2$

Ethisterone
17α-Ethynyl testosterone
Ethynyl testosterone
Gestoral
Pregneninolone
Anhydrohydroxyprogesterone

RN: 434-03-7   MP (°C):   269
MW: 312.46   BP (°C):

Solubility (Moles/L)	Solubility (Grams/L)	Temp (°C)	Ref (#)	Evaluation (T P E A A)	Comments
1.920E-06	5.999E-04	20	G072	1 2 2 1 2	
1.600E-06	4.999E-04	20	L077	1 2 2 2 1	
1.280E-06	4.000E-04	25	K003	2 1 1 1 1	
2.200E-06	6.874E-04	27.34	L077	1 2 2 2 1	
3.200E-06	9.999E-04	35	L077	1 2 2 2 1	
3.500E-06	1.094E-03	42.34	L077	1 2 2 2 1	
4.200E-06	1.312E-03	50	L077	1 2 2 2 1	

4223. C$_{21}$H$_{28}$O$_2$

1,1,1-Trimethyl-2,2-bis(p-ethoxylphenyl)ethane

RN: 27955-87-9   **MP** (°C):
MW: 312.46   **BP** (°C):

Solubility (Moles/L)	Solubility (Grams/L)	Temp (°C)	Ref (#)	Evaluation (T P E A A)	Comments
4.481E-07	1.400E-04	rt	C122	0 0 0 0 0	

4224. C$_{21}$H$_{28}$O$_5$

Prednisolone

11β,17α,21-Trihydroxypregna-1,4-diene-3,20-dione

Ropredlone

Predonin

Hostacortin H

Nisolone

RN: 50-24-8   **MP** (°C):   240
MW: 360.45   **BP** (°C):

Solubility (Moles/L)	Solubility (Grams/L)	Temp (°C)	Ref (#)	Evaluation (T P E A A)	Comments
6.173E-03	2.225E+00	25	G008	1 2 1 1 2	*sic*
5.963E-04	2.150E-01	25	K003	2 1 1 1 1	
1.379E-03	4.970E-01	25	K021	1 2 2 2 1	
5.770E-04	2.080E-01	25	M457	0 0 0 0 0	
7.000E-04	2.523E-01	30	H016	2 2 2 2 0	EFG
1.268E-03	4.570E-01	30	T002	1 0 2 0 2	anhydrous, form A
1.398E-03	5.040E-01	30	T002	1 0 2 0 2	anhydrous, form B
6.658E-04	2.400E-01	30	T002	1 0 2 0 2	hydrate
6.658E-04	2.400E-01	30	W006	2 2 2 1 2	hydrate, form C
4.694E-04	1.692E-01	37	C400	2 0 2 2 2	
9.738E-04	3.510E-01	37	H004	0 0 0 0 0	
5.500E-04	1.982E-01	ns	F327	0 0 1 2 2	
2.774E-04	1.000E-01	ns	K444	0 0 0 0 0	
1.398E-03	5.040E-01	ns	W006	2 2 2 1 2	anhydrous, form B

4225. C$_{21}$H$_{28}$O$_5$

Aldosterone

18-Oxocorticosterone

Aldocortin

Electrocortin

18-Oxo-11β,21-dihydroxy-4-pregnene-3,20-dione

RN: 52-39-1   **MP** (°C):   108
MW: 360.45   **BP** (°C):

Solubility (Moles/L)	Solubility (Grams/L)	Temp (°C)	Ref (#)	Evaluation (T P E A A)	Comments
1.420E-04	5.118E-02	37	H034	1 0 2 1 2	pH 7.4
1.413E-04	5.092E-02	ns	R427	0 0 0 0 0	

4226. C$_{21}$H$_{28}$O$_5$
Cortisone
17-Hydroxy-11-dehydrocorticosterone
Cortate

RN: 53-06-5   **MP** (°C):   222
MW: 360.45   **BP** (°C):

Solubility (Moles/L)	Solubility (Grams/L)	Temp (°C)	Ref (#)	Evaluation (T P E A A)	Comments
7.766E-04	2.799E-01	20	D041	1 0 0 0 0	
6.379E-04	2.299E-01	25	K003	2 1 1 1 1	
7.768E-04	2.800E-01	25	M023	1 0 2 1 1	
7.500E-04	2.703E-01	30	L344	2 0 1 1 0	EFG
6.000E-04	2.163E-01	37	E014	2 2 2 1 2	pH 7.3
7.768E-04	2.800E-01	ns	B338	0 0 0 0 1	

4227. C$_{21}$H$_{29}$FO$_5$
Fludrocortisone
9α-Fluoro-17-hydroxycorticosterone
9α-Fluorohydrocortisone
Florinef

RN: 127-31-1   **MP** (°C):   260dec
MW: 380.46   **BP** (°C):

Solubility (Moles/L)	Solubility (Grams/L)	Temp (°C)	Ref (#)	Evaluation (T P E A A)	Comments
2.918E-04	1.110E-01	25	K021	1 2 2 2 1	
8.516E-04	3.240E-01	25	L009	1 0 0 1 1	
2.411E-04	9.172E-02	37	C400	2 0 2 2 2	

4228. C$_{21}$H$_{29}$NO
N,N-Dicyclohexylcinnamamide
N,N-Dicyclohexyl-3-phenyl2-propenamide

RN: 6631-21-6   **MP** (°C):
MW: 311.47   **BP** (°C):

Solubility (Moles/L)	Solubility (Grams/L)	Temp (°C)	Ref (#)	Evaluation (T P E A A)	Comments
5.680E-06	1.769E-03	ns	H350	0 0 0 0 0	

4229. C$_{21}$H$_{29}$N$_3$O
Disopyramide
α-(2-(Diisopropylamino)ethyl)-α-phenyl-2-pyridineacetamide

RN: 3737-09-5   **MP** (°C):
MW: 339.48   **BP** (°C):

Solubility (Moles/L)	Solubility (Grams/L)	Temp (°C)	Ref (#)	Evaluation (T P E A A)	Comments
1.817E-05	6.170E-03	22.5	B440	0 0 0 0 0	
1.995E-02	6.774E+00	ns	R427	0 0 0 0 0	

4230. $C_{21}H_{30}N_4O_{10}$
Methylol riboflavine
Methylol-riboflavin

RN:　　　　　　　　　　**MP** (°C):
MW:　　498.49　　　　　**BP** (°C):

Solubility (Moles/L)	Solubility (Grams/L)	Temp (°C)	Ref (#)	Evaluation (T P E A A)	Comments
2.387E-02	1.190E+01	20	F300	1 0 0 0 2	compound not stable

4231. $C_{21}H_{30}N_6O_4S$
Benzenesulfonamide, *N*-[2-(dimethylamino)ethyl]-4-(2,3,4,5,6,7-hexahydro-2,6-dioxo-1,3-dipropyl-1H-purin-8-yl)-

RN:　89073-58-5　　**MP** (°C):　270dec
MW:　462.58　　　　　**BP** (°C):

Solubility (Moles/L)	Solubility (Grams/L)	Temp (°C)	Ref (#)	Evaluation (T P E A A)	Comments
4.302E-02	1.990E+01	ns	H316	0 0 0 0 0	0.1N HCL
1.081E-04	5.000E-02	ns	H316	0 0 0 0 0	pH 7.4

4232. $C_{21}H_{30}O_2$
Tetrahydrocannabinol
THC
Dronabinol
δ9-Tetrahydrocannabinol

RN:　1972-08-3　　**MP** (°C):
MW:　314.47　　　　**BP** (°C):

Solubility (Moles/L)	Solubility (Grams/L)	Temp (°C)	Ref (#)	Evaluation (T P E A A)	Comments
8.904E-06	2.800E-03	23	G018	1 0 0 1 0	

4233. $C_{21}H_{30}O_2$
Progesterone
δ4-Pregnene-3,20-dione
Corlutin
Corlutina
Lutein
Pregn-4-ene-3,20-dione

RN:　57-83-0　　**MP** (°C):　121
MW:　314.47　　　**BP** (°C):

Solubility (Moles/L)	Solubility (Grams/L)	Temp (°C)	Ref (#)	Evaluation (T P E A A)	Comments
1.700E-05	5.346E-03	10	B012	2 0 1 1 0	
2.200E-05	6.918E-03	20	B012	2 0 1 1 0	
3.210E-05	1.009E-02	20	L077	1 2 2 2 2	
2.600E-05	8.176E-03	21.70	M108	1 2 1 1 2	form A

(continued)

4233. $C_{21}H_{30}O_2$ (continued)

Solubility (Moles/L)	Solubility (Grams/L)	Temp (°C)	Ref (#)	Evaluation (T P E A A)	Comments
4.837E-05	1.521E-02	23	B014	0 0 1 2 2	
3.720E-05	1.170E-02	24.00	M108	1 2 1 1 2	form B
2.800E-05	8.805E-03	25	B012	2 0 1 1 0	
2.512E-05	7.899E-03	25	B041	1 0 2 2 0	EFG
3.802E-05	1.196E-02	25	F312	1 1 2 2 2	units assumed
2.862E-05	9.000E-03	25	K003	2 1 1 1 1	
6.359E-04	2.000E-01	25	P324	0 0 0 0 0	
2.810E-05	8.837E-03	25.30	M108	1 2 1 1 2	form A
3.690E-05	1.160E-02	27.34	L077	1 2 2 2 2	
3.600E-05	1.132E-02	30	B012	2 0 1 1 0	
3.498E-05	1.100E-02	30	M007	2 2 1 2 2	average of 8
3.800E-05	1.195E-02	30.20	M108	1 2 1 1 2	form A
4.520E-05	1.421E-02	30.50	M108	1 2 1 1 2	form B
4.230E-05	1.330E-02	35	L077	1 2 2 2 2	
5.390E-05	1.695E-02	35.50	M108	1 2 1 1 2	form B
4.690E-05	1.475E-02	36.40	M108	1 2 1 1 2	form A
3.816E-05	1.200E-02	37	A086	1 0 1 1 2	
3.528E-05	1.109E-02	37	C400	2 0 2 2 2	
4.800E-05	1.509E-02	37	H034	1 0 2 1 2	pH 7.4
4.260E-05	1.340E-02	37	H035	1 1 1 1 2	pH 7.4
4.007E-05	1.260E-02	37	L010	2 0 2 1 1	
4.260E-05	1.340E-02	37.50	B041	1 0 2 2 2	
3.981E-05	1.252E-02	37.50	B041	1 0 2 2 0	EFG
3.800E-05	1.195E-02	40	B012	2 0 1 1 0	
6.750E-05	2.123E-02	40.70	M108	1 2 1 1 2	form B
6.370E-05	2.003E-02	41.30	M108	1 2 1 1 2	form A
4.580E-05	1.440E-02	42.34	L077	1 2 2 2 2	
6.500E-05	2.044E-02	46.10	M108	1 2 1 1 2	form A
4.900E-05	1.541E-02	50	B012	2 0 1 1 0	
4.930E-05	1.550E-02	50	L077	1 2 2 2 2	
		amb	L434	0 0 0 0 0	
1.908E-05	6.000E-03	ns	B404	0 2 1 1 0	

4234. $C_{21}H_{30}O_3$

Deoxycorticosterone
21-Hydroxyprogesterone
4-Pregnen-21-ol-3,20-dione
11-Deoxycorticosterone
21-Hydroxypregn-4-ene-3,20-dione

RN: 64-85-7   **MP (°C):**   141.5
MW: 330.47   **BP (°C):**

Solubility (Moles/L)	Solubility (Grams/L)	Temp (°C)	Ref (#)	Evaluation (T P E A A)	Comments
4.387E-04	1.450E-01	25	K003	2 1 1 1 1	
4.588E-04	1.516E-01	37	C400	2 0 2 2 2	
1.800E-04	5.948E-02	37	E014	2 2 2 1 2	pH 7.3
1.070E-04	3.536E-02	37	H034	1 0 2 1 2	pH 7.4

4235. C$_{21}$H$_{30}$O$_3$
11α-Hydroxyprogesterone
11α-Hydroxy-4-pregnene-3,20-dione
RN: 80-75-1   **MP** (°C):
MW: 330.47   **BP** (°C):   165–166

Solubility (Moles/L)	Solubility (Grams/L)	Temp (°C)	Ref (#)	Evaluation (T P E A A)	Comments
3.522E-04	1.164E-01	37	C400	2 0 2 2 2	

4236. C$_{21}$H$_{30}$O$_3$
11β-Hydroxyprogesterone
11β-Hydroxypregn-4-ene-3,20-dione
RN: 600-57-7   **MP** (°C):
MW: 330.47   **BP** (°C):

Solubility (Moles/L)	Solubility (Grams/L)	Temp (°C)	Ref (#)	Evaluation (T P E A A)	Comments
9.333E-05	3.084E-02	37	C400	2 0 2 2 2	

4237. C$_{21}$H$_{30}$O$_3$
5,6-Dehydroisoandrosterone acetate
Androst-5-en-17-one, 3-(acetyloxy)-, (3β)-
RN: 853-23-6   **MP** (°C):   166
MW: 330.47   **BP** (°C):

Solubility (Moles/L)	Solubility (Grams/L)	Temp (°C)	Ref (#)	Evaluation (T P E A A)	Comments
3.480E-05	1.150E-02	ns	B057	0 2 1 1 2	

4238. C$_{21}$H$_{30}$O$_3$
Testosterone acetate
17-O-Acetyltestosterone
Androst-4-en-3-one, 17-(acetyloxy)-, (17β)-
RN: 1045-69-8   **MP** (°C):   140
MW: 330.47   **BP** (°C):

Solubility (Moles/L)	Solubility (Grams/L)	Temp (°C)	Ref (#)	Evaluation (T P E A A)	Comments
7.111E-06	2.350E-03	25	J004	1 0 1 1 2	
7.111E-06	2.350E-03	ns	B057	0 2 1 1 2	

4239. C$_{21}$H$_{30}$O$_3$

17-α-Hydroxyprogesterone
Pregn-4-ene-3,20-dione, 17-hydroxy-
Prodix
Prodox
U 3096

RN: 68-96-2   **MP** (°C): 222
MW: 330.47   **BP** (°C):

Solubility (Moles/L)	Solubility (Grams/L)	Temp (°C)	Ref (#)	Evaluation (T P E A A)	Comments
1.530E-05	5.056E-03	20	L077	1 2 2 2 2	
1.960E-05	6.477E-03	27.34	L077	1 2 2 2 2	
2.760E-05	9.121E-03	35	L077	1 2 2 2 2	
3.580E-05	1.183E-02	42.34	L077	1 2 2 2 2	
4.290E-05	1.418E-02	50	L077	1 2 2 2 2	

4240. C$_{21}$H$_{30}$O$_4$

Corticosterone
11,21-Dihydroxyprogesterone
δ(4)-Pregnene-11β,21-diol-3,20-dione
11β,21-Dihydroxypregn-4-ene-3,20-dione

RN: 50-22-6   **MP** (°C): 182
MW: 346.47   **BP** (°C):

Solubility (Moles/L)	Solubility (Grams/L)	Temp (°C)	Ref (#)	Evaluation (T P E A A)	Comments
6.943E-04	2.405E-01	37	C400	2 0 2 2 2	

4241. C$_{21}$H$_{30}$O$_4$

11β,17α-Dihydroxy-4-pregnene-3,20-dione
Pregn-5-ene-3,20-dione, 11,17-dihydroxy-
Pregn-5-ene-3,20-dione, 11b,17-dihydroxy-

RN: 603-97-4   **MP** (°C):
MW: 346.47   **BP** (°C): 516.3

Solubility (Moles/L)	Solubility (Grams/L)	Temp (°C)	Ref (#)	Evaluation (T P E A A)	Comments
2.361E-04	8.180E-02	37	C400	2 0 2 2 2	

4242. $C_{21}H_{30}O_4$

Cortexolone
11-Deoxy-17-hydroxycorticosterone
11-Deoxycortisol
11-Desoxycortisone
17,21-Dihydroxy-4-pregnene-3,20-dione
17α,21-Dihydroxypregn-4-ene-3,20-dione

RN:	152-58-9	**MP** (°C):	208
MW:	346.47	**BP** (°C):	

Solubility (Moles/L)	Solubility (Grams/L)	Temp (°C)	Ref (#)	Evaluation (T P E A A)	Comments
1.272E-04	4.408E-02	37	C400	2 0 2 2 2	

4243. $C_{21}H_{30}O_5$

Hydrocortisone
11β,17,21-Trihydroxypregn-4-ene-3,20-dione
Colifoam
Cortaid
Cortef
Bactine

RN:	50-23-7	**MP** (°C):	218.5
MW:	362.47	**BP** (°C):	

Solubility (Moles/L)	Solubility (Grams/L)	Temp (°C)	Ref (#)	Evaluation (T P E A A)	Comments
4.780E-04	1.733E-01	10	B012	2 0 1 1 0	
7.725E-04	2.800E-01	20	A067	0 0 0 0 1	
7.430E-04	2.693E-01	20	B012	2 0 1 1 0	
8.109E-04	2.939E-01	22.5	B422	2 0 2 2 2	
8.820E-04	3.197E-01	25	B012	2 0 1 1 0	
9.932E-04	3.600E-01	25	C437	0 0 0 0 0	Average
7.725E-04	2.800E-01	25	H015	1 0 0 0 1	
8.194E-04	2.970E-01	25	H098	1 0 2 0 2	
8.190E-04	2.969E-01	25	H320	0 0 0 0 0	
8.194E-04	2.970E-01	25	H320	0 0 0 0 0	
7.860E-04	2.849E-01	25	K003	2 1 1 1 1	
1.614E-03	5.850E-01	25	K021	1 2 2 2 1	
7.725E-04	2.800E-01	25	M023	1 0 2 1 1	
9.896E-03	3.587E+00	25	P324	0 0 0 0 0	
1.034E-03	3.748E-01	30	B012	2 0 1 1 0	
1.000E-03	3.625E-01	30	L344	2 0 1 1 0	EFG
1.077E-03	3.905E-01	37	C400	2 0 2 2 2	
1.070E-03	3.878E-01	37	H036	1 0 2 2 2	EFG
1.265E-03	4.585E-01	40	B012	2 0 1 1 0	
1.519E-03	5.506E-01	50	B012	2 0 1 1 0	
7.725E-04	2.800E-01	298	F016	0 0 0 0 0	
1.159E-03	4.200E-01	amb	L434	0 0 0 0 0	
1.104E-03	4.000E-01	amb	L445	0 0 0 0 0	Intrinsic
7.116E-04	2.579E-01	ns	B404	0 2 1 1 0	

4244. C$_{21}$H$_{30}$O$_6$
Cortisone acetate
Pregn-4-ene-3,11,20-trione, 21-(acetyloxy)-17-hydroxy-

RN:	50-04-4	MP (°C):	235
MW:	378.47	BP (°C):	

Solubility (Moles/L)	Solubility (Grams/L)	Temp (°C)	Ref (#)	Evaluation (T P E A A)	Comments
5.284E-05	2.000E-02	22.5	G301	0 0 0 0 0	
5.020E-05	1.900E-02	25	K003	2 1 1 1 1	
5.284E-05	2.000E-02	25	M023	1 0 2 1 0	
7.398E-05	2.800E-02	25	P096	0 0 0 0 0	
1.000E-04	3.785E-02	30	L068	1 0 0 1 0	EFG

4245. C$_{21}$H$_{31}$NO
N-Cyclododecylcinnamamide
2-Propenamide, N-cyclododecyl-3-phenyl

RN:	59832-03-0	MP (°C):	
MW:	313.49	BP (°C):	

Solubility (Moles/L)	Solubility (Grams/L)	Temp (°C)	Ref (#)	Evaluation (T P E A A)	Comments
3.910E-08	1.226E-05	ns	H350	0 0 0 0 0	

4246. C$_{21}$H$_{31}$N$_3$O$_2$
2-Pentoxy-N-[2-(diethyl-amino)ethyl]-4-quinoline carboxamide
N-[2-(Diethylamino)ethyl]-2-pentoxyquinoline-4-carboxamide

RN:	2717-02-4	MP (°C):	
MW:	357.50	BP (°C):	

Solubility (Moles/L)	Solubility (Grams/L)	Temp (°C)	Ref (#)	Evaluation (T P E A A)	Comments
5.300E-05	1.895E-02	ns	B018	0 0 0 0 1	
5.300E-05	1.895E-02	ns	M066	0 0 0 0 1	

4247. C$_{21}$H$_{32}$O$_2$
3,20-Pregnanedione
7α-17-Dimethyltestosterone
Bolasterone

RN:	128-23-4	MP (°C):	
MW:	316.49	BP (°C):	

Solubility (Moles/L)	Solubility (Grams/L)	Temp (°C)	Ref (#)	Evaluation (T P E A A)	Comments
1.833E-04	5.800E-02	37	H004	0 0 0 0 0	

4248. C$_{21}$H$_{32}$O$_2$

7α,17-Dimethyl-19-nortestosterone

RN: **MP** (°C):
MW: 316.49   **BP** (°C):

Solubility (Moles/L)	Solubility (Grams/L)	Temp (°C)	Ref (#)	Evaluation (T P E A A)	Comments
1.434E-04	4.540E-02	37	H004	0 0 0 0 0	

4249. C$_{21}$H$_{32}$O$_2$

Pregnenolone
3β-Hydroxy-5-pregnen-20-one
5-Pregnen-3β-ol-20-one
3β-Hydroxypregn-5-en-20-one

RN: 145-13-1   **MP** (°C):   193
MW: 316.49   **BP** (°C):

Solubility (Moles/L)	Solubility (Grams/L)	Temp (°C)	Ref (#)	Evaluation (T P E A A)	Comments
2.230E-05	7.058E-03	37	H034	1 0 2 1 2	pH 7.4
9.479E-05	3.000E-02	amb	L434	0 0 0 0 0	
1.295E-04	4.100E-02	rt	B408	0 0 2 2 2	

4250. C$_{21}$H$_{32}$O$_3$

Androstanolone acetate
Androstan-3-one, 17-(acetyloxy)-, (5α,17β)-
Stanolone acetate

RN: 1164-91-6   **MP** (°C):
MW: 332.49   **BP** (°C):

Solubility (Moles/L)	Solubility (Grams/L)	Temp (°C)	Ref (#)	Evaluation (T P E A A)	Comments
2.672E-01	8.884E+01	ns	B057	0 2 1 1 2	

4251. C$_{21}$H$_{33}$NO

2-Propenamide, N-dodecyl-3-phenyl-

RN: 55125-24-1   **MP** (°C):
MW: 315.50   **BP** (°C):

Solubility (Moles/L)	Solubility (Grams/L)	Temp (°C)	Ref (#)	Evaluation (T P E A A)	Comments
2.100E-06	6.626E-04	ns	H350	0 0 0 0 0	

4252. $C_{21}H_{33}NO_7$
Lasiocarpine
(7α-Angelyloxy-5,6,7,8α-tetrahydro-3H-pyrrolizin-1-yl)methyl-2,3-dihydroxy-2-(1'-
   methoxyethyl)-3-methylbutyrate

RN:	303-34-4	MP (°C):	97
MW:	411.50	BP (°C):	

Solubility (Moles/L)	Solubility (Grams/L)	Temp (°C)	Ref (#)	Evaluation (T P E A A)	Comments
1.641E-02	6.754E+00	ns	I312	0 0 0 0 0	

4253. $C_{21}H_{34}O_2$
Pregnanolone
3-Deoxo-3a-hydroxy-5b-dihydroprogesterone
3a,5b-Tetrahydroprogesterone
3a-Hydroxy-5b-pregnan-20-one
Pregnan-3a-ol-20-one
3a,5b-Pregnanolone

RN:	128-20-1	MP (°C):	
MW:	318.50	BP (°C):	431.2

Solubility (Moles/L)	Solubility (Grams/L)	Temp (°C)	Ref (#)	Evaluation (T P E A A)	Comments
2.512E-05	8.000E-03	rt	B408	0 0 2 2 2	

4254. $C_{21}H_{34}O_3$
Tetradecyl p-hydroxybenzoate
Tetradecyl 4-hydroxybenzoate

RN:	71177-53-2	MP (°C):	
MW:	334.50	BP (°C):	

Solubility (Moles/L)	Solubility (Grams/L)	Temp (°C)	Ref (#)	Evaluation (T P E A A)	Comments
1.088E-03	3.639E-01	25	D081	1 2 2 1 2	

4255. $C_{21}H_{35}NO_3$
4-Octoxybenzoic acid-2-(diethyl-amino)ethyl ester

RN:	38973-76-1	MP (°C):	
MW:	349.52	BP (°C):	

Solubility (Moles/L)	Solubility (Grams/L)	Temp (°C)	Ref (#)	Evaluation (T P E A A)	Comments
4.000E-05	1.398E-02	ns	M066	0 0 0 0 1	

4256. C$_{21}$H$_{36}$O$_4$
4-Nonylphenol triethoxylate
Ethanol, 2-[2-[2-(4-nonylphenoxy)ethoxy]ethoxy]-
RN: 51437-95-7   **MP** (°C):
MW: 352.52   **BP** (°C):

Solubility (Moles/L)	Solubility (Grams/L)	Temp (°C)	Ref (#)	Evaluation (T P E A A)	Comments
1.668E-05	5.880E-03	20.5	A335	0 0 0 0 0	
1.670E-05	5.887E-03	20.5	A335	0 0 0 0 0	

4257. C$_{21}$H$_{40}$O$_4$
α-Monoolein
1-Monoolein
Glycerol monooleate
9-Octadecenoic acid (Z)-, monoester with 1,2,3-propanetriol
1-Oleoyl-*sn*-glycerol
RN: 25496-72-4   **MP** (°C):
MW: 356.55   **BP** (°C):

Solubility (Moles/L)	Solubility (Grams/L)	Temp (°C)	Ref (#)	Evaluation (T P E A A)	Comments
<1.00E-05	<3.57E-03	30	O321	0 0 0 0 0	

4258. C$_{21}$H$_{44}$
3-Methyleicosane
18-Methyleicosane
RN: 6418-46-8   **MP** (°C):
MW: 296.58   **BP** (°C):

Solubility (Moles/L)	Solubility (Grams/L)	Temp (°C)	Ref (#)	Evaluation (T P E A A)	Comments
5.294E-13	1.570E-10	23	C332	0 0 0 0 0	

4259. C$_{21}$H$_{44}$
2-Methyleicosane
19-Methyleicosane
RN: 1560-84-5   **MP** (°C):
MW: 296.58   **BP** (°C):

Solubility (Moles/L)	Solubility (Grams/L)	Temp (°C)	Ref (#)	Evaluation (T P E A A)	Comments
5.091E-13	1.510E-10	23	C332	0 0 0 0 0	

4260. C$_{22}$H$_{12}$

Indeno(1,2,3-cd)pyrene
Indeno[1,2,3-cd]pyrene
o-Phenylenepyrene

RN: 193-39-5   **MP** (°C): 162.5
MW: 276.34   **BP** (°C): 536

Solubility (Moles/L)	Solubility (Grams/L)	Temp (°C)	Ref (#)	Evaluation (T P E A A)	Comments
6.876E-10	1.900E-07	ns	W302	0 0 0 0 1	

4261. C$_{22}$H$_{12}$

Benzo[g,h,i]perylene
Benz[g,h,i]perylene

RN: 191-24-2   **MP** (°C): 279
MW: 276.34   **BP** (°C): >500

Solubility (Moles/L)	Solubility (Grams/L)	Temp (°C)	Ref (#)	Evaluation (T P E A A)	Comments
4.958E-10	1.370E-07	25	D406	1 2 2 2 2	
6.500E-10	1.796E-07	25	K123	1 0 2 2 1	
9.409E-10	2.600E-07	25	M064	1 1 2 2 1	
9.400E-10	2.598E-07	25	M342	1 0 1 1 1	
9.409E-10	2.600E-07	ns	M344	0 0 0 0 1	
2.533E-09	7.000E-07	ns	W302	0 0 0 0 0	

4262. C$_{22}$H$_{14}$

Picene
1,2,7,8-Dibenzphenanthrene
3,4-Benzchrysene

RN: 213-46-7   **MP** (°C): 366
MW: 278.36   **BP** (°C): 518

Solubility (Moles/L)	Solubility (Grams/L)	Temp (°C)	Ref (#)	Evaluation (T P E A A)	Comments
1.550E-08	4.315E-06	20	E009	1 0 0 1 2	
8.981E-09	2.500E-06	27	D003	1 0 0 1 1	

4263. C$_{22}$H$_{14}$

1,2:3,4-Dibenzanthracene

RN: 215-58-7   **MP** (°C): 205
MW: 278.36   **BP** (°C): 518

Solubility (Moles/L)	Solubility (Grams/L)	Temp (°C)	Ref (#)	Evaluation (T P E A A)	Comments
5.748E-09	1.600E-06	25	B319	2 0 1 2 1	
8.200E-08	2.283E-05	25	K123	1 0 2 2 1	

4264. C$_{22}$H$_{14}$

1,2:7,8-Dibenzanthracene
Dibenz[a,j]anthracene
Dinaphthanthracene

RN: 224-41-9   **MP** (°C):   196
MW: 278.36   **BP** (°C):

Solubility (Moles/L)	Solubility (Grams/L)	Temp (°C)	Ref (#)	Evaluation (T P E A A)	Comments
3.100E-08	8.629E-06	25	K123	1 0 2 2 1	
4.311E-08	1.200E-05	27	D003	1 0 0 1 1	

4265. C$_{22}$H$_{14}$

1,2:5,6-Dibenzanthracene
1,2,5,6-Dibenzanthracene

RN: 53-70-3   **MP** (°C):   266
MW: 278.36   **BP** (°C):   524

Solubility (Moles/L)	Solubility (Grams/L)	Temp (°C)	Ref (#)	Evaluation (T P E A A)	Comments
8.945E-09	2.490E-06	24	H106	1 0 2 2 2	
7.904E-09	2.200E-06	25	B319	2 0 1 2 2	
2.150E-09	5.985E-07	25	K001	2 2 2 2 2	
1.100E-07	3.062E-05	25	K123	1 0 2 2 1	*sic*
8.945E-09	2.490E-06	25	M156	1 2 1 1 2	
1.800E-09	5.010E-07	25	M342	1 0 1 1 2	
1.796E-09	5.000E-07	27	D003	1 0 0 1 1	

4266. C$_{22}$H$_{16}$F$_3$N$_3$

Fluotrimazole
1H-1,2,3-Triazole, 1-[diphenyl[3-(trifluoromethyl)phenyl]methyl]-

RN: 57381-79-0   **MP** (°C):   132
MW: 379.39   **BP** (°C):

Solubility (Moles/L)	Solubility (Grams/L)	Temp (°C)	Ref (#)	Evaluation (T P E A A)	Comments
3.954E-09	1.500E-06	20	M161	1 0 0 0 1	

4267. C$_{22}$H$_{16}$O$_8$

Ethyl biscoumacetate
Tromexan

RN: 548-00-5   **MP** (°C):   154
MW: 408.37   **BP** (°C):

Solubility (Moles/L)	Solubility (Grams/L)	Temp (°C)	Ref (#)	Evaluation (T P E A A)	Comments
2.179E-04	8.900E-02	20	K028	2 1 2 1 2	pH 3.8, form I
3.747E-04	1.530E-01	20	K028	2 1 2 1 2	pH 3.8, form II
2.179E-04	8.899E-02	20	M042	1 0 0 0 1	pH 3.8, form I, mp 172-182 C
3.761E-04	1.536E-01	20	M042	1 0 0 0 2	pH 3.8, form II, mp 153-160 C

4268. C$_{22}$H$_{17}$ClN$_2$

Clotrimazole
1-(o-Chloro-α,α-diphenylbenzyl)imidazole
1-[α-(2-Chlorophenyl)benzhydryl]imidazole
Lotrimin

RN:	23593-75-1	**MP** (°C):	147–149		
MW:	344.85	**BP** (°C):			

Solubility (Moles/L)	Solubility (Grams/L)	Temp (°C)	Ref (#)	Evaluation (T P E A A)	Comments
<2.90E-05	<1.00E-02	25	H328	0 0 0 0 0	
8.700E-05	3.000E-02	amb	L434	0 0 0 0 0	

4269. C$_{22}$H$_{18}$N$_2$O$_4$S

Hydantoin, 5,5-diphenyl-1-(o-tolylsulfonyl)-
1-(o-Methylbenzenesulfonyl)-5,5-diphenyl-hydantoin

RN:	24759-41-9	**MP** (°C):		
MW:	406.46	**BP** (°C):		

Solubility (Moles/L)	Solubility (Grams/L)	Temp (°C)	Ref (#)	Evaluation (T P E A A)	Comments
1.870E-06	7.600E-04	37	F183	1 0 1 1 2	intrinsic

4270. C$_{22}$H$_{18}$N$_2$O$_5$S

1-(p-Methoxylbenzenesulfonyl)-5,5-diphenyl-hydantoin

RN:	24759-37-3	**MP** (°C):		
MW:	422.46	**BP** (°C):		

Solubility (Moles/L)	Solubility (Grams/L)	Temp (°C)	Ref (#)	Evaluation (T P E A A)	Comments
1.207E-06	5.100E-04	37	F183	1 0 1 1 2	intrinsic

4271. C$_{22}$H$_{19}$Br$_2$NO$_3$

Deltamethrin
3-(2,2-Dibromoethenyl)-2,2-dimethylcyclopropanecarboxylic acid, cyano(3-phenoxyphenyl) methyl ester

RN:	52918-63-5	**MP** (°C):	98–101	
MW:	505.22	**BP** (°C):	300	

Solubility (Moles/L)	Solubility (Grams/L)	Temp (°C)	Ref (#)	Evaluation (T P E A A)	Comments
3.959E-09	2.000E-06	25	M364	1 0 0 0 1	
3.959E-09	2.000E-06	ns	V414	0 0 0 0 0	

4272. C_{22}H_{19}F_6NOS

α-Piperidyl-3,6-bis(trifluoromethyl)-9-phenanthrenemethanol

RN: 31817-24-0   **MP** (°C): 215
MW: 459.46   **BP** (°C):

Solubility (Moles/L)	Solubility (Grams/L)	Temp (°C)	Ref (#)	Evaluation (T P E A A)	Comments
1.632E-05	7.500E-03	25	A013	1 0 2 2 0	average

4273. C_{22}H_{20}

10-Butyl-1,2-benzanthracene

RN: 188124-94-9   **MP** (°C): 97
MW: 284.40   **BP** (°C):

Solubility (Moles/L)	Solubility (Grams/L)	Temp (°C)	Ref (#)	Evaluation (T P E A A)	Comments
2.813E-08	8.000E-06	27	D003	1 0 0 1 1	

4274. C_{22}H_{20}Cl_2N_2O_3

Benzofenap

2-((4-(2,4-Dichloro-3-methylbenzoyl)-1,3-dimethyl-1H-pyrazol-5-yl)oxy)-1-(4-methylphenyl)
   ethanone

RN: 82692-44-2   **MP** (°C):
MW: 431.32   **BP** (°C):

Solubility (Moles/L)	Solubility (Grams/L)	Temp (°C)	Ref (#)	Evaluation (T P E A A)	Comments
3.020E-07	1.303E-04	ns	R427	0 0 0 0 0	

4275. C_{22}H_{20}O_{13}

Carminic acid
Carmine
Carminsaeure

RN: 1260-17-9   **MP** (°C):
MW: 492.40   **BP** (°C):

Solubility (Moles/L)	Solubility (Grams/L)	Temp (°C)	Ref (#)	Evaluation (T P E A A)	Comments
2.637E-03	1.298E+00	rt	D021	0 0 1 1 1	

4276. C$_{22}$H$_{22}$ClN$_3$O$_5$

Propaquizafop

2-[(Isopropylideneamino)oxy]ethyl (*R*)-2-[*p*-[(6-chloro-2-quinoxalinyl)oxy]phenoxy]-propionate
(*R*)-2-{[(1 Methylethylidene)amino]oxy}ethyl 2-{4-[(6-chloro-2-quinoxalinyl)oxy]phenoxy}
    propanoate

Agil

Shogun

RO 17-3664

RN: 111479-05-1   **MP** (°C):

MW: 443.89   **BP** (°C):

Solubility (Moles/L)	Solubility (Grams/L)	Temp (°C)	Ref (#)	Evaluation (T P E A A)	Comments
1.413E-06	6.270E-04	ns	R427	0 0 0 0 0	

4277. C$_{22}$H$_{22}$FN$_3$O$_2$

Droperidol

2H-Benzimidazol-2-one, 1-[1-[4-(4-fluorophenyl)-4-oxobutyl]-1,2,3,6-tetrahydro-4-pyridinyl]-1,3-
    dihydro-

Sintodril

Neurolidol

R 4749

RN: 548-73-2   **MP** (°C):

MW: 379.44   **BP** (°C):

Solubility (Moles/L)	Solubility (Grams/L)	Temp (°C)	Ref (#)	Evaluation (T P E A A)	Comments
1.081E-05	4.100E-03	30	P044	0 0 0 0 0	

4278. C$_{22}$H$_{22}$N$_2$O$_4$

N,N'-Dibutyl-1,4,5,8-naphthalenediimide

Benzo[*lmn*][3,8]phenanthroline-1,3,6,8(2H,7H)-tetrone, 2,7-dibutyl-
1,4,5,8-Naphthalenetetracarboxylic 1,8:4,5-diimide, *N,N'*-dibutyl-

RN: 17655-95-7   **MP** (°C):

MW: 378.43   **BP** (°C):   572.6

Solubility (Moles/L)	Solubility (Grams/L)	Temp (°C)	Ref (#)	Evaluation (T P E A A)	Comments
5.000E-09	1.892E-06	23	B410	2 1 2 2 2	

4279. C$_{22}$H$_{22}$N$_2$O$_8$
Methacycline base
Oxytetracycline, 6-methylene-
Tri-methacycline
Rondomycin

RN:	914-00-1	MP (°C):
MW:	442.43	BP (°C):

Solubility (Moles/L)	Solubility (Grams/L)	Temp (°C)	Ref (#)	Evaluation (T P E A A)	Comments
1.706E-02	7.548E+00	21	M044	2 0 2 2 2	

4280. C$_{22}$H$_{22}$N$_4$O$_6$
Benzoyl-mitomycin C

RN:		MP (°C):
MW:	438.44	BP (°C):

Solubility (Moles/L)	Solubility (Grams/L)	Temp (°C)	Ref (#)	Evaluation (T P E A A)	Comments
1.000E-05	4.384E-03	25	M316	1 1 1 1 2	

4281. C$_{22}$H$_{23}$ClN$_2$O$_8$
Chlortetracycline
7-Chlortetracycline
Acronize PD
Acronize

RN:	57-62-5	MP (°C):
MW:	478.89	BP (°C):

Solubility (Moles/L)	Solubility (Grams/L)	Temp (°C)	Ref (#)	Evaluation (T P E A A)	Comments
1.316E-03	6.300E-01	25	B191	1 0 0 0 1	
2.297E-03	1.100E+00	37	M104	1 2 1 1 0	form II, EFG, recrystallized
1.566E-03	7.500E-01	37	M104	1 2 1 1 0	form I, EFG, recrystallized
2.088E-04	1.000E-01	37	M105	1 2 1 1 0	EFG

4282. C$_{22}$H$_{23}$NO$_3$

Fenpropanate
Danitol
Herald
WL 41706
Miothrin
2,2,3,3-Tetramethylcyclopropane carboxylic acid, cyano(3-phenoxyphenyl)methyl ester

| **RN:** | 39515-41-8 | **MP** (°C): |
| **MW:** | 349.43 | **BP** (°C): |

Solubility (Moles/L)	Solubility (Grams/L)	Temp (°C)	Ref (#)	Evaluation (T P E A A)	Comments
4.074E-08	1.424E-05	ns	R427	0 0 0 0 0	

4283. C$_{22}$H$_{23}$NO$_7$

Noscapine
Narcotine
O-Methylnarcotoline
Opianin
Opian

| **RN:** | 128-62-1 | **MP** (°C): | 176 |
| **MW:** | 413.43 | **BP** (°C): | |

Solubility (Moles/L)	Solubility (Grams/L)	Temp (°C)	Ref (#)	Evaluation (T P E A A)	Comments
4.000E-05	1.654E-02	15	K059	2 2 2 0 0	
7.327E-04	3.029E-01	25	D004	0 0 0 0 0	
7.256E-04	3.000E-01	30	A073	1 1 1 1 0	
1.693E-03	7.000E-01	40	A073	1 1 1 1 0	
2.419E-03	1.000E+00	50	A073	1 1 1 1 1	
2.419E-03	1.000E+00	60	A073	1 1 1 1 1	
2.419E-03	1.000E+00	70	A073	1 1 1 1 1	
2.419E-03	1.000E+00	80	A073	1 1 1 1 1	
3.628E-03	1.500E+00	90	A073	1 1 1 1 1	
4.838E-03	2.000E+00	100	A073	1 1 1 1 1	

4284. C$_{22}$H$_{24}$ClN$_5$O$_2$

Domperidone
5-Chloro-1-[1-[3-(2-oxo-1-benzimidazolinyl)propyl]-4-piperidyl]-2-benzimidazolinone

| **RN:** | 57808-66-9 | **MP** (°C): | 242.5 |
| **MW:** | 425.92 | **BP** (°C): | |

Solubility (Moles/L)	Solubility (Grams/L)	Temp (°C)	Ref (#)	Evaluation (T P E A A)	Comments
4.015E-05	1.710E-02	22	J420	0 0 0 0 0	pH6.5

4285. $C_{22}H_{24}N_2O_8$
Tetracycline
Achromycin V
Sumycin
Robitet
Panmycin
RN: 60-54-8 **MP** (°C): 176dec
MW: 444.45 **BP** (°C):

Solubility (Moles/L)	Solubility (Grams/L)	Temp (°C)	Ref (#)	Evaluation (T P E A A)	Comments
9.900E-04	4.400E-01	25	B191	1 0 0 0 1	neutral pH
5.200E-04	2.311E-01	25	G012	2 0 2 1 0	EFG, pH 5.0
5.700E-04	2.533E-01	25	H017	1 2 2 2 0	EFG, pH 5.0
2.655E-03	1.180E+00	29	N031	1 2 2 2 0	EFG, pH 5.0
7.600E-04	3.378E-01	30	L069	1 0 1 1 0	EFG
1.777E-03	7.900E-01	35	N031	1 2 2 2 0	EFG, pH 5.0
7.875E-02	3.500E+01	37	M104	1 2 1 1 2	form II, recrystallized
6.232E-02	2.770E+01	37	M104	1 2 1 1 2	form I, recrystallized
6.478E-04	2.879E-01	ns	N302	0 2 1 2 2	

4286. $C_{22}H_{24}N_2O_8 \cdot H_2O$
Doxycycline (monohydrate)
Doxylin
Monodox
Vibra-tabs
Doxy-caps
Vibramycin
RN: 564-25-0 **MP** (°C): 201dec
MW: 462.46 **BP** (°C):

Solubility (Moles/L)	Solubility (Grams/L)	Temp (°C)	Ref (#)	Evaluation (T P E A A)	Comments
1.362E-03	6.300E-01	25	B132	2 1 1 1 0	EFG

4287. $C_{22}H_{24}N_2O_9$
Oxytetracycline
Glomycin
Hydroxytetracycline
Riomitsin
Terrafungine
Stevacin
RN: 79-57-2 **MP** (°C): 184
MW: 460.44 **BP** (°C):

Solubility (Moles/L)	Solubility (Grams/L)	Temp (°C)	Ref (#)	Evaluation (T P E A A)	Comments
4.234E-04	1.950E-01	20	L051	1 0 0 0 2	
9.990E-04	4.600E-01	25	B191	1 0 0 0 1	neutral pH

(continued)

4287. C₂₂H₂₄N₂O₉ (continued)

Solubility (Moles/L)	Solubility (Grams/L)	Temp (°C)	Ref (#)	Evaluation (T P E A A)	Comments
4.800E-04	2.210E-01	25	G012	2 0 2 1 0	EFG, pH 5.0
6.798E-04	3.130E-01	25	H005	1 0 1 2 2	Ph 5.8
5.000E-04	2.302E-01	25	H017	1 2 2 2 0	EFG, pH 5.0
6.515E-04	3.000E-01	29	N031	1 2 2 2 0	EFG, pH 5.0
8.687E-04	4.000E-01	37	M104	1 2 1 1 0	form II, EFG, recrystallized
6.515E-04	3.000E-01	37	M104	1 2 1 1 0	form I, EFG, recrystallized

4288. C₂₂H₂₄N₄O₅

Benzyl-mitomycin C

RN: **MP (°C):**
MW: 424.46   **BP (°C):**

Solubility (Moles/L)	Solubility (Grams/L)	Temp (°C)	Ref (#)	Evaluation (T P E A A)	Comments
1.490E-03	6.324E-01	25	M316	1 1 1 1 2	

4289. C₂₂H₂₄N₄O₅S

Methanesulfonamide, *N*-[1′-[2-(2,1,3-benzoxadiazol-5-yl)ethyl]-3,4-dihydro-4-oxospiro[2H-1-benzopyran-2,4′-piperidin]-6-yl]-

Methanesulfonamide, *N*-[1′-[2-(5-benzofurazanyl)ethyl]-3,4-dihydro-4-oxospiro[2H-1-benzopyran-2,4′-piperidin]-6-yl]-

RN: **MP (°C):**
MW: 456.52   **BP (°C):**

Solubility (Moles/L)	Solubility (Grams/L)	Temp (°C)	Ref (#)	Evaluation (T P E A A)	Comments
1.752E-05	8.000E-03	22	D405	1 1 2 2 2	Intrinsic

4290. C₂₂H₂₅NO₆

Colchicine
Colchicin

RN: 64-86-8   **MP (°C):**
MW: 399.45   **BP (°C):**

Solubility (Moles/L)	Solubility (Grams/L)	Temp (°C)	Ref (#)	Evaluation (T P E A A)	Comments
9.629E-02	3.846E+01	20	D041	1 0 0 0 0	
1.088E-01	4.348E+01	25	D004	0 0 0 0 0	
8.261E-02	3.300E+01	ns	K444	0 0 0 0 0	

4291. C$_{22}$H$_{26}$F$_3$N$_3$OS

Fluphenazine
Permitil
Modecate
Prolixin

RN:	69-23-8	**MP** (°C):	<25	
MW:	437.53	**BP** (°C):	271	

Solubility (Moles/L)	Solubility (Grams/L)	Temp (°C)	Ref (#)	Evaluation (T P E A A)	Comments
7.100E-05	3.106E-02	37	F011	1 0 1 1 1	pH 7.4

4292. C$_{22}$H$_{26}$N$_2$O$_9$

Doxycycline
4-(Dimethylamino)-1,4,4a,5,5a,6,11,12a-octahydro-3,5,10,12,12a-pentahydroxy-6-methyl-1,11-
  dioxo-2-naphthacenecarboxamide monohydrate
Doryx
Doxylin
Monodox
Vibramycin

RN:	564-25-0	**MP** (°C):	
MW:	462.46	**BP** (°C):	

Solubility (Moles/L)	Solubility (Grams/L)	Temp (°C)	Ref (#)	Evaluation (T P E A A)	Comments
2.350E+00	1.087E+03	25	B443	0 0 0 0 0	
2.162E-04	1.000E-01	ns	K444	0 0 0 0 0	

4293. C$_{22}$H$_{27}$ClN$_2$O$_4$S

Diltiazem hydrochloride
1,5-Benzothiazepin-4(5H)one,3-(acetyloxy)-5-(2-(dimethylamino)ethyl)-2,3-dihydro-2-(4-
  methoxyphenyl)-,
Dilacor XR
Cardizem
Cardcal
Coras

RN:	33286-22-5	**MP** (°C):	
MW:	450.99	**BP** (°C):	

Solubility (Moles/L)	Solubility (Grams/L)	Temp (°C)	Ref (#)	Evaluation (T P E A A)	Comments
1.131E-03	5.100E-01	25	A412	1 0 2 2 1	int

4294. $C_{22}H_{27}NO_2$

Danazol

17α-Pregna-2,4-dien-20-yno[2,3-d]isoxazol-17-ol

Danocrine

Cyclomen

RN: 17230-88-5   **MP** (°C):

MW: 337.47   **BP** (°C):

Solubility (Moles/L)	Solubility (Grams/L)	Temp (°C)	Ref (#)	Evaluation (T P E A A)	Comments
1.719E-06	5.800E-04	25	E409	0 0 0 0 0	
1.245E-06	4.200E-04	37	S446	0 0 0 0 0	

4295. $C_{22}H_{28}F_2O_5$

Flumethasone

Flumethasonpivalate

RN: 2135-17-3   **MP** (°C):

MW: 410.46   **BP** (°C):

Solubility (Moles/L)	Solubility (Grams/L)	Temp (°C)	Ref (#)	Evaluation (T P E A A)	Comments
2.436E-06	1.000E-03	20	A067	0 0 0 0 0	

4296. $C_{22}H_{28}N_2O$

Fentanyl

1-Phenethyl-4-(phenylpropionylamino)piperidine

N-(1-Phenethyl-4-piperidyl)propionanilide

Duragesic

RN: 437-38-7   **MP** (°C):

MW: 336.48   **BP** (°C):

Solubility (Moles/L)	Solubility (Grams/L)	Temp (°C)	Ref (#)	Evaluation (T P E A A)	Comments
5.944E-04	2.000E-01	25	R338	0 0 0 0 0	
3.566E-05	1.200E-02	35	R418	0 0 0 0 0	Intrinsic

4297. $C_{22}H_{28}N_6O_3S$

Delavirdine

1-[3-[(1-Methylethyl)amino]-2-pyridinyl]-4-[[5-[(methylsulfonyl)amino]-1H-indol-2-yl]carbonyl]
   piperazine

1-(5-Methanesulfonamido-1H-indol-2-ylcarbonyl)-4-[3-(1-methylethylamino)pyridinyl]piperazine

1-[3-(Isopropylamino)-2-pyridyl]-4-[(5-methanesulfonamidoindol-2-yl)carbonyl]piperazine

RN: 136817-59-9   **MP** (°C):

MW: 456.57   **BP** (°C):

Solubility (Moles/L)	Solubility (Grams/L)	Temp (°C)	Ref (#)	Evaluation (T P E A A)	Comments
6.571E-02	3.000E+01	ns	A426	0 0 0 0 0	Intrinsic

4298. C$_{22}$H$_{28}$O$_3$
Canrenone
17-Hydroxy-3-oxo-17α-pregna-4,6-diene-21-carboxylic acid lactone
RN: 976-71-6   **MP** (°C): 149-151
MW: 340.47   **BP** (°C):

Solubility (Moles/L)	Solubility (Grams/L)	Temp (°C)	Ref (#)	Evaluation (T P E A A)	Comments
8.000E-07	2.724E-04	25	G017	1 0 1 0 0	EFG
8.100E-05	2.758E-02	37	C004	0 0 0 0 0	*sic*
8.958E-07	3.050E-04	37	O306	1 0 2 2 2	
6.374E-07	2.170E-04	rt	O306	0 0 2 2 2	

4299. C$_{22}$H$_{28}$O$_3$
Norethindrone acetate
Norethisterone acetate
RN: 51-98-9   **MP** (°C): 161
MW: 340.47   **BP** (°C):

Solubility (Moles/L)	Solubility (Grams/L)	Temp (°C)	Ref (#)	Evaluation (T P E A A)	Comments
9.288E-06	3.162E-03	10	L078	1 0 1 2 0	EFG
1.312E-05	4.467E-03	20	L078	1 0 1 2 0	EFG
1.570E-05	5.345E-03	25	H099	1 0 2 2 2	
1.652E-05	5.623E-03	25	L078	1 0 1 2 2	
1.853E-05	6.310E-03	30	L078	1 0 1 2 0	EFG
2.937E-05	1.000E-02	40	L078	1 0 1 2 0	EFG

4300. C$_{22}$H$_{29}$FO$_4$
Fluorometholone
9-Fluoro-11β,17-dihydroxy-6α-methylpregna-1,4-diene-3,20-dione
21-Desoxy-9α-fluoro-6α-methyl-prednisolone
RN: 426-13-1   **MP** (°C):
MW: 376.47   **BP** (°C):

Solubility (Moles/L)	Solubility (Grams/L)	Temp (°C)	Ref (#)	Evaluation (T P E A A)	Comments
7.968E-05	3.000E-02	25	G008	1 2 1 1 0	

4301. C$_{22}$H$_{29}$FO$_5$
Betamethasone
Pregna-1,4-diene-3,20-dione, 9-fluoro-11,17,21-trihydroxy-16-methyl-, (11β,16β)-
RN: 378-44-9   **MP** (°C): 230
MW: 392.47   **BP** (°C):

Solubility (Moles/L)	Solubility (Grams/L)	Temp (°C)	Ref (#)	Evaluation (T P E A A)	Comments
1.478E-04	5.800E-02	25	K003	2 1 1 1 1	
1.936E-04	7.599E-02	25	P096	0 0 0 0 0	

(continued)

4301. $C_{22}H_{29}FO_5$ (continued)

Solubility (Moles/L)	Solubility (Grams/L)	Temp (°C)	Ref (#)	Evaluation (T P E A A)	Comments
1.500E-04	5.887E-02	30	O321	0 0 0 0 0	
1.529E-04	6.000E-02	30	O321	0 0 0 0 0	
1.605E-04	6.301E-02	37	C400	2 0 2 2 2	
1.605E-04	6.300E-02	ns	B404	0 2 1 1 0	
1.575E-04	6.180E-02	rt	I404	0 0 0 0 0	Intrinsic, Average

4302. $C_{22}H_{29}FO_5$
Dexamethasone
Dexamethasone alcohol

RN: 50-02-2   **MP** (°C): 262
MW: 392.47   **BP** (°C):

Solubility (Moles/L)	Solubility (Grams/L)	Temp (°C)	Ref (#)	Evaluation (T P E A A)	Comments
8.200E-05	3.218E-02	10	B012	2 0 1 1 0	
1.580E-04	6.201E-02	20	B012	2 0 1 1 0	
2.800E-04	1.099E-01	23	L345	1 0 1 1 2	
2.270E-04	8.909E-02	25	B012	2 0 1 1 0	
2.140E-04	8.399E-02	25	K003	2 1 1 1 1	
3.083E-04	1.210E-01	25	K021	1 2 2 2 1	
2.548E-04	1.000E-01	25	P312	0 0 0 0 0	
2.520E-04	9.890E-02	30	B012	2 0 1 1 0	
2.344E-04	9.200E-02	37	C400	2 0 2 2 2	
2.955E-04	1.160E-01	37	D026	0 0 0 0 0	
3.560E-04	1.397E-01	40	B012	2 0 1 1 0	
4.600E-04	1.805E-01	50	B012	2 0 1 1 0	
4.077E-04	1.600E-01	amb	L434	0 0 0 0 0	
2.548E-04	1.000E-01	ns	K444	0 0 0 0 0	
1.707E-04	6.700E-02	ns	N302	0 2 1 2 1	

4303. $C_{22}H_{29}NO_7S_2$
Methyl O-acetyl-3-(acetyloxy)-N-{5-[(3R)-1,2-dithiolan-3-yl]-pentanoyl}-L-tyrosinate

RN: **MP** (°C):
MW: 483.61   **BP** (°C):

Solubility (Moles/L)	Solubility (Grams/L)	Temp (°C)	Ref (#)	Evaluation (T P E A A)	Comments
4.817E-04	2.329E-01	ns	S453	0 0 0 0 0	

4304. C$_{22}$H$_{30}$ClNO$_2$
Propoxyphene hydrochloride
D-Propoxyphene hydrochloride
RN: 1639-60-7   **MP** (°C):
MW: 375.94   **BP** (°C):

Solubility (Moles/L)	Solubility (Grams/L)	Temp (°C)	Ref (#)	Evaluation (T P E A A)	Comments
9.842E-06	3.700E-03	25	A412	1 0 2 2 1	int

4305. C$_{22}$H$_{30}$Cl$_2$N$_{10}$
Chlorhexidin
Chlorhexidine
bis(5-(*p*-Chlorophenyl)biguanidinio)hexane
RN: 55-56-1   **MP** (°C):
MW: 505.46   **BP** (°C):

Solubility (Moles/L)	Solubility (Grams/L)	Temp (°C)	Ref (#)	Evaluation (T P E A A)	Comments
1.583E-04	7.999E-02	20	D341	0 0 0 0 0	
8.309E-05	4.200E-02	22.5	G301	0 0 0 0 0	

4306. C$_{22}$H$_{30}$N$_2$O$_2$
Aspidospermine
Aspidospermidine, 1-acetyl-17-methoxy-
RN: 466-49-9   **MP** (°C):   208
MW: 354.50   **BP** (°C):

Solubility (Moles/L)	Solubility (Grams/L)	Temp (°C)	Ref (#)	Evaluation (T P E A A)	Comments
4.701E-04	1.666E-01	c	D004	0 0 0 0 0	

4307. C$_{22}$H$_{30}$N$_2$O$_2$S
Sufentanil
N-[4-(Methoxymethyl)-1-[2-(2-thienyl)ethyl]-4-piperidyl]propionanilide
Sufenta
RN: 56030-54-7   **MP** (°C):
MW: 386.56   **BP** (°C):

Solubility (Moles/L)	Solubility (Grams/L)	Temp (°C)	Ref (#)	Evaluation (T P E A A)	Comments
1.966E-04	7.600E-02	25	R338	0 0 0 0 0	
3.363E-06	1.300E-03	35	R418	0 0 0 0 0	Intrinsic

4308. C$_{22}$H$_{30}$O$_5$
Methylprednisolone
6α-Methylprednisolone
Medrol
Solumedrol
Metrisone
Promacortine

RN:	83-43-2	**MP (°C):**	232.5
MW:	374.48	**BP (°C):**	

Solubility (Moles/L)	Solubility (Grams/L)	Temp (°C)	Ref (#)	Evaluation (T P E A A)	Comments
3.204E-04	1.200E-01	25	A014	1 0 1 1 0	EFG
2.403E-04	9.000E-02	25	A014	1 0 1 1 0	EFG, pH 5.0
2.534E-03	9.491E-01	25	G008	1 2 1 1 1	
3.445E-04	1.290E-01	25	K021	1 2 2 2 1	
1.335E-04	5.000E-02	27.14	H026	1 0 2 1 0	EFG, form I
1.923E-04	7.199E-02	30.0	H010	2 2 1 1 1	
4.273E-04	1.600E-01	31.72	H026	1 0 2 1 0	EFG, form II
3.124E-04	1.170E-01	37	H004	0 0 0 0 0	polymorph I
3.765E-04	1.410E-01	37	H004	0 0 0 0 0	polymorph II
5.341E-04	2.000E-01	40.32	H026	1 0 2 1 0	EFG, form II
2.937E-04	1.100E-01	40.32	H026	1 0 2 1 0	EFG, form I
4.273E-04	1.600E-01	51.52	H026	1 0 2 1 0	EFG, form I
1.362E-03	5.100E-01	81.45	H026	1 0 2 1 0	EFG, form II
1.068E-03	4.000E-01	81.45	H026	1 0 2 1 0	EFG, form I
2.670E-04	1.000E-01	ns	M169	0 0 0 0 1	

4309. C$_{22}$H$_{30}$O$_6$
5,16-β-Dihydroxy-6-β-methyl-3,11-dioxo-5-α-pregn-17(20)-ene-*cis*-20-carboxylic acid methyl ester
U-20235

RN:		**MP (°C):**	
MW:	390.48	**BP (°C):**	

Solubility (Moles/L)	Solubility (Grams/L)	Temp (°C)	Ref (#)	Evaluation (T P E A A)	Comments
6.402E-04	2.500E-01	ns	K029	0 0 2 1 1	

4310. C$_{22}$H$_{32}$O$_3$
Nandrolone butyrate

RN:		**MP (°C):**	
MW:	344.50	**BP (°C):**	

Solubility (Moles/L)	Solubility (Grams/L)	Temp (°C)	Ref (#)	Evaluation (T P E A A)	Comments
1.460E-05	5.030E-03	37	C026	0 0 0 0 0	

4311. C$_{22}$H$_{32}$O$_3$
Methyltestosterone acetate
17-α-Methyltestosterone acetate
RN: 1099-79-2 **MP** (°C): 164
MW: 344.50 **BP** (°C):

Solubility (Moles/L)	Solubility (Grams/L)	Temp (°C)	Ref (#)	Evaluation (T P E A A)	Comments
1.430E-05	4.926E-03	25	H099	1 0 2 2 2	
5.196E-06	1.790E-03	ns	B057	0 2 1 1 2	

4312. C$_{22}$H$_{32}$O$_3$
5,6-Dehydroisoandrosterone propionate
RN: 1167-87-9 **MP** (°C):
MW: 344.50 **BP** (°C):

Solubility (Moles/L)	Solubility (Grams/L)	Temp (°C)	Ref (#)	Evaluation (T P E A A)	Comments
2.415E-05	8.320E-03	ns	B057	0 2 1 1 2	

4313. C$_{22}$H$_{32}$O$_3$
Testosterone propionate
17-(1-Oxopropoxy)-(17β)-androst-4-en-3-one
Testosterone-17-propionate
Agovirin
Androsan
Androgen
RN: 57-85-2 **MP** (°C): 120
MW: 344.50 **BP** (°C):

Solubility (Moles/L)	Solubility (Grams/L)	Temp (°C)	Ref (#)	Evaluation (T P E A A)	Comments
1.710E-04	5.891E-02	20	F012	1 0 1 1 1	
4.300E-06	1.481E-03	25	J004	1 0 1 1 2	
5.806E-06	2.000E-03	25	K003	2 1 1 1 1	
6.096E-06	2.100E-03	30	T005	2 0 2 2 1	
1.060E-05	3.652E-03	37.50	B054	1 0 1 1 2	
4.296E-06	1.480E-03	ns	B057	0 2 1 1 2	

4314. C$_{22}$H$_{33}$N$_3$O$_2$
2-Hexoxy-N-[2-(diethyl-amino)ethyl]-4-quinoline carboxamide
N-[2-(Diethylamino)ethyl]-2-hexoxyquinoline-4-carboxamide
RN: 2717-03-5 **MP** (°C):
MW: 371.53 **BP** (°C):

Solubility (Moles/L)	Solubility (Grams/L)	Temp (°C)	Ref (#)	Evaluation (T P E A A)	Comments
6.700E-06	2.489E-03	ns	B018	0 0 0 0 1	
6.700E-06	2.489E-03	ns	M066	0 0 0 0 1	

4315. C$_{22}$H$_{34}$Cl$_2$O$_3$

2,4-Dichlorophenoxyacetic acid *n*-tetradecyl ester

RN: 65267-96-1 **MP** (°C):
MW: 417.42 **BP** (°C):

Solubility (Moles/L)	Solubility (Grams/L)	Temp (°C)	Ref (#)	Evaluation (T P E A A)	Comments
1.161E-05	4.848E-03	ns	M120	0 0 1 1 2	

4316. C$_{22}$H$_{34}$N$_6$O$_4$

2,5-Diaziridinyl-3,6-di(1′-piperazineethanol)-1,4-benzoquinone

RN: 59886-40-7 **MP** (°C): 170
MW: 446.55 **BP** (°C):

Solubility (Moles/L)	Solubility (Grams/L)	Temp (°C)	Ref (#)	Evaluation (T P E A A)	Comments
4.479E-02	2.000E+01	rt	C317	0 0 0 0 0	

4317. C$_{22}$H$_{34}$O$_3$

Androstanolone propionate

Androstan-3-one, 17-(1-oxopropoxy)-, (5α,17β)-

RN: 855-22-1 **MP** (°C):
MW: 346.51 **BP** (°C):

Solubility (Moles/L)	Solubility (Grams/L)	Temp (°C)	Ref (#)	Evaluation (T P E A A)	Comments
1.789E-06	6.200E-04	ns	B057	0 2 1 1 2	

4318. C$_{22}$H$_{35}$NO$_3$

Acetaminophen myristate

Acetaminophen tetradecanoate

RN: 54942-39-1 **MP** (°C): 114
MW: 361.53 **BP** (°C):

Solubility (Moles/L)	Solubility (Grams/L)	Temp (°C)	Ref (#)	Evaluation (T P E A A)	Comments
1.660E-05	6.000E-03	25	B010	1 1 1 1 0	

4319. C$_{22}$H$_{37}$NO$_2$

Anandamide

Arachidonoylethanolamide

AEA

RN: 94421-68-8 **MP** (°C):
MW: 347.55 **BP** (°C):

Solubility (Moles/L)	Solubility (Grams/L)	Temp (°C)	Ref (#)	Evaluation (T P E A A)	Comments
1.188E-06	4.130E-04	25	J414	0 0 0 0 0	Intrinsic

4320. C₂₂H₃₈O₅

4-Octylphenol tetraethoxylate

Ethanol, 2-[2-[2-[2-(4-octylphenoxy)ethoxy]ethoxy]ethoxy]-

RN: 51437-92-4   **MP** (°C):

MW: 382.55   **BP** (°C):

Solubility (Moles/L)	Solubility (Grams/L)	Temp (°C)	Ref (#)	Evaluation (T P E A A)	Comments
6.404E-05	2.450E-02	20.5	A335	0 0 0 0 0	
6.410E-05	2.452E-02	20.5	A335	0 0 0 0 0	

4321. C₂₂H₃₉O₃P

Diisooctyl phenyl phosphonate

RN: **MP** (°C):

MW: 382.53   **BP** (°C):

Solubility (Moles/L)	Solubility (Grams/L)	Temp (°C)	Ref (#)	Evaluation (T P E A A)	Comments
<2.61E-04	<1.00E-01	25	B070	1 2 0 1 0	

4322. C₂₂H₃₉O₃P

Dioctyl phenyl phosphonate

Di-*n*-octyl phenylphosphonate

DOPP

RN: 1754-47-8   **MP** (°C):

MW: 382.53   **BP** (°C):

Solubility (Moles/L)	Solubility (Grams/L)	Temp (°C)	Ref (#)	Evaluation (T P E A A)	Comments
<5.23E-04	<2.00E-01	25	B070	1 2 0 1 0	

4323. C₂₂H₄₂O₄

Dioctyl adipate

bis(2-Ethylhexyl) adipate

RN: 103-23-1   **MP** (°C):

MW: 370.58   **BP** (°C):

Solubility (Moles/L)	Solubility (Grams/L)	Temp (°C)	Ref (#)	Evaluation (T P E A A)	Comments
8.095E-06	3.000E-03	25	F067	1 0 2 2 1	

4324. C₂₂H₄₃N₅O₁₃

Amikacin

Antibiotic BB-K8

RN: 37517-28-5   **MP** (°C):   203

MW: 585.61   **BP** (°C):

Solubility (Moles/L)	Solubility (Grams/L)	Temp (°C)	Ref (#)	Evaluation (T P E A A)	Comments
3.159E-01	1.850E+02	25	K044	1 0 0 0 2	pH 10.4

4325. $C_{23}H_{16}O_6$
Pamoic acid
4,4′-Methylenebis[3-hydroxy-2-naphthalenecarboxylic acid]
3,3′-Dihydroxy-4,4′-methylenedi-2-naphthoic acid
Embonic acid

RN: 130-85-8   **MP** (°C):
MW: 388.38   **BP** (°C):

Solubility (Moles/L)	Solubility (Grams/L)	Temp (°C)	Ref (#)	Evaluation (T P E A A)	Comments
2.800E-01	1.087E+02	ns	F007	0 0 0 0 1	

4326. $C_{23}H_{18}F_2N_4O$
α-(2,4-Difluorophenyl)-α-(1-2-(2-pyridyl)phenylethenyl)-1H-1,2,4-triazole-1-ethanol
XD405

RN: 124669-93-8   **MP** (°C):
MW: 404.42   **BP** (°C):

Solubility (Moles/L)	Solubility (Grams/L)	Temp (°C)	Ref (#)	Evaluation (T P E A A)	Comments
7.418E-06	3.000E-03	22	M372	1 2 1 1 1	intrinsic

4327. $C_{23}H_{20}N_2O_2S$
G-1
p-Phenylthioethylphenylbutazone
1,2-Diphenyl-4-(2-phenylthioethyl)-3,5-pyrazolidinedione

RN: 3736-92-3   **MP** (°C):
MW: 388.49   **BP** (°C):

Solubility (Moles/L)	Solubility (Grams/L)	Temp (°C)	Ref (#)	Evaluation (T P E A A)	Comments
4.118E-03	1.600E+00	ns	B158	0 0 0 0 1	pH 7.0

4328. $C_{23}H_{20}N_2O_3S$
Sulfinpyrazone
Sulfoxyphenyl pyrazolidine
Sulfinpyrazole
1,2-Diphenyl-4-(2-(phenylsulfinyl)ethyl)-3,5-pyrazolidinedione
Anturane

RN: 57-96-5   **MP** (°C):
MW: 404.49   **BP** (°C):

Solubility (Moles/L)	Solubility (Grams/L)	Temp (°C)	Ref (#)	Evaluation (T P E A A)	Comments
6.431E-03	2.601E+00	22	J420	0 0 0 0 0	pH6.5

4329. C$_{23}$H$_{22}$
10-Amyl-1,2-benzanthracene

RN: 188124-96-1   **MP** (°C):
MW: 298.43   **BP** (°C):

Solubility (Moles/L)	Solubility (Grams/L)	Temp (°C)	Ref (#)	Evaluation (T P E A A)	Comments
2.681E-09	8.000E-07	27	D003	1 0 0 1 0	

4330. C$_{23}$H$_{22}$O$_6$
Rotenone
Tubatoxin
Derris
1,2,12,12α-Tetrahydro-2α-isopropenyl-8,9-dimethoxy(1)benzopyrano(3,4-b)furo(2,3-h)(1)
    benzopyran-6(6α H)-one

RN: 83-79-4   **MP** (°C):   163
MW: 394.43   **BP** (°C):

Solubility (Moles/L)	Solubility (Grams/L)	Temp (°C)	Ref (#)	Evaluation (T P E A A)	Comments
4.310E-07	1.700E-04	25	C100	1 0 2 1 1	
3.803E-05	1.500E-02	100	M161	1 0 0 0 1	

4331. C$_{23}$H$_{23}$NO
Trifenmorph
Frescon
N-Tritylmorpholine

RN: 1420-06-0   **MP** (°C):   175
MW: 329.45   **BP** (°C):

Solubility (Moles/L)	Solubility (Grams/L)	Temp (°C)	Ref (#)	Evaluation (T P E A A)	Comments
6.071E-08	2.000E-05	20	M161	1 0 0 0 1	

4332. C$_{23}$H$_{24}$N$_4$O$_2$
Diantipyrylmethane
4,4′-Methylenediantipyrine
4,4′-Diantipyrylmethane

RN: 1251-85-0   **MP** (°C):   182
MW: 388.47   **BP** (°C):

Solubility (Moles/L)	Solubility (Grams/L)	Temp (°C)	Ref (#)	Evaluation (T P E A A)	Comments
1.130E-03	4.390E-01	20	P054	0 0 0 0 0	
1.132E-03	4.398E-01	20	P054	0 0 0 0 0	

4333. C$_{23}$H$_{24}$N$_4$O$_6$
Benzylcarbonyl-mitomycin C
RN: **MP** (°C):
MW: 452.47   **BP** (°C):

Solubility (Moles/L)	Solubility (Grams/L)	Temp (°C)	Ref (#)	Evaluation (T P E A A)	Comments
2.240E-03	1.014E+00	25	M316	1 1 1 1 2	

4334. C$_{23}$H$_{24}$N$_4$O$_7$
Benzyloxycarbonyl-mitomycin C
RN: **MP** (°C):
MW: 468.47   **BP** (°C):

Solubility (Moles/L)	Solubility (Grams/L)	Temp (°C)	Ref (#)	Evaluation (T P E A A)	Comments
5.200E-04	2.436E-01	25	M316	1 1 1 1 2	

4335. C$_{23}$H$_{24}$N$_4$S$_2$
Dithiodiantipyrinylmethane
3H-Pyrazole-3-thione, 4,4'-methylenebis[1,2-dihydro-1,5-dimethyl-2-phenyl-
RN: 53799-78-3   **MP** (°C):   166
MW: 420.60   **BP** (°C):

Solubility (Moles/L)	Solubility (Grams/L)	Temp (°C)	Ref (#)	Evaluation (T P E A A)	Comments
5.000E-04	2.103E-01	ns	D087	0 2 0 0 1	

4336. C$_{23}$H$_{25}$N
Fendiline
RN: 13042-18-7   **MP** (°C):
MW: 315.46   **BP** (°C):

Solubility (Moles/L)	Solubility (Grams/L)	Temp (°C)	Ref (#)	Evaluation (T P E A A)	Comments
7.389E-06	2.331E-03	22.5	B440	0 0 0 0 0	

4337. C$_{23}$H$_{26}$FN$_3$O$_2$
Spiperone
8-[4-(4-Fluorophenyl)-4-oxobutyl]-1-phenyl-1,3,8-triazaspiro[4.5]decan-4-one
RN: 749-02-0   **MP** (°C):   192 C
MW: 395.48   **BP** (°C):

Solubility (Moles/L)	Solubility (Grams/L)	Temp (°C)	Ref (#)	Evaluation (T P E A A)	Comments
8.091E-05	3.200E-02	22	J420	0 0 0 0 0	pH6.5

4338. C$_{23}$H$_{26}$N$_2$O$_4$

Brucine

Brucin

RN: 357-57-3   **MP** (°C):   178

MW: 394.47   **BP** (°C):

Solubility (Moles/L)	Solubility (Grams/L)	Temp (°C)	Ref (#)	Evaluation (T P E A A)	Comments
8.112E-03	3.200E+00	15	F300	1 0 0 0 1	
1.330E-03	5.247E-01	15	K059	2 2 2 0 2	
1.698E-02	6.700E+00	100	F300	1 0 0 0 1	
1.267E-03	4.998E-01	rt	D021	0 0 1 1 1	

4339. C$_{23}$H$_{26}$N$_2$O$_4$·4H$_2$O

Brucine (tetrahydrate)

Strychnidin-10-one, 2,3-dimethoxy-, tetrahydrate

RN: 5892-11-5   **MP** (°C):   105

MW: 466.54   **BP** (°C):

Solubility (Moles/L)	Solubility (Grams/L)	Temp (°C)	Ref (#)	Evaluation (T P E A A)	Comments
6.677E-03	3.115E+00	c	D004	0 0 0 0 0	
1.420E-02	6.623E+00	h	D004	0 0 0 0 0	

4340. C$_{23}$H$_{26}$O$_3$

Phenothrin

(3-Phenoxylphenyl)methyl 2,2-dimethyl-3-(2-methyl-1-propenyl)cyclopropanecarboxylate

Sumithrin

3-Phenoxybenzyl D-*cis* and *trans*-2,2-dimethyl-3-(2-methylpropenyl)cyclopropanecarboxylate

RN: 26002-80-2   **MP** (°C):   <25

MW: 350.46   **BP** (°C):

Solubility (Moles/L)	Solubility (Grams/L)	Temp (°C)	Ref (#)	Evaluation (T P E A A)	Comments
5.707E-06	2.000E-03	30	M161	1 0 0 0 0	

4341. C$_{23}$H$_{27}$ClO$_4$

Delmadinone acetate

Pregna-1,4,6-triene-3,20-dione, 17-(acetyloxy)-6-chloro-

RN: 13698-49-2   **MP** (°C):   168

MW: 402.92   **BP** (°C):

Solubility (Moles/L)	Solubility (Grams/L)	Temp (°C)	Ref (#)	Evaluation (T P E A A)	Comments
1.506E-05	6.070E-03	37	K070	1 0 0 1 2	
1.134E-05	4.570E-03	ns	K070	1 0 0 1 2	

4342. C₂₃H₂₇FN₄O₂
4342. $C_{23}H_{27}FN_4O_2$

Risperidal

3-(2-(4-(6-Fluoro-1,2-benzisoxazol-3-yl)-1-piperidinyl)ethyl)-6,7,8,9-tetrahydro-2-methyl-4H-
  pyrido[1,2-a]pyrimidin-4-one

Risperidone

RN:	106266-06-2	**MP** (°C):			
MW:	410.50	**BP** (°C):			

Solubility (Moles/L)	Solubility (Grams/L)	Temp (°C)	Ref (#)	Evaluation (T P E A A)	Comments
1.090E-04	4.474E-02	25	E406	0 0 0 0 0	
<2.44E-04	<1.00E-01	rt	B435	0 0 0 0 0	

4343. C₂₃H₂₇NO₈
4343. $C_{23}H_{27}NO_8$

Narceine

o-Veratric acid, 6-[[6-[2-(dimethylamino)ethyl]-2-methoxy-3,4-(methylenedioxy)phenyl]acetyl]-
NIH 10760

RN:	131-28-2	**MP** (°C):	138		
MW:	445.47	**BP** (°C):			

Solubility (Moles/L)	Solubility (Grams/L)	Temp (°C)	Ref (#)	Evaluation (T P E A A)	Comments
1.300E-03	5.791E-01	15	K059	2 2 2 0 1	
2.915E-03	1.299E+00	c	D004	0 0 0 0 0	
1.016E-02	4.525E+00	h	D004	0 0 0 0 0	

4344. C₂₃H₂₇N₃O₇
4344. $C_{23}H_{27}N_3O_7$

Minocycline

Dynacin

Minocin

RN:	10118-90-8	**MP** (°C):			
MW:	457.49	**BP** (°C):			

Solubility (Moles/L)	Solubility (Grams/L)	Temp (°C)	Ref (#)	Evaluation (T P E A A)	Comments
1.137E-01	5.200E+01	25	B191	1 0 0 0 1	neutral pH

4345. C₂₃H₂₈ClN₃O₂S
4345. $C_{23}H_{28}ClN_3O_2S$

Thiopropazate

1-(2-Acetoxyethyl)-4-[3-(2-chloro-10-phenothiazinyl)propyl]piperazine

RN:	84-06-0	**MP** (°C):			
MW:	446.02	**BP** (°C):			

Solubility (Moles/L)	Solubility (Grams/L)	Temp (°C)	Ref (#)	Evaluation (T P E A A)	Comments
2.000E-05	8.920E-03	24	G022	2 0 1 1 1	

4346. $C_{23}H_{28}ClN_3O_5S$

Glyburide
HB 419
Glibenclamide
Diabeta
1-((p-(2-(5-Chloro-o-anisamido)ethyl)phenyl)-sulfonyl)-3-cyclohexylurea

RN: 10238-21-8   **MP** (°C):   169
MW: 494.01   **BP** (°C):

Solubility (Moles/L)	Solubility (Grams/L)	Temp (°C)	Ref (#)	Evaluation (T P E A A)	Comments
1.137E-05	5.615E-03	22	M382	2 1 1 1 1	average of 2
6.275E-05	3.100E-02	25	G088	1 1 1 1 0	
1.000E-05	4.940E-03	25	Z410	0 0 0 0 0	EFG
8.097E-06	4.000E-03	27	H093	1 0 1 1 0	
2.024E-05	1.000E-02	ns	K444	0 0 0 0 0	

4347. $C_{23}H_{28}O_7$

Prednisone acetate
Pregna-1,4-diene-3,11,20-trione, 21-(acetyloxy)-17-hydroxy-

RN: 125-10-0   **MP** (°C):
MW: 416.48   **BP** (°C):

Solubility (Moles/L)	Solubility (Grams/L)	Temp (°C)	Ref (#)	Evaluation (T P E A A)	Comments
5.522E-05	2.300E-02	25	K003	2 1 1 1 1	

4348. $C_{23}H_{29}ClFN_3O_4$

Cisapride
4-Amino-5-chloro-N- [1-[3-(4-fluorophenoxy)propyl]-3-methoxy-4-piperidyl]-2-methoxy-
   benzamide

RN: 81098-60-4   **MP** (°C):
MW: 465.96   **BP** (°C):

Solubility (Moles/L)	Solubility (Grams/L)	Temp (°C)	Ref (#)	Evaluation (T P E A A)	Comments
2.000E-05	9.319E-03	30	A417	0 0 0 0 0	pH 8.2
4.000E-04	1.864E-01	30	A417	0 0 0 0 0	pH 3.6

4349. $C_{23}H_{31}Cl_2NO_3$

Estramustine
Estradiol 3-[bis(2-chloroethyl)carbamate]
3-[bis(2-Chloroethyl)carbamate]

RN: 2998-57-4   **MP** (°C):
MW: 440.41   **BP** (°C):

Solubility (Moles/L)	Solubility (Grams/L)	Temp (°C)	Ref (#)	Evaluation (T P E A A)	Comments
~2.27E-06	~1.00E-03	30	L334	1 0 1 1 0	

4350. C₂₃H₃₁FO₆

9α-Fluorohydrocortisone acetate
Pregn-4-ene-3,20-dione, 21-(acetyloxy)-9-fluoro-11,17-dihydroxy-, (11β)-

RN:	514-36-3	MP (°C):
MW:	422.50	BP (°C):

Solubility (Moles/L)	Solubility (Grams/L)	Temp (°C)	Ref (#)	Evaluation (T P E A A)	Comments
1.278E-04	5.400E-02	25	K021	1 2 2 2 1	

4351. C₂₃H₃₁N₅O₄

Benzoic acid, 3-[(dipropylamino)methyl]-, 2-[(6-amino-4,5-dihydro-4-oxo-1H-imidazo[4,5-c]pyridin-1-yl)methoxy]ethyl ester
1H-Imidazo[4,5-c]pyridine, benzoic acid deriv.

RN:	137605-71-1	MP (°C):	
MW:	441.53	BP (°C):	674.6

Solubility (Moles/L)	Solubility (Grams/L)	Temp (°C)	Ref (#)	Evaluation (T P E A A)	Comments
2.944E-04	1.300E-01	21	B419	1 1 2 2 1	int

4352. C₂₃H₃₁O₇

Cortisone-21-hemi-succinate

RN:		MP (°C):
MW:	419.50	BP (°C):

Solubility (Moles/L)	Solubility (Grams/L)	Temp (°C)	Ref (#)	Evaluation (T P F A A)	Comments
4.768E-04	2.000E-01	ns	E307	0 0 0 0 0	

4353. C₂₃H₃₂O₂

Medrogestone
Pregna-4,6-diene-3,20-dione, 6,17-dimethyl-

RN:	977-79-7	MP (°C):	144
MW:	340.51	BP (°C):	

Solubility (Moles/L)	Solubility (Grams/L)	Temp (°C)	Ref (#)	Evaluation (T P E A A)	Comments
5.345E-06	1.820E-03	25	L033	1 0 2 1 2	

4354. C₂₃H₃₂O₄

Deoxycorticosterone acetate
Pregn-4-ene-3,20-dione, 21-(acetyloxy)-

RN:	56-47-3	MP (°C):	156
MW:	372.51	BP (°C):	

Solubility (Moles/L)	Solubility (Grams/L)	Temp (°C)	Ref (#)	Evaluation (T P E A A)	Comments
1.074E-05	4.000E-03	25	K003	2 1 1 1 1	

4355. $C_{23}H_{32}O_6$

Hydrocortisone acetate
Hydrocortisone-21-acetate
Cortisol acetate
Cortisol 21-acetate

RN: 50-03-3   **MP** (°C):   223dec
MW: 404.51   **BP** (°C):

Solubility (Moles/L)	Solubility (Grams/L)	Temp (°C)	Ref (#)	Evaluation (T P E A A)	Comments
3.486E-05	1.410E-02	25	C037	2 1 2 2 2	
1.555E-05	6.290E-03	25	H098	1 0 2 0 2	
1.555E-05	6.290E-03	25	H320	0 0 0 0 0	
1.550E-05	6.270E-03	25	H320	0 0 0 0 0	
2.472E-05	1.000E-02	25	K003	2 1 1 1 1	
3.461E-05	1.400E-02	25	K021	1 2 2 2 1	
2.472E-05	1.000E-02	25	M023	1 0 2 1 0	
2.472E-05	1.000E-02	ns	M169	0 0 0 0 1	
1.904E-05	7.700E-03	ns	N323	0 0 0 0 0	

4356. $C_{23}H_{34}O_3$

Testosterone butyrate
Androst-4-en-3-one, 17-(1-oxobutoxy)-, (17bet)-

RN: 3410-54-6   **MP** (°C):
MW: 358.53   **BP** (°C):

Solubility (Moles/L)	Solubility (Grams/L)	Temp (°C)	Ref (#)	Evaluation (T P E A A)	Comments
1.406E-06	5.039E-04	25	J004	1 0 1 1 2	
1.403E-06	5.030E-04	ns	B057	0 2 1 1 2	

4357. $C_{23}H_{34}O_3$

5,6-Dehydroisoandrosterone butyrate
Androst-5-en-17-one, 3-(1-oxobutoxy)-, (3β)-

RN: 15253-51-7   **MP** (°C):
MW: 358.53   **BP** (°C):

Solubility (Moles/L)	Solubility (Grams/L)	Temp (°C)	Ref (#)	Evaluation (T P E A A)	Comments
1.231E+00	4.413E+02	ns	B057	0 2 1 1 2	

4358. $C_{23}H_{34}O_3$

17-α-Methyltestosterone propionate

RN: **MP** (°C):
MW: 358.53   **BP** (°C):

Solubility (Moles/L)	Solubility (Grams/L)	Temp (°C)	Ref (#)	Evaluation (T P E A A)	Comments
2.845E-06	1.020E-03	ns	B057	0 2 1 1 2	

4359. C$_{23}$H$_{34}$O$_4$
Digitoxigenin
Card-20(22)-enolide, 3,14-dihydroxy-, (3β,5β)-
RN: 143-62-4 **MP** (°C):
MW: 374.53 **BP** (°C):

Solubility (Moles/L)	Solubility (Grams/L)	Temp (°C)	Ref (#)	Evaluation (T P E A A)	Comments
3.000E-05	1.124E-02	30	O321	0 0 0 0 0	

4360. C$_{23}$H$_{35}$NOS
5-Pregnene-20-one-3-spiro-2′-(1′,2′-thiazolidine)
RN: **MP** (°C): 127–136
MW: 373.61 **BP** (°C):

Solubility (Moles/L)	Solubility (Grams/L)	Temp (°C)	Ref (#)	Evaluation (T P E A A)	Comments
~1.34E-05	~5.00E-03	ns	B199	0 0 0 0 0	

4361. C$_{23}$H$_{36}$N$_2$O$_2$
Finasteride
Proscar
RN: 98319-26-7 **MP** (°C):
MW: 372.56 **BP** (°C):

Solubility (Moles/L)	Solubility (Grams/L)	Temp (°C)	Ref (#)	Evaluation (T P E A A)	Comments
1.074E-04	4.000E-02	amb	L434	0 0 0 0 0	

4362. C$_{23}$H$_{36}$O$_3$
Androstanolone butyrate
Androstan-3-one, 17-(1-oxobutoxy)-, (5α,17β)-
RN: 18069-66-4 **MP** (°C):
MW: 360.54 **BP** (°C):

Solubility (Moles/L)	Solubility (Grams/L)	Temp (°C)	Ref (#)	Evaluation (T P E A A)	Comments
1.220E-06	4.400E-04	ns	B057	0 2 1 1 2	

4363. C$_{23}$H$_{38}$O$_3$
Hexadecyl p-hydroxybenzoate
Hexadecyl 4-hydroxybenzoate
RN: 71067-09-9 **MP** (°C):
MW: 362.56 **BP** (°C):

Solubility (Moles/L)	Solubility (Grams/L)	Temp (°C)	Ref (#)	Evaluation (T P E A A)	Comments
1.045E-03	3.789E-01	25	D081	1 2 2 1 2	

4364. $C_{23}H_{40}O_5$
4-Nonylphenol tetraethoxylate
p-Nonylphenol tetraethoxylate

RN: 7311-27-5   **MP** (°C):
MW: 396.57   **BP** (°C):

Solubility (Moles/L)	Solubility (Grams/L)	Temp (°C)	Ref (#)	Evaluation (T P E A A)	Comments
1.929E-05	7.650E-03	20.5	A335	0 0 0 0 0	
1.930E-05	7.654E-03	20.5	A335	0 0 0 0 0	

4365. $C_{24}H_{12}$
Coronene
Coronen

RN: 191-07-1   **MP** (°C): 438
MW: 300.36   **BP** (°C): 525

Solubility (Moles/L)	Solubility (Grams/L)	Temp (°C)	Ref (#)	Evaluation (T P E A A)	Comments
4.680E-09	1.406E-06	20	E009	1 0 0 1 2	
3.329E-10	1.000E-07	25	B319	2 0 1 2 1	
4.661E-10	1.400E-07	25	M064	1 1 2 2 1	
4.660E-10	1.400E-07	25	M342	1 0 1 1 2	

4366. $C_{24}H_{20}N_2$
N,N'-Diphenylbenzidine

RN: 531-91-9   **MP** (°C): 247
MW: 336.44   **BP** (°C):

Solubility (Moles/L)	Solubility (Grams/L)	Temp (°C)	Ref (#)	Evaluation (T P E A A)	Comments
1.783E-07	6.000E-05	50	K068	1 0 2 2 0	buffer
1.783E-07	6.000E-05	rt	K068	0 0 2 2 0	buffer

4367. $C_{24}H_{22}N_2O_2$
G-3
p-Phenylpropylphenylbutazone
3,5-Pyrazolidinedione, 1,2-diphenyl-4-(3-phenylpropyl)-

RN: 32060-78-9   **MP** (°C):
MW: 370.46   **BP** (°C):

Solubility (Moles/L)	Solubility (Grams/L)	Temp (°C)	Ref (#)	Evaluation (T P E A A)	Comments
3.779E-04	1.400E-01	ns	B158	0 0 0 0 1	pH 7.0

4368. $C_{24}H_{26}N_2O_4$
Carvedilol

RN: 72956-09-3   **MP** (°C):
MW: 406.49   **BP** (°C):

Solubility (Moles/L)	Solubility (Grams/L)	Temp (°C)	Ref (#)	Evaluation (T P E A A)	Comments
1.105E-06	4.492E-04	22.5	B440	0 0 0 0 0	
7.380E-05	3.000E-02	ns	S469	0 0 0 0 0	

4369. $C_{24}H_{26}N_4O_2$
Methyldiantipyrylmethane
MDAM

RN: 1606-56-0   **MP** (°C):
MW: 402.50   **BP** (°C):

Solubility (Moles/L)	Solubility (Grams/L)	Temp (°C)	Ref (#)	Evaluation (T P E A A)	Comments
1.118E-03	4.498E-01	20	P054	0 0 0 0 0	

4370. $C_{24}H_{26}N_4S_2$
Methyldithiopyrylmethane
3H-Pyrazole-3-thione, 4,4'-ethylidenebis[1,2-dihydro-1,5-dimethyl-2-phenyl-

RN: 74713-70-5   **MP** (°C):   229
MW: 434.63   **BP** (°C):

Solubility (Moles/L)	Solubility (Grams/L)	Temp (°C)	Ref (#)	Evaluation (T P E A A)	Comments
5.000E-04	2.173E-01	ns	D087	0 2 0 0 1	

4371. $C_{24}H_{27}BrN_6O_{10}$
C.I. Disperse blue 79
2'-Acetylamino-4'-[bis(acetoxyethyl)amino]-6-bromo-2,4-dinitro-5'-ethoxyazobenzene

RN: 12239-34-8   **MP** (°C):   146
MW: 639.43   **BP** (°C):

Solubility (Moles/L)	Solubility (Grams/L)	Temp (°C)	Ref (#)	Evaluation (T P E A A)	Comments
1.000E-09	6.394E-07	25	B333	0 0 0 0 0	

4372. $C_{24}H_{27}N$
Prenylamine
N-(3,3-Diphenylpropyl)-α-methylphenylethylamine

RN: 390-64-7   **MP** (°C):   36.5
MW: 329.49   **BP** (°C):

Solubility (Moles/L)	Solubility (Grams/L)	Temp (°C)	Ref (#)	Evaluation (T P E A A)	Comments
1.517E-04	5.000E-02	37	C054	2 0 2 1 0	

4373. $C_{24}H_{29}N_5O_3$

Valsartan

(2S)-3-Methyl-2-[pentanoyl-[[4-[2-(2H-tetrazol-5-yl)phenyl]phenyl]methyl]amino]butanoic acid

RN: 137862-53-4   **MP** (°C):

MW: 435.53   **BP** (°C):

Solubility (Moles/L)	Solubility (Grams/L)	Temp (°C)	Ref (#)	Evaluation (T P E A A)	Comments
1.951E-04	8.499E-02	25	C431	0 0 0 0 0	

4374. $C_{24}H_{30}F_2O_6$

Fluocinolone acetonide

6α,9α-Difluoro-16α hydroxyprednisolone-16,17-acetonide

6α,9α-Difluoro-16α,17α-isopropylidenedioxy-1,4-pregnadiene-3,20-dione

RN: 67-73-2   **MP** (°C):   260.5

MW: 452.50   **BP** (°C):

Solubility (Moles/L)	Solubility (Grams/L)	Temp (°C)	Ref (#)	Evaluation (T P E A A)	Comments
2.387E-04	1.080E-01	25	K021	1 2 2 2 1	
4.641E-05	2.100E-02	25	O001	2 0 2 2 2	
2.210E-04	1.000E-01	25	P008	0 0 0 0 0	EFG

4375. $C_{24}H_{31}ClO_7$

Loteprednol etabonate

Lenoxin

Androsta-1,4-diene-17-carboxylic acid

17-[(Ethoxycarbonyl)oxy]-11-hydroxy-3-oxo-chloromethyl ester, (11b,17a)-

RN: 82034-46-6   **MP** (°C):

MW: 466.96   **BP** (°C):   600.1

Solubility (Moles/L)	Solubility (Grams/L)	Temp (°C)	Ref (#)	Evaluation (T P E A A)	Comments
<2.14E-06	<1.00E-03	23	B409	1 0 0 0 1	

4376. $C_{24}H_{31}FO_5S$

Timobesone acetate

17-β-Methythiocarbonyl-9α-fluoro-11β

RN: 79578-14-6   **MP** (°C):

MW: 450.57   **BP** (°C):

Solubility (Moles/L)	Solubility (Grams/L)	Temp (°C)	Ref (#)	Evaluation (T P E A A)	Comments
6.000E-03	2.703E+00	25	O318	0 0 0 0 0	

4377. $C_{24}H_{31}FO_6$

Triamcinolone acetonide
9α-Fluoro-16α-hydroxyprednisolone acetonide
Triamcinolone 16α,17-acetonide
Aristoderm
Adcortyl-A

RN:	76-25-5	**MP** (°C):	293	
MW:	434.51	**BP** (°C):		

Solubility (Moles/L)	Solubility (Grams/L)	Temp (°C)	Ref (#)	Evaluation (T P E A A)	Comments
9.205E-05	4.000E-02	23	F025	1 0 0 0 0	
9.436E-05	4.100E-02	25	K021	1 2 2 2 1	
6.076E-04	2.640E-01	25	L009	1 0 0 1 1	
4.833E-05	2.100E-02	28	B055	2 0 2 2 2	
4.027E-05	1.750E-02	28	B056	1 2 1 1 2	
5.869E 05	2.550E-02	37	B055	2 0 2 2 2	
4.764E-05	2.070E-02	37	B056	1 2 1 1 2	
9.205E-05	4.000E-02	37	F025	1 0 0 0 0	
7.733E-05	3.360E-02	50	B055	2 0 2 2 2	
6.099E-05	2.650E-02	50	B056	1 2 1 1 2	
2.532E-04	1.100E-01	amb	L434	0 0 0 0 0	

4378. $C_{24}H_{31}FO_6$

Betamethasone acetate
Betamethasone-17-acetate
9α-Fluoro-16β-methylprednisolone-21-acetate

RN:	987-24-6	**MP** (°C):	200dec	
MW:	434.51	**BP** (°C):		

Solubility (Moles/L)	Solubility (Grams/L)	Temp (°C)	Ref (#)	Evaluation (T P E A A)	Comments
6.904E-05	3.000E-02	25	K003	2 1 1 1 1	

4379. $C_{24}H_{31}FO_6$

Dexamethasone acetate
Dexamethasone-17-acetate
Dexamethasone acetate

RN:	1177-87-3	**MP** (°C):	263	
MW:	434.51	**BP** (°C):		

Solubility (Moles/L)	Solubility (Grams/L)	Temp (°C)	Ref (#)	Evaluation (T P E A A)	Comments
2.992E-05	1.300E-02	25	K003	2 1 1 1 1	
6.214E-05	2.700E-02	37	D026	0 0 0 0 0	

4380. C$_{24}$H$_{31}$NO$_4$
Drotaverine
1-(3,4-Diethoxybenzylidene)-6,7-diethoxy-1,2,3,4-tetrahydroisoquinoline
RN: 14009-24-6   **MP** (°C):
MW: 397.52   **BP** (°C):

Solubility (Moles/L)	Solubility (Grams/L)	Temp (°C)	Ref (#)	Evaluation (T P E A A)	Comments
3.459E-02	1.375E+01	37	C054	2 0 2 1 2	

4381. C$_{24}$H$_{32}$N$_2$O$_9$
Enalapril maleate
L-Proline, 1-[N-[1-(ethoxycarbonyl)-3-phenylpropyl]-L-alanyl]-,
(S)-1-(N-(1-(Ethoxycarbonyl)-3-phenylpropyl)-L-alanyl)-L-proline, (Z)-2-butenedioate salt
RN: 76095-16-4   **MP** (°C):
MW: 492.53   **BP** (°C):

Solubility (Moles/L)	Solubility (Grams/L)	Temp (°C)	Ref (#)	Evaluation (T P E A A)	Comments
4.264E-02	2.100E+01	25	A412	1 0 2 2 1	int

4382. C$_{24}$H$_{32}$O$_4$
Ethynodiol diacetate
Ovulen-50
RN: 297-76-7   **MP** (°C):   126
MW: 384.52   **BP** (°C):

Solubility (Moles/L)	Solubility (Grams/L)	Temp (°C)	Ref (#)	Evaluation (T P E A A)	Comments
3.641E-06	1.400E-03	25	L027	1 0 0 0 2	

4383. C$_{24}$H$_{32}$O$_4$S
Spironolactone
17-Hydroxy-7α-mercapto-3-oxo-17α-pregn-4-ene-21-carboxylic acid γ-lactone acetate
Spiractin
RN: 52-01-7   **MP** (°C):   134
MW: 416.58   **BP** (°C):

Solubility (Moles/L)	Solubility (Grams/L)	Temp (°C)	Ref (#)	Evaluation (T P E A A)	Comments
7.200E-06	2.999E-03	25	A348	0 0 0 0 0	
5.281E-05	2.200E-02	25	C037	2 1 2 2 2	
5.281E-05	2.200E-02	25	G084	2 0 2 2 1	
4.801E-05	2.000E-02	25	G095	2 1 2 2 1	
6.649E-05	2.770E-02	37	K092	2 0 0 1 2	
2.400E-05	1.000E-02	ns	K444	0 0 0 0 0	

4384. C$_{24}$H$_{32}$O$_5$
7-Carboxylic acid methyl ester canrenone
RN: **MP** (°C):
MW: 400.52   **BP** (°C):

Solubility (Moles/L)	Solubility (Grams/L)	Temp (°C)	Ref (#)	Evaluation (T P E A A)	Comments
1.960E-04	7.850E-02	37	C004	0 0 0 0 0	EFG

4385. C$_{24}$H$_{32}$O$_6$
Cortisone 17-propionate
Pregn-4-ene-3,11,20-trione, 21-hydroxy-17-(1-oxopropoxy)-
RN: 136370-32-6   **MP** (°C):
MW: 416.52   **BP** (°C):

Solubility (Moles/L)	Solubility (Grams/L)	Temp (°C)	Ref (#)	Evaluation (T P E A A)	Comments
1.921E-05	8.000E-03	25	M023	1 0 2 1 0	

4386. C$_{24}$H$_{33}$FO$_6$
Flurandrenolone
Fludroxycortide
6-Fluoro-16α-hydroxyhydrocortisone-16,17-acetonide
RN: 1524-88-5   **MP** (°C):
MW: 436.53   **BP** (°C):

Solubility (Moles/l)	Solubility (Grams/L)	Temp (°C)	Ref (#)	Evaluation (T P E A A)	Comments
6.758E-04	2.950E-01	25	K021	1 2 2 2 1	

4387. C$_{24}$H$_{34}$N$_2$O
Bepridil
1-Isobutoxy-2-pyrrolidino-3-N-benzylanilino-propane
Bepadin
RN: 64706-54-3   **MP** (°C):
MW: 366.55   **BP** (°C):

Solubility (Moles/L)	Solubility (Grams/L)	Temp (°C)	Ref (#)	Evaluation (T P E A A)	Comments
2.027E-02	7.430E+00	37	N032	1 0 1 1 2	

4388. C$_{24}$H$_{34}$N$_2$O$_3$
Lysine estrone ester
RN: **MP** (°C):
MW: 398.55   **BP** (°C):

Solubility (Moles/L)	Solubility (Grams/L)	Temp (°C)	Ref (#)	Evaluation (T P E A A)	Comments
3.162E-01	1.260E+02	ns	A074	0 0 0 0 0	EFG

4389. $C_{24}H_{34}O_5$
Dehydrocholic acid
3,7,12-Trioxo-5β-cholanic acid

RN: 81-23-2 **MP** (°C): 237
MW: 402.54 **BP** (°C):

Solubility (Moles/L)	Solubility (Grams/L)	Temp (°C)	Ref (#)	Evaluation (T P E A A)	Comments
4.472E-04	1.800E-01	15	G081	1 0 1 1 1	
1.615E-04	6.500E-02	30	O321	0 0 0 0 0	
1.600E-04	6.441E-02	30	O321	0 0 0 0 0	

4390. $C_{24}H_{34}O_6$
Hydrocortisone propionate
Hydrocortisone-21-propionate

RN: 6677-98-1 **MP** (°C):
MW: 418.53 **BP** (°C):

Solubility (Moles/L)	Solubility (Grams/L)	Temp (°C)	Ref (#)	Evaluation (T P E A A)	Comments
2.772E-05	1.160E-02	25	H098	1 0 2 0 2	
2.772E-05	1.160E-02	25	H320	0 0 0 0 0	
2.770E-05	1.159E-02	25	H320	0 0 0 0 0	

4391. $C_{24}H_{36}O_3$
Testosterone valerate
Androst-4-en-3-one, 17-[(1-oxopentyl)oxy]-, (17β)-
Testosterone 17-valerate

RN: 3129-43-9 **MP** (°C):
MW: 372.55 **BP** (°C):

Solubility (Moles/L)	Solubility (Grams/L)	Temp (°C)	Ref (#)	Evaluation (T P E A A)	Comments
7.778E-07	2.898E-04	25	J004	1 0 1 1 2	
7.811E-07	2.910E-04	ns	B057	0 2 1 1 2	

4392. $C_{24}H_{36}O_3$
5,6-Dehydroisoandrosterone valerate
Androst-5-en-17-one, 3-[(1-oxopentyl)oxy]-, (3β)-

RN: 7642-68-4 **MP** (°C):
MW: 372.55 **BP** (°C):

Solubility (Moles/L)	Solubility (Grams/L)	Temp (°C)	Ref (#)	Evaluation (T P E A A)	Comments
2.061E-05	7.680E-03	ns	B057	0 2 1 1 2	

4393. $C_{24}H_{38}O_3$
Androstanolone valerate
Androstan-3-one, 17-[(1-oxopentyl)oxy]-, (5α,17β)-
RN: 26271-72-7 **MP** (°C):
MW: 374.57 **BP** (°C):

Solubility (Moles/L)	Solubility (Grams/L)	Temp (°C)	Ref (#)	Evaluation (T P E A A)	Comments
8.143E-07	3.050E-04	ns	B057	0 2 1 1 2	

4394. $C_{24}H_{38}O_4$
Di-2-ethylhexyl isophthalate
D-(2-Ethylhexyl) isophthalate
Dioctyl isophthalate
RN: 137-89-3 **MP** (°C):
MW: 390.57 **BP** (°C): 400

Solubility (Moles/L)	Solubility (Grams/L)	Temp (°C)	Ref (#)	Evaluation (T P E A A)	Comments
2.816E-08	1.100E-05	24	H116	2 1 0 0 2	

4395. $C_{24}H_{38}O_4$
Octyl phthalate
Di(2-ethylhexyl)phthalate
Di-(2-ethylhexyl)-phthalate
Di-*sec*-octyl phthalate
bis(2-Ethylhexyl) phthalate
bis-(2-Ethylhexyl) 1,2-benzenedicarboxylate
RN: 117-81-7 **MP** (°C): −50
MW: 390.57 **BP** (°C): 386.9

Solubility (Moles/L)	Solubility (Grams/L)	Temp (°C)	Ref (#)	Evaluation (T P E A A)	Comments
2.560E-04	9.999E-02	20	F070	1 0 0 0 1	*sic*
1.050E-07	4.101E-05	20	L300	2 1 0 2 2	
1.536E-06	6.000E-04	22.5	G301	0 0 0 0 0	
7.297E-07	2.850E-04	24	H116	2 1 0 0 2	
6.913E-07	2.700E-04	25	D336	0 0 0 0 0	
1.280E-06	5.000E-04	25	F067	1 0 2 2 0	

4396. $C_{24}H_{38}O_4$
Apocholic acid
RN: 641-81-6 **MP** (°C): 175.5
MW: 390.57 **BP** (°C):

Solubility (Moles/L)	Solubility (Grams/L)	Temp (°C)	Ref (#)	Evaluation (T P E A A)	Comments
2.048E-03	8.000E-01	15	G081	1 0 1 1 0	

4397. C$_{24}$H$_{38}$O$_4$
bis(Tereoctyl) phthalate

RN:		**MP** (°C):	
MW:	390.57	**BP** (°C):	

Solubility (Moles/L)	Solubility (Grams/L)	Temp (°C)	Ref (#)	Evaluation (T P E A A)	Comments
5.633E-08	2.200E-05	25	D336	0 0 0 0 0	

4398. C$_{24}$H$_{38}$O$_4$
bis(Isooctyl) phthalate
Diisooctyl phthalate
1,2-Benzenedicarboxylic acid diisooctyl ester

RN:	27554-26-3	**MP** (°C):	-4
MW:	390.57	**BP** (°C):	239

Solubility (Moles/L)	Solubility (Grams/L)	Temp (°C)	Ref (#)	Evaluation (T P E A A)	Comments
1.024E-07	4.000E-05	25	D336	0 0 0 0 0	

4399. C$_{24}$H$_{38}$O$_4$
bis(n-Octyl) phthalate
Di-n-octyl phthalate
1,2-Benzenedicarboxylic acid

RN:	117-84-0	**MP** (°C):	−25
MW:	390.57	**BP** (°C):	220

Solubility (Moles/L)	Solubility (Grams/L)	Temp (°C)	Ref (#)	Evaluation (T P E A A)	Comments
5.121E-08	2.000E-05	25	D336	0 0 0 0 0	

4400. C$_{24}$H$_{39}$NO$_3$
Acetaminophen palmitate
Acetaminophen hexadecanoate

RN:	54942-40-4	**MP** (°C):	117
MW:	389.58	**BP** (°C):	

Solubility (Moles/L)	Solubility (Grams/L)	Temp (°C)	Ref (#)	Evaluation (T P E A A)	Comments
1.283E-05	5.000E-03	25	B010	1 1 1 1 0	

4401. C$_{24}$H$_{40}$N$_8$O$_4$
Dypyridamole
2,6-bis(Diethanolamino)-4,8-dipiperidinopyrimido-[5,4-d]pyrimidin
Dipridacot
Dipryridamole
Persantin
Dipyridamol

RN: 58-32-2   **MP (°C):**
MW: 504.64   **BP (°C):**

Solubility (Moles/L)	Solubility (Grams/L)	Temp (°C)	Ref (#)	Evaluation (T P E A A)	Comments
1.649E-06	8.320E-04	22.5	B440	0 0 0 0 0	
7.000E-05	3.532E-02	30	A417	0 0 0 0 0	pH 5.2
3.200E-03	1.615E+00	30	A417	0 0 0 0 0	pH 3.7

4402. C$_{24}$H$_{40}$O$_3$
3β-Hydroxy-5β-cholanoic acid
7α-Hydroxy-5β-cholanoic acid

RN: **MP (°C):**
MW: 376.58   **BP (°C):**

Solubility (Moles/L)	Solubility (Grams/L)	Temp (°C)	Ref (#)	Evaluation (T P E A A)	Comments
1.800E-07	6.779E-05	10	F307	1 2 2 2 2	pH 3.0
4.400E-07	1.657E-04	10	F307	1 2 2 2 2	pH 3.0
5.200E-07	1.958E-04	15	F307	1 2 2 2 2	pH 3.0
2.200E-07	8.285E-05	15	F307	1 2 2 2 2	pH 3.0
2.400E-07	9.038E-05	20	F307	1 2 2 2 2	pH 3.0
6.500E-07	2.448E-04	20	F307	1 2 2 2 2	pH 3.0
2.800E-07	1.054E-04	25	F307	1 2 2 2 2	pH 3.0
7.900E-07	2.975E-04	25	F307	1 2 2 2 2	pH 3.0
3.500E-07	1.318E-04	30	F307	1 2 2 2 2	pH 3.0
9.700E-07	3.653E-04	30	F307	1 2 2 2 2	pH 3.0
5.300E-07	1.996E-04	35	F307	1 2 2 2 2	pH 3.0
1.190E-06	4.481E-04	35	F307	1 2 2 2 2	pH 3.0
8.200E-07	3.088E-04	40	F307	1 2 2 2 2	pH 3.0
1.490E-06	5.611E-04	40	F307	1 2 2 2 2	pH 3.0
1.770E-06	6.666E-04	45	F307	1 2 2 2 2	pH 3.0
1.280E-06	4.820E-04	45	F307	1 2 2 2 2	pH 3.0
1.500E-06	5.649E-04	50	F307	1 2 2 2 2	pH 3.0
2.150E-06	8.097E-04	50	F307	1 2 2 2 2	pH 3.0

4403. $C_{24}H_{40}O_3$

Lithocholic acid
3α-Hydroxy-5β-cholan-24-oic acid
3α-Hydroxycholanic acid

RN: 434-13-9   **MP** (°C): 184
MW: 376.58   **BP** (°C):

Solubility (Moles/L)	Solubility (Grams/L)	Temp (°C)	Ref (#)	Evaluation (T P E A A)	Comments
3.800E-08	1.431E-05	10	F307	1 2 2 2 2	pH 3.0
4.000E-08	1.506E-05	15	F307	1 2 2 2 2	pH 3.0
4.600E-08	1.732E-05	20	F307	1 2 2 2 2	pH 3.0
1.000E-06	3.766E-04	20	I012	1 2 2 1 0	pH 2.4
5.000E-08	1.883E-05	25	F307	1 2 2 2 2	pH 3.0
6.000E-08	2.260E-05	30	F307	1 2 2 2 2	pH 3.0
7.500E-08	2.824E-05	35	F307	1 2 2 2 2	pH 3.0
1.000E-06	3.766E-04	37	I012	1 2 2 1 0	pH 2.4
1.000E-07	3.766E-05	40	F307	1 2 2 2 2	pH 3.0
1.100E-07	4.142E-05	45	F307	1 2 2 2 2	pH 3.0
1.400E-07	5.272E-05	50	F307	1 2 2 2 2	pH 3.0

4404. $C_{24}H_{40}O_4$

Hyodeoxycholic acid
3α,6β-Dihydroxy-5α-cholanoic acid

RN: 83-49-8   **MP** (°C): 198
MW: 392.58   **BP** (°C):

Solubility (Moles/L)	Solubility (Grams/L)	Temp (°C)	Ref (#)	Evaluation (T P E A A)	Comments
1.000E-05	3.926E-03	10	F307	1 2 2 2 2	pH 3.0
1.200E-05	4.711E-03	15	F307	1 2 2 2 2	pH 3.0
1.300E-05	5.104E-03	20	F307	1 2 2 2 2	pH 3.0
1.500E-05	5.889E-03	25	F307	1 2 2 2 2	pH 3.0
1.700E-05	6.674E-03	30	F307	1 2 2 2 2	pH 3.0
1.800E-05	7.067E-03	35	F307	1 2 2 2 2	pH 3.0
2.000E-05	7.852E-03	40	F307	1 2 2 2 2	pH 3.0
2.200E-05	8.637E-03	45	F307	1 2 2 2 2	pH 3.0
2.600E-05	1.021E-02	50	F307	1 2 2 2 2	pH 3.0

4405. $C_{24}H_{40}O_4$

Deoxycholic acid
Cholan-24-oic acid, 3,12-dihydroxy-, (3α,5β,12α)-
3α,12α-Dihydroxy-5β-cholanoic acid

RN: 83-44-3   **MP** (°C): 176
MW: 392.58   **BP** (°C):

Solubility (Moles/L)	Solubility (Grams/L)	Temp (°C)	Ref (#)	Evaluation (T P E A A)	Comments
2.400E-05	9.422E-03	10	F307	1 2 2 2 2	pH 3.0
2.600E-05	1.021E-02	15	F307	1 2 2 2 2	pH 3.0
6.113E-04	2.400E-01	15	G081	1 0 1 1 1	

(continued)

4405. $C_{24}H_{40}O_4$ (continued)

Solubility (Moles/L)	Solubility (Grams/L)	Temp (°C)	Ref (#)	Evaluation (T P E A A)	Comments
5.093E-04	2.000E-01	20	D041	1 0 0 0 0	
2.700E-05	1.060E-02	20	F307	1 2 2 2 2	pH 3.0
1.110E-04	4.358E-02	20	I012	1 2 2 1 2	pH 2.4
2.800E-05	1.099E-02	25	F307	1 2 2 2 2	pH 3.0
2.800E-05	1.099E-02	30	F307	1 2 2 2 2	pH 3.0
2.900E-05	1.138E-02	35	F307	1 2 2 2 2	pH 3.0
1.140E-04	4.475E-02	37	I012	1 2 2 1 2	pH 2.4
2.900E-05	1.138E-02	40	F307	1 2 2 2 2	pH 3.0
3.000E-05	1.178E-02	45	F307	1 2 2 2 2	pH 3.0
3.200E-05	1.256E-02	50	F307	1 2 2 2 2	pH 3.0

4406. $C_{24}H_{40}O_4$
Chenodeoxycholic acid
CDCA

RN: 474-25-9   **MP (°C):** 119
MW: 392.58   **BP (°C):**

Solubility (Moles/L)	Solubility (Grams/L)	Temp (°C)	Ref (#)	Evaluation (T P E A A)	Comments
2.500E-05	9.815E-03	10	F307	1 2 2 2 2	pH 3.0
2.500E-05	9.815E-03	15	F307	1 2 2 2 2	pH 3.0
2.600E-05	1.021E-02	20	F307	1 2 2 2 2	pH 3.0
2.290E-04	8.990E-02	20	I012	1 2 2 1 2	pH 2.4
2.700E-05	1.060E-02	25	F307	1 2 2 2 2	pH 3.0
2.800E-05	1.099E-02	30	F307	1 2 2 2 2	pH 3.0
3.000E-05	1.178E-02	35	F307	1 2 2 2 2	pH 3.0
2.560E-04	1.005E-01	37	I008	1 0 0 1 2	
2.560E-04	1.005E-01	37	I012	1 2 2 1 2	pH 2.4
3.150E-05	1.237E-02	40	F307	1 2 2 2 2	pH 3.0
3.400E-05	1.335E-02	45	F307	1 2 2 2 2	pH 3.0
3.600E-05	1.413E-02	50	F307	1 2 2 2 2	pH 3.0
2.291E-04	8.994E-02	ns	R427	0 0 0 0 0	

4407. $C_{24}H_{40}O_4$
Ursodeoxycholic acid
UDCA

RN: 128-13-2   **MP (°C):** 203
MW: 392.58   **BP (°C):**

Solubility (Moles/L)	Solubility (Grams/L)	Temp (°C)	Ref (#)	Evaluation (T P E A A)	Comments
7.000E-06	2.748E-03	10	F307	1 2 2 2 2	pH 3.0
7.500E-06	2.944E-03	15	F307	1 2 2 2 2	pH 3.0
8.000E-06	3.141E-03	20	F307	1 2 2 2 2	pH 3.0
5.100E-05	2.002E-02	20	I012	1 2 2 1 1	pH 2.4
9.000E-06	3.533E-03	25	F307	1 2 2 2 2	pH 3.0
1.000E-05	3.926E-03	30	F307	1 2 2 2 2	pH 3.0

(continued)

4407. C$_{24}$H$_{40}$O$_4$ (continued)

Solubility (Moles/L)	Solubility (Grams/L)	Temp (°C)	Ref (#)	Evaluation (T P E A A)	Comments
1.150E-05	4.515E-03	35	F307	1 2 2 2 2	pH 3.0
5.300E-05	2.081E-02	37	I008	1 0 0 1 1	
5.300E-05	2.081E-02	37	I012	1 2 2 1 1	pH 2.4
1.200E-05	4.711E-03	40	F307	1 2 2 2 2	pH 3.0
1.300E-05	5.104E-03	45	F307	1 2 2 2 2	pH 3.0
1.400E-05	5.496E-03	50	F307	1 2 2 2 2	pH 3.0
8.556E-04	3.359E-01	ns	K446	0 0 0 0 0	

4408. C$_{24}$H$_{40}$O$_5$
3α, 6α, 7α -Trihydroxy-5β-cholanate
RN: **MP** (°C):
MW: 408.58   **BP** (°C):

Solubility (Moles/L)	Solubility (Grams/L)	Temp (°C)	Ref (#)	Evaluation (T P E A A)	Comments
3.700E-05	1.512E-02	10	F307	1 2 2 2 2	pH 3.0
3.800E-05	1.553E-02	15	F307	1 2 2 2 2	pH 3.0
4.100E-05	1.675E-02	20	F307	1 2 2 2 2	pH 3.0
4.500E-05	1.839E-02	25	F307	1 2 2 2 2	pH 3.0
5.500E-05	2.247E-02	30	F307	1 2 2 2 2	pH 3.0
6.900E-05	2.819E-02	35	F307	1 2 2 2 2	pH 3.0
8.600E-05	3.514E-02	40	F307	1 2 2 2 2	pH 3.0
1.160E-04	4.740E-02	45	F307	1 2 2 2 2	pH 3.0
1.600E-04	6.537E-02	50	F307	1 2 2 2 2	pH 3.0

4409. C$_{24}$H$_{40}$O$_5$
Ursocholic acid
3α,7β,12α-Trihydroxy-5β-cholanoic acid
RN: 2955-27-3   **MP** (°C):
MW: 408.58   **BP** (°C):

Solubility (Moles/L)	Solubility (Grams/L)	Temp (°C)	Ref (#)	Evaluation (T P E A A)	Comments
1.590E-03	6.496E-01	10	F307	1 2 2 2 2	pH 3.0
1.610E-03	6.578E-01	15	F307	1 2 2 2 2	pH 3.0
1.640E-03	6.701E-01	20	F307	1 2 2 2 2	pH 3.0
1.670E-03	6.823E-01	25	F307	1 2 2 2 2	pH 3.0
1.710E-03	6.987E-01	30	F307	1 2 2 2 2	pH 3.0
1.762E-03	7.199E-01	35	F307	1 2 2 2 2	pH 3.0
1.828E-03	7.469E-01	40	F307	1 2 2 2 2	pH 3.0
1.872E-03	7.649E-01	45	F307	1 2 2 2 2	pH 3.0
2.000E-03	8.172E-01	50	F307	1 2 2 2 2	pH 3.0

4410. $C_{24}H_{40}O_5$
Cholic acid
Cholsaeure

RN: 81-25-4   **MP** (°C): 198
MW: 408.58   **BP** (°C):

Solubility (Moles/L)	Solubility (Grams/L)	Temp (°C)	Ref (#)	Evaluation (T P E A A)	Comments
2.210E-04	9.030E-02	10	F307	1 2 2 2 2	pH 3.0
6.486E-04	2.650E-01	15	F300	1 0 0 0 0	
2.140E-04	8.744E-02	15	F307	1 2 2 2 2	pH 3.0
6.853E-04	2.800E-01	15	G081	1 0 1 1 1	
6.851E-04	2.799E-01	20	D041	1 0 0 0 1	
2.247E-04	9.180E-02	20	E008	1 0 2 0 2	average of 3
2.200E-04	8.989E-02	20	F307	1 2 2 2 2	pH 3.0
4.280E-04	1.749E-01	20	I012	1 2 2 1 2	pH 2.4
2.350E-04	9.602E-02	25	F307	1 2 2 2 2	pH 3.0
2.670E-04	1.091E-01	30	F307	1 2 2 2 2	pH 3.0
3.240E-04	1.324E-01	35	F307	1 2 2 2 2	pH 3.0
4.600E-04	1.879E-01	37	I012	1 2 2 1 2	pH 2.4
3.830E-04	1.565E-01	40	F307	1 2 2 2 2	pH 3.0
4.830E-04	1.973E-01	45	F307	1 2 2 2 2	pH 3.0
6.390E-04	2.611E-01	50	F307	1 2 2 2 2	pH 3.0

4411. $C_{24}H_{50}$
Tetracosane
n-Tetracosane
Alkane C(24)

RN: 646-31-1   **MP** (°C): 54
MW: 338.67   **BP** (°C): 391.3

Solubility (Moles/L)	Solubility (Grams/L)	Temp (°C)	Ref (#)	Evaluation (T P E A A)	Comments
1.264E-02	4.282E+00	321	S355	1 1 1 2 0	EFG
8.878E-02	3.007E+01	369	S355	1 1 1 2 0	EFG

4412. $C_{24}H_{51}OP$
tri-*n*-Octylphosphine oxide
TOPO
Trioctylphosphine oxide

RN: 78-50-2   **MP** (°C):
MW: 386.65   **BP** (°C):

Solubility (Moles/L)	Solubility (Grams/L)	Temp (°C)	Ref (#)	Evaluation (T P E A A)	Comments
7.242E-06	2.800E-03	0	O002	2 0 2 2 1	
3.880E-06	1.500E-03	25	O002	2 0 2 2 1	

4413. $C_{24}H_{51}O_3P$
Dibutyl hexadecyl phosphonate
Phosphonic acid, hexadecyl-, dibutyl ester
RN: 84869-93-2   **MP** (°C):
MW: 418.65   **BP** (°C):

Solubility (Moles/L)	Solubility (Grams/L)	Temp (°C)	Ref (#)	Evaluation (T P E A A)	Comments
<4.78E-04	<2.00E-01	25	B070	1 2 0 1 0	

4414. $C_{24}H_{51}O_4P$
tris-(2-Ethylhexyl) phosphate
Disflamoll TOF
TEHP
Flexol TOF
RN: 78-42-2   **MP** (°C):
MW: 434.65   **BP** (°C):

Solubility (Moles/L)	Solubility (Grams/L)	Temp (°C)	Ref (#)	Evaluation (T P E A A)	Comments
1.380E-06	6.000E-04	24	H116	2 1 0 0 2	

4415. $C_{24}H_{54}OSn_2$
bis(Tributyltin) oxide
6-Oxa-5,7-distannaundecane, 5,5,7,7-tetrabutyl-
RN: 56-35-9   **MP** (°C):
MW: 596.08   **BP** (°C):   180

Solubility (Moles/L)	Solubility (Grams/L)	Temp (°C)	Ref (#)	Evaluation (T P E A A)	Comments
1.678E-04	1.000E-01	rt	M161	0 0 0 0 2	

4416. $C_{25}H_{22}O_{10}$
Silybin
Silibinin
Silybum substance E6
Silymarin I
RN: 22888-70-6   **MP** (°C):
MW: 482.45   **BP** (°C):

Solubility (Moles/L)	Solubility (Grams/L)	Temp (°C)	Ref (#)	Evaluation (T P E A A)	Comments
8.788E-05	4.240E-02	19.99	B439	0 0 0 0 0	
1.119E-04	5.400E-02	24.99	B439	0 0 0 0 0	
1.432E-04	6.910E-02	29.99	B439	0 0 0 0 0	
1.726E-04	8.329E-02	34.99	B439	0 0 0 0 0	
2.066E-04	9.969E-02	39.99	B439	0 0 0 0 0	

4417. C$_{25}$H$_{24}$F$_6$N$_4$

Hydramethylnon
Amdro
Comat
Amidinohydrazone;
Wipeout
Tetrahydro-5,5-dimethyl-2(1H)-pyrimidinone[3-[4-(trifluoromethyl)phenyl]-1-[2-[4-
   (trifluoromethyl)phenyl]ethenyl]-2-propenylidene]hydrazone

RN: 67485-29-4   **MP** (°C): 185–190
MW: 494.49   **BP** (°C):

Solubility (Moles/L)	Solubility (Grams/L)	Temp (°C)	Ref (#)	Evaluation (T P E A A)	Comments
1.202E-08	5.945E-06	ns	R427	0 0 0 0 0	

4418. C$_{25}$H$_{24}$N$_2$O$_2$S

G-8
o,p-Dimethylphenylthioethylphenylbutazone
3,5-Pyrazolidinedione, 1,2-diphenyl-4-[2-(2,4-xylylthio)ethyl]-

RN: 102892-46-6   **MP** (°C):
MW: 416.55   **BP** (°C):

Solubility (Moles/L)	Solubility (Grams/L)	Temp (°C)	Ref (#)	Evaluation (T P E A A)	Comments
3.121E-04	1.300E-01	ns	B158	0 0 0 0 1	pH 7.0

4419. C$_{25}$H$_{28}$N$_4$O$_2$

Ethyldiantipyrylmethane
EDAM

RN: 61358-28-9   **MP** (°C):
MW: 416.53   **BP** (°C):

Solubility (Moles/L)	Solubility (Grams/L)	Temp (°C)	Ref (#)	Evaluation (T P E A A)	Comments
3.601E-04	1.500E-01	20	P054	0 0 0 0 0	

4420. C$_{25}$H$_{28}$O$_3$

Estradiol benzoate
Estradiol monobenzoate
7β-Estradiol-3-benzoate

RN: 50-50-0   **MP** (°C): 190
MW: 376.50   **BP** (°C):

Solubility (Moles/L)	Solubility (Grams/L)	Temp (°C)	Ref (#)	Evaluation (T P E A A)	Comments
1.062E-06	4.000E-04	25	K003	2 1 1 1 1	
1.072E-06	4.034E-04	ns	R427	0 0 0 0 0	

4421. C₂₅H₂₈O₃

Ethofenprox
1-((2-(4-Ethoxyphenyl)-2-methylpropoxy)methyl)-3-phenoxybenzene
Etofenprox
Zoecon
MTI-500
Trebon
RN: 80844-07-1 **MP** (°C):
MW: 376.50 **BP** (°C):

Solubility (Moles/L)	Solubility (Grams/L)	Temp (°C)	Ref (#)	Evaluation (T P E A A)	Comments
2.512E-09	9.457E-07	ns	R427	0 0 0 0 0	

4422. C₂₅H₂₉I₂NO₃

Amiodarone
Cordarone
Aratac
RN: 1951-25-3 **MP** (°C):
MW: 645.32 **BP** (°C):

Solubility (Moles/L)	Solubility (Grams/L)	Temp (°C)	Ref (#)	Evaluation (T P E A A)	Comments
<6.72E-08	<4.34E-05	22.5	B440	0 0 0 0 0	
1.110E-03	7.164E-01	25	B337	2 2 2 1 2	

4423. C₂₅H₃₁FO₈

Triamcinolone 16, 21-diacetate
Pregna-1,4-diene-3,20-dione, 16,21-bis(acetyloxy)-9-fluoro-11,17-dihydroxy-, (11β,16apha)-
RN: 67-78-7 **MP** (°C): 235
MW: 478.52 **BP** (°C):

Solubility (Moles/L)	Solubility (Grams/L)	Temp (°C)	Ref (#)	Evaluation (T P E A A)	Comments
1.003E-04	4.800E-02	25	F026	0 0 0 0 0	

4424. C₂₅H₃₁NO₂

3-Hydroxy-17β-{[(1-methyl-1,4-dihydropyridin-3-yl)-carbonyl]oxy}-estra-1,3,5(10)-triene
RN: **MP** (°C):
MW: 377.53 **BP** (°C):

Solubility (Moles/L)	Solubility (Grams/L)	Temp (°C)	Ref (#)	Evaluation (T P E A A)	Comments
1.743E-07	6.580E-05	25	B366	0 0 0 0 0	

4425. C₂₅H₃₄O₃

Norethindrone dimethylpropionate
19-Norpregn-4-en-20-yn-3-one, 17-(2,2-dimethyl-1-oxopropoxy)-, (17α)-
RN: 65445-09-2   **MP** (°C):
MW: 382.55   **BP** (°C):

Solubility (Moles/L)	Solubility (Grams/L)	Temp (°C)	Ref (#)	Evaluation (T P E A A)	Comments
7.894E-08	3.020E-05	25	L078	1 0 1 2 2	

4426. C₂₅H₃₄O₆

Budesonide
16,17-Butylidenebis(oxy)-11-,21-dihydroxypregna-1,4-diene-3,20-dione
Rhinocort
RN: 51333-22-3   **MP** (°C):
MW: 430.55   **BP** (°C):

Solubility (Moles/L)	Solubility (Grams/L)	Temp (°C)	Ref (#)	Evaluation (T P E A A)	Comments
5.000E-05	2.153E-02	ns	F327	0 0 1 2 2	

4427. C₂₅H₃₄O₉

6-(1,3-Dihydro-7-acetate-5-methoxy-4-methyl-1-oxoisobenzofuran-6-yl)-4-methyl-4-hexanoic
    solketal ester
RN: **MP** (°C):
MW: 478.54   **BP** (°C):

Solubility (Moles/L)	Solubility (Grams/L)	Temp (°C)	Ref (#)	Evaluation (T P E A A)	Comments
1.881E-05	9.000E-03	25	L333	1 1 1 1 0	

4428. C₂₅H₃₅N₅O₄

Benzoic acid, 3-[(dibutylamino)methyl]-, 2-[(6-amino-4,5-dihydro-4-oxo-1H-imidazo[4,5-c]
    pyridin-1-yl)methoxy]ethyl ester
1H-Imidazo[4,5-c]pyridine, benzoic acid deriv.
RN: 137605-73-3   **MP** (°C):
MW: 469.59   **BP** (°C):   688.2

Solubility (Moles/L)	Solubility (Grams/L)	Temp (°C)	Ref (#)	Evaluation (T P E A A)	Comments
1.278E-05	6.000E-03	21	B419	1 1 2 2 1	int

4429. C$_{25}$H$_{36}$N$_4$O$_7$

Nonyloxycarbonyl-mitomycin C
2′-(2-Hexanoyl-2-pentanyl-acetyl)-6-methoxypurine arabinoside
RN: **MP** (°C):
MW: 504.59   **BP** (°C):

Solubility (Moles/L)	Solubility (Grams/L)	Temp (°C)	Ref (#)	Evaluation (T P E A A)	Comments
2.500E-07	1.261E-04	25	M316	1 1 1 1 2	
2.020E-03	1.019E+00	37	C348	0 0 0 0 0	pH 7.00

4430. C$_{25}$H$_{36}$O$_6$

Hydrocortisone butyrate
Hydrocortisone-21-butyrate
11,17-Dihydroxy-21-(1-oxobutoxy)-pregn-4-ene-3,20-dione
RN: 6677-99-2   **MP** (°C):
MW: 432.56   **BP** (°C):

Solubility (Moles/L)	Solubility (Grams/L)	Temp (°C)	Ref (#)	Evaluation (T P E A A)	Comments
1.787E-05	7.730E-03	25	H098	1 0 2 0 2	
1.787E-05	7.730E-03	25	H320	0 0 0 0 0	
1.780E-05	7.700E-03	25	H320	0 0 0 0 0	

4431. C$_{25}$H$_{36}$O$_7$

5,16-β-Dihydroxy-6-β-methyl-3,11-dioxo-5-α-pregn-17(20)-ene-*cis*-20-carboxylic acid methyl ester cycl
RN: **MP** (°C):
MW: 448.56   **BP** (°C):

Solubility (Moles/L)	Solubility (Grams/L)	Temp (°C)	Ref (#)	Evaluation (T P E A A)	Comments
1.672E-04	7.500E-02	ns	K029	0 0 2 1 1	

4432. C$_{25}$H$_{40}$O$_3$Si$_2$

Norethindrone pentamethyldisiloxyl ether
RN: **MP** (°C):
MW: 444.77   **BP** (°C):

Solubility (Moles/L)	Solubility (Grams/L)	Temp (°C)	Ref (#)	Evaluation (T P E A A)	Comments
2.301E-07	1.023E-04	25	L078	1 0 1 2 2	

4433. C$_{25}$H$_{42}$O$_3$
Octadecyl-*p*-hydroxybenzoate
RN: 71067-10-2   **MP** (°C):
MW: 390.61   **BP** (°C):

Solubility (Moles/L)	Solubility (Grams/L)	Temp (°C)	Ref (#)	Evaluation (T P E A A)	Comments
8.343E-04	3.259E-01	25	D081	1 2 2 1 2	

4434. C$_{25}$H$_{44}$
Nonadecylbenzene
1-Phenylnonadecane
RN: 29136-19-4   **MP** (°C):
MW: 344.63   **BP** (°C):   419

Solubility (Moles/L)	Solubility (Grams/L)	Temp (°C)	Ref (#)	Evaluation (T P E A A)	Comments
1.530E-02	5.272E+00	328	S355	1 1 1 2 0	EFG
2.396E-01	8.257E+01	363	S355	1 1 1 2 0	EFG

4435. C$_{25}$H$_{44}$O$_6$
4-Nonylphenol pentaethoxylate
RN: 20636-48-0   **MP** (°C):
MW: 440.63   **BP** (°C):

Solubility (Moles/L)	Solubility (Grams/L)	Temp (°C)	Ref (#)	Evaluation (T P E A A)	Comments
2.151E-05	9.480E-03	20.5	A335	0 0 0 0 0	
2.150E-05	9.473E-03	20.5	A335	0 0 0 0 0	

4436. C$_{25}$H$_{48}$O$_4$
Dioctyl azelate
Di(2-ethylhexyl) azelate
RN: 103-24-2   **MP** (°C):
MW: 412.66   **BP** (°C):

Solubility (Moles/L)	Solubility (Grams/L)	Temp (°C)	Ref (#)	Evaluation (T P E A A)	Comments
2.423E-07	1.000E-04	25	F067	1 0 2 2 0	

4437. C$_{25}$H$_{54}$O$_2$P$_2$
bis(Di-n-hexyl-phosphinyl)methane
HDPM

RN: 2785-33-3   **MP** (°C):
MW: 448.66   **BP** (°C):

Solubility (Moles/L)	Solubility (Grams/L)	Temp (°C)	Ref (#)	Evaluation (T P E A A)	Comments
1.426E-04	6.400E-02	0	O002	2 0 2 2 0	EFG
8.849E-05	3.970E-02	25	O002	2 0 2 2 1	average of 2
6.241E-05	2.800E-02	35	O002	2 0 2 2 0	EFG
4.458E-05	2.000E-02	40	O002	2 0 2 2 0	EFG
3.377E-03	1.515E+00	45	O002	2 0 2 2 0	EFG

4438. C$_{26}$H$_{18}$N$_2$O$_4$
Samaron violet
Mowilith red 3B(IG)

RN: 6408-72-6   **MP** (°C):
MW: 422.44   **BP** (°C):

Solubility (Moles/L)	Solubility (Grams/L)	Temp (°C)	Ref (#)	Evaluation (T P E A A)	Comments
3.000E-06	1.267E-03	98.59	M180	0 0 2 2 0	EFG
4.000E-06	1.690E-03	109.98	M180	0 0 2 1 0	EFG
4.500E-06	1.901E-03	120.54	M180	0 0 2 2 0	EFG
6.000E-06	2.535E-03	133.34	M180	0 0 2 2 0	EFG
8.000E-06	3.380E-03	141.78	M180	0 0 2 2 0	EFG

4439. C$_{26}$H$_{20}$N$_2$O$_8$S$_2$
1,5-Anthraquinone disulfonic acid anilide

RN: **MP** (°C):
MW: 552.59   **BP** (°C):

Solubility (Moles/L)	Solubility (Grams/L)	Temp (°C)	Ref (#)	Evaluation (T P E A A)	Comments
7.210E-03	3.984E+00	18	F047	1 2 1 1 1	

4440. C$_{26}$H$_{20}$N$_2$O$_8$S$_2$
1,8-Anthraquinone disulfonic acid anilide

RN: **MP** (°C):
MW: 552.59   **BP** (°C):

Solubility (Moles/L)	Solubility (Grams/L)	Temp (°C)	Ref (#)	Evaluation (T P E A A)	Comments
4.209E-02	2.326E+01	18	F047	1 2 1 1 1	

4441. $C_{26}H_{28}Cl_2N_4O_4$

Ketoconazole

(±)-cis-1-Acetyl-4-(4-[(2-[2,4-dichlorophenyl]-2-[1H-imidazol-1-ylmethyl]-1,3-dioxolan-4-yl)-
   methoxy]phenyl)piperazine

RN: 65277-42-1   **MP** (°C):
MW: 531.44   **BP** (°C):

Solubility (Moles/L)	Solubility (Grams/L)	Temp (°C)	Ref (#)	Evaluation (T P E A A)	Comments
1.505E-04	8.000E-02	37	C323	0 0 0 0 0	EFG
1.882E-05	1.000E-02	amb	L434	0 0 0 0 0	

4442. $C_{26}H_{28}N_2$

Cinnarizine

Stugeron

RN: 298-57-7   **MP** (°C):
MW: 368.53   **BP** (°C):

Solubility (Moles/L)	Solubility (Grams/L)	Temp (°C)	Ref (#)	Evaluation (T P E A A)	Comments
2.035E-03	7.500E-01	ns	B155	0 0 1 1 0	EFG, pH 3.0

4443. $C_{26}H_{28}N_4O_2$

Propyldiantipyrylmethane

PDAM

RN: 1461-17-2   **MP** (°C):
MW: 428.54   **BP** (°C):

Solubility (Moles/L)	Solubility (Grams/L)	Temp (°C)	Ref (#)	Evaluation (T P E A A)	Comments
1.400E-04	6.000E-02	20	P054	0 0 0 0 0	

4444. $C_{26}H_{29}NO$

Tamoxifen

Genox

Kessar

Nolvadex

(Z)-2-[4-(1,2-Diphenyl-1-butenyl)phenoxy]-N,N-dimethylethanamine

Tamoxen

RN: 10540-29-1   **MP** (°C):
MW: 371.53   **BP** (°C):

Solubility (Moles/L)	Solubility (Grams/L)	Temp (°C)	Ref (#)	Evaluation (T P E A A)	Comments
7.550E+00	2.805E+03	25	B443	0 0 0 0 0	extrapolated
		amb	L434	0 0 0 0 0	

4445. $C_{26}H_{30}Cl_2F_3NO$
Halofantrine
1-(1,3-Dichloro-6-trifluoromethyl-9-phenanthryl)-3-di(n-butyl)aminopropanol
RN: 69756-53-2   **MP** (°C):
MW: 500.44   **BP** (°C):

Solubility (Moles/L)	Solubility (Grams/L)	Temp (°C)	Ref (#)	Evaluation (T P E A A)	Comments
1.179E-06	5.900E-04	37	A423	0 0 0 0 0	

4446. $C_{26}H_{30}N_4O_2$
Isopropyldiantipyrylmethane
IPDAM
RN: 15536-49-9   **MP** (°C):
MW: 430.55   **BP** (°C):

Solubility (Moles/L)	Solubility (Grams/L)	Temp (°C)	Ref (#)	Evaluation (T P E A A)	Comments
4.644E-04	2.000E-01	20	P054	0 0 0 0 0	

4447. $C_{26}H_{30}N_4S_2$
Propyldithiopyrylmethane
3H-Pyrazole-3-thione, 4,4'-butylidenebis[1,2-dihydro-1,5-dimethyl-2-phenyl-
RN: 57094-83-4   **MP** (°C):   222
MW: 462.68   **BP** (°C):

Solubility (Moles/L)	Solubility (Grams/L)	Temp (°C)	Ref (#)	Evaluation (T P E A A)	Comments
2.400E-04	1.110E-01	ns	D087	0 2 0 0 1	

4448. $C_{26}H_{31}ClN_2O_8S$
Amlodipine
Amlodipine besylate
Norvasc
(RS)-3-Ethyl-5-methyl-2-(2-aminoethoxymethyl)-4-(2-chlorophenyl)-1,4-dihydro-6-methyl-3,5-
   pyridinedicarboxylate benzenesulfonate
RN: 88150-42-9   **MP** (°C):
MW: 567.06   **BP** (°C):

Solubility (Moles/L)	Solubility (Grams/L)	Temp (°C)	Ref (#)	Evaluation (T P E A A)	Comments
<1.76E-05	<1.00E-02	rt	B435	0 0 0 0 0	

4449. $C_{26}H_{32}F_2O_7$
Diflorasone diacetate
U-34865

RN: 33564-31-7 **MP** (°C):
MW: 494.54 **BP** (°C):

Solubility (Moles/L)	Solubility (Grams/L)	Temp (°C)	Ref (#)	Evaluation (T P E A A)	Comments
1.314E-05	6.500E-03	25	F003	0 0 0 0 0	
1.254E-05	6.200E-03	37	F003	0 0 0 0 0	
2.629E-05	1.300E-02	50	F003	0 0 0 0 0	

4450. $C_{26}H_{32}F_2O_7$
Fluocinolide
Fluocinonide
Fluocinolone acetonide acetate

RN: 356-12-7 **MP** (°C):
MW: 494.54 **BP** (°C):

Solubility (Moles/L)	Solubility (Grams/L)	Temp (°C)	Ref (#)	Evaluation (T P E A A)	Comments
1.072E-06	5.300E-04	25	O001	2 0 2 2 2	
2.022E-05	1.000E-02	25	P008	0 0 0 0 0	EFG

4451. $C_{26}H_{32}O_3$
Testosterone benzoate
Androst-4-en-3-one, 17-(benzoyloxy)-, (17β)-

RN: 2088-71-3 **MP** (°C):
MW: 392.54 **BP** (°C):

Solubility (Moles/L)	Solubility (Grams/L)	Temp (°C)	Ref (#)	Evaluation (T P E A A)	Comments
3.312E-05	1.300E-02	25	L342	1 0 1 1 2	

4452. $C_{26}H_{36}O_3$
Norethisterone heptanoate

RN: **MP** (°C):
MW: 396.58 **BP** (°C):

Solubility (Moles/L)	Solubility (Grams/L)	Temp (°C)	Ref (#)	Evaluation (T P E A A)	Comments
1.521E-07	6.030E-05	25	E301	1 0 1 1 2	

4453. $C_{26}H_{36}O_6$
Prednisolone 21-trimethylacetate
Prednisolone acetate

| **RN:** | 52-21-1 | **MP** (°C): | 233 |
| **MW:** | 444.57 | **BP** (°C): | |

Solubility (Moles/L)	Solubility (Grams/L)	Temp (°C)	Ref (#)	Evaluation (T P E A A)	Comments
2.609E-05	1.160E-02	25	C037	2 1 2 2 2	
6.298E-05	2.800E-02	25	K021	1 2 2 2 1	
2.699E-05	1.200E-02	ns	N302	0 2 1 2 1	

4454. $C_{26}H_{37}FO_5$
Dexamethasone TBA

| **RN:** | | **MP** (°C): | |
| **MW:** | 448.58 | **BP** (°C): | |

Solubility (Moles/L)	Solubility (Grams/L)	Temp (°C)	Ref (#)	Evaluation (T P E A A)	Comments
2.229E-05	1.000E-02	37	D026	0 0 0 0 0	

4455. $C_{26}H_{38}NO_8$
Glucosamine testosterone
17-β-(4-Androsten-3-one)-N-2-(2-desoxyglucosyl)

| **RN:** | | **MP** (°C): | 185–190 |
| **MW:** | 492.59 | **BP** (°C): | |

Solubility (Moles/L)	Solubility (Grams/L)	Temp (°C)	Ref (#)	Evaluation (T P E A A)	Comments
1.332E-03	6.560E-01	25	L009	1 0 0 1 1	

4456. $C_{26}H_{38}O_4$
Trimethylcyclohexyl phthalate
bis(cis-3,3,5-Trimethylcyclohexyl) phthalate

| **RN:** | 245652-81-7 | **MP** (°C): | 93 |
| **MW:** | 414.59 | **BP** (°C): | |

Solubility (Moles/L)	Solubility (Grams/L)	Temp (°C)	Ref (#)	Evaluation (T P E A A)	Comments
2.894E-07	1.200E-04	24	H116	2 1 0 0 2	

4457. C$_{26}$H$_{38}$O$_6$
Hydrocortisone valerate
Hydrocortisone-21-valerate
RN: 6678-00-8   **MP** (°C):
MW: 446.59   **BP** (°C):

Solubility (Moles/L)	Solubility (Grams/L)	Temp (°C)	Ref (#)	Evaluation (T P E A A)	Comments
6.830E-06	3.050E-03	25	H098	1 0 2 0 2	
6.830E-06	3.050E-03	25	H320	0 0 0 0 0	
6.780E-06	3.028E-03	25	H320	0 0 0 0 0	

4458. C$_{26}$H$_{39}$NO$_3$S
4-Pregnene-20-one-3-spiro-2′-(4′-ethoxycarbonyl-1′,3′-thiazolidine)
RN: **MP** (°C):   131–135
MW: 445.67   **BP** (°C):

Solubility (Moles/L)	Solubility (Grams/L)	Temp (°C)	Ref (#)	Evaluation (T P E A A)	Comments
~3.81E-06	~1.70E-03	ns	B199	0 0 0 0 0	

4459. C$_{26}$H$_{43}$NO$_3$
Acetaminophen stearate
Acetaminophen octadecanoate
Stearoyl acetaminophen
Octadecanoic acid, 4-(acetylamino)phenyl ester
Acetanilide, 4′-hydroxy-, stearate (ester)
RN: 20675-22-3   **MP** (°C):   117
MW: 417.64   **BP** (°C):

Solubility (Moles/L)	Solubility (Grams/L)	Temp (°C)	Ref (#)	Evaluation (T P E A A)	Comments
1.197E-05	5.000E-03	25	B010	1 1 1 1 0	
3.592E-05	1.500E-02	37	D029	0 0 0 0 0	

4460. C$_{26}$H$_{43}$NO$_6$
Glycocholic acid
Glycine, N-[(3α,5β,7α,12α)-3,7,12-trihydroxy-24-oxocholan-24-yl]-
RN: 475-31-0   **MP** (°C):   130
MW: 465.64   **BP** (°C):

Solubility (Moles/L)	Solubility (Grams/L)	Temp (°C)	Ref (#)	Evaluation (T P E A A)	Comments
7.085E-04	3.299E-01	20	E035	1 2 0 0 1	
2.188E-03	1.019E+00	60	E035	1 2 0 0 2	
5.035E-03	2.344E+00	80	E035	1 2 0 0 2	
1.810E-02	8.428E+00	100	E035	1 2 0 0 1	

4461. C$_{26}$H$_{50}$O$_4$
Dioctyl sebacate
Sebacic acid bis(2-ethylhexyl) ester
RN: 122-62-3   **MP** (°C):   −67
MW: 426.69   **BP** (°C):   248

Solubility (Moles/L)	Solubility (Grams/L)	Temp (°C)	Ref (#)	Evaluation (T P E A A)	Comments
2.344E-07	1.000E-04	25	F067	1 0 2 2 0	

4462. C$_{26}$H$_{56}$O$_2$P$_2$
bis(Di-*n*-hexyl-phosphinyl)ethane
HDPE
RN: 2785-34-4   **MP** (°C):
MW: 462.68   **BP** (°C):

Solubility (Moles/L)	Solubility (Grams/L)	Temp (°C)	Ref (#)	Evaluation (T P E A A)	Comments
2.810E-05	1.300E-02	0	O002	2 0 2 2 1	EFG
6.484E-06	3.000E-03	25	O002	2 0 2 2 1	
6.484E-06	3.000E-03	60	O002	2 0 2 2 1	EFG

4463. C$_{27}$H$_{22}$Cl$_2$N$_4$
Clofazimine
Lamprene
N,5-bis(4-Chlorophenyl)-3,4-dihydro-3-((1-methylethyl)imino)-2-phenazinamine
3-(*p*-Chloroanilino)-10-(*p*-chlorophenyl)-2,10-dihydro-2-(isopropylimino)phenazine
RN: 2030-63-9   **MP** (°C):   211
MW: 473.41   **BP** (°C):

Solubility (Moles/L)	Solubility (Grams/L)	Temp (°C)	Ref (#)	Evaluation (T P E A A)	Comments
<2.11E-06	<1.00E-03	ns	B404	0 2 1 1 0	
2.112E-05	1.000E-02	ns	K444	0 0 0 0 0	
2.000E-04	9.468E-02	ns	O322	0 0 0 0 0	EFG

4464. C$_{27}$H$_{29}$NO$_{11}$
Adriamycin
Adriblastin
RN: 23214-92-8   **MP** (°C):   205
MW: 543.53   **BP** (°C):

Solubility (Moles/L)	Solubility (Grams/L)	Temp (°C)	Ref (#)	Evaluation (T P E A A)	Comments
3.607E-02	1.961E+01	ns	I312	0 0 0 0 0	

4465. C₂₇H₃₀O₃

Norethindrone benzoate

RN: **MP** (°C):

MW: 402.54   **BP** (°C):

Solubility (Moles/L)	Solubility (Grams/L)	Temp (°C)	Ref (#)	Evaluation (T P E A A)	Comments
2.019E-08	8.128E-06	25	L078	1 0 1 2 2	

4466. C₂₇H₃₂N₄O₂

Butyldiantipyrylmethane

BDAM

RN: 61358-30-3   **MP** (°C):

MW: 444.58   **BP** (°C):

Solubility (Moles/L)	Solubility (Grams/L)	Temp (°C)	Ref (#)	Evaluation (T P E A A)	Comments
6.748E-05	3.000E-02	20	P054	0 0 0 0 0	

4467. C₂₇H₃₂N₄O₂

Isobutyldiantipyrylmethane

IBDAM

RN: 16671-34-4   **MP** (°C):

MW: 444.58   **BP** (°C):

Solubility (Moles/L)	Solubility (Grams/L)	Temp (°C)	Ref (#)	Evaluation (T P E A A)	Comments
1.350E-04	6.000E-02	20	P054	0 0 0 0 0	

4468. C₂₇H₃₂N₄S₂

Isobutyldithiopyrylmethane

3H-Pyrazole-3-thione, 4,4′-(3-methylbutylidene)bis[1,2-dihydro-1,5-dimethyl-2-phenyl-

RN: 73429-89-7   **MP** (°C):   209

MW: 476.71   **BP** (°C):

Solubility (Moles/L)	Solubility (Grams/L)	Temp (°C)	Ref (#)	Evaluation (T P E A A)	Comments
1.600E-04	7.627E-02	ns	D087	0 2 0 0 1	

4469. $C_{27}H_{32}O_{14}$

Naringin

4H-1-Benzopyran-4-one, 7-[[2-O-(6-deoxy-α-L-mannopyranosyl)-β-D-glucopyranosyl]oxy]-2,3-
   dihydro-5-hydroxy-2-(4-hydroxyphenyl)-, (S)-

RN: 10236-47-2 **MP** (°C):

MW: 580.55 **BP** (°C):

Solubility (Moles/L)	Solubility (Grams/L)	Temp (°C)	Ref (#)	Evaluation (T P E A A)	Comments
2.928E-04	1.700E-01	6	P070	1 2 1 1 1	
8.613E-04	5.000E-01	20	P070	1 2 1 1 1	
1.361E-03	7.900E-01	35	P070	1 2 1 1 1	
3.376E-03	1.960E+00	45	P070	1 2 1 1 2	
1.233E-02	7.160E+00	55	P070	1 2 1 1 2	
7.271E-02	4.221E+01	65	P070	1 2 1 1 2	
1.864E-01	1.082E+02	75	P070	1 2 1 1 2	

4470. $C_{27}H_{33}N_3O_8$

Rolitetracycline

N-(1-Pyrrolidinylmethyl)tetracycline

Syntetrin

Tetraverin

Synotodecin

RN: 751-97-3 **MP** (°C): 162dec

MW: 527.58 **BP** (°C):

Solubility (Moles/L)	Solubility (Grams/L)	Temp (°C)	Ref (#)	Evaluation (T P E A A)	Comments
>3.79E-02	>2.00E+01	21	M044	2 0 2 2 0	

4471. $C_{27}H_{34}O_3$

Testosterone phenylacetate

Androst-4-en-3-one, 17-[(phenylacetyl)oxy]-, (17β)-

RN: 5704-03-0 **MP** (°C):

MW: 406.57 **BP** (°C):

Solubility (Moles/L)	Solubility (Grams/L)	Temp (°C)	Ref (#)	Evaluation (T P E A A)	Comments
2.206E-05	8.970E-03	25	L342	1 0 1 1 2	
2.188E-05	8.895E-03	ns	R427	0 0 0 0 0	

4472. $C_{27}H_{34}O_{10}$

Cortisone tricarballylate

RN: **MP** (°C):

MW: 518.57 **BP** (°C):

Solubility (Moles/L)	Solubility (Grams/L)	Temp (°C)	Ref (#)	Evaluation (T P E A A)	Comments
1.350E-04	7.000E-02	25	M023	1 0 2 1 0	

4473. $C_{27}H_{36}N_2O_4$
Repaglinide
RN: 135062-02-1 **MP** (°C):
MW: 452.60 **BP** (°C):

Solubility (Moles/L)	Solubility (Grams/L)	Temp (°C)	Ref (#)	Evaluation (T P E A A)	Comments
1.988E-04	8.999E-02	25	M448	0 0 0 0 0	Intrinsic, EFG

4474. $C_{27}H_{38}N_2O_6$
p-Ureidophenyl prostaglandin E2
RN: **MP** (°C):
MW: 486.61 **BP** (°C):

Solubility (Moles/L)	Solubility (Grams/L)	Temp (°C)	Ref (#)	Evaluation (T P E A A)	Comments
2.800E-05	1.363E-02	25	A066	1 0 1 1 1	

4475. $C_{27}H_{38}O_3$
Norethindrone heptanoate
RN: **MP** (°C):
MW: 410.60 **BP** (°C):

Solubility (Moles/L)	Solubility (Grams/L)	Temp (°C)	Ref (#)	Evaluation (T P E A A)	Comments
1.468E-07	6.026E-05	25	L078	1 0 1 2 2	

4476. $C_{27}H_{40}N_2O_6$
p-Ureidophenyl prostaglandin F2 α
RN: **MP** (°C):
MW: 488.63 **BP** (°C):

Solubility (Moles/L)	Solubility (Grams/L)	Temp (°C)	Ref (#)	Evaluation (T P E A A)	Comments
6.900E-05	3.372E-02	25	A066	1 0 1 1 1	

4477. $C_{27}H_{40}O_6$
Hydrocortisone tebutate
Hydrocortisone-21-hexanoate
Hydrocortisone-21-caproate
RN: 508-96-3 **MP** (°C): 168
MW: 460.62 **BP** (°C):

Solubility (Moles/L)	Solubility (Grams/L)	Temp (°C)	Ref (#)	Evaluation (T P E A A)	Comments
3.083E-06	1.420E-03	25	H098	1 0 2 0 2	
3.083E-06	1.420E-03	25	H320	0 0 0 0 0	
3.060E-06	1.409E-03	25	H320	0 0 0 0 0	

4478. C$_{27}$H$_{42}$Cl$_2$N$_2$O$_6$

α-Chloramphenicol palmitate
β-Chloramphenicol palmitate
Chloramphenicol palmitate

RN: 530-43-8   **MP** (°C):   359
MW: 561.55   **BP** (°C):

Solubility (Moles/L)	Solubility (Grams/L)	Temp (°C)	Ref (#)	Evaluation (T P E A A)	Comments
1.100E-08	6.177E-06	20	M006	2 2 1 2 1	
8.500E-08	4.773E-05	20	M006	2 2 1 2 1	
1.500E-08	8.423E-06	25	M006	2 2 1 2 1	
9.600E-08	5.391E-05	25	M006	2 2 1 2 1	
7.123E-06	4.000E-03	28	R004	0 0 0 0 0	
1.800E-08	1.011E-05	29	M006	2 2 1 2 1	
1.440E-07	8.086E-05	29	M006	2 2 1 2 2	
2.700E-08	1.516E-05	32	M006	2 2 1 2 1	
2.600E-07	1.460E-04	32	M006	2 2 1 2 2	
3.100E-08	1.741E-05	35	M006	2 2 1 2 1	
3.800E-07	2.134E-04	35	M006	2 2 1 2 2	

4479. C$_{27}$H$_{42}$N$_4$O$_7$.0.3H$_2$O

2′-(2-Heptanoyl-2-hexanyl-acetyl)-6-methoxypurine arabinoside (0.3 hydrate)

RN: 145913-52-6   **MP** (°C):
MW: 540.06   **BP** (°C):

Solubility (Moles/L)	Solubility (Grams/L)	Temp (°C)	Ref (#)	Evaluation (T P E A A)	Comments
2.990E-04	1.615E-01	37	C348	0 0 0 0 0	pH 7.00

4480. C$_{27}$H$_{42}$O$_3$

Diosgenin
(25R)-Spirost-5-en-3β-ol

RN: 512-04-9   **MP** (°C):   204
MW: 414.63   **BP** (°C):

Solubility (Moles/L)	Solubility (Grams/L)	Temp (°C)	Ref (#)	Evaluation (T P E A A)	Comments
4.824E-08	2.000E-05	25	L033	1 0 2 1 0	

4481. C$_{27}$H$_{42}$O$_3$

Nandrolone nonanoate

RN: **MP** (°C):
MW: 414.63   **BP** (°C):

Solubility (Moles/L)	Solubility (Grams/L)	Temp (°C)	Ref (#)	Evaluation (T P E A A)	Comments
2.233E-06	9.260E-04	37	C026	0 0 0 0 0	

4482. C$_{27}$H$_{43}$NO$_8$
N-Methylglucamine testosterone
17-β-(4-Androsten-3-one)-N-methyl-N-1-(1-desoxyglucosyl) carbamate

RN: **MP (°C):**   183–185
MW: 509.65   **BP (°C):**

Solubility (Moles/L)	Solubility (Grams/L)	Temp (°C)	Ref (#)	Evaluation (T P E A A)	Comments
8.633E-05	4.400E-02	25	L009	1 0 0 1 1	

4483. C$_{27}$H$_{44}$N$_4$O$_6$
2′-Hexadecyl-6-methoxypurine arabinoside

RN: 145913-43-5   **MP (°C):**   97–99
MW: 520.67   **BP (°C):**

Solubility (Moles/L)	Solubility (Grams/L)	Temp (°C)	Ref (#)	Evaluation (T P E A A)	Comments
1.900E-05	9.893E-03	37	C348	0 0 0 0 0	pH 7.00

4484. C$_{27}$H$_{44}$O
Vitamin D3
Cholecalciferol
Activated 7-dehydrocholesterol
Oleovitamin D3

RN: 67-97-0   **MP (°C):**   85
MW: 384.65   **BP (°C):**

Solubility (Moles/L)	Solubility (Grams/L)	Temp (°C)	Ref (#)	Evaluation (T P E A A)	Comments
<5.98E-04	<2.30E-01	25	P312	0 0 0 0 0	

4485. C$_{27}$H$_{58}$O$_2$P$_2$
bis(Di-n-hexyl-phosphinyl)propane
HDPP

RN: 2896-56-2   **MP (°C):**
MW: 476.71   **BP (°C):**

Solubility (Moles/L)	Solubility (Grams/L)	Temp (°C)	Ref (#)	Evaluation (T P E A A)	Comments
2.727E-04	1.300E-01	0	O002	2 0 2 2 0	EFG
1.154E-04	5.500E-02	15	O002	2 0 2 2 0	EFG
3.566E-05	1.700E-02	25	O002	2 0 2 2 0	

4486. C₂₈H₂₉F₂N₃O

Pimozide

2-Benzimidazolinone, 1-[1-[4,4-bis(*p*-fluorophenyl)butyl]-4-piperidyl]-

Orap

RN:　　2062-78-4　　**MP** (°C):

MW:　　461.56　　　**BP** (°C):

Solubility (Moles/L)	Solubility (Grams/L)	Temp (°C)	Ref (#)	Evaluation (T P E A A)	Comments
6.283E-06	2.900E-03	30	P044	0 0 0 0 0	

4487. C₂₈H₃₁FN₄O

Astemizole

1-((4-Fluorophenyl)-methyl)-*N*-(1-(2-(4-methoxyphenyl)ethyl)-4-piperidinyl)-1H-benzimidazol-2-
amine

Hismanal

RN:　　68844-77-9　　**MP** (°C):

MW:　　458.58　　　**BP** (°C):

Solubility (Moles/L)	Solubility (Grams/L)	Temp (°C)	Ref (#)	Evaluation (T P E A A)	Comments
7.000E-04	3.210E-01	30	A417	0 0 0 0 0	pH 5.8
3.700E-03	1.697E+00	30	A417	0 0 0 0 0	pH 3.8

4488. C₂₈H₃₆O₃

Testosterone phenyl propionate

Androst-4-en-3-one, 17-(1-oxo-3-phenylpropoxy)-, (17β)-

RN:　　1255-49-8　　**MP** (°C):

MW:　　420.60　　　**BP** (°C):

Solubility (Moles/L)	Solubility (Grams/L)	Temp (°C)	Ref (#)	Evaluation (T P E A A)	Comments
5.350E-06	2.250E-03	25	L342	1 0 1 1 2	

4489. C₂₈H₃₆O₁₅

Neohesperidin dihydrochalcone

1-Propanone, 1-[4-[[2-*O*-(6-deoxy-a-L-mannopyranosyl)-b-D-glucopyranosyl]oxy]-2,6-
dihydroxyphenyl]-3-(3-hydroxy-4-methoxyphenyl)-

Glucopyranoside, 3,5-dihydroxy-4-(3-hydroxy-4-methoxyhydrocinnamoyl)phenyl 2-*O*-(6-deoxy-
a-L-mannopyranosyl)-, b-D-

Glucopyranoside, 3,5-dihydroxy-4-(3-hydroxy-4-methoxyhydrocinnamoyl)phenyl 2-*O*-a-L-
rhamnopyranosyl-

Neohesperidin DHC

NHDC

RN:　　20702-77-6　　**MP** (°C):

MW:　　612.59　　　**BP** (°C):　　927.1

Solubility (Moles/L)	Solubility (Grams/L)	Temp (°C)	Ref (#)	Evaluation (T P E A A)	Comments
6.530E-06	4.000E-03	rt	B417	0 0 1 2 1	

4490. $C_{28}H_{38}N_6O_{11}S$

Sildenafil citrate

1-[[3-(6,7-Dihydro-1-methyl-7-oxo-3-propyl-1H-pyrazolo [4,3-d]pyrimidin-5-yl)-4-ethoxyphenyl] sulfonyl]-4-methylpiperazine citrate

Viagra

RN: 171599-83-0 **MP** (°C):
MW: 666.71 **BP** (°C):

Solubility (Moles/L)	Solubility (Grams/L)	Temp (°C)	Ref (#)	Evaluation (T P E A A)	Comments
5.231E-03	3.488E+00	ns	S469	0 0 0 0 0	

4491. $C_{28}H_{39}NO_6$

p-Acetamidophenyl prostaglandin E2

RN: **MP** (°C):
MW: 485.63 **BP** (°C):

Solubility (Moles/L)	Solubility (Grams/L)	Temp (°C)	Ref (#)	Evaluation (T P E A A)	Comments
5.400E-05	2.622E-02	25	A066	1 0 1 1 1	

4492. $C_{28}H_{39}NO_6$

2-Oxo-5-indolinyl prostaglandin F2α

Prosta-5,13-dien-1-oic acid, 9,11,15-trihydroxy-, 2,3-dihydro-2-oxo-1H-indol-5-yl ester, (5Z,9α,11α,13E,15S)-

RN: 74973-22-1 **MP** (°C):
MW: 485.63 **BP** (°C):

Solubility (Moles/L)	Solubility (Grams/L)	Temp (°C)	Ref (#)	Evaluation (T P E A A)	Comments
6.000E 05	2.914E-02	25	Λ066	1 0 1 1 1	

4493. $C_{28}H_{39}N_3O_6$

α-Semicarbazono-p-tolyl prostaglandin E2

RN: **MP** (°C):
MW: 513.64 **BP** (°C):

Solubility (Moles/L)	Solubility (Grams/L)	Temp (°C)	Ref (#)	Evaluation (T P E A A)	Comments
2.500E-06	1.284E-03	25	A066	1 0 1 1 1	

4494. $C_{28}H_{40}FNO_{11} \cdot H_2O$

Glucosamine 9-α-fluorohyfrocortisome (monohydrate)

21-(9-α-Fluoro-11α, 17α-dihydroxy-4-pregnen-3,20-dione)-N-2-(2-desoxyglucosyl) carbamate

RN: **MP** (°C): 176-178
MW: 603.64 **BP** (°C):

Solubility (Moles/L)	Solubility (Grams/L)	Temp (°C)	Ref (#)	Evaluation (T P E A A)	Comments
5.964E-04	3.600E-01	25	L009	1 0 0 1 1	

4495. C_{28}H_{41}N_3O_6
α-Semicarbazono-*p*-tolyl prostaglandin F2 α
RN: **MP** (°C):
MW: 515.66   **BP** (°C):

Solubility (Moles/L)	Solubility (Grams/L)	Temp (°C)	Ref (#)	Evaluation (T P E A A)	Comments
1.600E-05	8.250E-03	25	A066	1 0 1 1 1	

4496. C_{28}H_{42}FNO_{11}.H_2O
Glucamine 9-α-fluorohyfrocortisome (monohydrate)
RN: **MP** (°C):   105–110
MW: 605.66   **BP** (°C):

Solubility (Moles/L)	Solubility (Grams/L)	Temp (°C)	Ref (#)	Evaluation (T P E A A)	Comments
4.456E-03	2.699E+00	25	L009	1 0 0 1 1	

4497. C_{28}H_{42}O_6
Hydrocortisone heptanoate
Hydrocortisone-21-heptanoate
RN: **MP** (°C):
MW: 474.64   **BP** (°C):

Solubility (Moles/L)	Solubility (Grams/L)	Temp (°C)	Ref (#)	Evaluation (T P E A A)	Comments
2.082E-06	9.880E-04	25	H098	1 0 2 0 2	
2.082E-06	9.880E-04	25	H320	0 0 0 0 0	
2.060E-06	9.778E-04	25	H320	0 0 0 0 0	

4498. C_{28}H_{44}O_3
Nandrolone decanoate
Deca-durabolin
Norandrostenolone decanoate
RN: 360-70-3   **MP** (°C):
MW: 428.66   **BP** (°C):

Solubility (Moles/L)	Solubility (Grams/L)	Temp (°C)	Ref (#)	Evaluation (T P E A A)	Comments
1.549E-06	6.640E-04	37	C026	0 0 0 0 0	

4499. C_{28}H_{46}O_4
Di-*n*-decyl phthalate
RN: 84-77-5   **MP** (°C):
MW: 446.68   **BP** (°C):

Solubility (Moles/L)	Solubility (Grams/L)	Temp (°C)	Ref (#)	Evaluation (T P E A A)	Comments
7.388E-07	3.300E-04	24	H116	2 1 0 0 2	

4500. C₂₈H₄₆O₄

Diisodecyl phthalate

RN:	26761-40-0	**MP** (°C):
MW:	446.68	**BP** (°C):

Solubility (Moles/L)	Solubility (Grams/L)	Temp (°C)	Ref (#)	Evaluation (T P E A A)	Comments
6.269E-07	2.800E-04	24	H116	2 1 0 0 2	

4501. C₂₈H₆₀O₂P₂

bis(Di-*n*-hexyl-phosphinyl)butane

HDPB

RN:	2785-35-5	**MP** (°C):
MW:	490.74	**BP** (°C):

Solubility (Moles/L)	Solubility (Grams/L)	Temp (°C)	Ref (#)	Evaluation (T P E A A)	Comments
3.627E-04	1.780E-01	0	O002	2 0 2 2 0	EFG
1.284E-04	6.300E-02	15	O002	2 0 2 2 0	EFG
4.076E-05	2.000E-02	25	O002	2 0 2 2 0	

4502. C₂₉H₂₀N₂O₄

1,4-Dibenzoylaminoanthraquinone

Benzamide, *N,N'*-(9,10-dihydro-3-methyl-9,10-dioxo-1,8-anthracenediyl)bis

RN:	4627-15-0	**MP** (°C):
MW:	460.49	**BP** (°C):

Solubility (Moles/L)	Solubility (Grams/L)	Temp (°C)	Ref (#)	Evaluation (T P E A A)	Comments
2.200E-05	1.013E-02	50	G077	1 0 0 0 1	

4503. C₂₉H₂₇N₅O₄

m-Nitrophenyldiantipyrylmethane

m-NPhDAM

RN:	1606-53-7	**MP** (°C):
MW:	509.57	**BP** (°C):

Solubility (Moles/L)	Solubility (Grams/L)	Temp (°C)	Ref (#)	Evaluation (T P E A A)	Comments
5.887E-05	3.000E-02	20	P054	0 0 0 0 0	

4504. C₂₉H₂₇N₅O₄

o-Nitrophenyldiantipyrylmethane

o-NPhDAM

RN:	14957-18-7	**MP** (°C):
MW:	509.57	**BP** (°C):

Solubility (Moles/L)	Solubility (Grams/L)	Temp (°C)	Ref (#)	Evaluation (T P E A A)	Comments
3.925E-05	2.000E-02	20	P054	0 0 0 0 0	

4505. C$_{29}$H$_{27}$N$_5$O$_4$
p-Nitrophenyldiantipyrylmethane
p-NPhDAM

RN: 55774-19-1   **MP** (°C):
MW: 509.57   **BP** (°C):

Solubility (Moles/L)	Solubility (Grams/L)	Temp (°C)	Ref (#)	Evaluation (T P E A A)	Comments
3.925E-05	2.000E-02	20	P054	0 0 0 0 0	

4506. C$_{29}$H$_{28}$N$_4$O$_2$
Phenyldiantipyrylmethane
PhDAM

RN: 1861-84-3   **MP** (°C):
MW: 464.57   **BP** (°C):

Solubility (Moles/L)	Solubility (Grams/L)	Temp (°C)	Ref (#)	Evaluation (T P E A A)	Comments
5.165E-04	2.399E-01	20	P054	0 0 0 0 0	

4507. C$_{29}$H$_{28}$N$_4$O$_3$
o-Hydroxylphenyldiantipyrylmethane
o-HPhDAM

RN: 1606-55-9   **MP** (°C):
MW: 480.57   **BP** (°C):

Solubility (Moles/L)	Solubility (Grams/L)	Temp (°C)	Ref (#)	Evaluation (T P E A A)	Comments
<2.08E-05	<1.00E-02	20	P054	0 0 0 0 0	

4508. C$_{29}$H$_{28}$N$_4$S$_2$
Phenyldithiopyrylmethane
3H-Pyrazole-3-thione, 4,4′-(phenylmethylene)bis[1,2-dihydro-1,5-dimethyl-2-phenyl-

RN: 74713-68-1   **MP** (°C): 160
MW: 496.70   **BP** (°C):

Solubility (Moles/L)	Solubility (Grams/L)	Temp (°C)	Ref (#)	Evaluation (T P E A A)	Comments
4.200E-05	2.086E-02	ns	D087	0 2 0 0 1	

4509. $C_{29}H_{32}O_{13}$

Etoposide
4'-Demethylepipodophyllotoxin ethylidene-β-D-glucoside
Vepesid
VP-16

RN: 33419-42-0 **MP** (°C): 236–251
MW: 588.57 **BP** (°C):

Solubility (Moles/L)	Solubility (Grams/L)	Temp (°C)	Ref (#)	Evaluation (T P E A A)	Comments
1.945E-04	1.145E-01	25	S466	0 0 0 0 0	
3.398E-04	2.000E-01	ns	D347	0 0 0 0 0	
3.388E-04	1.994E-01	ns	R427	0 0 0 0 0	

4510. $C_{29}H_{35}NO_2$

Mifepristone
RU-486

RN: 84371-65-3 **MP** (°C):
MW: 429.61 **BP** (°C):

Solubility (Moles/L)	Solubility (Grams/L)	Temp (°C)	Ref (#)	Evaluation (T P E A A)	Comments
1.105E-06	4.748E-04	22.5	B440	0 0 0 0 0	

4511. $C_{29}H_{36}N_4O_2$

Hexyldiantipyrylmethane
HDAM

RN: 7660-44-8 **MP** (°C):
MW: 472.64 **BP** (°C):

Solubility (Moles/L)	Solubility (Grams/L)	Temp (°C)	Ref (#)	Evaluation (T P E A A)	Comments
4.230E-05	1.999E-02	20	P054	0 0 0 0 0	
4.232E-05	2.000E-02	20	P054	0 0 0 0 0	

4512. $C_{29}H_{36}N_4S_2$

Hexyldithiopyrylmethane
3H-Pyrazole-3-thione, 4,4'-heptylidenebis[1,2-dihydro-1,5-dimethyl-2-phenyl-

RN: 74713-69-2 **MP** (°C): 169
MW: 504.77 **BP** (°C):

Solubility (Moles/L)	Solubility (Grams/L)	Temp (°C)	Ref (#)	Evaluation (T P E A A)	Comments
4.100E-05	2.070E-02	0	D087	0 2 0 0 1	

4513. $C_{29}H_{38}Cl_2N_2O_3$

3β-Hydroxy-13α-amino-13,17-seco-5α-androstan-17-oic-13,17-lactam-4-*N,N-bis*-(chloroethyl)
    amino phenyl-acetate

RN:　　　　　　　　　　**MP** (°C):
MW:　　533.54　　　　**BP** (°C):

Solubility (Moles/L)	Solubility (Grams/L)	Temp (°C)	Ref (#)	Evaluation (T P E A A)	Comments
3.186E-07	1.700E-04	25	P022	0 0 0 0 0	
3.599E-07	1.920E-04	30	P022	0 0 0 0 0	
4.517E-07	2.410E-04	44	P022	0 0 0 0 0	
6.110E-07	3.260E-04	73	P022	0 0 0 0 0	

4514. $C_{29}H_{38}O_3$

Testosterone phenylbutyrate

RN:　　　　　　　　　　**MP** (°C):
MW:　　434.62　　　　**BP** (°C):

Solubility (Moles/L)	Solubility (Grams/L)	Temp (°C)	Ref (#)	Evaluation (T P E A A)	Comments
3.681E-06	1.600E-03	25	L342	1 0 1 1 2	

4515. $C_{29}H_{40}N_2O_4$

Emetine
Emetan, 6',7',10,11-tetramethoxy-
NSC 33669

RN:　　483-18-1　　**MP** (°C):　　74
MW:　　480.65　　　　**BP** (°C):

Solubility (Moles/L)	Solubility (Grams/L)	Temp (°C)	Ref (#)	Evaluation (T P E A A)	Comments
2.000E-03	9.613E-01	15	K059	2 2 2 0 0	
2.078E-03	9.990E-01	c	D004	0 0 0 0 0	

4516. $C_{29}H_{42}O_6$

Cortisone caprylate

RN:　　　　　　　　　　**MP** (°C):
MW:　　486.65　　　　**BP** (°C):

Solubility (Moles/L)	Solubility (Grams/L)	Temp (°C)	Ref (#)	Evaluation (T P E A A)	Comments
4.110E-06	2.000E-03	25	M023	1 0 2 1 0	

4517. $C_{29}H_{44}FNO_{11} \cdot H_2O$
N-Methylglucamine 9-α-fluorohyfrocortisome (monohydrate)
21-(9-α-Fluoro-11β, 17α-dihydroxy-4-pregnen-3,20-dione)-N-methyl-N-1-(1-desoxyglucosyl)
   carbamate

RN: **MP** (°C):   120
MW: 619.69   **BP** (°C):

Solubility (Moles/L)	Solubility (Grams/L)	Temp (°C)	Ref (#)	Evaluation (T P E A A)	Comments
6.358E-03	3.940E+00	25	L009	1 0 0 1 1	

4518. $C_{29}H_{44}O_{12}$
Oubain
γ-Strophanthin
Ouabain
Quabain

RN: 630-60-4   **MP** (°C):   185
MW: 584.67   **BP** (°C):

Solubility (Moles/L)	Solubility (Grams/L)	Temp (°C)	Ref (#)	Evaluation (T P E A A)	Comments
2.223E-02	1.300E+01	25	P312	0 0 0 0 0	
1.693E-02	9.901E+00	c	D004	0 0 0 0 0	
2.851E-01	1.667E+02	h	D004	0 0 0 0 0	

4519. $C_{29}H_{46}N_4O_7 \cdot 0.4H_2O$
2'-(2-Octanoyl-2-heptanyl-acetyl)-6-methoxypurine arabinoside (0.4 hydrate)

RN: 145913-53-7   **MP** (°C):
MW: 569.92   **BP** (°C):

Solubility (Moles/L)	Solubility (Grams/L)	Temp (°C)	Ref (#)	Evaluation (T P E A A)	Comments
2.810E-05	1.601E-02	37	C348	0 0 0 0 0	pH 7.00

4520. $C_{29}H_{46}O_3$
Nandrolone undecanoate

RN: **MP** (°C):
MW: 442.69   **BP** (°C):

Solubility (Moles/L)	Solubility (Grams/L)	Temp (°C)	Ref (#)	Evaluation (T P E A A)	Comments
1.360E-06	6.020E-04	37	C026	0 0 0 0 0	

4521. C$_{29}$H$_{50}$O$_2$
Vitamin E
α-Tocopherol
RN: 59-02-9   **MP** (°C):
MW: 430.72   **BP** (°C):

Solubility (Moles/L)	Solubility (Grams/L)	Temp (°C)	Ref (#)	Evaluation (T P E A A)	Comments
4.833E-05	2.082E-02	33	D404	2 1 2 2 2	
4.852E-05	2.090E-02	33	D404	2 1 2 2 2	

4522. C$_{30}$H$_{28}$N$_4$O$_3$
Benzoyldiantipyrylmethane
BenzDAM
RN: 55774-17-9   **MP** (°C):
MW: 492.58   **BP** (°C):

Solubility (Moles/L)	Solubility (Grams/L)	Temp (°C)	Ref (#)	Evaluation (T P E A A)	Comments
<2.03E-05	<1.00E-02	20	P054	0 0 0 0 0	

4523. C$_{30}$H$_{30}$N$_{20}$O$_{10}$
Cucurbit[5]uril
RN: 259886-49-2   **MP** (°C):
MW: 830.70   **BP** (°C):

Solubility (Moles/L)	Solubility (Grams/L)	Temp (°C)	Ref (#)	Evaluation (T P E A A)	Comments
4.100E-04	3.406E-01	25	B424	1 0 1 2 2	

4524. C$_{30}$H$_{34}$O$_{13}$
Picrotoxin
Picrotoxine
RN: 124-87-8   **MP** (°C):
MW: 602.60   **BP** (°C):

Solubility (Moles/L)	Solubility (Grams/L)	Temp (°C)	Ref (#)	Evaluation (T P E A A)	Comments
4.964E-03	2.991E+00	20	D041	1 0 0 0 0	
6.776E-03	4.083E+00	rt	D021	0 0 1 1 1	

4525. C$_{30}$H$_{48}$O$_3$
β-Boswellic acid
RN: **MP** (°C):
MW: 456.72   **BP** (°C):

Solubility (Moles/L)	Solubility (Grams/L)	Temp (°C)	Ref (#)	Evaluation (T P E A A)	Comments
1.700E-02	7.764E+00	ns	R422	0 0 0 0 0	

4526. $C_{30}H_{48}O_{12}$
Periplocin
Card-20(22)-enolide, 3-[(2,6-dideoxy-4-O-β-D-glucopyranosyl-3-O-methyl-β-D-ribo-
   hexopyranosyl)oxy]-5,14-dihydroxy-, (3β,5β)-
Periplocoside

RN:	13137-64-9	**MP (°C):**	205		
MW:	600.71	**BP (°C):**			

Solubility (Moles/L)	Solubility (Grams/L)	Temp (°C)	Ref (#)	Evaluation (T P E A A)	Comments
1.321E-02	7.937E+00	c	D004	0 0 0 0 0	

4527. $C_{31}H_{33}N_5O_2$
p-Dimethylaminophenyldiantipyrylmethane
p-DMAPhDAM

RN:	2088-76-8	**MP (°C):**			
MW:	507.64	**BP (°C):**			

Solubility (Moles/L)	Solubility (Grams/L)	Temp (°C)	Ref (#)	Evaluation (T P E A A)	Comments
1.576E-04	7.999E-02	20	P054	0 0 0 0 0	

4528. $C_{31}H_{38}N_2O_{11}$
Dihydronovobiocin
Benzamide, N-[7-[[3-O-(aminocarbonyl)-6-deoxy-5-C-methyl-4-O-methyl-β-L-lyxo-
   hexopyranosyl]oxy]-4-hydroxy-8-methyl-2-oxo-2H-1-benzopyran-3-yl]-4-hydroxy-3-(3-
   methylbutyl)-

RN:	29826-16-2	**MP (°C):**			
MW:	614.66	**BP (°C):**			

Solubility (Moles/L)	Solubility (Grams/L)	Temp (°C)	Ref (#)	Evaluation (T P E A A)	Comments
2.928E-04	1.800E-01	28	A038	2 0 1 1 2	

4529. $C_{31}H_{42}FNO_{12}.H_2O$
Glucosamine triamcinolone acetonide (monohydrate)
21-(9-α-Fluoro-11β-hydroxy-16α, 17α-isopropylidenedioxy-1,4-pregnadien-3,20-dione)-N-2-(2-
   desoxyglucosyl) carbamate

RN:		**MP (°C):**	250–255		
MW:	657.69	**BP (°C):**			

Solubility (Moles/L)	Solubility (Grams/L)	Temp (°C)	Ref (#)	Evaluation (T P E A A)	Comments
5.717E-04	3.760E-01	25	L009	1 0 0 1 1	

4530. $C_{31}H_{44}FNO_{12}.H_2O$

Glucaminetriamcinolone acetonide (monohydrate)

21-(9-α-Fluoro-11β-hydroxy-16α, 17α-isopropylidenedioxy-1,4-pregnadien-3,20-dione)-N-1-(1-desoxyglucosyl) carbamate

RN:		MP (°C):	150
MW:	659.71	BP (°C):	

Solubility (Moles/L)	Solubility (Grams/L)	Temp (°C)	Ref (#)	Evaluation (T P E A A)	Comments
5.366E-03	3.540E+00	25	L009	1 0 0 1 1	

4531. $C_{31}H_{44}N_2O_7$

N-Acetyl-L-tyrosinamide prostaglandin E2

RN:		MP (°C):	
MW:	556.71	BP (°C):	

Solubility (Moles/L)	Solubility (Grams/L)	Temp (°C)	Ref (#)	Evaluation (T P E A A)	Comments
1.700E-04	9.464E-02	25	A066	1 0 1 1 1	

4532. $C_{31}H_{46}N_2O_7$

N-Acetyl-L-tyrosinamide prostaglandin F2 α

RN:		MP (°C):	
MW:	558.72	BP (°C):	

Solubility (Moles/L)	Solubility (Grams/L)	Temp (°C)	Ref (#)	Evaluation (T P E A A)	Comments
1.400E-04	7.822E-02	25	A066	1 0 1 1 1	

4533. $C_{31}H_{48}O_{12}$

Strophanthin

k-Strophanthin

RN:	11005-63-3	MP (°C):	179
MW:	612.72	BP (°C):	

Solubility (Moles/L)	Solubility (Grams/L)	Temp (°C)	Ref (#)	Evaluation (T P E A A)	Comments
3.709E-02	2.273E+01	25	D004	0 0 0 0 0	

4534. $C_{32}H_{32}O_{14}$

Chartreusin

Lambdamycin

NSC 5159

Antibiotic X 465A

RN:	6377-18-0	MP (°C):	246–249
MW:	640.60	BP (°C):	

Solubility (Moles/L)	Solubility (Grams/L)	Temp (°C)	Ref (#)	Evaluation (T P E A A)	Comments
2.342E-05	1.500E-02	25	P067	0 0 0 0 0	

4535. $C_{32}H_{37}NO_5S$

Dextropropoxyphene napsylate
Darvocet N-50
Darvocet N-100
Darvon-N

RN: 17140-78-2 **MP (°C):**
MW: 547.72 **BP (°C):**

Solubility (Moles/L)	Solubility (Grams/L)	Temp (°C)	Ref (#)	Evaluation (T P E A A)	Comments
2.556E-03	1.400E+00	22	N319	0 0 0 0 0	

4536. $C_{32}H_{40}BrN_5O_5$

Bromocriptine
2-Bromo-α-ergocryptine
Parlodel
Kripton
(5′α)-2-Bromo-12′-hydroxy-2′-(1-methylethyl)-5′-(2-methylpropyl)ergotaman-3′,6′,18-trione

RN: 25614-03-3 **MP (°C):**
MW: 654.62 **BP (°C):**

Solubility (Moles/L)	Solubility (Grams/L)	Temp (°C)	Ref (#)	Evaluation (T P E A A)	Comments
3.162E-06	2.070E-03	ns	R427	0 0 0 0 0	

4537. $C_{32}H_{41}NO_2$

Terfenadine
Seldane
Teldane

RN: 50679-08-8 **MP (°C):**
MW: 471.69 **BP (°C):**

Solubility (Moles/L)	Solubility (Grams/L)	Temp (°C)	Ref (#)	Evaluation (T P E A A)	Comments
2.056E-10	9.700E-08	25	A412	1 0 2 2 1	
		amb	L434	0 0 0 0 0	
2.138E-07	1.008E-04	ns	R427	0 0 0 0 0	

4538. $C_{32}H_{45}N_3O_4S$

Nelfinavir mesylate
Nelfinavir
NFV
Viracept

RN: 159989-65-8 **MP (°C):**
MW: 567.80 **BP (°C):**

Solubility (Moles/L)	Solubility (Grams/L)	Temp (°C)	Ref (#)	Evaluation (T P E A A)	Comments
7.925E-03	4.500E+00	ns	W424	0 0 0 0 0	

4539. C$_{32}$H$_{45}$N$_3$O$_4$S

Nelfinavir

(3S,4aS,8aS)-N-(1,1-Dimethylethyl)decahydro-2-[(2R,3R)-2-hydroxy-3-[(3-hydroxy-2-
methylbenzoyl)amino]-4-(phenylthio)butyl]-3-isoquinolinecarboxamide

RN:	159989-64-7	MP (°C):
MW:	567.80	BP (°C):

Solubility (Moles/L)	Solubility (Grams/L)	Temp (°C)	Ref (#)	Evaluation (T P E A A)	Comments
1.233E-02	7.000E+00	ns	A426	0 0 0 0 0	Intrinsic

4540. C$_{32}$H$_{46}$FNO$_{12}$.H$_2$O

N-Methylglucamine triamcinolone acetonide (monohydrate)

21-(9-α-Fluoro-11β-hydroxy-16α, 17α-isopropylidenedioxy-1,4-pregnadien-3,20-dione)-N-methyl-
N-1-(1-desoxyglucosyl) carbamate

RN:		MP (°C):	152
MW:	673.74	BP (°C):	

Solubility (Moles/L)	Solubility (Grams/L)	Temp (°C)	Ref (#)	Evaluation (T P E A A)	Comments
4.744E-03	3.196E+00	25	L009	1 0 0 1 1	

4541. C$_{32}$H$_{49}$NO$_9$

Cevadine

Cevane-3,4,12,14,16,17,20-heptol, 4,9-epoxy-, 3-[(2Z)-2-methyl-2-butenoate], (3β,4α,16β)-

Veratrine

RN:	62-59-9	MP (°C):	213.5
MW:	591.75	BP (°C):	

Solubility (Moles/L)	Solubility (Grams/L)	Temp (°C)	Ref (#)	Evaluation (T P E A A)	Comments
8.000E-03	4.734E+00	15	K059	2 2 2 0 0	

4542. C$_{32}$H$_{54}$O$_4$

Didodecyl phthalate

1,2-Benzenedicarboxylic acid, didodecyl ester

RN:	2432-90-8	MP (°C):
MW:	502.78	BP (°C):

Solubility (Moles/L)	Solubility (Grams/L)	Temp (°C)	Ref (#)	Evaluation (T P E A A)	Comments
2.784E-07	1.400E-04	24	H116	2 1 0 0 2	

4543. $C_{33}H_{25}N_3O_3$
Norbormide
5-(α-Hydroxy-α-2-pyridylbenzyl)-7-(α-2-pyridylbenzylidene)-5-norbornene-2,3-dicaboximide
Shoxin

RN:	991-42-4	**MP** (°C):	>160		
MW:	511.59	**BP** (°C):			

Solubility (Moles/L)	Solubility (Grams/L)	Temp (°C)	Ref (#)	Evaluation (T P E A A)	Comments
1.173E-04	6.000E-02	rt	M161	0 0 0 0 1	

4544. $C_{33}H_{34}O_3$
Norethindrone biphenyl-4-carboxylate

RN:		**MP** (°C):	
MW:	478.64	**BP** (°C):	

Solubility (Moles/L)	Solubility (Grams/L)	Temp (°C)	Ref (#)	Evaluation (T P E A A)	Comments
7.762E-09	3.715E-06	25	L078	1 0 1 2 2	

4545. $C_{33}H_{34}O_4$
Norethindrone 4-phenoxybenzoate

RN:		**MP** (°C):	
MW:	494.64	**BP** (°C):	

Solubility (Moles/L)	Solubility (Grams/L)	Temp (°C)	Ref (#)	Evaluation (T P E A A)	Comments
1.431E-07	7.079E-05	25	L078	1 0 1 2 2	

4546. $C_{33}H_{36}N_4O_6$
Bilirubin
21H-Biline-8,12-dipropanoic acid, 2,17-diethenyl-1,10,19,22,23,24-hexahydro-3,7,13,18-
   tetramethyl-1,19-dioxo-

RN:	635-65-4	**MP** (°C):	
MW:	584.68	**BP** (°C):	

Solubility (Moles/L)	Solubility (Grams/L)	Temp (°C)	Ref (#)	Evaluation (T P E A A)	Comments
7.000E-09	4.093E-06	18	K104	1 0 0 0 2	intrinsic

4547. C₃₃H₄₀N₂O₉

$C_{33}H_{40}N_2O_9$

Reserpine

3,4,5-Trimethoxybenzoyl methyl reserpate

Rauwilid

Rauwiloid

RN: 50-55-5   **MP** (°C):
MW: 608.69   **BP** (°C):

Solubility (Moles/L)	Solubility (Grams/L)	Temp (°C)	Ref (#)	Evaluation (T P E A A)	Comments
1.200E-04	7.304E-02	30	L068	1 0 0 1 0	EFG
1.643E-05	1.000E-02	ns	K444	0 0 0 0 0	

4548. C₃₃H₄₁N₅O₆S₂

$C_{33}H_{41}N_5O_6S_2$

Kynostatin

KNI-272

4-Thiazolidinecarboxamide, N-(1,1-dimethylethyl)-3-[(2S,3S)-2-hydroxy-3-[[(2R)-2-[[(5-isoquinolinyloxy)acetyl]amino]-3-(methylthio)-1-oxopropyl]amino]-1-oxo-4-phenylbutyl]-, (4R)-

RN: 147318-81-8   **MP** (°C):
MW: 667.85   **BP** (°C):

Solubility (Moles/L)	Solubility (Grams/L)	Temp (°C)	Ref (#)	Evaluation (T P E A A)	Comments
6.289E-06	4.200E-03	25	J308	0 0 0 0 0	

4549. C₃₃H₄₅NO₉

$C_{33}H_{45}NO_9$

Delphinine

Indaconitine, N-deethyl-3-deoxy-N-methyl-

RN: 561-07-9   **MP** (°C): 198–200
MW: 599.73   **BP** (°C):

Solubility (Moles/L)	Solubility (Grams/L)	Temp (°C)	Ref (#)	Evaluation (T P E A A)	Comments
3.335E-05	2.000E-02	25	D004	0 0 0 0 0	

4550. C₃₃H₄₇NO₁₃

$C_{33}H_{47}NO_{13}$

Natamycin

Pimafucin

RN: 7681-93-8   **MP** (°C):
MW: 665.74   **BP** (°C):

Solubility (Moles/L)	Solubility (Grams/L)	Temp (°C)	Ref (#)	Evaluation (T P E A A)	Comments
4.506E-05	3.000E-02	20	B190	1 2 1 1 0	
6.159E-04	4.100E-01	21	M044	2 0 2 2 2	*sic*

4551. $C_{34}H_{30}N_2O_6S$
Pyrantel pamoate
Pirantel pamoate
Dog Wormer
Helmex
Lombriareu
Trilombrin

RN: 22204-24-6 **MP** (°C): 266–267
MW: 594.69 **BP** (°C):

Solubility (Moles/L)	Solubility (Grams/L)	Temp (°C)	Ref (#)	Evaluation (T P E A A)	Comments
1.682E-05	1.000E-02	ns	K444	0 0 0 0 0	

4552. $C_{34}H_{34}N_4O_4$
Protoprophyrin IX
Protoporphyrin IX

RN: 553-12-8 **MP** (°C):
MW: 562.67 **BP** (°C):

Solubility (Moles/L)	Solubility (Grams/L)	Temp (°C)	Ref (#)	Evaluation (T P E A A)	Comments
1.900E-04	1.069E-01	25	C097	2 0 1 1 1	EFG

4553. $C_{34}H_{47}NO_{11}$
Aconitine
Acetylbenzoylaconine

RN: 302-27-2 **MP** (°C): 204
MW: 645.75 **BP** (°C):

Solubility (Moles/L)	Solubility (Grams/L)	Temp (°C)	Ref (#)	Evaluation (T P E A A)	Comments
4.691E-04	3.029E-01	25	D004	0 0 0 0 0	

4554. $C_{34}H_{50}O_7$
Carbenoxolone
Olean-12-en-29-oic acid, 3-(3-carboxy-1-oxopropoxy)-11-oxo-, (3β,20β)-

RN: 5697-56-3 **MP** (°C):
MW: 570.77 **BP** (°C):

Solubility (Moles/L)	Solubility (Grams/L)	Temp (°C)	Ref (#)	Evaluation (T P E A A)	Comments
1.160E-05	6.621E-03	24	B363	0 0 0 0 0	
1.630E-05	9.304E-03	37	B363	0 0 0 0 0	

4555. C_{34}H_{57}NO_7
Glucosamine cholesterol
3-β-(5-Cholestenyl)-N-2-(2-desoxyglucosyl) carbamate
RN: **MP (°C):**   155–158
MW: 591.84   **BP (°C):**

Solubility (Moles/L)	Solubility (Grams/L)	Temp (°C)	Ref (#)	Evaluation (T P E A A)	Comments
9.530E-04	5.640E-01	25	L009	1 0 0 1 1	

4556. C_{34}H_{58}O_4
Ditridecyl phthalate
Staflex DTDP
Truflex DTDP
Hexaplas DTDP
Jayflex DTDP
Polycizer 962BPA
RN: 119-06-2   **MP (°C):**
MW: 530.84   **BP (°C):**

Solubility (Moles/L)	Solubility (Grams/L)	Temp (°C)	Ref (#)	Evaluation (T P E A A)	Comments
6.405E-07	3.400E-04	24	H116	2 1 0 0 2	

4557. C_{34}H_{68}N_3O_8S_2
Lincomycin hexadecylsulfamate
RN: **MP (°C):**
MW: 711.06   **BP (°C):**

Solubility (Moles/L)	Solubility (Grams/L)	Temp (°C)	Ref (#)	Evaluation (T P E A A)	Comments
5.738E-04	4.080E-01	21	M044	2 0 2 2 2	

4558. C_{35}H_{44}N_2O_7
p-(p-Acetamidobenzamido)phenyl prostaglandin E2
RN: **MP (°C):**
MW: 604.75   **BP (°C):**

Solubility (Moles/L)	Solubility (Grams/L)	Temp (°C)	Ref (#)	Evaluation (T P E A A)	Comments
9.800E-08	5.927E-05	25	A066	1 0 1 1 1	

4559. C₃₅H₄₆N₂O₇

p-(*p*-Acetamidobenzamido)phenyl prostaglandin F2 α

RN: **MP** (°C):

MW: 606.77   **BP** (°C):

Solubility (Moles/L)	Solubility (Grams/L)	Temp (°C)	Ref (#)	Evaluation (T P E A A)	Comments
2.800E-07	1.699E-04	25	A066	1 0 1 1 1	

4560. C₃₅H₄₇NO₉

Rhizoxin

RN: 90996-54-6   **MP** (°C):

MW: 625.77   **BP** (°C):

Solubility (Moles/L)	Solubility (Grams/L)	Temp (°C)	Ref (#)	Evaluation (T P E A A)	Comments
1.918E-05	1.200E-02	25	P336	0 0 0 0 0	

4561. C₃₅H₆₁NO₇

N-Methylglucamine cholesterol

3-β-(5-Cholestenyl)-*N*-methyl-*N*-1-(1-desoxyglucosyl) carbamate

RN: **MP** (°C):   131–133

MW: 607.88   **BP** (°C):

Solubility (Moles/L)	Solubility (Grams/L)	Temp (°C)	Ref (#)	Evaluation (T P E A A)	Comments
1.842E-04	1.120E-01	25	L009	1 0 0 1 1	

4562. C₃₆H₄₇N₂O₇

N-Benzoyl-L-tyrosinamide prostaglandin E2

RN: **MP** (°C):

MW: 619.79   **BP** (°C):

Solubility (Moles/L)	Solubility (Grams/L)	Temp (°C)	Ref (#)	Evaluation (T P E A A)	Comments
4.700E-07	2.913E-04	25	A066	1 0 1 1 1	

4563. C$_{36}$H$_{47}$N$_5$O$_4$

Indinavir sulfate
Crixivan
IDV
Indinavir
Indinavir sulfate
MK-639

RN: 157810-81-6   **MP** (°C):
MW: 613.81   **BP** (°C):

Solubility (Moles/L)	Solubility (Grams/L)	Temp (°C)	Ref (#)	Evaluation (T P E A A)	Comments
>1.63E-01	>1.00E+02	ns	W424	0 0 0 0 0	

4564. C$_{36}$H$_{47}$N$_5$O$_4$

Indinavir
2,3,5-Trideoxy-N-[(1S,2R)-2,3-dihydro-2-hydroxy-1H-inden-1-yl]-5-[(2S)-2-[[(1,1-dimethylethyl)
  amino]carbonyl]-4-(3-pyridinylmethyl)-1-piperazinyl]-2-(phenylmethyl)-D-erythro-pentonamide
N-(2-hydroxy-1(S)-indanyl)-2-(phenylmethyl)-4(S)-hydroxy-5-[1-[4-(3-pyridylmethyl)-2(S)-(N-tert-
  butylcarbamoyl)piperazinyl]]pentanamide

RN: 150378-17-9   **MP** (°C):
MW: 613.81   **BP** (°C):

Solubility (Moles/L)	Solubility (Grams/L)	Temp (°C)	Ref (#)	Evaluation (T P E A A)	Comments
1.140E-04	7.000E-02	ns	A426	0 0 0 0 0	Intrinsic

4565. C$_{36}$H$_{49}$N$_2$O$_7$

N-Benzoyl-L-tyrosinamide prostaglandin F2 α

RN: **MP** (°C):
MW: 621.80   **BP** (°C):

Solubility (Moles/L)	Solubility (Grams/L)	Temp (°C)	Ref (#)	Evaluation (T P E A A)	Comments
1.800E-06	1.119E-03	25	A066	1 0 1 1 1	

4566. C$_{36}$H$_{56}$O$_{14}$

Digitalin
Card-20(22)-enolide, 3-[(6-deoxy-4-O-β-D-glucopyranosyl-3-O-methyl-β-D-galactopyranosyl)
  oxy]-14,16-dihydroxy-, (3β,5β,16β)-
Digitalinum verum

RN: 752-61-4   **MP** (°C): 229
MW: 712.84   **BP** (°C):

Solubility (Moles/L)	Solubility (Grams/L)	Temp (°C)	Ref (#)	Evaluation (T P E A A)	Comments
1.401E-03	9.990E-01	25	D004	0 0 0 0 0	

4567. $C_{36}H_{57}N_7O_{10}S$

L-Histidinamide, N-[(1,1-dimethylethoxy)carbonyl]-L-phenylalanyl-N-[2-hydroxy-
1-(2- methylpropyl)-4-[[3-methyl-1-[[(2-sulfoethyl)amino]carbonyl]butyl]amino]-4-oxobutyl]-,
[1S-[1R*,2R*,4(R*)]]-

RN: 100902-06-5	**MP** (°C):	
MW: 779.96	**BP** (°C):	

Solubility (Moles/L)	Solubility (Grams/L)	Temp (°C)	Ref (#)	Evaluation (T P E A A)	Comments
>1.10E-02	>8.58E+00	ns	B425	0 0 0 1 0	

4568. $C_{36}H_{58}N_8O_7$

L-Leucinamide, N-[(1,1-dimethylethoxy)carbonyl]-L-phenylalanyl-L-histidyl-(3S,4S)-4-amino-3-
hydroxy-6-methylheptanoyl-N-(2-aminoethyl)-

L-Histidinamide, N-[(1,1-dimethylethoxy)carbonyl]-L-phenylalanyl-N-[4-[[1-[[(2-aminoethyl)
amino]carbonyl]-3-methylbutyl]amino]-2-hydroxy-1-(2-methylpropyl)-4-oxobutyl]-

RN: 105192-87-8	**MP** (°C):	
MW: 714.91	**BP** (°C):	

Solubility (Moles/L)	Solubility (Grams/L)	Temp (°C)	Ref (#)	Evaluation (T P E A A)	Comments
4.900E-03	3.503E+00	ns	B425	0 0 0 1 0	pH 7.4

4569. $C_{36}H_{60}O_2$

Vitamin A palmitate
Retinol, hexadecanoate
Retinyl palmitate

RN: 79-81-2	**MP** (°C):	
MW: 524.88	**BP** (°C):	

Solubility (Moles/L)	Solubility (Grams/L)	Temp (°C)	Ref (#)	Evaluation (T P E A A)	Comments
5.000E-07	2.624E-04	25	P343	0 0 0 0 0	
1.905E-05	1.000E-02	ns	K444	0 0 0 0 0	

4570. $C_{36}H_{60}O_{30}$

α-Cyclodextrin
β-Hexaamylose
(C6H10O5)6
α-Dextrin

RN: 10016-20-3	**MP** (°C):	
MW: 972.86	**BP** (°C):	

Solubility (Moles/L)	Solubility (Grams/L)	Temp (°C)	Ref (#)	Evaluation (T P E A A)	Comments
9.345E-02	9.091E+01	20	F186	1 2 1 1 1	
2.409E-02	2.344E+01	20	P048	1 0 1 1 1	*sic*
1.118E-01	1.088E+02	23.7	J305	0 0 0 0 0	
1.204E-01	1.171E+02	23.7	J305	0 0 0 0 0	
1.460E-01	1.420E+02	25	B396	0 0 0 0 0	

(*continued*)

4570. C$_{36}$H$_{60}$O$_{30}$ (continued)

Solubility (Moles/L)	Solubility (Grams/L)	Temp (°C)	Ref (#)	Evaluation (T P E A A)	Comments
1.490E-01	1.450E+02	25	L432	0 0 0 0 0	
1.800E-01	1.751E+02	25	O321	0 0 0 0 0	
1.331E-01	1.295E+02	25	S462	0 0 0 0 0	
1.211E-01	1.178E+02	25.0	J305	0 0 0 0 0	
1.318E-01	1.282E+02	25.0	J305	0 0 0 0 0	
1.678E-01	1.632E+02	30.0	J305	0 0 0 0 0	
1.501E-01	1.460E+02	30.0	J305	0 0 0 0 0	
1.696E-01	1.650E+02	33.0	J305	0 0 0 0 0	
1.912E-01	1.860E+02	33.0	J305	0 0 0 0 0	
2.161E-01	2.102E+02	35.0	J305	0 0 0 0 0	
1.885E-01	1.834E+02	35.0	J305	0 0 0 0 0	
2.331E-01	2.268E+02	38.0	J305	0 0 0 0 0	
2.023E-01	1.968E+02	38.0	J305	0 0 0 0 0	
2.100E-01	2.043E+02	40	O321	0 0 0 0 0	
2.171E-01	2.112E+02	40.0	J305	0 0 0 0 0	
2.532E-01	2.463E+02	40.0	J305	0 0 0 0 0	
2.229E-01	2.169E+02	42.0	J305	0 0 0 0 0	
2.616E-01	2.545E+02	42.0	J305	0 0 0 0 0	
2.677E-01	2.604E+02	43.0	J305	0 0 0 0 0	
2.283E-01	2.221E+02	43.0	J305	0 0 0 0 0	
2.492E-01	2.424E+02	45.0	J305	0 0 0 0 0	
2.982E-01	2.901E+02	45.0	J305	0 0 0 0 0	
3.397E-01	3.305E+02	48.0	J305	0 0 0 0 0	
2.773E-01	2.698E+02	48.0	J305	0 0 0 0 0	
4.700E-01	4.572E+02	55	O321	0 0 0 0 0	
1.302E-01	1.266E+02	ns	M335	0 0 2 0 1	
1.490E-01	1.450E+02	rt	F041	0 2 2 0 2	

4571. C$_{36}$H$_{72}$N$_3$O$_8$S$_2$
Lincomycin octadecylsulfamate

RN: **MP** (°C):
MW: 739.12   **BP** (°C):

Solubility (Moles/L)	Solubility (Grams/L)	Temp (°C)	Ref (#)	Evaluation (T P E A A)	Comments
3.897E-04	2.880E-01	21	M044	2 0 2 2 2	

4572. C$_{36}$H$_{74}$
n-Hexatriacontane
Hexatriacontane

RN: 630-06-8   **MP** (°C):   75.0
MW: 506.99   **BP** (°C):

Solubility (Moles/L)	Solubility (Grams/L)	Temp (°C)	Ref (#)	Evaluation (T P E A A)	Comments
3.353E-09	1.700E-06	25	B069	1 0 1 1 1	
4.122E-09	2.090E-06	ns	B033	0 0 0 0 2	
4.122E-09	2.090E-06	ns	B033	0 0 0 0 0	

4573. C₃₇H₄₈N₆O₅S₂

Ritonavir
ABT-538
Norvir
Ritonavir

RN: 155213-67-5   **MP** (°C):
MW: 720.96   **BP** (°C):

Solubility (Moles/L)	Solubility (Grams/L)	Temp (°C)	Ref (#)	Evaluation (T P E A A)	Comments
6.935E-06	5.000E-03	ns	A426	0 0 0 0 0	intrinsic
1.387E-05	1.000E-02	ns	K444	0 0 0 0 0	
~1.39E+00	~9.99E+02	ns	W424	0 0 0 0 0	

4574. C₃₇H₆₇NO₁₃·2H₂O

Erythromycin (dihydrate)

RN: 114-07-8   **MP** (°C):
MW: 769.98   **BP** (°C):

Solubility (Moles/L)	Solubility (Grams/L)	Temp (°C)	Ref (#)	Evaluation (T P E A A)	Comments
6.857E-04	5.280E-01	30	F310	1 0 2 2 2	
4.922E-04	3.790E-01	40	F310	1 0 2 2 2	
4.377E-04	3.370E-01	50	F310	1 0 2 2 2	
4.143E-04	3.190E-01	60	F310	1 0 2 2 2	
4.598E-04	3.540E-01	70	F310	1 0 2 2 2	
5.688E-04	4.380E-01	80	F310	1 0 2 2 2	

4575. C₃₈H₅₀N₆O₅

Squinavir
Butanediamide, N1-[(1S,2R)-3-[(3S,4aS,8aS)-3-[[(1,1-dimethylethyl)amino]carbonyl]octahydro-2-
(1H)-isoquinolinyl]-2-hydroxy-1-(phenylmethyl)propyl]-2-[(2-quinolinylcarbonyl)amino]-
Saquinavir mesylate
Fortovase
Invirase
(S)-N-[(aS)-a-[(1R)-2-[(3S,4aS,8aS)-3-(tert-Butylcarbamoyl)octahydro-2(1H)-
isoquinolyl]-1-hydroxyethyl]phenethyl]-2-quinaldamidosuccinamide

RN: 127779-20-8   **MP** (°C):
MW: 670.86   **BP** (°C):

Solubility (Moles/L)	Solubility (Grams/L)	Temp (°C)	Ref (#)	Evaluation (T P E A A)	Comments
5.336E-05	3.580E-02	25	B431	1 0 1 1 0	
8.198E-05	5.500E-02	25	C437	0 0 0 0 0	Average
3.309E-03	2.220E+00	ns	W424	0 0 0 0 0	

4576. C$_{38}$H$_{60}$N$_8$O$_9$

Butanoic acid, N4-[N-[4-[[N-[N-[(1,1-dimethylethoxy)carbonyl]-L-phenylalanyl]-L-histidyl]amino]-
3-hydroxy-6-methyl-1-oxoheptyl]-L-leucyl]-2,4-diamino-

RN: 115511-05-2   **MP** (°C):
MW: 772.95   **BP** (°C):

Solubility (Moles/L)	Solubility (Grams/L)	Temp (°C)	Ref (#)	Evaluation (T P E A A)	Comments
2.300E-03	1.778E+00	ns	B425	0 0 0 1 0	pH 7.4

4577. C$_{38}$H$_{69}$NO$_{13}$

Clarithromycin
Biaxin
A-56268
TE-031

RN: 81103-11-9   **MP** (°C):   218.5
MW: 747.97   **BP** (°C):

Solubility (Moles/L)	Solubility (Grams/L)	Temp (°C)	Ref (#)	Evaluation (T P E A A)	Comments
1.330E-04	9.948E-02	20	N334	0 0 0 0 0	EFG
1.089E-04	8.145E-02	37	N334	0 0 0 0 0	EFG
4.893E-05	3.660E-02	50	N334	0 0 0 0 0	EFG

4578. C$_{40}$H$_{51}$NO$_{14}$

Streptovaricin C
Streptovaricin

RN: 1404-74-6   **MP** (°C):   189
MW: 769.85   **BP** (°C):

Solubility (Moles/L)	Solubility (Grams/L)	Temp (°C)	Ref (#)	Evaluation (T P E A A)	Comments
1.604E-03	1.235E+00	21	M044	2 0 2 2 2	

4579. C$_{40}$H$_{58}$N$_8$O$_7$

L-Histidinamide, N-[(1,1-dimethylethoxy)carbonyl]-L-phenylalanyl-N-[2-hydroxy-
1-(2-methylpropyl)-4-[[3-methyl-1-[[(2-pyridinylmethyl)amino]carbonyl]butyl]
amino]-4-oxobutyl]-

RN: 87691-49-4   **MP** (°C):
MW: 762.96   **BP** (°C):

Solubility (Moles/L)	Solubility (Grams/L)	Temp (°C)	Ref (#)	Evaluation (T P E A A)	Comments
1.800E-04	1.373E-01	ns	B425	0 0 0 1 0	pH 7.4

4580. $C_{40}H_{58}N_8O_7$

L-Histidinamide, N-[(1,1-dimethylethoxy)carbonyl]-L-phenylalanyl-N-[2-hydroxy-
1-(2- methylpropyl)-4-[[3-methyl-1-[[(4-pyridinylmethyl)amino]carbonyl]butyl]
amino]-4-oxobutyl]-

RN: 87691-50-7   **MP** (°C):
MW: 762.96   **BP** (°C):

Solubility (Moles/L)	Solubility (Grams/L)	Temp (°C)	Ref (#)	Evaluation (T P E A A)	Comments
3.400E-04	2.594E-01	ns	B425	0 0 0 1 0	pH 7.4

4581. $C_{40}H_{58}N_8O_8$

L-Histidinamide, N-[(1,1-dimethylethoxy)carbonyl]-L-phenylalanyl-N-[2-hydroxy-4-[[3-methyl-1-
[[[(1-oxido-4-pyridinyl)methyl]amino]carbonyl]butyl]amino]-1-(2-methylpropyl)-4-oxobutyl]-

RN: 100902-03-2   **MP** (°C):
MW: 778.96   **BP** (°C):

Solubility (Moles/L)	Solubility (Grams/L)	Temp (°C)	Ref (#)	Evaluation (T P E A A)	Comments
4.200E-03	3.272E+00	ns	B425	0 0 0 1 0	pH 7.4

4582. $C_{41}H_{59}N_7O_7$

L-Histidinamide, N-[(1,1-dimethylethoxy)carbonyl]-L-phenylalanyl-N-[2-hydroxy-4-[[3-methyl-1-
[[(phenylmethyl)amino]carbonyl]butyl]amino]-1-(2-methylpropyl)-4-oxobutyl]-

RN: 109585-11-7   **MP** (°C):
MW: 761.97   **BP** (°C):

Solubility (Moles/L)	Solubility (Grams/L)	Temp (°C)	Ref (#)	Evaluation (T P E A A)	Comments
<1.00E-05	<7.62E-03	ns	B425	0 0 0 1 0	pH 7.4

4583. $C_{41}H_{61}N_9O_7$

L-Histidinamide, N-[(1,1-dimethylethoxy)carbonyl]-L-phenylalanyl-N-[4-[[1-[[[2-amino-
2-(2- pyridinyl)ethyl]amino]carbonyl]-3-methylbutyl]amino]-2-hydroxy-1-(2-methylpropyl)-4-
oxobutyl]-

RN: 100901-99-3   **MP** (°C):
MW: 792.00   **BP** (°C):

Solubility (Moles/L)	Solubility (Grams/L)	Temp (°C)	Ref (#)	Evaluation (T P E A A)	Comments
4.000E-04	3.168E-01	ns	B425	0 0 0 1 0	pH 7.4

4584. $C_{41}H_{64}O_{13}$

Digitoxin

(3β,5β)-3-[(0-2,6-Dideoxy-β-D-ribo-hexopyranosyl-(1->4)-O-2,6-dideoxy-β-D-ribo-hexopyranosyl-(1->4)-2,6-dideoxy-β-D-ribo-hexopyranosyl)oxy]-14-hydroxycard-20(22)-enolide

Crystodigin

Digifortis

RN:	71-63-6	MP (°C):	256
MW:	764.96	BP (°C):	

Solubility (Moles/L)	Solubility (Grams/L)	Temp (°C)	Ref (#)	Evaluation (T P E A A)	Comments
1.307E-05	1.000E-02	20	J010	1 0 0 0 0	
5.098E-06	3.900E-03	25	M301	1 1 2 2 1	anhydrate
2.000E-05	1.530E-02	30	O321	0 0 0 0 0	
2.222E-05	1.700E-02	30	O321	0 0 0 0 0	
1.447E-05	1.107E-02	37	C303	2 2 2 2 2	average of 3
3.255E-06	2.490E-03	37	M301	1 1 2 2 1	anhydrate
1.300E-05	9.944E-03	ns	M070	0 0 0 0 1	
9.151E-06	7.000E-03	ns	N302	0 2 1 2 0	

4585. $C_{41}H_{64}O_{14}$

Digoxin

3β-((O-2,6-Dideoxy-β-D-ribo-hexopyranosyl-(1->4)-O-2,6-dideoxy-β-D-ribo-hexopyranosyl-(1->4)-2,6-dideoxy-β-D-ribo-hexopyranosyl)oxy)-12β,14-dihydroxy-5β-card-20(22)-enolide

Lanoxicaps

Lanoxin

RN:	20830-75-5	MP (°C):	260
MW:	780.96	BP (°C):	

Solubility (Moles/L)	Solubility (Grams/L)	Temp (°C)	Ref (#)	Evaluation (T P E A A)	Comments
1.253E-04	9.789E-02	25	F010	2 1 2 2 2	Swiss micron
6.786E-05	5.300E-02	25	F010	2 1 2 2 2	
7.375E-05	5.760E-02	25	F010	2 1 2 2 2	Swiss standard
8.297E-05	6.480E-02	25	F010	2 1 2 2 2	
1.000E-04	7.810E-02	25	H066	1 0 0 0 0	EFG
3.585E-05	2.800E-02	25	M301	1 1 2 2 1	
3.675E-05	2.870E-02	25	N301	2 0 2 2 2	
3.841E-05	3.000E-02	27	E052	2 0 2 2 0	EFG
3.585E-05	2.800E-02	30	O321	0 0 0 0 0	
4.000E-05	3.124E-02	30	O321	0 0 0 0 0	
6.312E-05	4.930E-02	37	C303	2 2 2 2 2	average of 6
3.457E-05	2.700E-02	37	M301	1 1 2 2 1	
3.483E-05	2.720E-02	37	N301	2 0 2 2 2	
4.443E-05	3.470E-02	37	R009	1 0 0 0 2	
2.817E-05	2.200E-02	100	D027	1 2 0 0 1	
1.268E-03	9.900E-01	amb	L434	0 0 0 0 0	
7.363E-06	5.750E-03	ns	F037	0 0 2 0 2	mp 225.5 C
8.963E-06	7.000E-03	ns	F037	0 0 2 0 2	mp 225.5 C
5.570E-06	4.350E-03	ns	F037	0 0 2 0 2	mp 228.5 C

(continued)

4585. C$_{41}$H$_{64}$O$_{14}$ (continued)

Solubility (Moles/L)	Solubility (Grams/L)	Temp (°C)	Ref (#)	Evaluation (T P E A A)	Comments
6.915E-06	5.400E-03	ns	F037	0 0 2 0 2	mp 235.5 C
1.280E-05	1.000E-02	ns	K444	0 0 0 0 0	
4.097E-05	3.200E-02	ns	N302	0 2 1 2 1	
5.900E-05	4.608E-02	rt	J034	0 0 0 0 0	

4586. C$_{41}$H$_{64}$O$_{14}$
Gitoxin
Anhydrogitalin
Pseudodigitoxin
Bigitalin

RN:	4562-36-1	MP (°C):	
MW:	780.96	BP (°C):	

Solubility (Moles/L)	Solubility (Grams/L)	Temp (°C)	Ref (#)	Evaluation (T P E A A)	Comments
3.000E-06	2.343E-03	ns	M070	0 0 0 0 0	

4587. C$_{41}$H$_{67}$NO$_{15}$
Troleandomycin
Triacetyloleandomycin

RN:	2751-09-9	MP (°C):	
MW:	813.99	BP (°C):	

Solubility (Moles/L)	Solubility (Grams/L)	Temp (°C)	Ref (#)	Evaluation (T P E A A)	Comments
3.071E-04	2.500E-01	28	A038	2 0 1 1 1	

4588. C$_{41}$H$_{68}$N$_8$O$_9$
L-Histidinamide, N-[(1,1-dimethylethoxy)carbonyl]-L-phenylalanyl-N-[4-[[1-[[[3-[bis(2-hydroxy-ethyl)amino]propyl]amino]carbonyl]-3-methylbutyl]amino]-2-hydroxy-1-(2-methylpropyl)-4-oxobutyl]-

RN:	87691-52-9	MP (°C):	
MW:	817.05	BP (°C):	

Solubility (Moles/L)	Solubility (Grams/L)	Temp (°C)	Ref (#)	Evaluation (T P E A A)	Comments
4.300E-03	3.513E+00	ns	B425	0 0 0 1 0	

4589. C$_{42}$H$_{59}$N$_7$O$_9$

Glycine, N-[N-[4-[[N-[N-[(1,1-dimethylethoxy)carbonyl]-L-phenylalanyl]-L-histidyl]amino]-3-hydroxy-6-methyl-1-oxoheptyl]-L-leucyl]-D-2-phenyl-

RN: 115511-06-3   **MP** (°C):
MW: 805.98   **BP** (°C):

Solubility (Moles/L)	Solubility (Grams/L)	Temp (°C)	Ref (#)	Evaluation (T P E A A)	Comments
7.400E-04	5.964E-01	ns	B425	0 0 0 1 0	

4590. C$_{42}$H$_{62}$N$_8$O$_7$

L-Histidinamide, N-[(1,1-dimethylethoxy)carbonyl]-L-phenylalanyl-N-[4-[[1-[[[[3-(aminomethyl)phenyl]methyl]amino]carbonyl]-3-methylbutyl]amino]-2-hydroxy-1-(2-methylpropyl)-4-oxobutyl]-

RN: 100901-98-2   **MP** (°C):
MW: 791.01   **BP** (°C):

Solubility (Moles/L)	Solubility (Grams/L)	Temp (°C)	Ref (#)	Evaluation (T P E A A)	Comments
1.400E-03	1.107E+00	ns	B425	0 0 0 1 0	pH 7.4

4591. C$_{42}$H$_{70}$O$_{35}$

β-Cyclodextrin
β-Cyclodextrin hydrate
Cycloheptaamylose hydrate
Cyclodextrin hydrate

RN: 7585-39-9   **MP** (°C): 298–300
MW: 1135.01   **BP** (°C):

Solubility (Moles/L)	Solubility (Grams/L)	Temp (°C)	Ref (#)	Evaluation (T P E A A)	Comments
1.044E-02	1.185E+01	15	W317	2 2 1 0 2	
1.216E-02	1.381E+01	20	F186	1 2 1 1 1	
1.282E-02	1.455E+01	20	W317	2 2 1 0 2	
1.410E-02	1.600E+01	21	C407	1 0 1 2 1	
1.540E-02	1.748E+01	23.7	J305	0 0 0 0 0	
1.630E-02	1.850E+01	25	B396	0 0 0 0 0	
1.586E-02	1.800E+01	25	C407	1 0 1 2 1	
1.558E-02	1.768E+01	25	H319	0 0 0 0 0	
1.600E-02	1.816E+01	25	O304	1 2 2 2 2	
1.600E-02	1.816E+01	25	O321	0 0 0 0 0	
1.621E-02	1.840E+01	25	S462	0 0 0 0 0	
1.674E-02	1.900E+01	25	T425	0 0 0 0 0	
1.551E-02	1.760E+01	25	W317	2 2 1 0 2	
1.630E-02	1.850E+01	25.0	J305	0 0 0 0 0	
2.026E-02	2.300E+01	30	C407	1 0 1 2 1	
1.895E-02	2.151E+01	30	W317	2 2 1 0 2	
2.203E-02	2.500E+01	35	C407	1 0 1 2 1	
2.440E-02	2.769E+01	35.0	J305	0 0 0 0 0	
3.100E-02	3.519E+01	40	O321	0 0 0 0 0	
2.980E-02	3.382E+01	40.0	J305	0 0 0 0 0	

(continued)

4591. $C_{42}H_{70}O_{35}$ (continued)

Solubility (Moles/L)	Solubility (Grams/L)	Temp (°C)	Ref (#)	Evaluation (T P E A A)	Comments
3.850E-02	4.370E+01	45.0	J305	0 0 0 0 0	
4.430E-02	5.028E+01	48.0	J305	0 0 0 0 0	
4.400E-02	4.994E+01	55	O321	0 0 0 0 0	
1.558E-02	1.768E+01	ns	M335	0 0 2 0 1	

4592. $C_{42}H_{70}O_{35}$
6-O-α-D-Glucosyl-α-cyclodextrin

RN: **MP** (°C):
MW: 1135.01   **BP** (°C):

Solubility (Moles/L)	Solubility (Grams/L)	Temp (°C)	Ref (#)	Evaluation (T P E A A)	Comments
8.000E-01	9.080E+02	25	O321	0 0 0 0 0	
1.030E+00	1.169E+03	40	O321	0 0 0 0 0	
1.190E+00	1.351E+03	55	O321	0 0 0 0 0	

4593. $C_{43}H_{55}NO_{13}$
Docetaxel
Taxotere
N-Debenzoyl-N-$tert$-butoxycarbonyl-10-deacetyl taxol

RN: 114977-28-5   **MP** (°C):
MW: 793.92   **BP** (°C):

Solubility (Moles/L)	Solubility (Grams/L)	Temp (°C)	Ref (#)	Evaluation (T P E A A)	Comments
7.557E-06	6.000E-03	22.5	C438	0 0 0 0 0	

4594. $C_{43}H_{58}N_4O_{12}$
Rifampin
Rifampicin

RN: 13292-46-1   **MP** (°C):
MW: 822.96   **BP** (°C):

Solubility (Moles/L)	Solubility (Grams/L)	Temp (°C)	Ref (#)	Evaluation (T P E A A)	Comments
1.300E-01	1.070E+02	25	B073	2 1 2 2 2	pH 2.12,*sic*
4.374E-03	3.600E+00	25	B073	2 1 2 2 1	pH 2.5
1.701E-03	1.400E+00	25	B073	2 1 2 2 1	pH 5.33
1.215E-03	1.000E+00	25	B073	2 1 2 2 1	pH 3.99
1.215E-03	1.000E+00	25	B073	2 1 2 2 1	pH 3.03
1.580E-03	1.300E+00	25	G096	1 0 0 0 0	pH 4.3
1.215E-04	1.000E-01	ns	K444	0 0 0 0 0	
3.393E-03	2.792E+00	rt	F182	0 0 0 0 1	pH 7.5

4595. C₄₃H₆₁N₇O₁₀

$C_{43}H_{61}N_7O_{10}$

L-Histidinamide, *N*-[(1,1-dimethylethoxy)carbonyl]-L-phenylalanyl-*N*-[4-[[1-[[[[3-(carboxymethoxy)
   phenyl]methyl]amino]carbonyl]-3-methylbutyl]amino]-2-hydroxy-1-(2-methylpropyl)-4-
   oxobutyl]-, [1S-[1R*,2R*,4(R*)]]-

RN:	100902-05-4	MP (°C):
MW:	836.01	BP (°C):

Solubility (Moles/L)	Solubility (Grams/L)	Temp (°C)	Ref (#)	Evaluation (T P E A A)	Comments
2.700E-04	2.257E-01	ns	B425	0 0 0 1 0	

4596. C₄₃H₆₂N₈O₇

$C_{43}H_{62}N_8O_7$

L-threo-Pentonamide, 5-cyclohexyl-2,4,5-trideoxy-4-[[*N*-[*N*-[(1,1-dimethylethoxy)carbonyl]-L-
   phenylalanyl]-L-histidyl]amino]-*N*-[3-methyl-1-[[(4-pyridinylmethyl)amino]carbonyl]butyl]-

RN:	105192-86-7	MP (°C):
MW:	803.02	BP (°C):

Solubility (Moles/L)	Solubility (Grams/L)	Temp (°C)	Ref (#)	Evaluation (T P E A A)	Comments
<1.00E-05	<8.03E-03	ns	B425	0 0 0 1 0	pH 7.4

4597. C₄₃H₆₂N₈O₈

$C_{43}H_{62}N_8O_8$

L-Phenylalaninamide, *N*-[(1,1-dimethylethoxy)carbonyl]-L-phenylalanyl-L-histidyl-(3S,4S)-
   4-amino-3-hydroxy-6-methylheptanoyl-L-leucyl-

L-Phenylalaninamide, *N*-[4-[[*N*-[*N*-[(1,1-dimethylethoxy)carbonyl]-L-phenylalanyl]-L-histidyl]
   amino]-3-hydroxy-6-methyl-1-oxoheptyl]-L-leucyl-,

RN:		MP (°C):	
MW:	819.02	BP (°C):	1171.2

Solubility (Moles/L)	Solubility (Grams/L)	Temp (°C)	Ref (#)	Evaluation (T P E A A)	Comments
<2.00E-05	<1.64E-02	ns	B425	0 0 0 1 0	pH 7.4

4598. C₄₃H₇₅NO₁₆

$C_{43}H_{75}NO_{16}$

Erythromycin ethyl succinate

RN:	1264-62-6	MP (°C):
MW:	862.07	BP (°C):

Solubility (Moles/L)	Solubility (Grams/L)	Temp (°C)	Ref (#)	Evaluation (T P E A A)	Comments
2.262E-04	1.950E-01	21	M044	2 0 2 2 2	

4599. $C_{44}H_{56}O_4$

p-tert-Butylcalix[4]arenetetrol
Tetra-*p-tert*-butyltetracalix[4]arene
p-tert-Butylcalix[4]arene
p-tert-Butylcalix[4]arene-25,26,27,28-tetrol
Formaldehyde-*p-tert*-butylphenyl cyclic tetramer
5,11,17,23-Tetra-*p-tert*-butyl-25,26,27,28-tetrahydroxycalix(4)arene

RN: 60705-62-6 **MP** (°C): 342–346
MW: 648.93 **BP** (°C): 683.1

Solubility (Moles/L)	Solubility (Grams/L)	Temp (°C)	Ref (#)	Evaluation (T P E A A)	Comments
<1.00E-05	<6.49E-03	25	B424	1 0 1 2 2	

4600. $C_{44}H_{64}N_8O_9$

D-Phenylalanine, 3-(aminomethyl)-*N*-[*N*-[4-[[*N*-[*N*-[(1,1-dimethylethoxy)carbonyl]-L-phenylalanyl]-L-histidyl]amino]-3-hydroxy-6-methyl-1-oxoheptyl]-L-leucyl]-

RN: 115511-03-0 **MP** (°C):
MW: 849.05 **BP** (°C):

Solubility (Moles/L)	Solubility (Grams/L)	Temp (°C)	Ref (#)	Evaluation (T P E A A)	Comments
3.900E-04	3.311E-01	ns	B425	0 0 0 1 0	pH 7.4

4601. $C_{44}H_{69}NO_{10}$

Tacrolimus
FK506

RN: 104987-11-3 **MP** (°C):
MW: 772.04 **BP** (°C):

Solubility (Moles/L)	Solubility (Grams/L)	Temp (°C)	Ref (#)	Evaluation (T P E A A)	Comments
1.580E-06	1.220E-03	25	A410	1 0 2 2 1	

4602. $C_{44}H_{74}O_{34}$

n-Ethyl-paba-β-cyclodextrin

RN: **MP** (°C):
MW: 1147.06 **BP** (°C):

Solubility (Moles/L)	Solubility (Grams/L)	Temp (°C)	Ref (#)	Evaluation (T P E A A)	Comments
5.100E-03	5.850E+00	ns	F327	0 0 1 2 2	

4603. C₄₄H₇₄O₃₅

Hydroxyethyl-β-cyclodextrin

RN: **MP** (°C):
MW: 1163.06   **BP** (°C):

Solubility (Moles/L)	Solubility (Grams/L)	Temp (°C)	Ref (#)	Evaluation (T P E A A)	Comments
3.224E-01	3.750E+02	ns	M335	0 0 2 0 1	

4604. C₄₅H₆₃Cl₂NO₆

Cosalane

RN: 154212-56-3   **MP** (°C):   262 C
MW: 784.91   **BP** (°C):

Solubility (Moles/L)	Solubility (Grams/L)	Temp (°C)	Ref (#)	Evaluation (T P E A A)	Comments
1.784E-09	1.400E-06	ns	V417	0 0 0 0 0	

4605. C₄₅H₆₆N₈O₇

L-Histidinamide, N-[(1,1-dimethylethoxy)carbonyl]-L-phenylalanyl-N-[2-hydroxy-4-[[3-methyl-1-
    [[4-(phenylmethyl)-1-piperazinyl]carbonyl]butyl]amino]-1-(2-methylpropyl)-4-oxobutyl]-

RN: 105192-85-6   **MP** (°C):
MW: 831.08   **BP** (°C):

Solubility (Moles/L)	Solubility (Grams/L)	Temp (°C)	Ref (#)	Evaluation (T P E A A)	Comments
<1.00E-05	<8.31E-03	ns	B425	0 0 0 1 0	pH 7.4

4606. C₄₅H₆₆N₈O₇

L-threo-Pentonamide, N-[1-[[[[3-(aminomethyl)phenyl]methyl]amino]carbonyl]-
    3-methylbutyl]-5- cyclohexyl-2,4,5-trideoxy-4-[[N-[N-[(1,1-dimethylethoxy)
    carbonyl]-L-phenylalanyl]-L-histidyl]amino]-

RN: 100902-07-6   **MP** (°C):
MW: 831.08   **BP** (°C):

Solubility (Moles/L)	Solubility (Grams/L)	Temp (°C)	Ref (#)	Evaluation (T P E A A)	Comments
2.000E-05	1.662E-02	ns	B425	0 0 0 1 0	pH 7.4

4607. $C_{45}H_{73}NO_{15}$

Solanine

β-D-Galactopyranoside, (3β)-solanid-5-en-3-yl O-6-deoxy-α-L-mannopyranosyl-(1®2)-O-[β-D-glucopyranosyl-(1-3)]-

Solanidane, β-D-galactopyranoside deriv

RN: 20562-02-1 **MP** (°C):
MW: 868.08 **BP** (°C):

Solubility (Moles/L)	Solubility (Grams/L)	Temp (°C)	Ref (#)	Evaluation (T P E A A)	Comments
3.000E-05	2.604E-02	15	K059	2 2 2 0 0	

4608. $C_{45}H_{76}O_{35}$

n-Propyl-paba-β-cyclodextrin

RN: **MP** (°C):
MW: 1177.09 **BP** (°C):

Solubility (Moles/L)	Solubility (Grams/L)	Temp (°C)	Ref (#)	Evaluation (T P E A A)	Comments
2.100E-03	2.472E+00	ns	F327	0 0 1 2 2	

4609. $C_{46}H_{62}N_4O_{11}$

Rifabutin

1',4-Didehydro-1-deoxy-1,4-dihydro-5'-(2-methylpropyl)-1-oxo

Ansamycin

Antibiotic LM 427

LM 427

Mycobutin

RN: 72559-06-9 **MP** (°C):
MW: 847.03 **BP** (°C):

Solubility (Moles/L)	Solubility (Grams/L)	Temp (°C)	Ref (#)	Evaluation (T P E A A)	Comments
2.243E-04	1.900E-01	ns	S469	0 0 0 0 0	

4610. $C_{46}H_{65}N_7O_{10}$

Acetic acid, [3-[[[2-[[5-cyclohexyl-2,4,5-trideoxy-4-[[N-[N-[(1,1-dimethylethoxy)carbonyl]-L-phenylalanyl]-L-histidyl]amino]-L-threo-pentonoyl]amino]-4-methyl-1-oxopentyl]amino]methyl]phenoxy]-

RN: 100902-09-8 **MP** (°C):
MW: 876.07 **BP** (°C):

Solubility (Moles/L)	Solubility (Grams/L)	Temp (°C)	Ref (#)	Evaluation (T P E A A)	Comments
3.000E-05	2.628E-02	ns	B425	0 0 0 1 0	

4611. $C_{46}H_{77}NO_{17}$
Tylosin
Vubityl 200
Vetil(R)

RN:	1401-69-0	**MP (°C):**	128
MW:	916.12	**BP (°C):**	

Solubility (Moles/L)	Solubility (Grams/L)	Temp (°C)	Ref (#)	Evaluation (T P E A A)	Comments
8.195E-03	7.508E+00	21	M044	2 0 2 2 2	

4612. $C_{46}H_{78}O_{35}$
n-Butyl-paba-β-cyclodextrin

RN:		**MP (°C):**	
MW:	1191.11	**BP (°C):**	

Solubility (Moles/L)	Solubility (Grams/L)	Temp (°C)	Ref (#)	Evaluation (T P E A A)	Comments
7.000E-04	8.338E-01	ns	F327	0 0 1 2 2	

4613. $C_{47}H_{51}NO_{14}$
Paclitaxel
5-β,20-Epoxy-1,2-α,4,7-β,10-β,13-α-hexahydroxy-tax-11-en-9-one 4,10-diacetate 2-benzoate 13-
　　ester with (2R,3S)-N-benzoyl-3-phenyl-isoserine
TAX
Taxal
Taxol
Taxol A

RN:	33069-62-4	**MP (°C):**	213–216
MW:	853.93	**BP (°C):**	

Solubility (Moles/L)	Solubility (Grams/L)	Temp (°C)	Ref (#)	Evaluation (T P E A A)	Comments
3.513E-07	3.000E-04	37	L435	0 0 0 0 0	
1.569E-06	1.340E-03	37	V412	0 0 0 0 0	

4614. $C_{47}H_{73}NO_{17}$
Amphotericin B

RN:	1397-89-3	**MP (°C):**	
MW:	924.10	**BP (°C):**	

Solubility (Moles/L)	Solubility (Grams/L)	Temp (°C)	Ref (#)	Evaluation (T P E A A)	Comments
8.116E-04	7.500E-01	28	A038	2 0 1 1 1	
3.246E-06	3.000E-03	ns	K067	0 0 2 1 0	intrinsic

4615. C_{47}H_{75}NO_{17}
Nystatin
Mycostatin
Biofanal
Nystex
Fungicidin

RN: 1400-61-9	**MP** (°C):	
MW: 926.12	**BP** (°C):	

Solubility (Moles/L)	Solubility (Grams/L)	Temp (°C)	Ref (#)	Evaluation (T P E A A)	Comments
3.887E-04	3.600E-01	24	M166	2 0 0 0 1	
4.319E-03	4.000E+00	ns	K444	0 0 0 0 0	

4616. C_{48}H_{72}O_{14}
Ivermectin
Heartgard-30
Ivomec

RN: 70288-86-7	**MP** (°C):	
MW: 873.10	**BP** (°C):	

Solubility (Moles/L)	Solubility (Grams/L)	Temp (°C)	Ref (#)	Evaluation (T P E A A)	Comments
4.581E-06	4.000E-03	ns	K444	0 0 0 0 0	

4617. C_{48}H_{80}O_{40}
6-*O*-α-D-Maltosyl-α-cyclodextrin
6-*O*-α-Maltosyl-α-cyclodextrin

RN:	**MP** (°C):	
MW: 1297.15	**BP** (°C):	

Solubility (Moles/L)	Solubility (Grams/L)	Temp (°C)	Ref (#)	Evaluation (T P E A A)	Comments
7.700E-01	9.988E+02	25	O321	0 0 0 0 0	
2.400E-01	3.113E+02	25	O321	0 0 0 0 0	
7.700E-01	9.988E+02	40	O321	0 0 0 0 0	
3.500E-01	4.540E+02	40	O321	0 0 0 0 0	
1.330E+00	1.725E+03	55	O321	0 0 0 0 0	
5.400E-01	7.005E+02	55	O321	0 0 0 0 0	

4618. $C_{48}H_{80}O_{40}$

γ-Cyclodextrin
Cyclooctaamylose
Ringdex C
Cyclomaltooctaose
Dexy Pearl γ-100

RN:	17465-86-0	**MP** (°C):		
MW:	1297.15	**BP** (°C):		

Solubility (Moles/L)	Solubility (Grams/L)	Temp (°C)	Ref (#)	Evaluation (T P E A A)	Comments
1.338E-01	1.736E+02	20	F186	1 2 1 1 1	
1.789E-01	2.320E+02	25	B396	0 0 0 0 0	
2.000E-01	2.594E+02	25	O321	0 0 0 0 0	
1.921E-01	2.492E+02	25	S462	0 0 0 0 0	
1.680E-01	2.179E+02	25.0	J305	0 0 0 0 0	
2.040E-01	2.646E+02	30.0	J305	0 0 0 0 0	
2.430E-01	3.152E+02	35.0	J305	0 0 0 0 0	
4.300E-01	5.578E+02	40	O321	0 0 0 0 0	
2.680E-01	3.476E+02	40.0	J305	0 0 0 0 0	
3.110E-01	4.034E+02	42.0	J305	0 0 0 0 0	
6.400E-01	8.302E+02	55	O321	0 0 0 0 0	
1.452E-01	1.883E+02	ns	M335	0 0 2 0 1	

4619. $C_{49}H_{87}NS$

Erythromycin lactobionate

RN:	3847-29-8	**MP** (°C):	145	
MW:	722.31	**BP** (°C):		

Solubility (Moles/L)	Solubility (Grams/L)	Temp (°C)	Ref (#)	Evaluation (T P E A A)	Comments
>2.77E-02	>2.00E+01	21	M044	2 0 2 2 0	

4620. $C_{50}H_{82}N_{10}O_{31}S_{10}$

Decane(S-(carboxymethyl)-L-cysteine))

RN:		**MP** (°C):		
MW:	1639.90	**BP** (°C):		

Solubility (Moles/L)	Solubility (Grams/L)	Temp (°C)	Ref (#)	Evaluation (T P E A A)	Comments
5.820E-05	9.544E-02	15	N331	0 0 0 0 0	
5.730E-04	9.397E-01	25	N331	0 0 0 0 0	

4621. $C_{51}H_{55}NO_{18}$
7-Malyl paclitaxel

RN:	265659-44-7	**MP** (°C):	166–168
MW:	970.00	**BP** (°C):	

Solubility (Moles/L)	Solubility (Grams/L)	Temp (°C)	Ref (#)	Evaluation (T P E A A)	Comments
3.093E-04	3.000E-01	ns	D401	0 2 2 2 0	

4622. $C_{51}H_{55}NO_{18}$
2'-Malyl paclitaxel

RN:	265659-38-9	**MP** (°C):	148–151
MW:	970.00	**BP** (°C):	

Solubility (Moles/L)	Solubility (Grams/L)	Temp (°C)	Ref (#)	Evaluation (T P E A A)	Comments
2.062E-04	2.000E-01	ns	D401	0 2 2 2 0	

4623. $C_{51}H_{70}N_{12}O_{11}$
His-pro-D-phe-his-leu-leu-thr-tyr

RN:		**MP** (°C):	
MW:	1027.20	**BP** (°C):	

Solubility (Moles/L)	Solubility (Grams/L)	Temp (°C)	Ref (#)	Evaluation (T P E A A)	Comments
8.100E-05	8.320E-02	20	B141	1 2 0 0 1	pH 7.5

4624. $C_{51}H_{74}O_{19}$
Penta-acetyl-gitoxin

RN:	7242-04-8	**MP** (°C):	
MW:	991.15	**BP** (°C):	

Solubility (Moles/L)	Solubility (Grams/L)	Temp (°C)	Ref (#)	Evaluation (T P E A A)	Comments
1.200E-05	1.189E-02	ns	M070	0 0 0 0 1	

4625. $C_{52}H_{72}N_{12}O_{10}$
His-pro-phe-his-leu-leu-val-tyr

RN:		**MP** (°C):	
MW:	1025.23	**BP** (°C):	

Solubility (Moles/L)	Solubility (Grams/L)	Temp (°C)	Ref (#)	Evaluation (T P E A A)	Comments
1.610E-04	1.651E-01	ns	B141	0 2 0 0 2	pH 7.5

4626. $C_{52}H_{72}N_{12}O_{10}$
His-pro-phe-his-leu-D-leu-val-tyr
RN: **MP** (°C):
MW: 1025.23   **BP** (°C):

Solubility (Moles/L)	Solubility (Grams/L)	Temp (°C)	Ref (#)	Evaluation (T P E A A)	Comments
1.370E-04	1.405E-01	ns	B141	0 2 0 0 2	pH 7.5

4627. $C_{52}H_{88}O_{39}$
n-Butyl-paba-γ-cyclodextrin
RN: **MP** (°C):
MW: 1337.26   **BP** (°C):

Solubility (Moles/L)	Solubility (Grams/L)	Temp (°C)	Ref (#)	Evaluation (T P E A A)	Comments
7.000E-04	9.361E-01	ns	F327	0 0 1 2 2	

4628. $C_{52}H_{97}NO_{18}S$
Erythromycin estolate
RN: 3521-62-8   **MP** (°C):   135
MW: 1056.41   **BP** (°C):

Solubility (Moles/L)	Solubility (Grams/L)	Temp (°C)	Ref (#)	Evaluation (T P E A A)	Comments
1.515E-04	1.600E-01	21	M044	2 0 2 2 2	

4629. $C_{54}H_{90}O_{45}$
6-O-α-D-Glucosyl-γ-cyclodextrin
RN: **MP** (°C):
MW: 1459.29   **BP** (°C):

Solubility (Moles/L)	Solubility (Grams/L)	Temp (°C)	Ref (#)	Evaluation (T P E A A)	Comments
9.800E-01	1.430E+03	25	O321	0 0 0 0 0	
1.010E+00	1.474E+03	40	O321	0 0 0 0 0	
1.180E+00	1.722E+03	55	O321	0 0 0 0 0	

4630. $C_{54}H_{90}O_{45}$
6-O-α-D-Maltosyl-β-cyclodextrin
6-O-α-Maltosyl-β-cyclodextrin
RN: **MP** (°C):
MW: 1459.29   **BP** (°C):

Solubility (Moles/L)	Solubility (Grams/L)	Temp (°C)	Ref (#)	Evaluation (T P E A A)	Comments
1.040E+00	1.518E+03	25	O321	0 0 0 0 0	
1.040E+00	1.518E+03	40	O321	0 0 0 0 0	
1.220E+00	1.780E+03	55	O321	0 0 0 0 0	

4631. C$_{54}$H$_{90}$O$_{45}$

6-*O*-α-D-Maltotriosyl-α-cyclodextrin
6-*O*-α-Maltotriosyl-α-cyclodextrin

RN: **MP** (°C):
MW: 1459.29 **BP** (°C):

Solubility (Moles/L)	Solubility (Grams/L)	Temp (°C)	Ref (#)	Evaluation (T P E A A)	Comments
1.070E+00	1.561E+03	25	O321	0 0 0 0 0	
1.220E+00	1.780E+03	40	O321	0 0 0 0 0	
1.370E+00	1.999E+03	55	O321	0 0 0 0 0	

4632. C$_{55}$H$_{59}$NO$_{22}$

2′,7-*bis*-(Malyl) paclitaxel

RN: 265659-41-4 **MP** (°C): 166–168
MW: 1086.08 **BP** (°C):

Solubility (Moles/L)	Solubility (Grams/L)	Temp (°C)	Ref (#)	Evaluation (T P E A A)	Comments
4.604E-04	5.000E-01	ns	D401	0 2 2 2 0	

4633. C$_{55}$H$_{70}$N$_{12}$O$_{10}$

His-pro-phe-his-leu-phe-val-tyr

RN: **MP** (°C):
MW: 1059.25 **BP** (°C):

Solubility (Moles/L)	Solubility (Grams/L)	Temp (°C)	Ref (#)	Evaluation (T P E A A)	Comments
1.760E-04	1.864E-01	ns	B141	0 2 0 0 2	pH 7.5

4634. C$_{55}$H$_{79}$N$_{13}$O$_{11}$

His-pro-D-phe-his-leu-leu-val-tyr-serinol

RN: **MP** (°C):
MW: 1098.32 **BP** (°C):

Solubility (Moles/L)	Solubility (Grams/L)	Temp (°C)	Ref (#)	Evaluation (T P E A A)	Comments
3.000E-04	3.295E-01	20	B141	1 2 0 0 2	pH 7.5

4635. C$_{55}$H$_{90}$N$_{11}$O$_{34}$S$_{11}$

Undecane(*S*-(carboxymethyl)-L-cysteine))

RN: **MP** (°C):
MW: 1802.09 **BP** (°C):

Solubility (Moles/L)	Solubility (Grams/L)	Temp (°C)	Ref (#)	Evaluation (T P E A A)	Comments
9.200E-06	1.658E-02	15	N331	0 0 0 0 0	
1.340E-04	2.415E-01	25	N331	0 0 0 0 0	
2.900E-04	5.226E-01	35	N331	0 0 0 0 0	

4636. C$_{56}$H$_{98}$O$_{35}$

β-Cyclodextrin, tetradeca-O-methyl-
Heptakis(2,6-di-O-methyl)-β-cyclodextrin

RN: 188367-19-3 **MP** (°C):
MW: 1331.38 **BP** (°C):

Solubility (Moles/L)	Solubility (Grams/L)	Temp (°C)	Ref (#)	Evaluation (T P E A A)	Comments
2.727E-01	3.631E+02	25	H319	0 0 0 0 0	

4637. C$_{57}$H$_{79}$N$_{13}$O$_{11}$

Pro-his-pro-phe-his-leu-leu-val-tyr

RN: **MP** (°C):
MW: 1122.35 **BP** (°C):

Solubility (Moles/L)	Solubility (Grams/L)	Temp (°C)	Ref (#)	Evaluation (T P E A A)	Comments
3.240E-04	3.636E-01	ns	B141	0 2 0 0 2	pH 7.5

4638. C$_{57}$H$_{79}$N$_{13}$O$_{11}$

Pro-his-pro-phe-his-leu-D-leu-val-tyr

RN: **MP** (°C):
MW: 1122.35 **BP** (°C):

Solubility (Moles/L)	Solubility (Grams/L)	Temp (°C)	Ref (#)	Evaluation (T P E A A)	Comments
4.100E-05	4.602E-02	ns	B141	0 2 0 0 1	pH 7.5

4639. C$_{60}$H$_{77}$N$_{13}$O$_{11}$

Pro-his-pro-phe-his-leu-phe-val-tyr

RN: **MP** (°C):
MW: 1156.36 **BP** (°C):

Solubility (Moles/L)	Solubility (Grams/L)	Temp (°C)	Ref (#)	Evaluation (T P E A A)	Comments
3.430E-04	3.966E-01	ns	B141	0 2 0 0 2	pH 7.5

4640. C$_{60}$H$_{92}$N$_{12}$O$_{10}$

Gramicidin S
Gramicidin
Cyclo(L-leucyl-D-phenylalanyl-L-prolyl-L-valyl-L-ornithyl-L-leucyl-D-phenylalanyl-L-prolyl-L-
   valyl-L-ornithyl)
Gramicidin S-A

RN: 113-73-5 **MP** (°C):
MW: 1141.48 **BP** (°C):

Solubility (Moles/L)	Solubility (Grams/L)	Temp (°C)	Ref (#)	Evaluation (T P E A A)	Comments
1.226E-04	1.400E-01	28	A038	2 0 1 1 2	

4641. $C_{60}H_{98}N_{12}O_{37}S_{12}$
Dodecane(S-(carboxymethyl)-L-cystein))

RN: **MP** (°C):
MW: 1964.28   **BP** (°C):

Solubility (Moles/L)	Solubility (Grams/L)	Temp (°C)	Ref (#)	Evaluation (T P E A A)	Comments
2.300E-06	4.518E-03	15	N331	0 0 0 0 0	
2.400E-05	4.714E-02	25	N331	0 0 0 0 0	
5.880E-05	1.155E-01	35	N331	0 0 0 0 0	

4642. $C_{60}H_{100}O_{50}$
6-O-α-D-Maltotriosyl-β-cyclodextrin
6-O-α-Maltotriosyl-β-cyclodextrin
6-O-α-D-Maltosyl-γ-cyclodextrin
6-O-α-Maltosyl-γ-cyclodextrin

RN: **MP** (°C):
MW: 1621.44   **BP** (°C):

Solubility (Moles/L)	Solubility (Grams/L)	Temp (°C)	Ref (#)	Evaluation (T P E A A)	Comments
9.400E-01	1.524E+03	25	O321	0 0 0 0 0	
9.400E-01	1.524E+03	40	O321	0 0 0 0 0	
1.140E+00	1.848E+03	55	O321	0 0 0 0 0	
1.100E+00	1.784E+03	55	O321	0 0 0 0 0	

4643. $C_{62}H_{86}N_{12}O_{16}$
Actinomycin D
Actactinomycin A IV
Actinomycin AIV
Actinomycin I1

RN: 50-76-0   **MP** (°C):
MW: 1255.45   **BP** (°C):

Solubility (Moles/L)	Solubility (Grams/L)	Temp (°C)	Ref (#)	Evaluation (T P E A A)	Comments
3.983E-04	5.000E-01	37	G025	1 0 0 0 1	
7.965E-04	1.000E+00	rt	G025	0 0 0 0 1	

4644. $C_{62}H_{111}N_{11}O_{12}$

Cyclosporin A
1,4,7,10,13,16,19,22,25,28,31-Undecaazacyclotritriacontane, cyclic peptide deriv.
Sandimmun neoral
Sandimmun
Sang-35
SDZ-OXL 400

RN:	59865-13-3	**MP (°C):**	148–151	
MW:	1202.64	**BP (°C):**		

Solubility (Moles/L)	Solubility (Grams/L)	Temp (°C)	Ref (#)	Evaluation (T P E A A)	Comments
3.326E-05	4.000E-02	25	B376	0 0 0 0 0	
8.315E-06	1.000E-02	amb	L434	0 0 0 0 0	

4645. $C_{63}H_{85}N_{21}O_{19}$

Candicidin
Candeptin
Vanobid

RN:	1403-17-4	**MP (°C):**	
MW:	1440.51	**BP (°C):**	

Solubility (Moles/L)	Solubility (Grams/L)	Temp (°C)	Ref (#)	Evaluation (T P E A A)	Comments
9.349E-03	1.347E+01	21	M044	2 0 2 2 2	

4646. $C_{63}H_{88}N_{14}O_{14}PCo$

Vitamin B12
Cyanoject
Hydrobexan
Alphamine
Crystamine
Cyomin

RN:	68-19-9	**MP (°C):**	
MW:	1355.40	**BP (°C):**	

Solubility (Moles/L)	Solubility (Grams/L)	Temp (°C)	Ref (#)	Evaluation (T P E A A)	Comments
9.149E-03	1.240E+01	20	F300	1 0 0 0 2	

4647. C$_{64}$H$_{112}$O$_{40}$

Dimethyl-β-cyclodextrin

β-Cyclodextrin, 2A,2B,2C,2D,2E,2F,2G,6A,6B,6C,6D,6E,6F,6G-Tetradeca-O-methyl-

Heptakis(2,6-di-O-methyl)-β-cyclodextrin

Tetradeca-O-methyl-β-cyclodextrin

Tetradecakis-2,6-O-methylcycloheptaamylose

RN: 51166-71-3   **MP** (°C):   298–300

MW: 1521.58   **BP** (°C):

Solubility (Moles/L)	Solubility (Grams/L)	Temp (°C)	Ref (#)	Evaluation (T P E A A)	Comments
1.397E-01	2.126E+02	c	D316	0 0 0 0 0	

4648. C$_{65}$H$_{106}$N$_{13}$O$_{40}$S$_{13}$

Tridecane(S-(carboxymethyl)-L-cyateine))

RN: **MP** (°C):

MW: 2126.46   **BP** (°C):

Solubility (Moles/L)	Solubility (Grams/L)	Temp (°C)	Ref (#)	Evaluation (T P E A A)	Comments
6.200E-06	1.318E-02	25	N331	0 0 0 0 0	
1.600E-05	3.402E-02	35	N331	0 0 0 0 0	

4649. C$_{66}$H$_{84}$O$_6$

4-tert-Butylcalix[6]arene

5,11,17,23,29,35-Hexa-tert-butyl-37,38,39,40,41,42-hexahydroxycalix[6]arene

RN: 78092-53-2   **MP** (°C):   380–381

MW: 973.40   **BP** (°C):   890.5

Solubility (Moles/L)	Solubility (Grams/L)	Temp (°C)	Ref (#)	Evaluation (T P E A A)	Comments
<1.00E-05	<9.73E-03	25	B424	1 0 1 2 2	

4650. C$_{66}$H$_{110}$O$_{55}$

6-O-α-D-Maltotriosyl-γ-cyclodextrin

6-O-α-Maltotriosyl-γ-cyclodextrin

RN: **MP** (°C):

MW: 1783.58   **BP** (°C):

Solubility (Moles/L)	Solubility (Grams/L)	Temp (°C)	Ref (#)	Evaluation (T P E A A)	Comments
8.500E-01	1.516E+03	25	O321	0 0 0 0 0	
8.500E-01	1.516E+03	40	O321	0 0 0 0 0	
1.040E+00	1.855E+03	55	O321	0 0 0 0 0	

4651. $C_{67}H_{93}N_{15}O_{13}$
Pro-pro-pro-his-pro-phe-his-leu-D-leu-val-tyr
RN: **MP** (°C):
MW: 1316.58   **BP** (°C):

Solubility (Moles/L)	Solubility (Grams/L)	Temp (°C)	Ref (#)	Evaluation (T P E A A)	Comments
3.650E-04	4.806E-01	ns	B141	0 2 0 0 2	pH 7.5

4652. $C_{67}H_{93}N_{15}O_{13}$
Pro-pro-pro-his-pro-phe-his-leu-leu-val-tyr
RN: **MP** (°C):
MW: 1316.58   **BP** (°C):

Solubility (Moles/L)	Solubility (Grams/L)	Temp (°C)	Ref (#)	Evaluation (T P E A A)	Comments
3.750E-04	4.937E-01	ns	B141	0 2 0 0 2	pH 7.5

4653. $C_{70}H_{89}N_{15}O_{13}$
Pro-pro-pro-his-pro-phe-his-leu-phe-val-tyr
RN: **MP** (°C):
MW: 1348.58   **BP** (°C):

Solubility (Moles/L)	Solubility (Grams/L)	Temp (°C)	Ref (#)	Evaluation (T P E A A)	Comments
2.240E-04	3.021E-01	ns	B141	0 2 0 0 2	pH 7.5

4654. $C_{70}H_{126}O_{35}$
β-Cyclodextrin, tetradeca-O-ethyl-
Heptakis(2,6-di-O-ethyl)-β-cyclodextrin
RN: 194715-43-0   **MP** (°C):
MW: 1527.76   **BP** (°C):

Solubility (Moles/L)	Solubility (Grams/L)	Temp (°C)	Ref (#)	Evaluation (T P E A A)	Comments
3.273E-05	5.000E-02	25	H319	0 0 0 0 0	

4655. $C_{72}H_{85}N_{19}O_{18}S_5$
Thiostrepton
Bryamycin
RN: 1393-48-2   **MP** (°C):   210
MW: 1664.92   **BP** (°C):

Solubility (Moles/L)	Solubility (Grams/L)	Temp (°C)	Ref (#)	Evaluation (T P E A A)	Comments
5.286E-05	8.800E-02	21	M044	2 0 2 2 1	
1.442E-04	2.400E-01	28	A038	2 0 1 1 1	

4656. $C_{72}H_{100}N_{18}O_{17}PCo$
Coenzyme B12
Cobamamide
Cobalamin, Co-(5′-deoxy-5′-adenosyl)-
Dibencozide
Funacomide
Deoxyadenosylcobalamin
RN: 13870-90-1   **MP** (°C):
MW: 1579.62   **BP** (°C):

Solubility (Moles/L)	Solubility (Grams/L)	Temp (°C)	Ref (#)	Evaluation (T P E A A)	Comments
1.646E-02	2.600E+01	24	M054	1 0 0 0 1	

4657. $C_{74}H_{100}ClN_{15}O_{14}$
Antarelix
AcDNal-Dcpa-ser-tyr-dhai-leu-lys(ipr)-pro-dala-NH2
RN: 151272-78-5   **MP** (°C):
MW: 1459.17   **BP** (°C):

Solubility (Moles/L)	Solubility (Grams/L)	Temp (°C)	Ref (#)	Evaluation (T P E A A)	Comments
>6.85E-03	>1.00E+01	ns	D350	0 1 0 1 1	

4658. $C_{75}H_{122}N_{15}O_{46}S_{15}$
Pendecane(S-(carboxymethyl)-L-cysteine))
RN: **MP** (°C):
MW: 2450.84   **BP** (°C):

Solubility (Moles/L)	Solubility (Grams/L)	Temp (°C)	Ref (#)	Evaluation (T P E A A)	Comments
3.400E-07	8.333E-04	25	N331	0 0 0 0 0	

4659. $C_{77}H_{107}N_{17}O_{15}$
Pro-pro-pro-pro-pro-his-pro-phe-his-leu-leu-val-tyr
RN: **MP** (°C):
MW: 1510.82   **BP** (°C):

Solubility (Moles/L)	Solubility (Grams/L)	Temp (°C)	Ref (#)	Evaluation (T P E A A)	Comments
1.328E-03	2.006E+00	ns	B141	0 2 0 0 2	pH 7.5

4660. C$_{80}$H$_{105}$N$_{17}$O$_{15}$

Pro-pro-pro-pro-pro-his-pro-phe-his-leu-phe-val-tyr

RN: **MP** (°C):
MW: 1544.83   **BP** (°C):

Solubility (Moles/L)	Solubility (Grams/L)	Temp (°C)	Ref (#)	Evaluation (T P E A A)	Comments
8.400E-04	1.298E+00	ns	B141	0 2 0 0 2	pH 7.5

4661. C$_{85}$H$_{117}$N$_{20}$O$_{18}$

Asp-arg-val-tyr-ile-his-pro-D-phe-his-leu-phe-val-tyr

RN: **MP** (°C):
MW: 1707.00   **BP** (°C):

Solubility (Moles/L)	Solubility (Grams/L)	Temp (°C)	Ref (#)	Evaluation (T P E A A)	Comments
6.200E-05	1.058E-01	20	B141	1 2 0 0 1	pH 7.5

References

A001 Andrews, L.J. and R.M. Keefer, Cation complexes of compounds containing carbon–carbon double bonds. IV. The argentation of aromatic hydrocarbons, *Journal of the American Chemical Society*, 71, 3644–3647, 1949.

A002 Andrews, L.J. and R.M. Keefer, Cation complexes of compounds containing carbon–carbon double bonds. VII. Further studies on the argentation of substituted benzenes, *Journal of the American Chemical Society*, 72, 5034–5037, 1950.

A003 Andrews, L.J. and R.M. Keefer, Cation complexes of compounds containing carbon–carbon double bonds. VI. The argentation of substituted benzenes, *Journal of the American Chemical Society*, 72, 3113–3116, 1950.

A004 Alexander, D.M., The solubility of benzene in water, *Journal of Physical Chemistry*, 63, 1021–1022, 1959.

A008 Allawala, N.A. and S. Riegelman, The release of antimicrobial agents from solutions of surface-active agents, *Journal of the American Pharmaceutical Association, Scientific Edition*, 42, 267–275, 1953.

A009 Andon, R.J.L. and J.D. Cox, Phase relationships in the pyridine series. Part I. The miscibility of some pyridine homologues with water, *Journal of the Chemical Society* (London), 4601–4606, 1952.

A010 Anderson, R.A. and K.J. Morgan, Effect of solubilisation on the antibacterial activity of hexachlorophane, *Journal of Pharmacy and Pharmacology*, 18, 449–456, 1966.

A011 Addison, C.C., The properties of freshly formed surfaces. Part VI. The influence of temperature and concentration on the dynamic and static surface tensions of aqueous decoic acid solutions, *Journal of the Chemical Society* (London), 579–585, 1946.

A012 Altwein, D.M., J.N. Delgado, and F.P. Cosgrove, effect of urea concentrations on the solubility of the isomeric monohydroxybenzoic acids, *Journal of Pharmaceutical Sciences*, 54, 603–606, 1965.

A013 Agharkar, S., S. Lindenbaum, and T. Higuchi, Enhancement of solubility of drug salts by hydrophilic counterions: properties of organic salts of an antimalarial drug, *Journal of Pharmaceutical Sciences*, 65, 747–749, 1976.

A014 Amin, M.I. and J.T. Bryan, Kinetics and factors affecting stability of methylprednisolone in aqueous formulation, *Journal of Pharmaceutical Sciences*, 62, 1768–1771, 1973.

A015 Addison, C.C., The properties of freshly formed surfaces. Part IV. The influence of chain length and structure on the static and the dynamic surface tensions of aqueous-alcoholic solutions, *Journal of the Chemical Society* (London), 98–106, 1945.

A016 Altshuller, A.P. and H.E. Everson, The solubility of ethyl acetate in water, *Journal of the American Chemical Society*, 75, 1727–1727, 1953.

A017 Andrews, L.J. and R.M. Keefer, Cation complexes of compounds containing carbon–carbon double bonds. IV. The argentation of aromatic hydrocarbons, *Journal of the American Chemical Society*, 72, 5801–5801, 1950.

A018 Albert, A. and D.J. Brown, Purine studies. Part I. Stability to acid and alkali. Solubility. Ionization. Comparison with pteridines, *Journal of the Chemical Society* (London), 2060–2071, 1954.

A019 Albert, A., D.J. Brown, and G. Cheeseman, Pteridine studies. Part III. The solubility and the stability to hydrolysis of pteridines, *Journal of the Chemical Society* (London), 4219–4232, 1952.

A020 Albert, A., J.H. Lister, and C. Pedersen, Pteridine studies. Part X. Pteridines with more than one hydroxy- or amino-group, *Journal of the Chemical Society* (London), 4621–4628, 1956.

A021 Azaz, E. and M. Donbrow, Solubilizaiton of phenolic compounds in nonionic surface-active agents, *Journal of Colloid and Interface Science*, 57, 11–15, 1976.

A022 Albert, A., The solubility of 8-hydroxymethylpurine, *Chemistry and Industry* (London), 202–202, 1955.

A023 Avico, U., E. Ciranni Signoretti, R. Di Francesco, and E. Cingolani, Physical parameters and biological activity of organic compounds, *Bollettino Chimico Farmaceutico*, 115, 242–253, 1976.

A025 Abelson, D., C. Depatie, and V. Craddock, Interactions of testosterone with amino acids, *Archives of Biochemistry and Biophysics*, 91, 71–74, 1960.

A027 Abidova, Z.K. and G.K. Khodzhaev, A separation method for a mixture of acids: benzoic, phthalic (o-, m-, p-), trimellitic trimesic, and hemimellitic, *Uzbekskii Khimicheskii Zhurnal*, 1, 69–76, 1960.

A028 Alric, R. and R. Puech, Coefficient de Partage, Solubilite intrinseque et puissance relative de sulfonyl-urees hypoglycemiantes, *Journal of Pharmacology*, 3, 435–447, 1972.

A029 Albert, A., Surface activity and association. Ionization. *Dipole Moments*, 147–261, 1966.

A031 Attane, E.C. and T.F. Doumani, Solubilities of aliphatic dicarboxylic acids in water, *Industrial and Engineering Chemistry*, 41, 2015–2017, 1949.

A032 Andrews, L.J. and R.M. Keefer, The argentation of organic iodides, *Journal of the American Chemical Society*, 73, 5733–5736, 1951.

A034 Arakawa, Y., M. Nakano, K. Juni, and T. Arita, Physical properties of pyrimidine and purine antimetabolites. I. The effects of salts and temperature on the solubility of 5-fluorouracil, 1-(2-tetrahydrofuryl)-5-fluorouracil, 6-mercaptopurine, and thioinosine, *Chemical and Pharmaceutical Bulletin*, 24, 1654–1657, 1976.

A035 A. Albert, The solubility of quinoline and the hydroxyquinolines, *Chemistry and Industry* (London), 252–252, 1956.

A037 Alberty, R.A. and E.R. Washburn, The ternary system isobutyl alcohol–benzene–water at 25°C, *Journal of Physical Chemistry*, 49, 4–8, 1945.

A038 Andrew, M.L. and P.J. Weiss, Solubility of antibiotics in twenty-four solvents. II, *Antibiotics and Chemotherapy* (Washington, D.C.), 9, 277–279, 1959.

A039 Abel, A.L., The substituted urea herbicides, *Chemistry and Industry* (London), 1106–1112, 1957.

A040 Audrieth, L.F. and A.D.F. Toy, The aquo ammono phosphoric acids. III. The *N*-substituted derivatives of phosphoryl and thiophosphoryl triamide as hydrogen bonding agents, *Journal of the American Chemical Society*, 64, 1553–1555, 1942.

A043 Angus, W.R. and R.P. Owen, Aqueous solubilities of R- and L-mandelic acids and three *o*-acyl-R-mandelic acids, *Journal of the Chemical Society* (London), 231–232, 1943.

A044 Apelblat, A., Extraction of nitric acid and hydrochloric acid by methyl diphenyl phosphate, *Journal of the Chemical Society A: Inorganic, Physical, Theoretical*, 3459–3463, 1971.

A045 Ando, K., Der Einfluß der Salze auf die Loslichkeit des Glykokolls und des Tyrosins, *Biochemische Zeitschrift* (1948–1967), 173, 426–432, 1926.

A046 Alric, R. and R. Puech, Mesure de la solubilite intrinseque et de la constante apparente d'ionisation acide de sulfamidothiodiazols et de sulfonylurees hypoglycemiants en solution aqueuse a 37°C, *Journal of Pharmacology*, 2, 141–154, 1971.

A047 Allport, N.L., *p*-Aminobenzenesulphonamide—research paper, *Quarterly Journal of Pharmacy and Pharmacology*, 9, 560–566, 1936.

A048 Abraham, M.H., E. Ah-Sing, R.E. Marks, and R.A. Schulz, Thermodynamics of solution of two forms of DL-alpha-amino-*n*-butyric acid in water, *Journal of the Chemical Society, Faraday Transactions*, 1, 181–185, 1977.

A049 Amoore, J.E. and R.G. Buttery, Partition coefficients and comparative olfactometry, *Chemical Senses and Flavor*, 3, 57–71, 1978.

A050 Albersmeyer, W., Quantitative determination of aromatic hydrocarbons in aqueous solution, *Gas-u. Wasserfach*, 99, 269, 1958.

A052 Azarnoosh, A. and J.J. McKetta, Solubility of propylene in water, *Journal of Chemical and Engineering Data*, 4, 211–212, 1959.

A055 Audrieth, L.F. and A.W. Browne, Azido-carbondisulfide. IV. Preparation and properties of the new interhalogenoid, cyanogen azido-dithiocarbonate, *Journal of the American Chemical Society*, 52, 2799–2805, 1930.

A056 Angelescu, E., Uber Loslichkeit in losungsmittelgemischen. II. Die loslichkeit eines stoffes, der in jedem verhaltnis mit einem der losungsmittel mischbar ist, *Zeitschrift fuer Physikalische Chemie, Stoechiometrie und Verwandschaftslehre*, 138, 300–310, 1928.

A057 Almgren, M., F. Grieser, and J.K. Thomas, Dynamic and static aspects of solubilization of neutral arenes in ionic micellar solutions, *Journal of the American Chemical Society*, 101, 279–291, 1979.

A058 Aquan-Yuen, M., D. MacKay, and W.Y. Shiu, Solubility of hexane, phenanthrene, chlorobenzene, and *p*-dichlorobenzene in aqueous electrolyte solutions, *Journal of Chemical and Engineering Data*, 24, 30–34, 1979.

A059 Alexander, K.S., B. Laprade, J.W. Mauger, and A.N. Paruta, Thermodynamics of aqueous solutions of parabens, *Journal of Pharmaceutical Sciences*, 67, 624–627, 1978.

A064 Austin, D.J., K.A. Lord, and I.H. Williams, High pressure liquid chromatography of benzimidazoles, *Pesticide Science*, 7, 211–222, 1976.

A065 Ahsan, S.S. and S.M. Blaug, Interactions of tweens with some pharmaceuticals, *Drug Standards*, 28, 95–100, 1960.

A066 Anderson, B.D. and R.A. Conradi, Prostaglandin prodrugs VI: Structure–thermodynamic activity and structure–aqueous solubility relationships, *Journal of Pharmaceutical Sciences*, 69, 424–430, 1980.

A067 Asche, H., Wirkstofffreigabe aus externa, *Fette, Seifen, Anstrichmittel*, 81, 370–373, 1979.

A068 Adjei, A., J. Newburger, and A. Martin, Extended Hildebrand approach: solubility of caffeine in dioxane–water mixtures, *Journal of Pharmaceutical Sciences*, 69, 659–661, 1980.

A069 Ali, S., Degradation and environmental fate of endosulfan isomers and endosulfan sulfate in mouse, insect and laboratory model ecosystem, Unpublished, 1978.

A070 Altsybeeva, A.I., V.P. Belousov, N.V. Ovtrakht, and A.G. Morachevskii, Phase equilibria in and thermodynamic properties of the *s*-butanol water system, *Russian Journal of Physical Chemistry*, 38, 676–679, 1964.

A072 Aiello, G., Uber die Verteilungskoeffizienten der Diuretica und Narkotica und die Theorie der Narkose, *Biochemische Zeitschrift* (1948–1967), 124, 192–205, 1921.

A073 Rakshit, J.N., Morphine, codeine, and narcotine in Indian opium, *Analyst* (London), 46, 481–492, 1921.

A074 Amidon, G.L., G.D. Leesman, and R.L. Elliott, Improving intestinal absorption of water-insoluble compounds: a membrane metabolism strategy, *Journal of Pharmaceutical Sciences*, 69, 1363–1367, 1980.

A075 Altsybeeva, A.I. and A.G. Morachevskii, Phase equilibria in the ternary system *sec*-butanol-methylethylketone–water, *Zhurnal Fizicheskoi Khimii* (Moscow), 38, 1574–1579, 1964.

A076 Ammar, H.O. and H.A. Salama, Solubilization of benzothiadiazide diuretics by cetomacrogol, *Pharmazeutische Industrie*, 42, 849–851, 1980.

A078 Ammar, H.O., S.A. Ibrahim, A.A. Kassem, and S.S. Abu-Zaid, Interaction of aromatic monocarboxylic acid derivatives with amidopyrine, *Pharmazeutische Industrie*, 42, 1312–1315, 1980.

A079 Ammar, H.O. and H.A. Salama, Effect of sodium salts of toluic acids on the water-solubility of riboflavine, *Pharmazeutische Industrie*, 43, 194–197, 1981.

A080 Ammar, H.O. and H.A. Salama, Interaction between bendroflumethiazide and caffeine, *Pharmazie* (Berlin), 36, 265–266, 1981.

A081 Aboutaleb, A.E., A.A. Ali, and R.B. Salama, Micellar solubilization of quinethazone, levomepromazine, and niridazole, *Pharmazie* (Berlin), 36, 35–37, 1981.

A082 Albert, A., Pteridine studies. Part VII. The degradation of 4-, 6-, and 7-hydroxypteridine by acid and alkali, *Journal of the Chemical Society* (London), 2690–2699, 1955.

A083 Albert, A., D.J. Brown, and H.C.S. Wood, Pteridine studies. Part V. The monosubstituted pteridines, *Journal of the Chemical Society* (London), 3832–3839, 1954.

A085 Albert, A., D.J. Brown, and G. Cheeseman, Pteridine studies. Part I. Pteridine, and 2- and 4-amino- and 2- and 4-hydroxy-pteridines, *Journal of the Chemical Society* (London), 474–485, 1951.

A086 Amidon, G.E., W.I. Higuchi, and N.F.H. Ho, Theoretical and experimental studies of transport of micelle-solubilized solutes, *Journal of Pharmaceutical Sciences*, 71, 77–84, 1982.

A087 Anderson, C.A., J.C. Cavagnol, C.J. Cohen, and J.W. Young, Guthion (azinphosmethyl): organophosphorus insecticide, *Residue Reviews*, 51, 123–130, 1974.

A088 Azmin, M.N., A. Setanoians, R.G.G. Blackie, and J.F.B. Stuart, Formulation of 1,3,5-triglycidyl-*s*-triazinetrione (alpha-TGT) for intravenous injection, *International Journal of Pharmaceutics*, 10, 109–118, 1982.

A089 Ammar, H.O., S.A. Ibrahim, and T.H. El-Faham, Interaction of chlorothiazide and hydrochlorothiazide with certain amides, imides and xanthines, *Pharmazeutische Industrie*, 43, 292–295, 1981.

A090 Aboul-Enein, H.Y., Glutethimide, *Analytical Profiles of Drug Substances*, 5, 142–149, 1976.

A091 Aboul-Enein, H.Y., Propylthiouracil, *Analytical Profiles of Drug Substances*, 6, 458–463, 1977.

A092 Aboul-Enein, H.Y., A.A. Al-Badr, and S.E. Irahim, salbutamol, *Analytical Profiles of Drug Substances*, 10, 665–669, 1981.

A093 Ammar, H.O., S.A. Ibrahim, and T.H. El-Faham, Effect of aromatic hydrotropes on the solubility of some benzothiadiazines, *Pharmazie* (Berlin), 37, 36–40, 1982.

A094 Archer, W.L. and V.L. Stevens, Comparison of chlorinated, aliphatic, aromatic, and oxygenated hydrocarbons as solvents, *Industrial and Engineering Chemistry, Product Research and Development*, 16, 319–325, 1977.

A095 Agrawal, D.K. and A.V. Deshpande, Spectrophotometric determination and solubility studies of some benzothiadiazine derivatives, *Pharmazie* (Berlin), 37, 150–150, 1982.

A096 Arbuckle, W.B., Estimating activity coefficients for use in calculating environmental parameters, *Environmental Science and Technology*, 17, 537–542, 1983.

A305 Amadori, E. and W. Heupt, Dithianon, *Analytical Methods for Pesticides and Plant Growth Regulators*, 10, 181–187, 1978.

A306 Asshauer, J., K. Hommel, and T. Hoppe, Pyrazophos, *Analytical Methods for Pesticides and Plant Growth Regulators*, 10, 237–241, 1978.

A308 Amadori, E. and W. Heupt, Chlorflurecol-methyl, *Analytical Methods for Pesticides and Plant Growth Regulators*, 10, 525–532, 1978.

A314 Anliker, R. and P. Moser, The limits of bioaccumulation of organic pigments in fish: their relation to the partition coefficient and the solubility in water and octanol, *Ecotoxicology and Environmental Safety*, 13, 43–52, 1987.

A323 Adams, W.J. and K.M. Blaine, A water solubility determination of 2,3,7,8-TCDD, *Chemosphere*, 15, 1397–1400, 1986.

A324 Alonso, M. and F. Recasens, Liquid–liquid equilibrium for the system acrylonitrile–styrene–water at 338 °K, *Journal of Chemical and Engineering Data*, 31, 164–166, 1986.

A325 Akiyoshi, M., T. Deguchi, and I. Sanemasa, The vapor saturation method for preparing aqueous solutions of solid aromatic hydrocarbons, *Bulletin of the Chemical Society of Japan* (Nippon Kagakukai Bulletin), 60, 3935–3939, 1987.

A326 Akita, K. and F. Yoshida, Phase-equilibria in methanol–ethyl acetate–water system, *Journal of Chemical and Engineering Data*, 8, 484–490, 1963.

A328 Arnold, V.W. and E.R. Washburn, Ternary system isoamyl alcohol–isopropyl alcohol–water at 10, 25 and 40 degrees, *Journal of Physical Chemistry*, 62, 1088–1090, 1958.

A330 Akade, M.A., D.K. Agrawal, and J.A.K. Lauwo, Influence of polyethylene glycol 6000 and mannitol on the in-vitro dissolution properties of nitrofurantoin by the dispersion technique, *Pharmazie* (Berlin), 41, 849–851, 1986.

A335 Ahel, M. and W. Giger, Aqueous solubility of alkylphenols and alkylphenol polyethoxylates, *Chemosphere*, 26, 1461–1470, 1993.

A336 Al-Razzak, L.A. and V.J. Stella, Stability and solubility of 2-chloro-2',3'-dideoxyadenosine, *International Journal of Pharmaceutics*, 60, 53–60, 1990.

A337 Anderson, B.D., M.B. Wygant, T.X. Xiang, and V.J. Stella, Preformulation soslubility and kinetic studies of 2',3'-dideoxypurine nucleosides: potential anti-AIDS agents, *International Journal of Pharmaceutics*, 45, 27–37, 1988.

A338 Anderson, B.D. and C.Y. Chiang, Physicochemical properties of carbovir, a potential anti-HIV agent, *Journal of Pharmaceutical Sciences*, 79, 787–790, 1990.

A339 Apelblat, A. and E. Manzurola, Solubility of oxalic, malonic, succinic, adipic, maleic, malic, citric, and tartaric acids in water from 278.15 to 338.15 K, *Journal of Chemical Thermodynamics*, 19, 317–320, 1987.

A340 Apelblat, A. and E. Manzurola, Solubility of suberic, azelaic, levulinic, glycolic, and diglycolic acids in water from 278.25 to 361.35 K, *Journal of Chemical Thermodynamics*, 22, 289–292, 1990.

A341 Apelblat, A. and E. Manzurola, Solubility of ascorbic, 2-furancarboxylic, glutaric, pimelic, salicylic, and *o*-phthalic acids in water from 279.15 to 342.15 K, and apparent molar volumes of ascorbic, glutaric, and pimelic acids in water at 298.15 K, *Journal of Chemical Thermodynamics*, 21, 1005–1008, 1989.

A346 Ahmed, S.M., A.A. Abdel-Rahman, S.I. Saleh, and M.O. Ahmed, Comparative dissolution characteristics of bropirimine–beta-cyclodextrin inclusion complex and its solid dispersion with PEG 6000, *International Journal of Pharmaceutics*, 96, 5–11, 1993.

A348 Acarturk, A., A. Sencan, and N. Celebi, Evaluation of the effect of low-molecular chitosan on the solubility and dissolution characteristics of spironolactone, *Pharmazie* (Berlin), 48, 605–607, 1993.

A350 Ammar, H.O. and S.A. El-Nahhas, Effect of aromatic hydrotropes on the solubility of allopurinol. Part 2: effect of nicotinamide and sodium salts of benzoic, naphthoic and nicotinic acids, *Pharmazie* (Berlin), 48, 534–536, 1993.

A351 Ammar, H.O. and S.A. El-Nahhas, Effect of aromatic hydrotropes on the solubility of allopurinol. Part 3: sodium salts of toluic acids, *Pharmazie* (Berlin), 48, 751–754, 1993.

A352 Aboutaleb, A.E., A.A.A. Rahman, and S. Ismail, Studies of cyclodextrin inclusion complexes: I. Inclusion complexes between alpha- and beta-cyclodextrins and chloramphenicol in aqueous solutions, *Drug Development and Industrial Pharmacy*, 12, 2259–2265, 1986.

A355 Alschaibani, H.A. and E.-E.A. Abu-Gharib, Transfer chemical potentials, solubility and reactivity of psoralen and 8-hydroxypsoralen in binary aqueous methanol mixtures, *Journal of the Chinese Society*, 42, 37–42, 1995.

A356 Arce, A., A. Blanco, P. Souza, and I. Vidal, Liquid–liquid equilibria of the ternary mixture water + propanoic acid + methyl ethyl ketone and water + propanoic acid + methyl propyl ketone, *Journal of Chemical and Engineering Data*, 40, 225–229, 1995.

A400 Achard, C., M. Jaoui, M. Schwing, and M. Rogalski, Aqueous solubilities of phenol derivatives by conductivity measurements, *Journal of Chemical and Engineering Data*, 41, 504–507, 1996.

A401 Ajisaka, N., K. Hara, K. Mikuni, and K. Hara, Effects of branched cyckodextrins on the solubility and stability of terpenes, *Bioscience, Biotechnology, and Biochemistry*, 64, 731–734, 2000.

A404 Andersin, R., Solubility and acid–base behaviour of midazolam in media of different pH, studied by ultraviolet spectrophotometry with multicomponent software, *Journal of Pharmaceutical and Biomedical Analysis*, 9, 451–455, 1991.

A405 Apelblat, A. and E. Manzurola, Solubilities of L-aspartic, DL-aspartic, DL-glutamic, *p*-hydroxybenzoic, *o*-anistic, *p*-anisic, and itaconic acids in water from T = 278 K to T = 345 K, *Journal of Chemical Thermodynamics*, 29, 1527–1533, 1997.

A406 Aranovich, G. and M. Donohue, Multilayer adsorption of slightly soluble organic compounds from aqueous solutions, *Journal of Colloid and Interface Science*, 178, 764–769, 1996.

A407 Armstrong, N., K. James, and C. Wong, Inter-relationships between solubilities, distribution coefficients and melting points of some substituted benzoic and phenylacetic acids, *Journal of Pharmacy and Pharmacology*, 31, 627–631, 1979.

A408 Avdeef, A., C. Berger, and C. Brownell, pH-metric solubility. 2: correlation between and acid–base titration and the saturation shake-flask solubility pH methods, *Pharmaceutical Research*, 17, 85–89, 2000.

A410 Arima, H., K. Yunomae, K. Miyake, and K. Uekama, Comparative studies of the enhancing effects of cyclodextrins on the solubility and oral bioavailability of tacrolimas in rats, *Journal of Pharmaceutical Sciences*, 90, 690–701, 2001.

A411 Anderson, B.D. and R.A. Conradi, Predictive relationships in the water solubility of salts of a nonsteroidal anti-inflammatory drug, *Journal of Pharmaceutical Sciences*, 74, 815–820, 1985.

A412 Avdeef, A. and C.M. Berger, pH-metric solubility. 3. Dissolution titration template method for solubility determination, *European Journal of Pharmaceutical Sciences*, 14, 281–291, 2001.

A413 An, Y.-J., E.R. Carraway, and M.A. Schlautman, Solubilization of polycyclic aromatic hydrocarbons by perfluorinated surfactant micelles, *Water Research*, 36, 300–308, 2002.

A414 Alexander, J., R.A. Fromtling, J.A. Bland, and E.C. Gilfillan, (Acylaxo)alkyl carbamate prodrugs of norfloxacin, *Journal of Medicinal Chemistry*, 34, 78–81, 1991.

A417 Avila, C.M. and F. Martinez, Themodynamic study of the solubility of benzocaine in some organic and aqueous solvents, *Journal of Solution Chemistry*, 31, 975–985, 2002.

A418 Al Omari, M.M., M.B. Zughul, J.E.D. Davies, and A.A. Badwan, Factors contributing to solubility synergism of some basic drugs with B-cyclodextrin in ternary molecular complexes, *Journal of Inclusion Phenomena and Macrocyclic Chemistry*, 54, 159–164, 2006.

A420 Abed, Y., N. Gabas, M.L. Delia, and T. Bounahmidi, Measurement of liquid–solid phase equilibrium in ternary systems of water–sucrose–glouse and water–sucrose–fructose, and predictions with UNIFAC, *Fluid Phase Equilibria*, 73, 175–184, 1992.

A423 Abdul-Fattah, A.M. and H.N. Bhargava, Preparation and in vitro evaluation of solid dispersions of halofantrine, *International Journal of Pharmaceutics*, 235, 17–33, 2002.

A426 Aungst, B.J., P-glycoprotein, secretory transport, and other barriers to the oral delivery of anti-HIV drugs, *Advanced Drug Delivery Reviews*, 39, 105–116, 1999.

A427 Antonic, J. and E. Heath, Determination of NSAIDs in river sediment samples, *Analytical and Bioanalytical Chemistry*, 387, 1337–1342, 2007.

B001 Bennett, G.M. and W.G. Philip, The influence of structure on the solubilities of ethers. Part II. Some cyclic ethers, *Journal of the Chemical Society* (London), 1937–1942, 1928.

B002 Bennett, G.M. and W.G. Philip, The influence of structure on the solubilities of ethers. Part I. Aliphatic ethers, *Journal of the Chemical Society* (London), 1930–1937, 1928.

B003 Bohon, R.L. and W.F. Claussen, The solubility of aromatic hydrocarbons in water, *Journal of the American Chemical Society*, 73, 1571–1578, 1951.

B004 Butler, J.A.V. and C.N. Ramchandani, The solubility of non-electrolytes. Part II. The influence of the polar group on the free energy of hydration of aliphatic compounds, *Journal of the Chemical Society* (London), 952–955, 1935.

B009 Back, E. and B. Steenberg, Simultaneous determination of ionization constant, solubility product and solubility for slightly soluble acids and bases. Electrolytic constants for abietic acid, *Acta Chemica Scandinavica*, 4, 810–815, 1950.

B010 Bauguess, C.T., F. Sadik, J.H. Fincher, and C.W. Hartman, Hydrolysis of fatty acid esters of acetaminophen in buffered pancreatic lipase systems I, *Journal of Pharmaceutical Sciences*, 64, 117–120, 1975.

B011 Breon, T.L. and A.N. Paruta, Solubility profiles for several barbiturates in hydroalcoholic mixtures, *Journal of Pharmaceutical Sciences*, 59, 1306–1313, 1970.

B012 Barry, B.W. and D.I.D. El Eini, Solubilization of hydrocortisone, dexamethasone, testosterone and progesterone by long-chain polyoxyethylene surfactants, *Journal of Pharmacy and Pharmacology*, 28, 210–218, 1976.

B013 Bischoff, F. and R.D. Stauffer, The dispersion of testosterone in aqueous bovine serum albumin solution, *Journal of the American Chemical Society*, 76, 1962–1965, 1954.

B014 Batra, S., Aqueous solubility of steroid hormones: an explanation for the discrepancy in the published data, *Journal of Pharmacy and Pharmacology*, 27, 777–779, 1975.

B016 Buchi, J., X. Perlia, and A. Strassle, Beziehungen zwischen den physikalisch-chemischen eigenschaften, der chemischen reaktivitat und der lokalanasthetischen wirkung in der reihe der 4-aminobenzoesaure-alkylester, *Arzneimittel-Forschung*, 16, 1657–1668, 1966.

B018 Buchi, J. and X. Perlia, Water-solubility and turbidity-ph of local anesthetic bases in homologous series, *Arzneimittel-Forschung*, 10, 544–549, 1960.

B019 Booth, H.S. and H.E. Everson, Hydrotropic solubilities-solubilities in 40 per cent sodium xylenesul-fonate, *Industrial and Engineering Chemistry*, 40, 1491–1493, 1948.

B028 Borsook, H. and J.W. Dubnoff, The biological synthesis of hippuric acid in vitro, *Journal of Biological Chemistry*, 132, 307–324, 1940.

B030 Burger, A., Dissolution and polymorphism of metolazone, *Arzneimittel-Forschung*, 25, 24–27, 1975.

B031 Bailey, C.R., The increased solubility of phenolic substances in water on addition of a third substance, *Journal of the Chemical Society* (London), 123, 2579–2589, 1923.

B032 Bull, H.B., K. Breese, and C.A. Swenson, Solubilities of amino acids, *Biopolymers*, 17, 1091–1100, 1978.

B033 Baker, E.G., Origin and migration of oil, *Science*, 129, 871–874, 1959.

B036 Bowman, M.C., F. Acree, Jr., and M.K. Corbett, Solubility of carbon-14 DDT in water, *Journal of Agricultural and Food Chemistry*, 8, 406–408, 1960.

B038 Butler, J.A.V., D.W. Thomson, and W.H. Maclennan, The free energy of the normal aliphatic alcohols in aqueous solution. Part I. The partial vapour pressures of aqueous solutions of methyl, *n*-propyl, and *n*-butyl alcohols. Part II. The solubilities of some normal alipathic alocoholsin water. Part III. The theory of binary solutions and its application to aqueous solutions, *Journal of the Chemical Society* (London), 674–686, 1933.

B039 Bates, T.R., S.-L. Lin, and M. Gibaldi, Solubilization and rate of dissolution of drugs in the presence of physiologic concentrations of lysolecithin, *Journal of Pharmaceutical Sciences*, 56, 1492–1495, 1967.

B040 Breusch, F.L. and E. Ulusoy, Physikalische Eigenschaften Homologer Kristallisierter Reihen mit Alter-nierendem und Nicht Alternierendem Schmelzpunkt, *Fette, Seifen, Anstrichmittel*, 66, 739–742, 1964.

B041 Bischoff, F. and H.R. Pilhorn, The state and distribution of steroid hormones in biologic systems. III. Solubilities of testosterone, progesterone, and alpha-estradiol in aqueous salt and protein solution and in serum, *Journal of Biological Chemistry*, 174, 663–682, 1948.

B042 Bolton, S., D. Guttman, and T. Higuchi, Complexes formed in solution by homologs of caffeine, *Journal of the American Pharmaceutical Association, Scientific Edition*, 46, 38–41, 1957.

B043 Bates, T.R., M. Gibaldi, and J.L. Kanig, Solubilizing properties of bile salt solutions I, *Journal of Pharmaceutical Sciences*, 55, 191–199, 1966.

B044 Bates, T.R., J.M. Young, C.M. Wu, and H.A. Rosenberg, pH-dependent dissolution rate of nitrofurantoin from commercial suspensions, tablets, and capsules, *Journal of Pharmaceutical Sciences*, 63, 643–645, 1974.

B045 Bates, T.R., M. Gibaldi, and J.L. Kanig, Solubilizing properties of bile salt solutions II, *Journal of Pharmaceutical Sciences*, 55, 901–906, 1966.

B046 Bandelin, F.J. and W. Malesh, The solubility of various sulfonamides employed in urinary tract infec-tions, *Journal of the American Pharmaceutical Association, Scientific Edition*, 48, 177–181, 1959.

B047 Bean, H.S. and H. Berry, The bactericidal activity of phenols in aqueous solutions of soap, *Journal of Pharmacy and Pharmacology*, 3, 639–649, 1951.

B048 Blaedel, W.J. and M.A. Evenson, The solubility of *p*-iodobenzenesulfonyl chloride, *Journal of Chemical and Engineering Data*, 9, 138–139, 1964.

B049 Bendich, A., G.B. Brown, F.S. Philips, and J.B. Thiersch, The direct oxidation of adenine in vivo, *Journal of Biological Chemistry*, 183, 267–277, 1950.

B050 Albert, A., Six-membered heteroaromatic rings containing nitrogen: Correlation of structure, in *Recent Work on Naturally Occurring Nitrogen Heterocyclic Compounds*, ed. K. Schofield, Chemical Society, London, 124–133, 1955.

B052 Bidner, M.S. and M. de Santiago, Solubilite de liquides non-electrolytes dans des solution aqueuses d'electrolytes, *Chemical Engineering Science*, 26, 1484–1488, 1971.

B054 Bischoff, F., R.E. Katherman, Y.S. Yee, and J.J. Moran, Solubilities of testosterone and estradiol esters in biologic systems, *Federation Proceedings, Federation of American Societies for Experimental Biology*, 11, 189–189, 1952.

B055 Block, L.H. and R.N. Patel, Solubility and dissolution of triamcinolone acetonide, *Journal of Pharmaceutical Sciences*, 62, 617–621, 1973.

B056 Behl, C.R., L.H. Block, and M.L. Borke, Aqueous solubility of 14C-triamcinolone acetonide, *Journal of Pharmaceutical Sciences*, 65, 429–430, 1976.

B057 Bowen, D.B., K.C. James, and M. Roberts, An investigation of the distribution coefficients of some androgen esters using paper chromatography, *Journal of Pharmacy and Pharmacology*, 22, 518–522, 1970.

B058 Baldeschwieler, E.L. and H.A. Cassar, A new petroleum by-product: octane-sultone, *Journal of the American Chemical Society*, 51, 2969–2975, 1929.

B059 Buffington, C. and H. Turndorf, Anesthetics alter the solubility of nonpolar compounds in water, *Bulletin of New York Academy Medicine*, 52, 838–841, 1976.

B060 Belfort, G., *Selective Adsorption of Organic Homologs onto Activated Carbon from Dilute Aqueo*, 2, 207–241, 1981.

B061 Baggesgaard-Rasmussen, H. and F. Reimers, Die loslichkeit des morphins in verschiedenen losungsmitteln, *Archiv der Pharmazie und Berichte der Deutschen Pharmazeutischen Gesellschaft*, 273, 129–139, 1935.

B062 Boyland, E. and B. Green, The interaction of polycyclic hydrocarbons and purines, *British Journal of Cancer*, 16, 347–360, 1962.

B063 Bowden, S.T. and J.H. Purnell, The influence of uranyl and thorium salts on the miscibility of phenol and water, *Journal of the Chemical Society* (London), 535–538, 1954.

B064 Bourne, D.W.A., T. Higuchi, and A.J. Repta, Acetylacroninium salts as soluble prodrugs of the antineoplastic agent acronine, *Journal of Pharmaceutical Sciences*, 66, 628–631, 1977.

B065 Breon, T.L., J.W. Mauger, G.E. Osborne, and A.N. Paruta, The aqueous solubility of variously substituted barbituric acids. I. Chemical effects, *Drug Development Communications*, 2, 521–535, 1976.

B066 Bamford, D.G., D.F. Biggs, M.F. Cuthbert, and W.R. Wragg, The preparation and intravenous anaesthetic activity of tetrahydrofuran-3-ols, *Journal of Pharmacy and Pharmacology*, 22, 694–699, 1970.

B068 Bowman, M.C., J.R. King, and C.L. Holder, Benzidine and congeners: analytical chemical properties and trace analysis in five substrates, *International Journal of Environmental Analytical Chemistry*, 4, 205–223, 1976.

B069 Baker, E.G., Crude oil composition and hydrocarbon solubility, *American Chemical Society, Division of Petroleum Chemistry, Preprints*, 3, 61–69, 1958.

B070 Burger, L.L. and R.M. Wagner, Preparation and properties of some organophosphorus compounds, *Industrial and Engineering Chemistry*, Chemical & Engineering Data Series, 3, 310–313, 1958.

B071 Bottari, F., G. Di Colo, E. Nannipieri, and M.F. Serafini, Release of drugs from ointment bases. II: In vitro release of benzocaine from suspension-type aqueous gels, *Journal of Pharmaceutical Sciences*, 66, 926–928, 1977.

B073 Boman, G., P. Lundgren, and G. Stjernstrom, Mechanism of the inhibitory effects of PAS granules on the absorption of rifampicin: adsorption of rifampicin by an excipient, bentonite, *European Journal of Clinical Pharmacology*, 8, 293–299, 1975.

B074 Bailey, C.R., The condensed ternary system phenol–water–salicylic acid, *Journal of the Chemical Society* (London), 126, 1951–1965, 1925.

B075 Backer, H.J., L'acide chloromethionique, *Recueil des Travaux Chimiques des Pays-Bas*, 49, 729–734, 1930.

B076 Backer, H.J., Preparation simple de l'acide methionique, *Recueil des Travaux Chimiques des Pays-Bas*, 48, 949–935, 1929.

B077 Backer, H.J., Quelques syntheses de l'acide bromomethionique, *Recueil des Travaux Chimiques des Pays-Bas*, 48, 616–621, 1929.

B078 Boulin, C. and L.-J. Simon, Action de l'eau sur le sulfate dimethylique, *Comptes Rendus Hebdomadaires des Seances de l'Academie des Sciences*, 170, 392–394, 1920.

B079 Boulin, C. and L.-J. Simon, Action de l'eau sur le sulfure d'ethyle dichlore, *Comptes Rendus Hebdomadaires des Seances de l'Academie des Sciences*, 170, 845–848, 1920.

B080 Balykova, L.A., G.P. Verkholetova, N.S. Lebedeva, and A.V. Starkov, Solubility and bactericidal activity of 1,3-dichlorohydantoin, 1,3-dichloro-5-methylhydantoin and trichloroisocyanuric Acid, *Zhurnal Mikrobiologii, Epidemiologii i Immunobiologii*, 44, 14–18, 1967.

B083 Biggar, J.W. and R.L. Riggs, Apparent solubility of organochlorine insecticides in water at various temperatures, *Hilgardia*, 42, 383–391, 1974.

B085 Bockris, J.O. and H. Egan, The salting-out effect and dielectric constant, *Transactions of the Faraday Society*, 43, 151–159, 1947.

B086 Brown, R.L. and S.P. Wasik, A method of measuring the solubilities of hydrocarbons in aqueous solutions, *Journal of Research of the National Bureau of Standards*, 78, 453–460, 1974.

B088 Bancroft, W.D. and F.J.C. Butler, Solubility of succinic acid in binary mixtures, *Journal of Physical Chemistry*, 36, 2515–2520, 1932.

B090 Bergen, Jr., R.L. and F.A. Long, The salting in of substituted benzenes by large ion salts, *Journal of Physical Chemistry*, 60, 1131–1135, 1956.

B092 Booth, H.S. and H.E. Everson, Hydrotropic solubilities–solubilities in aqueous sodium *o*-, *m*-, and *p*-xylenesulfonate solutions, *Industrial and Engineering Chemistry*, 42, 1536–1537, 1950.

B093 Biggar, J.W., G.R. Dutt, and R.L. Riggs, Predicting and measuring the solubility of p,p'-DDT in water, *Bulletin of Environmental Contamination and Toxicology*, 2, 90–100, 1967.

B094 Burgess, J. and R.I. Haines, Solubilities of 1,10-phenanthroline and substituted derivatives in water and in aqueous methanol, *Journal of Chemical and Engineering Data*, 23, 196–197, 1978.

B095 Bennetto, H.P. and J.W. Letcher, Solubility and solvation of bipyridyls and biphenyl in water, *Chemistry and Industry* (London), 847–848, 1972.

B096 Brust, H.F., A summary of chemical and physical properties of DURSBAN, *Down to Earth*, 22, 21–22, 1966.

B097 Bockris, J.O., J. Bowler-Reed, and J.A. Kitchener, The salting-in effect, *Transactions of the Faraday Society*, 47, 184–192, 1951.

B099 Blicke, F.F. and E.S. Blake, Local anesthetics in the pyrrole series, *Journal of the American Chemical Society*, 52, 235–240, 1930.

B100 Brian, R.C., *The History and Classification of Herbicides*, Book Chapter, 1, Academic Press, New York, 13–54, 1976.

B101 Bradfield, A.E. and A.F. Williams, The solubility of certain anilides in water–acetic acid mixtures, *Journal of the Chemical Society* (London), 2542–2544, 1929.

B102 Bogert, M.T. and J. Ehrlich, A unique case of a liquid that exhibits a minimum solubility in an unstable region, *Journal of the American Chemical Society*, 41, 741–745, 1919.

B103 Blix, G., Uber die loslichkeitsverhaltnisse von cystin im harn, *Hospodarsky Zpravodaj*, 178, 109–115, 1928.

B104 Buchowski, H., W. Jodzewicz, R. Milek, and A. Maczynski, Solubility and hydrogen bonding. Part I. Solubility of 4-nitro-5-methylphenol in one-component solvents, *Roczniki Chemii*, 49, 1879–1887, 1975.

B106 Braham, J.M., Some physical properties of mannite and its aqueous solutions, *Journal of the American Chemical Society*, 41, 1707–1719, 1919.

B107 Boyle, M., The Conductivities of iodoanilinesulphonic acids, *Journal of the Chemical Society* (London), 115, 1505–1517, 1919.

B108 Beech, D.G. and S. Glasstone, Solubility influences. Part V. The influence of aliphatic alcohols on the solubility of ethyl acetate in water, *Journal of the Chemical Society* (London), 67–70, 1938.

B109 Bhagwat, W.V. and N.R. Dhar, Dissociation constants of some inorganic acids from solubility measurements, *Journal of the Indian Chemical Society*, 6, 807–822, 1929.

B110 Biamonte, A.R. and G.H. Schneller, Observations on the solubility of certain sulfonamides, *Journal of the American Pharmaceutical Association, Scientific Edition*, 41, 341–345, 1952.

B111 Boutaric, A. and G. Corbet, Sur la temperature critique de dissolution de l'acroleine et de l'eau et sur la masse moleculaire de la resine d'acroleine soluble, *Comptes Rendus Hebdomadaires des Seances de l'Academie des Sciences*, 183, 42–44, 1926.

B112 Bell, E.V. and G.M. Bennett, Stereoisomerism of disulphoxides and related substances. Part IV. Di- and tri-sulphoxides of trimethylene trisulphide, *Journal of the Chemical Society* (London), 15–19, 1929.

B113 Brodsky, A.E. and M.I. Alferow, Uber die loslichkeit des benzochinhydrons in wasserigem alkohol, *Berichte der Deutschen Chemischen Gesellschaft*, 62, 2132–2133, 1929.

B114 Bastami, S.M. and M.J. Groves, Some factors influencing the in vitro release of phenytoin from formulations, *International Journal of Pharmaceutics*, 1, 151–153, 1978.

B115 Biltz, H. and M. Heyn, Alpha, zeta, und delta-methylharnsaure, *Annalen der Chemie*, 98–162, 1917.

B116 Biltz, H. and L. Herrmann, Uber die loslichkeit von harnsaure in wasser, *Annalen der Chemie*, 104–111, 1923.

B117 Bourgom, A., Contribution a l'etude du methylal comme solvant, *Bulletin des Societes Chimiques Belges*, 33, 101–115, 1924.

B118 Bhagwat, W.V. and S.S. Doosaj, Limitations of solubility method for determining dissociation constant, *Journal of the Indian Chemical Society*, 10, 477–490, 1933.

B119 Biilmann, E. and J. Bentzon, Uber alloxan und alloxanthin, *Berichte der Deutschen Chemischen Gesellschaft*, 51, 522–532, 1918.

B121 Berthoud, A. and S. Kunz, Solubilite et dissociation de la quinhydrone, *Helvetica Chimica Acta*, 21, 17–21, 1938.

B123 Brooks, W.B. and J.J. McKetta, The solubility of 1-butene in water, *Petroleum Refiner*, 34, 143–144, 1955.

B124 Barbaudy, J., Contribution a l'etude de la distillation des melanges ternaires heterogenes-le systeme eau-benzene-toluene, *Journal de Chimie Physique et de Physico-Chimie Biologique*, 23, 290–298, 1926.

B125 Bucherer, H.T. and R. Wahl, Uber die 2,5,1-aminonaphtolsulfonsaure (A-Saure) und ihre derivate, *Journal fuer Praktische Chemie*, 103, 129–150, 1921.

B126 Barbaudy, J., Systeme alcool ethylique-benzene-eau. I. Etude de la Surface de Trouble, *Recueil des Travaux Chimiques des Pays-Bas*, 45, 207–213, 1926.

B128 Braun, R.J. and E.L. Parrott, Influence of viscosity and solubilization on dissolution rate, *Journal of Pharmaceutical Sciences*, 61, 175–178, 1972.

B129 Blaug, S.M. and S.S. Ahsan, Interaction of parabens with nonionic macromolecules, *Journal of Pharmaceutical Sciences*, 50, 441–443, 1961.

B130 Burger, A., Zur polymorphie oraler antidiabetika, *Scientia Pharmaceutica*, 46, 207–222, 1978.

B131 Branson, D.R., *A new capacitor fluid—a case study in product stewardship*, Unpublished, 44–61, 1977.

B132 Bogardus, J.B. and R.K. Blackwood, Jr., Solubility of doxycycline in aqueous solution, *Journal of Pharmaceutical Sciences*, 68, 188–194, 1979.

B133 Bridges, J.W., S.R. Walker, and R.T. Williams, Species differences in the metabolism and excretion of sulphasomidine and sulphamethomidine, *Biochemical Journal*, 111, 173–179, 1969.

B134 Baker, D.C., T.H. Haskell, and S.R. Putt, Prodrugs of 9-beta-D-arabinofuranosyladenine. 1. Synthesis and evaluation of some 5-(o-acyl) derivatives, *Journal of Medicinal Chemistry*, 21, 1218–1221, 1978.

B135 Bachstez, M., Uber die konstitution der orotsaure, *Chemische Berichte*, 63, 1000–1007, 1930.

B136 Baykut, S., The solubility of the higher fatty acids in water, *Chemie-Physique Serie C*, 21, 36–45, 1956.

B138 Burger, A., Zur polymorphie oraler antidiabetika, *Scientia Pharmaceutica*, 43, 161–168, 1975.

B139 Beremzhanov, B.A., N.N. Nura, and R.S. Erkasov, Solubility of benzamide in aqueous solutions of sulfuric, selenic, and phophoric acids at 20°C, *Journal of General Chemistry of the USSR*, 45, 1191–1194, 1975.

B140 Burger, A. ZurPolymorphie oraler antidiabetika. I. Mitteilung: chlorpropamid, *Scientia Pharmaceutica*, 43, 152–161, 1975.

B141 Burton, J., K. Poulsen, and E. Haber, Solubility and lipophilicity relationships in the design of renin inhibitors, in *Polymeric Drugs*, eds. L. G. Donaruma, O. Vogl, Academic Press, New York, 219–237, 1978.

B142 Bolton, S., Interaction of urea and thiourea with benzoic and salicylic acids, *Journal of Pharmaceutical Sciences*, 52, 1071–1074, 1963.

B144 Baveja, S.K., V.S. Raju, M.P. Pakhetra, and S. Kaur, Formulation of intravenous osthole solution, *Indian Journal of Pharmacy*, 40, 230, 1978.

B147 Burger, A., Das Auflosungsverhalten von sulfanilamid in wasser, *Pharmazeutische Industrie*, 35, 626–633, 1973.

B148 Bogardus, J.B. and N.R. Palepu, Ionization and solubility of an amphoteric beta-lactam antibiotic, *International Journal of Pharmaceutics*, 4, 159–170, 1979.

B149 Ben-Naim, A. and J. Wilf, A direct measurement of intramolecular hydrophobic interactions, *Journal of Chemical Physics*, 70, 771–777, 1979.

B150 Brooker, P.J. and M. Ellison, The determination of the water solubility of organic compounds by a rapid turbidimetric method, *Chemistry and Industry* (London), 5, 785–787, 1974.

B151 Bittrich, H.-J., H. Gedan, and G. Feix, Zur Loslichkeitsbeeinflussung von kohlenwasserstoffen in wasser (Effects on the solubility of hydrocarbons in water), *Zeitschrift fuer Physikalische Chemie* (Leipzig), 260, 1009–1013, 1979.

B152 Bundgaard, H., A.B. Hansen, and C. Larsen, Pro-drugs as drug delivery systems. III. Esters of malonuric acids as novel pro-drug types for barbituric acids, *International Journal of Pharmaceutics*, 3, 341–353, 1979.

B153 Ben-Naim, A. and J. Wilf, Solubilities and hydrophobic interactions in aqueous solutions of monoalkylbenzene molecules, *Journal of Physical Chemistry*, 84, 583–586, 1980.

B154 Bundgaard, H. and M. Johansen, Pro-drugs as drug delivery systems VIII. Bioreversible derivatization of hydantoins by n-hydroxymethylation, *International Journal of Pharmaceutics*, 5, 67–77, 1980.

B155 Bogdanova, S.V., N. Lambov, and E. Minkov, Physicochemical studies of cinnarizine-polyvinylpyrrolidone solid dispersion, *Pharmazie* (Berlin), 36, 197–199, 1981.

B156 Button, D.K., The influence of clay and bacteria on the concentration of dissolved hydrocarbon in saline solution, *Geochimica et Cosmochimica Acta*, 40, 435–440, 1976.

B157 Bugaevskii, A.A., N.R. Sumskaya, and V.O. Kruglov, The salting-out of *p*-nitrophenol in aqueous sodium chloride solutions, *Russian Journal of Physical Chemistry*, 51, 1072–1073, 1977.

B158 Brodie, B.B. and C.A.M. Hogben, Some physico-chemical factors in drug action, *Journal of Pharmacy and Pharmacology*, 9, 345–380, 1957.

B160 Burchfield, H.P., Performance of fungicides on plants and in soil—physical, chemical, and biological considerations, *Plant Pathology*, 3, 447–520, 1960.

B161 Bhavnagary, H.M. and M. Jayaram, Determination of water solubilities of lindane and dieldrin at different temperatures, *Bulletin of Grain Technology*, 12, 95–99, 1974.

B162 Biggar, J.W., L.D. Donnen, and R.L. Riggs, Soil interaction with organically polluted water. Summary report Dept. of Water Science and Engineering, University of California (1966) cf. F.A. Gunther, W.E. Westlake, and P.S. Jaglan, *Residue Reviews*, 20, 1, 1968.

B164 Behrens, R. and H.L. Morton, Some factors influencing activity of 12 phenoxy acids on mesquite root inhibition, *Plant Physiology*, 38, 165–170, 1963.

B165 Becke, H. and G. Quitzsch, Das phasengleichgewichtsverhalten ternarer systeme der art c4–alkohol–wasser–kohlenwasserstoff, *Chemische Technik* (Leipzig), 29, 49–51, 1977.

B166 Ballard, B.E., The physicochemical properties of drugs that control their absorption rate after implantation, PhD Thesis, 210–239, 1961.

B167 Barnes, Jr., F.W. and W.F. Seip, Hollow crystals from buffer solutions of sodium diethyl barbiturate, *Science*, 131, 161–161, 1960.

B169 Bowman, B.T. and W.W. Sans, The aqueous solubility of twenty-seven insecticides and related compounds, *Journal of Environmental Science & Health, Series B*, B14, 625–634, 1979.

B170 Baker, E.G., A hypothesis concerning the accumulation of sediment hydrocarbons to form crude oil, *Geochimica et Cosmochimica Acta*, 19, 309–317, 1960.

B171 Banerjee, D. and B.K. Gupta, The estimation of compound hydrophobicities and their relevance to partition coefficients, *Canadian Journal of Pharmaceutical Science*, 15, 61–63, 1980.

B173 Banerjee, S., S.H. Yalkowsky, and S.C. Valvani, Water solubility and octanol/water partition coefficients of organics. Limitations of the solubility-partition coefficient correlation, *Environmental Science and Technology*, 14, 1227–1229, 1980.

B174 Bohm, R., Physico-chemical properties of the cyclic ketals and thioketals, *Pharmazie* (Berlin), 35, 802–803, 1980.

B175 Booth, J. and E. Boyland, The reaction of the carcinogenic dibenzcarbazoles and dibenzacridines with purines and nucleic acid, *Biochimica et Biophysica Acta*, 12, 75–87, 1953.

B177 Bansal, P.C., I.H. Pitman, J.N.S. Tam, and J.J. Kaminski, N-hydroxymethyl derivatives of nitrogen heterocycles as possible prodrugs I: *n*-hydroxymethylation of uracils, *Journal of Pharmaceutical Sciences*, 70, 850–854, 1981.

B178 Brodin, A., B. Sandin, and B. Faijerson, Rates of transfer of organic protolytic solutes between an aqueous and an organic phase. V. The thermodynamics of mass transfer, *Acta Pharmaceutica Suecica*, 13, 331–352, 1976.

B179 Briggs, G.G., Theoretical and experimental relationships between soil adsorption, octanol–water partition coefficients, water solubilities, bioconcentration factors, and the parachor, *Journal of Agricultural and Food Chemistry*, 29, 1050–1059, 1981.

B181 Baer, J.E., H.L. Leidy, A.V. Brooks, and K.H. Beyer, The physiological disposition of chlorothiazide (diuril) in the dog, *Journal of Pharmacology and Experimental Therapeutics*, 125, 295–302, 1959.

B182 Banerjee, P.K. and G.L. Amidon, Physicochemical property modification strategies based on enzyme substrate specificities I: rationale, synthesis, and pharmaceutical properties of aspirin derivatives, *Journal of Pharmaceutical Sciences*, 70, 1299–1309, 1981.

B183 Bevenue, A. and H. Beckman, Pentachlorophenol: a discussion of its properties and its occurrence as a residue in human and animal tissues, *Residue Reviews*, 19, 83–87, 1967.

B185 Bailey, G.W. and J.L. White, Herbicides: A compilation of their physical, chemical, and biological properties, *Residue Reviews*, 10, 97–120, 1965.

B186 Babers, F.H., The solubility of DDT in water determined radiometrically, *Journal of the American Chemical Society*, 77, 4666–4666, 1955.

B187 Brewer, G.A., Isoniazid, *Analytical Profiles of Drug Substances*, 6, 183–229, 1977.

B188 Bhattacharyya, P.K. and W.M. Cort, Amoxicillin, *Analytical Profiles of Drug Substances*, 7, 19–35, 1978.

B189 Benezra, S.A. and T.R. Bennett, Allopurinol, *Analytical Profiles of Drug Substances*, 7, 1–4, 1978.

B190 Brik, H., Natamycin, *Analytical Profiles of Drug Substances*, 10, 513–541, 1981.

B191 Barringer, W.C., W. Shultz, G.M. Sieger, and R.A. Nash, Minocycline hydrochloride and its relationship to other tetracycline antibiotics, *American Journal of Pharmacy*, 146, 179–191, 1974.

B192 Bartley, C.E., Triazine compounds, *Farm Chemicals*, 122, 28–34, 1959.

B193 Beilstein, P., A.M. Cook, and R. Hutter, Determination of seventeen s-triazine herbicides and derivatives by high-pressure liquid chromatography, *Journal of Agricultural and Food Chemistry*, 29, 1132–1135, 1981.

B194 Bhalla, H.L., Preformulation studies on vasicinone—a bronchodilatory alkaloid (study of some physicochemical aspects), *Drug Development and Industrial Pharmacy*, 7, 755–768, 1981.

B196 Bundgaard, H., M. Johansen, V. Stella, and M. Cortese, Pro-drugs as drug delivery systems xxi. preparation, physicochemical properties and bioavailability of a novel water-soluble pro-drug type for carbamazepine, *International Journal of Pharmaceutics*, 10, 181–192, 1982.

B197 Baranaev, M.K., I.S. Gilman, L.M. Kogan, and N.P. Rodinova, Separating dichloroethane from its aqueous solutions, *Journal of Applied Chemistry of the USSR*, 27, 1031–1036, 1954.

B198 Biedermann, W. and A. Datyner, The interaction of nonionic dyestuffs with sodium dodecyl sulfate and its correlation with lipophilic parameters, *Journal of Colloid and Interface Science*, 82, 276–285, 1981.

B199 Bodor, N. and K.B. Sloan, Soft Drugs V: Thiazolidine-type derivatives of progesterone and testosterone, *Journal of Pharmaceutical Sciences*, 71, 514–520, 1982.

B200 Barrier, G.E., J.L. Hilton, R.E. Frans, and D.E. Moreland, *Herbicide Handbook of the Weed Science Society of America*, Humphrey Press, Geneva, New York, 1–353, 1970.

B201 Barduhn, A.J. and M. Handley, Low-temperature solubility of caprolactam in water, *Journal of Chemical and Engineering Data*, 27, 306–308, 1982.

B300 Bowman, B.T. and W.W. Sans, Further water solubility determinations of insecticidal compounds, *Journal of Environmental Science & Health, Series B*, B18, 221–227, 1983.

B301 Bruggeman, W.A., L.B.J. Martron, D. Kooiman, and O. Hutzinger, Accumulation and elimination kinetics of di-, tri-, and tetra-chlorobiphenyls by goldfish after dietary and aqueous exposure, *Chemosphere*, 10, 811–832, 1981.

B302 Beerbower, A., P.L. Wu, and A. Martin, Expanded solubility parameter approach i: naphthalene and benzoic acid in individual solvents, *Journal of Pharmaceutical Sciences*, 73, 179–188, 1984.

B304 Banerjee, S., Solubility of organic mixtures in water, *Environmental Science and Technology*, 18, 587–591, 1984.

B305 Bengtsson, T.A., 4-Amino-4'-chlorodiphenyl as analytical reagent for sulphate, *Analytica Chimica Acta*, 18, 353–359, 1958.

B306 Baker, D.C., S.D. Kumar, W.J. Waites, and W.J. Lambert, Synthesis and evaluation of a series of 2'-o-acyl derivatives of 9-beta-D-arabinofuranosyladenine as antiherpes agents, *Journal of Medicinal Chemistry*, 27, 270–274, 1984.

B309 Bleidner, W.E., R. Morales, and R.F. Holt, Benomyl, *Analytical Methods for Pesticides and Plant Growth Regulators*, 10, 157–171, 1978.

B310 Brady, S.S., C. Van Hoek, and V.F. Boyd, Norflurazon, *Analytical Methods for Pesticides and Plant Growth Regulators*, 10, 415–435, 1978.

B314 Beezer, A.E., P.L.O. Volpe, M.C.P. Lima, and W.H. Hunter, Solution thermodynamics for alkoxy phenols in alcohol and in water–alcohol systems, *Journal of Solution Chemistry*, 15, 341–363, 1986.

B315 Beezer, A.E., W.H. Hunter, and D.E. Storey, Enthalpies of solution of a series of *m*-alkoxy phenols in water, *n*-octanol and water–*n*-octanol mutually saturated: derivation of the thermodynamic parameters for solute transfer between these solvents, *Journal of Pharmacy and Pharmacology*, 35, 350–357, 1983.

B316 Blackmann, G.E., M.H. Parke, and G. Garton, The physiological activity of substituted phenols. II. Relationships between physical properties and physiological activity, *Archives of Biochemistry and Biophysics*, 54, 55–70, 1955.

B317 Bobra, A., W.Y. Shiu, and D. Mackay, Quantitative structure–activity relationships for the acute toxicity of chlorobenzenes to daphnia magna, *Environmental Toxicology and Chemistry*, 4, 297–305, 1985.

B318 Burris, D.R. and W.G. MacIntyre, Water solubility behavior of binary hydrocarbon mixtures, *Environmental Toxicology and Chemistry*, 4, 371–377, 1985.

B319 Billington, J.W., G.-L. Huang, F. Szeto, and D. MacKay, Preparation of aqueous solutions of sparingly soluble organic substances: I. Single component systems, *Environmental Toxicology and Chemistry*, 7, 117–124, 1988.

B321 Buur, A., H. Bundgaard, and E. Falch, Prodrugs of 5-fluorouracil. IV. Hydrolysis kinetics, bioactivation and physicochemical properties of various *n*-acyloxymethyl derivatives of 5-fluorouracil, *International Journal of Pharmaceutics*, 24, 43–60, 1985.

B322 Bundgaard, H. and E. Falch, Allopurinol prodrugs. II. Synthesis, hydrolysis kinetics and physicochemical properties of various n-acyloxymethyl allopurinol Derivatives, *International Journal of Pharmaceutics*, 24, 307–325, 1985.

B323 Bundgaard, H. and E. Falch, Allopurinol Prodrugs. III. Water-soluble N-acyloxymethyl allopurinol derivatives for rectal or parenteral use, *International Journal of Pharmaceutics*, 25, 27–39, 1985.

B324 Bowman, B.T. and W.W. Sans, Effect of temperature on the water solubility of insecticides, *Journal of Environmental Science & Health, Series B*, 20, 625–631, 1985.

B325 Bowman, B.T. and W.W. Sans, Adsorption, desorption, soil mobility, aqueous persistence and octanol–water partitioning coefficients of terbufos, terbufos sulfoxide and terbufos sulfone, *Journal of Environmental Science & Health, Series B*, 17, 447–462, 1982.

B328 Bundgaard, H., U. Klixbull, and E. Falch, Prodrugs as drug delivery systems. 44. *O*-acyloxymethyl, *o*-acyl and *n*-acyl salicylamide derivatives as possible prodrugs for salicylamide, *International Journal of Pharmaceutics*, 30, 111–121, 1986.

B331 Bundgaard, H. and N.M. Nielsen, Glycolamide esters as a novel biolabile prodrug type for non-steroidal anti-inflammatory carboxylic acid drugs, *International Journal of Pharmaceutics*, 43, 101–110, 1988.

B332 Buur, A. and H. Bundgaard, Prodrugs of 5-Fluorouracil VIII. Improved rectal and oral delivery of 5-fluorouracil via various prodrugs. Structure–rectal absorption relationships, *International Journal of Pharmaceutics*, 36, 41–49, 1987.

B333 Baughman, G.L. and T.A. Perenich, Fate of dyes in aquatic systems: I. Solubility and partitioning of some hydrophobic dyes and related compounds, *Environmental Toxicology and Chemistry*, 7, 183–199, 1988.

B335 Brisset, J.L., Solubilities of various nitroanilines in water–pyridine, water–acetonitrile, and water–ethylene glycol solvents, *Journal of Chemical and Engineering Data*, 30, 381–383, 1985.

B337 Bonati, M., F. Gaspari, V. D'Aranno, and G. Tognoni, Physicochemical and analytical characteristics of amiodarone, *Journal of Pharmaceutical Sciences*, 73, 829–831, 1984.

B338 Bronaugh, R.L. and R.F. Stewart, Methods for in vitro percutaneous absorption studies iii: hydrophobic compounds, *Journal of Pharmaceutical Sciences*, 73, 1255–1258, 1984.

B340 Bundgaard, H., E. Falch, C. Larsen, and T.J. Mikkelson, Pilocarpine prodrugs. II. Synthesis, stability, bioconversion, and physicochemical properties of sequentially labile pilocarpine acid diesters, *Journal of Pharmaceutical Sciences*, 75, 775–783, 1986.

B342 Bolden, P.L., J.C. Hoskins, and A.D. King, Jr., The solubility of gases in solutions containing sodium alkylsulfates of various chain lengths, *Journal of Colloid and Interface Science*, 91, 454–463, 1983.

B348 Burris, D.R. and W.G. MacIntyre, Solution of hydrocarbons in a hydrocarbon-water system with changing phase composition due to evaporation, *Environmental Science and Technology*, 20, 296–299, 1986.

B349 Baker, R.J., B.J. Donelan, L.J. Peterson, and C.-C. Tsai, Correlation and estimation of aqueous solubilities of halogenated benzenes, *Physics and Chemistry of Liquids*, 16, 279–292, 1987.

B350 Billington, J.B., *Journal of Physical Chemical Reference Data*, 15, 7–9, 1986.

B351 Bailey, R.E., W.L. Rhinehart, S.J. Gonsior, and W.B. Neely, Hazard assessment of monochloro biphenyl in the aquatic environment. A Case History, Presentation at a Meeting, 1981.

B352 Barton, A.F.M. and J. Tjandra, Ternary phase equilibrium studies of the water–ethanol-1,8-cineole system, *Fluid Phase Equilibria*, 44, 117–123, 1988.

B353 Beezer, A.E., M.C.P. Lima, G.G. Fox, and B.V. Smith, Solution thermodynamics for *o*-alkoxyphenols in water and in water–alcohol systems, *Thermochimica Acta*, 116, 329–335, 1987.

B354 Bockstanz, G.L., M. Buffa, and C.T. Lira, Solubilities of alpha-anhydrous glucose in ethanol/water mixtures, *Journal of Chemical and Engineering Data*, 34, 426–429, 1989.

B355 Beezer, A.E., S. Forster, W.B. Park, and G.J. Rimmer, Solution thermodynamics of 4-hydroxybenzoates in water, 95% ethanol–water, 1-octanol and hexane, *Thermochimica Acta*, 178, 59–65, 1991.

B356 Burris, D.R. and W.G. MacIntyre, Water solubility behavior of hydrocarbon mixtures—implications for petroleum dissolution, J.H., Vandermeulen, S.E., Hrudrey, eds., Pergamon, New York, 85–113, 1987.

B360 Benjamin, E.J., B.A. Firestone, R. Bergstrom, and Y.Y.T. Lin, Selection of a derivative of the antiviral agent 9-[(1,3-dihydroxy-2-propoxy)-methyl]guanine (DHPG) with improved oral absorption, *Pharmaceutical Research*, 4, 120–123, 1987.

B361 Berner, B., D.R. Wilson, R.H. Guy, and H.I. Maibach, The relationship of pK_a and acute skin irritation in man, *Pharmaceutical Research*, 5, 660–663, 1988.

B362 Bisrat, M., E.K. Anderberg, M.I. Barnett, and C. Nystrom, Physicochemical Aspects of Drug Release. XV. Investigation of diffusional transport in dissolution of suspended, sparingly soluble drugs, *International Journal of Pharmaceutics*, 80, 191–201, 1992.

B363 Blanchard, J., J.O. Boyle, and S.V. Wagenen, Determination of the partition coefficients, acid disso-ciation constants, and intrinsic solubility of carbenoxolone, *Journal of Pharmaceutical Sciences*, 77, 548–552, 1988.

B366 Brewster, M.E., K.S. Estes, T. Loftsson, and N. Bodor, Improved delivery through biological mem-branes. XXXI: solubilization and stabilization of an estradiol chemical delivery system by modified B-cyclodextrins, *Journal of Pharmaceutical Sciences*, 77, 981–985, 1988.

B376 Borel, J.F., *Ciclosporin*, Karger, Basel, 1986.

B384 Belaj, F., R. Tripolt, and E. Nachbaur, Kristallstruktur und thermisches verhalten der additionsverbindun-gen von trithiocyanursaure mit tetrahydrofuran und 1,4-dioxan, *Monatshefte fuer Chemie*, 121, 99–108, 1990.

B385 Blyshak, L.A., K.Y. Dodson, G. Patonay, and W.E. May, Determination of cyclodextrin formation con-stants using dynamic coupled-column liquid chromatography, *Analytical Chemistry*, 61, 955–960, 1989.

B386 Braxton, B.K. and J.H. Rytting, Solubilities and solution thermodynamics of several substituted melamines, *Thermochimica Acta*, 154, 27–47, 1989.

B387 Boje, K.M., M. Sak, and H.-L. Fung, Complexation of nifedipine with substituted phenolic ligands, *Pharmaceutical Research*, 5, 655–659, 1988.

B388 Buur, A. and H. Bundgaard, Prodrugs of 5-fluorouracil. III. Hydrolysis kinetics in aqueous solution and biological media, lipophilicity and solubility of various 1-carbamoyl derivatives of 5-fluorouracil, *International Journal of Pharmaceutics*, 23, 209–222, 1985.

B390 Baba, K., Y. Takeichi, and Y. Nakai, Molecular behavior and dissolution characteristics of uracil in ground mixtures, *Chemical and Pharmaceutical Bulletin*, 38, 2542–2546, 1990.

B391 Biswas, P.K., S.C. Lahiri, and B.P. Dey, Solvational behavior of some substituted benzoic acids in etha-nol–water mixtures at 298.15 K, *Bulletin of the Chemical Society of Japan* (Nippon Kagakukai Bulletin), 66, 2785–2789, 1993.

B393 Bharath, A., C. Mallard, D. Orr, and A. Smith, Problems in determining the water solubility of organic compounds, *Bulletin of Environmental Contamination and Toxicology*, 33, 133–137, 1984.

B394 Brandani, S., V. Brandani, and D. Flammini, Solubility of trioxane in water, *Journal of Chemical and Engineering Data*, 39, 201–201, 1994.

B396 Brewster, M.E., J.W. Simpkins, M.S. Hora, and N. Bodor, The potential use of cyclodextrins in parenteral formulations, *Journal of Parenteral Science and Technology*, 43, 231–240, 1989.

B397 Brewster, M.E., W.R. Anderson, K.S. Estes, and N. Bodor, Development of aqueous parenteral formula-tions for carbamazepine through the use of modifies cyclodextrins, *Journal of Parenteral Science and Technology*, 80, 380–383, 1991.

B402 Becket, G., L. Schep, and M. Tan, Improvement of the in vitro dissolution of praziquantel by complex-ation with alpha-, beta- and gamma-cyclodextrins, *International Journal of Pharmaceutics*, 179, 65–71, 1999.

B403 Benes, M. and V. Dohnal, Limiting activity coefficients of some aromatic and aliphatic nitro compounds in water, *Journal of Chemical and Engineering Data*, 44, 1097–1102, 1999.

B404 Bevan, C. and R. Lloyd, A high-throughput screening method for the determination of aqueous drug solubility using laser nephelometry in microtiter plates, *Analytical Chemistry*, 72, 1781–1787, 2000.

B405 Beyers, H., S.F. Malan, and J.G. van der Watt, Structure–solubility relationship and thermal decomposi-tion of furosemide, *Drug Development and Industrial Pharmacy*, 26, 1077–1083, 2000.

B406 Bhardwaj, R., R. Dorr, and J Blanchard, Approaches to reducing toxicity of parenteral anticancer drug formulations using cyclodextrins, *PDA Journal of Pharmaceutical Science and Technology*, 54, 233–239, 2000.

B407 Breslow, R. and S. Halfon, Quantitative effects of anithydrophobic agents on binding constants and solu-bilities in water, *Proc. Natl. Acad. Sci. USA*, 89, 6916–6918, 1992.

B408 Brewster, M., W. Anderson, T. Loftsson, and E. Pop, Preparation, characterization, and anesthetic prop-erties of 2-hydroxypropyl-beta-cyclodextirn complexes of pregnanolone and pregnenolone in rat and mouse, *Journal of Pharmaceutical Sciences*, 84, 1154–1159, 1995.

B409 Bodor, N., J. Drustrup, and W. Wu, Effect of cyclodextrins on the solubility and stability of a novel soft corticosteroid, loteprednol etabonate, *Pharmazie*, 55, 206–209, 2000.

B410 Brochsztain, S. and M. Politi, Solubilization of 1, 4, 5, 8-naphthalenediimides and 1,8-maphthalimides through the formation of novel host–guest complexes with alpha, *Langmuir*, 15, 4486–4494, 1999.

B411 Burger, A., K. Koller, and W. Schiermeier, RS-ibuprofen and S-ibuprofen (dexibuprofen)-binary sys-tem and unusual solubility behaviour, *European Journal of Pharmaceutis and Biopharmaceutics*, 43, 142–147, 1996.

B412 Burger, A. and A. Lettenbichler, Polymorphism and preformulation studies of lifibrol, *European Journal of Pharmaceutics and Biophamaceutics*, 49, 65–72, 2000.

B413 Burrows, H., M. Miguel, A. Varela, and R. Becker, The aqueous solubility and thermal behaviour of some beta-carbolines, *Thermochimica Acta*, 279, 77–82, 1996.

B414 Butvin, P., J. Al-Ja'afreh, J. Svetlik, and E. Havranek, Solubility, stability, and dissociation constants of (2rs,4r)-2-substituted thiazolidine-4-carboxylic acids in aqueous solutions, *Chemistry Papers*, 53, 315–322, 1999.

B416 Beall, H. and K. Sloan, Topical delivery of 5-fluorouracil (5-FU) by 3-alkylcarbonyl-5-FU prodrugs, *International Journal of Pharmaceutics*, 217, 127–137, 2001.

B417 Benavente-Garcia, O., J. Castillo, M.J. Del Bano, and J. Lorente, Improved water solubility of neohesperidin dihydrochalcone in sweetener blends, *Journal of Agricultural and Food Chemistry*, 49, 189–191, 2001.

B418 Berberich, K.A., V.C. Reinsborough, and C.N. Shaw, Kinetic and solubility studies in zwitterionic surfactant solutions, *Journal of Solution Chemistry*, 29, 1017–1026, 2000.

B419 Bundgaard, H., E. Jensen, and E. Falch, Water-soluble, solution-stable, and biolabile n-substituted (aminomethyl) benzoate ester prodrugs of acyclovir, *Pharmaceutical Research*, 8, 1087–1093, 1991.

B420 Brix, R., S. Hvidt, and L. Carlsen, Solubility of nonylphenol and nonylphenol ethoxylates. On the possible role of micelles, *Chemosphere*, 44, 759–763, 2001.

B421 Brown, J.R., J.H. Collett, D. Attwood, and E.E. Sims, Physicochemical and biopharmaceutical characterization of BTA-243, a diacidic drug with low oral bioavailability, *International Journal of Pharmaceutics*, 213, 127–134, 2001.

B422 Bergstrom, C.A.S., U. Norinder, K. Luthman, and P. Artursson, Experimental and computational screening models for prediction of aqueous drug solubility, *Pharmaceutical Research*, 19, 182–188, 2002.

B423 Barton, A.F.M. and J. Tjandra, Ternary phase equilibrium studies of the water–ethanol–1,8-cineole system, *Fluid Phase Equilibria*, 44, 117–123, 1988.

B424 Buschmann, H.-J., E. Cleve, K. Jansen, and E. Schollmeyer, The determination of complex stabilities between different cyclodextrins and dibenzo-18-crown-6, cucurbit[6]uril, decamethylcucurbit[5]uril, cucurbit[5]uril, p-tert-butylcalix[4]arene and p-tert-butylcalix[6]arene in aqueous solutions using a spectrosphometric method, *Materials Science and Engineering C*, 14, 35–39, 2001.

B425 Bock, M.G., R.M. DiPardo, B.E. Evans, and D.F. Veber, Renin inhibitors containing hydrophilic groups. Tetrapeptides with enhanced aqueous solubility and nanomolar potency, *Journal of Medicinal Chemistry*, 31, 1918–1923, 1988.

B426 Bongiorno, D., L. Ceraulo, A. Mele, and V.T. Liveri, Structural and physicochemical characterization of the inclusion complexes of cyclomaltooligosaccharides (cyclodextrins) with melatonin, *Carbohydrate Research*, 337, 743–754, 2002.

B427 Bundgaard, H. and N.M. Nielsen, Esters of N,N-disubstituted 2-hydroxyacetamides as a novel highly biolabile prodrug type for carboxylic acid agents, *Journal of Medicinal Chemistry*, 30, 451–454, 1987.

B428 Bundgaard, H. and E. Falch, Allopurinol prodrugs. I. Synthesis, stability and physicochemical properties of various N1-acyl allopurinol derivatives, *International Journal of Pharmaceutics*, 23, 223–237, 1985.

B429 Bustamante, P., J. Navarro, S. Romero, and B. Escalera, Thermodynamic origin of the solubility profile of drugs showing one or two maxima against the polarity of aqueous and nonaqueous mixtures: niflumic acid and caffeine, *Journal of Pharmaceutical Sciences*, 91, 874–883, 2002.

B431 Boudad, H., P. Legrand, G. Lebas, and G. Ponchel, Combined hydroxypropyl-beta-cyclodextrin and poly(alkylcyanoacrylate) nanoparticles intended for oral administration of saquinavir, *International Journal of Pharmaceutics*, 218, 113–124, 2001.

B433 Bertau, M. and G. Jorg, Saccharidesas efficacious solubilisers for highly lipophilic compounds in aqueous media, *Bioorganic & Medicinal Chemistry*, 12, 2973–2983, 2004.

B434 Baena, Y., J.A. Pinzon, H.J. Barbosa, and F. Martinez, Temperature-dependance of the solubility of some acetanilide derivatives in several organic and aqueous solvents, *Physics and Chemistry of Liquids*, 42, 603–613, 2004.

B435 Banerjee, R., P.M. Bhatt, N.V. Ravindra, and G.R. Desiraju, Saccharin salts of active pharmaceutical ingredients, their crystal structures, and increased water solubilities, *Crystal Growth and Design*, 5, 2299–2309, 2005.

B438 Bala, I., V. Bhardwaj, S. Hariharan, and M.N.V Ravi Kumar, Analytical methods for assay of ellagic acid and its solubility studies, *International Journal of Pharmaceutics*, 40, 206–210, 2006.

B439 Bai, T.-C., G.-B. Yan, J. Hu, and C.-G. Huang, Solubility of silybin in aqueous poly(ethylene glycol) solution, *International Journal of Pharmaceutics*, 308, 100–106, 2006.

B440 Bergstrom, C.A.S., K. Luthman, and P. Artursson, Accuracy of calculated pH-dependent aqueous drug solubility, *European Journal of Pharmaceutical Science*, 22, 387–398, 2004.

B441 Breil, M.P., J.M. Mollerup, E.S.J. Rudolph, and L.A.M. van der Wielen, Densities and solubilities of gycylgylycine and glycyl-L-alanine in aqueous electrolyte solutions, *Fluid Phase Equilibria*, 215, 221–225, 2004.

B442 Blanco, L.H., N.R. Sanabria, and M.T. Davila, Solubility of 1,3,5,7-tetra azatricyclo[3.3.1.1 3,7] decane(HMT) in water from 275.15K to 313.15K, *Thermochimica Acta*, 450, 73–75, 2006.

B443 Bergstrom, A.S.C., M. Strafford, L. Lazorova, and P. Artursson, Absorption classification of oral drugs based on molecular surface properties, *Journal of Medicinal Chemistry*, 46, 558–570, 2003.

B444 Bretti, C., F. Crea, C.D. Stefano, and S. Sammartano, Solubility and activity coefficients of 2,2'-bipyridyl, 1,10-phenanthroline and 2,2',6',2''-terpyridine in NaCl(aq) at different ionic strengths and T=298.15 K, *Fluid Phase Equilibria*, 272, 47–52, 2008.

C004 Chien, Y.W. and H.J. Lambert, Solubilization of steroids by multiple co-solvent systems, *Chemical and Pharmaceutical Bulletin*, 23, 1085–1090, 1975.

C005 Carless, J.E. and J. Swarbrick, The solubility of benzaldehyde in water as determined by reactive index measurements, *Journal of Pharmacy and Pharmacology*, 16, 633–634, 1964.

C006 Corby, T.C. and P.H. Elworthy, The solubility of some compounds in hexadecylpolyoxyethylene monoethers, polyethylene glycols, water and hexane, *Journal of Pharmacy and Pharmacology*, 23, 39–48, 1971.

C008 Carless, J.E. and J.R. Nixon, The oxidation of solubilised and emulsified oils (research paper), *Journal of Pharmacy and Pharmacology*, 12, 340–347, 1960.

C011 Cadwallader, D.E., H.W. Jun, and L.K. Chen, Nitrofurantoin solubility in aqueous urea solutions, *Journal of Pharmaceutical Sciences*, 64, 886–887, 1975.

C014 Chantooni, Jr., M.K. and I.M. Kolthoff, Transfer activity coefficients of *ortho*-substituted and non-*ortho*-substituted benzoates between water, methanol, and polar aprotic solvents, *Journal of Physical Chemistry*, 78, 839–846, 1974.

C017 Costantino, L. and V. Vitagliano, The influence of solvation of purinic and pyrimidinic bases on the conformational stability of DNA solutions, *Biochimica et Biophysica Acta*, 134, 204–206, 1967.

C018 Cohn, E., T.L. McMeekin, J.T. Edsall, and J.H. Weare, Studies in the physical chemistry of amino acids, peptides and related substances. II. The solubility of alpha-amino acids in water and in alcohol–water mixtures, *Journal of the American Chemical Society*, 56, 2270–2282, 1934.

C020 Corson, B.B., N.E. Sanborn, and P.R. Van Ess, Some observations on benzoylformic acid, *Journal of the American Chemical Society*, 52, 1623–1626, 1930.

C022 Copley, M.J., E. Ginsberg, G.F. Zellhoefer, and C.S. Marvel, Hydrogen bonding and the solubility of alcohols and amines in organic solvents XIII, *Journal of the American Chemical Society*, 63, 254–256, 1941.

C023 Chapman, R.P., P.R. Averell, and R.R. Harris, Solubility of melamine in water, *Industrial and Engineering Chemistry*, 35, 137–138, 1943.

C024 Chey, W. and G.V. Calder, Method for determining solubility of slightly soluble organic compounds, *Journal of Chemical and Engineering Data*, 17, 199–200, 1972.

C025 Charonnat, R., La Solubilite de la 1-Phenyl-2,3-dimethyl-4-dimethylamino-5-pyrazolone dans L'eau, *Comptes Rendus Hebdomadaires des Seances de l'Academie des Sciences*, 185, 284–286, 1927.

C026 Chaudry, M.A.Q. and K.C. James, A Hansch analysis of the anabolic activities of some nandrolone esters, *Journal of Medicinal Chemistry*, 17, 157–161, 1974.

C031 Cheung, M.W. and J.W. Biggar, Solubility and molecular structure of 4-amino-3,5,6-trichloropicolinic acid in relation to pH and temperature, *Journal of Agricultural and Food Chemistry*, 22, 202–206, 1974.

C032 Chow, Y.P. and A.J. Repta, Complexation of acetaminophen with methyl xanthines, *Journal of Pharmaceutical Sciences*, 61, 1454–1458, 1972.

C033 Caronna, G., Antagonismo Batterico e influenze di solubilita, *Gazzetta Chimica Italiana*, 78, 827–835, 1948.

C034 Chen, L.-K., D.E. Cadwallader, and H.W. Jun, Nitrofurantoin solubility in aqueous urea and creatinine solutions, *Journal of Pharmaceutical Sciences*, 65, 868–872, 1976.

C035 Cone, N.M., S.E. Forman, and J.C. Krantz, Jr., Relationship between anesthetic potency and physical properties, *Proceedings of the Society for Experimental Biology and Medicine*, 48, 461–463, 1941.

C037 Chiou, W.L., Possibility of Errors in using filter paper for solubility determination, *Canadian Journal of Pharmaceutical Science*, 10, 112–114, 1975.

C038 Carstensen, J.T. and M. Patel, Dissolution patterns of polydisperse powders: oxalic acid dihydrate, *Journal of Pharmaceutical Sciences*, 64, 1770–1776, 1975.

C039 Clements, J.A. and S.D. Popli, The preparation and properties of crystal modifications of meprobamate, *Canadian Journal of Pharmaceutical Science*, 8, 88–92, 1973.

C040 Chiou, W.L., Pharmaceutical applications of solid dispersion systems: x-ray diffraction and aqueous solubility studies on griseofulvin-polyethylene glycol 6000 systems, *Journal of Pharmaceutical Sciences*, 66, 989–991, 1977.

C042 Clough, W.W. and C.O. Johns, Higher alcohols from petroleum olefins, *Industrial and Engineering Chemistry*, 15, 1030–1032, 1923.

C045 Campbell, A.N. and F.C. Garrow, The physical identity of enantiomers, *Transactions of the Faraday Society*, 26, 560–565, 1930.

C046 Cho, M.J. and M.J. Peterman, Pre-formulation studies of 7-benzoylindoline (U-26,952) and possible utilization of molecular interaction with beta-cyclodextrin in development of an oral dosage form, *Pharmaceutical Research and Development Technical Report*, 7271, 2–21, 1978.

C047 Cox, J.D., Phase relationships in the pyridine series. Part IV. The miscibility of the ethylpyridines and dimethylpyridines with water, *Journal of the Chemical Society* (London), 3183–3187, 1954.

C048 Calvet, R., M. Terce, J. Le Renard, Cinetique de dissolution dans l'eau de l'atrazine, de la propazine et de la simazine, *Weed Research*, 15, 387–392, 1975.

C049 Caldwell, W.T., E.C. Kornfeld, and C.K. Donnell, Substituted 2-sulfanilamidopyrimidines, *Journal of the American Chemical Society*, 63, 2188–2190, 1941.

C051 Cumber, A.J. and W.C.J. Ross, Analogues of hexamethylmelamine. The anti-neoplastic activity of derivatives with enhanced water solubility, *Chemico-Biological Interactions*, 17, 349–357, 1977.

C052 Campbell, A.N. and A.J.R. Campbell, The heats of solution, heats of formation, specific heats and equilibrium diagrams of certain molecular compounds, *Journal of the American Chemical Society*, 62, 291–297, 1940.

C053 Chiou, C.T., V.H. Freed, D.W. Schmedding, and R.L. Kohnert, Partition coefficient and bioaccumulation of selected organic chemicals, *Environmental Science and Technology*, 11, 475–478, 1977.

C054 Csontos, A., I. Racz, and L. Gyarmati, A contribution to the kinetics of dissolution of some modern drugs, *Pharmazie* (Berlin), 32, 498–500, 1977.

C055 Conti, J.J., D.F. Othmer, and R. Gilmont, Composition of vapors from boiling binary solutions, *Journal of Chemical and Engineering Data*, 5, 301–307, 1960.

C056 Conway, J.B. and J.J. Norton, Ternary system furfural–ethylene glycol–water, *Industrial and Engineering Chemistry*, 43, 1433–1435, 1951.

C057 Carlisle, P.J. and A.A. Levine, Stability of chlorohydrocarbons. I. Methylene chloride, *Industrial and Engineering Chemistry*, 24, 146–147, 1932.

C058 Campbell, A.N., The system aniline–phenol–water, *Journal of the American Chemical Society*, 67, 981–987, 1945.

C059 Chowhan, Z.T., pH-solubility profiles of organic carboxylic acids and their salts, *Journal of Pharmaceutical Sciences*, 67, 1257–1260, 1978.

C060 Carter, J.S. and R.K. Hardy, The salting-out effect. Influence of electrolytes on the solubility of *m*-cresol in water, *Journal of the Chemical Society* (London), 131, 127–129, 1928.

C061 Creighton, H.J.M. Solubility and electrolytic conductance of mesitylene phosphinous acid, *Journal of Physical Chemistry*, 30, 1207–1208, 1926.

C062 Cho, M.J. and J.J. Biermacher, Water-soluble prodrug of metronidazole: synthesis and serum hydrolysis of metronidazole phosphate, technical report, 5–10, 1976.

C064 Clarke, G.A., T.R. Williams, and R.W. Taft, A manometric determination of the solvolysis rate of gaseous t-butyl chloride in aqueous solution, *Journal of the American Chemical Society*, 84, 2292–2295, 1962.

C065 Chambon, M., J. Bouvier, and P. Duron, Etude physico-chimique du phenomene de solubilisation de la cafeine par le benzoate de soude, *Journal de Pharmacie et de Chimie*, 26, 216–231, 1937.

C066 Chapin, E.M. and J.M. Bell, The solubility of oxalic acid in aqueous solutions of hydrochloric acid, *Journal of the American Chemical Society*, 53, 3284–3287, 1931.

C068 Coull, J. and H.B. Hope, The ternary system isoamyl alcohol–propyl alcohol–water, *Journal of Physical Chemistry*, 39, 967–971, 1935.

C069 Casale, L., Amidi ed imidi tartariche. Nota III, *Gazzetta Chimica Italiana*, 47, 63–68, 1917.

C070 Casale, L., Amidi ed imidi tartariche. Nota I, *Gazzetta Chimica Italiana*, 47, 272–285, 1917.

C071 Casale, L., Amidi ed imidi tartariche. Nota II, *Gazzetta Chimica Italiana*, 48, 114–120, 1918.

C072 Cofman, V., Sulla preparazione dell'acido diiodosalicilico e la sua solubilita nell' acqua, *Gazzetta Chimica Italiana*, 50, 296–299, 1920.

C073 Creighton, H.J.M. and D.S. Klauder, Jr., Solubility of mannite in mixtures of ethyl alcohol and water, *Journal of the Franklin Institute*, 195, 687–691, 1923.

C074 Chambon, M., J. Bouvier, and P. Duron, Etude du systeme cafeine-benzoate de sodium-eau, *Bulletin Society of Chemistry, French*, 4, 1401–1407, 1937.

C075 Czerski, L. and A. Czaplinski, Solubility of ethane in water and NaCl and CaCl solutions at .0 and pressures above 1 atmosphere, *Roczniki Chemii*, 36, 1827–1834, 1962.

C076 Collett, A.R. and C.L. Lazzell, Solubility relations of the isomeric nitro benzoic acids, *Journal of Physical Chemistry*, 34, 1838–1847, 1930.

C077 Cohen, J. and J.L. Lach, Interaction of pharmaceuticals with schardinger dextrins. I—interaction with hydroxybenzoic acids and *p*-hydroxybenzoates, *Journal of Pharmaceutical Sciences*, 52, 132–136, 1963.

C079 Collett, J.H. and B.L. Flood, Some effects of urea on drug dissolution, *Journal of Pharmacy and Pharmacology*, 28, 206–209, 1976.

C081 Caramella, C., P. Colombo, U. Conte, and A. La Manna, On the direct compression of sulfamethoxydiazine polymorphic forms—II, *Farmaco, Edizione Pratica* (PAVIA), 30, 496–501, 1975.

C083 Cohen, E. and H. Goedhart, Die metastabilitat der materie und deren bedeutung fur unsere kalorimetrischen standarde, *Proceedings of the Koninklijke Nederlandse Akadamie van Wetenschappen*, 34, 3–14, 1931.

C086 Copp, J.L., Thermodynamics of binary systems containing amines—Part 2, *Transactions of the Faraday Society*, 51, 1056–1061, 1955.

C087 Corrigan, O.I., C.A. Murphy, and R.F. Timoney, Dissolution properties of polyethylene glycols and polyethylene glycol–drug systems, *International Journal of Pharmaceutics*, 4, 67–74, 1979.

C088 Copp, J.L. and D.H. Everett, Thermodynamics of binary mixtures containing amines, *Discussions of the Faraday Society*, 15, 174–188, 1953.

C090 Connors, K.A. and T.W. Rosanske, *trans*-cinnamic acid-alpha-cyclodextrin system as studied by solubility, spectral, and potentiometric techniques, *Journal of Pharmaceutical Sciences*, 69, 173–179, 1980.

C091 Chandy, C.A. and M.R. Rao, Ternary liquid equilibria: 1-Hexanol–water–fatty acids, *Journal of Chemical and Engineering Data*, 7, 473–475, 1962.

C092 Charykov, A.K. and T.V. Tal'nikova, pH-metric method of determining the solubility and distribution ratios of some organic compounds in extraction systems, *Journal of Analytical Chemistry of the USSR*, 29, 818–822, 1974.

C093 Crittenden, Jr., E.D. and A.N. Hixson, Extraction of hydrogen chloride from aqueous solutions, *Industrial and Engineering, Process Design and Development*, 46, 265–274, 1954.

C094 Chiou, C.T., L.J. Peters, and V.H. Freed, A physical concept of soil–water equilibria for nonionic organic compounds, *Science*, 206, 831–832, 1979.

C095 Collett, J.H. and G. Kesteven, The solubility of allopurinol in aqueous solutions of polyvinylpyrrolidone, *Drug Development and Industrial Pharmacy*, 4, 555–568, 1978.

C096 Chase, E.F. and M. Kilpatrick, Jr., The classical dissociation constant of benzoic acid and the activity coefficient of molecular benzoic acid in potassium chloride solutions, *Journal of the American Chemical Society*, 53, 2589–2597, 1931.

C097 Carlotti, M.E., M. Trotta, and M.R. Gasco, Behaviour of hematoporphyrin and protoporphyrin with antidepressant drugs, *Pharmazie* (Berlin), 37, 194–196, 1982.

C098 Crummett, W.B. and R.H. Stehl, Determination of chlorinated dibenzo-p-dioxines and dibenzofurans in various materials, *Environmental Health Perspectives*, 5, 15–25, 1973.

C099 Chamlin, G.R., The chemistry of benzene hexachloride and its insecticidal properties, *Journal of Chemical Education*, 23, 283–284, 1946.

C100 Cohen, J.M., L.J. Kamphake, A.E. Lemke, and R.L. Woodward, Effect of fish poisons on water supplies. Part 1. Removal of toxic materials, *Journal of the American Water Works Association*, 52, 1151–1566, 1960.

C101 Cox, J.R., Triazine derivatives as non-selective herbicides, *Journal of the Science of Food and Agriculture*, 13, 99–103, 1962.

C102 Clark, W.G., E.A. Strakosch, and N.I. Levitan, Solubility and pH data of some of the commonly used sulfonamides, *Journal of Laboratory and Clinical Medicine*, 28, 188–189, 1942.

C103 Crossley, M.L., E.H. Northey, and M.E. Hultquist, Sulfanilamide derivatives. V. Constitution and properties of 2-sulfanilamidopyridine, *Journal of the American Chemical Society*, 62, 372–374, 1940.

C104 Campanella, L., T. Ferri, and P. Mazzoni, Solubility of pyridinedicarboxylic acids, *Journal of Inorganic and Nuclear Chemistry*, 41, 1054–1055, 1979.

C105 Carringer, R.D., J.B. Weber, and T.J. Monaco, Adsorption–desorption of selected pesticides by organic matter and montmorillonite, *Journal of Agricultural and Food Chemistry*, 23, 568–572, 1975.

C108 Castaneda, J.M., F.J. Lozano, and S. Trejo, Ternary equilibrium for the system water/cyclohexanol/2-ethyl-2-(hydroxymethyl)-1,3-propanediol, *Journal of Chemical and Engineering Data*, 26, 133–135, 1981.

C109 Chandrasekaran, S.K., P.S. Campbell, and A.S. Michaels, Effect of dimethyl sulfoxide on drug permeation through human skin, *American Institute of Chemical Engineers Journal*, 23, 810–816, 1977.

C111 Chiou, C.T., D.W. Schmedding, and J.H. Block, Correlation of water solubility with octanol–water partition coefficient, *Journal of Pharmaceutical Sciences*, 70, 1176–1177, 1981.

C112 Cousse, H., G. Mouzin, J.-P. Ribet, and J.-C. Vezin, Physicochemical and analytical characteristics of itanoxone, *Journal of Pharmaceutical Sciences*, 70, 1245–1248, 1981.

C113 Chiou, C.T., D.W. Schmedding, and M. Manes, Partitioning of organic compounds in octanol–water systems, *Environmental Science and Technology*, 16, 4–10, 1982.

C114 Charnicki, W.F., F.A. Bacher, S.A. Freeman, and D.H. DeCesare, The pharmacy of chlorothiazide (6-chloro-7-sulfamyl-1,2,4-benzothiadiazine-1,1-dioxide): a new orally effective diuretic agent, *Journal of the American Pharmaceutical Association, Scientific Edition*, 48, 656–659, 1959.

C115 Cosgrove, B.A. and J. Walkley, Solubilities of gases in H_2O and $2H_2O$, *Journal of Chromatography*, 216, 161–167.

C116 Clever, H., E.R. Baker, and W.R. Hale, Solubility of ethylene in aqueous silver nitrate and potassium nitrate solutions, *Journal of Chemical and Engineering Data*, 15, 411–413, 1970.

C117 Coffin, D.E., Residues of parathion, methyl parathion, EPN, and their oxons in Canadian fruits and vegetables, *Residue Reviews*, 7, 61–63, 1967.

C118 Cadwallader, D.E. and H.W. Jun, Nitrofurantoin, *Analytical Profiles of Drug Substances*, 5, 348–369, 1976.

C119 Coca, J. and R. Diaz, Extraction of furfural from aqueous solutions with chlorinated hydrocarbons, *Journal of Chemical and Engineering Data*, 25, 80–83, 1980.

C120 Chiu, C.C. and L.T. Grady, Penicillamine, *Analytical Profiles of Drug Substances*, 10, 602–613, 1981.

C121 Chitwood, B.G., Nematocidal action of halogenated hydrocarbons, *Agricultural Applications of Petroleum Products*, Advances in Chemistry Series, 7, ACS, Washington, DC, 91–99, 1952.

C122 Coats, J.R., R.L. Metcalf, I.P. Kapoor, and P.A. Boyle, Physical–chemical and biological degradation studies on DDT analogues with altered aliphatic moieties, *Journal of Agricultural and Food Chemistry*, 27, 1016–1022, 1979.

C124 Cho, M.J., R.R. Kurtz, C. Lewis, and D.J. Houser, Metronidazole Phosphate-A Water-soluble prodrug for parenteral solutions of metronidazole, *Journal of Pharmaceutical Sciences*, 71, 410–414, 1982.

C302 Chiou, C.T., P.E. Porter, and D.W. Schmedding, Partition equilibria of nonionic organic compounds between soil organic matter and water, *Environmental Science and Technology*, 17, 227–231, 1983.

C303 Chiou, W.L. and L.E. Kyle, Differential thermal, solubility, and aging studies on various sources of digoxin and digitoxin powder: biopharmaceutical implications, *Journal of Pharmaceutical Sciences*, 68, 1224–1229, 1979.

C305 Chlou, C.T., Partition Coefficients of organic compounds in lipid–water systems and correlations with fish bioconcentration factors, *Environmental Science and Technology*, 19, 57–62, 1985.

C307 Carlson, R., R. Whitaker, and A. Landskov, Endothall, *Analytical Methods for Pesticides and Plant Growth Regulators*, 10, 327–340, 1978.

C309 Correa, J.M., A. Arce, A. Blanco, and A. Correa, Liquid–liquid equilibria of the system water + acetic acid + methyl ethyl ketone at several temperatures, *Fluid Phase Equilibria*, 32, 151–162, 1987.

C310 Carswell, T.S. and H.K. Nason, Properties and uses of pentachlorophenol, *Industrial and Engineering Chemistry*, 30, 622–626, 1938.

C311 Chiou, C.T., D.E. Kile, T.I. Brinton, and J.A. Leenheer, A comparison of water solubility enhancements of organic solutes by aquatic humic materials and commercial humic acids, *Environmental Science and Technology*, 21, 1231–1234, 1987.

C313 Chiou, C.T., R.L. Macolm, T.I. Brinton, and D.E. Kile, Water solubility enhancement of some organic pollutants and pesticides by dissolved humic and fulvic acids, *Environmental Science and Technology*, 20, 502–508, 1986.

C314 Chiarini, A. and A. Tartarini, pH-solubility relationship and partition coefficients for some anti-inflammatory arylaliphatic acids, *Archiv der Pharmazie*, 317, 268–273, 1984.

C315 Corrigan, O.I. and R.F. Timoney, Anomalous behaviour of some hydroflumethiazide crystal samples, *Journal of Pharmacy and Pharmacology*, 26, 838–840, 1974.

C316 Connors, K.A. and D.D. Pendergast, Microscopic binding constants in cyclodextrin systems: complexation of alpha-cyclodextrin with sym-1,4-disubstituted benzenes, *Journal of the American Chemical Society*, 106, 7607–7614, 1984.

C317 Chou, F.T., A.H. Khan, and J.S. Driscoll, Potential central nervous system antitumor agents. Aziridinylbenzoquinones, *Journal of Medicinal Chemistry*, 19, 1302–1308, 1976.

C318 Caron, G., I.H. Suffet, and T. Belton, Effect of dissolved organic carbon on the environmental distribution of nonpolar organic compounds, *Chemosphere*, 14, 993–1000, 1985.

C323 Carlson, J.A., H.J. Mann, and D.M. Canafax, Effect of pH on disintegration and dissolution of ketoconazole tablets, *American Journal of Hospital Pharmacy*, 40, 1334–1336, 1983.

C324 Chien, Y.E.W., Solubilization of metronidazole by water-miscible multi-cosolvents and water-soluble vitamins, *Journal of Parenteral Science and Technology*, 38, 32–36, 1984.

C329 Conway, J.B. and J.B. Philip, Ternary system: furfural-methyl isobutyl ketone-water at 25°C, *Industrial and Engineering Chemistry*, 45, 1083–1085, 1953.

C331 Chang, S.-S., J.R. Maurey, and W.J. Pummer, Solubilities of two *n*-alkanes in various solvents, *Journal of Chemical and Engineering Data*, 28, 187–189, 1983.

C332 Coates, M., D.W. Connell, and D.M. Barron, Aqueous solubility and octan-1-ol to water partition coefficients of aliphatic hydrocarbons, *Environmental Science and Technology*, 19, 628–632, 1985.

C333 Correa, J.M., A. Blanco, and A. Arce, Liquid–liquid equilibria of the system water + acetic acid + methyl isopropyl ketone between 25 and 55°C, *Journal of Chemical and Engineering Data*, 34, 415–419, 1989.

C340 Chan, H.K., S. Venkataram, D.J.W. Grant, and Y.E. Rahman, Solid state properties of an oral iron chelator, 1,2-dimethyl-3-hydroxy-4-pyridone, and its acetic acid solvate. I: Physicochemical characterization, intrinsic dissolution rate, and solution thermodynamics, *Journal of Pharmaceutical Sciences*, 80, 677–685, 1991.

C342 Chellquist, E.M. and W.G. Gorman, Benzoyl peroxide solubility and stability in hydric solvents, *Pharmaceutical Research*, 9, 1341–1345, 1992.

C346 Cohen, S., Y. Marcus, Y. Migron, and A. Shafran, Water sorption, binding and solubility of polyols, *Journal of the Chemical Society, Faraday Transactions 1*, 89, 3271–3275, 1993.

C347 Chen, C.-C., Y. Zhu, and L.B. Evans, Phase partitioning of biomolecules: solubilities of amino acids, *Biotechnology Progress*, 5, 111–118, 1989.

C348 Chamberlain, S.D., A.R. Moorman, L.A. Jones, and T.A. Krenitsky, 2'-Ester Prodrugs of the varicella-zoster antiviral agent, 6-methoxypurine arabinoside, *Antiviral Chemistry and Chemotherapy*, 3, 371–378, 1992.

C349 Chen, H. and J. Wagner, An apparatus and procedure for measuring mutual solubilities of hydrocarbons + water: benzene + water from 303 to 373 K, *Journal of Chemical and Engineering Data*, 39, 470–474, 1994.

C350 Chen, H. and J. Wagner, Mutual solubilities of alkylbenzene + water systems at temperatures from 303 to 373 k: ethylbenzene, *p*-xylene, 1,3,5-trimethylbenzene, and butylbenzene, *Journal of Chemical and Engineering Data*, 39, 679–684, 1994.

C400 Cai, X., D. Grant, and T. Wiedmann, Analysis of the solubilization of steroids by bile salt micelles, *Journal of Pharmaceutical Sciences*, 86, 372–377, 1997.

C401 Cakebread, S.H., Confectionery ingredients—composition and properties, *Confectionery Productions*, 274–278, 1971.

C404 Carta, R., Solubilities of L-cystine, L-tyrosine, L-leucine, and glycine in their water solutions, *Journal of Chemical and Engineering Data*, 44, 563–567, 1999.

C405 Carta, R. and G. Tola, Solubilities of L-cystine, L-tyrosine, L-leucine, and glycine in aqueous solutions at various pHs and NaCl concentrations, *Journal of Chemical and Engineering Data*, 41, 414–417, 1996.

C407 Chatjigakis, A., C. Donze, and A. Coleman, Solubility behavior of beta-cyclodextrin in water/cosolvent mixtures, *Analytical Chemistry*, 64, 1632–1634, 1992.

C409 Coffman, R.E. and D.O. Kildsig, Effect of nicotinamide and urea on the solubility of riboflavin in various solvents, *Journal of Pharmaceutical Sciences*, 85, 951–954, 1996.

C410 Colonia, E., A. Dixit, and N. Tavare, Phase relations of *o*- and *p*-chlorobenzoic acids in hydrotrope solutions, *Journal of Chemical and Engineering Data*, 43, 220–225, 1998.

C411 Cordero, J.A., L. Alarcon, E. Escribano, and J. Domenech, A comparative study of the transdermal penetration of a series of nonsteroidal antiinflammatory drugs, *Journal of Pharmaceutical Sciences*, 86, 503–507, 1997.

C413 Coyle, G.T., T.C. Harmon, and I.H. Suffet, Aqueous solubility depression for hydrophobic organic chemicals in the presence of partially miscible organic solvents, *Environmental Science and Technology*, 31, 384–389, 1997.

C414 Cappello, B., C. Carmignani, M. Iervolino, and M. Saettone, Solubilization of tropicamide by hydroxypropyl-beta-cyclodextrin and water-soluble polymers: in vitro/in vivo studies, *International Journal of Pharmaceutics*, 213, 75–81, 2001.

C415 Casini, A., A. Scozzafava, F. Mincione, and C. Supuran, Carbonic anhydrase inhibitors: water-soluble 4-sulfamoylphenylthioureas as topical intraocular pressure-lowering agents with long-lasting effects, *Journal of Medicinal Chemistry*, 43, 4884–4892, 2000.

C416 Chen, A., S. Zito, and R. Nash, Solubility Enhancement of nucleosides and structurally related compounds by complex formation, *Pharmaceutical Research*, 11, 398–401, 1994.

C423 Cervantes, M.C., S. Bongard, D. Champion, and A. Voilley, Temperature effect on solubility of aroma compounds in various aqueous solutions, *LWT*, 38, 371–378, 2005.

C431 Cappello, B., C. Di Maio, M. Iervolino, and A. Miro, Improvement of solubility and stability of valsartan by Hydroxypropyl-*B*-Cyclodextrin, *Journal of Inclusion Phenomena and Macrocyclic Chemistry*, 54, 289–294, 2006.

C434 Charumanee, S., A. Titwan, J. Sirithunyalug, and S. Okonogi , Thermodynamics of the encapsulation by cyclodextrins, *Journal of Chemical Technology and Biotechnology*, 81, 523–529, 2006.

C435 Covarrubias-Cervantes, M., S. Bongard, D. Champion, and A. Voilley, Effects of the nature and concentration of substrates in aqueous solutions on the solubility of aroma compounds, *Flavour and fragrance Journal*, 20, 265–273, 2005.

C437 Chen, X.-Q. and S. Venkatesh, Miniature device for aqueous and non-aqueous solubility measurements during drug discovery, *Pharmaceutical Research*, 21, 1758–1761, 2004.

C438 Crothers, M., N.M.P.S. Ricardo, F. Heatley, and C. Booth, Solubilisation of drugs in micellar solutions of diblock copolymers of ethylene oxide and styrene oxide, *International Journal of Pharmaceutics*, 358, 303–306, 2008.

D001 Deno, N.C. and H.E. Berkheimer, Phase equilibria molecular transport thermodynamics, *Journal of Chemical and Engineering Data*, 5, 1–5, 1960.

D002 Dixon, M.R., C.E. Rehberg, and C.H. Fisher, Preparation and physical properties of *n*-alkyl beta-ethoxypropionates, *Journal of the American Chemical Society*, 70, 3733–3738, 1948.

D003 Davis, W.W., M.E. Krahl, and G.H.A. Clowes, Solubility of carcinogenic and related hydrocarbons in water, *Journal of the American Chemical Society*, 64, 108–110, 1942.

D004 Dean, J.A., Physical constants of alkaloids, in *Lange's Handbook of Chemistry*, McGraw-Hill, New York, 394–417, 1973.

D005 Dunn, M.S., M.P. Stoddard, L.B. Rubin, and R.C. Bovie, Investigations of amino acids and peptides, *Journal of Biological Chemistry*, 151, 241–258, 1943.

D006 Duff, J.C. and E.J. Bills, The solubilities of nitrophenols in aqueous ethyl-alcoholic solutions, *Journal of the Chemical Society* (London), 1331–1338, 1930.

D007 Deno, N.C. and C. Perizzolo, The application of activity coefficient data to the relations between kinetics and acidity functions, *Journal of the American Chemical Society*, 79, 1345–1348, 1957.

D008 Donbrow, M. and H. Ben-Shalom, Molecular interactions of caffeine with *o*-, *m*-, and *p*-iodobenzoic acids and *o*-, *m*-, and *p*-fluorobenzoic acids, *Journal of Pharmacy and Pharmacology*, 19, 495–501, 1967.

D009 Dittert, L.W., H.C. Caldwell, T. Ellison, and J.V. Swintosky, Carbonate ester prodrugs of salicylic acid, *Journal of Pharmaceutical Sciences*, 57, 828–831, 1968.

D010 DeLuca, P.P., L. Lachman, and H.G. Schroeder, Physical–chemical properties of substituted amides in aqueous solution and evaluation of their potential use as solubilizing agents, *Journal of Pharmaceutical Sciences*, 62, 1320–1327, 1973.

D011 DeLuca, P.P. and L. Lachman, Lyophilization of pharmaceuticals. I. Effect of certain physical–chemical properties, *Journal of Pharmaceutical Sciences*, 54, 617–624, 1965.

D012 Donbrow, M., E. Touitou, and H. Ben-Shalom, Stability of salicylamide–caffeine complex at different temperatures and its thermodynamic parameters, *Journal of Pharmacy and Pharmacology*, 28, 766–769, 1976.

D013 Dunstan, I., J.V. Griffiths, and S.A. Harvey, Nitric esters. Part I. Characterisation of the isomeric glycerol dinitrates, *Journal of the Chemical Society* (London), 1319–1324, 1965.

D014 Drobnica, L., M. Zemanova, P. Nemec, and P. Nemec, Jr., Antifungal activity of isothiocyanates and related compounds, *Applied Microbiology*, 15, 701–709, 1967.

D015 Daabis, N.A., S.A. Khalil, and V.F. Naggar, The effect of urea, amidopyrine, phenazone and paracetamol on the solubility of some sparingly soluble antirheumatics, *Canadian Journal of Pharmaceutical Science*, 11, 114–117, 1976.

D016 Dalton, J.B. and C.L.A. Schmidt, The solubilities of certain amino acids in water, the densities of their solutions at twenty-five degrees, and the calculated heats of solution and partial molal volumes, *Journal of Biological Chemistry*, 103, 549–575, 1933.

D017 Dalton, J.B. and C.L.A. Schmidt, The solubilities of certain amino acids and related compounds in water, the densities of their solutions at twenty-five degrees, and the calculated heats of solution and partial molal volumes. II, *Journal of Biological Chemistry*, 109, 241–248, 1935.

D018 Dunn, M.S., F.J. Ross, and L.S. Read, The solubility of the amino acids in water, *Journal of Biological Chemistry*, 103, 579–595, 1933.

D019 Drobnica, L. and J. Augustin, Reaction of isothiocyanates with amino acids, peptides and proteins. I. Kinetics of the reaction of aromatic isothiocyanates with glycine, *Collection of Czechoslovak Chemical Communications*, 30, 99–105, 1965.

D020 Dalman, L.H., Ternary systems of urea and acids. iv. urea, citric acid and water. v. urea, acetic acid and water. VI. Urea, tartaric acid and water, *Journal of the American Chemical Society*, 59, 775–779, 1937.

D021 Dehn, W.M., Comparative solubilities in water, in pyridine and in aqueous pyridine, *Journal of the American Chemical Society*, 39, 1399–1404, 1917.

D022 Davis, T.L., A.A. Ashdown, and H.R. Couch, Two forms of nitroguanidine, *Journal of the American Chemical Society*, 47, 1063–1066, 1925.

D025 Daley, R.D., Primidone, *Analytical Profiles of Drug Substances*, 2, 409–421, 1973.

D026 Dempski, R.E., J.B. Portnoff, and A.W. Wase, In vitro release and in vitro penetration studies of a topical steroid from nonaqueous vehicles, *Journal of Pharmaceutical Sciences*, 58, 579–582, 1969.

D027 Desvergnes, L., Sur quelques proprietes physiques de certains derives nitres, *The Reviews of Chemical Industry*, 38, 265–266, 1929.

D029 Dittert, L.W., H.C. Caldwell, H.J. Adams, and J.V. Swintosky, Acetaminophen prodrugs. I—synthesis, physicochemical properties, and analgesic activity, *Journal of Pharmaceutical Sciences*, 57, 774–780, 1968.

D031 Daniel, R.J. and W. Doran, Some chemical constituents of the mussel (*Mytilus edulis*), *Biochemical Journal*, 20, 676–684, 1926.

D033 Daniels, T.C. and R.E. Lyons, Concerning the physical properties of solutions of certain phenyl-substituted acids in relation to their bactericidal power, *Journal of Physical Chemistry*, 35, 2049–2060, 1931.

D034 Dworkin, M. and K.H. Keller, Solubility and diffusion coefficient of adenosine 3':5'-monophosphate, *Journal of Biological Chemistry*, 252, 864–865, 1977.

D035 Desai, S.J., Quantitative mechanistic studies of drug release from inert matrices, PhD Thesis, 80–80, 1966.

D036 Dooley, K.H. and F.J. Castellino, Solubility of amino acids in aqueous guanidinium thiocyanate solutions, *Biochemistry*, 11, 1870–1874, 1972.

D037 Donbrow, M. and C.T. Rhodes, Potentiometric studies on solubilisation in non-ionic micellar solutions. Part I. Interpretation of pH changes and mechanism of solubilisation of benzoic acid, *Journal of the Chemical Society* (London), 6166–6171, 1964.

D038 Dalton, J.B. and C.L.A. Schmidt, The solubility of D-valine in water, *Journal of General Physiology*, 19, 767–771, 1936.

D039 Dalman, L.H., The Solubility of citric and tartaric acids in water, *Journal of the American Chemical Society*, 59, 2547–2549, 1937.

D040 De Santis, R., L. Marrelli, and P.N. Muscetta, Influence of temperature on the liquid–liquid equilibrium of the water–n-butyl alcohol–sodium chloride system, *Journal of Chemical and Engineering Data*, 21, 324–327, 1976.

D041 Dawson, R.M.C., D.C. Elliott, W.H. Elliott, and K.M. Jones, *Data for Biochemical Research*, 1, 1, Oxford University Press, Pergamon, 1969.

D043 Davis, W.W. and T.V. Parke, Jr., A Nephelometric method for determination of solubilities of extremely low order, *Journal of the American Chemical Society*, 64, 101–107, 1942.

D044 Dearden, J.C., J.H. Collett, and E. Tomlinson, In vitro dissolution rate as a parameter in structure–activity studies, *Experientia*, 23, 37–40, 1976.

D046 Davis, H.S. and O.F. Wiedeman, Physical properties of acrylonitrile, *Industrial and Engineering Chemistry*, 37, 482–485, 1945.

D047 Durand, R., Recherches sur l'hydrotropie. etude de la solubilite de l'heptane, de l'hexane et du cyclohexane dans les solutions aqueuses de quelques sels d'acides gras, *Comptes Rendus Hebdomadaires des Seances de l'Academie des Sciences*, 226, 409–410, 1948.

D049 Druckrey, H., R. Preussmann, N. Nashed, and S. Ivankovic, Carcinogene alkylierende substanzen I. Dimethylsulfat, carcinogene wirkung an ratten und wahrscheinliche ursache von berufskrebs, *Zeitschrift fuer Krebsforschung*, 68, 103–111, 1966.

D050 Desai, P.G. and A.M. Patel, Effect of polarity on the solubilities of some organic acids, *Journal of the Indian Chemical Society*, 12, 131–136, 1935.

D051 Desbarres, J. and H.O. El Sayed, Determination des pK et des solubilites d'acides peu solubles (acide pipemidique, acide undecanoique et derives), *Comptes Rendus Hebdomadaires des Seances de l'Academie des Sciences, Serie C: Sciences Chimiques*, 285, 431–434, 1977.

D052 Doolittle, A.K., Lacquer Solvents in Commercial Use, *Industrial and Engineering Chemistry*, 27, 1169–1179, 1935.

D055 Dawe, R.A., PhD thesis, Oxford University, 1965.

D056 Dyer, D.L., The effect of ph on solubilization of weak acids and bases, *Journal of Colloid Science*, 14, 640–645, 1959.

D058 Drucker, C., Experimentelle beitrage zur frage der elektrolytischen dissoziation, *Monatshefte fuer Chemie*, 53, 62–68, 1929.

D059 Duff, J.C., The solubilities of *o*- and *p*-nitrophenols in aqueous methyl-alcoholic solutions at 25 and 40 degrees. Formation of beta-p-nitrophenol, *Journal of the Chemical Society* (London), 2789–2796, 1929.

D060 Dalman, L.H., Ternary systems of urea and acids. I. Urea, nitric acid and water. II. Urea, sulfuric acid and water. III. Urea, oxalic acid and water, *Journal of the American Chemical Society*, 56, 549–553, 1934.

D061 Doosaj, S.S. and W.V. Bhagwat, Solubilities of weak acids in salts of weak acids at very high concentrations, *Journal of the Indian Chemical Society*, 10, 225–232, 1933.

D062 Dittmar, H.R., The decomposition of malic acid by sulfuric acid, *Journal of the American Chemical Society*, 52, 2746–2754, 1930.

D063 Drouillon, F., Etude du melange ternaire: eau, alcool ethylique, alcool butylique normal, *Journal de Chimie Physique et de Physico-Chimie Biologique*, 22, 149–160, 1925.

D064 De Brouwer, S., Sur l'acide orthotrifluortoluique et le nitrotrifluorcresol 1-3-6, *Bulletin des Societes Chimiques Belges*, 39, 298–308, 1930.

D065 Desvergnes, L., Sur la solubilite du 2-4-6-trinitrotoluene du tetryl et de la tetranitraniline dans les solvants organiques, *Moniteur Scientifique*, 14, 121–130, 1924.

D066 Desvergnes, L., Le 1-3-5-trinitrobenzene ou benzite, *Chimica e l'Industria* (Milan), 25, 3–16, 1931.

D067 Desvergnes, L., Sur quelques proprietes physiques des derives nitres: l'acide 2-4-6-trinitrobenzoique, *Moniteur Scientifique*, 16, 201–208, 1926.

D068 Doucet, M.A., Travaux originaux-de l'action de l'iode sur quelques semi-carbazides substituees en (1); Application a leur dosage, *Journal de Pharmacie et de Chimie*, 27, 361–365, 1923.

D069 Desvergnes, L., Sur quelques proprietes physiques des nitrophenols-orthonitrophenol, *The Reviews of Chemical Industry*, 36, 194–196, 1927.

D070 Desvergnes, L., Sur quelques proprietes physiques des derives nitres: 1-3-dinitrobenzene, *Moniteur Scientifique*, 15, 149–158, 1925.

D071 Desvergnes, L., Sur quelques proprietes physiques des derives nitres: 4-nitro-4-chlorobenzene, *Moniteur Scientifique*, 15, 73–78, 1925.

D072 Duclaux, J. and A. Durand-Gasselin, Les perchlorates et la serie lyotrope—II, *Journal de Chimie Physique et de Physico-Chimie Biologique*, 35, 189–192, 1938.

D073 De Groote, M., The solubility of vanillin and coumarin in glycerine solutions, *American Perfumer* (APRFA), 15, 372–374, 1920.

D077 Dolinski, J.H., Ueber die loslichkeit einiger organischer verbindungen in wasser bei verschiedenen temperaturen, *Berichte der Deutschen Chemischen Gesellschaft*, 38, 1835–1837, 1905.

D078 Dearden, J.C. and N.C. Patel, Dissolution kinetics of some alkyl derivatives of acetaminophen, *Drug Development and Industrial Pharmacy*, 4, 529–535, 1978.

D079 Desvergnes, L., Sur quelques proprietes physiques des derives nitres: 1-2-3 dinitranisol, *Moniteur Scientifique*, 14, 249–257, 1924.

D080 Desvergnes, L., Sur quelques proprietes physiques des nitrophenols-2.6-dinitrophenol, *The Reviews of Chemical Industry*, 36, 224–226, 1927.

D081 Dymicky, M. and C.N. Huhtanen, Inhibition of clostridium botulinum by *p*-hydroxybenzoic acid *n*-alkyl esters, *Antimicrobial Agents & Chemotherapy*, 15, 798–801, 1979.

D082 Draguet-Brughmans, M., M. Azibi, and R. Bouche, Solubilite et vitesse de dissolution du meprobamate: des cas significatifs, *Journa de Pharmacie de Belgique*, 34, 267–271, 1979.

D083 Organotrope carcinogene wirkungen bei 65 Verschiedenen *N*-nitroso-verbindungen an BD-Ratten, *Zeitschrift fuer Krebsforschung*, 69, 103–201, 1967.

D084 Durel, P. and M. Allinne, Sur la precipitation des produits sulfamides dans l'Urine, *Bulletins et Memoires de la Societe Medicale des Hopitaux de Paris*, 251–259, 1941.

D085 Dexter, R.N. and S.P. Pavlou, Mass solubility and aqueous activity coefficients of stable organic chemicals in the marine environment: polychlorinated biphenyls, *Marine Chemistry*, 6, 41–53, 1978.

D086 DeLassus, P.T. and D.D. Schmidt, Solubilities of vinyl chloride and vinylidene chloride in water, *Journal of Chemical and Engineering Data*, 26, 274–276, 1981.

D087 Dolgorev, A.V., Y.G. Lysak, Y.F. Zibarova, and A.P. Lukoyanov, Dithiopyrylmethane and its analogs as analytical reagents. Synthesis and properties, *Journal of Analytical Chemistry of the USSR*, 35, 560–567, 1980.

D088 Donbrow, M. and P. Sax, Thermodynamic parameters of molecular complexes in aqueous solution: enthalpy–entropy compensation in a series of complexes of caffeine with beta-naphthoxyacetic acid and drug-related aromatic compounds, *Journal of Pharmacy and Pharmacology*, 34, 215–224, 1982.

D089 David, W.A.L., R.L. Metcalf, and M. Winton, The systemic insecticidal properties of certain carbamates, *Journal of Economic Entomology*, 53, 1021–1025, 1960.

D091 Deppeler, H.P., Hydrochlorothiazide, *Analytical Profiles of Drug Substances*, 10, 406–423, 1981.

D092 Draguet-Brughmans, M., P. Draux, and R. Bouche, Polymorphisme du butobarbital, *Journa de Pharmacie de Belgique*, 36, 397–403, 1981.

D093 Datyner, A., The solubilization of nonionic dyestuffs at elevated temperatures in aqueous solutions of soldium dodecyl sulfate, *Journal of Colloid and Interface Science*, 65, 527–532, 1978.

D300 David, C., E. Szalai, and D. Baeyens-Volant, Photophysical processes in cyanobiphenyl derivatives. II. Cyanobiphenyl derivatives as fluorescent probes in micellar environment, *Berichte der Bunsengesellschaft fuer Physikalische Chemie*, 86, 710–716, 1982.

D302 Dunham, L.L. and W.W. Miller, Methoprene, *Analytical Methods for Pesticides and Plant Growth Regulators*, 10, 95–109, 1978.

D303 Darskus, R. and D. Eichler, Triforine, *Analytical Methods for Pesticides and Plant Growth Regulators*, 10, 243–253, 1978.

D304 Day, E.W., Ethalfluralin, *Analytical Methods for Pesticides and Plant Growth Regulators*, 10, 341–352, 1978.

D305 Dietz, Jr., E.A. and L.O. Moore, Monomethylarsonic acid, cacodylic acid, and their sodium salts, *Analytical Methods for Pesticides and Plant Growth Regulators*, 10, 385–401, 1978.

D306 Dunnivant, F.M. and A.W. Elzerman, Aqueous solubility and Henry's law constant data for PCB congeners for evaluation of quantitative structure–property relationships (QSPRs), *Chemosphere*, 17, 525–541, 1988.

D307 DeVoe, H. and S.P. Wasik, Aqueous solubilities and enthalpies of solution of adenine and guanine, *Journal of Solution Chemistry*, 13, 51–61, 1984.

D308 Dahlan, R., C. McDonald, and V.B. Sunderland, Solubilities and intrinsic dissolution rates of sulphamethoxazole and trimethoprim, *Journal of Pharmacy and Pharmacology*, 39, 246–251, 1987.

D311 Dahlund, M. and A. Olin, Chemical equilibria in aqueous solutions of olsalazine, 3,3'-azo-*bis*(6-hydroxybenzoic acid), *Acta Pharmaceutica Suecica*, 24, 219–232, 1987.

D315 DeRitter, E., Vitamins in pharmaceutical formulations, *Journal of Pharmaceutical Sciences*, 71, 1073–1075, 1982.

D316 Drugs, dimethyl-beta-cyclodextrin, *Drugs of the Future*, 9, 576–578, 1984.

D319 Drugs, triglycidulurazol, *Drugs of the Future*, 9, 209–210, 1983.

D321 Drugs, flupirtine, *Drugs of the Future*, 8, 773–775, 1983.

D330 Doucette, W.J. and A.W. Andren, Aqueous solubility of selected biphenyl, furan, and dioxin congeners, *Chemosphere*, 17, 243–252, 1988.

D331 Dickhut, R.M., A.W. Andren, and D.E. Armstrong, Aqueous solubilities of six polychlorinated biphenyl congeners at four temperatures, *Environmental Science and Technology*, 20, 807–810, 1986.

D332 Davison, R.R. and W.H. Smith, Vapor–liquid equilibrium of *n*-ethyl-*n*-butylamine–water and *n*-ethyl-*sec*-butylamine–water, *Journal of Chemical and Engineering Data*, 14, 296–298, 1969.

D335 Dickhut, R.M., Dissertation or Masters Thesis, 1985.

D336 DeFoe, D.L., G.W. Holcombe, D.E. Hammermeister, and K.E. Biesinger, Solubility and toxicity of eight phthalate esters to four aquatic organisms, *Environmental Toxicology and Chemistry*, 9, 623–636, 1990.

D337 Dickhut, R.M., A.W. Andren, and D.E. Armstrong, naphthalene solubility in selected organic solvent/water mixtures, *Journal of Chemical and Engineering Data*, 34, 438–443, 1989.

D339 Dempsey, G. and P. Molyneux, Quantitative investigations of amino acids and peptides. IV. The solubilities of the amino acids in water–ethyl alcohol mixtures, *Journal of the Chemical Society, Faraday Transactions 1*, 88, 971–977, 1992.

D340 De Smidt, J.H., J.C.A. Offringa, and D.J.A. Crommelin, Dissolution rate of griseofulvin in bile salt solutions, *Journal of Pharmaceutical Sciences*, 80, 399–401, 1991.

D341 Denton, G.W., *Chlohexidine*, 274–275, 1991.

D343 Dramur, U. and B. Tatli, Liquid–liquid equilibria of water + acetic acid + phthalic esters (dimethyl phthalate and diethyl phthalate) ternaries, *Journal of Chemical and Engineering Data*, 38, 23–25, 1993.

D344 Dumanovic, D., J. Jovanovic, S. Popovic, and D. Kosanovic, The solubility of some 4(5)- and 5-nitroimidazoles in water and twenty common organic solvents, *Journal of the Serbian Chemical Society*, 51, 411–416, 1986.

D345 Dwivedi, S.K., S. Sattari, F. Jamali, and A.G. Mitchell, Ibuprofen racemate and enantiomers: phase diagram, solubility and thermodynamic studies, *International Journal of Pharmaceutics*, 87, 95–104, 1992.

D346 De Ligny, C.L. and N.G. Van Der Veen, Solubilities of some tetra-alkyl-carbon, -silicon, -germanium and -tin compounds in mixtures of water with methanol, ethanol, dioxane, acetone and acetic acid and differences between the standard chemical potentials of these solutes, *Recueil des Travaux Chimiques des Pays-Bas*, 90, 984–1009, 1971.

D347 Darwish, A., A.T. Florence, and A.M. Saleh, Effects of hydrotropic agents on the solubility, precipitation, and protein binding of etoposide, *Journal of Pharmaceutical Sciences*, 78, 577–581, 1989.

D348 Demond, A.H. and A.S. Lindner, Estimation of interfacial tension between organic liquids and water, *Environmental Science and Technology*, 27, 2318–2331, 1993.

D349 Dey, B.P. and S.C. Lahiri, Solubilities of amino acids in methanol + water mixtures at different temperatures, *Indian Journal of Chemistry*, 27A, 297–302, 1988.

D350 Deghenghi, R., F. Boutignon, P. Wuthrich, and V. Lenaerts, Antarelix (EP 24332) a novel water soluble LHRH antagonist, *Biomedical and Pharmacotherapy*, 47, 107–110, 1993.

D351 Dickhut, R.M., K.E. Miller, and A.W. Andren, Evaluation of total molecular surface area for predicting air-water partitioning properties of hydrophobic aromatic chemicals, *Chemosphere*, 29, 283–297, 1994.

D400 Dai, J., L. Jin, L. Wang, and Z. Zhang, Determination and estimation of water solubilities and octanol/water partition coefficients for derivates of benzanilides, *Chemosphere*, 37, 1419–1426, 1998.

D401 Damen, E., P. Wiegerinck, L. Braamer, and H. Scheeren, Paclitaxel esters of malic acid as prodrugs with improved water solubility, *Bioorganic and Medicinal Chemistry*, 8, 427–432, 2000.

D402 Dollo, G., P. Le Corre, F. Chevanne, and R. Le Verge, Inclusion complexation of amide-typed local anaesthetics with beta-cyclodextrin and its derivatives. II. Evaluation of affinity constants and in vitro transfer rate constants, *International Journal of Pharmaceutics*, 136, 165–174, 1996.

D404 Dubbs, M. and R. Gupta, Solubility of vitamin E (alpha-tocopherol) and vitamin K3 (Menadione) in ethanol–water mixture, *Journal of Chemical and Engineering Data*, 43, 590–591, 1998.

D405 Dubost, D., M. Kaufman, H. Jahansouz, and G. Brenner, Physicochemical characterization of L-691, 121, a potent and selective class III antiarrhythmic agent, *Drug Development and Industrial Pharmacy*, 22, 873–880, 1996.

D406 De Maagd, P., D. Ten Hulscher, H. Van Den Heuvel, and D. Sijm, Physicochemical properties of polycyclic aromatice hydrocarbons: aqueous solubilities, *n*-octanol/water partition coefficients, and Henry's law constants, *Environmental Toxicology and Chemistry*, 17, 251–257, 1998.

D407 Demian, B., Correlation of the solubility of several aromatics and terpenes in aqueous hydroxypropyl-beta-cyclodextrin with steric and hydrophobicity parameters, *Carbohydrate Research*, 328, 635–639, 2000.

D414 Desai, K.G.H., A.R. Kulkarni, and T.M. Aminabhavi, Solubility of rofecoxib in the presence of methanol, ethanol, and sodium lauryl sulfate at (298.15, 303.15, and 308.15) K, *Journal of Chemical Engineering Data*, 48, 942–945, 2003.

D415 Dohanyosova, P., V. Dohnal, and D. Fenclova, Temperature dependence of aqueous solubility of anthracenes: accurate determination by a new generator column apparatus, *Fluid Phase Equilibria*, 214, 151–167, 2003.

D416 Dorn, P.B., C. Chou, and J.J. Gentempo, Degradation of bisphenol a in natural waters, *Chemosphere*, 16, 1501–1507, 1987.

D425 Druaux, C., M. Le Thanh, A.M. Seuvre, and A. Voilley, Application of headspace analysis to the study of aroma compounds–lipids interactions, *The Journal of the American Oil Chemists' Society*, 75, 127–130, 1998.

D426 Daniel_Mwambete, K., S. Torrado, C. Cuesta-Bandera, and J.J. Torrado, The effect of solubilization on the oral bioavailability of three benzimidazole carbamate drugs, *International Journal of Pharmaceutics*, 272, 29–36, 2004.

D428 Dong, Y., W.K. Ng, U. Surana, and R.B.H. Tan, Solubilization and preformulation of poorly water soluble and hydrolysis susceptible *N*-epoxymethyl-1,8-naphthalimide (ENA) compound, *International Journal of Pharmaceutics*, 356, 130–136, 2008.

E002 Erichsen, L.V., Das loslichkeitsdekrement der methylengruppe und die funktionsloslichkeit in homologen reihen, *Naturwissenschaften*, 39, 189–189, 1952.

E003 Elworthy, P.H. and H.E.C. Worthington, The solubility of sulphadiazine in water–dimethyl-formamide mixtures, *Journal of Pharmacy and Pharmacology*, 20, 830–835, 1968.

E004 Eganhouse, R.P. and J.A. Calder, The solubility of medium molecular weight aromatic hydrocarbons and the effects of hydrocarbon co-solutes and salinity, *Geochimica et Cosmochimica Acta*, 40, 555–561, 1976.

E005 Eggenberger, D.N., F.K. Broome, A.W. Ralston, and H.J. Harwood, The solubilities of the normal saturated fatty acids in water, *Journal of Organic Chemistry*, 14, 1108–1110, 1949.

E008 Ekwall, P., T. Rosendahl, and A. Sten, Solubility of bile acids, *Acta Chemica Scandinavica*, 12, 1622, 1958.

E009 Eisenbrand, J. and K. Baumann, Uber die bestimmung der wasserloslichkeit von coronen, fluoranthen, perylen, picen, tetracen, und triphenylen und uber die bildung wasserloslicher komplexe dieser kohlenwasserstoffe mit coffein, *Zeitschrift fuer Lebensmittel-Untersuchung und -Forschung*, 144, 312–317, 1970.

E010 Elworthy, P.H. and F.J. Lipscomb, A note on the solubility of griseofulvin, *Journal of Pharmacy and Pharmacology*, 20, 790–792, 1968.

E011 Edmonson, T.D. and J.E. Goyan, The effect of hydrogen ion and alcohol concentration on the solubility of phenobarbital, *Journal of the American Pharmaceutical Association, Scientific Edition*, 47, 810–812, 1958.

E012 El-Shibini, H.A.M., S. Abd-Elfattah, and M.M. Motawi, Die solubilisation des khellins durch coffein und dihydroxypropyltheophyllin, *Pharmazie* (Berlin), 27, 570–573, 1972.

E014 Eik-Nes, K., J.A. Schellman, R. Lumry, and L.T. Samuels, The binding of steroids to protein. i. solubility determinations, *Journal of Biological Chemistry*, 206, 411–419, 1954.

E015 England, Jr., A. and E.J. Cohn, Studies in the physical chemistry of amino acids, peptides and related substances. IV. The distribution coefficients of amino acids between water and butyl alcohol, *Journal of the American Chemical Society*, 57, 634–637, 1935.

E016 Emery, W.O. and C.D. Wright, Distribution of certain drugs between immiscible solvents, *Journal of the American Chemical Society*, 43, 2323–2335, 1921.

E017 Edwards, L.J., The dissolution and diffusion of aspirin in aqueous media, *Transactions of the Faraday Society*, 47, 1191–1210, 1951.

E018 El-Gindy, N.A. and F. El-Khawas, Solubility and dissolution enhancement of riboflavine by solid dispersion systems, *Pharmazeutische Industrie*, 39, 84–86, 1977.

E019 Evans, T.W., The Hill method for solubility determinations, *Industrial and Engineering Chemistry, Analytical Edition*, 8, 206–208, 1936.

E022 Eger II, E.I., R. Shargel, and G. Merkel, Solubility of diethyl ether in water, blood and oil, *Anesthesiology*, 24, 676–678, 1963.

E025 Eisenbrand, J. and K. Baumann, Uber die bestimmung der wasserloslichkeit von benzol, naphthalin, anthracen und pyren und uber die bildung wasserloslicher komplexe dieser kohlenwasserstoffe mit coffein, *Zeitschrift fuer Lebensmittel-Untersuchung und -Forschung*, 140, 210–216, 1969.

E028 Evans, B.K., K.C. James, and D.K. Luscombe, Quantitative structure–activity relationships and carminative activity, *Journal of Pharmaceutical Sciences*, 67, 277–278, 1978.

E029 Erichsen, L.V., Die kritischen losungstemperaturen in der homologen reihe der primaren normalen alkohole, *Brennstoff-Chemie*, 33, 166–172, 1952.

E031 Eisenbrand, J. and H. Picher, Bestimmung der dissoziationskonstanten, loslichkeiten und verteilungskoeffizienten von pantokain- und novokainbase, *Archiv der Pharmazie und Berichte der Deutschen Pharmazeutischen Gesellschaft*, 276, 1–17, 1938.

E032 Efremov, N.N., On the solubility in water of the nitro derivatives of phenol and of the dihydroxybenzenes, *Bulletin Academic Science USSR Division of Chemical Science*, 1–29, 1940.

E033 El-Fattah, S.A. and N.A. Daabis, The effect of dihydroxypropyl theophylline on the solubility and stability of menadione (vitamin K3), *Pharmazie* (Berlin), 32, 232–234, 1977.

E035 Emich, Loslichkeit der glycocholsaure, *Monatshefte fuer Chemie*, 3, 336–340, 1882.

E036 Earle, J.C., Notes on cineol, *Journal of the Society of Chemical Industry, London*, 37, 274–274, 1918.

E037 Evans, W.V. and M.B. Aylesworth, Some critical constants of furfural, *Industrial and Engineering Chemistry*, 18, 24–27, 1926.

E039 Efremov, Y.V. and I.F. Golubev, The solubility of omega-aminoundecanoic acid in aqueous alcoholic solutions, *Russian Journal of Physical Chemistry*, 36, 516–516, 1962.

E041 Ebian, A.R. and N.A. El-Gindy, Codeine crystal forms. I. Preparation, indentification, and characterization, *Scientia Pharmaceutica*, 46, 1–7, 1978.

E044 El Gamal, S., N. Borie, and Y. Hammouda, The influence of urea, polyethylene glycol 6000 and polyvinyl pyrrolidone on the dissolution properties of nitrofurantoin, *Pharmazeutische Industrie*, 40, 1373–1376, 1978.

E045 Eisenberg, M., P. Chang, C.W. Tobias, and C.R. Wilke, Physical properties of organic acids, *American Institute of Chemical Engineers Journal*, 1, 558–558, 1955.

E046 Edwards, L.J., Salicylamide: thermodynamic dissociation constant. Solubility and quantitative estimation by U.V. absorption spectrophotometry, *Transactions of the Faraday Society*, 49, 234–236, 1953.

E047 El-Banna, H.M. and O.Y. Abdallah, Physicochemical and dissolution studies of phenylbutazone binary systems, *Pharmaceutica Acta Helvetiae*, 55, 256–260, 1980.

E048 Ellgehausen, H., C. D'Hondt, and R. Fuerer, Reversed-phase chromatography as a general method for determining octan-1-ol/water partition coefficients, *Pesticide Science*, 12, 219–227, 1981.

E049 El Gholmy, Z.A., Effect of urea and sodium chloride on the aqueous solubility of acetazolamide, hydrochlorothiazide and frusemide, *Journal of Drug Research*, 11, 181–189, 1979.

E050 Eichler, D., Bromophos and bromophos-ethyl residues, *Residue Reviews*, 41, 65–67, 1972.

E051 Embil, K. and G. Torosian, Solubility and ionization characteristics of doxepin and desmethyldoxepin, *Journal of Pharmaceutical Sciences*, 71, 191–193, 1982.

E052 El-Nimr, A.E.M., S.M. Omar, and M.A. Kassem, Effect of bile salt–polyvinylpyrrolidone complexes on the solubilization profiles of diethylstilbestrol and digoxin, *Pharmazeutische Industrie*, 42, 311–314, 1980.

E301 Enever, R.P., K. Fotherby, S. Naderi, and G.A. Lewis, Long-acting contraceptive agents: the influence of physico-chemical properties of some esters of norethisterone upon the plasma levels of free norethisterone, *Steriods, An International Journal* (San Francisco), 41, 381–396, 1983.

E305 El-Harakany, A.A. and A.O. Barakat, Solubility of *tris*-(hydroxymethyl)-aminomethane in water–2-methoxyethanol solvent mixtures and the solvent effect on the dissociation of the protonated base, *Journal of Solution Chemistry*, 14, 263–269, 1985.

E307 Ekwall, P., L. Sjoblom, and J. Olsen, The spectrophotometric determination of steroid hormones solubilized in aqueous solutions of association colloids, *Acta Chemica Scandinavica*, 7, 347–351, 1953.

E308 Eckert, J.W., Fungistatic and phytotoxic properties of some derivatives of nitrobenzene, *Phytopathology*, 52, 642–649, 1962.

E311 Egawa, H., S. Maeda, E. Yonemochi, and Y. Nakai, Solubility parameter and dissolution behavior of cefalexin powders with different crystallinity, *Chemical and Pharmaceutical Bulletin*, 40, 819–820, 1992.

E312 El-Mahrouk, G.M., S.Y. Amin, and R.A. Shoukry, Complexation of khellin with different cyclodextrins, *Journal of Drug Research. Egypt*, 20, 91–101, 1991.

E314 Escalera, J.B., P. Bustamante, and A. Martin, Predicting the solubility of drugs in solvent mixtures: multiple solubility maxima and the chameleonic effect, *Journal of Pharmacy and Pharmacology*, 46, 172–176, 1994.

E316 Elamin, A.A., C. Ahlneck, G. Alderborn, and C. Nystrom, Increased |metastable solubility of milled griseofulvin, depending on the formation of a disordered surface structure, *International Journal of Pharmaceutics*, 111, 159–170, 1994.

E406 El-Barghouthi, M.I., N.A. Masoud, J.K. Kafawein, and A.A. Badwan, Host–guest interactions of risperidone with natural and modified cyclodextrins: phase solubility, thermodynamics and molecular modeling studies, *Chemical and Pharmaceutical Bulletin*, 53, 15–22, 2005.

E409 Erlich, L., D. Yu, D.A. Pallister, and T.X. Viegas, Relative bioavailability of danazol in dogs from liquid-filled hard gelatin capsules, *International Journal of Pharmaceutics*, 179, 49–53, 1999.

F001 Fuhner, H., Die Wasserloslichkeit in Homologen Reihen, *Berichte der Deutschen Chemischen Gesellschaft*, 57, 510–515, 1924.

F002 Franks, F., M. Gent, and H.H. Johnson, The solubility of benzene in water, *Journal of the Chemical Society* (London), 2716–2723, 1963.

F003 Flynn, G.L., R.W. Smith, and S.H. Yalkowsky, Solubility of hydrophobic species in aqueous systems i. solubility of u-34,865 in propylene glycol-water mixtures as a function of solvent composition and temperature, Technical Report, 1972.

F004 Franks, F., Solute–water interactions and the solubility behaviour of long-chain paraffin hydrocarbons, *Nature* (London), 210, 87–88, 1966.

F005 Feldman, S. and M. Gibaldi, Effect of urea on solubility—Role of water structure, *Journal of Pharmaceutical Sciences*, 56, 370–375, 1967.

F006 Flynn, G.L. and S.H. Yalkowsky, Correlation and prediction of mass transport across membranes I: influence of alkyl chain length on flux-determining properties of barrier and diffusant, *Journal of Pharmaceutical Sciences*, 61, 838–852, 1972.

F007 Florence, A.T. and A. Rahman, Polyvinylpyrrolidones and their influence on the dissolution rates of compounds of varying aqueous solubilities, *Journal of Pharmacy and Pharmacology*, 27, 55–55, 1975.

F008 Farhadieh, B., S. Borodkin, and J.D. Buddenhagen, Drug release from methyl acrylate-methyl methacrylate copolymer matrix I: kinetics of release, *Journal of Pharmaceutical Sciences*, 60, 209–212, 1971.

F009 Fritz, A., J.L. Lach, and L.D. Bighley, Solubility analysis of multicomponent systems capable of interacting in solution, *Journal of Pharmaceutical Sciences*, 60, 1617–1619, 1971.

F010 Florence, A.T. and E.G. Salole, Changes in crystallinity and solubility on comminution of digoxin and observations on spironolactone and oestradiol, *Journal of Pharmacy and Pharmacology*, 28, 637–642, 1976.

F011 Florence, A.T., A.W. Jenkins, and A.H. Loveless, Effect of formulation of intramuscular injections of phenothiazines on duration of activity, *Journal of Pharmaceutical Sciences*, 65, 1665–1668, 1976.

F012 Fedorova, E.A., L.F. Shashkina, and V.K. Fedorov, On the method of determination of the solubility of derivative testosterone, *Khimiko-Farmatsevticheskii Zhurnal*, 10, 139–142, 1976.

F013 Felder, E., D. Pitre, and M. Grandi, Radiopaque contrast media: XXXV. Physical properties of iopronic acid, a new oral cholecystografic agent, *Farmaco, Edizione Scientifica*, 31, 426–437, 1976.

F014 Fondyce, C.F. and L.W.A. Meyer, *Industrial and Engineering Chemistry*, 32, 1053–1053, 1940.

F015 Forist, A.A. and T. Chulski, pH–solubility relationships for 1-butyl-3-p-tolylsulfonylurea (orinase) and its metabolite, 1-butyl-3-p-carboxy-phenylsulfonylurea, *Metabolism, Clinical and Experimental*, 5, 807–812, 1956.

F016 Flynn, G.L., N.F.H. Ho, S. Hwang, and J. Park, Interfacing matrix release and membrane absorption-analysis of steroid absorption from a vaginal device in the rabbit doe, *Controlled Release Polymeric Formulations*, 87–122, 1976.

F017 Findlay, A. and A.N. Campbell, The influence of constitution on the stability of racemates, *Journal of the Chemical Society* (London), 1768–1775, 1928.

F018 Foy, C.L., *The Chorinated Aliphatic Acids*, Book Chapter, 207–211, Marcel Dekker, 1969.

F019 Fang, S.C., *Thiolcarbamates*, Book Chapter, 9, 147–149, Marcel Dekker, 1969.

F023 Feofilaktow, W.W., Uber die kondensation von brenztraubensaure mit paraformaldehyd unter zusatz von schwefelsaure, *Berichte der Deutschen Chemischen Gesellschaft*, 59, 2765–2777, 1926.

F024 Florey, K., Triamcinolone, *Analytical Profiles of Drug Substances*, 1, 378–381, 1972.

F025 Florey, K., Triamcinolone acetonide, *Analytical Profiles of Drug Substances*, 1, 398–409, 1972.

F026 Florey, K., Triamcinolone diacetate, *Analytical Profiles of Drug Substances*, 1, 423–433, 1972.

F027 Florey, K., Triflupromazine hydrochloride, *Analytical Profiles of Drug Substances*, 2, 523–546, 1973.

F029 Farmer, R.C., The decomposition of nitric esters, *Journal of the Chemical Society* (London), 117, 806–818, 1920.

F030 Flatt, R. and A. Jordan, Contribution au probleme de la solvatation: determination des rayons d'ions dissous, *Helvetica Chimica Acta*, 16, 37–53, 1933.

F033 Fischer, L.J. and S. Riegelman, Absorption and activity of some derivatives of Griseofulvin, *Journal of Pharmaceutical Sciences*, 56, 469–476, 1967.

F035 Freed, V.H., Mode of action other than aryl axyalkyl Acids, *Journal of Agricultural and Food Chemistry*, 1, 47–50, 1953.

F037 Florence, A.T., E.G. Salole, and J.B. Stenlake, The effect of particle size reduction on digoxin crystal properties, *Journal of Pharmacy and Pharmacology*, 26, 479–480, 1974.

F040 Freed, V.H., R. Haque, D. Schmedding, and R. Kohnert, Physicochemical properties of some organo-phosphates in relation to their chronic toxicity, *Environmental Health Perspectives*, 13, 77–81, 1976.

F041 French, D., M.L. Levine, J.H. Pazur, and E. Norberg, Studies on the Schardinger dextrins. The preparation and solubility characteristics of alpha, beta and gamma dextrins, *Journal of the American Chemical Society*, 71, 353–356, 1949.

F042 Frokjoer, S. and V.S. Andersen, Application of differential scanning calorimetry to the determination of the solubility of a metastable drug, *Archive of Pharmacy Chemistry Science*, 2, 50–59, 1974.

F043 Franzen, H. and E. Engel, Mitteilung aus dem chemischen institut der technischen hochschule zu karlsruhe, *journal fuer praktische chemie*, 102, 156–186, 1921.

F044 Ferguson, J., The use of chemical potentials as indices of toxicity, *Proceedings of the Royal Society of London, Series B: Biological Sciences*, 127, 387–404, 1927.

F045 Felder, E., D. Pitre, and M. Grandi, Radiopaque contrast media, *Farmaco, Edizione Scientifica*, 32, 755–766, 1977.

F047 Fierz-David, H.E., Uber die Anthrachinon-sulfosauren, *Helvetica Chimica Acta*, 10, 197–227, 1927.

F048 Frere, F.J., Ternary system diisopropyl ether–isopropyl alcohol–water at 25°C, *Industrial and Engineering Chemistry*, 41, 2365–2367, 1949.

F049 Fowler, A.R. and H. Hunt, The system nitromethane-n-propanol–water: vapor–liquid equilibria in the ternary and the three binary systems, *Industrial and Engineering Chemistry*, 33, 90–95, 1941.

F050 Fritzsche, R.H. and D.L. Stockton, Systems containing isobutanol and tetrachloroethane : liquid–vapor and liquid–liquid equilibria, *Industrial and Engineering Chemistry*, 38, 737–740, 1946.

F051 Fontein, F., Gleichgewichte in ternaren und quaternaren systemen, wobei zwei flussige schichten auftreten konnen, *Zeitschrift fuer Physikalische Chemie, Abteilung A: Chemische Thermodynamik, Kinetik, Electrochemie, Eigenschaftslehre*, 73, 212–251, 1910.

F052 Forcrand, M.D., Sur quelques constantes physiques du cyclohexanol, *Comptes Rendus Hebdomadaires des Seances de l'Academie des Sciences*, 154, 1327–1330, 1912.

F053 Fuoss, R.M., The system water–n-butanol–toluene at 30 degrees, *Journal of the American Chemical Society*, 65, 78–81, 1943.

F054 Frisch, F., Uber das naphthalingrun V, *Helvetica Chimica Acta*, 13, 768–785, 1930.

F055 Forbes, G.S. and A.S. Coolidge, Relations between distribution ratio, temperature and concentration in system: water, ether, succinic acid, *Journal of the American Chemical Society*, 41, 150–167, 1919.

F056 Fourneau, E. and G. Florence, Contribution a l'etude des ureides des acides bromo-valerianiques. iii.-influences sur les proprietes physiologiques de la migration de l'halogene dans la chaine de l'acide, *Bulletin Society of Chemistry, French*, 43, 1027–1040, 1928.

F057 Fourneau, E. and G. Florence, Contribution a l'etude des ureides des acides bromo-valerianiques. ii. influence de la ramification de la chaine sur les proprietes physiologiques, *Bulletin Society of Chemistry, French*, 43, 211–216, 1928.

F059 Finholt, P. and S. Solvang, Dissolution kinetics of drugs in human gastric juice-the role of surface tension, *Journal of Pharmaceutical Sciences*, 57, 1322–1326, 1968.

F062 Falck, V.A., Beitrag zur kenntnis des Sulfonal, *Pharmazeutische Zentralhalle*, 60, 409–416, 1919.

F063 Fendler, J.H., E.J. Fendler, G.A. Infante, and L.K. Patterson, Absorption and proton magnetic resonance spectroscopic investigation of the environment of acetophenone and benzophenone in aqueous micellar solutions, *Journal of the American Chemical Society*, 97, 89–95, 1975.

F064 Funasaki, N., S. Hada, and K. Kawamura, The surface tension of aqueous solutions of ethylene glycol diesters, *Nippon Kagaku Kaishi*, 12, 1944–1946, 1976.

F066 Flottmann, F., Uber loslichkeitsgleichgewichte, *Ztschrft fur Analytische Chemie*, 73, 1–39, 1928.

F067 Fukano, I. and Y. Obata, Solubility of Phthalates in Water, *Purosuchikkusu*, 27, 48–49, 1976.

F068 Frank, S.G. and M.J. Cho, Phase solubility analysis and pmr study of complexing behavior of dinoprostone with beta-cyclodextrin in water, *Journal of Pharmaceutical Sciences*, 67, 1665–1668, 1978.

F069 Freundlich, H. and G.V. Slottman, Uber das gelten der traubeschen regel bei der hydrotropie, *Biochemische Zeitschrift* (1948–1967), 188, 101–111, 1927.

F070 Fishbein, L. and P.W. Albro, Chromatographic and biological aspects of the phthalate esters, *Journal of Chromatography*, 70, 365–412, 1972.

F071 Freed, V.H., C.T. Chiou, D. Schmedding, and R. Kohnert, Some physical factors in toxicological assessment tests, *Environmental Health Perspectives*, 30, 75–80, 1979.

F072 Fox, Jr., C.L. and H.M. Rose, Ionization of sulfonamides, *Proceedings of the Society for Experimental Biology and Medicine*, 50, 142–145, 1942.

F073 Frisk, A.R., Sulfanilamide derivatives—chemotherapeutic evaluation of N1-substituted sulfanilamides, *Acta Medical Scandanavia*, 142, 1–20, 1943.

F074 Frisk, A.R., Blood concentration, acetylation and urinary excretion of sulfapyridine and sulfathiazole after various sulfapyridine and sulfathiazole derivatives administered by different routes, *Acta Medica Scandinavica*, 106, 369–403, 1941.

F075 Feinstone, W.J., R.D. Williams, R.T. Wolff, and M.L. Crossley, The toxicity, absorption and chemotherapeutic activity of 2-sulfanilamidopyrimidine (sulfadiazine), *Bulletin of the John Hopkins Hospital*, 67, 427–456, 1940.

F076 Fessi, H., J.-P. Marty, F. Puisieux, and J.T. Carstensen, Square root of time dependence of matrix formulations with low drug content, *Journal of Pharmaceutical Sciences*, 71, 749–752, 1982.

F176 Friberger, P. and G. Aberg, Some physicochemical properties of the racemates and the optically active isomers of two local anaesthetic compounds, *Acta Pharmaceutica Suecica*, 8, 361–364, 1971.

F178 Felder, E., D. Pitre, and P. Tirone, Radiopaque contrast media. XLIV. Preclinical studies with a new nonionic contrast agent, *Farmaco, Edizione Scientifica*, 32, 835–844, 1977.

F179 Felsot, A. and P.A. Dahm, Sorption of organophosphorus and carbamate insecticides by soil, *Journal of Agricultural and Food Chemistry*, 27, 557–563, 1979.

F181 Florey, K., Cephradine, *Analytical Profiles of Drug Substances*, 5, 22–37, 1976.

F182 Furesz, S., Chemical and biological properties of rifampicin, *Antibiotica et Chemotherapia* (1954–68), 16, 316–351, 1970.

F183 Fujioka, H. and T. Tan, Biopharmaceutical studies on hydantoin derivatives. I. Physico-chemical properties of hydantoin derivatives and their intestinal absorption, *Journal of Pharmaceutics Dynamics*, 4, 759–770, 1981.

F184 Freed, V.H. and P. Burschel, The relationship of water solubility to dosage of herbicides, *Zeitschrift fuer Pflanzenkrankheiten und Pflanzenschutz*, 64, 477–479, 1957.

F185 Filippov, T.C. and A.A. Firman, *Zhurnal Prikladnoi Khimii* (Leningrad), 25, 895–897, 1952.

F186 Freudenberg, K., E. Plankenhorn, and H. Knauber, Schardinger's dextrins—derived from starch, *Chemistry and Industry* (London), 731–735, 1947.

F300 Freier, R.K., *Aqueous Solutions Volume 1: Data for Inorganic and Organic Compounds*, Walter de Gruyter, New York, 1, 1976.

F301 Fini, A., V. Zecchi, L. Rodriguez, and A. Tartarini, Solubility–dissolution relationship for ibuprofen, fenbufen and their sodium salts in acid medium, *Pharmaceutica Acta Helvetiae*, 59, 106–108, 1984.

F302 Freier, R.K., *Book*, 2, 17–443, Walter de Gruyter, 1978.

F303 Friesen, K.J., L.P. Sarna, and G.R.B. Webster, Aqueous solubility of polychlorinated dibenzo-*p*-dioxins determined by high pressure liquid chromatography, *Chemosphere*, 14, 1267–1274, 1985.

F306 Fini, A., M. Laus, I. Orienti, and V. Zecchi, Dissolution and partition thermodynamic functions of some nonsteroidal anti-inflammatory drugs, *Journal of Pharmaceutical Sciences*, 75, 23–25, 1986.

F307 Fini, A., A. Roda, R. Fugazza, and B. Grigolo, Chemical properties of bile acids: III. Bile acid structure and solubility in water, *Journal of Solution Chemistry*, 14, 595–603, 1985.

F309 Farraj, N.F., S.S. Davis, G.D. Parr, and H.N.E. Stevens, The stability and solubility of progabide and its related metabolic derivatives, *Pharmaceutical Research*, 5, 226–231, 1988.

F310 Fukumori, Y., T. Fukuda, Y. Yamamoto, and N. Sato, Physical characterization of erythromycin dihydrate, anhydrate and amorphous solid and their dissolution properties, *Chemical and Pharmaceutical Bulletin*, 31, 4029–4039, 1983.

F311 Furer, R. and M. Geiger, A simple method of determining the aqueous solubility of organic substances, *Pesticide Science*, 8, 337–344, 1977.

F312 Fulford, M.D., J.E. Slonek, and M.J. Groves, A note on the solubility of progesterone in aqueous polyethylene gylcol 400, *Drug Development and Industrial Pharmacy*, 12, 631–635, 1986.

F314 Friesen, K.J., J. Vilk, and D.C.G. Muir, Aqueous solubilities of selected 2,3,7,8-substituted polychlorinated dibenzofurans (PCDFs), *Chemosphere*, 20, 27–32, 1990.

F315 Friesen, K.J. and G.R.B. Webster, Temperature dependence of the aqueous solubilities of highly chlorinated dibenzo-*p*-dioxins, *Environmental Science and Technology*, 24, 97–101, 1990.

F317 Faizal, M., F.J. Smagghe, G.H. Malmary, and J.R. Molinier, Equilibrium diagrams at 25 degrees C of water–oxalic acid–2-methyl-1–propanol, water–oxalic acid-1–pentanol, and water–oxalic acid-3-methyl-1-butanol ternary systems, *Journal of Chemical and Engineering Data*, 35, 352–354, 1990.

F318 Farraj, N.F., S.S. Davis, G.D. Parr, and H.N.E. Stevens, Modification of the aqueous solubility and stability of progabide, *International Journal of Pharmaceutics*, 52, 11–18, 1989.

F322 Forster, S., G. Buckton, and A.E. Beezer, The Importance of chain length on the wettability and solubility of organic homologs, *International Journal of Pharmaceutics*, 72, 29–34, 1991.

F325 Frolov, A.F., M.A. Loginova, and A.P. Karaseva, Mutual solubility in the system butyl alcohol–ethyl alcohol–methyl alcohol–water, *Journal of General Chemistry of the USSR*, 38, 1164–1166, 1968.

F327 Frilink, H.W., A.C. Eissens, A.J.M. Schoonen, and C.F. Lerk, The effects of cyclodextrins on drug release from fatty suppository bases. I: in vitro observations, *European Journal of Pharmaceutics and Biopharmaceutics*, 37, 178–182, 1991.

F415 Fioritto, A.F., S.N. Bhattachar, and J.A. Wesley, Solubility measurement of polymorphic compounds via the pH-metric titration technique, *International Journal of Pharmaceutics*, 330, 105–113, 2007.

F418 Freire, M.G., A. Razzouk, I. Mokbel, and J.A.P. Coutinho, Solubility of hexafluorobenzene in aqueous salt solutions from (280 to 340) K, *Journal of Chemical and Engineering Data*, 50, 237–242, 2005.

F419 Fuchs, D., J. Fischer, F. Tumakaka, and G. Sadowski, Solubility of amino acids: influence of the ph value and the addition of alcoholic cosolvents on aqueous solubility, *Industrial and Engineering Chemistry Research*, 45, 6578–6584, 2006.

F425 Fourie, L., J.C. Breytehbach, J.D. Plessis, and J. Hadgraft, Percutaneous delivery of carbamazepine and selected N-alkyl and N-hydroxyalkyl analogues, *International Journal of Pharmaceutics*, 279, 59–66, 2004.

G001 Getzen, F.W. and T.M. Ward, Influence of water structure on aqueous solubility, *Industrial and Engineering Chemistry, Product Research and Development*, 10, 122–132, 1971.

G002 Garrett, E.R. and P.B. Chemburkar, Evaluation, control, and prediction of drug diffusion through polymeric membranes. II, *Journal of Pharmaceutical Sciences*, 57, 949–959, 1968.

G003 Garrett, E.R. and P.B. Chemburkar, Evaluation, control, and prediction of drug diffusion through polymeric membranes. III, *Journal of Pharmaceutical Sciences*, 57, 1401–1409, 1968.

G004 Ginnings, P.M. and R. Baum, Aqueous solubilities of the isomeric pentanols, *Journal of the American Chemical Society*, 59, 1111–1113, 1937.

G005 Ginnings, P.M. and R. Webb, Aqueous solubilities of some isomeric hexanols, *Journal of the American Chemical Society*, 60, 1388–1389, 1938.

G006 Ginnings, P.M. and M. Hauser, Aqueous solubilities of some isomeric heptanols, *Journal of the American Chemical Society*, 60, 2581–2582, 1938.

G007 Ginnings, P.M. and D. Coltrane, Aqueous solubility of 2,2,3-trimethylpentanol-3, *Journal of the American Chemical Society*, 61, 525–525, 1939.

G008 Guttman, D.E., W.E. Hamlin, J.W. Shell, and J.G. Wagner, Solubilization of anti-inflammatory steroids by aqueous solutions of triton WR-1339, *Journal of Pharmaceutical Sciences*, 50, 305–307, 1961.

G009 Gabaldon, M., J. Sanchez, and A. Llombart, Jr., In vitro utilization of diethylstilbestrol by rat liver, *Journal of Pharmaceutical Sciences*, 57, 1744–1747, 1968.

G010 Goldberg, A.H., M. Gibaldi, J.L. Kanig, and M. Mayersohn, Increasing dissolution rates and gastrointestinal absorption of drugs via solid solutions and eutectic mixtures. IV, *Journal of Pharmaceutical Sciences*, 55, 581–583, 1966.

G011 Goldberg, A.H., M. Gibaldi, and J.L. Kanig, Increasing dissolution rates and gastrointestinal absorption of drugs via solid solutions and eutectic mixtures. III, *Journal of Pharmaceutical Sciences*, 55, 487–492, 1966.

G012 Gans, E.H. and T. Higuchi, The solubility and complexing properties of oxytetracycline and tetracycline I, *Journal of the American Pharmaceutical Association, Scientific Edition*, 46, 458–466, 1957.

G014 Gouda, M.W., A.A. Ismail, and M.M. Motawi, Micellar solubilization of barbiturates ii: solubilities of certain barbiturates in polyoxyethylene stearates of varying hydrophilic chain length, *Journal of Pharmaceutical Sciences*, 59, 1402–1405, 1970.

G015 Garrett, E.R., Prediction of stability in pharmaceutical preparations. IV, *Journal of the American Pharmaceutical Association, Scientific Edition*, 46, 584–586, 1957.

G016 Garrett, E.R. and D.J. Weber, Metal complexes of thiouracils i: stability constants by potentiometric titration studies and structures of complexes, *Journal of Pharmaceutical Sciences*, 59, 1383–1391, 1970.

G017 Garrett, E.R. and C.M. Won, Prediction of stability in pharmaceutical preparations. XVI: kinetics of hydrolysis of canrenone and lactonization of canrenoic acid, *Journal of Pharmaceutical Sciences*, 60, 1801–1809, 1971.

G018 Garrett, E.R. and C.A. Hunt, Physicochemical properties, solubility, and protein binding of delta-9-tetrahydrocannabinol, *Journal of Pharmaceutical Sciences*, 63, 1056–1064, 1974.

G020 Gibbs, I.S., A. Heald, H. Jacobson, and I. Weliky, Physical characterization and activity in vivo of polymorphic forms of 7-chloro-5,11-dihydrodibenz[b,e][1,4]oxazepine-5-carboxamide, a potential tricyclic antidepressant, *Journal of Pharmaceutical Sciences*, 65, 1380–1385, 1976.

G021 Guttman, D. and T. Higuchi, Reversible association of caffeine and of some caffeine homologs in aqueous solution, *Journal of the American Pharmaceutical Association, Scientific Edition*, 46, 4–9, 1957.

G022 Green, A.L., Ionization constants and water solubilities of some aminoalkylphenothiazine tranquillizers and related compounds, *Journal of Pharmacy and Pharmacology*, 19, 10–16, 1967.

G023 Green, A.L., Activity correlations and the mode of action of aminoalkylphenothiazine tranquillizers, *Journal of Pharmacy and Pharmacology*, 19, 207–208, 1967.

G024 Gadalla, M.A.F., A.M. Saleh, and M.M. Motawi, Effect of electrolytes on the solubility and solubilization of chlorocresol, *Pharmazie* (Berlin), 29, 105–107, 1974.

G025 Giri, S.N. and L.R. Kartt, Temperature dependent aqueous solubility of actinomycin D, *Specialia*, 31, 482–483, 1975.

G026 Gilligan, D.R. and M.N. Plummer, Comparative solubilities of sulfadiazine, sulfamerizine and sulfamethazine and their n4-acetyl derivatives at varying pH levels, *Proceedings of the Society for Experimental Biology and Medicine*, 53, 142–145, 1944.

G028 Guillory, J.K. and H.O. Lin, Some properties of sulfanilamide monohydrate, *Chemical and Pharmaceutical Bulletin*, 24, 1675–1678, 1976.

G029 Gross, P.M. and J.H. Saylor, The solubilities of certain slightly soluble organic compounds in water, *Journal of the American Chemical Society*, 53, 1744–1751, 1931.

G030 Ginnings, P.M., D. Plonk, and E. Carter, Aqueous solubilities of some aliphatic ketones, *Journal of the American Chemical Society*, 62, 1923–1924, 1940.

G031 Ginnings, P.M., E. Herring, and D. Coltrane, Aqueous solubilities of some unsaturated alcohols, *Journal of the American Chemical Society*, 61, 807–808, 1939.

G032 Gross, P., J.C. Rintelen, and J.H. Saylor, Energy and volume relations in the solubilities of some ketones in water, *Journal of Physical Chemistry*, 43, 197–205, 1939.

G033 Granger, F.S. and J.M. Nelson, Oxidation and reduction of hydroquinone and quinone from the standpoint of electromotive-force measurements, *Journal of the American Chemical Society*, 43, 1401–1415, 1921.

G034 Glew, D.N. and R.E. Robertson, The spectrophotometric determination of the solubility of cumene in water by a kinetic method, *Journal of Physical Chemistry*, 60, 332–337, 1956.

G035 Guseva, A.N. and E.I. Parnov, Isothermal sections of monocyclic arene–water binary systems at 25, 100, and 200 degrees [ethylbenzene and propylbenzene (ed. of translation)], *Russian Journal of Physical Chemistry*, 38, 439–440, 1964.

G036 Geissbuhler, H., The substituted ureas, in *Degradation of Herbicides*, Marcel Dekker, New York, 79–83, 1969.

G037 Gross, P.M., J.H. Saylor, and M.A. Gorman, Solubility studies. IV. The solubilities of certain slightly soluble organic compounds in water, *Journal of the American Chemical Society*, 55, 650–652, 1933.

G038 Gross, P., The determination of the solubility of slightly soluble liquids in water and the solubilities of the dichloro-ethanes and -propanes, *Journal of the American Chemical Society*, 51, 2362–2366, 1929.

G039 Gorman, W.G. and G.D. Hall, Dielectric constant correlations with solubility and solubility parameters, *Journal of Pharmaceutical Sciences*, 53, 1017–1020, 1964.

G040 Gilbert, E.C. and B.E. Lauer, A study of the ternary system methyl benzoate, methanol, water, *Journal of Physical Chemistry*, 31, 1050–1052, 1927.

G041 Gysin, H., Triazine herbicides their chemistry, biological properties and mode of action, *Chemistry and Industry* (London), 31, 1393–1400, 1962.

G042 Griffiths, R.V. and A.G. Mitchell, Surface transformation during dissolution of aspirin, *Journal of Pharmaceutical Sciences*, 60, 267–270, 1971.

G043 Gibbs, H.D., Phenol tests, *Journal of Biological Chemistry*, 72, 649–655, 1927.

G046 Grube, V.G. and M. Nubaum, Phasenthcorctische untersuchungen uber die entzuckerung der melasse, *Zeitschrift fuer Elektrochemie*, 34, 91–98, 1928.

G047 Gordon, J.E. and R.L. Thorne, Salt effects on the activity coefficient of naphthalene in mixed aqueous electrolyte solutions. I. Mixtures of two salts, *Journal of Physical Chemistry*, 71, 4390–4392, 1967.

G050 Griswold, J., M.E. Klecka, and R.V. West, Jr., Conjugate liquid phase equilibria—c4-hydrocarbon–furfural–water systems, *Chemical Engineering Progress Symposium Series—"Phase-Equilibria"*, 44, 839–846, 1948.

G051 Gold, B. and S.S. Mirvish, *N*-nitroso derivatives of hydrochlorothiazide, niridazole, and tolbutamide, *Toxicology and Applied Pharmacology*, 40, 131–136, 1977.

G052 Goeller, G.M. and A. Osol, The salting-out of molecular benzoic acid in aqueous salt solutions at 35 degrees, *Journal of the American Chemical Society*, 59, 2132–2134, 1937.

G053 Gross, P., Uber den aussalzeffekt an dichlorathanen und-propanen, *Zeitschrift fur Physical Chemistry Abt B*, 6, 215–220, 1929.

G054 Gavaudan, P. and H. Poussel, Le mecanisme de l'action insecticide du dichlorodiphenyl-trichlorethane (DDT) et la regle thermodynamique des narcotiques indifferents, *Comptes Rendus Hebdomadaires des Seances de l'Academie des Sciences*, 224, 683–685, 1947.

G055 Glew, D.N., The gas hydrate of bromochlorodifluoromethane, *Canadian Journal of Chemistry*, 38, 208–221, 1960.

G056 Gladis, G.P., Effects of moisture on corrosion in petrochemical environments, *Chemical Engineering Progress Symposium Series—"Phase-Equilibria"*, 56, 43–51, 1960.

G058 Gerrard, W., Significance of the solubility of hydrocarbon gases in liquids in relation to the intermolecular structure of liquids, *Chemistry and Industry* (London), 21, 804–805, 1972.

G060 Grut, D., The solubility of sucrose, *Zeitschrift fuer die Zuckerindustrie Czcchoslov*, 61, 345, 1937.

G061 Glew, D.N. and E.A. Moelwyn-Hughes, Chemical statics of the methyl halides in water, *Discussions of the Faraday Society*, 15, 150–161, 1953.

G062 Glasstone, S. and A. Pound, Solubility influences. Part I. The effect of some salts, sugars, and temperature on the solubility of ethyl acetate in water, *Journal of the Chemical Society* (London), 107, 2660–2667, 1925.

G063 Giacalone, A., Solubilita dell'acido 6-nitro-3-metilbenzoico in benzene, toluene ed acqua, *Gazzetta Chimica Italiana*, 65, 844–850, 1935.

G066 Gibbs, H.D., Phenol tests, *Journal of Physical Chemistry*, 31, 1057–1081, 1927.

G067 Goto, A., F. Endo, and K. Ito, Gel filtration of solubilized systems. I. On the gel filtration of aqueous sodium lauryl sulfate solution solubilizing alkyl paraben on sephadex G-50, *Chemical and Pharmaceutical Bulletin*, 25, 1165–1173, 1977.

G068 Guseva, A.N. and E.I. Parnov, The solubility of cyclohexane in water, *Russian Journal of Physical Chemistry*, 37, 1494–1494, 1963.

G072 Gale, M.M. and L. Saunders, The solubilisation of steroids by lysophosphatidylcholine testosterone, estradiol and their 17-alpha-ethinyl derivations, *Biochimica et Biophysica Acta*, 248, 466–470, 1971.

G073 Gouda, M.W., A.R. Ebian, M.A. Moustafa, and S.A. Khalil, Sulphapyridine crystal forms, *Drug Development and Industrial Pharmacy*, 3, 273–290, 1977.

G075 Gettins, J., D. Hall, P.L. Jobling, and E. Wyn-Jones, Thermodynamic and kinetic parameters associated with the exchange process involving alcohols and micelles, *Journal of the Chemistry Society, Faraday Transactions II*, 71, 1957–1964, 1978.

G076 Gasparini, G.M., The preparation and properties of trialkylacetohydroxamic acids: effect of the neoalkyl structure with regard to the solubility, the stability and some extractive capacities, *Gazzetta Chimica Italiana*, 109, 357–363, 1979.

G077 Geake, A. and J.T. Lemon, Semiquinone formation by anthraquinone and some simple derivatives, *Transactions of the Faraday Society*, 34, 1409–1427, 1938.

G078 Goto, A., R. Sakura, and F. Endo, Gel filtration of solubilized systems. V. effects of sodium chloride on micellar sodium lauryl sulfate solutions solubilizing alkylparabens, *Chemical and Pharmaceutical Bulletin*, 28, 14–22, 1980.

G079 Goring, C.A.I., Control of nitrification by 2-chloro-6-(trichloromethyl) pyridine, *Soil Science*, 93, 211–218, 1962.

G080 Goring, C.A.I., Theory and principles of soil fumigation, *Advances Pest Control Research*, 5, 47–84, 1962.

G081 Gillert, E., Cholerese und choleretica, ein beitrag zur physiologie der galle, *Zfrieidie Gesamte Experimental Medizin*, 48, 255–275, 1926.

G083 Gilligan, D.R., S. Garb, C. Wheeler, and M.N. Plummer, Adjuvant alkali therapy in the prevention of renal complications from sulfadiazine, *Journal of the American Medical Association*, 122, 1160–1165, 1943.

G084 Geneidi, A.S. and H. Hamacher, Enhancement of dissolution rates of spironolactone and diazepam via polyols and PEG solid dispersion systems, *Pharmazeutische Industrie*, 42, 401–404, 1980.

G085 Ghanem, A.H., H. El-Sabbagh, and H. Abdel-Alim, Solubilization of flufenamic acid, *Pharmazeutische Industrie*, 42, 854–856, 1980.

G086 Ghanem, A., M. Meshali, and I. Ramadaan, Dissolution rate of trimethoprim polyvinylpyrrolidone coprecipitate, *Pharmazie* (Berlin), 35, 689–690, 1980.

G088 Geneidi, A.S., M.S. Adel, and E. Shehata, Enhanced dissolution of gilbenclamide from gilbenclamide-poloxamer and glibenclamide-PVP coprecipitates, *Canadian Journal of Pharmaceutical Science*, 15, 81–84, 1980.

G089 Gallego, M., M. Garcia-Vargas, F. Pino, and M. Valcarcel, Analytical applications of picolinealdehyde salicyloylhydrazone, *Microchemical Journal*, 23, 353–359, 1978.

G090 Grabovskaya, Z.E. and M.I. Vinnik, Activity coefficients of certain aromatic compounds in concentrated sulphuric acid solutions, *Russian Journal of Physical Chemistry*, 40, 1221–1223, 1966.

G091 Gaynor, J.D. and V. Van Volk, s-Triazine solubility in chloride salt solutions, *Journal of Agricultural and Food Chemistry*, 29, 1143–1146, 1981.

G092 Gekko, K., Mechanism of poly-induced protein stabilization: solubility of amino acids and diglycine in aqueous polyol solutions, *Journal of Biochemistry* (Tokyo), 90, 1633–1641, 1981.

G093 Garrett, E.R. and M.R. Gardner, Prediction of stability in pharmaceutical preparations. XIX: stability evaluation and bioanalysis of clofibric acid esters by high-pressure liquid chromatography, *Journal of Pharmaceutical Sciences*, 71, 14–25, 1982.

G095 Geneidi, A.S. and H. Hamacher, Physical characterization and dissolution profiles of spironolactone and diazepam coprecipitates, *Pharmazeutische Industrie*, 42, 315–319, 1980.

G096 Gallo, G.G. and P. Radaelli, Rifampin, *Analytical Profiles of Drug Substances*, 5, 468–509, 1976.

G098 Grubb, P.E., Nalidixic acid, *Analytical Profiles of Drug Substances*, 8, 371–381, 979.

G099 Gysin, H. and E. Knusli, Chemistry and herbicidal properties of triazine derivatives, *Advances Pest Control Research*, 3, 289–355, 1960.

G101 Griswold, J., P.L. Chu, and W.O. Winsauer, Phase equilibria in ethyl alcohol–ethyl acetate–water system, *Industrial and Engineering Chemistry*, 41, 2352–2358, 1949.

G300 Geyer, H., P. Sheehan, D. Kotzias, and F. Korte, Prediction of ecotoxicological behaviour of chemicals: relationship between physico-chemical properties and bioaccumulation of organic chemicals in the mussel *Mytilus edulis*, *Chemosphere*, 11, 1121–1134, 1982.

G301 Geyer, H., R. Viswanathan, D. Freitag, and F. Korte, Relationship between water solubility of organic chemicals and their bioaccumulation by the alga chlorella, *Chemosphere*, 10, 1307–1313, 1981.

G302 Grant, D.J.W., M. Mehdizadeh, A.H.-L. Chow, and J.E. Fairbrother, Non-linear van't Hoff solubility–temperature plots and their pharmaceutical interpretation, *International Journal of Pharmaceutics*, 18, 25–38, 1984.

G306 Green, C.D., Perfluidone, *Analytical Methods for Pesticides and Plant Growth Regulators*, 10, 437–450, 1978.

G307 Green, C.D., Fluoridamid, *Analytical Methods for Pesticides and Plant Growth Regulators*, 10, 533–543, 1978.

G310 Gledhill, W.E., R.G. Kaley, W.J. Adams, and V.W. Saeger, An environmental safety assessment of butyl benzyl phthalate, *Environmental Science and Technology*, 14, 301–305, 1980.

G312 Gobas, F.A.P., J.M. Lahittete, G. Garofalo, and D. Mackay, A novel method for measuring membrane–water partition coefficients of hydrophobic organic chemicals: comparison with 1-octanol–water partitioning, *Journal of Pharmaceutical Sciences*, 77, 265–272, 1988.

G313 Groves, F.R., Solubility of cycloparaffins in distilled water and salt water, *Journal of Chemical and Engineering Data*, 33, 136–138, 1988.

G315 Gekko, K. and S. Koga, The stability of protein structure in aqueous propylene glycol amino acid solubility and preferential solvation of protein, *Biochimica et Biophysica Acta*, 786, 151–160, 1984.

G317 Gabas, N., T. Carillon, and N. Hiquily, Solubilities of D-xylose and D -mannose in water–ethanol mixtures at 25°C, *Journal of Chemical and Engineering Data*, 33, 128–130, 1988.

G318 Gandhi, R.B. and A.H. Karara, Characterization, dissolution and diffusion properties of tolbutamide-beta-cyclodextrin, *Drug Development and Industrial Pharmacy*, 14, 657–682, 1988.

G319 Hamaker, J.W., Decomposition: Quantitative aspects, in *Organic Chemicals in the Soil Environment*, eds. C.A.I. Goring and J.W. Hamaker, Dekker, New York, 384–385, 1972.

G323 Griswold, J., J.-N. Chew, and M.E. Klecka, Pure hydrocarbons from petroleum: recovery of aniline solvent from distex hydrocarbon products by water extraction, *Industrial and Engineering Chemistry*, 42, 1246–1251, 1950.

G430 Gerber, M., J.C. Breytenbach, J. Hadgraft, and J. du Plessis, Synthesis and transdermal properties of acetylsalicylic, *International Journal of Pharmaceutics*, 310, 31–36, 2006.

G431 Garzon, L.C. and F. Martinez, Temperature dependence of solubility for ibuprofen in some organic and aqueous solvents, *Journal of Solution Chemistry*, 33, 1379–1395, 2004.

G433 Gude, M.T., H.H.J. Meuwissen, L.A..M van der Wielen, and K.A.M. Luyben, Partition coefficients and solubilities of α-amino acids in aqueous 1-butanol solutions, *Industrial & Engineering Chemistry Research*, 35, 4700–4712, 1996.

G434 Green, A.R. and J.K. Guillory, Heptakis(2,6-di-o-methyl)-b-cyclodextrin complexation with the antitumor agent chlorambucil, *Journal of Pharmaceutical Sciences*, 78, 427–431, 1989.

H002 Hill, A.E., The mutual solubility of liquids. I. The mutual solubility of ethyl ether and water. II. The solubility of water in benzene, *Journal of the American Chemical Society*, 45, 1143–1155, 1923.

H003 Hill, A.E. and W.M. Malisoff, The mutual solubility of liquids. III. The mutual solubility of phenol and water. iv. the mutual solubility of normal butyl alcohol and water, *Journal of the American Chemical Society*, 48, 918–927, 1926.

H004 Hamlin, W.E., J.I. Northam, and J.G. Wagner, Relationship between in vitro dissolution rates and solubilities of numerous compounds representative of various chemical species, *Journal of Pharmaceutical Sciences*, 54, 1651–1653, 1965.

H005 Higuchi, T., M. Gupta, and L.W. Busse, Influence of electrolytes, pH, and alcohol concentration on the solubilities of acidic drugs, *Journal of the American Pharmaceutical Association, Scientific Edition*, 42, 157–161, 1953.

H006 Houston, J.B., D.G. Upshall, and J.W. Bridges, A re-evaluation of the importance of partition coefficients in the gastrointestinal absorption of anutrients, *Journal of Pharmacology and Experimental Therapeutics*, 189, 244–254, 1974.

H007 Herzog, K.A. and J. Swarbrick, Drug permeation through thin-model membranes. III: correlations between in vitro transfer, in vivo absorption, and physicochemical parameters of substituted benzoic acids, *Journal of Pharmaceutical Sciences*, 60, 1666–1668, 1971.

H008 Hunt, M.J. and L. Saunders, The solubilization of some local anaesthetic esters of *p*-aminobenzoic acid by lysophosphatidylcholine, *Journal of Pharmacy and Pharmacology*, 27, 119–124, 1975.

H009 Humphreys, K.J. and C.T. Rhodes, Effect of temperature upon solubilization by a series of nonionic surfactants, *Journal of Pharmaceutical Sciences*, 57, 79–83, 1968.

H010 Higuchi, W.I., P.D. Bernardo, and S.C. Mehta, Polymorphism and drug availability II. Dissolution rate behavior of the polymorphic forms of sulfathiazole and methylpredisolone, *Journal of Pharmaceutical Sciences*, 56, 200–207, 1967.

H011 Hamlin, W.E. and W.I. Higuchi, Dissolution rate-solubility behavior of 3-(1-methyl-2-pyrrolidinyl)-indole as a function of hydrogen-ion concentration, *Journal of Pharmaceutical Sciences*, 55, 205–207, 1966.

H012 Harkins, W.D. and H. Oppenheimer, Solubilization of polar-non-polar substances in solutions of long chain electrolytes, *Journal of the American Chemical Society*, 71, 808–811, 1949.

H013 Hinsvark, O.N., W. Zazulak, and A.I. Cohen, Liquid chromatography: its use in the biological characterization and study of metolazone-a new diuretic, *Journal of Chromatographic Science*, 10, 379–382, 1972.

H015 Hussain, A., Prediction of dissolution rates of slightly water-soluble powders from simple mathematical relationships, *Journal of Pharmaceutical Sciences*, 61, 811–813, 1972.

H016 Higuchi, T. and A. Drubulis, Complexation of organic substances in aqueous solution by hydroxyaromatic acids and their salts, *Journal of Pharmaceutical Sciences*, 50, 905–909, 1961.

H017 Higuchi, T. and S. Bolton, The solubility and complexing properties of oxytetracycline and tetracycline III, *Journal of the American Pharmaceutical Association, Scientific Edition*, 48, 557–564, 1959.

H018 Higuchi, T. and J.L. Lach, Investigation of Some complexes formed in solution by caffeine IV. Interactions between caffeine and sulfathiazole, sulfadiazine, *p*-aminobenzoic acid, benzocaine, phenobarbital, and barbital, *Journal of the American Pharmaceutical Association, Scientific Edition*, 43, 349–354, 1954.

H019 Higuchi, T. and J.L. Lach, Study of possible complex formation between macromolecules and certain pharmaceuticals. III. Interaction of polyethylene glycols with several organic acids, *Journal of the American Pharmaceutical Association, Scientific Edition*, 43, 465–470, 1954.

H020 Higuchi, T. and J.L. Lach, Investigation of complexes formed in solution by caffeine. VI. Comparison of complexing behaviors of methylated xanthines with *p*-aminobenzoic acid, salicylic acid, acetylsalicylic acid, and p-hydroxybenzoic acid, *Journal of the American Pharmaceutical Association, Scientific Edition*, 43, 527–530, 1954.

H021 Higuchi, T. and J.L. Lach, Investigation of some complexes formed in solution by caffeine. V. Interactions between caffeine and *p*-aminobenzoic acid, *m*-hydroxybenzoic acid, picric acid, *o*-phthalic acid, suberic acid, and valeric acid, *Journal of the American Pharmaceutical Association, Scientific Edition*, 43, 524–527, 1954.

H022 Higuchi, T. and D.A. Zuck, Investigation of some complexes formed in solution by caffeine. III. Interactions between caffeine and aspirin, *p*-hydroxybenzoic acid, *m*-hydroxybenzoic acid, salicylic acid, salicylate ion, and butyl paraben, *Journal of the American Pharmaceutical Association, Scientific Edition*, 42, 138–145, 1953.

H023 Higuchi, T. and D.A. Zuck, Investigation of some complexes formed in solution by caffeine. II. Benzoic acid and benzoate ion, *Journal of the American Pharmaceutical Association, Scientific Edition*, 42, 132–138, 1953.

H024 Hormann, W.D. and D.O. Eberle, The aqueous solubility of 2-chloro-4-ethylamino-6-isopropylamino-1,3,5-triazine (atrazine) obtained by an improved analytical method, *Weed Research*, 12, 199–202, 1972.

H025 Higuchi, W.I., N.A. Mir, A.P. Parker, and W.E. Hamlin, Dissolution kinetics of a weak acid, 1,1-hexamethylene *p*-tolylsulfonylsemicarbazide, and its sodium salt, *Journal of Pharmaceutical Sciences*, 54, 8–11, 1965.

H026 Higuchi, W.I., P.K. Lau, T. Higuchi, and J.W. Shell, Polymorphism and drug availability: solubility relationships in the methylprednisolone system, *Journal of Pharmaceutical Sciences*, 52, 150–153, 1963.

H027 Higgins, C.E., W.H. Balwin, and B.A. Soldano, Effects of electrolytes and temperature on the solubility of tributyl phosphate in water, *Journal of Physical Chemistry*, 63, 113–118, 1959.

H028 Hansen, R.S., Y. Fu, and F.E. Bartell, Multimolecular adsorption from binary liquid solutions, *Journal of Physical Chemistry*, 53, 769–785, 1949.

H030 Hoffman, Jr., C.S., Water solubilities of tetradecanol and hexadecanol by gas-liquid chromatography, PhD Thesis, 1967.

H031 Higgins, C.E. and W.H. Baldwin, Refractometric determination of mutual solubility as a function of temperature: tributyl phosphine oxide and water, *Analytical Chemistry*, 32, 233–236, 1960.

H032 Higgins, C.E. and W.H. Baldwin, Effect of centrifugation on solution temperature and solubility of tributyl phosphate and tributyl phosphine oxide in water, *Analytical Chemistry*, 32, 236–238, 1960.

H033 Haussler, A. and P. Hajdu, Mitteilung uber die dissoziationskonstante und loslichkeit von rastinon 'hoechst', *Archiv der Pharmazie*, 291, 531–536, 1958.

H034 Heap, R.B., A.M. Symons, and J.C. Watkins, Steroids and their interactions with phospholipids: solubility, distribution coefficient and effect on potassium permeability of liposomes, *Biochimica et Biophysica Acta*, 218, 482–495, 1970.

H035 Heap, R.B., A.M. Symons, and J.C. Watkins, An interaction between oestradiol and progesterone in aqueous solutions and in a model membrane system, *Biochimica et Biophysica Acta*, 233, 307–314, 1971.

H036 Hajratwala, B.R. and H. Taylor, Effect of non-ionic surfactants on the dissolution and solubility of hydrocortisone, *Journal of Pharmacy and Pharmacology*, 28, 934–935, 1976.

H037 Hetherington, H.C. and J.M. Braham, Preparation of dicyanodiamide from calcium cyanamide, *Industrial and Engineering Chemistry*, 15, 1060–1063, 1923.

H038 Hobson, R.W., R.J. Hartman, and E.W. Kanning, A solubility study of di-n-propylamine, *Journal of the American Chemical Society*, 63, 2094–2095, 1941.

H039 Helmkamp, G.K., F.L. Carter, and H.J. Lucas, Coordination of silver ion with unsaturated compounds. VIII. Alkynes, *Journal of the American Chemical Society*, 79, 1306–1310, 1957.

H040 Hungerford, E.H. and A.R. Nees, Raffinose preparation and properties, *Industrial and Engineering Chemistry*, 26, 462–464, 1934.

H041 Horvath, A.L., Temperature and pressure effects on solubility: selection and consistency, *Chemie-Ingenieur-Technik*, 48, 144–146, 1976.

H042 Herrett, R.A., *Methyl- and Phenylcarbamates*, 113–141, 1969.

H043 Hayashi, M. and T. Sasaki, Measurement of solubilities of sparingly soluble liquids in water and aqueous detergent solutions using non-ionic surfactant, *Bulletin of the Chemical Society of Japan* (Nippon Kagakukai Bulletin), 29, 857–859, 1956.

H044 Huckel, W., M.-T. Niesel, and L. Buchs, Die anomalitaten des benzylalkohols und seiner losungen, *Chemische Berichte*, 77, 334–337, 1944.

H046 Huang, M.L. and S. Niazi, Polymorphic and dissolution properties of mercaptopurine, *Journal of Pharmaceutical Sciences*, 66, 608–609, 1977.

H048 Halban, H.V. and H. Kortschak, Uber die loslichkeit der pikrinsaure in wasser und wasserigen elektrolyt-losungen, *Helvetica Chimica Acta*, 21, 392–401, 1938.

H049 Hurwitz, A.R. and S.T. Liu, Determination of aqueous solubility and pKa values of estrogens, *Journal of Pharmaceutical Sciences*, 66, 624–627, 1977.

H051 Hou, J.P. and J.W. Poole, The amino acid nature of ampicillin and related penicillins, *Journal of Pharmaceutical Sciences*, 58, 1510–1515, 1969.

H053 Hancock, W. and E.Q. Laws, The determination of traces of benzene hexachloride in water and sewage effluents, *Analyst* (London), 80, 665–674, 1955.

H054 Hahnel, R., Interactions of estradiol-17-beta with amino acids, *Journal of Steroid Biochemistry*, 2, 61–65, 1971.

H055 Hoevel-Kestermann, H. and H. Muhlemann, Analytische untersuchungen einiger jodhaltiger rontgen-kontrastmittel im hinblick auf die ph. helv. VI, *Pharmaceutica Acta Helvetiae*, 47, 394–423, 1972.

H056 Huttenrauch, R. and I. Keiner, Molekulargalenik, *Pharmazie* (Berlin), 31, 489–491, 1976.

H058 Haque, R. and D. Schmedding, A method of measuring the water solubility of hydrophobic chemicals: solubility of five polychlorinated biphenyls, *Bulletin of Environmental Contamination and Toxicology*, 14, 13–17, 1975.

H059 Hamabata, A., S. Chang, and P.H. von Hippel, Model studies on the effects of neutral salts on the conformational stability of biological macromolecules. III. Solubility of fatty acid amides in ionic solutions, *Biochemistry*, 12, 1271–1277, 1973.

H060 Halford, J.O., Relative strength of benzoic and salicylic acids in alcohol–water solutions, *Journal of the American Chemical Society*, 55, 2272–2278, 1933.

H061 Herskovits, T.T. and J.P. Harrington, Solution studies of the nucleic acid bases and related model compounds. solubility in aqueous alcohol and glycol solutions, *Biochemistry*, 11, 4800–4810, 1972.

H062 Hayashi, K., T. Matsuda, T. Takeyama, and T. Hino, Solubilities studies of basic amino acids, *Agricultural and Biological Chemistry*, 30, 378–384, 1966.

H063 Hruby, R. and V. Kasjanov, Solubility of sucrose, *The International Sugar Journal*, 42, 21–24, 1940.

H064 Hubacher, M.H., Solubility, density and melting point of phenolphthalein, *Journal of the American Pharmaceutical Association, Scientific Edition*, 34, 76–78, 1945.

H066 Higuchi, T. and M. Ikeda, Rapidly dissolving forms of digoxin: hydroquinone complex, *Journal of Pharmaceutical Sciences*, 63, 809–811, 1974.

H067 Hussain, A. and J.H. Rytting, Prodrug approach to enhancement of rate of dissolution of allopurinol, *Journal of Pharmaceutical Sciences*, 63, 798–799, 1974.

H068 Howard, J.E. and W.H. Patterson, Miscibility tests of dilute solutions of chromic chloride hexahydrates, *Journal of the Chemical Society* (London), 129, 2791–2796, 1926.

H069 Haward, R.N., Determination of the solubility of plasticisers in water, *Analyst* (London), 68, 303–305, 1943.

H070 Harte, R.A. and J.L. Chen, Tryptophan as solubilizing agent for riboflavin, *Journal of the American Pharmaceutical Association, Scientific Edition*, 38, 568–570, 1949.

H071 Hammett, L.P. and R.P. Chapman, The solubilities of some organic oxygen compounds in sulfuric acid–water mixtures, *Journal of the American Chemical Society*, 56, 1282–1285, 1934.

H072 Herz, W., Ueber die loslichkeit einiger mit wasser schwer mischbarer flussigkeiten, *Berichte der Deutschen Chemischen Gesellschaft*, 31, 2668–2673, 1898.

H073 Hurle, K.B. and V.H. Freed, Effect of electrolytes on the solubility of some 1,3,5-triazines and substituted ureas and their adsorption on soil, *Weed Research*, 12, 1–10, 1972.

H074 Hempelmann, F.W., Studies on xipamid (4-chlor-5-sulfamoyl-2′,6′-salicyloxylidide), *Arzneimittel-Forschung*, 27, 2140–2143, 1977.

H075 Himmelreich, M., B.J. Rawson, and T.R. Watson, Polymorphic forms of mebendazole, *Australian Journal of Pharmaceutical Sciences*, 6, 123–125, 1977.

H076 Hussain, A., P. Kulkarni, and D. Perrier, Prodrug approaches to enhancement of physicochemical properties of drugs. IX: acetaminophen prodrug, *Journal of Pharmaceutical Sciences*, 67, 545–546, 1978.

H077 Hine, J., H.W. Haworth, and O.B. Ramsay, Polar effects on rates and equilibria. vi. the effect of solvent on the transmission of polar effects, *Journal of the American Chemical Society*, 85, 1473–1476, 1963.

H078 Hill, A.E. and R. Macy, Ternary Systems. II. Silver perchlorate, aniline and water, *Journal of the American Chemical Society*, 46, 1132–1150, 1924.

H080 Hodgman, C.R., *Chemodynamics: Transport and Behavior of Chemicals in the Environment—A Problem in Environmental Health*, Book, 59–59, 1952.

H081 Haight, Jr., G.P., Solubility of methyl bromide in water and in some fruit juices, *Industrial and Engineering Chemistry*, 43, 1827–1828, 1951.

H082 Hoffman, W.F. and R.A. Gortner, Sulfur in proteins. I. The effect of acid hydrolysis upon cystine, *Journal of the American Chemical Society*, 44, 341–360, 1922.

H083 Hertelendi, L., Zur loslichkeit der nicotinsaure (beta-pyridincarbonsaure, vitamin b faktor), *Zeitschrift fuer Physikalische Chemie, Abteilung A: Chemische Thermodynamik, Kinetik, Electrochemie, Eigenschaftslehre*, 192, 379–380, 1943.

H084 Herz, W. and F. Hiebenthal, Uber loslichkeitsbeeinflussungen, *Zeitschrift fuer Anorganische Chemie*, 177, 363–380, 1928.

H085 Halban, H.V., G. Kortum, and M. Seiler, Die dissoziationskonstanten schwacher und mittelstarker elektrolyte, *Zeitschrift fuer Physikalische Chemie, Abteilung A: Chemische Thermodynamik, Kinetik, Electrochemie, Eigenschaftslehre*, 173, 449–463, 1935.

H087 Hermans, P.H., Die loslichkeitskurven der systeme mannit–borsaure–wasser und *cis*-tetrahydronaphthalin 1,2-diol-borsaure–wasser bei 25, *Zeitschrift fuer Anorganische Chemie*, 142, 111–114, 1925.

H089 Holleman, A.F., Sur l'analyse quantitative des produits de la nitration des acides metachloro- et metabromo-benzoiques, *Recueil des Travaux Chimiques des Pays-Bas et de la Belgique*, 29, 394–402, 1910.

H090 Holmberg, B., Stereokemiska studier. V. Diklorbarnstensyrornas sterokemi, *Arkiv for Kemi, Mineralogi och Geologi*, 8, 1–35, 1921.

H091 Hamada, Y., N. Nambu, and T. Nagai, Interactions of alpha- and beta-cyclodextrin with several non-steroidal antiinflammatory drugs in aqueous solution, *Chemical and Pharmaceutical Bulletin*, 23, 1205–1211, 1975.

H092 Ho, P.C., C.-H. Ho, and K.A. Kraus, Solubility of toluene in aqueous sodium alkylbenzenesulfonate solutions, *Journal of Chemical and Engineering Data*, 24, 115–118, 1979.

H093 Hajdu, P., K.F. Kohler, F.H. Schmidt, and H. Spingler, Physico-chemical and analytical studies with HB 419, *Arzneimittel-Forschung*, 19, 1381–1386, 1969.

H094 Horiba, S., Sucrose–water, *Memoirs of the College of Engineering, Kyoto Imperial University*, 2, 519–519, 1917.

H096 Hajdu, P. and D. Damm, Physico-chemical and analytical studies of penbutolol, *Arzneimittel-Forschung*, 29, 602–606, 1979.

H097 Hitchcock, D.I., The solubility of tyrosine in acid and in alkali, *Journal of General Physiology*, 6, 747–756, 1924.

H098 Hagen, T.A., Physicochemical study of hydrocortisone and hydrocortison *n*-alkyl-21-esters, PhD Thesis, 1979.

H099 Higuchi, T., F.-M.L. Shih, T. Kimura, and J.H. Rytting, Solubility determination of barely aqueous-soluble organic solids, *Journal of Pharmaceutical Sciences*, 68, 1267–1272, 1979.

H100 Hoover, T.B., Water Solubilities of PCB isomers, *PCB Newsletter*, 3, 4–5, 1971.

H101 Horiba, S., Studies of solution. I. The change of molecular solution volumes in solutions, *Memoirs of the College of Science and Engineering, Kyoto Imperial University*, 2, 1–43, 1917.

H102 Hamilton, I.C. and R. Woods, The effect of alkyl chain length on the aqueous solubility and redox properties of symmetrical dixanthogens, *Australian Journal of Chemistry*, 32, 2171–2179, 1979.

H103 Hoffman, C.S. and E.W. Anacker, Water solubilities of tetradecanol and hexadecanol, *Journal of Chromatography*, 30, 390–396, 1967.

H104 Hakala, M.-R. and J.B. Rosenholm, Thermodynamics of micellization and solubilization in the system water + sodium n-octanoate + n-pentanol at 25°C, *Journal of the Chemical Society, Faraday Transactions 1*, 76, 473–488, 1980.

H105 Holleman, A.F. and P. Caland, Quantitative untersuchungen uber die sulfonierung des toluols, *Berichte der Deutschen Chemischen Gesellschaft*, 44, 162–163, 1911.

H106 Hassett, J.J., J.C. Means, W.L. Banwart, and S.G. Wood, *Sorption Properties of Sediments and Energy-related Pollutants*, Environmental Protection Agency, Athens, Georgia, 103–103, 1980.

H107 Hassett, J.J., J.C. Means, W.L. Banwart, and A. Khan, Sorption of dibenzothiophene by soils and sediments, *Journal of Environmental Quality*, 9, 184–186, 1980.

H109 Hughes, Jr., R.E. and V.H. Freed, The determination of ethyl *N,N*-di-*n*-propylthiolcarbamate (EPTC) in soil by gas chromatography, *Journal of Agricultural and Food Chemistry*, 9, 381–382, 1961.

H110 Hill, D.J.T. and L.R. White, The enthalpies of solution of hexan-1-ol and heptan-1-ol in water, *Australian Journal of Chemistry*, 27, 1905–1916, 1974.

H111 Hyde, A.J., D.M. Langbridge, and A.S.C. Lawrence, Soap + water + amphiphile systems, *Discussions of the Faraday Society*, 18, 239–258, 1954.

H112 Harris, C.I., Adsorption, movement, and phytotoxicity of monuron and s-triazine herbicides in soil, *Weed Science*, 14, 6–10, 1966.

H114 Hawking, F., The rate of diffusion of sulphonamide compounds, *Quarterly Journal of Pharmacy and Pharmacology*, 14, 226–233, 1941.

H116 Hollifield, H.C., Rapid nephelometric estimate of water solubility of highly insoluble organic chemicals of environmental interest, *Bulletin of Environmental Contamination and Toxicology*, 23, 579–586, 1979.

H117 Haque, R., D.W. Schmedding, and V.H. Freed, Aqueous solubility, adsorption, and vapor behavior of polychlorinated biphenyl aroclor 1254, *Environmental Science and Technology*, 8, 139–142, 1974.

H118 Hafkenscheid, T.L. and E. Tomlinson, Estimation of aqueous solubilities of organic non-electrolytes using liquid chromatographic retention data, *Journal of Chromatography*, 218, 409–425, 1981.

H120 Hirano, K., T. Ichihashi, and H. Yamada, Studies on the absorption of practically water-insoluble drugs following injection. II. Intramuscular absorption from aqueous suspensions in rats, *Chemical and Pharmaceutical Bulletin*, 29, 817–827, 1981.

H121 Hlavaty, K. and J. Linek, Liquid–liquid equilibria in four ternary acetic acid–organic solvent–water systems at 24.6°C, *Collection of Czechoslovak Chemical Communications*, 38, 374–378, 1973.

H122 Hansen, R.S., F.A. Miller, and S.D. Christian, Activity coefficients of components in the systems water–acetic acid, water–propionic acid and water–n-butyric acid at 25°C, *Journal of Physical Chemistry*, 59, 391–395, 1955.

H123 Hutchinson, T.C., J.A. Hellebust, D. Tam, and W.Y. Shiu, The correlation of the toxicity to algae of hydrocarbons and halogenated hydrocarbons with their physical-chemical properties, *Environment Science Research*, 13, 577–586, 1978.

H124 Hayduk, W. and V.K. Malik, Density, viscosity, and carbon dioxide solubility and diffusivity in aqueous ethylene glycol solutions, *Journal of Chemical and Engineering Data*, 16, 143–146, 1971.

H125 Hassan, M.M.A., A.I. Jado, and M.U. Zubair, Aminosalicylic acid, *Analytical Profiles of Drug Substances*, 10, 1–27, 1981.

H127 Harms, H., Uber die energieverhaltnisse der OH–OH-bindung, *Zeitschrift fur Physical Chemistry Abt B*, 43, 257–270, 1939.

H129 Huttenrauch, R. and S. Fricke, Zur beziehung zwischen ordnungsgrad und losungsvermogen des wassers, *Pharmazie* (Berlin), 37, 147–148, 1982.

H300 Hashimoto, Y., K. Tokura, K. Ozaki, and W.M.J. Strachan, A comparison of water solubilities by the flask and micro-column methods, *Chemosphere*, 11, 991–1001, 1982.

H301 Hafkenscheid, T.L. and E. Tomlinson, Isocratic chromatographic retention data for estimating aqueous solubilities of acidic, basic and neutral drugs, *International Journal of Pharmaceutics*, 17, 1–21, 1983.

H302 Herzfeldt, C.D. and R. Kummel, Dissociation constants, solubilities and dissolution rates of some selected nonsteroidal antiinflammatories, *Drug Development and Industrial Pharmacy*, 9, 767–793, 1983.

H303 Hesse, J.L. and R.A. Powers, polybrominated biphenyl (PBB) contamination of the Pine river, Gratiot, and Midland Counties, Michigan, *Environmental Health Perspectives*, 23, 19–25, 1978.

H306 Hashimoto, Y., K. Tokura, H. Kishi, and W.M.J. Strachan, Prediction of seawater solubility of aromatic compounds, *Chemosphere*, 13, 881–888, 1984.

H307 Hiller, K.O., B. Masloch, and H.J. Mockel, Zusammenhang zwischen loslichkeit und kapazitatsfaktor bei der reverse-phase-bonded-phase-chromatographie von alkylbenzolen, alkylbromiden und alkyldisulfiden, *Ztschrft fur Analytische Chemie*, 283, 109–113, 1977.

H308 Holt, R.F. and R.E. Leitch, Oxamyl, *Analytical Methods for Pesticides and Plant Growth Regulators*, 10, 111–118, 1978.

H309 Hattori, T. and M. Kanauchi, Isoprothiolane, *Analytical Methods for Pesticides and Plant Growth Regulators*, 10, 229–236, 1978.

H313 Ho, P.C., Solubilities of toluene and n-octane in aqueous protosurfactant and surfactant solutions, *Journal of Chemical and Engineering Data*, 30, 88–90, 1985.

H316 Hamilton, H.W., D.F. Ortwine, D.F. Worth, and R.P. Steffen, Synthesis of xanthines as adenosine antagonists, a practical quantitative structure–activity relationship application, *Journal of Medicinal Chemistry*, 28, 1071–1079, 1985.

H319 Hirayama, T., N. Hirashima, K. Abe, and M. Ueno, Utilization of diethyl-beta-cyclodextrin as a sustained-release carrier for isosorbide dinitrate, *Journal of Pharmaceutical Sciences*, 77, 233–236, 1988.

H320 Hagen, T.A. and G.L. Flynn, Permeation of hydrocortisone and hydrocortisone 21-alkyl esters through silicone rubber membranes—relationship to regular solution solubility behavior, *Journal of Membrane Science*, 30, 47–65, 1987.

H322 Herzel, F. and A.S. Murty, Do carrier solvents enhance the water solubility of hydrophobic compounds? *Bulletin of Environmental Contamination and Toxicology*, 32, 53–58, 1984.

H324 Hughes, J., P. Tenni, C. McDonald, and V.B. Sunderland, Solubility of metronidazole for topical application, *Australian Journal of Hospital Pharmacy*, 12, 58–58, 1982.

H328 Hoogerheide, J.G. and B.E. Wyka, Clotrimazole, *Analytical Profiles of Drug Substances*, 11, 225–229, 1982.

H330 Hommelen, J.R., The elimination of errors due to evaporation of the solute in the determination of surface tensions, *Journal of Colloid Science*, 14, 385–400, 1959.

H332 Hogfeldt, E. and B. Bolander, On the extraction of water and nitric acid by aromatic hydrocarbons, *Arkiv foer Kemi*, 21, 161–186, 1963.

H333 Henriksons, U., T. Klason, L. Odberg, and J.C. Eriksson, Solubilization of benzene and cyclohexane in aqueous solutions of hexadecyltrimethylammonium bromide: a deuterium magnetic resonance study, *Chemical Physics Letters*, 52, 554–558, 1977.

H337 Haggard, H.W., An accurate method of determining small amounts of ethyl ether in air, blood, and other fluids, together with a determination of the coefficient of distribution of ether between air and blood at various temperatures, *Journal of Biological Chemistry*, 55, 131–143, 1923.

H338 Heric, E.L. and R.E. Langford, System furfural–water–valeric acid at 25 and 35°C, *Journal of Chemical and Engineering Data*, 17, 209–211, 1972.

H339 Heric, E.L. and R.E. Langford, System furfural–water–caproic acid at 25 and 35°C, *Journal of Chemical and Engineering Data*, 17, 471–473, 1972.

H340 Heric, E.L., B.H. Blackwell, L.J. Gaissert, and J.W. Pierce, Distribution of butyric acid between furfural and water at 25 and 35°C, *Journal of Chemical and Engineering Data*, 11, 38–40, 1966.

H341 Huang, G.L., Dissertation or Masters Thesis, 1983.

H342 Hardaway, L.A. and S.H. Yalkowsky, Cosolvent effects on diuron solubility, *Journal of Pharmaceutical Sciences*, 80, 197–198, 1991.

H343 Huerta-Diaz, M.A. and S. Rodriguez, Solubility measurements and determination of Setschenow constants for the pesticide carbaryl in seawater and other electrolyte solutions, *Canadian Journal of Chemistry*, 70, 2864–2868, 1992.

H345 Harva, O., The effect of long-chain alcohols on the properties of sodium laurate solutions, *Recueil des Travaux Chimiques des Pays-Bas*, 75, 101–111, 1956.

H347 Haj-Yehia, A. and M. Bialer, Structure–pharmacokinetic relationships in a series of valpromide derivatives with antiepileptic activity, *Pharmaceutical Research*, 6, 683–689, 1989.

H348 Haj-Yehia, A. and M. Bialer, Structure–pharmacokinetic relationships in a series of short fatty acid amides that possess anticonvulsant activity, *Journal of Pharmaceutical Sciences*, 79, 719–724, 1990.

H350 Holmes, H.L., *Book*, 2, Defence research establishment suffield, 1975.

H430 Hyvarinen, A.-P., H. Lihavainen, A. Gamma, and Y. Viisanen, Surface tensions and densities of oxalic, malonic, succinic, maleic, malic, and cis-pinonic acids, *Journal of Chemical and Engineering Data*, 51, 255–260, 2006.

H431 Huff Hartz, K.E., J.E. Tischuk, M.N. Chan, and S.N. Pandis, Cloud condensation nuclei activation of limited solubility organic aerosol, *Atmospheric Environment*, 40, 605–617, 2006.

H434 Hilal, S.H. and S.W. Karickhoff, Prediction of the solubility, activity coefficient and liquid/liquid partition coefficient of organic compounds, *QSAR & Combinatorial Science*, 23, 709–720, 2004.

I001 Ismail, A.A., M.W. Gouda, and M.M. Motawi, Micellar solubilization of barbiturates i: solubilities of certain barbiturates in polysorbates of varying hydrophobic chain length, *Journal of Pharmaceutical Sciences*, 59, 220–224, 1970.

I002 Ibrahim, H.G., F. Pisano, and A. Bruno, Polymorphism of phenylbutazone: properties and compressional behavior of crystals, *Journal of Pharmaceutical Sciences*, 66, 669–673, 1977.

I006 Irrera, L., Influenze di solubilita, *Gazzetta Chimica Italiana*, 61, 614–618, 1931.

I007 Ikeda, K., K. Uekama, and M. Otagiri, Inclusion complexes of beta-cyclodextrin with antiinflammatory drugs fenamates in aqueous solution, *Chemical and Pharmaceutical Bulletin*, 23, 201–208, 1975.

I008 Igimi, H. and M.C. Carey, Dissimilar pH-solubility relations of chenodeoxycholic (CDCA) and ursodeoxycholic (UDCA) acids, *Gastroenterology*, 76, 1159–1159, 1979.

I009 Ikeda, K., K. Kato, and T. Tukamoto, Solubilization of barbiturates by polyoxyethylene lauryl ether, *Chemical and Pharmaceutical Bulletin*, 19, 2510–2517, 1971.

I010 Ibrahim, S.A., H.O. Ammar, A.A. Kassem, and S.S. Abu-Zaid, Effect of some hydrotropic agents on the water solubility of aminophenazone, *Pharmazie* (Berlin), 34, 809–812, 1979.

I011 Inga, R.F. and J.J. McKetta, Solubility of propyne in water, *Journal of Chemical and Engineering Data*, 6, 337–338, 1961.

I012 Igimi, H. and M.C. Carey, pH-solubility relations of chenodeoxycholic and ursodeoxycholic acids: physical–chemical basis for dissimilar solution and membrane phenomena, *Journal of Lipid Research*, 21, 72–90, 1980.

I015 Ikekawa, A. and S. Hayakawa, Mechanochemical change in the solid state and the solubility of amobarbital, *Bulletin of the Chemical Society of Japan* (Nippon Kagakukai Bulletin), 54, 2587–2591, 1981.

I017 Inga, R.F. and J.J. McKetta, Solubility of cyclopropane in water, *Petroleum Refiner*, 40, 191–192, 1961.

I018 Ivanov, K.A., Solubility of lindane in H_2O, *Gigiena i Sanitariya*, 21, 82–83, 1956.

I019 Ikeda, Y., K. Matsumoto, K. Kunihiro, and K. Uekama, Inclusion complexation of essential oils with alpha- and beta-cyclodextrins, *Yakugaku Zasshi* (Tokyo), 102, 83–88, 1982.

I300 Irmann, F., Eine Einfache Korrelation zwischen wasserloslichkeit und struktur von kohlenwasserstoffen und halogenkohlenwasserstoffen, *Chemie-Ingenieur-Technik*, 37, 789–798, 1965.

I304 Ibrahim, S.A. and S. Shawky, Effect of some aliphatic acids, their sodium and potassium salts on aqueous solubility of certain diuretics, *Pharmazeutische Industrie*, 45, 207–212, 1983.

I306 IARC Committee, *Iarc Mongraphs on the Evaluation of the Carcinogenic Risk of chemicals to Man*, 1, 1–181, 1971.

I307 IARC Committee, Some aromatic amines, hydrazine and related substances, *n*-nitroso compounds and miscellaneous alkylating agents, *Iarc Mongraphs on the Evaluation of the Carcinogenic Risk of chemicals to Man*, 4, 1–259, 1973.

I308 IARC Committee, Some organochlorine pesticides, *Iarc Mongraphs on the Evaluation of the Carcinogenic Risk of chemicals to Man*, 5, 1–211, 1973.

I309 IARC Committee, Sex hormones, *Iarc Mongraphs on the Evaluation of the Carcinogenic Risk of chemicals to Man*, 6, 1–210, 1974.

I310 IARC Committee, Some anti-thyroid and related substances, nitrofurans and industrial chemicals, *Iarc Mongraphs on the Evaluation of the Carcinogenic Risk of chemicals to Man*, 7, 1–291, 1974.

I312 IARC Committee, Some naturally occurring substances, *Iarc Mongraphs on the Evaluation of the Carcinogenic Risk of chemicals to Man*, 10, 1–328, 1975.

I313 IARC Committee, Cadmium, nickel, Some epoxides, miscellaneous industrial chemicals and general considerations on volatile anaesthetics, *Iarc Mongraphs on the Evaluation of the Carcinogenic Risk of chemicals to Man*, 11, 1–277, 1976.

I314 IARC Committee, Some carbamated, thiocarbamates and carbazides, *Iarc Mongraphs on the Evaluation of the Carcinogenic Risk of chemicals to Man*, 12, 1–260, 1976.

I315 IARC Committee, Some miscellaneous pharmaceutical substances, *Iarc Mongraphs on the Evaluation of the Carcinogenic Risk of chemicals to Man*, 13, 1–233, 1976.

I316 IARC Committee, Some fumigants, the herbicides 2,4-D and 2,4,5-T, chlorinated dibenzodioxins and miscellaneous industrial chemicals, *Iarc Mongraphs on the Evaluation of the Carcinogenic Risk of chemicals to Man*, 15, 1–265, 1977.

I332 Isnard, P. and S. Lambert, Estimating bioconcentration factors from octanol-water partition coefficient and aqueous solubility, *Chemosphere*, 17, 21–34, 1988.

I333 Iwamoto, E., Y. Tanaka, H. Kimura, and Y. Yamamoto, Solute–solvent interactions with metal chelate electrolytes. Part III. Salting in of *tris*(acetylacetonato)cobalt (III) and benzene by aromatic and aliphatic ions, *Journal of Solution Chemistry*, 9, 841–856, 1980.

I334 Iwamoto, E., Y. Hiyama, and Y. Yamamoto, Hydrophobic and charge-dipole interactions in aqueous solutions of highly charged metal chelate cations and nitrobenzene, dinitrobenzenes, and toluene at 25°C, *Journal of Solution Chemistry*, 6, 371–383, 1977.

I335 Iwamoto, E., M. Yamamoto, and Y. Yamamoto, Salting-in of nitrobenzene and toluene by metal chelate electrolytes, *Inorganic Nuclear Chemistry Letters*, 10, 1069–1076, 1974.

I404 Igo, D.H., T.D. Brennan, and E.E. Pullen, Development of an automated in-line microfiltration system coupled to an HPLC for the determination of solubility, *Journal of Pharmaceutical and Biomedical Analysis*, 26, 495–500, 2001.

J001 John, L.M. and J.W. McBain, The hydrolysis of soap solutions. II. The solubilities of higher fatty acids, *Journal of the American Oil Chemists' Society*, 25, 40–41, 1948.

J003 Jacobsohn, G.M. and D. Levenberg, Solubilities of 17-ketosteroids in water, *Steriods, An International Journal* (San Francisco), 4, 849–853, 1964.

J004 James, K.C. and M. Roberts, The solubilities of the lower testosterone esters, *Journal of Pharmacy and Pharmacology*, 20, 709–714, 1968.

J005 Jeffreys, G.V., Phase equilibrium for system methylethylketone, cyclohexane, and water, *Journal of Chemical and Engineering Data*, 8, 320–323, 1963.

J007 Jones, W.J. and J.B. Speakman, Some physical properties of aqueous solutions of certain pyridine bases, *Journal of the American Chemical Society*, 43, 1867–1870, 1921.

J008 Jaworski, E.G., *Chloroacetamides*, Book Chapter, 165–167, Marcel Dekker, New York, 1969.

J009 Joshi, R.K., L. Krasnec, and I. Lacko, Studies on solubilization Part II, *Pharmaceutica Acta Helvetiae*, 46, 570–582, 1971.

J010 Jakovljevic, I.M., Digitoxin, *Analytical Profiles of Drug Substances*, 3, 149–159, 1974.

J011 James, K.C. and R.H. Leach, A Borax–chloramphenicol complex in aqueous solution, *Journal of Pharmacy and Pharmacology*, 22, 612–614, 1970.

J012 Jones, D.C., The systems *n*-butyl alcohol–water and *n*-butyl alcohol–acetone–water, *Journal of the Chemical Society* (London), 799–813, 1929.

J016 Janes, J.O., P.M. Loeb, R.N. Berk, and J.M. Dietschy, Intestinal absorption of oral cholecystographic agents, *Clinical Research*, 25, 312–312, 1977.

J017 Janecke, E., Uber das system methylalkohol–isobutylalkohol–wasser, *Zeitschrift fuer Physikalische Chemie, Abteilung A: Chemische Thermodynamik, Kinetik, Electrochemie, Eigenschaftslehre*, 164, 401–416, 1933.

J018 Juni, K., M. Nakano, and T. Arita, Controlled drug permeation. II. Comparative permeability and stability of butamben and benzocaine, *Chemical and Pharmaceutical Bulletin*, 25, 1098–1100, 1977.

J019 Jackson, R.F. and C.G. Silsbee, *The Solubility of Dextrose in Water*, Scientific Papers, National Bureau of Standards, 17, 715–724, 1922.

J020 Jones, J.H. and J.F. McCants, Ternary solubility data: 1-butanol-methyl 1 butyl ketone–water, 1-butyraldehyde-ethyl acetate–water, 1-hexane-methyl ethyl ketone–water, *Industrial and Engineering Chemistry*, 46, 1956–1958, 1954.

J021 Janecke-Heidelberg, E., Einzelvortrage, *Zeitschrift fuer Elektrochemie*, 36, 645–654, 1930.

J022 Juni, K., T. Tomitsuka, M. Nakano, and T. Arita, Analysis of permeation profiles of drugs from systems containing micelles, *Chemical and Pharmaceutical Bulletin*, 26, 837–841, 1978.

J023 Jaeger, A., Uber die loslichkeit von flussigen kohlenwasserstoffen in uberhitztem wasser, *Brennstoff-Chemie*, 4, 259–260, 1923.

J025 Jacobi, H., A. Lange, and K. Pfleger, Vergleichende Untersuchungen Wasserloslicher Theophyllinderivate, *Arzneimittel-Forschung*, 6, 41–43, 1956.

J026 Juni, K., K. Nomoto, M. Nakano, and T. Arita, Drug release through a silicone capsular membrane from micellar solution, emulsion, and cosolvent systems and the correlation of release data in vivo with release profile in vitro, *Journal of Membrane Science*, 5, 295–304, 1979.

J027 Janado, M., K. Takenaka, H. Nakamori, and Y. Yano, Solubilities of water-insoluble dyes in internal water of swollen sephadex gels, *Journal of Biochemistry* (Tokyo), 87, 57–62, 1980.

J028 Jachowicz, R., L. Krowczynski, and Z. Kubiak, Increasing the solubility of khellia for the preparation of injection solutions, *Farmagia Polska*, 33, 419–422, 1977.

J030 Jespersen, J.C. and K.T. Larsen, Identificering af terapeutisk vigtige barbitursyrer, *Yakugaku Zasshi*, 8, 212–226, 1934.

J031 Johansen, M. and H. Bundgaard, Pro-drugs as drug delivery systems. XIII. Kinetics of decomposition of *n*-mannich bases of salicylamide and assessment of their suitability as possible pro-drugs for amines, *International Journal of Pharmaceutics*, 7, 119–127, 1980.

J033 Jordan, L.S., Residue reviews, *Residue Reviews*, 32, 7–13, 1970.

J034 Jones, H., Complex formation between digoxin and beta-cyclodextrin, *Journal of Pharmacy and Pharmacology*, 33, 27–27, 1981.

J035 Janado, M. and T. Nishida, Effect of sugars on the solubility of hydrophobic solutes in water, *Journal of Solution Chemistry*, 10, 489–500, 1981.

J036 Jones, D.C., R.H. Ottewill, and A.P.J. Chater, The adsorption of insoluble vapours on water surfaces, in *Second International Congress of Surface Activity*, 188–199, 1957.

J037 Johansen, M. and H. Bundgaard, Pro-drugs as drug delivery systems. XII. Solubility, dissolution and partitioning behaviour of *n*-mannich bases and *n*-hydroxymethyl derivatives, *Archive of Pharmacy Chemistry Science*, 8, 717–727, 1980.

J300 James, K.C., Calculation of molecular surface areas and aqueous solubilities at ambient temperatures, *International Journal of Pharmaceutics*, 21, 123–128, 1984.

J302 Janado, M. and Y. Yano, The nature of the cosolvent effects of sugars on the aqueous solubilities of hydrocarbons, *Bulletin of the Chemical Society of Japan* (Nippon Kagakukai Bulletin), 58, 1913–1917, 1985.

J303 Jin, X.Z. and K.C. Chao, Solubility of four amino acids in water and of four pairs of amino acids in their water solutions, *Journal of Chemical and Engineering Data*, 37, 199–203, 1992.

J305 Jozwiakowski, M.J. and K.A. Connors, Aqueous solubility behavior of three cyclodextrins, *Carbohydrate Research*, 143, 51–59, 1985.

J308 Johnson, M.D., B.L. Hoesterey, and B.D. Anderson, Solubilization of a tripeptide hiv protease inhibitor using a combination of ionization and complexation with chemically modified cyclodextrins, *Journal of Pharmaceutical Sciences*, 83, 1142–1146, 1984.

J414 Jarho, P., A. Urtti, D.W. Pate, and T. Jarvinen, Hydroxypropyl-beta-cyclodextrin increases aqueous solubility and stability of anandamide, *Life Sciences*, 58, 181–185, 1996.

J415 Jamzad, S. and R. Fassihi, Role of surfactant and pH on dissolution properties of fenofibrate and glipizide—a technical note, *AAPS PharmScitech*, 7, E1–E6, 2006.

J417 Jennifer Luk, C.-W. and R.W. Rousseau, Solubilities of and transformations between the anhydrous and hydrated froms of L-serine in water–methanol solutions, *Crystal growth & Design*, 6, 1808–1812, 2006.

J418 Jouquand, C., V. Ducruet, and P. Le Bail, Formation of amylose complexes with C6-aroma compounds in starch dispersions and its impact on retention, *Food Chemistry*, 96, 461–470, 2006.

J420 Johnson, S.R., X.-Q. Chen, D. Murphy, and O. Gudmundsson, A computational model for the prediction of aqueous solubility that includes crystal packing, intrinsic solubility, and ionization effects, *Molecular Pharmaceutics*, 2007.

K001 Klevens, H.B., Solubilization of polycyclic hydrocarbons, *Journal of Physical and Colloid Chemistry*, 54, 283–297, 1950.

K002 Kablukov, I.A. and V.T. Malischeva, The volumetric method of measurement of the mutual solubility of liquids. The mutual solubility of the systems ethyl ether–water and iso-amyl alcohol–water, *Journal of the American Chemical Society*, 47, 1553–1561, 1925.

K003 Kabasakalian, P., E. Britt, and M.D. Yudis, Solubility of some steroids in water, *Journal of Pharmaceutical Sciences*, 55, 642–642, 1966.

K004 Kostenbauder, H.B. and T. Higuchi, A note on the water solubility of some *N*,*N*-dialkylamides, *Journal of the American Pharmaceutical Association, Scientific Edition*, 46, 205–206, 1957.

K005 Korenman, I.M., Hydrotropic dissolution, *Russian Journal of Physical Chemistry*, 45, 1011–1011, 1971.

K006 Krebs, H.A. and J.C. Speakman, The solubility of sulphonamides in relation to hydrogen-ion concentration, *British Medical Journal*, 1, 47–50, 1946.

K007 Kirkland, J.J., Columns for modern analytical liquid chromatography, *Analytical Chemistry*, 43, 36–48, 1971.

K008 Kakemi, K., H. Sezaki, M. Nakano, and K. Ohsuga, Effect of structure of pyridinecarboxylic acids and hydroxypyridines on molecular interaction in water, *Journal of Pharmaceutical Sciences*, 58, 699–702, 1969.

K009 Kakemi, K., H. Sezaki, T. Mitsunaga, and M. Nakano, Effect of structural similarity on molecular interaction in aqueous solution: interaction of phenazine and tetramethylpyrimido-pteridinetetrone with alkylxanthines and benzene derivatives, *Journal of Pharmaceutical Sciences*, 59, 1597–1601, 1970.

K010 Krause, G.M. and J.M. Cross, Solubility of phenobarbital in alcohol–glycerin–water systems, *Journal of the American Pharmaceutical Association, Scientific Edition*, 40, 137–139, 1951.

K011 Krause, F.P. and W. Lange, Aqueous solubilities of *n*-dodecanol, *n*-hexadecanol, and *n*-octadecanol by a new method, *Journal of Physical Chemistry*, 69, 3171–3173, 1965.

K012 Kakovsky, I.A., Physicochemical properties of some flotation reagents and their salts with ions of heavy non-ferrous metals, *Solubilization and Micelles*, 4, 225–237, 1957.

K013 Kakinuma, H., The solubility of urea in water, *Journal of Physical Chemistry*, 45, 1045–1046, 1941.

K017 Kramer, S.F. and G.L. Flynn, Solubility of organic hydrochlorides, *Journal of Pharmaceutical Sciences*, 61, 1896–1903, 1972.

K018 Katchen, B. and S. Symchowicz, Correlation of dissolution rate and griseofulvin absorption in man, *Journal of Pharmaceutical Sciences*, 56, 1108–1111, 1967.

K019 Kostenbauder, H.B. and T. Higuchi, Formation of molecular complexes by some water-soluble amides. II. Effect of decreasing water solubility on degree of complex formation, *Journal of the American Pharmaceutical Association, Scientific Edition*, 45, 810–813, 1956.

K020 Kostenbauder, H.B. and T. Higuchi, Formation of molecular complexes by some water-soluble amides. I. Interaction of several amides with p-hydroxybenzoic acid, salicylic acid, chloramphenicol, and phenol, *Journal of the American Pharmaceutical Association, Scientific Edition*, 45, 518–522, 1956.

K021 Katz, M. and Z.I. Shaikh, Percutaneous corticosteroid absorption correlated to partition coefficient, *Journal of Pharmaceutical Sciences*, 54, 591–594, 1965.

K022 Kaplan, S.A., R.E. Weinfeld, C.W. Abruzzo, and M. Lewis, Pharmacokinetic profile of sulfisoxazole following intravenous, intramuscular, and oral administration to man, *Journal of Pharmaceutical Sciences*, 61, 773–778, 1972.

K023 Khalafallah, N. and Y. Hammouda, The solubility and complexing properties of acetohexamide in the presence of hydrotropic agents, *Pharmazie* (Berlin), 28, 452–454, 1973.

K024 Kakeya, N., M. Aoki, A. Kamada, and N. Yata, Biological activities of drugs. vi. structure-activity relationship of sulfonamide carbonic anhydrase inhibitors, *Chemical and Pharmaceutical Bulletin*, 17, 1010–1018, 1969.

K025 Kinoshita, K., H. Ishikawa, and K. Shinoda, Solubility of alcohols in water determined by the surface tension measurements, *Bulletin of the Chemical Society of Japan* (Nippon Kagakukai Bulletin), 31, 1081–1083, 1958.

K027 Krasowska, H., Solubilization of indomethacin and cinmetacin by non-ionic surfactants of the polyoxy-ethylene type, *Farmaco, Edizione Pratica* (PAVIA), 31, 463–472, 1976.

K028 Kuhnert-Brandstatter, M. and A. Martinek, Uber den einfluss der polymorphie auf die loslichkeit von arzneimitteln, *Mikrochimica et Ichnoanalytica Acta*, 909–919, 1965.

K029 Kaiser, D.G., W.C. Krueger, L.M. Pschigoda, and B.F. Zimmer, *Aqueous Solubilities and Distribution Coefficients of U-20,235, U-25,312, U-2726 and U-22,338*, Technical Report, 1–7, 1969.

K031 Kresheck, G.C., H. Schneider, and H.A. Scheraga, The effect of D_2O on the thermal stability of proteins. thermodynamic parameters for the transfer of model compounds from H_2O to D_2O, *Journal of Physical Chemistry*, 69, 3132–3144, 1965.

K032 Kristian, P. and L. Drobnica, Reactions of isothiocyanates with amino acids, peptides and proteins. IV. Kinetics of the reaction of substituted phenylisothiocyanates with glycine, *Collection of Czechoslovak Chemical Communications*, 31, 1333–1339, 1966.

K033 Korman, S. and V.K. La Mer, Deuterium exchange equilibria in solution and the quinhydrone electrode, *Journal of the American Chemical Society*, 58, 1396–1403, 1936.

K034 Kienle, R.H. and J.M. Sayward, Solubilities of orthanilamide, metanilamide and sulfanilamide, *Journal of the American Chemical Society*, 64, 2464–2468, 1942.

K035 Komar, N.P., V.V. Mel'nik, K.V. Zimina, and A.G. Kozachenko, Solubility of adipic acid, *Vestnik Khar'Kovskogo Universiteta*, 67–71, 1971.

K036 Kuroki, A.K.N. and K. Konishi, Distribution of disperse dye between water and benzene phases, *Sen'i Kikai Gakkaishi*, 20, 256–261, 1964.

K040 Knox, J. and M.B. Richards, The basic properties of oxygen in organic acids and phenols; and the quadrivalency of oxygen, *Journal of the Chemical Society* (London), 115, 508–531, 1919.

K041 Kovach, I.M., I.H. Pitman, and T. Higuchi, Amino acid esters of phenolic drugs as potentially useful prodrugs, *Journal of Pharmaceutical Sciences*, 64, 1070–1071, 1975.

K042 King, J.R. and M.C. Bowman, 4-Ethylsulfonylnaphthalene-1-sulfonamide (ENS): analytical chemical behavior and trace analysis in five substrates, *Biochemical Medicine*, 12, 313–330, 1975.

K043 Kreilgard, B., T. Higuchi, and A.J. Repta, Complexation in formulation of parenteral solutions: solubilization of the cytotoxic agent hexamethylmelamine by complexation with gentisic acid species, *Journal of Pharmaceutical Sciences*, 64, 1850–1855, 1975.

K044 Kaplan, M.A., W.P. Coppola, B.C. Nunning, and A.P. Granatek, Pharmaceutical properties and stability of amikacin—Part I, *Current Therapeutic Research, Clinical and Experimental*, 20, 352–358, 1976.

K046 Kawashima, Y., M. Saito, and H. Takenaka, Improvement of solubility and dissolution rate of poorly water-soluble salicylic acid by a spray-drying technique, *Journal of Pharmacy and Pharmacology*, 27, 1–5, 1975.

K047 Kanke, M. and K. Sekiguchi, Dissolution behavior of solid drugs. I. Improvement and simplification of dissolution rate measurement, and its application to solubility determinations, *Chemical and Pharmaceutical Bulletin*, 21, 871–877, 1973.

K048 Krebs, H.A. and J.C. Speakman, Dissociation constant, solubility, and the pH value of the solvent, *Journal of the Chemical Society* (London), 593–595, 1945.

K049 Kodama, M., Y. Tagashira, A. Imamura, and C. Nagata, Effect of secondary structure of DNA upon solubility of aromatic hydrocarbons, *Journal of Biochemistry* (Tokyo), 59, 257–264, 1966.

K050 Koch, H. and R. Bodmann, Akute toxizitat von endomid und seinen metaboliten korrelation zwischen biologischer aktivitat und lipophilen eigenschaften, *Archiv der Pharmazie*, 309, 812–822, 1976.

K051 Kolthoff, I.M. and A.I. Medalia, The reaction between ferrous iron and peroxides. iii. reaction with cumene hydroperoxide, in aqueous solution, *Journal of the American Chemical Society*, 71, 3789–3792, 1949.

K052 Kuttel, D., Die solubilisierungsmoglichkeit einiger alkaloidbasen mit Tween 80, *Pharmazeutische Zentralhalle*, 107, 593–600, 1968.

K053 Kendall, J., On the ionic solubility-product, *Proceedings Royal Society*, 85, 200–219, 1911.

K055 Kudchadker, A.P. and J.J. McKetta, Solubility of cyclohexane in water, *American Institute of Chemical Engineers Journal*, 7, 707–707, 1961.

K056 Klemenc, A. and M. Low, Die loslichkeit in wasser und ihr zusammenhang der drei dichlorbenzole. eine methode zur bestimmung der loslichkeit sehr wenig loslicher und zugleich sehr fluchtiger stoffe, *Recueil des Travaux Chimiques des Pays-Bas*, 49, 629–640, 1930.

K057 Kendall, J. and J.C. Andrews, The solubilities of acids in aqueous solutions of other acids, *Journal of the American Chemical Society*, 43, 1545–1560, 1921.

K058 Krantz, Jr., J.C., W.E. Evens, Jr., S.E. Forman, and H.L. Wollenweber, Anesthesia VI. The anesthetic action of cyclopropyl vinyl ether, *Journal of Pharmacology and Experimental Therapeutics*, 75, 30–37, 1942.

K059 Kolthoff, J.M., Die dissoziationskonstante, das loslichkeitsprodukt und die titrierbarkeit von alkaloiden, *Biochemische Zeitschrift* (1948–1967), 162, 289–353, 1925.

K060 Kudielka, H., Zur Kenntnis der alpha-amino-n-capronsaure, *Monatshefte fuer Chemie*, 29, 351–358, 1908.

K061 Krantz, J.C., Jr., C.J. Carr, S.E. Forman, and H. Wollenweber, Anesthesia. IV. The anesthetic action of cyclopropyl ethyl ether, *Journal of Pharmacology and Experimental Therapeutics*, 72, 233–244, 1941.

K062 Kumar, S., S.N. Upadhyay, and V.K. Mathur, On the solubility of benzoic acid in aqueous carboxymethylcellulose solutions, *Journal of Chemical and Engineering Data*, 23, 139–141, 1978.

K063 Komar, N.P. and G.S. Zaslavskaya, The solubility of alpha-alpha'-bipyridyl in aqueous salt solutions, *Russian Journal of Physical Chemistry*, 47, 1642–1643, 1973.

K064 Kolthoff, I.M. and W. Bosch, The activity coefficient of benzoic acid in solutions of neutral salts and of sodium benzoate, *Journal of Physical Chemistry*, 36, 1685–1694, 1932.

K065 Kisarov, V.M., Solubility of chlorobenzene in water, *Journal of Applied Chemistry of the USSR*, 35, 2252–2253, 1962.

K067 Kral, F. and G. Strauss, A biologically active borate derivative of amphotericin b soluble in saline solution, *The Journal of Antibiotics*, 31, 257–259, 1978.

K068 Kolthoff, I.M. and L.A. Sarver, Properties of diphenylamine and diphenylbenzidine as oxidation–reduction indicators, *Journal of the American Chemical Society*, 52, 4179–4191, 1930.

K069 Kenaga, E.E., Some physical, chemical, and insecticidal properties of some o,o-dialkyl o-(3,5,6-trichloro-2-pyridyl) phosphates and phosphorothioates, *Bulletin of the World Health Organization*, 44, 225–228, 1971.

K070 Kent, J.S., Controlled release of delmadinone acetate from silicone polymer tubing: in vitro-in vivo correlations, *American Chemical Society, Division of Organic Coating and Plastics Chemistry*, 36, 356–361, 1976.

K072 Klevens, H.B., Solubilization, *Chemical Reviews*, 47, 1–73, 1950.

K075 Kremann, R. and E. Janetzky, Das ternare system antipyrin–coffein–wasser ein beitrag zur kenntnis des migranins, *Monatshefte fuer Chemie*, 44, 49–63, 1923.

K076 Kolthoff, I.M., The hydration of dissolved saccharose and the expression of the concentration in measuring the activity of ions, *Proceedings Royal Academy of Science of Amsterdam*, 29, 885–898, 1926.

K077 Klobbie, E.A., Gleichgewichte in den systemen aether–wasser und aether–wasser–malonsaure, *Zeitschrift fuer Physikalische Chemie* (Leipzig), 14, 615–632, 1894.

K078 Kuttel, D., Die solubilisationsfahigkeit des Tween 20, 60, 80 bei Einigen in wasser schlecht loslichen medikamenten, *Pharmazeutische Zentralhalle*, 103, 10–16, 1964.

K079 Kendall, J. and L.E. Harrison, Compound formation in ester–water systems, *Transactions of the Faraday Society*, 24, 588–596, 1928.

K084 Kremann, R. and H. Eitel, Das ternare system zucker-zitronensaure–wasser ein beitrag zur theorie der speiseeise vom standpunkt der phasenlehre, *Recueil des Travaux Chimiques des Pays-Bas*, 42, 539–546, 1923.

K085 Krupatkin, I.L., Ternary systems with layering without formation of chemical compounds, *Journal of General Chemistry of the USSR*, 26, 1815–1819, 1956.

K086 Kaneniwa, N., N. Watari, and H. Iijima, Dissolution of slightly soluble drugs. V. Effect of particle size on gastrointestinal drug absorption and its relation to solubility, *Chemical and Pharmaceutical Bulletin*, 26, 2603–2614, 1978.

K087 Khazanova, N.E., *Trudy Gosudarstevennoy Institute Azotnoi Promyshlennosi*, 4, 5–12, 1954.

K090 Kaneniwa, N. and A. Ikekawa, Solubilization of water-insoluble organic powders by ball-milling in the presence of polyvinylpyrrolidone, *Chemical and Pharmaceutical Bulletin*, 23, 2973–2986, 1975.

K091 Kaneniwa, N. and N. Watari, Dissolution of slightly soluble drugs. IV. Effect of particle size of sulfonamides on in vitro dissolution rate and in vivo absorption rate, and their relation to solubility, *Chemical and Pharmaceutical Bulletin*, 26, 813–826, 1978.

K092 Kata, M. and L. Haragh, Spray-embedding of spironolactone with beta-cyclodextrin, *Pharmazie* (Berlin), 36, 784–785, 1981.

K093 Klemm, K., W. Krastinat, and U. Kruger, Synthese und physikalisch-chemische eigenschaften von clanobutin, *Arzneimittel-Forschung*, 29, 1–2, 1979.

K095 Kitao, K., K. Kubo, T. Morishita, and A. Kamada, Studies on absorption of drugs. VII. Absorption of isomeric *n*-heterocyclic sulfonamides from the rat small intestines and relations between physicochemical property and absorption of unionized sulfonamides, *Chemical and Pharmaceutical Bulletin*, 21, 2417–2426, 1973.

K096 Kanke, M. and K. Sekiguchi, Dissolution behavior of solid drugs. II. Determination of the transition temperature of sulfathiazole polymorphs by measuring the initial dissolution rates, *Chemical and Pharmaceutical Bulletin*, 21, 878–884, 1973.

K097 Karlsson, K.G., Uber die zersetzungsgeschwindigkeit einiger ester in ihrer abhangigkeit von der was-serstoffionenkonzentration, *Zeitschrift Fur anorganische und Allgemeine Chemie*, 145, 1–57, 1925.

K103 Kanal, H., V. Inouye, R. Goo, and H. Wakatsuki, Solubility of 4-methyl-2-pentanone in aqueous phase of various salt concentrations, *Analytical Chemistry*, 51, 1019–1021, 1979.

K104 Kolosov, I.V. and E.P. Shapovalenko, Study of acid–base equilibria in aqueous solutions of bilirubin by the solubility method, *Journal of General Chemistry of the USSR*, 47, 1967–1967, 1977.

K105 Krupatkin, I.L., L.D. Vorob'eva, V.P. Maskhuliya, and M.E. Veselova, Liquid-phase equilibria in the systems water–2-furaldehyde-thiocyanates and water–butanone–thiocyanates, *Journal of General Chemistry of the USSR*, 45, 973–977, 1975.

K106 Kuroda, K., T. Yokoyama, T. Umeda, and Y. Takagishi, Studies on drug nonequivalence. VI. Physico-chemical studies on polymorphism of acetohexamide, *Chemical and Pharmaceutical Bulletin*, 26, 2565–2568, 1978.

K108 Koizumi, K., K. Mitsui, and K. Higuchi, Comparison between interactions of alpha- and beta-cyclodextrin with barbituric acid derivatives, *Yakugaku Zasshi* (Tokyo), 94, 1515–1519, 1974.

K112 Korenman, I.M. and R.P. Arefeva, Determination of the solubility of hydrocarbons in water, *Zhurnal Prikladnoi Khimii* (Leningrad), 51, 957–958, 1978.

K114 Krumholz, P., Structural studies on polynuclear pyridine compounds, *Journal of the American Chemical Society*, 73, 3487–3492, 1951.

K117 Kapoor, I.P., R.L. Metcalf, R.F. Nystrom, and G.K. Sangha, Comparative metabolism of methoxychlor, methiochlor, and DDT in mouse, insects, and in a model ecosystem, *Journal of Agricultural and Food Chemistry*, 18, 1145–1152, 1970.

K119 Krzyzanowska, T. and J. Szeliga, Determination of the solubility of individual hydrocarbons, *Nafta* (Katowice, Poland), 34, 413–417, 1978.

K120 Kurihara, N., M. Uchida, T. Fujita, and M. Nakajima, Studies on BHC isomers and related compounds. V. Some physicochemical properties of BHC isomers, *Pesticide Biochemistry and Physiology*, 2, 383–390, 1973.

K121 Koizumi, K., H. Miki, and Y. Kubota, Enhancement of the hypnotic potency of barbiturates by inclusion complexation with beta-cyclodextrin, *Chemical and Pharmaceutical Bulletin*, 28, 319–322, 1980.

K122 Kelly, F.H.C., Phase equilibria in sugar solutions. II. Ternary systems of water–sucrose–hexose, *Journal of Applied Chemistry*, 4, 405–406, 1954.

K123 Krasnoschekova, R. and M. Gubergrits, The relationship between reactivity and hydrophobicity of poly-cyclic aromatic hydrocarbons, *Organic Reaction*, 13, 432–439, 1976.

K129 Kralj, F. and D. Sincic, Mutual solubilities of phenol, salicylaldehyde, phenol-salicylaldehyde mixture, and water with and without the presence of sodium chloride or sodium chloride plus sodium sulfate, *Journal of Chemical and Engineering Data*, 25, 335–338, 1980.

K130 Korenman, Y.I., E.I. Polumestnaya, and E.V. Lyubeznykh, The extraction and solubility of naphthols in the presence of neutral salts, *Russian Journal of Physical Chemistry*, 53, 1663–1665, 1979.

K132 Korenman, Y.I. and V.S. Smirnov, The solubilities and distribution constants of xylenols in water-organic solvent systems, *Russian Journal of Physical Chemistry*, 54, 1553–1554, 1980.

K135 Kelly, F.H.C., Phase equilibria in sugar solutions. III. Ternary systems of water–hexose–inorganic salt, *Journal of Applied Chemistry*, 4, 407–408, 1954.

K136 Kelly, F.H.C., Phase equilibria in sugar solutions. IV. Ternary system of water–glucose–fructose, *Journal of Applied Chemistry*, 4, 409–411, 1954.

K137 Kanazawa, J., Measurement of the bioconcentration factors of pesticides by freshwater fish and their cor-relation with physicochemical properties or acute toxicities, *Pesticide Science*, 12, 417–424, 1981.

K138 Kenaga, E.E., Correlation of bioconcentration factors of chemicals in aquatic and terrestrial organisms with their physical and chemical properties, *Environmental Science and Technology*, 14, 553–556, 1980.

K142 Kotel'nikov, V.V. and V.P. Skripov, Isotope effect in the mutual solubility of water and triethylamine, *Nauchnaya Doklady Vysshei Shkoly Khimiya i Khimicheskaya Teknologiya*, 53, 248–249, 1959.

K143 Kato, Y., Y. Okamoto, S. Nagasawa, and T. Ueki, Solubility of a new polymorph of phenobarbital obtained by crystallization in the presence of phenytoin, *Chemical and Pharmaceutical Bulletin*, 29, 3410–3413, 1981.

K144 Kobinger, W. and F.J. Lund, Investigations into a new oral diuretic, rontyl (6-trifluoromethyl-7-sulfamyl-3,4-dihydro-1,2,4-benzothiadiazine-1,1-dioxide), *Acta Pharmacology et Toxicology*, 15, 265–274, 1959.

K148 Kulakov, V.N., A.G. Artyukh, I.M. Nikiforov, and T.V. Barinova, Solubility of 2-naphthoic acid in aque-ous solutions of *n*-methylpyrrolidone, *Journal of Applied Chemistry of the USSR*, 54, 335–338, 1981.

K301 Korenman, Y.I., Correlation between the partition constants of organic substances and their solubilities in water and extractants, *Russian Journal of Physical Chemistry*, 57, 382–384, 1983.

K304 Karickhoff, S.W., Semi-empirical estimation of sorption of hydrophobic pollutants on natural sediments and soils, *Chemosphere*, 10, 833–846, 1981.

K305 Konemann, H., Quantitative structure–activity relationships in fish toxicity studies. part i. relationship for 50 industrial pollutants, *Toxicology*, 19, 209–221, 1981.

K307 Korenman, Y.I. and E.I. Polumestnaya, The solubility of 1-naphthol in water at different temperatures, *Russian Journal of Physical Chemistry*, 51, 1392–1392, 1977.

K308 Korenman, Y.I., E.I. Polumestnaya, and L.I. Shestakova, The solubility of 2-naphthol in water at different temperatures, *Russian Journal of Physical Chemistry*, 51, 608–608, 1977.

K309 Korenman, Y.I., I.V. Karmaeva, and L.N. Sergeeva, The solubility of 2,4-xylenol in water at different temperatures, *Russian Journal of Physical Chemistry*, 51, 165–165, 1977.

K310 Korenman, Y.I., S.N. Taldykina, and T.N. Bogomolova, The solubility of picric acid in water at different temperatures, *Russian Journal of Physical Chemistry*, 50, 1780–1781, 1976.

K315 Kahrs, R.A., Profluralin, *Analytical Methods for Pesticides and Plant Growth Regulators*, 10, 451–459, 1978.

K316 Keeley, D.F., M.A. Hoffpauir, and J.R. Meriwether, Solubility of aromatic hydrocarbons in water and sodium chloride solutions of different ionic strengths: benzene and toluene, *Journal of Chemical and Engineering Data*, 33, 87–89, 1988.

K337 Kishii, H., M. Nakamura, and Y. Hashimoto, Prediction of solubility of aromatic compounds in water by using total molecular surface area, *Nippon Kagaku Kaishi*, 8, 1615–1622, 1987.

K431 Kuramochi, H., K. Maeda, and K. Kawamoto, Physicochemical properties of select polybrominated diphenyl ethers and extension of the UNIFAC model to brominated aromatic compounds, *Chemosphere*, 67, 1858–1865, 2007.

K436 Kuramochi, H. and K. Kawamoto, Modification of UNIFAC parameter table revision 5 for representation of aqueous solubility and 1-octanol/water partition coefficient for POPs, *Chemosphere*, 63, 698–706, 2006.

K437 Kurkov, S.V., G.L. Perlovich, and W. Zielenkiewicz, Thermodynamic Investigations of sublimation, solu-bility and slovation 0f [4-(benzyloxy)-phenyl] acetic acid, *Journal of Thermal Analysis and Calorimetry*, 83, 549–556, 2006.

K440 Kuramochi, H., K. Maeda, and K. Kawamoto, Measurements of water solubilities and 1-octanol/water partition coefficients and estimations of Henry's law constants for brominated benzenes, *Journal of Chemical and Engineering Data*, 49, 720–724, 2004.

K441 Koparkar, Y.P. and V.G. Gaikar, Solubility of o-/p-hydroxyacetonphenones in aqueous solutions of sodium alkyl benzene sulfonate hydrotropes, *Journal of Chemical and Engineering Data*, 49, 800–803, 2004.

K443 Kristl, A. and G. Vesnaver, Thermodynamic investigation of the effect of octanol-water mutual miscibility on the partitioning and solubility of some guanine derivatives, *Journal of Chemical Society*, 91, 995–998, 1995.

K444 Kasim, N.A., M. Whitehouse, C. Ramachandran, and G.L. Amimdon, Molecular properties of WHO essential drugs and provisional biopharmaceutical classification, *Molecular Pharmaceutics*, 1, 85–96, 2004.

K445 Kristl, A., Estimation of aqueous solubility for some guanine derivatives using partition coefficient and melting temperatures, *Journal of Pharmaceutical Sciences*, 88, 109–110, 1999.

K446 Kim, J., D.H. Jung, H. Rhee, and W.S. Choi, Improvement of bioavailability of water insoluble drugs: estimation of Intrinsic bioavailability, *Korean Journal of Chemical Engineering*, 25, 171–175, 2008.

K447 Kurkov, S.V. and G.L. Perlovich, Thermodynamic studies of fenbufen, diflunisal, and flurbiprofen: sublimation, solution and solvation of biphenyl substituted drugs, *International Journal of Pharmaceutics*, 357, 100–107, 2008.

L001 Loring, H.S. and V. Du Vigneaud, The solubility of the stereoisomers of cystine with a note on the identity of stone and hair cystine, *Journal of Biological Chemistry*, 107, 267–274, 1934.

L002 Leinonen, P.J. and D. MacKay, The multicomponent solubility of hydrocarbons in water, *Canadian Journal of Chemical Engineering*, 51, 230–233, 1973.

L003 Lindenberg, B.A., Sur la solubilite des substances organiques amphipatiques dans les glycerides neutres et hydroxyles, *Journal de Chimie Physique et de Physico-Chimie Biologique*, 48, 350–355, 1951.

L006 Lumsden, J.S., The physical properties of heptoic, hexahydrobenzoic, and benzoic acids and their derivatives, *Journal of the Chemical Society* (London), 87, 90–99, 1905.

L007 Liabastre, A.A., Experimental determination of the solubility of small organic molecules in H_2O and D_2O and the application of the scaled particle theory to aqueous and nonaqueous solutions, PhD Thesis, 48–167, 1974.

L008 Leiga, A.G. and J.N. Sarmousakis, The effect of certain salts on the aqueous solubilities of o-, m-, and p-dinitrobenzene, *Journal of Physical Chemistry*, 70, 3544–3549, 1966.

L009 Lange, W.E. and M.E. Amundson, Soluble steroids. I—sugar derivatives, *Journal of Pharmaceutical Sciences*, 51, 1102–1106, 1962.

L010 Lacey, R.E. and D.R. Cowsar, *Controlled Release of Biologically Active Agents*, 117–136, Plenum Press, New York, 1973.

L011 Le Petit, G.F., Medazepam pKa determined by Spectrophotometric and solubility methods, *Journal of Pharmaceutical Sciences*, 65, 1094–1095, 1976.

L012 Lindstrom, R.E. and A.R. Giaquinto, Salt effects in aqueous solutions of urea, *Journal of Pharmaceutical Sciences*, 59, 1625–1630, 1970.

L013 Lappas, L.C., C.A. Hirsch, and C.L. Winely, Substituted 5-nitro-1,3-dioxanes: correlation of chemical structure and antimicrobial activity, *Journal of Pharmaceutical Sciences*, 65, 1301–1305, 1976.

L014 Lang, B., Solubility and solubilization studies on reseptyl, *Pharmazie* (Berlin), 26, 689–691, 1971.

L015 Lister, J.H. and D.S. Caldbick, An investigation into the factors governing the aqueous solubility of xanthine (purine-2,6-dione), *Journal of Applied Chemistry* Biotechnology, 26, 351–354, 1976.

L016 Letellier, P., Influence du solvant sur les solubilites de composes organiques. correlations avec les variations des effects de substituants, *Bulletin de la Societe Chimique de France*, 5, 1569–1575, 1973.

L017 Lata, G.F. and L.K. Dac, Steroid solubility studies with aqueous solutions of urea and ureides', *Archives of Biochemistry and Biophysics*, 109, 434–441, 1965.

L020 Lamson, P.D., H.W. Brown, R.W. Stoughton, and A. Bass, Anthelmintic studies on alkyl-hydroxybenzenes. IV. Isomerism in polyalkylphenols, *Journal of Pharmacology and Experimental Therapeutics*, 53, 234–238, 1935.

L021 Lamson, P.D., H.W. Brown, R.W. Stoughton, and A.D. Bass, Anthelmintic studies on alkyl-hydroxybenzenes. V. Phenols with other than normal alkyl side chains, *Journal of Pharmacology and Experimental Therapeutics*, 53, 239–241, 1935.

L022 Lamson, P.D., H.W. Brown, R.W. Stoughton, and A. Bass, Anthelmintic studies on alkyl-hydroxybenzenes. II. *Ortho-* and *para*-n-akylphenols, *Journal of Pharmacology and Experimental Therapeutics*, 53, 218–226, 1935.

L024 Loos, M.A., *Phenoxyalkanoic Acids*, 1–5, 1969.

L025 Langecker, H., A. Harwart, and K. Junkmann, 2,4,6-Trijod-3-acetaminobenzoesaure-abkommlinge als kntrastmittel, *Archiv fuer Experimentelle Pathologie und Pharmakologie*, 220, 195–206., 1953

L027 Lau, E.P.K. and J.L. Sutter, Ethynodiol dacetate, *Analytical Profiles of Drug Substances*, 3, 254–277, 1974.

L028 Lane, W.H., Determination of the solubility of styrene in water and of water in styrene, *Industrial and Engineering Chemistry, Analytical Edition*, 18, 295–296, 1946.

L029 Logan, T.S., The aqueous solubility of acetanilide, *Journal of the American Chemical Society*, 67, 1182–1184, 1945.

L030 Leopold, A.C., P. Van Schaik, and M. Neal, Molecular structure and herbicide adsorption, *Weeds*, 8, 48–54, 1960.

L031 Laguerie, C., M. Aubry, and J.-P. Couderc, Some physicochemical data on monohydrate citric acid solutions in water: solubility, density, viscosity, diffusivity, pH of standard solution, and refractive index, *Journal of Chemical and Engineering Data*, 21, 85–87, 1976.

L032 Leuallen, E.E., Solubility of phenobarbital in ethanol–water systems, *Journal of the American Pharmaceutical Association, Practical Pharmacy Edition*, 10, 722–724, 1949.

L033 Liu, S.-T., C.F. Carney, and A.R. Hurwitz, Adsorption as a possible limitation in solubility determination, *Journal of Pharmacy and Pharmacology*, 29, 319–321, 1977.

L034 Lordi, N.G. and J.E. Christian, Physical properties and pharmacological activity: antihistaminics, *Journal of the American Pharmaceutical Association, Scientific Edition*, 45, 300–305, 1956.

L035 Lewkowitsch, Aqueous solubilities of R- and L-mandelic acids and three *o*-acyl-R-mandelic acids, *Berichte der Deutschen Chemischen Gesellschaft*, 16, 1566, 1883.

L037 Linderstrom-Lang, C.U. and R.F. Naylor, 4,4′-diaminodiphenyl sulphone: solubility and distribution in blood, *Biochemical Journal*, 83, 417–420, 1962.

L038 Logan, T.S., The effect of KCl, NaCl and Na$_2$SO$_4$ on the aqueous solubility of acetanilide, *Journal of the American Chemical Society*, 68, 1660–1661, 1946.

L039 Lutz, O., Ueber einige falle von sauerstoffwanderung in der molekel. I, *Berichte der Deutschen Chemischen Gesellschaft*, 35, 2460–2466, 1902.

L041 Lamouroux, F., Sur la solubilite dans l'eau des acides normaux de la serie oxalique, *Comptes Rendus Hebdomadaires des Seances de l'Academie des Sciences*, 128, 998–1000, 1899.

L042 Levine, L., J.A. Gordon, and W.P. Jencks, The relationship of structure to the effectiveness of denaturing agents of deoxyribonucleic acid, *Biochemistry*, 2, 168–175, 1963.

L044 Lykhol'ot, N.M., Studies of the solubility of sulfanilamide preparations. V. The solubility of sulfanilamide in phosphate-citrate buffer mixtures, *Farmatsevty Zhurnal*, 20, 44–46, 1965.

L045 Liu, S. and A. Hurwitz, The effect of micelle formation on solubility and pKa determination of acetylpromazine maleate, *Journal of Colloid and Interface Science*, 60, 410–413, 1977.

L047 Licht, Jr., W. and L.D. Wiener, Hydrotropic solvents for benzoic acid, *Industrial and Engineering Chemistry*, 42, 1538–1542, 1950.

L048 Larsson, E., Zur Elektrolytischen dissoziation der zweibasischen sauren iii. bestimmung zweiter dissoziationskonstanten aus loslichkeitsversuchen, *Zeitschrift fuer Anorganische Chemie*, 115, 247–254, 1926.

L049 Laddha, G.S. and J.M. Smith, The systems: glycol–n-amyl alcohol–water and glycol–n-hexyl alcohol–water, *Industrial and Engineering Chemistry*, 40, 494–496, 1948.

L050 Larsson, E., Die loslichkeit von sauren in salzlosungen. I, *Zeitschrift fuer Physikalische Chemie* (Leipzig), 127, 233–248, 1927.

L051 Linkov, G.I., *Antibiotiki* (Moscow), 20, 53–58, 1975.

L052 Lott, W.A. and F.B. Bergeim, 2-(p-aminobenzenesulfonamido)-thiazole: a new chemotherapeutic agent, *Journal of the American Chemical Society*, 61, 3593–3594, 1939.

L053 Leikola, E. and I. Suihkonen, Amino-, metyyli-ja nitroryhmien aseman vaikutus substituoidun bentseenimolekyylin liukoisuuteen, *Farmaseuttinen Aikakauslehti*, 69, 193–201, 1960.

L055 Lee, J.H., Hydrophilics, *School Science Review*, 43, 391–393, 1962.

L058 Lehr, D., Inhibition of drug precipitation in the urinary tract by the use of sulfonamide mixtures. I. Sulfathiazole–sulfadiazine mixture, *Proceedings of the Society for Experimental Biology and Medicine*, 58, 11–14, 1945.

L059 Leone, P. and E. Angelescu, Variazioni di solubilita di un corpo per la presenza di altri corpi. I. Acquafenolo-difenoli, *Gazzetta Chimica Italiana*, 52, 61–74, 1922.

L060 Levan, A. and G. Ostergren, The mechanism of c-mitotic action observations on the naphthalene series, *Hereditas*, 29, 381–432, 1943.

L061 Leone, P. and M. Benelli, Variazioni di solubilita di un corpo per presenza di altri corpi (II). Acquaepicloridrina-acido acetico, *Gazzetta Chimica Italiana*, 52, 75–86, 1922.

L062 Lloyd, B.A., S.O. Thompson, and J.B. Ferguson, Equilibria in liquid systems containing furfural, *Canadian Journal of Research, Section B: Chemical Sciences*, 15, 98–102, 1937.

L063 Ledbury, W. and C.W. Frost, The solubility of nitroglycerol in water, *Journal of the Society of Chemical Industry, London*, 46, 120–120, 1927.

L064 Linderstrom-Lang, K., Solubility of hydroquinone, *Comptes Rendus des Travaux du Laboratoire Carlsbreg*, 15, 4–28, 1924.

L065 Linderstrom-Lang, K., On the relation between the sizes of ions and the salting-out of hydroquinone and quinone, *Comptes Rendus des Travaux du Laboratoire Carlsbreg*, 17, 1–6, 1929.

L067 Lee, H.K., H. Lambert, V.J. Stella, and T. Higuchi, Hydrolysis and dissolution behavior of a prolonged-release prodrug of theophylline: 7,7'-succinyldetheophylline, *Journal of Pharmaceutical Sciences*, 68, 288–293, 1979.

L068 Lach, J.L. and W.A. Pauli, Interaction of pharmaceuticals with schardinger dextrins vi: interactions of beta-cyclodextrin, sodium deoxycholate, and deoxycholic acid with amines and pharmaceutical agents, *Journal of Pharmaceutical Sciences*, 55, 32–38, 1966.

L069 Lach, J.L. and J. Cohen, Interaction of pharmaceuticals with schardinger dextrins ii: interaction with selected compounds, *Journal of Pharmaceutical Sciences*, 52, 137–142, 1963.

L070 Lundberg, B., T. Lovgren, and B. Heikius, Simultaneous solubilization of steroid hormones. II: androgens and estrogens, *Journal of Pharmaceutical Sciences*, 68, 542–545, 1979.

L071 Lu, P.-Y., R.L. Metcalf, A.S. Hirwe, and J.W. Williams, Evaluation of environmental distribution and fate of hexachlorocyclopentadiene, chlordene, heptachlor, and heptachlor epoxide in a laboratory model ecosystem, *Journal of Agricultural and Food Chemistry*, 23, 967–973, 1975.

L072 Laseter, J.L., C.K. Bartell, A.L. Laska, and R.L. Evans, An ecological study of hexachlorobenzene, Unpublished, 77–77, 1976.

L073 Leo, H. and E. Rimbach, Uber die wasserloslichkeit des camphers, *Biochemische Zeitschrift* (1948–1967), 95, 306–312, 1919.

L074 Leech, P.N., W. Rabak, and A.H. Clark, American-made synthetic drugs—II, *Journal of the American Medical Association*, 73, 754–759, 1919.

L075 Lindstrom, R.E., Solubility in amide–water cosolvent systems. II: cosolvent excess at solute surface, *Journal of Pharmaceutical Sciences*, 68, 1141–1143, 1979.

L076 Labarre, J.-F., J.-P. Faucher, G. Levy, and G. Francois, Antitumour activity of some cyclophosphazenes, *European Journal of Cancer*, 15, 637–643, 1979.

L077 Lundberg, B., Temperature effect on the water solubility and water–octanol partition of some steroids, *Acta Pharmaceutica Suecica*, 16, 151–159, 1979.

L078 Lewis, G.A. and R.P. Enever, Solution thermodynamics of some potentially long-acting norethinedrone derivatives. III. Measurement of aqueous solubilities and the use of group free energy contributions in predicting partition coefficients, *International Journal of Pharmaceutics*, 3, 319–333, 1979.

L079 Lin, S.-L., L. Lachman, C.J. Swartz, and C.F. Huebner, Preformulation investigation. I: relation of salt forms and biological activity of an experimental antihypertensive, *Journal of Pharmaceutical Sciences*, 61, 1418–1422, 1972.

L080 Lakshmi, T.S. and P.K. Nandi, Interaction of adenine and thymine with aqueous sugar solutions, *Journal of Solution Chemistry*, 7, 283–289, 1978.

L081 Lakshmi, T.S. and P.K. Nandi, Effects of sugar solutions on the activity coefficients of aromatic amino acids and their *n*-acetyl ethyl esters, *Journal of Physical Chemistry*, 80, 249–252, 1976.

L082 Lozano, F.J., Ternary equilibrium for the system water/methyl isobutyl ketone/2-ethyl-2-(hydroxymethyl)-1,3-propanediol, *Journal of Chemical and Engineering Data*, 26, 131–133, 1981.

L083 Leja, J., Some electrochemical and chemical studies related to froth flotation with xanthates, *Minerals Science and Engineering*, 5, 278–286, 1973.

L084 Lesteva, T.M., S.K. Orgorodnikov, and T.N. Tyvina, Liquid–liquid phase equilibria in the system 3-methylbutanediol-1,3-n-butanol-water, *Journal of Applied Chemistry of the USSR*, 41, 1103–1105, 1968.

L085 Lauer, K., Der einfluss des losungsmittels auf den ablauf chemischer reaktionen, XIV. Mitteil: zur kenntnis der aromatischen kohlenwasserstoffe, *Berichte der Deutschen Chemischen Gesellschaft*, 70, 1127–1133, 1937.

L086 Lindstrom, R.E. and C.H. Lee, Solubility in amide–water cosolvent systems: a thermodynamic view, *Journal of Pharmacy and Pharmacology*, 32, 245–247, 1980.

L087 Lundberg, B., T. Lovgren, and C. Blomqvist, The effect of salt on the solubility of steroids in tetradecyltrimethylammonium bromide, *Acta Pharmaceutica Suecica*, 16, 144–150, 1979.

L088 Long, F.A. and W.F. McDevit, Activity coefficients of nonelectrolyte solutes in aqueous salt solutions, *Chemical Reviews*, 51, 119–169, 1952.

L089 Lantsman, M.C., Izvestiya tomskogo politekhnicheskogo instituta, 257, 202–204, 1973.

L090 Loskit, K., Uber polymorphie, Zeitschrift fuer Physikalische Chemie (Leipzig), 134, 156–159, 1924.

L091 Lehr, D., Choice of sulphonamides for mixture therapy, British Medical Journal, 2, 601–604, 1950.

L094 Lippold, B.H. and J.F. Lichey, Loslichkeits-pH-profile mehrprotoniger arzneistoffe am beispiel des assozi-ierenden dimetindens und des zwitterionischen liothyronins, Archiv der Pharmazie, 314, 541–556, 1981.

L095 Lu, P.-Y. and R.L. Metcalf, Environmental Fate and biodegradability of benzene derivatives as studied in a model aquatic ecosystem, Environmental Health Perspectives, 10, 269–284, 1975.

L096 Ley, G.J.M., D.O. Hummel, and C. Schneider, Gamma-radiation-induced polymerization of some vinyl monomers in emulsion systems, Irradiation of Polymers, Advances in Chemistry Series, 66, ACS, Washington, DC, 184–202, 1967.

L097 Lohr, G., Estimation of the uptake rate of solvents into latex particles, Polymers as Colloid Systems, 2, 71–81, 1978.

L099 Lorenz, L.J., Cefaclor, Analytical Profiles of Drug Substances, 9, 107–115, 1980.

L100 Langecker, H., A. Harwart, and K. Junkmann, 3,5-Diacetylamino-2,4,6-trijodbenzoesaure als rontgen-kontrastmittel, Archiv fuer Experimentelle Pathologie und Pharmakologie, 222, 584–590, 1954.

L103 Likhosherstov, M.C., S.V. Alekseev, and T.V. Shalaeva, Zhurnal Khimicheskoi Promyshlennosti, 12, 705–709, 1935.

L106 Lawrence, J. and H.M. Tosine, Adsorption of polychlorinated biphenyls from aqueous solutions and sewage, Environmental Science and Technology, 10, 381–383, 1976.

L300 Leyder, F. and P. Boulanger, Ultraviolet absorption, aqueous solubility, and octanol-water partition for several phthalates, Bulletin of Environmental Contamination and Toxicology, 30, 152–157, 1983.

L301 Lu, P.-Y., R.L. Metcalf, and E.M. Carlson, Environmental fate of five radiolabeled coal conversion by-products evaluated on a laboratory model ecosystem, Environmental Health Perspectives, 24, 201–208, 1978.

L303 Lynch, V.P., Chlormephos, Analytical Methods for Pesticides and Plant Growth Regulators, 10, 49–55, 1978.

L310 Linek, J., Liquid–liquid equilibrium in the isobutyl acetate–water system, Collection of Czechoslovak Chemical Communications, 41, 1714–1717, 1976.

L311 Lu, P.Y., R.L. Metcalf, and L.K. Cole, Pentachlorophenol: Chemistry, Pharmacology, and Environmental Toxicology, 53, Plenum Press, New York, 1977.

L319 Lo, J.M., C.L. Tseng, and J.Y. Yang, Radiometric method for determining solubility of organic solvents in water, Analytical Chemistry, 58, 1596–1597, 1986.

L320 Langford, R.E. and E.L. Heric, Furfural–water–formic acid system at 25 and 35°C, Journal of Chemical and Engineering Data, 17, 87–89, 1972.

L321 Lodge, K.B., Solubility studies using a generator column for 2,3,7,8-tetrachlorodibenzo-p-dioxin, Chemosphere, 18, 933–940, 1989.

L322 Li, A., W.J. Doucette, and A.W. Andren, Solubility of polychlorinated biphenyls in binary water/organic solvent systems, Chemosphere, 24, 1347–1360, 1992.

L329 Lilley, T.H., H. Linsdell, and A. Maestre, Association of caffeine in water and in aqueous solutions of sucrose, Journal of the Chemical Society, Faraday Transactions 1, 88, 2865–2870, 1992.

L332 Lee, L.S., P.S.C. Rao, and I. Okuda, Equilibrium partitioning of polycyclic aromatic hydrocarbons from coal tar into water, Environmental Science and Technology, 26, 2110–2115, 1992.

L333 Lee, W.A., L. Gu, A.R. Miksztal, and P.H. Nelson, Bioavailability improvement of mycophenolic acid through amino ester derivatization, Pharmaceutical Research, 7, 161–166, 1990.

L334 Loftsson, T., B.J. Olafsdottir, and J. Baldvinsdottir, Estramustine: hydrolysis, solubilization, and stabilization in aqueous solutions, International Journal of Pharmaceutics, 79, 107–112, 1992.

L335 Lipnick, R.L., D.E. Johnson, J.H. Gilford, and L.D. Newsome, Comparison of fish toxicity screening data for 55 alcohols with the quantitative structure–activity relationship predictions of minimum toxicity for non-reactive nonelectrolyte organic compounds, Environmental Toxicology and Chemistry, 4, 281–296, 1985.

L338 Lin, H.M. and R.A. Nash, An experimental method for determining the hildebrand solubility parameter of organic nonelectrolytes, Journal of Pharmaceutical Sciences, 82, 1018–1026, 1993.

L339 Leet, W.A., H.-M. Lin, and K.-C. Chao, Mutual solubilities in six binary mixtures of water + a heavy hydrocarbon or a derivative, Journal of Chemical and Engineering Data, 32, 37–40, 1987.

L342 Leung, S.L., G. Becker, R. Karunanithy, and J.T. Fell, Studies on long-acting aryl carboxylic acid esters of testosterone, Pharmaceutica Acta Helvetiae, 64, 121–124, 1989.

L343 Loftsson, T., S. Bjornsdottir, G. Palsdottir, and N. Bodor, The effects of 2-hydroxypropyl-beta-cyclodextrin on the solubility and stability of chlorambucil and melphalan in aqueous solution, International Journal of Pharmaceutics, 57, 63–72, 1989.

L344   Liu, F., D.O. Kildsig, and A.K. Mitra, beta-Cyclodextrin/steroid complexation: effect of steroid structure on association equilibria, *Pharmaceutical Research*, 7, 869–873, 1990.

L345   Loftsson, T., H. Frioriksdottir, S. Thorisdottir, and E. Stefansson, The effect of hydroxypropyl methyl-cellulose on the release of dexamethasone from aqueous 2-hydroxypropyl-beta-cyclodextrin formulations, *International Journal of Pharmaceutics*, 104, 181–184, 1994.

L346   Le, V.P. and B.C. Lippold, Influence of physicochemical properties of homologous esters of nicotinic acid on skin permeability and maximum flux, *International Journal of Pharmaceutics*, 124, 285–292, 1995.

L348   Lun, R., W. Shiu, and D. Mackay, Aqueous solubilities and octanol–water partition coefficients of chloroveratroles and chloroanisoles, *Journal of Chemical Engineering Data*, 40, 959–962, 1995.

L430   Lu, L. and X. Lu, Solubilities of gallic acid and its esters in water, *Journal of Chemical and Engineering Data*, 52, 37–39, 2007.

L432   Loftsson, T. and D. Duchene, Cyclodextrins and their pharmaceutical applications, *International Journal of Pharmaceutics*, 329, 1–11, 2007.

L434   Loftsson, T., D. Hreinsdottir, and M. Masson, Evaluation of cyclodextrin solubilization of drugs, *International Journal of Pharmaceutics*, 302, 18–28, 2005.

L435   Lee, J., S.C. Lee, G. Acharya, and K. Park, Hydrotropic solubilization of paclitaxel: analysis of chemical structures for hydrotropical property, *Pharmaceutical Research*, 20, 1022–1029, 2003.

L437   Long, B.W., L.S. Wang, and J.S. Wu, Solubilities of 1,3-benzenedicarboxylic acid in water + acetic acid solutions, *Journal of Chemical and Engineering Data*, 50, 136–137, 2005.

L441   Luszczyk, M. and S.K. Malanowski, Vapor–liquid equilibrium in α-methylbenzenmethanol + water, *Journal of Chemical and Engineering Data*, 51, 1735–1739, 2006.

L445   Loftson, T., K. Matthiasson, and M. Masson, The effect of organic salts on the cyclodextrin solubilization of drugs, *International Journal of Pharmaceutics*, 262, 101–107, 2003.

L450   Lebosse, R. and V. Ducruet, Aqueous solubility determination of volatile organic compounds by capillary gas chromatography, *Journal of High Resolution Chromatography*, 19, 413–416, 1996.

L451   Wang, W., X. Lu, X. Qin, and Y. Xu, Solubility of pyoluterin in water, dichloromethane, chloroform, and carbon tetrachloride from (278.2 to 333.2) K, *Journal of Chemical Engineering Data*, 53, 2241–2243, 2008.

L452   Lu, L. and X. Lu, Solubilities of polyhydroxybenzophenones in an ethanol + water mixture from (293.15 to 343.15) °K, *Journal of Chemical Engineering Data*, 53, 1996–1998, 2008.

M001   McAuliffe, C., Solubility in water of paraffin, cycloparaffin, olefin, acetylene, cycloolefin, and aromatic hydrocarbons, *Journal of Physical Chemistry*, 70, 1267–1275, 1966.

M002   McAuliffe, C., Solubility in water of C1–C9 hydrocarbons, *Nature* (London), 200, 1092–1093, 1963.

M003   McAuliffe, C., Solubility in water of normal C9 and C10 alkane hydrocarbons, *Science*, 163, 478–479, 1969.

M004   Mukherjee, S. and N.P. Datta, Electrochemical properties of stearic acid hydrosol. Part I, *Journal of the Indian Chemical Society*, 16, 563–582, 1939.

M006   Muramatsu, M., M. Iwahashi, and K. Masumoto, Polymorphic effects of chloramphenicol palmitate on thermodynamic stability in crystals and solubilities in water and in aqueous urea solution, *Journal of Chemical and Engineering Data*, 20, 6–9, 1975.

M007   Madan, D.K. and D.E. Cadwallader, Solubility of cholesterol and hormone drugs in water, *Journal of Pharmaceutical Sciences*, 62, 1567–1569, 1973.

M008   McMeekin, T.L., Unpublished data cited in C007, 201–202, 1943.

M010   McBain, J.W. and K.J. Lissant, The solubilization of four typical hydrocarbons in aqueous solution by three typical detergents, *Journal of Physical Chemistry*, 65, 655–658, 1951.

M011   Morozowich, W., T. Chulski, W.E. Hamlin, and J.G. Wagner, Relationship between in vitro dissolution rates, solubilities, and lt50's in mice of some salts of benzphetamine and etryptamine, *Journal of Pharmaceutical Sciences*, 51, 993–996, 1962.

M012   Mascardo, L.B. and M. Barr, The solubility of terpin hydrate in hydroalcoholic solutions, *Journal of the American Pharmaceutical Association, Practical Pharmacy Edition*, 14, 772–773, 1953.

M013   Merrill, E.J., Solubility of pentaerythritol tetranitrate-1,2-14C in water and saline, *Journal of Pharmaceutical Sciences*, 54, 1670–1671, 1965.

M014   McDonald, C. and R.E. Lindstrom, The effect of urea on the solubility of methyl *p*-hydroxybenzoate in aqueous sodium chloride solution, *Journal of Pharmacy and Pharmacology*, 26, 39–45, 1974.

M015   Marvel, J.R. and A.P. Lemberger, Complexing tendencies of saccharin in aqueous solutions, *Journal of the American Pharmaceutical Association, Scientific Edition*, 49, 417–419, 1960.

M017 Mitchell, A.G. and L.S.C. Wan, Oxidation of aldehydes solubilized in nonionic surfactants. I: solubility of benzaldehyde and methylbenzaldehyde in aqueous solutions of polyoxyethylene glycol ethers, *Journal of Pharmaceutical Sciences*, 53, 1467–1470, 1964.

M018 Mulley, B.A. and A.D. Metcalf, Non-ionic surface-active agents. Part I. The solubility of chloroxylenol in aqueous solutions of polyethylene glycol 1000 monocetyl ether, *Journal of Pharmacy and Pharmacology*, 8, 774–780, 1956.

M020 Marshall, H. and D. Bain, Sodium succinates, *Journal of the Chemical Society* (London), 97, 1074–1085, 1910.

M021 McDevit, W.F. and F.A. Long, The activity coefficient of benzene in aqueous salt solutions, *Journal of the American Chemical Society*, 74, 1773–1773, 1952.

M022 Mortimer, F.S., The solubility relations in mixtures containing polar components, *Journal of the American Chemical Society*, 45, 633–641, 1923.

M023 Macek, T.J., W.H. Baade, A. Bornn, and F.A. Bacher, Observations on the solubility of some cortical hormones, *Science*, 116, 399–399, 1952.

M024 McMeekin, T.L., E.J. Cohn, and J.H. Weare, Studies in the physical chemistry of amino acids, peptides and related substances. A comparison of the solubility of amino acids, peptides and their derivatives, *Journal of the American Chemical Society*, 58, 2173–2181, 1936.

M025 McMeekin, T.L., E.J. Cohn, and J.H. Weare, Studies in the physical chemistry of amino acids, peptides and related substances. III. The solubility of derivatives of the amino acids in alcohol–water mixtures, *Journal of the American Chemical Society*, 57, 626–633, 1935.

M026 Moyle, M.P. and M. Tyner, Solubility and diffusivity of 2-naphthol in water, *Industrial and Engineering Chemistry*, 45, 1794–1797, 1953.

M027 McCune, L.K. and R.H. Wilhelm, Mass and momentum transfer in solid–liquid system, *Industrial and Engineering Chemistry*, 41, 1124–1127, 1949.

M028 McBride, W., R.A. Henry, J. Cohen, and S. Skolnik, Solubility of nitroguanidine in water, *Journal of the American Chemical Society*, 73, 485–486, 1951.

M029 Mason, L.S., The solubilities of four amino butyric acids and the densities of aqueous solutions of the acids at 25 degrees, *Journal of the American Chemical Society*, 69, 3000–3002, 1947.

M030 McMaster, L., E. Bender, and E. Weil, The solubility of phthalic acid in water and sodium sulfate solutions, *Journal of the American Chemical Society*, 43, 1205–1207, 1921.

M031 Mirvish, S.S., P. Issenberg, and H.C. Sornson, Air–water and ether–water distribution of *n*-nitroso compounds: implications for laboratory safety, analytic methodology, and carcinogenicity for the rat esophagus, nose, and liver, *Journal of the National Cancer Institute*, 56, 1125–1129, 1976.

M032 Maren, T.H., Renal carbonic anhydrase and the pharmacology of sulfonamide inhibitors, in *Heffter's Handbook of Experimental Pharmacology*, ed. H. Herken, Springer-Verlag, Berlin, 24, 225–228, 1969.

M035 MacDonald, A., A.F. Michaelis, and B.Z. Senkowski, Chlordiazepoxide, *Analytical Profiles of Drug Substances*, 1, 15–25, 1972.

M036 MacDonald, A., A.F. Michaelis, and B.Z. Senkowski, Diazepam, *Analytical Profiles of Drug Substances*, 1, 79–89, 1972.

M037 McGovern, E.W., Chlorohydrocarbon solvents, *Industrial and Engineering Chemistry*, 35, 1230–1239, 1943.

M038 Mauger, J.W. and A.N. Paruta, Entropy of transfer of molecular benzoic acid from a pure liquid to an aqueous solution, *Journal of Pharmaceutical Sciences*, 63, 576–579, 1974.

M040 Mackay, D., Environmental and laboratory rates of volatilization of toxic chemicals from water, in *Hazard Assessment of Chemicals: Current Developments*, Vol. 1, eds. J. Saxena and F. Fisher, Academic Press, New York, 1, 303–319, 1981.

M041 Megson, N.J.L., The solubility of phenols in formalin, *Transactions of the Faraday Society*, 34, 525–532, 1938.

M042 Munzel, K., *Galenische Formgebung und Arzneimittelwirkung Neue Erkenntnisse und Feststellung*, 14, 269–337, 1970.

M043 Mullin, J.W., *Crystallisation*, Butterworths, London, 425–426, 1972.

M044 Marsh, J.R. and P.J. Weiss, Solubility of antibiotics in twenty-six solvents. III, *Journal of the Association of Official Analytical Chemists*, 50, 457–462, 1967.

M045 Marvel, J.R., D.A. Schlichting, C. Denton, and M.M. Cahn, The effect of a surfactant and of particle size on griseofulvin plasma levels, *Journal of Investigative Dermatology*, 42, 197–203, 1964.

M046 Mehta, S.C., Mechanistic Studies of: I. Crystal growth of sulfathiazole and its inhibition by PVP; II. dissolution of high-energy sulfathiazole-PVP coprecipitates, PhD Thesis, 19–19, 1969.

M047 McBride, W., R.A. Henry, and G.B.L. Smith, Solubility of nitroaminoguanidine, *Journal of the American Chemical Society*, 71, 2937–2938, 1949.

M048 Minkov, E., M. Zahariewa, B. Botev, and T. Trandafilov, solubilisierung und emulgierung von hexachlorophen, *Pharmazie* (Berlin), 24, 353–356, 1969.

M049 Mulley, B.A. and A.J. Winfield, Non-ionic surface-active agents. Part VII. Solubility of benzoic acid in aqueous solutions of some monododecyl polyoxyalkanols, *Journal of the Chemical Society A: Inorganic, Physical, Theoretical*, 1459–1464, 1970.

M051 Massol, G. and F. Lamouroux, Sur la solubilite dans l'eau des acides maloniques substitues, *Comptes Rendus Hebdomadaires des Seances de l'Academie des Sciences*, 128, 1000–1002, 1899.

M053 Mitchell, A.G. and D.J. Saville, The dissolution of commercial aspirin, *Journal of Pharmacy and Pharmacology*, 21, 28–34, 1969.

M054 Morris, J.G. and E.R. Redfearn, Vitamins and coenzymes, in *Data for Biochemical Research*, eds. R.M.C. Dawson, D.C. Elliot and K.M. Jones, Oxford University Press, New York, 191–215, 1969.

M056 Maher, N. and G. Sirois, Solubilite du phenobarbital et de l'hexobarbital en relation avec leur activite biologique (solubility of phenobarbital and hexobarbital in relation with their biological activity), *Revue Canadienne de Biologie*, 30, 45–49, 1971.

M057 Macartney, D.W., R.W. Luxton, G.S. Smith, and J. Goldman, Sulphamethazine trial of a new sulphonamide, *Lancet*, 1, 639–642, 1942.

M058 Mukerjee, P. and J.R. Cardinal, Solubilization as a method for studying self-association: solubility of naphthalene in the bile salt sodium cholate and the complex pattern of its aggregation, *Journal of Pharmaceutical Sciences*, 65, 882–885, 1976.

M059 Miller, Jr., F.W. and H.R. Dittmar, The solubility of urea in water. The heat of fusion of urea, *Journal of the American Chemical Society*, 56, 848–849, 1934.

M060 Masterton, W.L. and T.P. Lee, Effect of dissolved salts on water solubility of lindane, *Environmental Science and Technology*, 6, 919–921, 1972.

M061 Melnikov, N.N., F.A. Gunther, and J.D. Gunther, Book Chapter, 36, 36–439, Springer-Verlag, New York, 1971.

M062 McClure, H.B., Industrial applications of the glycols, *Industrial and Engineering Chemistry*, News Edition, 17, 149–153, 1939.

M063 May, W.E., S.P. Wasik, and D.H. Freeman, Determination of the aqueous solubility of polynuclear aromatic hydrocarbons by a coupled column liquid chromatographic technique, *Analytical Chemistry*, 50, 175–179, 1978.

M064 Mackay, D. and W.Y. Shiu, Aqueous solubility of polynuclear aromatic hydrocarbons, *Journal of Chemical and Engineering Data*, 22, 399–402, 1977.

M065 Mannheimer, M., Mutual solubility of liquefied gases and water at room temperature, *Chemist-Analyst*, 45, 8–10, 1956.

M066 Meyer, O., Dissertation or Masters Thesis, 1967.

M067 Mizuno, N., Y. Iwayama, H. Takagi, and A. Kamada, Stability and intestinal absorption of alpha,3-o-isopropylidene pyridoxine, *Yakuzaigaku*, 33, 172–178, 1973.

M068 McBee, E.T. and R.E. Hatton, Production of hexachlorobutadiene, *Industrial and Engineering Chemistry*, 41, 809–812, 1949.

M069 McKeown, A., The influence of electrolytes on the solubility of non-electrolytes, *Journal of the American Chemical Society*, 44, 1203–1209, 1922.

M070 Megges, R., H.J. Portius, and K.R.H. Repke, Penta-acetyl-gitoxin: the prototype of a prodrug in the cardiac glycoside series, *Pharmazie* (Berlin), 32, 665–667, 1977.

M071 May, W.E., S.P. Wasik, and D.H. Freeman, Determination of the solubility behavior of some polycyclic aromatic hydrocarbons in water, *Analytical Chemistry*, 50, 997–1000, 1978.

M072 Muller, F. and R. Suverkrup, Die zersetzung von *p*-aminosalicylsaure in gegenwart begrenzter wassermengen, *Pharmazeutische Industrie*, 39, 1115–1122, 1977.

M073 Mitchell, S., A method for determining the solubility of sparingly soluble substances, *Journal of the Chemical Society* (London), 1333–1336, 1926.

M074 Mondain-Monval, P., Sur la solubilite du saccharose, *Comptes Rendus Hebdomadaires des Seances de l'Academie des Sciences*, 181, 37–40, 1925.

M075 Meyer, K.H. and O. Klemm, La solubilite de l'anhydride du glycocolle, *Helvetica Chimica Acta*, 23, 25–27, 1940.

M077 Meyer, J., Die polymorphie der allozimtsaure, *Zeitschrift fuer Elektrochemie*, 17, 976–984, 1911.

M078 Moiseeva, L.M. and N.M. Kuznetsova, Determination of the solubility of aliphatic beta-diketones and their compounds with beryllium in water, *Zhurnal Analiticheskoi Khimii*, 26, 2094–2096, 1971.

M081 Mehl, I.W., Ein uebersichtsdiagramm log p-1/T fuer das stoffpaar methylamin–wasser, *Zeitschrift fuer die Gesamte Kaelte-Industrie*, 42, 13–14, 1935.

M082 May,W.E., The solubility behavior of polynuclear aromatic hydrocarbons in aqueous systems, PhD Thesis, 113–135, 1977.

M083 McBain, J.W. and M. Eaton, Hydrolysis in solutions of potassium laurate as measured by extraction with benzene, *Journal of the Chemical Society* (London), 131, 2166–2179, 1928.

M087 McBain, J.W. and P.H. Richards, Solubilization of insoluble organic liquids by detergents, *Industrial and Engineering Chemistry*, 38, 642–646, 1946.

M088 Mion, M. and G. Urbain, Contribution a l'etude du systeme eau, alcool ethylique, acide acetique, acetate d'ethyle, *Comptes Rendus Hebdomadaires des Seances de l'Academie des Sciences*, 193, 1330–1333, 1931.

M091 Masterton, W.L., Partial molal volumes of hydrocarbons in water solution, *Journal of Chemical Physics*, 22, 1830–1833, 1954.

M093 Mitchell, A.G., Bactericidal activity of chloroxylenol in aqueous solutions of cetomacrogol, *Journal of Pharmacy and Pharmacology*, 16, 533–537, 1964.

M094 Mahieu, J., La solubilite dans les melanges de deux solvants miscibles, *Bulletin des Societes Chimiques Belges*, 45, 667–674, 1936.

M095 Mueller, A.J., L.I. Pugsley, and J.B. Ferguson, The system normal butyl alcohol–methyl alcohol–water, *Journal of Physical Chemistry*, 35, 1314–1327, 1931.

M096 Mange, C.E. and O. Ehler, Solubilities of vanillin, *Industrial and Engineering Chemistry*, 16, 1258–1260, 1924.

M097 Monblanova, V.V. and V.M. Rodinov, Solubilities of some beta-amino acids, *Journal of General Chemistry of the USSR*, 23, 1899–1901, 1953.

M098 Michels, A. and E.C.F. Ten Haaf, The three-phase-lines of the systems: water–*ortho*-cresol, water–*meta*-cresol, and water–paracresol, *Proceedings of the Koninklijke Nederlandse Akadamie van Wetenschappen*, 30, 52–54, 1927.

M099 Mains, G.H., The system furfural–water. I. A study of its properties with reference to their commercial application in the production of furfural, *Chemical and Metallurgical Engineering*, 26, 779–784, 1922.

M101 Manchot, W., M. Jahrstorfer, and H. Zepter, Untersuchungen uber gasloslichkeit und hydratation, *Zeitschrift fuer Anorganische Chemie*, 141, 45–81, 1924.

M102 Matheson, I.B.C. and A.D. King, Jr., Solubility of gases in micellar solutions, *Journal of Colloid and Interface Science*, 66, 464–469, 1978.

M104 Miyazaki, S., M. Nakano, and T. Arita, Effect of crystal forms on the dissolution behavior and bio-availability of tetracycline, chlortetracycline, and oxytetracycline bases, *Chemical and Pharmaceutical Bulletin*, 23, 552–558, 1975.

M105 Miyazaki, S., M. Nakano, and T. Arita, A comparison of solubility characteristics of free bases and hydro chloride salts of tetracycline antibiotics in hydrochloric acid solutions, *Chemical and Pharmaceutical Bulletin*, 23, 1197–1204, 1975.

M106 Matsumaru, H., S. Tsuchiya, and T. Hosono, Interaction and dissolution characteristics of ajmaline–PVP coprecipitate, *Chemical and Pharmaceutical Bulletin*, 25, 2504–2509, 1977.

M107 Mullin, J.W. and T.P. Cook, Diffusion and dissolution of the hydroxybenzoic acids in water, *Journal of Applied Chemistry*, 15, 145–151, 1965.

M108 Muramatsu, M., M. Iwahashi, and U. Takeuchi, Thermodynamic relationship between alpha- and beta-forms of crystalline progesterone, *Journal of Pharmaceutical Sciences*, 68, 175–177, 1979.

M109 Morimoto, Y., R. Hori, and T. Arita, Solubilities of antipyrine derivatives in water and non-polar solvents, *Chemical and Pharmaceutical Bulletin*, 22, 2217–2222, 1974.

M110 Metcalf, R.L. and J.R. Sanborn, Pesticides and environmental quality in Illinois, *Illinois, Natural History Survey, Bulletin*, 31, 381–438, 1975.

M111 Merriman, R.W., The mutual solubilities of ethyl acetate and water and the densities of mixtures of ethyl acetate and ethyl alcohol, *Journal of the Chemical Society* (London), 103, 1774–1789, 1913.

M112 Morachevskii, A.G. and Z.P. Popovich, Liquid–vapor equilibrium and mutual solubility of components in the system tertiary-butyl alcohol-*sec*-butyl alcohol–water, *Journal of Applied Chemistry of the USSR*, 38, 2085–2088, 1965.

M113 Moustafa, M.A., A.R. Ebian, S.A. Khalil, and M.M. Motawi, Sulphamethoxydiazine crystal forms, *Journal of Pharmacy and Pharmacology*, 23, 868–874, 1971.

M114 Moriyoshi, T., Y. Aoki, and H. Kamiyama, Mutual solubility of i-butanol + water under high pressure, *Journal of Chemical Thermodynamics*, 9, 495–502, 1977.

M115 Mitchell, A.G. and J.F. Broadhead, Hydrolysis of solubilized aspirin, *Journal of Pharmaceutical Sciences*, 56, 1261–1266, 1967.

M116 Matsuura, H. and K. Sekiguchi, Studies on the effect of inorganic salts on the solubility of organic pharmaceutical compounds, *Yakuzaigaku*, 20, 213–218, 1960.

M117 Mizukami, S. and K. Nagata, On the solubility of N1-acetyl-sulfaisoxazole and its decomposed rate in the simulated intestinal fluid, *Shionogi Kenkyusho Nempo*, 6, 58–64, 1956.

M118 Metcalf, R.L., J.R. Sanborn, P.-Y. Lu, and D. Nye, Laboratory model ecosystem studies of the degradation and fate of radiolabeled tri-, tetra-, and pentachlorobiphenyl compared with DDE, *Archives of Environmental Contamination and Toxicology*, 3, 151–165, 1975.

M119 Moriyoshi, T., S. Kaneshina, K. Aihara, and K. Yabumoto, Mutual solubility of 2-butanol + water under high pressure, *Journal of Chemical Thermodynamics*, 7, 537–545, 1975.

M120 Mitzner, R. and C.-R. Kramer, Zur loslichkeit einiger 2,4-dichlorphenoxyessigsaurealkylester in wasser, *Zeitschrift fuer Chemie*, 17, 379–380, 1977.

M122 Markowski, W., E. Soczewinski, and K. Czapinska, Analogy of solubility and chromatographic parameters in reversed phase partition chromatography, *Polish Journal of Chemistry*, 52, 1775–1780, 1978.

M123 Mirgorod, Y.A., Assessing the initiation of submicelle formation in aqueous solutions of diphilic molecules, *Kolloid Zhurnal*, 40, 483–488, 1978.

M124 Massaldi, H.A. and C.J. King, Simple technique to determine solubilities of sparingly soluble organics: solubility and activity coefficients of D-limonene, *n*-butylbenzene, and *n*-hexyl acetate in water and sucrose solutions, *Journal of Chemical and Engineering Data*, 18, 393–397, 1973.

M125 McCants, J.F., J.H. Jones, and W.H. Hopson, Ternary solubility data for systems involving 1-propanol and water, *Industrial and Engineering Chemistry*, 45, 454–456, 1953.

M127 Marcus, Y., L.E. Asher, J. Hormadaly, and E. Pross, Selective extraction of potassium chloride by crown ethers in substituted phenol solvents, *Hydrometallurgy*, 7, 27–39, 1981.

M128 Martin, A., J. Newburger, and A. Adjei, Extended hildebrand solubility approach: solubility of theophylline in polar binary solvents, *Journal of Pharmaceutical Sciences*, 69, 487–491, 1980.

M129 Means, J.C., J.J. Hassett, S.G. Wood, and W.L. Banwart, Sorption properties of energy-related pollutants and sediments, in *Polynuclear Aromatic Hydrocarbons*, eds. P.W. Jones and P. Leber, Ann Arbor Science, Inc, Ann Arbor, Michigan, 327–339, 1979.

M130 Mackay, D. and P.J. Leinonen, Rate of evaporation of low-solubility contaminants from water bodies to atmosphere, *Environmental Science and Technology*, 9, 1178–1180, 1975.

M131 Mortland, M.M. and W.F. Meggitt, Interaction of ethyl *N,N*-di-n-propylthiolcarbamate (EPTC) with montmorillonite, *Journal of Agricultural and Food Chemistry*, 14, 126–129, 1966.

M132 Mackay, D. and W.Y. Shiu, The determination of the solubility of hydrocarbons in aqueous sodium chloride solutions, *Canadian Journal of Chemical Engineering*, 53, 239–242, 1975.

M133 McConnell, G., D.M. Ferguson, and C.R. Pearson, Chlorinated hydrocarbons and the environment, *Endeavour*, 34, 13–18, 1975.

M134 Metcalf, R.L., I.P. Kapoor, P.-Y. Lu, and P. Sherman, Model ecosystem studies of the environmental fate of six organochlorine pesticides, *Environmental Health Perspectives*, 4, 35–44, 1973.

M135 Morrison, T.J., The salting-out effect, *Transactions of the Faraday Society*, 40, 43–48, 1944.

M136 McGuire, M.J. and I.H. Suffet, *The Calculated Net Adsorption Energy Concept*, 1, 91–115, 1980.

M137 Mistry, F.R. and S.M. Barnett, An equilibrium phase diagram for the glucose-cellobiose-water system at 30.5°C, *Journal of Chemical and Engineering Data*, 25, 223–226, 1980.

M138 Metcalf, R.L., DDT substitutes, *Critical Reviews in Environmental Control*, 3, 25–59, 1972.

M139 McGuire, M.J., I.H. Suffet, and J.V. Radziul, Assessment of unit processes for the removal of trace organic compounds from drinking water, *Journal of the American Water Works Association*, 70, 565–572, 1978.

M140 Mitterhauszerova, L., K. Kralova, and L. Krasnec, Wechselwirkung zwischen kebuzon (ketophenylbutazon) und modifizierten starken, *Pharmazie* (Berlin), 35, 159–160, 1980.

M141 Meier, R., O. Allemann, and H.V. Meyenburg, 6-sulfanilamido-2,4-dimethylpyrimidin, *Schweizerische Medizinische Wochenschrift*, 74, 1091–1095, 1944.

M142 Marshall, E.K., Jr., A.C. Bratton, H.J. White, and J.T. Litchfield, Jr., Sulfanilylguanidine: a chemotherapeutic agent for intestinal infections, *Bulletin of the John Hopkins Hospital*, 67, 163–188, 1940.

M143 Manov, G.G., K.E. Schuette, and F.S. Kirk, Ionization constant of 5-5'-diethylbarbituric acid from to 60°C, *Journal of Research of the National Bureau of Standards*, 48, 84–91, 1952.

M145 McNabb, R.A. and F.M.A. McNabb, Physiological chemistry of uric acid: solubility, colloid and ion-binding properties, *Comparative Biochemistry Physiology*, 67, 27–34, 1980.

M146 Morachevskii, A.G., N.A. Smirnova, and R.V. Lyzlova, Phase equilibria in the ternary systems isobutyraldehyde-isobutyl alcohol–water and isovaleraldehyde–isobutyl alcohol–water, *Journal of Applied Chemistry of the USSR*, 38, 1245–1248, 1965.

M147 Matin, N.B., E.N. Zil'berman, V.I. Trachenko, and V.A. Afanas'ev, Solubility of acrylamide in water and some organic solvents, *Journal of Applied Chemistry of the USSR*, 52, 2228–2229, 1980.

M148 Mannelli, M., P. Gigli, G. Orzalesi, and T. Bisagno, Physico-chemical properties of ibuproxam, a new non-steroideal anti-inflammatory agent, *Bollettino Chimico Farmaceutico*, 119, 203–208, 1980.

M149 Mooney, K.G., M.A. Mintun, K.J. Himmelstein, and V.J. Stella, Dissolution kinetics of carboxylic acids. I: effect of ph under unbuffered conditions, *Journal of Pharmaceutical Sciences*, 70, 13–22, 1981.

M151 May, W.E., The solubility behavior of polycyclic aromatic hydrocarbons in aqueous systems, *Petroleum in the Marine Environment*, eds. L. Petrakis and F.T. Weiss, Advances in Chemistry Series, 185, ACS, Washington, DC, 143–192, 1980.

M152 Moolenaar, F., J. Visser, and T. Huizinga, Biopharmaceutics of rectal administration of drugs in man. absorption rate and bioavailability of glafenine after oral and rectal administration, *International Journal of Pharmaceutics*, 4, 195–203, 1980.

M153 Mullin, J.W. and M.J.L. Whiting, Succinic acid crystal growth rates in aqueous solution, *Industrial and Engineering Chemistry, Fundamentals*, 19, 117–121, 1980.

M155 Montagu-Bourin, M., P. Levillain, R. Ceolin, and C. Souleau, le systeme ternaire eau-phenanthroline-1, 10-acide perchlorique: I. Etude de la solubilite a 25°C, dans la region riche en phenanthroline-1,10 et cristallochimie du perchlorate d'hydrogenato-*bis* (phenanthroline-1,10), *Bulletin de la Societe Chimique de France*, 1, 109–112, 1981.

M156 Means, J.C., S.G. Wood, J. Hassett, and W.L. Banwart, Sorption of polynuclear aromatic hydrocarbons by sediments and soils, *Environmental Science and Technology*, 14, 1524–1528, 1980.

M157 Moustafa, M.A., A.M. Molokhia, and M.W. Gouda, Phenobarbital solubility in propylene glycol–glycerol–water systems, *Journal of Pharmaceutical Sciences*, 70, 1172–1174, 1981.

M158 Martin, A., A.M. Paruta, and A. Adjei, Extended Hildebrand solubility approach: methylxanthines in mixed solvents, *Journal of Pharmaceutical Sciences*, 70, 1115–1120, 1981.

M159 Mason, N.A., S. Cline, M.L. Hyneck, and G.L. Flynn, Factors affecting diazepam infusion: solubility, administration-set composition, and flow rate, *American Journal of Hospital Pharmacy*, 38, 1449–1454, 1981.

M160 Messer, C.E., G. Malakoff, J. Well, and S. Labib, Phase equilibrium behavior of certain pairs of amino acids in aqueous solution, *Journal of Physical Chemistry*, 85, 3533–3540, 1981.

M161 Martin, H. and C.R. Worthing, *Pesticide Manual: Basic Information on the Chemicals Used as Active Components of Pesticides*, British Crop Protection Council, London, 1977.

M162 Maier-Bode, H. and K. Hartel, Linuron and monolinuron, *Residue Reviews*, 77, 4–6, 1981.

M163 Mysels, K.J., Contribution of micelles to the transport of a water-insoluble substance through a membrane, *Pesticidal Formulations Research*, ed. J. W. Van Valkenburg, Advances in Chemistry Series, 86, ACS, Washington, DC, 24–77, 1969.

M164 Maier-Bode, H., Book Chapter, 63, 44–48, Springer-Verlag, 1976.

M165 Marrelli, L.P., Cephalexin, *Analytical Profiles of Drug Substances*, 4, 21, 1975.

M166 Michel, G., Nystatin, *Analytical Profiles of Drug Substances*, 6, 341–371, 1977.

M167 Manius, G.J., Trimethoprim, *Analytical Profiles of Drug Substances*, 7, 445, 1978.

M168 Moriyama, M., A. Inoue, M. Isoya, and M. Hanano, Dissolution properties and gastrointestinal absorption of chloramphenicol from hydrophilic high molecular compound correcipitates, *Yakugaku Zasshi* (Tokyo), 98, 1012–1018, 1978.

M169 Mauger, J.W., S.A. Howard, and E.Z. Damewood, A simulation experiment for the dissolution of mono-sized and multisized drug particles, *American Journal of Pharmaceutical Education*, 42, 60–63, 1978.

M171 Merckel, J.H.C., Die loslichkeit der dicarbonsauren, *Recueil des Travaux Chimiques des Pays-Bas*, 56, 811–814, 1937.

M172 Meleschenko, A.F., *Gigiena i Sanitariya*, 25, 54–57, 1960.

M175 Mackay, D., W.Y. Shiu, and A.W. Wolkoff, Gas chromatographic determination of low concentrations of hydrocarbons in water by vapor phase extraction, *American Society for Testing and Materials*, 573, 251–258, 1975.

M177 Moroi, Y., K. Sato, and R. Matuura, Solubilization of phenothiazine in aqueous surfactant micelles, *Journal of Physical Chemistry*, 86, 2463–2468, 1982.

M180 McDowell, W. and R. Weingarten, New experimental evidence about the dyeing of polyester materials with disperse dyes, *Journal of the Society of Dyers and Colourists*, 85, 589–597, 1969.

M183 May, W.E., S.P. Wasik, M.M. Miller, and R.N. Goldberg, Solution thermodynamics of some slightly soluble hydrocarbons in water, *Journal of Chemical and Engineering Data*, 28, 197–200, 1983.

M184 Mieure, J.P., O. Hicks, R.G. Kaley, and V.W. Saeger, Characterization of polychlorinated biphenyls, *Proceedings of National Conference on Polychlorinated Biphenyls*, 84–87, 1976.

M300 McNally, M.E. and R.L. Grob, Determination of the solubility limits of organic priority pollutants by gas chromatographic headspace analysis, *Journal of Chromatography*, 260, 23–32, 1983.

M301 Molin, L., G. Dahlstrom, M.-I. Nilsson, and L. Tekenbergs, Solubility, partition, and adsorption of digitalis glycosides, *Acta Pharmaceutica Suecica*, 20, 129–144, 1983.

M303 Means, J.C., J.J. Hassett, S.G. Wood, and A. Khan, *Sorption Properties of Polynuclear Aromatic Hydrocarbons and Sediments: Heterocy*, 395–404, 1980.

M308 Miller, M.M., S. Ghodbane, S.P. Wasik, and D.E. Martire, Aqueous solubilities, octanol/water partition coefficients, and entropies of melting of chlorinated benzenes and biphenyls, *Journal of Chemical and Engineering Data*, 29, 184–190, 1984.

M310 Mayer, J.M. and M. Rowland, Determination of Aqueous Solubilities of a Series of 5-Ethyl-5-alkylbarbituric Acids and Their Correlation with Log P and Melting Points, *Drug Development and Industrial Pharmacy*, 10, 69–83, 1984.

M311 McNally, M.E. and R.L. Grob, Headspace Determination of Solubility Limits of the Base Neutral and Volatile Components From the Environmental Protection Agency's List of Priority Pollutants, *Journal of Chromatography*, 284, 105–116, 1984.

M312 Mackay, D. and A.T.K. Yeun, Mass Transfer Coefficient Correlations for Volatilization of Organic Solutes from Water, *Environmental Science and Technology*, 17, 211–217, 1983.

M314 Meeussen, E. and P. Huyskens, Etude de la structure du butanol-n en solution par les coefficients de partage, *Journal de Chimie Physique et de Physico-Chimie Biologique*, 63, 845–854, 1966.

M315 Mentasti, E., C. Rinaudo, and R. Boistelle, Solubility and mechanism of dissolution of dihydrated and anhydrous uric acid, *Journal of Chemical and Engineering Data*, 28, 247–251, 1983.

M316 Mukai, E., K. Arase, M. Hashida, and H. Sezaki, Enhanced delivery of mitomycin C prodrugs through the skin, *International Journal of Pharmaceutics*, 25, 95–103, 1985.

M317 Mollgaard, B., A. Hoelgaard, and H. Bundgaard, Prodrugs as drug delivery systems. XXIII. Improved dermal delivery of 5-fluorouracil through human skin via *N*-acyloxymethyl pro-drug derivatives, *International Journal of Pharmaceutics*, 12, 153–162, 1982.

M318 Miles, J.R.W., B.T. Bowman, and C.R. Harris, Adsorption, desorption, soil mobility and aqueous persistence of fensulfothion and its sulfide and sulfone metabolites, *Journal of Environmental Science & Health, Series B*, 16, 309–324, 1981.

M319 Martin, A., P.L. Wu, and T. Velasquez, Extended Hildebrand solubility approach: sulfonamides in binary and ternary solvents, *Journal of Pharmaceutical Sciences*, 74, 277–282, 1985.

M320 Moro, M.E., M.M. Velazquez, J.M. Cachaza, and L.J. Rodriguez, Solubility of diazepam and prazepam in aqueous non-ionic surfactants, *Journal of Pharmacy and Pharmacology*, 38, 294–296, 1986.

M321 Meshali, M., H. El-Sabbagh, and I. Ramadan, Simultaneous solubility and dissolution rate of sulfamethoxazole and trimethoprim in binary mixture, *Pharmazie* (Berlin), 39, 407–408, 1984.

M323 Marco, J.M., M.I. Galan, and J. Costa, Liquid–liquid equilibria for the quaternary system water-phosphoric acid-1-hexanol-cyclohexanone at 25°C, *Journal of Chemical and Engineering Data*, 33, 211–214, 1988.

M324 Mackay, D., S. Paterson, and W.H. Schroeder, Model describing the rates of transfer processes of organic chemicals between atmosphere and water, *Environmental Science and Technology*, 20, 807–810, 1986.

M325 McDonald, C. and C. Richardson, The effect of added salts on solubilization by a non-ionic surfactant, *Journal of Pharmacy and Pharmacology*, 33, 38–39, 1981.

M327 McGowan, J.C., P.N. Atkinson, and L.H. Ruddle, The physical toxicity of chemicals. V. interaction terms for solubilities and partition coefficients, *Journal of Applied Chemistry*, 16, 99–102, 1966.

M333 Manzo, R.H., A.A. Ahumada, and E. Luna, Effects of solvent medium on solubility. IV: comparison of the hydrophilic–lipophilic character exhibited by functional groups in ethanol–water and ethanol–cyclohexane mixtures, *Journal of Pharmaceutical Sciences*, 73, 1869–1871, 1984.

M334 Martin, A., P.L. Wu, and A. Beerbower, Expanded solubility parameter approach. II: *p*-hydroxybenzoic acid and methyl *p*-hydroxybenzoate in individual solvents, *Journal of Pharmaceutical Sciences*, 73, 188–194, 1984.

M335 Menard, F.A., M.G. Dedhiya, and C.T. Rhodes, Potential pharmaceutical applications of a new beta-cyclodextrin derivative, *Drug Development and Industrial Pharmacy*, 14, 1529–1547, 1988.

M336 Murphy, T.J., M.D. Mullin, and J.A. Meyer, Equilibration of polychlorinated biphenyls and toxaphene with air and water, *Environmental Science and Technology*, 21, 155–162, 1987.

M337 Mohammadzadeh-K., A., R.E. Feeney, and L.M. Smith, Hydrophobic binding of hydrocarbons by proteins. I. Relationship of hydrocarbon structure, *Biochimica et Biophysica Acta*, 194, 246–255, 1969.

M339 Munz, C. and P.V. Roberts, Effects of solute concentration and cosolvents on the aqueous activity coefficient of halogenated hydrocarbons, *Environmental Science and Technology*, 20, 830–836, 1986.

M340 Marple, L., R. Brunck, and L. Throop, Water solubility of 2,3,7,8-tetrachlorodibenzo-p-dioxin, *Environmental Science and Technology*, 20, 180–182, 1986.

M342 Miller, M.M., S.P. Wasik, G.-L. Huang, and D. Mackay, Relationships between octanol–water partition coefficient and aqueous solubility, *Environmental Science and Technology*, 19, 522–529, 1985.

M344 Mackay, D., A. Bobra, W.Y. Shiu, and S.H. Yalkowsky, Relationships between aqueous solubility and octanol–water partition coefficients, *Chemosphere*, 9, 701–711, 1980.

M345 Major, C.J. and O.J. Swenson, Acetic acid–ethyl ether–water system. Mutual solubility and tie line data, *Industrial and Engineering Chemistry*, 38, 834–836, 1946.

M346 Malone, J.W. and R.W. Vining, Phase equilibria data for the system *n*-propyl alcohol–water–nitroethane, *Journal of Chemical and Engineering Data*, 12, 387–389, 1967.

M347 Matous, J., J.P. Novak, J. Sobr, and J. Pick, Phase equilibria in the system tetrahydrofuran(1)–water(2), *Collection of Czechoslovak Chemical Communications*, 37, 2653–2663, 1972.

M348 Mertl, I., Liquid–vapour equilibrium. II. Phase equilibria in the ternary system ethyl acetate–ethanol–water, *Collection of Czechoslovak Chemical Communications*, 37, 366–374, 1972.

M349 Means, J.C., S.G. Wood, J.J. Hassett, and W.L. Banwart, Sorption of amino- and carboxy-substituted polynuclear aromatic hydrocarbons by sediments and soils, *Environmental Science and Technology*, 16, 93–98, 1982.

M350 Matsuda, H., K. Ito, M. Tanaka, and H. Sumiyoshi, Inclusion complexes of various fragrance materials with 2-hydroxypropyl-beta-cyclodextrin, *STP Pharma Science*, 1, 211–215, 1991.

M352 Manzo, R.H. and A.A. Ahumada, Effects of solvent medium on solubility. V: enthalpic and entropic contributions to the free energy changes of di-substituted benzene derivatives in ethanol:water and ethanol:cyclohexane mixtures, *Journal of Pharmaceutical Sciences*, 79, 1109–1115, 1990.

M360 Maurin, M.B., L.W. Dittert, and A.A. Hussain, Mechanism of diffusion of monosubstituted benzoic acids through ethylene-vinyl acetate copolymers, *Journal of Pharmaceutical Sciences*, 81, 79–84, 1992.

M362 Maurin, M.B., W.J. Addicks, S.M. Rowe, and R. Hogan, Physical–chemical properties of alpha styryl carbinol antifungal agents, *Pharmaceutical Research*, 10, 309–312, 1993.

M364 Mestres, R. and G. Mestres, Deltamethrin: uses and environmental safety, *Reviews of Environmental Contamination and Toxicology*, 124, 1–3, 1992.

M368 Muraoka, K. and T. Hirata, Hydraulic behaviour of chlorinated organic compounds in water, *Water Research*, 22, 485–489, 1988.

M370 Morris, C.E., Solubility of meso-1,2,3,4-butanetetracarboxylic acid and some of its salts in water, *Journal of Chemical and Engineering Data*, 37, 330–331, 1992.

M372 Maurin, M.B., R.D. Vickery, C.A. Gerard, and M. Hussain, Solubility of ionization behavior of the antifungal alpha-(2,4-difluorophenyl)-alpha-[(1-(2-(2-pyridly)phenylethenyl)]-1H-1,2,4-triazole-1-ethanol bismesylate (XD405), *International Journal of Pharmaceutics*, 94, 11–14, 1993.

M373 Ma, K.-C., W.-Y. Shiu, and D. Mackay, Aqueous solubilities of chlorinated phenol at 25°C, *Journal of Chemical and Engineering Data*, 38, 364–366, 1993.

M374 Mazzenga, G.C. and B. Berner, The transdermal delivery of zwitterionic drugs. I: the solubility of zwitterion salts, *Journal of Controlled Release*, 16, 77–88, 1991.

M375 Marcilla, A.F., F. Ruiz, and M.C. Sabater, Two-phase and three-phase liquid–liquid equilibrium for *bis*(2-methylpropyl) ester + phosphoric acid + water, *Journal of Chemical and Engineering Data*, 39, 14–18, 1994.

M378 Moorman, A.R., S.D. Chamberlain, L.A. Jones, and T.A. Krenitsky, 5'-ester Prodrugs of the varicella-zoster antiviral agent, 6-methoxypurine arabinoside, *Antiviral Chemistry and Chemotherapy*, 3, 141–146, 1992.

M379 Muller, B.W. and E. Albers, Effect of hydrotropic substances on the complexation of sparingly soluble drugs with cyclodextrin derivatives and the influence of cyclodextrin complexation on the pharmacokinetics of the drugs, *Journal of Pharmaceutical Sciences*, 80, 599–604, 1991.

M380 Marques, H.M.C., J. Hadgraft, and I.W. Kellaway, Studies of cyclodextrin inclusion complexes. I. The salbutamol–cyclodextrin complex as studied by phase solubility and DSC, *International Journal of Pharmaceutics*, 63, 259–266, 1990.

M381 Morelock, M.M., L.L. Choi, G.L. Bell, and J.L. Wright, Estimation and correlation of drug water solubility with pharmacological parameters required for biological activity, *Journal of Pharmaceutical Sciences*, 83, 948–951, 1994.

M382 Mosharraf, M. and C. Nystrom, Solubility characterization of practically insoluble drugs using the coulter counter principle, *International Journal of Pharmaceutics*, 122, 57–67, 1995.

M383 Munoz de la Pena, A., T. Ndou, J.B. Zung, and I.M. Warner, Stoichiometry and formation constants of pyrene inclusion complexes with beta-and gamma-cyclodextrin, *Journal of Physical Chemistry*, 95, 3330–3334, 1991.

M402 Marsella, A., M. Jackolka, and S. Mabury, Aqueous solubilities, photolysis rates and partition coefficients of benzoylphenylurea insecticides, *Pest Management Science*, 56, 789–794, 2000.

M418 Modamio, P., C.F. Lastra, and E.L. Marino, Transdermal absorption of celiprolol and bisoprolol in human skin in vitro, *International Journal of Pharmaceutics*, 173, 141–148, 1998.

M438 Monene, L.M., C. Goosen, J.C. Breytenbach, and J. Plessis, Percutaneous absorption of cyclizine and its alkyl analogues, *European Journal of Pharmaceutical Sciences*, 24, 239–244, 2005.

M439 Mirmehrabi, M., S. Rohani, and L. Perry, Thermodynamic modeling of activity coefficient and prediction of solubility: Part 2. Semipredictive or semiempirical models, *Journal of Pharmaceutical Sciences*, 95, 798–809, 2006.

M440 Martinez, F. and A. Gomez, Estimation of the solubility of sulfonamidaes in aqueous media from partition coefficients and entropies of fusion, *Physics and Chemistry of Liquid*, 40, 411–420, 2002.

M444 Meyyappan, N. and N.N. Gandhi, Effect of hydrotropes on the solubility and mass transfer coefficient of benzyl benzoate in water, *Journal of Chemical and Engineering Data*, 50, 796–800, 2005.

M447 Marche, C., C. Ferronato, and J. Jose, Solubilities of alkycyclohexanes in water from 30°C to 180°C, *Journal of Chemical and Engineering Data*, 49, 937–940, 2004.

M448 Mandic, Z. and V. Gabelica, Ionization, lipophilicity and solubility properties of repaglinide, *Journal of Pharmaceutical and Biomedical Analysis*, 41, 866–871, 2006.

M456 Monnot, E.A., C.G. Kingberg, T.S. Johnson, and M. Slavik, Stability of intravenous admixtures of 5-fluoracil and spirogermanium, a novel combination of cytotoxic agents, *International Journal of Pharmaceutics*, 60, 41–52, 1990.

M457 Macheras, P.E., M.A. Koupparis, and S.G. Antimisiaris, Drug Binding and Solubility in Milk, *Pharmaceutical Research*, 7, 537–541, 1990.

M458 Modamio, P., C.F. Lastra, and E.L. Marino, A comparative in vitro study of percutaneous penetration of B-blockers in human skin, *International Journal of Pharmaceutics*, 194, 249–259, 2000.

M459 Ma, Z. and F. Zaera, Role of the solvent in the adsorption-desorption equilibrium of cinchona alkaloids between solution and a platinum surface: correlations among solvent polarity, cinchona solubility, and catalytic performance, *Journal of Physical Chemistry*, 109, 406–414, 2005.

M461 Mota, F.L., A.J. Queimada, S.P. Pinho, and E.A. Macedo, Aqueous solubility of some natural phenolic compounds, *Industrial and Engineering Chemistry Research*, 47, 5182–5189, 2008.

N001 Needham, Jr., T.E., A.N. Paruta, and R.J. Gerraughty, Solubility of amino acids in mixed solvent systems, *Journal of Pharmaceutical Sciences*, 60, 258–260, 1971.

N002 Ng, W.F. and C.F. Poe, A note on the solubility of strychnine in alcohol and water mixtures, *Journal of the American Pharmaceutical Association, Scientific Edition*, 45, 351–353, 1956.

N003 Nixon, J.R. and B.P.S. Chawla, Solubilization and rheology of the system ascorbic acid–water–polysorbate 80: temperature effects, *Journal of Pharmacy and Pharmacology*, 21, 79–84, 1969.

N004 Nelson, H.D. and C.L. De Ligny, The determination of the solubilities of some *n*-alkanes in water at different temperatures, by means of gas chromatography, *Recueil des Travaux Chimiques des Pays-Bas*, 87, 528–544, 1968.

N006 Nogami, H., T. Nagai, and H. Uchida, Physico-chemical approach to biopharmaceutical phenomena. II. Hydrophobic hydration of tryptophan in aqueous solution, *Chemical and Pharmaceutical Bulletin*, 16, 2257–2262, 1968.

N007 Nogami, H., T. Nagai, E. Fukuoka, and T. Yotsuyanagi, Dissolution kinetics of barbital polymorphs, *Chemical and Pharmaceutical Bulletin*, 17, 23–31, 1969.

N008 Nogami, H., T. Nagai, and H. Uchida, Physico-chemical approach to biopharmaceutical phenomena. IV. Adsorption of barbituric acid derivatives by carbon black from aqueous solution, *Chemical and Pharmaceutical Bulletin*, 17, 168–175, 1969.

N009 Nogami, H., T. Nagai, and H. Umeyama, Effect of third component on water structure around tryptophan in aqueous solution, *Chemical and Pharmaceutical Bulletin*, 18, 328–334, 1970.

N012 Needham, Jr., T.E., A.N. Paruta, and R.J. Gerraughty, solubility of amino acids in pure solvent systems, *Journal of Pharmaceutical Sciences*, 60, 565–567, 1971.

N013 Nex, R.W. and A.W. Swezey, Some chemical and physical properties of weed killers, *Weeds*, 3, 241–253, 1954.

N014 Nanda, A.K. and M.M. Sharma, Effective interfacial area in liquid–liquid extraction, *Chemical Engineering Science*, 21, 707–714, 1966.

N015 Neish, W.J.P., On the solubilisation of aromatic amines by purines, *Recueil des Travaux Chimiques des Pays-Bas*, 67, 361–373, 1948.

N016 Newton, D.W., W.J. Murray, and S. Ratanamaneichatara, Evaluation of solubility data useful for phase solubility determination of azathioprine, *Analytica Chemica Acta*, 135, 343–346, 1982.

N017 Nakano, M., Y. Arakawa, K. Juni, and T. Arita, Physical properties of pyrimidine and purine antimetabolites. II. Permeation of 5-fluorouracil and 1-(2-tetrahydrofuryl)-5-fluorouracil through cellophane, collagen, and silicone membranes, *Chemical and Pharmaceutical Bulletin*, 24, 2716–2722, 1976.

N018 Nakatani, H., Studies on pharmaceutical preparations of orotic acid. II. Isolation of reaction products of orotic acid and amines, and their solubility in water, *Yakugaku Zasshi* (Tokyo), 83, 6–9, 1963.

N019 Nakatani, H., Studies on pharmaceutical preparations of orotic acid. VI. Water soluble properties of orotic acid salts, *Yakugaku Zasshi* (Tokyo), 84, 1057–1061, 1964.

N021 Nasipuri, R.N. and S.A.H. Khalil, Adsorption–dissolution relationship in sulfamethazine–benzoic acid system, *Journal of Pharmaceutical Sciences*, 62, 473–475, 1973.

N023 Nogami, H., T. Nagai, and T. Yotsuyanagi, Dissolution phenomena of organic medicinals involving simultaneous phase changes, *Chemical and Pharmaceutical Bulletin*, 17, 499–509, 1969.

N024 Nozaki, Y. and C. Tanford, The solubility of amino acids, diglycine, and triglycine in aqueous guanidine hydrochloride solutions, *Journal of Biological Chemistry*, 245, 1648–1652, 1970.

N025 Nozaki, Y. and C. Tanford, The solubility of amino acids and two glycine peptides in aqueous ethanol and dioxane solutions, *Journal of Biological Chemistry*, 246, 2211–2217, 1971.

N026 Nozaki, Y. and C. Tanford, The solubility of amino acids and related compounds in aqueous ethylene glycol solutions, *Journal of Biological Chemistry*, 240, 3568–3573, 1965.

N027 Nozaki, Y. and C. Tanford, The solubility of amino acids and related compounds in aqueous urea solutions, *Journal of Biological Chemistry*, 238, 4074–4081, 1963.

N028 Nylen, P., Zur kenntnis der organischen phosphorverbindungen, II: uber beta-phosphon-propionsaure und gamma-phosphon-n-buttersaure, *Berichte der Deutschen Chemischen Gesellschaft*, 59, 1119–1123, 1926.

N031 Naggar, V., N.A. Daabis, and M.M. Motawi, Solubilization of tetracycline and oxytetracycline, *Pharmazie* (Berlin), 29, 122–125, 1974.

N032 Nang, L.S., D. Cosnier, G. Terrie, and J. Moleyre, Consequence of solubility alteration by salt effect on dissolution enhancement and biological response of a solid dispersion of an experimental antianginal drug, *Pharmacology*, 15, 545–550, 1977.

N034 Nathan, M.F., Choosing a process for chloride removal, *Chemical Engineering* (New York), 85, 93–100, 1978.

N035 Neudert, W. and H. Ropke, Uber das physikalischchemische verhalten des dinatriumsalzes des adipinsaure-bis-[2.4.6-trijod-3-carboxy-anilids] und anderer trijodbenzolderivate, *Chemische Berichte*, 87, 659–667, 1954.

N038 Niini, A., The determination of the reciprocal solubilities of water and certain organic liquids by means of a pycnometer and refractometer, *Suomen Kemistilehti A*, 11, 19–20, 1938.

N041 Negoro, H., T. Miki, and S. Ueda, Interaction between pyrazinamide and sodium *p*-aminosalicylate or sodium hydroxybenzoates in aqueous solution, *Chemical and Pharmaceutical Bulletin*, 7, 91–95, 1959.

N042 Noorduyn, A.C., Sur des hydrates d'oenanthol, *Recueil des Travaux Chimiques des Pays-Bas et de la Belgique*, 38, 345–350, 1919.

N043 Nelson, H.D. and J.H. Smit, Gas chromatographic determination of the water solubility of the halogenobenzenes, *Suid-Afrikaanse Tydskrf vir Chemie*, 31, 76–76, 1978.

N044 Nakano, N.I., N. Kawahara, T. Amiya, and D. Furukawa, 3-Hydroxy-2-naphthoates of lidocaine, mepivacaine, and bupivacaine and their dissolution characteristics, *Chemical and Pharmaceutical Bulletin*, 26, 936–941, 1978.

N045 Nakano, M., K. Juni, and T. Arita, Controlled drug permeation. I: controlled release of butamben through silicone membrane by complexation, *Journal of Pharmaceutical Sciences*, 65, 709–712, 1976.

N046 Nakano, N.I., Temperature-dependent aqueous solubilities of lidocaine, mepivacaine, and bupivacaine, *Journal of Pharmaceutical Sciences*, 68, 667–668, 1979.

N050 Niazi, S., Thermodynamics of mercaptopurine dehydration, *Journal of Pharmaceutical Sciences*, 67, 488–491, 1978.

N051 Nachod, F.C., Keto-enoltautomerien in leichten und schweren losungsmitteln, *Zeitschrift fuer Physikalische Chemie, Abteilung A: Chemische Thermodynamik, Kinetik, Electrochemie, Eigenschaftslehre*, 182, 193–219, 1938.

N053 Nango, M., H. Yamamoto, K. Joukou, and N. Kuroki, Solubility of aromatic hydrocarbons in water and aqueous solutions of sugars, *Journal of the Chemical Society, Chemical Communications*, 104–105, 1980.

N055 Newton, D.W., D.F. Driscoll, J.L. Goudreau, and S. Ratanamaneichatara, Solubility characteristics of diazepam in aqueous admixture solutions: theory and practice, *American Journal of Hospital Pharmacy*, 38, 179–182, 1981.

N056 Naggar, V.F.B., S. El-Gamal, and M.A. Shams-Eldeen, Physicochemical studies of phenylbutazone recrystallized from polysorbate 80, *Scientia Pharmaceutica*, 48, 335–343, 1980.

N057 Nicklasson, M., A. Brodin, and H. Nyqvist, Studies on the relationship between solubility and intrinsic rate of dissolution as a function of pH, *Acta Pharmaceutica Suecica*, 18, 119–128, 1981.

N059 Neira, M.C.O., J.M. Fernando, P. de Leon, and L. Fernanda, Effect of dielectric constants on the solubility of diazepam, *Revista Colombiana de Ciencias Quimico-Farmaceuticas*, 3, 37–61, 1980.

N061 Nash, N.G. and R.D. Daley, Disulfiram, *Analytical Profiles of Drug Substances*, 4, 170–177, 1975.

N062 Nishijo, J., K. Ohno, K. Nishimura, and I. Yonetani, Soluble complex formation of theophylline with aliphatic di- and monoamines in aqueous solution, *Chemical and Pharmaceutical Bulletin*, 30, 771–776, 1982.

N063 Newton, D.W., S. Ratanamaneichatara, and W.J. Murray, Dissociation, solubility and lipophilicity of azathioprine, *International Journal of Pharmaceutics*, 11, 209–213, 1982.

N301 Nyberg, L., L. Bratt, A. Forsgren, and S. Hugosson, Bioavailability of digoxin from tablets. I. In vitro characterization of digoxin tablets, *Acta Pharmaceutica Suecica*, 11, 447–458, 1974.

N302 Nurnberg, E., Neuere untersuchungsergebnisse von pharmazeutischen spruhtrocknungsprodukten, *Pharmazeutische Industrie*, 38, 228–232, 1976.

N304 Nakagawa, T. and M. Kanauchi, Isothioate, *Analytical Methods for Pesticides and Plant Growth Regulators*, 10, 75–82, 1978.

N305 Nakamura, T. and K. Yamaoka, Isoxathion, *Analytical Methods for Pesticides and Plant Growth Regulators*, 10, 83–94, 1978.

N306 Nakamura, T., K. Yamaoka, and M. Kotakemori, Hymexazol, *Analytical Methods for Pesticides and Plant Growth Regulators*, 10, 215–228, 1978.

N309 Newman, M., C.B. Hayworth, and R.E. Treybal, Dehydration of aqueous methyl ethyl ketone: equilibrium data for extractive distillation and solvent extraction, *Industrial and Engineering Chemistry*, 41, 2039–2043, 1949.

N311 Nyssen, G.A., E.T. Miller, T.F. Glass, and C.R. Quinn II, solubilities of hydrophobic compounds in aqueous–organic solvent mixtures, *Environmental Monitoring and Assessment*, 9, 1–11, 1987.

N312 Nishijo, J. and I. Yonetani, The interaction of theophylline with benzylamine in aqueous solution, *Chemical and Pharmaceutical Bulletin*, 30, 4507–4511, 1982.

N315 Newton, D.W., W.A. Narducci, W.A. Leet, and C.T. Ueda, Lorazepam solubility in and sorption from intravenous admixture solutions, *American Journal of Hospital Pharmacy*, 40, 424–427, 1983.

N316 Najib, N.M. and M.A.S. Salem, Release of ibuprofen from polyethylene glycol solid dispersions: equilibrium solubility approach, *Drug Development and Industrial Pharmacy*, 13, 2263–2275, 1987.

N317 Nielsen, N.M. and H. Bundgaard, Gycolamide esters as biolabile prodrugs of carboxylic acid agents: synthesis, stability, bioconversion, and physicochemical properties, *Journal of Pharmaceutical Sciences*, 77, 285–298, 1988.

N319 Nicklasson, M. and A. Brodin, The relationship between intrinsic dissolution rates and solubilities in the water–ethanol binary solvent system, *International Journal of Pharmaceutics*, 18, 149–156, 1984.

N320 Neely, W.B., Estimating rate constants for the uptake and clearance of chemicals by fish, *Environmental Science and Technology*, 13, 1506–1508, 1979.

N322 Nystrom, C. and M. Bisrat, Coulter counter measurements of solubility and dissolution rate of sparingly soluble compounds using micellar solutions, *Journal of Pharmacy and Pharmacology*, 38, 420–425, 1986.

N323 Nystrom, C., J. Mazur, M.I. Barnett, and M. Glazer, Dissolution rate measurements of sparingly soluble compounds with the coulter counter model TAII, *Journal of Pharmacy and Pharmacology*, 37, 217–221, 1985.

N326 Norris, J.M., J.W. Ehrmantraut, C.L. Gibbons, and J.S. Brosier, Toxicological and environmental factors involved in the selection of decabromodiphenyl oxide as a fire retardant chemical, *Applied Polymer Symposia*, 22, 195–219, 1973.

N330 Narasimhan, K.S., C.C. Reddy, and K.S. Chari, Solubility and equilibrium data of phenol–water–n-butyl acetate system at 30°C, *Journal of Chemical and Engineering Data*, 7, 340–343, 1962.

N331 Nakaishi, A., H. Maeda, T. Tomiyama, and Y. Kyogoku, Chain length dependence of solubility of monodisperse polypeptides in aqueous solutions and the stability of the beta-structure, *Journal of Physical Chemistry*, 92, 6161–6166, 1988.

N332 Narurkar, M.M. and A.K. Mitra, Synthesis, physicochemical properties, and cytotoxicity of a series of 5'-ester prodrugs of 5-iodo-2'-deoxyuridine, *Pharmaceutical Research*, 5, 734–737, 1988.

N333 Nkedi-Kizza, P., M.L. Brusseau, P.S.C. Rao, and A.G. Hornsby, nonequilibrium sorption during displacement of hydrophobic organic chemicals and 45-ca through soil columns with aqueous and mixed solvents, *Environmental Science and Technology*, 23, 814–820, 1989.

N334 Nakagawa, Y., S. Itai, T. Yoshida, and T. Nagai, Physicochemical properties and stability in the acidic solution of a new macrolide antibiotic, clarithromycin, in comparison with erythromycin, *Chemical and Pharmaceutical Bulletin*, 40, 725–728, 1992.

N335 Nielsen, N.M. and H. Bundgaard, Evaluation of glycolamide esters and various other esters of aspirin as true aspirin prodrugs, *Journal of Medicinal Chemistry*, 32, 727–734, 1989.

N336 Natarajan, A., V. Sapre, U.B. Hadkar, and P.Y. Shirodkar, The correlation of pK_a with biological activity of substituted salicylanide derivatives, *Indian Drugs*, 29, 545–552, 1992.

N337 Nielsen, L.S., F. Slok, and H. Bundgaard, N-alkoxycarbonyl prodrugs of mebendazole with increased water solubility, *International Journal of Pharmaceutics*, 102, 231–239, 1994.

N417 Namor, A.F.D., J.A. Zvietcovich-Guerra, V. Grachev, and F.J. Suerros-Velarde, Partition and transfer of chlorophenoxy acids (herbicides) in water-non-aqueous media, *New Journal of Chemistry*, 29, 1072–1076, 2005.

N419 Norstrom, F.L. and A.C. Rasmuson, Solubility and melting properties of salicylamide, *Journal of Chemical and Engineering Data*, 51, 1775–1777, 2006.

N420 Norstrom, F.L. and A.C. Rasmuson, Solubility and melting properties of salicylic acid, *Journal of Chemical and Engineering Data*, 51, 1668–1671, 2006.

O001 Ostrenga, J.A. and C. Steinmetz, Estimation of steroid solubility: use of fractional molar attraction constants, *Journal of Pharmaceutical Sciences*, 59, 414–416, 1970.

O002 O'Laughlin, J.W., F.W. Sealock, and C.V. Banks, Determination of solubility of several phosphine oxides in aqueous solutions using a new spectrophotometric procedure, *Analytical Chemistry*, 34, 224–226, 1964.

O003 Osol, A. and M. Kilpatrick, 55, 4440–4444, 1933.

O004 Osol, A. and M. Kilpatrick, 55, 4430–4440, 1933.

O005 Othmer, D.F., R.E. White, and E. Trueger, Liquid–liquid extraction data, *Industrial and Engineering Chemistry*, 33, 1240–1248, 1941.

O006 O'Connell, W.L., Properties of heavy liquids, *Transactions of Society of Mining Engineers*, 226, 126–132, 1963.

O007 Okano, T., K. Uekama, and K. Ikeda, Electronic properties of *n*-heteroaromatics XVI. Charge transfer properties of pyrazolone antipyretics. On the complex formation of aminopyrine with benzoic acid and salicylic acid, *Chemical and Pharmaceutical Bulletin*, 16, 6–12, 1968.

O009 Ouellette, R.P. and J.A. King, *Chemical Week Pesticides Register*, 1977.

O011 Othmer, D.F. and J. Serrano, Jr., Solubility data for ternary liquid systems: systems of acetic acid, higher boiling homologous acids, and water, *Industrial and Engineering Chemistry*, 41, 1030–1032, 1949.

O012 Othmer, D.F., W.S. Bergen, N. Shlechter, and P.F. Bruins, Liquid–liquid extraction data: systems used in butadien manufacture from butylene glycol, *Industrial and Engineering Chemistry*, 37, 890–894, 1945.

O013 Ostergren, G. and A. Levan, The connection between c-mitotic activity and water solubility in some monocyclic compounds, *Hereditas*, 29, 496–498, 1943.

O015 Olsen, A.L. and E.R. Washburn, Study of solutions of isopropyl alcohol in benzene, in water and in benzene and water, *Journal of the American Chemical Society*, 57, 303–305, 1935.

O016 Oliveri-Mandala, E. and F. Forni, Influenze di solubilita. (acetanilide-antipirina, acetanilide-piramidone). Nota IV, *Gazzetta Chimica Italiana*, 55, 783–788, 1925.

O017 Oliveri-Mandala, E. and L. Irrera, Influenze di solubilitia (coppie: tiourea-antipirina, caffeina-antipirina). Nota VII, *Gazzetta Chimica Italiana*, 60, 872–877, 1930.

O018 Oliveri-Mandala, E., Influenze di solubilita (coppie: cloralio-caffeina, urotropina-antipirina, urotropina-cloralio). Nota V, *Gazzetta Chimica Italiana*, 56, 889–896, 1926.

O019 Oliveri-Mandala, E., Influenze di solubilita (constituzione chimica e solubilita). Nota VI, *Gazzetta Chimica Italiana*, 56, 896–901, 1926.

O021 Odaira, I., Synthesis of the different acylurethanes and some allied compounds, *The Influence of the Acid Radicals*, 1, 324–330, 1915.

O025 Osteryoung, J. and J.W. Whittaker, Picloram: solubility and acid–base equilibria determined by normal pulse polarography, *Journal of Agricultural and Food Chemistry*, 28, 95–97, 1980.

O026 Othmer, D.F. and P.L. Ku, Solubility data for ternary liquid systems acetic acid and formic acid distributed between chloroform and water, *Journal of Chemical and Engineering Data*, 5, 42–44, 1960.

O027 O'Neill, J.J., P.L. Peelor, A.F. Peterson, and C.H. Strube, selection of parabens as preservatives for cosmetics and toiletries, *Journal of the Society of Cosmetic Chemistry*, 30, 25–38, 1979.

O028 Othmer, D.F., M.M. Chudgar, and S.L. Levy, Binary and ternary systems of acetone, methyl ethyl ketone, and water, *Industrial and Engineering Chemistry*, 44, 1872–1881, 1952.

O032 Ojile, J.E., C.B. Macfarlane, and A.B. Selkirk, Drug distribution during massing and its effect on dose uniformity in granules, *International Journal of Pharmaceutics*, 10, 99–107, 1982.

O300 Ottnad, M., N.A. Jenny, and C.-H. Roder, methyl isothiocyanate, *Analytical Methods for Pesticides and Plant Growth Regulators*, 10, 563–573, 1978.

O302 Ochsner, A.B., R.J. Belloto, Jr., and T.D. Sokoloski, prediction of xanthine solubilities using statistical techniques, *Journal of Pharmaceutical Sciences*, 74, 132–135, 1985.

O303 Otagiri, M., T. Imai, F. Hirayama, and K. Uekama, Inclusion complex formations of the antiinflammatory drug flurbiprofen with cyclodextrins in aqueous solution and in solid state, *Acta Pharmaceutica Suecica*, 20, 11–20, 1983.

O304 Okada, Y., S. Horiyama, and K. Koizumi, Studies on inclusion complexes of non-steroidal anti-inflammatory agents with cyclosophoraose-A, *Yakugaku Zasshi* (Tokyo), 106, 240–247, 1986.

O305 Otsuka, M. and N. Kaneniwa, Hygroscopicity and solubility of noncrystalline cephalexin, *Chemical and Pharmaceutical Bulletin*, 31, 230–236, 1983.

O306 Obikili, A., M. Deyme, D. Wouessidjewe, and D. Duchene, Improvement of aqueous solubility and dissolution kinetics of canrenone by solid dispersion in sucroester, *Drug Development and Industrial Pharmacy*, 14, 791–803, 1988.

O307 Othman, S., H. Muti, O. Shaheen, and W.A. Al-Turk, Studies on the adsorption and solubility of nalidixic acid, *International Journal of Pharmaceutics*, 41, 197–203, 1988.

O310 Ogston, A.G., Some dissociation constants, *Journal of the Chemical Society* (London), 1713–1713, 1936.

O311 Opperhuizen, A., F.A.P. Gobas, J.M.D. Van der Steen, and O. Hutzinger, Aqueous solubility of polychlorinated biphenyls related to molecular structure, *Environmental Science and Technology*, 22, 638–646, 1988.

O312 Owens, J.W., S.P. Wasik, and H. DeVoe, Aqueous solubilities and enthalpies of solution of *n*-alkylbenzenes, *Journal of Chemical and Engineering Data*, 31, 47–51, 1986.

O316 Orella, C.J. and D.J. Kirwan, The solubility of amino acids in mixtures of water and aliphatic alcohols, *Biotechnology Progress*, 5, 89–91, 1989.

O317 Orella, C.J. and D.J. Kirwan, Correlation of amino acid solubilities in aqueous aliphatic alcohol solutions, *Industrial and Engineering Chemistry, Product Research and Development*, 30, 1040–1045, 1991.

O318 Ong, J.T.H. and E. Manoukian, Micellar solubilization of timobesone acetate in aqueous and aqueous propylene glycol solutions of nonionic surfactants, *Pharmaceutical Research*, 5, 704–708, 1988.

O319 Oh, I., S.C. Chi, B.R. Vishnuvajjala, and B.D. Anderson, Stability and solubilization of oxathiin carboxanilide, a novel anti-HIV agent, *International Journal of Pharmaceutics*, 73, 23–31, 1991.

O320 Orazio, C.E., S. Kapila, R.K. Puri, and A.F. Yanders, Persistence of chlorinated dioxins and furans in the soil environment, *Chemosphere*, 25, 1469–1474, 1992.

O321 Okada, Y., Y. Kubota, K. Koizumi, and K. Ogata, Some properties and the inclusion behavior of branched cyclodextrins, *Chemical and Pharmaceutical Bulletin*, 36, 2176–2185, 1988.

O322 O'Reilly, J.R., O.I. Corrigan, and C.M. O'Driscoll, The effect of mixed micellar systems, bile salt/fatty acids, on the solubility and intestinal absorption of clofazimine (B663) in the anaesthetised rat, *International Journal of Pharmaceutics*, 109, 147–154, 1994.

O406 Oleszek-Kudlak, S., E. Shibata, and T. Nakamura, Solubilities of selected PCDDs and PCDFs in water and various chloride solutions, *Journal of Chemical and Engineering Data*, 2007.

P003 Polak, J. and B. Lu, Mutual solubilities of hydrocarbons and water at and 25°C, *Canadian Journal of Chemistry*, 51, 4018–4023, 1973.

P004 Parsons, G.H., C.H. Rochester, A. Rostron, and P.C. Sykes, The thermodynamics of hydration of phenols, *Journal of Chemical Social Perkin*, 2, 136–138, 1972.

P005 Pryor, W.A. and R.E. Jentoft, Jr., Solubility of *m*- and *p*-xylene in water and in aqueous ammonia from to 300°C, *Journal of Chemical and Engineering Data*, 6, 36–37, 1961.

P006 Peterson, C.F. and R.E. Hopponen, Solubility of phenobarbital in propylene glycol–alcohol–water systems, *Journal of the American Pharmaceutical Association, Scientific Edition*, 42, 540–541, 1953.

P007 Price, L., The solubility of hydrocarbons and petroleum in water as applied to the primary migration of petroleum, PhD Thesis, 60–261, 1973.

P008 Poulsen, B.J., E. Young, V. Coquilla, and M. Katz, Effect of topical vehicle composition on the in vitro release of fluocinolone acetonide and its acetate ester, *Journal of Pharmaceutical Sciences*, 57, 928–933, 1968.

P009 Poole, J.W. and C.K. Bahal, Dissolution behavior and solubility of anhydrous and trihydrate forms of ampicillin, *Journal of Pharmaceutical Sciences*, 57, 1945–1948, 1968.

P010 Paruta, A.N. and S.A. Irani, Solubility profiles for the xanthines in aqueous solutions of a glycol ether. II, *Journal of Pharmaceutical Sciences*, 55, 1060–1064, 1966.

P011 Paruta, A.N. and S.A. Irani, Solubility profiles for the xanthines in aqueous alcoholic mixtures. I. Ethanol and methanol, *Journal of Pharmaceutical Sciences*, 55, 1055–1059, 1966.

P012 Paruta, A.N., Solubility profiles for antipyrine and aminopyrine in hydroalcoholic solutions, *Journal of Pharmaceutical Sciences*, 56, 1565–1569, 1967.

P013 Paruta, A.N. and B.B. Sheth, Solubility of parabens in syrup vehicles, *Journal of Pharmaceutical Sciences*, 55, 1208–1211, 1966.

P014 Paruta, A.N., B.J. Sciarrone, and N.G. Lordi, Solubility of salicylic acid as a function of dielectric constant, *Journal of Pharmaceutical Sciences*, 53, 1349–1353, 1964.

P015 Paruta, A.N., Solubility of several solutes as a function of the dielectric constant of sugar solutions, *Journal of Pharmaceutical Sciences*, 53, 1252–1254, 1964.

P016 Paruta, A.N. and S.A. Irani, Dielectric solubility profiles in dioxane–water mixtures for several antipyretic drugs, *Journal of Pharmaceutical Sciences*, 54, 1334–1338, 1965.

P018 Paruta, A.N., B.J. Sciarrone, and N.G. Lordi, Solubility profiles for the xanthines in dioxane–water mixtures, *Journal of Pharmaceutical Sciences*, 54, 838–841, 1965.

P019 Patel, N.K. and H.B. Kostenbauder, Interaction of preservatives with macromolecules. I. Binding of parahydroxyethylene 80 sorbitan monooleate (Tween 80), *Journal of the American Pharmaceutical Association, Scientific Edition*, 47, 289–293, 1958.

P020 Paruta, A.N. and B.B. Sheth, Solubility of the xanthines, antipyrine, and several derivatives in syrup vehicles, *Journal of Pharmaceutical Sciences*, 55, 896–901, 1966.

P022 Plakogiannis, F.M. and P. Catsoulakos, Solubility behavior of 3-beta-hydroxy-13-alpha-amino-13, 17-seco-5-alpha-androstan-17-oic-13,17-lactam-4-*n,n-bis*-(chloroethyl) amino phenyl-acetate, a new anti-cancer agent, *Pharmaceutica Acta Helvetiae*, 51, 249–252, 1976.

P023 Pinck, L.A. and M.A. Kelly, The solubility of urea in water, *Journal of the American Chemical Society*, 47, 2170–2172, 1925.

P024 Phillips, J.P. and H.P. Price, Spectrophotometric study of 8-hydroxyquinaldine chelates. notes on 8-quinolinol chelates, *Journal of the American Chemical Society*, 73, 4414–4415, 1951.

P026 Pulver, R., B. Exer, and B. Herrmann, Uber die beeinflussung enzymatischer reaktionen durch phenylbutazon und die ubertragbarkeit fermentchemischer befunde auf die stoffwechselprozesse der zelle, *Schweizerische Medizinische Wochenschrift*, 86, 1080–1085, 1956.

P027 Park, K.S. and W.N. Bruce, The determination of the water solubility of aldrin, dieldrin, heptachlor, and heptachlor epoxide, *Journal of Economic Entomology*, 61, 770–774, 1968.

P028 Probst, G.W. and J.B. Tepe, Trifluralin and related compounds, in *Degradation of Herbicides*, eds. P.C. Kearney and D.D. Kaufman, Dekker, New York, 225–257, 1969.

P029 Pedersen, C.J., Cyclic polyethers and their complexes with metal salts, *Journal of the American Chemical Society*, 89, 7017–7036, 1967.

P031 Pinney, R.J. and V. Walters, The relation between the bactericidal activities and certain physico-chemical properties of some fluorophenols, *Journal of Pharmacy and Pharmacology*, 21, 415–422, 1969.

P033 Paul, M.F., R.C. Bender, and E.G. Nohle, Renal excretion of nitrofurantoin (furadantin), *American Journal of Physiology*, 197, 580–584, 1959.

P034 Paul, M.F., C. Harrington, R.C. Bender, and M.H. Bryson, Effect of pH and of urea on nitrofurantoin activity, *Proceedings of the Society for Experimental Biology and Medicine*, 125, 941–947, 1967.

P035 Poole, J.W. and C.K. Bahal, Dissolution behavior and solubility of anhydrous and dihydrate forms of Wy-4508, and aminoalicyclic penicillin, *Journal of Pharmaceutical Sciences*, 59, 1265–1267, 1970.

P036 Pearson, J.T. and G. Varney, The anomalous behaviour of some oxyclozanide polymorphs, *Journal of Pharmacy and Pharmacology*, 25, 62–70, 1973.

P037 Philip, J.C. and F.B. Garner, Influence of various sodium salts on the solubility of sparingly soluble acids. Part II, *Journal of the Chemical Society* (London), 95, 1466–1473, 1909.

P038 Philip, J.C. and R.S. Colborne, The solubility of anilinesulphonic acids, *Journal of the Chemical Society* (London), 125, 492–500, 1924.

P040 Pasquinelli, E.A., Correlation of the mutual solubilities of two liquids with the electric and magnetic properties of the pure components, *Transactions of the Faraday Society*, 53, 932–938, 1956.

P041 Powell, J.F., The solubility or distribution coefficient of trichlorethylene in water, whole blood, and plasma, *British Journal of Industrial Medicine*, 4, 233–236, 1947.

P043 Philip, J.C., Influence of various sodium salts on the solubility of sparingly soluble acids, *Journal of the Chemical Society* (London), 87, 987–1003, 1905.

P044 Parker, M.D., The influence of anions on the physical properties of butyrophenone-type molecules, *Dissertation Abstracts of International*, 35, 34–68, 1974.

P045 Pfeiffer, P. and O. Angern, Das aussalzen der aminosaeuren, *Zeitschrift fuer Physiologische Chemie*, 133, 180–192, 1924.

P046 Pearson, C.R. and G. McConnell, Chlorinated C1 and C2 hydrocarbons in the marine environment, *Proceedings of the Royal Society of London, Series B:Biological Sciences*, 189, 305–332, 1975.

P048 Pringsheim, H. and D. Dernikos, Weiteres uber die polyamylosen. (Beitrage zur chemie der starke, VI.), *Berichte der Deutschen Chemischen Gesellschaft*, 55, 1433–1449, 1922.

P049 Pucher, G. and W.M. Dehn, Solubilities in mixtures of two solvents, *Journal of the American Chemical Society*, 43, 1753–1758, 1921.

P051 Price, L.C., Aqueous solubility of petroleum as applied to its origin and primary migration, *The American Association of Petroleum Geologists Bulletin*, 60, 213–244, 1976.

P052 Peter, P.N., Solubility relationships of lactose-sucrose solutions. I. Lactose-sucrose solubilities at low temperatures, *Journal of Physical Chemistry*, 32, 1856–1864, 1928.

P053 Poelman, M.-C., F. Puisieux, and J.-C. Chaumeil, Interactions between antiseptics and surfactants. II. Study by the solubility method of the interaction of methyl-parahydroxy-benzoate with polyoxyethylene fatty alcohol ethers, *Annals Pharmaceutiques Francaises*, 33, 551–557, 1975.

P054 Petrov, B.I., V.P. Zhivopistsev, I.A. Kislitsyn, and M.A. Volkova, Solubility of diantipyrylmethanes in aqueous solution and in organic solvents, *Journal of Analytical Chemistry of the USSR*, 32, 1180–1186, 1977.

P055 Park, J.G. and H.E. Hofmann, Aliphatic ketones as solvents, *Industrial and Engineering Chemistry*, 24, 132–134, 1932.

P057 Pedersen, K.J., Studies of complex formation between aniline and picrate ion by solubility measurements, *Journal of the American Chemical Society*, 56, 2615–2619, 1934.

P059 Etude physico-chimique des melanges d'eau, d'aldehyde et de paraldehyde, *Bulletin de la Societe Chimique de France*, 27, 353–362, 1920.

P060 Patterson, W.H., Estimation of deuterium oxide–water mixtures. Part II. The solubility curves with *n*-butyric acid and with isobutyric acid, *Journal of the Chemical Society* (London), 1559–1561, 1938.

P061 Pippenger, C.E., Physiochemical and pharmacological properties of antiepileptic drugs (Appendix 1), in *Antiepileptic Drugs: Quantitative Analysis and Interpretation*, eds. C.E. Pippenger, J.K. Penry, and H. Kutt, Raven Press, New York, 321-333, 1978.

P064 Pleuger, G., Beitrage zur kenntnis der loslichkeit in flussigkeitsgemischen, *Pleuger, Zur Kenntnis der Loslichkeit in Flussigkeisgemischen*, 26, 167–170, 1925.

P065 Palitzsch, T., Studien uber die oberflachenspannung von losungen. IV. Uber den gegenseitigen einfluss von urethan und salzen auf ihr losungsvolum und ihre loslichkeit in wasser, *Zeitschrift fuer Physikalische Chemie, Abteilung A: Chemische Thermodynamik, Kinetik, Electrochemie, Eigenschaftslehre*, 145, 97–108, 1929.

P067 Poochikian, G.K. and J.C. Cradock, Enhanced chartreusin solubility by hydroxybenzoate hydrotropy, *Journal of Pharmaceutical Sciences*, 68, 728–732, 1979.

P068 Peck, C.C. and L.Z. Benet, General method for determining macrodissociation constants of polyprotic, amphoteric compounds from solubility measurements, *Journal of Pharmaceutical Sciences*, 67, 12–16, 1978.

P070 Pulley, G.N., Solubility of naringin in water, *Industrial and Engineering Chemistry, Analytical Edition*, 8, 360–360, 1936.

P073 Pavlovskaya, E.M., A.K. Charykov, and V.I. Tikhomirov, pH-potentiometric determination of the solubility of sparingly soluble organic extractants in water and aqueous solutions of neutral salts, *Journal of General Chemistry of the USSR*, 47, 2230–2234, 1977.

P076 Pomianowski, A. and J. Leja, Spectrophotometric study of xanthate and dixanthogen solutions, *Canadian Journal of Chemistry*, 41, 2219–2228, 1963.

P077 Petritis, V.E. and C.J. Geankoplis, Phase equilibria in 1-butanol–water–lactic acid system, *Journal of Chemical and Engineering Data*, 4, 197–198, 1959.

P081 Peachey, J.E., Chemical control of plant parasitic nematodes in the United Kingdom, *Chemistry and Industry* (London), 1736–1740, 1963.

P085 Paris, D.F., D.L. Lewis, and J.T. Barnett, Bioconcentration of toxaphene by microorganisms, *Bulletin of Environmental Contamination and Toxicology*, 17, 564–572, 1977.

P089 Plakogiannis, F.M., C. Iordanides, and C. Siakali-Kiolafa, Solubility studies of certain 2,3 quinolinophthalides, *Drug Development and Industrial Pharmacy*, 6, 61–75, 1980.

P091 Pitre, D. and E. Felder, Development, chemistry, and physical properties of iopamidol and its analogues, *Investigative Radiology*, 15, 301–309, 1980.

P094 Patel, D.M., A.J. Visalli, J.J. Zalipsky, and N.H. Reavey-Cantwell, Methaqualone, *Analytical Profiles of Drug Substances*, 4, 245–255, 1975.

P096 Patel, M.S., P.H. Elworthy, and A.K. Dewsnup, Solubilisation of drugs in nonionic surfactants, *Journal of Pharmacy and Pharmacology*, 63, 64–64, 1981.

P303 Paruta, A.N., Thermodynamics of aqueous solutions of alkyl *p*-aminobenzoate, *Drug Development and Industrial Pharmacy*, 10, 453–465, 1984.

P307 Pease, H.L., R.E. Leitch, and O.R. Hunt, Terbacil, *Analytical Methods for Pesticides and Plant Growth Regulators*, 10, 483–492, 1978.

P311 Pfeiffer, R.R., K.S. Yang, and M.A. Tucker, Crystal pseudopolymorphism of cephaloglycin and cephalexin, *Journal of Pharmaceutical Sciences*, 59, 1809–1814, 1970.

P312 Pitha, J., J. Milecki, H. Fales, and K. Uekama, Hydroxypropyl-beta-cyclodextrin: preparation and characterization; effects on solubility of drugs, *International Journal of Pharmaceutics*, 29, 73–82, 1986.

P313 Patterson, D. and R.P. Sheldon, The dyeing of polyester fibers with disperse dyes: mechanism and kinetics of the process for purified dyes, *Transactions of the Faraday Society*, 55, 1254–1264, 1959.

P314 Parrott, E.L., M. Simpson, and D.R. Flanagan, Dissolution kinetics of a three-component solid. II: benzoic acid, salicylic acid, and salicylamide, *Journal of Pharmaceutical Sciences*, 72, 765–766, 1983.

P315 Pascoe, P.F. and W.A. Sherbrock-Cox, The reaction between anhydrous ethyleneimine and water, *Journal of Applied Chemistry*, 13, 564–572, 1963.

P321 Pryanikova, R.O. and I.A. Markina, Liquid–liquid equilibrium in the benzonitrile–water and benzonitrile–water–ammonia systems. Generalised Sechenov equation, *Russian Journal of Physical Chemistry*, 59, 1306–1308, 1985.

P323 Platford, R.F., The octanol–water partitioning of some hydrophobic and hydrophilic compounds, *Chemosphere*, 12, 1107–1111, 1983.

P324 Bruck, S.D., Chapter 5. Molecular encapsulation of drugs by cyclodextrins and congeners, *Controlled Drug Delivery*, 1, 125–148, CRC Press, Boca Raton, 1983.

P325 Prankerd, R.J., S.G. Frank, and V.J. Stella, Preliminary development and evaluation of a parenteral emulsion formulation of penclomedine (NSC-338720; 3,5-dichloro-2,4-dimethoxy-6-trichloromethylpyridine): a novel, practically water insoluble cytotoxic agent, *Journal of Parenteral Science and Technology*, 42, 76–81, 1988.

P326 Pitman, I.H., Three chemical approaches towards the solubilisation of drugs: control of enantiomer composition, salt selection, and pro-drug formation, *Australian Journal of Pharmaceutical Sciences*, 17–19, 1976.

P329 Pal, A. and S.C. Lahiri, Solubility and the thermodynamics of transfer of benzoic acid in mixed solvents, *Indian Journal of Chemistry*, 28A, 276–279, 1989.

P331 Perez-Tejeda, P., C. Yanes, and A. Maestre, Solubility of naphthalene in water + alcohol solutions at various temperatures, *Journal of Chemical and Engineering Data*, 35, 244–246, 1990.

P332 Pinal, R., P.S.C. Rao, L.S. Lee, and P.V. Cline, Cosolvency of partially miscible organic solvents on the solubility of hydrophobic organic chemicals, *Environmental Science and Technology*, 24, 639–647, 1990.

P335 Panneman, H.J. and A.A.C. Beenackers, Solvent effects on the hydration of cyclohexene catalyzed by a strong acid ion exchange resin. I. Solubility of cyclohexene in aqueous sulfolane mixtures, *Industrial and Engineering Chemistry, Product Research and Development*, 31, 1227–1231, 1992.

P336 Prankerd, R.J. and V.J. Stella, The use of oil-in-water emulsions as a vehicle for parenteral drug administration, *Journal of Parenteral Science and Technology*, 44, 139–149, 1990.

P339 Pinal, R., L.S. Lee, and P.S.C. Rao, Prediction of the solubility of hydrophobic compounds in nonideal solvent mixtures, *Chemosphere*, 22, 939–951, 1991.

P340 Peters, C.A. and R.G. Luthy, Coal tar in water-miscible solvents: experimental evaluation, *Environmental Science and Technology*, 27, 2831–2843, 1993.

P342 Piasecki, A., Chemical structure and surface activity. Part XXI. The amphiphilic properties of 2-(2-alkoxyethyl)-2-methyl-4-hydroxymethyl-1-1,3-dioxolanes, *Colloids and Surfaces*, 36, 383–390, 1989.

P343 Palmieri, G.F., P. Wehrle, G. Duportail, and A. Stamm, Inclusion complexation of vitamin A palmitate with b-cyclodextrin in aqueous solution, *Drug Development and Industrial Pharmacy*, 18, 2117–2121, 1992.

P344 Puglisi, G., N.A. Santagati, R. Pignatello, and G. Mazzone, Inclusion complexation of 4-biphenylacetic acid with b-cyclodextrin, *Drug Development and Industrial Pharmacy*, 16, 395–413, 1990.

P348 Pedersen, M., M. Edelsten, V.F. Nielsen, and C. Slot, Formation and antimycotic effect of cyclodextrin inclusion complexes of econazole and miconazole, *International Journal of Pharmaceutics*, 90, 247–254, 1993.

P349 Pinho, S.P., C.M. Silva, and E.A. Macedo, Solubility of amino acids: a group-contribution model involving phase and chemical equilibria, *Industrial and Engineering Chemistry, Product Research and Development*, 33, 1341–1347, 1994.

P350 Prankerd, R.J. and R.H. McKeown, Physico-chemical properties of barbituric acid derivatives: IV. Solubilities of 5,5-disubstituted barbituric acids in water, *International Journal of Pharmaceutics*, 112, 1–15, 1994.

P351 Plchopin, N.L., W.N. Charman, and V.J. Stella, Amino acid derivatives of dapsone as water-soluble prodrugs, *International Journal of Pharmaceutics*, 121, 157–167, 1995.

P430 Perdue, J.E., S.G. Pavlostathis, and R. Araujo, Physicochemical properties of selected monoterpenes, *Environment International*, 24, 353–358, 1998.

P432 Park, S.-H. and H.-K. Choi, The effects of surfactants on the dissolution profiles of poorly water-soluble acidic drugs, *International Journal of Pharmaceutics*, 321, 35–41, 2006.

P433 Perlovich, G.L., N.N. Strakhova, V.P. Kazachenko, and O.A. Raevsky, Studying thermodynamic aspects of sublimation, solubility and solvation processes and crystal structure analysis of some sulfonamides, *International Journal of Pharmaceutics*, 334, 115–124, 2007.

P434 Phyu, Y.L., M.St.J Warne, and R.P. LiM, Toxicity and bioavailability of atrazine and molinate to the freshwater fish (*Melonotenia fluviatilis*) under laboratoty and simulated field conditions, *Science of the Total Environment*, 356, 86–99, 2006.

P438 Perlovich, G.L., A.O. Surov, L.K.R. Hansen, and A. Bauer-Brandl, Energetics aspects of diclofenac acid in crystal modifications and in solutions—mechanism of solvation, partitioning and distribution, *Journal of Pharmaceutical Sciences*, 96, 1031–1042, 2007.

R001 Ralston, A.W. and C.W. Hoerr, The solubilities of the normal saturated fatty acids, *Journal of Organic Chemistry*, 7, 546–555, 1942.

R002 Robb, I.D., Determination of the aqueous solubility of fatty acids and alcohols, *Australian Journal of Chemistry*, 19, 2281–2284, 1966.

R003 Richardson, N.E. and B.J. Meakin, The influence of cosolvents and substrate substituents on the sorption of benzoic acid derivatives by polyamides, *Journal of Pharmacy and Pharmacology*, 27, 145–151, 1975.

R004 Rogers, J.A. and J.G. Nairn, Solubility and dielectric constant correlations of the systems chloramphenicol palmitate–propylene glycol–water, *Canadian Journal of Pharmaceutical Science*, 8, 75–77, 1973.

R006 Rehberg, C.E. and M.B. Dixon, n-Alkyl lactates and their acetates, *Journal of the American Chemical Society*, 72, 1918–1922, 1950.

R007 Rehberg, C.E. and M.B. Dixon, Mixed esters of lactic and carbonic acids, n-alkyl carbonates of methyl and butyl lactates, and butyl carbonates of n-alkyl lactates, *Journal of Organic Chemistry*, 15, 565–571, 1950.

R009 Reddy, R.K., S.A. Khalil, and M.W. Gouda, Dissolution characteristics and oral absorption of digitoxin and digoxin coprecipitates, *Journal of Pharmaceutical Sciences*, 65, 1753–1758, 1976.

R010 Rebagay, T. and P. DeLuca, Correlation of dielectric constant and solubilizing properties of tetramethyldicarboxamides, *Journal of Pharmaceutical Sciences*, 65, 1645–1647, 1976.

R016 Randall, M. and C.F. Failey, The activity coefficient of the undissociated part of weak electrolytes, *Chemical Reviews*, 4, 291–318, 1927.

R017 Repta, A.J., M.J. Baltezor, and P.C. Bansal, Utilization of an enantiomer as a solution to a pharmaceutical problem: application to solubilization of 1,2-di(4-piperazine-2,6-dione)propane, *Journal of Pharmaceutical Sciences*, 65, 238–242, 1976.

R023 Rudy, B.C. and B.Z. Senkowski, Fluorouracil, *Analytical Profiles of Drug Substances*, 2, 221–242, 1973.

R024 Rudy, B.C. and B.Z. Senkowski, Isocarboxazid, *Analytical Profiles of Drug Substances*, 2, 295–314, 1973.

R025 Rudy, B.C. and B.Z. Senkowski, Sulfamethoxazole, *Analytical Profiles of Drug Substances*, 2, 467–486, 1973.

R027 Rudy, B.C. and B.Z. Senkowski, Methyprylon: an analytical profile, *Analytical Profiles of Drug Substances*, 2, 363–382, 1973.

R028 Raventos, J., The action of fluothane—a new volatile anaesthetic, *British Journal of Pharmacology*, 11, 394–410, 1956.

R030 Repta, A.J., B.J. Rawson, R.D. Shaffer, and T. Higuchi, Rational development of a soluble prodrug of a cytotoxic nucleoside: preparation and properties of arabinosyladenine 5'-formate, *Journal of Pharmaceutical Sciences*, 64, 392–396, 1975.

R034 Rehberg, C.E., M.B. Dixon, and C.H. Fisher, Preparation and physical properties of n-alkyl beta-n-alkoxypropionates, *Journal of the American Chemical Society*, 69, 2966–2970, 1947.

R036 Robinson, D.R. and W.P. Jencks, The effect of compounds of the urea-guanidinium class on the activity coefficient of acetyltetraglycine ethyl ester and related compounds, *Journal of the American Chemical Society*, 87, 2462–2470, 1965.

R037 Ralston, A.W., C.W. Hoerr, and E.J. Hoffman, Studies on high molecular weight aliphatic amines and their salts. VII. The systems octylamine–, dodecylamine– and octadecylamine–water, *Journal of the American Chemical Society*, 64, 1516–1523, 1942.

R039 Robinson, D.R. and M.E. Grant, The effects of aqueous salt solutions on the activity coefficients of purine and pyrimidine bases and their relation to the denaturation of deoxyribonucleic acid by salts, *Journal of Biological Chemistry*, 241, 4030–4042, 1966.

R041 Roberts, M.S., R.A. Anderson, and J. Swarbrick, Permeability of human epidermis to phenolic compounds, *Journal of Pharmacy and Pharmacology*, 29, 677–683, 1977.

R042 Roseman, M. and W.P. Jencks, Interactions of urea and other polar compounds in water, *Journal of the American Chemical Society*, 97, 631–640, 1975.

R044 Rose, F.L., A.R. Martin, and H.G.L. Bevan, Sulphamethazine (2-4'-aminobenzenesulphonylamino-4:6-dimethylpyrimidine) a new heterocyclic derivative of sulphanilamide, *Journal of Pharmacology and Experimental Therapeutics*, 77, 127–141, 1943.

R045 Roblin, Jr., R.O., J.H. Williams, P.S. Winnek, and J.P. English, Chemotherapy. II. Some sulfanilamido heterocycles, *Journal of the American Chemical Society*, 62, 2002–2005, 1940.

R046 Roblin, Jr., R.O., P.S. Winnek, and J.P. English, Studies in chemotherapy. IV. Sulfanilamidopyrimidines, *Journal of the American Chemical Society*, 64, 567–570, 1942.

R047 Ruigh, W.L. and A.E. Erickson, The variation of the oil–water distribution ratio of divinyl ether with concentration, *Anesthesiology*, 2, 546–551, 1941.

R048 Rauws, A.G., M. Olling, and A.E. Wibowo, The determination of fluorochlorocarbons in air and body fluids, *Journal of Pharmacy and Pharmacology*, 25, 718–722, 1973.

R049 Ross, J.D.M. and T.J. Morrison, Acid salts of monobasic organic acids. Part II, *Journal of the Chemical Society* (London), 867–872, 1936.

R058 Roblin, Jr., R.O. and P.S. Winnek, Chemotherapy. I. Substituted sulfanilamidopyridines, *Journal of the American Chemical Society*, 62, 1999–2001, 1940.

R060 Robbins, B.H., Studies of Cyclopropane I. The quantitative determination of cyclopropane in air, water, and blood by means of iodine pentoxide, *Journal of Pharmacology and Experimental Therapeutics*, 58, 243–259, 1936.

R063 Rice, P.A., R.P. Gale, and A.J. Barduhn, Solubility of butane in water and salt solutions at low temperatures, *Journal of Chemical and Engineering Data*, 21, 204–206, 1976.

R067 Rao, K.S., M.V.R. Rao, and C.V. Rao, Ternary liquid equilibria: acetone–water-n-heptanol and acetone–water-n-octanol systems, *Journal of Scientific and Industrial Research*, 20B, 283–286, 1961.

R069 Resetarits, D.E., K.C. Cheng, B.A. Bolton, and T.R. Bates, Dissolution behavior of 17-beta-estradiol (E2) from povidone coprecipitates, comparison with microcrystalline and macrocrystalline E2, *International Journal of Pharmaceutics*, 2, 113–123, 1979.

R070 Rosanske, T.W. and K.A. Connors, Stoichiometric model of alpha-cyclodextrin complex formation, *Journal of Pharmaceutical Sciences*, 69, 564–567, 1980.

R071 Repta, A.J. and A.A. Hincal, Complexation and solubilization of acronine with alkylgentisates, *International Journal of Pharmaceutics*, 5, 149–155, 1980.

R072 Reber, L.A., W.M. McNabb, and W.W. Lucasse, The effect of salts on the mutual miscibility of normal butyl alcohol and water, *Journal of Physical Chemistry*, 46, 500–515, 1942.

R075 Rose, F.L. and G. Swain, 2-p-Aminobenzenesulphonamidopyrimidines. preparation by a novel route, *Journal of the Chemical Society* (London), 689–692, 1945.

R076 Raiziss, G.W. and M. Freifelder, N1-Sulfanilylamino-alkyl-pyrimidines, *Journal of the American Chemical Society*, 64, 2340–2342, 1942.

R078 Reamer, H.H., B.H. Sage, and W.N. Lacey, Phase equilibria in hydrocarbon systems n-butane–water system in the two-phase region, *Industrial and Engineering Chemistry*, 44, 609–615, 1952.

R080 Rodriguez, M.M. and A.G. Asuero, Studies on pyridylhydrazones derived from biacetyl as analytical reagents, *Microchemical Journal*, 25, 309–322, 1980.

R081 Rosoff, M. and A.T.M. Serajuddin, Solubilization of diazepam in bile salts and in sodium cholate–lecithin–water phases, *International Journal of Pharmaceutics*, 6, 137–146, 1980.

R082 Ritschel, W.A., K.W. Grummich, S. Kaul, and T.J. Hardt, Biopharmaceutical parameters of coumarin and 7-hydroxycoumarin, *Pharmazeutische Industrie*, 43, 271–276, 1981.

R084 Rossi, S.S. and W.H. Thomas, Solubility behavior of three aromatic hydrocarbons in distilled water and natural seawater, *Environmental Science and Technology*, 15, 715–716, 1981.

R087 Rogers, J.A., Solution thermodynamics of phenols, *International Journal of Pharmaceutics*, 10, 89–97, 1982.

R302 Roland, B., K. Kimura, and J. Smid, interaction of neutral arenes with poly(vinylbenzo-18-crown-6) and poly(vinylbenzoglyme) in aqueous media, *Journal of Colloid and Interface Science*, 97, 392–400, 1984.

R303 Rabenort, B., P.C. DeWilde, F.G. DeBoer, and R.D. Cannizzaro, Diflubenzuron, *Analytical Methods for Pesticides and Plant Growth Regulators*, 10, 57–72, 1978.

R304 Roder, C.-H., N.A. Jenny, and M. Ottnad, Desmedipham, *Analytical Methods for Pesticides and Plant Growth Regulators*, 10, 293–303, 1978.

R308 Rawat, B.S. and S. Krishna, Isobaric vapor–liquid equilibria for the partially miscible system of water–methyl isobutyl ketone, *Journal of Chemical and Engineering Data*, 29, 403–406, 1984.

R318 Rao, R.J. and C.V. Rao, Ternary liquid equilibria systems: *n*-propanol–water–esters, *Journal of Applied Chemistry*, 9, 69–73, 1959.

R319 Rao, M.R. and C.V. Rao, Ternary liquid equilibria. IV. Various systems, *Journal of Applied Chemistry*, 7, 659–666, 1957.

R320 Regna, E.A. and P.F. Bruins, Recovery of aconitic acid from molasses, *Industrial and Engineering Chemistry*, 48, 1268–1277, 1956.

R321 Reinders, W. and C.H. De Minjer, Vapour–liquid equilibria in ternary systems. VI. The System water–acetone–chloroform, *Recueil des Travaux Chimiques des Pays-Bas*, 66, 564–604, 1947.

R338 Roy, S.D. and G.L. Flynn, Solubility and related physicochemical properties of narcotic analgesics, *Pharmaceutical Research*, 5, 580–586, 1988.

R418 Roy, S. and G. Flynn, Solubility behavior of narcotic analgesics in aqueous media: solubilities and dissociation constants of morphine, fentanyl, and sufentanil, *Pharmaceutical Research*, 6, 147–151, 1989.

R419 Rudolph, E.S.J., M. Zomerdijk, M. Ottens, and L.A.M. Van der Wielen, Solubilities and partition coefficients of semi-synthetic antibiotics in water + 1-butanol systems, *Industrial and Engineering Chemical Research*, 40, 398–406, 2001.

R422 Raman, G. and V.G. Gaikar, Hydrotropic solubilization of boswellic acids from boswellia serrata resin, *Langmuir*, 19, 8026–8032, 2003.

R423 Reza, J. and A. Trejo, Temperature dependence of the infinite dilution activity coefficient and Henry's law constant of polycyclic aromatic hydrocarbons in water, *Chemosphere*, 56, 537–547, 2004.

R424 Raevsky, O., E. Andreeva, O. Raevskaja, and K. Schaper, QSPR analysis of the partitioning of vaporous chemicals in a water–gas phase system and the water solubility of liquid and solid chemicals on the basis of fragment and physicochemical similarity and hybot descriptors, *SAR and QSAR in Environmental Research*, 16, 1–2, 2005.

R426 Remko, M., M. Swart, and F.M. Bickelhaupt, Theoretical study of structure, pKa, lipophilicity, solubility, absorption, and polar surface area of some centrallyl acting antihypertensives, *Bioorganic & Medical Chemistry*, 14, 1715–1728, 2006.

R427 Raevsky, O.A., O.E. Raevskaja, and K.-J. Schaper, Analysis of Water Solubility Data on the Basis of HYBOT Descriptors. Part 3. Solubility of solid neutral chemicals and drugs, *QSAR & Combinatorial Science*, 23, 327–343, 2004.

R428 Remko, M. and C.-W. Lieth, Theoretical study of gas-phase acidity, pKa, lipophilicity, and solubility of some biologically active sulfonamides, *Bioorganic & Medicinal Chemistry*, 12, 5395–5403, 2004.

R430 Rigg, M.W. and F.W.J. Liu, Solubilization of orange OT and dimethylaminoazobenzene, *The Journal of the American Oil Chemists' Society*, 30, 14–17, 1953.

R431 Roy, D., F. Ducher, A. Laumain, and J.Y. Legendre, Determination of the aqueous solubility of drug using a convenient 96-well plate-based assay, *Drug Development and Industrial Pharmacy*, 27, 107–109, 2001.

S005 Smith, C. and J.A. Calder, Solubility of alkylbenzenes in distilled water and seawater at 25.0°C, *Journal of Chemical and Engineering Data*, 20, 320–322, 1975.

S006 Saracco, G. and E. Spaccamela-Marchetti, Influenza della catena idrocarburica sulla solubilita in acqua di serie omologhe, *Annales de Chimica*, 48, 1357–1370, 1958.

S010 Sada, E., S. Kito, and Y. Ito, Solubility of toluene in aqueous salt solutions, *Journal of Chemical and Engineering Data*, 20, 373–375, 1975.

S012 Stearns, R.S., H. Oppenheimer, E. Simon, and W.D. Harkins, Solubilization by solutions of long-chain colloidal electrolytes, *Journal of Chemical Physics*, 15, 496–507, 1947.

S076 Schwarz, F.P. and S.P. Wasik, A fluorescence method for the measurement of the partition coefficients of naphthalene, 1-methylnaphthalene, and 1-ethylnaphthalene in water, *Journal of Chemical and Engineering Data*, 22, 270–273, 1977.

S115 Sidgwick, N.V. and J.A. Neill, The solubility of the phenylenediamines and of their monoacetyl derivatives, *Journal of the Chemical Society* (London), 123, 2813–2819, 1923.

S117 Sidgwick, N.V. and T.W.J. Taylor, The solubility and volatility of 3:5-dinitrophenol, *Journal of the Chemical Society* (London), 121, 1853–1859, 1922.

S118 Sidgwick, N.V. and W.M. Dash, The solubility and volatility of the nitrobenzaldehydes, *Journal of the Chemical Society* (London), 121, 2586–2592, 1922.

S119 Sidgwick, N.V. and W.J. Spurrell, The system benzene–ethyl alcohol–water between +25 and -5 degrees, *Journal of the Chemical Society* (London), 117, 1397–1404, 1920.

S120 Sidgwick, N.V. and R.K. Callow, The solubility of the aminophenols, *Journal of the Chemical Society* (London), 125, 522–527, 1924.

S124 Sergeeva, V.F. and M.I. Usanovich, Effect of some electrolytes on solubility of benzoic acid in water, *Journal of General Chemistry of the USSR*, 29, 1369–1372, 1959.

S131 Shnidman, L. and A.A. Sunier, The Solubility of Urea in Water, *Journal of Physical Chemistry*, 30, 1232–1240, 1932.

S133 Smith, H.A. and M. Berman, The solubility curves of the systems carbon tetrachloride–n-alkyl acids–water at 25 degrees, *Journal of the American Chemical Society*, 59, 2390–2391, 1937.

S146 Sekiguchi, K., M. Kanke, N. Nakamura, and Y. Tsuda, Dissolution behavior of solid drugs. V. Determination of the transition temperature and heat of transition between barbital polymorphs by initial dissolution rate measurements, *Chemical and Pharmaceutical Bulletin*, 23, 1347–1352, 1975.

S147 Sekiguchi, K., Y. Tsuda, and M. Kanke, Dissolution behavior of solid drugs. VI. Determination of transition temperatures of various physical forms of sulfanilamide by initial dissolution rate measurements, *Chemical and Pharmaceutical Bulletin*, 23, 1353–1362, 1975.

S149 Sekiguchi, K., M. Kanke, Y. Tsuda, and Y. Tsuda, Dissolution behavior of solid drugs. III. Determination of the transition temperature between the hydrate and anhydrous forms of phenobarbital by measuring their dissolution rates, *Chemical and Pharmaceutical Bulletin*, 21, 1592–1600, 1973.

S171 Schwarz, F.P., Measurement of the solubilities of slightly soluble organic liquids in water by elution chromatography, *Analytical Chemistry*, 52, 10–15, 1980.

S187 Sanborn, J.R., R.L. Metcalf, W.N. Bruce, and P.-Y. Lu, The fate of chlordane and toxaphene in a terrestrial-aquatic model ecosystem, *Environmental Entomology*, 5, 533–538, 1976.

S191 Sutton, C., The solubility of aromatic hydrocarbons and the geochemistry of hydrocarbons in the Eastern Gulf of Mexico, PhD Thesis, 1–198, 1974.

S192 Schmidt, L.H., H.B. Hughes, E.A. Badger, and I.G. Schmidt, The toxicity of sulfamerazine and sulfamethazine, *Journal of Pharmacology and Experimental Therapeutics*, 81, 17–42, 1944.

S198 Schwarz, F.P. and J. Miller, Determination of the aqueous solubilities of organic liquids at 10.0, 20.0, and 30.0°C by elution chromatography, *Analytical Chemistry*, 52, 2162–2164, 1980.

S200 Shenkin, Y.S. and L.N. Zaikina, The temperature dependence of the solubility of urea in water, *Russian Journal of Physical Chemistry*, 52, 1017–1018, 1978.

S203 Sanemasa, I., M. Araki, T. Deguchi, and H. Nagai, Solubilities of benzene and the alkylbenzenes in water—method for obtaining aqueous solutions saturated with vapours in equilibrium with organic liquids, *Chemistry Letters*, 225–228, 1981.

S204 Sahay, H., S. Kumar, S.N. Upadhyay, and Y.D. Upadhya, Solubility of benzoic acid in aqueous polymeric solutions, *Journal of Chemical and Engineering Data*, 26, 181–183, 1981.

S207 Salem, A.-B., Solubility Data of the system acetic acid–toluene-water at different temperatures, *Journal of Chemical Engineering of Japan*, 12, 236–238, 1979.

S212 Shoor, S.K., R.D. Walker, Jr., and K.E. Gubbins, Salting out of nonpolar gases in aqueous potassium hydroxide solutions, *Journal of Physical Chemistry*, 73, 312–317, 1969.

S227 Schwarz, F.P. and S.P. Wasik, Fluorescence measurements of benzene, naphthalene, anthracene, pyrene, fluoranthene, and benzo[e]pyrene in water, *Analytical Chemistry*, 48, 524–528, 1976.

S304 Summers, M.P., R.P. Enever, and J.E. Carless, Studies of the dissolution characteristics of three crystal forms of aspirin, *Symposium on Particle Growth in Suspensions*, 247–259, 1972.

S306 Spencer, J.N. and T.A. Judge, Hydrophobic hydration of thymine, *Journal of Solution Chemistry*, 12, 847–853, 1983.

S307 Stephenson, R., J. Stuart, and M. Tabak, Mutual solubility of water and aliphatic alcohols, *Journal of Chemical and Engineering Data*, 29, 287–290, 1984.

S309 Shell Development Company Analytical Department Biological Sci Res Center, Cyanazine, *Analytical Methods for Pesticides and Plant Growth Regulators*, 10, 275–292, 1978.

S314 Sarker, M. and D. Wilson, Solubilities of p-dichlorobenzene and naphthalene in several aqueous-organic solvent mixtures, *Journal of the Tennessee Academy of Science*, 69–74, 1987.

S352 Shiu, W.Y., W. Doucette, F.A.P. Gobas, and D. Mackay, Physical–chemical properties of chlorinated dibenzo-p-dioxins, *Environmental Science and Technology*, 22, 651–658, 1988.

S355 Sanders, N.D., Visual observation of the solubility of heavy hydrocarbons in near-critical water, *Industrial and Engineering Chemistry, Fundamentals*, 25, 169–171, 1986.

S357 Sohoni, V.R. and U.R. Warhadpande, system ethyl acetate–acetic acid–water at 30°C, *Industrial and Engineering Chemistry*, 44, 1428–1429, 1952.

S358 Sutton, C. and J.A. Calder, Solubility of alkylbenzenes in distilled water and seawater at 25.0°C, *Journal of Chemical and Engineering Data*, 20, 320–322, 1975.

S359 Sanemasa, I., Y. Miyazaki, S. Arakawa, and T. Deguchi, The Solubility of benzene-hydrocarbon binary mixtures in water, *Bulletin of the Chemical Society of Japan* (Nippon Kagakukai Bulletin), 60, 517–523, 1987.

S377 Sherblom, P.M., P.M. Gschwend, and R.P. Eganhouse, Aqueous solubilities, vapor pressures, and 1-octanol–water partition coefficients for C9-C14 linear alkylbenzenes, *Journal of Chemical and Engineering Data*, 37, 394–399, 1992.

S415 Sanemasa, I., J. Wu, and K. Toda, Solubility product and solubility of cyclodextrin inclusion complex precipitates in an aqueous medium, *Bulletin of the Chemical Society of Japan*, 70, 365–369, 1997.

S415 Seedher, N. and S. Bhatia, Solubility enhancement of Cox-2 inhibitors using various solvent systems, *AAPS PharmSciTech*, 4, 281–289, 2003.

S417 Schulze, M., H. Wilkes, and H. Vereecken, Direct determination of hydrophobic organic compounds in aqueous solution in the presence of dissolved organic carbon by high-performance liquid chromatography, *Chemosphere*, 39, 2365–2374, 1999.

S417 Song, W., A. Li, and X. Xu, Water solubility of phthalates by cetyltrimethylammonium bromide and beta-cyclodextrin, *Industrial Engineering Chemical Research*, 42, 949–955, 2003.

S420 Shah, S., K. Naeem, S.W.H. Shah, and H. Hussain, Solubilization of short chain phenylalkanoic acids by a cationic surfactant, cetyltrimethylammonium bromide, *Physicochemical and Engineering Aspects*, 148, 299–304, 1999.

S446 Sunesen, V.H., B.L. Pedersen, H.G. Kristensen, and A. Mullertz, In vivo in vitro correlations for a poorly soluble drug, danazol, using the flow-through dissolution method with biorelevant media, *European Journal of Pharmaceutical Sciences*, 24, 305–313, 2005.

S448 Staples, C.A., P.B. Dorn, G.M. Klecka, and L.R. Harris, A Review of the environmental fate, effects, and exposures of bisphenola, *Chemosphere*, 36, 2149–2173, 1998.

S450 Stuart, M. and K. Box, Chasing equilibrium: measuring the intrinsic solubility of weak acids and bases, *Analytical Chemistry*, 77, 983–990, 2005.

S453 Stefano, A.D., P. Sozio, A. Cocco, and F. Pinnen, L-Dopa and dopamine (R)-α-lipoic acid conjugates as multifunctional codrugs with antioxidant properties, *Journal of Medicinal Chemistry*, 49, 1486–1493, 2006.

S454 Sarraute, S., H. Delepine, M.F.C. Gomes, and V. Majer, Aqueous solubility, Henry's law constant and air/water partition coefficients of *n*-octane and two halogenated octanes, *Chemosphere*, 57, 1543–1551, 2004.

S460 Schaper, K.J., B. Kunz, and O.A. Raesky, Analysis of water solubility data on the basis of HYBOT descriptors, *QSAR & Combinatorial Science*, 22, 943–958, 2003.

S461 Susilo, R., J.D. Lee, and P. Englezos, Liquid–liquid equilibrium data of water with neohexane, methyl-cycloxane, *tert*-butyl methyl ether, *n*-heptane and vapor–liquid–liquid equilibrium with methane, *Fluid Phase Equilibria*, 231, 20–26, 2005.

S462 Sabadini, E., T. Cosgrove, and F. Egidio, Solubility of cyclomaltooligosaccharides (cyclodextrins) in H_2O and D_2O: a comparative study, *Carbohydrate Research*, 341, 270–274, 2006.

S464 Sarraute, S., I. Mkobel, M.F.C. Gomes, and J. Jose, Vapour pressures, aqueous solubility, Henry's Law constants and air/water partition coefficients of 1,8-dichlorooctane and 1,8-dibromooctane, *Chemosphere*, 64, 1829–1836, 2006.

S466 Shah, J.C., J.R. Chen, and D. Chow, Metastable polymorph of etoposide with higher dissolution rate, *Drug Development and Industrial Pharmacy*, 25, 63–67, 1999.

S468 Shareef, A., M.J. Angove, J.D. Wells, and B.B. Johnson, Aqueous solubilities of estrone, 17b-estradiol, 17α-ethynylestradiol, and bisphenol A, *Journal of Chemical and Engineering Data*, 51, 879–881, 2006.

S469 Singh, B.N., A quantitative approach to probe the dependence and correlation of food-effect with aqueous solubility, dose/solubility ratio, and partition coefficient (Log P) for orally active drugs administered as immediate-release formulations, *Drug Development Research*, 65, 55–75, 2005.

S470 Schaefer, H.G., D. Beermann, R. Horstmann, and J. Kuhlmann, Effect of food on the pharmacokinetics of the active metabolite of the prodrug repirinast, *Journal of Pharmaceutical Sciences*, 82, 107–109, 1993.

S471 Szterner, P., Solubilities in water of uracil and its halogenated derivatives, *Journal of Chemical Engineering Data*, 53, 1738–1744, 2008.

S472 Shalmashi, A. and A. Eliassi, Solubility of L-(+)-Ascorbic acid in water, ethanol, methanol, propan-2-ol, acetone, acetonitrile, ethyl acetate and tetrahydrofuran from (293 to 323) °K, *Journal of Chemical Engineering Data*, 53, 1332–1334, 2008.

T002 Taylor, Jr., P.W. and D.E. Wurster, Dissolution kinetics of certain crystalline forms of prednisolone. II. Influence of low concentrations of sodium lauryl sulfate, *Journal of Pharmaceutical Sciences*, 54, 1654–1658, 1965.

T003 Tabern, D.L. and E.F. Shelberg, Physico-chemical properties and hypnotic action of substituted barbituric acids, *Journal of the American Chemical Society*, 55, 328–332, 1933.

T005 Thakkar, A.L. and P.B. Kuehn, Solubilization of some steroids in aqueous solutions of a steroidal nonionic surfactant, *Journal of Pharmaceutical Sciences*, 58, 850–852, 1969.

T008 Ts'o, P.O.P., I.S. Melvin, and A.C. Olson, Interaction and association of bases and nucleosides in aqueous solutions, *Journal of the American Chemical Society*, 85, 1289–1296, 1963.

T015 Taylor, C.A. and W.H. Rinkenbach, The solubility of trinitro-phenylmethyl-nitramine (tetryl) in organic solvents, *Journal of the American Chemical Society*, 45, 104–107, 1923.

T020 Taylor, C.A. and W.H. Rinkenbach, The solubility of trinitrotoluene in organic solvents, *Journal of the American Chemical Society*, 45, 44–59, 1923.

T023 Taylor, D. and G.C. Vincent, Phase equilibria in sulphonic acid–water systems, *Journal of the Chemical Society* (London), 3218–3224, 1952.

T025 Ts'o, P.O.P. and P. Lu, Interaction of nucleic acids, II. Chemical linkage of the carcinogen 3,4-benzpyrene to DNA induced by photoradiation, *Proceedings of the National Academy of Science of the United States of America*, 51, 272–280, 1964.

T033 Thorne, P.C.L., The solubility of ethyl ether in solutions of sodium chloride, *Journal of the Chemical Society* (London), 119, 262–268, 1921.

T066 Tsonopoulos, C. and J.M. Prausnitz, Activity coefficients of aromatic solutes in dilute aqueous solutions, *Industrial and Engineering Chemistry, Fundamentals*, 10, 593–599, 1971.

T067 Tewari, Y.B., M.M. Miller, S.P. Wasik, and D.E. Martire, Aqueous solubility and octanol/water partition coefficient of organic compounds at 25.0°C, *Journal of Chemical and Engineering Data*, 27, 451–454, 1982.

T301 Takano, J., T. Yauoka, and S. Mitsuzawa, Solubility measurements of solid organic compounds in water by TOC method, *Nippon Kagaku Kaishi*, 1830–1834, 1982.

T303 Treiner, C., C. Vaution, and G.N. Cave, Correlations between solubilities, heats of fusion and partition coefficients for barbituric acids in octanol + water and in aqueous micellar solutions, *Journal of Pharmacy and Pharmacology*, 34, 539–540, 1982.

T305 Thier, W.G., K. Hommel, and T. Hoppe, Triazophos, *Analytical Methods for Pesticides and Plant Growth Regulators*, 10, 127–137, 1978.

T418 Tewari, Y.B., P.D. Gery, M.D. Vaudin, and R.N. Goldberg, Saturation molalities and standard molar enthalpies of solution of cytidine(cr), hypoxanthine(cr), thymidine(cr), thymine(cr), uridine(cr), and xanthine(cr) in H2O(1), *Journal of Chemical and Thermodynamics*, 36, 645–658, 2004.

T420 Tewari, Y.B., P.D. Gery, M.D. Vaudin, and R.N. Glodberg, Saturation molaties and standard molar enthalpies of solution of 2'-deoxyadensine . H$_2$O(cr), 2'-deoxycytidine . H$_2$O(cr), 2'-dexyguanosine . H$_2$O(cr), 2'-deoxyinosine(cr), and 2'-deoxyuridine(cr) in H$_2$O(1), *Journal of Chemical and Thermodynamics*, 37, 233–241, 2005.

T423 Tolls, J., J.V. Dijk, E.J.M. Verbruggen, and G. Schuurmann, Aqueous solubility–molecular size relationships: a mechanistic case study Using C10⁻ to C19⁻ alkanes, *Journal of Physical Chemistry*, 106, 2760–2765, 2002.

T424 Tamura, K. and H. Li, Mutual Solubilities of terpene in methanol and water and their multicomponent liquid–liquid equilibria, *Journal of Chemical and Engineering Data*, 50, 2013–2018, 2005.

T425 Tavornvips, S., F. Hirayama, H. Arima, and H. Hashimoto, 6-o-α(4-o-α-D-glucuronyl)-D-glucosyl-b-cyclodextrin: solubilizing ability and some cellular effects, *International Journal of Pharmaceutics*, 249, 199–209, 2002.

T426 Maleka, T.S.P., W. Liebenberg, M. Song, and M.M. de Villiers, Preparation and physicochemical properties of niclosamide anhydrate and two monohydrates, *International Journal of Pharmaceutics*, 269, 417–432, 2004.

T428 Terashima, M., M. Fukushima, and S. Tanaka, Evaluation of solubilizing ability of humic aggregate basing on the phase-separation model, *Chemosphere*, 57, 439–445, 2004.

U001 Udani, J.H. and J. Autian, Study of the stability of secobarbital sodium solutions. I. Determination of solubility of secobarbital as a function of solvent and hydrogen ion concentration, *Journal of the American Pharmaceutical Association, Scientific Edition*, 49, 376–380, 1960.

U001 Ulijn, R.V., L. De Martin, P.J. Halling, and A.E.M. Janssen, Enzymatic synthesis of B-lactam antibiotics via direct condensation, *Journal of Biotechnology*, 99, 215–222, 2002.

U010 Ueda, M., A. Katayama, N. Kuroki, and T. Urahata, Effect of urea on the solubility of benzene and toluene in water, *Progress in Colloid and Polymer Science*, 63, 116–119, 1978.

U013 Ueda, M., A. Katayama, T. Urahata, and N. Kuroki, Effect of alcohols on the solubilities of aromatic hydrocarbons in water, *Kagaku To Kogyo* (Tokyo), 54, 252–258, 1980.

V004 Vermillion, H.E., B. Werbel, J.H. Saylor, and P.M. Gross, Solubility studies. VI. The solubility of nitrobenzene in deuterium water and in ordinary water, *Journal of the American Chemical Society*, 63, 1346–1347, 1941.

V009 Van Arkel, A.E. and S.E. Vles, Loslichkeit von organischen verbindungen in wasser, *Recueil des Travaux Chimiques des Pays-Bas*, 55, 407–411, 1936.

V013 Vesala, A., Thermodynamics of transfer of nonelectrolytes from light to heavy water. I. Linear free energy correlations of free energy of transfer with solubility and heat of melting of a nonelectrolyte, *Acta Chemica Scandinavica, Series A: Physical and Inorganic Chemistry*, 28, 839–845, 1974.

V033 Vaution, C., C. Treiner, F. Puisieux, and J.T. Carstensen, Solubility behavior of barbituric acids in aqueous solution of sodium alkyl sulfonate as a function of concentration and temperature, *Journal of Pharmaceutical Sciences*, 70, 1238–1242, 1981.

V300 Valko, E.I. and M.B. Epstein, Comicellization, *Solubilization and Micelles*, 1, 334–339, 1957.

V301 Van Rossum, A., P.C. DeWilde, F.G. DeBoer, and P.K. Korver, Tetradifon, *Analytical Methods for Pesticides and Plant Growth Regulators*, 10, 119–126, 1978.

V302 Van Auken, O.W. and M. Hulse, Hexachlorophene, *Analytical Methods for Pesticides and Plant Growth Regulators*, 10, 189–214, 1978.

V303 Van Rossum, A., P.C. DeWilde, F.G. DeBoer, and P.K. Korver, Dichlobenil, *Analytical Methods for Pesticides and Plant Growth Regulators*, 10, 311–320, 1978.

V410 Viernstein, H., P. Weiss-Greiler, and P. Wolschann, Solubility enhancement of law soluble biologically active compounds by b-cyclodextrin and dimethyl-b-cyclodextrin, *Journal of Inclusion Phenomena and Macrocyclic Chemistry*, 44, 235–239, 2002.

V412 Varma, M.V.S. and R. Panchagnula, Enhanced oral paclitaxel absorption with vitamin E-TPGS: effect on solubility and permeability in vitro, in situ and in vivo, *European Journal of Pharmaceutical Sciences*, 25, 445–453, 2005.

V414 Voutsas, E., C. Vavva, K. Magoulas, and D. Tassios, Estimation of the volatilization of organic compounds from soil surfaces, *Chemosphere*, 58, 751–758, 2005.

V416 Viamajala, S., B.M. Peyton, L.A. Richards, and J.N. Petersen, Solubilization, solution equilibria, and biodegradation of PAH's under thermophilic conditions, *Chemosphere*, 66, 1094–1106, 2007.

V417 Venkatesh, S., J. Li, Y. Xu, and B.D. Anderson, intrinsic solubility estimation and pH-solubility behavior of cosalane (NSC 658586), an extremely hydrophobic diprotic acid, *Pharmaceutical Research*, 13, 1453–1459, 1996.

W003 Wauchope, D. and F.W. Getzen, Temperature dependence of solubilities in water and heats of fusion of solid aromatic hydrocarbons, *Journal of Chemical and Engineering Data*, 17, 38–41, 1972.

W005 Weil-Malherbe, H., The solubilization of polycyclic aromatic hydrocarbons by purines, *Biochemical Journal*, 40, 351–363, 1946.

W006 Wurster, D.E. and P.W. Taylor, Jr., Dissolution kinetics of certain crystalline forms of prednisolone, *Journal of Pharmaceutical Sciences*, 54, 670–676, 1965.

W007 Wurster, D.E. and D.O. Kildsig, Effect of complex formation on dissolution kinetics of m-aminobenzoic acid, *Journal of Pharmaceutical Sciences*, 54, 1491–1494, 1965.

W011 Weiss, J.M. and C.R. Downs, The physical properties of maleic, fumaric and malic acids, *Journal of the American Chemical Society*, 45, 1003–1008, 1923.

W013 Wise, W.S. and E.B. Nicholson, The solubilities and heats of crystallisation of sucrose and methyl alpha-D-glucoside in aqueous solution, *Journal of the Chemical Society* (London), 2714–2716, 1955.

W016 Watari, N. and N. Kaneniwa, Dissolution of slightly soluble drugs. ii. effect of particle size on dissolution behavior in sodium lauryl sulfate solutions, *Chemical and Pharmaceutical Bulletin*, 24, 2577–2584, 1976.

W019 Walters, V., The dissolution of paracetamol tablets and the in vitro transfer of paracetamol with and without sorbitol, *Journal of Pharmacy and Pharmacology*, 20, 228–231, 1968.

W022 Wiley, R.H. and N.R. Smith, Reciprocal solubility of 4,6-dimethyl-1,2-pyrone and water, *Journal of the American Chemical Society*, 73, 1383–1384, 1951.

W024 Wallnofer, P.R., M. Koniger, and O. Hutzinger, ANRNX, 13, 14, 1973.

W025 Weil, L., G. Dure, and K.E. Quentin, Wasserloslichkeit von insektiziden chlorierten kohlenwasserstoffen und polychlorierten biphenylen im hinblick auf eine gewasserbelastung mit diesen stoffen, *Zeitschrift fuer Wasser und Abwasser Forschung*, 7, 169–175, 1974.

W026 Wright, R., Selective solvent action. Part VI. The effect of temperature on the solubilities of semisolutes in aqueous alcohol, *Journal of the Chemical Society* (London), 130, 1334–1337, 1927.

W029 Ward, H.L. and S.S. Cooper, The system, benzoic acid, *ortho* phthalic acid, water, *Journal of Physical Chemistry*, 34, 1484–1493, 1930.

W033 Wiese, C.S. and D.A. Griffin, The solubility of aroclor 1254 in seawater, *Bulletin of Environmental Contamination and Toxicology*, 19, 403–411, 1978.

W038 Walker, W.H., A.R. Collett, and C.L. Lazzell, The solubility relations of the isomeric dihydroxyben-zenes, *Journal of Physical Chemistry*, 35, 3259–3271, 1931.

W044 Walker, J. and J.K. Wood, solubility of isomeric substances, *Journal of the Chemical Society* (London), 73, 618–627, 1898.

W053 Wilkerson, A.S., Optical properties of 2-sulfanilamidopyrimidine (sulfadiazine), *Journal of the American Chemical Society*, 64, 2230–2230, 1942.

W055 Watari, N., M. Hanano, and N. Kaneniwa, Dissolution of slightly soluble drugs. VI. Effect of particle size of sulfadimethoxine on the oral bioavailability, *Chemical and Pharmaceutical Bulletin*, 28, 2221–2225, 1980.

W057 Worley, J.D., Benzene as a solute in water, *Canadian Journal of Chemistry*, 45, 2465–2467, 1967.

W300 Wasik, S.P., M.M. Miller, Y.B. Tewari, and W.H. Zoller, Determination of the vapor pressure, aqueous solubility, and octanol/water partition coefficient of hydrophobic substances by coupled generator col-umn/liquid chromatographic methods, *Residue Reviews*, 85, 29–42, 1983.

W302 Wise, S.A., W.J. Bonnett, F.R. Guenther, and W.E. May, A relationship between reversed-phase C18 liquid chromatographic retention and the shape of polycyclic aromatic hydrocarbons, *Journal of Chromatographic Science*, 19, 457–465, 1981.

W305 Ward, A.F.H. and A.G. Chitale, A study of solubilization by electrical conductivity measurements, *Solubilization and Micelles*, 1, 405–409, 1957.

W310 Whiteoak, R.J., J.B. Reary, and K.C. Overton, bendiocarb, *Analytical Methods for Pesticides and Plant Growth Regulators*, 10, 3–17, 1978.

W311 Weeren, R.D. and D. Eichler, Bromophos, *Analytical Methods for Pesticides and Plant Growth Regulators*, 10, 31–40, 1978.

W312 Weeren, R.D. and D. Eichler, Bromophos-ethyl, *Analytical Methods for Pesticides and Plant Growth Regulators*, 10, 41–43, 1978.

W313 Whiteoak, R.J., M. Crofts, R.J. Harris, and K.C. Overton, Ethofumesate, *Analytical Methods for Pesticides and Plant Growth Regulators*, 10, 353–366, 1978.

W314 Whitacre, D.M., Y.H. Atallah, J.E. Forrette, and H.K. Suzuki, Methazole, *Analytical Methods for Pesticides and Plant Growth Regulators*, 10, 367–384, 1978.

W317 Wiedenhof, N. and Lammers, J.N.J., Properties of cyclodextrins. Part II. Preparation of a stable beta-cyclodextrin hydrate and determination of its water content and enthalpy of solution in water from 15–30°C, *Carbohydrate Research*, 7, 1–6, 1968.

W319 Webster, G.R.B., K.J. Friesen, L.P. Sarna, and D.C.G. Muir, Environmental fate modelling of chlo-rodioxins: determination of physical constants, *Chemosphere*, 14, 609–622, 1985.

W332 Webster, G.R.B., L.P. Sarna, and D.C. Muir, *Octanol–Water Partition Coefficient of 1,3,6,8-TCDD and OCDD by Reverse Phase HPLC*, Book Chapter, 79–87, Butterworth, 1985.

W402 Wang, L. and X. Wang, Solubilities of dichlorophenylphosphine sulfide and *bis*(4-carboxyphenyl) phen-ylphosphine oxide in water, *Journal of Chemical and Engineering Data*, 45, 743–745, 2000.

W412 Wang, Z.-W., Q.-X. Sun, J.-S. Wu, and L.-S. Wang, Solubilities of 2-carboxyethylphenylphosphinic acid and 4-carboxyphenylphosphinic acid in water, *Journal of Chemical and Engineering Data*, 48, 1073–1075, 2003.

W414 Wang, S., Q. Li, Z. Li, and M. Su, Solubility of xylitol in ethanol acetone, *N,N*-dimethylformamide, 1-butanol, 1-pentanol, toluene, 2-propanol, and water, *Journal of Chemical and Engineering Data*, 52, 186–188, 2007.

W416 Wu, Z., M. Razzak, I.G. Tucker, and N.J. Medlicott, Physicochemical characterization of ricobenda-zole: I. solubility, lipophilicity, and ionization characteristics, *Journal of Pharmaceutical Sciences*, 94, 983–993, 2005.

W417 Wang, S., J. Wang, and Q. Yin, Measurement and correlation of solubility of 7-aminocephalosporanic acid in aqueous acetone mixture, *Industrial and Engineering Chemistry Research*, 44, 3783–3787, 2005.

W418 Wang, L.-H. and Y.-Y. Chang, Solubility of puerarin in water, ethanol, and acetone from (288.2 to 328.2) °K, *Journal of Chemical and Engineering Data*, 50, 1375–1376, 2005.

W422 Wang, L.-S., M. Yang, S.-B. Wang, and X. Ouyang, Solubilities of phenylphosphinic acid, hydroxymethylphenylphinic acid, *p*-methoxyphenylphosphinic acid, *p*-methoxyphenylhydroxymethylphosphinic acid, triphenylphosphine, tri(p-methoxyphenyl)phosphine, and tri(*p*-methoxyphenyl)phosphine oxide, *Journal of Chemical and Engineering Data*, 51, 462–466, 2006.

W424 Williams, G.C. and P.J. Sinko, Oral absorption of the HIV protease inhibitors: a current update, *Advance Drug Delivery Reviews*, 39, 211–238, 1999.

Y020 Young, F.E., D-Glucose–water phase diagram, *Journal of Physical Chemistry*, 61, 616–619, 1957.

Y409 Yang, Z., H. Hu, X. Zhang, and Y Xu, Solubility of phenazine-1-carboxylic acid in water, methanol, and ethanol from (278.2 to 328.2) °K, *Journal of Chemical and Engineering Data*, 52, 184–185, 2007.

Y410 Yurquina, A., M.E. Manzur, P. Brito, and M.A.A. Molina, Physicochemical studies of acetaminophen in Water-Peg 400 systems, *Journal of Molecular Liquids*, 133, 47–53, 2007.

Y412 Yi, Y., D. Hatziavramidis, and A.S. Myerson, Development of a small-scale automated solubility measurement apparatus, *Industrial and Engineering Chemistry Research*, 44, 5427–5433, 2005.

Y414 Yu, Y.L., X.M. Wu, S.N. Li, and J.Q. Yu, An exploration of the relationship between adsorption and bioavailability of pesticides in soil to earthworm, *Environmental Pollution*, 141, 428–433, 2006.

Y418 Yang, G.-F., H.-B. Wang, W.-C. Yang, and C.-G. Zhan, Bioactive permethrin/B-cyclodextrin inclusion complex, *Journal of Physical Chemistry*, 110, 7044–7048, 2006.

Y419 Yamamoto, H. and H.M. Linjesrand, Partitioning of selected estrogenic compounds between synthetic membrane vesicles and water: effects of lipid components, *Environmental Science and Technology*, 38, 1139–1147, 2004.

Y421 Yu, X., G.L. Zipp, and G.W.R. Davidson III, The effect of temperature and pH on the solubility of quinolone compounds: estimation of heat of fusion, *Pharmaceutical Research*, 11, 522–527, 1994.

Y421 Yazadanian, M., K. Briggs, C. Jankovsky, and A. Hawi, The "high solubility" definition of the current FDA guidance on biopharmaceutical classification system may be too strict for acidic drugs, *Pharmaceutical Research*, 21, 293–299, 2004.

Z008 Zerpa, C.O., P.B. Dharmawardhana, W.R. Parrish, and E.D. Sloan, Solubility of cyclopropane in aqueous solutions of potassium chloride, *Journal of Chemical and Engineering Data*, 24, 26–28, 1979.

Z407 Zielenkiewicz, W., B. Golankiewicz, G.L. Perlovich, and M. Kozbial, Aqueous solubilities, infinite dilution activity coefficients and octanol–water partition coefficients of tricyclic analogs of acyclovir, *Journal of Solution Chemistry*, 28, 731–745, 1999.

Z408 Zielenkiewicz, W., J. Poznanski, and A. Zielenkiewicz, Partial molar volumes of aqueous solutions of some halo and amino derivatives of uracil, *Journal of Solution Chemistry*, 29, 757–769, 2000.

Z409 Zeng, Q.-R., H.-X. Tang, B.-H. Liao, and C. Tang, Solubilization and desorption of methyl-parathion from porous media: a comparison of hydroxypropyl-B-cyclodextrin and two nonionic surfactants, *Water Research*, 40, 1351–1358, 2006.

Z410 Zerrouk, N., G. Corti, S. Ancillotti, and P. Mura, Influence of cyclodextrins and chitosan, separately or in combination, on glyburide solubility and permeability, *European Journal of Pharmaceutics and Biopharmaceutics*, 62, 241–246, 2006.

Z411 Zimmermann, T., R.A. Yeates, H. Laufen, and A. Wildfeuer, Influence of concomitant food intake on the oral absorption of two triazole antifungal agents, itraconazole and fluconazole, *European Journal of Clinical Pharmacology*, 46, 147–150, 1994.

Z412 Zi, P., X. Yang, H. Kuang, and L. Yu, Effect of HPBetaCD on solubility and transdermal delivery of capsaicin through rat skin, *International Journal of Pharmaceutics*, 258, 151–158, 2008.

Index 1: Molecular Formula

Index 2: Names and Synonyms

O

Q

R

Index 3: Chemical Abstracts Service Registry Number (RN)

50-02-2	4302	52-44-8	1836
50-03-3	4355	52-45-9	1834
50-04-4	4244	52-46-0	2967
50-06-6	2794	52-49-3	4159
50-07-7	3491	52-51-7	156
50-11-3	1875	52-67-5	521
50-18-0	1294	52-68-6	324
50-22-6	4240	52-86-8	4202
50-23-7	4243	53-03-2	4214
50-24-8	4224	53-06-5	4226
50-27-1	3918	53-16-7	3907
50-28-2	3916	53-19-0	3215
50-29-3	3203	53-34-9	4215, 4216
50-30-6	1008	53-41-8	4057
50-31-7	995	53-42-9	4059
50-32-8	4076	53-43-0	4047
50-33-9	3992	53-70-3	4265
50-36-2	3785	54-12-6	2413
50-44-2	404	54-20-6	391
50-49-7	4026	54-31-9	2770
50-50-0	4420	54-42-2	1793
50-52-2	4211	54-85-3	775
50-55-5	4547	55-18-5	372
50-59-9	3981	55-21-0	1133
50-65-7	2995	55-38-9	2232
50-70-4	973	55-56-1	4305
50-71-5	229	55-63-0	151
50-76-0	4643	55-91-4	939
50-78-2	1731	56-04-2	422
50-81-7	801	56-12-2	355
50-84-0	1009	56-23-5	40
50-85-1	1453	56-25-7	2150
50-89-5	2205	56-29-1	2878
50-99-7	907, 914	56-35-9	4415
51-20-7	230	56-38-2	2197
51-21-8	233	56-40-6	89
51-28-5	626	56-41-7	198
51-34-3	3784	56-45-1	206
51-35-4	462	56-47-3	4354
51-36-5	1007	56-49-5	4184
51-43-4	1862	56-53-1	3895
51-44-5	1010	56-54-2	4122
51-52-5	1208	56-55-3	3843
51-55-8	3804	56-56-4	4095
51-66-1	1801	56-72-4	3288
51-67-2	1547	56-75-7	2408
51-79-6	204	56-81-5	220
51-98-9	4299	56-82-6	182
52-01-7	4383	56-84-8	313
52-21-1	4453	56-85-9	477
52-31-3	2880	56-86-0	466
52-39-1	4225	56-87-1	946
52-43-7	2118	56-89-3	868

RN	Page		RN	Page
57-00-1	367		60-11-7	3282
57-06-7	265		60-12-8	1520
57-10-3	3709		60-18-4	1807
57-11-4	3957		60-27-5	316
57-13-6	22		60-29-7	378
57-15-8	306		60-32-2	927
57-24-9	4199		60-35-5	88
57-27-2	3767		60-51-5	527
57-41-0	3427		60-54-8	4285
57-42-1	3510		60-57-1	2719
57-43-2	2522		60-79-7	4023
57-44-3	1556		60-80-0	2412
57-48-7	908		60-87-7	3775
57-50-1	2960		60-89-9	4017
57-53-4	1908		60-92-4	2137
57-62-5	4281		60-99-1	4028
57-66-9	3143		61-00-7	4015
57-67-0	1214		61-33-6	3630
57-68-1	2832, 2833		61-57-4	742
57-74-9	1995		61-68-7	3463
57-83-0	4233		61-73-4	3656
57-85-2	4313		61-82-5	78
57-91-0	3917		61-90-5	920
57-96-5	4328		62-23-7	1053
58-00-4	3750		62-38-4	1448
58-05-9	4118		62-44-2	2162
58-08-2	1515		62-53-3	763
58-14-0	2803		62-55-5	93
58-15-1	3118		62-56-6	23
58-18-4	4156		62-57-7	359
58-22-0	4046, 4048		62-59-9	4541
58-25-3	3579		62-73-7	305
58-27-5	2372		63-05-8	4039
58-32-2	4401		63-25-2	2775
58-38-8	4119		63-42-3	2959
58-39-9	4208		63-68-3	520
58-40-2	3776		63-74-1	792
58-55-9	1160		63-84-3	1810
58-61-7	2183		63-91-2	1799
58-63-9	2135		63-98-9	1760
58-74-2	4109		64-00-6	2477
58-85-5	2248		64-19-7	81
58-86-6	503		64-77-7	2914
58-89-9	718		64-85-7	4234
58-90-2	565		64-86-8	4290
58-93-5	1150, 1151		64-95-9	4131
58-94-6	1073		65-28-1	3910
58-96-8	1838		65-45-2	1134
59-02-9	4521		65-46-3	1868
59-05-2	4115		65-49-6	1143
59-23-4	910		65-71-4	424
59-30-3	3988		65-85-0	1099
59-31-4	1707		65-86-1	393
59-46-1	3147		66-02-4	1739
59-50-7	1120		66-22-8	245
59-51-8	519		66-25-1	882
59-52-9	218		66-27-3	103
59-63-2	2817		66-71-7	2722
59-66-5	278		66-76-2	3964
59-67-6	676		66-81-9	3538
59-87-0	743		66-97-7	2362
59-92-7	1811		67-20-9	1371
60-01-5	3555		67-52-7	249
60-09-3	2776		67-56-1	25

67-63-0	217	74-98-6	214
67-64-1	175	74-99-7	125
67-66-3	5	75-00-3	85
67-72-1	112	75-01-4	61
67-73-2	4374	75-03-6	86
67-78-7	4423	75-04-7	105
67-97-0	4484	75-05-8	66
68-19-9	4646	75-09-2	9
68-22-4	4139	75-11-6	10
68-26-8	4154	75-15-0	44
68-35-9	2064	75-17-2	18
68-90-6	3726	75-19-4	153
68-94-0	397	75-25-2	3
68-96-2	4239	75-26-3	188
69-23-8	4291	75-27-4	1
69-53-4	3641	75-28-5	369
69-65-8	971	75-29-6	191
69-72-7	1103	75-30-9	195
69-79-4	2956	75-34-3	74
69-89-6	400, 401	75-35-4	52
69-93-2	402, 403	75-37-6	76
70-18-8	2268	75-45-6	4
70-25-7	97	75-47-8	6
70-30-4	2989	75-50-3	222
70-34-8	592	75-52-5	19
70-47-3	331	75-56-9	174
70-55-3	1194	75-60-5	107
70-69-9	1796	75-65-0	374
71-00-1	811	75-69-4	38
71-23-8	216	75-71-8	37
71-30-7	266	75-73-0	41
71-36-3	375	75-80-9	60
71-41-0	537	75-83-2	935
71-43-2	702	75-84-3	532
71-55-6	63	75-85-4	539
71-63-6	4584	75-97-8	878
72-14-0	1750	75-98-9	496
72-18-4	517	75-99-0	131
72-19-5	363	76-01-7	49
72-20-8	2718	76-03-9	48
72-43-5	3593	76-06-2	39
72-44-6	3584	76-13-1	110
72-48-0	3192	76-14-2	109
72-54-8	3214	76-20-0	1668
72-55-9	3190	76-22-2	2254
73-22-3	2414	76-24-4	1372
73-24-5	415	76-25-5	4377
73-32-5	929	76-44-8	1980
73-40-5	416	76-57-3	3899
73-48-3	3447	76-58-4	4019
73-49-4	2112	76-73-3	2913
74-11-3	1029	76-74-4	2523
74-79-3	948	76-75-5	2521
74-82-8	21	76-76-6	1876
74-83-9	12	76-84-6	3975
74-84-0	100	76-93-7	3255
74-85-1	69	76-94-8	2389
74-86-2	50	77-02-1	2204
74-87-3	14	77-09-8	4090
74-88-4	17	77-21-4	3068
74-89-5	28	77-26-9	2496
74-95-3	8	77-27-0	2911
74-96-4	84	77-28-1	2247
74-97-5	7	77-30-5	2937

86-57-7	2005	90-05-1	1168
86-60-2	2036	90-11-9	2000
86-73-7	3008	90-12-0	2380
86-74-8	2742	90-13-1	2003
86-87-3	2766	90-14-2	2004
86-88-4	2390	90-15-3	2017
87-17-2	3033	90-34-6	3522
87-20-7	2892	90-39-1	3552
87-26-3	2508	90-43-7	2764
87-29-6	3600	90-45-9	3015
87-33-2	796	90-51-7	2046
87-40-1	1036	90-64-2	1455
87-47-8	3075	90-82-4	2228
87-58-1	227	90-94-8	3773
87-61-6	582	91-01-0	3050
87-64-9	1122	91-10-1	1537
87-65-0	616	91-16-7	1531
87-66-1	755	91-17-8	2272
87-68-3	388	91-18-9		629
87-69-4	299	91-20-3		2010
87-72-9	504	91-22-5		1701
87-78-5	974	91-33-8		3444
87-79-6	911	91-57-6		2379
87-82-1	987	91-58-7		2002
87-86-5	553	91-59-8		2030
87-87-6	568	91-64-5		1693
87-88-7	558	91-75-8		3770
87-89-8	912	91-79-2		3341
87-90-1	225	91-80-5		3340
87-92-3	2954	91-81-6		3665
87-99-0	543	91-82-7		4112
88-04-0	1470	91-84-9		3807
88-06-2	589	91-85-0		3667
88-09-5	891	91-94-1		2755
88-13-1	406	92-06-8		3853
88-14-2	407	92-24-0		3841
88-19-7	1195	92-36-4		3249
88-20-0	1216	92-44-4		2019
88-21-1	769	92-50-2		2226
88-29-9	3929	92-52-4		2750
88-43-7	713, 715	92-66-0		2730
88-44-8	1199	92-69-3		2763
88-67-5	1042	92-82-0		2721
88-72-2	1136	92-84-2		2744
88-73-3	608	92-86-4		2705
88-74-4	732	92-87-5		2788
88-75-5	677	92-94-4		3854
88-85-7	2123	93-00-5		2043
88-88-0	557	93-07-2		1781
88-89-1	595	93-09-4		2371
88-96-0	1421	93-35-6		1694
88-99-3	1379	93-37-8		2394
89-00-9	1052	93-55-0		1769
89-05-4	1999	93-58-3		1447
89-51-0	1728	93-60-7		1140
89-56-5	1462	93-65-2		2081
89-61-2	579	93-71-0		1553
89-69-0	559	93-72-1		1700
89-78-1	2323	93-76-5		1338
89-83-8	2218	93-80-1		2024
89-86-1	1110	93-89-0		1775
90-01-7	1167	94-09-7		1806
90-02-8	1102	94-11-1		2409
90-04-0	1192	94-12-2		2165

352-11-4	1072	470-82-6	2288
352-34-1	619	470-90-6	2824
352-93-2	383	471-03-4	322
352-97-6	211	471-46-5	77
353-54-8	33	472-54-8	4147
353-59-3	32	473-18-7	2263
356-12-7	4450	473-55-2	2271
357-57-3	4338	473-72-3	2260
360-70-3	4498	474-25-9	4406
363-24-6	4162	474-86-2	3894
366-18-7	2015	475-25-2	3570
367-12-4	666	475-31-0	4460
368-88-7	669, 670, 671, 672	476-66-4	3184
371-41-5	668	479-18-5	2210
371-86-8	983	479-23-2	4086
372-20-3	667	479-45-8	1069
378-44-9	4301	479-92-5	3316
382-45-6	4034	480-16-0	3409
389-08-2	2793	480-63-7	2146
390-64-7	4372	481-06-1	3492
392-56-3	991	481-29-8	4056
404-86-4	3935	481-37-8	1882
406-90-6	260	482-05-3	3222
420-04-2	11	483-18-1	4515
426-13-1	4300	484-12-8	3479
431-03-8	283	484-19-5	3760
433-97-6	1345	485-35-8	2458
434-03-7	4222	485-65-4	4027
434-13-9	4403	485-71-2	4013
434-22-0	3930	486-64-6	2388
434-45-7	1344	486-84-0	2756
437-38-7	4296	487-21-8	634
437-50-3	3224	487-65-0	1488
439-14-5	3572	488-59-5	915
440-58-4	2772	488-73-3	903
443-48-1	813	488-81-3	542
443-79-8	932	489-98-5	642
444-27-9	308	490-11-9	1057
445-29-4	1038	490-26-6	1698
446-86-6	1715	490-79-9	1111
451-13-8	1464	492-11-5	700
452-35-7	1764	492-27-3	2007
453-20-3	342	492-38-6	1724
455-38-9	1037	492-62-6	913
456-22-4	1039	493-01-6	2273
456-42-8	1071	494-44-0	2042
458-88-8	1640	495-69-2	1744
460-12-8	228	495-73-8	3037
460-19-5	113	496-11-7	1751
461-58-5	79	496-15-1	1475
461-72-3	133	496-64-0	408
461-98-3	809	496-67-3	839
462-02-2	123	497-59-6	1024
462-06-6	663	498-21-5	453
462-60-2	167	498-23-7	428
462-95-3	540	498-24-8	429
463-58-1	42	498-59-9	1263
463-82-1	525	498-71-5	2291
464-07-3	955	499-71-8	2255
464-43-7	2279	499-75-2	2219
464-88-0	2292	499-78-5	648
466-49-9	4306	499-80-9	1060
466-99-9	3766	499-81-0	1058
467-98-1	3900	500-28-7	1466

1988-14-3	3357	2270-20-4	2466
2008-41-5	2561	2275-18-5	1960
2008-58-4	1031	2275-23-2	1659
2021-28-5	2468	2276-96-2	2835
2029-64-3	1939	2277-92-1	2988
2030-63-9	4463	2284-20-0	1398
2032-59-9	2492	2303-16-4	2264
2040-96-2	1601	2303-17-5	2240
2043-43-8	202	2305-32-0	1558
2049-95-8	2487	2307-68-8	3124
2050-47-7	2706	2309-49-1	1845
2050-67-1	2715	2310-17-0	2844
2050-68-2	2714	2312-15-4	4060
2050-89-7	2789	2314-09-2	3881
2051-24-3	2987	2315-36-8	856
2051-30-1	2346	2315-68-6	2147
2051-60-7	2731	2340-66-1	4004
2051-61-8	2733	2348-82-5	2374
2051-62-9	2732	2350-32-5	3822
2051-85-6	2759	2361-96-8	3103
2052-14-4	2469	2363-58-8	4055
2059-76-9	1014	2364-46-7	1982
2062-78-4	4486	2367-82-0	570
2076-56-4	1354	2373-84-4	1835
2078-54-8	2918	2381-16-0	3971
2086-83-1	4102	2382-80-1	3488
2088-71-3	4451	2385-74-2	1878
2088-76-8	4527	2385-85-5	2360
2095-24-1	3617	2392-67-8	1766
2104-64-5	3265	2392-68-9	1006
2104-96-3	1410	2392-80-5	2361
2114-20-7	1297	2396-63-6	1148
2131-41-1	3058	2396-65-8	1819
2131-57-9	1710	2401-85-6	1979
2131-59-1	1002	2404-58-2	2357
2131-60-4	1046	2404-73-1	1970
2131-62-6	1348	2417-10-9	2767
2131-63-7	1347	2425-10-7	2166
2131-64-8	1765	2432-12-4	1078
2135-17-3	4295	2432-20-4	1086
2136-99-4	2573	2432-21-5	1084
2138-20-7	2891	2432-26-0	631
2152-56-9	544	2432-27-1	632
2162-99-4	1608	2432-90-8	4542
2163-68-0	1585	2432-99-7	2562
2163-69-1	2550	2437-79-8	2667
2164-08-1	3127	2439-99-8	387
2164-09-2	2023	2446-69-7	2927
2164-17-2	2085	2447-57-6	2840
2180-92-9	3937	2451-01-6	2336
2205-27-8	1539	2451-62-9	2865
2207-01-4	1600	2452-84-8	3639
2207-04-7	1603	2463-84-5	1467
2212-67-1	1900	2475-44-7	3585
2213-23-2	1946	2475-45-8	3250
2216-30-0	1942	2475-46-9	3743
2216-32-2	1954	2487-01-6	3090
2216-33-3	1943	2491-15-8	149
2216-34-4	1959	2491-52-3	2745
2216-51-5	2324	2491-74-9	3267
2217-08-5	2245	2497-06-5	1678
2235-90-7	2872	2498-77-3	3969
2236-60-4	697	2516-95-2	1005
2243-95-0	1983	2536-31-4	3414

16488-56-5	3947	19167-62-5	778
16488-57-6	4071	19167-63-6	798
16533-50-9	1855	19216-56-9	4010
16606-02-3	2692	19549-73-6	1964
16671-34-4	4467	19578-81-5	3012
16747-25-4	1945	19666-30-9	3486
16752-77-5	473, 867	19679-38-0	3596
16766-29-3	1394	19780-41-7	1963
16789-46-1	1957	19810-30-1	2025
16805-99-5	2382	19872-68-5	3110
16806-29-4	1430	19872-70-9	3815
16840-28-1	2417	19872-72-1	3747
16878-76-5	644	19922-87-3	240
16878-77-6	1097	19937-59-8	2156
16891-79-5	1209	20043-94-1	4063
16909-11-8	2429	20168-99-4	4190
17103-43-4	2381	20195-08-8	1975
17103-48-9	2066	20195-16-8	3401
17103-49-0	2022	20203-81-0	793
17109-49-8	3286	20279-51-0	1937
17140-78-2	4535	20354-26-1	1687
17199-29-0	1452	20427-84-3	4072
17230-88-5	4294	20460-96-2	3111
17239-22-4	3517	20473-73-8	1582
17239-23-5	3441	20473-77-2	2295
17239-27-9	2854	20562-02-1	4607
17243-26-4	2448	20629-50-9	3967
17243-29-7	2407	20636-48-0	4435
17260-71-8	710	20651-71-2	2467
17296-50-3	3437	20675-21-2	3337
17301-94-9	2349	20675-22-3	4459
17321-62-9	2094	20675-23-4	3105
17321-63-0	2056	20675-24-5	2810
17348-59-3	1317	20675-25-6	2815
17418-58-5	4185	20682-28-4	3737
17465-86-0	4618	20702-77-6	4489
17530-23-3	1884	20830-75-5	4585
17530-24-4	2270	20923-67-5	1646
17560-51-9	3606	21035-44-9	976
17598-81-1	909	21087-64-9	1569
17629-30-0	3954	21149-88-2	2498
17655-95-7	4278	21256-18-8	3861
17698-03-2	3393	21312-10-7	2820
17700-09-3	560	21321-07-3	2016
17804-35-2	3320	21413-25-2	3746
18031-40-8	2214	21413-28-5	4186
18046-21-4	3722	21413-53-6	4183
18069-66-4	4362	21416-67-1	2503
18181-70-9	1413	21452-14-2	2402
18181-80-1	3739	21452-18-6	2790
18259-05-7	2639	21540-35-2	2206
18264-75-0	29	21548-32-3	860
18530-56-8	3165	21609-90-5	3009
18559-94-9	3160	21616-46-6	3586
18691-97-9	2099	21725-46-2	1860
18854-01-8	3085	21829-25-4	3757
18883-66-4	1591	21885-31-4	848
18979-72-1	2221	21914-07-8	3568
18979-73-2	2512	21988-05-6	1040
19044-88-3	2917	22005-65-8	696
19077-97-5	1846	22071-15-4	3588
19077-98-6	3055	22131-79-9	2392
19167-57-8	1146	22204-24-6	4551
19167-60-3	745	22204-53-1	3275

29091-05-2	2438	32407-99-1	1812
29104-30-1	3873	32451-19-7	1485
29122-68-7	3370	32598-10-0	2663
29136-19-4	4434	32598-11-1	2664
29232-93-7	2532	32598-12-2	2657
29446-15-9	2648	32598-13-3	2656
29826-16-2	4528	32620-68-1	2801
29839-52-9	1628	32620-70-5	3066
29839-54-1	2333	32620-72-7	3259
29839-55-2	2334	32690-93-0	2658
29839-58-5	1926	32861-85-1	3001
29839-59-6	1626	33025-41-1	2660
29839-60-9	1627	33069-62-4	4613
29839-61-0	1893	33089-61-1	4022
29839-62-1	1895	33146-45-1	2711
29839-63-2	3380	33213-65-9	1690
29839-64-3	1927	33244-57-4	3436
29839-66-5	1920	33284-50-3	2708
29839-67-6	1636	33284-52-5	2655
29839-68-7	1902	33284-53-6	2659
29839-70-1	2327	33286-22-5	4293
29839-71-2	2329	33376-25-9	1874
29839-72-3	2331	33419-42-0	4509
29839-73-4	2325	33423-92-6	2597
29839-74-5	1620	33433-95-3	1343
29839-76-7	1928	33439-45-1	2057
29839-77-8	2326	33564-31-7	4449
29839-78-9	2969	33669-70-4	637
29848-44-0	494	33693-04-8	2307
29848-45-1	2330	33820-53-0	3539
29883-15-6	4143, 4144	33857-26-0	2646
29973-13-5	2479	33979-03-2	2618
30003-26-0	897	34006-76-3	3332
30003-27-1	2328	34014-18-1	1891
30007-47-7	273	34123-59-6	2909
30010-08-3	1275	34197-26-7	3769
30010-09-4	2335	34244-80-9	1089
30091-04-4	3271	34256-82-1	3346
30097-06-4	433	34392-82-0	2384
30377-37-8	418	34462-96-9	2679
30448-43-2	3674	34491-29-7	2054
30516-87-1	2182	34622-58-7	2869
30544-61-7	3871	34645-84-6	3213
30560-19-1	371	34801-09-7	1502
30564-38-6	1090	34883-39-1	2707
30565-25-4	327	34883-43-7	2709
30652-11-0	1193	34968-90-6	701
30665-94-2	3595	35065-27-1	2614
30746-58-8	2595	35065-28-2	2615
30979-48-7	1590	35065-29-3	2587
31127-54-5	3027	35065-30-6	2576
31218-83-4	2317	35075-35-5	1568
31220-35-6	3636	35367-38-5	3199
31431-39-7	3577	35400-43-2	2935
31482-09-4	1913	35693-92-6	2693
31508-00-6	2636	35693-99-3	2666
31637-97-5	3872	35694-06-5	2617
31647-36-6	4181	35694-08-7	2572
31807-55-3	2972	35801-62-8	1251
31817-24-0	4272	35801-67-3	3511
31879-05-7	3457	35822-46-9	2565
31889-35-7	1932	35846-47-0	2787
32060-78-9	4367	35858-24-3	3518
32365-02-9	1814	36322-90-4	3442

36330-85-5	3589	40186-70-7	2583
36355-01-8	2591	40186-72-9	2566
36370-80-6	2936	40326-33-8	1633
36378-49-1	4054	40363-76-6	3406
36378-71-9	3961	40487-42-1	3144
36559-22-5	2671	40596-69-8	4073
36566-80-0	1228	40691-50-7	3407
36614-38-7	1324	41047-52-3	1491
36653-71-1	645	41198-08-7	2474
36653-82-4	3714	41295-28-7	3614
36729-58-5	2354, 2355	41340-25-4	3783
36734-19-7	3052	41394-05-2	2063
36756-79-3	3696	41464-39-5	2670
36765-01-2	1075	41464-40-8	2676
37076-68-9	1472	41464-41-9	2669
37139-21-2	3698	41464-43-1	2662
37517-28-5	4324	41464-47-5	2672
37680-65-2	2690	41464-51-1	2627
37680-66-3	2683	41483-43-6	3169
37680-73-2	2640	41492-69-7	2535
37723-78-7	3487	41541-11-1	3905
37764-25-3	1544	41541-13-3	3771
37809-02-2	159	41541-14-4	3778
37924-13-3	3243	41814-78-2	1714
38026-46-9	1157	41992-23-8	3833
38194-50-2	4098	42175-34-8	3178
38260-54-7	2266	42200-33-9	3823
38363-40-5	3940	42509-80-8	1899
38379-99-6	2634	42576-02-3	3201
38380-01-7	2635	42607-20-5	846
38380-02-8	2628	42607-21-6	2089
38380-03-9	2629	42782-57-0	3752
38380-07-3	2600	42874-03-3	3412
38380-08-4	2602	42924-53-8	3478
38411-25-5	2581	43121-43-3	3287
38434-77-4	150	45376-90-7	488
38444-73-4	2684	47000-92-0	2086
38444-76-7	2685	49562-28-9	4108
38444-78-9	2689	50471-44-8	2739
38444-81-4	2697	50512-35-1	2930
38444-85-8	2687	50679-08-8	4537
38444-86-9	2696	50785-22-3	2401
38444-90-5	2691	51146-56-6	3134
38444-93-8	2677	51146-57-7	3135
38713-56-3	3694	51166-71-3	4647
38765-78-5	1540	51203-19-1	2103
38821-53-3	3640	51203-20-4	2173
38964-22-6	2647	51207-31-9	2594
38973-73-8	3941	51218-45-2	3528
38973-74-9	4064	51235-04-2	2938
38973-75-0	4164	51333-22-3	4426
38973-76-1	4255	51337-71-4	4003
39075-90-6	2223	51338-27-3	3582
39196-18-4	1907	51437-89-9	3705
39227-28-6	2571	51437-90-2	3948
39227-53-7	2680	51437-91-3	4172
39227-54-8	2681	51437-92-4	4320
39227-58-2	2621	51437-95-7	4256
39227-61-7	2575	51453-65-7	3104
39515-41-8	4282	51481-61-9	2252
39588-36-8	2391	51527-19-6	2375
39719-87-4	2420	51543-29-4	2818
39832-36-5	1840	51543-30-7	2861
40038-46-8	2760	51543-31-8	2454

51736-24-4	2902	55312-69-1	2632
51803-78-2	3047	55335-06-3	1011
51908-16-8	2599	55380-34-2	876
51940-44-4	3313	55441-71-9	740
51953-05-0	399	55511-98-3	2251
51953-13-0	243	55648-40-3	2456
51953-14-1	247	55702-45-9	2688
51953-17-4	242	55774-17-9	4522
52061-82-2	3392	55774-19-1	4505
52222-35-2	3534	55791-76-9	3140
52645-53-1	4191	56030-54-7	4307
52663-58-8	2661	56030-56-9	2616
52663-59-9	2675	56070-16-7	1976
52663-60-2	2624	56209-30-4	1510
52663-62-4	2631	56265-21-5	4140
52663-63-5	2609	56265-22-6	4018
52663-65-7	2579	56265-23-7	4012
52663-67-9	2578	56265-24-8	4130
52663-68-0	2577	56265-26-0	4032
52663-69-1	2582	56265-27-1	3898
52663-70-4	2585	56288-27-8	3494
52663-71-5	2584	56392-17-7	4065
52663-74-8	2586	56529-85-2	1699
52663-77-1	2567	56558-16-8	2623
52704-70-8	2611	56563-18-9	720
52709-86-1	3598	56595-20-1	1249
52712-04-6	2612	56741-95-8	2011
52712-05-7	2589	56980-93-9	4165
52717-51-8	1841	56995-20-1	3481
52717-52-9	2129	57045-86-0	1386
52744-13-5	2604	57094-83-4	4447
52756-22-6	3987	57117-31-4	2574
52918-63-5	4271	57117-44-9	2569
53162-40-6	2395	57229-74-0	314
53469-21-9	2695	57381-79-0	4266
53496-15-4	1281	57440-16-1	2848
53535-33-4	1318	57808-66-9	4284
53648-05-8	3141	57837-19-1	3516
53744-47-1	134	58182-63-1	3730
53780-34-0	2437	58447-18-0	3878
53799-78-3	4335	58471-47-9	1152
53982-07-3	3765	58522-87-5	2052
53983-00-9	441	58528-60-2	3755
53983-01-0	1229	58947-88-9	640
53994-73-3	3443	58965-05-2	2385
54010-85-4	818	59080-33-0	574
54029-12-8	2864	59080-37-4	2644
54063-53-5	4218	59277-89-3	1551
54135-80-7	1034	59467-70-8	3847
54612-53-2	3153	59536-65-1	2592
54942-37-9	3934	59602-16-3	2224
54942-38-0	4160	59746-11-1	1762
54942-39-1	4318	59804-37-4	3039
54942-40-4	4400	59831-97-9	3308
54942-41-5	3684	59831-98-0	3658
54942-42-6	2442	59831-99-1	3682
54965-21-8	2862	59832-00-7	3802
55030-48-3	3812	59832-01-8	3933
55125-24-1	4251	59832-02-9	4051
55205-89-5	2894	59832-03-0	4245
55215-17-3	2633	59832-05-2	3495
55215-18-4	2601	59832-06-3	3657
55268-74-1	4029	59832-07-4	3801
55283-68-6	3059	59832-08-5	4035

69477-71-0	2169	73632-82-3	2249
69570-81-6	2062	73632-83-4	2500
69577-07-7	1890	74103-06-3	3439
69588-11-0	2092	74103-08-5	3850
69655-05-6	2132	74103-09-6	3717
69679-30-7	3818	74103-11-0	2743
69679-31-8	3939	74103-12-1	3034
69679-32-9	4161	74115-24-5	3188
69698-47-1	3797	74158-10-4	1813
69756-53-2	4445	74223-64-6	3285
69806-50-4	3990	74472-34-7	2668
70136-02-6	3358	74472-37-0	2638
70288-86-7	4616	74472-42-7	2606
70362-47-9	2651	74472-44-9	2605
70362-48-0	2674	74692-14-1	15
70424-68-9	2637	74713-68-1	4508
70424-70-3	2626	74713-69-2	4512
70458-96-7	3627	74713-70-5	4370
70648-26-9	2570	74973-14-1	1758
71067-09-9	4363	74973-19-6	3745
71067-10-2	4433	74973-22-1	4492
71119-16-9	3758	75121-79-8	2509
71119-19-2	3076	75410-15-0	664
71119-20-5	2405	75410-16-1	1129
71119-21-6	2386	75410-27-4	1132
71119-29-4	3132	75410-28-5	2366
71119-30-7	3131	75438-57-2	1825
71119-31-8	2885	75922-48-4	3474
71119-32-9	2461	76006-86-5	1373
71119-34-1	3298	76095-16-4	4381
71119-35-2	3668	76432-30-9	2432
71119-36-3	3913	76432-33-2	2433
71119-37-4	2423	76432-35-4	2434
71119-38-5	2106	76466-16-5	1732
71125-38-7	3262	76578-14-8	3977
71138-72-2	2838	76748-72-6	4110
71177-53-2	4254	76824-35-6	1596
71283-80-2	3864	76842-07-4	2630
71422-67-8	4075	76963-41-2	2945
71720-65-5	3077	77340-50-2	330
71759-43-8	665	77501-90-7	3848
71759-45-0	1791	77632-11-2	747
71963-77-4	3706	77658-97-0	2485
72402-00-7	2287	77868-40-7	3650
72487-80-0	1388	77868-41-8	3651
72509-76-3	3882	77868-43-0	3649
72559-06-9	4609	77868-44-1	3648
72678-81-0	2055	77868-45-2	3791
72678-82-1	2093	77868-46-3	4036
72678-86-5	2877	77868-48-5	3911
72678-93-4	2450	77942-93-9	3790
72762-00-6	412	78002-88-7	1253
72956-09-3	4368	78092-53-2	4649
73042-04-3	1130	79578-14-6	4376
73080-51-0	4111	79617-96-2	3749
73239-20-0	2200	80421-90-5	3400
73383-40-1	1232	80496-87-3	1339
73429-89-7	4468	80702-24-5	4126
73544-88-4	3229	80844-07-1	4421
73632-76-5	2275	81098-60-4	4348
73632-77-6	2531	81103-11-9	4577
73632-78-7	3500	81777-89-1	2821
73632-79-8	2276	82034-46-6	4375
73632-81-2	3089	82310-91-6	310